FUNDAMENTALS OF ZOOLOGY

FUNDAMENTALS OF ZOOLOGY

[for Undergraduate General and Honours Courses; Entrance and Public Examinations]

Dr K C Ghose DSc, FZS, FAZ

Professor (Retd.) and Ex-Head, Department of Zoology, Calcutta Univesity
Jubilee Postgraduate Scholar; University Gold Medallist & Prize Winner
Hem Chandra Gossawin Gold Medallist and
Prize Winner; Matilal Mullick Gold Medallist

and

Dr Buddhadeb Manna MSc, PhD, FZS, FHSI, FASc, MIPHA

Professor Emeritus (UGC), Department of Zoology, Calcutta University
Professor and former Head, Department of Zoology, Calcutta Univesity
Formerly, Assistant Professor in Zoology,
City College, Kolkata and Kalyani University,
Kalyani, Nadia, West Bengal
Recipient: 20th Century Gold Medal, Zoological Society of India
Zoological Congress Gold Medal, Zoological Society of Eastern India

New Central Book Agency (P) Ltd

LONDON
HYDERABAD ERNAKULAM BHUBANESWAR
NEW DELHI KOLKATA PUNE GUWAHATI

NCBA

REGD OFFICE
8/1 Chintamoni Das Lane, Kolkata 700 009, India
email: ncbapvtltd@eth.net

OVERSEAS
NCBA (UK) Ltd, 149 Park Avenue North, Northampton, NN3 2HY, UK
email: ncbauk@yahoo.com

EXPORT
NCBA Exports Pvt. Ltd, 212 Shahpur Jat, New Delhi 110 049
email: ncbaexports@ncbapvtltd.com

BRANCHES

2/27 Ansari Road, 1st Floor
Daryaganj, **New Delhi** 110 002
email: ncbadel@ncbapvtltd.com

House No. 3-1-315, 1st Floor
Nimboliadda, Kachiguda, **Hyderabad** 500 027
email: ncbahydb@ncbapvtltd.com

Shop Nos. 3 & 4, Vinayak Towers
681/B Budhwar Peth, Appa Balwant Chowk
Pune 411 002
email: ncbapune@ncbapvtltd.com

GSS Shopping Complex, 1st Floor
Opposite College Post Office, Convent Road
Ernakulam 682 035
0484-4021054/53

Shop No. 15, Bharati Towers, Ground Floor
Forest Park
Bhubaneswar 751 001
email: ncbabub@ncbapvtltd.com

Radhamayee Complex (Second Floor)
Opp. SBI, Panbazar, Jaswanta Road
Guwahati 781 001
email: ncbaguwahati@ncbapvtltd.com

FUNDAMENTALS OF ZOOLOGY • Dr K C Ghose and Dr B Manna

First Published: May 2003
Reprinted: April 2004, March 2007, March 2012
Thoroughly Revised, Enlarged and Updated Second Edition: August 2017

PUBLISHER AND TYPESETTER
New Central Book Agency (P) Ltd
8/1 Chintamoni Das Lane, Kolkata 700 009

PRINTER
New Central Book Agency (P) Ltd
Web-Offset Division, Dhulagarh, Sankrail, Howrah

COVER DESIGNER
Ankit, Kolkata

ILLUSTRATIONS
Narayan Sur, Gobinda Jana, Niranjan Giri and Sunil Pal

PROOF CHECKING
Kakoly Banerjee

PROJECT TEAM
Srimanta Kr. Hazra, Swapan Sarkar, Pradip Biswas, and Prodip Baidya

ISBN: 978 93 5255 042 5

Price ₹ 735.00

Dedicated to

The Almighty

CONTENTS

PART ONE

SCOPE AND HISTORY OF ZOOLOGY: CELLS AND TISSUES: ANIMAL FORMS AND FUNCTIONS

PART TWO

DIVERSITY IN ANIMAL FORMS

PART THREE

ORIGIN AND CONTINUITY OF ANIMAL LIFE: ENVIRONMENT: HUMAN INTERFERENCE: DISEASES AND ANIMALS

Preface to the Second Edition

Since publication of the first edition of this book *Fundamentals of Zoology*, I have received many encouraging letters and advice from a large number of teachers, experienced and young, teaching in undergraduate and postgraduate colleges of universities, research associates from different corners of the country—for further additions to the book. In the present edition I have tried to comply with them.

Maintaining the pattern and uniqueness of the book I have tried to incorporate all the suggestions. Each and every chapter has been given special attention.

The Part One deals with the fundamentals of life processes and the basic structures of the organism. In this part I have included three new chapters: (i) Regulation of Body Temperature, (ii) Regeneration, and (iii) Biological Luminescence.

Part Two of the book deals with diversity in animal forms and their morphology. Each chapter has been reoriented with addition of relevant necessary information. Few special topics, like, Leishmania and Kala-azar, polymorphism in Cnidaria, corals and coral reef, *Wuchereria* and filariasis, metamerism in annelids, life of sacculina, biology of few important vectors, communication in insects, foot-modification and nervous system in echinoderms, migration of fishes and birds, parental care in amphibians and birds, Axolotl and Neoteny, poison apparatus and biting mechanism in snake, dentition, echolocation, ruminant stomach in mammals, and biological status of Monotremata, are added in this part.

Similarly, in Part Three of this book, two new chapters Microbiology and Biotechnology have been added in view of exposure of the readers, although in a preliminary state. The additional topics in this part are morphogenetic movement, fate map, organisers, development of brain and eye in chick, placenta in mammals. The chapter Economic Zoology has been reoriented with up-to-date need and information.

Now, we think, in this form, the book will be more useful and utilized by a wider group of readers to meet their basic needs.

I have made every endeavour to present a simple but very useful textbook, written in an easily understandable language. It is for the readers to judge its usefulness. Suggestions from the readers are gratefully welcome.

I gratefully acknowledge all the teachers from different graduate and postgraduate colleges and universities for their kind help and suggestions received for revising this textbook. I gratefully remember the inspiration and teaching of my teacher Professor Krisna Chandra Ghose, who is no more in this world.

I sincerely thank Sri Amitabha Sen, Director, New Central Book Agency (P) Ltd, for his interest and unstinted continuous effort to publish this edition.

1 February 2017 **B Manna**
Kolkata

Preface to the First Edition

The factors governing life on earth are changing constantly and the same is true for life too. The unique property of the living forms is their ability to change themselves, accepting the challenges caused by changes in the surroundings—and this has enabled them to exploit the environment successfully, leading to their survival, multiplication, and continuation on earth since first appearance.

The tree of life has grown steadily for millions of years, diversifying itself in many branches—barring the termination of a few, owing to certain drastic changes, which may be called catastrophes. The continuous exposure to changing environment has slowly led to the explosion of myriad forms of life on earth. It is for the authors of books to pass the information to their readers.

To maintain the standard of a book some well-established themes, presently modified or discarded, are rewritten or omitted altogether, and new matters are incorporated after proper screening and judgement. For this, the author has to collect, collate and analyse enormous data, screen out information from noises and present those in logical sequence.

The progressive changes in the animal world are reflected in an ideal textbook on life sciences, irrespective of its standard. The students should be made to understand that new discoveries create the need to forsake many established concepts, as the firm belief 'Anatomy does not change' is not tenable today, though it cannot be denied that the changes in anatomy are extremely slow and not detectable within a short period of time, but seasonal, temporary changes in anatomical structures—particularly in relation to food and feeding habits—are not rare in the animal world.

Attempts have been made to create interest for Zoology in young minds and to help them equip with current information. Unfortunately, the students are not taught how to acquire reading skill, spoon-feeding has made them dull and passive readers. They want everything in tabloid form and shun reading books. They are not enthusiastic in reading even very attractive and interesting books. Students must understand that knowledge cannot be acquired without toils and tears.

· The syllabi in Zoology for undergradutate students in a number of Indian universities have been prepared in a haphazard manner, not in sequence. To climb a mountain one has to go up by steps. And, before moving to the next step, the mountaineer has to acquire certain ideas about the region to which he is moving. Likewise, one has to acquire certain basic knowledge to understand the topic. For this, the chapter dealing with the required basic information should precede the chapter dealing with a particular topic. Unfortunately, in the order of arrangement of the topics in the syllabi of these universities, no attention has been paid to this norm. This is something like teaching 'formation of three germinal layers' in an embryology class to students who do not possess even elementary knowledge on fertilisation, cleavage and gastrulation.

Taking the risk of criticism, we have arranged the topics in a fashion, following which properly, the students would be armed with the necessary knowledge before passing over to the next more complex topic.

Part One of the book deals with the fundamentals of life processes and basic structures of ogranisms; Part Two deals with diversity in animals; Part Three with development, evolution, genetics, economic zoology, ecology, wildlife and environmental pollution. Chapter 52 and Annexures I and II, are meant to help both students and persons interested in life sciences.

We are aware of our limitations and also the limitations imposed by factors beyond our control. We have made every endeavour to present a simple but useful textbook, written in an easily understandable language. It is for the readers to judge its usefulness.

In the preparation of the manuscript we received enormous cooperation and encouragement from a large number of teachers, experienced and young, teaching in undergraduate and postgraduate colleges of a number of universities. Quite a good number of research associates, fellows and scholars volunteered their cooperation enthusiastically. We owe much to them and we gratefully acknowledge their valuable help with all sincerity. We deeply regret our inability to ackowledge indebtedness individually, for which we beg to be excused.

We will be failing in our duty if we do not specially mention the help rendered by Prof. D. K. Nanda, Prof. N. C. Dutta, Prof. S. Banerjee and Dr. S. K. Raut, Reader, Department of Zoology, Calcutta University; Dr. Badal Chandra Bharati Goswami, Sree Gopal Banerjee College, Mogra, Hooghly; Dr. Ashoke Das, Sri Prafulla Kr. Das, Zoological Survey of India; Sri Saibal Roychowdhury, Head, Department of Zoology, Rishi Bankim Chandra College, Naihati; Dr. Narayan Chandra Das, Scottish Church College, Kolkata, and Zoological Survey of India. We owe much to Sri Ajoy Kr. Mondal, Research Scholar, Department of Zoology, Calcutta University, for his help in the preparation of the manuscript in a number of ways including dissection of circulatory system of guinea pig (Cavia sp.) and making sketches from dissected specimens.

We sincerely thank Sri. J. N. Sen, Director, New Central Book Agency (P) Ltd, for his interest and continuous efforts to publish the book.

1 January 2003
Kolkata

K C Ghose
B Manna

PART ONE

SCOPE AND HISTORY OF ZOOLOGY:

CELLS AND TISSUES: ANIMAL FORMS AND FUNCTIONS

1 ZOOLOGY: SCOPE AND HISTORY

Science is the analytical study of facts. Biology embraces the study of all living organisms and zoology is concerned with that of animals.

DIVISIONS OF ZOOLOGY

Morphology or **Anatomy.** The study of gross structures of organisms—both external and internal, and the relation of one organ with the other.

- **Microanatomy.** The study of histological structures or tissues under magnification.
- **Physiology.** It deals with the functions of organs in relation to structure.
- **Cytology.** The study of structural details of the cells, their modifications and divisions.
- **Embryology.** The science of development of animals from zygote stage to fully formed young individual. Nowadays it has been replaced by the term morphogenesis or embryogeny.
- **Parasitology.** The structure, modification and the influence of parasitic life, both on parasite and host.
- **Entomology.** Various aspects of insects including their influences on other animals.
- **Taxonomy.** Nomenclature of animals and classification or assigning them proper position in the animal kingdom.
- **Ecology.** Relation of animals with the environment and the influence of the latter on the former.
- **Zoogeography.** Distribution of animals on the surface of the earth.
- **Evolution.** Descent with modifications or the appearance of new species from preexisting species.
- **Genetics.** Genetics is the science of inheritance. It is concerned with the transmission of potentialities of characters from generation to generation.
- **Molecular Biology.** Quantitative studies of the molecular structures and the factors responsible for biological processes. It includes analysis of gene structures and functions and genetic control of synthesis of biological molecules and their regulatory processes in the cell.

PRACTICAL APPLICATION OF ZOOLOGY

Our ever-increasing knowledge in zoology has enabled us to apply this science in human benefit, ranging from prevention of diseases to production of various items for our use, introduction and stabilisation of new hybrids, and in many other fields.

• **Medicine:** A knowledge of animals producing various disease, viz. malaria, filaria, nagana, liver rot, etc., is essential for proper treatment. Further knowledge on anatomy and physiology of experiment animals like rabbit, monkey and others enables us not only to test the drugs but in the manufacture of hormones, enzymes, vaccines, etc.

• **Fibre.** The beautiful silk is really the secretion from the silk glands of the silkworm *Bombyx mori* and others. The larva builds a hard protective cocoon around it, within which it pupates. The silk thread is obtained from this cocoon.

• **Lac.** The lac is a resinous substance secreted by the female lac insect *Tachardia lacca* to build their nests on the branches of food tree. The dye obtained in the processing of lac is used for colouring and the shellac in various industries, viz. pottery, varnish, toy, electric insulation, shoe polish, etc.

• **Pearl.** The pearl is an important item of commerce. It is extensively used in the manufacture of ornaments and decorative articles. Large scale pearl cultivation is carried out in a number of countries including India, of which Japan tops the list.

• **Food.** The animals themselves or their products are very important items in the list of our food. With increased demand for animal and animal products for food—fishery, poultry, piggery and dairy have been started on commercial basis.

a. **Fishery.** The knowledge on anatomy, physiology, breeding, growth, migration and behaviour of marine, estuarine and fresh water fishes is being utilised in fisheries and it has been possible to produce a number of high-yielding fishes. Sea is a vast reservoir of fish and sea fishing is encouraged by a number of countries.

b. **Dairy.** Healthy cattle and buffaloes are a good source of protein and milk. In dairy, in addition to maintaining proper health of the livestock, efforts have been made with success to improve stock by hybridization.

c. **Poultry.** Birds are a good source of most tasty meat and egg. Fowl and duck are the common animals of a poultry. The maintenance of the health of the birds and improvement of stock by hybridization and other means go side by side in an ideal poultry.

HISTORY OF ZOOLOGY

The association of man and animals dates back to the prehistoric period. The prehistoric men knew animals, they could distinguish them from one another, from different angles, primarily from their daily needs and safety. The early Egyptians knew quite a lot about animals, and domesticated cattle, sheep, pigs, cats and ducks.

Attempts were made by Greek philosophers of fifth and sixth century BC to explain the origin of life. **Aristotle** (384-322 BC) wrote about animals of Greece and the nearby region of Asia Minor in his **Historia animalium**. Greek physician **Galen** (131-201 AD) the founder of experimental physiology, dissected pigs and monkeys and made notable discoveries on the functions of brain and nerves and demonstrated that arteries carry blood and not air, the idea held till then.

Belgian anatomist **Andreas Vesalius** (1514-64), Professor in the University of Padua in Italy, published his observations based on the dissection of human body '**On the structure of the Human Body**' in 1543. Galen's description of human anatomy had many errors. In spite of that, Vesalius' challenge to unquestioned authority of Galen for 1,300 years invited severe adverse criticism and he was finally forced to resign from Professorship.

Since the domestication of animals by early Egyptians, our knowledge in zoology has been enriched from the contributions of a large number of workers, the names of the majority of whom are unheard to the average students of zoology. From the time of Vesalius, till the appearance of **Charles Darwin**, the progress was not that fast. Darwin, a man of vision, brought the science of animals from the laboratories and libraries of a few people or organisations to the common man. The **father of zoology** helped at large to get an idea about the vastness and diversity of the animal world.

The progress of zoology has been spectacular from the early twentieth century. Naturalists, systematists, anatomists, physiologists, cytologists and geneticists, embryologists, evolutionists, biophysicists and biochemists, ecologists, mathematicians and statisticians, all nurtured the giant tree of zoology with utmost care and attention. Today, molecular biology and genetics have developed to such an extent that different disciplines have been created for them.

CONTRIBUTIONS OF HARVEY, LINNAEUS, SCHLEIDEN, SCHWANN, LAMARCK, DARWIN, de VRIES, MENDEL, PASTEUR, SPEMANN AND MORGAN

W. Harvey (1578-1657)

William Harvey was born in a wealthy family in Folkstone, England. He had his schooling in the King's School at Canterbury and at the age of sixteen entered Cains College. Here he studied Latin and Greek to become a graduate with arts degree. In 1598 he came to University of Padua, Italy, and studied Anatomy and Physiology from Fabricius. Fabricius was the most learned and honoured teacher in Italy at that time. He came back to England in 1602 and obtained M.D. degree from Cambridge. In 1615 he became lecturer of anatomy in the Royal College of Physicians in London.

Harvey's book "**Anatomical dissertation concerning the motion of the heart and blood**" is still considered as one of the greatest monuments in Biology. He was the first amongst the biologists who gave a really reasonable explanation, in physical terms, of any bodily process. His investigation on the anatomy of about sixty animals and embryology of insects and vertebrates had a tremendous influence in promoting further biological research.

C. Linnaeus (1707-78)

Caroleus von Linnaeus, son of a poor Sweedish clergyman, was an eminent naturalist. He was a student of medicine in Upsala. In 1732 the Royal Society of Upsala sent him to Lapland as a collector and observer. On returning he completed his studies in medicine from the University of Hardewyk in 1735. Within a short time he was appointed as Professor of Natural History of Upsala.

Linnaeus had a fascination to classify animals and plants. He introduced the system of classification which is known as **Binomial nomenclature** and placed all the known plants and animals in his classificatory system. He classified not only plants and animals but also the minerals and even diseases.

His great contributions were the publications of **Systema Naturae** (1735)*, **Fundamenta Botanica** (1735), **Bibliotheca Botanica** (1735) and **Genera Plantarum** (1737).

The Systema Naturae is not only a treatise but also a catalogue of the production of nature, most methodically arranged. His binomial nomenclature system is still in practice all over the world.

M. J. Schleiden (1804-81)

Schleiden, a German botanist, was a graduate in medicine but his studies centred mainly around Botany. The nucleus in the plant cells was discovered by **Robert Brown** in 1831 and this greatly influenced the studies of Schleiden. In 1838, he suggested that the cells were the structural units of plants. His co-worker **Theodor Schwann** concluded that Schleiden's idea is applicable to animals also. The cell

* Systema Naturae written in 1735, published in 1758.

theory, though really postulated by Schwann, is based on Schleiden's observations also. The cell theory has played a very important role in the understanding of structures and functions of different organs of the living organism. The present-day modification of the cell theory of Schleiden and Schwann stands as:

1. Cells are the structural and functional units of the body of an organism.
2. Every individual begins its life from a single cell.
3. New cells are produced only by the division of preexisting cells.

T. Schwann (1810-82)

Theodor Schwann, a German zoologist postulated cell theory in 1839 from his detailed study of the animal tissue. The contributions of Schleiden on the structure of plant tissues influenced his idea greatly. The cells were regarded as the structural units of a living body and tissues were made up of cells. He first used the term **'Cell theory'** and presented it to the scientific world. His important contributions are:

1. Every organism is made up of organic units, the cells.
2. Every organism begins life from a single cell.
3. New cells are always produced by the division of preexisting cells.
4. Experiments on spontaneous origin of life.
5. Fermentation is carried by microorganisms (bacteria).
6. Discovery of a sheath (Sheath of Schwann) in nerve fibre.

J. B. Lamarck (1774-1829)

Jean Baptiste de Lamarck, a French scientist, is the most pathetic figure in the evolutionary history. He was a man of vision, with the power of keen observation and logical deduction but suffered a setback mainly for his struggle with poverty and bitter criticism from his contemporaries.

He came from an army family and fought many battles. He was meant for church but religious studies could not interest him. He began his carrier as a botanist, but became a zoologist when he was offered an appointment in zoology at the **Jardin des Plantes**.

Lamarck's contributions to general zoology are manifold, although his name is usually associated with the theory of organic evolution. He came very close to the cell theory of Schleiden and Schwann, thirtynine years before they postulated it. He extensively studied invertebrates, greatly improved classification and reorganised vertebrates and invertebrates as distinct sections of the animal kingdom. He was also a distinguished anatomist and palaeontologist.

Lamarck conceived that species were not constant, but rather were derived from preexisting species. To account for this he advocated the theory of **inheritance of acquired characters** and **the laws of use and disuse**. He was the first to give the idea of the **'tree of life'**.

C. Darwin (1809-82)

Charles Robert Darwin, an English scientist is famous for his theory of **'Natural Selection'**. He closely studied for long years a large number of animals belonging to different classes, plants and fossils and collected a huge volume of data. He published his famous book **"The Origin of Species by Natural Selection or the Preservation of Favoured Races in the Struggle for Life"**, in the year 1859. Evolution by **natural selection**, as conceived by Darwin, can be fully summarized in three points: (i) All plants and animals reproduce in excess of the numbers which can actually survive, and this leads to a struggle for survival. (ii) The members of the same species vary, some of the variations are neutral, while others help or hinder the organism in its struggle for survival. (iii) As a consequence, **'the sur-**

vival of the fittest' variants is expected, with the less fit variants being destroyed either by their physical or biotic environments.

Darwin postulated two other theories of much less importance, **'Artificial Selection'** and **'Sexual Selection'**. For his notable contributions and popularisation of biology he is also known as **'Father of Biology'**.

H. de Vries (1848-1935)

Hugo de Vries, a Dutch botanist, published **'Mutation theory'** in 1901. The theory is based mainly on his experiment on the plant, evening primrose, *Oenothera lamarckiana*.

Certain strikingly different forms appeared suddenly among a population of normal type of primrose. On cultivation he recognised appearance of seven types that breed true. He called them **mutants**. Starting with nine plants from the field and growing them for eight years de Vries found that the same seven mutants, appeared from the stock parental forms, year after year. Today, it is well-known that a mutant is a true breeding variant which arises abruptly among normal forms and the mutation is a sudden measurable change in organisms which can be transmitted to the offsprings. de Vries opened a new period of investigation on the phenomenon of heredity. He made important contributions to the study of origin of species.

Although, de Vries formulated an extremely important theory which stands well established today, the material on which his theory was based, was later found to be rather abnormal. Even de Vries had noticed that *O. lamarckiana* was an inconsistent species. de Vries proved his theory with a very unnatural species but the facts do not anyway diminish the importance of the theory.

He also developed the **Theory of Pangenesis**. Pangenes are the materials that remain within the nucleus, multiply and come out of the nucleus to determine the form and activity of tissue.

G. J. Mendel (1822-84)

Gregor Johan Mendel, an Austrian monk, and abbot of the Augustinian Monastery at Brünn, Austria, is pioneer in the field of **Heredity**. Painstakingly for years he carried experiments on inheritance of characters with pure races of garden peas and from the results he deduced certain fundamental laws of heredity. These are known as **'Law of dominance'**, **'Law of segregation'** and **'Law of independent assortment'**. His findings were published in 1886 in an obscure journal and were made known to the world only in 1900, sixteen years after his death by three scientists Correns, de Vries and Tschermak who had independently reached the same conclusion.

The discovery of Mendel's work gave tremendous impetus to the study of heredity and many new fields have been explored. Genetics, from a humble beginning in the hands of Mendel, **'Father of genetics'**, has risen to a tremendous height and it has helped us to understand many of the problems of inheritance, and explain the complicated process of organic evolution.

L. Pasteur (1822-95)

The French scientist, Louis Pasteur is regarded as the founder of modern bacteriology. He discovered causative organisms of wine fermentation and the theory of **'Spontaneous Origin'** was finally disposed. He and Joubert identified first pathogenic anaerobe *Vibrion septique*, and established distinction between aerobic and anaerobic bacteria. He experimented with attenuated virus in chicken cholera, confirmed Koch's work on *Anthrax*, and was pioneer in developing attenuated bacterial culture with Chamberland and Roux for therapeutical use. They discovered rabies virus in the blood and first used attenuated vaccine as a prophylactic (1885). The Pasteur Institute' in Paris was established for his investigations.

H. Spemann (1860-1941)

Hans Spemann, an eminent German experimental embryologist performed a series of experiments on differentiation of eggs during cleavage and gastrulation. He tightly tied newt's egg through its first cleavage furrow with human hair (1901-1903). During subsequent division the upper half produced a miniature whole embryo whereas the other half developed into an undifferentiated ball of living cells. Spemann concluded that during first cleavage, the upper half of the egg becomes equipped with quality to form an embryo whereas the other half lacks it. In 1918, he made another experiment on triton egg. He transplanted a small piece of ectoderm of one early gastrula to another gastrula of same age. The transplanted piece developed in accordance with at the new position. But the dorsal lip of the blastopore of amphibia grafted in similar manner, developed an embryo-like body with neural tube, notochord and somites. Spemann inferred that the dorsal lip of the blastopore was already determined as centre of differentiation. He called the dorsal lip of blastopore as **'organizer'**. The discovery of concept of organizer action helps the experimental embryologists in many ways in recent times.

T. H. Morgan (1866-1945)

Thomas Hunt Morgan, a Nobel Laureate was born at Kentucky. He entered Johan Hopkins University—named after the great biochemist—in 1886, after completing his undergraduate studies at Kentucky College and studied Embryology. He got Ph.D. degree in 1890 and then joined Brym Mowr as Associate Professor of Biology in 1891. Then he joined as a Professor of Experimental Biology at Columbia University in 1909 and maintained there till his death in 1945.

Morgan was interested in embryology and the process of differentiation at the beginning of his research career but later he turned to genetics. The work of Morgan and his distinguished associates are major contribution in heredity after Mendel. Morgan formulated many genetic principles, which accelerated investigations on hereditary mechanism.

2 LIVING FORMS: PHYSICAL AND CHEMICAL BASIS OF LIFE

An animal body (plant also) is a very complex unit. In the long process of evolution, unicellular organisms have been transformed into highly specialised, multicellular ones. In higher forms, where the body of an individual is made of numerous cells, a division of labour has set in, and different functions are carried out by different organs—all leading to the survival and perpetuation of the individual. The cells, which built up the body are considered as **structural and functional units** of the whole, and all activities of the individual are traceable to them. The cell, which constitutes the body of an acellular (unicellular) organism, is necessarily much more complex than that of a multicellular one, since it has to perform all the activities of life, which are carried out collectively by numerous cells present in the latter.

An intimate relationship exists between every organism and its environment, be it a free living or commensal, symbiont or parasitic. It is a two-way process; the environment influences the organism and the organism influences the environment. It is from the environment, the organism obtains its total requirements and also to some extent protection. Adaptation in all living objects are aimed to successful exploitation of its environment for its betterment. Utilisation of matters available in the surroundings, and the myriad activities associated with life are entirely dependent on certain chemical reactions.

☒ Metabolism

The wide variety of chemical reactions which occur in a living body are collectively called metabolism. All cells are being constantly supplied with new matters, changing them chemically in a number of ways (chemical reactions). New cells are built, the large molecules of carbohydrates, fats and proteins are converted into simpler ones, and the potential energy in the complex molecules is converted into kinetic energy and heat (heat is also a form of kinetic energy, but it deserves special mention). The energy flow is a continuous process within a cell, from one cell to other and from one organism to other. The study of energy transformation in living organisms is called **bioenergetics**.

Two opposite processes constitute the metabolism in living organisms:

1. **Anabolism** (Gr. *ana* = up + *bole* = throw). Simpler substances combine to form complex compounds, resulting in storage of energy and production of new

cellular materials. It is a **synthetic process.**

2. **Catabolism** (Gr. *cata* = down + *bole* = throw). The complex structures are broken down to simpler forms and the energy is released. This is a **biological degradation process.**

Both the processes are continuous and interdependent. Catabolism releases energy, which is essential for the synthesis of most molecules. In general, the anabolic reactions are more in the period of growth, while catabolic reactions predominate in starvation, diseases and senescence.

☒ Atoms and molecules

To understand the living matter and its requirements some knowledge on the basic principles of physics and chemistry is necessary. The constituents of the universe are **matter** and **energy**.

☒ Matter and energy

Matter and energy are related by Einstein equation, $E = mc^2$, where E = energy, m = mass and c = the velocity of light, which is a constant. Matter and energy are separate and distinguishable*. Matter occupies space and has mass, and the energy, which lacks both—is capable to do work. It can produce a change or motion in matter.

ATOMS

Regardless of the states—solid, liquid and gas, a matter is composed of units, the atoms. An atom is the smallest particle of an element which can take part in chemical reactions. Till now, 112 elements have been identified. An element is composed of atoms, all of which have the same numer of **protons** in the nucleus and the same number of **electrons** circling the orbitals. The hydrogen atom is smallest and the uranium atom is largest. In an atom a number of subatomic particles are present, of which three are important. These particles are arranged in a central core and in one or more orbits (Fig. 2.1). Electrons are extremely small and negatively (–) charged, whirling in the orbits (Fig. 2.2) around the nucleus. Protons, extremely heavy and with a positve (+) charge, and **neutrons** with no electric charge are present in the nucleus. Each type of atom has characteristic number of electrons and characteristic number of protons and neutrons. The number of protons in the nucleus equals the number of electrons in the orbit, and an atom is neutral. The number of

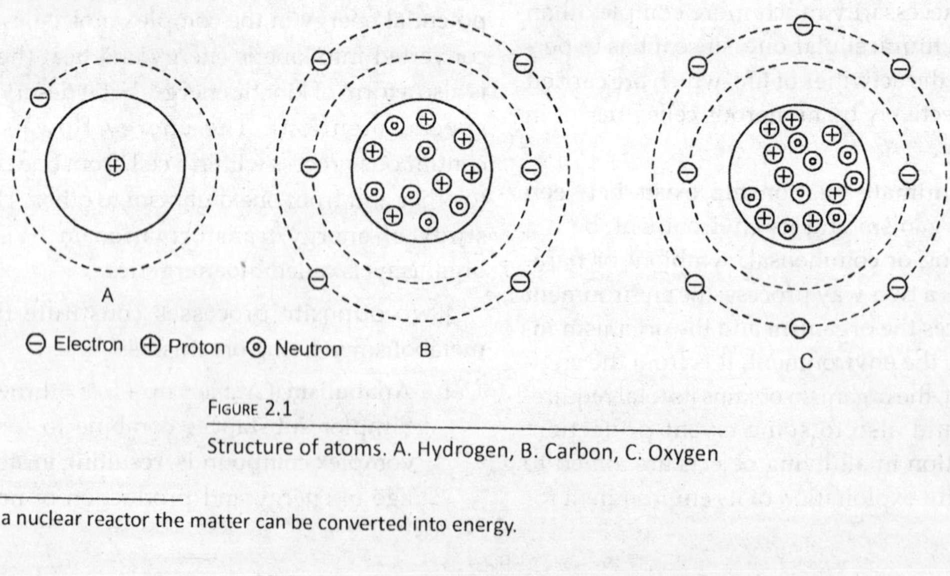

FIGURE 2.1
Structure of atoms. A. Hydrogen, B. Carbon, C. Oxygen

* In a nuclear reactor the matter can be converted into energy.

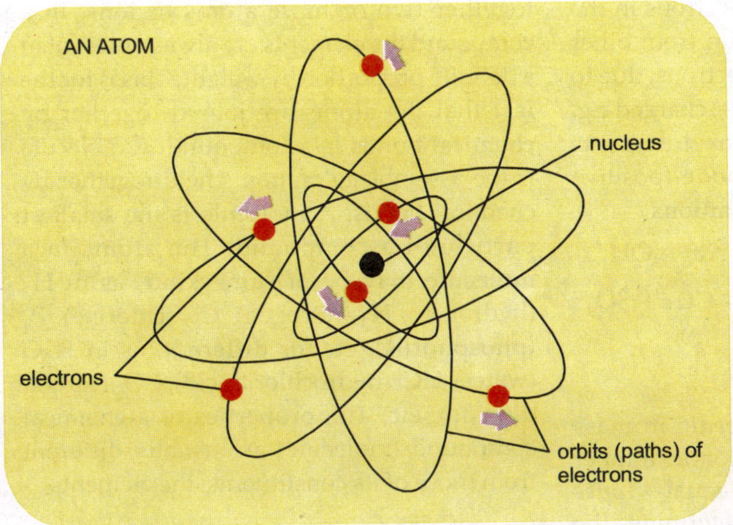

FIGURE 2.2
An atom. Orbits (paths) of electron

electrons is the **atomic number** and the sum of the numbers of protons and neutrons is the **mass number*** or **atomic** weight.

Element symbol	Atomic weight	Mass number	Atomic number	Number of protons	Number of neutrons	Number of electrons
Hydrogen $_1H^1$	1.008	1	1	1	0	1
Carbon $_6C^{12}$	12.011	12	6	6	6	6
Sulphur $_{16}S^{32}$	32.066	32	16	16	16	16

ELEMENTS

An element is composed of atoms, all of which have the same number of protons and electrons. For convenience, a symbol has been assigned to each of the elements, which usually is the first letter of the name of the element—H for hydrogen, C for carbon, O for oxygen, S for sulpur, and so on. In the symbols of elements, the first letter of which is the same as in other element, a second letter is added, Ca for calcium, Co for cobalt, Cl for chlorine, etc.

IONS

Ions are electrically charged particles. The chemical properties of an element depend primarily on the number and arrangement of electrons in the outer orbit of the atom and to some extent on the number and arrangement of electrons in the inner orbit. These, in turn, depend on the number of protons and neutrons in the nucleus of the atom.

The number of electrons in the outer orbit varies from zero to eight. The element is chemically inert if the number is zero or eight, e.g., helium and argon.

When the number of electrons in the outer orbit is less than eight, the atom tends to lose or gain electrons. Since the number of proton does not change, the loss or gain of electrons results in an atom with a net positive (+) or negative (–) charge. Such electrically charged atoms are termed ions.

Atoms with one to three electrons in the outer orbit tend to lose electron to other atoms. Because of the excess protons, due to this loss, the atom is positively (+) charged, e.g. Na^+ (sodium ion), Ca^+ (Calcium ion), etc. Since these atoms migrate to the **Cathode** (negative electrode) of an electrolytic cell, they are **Cations.**

* The mass number is a whole number, since the number of protons or neutrons cannot be fractions.

Atoms with five to seven electrons in the outer orbit tend to gain electron from other atoms. Because of the excess electrons, due to the gain, the atom is negatively (–) charged e.g. Cl^- (Chlorine ion), OH^- (hydroxyl ion), etc. These atoms migrate to the **anode** (positive electrode) and they are termed **anions**.

$$NaCl \rightleftharpoons Na^+ + Cl^-, \; NaOH \rightleftharpoons Na^+ + OH^-$$

$$H_2SO_4 \rightleftharpoons 2H^+ + SO_4^-, CuSO_4 \rightleftharpoons CU^+ + SO_4^-$$

ISOTOPES

The atoms of an element differing in mass number but having the same atomic number are termed Isotopes (Gr. ***isos*** = equal + ***topes*** = place). The atomic number determines the chemical properties of an element. The isotopes of an element are identical in chemical properties but differ in physical properties, viz., density, rate of diffusion, etc. The isotopes of an element can be identified by differences in physical properties.

The differences in the atomic masses of various isotopes of an element is due to unequal number of neutrons in the nucleus. The number of proton is constant. Chlorine has two isotopes of atomic masses 35 and 37, one containing 17 protons and 18 neutrons in the nucleus, while the other contains 17 protons and 20 neutrons. The atomic number in each case is 17.

Isotopes are extensively used in the study of metabolism. Ordinary carbon (^{12}C) in sugar has been replaced by radioactive carbons (^{11}C or (^{14}C) or heavy carbon (^{13}C). The labelled substances (tracers) are fed or injected into animal or a plant or cells are incubated in solution containing the **tracer**. The course of the labelled substance is traced in the organism or in the cell and the sequence of chemical reactions are determined.

CHEMICAL COMPOUNDS

Substances present in living matters are mostly chemical compounds, not elements. A chemical compound is formed by joining together two or more atoms or ions. In a compound the elements are always present in a definite proportion by weight. This is for the fact that the atoms are joined together by **chemical bonds** in a compound. A molecule is the assemblage of atoms held together by chemical bonds. A molecule is the smallest particle of a compound. The atoms in a molecule may be of same kind, as in H_2 (hydrogen), N_2 (nitrogen), Cl_2 (chlorine), P_4 (phosphorus), etc. or different, as in H_2O (water), HCl (hydrochloric acid), CO_2 (carbon dioxide), etc. The properties of a chemical compound (molecule) are usually different from those of its constituents, the elements.

⊠ Chemical bonds

The force which holds the atoms of elements together in a molecule is termed chemical bond. **Ionic, covalent** and **hydrogen bonds** are present in organic molecules. Ionic and hydrogen bonds are weak, formed and cleave easily, while the covalent bonds are strong.

Ionic bond. It results from the transfer of electrons from electropositive elements to electronegative elements.

$2Na^+ + Cl_2^- = 2NaCl \,||\, Ca^+ + Cl_2^- = CaCl_2$
(sodium chloride) (calcium chloride)

Covalent bond. It consists of mutual sharing of a pair of electrons between two electronegative (deficit of electrons) elements. The bond is strong. Its formation is endergonic and requires the input of energy. Both the formation and cleavage of the bond require enzymatic reaction within the cell.

```
  H
  :
H..N..H ‖ H..C::C..H
(Ammonia)  (Acetylene)
```

Hydrogen bond. In it a hydrogen atom is shared between two atoms. Under certain conditions an hydrogen atom is attracted by strong forces of two atoms and it acts as a bond between them. This is hydrogen bridge or hydrogen bond. It is really a covalent bond.

Hydrogen bond is much weaker than ionic or covalent bond having a bond energy of a few kcal, only; 0.7 kcal for F–H–F, 4.5 kcal for O–H–O as against 100 kcal for covalent bond.

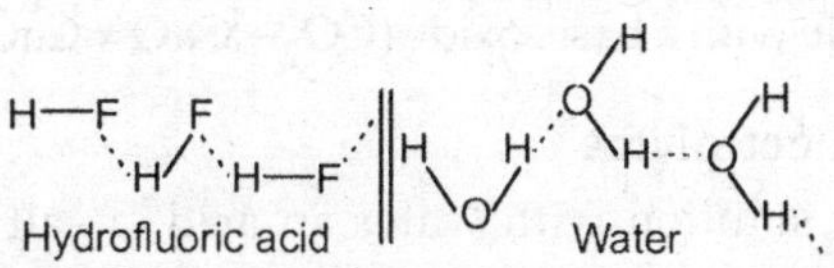

INORGANIC AND ORGANIC COMPOUNDS IN LIVING SYSTEM

For the normal functioning of animal body a constant in and outflow of substance is a necessity. The animal receives water, oxygen, minerals, carbohydrates, proteins, lipids and vitamins from outside and releases toxic wastes, chiefly carbon dioxide and nitrogenous wastes, the products of metabolism. The proportion of the substances involved in the function, of cells is always nearly constant. In addition, certain substances, taking part in chemical reactions, are present in small amounts in the body.

⊠ Inorganic Compounds

The inorganic compounds—water, carbon dioxide, acids, bases and salts are always present in living body.

⊠ Water

Water (H_2O) is an important constituent of a cell. The percentage of water in a vertebrate cell varies from 30 (bones) to 80 in brain cells. The water content is higher in embryonic and young cells and decreases with age. About 75% of our body weight is water. It is about 90% in jelly fish. The important functions of water are:

(a) Most of the chemicals present in cells dissolve in water. Chemical reactions in cells take place only in a soluble state.
(b) Removal of waste products of metabolism is possible only in solution.
(c) Heat absorbing capacity of water with a minimal change in its temperature is great and thus protects the organism against sudden thermal change.
(d) The latent heat of evaporation of water is quite high and the mammalian body can dissipate excess heat by sweating.
(e) High heat conductivity of water helps in even distribution of heat throughout the body.
(f) Water acts as a lubricant and facilitates rubbing of one organ against another and movement of bones in joints.

⊠ Carbon dioxide

Carbon dioxide, a harmful gas, is released in tissue in respiration. Solubility of carbon dioxide (CO_2) in water is about 30 times greater than that of oxygen (O_2). When dissolved in water it forms carbonic acid (H_2CO_3). Carbon dioxide combines with cations—sodium or calcium, to form salts, carbonates and bicarbonates. The gas is of considerable importance in body fluids and blood (p. 104).

⊠ Acid

An acid is a compound having replaceable hydrogen, a part or whole of which may be replaced by a metal or a group of elements resulting in the formation of a salt. An acid in solution turns blue litmus red; is sour in taste; releases hydrogen ions (H^+) in aqueous solution and reacts vigorously with bases (alkalis). Hydrochloric acid (HCl), an inorganic acid, is secreted by the glands in the stomach of vertebrates. Lactic acid ($C_3H_6O_3$), an organic acid, is present in sour milk.

⊠ Base

A base is a compound, usually oxide or hydroxide of a metal or a group of elements equivalent to the metal. A base can neutralize an acid producing a salt and water only. A base is soluble in water and release hydroxyl ions (OH^-); aqueous solution conducts electricity; of soapy nature; turns red litmus blue, phenlophthalein pink, methyl orange yellow. Common inorganic bases are sodium hydroxide (NaOH) and ammonium hydroxide (NH_4OH).

☒ Hydrogen ion concentration: pH

pH is the most convenient method of expressing hydrogen ion concentration in a solution. It is a **figure**, not the state of the substance, **i.e. neutral, acid** or **alkaline**.

Though commonly expressed as $2H_2O \rightleftharpoons H_3O^+ + OH^-$, the equation for dissociation of water is usually written $H_2O = H^+ + OH^-$, from which we have $\frac{[H^+][OH^-]}{[H_2O]} = K_1$.

H_2O is constant so $[H^+]\,[OH^-] = K_2$. This constant is 10^{-14}. At the neutral point $[H^+] = [OH^-]$ each equals 10^{-7}, equivalents per litre ($[H^+] \times [OH^-] = 10^{-7} \times 10^{-7} = 10^{-14}$). When the concentration of hydrogen ions is greater than this, the solution is **acid**, when less, **alkaline**.

The term most used to indicate the hydrogen ion concentration is the pH, its negative logarithm.

Thus, at neutrality this is 7 and the pH ranges from O (N) to 14. For a fully dissociated strong acid, viz. hydrochloric, a 0.1 N solution has a pH nearly 1.0; 0.01 N about 2.0. For a strong base, viz. sodium hydroxide, a 0.1 N solution has a pH 13.0; a 0.01 N almost 12.0. The pH of animal cells vary with the groups, within a narrow limit. In mammals it is normally around 7.0; in fishes and birds around 6.0. A large change in pH is inconsistent with life and such changes are prevented by chemical reactions with certain substances know as **buffers**.

☒ Salt

A salt is a compound formed in chemical reaction between an acid and a base. In the reaction the hydrogen atoms present in an acid is partially or completely replaced by a metal or a group of elements acting like a metal. Salts are of three types—**Normal salts** (all the replaceable hydrogen atoms present in the molecule of an acid are replaced); **Acid salts** (partial displacement of replaceable hydrogen atoms) and **Basic salts** (produced by the reaction of an acid with higher proportion of base than necessary). Sodium chloride is formed by the direct union of a metal and a nonmetal ($2Na + Cl_2 = 2NaCl$); calcium carbonate is formed by the union of an acidic oxide with a basic oxide ($CO_2 + CaO = CaCO_3$).

☒ Electrolytes

In a solution with water an acid, a salt or a base separates into its constituent ions. The charged particles can conduct electric current and have been named as electrolytes. Substances which do not dissociate in solution with water, e.g. sugar, alcohol, etc. cannot conduct electric current and are known as **non-electrolytes**.

☒ Organic Compounds

The major types of organic matters present in living cells are carbohydrates, proteins, lipids and nucleic acids. An animal cell contains about 80 per cent water, 1 per cent carbohydrates, 12 per cent proteins, 2 per cent nucleic acids, 5 per cent lipids and a fraction of 1 per cent of steroids and other substances.

☒ Carbohydrates

The elements carbon (C), hydrogen (H) and oxygen (O) combine in the ratio 1C : 2H : 1O to form carbohydrate. In living organisms the carbohydrates are present in the form of sugar (plants and animals), glycogen (animals), starch and cellulose (plants). Carbohydrates are principal source of energy in the majority of organisms but not essential to the body or the cells, as the cells could obtain their energy from a mixture of protein or fat. In addition to being primary energy source carbohydrates provide carbon skeleton necessary in the manufacture of basic components of protoplasm.

Carbohydrates in its simplest form, monosaccharides are water soluble and present in living cells in small amount, is a ready source of energy and for this it is considered as first fuel.

The chemical structure of glucose, galactose and fructose, all monosaccharides are different though their chemical formula ($C_6H_{12}O_6$) is same. Their difference in chemical structure (Fig. 2.3) and properties are little but they differ considerably in intensities of sweetness and physiological activities in the animal body.

Simple sugars unite to form large molecules of carbohydrates. They are disaccharides (two monosaccharides), trisaccharides (3), tetrasaccharides (4), penta, hexa ... polysaccharides. The phenomenon is known as **polymerisation**. The polysaccharide may be a single long chain or a branched one. The carbohydrates, di- to poly are collectively known as **oligosaccharides**.

The joining of the monosaccharides is effected by losing one molecule of **water**. The formula of a disaccharide is $C_{12}H_{22}O_{11}$; that of a polysaccharide is ($C_6H_{10}H_5$)n. The most common double sugar, **sucrose** (table sugar), is formed by joining one molecule of glucose and one molecule of fructose (Fig. 2.4). Two molecules of glucose join to form **maltose; lactose** is composed of glucose and galactose.

Starch $(C_6H_{10}O_5 . H_2O)_n$ is normally insoluble in water and is formed by joining many molecules of single sugars.

A molecule of **glycogen**, the animal starch, is made up of about thirty thousands of D-glucose molecules joined together by glycosidic bonds, the majority of which are α-1,4 glycosidic bonds. There are branches in the chain at intervals of about 10 units involving α-1,6 linkages. Animals store carbohydrate in the form of glycogen, which is transformed into glucose in needs and is used as a source of energy. All cells can store glycogen. In mammals the liver and muscle cells can store **glycogen** up to 15-20 per cent.

Cellulose is insoluble in water and is formed by the joining of thousands of molecules of glucose. It is very tough and in nature hydrolysed with the enzyme **cellulase** produced by microorganisms. Due to very compact arrangement of microfibrils in a cellulose fibre, enzymic hydrolysis is restricted only to the ends of the fibres and consequently the rate of biological degradation is very slow. The major part of plant cell wall is made up of cellulose.

⊠ Metabolism of Carbohydrate

Carbohydrate metabolism is the metabolism of simple sugar–glucose and substances related to glucose. Glucose is a ready source of chemical energy for animals. More than 50% of energy requirements of animal body is obtained from carbohydrate. Oxidation of glucose for energy production occurs through the pathways–glycolysis and tricarboxylic acid cycle (Citric acid or Krebs cycle).

⊠ Glycolysis

The conversion of glucose or glycogen to pyruvate and lactate is known as glycolysis or Embden-Meyarhof Pathway (EMP). Under anaerobic condition pyruvate is converted into

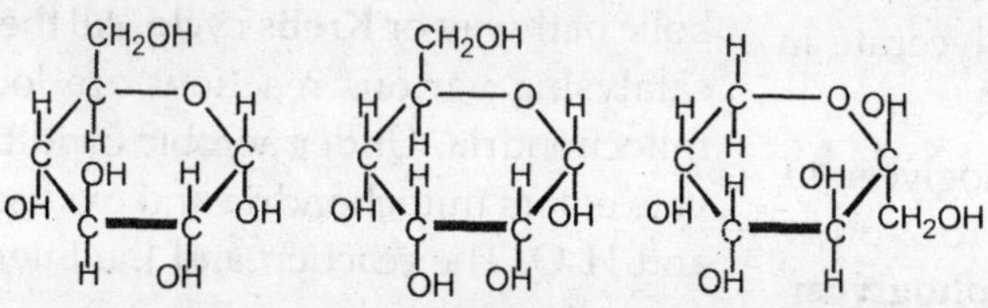

D-GLUCOSE D-GALACTOSE D-FRUCTOSE

FIGURE 2.3

Structural formula of Glucose, Galactose, Fructose

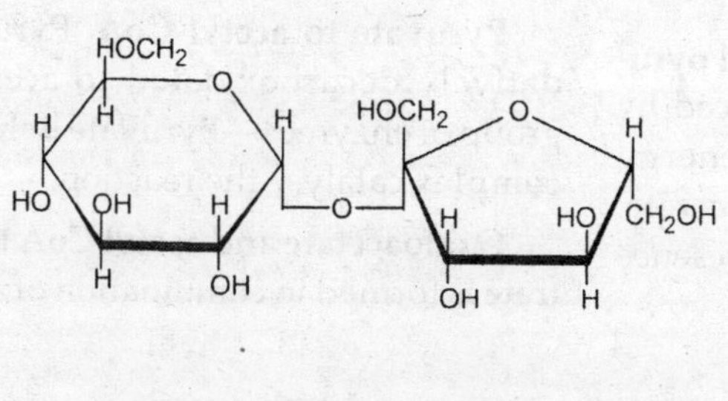

FIGURE 2.4

Sucrose (double sugar; table sugar), Structural formula. A glucoside linkage joins a glucose and a fructose molecule

lactic acid; under aerobic condition pyruvate enters mitochondria and completely oxidised to CO_2 and H_2O.

☒ Embden-Meyerhof Pathway

The reactions and the intermediates formed (Fig. 2.5) in the cycle are:

Glucose to glucose-6-phosphate. Phosphorylation of glucose to glucose-6-phosphate is effected by the enzyme **hexokinase.** ATP acts as phosphate donor in the presence of magnesium ion.

Glucose-6-phosphate to fructose-1, 6-diphosphate. The enzyme **phosphohexose isomerase** catalyze the reaction.

Fructose-6-phosphate to fructose-1, 6-diphosphate. In the presence of magnesium ions and the enzyme **phosphofructokinase,** fructose-6-phosphate is further phosphorylated to fructose-1, 6-diphosphate by ATP.

Fructose-1, 6-diphosphate to triose phosphates. The enzyme **aldolase splits** Fructose-1, 6-diphosphate to two triose phosphates-glyceraldehyde -3- phosphate and dihydroxy-acetone phosphate.

Glyceraldehyde-3-phosphate to 1, 3-diphosphoglycerate. Glyceraldehyde -3-phosphate is oxidised to 1, 3-diphosphate glycarate, containing a high energy phosphate bond. The **Glyceraldehyde-3-Phosphate** dehydrogenase catalyse the reaction.

1, 3-diphosphoglycerate to 3-phosphoglycerate. Phosphoglycerate Kinase catalyse the reaction and 1, 3-diphosphoglycerate in converted into 3-phosphoglycerate.

3-phosphoglycerate to 2-phosphoglycerate. 3-phosphoglycerate is converted to 2-phosphoglycerate by the enzyme **phosphoglyceromutase.**

2-phosphoglycerate to phosphoenol pyruvate. phosphoenol pyruvate is produced by dehydration and redistribution of energy within the molecule of 2-phosphoglycerate. **Enolase** catalyse the reaction in the presence of magnesium ions.

Phosphoenol pyruvate to pyruvate. Phosphoenol pyruvate is inversably converted to pyruvate. **Pyruvate Kinase** in presence of magnesium ions transfer the high energy phosphate to ADP.

Pyruvate to lactate. Pyruvate is the normal end product of glycolysis in aerobic condition. In anaerobic condition pyruvate is reduced to lactate by NADH in the reaction catalysed by **lactate dehydrogenase.**

☒ Energetics

(One mole of glucose gives two moles of pyruvate)

Reaction where ATP generated. (under aerobic condition) Glyc-3 phosphate —

→ 1, 3-Diphosphoglycerate	2 × 3 = 6
1, 3-Diphosphoglycerate	
→ 3-phosphoglycerate	2 × 2 = 4
Phosphoenol pyruvate	
→ Pyruvate	2 × 1 =2
Total	**12**

Reaction where ATP is consumed:

Glucose to glucose-6-phosphate	1
Fructose-6-phosphate to fructose-1, 6-**Diphosphate**	1
Total	2
Net ATP formed	8

☒ Citric acid cycle (Tricarboxylic acid cycle)

Citric acid cycle is the common pathway for the metabolism of carbohydrates, fats and proteins. The cycle is also known as common metabolic pathway or **Krebs cycle.** All the enzymes catalysing various reactions are localised in mitochondria. Under aerobic condition pyruvate enters mitochondria and oxidised to CO_2 and H_2O. The reaction and the intermediates formed in the cycle are (Fig. 2.6)—

Pyruvate to acetyl-CoA. Pyruvate is oxidatively decarboxylated to acetyl–CoA. A group of enzymes—**Pyruvate dehydrogenase complex** catalyse the reaction.

Oxaloacetate and acetyl CoA to citrate. Citrate is formed in combination of a acetyl CoA

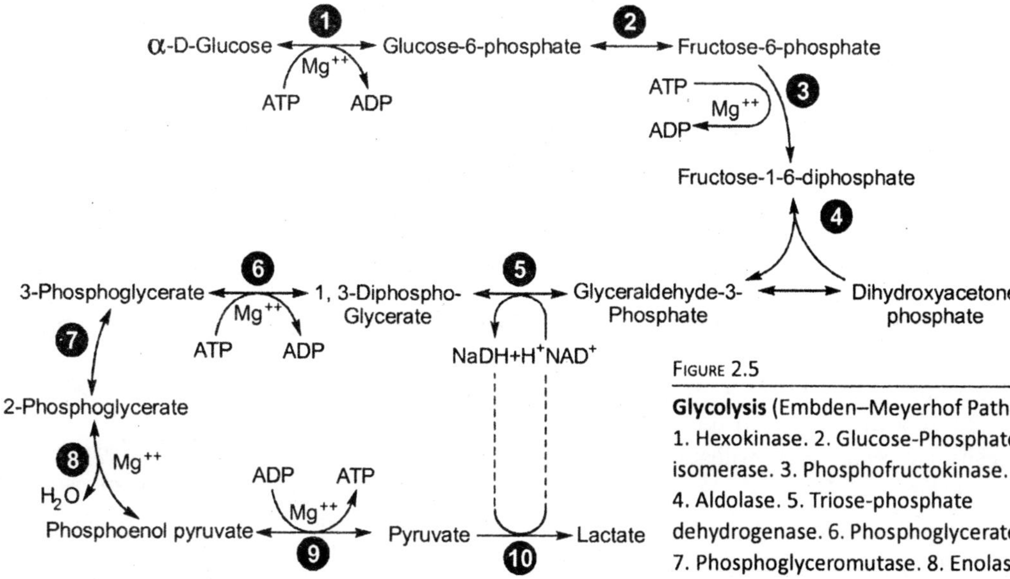

FIGURE 2.5

Glycolysis (Embden–Meyerhof Pathway). 1. Hexokinase. 2. Glucose-Phosphate isomerase. 3. Phosphofructokinase. 4. Aldolase. 5. Triose-phosphate dehydrogenase. 6. Phosphoglycerate. 7. Phosphoglyceromutase. 8. Enolase. 9. Pyruvate Kinase. 10. Lactate dehydrogenase.

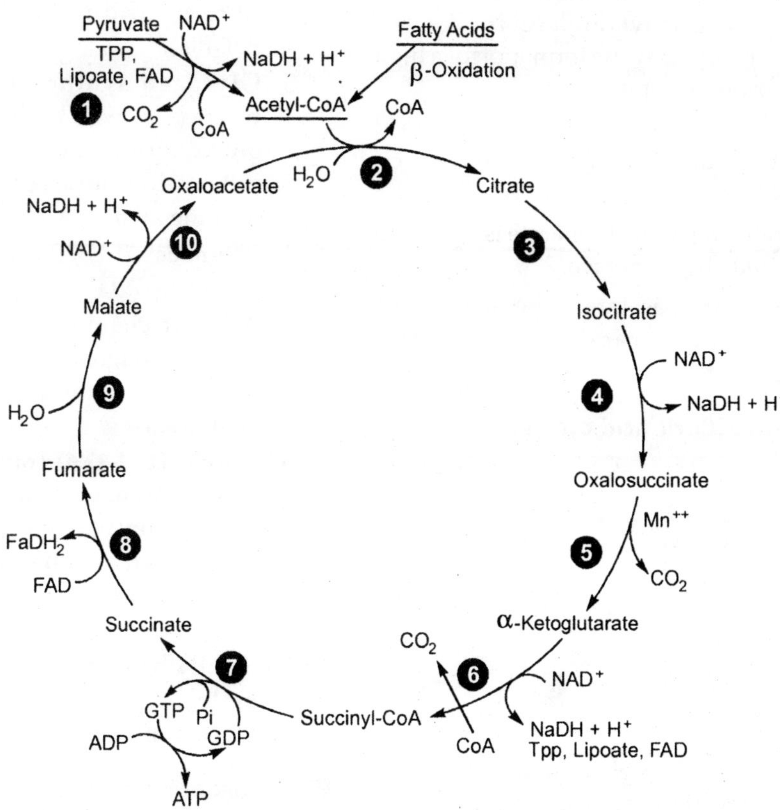

FIGURE 2.6

Citric acid cycle. 1. Pyruvate dehydrogenase complex. 2. Citrate synthatase. 3. Aconitase. 4. Isocitrate dehydrogenase. 5. Isocitrate dehydrogenase. 6. Ketoglutarate dehydrogenase complex. 7. Succinic thiokinase. 8. Succinyl thiolase. 9. Fumarase. 10. Malate dehydrogenase.

and oxaloacetate. **Citrate synthetase** catalyses the reaction. A molecule of water is consumed and CoA released.

Citrate to isocitrate. The conversion of citrate to isocitrate takes place in the steps–catalysed by **aconitase**. First the citrate is dehydrated to cis-aconite, which is rehydrated to isocitrate.

Isocitrate to oxalosuccinate. Dehydrogenation by **isocitrate dehydrogenase** causes the conversion.

Oxalosuccinate to α-Ketoglutarate. **Isocitrate dehydrogenase** in the presence of magnesium ions converts oxalosuccinate to α-Ketoglutarate.

α-Ketoglutarate to succinyl CoA. α Ketoglutarate dehydrogenase catalyze the reaction oxidatively.

Succinyl CoA to succinate. The reaction is catalysed by **succinyl thiolase**. The conversion is accompanied by the formation of a high energy phosphate bond.

Succinate to fumerate. Fumerate is formed by dehydrogenation in the presence of **succinate dehydrogenase.**

Fumerate to malate. Fumerate is hydrated to form malate by **fumerase.**

Malate to oxaloacetates. The dehydrogenation reaction is catalysed by **malate dehydroganase.**

⊠ Energetics (Citric acid cycle)

Reaction	ATP formed through oxidation of pyruvate
Pyruvate → Acetyl-CoA	2 × 3 = 6
Isocitrate → Oxalosuccinate	2 × 3 = 6
α-Ketoglutarate → succinyl-CoA	2 × 3 = 6
Succinyl-CoA → succinate	2 × 1 = 2
Succinate → fumerate	2 × 2 = 4
Malate → Oxaloacetate	2 × 3 = 6
Net ATP formed	30

Total number of ATP formed per mole of glucose:

Glycolysis	8
Citric acid cycle	30
Total	38

⊠ Glycogenesis

The formation of glycogen in the animal body is glycogenesis. It is synthesized chiefly in liver and muscles and stored there. On an average in a human body the amount of glucose stored is 108g in liver and 245g in muscles. Glucose units are added to the nonreducing units of the preexisting glycogen by α-(1, 4) and α-(1, 6) linkage to form glycogen under combined act of glycogen synthesis and branching enzymes.

⊠ Synthesis of glycogen

1. Glucose-6-phosphate is formed by phosphorylation of glucose by **hexokinase**, which isomerizes to form glucose-1-phosphate.
2. **UDP-glucose pyrophosphorylase** combines UTP with glucose-1-phosphate to form uridine diphosphoglucose (UDPG).
3. **Glycogen synthetase** catalyses the reaction in which UDPG transfer its glucose to the nonreducing end of a preexisting polysaccharides **primer**, forming an α -1, 4-glucosidic bond.
4. The released UDP can be reconverted to UTP by ATP (Fig. 2.7).
5. A linear chain of glucose molecules linked through a α-1, 4 glucosidic bond is formed by the action of **glycogen synthases.**
6. **Amylo (1, 4-1, 6) transglycerylase**, a branching enzyme catalyses cleavage of an α-1, 4-linkage near the growing end of the chain and the fragment is transferred to form an α-1, 6-linked branch point on the lower chain (Fig. 2.8).
7. Finally a highly branched glycogen molecule containing thousand of glucose units is formed.

⊠ Glycogenolysis

Breakdown of glycogen to glucose is glycogenolysis. Glycogenolysis occurs in the storage sites of glycogen liver and muscles. In liver

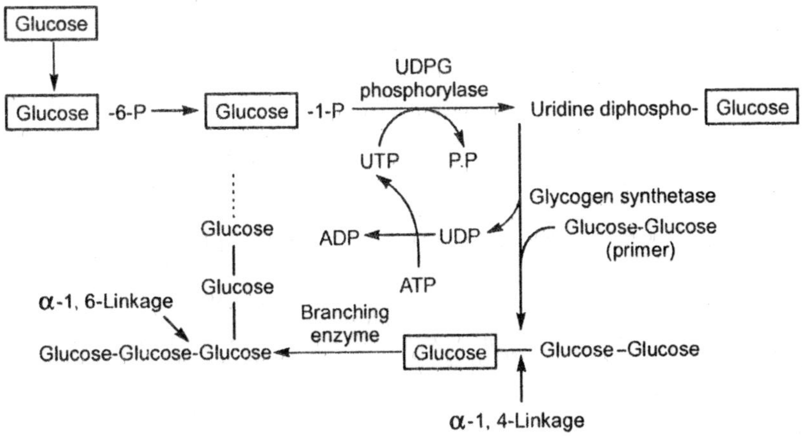

FIGURE 2.7

Glycogenesis

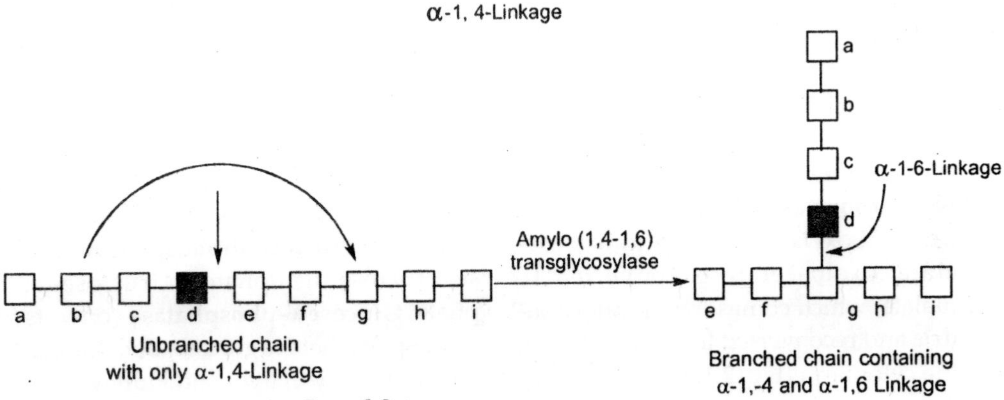

FIGURE 2.8

Action of branching enzymes

the end product is glucose and in muscles the end product is lactic acid.

Breakdown of Glycogen :

1. Breakdown of glycogen is catalysed by **phosphorylase** and a debranching enzyme **amylo-1, 6-glucosidase**.
2. The terminal glucose unit from the nonreducing end of the glycogen molecule is cleaved from α-1, 4 glucosidic linkage by the enzyme **phosphorylase** and converts it to glucose-1-phosphate.
3. The branching enzyme catalyses the hydrolysis of α-1, 6-glucosidic linkage and the glucose is released (Fig. 2.9).

☒ Gluconeogenesis

Gluconeogenesis occurs when carbohydrate reserve in the body is exhausted. Glucose formed by the process maintains normal blood glucose level. The formation of glucose from noncarbohydrate precursors—glycogenic amino acid, intermediates of TCA cycle, glycerol, pyruvate, lactate etc. is gluconeogenesis, Gluconeogenesis occurs in liver and kidneys.

1. **Lactate to glucose.** In aerobic glycolysis in muscles, lactic acid is the major breakdown product. Muscle lactate is transported to the liver by blood, where it is converted to glucose and glycogen by enzymes involved in gluconeogenesis. The conversion of muscle lactic acid to glucose in liver and its reentry into the muscles is known as **Cori cycle**.

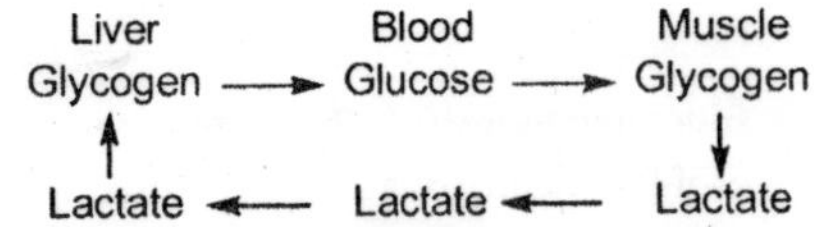

Cori cycle

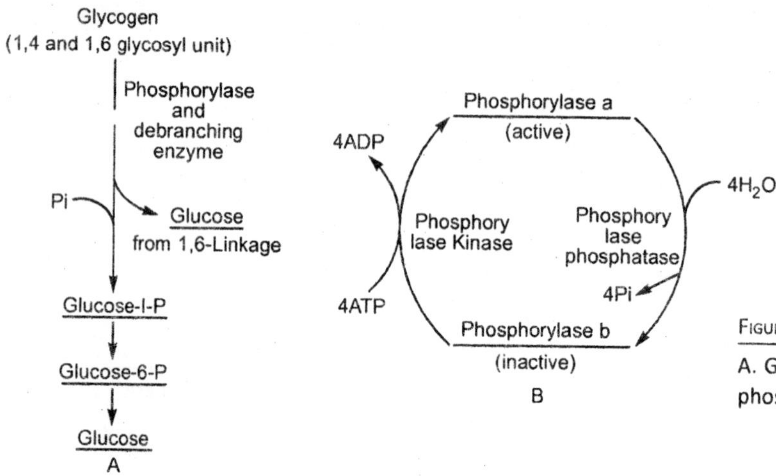

FIGURE 2.9

A. Glycogenolysis. B. Activation of phosphorylase

2. **Amino acid to glucose.** Glucogenic, amino acids are converted to citric acid cycle intermediates or **pyruvate. Pyruvate carboxylase** and ATP in mitochondria carboxylate pyruvate to oxaloacetate; oxaloacetate is converted to malate, which comes out of mitochondria and reconverted to oxaloacetate in the cytoplasm (Fig. 2.10).

Phosphoenol pyruvate carboxylase and GTP decarboxylate oxaloacetate and phosphoenol pyruvate is formed. Phosphoenol pyruvate is converted to fructose-1, 6-diphosphate in tissues. **Fructose diphosphatase** hydrolyses fructose disulphate to fructose-6-phosphate. **Glucose-6-phosphatase** converts fructose-6-phosphate to glucose. Both the enzymes are found only in liver and kidneys.

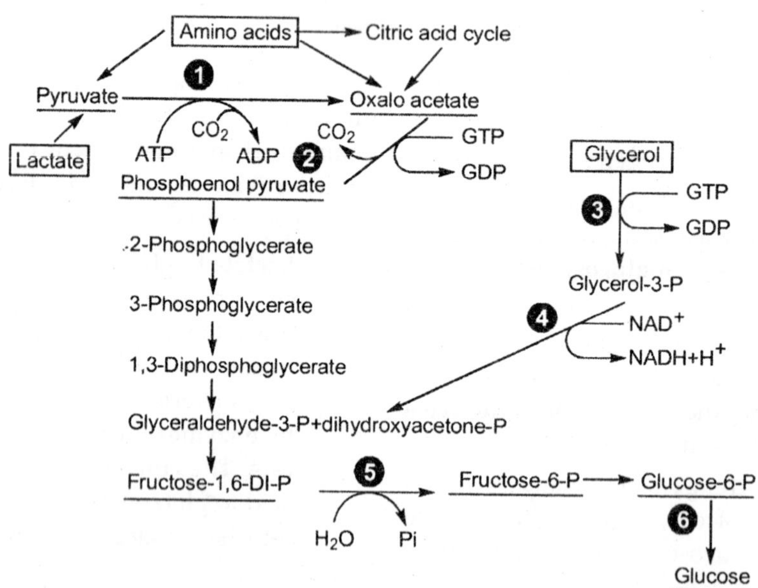

FIGURE 2.10

Gluconeogenesis. 1. Pyruvate carboxylase. 2. PEP carboxy kinase. 3. Glycerol kinase. 4. Glycerol-P-dehydrogenase. 5. Fructose diphosphatase. 6. Glucose-6-Phosphatase

Glycerol to glucose. Glycarol, the breakdown product of triglycerides in liver yields glucose. Glycerol is initially phosphorylated by ATP followed by reduction to form dehydroxyacetonephosphate. The product enters the pathway of gluconeogenesis (Fig. 2.10)

Electron transfer

NADH and FADH are reoxidised by electron transport and oxidative phosphorylation, and trap the energy released as ATP.

The oxidation of a molecule involves the loss of electrons. The reduction of a molecule causes a gain of electrons. Since electrons can neither be created nor destroyed in a chemical reaction, oxidation of one molecule causes reduction in the other (**oxidation-reduction reaction**). The Oxidation-reduction process involves the transfer of electrons.

$$NADH + H^{+} + \frac{1}{2}O_2 \rightleftharpoons NAD^{+}H_2O$$

In oxidation of NADH to NAD^+, NADH loses one electron. When the molecular oxygen is reduced to water it gains electrons. Oxidative phosphorylation is the major source of ATP in living cells. In eukaryotes, electron transport and oxidative phosphorylation occur in the inner membrane of mitochondria. In the process the NADH and FADH produced in the citric acid cycle (glycolysis) are reoxidised and trap the energy released as ATP. In prokaryotes the electron transport and oxidative phosphorylations are located in the plasma membrane.

Abbreviations

Acyl-CoA = Acyl Coenzyme A
Acetyl-CoA = Acetyl Coenzyme A
ADP = Adenosine diphosphate
AMP = Adenosine monophosphate
ATP = Adenosine Triphosphate
FAD = Flavin adenine dinucleotide
FADH = Flavine adenine dinucleotide (reduced form)
GTP = Guanosin Triphosphate
NAD^+ = Nicotinamide adenine dinucleotide
NADH = Nicotinamide adenine dinucleate (reduced form)
NADP = Nicotinamide adenine dinucleate phosphate
NADPH = Nicotinamide adenine dinucleate phosphate (reduced form).
TCA = Tricarboxylic acid cycle.
UDP = Uridine diphosphate.
UTP = Uridine triphosphate.

Protein

Proteins are polymers of amino acids linked by peptide bonds. Carbon, hydrogen, oxygen, nitrogen, sulphur and phosphorus are constituents of protein. The protein bond is a covalent bond between the amino group of one amino acid and the α-carboxyl group of another. Protein molecules are largest and most complicated biological molecules. The formula of **haemoglobin** in red blood corpuscles of vertebrates, which is a small to medium-sized protein molecule is $C_{3032}H_{4816}O_{872}N_{780}S_8Fe_4$.

Over fifty per cent of the dry weight of animal cells is protein. They are integral part of cell membrane, skin, bone, tendon and muscles. All enzymes are protein in nature. Proteins of even two organs of the same individual vary in their amino acid sequence. To synthesize new proteins, the dietary proteins must be degraded (broken down) to their constituent amino acids.

Protein molecules are made of simple units, the **amino acids**. A protein molecule is formed by joining of hundreds of amino acids in definite proportion and particular sequence. Innumerable types of protein molecules can be formed in this way.

The structure of protein molecules is described in terms of primary, secondary, tertiary and quaternary levels.

Primary structure. Proteins are made up of one or more linear, unbranched chain (Fig. 2.11) of amino acids depending on the sequence of amino acids. The chain is some

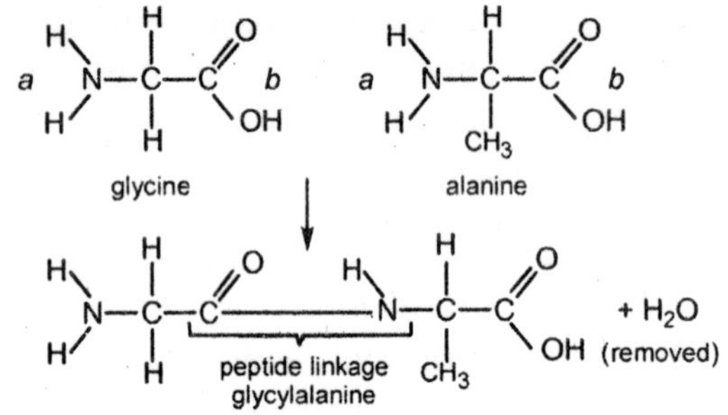

FIGURE 2.11

Synthesis of glycylalanine. The amino group (a) and the acid (carboxyl) group (b) of amino acids glycine and alanine joined in a peptide linkage. One molecule of water removed

hundreds of amino acid long. This is known as primary structure.

Secondary structure. Hydrogen bonds between two amino acids and another amino acid and four residues along the primary sequence, leads to a coiling (Fig. 2.12) of the primary structure resulting in a coil (α-helix) or zigzag chain (pleated sheet) or a mixture of the two. This physical structure of the polypeptide is termed secondary structure.

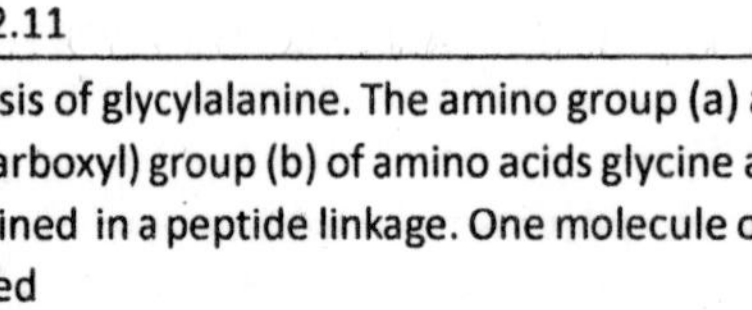

FIGURE 2.12

A tripeptide

Tertiary structure. A strand of amino acids, whether in helix or not, tends to fold back on itself or coil over itself to produce various three-dimensional configuration (e.g. a globular form) producing tertiary structure. Covalent bonds, hydrogen bonds, electrostatic bonds and hydrophobic bonds are involved in tertiary structure. Covalent disulfide bonds (—S—S—) occur in tertiary structure of many proteins (Fig. 2.13).

Quaternary structure. Certain proteins are made up of a specific number of distinct polypeptide chains (e.g. insulin—A and B chains; chymotrypsin—A, B and C chains). Interaction of individual chains of protein with each other forms the quaternary structure.

Diversity of protein molecules

A single cell contains hundreds of different kinds of proteins. The proteins of one type of cell differ from those in other types. The proteins of each species of organisms are unique for the species and they differ from those in all other species. Proteins of one individual may differ from the proteins of another individual. Differences in the kind of proteins are present even in the different organs of the same individual. The proteins present in the nerve cells of an individual are different from those in the cells of other organs, viz. muscles, liver, kidney, heart, etc. Each organ has its specific proteins. It is really the differences in the kinds of proteins, which has led to the differentiation of the organs during development.

Unlike proteins **interact.** Precisely for this, grafts of tissues or organs from one animal usually do not grow when implanted on a host of different species but degenerate and rejected by the host.

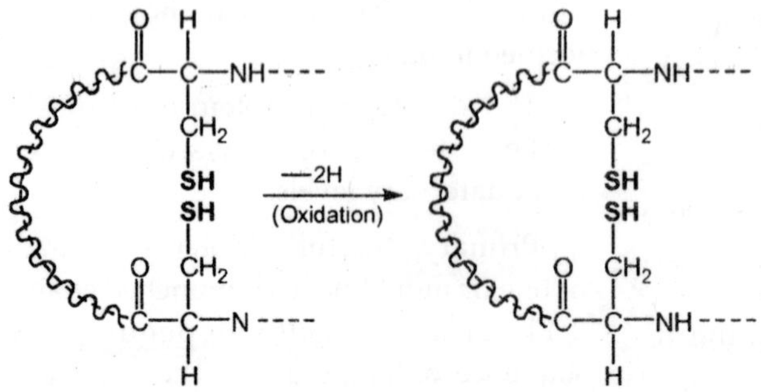

FIGURE 2.13

The formation of disulphide bond (cross links) in a polypeptide.

⊠ Denaturation and renaturation

Covalent bonds, hydrogen bonds, electrostatic bonds and hydrophobic bonds stabilize specific arrangement and pattern of the three dimensional tertiary structure of protein molecules. The specific tertiary structure of the protein is modified or disrupted by certain factors, viz. mild heat, urea, extremes of pH leading to a structural change and the coiled peptide chain unflods. This is known as denaturation. The change in the configuration of the tertirary structure leads to the loss of biological activity.

Denaturation may be reversible or irreversible. If the denaturation is carried out in dilute solution or by mild heat and gently enough, the process can be reversed by withdrawing the denaturation causing factors. This is **renaturation.** When denaturation is extreme as caused by strong solution or high temperature, the denaturation process cannot be reversed and the original condition of the tertiary structure cannot be restored.

⊠ Amino acids

Amino acids, the structural units of protein molecules are a mixture of simple nitrogen containing organic molecules. The chemical composition of an amino acid is:

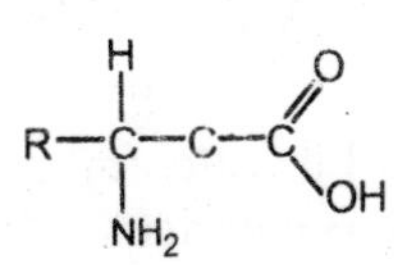

The NH_2 group is the amino group (base); the COOH group (an alpha carbon) is acid or carboxyl group. In addition the α-carbon is bound by a **side chain** (R), which is different in each amino acid. R-groups vary considerably. There are 30 or more different R groups, and therefore, 30 or more different amino acids. The amino group (base) of one amino acid combines with carboxyl group (acid) of another amino acid.

Animal cells can synthesize some of the amino acids. Humans cannot synthesize eight of the identified amino acids. They obtain those through their food. The ones that the animal cannot synthesize, but must be supplied with are known as **essential amino acids**.

⊠ Classification of protein

On the basis of solubility and composition proteins have been classified into three major groups.

I. Simple proteins

On hydrolysis simple proteins yield α-amino acids or their derivatives. Five simple proteins have been identified.

- **Albumins.** Water soluble, coagulate on heating and precipitate at high salt concentration. Examples: Serum albumin, egg albumin, etc.
- **Globulins.** Insoluble in water; soluble in salt solution; precipitate in salt solution of higher concentration. Example: Serum globulin, myasinogen (muscle), Ovoglobulin (egg), etc.
- **Glutenils.** Present mostly in plants; insoluble in water and dilute salt solution; soluble in dilute acids and alkaline. Examples: Oxyzenin (rice), Glutenin (wheat) etc.
- **Prolamines.** Present mostly in plants. Insoluble in water and absolute alcohol but soluble in 70% alcohol. Example: Zein (corn), gliadin (wheat), etc.
- **Scleroproteins.** Insoluble in all solvents. Function-maintenance of structure and protection. Examples: keratin (nail, hoof), collagen (bone, skin), elastin (ligament).

II. Conjugated proteins

On hydrolysis yield a nonprotein substance (prosthetic group). 7 conjugated proteins have been identified.

- **Nucleoprotein.** Water soluble and basic in nature; occur in combination with nucleic acids. Examples: Nucleohistones, nucleoprotamins.
- **Phosphoproteins.** Phosphorus radical (a prosthetic group) is usually attached as a monoester to the hydroxyl group of serine. Examples: Vitelline (egg yolk), Casein (milk) etc.

• **Glycoproteins.** Contain mucopolysaccharide (carbohydrate) as a prosthetic group. Examples: Ovomucoid (egg white), mucin (saliva).

• **Porphyrin proteins.** Contain porphyrin (bearing a metal ion) as a prosthetic group. Examples: Haemoglobin, myoglobin, etc.

• **Lipoproteins.** A combination of protein with lipids—cholesterol, fatty acid, phospholipid, etc. Examples: α and β lipoprotein of blood, lipoproteins of cell membrane, egg yolk, milk, etc.

• **Flavoproteins.** Contain riboflavin as prosthetic group. Examples: yeast, milk, egg, etc.

• **Metaloproteins.** Contain metal ions–Fe, Mn, Zn, Cu. Mg, etc. as their prosthetic group. Examples: ceruloplasmin (Cu), siderophilin (Fe), etc.

III. Derived proteins

Derived proteins are formed due to partial hydrolysis of protein, or removal of prosthetic group, or changes in their molecules. Two major groups of derived proteins are—

• **Primary derived proteins.** Denatured form of proteins, no hydrolytic cleavage of protein molecules.

• **Metaproteins.** Formed by the action of slightly strong acids or alkalies. Examples: Acid and alkali albuminates.

• **Coagulated proteins.** Denatured due to extrinsic factors heat, ultraviolet light, alcohol, etc. Examples: Alcohol precipitated protein, coagulated albumin, coaken, protein, etc.

• **Secondary derived proteins.** Formed due to hydrolytic cleavage of peptide bonds. Examples: Peptones, peptides, proteoses, etc.

⊠ Conversion of food protein to body protein

Protein molecules being large cannot be absorbed in animal body. In digestion protein molecules are hydrolysed by the enzymes hydrolases to their constituents amino acids. Amino acid absorbed from the small intestine are transported to tissues through blood and interstitial fluid. The cells synthesise their protein from the amino acids.

Naturally occurring proteins contain two forms of amino acids—L-amino acid and D-amino acid. Majority of the proteins contain L-amino acid, though D-amino acid is found in some natural peptides.

⊠ Protein hydrolysing enzymes-hydrolases

Pepsin. Catalyses protein to proteoses and peptones–stomach (pH 1.5 to 2.5).

Trypsin. Splits peptide bonds involving carboxyl group of arginine and lysine—small intestine (optimum pH 5.2 to 6.60)

Chymotrypsin. Cleaves peptide bonds involving carboxyl groups of phenylalanine, tyrosine and tryptophan—small intestine (optimum pH 8.9).

Carboxypeptidases. Proteins and peptides are cleaved from the carboxyl end in small intestine (optimum pH 7.4).

Aminopeptidases. Catalyses the removal of the terminal residue with a free amino group—succus entericus*.

Tripeptidases. Acts on tripeptides and liberate amino acids—succus entericus.

dipeptidases. Liberate amino acids from dipeptides—succus entericus.

⊠ Absorption

Amino acids liberated by hydrolysis of dietary proteins are rapidly absorbed from the lumen of the small intestine by active transport to portal blood. Separate transport mechanisms exist for absorption of neutral and basic amino acids. Pyridoxine (vitamin B_6) plays important role in absorption of amino acids.

⊠ Urea cycle

Ammonia released in deamination of amino

* Succus entericus. The intestinal juice, secreted by the glands of the intestinal mucous membrane.

acids is toxic and it is quickly removed from the body. Urea is a nontoxic compound and formed mainly in the liver by Krebs cycle. Two molecules of ammonia and one molecule of CO_2 are converted to urea in each turn of the cycle. The urea formed in the liver is transported to the kidneys via the blood. In human about 25 to 30g urea in secreted by urine during 24 hours.

Enzymes required in the formation of intermediates in the urea cycle are present in mitochondria.

Carbamyl phosphate formation. In the presence of biotin N-acetylglutamate and 2 molecules of ATP, ammonia combines with CO_2 derived from decarboxylation reaction to form carbamyl phosphate. The reaction is catalyzed by **carbamyl phosphate synthetase**.

Citrulin formation. Transfer of carbamyl group (—CO—NH_2) from carbamyl phosphate to ornithine results in the formation of citrulin. The reaction is catalysed by **Ornithine transcarbamylase.**

Arginine succinate formation. The amino group of aspartate provide the second molecule of nitrogen required for the synthesis of urea, condenses with citrulline to form argino-succinate. The reaction is catalysed by **argino-succinate synthetase** and requires one molecule of ATP.

Arginine formation. Arginosuccinate is reversibly cleaved and form arginine and fumarate. catalysed by **arginosuccinase** (Fig. 2.14).

Urea formation. Enzyme arginase cleaves arginine to urea and ornithine. Ornithine again enters the cycle at the second step and the cycle continues.

☒ Transamination

A new keto and an amino acid is formed when a α-amino group from an amino acid is transferred to a keto acid. The process is transamination. Transamination reactions are catalysed by specific transaminases in which pyridoxal phosphate is involved as a coenzyme (Fig. 2.15).

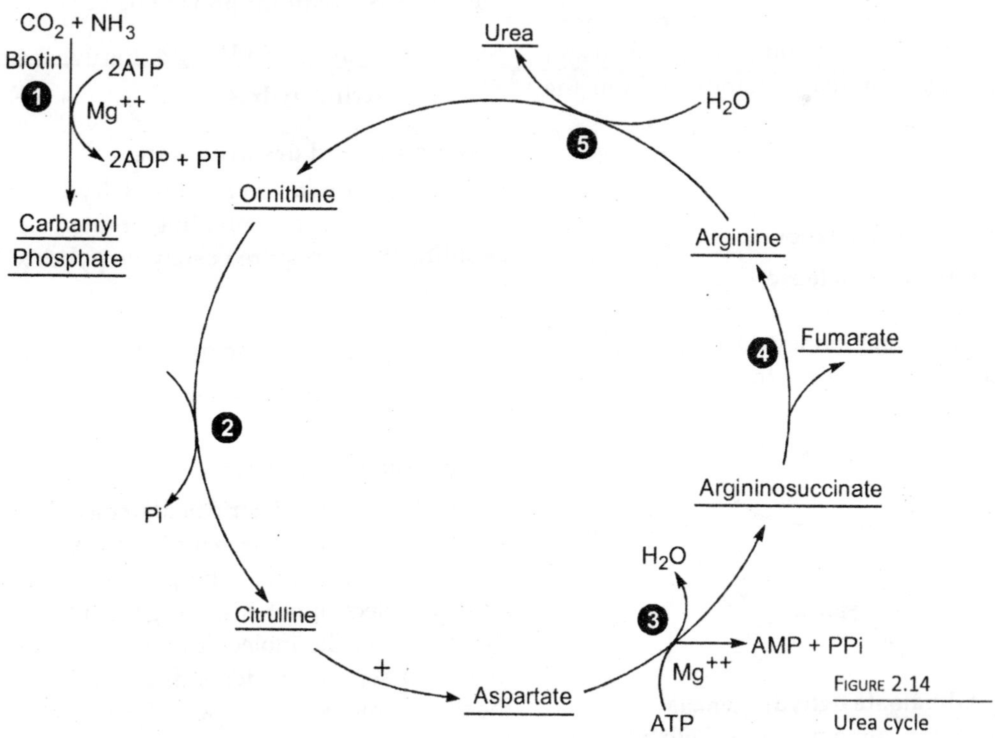

FIGURE 2.14
Urea cycle

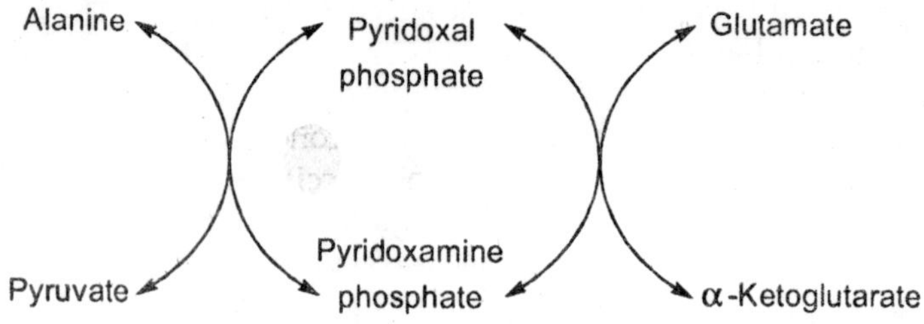

FIGURE 2.15
Transamination

Two most active transaminases in the liver are—glutamic-pyruvate transaminase (GPT) and glutamic-oxaloacetate transaminase (GOT). GTP is also found in cardiac and skeletal muscles.

ALANINE + α-KETOGLUTARATE $\xrightarrow{GPT}$ PYRUVATE + GLUTAMATE ASPARTATE + α-KETOGLUTARATE $\xrightarrow{GOT}$ OXALOACETATE + GLUTAMATE.

Deamination

The amino acids in addition to their role in the formation of protein can be used for energy, which also can be stored. The nitrogen is removed by a process, deamination catalysed by enzyme, mainly in liver and kidneys producing the corresponding α-keto acid and ammonia. Ammonia in removed from the body.

Deamination occurs in two ways—oxidative and non-oxidative.

1. Oxidative deamination

- **Amino acid oxidase**

Both L and D amino acid specific **oxidises** occur in liver and kidney. D-amino acids are not functionally important. The L-amino oxidases are FAD

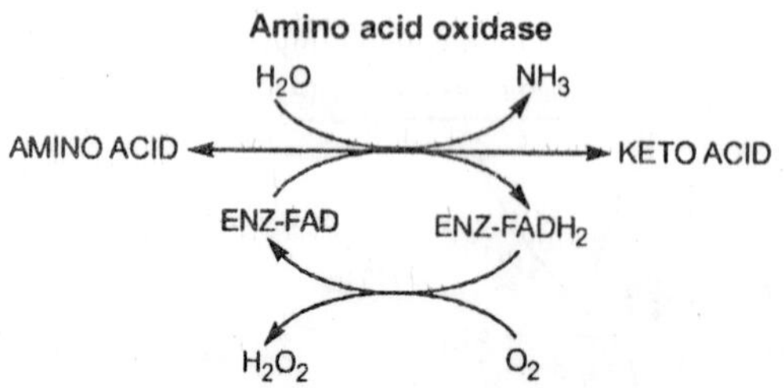

- **L-Glutamate dehydrogenase**

It catalyses glutamate to α-keto glutarate liberating ammonia which is utilised in urea formation. The enzyme utilises NAD^+ as co-enzyme.

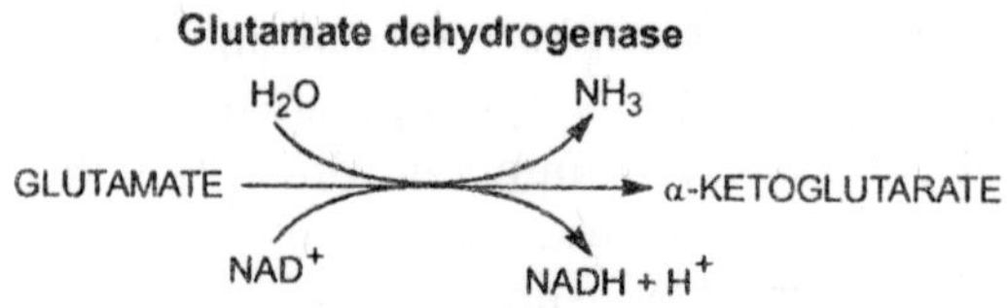

2. Non-Oxidative deamination

- **Amino acid hydrases**

Deamination reactions are catalysed by specific **amino acid hydrases** on amino acids are—serine, homoserine, threonine. The hydrases occur mainly in liver, and the coenzyme is **pyridoxal phosphate**

Serine $\xrightarrow{-H_2O}$ [Imino acid] $\xrightarrow{+H_2O}$ Pyruvic acid $+ NH_3$
Serine hydrase

- **Amino acid desulfhydrases**

Specific **amino acid desulfhydrases** deaminate sulphur containing amino acids. Desulfhydrases require coenzyme pyridoxal phosphate.

Cysteine $\xrightarrow{-H_2S}$ [Imino acid] $\xrightarrow{+H_2O}$ Pyruvic acid $+NH_3$
Cysteine desulfhydrase

Electrophoresis of protein

When placed in an electric field, molecules of proteins with a net charge move towards one electrode or the other. The phenomenon is known as electrophoresis. The greater the net change, faster the molecules will move. The molecular separation depends on the size of the protein molecules and also the net charge. In the process the gel serves as a molecular sieve. The gel is made of an inert chemical polyacrylamide. Small molecules pass readily through the pores in the gel, while the movement of larger molecules are retarded. The gel is readily formed by polymerization of acrylamide. With appropriate concentrations of acrylamide and the cross-linking reagent methylene bisacrylamide, the **pore sizes** in the gel can be controlled. Higher the concentration of acrylamide, the smaller will be the pore size. Usually the gel is cast between two glass plates, 7-20 cm^2, separated by a distance of 0.5-1.0 mm. The protein samples mixed with a blue dye (bromophenol) are poured in the cells in the top of the gel (Fig. 2.16). The molecules of the bromophenol being small migrate quickly through the gel during electrophoresis and indicates the progress at electrophoresis. The buffer-pH approximately 9.0 is used in both the upper and lower reservoirs and the gel. An electric current-approximately 300 V– is applied across the gel from top to bottom for 30-90 minutes. Most proteins have **net negative charges** and will move towards the anode in the lower reservoir. The gel is removed from electrophoresis apparatus and the protein within it can be observed.

☒ Lipids

The heterogeneous group of compounds, sparingly soluble in water but soluble in organic solvents, viz. methanol, chloroform and carbon tetrachloride and made up of carbon, hydrogen and oxygen are called lipids or fats.

Each molecule of fat contains one molecule of **glycerol** and three molecules of **fatty acids**, and termed **triglyceride**. Fatty acids are long chains of carbon atoms with a carboxyl group (—COOH) at one end (Fig. 2.17). All natural fatty acids contain an even number of carbon atoms. Fatty acids with one or more double bonds between carbon atoms are **unsaturated**, others are **saturated**. Fats containing unsaturated fatty acids are usually liquid at room temperature, while saturated fats are solids.

Fats are important structural components of cells and also biological fuels. Fats yield more than double the amount of energy produced by carbohydrates per unit. Excess carbohydrates in the body are transformed into fats and stored. Fats are stored as globules in the adipose tissue. The layer of adipose tissue under the skin serves as an insulator against the loss of body heat.

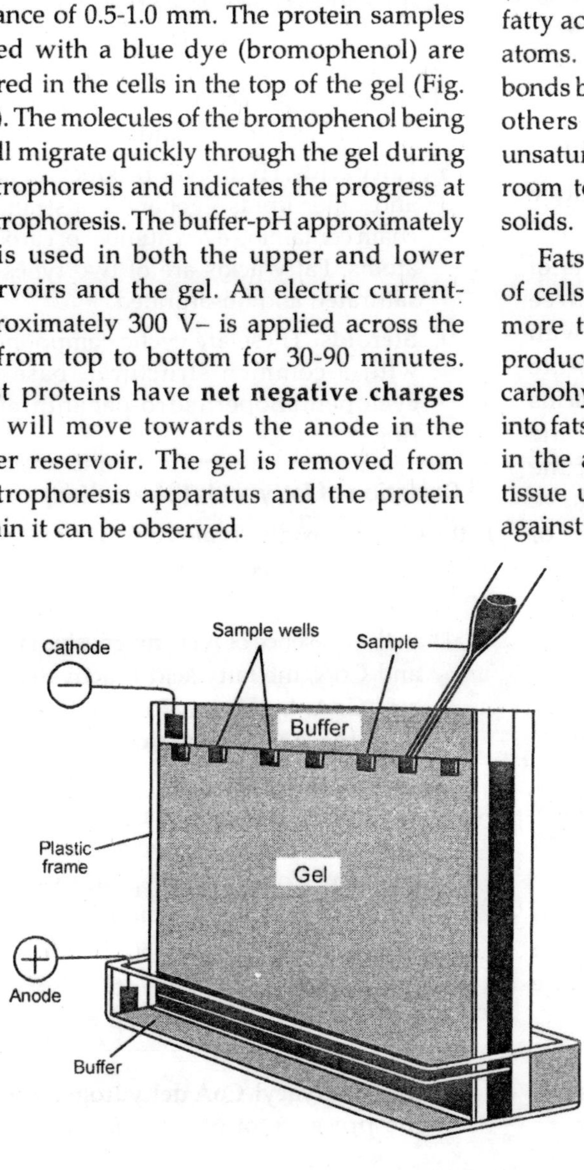

FIGURE 2.16

Gel electrophoresis apparatus

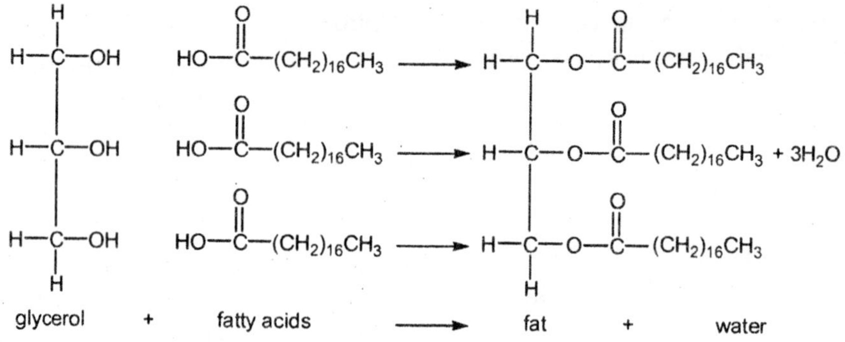

FIGURE 2.17

Synthesis of fat (triglyceride). Joining a molecule of glycerol and three molecules of stearic acid. Three molecules of water removed.

☒ Classification of lipids

Lipids have been divided into three groups.

I. Simple lipids. Fats and waxes constitute simple lipids.

i. **Fats.** Fats (triglycerides) are esters of fatty acids with glycerol. Three molecules of fatty acids esterified with one molecule of glycerol constitute triglycerides. Three different fatty acids are esterified with glycerol in the majority of neutral fats. Most of the naturally occurring fatty acids have straight chains with even number of carbon.

ii. **Waxes.** Waxes are esters of fatty acids with alcohols other than glycerol.

II. Compound lipids. These are esters of fatty acids with alcohol having additional groups. Compound lipids are of three major types.

i. **Phospholipids.** These are esters of fatty acids with glycerol bearing an esterified phosphoric acid and a nitrogen base.

ii. **Glycolipids.** Complex lipids–an amino alcohol attached to a fatty acids with an amino linkage and to a carbohydrate moiety glycosidically.

iii. **Lipoproteins.** These are lipid-protein complexes. The lipid compounds are mostly esterified cholesterol.

III. Derived lipids. Fatty acids and steroids are considered as derived lipids.

i. **Fatty acids.** Hydrolytic products of fats and other lipids—generally a straight chain containing even number of carbon atoms. Fatty acids are of two types—saturated and unsaturated.

ii. **Steroids.** These are cyclic compounds with a common structural base of cyclopentanoperhydro-penanthrene ring.

☒ Oxidation of fatty acids (β-oxidation)

Fatty acids are oxidised chiefly by β-oxidation, which occurs in mitochondria. It is a stepwise process.

1. In the presence of ATP, magnesium ions and CoA, the fatty acid is activated to acyl-CoA. In the reaction catalysed by **thiokinase** ATP breaks down to AMP, and pyrophosphate.
2. **Acyl-CoA dehydrogenase** in the presence of FAD remove two α, β-unsaturated acyl-CoA. In the respiratory chain $FADH_2$ is reoxidised yielding two high energy phosphate bonds.
3. **Enoyl hydrase** catalyse addition of one molecule of water to form β-hydroxyacyl-CoA.
4. **β-hydroxyacyl-CoA dehydrogenase** in the presence of NAD^+ removes two

hydrogen atoms in which β-hydroxy-acyl-CoA is converted to β-Ketoacyl-CoA.

5. β-Ketoacyl-CoA is cleaved by **thiolase** into one molecule acyl-CoA and one molecule acetyl-CoA. The acyl-CoA formed at the end of each cycle is the carbon short of the starting molecule. Acyl-CoA reenters the cycle and the fatty acid is fully degraded to acetyl-CoA. Acetyl-CoA enters the citric acid cycle for further oxidation (Fig. 2.18).

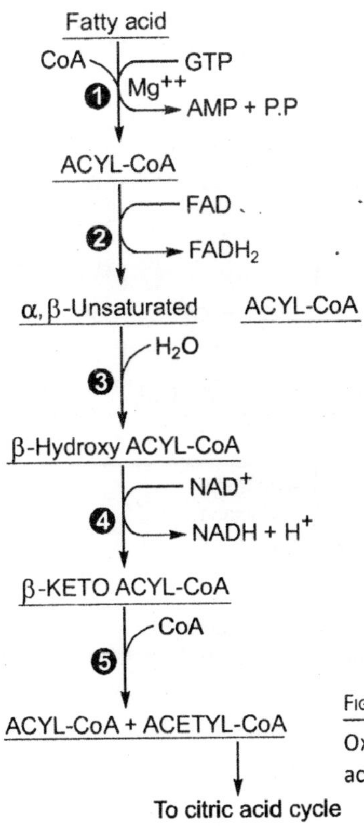

FIGURE 2.18

Oxidation of fatty acids

⊠ Energy yield of Palmitic acid through β-oxidation

Palmitic acid is C_{16} acid.

1. Palmitic acid undergoes 7 steps of β-oxidation and 8 molecules of acetyl-CoA are produced.
2. 5 ATP molecules are generated (2 from $FADH_2$ and 3 from NADH + H^+) in each step. In 7 steps of oxidation a total (7 × 5) – 2 = 33 molecules are formed.
3. 12 ATP molecules are produced by complete oxidation of one acetyl-CoA molecule in cytric acid cycle. So a total of 12 × 8 = 96 molecules are formed by complete oxidation of acetyl-CoA.
4. Total high energy bonds formed on complete oxidation of Palmitate = 33 + 96 = 129.

⊠ Cholesterol metabolism

The sources of cholesterol are meat, brain, liver and egg-yolk.

Cholesterol (Fig. 2.19) is present in large amounts in brain and nervous tissue. It is also found in liver, kidney, spleen, and skin.

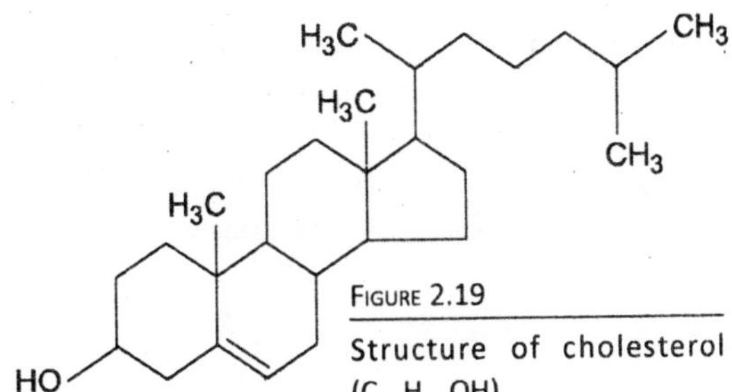

FIGURE 2.19

Structure of cholesterol ($C_{27}H_{15}OH$)

⊠ Biosynthesis of Cholesterol

Acetyl-CoA molecules generated from breakdown of carbohydrates or fatty acids are precursors of cholesterol.

1. A **synthetase** enzyme catalyse reaction in which acetyl-CoA combines with acetoacetyl-CoA and β-hydroxy-methyl glutaryl-CoA (HMG-CoA) is formed.
2. **HMG-CoA reductase** catalyse reduction of HMG-CoA by NADPH to produce mevalonic acid.
3. In a stepwise reaction mevalonic acid is phosphorylated to form isoprenyl pyrophosphate a 5-carbon moiety.
4. Two molecules of farnesyl pyrophosphate (15 carbon each) are formed by condensation of six molecules of isoprenyl pyrophosphate.

5. Two molecules of farnesyl pyrophosphate condense to from squalene ($C_{30}H_{50}$).
6. Undergoing a ring closure on oxidation a squalene forms lanosterol. Two methyl groups are removed resulting in zymosterol.
7. Desmosterol is formed from zymosterol by shifling the double bond in ring 11 to the position between C-5 and C-6. Subsequently the double bond in the side chain is reduced and cholesterol is formed (Fig 2.20).

Cholesterol is transported in blood in association with β-lipoprotein. In a normal human, plasma cholesterol level ranges between 100 to 200 mg/100 ml blood. High cholesterol levels are believed to be associated with coronary heart diseases and arthrosclerosis. The disease is due to deposition of esterified cholesterol and other lipids in connective tissue of arterial wall.

Nucleic acids

Nucleic acids contain carbon, hydrogen, nitrogen, oxygen and phosphorus. These are complex molecules, larger than most protein molecules. Nucleic acids are of two types (Fig. 2.21).

a. **Ribonucleic acid** or **RNA** contains a five-carbon sugar, ribose. A RNA molecule has only a single polynucleotide chain.
b. **Deoxyribonucleic acid** or **DNA** contains a five-carbon sugar, with one less oxygen atom, **deoxyribose**. A DNA molecule has two right-handed helical polynucleotide chains that form a **double helix** around the same central axis.

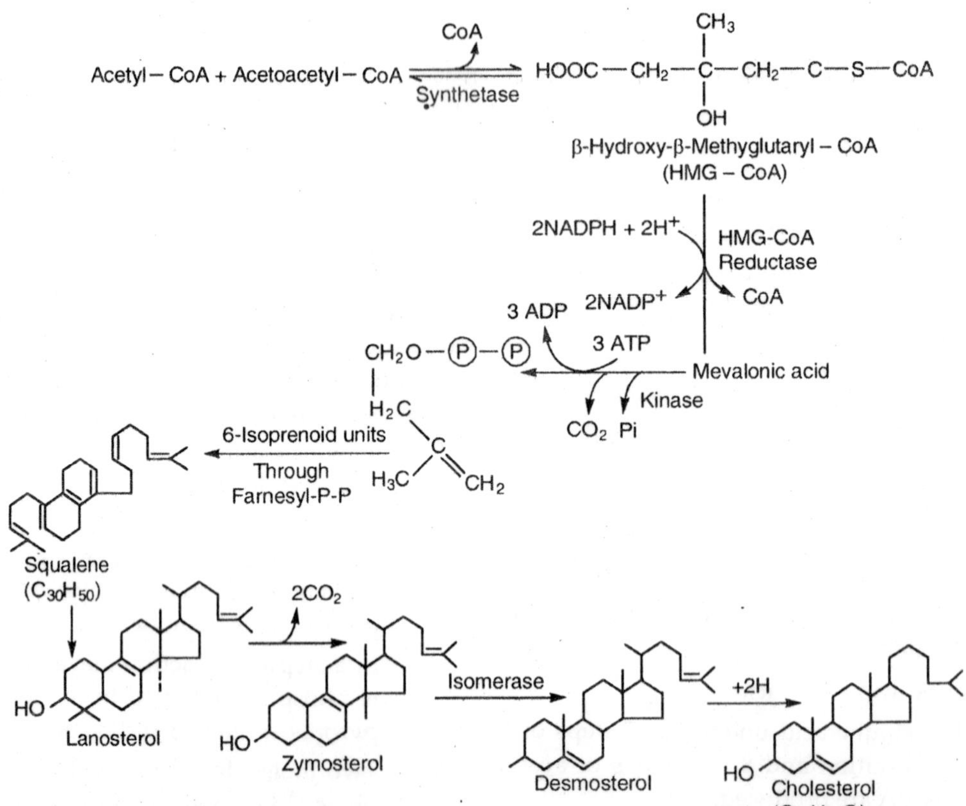

FIGURE 2.20

Biosynthesis of cholesterol

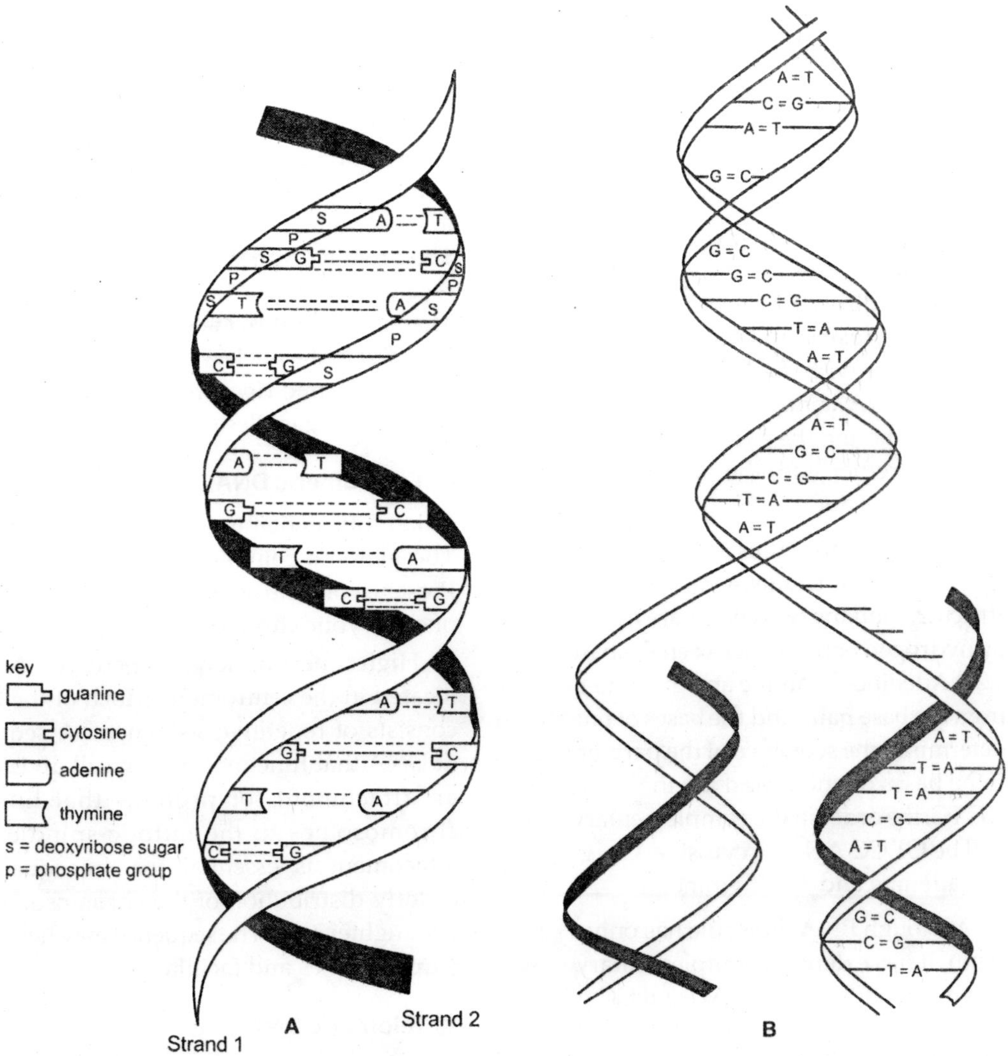

FIGURE 2.21

Deoxyribonucleic acid (DNA) molecule: A. A DNA molecule, B. Replication of a DNA molecule

The major bulk of DNA is present in the chromosomes and only a small amount in the mitochondria. It is the primary carrier of biological information.

RNA is synthesized by DNA. It is stored in the nucleolus as nuclear RNA or move out to cytoplasm.

There are different types of DNA and RNA. Nucleic acids differ in their structural details and their metabolic functions.

Three different types of RNA have been identified.

a. **Messenger RNA (*m*RNA).** Constitute about 5 per cent of total RNA. Carry coded information from DNA to specific protein synthesis sites of the cytoplasm.

b. **Transfer RNA (*t*RNA).** They make up about 15 per cent of total RNA and carry specific amino acids to mRNA to help synthesize specific proteins.

c. **Ribosomal RNA (*r*RNA).** Constitute approximately 80 per cent of total RNA, and found in ribosomes. Their function is to catalyse protein synthesis.

Units called **nucleotides** constitute nucleic acids. Each nucleotide contains a nitrogenous base, a five-carbon sugar and phosphoric acid. Nitrogenous bases are of two types, **purines** and **pyrimidines**. DNA contains **adenine** and **guanine**, **cytosine** and the pyrimidine-thymine, together with deoxyribose and phosphoric acid. RNA contains the purines, adenine and guanine and the pyrimidine, cytosine and **uracil**, together with the pentose, ribose and phosphoric acid.

The DNA molecules are composed of two polynucleotide chains arranged in a helical structure, and the two chains are held together by hydrogen bonding between the bases. (Fig. 2.21) Adenine-thyamine and guanine-cytosine from the base pairs and the bases of one strand determines the sequence of the partner strand. If the bases of one strand are in the order GA GG TC, the order in the complementary strand will be CT CC AG; C = cytosine, G = guanine, A = adenine and T = thymine.

Although RNA molecule has only a single chain, it has extensive complementary region in which bonds between AU and GC pairs are formed between different regions of the molecule. U = uracil.

MITOCHONDRIAL AND CHLOROPLAST DNA

Mitochondria and chloroplast probably evolved from endosymbiotic bacteria. Similer to bacterial multiplication by fission mitochondria and chloroplasts are derived from preexisting mitochondria and chloroplasts by simple fission. It is presumed that in course of evolution mitochondria and chloroplast of present day eukaryotes were derived from aerobic bacteria and cyanobacteria that lived in early eukaryotic cells. Both mitochondria and chloroplast contain DNA ribosomes and the enzyme system to synthesize proteins encoded in their DNA. Sequences in mitochondrial DNA and chloroplast DNA are similar to sequences in certain aerobic bacteria and cyanobacteria respectively. The ribosomes of mitochondria and chloroplasts are similar in size and structure and RNA sequences to those in bacteria than those in the eukaryotic cells. The protein synthesize enzymes more closely resemble bacteria than in eukaryotic cells. Most mitochondrial and chloroplast proteins encoded in the nuclear genes, are transformed to cytoplasmic ribosomes and subsequently reach the organelles.

⊠ Centromeric DNA

Eukaryotic genomes are usually much large. The genome contains repetitive sequences. Presence of centromere is a definitive feature of eukaryotic chromo-somes.

High concentration of repetitive DNA are located in the centromere. Much of the DNA consists of repetitive sequences. A sequence of DNA that functions during cell division as an attachment for proteins that link the chromosomes to the mitotic spindle. The attachment is essential for the equal and orderly distribution of the chromosome sets to daughter cells. The sequence may help align chromosomes and facilitate recombination.

⊠ Telomeric DNA

Telomeres are sequences at the end of the eukaryote chromosomes that help to stabilize the chromosome. Telomeric DNA consists of more than 1000 or more tandem repeats of a short G-rich sequence (TTGGGG) in ***Tetrahymena*** (a protozoon) and TTAGGG in human on the 3′-ending strand of each chromosome end. Further the strand has a 12-16 nt single-strand overhang.

SELFISH OR JUNK DNA

There are certain moderately repetitive DNA, functions of which are not known. These have been termed as selfish DNA. The DNA has disseminated itself throughout the genome for generations and apparently is a molecular parasite. Though very little, the harmless DNA is burden to increased metabolism for slowly growing eukaryotes.

☒ C-value paradox

How many genes are present in an organism can be determined from the genome. Correlation between quantity of unique genetic material (C-value) and the complexity of its morphology and metabolism has many exceptions. This is known as **C-value paradox.** The complete genome sequences of several prokaryotes and one eukaryote along with preliminary data on more complicated genome indicates that the apparent number of genes, like the overall quantity of DNA parallels the organisms complexity.

TRANSCRIPTION

Transcription involves the synthesis of RNA from a DNA template by RNA polymerase. This is the first stage of DNA expression. Genes are classified by their promoters, and each class is transcribed by a different RNA polymerase.

The enzyme RNA polymerases synthesize RNA using DNA as a template. One strand of the double helix is template and the other nontemplate. The RNA is produced on the template strand and the molecule synthesized is a copy of the nontemplate strand. The RNA molecule synthesized is known as **transcript**. The RNA molecule is capable of synthesizing a protein or act as ribosomal RNA (rRNA) or tRNA (transfer RNA).

Ribosomal RNA. Ribosomes are large molecules made of ribosomal RNA (rRNA) and protein. Large numbers of ribosomes are present in the cytoplasm and they transfer messenger RNA to protein.

Messenger RNA. It (mRNA) acts as a template for protein synthesis.

Transfer RNA. They (tRNA) are small molecules and carry amino acids to synthesize protein directed by a messenger RNA.

DNA FINGER PRINTING

Hypervariability in the satellite DNA is the basis of DNA finger printing, the technique, which identify individual and their family relationship (blood relation).

The eukaryotic genome contains highly repetitive DNA with very short sequences from 2 to 30 base pairs in length. Many thousand such copies are arranged in tandem. Such assays are known as satellite DNA. On the length of repeating sequence, the satellite DNAs have been placed in two groups, mini satellites (VNTRs)—variable number of tandem repeats and micro satellites (STRs)—simple tandem repeats.

Mini satellite repeats are the tool for DNA finger printing. The number of repeats in some satellite sequences vary significantly between individuals. The precise length of a set of different minisatallite assays in the genome are diagnostic of an individual. The probability of another individual having the name set of assay length (except identical twin clone) is almost nil.

Eukaryotic nuclei have three polymerases:				
Enzyme	**Location**	**Product**	**Relative activity**	**α Amanitin sensitivity**
RNA Polymerase I	Nucleolus	Ribosomal RNA (rRNA)	50-70%	Not sensitive
RNA polymerase II	Nucleoplasm	Messenger RNA (mRNA)	20-40%	Sensitive
RNA polymerase III	Nucleoplasm	Transfer RNA (tRNA)	10%	Species-specific

Family relationship can be determined by the matching of lengths from different individuals, since half of these are inherited from each parent.

GENE

A gene is a self duplicating unit located on a chromosome. It is an unit of information and corresponds to a discrete segment of DNA with a base sequence encoding the amino acid sequence of a polypeptide. Genes vary greatly in size—from less than 100 base pairs to several million base pains. The genes are dispersed, and separated by noncoding intergenic DNA. Information is encoded on the template strand of DNA, which directs the synthesis of RNA molecule.

ALLELES

In a population of the members of the same species many different alternative forms of the same gene are called alleles. In sexually reproducing organism one member of each pair of chromosomes is maternal and the other paternal. Each diploid organism contains two copies of each gene as a consequence of biparental inheritance. The members of each pair of genes influencing the character need not be identical. Therefore phenotype traits may be under the control of more than one gene and the allele forms of each gene is involved. Mendel symbolised alleles by using uppercase and lower case of initial letter—'T' for dominant tall and 't' for recessive dwarf.

GENE EXPRESSION

Genetic information is encoded in the base sequence of DNA as a series of genes. The process by which the cells decode and use the information to synthesize proteins, which control cellular functions, is known as gene expression. In gene expression the information in the DNA is copied to RNA molecule. The base sequence of the synthesized RNA molecule is complementary to that of the DNA template. The RNA synthesize a protein, the amino acid sequence of which is specified by the base sequence of the RNA. The DNA sequence of the gene is colinear with the amino acid sequence of the polypeptide.

ONE GENE ONE POLYPEPTIDE CHAIN

Protein may consists of one or more polypeptide chains. Gene is the unit of distinct information. Genes encode proteins, which in turn synthesizes other types of materials including nucleic acids. The sequence of a protein is regulated by the sequence of a gene. **One gene: One enzyme hypothesis** proposed that each particular enzyme catalyses each metabolic step and the enzyme is produced by a single gene. Some proteins contain more than one subunit. The protein is **homomultimer** if all the subunits are same. It is **heteromultimer** if the subunits are different. In haemoglobin a heme group (iron bounding group) is associated with two α subunits and two β subunits. Each type of subunit has a different polypeptide chain, which is represented by its own gene. If the gene coding either the α or β polypeptide mutate, the function of haemoglobin is inhibited. In general, in any heteromultimeric protein the one gene: one enzyme hypothesis is more precisely expressed as one gene: one poly-peptide chain.

3 THE ANIMAL CELL

A cell is a fundamental morphologic and physiologic unit in living beings. It consists of protoplasm (cytoplasm and nucleus together) and a cell membrane, the plasma membrane.

It was Robert Hook (1665), who first observed that a thin slice of a cork, when seen under magnifying lenses, appears to be porous, much like a honeycomb. He called these pores, **cells**. Later, the works of Mirbel (1808), Lamarck (1809) confirmed Hook's idea and in 1838-39, Schleiden and Schwann advanced their '**Cell theory**', stating that all plants and animals are aggregates of these minute structures, the **cells**.

Cells vary in shape and size immensely. Their sizes may vary from 0.001 mm (bacteria) to several cm (eggs of some birds). Shape too, may be spherical, columnar, polygonal, or amorphous and so on. Various types of cells and tissues show chemical differences, like their morphological variations. The following is the average composition of an animal cell.

Water	...	85-90%
Protein	...	7-10%
Fatty substances	...	1-2%
Other organic substances	...	1-1.5%
Inorganic materials	...	1-1.5%

A TYPICAL ANIMAL CELL

A cell usually has several structural components (Fig. 3.1) embedded in the cytoplasm, and the whole structure is surrounded by a cell membrane. The different structures—nucleus, cytoplasm, mitochon-dria, Golgi complex, centrosome, endoplasmic reticulum, ribosome, lysosome, cytoplasmic inclusions and cell membrane or plasma membrane have some specific functions.

☒ Nucleus

Generally it lies in the centre of the cytoplasm, and is usually spherical. A nondividing nucleus consists of (a) a nuclear membrane, (b) nucleoplasm, (c) chromosomes and (d) nucleolus.

The nuclear membrane surrounds the nucleoplasm in which the chromosomes lie dispersed. Under high magnification it is seen to be double layered with frequent passages or pores through them. Through these pores the nuclear materials and the cytoplasm are continuous. RNA molecules pass out through these pores. The membrane has a certain degree of mechanical strength.

The nucleoplasm is **granular** and composed of **glycoproteins** and some amount of ribonucleic acid (RNA).

The most important constituent of the nucleus is the chromatin reticulum which becomes differentiated into chromosome during cell division. Two types of chromatins are recognised, **euchromatin** and **heterochromatin**. The **genes** are thought to be in association with the euchromatin. The chromosomes can only be seen in a dividing cell. Usually they are more or less cylindrical bodies constricted at one or more points. These regions are called **centromere** or **kinetochore**. A chromosome is composed of one or more **helically coiled** threads called **chromonemata**, embedded in matrix. The most common chemical fraction found in a chromosome is **deoxyribonucleic acid** (DNA).

The **nucleolus** may be one or more and is formed during telophase of mitosis in some specific regions. Under high magnification it is seen to be composed of a coiled thread with some granular structures. These structures seem to be in intimate contact with the surrounding nucleoplasm.

Cytoplasm

The cytoplasm is a highly complex substance and is considered as an aqueous crystalloid and colloidal solution of a number of chemical substances. Its viscosity is about four times greater than that of water but it does not flow freely due to certain amount of cohesion. This is due to large organic molecules, chiefly protein, and the inorganic ions are insufficient to materially alter the viscosity.

Cytosol

Cytosol the cytoplasm minus its membrane bound organelles—is a major site of metabolism in the cell. It contains within it the cytoskeleton and microtubules.

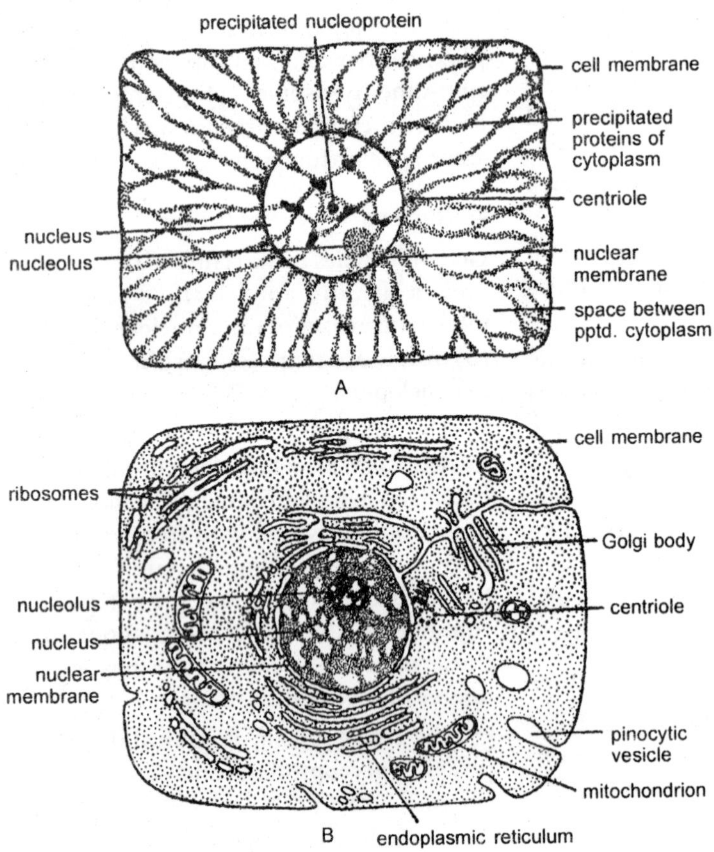

FIGURE 3.1

A typical animal cell: A. Through an optical microscope, B. Through an electron microscope

⊠ Cytoskeleton

The cytoskeleton is made up of two types of filaments—microfilaments, intermediate filaments and microtubules.

Filaments. The microfilaments are made of actin with an approximate diameter of 7nm. Its function is mechanical support. Microfilaments through their interaction with myosin become contractile and are responsible for cytoplasm streaming and formation of membrane invaginations.

The intermediate filaments are possibly of load bearing function.

The cytoskeleton alters the shape of the cell and helps in transporting intracellular vesicles.

⊠ Microtubules

These are hollow tubular structures made up of the protein—tubulin with an outer diameter of 30 nm. The rigid wall of the tubule is made of a helical array of alternating α- and β-tubulin. Thirteen tubulin subunits are present per turn of the filament. Microtubules form a supportive network that guides the movement of subcellular organelles within the cell. The spindle separating the daughter chromosomes during cell division is formed of microtubules.

⊠ Mitochondria

These are some distinct cytoplasmic bodies with extremely variable morphologic forms. They are either granular or rod-shaped or thread-like and vary in size, number and position. Under high magnification a mitochondrion is found to be composed of a double osmiophilic membrane enclosing an inner matrix of variable diameter. A series of septation formed by the inner membrane called **cristae mitochondriales**, which run at right angles to the long axis of the mitochondrion are present in the matrix (Fig. 3.2). Chemically, mitochondira are composed of protein, fat, phospholipid, and a small amount of ribonucleic acid (RNA). Additional substances like many essential enzymes including cytochrome, succinoxidase, and various phosphatases as well as enzymes involved in the tricarboxylic acid cycle are also present.

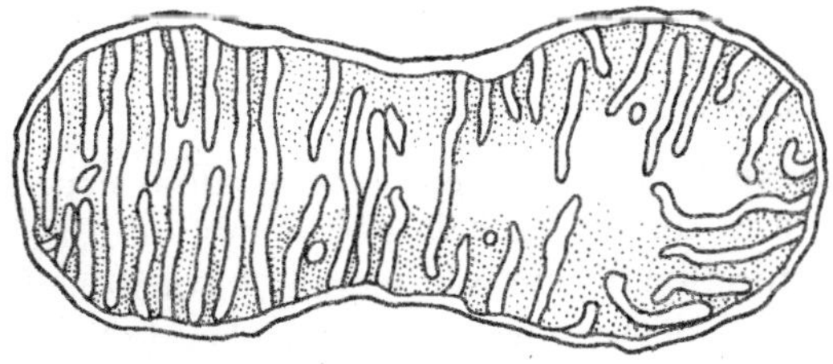

FIGURE 3.2

A mitochondrion

Mitochondria play a major role in cell respiration. Because of the great number of essential enzymes present in the mitochondria and due to the fact that mitochondria apparently enter into all basic metabolic processes, they have been considered to be the 'principal power plant of the cells'. Mitochondria often change their shape and position along with the development of the organism.

⊠ Golgi complex

The Golgi complex is composed of a system of membrane bound vesicles, associated with membrane elements arranged in more or less parallel fashion (Fig. 3.3). It is composed of lipoprotein, RNA, and an alkaline phosphatase. Like mitochondria, Golgi complex

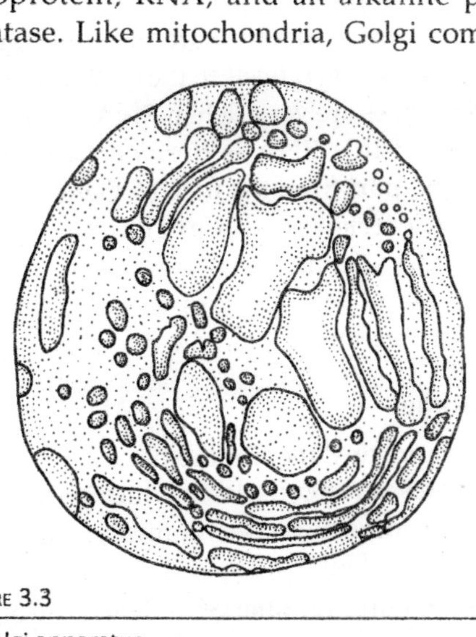

FIGURE 3.3

A Golgi apparatus

also reveals consistent changes associated with the development sequence.

In secretory cells there seems to exist a relationship between Golgi complex and secretion.

☒ Centrosome

The centrosome is a highly conspicuous structure in certain animal cells, notably during maturation divisions. Generally it consists of a small, dense, refractile body embedded in a clear area from which fibrillar rays radiate. The central body—usually called **centriole,** is situated in the cytoplasm just outside the nuclear membrane and divides during cell division (Fig. 3.5).

☒ Endoplasmic reticulum

The cytoplasmic portion of the membrane system is endoplasmic reticulum. It is a continuous and morphologically highly variable system of flattened sacs and tubules separated from the matrix by a membrane 40 Å thick [1 μm = 10,000 Angstrom (Å) units].

It has a relationship with protein synthesis, intracellular transport and intracellular impulse conduction in striated muscle cells.

☒ Ribosome

The ribosomal particles or the rough form of the endoplasmic reticulum are thought to be the site of protein synthesis in the cells. Chemically, ribosomes are very rich in RNA and proteins; part of this RNA is derived from the nucleus and part from the matrix. Genetic information carried in the nucleus is transcribed into specific proteins outside the nucleus, the nuclear RNA of the ribosomes acting as a messenger from the nucleus.

☒ Lysosome

Lysosomes are tiny, membrane-bound, vesicular structures in the cytoplasm which perform intracellular digestion. Lysosomes were first reported by de Duve in 1955.

They contain a number of hydrolyzing enzymes capable of degrading macromolecules in the cells. In need, the lysosome membrane ruptures, releasing the enzymes, which presumably hydrolyse macromolecules of dead cells and the contents are reabsorbed.

☒ Cytoplasmic inclusion

Besides the above structures, cytoplasm often includes reserve food like fat, protein, carbohydrates, etc. Presence of pigment is also common.

☒ Cell membrane or Plasma membrane

Each cell is completely enveloped by a three-layered, dense-light-dense cell membrane and is separated from other cells. It consists of a double structure having a lipoidal and protein component.

Plasma membrane bears a number of pores and is permeable to water and certain solutes, but the passage of different solutes does not get the same facility. Osmotic concentration is maintained by this mechanism which regulates the concentration of the dissolved substances in the cells. It receives information through receptor proteins and also involved in exocytosis (secretion) and endocytosis (internalisation).

☒ Cell surface receptors

Cell surface receptors are hydrophilic and some lipophilic hormones bound to cell surface. These integral membrane proteins in the plasma membrane bind the hormone (ligand) —the signalling molecule. A conformational change in the receptor due to binding of the ligand to the receptor initiates a sequence of reactions in the target cell the signal transduction, which initiates a change in cell function. The distribution of receptors often varies in different cells and there may be more than one type of receptor for a particular ligand. This permits different target cells to respond differently to the same signalling molecule. Cell surface receptors are of three types.

Enzyme-linked receptors. The ligand is bound to the extracellular face of the cell. This causes a conformational change in the cell surface at the receptor and activates an intrinsic enzyme activity.

Ion-channel-linked receptors. Conformational change occurs in the protein due to binding of the ligand is such that a specific ion channel is opened which allows certain ion to flow through, that alters the electric potential across the membrane.

G protein-linked receptors. On binding its ligand the G-protein-linked receptors activate G proteins [Guanosine triphosphate (GTP) binding proteins] that in turn produces an intracellular second messenger.

☒ Cell signalling

Communication between cell is a necessity to coordinate growth and metabolism in multicellular organisms. Extracellular signalling molecules (hormones) perform the function. The hormone is secreted by the signalling cell, which is bound to a receptor in the target cell, initiating a response in that cell. In endocrine signalling the hormone may act on the secretary cell itself, or on nearby cells or cells at a distance. The hormones are carried to distant target cells through blood (chordate) or nerve (nonchordate).

PROTOPLASM: COMPOSITION AND FUNCTION

Protoplasm is a complex, alterable substance with myriad functions. Protoplasm of different organisms differs, but the fundamental features are strikingly similar everywhere and this led T. H. Huxley (1868) to call protoplasm "**the physical basis of life**".

☒ Physical nature of protoplasm

It is a viscous, elastic; colloidal system with inherent structural bonds, and the ability to split rays of light.

Viscosity and elasticity. The specific gravity of protoplasm is a little more than that of water but it is highly viscous. The range of viscosity varies from 2 to 20 centipoises. The elasticity of protoplasm has been demonstrated by micromanipulator.

Inherent structural bonds. Protoplasm has very definite structural bonds that resist the movement of objects through it.

Light refraction. Movement of light in the protoplasm is slower than that in air. Protoplasm splits rays of light into two beams, only one of which is bent. This is known as birefringence.

Colloidal nature. Protoplasm is a complex colloidal system. It may exist as a sol or a gel. In many cells the protoplasm in the centre has the properties of a sol, surrounded by a stiff gel, the cortex.

☒ Chemical nature of protoplasm

The protoplasm is composed of about 80 per cent water, 15 per cent protein, 3 per cent lipid, 1 per cent carbohydrate and 1 per cent salt.

Water. Water, the major constituent of the protoplasm, is bound with other substances of the protoplasm, mainly protein. Some workers belive that in normal protoplasm water remains in a free state.

Protein. Protein is the most important component of protoplasm. Fundamentally a protein molecule consists of carbon, oxygen, hydrogen, nitrogen, usually with a small amount of sulphur woven together in an intricate pattern. Nucleoproteins are most important since they play a major role in the life of a cell. Nucleotides the building blocks of nucleoproteins, consist of a sugar and phosphoric acid in combination either with a pyrimidine or purine base. The nucleic acids are of two types—ribonuleic acid (RNA) and deoxyribonucleic acid (DNA).

Lipids. Lipid includes natural fats, fatty acids, fatty oils, phosphatides, waxes and steriods. The natural fat composed of fatty acid and glycerol constitutes the most common type of lipid.

Carbohydrates. In protoplasm, carbohydrates exist both in bound and free state, and take part in vital cellular functions. The simple sugars constitute an immediate source of fuel, that is energy for cell, whereas the polysaccharides represent reserve sources.

Inorganic salts. The inorganic salts of calcium (Ca), potassium (K), magnesium (Mg), iron (Fe) and sodium (Na) are present in small quantities in the protoplasm. Traces of salts of other metals like zinc (Zn), manganese (Mn), aluminium (Al), boron (B) and copper (Cu) are also present.

⊠ Functions of protoplasm

The following physiological activities are associated with protoplasm.

1. **Irritability.** It is essential for response to stimuli.
2. **Conductivity.** The protoplasm can transmit stimuli from one place to another.
3. **Contractility.** It can contract in any direction.
4. **Intake of foreign substances.** The protoplasm exhibits selective absorption, phagocytosis, and it is capable of assimilating certain substances.
5. **Secretion and excretion.** It can synthesize useful substance, e.g. enzymes and others and secrete those. It is capable of removing or excrete harmful or unwanted substances.
6. **Respiration.** The protoplasm carries on respiration to meet the need of energy with the help of mitochondria.

MULTIPLICATION OF CELL

New cells are produced from the preexisting cells by division. Ordinarily, cells divide by two means—Mitosis and Meiosis. The former is meant for multiplication of cells and is found in somatic cells, while the latter produces gametes, and is naturally restricted to germ cells. A cell division consists of karyokinesis or nuclear division followed by cytokinesis or cytoplasmic division.

CELL CYCLE

The life of an eukaryotic cell can be defined as a cell cycle (Fig. 3.4). The sequence of events between eukaryotic cell divisions includes mitosis and cell division (**M phase**), a gap stage (**G_1 phase**), a period of DNA synthesis (**S phase**) and a second gap stage (**G_2 phase**) before the next M phase. Some cells-neurons, muscle cells never divide and this state is known as **G_0 phase.**

The cell cycle for cells in culture takes about 16-24 hours. Cell cycle for different types of cells in multicellular animals vary from 3 hours to 100 days. Mitosis and cell division occur in M-phase, which is about one hour. G_1 phase is longest part of the cycle. Most of the variations occur in this phase. The chromosomal DNA is replicated in the S phase. In G_2 phase the cells prepare for mitosis. **Cyclins,** a class of protein accumulate throughout the cell cycle and then abruptly disappear during mitosis. Cyclins combine with a 34-kD protein and form an active Ser/Thr protein kinase, **Cyclin dependant protein Kinase** (CDK), whose full activity may require phosphorylation.

GENETIC CONTROL OF CELL CYCLE

The cell cycle is controlled through protein phosphorylation, catalysed by multiple **protein kinase complexes**. The protein kinases are specific and consists of regulatory and catalytic subunits. The regulatory subunits are cyclins and the catalytic submits are **cyclin dependent kinases** (CDKs). The CDKs are active only in association with cyclins. A specific CDK-cyclin complex determine the proteins to be phosphorylated by the protein kinase.

Three classes of cyclin-CDK complexes exist. The G_1 phase CDK complexes activate transcription factors for enzymes needed for DNA synthesis and the genes encoding S-phase CDK complexes. The S-phase CDK

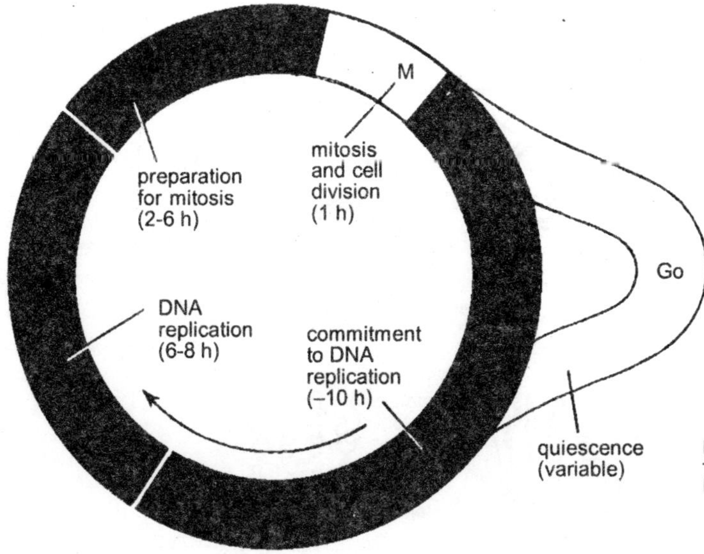

FIGURE 3.4

Eukaryotic Cell cycle stages

complexes stimulate organised DNA synthesis to a fashion that each chromosome is replicated only once. The mitotic CDK complexes induce condensation of chromosomes and separation of chromatins to daughter cells.

ONCOGENES AND CANCER

In an animal body growth and differentiation of cells are normally strictly controlled, with a few exceptions, viz. hair forming and blood forming cell. Cells, however under certain conditions may multiply uncontrollably and form a **tumor**. Tumours are of two types **benign** and **malignant** (cancerous). Benign tumours stop growing sometimes after its formation and are harmless. Malignant tumors grow in an invasive manner. Those are incurable and eventually lead to the death of the patient. Uncontrollable proliferation of cells is caused by oncogene.

☒ Oncogene

It is a mutant version of a normal gene (a protooncogene). Oncogene interferes with the mechanisms that normally control cell growth and differntiation and thereby contributes to uncontrollable proliferation (cancer).

Cancer. Among the many causes of cancer are viruses that carry oncogenes.

1. The Rous sarcoma virus (RSV) induces the formation of **sarcomas** (cancer arising from connective tissue) in chickens. Oncogenes are not necessarily of viral origin. In human, cancer is caused by proto-oncogenes that have mutated to form oncogenes. To date 50 oncogenes have been identified. The subversive effects of oncogene products are through their differences from the corresponding cellular proteins.

MITOSIS

It is a type of cell division in which a constant chromosome number is maintained in the parent (dividing) and daughter (divided) cells (Fig. 3.5). Mitosis effects a qualitative and quantitative distribution of the essential particles of heredity, the genes to dauther cells. It is a dynamic mechanical process, a number of push and pull, stress and strain.

☒ Karyokinesis

The whole cycle of karyokinesis is a continuous one and its division to prophase, prometaphase, metaphase, anaphase and telophase is only conventional.

Prophase. The prophase is initiated with the division of the **centrosome** into two and

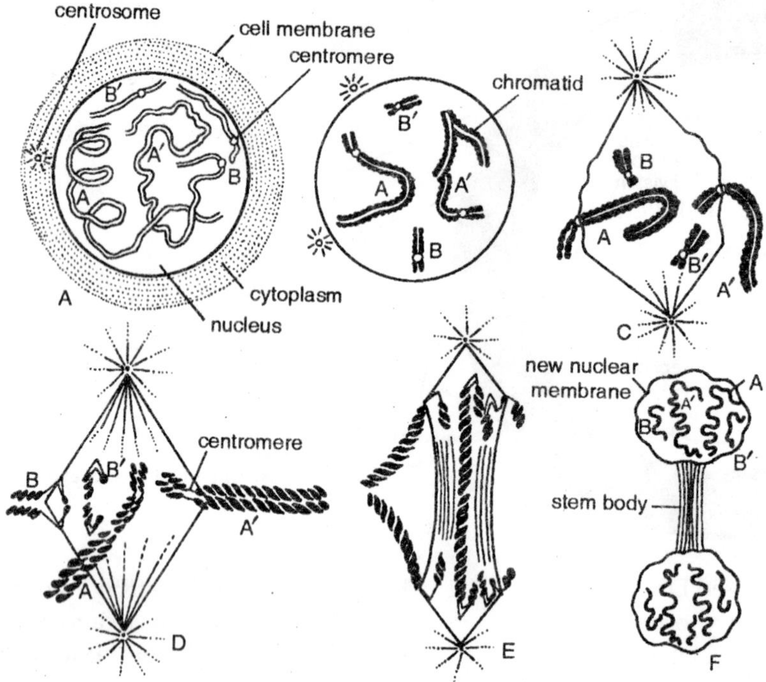

FIGURE 3.5

Mitosis. A-B, Prophase; C. Metaphase; D-E. Anaphase; F. Telophase

the appearance of the chromosomes. During this stage, two halves of the centrosome travel in opposite direction around the nucleus until they are 180° apart and **aster** is formed around each daughter centrosome.

Each chromosome is composed of two closely parallel **chromatids**. The division of a chromosome into two chromatids takes place before the beginning of mitosis and the cell division is really concerned not with the multiplication of the chromosomes but only with the distribution of the products.

Each chromosome has a non-staining gap, the **centromere**, which consists mainly or entirely of protein, with little or no DNA or RNA, and plays an important role in attaching the chromosome to the spindle. Throughout the prophase stage the chromatids are slightly irregular, woolly or hairy in appearance and by the end of the prophase they become smooth in outline. The nuclear membrane usually disappears at the end of this stage.

Prometaphase. The spindle is formed. It is a relatively solid gelatinous body, apparently composed entirely of protein, with a very small amount of RNA. Centrosomes possibly act as organizers of the spindle and are placed at the two poles. The chromosomes arrange themselves with their centromeres on the equatorial plane of the spindle.

Metaphase. The chromosomes reach maximal condensation and the chromatids are united only at the centromere which is still undivided. **Metacentric** chromosomes usually appear V-shaped at metaphase, while **acrocentrics** appear as straight or slightly curved rods.

Anaphase. The centromere divides into two daughter centromeres and separate from one another. At this time the equatorial region of the spindle between the daughter centrosomes elongates and the poles of the spindle become further apart in the late anaphase. The equatorial part of the spindle narrows and appears as a thin bundle, made up of thread-like structure, the stem body, between the two separating groups of daughter chromosomes.

Telophase. The two groups of chromosomes do not actually reach the poles of the spindle but are at some distance from the equator. The polar caps of the spindle disappear, the chromosomes lose their smooth outline, undergo decondensation and form a tangled mass. A nuclear membrane appears around the chromosomes, the stem body disappears, and two daughter nuclei are formed.

⊠ Cytokinesis

The cytoplasmic division starts in the telophase or even in the anaphase in some cases. A furrow appears around the cytoplasm which gradually moves inward and finally cleaves the cytoplasm into two halves, each receiving a daughter nucleus, and two daughter cells are formed.

⊠ Significance of Mitosis

1. The exact longitudinal division of the chromosomes ensures equal distribution of genetic constitution in the daughter cells.
2. Lineal heredity is established whether it be from cell to cell or organism to organism.
3. It ensures the maintenance of constancy of a species which is essential in the perpetuation of the race since its inception.

MEIOSIS

Meiosis is the antithesis of fertilization. It occurs mainly during the formation of gametes. In diploid organisms it results in the haploid number of chromosomes. In meiosis the nucleus divides twice, the first and second meiotic divisions, but chromosomes divide only once (Fig. 3.6).

A gap of interphase of uneven duration is present between the successive division. Each division is broadly separated into five stages—prophase, prometaphase, metaphase, anaphase and telophase.

FIRST MEIOSIS

⊠ Prophase

It is a very long stage and may be divided into the following subdivisions:

Leptotene

1. The chromosomes are not longitudinally divided and consist of a single chromatid.
2. The threads are composed of a series of granules (chromomeres) connected by non-staining gaps.
3. Leptotene chromosomes preserve the same structure and number as in the somatic tissue.

Zygotene

1. The homologous chromosomes (paternal and maternal) come to pair which begin at one or more points, usually at centromere.
2. Pairing is not merely between homologous chromosomes but strictly between homologous regions, which is revealed from various sorts of loops, and it seems that the force of attraction is a mutual one between homologous **chromomeres** (or genes), resulting in a delay in the spliting of the chromosomes.

Pachytene

1. As a result of pairing the number of chromosomes is reduced to half.
2. The paired chromosomes are called **bivalents**, each of which now splits longitudinally into two chromatids. The bivalents have two distinct centromeres.
3. In pachytene, the threads are strictly parallel in the beginning but in some they are coiled and, on splitting, it seems that the two threads are wound around each other.

Diplotene

1. With the splitting of the pachytene chromosomes the force of attraction between the paternal and maternal chromosomes is replaced by a force of attraction between the chromatids.
2. The moment the two homologous chromosomes begin to separate, marks the transition from pachytene to diplotene.

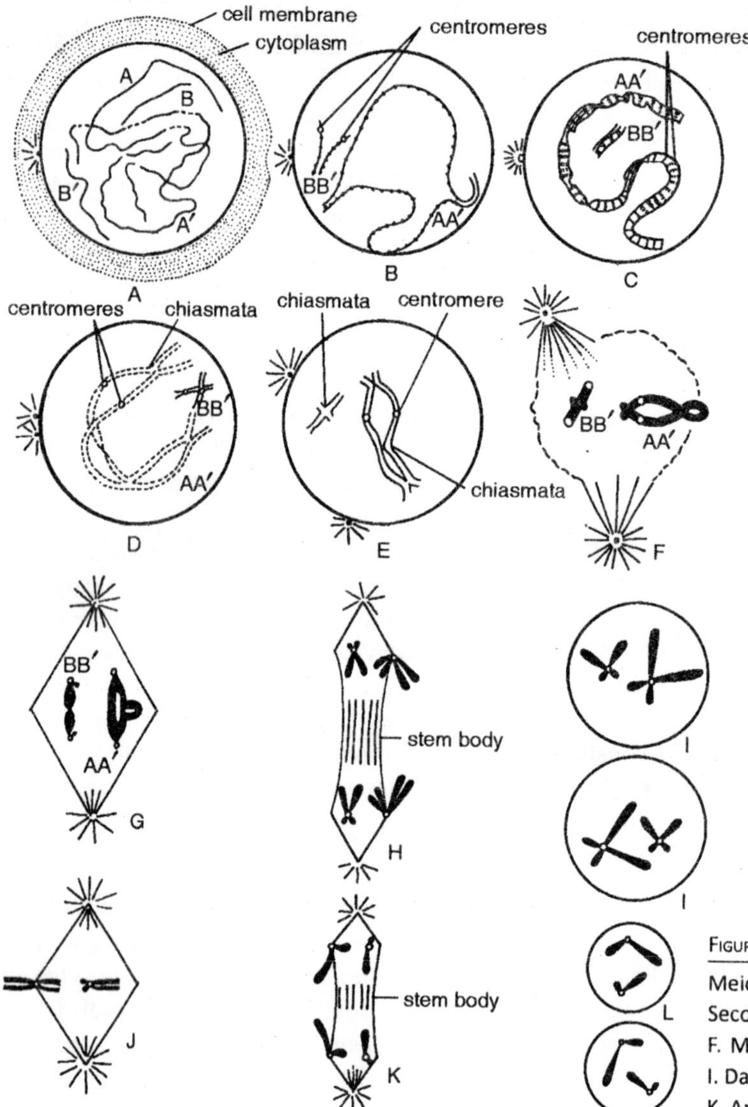

FIGURE 3.6

Meiosis: A-I. First meiosis; J-L. Second meiosis. A-E. Prophase; F. Metaphase; G-H. Anaphase; I. Daughter cells; J. Metaphase; K. Anaphase; L. Daughter cells

3. They do not actually separate completely but remain held together at some points called **chiasmata**. There is usually at least one chiasma in each bivalent.
4. Chiasma in a bivalent is accompanied by three changes:
   a. Shortening and thickening of the threads.
   b. Rotation of 180° is most marked in bivalents with a single chiasma.
   c. Actual moving of the chiasma towards the end or terminalisation.

Diakinesis

1. The bivalents become much thicker and shorter.
2. The split in some cases becomes invisible.
3. Terminalization may still continue.

Prometaphase. It is the period between the disappearance of the nuclear membrane and the moment when the spindle is fully formed.

Here the diakinesis bivalents have contracted still further and begin to be associated with the developing spindle.

Metaphase

1. Each bivalent possesses two centromeres, one maternal and the other paternal.
2. They are generally far apart and they arrange themselves at equal distances above and below the equatorial plane.
3. The centromeres have not divided nor they do so until the second meiotic division.

Anaphase

1. As the centromeres separate, they drag after them the chromatids which are attached to them.
2. The unterminalised chiasmata move to the end, away from the centromere and slip off at the ends.
3. Each centromere, as it moves to the pole, drags after it two chromatids.
4. The result of this stage is often said to be a separation of whole chromosomes.
5. The chromosomes which separate at the first anaphase are not the same as the maternal and paternal chromosomes which came together at **zygotene** but these have interchanged parts of their length by **crossing over**, so that the actual chromosomes which separate at first division are new combinations of segments of the maternal and paternal chromosomes.
6. There is no correlation between the mode of orientation of one bivalent and another in the same cell.
7. Alike mitosis, here also the spindle elongates to form the stem body and completes the anaphase separation of two groups of chromosomes.

Telophase

1. The telophase nuclei are formed.
2. They pass into a more or less complete resting* stage.

SECOND MEIOSIS

⊠ Prophase

If the chromosomes have not gone into a resting stage during interkinesis, there will naturally be no prophase to the second meiosis. In this case, the telophase nuclei of the first meiosis will pass directly into prometaphase of the second meiosis. Prophase, if present, is too short.

⊠ Prometaphase

It differs from the ordinary mitotic division in the following:

1. The number of chromosome is half the somatic number.
2. The chromatids diverge widely, being only held together at the centromere and not approximated throughout their length.

⊠ Metaphase

1. The chromatids tend, however, to come in closer contact.
2. The chromosomes have only one spiral.

⊠ Anaphase

1. Centromeres divide.
2. Two sister chromatids move to opposite poles.

⊠ Telophase

1. Four daughter nuclei are formed with haploid number of chromosomes.
2. The nuclear division is followed by cytokinesis and four cells are formed at the end of meiosis.

---

* The cells are continuously active and there is no resting stage. The term 'resting stage' is used to mean the period between two successive divisions.

⊠ Significance of meiosis

At fertilization two gametes unite but their chromosomes remain separate. In meiosis, the chromosome number is reduced to half (haploid) and in this way the original number (diploid) is restored in the zygote.

THE STRUCTURE OF A CHROMOSOME

In 1888, Waldeyar introduced the term chromosome (Gr. *Chroma* = Colour + *Soma* = body), because of its affinity towards dyes.

1. The chromosomes are darkly stained, individualised, protoplasmic bodies located in the nucleus.
2. In an interphase nucleus, these form an interwoven network of fine chromatin.
3. They appear as well-defined structures during cell division and are most distinct in metaphase only.
4. The chromosomes are self-duplicating structures and maintain their morphological and physiological characteristics all along.

⊠ Microscopic structure

A chromosome has three distinct zones—pellicle, matrix and chromonemata.

1. **Pellicle.** The very thin outer achromatic envelope.
2. **Matrix.** The ground substance.
3. **Chromonemata.** Two identical, spirally coiled thread-like structures embedded in the matrix. They are tightly coiled together and appear as a single thread of about 800 Å thickness.

a. A **chromonemata** consists of about 8 microfibrils, each of which is formed of double helix of DNA.

b. **Chromomeres.** These are small, dense masses at regular intervals on the chromonemata, and found only in leptotene and zygotene stages of meiosis.

c. **Interchromomeres.** The highly stained parts of the chromonemata in between the chromomeres.

d. **Centromere** or **Kinetochore.** A constricted zone where two chromonemata are joined together. This is also called **primary constriction.**

i. The chromosome attaches itself to the spindle fibre with the centromere during cell division.

ii. The centromere helps in the orientation of the chromosomes on the spindle. It divides during cell division and play an important role in the segregation of chromosomes to the poles.

iii. The centromere is usually one (monocentric); the number may be two (dicentric), many (polycentric) or none (acentric). The acentric chromosomes do not survive.

iv. The centromere may be located in different sites of a chromosome. **Telocentric**, occupying terminal position; **acrocentric**, subterminal position; **metacentric**, middle position. In metaphase the telocentric chromosome assumes rod-shape, with one small anterior arm and metacentric, V-shape.

4. **Arms.** The part of the chromosome on either side of centromere represents arms.
5. **Nucleolus organizer region or secondary constriction.** This is sometimes present in one or both the arms of a chromosome. This area is associated with the nucleolus and participates in its formation.
6. **Satellite.** Sometimes the secondary constriction is subterminal and the small part at the terminal end beyond the constriction appears as a knob. The shape and size of the satellite are constant for a particular chromosome.

AUTOSOMES AND SEX CHROMOSOMES

The sex of an animal is genetically determined. The chromosome complement in animals is grouped under two categories—**autosomes** and **sex chromosomes** (Fig. 3.7).

FIGURE 3.7

Autosomes and Sex chromosomes in Drosophila

☒ Autosomes

Autosomes are somatic chromosomes. They are homologous pairs, identical in shape and size.

☒ Sex Chromosomes

X and Y are sex chromosomes. They are smaller and have a tendency to divide slightly later than the autosomes. In females, the sex chromosomes, called X chromosomes are identical. The males may have only a single X chromosome (fruit fly) or an odd pair, one X chromosome and a somewhat smaller Y chromosome. The situation is reverse in butterflies and birds, the male has two X chromosomes and the female has one X and Y chromosome. In certain insects the male is XO and the female is XX.

In vertebrates, the male of most species is characterised by XY chromosomes and the female by XX chromosomes.

☒ Barr bodies

The cells of a normal male do not have Barr bodies. A cell of a normal female has one Barr body.

a. Only one X chromosome is functionally active in an interphase cell. If the cell contains more than one, all but one of them condense into small, discrete masses, the **Barr bodies**.
b. The total number of Barr bodies in any cell is always one less than the number of X chromosomes present (Fig. 3.8).

Telomeres. The two terminal ends or telomeres of a chromosome exibit physiological differentiation and polarity. Telomere prevents fusion of other chromosome segments to it.

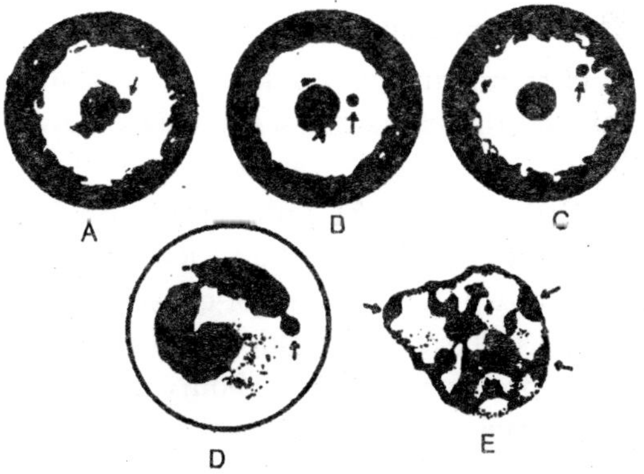

FIGURE 3.8

Sex chromosome and Barr body relation, A-C. Barr body in a nerve cell of a female cat. A. Near the nucleolus, B. In the nucleoplasm, C. Under the nuclear membrane. D. A drumstick nuclear appendage, Human female, E. Three Barr bodies in a nucleus of a female with XXXX

ULTRASTRUCTURE OF A CHROMOSOME

Two concepts—multistranded and unistranded, are currently held on the fine structure of a chromosome.

☒ Multistrand concept

1. Each strand of a chromosome is composed of a repeated linked nucleotides held together by phosphate-sugar groups.
2. The nucleotides of two strands are joined together through nitrogenous bases, two DNA molecules with their associated proteins constitute a microfibril.
3. Each microfibril is about 60-100 Å thick. Four microfibrils (8 double helices of DNA) collectively constitute a half chromatid.
4. Two such half chromatids form a chromatid.
5. A chromosome contains two chromatids or chromonemal threads. A chromosome if formed of 32 helices of DNA and is about 1,600 Å in diameter.

Unistrand concept

A chromosome is composed of a single microfibril made up of a single DNA double helix molecule and its associated proteins.

DuPraw (1965) holds that the DNA molecule is wrapped in the nucleoproteins and the DNA-protein fibres get coiled tightly to form 250-300 Å fibres (folded fibre model). The DNA-protein molecules of both chromatids of a chromosome remain held together by an unreplicated thread of chromosome in centromere region.

DNA: The genetic material

Avery, Mcleod and **McCarty** in USA proved that DNA is the carrier of genetic information from generation to generation.

1. The capsulated bacteria *Diplococcus pneumonia* are virulent and cause pneumonia. On the specificity of the bacteria they are specified as **type II, type III** and so on.
2. The different types injected separately in the blood stream of rabbits produced different forms of antibodies.
3. It proves that the capsule types are distinct properties of the bacterial strains.
4. When the different types are cultured separately, their respective characteristics are transmitted to their progeny.
5. In a culture, non-capsulated strain may develop by mutation which is non-virulent.
6. The capsulated ones are smooth in appearance (S) and the non-capsulated form are rough (R).
7. DNA extracted from heat killed type III S was mixed with vitro culture of type II R.
8. Type II R cells were precipitated with the addition of its antibodies.
9. In cases where transformation occurred, type III S grew diffusedly in the medium as they could not be precipitated by the antibodies added.
10. It was concluded that heat-killed type III S DNA was incapable of producing live III S. It also could not mutate II R to III S. But the transformation of type II R to type III S occurred by induced transformation with DNA of type III S.

DNA and RNA

The nucleic acids are of two types (p. 30). **Deoxyribonucleic acid** (DNA) and **Ribonucleic acid** (RNA). DNA is Feulgen positive, restricted to chromosomes while RNA is Feulgen negative and present mostly in the cytoplasm and only a small part of it in the chromosome. The differences are due to difference in sugar, pyrimidine bases and also the molecular organisation.

Constituents	DNA	RNA
Nitrogenous base		
A. Purines	a. Adenine b. Guanine	a. Adenine b. Guanine
B. Pyrimidines	a. Cytosine b. Thymine	a. Cytosine b. Uracil
Sugar	**Deoxyribose**	**Ribose**
Phosphoric acid	Present	Present
Dye reaction	Feulgen positive	Feulgen negative

4 CELL FUNCTIONS

Living units constituting an animal and plants are called cells. **Protoplasm**, the **cytoplasm** and the nucleus together, is surrounded by a thin membrane, the **plasma membrane**. As a rule, the number of nucleus in a cell is one. Several nuclei are present in a cell of the skeletal muscles and the nucleus is lost in mammalian red blood corpuscles (**Erythrocytes**) during maturation.

Functions of cells in different organs in an animal body are different; they also differ in their shapes and sizes, but cells in the living world are remarkably similar in their basic composition.

The cells are not static. From the new materials obtained from food substances—molecules of carbohydrates, proteins and fats are readily synthesized and incorporated into the cells but the total amount of these substances in the body are nearly constant. This is due to the fact that carbohydrates, proteins and fats and even the bone materials are being constantly synthesized and broken down. In young, the synthesis is greater than breakdown and an increase in the body volume occurs. In adults, the two processes are nearly equal and there is a little or no change in the body volume.

Against this background, the deoxyribonucleic acid (DNA) molecules, which constitute the genes in the nucleus, are remarkably constant. The synthesis and breakdown of DNA molecules are very slow and occurs only occasionally. New DNA molecules are synthesized only before the division of a cell. DNA is responsible for transmission of potentialities of characters to succeeding generations, and the stability of DNA is to ensure minimum deviation* in the transmission. On the other hand the amount of ribonucleic acid (RNA) per cell varies greatly, as the RNA molecules are constantly synthesized and broken down.

ENERGY

The ability to do work or to produce a change in the matter is called energy. All the different forms of energy have been placed under two heads—potential energy and kinetic energy.

⊠ Position or potential energy

It is considered as stored energy.

* Recent investigations suggest that, of the genes in action, only forty per cent are involved in the replication of DNA, while sixty per cent act as brakes to prevent any deviation from the original code in the replicated molecules.

It cannot produce a change but it has latent power, which, on release, is capable of doing work. The water in a lake on a mountain or in a dam does not do any work but it is under gravitational pull. If the water is allowed to flow downwards it can move a machine or a turbine, the stored energy being released in downflow. A piece of coal is inert but on burning the energy stored in it is converted into kinetic energy—heat and light.

☒ Motion or kinetic energy

The energy may be in the form of mechanical energy of moving body. Heat and light are placed under this category.

The energy source of earth is **radiant energy*** from the sun. The green plants can synthesize organic molecules, sugar in the form of chemical energy, a kind of potential energy. In the process, carbon dioxide and water are joined by using sunlight as a source of energy. Light is absorbed in the chlorophyll pigment and transformed by the cells into the chemical energy needed for the synthesis of carbohydrate. The radiant (kinetic) energy is transformed into chemical (potential) energy, which is stored in carbohydrate molecules. From simple inorganic compounds, plants can synthesize highly complex organic compounds, such as carbohydrates, fats and proteins and a considerable amount of energy is stored in these molecules.

In the living cells, the food molecules are subjected to chemical reactions and the potential energy in the molecules change into heat, light, motion or some other kinetic energy. Energy is neither created nor destroyed but simply transformed from one form to another. Living cells are unique in their ability to transform chemical energy directly into heat energy, without being initiated by some other energy.

☒ Energy transformation in living cells

Type of cell	Transformation
1. Majority of the cells	Chemical energy to heat energy
2. Muscle cells	Chemical energy to mechanical energy
3. Cells of excretory organs	Chemical energy to osmotic energy
4. Nerve and brain cells	Sound to electrical energy
5. Cells of inner ear	Chemical energy to electrical energy
6. Retinal cells of the eye	Light to chemical energy
7. Sense organs—taste and smell	Chemical energy to electrical energy
8. Cells of the bioluminescent organs of fire fly	Chemical energy to radiant energy

MOLECULAR MOTION

The three states of matters are solid, liquid and gas. The number of molecules in a given volume of matter at normal temparature and pressure, irrespective of the state, is constant.

$$\text{It is } \frac{6.023 \times 10^{23}}{22.4} \text{ in 1 litre of liquid or gas.}$$

The molecules constituting a matter are in constant motion. The freedom of molecular movement differs with the state of the matter.

☒ Solid

In a solid the molecules are closely packed together and the **cohesive force** between the molecules permit them only to vibrate and not to move around. This helps the solid to retain a definite shape.

☒ Liquid

In a liquid the distance between the molecules is more. Cohesive force is less and they can freely move around. For this a mass of liqid attains the shape of the container.

* A tremendous amount of energy is received by earth from the sun in the form of sunlight or light particles called photons. The amount of energy received by the earth in every three minutes in sunny days is equal to the heat produced by burning 1,00,000 tons of coat. Only a small part of the radiant energy is utilised in the synthesis of plant materials. It is estimated that microorganisms degrade about 800 billion ton plant materials every year and those are returned to earth.

⊠ Gas

The molecules are far apart, the cohesive force is negligible and the molecular movement is restricted only by external barriers. For this, even a small amount of gas fills up a large container.

⊠ Surface tension

Due to the movement of the molecules, some of them in a fluid tend to move away from the exposed surface. They are held back by a force called surface tension (Fig. 4.1).

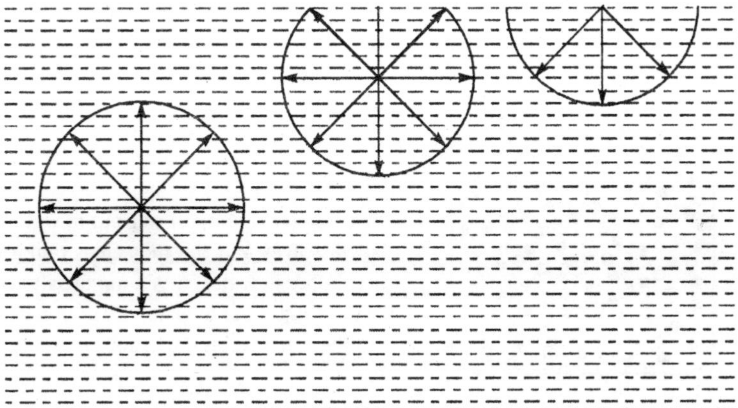

FIGURE 4.1
Surface tension

⊠ Heat

With heating, the rate of movement of the molecules in a matter increases, irrespective of the state. The increase is directly proportional to the heat applied. The rate of movement of the particles is the **heat** and the measurement of the heat is **temperature**. At boiling point, the movement of the molecules in a liquid is so fast that a large number of them cross the threshold of surface tension and escape from the surface.

DIFFUSION

The movement of molecules of a solute in a solvent from higher to lower concentration is diffusion. The concentration of a solute in a solvent depends on the number of its molecule in a given volume of solvent. The solute molecules move from higher to lower concentration and spread gradually till they are uniformly distributed in the whole solution and an equilibrium is reached.

Molecules in a matter are in constant motion. The movement is a manifestation of the kinetic energy in the molecules.

The molecules tend to disperse in a straight line but the direction of movement at any particular moment is unpredictable, because its direction depends on its chance collision with other molecules or the wall of the container (Fig. 4.2). This retards the rate of movement of the molecules but facilitates the even distribution of the substance in a short period.

The diffusion of the molecules of one substance in a solution is independent of the diffusion of the molecules of other substances. In a mixed solution the relative concentration of each solute is critical in determining the direction of diffusion of the solute. No two molecules can occupy the same space at a time. Increase in the concentration of one will influence the concentration of the other. In high solute concentration, the solvent concentration must be low. An increase in concentration of one solute results in the corresponding decrease in the concentration of other solutes or the solvent.

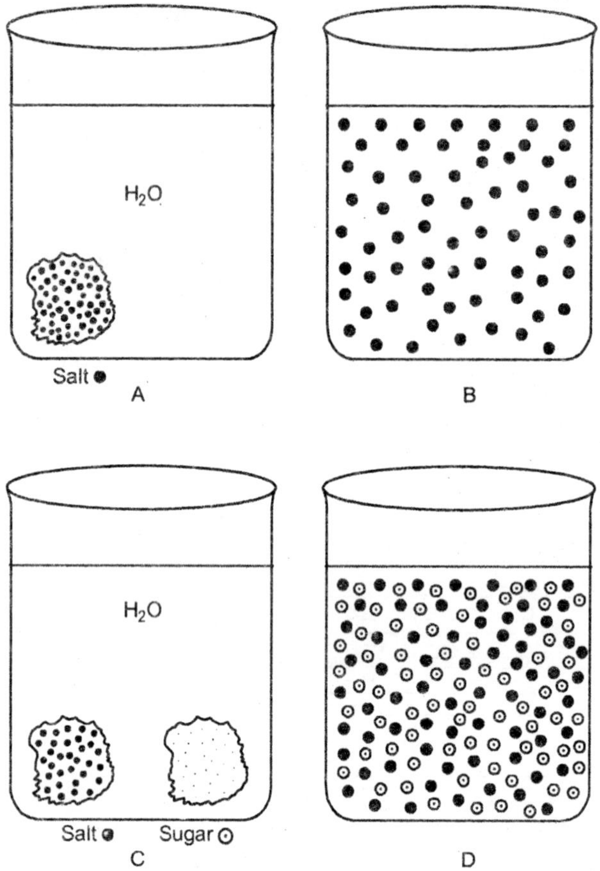

FIGURE 4.2

Diffusion: A. A lump of salt put in water; B. Salt dissolves and the molecules uniformly distributed in water; C. A lump of salt and a lump of sugar put in water; D. Both dissolve and their molecules uniformly distributed in water

Osmosis

The diffusion of water or a solvent or solute through a semipermeable membrane is osmosis (Gr. *osmes* = action of pushing). It is passive. When a solution of salt or sugar is separated from water by a semipermeable membrane that allows only the water (small molecules) to pass but not the large molecules of sugar or salt, the movement will be one way, i.e. the water molecules will pass to the solution (Fig. 4.3). The water molecules diffuse from higher (pure water) to lower concentration (solution). A reverse movement also occurs but at a much slow rate. As a result, the level of the solution rises till an equilibrium is reached, i.e. the weight of the solution exerts a hydrostatic pressure just equal to the pressure of water to enter the solution.* The pressure exerted by the water column is called **Osmotic pressure**. Substances dissolved in the intracellular fluid exerts osmotic pressure on the interstitial fluid and vice versa. When the two processes are equal, there is no entry or exit of water from the cell.

* A number of factors regulate the movement of molecules through the semipermeable membrane.

1. **Pore diameter.** The membrane is provided with pores. The molecules larger than the pore diameter cannot pass through the membrane.
2. **Solutes.** Concentration (number of molecules) of solutes, one or more, in each of the separated solution.
3. **Other factors.** (a) Electric charge of the molecules (ions), (b) Number of bound water molecules, (c) Fat solubility of diffusing molecules.

⊠ Isoosmotic solution

A solution in which a cell neither shrinks nor swells, such a solution is called isoosmotic or isotonic with the intracellular fluid of the cell.

⊠ Hyperosmotic solution

A cell placed in a solution containing more dissolved substances than in the intracellular fluid, will shrink, as the water from the cell would move out to the solution. Such a solution is known as hyperosmotic with the intracellular fluid of the cell.

⊠ Hypoosmotic solution

A cell placed in a solution containing less dissolved substances than in the intracellular fluid, swells as the water from the solution moves into the cell. Such a solution is called hypoosmotic to the cell.

Erythrocyte is an ideal material for demonstration of osmosis. Plasma membrane of R.B.C. is permeable to water but seemingly impermeable to salts. 160 milimolar NaCl solution is isoosmotic with plasma. If the R.B.C. is placed in a more concentrated (hyperos-motic) solution, water moves out of the cell and the cell shrinks; in isoosmotic solution water does not move in any way and the cell retains its shape; in a more dilute (hypoos-motic) solution, water enters the cell when placed in pure water, the cell swells and bursts.

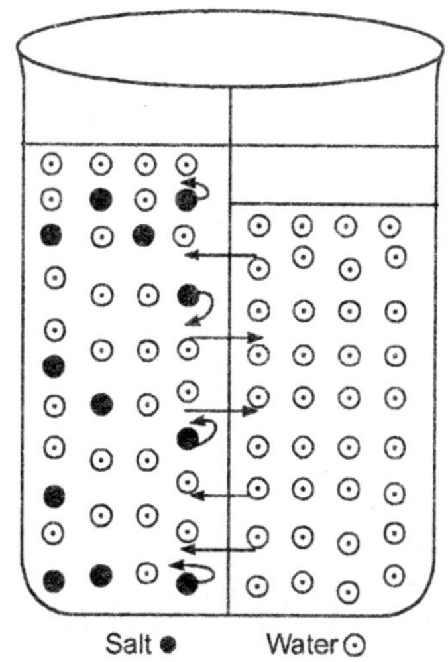

FIGURE 4.3

Osmosis: Semipermeable membrane permits small water molecules to pass but not the larger salt molecules. The water molecules diffuse from higher concentration (pure) to lower concentration (salt solution).

TRANSPORT ACROSS THE PLASMA MEMBRANE

All substances moving from outside to inside a cell, be it from external environment to epithelial cells of the body surface or from interstitial fluid* to cell inside the body, must pass through the plasma membrane surrounding the cell. The plasma membrane is provided with minute pores, the diameters of which determine the maximal size of the molecules allowed to pass through the membrane. The transport of substances from the surroundings to inside the cells and from the cells to the surroundings take place by two means.

⊠ Passive transport

In this type of transport no energy is required. Movement of molecules entirely depends on their relative sizes and concentrations on either side of the plasma membrane. The flow takes place from higher to lower concentration (Fig. 4.4A-C).

Movement of ions across the membrane follows the principle of passive transport. Particles with an electric charge will move to the particles having opposite charge, provided the flow is from higher to lower concentration.

⊠ Active transport

Movement of molecules against concentration gradient is known as active transport.

* The fluid filling up the spaces between the cells. Plasma water in blood, lymph fluid, transcellular water (fluid secreted in the coelom and digestive system), and interstitial fluid together constitute **extracellular fluid**.

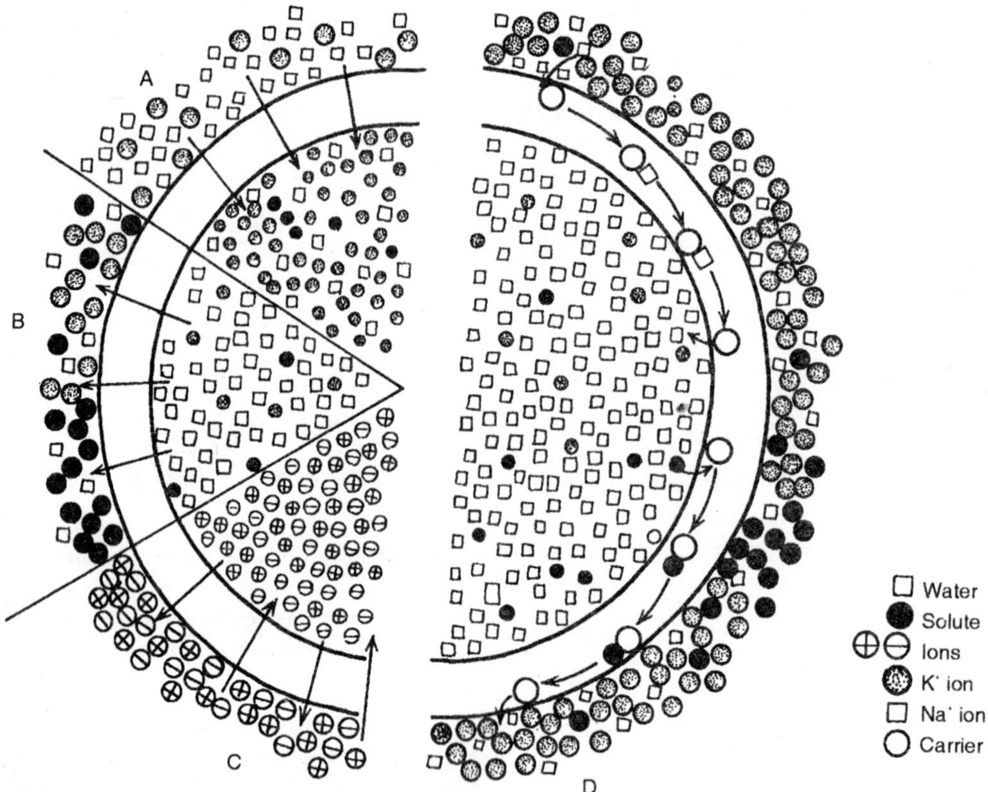

FIGURE 4.4

Movement of particles through membrane (schematic): A-C. **Passive transport**, D. **Active transport**.

A. Water molecules diffuse from higher to lower concentration (inside)

B. Water molecules diffuse from higher to lower concentration (outside)

C. Diffusion of ions from higher to lower concentration (flow of ion regulated by concentration)

Energy is required to move the molecules in a direction opposite to that which would have taken place in passive transport (higher to lower concentration) (Fig. 4.4D).

In active transport an important role is played by **carrier molecules** in the membrane. The carrier molecules combine with the molecules of the substance to be transported from outside, form temporary compound, carry those inside the cell and released there. Alternately, it may operate in the same way to carry molecules from inside to outside the cell.

☒ Cell and environment: Exchange of materials

An organism is intimately associated with its surroundings. Exchange of materials are constantly taking place between the cell and its surroundings. Exchange takes place betweeen cells on the body surface and environment and also between the cells and its surroundings inside the body.

The chemical factors in outside environment, oxygen and carbon dioxide concentration, acidity, alkalinity and temperature; in the internal environment, the concentrations of different substance, viz., carbohydrate, amino acid and other nutrients and carbon dioxide in the interstitial fluid greatly influence the activities of an organism. In terrestrial forms, the chief requirements of organisms are received from the lumen of the gut (products of digestion) and lung alveoli (oxygen). In aquatic forms, in addition to those, substances from outside are taken in and given out from the body through the skin, gills, etc.

The plasma membrane, surrounding the cell, plays a vital role in the process of exchange of materials. All nutrients and waste products must pass through the plasma membrane to enter or leave the cell. The plasma membrane is surrounded by interstitial fluid. Substances only in solution can pass through the membrane. The membrane bears ultramicroscopic pores. These pores determine the maximal size of the molecules that can pass through them. Electric charge, if any, the number of water molecule bonds to different molecules, and its solubility in fatty substances, largely govern the diffusion.

The membranes are of different types: (a) **permeable**, allows any substance to pass through it; (b) **nonpermeable**, allows no substance to pass and (c) **semipermeable** or **differently permeable**, allows certain but not all subtances to diffuse through it. The plasma membrane, nuclear membrane and vacuoles are semipermeable.

CATALYST

A catalyst is an agent, which accelerates the velocity of a chemical reaction without altering its end point, and not being used in the chemical reaction. Enzymes are organic catalysts.

BIOSYNTHESIS

Synthesis of a complex biological molecule from simple substances is biosynthesis. A cell is a complete unit with all its constituents. Unlike energy flow, biological compounds do not pass from one cell to another or from one organ to the other and from one to another organism. Each cell synthesizes its own requirements from materials supplied to it. Several steps are involved in the synthesis of a compound and these are controlled by enzymatic reactions, which, in turn are genetically determined. The major constitutents of a living cell are proteins, enzymes (also protein), carbohydrates, lipids and nucleic acids. Protein, the structural material, varies in the cells of different organs of an individual. Glycogen molecules present in liver and muscle cells are similar but not obtained in preformed state from a common pool but synthesized in each cell. Same is true for lipids in adipose tissue, subcutaneous tissue and other organs. Synthesis of nucleic acids is carried independently in each cell.

ENZYMES

The enzymes are organic catalysts synthesized in the cells. They are protein in nature, accelerates the rate of reaction but does not determine direction of reaction. A small quantity of enzyme can bring reactions in a large amount of substrate. Being protein in nature, they lose activity within a short period in a reaction system and this needs a continuous replacement by further synthesis.

⊠ Naming of enzymes

An enzyme is named by adding suffix **ase** after the name of the substrate, amylum- **amylase**, sucrose–**sucrase**, protein–**proteinase**, lipid–**lipase**, urea–**urease**, etc. Some enzymes were named before the introduction of present naming system and those are retained following the law of priority. **Trypsin**, **ptyalin**, **renin**, **pepsin**, and some others belong to this category.

⊠ Substrate

The compound on which an enzyme acts—amylum, sucrose, protein, lipid, urea, etc. are common substrates.

Specificity. An enzyme reacts only with a particular substrate. The degree of specificity, however, differs in some cases.

⊠ Enzyme-substrate complex

An enzyme combines with its substrate to form a reactive intermediate, the enzyme-substrate complex. In the complex; two molecules must

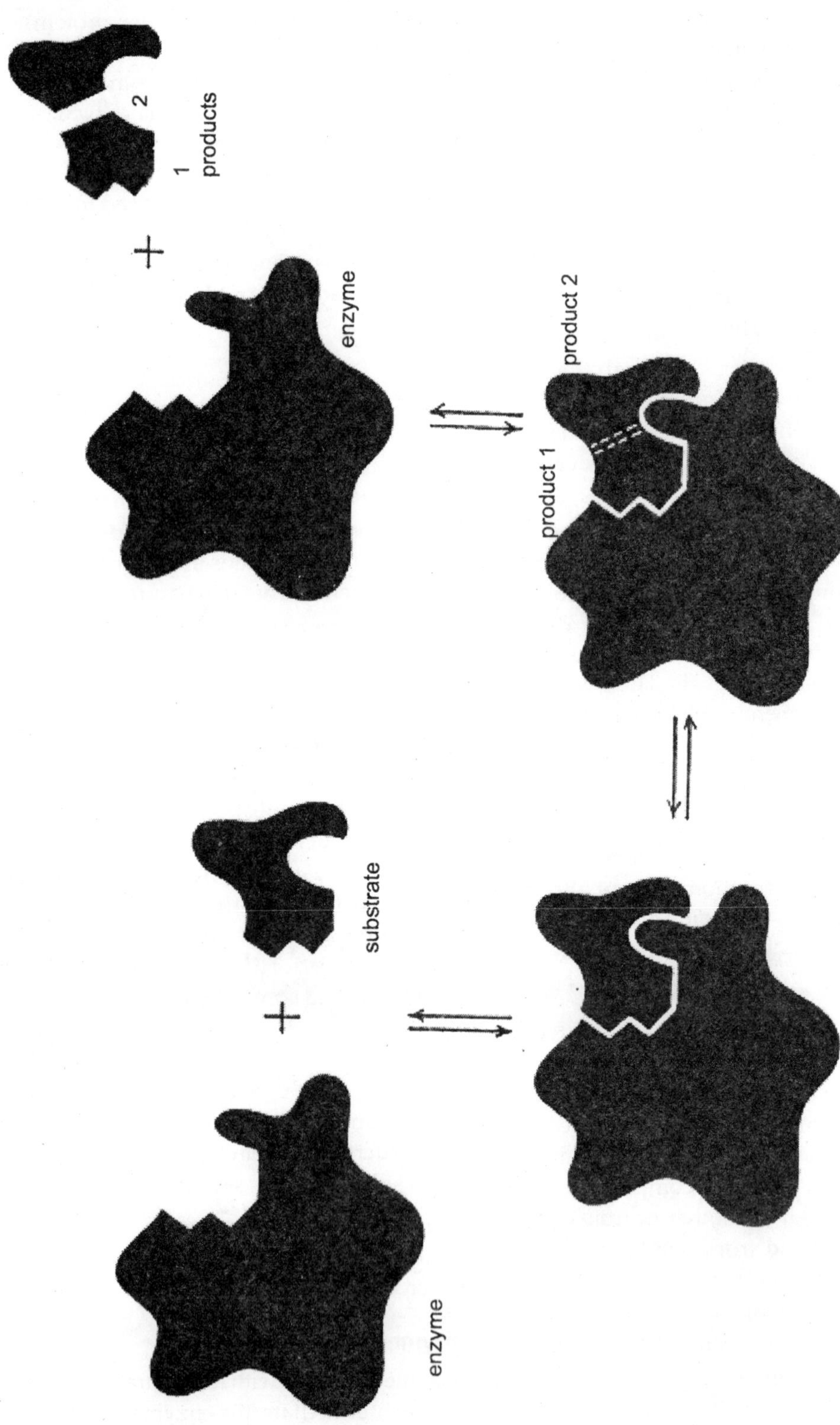

FIGURE 4.5
Enzyme reaction. Formation of an enzyme-substrate complex, followed by catalysis

fit together closely like a lock and key. The molecules are attached at a particular site, the **active site**, for specific configuration of the molecules. The specificity of an enzyme is due to this configuration in the combining molecules (Fig. 4.5).

⊠ Turnover

The number of substrate molecules acted upon by one molecule of enzyme in one minute in optimum conditions. This figure is very high in most cases. This is also known as **enzyme activity**. The enzyme molecule is released at the breakdown of the substrate molecule and it is free to combine immediately with another substrate molecule.

⊠ Reversible reaction

The enzyme-substrate reaction is reversible. When the reaction reaches equilibrium, the direction of reaction is reversed.

$$\underset{\text{ethyl alcohol}}{CH_3CH_2OH} + \underset{\text{acetic acid}}{CH_3COOH} \rightleftharpoons \underset{\text{ethyl acetate}}{CH_3COOC_2H_5} + \underset{\text{water}}{H_2O}$$

⊠ Enzyme activator

Molecules stimulating enzyme reaction. Chlorine ions Cl^- and bile salts activate salivary amylase and pancreatic juice, respectively.

⊠ Enzyme inhibitor

Substance inhibiting enzyme reaction. Inhibitors are **noncompetitive** and **competitive.** Noncompetitive inhibitor combines with the active site of the enzyme molecule. It is irreversible. Competitive inhibitor competes with other substrates for the available enzyme molecules. It is reversible.

⊠ Factors affecting rate of enzyme reaction

pH, temperature, enzyme and substrate concentration, end products and period of reaction, all interfere with enzyme reactions. The range in which an enzyme is most active is called **optimum range** and this is very narrow. All enzymes are **thermolabile.***

⊠ The mechanism of enzyme reaction

An enzyme molecule combines with a substrate molecule at their active sites and a reactive enzyme-substrate complex is formed. The bonds involved in the complex are ionic bonds, hydrogen bonds and hydrophobic bonds. The complex breakdown quickly releasing the free enzyme and the reaction products.

⊠ Classification of enzymes

On the basis of reaction, enzymes have been classified into six major groups (a) **oxidoreductases**-involved in biological oxidation and reduction, (b) **transferases**-transfer of one radical from one compound to another, (c) **hydrolases**-break up complex molecules into simple ones by adding of water molecule, (d) **lyases**-cleave the substance without a reactant, (e) **isomerases**-intramolecular rearrangement of atoms in the substrate, (f) **ligases (synthetases)** linking of two substrate moleules.

⊠ Michaelis constant (Km)

The substrate concentration at which an enzyme catalyses reaction proceeds at one-half its maximum velocity.

---

* Thermolabile means the substance is changed or loses its activity above certain temperature. The activity of an enzyme does not come to a halt at the highest level of optimum temperature range. The activity continues, though at a reduced rate, at temperature slightly above the optimum range. **Cellulase** is the most thermostable enzyme and its activity has been recorded at a temperature as high as 50°C.

5 TISSUE AND TISSUE SYSTEM

A tissue is a group of similar cells forming a continuous mass or layer to carry on similar functions. Two or more kinds of tissues in combination form an **organ**. The different tissues of an organ work together and an organ performs a particular function. Typical organs are liver, heart, eye, etc. A few organs in combination form a **system**, such as digestive system, circulatory system, nervous system, etc. A few systems in combination form the body of an individual.

TISSUES

Broadly, six types of tissues have been identified. They are classified according to their functions.

⊠ Epithelial tissue

It forms a continuous layer covering the body externally or internal surfaces. It is protective in function. Epithelial tissue covering the external body surface is external epithelium and that covering the internal surface, viz. body cavity and other organs are internal epithelium. Epithelial cells are usually supported by a basement membrane. The outer epithelium, to some extent, protects the body from mechanical injury, from chemicals and from the invasion of microorganisms. The epithelium in the intestine absorbs water and nutrients; in the kidney glucose, amino acids, salts and, in some cases, water. Epithelial tissues (Fig. 5.1) are classified according to their shape.

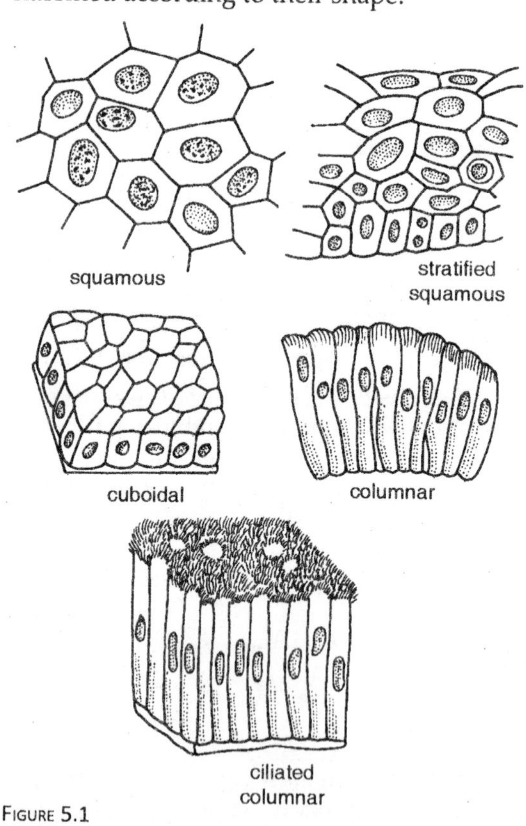

FIGURE 5.1
Epithelial tissue

a. **Simple** or **Nonstratified epithelium.** The tissue is made of a single layer of cells. Present in the intestine.

b. **Stratified epithelium.** The tissue is made of many layers. Present in the palate of rabbit.

c. **Pavement** or **Squamous epithelium.** The cells of the tissue are thin and flat. Present in outer skin of frog.

d. **Cuboidal epithelium.** The cells are cuboid in outline. Present in liver, kidney tubules etc.

e. **Columnar epithelium.** The cells are elongated and supported by a basement membrane. The length is more than the width. Present in intestinal mucous membrane.

f. **Ciliated epithelium.** It is modified columnar epithelium. The cells are provided with cilia at the free end. Present in oral cavity and genital ducts.

Some epithelial cells, both external and internal, respond to stimuli. These are sensory cells. Some cells are capable of secreting certain substances. These are glandular cells.

Connective tissues

These tissues form supporting structures of the body and connect various parts of the body. Large amount of nonliving intercellular substance, the **matrix,** is present in these tissues holding the cells together. The nature and function of the connective tissue is determined by its intercellular matrix. Five major types of connective tissues are found.

a. **Fibrous connective tissue.** The cellular part of the tissue consists of small, scattered cells. The intercellular structure is constituted by unbranched, wavy fibres. The white fibres or collagen fibres are arranged in bundles and embedded in a clear homogeneous matrix. The yellow elastic fibres or elastin are straight and arranged closer (Fig. 5.2). Other types of fibrous tissues are present in muscular sheath, tendon and ligament.

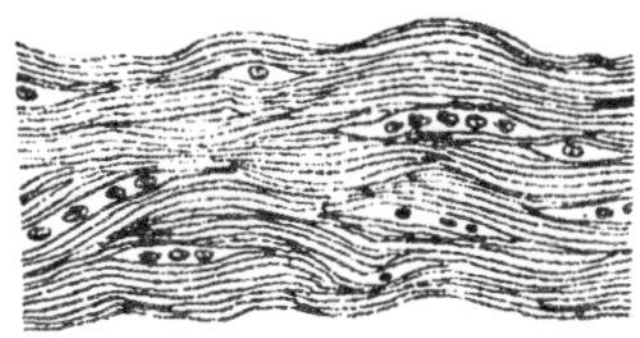

FIGURE 5.2
Fibrous tissue

b. **Tendon.** Thick, closely packed bundles of collagen fibres constitute a tendon. Tendons are flexible and join a muscle to a bone or to another muscle.

c. **Ligament.** Ligament is similar to tendon, it connects one bone to another.

d. **Cartilage.** The cartilaginous tissue is semitransparent. Oval cells, singly or in groups, are embedded in the semitransparent intercellular matrix which forms the main bulk of the cartilage. Cartilage can grow from within by divisions of cartilage cells, the modified **fibroblasts.** They secrete more matrix around themselves (Fig. 5.3). Hyaline or semitransparent cartilage is present in skull, limb girdles, between the vertebrae and other parts of the body. The cartilage is a firm but elastic structure and most bones develop from cartilage by calcification.

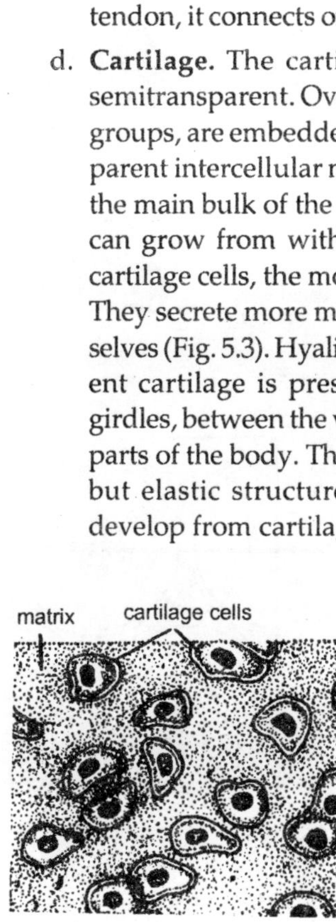

FIGURE 5.3
Cartilage (section)

e. **Bone.** The bones are of two types—the long bone and membrane bone. The bone is almost similar to cartilage but the matrix is rendered hard by a rich deposition of lime salts. Membrane bone is formed directly from mesenchyme. A long bone is surrounded by a thin, fibrous **periosteum** (Gr *peri* = around + *osteum* = bone), except articular surfaces. A bone consists of an outer layer of **compact bone**, and a **medullary cavity** at the centre (Fig. 5.4A). The compact bony layer is made of many **Haversian systems**, closely packed. A system has a narrow, central Haversian canal, around which are concentric bony rings or lamellae, having narrow lacunae in between them. Osteoblasts, osteocytes and osteoclasts are present in the lacunae. Narrow canaliculi connect the lacunae, with each other and also with the Haversian canal. Interstitial lamallae separate the Haversian systems. The Haversian systems are enclosed in a secondary lamella. (Fig. 5.4 B, C). The medullary cavity is occupied by **bone marrow**, which is the site for production of blood cells. The bones support and add rigidity to the body.

f. **Adipose tissue.** In this tissue the cells are enlarged and heavily laden with fat in the form of globules. The fat is secreted by the cytoplasm and due to accumulation of huge quantity of fat the cytoplasm and the nucleus are pushed to one side (Fig. 5.5). Large amount of adipose tissues are present under the skin.

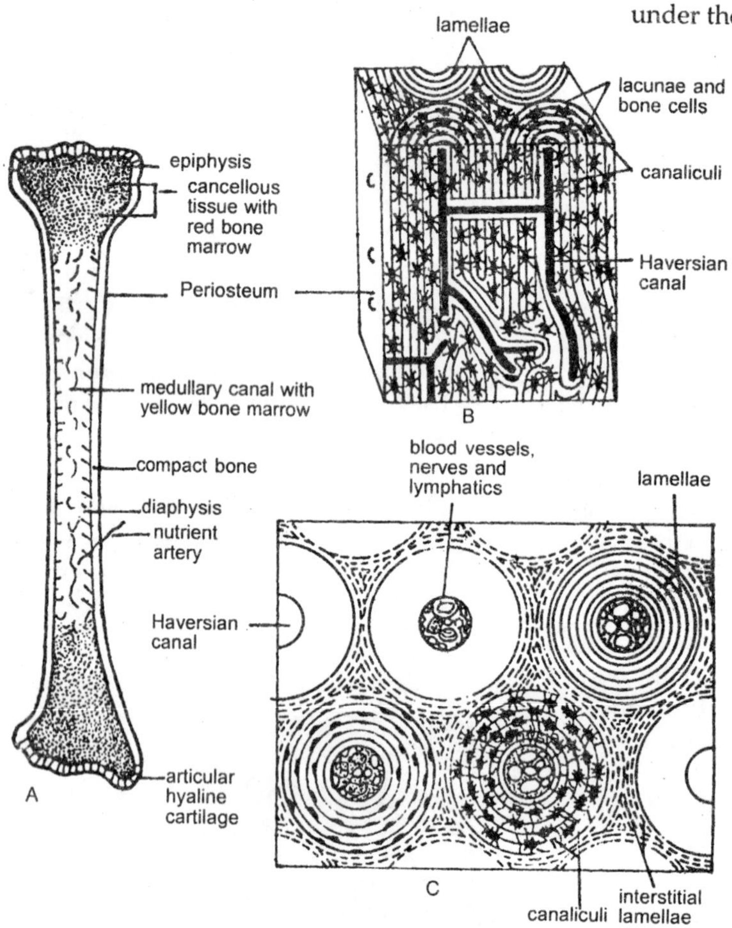

FIGURE 5.4

Long bone: A. Longitudinal section; B. Longitudinal section (magnified); C. Transverse section (magnified)

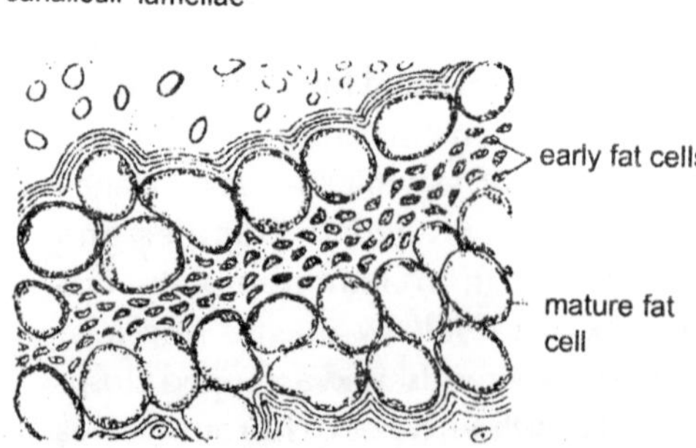

FIGURE 5.5

Adipose tissue

Muscular tissue

All the movements of animal body are performed with the help of muscular tissue. The cells of muscular tissues are elongated and slender (Fig. 5.6-5.9). All muscle cells—whether skeletal, cardiac or smooth muscle, contain contractile protein arranged as **myofibrils**. In **striated** muscle (skeletal and cardiac) the filaments are arranged **symmetrically** in an orderly fashion. In a skeletal muscle groups of muscle fibres are bound together and invested in a connective tissue sheath, through which blood vessels and nerves pass. The cells in a muscle develop tension along their long axis and the muscle contracts. Muscles only contract and relax, they never elongate. In **smooth** muscle the arrangement is random. Proteins, **actin** and **myosin** are present in all myofilaments.

a. **Skeletal (Striated voluntary) muscle.** The muscle cells **(sarcomeres)** are large (Fig. 5.6). The plasma membrane is known as **sarcolemma**. The cytoplasm or the sarcoplasm. is composed of longitudinally arranged myofibrils, the essential contractile component of the cell. The cells are multinucleated. Cross striations run at right angles to the presence of alternating thin bands or discs. The bands (Fig. 5.7) are:

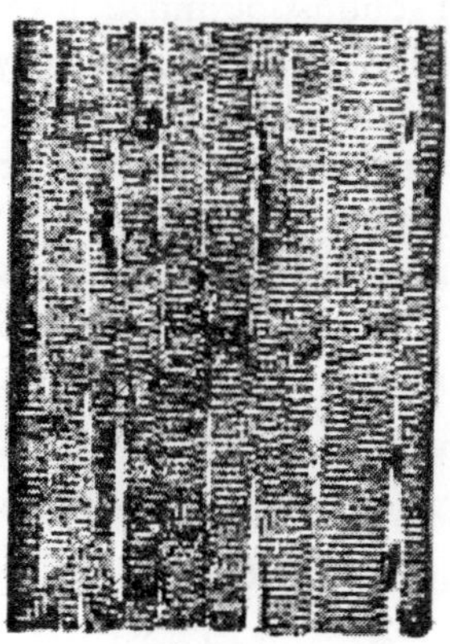

FIGURE 5.6

Skeletal muscle

i. I (Isotropic) bands appear light. A narrow central dark disc, the Z (Zwischenscheibe) band is present in a I band.

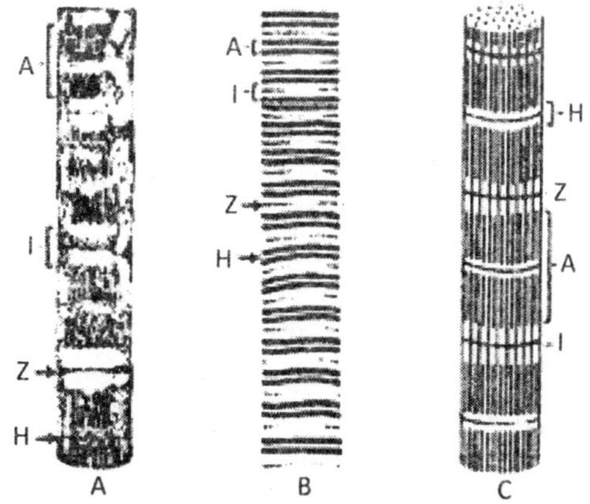

FIGURE 5.7

Myofibril: Skeletal muscle cell

A. An electron microscopic view of a myofibril
B. An explanatory diagram pointing the various bands in a myofibril.
C. A three dimensional view of a myofibril

ii. A (Anisotropic) bands appear dark. A narrow, less dense central portion is known as H (Heller) band.

The functional unit of the sarcomeres is the zone of myofibril between two Z bands.

b. **Cardiac muscle.** The sarcomeres of cardiac muscle are similar to those of skeletal muscle. In addition, cardiac muscle cells are joined by dense **intercalated** discs (Fig. 5.8), which appear to replace Z bands. Due to irregular arrangement of membranes, the firm contacts between adjacent sarcolemmal membranes at the intercalated discs extend over large areas.

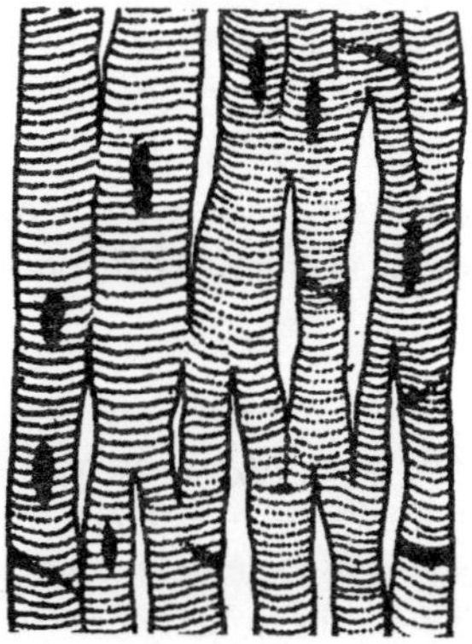

FIGURE 5.8

Cardiac muscle

c. **Smooth (Involuntary) muscle.** Smooth muscle cells are generally small. The sarcoplasm contains many longitudinally arranged myofibrils (Fig. 5.9). The single nucleus is eccentric. Unlike the muscle fibres of the skeletal muscles in which longitudinal myofibrils are packed, the myofibrils are arranged at random in a smooth muscle fibre. The fibres contain actin and myosin and the mechanism of contraction is same as in skeletal muscle. Smooth muscle has two functions, sustained and intermittent contraction.

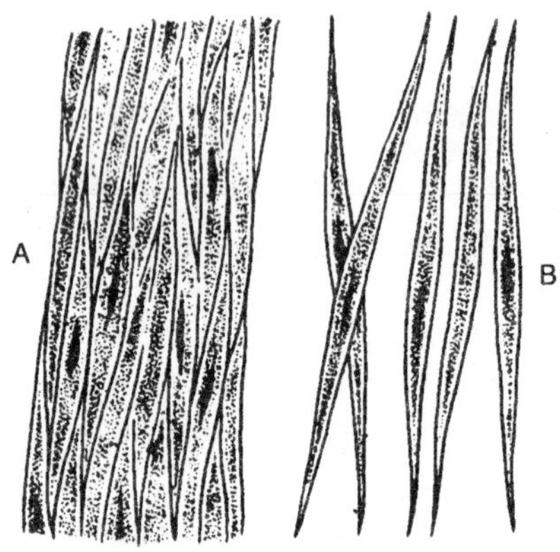

FIGURE 5.9

Smooth muscle: A. A muscle, B. Smooth muscle fibres (cells)

The smooth muscles in vertebrates are of two types: (a) **multiunit smooth muscle** present in iris and ciliary body of the eye and in the walls of the blood vessels and (b) **visceral smooth muscle** in the walls of the gut, uterus and visceral organs. Like other muscles, the visceral smooth muscles also possess certain spontaneous activities and contract automatically when stretched, which may happen when an organ is filled with food or liquid.

Blood and Lymph tissue

The blood and lymph are considered liquid tissues, as loose cells float in the plasma which is comparable to the ground substance of connective tissue.

Blood is essential for the various activities of an animal, because this is the important transport medium in the body. Blood consists of a liquid portion,—the **plasma** and the blood cells or **corpuscles** floating in it. Besides the corpuscles, fibrinogen remains in the plasma in a soluble state.

In most of the invertebrates the blood corpuscles are amoebocytes but oval and nucleated in nemertines, oval and biconcave in some annelids and phoronis. In invertebrates, usually the outflow of blood (haemolymph) through injury is prevented by agglutination of blood cells.

Human plasma constitutes about 55 per cent of the blood volume and 92 per cent of the plasma is water. Plasma proteins, amino acids, glucose, lipids and nitrogenous wastes constitute 8 per cent solid. Albumin, α, β and γ globulins, euglobulin and fibrinogen helps in coagulation of blood. The blood cells in vertebrates are of three types—**erythrocytes, leucocytes** and **thrombocytes**. All cells except lymphocytes* originate in red bone marrow and are temporarily stored in spleen, which is also a site of origin of the cells. Lymphocytes (Fig. 5.10) originate in the lymph nodes and also stored there.

Red blood corpuscles or R.B.C. or erythrocytes

These are much more in number and an iron compound, **haemoglobin** is present in the cytoplasm of the cell. This is a respiratory pigment. In most of the invertebrates the respiratory pigment is **haemocyanin**, a copper com-

* Lymphocytes are of two types, ***T-lymphocyte*** and ***B-lymphocyte***. Both owe their origin to common stem cells in the bone marrow but mature in different tissues. T-lymphocytes undergo maturation in thymus and B-lymphocytes possibly in the foetal liver, tonsil, appendix and lymphoid tissue of the intestinal wall in man. Mature lymphocytes enter the lymph nodes (Fig. 5.10) and general circulation.

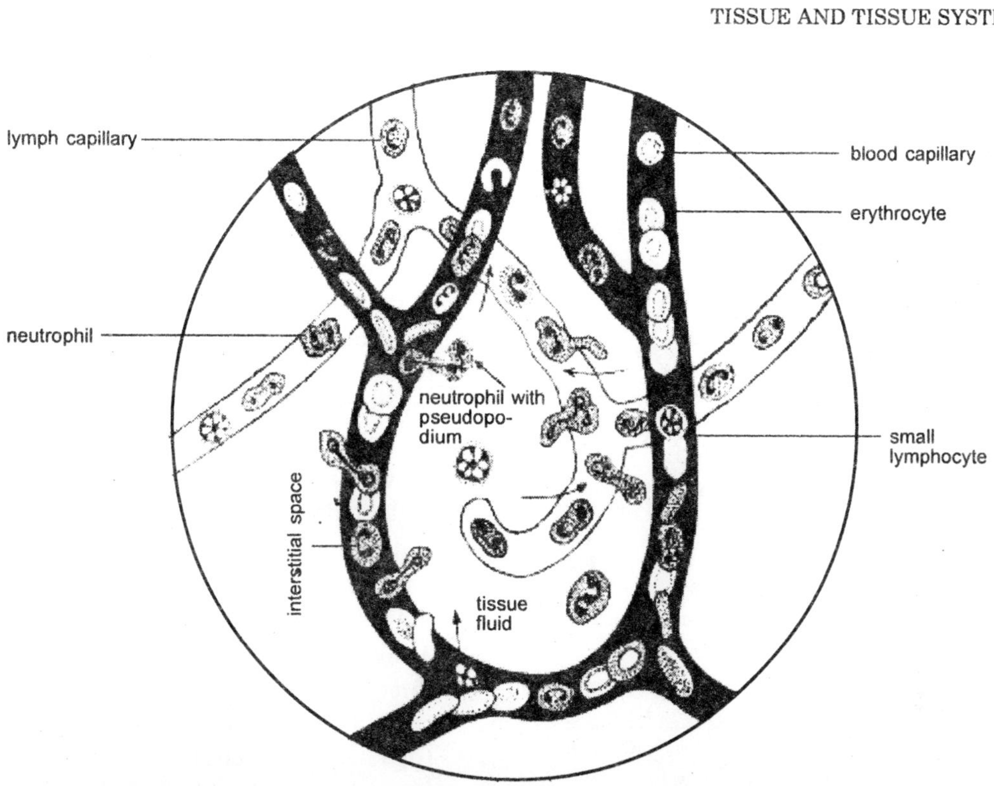

FIGURE 5.10

Lymphocytes: Amoeboid movement from blood capillary to lymph capillary

pound. The colour of vertebrate blood is red due to haemoglobin, while the haemocyanin content blood is almost colourless. In all vertebrate R.B.C. except that of mammals, a nucleus is present, the shape of which differs in different classes of vertebrates (Fig. 5.11).

a. **Fish.** Almost round.
b. **Amphibia, reptiles** and **birds.** Biconvex and oval.
c. **Mammal.** Round and biconcave.

⊠ White blood corpuscles or W.B.C. or leucocytes

These are less in number, large in size and the nucleus is not round (Fig. 5.11). Some of the cells are phagocytic and can engulf bacteria and other foreign particles with pseudopodia.

Leucocytes are of the following types.

a. **Monocytes.** Large size, with a single nucleus. In a young monocyte the nucleus is round or elliptical but becomes kidney-shaped later.
b. **Lymphocytes.** Nearly circular with a large nucleus.
c. **Granulocytes.** Capable of amoeboid movement. The cytoplasm is granular. Granulocytes are of three types.
   i. **Eosinophil.** Take up only acid stains. The nucleus bears 2-3 segments.
   ii. **Basophil.** Take up only basic stains. The nucleus is fragmented.
   iii. **Neutrophils.** Take up only neutral stains. The nucleus bears 2-7 segments.

⊠ Thrombocytes or blood platelets

These are small cells of irregular shape and help in coagulation of blood.

⊠ Blood Volume

The amount of blood present in the body or the blood volume differ in animals. It is 7 to 10 per cent of the body weight in man; 5 per

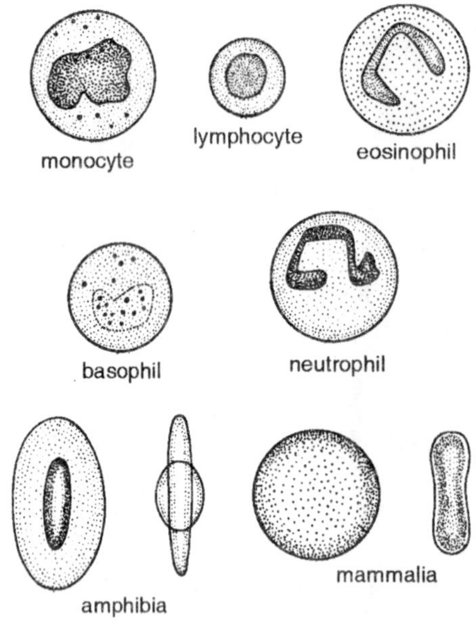

FIGURE 5.11

Blood cells (First two rows leucocytes, last row erythrocytes).

cent in bony fishes; 5-8 per cent in *Octopus;* 5-25 per cent in insects.

Functions of blood

a. **Transport of respiratory gases.** Haemoglobin combines with O_2 to form unstable oxyhaemoglobin. The oxygen is taken up in the respiratory organs, skin, gill, lung as the case may be and released in the tissue. In the plasma, haemoglobin directly combines with CO_2 to form carbamino compounds and the CO_2 is released in the respiratory organ. Haemocyanin has similar function.

b. **Transport of nutrients.** It carries nutrients, the products of digestion of food and water, minerals, etc. absorbed from the intestine, to the cells. It also carries substances from one place of the body to other, i.e. from storage site to other parts.

c. **Removal of wastes.** Waste products of metabolism are carried from the site of origin to the organ of excretion.

d. **Carrier of chemicals.** It transports hormones, vitamins and other essential chemicals from the site of origin or point of entrance to the site of activity.

e. **Acid-base equilibrium.** It is a powerful buffer and helps to maintain a constant reaction in the body with the help of kidney, lung and skin.

f. **Water balance.** It helps in maintaining water balance in the body.

g. **Ion balance.** It helps to maintain ion balance between the cells and surrounding tissue fluid (interstitial fluid).

h. **Temperature regulation.** Water has three important qualities in temperature regulation—(i) High specific heat, (ii) High conductivity and (iii) High latent heat of evaporation. The water present in blood helps in temperature regulation.

i. **Defensive action.** (i) The white cells engulf bacteria by phagocytosis. (ii) Antigen develop antibody to combat toxin.

Blood clot

Whenever a blood vessel is ruptured or severed the out flow of blood is prevented by (1) muscular spasm of the vessel, (2) formation of a platelet plug and (3) clotting in the ruptured vessel.

The clot is a meshwork of fibrin threads with entrapped blood cells, platelets and plasma. The fibrin threads adhere to the damaged surfaces of the blood vessel. The clot prevents blood loss.

Clotting occurs in three steps:

1. In response to the rupture of the blood vessel or damage to blood itself, a **prothrombin activator** is formed.
2. Prothrombin is catalysed into **thrombin** by prothrombin activator.

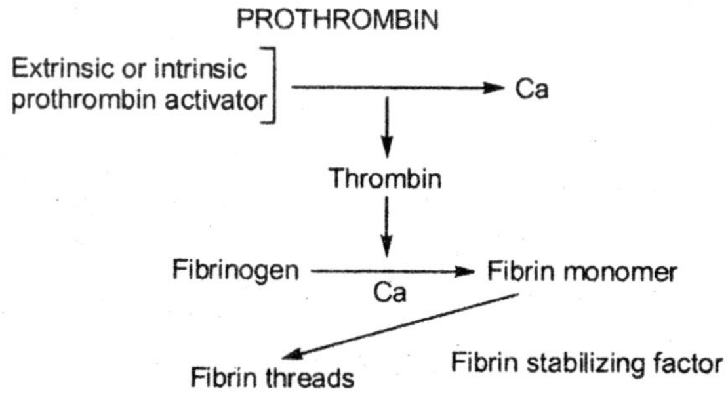

[Conversion of prothrombin to thrombin, and polymerisation of fibrinogen to form fibrin threads]

3. Thrombin converts fibrinogen into **fibrin threads**, which enmesh red blood cells and plasma to form the clot.

⊠ Rh Agglutinogens

At least eight different types of Rh agglutinogens called **Rh factors** have been identified. Certain Rh factors cause transfusion reaction—agglutination. If the factor is present in the blood cells, the person is **Rh positive.** When the red cells do not contain one of the Rh agglutinins, the person is **Rh negative**.

⊠ Lymph

Excess interstitial fluid passes into lymph capillaries and it is called lymph. The lymph is returned to the heart and thereby to the blood through a set of vessels, the **lymphatic system**. Certain glands known as **lymph nodes** are associated with lymph vessels. The lymph nodes act as filters to remove bacteria and harmful substances. The lymph probably removes proteins, cell fragments and other materials from interstitial fluid (Fig. 9.12).

⊠ Defence System

The first line of defence in animals has evolved by the capacity of some leucocytes—the **lymphocytes**, to engulf bacteria, virus, etc. In vertebrates, a second line of defence has evolved by their ability to produce a type of special proteins, the **antibodies**. Antibody is a chemical synthesized in the body in response to a foreign substance, the **antigen** and it can combine with the antigen and destroy it. The antigen-antibody reaction is specific. Antigens are proteins introduced by invading bacteria or virus, but any protein or even some polysaccharides may act as antigen.

⊠ Immunity

Once an animal produces antibody in response to **infection** or **vaccination**, the capacity to synthesize antibody in response to fresh invasion is retained for a long time. This is known as **immunity**. As the capacity is acquired, it is also known as **acquired immunity**.

A serum containing a specific antibody prepared from animals with acquired immunity is injected to a person exposed to a new or dangerous antigen, and the immunity developed there from in the recepient is known as **passive immunity**. Injection of **antivenom** or **antirabies serum** induces passive immunity against snake venom or rabies virus.

⊠ Immunoglobin

The immune response by the formation of **antibody** is initiated by the presence of a foreign macromolecule—the antigen. The antibodies are gamma globulins called **immunoglobins** with a molecular weight 150,000 and 900,000. All the immunoglobins are composed of light and heavy polypeptide chains. In most cases two light and two heavy chains are present.

Two types of immunity have been identified.

1. **Cellular immunity.** It guards cells from infection by virus, fungi, parasites and foreign tissue. The cellular immune response is mediated by **T lymphocytes** or **T cells**. The name is given because they develop in the thymus. T lymphocytes recognise and kill the foreign invaders. The cells are cytotoxic T lymphocytes (CTL), also called killer T cells.
2. **Humoral immunity.** It is most effective against bacterial infection and the extracellular phases of virus infection. The immune response is mediated by diverse soluble proteins called **antibodies** or **immunoglo-bulins**. Antibodies are produce by B lymphocytes or B cells, which mature in bone marrow in mammals.

   Recognition of **antigen**, the foreign component of macromolecules (carbohydrates protein, nucleic acids) in both cellular and humoral immunity depands upon the recognitions of foreign macromolecules.

   Human immunoglobins have been named as IgA, IgD, IgE, IgG and IgM classes which contain α, β, ε, γ and μ heavy chains respectively.

⊠ Blood group factors

Besides the capability of the lymphocytes to produce antibodies, some antigen and antibody systems are present in human blood. They have been designated as blood group factors OAB.

Agglutinogens. Two specific antigenic proteins are present for all time in the plasma membrane of erythrocytes. They have been named A and B.

Agglutinins. The antibodies α and β, specific to antigens A and B may be present in the plasma of blood.

Agglutination. It is the phenomenon of clumping of blood cells.

In human, some individuals have antigen A, some the antigen B, some both A and B and some none of them.

An agglutinogen (say A) and its specific agglutinin (α) are never present in the same person, since its own blood cells would be coagulated. Similarly, B and β cannot be present in the same person. Individuals having no agglutinogen may have both the agglutinins α and β.

On the basis of the presence or absence of the blood group factors, people have been divided broadly into four groups, O, A, B and AB.

To determine the blood group of a person, the test tube method of blood grouping is routinely followed.

⊠ Test tube method of blood grouping

Two drops of blood are collected in a test tube containing 1 to 2 ml physiological saline solution.

In two small test tubes, one drop of corpuscle suspension and one drop of saline solution are put in each. One drop of anti A (α) serum is added to one tube and one drop of anti B (β) serum to the other tube. Four such sets are prepared

The contents of each tube are gently mixed and left for one hour in room temperature. Positive reactions (Fig. 5.12) can be detected within a few minutes, and the extra time is given as a precautionary measure. The contents of each tube is examined for agglutination under a low power microscope.

⊠ Results

a. No agglutination by α and β = group O.
b. Agglutination by α but not by β = group A.
c. Agglutination by β but not by α = group B.
d. Agglutination by both α and β = group AB.

blood group	agglutinogen in erythrocyte	agglutinins in plasma	reaction with	
			anti-A serum	anti-B serum
O universal donor	none	a and b		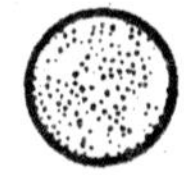
A	A	b	 agglutination	
B	B	a	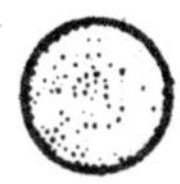	 agglutination
AB universal recipient	A and B	none	 agglutination	 agglutination

FIGURE 5.12

The ABO blood groups. The appearance of the blood of each group when exposed to anti-A and anti-B serum is shown in the two right hand columns. Clumping of cells indicates reaction of serum antibodies from recipient with cell antigens from donor.

☒ Nerve tissue

Response to stimuli is the characteristic of living cells. A number of controlling processes, viz. nerves, hormones and metabolic products, operate in several ways to regulate and integrate functions of different organs in an animal body. But instantaneous response is possible through nerves alone, as the waves of excitement or impules can move at a rapid rate through nerves only.

The tissue made up of specialized cells—the nerve cells or **neurons**, capable of quick transmission of impulses is termed nerve tissue. Waves of excitement appear in the cells of receptors and sense organs due to stimuli in the surroundings, both external and internal. These are transmitted to the controlling centres by one set of neurons (sensory) and impulses from the controlling centres to sites of action or effectors are transmitted through another set of neurons (motor). The ability of the neurons to carry the waves of excitement rapidly enable the animals to integrate their internal functions and also to adjust with the environment. A nerve is a bundle of hundreds or thousands of nerve fibres. A ganglion is an aggregation of a number of cell bodies of neurons. Nerves and nerve ganglia constitute the nervous system, which is present in all metazoons except sponges.

☒ Neurons

A typical neuron is enclosed in a plasma membrane and consists of three parts.

a. **Cell body.** The cell body, **soma** or perikaryon is oval or irregular in outline and bears a prominent nucleus, chromatin granules, and the **Nissl bodies** (Fig. 5.13). Centrosome is absent. Nerve cells do not multiply. The cell bodies are restricted to the periphery of the brain, known as **nuclei**; the central portion of the spinal cord and form the **grey matter**, and in the ganglia.

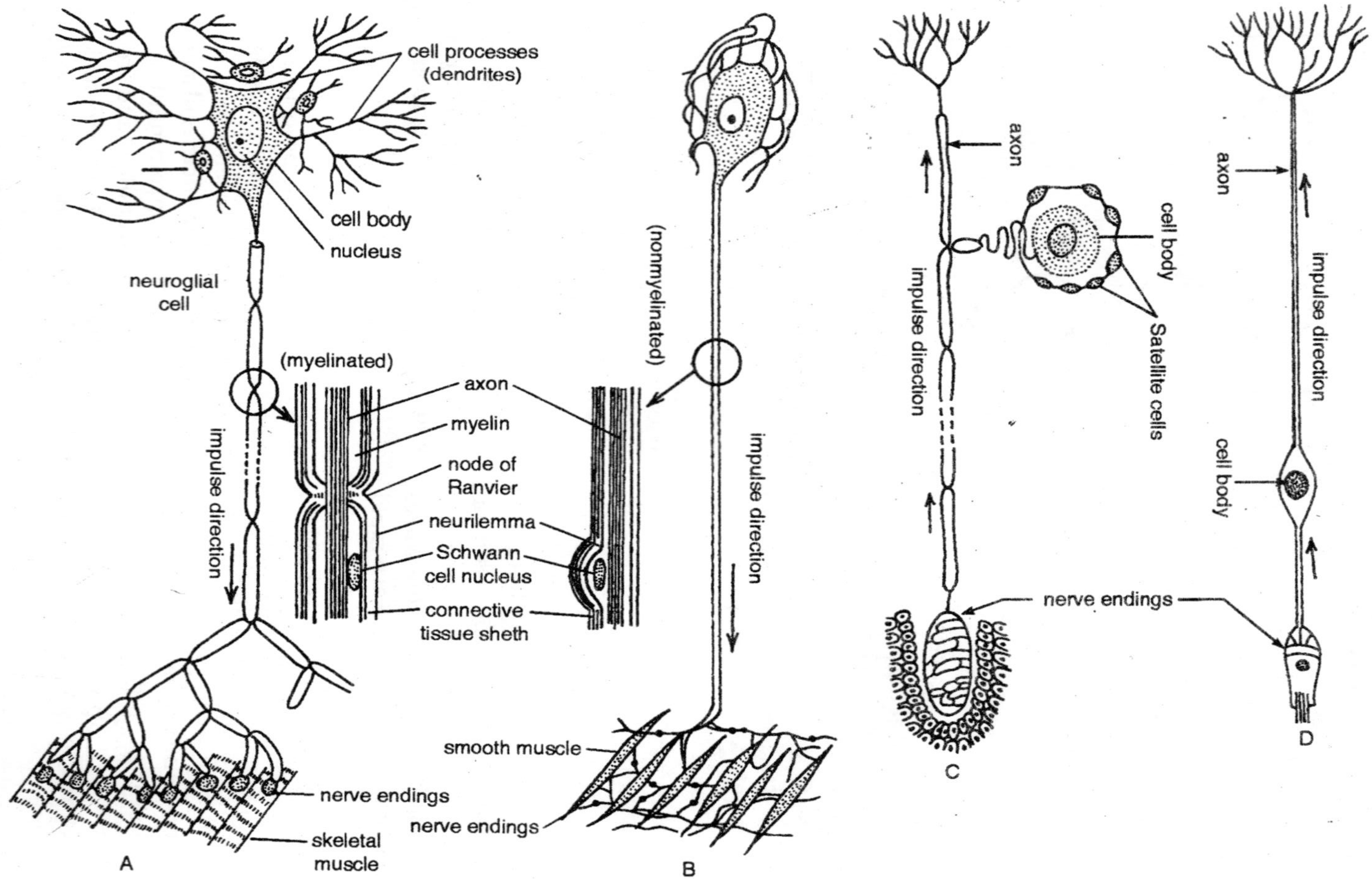

FIGURE 5.13

Neurons: A-B. Multipolar neurons, A. Anterior horn cell of spinal cord, B. Synaptic ganglion cell, C. Unipolar neuron, D. Bipolar neuron.

The cells vary in size from the small granule cells of the cerebellum of about 5.0 μm to the large motor cells of the spinal grey matter.

b. **Dendrites.** A number of branched protoplasmic outgrowths from one end of the cell body are dendrites. Dendrites carry impulses towards the cell body.
c. **Axon.** A single, long extension of the cell body from the opposite end, ending in brush-like filaments, the **telodendria.** Axon carries waves away from the cell body. Damaged axon can regenerate if the cell body of the neuron is not injured.

The processes of the nerve cells are present deep within the brain, at the periphery of the spinal cord and form **white matter.** They form the peripheral nerves.

Some of the neurons are enveloped by a cellular sheath, the **neurilemma.** From the presence or absence of a sheath, the **neurilemma.** From the presence or absence of a sheath around the axon, the neurons are called **myelinated** or **nonmyelinated.** The first group is present in the peripheral nervous system, while the second group is found in the central nervous system. The neurilemma* is constituted by Schwann cells. Concentric lamellae of fatty substances, the myelin, often called myelin sheath, is present between the axon and the neurilemma (Fig. 5.14). The myelin sheath is absent in the gaps are known as **nodes of Ranvier.**

Nerve cells vary greatly both in length and thickness. The average neuron is slightly less than 10.0 μm in diameter but the length may be more than a metre. According to the number of processess they bear, the neurons are of four types:

a. **Apolar**—cell body bears no process.
b. **Multipolar**—cell body with one axon and a number of dendrites.

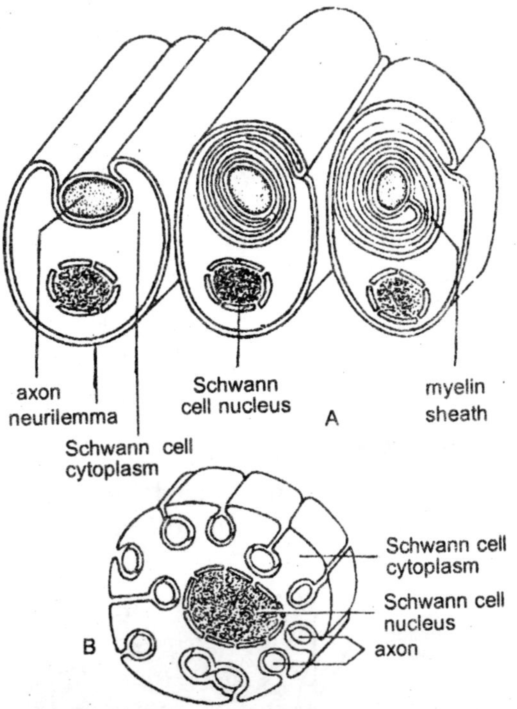

FIGURE 5.14

Neurons (nerve cells) cross section: A. Myelinated. B. Nonmyelinated

c. **Bipolar**—the axon and the dendron are placed symmetrically, one on either side of the cell body.
d. **Unipolar**—axon and dendron arising from a common stem.

☒ Nerve impulse

Nerve impulses in a neuron always travel in the same direction. The speed of the nerve impulse is about four times greater in the myelinated fibres. Thick nerves carry impulses at a faster rate. A nerve impulse may move over 100 metres per second and only milliseconds will elapse before the effect takes place. The speed of nerve impulse varies with homeotherms and poikilotherms, slow and first moving animals and also with ordinary and giant fibres.

* **Neuroglia.** A general term for the fibrous and cellular, nonneurons, supporting cells of the nervous system. Neuroglia proper comprises the **astrocytes** and **oligodendroglia** cells of the central nervous system.

Speed of nerve impulse in the motor nerves of some animals—metre/second.

Snail	0.4
Earthworm	0.6
Earthworm (giant fibres)	10-30
Cockroach	1.5-6
Cockroach (giant fibres)	9-12
Crab	3.5
Fish	50-60
Snake	10-35
Mammals	8-120

☒ Reproductive tissue

New individuals are formed from the preexisting individuals and the species is perpetuated. No specialised reproductive tissues are involved in asexual reproduction. Sexual reproduction involves gametes. In majority of the metazoons, the gametes are produced in definite structures,* the **gonads, the sperms** in the **testis** and the **ova** in the **ovary**.

The testes of vertebrates are made of thousands of sperm tubules (Fig. 5.15). The walls of the tubules are lined with primordial germ cells, the **spermatogonia**. The ovaries bear many primordial germ cells, the **oogonia**, which are enclosed in follicles in mammals (Fig. 5.16). The number of eggs produced by a female is much less than the number of sperms produced by a male.

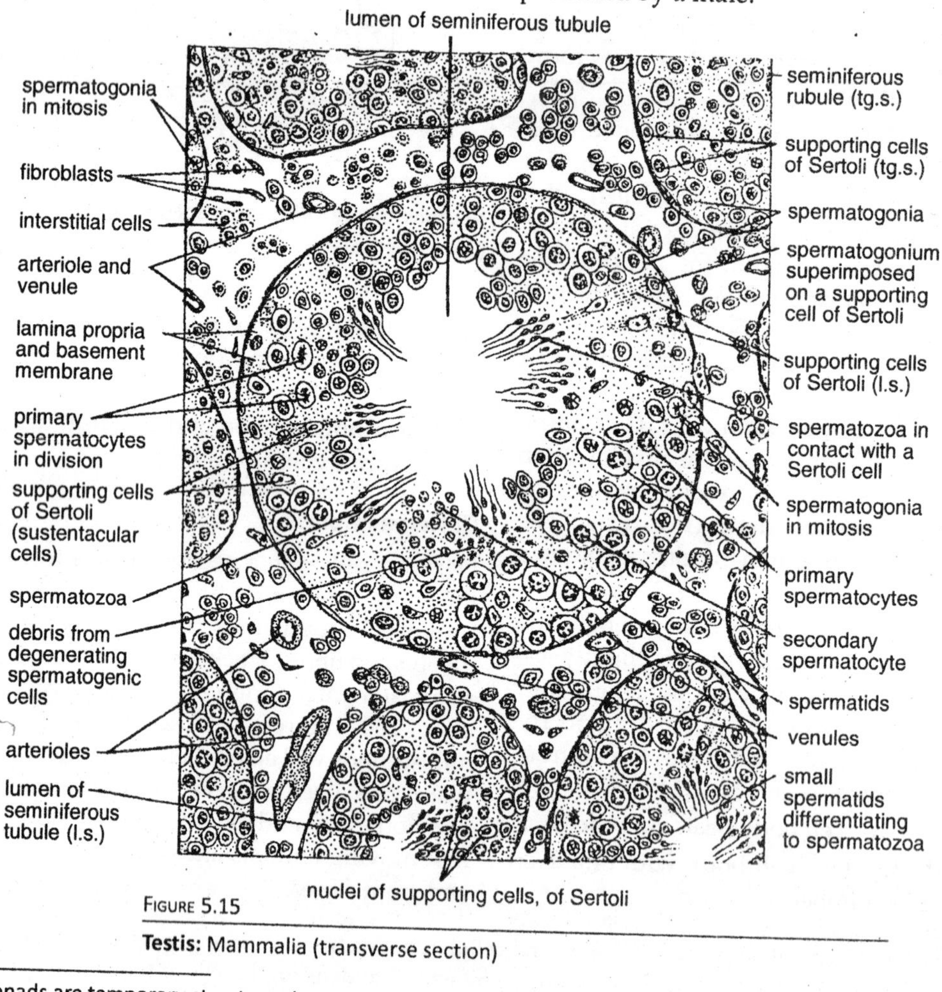

FIGURE 5.15

Testis: Mammalia (transverse section)

* Gonads are temporary structures in sponges and some annelids (*Neanthes sp.*).

The spermatogonia and oogonia divide many times and **primary spermatocytes** and **primary oocytes** are produced. Each of them undergoes two divisions, one of which is meiotic, and four sperms and one ovum with three nonfunctional polar bodies are formed.

⊠ Sperms

A spermotozoon is an intricate motile cell, which moves to the egg to fetilize it. The sperms are minute in size bearing an ovoid head piece connected to a narrow, long tail piece by a middle piece (Fig. 5.16). Deviation from this is found in some animals. They vary extremely in shape and size but their essential constituents remain unaltered. Some gastropods produce two types of sperms—pear-shaped, functional eupyrene and worm-like nonfunctional oligopyrene forms. The sperm is spirally twisted in passerine birds.

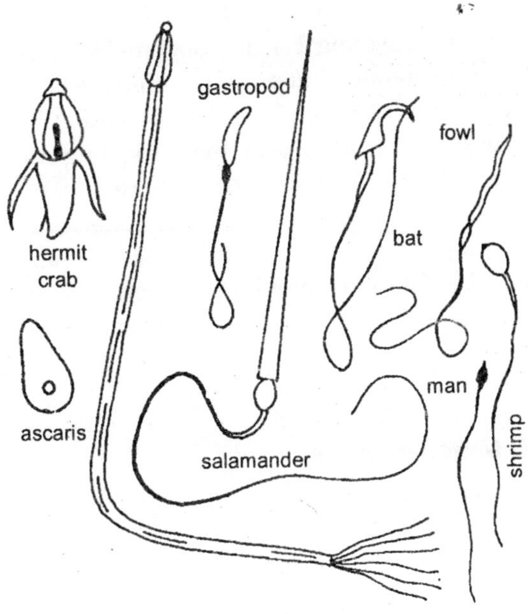

FIGURE 5.16

Sperms of animals belonging to different groups

The head piece is nearly round in shrimp and hook-shaped in bats and mouse. The tail is specialised for movement. The terminal end of the tail piece is termed flagella. The structure of the tail piece vary in different animals. It is lacking in some parasitic round worms and crustaceans. In sea-urchin tail piece, nine fibres are arranged in a circle with two more in the middle extending further. In some animals, viz. rabbit and others, the fibres with similar arrangement has been termed axial filament.

⊠ Ova

The ova are large, usually round (Fig. 5.17) and contain yolk for the development of embryo. The shape and size are variable, the latter depending on the amount of yolk. In addition to plasma membrane and vitelline membrane, in most cases, the ova are protected with additional coverings, shell membrane and an outer softer leathery and hard or calcareous shell depending on the hazards they are expected to meet after laying. The additional membranes are secreted by accessory glands following fertilization.

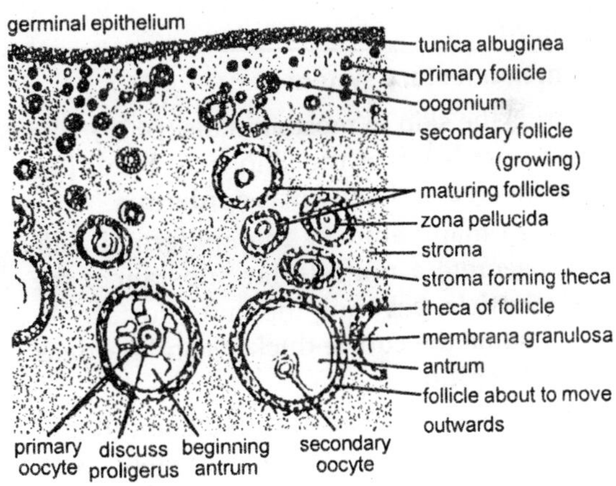

FIGURE 5.17

Ovary : Mammalia (Transverse section)

6 ORGANS AND SYSTEMS

ORGANS

Organs are specialised parts of the body to carry out particular functions. In combination a few organs form a system. Heart, stomach, liver, kidney etc. are typical organs. Organs—gut, liver, pancreas, salivary glands constitute the digestive system. Following systems are present in an organised animal body and they collectively maintain the body and help in propagation.

SYSTEMS

⊠ Integumentary system

It forms the skin in most animals (p. 79). The outermost layer of the skin may be uni or multilayered and known as **epidermis**.

a. In many animals the epidermal cells secrete a noncellular substance, the **cuticle**, covering the outer surface. In some the cuticle thickens and hardens to **chitin**.
b. Protective structures like hairs (mammals), feathers (birds), scales (snakes, lizards) develop from the epidermis.
c. Sensory cells present in the epidermis act as receptors for stimuli.
d. Glands—protective and of vital functions, develop from the **dermis**, the layer below epidermis. In fishes the scales are dermal in origin.

⊠ Skeletal system

It includes all the hard structures of animal body, either protecting the soft parts or adding rigidity to the body or serving for attachment of muscles. The skeleton situated outside the body and covering it is called external skeleton or **exoskeleton**. That present within the body is internal skeleton or **endoskeleton**. The exoskeleton is formed by the thickening and hardening of cuticle secreted by the epidermis or dermis. The endoskeleton consists of cartilage and bone. Both the exo-and endoskeleton are movably jointed and the muscles are attached to them for movement (p. 78).

⊠ Muscular system

A muscle is the aggregate of a large number of muscle fibres into a bundle which has the capacity to contract lengthwise. In some cases the two ends of a muscle are attached to skeletal system and with the contraction of muscles corresponding parts are drawn nearer. In others the muscles are not connected with the skeleton but arranged one layer after another in such a fashion that the direction of contraction of different muscular layers are different (p. 61, 78).

☒ Digestive or alimentary system

It is a system of cavities in which food is received, digested, nutritive substances absorbed and the faeces thrown out. The gastrointestinal tract is lined with epithelial secretory cells, and it can be divided into three distinct regions—**ingestive, digestive** and **absorptive** and **egestive**. Mouth cavity to gullet which receives food and prepare it for digestion is the ingestive part. Stomach to small intestine in which the food is digested and nutritive materials absorbed is the digestive and absorptive part. Various digestive glands—liver, pancreas open into this region. The posterior portion of the digestive tract or rectum, in which undigested portion of food or faeces is temporarily stored in its passage to outside through the **anus**, is the egestive part (p. 85).

☒ Respiratory system

The system responsible for gaseous exchange between the animal body and the environment is called respiratory system. The phenomenon is **respiration**. The gaseous exchange i.e., absorption of O_2 and release of CO_2, takes place in respiratory organs. In aquatic forms, gills and in terrestrial animals, lungs are the chief respiratory organs. In addition, the skin and other accessory structures take part in respiration in some animals. Animals without definite respiratory structures respire through body surface (p. 92).

☒ Blood vascular system

It consists of a heart or central pumping station and narrow tube-like structures, the blood vessels. Vessels carrying blood to different body parts from the heart are **arteries** and collectively named **arterial system**. The vessels carrying blood from body to the heart are **veins** and collectively called **venous system**. Such organised vascular system is absent in lower animals (p. 117).

☒ Excretory system

Organs meant for throwing out nitroge-nous wastes from the animal body are excretory organs, and the system is excretory system. The system varies widely in structure in different animals, starting from simple tubes, one end opening inside and the other end outside the body, to a complex kidney in vertebrates (p. 127).

☒ Nervous system

It is composed of nerve cells (**neurons**) and supporting cells (**neuroglial cells**). Nervous system maintains communication between different parts of the body and also between the body and the environment. Nervous system assumes a net-like structure in lower animals (*Hydra*) but it is highly organised in higher animals. It consists of a **central nervous system** (brain and spinal cord), and a peripheral nervous system with **afferent** and **efferent nerve fibres**. The brain is the controlling centre. Messages are sent to the brain through afferent (**receptor**) nerve fibres and impulses are carried to different body parts through efferent (**motor**) neurons (p. 131). Eyes, nose, tongue and ears are highly specialised sense organs. They are intimately associated with nervous system (p. 141).

☒ Endocrine system

Hormones secreted by endocrine glands constitute endocrine system in animals. Endocrine glands are highly specialized structures and ductless and the hormones are extremely complex chemical compounds. Hormones are capable of inducing physiological changes in the target organs located at a distance. They are moved from the sites of origin to targets through blood or through axons of the neurons or through diffusion (p. 156).

☒ Reproductive system

The system or set of organs by which reproduction takes place in animals. The organs are ovaries and testes, associated structures, and their ducts. Animals having both the sexes in an individual are **hermaphrodite** and those bearing only one sex are **unisexual** (p. 159).

7 THE ANIMAL: STRUCTURES AND FUNCTIONS

The earth is inhabited by million or more species of animals. They exhibit a tremendous diversity, both in their sizes and structures. The body size ranges from a few microns in an acellular animal to the giant dinosaurs and mammoths of the past and present day elephants and baleen whales. The entire surface of the planet earth is teeming with animals—from high, ice-covered mountains to deepest seas; from freezing poles to the scorching deserts. All live in their own ways; they have changed themselves (adaptation) to adjust to the conditions in which they live. In the long history of evolution, many species have become extinct and still more new species have appeared.

To meet the two prime needs of life—food and safety, animals migrate in all possible directions leading to a wide distribution. Gradually, over a long period of time, their body forms undergo a sea-change. Animals belonging to the same group assumed different body forms due to their living in different conditions. This is known as **divergent evolution** or **adaptive radiation**. Mammals exhibit a remarkable adaptive radiation. On the other hand, animals belonging to widely different groups developed similar body forms following their living in similar environment. This is **convergent evolution**. Aquatic animals have assumed the shape of fishes; desert animals exhibit remarkable convergent evolution. In the first category they were subjected to different types of selection pressures, while in the second category similar selection pressures operated upon them.

Adaptations are restricted not only to body forms. Selection pressure has influenced even the biochemical reactions in the cells. This is known as **biochemical convergence** or **divergence**.

Life is dynamic. And this is evidenced from its response to all types of agents of selection, which has culminated in the present day diversity in the animal world.

In spite of the wild divergences, a basic plan is common to all, since life evolved on the planet earth only for once, and the diversity of today is the result of evolution.

SYMMETRY

Symmetry may be defined as equivalence among constituents of a system or between different systems. Except a few asymmetrical species, the vast majority of the animals exhibit some sort of symmetry. From the point of view of symmetry, animals may be placed under the following categories.

⊠ Asymmetrical

Absence of equivalence among different organs in any two halves of the body of an animal (Fig. 7.1). In other words, the organs in one half of the body are not similar to those in the other half, in any case. Examples; Most sponges, adult gastropod molluscs.

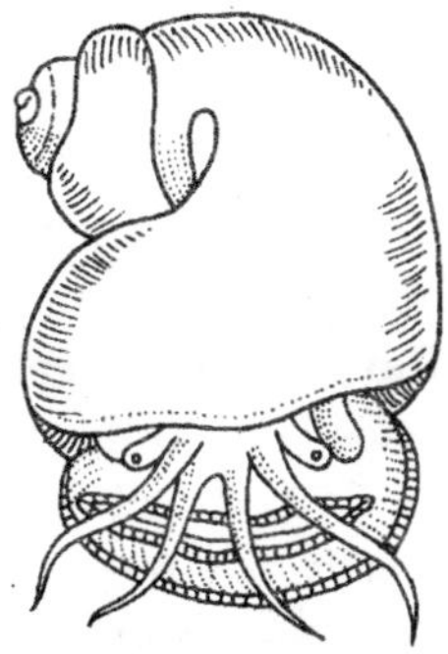

FIGURE 7.1
Asymmetry (*Pila* sp.)

⊠ Bilaterally symmetrical

Organs in one half of the body are similar to those in the other half in an animal (Fig. 7.2).

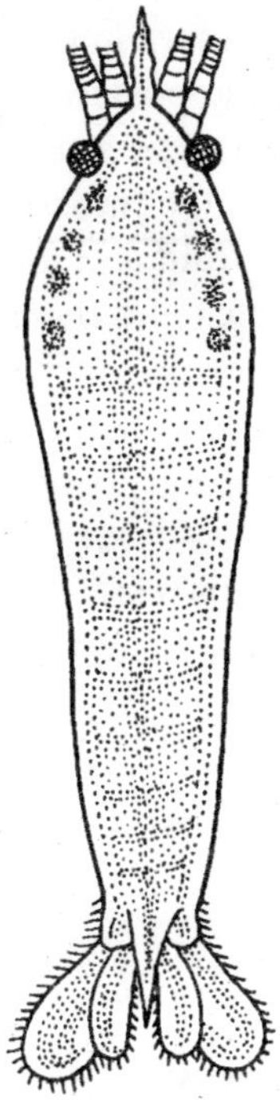

FIGURE 7.2
Bilateral symmetry (*Macrobrachium* sp.)

Majority of the animals exhibit bilateral symmetry.

⊠ Radially symmetrical

In these animals similar organs are located around the central body axis. Organs in one half of the body are similar to those in the other half, when the line of division passes through a particular diameter (Fig. 7.3). Examples: Cnidaria, Echinodermata.

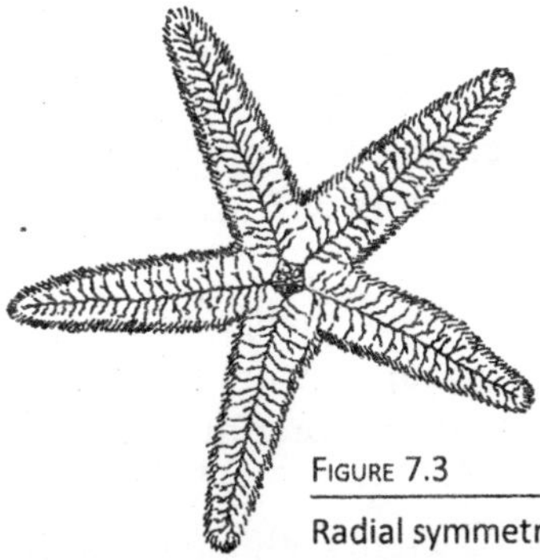

FIGURE 7.3
Radial symmetry (*Asterias* sp.)

In bilaterally symmetrical and asymmetrical animals, anterior, posterior, dorsal, ventral and lateral surfaces are distinguished. The anterior end is directed forward and first to meet the environment. The posterior end is opposite to the anterior end. The dorsal surface is directed upward and well exposed to the surroundings; the ventral surface is in contact with the substratum. The lateral surfaces are exposed to both the surroundings and the substratum. Radially symmetrical animals are usually sessile or slow moving. In the first group, only the anterior and posterior ends and in the second group only the dorsal and ventral surfaces are distinguished.

All animals are bilaterally symmetrical in their early developmental stages. Asymmetry or radial symmetry are secondary* adaptations in course of evolution.

* Many zoologists believe first multicellular animals were radial but motile. They developed anterior and posterior ends but not lateral surfaces. Hydras, sea anemones and corals are descendants of such ancestors and their radial symmetry is **primary**. On the other hand, echinoderms are radial but motile and their radial symmetry is **secondary**.

LOCOMOTION

Change of place is locomotion and change of position of parts or organs, the body remaining fixed to one point is **movement**.

Locomotion is essential for procurement of food and moving away from enemies and environmental hazards. Food supply in an area is limited; it is consumed within a period of time and the animals must move to a new pasture. Sedentary animals cannot change place, but their body parts always move in search of food. Such animals usually live in places where a regular food supply is ensured by certain abiotic factors, viz. water current tidal waves, etc.

Locomotion involves change of place. It requires a means for **support**, development of a **thrust** against surrounding medium and **stability**.

Some sort of skeleton, within (endoskeleton) or outside (exoskeleton) the body is present in most animals. It supports the animal and helps to maintain body shape.

Most of the aquatic animals have a **neutral buoyancy** and neither sink nor rise. In some cases, gas chembers (cnidarians, molluscs), swim bladder (fishes), oil in tissues (planktons) and certain other materials help in maintaining neutral buoyancy. Land animals are supported by skeleton.

The thrust is derived by pushing against the substratum or surrounding medium. This is effected by forming a jet of water (squids), undulation of the body (fishes and both aquatic and terrestrial snakes), air moving over wings (binds and some mammals) and pushing of legs (most land forms).

In protozoon, the thrust is derived from a change in the body shape in forming pseudopodia (*Amoeba*) (Fig. 7.4), beating movement of cilia (*Paramoecium*) (Fig. 7.5) and contraction of contractile fibres (*Monocystis*). In cnidaria, body flexion and heating movement of the tentacles lead to the same result. In aquatic annelids and arthropods (crustaceans), the thrust is created by the beating movement of parapodia and swimming appendages, due to the contraction of the muscles attached to them. In land forms, contraction of the muscles in the body wall plays the same role. In earthworm, anchoring structures like setae are present in the ventral body wall. The setae of the posterior part of the body are fixed to the substratum; contraction of the circular muscles put pressure on the coelomic fluid and forces the longitudinal muscles stretch and the anterior part of the body to elongate.

In the second step, the setae in the anterior part are anchored; those of the posterior part

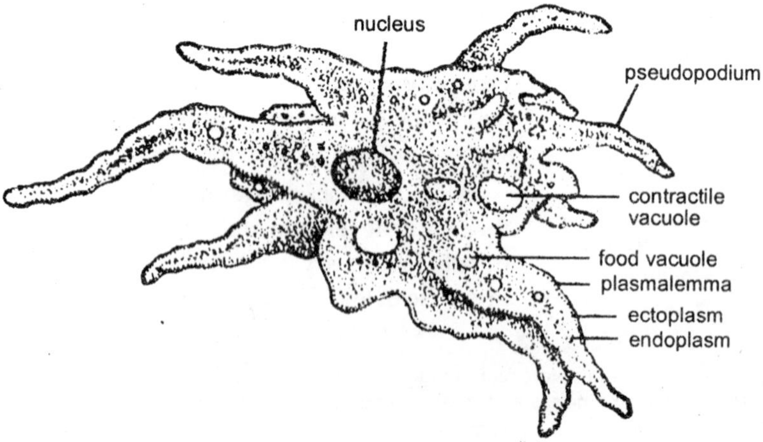

FIGURE 7.4

Pseudopodial movement (***Amoeba proteus***)

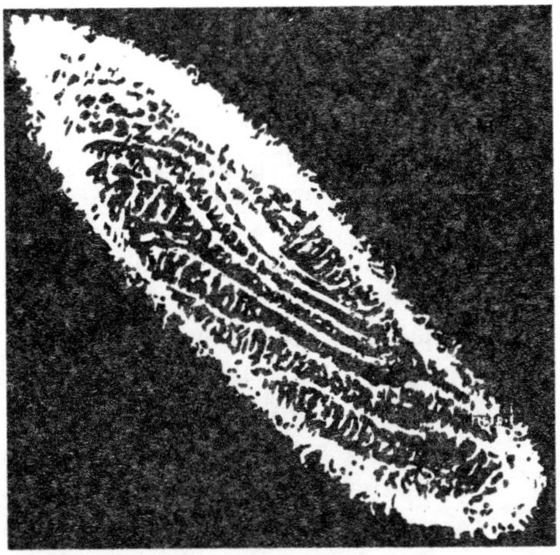

FIGURE 7.5

Ciliary movement (***Paramoecium*** sp.): Waves of ciliary beating passing over the body surface (electron photomicrograph)

released; the contraction of longitudinal muscles also increases the coelomic pressure, the circular muscles are stretched, leading to the shortening of the body and the animal moves forward.

In arthropods, the thrust is created by contraction of muscles attached to the legs and wings. In molluscs, pressure of haemolymp forces forward extension of the foot. This is followed by changing the shape of the foot or by creating a series of waves appearing on the sole of the foot due to rhythmic contraction of the fibres of the foot muscle.

In echinoderms, the muscular contraction of the ampullae of the water vascular system creates a hydrostatic pressure and forces the tube feet to elongate. The tube feet are fixed to the substratum, muscular relaxation causes the fall of pressure, the tube feet shorten and the animal moves forward with a jerking fashion.

Muscles work only in contracting and against antagonistic forces. All muscles are not attached to hard structures or bones. Such structures are foot muscles of molluscs and all smooth muscles.

In vertebrates, the somatic muscles have definable origin and insertion. The insertion is usually through a tendon and the attachment may be direct to the bone or through a tendon.

In organised invertebrates and in all vertebrates, the thrust for movement and locomotion is derived from contraction of the muscles.

Fishes are animals with primary aquatic adaptation. The musculature in fishes is some what different from that in tetrapods, primarily adapted for land life. Most of the somatic muscles in fishes are segmentally arranged (myomeres) (Fig. 7.6). The movement in fishes

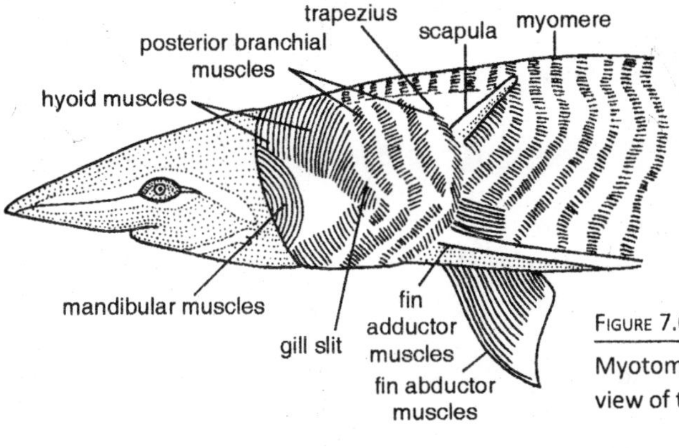

FIGURE 7.6

Myotomes (***Scoliodon*** sp.): Lateral view of the anterior region

evolved from fish adbuctor and **pectoralis** and **biceps** from the adductor.

BODY COVERING

From acellular to large bodied, all animals have a body covering. It ranges from a thin plasma membrane in acellular forms, some what tough in some annelids and arthropods to very hard and tough in arthropods, molluscs and echinoderms. In most vertebrates, the covering is thick but flexible and may be provided with additional armaments like scales, spines, plates, feathers, hairs, etc.

The skin or **integument** (La. *integumentum*) protects the animals from dangers in the environment and invasion of bacteria, para-

FIGURE 7.7

Movement by lateral undulations of body and tail in aquatic medium, A. Scoliodon sp., B. Water snake

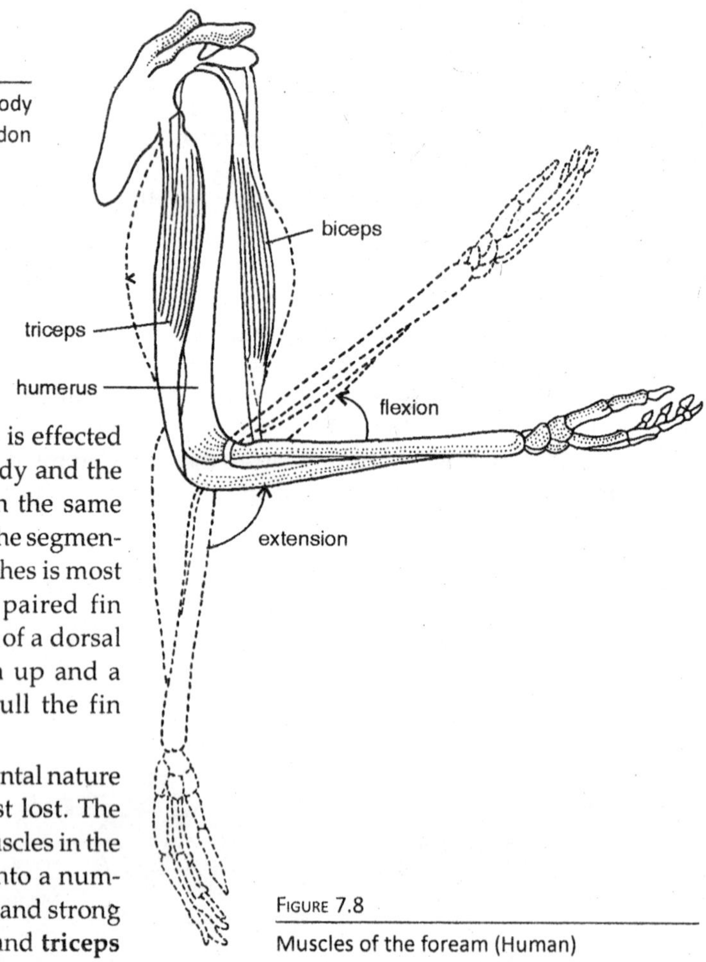

FIGURE 7.8

Muscles of the foream (Human)

and in some aquatic vertebrates is effected by lateral undulations of the body and the tail (Fig. 7.7), the latter being in the same long axis with that of the body. The segmental arrangement of muscles in fishes is most efficient for the purpose. The paired fin muscles are simple and consists of a dorsal abductor muscle to pull the fin up and a ventral adductor muscle, to pull the fin downwards.

In terrestrial forms, the segmental nature of the trunk muscles was almost lost. The single adbuctor and adductor muscles in the fins of the fishes have divided into a number of muscles and became large and strong (Fig. 7.8). The **latissimus dorsi** and **triceps**

sites, predators and ultraviolet radiations. In no case, the animal is totally isolated from its surroundings and in cases where the exoskeleton is hard—as in crabs and certain molluscs there are openings through which contact between the body and the environment is maintained. In many animals the integument has a regulatory role and exchange of water, gases and ions takes place across it. Receptors and external glands are present in the skin. It also helps in heat regulation in most animals, pigment cells present in the skin imparts a particular colour to the animal.

⊠ Invertebrate integument

The integument is usually made up of a basement membrane-supported simple, columnar epithelium, termed **epidermis** (Gr. *epi* = upon + *derma* = skin).

An exposed epidermis may contain many ciliated and gland cells. In some cases overlying noncellular **cuticle** (annelids, arthropods) or a **shell** (molluscs) is secreted by the gland cells in the epithelium, where the latter is termed **hypodermis**.

⊠ Vertebrate integument

The vertebrate integument consist of an **epidermis** of **stratified epithelium** supported by an underlying **dermis** made up of a thick layer of connective tissue. In some animals, the epithelial cells synthesize and accumulate **keratin** (Gr. *Keras* = horn) a protective and water insoluble layer.

In fishes, the epidermis contains mucous cells and multicellular glands. Their secretion covers the body surface with a layer of mucus and helps in reduction of friction with water and also prevents attachment of parasites and

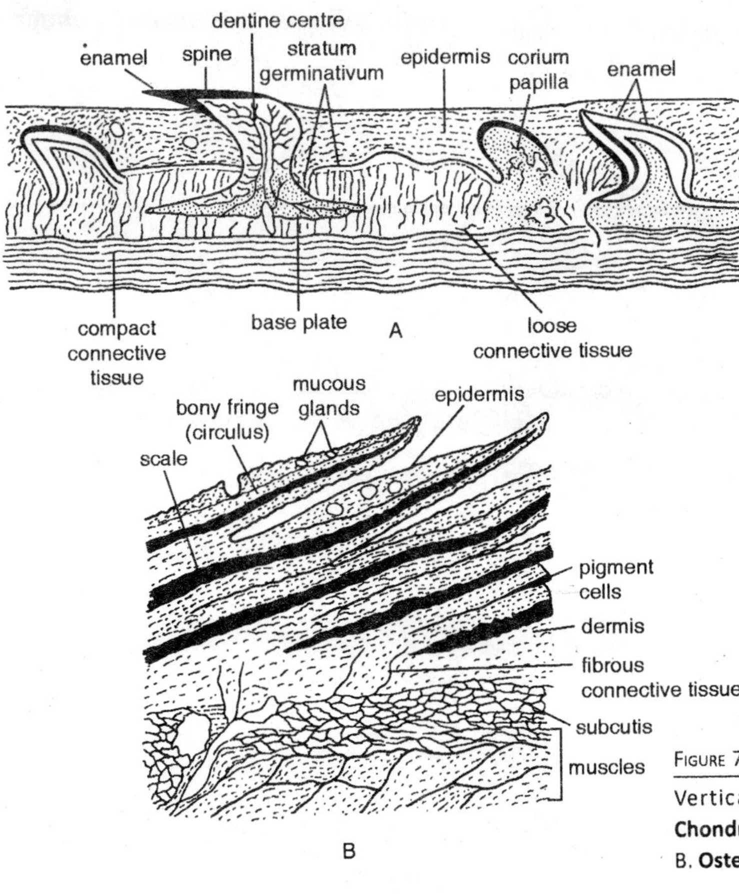

FIGURE 7.9

Vertical section of skin: A. **Chondrichthyes** (elasmobranch), B. **Osteichthyes** (teleosts)

predators. In some fishes, horny **dermal scales** (Fig. 7.9) develop, the posterior parts of which overlap the scales of the next row and protect the fish. Spines and plates are modified scales.

Amphibian skin (Fig. 39.2) is basically similar to that of the fishes, except that it does not synthesizes keratin and may contain poison glands.

In terrestrial vertebrates, the skin is greatly modified to reduce the loss of body water and to make it more protective. The **epidermis** is highly keratinized and forms a distinct horny layer, the **stratum corneum**. This layer is quite thick in reptiles and forms horny scales and plates (Fig. 7.10A). The **feathers** of birds are basically elongated and flattened horny scales (Fig. 7.10B). In mammals, a different type of epidermal derivative is hair, the follicle of which extends to the dermis (Fig. 42.2). A layer of air entrapped in the feathers and hairs acts as an insulator against heat loss.

The skin of reptiles and birds is practically without glands except a few **scent** (odoriferous) **glands**. On the other hand, a large number of glands are present in mammals (Fig. 39.2). The **sebaceous** (L. *sebum* = tallow) **glands**, the **sweat glands** are ingrowths of epidermis into the dermis.

SKIN COLOURATION

Pigment in the epithelial cells, vascularity of skin, presence of surface striations and refractive granules produce colouration in animals. Prawns, crabs, squids, frogs, snakes, lizards may have several kind of pigments in special cells known as **chromatophores** (Gr. *khroma* = colour + *pheros* = bearing). Colour pattern helps prey animals to conceal from predators, and the predators to lie unseen when waiting for prey. In human, **melanin** (Gr. *meles* = black) pigment is present both within and between basal epidermal cells, except **albinos**. It is specially abundant in black skin.

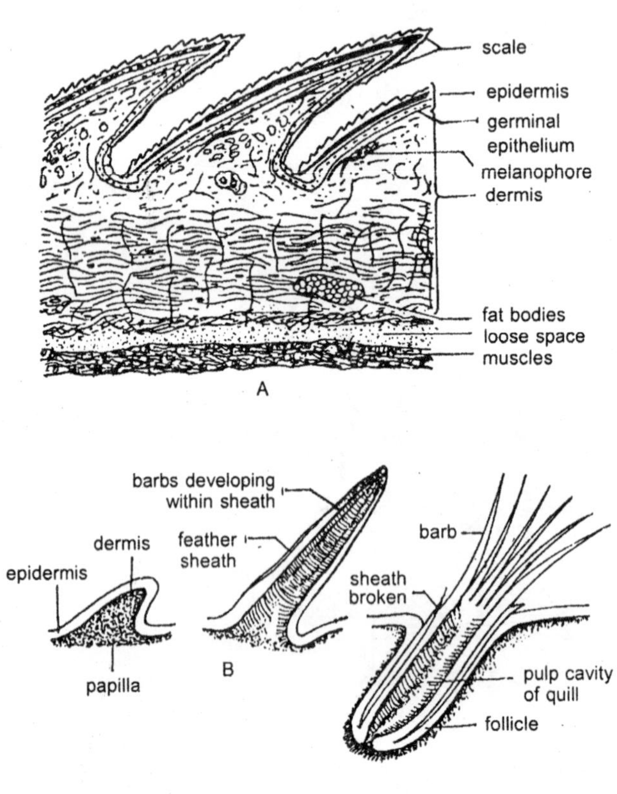

FIGURE 7.10

Exoskeleton: Vertical section of skin (slightly diagrammatic) A. ***Calotes*** sp., B. Development of down feather (**Bird**)

SUPPORT: SKELETON

A support to maintain the shape and position, essential for carrying out normal functions of the body, is present in most animals. The different types of supports met with in the animal kingdom are called skeleton.

Protection of the body or some internal organs and support are important functions of the skeleton. But in certain animals, the skeleton does not support the body, as in minute arthropods and molluscs. In addition to support and taking part in action like locomotion, biting, etc. it acts as a store-house for calcium, phosphorus, minerals, and as the site for manufacture of blood cells. In invertebrate skeleton, calcium is present in the form of carbonate and in the form of calcium phosphate in vertebrates. The skeletons are of three types, **exoskeleton, hydroskeleton** and **endoskeleton**.

• Exoskeleton

Hard structures derived from epidermis or dermis, present outside the body and partly or almost fully enclosing it are called exoskeleton.

Many fresh water protozoon secrete a chitinoid (*Arcella*), calcic (*Elphidium*) shell (Fig. 7.11); some protozoon living in damp soil secrete a cementing material in which sand and soil particles are embedded and acts as a shell.

The polyps of stony corals secrete calcium carbonate, which forms a skeleton beneath and around them. With the growth and multiplication of polyps, new materials are added and the coral grows in size (Fig. 7.12).

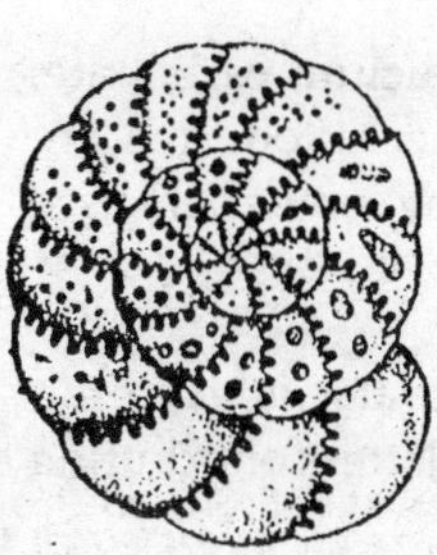

FIGURE 7.11
Exoskeleton (***Elphidium*** sp.)

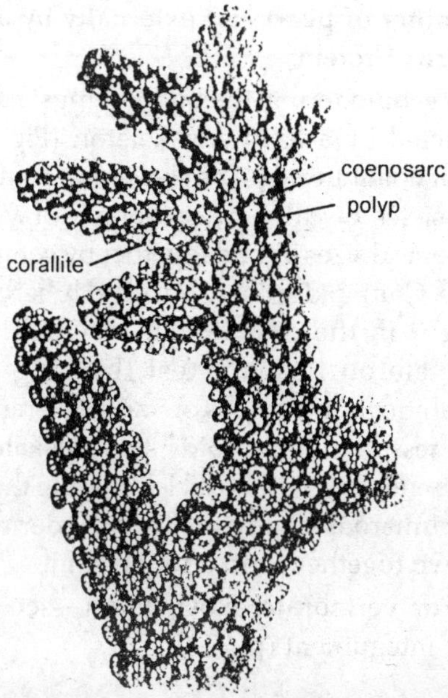

FIGURE 7.12
Madrepora oculata

The corals may be flat, upright and branched or of any other shape. The chief constituent of the coral reef is the exoskeleton of stony corals of many generations.

In annelids, the exoskeleton is a thin **cuticle** or **chitin** around the body and bears a number of pores for communication of internal organs with the exterior.

The exoskeleton in arthropods is a complex glycoprotein secreted by the epidermis. It encloses the body almost completely (Fig. 7.2). It is not a complete tube but thick pieces are joined together to allow free movement. In appendages, the exoskeleton is tubular. The skeleton may extend inwards to line parts of the gut (crustacea) or trachaea (insects).

The hard covering made of several pieces (*Chiton*), two valves (mussels) or a single piece (gastropods) enclosing the body is secreted by the mantle in molluscs. The exoskeleton, however, is absent in slugs (Gastropoda: Pulmonata) and cephalopods. The chief constituent of molluscan shell is calcium carbonate lined internally with a layer of **nacre** or **mother of pearl** and externally by a layer of **horny** protein.

In echinoderms, the body in most cases, is protected by a hard exoskeleton (Fig. 7.13), except holothuroids. The skeleton is composed of ossicles of calcium carbonate of various shapes and sizes bound together by connective tissue. Chitinoid spines, wart and tubercles are present in the ossicles. In sea urchin, the exoskeleton forms a test, bearing many perforations and spines of varying length and thickness. In holothuroids, the exoskeleton is represented by a thin cuticle covering the body and numerous small ossicles in the dermis and the two together form a tough skin.

[For vertebrate exoskeleton—see vertebrate integument (p. 72.)]

Hydroskeleton

The hydroskeleton is not that important in higher forms, but it has useful functions in lowly organised animals and in larval forms. The fluid in the body under pressure—helping in locomotion, is considered as hydroskeleton.

The coelomic fluid in earthworm, the haemocoelomic fluid in the foot of arthropods, molluscs and podocyst of mollusc larvae, the notochord in *Branchiostoma* are considered as hydroskeleton.

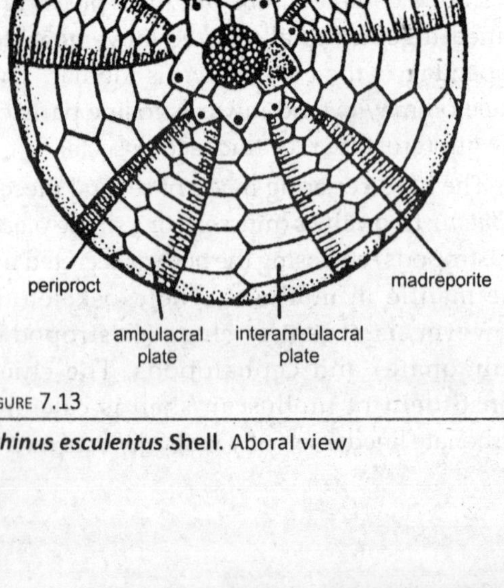

FIGURE 7.13

***Echinus esculentus* Shell.** Aboral view

Contraction of circular and longitudinal muscles in the body wall of an earthworm exerts pressure on the coelomic fluid and forces the animal to elongate or shorten (widen). Increased pressure on he haemo-coelomic fluid (blood) in the foot due to contraction of foot muscle in mussels, forces the foot to elongate and it is pushed forward into the substratum. A sudden rush of haemocoelomic fluid causes forceful extension of the legs during a jump in jumping spiders. The notochord in *Branchiostoma* is an elongated cord of cells filled with a fluid. The firm connective tissue sheath surrounding it helps to maintain the turgidity. The notochord is flexible but firm and prevent shortening of the body during movement, due to contraction of the longitudinal segmental muscles.

Endoskeleton

The skeletal structures present in the body wall or in deep body tissue are collectively known as endoskeleton.

Invertebrate endoskeleton

Radiolarians, a group of large planktonic protozoa, have a very delicate endoskeleton made of elaborate silica lattices (Fig. 7.14). Variously shaped spicules made of spongin fibres or calcium carbonate or both present in the body, support the sponges and help to maintain shape. Portions of some of the spicules project beyond body surface.

In cephalopods, internal shell of calcareous particles arranged in the form of lamellae and a cartilaginous cranium enclosing the nerve ganglia are present.

The ossicles made of calcium carbonate, situated in the dermis of echinoderms are considered as endoskeleton by some zoologists.

Vertebrate endoskeleton

The endoskeleton of vertebrates is made up of bones and cartilages, the latter contributing only a small fraction.

The entire endoskeleton of Agnatha (jawless) and Chondrichthyes (sharks) is made of cartilage. The chief constituents of bones is calcium phosphate. Vertebrate bones develop from mesoderm cells, and are of two types—the **membrane bone** (intramembranous ossification) and **cartilage bone** or **long bone** (intracartilaginous ossification).

In both the cases, prior to deposition of bony or cartilaginous materials, the mesoderm cells are transformed into mesenchymes. In the membrane bone, the materials contributing to the formation of bone are deposited directly in between the cells and around blood vessels. In the long bone, in the foetal stage, a part of condensed mesenchyme is replaced by hyaline cartilage. In the adult, the cartilage is replaced by bone. The membrane bones are usually flat and light but quite strong, and present in the roof and sides of the skull. The long bones are thick and heavy.

The vertebrate skeleton is divided into a **somatic skeleton** (Gr. *soma* = body), present in the body wall and **visceral skeleton** (Gr. *viscera* = bowels) restricted to primitive pharyngeal wall and gills. The somatic skeleton is subdivided into an **axial skeleton**, comprising vertebral column, sternum, ribs and most of the skull and an **appendicular skeleton** consists of girdles, both pectoral and pelvic and limb bones.

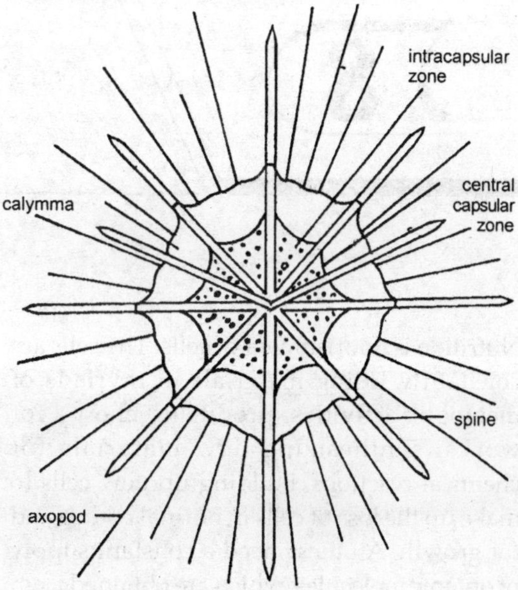

FIGURE 7.14

Radiolarian endoskeleton

8 NUTRITION

Nutrition is nourishment of cells. The cells are constantly losing materials in myriads of metabolic activities, producing energy for works, synthesizing new materials for chemical reactions, building up new cells to make up the loss of cells in normal course and for growth. All these need a constant supply of organic molecules, which are obtained from external sources, the **food**. The food consumed cannot be used directly and are subjected to chemical reactions to break or hydrolyse them down to their basic units. The process of transformation of food to their units is **digestion** and the products are **nutrients**.

Barring a few, the systematic positions of which are uncertain, animals depend directly or indirectly on plants. The organic materials used as food are same for all animals, though the proportion of different compounds vary. The modes of procurement of food are different and they have been named accordingly. All animals are **heterotrophs**. Heterotrophs (Gr. *heteres* = other + *trophe* = nourish) obtain their food from plants; other animals or from plant and animal products. A heterotroph may be (a) **holozoic** (Gr. *holo* = whole + *zoon* = animal), consuming organisms or their products directly, (b) **parasitic**, absorbing food from the body of the host and (c) **saprozoic**, digesting organic matters outside the body and then absorbing them.

DIET AND FOOD

Diet and food are different. Food provides materials for energy, maintenance and growth of the body. Other substances help in the utilisation of food and maintaining normal functioning of the cells. A diet consists of **energy food**, carbohydrate, protein and lipid and nonenergy food, water, inorganic ions, minerals, vitamins, etc.

⊠ Diet and feeding habits

The nature of diet has greatly influenced the feeding habits of animals. Animals have been named after the type of diet on which they depend.

Herbivores. Animals feeding on plants. They may be **microphagous** herbivores feeding on microscopic plants, e.g. copepods, daphnia; **macrophagous** herbivores, feeding on macroscopic plants, e.g. cattle, deer, etc.

Carnivores. Feeding on freshly killed animals, e.g. lions, tigers. Animals feeding on leftovers are **scavengers**, e.g. crabs, jackals; those feeding on dead animals are **carrion** feeders, e.g. vultures, hyaenas.

Filter feeders. Procure food from water by filtration, e.g. mussels, gastropods.

Detritus feeders. Feeding on decomposing organic matters, e.g. mussels, sedentary bottom dwellers.

DIGESTIVE SYSTEM

The organs in which food substances are broken down to simple absorbable forms, the products or **nutrients** absorbed and the undigested wastes thrown out are collectively called digestive system (Fig. 8.1).

Except acellular animals, sponges and certain parasites, some sort of digestive system is present in all animals. Many of the invertebrates have well organised digestive system but it has attained perfection in vertebrates. The system consists of a tube, the **alimentary canal** or gut of varying length and different zonations, starting at mouth and ending in a **cloaca** or **anus** and digestive glands.

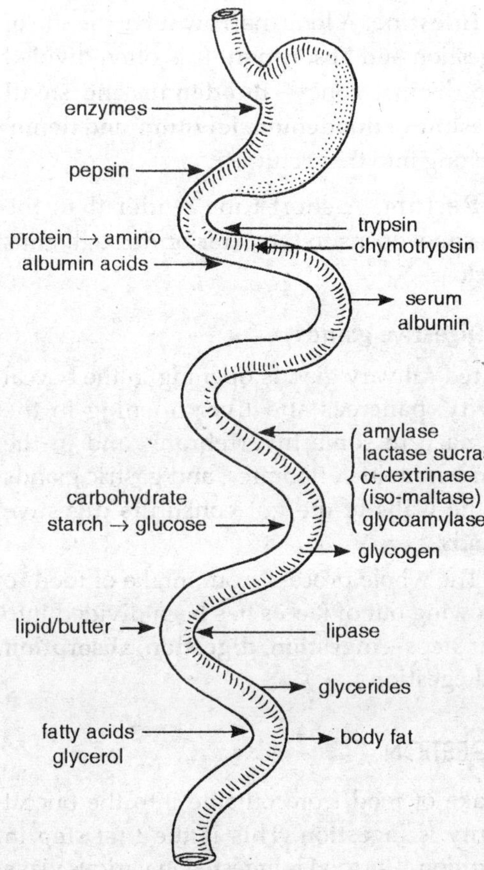

FIGURE 8.1

Conversion of food substances to body substances

⊠ Alimentary canal

In majority of the animals the alimentary canal has five distinct zones, one or more of which may again be divided into subzones.

Buccal or mouth cavity. The anterior part of the gut in which the mouth opens. It is a spacious chamber to receive food (ingestion) and guarded by a pair of jaws, which may or may not bear teeth. Denticles, glands and tongue may be present in the chamber.

Pharynx. A small, muscular chamber connecting the buccal cavity with oesophagus. It guides food to next part of the gut.

Oesophagus. A narrow, muscular tube joining pharynx with the stomach. In quite a number of animals, belonging to different groups, a crop, is present between the oesophagus and the stomach. Usually the posterior part of the oesophagus is transformed into a crop.

Crop*. A thin walled sac for temporary storage of food, e.g. pulmonate molluscs, birds.

Gizzard*. A muscular structure for trituration of food, the inner surface of which is thrown into folds, often bearing hard, chitin or teeth, e.g. insects, birds.

Stomach. A fairly large, muscular sac used for a short period storage of food and digestion. Posteriorly, it opens into the intestine. In general, the stomach is absent in animals having a gizzard.

* The crop is temporary food storage organ and the gizzard is a food crushing organ. These structures are absent in many animals.

Intestine. A long, narrow tube, the site of digestion and absorption. It is often divided into distinct zones—**duodenum** and **small intestine** or **duodenum, jejunum**, and **ileum**, opening into the rectum.

Rectum. A short tube, wider than the intestine, opens to the exterior through anus or cloaca.

⊠ Digestive glands

Paired salivary glands opening in the buccal cavity, pancreas and liver opening in the stomach in some invertebrates and in the duodenum in vertebrates, and gastric glands in the walls of the gut constitute digestive glands.

The whole process from intake of food to throwing out of faeces has been divided into four steps—**ingestion, digestion, absorption** and **egestion**.

INGESTION

Intake of food from outside into the buccal cavity is ingestion. This is the first step in nutrition. The food is ingested in various ways in different animals.

DIGESTION

It is a series of chemical reactions by which the complex compounds constituting food are broken down to compounds of lower order so that those can be absorbed. The chemical reaction is brought about by organic catalysts known as **enzymes** (Fig. 8.1) produced in specialised exocrine glands.* The enzymes are protein in nature and specific in reaction. The substance on which the enzyme acts is known as **substrate**. An enzyme is named after adding the suffix *-ase* after the name of the substrate, viz. Protein-Protein-ase. Being protein in nature, enzyme is denatured with both physical and chemical factors.

In majority of the animals food is crushed to small particles by mechanical means. This increases surface of exposure to enzyme molecules and the reaction rate becomes faster. The enzyme which splits substrate molecules by adding H_2O molecules is called **hydrolases** or hydrolysing enzyme. Digestive enzymes belong to this category.

Two types of digestion are found in animals.

⊠ Intracellular digestion

Digestion takes place inside the cells. The food particles are engulfed by the cell, digestive enzymes are secreted there and the food is digested. In lower invertebrates this is the only mode of digestion. Even in organised invertebrates, where extracellular digestion is well-developed, intracellular digestion persists and is of considerable importance.

⊠ Extracellular digestion

Food is digested outside the cells. Food reaches the lumen of the gut. Enzymes from exocrine glands and gastric glands are released in the lumen. By the contraction and relaxation of smooth muscles in the gut wall, the enzymes are thoroughly mixed with food and digestion occurs. Extracellular digestion is much more efficient than intracellular digestion.

⊠ Carbohydrate digestion

The need for carbohydrate food in animals is met with different forms of carbohydrates including cellulose of plant origin and glycogen of animal source. The unit of a carbohydrate molecule is monosaccharide. A polysaccharide molecule is made up of many monosaccharides. The intermediates in size are di-, tri-, tetra-, penta-... collectively called oligosaccharides. Digestion of oligosaccharides could be accounted for by five enzymes, α-glucosidases, β-glucosidases, α-galactosidases, β-galactosides and β-fructosidases.

---

* The gland provided with a duct, through which the secretion is released at a particular site is known as exocrine gland.

The bulk of carbohydrate food of animals is starch of plant origin. The starch is composed of amylose and amylopectin. Glycogen is the polysaccharide of animal origin. In general, amylose is broken by α amylase to maltose (87 per cent) and glucose (13 per cent). Amylopectin is cleaved to limit by dextrin glucosidase, and maltose is broken by maltase (an α-glucosidase) to two α-glucose molecules. Thus, three enzymes are needed in the digestion of starch and glycogen.

In carbohydrate digestion, polysaccharides are cleaved to monosaccharides, the form in which it is absorbed. In some cases disaccharides are also absorbed.

⊠ Protein digestion

The components of protein are amino acids. In protein digestion, the protein molecules are degraded to amino acid molecules, the form in which they are absorbed. Protein exists in innumerable forms. The protein in one organ differs from the protein of another organ of the same animal in structure, i.e., in the sequence of amino acids. To synthesise the required type of protein it is essential that the protein in food must be split to amino acids.

There are three steps in protein hydrolysis. Protein → polypeptide → dipeptide → amino acids. Different enzyme is needed in each step.

Proteases are classified into two groups.

⊠ Endopeptidases (Proteinases)

These enzymes attack central bonds as well as certain specific peptides.

a. ***Pepsin.** In the stomach of animals it is secreted as pepsinogen, which is activated autocatalytically in an acid medium, pH 6.0. It acts in acid medium, pH 1.0-5.0, and inactivated in a neutral or alkaline one. The enzyme attacks peptide bonds with an adjacent aromatic amino acid.

b. ***Trypsin.** It is secreted as trypsinogen, which may be activated by a specific enzyme, intestinal enterokinase or it may be activated autocatalytically. Trypsin acts on peptide links adjacent to arginine or lysine. It acts in an alkaline medium, pH 7.0-9.0.

c. ***Chymotrypsin.** It is pancreatic in origin and attacks peptide links with adjacent aromatic amino acids of a particular nature in alkaline medium, pH 7.0-9.0.

d. ***Cathepsins.** These are extracellular in many invertebrates, and intracellular in the liver, kidney and spleen of mammals. Substrate requirement of different cathepsins are different. They act in a weak acid medium, pH 4.0-6.0 and their substrate requirements resemble pepsin.

⊠ Exopeptidases

These enzymes attack terminal peptide bonds, i.e. substrates with free polar groups, and they normally act on polypeptide products of proteinase action.

a. **Carboxypeptidases.** A carboxypeptidase removes a terminal amino acid with a free carboxyl group. Its pH optima is in the neutral of weak alkali range. It acts best on peptides containing tyrosine, phenylalanine or tryptophan.

b. **Aminopeptidases.** An aminopeptidase acts on a peptide bond next to terminal amino acid with a free amino group. Pancreatic aminopeptidase has an optimum pH 8.0.

c. **Dipeptidases.** There are numerous dipeptidases. It breaks dipeptide into amino acids. The pH is 7.6-7.7.

---

* The terminology do not follow the rules for naming an enzyme. These were coined before the adoption of the rule and are being retained.

☒ Lipid digestion

Fat or triglycerides may be partially hydrolyzed to monoglyceride and diglyceride or fully hydrolyzed to the constituents glycerol (alcohol) and fatty acid.

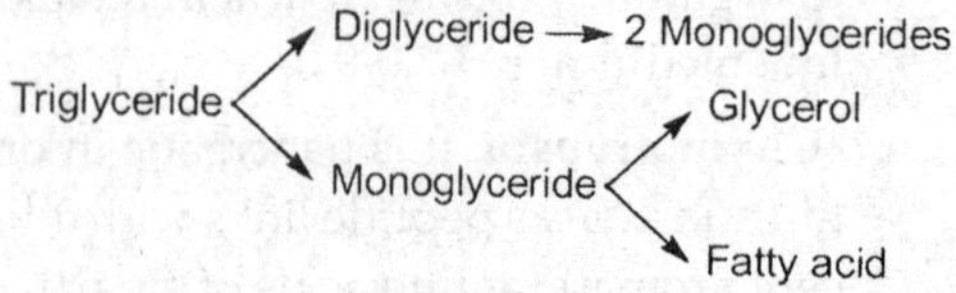

Two groups of fat splitting enzymes, lipases and esterases have been identified though the distinction between the two is not sharp. pH range of both is broad 5.0-9.6.

Lipases. Hydrolyse esters of higher fatty acids. Pancreatic lipase is highly active on olive oil.

Esterases. Hydrolyse esters of shorter fatty acids. Liver esterase is highly active on methyl butyrate.

Bile salts activate pancreactic lipase and also emulsify fat to droplets as small as 0.5 μm in diameter. This increases the surface of the substrate for enzyme reaction, and also favours absorption of fat droplets.

ABSORPTION

Nutrients, the end products of digestion are to be moved to the chief internal transport medium, the blood, for distribution to different parts of the body. The transfer of nutrients from the lumen of the gut to the blood through the lining epithelium of the gut wall is absorption. In general, the small intestine is the chief site of absorption (Fig. 8.1) but absorption from different zones of the gut except rectum has been reported in many invertebrates.

In vertebrates alcohol and only a minute fraction of glucose are absorbed in the stomach. All other materials are absorbed from the intestine.

Absorption surface greatly influences the rate of absorption. A large surface is a prerequisite for rapid absorption. To increase the absorption surface the mucous membrane of the intestine is thrown into a large number of finger-like folds, the **villi**. The free borders of the columnar epithelium of the villi bear numerous narrow projections, the **microvilli**, resulting about 6,000 fold increase in the absorption surface. In man, the surface area of small intestine is 550 m^2 approximately.

The villus is richly supplied with blood capillaries. The space, the **lacteal** in the villus is filled with lymph. The villus is bathed in **chyme**, the fluid in the lumen of the gut. Due to rapid blood flow in the capillaries, the absorbable materials are drawn towards the blood. In all probability, the nutrients pass to the capillaries following concentration gradient. Since absorption in the small intestine is more rapid than expected of simple diffusion, it is believed that active transport is involved (p. 53). Monosaccharides, to a certain extent disaccharides, amino acids, salts, etc. are quickly absorbed as their molecule sizes are rather small.

Emulsified and hydrolysed triglycerides are absorbed. Long chain fatty acids have been detected in the lymph and short chain ones in the portal vein. Incomplete walls of the capillaries (p. 101) are the chief passages for large molecules of fatty acids. Natural fats or triglycerides are absorbed as such by **pinocytosis** and digested intracellularly.

VITAMINS

Vitamins are organic compounds mostly obtained from external sources and are essential for specific physiological functions in an animal. In their chemical structure the vitamins differ greatly. Animals are incapable of synthesizing the required amount of vitamin and they depend on the variable and minute quantity of vitamin present in the food on which they live. Again, requirements of specific vitamins are not same for all animals. Except, guinea pig, monkey, man and a few such animals, others can synthesize vitamin C (ascorbic acid) from glucose.

⊠ Vitamin deficiency

Deficiency of vitamins causes certain physiological conditions or diseases. Deficiency in vitamins are two types—primary and conditioned.

Primary deficiency is the result of inadequate intake of the vitamin or its precursor for a long time.

Conditioned deficiency where adequate amount of vitamin is available but it is not fully utilised due to deficiency in absorption, passing out with urine and even greater absorption. Normally, vitamin requirements are fully met with in a balanced diet.

Occurrence. It is present in common foodstuff of animals. Plants can manufacture all the vitamins while animals can synthesize only a few.

Daily requirement. Only a small amount is needed for routine activity. The daily need varies with the rate of metabolism.

Store. These can be stored in the body to a certain extent.

Fate. Partly excreted, partly destroyed.

• Synthesis by animals

Only a few are synthesised by animals. Vitamin A is synthesised from carotene and vitamin D by ultraviolet irradiation of vitamin D-precursor ergosterol.

Solubility. Some vitamins are water soluble while others are fat soluble.

Nonantigenic. Vitamins are not antigenic.

Administration. Vitamins are unaffected by digestive process and can be taken orally.

Name	Source	Functions
Fat Soluble vitamins		
A.		
A_1.	Carrot, green vegetables, milk, fish liver, eggs, etc.	Prevents night blindness, help in regeneration of visual purple.
A_2.	As above.	Prevents keratinisation of epithelial tissue.
D.		
D_2. Calciferol	Milk, fish liver, egg, etc.	Prevents rickets.
D_3.	As above	Prevents rickets, essential for formation of bone.
E. Tocopherol	Green vegetables, wheat germ oil, egg, liver, meat, etc.	Prevents sterility.
K.	Green leafy vegetables, soyabean, eggs, liver etc.	Blood prothrombin formation, antihaemorrhagic.
Water soluble vitamins		
B Complex		
Thiamine (B_1)	Cereals, peas, beans, yeast, egg, liver, pork, etc.	Antineuritic.
Riboflavin (B_2)	Green vegetables, yeast, milk, egg, liver, meat, etc.	Prevents cheilosis, associated with oxidation in the cells.
Nicotinic acid	Pea nuts, yeast, chicken, meat, liver etc.	Prevents human pellegra.
Pantothenic acid (B_4)	Sweet potato, yeast, cane molasses, milk, egg, liver, meat, etc.	Check pellegra factor.
B_5	Cereals, germinating wheat, nuts, yeast, milk, meat, liver, etc.	Growth factor in birds, not known in mammals.

Contd.

Name	Source	Functions
Pyridoxine (B_6)	Vegetables, wheat bread, yeast, milk, lamb leg, pork, etc.	Antidermatitic.
Folic acid	Green vegetables, soyabean, yeast, egg, liver, etc.	Prevents macrocytic anaemia.
Cyanocobalamin (B_{12})	Soyabean, milk, egg, meat, liver, kidney, etc.	Antipernicious anaemia factor.
Biotin (vitamin H)	Fresh vegetables and fruits, tomato, yeast, liver, kidney, etc.	Prevents dermatitis in dogs and rats.
Choline	Leafy vegetables, germinating cereals, peas, egg, kidney, liver, brain, etc.	Methyl donor acts via betaine and methionine. Factor in synthesis of acetylcholine, phospholipids, etc.
Para-amino-benzoic acid (PABA)	Peas and beans, germinating cereals, yeast, liver, brain, etc.	Prevents premature hair greying in rats.
Inositol	Cereals, yeast, brain, etc.	Antialopecia factor.
Vitamin C (ascorbic acid)	Fresh vegetables and fruits, citrus fruits, tomato, etc.	Prevents human scurvy. Associated with many physiologic functions.

9 GASES: RESPIRATORY STRUCTURES: INTERNAL TRANSPORT: GAS TRANSPORT

GASES

A constant supply of energy is required to carry out the myriad activities of the cells. The energy is derived from the biological oxidation of nutrients obtained from food the animal consumes. Respiration is a broad term meaning intake of oxygen, chemical reaction of the oxygen with carbohydrates, fatty acids and amino acids—producing energy, carbon dioxide and water, and the expulsion of carbon dioxide from the body.

Exchange of gases in respiration takes place in several steps.

1. Transport of oxygen in the air or the oxygen dissolved in water to organs of respiration, gills or lungs. In cases, the body surface functioning as the respiratory organ, it is in direct contact with the oxygen in water or air.
2. Diffusion of oxygen through a moist respiratory membrane to reach the transport medium.
3. Transport of oxygen from the respiratory membrane to the cells and its participation in biological oxidation.
4. Transport of carbon dioxide produced in the cells during chemical reactions, to the exterior following the reverse route.

The gas can take part in a chemical reaction only in a dissolve state. The quantity of a gas available for respiration varies with the solubility of the gas in water, which in turn depends on the (a) concentration (pressure) of the gas, (b) availability of the gas and (c) temperature. The amount of gas dissolved in water is important in respiration of aquatic animals. Dry air contains about 21 per cent O_2 and 0.04 per cent CO_2. Since O_2 is one of the constituents of the air, the availability of O_2 is expressed in terms of **partial pressure** (Dalton's law of partial pressure). The atmospheric pressure at sea level is 760 mm mercury (Hg). The percentage of O_2 in air being 21, the partial pressure of O_2 at sea level is $760 \times 0.21 = 159.60$ mm Hg. It may be said that 159.60 mm Hg at sea level is the tension of O_2 in water. Presence of water vapour reduces the percentage of O_2 and consequently O_2 pressure. In the lung, saturated with water vapour, the partial pressure of O_2 is 104 mm Hg. The partial pressure determines the gas to be dissolved in water. O_2 is sparingly soluble in water, the amount of O_2 going to solution in fresh water at 0 °C is 1 ml in 100 ml; at 15 °C 0.7 ml; at 20 °C 0.6 ml and at 37 °C about 0.5 ml. The solubility for gases decreases with the increase in temperature. The solubility decreases also with the fall of partial pressure.

Decrease in atmospheric pressure with higher altitude is associated with reduced solubility of a gas. At 6,000 m the atmospheric pressure is half of that at sea level and the amount of O_2 going in solution is also one-half.

CO_2 is much more soluble in water, about 30 times more than O_2. It forms carbonic acid (H_2CO_3), which form salts—carbonates and bicarbonates. Since the solubility of O_2 in water is very low, the amount of gas received by aquatic animals (water-breathers) is less than that received by those breathing air.

Both water-breathing and air-breathing animals need a constant renewal of O_2 supply for respiration. Aerial respiration is superior to aquatic respiration, since O_2 concentration is high in air. One fifth of the air is O_2 and the amount of air to be moved in and out of respiratory structures is five times the weight of O_2. Contrast to this, a large amount of water is to be moved in aquatic animals. At 15 °C, 100 ml water saturated with O_2 contains only 0.7 ml O_2 (about 1 mg) and the amount of water to be moved is 1,00,000 times the weight of O_2.

In land forms, O_2 first dissolves in moist surface of the body or the respiratory organ. In aquatic forms, O_2 dissolved in water is used. Passing of a gas across the respiratory membrane and the plasma membrane involved in the supply and removal of the gases to and from the membrane is known as **ventilation**.

From the respiratory organ the O_2 enters the body fluid till an equilibrium is reached. The state in which the number of gas molecules passing to the liquid equals the number of gas molecules moving out from liquid is known as **equilibrium**. Oxygen combined with respiratory pigment is carried by the blood or haemolymph (haemocoelomic fluid) and released in the intrastitial (intercellular) fluid, from where it enters the cells through diffusion across the plasma membrane.

Respiration is a continuous process. Since the oxygen used in respiration is obtained from the surroundings, air or water, the concentration of the gas in the cells is always less than that in the source, i.e. the medium from where the gas is received. On the other hand, continuous release of carbon dioxide in the cells during respiration always leads to much higher concentration of the gas in the cells than that in the surroundings.

RESPIRATORY STRUCTURES

A large ventilation surface is a very effective arrangement for both acceptance and release of gases. This has been achieved with the development of respiratory structures like **ctenidia** and **gills** in aquatic and **lungs** in terrestrial animals.

The aquatic respiratory structures of the invertebrates, the ctenidia, also called gills, differ structurally from the gills of aquatic vertebrates, though, functionally, the two are alike. Similarly, the lungs in some terrestrial invertebrates, as in pulmonate molluscs differ basically from the lungs of vertebrates but both are functionally alike. Some tetrapods, with lungs as the respiratory organs in the adult, bear functional gills in the larval stages, which they spend in water. In all cases, however, the respiratory structures are richly supplied with blood.

OXYGEN CONFORMERS AND REGULATORS

The organisms in which some sort of equilibrium is maintained between the amount of oxygen available and the metabolism are known as oxygen conformers. During low tide, in response to the less availability of oxygen, the burrowing annelid *Glycera* lowers down its metabolism. In high tide, the metabolism goes up with the availability of more oxygen. Other examples are *Spirotomum, Cassiopea, Sipunaulus*, etc.

The animal in which metabolism is independent of oxygen supply up to a certain lower level, the **critical level**, is known as oxygen regulator. Below the critical level, the oxygen consumption goes down rapidly and this may lead to the death of the organism. Most animals belong to this category.

GAS EXCHANGE IN AQUATIC ANIMALS

Small aquatic animals, protozoa and somewhat large forms, viz. sponges and hydras, in which water current bathes the cells, a definite respiratory structure is absent. The situation is similar in forms in which the body surface acts as the respiratory membrane (Fig. 9.1). The oxygen dissolved in water directly moves into the cells through plasma membrane and carbon dioxide diffuses out. In stagnant water, organelles like cilia develop to maintain a water current, which brings oxygen supply and remove carbon dioxide from the surface.

In larger invertebrates, specialised respiratory structures, ctenidia have developed. The ctenidia may be located in different regions of the body in different animals, but, as a rule, they are restricted to the anterior part of the body, in the majority. The ctenidia are present in association with parapodia in annelids, on both sides of the cephalothorax in crustaceans and in the mantle cavity in molluscs.

⊠ Invertebrate gill

A ctenidium (gill) is made of a number of lamellae. Each lamella is composed of a large number of gill filaments. The lamellae are attached to a central axis. Channels in the central axis carry blood to and from spaces in the lamellae. Gas exchange occurs in the gill filaments. In crustacea, the ctenidia (Figure 9.2) on the sides of the cephalothorax, lie in the branchial chambers formed by the sides of the carapace. In prawn the chambers communicate with the exterior along their anterior, posterior and ventral borders. Movements of the scaphognathites of the second maxillae set a constant backward to forward water current in the gill chambers. In crabs, the carapace is sealed along the ventral margin and the water enters through apertures at the bases of the great chela, the first pair of walking legs. In molluscs, the gills are lodged in the mantle cavity. In a bivalve, by the beating movement of cilia present in the gill filaments, a current of water enters the mantle cavity through the inhalant siphon and escapes to the exterior through exhalant siphon. In prosobranchs, contraction of muscles in the mantle and nuchal lobes enlarges the mantle cavity and water enters through the left nuchal lobe or the siphon and goes out through the right siphon due to relaxation of muscles in the mantle.

⊠ Vertebrate gill

The respiratory organs of adult cyclostomes and fishes are internal gills. Some amphibians

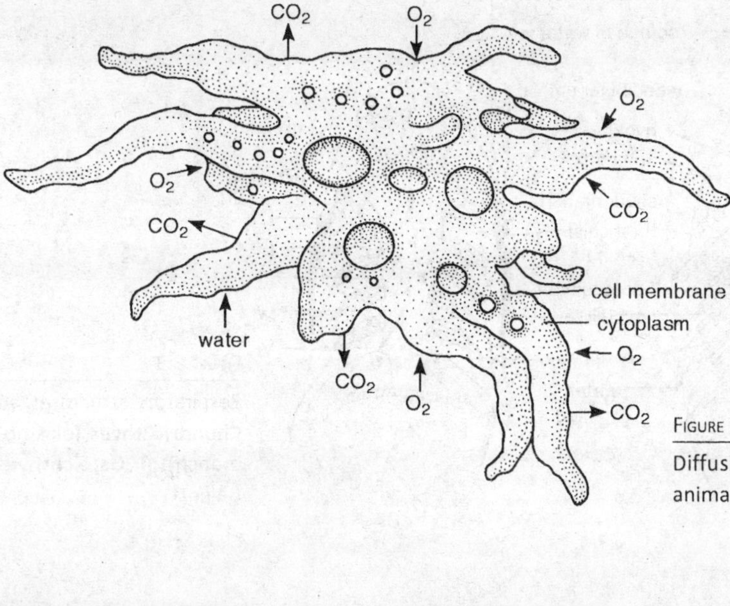

FIGURE 9.1

Diffusion of gases in acellular animals. ***Amoeba*** sp.

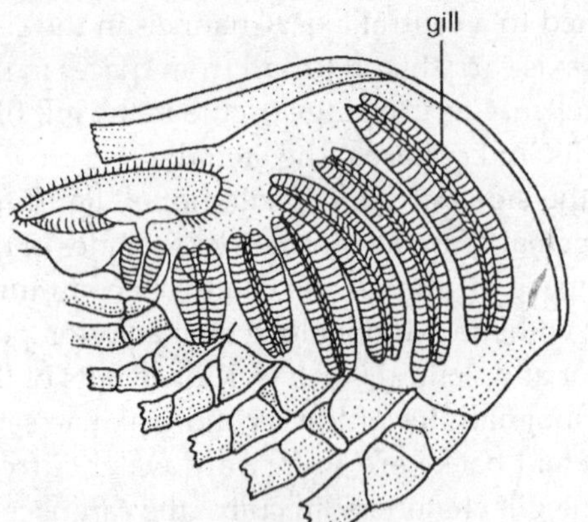

FIGURE 9.2

Respiratory structures (***Macrobrachium*** sp.)

respire with larval gills during development, which takes place in water.

Structurally, the respiratory organs of cyclostomes and Chondrichthyes (Elasmobranchii) exhibit some differences between them and also with osteichthyes (Teleostomi), but in all cases, the gills are located in the lateral walls of the pharynx, hence called **pharyngeal gills**.

During embryonic development a series of paired outgrowths appear in the lateral walls of the pharynx, extending from inside to outside. These are **gill pouches.** The out pushings or gill pouches finally open to the exterior by narrow slits, the **gill slits** or **gill clefts**. Thus, each gill pouch is in communication with the pharyngeal cavity by an **internal** and with the outside by an **external branchial aperture**. Two successive gill pouches are separated by a fibrous partition (the original pharyngeal wall), the **interbranchial septum**. The membrane lining the anterior and posterior walls of each gill pouch is folded into a number of horizontal ridges, the **branchial filaments**. The filaments are richly supplied with blood, where gaseous exchange takes place.

The visceral arches supporting the pharynx are a series of U-shaped rods. The visceral bar or each half of the arch is lodged in the pharyngeal side of an interbranchial septum, alternate with the gill pouches. As a result, a visceral arch bears the posterior set of filaments of one pouch and the anterior set of the filaments of the next one.

In Osteichthyes (Teleostomi) the inter-branchial septa are reduced to narrow bars enclosing the visceral arches of the pharyngeal wall and a double row of gill filaments develop from each visceral bar.

A gill made of one set of gill filaments is **hemibranch** or half gill and a gill with two hemibranchs is **holobranch**.

In cyclostomes (Fig. 35.5) and elasmobranchs there are several pairs of gill pouches (Fig. 9.3). A gill pouch is a biconvex sac, containing numerous highly vascular gill lamellae. The gill pouches open into the pharynx.

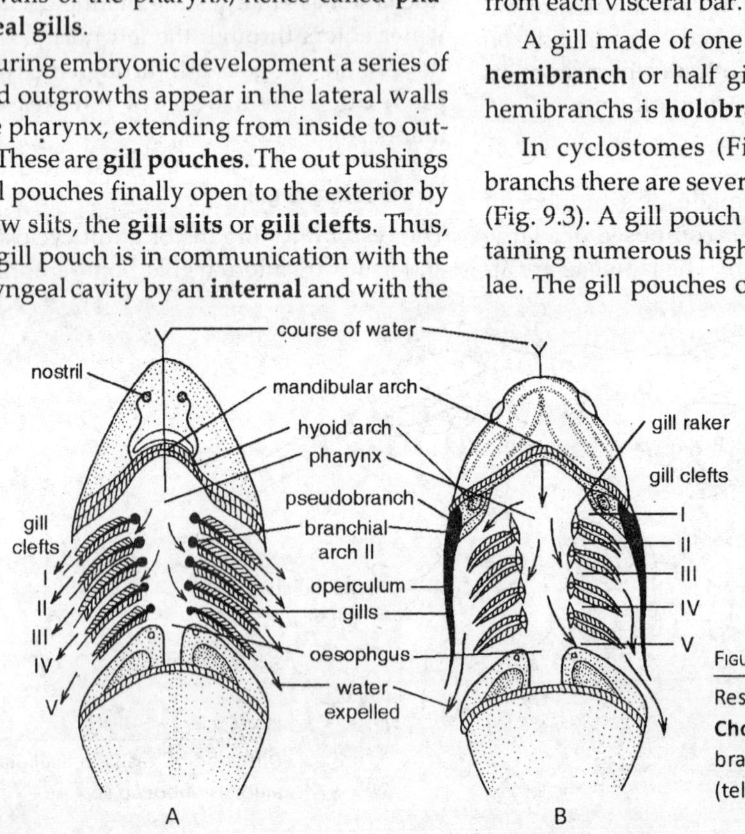

FIGURE 9.3

Respiratory structures: A. **Chondrichthyes** (elasmobranch), B. **Osteichthyes** (teleost)

In lampreys, several pairs of gill slits are present, while in hag fishes, the gill slits are separated by partition. The number of paired gill slits vary in different Chondricthyes.

In teleosts, the partition between two gill slits is reduced to a branchial arch bearing two rows of gill filaments (Fig. 9.3). All the gills of each side are placed in a gill chamber, covered externally by a bony operculum (L. *operculum* = lid). A thin branchiostegal membrane is present at the posterior margin of the operculum. During respiration the gill chamber becomes tightly closed by pressing the membrane against the body wall.

A respiratory current is established by lowering the floor of the buccal cavity. The water rushes in the cavity through the open mouth (Fig. 9.4A). With the closure of the mouth and the oesophagus and raising of the buccal floor and contraction of the pharyngeal walls, water enters the gill pouches and passes out through gill slits in cyclostomes and elasmobranchi (Fig. 9.3A). In teleosts with raising of the büccal floor and contraction of the pharyngeal walls, the branchiostegal membranes are forced open and water from the buccal cavity is driven to the gill chamber through the gill slits and thence to the exterior (Figs. 9.3B, 9.4B).

Afferent and efferent branchial arteries are lodged in the gill arch. The former carries deoxygenated blood to the gills and the latter drains away oxygenated blood from the gills (Fig. 9.5).

For an active respiration the gills must have a large surface area and a large volume of water must move across the gill lamellae or filaments. Oxygen is only slightly soluble in water and the weight of the water passing through the gill slits is about 0.1 million times the weight of available oxygen. Usually, water moves in one direction across the gills. Diffusion of gases between the water and the blood or haemolymph in the gills is rather rapid. The diffusion is accelerated by a mechanism termed **countercurrent exchange** in some crustaceans, molluscs and fishes. In the process, the direction of water flow across the gills is opposite to the flow of blood or fluid (Fig. 9.6). About 90 per cent of dissolved oxygen may be removed from water by the counter current exchange.

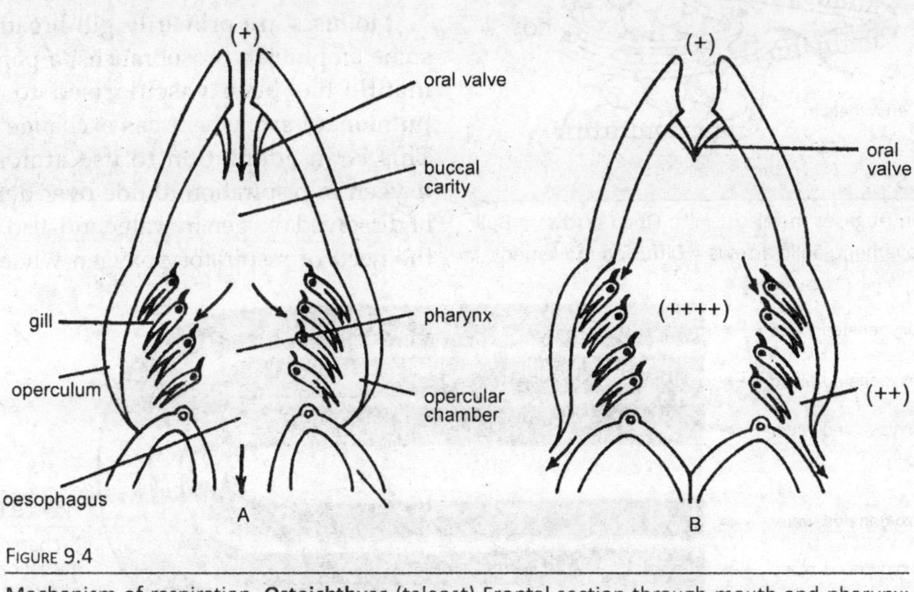

FIGURE 9.4

Mechanism of respiration, **Osteichthyes** (teleost) Frontal section through mouth and pharynx: A. Inspiration, B. Expiration

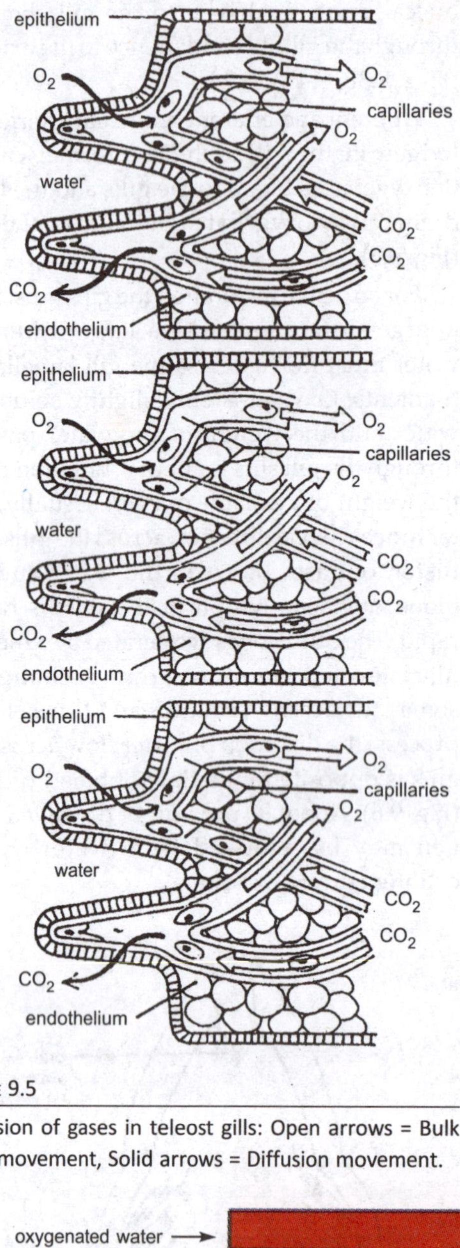

FIGURE 9.5

Diffusion of gases in teleost gills: Open arrows = Bulk flow movement, Solid arrows = Diffusion movement.

SPECIALIZED RESPIRATORY STRUCTURES

In some land invertebrates and some aquatic invertebrates and vertebrates, respiratory structures to use atmospheric oxygen developed. These are **tracheae, book-lung** in arthropods, **pulmonary sac** in prosobranch and **lung** in pulmonate molluscs and **accessory respiratory organs** in fishes.

Tracheae and book-lungs

Unlike other terrestrial animals, the cells of insects obtain their oxygen supply directly from the air through a series of branching tubes, the **tracheae** and tracheoles. A trachea is communicated to the exterior through an opening, the **stigmata** or **spiracle**. The other end is branched repeatedly and the tracheoles, less than 1 μm in diameter, open directly in the tissue fluid (Fig. 9.7). In arachnids, the respiratory organs, **book-lungs** are directly connected to the stigmata. The lining membrane of the book-lung is the respiratory surface. It is folded into numerous delicate lamellae, like the pages of a book and the respiratory surface is greatly increased (Fig. 9.8).

• Pulmonary sac and lungs

Molluscs are primarily gill breathers. In some amphibian prosobranchs, a part of the mantle has been vascularised to form a **pulmonary sac**, where gas exchange occurs. This is an adaptation to use atmospheric oxygen in respiration to tide over deficiency of dissolved oxygen in water and also to meet the need of respiratory oxygen when out of

FIGURE 9.6

Exchange of gases: counter-current exchange mechanism

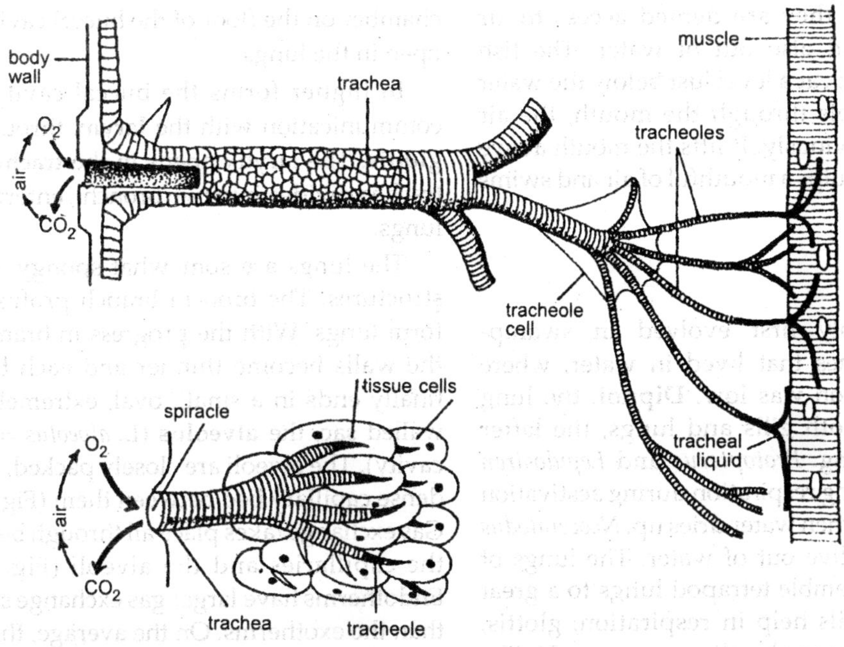

FIGURE 9.7

Diffusion of gases in insects: A. The tracheal system, B. Trachea and tracheoles

water or in aestivation. The pulmonary sac communicates with the exterior. Air is sucked in through the expanded left siphon and expelled through the opening of the sac. The sac may be considered, at best, as an **accessory respiratory structure**.

The pulmonate lung is not homologous with the pulmonary sac of prosobranchs. It is not a vascularised mantle. It develops from a separate invagination and maintains its identity till late larval life. In the adult, it fuses with the mantle and loses its separate identity (p. 489). Presumably for this the pulmonate lung has been confused with the pulmonary sac of prosobranchs. The mantle cavity enlarges and air is sucked into it through the **pneumostome**. The expired air is expelled out due to contraction of the mantle cavity. The air flow in and out of the mantle cavity is regulated by opening and closing of the pneumostome. The lung of pulmonate is called **diffusion lung**.

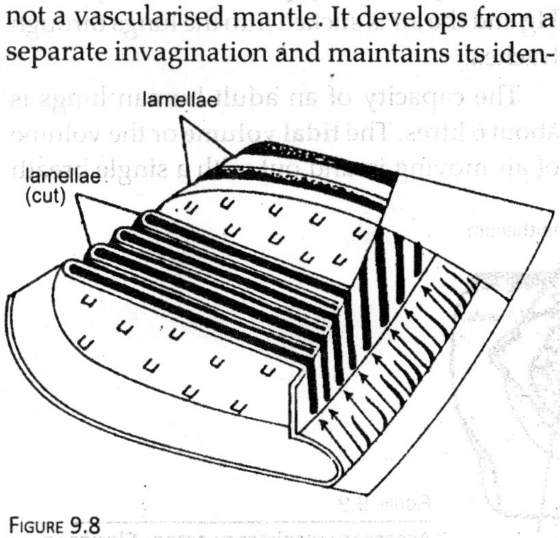

FIGURE 9.8

Arachnid book lung (section), a part (diagrammatic)

☒ Vertebrate accessory respiratory organs

The gills are primary respiratory structures in fishes. The accessory respiratory organs, capable to make use of oxygen in the air, have developed in some bony fishes, probably in response to low oxygen tension in the water body in which they lived, to supplement gill respiration to a certain extent. The structures are paired, highly vascular, in association with the gills and lodged in gill chambers in most cases (Fig. 9.9). These fishes are unable to survive if they are denied access to air breathing and also out of water. The fish moves up and, at a level just below the water surface, expels through the mouth, the air engulfed previously. It lifts the mouth above water level, gulps a mouthful of air and swims **downward**.

FISH LUNGS

Lung possibly first evolved in swamp-dwelling fishes that lived in water, where oxygen tension was low. **Dipnoi**, the lung fishes, use both gills and lungs, the latter predominating. *Protopterus* and *Lepidosiren* survive on lung respiration during aestivation in summer, when water dries up. *Neoceratodus* is unable to live out of water. The lungs of dipnoans resemble tetrapod lungs to a great extent; nostrils help in respiration; glottis, trachea and lung alveoli are present. Unlike tetrapod lungs they are situated ventrally.

TETRAPOD LUNGS

Since the air is moved in and out of the lungs by a pumping action in vertebrates, they are called **ventilation lungs**.

Lungs are highly vascularised, paired structures located in the thorax. The lungs develop as bilobed outgrowths from the floor of the pharynx and extend backward. This state is almost retained in amphibians. The buccal cavity communicates with the **laryngotracheal chamber** through glottis (Gr. *glotta* = tongue). Two extremely short **bronchi** run backward from the laryngotracheal chamber on the floor of the buccal cavity and open in the lungs.

In higher forms the buccal cavity is in communication with the **larynx** through the **glottis**. The larynx opens in the **trachea**, the trachea divides into two bronchi, entering the lungs.

The lungs are somewhat spongy, elastic structures. The bronchi branch profusely to form lungs. With the progress in branching, the walls become thinner and each branch finally ends in a small, oval, extremely thin walled sac, the **alveolus** (L. *alveolus* = small cavity). The alveoli are closely packed, with a dense capillary bed between them (Fig. 9.10). Gas exchange takes place all through between the capillaries and the alveoli (Fig. 9.11). Endotherms have larger gas exchange surface than the exotherms. On the average, the lung volume in mammals is about 6 per cent of the total body volume.

Lung ventilation in vertebrates is effected by changing the volume of the thoracic cavity. In mammals the thoracic cavity is bounded by the ribs on the sides and by the diaphragm towards the abdomen. In inspiration, contraction of **external intercostal muscles** pulls the ribs up and out, the muscular **diaphragm** contracts and flattens and the chest expands. The **intrapulmonary pressure** falls by 1 mm Hg and the air is sucked into the lungs through trachea.

The capacity of an adult human lungs is about 6 litres. The **tidal volume** or the volume of air moving in and out with a single breath

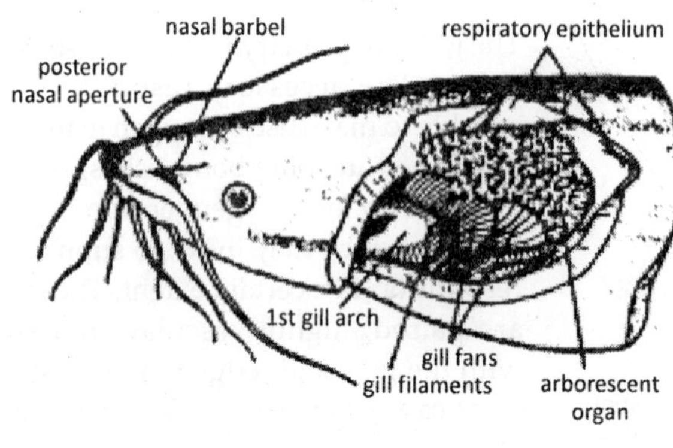

FIGURE 9.9

Accessory respiratory organ: ***Clarias*** sp. (Magur fish)

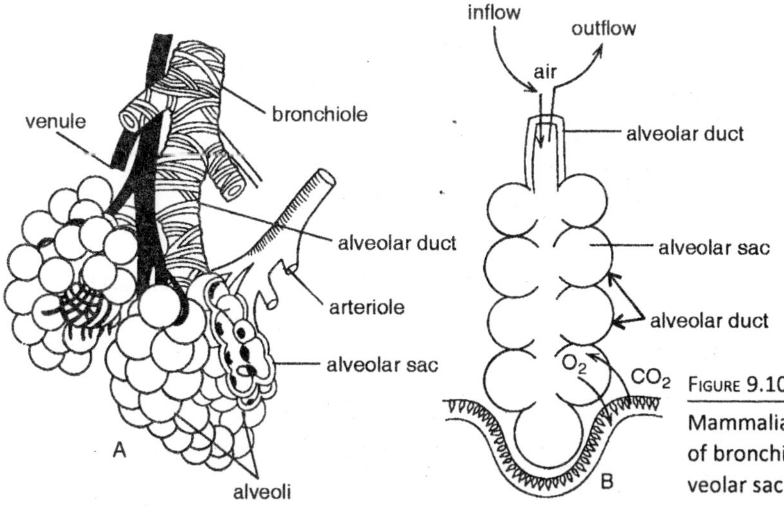

FIGURE 9.10

Mammalian lung: A. Termination of bronchiole in alveoli, B. An alveolar sac cut open (schematic)

is about 0.5 L at rest. The total breathing capacity, i.e., the amount of air that can be expelled after the lungs filled in capacity, is about 4 L. This is known as **vital capacity**. The vital capacity is usually higher in man than that in woman. Athletes have still higher capacity, 5 to 6 L or more.

During expiration, inspiratory muscles relax, the elastic lungs recoil, the intrapulmonary pressure increases and the air is forced out. In all vertebrates the inspiration is **passive** and expiration is **active**.

In an average human male, the gas exchange surface is 70 m^2. Since the major constituent of air is inert nitrogen, the weight of the air to be moved through the lungs in respiration is about 5.0 times the weight of the available oxygen. About 25 per cent of oxygen is extracted by mammals from the air present in the lungs. All the lung alveoli do not collapse in respiratory movement and a certain amount of residual air, about 1.5 L, is always left in the lungs. This is called **residual volume**, which varies in quiet and fast breathing. For this the lungs never contain fresh air as the air inhaled gets mixed with air already present in the lungs. The oxygen content of the mixture is in equilibrium with the blood. At rest, in human, about 10 L air is taken into the lungs per minute. In heavy exercise the amount may reach 125 L.

INTERNAL TRANSPORT

The cells need a constant supply of water, oxygen, nutrients, minerals, vitamins and other materials and removal of nitrogenous wastes, carbon dioxide and other products of metabolism. The cells of the metazoons are bathed in an **interstitial fluid**.

The systems which carry materials from the sites of **intake** or sites or **origin** in the cells to the interstitial fluid or other parts of the body, respectively, are collectively called internal transport system. Materials from one to other part within the body are carried only through fluid media.

Very small animals as the protozoon and certain metazoons in which the cells are exposed to water current, as in sponges and some cnidarians, do not need a transport system. With the progress of evolution and enlargement of body size, the problem of transporting materials from one to another part of the body became complex and a well knit transport system developed.

ROLE OF TRANSPORT SYSTEMS

The fluids transporting materials within the body are **plasma water** in blood, **lymph fluid**, **transcellular water** (fluid secreted by cells in the coelom and digestive system) and **inter-**

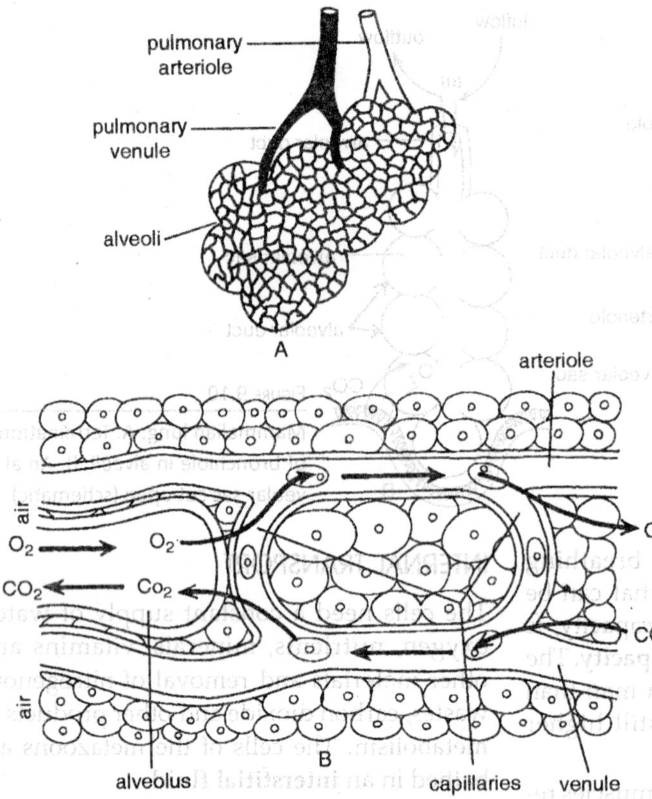

FIGURE 9.11

Vertebrate lung: A. Lung alveoli, B. Diffusion of gases in an alveolus (schematic), Open arrows = Bulk flow movement. Solid arrows = Diffusion movement

stitial fluid. Flow of liquids through body cavities or a circulatory system that carries different materials is known as **bulk flow**. The degree of perfection of fluid flow depends on the size of the animal and its metabolic rate, larger the body size of higher the metabolism, greater is the perfection of bulk flow.

Bulk flow helps in quick transport of oxygen, glucose, amino acids; minerals and other materials needed in the metabolism of cells, from the site of intake to the interstitial fluid, from where the materials enter the cells through diffusion across the plasma membrane. Conversely, the metabolic products of cells—carbon dioxide, nitrogenous wastes and other by-products diffuse out of the cells and bulk flow carry them quickly to their sites of disposal or sites of actions (hormones).

Most cells cannot withstand large changes in their immediate environment. Addition or removal of materials by different organs from the blood and interstitial fluid occurs in a regular way to maintain the steady state of the fluids. The maintenance of a constant internal environment is known as **homeostasis**.

☒ Modes of internal transport

Internal transport in animals involve moving the materials from the site of intake to interstitial fluid or from interstitial fluid to specific organs and from the organs to the site of expulsion. This is effected by a continuous movement of **haemolymph** or **blood** in the body in definite channels.

CIRCULATORY SYSTEM

The movement of the fluid, blood or haemolymph is known as circulation and the structures taking part in it are collectively called **circulatory system**. It involves a force pump, the heart, a set of vessels carrying blood away from the heart and another set returning blood to the heart. The circulatory systems are of two types, **open** and **closed**.*

☒ Open circulatory system

Blood moves out of the heart through **arteries, arterioles** and **capillaries**. In the tissues, the arterial capillaries are continuous with narrow spaces or **sinuses**. The sinuses are not true vessels and not lined with endothelium but by a very thin connective tissue. The sinuses join to form bigger sinuses and finally a large sinus, which sends deoxygenated blood to the gills through a number of narrow sinuses. Sinuses collecting blood from the gills join to form a large sinus, often assuming the shape of a blood vessel but not separated from the adjoining tissues, opens in the heart. These channels through which the returning blood flows are collectively called **haemocoel** and the circulatory fluid **haemolymph** (Fig. 12.1). Arthropods, most molluscs and lower chordates have open circulatory system.

☒ Closed circulatory system

In this type, the capillaries formed by the branching of an arteriole, travelling some distance through the tissues, join to form a **venule**. The venules form veins and finally one or more large veins open in the heart. Arteries and veins are distinct vessels. The closed circulatory system (Fig. 12.4) is found in some annelids, cephalopods and vertebrates.

☒ Open circulation

1. The haemolymph moves slowly and the animals are slow moving and sluggish.
2. The circulatory fluid and materials they carry are evenly distributed to all the tissues, which may not always be advantageous.

☒ Closed circulation

1. The process is more advanced and the outflow and return of blood are rapid.
2. The amount of flow can be regulated.
3. Blood can be shunted to particular active tissue by contracting and dilating the arteries.

CAPILLARY EXCHANGE

Blood is composed of a liquid component, the plasma, blood cells, protein and other materials. About 55 per cent of the blood volume is **plasma**. The plasma contains about 90 per cent water, free in solution are 7 to 8 per cent soluble proteins, 1 per cent electrolytes and 1 to 2 per cent other materials, viz. glucose, amino acids, lipids, vitamins, hormones, metabolic intermediates and nitrogenous wastes.

The blood in the arteries moving to the whole of the body, except respiratory organs, is rich in oxygen and low in carbon dioxide. The blood in the ventral aorta of fishes and pulmonary arteries of land and secondarily adapted aquatic vertebrates, moving, to the respiratory organs, is low in oxygen and rich in carbon dioxide.

On its way an artery branches to a number of arterioles which have a diameter of 0.1 mm or less and an arteriole subdivides into many capillaries, the average diameter being 8-10 μm; the total cross-sectional area increases, which causes a gradual fall of velocity of blood flow at each step. The total cross-sectional area of all the capillaries of an arteriole is greater than that of the arteriole and the result is decreased velocity of blood flow. Further, the friction of the moving blood with the blood vessels causes continuous fall of mean blood pressure. The fall of pressure continues in the capillaries and veins. The velocity of blood

* The blood never comes in direct contact with the tissues. In closed circulation, the capillaries connecting the arteriole with the venule are continuous tubes. For the convenience of studies the half adjacent to arteriole has been named **arterial capillary** and the other half **venous capillary**. In open circulation, the narrow sinuses with which the arterial capillaries are connected, are comparable to venous capillaries.

flow begins to increase with the flow of blood into the venule. Several venules join to form a vein, several veins to a larger vein, and at each step the cross-sectional area decreases, leading to faster flow of blood.

In animals with open circulatory system the haemolymph flows freely into sinuses from the arterioles. Cells are bathed in haemolymph and exchange of materials takes place across the plasma membrane of the cells.

In animals with closed circulatory system, the blood flows into the capillaries. The capillaries are numerous closely packed, extremely thin-walled, narrow tubes with a diameter 8.0 to 10.0 μm. The capillary walls are formed by membranes of the endothelium and bear numerous minute pores, the diameter of which may be quite large, as 100 nm in some glands. These openings are closed only by delicate protein fibres constituting the basement membrane. In addition, at many points the endothelial cells are not firmly united with one another and they overlap, leaving direct passages through the wall.

Following the concentration gradient, water and dissolved substances with small molecular weight pass from the capillaries to the interstitial fluid, *vice versa*, and from interstitial fluid to cells through plasma membrane and reverse. Oxygen, glucose, amino acids, vitamins, ions, etc. diffuse out through the capillary walls to the interstitial fluid and from there move to the cells across the plasma membrane. Carbon dioxide, nitrogenous wastes and other products of cell metabolism move out to the interstitial fluid through plasma membrane and from there to the capillaries through capillary walls (Fig. 9.12). Water moves each way through the capillary walls.

Several forces help in maintaining a constant pressure of interstitial fluid.

Most protein molecules are large and those with molecular weight of about 70,000 or more cannot pass through the capillary wall. The pressure of plasma in the capillaries, the **hydrostatic pressure** or **blood pressure**, which varies along the length of the capillaries, causes water and dissolved substances to filter out of the capillaries, while proteins remain in the capillaries and exerts an osmotic pressure, **colloid osmotic pressure**, also called **protein osmotic pressure** (protein's hunger for water), which is fairly constant along the length of the capillaries. The higher hydro-

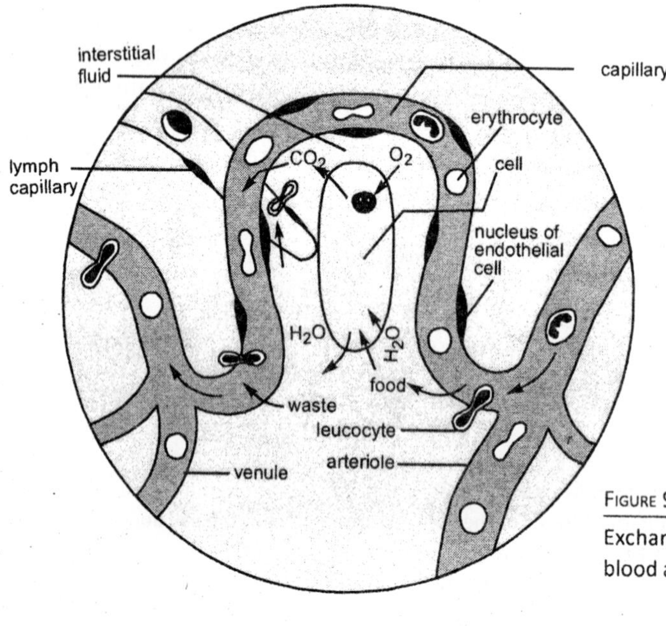

FIGURE 9.12

Exchange of materials between blood and interstitial fluid

static pressure at the arterial end of the capillaries gradually decreases towards the venous end with the flowing out of water.

In the exchange process, opposite forces are at work. At the arterial end the hydrostatic pressure is greater (35 mm) in the beginning of capillary than the protein osmotic pressure (25 mm Hg) and there is a net force, the **filtration pressure**, that forces out water and dissolved materials, including some plasma proteins out of the capillaries. With the fall of pressure in the capillaries, less fluid is filtered out and when the hydrostatic pressure is less than the colloid osmotic pressure, the process is reversed. The outflow of water causes higher protein osmotic pressure at the venous end, which is higher than the hydrostatic pressure and there is a net force, the **absorption pressure** that draws water and dissolved materials into the capillaries. In most cases, the filtration pressure is a bit higher than the absorption pressure. Absorption surface of the capillaries at the venous end is slightly more and more water can move in through it.

The amount of fluid exchange through the capillary wall fluctuates. Usually the outflow exceeds the inflow and the excess interstitial fluid passes into lymph capillaries and a constant pressure is maintained in the cell environment. The interstitial fluid entering the lymp capillaries is known as **lymph**. The lymph capillaries join to form lymph ducts, that drain into larger ducts, which finally open into the venous system and the fluid is returned to the blood.

GAS TRANSPORT

All animals, from annelids to mammals have large bodies and mechanisms other than simple diffusion have developed to carry gases from respiratory membrane to tissues and tissues to respiratory membrane. Both oxygen and carbon dioxide are soluble in water but they greatly differ in the degree of solubility.

The amount of gas going in solution depends on a number of abiotic factors.

RESPIRATORY PIGMENTS

Four types of respiratory pigments, **haemoglobin, haemocyanin, chlorocruorin** and **haemoerythrin**, are found in animals. All of them have metal containing proteins.

⊠ Haemoglobin

A haemoglobin molecule consists of two portions, iron containing portion, **haematin** and iron free portion (protein) **globin**. Each molecule of haem combines with one molecule of globin, and all haemoglobin contains four pairs of haem + globin with a total molecular weight 68,000. Normal haemoglobin (Hb) has two α globin chains and two β globin chains. Haemoglobin imparts red colour to the blood. In invertebrates, some annelids, arthropods and molluscs the pigment is in solution in blood or haemolymph. The molecules are much larger in size than those in blood cells. In vertebrates, the pigment is present in **erythrocytes** or R.B.C. This arrangement permits blood to carry much greater amount of the pigment and thereby transport efficiency. A single mammalian erythrocyte may have 265,000,000 haemoglobin molecules.

⊠ Bohr effect

Haemoglobin transports about 20% of the total H^+ and CO_2 formed in the tissues to lungs and kidneys. The binding of H^+ and CO_2 is inversely related to the binding of O_2. The affinity of haemoglobin for O_2 decreased as the H^+ and CO_2 are bound in the peripheral tissues with low pH and CO_2 concentration and the O_2 is released. On the other hand, the affinity of haemoglobin for O_2 increases in the lung capillaries as CO_2 is released and the blood pH rises, and the protein binds more O_2 for transport to the peripheral tissues. The effect of pH and CO_2 concentration on the binding and release of O_2 is known as Bohr effect.

$$Hb + O_2 \rightleftharpoons HBO_2$$

Due to the effect of H^+ concentration on the binding equilibrium, the reaction is

$$HHb + O_2 \rightleftharpoons HbO_2 + H^+$$

where HHb^+ denotes a pronoted form of haemoglobin.

☒ Haemocyanin

A haemocyanin molecule consists of two copper atoms as a part of the protein chain. Two atoms of copper are necessary to carry one molecule of oxygen. Its carrying capacity of oxygen is much less than that of the haemoglobin. Haemocyanin occurs in two forms, oxidised and reduced and remains dissolved in the haemolymph. In oxygenated stage it is shining blue but colourless in reduced state. Its molecular weight ranges from 1 to 7 millions. Haemocyanin occurs in some invertebrates—crustaceans, insects, gastro-pods, cephalopods and others.

☒ Gas transport in blood

Requirements of oxygen and removal of carbon dioxide can be met with by simple diffusion only in a very few animals. Gas in solution is also unable to meet the needs. In solution, the plasma can carry only about 0.2 ml oxygen and 0.3 ml carbon dioxide in each 100 ml blood. Presence of haemoglobin in blood increases the capacity by about 100 times in oxygen and 170 times in carbon dioxide.

☒ Oxygen transport

Following concentration gradient oxygen diffuses to the blood from the respiratory membrane. At high oxygen pressure the gas combines with haemoglobin (Hb) and form **oxyhaemoglobin** (HbO_2), an unstable compound. The reaction is reversible.

$$Hb + O_2 \underset{\text{tissues}}{\overset{\text{lung/gill}}{\rightleftharpoons}} HbO_2$$

Each iron atom of haemoglobin molecule can bind one oxygen molecule and when all the four sites are occupied it is called **saturated**. These are carried to the capillaries and at low oxygen pressure in the tissues the oxygen dissociates from its binding and the haemoglobin gives up all its oxygen. The proportion between haemoglobin and oxyhaemoglobin is regulated by oxygen concentration (pressure). This is known as **oxygen dissociation**. Haemoglobin is saturated with oxygen at the oxygen partial pressure 100 to 110 mm Hg. No more oxygen is bound by the haemoglobin even at still higher oxygen pressure. At lower pressure oxygen is released and at 30 mm oxygen pressure, half the haemoglobin remains as oxyhaemoglobin. At zero oxygen pressure all oxygen is given up (Fig. 9.13).

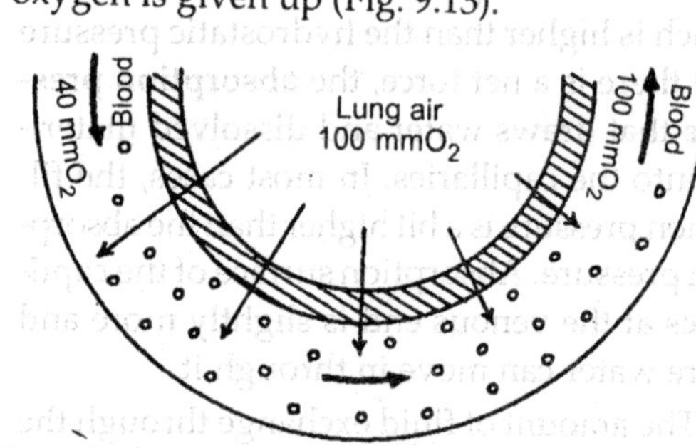

FIGURE 9.13

Oxygen diffusion in vertebrate lung

☒ Carbon dioxide transport

Only a small part of the carbon dioxide diffusing into the capillaries from the interstitial fluid goes into solution in the blood. Most of the carbon dioxide combines with water to form carbonic acid (H_2CO_3), which dissociates into hydrogen ions (H^+) and bicarbonate ions. The reaction is reversible.

$$CO_2 + H_2O \underset{\text{tissues}}{\overset{\text{lung/gill}}{\rightleftharpoons}} H_2CO_3 = H^+ + HCO_3^-$$

15 to 20 per cent carbon dioxide in the blood binds with the amino group of globin in haemoglobin and form carboamino haemoglobin. By reversible reaction the carbon dioxide is released in the lungs (Fig. 9.14).

Most of the resulting carbon dioxide in the blood is in the form of sodium bicarbonate. HCO_3^- combines with sodium ions (Na^+) in plasma and form sodium bicarbonate ($NaHCO_3$).

Carbonic acid and sodium bicarbonate together form a buffer system, which helps in maintaining a constant pH in the blood. The oxygen dissociation is influenced by carbon dioxide. A higher carbon dioxide concentration leads to release of more oxygen from the haemoglobin at any given oxygen pressure. This is important in removal of carbon dioxide from the tissues. With the release of oxygen from haemoglobin in the capillaries carbon dioxide enters from the tissues and the amount of oxygen to be given up from the haemoglobin is increased, leading to additional delivery of oxygen to the tissues.

The total amount of carbon dioxide in venous blood is about 60 ml per 100 ml blood. About 10 ml carbon dioxide is released by reverse reaction in the lungs. The arterial blood contains about 50 ml carbon dioxide per 100 ml blood. This is known as **carbon dioxide tolerance**. About 10 per cent carbon dioxide in blood is in solution; about 20 per cent in combination with haemoglobin and about 70 per cent in the form of sodium bicarbonate ($NaHCO_3$).

RESPIRATION: OXIDATION OF FOODSTUFFS

The oxygen is transported to cells since it is essential for oxidation of foodstuffs for liberation of energy. Among the food substances oxidised, glucose tops the list and hence it is considered as first fuel.

By oxidation of glucose ($C_6H_{12}O_6 + 6O_2 \rightarrow 6CO_2 + 6H_2O$ + 673 Cal. energy), carbon dioxide, water and energy are liberated. The breakdown of glucose involves a number of enzymes and 30 to 40 steps of chemical reactions. The whole reaction may be summed up in two major stages.

1. Glycogen and glucose breakdown into pyruvic and lactic acids in the absence of oxygen. This is **anaerobic stage** or glycolysis, taking place in the cytosol or soluble fraction of cells.

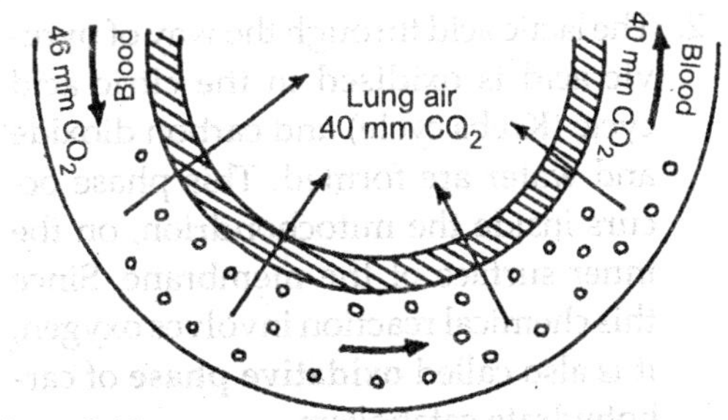

FIGURE 9.14

Carbon dioxide release in vertebrate lung

Respiration involves two distinct steps, external and internal respiration

External respiration	Internal respiration
1. It is a physical process in which oxygen (O_2) is absorbed and carbon dioxide (CO_2) is given out	1. It is a chemical reaction in which oxygen (O_2) combines with carbohydrates, fatty acids and amino acids (biological oxidation), producing carbon dioxide (CO_2), water and energy.
2. Enzymes do not play any role in the process. out.	2. The whole process is enzyme controlled.
3. Restricted to respiratory organs only.	3. It takes place in all living cells.
4. Respiratory structures are body surface, gills, lungs, etc. exposed to the surroundings.	4. Respiration occurs within the cells.
5. Sites of gaseous exchange—moist respiratory surfaces; inflow of O_2 and outflow of CO_2 occurs through plasma membrane by passive diffusion.	5. Sites of reaction—cystosole of the cell and mitochondria; O_2 and CO_2 move in and out of the cell through plasma membrane by passive diffusion.
6. Partial pressure of O_2 in the surroundings determines the rate of O_2 diffusion.	6. Pressures of O_2 and CO_2 inside the cell and interstitial fluid determine the rate of diffusion of the gases and CO_2 concentration in blood influences the release of O_2 from oxyhaemoglobin (HbO_2).

2. The lactic acid through the way of pyruvic acid is oxidised in the citric acid cycle (Krebs cycle) and carbon dioxide and water are formed. This phase occurs inside the mitochondrion, on the inner surface of the membrane. Since this chemical reaction involves oxygen, it is also called **oxidative phase** of carbohydrate catabolism.

HOMEOSTASIS

Maintenance of a constant internal environment is known as homeostasis. For normal functioning cells need a surrounding which does not change much. The cells are bathed in the interstitial or tissue fluid, consists of plasma fluid, glucose, salts, respiratory gases, etc. To maintain a constant state of the fluid, the amount and ratio of all the constituents have to be ensured. The factors effecting these are:

1. **Sugar level.** The human blood normally contains 90 mg/100 ml of glucose. The excess glucose, if any, is transported to the muscles and liver, converted into glycogen and stored there. In case the glucose level in blood falls, the glycogen is converted into glucose and the level is maintained. Both the processes are controlled by insulin, secreted by the pancreas.
2. **Osmotic pressure.** The osmotic pressure of the tissue fluid is maintained (p. 128) by kidney (osmoregulation). Antidiuritic hormone (ADH) secreted by the posterior lobe of the pituitary regulates reabsorption of water in the kidney tubules.
3. **Respiratory gases.** In the event of oxygen depletion and carbon dioxide increase in the interstitial fluid, increased heart rate helps in greater absorption of oxygen by the respiratory membrane and the respiration rate also increases. Further, with more carbon dioxide entering into it, the blood becomes more acid and extra oxygen is given up by the haemoglobin (p.108, 105) than that in the normal state.
4. **Temperature.** In animals, the normal metabolism is carried only within a range of temperature, which varies in different groups. The homeotherms maintain a near constant body temperature by their efficient metabolic activities and insulation against heat loss by the deposition of a layer of fat beneath the skin and feathers or furs covering the body. The poikilotherms are unable to maintain the optimal range of temperature by metabolism. Some of them collect heat from sun rays, while the majority shelter them in burrows or under some covers or even hibernate to bring down metabolism to a minimum.

10 REGULATION OF BODY FLUIDS

OSMOREGULATION

The mechanisms by which internal body fluids are maintained relatively constant, both in its volume and its contents, is known as **osmoregulation**. The interstitial fluid, the coelomic fluid, the water in blood are collectively named internal body fluid or **extracellular fluid**. The chief constituents of extracellular fluid are ions, glucose, phosphatase, amino acids, etc. and they are in osmotic balance with the intracellular fluid. To maintain such an environment, there must be a regulatory mechanism to balance water loss against water gain and in and outflow of other materials.

Water is gained through osmosis following concentration gradient in aquatic animals through body surface, gills, from the water they drink, from the food they consume, and in land animals—through the water they drink, food taken, absorption of atmospheric moisture and in both groups from water-yielding metabolic reactions. Water is lost in urine (Chapter 12), through respiratory membranes, sweat glands, surface evaporation, etc.

Animals inhabit four different environments—fresh water, brackish water, sea water (marine) and land.

The salinity of open sea water is about 350‰ (35 parts per thousand), that of brackish water far low and variable, as the rivers and streams discharge fresh water at the estuaries. The degree of capability in maintaining a constant salt concentration in the extracellular fluid and also the body volume, varies greatly in aquatic animals. Accordingly they may be placed in the following categories.

⊠ Osmoconformers

The salt concentration in the body fluid changes with the salt concentration in the medium, the salt concentration of the body fluid conforms with that of the medium.

⊠ Osmoregulators

The salt concentration in the body fluid does not change with the salt concentration in the medium; the salt concentration of the body fluid remains near constant, irrespective of changes in the medium.

⊠ Volume conformers

The body volume changes with the change in salt concentration in the medium. In lower concentration in the medium, water enters the body and the volume increases, in higher concentration it decreases.

☒ Volume regulators

Animals capable of maintaining a constant body volume irrespective of salt concentration in the medium.

☒ Ionic conformers

Animals incapable of maintaining a constant ionic level in the body against changes in ionic concentration in the medium. Ionic concentration in the body fluid changes directly with the ionic concentration in the medium.

☒ Ionic regulators

Animals capable of maintaining a constant ionic level in the body against changes in ionic concentration in the medium.

Capability of animals to survive in different aquatic environments also vary greatly.

☒ Stenohaline (Gr. *stenos* = narrow + *halinos* = saline)

Animals with a limited tolerance of changes in the salinity of the medium. A minimal concentration of salt is needed for survival and this phenomenon is present in osmoconformers. Stenohalines usually live in the open seas.

☒ Euryhaline (Gr. *eurys* = wide)

Animals able to tolerate a wide range of salinty. These marine organisms can move into brackish or even fresh water and survive there.

☒ Osmoregulatory organs and nitrogenous wastes

The chief osmoregulatory organs in aquatic animals are gills and excretory organs. In land forms, the sites for water loss are excretory organs, respiratory membranes and skin.

Animal living in water or having access to plenty of water, secrete major bulk of nitrogenous wastes in the form of ammonia, which is highly toxic and needs immediate removal (**ammonotelic**). In a situation where water availability is somewhat limited, ammonia is largely converted into urea, which is highly soluble but less toxic (**ureotelic**). The urea is converted to uric acid, when is only sparingly soluble in water and almost nontoxic, in animals where reduction in volume of urine and thereby conservation of water is important (**uricotelic**).

☒ Osmoregulation in invertebrates

The salt concentration in the body fluid of invertebrates ranges from 0.1 to 0.4 per cent.

In protozoon and some small animals water entering the body by diffusion is expelled to the exterior through contractile vacuole (Fig. 10.1) or some other means. Cnidaria and ctenophores are osmotic conformers.

Very little is known about the osmoregulation in marine invertebrates, though a host of them live in brackish water and in open sea.

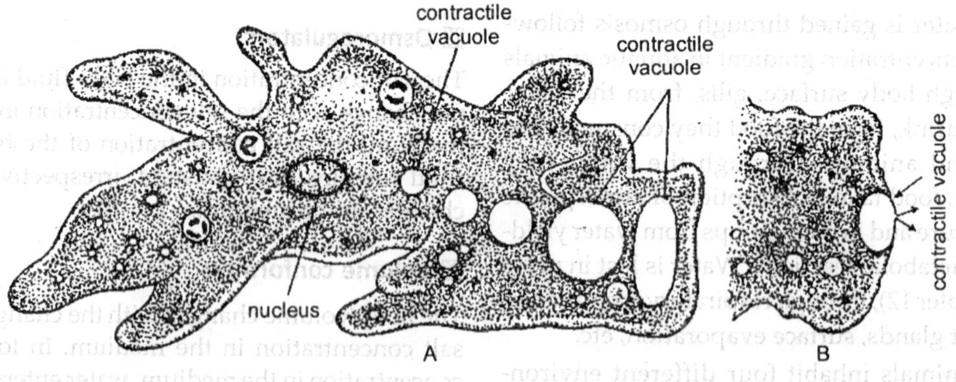

FIGURE 10.1

A. Origin and movement of contractile vacuole, B. A part of A, contents of contractile vacuole being discharged

Experiments on ***Neanthes, Arenicola, Maia, Carcinus*** and others show that osomoregulation in these animals is restricted within a very narrow range.

Excretory organs, the chief osmoregulatory structures, exhibit a wide diversity in invertebrates, but basically, they are similar. The organ consists of a **tubule** or a sac, capable of concentrating wastes. One end of the organ usually opens in the coelom and the other end opens outside. Excretory molecules filtered into the coelomic fluid from the haemolymph or blood, move into the tubule through the opening, the **nephrostome** by ciliary action, while in the tubule, glucose, amino acids, salts, minerals and even water in some cases, are reabsorbed. The cells of the tubules also secrete wastes and the fluid in the tubule, rich in wastes, the urine passes to the exterior through an opening, the **nephridiopore** (Nereis). Such an excretory tubule (Fig. 27.7) has been named **nephridium**. In the absence of a nephrostome (Fig. 27.33) the blood filtrate is directly received by the tubule and due to secretion of wastes and selective reabsorption, the urine is formed (leech). The nephridiopore may be absent (earthworm) and the nephridium open directly to the gut singly or many of them joining to form a common passage before opening in the gut (Fig. 27.20).

Nephridia are segmentally arranged in annelids. In platyhelminthes, rotifers and some others the nephridium assumes a different shape, the flame cell (Fig. 25.6), opening to the exterior.

A molluscan kidney is really a nephridium. Since the number is one or two, the size has greatly enlarged. It is spongy in nature, enclosing a space within, one end opening into the pericardial chamber through a renopericardial aperture and the other end in the mantle cavity through a renal aperture. In molluscs, other than pulmonates, living in water throughout the year, the nitrogenous wastes excreted are chiefly in the form of ammonia.

The salt concentration in the body fluid of fresh water mussels and gastropods is as low as 0.1 to 0.2 per cent. Salt level is maintained by selective reabsorption of salts in the kidneys and producing a very hypoosmotic urine and also by active absorption of salts through the mantle and body surface.

The amphibian gastropod molluscs and pulmonates produce uric acid to conserve water during the long period of aestivation; no urine is formed and the uric acid is stored in the digestive gland, to be released slowly after regaining activity in the next monsoon. To prevent water loss, the aestivating non-operculate snails secrete a calcic **epiphragm** closing the mouth of the shell, leaving only a small opening for movement of air for respiration. The land snails can withstand a water loss equivalent to the 50 per cent of their body weight, the slugs can withstand 80 per cent water loss.

The paired, saccular excretory glands, the green glands are present in crustacea. They are located in the cephalothorax and richly supplied with blood. Filtrate is subjected to selective reabsorption in the glandular portion, stored temporarily in the sac and a hypotonic urine is passed to the exterior through the renal pores at the base of the antenna (Fig. 28.11). The gills have the ability to collect salts.

The excretory organs of insects (Fig. 28.28) and arachnids (Fig. 28.43) are narrow thread-like Malpighian tubules at the junction of intestine and rectum. The number of tubules vary from two in trochophore larva, *Scolopendra* to four in scorpion and many in cockroach. Wastes are absorbed from the haemolymph, concentrated in the tubules and discharged in the gut. Water is absorbed in the rectum and the urine is passed out in the form of a paste. In terrestrial arthropods the excretory organs are modified to secrete hyperosmotic urine and thereby minimise water loss.

⊠ Osmoregulation in vertebrates

The salinity of vertebrate body fluids ranges from 0.5 to 0.9 per cent. As a rule, the vertebrates are osmoregulators and the hazards of maintaining a constant osmotic concentration in the body fluid are somewhat pronounced.

The fresh water forms have adaptations to prevent inflow of water, outflow of salts and also to collect salts from the medium. A certain amount of salt and water are received through food.

Against this, the animals living in brackish water and in open seas must have a different kind of adaptation. The salinity of even brackish water is much higher than the salinity of the body fluid, the salinity shooting up to about 400 per cent greater in open seas. In these forms, mechanisms have developed to prevent outflow of body fluid water and inflow of salts and also to excrete salts, which enter the body through food and the water.

In land forms the problem chiefly centres about conservation of water. Water enters the body only through extraneous sources like food, including plant sap and animal body fluid and the water they drink. On the other hand, they lose water in different ways, namely excretion, expiration, perspiration, etc. Of the various adaptations for conservation of water, formation of insoluble nitrogenous wastes and hypoosmotic urine are most important.

⊠ Osmoregulation in fresh water vertebrates

Fresh water vertebrates live in a very hypoosmotic medium. To maintain an equilibrium with the body fluid, water tends to move in through body surface and gills and salts tend to go out in excretion.

These are overcome by maintaining a minimum salt concentration in the body fluid to permit metabolism; reabsorption of salts in the excretory organs and formation of hypoosmotic urine and expulsion of nitrogenous wastes in the form of ammonia. Selective absorption of salts by active transport through gills and skin is present. The fresh water teleosts absorb Na^+ by active transport through gills.

Frogs take up Na^+ and Cl^- through skin, water and salts are reabsorbed from the urinary bladder and urine discharge is usually stopped in dry season.

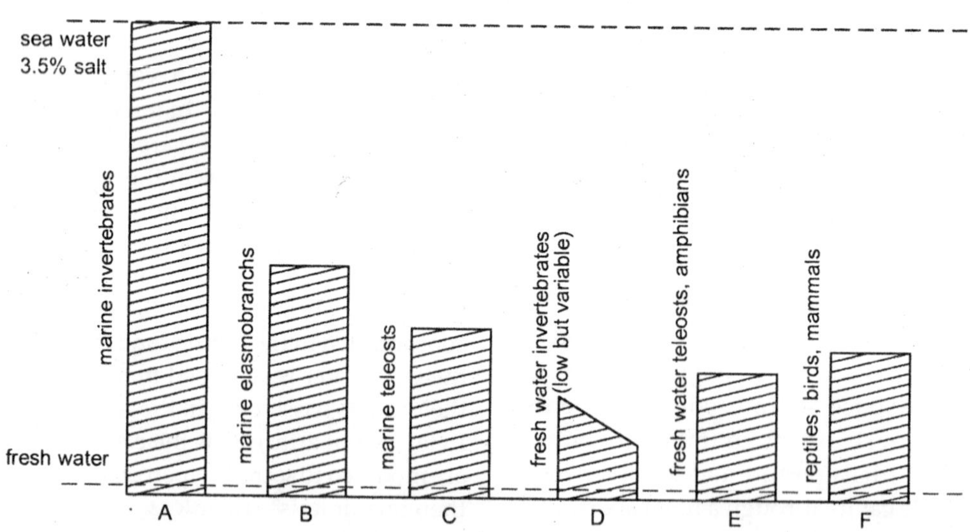

FIGURE 10.2

Osmotic concentration in marine, fresh water and land animals in relation to the concentration of sea water

The skin being impermeable, fresh water reptiles receive water through diet. To overcome deficit of water, the filtration rate is low in the kidneys, urine is very concentrated and may be even solid. Nitrogenous wastes are converted into uric acid.

☒ Osmoregulation in marine vertebrates

Marine vertebrates live in a very hyper-osmotic environment, the salinity in open sea being about four times higher than that in their body fluids. To maintain an equilibrium with the surrounding medium, body water tends to move out and salts tend to move in through structures like gills and body surface. Water conservation in marine vertebrates is hazardous and as they cannot use sea water freely, the sea has been described as a physiological desert to its inhabitants.

The adaptations that helped marine teleosts to live successfully in sea are (a) an impermeable body surface and a mucous covering preventing the in and outflow of materials. (b) Capability to drink sea water and make use of water and monovalent ions in it and passing out the divalent ions from the gut. (c) Urine volume is minimal and only some nitrogenous wastes are removed through it. (d) Ammonia, constituting the main bulk of the nitrogenous wastes is eliminated through gills. (e) Removal of excess Na^+ and Cl^- ions through gills.

Presence of a large amount of urea and trimethyl amine oxide in blood and tissues make the elasmobranchs isoosmotic in sea water. Like other fishes, elasmobranch gills are not permeable to urea. The accumulated salts through food and inward diffusion across the gills are excreted through kidneys and a **rectal gland** (Fig. 37.9) in communication with the rectum.

Marine reptiles and birds exhibit similar osmoregulatory mechanism. Their internal osmotic pressure is less than that of their environment. Salt intake occurs through food. The water is reabsorbed from the kidneys and the nitrogenous wastes are converted to uric acid. The excess salt is secreted in turtles through **orbital glands**, the modified lacrimal glands; through **nasal glands** in marine iguana and through **salt glands** (Fig. 10.3) in marine birds.

The marine mammals eat both invertebrates and vertebrates living in sea. The invertebrates are isoosmotic with sea water and the fishes are hypoosmotic with sea water. Marine mammals maintain their body salt concentration by secreting urine having greater salt concentration than sea water.

☒ Osmoregulation in land vertebrates

Of the different modes of water losses in terrestrial vertebrates, viz. urine, faeces and evaporation, the last one accounts for maximum loss. The air is capable of holding a certain amount of water vapour. This is constant for a volume of air at a given temperature and pressure. **Saturation deficit** is the amount of

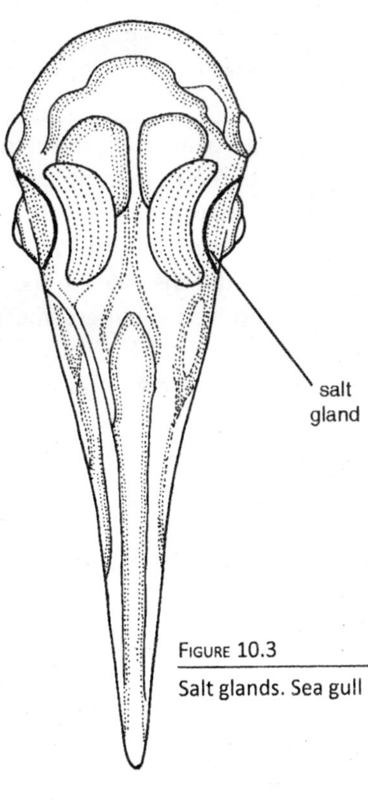

FIGURE 10.3

Salt glands. Sea gull

water vapour required to saturate the air in addition to the water vapour already present in it. The saturation deficit is the amount of water to be evaporated from the organisms to the environment. This again is regulated by different biotic and abiotic factors in the area. The principal sites of evaporation in land animals are the body surface and the respiratory surface. Water lost is made up by drinking water, through food and absorption of moisture through body surface.

Frogs and toads are greatest victims of water loss by evaporation. The loss is minimised by venturing out only in nights and excreting urea, which is less toxic but highly soluble in water. Certain amount of water is obtained from food.

More resistant skin to water loss and excretion of nitrogenous wastes in the form of uric acid play important role in water conservation in reptiles.

In birds, water loss from body surface is prevented by feathers. Urine is hyperosmotic and nitrogenous wastes are eliminated in the form of uric acid.

Of all the groups of animals, the mammals are best adapted to their environments. With one exception (desert) they live close to sources of water. The skin is resistant to water loss, hairs have developed to insulate the body from outside, urine produced is hyperosmotic, the uric acid concentration is as high as 6 per cent. In addition, they procure a good amount of water with food.

Extreme modifications are found in desert animals. The Kangaroo rats (*Dipodomys*) of South Western North American desert, practically with no access to water, live on dry seeds. But their bodies contain about 66 per cent water, as in other mammals. The water loss does not exceed the gain, even if they live on dry food alone. Most of their water requirement comes from the water produced in oxidation of the dry food and from a small percentage of water present in the grains. Further, they lack sweat glands, as they do not need water for temperature regulation, because of their natural habits help them to avoid the heat of the day. Their faeces is almost dry and they produce highly concentrated urine containing as much as 23 per cent urea and 7 per cent salt.

Camels, the ship of the desert, can survive great water loss. They can drink water as much as 150 litres, one-third of their total body weight, within 10 minutes. The water loss is gradual and the animal can survive for at least 17 days after a drink. Sweat glands are present but seating begins only at 41°C. The desert nights are cool. The body temperature of camel drops as low as 34°C at dawn and much of the day elapses before the temperature rises to 41°C, at which sweating begins. The remarkable fluctuation of 7°C in contrast to only about 1.1°C in man, helps to minimise loss of water through sweating. The significant adaption in camels is that the **microorganisms in the gut convert urea back to protein** and thereby save water loss from the kidneys to excrete urea.

11 REGULATION OF BODY TEMPERATURE

Active animal life is limited to a narrow range of temperature—a few degrees below – 0 °C to about +50 °C. Compared to cosmic temperatures the temperature range for animal life is very narrow. Livable temperature, however, exists over most of the earth surface and throughout the oceans, at least during a part of the year.

Depending on the inability and ability to maintain a constant body temperature, independent of the surrounding temperature, the animals have been placed under two divisions.

1. **Poikilotherms** or **Cold blooded.** The majority of the animals living on earth are poikilotherms. The poikilotherms are not necessarily cold blooded. Their body temperature is same as that of the surroundings. An earthworm has a temperature of the soil in which it lives. A fish has the temperature of the water it swims in.
2. **Homoiotherm** or **Warm blooded.** Only birds and mammals are homothermic animals. They can maintain a quite constant body temperature inspite of wide variations in the surrounding temperature. A tropical monkey, a walrus and polar bear have the same body temperature about 38 °C.

TEMPERATURE REGULATION IN POIKILOTHERMS

The poikilotherms having same body temperature with the surroundings does not mean that they have no control over their body temperature. The animals can to a great extent influence their body temperature by seeking out environments with certain temperatures to keep the body temperature at a desired level and thus maintain their level of metabolic activity.

Poikilotherms exhibit diverse adaptation to prevent both cooling and overheating.

⊠ Adaptation to raise body temperature

In cold days snakes, lizards, crocodiles and a host of poikilotherms busk in the sun in early morning hours. The radiant energy from the sun makes their temperature rise above that of the surroundings. After necessary warming up of the body they move away from the exposure to sun rays.

The high rate of metabolism during flight is a special thermal problem in insects. The muscles of a cold insect cannot contract fast enough to carry it in flight. The insects, however, can increase the temperature of the flight muscles considerably by contractions similar to shivering in man. In a cool day the spread wings of a butterfly vibrate for a few minutes to warm up before it takes off. It has been measured that the temperature in the muscles of butterfly wings increased 20 °C to 35 °C in a six-minute period of warm up.

☒ Adoptation to keep body temperature cool

In a hot day the poikilotherms hide under rocks, in cool burrows, under the fallen leaves in the shade of trees, etc. to prevent rise of body temperature. A desert lizard dies if exposed for 15 minutes at about 45 °C under the open sun. It survives by keeping its body cool in the same way as in others.

TEMPERATURE REGULATION IN HOMOIOTHERM

In the animal kingdom only two groups—birds and mammals are homotherms. They maintain an almost constant body temperature—with a slight variation. Both lowering and rising of a few degree temperature have adverse effects on them. The variation in body temperature in homotherms is not much at best a few degrees. The body temperature is about 37 °C in human, 37.7 °C in horse, 36.2 °C in elephant, 36.5 °C in mouse. The body temperature of birds is a little high, 40 °C to 45 °C.

The body temperature in human shows a slight diurnal change, the night temperature is 1°C less. Variation of a few degrees of temperature is dangerous to homotherms. In human a few degrees higher temperature causes serious fever, and a few degrees lower causes dangerous cooling.

Thermoneutrality. The range of temperature in which normal activity of homoiotherms continue is rather narrow. It is best known in human. In human (without clothing) the comfortable range is 27 °C to 31 °C. At temperature higher than 31 °C the air is not sufficiently cool to remove the heat produced in metabolism and the only way to cool the body is to evaporate water by sweating. Hotter the surroundings the more the person is to sweat. On the other hand if the surrounding temperature is less than 27 °C the body loses heat faster than it is produced and the only way to produce more heat is by involuntary muscle contraction called shivering. The cooler it is, the more the body shivers.

☒ Temperature regulation in mammals

In mammals other than human—some sort of arrangements are found for retaining and losing heat as the case may be.

Prevention of heat loss

To prevent heat loss some sort of insulation—like fur, layer of fat ensheathing the body is necessary.

1. In artic, Alaskan ground squirrels the lower limit of thermoneutrality is about 10 °C. Below this they must increase their metabolism to maintain high body temperature.
2. The Eskimo dog has much heavier fur insulation. The animal curls up in the snow and sleeps. The insulation is so great that the dog is to increase metabolism when the temperature drops below 10 °C.
3. The artic seals, walrushes have both hair and a heavy layer of blubber under the skin for insulation. They swim in water close to 0 °C. Wet fur cannot prevent heat loss, still it is necessary to protect outer part of the skin from damage, when they are out of water and rest on ice in subzero temperature.
4. In whales a heavy layer of blubber insulates them from heat loss. They do not leave water and hairs are lost due to lack of any function.

Prevention of overheating

1. **Sweating.** Evaporation of water needs a considerable amount of heat. Overheating is prevented by inducing

production of sweat by sweat glands. The sweat evaporates from the skin and the body is prevented from over-heating. However all mammals do not passes sweat glands and in some cases the amount of sweat produced is not sufficient to keep the body cool.

2. **Panting.** A rapid rate of respiration is called panting. Panting increases the passage of water over the wet surface. In dog the number of breaths per minute 15-30 normally, but in panting the rate rises to over 300 per minute. Water evaporates from the tongue and upper respiratory tract and the body is cooled. Cattle and sheep both sweat and pant but panting is not so efficient as in dog.
3. **Wetting the fur.** When overheated some animals–cats, rabbits, kangaroos wet down the fur of their limbs and belly by licking. The moisture evaporates from the wet fur and the body is cooled. This, however, is not a very efficient method of cooling.
4. **Living in underground burrows.** Cooling by evaporation dehydrates the body. The body surface is relatively large in small animals than in large-bodied ones. In hot surroundings the small animals will heat up faster and metabolic heat production is also higher in them. Such animals have not developed any special mechanism for heat regulation. The small animals such as rodents have no sweat gland and do not pant. They avoid the heat by living in underground burrows.

☒ Temperature regulation in birds

1. **Prevention of heat loss.** Feathers are excellent insulators. The whole body of the bird except beaks and extremities of the legs and toes are completely covered with neatly arranged feathers that do not permit heat loss even in cold surroundings. The beaks are cornified. The legs are ensheathed by an impervious skin, which do not permit escape of heat. In arctic birds the amount of body fat is very high.
2. Chicks of emperor penguines huddle together in the ice-covered Antarctic breeding ground and thereby reduce the surface area exposed to cold.

Prevention of overheating

Birds have no sweat glands. They pant to some extent. The extensive air-sac system connected with the lungs possibly help in heat regulation. The process, however, is not clearly understood.

HIBERNATION

Hibernation is not a temperature regulating adaptation. It is an adaptation to tide over the cold winter period when more food is required for high degree of metabolism to generate more heat, but the amount of food available is less than that in other seasons. In the winter the animals become lethargic and torpid. To survive in cold surroundings they shut them from outside and take shelter in cracks and holes in soil, in burrows, under stones, rocks, debris, in holes of trees and some such places for the whole winter.

In a hibernator the body temperature is greatly decreased, and the metabolism, respiration, heart rate, etc. are considerably lowered.

Mammals of different groups—insectivores, bats, rodents, wood bucks, hamsters, hedge-hogs, ground squirrels undergo hibernation. Hibernation has also been reported in birds. It is most common in terrestrial poikilothermic vertebrates.

In winter a similar situation is found in a number of invertebrates. The phenomenon, however, is not considered as hibernation. It has been described as a torpor or dormant state—**aestivation**.

Some terrestrial invertebrates undergo aestivation in summer. To prevent desiccation they keep themselves away from sun and hot wind for the whole summer. Some non-operculate pulmonats secrete a calcareous membrane, epiphragm, which tightly fits in the month of the shell to insulate the body from hot surroundings. A narrow slit-like opening corresponding to the pneumostome is present to permit in and outflow of air for respiration.

AROUSING

In hiberation arousing or the resumption of normal activity from a torpor state takes a longer period. In aestivating animals arousing occurs within minutes if the favourable condition is restored by natural or artificial means.

12 THE CIRCULATORY SYSTEM

The organs taking part in circulating blood through the body of an animal are collectively called circulatory or blood vascular system. In a typical vertebrate circulatory system, the structures involved are—a heart, which pumps out blood and receives it back, a set of vessels, the arteries, carrying blood away from the heart and another set of vessels, the veins, returning blood back to the heart. The arterioles are the ultimate branches of an artery. Several venules join to form a vein. An arteriole and a venule are joined by a number of capillaries.

INVERTEBRATE CIRCULATORY SYSTEM

The structures are somewhat different in different groups of invertebrates. A circulatory system with distinct vessels is first met with in annelids. The skin serves as the respiratory structure and as a result the circulatory system is simplified (Fig. 27.17). A contractile dorsal vessel, provided with valves at regular intervals to prevent back flow of blood, acts as a collecting vessel and runs forward from the posterior end. At a short distance from the anterior end, it sends major portion of the blood to a **ventral vessel** through a number of

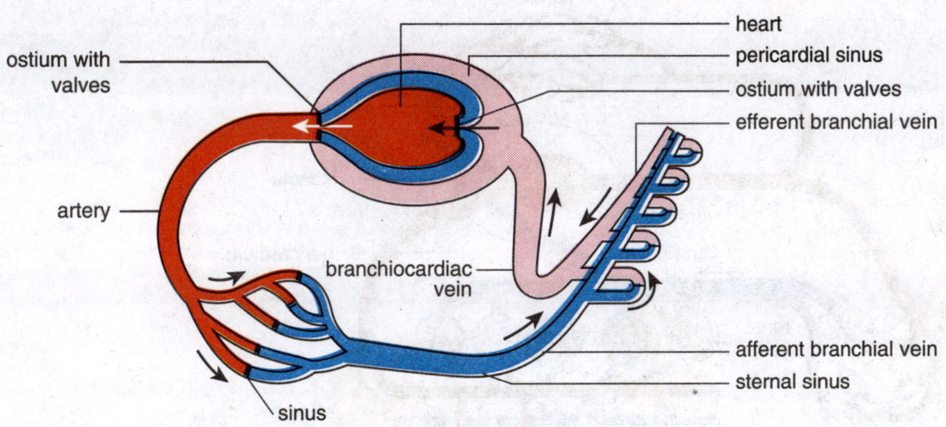

FIGURE 12.1

Circulation. Crustacea, Arthropoda (schematic)

pulsating lateral vessels, the **lateral hearts**. Anterior to this the dorsal vessel acts as a distributing vessel. The ventral vessel is a distributing vessel and sends blood from anterior to posterior region, supplying blood to different organs through narrow branches. In annelids, the blood flows through distinct vessels but they have not been designated as arteries and veins, since they lack an endothelial lining.

In arthropods (Fig. 12.1) and molluscs (Fig. 12.2) the arrangement has undergone much change due to interposition of respiratory structures in the course of circulation. The heart is enclosed in a **pericardium**. The pericardial chamber, named **pericardial sinus** contains **haemolymph**. The sinus is haemo coelomic in arthropods and coelomic in origin in molluscs. The heart consists of one atrium (auricle) and one **ventricle**, lying in the same long axis. In some molluscs, the artria are two in number. Oxygenated blood is moved through the arteries, the final branches of which communicate with the sinuses in the tissues. The sinuses join to form a large sinus, which receives the deoxygenated blood from the tissues. From there it is sent to the respiratory organs and the oxygenated blood is returned to the heart, through another set of channels. The sinuses, through which the blood flows are not provided with a distinct wall. They are lined only with a thin layer of connective tissue and are not considered as true vessels. This type of circulation is known as **open circulation** (p. 101).

VERTEBRATE CIRCULATORY SYSTEM

The circulatory system in craniates differs basically from that of the invertebrates. A progressive evolution in this group had led to a gradual change, leading to the appearance of an efficient heart, since activities of animals are directly related to a rapid flow of oxygenated blood.

The general scheme of circulation, sending blood through arteries and returning it to the heart through veins, has been complicated by the interposition of gills in the course of outflow of blood from the heart, and the arteries are to function as both **afferent** (sending away) and **efferent** (returning) vessels in branchiate craniate (Fig. 12.3). Flow of oxygenated blood to the tissues directly from the gills and returning deoxygenated blood from tissues to the heart created the necessity of imposition of some blood

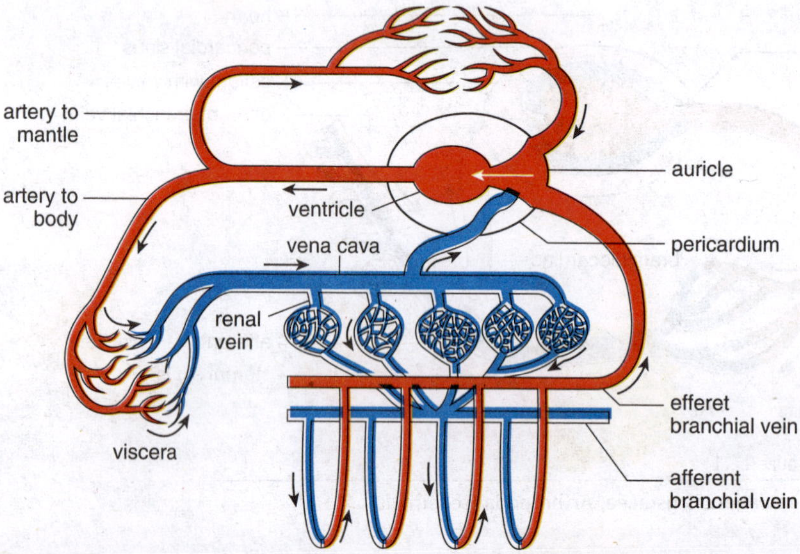

FIGURE 12.2

Circulation. Bivalvia, Mollusca (schematic)

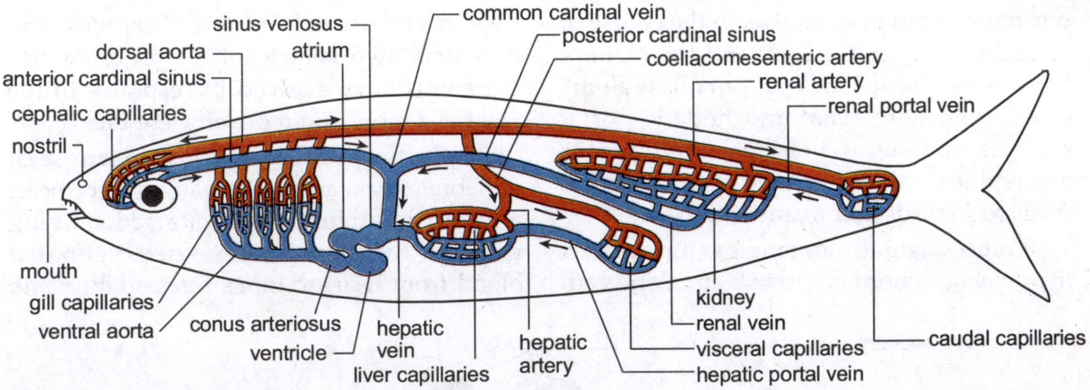

FIGURE 12.3

Fish. Circulatory system (schematic)

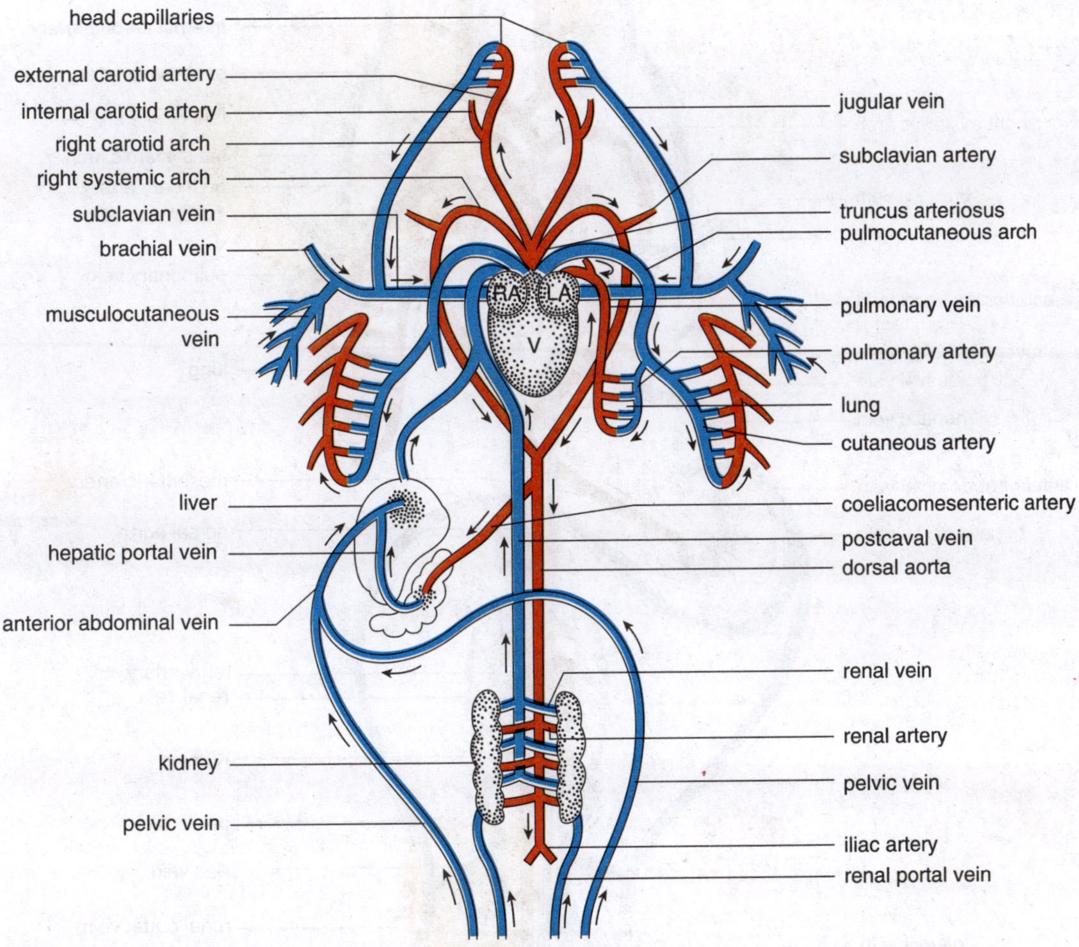

FIGURE 12.4

Amphibia. Circulatory system (schematic). Heart magnified

purifying organ in its course. In the course of returning current, the liver and the kidneys have been imposed and portal systems developed. Both renal and hepatic portal systems continue to present in amphibians and reptiles; only the hepatic portal system is retained in birds and mammals.

From a single atrium, single ventricle heart in branchiate chordates to two atria, one ventricle heart in amphibians and reptiles, and two atria, two ventricle heart in birds and mammals have evolved in response to the need of a more and more efficient heart.

Replacement of gills by lungs in terrestrial vertebrates has caused big changes in the heart and aortic arches. Blood started moving through different routes—deoxygenated blood from heart to lungs for oxidation and

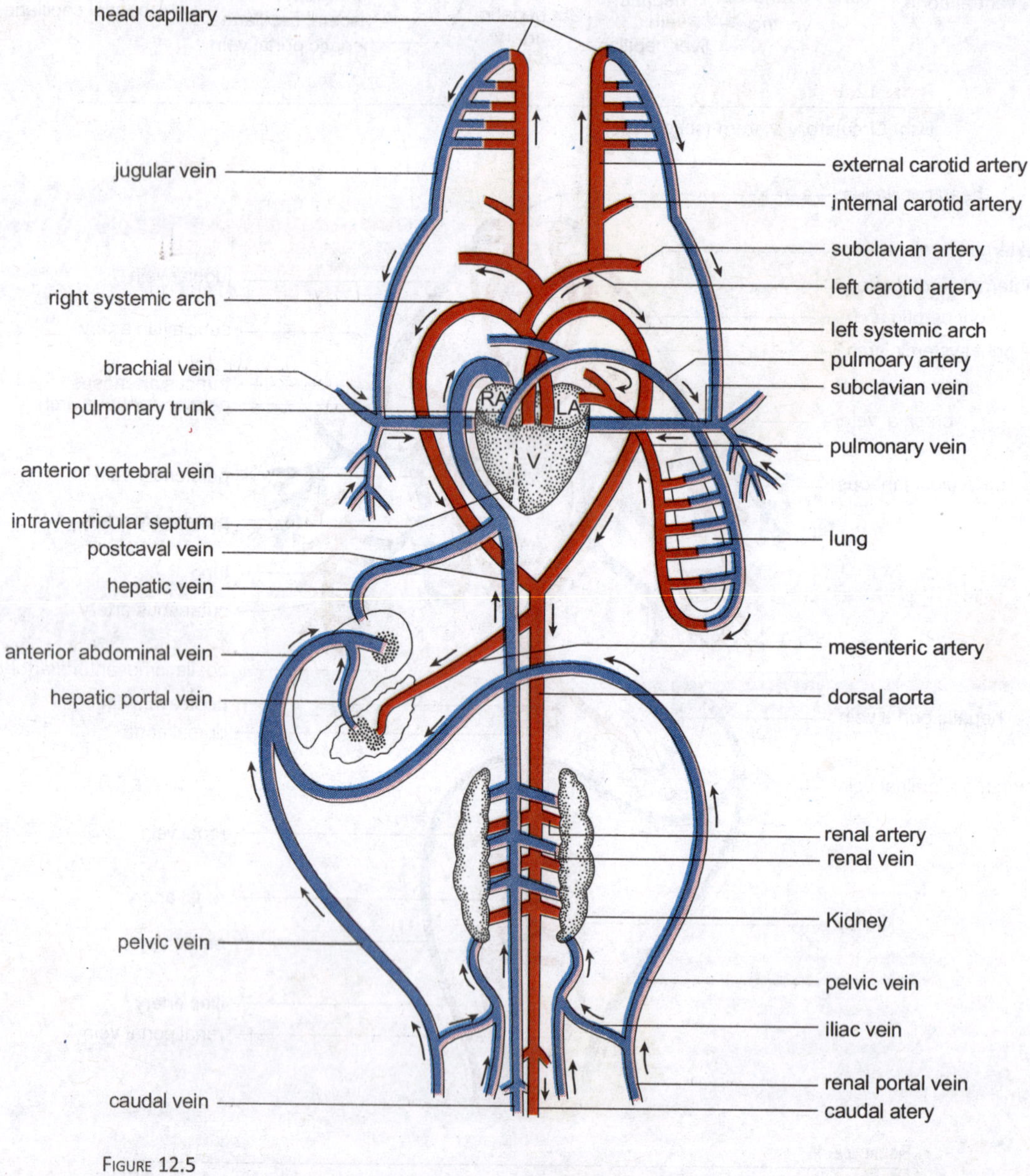

FIGURE 12.5

Reptilia. Circulatory system (schematic). Heart magnified

back to heart, and oxygenated blood from heart to tissues (where it loses oxygen) and back to heart, the latter being (Fig.12.4) intervened by both the liver and the kidneys or by the liver alone. Absence of any partition in the ventricle in three-chambered heart of amphibia permits mixing of deoxygenated and oxygenated blood and this stands in the way of leading an active life in this group. In reptiles, an incomplete partition in the ventricle (Fig.12.5) restricts mixing of two bloods to a certain extent. In crocodiles, a group of reptiles, the partition is near complete and the mixing of blood is prevented to a great extent, enabling them to lead more active life. In birds and mammals, a four-chambered heart (Fig. 12.6) has ensured complete separation of oxygenated and deoxygenated blood and this has helped them to lead very active life.

With the progress of evolution a gradual

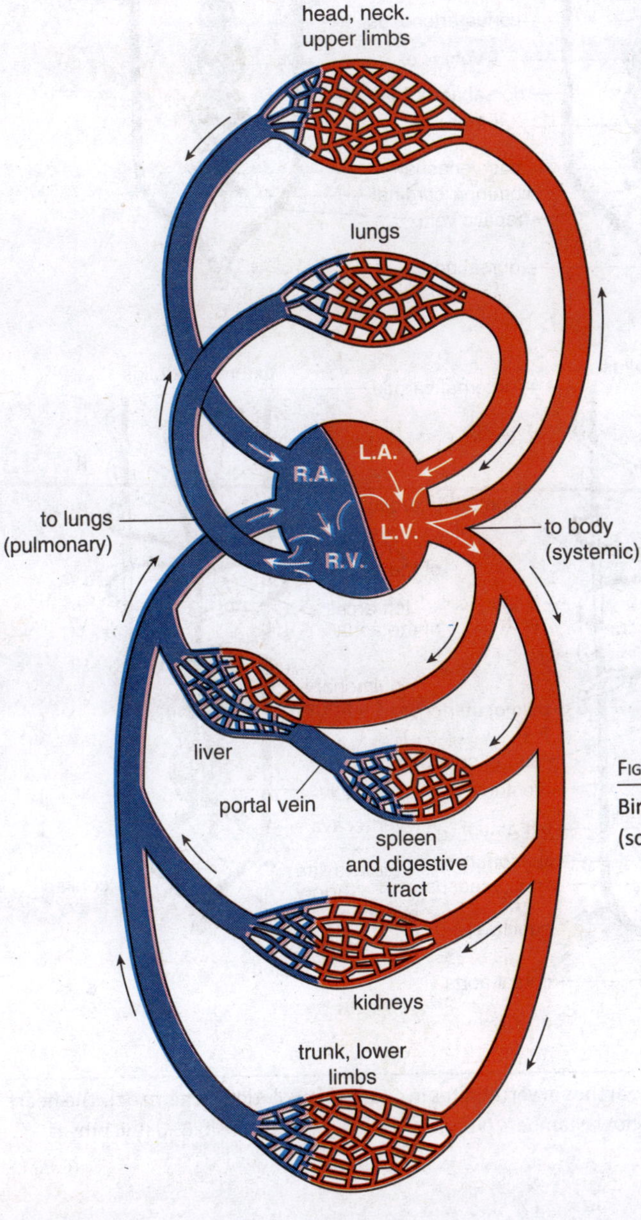

FIGURE 12.6

Birds and mammals. Circulation (schematic)

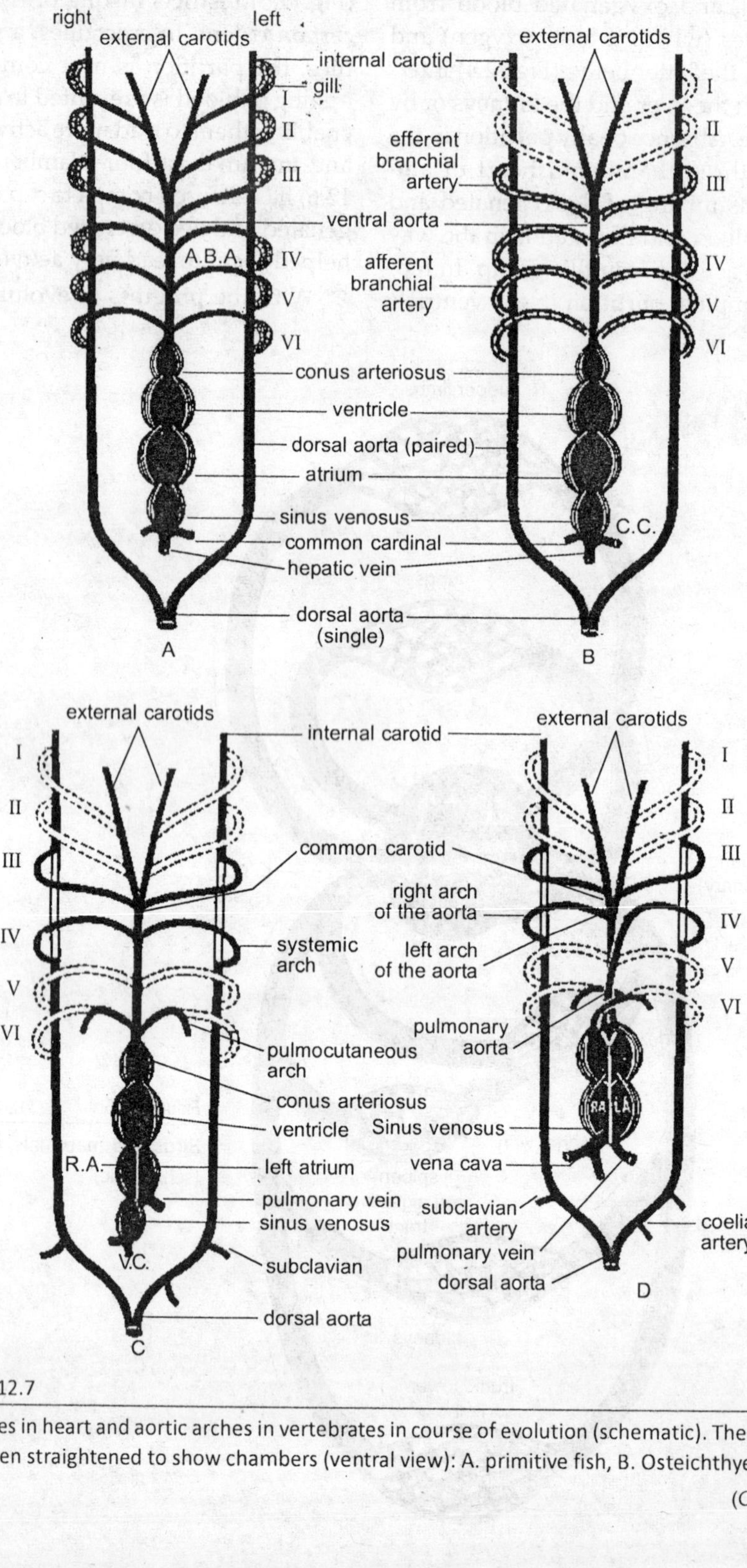

FIGURE 12.7

Changes in heart and aortic arches in vertebrates in course of evolution (schematic). The heart has been straightened to show chambers (ventral view): A. primitive fish, B. Osteichthyes,

(*Contd.*)

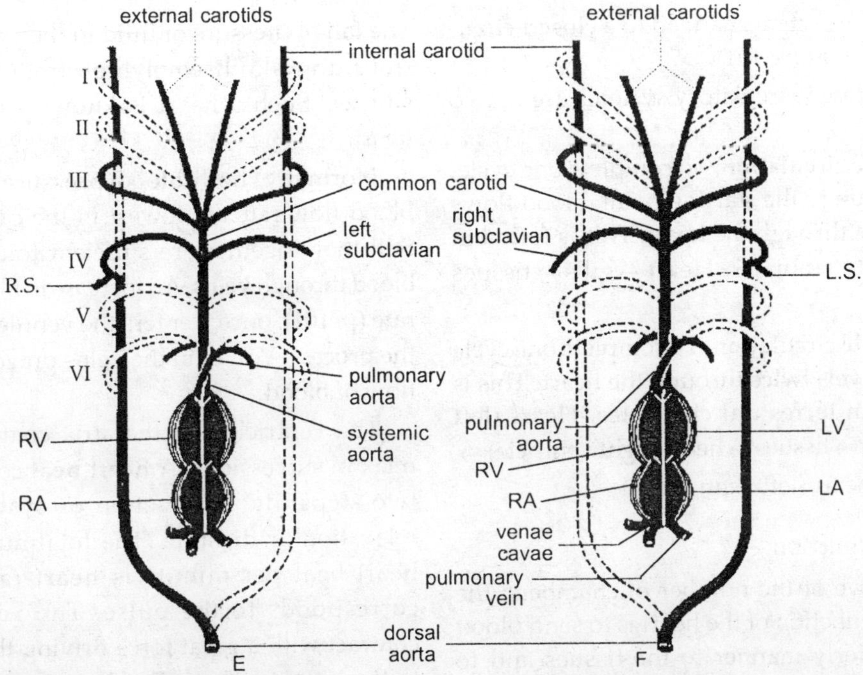

FIGURE 12.7

C. Amphibia, D. Reptilia. E. Aves, F. Mammalia. A.B.A. Afferent branchial artery, C.C. Common cardinal, L.S. Left subclavian, R.A. Right atrium, R.S. Right subclavian, S.A. Systemic aorta, V.C. Venae cavae.

reduction in the number of aortic arches occurred in craniata. From eight in cyclostomes it came down to six in primitive fishes, to four in Osteichthyes. With the adaptation to land life the number of arches drastically reduced to three in tetrapods (Fig. 12.7). Their embryos, however, always possess six arches.

Twenty per cent of the atmospheric air being oxygen, the land forms had to move a far less amount of air to respiratory membranes, than the amount of water to be moved in aquatic forms. In primitive tetrapods (frogs and toads) the sixth pair formed the **pulmonary arteries**, their bases constituting the **pulmonary arch**, took over the function of carrying deoxygenated blood from the heart to lungs and skin. This is maintained (skin excluded) in higher groups of vertebrates. The anterior two pairs degenerated greatly, the rudiments of their ventral portions formed a part of the ventral branches of the external carotids. The anterior portion of the ventral aortae became **external carotids**. The third pair formed the **carotid arches**. The anterior portion of the dosal aortae formed **internal carotids**. The part of the ventral aortae between the bases of the third and fourth arches became **common carotids**. The fourth pair formed systemic arches; the fifth pair degenerated. Amphibians and reptiles retained paired systemic arches. Asymmetry in the systemic arches occurred in birds and mammals. The left systemic arch degenerates in mammals. In mammals the right subclavian artery arises from the right systemic arch before it is reduced during development.

The change from somewhat sluggish life in most invertebrates to active life of vertabrates created a demand for separation of blood from coelomic fluid and also a rapid circulation. In response to this, the open circulatory system changed to a **closed circulatory system** (p. 101).

The closed circulatory systems are of two types:

Single circulation. To complete one cycle, i.e. to retun to the starting point, blood flows only once through the heart. This is found in branchiate craniates. Heart → gills → tissues → heart.

Double circulation. To complete one cycle blood travels twice through the heart. This is present in terrestrial chordates. Heart (left ventricle) → tissues → heart (right ventricle) → lungs → heart (left ventricle).

⊠ Heart function

Irrespective of the number of chambers, the primary function of the heart is to send blood in an orderly manner to the tissues and to receive the blood back. This is met with in vast majority of animals by having a muscular heart and vessels. The blood is received by the atrium/atria and pumped out by the ventricle(s). Valves are present at the openings in the heart and the opening of the aorta to prevent backflow of blood and thereby ensure a steady, one-way movement. In addition the thick walls of the arteries can contract and the veins, being thin walled, are provided with valves at regular intervals.

The pressure exerted by the ventricle is sufficient to drive blood through arteries and its return to the heart. In some arthropods the blood pressure falls below the thrust required to drive the haemolymph into the heart. The pericardial sinus enclosed in the pericardium is filled with haemolymph. Contraction of muscle stands extending from the heart to the pericardium causes expansion of the heart and the haemolymph in pericardial sinus is sucked into the heart through ostia in the heart wall. The fall of pressure of fluid in the pericardial sinus draws in haemolymph from the large sinuses. Such a heart is known as **suction heart**.

Normally, both blood pressure and the blood flow rate are lowest in the capillaries. Still they together are suffcient to drive the blood through veins. An increase in blood flow rate (p. 101), once it enters the venules help in the process. Valves in the veins prevent backflow of blood.

The ventricle and the atria contract and relax in successions. A **heart beat** consists of two steps, the contraction or **systole** and relaxation or **diastole**. The total number of heart beat per minute is **heart rate**. This corresponds to the pulse. The ventricles contract with a great force driving the blood to the arteries. The ventricles relax, pressure within it falls and blood from the atria enters it. The atria contract when the ventricles are near full with blood. This is again followed by the contraction of the ventricles. The alternate stretching and contraction of arteries are associated with the contraction and relaxation of the ventricles.* This can be felt as **pulse**.

⊠ Cardiac output

Outflow of blood or haemolymph from the heart in per unit time is called cardiac output. Small animals, the animals under exercise need increased cardiac output. Since the weight of heart in most mammals is about 0.6 per cent of the body weight, the cardiac output is increased by increasing heart rate. But the increased rate is not proportional as the blood carries more oxygen at such time. The pulse rate of an elephant is 25 beats per minute, of an adult man 70 and that of a shrew 600.

---

* **Systolic pressure.** The amount of pressure in the blood vessels when the heart contracts and pushes blood through the circulatory system.

Diastolic pressure. The pressure in the blood vessels between beats of the heart.

HAEMOSTASIS (In blood circulation)

The mechanism of maintaining a near constant blood composition and blood volume to retain an active balance of body processes is known as haemostasis. This is effected in two steps :

1. The blood composition is kept near constant by receiving fresh materials to replace those released in the process of circulation and eliminating the by products of cell metabolism passing to it.
2. The blood volume is maintained by a short term closure of damaged blood vessels by constrictions in case of a wound and thereby reducing blood loss, followed by the formation of a blood clot at the broken ends of the blood vessels.

13 THE EXCRETORY SYSTEM

EXCRETION

Wastes* are the toxic products of cell metabolism. The chief wastes formed in the body belong to two distinct categories. The nitrogenous wastes,—**ammonia, urea, uric acid** and in some cases guanine, are formed by deamination of amino acids, and the carbon dioxide formed by the biological oxidation of organic molecules. Being toxic, they are harmful and to be moved out of the body at the earliest. Of the nitrogenous wastes, ammonia being highly toxic has to be removed immediately and it occurs in animals living in water or having access to plenty of water. These animals are called **ammonotelic**. Examples: *Neanthes, Arenicola, Carcinus, Astacus,* mussels, cuttle fishes, human, etc. Urea is less toxic but highly soluble in water. It is found in animals where water conservation is necessary. They are known as **Ureotelic**. Examples: Some earthworms, shore gastropods, elasmobranchs, lung fishes, tortoises and turtles, human, etc. In extreme case of water shortage, the waste is uric acid. Uric acid is least toxic, sparingly soluble in water and can be retained in the body for a long time without any harm. These animals are named **uricotelic**. Examples: Some insects, pulmonate gastropods, land reptiles, birds, desert animals, viz. kangaroo rat, etc. Some animals produce more than one type of nitrogenous wastes, which is dependent on the habitat in which they live at a particular period.

The removal of wastes from the body is known as **excretion**. Since the carbon dioxide produced in respiration is transported by the blood to the respiratory membranes (body surface, gills or lungs) and released (p. 108) by diffusion, the term excretion is used to mean removal of nitrogenous wastes only. The organs involved in the process are collectively called excretory system. Excretory organs do not secrete wastes. They only concentrate the wastes and discharge those to the exterior.

In addition to the removal of nitrogenous wastes the organs help in removal or retention by selective absorption of substances, viz, water, salts, ions, minerals, etc. as per need, and for their role in maintaining osmotic balance they are considered also as osmoregulatory organs (p. 110).

* Since faeces, the undigested portion of the food, also called wastes, is not the product of cell metabolism, it is not discussed here.

INVERTEBRATE EXCRETORY ORGANS

Excretory organs in different degress of specialisation are present in all cases except protozoons, diploblastic animals and echinoderms. They are paired, the number varying from one to many pairs and made of narrow excretory tubules or nephrons, with or without internal and external openings and richly supplied with blood.

A series of flame cells (Fig. 25.5) constitute excretory organs in platyhelminthes; segmentally arranged nephridia with both nephrostome and nephridiopore (*Neanthes*) (Fig. 27.7); without nephrostome but with nephridiopore (Fig. 27.33) (*Hirudinaria*) or a number of them opening in the gut through a common duct (*Pheretima*) (Fig. 24.20) are found in annelids. A pair of green glands with bladders and a median renal sac in crustaceans (Fig. 28.11); fine thread-like Malpighian tubules (Fig. 28.28), from two to many in insects and arachnoids; a single pair or one large, spongy kidneys, the modified nephridia in molluscs (Fig. 29.20) in communication with the pericardial chamber and mantle cavity, constitute the excretory system in highly organised invertebrates (p. 9).

VERTEBRATE EXCRETORY ORGANS

In vertebrates the excretory organs are a pair of kidneys, located dorsally by the sides of the vertebral column and the ureters opening in the cloaca or to the exterior directly through renal apertures.

From the sites of their origin in the embryos the vertebrate kidneys have been named as **pronephros** or **pronephric** (head kidney), **mesonephros** or **mesonephric** (middle kidney) and **metanephros** or **metanephric** (tail kidney). The kidneys of cyclostomes are pronephric (develop from pronephros), those of fishes and amphibians mesonephric (from mesonephros), and metanephric (from metanephros) in amniota (reptiles, birds and mammals). The urinary bladder is a dilatation of the mesonephric duct in fishes and amphibia; in reptiles and birds it is a diverticulum of the cloaca; in mammals the cloaca is divided into two parts, the dorsal part forming the rectum and the ventral part the urethra and the ureters open in the bladder.

A MAMMALIAN KIDNEY

The kidneys are bean-shaped (Fig. 13.1), enclosed in a smooth, tough capsule. The renal tissues consist of an outer **renal cortex** and an inner renal medulla. Several renal pyramids constitute the medulla. The base of a pyramid is placed outward and the apex or papilla, inward. The cortical tissues extend in the spaces between the pyramids. The anterior end of the ureter enlarges to form a large space in the medially directed hilum of the kidney, **Calyces**, the narrow spaces, are present between the papillae (Fig. 13.2A).

A human kidney contains about 45,00,000 excretory or filtration units, the **nephrons**. A nephron consists of a **Bowman's capsule** at the proximal end, the rest is a long narrow tubule. The capsule is lodged in the renal cortex. The Bowman's capsule is a two layered, hollow sphere, a part of which is pushed inside by a knot of capillaries, the **glomerulus** (L. *glomus* = ball). The capsule and the glomerulus together are known as **renal corpuscle** (Fig. 13.2B and C). The walls of the glomerular capillaries are made of a single layer of endothelium bearing pores about 10 nm* in diameter and a layer of large cells, the podocytes (Gr. *podos* = foot + Fr. *cyte* = cell) bearing primary and secondary folds, the **pedicels** that interlaced with each other and rest on a basement membrane lying against the endothelium (Fig. 13.2D). The endothelial pores and the podocyte slits make the glomerular capillary wall manifold permeable than other capillaries but does not permit

* 1 Nanometre (nm) = 10 angstrom; 1 Micrometre (mm) = 10,000Å

molecules larger than 40,000 molecular weight to pass. The walls of the Bowman's capsule are made of a single layer of flattened cells. The inner or the **visceral** layer lie against the glomerulus and the outer or **parietal** layer is continuous with the walls of the tubule.

The walls of the tubule are constituted by cuboidal and flat cells. The tubule has three distinct zones, the **proximal convoluted tubule**, with the wall made of a single layer of cuboidal cells with brush borders, the middle or U-shaped **loop of Henley** and the last part, **distal convoluted** tubule joining the **collecting tubule**. The cells of the walls of the tubule beyond the proximal convoluted tubule are made of cuboidal cells with a few microvilli in the first part of the descending loop of Henley and of squamous epithelium in the rest. A collecting tubule receives several nephrons and ends in the renal pelvis.

The renal corpuscles and the convoluted tubules lie in the cortex, a major part of the collecting tubules and the loops of Henley extend into the medulla.

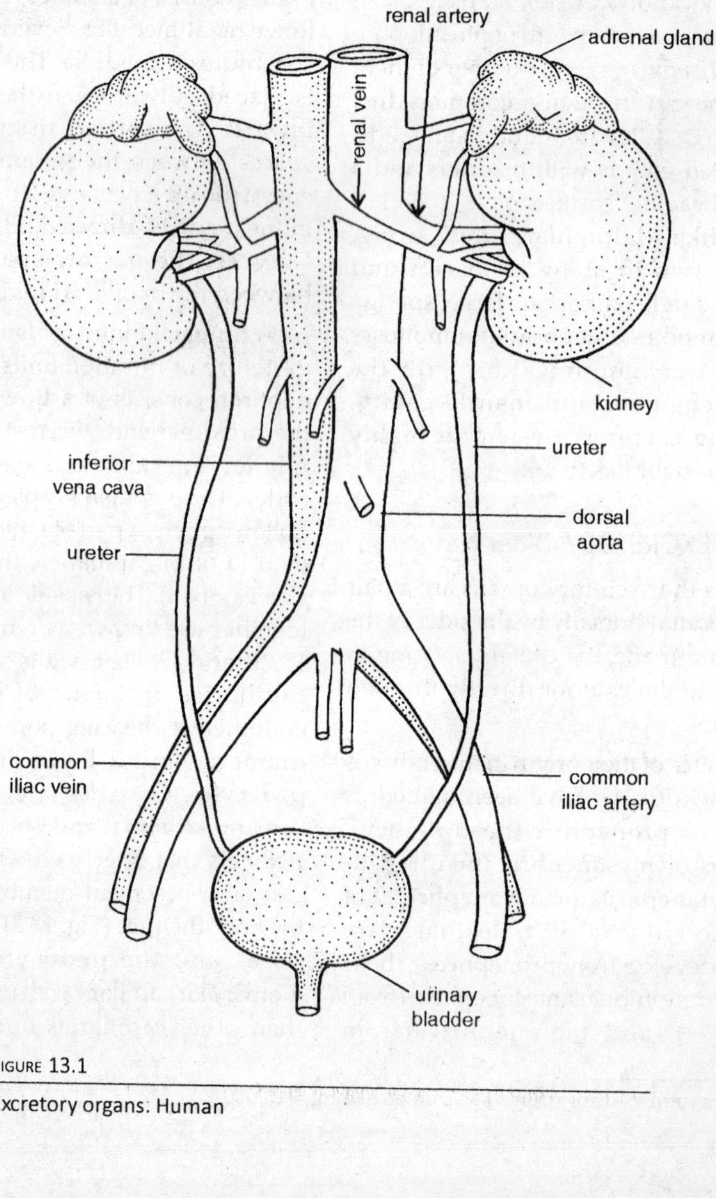

FIGURE 13.1

Excretory organs: Human

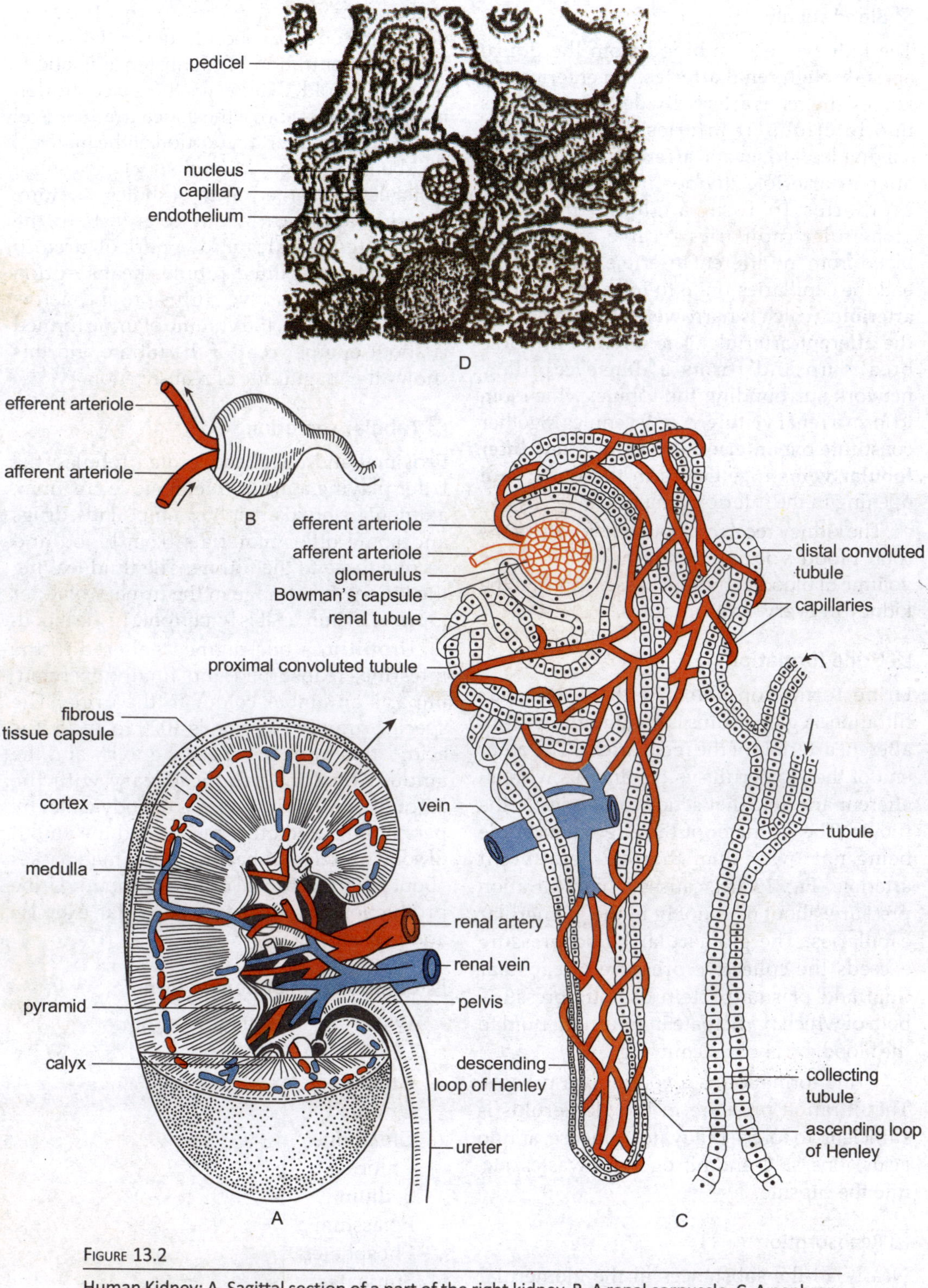

FIGURE 13.2

Human Kidney: A. Sagittal section of a part of the right kidney. B. A renal corpuscle. C. A nephron. D. Renal corpuscle (section) showing pedicel

Blood supply

The kidneys receive blood from the dorsal aorta through **renal arteries**. On entering the kidney the renal artery divides several times into **interlobular** arteries, each of which subdivides into several **afferent arterioles**. An afferent arteriole divides into a number of capillaries to form a glomerulus. The glomerular capillaries receive oxygenated blood from the afferent arteriole. At the other end the capillaries unite to form the **efferent arteriole**, which is narrower in diameter than the afferent arteriole. The efferent arteriole breaks up and forms a dense capillary network surrounding the tubule, which join to form a **renal venule**. A few venules together constitute one **interlobular vein**. A few **inter lobular veins** together form the **renal vein** opening in the inferior vena cava.

The kidney receives about one-fourth of the total blood volume per minute. The total volume of blood in human passes through the kidney in every 4 to 5 minutes.

Urine formation

Urine formation starts with glomerular filtration. A glomerulus is continuous with an afferent arteriole of the renal artery. The other end of the glomerulus is continuous with an efferent arteriole that sends capillaries to the tubule. The diameter of the efferent arteriole being narrower than that of the afferent arteriole (Fig. 13.2B), causes a high filtration pressure, about 60 mm Hg in the glomerular capillaries. The glomerular blood pressure exceeds the collective pressure of capsular fluid and plasma protein osmotic pressure, both of which participate in returning fluid to the blood, by about 20 mm Hg.

Urine formation is a continuous process. The filtration pressure in the glomerulus is sufficient to forece out water, glucose, amino acids, ions, salts and nitrogenous wastes, etc. into the capsule.

Reabsorption

Nearly all the substances in the glomerular filtrate, except nitrogenous wastes are reabsorbed in varying degrees in the tubules, depending on their concentration in the tubule. If the concentration of a substance is above renal threshold, that means the concentration in the filtrate is above the concentration level in the plasma, a certain portion of the material will be reabsorbed. More than 90 per cent of water is reabsorbed in the tubules, sodium, chloride and bicarbonate ions in both the convoluted tubules and a part of urea in proximal convoluted tubule. Reabsorption involves both passive diffusion and active transport. In man, the volume of urine formed is about one per cent of the filtrate and this shows the magnitude of reabsorption.

Tubular secretion

Proximal and distal convoluted tubules, the latter playing a major role, remove creatinin, ammonia, steroids, sulphur compounds, drugs and some other materials from blood and secrete these to the filtrate. The fluid leaving the kidney for storage in the urinary bladder is urine. In human it is hypertonic to the blood.

Urobilin, a bile pigment, altered in the intestine, reabsorbed and finally excreted, imparts an amber colour to the urine. The specific gravity of urine is 1020 to 1030; it is acidic in nature. The specific gravity and the amount of urine produced vary with the amount of water intake, loss of body water by perspiration and other means and the amout of solid excreted. A healthy, adult man passes about 1,000 to 1,500 ml urine in 24 hours. Urine production is reduced in muscular exercise and in sleep.

Composition of human urine

Constituent	Percentage
Water	96%
Urea	2%
Ammonia, Uric acid, Creatinine, Chloride, Sodium, Potassium, Phosphates, Sulphates, Oxalates	2%

14 THE NERVOUS SYSTEM

Response to stimuli and conduction of waves of excitement, caused by the stimuli are inherent properties of living cells. With the progress of evolution and appearance of metazoon animals a quick conduction of waves of excitements was required and specialised cells, the nerve cells or **neurons**; presumably developed in response to it.

NEURON

The neurons are units of the nervous system. The nervous system receives both external and internal stimuli, analyse and integrate them and produces coordinated responses, implementing the actions of the effectors.

A nervous system, ectodermal in origin, is present in all groups of animals except protozoons and sponges. The structure of a neuron is essentially same in all animals, the average diameter of an axon is a little less than 10.0 μm but it may be a few feet long in cephalopods and more than a metre, running from toe to spinal cord in a giraffe. A neuron consists of three parts (Fig. 5.13), the **cell body**, the **dendrites** and the **axon** (p. 67).

From the functional point of view the neurons are of three types.

⊠ Sensory or afferent neurons

Carry impulses from the receptor cells of the sense organs (external environment) or internal environment to the brain or the interneurons (connecting neurons) of the reflex arc.

⊠ Motor or efferent neurons

Originate in the brain, spinal cord and autonomic ganglia and transmit impulses to the effectors, the different organs.

⊠ Interneuron or connecting neurons

Connect two or more neurons. They are restricted to the central nervous system.

NERVE

A nerve is a bundle of neurons enclosed in a tough tissue sheath. The neurons in a nerve are arranged in a fashion in which the **telodendria** or an axon are placed close to the **dendrites** of another neuron. A nerve constituted entirely of sensory neurons is called **sensory nerve**, entirely with motor neurons, **motor nerve** and with sensory and motor neurons, the **mixed nerve**.

⊠ Nerve ganglion

An aggregation of cell bodies of a number of neurons is known as nerve ganglion. The ganglia may be connected with one another by the processes of the neurons (nerve net system) or by nerves (majority of the invertebrates).

☒ Nerve impulse

The nerve impulse involves both physical and chemical actions. Stimulus in any form excites the nerve. The excitation wave leads to certain physical and chemical actions which create waves of impulses. The waves of impulses pass along the nerve fibres.

☒ Transmission of nerve impulse

The receptor cells in the sense organs are sensitive to stimuli. Any perceptible change in the environment, external or internal is a **stimulus.** The dendrites which penetrate deep into the receptor cells receive the impulses or the neural message in the receptor cells, caused by the stimulus. The impulses are transmitted to the axon via the cell body and finally to the brain or spinal cord. In a few cases the axon reaches the brain or spinal cord directly. In most cases, a number of neurons are involved in the transmission.

The neurons are discontinuous, i.e. they are not in direct contact with one another. This needs transfer of neural messages from axon of one neuron to the dendrites of the next neuron. The junction between the axon of one neuron and the dendrites of the next neuron in known as a **synapse** (Gr. *synapsis* = conjunction). The axon at the free end divides into a number of branches, the telodendria, each with a terminal swelling, the **presynaptic knob** or **bouton**. The boutons are separated from the dendrites or the cell body of the next neuron by a **synaptic cleft** (Fig. 14.1) or gap about 20 nm. **Synaptic vesicles** are present at the ends of presynaptic knobs. Short-lived chemical transmitters, which carry neural messages across the synaptic cleft reach the synaptic vesicle. These chemicals are secreted by the nerve cells, transported along axons and stored in the vesicles. **Noradrenaline, gammaquinobutyric acid** (GABA) and **acetylcholine** are considered as the chemical transmitters.

INVERTEBRATE NERVOUS SYSTEM

Beginning from a primitive nerve net system, the evolution of nervous system has progressed towards aggregation of ganglia (cephalisation) giving rise to a controlling and coordinating centre, the **brain**, towards the anterior end of the body, which is first to meet the environment and also is the site where important organs are concentrated. Along with this, different types of neurons, transmitting waves of excitement in one direction and transferring the same from one to the other neurons have developed, which has culminated in cephalopods, the most active invertebrate.

☒ Nerve net system

A nervous system is first met with in the cnidarians. The nervous system of cnidarians and some flatworms is a network formed by

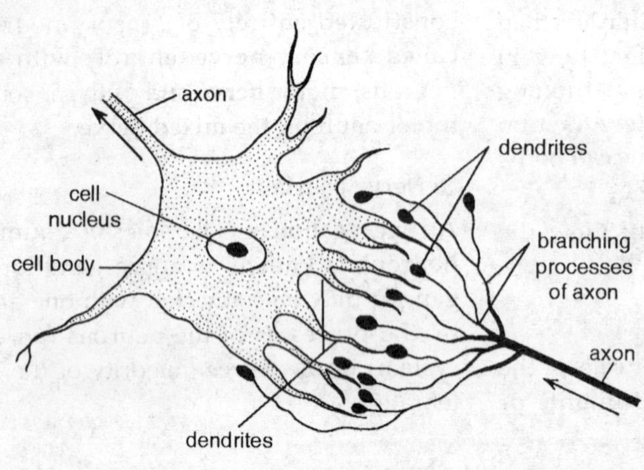

FIGURE 14.1
A synapse

ganglia and their processes. Such a type is called nerve net system (Fig. 23.7).

The ganglia are constituted by cell bodies and the connections by the processes. Sensory, motor and interneurons are present, but they cannot be differentiated. Usually the processes are connected through interneurons, though connection between sensory and motor neurons through synapses are not wanting. Impulses are conducted across in either direction. A controlling system is lacking. The conduction is diffuse and tends to spread in all directions from the starting point.

⊠ Cephalic dominance system

For the first time the **nerve ganglia** are formed by the aggregation of cell bodies in the flatworms. The ganglia are close to one another at the anterior end of the body and form the **brain**, the site for sensory integration and motor command. The nerve cords are **solid** and run backward up to the posterior end of the body. With the progress of evolution the number of nerve cords reduced to two as in turbellaria (Fig. 14.2) and finally one solid, double ventral nerve cord in annelids (Fig. 27.8). A pair of ganglia in each segment of the body help the segments to act in a coorndinated way. The motor centre is present in the suboesophageal ganglion and the brain has an inhibitory control over it. In the arthropod brain **protocerebrum** associated with vision centre and associated area, **deuterocerebrum** with olfactory and antennal centres and **tritocerebrum** innervating the mouth and alimentary canal are distinguishable (Fig. 28.32). The motor activity is controlled by the suboesophageal ganglion. In molluscs the nerve ganglia are paired like other invertebrates, except the visceral ganglion. They are located at different points along the nerve cords but they tend to aggregate at one point, which is first found in pulmonate (Fig. 29.31). Extreme condensation of the nerve ganglia forming a central brain, associated with the appearance of a cartilaginous **cranium** is present in cephalopods (Fig. 29.41).

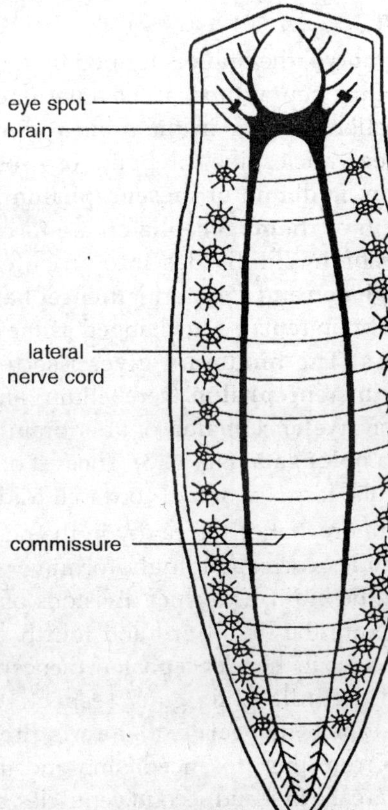

FIGURE 14.2

Planaria **sp.** Nervous system

VERTEBRATE NERVOUS SYSTEM

The nervous system in the vertebrates is unique in its structural complexity and functional coordination. The nervous system in vertebrates has been divided into the central nervous system and the peripheral nervous system. The autonomous nervous system is a special part of the peripheral nervous system.

CENTRAL NERVOUS SYSTEM

The central nervous system is composed of the **brain** and the **spinal cord**. Impulses are transmitted to the central nervous system through afferent (sensory) fibres of the peripheral nerves. The brain interpretes the messages and despatch them to the effectors through efferent (motor) neurons.

Brain

In the embryo, the brain is formed by the dilation of the neural tube, at the anterior end. Constrictions appear in the brain rudiment and three vesicles, the forebrain, or **prosencephalon,** midbrain or **mesencephalon** and hindbrain or **rhombencephalon** are formed. The forebrain subdivides into an anterior **telencephalon** and a posterior **diencephalon**. The midbrain remains unchanged as **mesencephalon**. The hindbrain gives rise to an anterior **metencephalon** (cerebellum) and a posterior **myelencephalon** or after brain, the **medulla oblongata** (Fig. 14.3). The rest of the neural tube form the spinal cord in the adult.

The **neurocoel** or the cavity in the neural tube divides correspondingly forming **ventricle** in the brain. The chief divisions of the ventricle are the first, third and fourth ventricles located in the telencephalon, diencephalon and medulla oblongata, respectively. Paired lobes, the **cerebral hemispheres** develop from the prosencephalon and their cavities are the first and second ventricles, also known as **lateral ventricles**, communicate with the third ventricle in the diencephalon by a narrow **foramen of Monro.** The **aqueduct of Sylvius** joins the third ventricle with the fourth ventricle in the metencephalon and myelencephalon (Fig. 29.18). The fourth ventricle is continuous with the **central canal** (neurocoel), which extends up to the end of the spinal cord. The canal is lined by a ciliated epithelium. The ventricles and the central canal are filled with a cerebrospinal fluid

The solid portion of the central nervous system enclosing ventricles in the brain and central neural canal in the spinal cord is made of grey and white matter.

Grey matter

Aggregation of the cell bodies of the neurons gives a grey appearance to the region formed by them and the portion is called grey matter.

White matter

The portion formed by the aggregation of nerve processes is whitish in colour and termed white matter.

In brain, the grey matter is restricted to the outer zone, while in the spinal cord it lies inwards, towards the central canal and vice versa.

The brain and the spinal cord are ensheathed in **meninges**, consists of three lay-

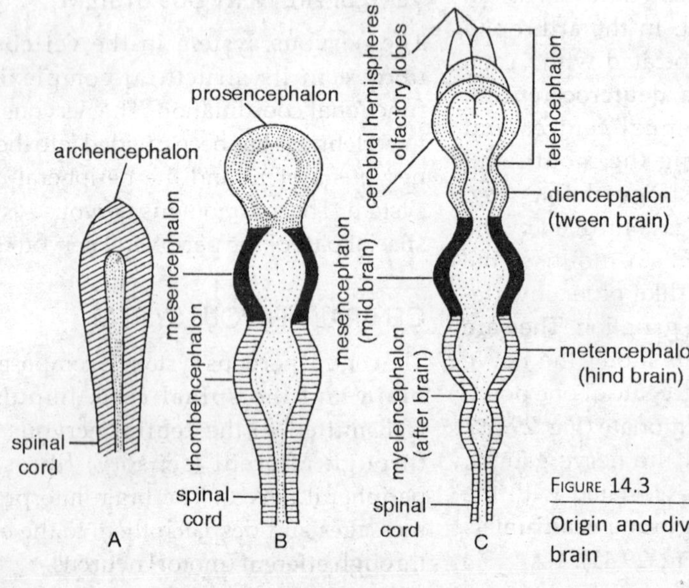

FIGURE 14.3
Origin and divisions of vertebrate brain

ers. The outermost layer, **dura mater** lies against the cranium bone, the innermost layer, the **pia mater** lies on the surface of the brain and the middle layer, the **arachnoid** lies in between the two. The spaces between the piamater and the arachnoid are filled with cerebrospinal fluid.

⊠ Modifications of vertebrate brain

Starting from distinct five divisions which is found in fishes, the vertebrate brain undergoes a series of modifications resulting in the most complex brain in mammals (Figs. 37.15, 38.9, 39.17, 40.21, 41.27, 42.29).

The primitive brain becomes complicated due to unequal thickening of the walls. The floor of the medulla thickens greatly but the roof is made of a single layer of nonnervous epithelium. The great thickness of the **cerebellum** obliterates the epicoele. The ventral wall of the mesencephalon thickens to form two longitudinal bands, the **crura cerebri**; the dorsal wall forming a pair of large **optic lobes**. The sides of the diencephalon thickens to give rise paired **optic thalami**, the roof remaining a single layer thin membrane, but a part of it forms **pineal apparatus**. The floor of the diencephalon grows downward to form **infundibulum** to which the pituitary gland is attached. The pituitary gland is formed partly from a diverticulum of the pharynx and partly from the extremity of the infundibulum. The telencephalon grows forward, enlarges greatly and due to fusion of the internal dorsal and ventral edges of the lateral walls, the two **cerebral hemispheres (cerebrum)** are formed.

With the progress of evolution, the horizontal plain of the vertebrate brain underwent drastic changes. Owing to the bending down of the anterior part (cerebral flexure) the brain assumed the shape of a retort, the axis of the forebrain inclined to that of the hind brain. The cerebral hemispheres grew forward, parallel to the hindbrain. The thickening of the floor of the mid and hindbrain made them obscured in the adult.

The grey column of the spinal cord extend up to the medulla. In the brain, the column breaks up into aggregated cell bodies, the **nuclei**. The dorsal nuclei are sensory nuclei and contain cell bodies of afferent neurons from the cranial nerves. The ventral nuclei are motor nuclei, as they contain the cell bodies of efferent neurons.

⊠ Cerebral hemispheres

The olfactory lobes are associated with the anterior region of the cerebral hemispheres. The cortex of the cerebral hemispheres known as **cerebral cortex** exhibits zonal specialisation to carry specialised functions. These may be placed in three groups:

a. Memory, intelligence, thinking, reasoning, sense of responsibility and moral sense, word perception and writing in man.
b. Sensory perception. Touch, sight, hearing, taste, smell, temperature, pain, etc.
c. Voluntary muscle initiation and control.

⊠ Mesencephalon

Groups of nerve cells and nerve fibres in the mesencephalon connect the **cerebrum** (cerebral hemispheres) with the posterior part of the brain and the spinal cord. The nerve cells function as relay centres for ascending (afferent) and descending (efferent) nerve fibres.

⊠ Cerebellum

The cerebellum is concerned with coordination of voluntary muscles, posture, movement, balance and equilibrium of the body.

⊠ Medulla oblongata

In the deeper structure lie groups of cells associated with reflex activity. The medulla acts as cardiac centre, respiratory centre, vasomotor centre and reflex centre.

⊠ Reticular formation

An aggregation of neurons in the core of the brain stem* is known as reticular formation.

It is surrounded by neural pathways conducting impulses between the brain and the spinal cord. Numerous synapses connect it with other parts of the brain and enable it to constantly receive both sensory and motor impulses. The reticular formation is considered to be a primitive part of the brain. Reticular formation is concerned with conciousness or awareness of the evironment.

⊠ Spinal cord

The spinal cord is hollow, cylindrical tube, suspended in the neural canal of the vertebral column and surrounded by meninges and cerebrospinal fluid. It runs backward from the medulla oblongata and extends from the upper border of atlas to very close to the posterior of the vertebral column (lower border of the first lumbar vertebra in man).

Two types of tissues constitute the spinal cord. The grey matter surrounding the central canal is λ or X-shaped in transverse section. Nerve cell nuclei and synapses are present in the grey matter. The white matter surrounds the grey matter. Myelinated nerve fibres running longitudinally and so connecting different levels of the spinal cord are present in the white matter.

The spinal cord can carry out certain functions, e.g. spinal reflexes, independent of the brain. Extensive connections between sensory and motor nerves in the spinal cord enable it to carry such activities.

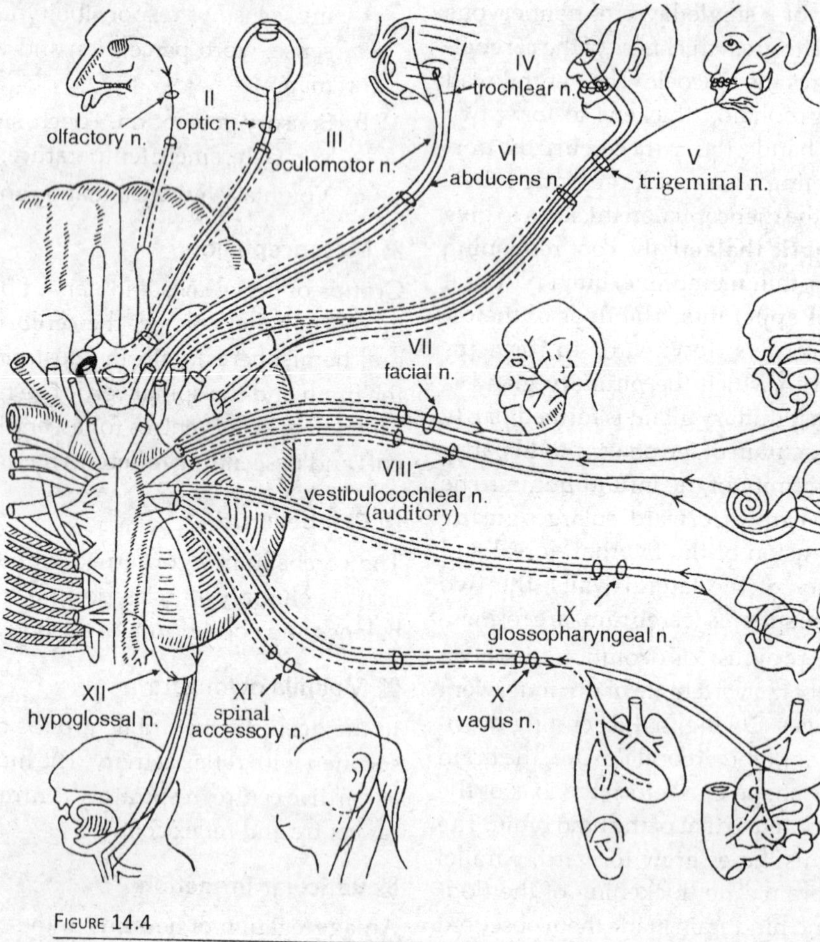

FIGURE 14.4

Cranial nerves. Primate (schematic)

PERIPHERAL NERVOUS SYSTEM

The paired cranial and spinal nerves constitute the peripheral nervous system. The autonomic nervous system is a part of the peripheral nervous system that controls the involuntary functions of the body.

☒ Cranial nerves

The cranial or cerebral nerves are paired, originate in the nuclei in the brain, pass out of the cranium through passages in it and terminate in the effectors. The cranial nerves are ten pairs in fishes and amphibia and twelve pairs in Amniota, i.e. reptiles, birds and mammals. Some of the nerves are sensory, some are motor and others are mixed. The cranial nerves are referred to as nerves I to XII (Fig. 14.4). The number, name, origin in the brain, peripheral termination, nature and function of vertebrate cranial nerves are given in the table.

☒ Spinal nerves

Paired nerves arising from the spinal cord are known as spinal nerves. One pair of nerves from each somite emanate laterally from the spinal cord. Each nerve has two roots, the **dorsal root** bears a ganglion, the **dorsal ganglion** and a nonganglionated **ventral root** (Fig. 14.5). Each root originates from one horn of the grey

Number Name	Origin in the brain	Peripheral termination	Nature	Function
i. Olfactory	Smell area in temporal lobe of cerebrum	Mucous membrane in roof of nose	Sensory	Sense of smell
ii. Optic	Sight area in occipital lobe of cerebrum	Retina of the eye	Sensory	Sight and balance
iii. Oculomotor	Floor of midbrain	Superior, inferior, medial rectus and ciliary muscles of the eye, circular fibres of the iris	Motor	Eye ball movement, focusing, pupil size regulation
iv. Trochlear	Floor of midbrain	Superior oblique muscle of the eye	Motor	Eye ball movement
v. Trigeminal	Motor fibres pons varoli; sensory fibres trigeminal ganglion	Mastication muscles; sensory—iris, cornea, cheek, lower jaw, gums	Mixed	Chewing, facial sensation
vi. Abducens	Floor, fourth ventricle	Lateral rectus muscle of the eye	Motor	Eye movement
vii. Facial	Pons Varoli	Tongue, facial muscles	Mixed	Taste, facial expression
viii. Auditory	Hearing area of cerebrum, cerebellum	Semicircular canals, organ of corti	Sensory	Hearing, balance
ix. Glossopharyngeal	Medulla oblongata	Back of the tongue, parotid glands, pharynx	Mixed	Taste, saliva secretion, pharynx movement
x. Vagus	Medulla oblongata	Larynx, pharynx, glands, ducts, heart, blood vessels, lung, alimentary canal	Mixed	Movement of muscles and secretion in glands
xi. Spinal accessory	Medulla oblongata	Sternocleidomastoid, trapezius, laryngeal and pharyngeal muscles	Motor	Movement of head, shoulder; larynx and pharynx
xii. Hypoglossal	Medulla oblongata	Tongue	Motor	Tongue movement

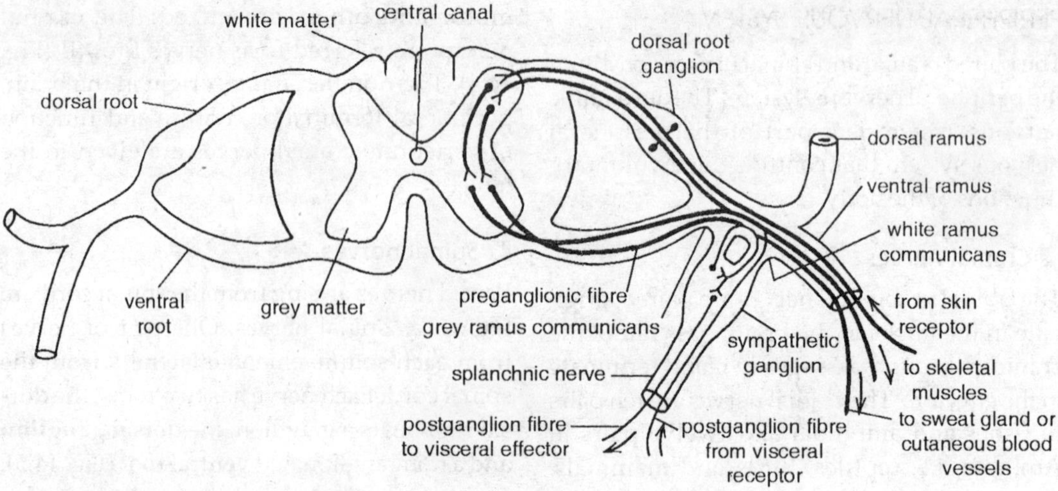

FIGURE 14.5

Cross section of spinal cord and spinal nerves (A single reflex arc)

matter and the two join to form the trunk of the nerve. The nerve comes out of the spinal canal, the space between the spinal cord and the vertebrae, through the passage between the arches of the adjacent vertebrae. The nerve roots have a covering of dura and arachnoid maters, which terminate before the formation of the trunk.

The dorsal root consists of sensory neurons, carries sensory impulses to the spinal cord from the skin, body somites and viscera. Impulses are interrupted by a synapse in the ganglion. The ventral root consists of motor neurons, carries impulses from the spinal cord to somatic and visceral components. The dorsal root may possess motor neurons but the ventral root never possesses sensory neurons.

Immediately after leaving the vertebral column each spinal nerve divides into a posterior ramus and an anterior ramus. The posterior (dorsal) ramus innervates the dorsal part of the axial musculature, skin of the back and lateral and ventral parts of the skin and musculature of the body wall. The anterior (ventral) ramus innervates internal organs, smooth muscles, blood vessels and glands.

The visceral motor component enters a **sympathetic ganglion**. The sympathetic ganglia of one side are connected to a longitudinal **sympathetic trunk** and a trunk lies on each side of the vertebral column. The visceromotor and viscerosensory systems together form the sympathetic nervous system.

⊠ Reflex action

Not all activities of animals are the results of judgement and deliberations. Certain actions occur without even the animal being aware of it, e.g. sudden withdrawal of hand or feet from hot, cold, sharp and pointed objects; sudden closure of eyelids in strong light, etc. Such actions are termed reflex actions. These are **protective** or **inherited reflexes** (present at birth). Reflex actions which were not initially present but developed due to constant practice for a long time or learning by habit are known as **conditioned reflexes**. Walking, cycling, secretion of saliva in dogs with the ringing of a bell at feeding time, etc. are conditioned reflexes.

⊠ Reflex arc

The neural pathway along which the impulses travel to execute the reflex action is called reflex arc. Each spinal nerve has two roots. Sensory neurons are present in the dorsal root and only motor neurons are present in the ventral root. In a single reflex arc (Fig. 14.5) the impulse received from the receptor passes

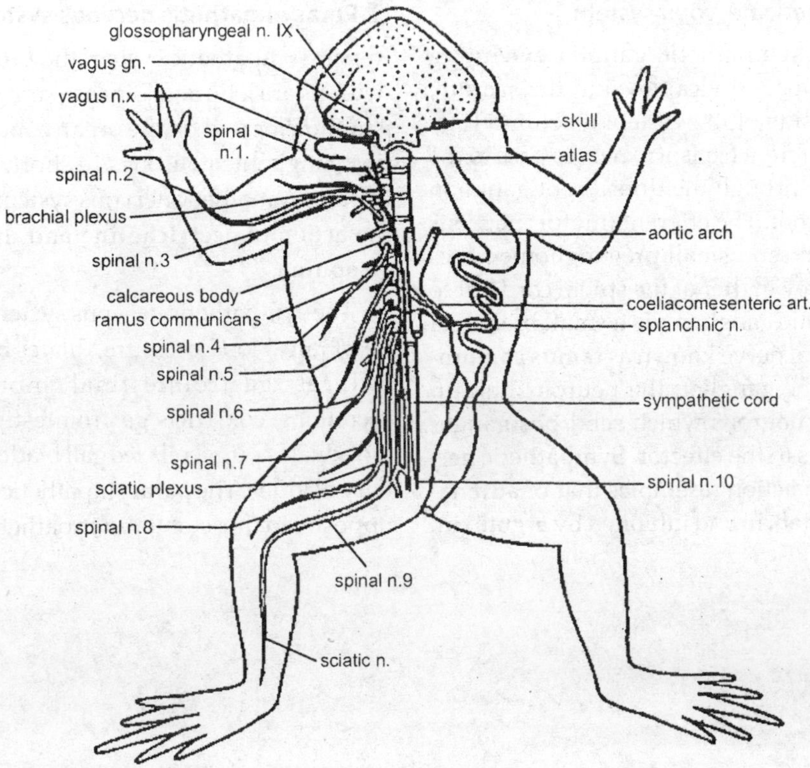

FIGURE 14.6

Spinal and sympathetic nerves. ***Bufo sp.*** Ventral view

to the spinal cord through afferent neurons via the dorsal root ganglion and dorsal root. The impulse from the spinal cord passes to the motor neurons in the ventral root via the synapse between the two and transmitted to the effector. The **knee jerk** is a typical simple reflex. One or more interneurons between the sensory and motor neurons are involved in the majority of the reflexes. Other reflexes are **righting reflex** and **antigravity reflex**.

☒ Autonomic nervous system

The part of the nervous system controlling the functions of certain organs in an animal body, automatically or involuntarily, is known as autonomic nervous system. A number of organs, viz. eye, nose, heart, lung, gut, ciliary and iris muscles of the eye, muscles related to hairs, glands, gall bladder, urinary bladder and many other organs are supplied with a special set of peripheral nerves made of self-operating units, and their functions cannot be controlled voluntarily. The autonomic system differs from the rest of the nervous system in the following features:

1. It has a double set of nerves, both sensory and motor.
2. The axon endings of each set secrete a different transmitter substance to the effectors.
3. Cerebrum has no voluntary control over it.

The autonomic nervous system has two components, the **sympathetic** and **parasympathetic**. Sympathetic and parasympathetic systems usually have antagonistic effects on the organs involved.

☒ Sympathetic nervous system

A chain of sympathetic ganglia extending from the upper cervical level to the sacrum, lie on each side of the vertebral column (Fig. 14.6). The afferent (sensory) nerves pass to the spinal cord through the dorsal root ganglion and dorsal root. The efferent (motor) neurons are long and send small preganglionic fibres not to the grey matter of the spinal cord but to the autonomic ganglion via a small branch of the segmental nerve known as **ramus communicans**. In the ganglion the neurons link up with motor neurons which send postganglionic neurons to the effector. Sympathetic nervous system action resembles that of **adrenaline** and **ephedrin** and inhibited by **ergotoxin**.

☒ Parasympathetic nervous system

In parasympathetic system the preganglionic neurons are long and join the parasympathetic ganglion located on the organ concerned. The postganglionic neurons are short. The action of parasympathetic nervous system resembles the action of **acetylcholine** and inhibited by **atrophine**.

The sympathetic nervous system increases heart rate, blood pressure, dilates bronchioles reduces gastrointestinal motility and secretions, constricts gastrointestinal sphincters, dilates pupil, relaxes gall bladder and urinary bladder. The parasympathetic actions are opposite to those of the sympathetic system.

15 RECEPTORS AND SENSE ORGANS

AWARENESS IN ANIMALS

The animals are in a state of continuous alertness for their very survival. They are to face environmental assaults, to move to safety, to procure food, to find mates for perpetuation of the species and to be aware of the multitude of factors constituting the surroundings.

The awarness of environment in animals come from sense organs and their receptor cells. The sense organs stand between the organism and the environment, i.e. they are exposed to the surroundings. The receptor cells are the sensory constituents of the sense organs, which receive different types of environmental energy, transform them into electrical energy and transmit the information to the central nervous system through nerve impulses. The part of the brain receiving the impulse is responsible for the awareness of the sensation. Certain accessory structures are almost always present in a sense organ. The supporting cells in taste buds, cornea, ciliary muscles and iris in mammalian eyes are such structures, which help the receptor cells to function properly.

Animals respond to stimuli both through the sense organs located in particular positions on the body and also through receptor cells scattered on the body surface or on internal organs.

Receptor cells in most cases are specific, i.e. specialised to respond to a particular type of stimulus (energy). Different stimuli, viz. touch, light, heat, chemicals, activate different types of receptor cells. Usually one type of receptor cells constitute a sense organ. The receptor cells are not only specific but they have a limitation to receive only a small part of a long range of stimulus and that also varies in different animals. Human eyes can receive light in the range of wave lengths 400 to 700 nm, while the same is 275 to 690 nm for insects. The numerous impulses received by the receptor cells are transmitted to the central nervous system directly or through synaptic connections between one or more interneurons. Only a few are accepted by the central nervous system named as **information**, others are **noise** and rejected.

RECEPTORS

Receptors are classified according to their ability to respond to the kinds of stimuli. Broadly, they have been placed under five groups—**mechanoreceptors, chemoreceptors, photoreceptors, thermoreceptors** and **electroreceptors**.

⊠ Invertebrate receptors

Sense organs in invertebrates exhibit wide diversity in their morphology and orientration.

And it may be presumed that the sense organs in different phyla of invertebrates have developed independently. Specialised receptor cells are absent in protozoon and sponges.

☒ Mechanoreceptors

Four types of mechanoreceptors are present in invertebrates:

☒ Tactile receptors

The tactile (touch) receptor is a **phasic** receptor. In cnidaria the tactile receptors **cnidocils** are present (Fig. 15.1) in the cnidocytes (cnidoblasts). The cnidocil being touched, the cnidocyte contracts violently discharging the nematocyst. This effects both defence and capture of prey. Sensory cells are scattered on the body surface of turbellaria, earthworm; on the tentacles and palpi of *Neanthes*, on the antennae and palpi in arthropods and on the tentacles, labial palps, head and visceral stalk in molluscs.

☒ Proprioreceptors

The proprioreceptors are tonic receptors, and register stress changes in the exoskeleton. In the invertebrates they are concerned with transmitting postural information and are present in crustacea and insects. In insects they are modified cuticular structures in the form of hair, plate, etc. and located on the stretchable area of the cuticle. Action of wings in flight is controlled by proprioreceptor cells present at the base of the halteres.

☒ Equilibrium receptor

Maintenance of body equilibrium is essential for the normal activities in an animal. Equilibrium receptors respond to loss of balance and help the animal to correct it. **Lithocysts** (Fig. 15.2), the marginal sense organs, ensure horizontal swimming in cnidarians. **Statocysts** (Gr. *statas* = standing + *kystis* = sac) are organs of balance in arthropods and molluscs. A statocyst is usually an oval

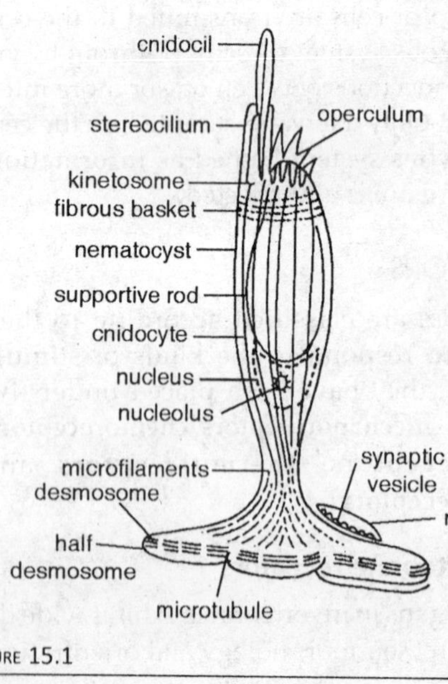

FIGURE 15.1

Epitheliomuscular cell of a hydrozoan jelly fish

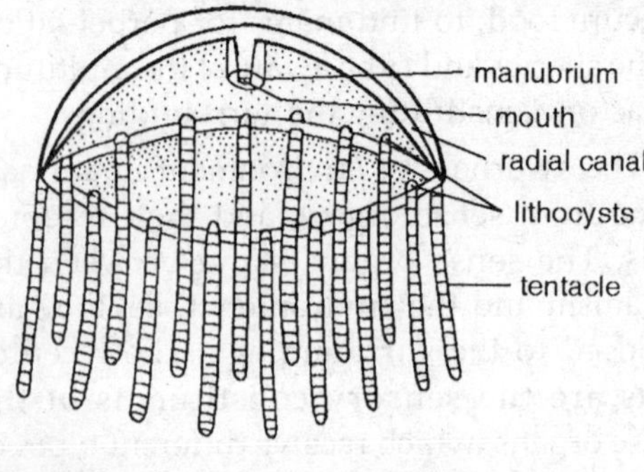

FIGURE 15.2

Lithocyst, obelia medusa

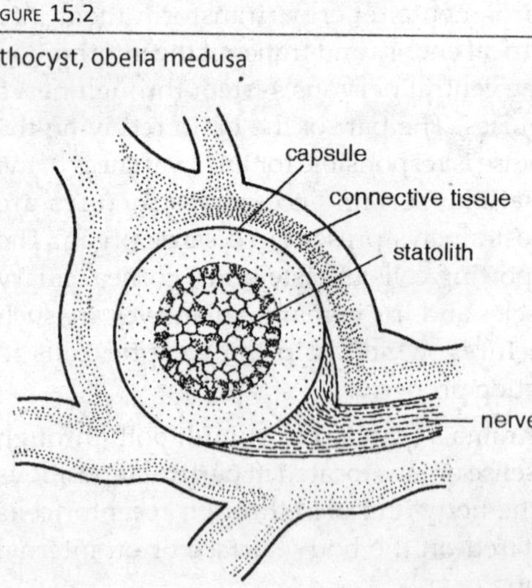

FIGURE 15.3

A statocyst. ***Pila sp.***

sac, may or may not be in communication with the exterior and lined internally by a sensory membrane bearing processes or hairs. A single calcareous particle or **statolith** or a number of small particles, **statoconia** (Fig. 15.3) are present at the centre. In some cases, sand particles (Fig. 15.4) occupy the place of statolith. With the change in the body position, the particles exert pressure on certain sensory cells and this signals the animal to correct its position with respect to the gravity*. A pair of **halteres**, the modified hind wings in flies and mosquitoes take care of maintaining stability by controlling lift, rotation and landing in their flight.

⊠ Phonoreceptors

Phonoreceptors or sense of hearing can detect pressure waves caused by mechanical disturbances at some distance from the body. Three types of phonoreceptors have been identified in arthropods:

1. **Sensillae hairs.** The hairs can perceive low frequency vibrations. They react with the sound waves and are located on the body surface. Frequency of sound waves below 20 hertz is called **vibration**.
2. **Chordotonal organs.** They differ in structures in different arthropods. The type known as **tymponal organ**, the most complex and efficient organ of sound perception is found in moths. The sensory neurons attached to the apical bodies, the **scolopodia** are bipolar.
3. **Lyriform organs.** It consists of parallel slits of varying length in the cuticle, resembling the strings of a lyre. Nerve endings of sensory neurons are present in the cuticle covering it.

   Hearing, in true sense of the term, is absent in invertebrates. They can only perceive the disturbances in the medium—air, water or solids.

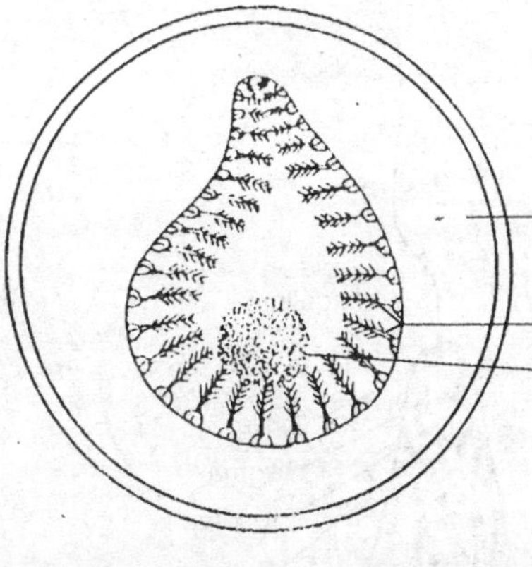

FIGURE 15.4

A statocyst with sand particles, ***Macrobrachium* sp.**

⊠ Chemoreceptors

Receptors for smell and taste are chemoreceptors (p. 128). These sense organs are capable to differentiate between chemicals and thereby help in controlling the activities of animals in relation to its surroundings, in the selection of food, in social behaviour, in defence from predators, in selection of mate and others.

The mouth parts and terminal segment of butterflies and moths bear hairs and pegs. Receptors for taste and touch are present in these structures. The taste receptors in flies are specific for sugar, water and salts.

Not much is known about sense of smell in the invertebrates. The antennae and palpi in insects bear receptors for smell. Olfactory receptors in moths and ants are quite powerful. The pulmonate molluscs move towards the food, even after removal of both the eyes by surgery. This suggests some sort of sense of smell are present in them.

Osphradium in molluscs can test the quality of water and it is considered as a chemoreceptor.

⊠ Photoreceptors

Sense organs capable to respond to light stimulus are photoreceptors. Photoreceptors in different degrees of perfection are present in

* The statocyst is innervated by a nerve emanating from the cerebral ganglion of the side.

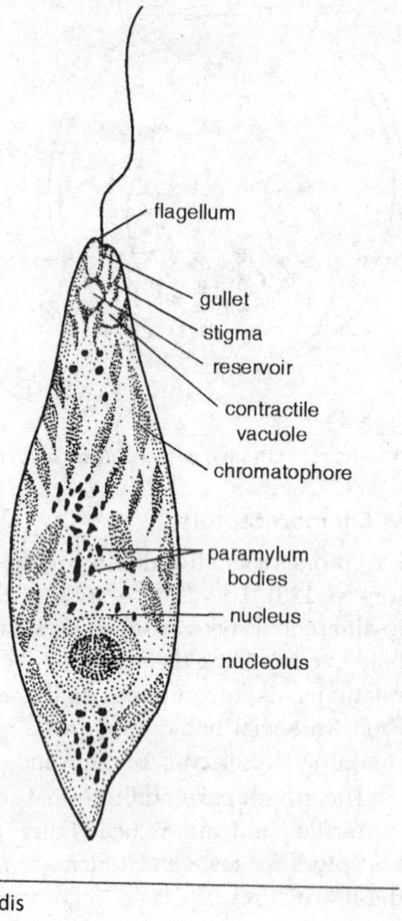

FIGURE 15.5

Euglena viridis

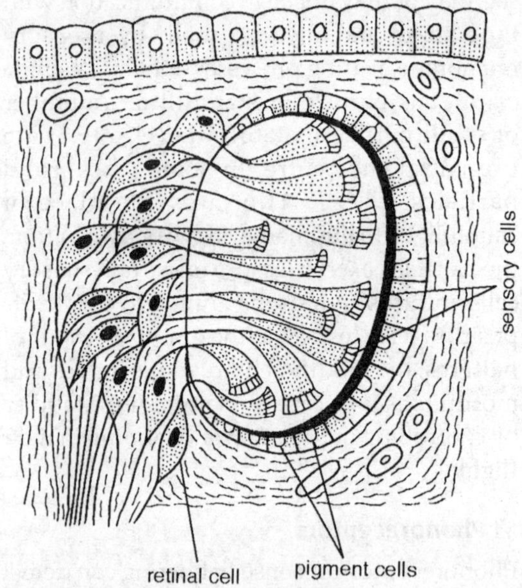

FIGURE 15.6

An ocellus. (vertical section) ***Planaria***

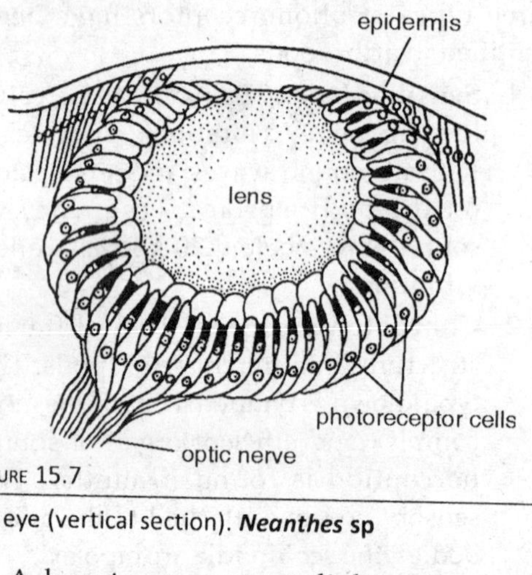

FIGURE 15.7

An eye (vertical section). ***Neanthes* sp**

the majority of the animals. Even the protozoon, in which such an organelle is lacking, respond to light and move away from bright light. In *Euglena*, the photoreceptor, **stigma**, simply a mass of granules (Fig. 15.5) is present near the base of the flagellum.

In the metazoon, portions of the skin are specialised to form eyes. Clusters of simple eyes of **ocelli** in the jelly fishes help them to orient to light. Isolated light sensitive cells are scattered over the body surface in earthworm. Bowl-shaped, black pigmented ocelli containing clusters of photosensitive cells are present in planaria (Fig. 15.6). The pigment allows light to reach the receptors only from above and slightly to the front, but image formation does not occur. **Eye spots** are present in leech. The eyes of neries (Fig. 15.7) are more specialised than those of planaria and leech.

A lens to concentrate light and a great increase in the number of photoreceptor cells led to the evolution of image forming eyes.

Two types of eyes—**ocelli** or simple eyes and **compound** eyes, are found in arthropods.

⊠ Simple eye

The ocelli type eyes are constituted by a single **ommatidium**, as in arachnids (Fig. 15.8) and compound eyes in crustacea and insects. The eyes of cephalopods are **camera type** like those of the vertebrates.

☒ Compound eye

A compound eye is formed of a number of visual elements, **ommatidia** (Fig. 25.13) and visions are of two types, **mosaic** and **superimposed** (p. 380).

☒ Thermoreceptors

All animals are sensitive to variations in temperature. No specialised thermoreceptors are on record in invertebrates. The thermoreceptor cells are usually scattered all over the body surface and may be concentrated in certain regions. e.g. in the antennae of some insects, particularly warm blood sucking forms.

☒ Vertebrate receptors

As a rule, the sense organs in vertebrates are highly specialized and a gradation is present along the evolutionary scale from fishes upwards, attending perfection in mammals.

☒ Mechanoreceptors

Five types of mechanoreceptors are found in vertebrates.

☒ Tactile receptors

The whole body surface forms an organ of touch. Special tactile receptors are widely distributed on the body surface. In human, extreme concentration of touch receptors are in the fingertips and lips. Different types of touch receptors (Fig. 15.9) are:

a. **End bulbs** (Fig. 15.9A) consist of ovoidal groups of sensory cells innervated by a special nerve.
b. **Touch cells** (Fig. 15.9B) formed by termination of a sensory nerve in the dermis.
c. **Touch corpuscles** (Fig. 15.9C) ramified sensory nerve in a mass of connective tissue, the terminal branches ending in touch cells.
d. **Pecinian corpuscle** (Fig. 15.9D) the naked axon is surrounded by a complex laminated sheath containing fluid.

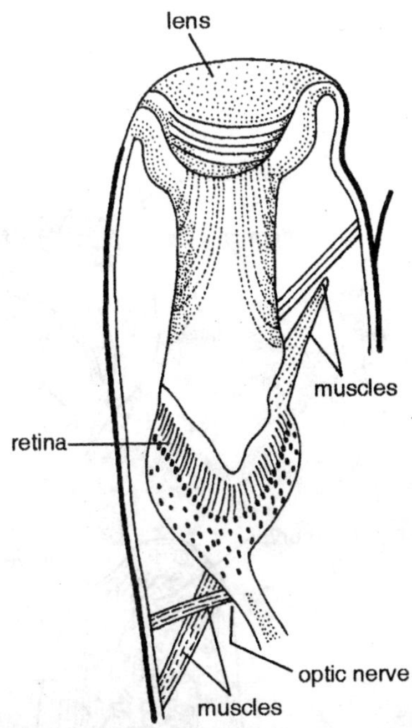

FIGURE 15.8

An eye (vertical section). Spider

☒ Proprioreceptors

These are internal receptor organs present in skeletal muscles, tendons and joints. They are capable of registering muscle tension or stretch and ensure harmonic contraction of all the muscles involved in a single movement. Proprioreceptors help us to work in dark. Nonfunctioning of proprioreceptors causes numbness of an organ.

☒ Equilibrium receptors

The ability to differentiate the orientation of the body in relation to its surroundings and to maintain an equilibrium is present in all vertebrates. The internal ears or **membranous labyrinths** control these functions. The membranous labyrinth is enclosed in a cavity, the **bony labyrinth** and bathed in a fluid, the **perilymph**, which is protective and acts as a buffer.

The membranous labyrinth (Figs. 15.10, 15.11) consists of three **semicircular canals** (except agnatha*) and two small, hollow sacs,

* Semicircular canal—one in *Myxine*, two in *Petromyzon*.

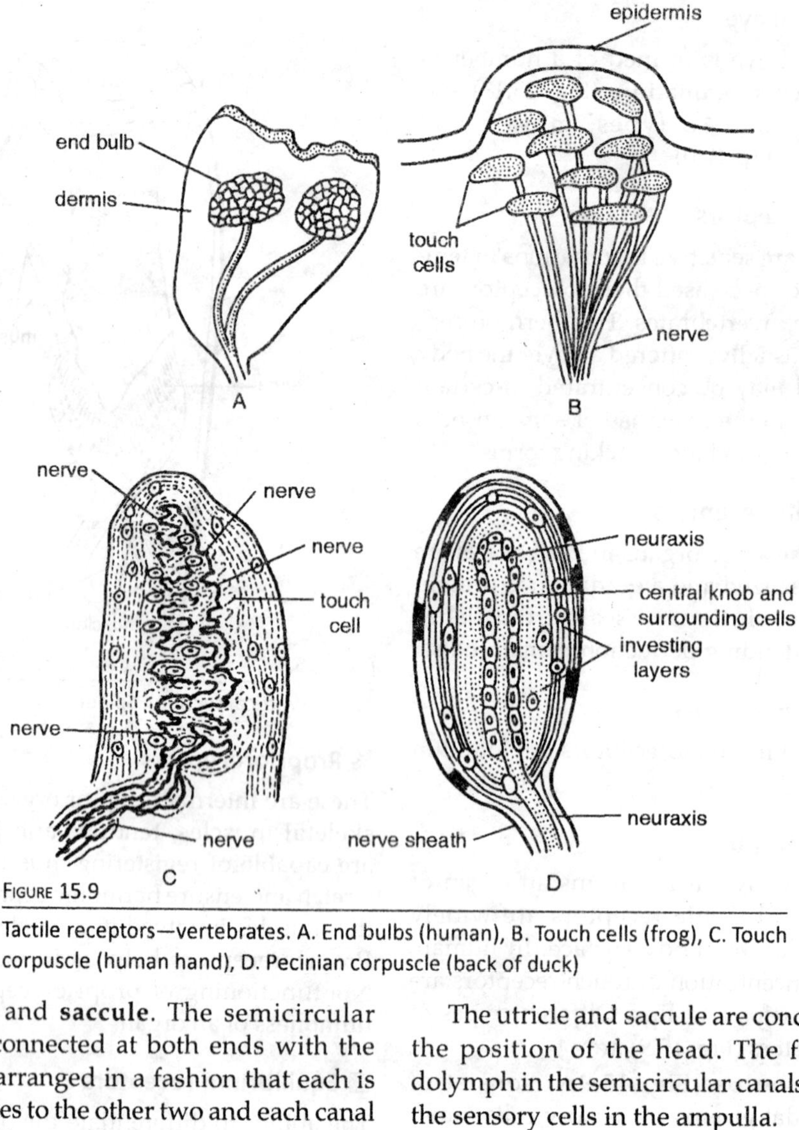

FIGURE 15.9

Tactile receptors—vertebrates. A. End bulbs (human), B. Touch cells (frog), C. Touch corpuscle (human hand), D. Pecinian corpuscle (back of duck)

the **utricle** and **saccule**. The semicircular canals are connected at both ends with the utricle and arranged in a fashion that each is at right angles to the other two and each canal bears a small **ampulla** at the lower end. The anterior and posterior canals are vertical in position while the external one is in horizontal plane. Patches of elongated sensory cells produced into hair-like processes are present in the ampullae, utricle and saccule. They are known as **cristae acoustica** in utricle and saccule and **cristae ampularis** in the ampulla. The sensory cells are innervated by the branches of the auditory nerve. The utricle and saccule contain small, calcareous **otoliths**. The membranous labyrinth is filled with a fluid, the endolymph.

The utricle and saccule are concerned with the position of the head. The flow of endolymph in the semicircular canals stimulates the sensory cells in the ampulla.

☒ Maintenance of equilibrium

Change of head position (**static equilibrium**) affects the gravitational pull on the otoliths and hence the pressure on the sensory hair cells in the utricle and saccule in relation to gravity.

Otoliths are pushed back upon certain hair cells of the utricle and saccule during rapid forward movement (**linear accelaration**).

The hair cells in the ampulla move faster than the endolymph due to the inertia of the latter in sudden turns of the head (**angular

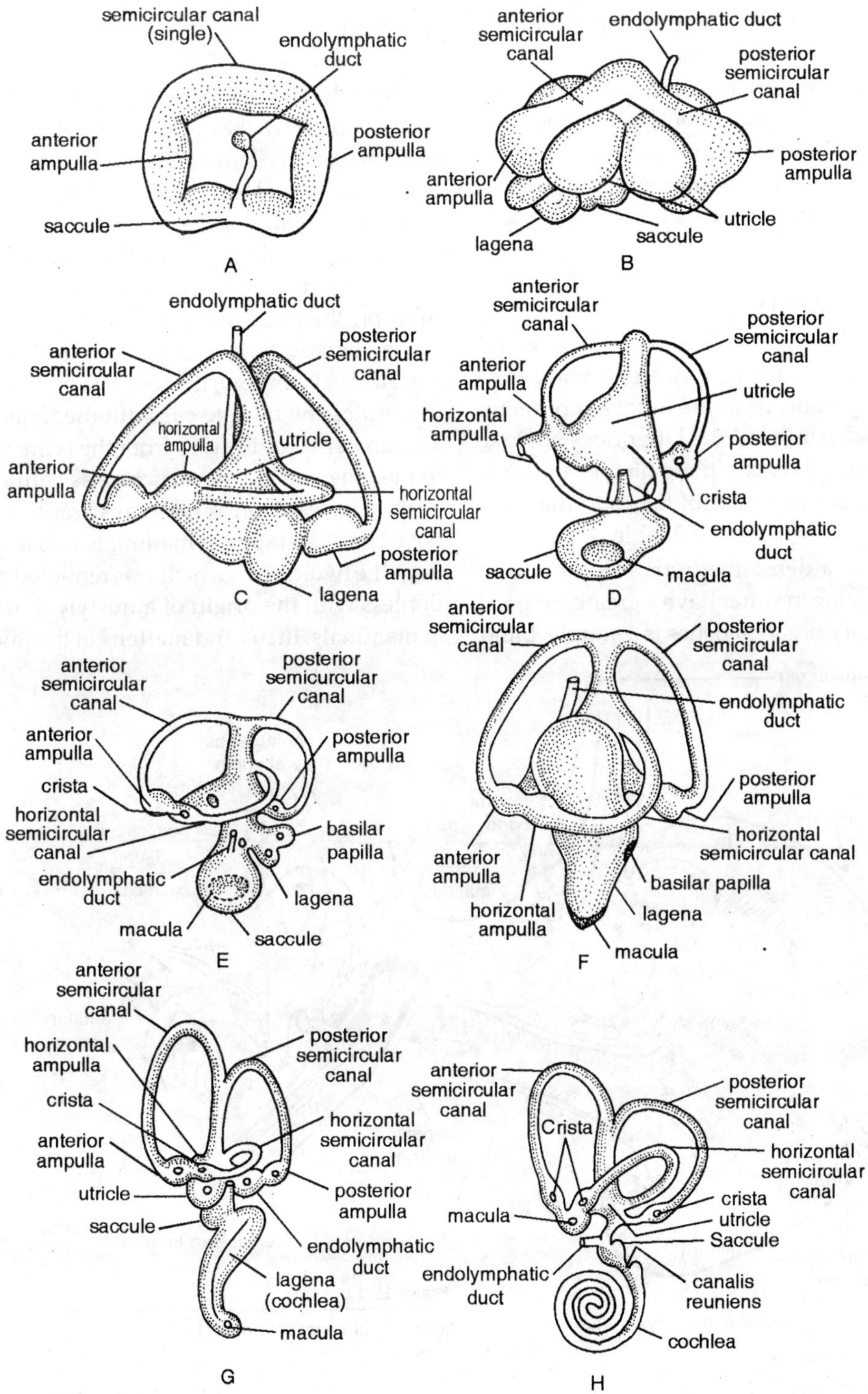

FIGURE 15.10

Membranous Labyrinth of Vertebrates: A. Myxine, B. Petromyzon, C. Scoliodon, D. Osteichthyes, E. Amphibia, F. Reptilia, G. Aves, H. Mammalia

accelaration) and the hair cells are stimulated due to differential movement.

The impulses are transmitted to the brain and the person becomes conscious of the body position.

In addition to the membranous labyrinths, eyes and proprioreceptors help in maintaining equilibrium.

☒ Phonoreceptors

The ear is the organ of hearing. The phonoreceptor or the structure concerned with hearing is inseparable from the inner ear or membranous labyrinth but the latter does not take part in hearing. A complete history of transformation of simple ear of fishes to the complex ear of mammals is available.

Water is a dense medium and the sound waves moving in water have a greater impact. The auditory organ in fishes is a simple, blind diverticulum, the lagena at the posterior end of the saccule (Fig. 15.10).

With the change of surroundings from water to air on land, the organs of hearing underwent profound changes. The air being a lighter medium the need to amplify the force of sound waves was a must and this led to the development of new structures. The first new acquisition in amphibia was the ear drum or **tympanum** on the body surface, which enclosed a space, the middle ear, external to the internal ear. An oval window, the **fenestra ovalis** communicates the middle ear with the inner ear. To transfer sound waves from the tympanum to the inner ear, a new structure as **columella auris** (Fig. 39.21) in amphibians, reptiles and birds* and as **stapes** in mammals was formed from the hyoid arch, which was rendered functionless with the origin of autostyly. Further, in mammals, **incus** and **malleus** in the middle

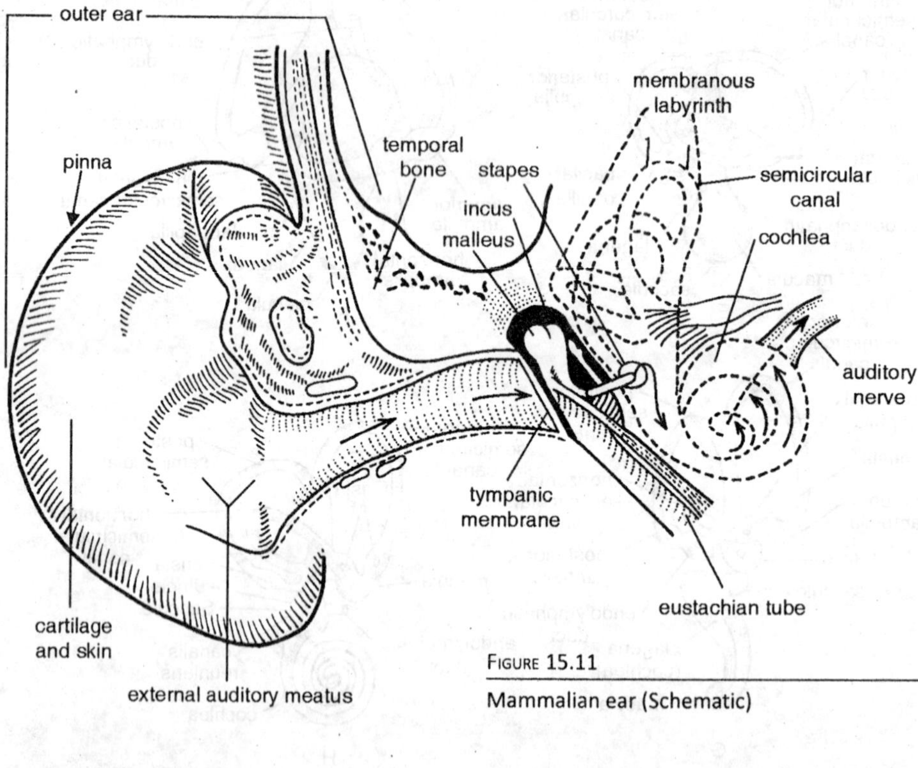

FIGURE 15.11

Mammalian ear (Schematic)

* True hearing is absent in present day snakes due to complete degeneration of the hearing organs. A large part of their body surface is in contact with the dense medium, water or solid and they perceive sound waves as vibrations through the skin.

ear developed from the quadrate and articular, respectively.

The ear in mammals though basically similar to that of other tetrapods, is more elaborate. The tympanam shifted from surface to inside and the passage from it to the exterior, the **external ear** appeared. The **cochlea**, the part of the membranous labyrinth concerned with hearing is an outgrowth from the saccule.

The capability to detect sound waves at different frequencies varies in different animals. It is 20 to 20,000 hertz in man, over 100,000 in bats.

⊠ The mammalian ear

The mammalian ear consists of three distinct parts—external ear, middle ear or tympanic cavity and internal ear (Fig. 15.11).

⊠ The external ear

It consists of two components, the skin covered, cartilagenous **pinna** or **auricle** and the slightly S-shaped auditory canal or **external auditory meatus** extending from auricle to tympanum. The tympanum completely separates the external auditory meatus from the middle ear.

⊠ The middle ear or tympanic cavity

The cavity is lined with either a simple squamous or a cubical epithelium. It is filled with air coming from the **nasopharynx** through the **auditory** or **eustachian tube** lined with a ciliated epithelium. Atmospheric air pressure on both the sides of the tympanum allows it to vibrate in response to sound waves. The medial wall of the tympanic cavity bears two openings, the **fenestra ovalis** (already referred to) or oval window and **fenestra cochlea** (round window). The stapes and a fine sheet of fibrous tissue occlude the fenestra ovalis and the fenestra cochlea, respectively.

⊠ Auditory ossicles

Three small bones from outer to inside—**malleus, incus** and **stapes**, the last one fitting into the fenestra ovalis, constitute auditory ossicles.

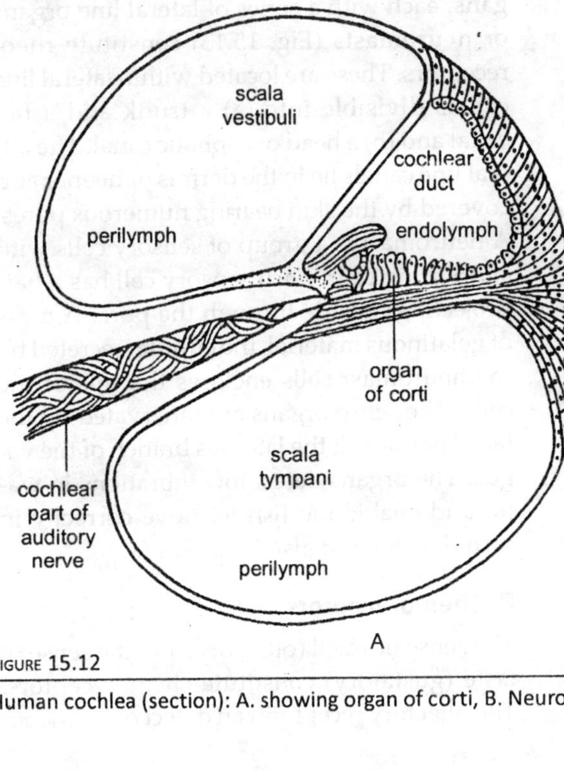

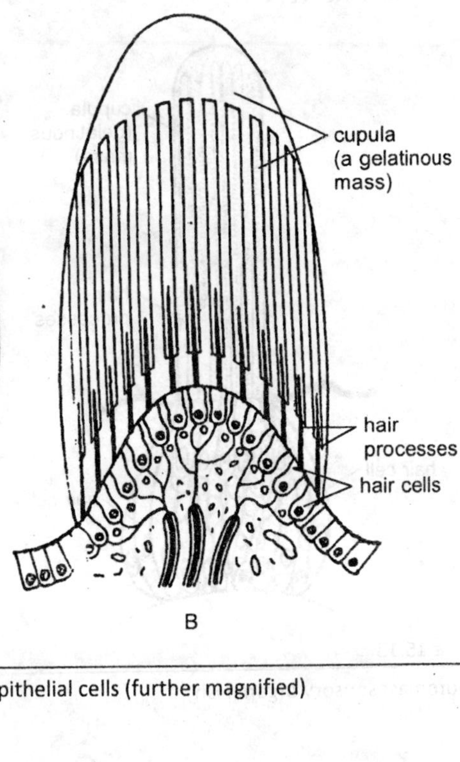

FIGURE 15.12

Human cochlea (section): A. showing organ of corti, B. Neuroepithelial cells (further magnified)

☒ The internal ear

Three structures—the vestibule, the membranous labyrinth and cochlea lodged in a cavity, the bony labyrinth constitute internal ear.

a. The expanded part adjacent to the middle ear and containing the oval and round windows constitute the **vestibule**.
b. The **membranous labyrinth** consists of three semicircular canals, utricle and saccule (p. 146-147).
c. The **cochlea** (Gr. *kochlias* = snail), an outgrowth of the saccule is a spirally coiled tube of two and half turns. In a cross section, the cochlea is triangle-shaped. The base of the triangle or the **basilar membrane** is lined by **neuroepithelial cells** and nerve fibres. The neuroepithelial cells are arranged side by side and a majority of them are long and narrow (Fig. 15.12). These cells and their nerve fibres constitute the **organ of corti**, the hearing organ proper. The organ of corti is innervated by the auditory nerve.

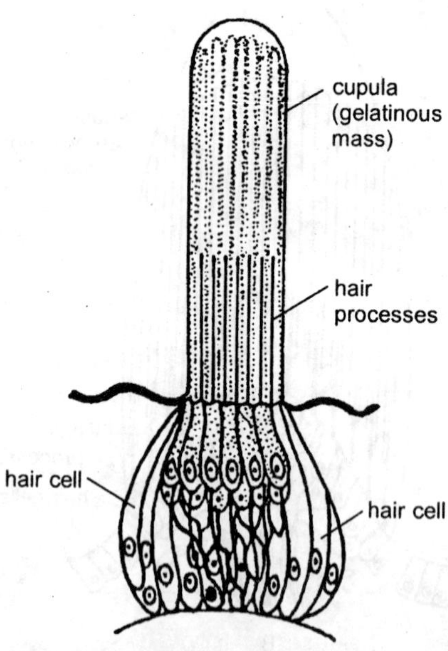

FIGURE 15.13

Neuromast sensory organ. Fish

☒ Machanism of hearing

Sound waves travel at about 1,088 feet (332 metres) per second. The sound waves are concentrated and directed by the auricle through the external auditory meatus. The tympanic membrane vibrates in response to the pressure of the sound waves. The vibration of the tympanum is transmitted to the oval window through the middle ear by the ossicles. To and fro movements of the stapes create pressure waves in the perilymph, which, in turn, causes wave motion in the endolymph. The neuroepithelial cells of the organ of corti are stimulated by the motion and the nerve impulses are send to the hearing area of the cerebrum in the brain through the auditory nerve. Vibrations of the membrane of the round window neutralise the wave motion in the endolymph.

☒ Rheoreceptors

Rheoreceptors or detectors of low vibrations in water, are present in fishes and aquatic amphibians. These are also known as lateral line sense organs. Two **lateral line sense organs**, each with a series of lateral line organs or **neuromasts** (Fig. 15.13) constitute rheoreceptors. These are located within lateral line canals divisible into (a) a trunk and a tail canal and (b) a head or cephalic canal. The lateral line canals lie in the dermis or deeper and covered by the skin bearing numerous pores. A neuromast is a group of sensory cells with supporting cells. Each sensory cell has a hair projecting outside through the pore. A mass of gelatinous material, the **cupula**, secreted by the neuromast cells encloses the tips of the cells. The sense organs are innervated by the facial nerve and the lateralis branch of the vagus. The organs detect low vibrations in water and enable the fish to move correctly in turbid water and also in darkness.

☒ Chemoreceptors

The sense of smell (olfactory) and the sense of taste (gustatory) constitute chemoreceptors. The olfactory receptors can detect objects from

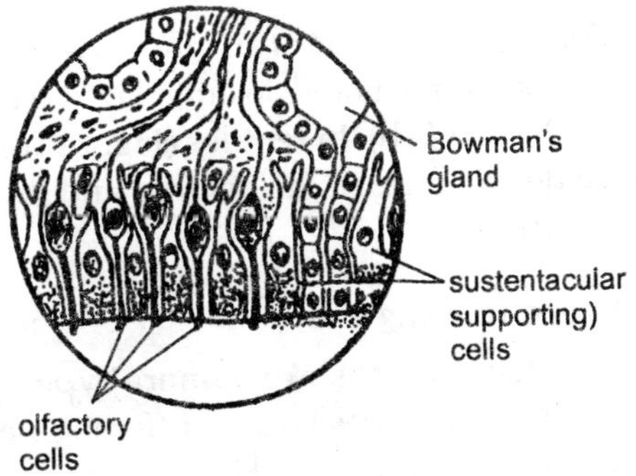

FIGURE 15.14

Olfactory epithelial cells. Human

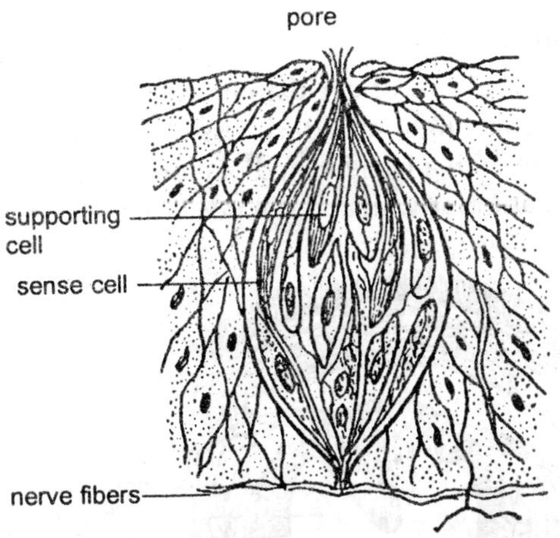

FIGURE 15.15

A taste bud

a distance while the gustotory receptors are active only in contact with the substance.

⊠ Olfactory receptors

The olfactory receptors are located in the posterior part of the nasal cavities. The receptors or the olfactory epithelium (Fig. 15.14) is made of **olfactory cells**, sensitive to smell. One end of the cell responds to the chemical stimuli while the other end is continuous with a neuron. The nerve fibres join to form the olfactory nerve, which passes directly to the olfactory lobe of the brain.

It is believed that substances containing volatile particles when dissolved in the fluid of the nasal cavity, releases the volatile particles, which stimulate the receptor cells.

This initiates impulses, which are transmitted to the site of smell in the brain.

In many air breathing vertebrates (amphibians, reptiles, mammals) an offshoot of the olfactory organ separated from it, forms a distinct sac lined by sensory epithelium and innervated by the olfactory and trigeminal nerve. It opens into the mouth and known as **Jacobson's organ** (Fig. 40.23).

⊠ Gustatory receptors

The sense organs for taste are bud-like structures, known as **taste buds** (Fig. 15.15). In fishes the taste buds are numerous and located in the mouth cavity, pharynx, skin of the head, tentacles, branchial chambers and, in some cases, scattered on whole surface of the body. In mammals, the buds are located on the papillae in the tongue and soft palate.

The taste buds are derived from the epithelium and oval in shape. Each bud contains two types of narrow cells, the taste cells and supporting cells. The free ends of the taste cells bear microvilli. The bud is in communication with the surroundings through a minute pore and many microvilli project into the pore. The taste bud is innervated by glossopharyngeal nerve.

A taste cell may be connected with only one neuron or more than one. The taste cells are stimulated by substances in solution. Sweet, sour, salt and bitter are four basic tastes. Some buds are specific while others may be stimulated by more than one tastes. This makes the detection and processing of information in the tongue very complex. But the messages finally transmitted to the brain are different for different tastes.

⊠ Photoreceptors

Two types of photoreceptor organs or eyes are found in vertebrates. Present day vertebrates possess a pair of image forming, simple, lateral eyes. The ancestral forms had another type of eye, the median pineal eye, which still persists in cyclostomes and some reptiles.

☒ Pineal eye

The pineal eye of cyclostomes can possibly distinguish light and darkness. It is a round capsule with a lens formed of elongated cells at the anterior wall and the posterior wall resembles a retina, made of an inner layer of rod-like visual elements and an outer layer of nerve fibres (Fig. 35.12B). In reptiles (Fig. 15.16), the nerve degenerates in the embryo and the eye lost its function.

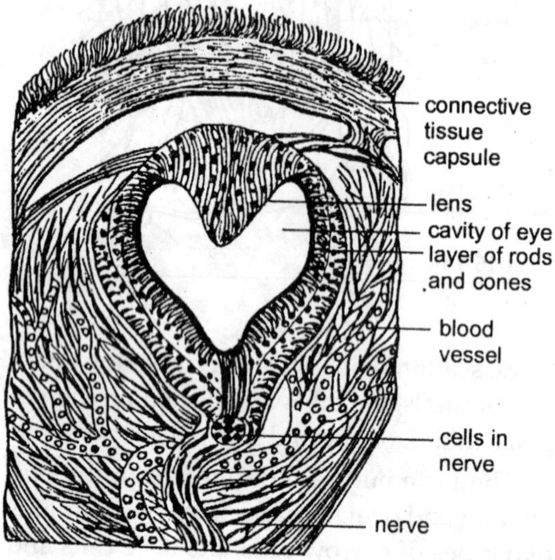

FIGURE 15.16
Pineal eye (vertical section). ***Sphenodon* sp.**

☒ Image forming eye

The vertebrate eyes are camera-type eyes. Since the vertebrates live in different media, their eyes are also adapted to see under water, in bright light, in dim light and in darkness (in relation to human vision). Still the basic features and functions of different components are similar in all groups of vertebrates.

☒ Mammalian eye

Eyes or organs of vision are two in number in a mammal, like that of any other vertebrate. The eye lies safely in a bony orbit of the skull, excepting a small exposure in front bounded by the eyelids. It is a hollow ball-like structure with a lens in front and a sensitive screen behind and largely corresponds to an ordinary photographic camera (Fig. 15.17). The eye is

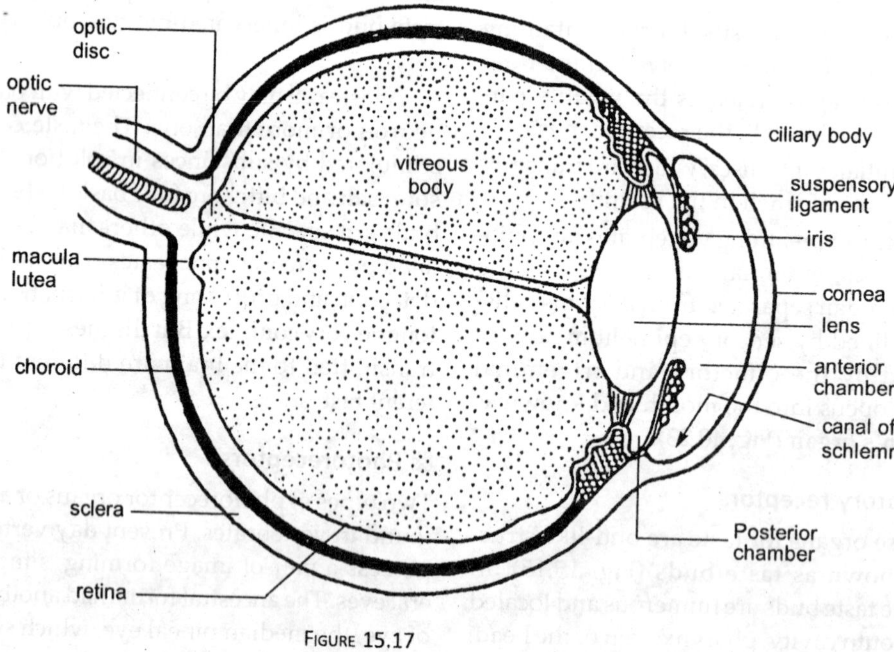

FIGURE 15.17
A human eye (vertical section)

moved by six muscles—superior oblique, inferior oblique, superior rectus, inferior rectus, anterior rectus and posterior rectus.

⊠ Structure

The wall of the eye-ball has three concentric layers—outer tunica fibrosa, middle tunica vasculosa and inner retina.

⊠ Tunica fibrosa

The outermost layer of the eye-ball, tunica fibrosa, skeletal in function, and maintains the rigidity of the eye-ball. It consists of **sclera** and **cornea**. The sclera is opaque, and is supplied by optic nerve and blood vessels. In the front, the sclera continues as a clear, transparent membrane, the cornea. A thin mucous membrane, in continuation with the eyelids, is present in front of the cornea. This is the **conjunctiva**.

⊠ Tunica vasculosa

The layer beneath the sclera is called tunica vasculosa. It is composed of **choroid, ciliary body** and **iris**, rich in blood vessels and pigment cells.

Choroid. The inner posterior five-sixths of the sclera is lined by choroid. The choroid is pigmented and richly supplied with blood.

Ciliary body. The anterior continuation of the choroid is made of nonstriated muscle fibres, the ciliary muscle and known as ciliary body. It is attached to the suspensory ligament, the other end of which is continuous with the capsule enclosing the **lens.**

Iris. The iris is a circular body behind the cornea, in front of the lens and with an aperture at the centre, the pupil. The iris is composed of two layers of muscle fibres, one circular and the other radiating and contain pigment cells. Contraction of the circular and radiating muscle fibres constrict and dilate the pupil. The space between the cornea and the lens is filled with a watery fluid, the **aqueous humor**. The iris incompletely divides the space into a large anterior and a very narrow posteior chamber.

Lens. The lens lies just behind the pupil. It is a transparent, circular, biconvex, highly elastic body. The lens is enclosed in a transparent capsule and suspended from the ciliary body by suspensory ligament. Contraction of the ciliary muscles release the pull on the lens and it thickens and moves forward.

⊠ Retina

It is the innermost layer of the eye wall. The retina lines about posterior three-fourths of the eye-ball. It is thickest at the posterior end, becomes thinner anteriorly and terminates just behind the level of the ciliary body. The retina is composed of a layer of pigmented epithelium constituted by receptor cells, the **rods** and **cones** (Fig. 15.18), against the choroid and several layers of sensory and connector neurons and their axons on the inner side. Due to the back position of the light sensitive cells, the rods and cones, the light rays must pass through several layers of neurons to reach them. The rods are more numerous and sensitive to dim light and colours while the cones are less numerous and respond to bright light. The retina acts as a sensitive screen on which the images of objects are focussed. The rods and cones receive the visual impulses. The retinal layer is highly innervated by the optic nerve. The point where the optic nerve enters the retina is known as **blind spot**. Situated close to the blind spot is the **macula lutea** (yellow spot). A small depression, the **fovea centralis** at the centre of this region is the site of greatest visual activity. Fovea centralis contains only cones. The high activity is due to the presence of maximal number of receptor cells, all cones, in the area.

⊠ Vitreous body

The whole space between the lens and the retina, about four-fifths of the eye-ball, is occupied by a thin jelly, the vitreous humor, enclosed in a delicate, transparent membrane, the

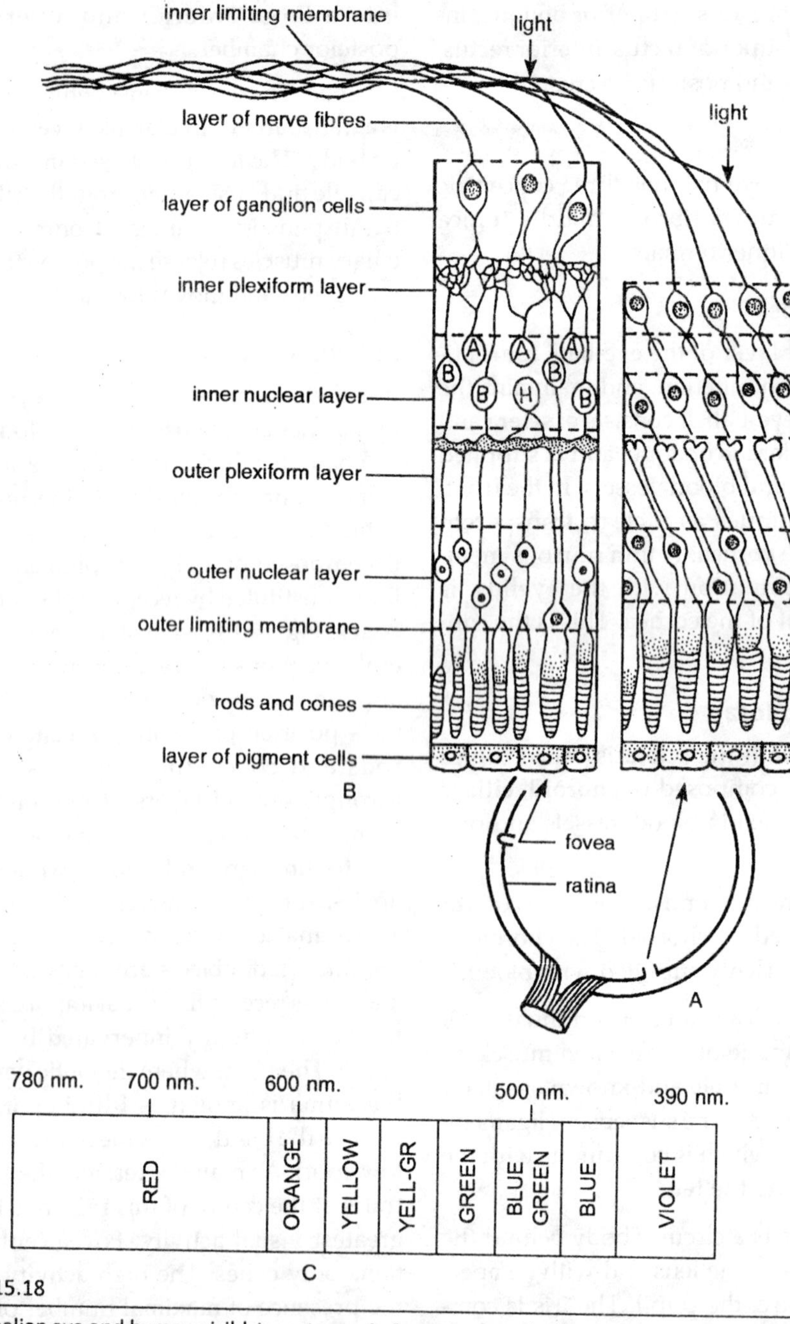

FIGURE 15.18

Mammalian eye and human visible spectrum, A. Section of eye-ball. B. Section of retina showing ten layers, C. Human visible spectrum

hyaloid membrane. A narrow canal, the **hyaloid canal**, filled with lymph runs forward through the vitreous body from the entrance of the optic nerve to the posterior surface of the lens.

Functions

Light rays enter the eye-ball through the transparent cornea and are focussed by the lens on the retina. An inverted image of the object is formed on the retina and this is conveyed to the visual centres of the optic lobes in the brain by the optic nerves. The inverted image is corrected and a direct image is formed. The visible spectrum of man ranges from violet (390 nm) to red (789 nm).

⊠ Accommodation

Normally, the eye is focussed for distant vision. It can be adjusted for near vision by increasing the convexity of the lens. Due to the tendency of the eye-ball to remain globular normally, there is a pull on the lens which keeps it flat. Contraction of the ciliary muscles releases this pull on the lens and the elastic lens becomes more convex and moves forward. The divergent rays from the near object are thus focussed to form a clear image on the retina.

⊠ Thermoreceptors

Cells sensitive to temperature variations are present in all vertebrates. In most cases, they are scattered all over the body. Fishes are highly sensitive to temperature variations and can detect variation as low as 0.1 °C. This enables the fishes to know that they are entering a different body of water and to regulate their migration. In snakes, the thermoreceptors are usually located in pits on the sides of the head. This helps them to locate the presence of warm-blooded prey.

⊠ Electroreceptors

Electroreceptors are sensory cells which can detect changes in the electric field of the surroundings. The fishes, electric ray or electric eel can generate electric pulses exceeding 500 volts. The majority of the fishes having electric organs, however, give off very weak pulses. These are believed to be a part of the **electric guidance system**. It is presumed that with their specific electric organs the fishes create a pulsating electric field around them and any object in the surroundings with a different electrical conductivity than that of the water will disturb the field. With their special receptor cells these fishes can detect the disturbance in the electric field and alter their course in water.

16 THE ENDOCRINE SYSTEM

HORMONES

Chemical messengers, the **hormones**, (Gr. *hormaein* = to set in motion) are secreted by certain glands without ducts, the **endocrine glands**. The hormones induce physiological action in organs, elsewhere in the body. Hormones constituting the endocrine system coordinate and control activities of many organs, the **targets** in multicellular animals. Some of the targets are located at quite a distance from the site of secretion. Since the glands secreting hormones are without any duct, the secretions are transported to the targets in different ways.

The functioning of the endocrine system involves three components, the **endocrine glands,** the **transport media** and the **target organs**.

The cells of an endocrine gland are specialised in one direction and secrete only one type of chemical, the hormone, the composition of which is very complex. The hormones are transported to the target organs through (a) diffusion, (b) blood and, (c) axons of the neurons. In transport through blood some hormones are in solution in plasma, some are bound to albumin, some others nonspecifically bound to albumin and still others specifically bound to certain proteins. The major bulk of hormones in transport through blood are bound to serum protein.

The vertebrate hormones have been extensively studied. The location of the endocrine glands in the body is almost same in relative position in all vertebrates.

The hormones secreted by the neurons (nerve cells) are **neurohormones**. Neurosecretory cells in vertebrates are usually located in the brain or some nerve ganglia. The neurohormones are moved along the axon and released at the axonal tip.

Hormones are effective in a very low concentration and a minute quantity is capable of bringing about marked reactions in the cells of the target organs. Presumably, they have serveral independent mechanisms of action to regulate cellular activities. In general, the response of the targets to the hormones is slow, which may extend even for weeks and the control is long lasting.

Hormone action is highly specific. Many hormones are being carried at a time by the blood. Each type of target cells possess a specific protein, the **receptor** and only this can form a specific combination with a particular hormone. This is known as **target selection**.

Vertebrate Hormones

Source	Hormone	Function
Thyroid	Thyroxine	1. Growth and differentiation. 2. Facilitates utilization of carbohydrates. 3. Stimulates adrenal cortex.
Parathyroid	1. Parathormone	Act on kidneys and gut to keep blood ionized calcium constant.
Parathyroid	2. 1, 25 dihydroxy-cholecalciferol (1, 25-DHCC)	Activate adenylate cyclase in bones and kidneys.
	3. Calcitonin	Supress mobilisation of calcium from bone.
Thymus (Regresses after puberty)	Thymosin.	Helps in development of T-lymphocytes.
Pancreas (Islet of Langerhans)	1. Insulin	1. Converts glucose into glycogen. 2. Conversion of carbohydrates to fat. 3. Protein synthesis.
	2. Glucagon	Stimulates glycogenolysis.
Adrenal medulla	1. Adrenaline 2. Noradrenaline	The effect of two hormones is qualitatively similar on heart, blood vessels, smooth muscles and carbohydrate metabolism.
Neurohypophysis (posterior pituitary)	1. Antidiuretic hormone	1. Increases water absorption in the kidney tubules. 2. Constriction of the arteriovenous capillaries.
	2. Oxytocin	1. Contraction of smooth muscles. 2. Lactation, ejection of milk.
Adenohypophysis (anterior pituitary)	1. Somatotropic hormones (STH)	1. Growth in young animals. 2. Increased bone growth.
	2. Thyroid stimulating hormones (TSH)	1. Stimulation of thyroid activity. 2. Swelling of the orbital content of the eye.
	3. Adrenocorticotropic hormones (ACTH)	Stimulation of adrenal cortex causing secretion of glucocorticoids.
	4. Follicle stimulating hormone (FSH)	Stimulates growth of ovarian follicles in female and promote spermatogenesis in male.
	5. Leuteinizing hormones (LH)	Stimulates ovarian secretion of oestrogen and progesterone and testicular secretion of androgens.
	6. Prolactin or Leutotropic hormones (LTH)	1. Maintenance of corpus leteum. 2. Mammary gland development. 3. Encourage maternal instinct and behaviour.
	7. Melanocyte stimulating hormones (MSH)	1. Stimulates pigment synthesis. 2. Dispersion of melanophores in lower vertebrates.

Contd.

Source	Hormone	Function
Ovary	1. Oestrogens	Normal sexual development of the accessory and secondary characters of female.
	2. Progesterone	Activation of the secretory phase of the uterine endometrium.
Testis	Androgens	1. Influence male accessory organs and secondary sex characters. 2. Stimulate red cell production and blood flow through the tissues.

Invertebrate hormones

Source	Hormone	Function
Intracerebral gland (brain surface)	Prothoraciotropic hormone	Initiates moulting
Prothoracic gland (Prothoracic ganglion)	Ecdysone	Regulates moulting
Corpora allata (Head)	Juvenile hormone	Metamorphosis

17 THE REPRODUCTIVE SYSTEM

REPRODUCTION

The appearance of new individuals, the offsprings, from the preexisting organisms, the parents and resembling them but not identical, is known as reproduction.

The term reproduction has got a wide range of meaning and does not merely mean multiplication of individuals for generations. It is a biological dictum, a natural prophacy to represent the self in a newer but heritable conformity that may establish a new pattern in the model of evolution. The organs involved in reproduction constitute the reproductive system.

The similarities between the parents and the offsprings are due to the ability of self-duplication of the nucleic acids, the genetic materials carrying information from generation to generation; the variations in the genotype is due to mutation and the variations in the morphology of the offsprings (phenotypes) are the results of the operation of the agents of selection in the surroundings, modifying the epigenetic landscape during the period of embronic development.

MODES OF REPRODUCTION

The different modes of reproduction in the animal kingdom may be placed under two distinct head, asexual and sexual.

⊠ Asexual reproduction

It is most common in lower animals but occurs also in somewhat highly organised forms. Production of a generation of cercaria from a redia in liver fluke, budding in bryozoa and tunicates belong to this category. In asexual reproduction the sexual units, male and female gametes are not involved. A single individual is capable to produce offsprings. Binary fission, multiple fission and budding are the common modes of asexual reproduction.

⊠ Binary fission

It may be transverse, longitudinal and oblique. The individual divides into two equal halves, each half grows into an adult individual. The nucleus divides mitotically into two, which is followed by a division of the cytoplasm, each half receiving (Fig. 21.2) a daughter nucleus.

In transverse fission, the cytoplasm divides by a transverse constriction, e.g. *Amoeba*, *Paramoecium* (Fig. 21.13), etc. In longitudinal fission the split appears at the anterior end and proceeds backward, e.g. *Vorticella* (Fig. 21.16), *Euglena*, *Trypanosoma*, etc. The half lacking some asymmetrical structures constructs them and a full grown adult is formed. Oblique fission is similar to that of transverse fission but the cytoplasm divides obliquely, e.g. Dinoflagellates, *Ceratium*, etc.

⊠ Multiple fission

It is also known as **sporulation**. In sporulation the nucleus undergoes repeated mitotic divisions producing a large number of daughter nuclei. The cytoplasm breaks up into many fragments, a bit surrounds a daughter nucleus and a tiny individual is formed. On liberation from the covering, the individuals grow and reach the adult stage. Multiple fission is the normal mode of reproduction in sporozoa (Fig. 21.9). In unfavourable conditions *Amoeba* undergoes multiple fission.

⊠ Budding

Multiplication by budding involves formation of **gemmules** or internal buds in sponges and **lateral buds** in some invertebrates and chordates, viz. sponges, cnidarians, bryozoa, tunicates and others.

A lateral bud is given off from some part of the parent animal, it grows and assumes the shape of a complete parent animal (Fig. 23.8) and separated from the parent either before or after completion of the development. In some cases, the fully developed buds remain in a vital continuity with the parent and form a compound animal consisting of a number of units. Such a compound animal is termed a **colony**, e.g. *Volvox, Obelia* (Fig. 23.12), *Bugula*, etc.

⊠ Sexual reproduction

In sexual reproduction, two sexual units or gametes are involved. Prior to the formation of gametes the germ cells undergo a pregametic meiosis and the chromosome number is reduced to half (n) in the gametes. The original number (2n) is restored in the zygote by the union of two gametes, which produces a new individual. The gametes may be similar and the union is known as **isogamy** or they may be dimorphic or dissimilar and the union is named as **heterogamy**.

The sexual reproduction is much more advantageous than asexual reproduction. Formation of germ cells, which are small in size, not only economise the involvement of living materials but also less exhausting physiologically. More important than these is the union of two units, coming from different sources, bringing different materials, and their combination is a perenial source of variations, which are the raw materials for evolution.

Majority of the animals are male or female, and the evolution of sex occurred in the early period of animal history. The male products are tiny, nearly always motile spermatozoa (Figs. 5.17 and 43.2) and the female products are fairly large, passive ova (Fig. 43.3) often containing yolk.

Sexual reproduction occurs in all groups of animals. Starting from protozoon it becomes more and more complex, attaining perfection in mammals. Sexual reproduction is of two types, conjugation and fertilization.

⊠ Conjugation

Three types of conjugations are found in animals:

Simple conjugation. It is the simplest form of sexual reproduction. Two similar cells unite with the fusion of their nuclei, e.g. rhizopods, sporozoa, flagellates and other protozoa. The conjugation is temporary in certain ciliates, *Paramoecium* (Fig. 21.15), for exchange of nuclear materials.

Multiple Conjugation. More than two cells unite temporarily, as in some sporozoa.

Dimorphic conjugation. Two cells of different sizes fuse to form an individual. In *Vorticella,* the small active, free swimming unit (male) fuses with a normal size (Fig. 21.16) fixed individual (female).

Fertilization. Fusion of a microgamete with a macrogamete. Found in many Protozoa, *Eimeria, Plasmodium*, etc. More commonly, the male gametes or spermatozoa are formed in the male gonads or testes and the female gametes or ova in the ovaries of females in higher forms. Only one spermatozoon fuses with an ovum to form a zygote, the unicellular starting point of multicellular life.

☒ Unisexuality and bisexuality

A range of modifications occur in the invertebrates in the location, arrangement, number and types of gonads; their ducts and other accessory reproductive organs. Depending on the presence of one or both types of gonads they have been placed in two groups.

Unisexual or dioecious. The individual possesses either the testes or the ovaries. Examples: Round worms, bivalves, prosobranchs, cephalopods, vertebrates etc.

Bisexual or **hermaphrodite** or **monoecious.** The individual possesses both the testes and the ovaries. Examples: Sponges, *Hydra*, annelids, barnacles, pulmonate molluscs, etc. In majority of the bisexual animals, however, self-fertilization is prevented by producing sperms and ova at different periods of time (**dichogamy**). Though rarely, self fertilization (**autogamy**) occurs in some parasitic flatworms.

Protandry. In some animals the same gonad first produces male gametes and then female gametes. This is known as protandry and the animal **protandrous**. Examples: Some nematodes, bivalves, gastropods, star fishes, tunicates, cyclostomes etc.

Protogyny. The first female and then male condition is known as protogyny and the animals **protogynous**. Protogynous forms are rare and occur only in some fishes.

☒ Invertenrate reproductive system

A range of diversity is found in the components of the reproductive system in invertebrates. The gonads are temporary structures in some diploblastic animals and polychaetes. They are derived from amoebocytes, a type of mesenchyme cells in sponges; both from **ectoderm** (*Hydra*) and **endoderm** (*Aurelia*) in cnidaria and the gametes are shed outside or in the radial canals respectively. In all triploblastic forms gonads are mesodermal in origin.

The number of testes in an individual varies greatly. They are one pair in most flatworms, polychaetes, arthropods and molluscs; 2 pairs in some cnidarians (*Obelia, Aurelia*), earthworms; 5 to many pairs in leeches and only one in round worms and gastropod molluscs. In the latter, one of the testes disappears during torsion in the embryo.

The number of ovaries is more constant in the invertebrates. They are two pairs in some cnidarians (*Obelia, Aurelia*), one pair in roundworms, annelids (except nereis, where the ovaries are many and temporary), arthropods and one in flat worms and gastropods. One ovary degenerates during torsion in gastropods.

Both the testes and ovaries are connected with ducts. In males, the sperms are discharged to the exterior through a pair or one pore. In females, the ova are carried to a particular site in the body for fertilization or released to the exterior through a pore. Certain glands and specialised structures constitute a part of the reproductive systems of both males and females.

☒ Vertebrate reproductive system

In all craniata the gonads are mesodermal in origin, paired and located in the coelom, suspended to the dorsal wall by a fold of peritoneum. The single pair of gonads develop from the dorsal wall of coelom as ridges and covered by coelomic epithelium. The sperms and ova are derived from the primordial germ cells in the gonads (p. 851).

The crypts or **tubules** lined with epithelium constitute the **testes**. The sperms produced in the tubules are motile and carried through vasa efferentia and vasa deferentia to the storage site (**seminal receptacle**) or copulatory organ or discharged to the exterior. The mesonephric ducts in male function as spermiducts and retain their connections with the urinary system in the adult (Fig. 39.22).

The connective tissue or **stroma** covered by epithelium constitute the **ovaries**. The primordial germ cells in the ovaries enlarge and form ova. In most cases, the mature ova are shed into the coelom, close to the infundibula or the funnels of the oviducts. The oviducts carry the ova to the site of fertilization or to the exterior. The pronephric ducts function as oviducts in females and they are not connected with the urinary system in the adult (Fig. 39.23). In some fishes and amphibia the ova are shed into the coelom and pass out through the genital pore.

In most vertebrates, the posterior part of the oviducts dilate to form uteri (ovisacs) or the two ducts may join and dilates to form a single, large uterus (human). In birds only the left oviduct persists in the adult (Fig. 41.31).

⊠ Accessory reproductive organs

Certain accessory structures are always associated with the gonads in all animals where the gametes are to be transported to the genital atrium or genital pore or to the uteri of females of the same species.

The accessory structures and the glands in the walls of the ducts secrete fluid to facilitate movement of the gametes. The oviducts and certain accessory glands play a major role in vitellogenesis, i.e. contributing yolk to the eggs and providing protective coverings, the vitelline membrane, the shell, leathery (in majority of the cases) or calcareous (reptiles and birds) and also in secreting a protective capsule enclosing eggs (insects, molluscs). Spermathecae, small sacs for storing sperms after mating, are present in the females of many species. The sacs may or may not be connected with the oviducts.

⊠ Gonadal activities

Animals do not breed round the year. The period in which a particular species breed is the breeding season for the species. It is related to geographical area, climate, food supply, etc. This varies in different animals. Broadly, animals thriving well in wet season, breeds in monsoon, while the majority of the others breed in the spring. The period may be stretched in either way.

The ovaries are active only in the breeding season. This is probably due to the limited number of primordial germ cells (4,00,000 in human) in the ovaries, destined to form ova. On the other hand, the testes are active throughout the year but not at equal tempo. The peak period of sperm production is the breeding season. This is presumably due to the capacity of an individual to produce an enormous number of sperms.

As a rule, the number of sperms produced by males is always much more than the eggs produced by the females of the same species. This serves dual purpose, to compensate the death of a large number of sperms in course of their movement towards the ova, and though only a single spermatozoa is needed to fertilize an ovum, presence of a large number of them is required in the vicinity of the egg to produce the required amount of enzyme to dissolve the coverings of the egg. In mammals, the enzyme, hyaluronidase secreted by the sperms dissolves the hyaluronic acid, the binding material holding the follicle cells around the ovum.

⊠ Gamete production

A regulatory mechanism to control gamete production is present in all sexually reproducing animals. The gonads and the accessory structures are remarkably modified to meet environmental assaults. Transmission from host to host to complete life cycle is compulsory in endoparasites, with a few rare exceptions. Large number of embryos perish in the process. To compensate the loss, the gonads have enlarged enormously in size to produce a very large number of gametes. A female *Ascaris*, a round worm, produces 20,000 eggs a day, a single liver fluke, a flatworm, produces 50,000 eggs. In the tape worm, *Taenia*, a proglottid beyond 200 is a complete hermaphrodite unit. The proglottids from 400 onwards, i.e. the last 400 to 500 are fully occupied by a much branched uterus, heavily loaded with embryos.

Contrast to this, a drastic reduction in the number of ova produced is associated with internal fertilization and formation of a protective membrane, a leathery egg shell, or a calcareous egg shell or a similar structure around the egg after fertilization to protect the embryo. In animals with intrauterine development, the security of the foetus is ensured and the number of ova produced at a time has come down to one to only a few. To provide yolk to the eggs is a tremendous physiological stress on the mother and the number of eggs produced is reduced with the enlargement of egg size. In birds, one oviduct has atrophied to prevent the development of many eggs at a time.

⊠ Fertilization

The union of a sperm and an ovum (Fig. 43.4) is fertilization (p. 854). Most sperms are motile but the distance they can travel in their brief life span in the open environment is rather small. Fertilization takes place both outside and inside the female body and known as external and internal fertilization, respectively.

⊠ External fertilization

External fertilization is possible only in animals living or breeding in water*. Being attracted by chemical stimuli the motile sperms move to the passive ova and fertilize them. The union of the gametes is a matter of chance and many sperms die in their hazardous journey. Factors in the surroundings also kill many ova and above all the life span of the gametes is very short.

To ensure fertilization, millions of eggs are produced by a female in one spawn and still more number of sperms are released by the male, so that at least fertilization of a few eggs is ensured.

In addition, certain adaptations help in synchronising the release of sperms and ova and that close to one another. Internal neural and hormonal controls are activated by both abiotic and biotic factors—temperature, photo-period and lunar and tidal cycles, courtship, stimulation received from shedding of gametes (sperms) of a member of the opposite sex at close proximity of the partner during ovulation period, are adaptations to synchronise gamete release.

⊠ Internal fertilization

Fertilization of ova inside the body of a female is internal fertilization. This is advan-tageous because placing the sperms close to the ova greatly increases the chances of fertilization and thereby cut down the wasteful expenditure of living materials spent in the production of a large number of gametes. At the same time it needs several adaptations, viz. space to accommodate the ova in the body of the female, at least for a certain period both in the pre- and post-fertilization phases. The eggs receive yolk and protective coverings, the shell and others after fertilization.

Internal fertilization occurs almost in all terrestrial animals. A fluid medium is necessary for the movement of the sperms towards the ova and a certain amount of moisture is required to prevent desiccation of the embryos during development. Both are lacking on land. The ducts through which the sperms move to reach the eggs contain fluid secreted by the walls of the duct or certain accessory glands. The eggs laid by land animals contain sufficient amount of water required for the incubation period. The loss of water through evaporation is prevented by an impervious covering and where such effective coverings are absent, eggs are laid in burrows in moist soil or covered by soil.

Internal fertilization is not restricted only to higher vertebrates and terrestrial animals and occurs in many marine and fresh water invertebrates.

SPECIAL MODES OF REPRODUCTION

The apparent uncommon modes of reproduction are common in certain animals. Partheno-

* In some land forms (earthworm), fertilization is external but occurs only within a cocoon formed during mating.

genesis, paedogenesis and polyembryony belong to this category.

⊠ Parthenogenesis

Development of an unfertilized egg into an adult is parthenogenesis (Gr. *perthenos* = virgin + *gene* = birth). A number of abiotic factors, viz. pH, salinity of the water (in aquatic forms), temperature, chemicals and mechanical stimuli induce parthenogenetic development of eggs in many species. It is common in many insects, viz. ants, termites, aphids, bees, wasps, etc. In most of the parthenogenetic species, pregametic meiosis is absent and the eggs contain diploid number of chromosomes. Some species of rotifers produce both haploid and diploid eggs. In rotifers and some lizards males are unknown. Parthenogenesis occurs in aphids (insects) and water fleas (crustacea) for several generations, following which males are produced and mating occurs. Parthenogenesis may be sex related. In honey bee, queens and workers develop from fertilized eggs and drones (males) from unfertilized eggs.

⊠ Paedogenesis (Neoteny)

Sexual reproduction in the larval or preadult stage is paedogenesis. This has evolved presumably in response to the advent of unfavourable conditions threatening survival before attaining adulthood and reproducing sexually. The axolotl larva of salamander *Ambystoma tigrinum*, still possessing gills, develop sex organs and breeds regularly. Some of the diptera, gall midges, reproduce sexually in their larval and pupal stages.

⊠ Polyembryony

Appearance of a few individuals, instead of one, from a single fertilized ovum is polyembryony. It is formed both in invertebrates and vertebrates. The parasitic hymenoptera, *Ageniaspis fuscicollis* produces 10 to 15 and *Litomastix truncatellus* about 2,000 larvae from one fertilized egg. Polyembryony is common in bryozoa. The nine-banded armadillo, *Tatusia novemeineta* produces 4 to 5 and other species up to eight or nine embryos from one fertilized egg.

18 REGENERATION

The ability to construct a lost part of the body is regeneration.

Regeneration occurs in many animals belonging to widely diverse groups. A cell can grow the lost part provided the nucleus is not damaged. Damaged neurons not involving the cell body can reconstruct the lost portion. A hydra can reconstruct any of its lost part within a short period. Sea anemone can regenerate new individuals from fragments of tissues from the basal disc. The regeneration capability of planaria (Fig. 18.1) is remarkable. Any part of the body separated can build up the missing portion and forms a new individual. Gastropod molluscs can regenerate their both tentacles with eyes, even if they are lost at a time. A starfish can reconstruct the whole body from a broken arm.

Many crustacea can grow lost legs. Salamanders, lizards can built new legs and tails if the original one is lost. The victim escapes by detaching the body parts seized by an enemy or a predator, and this is considered as an adaptation for defence.

Union of broken bones and healing of wounds are viewed as regeneration. Though in a much less degree, regeneration occurs in birds and mammals. Large areas of damaged epidermal tissue, damaged blood vessels and liver tissues redevelop.

⊠ Type of Regeneration

Three types of regeneration occur in nature.

Morphalaxis. It involves reconstitution of missing parts of the body by the respecification of the existing tissues after the loss of a body part. There is no cell division or growth of new tissues in the process. Example: *Hydra*.

Hydra has a tubular body with a 'head' consists of a conical hypostome surrounded by a ring of tentacles at the distal end and a 'foot' or 'basal disc' at the proximal end.

When a Hydra is cut into two halves the half with the head regenerates a new basal disc and the other half containing the basal disc regenerates a new foot (Fig. 18.1).

If a hydra is cut into several pieces, the middle portions will regenerate both head and basal disc at their opposite ends. Since there is no multiplication of cells by division the new individuals are of smaller size.

Every portion of the hydra along the apical-basal axis is able to form an entire body, a head or a basal disc. This indicates some sort of polarity exists in hydra that controls the head at one place and the basal disc at another.

Planaria, a Platyhelminthes has unique power of regeneration. The body cut across or length wise, each part will regenerate the missing parts (Fig. 18.2). The regeneration

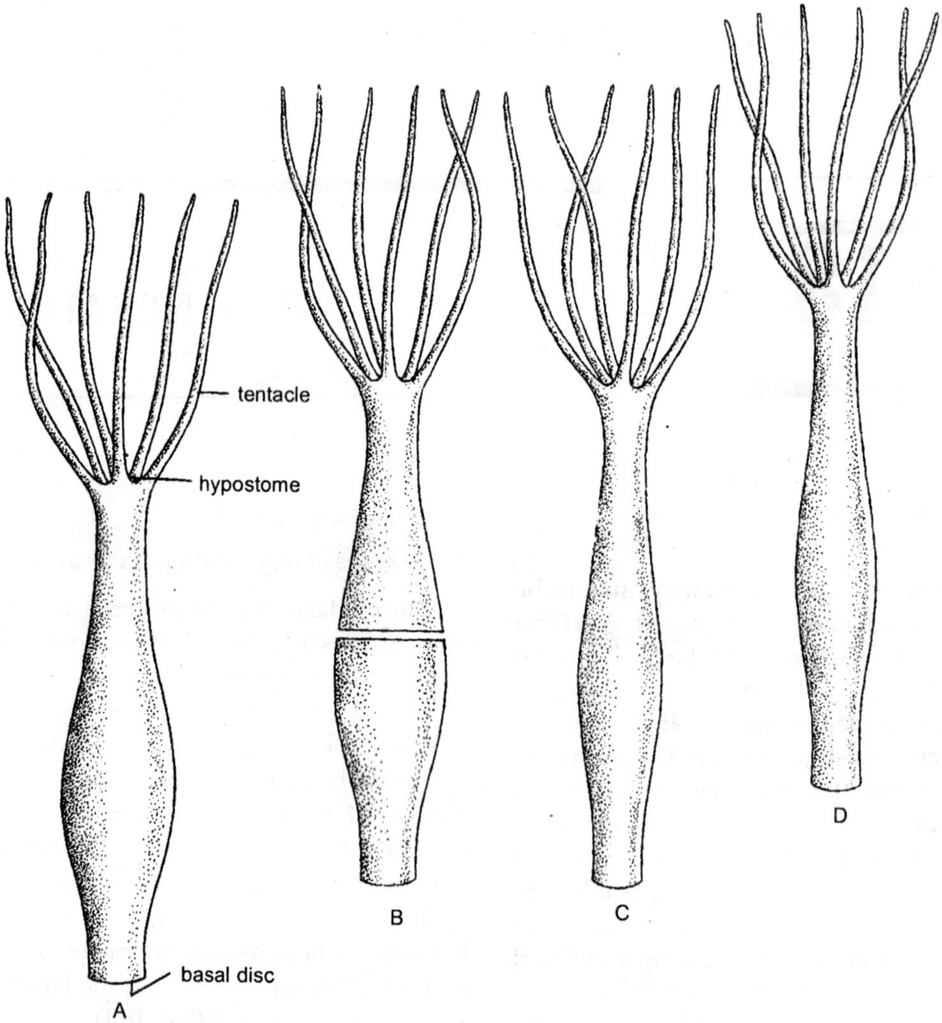

FIGURE 18.1

Hydra. Regeneration. A Entire hydra, B. Cut in the middle. C. Anterior portion regenerates posterior portion. D. Posterior portion regenerates anterior portion

power is maximum near the head and minimum near the posterior end.

When a cut is made in planeria, new undifferentiated cells derived from existing cells migrate to the cut end and a regeneration *blastema* is formed. The lost portion is regenerated from the blastema. The regeneration is a combination of morpholaxis and epimorphosis.

Epimorphosis. It involves the progressive replacement of missing parts by new cells formed by multiplication of existing cells and growth of new tissues. The mechanism involves the dedifferentiation of adult structure to form an undifferentiated mass of cells that becomes respecified to reconstruct the lost organ. Examples: Limb generation in insects, amphibians; tail regeneration in reptiles (house lizards), etc.

Epimorphic regeneration. Salamander limb. In an amputated salamander limb the remaining cells can reconstruct a complete

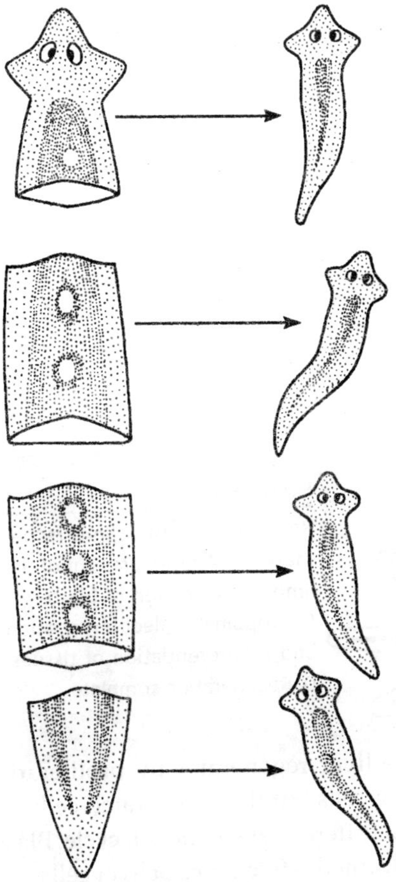

FIGURE 18.2

Regeneration. Planaria

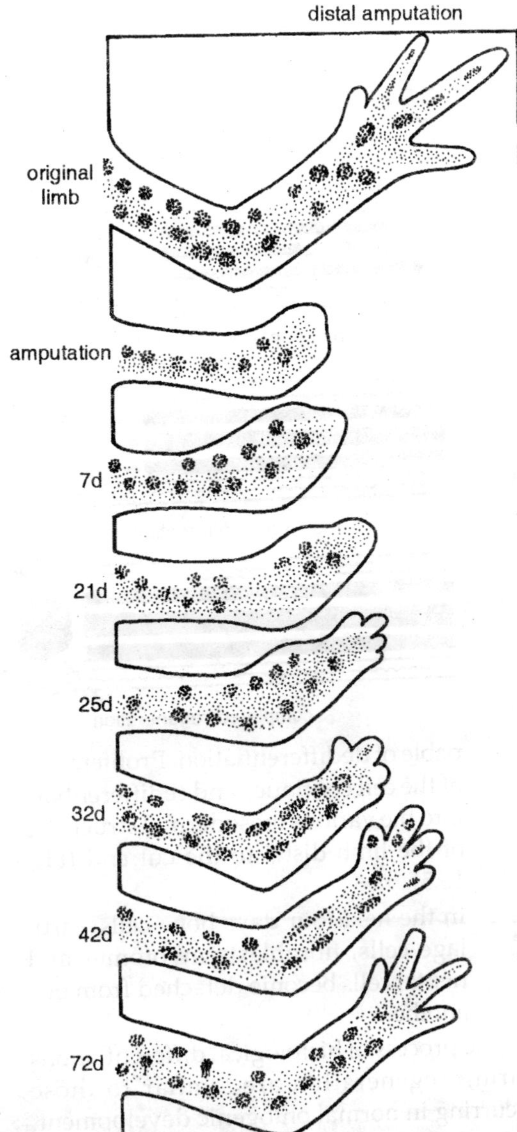

FIGURE 18.3

Regeneration of Salamander fore limb. Amputation of left limb made below elbow

limb similar to the previous one—with all its differentiated cells arranged in the proper order (Fig. 18.3, 18.4). The new cells formed construct only the missing structure. It may be cited that in a wrist amputation only a new wrist is formed and not a new elbow. The new structure is formed by differentiation and respecification.

1. Following amputation limb generation begins with the proliferation of ectodermal cells at the site of amputation. Within 6-12 hours the epidermal cells from the remaining stump migrate to cover the wound surface, forming the **wound epidermis**. The cells of the wound epithelium proliferate and a thickened **apical ectodermal cap** (AEC) is formed.
2. The underlying cells are induced by the apical ectodermal cap to produce undifferentiated mass of mesenchyme —**regeneration blastema**.
3. Blastema formation indicates that the differentiated cells in the limb are ca-

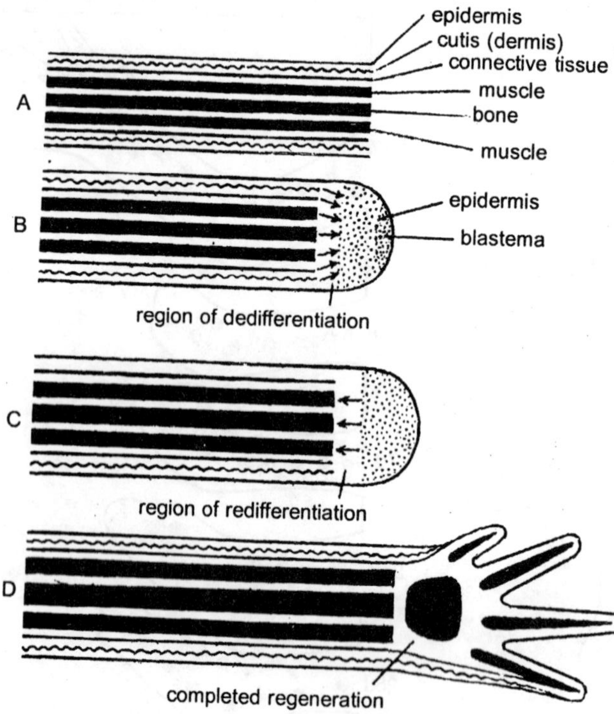

FIGURE 18.4

Regeneration of a limb (schematic). A. A transverse cut of the limb. B. Formation of blastema. C. Stopping of dedifferentiation and redifferentiation of tissues. D. Regeneration complete

pable of dedifferentiation. Proliferation of the cell continues and redifferentiate into the various mesodermal structures of the limb distal to the cut end (Fig. 18.5).

4. In the next four days bone cells, cartilage cells, fibroblasts myotomes and neutal cells become detached from one another.

The process of histological differentiations during regeneration are similar to those occurring in normal ontogenic development.

Compensatory regeneration. The cells divide but maintain their differentiated function. Cell proliferation occurs producing cells similar to themselves. This type of regeneration is found in mammalian liver.

The liver regenerates by proliferation of existing cells but the regenerated cells do not fully dedifferentiate in the cell cycle. Blastema is not formed—five types of liver cells—hepatocytes, duct cells, fat-storing cell, endothelial cells and Kupffer macrophages—each divide to produce more similar cells. Each type of cell retain their cellular entity and retains the ability to carry specific function.

☒ Regeneration in animals

Nonchordata

Protozoa. Many protozoon if cut into two pieces the part containing the nucleus regenerate the lost part while the enucleated part perish.

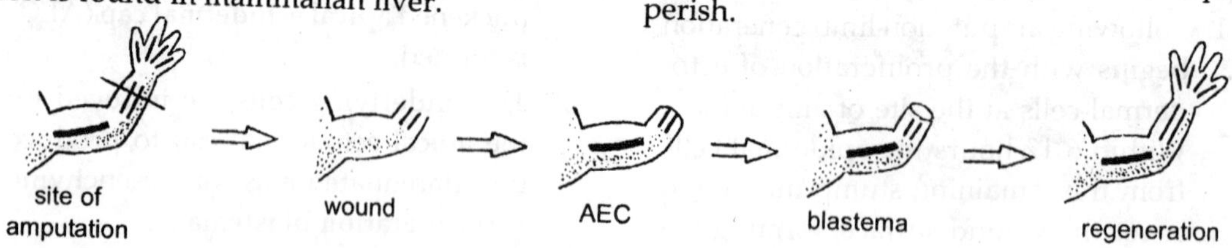

FIGURE 18.5

Limb regeneration in urodele amphibians. AEC = apical ectodermal cap

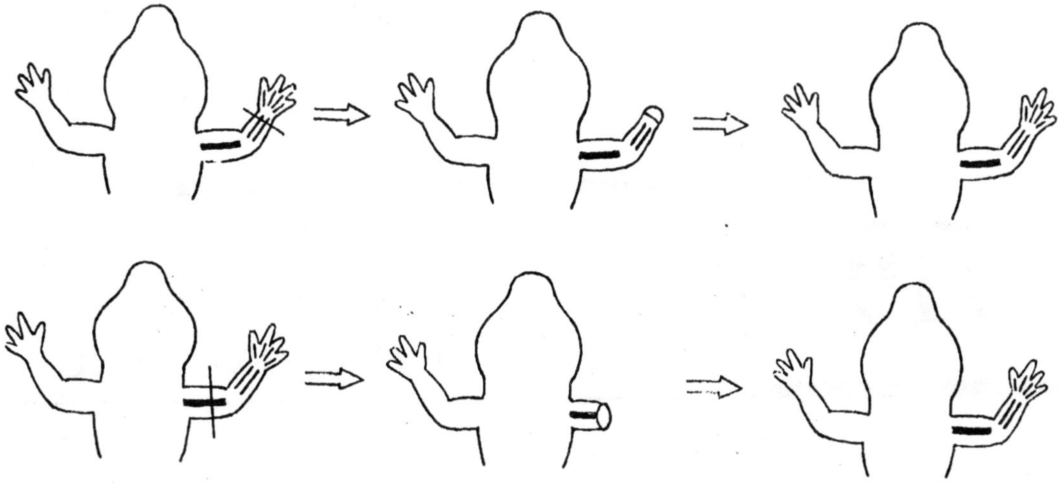

FIGURE 18.6

Regeneration in the distal direction irrespective of the level of amputation

Porifera. Regeneration ability is quite high in sponges. A sponge cut into several pieces, each piece grows into a complete sponge.

Cnidaria. Hydra regenerate through a morphalaxic process.

Platyhelminthes. Planaria regenerate through a combination of morphalaxic and epimorphic regeneration.

Annelida. The regenaration is somewhat restricted at the anterior end, since only a limited number of segments are regenerated. Such restriction is not present in the posterior end. The process of regeneration is epimorphic.

Arthropoda. Regeneration is limited to **autotomy** or loss of appendages.

Mollusca. Regeneration ability is quite high. The snails and slugs can even regenerate head or at least a greater part of it with all its components.

Echinodermata. Autotomy occurs in this group. A starfish can reconstruct its lost arm. A complete brittle star is regenerated from a broken arm.

⊠ Chordata

Urochordata. Under adverse condition dedifferentiation takes place, in some. The organs breakdown, lose their shape and in extreme condition become a mass of cells in a small ectodermal bag on return of favourable condition the cells rebuilt the organ by redifferentiation.

Fish. Regeneration is restricted only to the fins, except the caudal.

Amphibia. In tailed forms regeneration is quite high, e.g. Salamander, triton, etc. In tailless frogs and toads regeneration ability in adult is only slight.

Reptilia. In house lizard, *Hemidactylus* regeneration of tail takes place following autotomy.

Aves. Regeneration ability is poor. Healing of broken bone occurs; some birds are able to regenerate the beak; new feather develops if the old one is plucked off.

Mammalia. Regeneration is absent in limbs and other external parts. It is restricted to repairing of damaged skin, healing up of wounds and repairing broken bones. The salivary glands and liver regenerate to some extent.

Regeneration is quite high in majority of the lower animals. The ability to regenerate lost organ decreases with the animals moving up the upper rungs of evolutionary ladder.

19 BIOLOGICAL LUMINESCENCE: BIOLUMINESCENCE

Light emission can occur when a physical system undergoes a discrete change to free energy. The emission of light by organisms. Bioluminescence is an example of *chemoluminescence*. Bioluminescence is also termed as '*cold light*' to distinguish it from thermal luminescence.

In bioluminescence the energy released in a biological oxidation forces an electron of a molecule to a higher energy level. The oxidation energy serves the same purpose as the incident proton does in *photoluminescence*.

The emission of light by organic chemicals was first reported by French Physiologist Dubois from an edible luminous clam, *Pholas dactylus*, in 1987.

The luminescence is caused by a variety of organisms starting from bacteria, fungi (Fig. 19.1), protozoan (Fig. 19.2) to fishes (Fig. 19.3) in the animal kingdom.

Luminous organisms inhabit land (soil, vegetation, trees, etc.) and sea. In sea they are present from water surface to a great depth*. The burning sea observed by fishermen for centuries is due to the presence of a large number of dinoflagellates an sea surface, that laminate when disturbed. The only known fresh water luminous species is limpit (*Latia* sp.) of New Zealand.

The periodicity of light (flash) produced by luminous organism is extremely variable. It is of a short duration but repeated in a rhythmic fashion. The intensity of light varies from dim to bright. Bacteria omit a blue or blue-green light (wave length 480 to 500 nm), fungi (530 nm); larvae of South American rail road beetle give off a red light (640 nm) and firefly (562 nm).

☒ The luminous organisms

Fungi. *Panus* sp.

Protozoa. Dinoflagellate-Noctiluca sp., *Cenalium* sp., *Pyradinium* sp., *Gonyaulax* sp., etc.

Cnidaria. Jelley fish (Acquorea sp., *Campanularis* sp., Sea pen (*Pleurobrachia* sp., *Panatula* sp.)

Ctenophora. Comb jelly (*Hormiphora* sp.)

Annelida. *Odontosyllis* sp.

Arthropoda, Crustacea: *Cyprinida* sp.

* Some luminous bacteria live in symbiosis in squids and fishes. The host has a complicated luminous organ in which the light is supplied by the bacteria. The bacteria emits light continuously but the host has special device—such as a movable screen—that turns the light on and off.

FIGURE 19.1

Luminous toadstools

Insecta. Lantern fly (*Phturis* sp.), Glow worm (*Pyrophorus* sp.), Springtail (*Orchesella* sp.), Click beetle (*Pyrophorus* sp.).

Mollusca. Fresh water limpet (*Lotia* sp.), Squids (*Thaumeta* sp., *Heteroteuthis* sp.).

Echinodermata. Brittle star (*Ophioflurix* sp.), Sea lily (*Annachrinus* sp.), Crinoid (*Antedon* sp.).

Fish. *Photoblepharus* sp., Lantern fish-*Diephus* sp., Angler fish-*Linophryne* sp., Hatchet fish-*Argyropelecus* sp., Goase fish-*Lophius* sp., etc.

⊠ Physiology and Biochemistry of luminescence

Luminescence in living organisms is caused by an heat-stable oxidizable substance—*luciferin*. In the presence of *oxygen* and an enzyme *luciferase* it is destroyed by an oxidative reaction, which gives out light, luciferin is not a single compound common to all luminous organisms—apparently there are many different luciferius.

Luminescence are of two types—intracellular and extracellular.

FIGURE 19.2

***Noctiluca scintillans*.** Source of luminescence in the sea. Inset-*Noctiluca* magnified

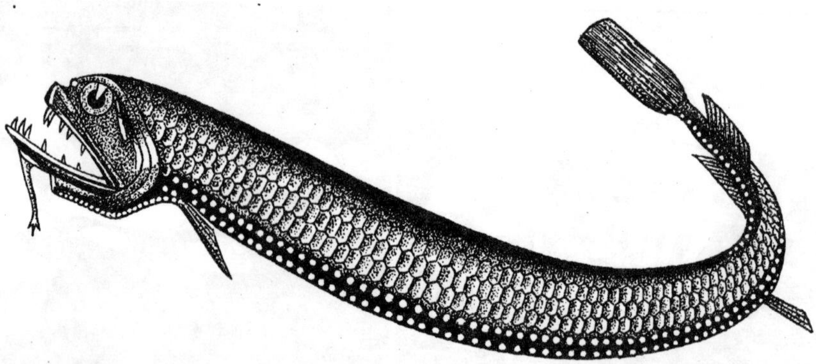

FIGURE 19.3

Stomias boa (deep sea fish). White dots one luminous organ

Intracellular luminescence. Produced from luminescent glands or *Photophores*—deep sea fishes.

Extracellular luminescence. Produced from unicellular or multicellular structure restricted to definite regions of the body—present in many organisms.

The flashes of luminescent organisms are controlled by nervous stimulation. The flash is probably triggered by a nerve impulse received by the luminous gland. A sequence chemical reactions then produces light. The chemicals involved are *ATP* (Adenosine triphosphate + LH_2 (Luciferin) + O_2 (Oxygen) $\rightarrow$ *AMP* (Adenosine monophosphate) + *L* = *O* (Oxidized luciferin) + *PP* (Inorganic pyrophosphate) + H_2O (Water) + *Light*. The reaction is catalyzed by the enzyme luciferase.

☒ Location of light organs

Luminescent structures occur in a variety of locations in an animal body—skin, ventral musculature, abdomen and other parts. In some cases they are distributed all over the body.

☒ Importance of luminescence

The ability to produce light plays an important role in the life of these organisms. In the depths of the ocean, about 750 meters and below there is no light. Organisms living there cannot see one another or other objects in the dark. Many organisms living there are capable of producing light (Fig. 19.3) So the depths of the ocean are full of light moving rhythmically around and continually turning off and on. The predators can identify the prey and the prey also can escape from the predator.

1. **Social signal. Lantern fishes** have a definite pattern of light organs varying from species to species by which they can identify members of their own and other species.
2. **Feeding**

a. The deep sea **Angler fish** bears a luminous organ (Fig. 19.4) on the tip of a retractile rod-like appendage on its head. The light lures preys into its jaws.

b. **Goose fishes** use their **illicium** (the light at the end of the first dorsal fin ray) to lure food animals.

3. **Defence.** Some deep sea squids spart a luminous cloud; Hatchet fish ejects luminescent cloud from the ventral photophores. The act help them in countershading and blinding the pursuing predator.
4. **Mating.** The reproductive cycle of many marine and land organisms are intimately tied to light emission.

a. In the sea off Bermuda the females of a species of marine annelid *Odontosyllis* sp. come to the surface three days following a full moon and secrete a glowing circle of luminous substance. The males move up from depth below the sea, emit puffs of light, straight to the circle of light and both eggs and sperms are discharged into the water.

b. The light of the firefly is a mating device to attract member of the opposite sex is universally accepted. The advertising flashing light of a flying male signals its presence to a responsively flashing female waiting in the grass.
c. In Myanmar (Burma) and Thailand all the fireflies of one tree may flash simultaneously, while these on another tree at some distance away may also flash in unison. The flashing of the fireflies in two different trees are not simultaneous but out of step with those on the first tree. It is presumed that all the fireflies of one tree are males and those of the other tree are females. However, this concept—flushing out of step in the two trees—a mating device is yet to be accepted by many workers.

Many different chemical types of bioluminescence reaction suggest that they occurred independently during evoluation and are an expression of the reverse process of photosynthesis. In photosynthesis organic molecules are produced by various reduction reactions and oxygen is evolved. In bioluminescence oxygen decomposes organic molecules in a oxidative process that leads to an excited state and light emission. Life appeared and evolved by using excited states, as plants are capable of carrying out photosynthesis.

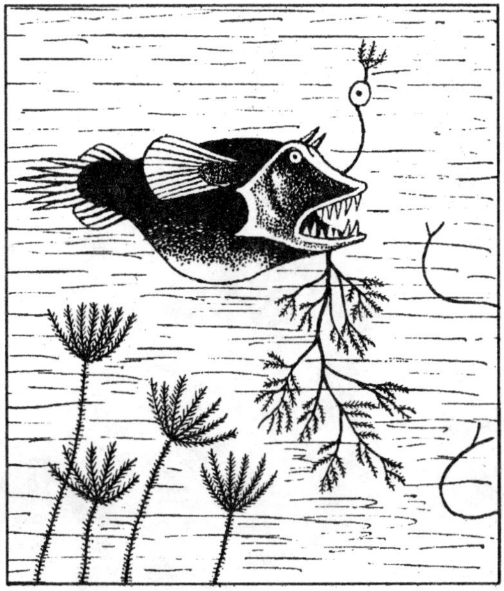

FIGURE 19.4

Angler fish of deep sea. White portions with luminous organ

Many more light-emitting organisms might have existed during the early evolution of life, than at the present time. Bioluminescence is possibly losing its importance as the organisms evolve effective means of utilizing the energy of the excited state. The utility of light emission for specific purpose may thus have been secondary in the evolutionary process.

PART TWO

DIVERSITY IN ANIMAL FORMS

20

WHO'S WHO AMONG THE ANIMALS

The planet earth is the home of numerous kinds of animals and their number is countless. Classification of animals and plants is an attempt to bring some sort of order in the vast population of living organisms, which would help a methodical study of the living world (Fig. 20.1).* Classification and naming of animals and plants go hand in hand and they cannot be viewed separately.

BINOMIAL NOMENCLATURE

Naming a kind of animal/plant is to put an identity tag on them. A man is identified by his first name, the surname indicates the family of his origin. A third name, if any, is just an ornament. A somewhat similar method has been adopted in naming animals/plants. An animal is known by different names in different countries and even in different areas of the same country. Human beings are named after the name of the country of their birth, where their ancestors lived; Indian (a native of India), British or English (England), French (France) and so on. A country again may have many clans or tribes. The human race has several distinct elements, viz. Nordic, Negrito, Mongoloid, Mongolo Dravidian, etc. They differ in many features but a close study reveals that their resemblances are greater than the differences and the human race is a single group, the *Homo sapiens* to a biologist. [*Sapiens* is the smallest group, a species, which includes all humans of today and *Homo* is the next large group, the genus.]

A name, designated as the specific name (**species**), is given to a kind of animal/plant for the convenience of study but the naming is not arbitrary and is based on strong evidences. In the old system, known as **artifical system**. Placing of an animal in its proper systematic position is determined chiefly on the basis of morphological, anatomical and histological, characters and, to same extent, on embryological and palaeontological evidences. The method is also known as **Pigeon hole system**. Dependence on incomplete information led to many confusions and quite a number of animals were misplaced. *Microfilaria* sp. an embryo stage of *Wuchenria* sp. (Aschelminthes) was considered a distinct species before its developmental history was known. There are many such examples in the living world.

In the **natural system** of classification of the present day, different characters—morphological, anatomical, histological and evidences from histology, embryology, palaeontology, genetics, physiology,

* Figures 20.1 to 20.26 at the end of the chapter.

biochemistry and others—are carefully considered before placing an animal in a particular species. Statistical analysis of the available data are carried in most cases.

But the attempt has not met with complete success in many cases. Even today, the artificial system of classification has not lost its importance and the systematists (those who study Taxonomy or classification) are forced to use this method for lack of required information. Palaeontological records of many animals are missing; we have no knowledge about the embryology of many, nor can we breed all known forms. Information on functions of organ-systems and their biochemical constituents are wanting in many cases. All these are strong hurdles in the way of full implementation of the natural system of classification.

The scientific name (Linnaeal name) given to an animal is not for the benefit of the common people but for the use of the scientific community, to identify an animal in any part of the globe, irrespective of its habital and local name. As *Homo sapiens* includes the total human population on earth, so *Columba* sp. embraces all the pigeons of the world and *Gallus* sp. includes all the fowls.

In naming, two names are assigned to one kind of animal. The first one is the generic name, the name of a genus and the second one is the specific name or the name of the species. The names are always written in *italics;* the first letter of the generic name is capital and the specific name—with a few rare exceptions — is written in small hands. This sort of double naming—known as **binomial nomenclature** —was introduced by a Swedish naturalist Linnaeus.

A **Species** is the smallest division in the classification of the animal kingdom (plant kingdom also). The members of a species resemble one another in great details and interbreed, producing fertile offsprings. The **gene pool** of all the members of a species is similar and there should be no gene flow from one species to another, i.e. a species is a **genetically closed system**. But, both in nature and through human interference, it has been possible to get offsprings by crossing different species and even different genera. The offsprings thus produced, if not fully, are greatly sterile, i.e. they cannot produce offsprings.

A few closely allied *species* together constitute a **genus**. A genus may be quite big—embracing a large number of species showing considerable diversity but not attaining the generic level, and is subdivided into a number of **subgenus**. The subgenus is written within parenthesis immediately after genus and in the style adopted for genus. Similarly, a species may contain a number of subspecies written immediately after the species, following the style for species and no parenthesis is used. Giving of three names—genus, species and subspecies—is known as **trinomial nomenclature**. The name of the author who first described the species is written after the species in unitalicised, Roman letters.

As a rule, a law of priority* is strictly followed in nomenclature. The name of a species or a genus can be changed only on the proof that the group was described previously under a different name and that too after following certain norms. The genus *Nereis* sp. has been changed to *Neanthes* sp. and *Ophicephalus* sp. to *Channa* sp. The name of the author is changed if recorded evidence is found that the species was described by

* *Law of Priority.* Priority is the basic principle of Zoological nomenclature. The International code at Zoological Nomenclature is the system of rules and recommendations authorized by the International Congress of Zoology. The object of the code is to promote stability and universality in the naming of animals. The purpose is to ensure that each name is unique and distinct.

Valid Name. The oldest available name of a Taxon is the valid name, provided it is not invalidated by any provision of the Code or not suppressed by the International commission of Zoological Nomenclature.

International Code of Zoological Nomenclature adapted by the International Congress of Zoology, London, July, 1958. Official Copy available: Trust for zoological Nomenclature. 14, Belgrave Square. London S.W.I.U.K.

another person prior to that of the present author. The credit is given to the original author by the law of priority. For quite a long time it was held, Ferussac first described the giant African land snail *Achatina fulica* Ferussac. Subsequently, it was found that the description of the same species recorded by Bowdich was older than that of Ferussac and Bowdich replaced Ferussac, *Achatina fulica* Bowdich.

While describing a new species it is mandatory for the author to select a typical specimen, the **holotype**, which is to be preserved for future reference. Other individuals of the species are **paratypes**. The same rule is applicable for a genus or a family.

PRINCIPLES OF CLASSIFICATION

Classification of animals is a stepwise system. The smallest division, the species occupies the bottom of the hieararchy and the animal kingdom or **Kingdom Animalia**, embracing all animals, is at the top. There are quite a few steps in between the two. A larger division is created with a number of closely allied smaller divisions. In each step, the number of animals in a division increases with an increase in diversity.

A few **species** constitute a **genus** (singular), a few genera (plural) make a **family**, a few families an **order**, a few orders a **class**, a few classes a **phylum** and several phyla the **subkingdom**; two subkingdoms form the animal kingdom. According to need subdivisions are interposed between two divisions, viz. **subphylum**, **superclass**, **subclass, superorder, suborder, superfamily** and **subfamily**. Further subdivisions like **supra** and **infra** are interposed, if necessary. The systematic position of two species are given below:

The giant African land snail

Kingdom	**Animalia**
Subkingdom	**Metazoa**
Phylum	**Mollusca**
Class	**Gastropoda**
Subclass	**Pulmonata**
Order	**Stylommatophora**
Family	**Achatinidae**
Subfamily	**Achatininae**
Genus	*Achatina* sp.
Subgenus	*Achatina* (*Lisachatina*) sp.
Species	*Achativa* (*Lisachatina*) *fulica* sp.
subspecies	*Achatina* (*Lisachatina*) *fulica fulica* Bowdich

The dog

Kingdom	**Animalia**
Subkingdom	**Metazoa**
Phylum	**Chordata**
Subphylum	**Vertebrata**
Superclass	**Gnathostomata**
Class	**Mammalia**
Subclass	**Theria**
Infraclass	**Eutheria**
Order	**Carnivora**
Suborder	**Fissipedia**
Superfamily	**Canoidea** or **Arteloidea**
Family	**Canidae**
Genus	*Canis* sp.
Species	*Canis familiaris* **Linn.***

☒ Subdivisions of the kingdom Animalia

The kingdom Animalia has been divided into two subkingdoms, **Protozoa** and **Metazoa**. Certain metazoans have been placed in a special category for their peculiarities. They are commonly known as **aberrant groups**.

☒ Kingdom Animalia

Animals are charactarised by the presence of locomotory structures in the majority; growth limited and not localised; inability to manufacture food; cells enclosed in a plasma membrane.

Subkingdom **Protozoa****

Small, unicellular (acellular); (Fig. 20.2) free-living or parasitic; terrestrial, fresh water, marine; locomotory organelles pseudopodia, flagella, cilia; reproduction both asexual and sexual.

---

* Linn. is the abbreviation of Linnaeus.

** For examples see detailed classification.

Phylum **Sarcomastigophora**

Locomotory organelles flagella, pseudopodia; nucleus single type; sexuality essentially syngamy.

Phylum **Labrynthomorpha**

Marine or estuarine; trophozoite stage; cytoplasmic network with spindle-shaped or spherical nonamoeboid cells; zoospores heterokont.

Phylum **Apicomplexa**

Apical complex present; cilia absent; syngamy occurs; all parasitic.

Phylum **Microspora**

Spores unicellular with imperforate wall; mitochondria absent; obligatory intracellular parasites.

Phylum **Ascetospora**

Spores multicellular; sporoplasm one or more; polar capsules or polar filaments absent; parasitic.

Phylum **Myxozoa**

Spores of multicellular origin with one or more polar capsules and sporoplasm; with 1, 2, 3 (rarely more) valves; parasitic.

Phylum **Ciliophora**

Simple cilia or compound ciliary organelles; nuclei two types; binary fission transverse; budding and multiple fission occur.

Subkingdom **Metazoa**

Multicellular; diplo- or triploblastic; acoelomate or coelomate; notochord and vertebral column may or may not be present.

Phylum **Porifera**

Sedentary (Fig. 20.3) aquatic, mostly marine; diploblastic; a single cavity within the body often forming a canal system; skeleton made of spongin fibres and siliceous or calcic spicules; reproduction by gemmation or sexual; larvae free swimming.

Phylum **Cnidaria**

Diploblastic; radially symmetrical (Fig. 20.4); presence of a coelenteron with a single aperture; cnidocytes present; reproduction both asexual and sexual.

Phylum **Ctenophora**

Radially symmetrical (Fig. 20.5); mesoderm present; coelenteron with a single aperture; cnidocytes absent; tentacles two and retractile; sense organ single, in the aboral pole; hermaphroditic.

Phylum **Platyhelminthes**

Bilaterally symmetrical; devoid of exoskeletal structures (Fig. 20.6); triploblastic; coelom absent; gut when present, bifid; excretion by flame cells; hermaphrodite with complex reproductive organs; both parasitic and free-living.

Phylum **Aschelminthes**

Triploblastic with a pseudocoel; body covered with a proteinous cuticle (Fig. 20.7); alimentary canal free with ectodermal fore and hindgut and endodermal midgut; sexes separate; female organs paired, male organ single; both parasitic and free-living.

Phylum **Annelida**

Body elongated; segmentation metameric (Fig. 20.8); locomotory organs setae or parapodia; head consists of a preoral prostomium and a postoral peristomium; alimentary canal straight with distinct zones; mouth and anus opposite; true coelom present; vascular system closed; excretion by segmental nephridia, nervous system well-developed; development direct or with a trochophore larva.

Phylum **Arthropoda**

Bilaterally symmetrical; appendages jointed; chitinous cuticle with flexible joints form exoskeleton (Fig. 20.9), haemocoel functions as vascular system; heart dorsal to gut; respiratory organs gills or tracheae or booklungs; sexes separate; metamorphosis often present.

Phylum **Mollusca**

Bilaterally symmetrical, at least in early stages (Fig. 20.10); an external shell often present; a mantle encloses the body; a radula present in the majority; coelom restricted to pericardium; respiratory organs—gills or lung or both gills and pulmonary sac; larvae—either trochophore or veliger.

Phylum **Echinodermata**

Coelomate, marine, radially symmetrical; exoskeleton of calcareous plates or ossicles with movable or immovable calcareous spines (Fig. 20.11); coelom converted into a water vascular system; alimentary and nervous system well-developed; larvae bilaterally symmetrical and free-swimming.

☒ Aberrant groups

Certain metazoons neither fit well in the hierarchy of invertebrates nor do they possess notochord. Since not much is known about their relationship with other groups, it is customary to assign them a special status—the aberrant groups. Based on the absence or presence of coelom they have been divided into two major groups (Fig. 20.12).

☒ Acoelomate groups (Fig. 20.12)

Coelom absent.

Phylum **Calyssozoa (Entoprocta)**

Sedentary; solitary or colonial; mostly marine; body cup-shaped, stalked with a circlet of tentacles; epistome overhangs the mouth. Examples: *Pedicellaria, Myosoma, Lexosoma*, etc.

Phylum **Rotifera (wheel animalcule)**

Microscopic; a retractile trochal disc at the anterior end; mastax present in pharynx; body cavity large; excretory organs a pair of flame cells; sexes separate. Examples: *Rotaria, Macrochaetus, Asplanchna, Ascomorpha, Philodina*, etc.

Phylum **Nematomorpha**

Body long, thread-like; both fresh water and marine; a pharyngeal ring constitutes nervous system; sexual stage free-living, asexual stage parasitic. Examples: *Gorchius, Nectonema*, etc.

Phylum **Acanthocephala**

Body cylindrical; proboscis at the anterior end; digestive system, excretory pore absent; no sense organs; intestinal parasites of vertebrates. Examples: *Echinorynchus, Acanthogyrus*, etc.

☒ Coelomate groups (Fig. 20.13)

Coelom present.

Phylum **Phoronida**

Body cylindrical, worm-like, unsegmented; enclosed in a membranous tube; horseshoe-shaped lophophore bears a crown of numerous ciliated tentacles; mouth and anus separated by a short space; nephridia two; Examples: *Phoronis, Phoronopsis*, etc.

Phylum **Brachiopoda**

Body enclosed in a shell with two valves—dorsal and ventral; lophophore large and complex; mouth crescentic; anus may be absent; sexes separate; marine. Examples: *Lingula, Cranin, Magellania*, etc.

Phylum **Chaetognatha**

Body small, slender, nearly cylindrical, transparent; head, trunk and tail distinct; locomotory organs caudal and horizontal lateral fines; marine. Examples: *Sagitta, Spadella*, etc.

Phylum **Bryozoa**

Branched, colonial, minute individuals located in separate chambers; ciliated tentacles around the mouth, alimentary canal U-shaped. Examples: *Plumatella, Bugula*, etc.

Phylum **Pogonophora**

Tube dwellers, inhabiting deep sea; body bilaterally symmetrical with an anterior protosome bearing one to many tentacles; a short mesosome and a long metasome or trunk with rings or adhesive papillae; no mouth, digestive tract and anus in the adult; coelom unlined, that of protosome extending into tentacles, the mesosome extending into tentacles, the mesosome and trunk, each with a pair of compartments; circulatory system closed, with a middorsal and a midventral vessel; vertical vessel enlarged in protosome as a heart; coelomoducts, two, in the protosome, opening to the exterior; nervous system in epidermis, enlarged into cephalic lobe from which arise tentacular nerves and a middorsal unpaired nerve cord; sexes separate, gonads paired. Example: *Lamellisabella*.

Phylum **Hemichordata***

Worm-like body with distinct proboscis, collar and trunk (Fig. 20.14); gill slits at the

* For examples see detailed classification

anterior end of the trunk; coelomic pouches, notochord, dorsal nerve cord present; marine, burrow dwellers.

Phylum **Chordata***

Bilaterally symmetrical triploblastic, coelomate; notochord at some stage of life; vertebral column in the majority; pharyngeal gill clefts either throughout life or in the embryo; dorsal spinal cord hollow; neurocoel present in central nervous system.

Subphylum **Urochordata**

Sedentary, enclosed in a cuticular test (Fig. 20.15); dorsal nervous system, notochord; gill slits only in larval stage; heart tubular, ventral; metamorphosis retrogressive.

Subphylum **Acrania (Cephalochordata)**

Body wall with over fifty myotomes; gill slits numerous (Fig. 20.16); notochord persistent; hollow nerve cord dorsal; larval form asymmetrical.

Subphylum **Craniata (Vertebrata)**

Presence of endoskeleton; vertebral column present; a cranium present; paired lateral appendages; a posterier tail usually present; vascular system closed.

Superclass **Agnatha**

Mouth round, suctorial, without jaws; paired fins absent; fins without fin rays (Fig. 20.17); notochord persistent; nasal organ unpaired; conus arteriosus, spleen, distinct pancreas, sympathetic nervous system, genital ducts absent.

Superclass **Gnathostomata**

Limbs two pairs; mouth with upper and lower jaws.

Class **Placodermi (Aphetohyoidea).** Extinct.

Primitive forms; hyoid arch acts as branchial arch; jaws attached to cranium by ligament (Figs. 20.18).

Class **Chondrichthyes (Elasmobranchii)**

Dermal scales placoid; fins with fin rays; gill slits open directly to the exterior; no operculum; skeleton cartilaginous; spiral valve in intestine (Fig. 20.19).

Class **Osteichthyes (Teleostomi)**

Bony fishes; dermal scales cycloid in the majority, ctenoid in some, cosmoid in others, fins with fin rays; operculum present; gill filaments attached to gill arches; lungs present in some in addition to gills; skull hyostylic; jaw suspended by mandibular arch.

Subclass **Actinopterygii**

Scales reduced to thin bony structures; the bony or cartilaginous supports of the fin reduced; fins outside the body wall supported by fin rays alone; dorsal fin single; tail usually homocercal; bulbous arteriosus prominent and elastic; pineal apparatus rarely occurs (Fig. 20.20).

Subclass **Crossopterygii**

Dermal scales cosmoid; pectoral fin supported by jointed median axis; caudal fin with epichordal lobe; nostrils within the mouth; lung present (Fig. 20.21).

Class **Amphibia**

Occipital condyles two; heart with a sinus venosus, two atria, one ventricle and conus arteriosus; gills in larva; lungs in the adult; metamorphosis present (Fig. 20.22).

Class **Reptilia**

Scales epidermal; occipital condyle one; heart with sinus venosus, two atria, one partially divided ventricle; land eggs large with calcareous shell; amnion and allantois present (Fig. 20.23).

Class **Aves**

Warm blooded, oviparous, feathered biped; forelimbs modified to wings (Fig. 20.24) pectoral muscles large and used in flight; air sace present; single occipital condyle; skull light without sutures; teeth absent.

Class **Mammalia**

Body covered with hair, at least in the early stage; mammary glands present (Figs. 20.25 and 20.26); diaphragm present; occipital condyles two; teeth heterodont, thecodont; viviparous; placenta for foetal nourishment.

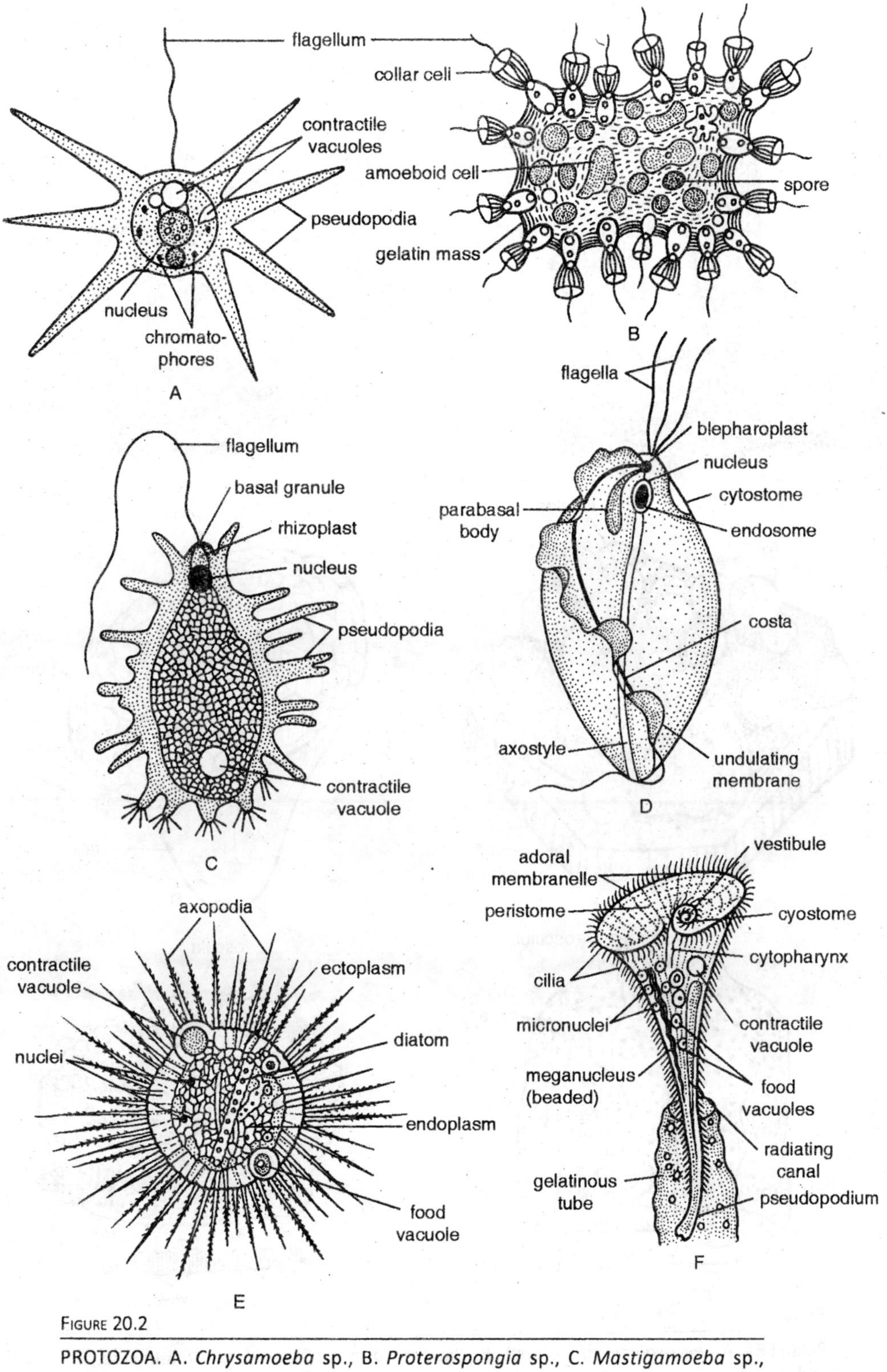

FIGURE 20.2

PROTOZOA. A. *Chrysamoeba* sp., B. *Proterospongia* sp., C. *Mastigamoeba* sp., D. *Trichomonas* sp., E. *Actinospherium* sp., F. *Stentor* sp.

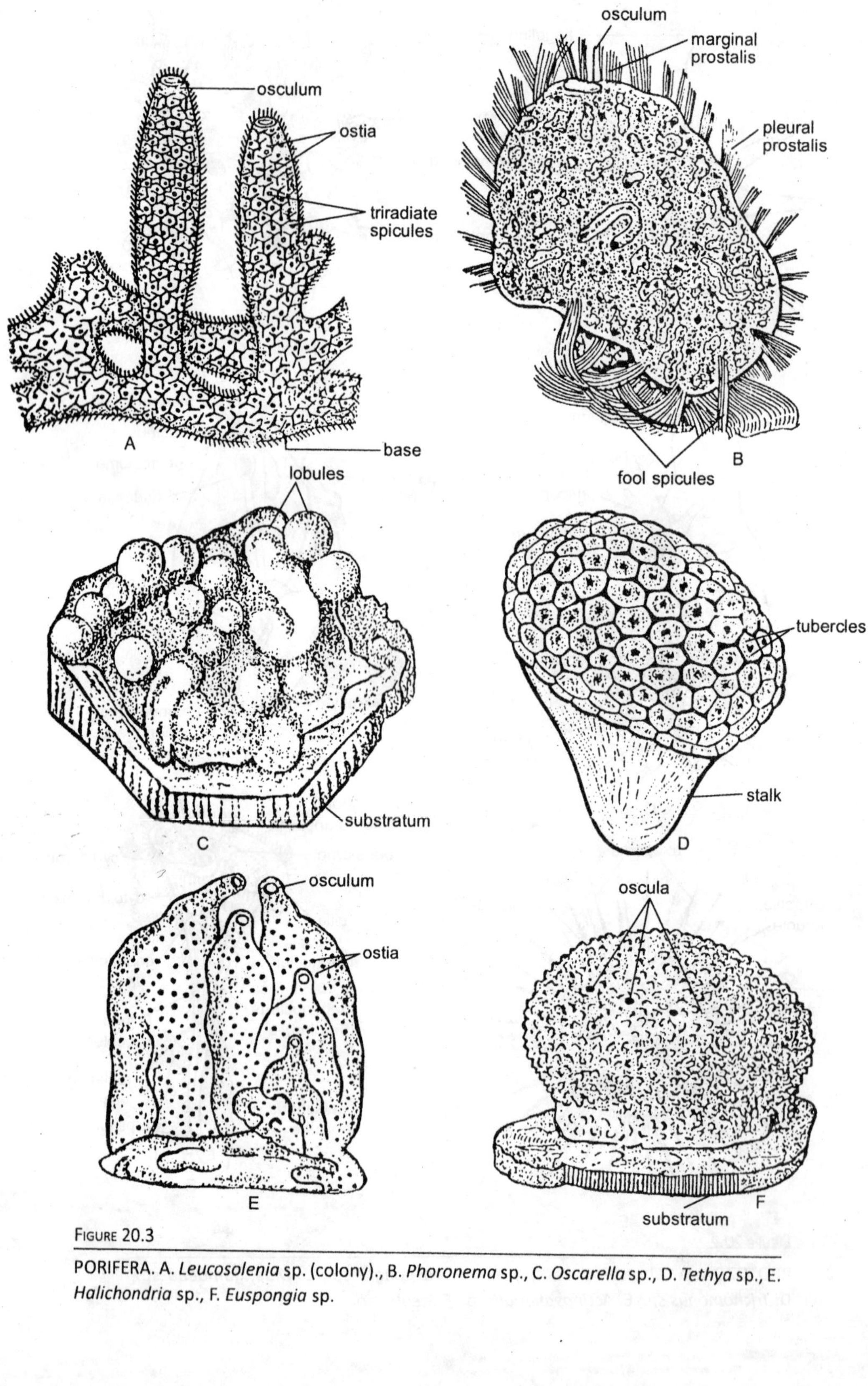

FIGURE 20.3

PORIFERA. A. *Leucosolenia* sp. (colony)., B. *Phoronema* sp., C. *Oscarella* sp., D. *Tethya* sp., E. *Halichondria* sp., F. *Euspongia* sp.

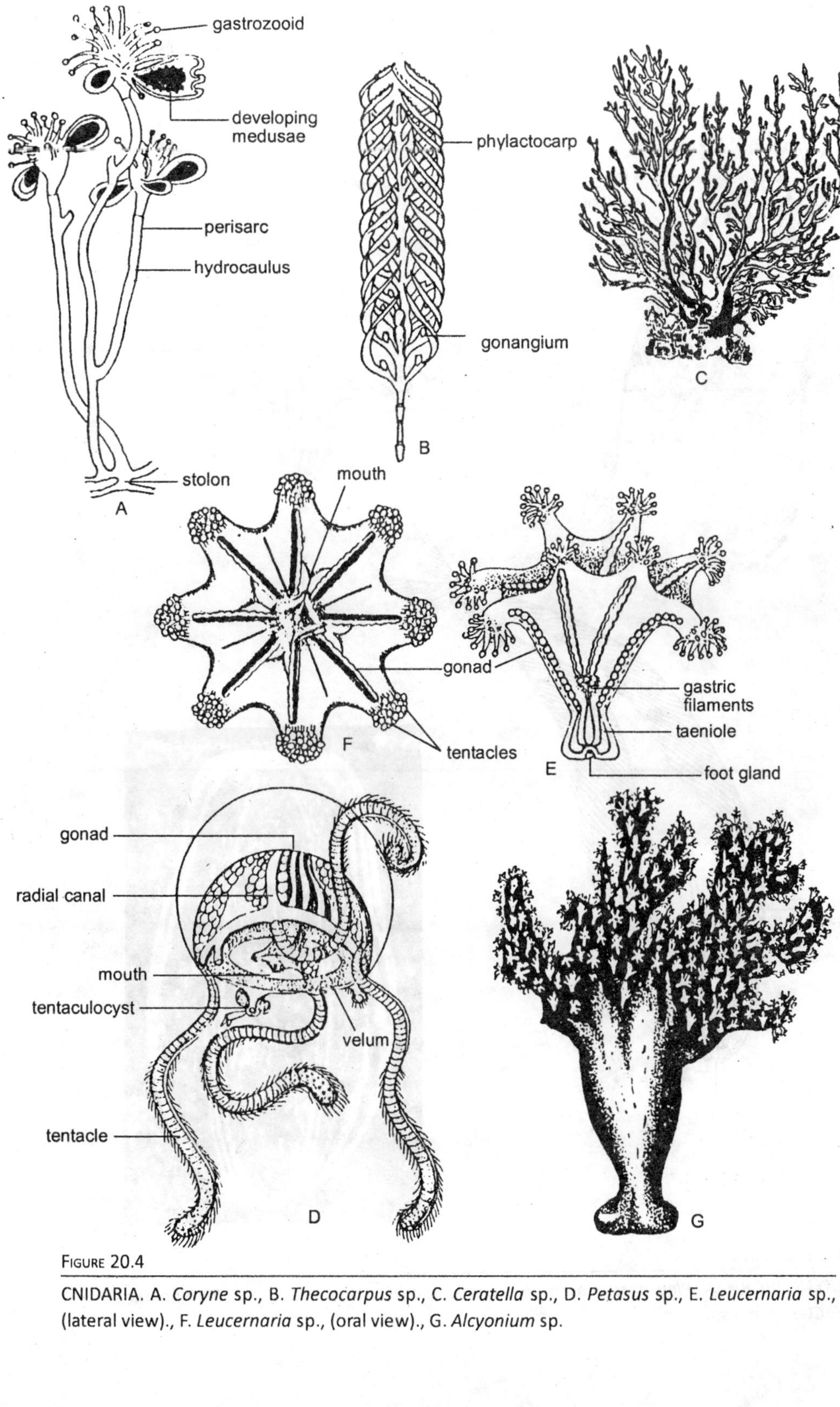

FIGURE 20.4

CNIDARIA. A. *Coryne* sp., B. *Thecocarpus* sp., C. *Ceratella* sp., D. *Petasus* sp., E. *Leucernaria* sp., (lateral view)., F. *Leucernaria* sp., (oral view)., G. *Alcyonium* sp.

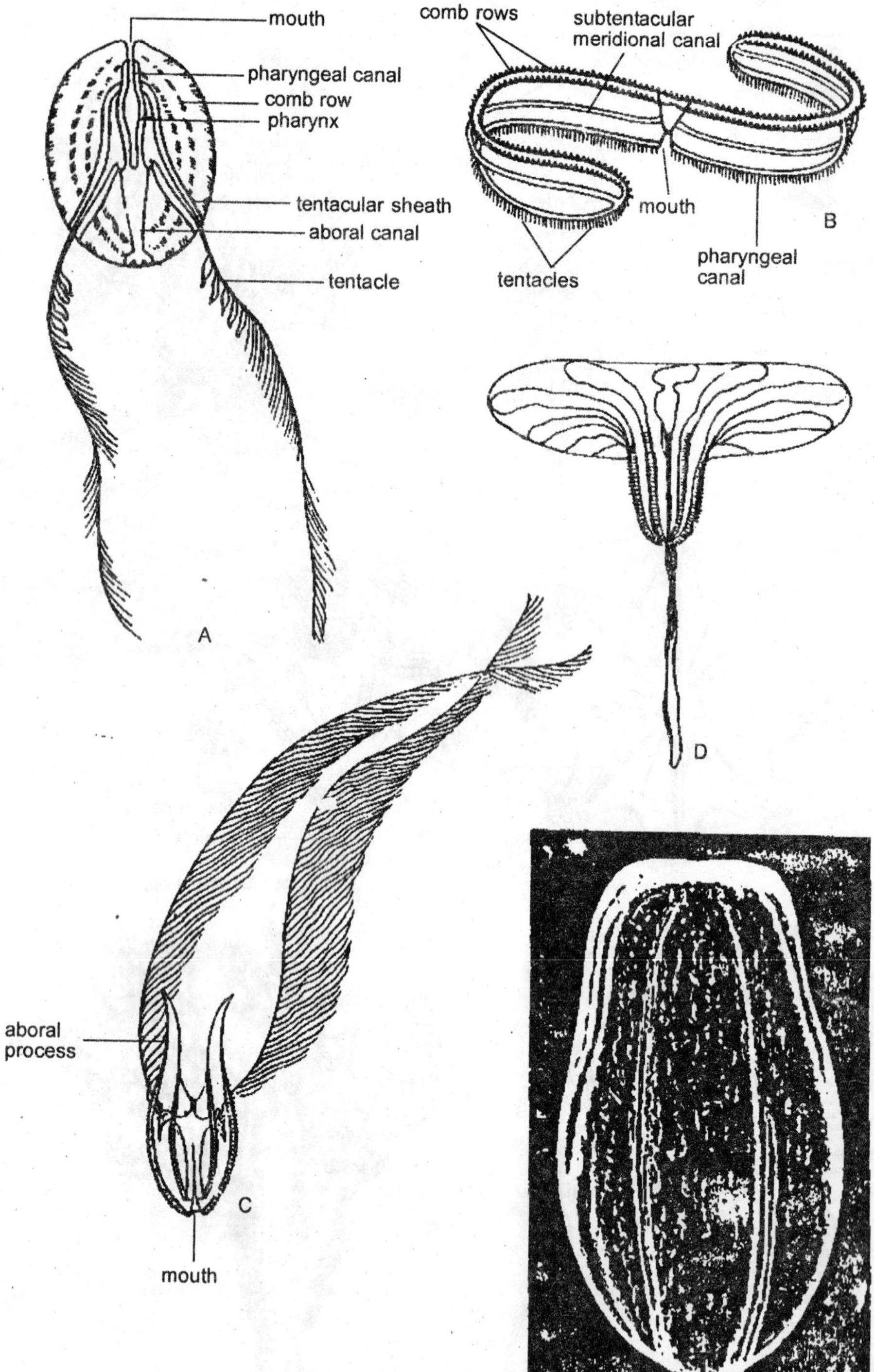

FIGURE 20.5

CTENOPHORA. A. *Pleurobrachia* sp., B. *Velamen* sp., C. *Callianira* sp., D. *Eulampetia* sp., E. *Ctenoplana* sp.

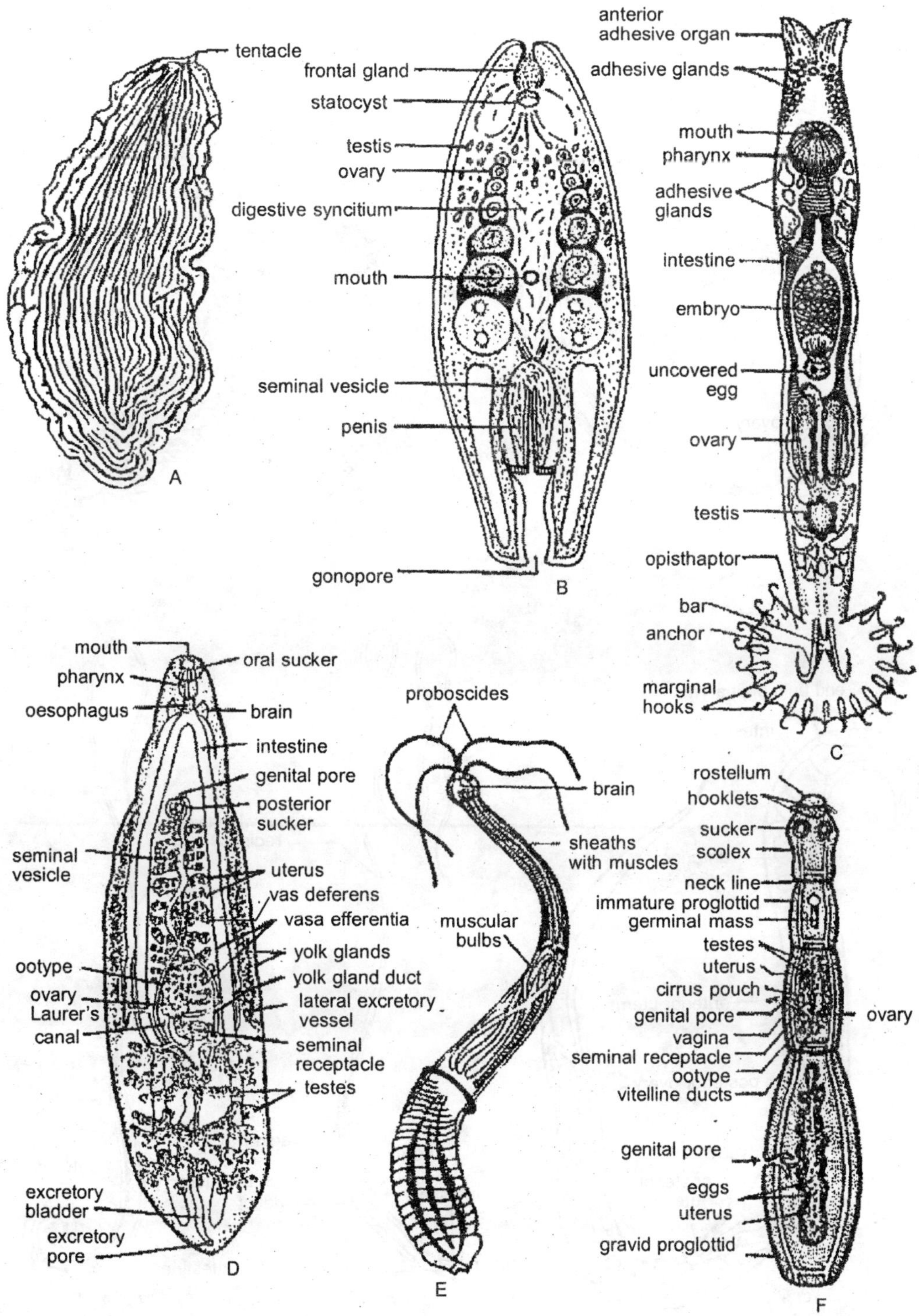

FIGURE 20.6

PLATYHELMINTHES. A. *Prosthaceracus* sp., B. *Pseudophanostoma* sp., C. *Gyrodactylus* sp., D. *Opisthorchis* sp., E. *Tetrarhynchus* sp., F. *Echinococcus* sp.

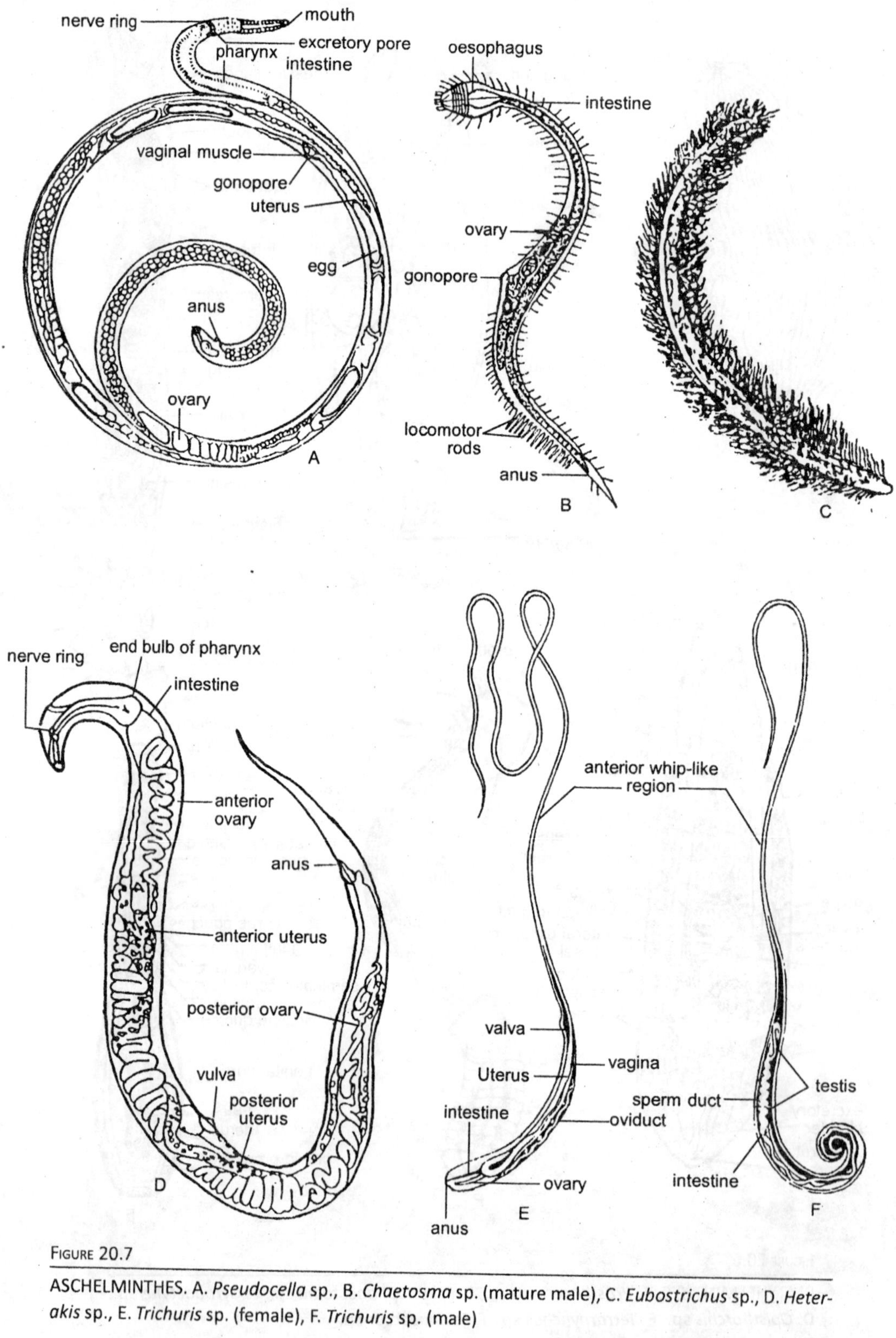

FIGURE 20.7

ASCHELMINTHES. A. *Pseudocella* sp., B. *Chaetosma* sp. (mature male), C. *Eubostrichus* sp., D. *Heterakis* sp., E. *Trichuris* sp. (female), F. *Trichuris* sp. (male)

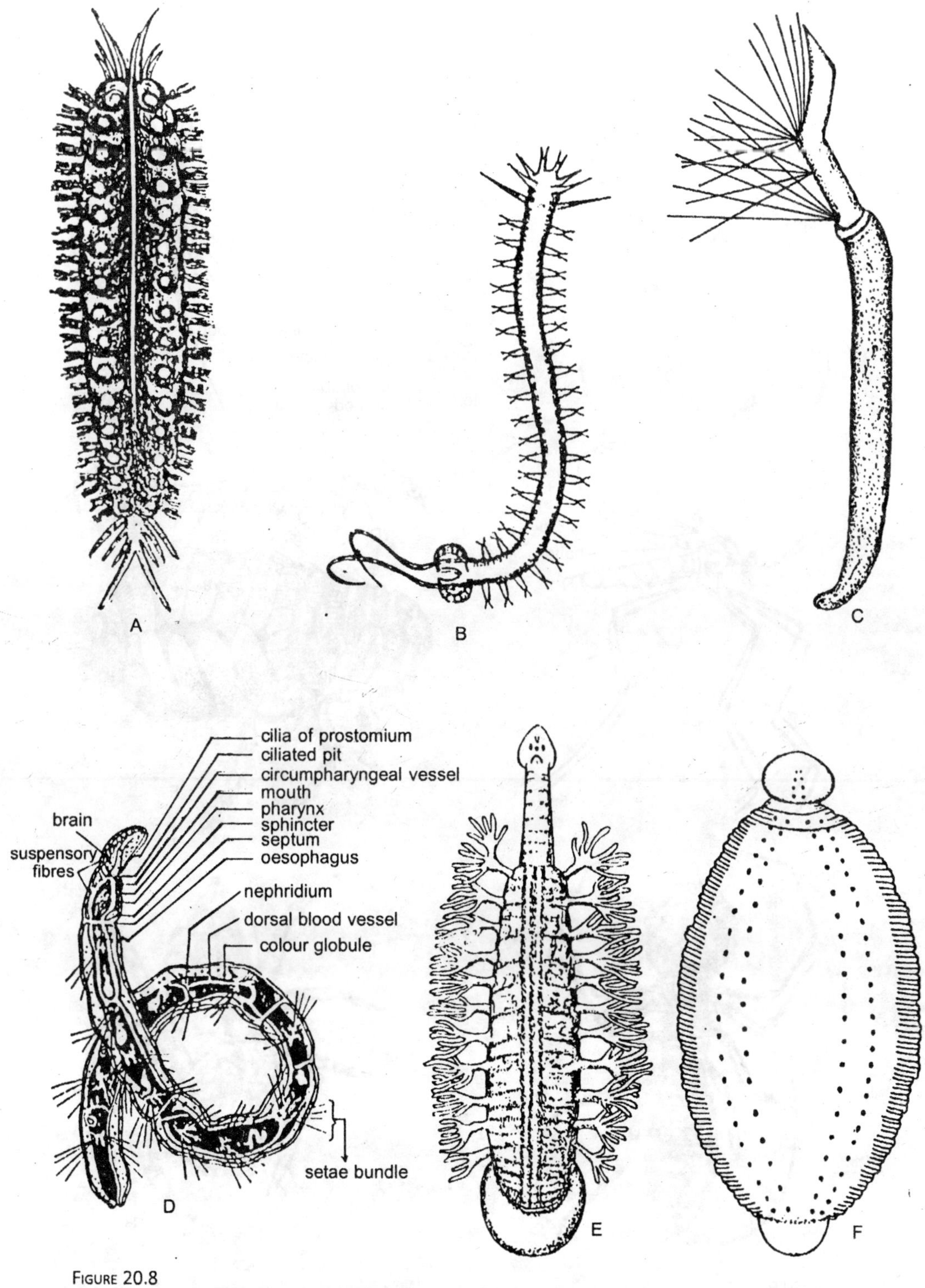

FIGURE 20.8

ANNELIDA. A. *Lagisca* sp., B. *Hesionides* sp., C. *Ripistis* sp., D. *Aelosoma* sp., E. *Ozobranchus* sp., F. *Theromyzon* sp.

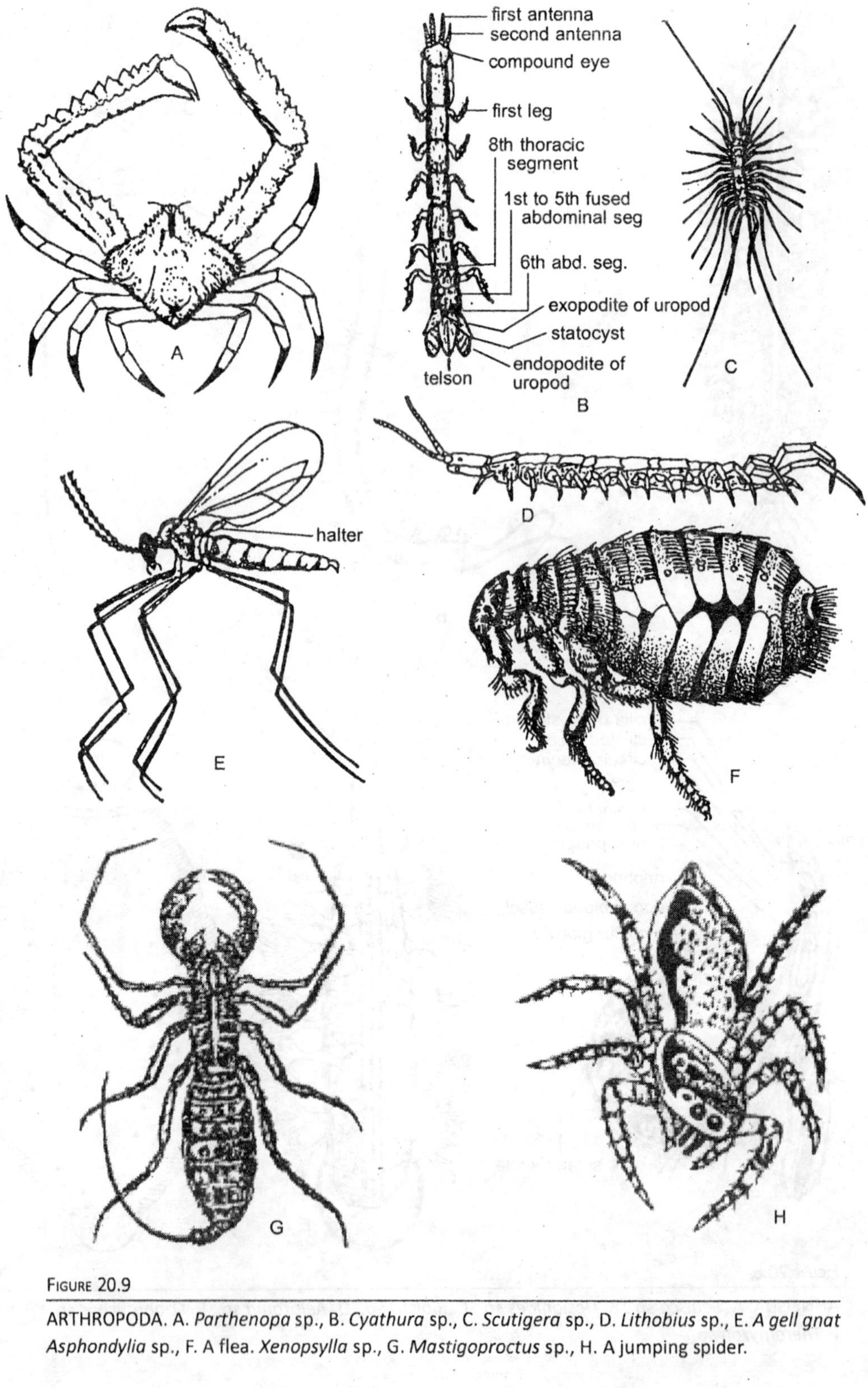

FIGURE 20.9

ARTHROPODA. A. *Parthenopa* sp., B. *Cyathura* sp., C. *Scutigera* sp., D. *Lithobius* sp., E. *A gell gnat Asphondylia* sp., F. A flea. *Xenopsylla* sp., G. *Mastigoproctus* sp., H. A jumping spider.

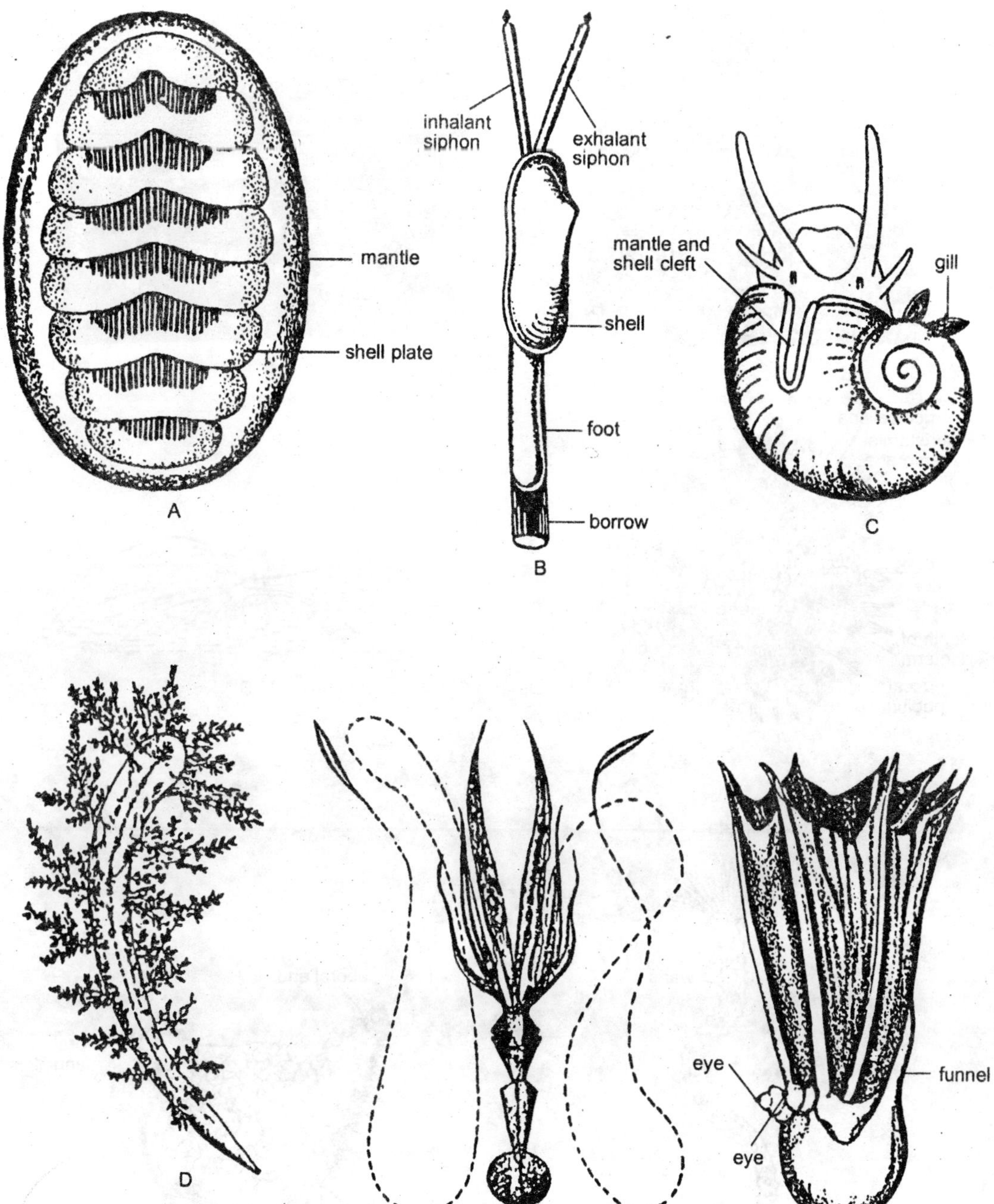

FIGURE 20.10

MOLLUSCA. A. *Chaetopleura* sp., B. *Tagelus* sp., C. *Scissurella* sp., D. *Dendronotus* sp., E. *Chiroteuthis* sp., F. *Amphitretus* sp.

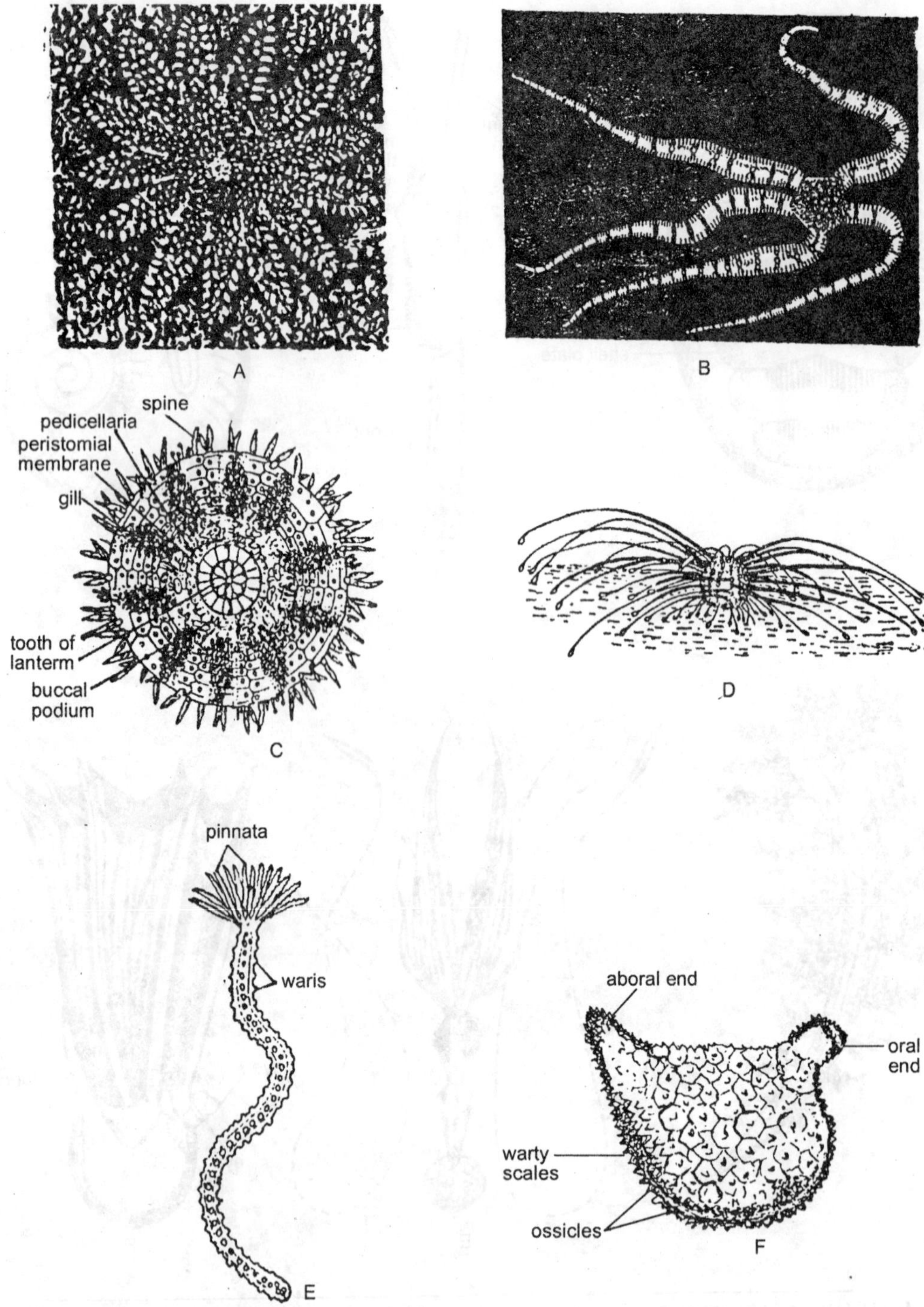

FIGURE 20.11

ECHINODERMATA. A. *Crossaster* sp. (aboral view). B. *Ophionereis* sp. (aboral view). C. *Arbacia* sp. (oral view). D. *Plesiodiodema* sp. E. *Leptosynapta* sp. F. *Sphaerothuria* sp.

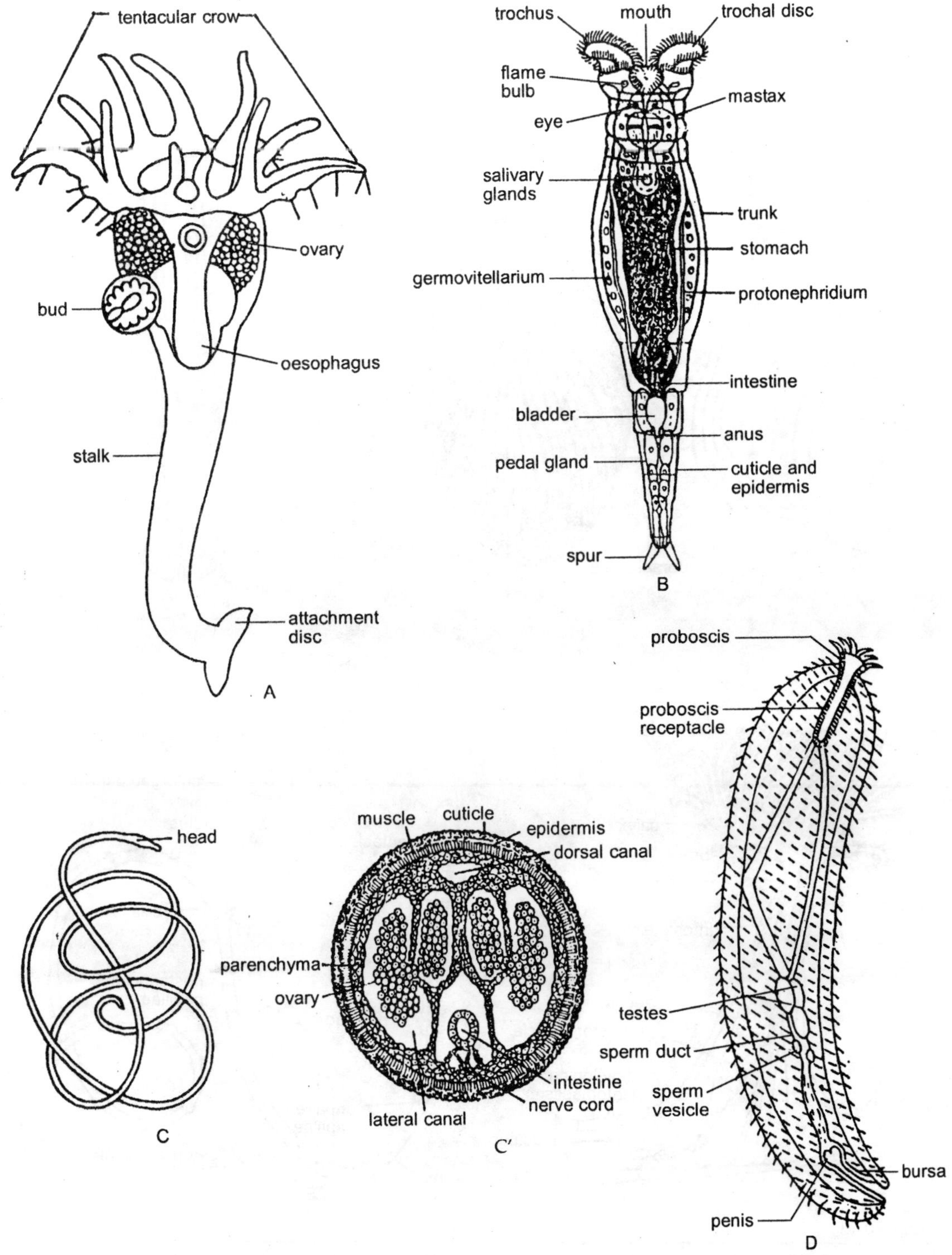

FIGURE 20.12

ABERRANT GROUPS. ACOELOMATA. A. *Lexosoma* sp. (CALYSOZOA). B. *Phylodina* sp. (ROTIFERA). C–C′. *Gordius* sp. (NEMATOMORPHA). C. An entire specimen. C′. Transverse section. D. Acanthogyrus sp. (ACANTHOCEPHALA)

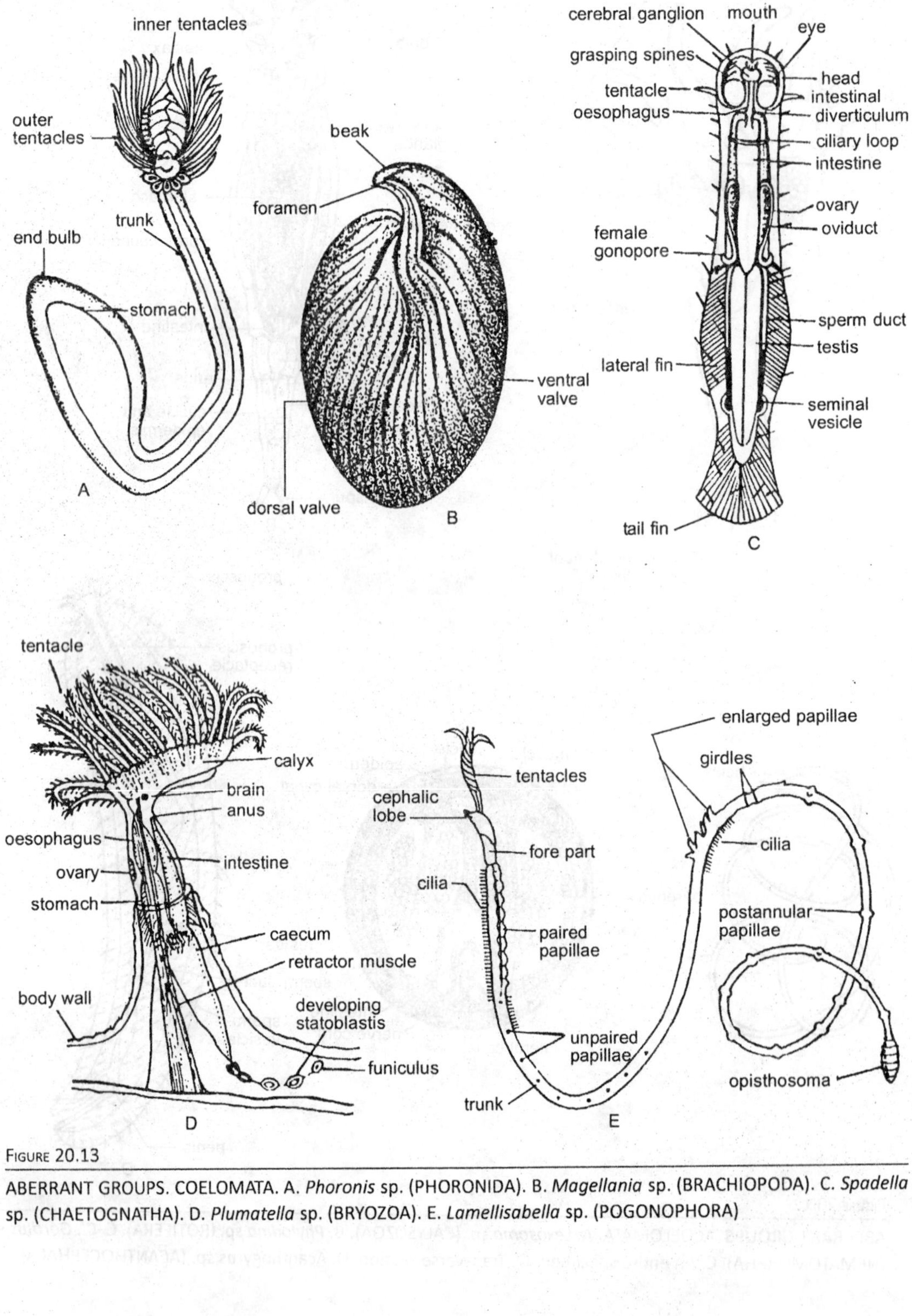

FIGURE 20.13

ABERRANT GROUPS. COELOMATA. A. *Phoronis* sp. (PHORONIDA). B. *Magellania* sp. (BRACHIOPODA). C. *Spadella* sp. (CHAETOGNATHA). D. *Plumatella* sp. (BRYOZOA). E. *Lamellisabella* sp. (POGONOPHORA)

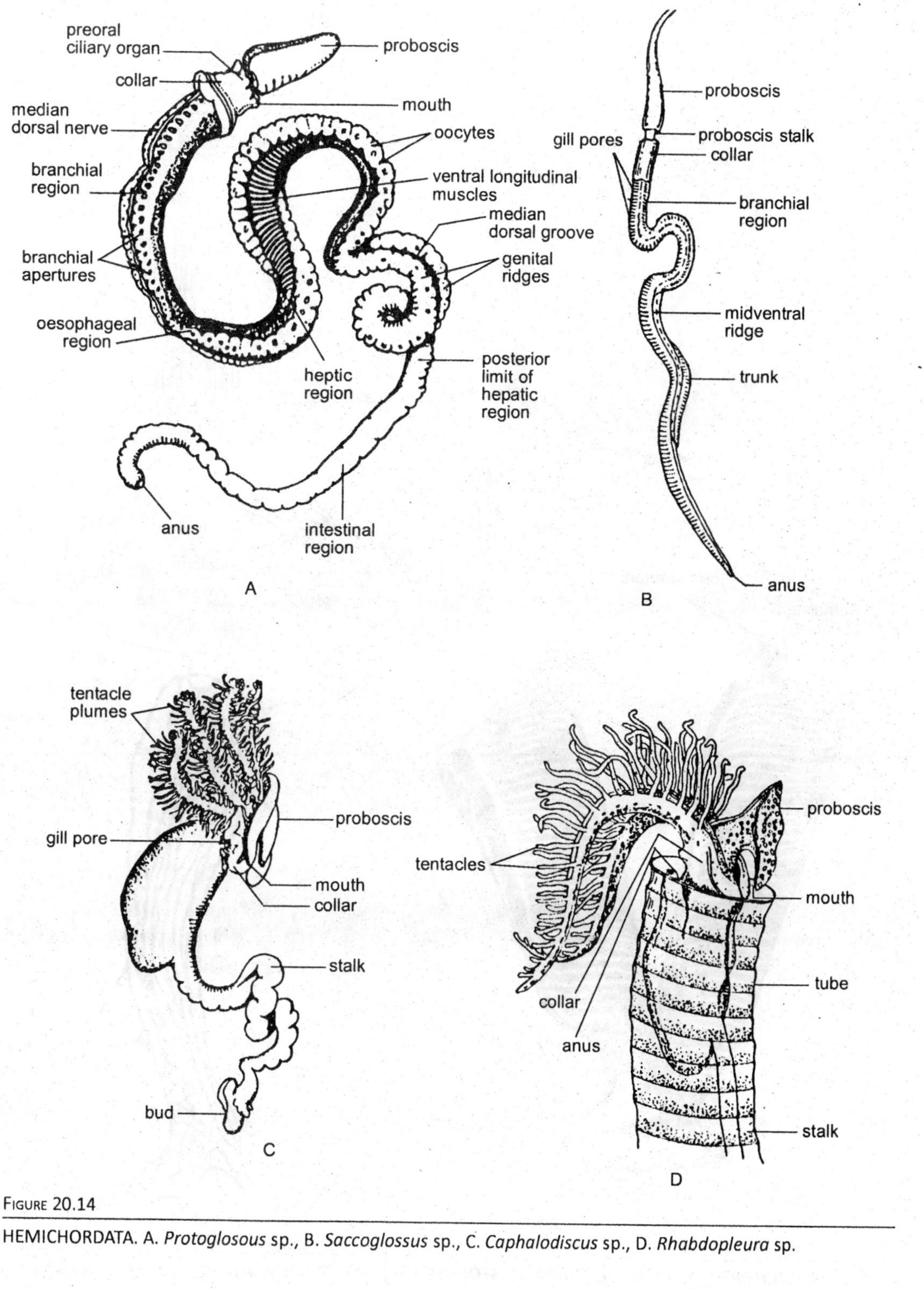

FIGURE 20.14

HEMICHORDATA. A. *Protoglosous* sp., B. *Saccoglossus* sp., C. *Caphalodiscus* sp., D. *Rhabdopleura* sp.

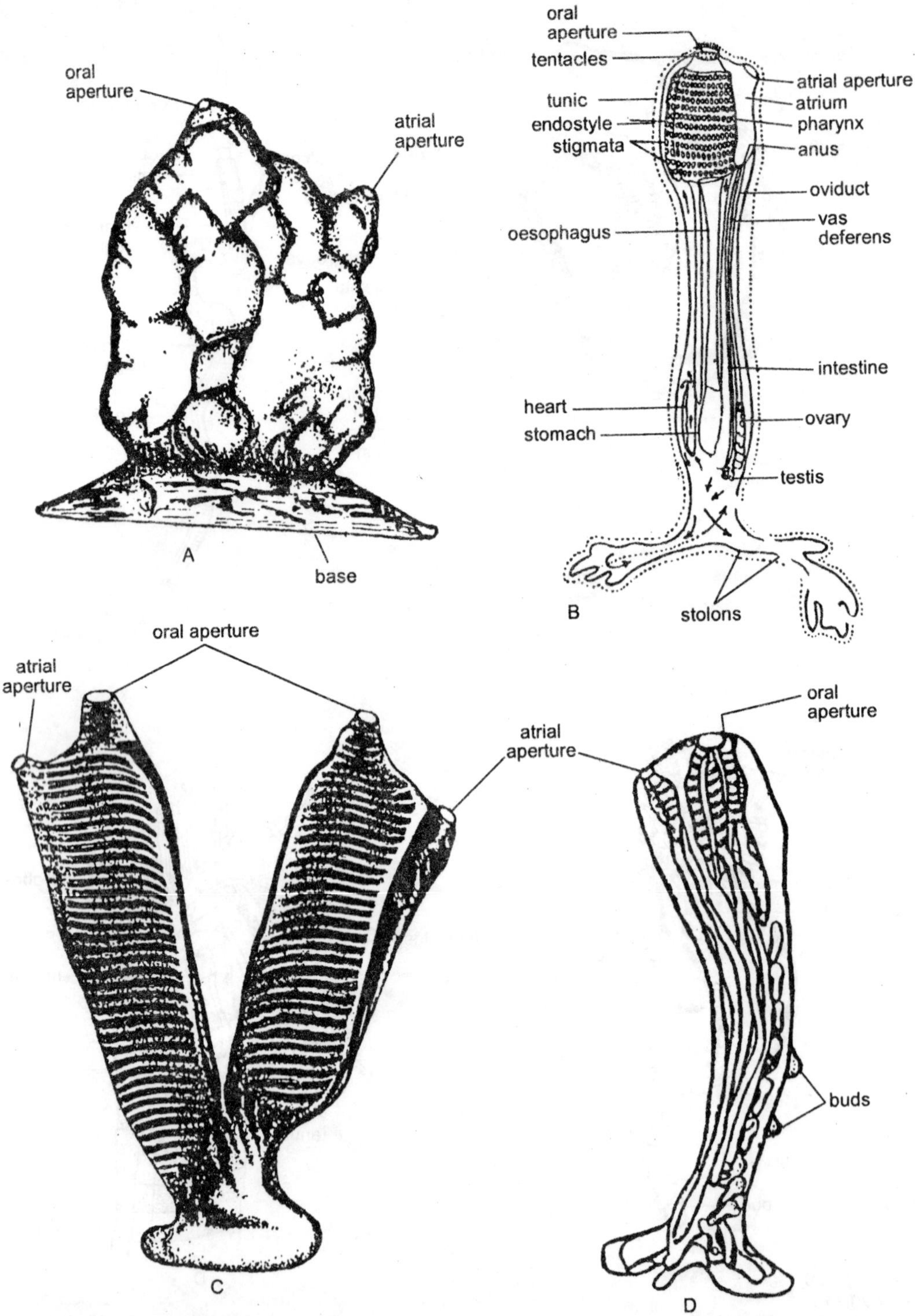

FIGURE 20.15

UROCHORDATA. (ADELOCHORDATA; STOMOCHORDATA) A. *Polycarpa* sp., B. *Clavelina* sp., C. A transparent ascidian, D. *Circinalium* sp.

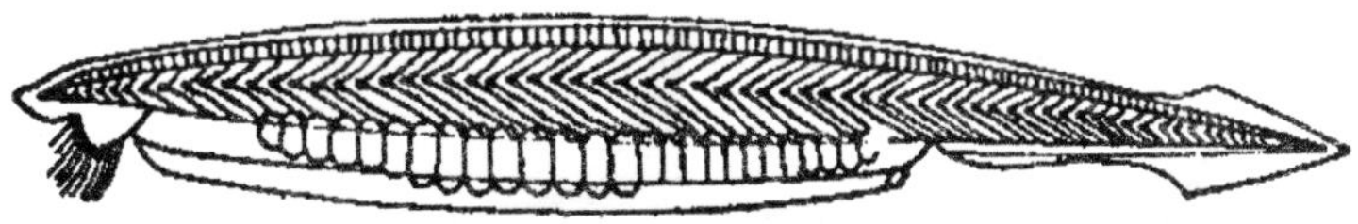

FIGURE 20.16

ACRANIA (CEPHALOCHORDATA). *Branchiostoma* sp. (Amphioxus)

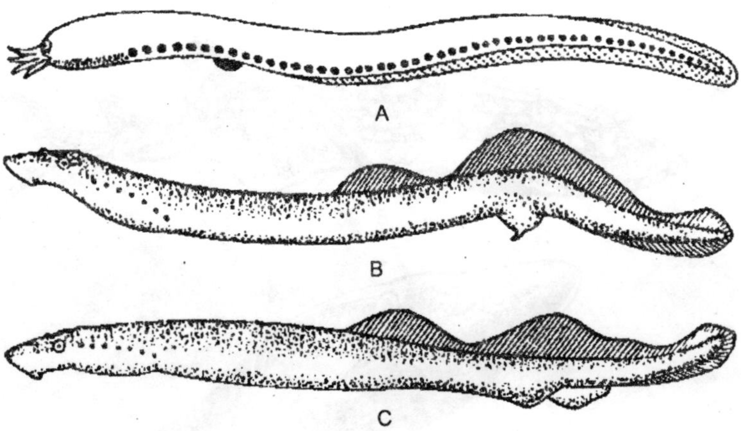

FIGURE 20.17

AGNATHA. A. Myxine sp., B–C. Brook Lamprey. *Lampetra* sp., B. Male. C. Female.

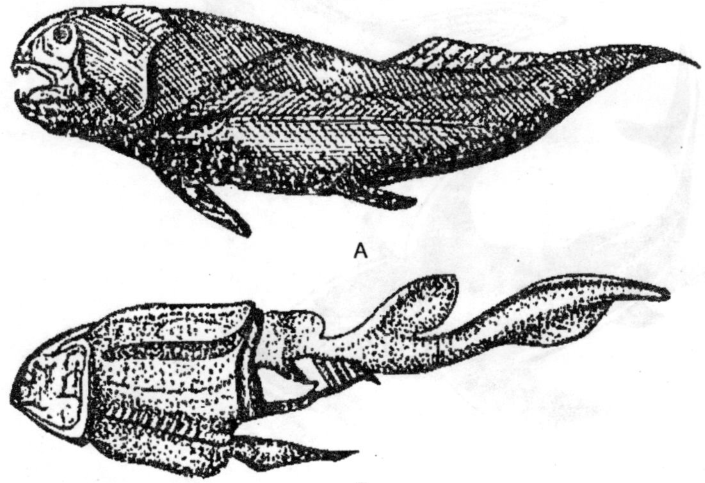

FIGURE 20.18

PLACODERMI (APHETOHYOIDEA) EXTINCT. A. *Dinichthys* sp., B. *Bothriolepis* sp.

FIGURE 20.19

CHONDRICHTHYES. (ELASMOBRANCHII: SELACHI). A. Basking shark (Cetorhinus sp.). B. Electric ray (*Torpedo nobilena*). C. Great white shark (*Carcharodon* sp.). D. Hammer-headed shark (*Sphyrna* sp.)

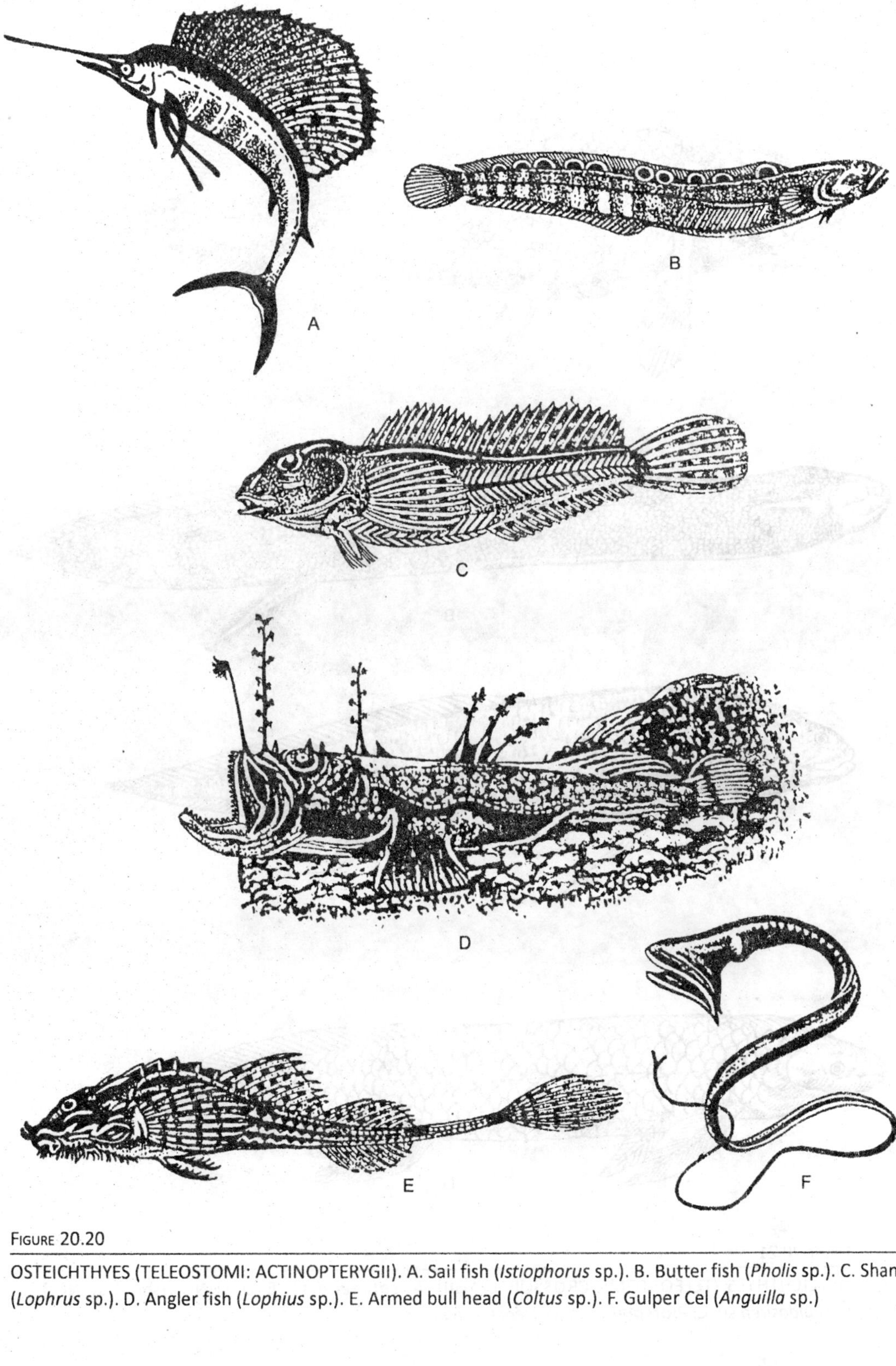

FIGURE 20.20

OSTEICHTHYES (TELEOSTOMI: ACTINOPTERYGII). A. Sail fish (*Istiophorus* sp.). B. Butter fish (*Pholis* sp.). C. Shanny (*Lophrus* sp.). D. Angler fish (*Lophius* sp.). E. Armed bull head (*Coltus* sp.). F. Gulper Cel (*Anguilla* sp.)

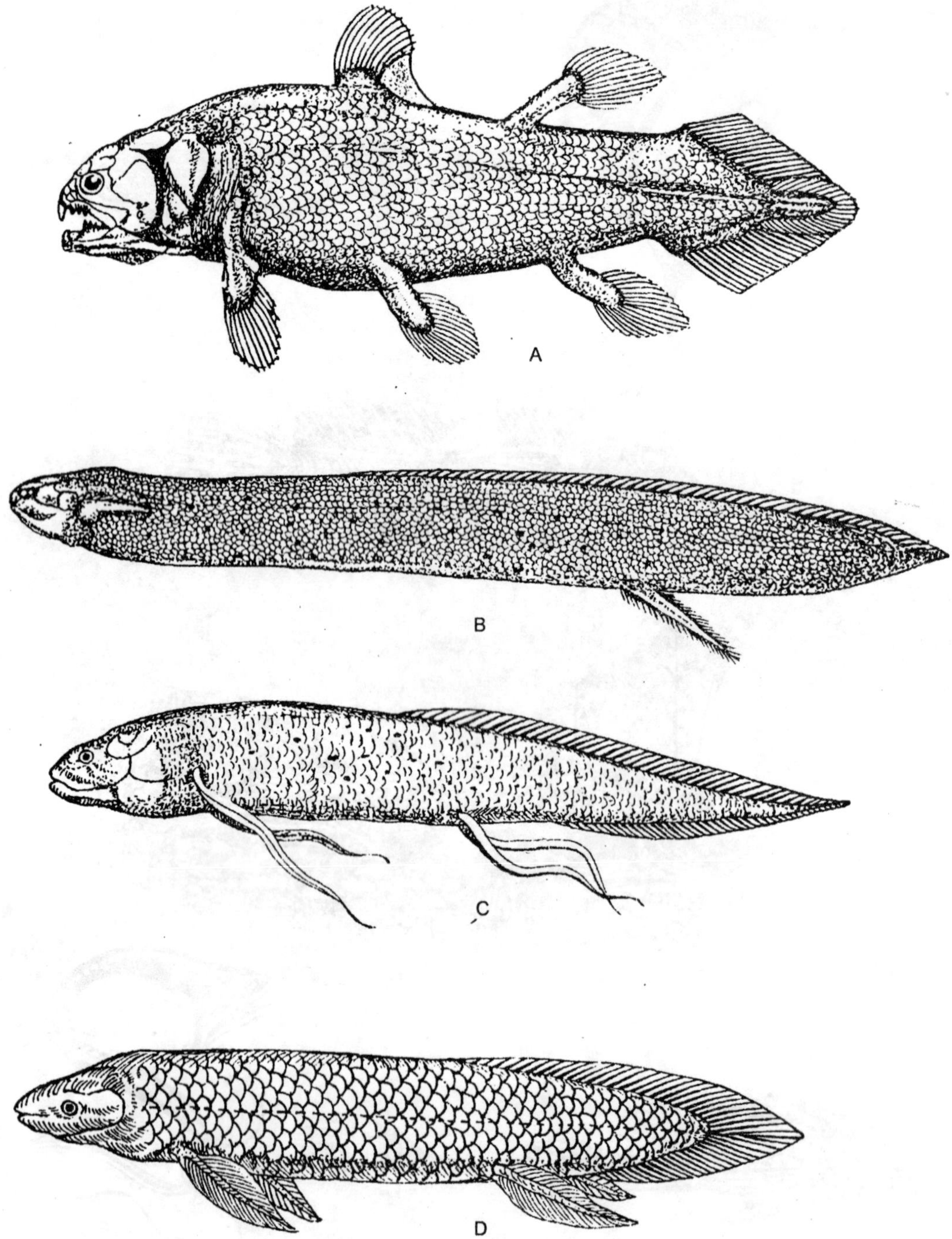

FIGURE 20.21

OSTEICHTHYES (TELEOSTOMI: CROSSOPTERYGII). A. *Latimeria* sp. B–D. LUNG fishes (DIPNOI). B. *Lepidosiren* sp. C. *Protepterus* sp. D. *Neoceratodus* sp.

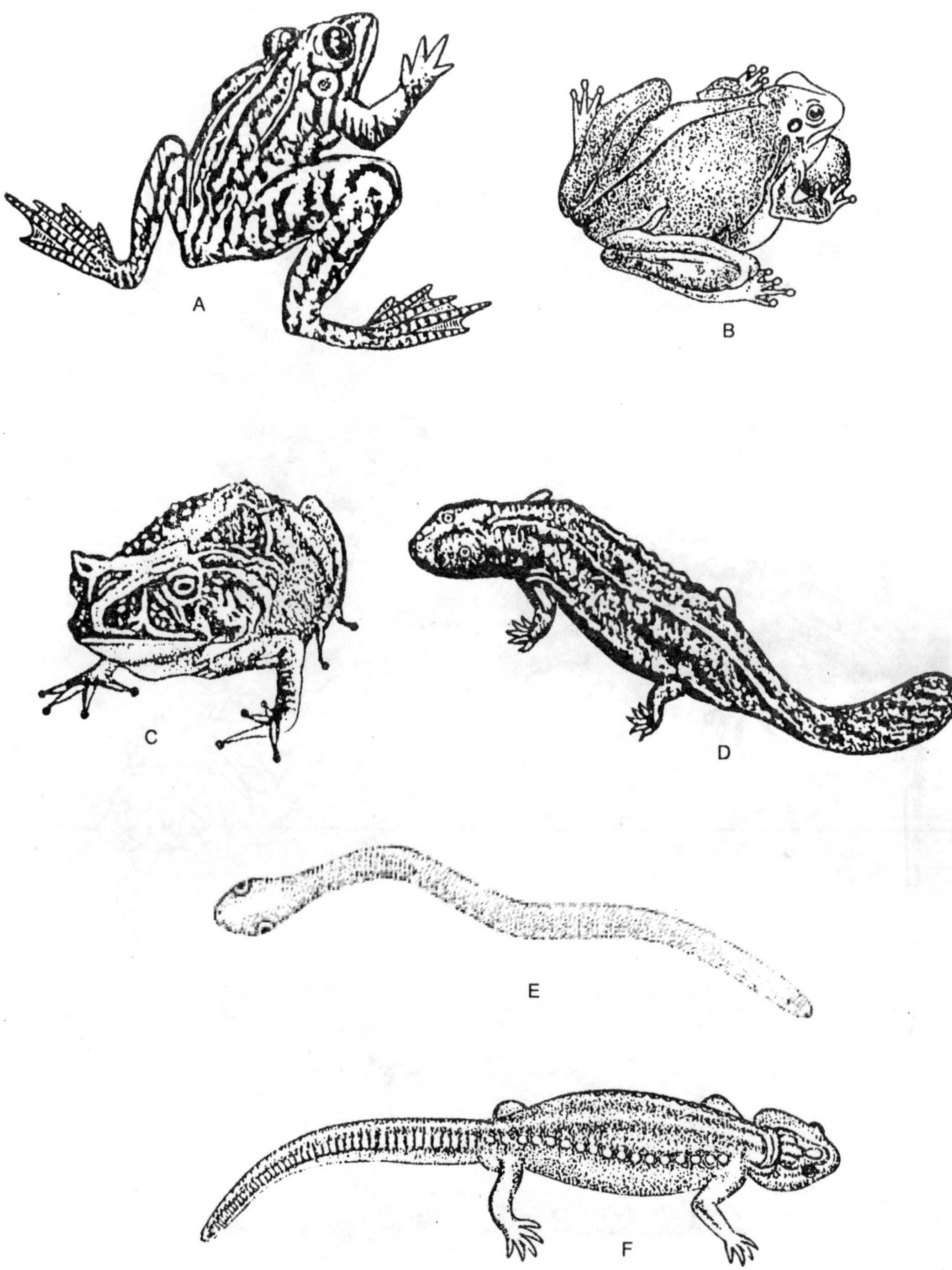

FIGURE 20.22

AMPHIBIA. A. Frog (*Rana tigrina*). B. Tree frog (*Hyla* sp.) Male during cry. C. Horned frog (*Ceratophrys orneta*). D. Japanese giant salamander (*Megalobatrachus* sp.). E. *Euraeotyphlus* sp. F. Darjeeling salamander (*Tylototriton* sp.)

FIGURE 20.23

REPTILIA. A. spectacled caiman (*Caiman* sp.). B. Tree Boa (*Corallus* sp.). C. Iguana sp. D. Ornate box turtle (*Geoclemys* sp.). E. Horned toad (*Phrynosoma* sp.)

FIGURE 20.24

AVES. A. Golden pheasant (*Chrysolophus* sp.). B. Harpy eagle (*Harpia* sp.). C. Golden eagle (*Aquila* sp.). D. Mallard drake (*Anas* sp.). E. Bull finch (*Pyrrhula* sp.). F. A carmine bee-eater (*Merops nupicus*)

FIGURE 20.25

MAMMALIA. PROTOTHERIA, THERIA (METATHERIA) A–B. PROTOTHERIA. A. Duck-billed platypus (Ornithorhyncus sp.). B. Five-toad Echidna (*Tachyglossus* sp.). C–F. THERIA (METATHERIA). C. Morsupial mole (Notoryctes sp.). D. Kangaroo (*Macropus* sp.). E. Cuocus (*Phalanger* sp.). F. Wombat (*Vombatus* sp.)

FIGURE 20.26

MAMMALIA. THERIA (EUTHERIA). A. Skunk (*Mephitis* sp.). B. Armadillo (*Dasypus* sp.). C. Kit fox (*Urocyon* sp.). D. Black bear—young (*Selenarctos* sp.). E. Crab-eater seal (*Lobodon* sp.). F. Coyote (*Canis latraus*)

21 SUBKINGDOM PROTOZOA

The Protozoa are essentially single-celled, acellular.* eukaryotic organisms. They are not a natural group, but have been placed together as a matter of convenience. The subkingdom Protozoa includes over 65,000 named species, of which over half are fossil, and 11,000 are parasitic. Among living species, sarcodines cover 250 parasitic and 11,300 free-living, of which 4,600 species are foramini-ferids; flagellates include, 1,800 parasitic and 5,100 free-living species; 'Sporozoa' (including Apicomplexa, Microspora, Myxospora and Ascetospora) contains 5,600 parasitic species, and the numbers of ciliate species are 2,500 parasitic and 4,700 free-living.

Recently, the subkingdom Protozoa has been classified in seven phyla: *viz.* Sarcomastigophora, Labyrinthomorpha, Apicomplexa, Microspora, Ascetospora, Myxospora and Ciliophora.

EUGLENA sp.

The *Euglena* is an acellular, fresh water organism placed in the order Euglenida, class Phytomastigophora, subphylum Mastigophora, phylum Sarcomastigophora, subkingdom Protozoa.

The genus *Euglena* comprises probably about fifty species and they vary considerably in shape, size and structural details. The most common species is *Euglena viridis* (Fig. 21.1).

⊠ Structure

1. The body is soft and stout, the anterior end being blunt and the posterior end pyriform.
2. It is covered by a pellicle, which permits characteristic euglenoid movement (metaboly).
3. At the anterior end, a depression, known as **gullet**, is present. A whip-like long **flagellum** arises by two roots from the **blepharoplast** in the gullet.
4. The cytoplasm is divisible into an outer, clear, flexible **ectoplasm** and an inner, semifluid, granular **endoplasm**.

   Following structures are found embedded in the endoplasm:

a. The **nucleus** is rounded, surrounded by a membrane and is placed near the centre of the body, with a small nucleolus.

* In acellular organisms the body of the individual is not subdivided into cells. The term unicellular is very often used opposed to multicellular. But the whole organism cannot be equated with a cell of a multicellular organism which is already a differentiated cell.

b. One or more **contractile vacuoles**—acting as water regulators—are present at the anterior end of the body.

c. A large noncontractile space—the **reservoir**—is situated close to the contractile vacuole and is connected with the gullet.

d. A red speck—the **stigma** or eye spot, a derivative of chlorophyll and sensitive to light—is placed close to the reservoir. The stigma is in close contact with a well-marked thickening or one of the two branches of the flagellum, which is suggested to be a specialised sensitive organ (Wager, 1900).

e. A green pigment, **chlorophyll**, which is characteristic of plants, is present, giving a green colour to *Euglena* and the organism can manufacture carbohydrate food materials—like plants by photosynthesis.

f. A single group of chloroplasts radiates from the centre. In the centre of each chloroplast is a **pyrenoid**, which may be enclosed in a paramylum sheath.

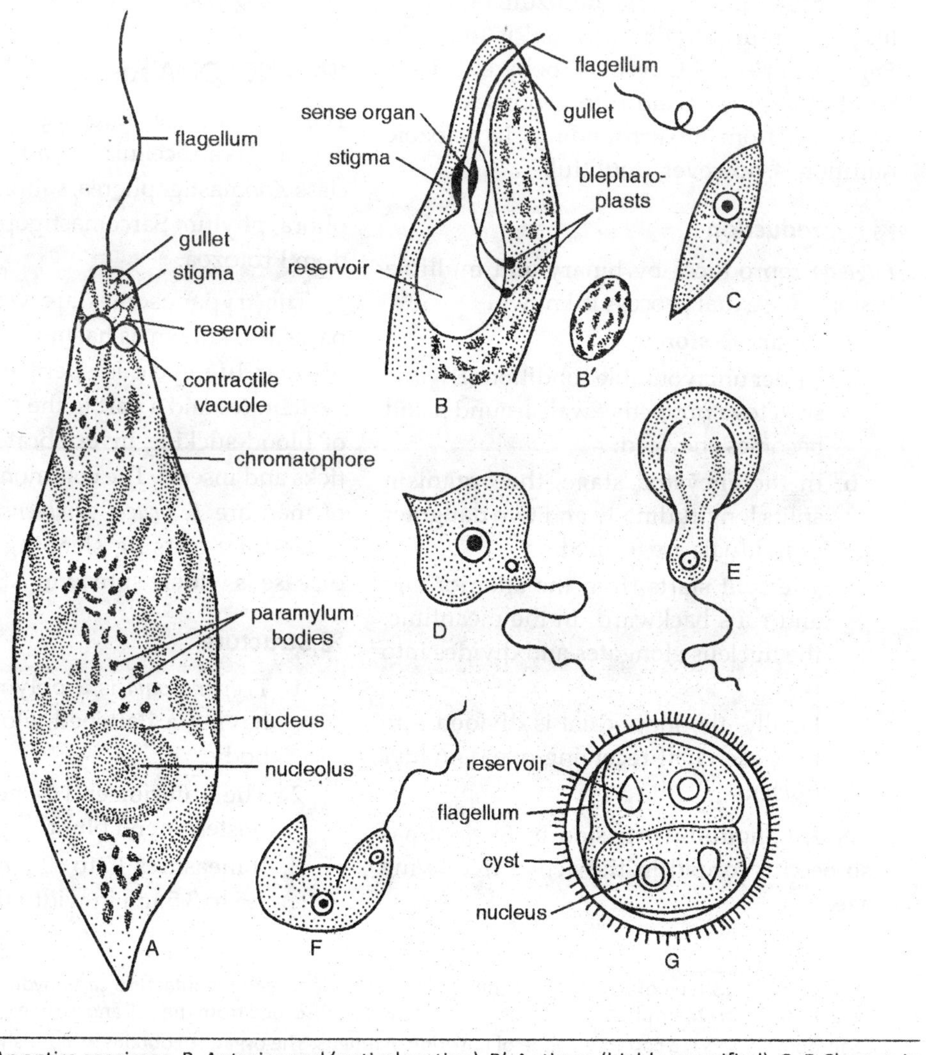

FIGURE 21.1

Euglena sp. A. An entire specimen. B. Anterior end (optical section). B′. A stigma (highly magnified). C–F. Changes in shape in movement. G. Resting stage after binary fission.

g. Paramylum are also found in cytoplasm in various shapes, mainly as rod-shaped grains allied to starch.

Locomotion

Euglena moves forward through the water by the lashing movement of the flagellum. It also exhibits a slow worm-like movement by alternate contraction and expansion of the body known as **euglenoid movement** or **metaboly**.

Nutrition

Holophytic (plant-like)*, **holozoic** (amimal-like), and **saprophytic** nutrition† is found in *Euglena*. *Euglena* obtains its carbohydrate food by photosynthesis and nitrogenous food by absorption from the surroundings. Its holozoic nutrition is, however, doubtful.

Reproduction

Euglena reproduces by binary and multiple fission. No sexual process is known.

1. **Binary fission**

a. Under unfavourable conditions *Euglena* secretes a protective wall around it and becomes encysted.

b. In the encysted stage, the organism splits longitudinally and two daughter individuals are formed.

c. The split starts from the anterior end and runs backward. In the meantime, the nucleus elongates and divides into two.

d. Finally, the individual is divided into two, each half receiving one daughter nucleus.

N.B. Longitudinal fission in *Euglena* has also been recorded in the active free-living stage.

2. **Multiple fission**

a. In the encysted stage the nucleus divides repeatedly and a large number of minute daughter nuclei are produced.

b. The cytoplasm breaks up and a small amount surrounds each daughter nucleus and many minute animals known as **flagellulae** are formed.

c. Under favourable circumstances the flagellulae come out of the cyst, and passing a short period through amoeboid stage develop into adult *Euglena*.

TRYPANOSOMA sp.

The *Trypanosoma*—causing sleeping sickness in man—is an acellular organism placed in the class Zoomastigophorea, subphylum Mastigophora, phylum Sarcomastigophora, subkingdom Protozoa.

The trypanosomes are vertebrate blood parasites, known as haemoflagellates. A part of their life cycle is spent in the blood of vertebrates and a part in the alimentary canal of blood-sucking invertebrates like leeches, ticks and insects. The common trypanosomes of man are *T. brucei gambiense* and *T. brucei rhodesiense* causing sleeping sickness. The disease is widely spread in Africa.

Structure

1. *Trypanosoma* is a polymorphic animal having a flattened, elongate-fusiform body.
2. The anterior end is pointed and the posterior end blunt.
3. It measures 15 to 60 μm‡ in length and 1.5 to 3.5 μm in width (Fig. 21.2A).

---

* Holophytic. This type of nutrition is common in plants. Plants can manufacture carbohydrate food material in the presence of chlorophyll and sunlight, with the water absorbed from the soil and carbon dioxide from the air.

† Saprophytic. This is a common mode of nutrition in fungus. The organism obtains its nourishment by absorbing food substances from decomposed organic matter.

‡ mm = 1/25,000 inch; 1/1,000 part of a millimetre = .001 mm.

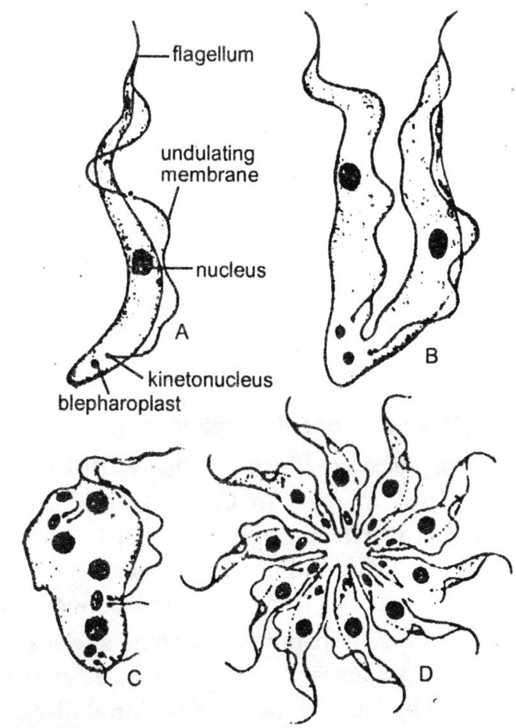

FIGURE 21.2

Trypanosoma sp. Structure and multiplication. A. Trypanosomal form. B. Longitudinal fission. C–D. Dividing individual (Rosette)

4. A strong, tough, elastic **pellicle** encloses the body. The pellicle is lined by a layer of ectoplasm and the central mass is **endoplasm**.
5. A large vesicular **nucleus** is present in the centre of the body.
6. Behind the nucleus and near the posterior end of the body is a spherical or rod-shaped or disc-shaped **parabasal body** or **kinetoplast**. A small **blepharoplast** is present in front of it. The two are connected by a delicate **rhizoplast**.
7. A **flagellum** arises from the posterior end of the blepharoplast and extends forward, towards the anterior end along the outer margin of the body, being connected to it by an undulating membrane.
8. A small portion of the flagellum extends freely beyond the anterior end.
9. The **contractile vacuoles** are absent.
10. **Volutine** granules in the form of refractile bodies are scattered in the endoplasm. *Trypanosoma* are polymorphic and occur in four morphologically different forms of which first three stages develop in the gut of vector and *trypanosome* stage are found in difinite host.

a. **Leishmanial stage.** Without flagellum and with anteriorly placed kinetoplast.
b. **Leptomonad stage.** Elongated body with a broad anterior end, a free flagellum and an anteriorly placed kinetoplast.
c. **Crithidial stage.** Kinetoplast and flagellum situated in front, anterior to the nucleus.
d. **Trypanosome stage.** Kinetoplast situated far behind the nucleus near the posterior end of the body.

⊠ Life Cycle

1. *Trypanosoma* completes its life cycle in two hosts (Fig. 21.3). The definitive host is a vertebrate and the intermediate or secondary host is an invertebrate, usually a Tsetse fly, *Glossina* sp.
2. In the vertebrate host, trypanosomes usually live in the blood or cerebrospinal fluid.
3. **Inoculation.** With the bite of Tsetse fly, the parasites enter the blood stream of man in the metacyclic form. In some cases the parasites may reach cerebrospinal fluid from the blood.
4. **Vertebrate host.** In the blood stream the parasites multiply by binary fission (Fig. 21.2B) and is represented in three morphological forms:

i. **Slender form.** With a large flagellum..
ii. **Intermediate form.** With a short flagellum.
iii. **Stumpy form.** With a broad body devoid of a flagellum.

5. **Secondary host.** The Tsetse fly (*Glossina palpalis*) sucks the trypano-somes along

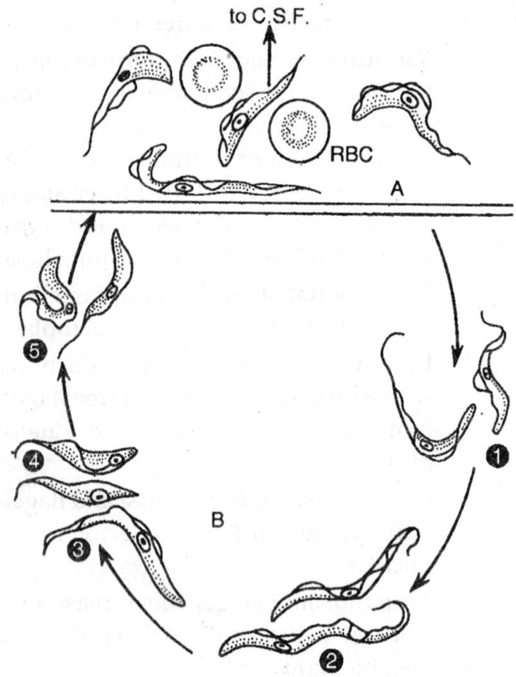

FIGURE 21.3

Trypanosoma sp. Life history. A. In man, antelope, etc. B. In Tsetse fly. 1. Forms in mid gut. 2. Long slender forms in proventriculus. 3. Freshly arrived form in salivary gland. 4. Crithidial form. 5. Infective form. C.S.F. (cerebrospinal fluid). R.B.C. (Red blood corpuscle)

with host's blood. The parasites remain unchanged in the midgut for a few hours.

6. In favourable conditions the parasites multiply by repeated binary fission (Fig. 21.2 C, D) and after about two weeks numerous slender forms appear. These gradually ascend the foregut.
7. By the end of the third week some of these slender forms reach the salivary gland through the labial cavity and hypopharynx.
8. In the salivary glands the parasites change into **crithidal** stage.
9. They multiply rapidly for about 2-3 days and finally metacyclic trypanosomes are formed.
10. These are the infective forms and are to be transferred to the primary host.

☒ Pathogenicity

The trypanosomes cause **gambian** fever, which persists for only a few days. Subsequently enlargement of lymph glands, anaemia and weakness are reported. These are followed by the enlargement of spleen and liver. On entering the cerebrospinal fluid the parasites cause sleeping sickness in man, in which the patient loses consciousness, first at regular intervals, leading into a state of coma, ending in death.

☒ Other pathogenic trypanosomes

Trypanosoma 'cruzi' causes '**Chagas**' disease or swelling in man in Central and South America. The parasites multiply in the muscles, heart and nervous tissues. *Triatoma megista*, bed bug is the intermediate host.

T. brucei transmitted by some species of ***Glossina*** causes '**Nagana**' fever. ***T. evansi*** carried by tabanid flies causes '**Surra**' disease in horse, cattle and camel in the Indian subcontinent.

☒ Clinical features

1. The incubation period of *Trypanosoma gambiensi* is on an average 6 to 14 days. The clinical course may be divided into (i) stage of **trypanosome fever** or febrile stage, lasting for several months, (ii) Stage of **sleeping sickness** or meningo-encephalitis.
2. The symptoms in the first stage are fever, headache, inability to sleep, asthenia, enlargement of all the superficial lymph glands, liver, spleen, localised oedema of eye lids, sheath of penis, neck and abdomen.
3. In the second stage the patient gradually becomes lethargic, sleep overcoming all the activities, becomes emaciated, urine and faeces pass involuntarily, bed sores develop, resulting in death, due to complication in urinary bladder infection and pneumonia.

Management of Trypanosoma

1. *Trypanosoma* spp. are transmitted from one individual to another by the vector tsetse fly *Glossina* sp. (Fig. 21.4) Factors influencing transmission of the disease are (i) reservoir host, (ii) vectors, and (iii) favourable environment for multiplication of vectors.

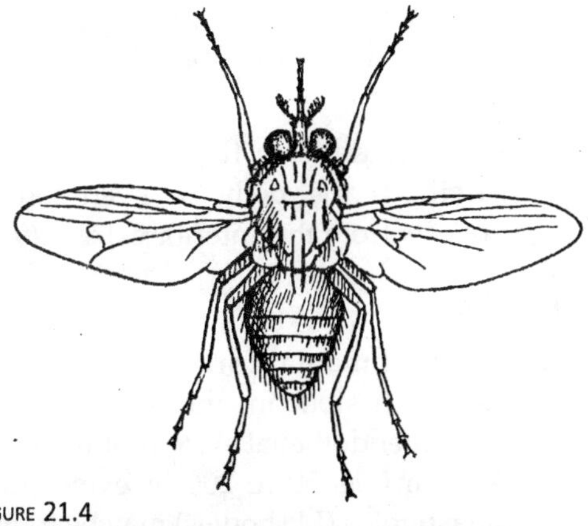

FIGURE 21.4

Tsetse fly (*Glossina* sp.)

2. *Glossina*, vector of *T. brucei* and *T. rhodesiense* occur in open, pupating in dry febrile earth, while vector of *T. gambiense* are riverine tsetse flies, breeding in shady, moist areas along the rivers. Tsetse flies are larviparous and deposit their larvae on the soil under bush.
3. The adults rest in bushes at certain heights above the ground and no higher. Bush removal and trimming are effective means of control.
4. The flies feed only in day time. As a preventive measure many farmers graze their livestocks at night.
5. The most satisfactory means of management is by spraying chemical insecticides DDT, BHC etc. from aircraft. But the chemical insecticides are very expensive and not ecofriendly. The residual effects of insecticides tells upon health of humans and other animals of the ecosystem.
6. Another method of control is use of different types of traps by which several kilograms of flies are killed every day.
7. The early detection of infection in patient's blood is necessary for proper treatment. Two drugs, Bayer 205 (also called Antrypol, Germanin, or Suramin) and Pentamidine (or Lomidine) are widely used for both treatment and prophylaxis. Early treatment helps in speedy recovery.

LEISHMANIA

Leishmania, haemoflagellate parasite causing leishmaniasis in man is placed in the class Zygomastigophora, subphylum Mastigophora, phylum Sarcomastigophora, subkingdom Protozoa.

It is a parasite in the macrophage cells of man. A part of the life cycle is spent in the gut of a small blood sucking sand fly *Phlebotomus spp*.

The common Leishmanias in man:

1. *Leishmania denovani* caused the deadly disease, ***Leishmaniasis*** or Kala-azar or **Dum Dum** fever.
2. *L. tropica* causes **oriental sore**.
3. *L.* braziliensis causes **nasopharyngeal Leishmaniasis or Espundia**.

Structure

The parasite exists in two stages:

1. The **aflagellar** (non flagellar) stage occurring in man, called the **leishmanial** or **amastigote** form. Leishmanial forms are sometimes called **L.D. bodies** (**Leishman-Donovan bodies**).
2. The flagellar stage occurring in the gut of sand flies, are **Leptomonad** or **promastigote** form.

Leishmanial or amastigote form (Fig. 21.5A).

It lives in the reticuloendothelial system of man.

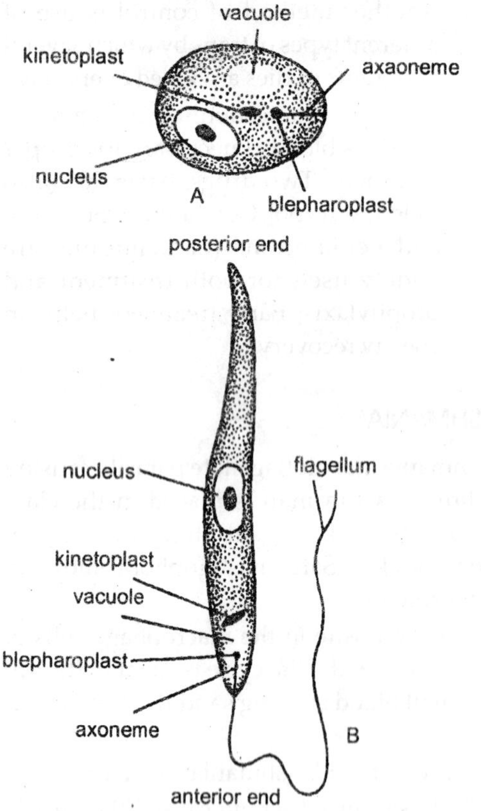

FIGURE 21.5

• **Leishmania donovani.** A. Amastigote form. B. Promastigote form

1. The body is round or oval, covered with a delicate cell membrane, measuring 2-4 micron along the longitudinal axis.
2. The nucleus is oval or round, less than one micron, usually situated in the middle.
3. Lying tangentially or at right angles to the nucleus, a tiny rod-shaped body, known as Kinetoplast is present. The Kinetoplast consists of a rodlike parabasal body and a dot like blepharoplast.
4. A delicate filament extends from the Kinetoplast to the margin of the body, called axoneme (rhizoplast) and represents the root of the flagellum.
5. A vacuole is present close to the axoneme. It is the smallest, nucleated cell known as Leishmanial form.

Leptomonad or promastigote forms (Fig. 21.5B)

1. It is long, slender, spindle-shaped measuring 15-20 × 1-2 microns.
2. The nucleus is centrally located and the kinetoplast lies transversely near the anterior end.
3. A light staining area, the eosinophilic vacuole lies in front of the kinetoplast over which runs the root of the flagellum.
4. A single delicate flagellum of same length as the body or even longer projects from the anterior extremity.

⊠ Life Cycle

1. The leishmanial form multiplies by binary fission in the cells of the reticuloendothelial system of the host. As many as 50 to 200 or even more amastigotes (LD bodies) may be found embedded in the cytoplasm of a host cell. The cell enlarges and eventually ruptures.
2. The parasites are liberated into the circulation and are again taken up by or invade, fresh cells and the cycle repeated, some remaining free.
3. *Phlebotomus* takes up the parasites along with its blood meal.
4. In the midgut of sand fly the parasites are transformed into leptomonad forms, multiply by longitudinal binary fission to enormous numbers.
5. Heavy infection in the pharynx and buccal cavity of Phlebotomus occur between 6th to 9th day of the infective blood-meal. Sometimes buccal cavity is blocked by the parasites.
6. The infection is transferred to the blood circulation of a healthy man when bitten by the infected fly.
7. Some of the flagellates enter the cells of the reticuloendothelial system and the cycle is repeated.

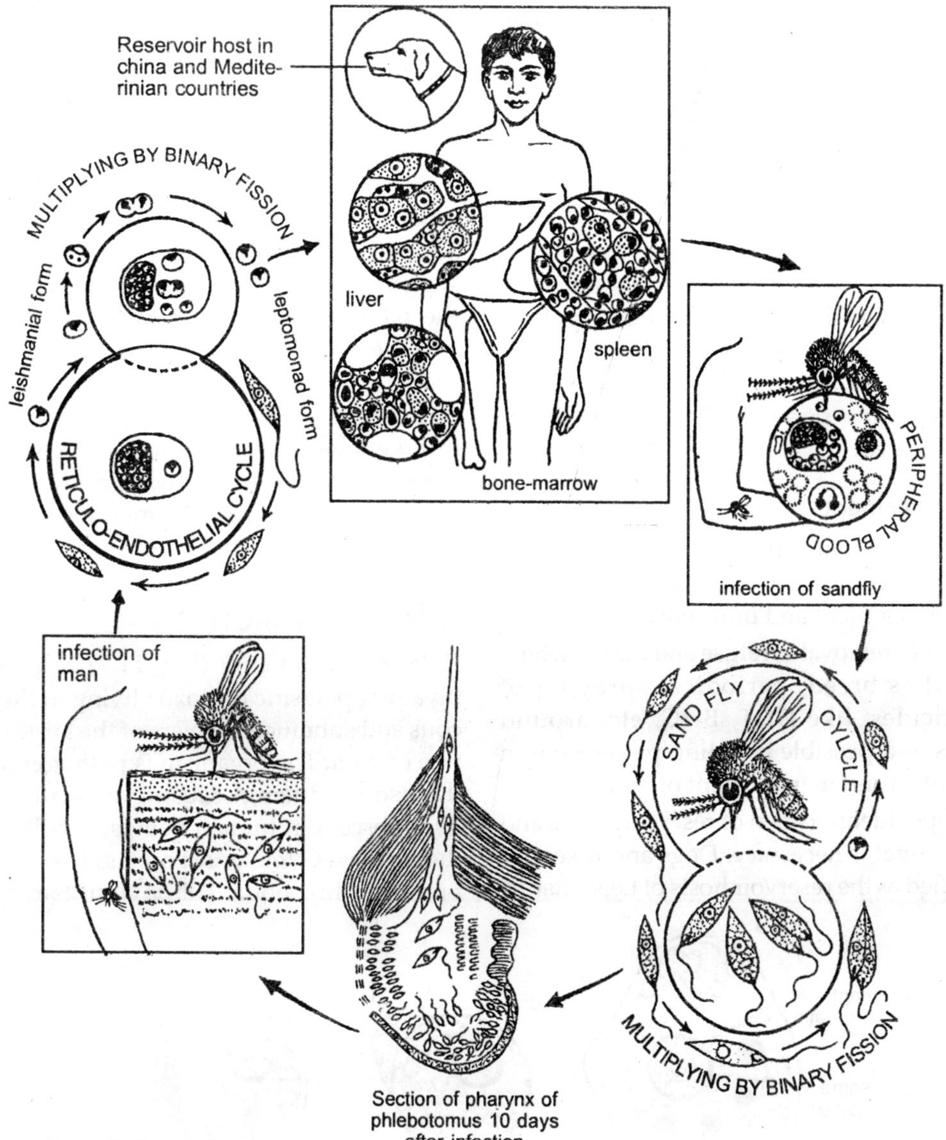

FIGURE 21.6

Leishmania donovani. Life cycle in man and sand fly

☒ Pathogenecity and clinical features

1. Infection with *L. donovani* produces visceral leishmaniasis or Kala-azar. The incubation period generally varies from 3 to 6 months.
2. Gradually all the reticuloendothelial tissue of spleen, liver, bone marrow and lymph nodes become heavily parasitised.
3. Spell of fever (pyrexia) and apyrexia occur for months. Profuse sweating common but rigors occur only exceptionally in the more chronic forms.
4. In well established cases, emaciation and anaemia together with enlargement of spleen (Splenomegaly) and liver, oedema of legs, and a generalised lymphadenopathy with enlargement of the inguinal and femoral glands occur.

5. In many cases, especially in Indian Kala-azar, the skin acquires a strange earth-grey colour, best seen on the feet, hands and abdomen in fair skinned people. The name Kala-azar (black, fever) has been derived from the black pigmentation.
6. Death in Kala-azar is always due to some complications, such as amoebic or bacillary dysentery, pneumonia, pulmonary tuberculosis, and other septic infections. About 75 to 95% of patient die within a period of 2 years, if left untreated.

Management

The management of the disease leishmaniasis involves vector control, reservoir control, disease treatment and immunization.

Timely removal of refuse and rubble where sand flies breed, periodical spraying of insecticides, like DDT, BHC, etc. around houses and possible breeding places help in the control and management of the vector.

Proper identification of **reservoir hosts** and their habitat is necessary. Dogs and foxes are identified as the reservoir hosts of Leishmania. Constant watch on reservoirs hosts and their treatment in time help to control spreading of the disease.

Timely diagnosis of the Leishmania parasite in the blood and spleen puncture materials of the patient and proper treatment with sodium stibagluconate (Pentosum, Solustibosan) or any other suitable drug, suggested by the physician help recovery of the patient. The disease may be managed by (i) eliminating the immediate primary source of infection of the vector animal, (ii) use of biodegradable pyrethroid insecticide-impregnated mosquito nets of 22 mesh as suggested by the WHO and periodic fumigation of the sleeping rooms are the best procedure for personal management.

ENTAMOEBA HISTOLYTICA

Entamoeba histolytica (Fig. 21.7) is a tissue-invading parasitic protozoa living in the mucous and submucous layers of the large intestine of man. The organism is **pathogenic** (i.e. disease producing) causing dysentery and liver abscess in human beings. It belongs to the subclass Gymnamoeba, class Lobosea, superclass Rhizopoda, subphylum Sarcodina,

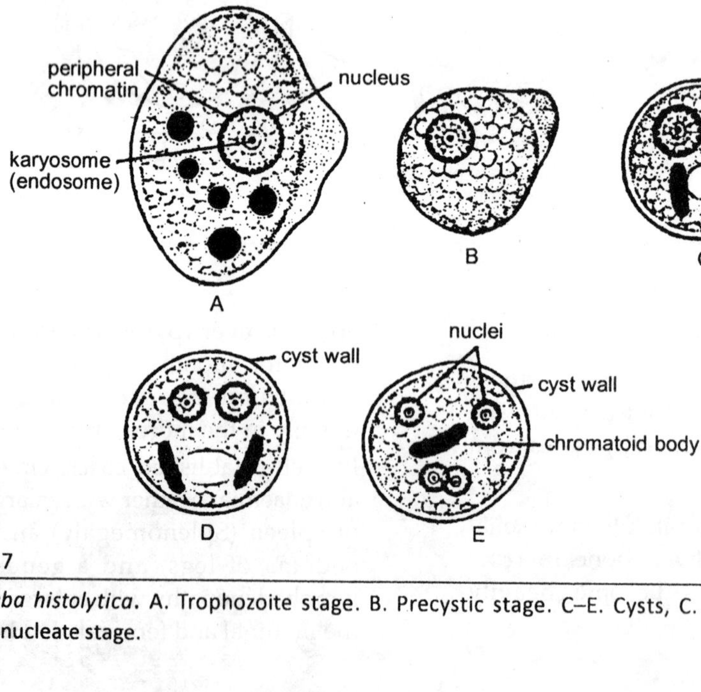

FIGURE 21.7

Entamoeba histolytica. A. Trophozoite stage. B. Precystic stage. C–E. Cysts, C. Uninucleate. D. Binucleate. E. Quadrinucleate stage.

phylum Sarcomastigophora and subkingdom Protozoa.

⊠ Structure

The amoeba has three stages in its life cycle, viz. The trophozoite stage, the precystic stage and the cystic stage.

A. The trophozoite amoeba

This is the growing or feeding stage of the parasite having the following features:

1. The trophozoite is somewhat elongated (Fig. 18.4A); measuring 18 to 40 μm, but the shape is not fixed as the animal constantly changes its contour.
2. The cytoplasm is divisible into a clear translucent **ectoplasm** and a granular **endoplasm**.
3. The nucleus is spherical and varies from 4 to 6 μm in size.
   a. The **nucleus** is surrounded by a delicate nuclear membrane lined with a single layer of chromatin granules.
   b. The centre of the nucleus is occupied by a karyosome surrounded by a clear halo.
   c. The space between the membrane and the karyosome is traversed by fine threads arranged radially in the form of spokes of a wheel.
4. Ingested blood cells are found in the cytoplasm.

⊠ Encystation

In some cases, the trophozoites are discharged into the lumen from the intestinal wall and are transformed into precystic forms, which produce cystic forms. This phenomenon is known as **encystation** and leads to the production of a different strain.

B. Precystic amoeba

1. This is slightly ovoid with a short and blunt pseudopodium projecting from one side (Fig. 21.7B).
2. Size varies from 10 to 20 μm.
3. The motility ceases and ingested particles are extruded.

C. Cystic amoeba

1. The parasite becomes rounded and surrounds itself by a highly refractile cyst wall (Fig. 21.7C-E).
2. Size varies from 7 to 20 μm.
3. Contains one or more chromatoid bodies with round ends.
4. The cyst is uninucleate at the beginning, but undergoes double fission so that four nuclei are produced. The quadrinucleate stage is considered as the mature cyst.

⊠ Excystation

The liberation of the parasite from the cyst is known as excystation. The mature cysts are infective forms. They reach the alimentary canal with contaminated food and drink. The cyst wall is digested by the action of **trypsin** in the intestine. When the cyst reaches the iliocaecal junction, particularly the caecum, vigorous amoeboid movement starts in the cytoplasm and a rent appears in the cyst wall. The parasite comes out through the rent. The four nuclei now divide into eight, the cytoplasm also undergoes division and **eight** minute amoebulae, also called **metacystic amoebae**, are formed. The young amoebae are active and invade the intestinal tissue to reach the mucous and submucous layers, their normal habitat.

⊠ Reproduction

Entamoeba reproduces (Fig. 21.8) by binary fission, the rate of multiplication being very high.

1. The nucleus slightly elongates and assumes an ovoid shape.
2. A constriction appears in the middle which grows deeper and the nucleus is divided into two by a modified type of mitosis.
3. The cyroplasmic mass elongates slightly, a constriction appears at the middle which grows deeper and the parasite is divided into two equal halves.

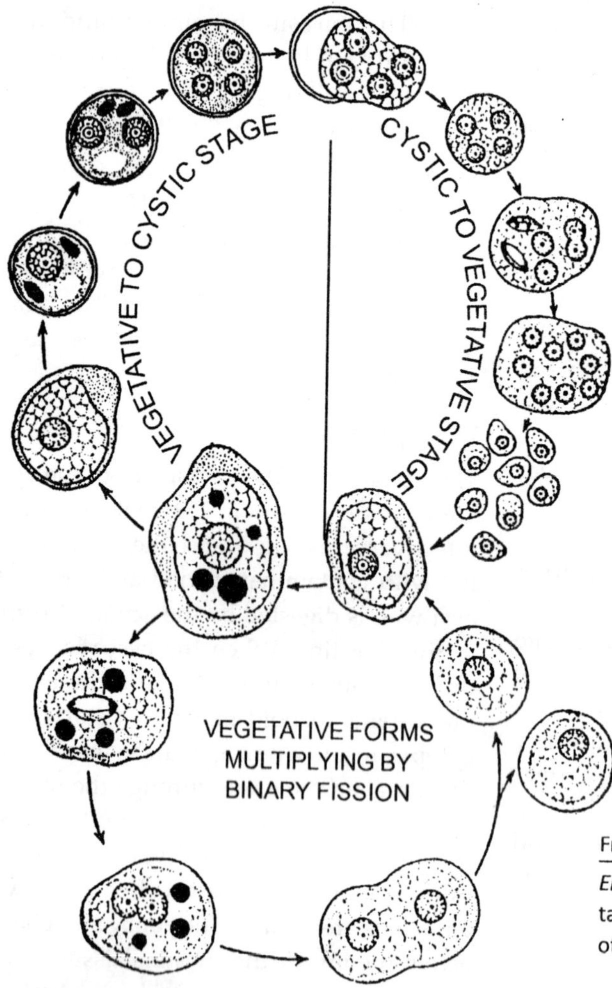

FIGURE 21.8

Entamoeba histolytica. Life history Top. Encystation and excystation. Below, Multiplication of Trophozoite

4. Each half receives one daughter nucleus and two individuals are thus produced.

☒ Modes of infection

The life cycle of *Entamoeba histolytica* passes only in one host, the man. Transmission from man to man is effected through faecal contamination of drinking water and vegetables directly or through the agency of flies or cockroaches. The mature quadrinucleate cysts are infective forms. Eating of uncooked vegetables and fruits and drinking of water contaminated with infective forms of parasites lead to the infection of a new host.

Cysts can pass through the intestine of the fly or cockroach in a viable condition, and infest new hosts through contaminated food. Cooks, mess boys and food handlers are important transmitters of infection in tropical countries.

☒ Pathogenicity and clinical features

1. ***Entamoeba histolytica*** is unique among the amoebas of humen in its ability to hydrolyse host tissues. The trophozoites have **active cysteine proteases** on their surfaces, and their pathogenic strains possess an additional form of this enzyme.
2. The onset may be sudden after an incubation period of 8 to 10 days. Amoebiasis is usually waterborne

disease. More commonly the disease develops slowly with intermittent diarrhoea, cramps, vomiting and general malaise. Infection in the caecal area may mimic the symptoms of appendicitis.

3. ***Entamoeba histolytica*** causes ulcer in the intestinal mucosa. Depending on the number and distribution of intestinal lesions the patient might experience pain in the entire abdomen, fulminating diarrhoea, dehydration and loss of blood. Amoebic diarrhoea is marked by bouts of abdominal discomfort with 4 to 6 stools per day but little fever.
4. In acute onset, there may be headache, fever, severe abdominal cramps. An average of 15 to 20 stools consisting of liquid faeces, flecked with bloody mucus are passed per day.
5. Initially, intestinal abscess may be formed, but in chronic cases the trophozoites enter the mesenteric venules and travel to the liver through hepatoportal system, enters the sinusoids and begin to form abscess in liver. This is the case of **hepatic amoebiasis.**
6. The trophozoites may invade lungs, kidney, adrenals, spleen, male and female genitalia, pericardium, even brain, forming abscess in these organs.
7. Death may occur from peritonitis, resulting from gut perforation or from cardiac failure and exhaustion.

⊠ Management

1. *Entamoeba histolytica* is found throughout the world. The higher prevalence in the tropics results from lower standards of sanitation and the greater longevity of cysts in a favourable environment.
2. *E. histolytica* is a faeces-borne organism and primary mode of infection is oral to anal (oro-faecal) contact. The manner of disposal of human excreta in a given area is the most important in the epidemiology of this organism.
3. Transmission depends heavily on contaminated food and water.
4. The carriers (cyst passers) handling food can infect the rest of their family groups or even hundreds of people if they work in eateries.
5. The use of human faeces as fertilizer contributes heavily in transmission of Entamoeba cysts.
6. Filth flies, particularly *Musca domestica* and cockroaches are the mechanical vectors.
7. For **integrated amoebiasis management** following measures should be taken simultaneously.
   i. good sanitary measures and quick disposal of faces.
   ii. prevention of faecal pollution of water by chlorination, and use of sand filter in water supply.
   iii. use of human excrement as fertilizer only after proper treatment.
   iv. detection, isolation and treatment of carrier and judicious use of insecticides in the breeding places of house fly and cockroaches.
8. For personal management, (i) Drinking of boiled/filtered water, (ii) Protection of foodstuff and drinks from contamination from flies, cockroaches and rats by keeping those under cover, (iii) avoiding eating of raw vegetables and fruits, (iv) washing of hand, personal cleanliness and elementary hygiene followed while taking meals and (v) early detection of parasites by stool examination and proper treatment with antiamoebic drugs to eliminate the family source of contamination.

ELPHIDIUM (*Polystomella*) sp.

Elphidium (*Polystomella*) is a marine animal placed in the class Granuloreticulosa, superclass Rhizopoda, subphylum Sarcodina, phylum Sarcomastigophora and subkingdom Protozoa (Fig. 21.9). It is holozoic and feeds on diatoms.

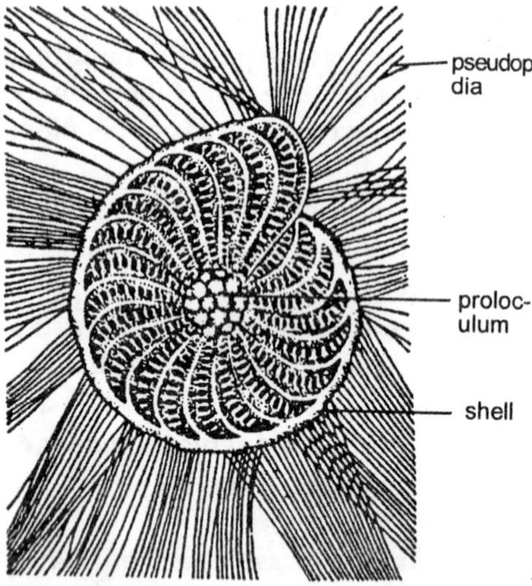

FIGURE 21.9

Elphidium crispa

Structure

1. The shape of the animal is more or less round.
2. It is surrounded with a many-chambered, multilocular and perforate calcareous shell consisting of 45 to 50 chambers.
3. The chambers are boat-shaped and spirally arranged and each completely covers its predecessor. They communicate with each other through minute pores or **foramen.**
4. The margins of the shells are provided with peg-like projections. The shell bears a keel and the last chamber has a terminal aperture.
5. The chambers are filled up with cytoplasm, which passes through the openings in the septa; the cytoplasm of the whole body of the animal is not divided into compartments but it is continuous. Nucleus may be one or many. The cytoplasm extends through the pores of the shell to form a layer over the shell.
6. Long, slender pseudopodia project outside through the minute pores present all over the shell and also through the terminal aperture; they often branch and anastomose with one another. Pseudopodia are of reticulate type, i.e. narrow, almost thread-like, branched and continually anastomose.
7. The adult *Elphidium* is **dimorphic**, some are *Schizonts* or **Agamonts**, commonly known as **microspheric** individual, characterised by an initial central chamber or proloculus of relatively minute size, and some others are **gamonts**, more commonly known as **macrospheric** individual, which give rise to gametes, characterised by a large centrally placed chamber.

Life history

Elphidium reproduces by both asexual and sexual modes. A well-marked alternation of generations is always present. The mode of reproduction in different species is not identical. The following description is applicable to *Elphidium crispa* (Fig. 21.10).

Formation of macrospheric forms

The microspheric form acts as a schizont or agamont which gives rise to the macrospheric form by a process of **schizogony**) or multiple fission.

1. With the growth of the animal, the number of nuclei in the microspheric individual increases by division and gives off chromidia to the body protoplasm.
2. With the commencement of reproduction the nuclei resolve into chromidia and the protoplasm leaves the shell, which is abandoned altogether.
3. The chromidia-laden protoplasm breaks up into a large number, about 200 amoeboid agametes or amoebulae or **pseudopodiospores.**
4. Each pseudopodiospore bears a number of chromidial granules, which fuse to

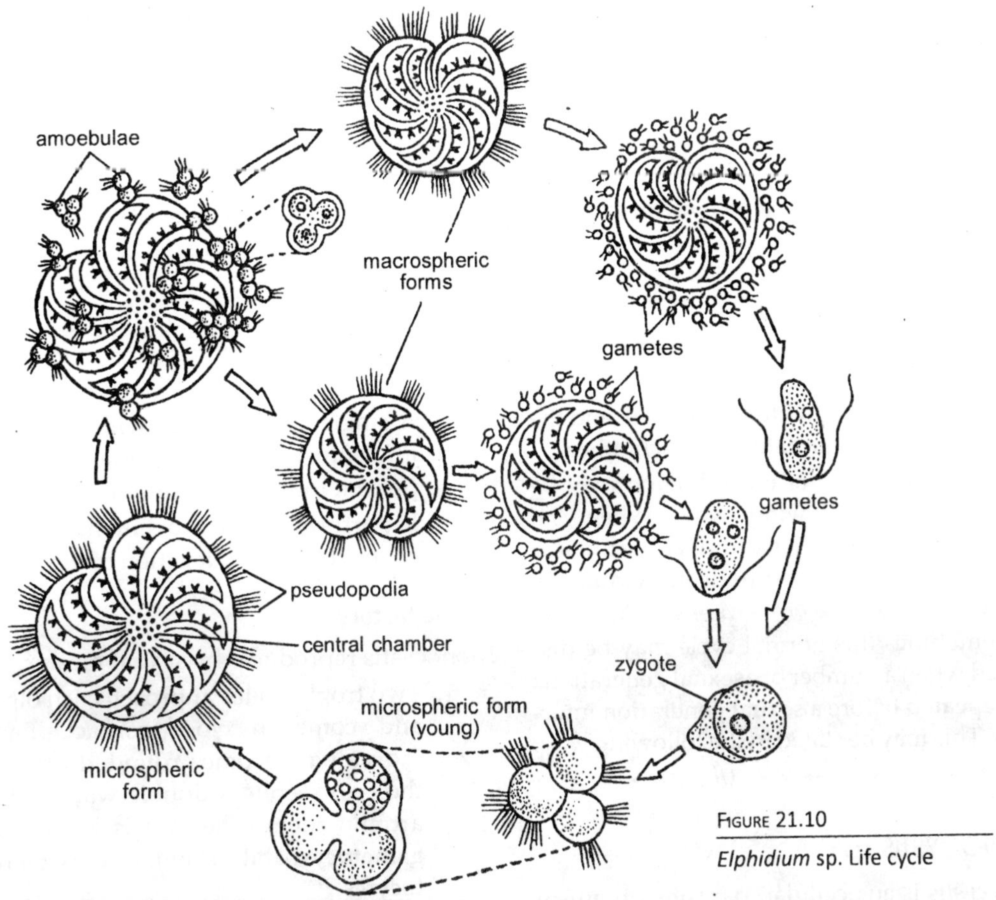

FIGURE 21.10
Elphidium sp. Life cycle

form a nucleus and secretes a single-chambered shell, the initial chamber of a macrospheric individual.

5. Each amoebula feeds, grows, secretes new chambers successively and becomes a macrospheric adult.

☒ Formation of microspheric forms

1. A single nucleus (the original nucleus of amoebula) is present in the full-grown macrospheric individual.
2. The nucleus of the gamont moves from chamber to chamber and then follows a complicated nuclear reorganization, when the primary nucleus is seen to have completely resolved into (chromidia?) a number of secondary nuclei. In some cases, it undergoes sporulation into a multitude of minute (2-5 μm) biflagellate isogametes, which are set free and escape outside the shell.
3. The gametes are of different sizes, produced by different individuals and bear unequal flagella.
4. The unequal gametes fuse in pairs, lose the flagella and the zygote is formed, which secretes a single chambered shell around it, and the microspheric individual is produced.

☒ Alternation of Generations

An alternation of generations in the life history of *Elphidium* is quite distinct. The microspheric form gives rise to the gametocytes (macro-spheric) by multiple fission. The gametocytes in turn produce gametes, which by their union form the zygotes and, thereby,

Macrospheric form (Fig. 21.10)	Microspheric form (Fig. 21.10)
1. The first formed chamber or proloculus is large.	1. The first formed chamber or proloculus is small.
2. Initial chamber is naked like other chambers.	2. Initial chamber is enclosed in a thick capsule-like cyst.
3. Nucleus one, large and laterally placed.	3. Nuclei many, small and scattered throughout the cytoplasmic mass.
4. Chromidia absent.	4. Chromidia present.
5. Produced asexually.	5. Produced sexually.

the microspheric individual is formed, and the two generations alternate with each other.

The usual mode of reproduction may be stated in the following way. When *'m'* represents microspheric individual; *am* = amoebula; 'M' = Macrospheric individual, *'fl'* = flagellula, $\rightarrow m \rightarrow am \rightarrow M \rightarrow (fl + fl) \rightarrow m...$

Sometimes this normal cycle may be disturbed when a number of asexual generations are repeated before a sexual generation interferes. This may be stated in the following way: $m \rightarrow am \rightarrow m \rightarrow am \rightarrow M \rightarrow (fl + fl) \rightarrow m ....$

MONOCYSTIS sp.

Monocystis is an acellular, parasitic organism placed in the subclass Gregarina, class Sporozoea, phylum Apicomplexa, subkingdom Protozoa (Fig. 21.11). It lives as a parasite in the vesicula seminalis of common earthworm (*Pheretima* sp.).

☒ Structure

The adult organism or the Trophozoite has the following features:

1. Body spindle-shaped and flat and covered by a thin layer of cuticle. The anterior end is broader and the posterior end is narrower.
2. The cytoplasm is divisible into a thin outer layer, the **ectoplasm** and an inner granular mass, the **endoplasm**.
3. Small thread-like structures, comparable to the muscle fibres of higher animals, known as **myonemes** are present in the innermost layer of cortex.
4. A single large nucleus with a distinct nucleolus and nuclear membrane is present towards the anterior end.
5. It performs slow movements by the contraction and relaxation of the body.

☒ Life history

Monocystis reproduces by **sexual** means:

1. Two trophozoites approach each other and come in close contact. Each individual becomes rounded off and the two secrete a double-walled cyst around them. The cyst is known as **gametocyst** and the individuals **gametocytes**.
2. The nucleus of each gametocyte undergoes repeated divisions, one of which is pregametic meiosis and many daughter nuclei are produced.
3. The nuclei arrange themselves along the periphery of the gametocyte, the cytoplasm breaks and a small bit surrounds each nucleus and thus many **gametes** are formed.
4. The gametes begin to move and the partition wall between the two gametocytes break down and the gametes are set free in the gametocyst.
5. Gametes produced from the two gametocytes unite in pairs and **zygotes** are formed.
6. Each zygote secretes a boat-shaped chitinoid case around itself. The **zygote** is now known as **spore** and the case as **sporocyst**—the whole structure is

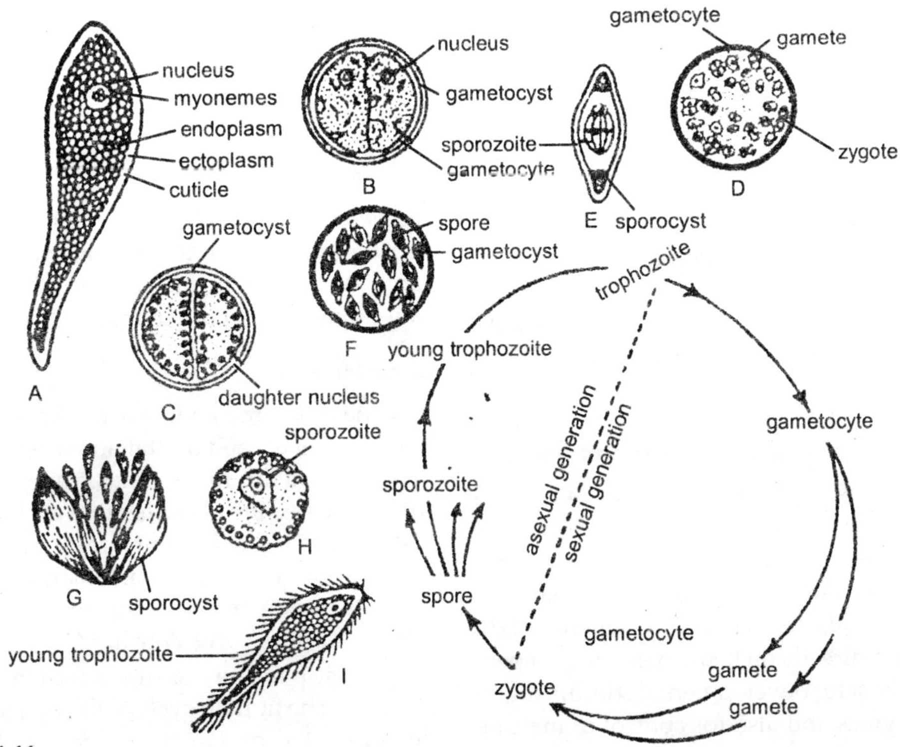

FIGURE 21.11

Monocystis sp. Structure and life cycle. A. Trophozoite. B–F. Stages of reproduction. G. Sporozoites liberating from sporocyst. H. A sporozoite in a sperm morula. I. Young trophozoite with tails of destroyed sperms attached to its body

commonly known as **pseudonavicella** (Fig. 21.11F).

7. The nucleus of the **spore** undergoes three successive fissions and 8 nuclei are produced. The cytoplasm aggregate around the nuclei and 8 sickle-shaped **sporozoites** are formed, which remain packed as a bundle within the sporocyst.

Further development does not take place if the sporocysts with the sporozoites are not transferred to a new *Pheretima*.

⊠ Transmission

The mode of transference of parasites from host to host is known as transmission. Sporocysts may be transferred by any of the following means:

1. The infested earthworm may die and, after the decomposition of the body, sporocysts get mixed with the soil and may reach the intestine of another *Pheretima* when it 'eats' soil.
2. The infested *pheretima* may be eaten by bird. The body is digested in the intestine of the bird and the undigested sporocysts coming out with the faecal matter get mixed with the soil, which may be eaten up by another *Pheretima*.
3. During mating certain amount of seminal fluid of earthworms falls on earth. Sporocysts—which remain floating in the seminal fluid—come out with it and get mixed with the soil and may be eaten up by another worm.

⊠ Changes in the new host

1. On reaching the intestine of the *Pheretima*, the sporocyst dissolves by the specific action of the enzymes and the sporozoites are liberated.

2. The sporozoites bore through the gut wall, reach the coelom and swim to the ciliated funnels of the male reproductive system to live temporarily as intracellular parasites.
3. The sporozoites leave the funnels and attack the cells of the sperm morulae and grow as intracellular parasites.
4. Young trophozoites leave the sperm morulae, come to the vesicula seminalis where they grow and become adult by absorbing seminal fluid.

⊠ Effect of parasitism on host and parasite

Parasitism is a heterospecies association (between two different species) in which one (the parasite) remains not only in continuous, intimate association with another organism (the host), but is also metabolically dependent directly or indirectly on it; the parasite is dependent on host for developmental stimuli, digestive enzymes and also for control of maturation.

On host. *Monocystis* destroys large number of sperm cells. But still the spermatozoa produced are much more than the demand and it cannot inflict any material harm on the host.

On parasite. Due to easy available ready food, all the vegetative organelles—like organelles of locomotion, digestion, etc.—are degenerated to a great extent. On the other hand, the chances of destruction during transmission from host to host are always great but this has been compensated by the production of numerous spores, so that at least a few may reach the host and the race is successfully continued.

MALARIA PARASITE (*PLASMODIUM* sp.)

The causative organisms of malaria fever in man, *Plasmodium*, are intraerythrocytic protozoan parasites, reported from all countries extending from 60° N to 40° S. The genus *Plasmodium* belongs to family Plasmodidae, suborder Haemosporina, order Eucoccidia, subclass Coccidia, class Sporozoea, phylum Apicomplexa, subkingdom Protozoa. The Plasmodia normally infecting men are *P.* (*Plasmodium*) *vivax*, *P* (*P.*) *ovale*, *P* (*P.*) *malariae* and *P.* (*Laverina*) *falciparum*. *Plasmodium vivax* causes benign tertian or vivax malaria, *P. falciparum* causes malignant tertian or cerebral or pernicious falciparum malaria, *P. ovale* cause ovale tertian malaria and *P. malariae* cause quartan malaria in man.

⊠ Malaria

It is one of the most vicious diseases of man. The word was used by Italian writer in 1873, meaning **bad air**. Malaria has played a major role in the decline of civilization and shaping history. The fall of ancient Greek and Roman empires was due to malaria. Malaria epidemic severely threatened both Japanese and Allied armies during World War II.

Malaria parasite was discovered in 1880 by Laveran. The human cycle of the parasite was demonstrated by Golgi in 1886. In 1898 Sir Ronald Ross described vector role of female mosquitoes and in the same year Bignani, Grassi and Bastianalli reported sexual cycle of *P. vivax* from the females of *Anopheles maculipennis*.

PLASMODIUM VIVAX

The malaria parasite *Plasmodium vivax* causes benign, tertian or vivax malaria in man. Fever occurs with chill every third day. An alternation of generations associated with alternation of hosts is present in the life cycle.

⊠ Life cycle and reproduction

Malaria parasites of man pass through two stages in their life cycle—an extrinsic phase in Anopheles mosquito and an intrinsic phase in man (Fig. 21.12). The extrinsic phase in which sexual reproduction occurs, is referred to as definitive and the intrinsic phase, with asexual reproduction, as intermediate.

⊠ Invertebrate phase

During its blood meal from an infected person, the female Anopheles ingests both asexual and sexual forms but only the sexual forms survive and develop, while the rest die immediately. In the gut of a susceptible mosquito the mature macrogametocytes (female) and microgametocytes (male) quickly mature into gametes. The first phase of development starts in the midgut (stomach) of the mosquito.

⊠ Formation of macrogamete

The macrogametocyte develops into macrogamete and its nucleus shifts to the surface where a projection is formed. Only one rounded macrogamete is formed from one macrogametocyte.

⊠ Formation of microgamete

Microgametocyte undergoes multiple nuclear division and mature by exflagellation. Within 10-12 minutes 5-8 thread-like microgametes break away from red blood cell sucked by the mosquito from man and become motile.

⊠ Zygote

In fertilization, a microgamete penetrates the projection of the macrogamete and the zygote is formed. This is completed within 20 minutes to 2 hours in the mosquito.

⊠ Formation of sporozoite

1. The zygote puts out a pseudopodium and forms a fusiform body, the **ookinete** or a **travelling vermicule stage**.
2. The oökinete grows and penetrates the brush border and body of a midgut epithelial cell and comes to rest, where it is transformed into a spherical mass or **oocyst**, surrounded by a cyst wall.
3. The oocyst is 6-12 μm in diameter, which increases from 6 to 60 μm.
4. The nucleus divides repeatedly and ultimately a large number of sickle-shaped sporozoites are formed from one oocyst.
5. The occyst ruptures on or about 10th day of infection and the sporozoites are released in the body cavity (haemocoel) of mosquito.
6. Vivax sporozoites are narrow, slightly curved and measures about 14 μm in length.
7. Sporozoites are distributed to various organs and tissues of mosquito, but due to their affinity for salivary glands, attack them in large number and numerous sporozoites reach the salivary glands.
8. Malarial infection may be transmitted to a new host by a single bite of such a mosquito.

• Human phase

Exoerythrocytic Schizogony or Preerythrocytic Schizogony and release of exoerythrocytic merozoites

1. Bite of an infected female Anopheles allows the entry of sporozoites in human blood. Sporozoites leave the blood stream within one hour and enter reticuloendothelial cells of liver.
2. The sporozoite becomes round or oval, about 40 μm in diameter, with somewhat clumped cytoplasm, irregular nucleus and relatively long vacuoles after 8 days.
3. The nucleus undergoes multiple division, vacuoles and clumps disappear and the cytoplasm condenses around nuclear particles, giving rise to as many as 10,000 **merozoites**.
4. The merozoites are spherical, 1-2 μm in diameter and, being released, they make their way to host's circulation.

A dormant or **secondary erythrocytic schizoite** or **hypnozoite** occurs in *P. vivax* and *P. ovale*, with one or two or more generations of exoerythrocytic merozoites. This hypnozoitic cycle may persist in liver with a period of latency for several years.

⊠ Erythrocytic stages

The exoerythrocytic merozoites enter erythrocytes and reticulocytes. The merozoite possesses a selectively adhesive surface for attachment to the erythrocyte (Bannister, 1977), and an apical complex consisting of rhoptries,

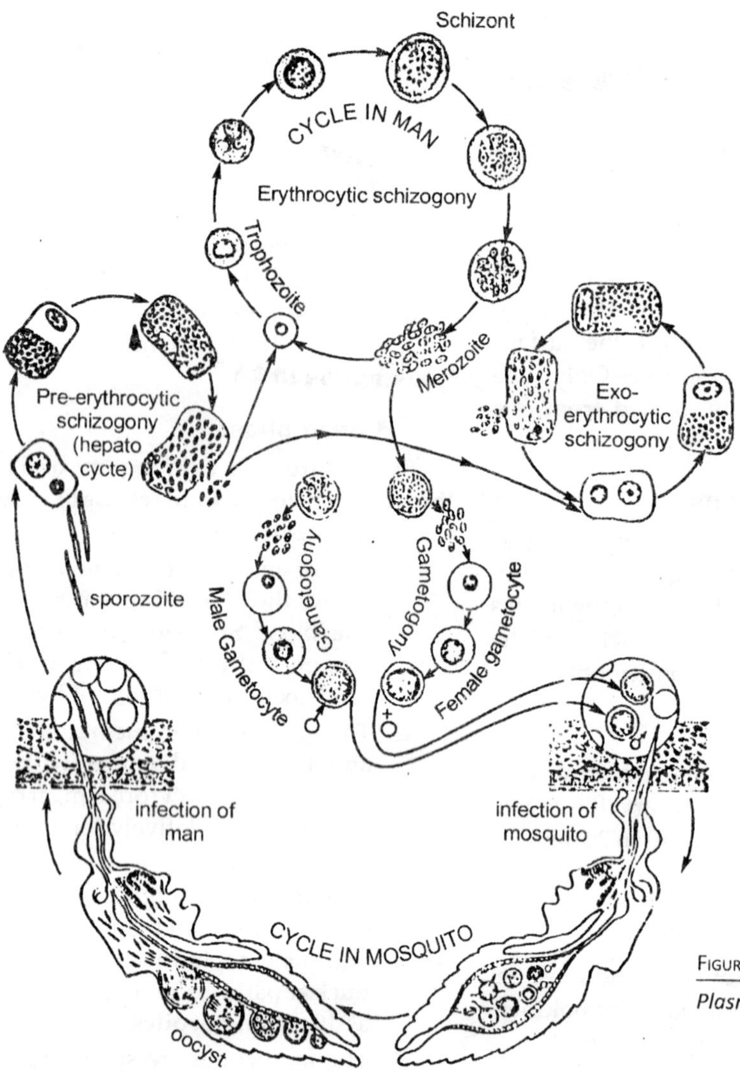

FIGURE 21.12
Plasmodium vivax. Life cycle

micronemes and a polar ring for invasion and usual organelles, such as rhibosomes, nuclear complex and other structures. As the apical complex of the merozoite is attached to the red blood cell membrane, a parasitophorous vacuole develops, where merozoite is engulfed. At this time, the outer coat of the merozoite disappears.

⊠ Trophozoite

In the red cell the merozoite is vacuolated, ring-shaped, more or less amoeboid and uninucleate. It is now called trophozoite, a name it retains until its nucleus begins to divide. The trophozoite measures 2.5 to 3 μm and consists of a cytoplasmic ring, thicker on one side, a chromatin dot (nucleus) and a vacuole. It is active and constantly sends out pseudopodia. It feeds on haemoglobin, which it does not metabolise. The residue of globin and an iron porphyrin haematin, which is called **malaria pigment** (a compound of haematin, ferrihaemic acid and protein). A coarse brick-red stripping of the cytoplasm, known as **Schüffner's dots,** appear. According to Aikawa. et. al. (1966) the parasite has a specific **cytostome** that ingests host cell cytoplasm, while Ladda (1966) considered the single food ingesting organelle resembling the **micropyle**.

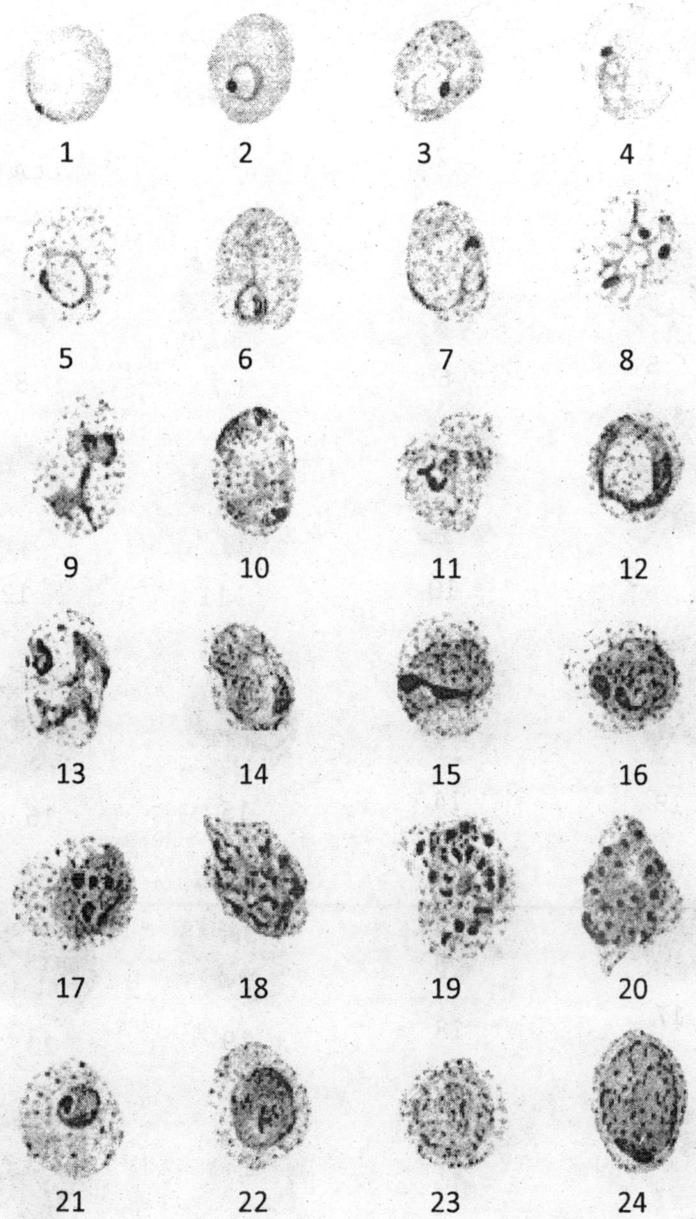

FIGURE 21.13

Plasmodium vivax. 1. Normal size erythrocyte with marginal ring-form trophozoite. 2. Young signet-ring form trophozoite in a macrocyte. 3. Slightly older ring-form trophozoite showing basophilic stippling. 4. Polychromatophilic erythrocyte containing young parasite with pseudopodia. 5. Ring-form trophozoite showing pigment in cytoplasm in an enlarged erythrocyte containing Schüffner's dots. (Schuffner's dots do not occur in all host cells containing growing and older forms of *P. vivax.*) 6,7. Medium trophozoite forms. 8. Three amoeboid trophozites with fused cytoplasm. 9,11,12,13. Older amoeboid trophozites in process of development. 10. Two amoeboid trophozites in one host erythrocyte. 14. Mature trophozoite. 15. Mature trophozite with chromatin in process of dividing. 16,17,18,10. Schizonts showing progressive stages of division. 20. Mature schizont. 21,22. Developing gametocytes. 23. Mature microgametocyte. 24. Mature macrogametocyte. **See color plates Fig. 21.14**).

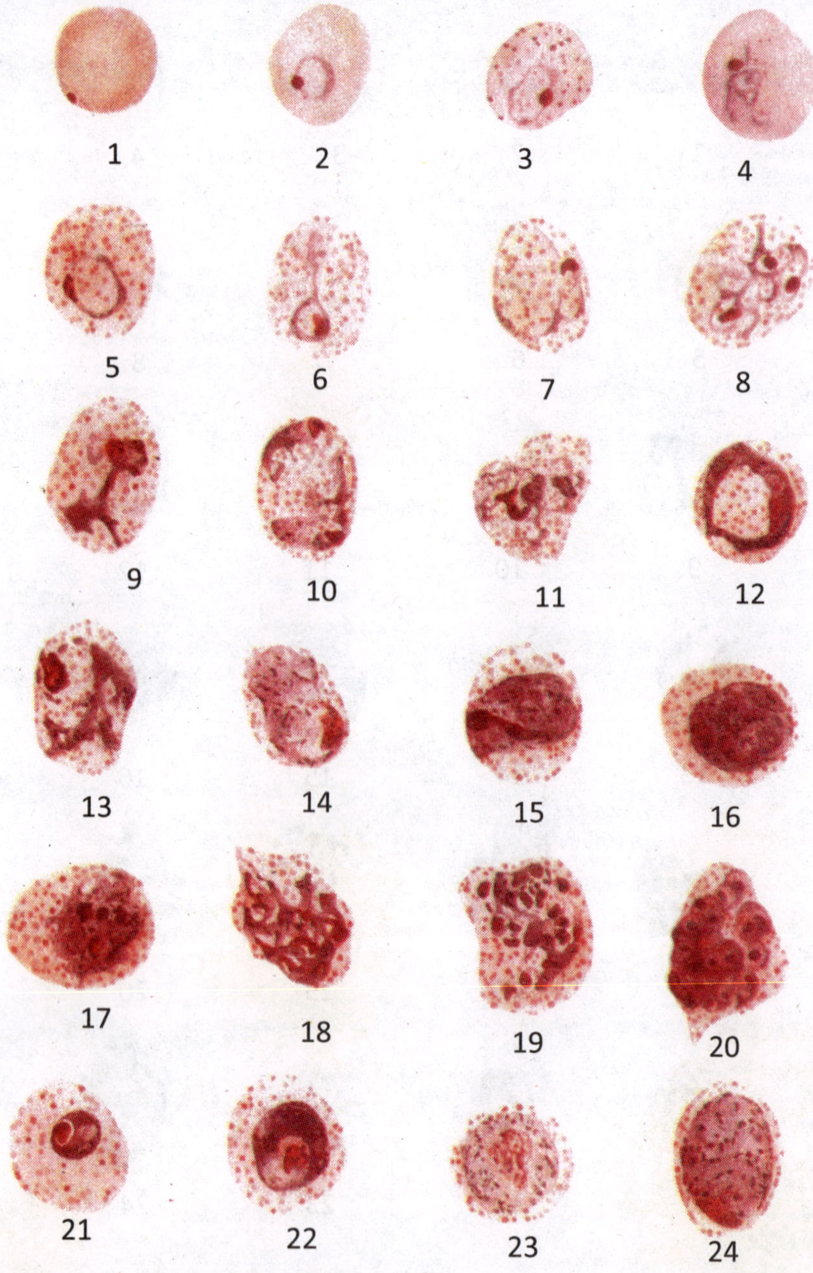

FIGURE 21.14

Plasmodium vivax. 1. Normal size erythrocyte with marginal ring-form trophozoite in a macrocyte. 2. Young signat-ring in a macrocyte. 3. Slightly older ring-form trophozoite in erythrocyte showing basophilic stippling. 4. Polychromatophilic erythrocyte containing young parasite with pseudopodia. 5. Ring-form trophozoite showing pigment in cytoplasm in an enlarged erythrocyte containing Schüffner's dots (Schüffner's dots do not occur in all host cells containing growing and older forms of *P. vivax.*). 6, 7. Medium trophozoite forms. 8. Three amoeboid trophozoites with fused cytoplasm. 9, 11, 12, 13. Older amoeboid trophozoites in process of development. 10. Two amoeboid trophozoites in one host erythrocyte. 14. Mature trophozite. 15. Mature trophozite with chromating in process of dividing. 16, 17, 18, 19. Schizonts showing progressive stages of division. 20. Mature schizont. 21, 22. Developing gametocytes. 23. Mature microgametocyte. 24. Mature macrogametocyte.

☒ Schizont

The trophozoite becomes full grown after about 36-40 hours. All amoeboid activities are lost, the parasite assumes round shape and measures 9-10 μm. The vacuole disappears and the mature schizont is formed. In the next 6-8 hours the mature schizont undergoes erythrocytic schizogony; the nucleus divides and, subsequently, 12-18 daughter individuals, the **erythrocytic merozoites** are produced. The merozoites arrange themselves in the form of a '**rosette**'. The red blood cell can no more hold the parasites, it bursts and merozoites released in the blood stream. The release of merozoites is associated with the rise of body temperature of the host.

☒ Erythrocytic merozoites

The parasite is oval in shape, measures 0.5 μm in width and contains a central chromatin. Following release in blood stream many are destroyed by the immune system of the host, but some invade new red blood cells and the cycle is repeated in every 48 hours.

☒ Gametogony

Certain schizonts are modified for sexual reproduction and the resulting merozoites instead of undergoing schizogony change into sexual form known as gametocytes.

1. The earliest gametocyte of *P. vivax* is round or oval with cytoplasm taking up blue stain.
2. The gametocyte bears a spherical mass of chromatin that stains deep red or violet.
3. The microgametocyte measures 9-10 μm in size and when fully developed fills the erythrocyte.

   It contains a large nucleus in which granules or fine fibrils of chromatin are arranged in the form a spindle.
4. The chromatin may be divided into several masses, collected near the periphery in those, which are about to undergo exflagellation.
5. The microgametocytes undergoing exflagellation are irregular, pale blue bodies, from which several long thread-like red or violet flagellated objects, the microgametes project. Pigment is in an irregular mass and deep red staining chromatin is abundant.
6. The macrogametocytes are 10-12 μm in size. It has a smaller nucleus than that of microgametocyte and chromatin is arranged in a compact mass near the periphery. The cytoplasm stains bright blue and chromatin deep red or violet.

   The micro- and macrogametocytes are sucked in by the feamle Anopheles during its blood meal and the life cycle of *P. vivax* is completed.

☒ Pathogenecity and Clinical features

The incubation period is 10-15 days but may be many months depending on a number of factors.

1. *Plasmodium vivax* causes **vivax malaria** or **Benign Tertian** (BT) malaria or **tertian ague.** Fever is at first remittent, going on to intermittent tertian paroxysm. Convulsion may occur in high fever. Each paroxysm of the fever shows a succession of three stages cold, hot and sweating. The cold stage lasts 20 minutes to an hour. The hot stage lasts 1 to 4 hours, and sweating stage 2 to 3 hours. Vomitting are usual. Fever paroxysms typically recur every 48 hours.
2. The spleen enlarges (splenomegaly) by the second week of fever. The liver uniformly enlarge in latter stages. All the organs become pigmented with slate grey pigments.
3. A microcytic or a normocytic anaemia develops due to enormous breakdown of RBC in long suffering.

☒ Management

Management of malaria involves protection against the vectors and proper treatment immediatly after detection.

1. Anopheles, the malaria vectors, usually bite at night. They enter houses at dusk

☒ General plan of life cycle of Plasmodium sp.

Sexual stage in mosquito

Macro- and microgametocytes enter the mosquito gut with RBC.

↓

Micro- and macrogamete unite and the zygote is formed.

↓

Oökinete pierces the stomach wall, undergoes sporulation and sporozoites are formed. (Infective to man)

↓

Asexual cycle in man

Undergoes preerythrocytic schizogony in the liver cells; the process is repeated to form numerous merozoites, merozoites invade RBC.

↓

Erythrocytic schizogony

↓

Exoerythrocytic schizogony

Macro- and microgametocytes in RBC of peripheral blood

and after a blood-meal rest on the walls or ceiling, leaving the room at dawn. Spraying a room with a short-acting insecticide from a flit gun is useful. All possible hiding places in the room should be sprayed.

2. Mosquitoes breed in water. All possible sites for stored water—holes, water containers, reservoirs, broken pots, rejected motor tyers, etc. should be periodically examined for mosquito larvae.
3. Edges of streams, and ponds with vegetation provide shelter for mosquito larvae and these should be cleaned regularly.
4. Rearing of larvivorus fishes like kholisa, tilapia, techokha, gambusia in ponds; treatment of breeding places with larvicidal oils or with insecticides in oil or water emulsions kill the mosquito larvae.
5. Spraying of DDT in and around the house reduces malaria cases. Spraying insecticides in and around the house before onset of rain and repeat after four months is helpful.
6. Regular resin smoking (doono) in the bedrooms and around the house in dusk are helpful.
7. Personal management like regular use of mosquito net and if possible treated with WHO specified pyrethroid insecticide, reimpregnated on a regular six-monthly or annual basis is highly effective.
8. Immediate appropriate drug treatment of infected persons following diagnosis is very much helpful. In addition to quinine products, artemisinin and its derivatives, the rapid acting Chinese drugs for the treatment of malaria is well known. Artemisinin is prepared from the Chinese herb *Artemisia annua*.

PARAMOECIUM sp.

Paramoecium is an acellular, fresh water animal placed in the subclass Hymenostomatia, class Oligohymenophorea, phylum Ciliophora, subkingdom Protozoa.

⊠ Structure

1. The body is like a crude slipper (Fig. 21.15), the anterior end blunt and the posterior end narrower.
2. The surface is covered with a tough and flexible pellicle and cilia arising from basal granules are arranged in longitudinal spiral rows.
3. A large depression known as **peristome** is present on the ventral surface. The peristome narrows down to form the **gullet** which ends in the soft endoplasmic mass.
4. The cytoplasm is divisible into an outer firmer **cortex** and an inner granular **medulla**.
5. The **nuclei** are two in number, one is larger, called the **macronucleus** and the other is smaller, the **micronucleus**.
6. The micronucleus lies close to the macronucleus.
7. Contractile vacuoles are two and placed at the opposite poles. To each of them are connected a number of radiating feeder canals. In some species of *Paramoecium* (*P. putrinum* and *P. trichium*) the radiating canals are absent.
8. Food vacuoles are many and move along a definite path.
9. Under certain conditions, the cortex is seen to project into minute radially arranged threads called **trichocysts**.

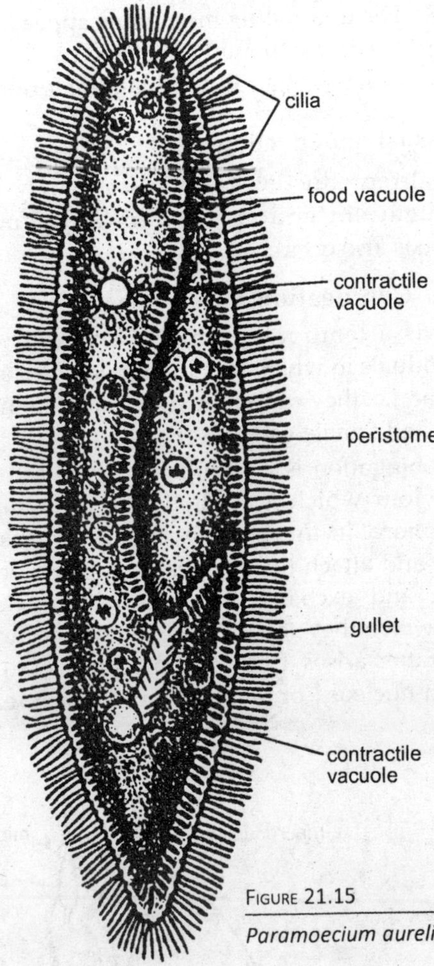

FIGURE 21.15

Paramoecium aurelia

⊠ Reproduction

Paramoecium reproduces both by asexual (Fig. 21.15) and sexual (Fig. 21.16) means. Asexual is the normal mode of reproduction in *Paramoecium*.

⊠ Asexual reproduction

1. It takes place by transverse binary fission. One fission takes about 2 hours and the next fission starts after 24 hours of the first fission.
2. Both the micro- and macronuclei elongate and constrictions appear in the middle.
3. A small, new gullet is formed from the posterior end of the old gullet and it moves backward.
4. A constriction appears at the middle of the cytoplasmic mass.
5. A new contractile vacuole appears on each side of the constriction.
6. The constrictions of both the cytoplasm and nuclei grow deeper. The micronucleus divides into two by **mitosis**, and the macronucleus by **vegetative division**.
7. Finally, the cytoplasm is divided into two equal halves and each half receives one micro- and one macronucleus, two contractile vaculoes and one gullet.

8. The undulating membrane appears in the newly formed gullet and two identical daughter individuals are formed.

Sexual reproduction

Sexual reproduction takes place mainly by conjugation; endomixis and autogamy. Hemixis also occurs.

A. Conjugation

It is a temporary union between two individuals in which the uniting gametes are similar, i.e. they cannot be differentiated into male and female gametes (Fig. 21.17).

Conjugation is a peculiar type of sexual behaviour which is limited to the phylum Ciliophora. In this process, two individuals come and attach themselves along their oral sides, and exchange their gamete nuclei, after which they separate and a new nuclear apparatus arises from the products of the fusion-nucleus, or synkaryon in each exconjugant, while the old macronucleus degenerates. Following are the main events in the process of conjugation:

1. Two individuals known as **conjugants**, approach each other and become applied along their oral sides.
2. The macronucleus remains unchanged but the micronucleus of each conjugant undergoes two divisions, one of which is **meiotic**, producing four haploid nuclei.
3. Three out of four disappear and the fourth one divides to produce two pronuclei. The pronuclei usually take a position near the oral region, where a cone-shaped protrusion, the **paroral** cone extends from each towards the other.
4. One of the pronuclei is known as the **active pronucleus** and the other as the **passive** or **stationary pronucleus**.

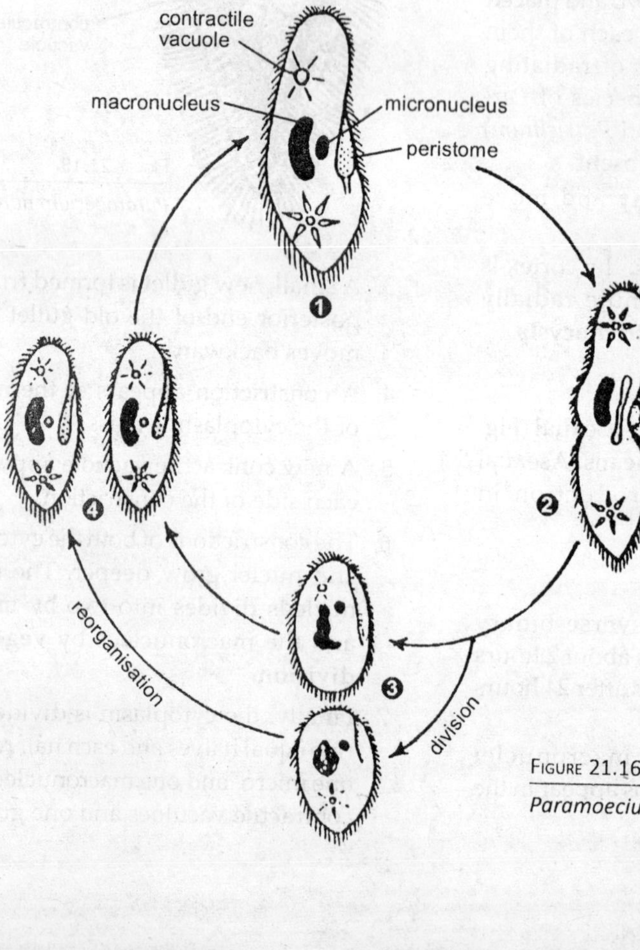

FIGURE 21.16
Paramoecium sp. Binary fission

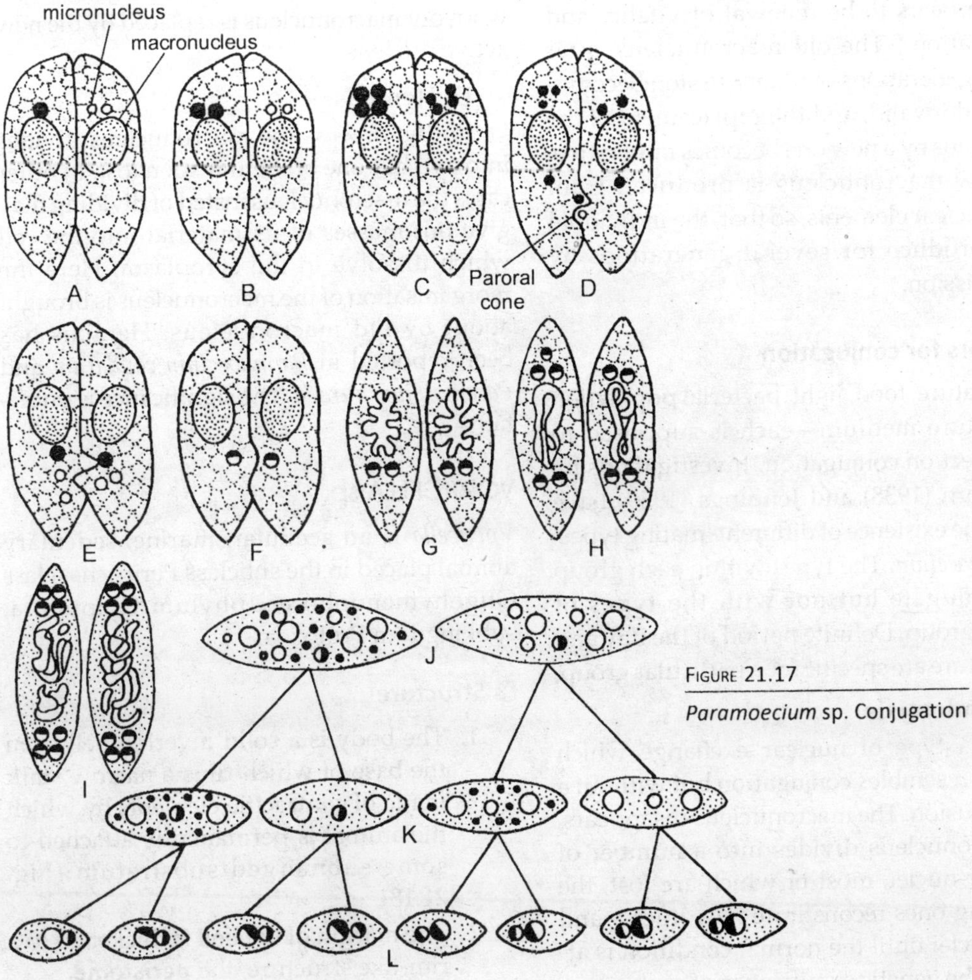

FIGURE 21.17

Paramoecium sp. Conjugation

5. The active pronucleus of each individual migrates through the paroral cones and fuses with the passive pronucleus of the other and the fusion nucleus is known as fusion or **zygote nucleus** or **synkaryon**.
6. The conjugants begin to separate, the macronucleus is gradually transformed into a skein, which breaks up and is ultimately lost. Each zygote nucleus divides thrice to form 8 nuclei. Four of these begin to grow and are known as **macronuclear anlagen**, since each of them will develop into a macronucleus.
7. As growth of the anlagen proceeds, three of the other four nuclei grow slightly, then begin to degenerate and ultimately disappear, while the eighth nucleus becomes the functional micronucleus.
8. At this stage binary fission starts. In fission, only the micronucleus divides and the macronuclear anlagen are simply distributed in the four descendants. Each daughter individual gets one macro- and one micronucleus. During these changes the fragments of the old macronucleus are gradually absorbed, while the anlagen attain their full size.

⊠ Effect of conjugation

The essence of conjugation is the exchange of nuclear materials between two individuals; its effect appears to be renewal of vitality and rejuvenation.* The old macronucleus, after several generations of binary fission becomes tired and invalid, and the replacement of the old nucleus by a new one becomes imperative. The new macronucleus is produced from micronuclear elements, so that, the individual can reproduce for several generations by binary fission.

☒ Factors for conjugation

Temperature, food, light, bacterial population and culture medium—each is supposed to have effect on conjugation. Investigations by Sonneborn (1938) and Jennings (1939) have shown the existence of different 'mating types' in *Paramoecium*. The types within each group will conjugate but not with the types of another group. Definite period of the day and temperature are specific for a particular group.

B. Endomixis

It is a type of nuclear exchange which strongly resembles conjugation but without a nuclear fusion. The macronucleus disappears, the micronucleus divides into a number of daughter nuclei, most of which are lost, the remaining ones reconstitute new micro- and macronuclei until the normal condition is attained. The genetic significance of such a process is uncertain. Conjugation retards the rate of endomixis. It seems to be a substitute for conjugation since it occurs in the absence of the latter. The main purpose is to replace periodically the old, worn out macronucleus which results in the rejuvenation of the cell.

C. Autogamy

This process occurs in single individuals and almost parallels endomixis. As the term signifies, the gametes produced from the same individual fuse and neclear reorganization takes place from this nucleus. The process is common in *Paramoecium aurelia*. Autogamy brings about rejuvenation of the race. The old worn out macronucleus is replaced by the new active nucleus.

D. Hemixis

In this process the macronucleus divides into two but one of them is soon lost. In other cases the macronucleus insted of dividing may give out masses of its material into the cell which dissolve in the cytoplasm. Here the reorganisation of the macronucleus is brought about by old macronucleus. Hemixis has been reported in *Paramoecium caudatum* and *Paramoecium aurelia*. The significance is uncertain.

VORTICELLA sp.

Vorticella is an acellular, marine, sedentary animal placed in the subclass Peritricha, class Oligohymenophorea, phylum Ciliophora, subkingdom Protozoa.

☒ Structure

1. The body is a solid inverted bell from the base of which runs a narrow **stalk** made of contractile elements by which the animal is permanently attached to some submerged substratum (Fig. 21.18).
2. The margin of the bell is thickened to a rim-like structure, the **peristome**.
3. A disc-like structure is present on the peristome at the top of the bell.
4. The mouth lies between the peristome and the disc and the tube connecting the gullet and the mouth is known as **vestibule**. In the vestibule opens the reservoir and it contains the **anal spot**.
5. The cilia are large and arranged in rings around the peristome and the disc. The cilia fuse to form an undulating membrane in the vestibule.
6. **Reservoir, food vacuoles** and a contractile vacuole are present.
7. Nuclei are two in number; the **micronucleus** is small rounded and the

* These ideas are seriously questioned now-a-days. Jennings' experiments have shown that the importance of conjugation is in the production of diversities.

macronucleus is large and horse shoe-shaped.

☒ Reproduction

Vorticella reproduces both by asexual and sexual means. Asexual reproduction is effected by fission and this is the normal mode of reproduction in Vorticella.

☒ Asexual reproduction

1. It takes place by **longitudinal** binary fission (Fig. 21.18).
2. Both the micro- and macronuclei become elongated and a constriction appears in the middle of each nucleus.
3. A constriction appears at the middle of the disc and a mouth, vestibule and gullet appear on that half which lacks these structures.
4. The constriction of the cytoplasm runs inward, each nucleus divides into two; the micronucleus by **mitosis** and the macronúcleus by **vegetative division**; one new contractile vacuole, reservoir and anal spot appear and two identical daughter *Vorticella* are produced on the original stalk.
5. One of the two daughter individuals gets detached from the stalk, develops a second circlet of cilia at the basal end, assumes the shape of a barrel and becomes free swimming.
6. After a brief period of free life the second circlet of cilia is lost, and developing a stalk from that end, the animal becomes permanently attached to some substratum.

   Vorticella is a sedentary animal. By developing a free living stage at the time of reproduction, the nonlocomotory species becomes locomotory and, by its dispersal, overcrowding of individuals in a limited area is avoided.

☒ Sexual reproduction

It takes place by conjugation (Fig. 21.18). After a number of generations produced by binary fission the individuals appear weakened and unhealthy and conjugation takes place at this stage. The conjugation is permanent in *Vorticella* and it resembles fertilization to some extent. The different types of gametes produced during conjugation are known as macrogamete and microgamete.*

☒ Formation of macrogamete

1. The micronucleus of some of the weakened, ordinary stalked individuals, undergoes one meiotic division and two daughter nuclei with half the original number of chromosomes are produced in each individual.
2. One of the two nuclei disappears and the individual is now known as macrogamete.

☒ Formation of microgamete

1. Vegetative structures of some of the weakened individuals disappear and the micronucleus divides thrice, of which one is meiotic and 8 daughter nuclei with half the original number of chromosomes are formed.
2. The cytoplasm fragments, a small bit surrounds each daughter nucleus and 8 microgametes are produced.
3. The microgamete is oval with one end pointed. At the pointed end, a circlet of cilia develops and the gametes come out leaving the mother animal empty.

☒ Conjugation

The microgametes swim in water and one comes to rest on a macrogamete at the junction of the bell and the stalk. A **protoplasmic bridge** is established between the two and the nucleus of the microgamete migrates to fuse with the nucleus of the macrogamete. The

* Conjugation in *Vorticella* differs from that of *Paramoecium* at least in two major features: *a.* the gametes are dissimilar, *b.* the union is permanent.

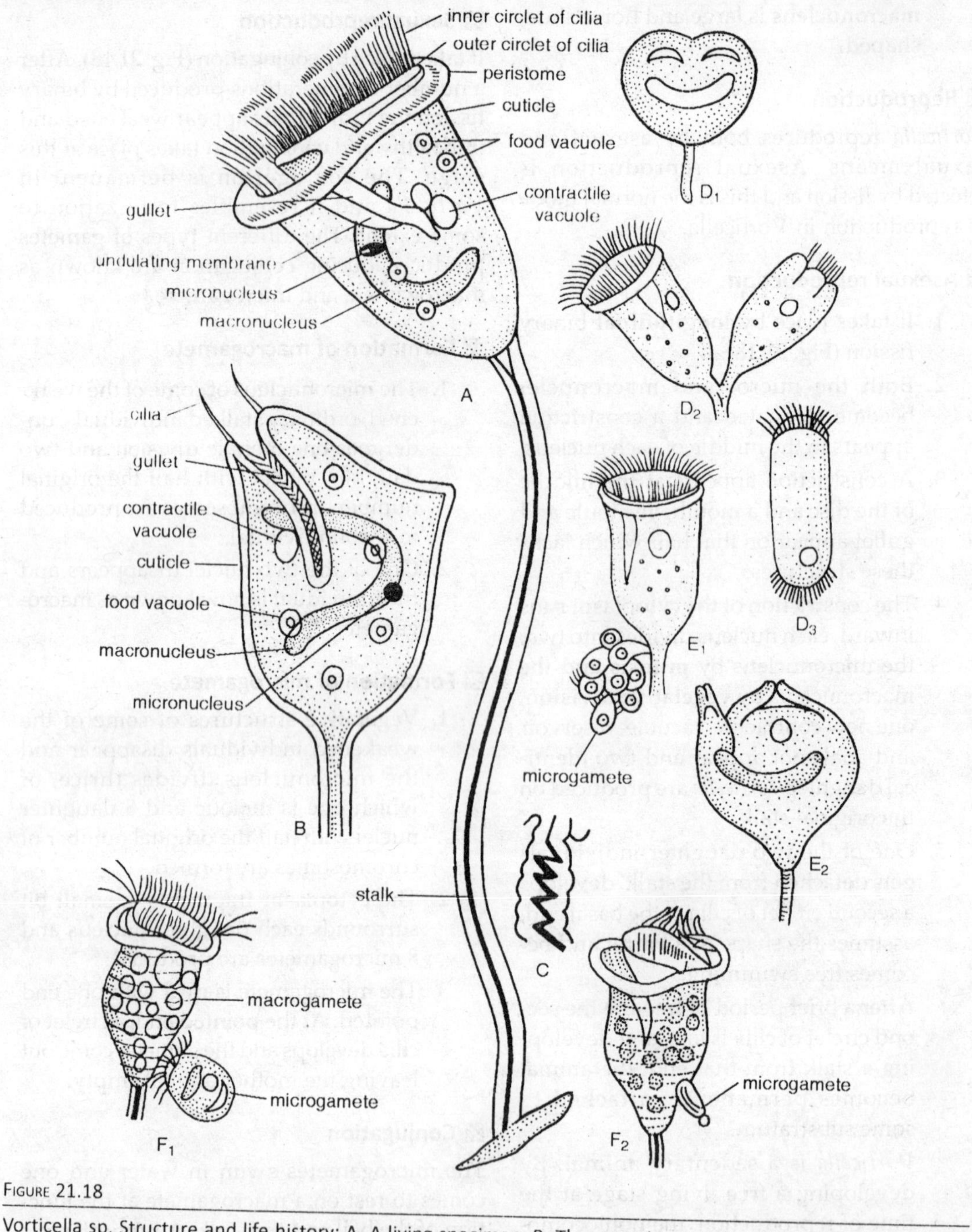

FIGURE 21.18

Vorticella sp. Structure and life history: A. A living specimen. B. The same in optical section. C. The coiled stalk. D_1–D_2. Stages of binary fission. D_3. Free daughter individual. E_1. Formation of microgamete. E_2. Formation of macrogamete. F_1–F_2. Conjugation

cytoplasm of the microgamete is lost and the macrogamete now with the fusion nucleus acts as a healthy ordinary *Vorticella*.

⊠ Effect of conjugation

The essence of conjugation is the reception of nuclear material from another individual: its effect appears to be a renewal of vitality which is essential for rapid reproduction by binary fission.

LOCOMOTION IN PROTOZOA

Change of place is known as locomotion. Power of locomotion is virtually an essential feature of free-living animals. For procuring food, shelter, for avoiding overcrowding and for the purpose of reproduction, locomotion seems to be indispensable. The members of the subkingdom Protozoa exhibit a great diversity both in structure and habitat and, consequently, varied forms of locomotory organelles are found in the group, possibly in response to different life conditions. The organelles, by which the change of place is effected, are known as **locomotory organelles**. Following different modes of locomotion are found in Protozoa:

1. **Pseudopodial movement.** Some protozoons move with the help of pseudopodia. Pseudopodia are blunt, finger-like temporary protrusions of the cytoplasm. These may be variously shaped. Large number of pseudopodia are present on the body surface of some individuals.

   In case of *Amoeba*, the pseudopodium is known as **lobopodium**. The pseudopodium, which is in the advancing direction and in contact with the substratum, helps in locomotion. The pseudopodium is fixed on the support by some adhesive secretion and the protoplasm of the body gradually flows into it. As to the mechanism of pseudopodial movement, the most accepted view is **change of viscosity** (Hyman). It states that **sol-gel** conversion initiates movement in the protoplasm. The ectoplasm —where a pseu-dopodium is formed— is dissolved by the formation of an acid and the endoplasm flows. New pseudopodia appear and, by the repetition of the process, the animal slowly creeps forward. The sol-gel phase transition of the cytoplasm during locomotion is the result of disassembly and assembly of the actin cytoskeleton and dynamic interaction between cytoskeletal actin and the motor molecule myosin. The myosin of ameoboid cells is monomeric (Fig. 21.19). Monomeric myosin forms cross-links with actin and in the presence of Ca^{2+} and ATP causes contraction. The ectoplasm of an amoeba is contractile and this contraction provides the force to move the cytoplasm forward during amoeboid movement. The rear contraction would force the cytoplasm forward by compressing it, while front contraction would pull the cytoplasm forward by exerting tension through the cytoskeleton on trailing part of the cell. In *Polystomella*, the pseudopodia are known as **reticulopodia** (branching).

2. **Flagellate movement.** Certain protozoons move with the help of flagella. Flagella are whip-like structures in the formation of which cytoplasm takes part. These are usually 2-4 in number. A flagellum has an inner stiff structure, known as **axoneme**, which is surrounded by a protoplasmic sheath. *Euglena* moves by the rapid lashing movement of the single flagellum, which is placed at the anterior end of the body and is known as **tractellum**. In many cases, flagella act as small oars. The most plausible mechanism, as suggested by Lowdens is that the basic function of a flagellum is to produce rotation of the organism on its major axis, along with a mechanism which directs the animal.

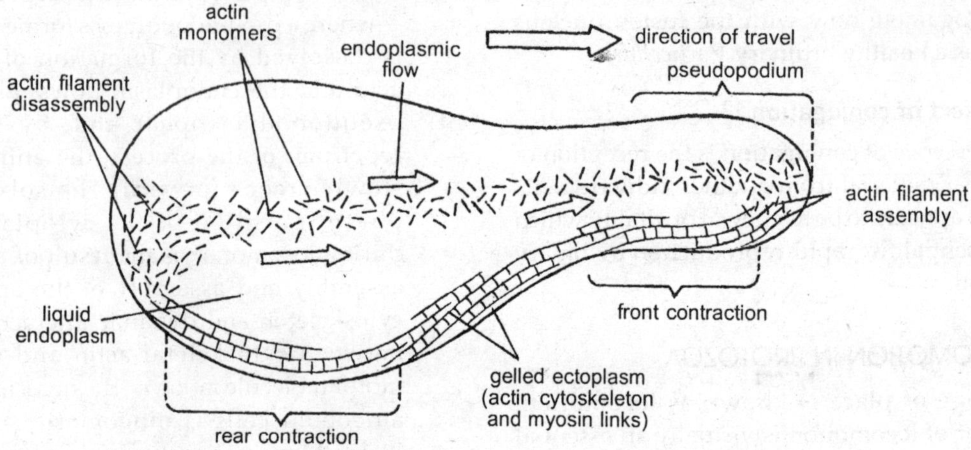

FIGURE 21.19

Amoeboid movement—rear contraction and front contraction shown

3. **Ciliary movment.** Some protozoons move with the help of cilia. Cilia are small hair-like structures, present usually in large numbers on body surface. In some cases (*Vorticella*), the cilia are restricted only in a concentric circlet at the distal end of the body. Cilia are usually arranged in definite rows. A cilium has practically the same histology as that of a flagellum. The *Paramoecium* moves forward in a rotating fashion by the beating movement of the cilia. The cilia act as small oars and the backward strokes are swift, which push the animal forward (Fig. 7.5).
4. **Peristaltic movement.** Some protozoons move with the help of myonemes. **Myonemes** are small thread-like contractile fibrils usually located in the inner layer of ectoplasm. In *Monocystis*, the contraction of the fibrils is responsible for the contraction of the pellicle and the animal slowly moves forward in a jerking fashion.

Protozoa embraces a large number of varied individuals. Different forms of locomotory organelles are present in different stages of the life of the same animal. The members of the subphylum Sarcodina possess pseudopodia in the adult stage but, in the early stages, some of them bear flagella, which is the characteristic locomotory organelles of the subphylum Mastigophora. There are some colourless mastigophores, which are permanently amoeboid, closely resembling amoebae but have a long flagellum at the same time. Such phenomenon indicates the possible ancestry of protozoa to a unicellular animal, which was capable of assuming either the amoeboid or flagellar phase.

NUTRITION IN PROTOZOA

Different modes of nutrition are found in protozoa. The principal modes of nutrition are holozoic, holophytic, saprozoic or saprophytic, parasitic, mixotrophic, etc.

1. **Holozoic.** (*holos* = whole + *zoon* = animal). Most of the free-living protozoons depend on engulfing the whole of the solid food particles. The food may be bacteria, protozoons or other small particles. The food capturing devices are of the following types:

a. **Circumvallation.** Encircling the active prey by pseudopodia. Exmaple: *Amoeba proteus* (Fig. 21.19A).

b. **Invagination.** The prey is immortile, food cup is formed and cytoplasm flows

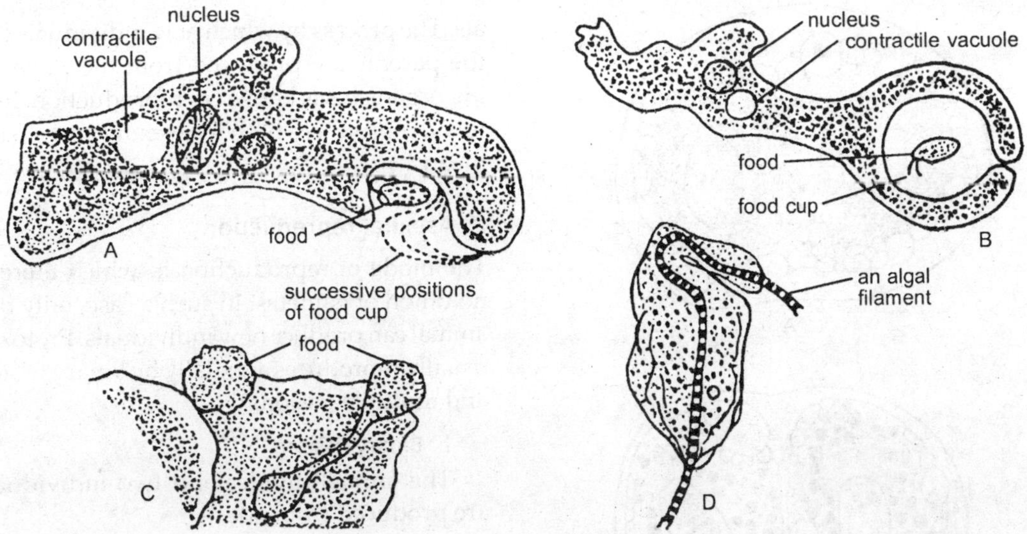

FIGURE 21.20

Feeding in lobose rhizopods. (*Amoeba* sp.): A. Circumvallation. B. Invagination (food cup). C. Import. D. Circumfluence.

around to engulf it. Example: *Amoeba dubia* (Fig. 21.20B).

c. **Import.** The prey is first killed by toxin and taken in the cytoplasm by invagination. Example. *Amoeba* sp. (Fig. 21.20C).

d. **Circumfluence.** Food like algae is simply drawn into the body. Example: *Amoeba* sp. (Fig. 21.20D).

The digestion mostly occurs in food vacuoles, in which acids or alkalis and digestive enzymes are secreted by the surrounding cytoplasm. Nutrients are absorbed by the cytoplasm and the undigested portion of the food in the vacuole is egested by repulsion (*Amoeba*), or through anal aperture or cytopyge (*Paramoecium, Nyctotherus,* etc.).

2. **Holophytic** (*holos* = whole + *phyton* = plant origin) or **autotrophic (photosynthesis)**.

The chlorophyll-bearing flagellates or phytomastigophores synthesize carbohydrates from raw materials, water and carbon dioxide using radiant energy from the sun. Example: *Euglena*.

3. **Saprophytic** (*sapros* = rotten) or **Saprozoic**. Some protozoons cause disintegration of food materials into smaller molecules outside their body and subsist entirely on the dissolved organic matters. The monosaccharides, amino acids and other molecules are absorbed through the body surface. Examples: *Entamoeba histolytica, Gratium* (a dinoflagellate), etc.

4. **Parasitic.** Parasitic protozoons are total or partial parasites in different organs of the hosts and obtain their food fully or partially from the organs. The parasites either rob the food of its host (*Opalina. Gregarina, Nyctotherus,* etc.) or feed upon the tissues of the hosts (*Plasmodium, Entamoeba,* etc.).

REPRODUCTION IN PROTOZOA

Every individual dies in course of time and the gap created by the death of the individual is filled up by the appearance of new individu-

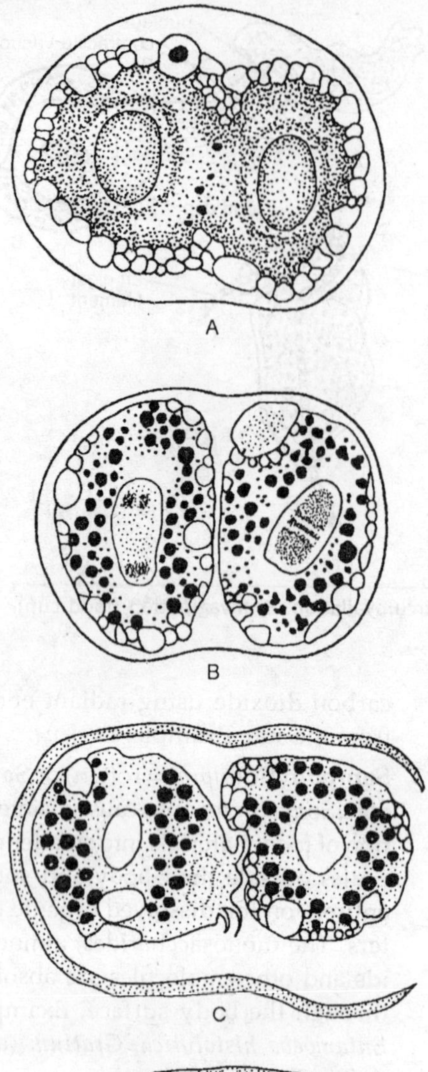

FIGURE 21.21

Paedogamy in *Actinophrys* sp. A. Encysted individual divides. B. Each daughter undergoes maturation division. C. Left daughter individual acts as male, right as female. D. Zygote

als. The process by which new individuals like the parents are produced from the preexisting organisms is known as reproduction. Protozoa reproduces both by asexual and sexual means.

⊠ Asexual reproduction

The mode of reproduction in which there is no union of gametes. In such a case, only one animal can produce new individuals. Protozoa usually reproduces asexually by binary fission and multiple fission.

I. Binary fission

The animal divides and two individuals are produced from one.

1. The micronucleus divides into two by a simplified form of mitosis.
2. The macronucleus divides into two by amitosis.
3. The cytoplasm divides into two equal halves by a constriction.
4. The daughter individuals can reconstruct the wanting structures which it does not obtain from the parent. Asymmetrical structures like gullet, peristome of *Paramoecium* cannot be equally shared by both the daughter individuals.

Binary fission is again of two types:

a. **Transverse fission.** The animal divides transversely into two Examples: *Amoeba, Paramoecium*, etc.

b. **Longitudinal fission.** The animal splits into two along the long axis of the body. Examples: *Euglena, Vorticella*, etc.

c. Oblique binary fission. The plane of fission is oblique. Examples: *Dinoflagellata, Ceratium, Cochliodinium*, etc.

II. Multiple fission or Sporulation

Many individuals are produced from one at a time. Examples: Some *Amoebae, Euglena, Polystomella*, etc.

1. The animal becomes encysted, the nucleus divides repeatedly and a large number of minute daughter nuclei are produced.

2. The cytoplasm fragments and a small bit of it surrounds each daughter nucleus and, thus, many minute animals are formed.
3. Under favourable circumstances the cyst bursts and these small animals come out and grow to the adult stage.

Multiple fission is common in sarcomastigophorans and apicomplexans. The process has been differently named according to the period and time of occurrence. Following types of multiple fission are found in protozoa.

a. **Gamogony.** Products are gametes. Examples: *Monocystis.*
b. **Scizogony** or **agamogony**. Occurs in asexual stages, the resulting individuals are known as agametes or merozoites. Example: *Plasmodium.*
c. **Sporogony.** It occurs following sexual fusion. The products are surrounded by a cyst or a resistant covering and termed as spores. Motile spores are known as **swarmers** or **swarmospores.** The swarmers are of two types:
   i. **Flagellospore.** Spores bearing flagella.
   ii. **Pseudopodiospore** or **amoebospore.** Spores bearing pseudopodia.

III. **Plasmotomy**

The multinucleate individual divides into many small multinucleate offsprings by simple division of cytoplasm independent of nuclear division. The daughter individuals regain the normal size and the number of nuclei is restored by further nuclear division.

IV. **Budding**

New individuals are produced by separation of a portion of the cytoplasm of the parent organism with a daughter nucleus. It may be simple or multiple, exogenous or endogenous. Budding is common in suctorians. Examples: *Noctiluca, Tokophrya*, etc.

⊠ Sexual reproduction

The modes of reproduction in which two gametes unite to form a new individual is known as sexual reproduction. The two units (male and female gametes) from two separate individuals unite by fusion of their cytoplasm, followed by the union of their nuclei. Most protists (protozoa) can continue to live, multiplying asexually for prolonged periods and may undergo sexual reproduction only at irregular intervals. However, there are many protozoans in which sexual reproduction is of regular occurrence. Sexual reproduction involves meiotic division reducing the chromosomes to haploid number. In majority, reduction division occurs shortly before syngamy. This is called **gametic meiosis**, in which gametes become haploid. But in some protozoans reduction division occurs in one of the subsequent divisions after formation of zygote. This is termed as **zygotic meiosis**, in which only zygote is diploid but rest of the life cycle is haploid. Of different types of sexual reproduction in protozoans syngamy, conjugation, automyxis are important.

I. **Syngamy or Sexual fusion**

Syngamy is the complete and permanent union or fusion of two specialised protozoan individuals or gametes resulting in the formation of a fertilized cell or **zygote** or **oospore**. The nuclei of the gametes fuse to form the zygote nucleus or **synkaryon**. The zygotes develop into adult, either directly or through encystment and fission of various types.

Depending upon the degree of differentiation of the fusing gametes syngamy may be of the following types.

a. **Autogamy.** The gametes derived from the same parent cell fuse. Examples: *Actinophrys, Actinosphaerium, Paramoecium aurelia*, etc.
b. **Paedogamy.** The fusing individuals are young. Example: *Actinophrys* (Fig. 21.21).
c. **Hologamy.** The two mature individualals behave as gametes and fuse. Example: *Copromonas* (21.22).

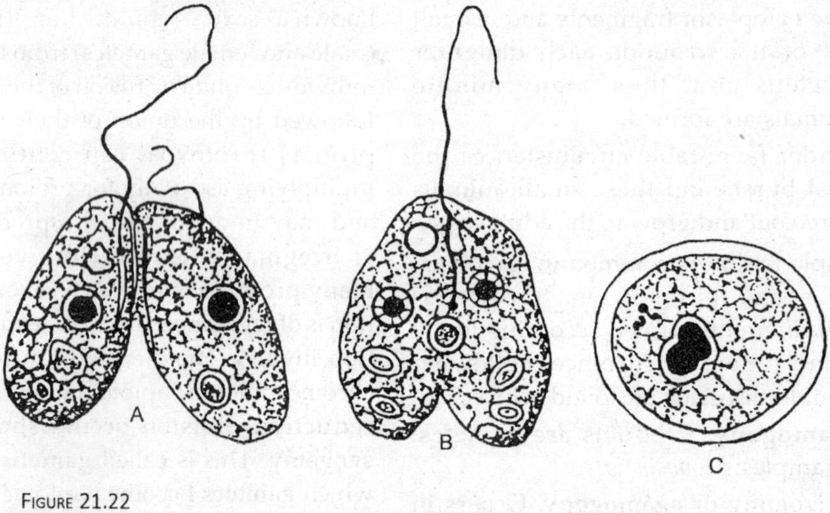

FIGURE 21.22

Hologamy in a flagellate *Capromonas* sp. A–B. Fusion of two individuals. C. Encysted zygote.

d. **Merogamy.** The uniting individuals are smaller than the ordinary vegetative individuals, called **Merogametes.**

e. **Isogamy.** Union of the gametes of similar size and shape. The isogametes are produced by multiple or repeated binary fission. Isogamy has been reported in Foraminifera (*Elphidium*), Phytomonadina (*Chlamydomonas, Copromonas*) and Gregarinida (*Monocystis*).

f. **Anisogamy.** The two fusing gametes differ in size, shape and behaviour. The gametes are termed as **heterogametes** or **anisogametes** and their fusion is known as **anisogamy** or **heterogamy**. The formation of morphologically different gametes, is the first indication of sex differentiation in Protozoa. The smaller gametes, the **microgametes**, or male gametes, are active, motile, generally flagellated and more numerous. They are produced by multiple or repeated fissions. The fusion of two microgametes is called **Microgamy**. Example: *Foraminifera, Arcella*, etc. The larger gametes, **macrogametes**, are immotile, volumi-nous, and referred to as female gametes. The fusion of two macrogametes is called **Macrogamy**. Examples: *Plasmodium, Eimeria, Volvox*, etc.

The syngamy brings about a combination of two different lines of hereditary characters. It increases the external differences in offsprings. It also renews the vigour which is lost due to repeated binary fissions. The fusion of two nuclei initiates the development of eggs.

II. Conjugation

The conjugation is the temporary union of two mating types of individuals of the same species to facilitate exchange of nuclear materials. They retain their distinct individuality and separate out after nuclear exchange. The pairing gametes are known as **conjugants**. The conjugants may be either isogamous (*Paramoecium*) or anisogamous (*Vorticella*).

Conjugation is considered to be an episode in reproduction and not a mode of multiplication. In conjugation (i) reorganization of a fresh meganucleus occurs to accelerate the metabolic activities, (ii) rejuvenation and revival of lost vigour, (iii) new nuclear combinations and new hereditary combinations arise.

III. **Automixis**

Automixis is the fusion of two gametic nuclei originating by the division of the single nucleus of an individual. Automixis may be of the following types.

a. **Autogamy.** The fusing nuclei come from the same cell as in *Paramoecium*. All the steps in nuclear changes are similar to conjugation but the union occurs between the pronuclei of the same individual.

b. **Paedogamy.** The fusion occurs between two nuclei coming from two different cells of a parent. A single organism encysts and then divides into two or more gametocytes. The nuclei of these gametocytes undergo meiosis and the gametes thus produced unite in pairs forming the zygotes. Examples: *Actinosphaerium, Actinophrys, Myxospo-ridians,* etc.

c. **Cytogamy.** In a number of species of *Paramoecium* the two individuals fuse with their oral surfaces. The nuclear changes occur as in conjugation but no nuclear exchange occurs. The two gametic nuclei in each individual fuse to form synkaryon. Cytogamy is said to be intermediate between conjugation and autogamy.

d. **Hemixis** (See p. 232).

☒ Other modes of reproduction

1. **Plasmogamy.** Two or more individuals may fuse by their cytoplasm to form a plasmodium and separate out unchanged with their distinct nuclei. This sexual phenomenon is known as Plasmogamy and occurs in certain Rhizopoda and Mycetozoa.
2. **Regeneration.** The regeneration and replacement of lost parts among free-living and few parasitic protists is widespread. A proper proportion of cytoplasm and nucleus can regenerate into an entire individual.
3. **Parthenogenesis.** The gametes which fail to fertilize start their development parthenogenetically. Examples: *Actinophrys, Chlamydomonas,* etc.

CLASSIFICATION: SUBKINGDOM PROTOZOA*

1. Protozoa are small, acellular animals living both in water (fresh water and marine) and on land. All the functions of life are performed by the single cell.
2. Organelles like pseudopodia, flagella, cilia are commonly present for locomotion and food capture.
3. Surface differentiation and skeletal secretions for protection and rigidity of shape are found in some cases.
4. No distinct respiratory structure, and excretory organ is detectable.
5. Reproduction takes place by both asexual and sexual means. Asexual reproduction by fission or budding and sexual reproduction by fusion of gametes or conjugation.
6. A complicated life history with alternation of sexual and asexual generations and some degree of embryonic development are present in some cases.
7. Free-living forms feed on small plants, other protozoa, bacteria, etc., while the parasites feed on tissues of the host.

The subkingdom **Protozoa** has been subdivided into seven phyla, Sarcomastigophora, Labrinthomorpha, Apicomplexa, Microspora, Acetospora, Myxozoa and Ciliophora.

Phylum I. SARCOMASTIGOPHORA

Locomotory organelles flagella or pseudopodia or both. Nucleus single type except in developmental stages of certain foraminifera.

* Society of the Systematics and Evolution of the Protozoologists (Lavine, *et. al.* 1980; J. Protozoology, 27: 37-58).

Spore formation absent. Sexuality often present, essentially syngamy.

Subphylum I. MASTIGOPHORA

Solitary or colonial; flagella one or more in trophozoites; asexual reproduction by symmetrogenic binary fission; nutrition phototrophic or heterotrophic.

Class 1. **Phytomastigophora** (plant-like flagellates)

Mostly free-living; some ectoparastitic and very few are endoparasitic; flagella one or two; amoeboid forms common in some groups; chromatophores present; sexual reproduction occurs in some. Examples: *Euglena, Cryptomonas, Colacium, Haplazoon, Volvox*, etc.

2. **Zoomastigophora** (animal-like flagellates)

Mostly parasitic; flagella one to many; amoeboid forms with or without flagella in some groups; chromatophores absent; sexual reproduction is known in some. Examples: *Trypanosoma, Leishmania, Trichomonas, Giardia, Trichonympha*, etc.

Subphylum II. OPALINATA

Parasitic; cilia-like locomotory organelles over the entire body surface in oblique rows; cytostome absent; nuclei single type and two or many; asexual reproduction usually by symmetrogenic binary fission; life cycles involve syngamy with anisogamous flagellated gametes. Examples: *Opalina, Protooplina*, etc.

Class **Opalinatea**

(Characters similar to those of the subphylum.

Subphylum III. SARCODINA

Mostly free-living; body naked or with external and internal tests or skeletons; typical pseudopodia present; flagella, when present, restricted to developmental stages; asexual reproduction by fission; sexual reproduction, if present with flagellate or, more rarely, amoeboid gametes.

Superclass I. RHIZOPODA

Locomotion by lobopodia, filopodia or reticulopodia or by protoplasmic flow without production of discrete pseudopodia.

Class 1. **Lobosa**

Pseudopodia lobose, rarely filiform or anastomosic; usually uninucleate.

Subclass i. **Gymnamoebia**

Test absent. Examples: *Amoeba, Entamoeba, Acanthamoeba*, etc.

Subclass ii. **Testacealobosia**

Body enclosed in a test, tectum or other complex membrane, external to plasma membrane. Examples: *Arcella, Trichosphaeriumor*, etc.

Class 2. **Acarpomyxea**

Small plasmodia or much expanded similar uninucleate forms, usually branched, sometimes forming reticulum of coarse branches; no test; no spore or fruiting bodies known. Examples: *Leptomyxa, Rhizamoeba*, etc.

Class 3. **Acrasea**

Uninucleate amoebae with eruptive, lobose pseudopodia; amoebae aggregating to form pseudoplasmodium which gives rise to fruiting bodies without stalk tube; sexuality unknown. Example: *Acrasis*.

Class 4. **Eumycetozoea**

Myxamoebae with filiform subpseudopodia; flagella, when present nonmastigonemate, in unequal, less often equal, apical pair; aerial fruiting bodies.

Subclass i. **Protosteliia**

Trophic stage varying from single amoeba to plasmodia which lack shuttle streaming; flagellate cells present or absent; fruiting bodies consist of one to several spores on narrow, hollow stalk. Example: *Ceratiomyxa*.

Subclass ii. **Dictyosteliia**

Amoebae aggregate to form multicellular pseudoplasmodium that gives rise to multispored fruiting body; stalk tube present; no flagellate cell. Example: *Dictostelium*.

Subclass iii. **Myxogastria**

Major trophic stage multinucleate plasmodium typically with shuttle streaming; fruiting bodies multispored; flagellate cells present; syngamy and meiosis in life cycle. Exampes: *Echinostelium. Licea*, etc.

Class 5. **Plasmodiophorea**

Obligate intracellular parasites with minute plasmodia; zoospores produced in zoosporangia and bearing an anterior pair of unequal, nonmastigonemate flagella; resting spores formed in compact sori or loose clusters within host cells. Examples: *Plasmodio-phora, Sorosphaera*, etc.

Class 6. **Filosea**

Hyaline, filiform pseudopodia, often branched, sometimes anastomosing; no spores or flagellate stages known. Examples: *Nuclearia, Euglypha*, etc.

Class 7. **Granuloreticulosea**

Delicate, finely granular or hyaline reticulopodia or, rarely, finely pointed, granular but anastomosing pseudopodia. Examples: *Allogromia, Ammonia, Elphidium*, etc.

Class 8. **Xenophyophorea**

Multinucleate plasmodium enclosed in branched tube system composed of transparent organic substance; numerous barite crystals in cytoplasm; faecal pellets retained outside organic tube system as conspicuous, dark masses; test of foreign matter surrounding tube system and faecal pellet masses; marine. Examples: *Psammina, Stamnophyllum*, etc.

Superclass II. ACTINOPODA

Typical floating forms; body spherical; naked or with test; pseudopodia delicate and radiose; reproduction by both asexual and sexual means; gametes usually flagellated.

Class 1. **Acantharea**

Strontium sulphate skeleton, usually composed of 20 radial or 10 diametrical spines; sometimes many more spines randomly oriented; extracellular outer (cortex) and inner envelopes usually present; inner envelope (capsular membrane) often close to central cell mass; marine, usually planktonic. Examples: *Acanthochiasma, Amphilithium, Actinelius*, etc.

Class 2. **Polycystinea**

Siliceous skeleton present in most species, made up usually of solid elements consisting of one or more latticed shells with or without radial spines or, of one or more isolated spicules; capsular membrane composed usually of grossly polygonal plates and many containing more than 3 pores; axonemes often originate from axoplast in endoplasm; marine, planktonic. Examples: *Orescena, Coccodisms. Plagonium*, etc.

Class 3. **Phaeodarea**

Skeleton (sometimes absent) of mixed silica and organic matter, consisting usually hollow spines and/or shells; very thick capsular membrane with astropyle (functioning as cytopharynx) at one pole; ectoplasm with phaeodium (group of dark corpuscles and debris); marine, planktonic. Examples: *Aulacontha, Castanella, Medusetta, Concharium, Coelodendrum*, etc.

Class 4. **Heliozoea**

Without central capsule; skeletal structures if present, siliceous or organic; axopodia radiating on all sides; most species fresh water, some marine. Examples: *Clathrulina, Actinophrys, Acanthocystis.* etc.

Phylum II. LABYRINTHOMORPHA

Trophic stage, ectoplasmic network with spindle-shaped or spherical, nonamoeboid cells; in some genera amoeboid cells move within network by gliding; heterokont zoospores produced by most species; saprobic and parasitic on algae, mostly in marine and estuarian waters.

Class 1. **Labyrinthulea**

Characters similar to those in the phylum. Examples: *Labyrinthula, Thravstochytrium*, etc.

Phylum III. APICOMPLEXA

Presence of apical complex (visible only in electron microscopy); the complex includes a polar ring(s), micronemes, rhoptries, subpellicular microtubules, micropore(s) and a conoid at the apical terminal; cilia absent; syngamy occurs; all are parasites.

Class 1. **Perkinsea**

Spherical enclosing several daughter cells known as zoospores; zoospore flagellated. Reproduction by multiple fission of the original single nucleus; spores with very large, partially eccentric vacuole; usually oval nucleus contains a compact, deep staining endosome. Example: *Parkinsus* (parasite in oyster).

Class 2. **Sporozoea**

Parasitic; nucleus single type; typical simple spores present; syngamy associated with sexual reproduction, when present; infective sporozoites result from sporogony; locomotion of adult by body flexion, gliding or, undulation of longitudinal ridges.

Subclass i. **Gregarinia**

Mature gametes large, extracellular; mucron or epimerite in mature organism; generally syzygy of gametes, isogametes or nearly so, with similar members of male and female gametes produced by gamonts; zygotes form oocysts within gametocytes; life cycle consists of gametogony and sporogony; parasites in digestive tract or body cavity of invertebrates and lower chordates. Examples: *Monocystis, Stenophora, Gregarina*, etc.

Subclass ii. **Coccidia**

Gamonts ordinarily present; mature gamonts small, typically intracellular, without mucron or epimerite; syzygy generally absent, if present, involves markedly anisogamous gametes; life cycle characteristically consisting of merogony, gametogony and sporogony; most species parasites in invertebrates; Examples: *Grellia, Haemogregarina, Eimerina, Isospora, Toxoplasma, Sarcocystis, Haemoproteus, Plasmodium, Leucocytozoon*, etc.

Subclass iii. **Piroplasmia**

Pyriform, round, rod-shaped or amoeboid; conoid absent; no oocysts, spores and pseudocysts; flagella absent; usually without subpellicular microtubules, with polar ring and rhoptries; locomotion by body flexion, gliding, or in sexual stages, by large axopodium-like strahlen; asexual and probably sexual reproduction; parasitic in erythrocytes and sometimes in other circulating and fixed cells; known vectors are ticks. Examples: *Babesia, Theileria*, etc.

Phylum IV. MICROSPORA

Unicellular spores, each with imperforate wall, containing one uninucleate or dinucleate sporoplasm and simple or complex extrusion apparatus always with polar tube and polar cap; without mitochondria; obligatory intracellular parasites in nearly all major animal groups.

Class 1. **Rudimicrosporea**

Spore with simple (rudimentary) extrusion apparatus consisting of polar cap and thick (manubroid) polar tube extending backward from cap, bending laterally and terminating in a funnel (infundibulum); sporulation sequence with dimorphism, occurring either in parasitophorous vacuole or in the thick-walled cyst; hyperparasites of gregarines in annelids. Examples: *Amphoacantha, Metehnikovella*, etc.

Class 2. **Microsporea**

Spore with a complex extrusion apparatus of Golgi origin, often including polaroplast and a posterior vacuole in addition to polar tube and polar cap; polar tube typically filamentous, extending backward from the polar cap and coiling around inside of spore wall; parasites of every major group of animals, from protozoons to man—especially abundant in arthropods, fish and mammals; undergo both schizogony and sporogony; spores are small, 3-6 μm, spherical, ovoid or cylindrical; hollow polar filament in resting spore remain tightly coiled surrounding the nucleus and most of the cytoplasm; in some two nuclei are associated intimately and known as diplokaryon. Examples: *Thelohania, Nosema*, etc.

Phylum V. ASCETOSPORA

Spore multicellular (or uninucleate?); with one or more sporoplasms; without polar capsules or polar filaments; all parasitic.

Class 1. **Stellatosporea**

Haplosporosomes present; spores with one or more sporoplasms. Examples: *Marteilia, Minchinia, Haplosporidium*, etc.

Class 2. **Paramyxea**

Spore bicellular, consisting of parietal cell and one sporoplasm; spore without orifice. Exaple: *Paramyxa*.

Phylum VI. MYXOZOA

Spores of multicellular origin, with one or more polar capsules and sporoplasms; with 1, 2 or 3 (rarely more) valves; all species parasitic.

Class 1. **Myxosporea**

Spores with one or two sporoplasms and 1-6 (typically two) polar capsules, each with a coiled polar filament; filament function probably anchoring; spore membrane generally with two, occasionally up to six valves; trophozoite, main stage of proliferation well developed; coelozoic or histozoic in cold blooded vertebrates. Examples: *Myxidium, Sphaeromyxa, Ceratomyxa, Henneguya, Myxobolus, Thelohanellus*, etc.

Class 2. **Actinosporea**

Spores with three polar capsules, each enclosing a coiled polar filament; membrane with three valves, several to many sporoplasms; trophozoite stage reduced; proliferation during sporogenesis; parasitic in invertebrates, specially in annelids.

Subclass i. **Actinomyxia**

Spores with three polar capsules each enclosing a polar filament; membrane with three valves; several to many sporoplasms. Example: *Triactinomyxon*.

Phylum VII. CILIOPHORA

Simple cilia or compound ciliary organelles typical, at least in one stage of life cycle; subpellicular infraciliature present even when cilia absent; two types of nuclei, with rare exception; binary fission transverse, basically homothelogenic and generally parakinetal; budding and multiple fission also occur; sexuality involves conjugation, autogamy and cytogamy; nutrition heterotrophic; typical contractile vacuole present; most species free-living but many commensal, some true parasitic, and a large number symphorionts on variety of hosts.

Class 1. **Kinetofragminophorea**

Oral infraciliature only slightly distinct from somatic infraciliature and differentiated from anterior parts or, other segments of all or some of somatic kineties; cytostome often apical (or subapical) or midventral, on surface of the body or at the bottom of the atrium or vestibulum; cytopharyngeal apparatus commonly prominent; compound ciliature, oral or somatic, typically absent.

Subclass i. **Gymnostomatia**

Cytostomal area superficial, apical or subapical; circumoral infraciliature without kinetosomal differentiation other than closer packing of kinetosomes, insertion of supplementary segments of kineties or, pairing (not as dyads) of kinetosomes; cytopharyngeal apparatus rhabdos type; toxcysts common, somatic ciliation usually uniform. Examples: *Bundleia, Didesmis, Metacystis Didinium, Amphileptus, Stephanopogon, Laxades*, etc.

Subclass ii. **Vestibulifera**

Apical or near apical vestibulum commonly present, equipped with cilia derived from anterior parts of somatic kineties and leading to cytostome; stomatogenesis, sometimes involving two anlagens; cytopharyngeal apparatus resemble rhabdos; free-living or parasitic, especially in digestive tract of vertebrates. Examples: *Balantidium, Isotricha, Entodinium, Colpoda*, etc.

Subclass iii. **Hypostomatia**

Cytostome nonpolar, on ventral surface; body cylindrical or flattened dorsoventrally, often with reduction of somatic ciliature; cytopharyngeal apparatus typical; oral area may be sunk into citrium; some species astomatous; free-living or ecto or endocommensals, principally of invertebrates. Examples: *Leptopharynx, Nossulopsis, Hartmannula, Isochona, Opalinopsis*, etc.

Subclass iv. **Suctoria**

Suctorial tentacles generally multiple (polystomy) containing haptocysts; adult sessile and sedentary, seldom with cilia; reproduction by budding; stalk commonly present, noncontractile, produced by scupoloid; conjugation often involving micro and macroconjugant; migratory larva ciliated, without tentacles or stalk; marine and fresh water, with wide distribution, occasionally endocommensal. Examples: *Podophyra, Acineta, Tokophyra, Discophyra, Heliophyra,* etc.

Class 2. **Oligohymenophorea**

Oral apparatus, at least partially in buccal cavity, generally well-defined, although absent in one group; oral ciliature clearly distinct from somatic ciliature, consisting of paroral membrane (stichodyad) on right side and a small number of compound organelles on left side; cytostome usually ventral and/or near anterior end at the bottom of buccal or infundibular cavity; colony formation common in some groups.

Subclass i. **Hymenostomatia**

Body ciliation often uniform and heavy; buccal cavity, when present, ventral; kinetodesmata regularly present, usually conspicuous; sessile forms with stalk; colony and cyst formation relatively rare; fresh water forms predominant. Examples: *Colpidium, Tetrahymena, Paramoecium, Conchophthirus, Pleuronema, Anoplophrya,* etc.

Subclass ii. **Peritrichia**

Oral ciliary field prominent, covering apical end of body and dipping into infundibulum; paroral membrane, generally called 'haplokinety' and adoral membranelles, 'polykineties', becoming 'peniculi' in infundibulum present; somatic ciliature reduced to temporary posterior circlet of locomotor cilia; widely distributed species, many stalked and sedentary, others mobile, all with aboral scopula; myonemes associated with strong contractility of stalk or parts of the body; conjugation total, involving fusion of macro and microconjugants. Examples. *Vorticella, Zoothamnium, Trichodina,* etc.

Class 3. **Polymenophorea**

Well-developed, conspicuous adoral zone with numerous buccal or peristomial organelles, often extending out on to body surfaces; on right side, one or several lines of paroral ciliature; stomatogenesis parakinetal or apokinetal; somatic ciliature complete or reduced or, appearing as cirri; cytostome at the bottom of buccal cavity or infundibulum; cytoproct often absent; large and commonly free-swimming forms in great variety of habitats.

Subclass 1. **Spirotricha**

Characters similar to those of the class. Examples: *Stentor, Nyctotherus, Folliculina, Saprodinium, Tintinnopsis, Stylanchia,* etc.

22 PHYLUM PORIFERA

The Porifera (*L. porous* = pore + *ferre* = bear) or sponges are lowly organised group of plant-like sessile animals remaining attached to the substratum. Of the described sponges, about 5,000 species are marine and very few are fresh water. They grow mainly in shallow waters but their habitat may extend up to a depth of 5,600 metres. Sponges are of varying shapes, sizes and colours. Sponges were once thought to be plants. Their animal nature was learned in 1765, but their place in the Animal kingdom was uncertain until about 1857. Sponges resemble some colonial protozoans (Proterospongia) in having groups of flagellated collar cells and intracellular digestion but differs in having some cells arranged as tissues, limited cellular division of labour and a body with many pores. Absence of true tissues and extensive system of pores and canals powered by the action of peculiar flagellated cells, the **choanocytes**, make them distinct from other metazoan. Very often they are considered as a subkingdom, **Parazoa**. Sponges are distributed in all seas—from the equator to the poles.

SCYPHA sp.

Sycon, recently named *Scypha*, is the simplest form of sponge, marine in habitat. It remains attached to sea shore rocks just below the tide line. The colour of a live specimen is a combination of shades of grey and light brown. Scypha belongs to subclass Calcaronea, class Calcarea, phylum Porifera.

⊠ Structure

The *Scypha* is either a simple solitary cylinder or a number of cylindrical branches (Fig. 22.1) connected together to form a colony. Solitary or colonial, it is anchored to rock by a structure, called **stolon** at the base. The cylinder is vase-shaped, swollen at the middle and narrows down at both the ends, being a little more

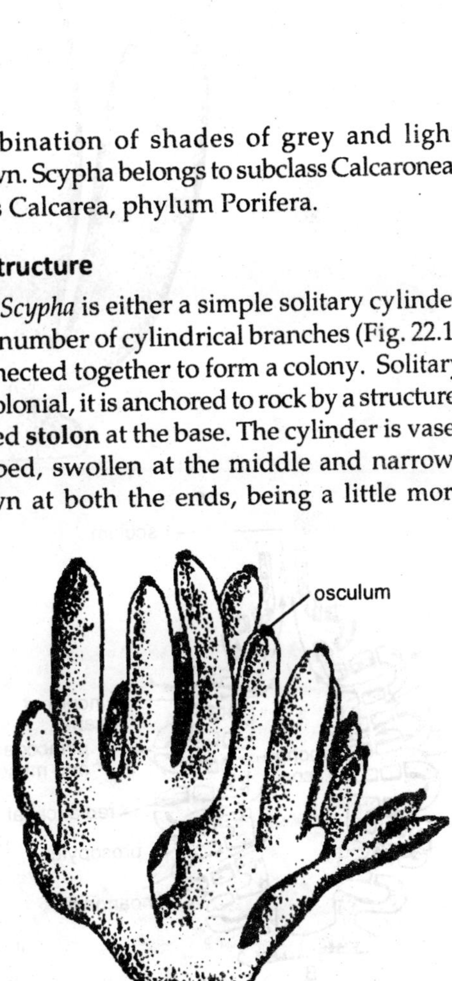

FIGURE 22.1

Scypha sp. An entire organism with a number of branching cylinders narrow at the free or distal end. Due to protrusion of numerous monaxon spicules from the surface, it appears bristly.

☒ Osculum

The **osculum** or the exhalant pore is a wide opening, present at the free end of the cylinder. It establishes direct communication between the paragastric cavity or the **spongocoel** and the exterior. The osculum is surrounded by numerous straight, monaxon, calcareous spicules arranged in a circlet, imparting the appearance of a delicate fringe to it.

☒ Dermal pores

The body surface bears numerous regularly

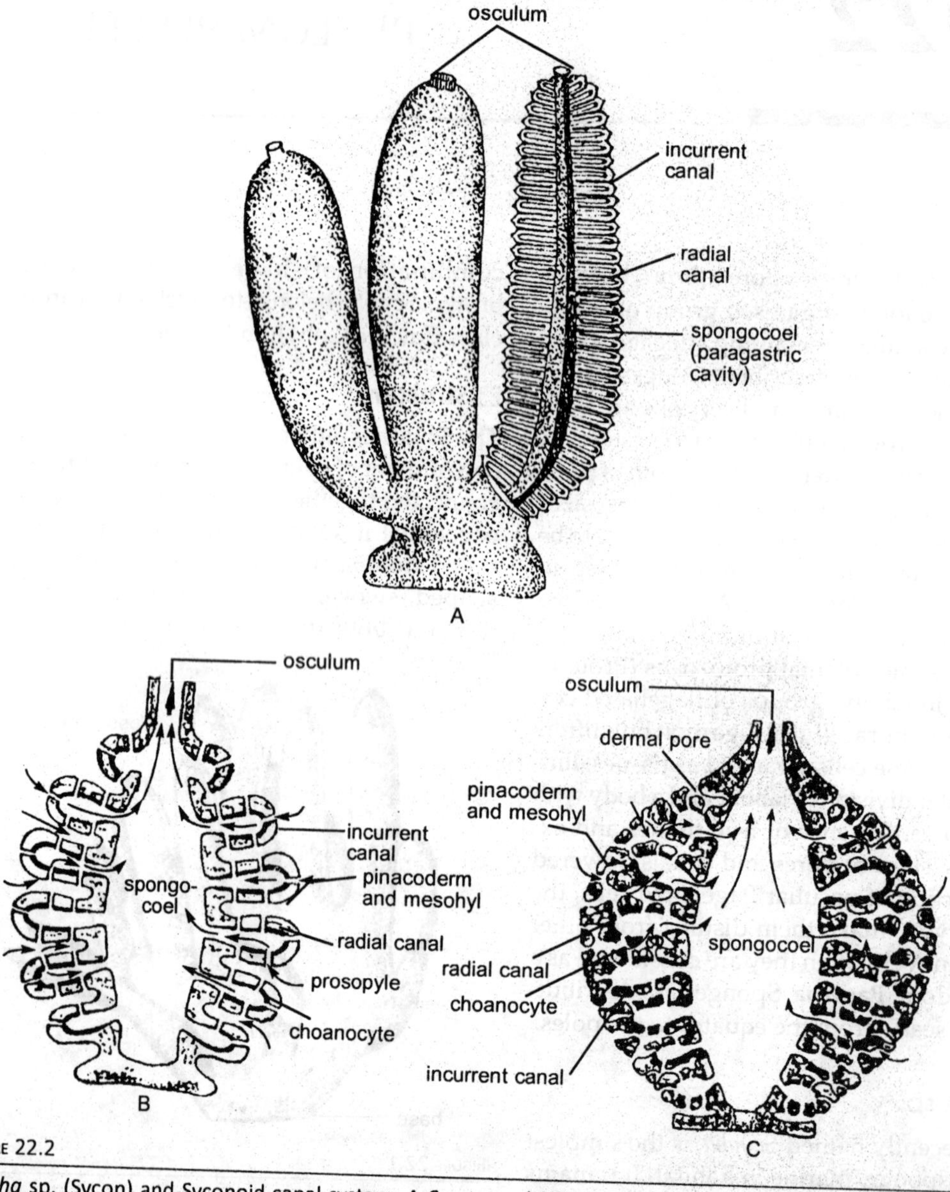

FIGURE 22.2

Scypha sp. (Sycon) and Syconoid canal system. A *Sycon* sp. (slightly magnified), a portion of a colony (the right cylinder (branch) bisected longitudinally). B. Syconoid canal system (early stage without cortex). C. Syconoid canal system (final stage with cortex)

arranged, polygonal elevations separated by depressed lines or furrows. Numerous microscopic apertures are present in the furrows. These are **ostia** or **inhalant pores** or **dermal pores** leading to the incurrent canals.

⊠ Canal system

It is evident from the term 'porifera' that the surface of the body bears a large number of pores, minute in size and inhalant in function. These pores open into a system of channels, which, after penetrating almost all the portions of the body, open to the exterior by an opening known as osculum, at the tip of the branch. All the canals are collectively called canal system. The canal system in *Scypha* (*Sycon*) is known as **syconoid** type. It establishes a continuous passage for the inflow and outflow of water within the body of a sponge (Fig. 22.2).

The following structures are found in association with the canal system of *Scypha*. (Fig. 22.3, 22.4).

1. **Spongocoel.** At the free end of a branch, a fairly large aperture known as osculum is present. Osculum leads internally into a canal, the spongocoel, which runs along the middle of the branch and receives a large number of excurrent canals of the branch. The paragastric cavities or spongocoels of different branches in a colonial form, are in communication with one another through a broad chamber at the base. The cavity is lined by flattened cells of ectodermal origin, the **pinacocytes**.
2. **Incurrent canal.** The surface of each branch of *Scypha* is provided with alternate elevations and depressions. At each depression a group of minute pores known as inhalant pores or **ostia** are present. Ostia lead internally into the incurrent canal. The canal is dilated towards the outer end and narrower and blind at the inner end. It is lined by flattened cells of ectodermal origin, the **pinacocytes**.
3. **Radial canal.** Lying alternate and parallel to the incurrent canals are the radial canals. The radial canals are situated opposite to the points of elevations on the surface. Each canal is narrower and blind towards the outer end but broad and open at the inner

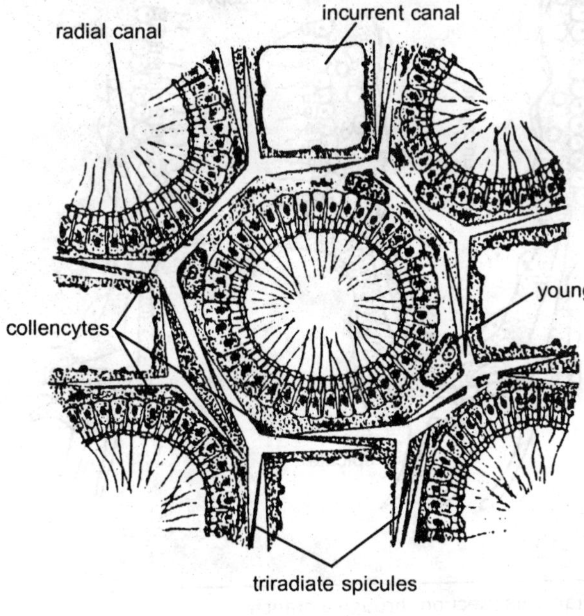

FIGURE 22.3

Scypha sp. Longitudinal section passing through the wall of a branch

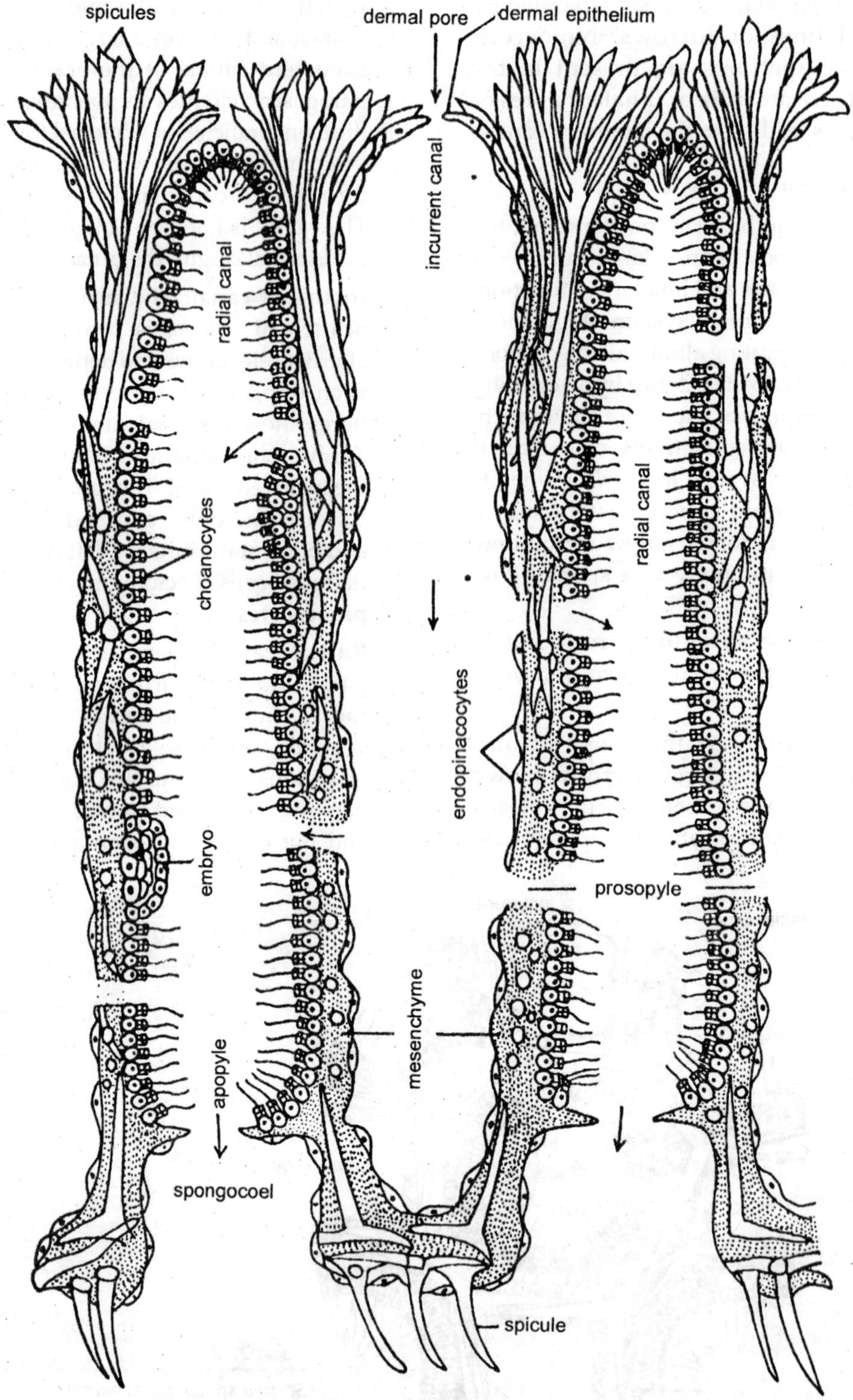

FIGURE 22.4

Scypha sp. Transverse section through a branch

end. It is lined by flagellated **choanocyte** cells of endodermal origin.

The radial canals are connected with the incurrent canals by narrow passages known as **prosopyles**.

4. **Excurrent canal.** These are short but wide canals. Each canal is situated along the same long axis of the radial canal. It is lined by flattened cells of ectodermal origin.

   The excurrent canal is connected at the outer end with the radial canal by an aperture, the apopyle, and at the inner end with the paragastric cavity by a large aperture, the gastric or internal ostium.

☒ Course of water current

Due to the rapid backward movement of the flagella of the cells lining the radial canals, a constant water current is maintained in the canal system. The rate of flow of water can, of course, be regulated by increasing or decreasing the diameter of the different apertures in the canal system.

Water from outside enters the incurrent canals through ostia and from there to the radial canals through the prosopyles. It passes from radial canals to excurrent canals through the apopyles and from there to the spongocoel through gastric ostia and, thence, to the exterior through the osculum.

☒ Role of the canal system

The canal system serves various functions.

1. It serves the purpose of nutrition. The food such as diatoms, protozoans, etc., are carried by the water current and reach the radial canals where they are picked up and digested by the flagellate cells.
2. Oxygen for respiration is carried by the streaming current of water.
3. It functions for excretion. Current of water, passing out of the osculum, also remove the carbonic acid and other nitorgenous waste matters.
4. The outward current takes away the reproductive units from the body of the sponge.
5. The complicated canal system also increases the surface area of the animal which is directly exposed to water.

☒ Histological elements

The different microscopic elements constituting the body wall of *Scypha* are known as histological elements. The elements may be divided into three groups—cell elements, skeletal elements and mesenchymal substance (Figs. 22.5 and 22.6).

1. Cell elements

The cells of the sponges are fairly differentiated, but, excepting choanocytes, others seem to be only modified forms of undifferentiated amoeboid cells corresponding to the primitive connective tissue cell of higher animals. Following different types of cells are found in the body wall of *Scypha* (Figs. 22.5 and 22.6).

a. **Ectodermal cells.** These are flattened scale-like cells with inconspicuous nuclei and their edges are closely cemented together to form an **epithelium**. The cells of the dermal layer covering the outer surface of the sponge are known as **pinacocytes**. Each pinacocyte has a thickened central bulging containing a nucleus. Some pinacocytes (endopinacocytes) also line

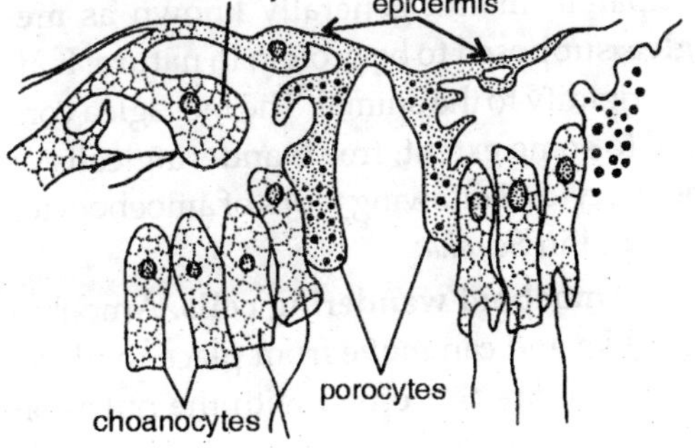

FIGURE 22.5

Scypha sp. Section of the body showing different types of cells

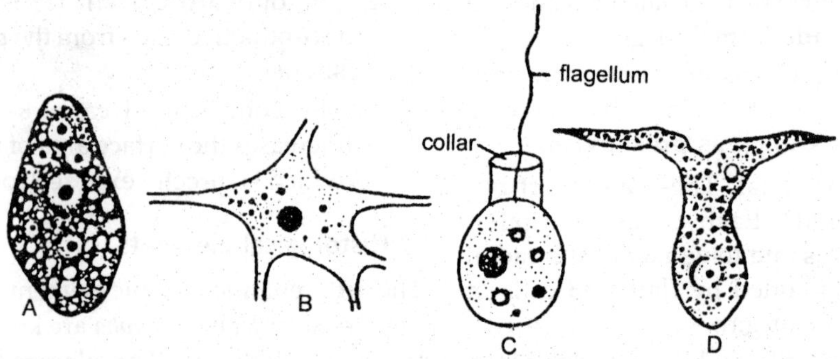

FIGURE 22.6

Scypha sp. Types of cells. A. Archaeocyte. B. Amoebocyte. C. Choanocyte. D. Epidermal

the incurrent canals and spongocoel. They are highly contractile and those lining the incurrent canals are called **skeletogenous** cells.

b. **Endodermal cells.** These are columnar cells, each with a large nucleus, one or more vacuoles and bears flagellum at the inner end, the base of which is surrounded by a delicate, transparent, collar-like upgrowth. Such cells are known as **choanocytes** and are restricted to the radial canals.

c. **Myocytes.** These are fusiform contractile cells around the ostia, prosopyles and apopyles effecting the closure of the apertures, and are usually arranged in a circular fashion to form a sphincter.

2. Mesenchyme substance

The mesenchyme consists of a gelatinous, transparent matrix generally known as **mesoglea**, supposed to be protein in nature. It affords rigidity to the animal. The mesoglea contains, to some extent, free wandering cells or **amoebocytes**. Following types of amoebocytes are found in *Scypha:*

a. **Amoeboid wandering cells.** Amoeba-like and can move from place to place. They are concerned with the nutrition of the animal.

b. **Collencytes.** The cells bear slender branching pseudopods and connect the different elements of the body. If bipolar, collencytes are termed **desmacytes** or fibre cells.

c. **Chromocytes.** These are with lobose pseudopods and contain pigments. The colouration of the sponge is dependent on these pigments.

d. **Thesocytes.** Cells with lobose pseudopods and store reserve food materials.

e. **Scleroblasts.** The cells secrete the skeletal elements or **spicules**. They are of two types:

i. **Calcoblasts.** Amoebocytes secreting calcareous spicules.

ii. **Silicoblasts.** Amoebocytes secreting siliceous spicules.

f. **Reproductive cells.** These are modified amoebocytes and possibly also modified choanocytes.

g. **Porocytes.** Also known as pore cells, they are tubular and with a central canal acting as an incurrent passage. The pores can be closed by a thin cytoplasmic sheet, the **pore diaphragm**. Porocytes are possibly transformed pinacocytes.

The mesenchyme is responsible, after all, for one of the most important characters in sponges, that is, secreting the skeletal elements.

3. **Skeletal elements.** These are hard structures connected together in such a way as to support and protect the soft parts of the body. These structures are

pointed and are made of calcium carbonate and known as **spicules**. Spicules develop from the **scleroblasts** and one scleroblast is required to form each arm of a spicule. The spicules may be needle-like or club-shaped and range from one to many axes,—or **monaxon** to **polyaxon**,—of which triaxon is the most abundant type (Fig. 22.4). There are spicules where the body is round and the growth is concentric. These are known as spheres while the crepis—on being deposited with layers of silica in an irregular fashion is termed **Desma**. The club-shaped spicules projecting on the outer surface beyond the ectoderm are known as **oxeote** spicules. Spicules are responsible for the framework of ***Scypha***. Although the primary function of the calcareous skeleton is to support, it also serves to buffer the mesenchyme against any drop in pH that could cause hardening of the ground substance.

⊠ Reproduction and development

Reproduction in sponges takes place both by asexual and sexual means (Fig. 22.7).

A. **Asexual reproduction**

It is effected by internal buds or **gemmules**. Some of the amoebocytes come to lie in the mesogleal layer around the lining cells of the radial canals, and after repeated division a mass of cells or the gemmule is formed. This process is known as **gemmation**. A gemmule cannot escape to the outside until the branch in which it grows is separated from the main body of the sponge and lost. For this reason, gemmules are formed in older branches. By the time of its liberation the gemmule is transformed into an **amphiblastula**, the subsequent development of which is like that of the same in sexual reproduction.

B. Sexual reproduction

Sponges are **monoecious**, i.e. ova and sperms develop in the same individual. Sexual reproduction is effected by the formation of the **archaeocytes** (specialized amoebocytes). Sex organs are absent. The mesenchyme around the gastral layer of the flagellated chamber is the seat of the sex cells.

Spermatozoon is with a round head and a long tail. Spermatozoa from another sponge are carried to the radial canal with water current. The ovum is large and round and remains attached to the maternal tissue. Fertilization takes place within the body. The fertilized egg of *Scypha* divides vertically and eight concial cells are formed. These cells then undergo horizontal cleavage, with the result the eight long cells and eight small cells are formed enclosing a **blastocoel**. The long cells are destined to form the future epidermis while the small cells form the future choanocytes. The eight small cells increase in size, elongate, and each acquires a flagellum on its inner side. The larger cells do not divide, instead, they assume a round form and become granular. The mouth opening appears at their middle which absorbs the neighbouring cells. This stage of the **blastula** is known as **stomoblastula**. The blastula undergoes inversion and the flagellated cells are brought outside.

The embryo at this stage is known as **amhiblastula** (Fig. 22.7F). This is a typical calcareous larva. It escapes from the parent and has two types of cells—the small narrow flagellated cells and the large round granular cells. After a brief period of free existence which may last for a few to several hours, the larva undergoes gastru-lation, in which the flagellated half is invaginated into or overgrown by the large granular cells (Fig. 22.7G.). It attaches itself to some solid object by the blastoporal end and is gradually transformed into a narrow cylinder and then to a small sponge. The flagellated cells become the collared cells, while the granular cells become numerous and form the dermal epithelium. The mesenchyme cells are formed from both the layers. Gradu-ally, the cylinder (larva) increases in thickness. With the growth of the intermediate layer the radial and other canals gradually appear.

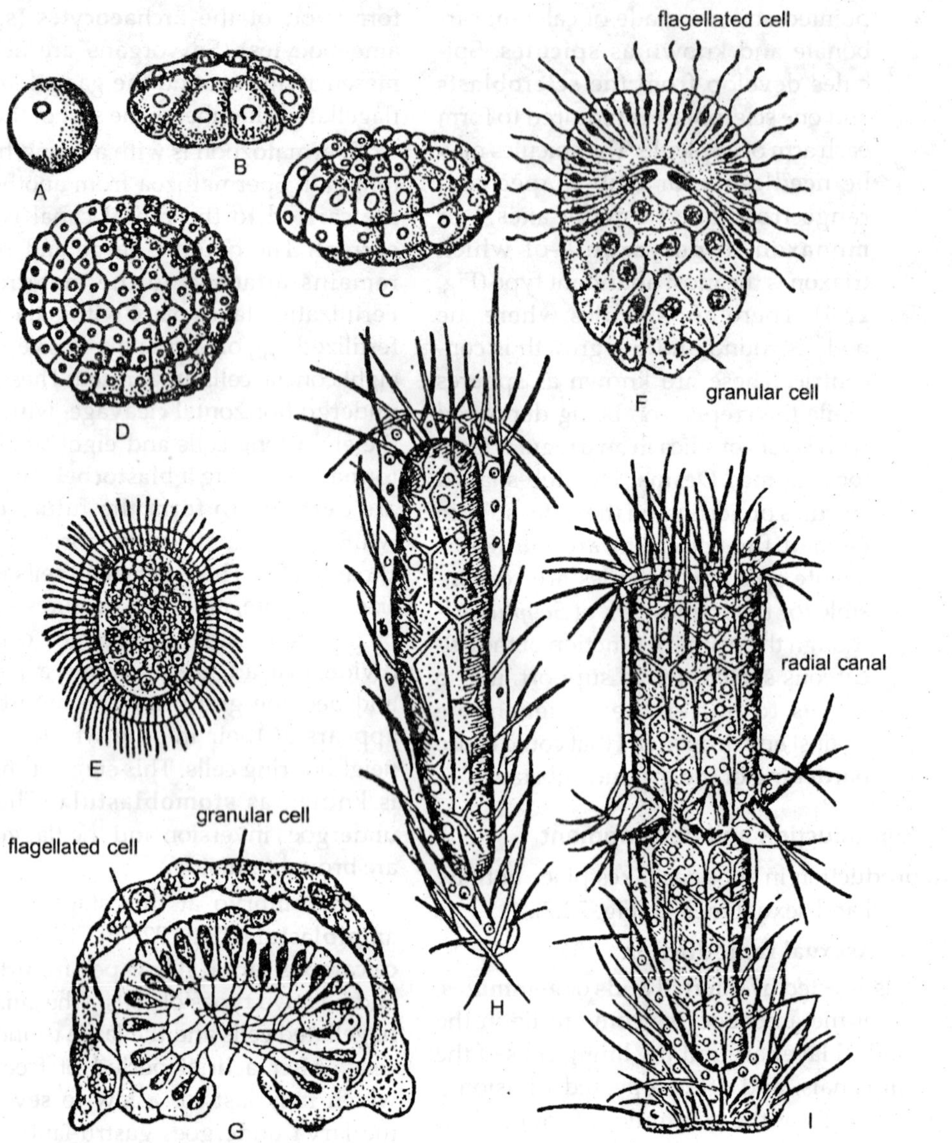

FIGURE 22.7

Scypha sp. Development. A. A zygote. B. B-cell stage. C. 32-cell stage. D. An early blastula. E. A parenchymula larva of *Leucosolenia* sp. F. An amphiblastula. G. A gastrula. H-I. A young sponge

⊠ Commercial Sponge

Sponges are mostly beneficial to man. Skeleton of some sponges are used to manufacture commercial sponge, which is of great economic importance. The spongin skeleton is treated with hydrochloric acid and the spicules dissolve. The residue left after the acid treatment is further subjected to chemical treatment, and the skeleton becomes marketable as the 'bath sponge'. Sponges are extensively used in bathrooms, laboratories and in surgical operations.

CLASSIFICATION: PHYLUM PORIFERA

1. Sponges are usually small, plant-like animals that are radially symmetrical

or irregular in shape, occurring singly or in colonies.

2. Multicellular, body with two loosely arranged and differentiated germ layers and an intermediate gelatinous layer containing amoeboid cells and skeletal pieces known as mesohyl or mesenchyme (not truly diploblastic), hardly formed in tissues.
3. There are no organs, head, mouth or gut. The body structure is organised around a system of canals and chambers through which water flows.
4. Flagellated coller cells, or choanocytes, which line the chambers, not only create the water current but also filter out fine food particles.
5. The supporting framework or the skeleton consists of fine, flexible fibres made of proteinous spongin or spongin fibres and siliceous spicules; or siliceous spicules alone; or spicules of calcium carbonate.
6. The body surface is provided with pores and, at the top, a large opening, the **osculum**. The interior cavity, the atrium or spongocoel opens to the outside through osculum.
7. The nerves and sensory cells are altogether absent; the cells act independently. Most cells can transmit stimuli slowly.
8. Both hermaphrodite and unisexual forms exist. Reproduction occurs asexually by the development of buds and gemmules, and sexually by development of typical ova and sperms; specialised gonads absent; the ova and sperms develop from the cell type archaeocytes.
9. Development direct; the embryo leaves the sponge body as an active, free-swimming ciliated larva, which finally attaches itself to a substratum and undergoes metamorphosis to reach the plant-like adult stage. The larvae are of two types: the Amphiblastula and Parenchymula.
10. The sponges have power of regeneration. Because of their highly variable characteristics, number of intermediate forms and immense collateral affinites, it is difficult to classify sponges. The peculiarities of the skeleton is the main basis of classification and the sponges are grouped into four classes accordingly.

Class 1. **Calcarea** or **Calcispongiae**

a. Calcareous sponges; skeleton solely of calcareous spicules which may be one, three or four-rayed and not distinguishable into mega and microscleres.
b. Body form is asconoid, syconoid or leuconoid.
c. Mosthy coloured, size not exceeding 10 cm in height.

Subclass i. **Calcaronea**

a. Triradiate spicules usually having one long ray.
b. Choanocyte nucleus is apical in position and the flagellum arises directly from the nucleus.
c. Larvae are ampliblastula.

Examples: *Leucosolenia*, *Scypha* (*Sycon*), *Grantia*, etc.

Subclass ii. **Calcinea**

a. Triradiate spicules with equal rays.
b. Choanocyte nucleus is basal in position and flagellum arises independently.
c. Larvae are parenchymula.

Examples: *Clathrina*, *Leucetta*, *Minchinella*, *Petrobiona*, etc.

Class 2. **Hexactinellida** or **Hyalospongiae**

a. Glass sponges; skeleton purely siliceous, composed of six-rayed spicules (hexactines). The skeleton is beautiful, glass-like, hence glass sponge.
b. The internal structure consists of a network, the trabecular net, made of thin strands forming open meshes.

c. Without jelly and osculum, may be covered by a sieve plate of silica.
d. Canal system simple and choanocytes small.
e. Mostly radially symmetrical; strongly individualised; may be cylindrical or shaped like a vase, urn, funnel, attached directly at the base or more usually by way of root tufts of spicules (basal stalk).
f. Exclusively marine, deep-sea sponges occurring at depths of 20-5,000 metres. Examples: *Euplectella aspergillum* (Venus flower basket of Philipines), *Staurocaptus* (cup-shaped glass sponge, without roof spicules), *Hyalonema* (glass-rope sponge), etc.

Class 3. **Demospongiae**

a. Either devoid of skeleton or with skeleton made up of spongin fibres alone (horny sponges) or in having combination of both siliceous spicules and spongin fibres.
b. Canal system mostly complicated with usually small rounded flagellated chambers having small choanocytes.

Subclass i. **Tetractinomorpha**

a. The body is often radially constructed and cortical or axial development is advanced.
b. Usually lack spongin fibres; having tetraxonid and monaxonid megascleres.

Examples: *Oscarella* (without skeleton), *Plakina, Geodia, Tethya, Axinella, Aciculitoes*, etc.

Subclass ii. **Ceractinomorpha**

a. Extremely variable quantity of spongin present; in most cases spongin fibres incorporate some spicules or detritus.
b. Monaxonid megascleres and sigmoid or chelate microscleres present.
c. Body form mostly flattened with growth and looks like crumb of bread.

Examples: *Haliochondria, Clathria, Spongilla* (fresh water type), *Spongia, Aplysilla, Dysidea*, etc.

Class 4. **Sclerospongiae** (Boring sponges)

a. A small group of tropical sponges with siliceous spicules and spongin fibres but encased within a solid external skeleton of calcium carbonate.
b. Found within marine caves and tunnels associated with coral reefs. Boring sponges are found throughout the world and play an important role in breaking down coral and shell.*

Example: *Cliona*.

* This group excavates tunnels in shell, coral and coralline rock by chemical process in which chips of calcium carbonate are etched out by simple sponge cell.

23 PHYLUM CNIDARIA

The phylum Cnidaria, (Gr. *cnidos* = thread) contains familiar fresh water hydras, marine jelly fishes, sea anemones and corals. These are lowest animals with definite tissues, with conspicuous radial symmetry, which in some become biradial, with one end of the body bearing the mouth and tentacles. The Cnidaria closely approximate an ideal diploblastic condition with two layers of cells in the body and highly diagnostic cnidocytes. Of about 10,000 known species 5,000 are fossils. Almost all are aquatic and the majority are marine.

HYDRA sp.

Due to its capability to regenerate, *Hydra* has derived its name from the Greek mythological may-headed sea serpant **Hydra**, which was supposed to have unlimited potentiality to regenerate its head, if cut off. Majority of the hydras live in clear fresh water, remaining submerged, being attached to an object in a vertical position.

⊠ Structure

Hydra belongs to the class Hydrozoa, phylum Cnidaria.

1. The body is more or less cylindrical (Fig. 23.1) and one end remains attached to a submerged object, wood or stone.
2. The attached end is known as the proximal end, the point of attachment is the **pedal disc**.

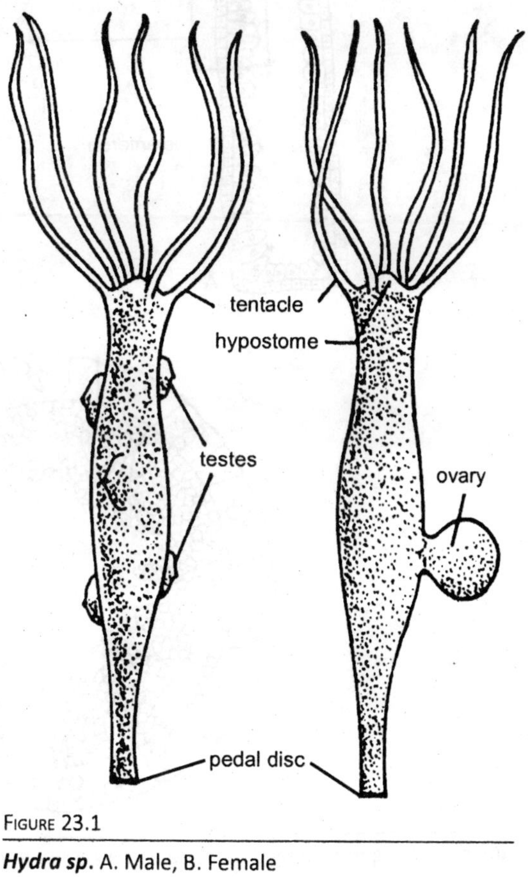

FIGURE 23.1

Hydra sp. A. Male, B. Female

3. The distal or the opposite end is provided with 6 or 8 tentacles radiating in all directions and arranged in a circle.
4. **Hypostome** is a small, elevated cone surrounded by the tentacles.
5. The **mouth** is round and situated at the centre of the hypostome.
6. The mouth communicates the **coelenteron** or the **gastrovascular cavity** to the exterior.
7. The whole surface of the body, except the pedal disc, is provided with a large number of stinging cells or **cnidocytes (cnidoblasts)**.

   The cnidocytes are crowded in the hypostome and the tentacles.
8. One or more **buds**, the tubular outgrowths, similar to the structure of the body are often present, as lateral outgrowths of the body.

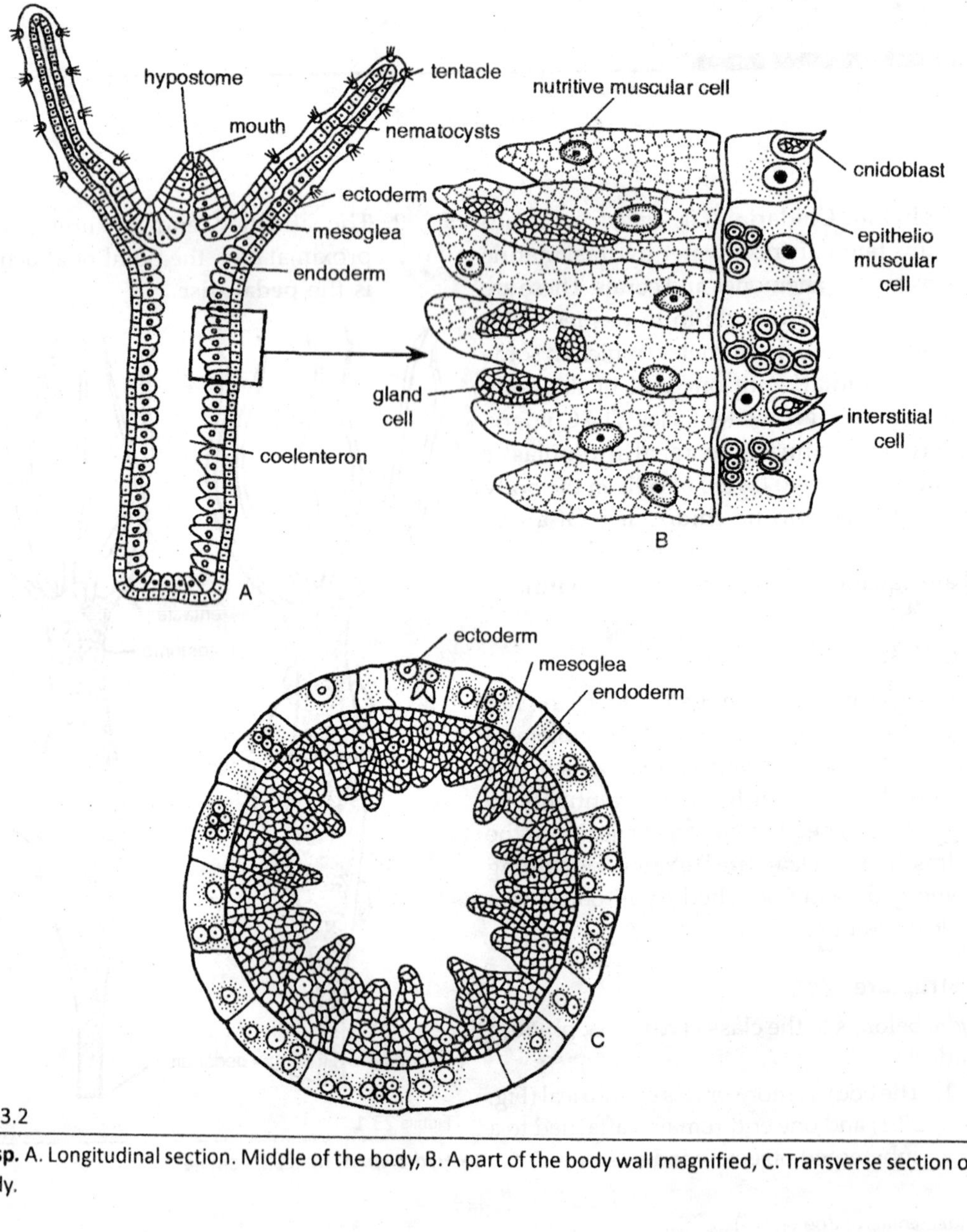

FIGURE 23.2

***Hydra* sp.** A. Longitudinal section. Middle of the body, B. A part of the body wall magnified, C. Transverse section of the body.

9. Small, conical outgrowths, male or female gonads, also called **testis** or **ovary**, are present on the body.
10. Indian Hydra is unisexual.

⊠ Body wall

Hydra is the simplest fresh water metazoa. It is the smallest and solitary cnidarian polyp.

The wall of the body, tentacles and buds are made up of two layers of cells (Fig. 23.2), the outer ectoderm and the inner endoderm separated by a thin layer of noncellular gelatinous mesoglöea. The ectodermal layer is protective, receptive and reproductive and the endodermal layer is mainly nutritive.

⊠ Cell elements

Of the different types of cells some are restricted to ectoderm, some to the endoderm and some are common to both ectoderm and endoderm (Fig. 23.3).

⊠ Cells confined to ectoderm

1. Epitheliomuscular cells

These are large elongated cells with oval nucleus. The free ends are smooth with rounded corners and being placed side by side form a continuous surface. Two basal strands, opposite each other, are present in each cell. A myoneme extends through them at the base of the cell. By the contraction of these processes the animal is shortened and can bend in various directions. The cells of the pedal disc can send pseudopodia by which hydra can move (Fig. 23.3A).

2. Cnidocytes

Cnidocytes (Cnidoblasts) are weapons of offence and food capture. It has the following components (Fig. 23.3B-F).

a. A tough, ovoid, fluid-filled capsule, the **nematocyst**, invaginated at the anterior end and continuing as a long, hollow thread, bearing minute **barbs** at the base.
b. A layer of specially contractile cytoplasm with a large nucleus.
c. A delicate protoplasmic process, the **cnidocil** or **trigger hair** at the anterior end, projecting beyond the surface.
d. Contact of the cnidocil with some small animals causes, sudden contraction of the cytoplasm, the nematocyst is everted and shot out with a great force.

Usually four types of cnidocytes are present and distributed in a characteristic regional pattern. They are most concentrated in the tentacles and anterior region, become less numerous towards the base and absent in the basal disc.

⊠ Types of cnidocytes

1. **Penetrant or stenoṭele.** Large, about 16 μm, spherical; the long thread remains coiled in a transverse plane. Three pairs of **barbs** and rows of small thorns or **nettles** are present at the base (Fig. 23.3C). The discharged thread pierces the animal body and the fluid hypotoxin, protein in nature, in the nematocyst, has a numbing effect on the animals.
2. **Volvent** or **desmoneme.** Pear-shaped, about 9 μm in size; thread short, in the form of a single loop (Fig. 23.3D). The thread coils round the prey after discharge.
3. **Glutinant steroline** or **atrichus isorhize.** Size about 7 μm; thread unarmed and straight (Fig. 23.3E).
4. **Glutinant streptoline** or **holotrichous.** Oval, about 9 μm. in size; thread long with three or four transverse coils and bears small nettles (Fig. 23.3F). The thread coils round the prey after discharge.

   The penetrant and volvent types help in capture of preys while the streptoline and steroline glutinant types help both in food capture and locomotion.

   Cnidocytes once used become nonfunctional and cast off.

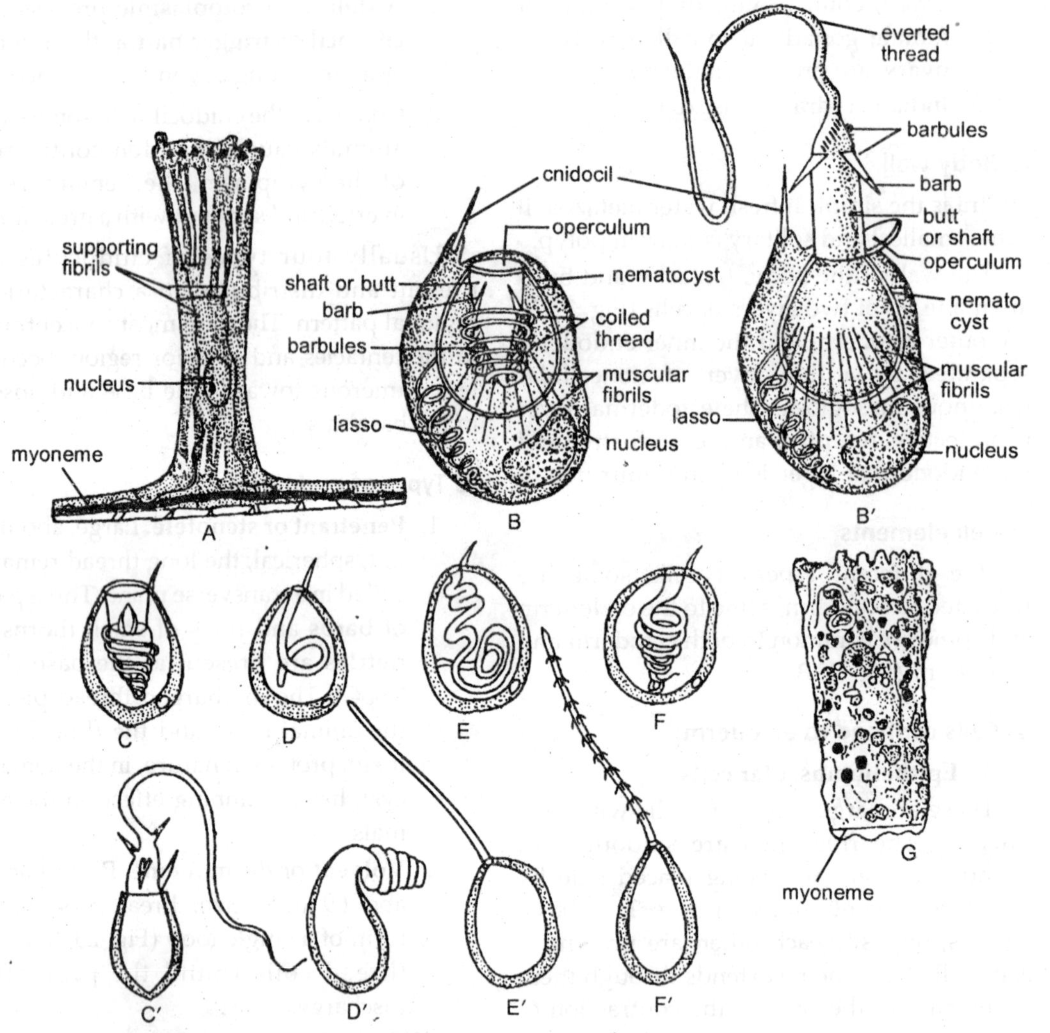

FIGURE 23.3

***Hydra* sp.** Cell elements... A. An Epitheliomuscular cell. B. A typical cnidocyte (at rest). B′. The nematocyst discharged (A, B. further magnified). C-F. Different types of cnidocytes (at rest). C′-F′. The nematocysts discharged. C. Penetrant. D. Volvent. E. Small glutinant. F. Large glutinant. G. An epithelionutritive cell. H. A gland cell. I. A sensory cell. J. A nerve cell. K. An interstitial cell

3. Reproductive cells

Hydra is dioecious or monoecious. Undifferentiated male and female germ cells are present in the ectodermal layer which, in the breeding season, multiply to form the gonads, testis or ovary (Figs. 23.9, 23.10).

☒ Cells confined to endoderm

1. Epithelionutritive or Nutritivemuscular cells

Large cells (Fig. 23.3G) with one or more vacuoles and a prominent nucleus. Two flagella are present at the free tip of some of them. The basal end is provided with two contractile extensions containing a myoneme. The extensions take a circular course next to mesogloea. The *Hydra* elongates when these processes contract. These cells carry out nutritive function.

2. Mucous cells

Modified gland cells situated in the oral gastrodermis and possess flagella. Their secretion helps in the lubrication of food for easy swallowing.

3. Albumen cells

Elongated cells with narrow bases and may or may not possess flagella. They supply albumen to the developing ovum.

4. Storage cells

Modified gland cells.

☒ Cells common to ectoderm and endoderm

1. Gland cells

Abundant in tentacles, oral region and pedal disc. In *Hydra,* the pedal disc is thickly populated (beset) with glandulomuscular cell, and they have muscular basal extensions (Fig. 23.3H). In the gastrodermis, they secrete digestive ferments.

2. Sensory cells

Elongated cells found in the tentacular and the oral region. Their bases are continuous into one or more fine fibrils passing into the general nerve plexuses (Fig. 23.3I).

3. Nerve cells

Commonly known as ganglion cells, which conduct functional impulses (23.3J).

4. Interstitial cells

Situated between the narrow bases of the epithelial cells and are considered as persistent undifferentiated embryonic cells. They are capable of forming cnidocytes, transforming into sex cells as well as other types of cells. They are also supposed to take part in building and repairing processes (Fig. 23.3K).

☒ Locomotion

Hydra usually remains attached to a submerged object or the substratum under water with the basal disc and remains in an upright position. The tentacles are in constant move-

FIGURE 23.4

***Hydra* sp.** Locomotion: Looping movement

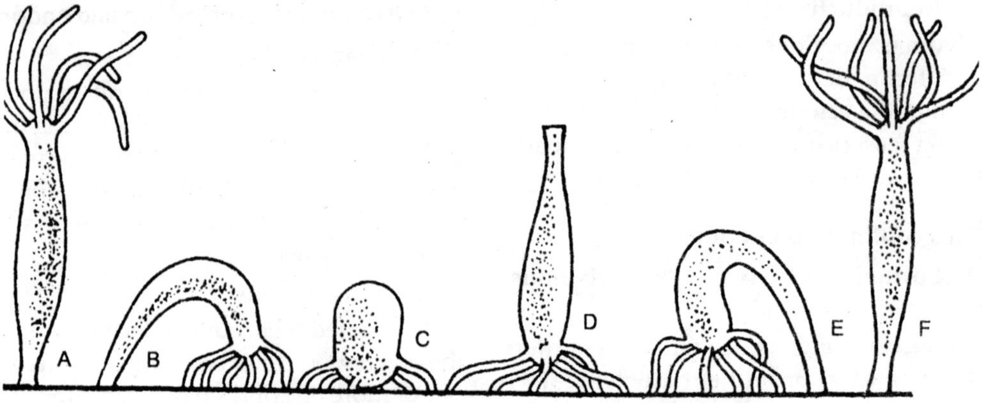

FIGURE 23.5

***Hydra* sp.** Locomotion: Somersault movement

ment in search of food and also in response to certain abiotic factors. Locomotion in Hydra (Figs. 23.4-23.6) is effected in the following ways.

1. **Gliding.** Hydra glides over the substratum with the amoeboid move-ment of the cells of the basal disc, without any change in the erect body position.
2. **Floating.** Occasionally the cells of the basal disc secrete gas bubble, which makes the body light and enables the Hydra to float on the water surface. In that state, it is carried away by wind or water current. The tentacles often help in swimming.
3. **Looping.** It is the most common type of locomotion in Hydra. It resembles the looping of a caterpillar (Fig. 23.4). In the normal erect position the body elongates, followed by bending of the distal part and the tentacles are attached to the substratum with the help of glutinant cnidocytes. The basal disc is released, the proximal end drawn forward and the basal disc reattached to the substratum, close to the tentacles, followed by assuming the erect position.
4. **Somersaulting.** The first phase of somersaulting resembles looping. The erect body extends and bends, the tentacles are attached to the substratum forming a loop. The basal disc is freed, the animal stands on the tentacles and the body contracts to form a small knob. This is followed by extension of the body and bending by which the basal disc is placed on the substratum in a forward position, forming a loop. The tentacles are released and the animal assumes an upright position (Fig. 23.5). By repeating the process, Hydra moves to a distant place. This is the normal mode of locomotion in Hydra.
5. **Inverted.** The tentacles are fixed to the substratum, the basal disc is up and the Hydra is in an inverted position (Fig. 23.6). By alternate contraction and

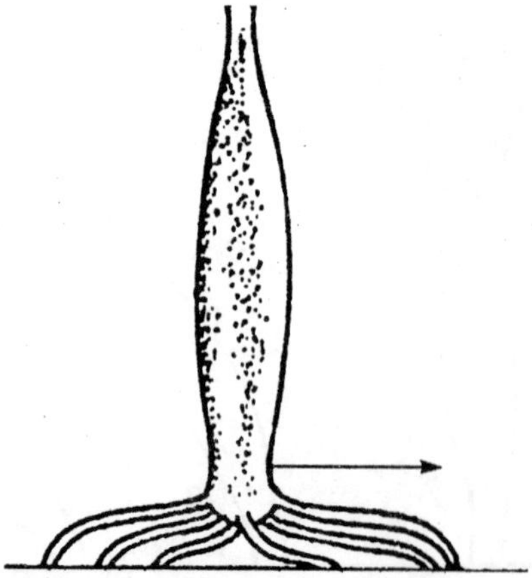

FIGURE 23.6

***Hydra* sp.** Locomotion: Inverted movement (Caterpillar like) relaxation of the tentacles, Hydra moves forward slowly.

6. **Climbing.** Hydra relaxes the tentacles and attaches them with aquatic vegetation or any object. The basal disc is released and tentacles contract, pulling the body to a new position.

☒ Nutrition

Hydra can digest proteins, fats, some carbohydrates, but not starch.

1. Small annelids, crustaceans, insect larvae are chief food of Hydra.
2. The nematocysts discharge in contact with the food animal and the toxin injected into the body of the victim, paralyses the prey.
3. The paralysed prey is wrapped about by tentacles and gradually drawn towards the mouth which opens widely to receive it.
4. Usually, Hydra swallows only living prey and it engulfs only those animals which contain **glutathione.**
5. Glutathione is present in the tissue fluids of most animals and is necessary to evoke the feeding reactions in Hydra.
6. The engulfed materials are covered by mucous secreted by the mucous gland cells in the first part of coelenteron.
7. Proteolytic enzyme is secreted in the enteron from the gland cells.
8. In the alkaline medium **extracellular digestion** of protein occurs.
9. Some endodermal cells produce pseudopodia and engulf the partly digested food into food vacuoles.
10. The rest of the digestion take place in food vacuoles, which is **intracellular digestion.**
11. The digested food is absorbed by the endoderm cells and then transferred to ectoderm and distributed to all parts of the body.
12. Undigested food materials are ejected through the mouth on contraction of the body.

☒ Respiration and Excretion

Hydra has no blood, no blood vessels, nor any organ of respiration and excretion. As the body wall is thin and the gastrovascular cavity usually circulated with water, most of the cells of the body remain freely exposed to the surrounding water. The exchange of oxygen and carbon dioxide gases—necessary for respiration and excretion of nitrogenous wastes (chiefly ammonia) occurs directly by diffusion through the cell membranes as in protozoans.

☒ Nervous system

Hydra is the first metazoan in which a nervous system is met with. A network formed by the nerve ganglia and the processes of the neurons constitute the nervous system (Fig. 23.7) and termed as **nerve net system** (p. 132).

☒ Reproduction

Hydra reproduces both by asexual and sexual means:

I. Asexual reproduction

Both budding and fission are present, the first one is most common.

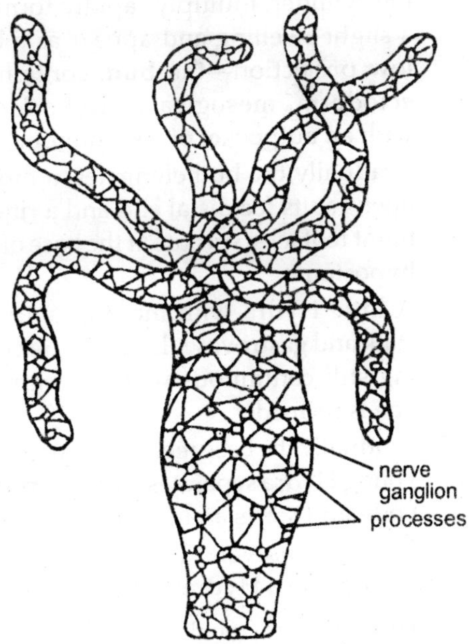

FIGURE 23.7

***Hydra* sp.** Nervous system—nerve net type

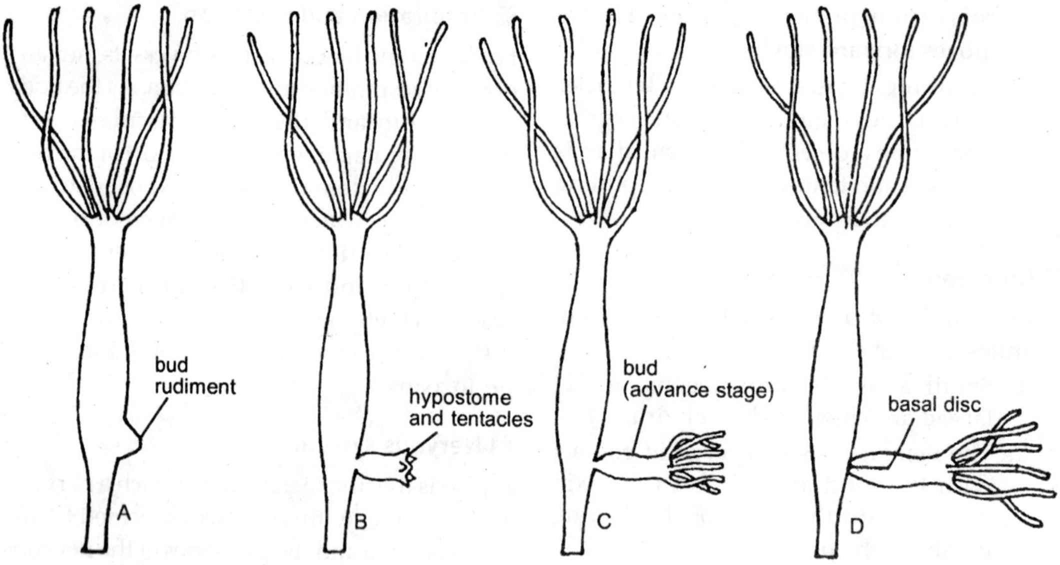

FIGURE 23.8

***Hydra* sp.** Bud formation: A. 3 hours, B. 24 hours, C. 48 hours, D. 72 hours

A. Budding

1. It occurs in summer in well-fed and healthy Hydra when food supply is abundant (Fig. 23.8).
2. The interstitial cells in the ectoderm, near the base or about the midway of the cylinder, multiply rapidly forming a slight swelling and appear as a hollow projection—the **bud**, containing ectoderm, mesoglea and endoderm with an outpocket of coelenteron.
3. Gradually the bud elongates, a mouth opens at its fee distal end and a ring of blunt tentacles sprouts at the base of the hypostome.
4. A little Hydra with mouth, tentacles, stalk and gastrovascular cavity, remaining still continuous with the parental one is formed.
5. Gradually the base region constricts, the young Hydra separates from the parent body and starts an independent life.
6. Sometimes several buds may develop at a time in a Hydra.

B. Fission. Occasionally both transeverse and longitudinal fission occur in Hydra, although it is not a normal mode of reproduction. Accidental breakdown, or for regaining normal shape from atypical condition, such as persistence of buds, two heads, two discs, etc. leads to such fission.

II. Sexual reproduction

1. Most hydras are **dioecios** but some species are **monoecious** (hermaphrodite) in which sperms mature before the eggs, and self-fertilization is prevented.
2. Sexual reproduction usually occurs in winter and spring months.
3. In some species (*Hydra oligactis*) male is smaller and bears one to eight testes, while the larger female bears one or two ovaries.
4. Sex reversal has also been observed where in one season a Hydra develops ovaries but in the next season it develops testes.
5. In hermaphrodite species usually testes develop towards the distal end and the ovaries towards the basal disc.
6. Lowering of temperature and increase in free carbon dioxide in water promote the formation of gonads in Hydra.

☒ Development of testes

1. A testes is a slight conical elevation, arising from ectodermal interstitial cells and having a number of elongate cysts.
2. At the base of a cyst is the primordial germ cell (interstitial cells), which divide repeatedly by **mitosis** to form a variable number of **spermatogonia**, which grow into spermatocytes (Fig. 23.9).
3. The spermatocytes undergo **meiotic division** forming spermatids which develop into spermatozoa.
4. A mature **spermatozoan** is minute, possessing three parts—a swollen head containing nuclues, a narrow mid-piece, and a long slender tail.
5. The mature sperms make their way through the rupture of outer ectodermal wall.
6. After release the sperms usually remain viable for 24 to 48 hrs.

☒ Development of ovary

1. The development of ovary is similar to that of testis. The interstitial cells multiply to form the germ mother cells or **oogonia**.
2. One centrally located cell, the **oocyte**, becomes larger and amoeboid with a large nucleus and yolk materials.

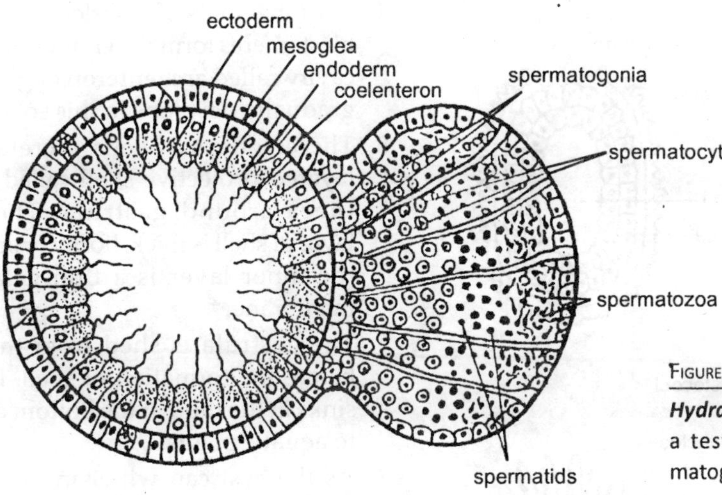

FIGURE 23.9

***Hydra* sp.** Section through a testis (stages of spermatogenesis)

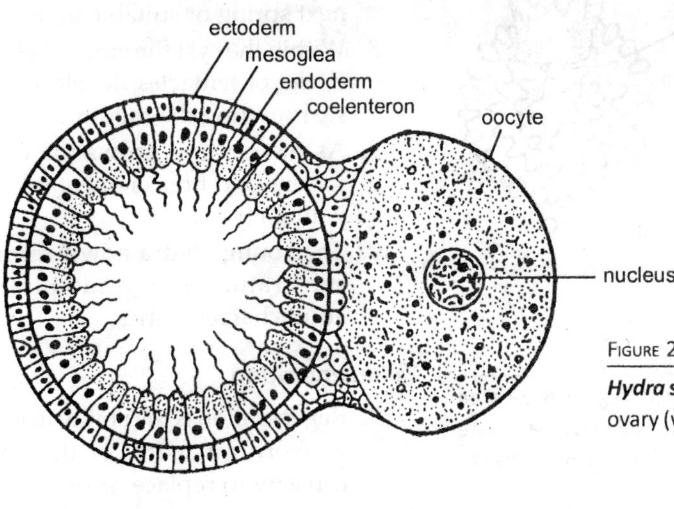

FIGURE 23.10

***Hydra* sp.** Section through an ovary (with a mature oocyte)

3. Subsequently, it undergoes a **meiotic** division to produce one **ovum** and two polar bodies.
4. One ovary usually contains a single ovum and remains attached to the parent by a broad base. Surrounding the ovum a gelatinous sheath is present (Fig. 23.10).

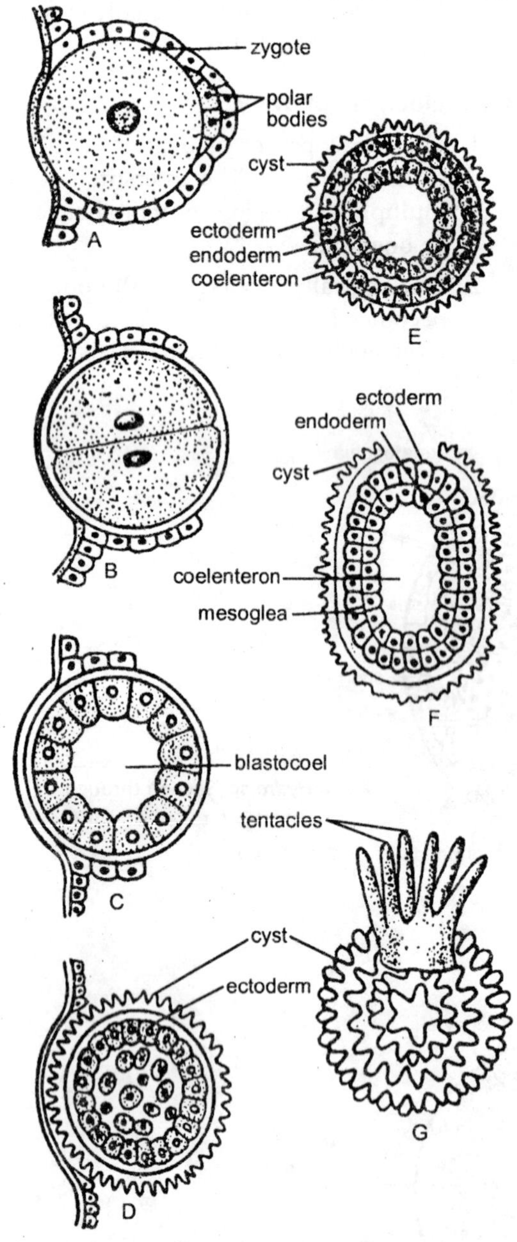

FIGURE 23.11

***Hydra* sp.** Stages of development. A. Zygote. B. 2-celled stage (blastomeres). C. Blastula. D. Early gastrula. E. Late gastrula. F. Fully formed embryo. G. Young Hydra hatching out

Fertilization and Development

1. Several sperms penetrate the gelatinous sheath but only one can successfully enter the ovum and fertilize it to form a **zygote** (Fig. 23.11).
2. The zygote cleaves immediately. The **cleavage** is holoblastic and results in a hollow spherical ball, the **blastula** or **coeloblastula**. The cavity inside it is the **blastocoel**.
3. Gradual development leads to a solid gastrula or **stereogastrula**.
4. The inner mass of cells derived from the blastomeres form the **endoderm**. A new cavity called **archenteron** or **gastrocoel** gradually develops in this solid mass.
5. The ectodermal cells secrete a two layered protective shell called **cyst** or **theca** around gastrula. The outer layer of shell is thick, horny or spiny but the inner layer is a thin gelatinous membrane.
6. The gastrula is shed from the body before the formation of shell. It either sinks in the mud at the bottom or sticks to aquatic plants.
7. As the cyst can withstand drying or freezing, it remains unchanged until the next spring or suitable time.
8. Within the cyst the embryo elongates, a circlet of tentacles develop at one end surrounding the mouth.
9. As the embryo grows the outer shell softens and breaks open, releasing the young Hydra.
10. The young Hydra now fixes itself to a substratum and gradually assumes adult characteristics.

Regeneration

1. Regeneration is the ability to restore lost or worn out parts of the body. The capacity to replace or regenerate parts

lost by injury or otherwise is allied to growth after fragmentation.

2. Young animals and species low in the evolutionary scale usually have greater regenerative power than the older or higher animals.
3. Regeneration power in Hydra was reported first by Trembly (1744). Hydra, if cut into several pieces, each will regenerate the lost parts and develop a whole new individual.
4. The body parts usually retain their polarity in original, with oral end developing tentacles and aboral end, basal disc.
5. If the head of Hydra is split into two and the parts are slightly separated, it results into a Y-shaped Hydra or two-headed individual, having two months and two sets of tentacles.
6. Brien (1953) and others found that in Hydra there is a growth zone just below the base of the tentacles. The interstitial cells give rise to new body cells of all types. As the new cells are formed, other cells are pushed towards the tentacles or the basal disc where the old cells are shed. All the cells may be renewed within 45 days.
7. If the internal cells of the growth zone are destroyed by X-rays, the Hydra lives only for a few days.

OBELIA sp

Obelia is a marine, sedentary, colonial cnidarian under the class Hydrozoa, phylum Cnidaria.

☒ Structure

1. It is a branched, fixed colony (Fig. 23.12). Some of the horizontal branches anchoring the colony on some support are called **Hydrorhiza** while other branches are vertical and known as **Hydrocaulus**.
2. Each branch consists of a granular coenosarc made of two cell layers enclosing the coelenteron and surrounded by a thin transparent horny **perisarc**.
3. The vertical branches towards the base are further branched and all the branches end in zooids.
4. Zooids are of three types:

a. **Polyps** or **gastrozooids** (vegetative zooids). Barrel-shaped and responsible for the nutrition of the colony. The

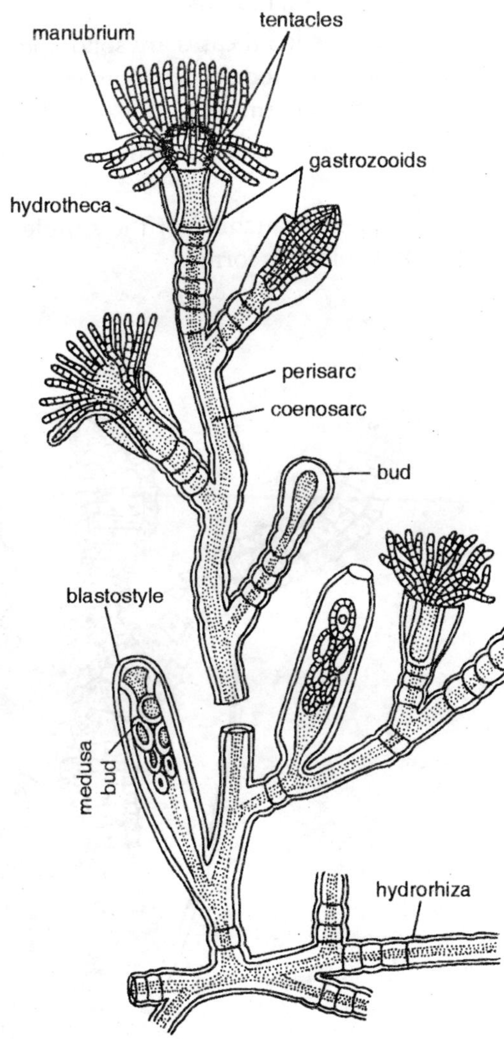

FIGURE 23.12

Obelia geniculata. Full colony

perisarc enclosing the polyp is termed **hydrotheca**.

b. **Blastostyles** or **gonozooids.** Club-shaped zooids, bearing the medusae buds.

c. **Medusae buds.** Umbrella-like reproductive zooids bearing gonads, enclosed in a gonotheca.

5. Medusae buds are unisexual and free-living at maturity. One medusa bears either four testes or four ovaries close to the four radial canals.
6. The tentacles in all cases are solid; the solid core or endoderm surrounded by a layer of ectoderm cells.
7. The larvae are ciliated and free-swimming.
8. By repeated branching of the simple polyp, a colony is formed.

☒ Polyps or Gastrozooids or Nutritive Zooids

1. The zooid is barrel-shaped, partially enclosed by a cup-shaped **hydrotheca**, a continuation of the perisarc.
2. At the distal end a conical projection the **hypostome** or manubrium, bearing a **mouth** is present (23.13).
3. A circlet of about twenty-four tentacles are present around the hypostome. The tentacles are solid with a core of endoderm cells surrounded by a layer of ectodermal cells.
4. The tentacles and the hypostome bear cnidocytes.
5. At the proximal end, the zooid is continuous with the coenosarc.

☒ Blastostyle or Gonozooid or Reproductive zooid

1. These are a few in number and restricted to the basal end of the hydrocaulus.

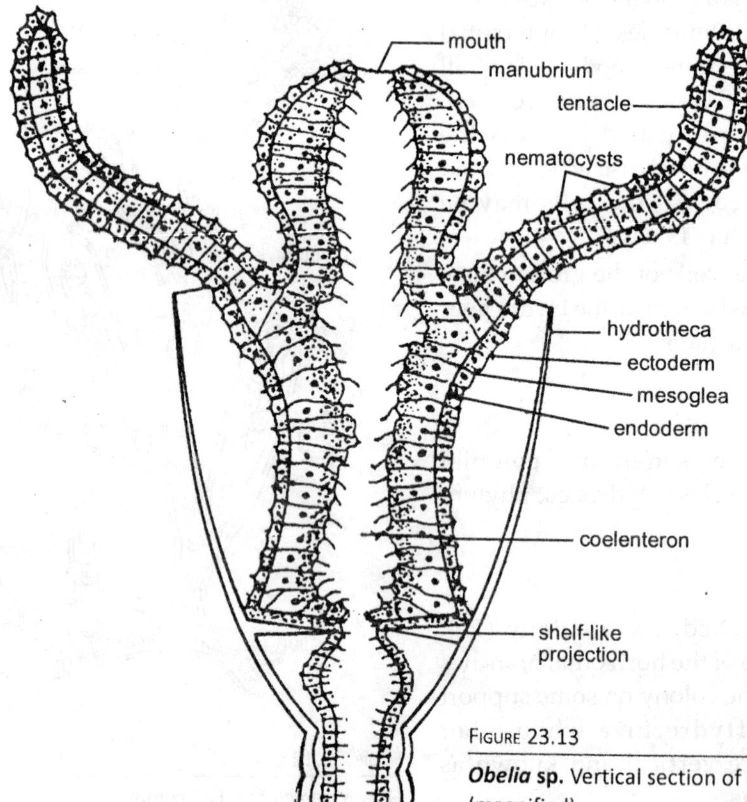

FIGURE 23.13

***Obelia* sp.** Vertical section of a polyp (magnified)

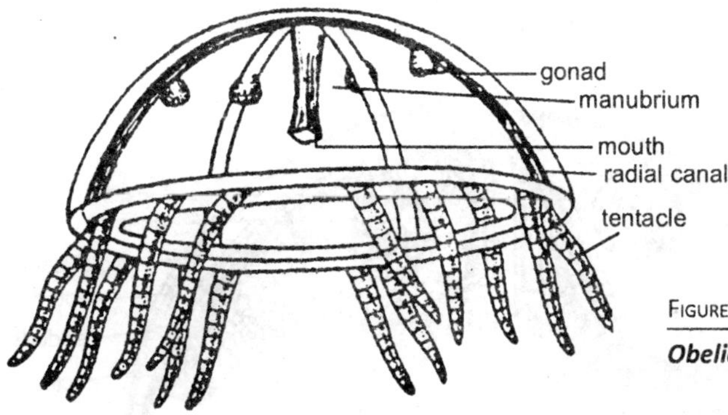

FIGURE 23.14

***Obelia* sp.** A medusa

2. The blastostyle is long, cylindrical and devoid of mouth and tentacles and enclosed in a transparent **gonotheca**, modified perisarce (Figs. 23.12, 23.16).
3. A number of small buds—the **medusae buds** develop on the walls of the blastostyle.

Medusa

1. At maturity, small umbrella-like medusae buds get detached from the blastostyle and escape to the sea water through an aperture at the free end of the gonotheca. A hydro-medusa (Fig. 23.14) is a solitary and free-swimming modified zooid.
2. A typical medusa consists of a bowl-shaped gelatinous disc or umbrella, the concave surface of which is described as **subumbrella** and the outer convex surface as **exumbrella**.
3. A cylindrical projection, the **manubrium**, hangs from the middle of the subumbrellar surface. A square mouth is present at the apex of the manubrium.
4. The mouth leads to a small, central, rounded or quadrangular chamber known as **gastric cavity** or **stomach**.
5. From gastric cavity four **gastrodermal canals** radiate to the margins of the bell. These are radial canals, opening into a **ring canal** or circular canal running in the margin of the bell. The radial canals mark out four principal radii.
6. From the middle of the radial canal four **gonads** project. Since sexes are separate, these are either testes or ovaries.
7. The gonads mature after the medusae escape from the **gonotheca**.
8. The edge of the bell is produced inwards as a thin fold called **velum**. The medusae with a velum are known as **craspedote** and those without a velum are **acraspedote**. The hydromedusa is craspedote.
9. From the edge of bell numerous small solid tentacles hang downwards.
10. The tentacles have swollen bases of accumulated interstitial cells, called vesicles or **bulbs**. Cnidocytes are formed continuously in the bulb and migrate to the tentacles. They are confined to manubrium and tentacles and a few on the bell margin.
11. Near the bulbs the ectoderm has pigment granules and nerve cells called **ocelli**. The pigments are accumulated excretory products.
12. Marginal sense organs, **statocysts** or **lithocysts** are eight in number and attached at regular intervals on the subumbrellar side to the bulbs of the tentacles.
13. A statocyst is a tiny, circular, closed vesicle lined with ectoderm and filled with a fluid containing calcareous granules called **otoliths**.
14. Otolith produces a stimulus on thin sensory processes of the sensory cells of ectodermal lining which are transmitted to the muscles. The muscles coordinate the snake-like swimming movements of the medusa.

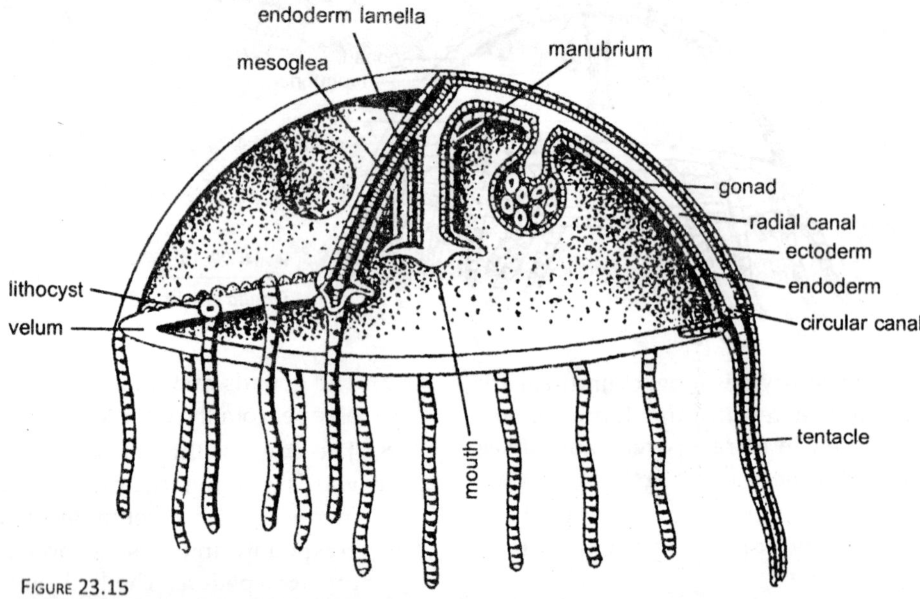

FIGURE 23.15

***Obelia* sp.** Medusa (more than one-quarter of the umbrella and manubrium removed)

⊠ Histology of hydromedusa (Fig. 23.15)

1. Both exumbrellar and subumbrellar sufaces of medusa are covered by ectoderm cells.
2. The whole canal system is lined by endoderm cells, continued, at the lip of the mouth, into the surface ectoderm.
3. A thin sheet of endodermal lamella, presumably formed by the fusion of an upper and a lower layer of endoderm, lie between the radial canals and between the ex- and subumbrellar layers of ectoderm.
4. The velum is composed of a double layer of ectoderm enclosing a ring canal and a strip of narrow mesoglea in between the canal and ectoderm.
5. The tentacles are solid, each containing a core of vacuolated endodermal cells covered by ectoderm.
6. The gelatinous mesoglea form the main bulk of the body and contains certain noncellular fibres.
7. Well-developed musculature with regularly arranged circular, longitudinal and radial tracts is present.
8. The interstitial cells are mainly accumulated at the bulbs or vesicles of tentacles and give rise to cnidocytes of only one type.

⊠ Gonads

1. The medusae are sexual reproductive zooids or gonozooids possessing gonads.
2. They are **dioecious**—testes and ovaries are borne by separate individuals.
3. Each medusa bears only four gonads situated on the subumbrellar surface, one on the middle of the course of each radial canal.
4. The gonads mature after the medusae escape from the gonotheca.
5. Each gonad (testis or ovary) is an ovoid, knob-like body; it has an outer covering of ectoderm, continuous with that of the subumbrella, and an inner lining of endoderm continuous with that of the radial canal (Fig. 23.14, 23.15).
6. The space between the two layers is filled with a mass of interstitial cells which become differentiated into ova or sperms, as the case may be.

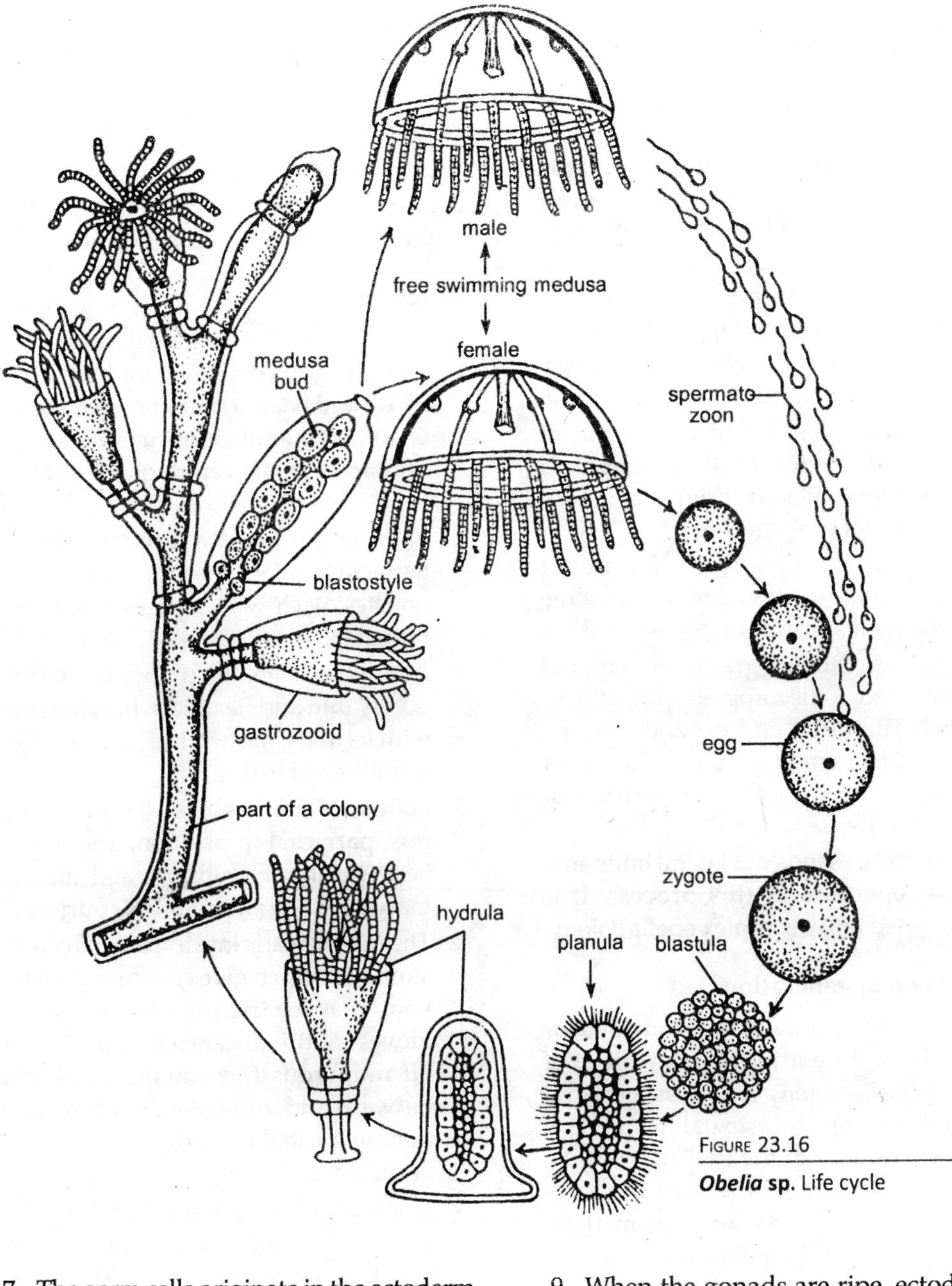

FIGURE 23.16

***Obelia* sp.** Life cycle

7. The germ cells originate in the ectoderm of the manubrium quite early when the medusa itself remains attached to the blastostyle.
8. Subsequently, they migrate to the gonads to lie between the mesoglea and the subumbrellar ectoderm and undergo maturation.
9. When the gonads are ripe, ectodermal covering ruptures and the germ cells are shed in water.
10. The ova are large rounded cells.
11. The sperms are minute, actively swimming flagellated cells.
12. The medusae die soon after liberating the gametes.

Fertilization and Development

1. Fertilization occurs either in the sea water where the germ cells are set free, or the spermatozoa may be carried by water currents to the female medusae and fertilize the ova in situ. Zygote formed after fertilization, immediately undergoes cleavage.
2. The cleavage is holoblastic and a **blastula** is formed (Fig. 23.16).
3. By invagination the blastula is converted into an oval, ciliated **planula** larva.
4. The planula consists of an outer layer of ciliated ectoderm and an inner mass of endoderm cells enclosing a space, the rudiment of coelenteron.
5. The planula swims freely for a brief period and settles down on some submerged substratum by one end.
6. The proximal end gradually narrows down and a disc appears for attachment. The distal end expands and by developing a manubrium and a circlet of tentacles, it turns to a **hydrula** or simple polyp.
7. The hydrula sends out lateral buds and, by a repetition of this process, it is converted into a complex obelia colony.

Alternation of generations

A distinct alternation of generations or **metagenesis** is present in the life history of *Obelia*. The Obelia colony is sexless, bears no gonads and develops by asexual process, i.e. by repeated budding of the hydrula. But the medusae buds, some of the zooids of the colony, develop gonads and, from their fertilized egg, new *Obelia* colony arises. The asexual generation is dependent on, and is alternated by the sexual generation.

Obelia is a permanently fixed colony but the planula larvae it produces are free swimming. The larva can swim from place to place with the help of cilia and, being aided by water current, it can travel a long distance. Thus, a nonlocomotory species becomes locomotory and an overcrowding of individuals within a limited area is avoided and, thereby, the species is successfully continued.

AURELIA sp.

Aurelia is a marine, free-swimming jelly fish placed in the class Scyphozoa, phylum Cnidaria. It is the medusoid stage of the species.

Structure

1. The body consists of a gelatinous disc with a slightly convex **exumbrella** (Fig. 23.17) and bears a four-cornered **mouth** on a short **manubrium** guarded by four **oral arms** at the centre of the **subumbrella**.
2. The margin bears a large number of tentacles and is smooth except eight notches, each with a pair of **marginal lappets**.
   Along the edge, the umbrellar portion is very thin and flexible, which together with its marginal notches and tentacles, form the **velarium**.
3. Four small depressions, the subgenital pits, parradial in position, are present between the manubrium and the bell margin on the subumbrellar surface.
4. The mouth leads into a spacious central **stomach**, which gives out four pouches known as **gastric pouch**. Each pouch moves a little distance and then gives off numerous fine canals, which run outward and, ultimately, unite to form a circular canal lodged within the edge of the umbrella.
5. **Gonads** are four in number, and are visible through the transparent umbrella as coloured horseshoe-shaped patches. They are frill-like structures lying on the floor of the gastric pouch, with the concavity facing inwards.
6. The main bulk of the umbrella is mesogleal in structure but it contains numerous amoeboid cells, which come from the ectoderm. It has also some branching fibres. The outer (ex-) and the

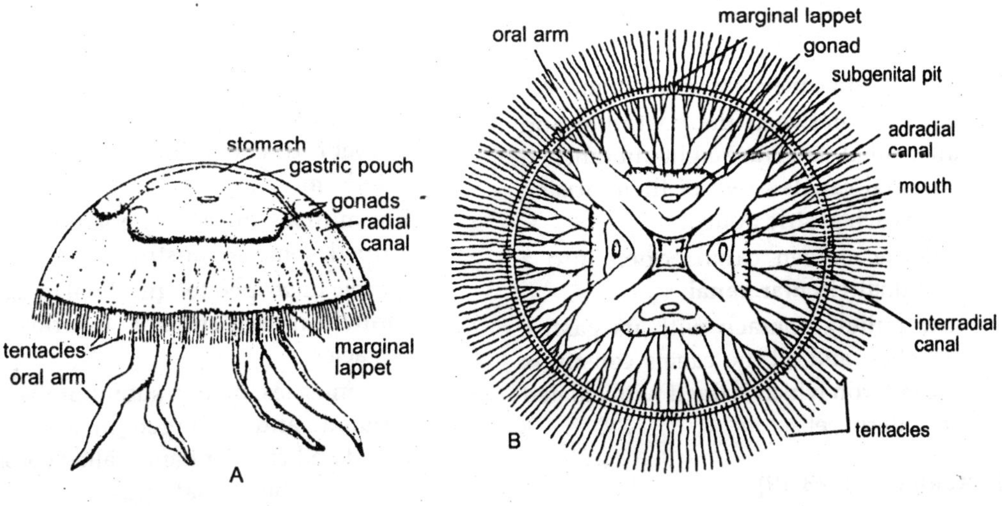

FIGURE 23.17

Aurelia aurita. A Floating position. Dorsolateral view, B. Subumbrellar view

inner (sub-) surfaces of the umbrella are covered with ectoderm. The stomach and the canal system are lined with endodermal cells and are ciliated throughout.

☒ Coelenteron

1. The coelenteron (Fig. 23.18) of *Aurelia* has been modified, and remains only along four meridional areas, the **radial canals** and as a **circular** canal.
2. The mouth leads into a short **gullet**, which joins a large rectangular stomach.
3. The **stomach** is produced laterally into four interradial **gastric pouches**. Numerous **gastric** filaments are present on the floor of the gastric pouches and also at the junction of the gullet and stomach.
4. At four corners of the mouth are four **parradial canals**, at right angles to each other.

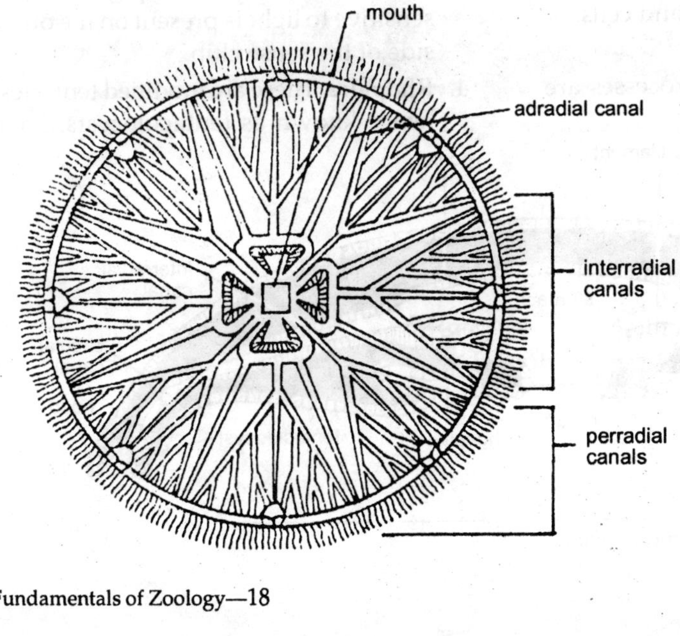

FIGURE 23.18

Aurelia aurita. Coelenteron and canal system (ventral view; oral arms removed)

5. Between these are four interradial canals.
6. These eight canals are much branched.
7. Halfway between the parradial and interradial canals are eight unbranched **adradial canals**—two of them arising from each gastric pouch.
8. All the sixteen canals open into a marginal **circular canal**.
9. The gullet, stomach with its gastric pouches, and all the canals constitute the **enteron** which is lined with ciliated endoderm cells.

⊠ Histology (Fig. 23.19)

1. The ectoderm covers the bell all around.
2. The gullet, and the subgenital pits are also lined by the invaginated epidermis. The epidermis contains epithelial, sensory, nerve, gland cells and cnidocytes.
3. The cnidocytes occur on the oral arms, the ex- and subumbrellar surfaces, the marginal tentacles as well as on the gastric filaments.
4. The gastrovascular cavity, its radiating canals and their extensions are lined by gastrodermis. The interspaces between the gastrovascular canals are filled with a thick solid layer of gastrodermal lamella. The cells are mainly flagellated columnar epithelial and gland cells.
5. Gonads are gastrodermal.
6. Nerve cells and muscle processes are wanting.
7. A thick gelatinous mesogloea, between the ecto- and endoderm, form the main mass of the umbrella. It contains branching fibres and wandering amoeboid cells derived from the ectoderm.
8. The muscles are usually epidermal and striped, some are unstriped.
9. Simple unstriped longitudinal muscles form the musculature of oral arms and tentacles.
10. The muscles in the subumbrella are striated, and arranged in a strong, broad and circular band, called coronal muscle. The radial muscles extend along the main radii from the manubrium to the coronal muscle.
11. Locomotion is the result of rhythmic contraction of muscles.

⊠ Sense Organs

1. The sense organs include the olfactory pits, the ocelli and the tentaculocysts or rhopalia.
2. **Olfactory pits.** Two, lined with sensory epithelium—one lie on the exumbrella and the other in the sense club. These are **chemoreceptors** and recognize the food.
3. **Ocellus.** An ectodermal pigment spot sensitive to light is present on the outer side of the sense club.
4. **Rhopalia.** These are modified tentacles, hence known as tentaculocysts. Four

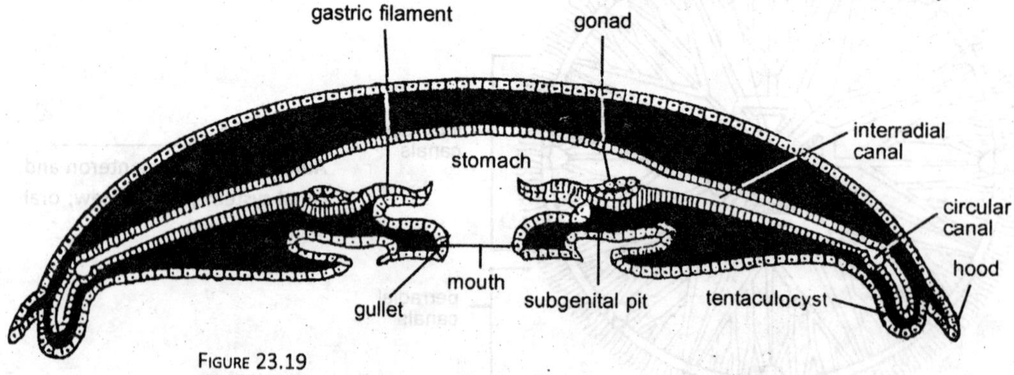

FIGURE 23.19

***Aurelia aurita*. Vertical section**

parradial and four interradial rhopalia (Fig. 23.20) are present in the eight marginal notches.

a. Each rhopalium consists of a hollow club-shaped projection, called **statocyst** or **lithostyle**.
b. The statocyst remains covered over by a hood-like process of the bell margin, connecting a pair of lateral marginal lappets.
c. The statocyst is a specialised hollow tentacle enclosing a prolongation of the circular canal lined by gastrodermis. The apical gastrodermal cells contain **otoliths** or **statoliths**, made of calcium sulphate.
d. The rhopalia are the organs of equilibrium. They also control the rhythmic swimming movements of the bell.

☒ Reproduction

The sexes are lodged in distinct individuals and the four horseshoe-shaped gonads are either testes or ovaries. The spermatozoa or ova are shed in the stomach. The spermatozoa come to the exterior with water current and enter the stomach of a female individual with water current. Fertilization is internal.

☒ Development and Life History (Fig. 23.21)

1. The zygote divides regularly and a **blastula** is formed.
2. By invagination of the blastula an oval, two-layered ciliated **planula** larva is formed, blastopore of which is not closed.
3. After a brief period or free life, the larva settles down by the end opposite to the blastopore, loses its cilia and gradually transforms into a **hydratuba**.
4. The definite **mouth** is formed at the distal end.
5. On the opposite side of the mouth two tentacles are formed by outpushing. Soon two others appear at right angles to the first pair; then four, and subsequently eight more tentacles appear. The larva is now provided with sixteen tentacles arranged in a circlet.
6. By this time, the proximal end narrows down to a stalk-like structure the **stolon**, and the distal end becomes expanded.
7. Interradial in position, the endoderm of the gastric cavity is folded into four longitudinal ridges, the **gastric ridges** or **taenioles**.
8. The mouth assumes a square shape and its sides are elevated to form a short **manubrium**.

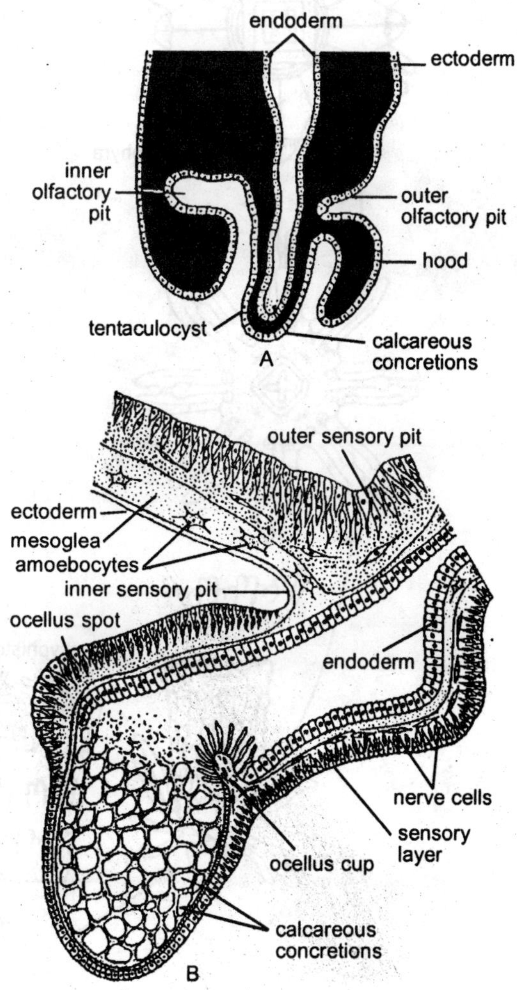

FIGURE 23.20

***Aurelia* sp.** A. Sense organ or rhopalium (vertical section). B. Part of A (further magnified)

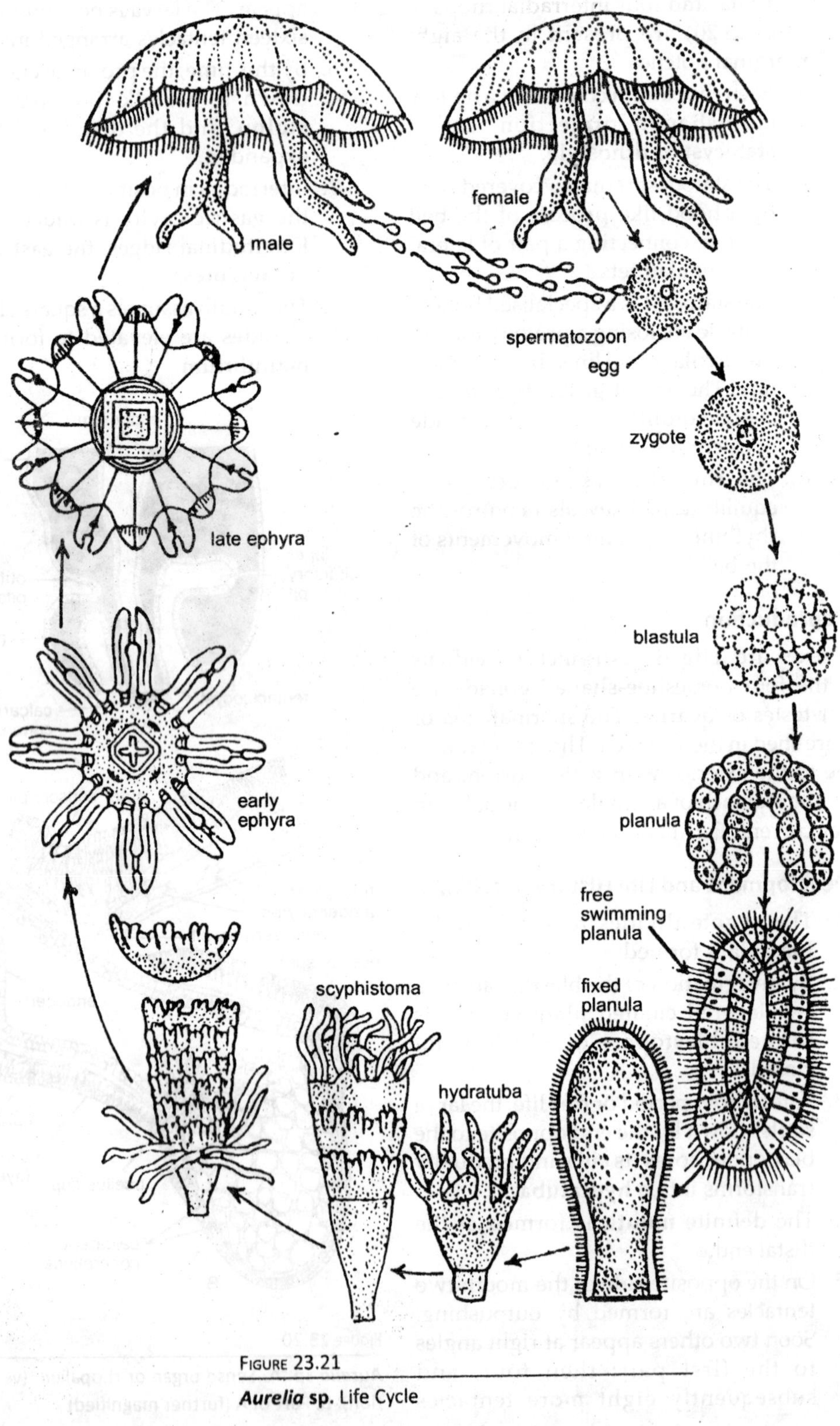

FIGURE 23.21

***Aurelia* sp.** Life Cycle

9. The ectoderm of the region between the mouth and the circlet of tentacles becomes invaginated in four interradii to form four septal funnels or **infundibula**, which sink in the four gastric ridges.
10. The hydratuba feeds and produces new hydratubae by budding from its stolon throughout the summer. The buds may separate from the mother hydratuba. Budding ceases at the end of the summer.
11. The hydratuba stage persists for a long period and undergoes **strobilation** or **transverse** fission in the next year.
12. This stage of a scyphozoan polyp is known as **scyphistoma larva**, and it is about 1.25 cm in length.
13. Several transverse constrictions appear on the larva and, as the constrictions grow deeper, the larva appears like a pile of discs, one placed above the other.
14. The edge of each is provided with eight bifid arms bearing deep notches. As the constriction proceeds, the stomach becomes closed on the lower surface of each disc, while on the upper surface it remains open and its outer edges are transformed into a manubrium.
15. Finally, the disc-like bodies separate form one another, turn upside down and are known as **ephyra.** Each ephyra carries with it a part of the stomach with the gastric ridges.
16. The mouth becomes elevated on the manubrium, tentaculocysts appear at the notches of the bifid arms, gastric filaments appear on the gastric ridges, soon followed by others and the stomach extends in the arms.
17. The swimming ephyra grows rapidly, notches are filled up and the umbrella becomes rounded; four oral arms and marginal tentacles appear and it is transformed into an adult *Aurelia*.

⊠ Alternation of Generations

In a sense, an alternation of generations is present in the life history of *Aurelia*. The sexual generation is represented by the adult *Aurelia*, which is produced by asexual means from the scyphistoma larva and the scyphistoma is produced sexually from the adult *Aurelia*. But unlike the budding of the hydrozoan polyp, the polyp undergoes several constrictions to produce a few adult individuals and it is not considered as a case of asexual generation.

The life history of *Aurelia* may better be described as a metamorphosis complicated by multiplication in the larval stage than a true alternation of generations.

It has also been seen that, under certain conditions, direct development from egg to scyphistoma, and that to the adult *Aurelia* occurs without the intervention of the socalled asexual reproduction.

TEALIA sp. (Sea anemone)

Sea anamones are members of the class Anthozoa under the phylum Cnidaria. This class as a whole was undergone high grade of specialization in an entirely different direction. All the members of the class remain as polyps, incapable of producing a free-swimming medusoid generation. Sea anemones are brilliantly coloured flower-like animals (Fig. 23.22) found attached to rocks at low water mark. Some are inserted in the sand.

⊠ Body form and mesentery

1. Column-like body, broader than high.
2. At the anterior end lies the **oral disc**.
3. In the middle of the oral disc lies the **mouth**.
4. From the edges of the **oral disc** are given off a number of short conical tentacles, arranged in five rings.
5. Outermost whorl carries the maximum number of tentacles, which is forty-eight. The subsequent whorls bear successively lesser number of tentacles.

Obelia medusa and Aurelia

Obelia	Aurelia
A. Medusa	
1. It has a small size, often called swimming bell and the medusa is produced by budding from blastostyles of polypoid obelia colony.	1. The size is large, often called jelly fish, and is formed by metamorphosis of ephyra, which is produced by strobilation of scyphistoma.
2. Medusa phase is reduced.	2. Medusa phase is dominant.
3. The bell margin is entire and is provided with few long, solid tentacles.	3. The bell margin is notched at eight radii having tentaculocysts and beset with many small tentacles.
4. The velum contains an endodermal ring canal. Craspedote medusa.	4. Velum is absent but the velarium containing endodermal canal is present. Acraspedote medusa.
5. Oral arms absent and the manubrium is comparatively elongated.	5. Four long oral arms, one at each corner of the four-sided mouth. The manubrium is comparatively short.
6. Gastric ridges and gastric filaments are absent.	6. Gastric ridges bearing endodermal gastric filaments are present.
7. Muscle strands are derived from both epidermis and gastrodermis	7. Muscle strands are derived only from the epidermis.
8. Ocelli and olfactory pits are absent.	8. Ocelli and olfactory pits are present.
9. Statocysts are ectodermal and unprotected by hood and marginal lappets.	9. Tentaculocysts are endodermal and protected by hood and marginal lappets.
10. Gonads lie on radial canals externally and the gametes are ectodermal.	10. Gonads lie on the floor of gastric pouches internally and the gametes are endodermal.
11. The subgenital pit is absent.	11. The subgenital pit is present.
B. Polyp	
12. Obelia colony or polypoid stage is the dominant phase; the adult form is trimorphic.	12. The scyphistoma or polypoid stage is very much reduced. It is solitary and a larval stage.
13. Hydrocaulus and hydrorhiza separable into perisarc and coenosarc.	13. No such structure.
14. Life history is remarkable for metagenesis	14. No true metagenesis but a case of metamorphosis complicated by multiplication in the larval stage.
C. Development	
15. Fertilization is external, occurring in water.	15. Fertilization internal, the union occurs inside the enteron of the female.
16. The cleavage is holoblastic and equal.	16. Cleavage holoblastic but unequal.
17. Planula larva is without a blastopore and coelenteron.	17. Planula possesses a blastopore and a coelenteron.
18. Planula finally grows into a simple hydrula stage which by repeated budding, produces obelia colony.	18. The planula is transformed into a hydra-tuba, which produces new hydratuba by budding. The hydratuba grows into a trumpet-like asexual polypoid scyphistoma which by strobilation or metamorphosis produces the aurelia medusa.

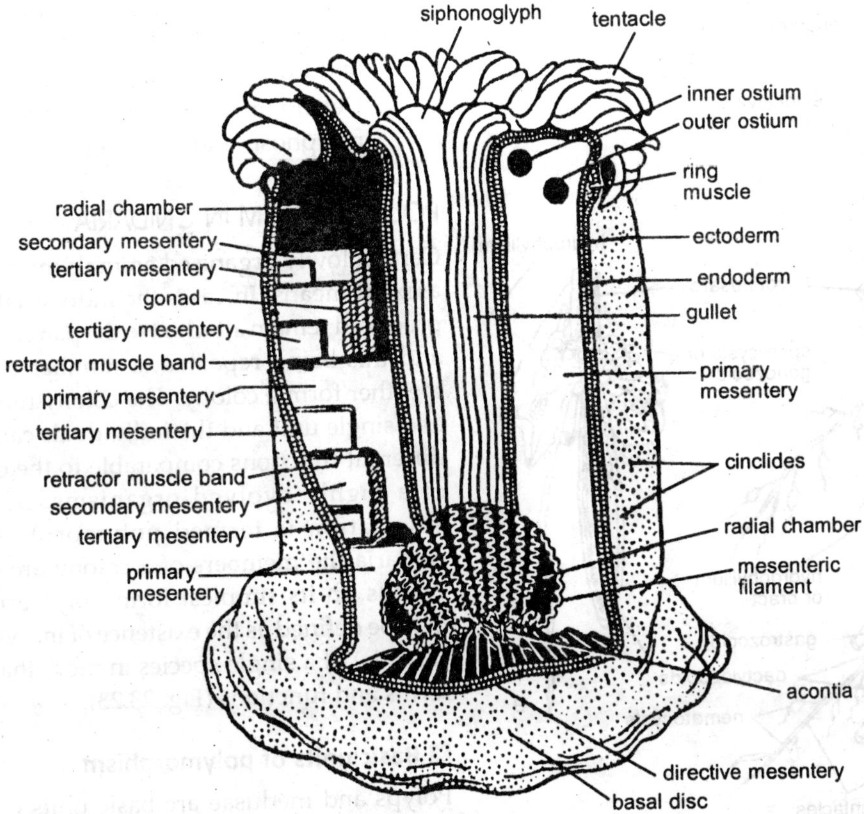

FIGURE 23.22

***Tealia* sp.** Vertical section (diagrammatic)

6. The posterior region of the body is flattened to form a **basal disc** for the purpose of attachment.
7. The oral disc invaginates to form a **gullet** or food tube, suspended into the general cavity.
8. The gullet has on its either side a pair of smooth, heavily ciliated furrows placed one at each end of the long diameter, through which the water passes to the enteron, the **siphonoglyph**.
9. The general cavity is subdivided into radial compartments by six distinct septa, each of which extends from the wall to the gullet. The septa are complete and known as **primary mesenteries**.
10. In between the primary mesenteries, two types of incomplete partition walls are found to be inserted; some are larger than the other. The larger partition wall is known as **secondary mesentery** and the smaller as the **tertiary mesentery**. Thus the gastrovascular cavity is subdivided into a number of vertical chambers by projecting partitions.
11. Each septum has a pair of ostia beneath the oral disc, for direct communication of the intermesenteric compartments.
12. The thick free inner margin of the septum convolutes to form mesenteric filament, and continues posteriorly as thread-like **acontium**. Acontium protrudes either through the mouth or through **cinclides** (openings in the body wall) for subduing prey.

⊠ Body wall

1. The epidermis and gastrodermis are separated by a very thick and tough

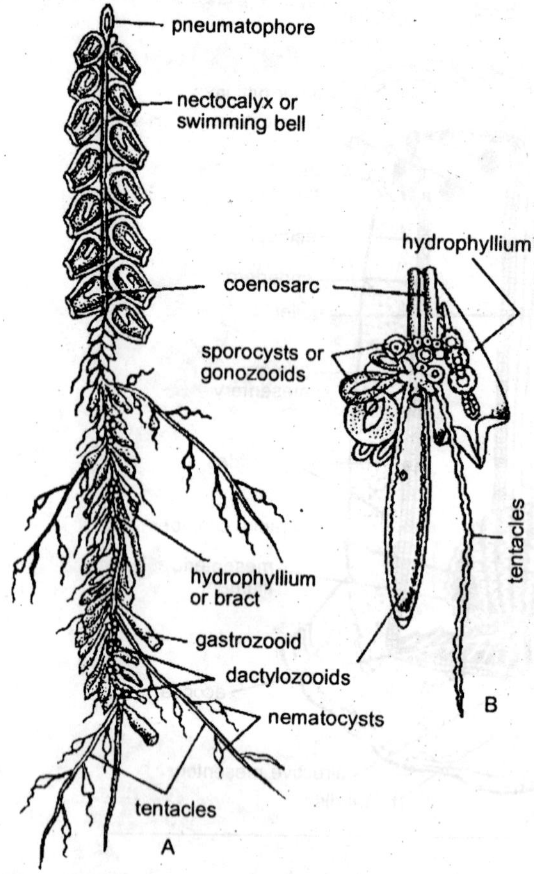

FIGURE 23.23

Halistemma. A. Entire Colony. B. Cormidium

layer of **mesoglea**. The former two consist mainly of very long, columnar, ciliated epitheliomuscular cells. The mesogloea, being traversed by a network of delicate fibres with interspersed cells, has assumed the characteristics of an intermediate cell layer.

2. Stinging capsules occur in the epidermis, and are also abundant in the mesenteric filaments.
3. Gland cells are abundant in the epidermal lining of the gullet and in the mesenteric filaments.
4. Each septum has got three sets of muscles, retractor (vertical), transverse (across) and parallel (oblique).
5. Nerve net occurs both in ectoderm and endoderm.
6. The gonads are endodermal in origin and the sexes are separate.
7. Development with complicated metamorphosis and planula stage.

POLYMORPHISM IN CNIDARIA

Certain lowly organised animals multiply by asexual means. In cases the individuals thus produced remain attached to the parent animal and this being repeated, a number of them together form a colony. The colony functions as a single unit and the individuals carry out different functions comparable to the organs of a highly evolved organisms. Such an association is termed polymorphism. In cnidaria the members of a colony are called **Zooids**. In its simplest form polymorphism may be defined as the existence of individuals (Zooids) of a single species in more than one forms and functions (Fig. 23.23).

☒ Basic units of polymorphism

Polyps and medusae are basic units of poly morphism in cnidaria. Polyps and medusae occur in a number of morphological variations, several of which may be found in a single species.

Dimorphism. Colony with a polyp or nutritive Zooid and medusa, the reproductive Zooid.

Trimorphism. In addition to polyp and medusa, the colony develops a modified polyp, the gonozooid or blastostyle. They produce medusae.

Polymorphism. Three kinds of polypoid Zooids and four kinds of medusoid cooids are present. *Siphonophora* belonging to Hydrozoa exhibit highest degree of polymorphisms (Fig. 23.24).

☒ Polypoid Zooids

1. **Gastrozooid or Nutritive Zooid.** Tubular or saccular with a mouth, usually a single, long and highly contractile tentacle arises from the base

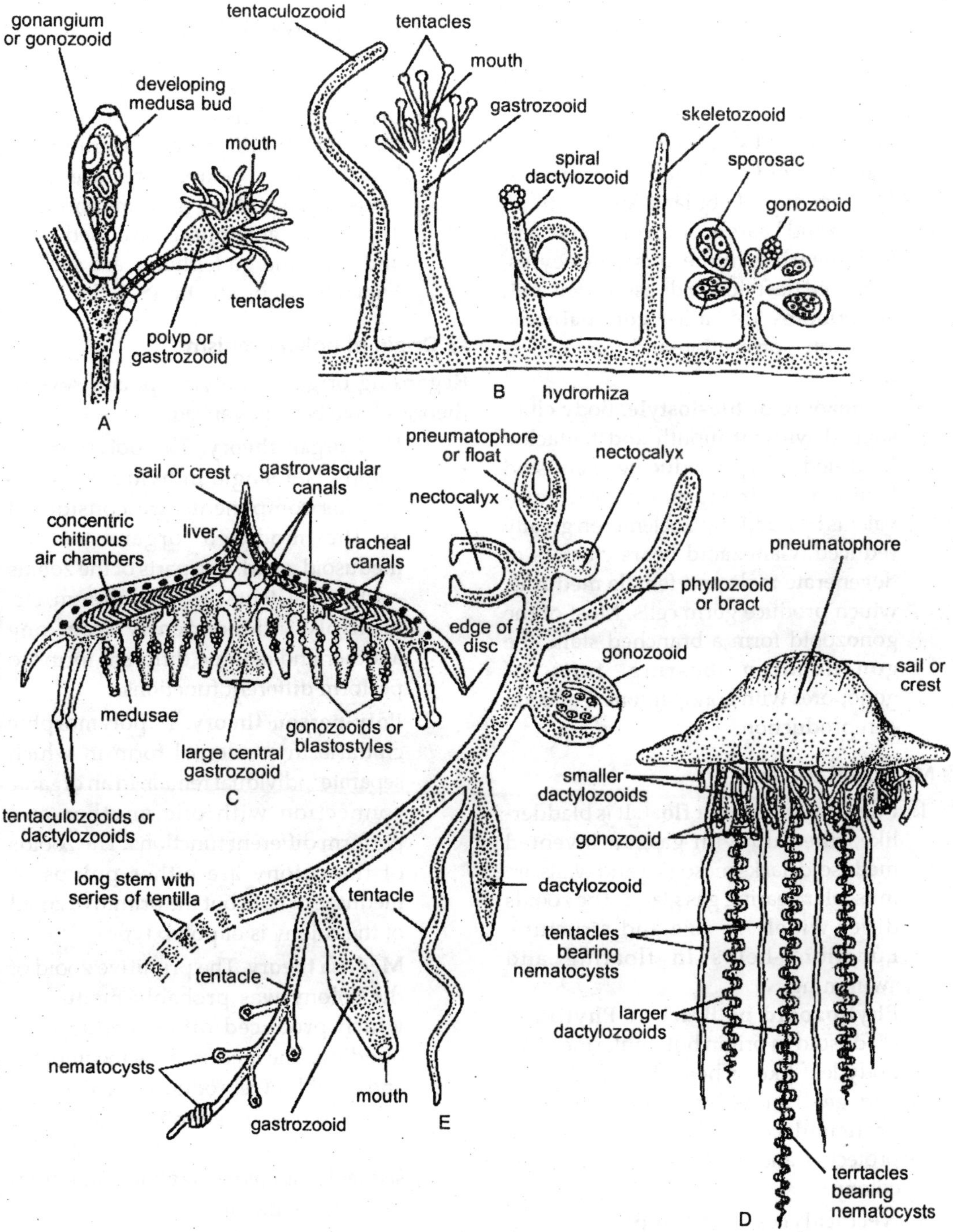

FIGURE 23.24

Polymorphic colonies of Hydrozoa. A. Obelia, B. *Hydractinia*, C. *Valella*, D. *Physalia*, showing single cormidium. E. Generalised calycophoron *Siphonophore* showing a single cormidium

and bears numerous fine lateral contractile branches, the **tantilla**. The tantilla terminates in a large knob or coil of nematocysts. Function—ingestion and digestion of food.

2. **Dactylozooid or Testers or Palpars or Macrozooids.** Tubular zooids with no mouth and bears a long, contractile tentacle bearing numerous cnidocytes. Modified dactylozooids associated with gonophores are called **gonopalpons**. Function—protection and food collection.
3. **Gonozooid or Blastostyle.** Body club-shaped without mouth and tentacles. Modified polyp producing medusoid forms or their morphological equivalents by budding. Coelenteron greatly reduced. Ganozaoid bears clusters of degenerate male and female medusae, which produce germ cells. Most of the gonozooid form a branched stalk, the **gonodendron** bearing tufts of gonopores with gonopalpons. Function—reproduction.

⊠ Medusoid zooids

1. **Pneumatophores or float.** It is bladder-like, filled up with gas; an invented medusoid, lacks mesoglea and walls are muscular bearing gas gland. The zooids differ widely in size and structure. Function—helps in floating and swimming.
2. **Phyllozooid or Bract or Phyllium.** Medusoid in origin but unlike medusa. Body leaf-like or helmet-shaped, thick and gelatinous having a simple or branched coelenteron. Function—protective and shield some zooids of the colony.
3. **Nectocalyces or Nectophores.** Bell-shaped medusoid with a velum, radial canals and a circular canal. Mouth, manubrium, tentacles and sense organs are lacking. The body is highly muscular. Function—helps in movement and swimming of the colony.
4. **Gonophores.** Occur on blastostyles singly or in clusters. Degenerate medusae lacking tentacles, mouth and sense organs. Velums, radial canals and manubrium are present. Manubrium bears gonads. Gonophores are dioecious but the colony is hermaphrodite. The female gonophores are medusoid and the male gonophores are saclike. Function—reproduction.

⊠ Origin of polymorphism

Regarding origin of polymorphism several theories have been advanced.

1. **Poly-organ theory.** The polymorphic colony is a single individual and its various components are considered as the modified organs of the medusoid zooid. The parts of the zooids —umbrella, tentacles, manubrium, etc. multiply independently from one another and assume different forms to perform different functions.
2. **Poly-person theory.** A polymorphic cnidaria in a colonial form in which separate individual remain in an organic connection with one another and perform different functions. The zooids of the colony are either polyps or medusae or both but the primitive zooid of the colony is of polyp type.
3. **Medusa theory.** The primitive zooid of the colony was probably medusoid, which produced other medusae by budding. The medusae are locomotory and capable of reproduction. The zooids of the colony are simply the organs of medusoid form, which have subsequently shifted their attachment from the original medusa.

CORALS

The corals are solitary or colonial polypoid forms. They secrete a skeleton of calcium carbonate in which they live and protect themselves from the hazards of the surround-

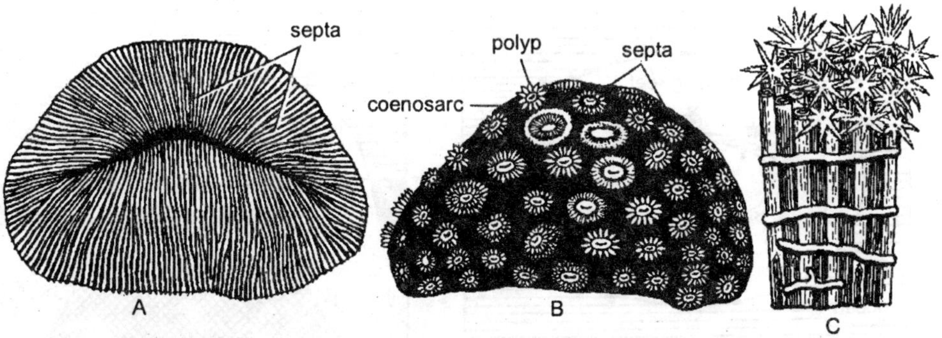

FIGURE 23.25

Some corals. A Fungia fungites, *B. Astrea calicularis*, C. *Tubipora purpurea*

ings. Corals are abundant in tropical seas; some grow in temperate and arctic seas.

The principle materials of coral reefs are stony corals. Other materials—certain branching coraline algae, impregnated with lime growing on coral colonies, foraminiferan shells, calcified alcyonarians, gorgonians, milliporan, tubiporan, helioporan, etc. play considerable role in their formation.

CORAL REEFS

In course of time corals (Fig. 23.25) form coral reefs and coral islands. The stony corals, Madreporaria, belonging to Anthozoa are principal builders of coral reefs. The reef building corals thrive in warm and shallow water, preferably 22°C and do not form reef below 50 metres from the sea level. The reef borders the coast closely.

The coral reefs are three primary types—Fringing reefs, Barrier reefs and Atolls.

⊠ Fringing reef (Fig. 23.26)

It extends from the shore up to about 40 metres, following the coast contour. No navigation channel is present between the reef and the shore. The front of the reef edge is the most active region of growth. A more or less flat surface known as **reef flat** is present between the shore and the front. Coral sand, dead corals, mud, coral colonies and other animals constitute the fringe coral. Sometimes the reef beds are broken resulting into irregular channels called **lagoons**. The depth of the water channels is 37-55 metres.

⊠ Barrier reef

It is similar to fringing reef but if does not follow the contour of the shore. like fringing reef. The barrier reef is located in the sea nearly 1-15 kilometer away from the shore. The navigable channels (lagoons) are about 90-100 metres deep. The lagoons usually contain inner islands and reefs. The great barrier reef on the north-eastern coast of Australia is 1900 kilometers long, 52-122 kilometers wide and situated nearly 144 kilometers away from the shore.

⊠ Atoll (Fig. 23.28)

The atoll is a more or less circular or horse shoe-shaped reef enclosing a lagoon varying from a few kilometres to 64-80 kilometres. (Fig. 23.28) The reef does not enclose any island. The depth varies from 32-112 metres. Atoll reef may be broken to form channels. Some of them suitable for navigation. The atoll reef is also called coral island or lagoon island. Thousands of atolls are found in South Pacific ocean.

The Aldebra atoll reef in the Indian ocean lies 416 kilometre north-east of Malagasy Republic and 640 kilometre from the coast of Africa.

⊠ Coral reef formation

The reef-building corals do not grow well below 45 metres and require optimum temperature about 22°C and sunlight.

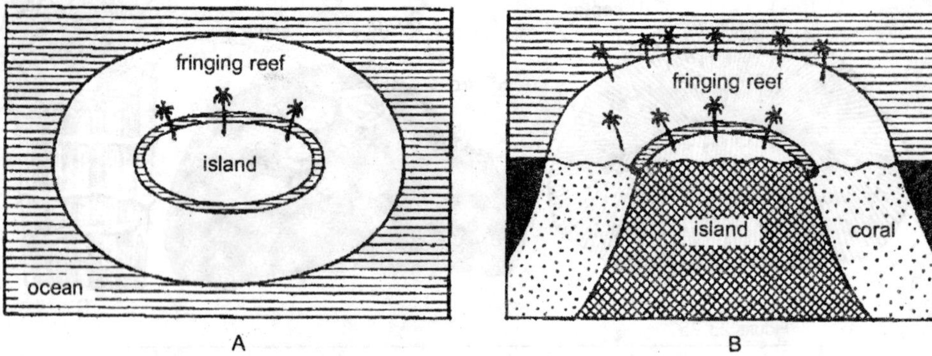

FIGURE 23.26

Coral reefs. A. Fringing reef, B. Fringing reef in section

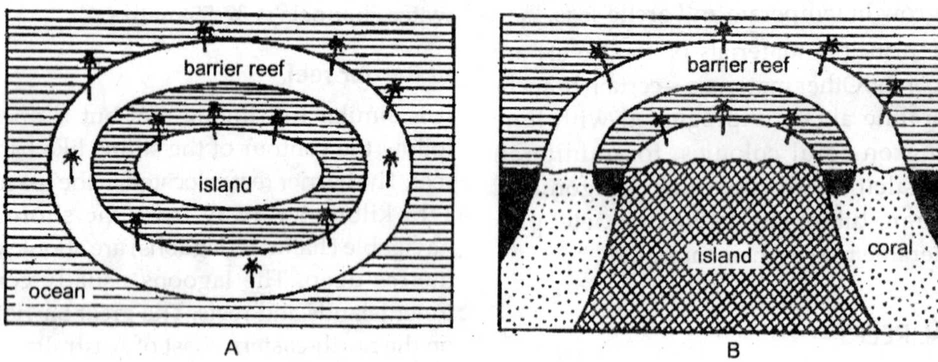

FIGURE 23.27

Coral reef. A. Barrier reef, B. Barrier reef in section

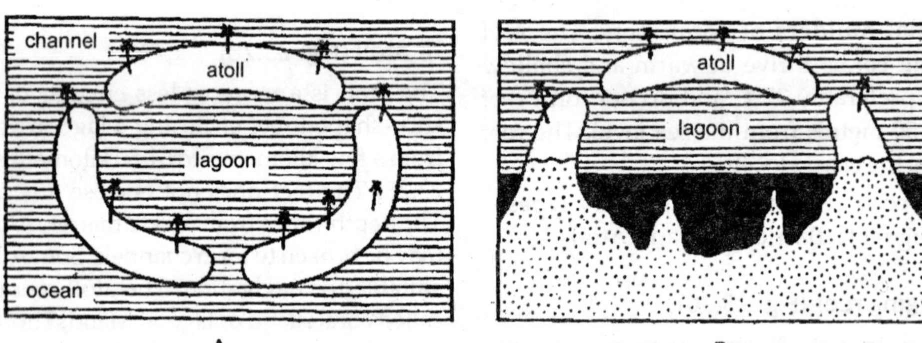

FIGURE 23.28

Coral reefs. A Atoll reef, B. Atoll reef in section

☒ Theories of coral formation

1. **Darwin-Dana Subsidence theory.** Different types of coral reefs arise from fringing reefs. Fringing reef corals grow on downward inclined slopes. Sometimes the shore between the developing fringing reef and the land mass sinks and the fringing reef turns to **barrier reef.** The barrier reef and the land mass are separated by sea water. If the main land mass is an island and it subsides under water, the reef turns into an atoll.

2. **Submerged Bank theory.** The coral reefs grow on surfaces in lower depths. Later, the surface with growing reefs subsides slowly and go down to greater depths. It supports independent origin of different types of reefs.
3. **Semper-Murray Solution theory.** Corals grow on the highest peaks of ocean bottom. Sediment deposition brings them to an optimum level. Better growth of the corals in the edges of coral deposition and by the solution of its inner part, the barrier reefs and atolls are formed.
4. **Daly Glacial-control theory.** Formation of ice caps during last glacial period lowered the water level of the ocean. With the welling of ice and raising of temperature corals started growing on flat ocean bottom. They grew with rising of ocean water bed.

   The coral reefs played a significant role to protect the ocean coast from erosion due to sea waves.

CLASSIFICATION: PHYLUM CNIDARIA

1. The Cnidaria are radially symmetrical in oral-aboral axis (biradial in some Anthozoa); no head and segmentation; simplest of all metazoans with a low grade of tissue formation.
2. The body wall is composed of two principal layers, an outer epidermis (ectoderrm) and an inner gastrodermis (endoderm), separated by a gelatinous noncellular layer, the mesoglea; there are cnidocytes and a few other structures in either or both layers.
3. The body wall encloses a single internal cavity lined by gastrodermis called **coelenteron** or gastrovascular cavity, which opens outside only by a single opening, the mouth.
4. Skeleton external or internal, calcic, horny or none; muscle fibres as a continuous or separate mass in epithelia.
5. A diffuse network of unpolarised nerve cells in body wall; but no central nervous system.
6. Reproduction—both asexual and sexual; asexual by budding, leading to extensive colony formation and sexual by production of ova and sperms.
7. The gonads are only aggregations of developing gametes, no gonoduct is present.
8. Fertilization usually external, and development leads to free-living, ciliated larva, the planula.
9. All are aquatic, mostly marine; individuals either solitary or colonial and sedentary or free-swimming.

The phylum Cnidaria has been divided into three classes.

Class 1. Hydrozoa

1. The individuals are colonial in most cases, some are polymorphic; a few solitary; marine and fresh water (Hydra); sessile or free; 2,700 species known.
2. The medusae develop from the sessile polyp colony by budding and are always free-swimming unisexual, with a true muscular and nonmuscular velum and nerve ring; reproduce sexually, forming hydroid colonies through fertilized eggs.
3. The gastrovascular cavity is simple, without stomodaeum or nematocyst bearing gastric filaments and not divided by vertical partitions and mesenteries.
4. Tentacles are usually solid.
5. The ectoderm secretes a cuticular horny layer called perisarc, which provides rigidity and support; perisarc is calcareous and enormously thickened forming a massive stony structure or coral.
6. Gonads commonly of ectodermal origin and discharged externally.

7. An alternation of generations or metagenesis is present in most cases; the dominant phase is a polypoid or hydroid stage, usually sessile, branched, colonial and often poly-morphic.
8. Embryo is a ciliated planula.

Examples: *Hydra, Obelia, Bougainvillea, Stylaster, Physalia, Sertularia, Valella, Porpita,* etc.

Class 2. Scyphozoa (True medusae)

1. The individuals are medusoid, polyp stage absent or inconspicuous; exclusively marine forms, commonly known as jelly-fish; about 200 known species.
2. Medusae phase dominant; medusae large, bell or umbrella-shaped and free-swimming or attached by an aboral stalk.
3. Gastric tentacles bearing cnidocytes surround the mouth are endodermal in origin and hollow.
4. Velum and nerve ring absent and, in many cases, sense organs are represented by tentaculocysts.
5. A skeleton is entirely absent.
6. Vertical partitions are present in the enteron of some forms and in the larval enteron of others.
7. Gonads are endodermal in origin and discharge their products into the digestive cavity.
8. Polyp when present is nonsexual, known as scyphistoma, gives rise to medusae by transverse fission (strobilation).
9. Development is direct or with an alternation of generations.

Example: *Lucernaria, Pericolpa, Aurelia, Pilema, Tamoya, Chironex, Rhizostoma, Cyanea,* etc.

Class 3. Actinozoa or Anthozoa

1. Solitary or colonial polyps without medusoid stage; exclusively marine; attached or fixed to the substratum; cylindrical body with biradial symmetry; about 6,100 species are known.
2. Oral end is expanded to form an oral disc, bearing six to several hundred hollow tentacles surrounding the mouth; ectoderm and endoderm are separated by a strong mesoglea containing fibres and cells.
3. Vertical partitions are present in the enteron.
4. A skeleton may or may not be present.
5. The free margins of the mesenteries bear coiled mesenteric filaments, which serve the function of gastric filaments.
6. The mesenteries are formed of a double layer of endoderm with a supporting plate of mesogloea.
7. The nervous system is a typical nerve net, central nervous system absent.
8. The sex cells are endodermal in origin and the sexual products are discharged into the coelenteron; asexual reproduction by budding.
9. The planula metamorphoses to adult form.

Subclass i. **Alcyonaria or Octocorallia**

a. Entirely colonial polyps in warm coastal waters, with eight pinnately, branched tentacles and eight complete septa.
b. One ventral siphonoglyph and an endoskeleton present.
c. Often brightly coloured and phosphorescent.

Example: *Clavularia, Telesto, Alocyonium, Tubipora, Heliopora, Gorgonia, Corallum, Renilla, Pennatula,* etc.

Subclass ii. **Zoantharia or Hexacorallia**

a. Colonial or solitary polyps with usually more than eight unbranched tentacles.
b. Skeleton present in some forms.

Examples: *Urticina, Adamsia, Metridium, Zoanthus, Cerianthus, Fungia, Madrepora, Antipathes,* etc.

24 PHYLUM CTENOPHORA

The members of the phylum Ctenophora called comb jellies, sea walnuts or more commonly ctenophores, are marine, pelagic, free-swimming, transparent, bioluminescent, biradially symmetrical, primarily triploblastic, acoelomate animals. The name Ctenophora, meaning comb-bearer, comes from the presence of eight longitudinal, ciliated bands, which propel the animal through water, oral end forward. Ctenophores are carnivores on other planktons, possess long aboral tentacles with specialised adhesive cells used in catching prey. About 100 species are known. These are eaten by many animals, even by whales.

HORMIPHORA sp.

It is a pelagic animal found in Mediterranean seas. Hormiphora belongs to class Tentaculata, phylum Ctenophora.

⊠ Structure

1. **Hormiphora** is pear-shaped, glossy, transparent, about 5-20 mm in diameter (Fig. 24.1).
2. From the opposite sides of the broad or **aboral end**, project two long, solid **tentacles**. The tentacles bear numerous lateral branches containing **adhesive cells**, and are completely retractile into deep sheaths.
3. **Mouth**, a slit-like opening present at the narrow **oral** end.
4. Eight adradially arranged, equidistant meridional bands, the **comb-ribs** or **costae** or **swimming plates**, starting from near the aboral pole and extending about two-thirds of the distance towards the oral pole are present on the surface. These help in swimming.

a. A row of transversely arranged comblike structures of narrow plates, frayed at their outer ends constitute a comb-rib.

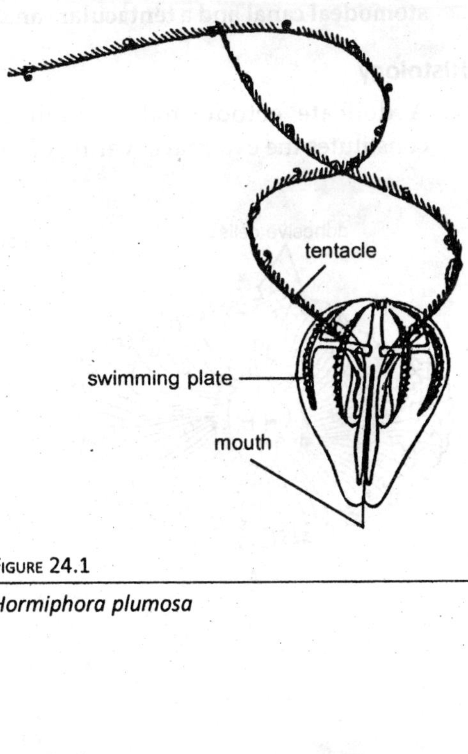

FIGURE 24.1

Hormiphora plumosa

b. The combs are formed by the fusion of rows of cilia at their proximal ends.

5. An aboral **sense organ** is present in the depression at the centre of the aboral pole.

☒ Enteron or gastrovascular system

1. The mouth leads to a flattened tube, the **stomodaeum**, which functions as stomach and extends to about two-thirds towards the aboral pole.
2. The walls of the stomodoeum are provided with internal ridges.
3. The stomodoeum narrows towards the aboral pole and opens in the **infundibulum**.
4. Running up to the aboral pole, the infundibulum divides into four short branches two of which open to the exterior as minute **excretory pores**.
5. The other two branches of the infundibulum are **parradial canals**, pass outwards in the transverse plane and each divides into two **adradial canals** opening in the **meridional canal**.
6. Each parradial canal gives out a **stomodeal canal** and a **tentacular canal**.

☒ Histology

1. A delicate ectodermal epithelium constitutes the external covering of the animal. The cells forming the comb are large.
2. The lining of the stomodaeum is ectodermal while that of the infundibulum and its canals is endodermal.
3. A jelly-like mesoglea fills the spaces between the external ectoderm and the canal system.
4. The tentacle sheath and the covering of the tentacles are ectodermal.
5. The longitudinal muscles constituting the core of the tentacles are of mesodermal origin.
6. Delicate muscle fibres are present beneath the external epithelium covering the body and lining the canal system and the mesoglea.
7. The branches of the tentacles are covered by **adhesive cells**, which readily adhere the animal to an object (Fig. 24.2).

a. The outer surface of an adhesive cell is convex, bearing a number of papillae, which help to attach the tentacle firmly to an object.

b. A spirally coiled filament, with its delicate inner end extending to the muscular axis of the branch of the tentacle, is present inside the cell.

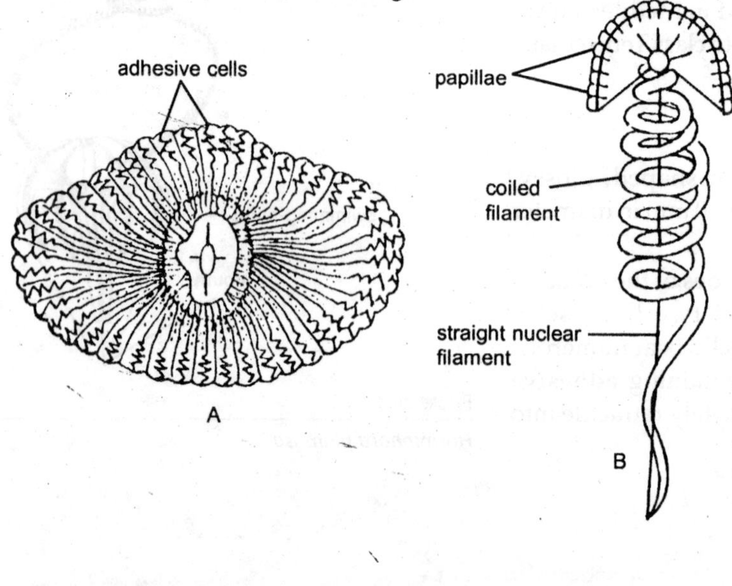

FIGURE 24.2
***Hormiphora* sp.** A. transverse section of one of the branches of a tentacle. B. Adhesive cell (highly magnified)

c. The filament acts as a spring to prevent the tearing of the adhesive cells, during struggle with the prey.
d. The nucleus of the cell is modified to a straight filament.

☒ The nervous system

The nervous system consists of subepithelial plexus of nerve fibres all over the body surface and constituted by multipolar ganglionic cells with their processes. The nerve fibres extend into the mesoglea.

☒ Sense organ

The principal sense organ is the **aboral sense organ**, which is a modified **statocyst** (Fig. 24.3) and help in maintaining equilibrium.

1. Two narrow areas of ciliated epithelium, the **polar plates**, are present in a depression at the centre of the aboral pole.
2. Four equidistant groups of very large S-shaped cilia, united to form as many springs, supporting a mass of calcareous particles, arise from the depression.
3. A pair of ciliated grooves from each spring run outwards and reach the swimming plates of the quadrant.

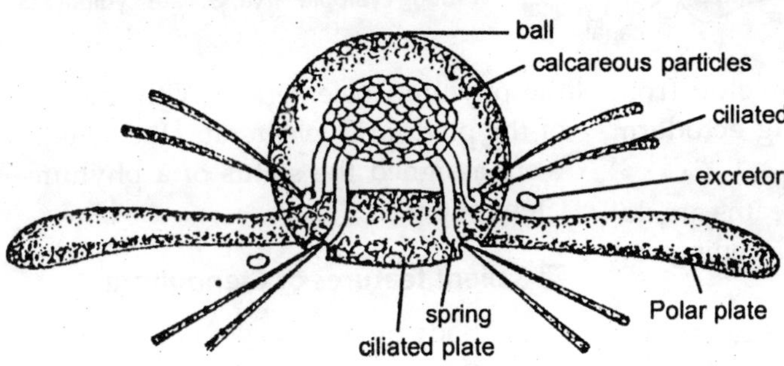

FIGURE 24.3
***Hormiphora* sp.** A statocyst

4. The calcareous mass with its springs is enclosed in a transparent case, formed of coalesced cilia.

☒ Reproductive system

Hormiphora is hermaphrodite, i.e., **testes** and **ovaries** are present in the same individual.

1. Gonads are endodermal in origin and lie against the wall of the meridional canals.
2. The gonads are continuous or discontinuous. A testis extends along the whole length of the canal on one side, while the ovary extends along the whole length of the opposite side.
3. The gametes are discharged into the canals and escape to the exterior.
4. Fertilization is external.

☒ Development

Development of Hormiphora is not very clearly understood.

1. The zygote undergoes cleavage and eight micromeres and eight macromeres are formed (Fig. 24.4A).
2. The micromeres give rise to ectoderm and the macromeres are the source of endoderm.
3. The micromeres divide rapidly, overspreads the macromeres, producing an embryo, having a central mass of large cells with yolk, partly surrounded by a layer of small cells, incomplete below.
4. The endoderm cells multiply, invagination starts and a gastrula formed by epiboly and emboly, without an archenteron.
5. Stomodaeum is formed by invagination of the ectoderm, while the gastrovascular cavities arise by rapid multiplication of endoderm cells.
6. The gelatinous mesoglea appears between the ectoderm and endoderm. The cells of the mesoglea are ectodermal in origin.
7. The complex canal system of the adult appears.

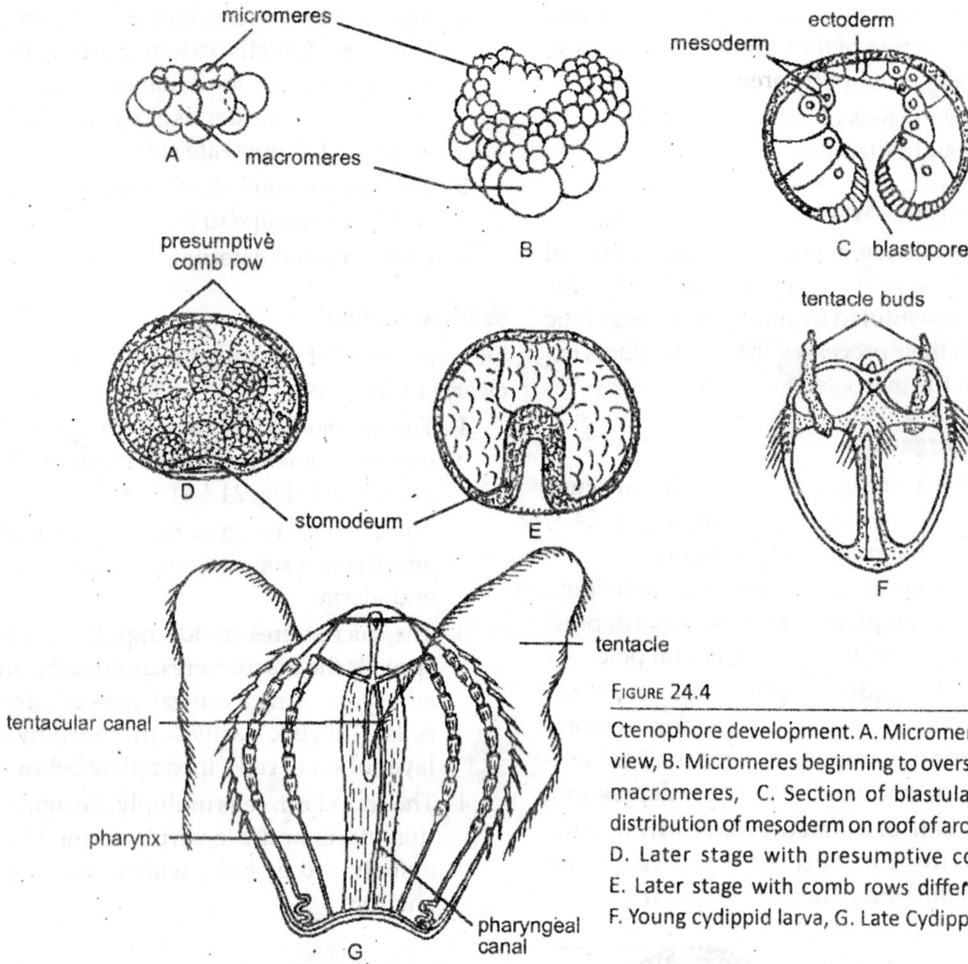

FIGURE 24.4

Ctenophore development. A. Micromeres, lateral view, B. Micromeres beginning to overspread the macromeres, C. Section of blastula showing distribution of mesoderm on roof of archenteron, D. Later stage with presumptive comb row, E. Later stage with comb rows differentiating, F. Young cydippid larva, G. Late Cydippid larva

8. The swimming plates develop from bands of rapidly dividing ectoderm cells.
9. The tentacles and their sheaths are derived from ectodermal thickenings on each side.
10. The muscle fibres and the calcareous bodies of the sense organs develop and the young hormiphora is formed.

BIOLOGICAL STATUS (SYSTEMATIC POSITION) OF CTENOPHORA

Ctenophora is a group of marine, pelagic animals, in which colony formation—common with its closely allied form Cnidaria—is entirely unknown. Previously these animals were placed in subphylum Acnidaria under the phylum Coelenterata. With the splitting of the phylum Coelenterata, the acnidarians were assigned the status of a phylum—the Ctenophora.

⊠ Salient features of Ctenophora

1. Body biradially symmetrical.
2. Presence of eight meridional rows of ciliary plates (comb plates).
3. Absence of cnidocytes.
4. Presence of adhesive cells (colloblasts) in the tentacles.
5. Mesenchymal muscle cells present in the mesoglea between the ecto- and endoderm layers.
6. Development mosaic with cydippid larva.

Owing to its peculiar anatomical organisation and the common features shared with different groups of invertebrates, attempts have been made by a number of workers to establish its relationship with other invertebrate phyla.

⊠ Affinities with Porifera

Similarities	Dissimilarities
1. Simple structural organisation.	1. Development of osculum in sponges and mouth in ctenophora are different.
2. Presence of mesoglea.	2. Inhalant pores absent in ctenophora.
3. Absence of distinct mesoderm.	3. Absence of collar cells (collenchymes) in ctenophora.
	4. Colloblasts (adhesive cells) absent in sponges
	5. Lack of nervous system and sense organs in sponges.

⊠ Affinities with Cnidaria

Similarities	Dissimilarities
	(Structures present in Ctenophora but lacking in Cnidaria)
1. Biradial symmetry.	1. Meridional comb plate.
2. Simple organisation—lack of organ systems.	2. Adult locomotory organ—cilia.
3. Arrangement of different structures in oral-aboral axis.	3. Presence of collobasts.
4. Presence of branched coelenteron.	4. Digestive structures organised.
5. Presence of gelatinous mesoglea.	5. Development mosaic and determinate type.
6. Net type nervous system.	6. Development of muscle cells directly from mesenchyme cells.
	(Structures present in Cnidaria but absent in Ctenophora)
7. Presence of statocyst (sense organ)	7. Cnidocytes or stinging cells.
	8. Formation of colony.

⊠ Affinities with (Anthomedusae), Hydrozoa (Cnidaria)

Similarities	Dissimilarities		
	Structure	**Ctenophora**	**Ctenaria**
1. A pair of opposite parradial tentacles.	1. Gullet	Ectodermal	Endodermal
2. Deep pouch at the base of the tentacles similar to tentacular sheath of *Hormiphora* (Ctenophora).	2. Tentacles	Possess Muscular base	Lacks Muscular base
3. Eight radial canals formed by bifurcation of four interradial pouches of the stomach.	3. Origin of Meridional gonad	Manubrium canals	
4. Subumbrellar cavity of Ctenaria comparable to the stomodaeum of Hormiphora.	4. Aboral sense organ	Present	Absent

⊠ Affinities with Hydroctena (Narcomedusae): Hydrozoa (Cnidaria)

Similarities	Dissimilarities		
	Structure	Ctenophora	Hydroctena
1. A pair of ensheathed tentacles.	1. Velum	Absent	Present
2. Presence of aboral sense organ.	2. Swimming plate	Present	Absent

⊠ Affinities with Actinozoa

Similarities	Dissimilarities
1. Homologous stomodaeum	1. Absence of aboral sense organ in Actinozoa.
2. Structural similarity in transverse section of embryos; *Actinia* (Actinozoa), *Beroe* (Nuda).	2. Presence of hollow tentacles in Actinozoa.
3. Cellular mesoglea.	
4. Gonads endodermal in origin.	

⊠ Affinities with Scyphozoa

Similarities	Dissimilarities		
	Structure	Ctenophora	Scyphozoa
1. Origin of stomodaeum similar.	1. Oral arms	Absent	Present
2. Coelenteron four-lobed.	2. Marginal tentacles	Absent	Present
3. Gonads endodermal in origin.	3. Comb plates	Present	Absent

⊠ Affinities with Platyhelminthes

Similarities	Dissimilarities
1. Arrangement of cilia on body surface.	1. Ctenophora biradially symmetrical; Platyhelminthes bilaterally symmetrical.
2. Locomotory organs of Muller's larva (Platyhelminthes), the eight ciliated ridges comparable to meridional comb plates of Ctenophora.	
3. Statocyst of Ctenophora comparable to dorsal polar nerve of Turbellaria.	
4. Origin of mesoderm more or less similar.	

The concept that bilaterally symmetrical Platyhelminthes evolved from biradial ctenophore Platyctenea is not accepted nowadays.

⊠ Affinities with Nemertines (Pilidium larva)

Similarities	Dissimilarities
1. Locomotory structure—lobed ciliary band.	1. Anatomical organisation differ from each other.
2. Cup-like sense organ in the aboral end. (Sense ogran of Pilidium larva and ctenophores differ structurally)	2. Absence of paired tentacles in nemertines.
	3. Absence of true mesoderm in ctenophore.

In spite of certain resemblances with Cnidaria, particularly Acanthomedusae (Hydrozoa). Actinozoa, and Scyphozoa, relationship of Ctenophora with Cnidaria is doubtful. In our present state of knowledge it is impossible to determine the affinity of Ctenophora with any group of Cnidaria. Ctenophores have certain resemblances with Turbellaria (Platyhelminthes) and it has been suggested that Ctenoplanidae (Ctenophora) possibly represent an intermediate stage between Ctenophora and Platyhelminthes.

CLASSIFICATION: PHYLUM CTENOPHORA

1. Exclusively marine, solitary, free-swimming, pelagic, carnivorous, phosphorescent, radially symmetrical, triploblastic, acoelomate metazoa.
2. Soft, delicate, transparent, gelatinous body, without segmentation.
3. In some, body typically spherical, pear-shaped or cylindrical.
4. Absence of cnidocytes and gastric filaments.
5. Skeletal, respiratory, circulatory and excretory systems absent.
6. Eight meridional rows of comb-like ciliary plates are present on the surface. The ciliary combs retain ciliary movement throughout life.
7. A pair of long, solid and contractile tentacles usually project from blind pouches in opposite sides of the body.
8. The tentacles bear special adhesive cells, the **lasso cells** or **colloblasts**, which help in food capture.
9. In mouth, lying at one end leads into a large ectodermal stomodaeum connected with a series of endodermal gastrovascular canals.
10. Diffused nervous system; aboral end bears a sensory organ of equilibrium, the statocyst.
11. All are hermaphroditic, the testes and ovaries formed side by side from the endoderm of gastrovascular canal.
12. Alternation of generations absent; development accompanied by metamorphosis with formation of a characteristic cydippid larva.

The phylum Ctenophora has been divided into two classes.

Class 1. **Tentaculata** or **Micropharyngea**

1. The body in section is either circular or slightly compressed in lateral plane.
2. Adults with two long aboral tentacles, retractile in sheaths and unbranched meridional and stomodaeal vessel; in certain cases only larva has tentacles, while the adult has oral lobes.

Examples: *Hormiphora*, *Pleurobrachia*, *Bolinopsis*, *Velamen*, *Ctenoplana*, etc.

Class 2. **Nuda** or **Macropharyngea**

1. Tentacles absent.
2. Mouth very wide, gullet occupies the greater part of the interior body.
3. The meridional vessels are produced into a complex system of anastomosing branches. Example: *Beroe*.

25 PHYLUM PLATYHELMINTHES

The Platyhelminthes (Gr. *Platy* = flat; *helmins* or *helminthos* = worm) are soft-bodied, dorsoventrally flat, free-living or parasitic, triploblastic acoelomate, lowly organized worms with an anterior end or **head** bearing sense organs, that move forward to "meet the environment" and a posterior end or **tail** behind. They travel or rest on the **ventral** or undersurface, that remains downward against the ground or bottom, while the opposite **dorsal surface** or back forms the upper surface.

The Platyhelminthes show many advances over Porifera and Coelenterata having—

1. Bilateral symmetry with anteroposterior and dorsoventral relations.
2. A nervous system with enlarged anterior ganglia and nerve cords, extending along the body.
3. Mesoderm as a third germ layer (in place of mesoglea and mesenchyme) producing muscles and other organs between the ectoderm and endoderm and the bundles of muscles make various movements possible.
4. Gonads internal, with permanent reproductive ducts and copulatory organs.

The Platyhelminthes differ from most higher animals in having: (i) no body cavity, (ii) the gut branched, extending to various parts of the body, (iii) usually no anus and (iv) united sexes. The phylum includes a group of free-living flatworms, the Turbellaria, a group of leaf-like external or internal parasitic flat worms; the Trematoda, and a group of tape-like strobilated worms, parasitic in the vertebrate intestine, the Cestoda.

FASCIOLA sp. (Liver fluke)

Fasciola hepatica is one of the largest fluke of the world reaching a length of 30 mm and width of 13 mm. *Fasciola*, a common endoparasite in the larger bile ducts of sheep, goat or cattle, and rarely human, is placed in the class Digenea, phylum Platyhelminthes. Its another host is an amphibious land snail, *Lymnaea truncatula*.

⊠ Structure

1. Body soft, flattened, leaf-like with a triangular head lobe (Figs. 25.1 and 25.2). It is brown to pale-grey in colour and measures 2.15-3.0 cm × 1.2-1.5 cm. The body is covered with the integument, the greater portion of which bears minute spines.
2. The mouth is anterior and terminal, surrounded by the oral sucker present on the oral cone.

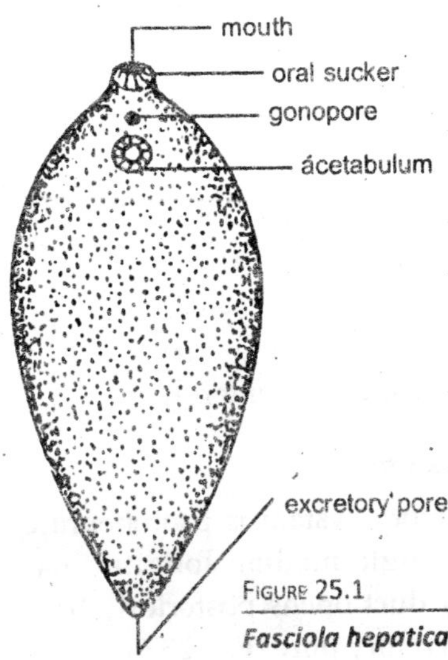

FIGURE 25.1

Fasciola hepatica (ventral view)

3. The posterior sucker or acetabulum is ventral and 2-4 mm behind the mouth.
4. The genital aperture is placed in between the two suckers.
5. The excretory pore is median and posterior terminal.

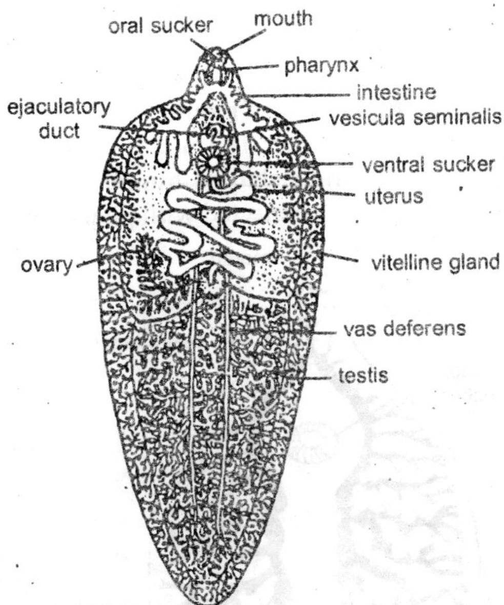

FIGURE 25.2

Fasciola hepatica Ventral view. showing internal organs

BODY WALL (Fig. 25.3)

1. The outer surface is coverted by a syncytium known as **tegument**.
2. In light microscopy the tegument appears as a noncellular homogeneous layer of about 7-16 μm thick.
3. Electron microscopic study reveals, it has two distinct zones. The outer zone consists of cytoplasmic syncytium; embedded in it are mitochondria, endoplasmic reticulum, various types of vacuoles, glycogen granules and other inclusions.
4. The outer surface is thrown into numerous microvilli (folds). Some tegumentary spines are also present which are overlayed by the superficial plasma membrane.
5. The outer syncytial zone is connected by cytoplasmic bridges to inner nucleated bodies, known as **cytons** or **perikarya**. Collectively the cytons form the inner integumentary zone.
6. In between the two layers lies a thin layer of connective tissue fibres known as **basal lamella**. Beneath this are a series of circular muscles under which lies the longitudinal muscle bands.

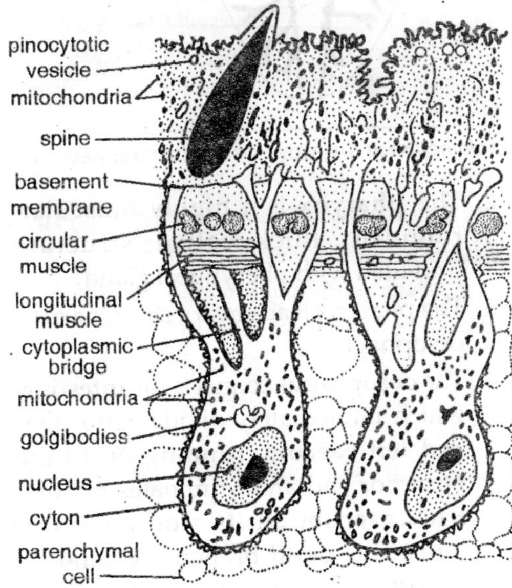

FIGURE 25.3

Fasciola hepatica. Section through integument (ultra-structure)

☒ Digestive System

1. Digestive canal (Fig. 25.4) begins from a ventrally situated **mouth**, surrounded by the oral sucker. The mouth leads into a small funnel-shaped **buccal cavity**. The latter is connected with pharynx.

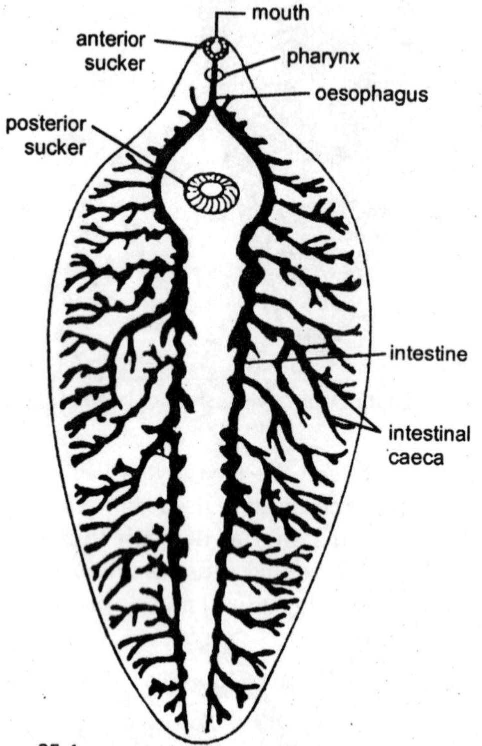

FIGURE 25.4

Fasciola hepatica**.** Digestive system (ventral view)

2. The **pharynx** is a highly muscular, rounded, thick-walled structure provided with pharyngeal gland.
3. The pharynx leads into a short, narrow oesophagus.
4. The **oesophagus** connects the **intestine**, which immediately divides into right and left **caecum**, running up to the posterior end and terminates blindly. Each caecum on its side gives out numerous blind branches extending almost to all parts of the body.
5. Liver fluke feeds on tissue elements and exudates including bile, blood, and lymph.
6. Fasciola lacks circulatory system, the ramification of digestive canal helps in distributing nutrients to different parts of the body.
7. Digestion is mostly extracellular. Intracellular digestion also occurs.
8. Undigested wastes are egested through the mouth. The alimentary canal is really a glorified coelenteron.

☒ Excretory System

1. The excretory system is protonephric type. A single median longitudinal **excretory duct** opens posteriorly by a single **excretory pore** (Fig. 25.5).
2. Anteriorly, the excretory duct branches into four trunks, two dorsal and two ventral, each of which divides repeatedly to form smaller vessels. These smaller vessels branch repeatedly ending into **flame cells**.

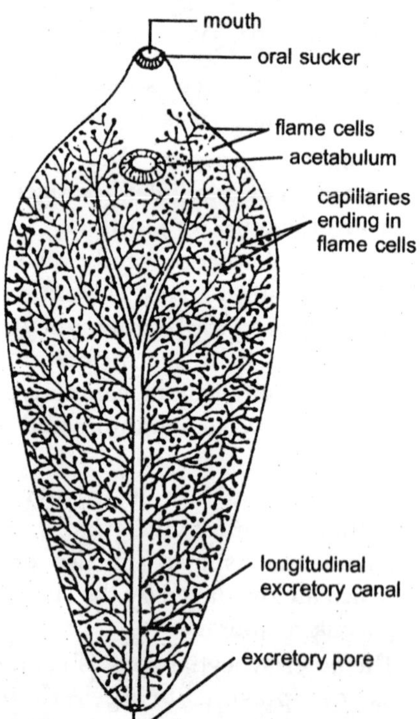

FIGURE 25.5

Fasciola hepatica**.** Excretory organ

3. A flame cell (Fig. 25.6) is a modified mesenchyme cell.

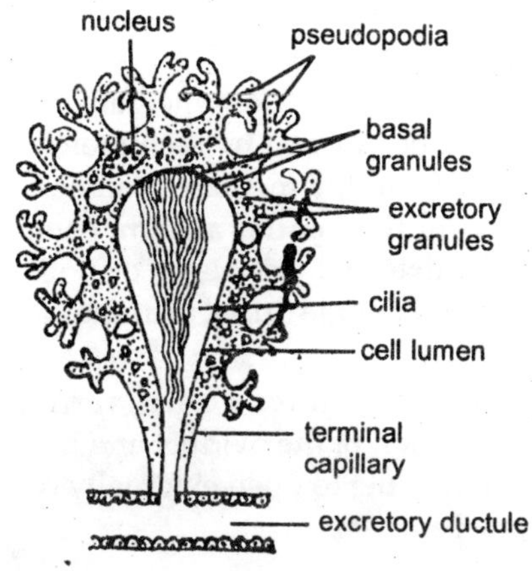

FIGURE 25.6

***Fasciola hepatica*.** A flame cell

a. It is of irregular shape, with granular cytoplasm and a nucleus.
b. A bundle of flagella arises near the nucleus.
c. The flagella are enclosed in a funnel, shaped space, formed by the end of a capillary, coming from the excretory duct and flicker constantly like a flame.
d. The flame cells maintain a constant hydrostatic pressure by which wastes are driven into the excretory duct.
e. The cells are scattered in the parenchyma in a definite pattern and remove metabolic wastes from nearby tissues.

☒ Nervous System and Sense Organs

1. The nervous system (Fig. 25.7) consists of a cerebral ring encircling the oesophagus.
2. A pair of lateral cerebral and a ventral ganglia are present over the cerebral ring.
3. A number of slender nerves arise from the ganglia, run forward and innervate the head lobe.
4. Three pairs of longitudinal nerve cords—dorsal, ventral and lateral—are given out from these ganglia, which run backward.
5. The lateral pairs are long and stout and run posteriorly as lateral nerves, sending branches in their way.
6. Special sense organs are unknown.

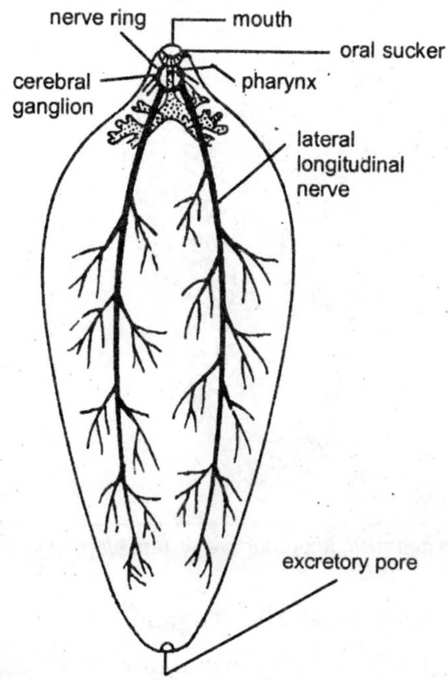

FIGURE 25.7

***Fasciola hepatica*.** Nervous system

☒ Reproductive System

Fasciola is hermaphrodite and bears organs of both the sexes (Fig. 25.8). The common genital aperture is situated on the midventral line, just anterior to the ventral sucker.

☒ Male Reproductive Organs

1. The **testes** are a pair, occupying the major part of the middle region of the body, one behind the other.
2. Each testis is a much branched-tubule, from the anterior end of which runs forward a narrow **vas deferens**.
3. Anteriorly, the two vasa deferentia join to form a median **vesicula seminis**, from which runs an **ejaculatory duct**, opening at the tip of the **cirrus** or **penis**, which, in turn, opens in the **genital atrium**.

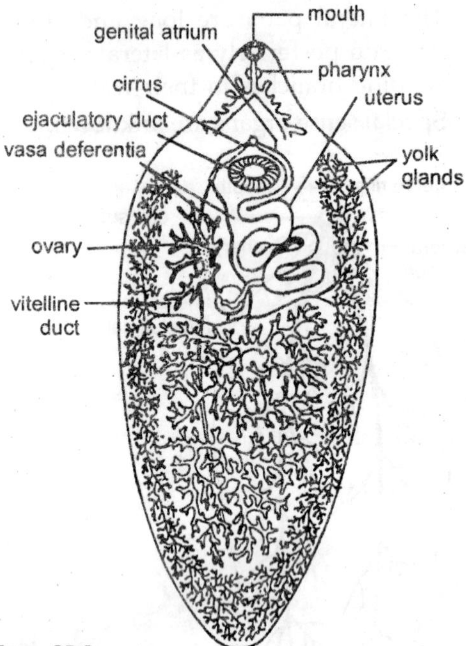

FIGURE 25.8

***Fasciola hepatica*.** Reproductive system (ventral view)

⊠ Female Reproductive Organs

1. **Ovary** or the **germarium** is single, consists of much-branched tubules and located in the right hand side, in front of the testes.
2. The main **oviduct** is formed by the joining of the tubules.
3. Numerous rounded follicles, constituting the **vitelline** or **yolk glands** occupy considerable zone in the lateral regions of the body. They are connected with the **vitelline ducts.**
4. On each side, two large ducts, anterior and posterior, unite to form a single main **lateral duct** and there are two lateral ducts—right and left.
5. The lateral ducts run inward, nearly transversely and open into a small sac, the **yolk reservoir.**
6. Arising from the reservoir, a short, median **vitelline duct** runs forward to open into the oviduct.
7. Groups of unicellular **shell glands** or Mehlis' gland or **accessory female glands**, opening by small ducts into the oviduct, are present around the junction.
8. The lumen of the oviduct at the region is termed **ootype.**
9. The **uterus** is a wide, median, convoluted tube, formed by the union of the oviduct and the median vitelline duct, opens in the **genital atrium,** common to the external apertures of both the male and female ducts, near the base of the cirrus.
10. A canal, the **Laurer's canal**, runs from the junction of the oviduct and median vitelline duct to open externally on the dorsal surface.

⊠ Fertilization and Development

1. The eggs are large ovoidal structures with brown colour due to the presence of bile pigment. On the average, they measure, 40-80 μm (microns). The eggs are fertilized in the uterus and self- or cross-fertilization may occur.
2. The **zygote** receives yolk and a chitinoid covering in the ootype, remains in the uterus for a short period and is discharged. Passing down the bile duct, the zygote reaches the intestine of the host and passes to the exterior with the faeces.
3. Active development starts and after 2-3 weeks the miracidium larva hatches out of the egg in suitable acquatic media.

⊠ Life Cycle (Fig. 25.9)

Miracidium larva

1. The body is conical with a triangular head lobe and is covered with vibratile cilia.
2. A pair of crossed eye spots are present anteriorly.
3. An imperfectly developed intestine and a pair of flame cells opening to the exterior are present.
4. The rest of the interior is filled up with a mass of germ cells.

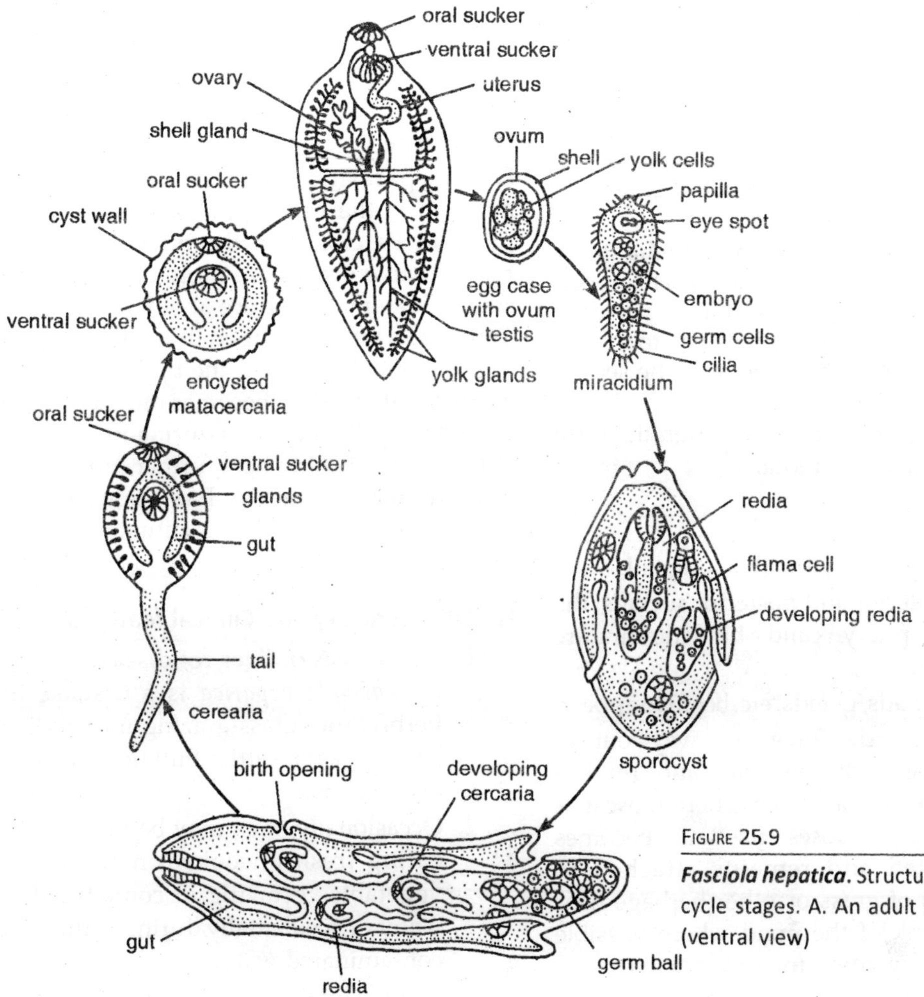

FIGURE 25.9

Fasciola hepatica. Structure and life cycle stages. A. An adult individual (ventral view)

5. The miracidium swims actively in water or moves on damp herbage and can survive only if it reaches an intermediate host snail *Lymnaea truncatula* approximately within eight hours time.
6. The embryo bores into the snail and comes to the pulmonary sac or other organs.
7. The ciliated ectoderm is lost, it grows to an elongated sac and forms the sporocyst.

☒ Sporocyst

1. The body wall is formed by a single layer of cells.
2. The flame cells and remnants of eye spots are present.
3. The internal cavity contains germ cells.
4. The germ cells of the sporocyst behave like parthenogenetic ova. Each cell divides to produce blastula, gastrula and finally a form of larva, the redia.
5. Five to eight rediae are usually formed in each sporocyst.

☒ Redia

1. The body is cylindrical, with a circular ridge near the anterior end and a pair of short processes, lappets or procruscula, near the posterior end.

2. The mouth leads into pharynx and a sac-like intestine is present.
3. A system of excretory vessel is present.
4. Germ cells are present in the internal cavity.
5. A birth opening is situated anteriorly near the circular edge.
6. From the germ cells in the redia, 14-20 cercariae develop in each redia. Cercariae are produced if the season is summer but redia gives rise to a fresh generation of 8-12 rediae if the season is winter.
7. The cercaria escapes through the birth opening of the redia.

☒ Cercaria

1. The body is oval with a long tail.
2. The anterior and posterior suckers, the mouth, pharynx and a bifid intestine are present.
3. The gonads, glands, etc. begin to appear.
4. The cercaria forces its way out of the organ of the snail and reaches pulmonary sac, from where it escapes outside. It loses its tail, becomes encysted and remains attached to blades of grass or other herbage near the bank of the pond where possible host may come for eating leaves.
5. The encysted cercaria* known as **metacercaria**, is taken by the herbivorous host with the grass and occasionally by man with vegetables. The metacercaria excysts in the duodenum and the young fluke, on escape, may reach the liver through bile ducts or hepatic portal vein, and grows rapidly to reach the adult stage.

The eggs come out with the faeces about 3 to 4 months (incubation period) after infection.

☒ Effect of parasitism

On host

The attack of liver fluke causes 'liver-rot', which is disastrous to the host and death has been recorded in most cases of liver-rot. Jaundice and adenomata have also been reported.

On parasite

Due to parasitic life, considerable degeneration of the vegetative organs has taken place in Fasciola. On the other hand, the reproductive organs are more developed. A single fluke may produce about 50,000 eggs. Twice in its life cycle, the embryos are exposed to the environment and the cycle—which is already full of risks—becomes more risky. To compensate the huge loss during its perilous journey from host to host further multiplication by asexual means has appeared, in addition to the already accentuated rate of multiplication.

☒ Pathogenecity and Clinical features

1. Fascioliasis or liver rot disease caused by *Fasciola hepatica* is a disease of herbivorous grass-grazing animals like sheep, goats, cattle, buffaloes, horses and rabbits.
2. Occasionally, man may be infected by eating fresh uncooked, half cooked vegetables growing in contaminated water or accidentally during bathing in contaminated water.
3. Human infection causes serious hepatic pathogenicity including biliary colic, with vomiting, persistent diarrhoea, and a tender hepatomegaly with peripheral eosinophilia.
4. During migration of the young flukes from the liver parenchyma to bile passage they cause extensive damage to the liver tissue. Flukes feed on liver cells and blood.
5. The adult flukes secrete toxin (possibly proline) which cause irritation, inflammation and oedema, which in turn stimulate production of fibrous tissue in the bile ducts and biliary epithelium proliferate to form adenomata.

* The cercaria is not infective until about 12 hours after encystment.

6. In heavy infection the gall bladder is damaged, and the walls of the bile ducts are eroded completely, the worms then reenter the parenchyma, causing large abscesses and severe anaemia.
7. Migrating juveniles frequently produce ulcers in ectopic locations, such as the eye, brain, skin and lungs.

⊠ Management

1. Mass anthelmintic therapy in the animal farm is necessary for protection of livestock. Several drugs are available in chemotherapy of fascioliasis, Rafoxamide is most effective.
2. Infection free nonpolluted drinking water is helpful in the management of fascioliasis.

⊠ Parasitic adaptation

1. Dorsoventrally flat body is suitable for living in the bile ducts of the hosts.
2. Two suckers help in proper attachment in the site, and also browsing the epithelium of the bile passage.
3. Flukes being hermaphroditic, the probability of self-fertilization increases in absence of other fluke.
4. Asexual multiplication in the larval stage in the intermediate host leads to an enormous increase in number of larvae. From one egg about 2,400 adults may be produced.
5. Gut is incomplete with only one opening, the mouth, and without digestive gland as they feed on semidigested materials from the host.
6. Respiratory and circulatory system absent in response to habitat.

TAENIA SOLIUM (Tape worm)

Taenia solium, an endoparasite living in the intestine of man, is placed under the class Cestoidea, phylum Platyhelminthes. Its definite host is man* and the intermediate or larval host is the pig. It is cosmopolitan in distribution.

⊠ Structure

1. Body ribbon-like, 2-3 metres long and consists of about 800-900 segments, known as **proglottids** (Fig. 25.10).
2. The **head** or **scolex** is small but distinct, pear-shaped and bears a conical, retractile elevation, the **rostellum** at the tip.
3. The rostellum bears 22 to 33 **hooks**, small and large, alternately placed and arranged in two rows.
4. Four cup-shaped **suckers** are present on the sides of the head at about the middle. The hooks and rostellum anchor the parasite to the intestine of the host.

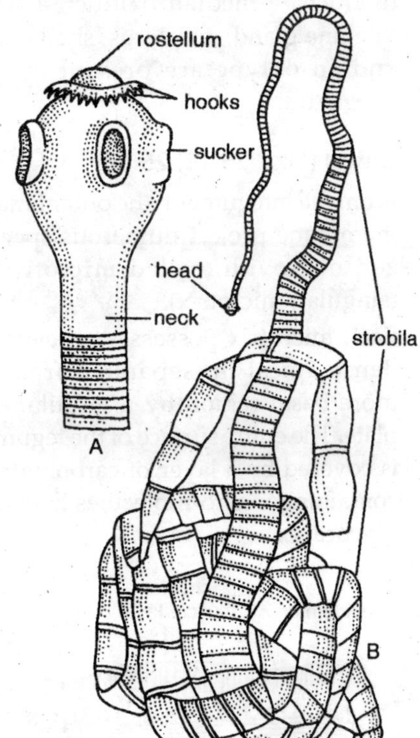

FIGURE 25.10

Taenia solium. A. Scolex with neck, B. An entire specimen

* Man may also serves as the larval host. in accidental case.

5. A small, unsegmented, narrow **neck** is present behind the head.
6. The rest of the body is ribbon-like, segmented and called **strobilla**. The proglottids progressively enlarge in size posteriorly.
   a. A proglottid at about the middle of the strobilla is roughly rectangular in outline (Fig. 25.15). It is thick and muscular.
   b. A longitudinal nerve and a pair of longitudinal excretory canals are present in each lateral side of a proglottid.
   c. A transverse excretory canal connects a pair of longitudinal canals at the posterior end of the proglottid.
   d. Numerous testes, their efferent ducts, a vas deferens, a pair of ovaries, two oviducts, a median oviduct, a single vitelline gland, an uterus, shell glands and an ootype are present in each proglottid.

⊠ Body wall (Figs. 25.11, 25.12)

1. From the integument, the outer limiting membrane project numerous specialized microvilli, called **microtriches** (singular, microtrix).
2. Each microtrix possesses an electron-dense apical tip separated from the more basal region by a multilaminar plate. The entire surface of the tegument is covered by a layer of carbohydrate-containing macro-molecules known as **glycocalyx**.
3. The cytoplasmic part of the microtriches are syncytial which are continuous with the external level of the tegument.
4. The external and internal levels of the tegument are joined by cytoplasmic bridges.
5. The external level of tegument contains numerons electron dense organelles like mitochondria, Golgi bodies, ribosomes, and endoplasmic reticulum. In addition, a number of vacuoles, vesicles and rhabdiform organelles, and glycogen granules occur.
6. The internal levels of the tegument consists of discrete cells known as **cytons** or **perikarya**.
7. The **cyton**, a large cell, includes a prominant nucleus, mitochondria, Golgi-bodies, endoplasmic reticulum, ribosomes and some vesicles.
8. The **basal lamina**, a layer of connective tissue, immediately beneath the external level of the tegument.
9. Beneath the basal lamina is the muscle layers. Musculatures are arranged in two layers. The outer layer is circularly or transversely arranged and the inner layer is longitudinally oriented.
10. A thin layer of neuromuscular cells lies underneath the muscles of the body wall; some cytoplasmic processes of these cells join with other similar processes of the cells while some other processes join with fine nerve fibres.
11. The space enclosed by the body wall except for that occupied by the repro-

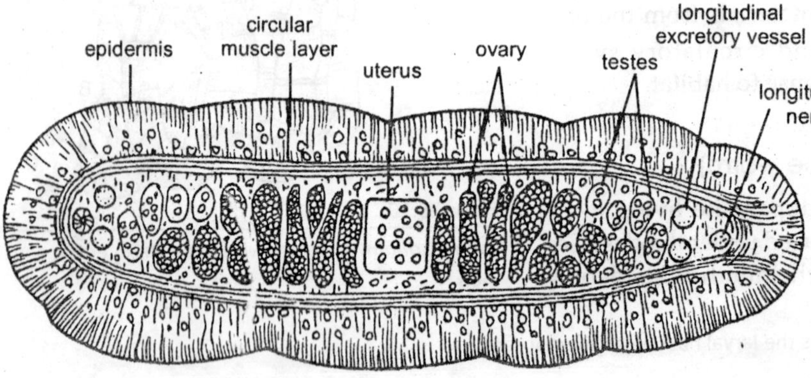

FIGURE 25.11
Taenia sp. Transverse section

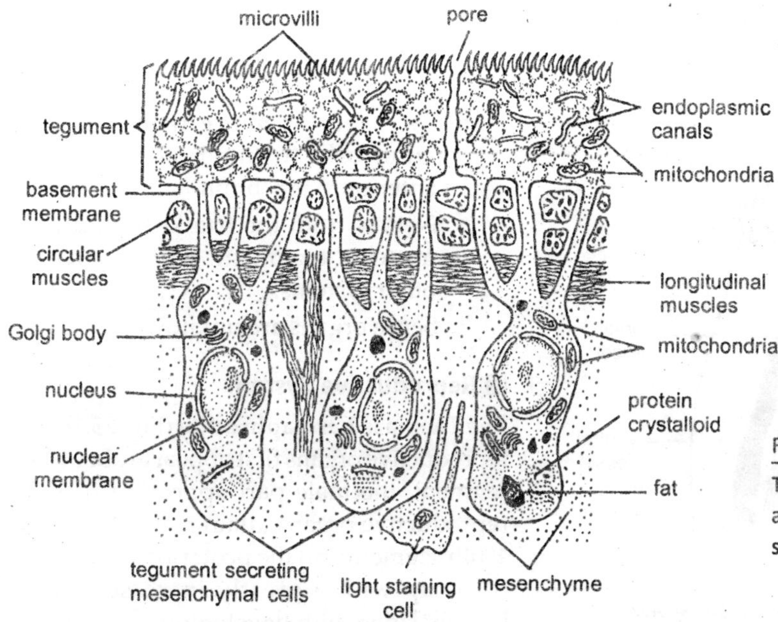

FIGURE 25.12

Taenia solium. T.S. of body wall as seen under an electron microscope

ductive organs, muscle fibres, nervous tissue and osmoregulatory structures, is filled with a spongy tissue called **parenchyma**. The parenchymal cells serve as sites for synthesis and storage of glycogen.

⊠ Digestive System

In Taenia the digestive organ is absent. They depend on the absorption of semidigested nutrients through the host's gut and also they depend on the membrane contact digestion.

⊠ Excretory and Osmoregulatory Systems

The excretory organs not only excrete metabolic wastes but also maintain hydrostatic pressure in the body. Thus, the excretory system also helps to maintain water balance of the body. In Taenia, the excretory system is protonephric type as in trematodes and consists of four longitudinal canals, transverse canals and flame cells. The flame cells resemble those of the liver fluke (Fig. 25.14).

⊠ Excretory Canals (Figs. 25.13, 25.14)

1. Four main **collecting canals** (ducts) extend the entire length of the strobilla. The two **ventral** canals ventrolaterally located and the two **dorsal canals** are dorsolateral. All the canals are situated in the peripheral zone of medullary parenchyma.
2. A single **transverse canal** connects the two ventral canals at the posterior end of each proglottid.
3. The ventral canals carry fluid away from scolex, the dorsal canals towards it.

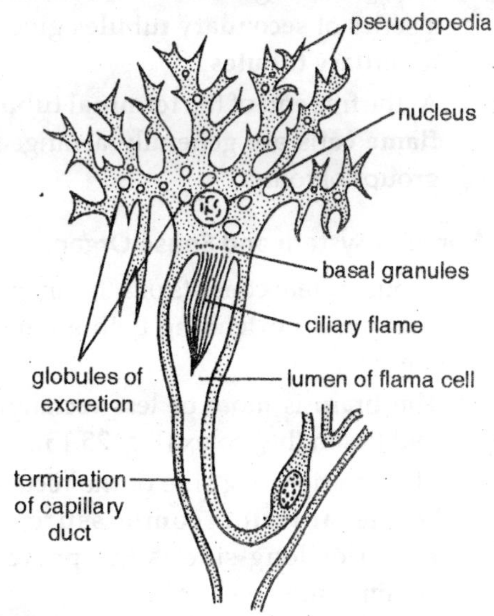

FIGURE 25.13

***Taenia* sp.:** A flame cell

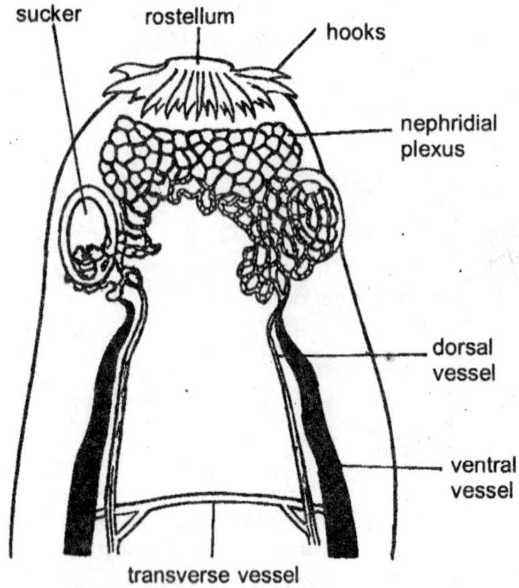

FIGURE 25.14

Taenia solium. Naphridial plexus in the scolex

4. Within the scolex, the four longitudinal canals may be joined by a network of canals called *nephridial plexus* (Fig. 25.15) or by a single ring canal or the dorsal and ventral canals of each side may be joined by a simple connection with no apparent exchange between the two sides.
5. Along the length of the ventral canals, a series of **secondary tubules** give rise to **tertiary tubules.**
6. At the free end of the **terminal tubules**, **flame cells** are generally arranged in groups of fours.

⊠ Nervous System and Sense Organs

The nervous system consists of a 'brain' in the scolex and nerves extending both anteriorly and posteriorly.

1. The **brain** is more or less rectangular and lies in the scolex (Fig. 25.15).
2. The anterior long side of the rectangle is the **anterior commissure**, the posterior long side is the **posterior commissure.**
3. The rectangle itself is the **cephalic ganglia.**
4. Two long **major trunks** extend posteriorly along the entire length of the strobilla.
5. Two shorter **nerve trunks** proceed anteriorly to innervate the tissues anterior to the 'brain'.
6. Nerves from the trunks extend to the muscles, tegument, reproductive and other systems.
7. Sense organs are unknown.

⊠ Reproductive System

The Taenia is **hermaphrodite** (Fig. 25.16). The reproductive organs first appear at about 200th segment, the male part appearing first. The reproductive organs attain maturity at about 400th segment and the posterior segments are filled up with a greatly increased uterus, heavily laden with developing embryos.

⊠ Male reproductive organs

1. The **testes** are numerous, present throughout the proglottid and each testis is connected with an **efferent duct**. The efferent ducts unite to form larger ducts and finally by joining the larger ducts a **vas deferens** is formed.
2. The convoluted vas deferens runs to the lateral margin of the proglottid, be-

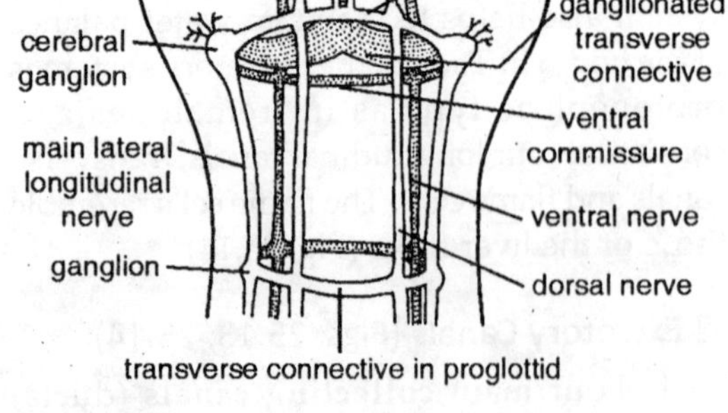

FIGURE 25.15

Central nervous system of a tapeworm

comes narrow and passing through a protrusible process, the **cirrus** opens in the **genital atrium**. At the base, the cirrus is enclosed in a muscular **cirrus sac**.

☒ Female reproductive organs

1. The **ovaries** or **germaria** are paired, bilobed and located at the posterior region of the proglottid.
2. Numerous branching tubules constituting the ovary, join to form an **oviduct**. The two oviducts join medially to form a **common oviduct**.
3. The **vitelline** or **yolk gland** is single, lobulated and situated at the posterior border of the proglottid, opening in the median oviduct through a **yolk duct**.
4. Numerous **shell glands** or **Mehlis' glands** are located around the yolk duct and open through their ducts into the oviduct at its junction with the yolk duct.
5. The part of the oviduct receiving the shell gland ducts and the yolk duct is termed **ootype**. Anteriorly, it is continuous with a median, elongated, blind **uterus**.
6. The female genital pore located in the genital atrium, leads to a narrow passage, the **vagina**, which runs backward to the middle of the segment and dilates at the end to form **receptaculum seminis**, which is usually filled with sperms.
7. The **fertilizing** or **spermatic duct**, runs from this duct to join the oviduct.
8. The simple, cylindrical **uterus** begins to branch from at about 600th segment. The other reproductive organs degenerate and the gravid uterus, heavily laden with developing embryos look like a longitudinal stem with 5-10 branches (Fig. 25.17) and occupies practically whole of the segment in the posterior part of the strobilla.

☒ Fertilization and Development

1. Eggs are fertilized in the oviduct and self- or cross-fertilization may take place.
2. The zygote becomes surrounded by

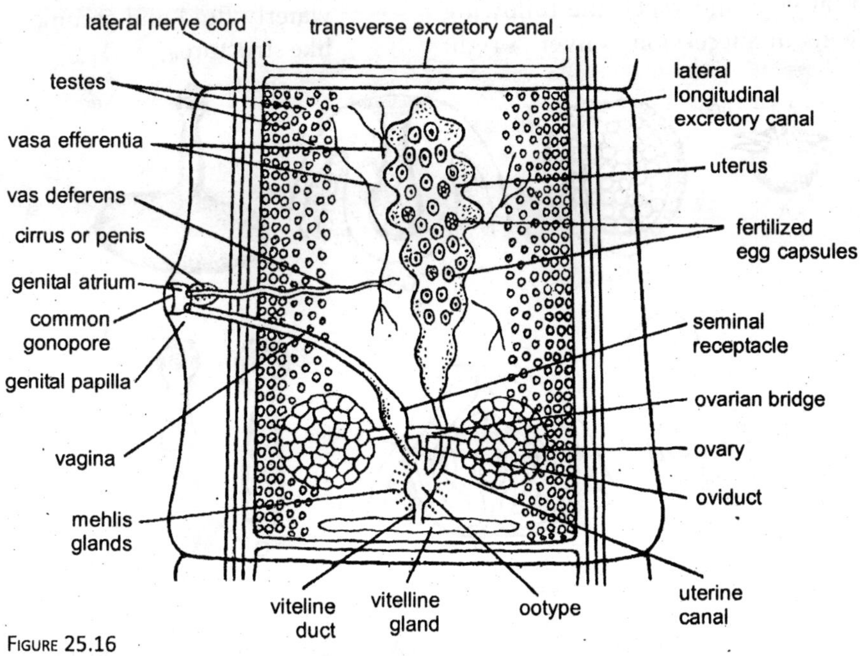

FIGURE 25.16

***Taenia solium*.** An immature proglottid

yolk and a chitinoid egg shell, and passes to the uterus.

3. The ripe proglottids loaded with developing embryos get detached in chains of 5 or 6, and pass to the exterior with the faeces of the host. This type of detachment of proglottids is known as **apolysis.**
4. The proglottids are eaten up by pig; muscles of the segments get digested and six-hooked **hexacanth** embryos are liberated.

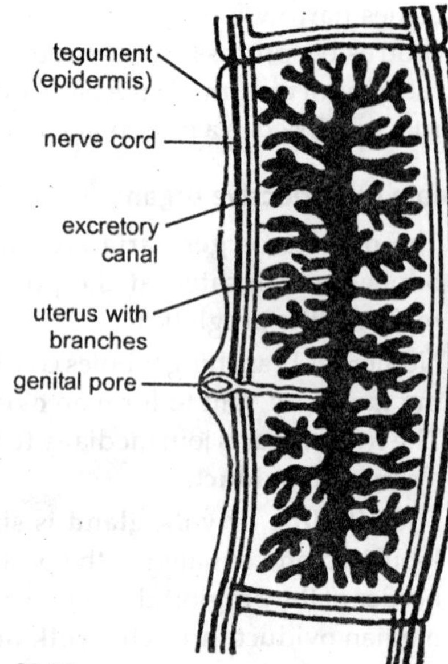

FIGURE 25.17

***Taenia* sp.** A gravid proglottid

⊠ Life Cycle

Hexacanth embryo or **onchosphere** (Fig. 25.18)

a. The body is rounded and enclosed in two membranes.
b. Six curved chitinoid hooks **appear** at one end and the hexacanth embryo is formed. The embryo, along with the membranes, is termed **onchosphere.**

5. The embryos bore the wall of the gut of the host with the hooks and enter blood vessels. From there they enter the general circulation, usually through the portal vein and reach the following organs in succession → liver → right side of the heart → lungs → left side of the heart → the systemic circulation. The embryo is filtered out and finally enters the muscular tissue.
6. Here the embryo increases greatly in size, having a large cavity filled up with waterly fluid and assumes a bladder-like structure.

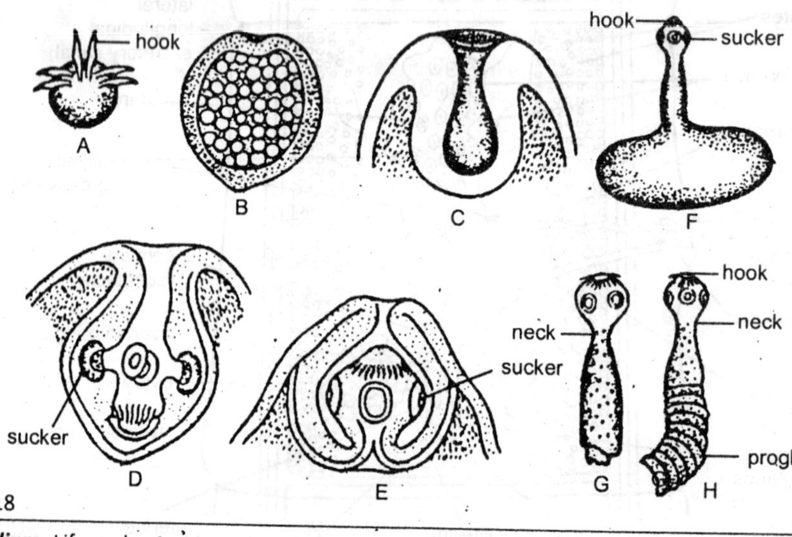

FIGURE 25.18

Taenia solium. Life cycle. A. An onchosphere (Hexacanth embryo). B. Bladder-like stage. C. Early invagination. D. Appearance of hooks and suckers. E. A cysticercus. F. A cysticercus with evaginated scolex. G. Scolex with remains of bladder. H. Anterior end of a young *Taenia*.

7. A hollow invagination or ingrowth takes place at one point of the bladder and on the inner surface of this invagination develop hooks and suckers, characteristic of the head of adult Taenia.
8. The hollow ingrowth becomes everted and the suckers and hooks come close to the surface.
9. The embryo now looks like a bladder with a head and neck of Taenia lying within it. This is known as bladder worm or **cysticercus** stage. It is rich in salt and albuminous material and when the pig's flesh is infected with the cysticercus cellulose, it is known as **measly pork**.
10. If a portion of measly pork is eaten by man, the bladder is dissolved by the gastric juice. The gastric juice causes the albuminous material to swell up and forces the fluid into the cavity, thereby the head comes out through the pore. The scolex attaches itself to the wall of the intestine by hooks and suckers and develops the series of proglottids to reach the adult stage (Fig. 25.18).
11. The worm gains sexual maturity by 2-3 months time, and survives for 25 years or more.

INFLUENCE OF PARASITIC MODE OF LIFE ON THE STRUCTURES OF *TAENIA SOLIUM* AND *FASCIOLA HEPATICA*

The tape worm and the liver fluke both are included in the same phylum Platyhelminthes, but they belong to different classes: **Cestoidea** and **Digenea**, respectively. Both are endoparasites with an intermediate host to spend an apprenticeship in a third kind before they are ready for their life of smooth and ease in their final hosts. Both are highly adapted to the parasitic mode of life with remarkable simplification of some organs and complication of certain parasitic mode of life with remarkable simplification of some organs and complication of certain others; both have to a large extent substituted self-fertilization for cross-fertilization (both hermaphrodite); both are efficient egg-making machines to overcome the chances of destruction.

Due to similar life conditions, resemblances are found in them, but, in spite of that, differences are also not wanting, which are the result of differential degree of adaptation to the parasitic mode of life.

The structures of the Taenia have been more influenced by the parasitic mode of life and it represents a more degenerated and specialized form than that of the Fasciola.

⊠ Pathogenecity and Clinical features

The adult Taenia in the intestine of man may cause vague abdominal discomfort, chronic indigestion, anaemia, diarrhoea alternating with constipation, weakness, loss of weight or excessive gain in weight. The disease due to infection of Taenia is known as Taeniasis.

⊠ Cysticercosis

1. Man may occasionally be infected with eggs of **Taenia** and the whole larval development occurs in the human body as in pigs. The cysticercus stage develops in different organs and the disease is known as **cysticercosis**.
2. In many countries, neurocysticercosis due to the presence of larvae in the brain is a major health hazard, which include cranial nerve palsy, stroke, paralysis, mental disturbances, etc.
3. The distribution of the cysticerci in the subcutaneous tissues and muscles cause visible nodules. Cysticercus in the eye cause blindness.
4. The cysts become Calcified in course of 5 to 6 years.

* Infinite increase of proglottids by means of splitting off (strobilation) segments from the neck region.

Table 25.1: A comprative chart of *Taenia solium* and *Fasciola hepatica*

Taenia solium	Fasciola hepatica
1. The body is elongated, divided into a head, a neck and proglottids.	1. Body is flat, unsegmented and leaf-like.
2. The head is equipped with hooks around the rostellum, and suckers, etc., to assure better attachment.	2. Head lobe with mouth and sucker (anterior sucker) and without hooks. A posteior sucker is also present.
3. Alimentary canal absent; already digested food of the host is absorbed through the body surface.	3. A bifid alimentary canal. Each caecum bears a number of blind branches.
4. Nervous system is composed of two undifferentiated ganglia and nerve cords.	4. Nervous system is composed of two differentiated ganglia and nerve cords.
5. Excretory organs are well-developed.	5. Excretory organs are more developed.
6. Each segment (proglottid) is a self-sufficient reproductive unit with male and female organs*	6. The animal, as a whole, is a single reproductive unit.
7. Self-fertilization either in the same segment or between different segments of the same individual takes place.	7. Usually cross-fertilization between different individuals takes place.
8. Only feeble movement of contraction and expansion is present in the larval form.	8. Larval forms bear well-developed cilia as locmotory organelles (miracidium, cercaria).
9. No known sense organ in the larval form.	9. Presence of eye spots.
10. No asexual method of larval reproduction.	10. Asexual method of larval reproduction present (rediae).
11. Larval life less complicated.	11. Larval life more complicated
12. Strobilation has less risk of loss or death as it takes place within the body of the host.	12. Asexual reproduction and larval life are more risky, specially as the presence of water is essential.
13. The life cycle is completed in two hosts and both the hosts belong to the class Mammalia.	13. Two hosts are involved in the life cycle, of which one is a vertebrate and the other an invertebrate (aquatic snail).

☒ Management

1. Taeniasis and cysticercosis are serious diseases in man. Its management involves proper sanitary arrangement and sewage disposal.
2. Effective anthelmintic treatment of infected individual to prevent infection of intermediate hosts.
3. Maintenance of personal hygiene to prevent autoinfection or infection from others infected with adult worm.
4. Cysticerci are killed if meat is kept frozen at –10 °C for ten days. Proper cooking of meat at high temperature kills the cysticerci.
5. Social awareness of such infection and sanitation also helps in management.

☒ Parasitic adaptation

1. Adult *Taenia* always harbours in the intestine. The body is flat, tape-like for accomodation in the intestinal lumen, without much obstruction.
2. Scolex is the hold fast organ, provided with hooks and blind suckers, which remain inserted in the mucus and submucus layers of the upper intestinal

wall and rest of the body floats in intestinal luman, to counteract the peristaltic movement of the host gut and better absorption of nutrition.

3. It lacks alimentary canal, respiratory and circulatory system. The integument is without cuticular covering and plays vital role in nutrient absorption.
4. Proglottids bear both male and female reproductive organs. Proglottids gradually become mature and larger in the posterior end. Enormous number of worms may be produced from a single proglottid. This helps the warm to survive and propagate in the severe adverse conditions.

CLASSIFICATION: PHYLUM PLATYHELMINTHES

1. Body bilaterally symmetrical, tape-like, triploblastic, acoelomate, dorsoventrally commpressed, devoid of any skeletal system and internal transport system.
2. The epidermis soft and ciliated. Scolex, external suckers, hooks or both, for attachment to the host present.
3. Digestive system incomplete, mouth but no anus and usually much-branched; none in some (Cestoidea).
4. Excretory system with many flame cells connected to excretory ducts.
5. Nervous system with a pair of anterior ganglia or a nerve ring connected to one to three pairs of longitudinal nerve cords with transverse commissures.
6. Sexes usually united (monoecius); hermaphrodite; reproductive system of each sex with gonads, ducts and accessory organs; fertilization internal; eggs microscopic, each enclosed with several yolk cells in a shell; development either direct (Turbellaria, monogenetic trematodes) or with one or more larval stages (digenetic trematodes and cestodes);
7. Asexual reproduction is also present in some.

The phylum has been divided into five classes:

Class 1. Turbellaria

1. Mostly free-living nonsegmented worms, anterior end often bears eyes and a triangular projection.
2. The ectoderm is ciliated, pigmented, often glandular and with peculiar rod-like bodies.
3. A pair of anteriorly placed ganglia give off lateral nerve cords.
4. Digestive cavity always blind with a protrusible, often branched muscular pharynx.
5. Excretory pores variable in number and position—either single posterior pore, or a pair of lateral pores, or numerous scattered pores or excretory system absent.
6. Some forms multiply by fission; regeneration also occurs.
7. Life cycle usually direct with juvenile worm, emerging from egg or cocoon.
8. Marine, fresh water and some terrestrial.
9. Size varis from microscopic to ones more than 60 cm. in length (land planarians), but most are less than 10 mm long.

Examples: *Convoluta, Temnocephala, Bipalium, Planaria, Dugesia, Notoplana*, etc.

Class 2. Monogenea

1. Typically ectoparasites of the skin and gills of fishes but some have become endoparasitic in the coelom and very rarely in the gut of fishes, others in the pharynx and urinary bladder of amphibians and reptiles.
2. Epidermis of adult devoid of cilia and rhabdites.
3. Gut rearly always present (except in Gyrocotylidea) and typically with pharynx and bifurcated intestine, with

mouth either simple or surrounded by an oral sucker.

4. Main attachment organ (the haptor) always posterior and terminal, typically armed.
5. Excretory system terminating in a pair of lateral pores at the anterior end.
6. Free-swimming larva provided with 10-16 posterior marginal hooks and, sometimes, one or two pairs of median hooks in addition.
7. Life cycle simple, no larval multiplication, confined to a single host (usually vertebrate).

Examples: *Entobdella, Gyrodactylus, Polystoma, Gyrocotyle (*endoparasite in intestine of holocephalan fishes) etc.

Class 3. Digenea (Flukes)

1. Body leaf-like or cylindrical; endoparasites with epidermis devoid of cilia in the adult.
2. Gut always present, provided typically with oral sucker, pharynx and bifurcate intestine.
3. A simple, unarmed acetabulate ventral sucker present as attachment organ.
4. Excretory system in adult terminates in a single posterior pore.
5. The life cycle complex involving more than one host; a free-swimming larva enters a gastropod or lamellibranch and multiply by polyembryony and develop free-swimming tailed cercaria larva which enters another host to form metacercaria; metacercaria enters the final host to attain adult stage.

Examples: *Fasciola, Echinostoma, Cotylophoron, Schistosoma* (blood fluke), etc.

Class 4. Aspidagastrea

1. Endoparasites; epidermis of adult devoid of cilia and rhabdites.
2. Gut always present, without oral sucker (funnel-like mouth) followed by pharynx and simple intestine.
3. Very large ventral sucker, subdivided into one or more longitudinal rows of alveoli or several distinct suckers.
4. Single pair of posterior excretory pores.
5. Development either direct or two hosts cycle in which the adult, parasite on a mollusc persists in the intestine of a fish or a reptile.

Example: *Aspidogaster.*

Class 5. Cestoidea (Tape worm)

1. Endoparasitic; epidermis of adult devoid of cilia and rhabdites, but with microtriches; body elongated, tape-like.
2. Gut completely absent.
3. Head or scolex usually round and bears blind suckers, hooks, bothria, etc.

Subclass i. **Cestodaria**

1. Body not divided into proglottids.
2. Only one set of gonad.
3. Ten hooks present in larva.

Examples: *Amphilina.*

Subclass ii. **Eucestoda**

1. Body long, ribbon-like, consists of a scolex (head), unsegmented neck, and a number of proglottids.
2. Six hooks are present in larva.
3. Each segment (proglottid) is provided with both sets of reproductive organs.

Examples: *Taenia, Diphylobothrium, Tetrarhynchus, Echinococcus,* etc.

26 PHYLUM ASCHELMINTHES/ NEMATHELMINTHES/NEMATA

Nemathelminthes (Gr. *nemat* = thread; *helminthos* = worm) or Nemata (Chitword, 1958) is bilaterally symmetrical, unsegmented, elongate, dimorphic, cylindrical worm-like pseudocoelomates and are the most abundant and widespread animals. Body covered with cuticle, mouth terminal, surrounded by lips, anterior body characteristically with papilliform sensory organs and two amphids. Approximately 12,000 known species exist in different habitats. Many are free-living in soil and waters and many others are parasites in the tissues or fluids of animals and plants. Some live in the roots of plants, seeds, gum of tree wounds, and the intestine, blood, and other organs of animals. In an acre of land several hundred billion nematodes may be present. Decomposing plants and animal bodies often contain large populations of nematodes.

Food crops, domestic animals and humans are parasitised by many species and, consequently, round worms are of tremendous economic and medical importance. There are few plants and animals that do not have nematode parasites. Different species of nematodes found in man cause different disease; nematodes in domestic animals cause losses amounting to hundreds of millions of rupees throughout the world; nematodes parasitic in plants attack all types of commercial crops, and about 10 per cent of all crops grown are consumed by nematodes. In addition, they greatly affect our interest by destruction of plant tissues, by causing poor growth and production of food animals or by direct deprivation within human body. Some nematodes are responsible for the destruction of dead plants and animal materials and thereby release organic substances locked there. Some soil nematodes are predatory.

Most free-living and plant parasitic nematodes are less than one millimetre long but species parasitic in the gut of vertebrates and muscles may attain larger sizes, even up to a metre (Guinea worm).

The nematodes differ from flatworm in the body shape, structure of integument, in presence of complete digestive tract, a body cavity and having separate sexes. The muscles of nematodes are all aligned with the body axis and permits only dorsoventral bending, thus less complex than those in flatworms, nemarteans, and annelids. The body cavity or pseudocoelom compares with that of rotifers, gastrotrichs, and acanthocephalans, lacking a mesodermal lining unlike the true coelom of annelids and vertebrates or the parenchyma-filled body of flatworm and nemarteans.

ASCARIS sp.

Ascaris belongs to order Ascaroidea, phylum Aschelminthes or Nemathelminthes or Nematoda. It is a parasitic nematode living in the intestine of man, and the largest of the common intestinal nematode parasites in man. The female is about 25-40 cm long and 6-8 mm in diameter. The male is considerably smaller, about 15-25 cm long with a maximum diameter of 3-4 mm.

☒ External features

1. The animal is narrow, elongated and light-yellowish brown in colour. Four white longitudinal streaks, one dorsal, one ventral and two laterals are present (Fig. 26.1).
2. The **mouth** is anterior and terminal, markedly constricted off, and is bounded by three **lips**; one median and dorsal and two ventrolateral.
3. **Excretory pore** is ventral and near the anterior end.
4. In male, the tail end is curved ventrally in the form of a hook with a conical tip. The **anus** opens at this tip and serves as a common aperture for **rectum** and **genital duct**. Two copulatory setae protrude from the aperture.
5. In female, the posterior extremity is conical and straight. The anus is situated slightly anterior to the posterior extremity on the ventral surface, guarded by one pair of postanal papillae. Tail present.
6. At about one-third of the entire length form the anterior end, the body is narrower in female and this region is known as **vulvar waist**. The **vulva** is situated on the ventral surface of the vulvar waist.

☒ Body Wall

1. The body is covered with a thick elastic transparent **cuticle** (Fig. 26.2) divisible into several layers with numerous vertical channels. The transverse wrinkles of the cuticle give the characteristic segmental appearance of the body. Cuticle also lines the buccal cavity, oesophagus, rectum, cloaca, vagina and excretory pore. The cuticle is constituted by the following layers:

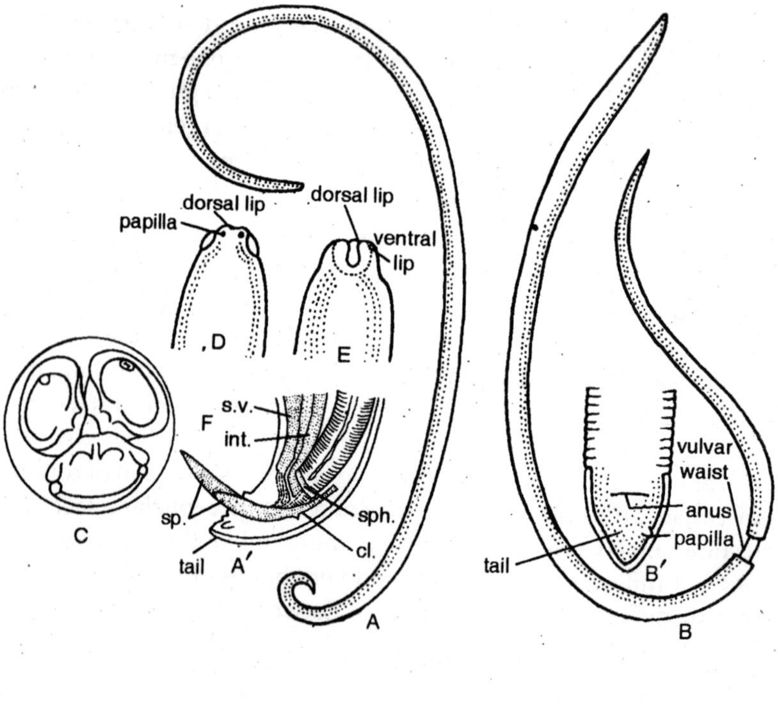

FIGURE 26.1

Ascaris lumbricoides. A. Male entire specimen. A′. Posterior end enlarged, Inset. Anterior end-dorsal and ventral-views, B. Female entire specimen, B′. Posterior end ventral-view enlarged, C. Enface view of aral region, D. Dorsal view of male anterior end and E. Ventral view of male anterior end.

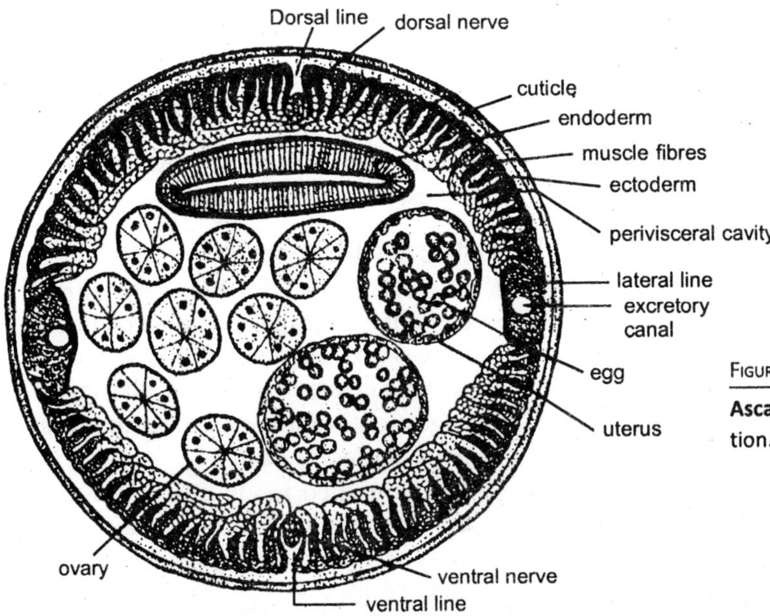

FIGURE 26.2

Ascaris sp. Female. Transverse section. Middle region

a. **Epicuticle.** It is trilaminate, consis-ting of two electron-dense layers separated by a less electron dense layer. Chemically, it is keratin, and also contains quinones and polyphenols.
b. **Exocuticle.** Immediately beneath the epicuticle is this layer.
c. **Mesocuticle.** This is a thick layer containing numerous obliquely ariented collagenous fibres.
d. **Endocuticle.** This is the innermost fibrillar layer.

2. The cuticle remains separated from the underlying tissues by a basal lamina.
3. The **hypodermis or epidermis** is a thin syncytial layer under the basal lamina. This layer projects along the middorsal, midventral, and lateral lines and form **chords**. Excretory canals and lateral nerves run along the lateral chords, while dorsal and ventral chords contain dorsal and ventral nerve, respectively (Fig. 26.2).
4. In four quadrants, a single layer of muscle cells remain attached to hypodermis. Spindle-shaped muscle cells contain two parts, a muscular and a protoplasmic part. The protoplasmic part remains inwards and connected with the neuromuscular process.

Pseudocoel. The body cavity is bounded externally by the muscle cells (mesoderm) and internally by the lining of the gut. Embryologically, it does not develop like a true coelom and is not lined with a layer of mesodermal cells on both the sides. The pseudocoel, in living nematodes, remains filled with a fluid containing mobile **phagocytic cells.**

☒ Digestive System

The alimentary canal with associated glands constitute digestive system.

☒ Alimentary Canal

It is a narrow straight tube divisible into three regions (Fig. 26.3) — the club-shaped foregut or oesophagus about 10-15 mm long and lined with cuticle; a nanow midgut of endodermal origin and the hindgut or rectum lined by cuticle.*

* The digestive and reproductive organs float in the body cavity containing an irritating fluid known as Ascaron.

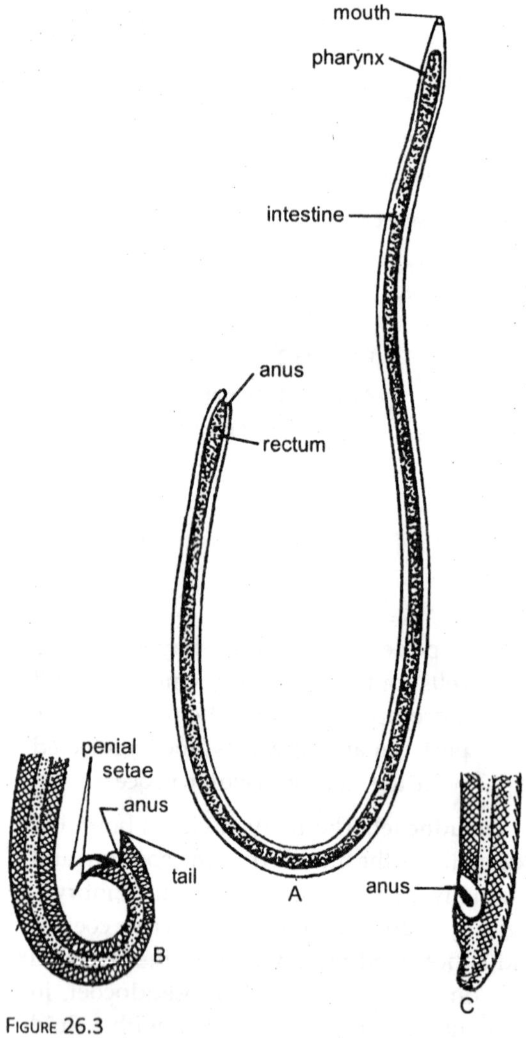

FIGURE 26.3

Ascaris lumbricoides. Digestive system (semidiagrammatic). A. Female. B. & C. Posterior part of male and female.

1. **Mouth** is anterior, terminal and bounded by three lips, one median and dorsal and two ventrolaterals. The mouth leads to a buccal cavity opening in the pharynx.
2. **Oesophagus** also called **pharynx** is a muscular bulb and acts as a suction tube. It leads to intestine.
3. **Intestine** is a straight tube of one-layered epithelium bounded externally by cuticle.
4. **Rectum** or the terminal portion of intestine opens to the exterior at the anus.
5. Three branched **osophageal glands**, one dorsal and two ventrolateral, remain embedded between the muscle fibres.

⊠ Feeding and Digestion

1. Ascaris feeds on hosts intestinal contents, predigested food of hosts, tissues and exudates or blood by the sucking action of muscular pharynx.
2. Digestion is partly extracellular and partly intracellular.
3. The simplified foods diffuse through the intestinal wall into the fluid filled body cavity and subsequently absorbed by the cells.
4. The excess food may remain stored as glycogen and fat in the epidermal, intestinal, epithelial and muscle cells.

⊠ Excretory System

It consists of two lateral longitudinal canals (Fig. 26.4) which meet together ventrally and open in the excretory pore on the ventral surface at the anterior end. Four or six big tuft-shaped cells with ramifying processes are in close contact with the canal. They absorb solid wastes and transfer the same to the canals in a dissolved state.

⊠ Nervous System

A nerve ring is present around the pharynx from which run six nerves forwards and six backwards (Fig. 26.5). Of the six backwards, two—one dorsal and another ventral—run up to the posterior end and they are connected by transverse nerve bands. The ventral nerve forms a ganglion just in front of the anus.

⊠ Reproductive System

The sexes are separate. Ascaris exhibits considerable sexual dimorphism.

Male. In the male, the genitalia (Fig. 26.6) consists of a single, coiled tubule, eight times the length of the body, differentiating into anterior **testis**, middle region as **vas deferens**

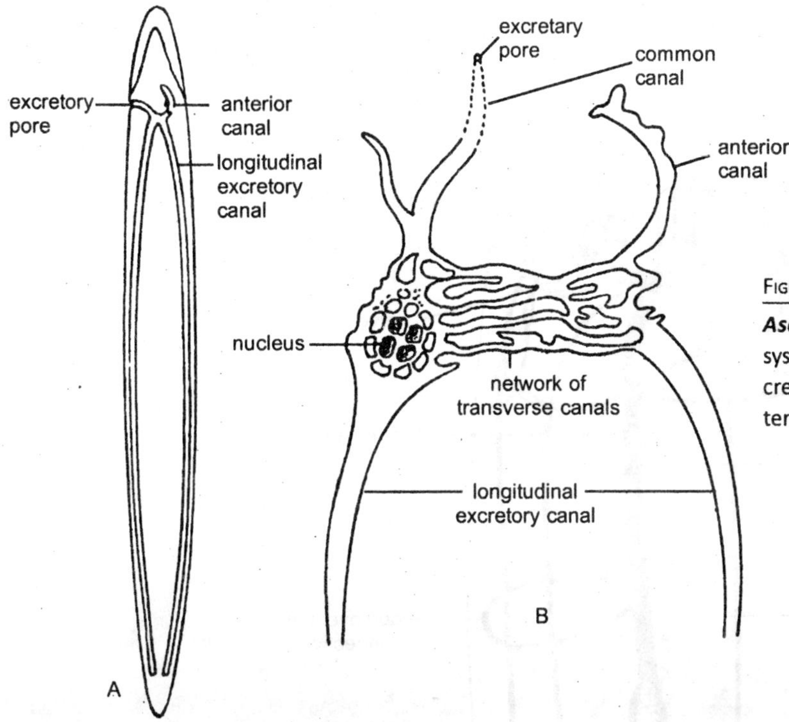

FIGURE 26.4

Ascaris lumbricoides. Excretory system (semidiagrammatic). A. Excretory system. B. Same as A, anterior part (magnified)

and the posterior region as **vesicula seminalis**, continuing as **ejaculatory duct** opening in the anus which is marked by the presence of a pair of **penial setae**.

Female. In the female (Fig. 26.7) **ovaries** are paired tubes, which pass gradually into **oviducts, seminal receptacles** and **uteri**, which join to form an unpaired conical **vagina** opening in the female **gonopore** on the ventral surface of the vulvar waist. The female genital organs also remain coiled in the body and, if extended, is found to be several times the length of the body.

☒ Development

The eggs are produced roughly at the rate of 20,000 a day. Fertilization takes place in the upper part of the uterus. Completed eggs being enclosed in chitinoid egg shell pass out of the body of the host with faeces.

Development and infection are direct. Infective embryos reach the intestine of man directly with water, soil, green vegetables, etc. The embryonated eggs move down to the duodenum, where the digestive juice weakens the egg shell. The larva emerges through a rent in the egg shell.

☒ Larval Wanderings

The newly hatched larvae measure about 0.2-0.3 mm and burrow their way through the mucous membrane of the small intestine to reach the lymphatics or venules. From there they pass to the liver through portal circulation, where they reach the right atrium of the heart and go to the lungs through the pulmonary artery. Here they leave the blood stream, attack the lung alveoli, and in mass infection the patient develops pneumonic symptoms*. In the lungs they moult twice. The larvae now crawl up the glottis and finally reach the intestine again. The cycle covers, on the average, a period of 10-15 days and the size of the larva increases from 0.2 to 2 mm.

* In massive infection, the worms may reach aberrant sites, viz., kidney, brain, spinal cord and other organs. In extreme cases they have been found to be excreted through urine.

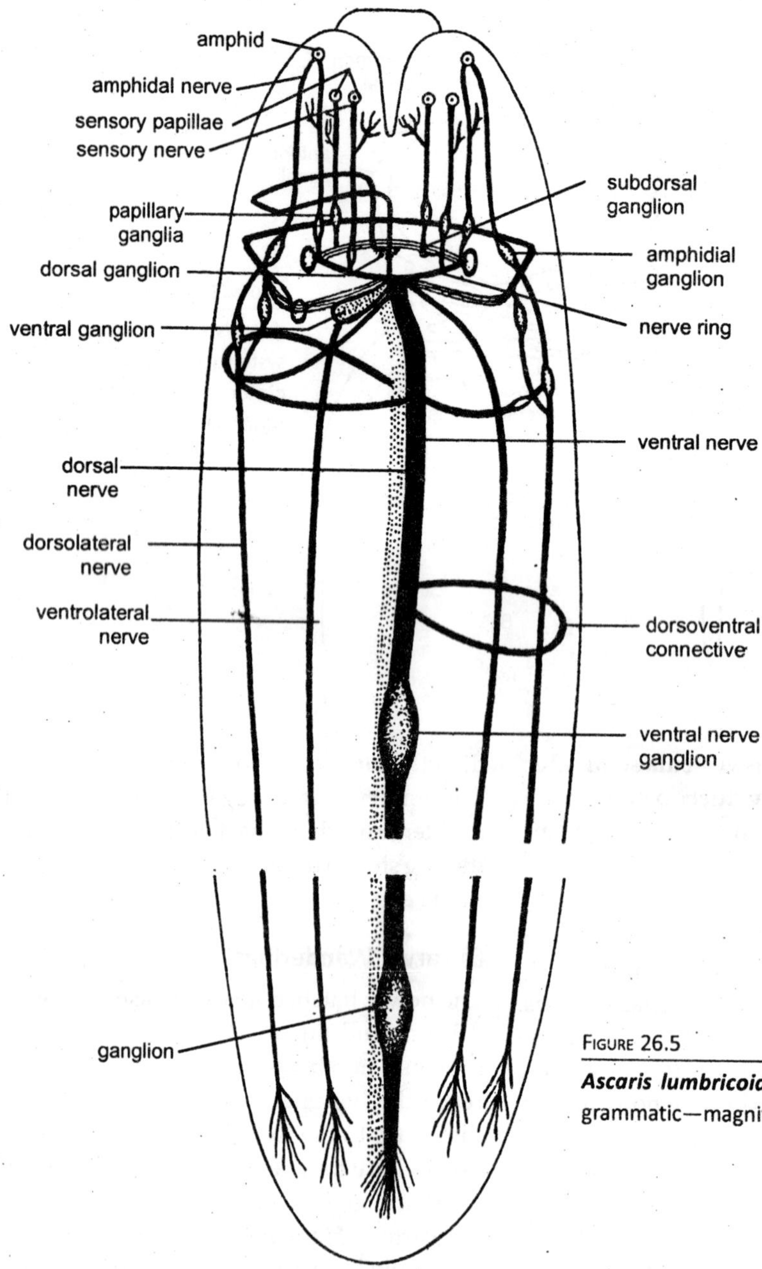

FIGURE 26.5

***Ascaris lumbricoides*.** Nervous system (diagrammatic—magnified)

Fourth moulting occurs in the upper intestine in between 25-29 days and the young continues to grow to attain the normal size. Sexual maturity is attained in about 6-10 weeks time.

☒ Parasitic Adaptation

Because of certain peculiarities in structure such as the reduction of some organs and specialisation of others *Ascaris* has been adapted to lead a normal life.

1. The body is long, cylindrical in shape with both the ends pointed.
2. The body wall is covered with cuticle, formed of albuminous proteins resistant to the digestive enzymes of the host. Physiological studies show

that the worm produces enzyme inhibitors that protect it from the host's digestive enzyme.

3. Locomotory organs are absent as protection from **enemies** and food supply are ensured by host.
4. Cilia are completely absent.
5. The alimentary canal is simple and poorly developed as the parasite feeds on the contents of the intestine of host which are in a semidigested condition.
6. Respiration is almost entirely **anaerobic**. There is a large accumulation of fatty acids within the parasite, and this settles the first step needed for anaerobic glycolysis of glycogen, richly present in the tissue of the parasite. The energy thus produced is adequate for the worm to carry on its vital processes. It possesses some cytochrome, thus it can respire **aerobically** also. For this it is sometimes correctly referred to as '**facultative anaerobe**' and not an **obligate anaerobe**.
7. Sense organs are poorly developed, being present only on lips in the form of papillae.
8. It is normally hypoosmotic to intestinal fluid.

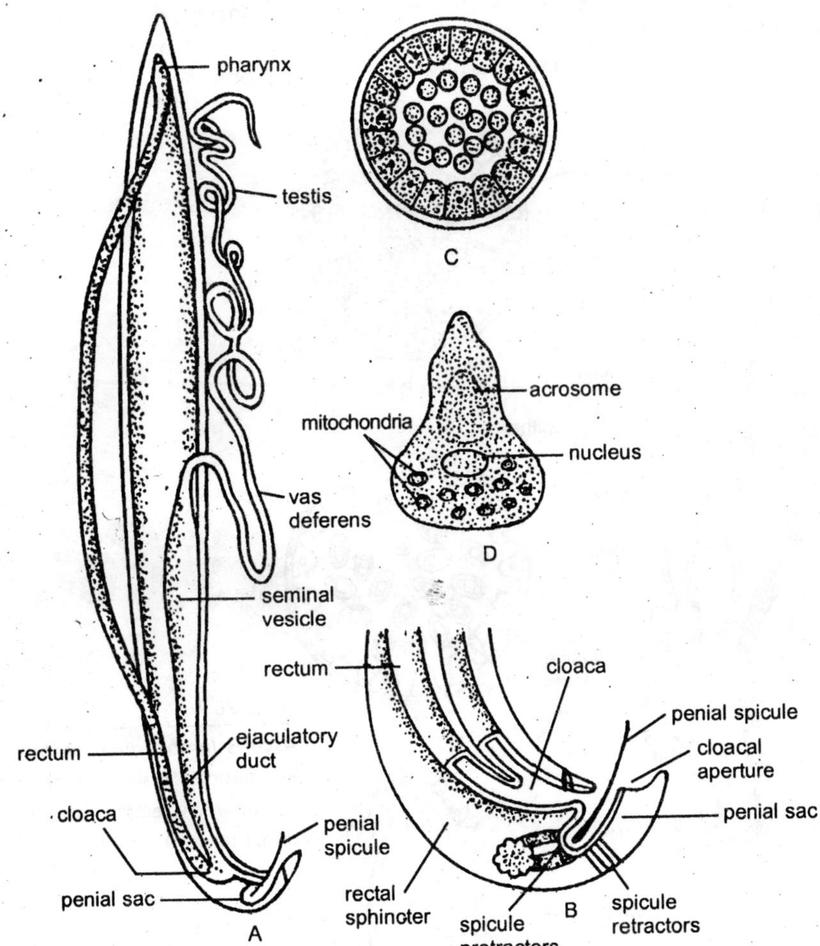

FIGURE 26.6

Ascaris lumbricoides. Reproductive system and gametes. A. Reproductive system (male). B. Posterior end (lateral view). C. Transverse section—vas deferens. D. A spermatozoon

9. The reproductive system is highly developed. Numerous eggs are produced and are covered with thick warty chitinous shell which protects them from the digestive enzyme of the host, prolonged dryness and cold for several days.

☒ Pathogenecity and Clinical features

1. Human infection occurs due to eating raw vegetables and drinks contaminated with embryonated eggs of Ascaris; children are more susceptible.
2. The larval stages are more harmful. The newly hatched rhabditiform larvae penetrate the intestinal wall and cause internal bleeding or haemorrhage, muscular spasms, fever and anaemia.
3. In lungs, the larvae damage the capillaries and inflammatory reaction occurs. In heavy infections entire lungs are affected with a marked exudate in air passages and desquamated epithelium. In extreme cases lungs become oedematous, haemorrhagic and completely consolidated leading to pneumonia with fatal result.
4. During migration of larvae through lungs, bronchial irritation, intense cough, and occasional hcemoptysis occur.
5. Sometimes, through general circulation, the larvae reach kidneys, brain, spinal cord and muscles and cause serious lesions.

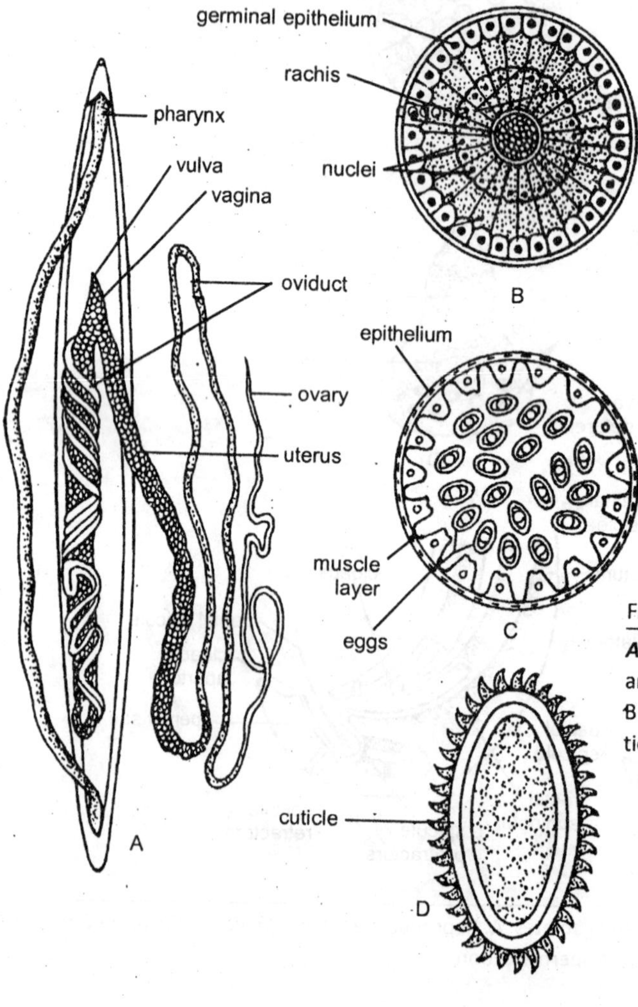

FIGURE 26.7

Ascaris lumbricoides. Reproductive system and gametes. A. Reproductive system (female). B. Transverse section, ovary. C. Transverse section, uterus. D. An egg.

6. The adult worms live in lumen of small intestine and in some cases cause temperature, colic pains, abdominal discomfort, indigestion, loss of appetite, appendicitis, peritonitis, abdominal tumour, diarrhoea, general nervousness, convulsions, vomiting, eosinophilia, etc. Allergic symptoms like articarial rashes, eye pain, asthma, insomnia and restlessness may appear.
7. Massive infection may cause fatal intestinal block. Even 100 worms are enough to plug the digestive tube completely, which may cause death.
8. Wandering *Ascaris* may cause great damage while passing through the bile duct or mechanical injury to the tissues of liver, resulting in serious illness.
9. Worms that reach the oesophagus, while the host is asleep, may crawl into the trachea, eustachian tube and middle ear, causing extensive damage.

⊠ Management

1. Ascariasis is very common in all tropical and oriental countries where disposal of human night soil is primitive. Before using it as fertilizer it should be properly treated.
2. Roots, stems, leaves and fruits growing in soil fertilized with human night soil should be thoroughly washed and cooked before consumption.
3. Proper sanitary toilet and awareness of the use of toilet by the people is necessary.
4. Before taking meal hand must be washed.
5. Treatment with anthelmintic drugs cures the patients and the source of infection is minimized.

WUCHERERIA BANCROFTI

Wuchereria bancrofti, a parasitic filaroid nematode living in the lymph nodes of man cause horrifying, debilitating disease, "Bancroft's filaria or elephantiasis. It belongs to the family Ochocercidae and order Filaroidea.

Both male and female worms live together in neibourhood of inguinal, abdominal and axillary lymph nodes of man. The embryoes (microfilariae) are found in circulating blood. The adult worms derive nourishment from the lymph.

⊠ Structure (Fig. 26.8)

1. The adults are long, creamy white, filiform nematodes, with both ends tapering; head end terminating in a slightly rounded swelling.
2. The tail end of male is curved ventrally and contains two spicules of unequal length measuring 100 and 200 microns respectively. The caudal end of female is narrow but abruptly pointed.
3. The male measures 2.5 to 4 cm in length and 0.1 mm in thickness, while female measures 8 to 10 cm in length and 0.2 to 0.3 mm in thickness.
4. Alimentary tract consists of oesophagus, intestine, and a short rectum. The oral aperture is unarmed and the anus is situated close to the posterior extremity of the worm.
5. Testis single, uncoiled and the gubernaculum is crescent-shaped. Circumanal papillae are 12—8 pre-anal and 4 postanal.
6. The female reproductive organ consists of 2 ovaries and their corresponding oviducts, situated posteriorly. Two uteri remain coiled, occupying the entire body. Vagina single; vulva situated on the ventral surface, about 0.6 to 1.3 mm from the head end.
7. Females are ovoviviparous laying eggs with well-developed embryos. Eggs measure 40 to 50 microns by 25 to 35 microns.
8. Life span of adult worms is long, 4 to 6 years or more.

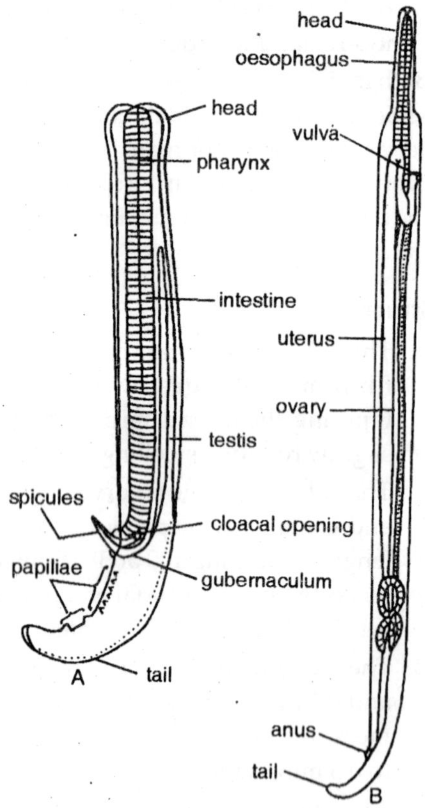

FIGURE 26.8

Wuehereria bancrofti. A. Male, B. Female

⊠ Microfilaria (Embryo) (Fig. 26.9)

1. Microfilaria is minute snake-like, (or Warm-like) transparent, with blunt head and pointed tail, measuring about 290 microns long and 6 to 7 microns thick.
2. The embryo is encased in a long hyaline sheath within which it move forwards and backwards.
3. The central axis of the body is composed of granules of somatic cells or nuclei. These granules extend from the head to the tail end and are broken at definite places.
4. At the anterior end of the embryo there is a space called **Cephalic** space. Two anterior shiny 'V-spots' called rudimentary excretory cell and excretory pore and one shiny V-spot or tail spot, at the tail end representing the terminal end of alimentary canal (anus or cloaca) are present.
5. In addition to the anterior and posterior 'V-spots' there are other spaces breaking the column of nuclei. These are (i) an oblique space, the nerve ring, (ii) a few G-cells (genital cells) posteriorly. G-cells 2, 3, 4 are in front of anal pore. G-cell 1 is further in front.
6. A dark stain region occupying one tenth the length of the worm extends from the anterior V-spot to the G-cell 1. This region is known as *Central (internal) body* of Manson or *Innerkörper* of Fülleborn,
7. The nuclei (granules) do not extend up to the tip of the tail (terminal 5%).

Nocturnal periodicity of Microfilaria. The microfilariae appear in peripheral blood of the host at night mostly from 10 p.m. to 4 a.m. The number rises to a maximum at midnight, when the patient is in deep sleep. Gradually, the number decreases and completely disappear in the morning.

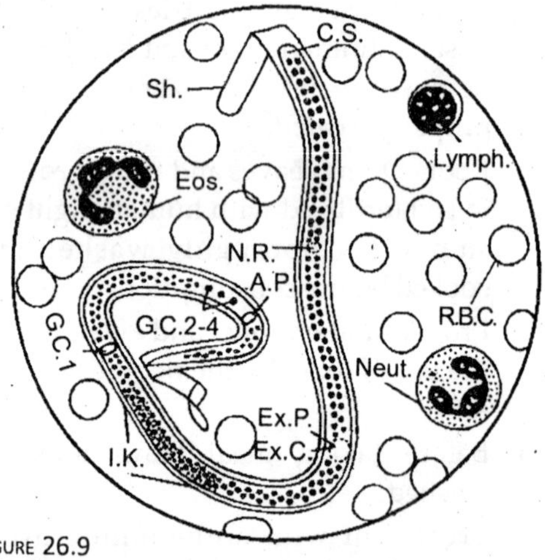

FIGURE 26.9

Microfilaria bancrofti in blood film. Sh—sheath, C.S.—Cephalic space, N.R.—Nerve ring, A.P.—Anal pore, G.C.—Genital Cell, Ex.P.—Excretory pore, Ex.C.—Exeratory cell, I.K.—Innenkörper

⊠ Life Cycle (Fig. 26.10)

Wuchereria bancrofti completes its life cyacle in two hosts—man and mosquito.

☒ Human cycle

1. Man is the definitive host. The adult worms harbour in the lymphatic system, particularly in lymph mode.
2. The embryos are discharged in the lymphatics find their way into the blood stream. They live in the peripheral blood for a considerable time without undergoing any developmental changes. The embryos are subsequently taken up by female culicine mosquitoes* with blood meal.

☒ Mosquito cycle

1. In the stomach of mosquito the embryo (sheathed microfilariae) lose the sheath within 2 to 6 hours, penetrate the gut wall and migrate to the thoracic muscles.
2. After sometimes the snake-like body changes to a thick, short, sausage-shaped one (*sausage stage*) with a spiky tail within 2 days and 1st cuticular moult occurs. It is the J_2 stage. It measures 124 to 250 × 10 to 17 microns and possesses rudimentary digestive tract.
3. In next 3 to 7 days the worm grows rapidly, moulting takes place measures 220 to 300 microns long and 10 to 20 microns thick. Within two weeks the metamorphosis is completed. The larva becomes filariform, measures 1500 to 2000 microns long and 18 to 23 microns thick. This is the J_3 stage. The body cavity, digestive system and genital organs are fully developed. Tail atrophies to a mere stump. This stage is infective to definitive host.

   The larvae migrate to the haemocoel and eventually to the labium or proboscis sheath and remain coiled up. The larvae enter human skin through the wound made by the mosquito during its blood meal.

☒ Human cycle

1. The worms migrate through peripheral lymphatics and settle in larger lymph vessels particularly inguinal, scrotal or abdominal lymphatics and grow to adult stage.
2. After a period of about 5 to 18 months they become sexually mature. Males and females remain coiled together and can only be separated with difficulty. The male fertilizes the female. Gravid females perturate.
3. A new generation of microfilariae is emitted, which passes either through the thoracic duct or the right lymphatic duct, to the venous system and pulmonary capillaries and then into the peripheral circulation.

☒ Pathogenecity and Clinical features

1. The incubation period of *W. bancrofit* is about 12 to 18 months and cause Bancroft's filariasis.
2. Microfilariae in the peripheral blood are not known to produce pathogenicity. Microfilariae in the tissue cause **occult filariasis** leading to Tropical Pulmonary Eosinophilia with febrile symptoms along with violent bout of dry cough emitting blood-tinged sputum in small amount.
3. The adult worms, living or dead cause different pathogenicity—inflammatory (acute) phase and obstructive phase in the host.
4. **The inflammatory responses** in the host is induced by the adult worms, particularly the female. Intense inflammation of lymphatics occurs, usually in the lower half of the body, with chills, fever, and toxemia. This is the allergic

* A large number of species of mosquito—*Culex, Aedes* and *Anopheles* act as intermediate host for *W. bancrofti* in different regions. In India and China *culex pipiens fatigans* is the usual intermediate host for *W. bancrofti*. as both the vector and the parasite are nocturnal.

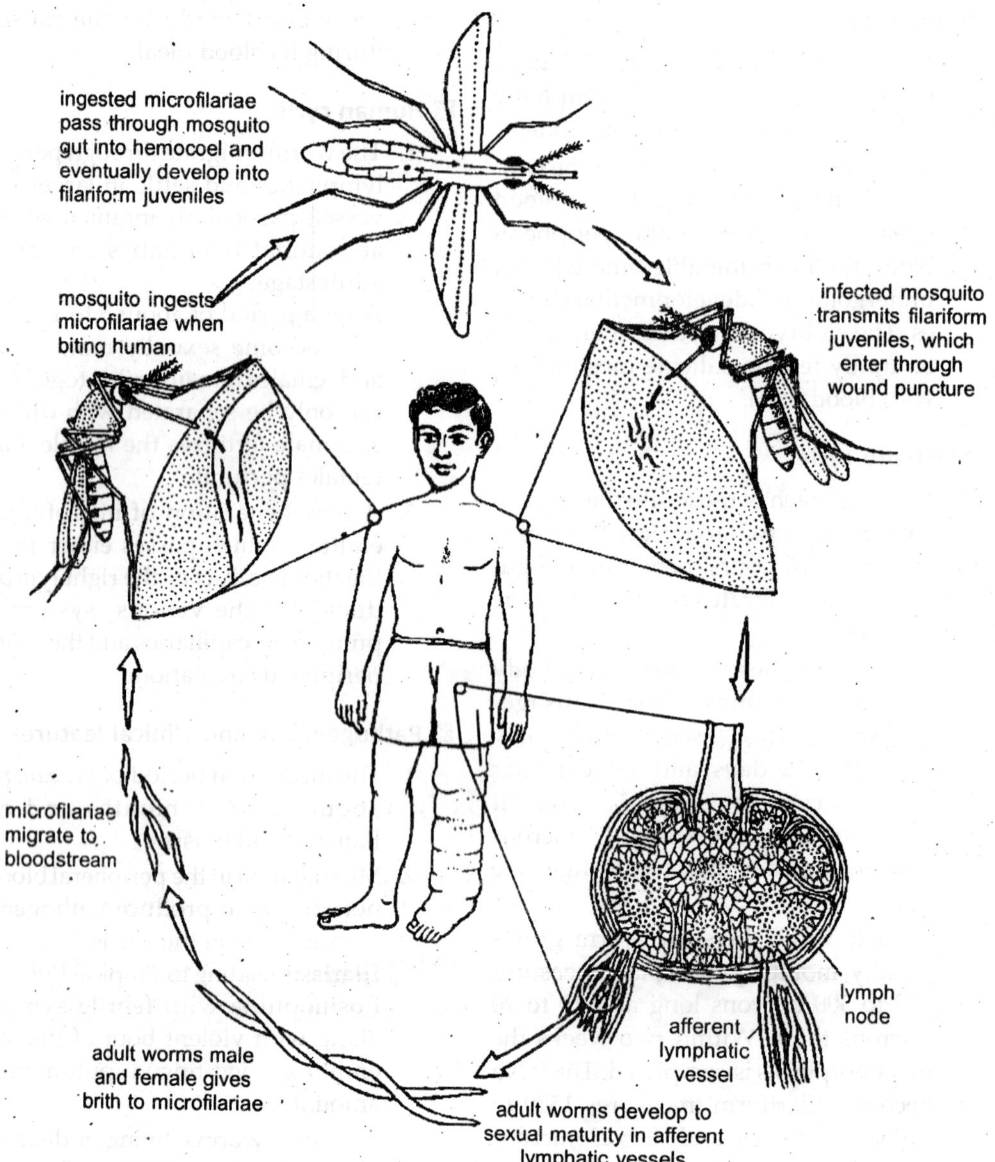

FIGURE 26.10

Wucheraria bancroti. Life cycle in man and mosquito; diagramatic representation

reaction caused by the toxic substances emanating from dead and degenerating worms or by discharge of metabolites by growing and gravid females. The area of the affected lymphatics is swollen and painful, and the overlying skin may be reddened. The common symptoms in this phase are (i) **lymphangitis** (inflammation in lymph channels), (ii) **lymphadenitis** (inflammation of lymph nodes), (iii) **Orchitis** (inflammation of testes usually with sudden enlargement and considerable pain), (iv) **hydrocele** (forcing of lymph into the tunica vaginalis of the testis or spermatic cord), and (v) **epididymitis** (inflammation of the spermatic cord). In some lymphangitis accompanied by acute febrile elephantoid fever recur frequently, at least once in a month coinciding with lunarphase. Sudden onset of fever with 104 °F temperature, rigors and sweating occurs. This fever may continue from a few hours to several days.

5. The **obstructive phase** is marked by lymph varices, lymph scrotum, hydrocele, chyluria and elephantiasis. One or more adult worms may obstruct the lymph channel. This physical obstruction of lymph channel leads to hyperplasia of endothelium leading to inflammatory thickening. Lymphatics gradually develop extensive fibrosis due to recurrent lymphangitis. This may lead to hypertrophy of the concerned part of the body. The hyperplasia and hypertrophy occur due to excess protein in lymph which stimulates growth.
6. In case the lymph returns obstructed, it accumulates in the duct causing dilation —the lymph varices.
7. Rupture of "varicose" lymph ducts causes discharge of chyle (lymph) through urine, which looks milk-white, containing albumen, fat and fibrinogen. This is common in lymphatic filariasis and is known as *Chyluria* or lymph in the urine.
8. In some cases, repeated attacks of acute lymphatic inflammation leads to a chronic lymphedema with much fibrous infiltration and thickening of the skin. This is known as elephantiasis. In men the organ most commonly afflicted with elephantiasis are scrotum, legs and arms, vulva, breasts are rarely affected. Elephantoid organs are composed mainly of fibrous connective tissues, granulomatous tissue and fat. The skin becomes thickened and cracked, and invasive bacteria and fungi further complicate the situation.

☒ Management

1. Proper management of the disease involves protection against the vector mosquito, control of mosquitoes and early treatment of the patient. Accurate diagnosis involves thick smear preparation of blood taken at mid night when the patient is in deep sleep. X-ray examination can defect dead, calcified worms.
2. The drug of choice for the past 40 years has been Hetrazan (DEC, diethyl carbamazines) which eliminates microfilariae from the blood. Careful administration of DEC for 7 to 12 days kills the adult worms. Ivermectin is also microfilaricidal and can be safely used.
3. Oedematous limbs are sometimes successfully treated by applying pressure bandages, which force the lymph out of the swollen area.
4. Long term protection from mosquito bite by use of mosquito net in the endemic areas, use of repellant, and management of vector mosquitoes and mass chemotherapy of indigenous people by microfilaricidal drug help in controlling bancroftian filaria.

ZOONOSIS

The Zoonoses* are the diseases and infections which are naturally transmitted between vertebrate animals and man, as defined by WHO (1959). These diseases often produce much more serious harm in man than they do in their nonhuman hosts. Different micro-organisms, protists and many helminths have Zoonotic potentiality. All the zoonotic diseases have reservoir hosts from which it may again spread and may form epidemic. So presence of man and reservoir hosts in the same locality makes control programme difficult.

☒ Types of Zoonoses

A. On the basis of reservoir hosts

1. **Anthropogenesis.** Infections transmitted from man to lower vertebrates.

* Singular Zoonosis

2. **Zooanthroponoses.** Infections transmitted from lower vertebrate to man.
3. **Amphixenoses.** Where men and animals act as reservoir hosts and transmission occur in either direction.

B. On the basis of life cycle of the infecting agent

1. **Direct Zoonoses.** Transmission occurs by direct contact, the agent undergoing little or no propagative changes or development during transmission. Example: *rabies, brucellosis*.
2. **Cyclo Zoonoses.** Transmission requires more than one species of vertebrate hosts, but not invertebrate host, to complete the developmental cycle of the agent. Example: **echinococcosis**.
3. **Metazoonoses.** Transmission by invertebrate vector in which the pathogen develops and multiplies. Incubation period is long e.g. plague, arbovirus infection, schistosomiasis etc.
4. **Saprozoonoses.** Transmission requires a vertebrate host and a non-animal developmental site or reservoir. Organic matters including food, soil and plants are non-animal reservoirs. Example: **mycoses, larva migrans,** etc.

Special group of people at risk

1. **Agricultural labour.** Industrial workers live stock inspectors, veterinarians and their family members.
2. **Animal product manufacturers.** Butchers, slaughter man, Handlers of meaes, milk, eggs, hide, fur, etc.
3. **Sylvian and Camp dwellers.** Wild life workers, foresters hunters, surveyors, explorers, tourists, etc.
4. **Animal handlers.** Zoogarden employee, wild life park employee, visitors of those places, pet or wild animal dealers.
5. **Clinical laboratory workers.** Animal autopsy performers, laboratory animal handlers, veterinary and public health workers.
6. **Overcrowding.** Refugees, disaster victims, pilgrimage participants.

Zoonotic factors

The Zoonotic diseases occur through contamination of the habitat due to high number of pathogens and unusual contact between different species, which may be direct or indirect through intermediate hosts.

The situation becomes alarming when parasites and pathogens increase in number with the increase in host number and consequently increase in infection in vector animals, vegetation, tree bark, etc.

ZOONOTIC HELMINTHS

1. ***Trichinella spiralis*** is a muscle parasite of man. The larvae form cysts in the stripped muscle, thousands in a single location and cause myositic disease, which is fatal and cannot be diagnosed till death of the host. There are three principal maintenance cycle of parasites (ii) pig to pig, rat to rat, and (iii) sylvatic. The three may act as reservoir also. Unprocessed or inadequately processed pork is the usual source of infection.
2. ***Trichuris trichiura*** (whip worm) grow well with Ascaris in man. Trichuris causes colitis, chronic diarrhoea, chronic dyscentery, and ultimately prolapsed rectum. Pig is the reservoir host for *Trichuris* and *Ascaris*.

   Hookworms are predominant human parasites. Simian, pangolin, dogs, etc. are the reservoir host. Brugia malayi, a filaroid worm is zoonotic and transmitted by mosquitoes.
3. ***Echinostoma*** spp. are trematode parasites of dog, rat, pig, etc. and are zoonotic. Dog, rat, pigs are reservoir hosts.
4. The cestodes—*Bertiella studeri*, a parasites of Orang utan and other primates; *Dipylidium caninum* a common tape worm of dog and cat, *Raillietina* spp. a

common tape worm of rat; *Vampirolepis nana* a common dwarf tape worm of mice; *Echinococcus granulosus*, a common tape worm of dog, fox, jackal, cat, etc. and its cysticercous larval stage in sheep, cattle, rodents etc; very often infect man and cause serious consequences.

BIOLOGY AND MANAGEMENT OF NEMATODE PEST

1. The nematodes are triploblastic (having three germinal layers), bilaterally symmetrical, pseudocoelomate, round, cylindrical, non segmented, thread-like worms with relatively simple structures.

   Male and females are separate. Posterior end of the body is frequently curved in males while pointed in females. The region behind the anus of female is called tail. Circulatory and respiratory organs and flame cells are absent. Nervous system consists of a circular nerve ring encircling oesophagus; several longitudinal nerve trunks are given out anteriorly and porteriorly.
2. Nematodes are enormous in number, may be free-living or parasitic in plants or animals.
3. They are widely distributed in all types of environments, from polar region to the tropics, in deserts, hot springs, high mountain elevations, great depths of oceans. Most free living species are aquatic, freshwater or salt water; found in marshes, on bottom of ponds, lakes and streams, and in the ocean at all the depths. Terrestrial forms live in soil and in decaying organic matter.
4. There are about 500,000 to 1000,000 living species of nematodes. The top six inches of an acre of arable soil may contain thousand of millions of soil nematodes and a single decomposing apple on the ground of an orchard may yield 90,000 or more round worms of different species.
5. Several hundred species of nematodes living in the soil become parasites of plants, causing a variety of plant diseases. Plant parasitic nematodes or **phytonematodes** are smaller in size, microscopic, and have a stylet or spear.
6. Nematodes show wide range of feeding habit. Many feed entirely on the microorganisms present in the decaying vegetables; others live on the surface of plants and suck their juices; still others venture within the plant and wander destructively through its tissues.
7. Many nematodes are parasitic in invertebrate animals and some cause varieties of diseases in higher animals including man. Many parasitise and destroy insects, spiders, crustaceans and snails. The hosts become infected by eating eggs of parasites with food or infective juvenile parasites penetrate into the body of the host.
8. All vertebrates are parasitised by nematodes. The nematodes in vertebrate body may harbour in the eye, mouth, tongue, alimentary canal, wall of oesophagus, salivary glands, liver, lungs, kidneys, heart, body cavity and muscles, causing various diseases.
9. Most of the free-living nematodes are microscopic. Many species invading blood or lymph channels of their hosts are microscopic. Nematode parasites living in the intestine are larger in size, while some in tissue habitats, e.g., guinea worm, kidney worm are still longer, a metre or more.
10. The nematodes are mostly dioecious, produce eggs with tough resistant covering.
11. The life cycle of nematodes ranges from vary simple to extremely complicated. Three different types of life-cycles are known among nematodes:

i. A life cycle involves no intermediate host is termed **monoxenous**.

ii. A life cycle involving an intermediate host is termed **heteroxenous**.

iii. A life cycle, where definitive host serves as an intermediate host and without a free larval stage as in **Trichinella spiralis**, is termed **autoheterogenous**.

12. The majority of nematodes are oviparous, some are ovoviviparous. All nematode larvae, whether hatch in water, soil or within animal or plant host undergo a series of four moults (ecdysis) before reaching maturity.

Management of Nematode pests

The parasitic nematodes are pestiferous and some of them are worst enemies of man. They are destructive and cause serious harm to health of man and their domesticated animals. Plant parasitic nematodes destroy tremendous amount of cash crop resulting a loss of millions rupees per year in India.

1. Control of **geohelminths** (transmitted through contaminated soil) like Ascaris, Trichuris, hookworms, etc., is possible by proper sanitation, cleanliness, washing of hands before eating, proper covering of prepared foods, and drinking filtered water.
2. Nematodes like, *Wuchereria, Loa, Onchocerca*, Guinea worm, are transmitted by vectors Culex mosquito, Chrysops fly, Simulium fly and Cylops respectively. Proper management of breeding sites by regular cleaning, spraying pesticides time to time, and use of bleaching powder in step wells help to control the vectors and diminish transmission of these nematodes.
3. The plant parasitic nematodes (Fig. 26.11) *Tylenchus, Ditylenchus, Anguina, Meloidogyne, Heterodera*, etc., attacks a number of crop plants including rye, oats, number of bulb lilies, onions, wheat, paddy, coffee, banana, sweet potato, etc. and cause devastating loss of crop by devouring plant tissues, forming galls and transmitting several pathogenic bacteria, virus and fungi. The spraying of systemic anthelmintic time to time and rotation of crops and fallowing of lands periodically help to control and manage phytonematodes.

ANIMAL ASSOCIATION

Animals live together in groups and for most effective utilization of the environment they are adopted to different types of intraspecific and interspecific competition, cooperation and exploitation. This phenomenon is known as **animal association**.

1. The majority of the animals live independently in their natural habitats procuring their own food, utilizing free water and oxygen for their metabolic processes.
2. Some animals develop a variety of patterns of associations which may broadly be divided in two groups—homogenetic and heterogenetic association.

Homogenetic Association

It is an association between individuals of the same genotype. The individuals of the same species may form loosely united communities, such as flocks of sheeps, while others with elaborately organised communities, as in honey bees where individual members exhibit considerable division of labour.

Heterogenetic association

Heterogenetic associations are in general much more complex. Animals that spend a part or whole of life intimately associated with members of different species is known as **symbiont** and the relationship **symbiosis**. Four subordinate categories of symbiosis are—phoresis, commensalism, parasitism and mutualism.

i. **Phoresis.** In this association no metabolic interaction of dependence occurs. In

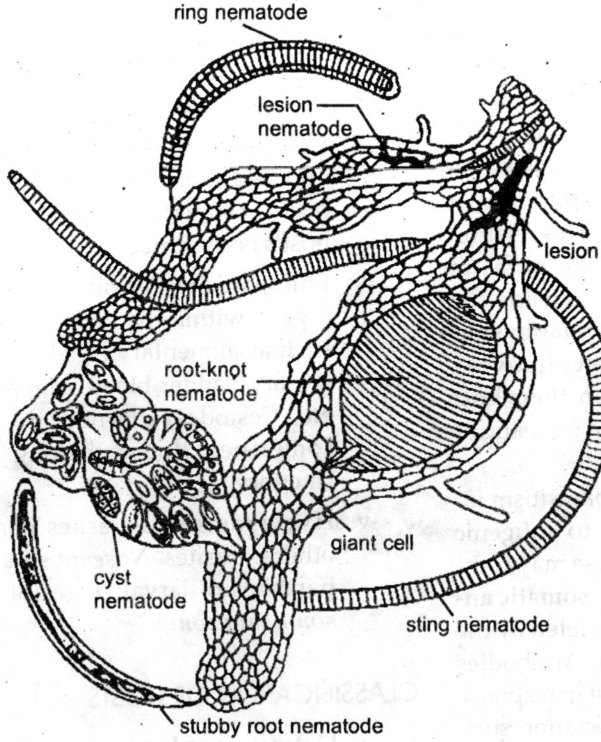

FIGURE 26.11 Different plant parasitic nematodes in different position of plant roots

this relation of the two, the smaller one the phoront is carried by the larger one, the **host**. *Fierasfier* sp., a fish live within the respiratory trees of sea cucumbers. These relatively helpless fishes are readily attacked and devoured by other species. The sea cucumbers remain undisturbed by the presence of this fish.

ii. **Commensalism.** Commensalism means "eating at the same table". It is a type of loose association in which two animals of different species live together without either being metabolically dependent on the other, although one or both organisms may receive some benefit from the association. A well-known example of this association is hermit crab and sea-anemone (Fig. 26.12). The hermit crab enters in an empty gastropod shell. Sea anemone is attached on the shell covering the exposed part of the crab. In the association the sea anemone benefits directly having access to new places and the remains of the food captured and not consumed by the crab, whereas the crab benefits by the presence of sea anemone which covers and assists in warding off undesirable predotors. (Fig. 26.12). The association is the coaction in which camouflage, food procurement, support, protection and locomotion of two species are assured without any harm to each other.

iii. **Parasitism.** Parasitism is an association with intimate and obligatory relationship between two hetero specific organisms during which the **parasite**, usually the smaller of the two partners, is metabolically dependent on the other, the **host**.

a. Parasites are metabolically dependent on their hosts. Ingestion of nutrition from the blood of the host (human) by *Schistosoma* spp, or ingestion of liver cells of the host by the common liver fluke ***Fasciola hepatica*** while within its bovine or sheep hosts are such examples. In case of blood sucking ticks, leeches and mosquitoes the nutrition is obtained directly from hosts.

b. The parasitic relationship may be permanent, as in case of tape worm in the intestine of mammals. This relation is obligatory, because the parasite cannot normally survive outside the host environment. The relationship may be temporary or facultative—leeches and ticks feed on their host blood but can live either a parasitic or a non-parasitic life.

The intimate relationship in parasitism involves the exposure of the host to antigenic substances of parasite origin. These may comprise the body of the parasite, i.e. **somatic antigen** or molecules secreted or excreted by the parasites i.e., **metabolic antigen**. Antibodies are usually synthesized by the host in response to the antigens. In no other association such response is observed.

CLASSIFICATION OF PARASITES

i. **Facultative.** Animals living temporarily as parasite, e.g., ectoparasites tick, mites, leeches, etc.

ii. **Obligatory.** Parasites unable to survive without host environment, e.g., cestodes, digenetic trematodes, etc.

iii. **Ectoparasite.** Parasites remaining attached to the outer surfaces of their hosts or superficially embedded in the host's body surface, e.g. louse, flee, etc.

iv. **Endoparasite.** Parasites live in internal organs within the body of their hosts such as alimentary canal, liver, lungs, urinary bladder, blood, lymphs, muscle, etc. Cestodes, digenetic trematodes, some nematodes belong to this category.

v. **Hyperparasite.** Parasites parasitising other parasites. ***Nosema*** sp are hyper parasites of larval stages of ***Schistosoma mansoni***.

CLASSIFICATION OF HOSTS

i. **Definitive or Final host** in the body of which the parasites attain sexual maturity. Sheep, Cattles are the final hosts for ***Fasciola hepatica***.

ii. **Intermediate host.** The animal which serves as essential and temporary host for completion of usually asexual part

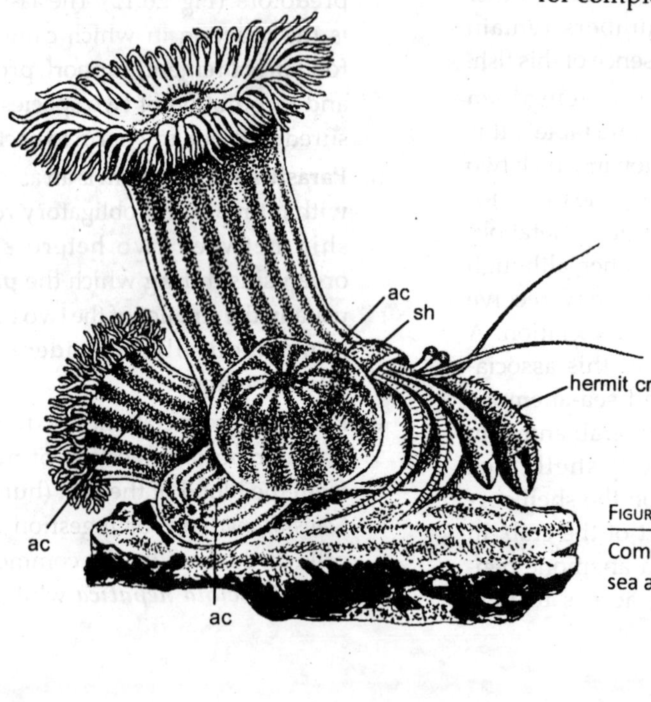

FIGURE 26.12 Commensalism of hermit crab, sea anemone

of life cycle. Snail serves as the intermediate host for ***Fasciola hepatica*** and mosquito serves as intermediate host of malarial parasites.

iii. **Vector.** It is the intermediate host that inject the parasites to the definitive or intermediate host. Mosquito acts as vector for malaria parasite ***Plasmodium***.

iv. **Reservoir hosts.** The animals that become infected and serve as a source from which other animals can be infected. Cat and rat act as the reservoir hosts for ***Toxoplasma gondii***.

☒ Mutualism

i. Mutualism is an obligatory interaction that is strongly beneficial to both species. Mutualist and the host are metabolically dependent on each other.

ii. The flagellated protozoans (*Trychonympha* spp and others) living in the gut of termites depend almost entirely on carbohydrate diet and acquire their nutrients from the wood chips ingested by their host. The flagellates produce cellulase (enzyme), which helps in biodegradation of cellulose to sugars. The host benefits the protozoans by providing anaerobic condition as well as wood chips, while the host benefits itself with the sugar produced from wood chips by the protozoan.

CLASSIFICATION: PHYLUM ASCHELMINTHES (NEMATHELMINTHES)

1. The nematodes are cylindrical, triploblastic roundworms, tapering at both the ends.
2. Body cavity is a pseudocoel, lies between the mesoderm and endoderm.
3. The body is generally covered with an elastic proteinous cuticle.
4. Alimentary canal is straight with ectodermal fore and hindgut, and an endodermal midgut without any muscles or glands.
5. Sexes separate; gonads tubular and continuous with their ducts.

a. Female organs are usually paired and open by a vulva.

b. Male organ single and opens into a cloaca.

The phylum Aschelminthes has been divided into 17 orders of which 7 are mostly free-living, 2 are free-living or parasitic and 8 are parasitic.

☒ Orders

Enoploidea. Cuticle ringed, often with bristles; anterior end bears 6 papillae.

Examples: *Anticoma, Enoplus*, etc.

Dorilaimoidea. Cuticle smooth, without bristles; anterior end with two circlet of papillae of 6 and 10 in each.

Examples: *Dorylaimus, Actionolaimus*, etc.

Mermithoidea. Smooth filiform parasites occurring in invertebrates, chiefly insects; head sense organs usually reduced to 16 papillae.

Examples: *Mermis, Aproctonema*, etc.

Chromadoroidea. Aquatic nematodes with smooth or ringed cuticle; usually heavily ornamented with bristles, knobs, punctuations etc. Pharynx with a posterior bulb.

Example: *Paracanthonchus*.

Araeolaimoidea. Cuticle smooth, sometimes with bristles; labial papillae present, anterior end usually with 4 cephalic bristles.

Examples: *Plectus, Odontophora*, etc.

Monhysteroidea. Cuticle smooth or slightly ringed, often bristly; anterior end with 4, 6 or 8 bristles.

Examples: *Siphonolaimus, Cylindrolaimus*, etc.

Desmoscelecoidea. Cuticle heavily ringed with prominent bristles throughout or, in restricted areas, anterior end with 4 bristles.

Examples: *Desmoscolex, Tricoma*, etc.

Rhabditoidea or **Anguilluloidea.** Free-living or parasitic nematodes with ringed or smooth cuticle; pharynx with a posterior bulb and also usually with a swelling, anterior to the nerve ring.

Examples: *Rhabditis, Diplogaster*, etc.

Rhabdiasoidea. Cuticle smooth, without definite pharyngeal bulb. Parasitic stage hermaphroditic or parthenogenetic. Free-living stages may develop into males and females.

Examples: *Rhabdias, Entomelas*, etc.

Oxyuroidea. Pharynx with a posterior bulb, usually valvulated; females with a long pointed tail; males with one or two equal spicules.

Examples: *Oxyuris, Enterobius*, etc.

Ascaroidea. Mouth surrounded by three prominent lips; pharynx without a posterior bulb; tail of females blunt, males with two spicules.

Example: *Ascaris.*

Strongyloidea. Mouth without conspicuous lips, but often with leaf crowns; pharynx without a bulb. Female usually with ovijectors; male with a copulatory bursa, supported by muscular rays.

Examples: *Ancylostoma, Strongylus, Necator*, etc.

Spiruroidea. Mouth usually with two lateral lips, sometimes 4 or 6. Pharynx without a bulb, anteriorly muscular, posteriorly glandular. Males without bursa; spicules unequal and dissimilar.

Examples: *Theliazia, Rictularia*, etc.

Dracunculoidea. Without definite lips or cuticularised buccal capsule. Vulva near or anterior to the middle of the body.

Examples: *Dracunculus, Micropleura*, etc.

Filaroidea. Filiform worms without lips. Buccal capsule small or rudimentary; no pharyngeal bulb; vulva anterior.

Examples: *Filaria, Loa*, etc.

Trichuroidea or **Trichinelloidea.** Worms with filiform anterior portion of the body. Mouth without lips. Pharynx slender, provided with a cirrus.

Examples: *Trichinella, Trichuris*, etc.

Dioctophymoidea. Moderate to very long nematodes; mouth without lips and surrounded by 6, 12, or 18 papillae; pharynx elongated, without bulb. Male with a muscular bursa.

Examples: *Dictyophyme, Hystrichis*, etc.

27 PHYLUM ANNELIDA

Phylum Annelida (L. *annelus* = little ring) includes elongated, vermiform, metamerically segmented* bilaterally symmetrical, triploblastic, true coelomate animals. They inhabit damp soil, moist grounds, burrows or tubes; may be aquatic, both fresh water and marine; very few are ectoparasites and endoparasites (leeches).

Annelids differ from arthropods—the next higher phylum—in having a large coelom, only simple unjointed appendages, little or no specialization of somites in different parts of the body and no moulting.

Annelids resemble arthropods in having a segmented body covered with cuticle, in the structure of nervous system and in the origin of mesoderm from special embryonic cell. More than 8,700 species ranging from 1 mm to 3.5 mm are known.

NEANTHES sp. (Nereis)

Neanthes (previous name *Nereis*) belongs to order Polychaeta, class Chaetopoda, phylum Annelida. It is commonly known as sand-worm because it is found buried in sand, and also clam worm, as it lives with clams. It is cosmopolitan in distribution.

Neanthes lives in shallow seas, in rock crevices or hidden under the stones or sea weeds. Some live in tubular burrows of loose tube made up of mucus in sand or mud at tide level.

It is a carnivore, feeding on dead fishes, molluscs, etc. It is active during night and remains passive in daytime.

⊠ External features

1. The body is long, narrow, dorsoventrally flattened, bilaterally symmetrical with a tapering posterior and a broader anterior end (Fig. 27.1).
2. The size varies from 300 to 400 mm in length and 2 to 6 mm in width.
3. Colour usually greenish-blue with tinges of red or yellowish brown.
4. A distinct **head** is present at the anterior end of the body. The rest of the body or the **trunk** is divided by ring-like narrow grooves into a series of segments, the **Metameres** or **somites**, arranged in a linear series.
5. The number of metameres varies from 80 to 200.

* Metameric segmentation is the segmentation in both external and internal features, including muscles, circulatory, excretory, reproductive organs and nerves.

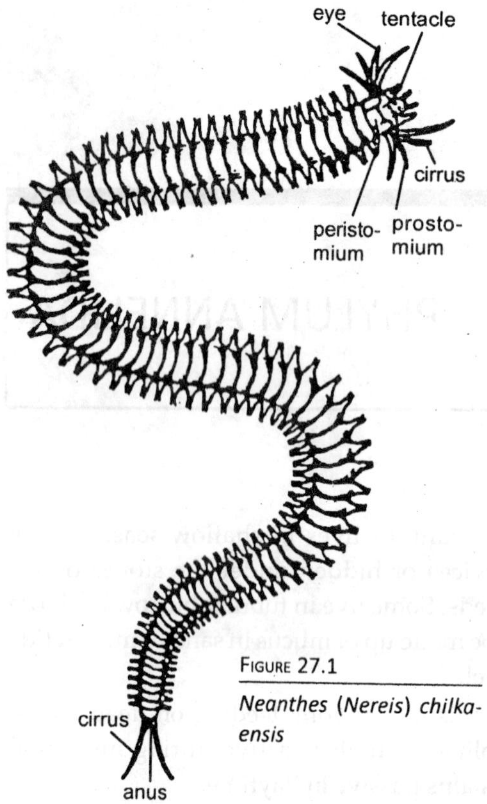

FIGURE 27.1

Neanthes (Nereis) chilkaensis

6. The head is differentiated into an anterior **prostomium** and a posterior **peristomium** (Fig. 27.2).
   a. The prostomium is a narrow, nearly triangular fleshy outgrowth situated middorsally in front of the mouth.
   b. The peristomium is a large ring-like structure with a transverse ventral **mouth** and is formed by the fusion of first two body segments. It forms the lateral and ventral margins of the mouth.
   c. The head bears several **sense organs**. The important sense organs are—prostomial eyes, prostomial tentacles, prostomial palps, nuchal organs and peristomial cirri.
   i. Prostomial eyes are four and dorsal in position. They are round, black and sensitive to light.
   ii. The **tentacles** are a pair of short, cylindrical tactile organs projecting from the anterior border of the prostomium.
   iii. The **palps** are paired, long and jointed and tactile in function, located on the ventral side of the prostomium.
   iv. The **nuchal organs** are ciliated pits of doubtful function, one on each side of the prostomium.
   v. Four long, slender **cirri**, tactile in function are present on the anterolateral border of the peristo-mium.
7. The segments of the trunk are almost similar in structure except the last one.
8. Each body segment except the anal bears on either lateral side a flat, fleshy, hollow and vertical flap-like outgrowth, the **parapodium** (Fig. 27.3).

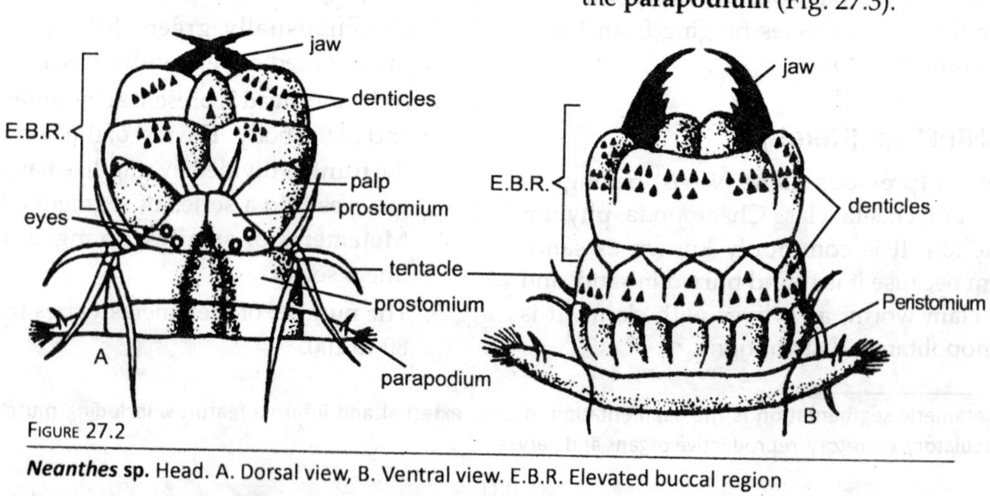

FIGURE 27.2

Neanthes sp. Head. A. Dorsal view, B. Ventral view. E.B.R. Elevated buccal region

i. A parapodium is a **biramous appendage** consisting of an upper or dorsal blade, the **notopodium** and a lower or ventral blade, the **neuropodium.** Each blade is further subdivided into two lobes, an upper and a lower.

ii. The dorsal margin of the notopo-dium is produced into a short, cylindrical, tactile appendage, the **dorsal cirrus**. A similar structure, **ventral cirrus** is present on the ventral margin of the neuropodium.

iii. Both the noto and neuropodium have a bundle of bristle-like setae or chaetae, lodged in **setigerous** or **chaetigerous sac** formed by the invagination of the epidermis. The setae are moved by strands of muscle fibres present in the parapodium.

iv. A stout, straight, thick and dark coloured chitinous rod called **aciculum**, deeply embedded in the parapodium is present in the middle of each bundle of setae. The two acicula constitute the endoskeleton of the parapodium and serve to support and for attachement of the setal muscles.

9. The papapodia of the mid-body segments, are largest and decrease in size towards both the anterior and posterior ends.

10. On the ventral side of the body, laterally and close to the base of each parapodium a minute opening, the excretory pore of **nephridiopore** is present.

11. The last segment is elongated, devoid of parapodia and bears a pair of long, ventrally placed anal **cirri** and the **anus**. The segment is called tail or **anal segment** or **pygidium**.

☒ Transverse section

1. The outermost layer is thin, porous **cuticle** secreted by the underlying epidermis. Two intersecting systems of

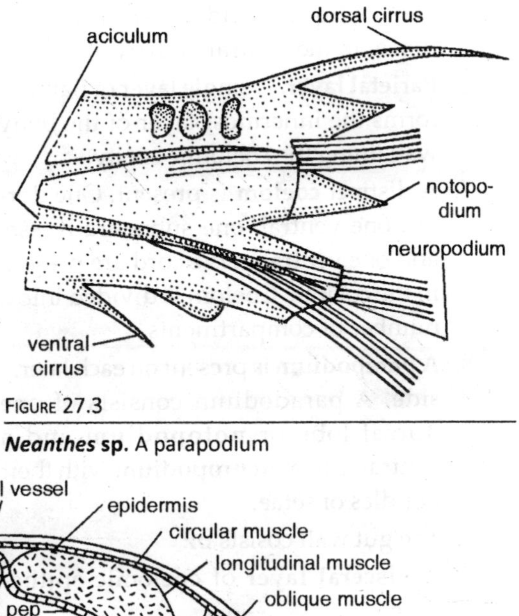

FIGURE 27.3

***Neanthes* sp.** A parapodium

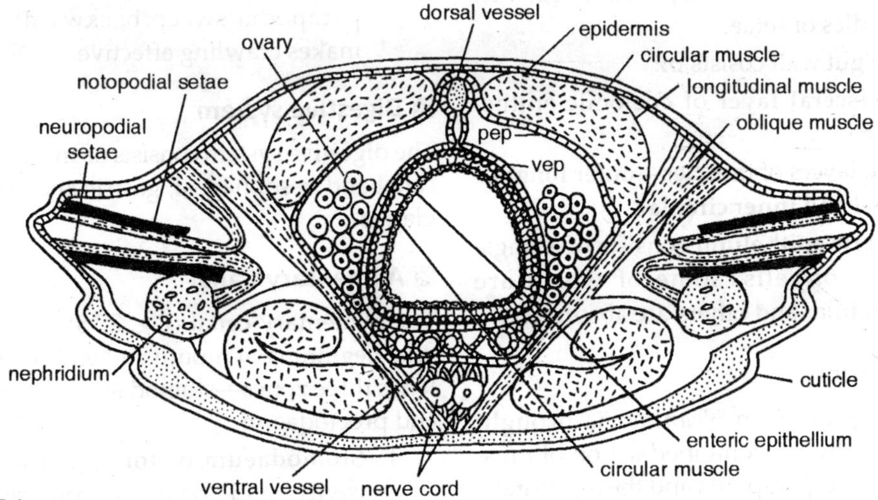

FIGURE 27.4

***Neanthes* sp.** Transverse section. Middle region. Vep. visceral epithelium. Pep. parietal epithelium

fine lines are present in the cuticle (Fig. 27.4).

2. The **epidermis** is a single layer of cells containing different types of numerous unicellular glands and plexuses of blood vessels on the dorsal surface. It is thick ventrally and thin dorsally.
3. **Dermis** absent.
4. Two layers of muscles are present. The outer **circular muscle layer** is a continuous one. The inner **longitudinal muscular layer** consists of four bundles of fibres, two dorsolateral and two ventrolateral.
5. Certain muscles run across the body wall obliquely and others run from dorsal to the ventral body wall.
6. **Parietal layer.** A single layer of flat cells forms the innermost layer of the body wall lining the coelomic cavity.
7. A distinct **coelom** is present. One dorsal, one ventral, one subneural vessel and one ventral nerve cord are present in the coelom. Coelom is divided into a number of compartments.
8. A **parapodium** is present on each lateral side. A paradodium consists of one dorsal lobe or **notopodium** and a ventral lobe or **neuropodium** with their bundles of setae.
9. The gut wall cosists of:

a. A **visceral layer** of coelomic epithelium.

b. Two layers of muscles—outer **longitudinal and inner circular.**

c. **Enteric epithelium** consists of a single layer of cells, some of which are glandular and others are absorptive.

☒ Locomotion

1. Locomotion in *Neanthes* is brought about by the combined action of musculature, parapodia and the hydrostatic action of the coelomic fluid.
2. Neanthes shows **creeping** and **crawling** movements but sometimes it **swims.**
3. The creeping movements are carried out by the action of parapodia only.

i. Each parapodium performs two types of strokes—**back stroke** or **effective stroke** and **recovery** or **forward stroke.**

ii. In the effective or back stroke the aciculum of parapodium is extended and lowered to come in contact with the substratum and moved back-wards against the substratum.

iii. In the recovery stroke, the aciculum is retracted, and the parapodium is lifted above and moved forward.

iv. The combined effective and reco-very strokes of numerous parapodia propel the worm forward.

v. The parapodia of two sides work alternately causing successive waves along each side of the worm.

4. The undulatory movement of the body musculature also helps the worm to crawl or swim rapidly.

i. The waves of contractions of the longitudinal muscles of the body wall cause undulation of the body.

ii. The longitudinal muscles of one side contract when the parapodia of that side are moved and the muscles relax when parapodia sweep backwards, which makes crawling effective.

☒ Digestive system

The digestive system consists of an alimen-tary canal and oesophageal, gastric and intestinal glands.

☒ Alimentary Canal

The alimentary canal (Fig. 27.5) is a straight tube beginning at mouth and ending in anus. It has three zones—stomodaeum, mesenteren and proctodaeum.

1. **Stomodaeum or foregut.** The buccal cavity and pharynx comprise stomodaeum. Internally it is lined by a layer of cuticle and ectoderm cells. An-

teriorly it bears the mouth and posteriorly ends in oesophagus. In the buccal cavity and pharynx the cuticle is thickened to form denticles. One pair of denticles in the posterior part of the pharynx enlarge to form claws.

2. **Mesenteron** or **midgut.** It is lined internally by endoderm and comprises **oesophagus,** and **stomach-intestine**.
3. **Proctodaeum** or **hindgut.** The posterior most part of alimentary canal, called **rectum,** is lined internally by ectoderm and cuticle. Rectum opens outside through anal opening.

☒ Feeding

1. Neanthes is carnivorous, feeding on petrified body and small crustaceans, sponges, molluscs, etc.
2. The contraction of protractor muscles and pressure of coelomic fluid cause protrusion of the pharynx.
3. Retraction of the muscles brings the jaws close to held the prey to carry it into the pharynx. This type of feeding on surface is known as **raptorial feeding.**
4. When in U-shaped burrow, it secretes mucus in front of mouth and creates a current by beating the parapodia. The food particles coming along with current and held in mucous cone are ingested by Neanthes. This is known as **filter feeding.**

☒ Digestion and Absorption

The ingested food undergoes mastication in the buccopharyngal region, pushed backwards by the peristaltic movement of gut.

1. Digestion is mainly extracellular and food is digested by the digestive juices secreted by the oesophageal glands and the gland cells of the epithelial lining of stomach-intestine.
2. Absorption of digested food occurs mainly in the stomach-intestine by diffusion.

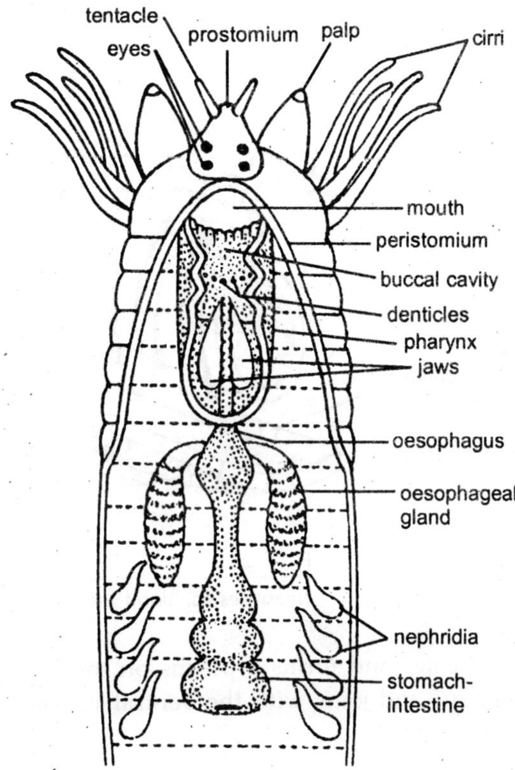

FIGURE 27.5

***Neanthes* sp.** Anterior end of the body. Dorsal wall removed to expose digestive system

3. Undigested materials pass on to the rectum and egested through the terminal anus.

☒ Blood Vascular System

The blood vascular system is well-developed and closed type and includes a system of blood vessels running to all parts of body.

1. The blood vessels have muscular walls and the constant circulation of blood is maintained by peristaltic movement of the vessels.
2. There are two main longitudinal vessels (Fig. 27.6) one median dorsal and other median ventral to the gut.

☒ Dorsal blood vessel

A collecting and contractile vessel, carries blood from behind forwards. Anteriorly, it

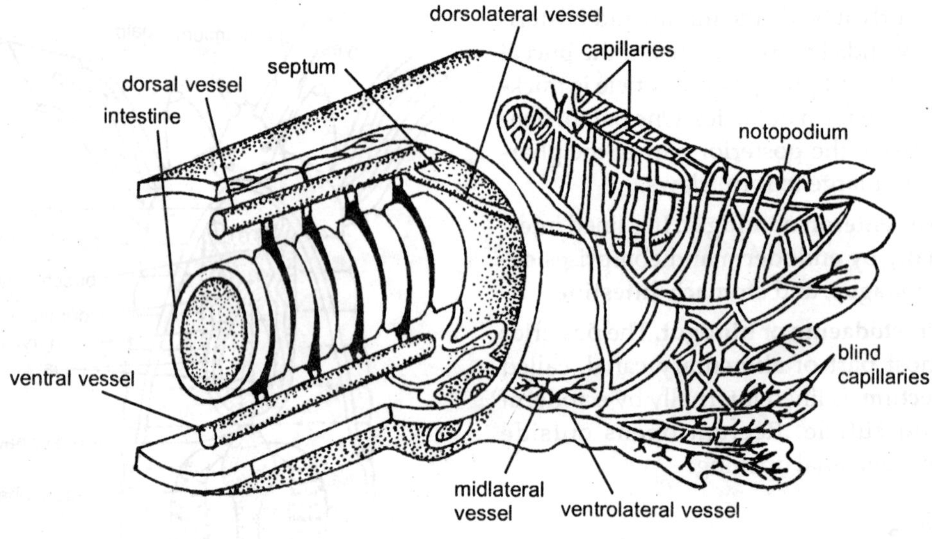

FIGURE 27.6

Neanthes sp. Blood vessels in a segment (diagrammatic)

bifurcates and forms a plexus on the oesophagus and joins with the ventral vessel in fifth segment.

In stomach region it receives blood from the stomach-intestine by means of two pairs of short dorso-intestinals or efferent-intestinals in each segment. It also collects blood from parapodia, nephridia and body wall through lateral vessels. In oesophageal region it supplies blood to oesophagus through branches.

⊠ Ventral blood vessel

A distributing and noncontractile vessel, carries blood from anterior to posterior end. It supplies blood to parapodia, nephridia and body wall and also to stomach-intestine through a pair of short efferent-intestinals in each segment.

The dorsal and ventral vessels are connected on either side in each segment, except the first five, by a pair of loop-like transverse **vessels or lateral vessels**. Blood from ventral vessel is sent to the skin, parapodia and other organs through afferent vessels from where efferent vessels arise and send blood to the dorsal vessel.

In each segment a pair of **ventrointestinal vessels** or **circumintestinal vessels** carry blood from the ventral vessel to the dorsal vessel. The circumintestinal vessel forms a network of capillaries in the stomach-intestine.

A thin **subneural vessel** runs below the nerve cord, collects blood from the lower body wall and supplies it to the ventral vessel.

The bright red colour of the blood is due to the presence of respiratory pigment **erythrocruorin**, similar to **haemoglobin**, in solution in plasma. The blood cells remain suspended in the plasma.

⊠ Respiratory organs

Special respiratory organ absent. Gaseous exchange takes place in the surface of parapodia and body wall. Blood is the carrier of oxygen and carbon dioxide.

⊠ Excretory System

Annelids are segmented animals. Their excretion is performed by certain organs known as **nephridia**. The nephridia with their ducts are usually repeated in each segment. Due to such arrangements, they are also called **segmental organs**. The nephridia of different annelids—though differ considerably in their structural details—are similar in their fundamental construction and extraction of nitrogenous wastes.

The nephridia of Neanthes are simplest (Fig. 27.7). It is a long, narrow, ciliated, folded or coiled tubule, with an opening, the **nephrostome**, lying in the coelom and a **nephridiopore** at the other end, opening outside.

1. Nephridia are one type and one pair occurs in all the segments except a few anterior and posterior.
2. The body of the nephridium is of an irregular oval shape, the anterior end of which is attached to the mesentery.
3. The body consists of a much coiled ciliated tubule richly supplied with blood.
4. The nephridium passes through the septum and the nephrostome, which is a ciliated funnel, opens in the coelom of the next anterior segment and is a continuation of the tubule of the body.
5. The tubule of the main body opens to the exterior by a fine circular pore, the nephridiopore, capable of dilatation and contraction and is placed on the ventral surface near the base of the ventral cirrus of the parapodium.
6. Wastes, collected by the nephridia are discharged to the exterior directly by the action of the cilia lining the excretory tube.

⊠ Nervous System

Nervous system is well-developed. It has three components—the central, peripheral and visceral nervous system (Fig. 27.8).

⊠ Central Nervous System

Cerebral ganglia or brain, circumoe-sophageal connectives, and ventral nerve cord constitute central nervous system.

a. The **brain** is a large bilobed mass, situated in the prostomium, above the

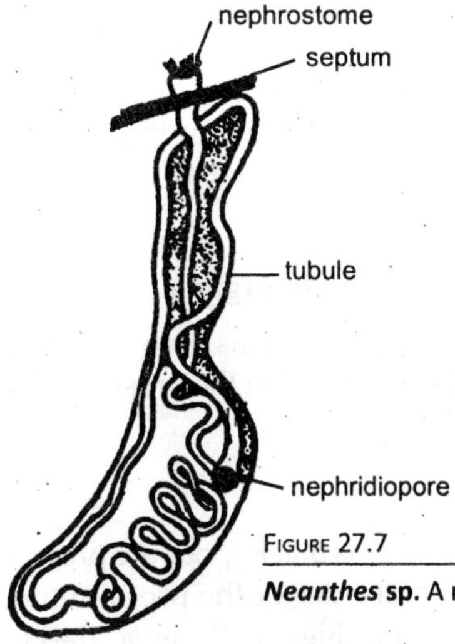

FIGURE 27.7

***Neanthes* sp.** A nephridium

the buccal cavity. Numerous large size neurons or nerve cells and nerve fibres are present.

b. A pair of stout **circumpharyngeal connectives** arise from the brain, encircle the pharynx and join with the **suboesophageal ganglion** in the third segment, ventral to the oesophagus.
c. The **ventral nerve** cord is made up of two solid nerve cords, enclosed in a common sheath. Starting from subpharyngeal ganglion, in the fourth segment it runs below the ventral blood vessel. The nerve chord is provided with a swelling, formed by a pair of ganglia in each segment.

⊠ Peripheral Nervous System

The nerves emanating directly from the brain and ventral nerve cord to supply different organs are included in peripheral nervous system.

a. The brain gives out four short **optic nerves** to the eyes, two **tentacular nerves**, and two **palpal nerves** to tentacles and palps, respectively.
b. Three pairs of nerves arise from each ganglion of the ventral nerve cord and innervate parapodia, body wall and other structures of that segment.

Visceral Nervous System

A set of fine nerves arising from the brain and innervating the anterior part of the gut and associated structures constitute visceral nervous system.

Receptors and Sense organs

Four eyes, two nuchal organs, two tentacles, two palpi, cirri, all located in the prostomium, constitute the sense organs.

Eyes

The eyes, appearing as black spots, are present on the dorsolateral sides of the prostomium. Each eye (Ch. 15. Fig. 15.7) is a darkly pigmented cup with a circular aperture, the **pupil**, a gelatinous **lens** behind it and a **retina** occupying the concave surface. The cuticle of the body surface and the epidermis, with somewhat flattened cells, pass over the eye and form the **cornea**. The cells of the retina are long and narrow, parallely arranged with one another in a radial direction. The inner part of the cells are hyaline and project towards the lens; the main body densely pigmented and the outer ends narrow into nerve fibres, forming part of the optic nerve.

Olfactory organs

The **nuchal organs** are olfacoty in function. They are located near the posterior part of the prostomium on the dorsal side in close contact with the posterior part of the brain. Each nuchal organ has two pits, lined with a ciliated epithelium and gland cells.

Tactile Sense organs

The **tentacles, palpi** and **cirri** are tactile sense organs and help the animals to detect changes in the surroundings.

Reproductive System

1. **Neanthes** is a dioecious animal. The gonads develop from the ventral coelomic epithelium only in breeding season.
2. In male only one pair of testes lie between nineteenth and twenty-fifth segments.
3. In female, ovaries lie in many segments around blood vessels.
4. The gonads have no ducts; the sex cells are discharged into the coelomic fluid and mature in the fluid.

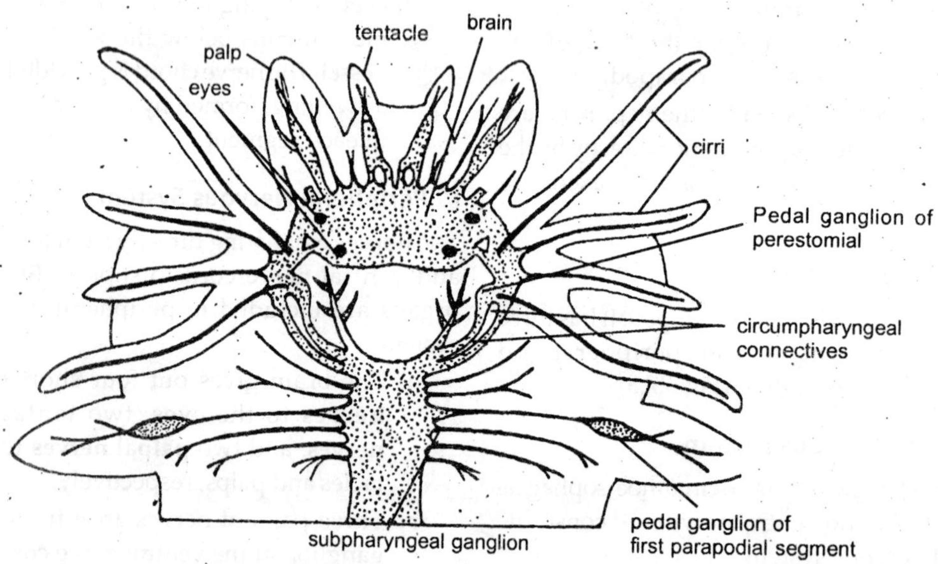

FIGURE 27.8

***Neanthes* sp.** Nervous system (anterior portion)

5. The mature gametes are discharged to the exterior by temporary rupture of the body wall and fertilization take place externally in sea water.
6. Neanthes exhibt polymorphism, a **Heteronereis** phase develops in the breeding seasons. Sometimes, during breeding seasons, some individuals leave their usual habitats, the burrows, and become free-swimming. They are called heteronereis. The body is differentiated into two distinct parts, an anterior **atoke** and a posterior **epitoke** (Fig. 27.9).
   - i. In the atoke eyes increase in size, peristomial cirri become longer.
   - ii. In the epitoke the parapodia become larger, leaf-like; dorsal and ventral cirri enlarge; additional foliaceous outgrowths are formed; setae become more numeours and longer with oar-shaped blades.
   - iii. The larger modified parapodia help in swimming and larger body surface serves for rapid respiration during swimming.
   - iv. The last segment or **pygidium** develop sensory papillae.
   - v. The gonads are abundant and confined to the epitoke.
   - vi. Heteroneries swim upwards and spawn near the surface by rupture of the posterior body wall.
   - vii. Fertilization takes place in the surface layer of the sea.

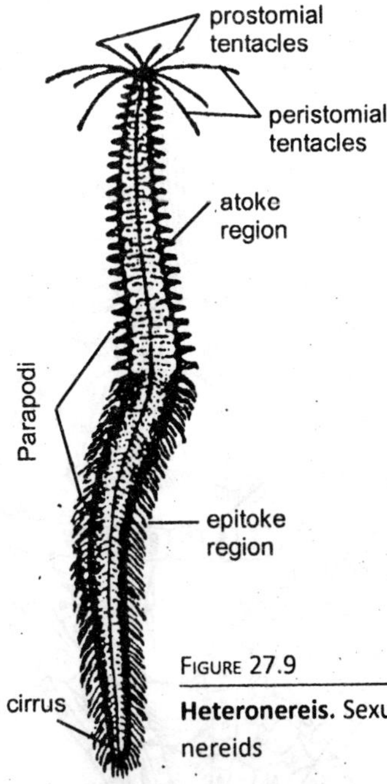

FIGURE 27.9

Heteronereis. Sexual phase, nereids

⊠ Development

The eggs are telolecithal; cleavage spiral; gastrula develops into a trochosphere or trochophore.

A larva is a transitional stage in the development of animals. In some animals the eggs, after fertilization, undergo development through a complicated series of stages which are not directly progressive towards the adult stages but assume certain structures which are unlike the parents; although, ultimately, the adult structure is assumed. The phenomenon is known as **metamorphosis** and the individual undergoing the process is known as **larva.** The larva is distinctly separate from the embryo in as much as it is capable of taking its own care and responsibility.

⊠ Structure of a Trochophore larva

1. The trochophore larva is somewhat conical or oval in shape with a protruding equator, having the general appearance of a spinning top (Fig. 27.10).
2. Externally, it has one-layered epithelium (ectoderm), which is thickened in the apical pole, forming a sensory plate, the **apical plate**, with a tuft of cilia.
3. Two main bands of cilia encircle the body. The ciliated band around the equator is known as **prototroch**, which serves as the chief organ of locomotion and food-current guide.
4. The second band, known as **metatroch** passes below the mouth.

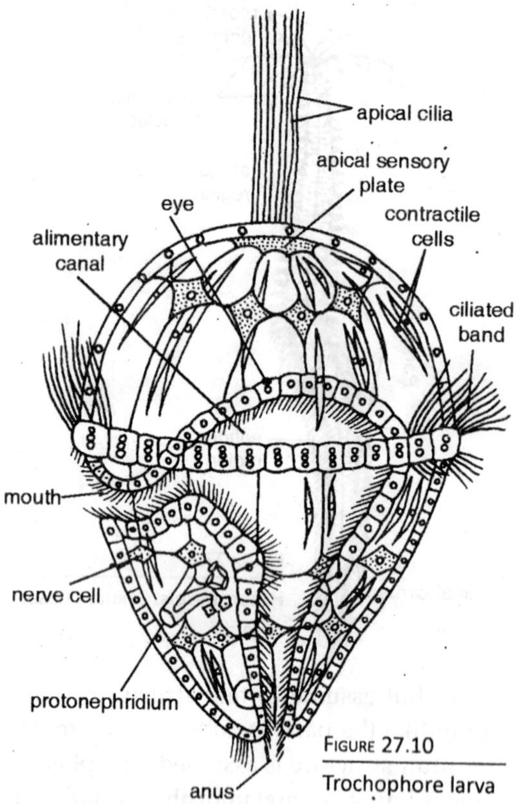

FIGURE 27.10
Trochophore larva

5. In some, a third girdle of a ciliated circlet is present, around the anus. This is known as **paratroch**.
6. There is no trace of **metamerism**, the rudiment of the adult trunk is represented only by a small region at the lower pole of the larva.
7. A spacious **blastocoel** is present, the larva is devoid of coelom at this stage. A pair of **protonephridia** are present along with **mesenchyme** and larval muscles.
8. The larva has a complete digestive-tract, ciliated all through, with mouth, stomach and intestine, which opens through an **anus** at the lower pole.
9. There may be present a ganglionic mass under the apical plate, known as **cerebral ganglia**.
10. The larval excretory organ consists of a **flame cell** and an excretory tube, opening to the exterior near the anus.

The trochophore larva is adapted to swimming and feeding on microscopic organisms the plankton. It undergoes successive changes—called metamorphosis—and transformed into adult form.

☒ Metamorphosis

1. With the start of metamorphosis, the posterior end of the trochophore larva beyond prototroch elongates, carrying the gut with it.
2. Traces of segmentation become visible; sometimes a series of segmentation occurs before appearance of parapodia.
3. Gradually, parapodial rudiments with their setae are developed.
4. The external segmentation is accompanied by the division of the mesoderm bands into a series of coelomic sacs.
5. The ventral ectoderm develops a median thickening to form the ventral nerve cords.
6. The larval nephridia, ectodermal muscle bands and ciliated bands of trochophore larva are lost.
7. The region anterior to prototroch gives rise to prostomium; the cells of the apical plate form the cerebral ganglia; mesodermal tissue grows forward to form a number of coelomic sacs within the prostomium.
8. After settlement, additional segments are added from the growing region.

TROCHOPHORE LARVA: BIOLOGICAL STATUS (SYSTEMATIC POSITION)

The trochophore is an important larva, showing phylogenetic relationship with annelids, and molluscs.

☒ Affinities of Trochophore larva

Trochophore is the free-swimming larval stage of polychaete, belonging to phylum Annelida. Since 1840, when the larva was first described by Loven, and the Trochophore theory based on the occurrence of Trochophore larva in a

☒ Affinities with Ctenophora

Similarities	Dissimilarities
1. Body pear-shaped. 2. Prototorch derived from the fourth group of ciliated cells. 3. Subectodermal radiating nerves. 4. Aboral sense organ—statocyst of ctenophore and apical sensory plate of trochophore.	1. The cleavage pattern is different. 2. Absence of anus in ctenophore.

☒ Affinities with Turbellaria (Platyhelminthes)

Similarities	Dissimilarities
1. Muller's larva of *Planocera* resembles Trochophore larva.	1. Embryonic differentiation of mesoderm.
2. Mode of development.	2. Absence of anus in Muller's larva (enteron with only one opening).
3. Disposition of ciliary bands.	3. Presence of a tuft of cilia at the caudal end of Muller's larva.
4. Presence of eye spots at the aboral end.	

☒ Affinities with Nemertina (Pilidium larva)

Similarities	Dissimilarities
1. Body helmet-shaped. 2. Ciliary ring between aboral and oral end of Pilidium corresponds to the prototorch of Trochophore. 3. Coelom schlizocoelic. 4. Disposition of nervous system. 5. Occurrence of stomodoeum.	1. Mode of development of mesoderm. 2. Absence of anus in Pilidium.

☒ Affinities with Rotifera

Similarities	Dissimilarities
1. Superficial resemblance between Trochophore of annelids and Trochosphorea larva of rotifers. 2. Disposition of ciliated girdles. 3. Shape of the intestine. 4. Structure of nephridia. 5. Structure of nervous system and sense organs.	1. The two differ in other structures.

⊠ Affinities with Mollusca (Trochophore of Patella; Trochosphore of Palludina)

Similarities	Dissimilarities
1. The preoral ciliated ring.	1. Molluscan larvae differ from Trochophore in many respects, viz. absence of protonephridium, mantle, foot, etc. in the latter.
2. Presence of an apical plate with a tuft of cilia (strong cilia along the border of velum in Veliger larva)	

⊠ Affinities with Echinodermata

Similarities	Dissimilarities
1. Body shape.	1. Cleavage spiral in annelids, radial in echinoderms.
2. Presence of similar ciliated bands.	2. Coelom schizocoelic in annelids, enterocoelic in echinoderms.

number of invertebrate phyla, postulated by Hatscheck (1878), attempts have been made to establish a relationship between different invertebrate phyla, often with a diverse group—like molluscs, though, in most cases, similarity does not extend beyond the early development stages.

⊠ Affinities with Hemichordata

Similarities. An apical plate with a tuft of cilia in Tornaria larva.

Trochophore occupies an important place in the hierarchy of invertebrate evolution. To a greater or lesser degree it resembles the larvae of a number of invertebrate phyla but in most cases marked differences begin to appear beyond the early larval stages. It has been suggested that the trochophore is the larva of an ancestral form—the **Trochozoon;** a common ancestor of most, if not all the bilateral phyla. Trochophore is considered as a transitional stage from the radial (ctenophores) to bilateral groups (rotifers) and it represents the larva of the ancestor, not that of the ancestor itself.

PHERETIMA sp. (Earthworm)

The earthworm (*Pheretima*) is a member of the order Oligochaeta, class Chaetopoda and phylum Annelida. It lives in self-made burrows in the upper layers of moist soil.

⊠ External features

1. The body of the earthworm is long, narrow, almost cylindrical and slightly narrows down at both the ends. The anterior end is pointed, while the posterior end is more or less blunt.
2. A full grown worm is about 150 mm long and 3 to 5 mm in diameter in the broadest portion of the body (Fig. 27.11A).
3. The dorsal surface is brown while the ventral surface is dull grey. A dark median line runs along the dorsal surface throughout the length of the body.
4. The surface bears a series of narrow ring-like grooves, and the body is divided into about 100 to 200 small **segments** or **metameres**.
5. The external segmentation corresponds to internal segmentation and the body is separated internally to a number of compartments.
6. The **mouth** is a slightly ventral, crescent-shaped opening at the anterior end.
7. A small, fleshy lobe, the **prostomium** (Fig. 27.11B) is situated on the dorsal surface of the first or the buccal segment and overhangs the mouth.

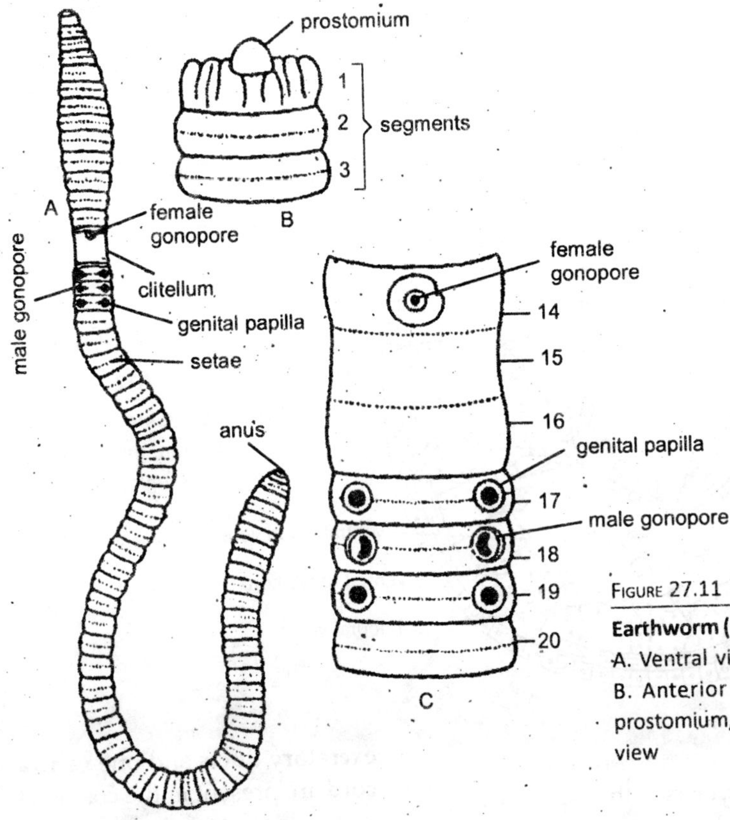

FIGURE 27.11

Earthworm (*Pheretima posthuma*). A. Ventral view (entire specimen), B. Anterior three segments and prostomium, C. Genital area, ventral view

8. At a distance of about 20 mm posterior to the anterior end, a prominent band known as **clitellum** is present.
9. The clitellum includes three segments 14 to 16, and divides the body into three distinct regions—the anterior or the **preclitellar**, the middle **clitellar** and the posterior **postclitellar** regions.
10. A median round aperture, the **female genital pore** is present on the ventral surface of the 14th segment (Fig. 27.11C).
11. A pair of crescentic apertures on raised papillae, the **male genital pores** are present on the ventral surface of the 18th segment.
12. In front and behind the male pores, on the 17th and 19th segments, a pair of papillae, the **genital papillae** are present in each segment. The ventral surface of the 17th to 19th segments is called the **genital area**.
13. **Spermathecal pores** or openings of the spermathecae are four pairs, placed ventrolaterally in the intersegmental grooves between the 5 and 6, 6 and 7, 7 and 8 and 8 and 9 segments.
14. At about the middle of each segment except the first, the last, and the clitellar segments, chitinous needle-like structures called **setae** are arranged in a circle. The setae help in locomotion.
15. The **dorsal pores** are minute openings in the middorsal line of intersegmental grooves. Starting from the groove between the 12 and 13 segments and except the last groove, they are present in all the intersegmental grooves.
16. The **anus** is a slit-like opening at the tip of the last segment.
17. The **nephridiopores** are minute, and scattered irregularly all over the skin except that of the first two segments.

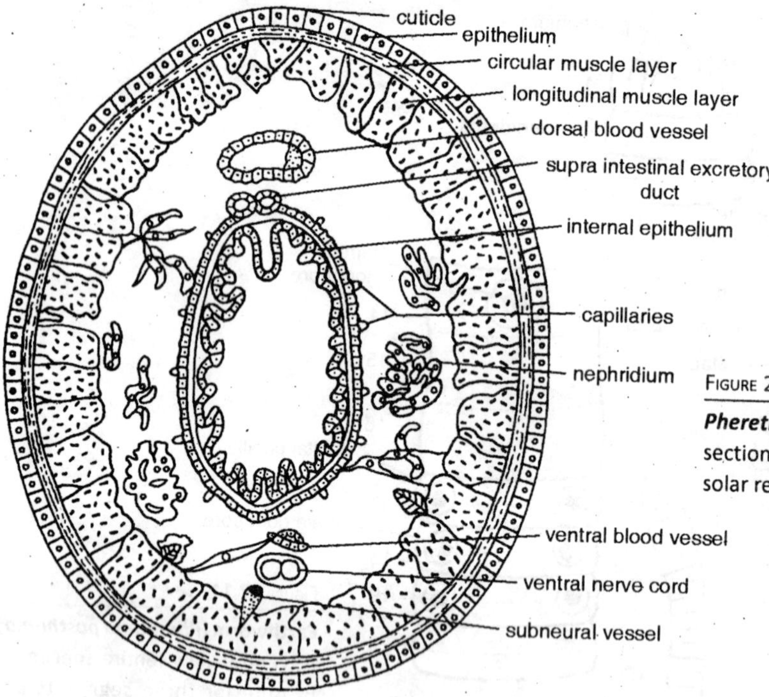

FIGURE 27.12

***Pheretima* sp.** Transverse section. Intestine. Typhlosolar region

⊠ Transverse Section

1. The outermost layer is a thin, porous cuticle secreted by the underlying epidermis (Fig. 27.12).
2. The **epidermis** is a single cell layer thick and consists of two kinds of cells, the large glandular cells with eccentric nucleus, and columnar cells placed between the glandular cells. A few cells functioning as sense cells are also present.
3. **Dermis** absent.
4. Two layers of muscles are present.
   The outer **circular muscular layer** is narrow and continuous. The inner **longitudinal muscular layer** is thick and consists of muscle fibres arranged in narrow bands.
5. Other muscles absent.
6. **Parietal layer.** A single layer of coelomic epithelium forms the innermost layer of the body wall lining the coelomic cavity.
7. A distinct **coelom** is present. One **dorsal**, one **ventral**, one **subneural blood vessel**, two **supraintestinal excretory ducts** and one **ventral nerve cord** are present in the coelom. Coelom is divided into a number of compartments.
8. **Parapodia** absent. **Chitionoid setae** embedded in the body wall may be found.
9. The gut-wall consists of:
   a. A **visceral layer** of coelomic epithelium.
   b. Two layers of muscles—**outer longitudinal** and **inner circular.**
   c. **Enteric epithelium** consists of a single layer of cells, some of which are glandular and others are absorptive. It is thrown into folds. A large median fold known as **typhlosole** in the region extending from 26 to 95 segments before anus, hangs in the lumen from the dorsal wall.

⊠ Locomotion

The locomotory structures are setae, circular muscle fibres arranged in the form of a ring, longitudinal muscle fibres in bands and the coelomic fluid.

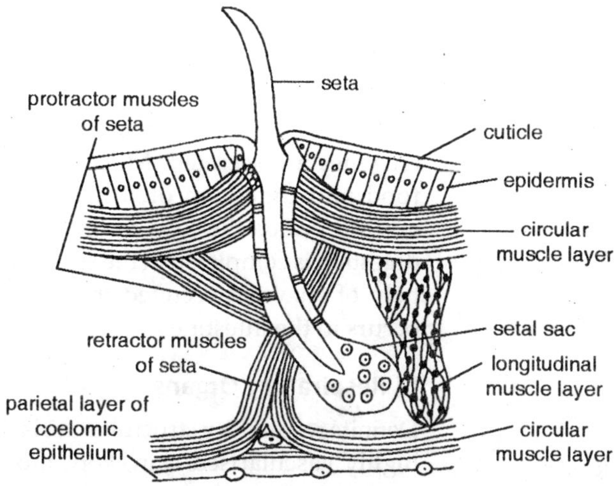

FIGURE 27.13

***Pheretima* sp.** A seta in body wall

Setae. The setae are arranged in the form of a ring in all segments, except the first, last and clitellar segments.

1. A seta is a naroow, elongated, chitinous structure hardened by sclerotised protein.
2. It is light yellow; the shape is in the form of a 'f' (Fig. 27.13), with a distal pointed end and a blunt proximal end embedded in setigerous sac in epidermal pit. A swelling, **nodule**, may be present in the middle.

⊠ Machanism of Locomotion

1. The setae of the posterior part of the body are fixed to the substratum; contraction of the circular muscles put pressure on the coelomic fluid and forces the longitudinal muscles to stretch and the anterior part of the body elongates.
2. In the next step, the setae of the anterior part are anchored; those of the posterior part released; the contraction of the longitudinal muscles increases the coelomic pressure, the circular muscles are stretched, leading to the shortening of the body and the earthworm moves forward.

⊠ Digestive System

The **alimentary canal** is a straight tube, starting at the mouth and ending in the anus (Fig. 27.14).

Mouth. The ventral mouth is crescent-shaped, situated in the peristomium. It is followed by a small **buccal cavity**, opening into the pharynx.

Pharynx. The thick-walled, pear-shaped, muscular structure in the 3-4 segments. A **pharyngeal bulb** is present on the dorsal wall.

Oesophagus. The muscular tube, extending from 5-8 segment and connects the pharynx with the stomach. The posterior part of the oesophagus is thick-walled, muscular with an inner cuticular lining, oval in shape and known as gizzard.

Gizzard. The food matters are grind to fine particles in it. It is a hard, muscular structure lined internally by a tough cuticle.

Stomach. It is highly vascular, glandular, lies in 9-14 segments and continues posteriorly as intestine.

Intestine. A thin walled, somewhat wide tube, running up to the anus in the last segment.

Intestinal coeca. A pair of small, conical outgrowths, directed anteriorly, project from the sides of the intestine in the 26th segment.

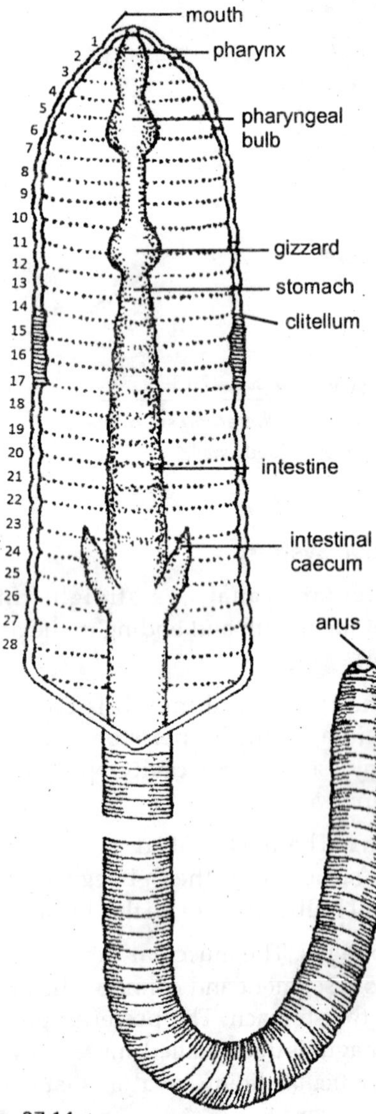

FIGURE 27.14

***Pheretima* sp.** Digestive system

Typhlosole. A fold of the dorsal intestinal wall in the 26-95 segments, hangs downward in the lumen of the intestine (Fig. 27.12).

The intestinal wall is made of an outer peritoneal epithelium, longitudinal muscle fibres, circular muscle fibres and an inner epithelium of ciliated and glandular cells.

⊠ Feeding and Digestion

Earthworm obtains organic materials from the soil it swallow, rejecting the rest as cast. It also feeds upon decaying matters, small dead animals, etc. The mouth and pharynx help in feeding.

In the pharynx mucin and proteolytic enzymes are secreted. Mucin lubricates the food and digestion of protein starts. The gizzard grinds the food to fine particles and digestion is completed in the stomach with the help of enzymes secreted there. Absorption occurs in the intestine.

⊠ Respiratory Organs

Definite respiratory structures are lacking. The highly vascularised skin, kept moist by the secretion of epidermal glands and coelomic fluid coming out through dorsal pores, serves the functions of respiratory membrane.

⊠ Blood Vascular System

The blood of *Pheretima* is red due to the presence of haemoglobin dissolved in plasma. The corpuscles are colourless and nucleated. An elaborate system of blood vessels is found in earthworm.

The arrangement of blood vessels is different in the intestinal region, i.e., posterior to the 13th segment, from that of the anterior region.

⊠ Blood vessels in the intestinal region (Figs. 27.15, 27.16)

Dorsal blood vessel

It runs along the middorsal line between the gut and the body wall. It has a thick wall and contracts rhythmically. A pair of valves are present in the lumen of a vessel in each segment, to prevent backflow of blood. It is a collecting vessel and receives blood from

a. Two pairs of **dorsointestinal vessels** from the two ring-like vessels surrounding the intestine in each segment, bringing blood from the intestine.
b. A pair of **commissural vessels** from the posterior face of each septum receiving capillaries from the nephridia and body wall. The dorsal vessel drives the collected blood forward.

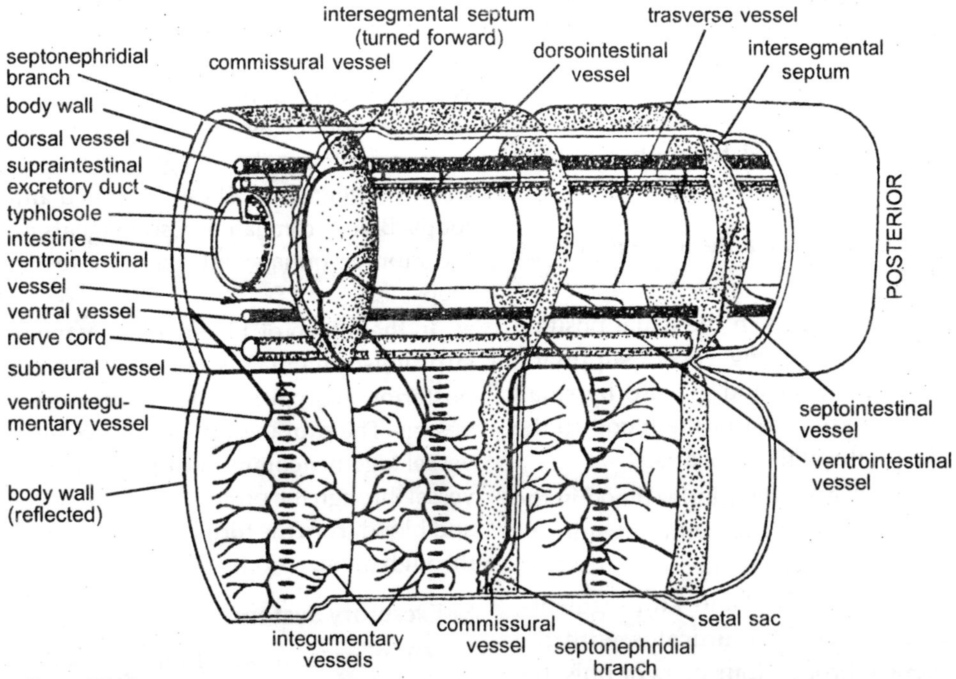

FIGURE 27.15

***Pheretima* sp.** Blood vascular system. Intestinal region (four segments shown)

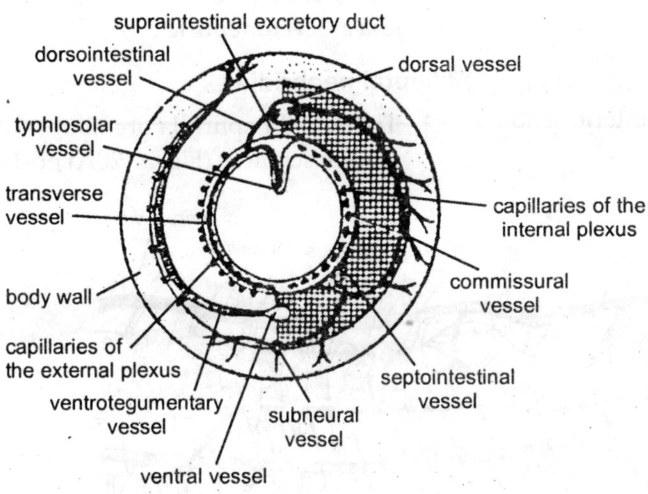

FIGURE 27.16

***Pheretima* sp.** Transverse section of the intestinal region. Section passing through the intersegmental region in the right half and through one of the dorsointestinal vessel in the left half

Ventral blood vessel

It runs along the midventral line just below the gut. It has no valves and acts as a distributing vessel by driving blood backward. It sends the following vessels—

a. A pair of **ventrointegumentary** in each segment, which piercing the septum posteriorly, send blood to the body wall, septal and integumentary nephridia.

b. A median **ventrointestinal**, to the ventral wall of the intestine in each segment.

Subneural blood vessel

It runs along the midventral line below the ventral nerve cord. It extends from the posterior end of the body up to the 14th segment. It is a collecting vessel and receives blood from—

a. A pair of small vessels in each segment bringing blood from the ventral region of the body wall.
b. The subneural vessel is connected with the dorsal vessel through a pair of commissurals in each septum.

⊠ Blood vessels in the anterior region (Figs. 27.17, 27.18)

The dorsal vessel runs in the usual position but acts as a distributing vessel, supplying blood to the pharynx and gizzard by small vessels and sending rest of the blood to the ventral vessel through the hearts.

The ventral vessel distributes blood in the usual fashion by a pair of ventrointegumentary vessels in each segment.

Hearts. In the 7, 9, 12 and 13th segments, the dorsal vessel communicates with the ventral vessel through four pairs of pulsatile loops known as hearts. The first two pairs are called **lateral hearts** and the last two pairs **laterooesophageal hearts** (Fig. 27.19).

The subneural vessel bends upward, bifurcates to two **laterooesophageals** and run by the sides of the gut up to the anterior end and act as collecting vessels.

In the region of 9 to 13th segments a short longitudinal vessel, the **supraoesophageal* vessel** is present dorsal to the oesophagus. The laterooesophageals are connected with the supraoesophageal by two pairs of loops in the 10 and 11 segments. These are called **anterior loops**. Blood from laterooesophageals pass to the supraoesophageal through these anterior loops.

In the region of 12 and 13th segments the supraoesophageal is connected with the posterior two pairs of hearts through small vessels. The supraoesophageal and the dorsal vessel together communicate with the ventral vessel through the posterior two pairs of hearts and blood from the former passes to the latter through them.

⊠ Excretory System

Excretion or 'removal of nitrogenous wastes is carried by nephridia. Nephridia are of 3 types. The septal and pharyngeal nephridia are '**enteronephric**', but the integumentary nephridia are '**ectonephric**'.

⊠ Septal nephridia

1. The septal nephridia are about 80-100 in each segment (Fig. 27.20A) posterior

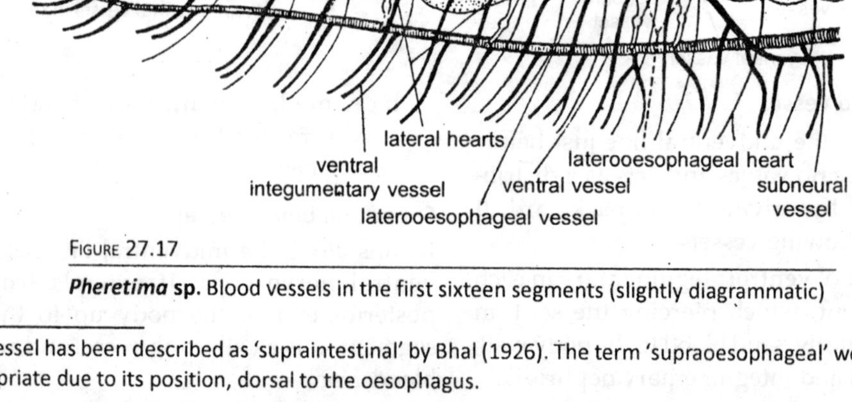

FIGURE 27.17

***Pheretima* sp.** Blood vessels in the first sixteen segments (slightly diagrammatic)

* This vessel has been described as 'supraintestinal' by Bhal (1926). The term 'supraoesophageal' would be more appropriate due to its position, dorsal to the oesophagus.

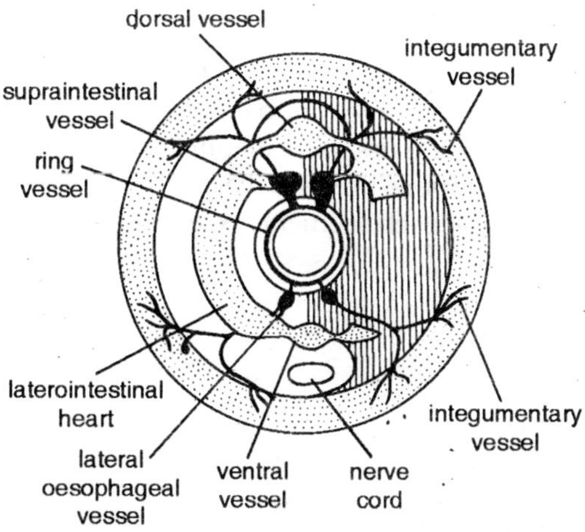

FIGURE 27.18

***Pheretima* sp.** Transverse section of the laterooesophageal heart region. The intersegmental septum behind the heart is shown on the right side.

to the 15. The body of the nephridium consists of a short **straight lobe** and a long, **spirally twisted loop** (Fig. 27.20B). The twisted loop is more than double the length of the straight lobe and is made up of two limbs; a **proximal** and a **distal**, the two spirally twisting each other.

The straight lobe continues into distal limb of the twisted loop at the base of the nephridium. The proximal limb of the twisted loop receives the ciliated ductule from the **funnel** and also gives out of the **terminal nephridial duct**.

2. The body consists of a much coiled ciliated tubule richly supplied with blood.
3. The funnel bears a ciliated slit-like opening, the **nephrostome**, by means of which the nephridium is communicated with the coelom. The funnel is lodged in the same segment in which the body of the nephridium lies. A narrow **ciliated ductule** connects the funnel with the body of the nephridium.
4. The **terminal duct** is a narrow continuation of the main body and the ducts of all the nephridia on half of each face of a septum unite to form a transverse duct, the **septal excretory canal**, which opens into the **supraintestinal excretory duct** of the side. A pair of supraintestinal excretory ducts are present on the dorsal surface of the gut and these ducts open in the gut by short vertical ducts near each intersegmental partition (Fig. 27.20A).

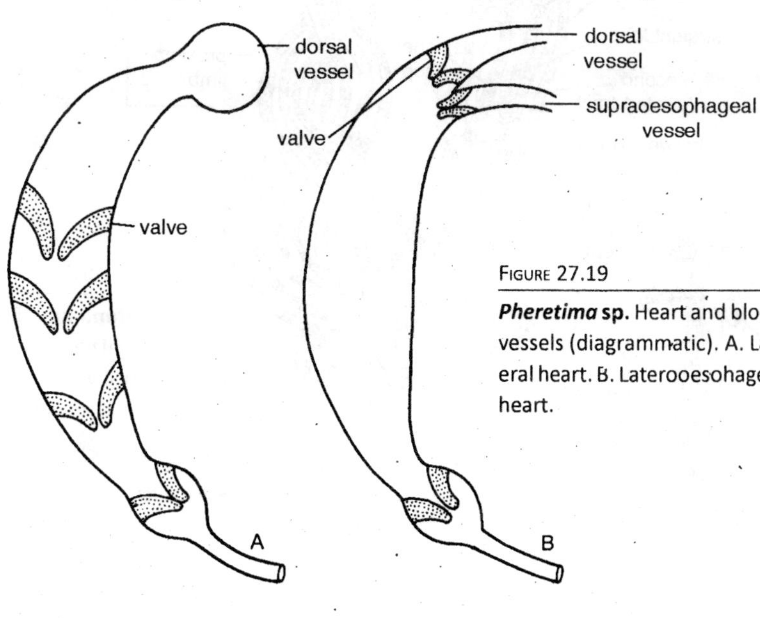

FIGURE 27.19

***Pheretima* sp.** Heart and blood vessels (diagrammatic). A. Lateral heart. B. Laterooesohageal heart.

5. Wastes, collected by the nephridia are discharged into the lumen of the intestine and pass to the exterior with the excreta. As the nephridia open into the gut, they are called **enteronephric** nephridia.

☒ Integumentary nephridia .

Smallest nephridia attached to the body wall and occur in large numbers in each segment, excepting the first two. Nephrostome absent, nephridiopore present. They are 200-500 in number in each segment, excepting the 14, 15

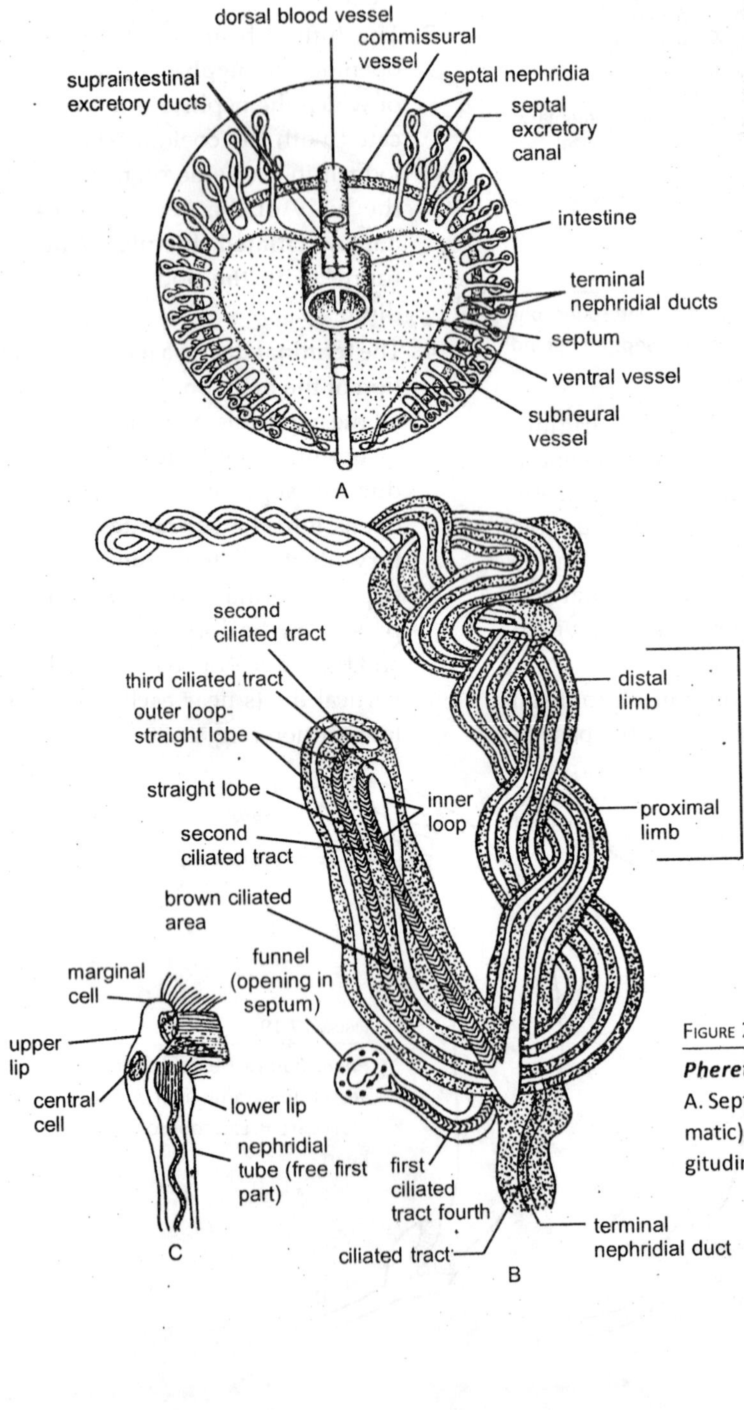

FIGURE 27.20

***Pheretima* sp.** Excretory system. A. Septal nephridial system (diagrammatic). B. A septal nephridium. C. Longitudinal section of a funel.

and 16th, where they are numerous. All of them open directly to the exterior and termed **ectonephric** nephridia.

⊠ Pharyngeal nephridia

Occur in large numbers in 4 to 6 segments as paired tufts associated with blood glands. Nephrostome absent. Terminal ducts unite to form a common duct and a pair of such ducts are present in each segment. These three pairs of ducts run forward in the fourth, fifth and sixth segments. The ducts of those in the 4 and 5 segments open in the lumen of the pharynx but of those in the 6 segment open in the buccal cavity.

⊠ Nervous System

Nervous sysem is well-developed and includes central, peripheral and visceral nerves (Fig. 27.21).

⊠ Central Nervous System

Cerebral ganglia or brain, peripharyngeal connectives and the ventral nerve cord constitute the central nervous system.

1. **Nerve ring.** Situated at the anterior end of the body and vertical in orientation. The pharynx passes through it.
   a. **Supraoesophageal** or **cerebral ganglia.** Two in number, fused to form the dorsal part of the ring or so-called brain.
   b. **Peripharyngeal connectives.** Two, constitute the lateral sides of the ring around pharynx and connect the brain with the subpharyngeal ganglia.
   c. **Subpharyngeal ganglia.** Two, fused to form the ventral portion of the nerve ring.
2. **Ventral nerve cord.** It is solid, consists of two nerves, right and left, enclosed in a sheath. It runs backward from subpharyngeal ganglia to the posterior end of the body, ventral to the ventral blood vessel. A ganglion in the ventral nerve cord is formed by the fusion of two ganglia.

⊠ Peripheral Nervous System

1. Nerves emanating directly from the brain and ventral nerve cord are in-

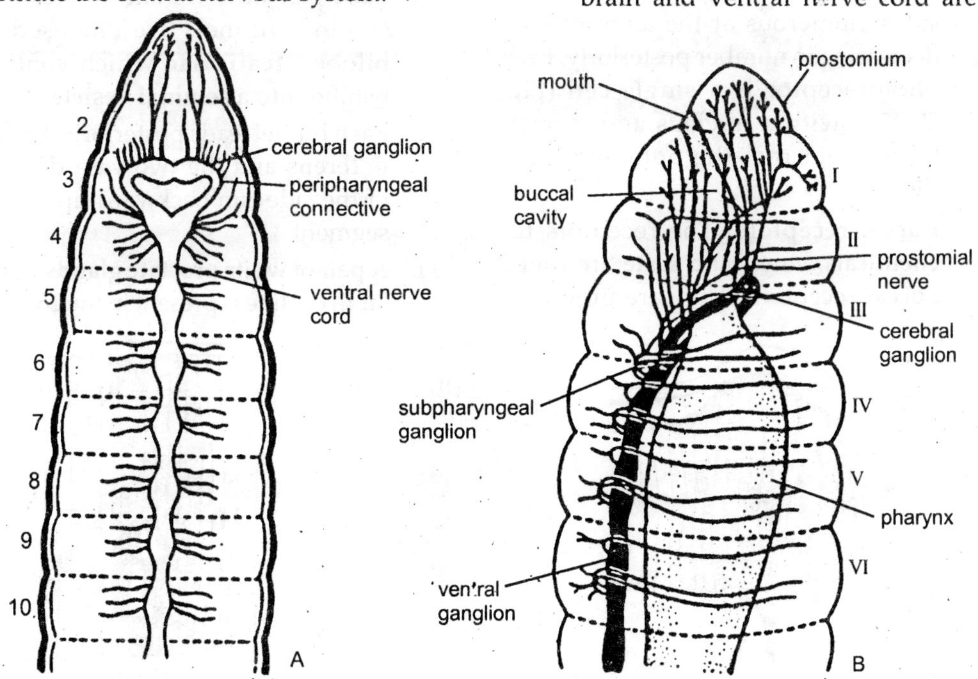

FIGURE 27.21

***Pheretima* sp.** Nervous system. Anterior end. A. Dorsal view. B. Lateral view

cluded in the peripheral nervous system.

2. Nerves from the brain innervate the prostomial region. A series of **stomatogastric** nerves from the peripharyngeal connectives pass to the pharynx and adjacent parts of the alimentary canal. The subpharyngeal ganglia send nerves to 2-4 segments.
3. Two pairs of nerves from each ganglion of the ventral nerve cord innervate the structures present in the segment.

Receptors and Sense organs

Three types of receptors have been identified in earthworm. A sensory fibre connects the base of each cells of the receptor to the central nervous system.

1. **Epidermal receptors.** Long, slender cells, terminating in small, hair-like processes; **tactile** in function, are located in groups in the epidermis (Fig. 27.22A).
2. **Photo receptors.** They are present in the epidermis of the dorsal and dorsolateral sides; numerous at the anterior end; decreasing in number posteriorly. Each photoreceptor is a single cell (Fig. 27.22C) with a nucleus and a small transparent rod, the optic organelle (**lens**).
3. **Buccal receptors.** The receptors for **chemicals, smell** and **taste** are called buccal receptors. They are groups of cells, present in large numbers in the wall of the buccal cavity. At the distal end the cells are broad and sensory hairs are better developed (Fig. 27.22B).

Reproductive System

Earthworm is **hermaphrodite** but **auto-fertilization** does not take place.

Male Reproductive System

1. The male reproductive organs consist of two pairs of testes, two pairs of vasa deferentia with seminal funnels, two pairs of seminal vesicles, a pair of prostate glands with ducts and a pair of genital apertures (Fig. 27.23A, B).
2. The **testes** are two pairs. They are small, white bodies composed of five to eight digitate processes coming out from a narrow base and are placed close to the middle line in the 10 and 11 segments.
3. A ciliated funnel, the **seminal funnel** is present just behind each testis and two pairs of funnels are present.
4. Each pair of testes and the respective seminal funnels are enclosed in a bilobed **testis sac** which continues behind into a **seminal vesicle**.
5. Each funnel leads posteriorly into a **vas deferens** and the two vasa deferentia of the side run backward up to the 18 segment.
6. A pair of white **prostate glands**, ir-regular in outline is present in the 16 to 22nd

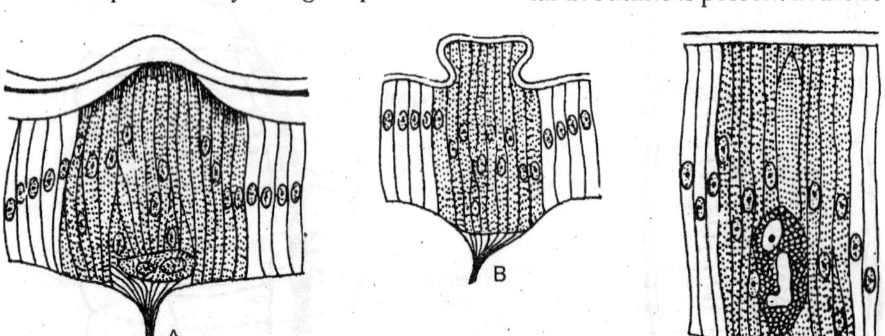

FIGURE 27.22

A-B. ***Pheretima* sp.** C. ***Lumbricus* sp.** Receptors and sense organs. A. An epidermal receptor. B. A buccal receptor. C. A photoreceptor

segmental regions. From each gland arises a duct, which is joined by the two vasa deferentia of the side and the three ducts are surrounded by a thick muscular sheath. This common structure is known as the **prostatic duct** (Fig. 27.23C). All the three ducts, however, run separately all along, and also open to the exterior separately through the **male genital aperture** situated on the ventral surface of the 18 segment.

7. **Spermatozoa** budded in the testes, mature in the seminal vesicles and pass to the exterior through the vasa deferentia.

Female Reproductive System

1. The female reproductive organs consist of a pair of ovaries, a pair of oviducts, a female gonopore and four pairs of spermathecae (Fig. 27.23A, B).
2. **Ovaries** are white bodies attached to the posterior face of the septum between the 12 and 13 segments. They resemble the testes very closely excepting the distal digitate processes, which are rather long (Fig. 27.24).
3. Just behind each ovary is a ciliated funnel, the **oviducal funnel**, which ends in the oviduct. The two oviducts open externally by a common aperture placed on the ventral surface of the 14 segment.

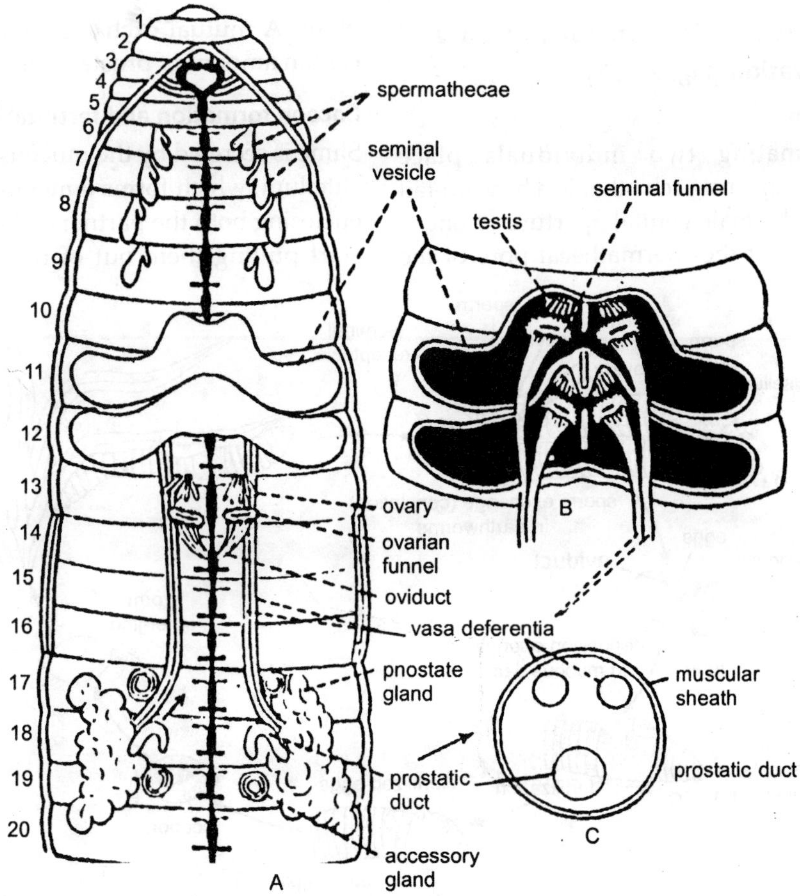

FIGURE 27.23

***Pheretima* sp.** Reproductive system. A. Entire reproductive system, B. A part of A (enlarged), the seminal vesicles cut open, C. A transverse section of a prostatic duct

4. **Ova** budded in the ovaries are shed in the coelom and being collected by the oviducal funnels pass to the exterior through the oviducts.
5. **Spermatheca** is a flask-shaped sac with an anterior blind diverticulum and opens to the exterior by a narrow duct. Sperms received from other partner during mating are stored up temporarily in these diverticula. Four pairs of spermathecae, one pair in each of the 6, 7, 8 and 9 segments are present.

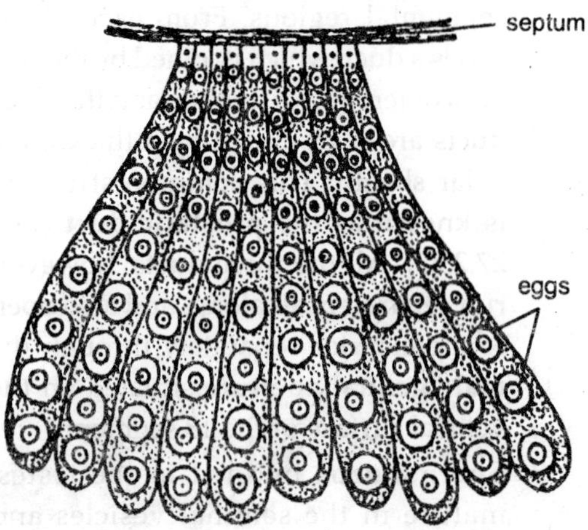

FIGURE 27.24

***Pheretima* sp.** An ovary

⊠ Reproduction

Earthworm reproduces in monsoon through cross fertilization.

⊠ Copulation, Cocoon formation and Fertilization (Fig. 27.25)

Copulation

During mating two individuals place themselves against each other in a head-to-tail position. The male genital aperture of one is placed against the spermathecal pore of the other. A mutual exchange of sperms and a certain amount of prostatic fluid takes place.

Cocoon formation and fertilization

Slime is secreted by the mucous glands in the clitellum, which forms a membranous girdle enclosing both the partners. The earthworms start pulling them out of the girdle. While

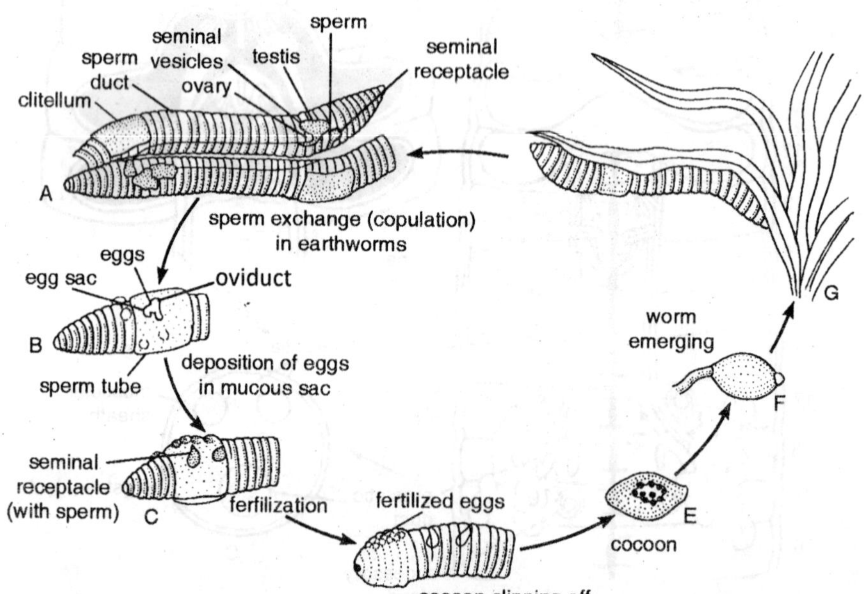

FIGURE 27.25

***Pheretima* sp.** mating and reproduction. A. Two individuals placed opposite to each other during copulation; B. Deposition of eggs in mucous sac; C. Fertilization; D. Coccon slipping off; E. Cocoon with eggs; F. Emerging worm; G. Photograph of mating

passing through the girdle, ova are discharged into it. Sperms are discharged from the spermathecae into the girdle when the anterior ends of the animals pass through it. With the moving out of the animals, the ends of the elastic girdle close and a cocoon is formed. Fertilization and development take place inside the cocoon. Development is direct.

HIRUDINARIA sp. (Leech)

Hirudinaria is placed in the order Gnathobdellida, class Hirudinea and phylum Annelida. It is commonly found in swamps and ponds. It lives on blood of other animals.

☒ External features

1. The body of the leech (*Hirudinaria*) is elongated, dorsoventrally flat, with a convex dorsal and a flat ventral surface (Fig. 27.26).
2. The body is broader at the middle and narrower towards both the ends.
3. A full grown leech varies in length from 300 to 350 mm with an olive green dorsal and an orange red ventral surface.
4. Each lateral line of the body bears an orange or yellow stripe, and ventral to each stripe is a broad black stripe.
5. Two suckers are present, one at the anterior and the other at the posterior end of the body.
 i. The anterior or **cephalic sucker** is oval shape and placed on the ventral surface at the anterior end of the body. It is formed by the fusion of a few anterior body segments and the prostomium. It bears a depression at the centre.
 ii. The posterior or **anal sucker** is circular in outline, large in size and is a highly muscular disc at the posterior end of the body.
6. The body is divided into 33 **segments**, the last 7 fused to form the posterior sucker.

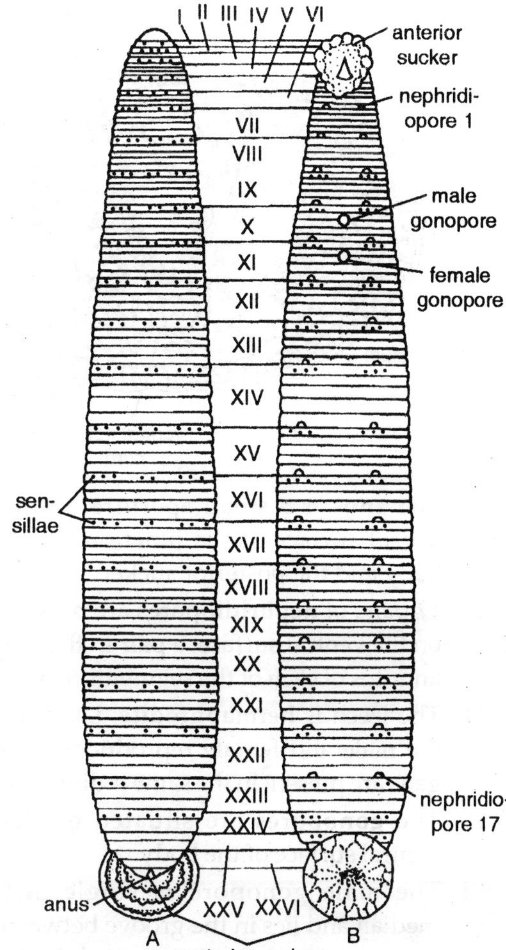

FIGURE 27.26

Leech (*Hirudinaria medicinalis*). A. Dorsal view, B. Ventral view

 a. Each segment is subdivided into several, usually 5 rings or **annuli** by a series of close set transverse grooves.
 b. The number of segments is determined by counting the seg-mental receptor organs or **sensillae**. These are dotlike elevations on the skin of both the dorsal and ventral surfaces and restricted to the first annulus of **somite**.
7. 5 pairs of sensillae (eyes), appearing as black dots on the dorsal surface of 2 to 6 segments, are called **eyes**.
8. The **mouth** is a narrow triradiate aperture at the centre of the depression of the anterior sucker.

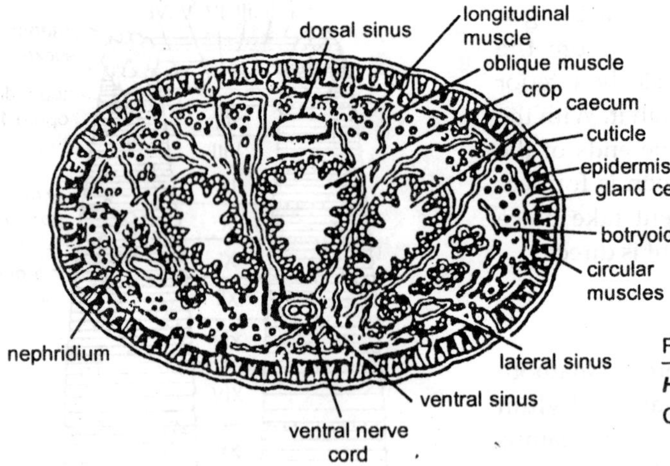

FIGURE 27.27
***Hirudinaria* sp.** Transverse section. Crop region

9. The **anus** is a small aperture on the dorsal surface at the junction of the body and the posterior sucker.
10. 17 pairs of **nephridiopores** are present on the ventral surface, a pair in the last annulus of each of the segments 6 to 22.
11. The leech is **hermaphrodite**, i.e., both the male and female reproductive organs are present in the same individual.
12. The **gonopores** are situated on the ventral surface of the body.
   i. The male gonopore is single and median and lies in the groove between the second and third annuli of the 10 segment.
   ii. The female gonopore is also single and median. It lies in the groove between the second and third annuli of the 11 segment.

⊠ Transverse section

1. The outermost layer is a thin, porous **cuticle** secreted by the underlying epidermis (Figs, 27.27, 27.28).
2. The **epidermis** is a single layer of cells which are wider at the outer and narrower at the inner curved ends. Some are modified as **unicellular glands** and other acts as **sense organs**. It is not a continuous layer but broken at intervals.

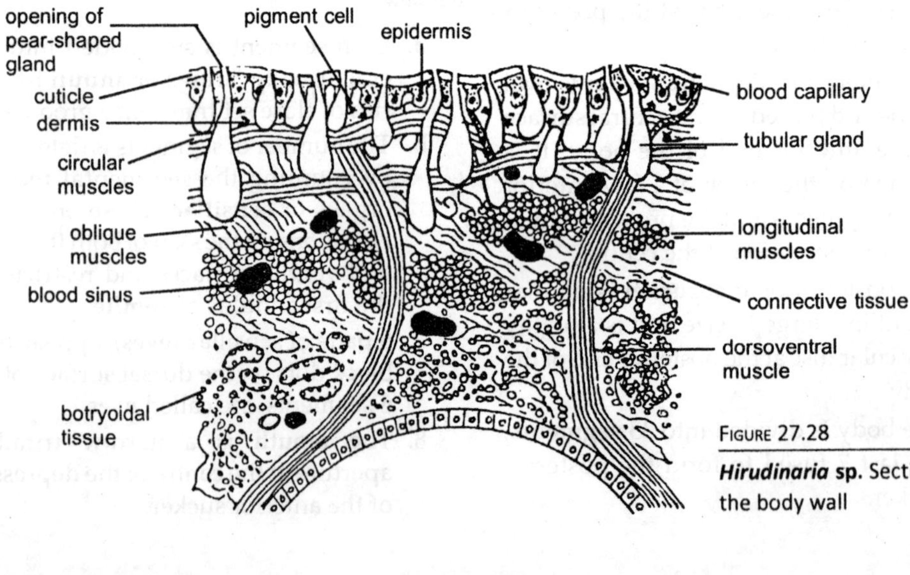

FIGURE 27.28
***Hirudinaria* sp.** Section through the body wall

3. Below the epidermis, there is a matrix of fibrous connective jelly containing connective tissue cells, pigment cells, fat cells, short muscle fibres and portions of various glands. This layer is often described as **dermis**.
4. Two layers of muscles are present. The outer **circular muscular layer** is a continuous one. The inner **longitudinal muscular layer** is also continuous but not uniformly thick and consists of several strata.
5. Certain muscles run across the body wall obliquely and dorsoventral muscles run from dorsal to ventral surfaces.
6. **Parietal** layer absent.
7. **Coelom** absent. The space is filled up by spongy connective tissue, embedded in which are two lateral blood vessels, one dorsal sinus and one ventral sinus, enclosing the nerve cord. **Botryoidal tissue** is present just below the longitudinal muscular layer.
8. **Parapodia** and **setae** absent.
9. The gut wall is made up of an **enteric epithelium**, consists of a single layer of cells, some of which are glandular and others are absorptive. The epithelium is thrown into folds.

☒ Locomotion

Change of place is effected by swimming and looping movements.

1. **Swimming.** The dorsoventrally flattened body moves in water by lateral undulations. The movement is produced by the contraction of the dorsoventral muscles and relaxation of the longitudinal muscles and this is repeated.
2. **Looping movement.** Leeches move on a substratum by looping movement. Both the anterior and posterior suckers, circular and longitudinal muscles help in it. It has two stages (Fig. 27.29).
   i. With the attachment of the posterior sucker to a surface, a simultaneous contraction of circular and relaxation of longitudinal muscles occur, resulting in lengthening of the body.
   ii. The anterior sucker is fixed on the substratum at a forward position, the posterior sucker released; longitudinal muscles contract, circular muscles relax; the posterior part is drawn forward and the body assumes the form of an inverted "U". The posterior sucker is fixed behind the anterior sucker, and the process is repeated.

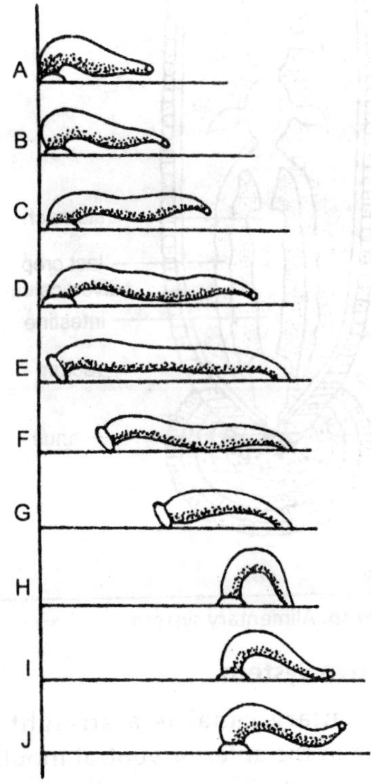

FIGURE 27.29

***Hirudinaria* sp.** Locomotion. A. Posterior sucker fixed to the substratum and the anterior portion extending forward. B-D. Extending the body further forward. E. Anterior sucker fixed to the substratum, posterior sucker released. F-H. The posterior part being drawn forward and the posterior sucker fixed. I-J. Repeats the process for progress.

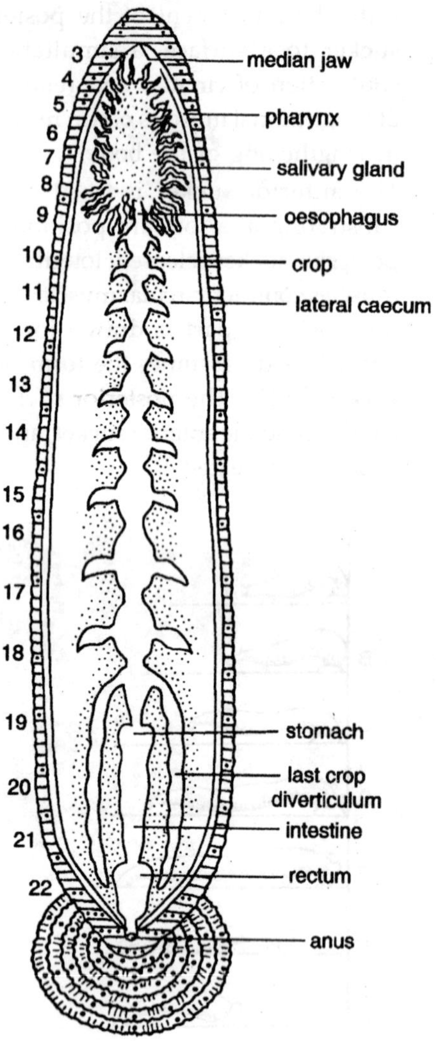

FIGURE 27.30

***Hirudinaria* sp.** Alimentary system

⊠ Digestive System

The alimentary canal is a straight tube, beginning in the anterior ventral mouth and ending in a posterior dorsomedian anus (Fig. 27.30).

Mouth. A triradiate opening at the base of the preoral chamber in the **anterior sucker** and guarded by the **velum.**

Buccal cavity. A small chamber behind the mouth, opening in the pharynx. Three jaws—one dorsomedian and two ventrolaterals—are present on the walls of the buccal cavity.

a. A jaw (Fig. 27.30) is a compressed muscular cushion, with a sharp, evenly curved free edge, covered with chitin, produced into numerous striations or teeth. Each jaw can be moved forward and backward through a certain arc, and the three acting together produce the typical triradiate wound in the skin of the prey.
b. The jaws bear papillae on both sides, bearing the openings of the unicellular salivary glands.

Pharynx. A small, muscular chamber between the buccal cavity and the oesophagus. Numérous unicellular **salivary glands** are lodged in the space between the pharyngeal wall and body wall. Their secretion is discharged close to the mouth. **Radial muscles** run from the pharyngeal wall to the body wall.

Oesophagus. Small and joins the pharynx with the crop.

Crop. It is the largest part of the alimentary canal. The crop is thin-walled and extends from 8 to 18 segments and produced into eleven pairs of lateral, blind pouches or **caeca.** The first ten pairs are laterally directed, corresponding to each segment, while the last pair runs backwards as far as the 24 segments.

a. The lumen of the crop is divided between the pairs of pouches, corresponding to each segment by a transverse septum with an opening, controlled by **sphincter.**
b. The crop is capable of great dilatation and communicates with the stomach by a minute aperture.

Stomach. A small, thin-walled chamber in the 19 segment, broad anteriorly and narrowing posteriorly, ending in the intestine. The wall bears transverse folds.

Intestine. It is a continuation of the stomach, located in 20 to 22 segments. The intestine is narrow, thin-walled and bears longitudinal and transverse folds.

It narrows posteriorly and opens in the rectum.

Rectum. Thin-walled, extends from 22 to 26 segments and opens through the anus on the dorsal surface of the 26 segment in the last annulus.

Feeding

The leech is a sanguinivorous animal, feeding on the blood of vertebrates. A cropfull of blood provides nourishment for several months.

1. The leech attaches itself to the victim's body with the posterior sucker.
2. The cup-shaped anterior sucker is pressed against victim's skin; the jaws are protruded and the serrated margins of the jaw come in contact with the skin.
3. With the operation of the muscles, the jaws move to and fro; a triradiate wound is made in the skin and blood starts flowing out.
4. Alternate contraction and expansion of pharynx due to the action of the radiating pharyngeal muscles, create a suction force and blood is sucked into the pharynx.
5. Secretion of the salivary glands containing **hirudin**, an anticoagulant, mixes with the blood, coagulation is prevented and the steady outflow of blood continues.
6. Blood is stored in crop and its caeca; haemolysed and transformed into a jelly-like mass.

Digestion

1. The sphincter between the last crop and the stomach controls blood flow and a small amount passes at a time from the crop to the stomach. The colour of the blood turns green from red.
2. The proteolytic enzymes secreted in the stomach digest blood.
3. Absorption takes place in the intestine and the residue is egested through the anus.

Respiratory System

No definite respiratory organs are present. The moist skin, due to aquatic life, is richly supplied with haemolymph and functions as the respiratory membrane.

Blood Vascular System

In leech, the blood containing spaces are of two types—**blood vessels** proper, with muscular wall and **blood sinuses**, without a definite wall and valves.

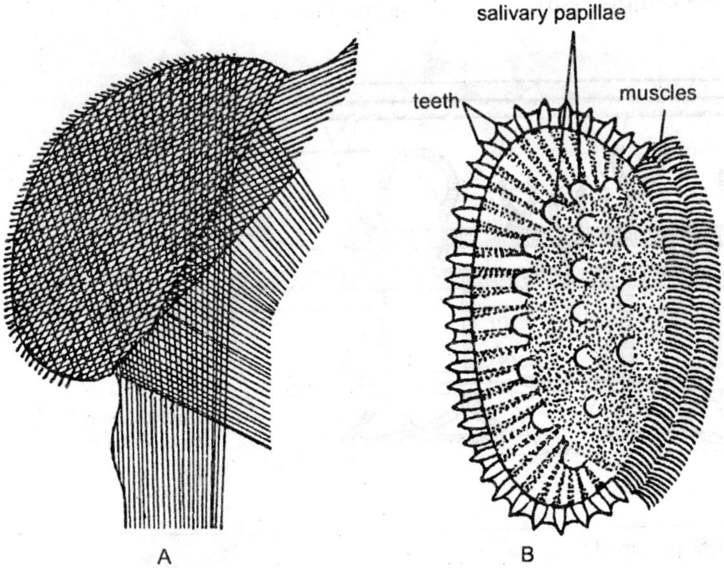

FIGURE 27.31

***Hirudinaria* sp.** Jaw. A. Frontal view, B. Lateral view

Two principal vessels, lateral in position and two principal sinuses—one dorsal and the other ventral, with branches and a number of spaces constitute the blood vascular system (Fig. 27.32). All the four principal channels are in communication at the posterior end of the body (Fig. 27.33). The longitudinal channels run from 6 to 22 segment.

☒ Dorsal sinus

1. Middorsal in position, runs beneath the body wall and dorsal to the alimentary canal; gives out two pairs of **dorsal channels** in each segment, ending in the dorsal and dorsolateral body wall and in the alimentary canal.
2. In the 6 segment, the dorsal sinus terminates in capillaries, forming **plexus** in the anterior five segments.
3. Posteriorly, the dorsal sinus bifurcates in the 22 segment. Running round the rectum, the branches join the posterior dilatation of the ventral sinus.
4. Haemolymph flows from posterior to anterior end.

☒ Ventral sinus

1. It encloses the nerve cord and runs ventral to the alimentary canal. It is largest in diameter and encloses cerebral ganglia, peripharyngeal connectives and subpharyngeal ganglia anteriorly and terminal ganglia posteriorly.
2. At the level of nerve ganglia in each segment arise a pair of branches, each dividing into two, the ventral branch ending in the ventrolateral wall and the anterodorsal branch in the dorsolateral body wall.
3. In segments between 6 to 22, ventral sinus gives out a pair of nephridial branches in each segment.
4. Haemolymph flows from anterior to posterior end.

☒ Lateral vessels

One on each side of the alimentary canal.

1. In the 6 to 22 segments, a **lateroventral** branch arises from each lateral vessel in each segment. The branch divides into two—anterior and posterior, and join with the corresponding branch from other lateral vessel, in the midventral line, below ventral sinus and form **ventral commissure** of the laterals. Three intersegmental and longitudinal commissural branches connect the commissures of the successive segments.

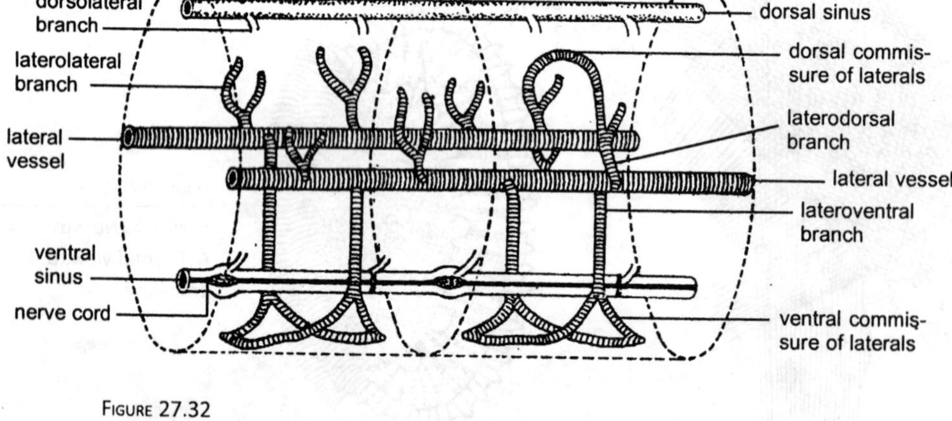

FIGURE 27.32

***Hirudinaria* sp.** Principal blood channels

2. Valves are absent at the origin of the lateroventrals but present at the origin of its branches.
3. Each lateral vessel receives slender vessels from each segment:
   a. A **laterolateral** from the lateral body wall and nephridium.
   b. A **laterodorsal** from the lateral and dorsal body wall, gut and nephri-dium.
4. The laterodorsals of both sides meet above the dorsal sinus and form a **dorsal commissure** of the lateral vessels in the segments 6 to 22.
5. The laterodorsals and laterolaterals are joined by a **lateral commissure** in each segment.
6. The laterodorsals, laterolaterals are collecting vessels and their junctions with the lateral vessels are guarded by valves. The haemolymph flow is from posterior to anterior.

Anteriorly, the lateral vessels branch into capillaries in the five first segments. Posteriorly, they end in a dilatation of the ventral sinus and a direct communication of all the four channels established.

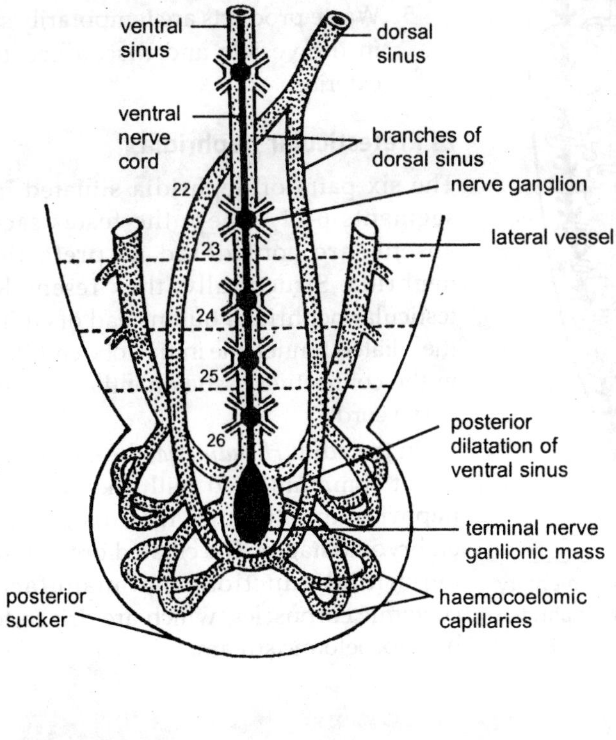

FIGURE 27.33

***Hirudinaria* sp.** Posterior end—union of four longitudinal sinuses

☒ Course of circulation

Presence of haemoglobin in the fluid gives red colour to the haemolymph. Colourless corpuscles are less in number.

1. The dorsal and ventral sinuses are distributory, the lateral vessels are both collecting and distributory in function.
2. The dorsal, dorsolateral body wall and the gut receive haemolymph from the dorsal sinus.
3. Haemolymph from the dorsal and dorsolateral body wall passes to the lateral vessels through dorsolaterals and from the gut through the laterointestinals via dorsal commissures of the laterals.
4. The ventral, ventrolateral body wall and the nephridia receive supplies from ventral sinus and the haemolymph from there goes to lateral vessels through laterolateral and laterodorsal branches.
5. In each segment, the ventral body wall, gut, nephridia and genital organs receive supply from two lateral vessels through lateroventrals and the haemolymph returns to lateral vessels

through laterodorsals and laterolaterals.

Excretory System

The nephridia are one type, seventeen pairs, and one pair occur in each segment from 6 to 22 segments (Fig. 27.34A). They are, however, absent in the first five and last four segments of the body and also in the segments forming the posterior sucker.

The nephridia may be placed under two categories.

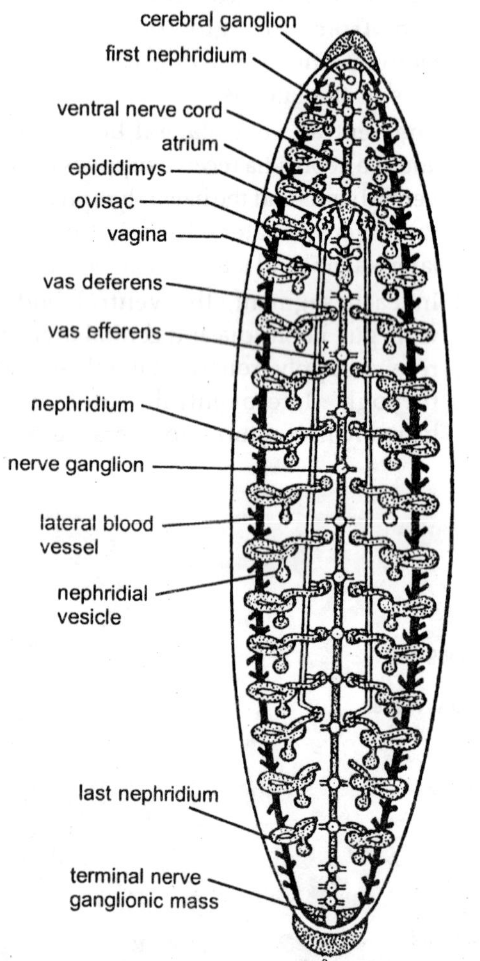

FIGURE 27.34

***Hirudinaria* sp.** A. Dorsal body wall and alimentary canal removed to expose lateral sinuses, nephridia, reproductive and nervous system.

Testicular nephridia

Eleven pairs, situated in 12 to 22 segments.

1. Body of the nephridim is attached to the ventral body wall and consists of a horseshoe-shaped **main lobe** and an **inner lobe** in the concavity of the main lobe (Fig. 27.34B).
2. The body consists of a much coiled ciliated tubule richly supplied with blood.
3. The **nephrostome** is absent. The posterior limb of the main lobe runs forward as a stout **apical lobe**. The inner lobe runs up to the half of the apical lobe. The **initial lobe** is a cord of cells and coiling the apical lobe joins the main lobe posteriorly. Anteriorly, it runs towards the testis sac and ends in the **ciliated organ** (Fig. 27.34C) corresponding to the nephrostome.
4. The anterior limb of the main lobe turns backward and ends in a narrow **vesicle duct**. The duct opens in a vesicle, which in turn opens to the exterior by the **nephridiopore** placed on the ventral surface.
5. Waste products are temporarily stored in the vesicle and discharged to the exterior.

Pretesticular nephridia

The six pairs of nephridia situated in the segments 6-11, where the testes sacs are absent, are considered as pretesticular nephrida. Structurally, they resemble the testicular nephridia, but instead of ending in the ciliated funnel, the initial lobe ends loosely in the connective tissue on either side of the nerve cord.

In the adult *Hirudinaria*, the ciliated organ is not communicated with the body of the nephridium, although it is distinct in the embryonic stages. The ciliated organ loses its nephridial function and manufactures coelomic corpuscles, which are driven to the haemocoelomic stream.

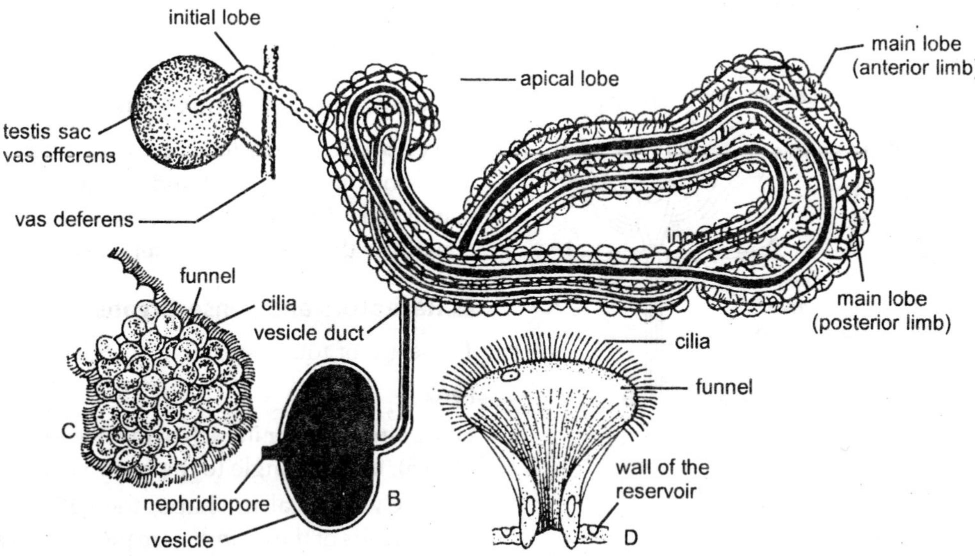

FIGURE 27.34 (CONTD.)

B. A testicular nephridium. C. A ciliated organ. D. A ciliated funnel

Some workers are inclined to believe that the botryoidal tissue has got certain excretory function.

⊠ Nervous System

The well-developed nervous system has three components—the central, the peripharyngeal and the sympathetic nervous system (Fig. 27.35).

⊠ Central Nervous System

The cerebral ganglia or the brain, peripharyngeal connectives, subpharyngeal ganglia and ventral nerve cord constitute central nervous system.

1. **Nerve ring.** Situated in the fifth segment, the nerve ring is small, somewhat elongated and oriented vertically. The pharynx passes through it.

a. **Cerebral ganglia or brain.** The two ganglia are fused to form the dorsal part of the ring or the **brain**.

b. **Peripharyngeal connectives.** Two, constitutes the lateral sides of the ring around the pharynx and connect the brain with the subpharyngeal ganglia.

c. **Subpharyngeal ganglia.** Large, triangular in shape, formed by the fusion of four pairs of embryonic ganglia and form the ventral part of the nerve ring.

2. **Ventral nerve cord.** It is solid, consists of two nerves, right and left, enclosed in a sheath. It runs backward from the subpharyngeal ganglia ventral to the gut, being enclosed in the ventral sinus. A ganglion in the ventral nerve cord is formed by the fusion of two ganglia and twenty-one such ganglia are situated, one in each, from 6 to 26 segmens, in the first, annulus. The terminal ganglionic mass is large and formed by the fusion of seven pairs of embryonic ganglia.

⊠ Peripheral Nervous System

Nerves emanating directly from the brain and ventral nerve cord are included in the peripheral nervous system.

1. Nerves from the brain innervate the first pair at eyes, prostomium and walls of the buccal chamber.

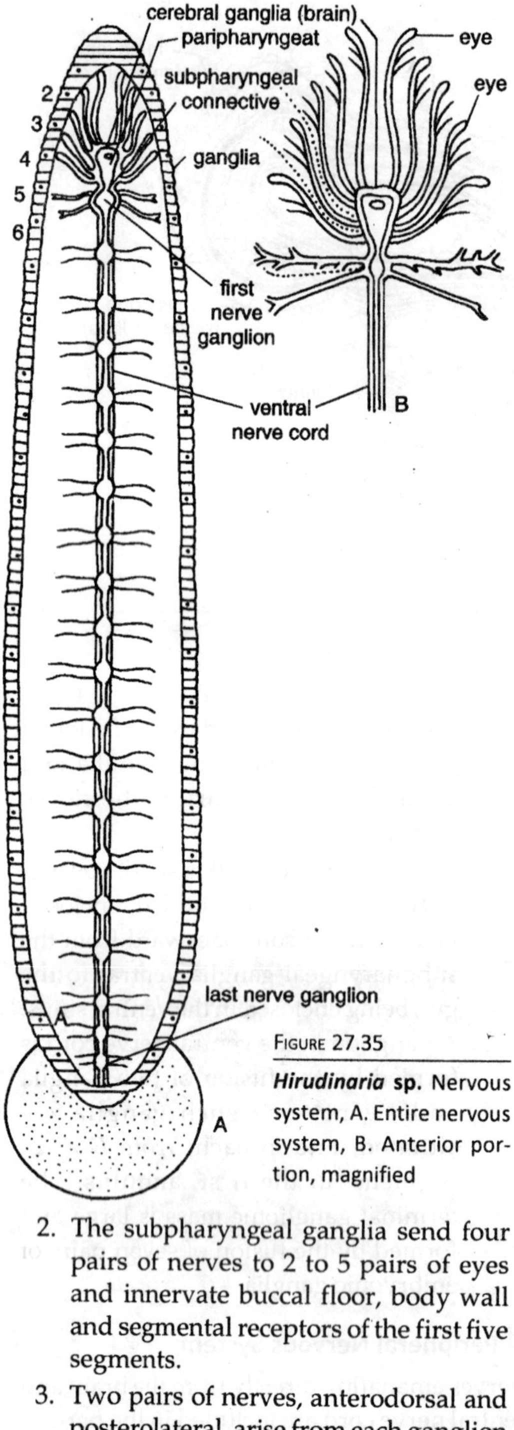

FIGURE 27.35

***Hirudinaria* sp.** Nervous system, A. Entire nervous system, B. Anterior portion, magnified

2. The subpharyngeal ganglia send four pairs of nerves to 2 to 5 pairs of eyes and innervate buccal floor, body wall and segmental receptors of the first five segments.
3. Two pairs of nerves, anterodorsal and posterolateral, arise from each ganglion of the ventral nerve cord.
   a. The anterolaterals supply the body wall, nephridia, reproductive organs and receptors.
   b. The posterodorsals innervate body wall, viscera, muscles and testis sacs.

☒ Sympathetic Nervous System

Nerve plexus below the epidermis, in muscle layers of the body wall and in the gut wall, connected with the peripharyngeal nerve ring, constitute the sympathetic nervous system.

☒ Receptors and Sense Organs

These are of four types (Fig. 27.36).

Eye

Five pairs, one pair in each of the five segments (2 to 5), at right angle to the body surface.

a. Cylindrical in shape, the cells of the walls of the cylinder are pigmented and the structure is called **pigment cup** (Fig. 27.36A).
b. A convex, transparent cuticle and epidermal layer form the broad convex surface of the eye.
c. Large, clear, refractile cells bearing a crescentic hyaline substance, the **lens** or optic organelle at the centre, arranged in longitudinal rows, are present in the cylinder. The cells are **photoreceptors** and possibly can distinguish between light and darkness.
d. An optic nerve enters the cylinder near its base and runs along the middle line.

Segmental receptor

a. The receptors are small, white areas, segmentally arranged on elevated, elliptical papillae and located in the first annulus of each segment (Fig. 27.36B).
b. Each segment bears four pairs of the receptors on the dorsal and three pairs on the ventral surface.
c. Two types of cells, **tactile or tangoreceptor** and **photoreceptor**, are present in a receptor.
d. The tactile cells are slender, with hair-like processes at the free ends. The optic organelles or the lens is present in the photoreceptor.

Annular receptor

These are thirty six in number, eighteen on the dorsal and eighteen on the ventral surface, arranged in a line across the middle annulus of each segment (Fig. 27.36C).

a. A receptor is formed of a group of flattened, overlapping cells, projecting beyond the integument.
b. The receptors are **tactile** in function.

Free nerve endings

The free nerve endings are present all over the body between epidermal cells. The ganglion cells of the nerve endings occur beneath the epidermis. The receptors help to detect **chemical changes** in water.

☒ Reproductive System

The leech is hermaphrodite but autofertilization does not take place.

Male Reproductive System

1. The male reproductive organs consist of testes enclosed in testes sacs, the vasa efferentia, vasa deferentia, the epididymis, the ejaculatory ducts, the atrium and the male genital pore (Fig. 27.37).
2. The **testes sacs** are 11 pairs, one pair in each segment, on either side of the nerve cord in the segments from 12th to 22nd.
3. A short duct, the **vas efferens** arises from the posterior border of each testis sac and runs outward to end in the vas deferens. 11 vasa efferentia from 11 testes sacs of the side end in the **longitudinal vas deferens** of the side. The two vasa deferentia run anteriorly from the 22 segment, and each forms a much convoluted **epididymis** in the 10 segment. A narrow muscular **ejaculatory** duct arises from each epididymis and leads to a median sac, the **atrium**.
4. The atrium is an oval sac consists of a large anteriorly placed base and a posteriorly directed neck. The base bears a thick wall having a large number of

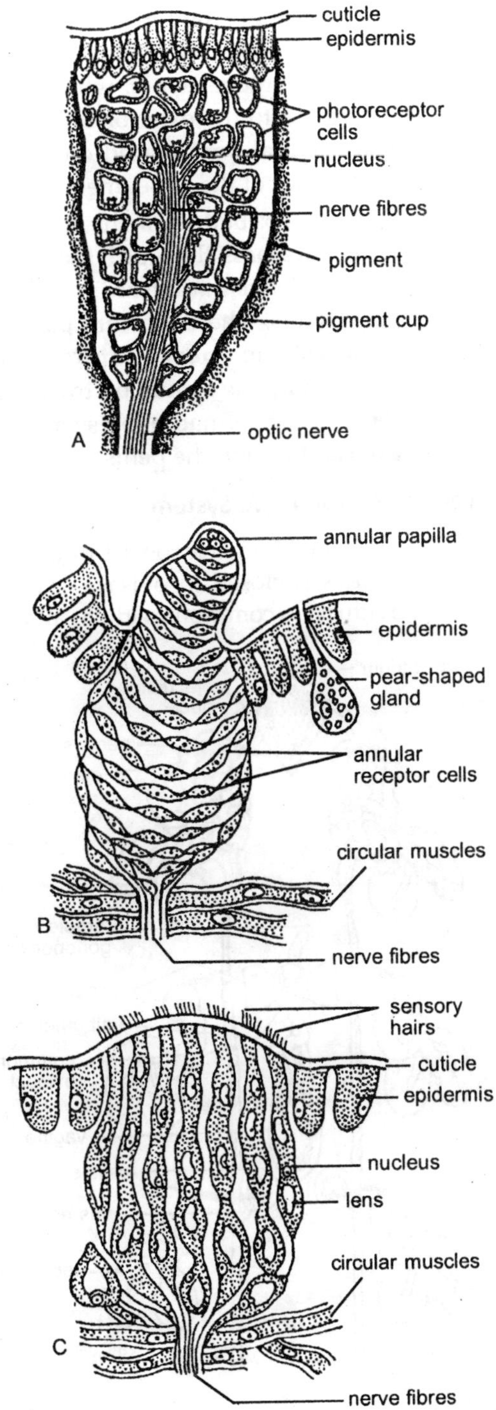

FIGURE 27.36

***Hirudinaria* sp.** Receptors and sense organs (vertical section). A. Eye. B. Dorsal segmental receptor. C. Annular receptor

unicellular **prostate glands**. The neck contains a **penis sac** bearing a long, hollow filamentous **penis**, which can be protruded to the exterior through the male **genital pore** placed on the ventral surface of the 10 segment.

5. Spermatozoa budded in the walls of the testes sacs pass through vasa efferentia, vasa deferentia and are stored in the epididymis, where they are packed in bundles, the **spermatophores** and thrown into the atrium by the ejaculatory ducts. Spermatophores pass to the exterior through the penis.

Female Reproductive System

1. The female organs consist of a pair of ovaries enclosed in ovisacs, a pair of oviducts, a common oviduct, albumen glands and a female gonopore (Fig. 27.37).
2. The **ovisacs** are white bodies lying in the 11 segment by the sides of the ventral nerve cord. Each ovary is a delicate filamentous cord made up of club-shaped cells.
3. A short duct, the **oviduct** arises from the base of each ovisac. The two oviducts unite below the ventral nerve cord and form a **common oviduct**, the region of union is covered by unicellular **albumen glands**. The common duct forms a 'S'-shaped structure and passes through a mass of albumen gland and opens in a large muscular sac, the **vagina**. The vagina bears a lateral diverticulum, the **vaginal caecum** and it opens to the exterior through the gonopore placed on the ventral surface of the 11 segment.
4. The **ova**, budded in the ovaries while passing through the anterior part of common oviduct, receive albumen from the albumen glands and remain stored in the vagina where they are fertilized during mating and then escape to the exterior.

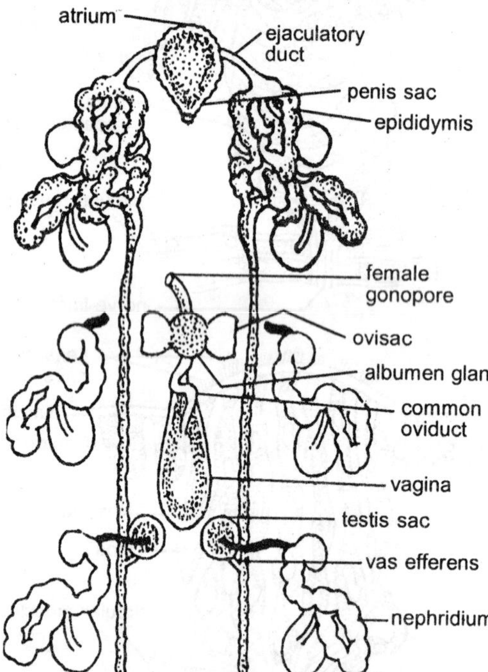

FIGURE 27.37

***Hirudinaria* sp.** Reproductive system, pretesticular and testicular nephridia. Male reproductive system (a part). Female reproductive system (whole). Three pairs of nephridia—two pairs pretesticular and one pair testicular

☒ Copulation and Cocoon formation

Leeches reproduce in early spring through cross-fertilization.

Copulation

During mating two individuals come close in head-to-tail position and ventral surface of one is placed against the ventral surface of the other. The male gonopore of one is placed against the female gonopore of the partner and the penis of one enters the vagina of the other and vice versa. Mating continues for about an hour, during which mutual exchange of **spermatophores** occur. Fertilization takes place in the vagina.

Cocoon formation

During mating a snow-white foamy girdle is secreted around the clitella by the glands of the clitellar segments. While the leeches pull out of the girdle, the fertilized ova from the vagina are forced out into the girdle. Before abandoning the girdle the leeches plug its both ends with the secretion of prostomial glands and the cocoon is formed.

Development

Development is direct and completed within fifteen days, following which the plugs of the cocoon drop off and young leeches come out.

METAMERISM IN ANNELIDA

1. The body of an annelid consists of a linear series of similar ring-like parts joined together to make the entire animal. An intricate connection between the segments is present. This type of segmentation with a linear repetition of similar body parts, arranged, along the antero-posterior axis is termed **metamerism**. Metamerism fundamentally effects the mesoderm.
2. These segments or body parts are known as **metameres** or **somites**. Each metamere is built on the same fundamental plan containing a part of all the systems of the body (nephridia, blood vessels, reproductive organs, muscles and nerves) which work in unison thus acting as a single integrated functional unit.
3. The youngest metamere occurs at the posterior end of the series.

⊠ Classification of metamerism

1. **External and internal.** In annelids metameric segmentation is conspicuously visible, both externally and internally. All the body organs, such as, musculature blood vessels, excretory organs, gonads, and nerves are repeated in each segment. Even the coelom is divided into segmental compartments by the intersegmental transverse mesenteries, called **septa**. Only the digestive tract remains unaffected, but it also extends through every segment.
   Metameric segmentation is arthropodous, chiefly external, while in vertebrates internal metamerism is present in the body muscle, vertebrae, some blood vessels and nerves.
2. **Complete and incomplete.** In annelids, segmentation is complete. Segments are essentially alike or **homonomous**, each having segmental nerves, blood vessels, nephridia and coelomoducts.
   In arthropods and chordates metamerism is incomplete and confined to certain organs only. The segments are unlike or **heteronomous**.

⊠ Significance of metamerism

1. Segmentation was initiated in the musculature of an elongated swimming worm. Breaking up of the body into metameres would facilitate swimming movements.
2. The coordination of muscular action and fluid-filled coelomic compartments cause efficient swimming and creeping which is an advancement over the simple ciliary and creeping movement of the lower invertebrates.
3. Fluid-filled coelomic compartments also provide hydrostatic skeleton for burrowing.
4. The flow of coelomic fluid from one part of the body to the other cause differential turgor pressure. This type of differential turgor pressure helps the animal in precise movements.
5. The opportunity for each metamere to become specialised for particular functions lead to a rapid evolution of high grade of organization.
6. Metamerism is a feature that has contributed towards greater complexity of animal bodies up to higher vertebrates.

CLASSIFICATION: PHYLUM ANNELIDA

1. Elongated, bilaterally symmetrical worms with metamerically segmented

body; segments are often expressed externally by ring-like constrictions, called annuli or metameres.

2. The head usually consists of a preoral prostomium and a postoral peristomium bearing sense organs in many cases.
3. The alimentary canal is straight and divisible into distinct regions; an anterior mouth and a posterior anus.
4. The outermost body covering, the cuticle is thin, secreted by underlying epidermal cells and not composed of chitin.
5. The paired appendages, when present, are never jointed.
6. The locomotory organs are segmentally repeated groups of chitinous bristles, called setae or chaetae, embedded in the skin.
7. The body cavity is a true perivisceral coelom lying between the two mesoderm layers.
8. The blood vascular system closed and well-developed with definite blood vessels.
9. The excretory organs are paired tubes, arranged metamerically, and are known as segmental organs or nephridia.
10. The nervous system consists of two dorsal cerebral ganglia connected with a ganglionated double-ventral nerve cord by two circumoesophageal connectives.
11. The sexes may be united when development is direct, as in Oligochaeta and Hirudinea or separate when development is indirect as in Polychaeta, Archiannelida.
12. The gonads develop from the coelomic epithelium. The reproductive elements may pass out either through the ectodermal nephridia or through a series of paired genital ducts, the mesodermal coelomoducts.

The phylum has been divided into three classes.

Class 1. **Chaetopoda**

1. Setae develop in skin-sacs and elevated on the parapodia.
2. Segments are usually numerous, well-marked and similar throughout.
3. A large coelom, divided into chambers by septa and not continuous with blood vascular system.
4. Coelomoducts function as reproductive ducts in many cases.

The class Chaetopoda has been subdivided into two orders.

Order i. **Polychaeta**

a. Marine chaetopods with parapodia as locomotory organs.
b. Head distinct and bears a number of appendages—eyes, tentacles, etc.
c. Clitellum is never present.
d. Sexes distinct, ovaries and testes are metameric in arrangement.
e. Development indirect, larva always trochophore.

Examples: *Neanthes*, *Aphrodite*, *Terebella Sabella*, etc.

Order ii. **Oligochaeta**

a. Land and fresh water chaetopods without appendages.
b. Simple setae for locomotion are present in each segment.
c. Head not distinct, clitellum usually present.
d. Hermaphrodite, gonads of each type nerver exceeds two pairs.
e. Development direct, metamorphosis always absent.

Examples: *Pheretima*, *Lumbricus*, *Tubifex*, etc.

Class 2. **Hirudinea**

1. Free-living or parasitic, live either on land or in fresh water or in sea.

2. Body elongated and flattened, devoid of setae and marked externally by more rings than the true segments.
3. The anterior end is modified into a ventrally directed sucker and several hindermost segment fuse to form a powerful posterior sucker, directed downwards.
4. Alimentary tract straight, the mouth and anus opposite and terminal.
5. Coelom is obliterated by connective tissue and is represented only in sinuses containing haemolymph.
6. Hermaphrodite, testes numerous and usually segmentally arranged; the ovaries are a single pair. Genital ducts always present.
7. The penis and vagina are unpaired and open to the exterior by respective median aperture.
8. Development always direct.

The class Hirudinea has been subdivided into four orders.

Order i. **Acanthobdellida**

a. A short proboscis is present.
b. Anterior sucker absent.

Example: *Acanthobdella* (a parasite on fish).

Order ii. **Rhynchobdellida**

A protrusible proboscis is present.

Examples: *Piscida, Pontobdella,* etc.

Order iii. **Gnathobdellida**

Mouth with more than one toothed jaws.

Example: *Hirudinaria.*

Order iv. **Herpobdellida**

Mouth without true jaws.

Example: *Trocheta.*

Class 3. **Archiannelida**

1. Primitive, marine worms with narrow, elongated, cylindrical body.
2. True segmentation present, but the segments may not always be distinct externally.
3. Head bears a distinct prostomium with a pair of tentacles and a peristomium with a ventral mouth.
4. Alimentary tract is a straight tube and the anus is teminal.
5. Sense organs probably in the form of paired ciliated grooves.
6. Sexes united or separate.
7. Development with a metamorphosis; larva trochophore.

Examples: *Polygordius, Protodrilus,* etc.

28

PHYLUM ARTHROPODA

The phylum Arthropoda (Gr. *arthros* = joint + *podos* = foot) contains a vast assemblage of animals found in every type of habitat showing diversity in life styles. More than 8,91,000 species are known and many of them having enormous population, occur at altitudes of over 6,096 metre on mountains and to depths of more than 9,753 metre in the sea. Different species are successfully adapted in the air, on land, in soil and in fresh, brackish and salt water. Some are parasites of plants and animals both ecto and endoparasites. Many are economically important to man. Some are food for man, some are food for fishes, some are pest of crops, stored foods, household goods and clothing; some are carriers of diseases of man, domestic and other animals as well as crops.

Arthropods are bilaterally symmetrical, triploblastic coelomate, metamerically segmented animals. The body is segmented externally in varying degrees having paired jointed appendages in each segment differentiated both in form and function to serve special purposes; covered by an organic exoskeleton containing chitin; nervous system, eyes and other sense organs proportionately large and well-developed to respond quickly to stimuli.

Arthropoda and Annelida are the only invertebrate phyla with conspicuous segmentation in the body, muscles and nervous system. The arthropods differ from annelids in the absence of cilia and intersegmental septa; reduction in coelom; presence of an exoskeleton with chitin; jointed appendages; a lacunar blood vascular system and compound eyes; concentration of excretory organs and gonads, and separation of sexes. This is the only major invertebrate phylum with many members adapted for life on land apart from moist surroundings and the insects are the only invertebrates capable of flight.

MACROBRACHIUM sp. (Prawn)

☒ External features

The fresh water prawn *Macrobrachium* (former *Palaemon*) belongs to subclass Branchiopoda, class Crustacea, phylum Arthropoda. The genus is widely distributed in tropical and temperate countries (Fig. 28.1).

1. The size is variable, the average being 15 to 20 cm.
2. The body is elongated and divisible into an anterior **cephalothorax** and a posterior **abdomen**.

☒ Cephalothorax

1. The cephalothorax is formed by the fusion of 5 cephalic and 8 thoracic segments, and covered externally by a hard cephalothoracic shield, the **carapace**, anteriorly drawn into a serrated and pointed **rostrum**.
   i. The carapace hangs freely on the sides and encloses, on either side, a narrow gill chamber containing gills or **branchiae**.
   ii. The portion of the carapace covering the gills are called **branchiostegite** or gill cover.
2. A pair of stalked **compound eyes** are present near the base of the rostrum.
3. 13 pairs of paired, **biramous** appendages are present in the cephalothorax. The appendages are two pairs of **antennae**, one pair of **jaws**, two pairs of **maxillae**, three pairs of **maxillipeds** and five pairs of **peraeopods or walking legs**. The first two pairs are **chelate**.
4. The **mouth** is a slit-like aperture situated on the ventral surface of the head. It is bounded by the **labrum** in front, **labium** behind, and jaws named **mandibles** on either side.
5. An **excretory pore** opens at the base on the inner surface of each of the second antenna.
6. **The male genital apertures** are present at the base of the last pair of walking legs and the **female genital apertures** at the base of the third pair of walking legs.

☒ Abdomen

1. The abdomen consists of six distinct, movable segments.
2. Each segment is enclosed in a hard cuticle divisible into a dorsal, convex **tergum** a pair of lateral **pleurons** and a ventral **sternum**.
3. A pair of biramous swimming legs called **pleopods** or **swimmerets** are present in each segment. The last pair is known as **uropod**.
4. At the end of the abdomen a median triangular piece called **telson** is present. It does not carry appendages.
5. The **anus** opens on the ventral side near the base of the telson.

☒ Appendages

Appendages are externally projected parts of the body in the formation of which various systems of the body take part. These develop generally at right angles to the long axis of the body and serve as locomotory organs, feelers, food seizers, manipulators and sex organs. The appendages are jointed in all arthropods (Figs.

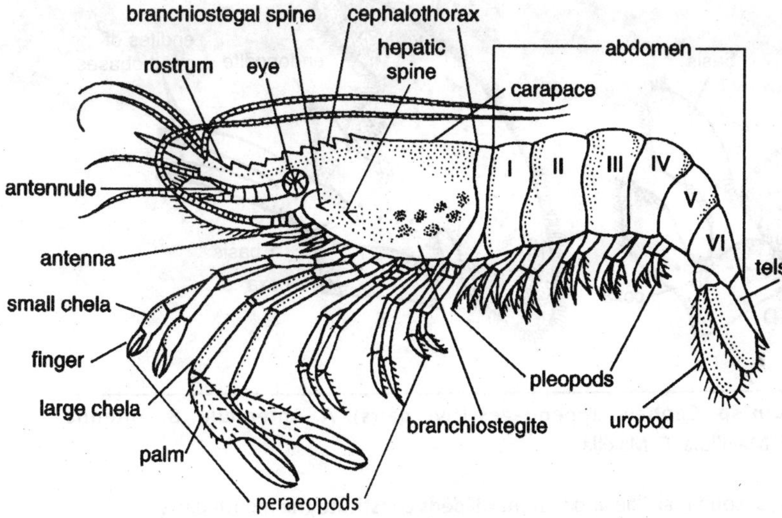

FIGURE 28.1 *Macrobrachium rosenbergii* (prawn)

28.2–28.4). Biramous appendages are nineteen pairs; five pairs in the head, eight pairs in the thorax and six pairs in the abdomen.

Cephalic appendages (Fig. 28.2)

1. **Antennules.** Each consists of a three-jointed protopodite bearing three many, jointed flagella at the distal end and a statocyst at the base.

   Function. Sensory and tactile.

2. **Antennae.** Each consists of a two-jointed protopodite bearing a flat squama and a many jointed flagellum.

   Functions. Tactile and balance.

3. **Mandibles***. The body is unjointed, bears teeth and masticatory lobes known as molar processes and a jointed mandibular palp on the outer surface.

   Functions: Cutting and crushing of the food.

4. **1st maxillae** (maxillulae). Leaf-like, with two inner lobes acting as gnathobases and an outer lobe.

   Functions: Tearing and passing the food to the mouth.

5. **2nd maxillae.** Leaf-like, with a flattened scaphognathite.

   Functions: Masticatory and respiratory.

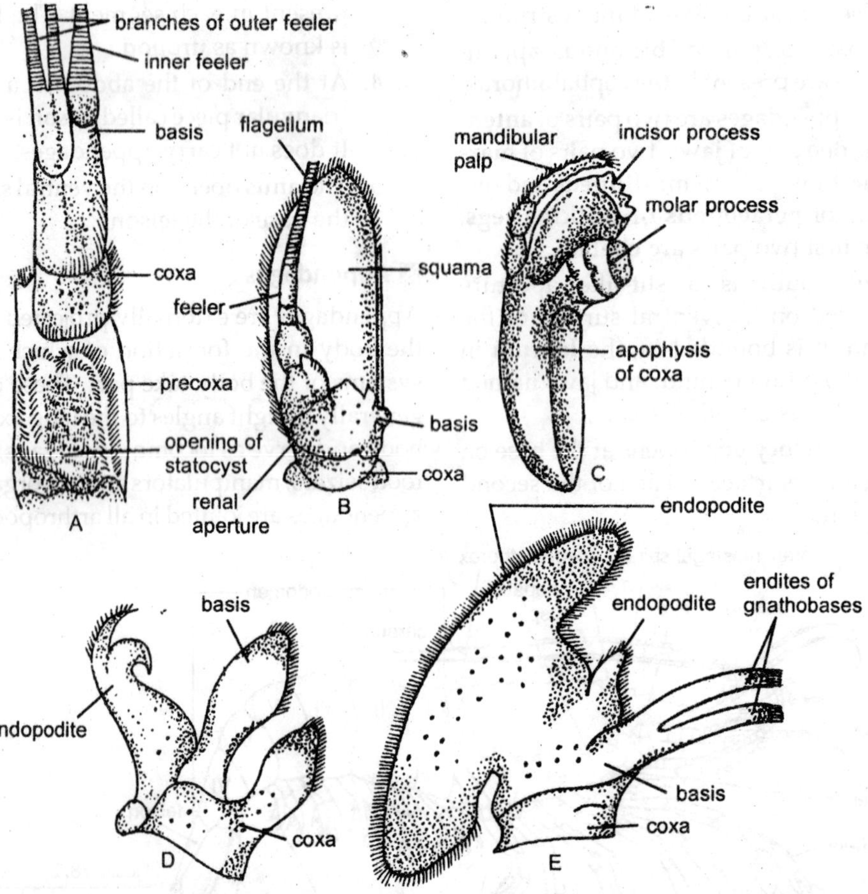

FIGURE 28.2

***Macrobrachium* sp.** Cephalic appendages (five pairs). A. Antennule. B. Antenna. C. Mandible. D. Maxillula. E. Maxilla

* Mandibles, first maxillae, second maxillae and first maxillipeds constitute the mouth parts.

☒ Thoracic appendages (Fig. 28.3)

6. **1st maxillipeds.** Leaf-like protopodite with a whip-like exopodite and a slender endopodite.

   Functions. Respiratory, masticatory and sending the food to the mouth.

7. **2nd maxillipeds.** The protopodite with an epipodite bears a five-jointed endopodite, a whip-like unjointed exopodite and a gill.

   Functions. Respiratory.

8. **3rd maxillipeds.** The protopodite bears a three-jointed leg-like endopodite and a slender unjointed exopodite. A small epipodite is present.

   Functions. Locomotory and respiratory.

9-13. **Peraeopdos** (walking legs). Five pairs, each leg consists of seven podomeres or segments.

   The first two legs end in chela and the second is the largest. The two basal segments represent the coxopodite and basipodite and the remaining five are ischium, merus, carpus, propodus and dactylus, respectively, in order of succession.

   Functions. Holding the prey and walking in first two and only walking in the last three.

☒ Abdominal appendages (14-19) (Fig. 28.4)

Six pairs; each pleopod consists of a two-jointed protopodite bearing expodite and endopodite. In the second pleopod of male, an **appendix masculina**, concerned with reproduction is found. In the second to fifth pleopods, **appendix internae** are present.

Functions. Swimming.

Appendix masculinae help in mating. Appendix internae form a basket in female to carry eggs.

Buccal cavity. A small, anteroposteriorly compressed chamber, next to mouth, bearing irregular internal folds.

Oesophagus. A wide, vertically oriented tube, joining the buccal cavity with the cardiac stomach. The inner lining bears one anterior, one posterior and two lateral folds.

☒ Stomach

A spacious, horizontally oriented sac, divided into two chambers.

A. **Cardiac stomach.** Large, bag-like, constitute the dorsal part, bearing following plates on its walls:

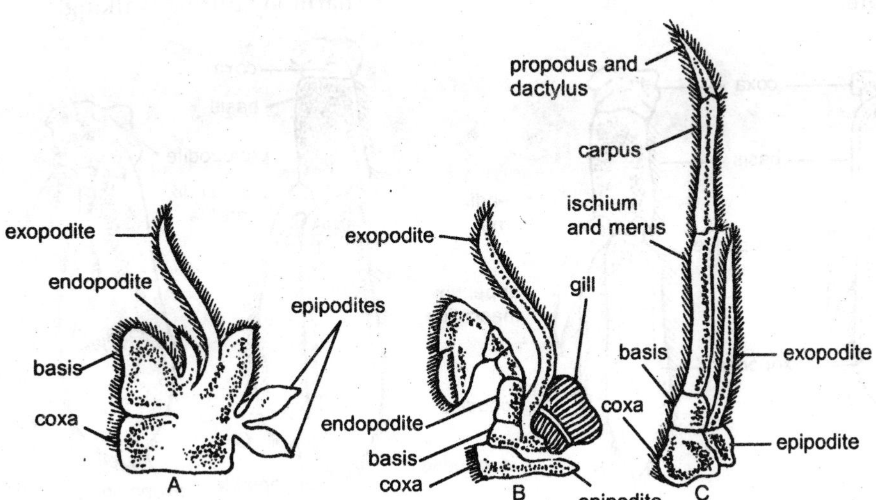

FIGURE 28.3

***Macrobrachium* sp.** Thoracic appendages: A. First maxilliped. B. Second maxillipeds. C. Third maxilliped

a. A **circular plate** anteriorly, just behind the oesophagus.
b. A **lanceolate plate** dorsally in the posterior part.
c. A **hastate plate** (Fig. 28.6) resembling the head of a spear in the midventral region.

i. The slightly convex upper part of the hastate plate gradually slopes laterally, forming a **median ridge** in the middle.
ii. Delicate **setae** are present on both the upper and posterior surfaces of the plate.
iii. An elongated **lateral groove** is present on either side of the plate.
iv. Each lateral groove is bounded by a **supporting rod** and a **ridged plate**, both cuticular, on the inner and outer side, respectively.
v. A **comb plate**; bearing rows of comblike setae is present on the inner side of each ridged plate.
vi. The comb plates join at the anterior end but remain free posteriorly, close to the cardio- pyloric opening.
vii. A longitudinal **guiding ridge** is formed by the folding of the inner wall of the cardiac stomach, lateral to each comb plate.
viii. The two guiding ridges guide the food to the pyloric stomach through cardiopyloric opening.

The cardiac stomach opens into the pyloric stomach through a narrow X-shaped **cardio-pyloric opening**, guarded by an **anterior**, one **posterior** and two **lateral valves**.

Uropod. Protopodite small, the exo- and endopodite are broad and oval. With the telson it forms the powerful tail fin.

Functions. Balance and back swimming with a jerk.

⊠ Locomotion

Change of place in prawn takes place into three ways.

1. **Swimming.** The body is kept straight in a horizontal position. The pleopods act like oars. The forward movement of the pleopods is slow, while the back stroke is fast and the animal moves forward. The uropod helps in guiding and the antennae move constantly, presumably feeling the surroundings.
2. **Walking.** The straightened body is supported by all the five pairs of pereopods. The walking legs move in harmony during walking.

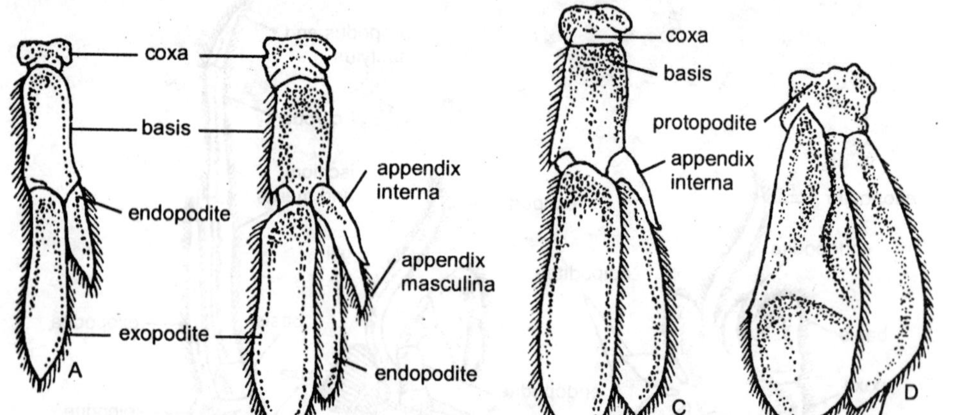

FIGURE 28.4

***Macrobrachium* sp.** Pleopods (abdominal appendages): A. First swimmerete. B. Second swimmerete. C. Third swimmerete. D. Uroped (last swimmerete)

3. **Darting.** This is resorted under emergency. With stretched pleopods and uropod the abdomen suddenly moves forward towards the cephalothorax and the animal swiftly moves backward with a jerk due to the sudden thrust.

☒ Digestive System

The digestive system consists of an alimentory canal and a hepatopancreas or digestive gland (Fig. 28.5).

☒ Alimentary canal

It has three distinct zones—an anterior **foregut** ending in **stomach**, a **midgut**, the constituent of which is **intestine** and **hindgut** or the **rectum**. The fore and hindgut are lined by a layer of thick cuticle.

Foregut

Following structure constitute the foregut.

Mouth. A slit-like opening situated ventrally in the head region. It is bounded by labrum anteriorly, mandibles laterally and a two-lobed labium posteriorly.

Pyloric stomach

a. Small in size, the lateral walls form prominent folds, imperfectly dividing the cavity into two—a small **dorsal** and a large **ventral chamber.**
b. The ventral chamber is subdivided into two lateral compartments and receive the ducts from the hepatopancreas.
c. A rectangular **filter plate** bearing alternate ridges and grooves is present on the floor of the ventral chamber.
d. The filter plate and the bristles of the lateral walls of the ventral chamber act as a **pyloric filtering apparatus** and permits only liquefied food to pass.

Midgut or Intestine

It is a long, slender tube. Arising from the posterior end of the pyloric stomach it runs backward, ascending between the two lobes of the hepatopancreas to reach the dorsal groove in the abdomen beyond cephalothorax and runs posteriorly to end in the rectum in the last segment.

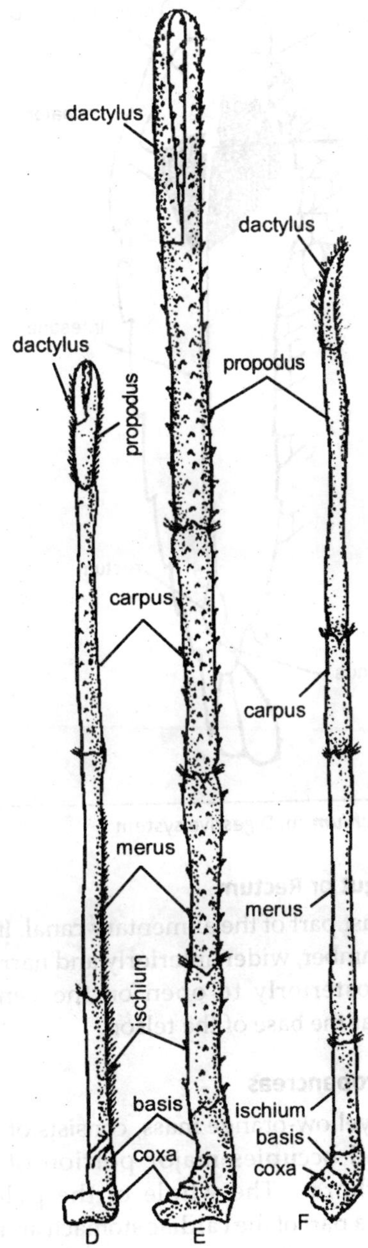

FIGURE 28.4 [Contd.]

***Macrobrachium* sp.** (contd.). Peraeopods (thoracic appendages). D. First walking leg. E. Second walking leg. F. Third walking leg (Third to fifth lack pincer)

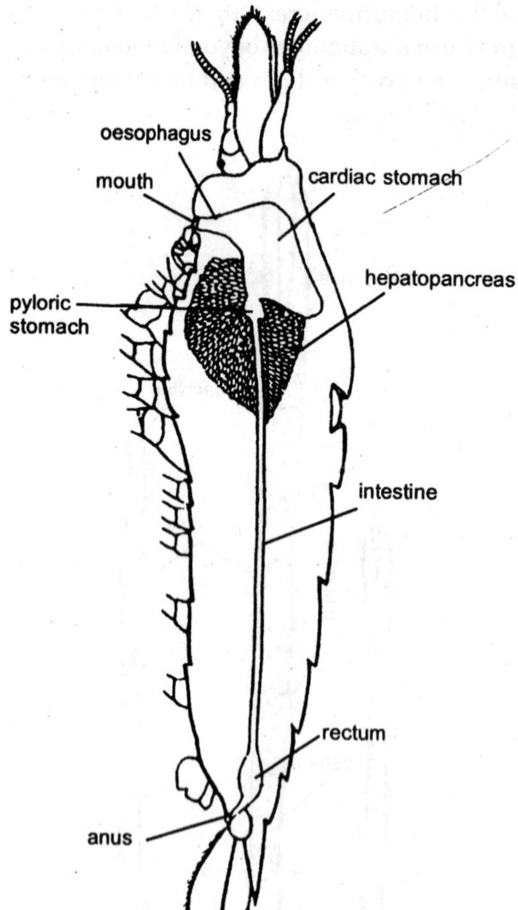

FIGURE 28.5

***Macrobrachium* sp. Digestive system**

Hindgut or Rectum

The last part of the alimentary canal. It is a small chamber, wider anteriorly and narrows down posteriorly to open on the ventral surface, at the base of the telson.

⊠ Hepatopancreas

A large, yellow-orange mass, consists of two lobes and occupies major portion of the cephalothorax. The whole of the pyloric stomach, a part of the cardiac stomach and the anterior part of the intestine are embedded in it. A duct arises from each lobe of the hepatopancreas and the two open separately into the ventral chamber of the pyloric stomach, just after the pyloric filter plate. The hepatopancreas plays an important role in digestion and also acts as a storage organ.

⊠ Feeding and Digestion

Prawn feeds upon small animals, eggs of other animals, algae and decaying plant matters.

Feeding

1. The food is captured by the chelate legs and brought to the mouth.
2. Maxillipeds, maxillulae and maxillae help in tearing it into pieces.
3. The mandibles crush the food and the crushed food is taken into the buccal cavity and from there to oesophagus.
4. Peristaltic movement of the oesophagus drives the food into the cardiac stomach.

⊠ Digestion

1. The food is churned by the cuticular plates of the cardiac stomach and the fine particles, filtered by the comb plates, reach the lateral grooves and finally guided to the ventral chamber of the pyloric stomach.
2. Enzymes secreted by the hepatopancreas digest proteins, carbohydrates and fats.
3. The filtering apparatus filters the nutrient, which is in a liquid state and passed to the hepatopancreas via the dorsal chamber of the pyloric stomach.
4. The undigested food is moved to the intestine, where certain amount of digestion and absorption take place.
5. The residue reaches the rectum and egested through the anus.
6. The digested food material that is absorbed through the intestinal wall is circulated to different parts of the body through lacunae or sinuses.

⊠ Respiratory System

Respiration is a mechanism by which gaseous exchange takes place between the organism and the environment, in which oxygen is taken in and carbon dioxide is given out.

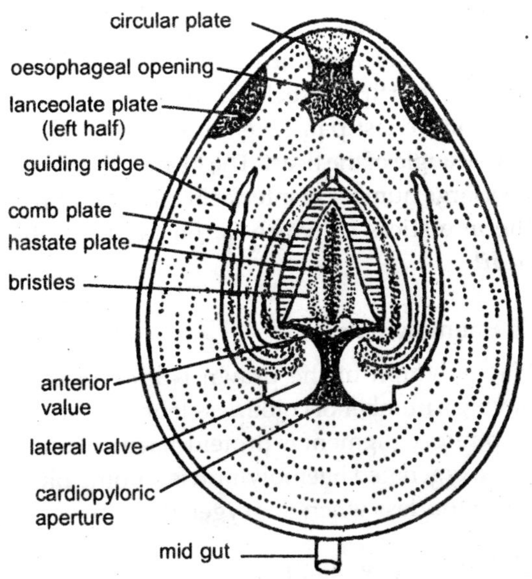

FIGURE 28.6

***Macrobrachium* sp.** Floor of the cardiac stomach (dorsal wall removed)

Macrobrachium lives in water and respire by gills, taking up oxygen dissolved in water.

1. The respiratory organs consist of the lining membrane of the branchiostegite, three pairs of epipodites and eight pairs of gills. The gills are lodged in gill chambers, which communicates with the exterior along its anterior, posterior and ventral borders.
2. The gills are crescent-shaped and increase in size anteroposteriorly (Fig. 28.7). The first gill is smallest and the last one the largest. From their point of origin, the first gill is podobranch, being attached to the coxa of the second maxilliped, the 2nd and 3rd are arthrobranchs being attached to the membrane articulating the 3rd maxilliped with the body and the last five are pleurobranchs, being attached to the body above the articulation of the walking legs.
3. Each gill consists of a long, narrow rachis supporting two rows of rhomboidal gill-plates diverging from each other at right angles to the elongated axis. The gill-plates are larger in size in the middle but smaller towards the ends. Each gill-plate is made up of a double layer of cuticle with a single layer of cells in between. The gill is attached to the body about the middle of its length, and is highly vascular.
4. The axis of the gill is roughly triangular in cross-section. Three longitudinal canals, two laterals and one median, run along the axis. The laterals are connected with each other by transverse channels and also with the median canal by marginal channels. Haemolymph enters through the transverse channels and traverses other channels. Gaseous exchange takes place in the gill filaments.
5. The lining membrane of the branchiostegite and the epipodites of the three maxillipeds are highly vascular and aid in the process of respiration.

⊠ Mechanism of respiration

6. The movements of scaphognathites maintain a constant backward to forward water current in the gill-chambers. The haemolymph in the respiratory organs gives up CO_2 and absorbs O_2.

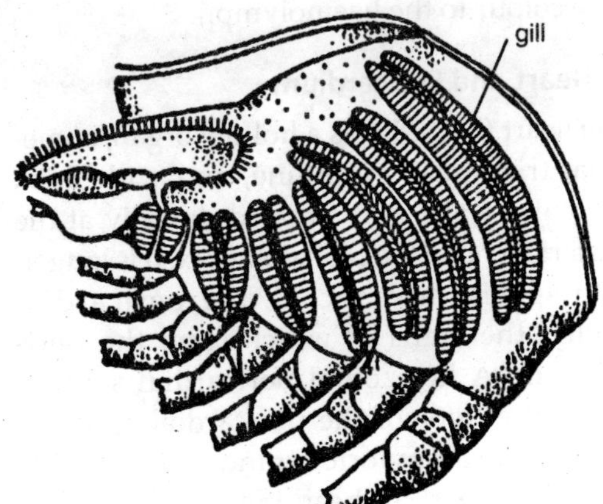

FIGURE 28.7

***Macrobrachium* sp.** Anterior region (lateral view), Carapace removed to expose gills

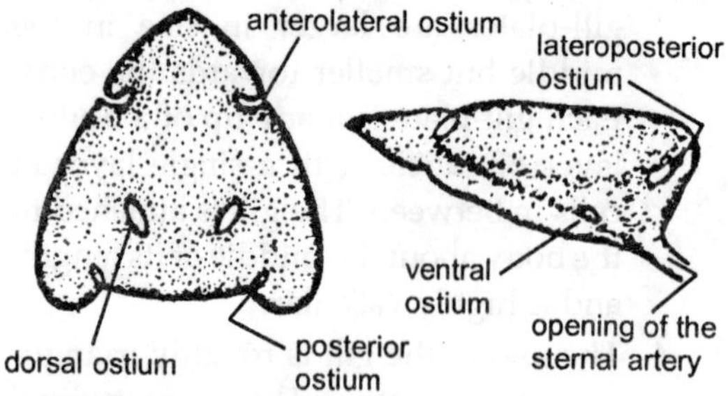

FIGURE 28.8

***Macrobrachium* sp.** Heart. A. Dorsal view. B. Lateral view

⊠ Circulatory System

In Arthropoda, a part of the blood vascular system is expanded to surround other organs; as the coelom is reduced, the other space in the mesoderm, the haemocoel, is elaborated (Fig. 12.1) and functions as the cavity of the blood vascular system.

The main functions of haemolymph (blood) are to transport food and oxygen and the elimination of respiratory wastes in general, although a number of other functions are complimentary to these. The blood of arthropods contain 87 to 98 per cent of water, and mainly 3 types of blood corpuscles are found; viz.—**amoebocytes, granulocytes** and **thrombocytes**. The respiratory pigment, **haemocyanin**, is a prosthetic group of copper, dissolved in the **haemolymph.** Haemocyanin is colourless but oxyhaemocyanin imparts blue colour to the haemolymph.

⊠ Heart and Pericardium

The heart (Fig. 28.8) is a hollow organ, somewhat triangular in outline, and with thick muscular walls. It is situated dorsally at the posterior end of the cephalothorax. It is lodged in a special haemocoel, the pericardial cavity, the walls of which form the pericardium. A horizontal pericardial septum forms the floor of the pericardial sinus. A median cardiopyloric strand and 2 lateral strands support the heart in the pericardium.

Five pairs of valved ostia are present on the walls of the heart; one pair a little behind the middle on the ventral surface, one on each side; second pair opposite to the first pair on the dorsal surface; third pair on the posterior border; fourth pair behind the apex and the fifth or the last pair, one on each side of the lateral angle of the heart. The heart is traversed by a large number of interlacing muscle fibres, the interstices of which is the cavity of the heart.

⊠ Haemolymph

Haemolymph of the prawn is a clear fluid having a number of colourless leucocytes. The respiratory pigment is proteid-haemocyanin. The oxygenated haemolymph is shining blue, but colourless when deoxygenated.

⊠ Arteries (Fig. 28.9)

These are thick-walled vessels, through which the heart pumps out its contents—the haemolymph.

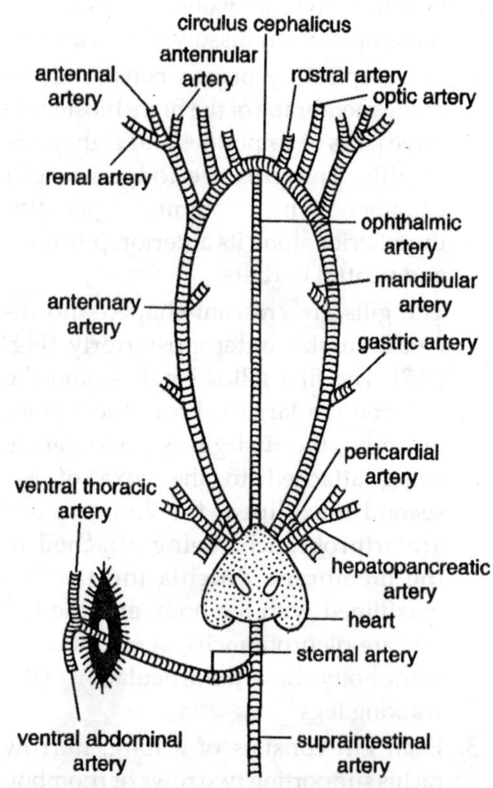

FIGURE 28.9

***Macrobrachium* sp.** Arterial system (anterior part)

From the apex of the heart proceeds anteriorly a slender, median **ophthalmic artery** up to the root of the oesophagus.

Two **antennary arteries** arise from the inner lateral sides of the heart and run anteriorly, slightly obliquely. Behind the eyes, the arteries of the two sides anastomose and form a loop, the **circulas cephalicus**, with which the median ophthalmic artery joins. From this loop comes off a **rostral artery** on each side.

Each antennary artery on its way gives off a **pericardial branch** to the pericardium, a **gastric branch** to the cardiac stomach, a **mandibular branch** to the mandibular muscles, and finally an **optic artery** supplying the eye of the side. Before giving off the optic artery, the antennary artery sends a common artery, which divides into **renal, antennal** and **antennular** branches and supply the respective organs.

A pair of small **hepatopancreatic arteries** arise from the heart, ventrolateral to the roots of the antennary arteries. The end in branches in the hepatopancreas.

A short and stout **dorsomedian artery** arises from the posterior and ventral region of the heart. It divides immediately into a **supraintestinal** and a **sternal artery**.

The supraintestinal artery runs up to the posterior tip of the adbomen lying dorsal to the alimentary canal. In its course, it gives off a number of small branches to the intestine.

The sternal artery is a large vessel. It runs obliquely to the ventral region of the body either through the right or left side of the midgut. It then pierces through the thoracic ganglionic mass of the ventral nerve cord and divides into two branches. One proceeds anteriorly lying below the nerve cord and is known as **ventral thoracic**, while the other the

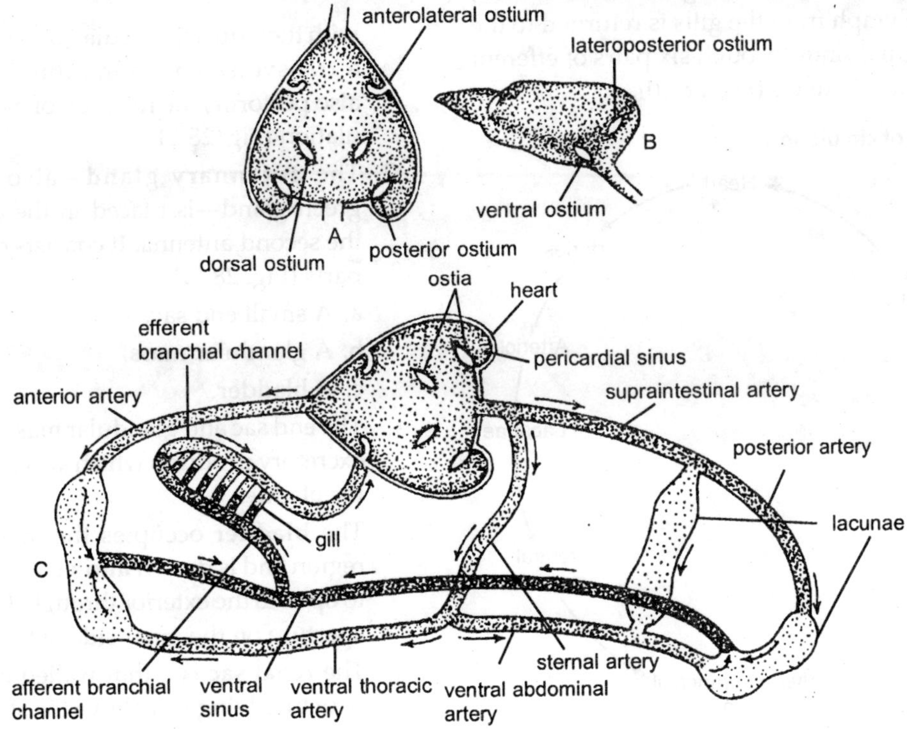

FIGURE 28.10

***Macrobrachium* sp.** Heart and circulation. A Heart (dorsal view). B. Heart (lateral view) pericardium removed. C. Circulation (schematic)

ventral abdominal, runs posteriorly below the nerve cord. They branch profusely, and the former supplies blood to the thorax, first three pairs of walking legs, the maxillipeds, maxillae and the maxillulae, while the latter supplies blood to the ventral region of the abdomen, fourth and fifth pairs of walking legs, the adbominal appendages and the midgut.

☒ Sinuses

Haemolymph from the arteries is finally received into minute intercommunicating body spaces, the lacunae, which ultimately open into two large spaces, the ventral sinuses, situated lengthwise in the ventral region, beneath the hepatopancreas. The two sinuses are connected with each other at several places.

For aeration, haemolymph from the ventral sinuses is sent through six pairs of afferent branchial channels to the gills. After aeration, haemolymph from the gills is returned to the pericardial sinus through six pairs of efferent branchial channels (Fig. 28.10).

Course of circulation

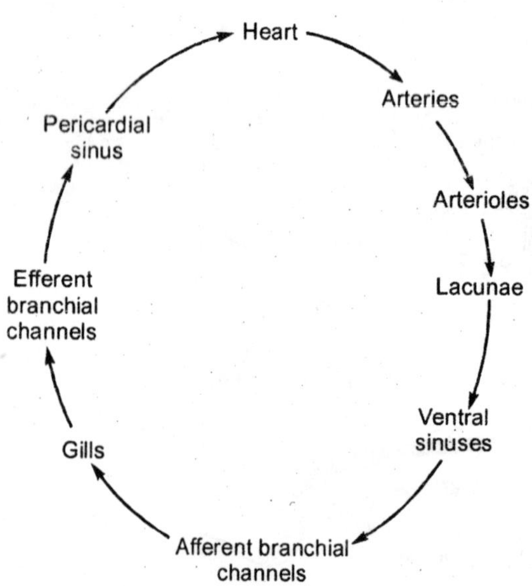

☒ Excretory System

1. The excretory organs consist of a pair of cream-coloured antennary glands with their ducts, a median renal sac and a transverse communicating duct. It also performs the function of osmoregulation (Fig. 28.11).

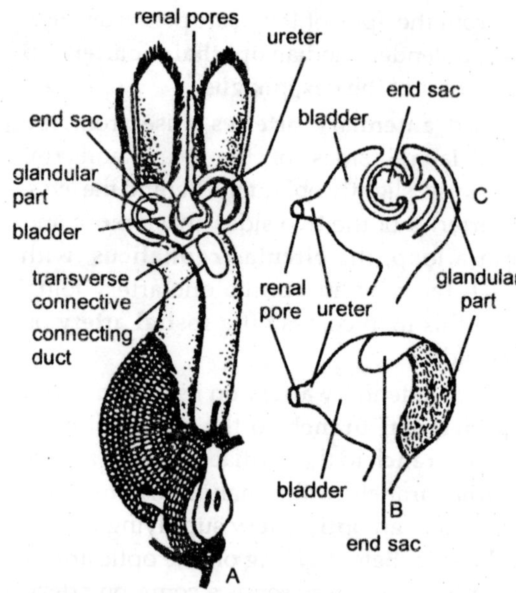

FIGURE 28.11

***Macrobrachium* sp.** Green (Antennary) gland. A. In situ, B. A gland, dissected out, C. Sectional view of a gland

2. The **antennary gland**—also called green gland—is placed in the coxa of the second antenna. It consists of three parts (Fig. 28.11):
   a. A small end sac.
   b. A glandular mass.
   c. A bladder.

   The end sac and glandular mass extract excretory products which are carried to the bladder.

   The **bladder** occupies the innermost region and is drawn into a narrow tube to open to the exterior through the renal aperture on the inner side of the coxa.
3. The **renal sac** is a thin-walled median structure lying just above the stomach. It is connected with each antennary gland by a narrow duct anteriorly. The two ducts are again connected by a narrow transverse duct. Posteriorly, the

renal sac ends blindly at the region of the gonad. The renal sac acts as a temporary reservoir for waste products.

4. No other renal organs.

The thick chitinous layer of the integument is a nitrogenous product secreted by the ectoderm and is cast off in each moult. **Moulting** is considered as a special mechanism to get rid of nitrogenous wastes.

⊠ Nervous System

The system which controls and regulates the various activities of an organism is known as nervous system.

The nervous system of prawn consists of a central nervous system, a peripheral nervous system and a visceral or sympathetic nervous system.

⊠ Central Nervous System

The central nervous system consists of a pair of supraoesophageal ganglia, a pair of circumoesophageal connectives, a suboesophageal or thoracic ganglionic mass and a double ventral ganglionated nerve cord. All these send out nerves which supply the respective organs (Fig. 28.12).

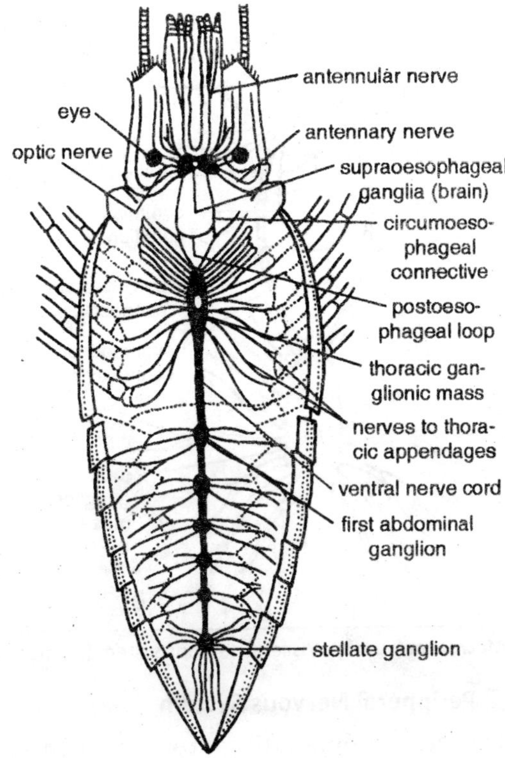

FIGURE 28.12

***Macrobrachium* sp.** Nervous system

1. **Brain or supraoesophageal ganglia.** It is a bilobed structure formed by the fusion of the right and left ganglia and is situated beneath the base of the rostrum just in front of the junction of the oesophagus with the cardiac stomach. The supraoesophageal ganglia is formed by the fusion of several pairs of ganglia.

2. **Circumoesophageal connectives.** Two stout nerve cords, arise from the hinder part of the supraoesophageal ganglia and run backward and downward round the oesophagus to meet ventrally in the thoracic ganglionic mass.

a. **Postoesophageal** or **transverse loop.** A loop embracing the oesophagus posteriorly and connects the two circumoesophageal connectives.

3. **Thoracic ganglionic mass.** A large ganglion formed by the fusion of several pairs of ganglia and form the anteriormost ganglion of the ventral nerve cord.

4. **Ventral nerve cord.** It lies beneath the mass of the abdominal muscles. The ventral nerve cord is formed by the fusion of two nerves and two ganglia unite to form each ganglion of the ventral nerve cord. There are six abdominal ganglia on the nerve cord corresponding to the six abdominal segments. The last ganglion is comparatively large and is known as **stellate ganglion**; it is possibly formed by the fusion of several ganglia.

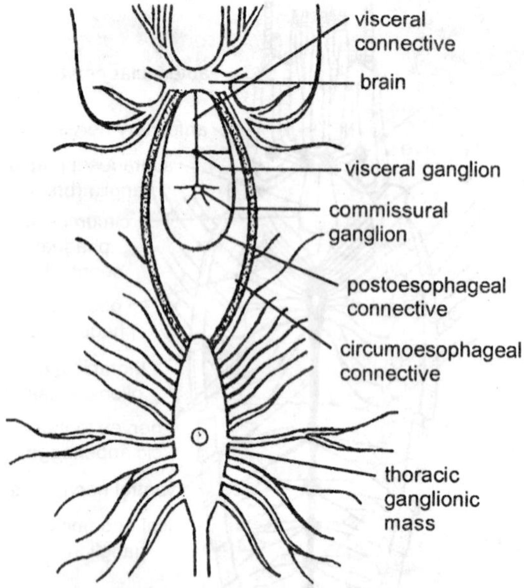

FIGURE 28.13

***Macrobrachium* sp.** Visceral nervous system (enlarged)

⊠ Peripheral Nervous System

The nerves emanating from the central nervous system constitute peripheral nervous system:

1. **Optic nerve.** Arising from the outer side of each supraoesophageal ganglion it runs forward and outward and innervate the eye of the side.
2. **Antennulary nerve.** Arises from the anterior part of the supraoesophageal ganglion, runs forward and sends branches to the antennule and the statocyst of the side.
3. **Antennary nerve.** Arising from the lower portion of the supraoesophageal ganglion and passing downwards and obliquely, curves forward to innervate the antenna.
4. The small nerve arises along the antennery nerve and innervate the labrum.
5. **Cephalothoracic nerves.** Eleven pairs of nerves arise from the thoracic ganglionic mass and innervate all the cephalothoracic appendages except the two pairs of antennae.
6. **Abdominal nerves.** Two pairs of nerves arise from each abdominal ganglion and innervate the corresponding muscles and appendages. Two additional pairs of nerves from the stellate ganglion send branches to rectum, telson and adjacent organs.

⊠ Visceral Nervous System

1. Two small **visceral** or oesophageal ganglia are present on the roof of the cardiac stomach, one behind the other (Fig. 28.13).
2. A small nerve arising from the posterior border of the brain connects the two ganglia behind.
3. Two delicate connectives join the anterior visceral ganglion with the two **commissural ganglia** on the circum-oesophageal connectives.

⊠ Receptors and sense organs

The sense organs include eyes, statocysts, tactile organs and olfactory setae.

Eyes

The prawn bears two compound eyes. Each eye is a collection of a large number of visual elements called **ommatidia** and is borne on a movable stalk.

Structure (Fig. 28.14)

1. The outer convex transparent cuticular covering of the eye is known as **cornea**. The cornea is divided into a large number of square facets, each corresponding to a single **ommatidium**. The **ommatidia** are arranged regularly along the radii of the eye.
2. Each ommatidium is a complete visual unit, made up of cells arranged in end-to-end position along the long axis. Beneath the corneal facet is a pair of flat **corneagen cells** of epidermal origin which secrete a new cornea when the old one is lost during moulting.
3. Beneath the corneagen cells lie four tall cells—the **cone cells**—the inner borders

of which give rise to a refractive crystalline cone.

4. The basal part of the ommatidium is made of spindle-shaped, transversely striated structure, the **rhabdome**, which is surrounded by seven elongated cells, the **retinular cells**. The rhabdome with the retinular cells are known as the **retinula** bearing a pigment, guanine, the nature of which is said to be of melanin.
5. The optic nerve breaks up into branches and innervate the retinular cells.
6. Each ommatidium is separated from its neighbours by a partition of pigment cells, the **chromatophores**. These are arranged in two groups.

a. The distal group surrounding the lens and the cone cells constitute the **irish sheath.**

b. The proximal group surrounding the retinula constitute the **retinal sheath**. The pigment sheaths can extend or retract under the influence of light. In bright light, they are extended and in weak light they are retracted.

⊠ Vision

In prawn, two types of visions are found. At daytime, when the intensity of light is high, the vision is of mosaic type. In the night, or dim light, when the intensity of light is less, the vision is of superimposed type.

1. **Mosaic vision.** In bright light, the pigment sheath is extended and any ray of light which falls obliquely on the ommatidium is absorbed by the pigment sheath. Almost parallel rays falling on each ommatidium from an object, reach the rhabdome and an image of a point of the object is formed. Therefore many images of many points of the object are formed. Such an image is known as **apposition** image. By the apposition of those points of images in a number of ommatidia an erect image of the object is formed. In such a vision, any slight change of the object is quickly detected.

   Here, the sharpness of the image is dependent upon the number of ommatidia involved and the degree of their separation.

2. **Superimposed vision.** In dim light, the pigment sheath is retracted and greater portion of the ommatidium is uncovered. Any ray of light striking obliquely on the sides of the ommatidium passes to the next and, in doing

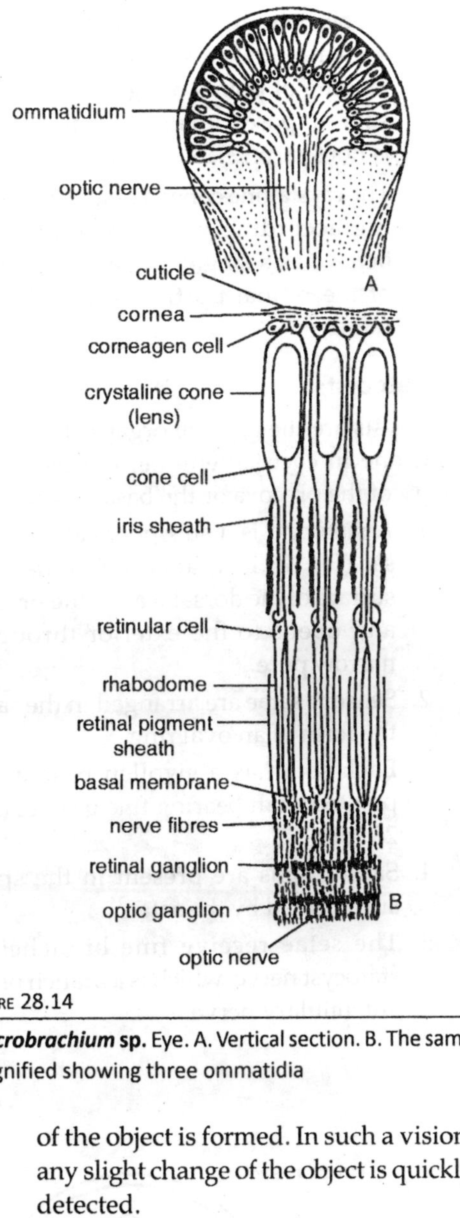

FIGURE 28.14

***Macrobrachium* sp.** Eye. A. Vertical section. B. The same magnified showing three ommatidia

so, becomes refracted to reach the next ommatidium. In such a case, an overlapping of points of lights occur and a superimposed image is formed, which is not sharp.

The crystalline cones, capable of adjusting accordingly, act in unison and behave as a single unit and the whole of the retinal portion act as a single retina.

⊠ Statocysts

Statocysts are the balance organs. They are a pair, one in each antennule, located in the cavity of the precoxa or the basal segment.

1. A statocyst is a subspherical cuticular sac (Fig. 28.15A) attached to the inner surface of the dorsal wall of the precoxa and opens to the exterior through a narrow pore.
2. Sensory setae are arranged in the sac in the form of an oval ring.
3. Each seta has a swollen base and a pointed shaft bearing fine bristles (Fig. 28.15B).
4. Sand grains are present in the space surrounded by the setae.
5. The setae receive fine branches of statocyst nerve, which is a branch of the antennulary nerve.

During movement the sand grains are displaced with the change of position and press against certain setae, which helps the animal to correct its position.

⊠ Tactile sense organs

They are present along the margins of the appendages, abundant in antennae and flattened portion of pleopods. A tactile seta (Fig. 28.15C) consists of:

1. A hollow base or shaft connected to the appendage.
2. A tapering blade bearing double linear rows of tiny barbs is attached to the distal end of the shaft.

The sense organs carry sense of touch.

⊠ Olfactory setae

A seta is located on the small, middle feeler, between the two long feelers of an antennule. It has:

1. A two-jointed shaft, proximally attached to the integument by a flexible membrane.
2. The free end of the distal segment is bluntly rounded and covered with a thin membrane (Fig. 28.15D).
3. Innervated by nerve fibres from the olfactory branch of the antennulary nerve.

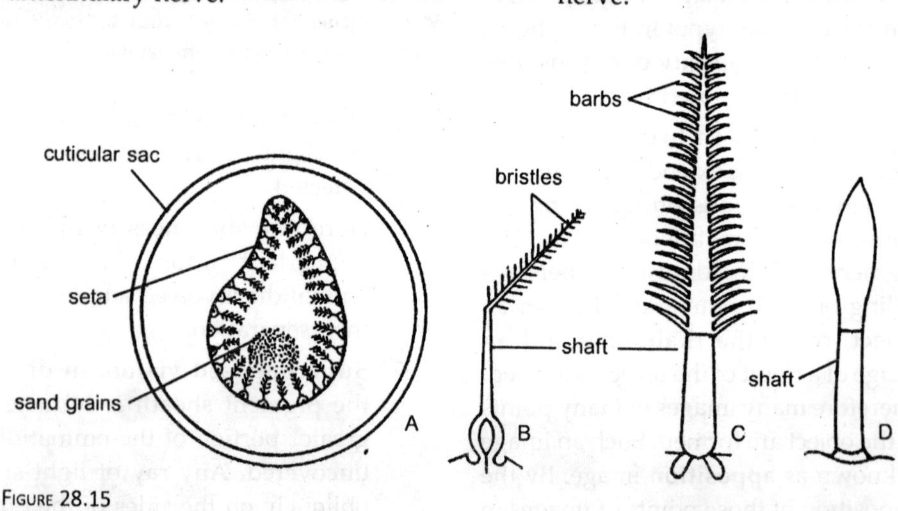

FIGURE 28.15

***Macrobrachium* sp.** A. Statocyst. B. A statocystic seta. C. A tactile seta. D. An olfactory seta

☒ Reproductive System

The sexes are separate.

Male reproductive system

1. The male reproductive system consist of a pair of testes, a pair of vasa deferentia, a pair of seminal vesicles and a pair of gonopores (Fig. 28.16A).
2. The **testes** are soft, white, elongated bodies, fused at both the ends and are situated in the cephalothorax, below the heart and above the hepatopancreas.
3. From each testis arises a narrow tube, the **vas deferens**, which is much coiled at first and then **descends** down towards the base of the fifth walking leg of the side.
4. The terminal end of each vas deferens forms a club-shaped swelling, known as **seminal vesicle**, which opens to the exterior by the male gonopore on the inner side of the coxa of the 5th walking leg.

Female reproductive system

1. The female reproductive system consist of a pair of ovaries, a pair of oviducts and a pair of female gonopores (Fig. 28.16B).
2. **Ovaries** are small and whitish in off-seasons but large and dark brown in the breeding season. They are fused at both the ends, larger in size than the testes and occupy same position as the testes in the male.
3. From the middle of the outer side of each ovary arises an **oviduct**, which narrows downwards to open in the gonopore on the inner side of the coxa of the 3rd walking leg of the side.

☒ Fertilization and Development

1. Prawns breed in rainy season.
2. The eggs are round and yolk-filled.
3. Fertilization external and the fertilized eggs are carried in the abdominal basket, formed by the appendix internae of the second to fifth pleopods in females. Development direct, the newly hatched young, resembling the adult, leave the abdominal basket to lead a free life.

SACCULINA CARCINI

Sacculina belongs to Grade Rhizocephala, sub class cirripedia, class crustacea, subphylum Mandibulata, phylum Arthropoda. Sacculina, a parasite on crab (*Carcinus* sp.), appears like a soft round tumour on the abdomen of the host (Fig. 28.17).

1. The adult Sacculina is characterised by—(i) loss of segmentation and appendages, (ii) no mouth and anus, (iii) numerous branched filaments from the body penetrating in all parts of the

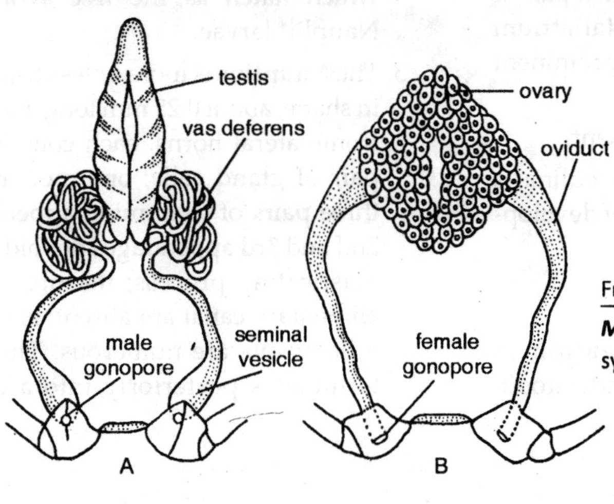

FIGURE 28.16

***Macrobrachium* sp.** Reproductive system. A. Male. B. Female

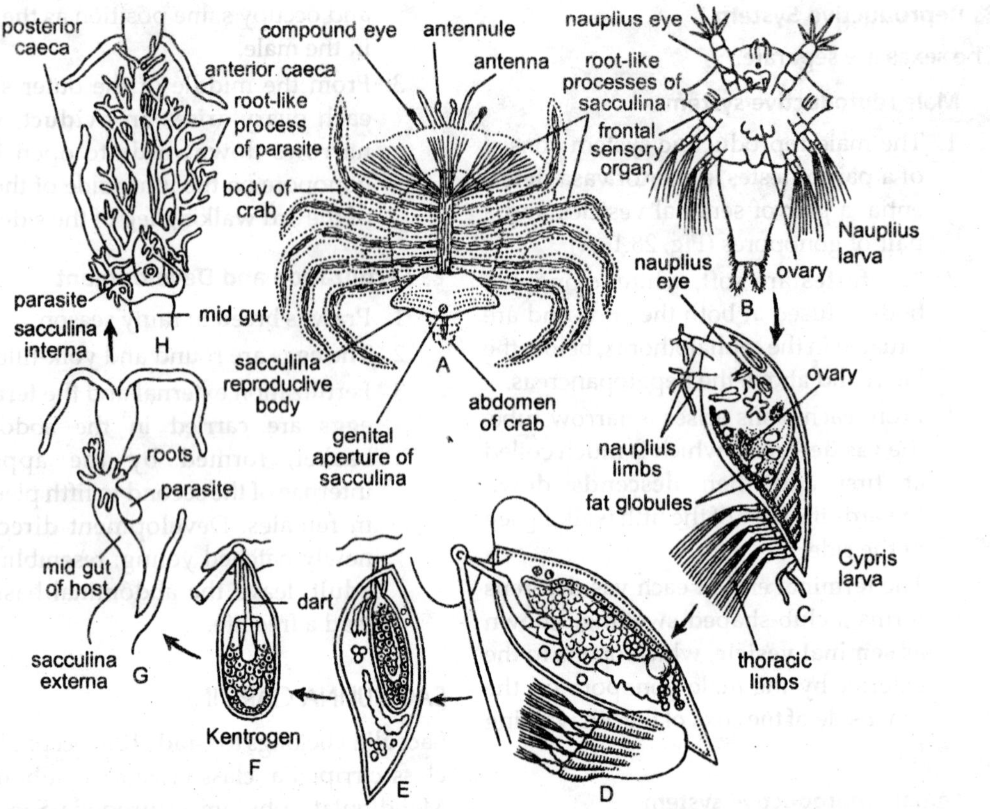

FIGURE 28.17

Sacculina carcini. Life cycle stages. A. *Sacculina* on the abdomen of crab. B-G. Life cycle stages. H. Branched treelike modification in the mid gut of crab (C. Cypris stage; F. Kentrogen stage; G. Sacculina externa, H. Sacculina interna)

body of the host except heart and gills (iv) all the organ systems are degenerated except the reproductive organs.

2. A pair of elongated testes and a pair of ovaries with accessory genital atrium and collaterial glands and a prominent genital aperture are present.
3. A single nerve ganglion present.
4. The arthropod nature of sacculina is known only from the study of developmental stages.

☒ Life history

1. The mechanism of fertilization in sacculina is not yet clearly understood. Some consider that the neotenic male cypris larva enters through the genital opening and fertilize the eggs.
2. An adult female lays numerous eggs, which hatch as the free swimming Nauplii* larvae.
3. The **Nauplius** is more or less triangular in shape, about 0.25 mm long, have two frontolateral horns, each containing a pair of gland cells; one median eye; three pairs of swimming appendages; 2nd and 3rd appendages devoid of any masticating process; the mouth and alimentary canal are absent; and germ cells or ova are numerous. The body terminates posteriorly into a caudal

* Singular Nauplius

furca. Nauplius is free living and does not feed until it moults into cypris form. Nauplius metamorphose into cypris larva in 3 to 4 months.

4. The cypris is free living, bears a bivalved shell, six pairs of thoracic biramous appendages and numerous germ cells. Abdomen extremely reduced, without any appendage and terminated in a pair of caudal ramii. Single eye persists. A pair of frontolateral glands open near the margin of the valves of the shell. A pair of 3-segmented antennules are present. The terminal segment of antennules bears backwardly curved structure, the organ for attachment.
5. The cypris larva attaches itself to the base of a bristle of a crab with the antennules, lose thoracic appendages and accessory muscles along with the bivalve shell, leaving only a mass of cells enclosed in a new sac in continuation with the antennules attached to the host. The old cuticle is replaced by a new one to enclose the rest of the body like a sac. The pointed end of the cuticle forms a hollow **stylet** or **dart**. This stage is known as **kentrogen** stage.
6. Once attached to the host, the protoplasm of the Kentrogen larva differentiates and becomes a ball of embryonic cells.
7. The stylet pierces the abdominal wall of the crab and the cell mass passes into the body of the host, through the hole made by the stylet.
8. The larval mass is carried by the host's blood to the region of alimentary canal in the thoracic zone.
9. In the thoracic zone the larval mass grows differentiating into two main parts (i) a complex root system, the internal **sacculina** (*Sacculina interna*), which gradually extends throughout the body of the crab, even into the limbs and serves to absorb nourishment, (ii) a second part is a rounded outgrowth which appears after the root system forming the tube-like body of the adult parasite or the **external succulina** (*Sacculina externa*).
10. As *Sacculina externa* grows it pushes through the ventral body wall of the abdomen of the crab and in about 8 weeks it emerges as a broad sac enclosing the reproductive organs. The enclosed brood chamber communicates with the exterior through a small genital opening.
11. Sacculina reproduces throughout the year but has a peak nauplius production between August and December.
12. If the host is a freshwater crab, the life cycle is somewhat altered. The nauplius stage is suppressed and the eggs produce **cypris** larvae.

☒ Host-parasite interaction of Sacculina

In association with crab and sacculina, some severe changes are observed in the body of both the host and the parasite.

☒ Effects on the host

I. Parasitic castration and sex reversal

1. The gonads of the hosts are affected in two ways. The multiplication and differentiation of the gametocytes are disturbed and destruction of sex cells takes place, resulting in incomplete atrophy of gonads.
2. Following removal of sacculina parasite the female host generally regenerates new ovarian tissue, but the male commonly develops complete or partial ovaries instead of testes causing sex reversal.
3. Such sex reversal may not be directly attributed to the parasite, because similar occurrences are known among nonparasitised crab also. The pathologic effects of sacculina are most severe in species of crabs other than *Carcinus*.

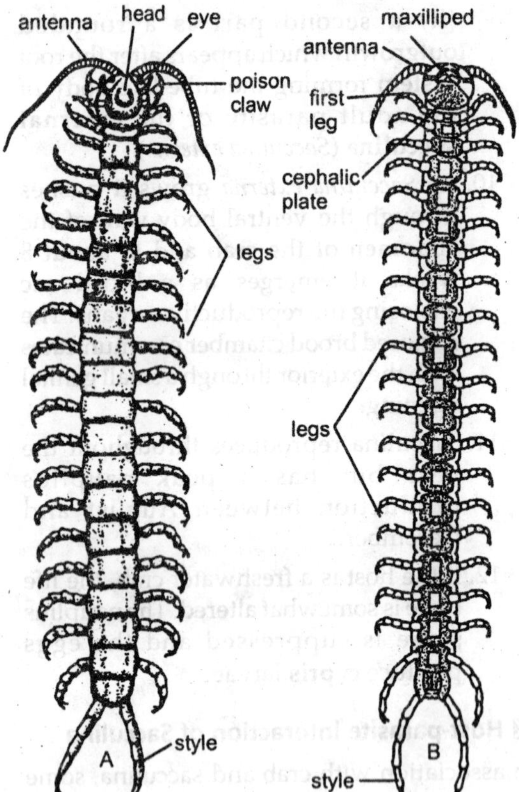

FIGURE 28.18

Scolopendra subspinipes. A. Dorsal view. B. Ventral view

4. Sacculina often induce modifications of external secondary sex characters in crab hosts. Such modifications include (i) the broadening of abdomen and feminization of certain appendages in males and narrowing of abdomen and degeneration of pleopods in females. These modifications are the result of damage to gonads and are manifested in sex reversal in males. (iii) The presence of mature parasite prevents further moulting of the crab.

II. Physiological alterations

1. In addition to the morphological alterations sacculina cause physiological changes also. Lipochromes in the blood increase, calcium and magnesium levels become almost double in the haemolymph, and the fat content in the haepatopancreas either increased or decreased and the glycogen reserves in the digestive gland and surface epithelium become depleted.

⊠ Effects on parasites

i. **Structures degenerated.** The parasite (sacculina) loses segmentation, appendages and even typical crustacean characteristics. It loses alimentary, circulatory and excretory organs.
ii. **Structures developed.** Organs for holding, piercing and absorbing nutritions from crab host.
`iii. Increased reproductive capability leading to prolific multiplication, and dispersal and successful continuation of the species.

SCOLOPENDRA sp.

Scolopendra belongs to class Chilopoda, subphylum Mandibulata, phylum Arthropoda. Several species of Scolopendra are found in India. These elongated, swift moving animals live in dark places, burrows, crevices, decaying logs of wood, under partially decomposed leaves in shades of trees. They occur in all tropical and subtropical countries.

⊠ External features

The multiped body is long, worm-like dorsoventrally flattened and divisible into a **head** and a **trunk**. The dorsal surface is reddish-yellow and the ventral surface yellowish in colour (Fig. 28.18).

⊠ Head

The head consists of head capsule, head lobes and head appendages.

Head Capsule

1. It is a somewhat roundish, flattened and well-sclerotized dorsal **cephalic** plate (Fig. 28.19).
2. The cephalic plate bears a pair of **antennae** in front and four pairs of **eyes** on the dorsolateral sides.
3. A pair, each of **accessory pleurites** and **chief pleurites**, two pairs of anterior

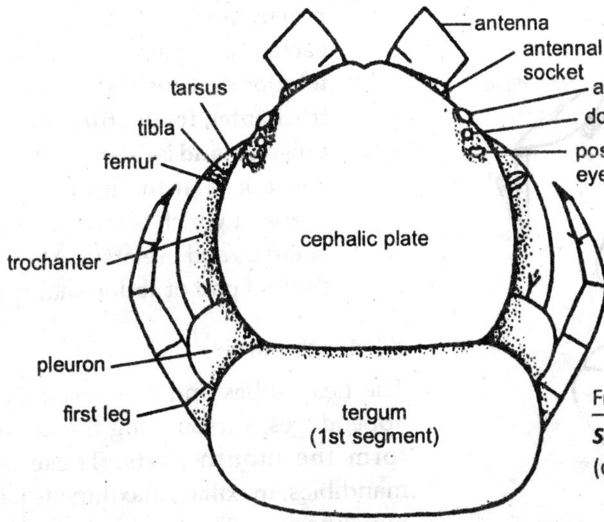

FIGURE 28.19

***Scolopendra* sp.** Cephalic end (dorsal view)

head pleurites, and three pairs of **mandibular pleurites** form to lateroventral sclerites.

4. **Hypopharyngeal suspensoria** constitute the ventral sclerites of the head wall. These support the hypopharynx.

Head lobes

Two median structures, the labrum and the hypopharynx (Fig. 28.20) form the head lobes.

1. **Labrum.** It is a preoral median cephalic lobe, the tip being projected posteriorly. The dorsal and ventral sclerites of the labrum are much wider than long and converge to form an extensive blade. The dorsal labral sclerite is provided with numerous sensory setae projecting into the preoral cavity.
2. **Hypopharynx.** It is a tongue-like median lobed structure associated with the ventral wall of the gnathal region of the head and divisible into a long median and two small lateral lobes supported by the hypopharyngeal suspensoria.

Head appendages

The head appendages comprise a pair of **antennae**, a pair of **mandibles**, two pairs of **maxillae** and a pair of **maxillipeds** (Fig. 28.20).

1. **Antennae.** Two slender many-jointed appendages narrowing towards the free end.
2. **Mandibles.** Two, oriented horizontally on the underface of the head.

   Anteriorly, the mandibles are broad and converge laterally while posteriorly they are curved and tapering. The mandible is made up of two units—the posterior unit and the anterior unit. The anterior lobed unit hinges on the posterior unit as a flap. The long incurved sclerite called the posterior sclerite, represents the posterior unit. The anterior unit is a gnathal lobe whose inner wall is short and membranous and the outer wall is more extensive and sclerotized. The free anterior part of the gnathal lobe bears five large, tricuspid teeth and ventrally crescentic, brush-like setae guarding the oral entrance.
3. **First Maxillae.** These are united to form labium-like structure, a dorsoventrally flattened body situated posteroventrally to the hypopharynx. The coxa is more extensive on the ventral side and bears medially a pair of seta-bearing endites. The telopodite is made of two segments of which the distal is longer and carries sensory setae.

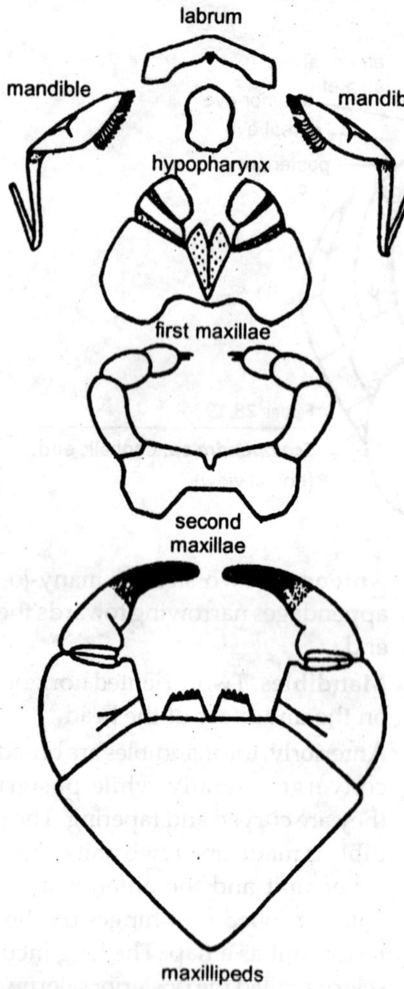

FIGURE 28.20

***Scolopendra* sp. Mouth parts (ventral view)**

4. **Second Maxillae.** These appendages are more leg-like. The coxa is dorsoventrally flattened and is more extensive on the ventral side. The telopodite is movable on its horizontal plane. The tibia bears a dorsal spur while the tarsus carries a comb-like longitudinal row of setae dorsally. Two strong spurs develop on the claw-like pretarsus.
5. **Maxillipeds.** Maxillipeds or the **posion-claws** are virtually the first pair of walking legs modifed as accessory feeding organs. Ventrally, the coxae fuse to form a single plate, dorsally they remain as separate plates and anteriorly each bears a pair of dental projects. The telopodite consists of four segments, **trochanter, femur, tibia** and **tarsus**. The **poison gland** is located in the telopodite and opens at the tip of the claw-like segment which is formed by the fusion of tarsus and pretarsus. Maxillipeds are the first pair of abdominal appendages.

☒ Mouth parts

The head lobes and the postantennal head appendages, surrounding the mouth region, form the mouth parts. These consist of mandibles, maxillae, maxillipeds labrum and hypopharynx. The head being **prognathous**, the mouth parts are directed anteriorly and underlap each other from behind forward.

☒ Trunk and trunk appendages

The trunk is segmented and several times longer than the head. It is divided into 24 identical segments (Fig. 28.18).

1. The first trunk segment is apposed to the head and the pair of appendages it bears, are maxillipeds or poison claws.
2. The legs in rest of the segments are seven-jointed, ending in a claw, bearing two ventral spines.
3. The legs become gradually longer posteriorly and the **anal legs** in the last segment are longest.
4. The **telson** forming the terminal part of the trunk bears **anus** and is not considered as a segment.
5. The last segment is the **genital segment**. In males, a pair of projections—the **genital styles**, are present in the genital segment.

☒ Locomotion

The legs support the body in a horizontal position. During locomotion, the legs of one side are lifted and moved foward, the body being supported by the legs of the opposite side. In the next step, the legs are put against the support in a forward position and the legs of the other side are lifted and moved forward. This is repeated and the animal can move very fast.

☒ Glands

Both unicellular and multicellular glands are present in the hypodermis, below the cuticle. The glands are of three categories—the poison glands, the salivary glands and the trunk glands.

1. **Poison glands.** A poison gland is located in each of the maxilliped.
   a. The gland is a somewhat cylindrical sac, lined by numerous unicellular glands.
   b. The secretion is carried through a highly cuticularised duct in the middle, opening at the centre of the maxilliped claw.
   c. The secretion is yellowish, acidic and helps in defence, lubrication and digestion of food.
2. **Salivary glands.** These glands are of three types and help in digestion. All the glands open in the head region.
   a. **Epipharyngeal glands**
      i. Two pairs, located ventral to the brain and open on the epipharyn-geal surface through ducts.
      ii. The anterior pair is larger and communicate laterally with each other.
   b. **Labial glands**
      i. A pair of lobular glands, located between fifth and seventh segments.
      ii. Numerous small ducts arising from the lobules join to form a common duct.
      iii. The common duct runs anteriorly along the gut and opens on the ventral surface of the head, between the hypopharynx and labium.
   c. **Maxillary glands**
      i. One pair, structurally similar to the labial glands and located in the third and fourth segments.
      ii. The ducts open on the ventral side of the head region, near the second maxillae.
3. **Trunk glands.** Present in the trunk and are of two types.
   a. **Coxal glands.** One pair in each of the second to fourth segments. The glands are vesicular and their ducts unite to form a common duct opening in the coxopleural plates.
   b. **Coxopleural glands.** Flask-shaped, many in number and located in the last leg bearing segment. The glands open separately on the coxopleural plates.

☒ Digestive System

The digestive system consists of a long, straight alimentary canal, starting from the preoral cavity and ending in anus (Fig. 28.21) in the telson.

Salivary glands (previously described) and unicellular glands in the lining epithelium of the gut are digestive glands.

Alimentary canal

The alimentary canal is divided into three zones—foregut, midgut and hindgut. The fore and hindgut are lined by cells of ectodermal origin, while the lining of the midgut is endodermal.

Foregut

It is the longest part, extends posteriorly up to 13 leg bearing segment and differentiated into pharynx, oesophagus, crop and proventriculus.

1. **Preoral cavity.** The cavity is bounded by the dorsal labral sclerite and membranous epipharyngeal anterodorsally, mandibles laterally and anterior portion of hypopharynx ventrally. The preoral cavity opens into the pharynx through a buccal chamber.
2. **Pharynx.** It is a broad chamber, narrows posteriorly into a short tube, the oesophagus. The internal living epithelium of the pharynx is lined by a cuticular lamina, which is largely folded into a dorsal and a ventral fold. Ven-

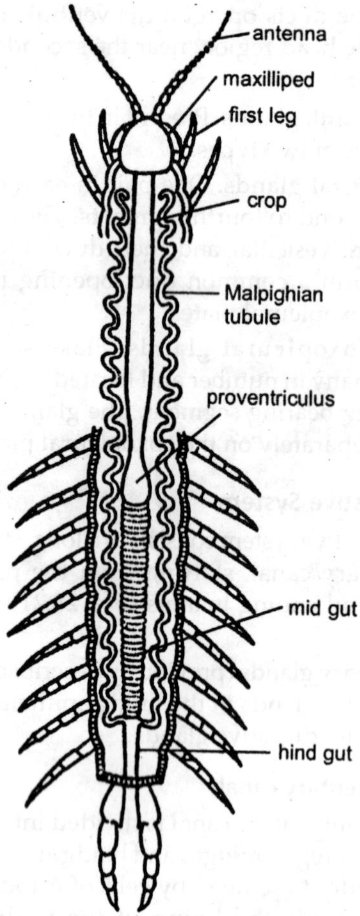

FIGURE 28.21

***Scolopendra* sp.** Postcephalic part of the alimentary canal and Malpighian tubules (dorsal view)

trally, the spindle-shaped transverse oral muscle marks the anteriormost limit of the pharynx. The extrinsic musculature are seven pairs and act as dilators of the pharynx.

3. **Oesophagus.** A short tube ending in crop.
4. **Crop.** The longest and thinnest part of the foregut which presents wrinkled appearance due to irregular folds of the cuticular intima internally. Crop extends up to the 12 leg bearing segments.
5. **Proventriculus.** A very highly specialised and important part. It is located between the crop and the ventriculus and confined to 12 and 13 segments. Proventriculus is divisible into anterior and posterior proventriculus. The anterior proventriculus acts as a chewing apparatus.

Midgut or Mesenteron or Ventriculus

It extends from the 13 to 19 segment and often bears many superficial transverse constrictions. The loss of cells during secretion (holocrine secretion) is compensated by regeneration of cells out of a group of cells called **nidi**, embedded in the **epithelium.** A protective envelope, the **peritrophic membrane**, protects the delicate epithelium from injury. The midgut epithelium secreting digestive enzymes are either completely or partially lost (holocrine and apocrine secretion).

Hindgut

It is the posterior ectodermal part of the alimentary canal. The bases of the Malpighian tubules mark the anterior limit of the hindgut. The extrinsic musculature, extending between anal sclerite and anal passage, helps in evacuation of the excreta through the anus.

☒ Respiratory System

Respiration is carried by tracheal system (see Cockroach).

Nine pairs of triradial, slit-like spiracles are present on the posterior and dorsolateral sides, one in each of the 3, 4, 5, 8, 12, 14, 16, 18 and 20 leg-bearing segments. The spiracles lead to tracheae, which branch profusely and anastomose, ending in tissues (Fig. 9.7).

☒ Circulatory System

In *Scolopendra* the haemolymph is carried through an open circulatory system formed by blood vessels and sinuses.

☒ Heart and Pericardium

Dorsal vessel

1. The **heart**, also called dorsal vessel is an elongated, muscular tube, extending

from 2 to 21 pedal segments, where it narrows abruptly. In front, it continues as an artery to the cephalic organ.

2. The heart lying in the 3 to 21 segments is enclosed in a **cardiac diaphragm** or **pericardium**.
3. The pericardium is drawn into two points in each segment and attached to the tergum of the segment.
4. A pair of openings, the **ostia**, communicate the heart with the pericardial cavity in each segment.
5. In each segment, the heart sends out a pair of vessels to the structures there.
6. Through a number of perforations, the pericardial cavity is in communication with the dorsal sinus.

Ventral vessel

7. The ventral vessel, also called **supraneural vessel**, is situated above the nerve cord and extends from 2 to 21 segment.
8. A ring-like **stomodaeal vessel** connects the dorsal and ventral vessel at the anterior end, which sends four **cephalic vessels** anteriorly—one dorsal, two laterals and one ventral, supplying haemolymph to the anterior part of the body. **Haemolymph** is a light yellow fluid with a few suspended corpuscles, the **haemocytes**.

☒ Sinuses

Two large sinuses, dorsal and perivisceral sinus and numerous small sinuses bring back haemolymph from different parts of the body to the heart.

1. **Dorsal sinus.** Located dorsally between the pericardium and the body wall. It is in communication with the pericardial cavity through narrow passages and also with the ventral part of the body.
2. **Perivisceral sinus.** It is a space around the gut surrounded by a perforated membrane.

☒ Circulation

Haemolymph from the heart is circulated to different parts of the body by pumping action of the heart. The fluid is finally collected in the dorsal sinus, from where it goes to the heart through the pericardial cavity.

☒ Excretory System

A number of organs (Fig. 28.22) which are concerned with the removal of intracellular wastes from the body and maintenance of a constant internal environment by retaining the necessary constituents constitute the excretory system. The Malpighian tubules are chief excretory organs. Other organs taking part in the process are **Kowalevsky's corpuscles**, lymphatic threads and fat bodies.

☒ Malpighian tubules

a. The most important excretory organs are the Malpighian tubules, which are two in number and originate at the junction of the mid and hindgut in the 19th segment.
b. The tubules run posteriorly for a short distance at its beginning and then turn abruptly forward along either side of the alimentary canal, pursuing a sinuous course and reaching as far as second segment where they terminate blindly (Fig. 28.22B).
c. The tubule consists of a columnar epithelium which lacks the striated border towards the lumen. Muscular elements of any sort are absent (Fig. 28.22B).
d. Haemolymph of centipedes contain a considerable concentration of uric acid and the Malpighian tubules eliminate this nitrogenous waste. The uric acid does not crystallise in the tubules. The crystals are formed in the hindgut due to reabsorption of water in the latter. The contents of the tubules are flushed out into hindgut by the continuous circulation of fluid in the lumen of the tubes.

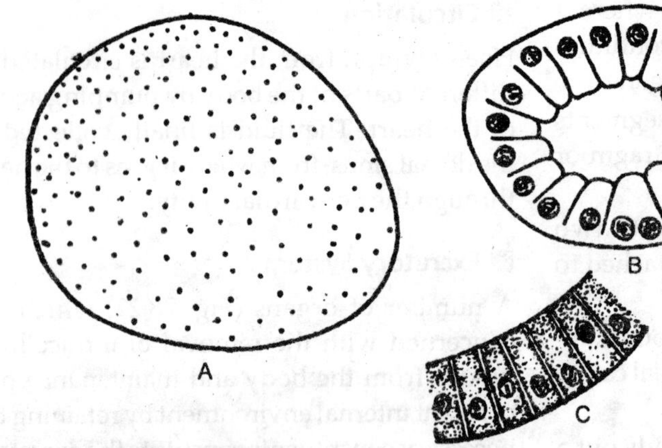

FIGURE 28.22

***Scolopendra* sp.** Excretory organs. A. Kowalevasky's corpuscle, B. Transverse section of a Malpighian tubule, C. Part of a lymphatic thread

⊠ Kowalevsky's corpuscles

The branches of the segmental arteries behind the 3rd segment terminate into somewhat rounded bodies known as Kowalevsky's corpuscles (Fig. 28.22A). The corpuscle is a syncytial mass and is responsible for building up blood cells in addition to its excretory function.

⊠ Lymphatic threads

These are filament-like, bluish in appearance, found round about the Malpighian tubules. Each filament is composed of a row of cells, capable of absorbing carmine particles (Fig. 28.22C).

Fat bodies are supposed to have an excretory function in addition to its storage of reserve materials.

⊠ Nervous System

The central nervous system, peripheral nervous system and visceral nervous system constitute the nervous system of *Scolopendra*.

⊠ Central Nervous System

It consists of cerebral ganglia, circumoesophageal connectives, suboesophageal ganglia and the ventral nerve cord.

1. **Cerebral ganglia or brain.** A pair of ganglia fused to form a bilobed structure, located dorsally in the head.
2. **Suboesophageal ganglia.** Located in the head, ventral to the oesophagus.
3. **Circumoesophageal connectives.** A pair, run ventrally from the cerebral ganglia around the oesophagus to join the suboesophageal ganglia.
4. **Ventral nerve cord.** It runs posteriorly along the midventral line, up to the posterior end and bears a ganglion in each segment. The genital ganglia lie in the last segment.

⊠ Peripheral Nervous System

The nerves arising from the cerebral ganglia, suboesophageal ganglia and ventral nerve cord constitute peripheral nervous system.

1. The cerebral ganglia send nerves to antennae, eyes and other organs in the head.
2. The jaws and maxillipeds are innervated by nerves emanating from suboesophageal ganglia.
3. Nerves from the ganglia in the ventral nerve cord innervate organs of the segment concerned.

⊠ Visceral Nervous System

A visceral nerve from the cerebral ganglia innervate the alimentary canal.

⊠ Sense Organs

Three types of sense organs are present:

1. **Eyes (Sensila optica).** One pair of simple ocellus are present on the dorsolateral sides of the head.

2. **Sensory hairs** and **spines.** These are tactile sense organs and scattered all over the body.
3. **Organs of Tomosuary.** Located in the base of the antennae and presumably auditory in function.

☒ Reproductive System

The male and the female can be distinguished from their external dimorphic characters. Males carry a pair of unsegmented genital appendages which are absent in females. Postgenital segment is absent in the females.

☒ Male Reproductive System

The male reproductive system (Fig. 28.23) consists of testes, vasa efferentia, vas deferens, accessory glands, copulatory organ and the external genitalia. The external genitalia consists of genital, postgenital and anal segments.

1. **Testes.** Ten closely juxtaposed pairs of spindle-shaped testes are arranged serially, the anterior end of the first pair being located in the 5th segment. The testis is covered by circular muscle fibres and fibrous tunica of connective tissue. Testis produces two types of sperm cells which differ in size of the head and tail.
2. **Vas efferens.** The testis is drawn into a narrow **vas efferens** at each end. The two vasa efferentia of a pair of testes are bound together in a common fibrous coat and open in a midian **vas deferens.** The posterior vasa efferentia of a testicular pair join the vas deferens slightly ahead of the anterior ones of the succeeding pair.
3. **Vas deferens.** It is differentiated into testicular and posttesticular zones. The testicular part is a narrow tube, concealed under the testes, and it continues as a very thin tube up to the 4th segment where it terminates blindly. The posttesticular part is a fairly transparent and prominent coiled structure. Anteriorly, it is narrow and highly coiled. Posteriorly, it is dilated at many places for accommodation of spermatophores. More posteriorly it divides into **right and left genital ducts** which occupy a position ventral to gut. The two then unite to form **terminal genital duct** communicating with the **ejaculatory canal.**
4. **Accessory glands.** These are two pairs, very closely set, white and lobular exocrin glands. The outer pair is longer and extend forward up to the 18th

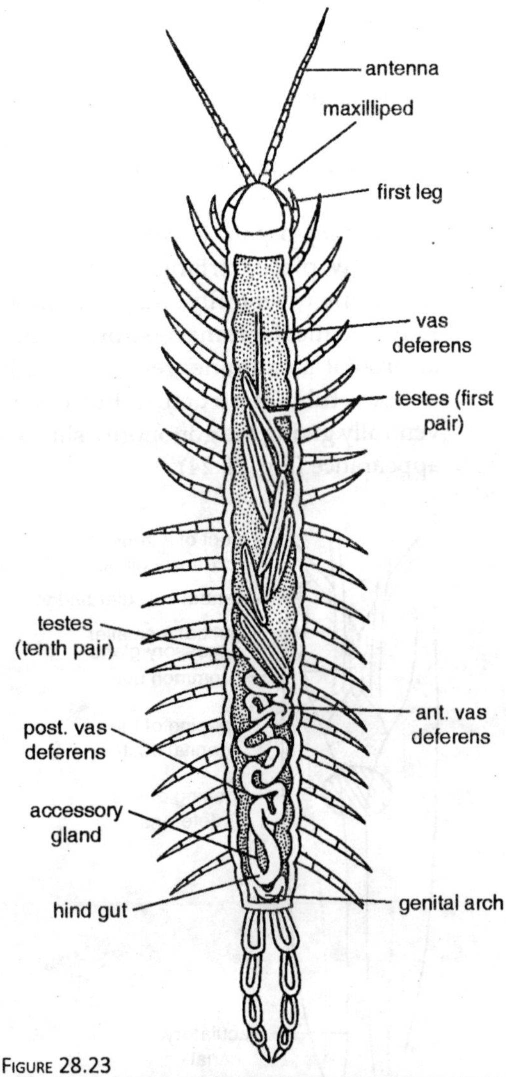

FIGURE 28.23
***Scolopendra* sp.** Male reproductive organs (ventral view)

segment. The ducts of this pair enter the copulatory organ to unite into a common duct, which communicates with the ejaculatory canal while the ducts of shorter pair enter the genital atrium.

5. **Copulatory organ.** It consists of two vertically placed triangular sclerites united anteroventrally and posterodorsally by cuticular transverse bridge and arthrodial membrane, respectively. It is not a tubular structure but opens ventrally giving the gonopore a slit-like appearance (Fig. 28.24).

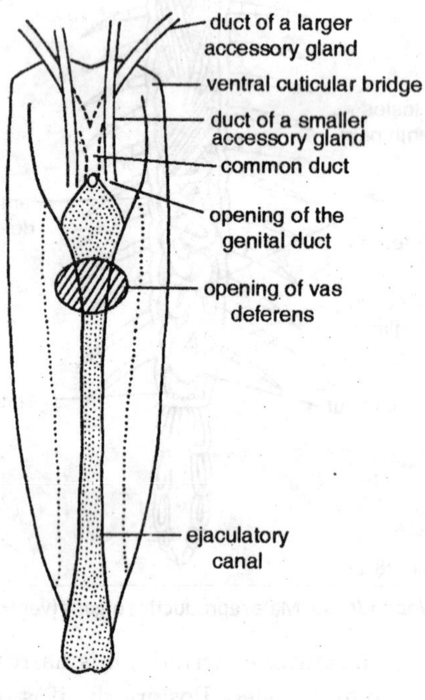

FIGURE 28.24

***Scolopendra* sp.** (Male). Copulatory organs, ducts of the accessory glands, vas deferens, ejaculatory canal (ventral view, semidiagrammatic)

⊠ Female Reproductive System

The female reproductive system (Fig. 28.25) consists of an ovary, an oviduct, a pair of receptacula seminis, accessory glands and the external genitalia, the last three being restricted to only genital and anal segments.

1. **Ovary.** It consists of an unpaired thinwalled egg tube situated on the middorsal line of the gut beneath the heart. It abruptly narrows down into a terminal filament in the 8th segment and extend up to the 5th segment. The ovarian epithelium is covered by longitudinal striated muscle fibres and con-

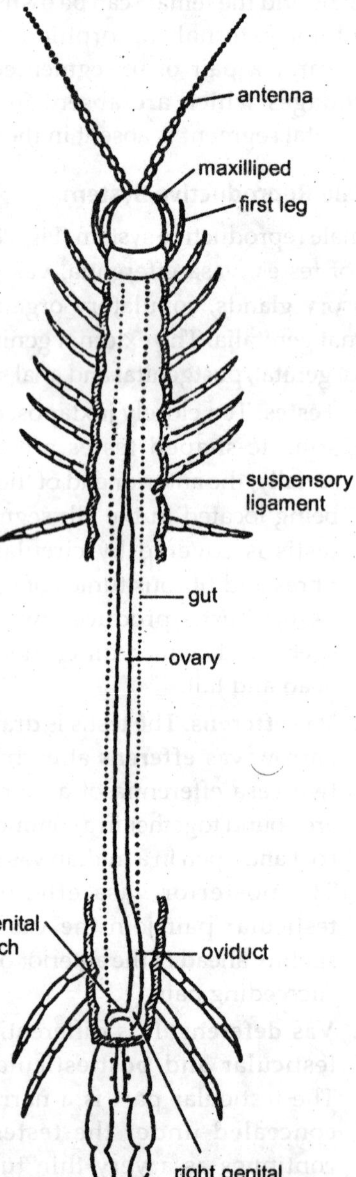

FIGURE 28.25

***Scolopendra* sp.** Female reproductive system (semidiagrammatic, ventral view)

nective tissue. Eggs are produced in the ovary. In the egg, vitelline membrane is apparently lacking although a periplasmic layer is recognisable. The egg contains protoplasmic reticulum, the meshes of which contain yolk globules.

2. **Oviduct.** Posteriorly the ovary passes imperceptibly into the oviduct. Intima is absent in the inner lining of the oviduct and the epithelium is thrown into folds which permits expansion of the lumen. The muscularis is comparatively thick and composed of longitudinal as well as a few oblique muscles. The oviduct divides in the last segment into **right** and **left genital ducts**. The ducts open separately into the genital atrium ahead of the openings of the ducts of the receptacula seminis and those of the accessory glands.
3. **Receptacula seminis.** Two in number, act as sperm storage organs in the last segment of the body (Fig. 28.26). Each receptaculum seminis is a tubular structure, closely bent on itself and communicating independently with the genital atrium by means of a duct which is highly coiled, particularly near the receptacle. The receptacle epithelium produces a secretion which is both protective and nutritive in function.
4. **Accessory glands.** A single pair of ectodermal origin, white in colour and lobular in appearance (Fig. 28.26). The ducts of the glands open posteriorly into the openings of the receptacular ducts which are ultimately connected with the genital atrium. They probably help in gluing the eggs by their secretion.

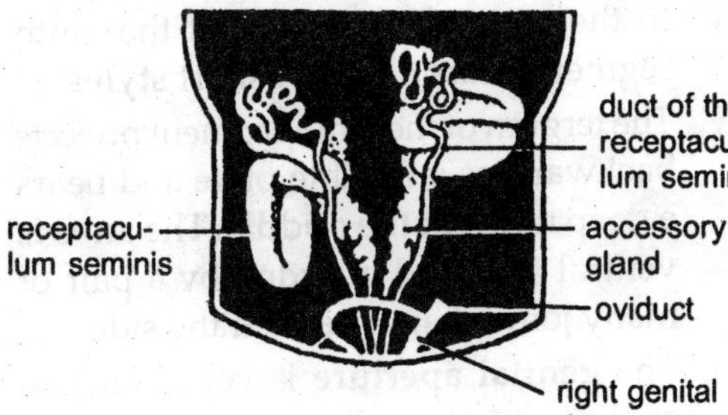

FIGURE 28.26

***Scolopendra* sp.** (Female). Receptacula seminis and accessory glands

☒ Reproduction

The sperms are packed in **spermatophores**, containing long, thread-like sperms twisted in bundles, and a nutritive fluid. The ovum contains scattered yolk granules. Fertilization and early development internal.

PERIPLANETA sp. (Cockroach)

The *Periplaneta* is a common household pest. It belongs to the subclass Pterygota, class Insecta, subphylum Mandibulata, phylum Arthropoda. It is nocturnal in habits, and almost omnivorous in diet.

☒ External features

The entire body surface is invested in a chitinous cuticle of brown colour.

The body is divisible into four distinct regions—the **head**, the **neck**, the **thorax** and the **abdomen** (Fig. 28.27).

Head

The head is triangular. The narrower end is directed downward and the broad end is attached to the body at right angle, through the neck.

i. The **mouth** is situated in front of the head. It is guarded by the **labrum** or upper lip in front, **labium** or lower lip behind and a pair of **mandibles** on the sides.
ii. A pair of compound **eyes**, appearing as reniform black patches, are present on the sides of the head.
iii. A pair of many jointed, long, slender feelers or **antennae** are present just below the eyes.
iv. A more flexible pair of jaws, the first **maxillae**, are situated behind the mandibles.

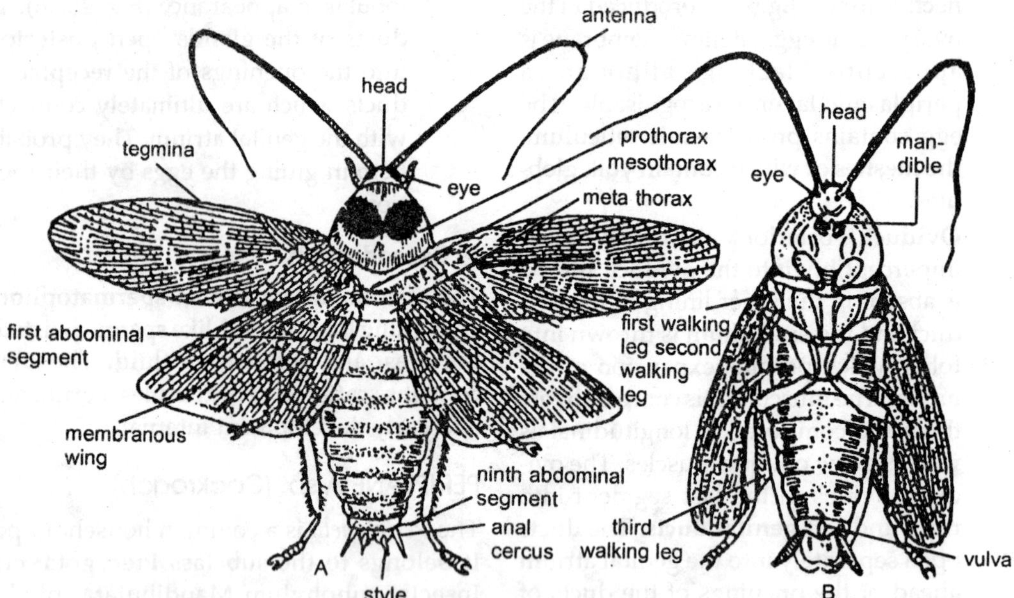

FIGURE 28.27

Cockroach (*Periplaneta americana*). A. Dorsal view, Male. B. Ventral view, Female

Neck

The neck is a small narrow region connecting the head with the thorax. It is covered by a thin, flexible cuticle supported by thickened and hardened patches, the **cervical sclerites.**

Thorax

The thorax consists of three segments, **prothorax, mesothorax** and **metathorax** covered by a chitinous dorsal **tergum** and a ventral **sternum.**

i. **Pronatum** is a shield-like hard structure present on the dorsal surface of the prothorax.
ii. A pair of thick, opaque, brown plates, the **elytra** or anterior wings are attached to the anterior border of the tergum of the mesothorax.
iii. A pair of delicate, membranous **posterior wings** are attached to the tergum of the metathorax.
iv. A pair of five **jointed legs** are attached to the sternum of each of the thoracic segments.

Abdomen

The abdomen gradually narrows down posteriorly. It consists of ten segments. Each segment is covered by a thin **tergum** and a **sternum.**

i. A segment is partially covered at the anterior end by its preceding segment.
ii. The eighth and ninth terga are hidden from view by being overlapped by the seventh.
iii. In the female, the sternum of the seventh segment is very much prominent.
iv. In the male, the sternum of the ninth segment bears a pair of short **styles.**
v. The tergum of the tenth segment projects backward as a flexible plate and bears a deep notch at the middle. The **anus** is ventral to it and guarded by a pair of many jointed **anal cerci** at the sides.
vi. The **genital aperture** is ventral to the anal opening.
vii. Ten pairs of respiratory apertures called **stigmata**, two in the thorax and eight in the abdomen are present on the sides of the body.

⊠ Appendages

Appendages are seven pairs; four pairs in the head and three pairs in the thorax.

Cephalic appendages (Figs. 28.27 and 28.28).

1. **Antennae.** Each antenna is long, slender and many-jointed (Fig. 28.27).

   Function. Sensory.

   Mandibles, 1st maxillae, 2nd maxillae and hypopharynx constitute the **mouth parts**.
2. **Mandibles.** The body is a strong chitinoid structure provided with teeth.
   Function: Cutting of food.
3. **1st maxillae.** Each consists of a protopodite bearing a double endopodite having lacinia and galea and a five-jointed exopodite. The maxillary palp is supposed to represent the endopodite of crustacea.

   Functions. Holding and masticating the food and sensory.
4. **2nd maxillae.** The pair fused to form a **labium.** The fused base form two sclerites, submentum and mentum. Mentum bears the prementum, which in turn carries an outer, paired and lobed structure, the paraglossae and an inner, paired and lobed glossae; together forming the ligula.

   Functions. Sending the food to the mouth and sensory.
5. **Thoracic appendages.** Three pairs of legs. Each leg consists of five segments, coxa, trochanter, femur, tibia and a six-jointed tarsus ending in a pair of claws (Fig. 28.29).

   Function. Walking.
6. **Abdominal appendages.** The **anal cerci** are possibly the relics of last abdominal appendages (Fig. 28.27).

   Function. Tactile.

The **styles** of male and the **gonapophyses** are possibly relics of appendages.

⊠ Muscles

Muscles concerned with movement of the appendages and feeding, in running and flight and abdominal movement for respiration are striated. Insertion of muscles occur through tendons in many cases.

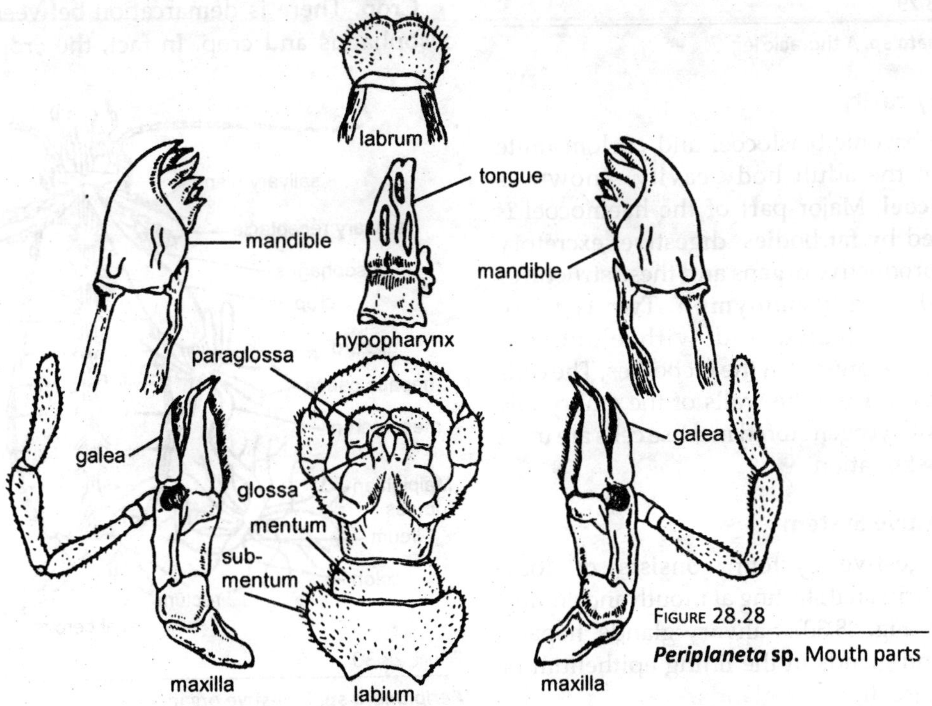

FIGURE 28.28

***Periplaneta* sp.** Mouth parts

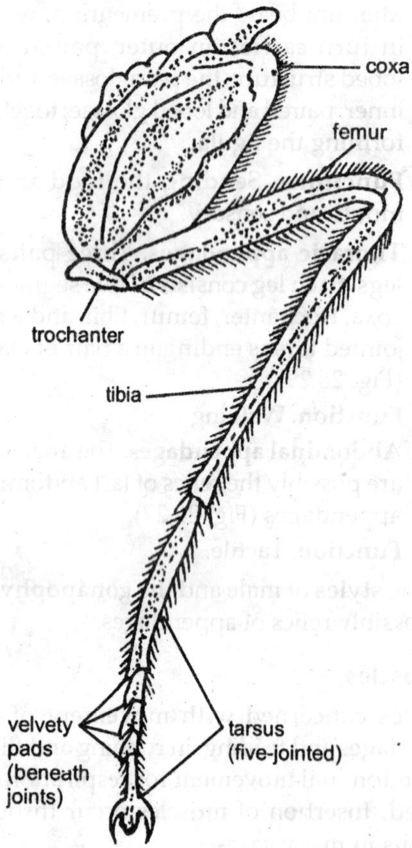

FIGURE 28.29

***Periplaneta* sp.** A thoracic leg

☒ Body cavity

The embryonic blastocoel and coelom unite to form the adult body cavity, known as **haemocoel**. Major part of the haemocoel is occupied by fat bodies, digestive, excretory and reproductive organs and these structures are bathed in **haemolymph**. Two types of cells—binucleated and with elongated nucleus, are present in the fat bodies. The cells are derived from the walls of the embryonic cavities. Glycogen stored in these cells are used during starvation.

☒ Digestive System

The digestive system consists of long alimentary canal starting at mouth and ending in anus (Fig. 28.30). Salivary glands, hepatic caeca and glands in the lining epithelium of the gut are digestive glands.

☒ Alimentary Canal

The alimentary canal is divided into three zones—foregut, midgut and hindgut. The fore and hindgut are lined by cells of ectodermal origin, while the lining of the midgut is endodermal.

Foregut

The foregut or stomodaeum starts from mouth, ending in gizzard. It is lined by cuticle.

Mouth. Ventral in position, located at the base of the buccal cavity.

Buccal cavity. An ill-defined chamber, bounded anteriorly by epipharynx and labrum, posteriorly by hypopharynx and labium and laterally by two mandibles.

Pharynx. A short vertically oriented tube, opening into the oesophagus. The wall of the opening between the pharynx and the oesophagus is thick, muscular and the opening is guarded by a sphincter.

Oesophagus. A short, narrow tube, extending up to the prothorax and ending in crop.

Crop. There is demarcation between the oesophagus and crop. In fact, the crop is a

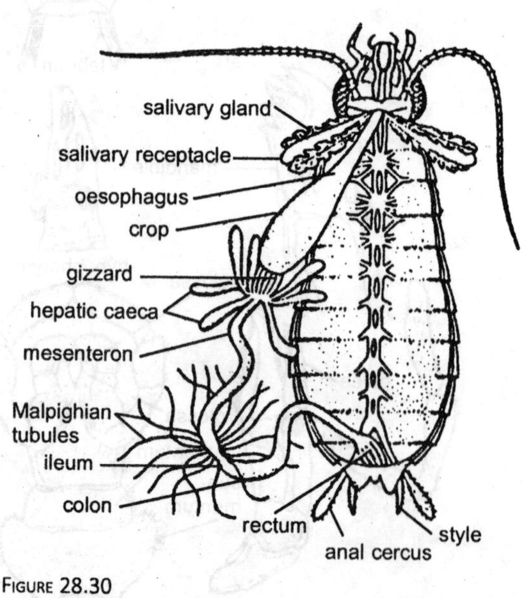

FIGURE 28.30

***Periplaneta* sp.** Digestive organ

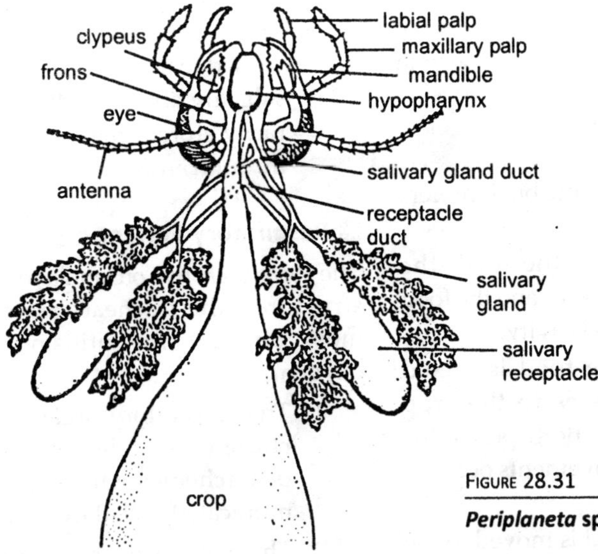

FIGURE 28.31

***Periplaneta* sp.** Salivary apparatus

large, sac-like dilatation of oesophagus. It extends into the abdomen. The crop acts as a temporary reservoir of food, where it can be retained for about 60 days.

Gizzard or proventriculus. A round, thick-walled, muscular structure, posterior to the crop. It has two parts — the anterior contains six chitinous teeth in the inner wall and the posterior, two circular hairy cushions. The food is crushed by the teeth and the hairy cushions permit only finer food particles pass to the midgut.

Midgut or mesenteron

A narrow tube, running from the gizzard to the hindgut. 7 to 8 **hepatic caeca** are present at the junction of the gizzard and the hindgut. The hepatic caeca presumably secrete digestive enzymes.

Hindgut

A narrow tube, divisible into three zones —**ileum, colon** and **rectum**. The junction of the mid and hindgut is marked by 60 to 70 extremely fine, yellowish Malpighian tubules. (The tubules are excretory in function). The ileum is narrow; colon broad and slightly coiled, with irregular folds of the lining epithelium, covered with a layer of chitin. The rectum is sac-like, bearing papillae on the inner wall. Rectal glands in the wall absorb water and salt and help in osmoregulation. The rectum opens to the exterior through anus, provided with a sphincter muscle, between two podical plates at the posterior end.

⊠ Digestive glands

1. **Salivary apparatus.** It consists of two salivary glands and two receptacles with their ducts (Fig. 28.30 and 28.31). The glands and the receptacles lie on the dorsolateral aspects of the crop. Each gland consists of two leaf-like glands, made of diffused lobes. The receptacles are elongateoval sacs, the anterior ends being narrower. The ducts of the glands and receptacles run forward by the sides of the crop. The ducts from the two glands unite and those from the receptacles also unite to form two common ducts, which again unite and form an **efferent salivary duct** opening on the ventral side of the hypopharynx.
2. **Hepatic caeca.** 7 to 8 in number, at the junction of the fore and midgut (Fig. 28.30). The caeca are about one-third of the midgut in diameter and open in it. The cells in the lining epithelium of the caeca are believed to secrete digestive enzymes.

3. **Glands of the midgut.** The cells of the lining epithelium of the midgut secrete digestive enzymes.

☒ Feeding and digestion

Cockroach is an almost omnivore but it prefers starchly food.

1. The maxillae procure the food, the mandibles cut it into pieces and the food is taken into the buccal cavity.
2. In buccal cavity, the food is mixed with saliva and passes to the crop through oesophagus. Both peristaltic and antiperistaltic movements occur in the crop.
3. From the crop the food is moved to the gizzard, where it is crushed with cuticular teeth and only the finer particles passing through hairy cushions enter the midgut.
4. Midgut is the site of both digestion and absorption. Carbohydrases, proteinases and lipases are secreted by digestive glands.
5. The undigested part of the food is passed to the hindgut, where water and salts are absorbed, the residue temporarily stored there and finally egested through anus.

☒ Respiratory System

The *Periplaneta* live on land and the respiratory structures, the tracheae, are considered as most efficient for utilising atmospheric oxygen.

1. The respiratory organs consist of **tracheae** or air tubes and their branches or **tracheoles** which are directly communicated with the exterior (Fig. 28.32).
2. The tracheae open to the exterior by ten pairs of apertures, the **spiracles** or **stigmata**. Two are placed on each side of the thorax; one between pro-and mesothorax, and the other between meso and metathorax. Eight occur on each side of the abdomen between terga and sterna of the first eight segments. The stigmata are guarded by hairs, etc.

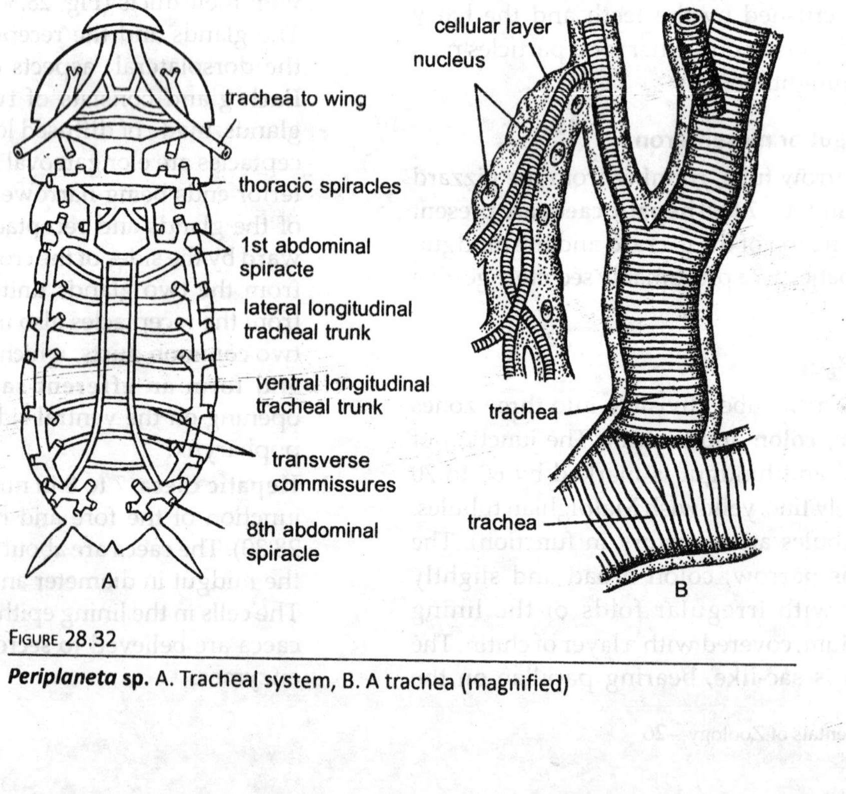

FIGURE 28.32

***Periplaneta* sp.** A. Tracheal system, B. A trachea (magnified)

to prevent the entrance of harmful particles. Each spiracle is controlled by muscular valves. From each spiracle leads inward a small tube which ends in a stout trachea and all of them are connected by longitudinal and transverse tracheae.

3. There are two lateral longitudinal tracheal trunks, one on each side of the body. They are connected by several transverse tracheae, and the tracheae form a system of intercommunicating network. Each trachea divides and subdivides into numerous branches and the ultimate, very fine branches entering the organs are called tracheoles. The tracheoles end in tissues (Fig. 9.7). The wall of the trachea is strengthened by chitinoid lining in the form of spiral threads.

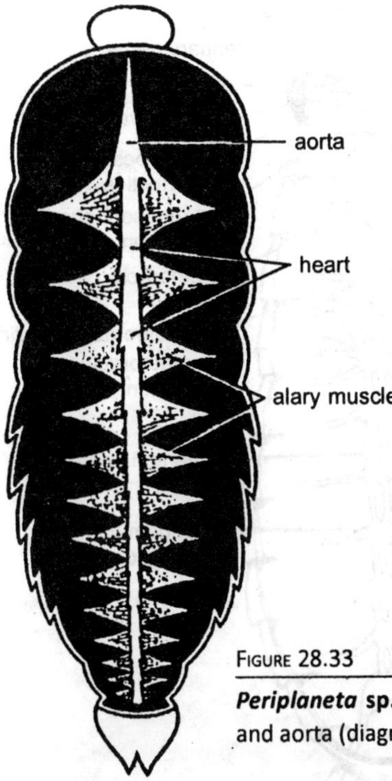

FIGURE 28.33

***Periplaneta* sp.** The heart and aorta (diagramatic)

⊠ Mechanism of Respiration

The alternate expansion and contraction of the abdomen brings inhalation and exhalation of air in the tracheal system. The opening and closing of the spiracles are also dependent on the CO_2 concentration of the inhaled air. The walls of the tracheoles allow gaseous exchange and diffusion takes place between the cell sap and the lumen of the tube. The tissues get direct supply of oxygen from the atmospheric air.

⊠ Circulatory System

The heart and pericardium

The heart (Fig. 28.33) is an elongated tube, which runs through the whole length of the thorax and abdomen as a dorsomedian contractile vessel. The cavity of the vessel is divided into 13 segmentally arranged chambers, of which 3 are thoracic and 10 are abdominal. Externally, each chamber is marked by constrictions and internally communicates with the adjacent chambers through narrow passages guarded by valves. The wall of each chamber is perforated by two openings known as **ostia**. The whole structure is enclosed in a space known as the **pericardial sinus** bounded by a thin membraneous **pericardium**. The pericardium also bears a number of perforations and as a result, the pericardial cavity is in communication with the perivisceral cavity or the haemocoel.

Arteries

The heart is anteriorly continued into the head as a slender tube, the aorta. The aorta runs forward along the dorsal surface of the oesophagus and ends in front of the peripharyngeal nerve ring in the form of a funnel. By the successive contraction of the chambers of the heart, haemolymph is driven forward and poured in the haemocoel through the aorta. Here, the blood bathes all the visceral organs.

Sinuses

The haemocoel remains filled up with haemolymph coming from the heart. From there it enters the pericardial sinus through apertures in the pericardium.

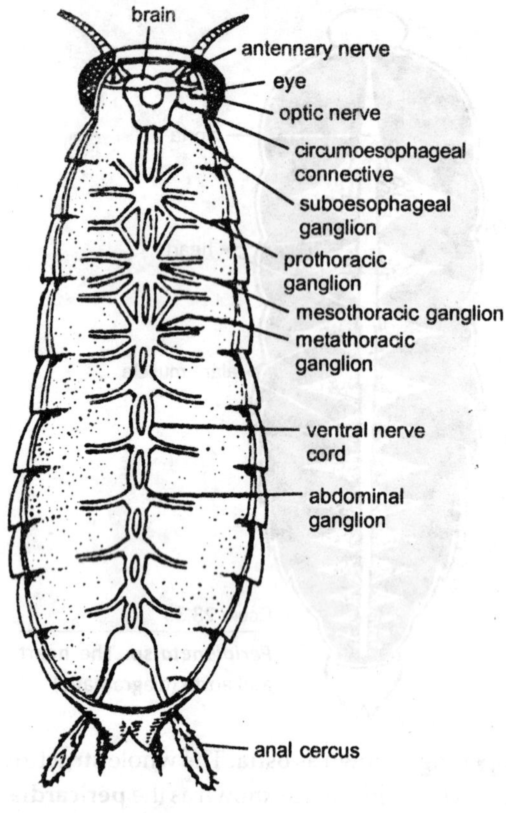

FIGURE 28.34

***Periplaneta* sp.** Nervous system

Haemolymph

Haemolymph is a watery fluid having no colour. White blood corpuscles, the amoebocytes float in it. The haemolymph bears no respiratory pigment and, as such, it is not concerned with respiration.

Course of circulation

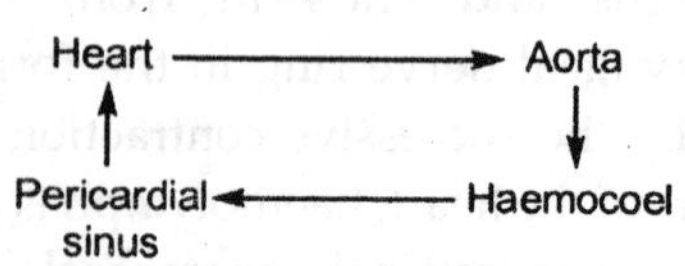

Excretory System

1. The excretory organs consist mainly of numerous Malpighian tubules and nephrocytes (Fig. 28.30).
2. The Malpighian tubules are thread-like yellow structures of variable number present at the junction of the mid and hindgut. Each tubule is a hollow structure opening into the lumen of the intestine and the wall is made of a single layer of glandular epithelium.

   Nitrogenous wastes are absorbed by the cells from the haemolymph and the same are discharged into the lumen of the tubule which in turn, convey the wastes into the lumen of the intestine.
3. Certain cells, the **nephrocytes** associated with the fat bodies and the hypodermis also serve as excretory structures.

⊠ Nervous System

The nervous system comprises of central nervous system, peripheral nervous system and visceral nervous system (Fig. 28.34).

Central Nervous System

It consists of a brain or supraoesophageal ganglia, suboesophageal ganglia, circumoesophageal connectives and a double ventral nerve cord with ganglia.

1. **Brain or supraoesophageal ganglia.** It is a large bilobed mass at the posterior part of the head, below the epicranial plates and above the oesophagus and formed by the fusion of three pairs of ganglia (Figs. 28.34, 28.35).
2. **Suboesophageal ganglia.** Formed by the fusion of three pairs of ganglia associated with the segments that produce mouth appendages and located in the head below the oesophagus.
3. **Circumoesophageal connectives.** The brain is connected with the suboesophageal ganglia by a pair of short, stout nerves, running around the oesophagus.
4. **Ventral nerve cord.** The nerve cord is formed by the fusion of two solid nerve cords and run backwards from the suboesophageal ganglia along the midventral line of the thorax and abdomen.

a. The nerve cord bears nine pairs of ganglia, three pairs in the thoracic segments and six pairs in the first six abdominal segments.

b. The thoracic ganglia are large in size. The fifth abdominal ganglia located at the junction of the fifth and sixth segments is largest and occupies the terminal position of the nerve cord.

☒ Peripheral Nervous System

Nerves arising from the ganglia of the central nervous system constitute the peripheral nervous system.

1. **Optic and antennary nerves.** Paired, given off from the brain and end in the compound eyes and the antennae.
2. **Mandibular, maxillary and labial nerves.** Emanate from suboesophageal ganglia and supply mandibles, maxillae and labium.
3. **Thoracic nerves.** Several pairs of nerves from each thoracic ganglion innervate the muscles, appendages and other organs in the corresponding segment. The first pair communicate with the ventral nerve cord.
4. **Abdominal nerves.** Paried nerves from each abdominal ganglion end in the organs of the corresponding segment.
5. The last abdominal ganglion supplies nerves to the sixth abdominal segment and other segments behind.

☒ Visceral or Sympathetic Nervous System

The system consists of a frontal ganglion, a visceral ganglion, a pair of oesophageal ganglia, a hypocerebral ganglion and their connectives, (Fig. 28.36).

1. **Frontal ganglion.** A median ganglion, just in front of the brain and connected to it by two bilateral connectives and sends a **recurrent nerve** posteriorly to the **visceral ganglion** on the dorsal surface of the crop.
2. **Oesophageal ganglia.** A pair, on the dorsolateral sides of the oesophagus, just behind the brain. The two are joined by connectives and also connected to a median **hypocerebral ganglion.**

☒ Receptors and Sense Organs

Photoreceptors, mechanoreceptors, thermoreceptors and chemoreceptors constitute sense organs.

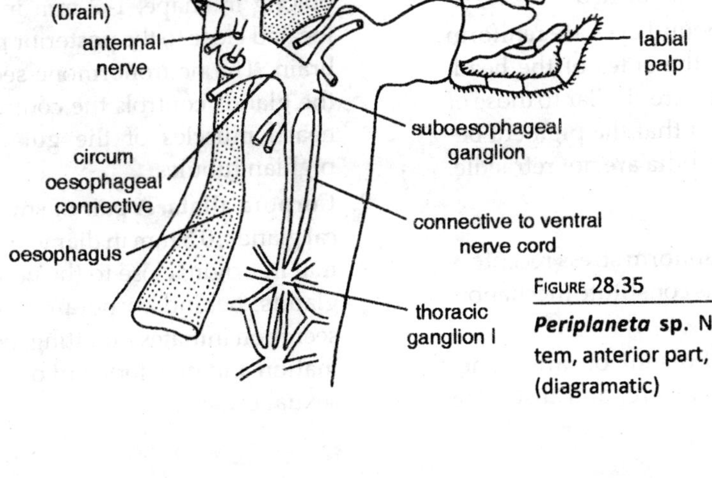

FIGURE 28.35

***Periplaneta* sp.** Nervous system, anterior part, Lateral view (diagramatic)

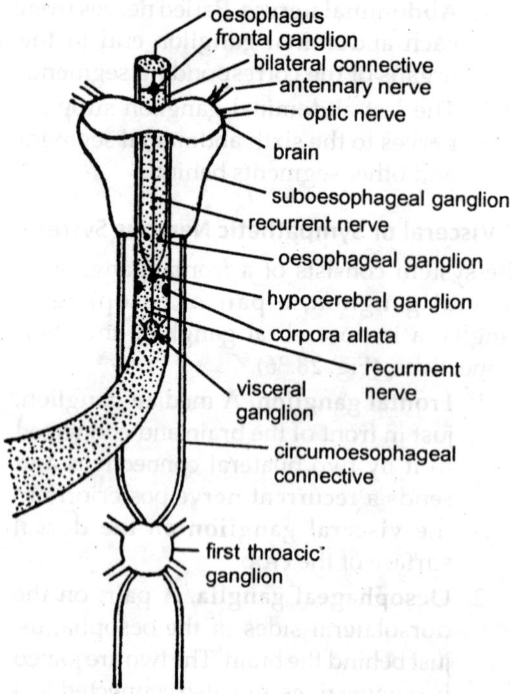

FIGURE 28.36

***Periplaneta* sp.** Visceral nervous system (diagramatic)

Photoreceptors

Ocelli and eyes are photoreceptors.

1. **Ocelli.** An ocellus appears as a white area near the base of each antenna. It consists of an outer transparent **corneagen layer** of flattened cells; 3 to 5 **retinular** cells; one **rhabdome** formed by the inner margins of the retinular cells. The corneal lens in an ocellus is single and a crude image of an object—at a close range—is formed.
2. **Eyes.** The compound eyes are reniform and located on the sides of the head. Structurally, they are similar to those of the prawn except that the pigment between the ommatidia are not retractile.

Mechanoreceptors

Tactile hairs, campaniform stress receptors and chordatonal organs constitute mechanoreceptors.

1. **Tactile hairs.** Present on antennae, mouth palpi, anal cerci and distal leg segments. These are in the form of hairs, bristles, plates, spines, etc.
2. **Campaniform stress receptors.** Present on the labial palps and their adjacent areas and ridges on the segments of the leg. They are sensitive to pressure difference on surface.
3. **Chordotonal organs.** Aggregates of specialised cells, arranged parallely or in the shape of a fan on the legs. They respond to vibration stimulus.

Thermoreceptors

Sensitive to temperature; located in pads between the first four tarsal segments of the leg.

Chemoreceptors

Sensitive to chemical stimuli—smell or touch.

1. **Olfactory organs.** Distributed on the antennae; represented in two forms—thick-walled bristles and thin-walled hairs.
2. **Taste receptors.** Distributed on the tips of the maxillary and labial palps, inner surface of the mouth parts and inner border of the mouth and pharynx.

Endocrine System

Endocrine organs, corpora cardiaca, corpora allata and prothoracic glands constitute the endocrine system.

1. **Corpora cardiaca.** Corpora cardiaca are oblong in shape, 1–2 mm in size and located close to the posterior part of the brain. The neurohormone secreted by the glands controls the contractility of heart, muscles of the gut and Malpighian tubules.
2. **Corpora allata.** A pair of small spherical glands, 0.5 mm in diameter, ectodermal in origin, close to the oesophageal glands, behind corpora cardiaca. The secretion initiates moulting, occyte formation and development of secondary sexual organs.

3. **Ecdysial gland (Prothoracic gland).** A semitransparent gland in the midventral position of the prothorax. It consists of two bands of tissues that cross in the form of 'X'. The anterior arms of the cross are relatively long, the overall length being 3.0 ± 2.0 mm. The breadth of the anterior arm is 0.025 ± 0.005 mm.

⊠ Reproductive System

The sexes are separate.

Male reproductive system

1. The male reproductive system consists of a pair of testes, a pair of vasa deferentia, a pair of seminal vesicles, an ejaculatory duct, a conglobate gland and a gonopore (Fig. 28.37).
2. The **testes** are small bodies made up of a number of vesicles and situated in the 4th and 5th abdominal segments below the terga.
3. Each testis is connected to a narrow duct, the **vas deferens**, which runs backwards, inwards and downwards.
4. Each vas deferens, before uniting with its fellow of the opposite side, dilates a little forming a **seminal vesicle**, from which arise a large number of blind pouches and the whole arrangement is known as **mushroom gland.**

   After forming the structure, the two vasa deferentia unite and form a muscular tube, the **ejaculatory duct**, which opens in the **gonopore**, below the anus, guarded by gonapophyses. The **conglobate gland** is associated with the ejaculatory duct.

Female Reproductive system

1. The female reproductive system consists of a pair of ovaries, a pair of oviducts, a pair of colleterial glands, a pair of spermathecae and a gonopore (vulva) (Fig. 28.38).

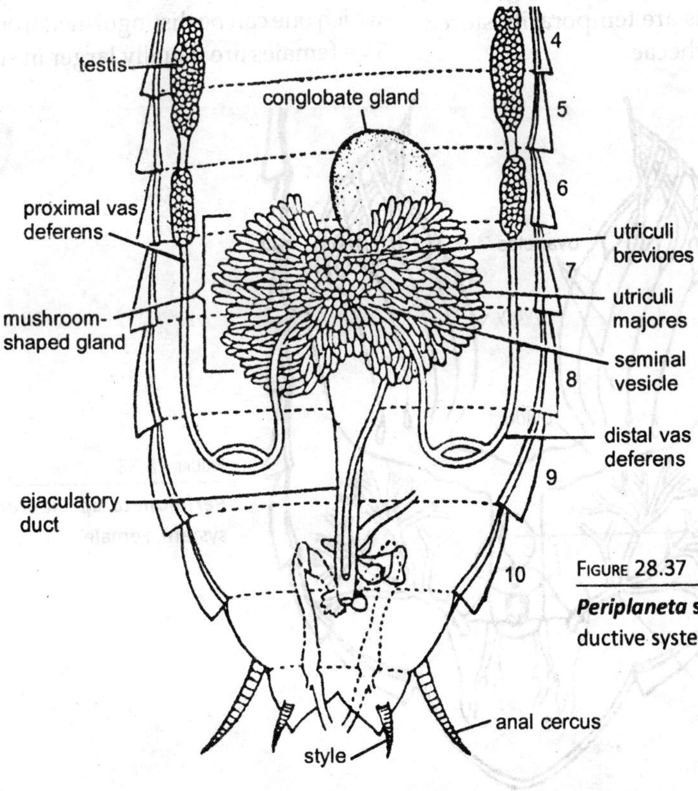

FIGURE 28.37

***Periplaneta* sp.** Reproductive system. Male

2. **Ovaries** are situated in the abdomen. Each ovary consists of 8 **ovarioles** con nected anteriorly with the body wall by a ligament.
3. From the posterior end of the ovariole runs a narrow tubule and by the union of tubules coming from all the ovarioles of an ovary, a wider tube, the **oviduct** is formed. The two oviducts unite to form a median chamber, sometimes called the **vagina**, which opens in the **gonopore (vulva)**, guarded by the gonapophyses on the sternum of the 8th abdominal segment. A pair of asymmetrical **spermathecae** of which only one is functional, open together in the genital pouch in the middle of the sternum of the 9th segment. A pair of **colleterial glands** open behind the spermathecae.

⊠ Fertilization and Development

1. **Spermatophores** packed with sperms are transferred to female during mating and the sperms are temporarily stored in the spermathecae.
2. Sixteen centrolecithel eggs from each ovary pass to the genital pouch through the oviduct and fertilized there.
3. Sixteen fertilized eggs are enclosed in a case, the **Ootheca**, formed by the secretion of the colleterial glands.
4. Development direct and the youngs come out of the ootheca as nymphs. The nymphs are soft-bodied, white, without gonads and resemble the adult, except being smaller in size and wings lacking
5. The nymphs undergo several **moultings** and the stage between two successive moultings is called **instar**. Wings appear with the completion of last moulting.

MELANOPUS sp. (Grasshopper)

⊠ Reproductive System

The grasshopper is an insect of a relatively generalised type. The male and the female has external dimorphic characters, by means of which one can be distinguished from the other. The females are usually larger in size than the

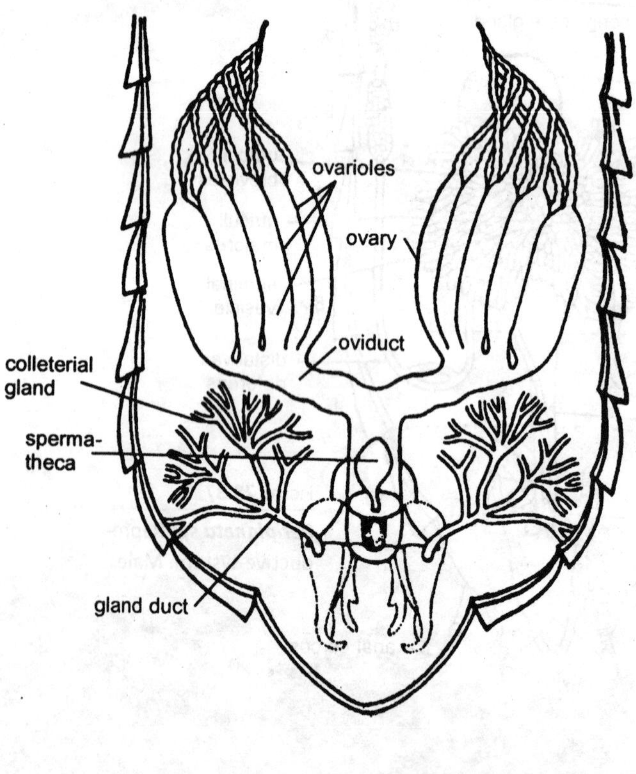

FIGURE 28.38

***Periplaneta* sp.** Reproductive system. Female

males and have a pair of hard ovipositors, the four-pronged terminus of the abdomen. Quite a few species of grasshoppers are common in India.

⊠ Male Reproductive System

The male reproductive system (Fig. 28.39B) consists of a pair of testes, a pair of vasa deferentia, an ejaculatory duct, a copulatory organ, accessory glands and a genital opening.

1. The **testes** are almost oval in shape, situated ventral to the alimentary canal and each is formed of **testicular follicles**, where the spermatozoa are formed. The testes are held together by terminal filaments.
2. From the two testes come out two sperm ducts or **vasa deferentia**.
3. The two vasa deferentia unite to form a short but broad **ejaculatory duct** which opens on the terminal segment of the abdomen.
4. A thick concial **penis** or copulatory organ is situated between the anus and the external genital opening.
5. **Accessory glands** are present at the anterior end of the ejaculatory duct in between the vasa deferentia. These apparently secrete a fluid that aids in the transfer of spermatozoa to the female.

⊠ Female Reproductive System

The female reproductive system (Fig. 28.39A) consists of a pair of ovaries lateral to the alimentary canal, a pair of oviducts, a vagina, a spermatheca, certain accessory structures and a genital opening.

1. The ovaries are a pair of slightly elongated bodies, each of which is composed of a number of egg tubules or ovarioles held together by terminal filaments. The **ovarioles** are devoid of lumen, contain oögonia and oöcytes arranged in a linear series with nurse cells, and other tissue cells.
2. The **oöcytes** grow in course of their way down.
3. The ovaries are connected with a pair of oviducts. They are swollen at the middle and unite to form a **vagina** posteriorly, which opens into a genital chamber, situated between the plates of the **ovipositor**, the organ for digging holes to place the fertilized eggs.

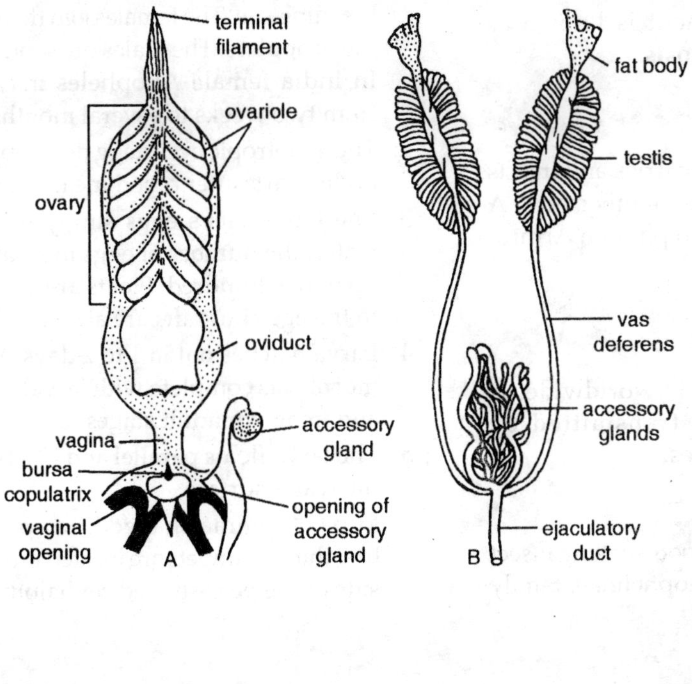

FIGURE 28.39

Grasshopper (*Melanopus* sp.) Reproductive system. A. Female. B. Male

4. A tubular **seminal receptacle** or **spermatheca** opens dorsal to the vaginal pore. The spermatheca receives spermatozoa and stores them. The sperms are released to fertilize eggs.
5. There is a small pore, the micropyle, which acts as the passage for the entrance of spermatozoon.

☒ Fertilization and Development

1. The matured, eggs, escape from the ovariole and pass down the oviducts. Yolk is deposited and the shell is secreted around the fertilized egg during this period.
2. The egg is surrounded by a delicate vitelline membrane and a brownish flexible shell. The fertilized eggs are deposited in holes or tunnels and are covered with a sticky secretion that adheres as an **egg pod**.
3. The young grasshopper, after hatching, looks like the adult excepting a large head and the conspicuous absence of wings. This stage is known as **nymph**. It is a stage of voracious feeding and rapid growth. The nymph undergoes five moultings, after which the adult condition is reached. During the process, the wings are formed and this mode of development is known as **gradual metamorphosis**.

BIOLOGY OF VECTORS

Vector is an organism that carries and injects pathogens from one host to another host. A number of vectors belong to phylum Arthropoda.

MOSQUITOES

Distribution of mosquito is worldwide. A number of diseases are transmitted by Anopheles, Culex and Aedes.

☒ Anopheles

Anopheles mosquito is a blood sucking insect belonging to subfamily Anophelinae, family Culicidae, suborder Nematocera, order Diptera. It transmit **malaria parasites** from host to host.

☒ Morphology (Fig. 28.40)

1. The wings are profusely spotted. The characteristic venation of the wings helps in identification of the species. In resting position the head, thorax and abdomen are in a straight line and at 45° angle, with the support.
2. The proboscis is elongated and that of female is specially adapted for piercing the skin and sucking blood. The palpi in both male and female are as long as the proboscis.
3. The character of the antennae differs in two sexes. In male antennae possess bunches of long hairs; in female the hairs are short and scanty.
4. Thorax bears three pairs of jointed legs and a pair of wings attached to the mesothorax.
5. Abdomen consists of 8 to 9 visible segments. The terminal segments are modified for sexual purposes.

☒ Breeding and life history

1. The males appear in swarms near the breeding site. The females join them and they copulate. The males die soon after. In India female Anopheles may live from two weeks to several months.
2. The genotropic cycle (egg development cycle) starts after blood meal.
3. The female lays eggs singly in fresh water, the number of eggs in a batch is up to one hundred. Floats are attached to the eggs but rafts are absent.
4. Larva hatches out in 1 to 2 days. Metamorphosis complete with larval, pupal and imago or adult stages.
5. The larva floats parallel and just below the water surface.
6. Pupa "comma-shaped", a pair of breathing trumpets are located on either side of the pear-shaped cephalothorax.

7. In favourable temperature the larval development may be completed within 9 days.

☒ Bionomics

1. Eggs, larvae and pupae are aquatic, hence presence of water is essential for the existence of mosquitoes.
2. The males are not blood feeder, feed on plant sap. Females prefer a blood meal from man or a vertebrate, which is essential for the maturation of ova.
3. Anopheles are mostly nocturnal, becoming active just after sun set. During day time they hide in shady places.
4. A large number of Indian species act as vectors of malaria parasites. *A. culifacies* and *A. fluviatilis* breeds in irrigation canals and rice fields; A. ***stephensi***, a notorious vector of ***Plasmodium*** in Indian cities breeds in cisterns, wells and temporary receptacles of water; A. ***philippinensis*** breeds in ditches, swamps, rice fields and lakes; A. ***annularis*** breeds in stagnant water, rice fields, ponds with weeds and in swamps and lakes; A. ***sundaicus*** breeds in brackish water and A. ***varuna*** breeds in stagnant fresh water tanks and ditches.

☒ Transmission of Parasites

1. Only a few species of *Anopheles* mosquito transmit malarial parasite *Plasmodium* from host to host.
2. During blood meal the **gametocytes** of plasmodium enter the stomach of mosquito with blood and undergo stages of development in the stomach wall and coelom and are transformed to infective **sporocysts.** The sporocysts crowd in the salivary glands of mosquito. During second blood meal from another host the sporocysts are transferred to the new host and infection continues.

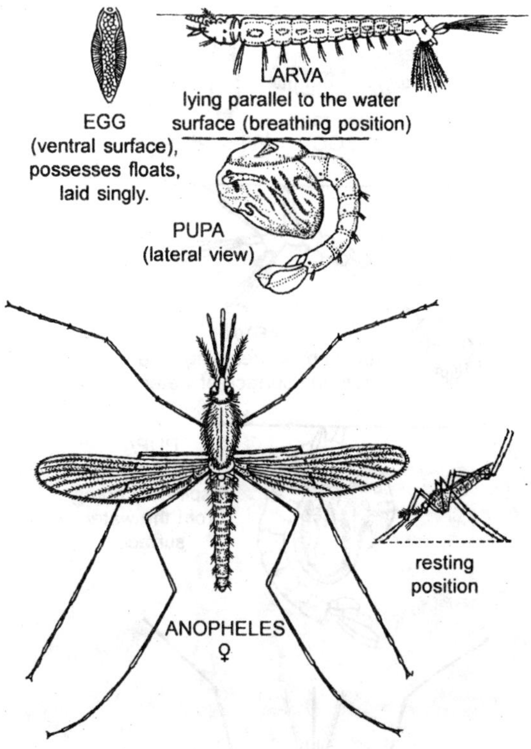

FIGURE 28.40

***Anopheles* sp.** Life cycle stages and resting position

CULEX

Culex mosquito is a blood sucking insect belonging to subfamily Culicinae, family Culicidae, suborder Nematocera, order Diptera. It transmits filaria worms *Wuchereria bancrofti* and *Dirofilaria immitis* and also bird malaria, avian pox and arbovirus encephalitides from host to host.

☒ Morphology (Fig. 28.41)

1. Body colour grey-black or brown with scales.
2. Wings smooth, narrow, long spotless and of uniform colour.
3. The maxillary palps are stiff and bear many bristles. The palpi in the female are short and three-jointed. In male they are as long or even larger than proboscis and five-jointed.

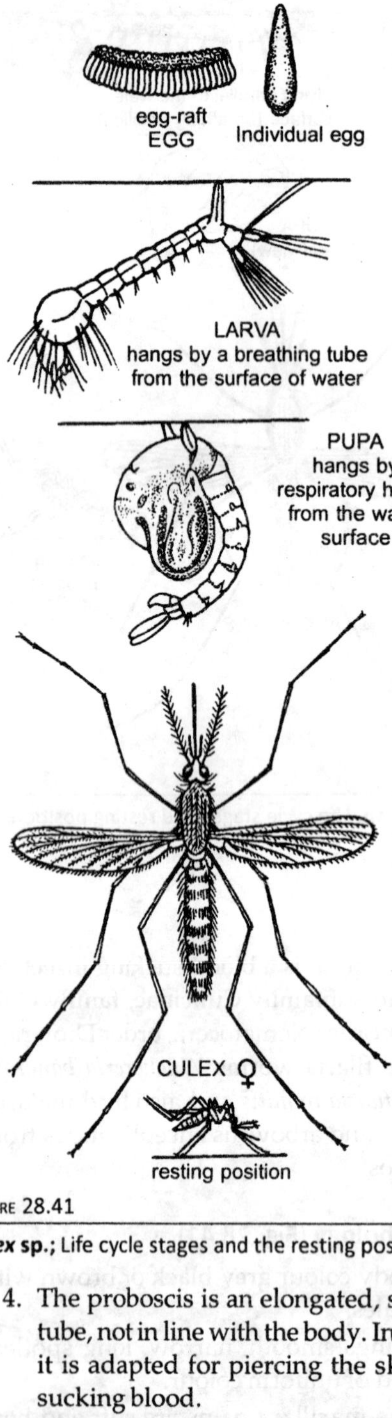

FIGURE 28.41

***Culex* sp.**; Life cycle stages and the resting position

4. The proboscis is an elongated, straight tube, not in line with the body. In female it is adapted for piercing the skin and sucking blood.
5. Thorax bears three pairs of jointed legs and a pair of wings attached to the mesothorax.
6. Brownish-yellow basal bands present in the abdomen.

☒ Breeding and life history

1. *Culex* mosquitoes, both male and females assemble in swarm near breeding places, and they copulate. The male dies after copulation. The females live four weeks to several months.
2. Following copulation the female takes blood meal.
3. After a blood meal the female lays eggs at night on still water of ponds, pools or rain-filled containers. The eggs are laid side by side standing erect and glued together to form a boat-shaped rafts which float on water. One female may lay upto 300 eggs.
4. The eggs hatch in 1 to 3 days. The larva wriggles after hatching, leads an active life, swims, feeds and grows up to 14 days. During this period it moults four times.
5. The larva possesses paired bristles on the head, thorax and abdomen, some of them forming bushy tufts. Abdomen is slender has nine segments. The eighth segment has a chitinous and tubular respiratory siphon. At the tip of the siphon are two spiracles leading into trachea.
6. At resting the larva hangs on the surface of water with its head downwards, inclined at an angle. The siphon projects just above the surface of water and draws in air.
7. The pupa is floating comma-shaped, with two respiratory trumpets on its cephalothorax.
8. Pupal stage lasts from two to seven days depending on temperature. Total larval period is 20 days or more depending on temperature.

☒ Bionomics

1. Eggs, larvae and pupac develops in the water so water is essential for their existence

2. The normal food of both sexes are nectar of flowers and juices of plants. The adult females prefers blood meal from man, domestic animals and even birds. It is a night feeder.

☒ Transmission of parasites

1. ***Culex pipiens*** is the major vector of filaria worms ***Wuchereria bancrofti, Brugia malayi*** and ***Dirofilaria immitis.***
2. They also transmit bird malaria, avian pox, and arbovirus, encephalitides.
3. Of the **Culex** species ***C. quinquefasciatus*** (= ***C. fatigans***), ***C. tarsalis*** transmit equine encephalitis (WEE) and St. Louis encephalitis.

PHLEBOTOMUS

Phlebotomus belongs to the subfamily Phlebotominae, family Psychodidae, suborder Nematocera, order Diptera. They are blood-sucking nocturnal hopping flies commonly known as **sand flies**, distributed mainly in the warmer parts of the globe.

☒ Morphology

1. Sand flies are minute hairy flies measuring 1.5 to 2.5 cm in length with conspicuous black eyes.
2. The greater part of the body and legs are covered with long hairs. The legs are silvery white.
3. Antennae with 12-16 segments.
4. Mouth parts rostrate, palpi with 4 segments.
5. The suctorial proboscis is as long as the head.
6. Thorax bears three pairs of jointed legs and a pair of lanceolate type hairy wings with prominent venation attached to mesothorax. At rest the wings are held at 60° angles with the body.
7. Abdomen segments 9, the last segment not clearly visible.
8. Claspers and substyles present in male.

☒ Breeding and life history

1. To lay egg phlebotomus require a combination of darkness, high humidity and organic debris on which the larvae feed. Oviposition takes place after a blood meal.
2. Eggs are long and ovoid, laid in batches of 50.
3. Incubation period 9 to 12 days. The larvae are minute, whitish in colour with long anal spines and chewing mouth parts (Fig. 28.42).
4. The larvae are free-living, feed on organic debris, like animal excreta.
5. There are four larval instars. The fourth larval instar metamorphoses into pupa, which is not enclosed in a cocoon.
6. The pupa develops into adult in about 10 days.
7. The whole life cycle is completed in about 30 days in Indian condition, but in winter it may be longer, even up to 3 months.
8. The average life of a sand fly is 14 to 16 days.
9. Egg to egg cycle requires 7 to 10 weeks.

☒ Bionomics

1. Because of their soft, delicate exoskeleton, the survival of sand flies depends on avoidance of hot and dry places. Their suitable living spaces are crevices, animal burrows, holes of trees, under logs and among fallen dead leaves.
2. Sand flies are weak fliers capable of navigating only short distances. They rarely take to wings and move from side to side by a series of short spasmodic hops.
3. Both male and female feed on plant juice, only the female take blood meal for maturation of their eggs.
4. They feed usually at twilight or night and in early morning.

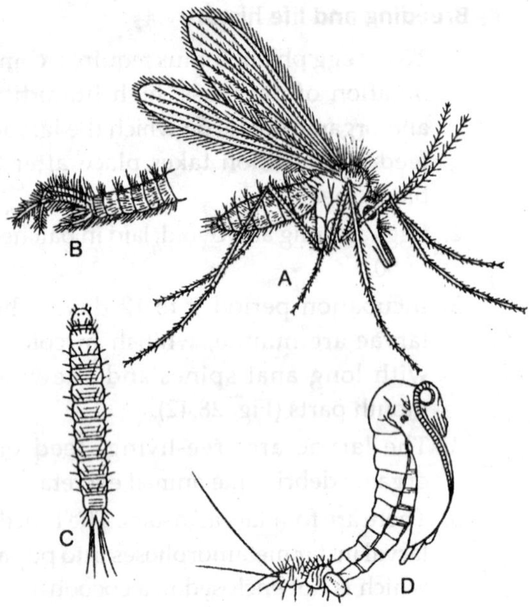

FIGURE 28.42
***Phlebotomus* sp.**; Life cycle stages. A. adult female, B. abdomen with claspers, C. Larval stage, D. Pupa

☒ Transmission of parasites

1. Sand flies are vectors of *Leishmania donovani* causing visceral leishmaniasis or Kala-azar. ***L. tropica*** causes oriental sores, a local lesion of skin and subcutaneous tissue. *L. braziliesis* cause espundia or uta, a mucocutaneous ulcer disease.
2. Sand flies also transmit bacterium *Bartonella bacilliformis* causing a disease known as carrion's disease and some other viral diseases.

INSECT AND MAN

About seventy-five per cent of total animal population of the world is covered by insects alone. On the basis of their activities they may be broadly classified into two groups: (a) Insects of economic importance and (b) Insects indifferent to human society. The first category can be subdivided again into two groups, (i) **Harmful insects**, which do remarkable damage to our interest by carrying or spreading many dreadful and contiguous diseases or by destroying the crops, stores, etc. and (ii) **Beneficial insects**, which do good to us, either by supplying us with very useful raw materials such as silk, honey, lac, etc. or by fixing nitrogen with the help of certain nitrogen-fixing bacteria and thus increasing the fertility of the soil. The following insects are important with reference to man.

☒ Mosquito *(Anopheles* sp., *Culex* sp., *Aedes* sp.)

Mosquitoes are quite common in the tropics and are found in abundance in marshy places. They are of sucking habits as well as carriers of germs of many diseases—**malaria, elephantiasis, yellow fever**, etc. and as such they are of considerable importance. Only the females are dangerous, since they require animal blood-meal for the proper growth and development of their eggs. Anopheles female causes malaria, culex carries the microfilaria, as well as the virus of dengue fever, Aedes is the carries of yellow fever and many other viral diseases.

☒ House Fly *(Musca domestica)*

House fly is a common nuisance to ordinary household. The house fly, *Musca domestica* is a menace to human welfare. Female flies are attracted by all kinds of dead and decomposing organic matters for laying their eggs. They are in the habit of regurgitating a part of the food they take and leave a vomit spot on the surface it sits on. Due to this ugly habit, the flies are the disseminators of various diseases, the enteric ones in particular. It may so happen that before sitting on exposed human food, they might have been feeding on some decomposed food materials or even faecal matter. Consequently, the pathogenic organisms, which are quite likely to be present there, contaminate the human food, being carried to them either through the stickly pad of the legs, or owing to the regurgitation of the previous diet. **Cholera, typhoid, paratyphoid, anthrax, tuberculosis, diarrhoea, dysentery** and a number of diseases are directly or indirectly caused by the house fly.

⊠ **Silkmoth** *(Bombyx mori)*

Silkmoth is one of the beneficial insects that supplies silk for human luxury and comfort. In India, silk is produced by two types of silkmoths of the family, Bombycidae and Saturnidae. The larvae of the moth spin the silk casings or *cocoons* to cover themselves during the *chrysalis stage* which is used for making the silk thread. It is the basis of large industries as well as cottage industry. Nowadays, besides the silk, the intestine of the silkmoth is used in the preparation of gut, which is used for fishing as well as surgical works.

⊠ **Honey Bee** *(Apis dorsata)*

The painful bite of the honey bee may possible be counter-balanced with the important and palatable honey they produce. They are very ancient creatures. The colonial insects live in combs made up of wax scales of real bee's wax, which is given off from the abdomen. From the honey-sacs, usually the nector is regurgitated and is stored in the cells of the comb. After proper processing, the pure, tasteful honey is obtained. Both wax and honey obtained from honey bees are of economic importance.

⊠ **Locust** *(Locusta migratoria)*

Locusts are migratory vegetable feeders. It passes through different phases of its life of which **gregaria** phase is one. This phase is harmful when great masses of locusts take long distance flight. In course of their journey, they stop at places and consume every bit of vegetation during their temporary settlement over that region. During flight, they run in thick swarms, which look like dense cloud. Due to their such habits they are considered as one of the greatest enemies of human welfare.

⊠ **Lac insect** *(Tachardia lacca)*

The lac insects are mostly recognised by females which have bag-like bodies covered up by resinous substances secreted by the animal. The lac insect *Tachardia lacca* gives off an exudation of a resinous nature which yield shellac of commercial importance and red dye. They are largely found in the forests of India and Myanmar (Burma). The shellac is used in making varnishes and polishes, insulation, as sealing wax, linoleum, buttons, pottery, shoe polishes, toys, bangles and various other items.

⊠ **Rice bug** (*Leptocorisa varicornis*)

In Bengal, it is popularly known as "Gondhi poka" because of its' nauseating smell. It belongs to the subclass Pterygota, division Endopterygota. It is very easily recognised by its green colour and slender form. The maximum size attained by it is about 25 mm. It is widely distributed and is reported to cause a serious damage to paddy crops in all the states of India. It feeds upon ripening seeds and also sucks sap from developing ears, which, as a result, die and turn white.

⊠ **Insects and diseases**

The relation between some of the important diseases and insects is surprisingly intimate. The insects are one of the most highly organised invertebrates. Some of the insects carry (**carriers**) disease-producing organisms only to transmit those to humans or other animals. Others harbour the causative organisms in them and are known as **vectors**. They are potential sources for continuous transmission of parasites. A great number of insects cause diseases while some others are entirely beneficial to human beings. Sometimes, the insects themselves are parasites or blood-suckers. It is in their capacity as carriers of germs or as intermediate hosts of parasites that they had to be reckoned with the foremost of human foes. With the advancement of the scientific knowledge in the current century, it has been clearly demonstrated that the most important human and animal diseases such as **Yaws, Pinkeye, Q fever, Tularemia** and **Anthrax**, are exclusively transmitted by insects. 'Arthropods' play an important role in the transmission of diseases due to bacterial, protozoan or helminthic infections.

☒ Transmission

It is chiefly effected in three ways:

1. Mechanical transmission
2. Biological transmission
3. Transovarial transmission

1. Mechanical transmission

This may be indirect. The simplest method is the indirect mechanical transmission where the insects take up the role of passive carriers of diseases. They only pick up the organisms causing disease, from the bodies or the excretions of man or animals and deposit them on food. The commonest carriers are the house flies and roaches.

A step further towards the specalization of transmission may be found in the direct mechanical transmission method. Here, the insect picks up the germs from the body of diseased individual and directly injects them into the blood of other animals through the skin or sores or wounds. Tsetse flies and certain others transmit the blood diseases namely, Anthrax, Fowl pox and trypanosomes causing some animal Trypanosomiases. The flies that feed on wounds transmit skin or eye diseases 'oriental sores', 'trachoma', etc. The organisms of all these diseases do not survive for more than a few minutes to a few days in the vectors.

Organs concerned with transmission

1. **Mouth parts**
   a. **Biting.** In the insects transmitting diseases, specially in plants, the mouth parts are modified for biting. Example: Thrips.
   b. **Sucking.** In mosquitos, the mouth parts are modified for sucking. Generally, the hypopharynx and epipharynx undergo modifications. The mandible is lost or rudimentary. In fleas, hypopharynx, mandible and maxilae are modified to form piercing organs.

   Another variety of sucking modification is found in the sponging type, where the organ concerned is popularly known as proboscis. Example: House fly.
2. **Legs.** In house flies, legs carry the germs of diseases.
3. **Antennae.** The antennae serve the same purpose in the insects in general.
4. **Body.** In house fly, the body as a whole acts as the carrier.

2. Biological transmission

When multiplication or cyclical changes or both take place within the body of the insects, it is known as biological transmission. It may be propagative, cyclopropagative, cyclodevelopmental, and transporting.

3. Transovarial transmission

This means transmission of disease through infected ovary and the egg, often referred to as hereditary transmission. The infection is effected specially by mites and ticks. Insect-borne infections are extremely insignificant, although some virus infections may be significant.

MANAGEMENT

The harmful insects are managed by various ways.

☒ Chemical management

Chemicals used to kill insects are known as insecticides. Injurious insects are killed by certain chemicals, viz. hydrocyanic acid, sulphur dioxide, arsenic, etc. But there are other chemicals which are more specially detrimental to the insects, namely pyrethrum and a number of chlorinated hydrocarbons, such as Dichloro Diphenyl Trichloroethane (D.D.T.), benzene hydrochloride (B.H.C.) Organophosphates are extensively used as systematic poison. The majority of these insecticides are believed to act primarily on the nervous system of insects and paralyse them, some very rapidly. The insecticides are absorbed either through the cuticle or the stomach.

☒ Modes of use of Insecticides

Insecticides are used in the form of dusts, solutions, emulsions, water suspensions and fumigants; or these may be mixed up with the food and used as bait.

1. **Dusting.** Applied to protect the health of man or animals against insects.
2. **Residual sprays.** Chlorinated hydrocarbons can be mixed up—usually with kerosense oil—or some other suitable medium and sprayed.
3. **Space sprays.** These are effective in small indoor disinfections. They may be sprayed in the form of fogs, mists, smokes, etc.
4. **Water treatment.** Applied to kill aquatic larvae of mosquitoes. Paris green dust is effective only in the case of Anopheles larvae.
5. **Skin application.** Application of insecticides, particularly D.D.T., is effective in case of animals, since it is not absorbed through the skin.
6. **Fumigation.** Hydrocarbonic acid gas is a powerful fumigant and is extensively used, but it is very toxic to all forms of life.

Repellants. These are chemicals which may be applied to the skin or clothing to keep off insects from biting. Indalone, is particularly effective against biting flies and some mosquitoes.

⊠ Biological management

Insects may be controlled by other insects; some insects devour the eggs and adults of others.

⊠ Common insect-borne diseases

a. **Sucking lice.** Ectoparasite on mammals, feed on blood; two species are parasites on man. Diseases are transmitted by continuation of scratches in the human body; or by excreta, or by the excreta of the infected louse.

b. **Mosquitoes.** (i) Anopheles. Females are carriers of malaria parasites. (ii) Culex. Intermediate hosts for filaria. They are also carrier for the causative agents of dengue fever.

c. **Flies.** Carrier of many pathogenic bacteria like those of cholrea and typhoid.

d. **Rat flea.** Carry bacilli from diseased rat to man. The plague bacilli rapidly multiply in the haemolymph of the flea and when the flea bites a man the bacilli are released in the blood of the victim.

COMMUNICATION IN INSECTS

Insects display different form of communications among themselves. A variety of stimuli and sensory channels—sound, visual stimuli, chemical stimuli and mechanical contact—serve as signals eliciting behavioural responses. The highest development of communication is observed in social insects which has been deciphered in honey bee by Von Frisch (1956) and Wenner (1964).

⊠ Communication in honey bee

Food related communication

The honey bees can very efficiently communicate amongst themselves the source of food, its direction and distance from the colony and richness of the source by dance and sound vibration of the scout bees.

1. **Round dance.** The successful foraging scout returns to the hive with nectar and pollen which she shares with other field bees.

   The scout bee then begin an excited dance circling to the right and then to the left, with a straight-line run between the two semicircles traced by the dance (Fig. 28.43) and repeating the same movement for about $1\frac{1}{2}$ minutes in one place. This dance is called **Round** or **Circling or Rundtanz dance.**

   This type of dancing indicates that the source of food is present at a distance between 50 and 100 metres away. The orientation of the circular movement shows the direction of the food.

2. **Wagging dance.** Sometimes a reporter bee after returning from the field runs

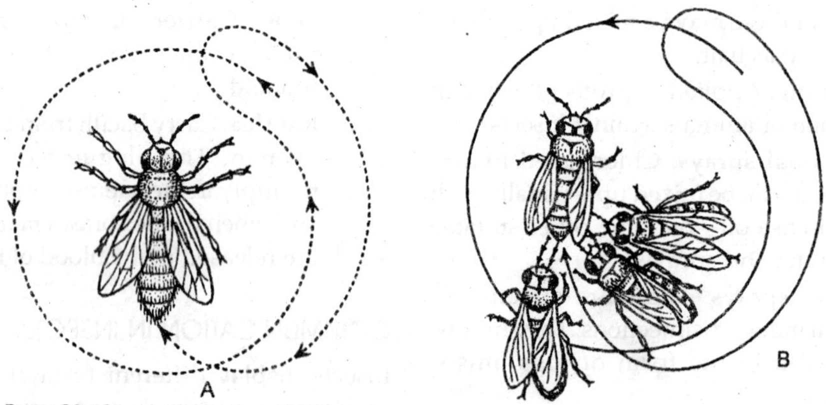

FIGURE 28.43

Round dance (Rundtanz) of honey bee (A, B). Arrows indicate the performance of dance in a series of circles alternately to the left and right every second

a certain short distance in a straight line, wagging its abdomen very swiftly from side to side and then makes a 360° turn towards left, run ahead once again and turns right and repeats for several times (Fig. 28.44). This dance is called **Wagging dance** or Schwänzeltanz.

i. This type of dance indicates the direction and distance of food source from 100 to 300 metres or more.
ii. The speed of the wagging dance carries distance instructions. The more rapidly the bee performs its wagging runs, the shorter is the distance.
iii. The wagging dance is performed not on a horizontal exposed platform but in the dark interior of the hive on the vertical surface of the honey comb.
iv. Here the dancer uses a remarkable method of informing her mates of the correct angle with respect to the sun. The direction to the sun is represented by the straight upward direction along the wall.
v. If the tail-wagging run is directed upward, the source of food is located towards the sun; if it is downward, the food is away from the sun; the inclination of the run to the right or to the left of vertical indicates the angle of the sun to the right or to the left of the food source.
vi. If the goal lies at an angle of 40° to the left of the sun, the wagging run points 40° to the left of the vertical. The angle to the sun is represented by an equal angle with respect to the upright.
vii. While flying to the feeding place, the bee observes the sun. During dance she orient herself so that in the straight run, she sees the sun on the same side and at the same angle.

3. On cloudy days the sky polarisation pattern acts as a reference point instead of the sun. If the food source is closer than 80 metres the clues provided by chemoreception are sufficient for finding the food, and the tail-wagging dance is not performed by the scout bee.
4. **Sound vibration.** The foraging bee communicates to other members of the hive with sound waves also.

i. During the straight run the dancing bee produces vibrations emitted as train of pulses. The vibrations are produced by wing movements.
ii. The surrounding bees apparently receive the dancer's vibration through their antennae, which touch the dancing bee's thorax, or through their legs.
iii. By means of sound the foraging bees indicate the distance of the food source from the hive.

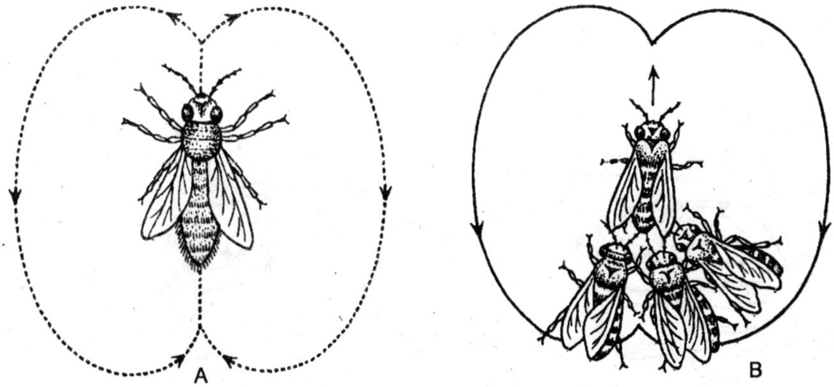

FIGURE 28.44

Waggle dance (Schwärzeltanz) of honey bee (A, B). Arrows indicate the dance begin up and down a vertical comb writhin the hive

iv. The average number of vibrations is proportional to the distance of the food from the hive.

v. There are indications that the sound signals of the forager may also provide clues about the richness of the food source and perhaps other information.

☒ Odour related communication in insects

1. **Queen secretion.** Odours also play a vital role in communication among honey bees. Healthy queen secretes an aromatic substance called "queen substance", which is licked off by her nurse bees. When the queen dies the secretion stops and the absence of "queen substance" is immediately relayed to almost 50,000 or more members of the colony.
2. **Pheromones.** Insects communicate among themselves by body secretion, which may function as sex-attractants, and irritants. This type of communications plays an important role in locating food and the nest, in attracting mates, in recogising a member of the same colony, and also in defence.

   Some of these chemicals, known as pheromones have been isolated and identified. The pheromones have been used in applied fields of insect control to lure males and females to their destruction.

PALAMNAEUS sp.

The Scorpion, *Palamnaeus* sp. belongs to the class Arachnida, phylum Arthropoda. Scorpions are one of the oldest terrestrial arthropods and the present day species differ very little from their fossil ancestors.

Scorpions are widely distributed and found in tropical and subtropical regions, from mountains to plains, rain forests to deserts. They live in holes, crevices, under stones, logs of wood, decaying leaves and organic matters, within sand and many other places.

☒ External features

1. The body is narrow, elongated, dorsoventrally flattened and divisible into an anterior Prosoma or Cephalothorax and a posterior Opisthosoma or abdomen (Fig. 28.45).
2. The colour of the exoskeletal covering harmonises to a considerable extent with the surroundings and ranges from yellowish orange to black.
3. **Prosoma.** it is formed by the fusion of seven segments.

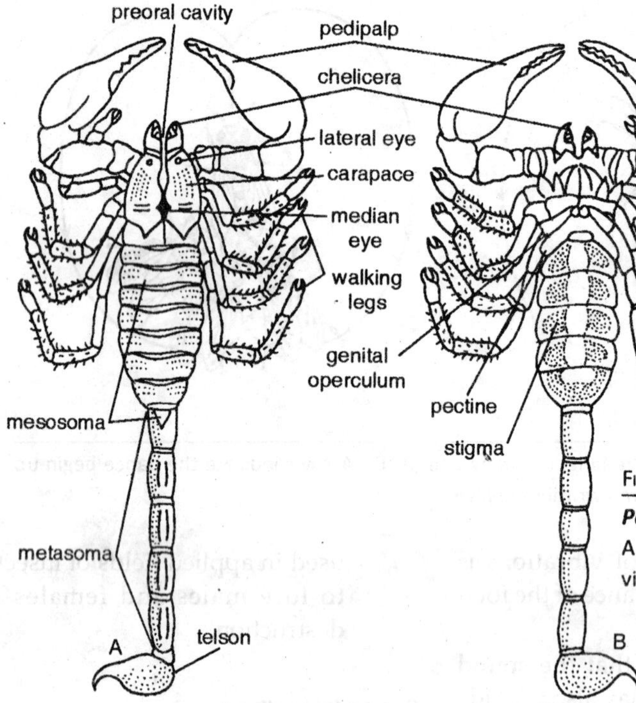

FIGURE 28.45

Palamnaeus bengalensis.
A. Dorsal view, B. Ventral view

a. Dorsally, the prosoma is covered by an unsegmented, more or less square **carapace**, formed by the fusion of the terga of the segments. The anterior border of the carapace is notched, forming a right and a left frontal lobe.

b. A single, **triangular plate** formed by the fusion of sternites, cover the ventral surface of the prosoma.

c. A pair of simple **median eyes** at the middle and 2-3 pairs of **lateral eyes** on the anterolateral margins are present in the carapace.

d. The **mouth** is a small, transverse, ventrally located opening at the anterior end of the cephalothorax. A small, fleshy labrum overhangs the mouth ventrally.

e. The cephalothorax bears six pairs of appendages.

i. A pair of three-jointed **chelicerae** anteriorly.

ii. A pair of six-jointed **pedipalpi** posterior to chelicerae.

iii. Four pairs of seven-jointed walking legs.

4. **Opisthosoma.** It is immovably attached to the cephalothorax and divisible into an anterior **mesosoma** or preabdomen and a posterior **metasoma** or postabdomen.

A. Mesosoma

It is broad, flat and consists of seven segments. Each segment has a dorsal **tergum**, a ventral **sternum**, connected by soft integument at the sides. The mesosoma bears two pairs of modified appendages, and four pairs of stigmata.

a. **Genital operculum.** A movable, bifid flap, covering the genital opening, at the middle of the narrow sternum of the first abdominal segment.

b. **Pectines.** A pair of comb-like structures attached to the sternum of the second adbominal segment.

The genital operculum and pectines are considered as modified appendages.

c. **Stigmata.** Oblique, slit-like openings on the ventrolateral borders of the 3, 4, 5 and 6 preabdominal somites.

B. **Metasoma**

It is narrow, tubular, five-segmented, enclosed in a complete ring of chitinous material and flexibly jointed. In live specimen, it is turned upward over the mesosoma. The metosoma bears:

a. **Poison bulb.** The terminal metasomal somite is bulb-like at the base and pointed at the apex, forming a poison sting. A pair of poison glands open in the tip of the sting.

b. **Anus.** Situated ventrally in the last metasomal somite, immediately in front of the sting.

⊠ Appendages

Appendages are six pairs, two pairs in the head and four pairs in the thorax (Fig. 28.46).

Cephalic appendages

1. **Chelicerae.** Paired, preoral in position. The appendage is small, three-jointed, ending in a chela.
   Function. Holding the prey.
2. **Pedipalpi.** Paired, post-oral in position. It is a large, six-jointed structure ending in a great chela.
   Functions. Seizure of prey and mastication by the bases.

Thoracic appendages

Four pairs of legs. Each leg consists of seven segments, the last one ending in a pair of curved, pointed, horny claws. The claws are absent in the 3rd and 4th pairs of legs.

Functions. Walking but in the first and second pairs, the bases help in mastication of food.

Abdominal appendages

The **operculum** bearing the genital pore and the comb-like tactile **pectines** are probably the remnants of the first and second pairs of abdominal appendages, respectively.

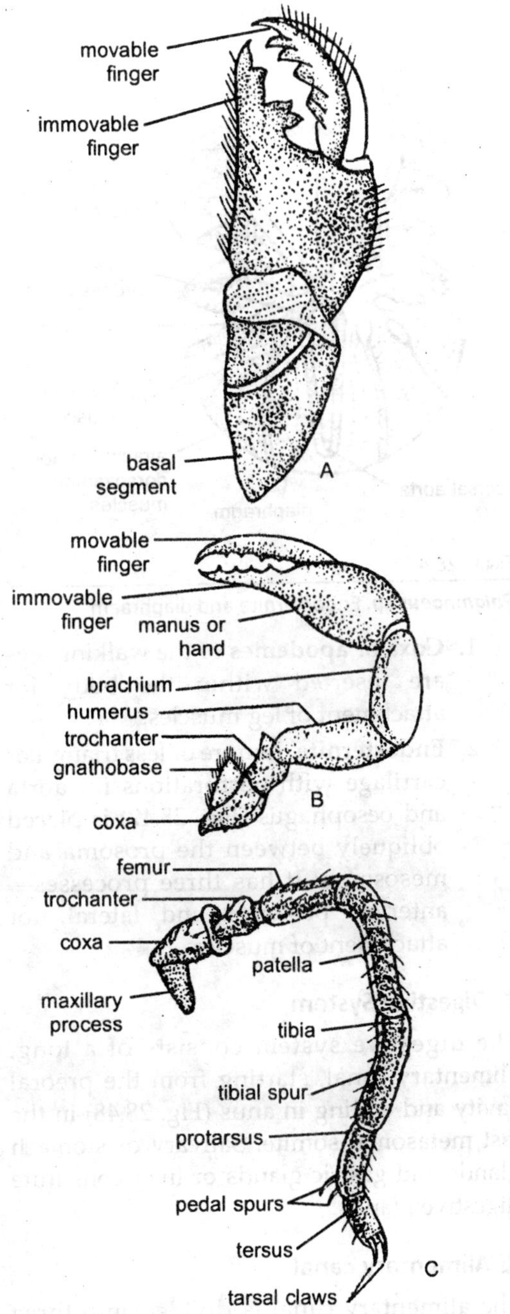

FIGURE 28.46

***Palamnaeus* sp.** Appendages. A. Chelicera. B. Pedipalp. C. First walking leg

⊠ Endoskeleton

The coxa of the walking legs and the endosternite constitute the endoskeleton.

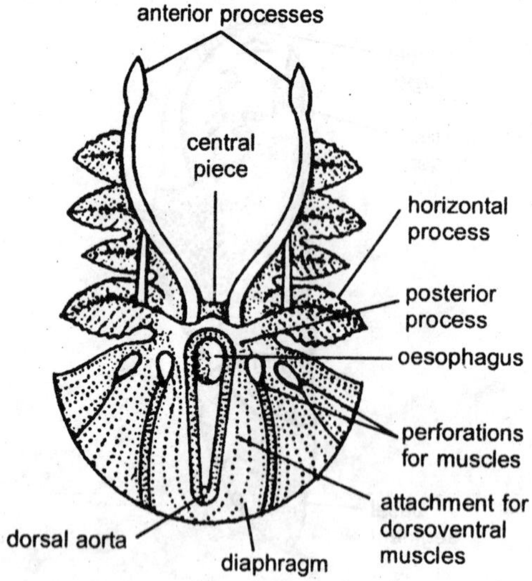

FIGURE 28.47

***Palamnaeus* sp.** Endosternite and diaphragm

1. **Coxa** or **apodemes** of the walking legs are inserted within the body for attachment of leg muscles.
2. **Endosternite.** A more or less triangular cartilage with perforations for aorta and oesophagus (Fig. 28.47) is placed obliquely between the prosoma and mesosoma. It has three processes—anterior, posterior and lateral, for attachment of muscles.

☒ Digestive System

The digestive system consists of a long, alimentary canal, starting from the preoral cavity and ending in anus (Fig. 28.48) in the last metasomal somite. Salivary or stomach glands and gastric glands or liver constitute digestive glands.

☒ Alimentary canal

The alimentary canal is divided into three zones—foregut, midgut and hindgut. The fore and hindgut are lined by cells of ectodermal origin, while the lining of the midgut is endodermal.

Foregut

The sclerotised foregut is divisible into a preoral cavity, a pharynx and an oesophagus.

1. **Preoral cavity.** It is located at the anterior end of the cephalothorax and bounded dorsally by two chelicerae, ventrally by gnathobases of the first two pairs of legs and laterally by the coxae of the pedipalpi. Labrum, a club-shaped structure hangs in the cavity.
2. **Mouth.** A small transverse opening inside the preoral cavity, near the base of the labrum. It leads to the pharynx.
3. **Pharynx.** A well-developed suctorial bulb, leading to the oesophagus. The pharyngeal wall is muscular and a number of radiating bundles of muscle fibres from it run to the walls of the cephalothorax.
4. **Oesophagus.** A short tube, opening posteriorly into the midgut.

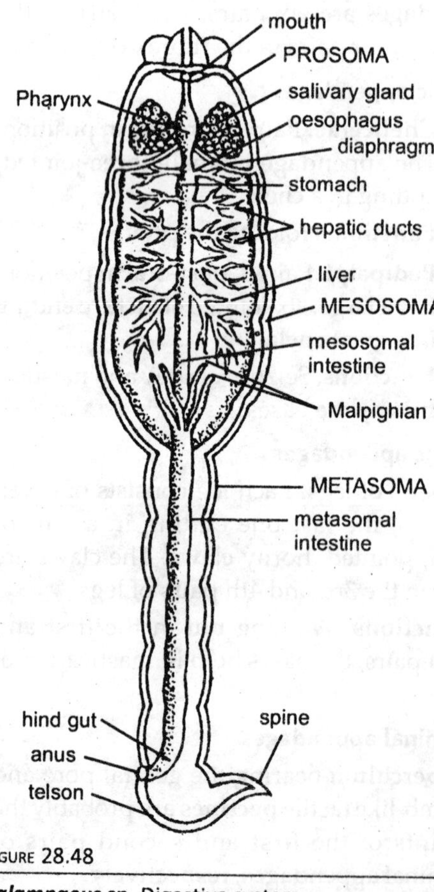

FIGURE 28.48

***Palamnaeus* sp.** Digestive system

Midgut

A long tube with a lining of glandular epithelium and divisible into a stomach and an intestine.

1. **Stomach.** A small sac, runs up to the diaphragm and continues posteriorly as intestine.
2. **Intestine.** A long, narrow, straight tube runs posteriorly along the middle line of the body and divisible into an anterior **mesosomal** or **preabdominal** and a posterior **metasomal** or **postabdominal** intestine. The junction of the two is demarcated by the presence of Malpighian tubules.

   The mesosomal part is embedded in the liver and receives five to six pairs of narrow **hepatic ducts**, one after another.

   The metasomal intestine ends posteriorly in the hindgut.

Hindgut

It is a small cuticularised tube in the last metasomal segment and opening to the exterior through ventral anus in front of the sting.

⊠ Digestive glands

Salivary or stomach glands, gastric glands or liver constitute digestive glands. Unicellular glands in the lining epihelium of the midgut also secrete digestive juice.

1. **Salivary glands.** Paired glands, located on either side of the oesophagus in the prosoma. Saliva, the secretion of the glands, is discharged into the oesophagus through separate ducts from the glands.
2. **Liver.** A large, paired glandular mass occupying the dorsal part of the mesosomatic cavity. The secretion of the gland is released in the lumen of the intestine through 5-6 pairs of **hapatic ducts.** Each duct is formed by the union of a number of smaller ducts arising in the mass of the liver.

⊠ Feeding and digestion

Scorpion is nocturnal and comes out in search of food only at night. It is a predator and depends chiefly on insects, spiders, myriapods, etc. for nourishment.

1. The prey, which comes in contact with the appendages in the dark is grabbed by the chela at the pedipalps and stung.
2. The prey is brought close to the preoral cavity and wounds are made on the body by the chelicerae.
3. The strong suctorial pharynx sucks fluid from the victim's body and saliva mixes with it in the preoral cavity.
4. The partially digested food is moved to the midgut, where further digestion occurs with the help of enzymes, released by the liver and unicellular glands in the gut wall, and the undigested part of the food is moved to the hindgut.
5. From the midgut food reaches the liver through hepatic ducts, where digestion is completed.
6. Residue collected in the hindgut is egested through the anus.

⊠ Respiratory System

The Palamnaeus live on land and the respira-tory structures, book-lungs can use atmospheric oxygen. Though physiologically less efficient, the book-lung is a more highly evolved respiratory organ, than those in other arthropods.

⊠ Respiratory organs

1. The respiratory organs consist of four pairs of **book-lungs**. The book-lungs open to the exterior by four pairs of apertures, the stigmata (Fig. 28.45B). Sternum of each of the 3rd, 4th, 5th and 6th preabdominal segments bears a pair of stigmata on its ventrolateral sides. These are narrow oblique slits. Each stigmata opens directly into a book-lung and they are not interconnected.

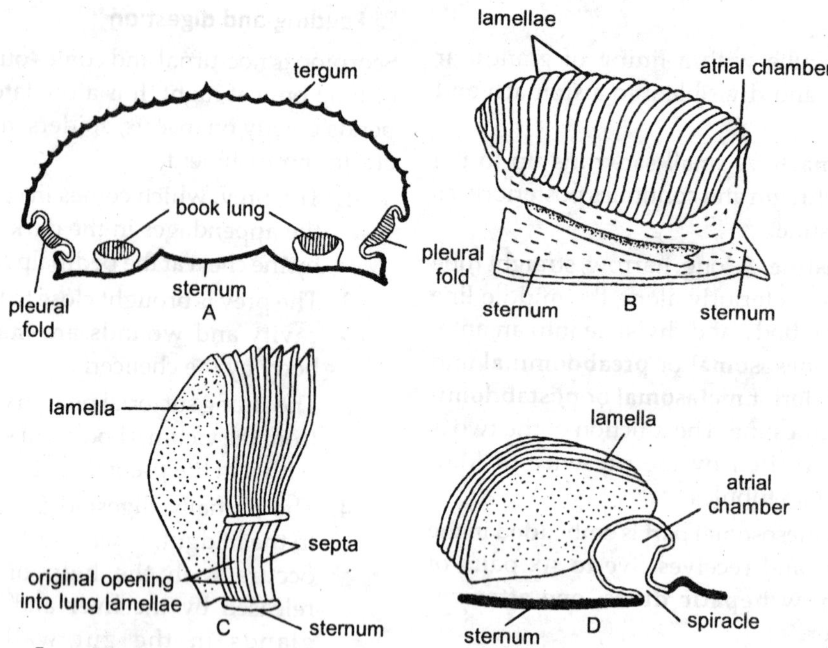

FIGURE 28.49

***Palamnaeus* sp.** Book-lung. A. Position of booklung in cross section, B. Posterodorsal view, C. Longitudinal view of a few lamella in natural condition, D. Vertical section

2. Each book-lung is a compressed sac lined with a thin cuticle. The lining membrane is folded up into numerous delicate laminae lying parallel to one another like the pages of a book, roughly 130-150 in number (Fig. 28.49).
3. Deoxygenated haemolymph flows through the narrow space in each lamina, separated from the air only by the membranous walls of the lamina (Fig. 9.8, Chapter 9).

☒ Mechanism of respiration

The alternate expansion and contraction of the abdomen brings inhalation and exhalation of air in the book-lungs. Gaseous exchange takes place between the haemolymph in the laminae and the air in between the laminae.

☒ Circulatory System

Heart and Pericardium

The heart is a dorsal elongated tube lying in a groove in the liver mass, entirely in the preabdominal region, extending from the seventh to the thirteenth segment. It is clearly demarcated in its anterior and posterior limits by valves. The heart is single-chambered. (Fig. 28.50). Seven pairs of ostia, guarded by valves, are present on the walls of the heart. The heart is suspended by eight groups of ligaments arranged metamerically (Fig. 28.51). The pericardium is thin, membraneous and horizontal in position. The pericardial cavity is divided into four sinuses, two laterals, one ventral and one dorsal. The cardiac nerve extends from one to the other end of the heart.

☒ Haemolymph

The freshly shed haemolymp is indigo in colour. White blood corpuscles, the amoebocytes float in it. The haemolymph contains a blue coloured respiratory pigment, proteid-haemocyanin. The oxygenated haemolymph is bright blue, but colourless when deoxygenated.

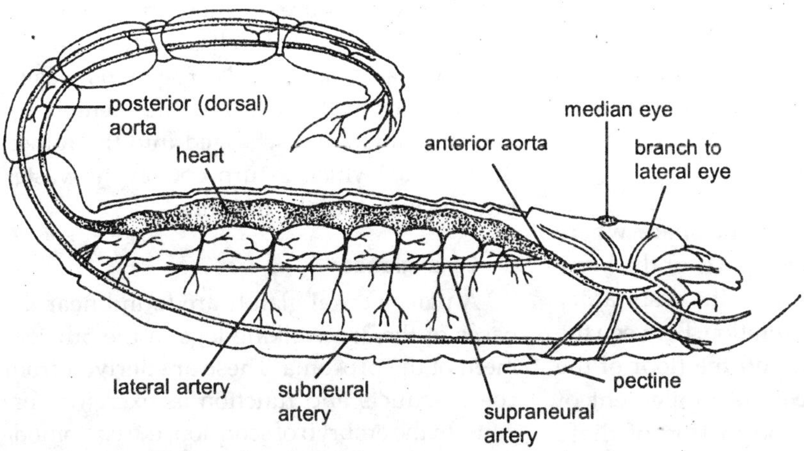

FIGURE 28.50

***Palamnaeus* sp.** Heart and arterial system (lateral view)

Arteries (Fig. 28.50)

The heart continues anteriorly as the **anterior aorta** for a considerable distance. Afterwards, it divides into two branches, which embrace the oesophagus on both the sides and unite below in the form of a short arch connecting the aorta with the right and left thoracic sinuses. In the ventral region, from the arch comes out a common vessel that runs over the nerve cord. This vessel is known as the **supraneural artery**.

On reaching the anterior end of the suboesophageal ganglionic mass it turns downward and then runs backward as a **subneural artery** ending behind the ganglionic mass.

COURSE OF CIRCULATION

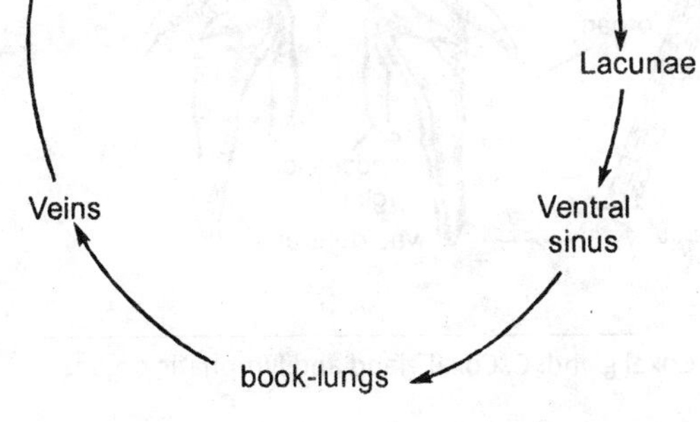

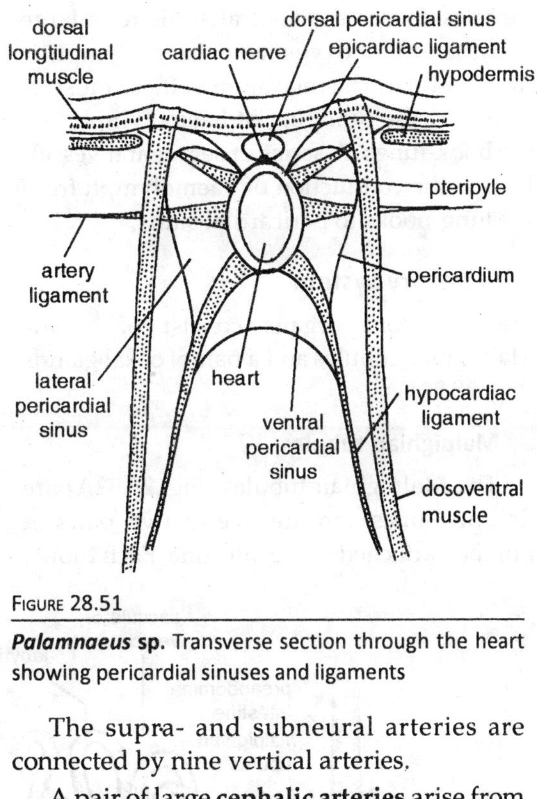

FIGURE 28.51

***Palamnaeus* sp.** Transverse section through the heart showing pericardial sinuses and ligaments

The supra- and subneural arteries are connected by nine vertical arteries.

A pair of large **cephalic arteries** arise from the aortic arch. Each gives off a number of branches, of which the **ophthalmic artery** and the **cheliceral artery** are more important. The ophthalmic supplies haemolymph to the eye and the cheliceral to the chelicera.

Each **thoracic sinus** sends a small vessel to the supraneural artery and four large vessels to the appendages supplying haemolymph to the pedipalp and legs of the side.

The heart continues posteriorly as **posterior aorta**. It runs straight backward up to the posterior tip and supplies to the intestine as well as various organs by a number of delicate branches.

Sinuses

The ventral sinus is a large space where haemolymph from various parts of the body is collected as well as sent to the respiratory organs or lung-books for aeration. Between the roof of the ventral sinus and the floor of the pericardial sinus a special arrangement of muscles is found. The contraction of these muscles causes the ventral sinus to enlarge and thereby deoxygenated haemolymph is drawn into it. The relaxation of these muscles results in the pumping of haemolymph into the book-lungs. A series of segmental vessels help in the conduction of haemolymph from the lung-books to pericardial sinus.

⊠ Excretory System

The excretory organs consist of a few Malpighian tubules and a pair of coxal glands (Fig. 28.52).

Malpighian tubules

The Malpighian tubules (Fig. 28.52A) are delicate tubes and are one or two pairs in number, attached to the intestine. Each tubule is a hollow structure opening into the lumen of the intestine and the wall is made of a single layer of glandular cells. Nitrogenous wastes are absorbed by the cells from the haemolymp and the same are discharged into the lumen of the tubule which, in turn, convey the wastes into the lumen of the intestine.

Coxal glands

A pair of coxal glands are found near the bases of the 3rd walking legs in the 5th segment of the prosoma. These are derived from coelomoducts and function as excretory organs. In the embryo of scorpion, the coelomoducts are found in the 3rd, 4th, 5th, 6th and 8th segments. All of them disappear in the adult except those in the 5th segment which persist as coxal glands. Each coxal gland consists of an end sac, a colied tube and a bladder opening to the exterior by a duct at the base of the 3rd walking leg (Fig. 28.52B).

Lymphatic organ

A diverticulum or lymphatic organ, possibly excretory (phagocytic) in function, lies on each side of the abdomen connected with the coxal gland of the side (Fig. 28.52C).

Nephrocytes

The nephrocytes are large cells, located in the walls of the mesosoma and are believed to be of excretory function.

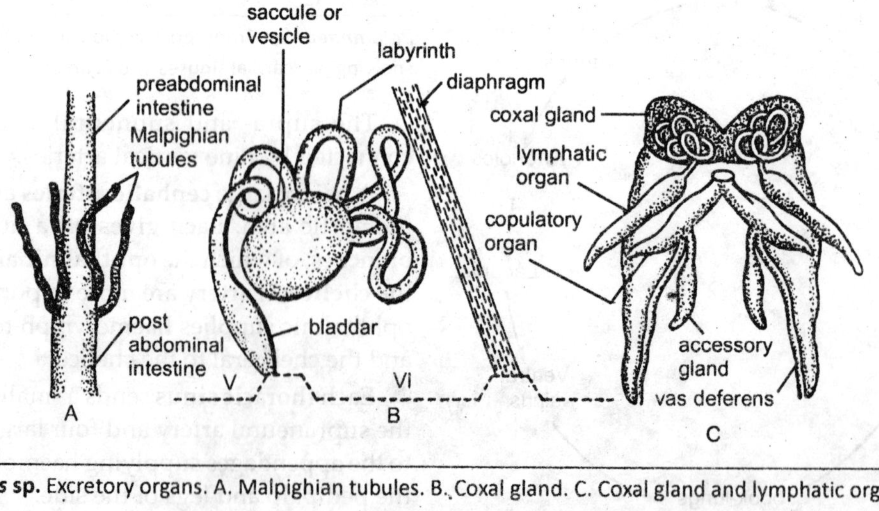

FIGURE 28.52

***Palamnaeus* sp.** Excretory organs. A. Malpighian tubules. B. Coxal gland. C. Coxal gland and lymphatic organs

☒ Nervous System

The nervous system consists of a central nervous system, a peripheral nervous system and a visceral nervous system (Fig. 28.53).

Central nervous system

The central nervous system includes a brain or supraoesophageal ganglia, suboesophageal ganglia, circumoesophageal connectives and a double ventral ganglionated nerve cord.

1. **Brain** or **Supraoesophageal ganglia.** It is a small bilobed mass, situated in the prosoma, just beneath the median eyes.
2. **Suboesophageal ganglia.** It is formed by the fusion of ten pairs of ganglia belonging to embryonic 2 to 11 segments and located ventromedially below the oesophagus.
3. **Circumoesophageal connectives.** A pair of short and stout nerves arise from the brain, one on each side, encircle the oesophagus and unite ventrally with the suboesophageal ganglia.
4. **Ventral nerve cord.** A double ventral nerve cord formed by the union of two solid nerve cords runs posteriorly from the suboesophageal ganglia up to the fourth segment of the metasoma.

The nerve cord bears three **preabdominal ganglia** in the mesosoma and four **postabdominal ganglia** in the metasoma, the last one formed by the fusion of two ganglia of the 18 and 19 segments and located in the fourth metasomal segment.

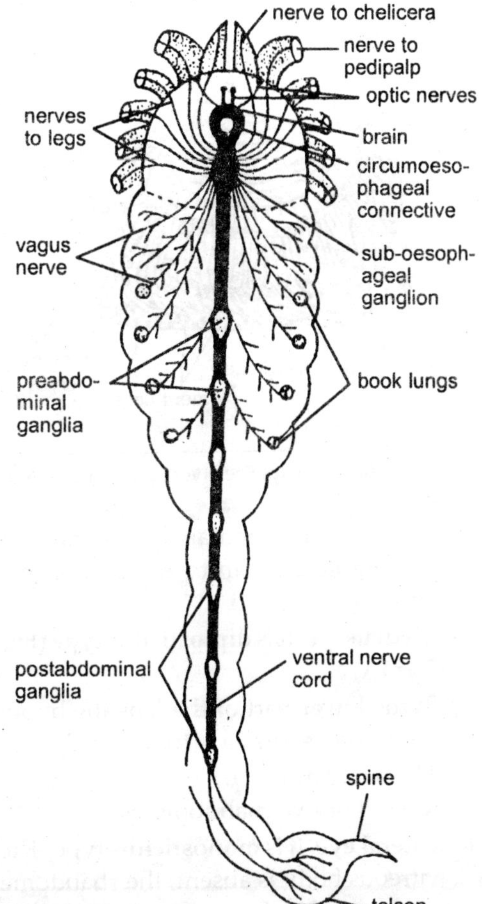

FIGURE 28.53

***Palamnaeus* sp.** Nervous system

☒ Peripheral nervous system

Nerves arising from the ganglia of the central nervous system:

1. The brain gives off paried nerves to median and lateral eyes and rostrum and a number of fine nerves to the walls of the preoral cavity, pharynx, oesophagus and adjoining structures.
2. Six pairs of lateral nerves emanating from the suboesophageal ganglia innervate the chelicerae, pedipalpi and four pairs of walking legs.
3. Two to four pairs of **vagus nerves** arising from the suboesophageal ganglia run posteriorly to supply the heart and alimentary canal, genital operculum, the pectines and the first two pairs of book-lungs in the mesosoma.
4. Paired nerves emanate from each ganglion of the ventral nerve cord and innervate the structures in the corresponding segments. Nerves from the first two preabdominal ganglia supply the last two pairs of book-lungs.

☒ Receptors and sense organs

The sense organs consist of eyes, pectines, sensory setae, trichobothria and slit sense organs.

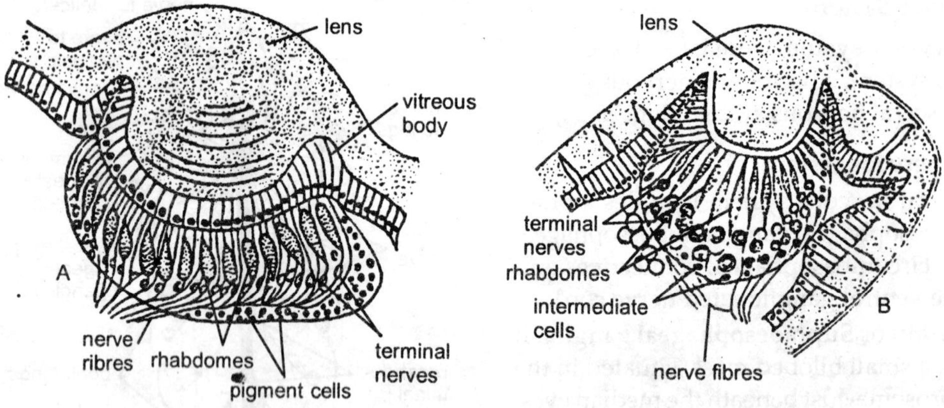

FIGURE 28.54

***Palamnaeus* sp.** Eyes (vertical section). A. Median eye. B. Lateral eye

1. **Eyes.** Both the median and lateral eyes are simple eyes and provided with a lens.
 a. **Median eye.** It is **diplostichus type** (Fig. 28.54A)
  i. In the lower part of the lens the hypodermal cells form a **vitreous** body.
  ii. The **rhabdome** of the retinulae is formed of five rhabdomeres.
 b. **Lateral eye.** It is **monostichus** type. The vitreous body is absent; the rhabdome is of irregular shape. (Fig. 28.54B).

Functions. Not properly understood.

2. **Pectines**
 a. Comb-like in appearance and consists of a shaft and movable processes (Fig. 28.55).
 b. The shaft is three segmented.
 c. Movable processes, also called leaflets are 4 to over 30 and attached in a row along the posterior margin of the shaft.

Functions. Food procurement and breeding.

3. **Sensory setae.** Fine hair-like structures scattered all over the body.
 Functions. Tactile sense organs.
4. **Trichobothria.** About 66-68 are present in each chela. Each organ consists of:
 a. A flask-shaped, cuticularised structure, the **bothrium.**
 b. The **trichobothrium** is a seta, inserted into the bothrium through the middle of cuticular outer membrane.
 c. The seta bends and narrows down within a fluid-filled cylinder supplied by nerve fibres.

Functions. Response to air current.

5. **Slit sense organs.** Scattered all over the body, being crowded on the appendages. A slit sense organ consists of:
 a. A slit on the integument, covered with a membranous epicuticular covering and measures about 0.005-0.16 × 0.002–0.003 mm.
 b. The slit leads to a narrow passage, which is a membrane-lined tube and penetrates in the procuticle.

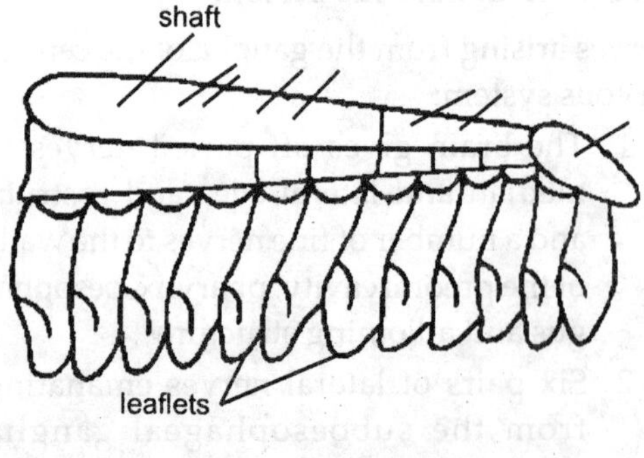

FIGURE 28.55

***Palamnaeus* sp.** A pectine.

c. The inner walls of the tube receive rich supply of nerve fibres.

Function. Concerned with the movement of appendages and also respond to vibration.

☒ Reproductive System

The sexes are separate.

Male reproductive system

1. The male reproductive system consists of a pair of testes, a pair of vasa deferentia, a pair of genital chambers, a pair of paraxial organs, a pair of flagella, a pair of seminal vesicles, a common genital atrium and a gonopore (Fig. 28.56A).
2. Each **testis** consists of a pair of longitudinal tubules placed in the preabdomen, and connected by transverse tubes.
3. The two longitudinal tubules of each testis unite anteriorly to form a **vas deferens**, which runs forward and downwards.
4. The terminal portion of each vas deferens opens into the inner side of the **genital chamber** along with an enlarged **seminal vesicle**. The genital chambers are muscular tubes and bear the **paraxial organ** having a chitinoid flagellum. It also receives the **accessory gland**. The genital chambers unite to form a median **genital atrium**, which opens to the exterior by the gonopore situated just behind the operculum.

Female reproductive system

1. The female reproductive system consist of a single ovary, two oviducts, a median genital chamber and a gonopore (Fig. 28.56B).
2. **Ovary** is situated in the preabdomen. It consists of three longitudinal tubules connected by cross branches. The median tubule is shorter than the other two. The ovarian tubules and the cross branches bear numerous diverticula (Fig. 28.56C).
3. From the two lateral tubules run two narrow **oviducts**. Anteriorly each oviduct is slightly swollen and form the seminal receptacle and the two open in the sac-like median **genital chamber**, the latter opening in the gonopore on the second adbominal segment below the bilobed operculum.

☒ Fertilization and development

Prior to breeding a nuptial dance is of universal occurrence in scorpion. The dance may continue for 30 to 90 minutes. During the dance, the male scraches a suitable surface with the help of its pectines for depositing the spermatophores.

1. The male deposits the spermatophores on a preselected site.
2. The male drags the female over the site.
3. The lever of the spermatophore being touched by the female genital operculum, the ejecting apparatus shots the sperms within the female genital duct. Fertilization is internal.
4. Palamnaeus is viviparous. Fertilized egg develops in the follicle of the ovary, which acts as placenta and provides nourishment to the embryo.
5. Young one resembling the adult is born. Immediately after birth, the young scorpion climbs on mother's back, remains there for 40 to 63 days and undergoes three moulting, before starting an independent life.

RESPIRATORY STRUCTURES AND RESPIRATION IN ARTHROPODA

Respiration is a process in which an organism receives oxygen from its surroundings and gives up carbon dioxide. It has two steps —external and internal respiration. The discussion is restricted to external respiration, i.e. the structures involved in absorption of

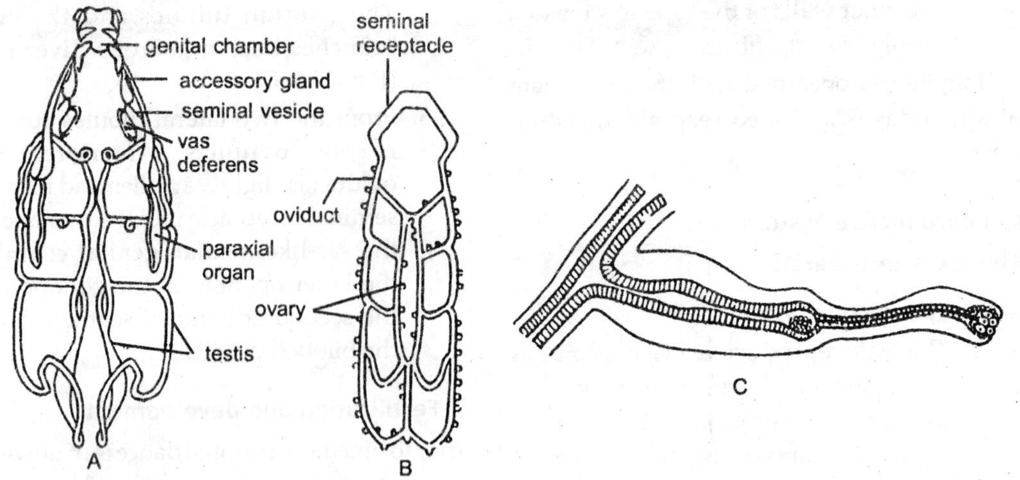

FIGURE 28.56

Palamnaeus **sp.** Reproductive system. A. Male. B. Female. C. Ovarian diverticulum of a viviparous species

oxygen from the surroundings—water or air. Both aquatic and aerial respiratory structure are found in Arthropoda.

⊠ Aquatic respiratory structures

Aquatic animals absorb dissolved oxygen in water. The respiratory organs in aquatic arthropods are chiefly gills. In some forms —brachiopods, Ostracoda, Copepoda and Cirripedia and some crustacean larvae, however, respire through skin.

⊠ Crustacean gills

A gill or ctenidium is made of a number of lamellae. Each lamella is composed of a large number of plate-like gill filaments. The lamellae are attached to a central axis. Channels in the central axis carry haemolymph to and from spaces in the lamellae.

In majority of the Crustacea the gills lie on the sides of the cephalothorax in the branchial chamber (Fig. 28.7) In most cases the gills are crescent-shaped. In podabranchi the gills are tufted and uncovered by carapace. In decapods and its allies the gills are plume-like or of soft cylindrical gill filaments.

In many land crabs the gills perform respiratory function from the moisture enclosed in nearly closed gill chamber. In Cocoanut crab (*Birgus*) the upper part of the gill chamber is separated from the rest and forms almost a closed cavity into which the vascular tufts project and function as the lung.

Origin of gills. Gills originate as out pushing of the thoracic wall. In Amphipoda the gills are out growths of the thoracic limbs. In Isopoda the endopodites of second and fifth pleopods are modified as gills. In stomatopod the gill filaments develop from the exopodites of the first to fifth pleopods. In some land forms a system of air-tubes develop for aerial respiration.

Insect gills

In aquatic larvae of some insects a series of soft external simple or divided processes, called **tracheal gills** are present. They are attached to the abdominal segments and richly supplied with tracheae. However, they have no connection with the exterior. In the nymph of some insects the inner surface of the rectum bears gills. These are termed **rectal gills**.

Arachnid gills

In Xiphosura the organs of respiration are external appendages or book-gills in the shape of delicate laminae attached to the abdominal appendages.

☒ Mechanism of respiration

The movements of scaphagnathites and second maxillae maintain a constant backward to forward water current in the gill-chambers. The haemolymph in the respiary structures gives up CO_2 and absorbs O_2. In crabs the carapace is sealed along the ventral margin and water enters and leaves through the apertures at the bases of the great chela.

☒ Aerial respiratory structures

The aerial respiratory structures of arthropods are trachea and book-lungs—the former is present in insects and chilopods and the latter in arachnids.

Tracheal system

The chief respiratory structure in insects is the tracheal system. It consists of tracheae or air-tubes and their branches, the tracheoles, which are directly communicated with the exterior (Fig. 28.32). The wall of the trachea is strengthened by chitinoid tissue in the form of spiral threads. These are, however, absent in tracheoles. The tracheae open to the exterior by small openings, the spiracle or stigmata. Each opening has two lids for opening and closing operated by muscles. The openings are guarded against the entry of foreign particles, either by being surrounded by bundles of hairs or by being provided with sieve-like membrane. In insects adapted for active flight, the tracheal system in certain parts of the body is dilated and form large air-sacs or air reservoirs. The tracheoles form network of air tubes within the tissue and thereby a direct passage for air from outside to the tissues is established. The tracheoles are blind and bathed in tissue fluid (Fig. 9.7).

Types of tracheae

The tracheal system in insects is classified on the basis of number of spiracles in the adult.

a. **Polypneustic.** Number of functional spiracles eight or more pairs.
b. **Oligopneustic.** Functional spiracles one or two pairs.
c. **Aponeustic.** No functional spiracle present.

☒ Mechanism of respiration

The alternate expansion and contraction of the abdomen brings inhalation and exhalation of air in the tracheal system. The openings and closing of the spiracles are also dependent on the CO_2 concentration of the inhaled air. The walls of the tracheoles allow gaseous exchange and diffusion takes place between the cell sap and the lumen of the tube. The tissues get direct supply of oxygen from the atmospheric air.

Book-lungs

Book-lungs are paired blind sacs located in the preabdominal segments of the opisthosoma of the arachnids. They are communicated with the exterior through narrow oblique slits—the stigmata. Book-lungs are independent structures and not intercommunicated like that in the tracheal system.

Each book-lung is a compressed sac lined with a thin cuticle. The lining membrane is folded up into numerous delicate laminae lying paratted to one another like the pages of a book, roughly 130-150 in number (Fig. 28.49) Deoxygenated haemolymph flows through the narrow spaces in each lamina, separated from the air only by membranous walls of the lamina.

☒ Mechanism of respiration

The alternate expansion and contraction of the abdomen causes inhalation and exhalation of air in the book-lungs. Gaseous exchange, takes place between the haemolymph in the laminae and the air in between the laminae.

BIOLOGICAL STATUS (SYSTEMATIC POSITION) OF PERIPATUS

Peripatus and its relatives are survivors of Archaic Middle Cambrian marine forms like **Aysheaia**, have problematic zoological position. They are ancestors of tracheate form

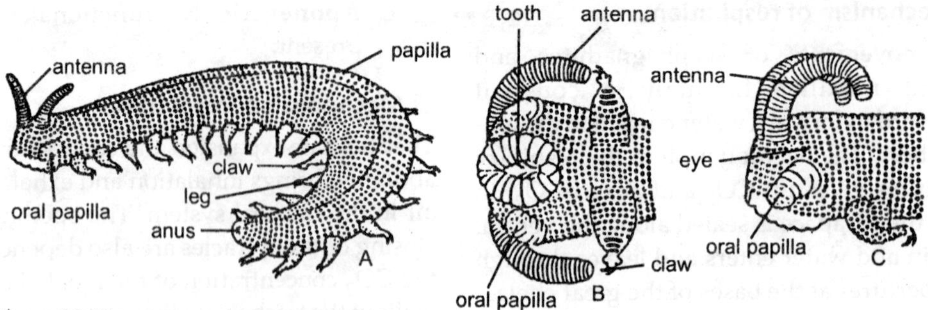

FIGURE 28.57

Peripatus capensis. A. Laterodorsal view, B. Ventral view, mouth region. C. Lateral view, head region

having ladder-like nervous system, being placed in between annelids and myriapods. They are represented by fourteen modern genera widely distributed but in restricted areas.

☒ Habits and habitats

Peripatus sp. belong to phylum Onychophora. They are shy, nocturnal creatures with strong dislike to light. They live in humid habitats, beneath logs of wood, leaves, and stones, along the banks of rivers, mostly in tropical rain forests. In low winter temperature they become inactive and remain in hinding.

☒ Structural peculiarities

1. Caterpillar-like, elongated, (Fig. 28.57) soft and nearly cylindrical body, enclosed in a thin chitinous cuticle.
2. Colouration varies from dark-grey to brown but red and blue colour are not rare.
3. Covering surface velvety with transverse wrinkles and numerous small papillae or tubercles.
4. Body divisible into an indistinct head and an elongated trunk.
5. A pair of eyes at the anterior end.
6. Antennae one pair, surrounded by spiny rings.
7. Mouth ventral, a pair of jaws and oral papillae present.
8. Walking legs 15 pairs.
9. Legs stumpy, unjointed, ringed by ridges with tubercles and terminate in a foot, bearing two curved claws.
10. Respiratory organs tracheae.

☒ Evolutionary status of Peripatus

It is presumed that Onychophora represents a phylogenetically intermediate stage between Annelida and Arthropoda. However, some of their special adaptations do not confirm the "missing link" status of Onychophora, as suggested by some authors. Kestner (1968) lists the following features in this regard.

1. Transformation of the claws of the 2nd and 3rd appendages into mandible.
2. The number of tracheal spiracles and their disposition on the body.
3. Fine structures of the eye.

The importance of onychophores in tracing the phylogeny of arthropods has been dis-

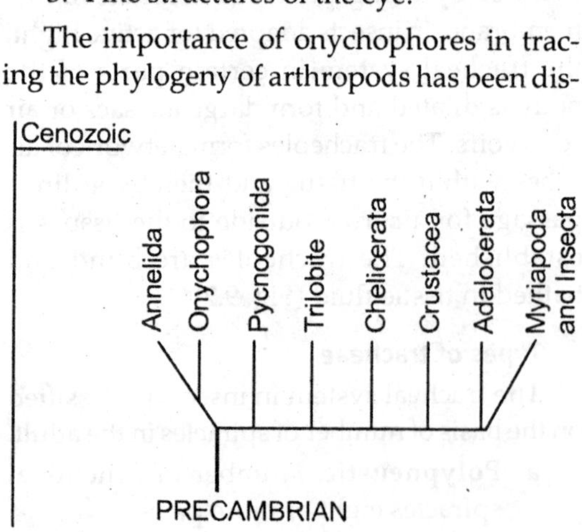

Phylogenetic tree of the major groups of arthropods (Sharov 1966)

cussed by Sharov (1966) and Manton (1973), who consider that arthropods are **monophyletic** and the onychophores represent an isolated line of evolution.

striking. *Peripatus* was assigned different systematic position by zoologists. It was placed in Annelida by some while others thought it was Myriapoda, (Mosely, 1874).

Affinities with Annelida

Similarities	Dissimilarities
1. A true head is absent; body segments homonomous 2. Hollow appendages resembling parapodia. 3. Paired segmental nephridia. 4. Slime and coxal glands are considered homologous with chaetopod glands. 5. Simple chaetopodian eyes with retinal cells. 6. Ciliated gonoducts. 7. Cholinergic motor neurons.	1. Exoskeleton very flexible, sclerotization weakly developed (Barrington, 1921). 2. Presence of tracheae in Peripatus. 3. Presence of 13-42 pairs of stumpy legs with two terminal chitinous claws

Affinities with Arthropoda

Similarities	Dissimilarities
1. Culticle chitinous 2. Appendages muscular and clawed. 3. Serpentine peristaltic motion like Scolopendra. It may even coil up when irritated. 4. Dependency of locomotion on leg musculature. 5. Heart ostiate and lies within pericardial sinus. 6. Haemocoelomic channels present. 7. Presence of tracheae and spiracles. 8. Transformation of nephridium to slivary glands. 9. Coxal vesicle and its function. 10. Cleavage superficial.	1. Skin with warty papillae. 2. Restriction of jaws to a single pair. 3. The jaws of peripatus are derived from the third leg segment, while in other arthropods a pair of limbs further back give rise to the jaws. 4. Short, atumpy legs. 5. Metamerically segmented nephridia. 6. Tracheae diffused in *Peripatus*. 7. Nervous system ladder-like. 8. Gonoducts ciliated.

☒ Affinities with mollusca

Similarities

1. Slug-like appearance.
2. Antennae tentacle-like.
3. Skin moist and thin.
4. Ladder-like nervous system in both Chiton and Peripatus.

The similarities are superficial having no phylogentic significance.

☒ Affinities of peripatus : A discussion

Peripatus possesses a number of characters common with groups of nonchordates but at the same time differences are also no less

Marshall and William (1972) and Storer (1983) considered *Peripatus* as an aberrant form and placed it in a separate subphylum Onychophora under phylum Arthropoda.

Tiegs and Manton (1958) and Manton (1970) believe the Arthropoda is **polyphyletic** and that the modern Onychophora, Myriapoda, Symphyla and Insecta have evolved from onychophore–like ancestral stock.

Kaestner (1968) Meglitsch (1972) and Barnes (1980) considered Onychophora as an independent phylum.

However, the Onychophora differs from other tracheates in the simplicity and diffuseness of tracheae; in having only one pair of jaws; in the absence of external segmentation; in the nature of body wall and the number of short rings bearing minute spines.

The ladder-like ventral nervous system is considered primitive for its resemblance to that of primitive molluscs, phyllopod crustaceans and nemerteans. It is not certain that the antennae, jaws and oral papillae of *Peripatus* precisely correspond to the antennae, mandibles and first maxillae of insects.

It may be concluded that *Peripatus* is a survivor of archaic type, related to the Middle Cambrian marine fossil *Aysheaia*, ancestral to tracheate and closely related to annelids, might have been derived from the type of lobopod annelids by the loss of eversible proboscis and subsequent development of jaws. However, the *Peripatus* and its relatives may be placed in a separate phylum Onychophora.

BIOLOGICAL STATUS (SYSTEMATIC POSITION) OF LIMULUS

The horse-shoe crabs—misnomerly called king crabs—are marine arthropods, which were abundant in Ordovician and Silurian periods. They have undergone little changes from the Silurian time and are represented by only five species today. This has earned the name "**living fossils**" for them. Extensive studies on these archaic creatures provide a good opportunity to know about a very primitive grade of arthropod organization.

⊠ Historical background

Straus and Surekhein (1821) were first to study the horse-shoe crabs and considered them as arachnids. Later, Lankester (1881) recorded the presence of book-gills and strengthened their arachnid affinity. Tiegs and Manton (1958) placed them under the class Merostomata and subsclass Xiphosaura along with subclass Eurypterida.

⊠ Systematic account

The five existing species of horse-shoe crabs are placed in three genera—*Limulus, Tachypleus* and *Carcinoscorpias*. The largest measuring 60 cm including the telson is *Limulus polyphemus*.

⊠ Distribution

Limulus first appeared about 200 million years ago, in the Triassic Period. A number of fossils have been discovered from the Jurassic rocks.

Today *Limulus* is restricted to the North-Western Atlantic coast of North America, while the other two genera are found along the coast of Asia (Japan, India, Korea and East Indies).

⊠ Ecology

Horse-shoe crabs are inhabitants of shallow seas but some members occasionally occur in fresh water, as well as brackish estuarine waters. They burrow and forage for molluscs and worms in sand and mud with a head-on shovelling action of the carapace. The sixth pair of appendages push behind dug-out mud and sand. In breeding season members of both the sexes come ashore and dig out holes at the upper limit of high tide for laying eggs. Adult individuals move by crawling while the young ones swim with help of their abdominal appendages.

⊠ Anatomical peculiarities

1. Body divisible into prosoma, opisthosoma and metasoma (telson).
2. Prosoma covered dorsally with a broad horse shoe-shaped carapace (Fig. 28.58).
3. A median longitudinal ridge and two lateral grooves; two median eyes and two lateral eyes are present on the carapace.
4. Opisthosoma with serrated border; movable spines are six pairs.
5. Carapace and opisthosoma joined by a transverse hinge.
6. Postoral caudal spine (telson) hinged and movable.

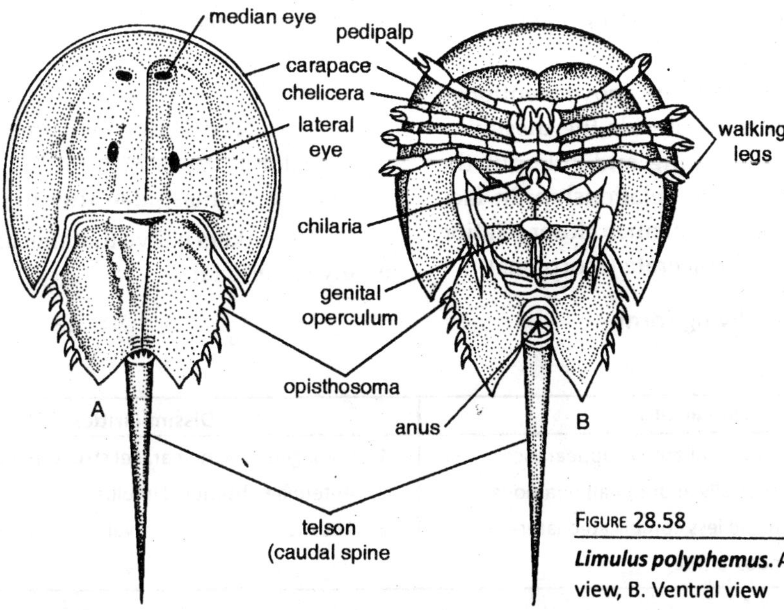

FIGURE 28.58
Limulus polyphemus. A. Dorsal view, B. Ventral view

7. Six pairs of cephalothoracic appendages —chelicerae, pedipalpi and four pairs of walking legs.
8. Mesosoma bears six pairs of flattened appendages. The first pair are without gills and united at the middle to form genital operculum. The rest five pairs are free and each carries a gill-book.
9. Muscles well-developed and attached to the abdomen and sternites.
10. Digestive tube bent, differentiated into zones; the anus ventral, at the base of the telson.
11. Heart tubular, long and with eight pairs of ostia.
12. Nervous system centralised and divisible into brain and a ganglionated, ventral nerve cord.

⊠ Affinities of Limulus

Limulus has certain features common with several groups of both extinct and living nonchordates but quite a number of differences have also been recorded.

⊠ Affinities with extinct forms

With Trilobite

Similarities

1. Cephalothorax with lateral eyes.
2. Appendages biramous.
3. Presence of lateral pleural spines.
4. Presence of trilobite stage in the life history of Limulus.
5. Certain fossils—Synxiphosura are very close to trilobites.

In spite of above similarities, Limulus differs from trilobites in having a telson, in the structure of genital operculum, abdominal appendages, and in the absence of antennae. Cephalothorax of trilobites is distinctly flattened.

With Eurypterida (fossile Merostomata)

Similarities

1. Three similar body segments—prosoma, mesosoma and metasoma.
2. Similar cephalothoracic appendages and telson.

3. Structures of median and lateral eyes similar.

Although *Limulus* bears above mentioned similarities with Eurypterida,the dissimilarities are no less striking and rule out the possibility of any close relationship. In fact, the Eurypterida exhibit a much closer relation with the modern Chelicerata (scorpions)

With Hemiaspidae (fossil crustacea)

1. Construction of carapace and telson similar.
2. Lateral compound eyes present in both.
3. Trilobite larva of Limulus resembles some Hemiaspidae after first moult.

Many workers belive that the affinities are merely superficial.

⊠ Affinities with living form

With crustacea

Similarities	Dissimilarities
1. Aquatic habits and similarity in appearance. 2. Appendages (specially abdominal) biramous. 3. Simple median and less complicated lateral eyes. 4. Presence of endosternite.	1. Book-gills have no parallel structure in Crustacea. 2. Antennae absent in Limulus. 3. Absence of Nauplius larval stage in Limulus.

With Arachnida

Similarities	Dissimilarities
1. Presence of a broad carapace.	1. Respiratory structures—book-gills and book-lungs are different.
2. Number and arrangement of cephalothoracic appendages.	2. Malpighian tubules absent in Limulus.
3. Caudal spine of Limulus resembles the postabdominal part of scorpion.	
4. Structure of genital operculum.	
5. Presence of median eyes.	
6. Presence of endosternite.	
7. Pharynx suctorial and symmetrical digestive glands.	
8. Presence of coxal glands.	

⊠ Affinities of Limulus : A discussion

Attempts have been made to establish affinities of this archaic creature with other forms including chordates. Gaskell (1908) tried to establish its affinities with **ammocoete larva** of Petromyzon, largely based on the structure of dermal skeleton, median and lateral eyes, gills and endosternum. But these similarities are only superficial. Patten (1912) attempted to establish affinity with **ostracoderms** (fossil armoured agnathans), but this gained a little support.

It appears that *Limulus* has much closer relationship with Arachnid and extinct Eurypterids than to any other arthropods. Formerly it was placed under the class Arachnida but Manton (1970) and others have assigned a separate class Merostomata for it along with eurypterids, and hold that Arthropoda is *polyphyletic*.

CLASSIFICATION: PHYLUM ARTHROPODA

1. Arthropoda is the largest phylum comprising bilaterally symmetrical, triploblastic, coelomate invertebrates.
2. Body segmented but most species exhibit some degree of reduction in metamerism.
3. The segments usually bear paired, lateral and jointed appendages, variously modified as jaws, gills, legs, etc.
4. The body is covered with an exoskeleton of chitin and protein, which is thick, tough and nonliving, covering each segment but remains soft and flexible between segments permitting free movement. Moulting or ecdysis of exoskeleton occurs very often.
5. The muscles are attached to the inside of the exoskeleton and the skeletomuscular system functions as a lever system.
6. The blood vascular system is open, capillaries are absent and the arteries open into irregular spaces, called lacunae or sinuses which collectively form haemocoel. The dorsal heart is a primitively pulsating tube.
7. Respiration is carried out on either through general body surface or through gills or book-gills in aquatic forms and tracheae or book-lungs in the terrestrial forms.
8. Excretion occurs through Malpighian tubules or by green glands (coelomoducts); nephridia absent.
9. The nervous system is annelidan type; the dorsal brain is connected around the oesophagus with a double ventral ganglionated nerve cord by a pair of circumoesophageal connectives.
10. A distinct cephalization is present.
11. Compound eyes with mosaic vision.
12. The sexes are usually separate, with sexual dimorphism; gonads and their ducts are usually paired; fertilization internal; development includes little or more metamorphosis; parental care often well-marked.

In present-day classification phylum Arthropoda has been subdivided into seven subphyla—Oncychophora, Tardigrada, Pentastomida, Trilobitomorpha, Chelicerata, Pycnogonida and Mandibulata.

Subphylum I. ONYCHOPHORA

1. The body is caterpillar-like, soft, covered with a thin layer of pigmented cuticle.
2. It is divided into rings, which do not correspond to the segmental arrangement of the appendages.
3. A pair of preoral antennae, a pair of jaws, a pair of oral papillae and many pairs of short, stumpy legs, each with two claws are present.
4. Respiration is carried by tracheae, opening into irregularly scattered pits.
5. The heart is elongated and dorsal and with valvular ostia.
6. Almost all are viviparous.

Examples: *Peripatus, Euperipatus, Ophisthopatus*, etc.

Subphylum II. TARDIGRADA

1. Soft-skinned animals of minute size, not exceeding a millimetre in length.
2. Body is vaguely segmented and not divisible into regions.
3. Four pairs of short unjointed legs, the last pair is terminal and each is provided with two or four claws.
4. Mouth is surrounded by papillae; the buccal cavity contains a pair of horny, sometimes partly calcified teeth.
5. A brain and a ventral nerve cord with ganglia present.
6. Two eye spots at the anterior and are the only sense organs.
7. The gonads in both sexes are saccular and open into the terminal part of the intestine.

8. Live among damp moss; others in fresh or marine water. Example: *Macrobiotus*.

Subphylum III. PENTASTOMIDA

1. Tongue or fluke-shaped or cylindrical worms with superficial annulation.
2. On each side of the mouth two pairs of short limbs or only claws present.
3. Alimentary canal straight and simple.
4. Nervous system reduced, consisting of a circumoesophageal ring, a sub-oesophageal ganglion and a simple nerve cord.
5. Heart, respiratory and excretory organs are lacking.
6. Larva with two pairs of jointed legs.
7. All are parasites in vertebrates.

Examples : *Railiatiella, Cephalobaena, Procephabis, Arumlifers*, etc.

Subphylum IV. TRILOBITOMORPHA or TRILOBITA (extinct group)

1. Body divisible into two distinct regions (tagmata)—prosoma and opisthosoma.
2. Two longitudinal furrows divide the oval-shaped body into three distinct longitudinal zones.
3. Antennae paired and unjointed.
4. Appendages similar and unspecialised.

Examples: *Agnostus, Trinuclaus*, etc.

Subphylum V. ARACHNOMORPHA or CHELICERATA

1. Body divided into cephalothorax and abdomen.
2. Cephalothorax consists of one preoral and five postoral segments, each bearing a pair of appendages.
3. Appendages of first postoral segment are known as chelicerae.
4. Abdomen consists of 12 to 13 segments and a telson.
5. Second abdominal segment bears genital aperture covered by operculum, which is a modified abdominal appendage.
6. Eyes compound, degenerated in most cases.
7. Median ocelli present.
8. Respiration by book-lungs, book-gills, tracheae and skin.

Class 1. **Merostomata**

1. Aquatic arthropods.
2. Abdomen subdivided into a 7-segmented mesosoma and a 5-segmented metasoma.
3. Mesosoma bears 4 to 5 pairs of similar appendages.
4. A caudal style present.
5. Respiratory organs gills.

Subclass i. **Xiphosura**

1. The prosoma has an unsegmented carapace, the opisthosoma has a dorsal shield with serrated edges; prosoma and opisthosoma are jointed by a transverse hinge.
2. The horse shoe-shaped carapace of the prosoma is marked by a median longitudinal ridge and lateral grooves.
3. The prosoma bears a pair of lateral compound eyes and a pair of anterior, median ocelli.
4. Three-jointed, chelate chelicera followed by five pairs of similar appendages, all of which are chelate except the fifth, which bear terminal spines.
5. Generally marine.

Examples: *Limulus* (distribution—North America), *Tachypleus, Carcinoscorpious* (South-East Asia).

Subclass ii. **Eurypterida** (Extinct)

Eample: *Eurypterus*.

Class 2. **Arachnida**

1. The body is usually divisible into cephalothorax and abdomen.
2. The cephalothorax bears sessile, usually simple eyes, two pairs of cephalic appendages, chelicerae and pedipalpi, and four pairs of legs.
3. Antennae are always absent.

4. Respiratory organs are either tracheae, or book-gills or book-lungs.
5. Heart is tubular and ordinarily not divided into chambers.
6. Sexes are almost always separate.
7. Development usually direct.

Eample: *Scorpio, Palamnaeus, Argiope, Mynossus*, etc.

Subphylum VI. PYCNOGONIDA

1. Marine, spider-like; chelicerae short, pedipalpi long and segmented.
2. At the anterior end a distinct proboscis is present.
3. Genital pores thoracic.
4. Third pair of appendages, the ovigers in male carry the eggs.
5. Reduced abdomen bears two to three segments.
6. Partially sedentary marine animals.
7. Young are parasites on soft bodied invertebrates.

Example: *Nymphon*.

Subphylum VII. MANDIBULATA

1. Body divided into head, thorax and abdomen.
2. Three preoral segments are fused and bear antennae, compound eyes and simple eyes.
3. The fourth segment bears a pair of mandibles.
4. Respiratory organs gill, tracheae or skin.

Class 1. **Crustacea**

1. The body is elongated and divisible into head, thorax and abdomen. In many cases, the head is fused with the thorax and form the cephalothorax.
2. Head bears two pairs of antennae and at least three pairs of masticatory appendages in addition to others.
3. The abdomen is usually segmented and often bears appendages.
4. Appendages perform varied functions like feeding, cutting, swimming, feeling, etc.
5. The appendages are typically biramous, each having a basal portion and two distal branches.
6. Respiration by gills.
7. Coelom is restricted to the gonads as gonocoel.
8. The chitinous cuticle is usually impregnated with calcium carbonate.

Subclass i. **Cephalocarida**

1. Ninetten trunk segments behind a horseshoe-shaped head; only anterior nine trunk segments bear appendages.
2. Both pairs of antennae are short.
3. Eyes absent.
4. Hermaphroditic with genital opening on ninth segment.

Examples: *Hutchinsoniella*

Subclass ii. **Branchiopoda**

1. Varying number of trunk segments provided with appendages of a more or less uniform character, usually foliaceous and rarely leg-like.
2. Abdomen without appendages; terminally a pair of unjointed or many-jointed caudal style.
3. A cephalic carapace is often present; it may be either shield-like or bivalved.
4. Paired compound eyes.
5. The antennules and second maxillae reduced.
6. Larva is a nauplius or metanauplius.

Examples: *Branchipus, Triops, Lepidurus, Daphnia, Polyphemus*, etc.

Subclass iii. **Ostracoda**

1. Indistinctly segmented body bearing usually up to two pairs of nonfoliaceous trunk appendages.
2. Posterior part of trunk is terminated by a pair of caudal styles.
3. Body is entirely enclosed in a bivalved carapace.

4. Both pairs of antennae well-developed and used for swimming.
5. The larvae typically nauplius, enclosed within a bivalve carapace.

Examples: *Cypris, Notodromus*, etc.

Subclass iv. **Mystacocorida**

1. Body divided into a head, five-segmented thorax, and six-segmented abdomen; the last four throcic and all adbominal segments are free.
2. No abdominal appendages except on the last, which carries a pair of caudal styles.
3. Sexes separate; earliest larval stage is Nauplius.

Example: *Derocheilocaris*.

Subclass v. **Copepoda**

1. Elongated, segmented body divided into head, thorax and abdomen.
2. Thorax six-segmented, but usually the first segment and sometimes second one too fuses with the head to form cephalothorax.
3. Each thoracic segment bears a pair of appendages; of these the first is always, and the last is often, uniramous and others are biramous.
4. Except for caudal rami, the abdominal segments do not possess appendages.
5. No carapace; no compound eye; a median nauplius eye is usually present.
6. The genital aperture is on special genital segment preceding the abdomen.
7. Larvae Nauplius.

Examples: *Cyclops, Calanus, Lernaea, Ergasilus*, etc.

Subclass vi. **Branchiura**

1. Body divided into head, thorax and abdomen.
2. A bilobed cephalic fold arises from the head and fuses with the sides of the first thoracic segment to form a carapace-like structure.
3. Thorax has four pairs of biramous appendages; abdomen unsegmented, without appendages, bilobed and ends with a minute pair of caudal claws.
4. Mouth suctorial; eyes compound.
5. Both pairs of antennae reduced; the first one clawed. Often the first maxillae are modified to form a pair of suckers.
6. The genital aperture opens at the basis of fourth thoracic legs.
7. Sometimes the eggs hatch to a nauplius larva, but commonly the young have all adult appendages on hatching.

Example: *Argulus*.

Subclass vii. **Cirripedia**

1. Almost exclusively marine, sessile, either parasitic or free-living. In parastic forms the structures may become so modified that almost all traces of arthropodan features disappear.
2. Body imperfectly segmented; carapace forms a pair of mantle folds, which usually enclose the animal entirely and is supported by a system of calcareous plates forming a hard shell.
3. Six similar pairs of biramous appendages are present in most cases, but the posterior region of the body is limbless and rudimentary.
4. The antennules of the larva give rise to the organs of attachment and then become vestigeal; the antennae usually disappear altogether.
5. Eyes absent.
6. Mostly hermaphrodite; the young hatch as nauplius larva.

Examples: *Lepas* (goose barnacle), *Balanus* (acorn barnacle), *Synagoga, Proteolepas, Sacculina*, etc.

Subclass viii. **Malacostraca**

1. Body divisible clearly into head, thorax and abdomen and a telson constituting a segmental region.
2. Distinct segmentation present; with a fixed number of 20 segments typically, rarely 21; the head constitutes six, thorax eight and abdomen six or seven segments.

3. Each segment bears appendages except the first.
4. Posterior margin of the head commonly fuses with one or more thoracic segments to form a cephalothorax; no typical caudal style.
5. Paired sessile or stalked compound eyes mostly present.
6. The antannules are often biramous.
7. The male gonopore always on the eighth, and female gonopore on sixth thoracic segment.
8. The nauplius stage often occurs before hatching.

Examples: *Nebalia, Palaeocaris, Anaspides, Bathynella, Squilla, Thermosbaena, Spelaeogriphus, Gnathophansia, Mysis, Leptocuma, Gammarus, Caprella, Asellus, Cryptoniscus, Euphausia, Thysanopoda, Macrobrachium, Penaeus, Palinurus, Scyllarus, Homarus, Astacus, Eupagurus, Cancer, Lithodes**, *Portunus, Uca* etc.

Class 2. **Chilopoda**

1. Body usually dorsoventrally flattened.
2. Number of legs vary from 15 to over 100.
3. First pair of thoracic appendages are maxillipeds and function as poison claws.
4. Each leg bears 7 segments.
5. Segment just anterior to telson is the genital segment.
6. A pair of gonopods are present.
7. Respiratory organs are tracheae.
8. Habitat humid areas; nocturnal and carnivore.

Exmaples: *Scolopendra, Geophilus, Scutigera, Lithobius*, etc.

Class 3. **Symphyla**

1. The number of body segments usually fourteen.
2. 12 pairs of legs are present.
3. A pair of dorsal cerci present on the thirteenth segment.
4. Second maxillae fused to form the labium.
5. Spiracles restricted to the head and the tracheae extend in the first three body segments.
6. Both herbivores and omnivores.

Examples: *Scolopendrella, Scutigerella*, etc.

Class 4. **Pauropoda**

1. Saprophytes, body length is about 1 mm or little more in some cases.
2. 5 segments constitute the head and ten segments are present in the trunk.
3. Antennae branched.
4. Gnathochilarium is formed by the union of first maxillae.
5. Second to tenth segment bear legs.

Example: *Pauropus*.

Class 5. **Diplopoda**

1. The body bears a distinct head but the remaining portion is not distinguishable into different regions.
2. The head bears a pair of many-jointed, short club-shaped antennae, a pair of eyes and 2 or 3 pairs of jaws.
3. The trunk consists of a number of similar segments, each bearing one or two pairs of legs.
4. Respiratory organs are tracheae.
5. Salivary glands are always present.
6. Excretion by Malpighian tubules.
7. Development usually accompanied by a metamorphosis.

Examples: *Julus, Polyxenus, Chordeuma*, etc.

Class 6. **Insecta or Hexapoda**

1. The body is divided into head, thorax and abdomen.
2. The head bears a pair of antennae, a pair of compound eyes and three pairs of mouth appendages.
3. The thorax bears a pair of legs on each of its three segments and generally a pair of wings on each of the posterior two.

---

* The king crab, *Lithodes* sp is a big edible crab. family Lithodidae. It's size after expansion of large chela (pedipalp) is about 240 cm.

4. The abdomen is devoid of appendages in the adult and has a variable number of segments.
5. A liver is absent but salivary glands are always present.
6. The heart is tubular, divided into eight chambers and situated in the abdomen.
7. Respiration by means of air tubes or tracheae.
8. Excretion by means of Malpighian tubules.
9. Sexes separate, development direct or indirect.

Subclass i. **Diptera**

1. Abdomen consists of eleven segments both in adult and juvenile.
2. Primarily wingless.
3. Segmented antennae.
4. Caeca present.
5. Malpighian tubules absent or vestigeal.

Examples: *Campodea, Heterojapyx*, etc.

Subclass ii. **Protura**

1. Abdominal segments added secondarily through postembryonic growth, making a final complement of eleven segments and a posterior telson.
2. Antennae absent.
3. Primarily wingless.
4. Malpighian tubules present but rudimentary.
   Minute, white insects widely distributed; habitat moist litter, bark, etc.

Examples: *Protura, Acerentomen, Eosentomon*, etc.

Subclass iii. **Collembola (Spring tail)**

1. Abdomen with six segments.
2. Usually with three sets of appendages —a ventral tube, a retaining hook on segment 3, which functions with a powerful springing organ on segment 4.
3. Primarily wingless.
4. Tracheael system absent in most species.
5. No Malpighian tubules.
6. Metamorphosis absent.

Known as springtail; live in soil, leaf litter, decaying vegetable materials, etc.

Examples: *Podura, Orchesella, Isotoma, Neanura*, etc.

Subclass iv. **Apterygota**

1. Primarily wingless.
2. Mandibles having only a single point of articulation.
3. Pregenital abdominal appendages usually present.
4. Metamorphosis from juvenile to adult slight or absent.

Example: *Lepisma, Pterobius*, etc.

Subclass v. **Pterygota**

1. Adults winged or secondarily wingless.
2. Pregenital abdominal appendages lacking in adult.
3. Mandibles articulated at two point; or modified; mouth parts highly variable.
4. Metamorphosis from juvenile slight, or gradual to extreme and complete.

Divison 1. **Exopterygota**

1. Wings when present form external pads in the late nymphal instars.
2. Extreme metamorphosis between juvenile (nymph) and adult generally lacking.

Example: *Ephemer Periplaneta, Aeschna* (Dragon fly), etc.

Divison 2. **Endopterygota**

1. Wings, when present, form internal pads.
2. Metamorphosis between juvenile (nymph) and adult more or less extreme, passing through a pupal stage.

Examples: Moths and butterflies, *Duomitus, Hypena, Noctua, Trichophaga, Tincola*, etc.

29 PHYLUM MOLLUSCA

The Mollusca is the second largest phylum of the animal kingdom containing some 50,000 living and 35,000 fossil species, which enjoy wide distribution in both time and space, having continuous record since Cambrian time.

The phylum includes diverse animals like snails, clams, chitons, squids and octopuses. The squids are one of the largest invertebrate animals. Most molluscs are free-living that creep slowly; some attach to rocks, woods or any hard object; some burrow, others float, and some others like octopuses and squids swim freely.

Molluscs (*L. Mollis* = Soft) have soft unsegmented bodies, consisting typically of an anterior head, a ventral soft foot (modified in different groups) and a dorsal visceral mass. The body is more or less surrounded by a thin fleshy mantle which secretes a protective calcareous shell. The shell has been lost in some but the mantle almost always remains.

The molluscs are vary much popular group of animals because of their colourful shells. Shell collection is an exciting hobby throughout the world. Attractive ornaments, decorative articles, buttons etc., are produced from shells. Pearls are produced by marine bivalve molluscs. Oysters, snails, squids, etc. are the chief constituents of delicious dishes for many.

Malacology is the science that deals with molluscs and conchology is the study of shells particularly those of molluscs.

LAMELLIDENS sp. (Fresh water mussel)

Lamellidens belongs to the class Pelecypoda or Bivalvia, phylum Mollusca. Fresh water mussels live in ponds, lakes, slow moving streams all over the world. The *Lamellidens* are distributed throughout Indian subcontinent. Pearls of small sizes have been reported from *Lamellidens* in West Bengal (India) and Bangladesh.

The mussels remain partially embedded in mud or sand at the bottom of a water body. On slight disturbance, the foot and siphons are withdrawn inside the shell and the gape between the valves of the shell is tightly closed.

⊠ External features

1. The mussel is enclosed in a hard, calcareous, horny-blackish, bivalve **shell** (Fig. 29.1A). The adult mussel is about 10 cm in length.
2. The body is light cream in colour, soft, elongateoval with a somewhat broad and a narrow posterior end.
3. A **mantle** or **pallium** having two equal halves, the **mantle lobes,** lines the inner surface of the shell and encloses a space,

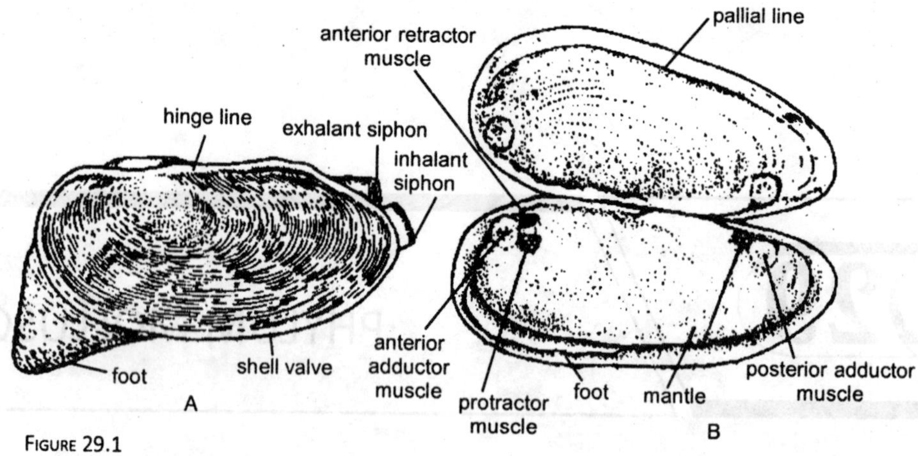

FIGURE 29.1

Fresh water mussel (*Lamellidens marginalis*). A. An entire specimen. B. Left valve of the shell open

the **mantle** or **pallial cavity**. The organs in the cavity are collectively termed **pallial complex** (Fig. 29.2).

4. Two siphons—the dorsal **exhalant siphon** with smooth margin and the ventral **inhalant siphon** with fimbriated margin—formed by the mantle protrude beyond the shell at the posterior end.
5. In natural state, a narrow gape between the valves of the shell allows the muscular foot to project out from the anteroventral end of the body.
6. The viscerel mass, a pair of gills or ctenidia, two renal pores and the anus are lodged in the mantle cavity.

⊠ Shell

The shell consists of two equal halves or **valves** is made of conchiolin and calcic sub-stances. It grows along the margins with the growth of the mussel.

1. The two valves are hinged along a straight, dorsal **hinge line** by a tough, elastic **hinge ligament**, running transversely from one to another valve.
2. Each valve bears teeth-like projections, the **hinge-teeth** along the hinge line. The teeth of one valve fit with the sockets of the other.
3. Series of concentric rings, the **lines of growth**, widening towards the margin are present on the outer surface of the valve.
4. A small dorsomedian elevated region, the **umbo** is the thickest and oldest part of the shell. The lines of growth start from the umbo.
5. The inner surface of the shell is smooth, pearly in colour with a faint bluish tinge.
6. A narrow streak, the **pallial line**, marking the outer border of the mantle lobe is present on the inner surface of the shell, close to the margin.
7. Some scars, indicating the attachment of muscles are present on the inner surface of the shell (Fig. 29.1B).

a. **Anterodorsal scars**

i. Anterior adductor muscle—large, nearly round.

ii. Anterior retractor muscle—small, and posterodorsal to the adductor.

iii. Protractor muscle—small and posterior to the adductor.

b. **Posterodorsal scars**

i. Posterior adductor muscle—large and oval.

ii. Posterior retractor muscle—small, close to the posterior adductor.

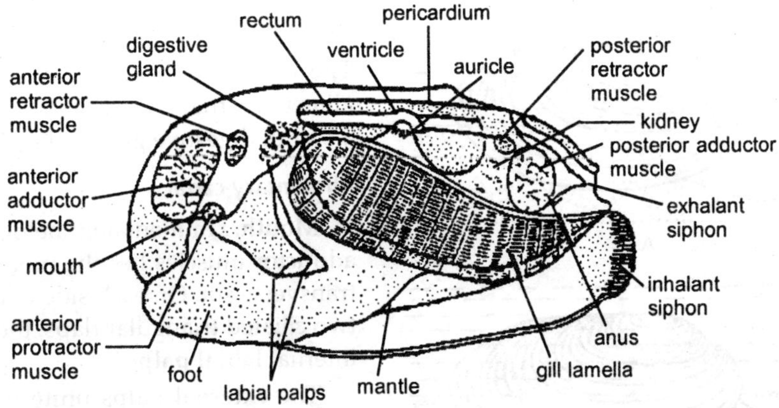

FIGURE 29.2

***Lamellidens* sp.** Pallial complex. Left half of the mantle removed

☒ Microscopic Structure

The shell consists of three distinct layers:

1. **Periostracum.** The thin outermost layer made of **conchiolin**, a substance related to chitin, and imparts colour to the shell.
2. **Ostracum** or **Prismatic layer.** The thick middle layer, composed of alternate layers of conchiolin and prisms of calcic substances.
3. **Nacreous layer** or **mother of pearl.** The innermost layer composed of alternate layers of conchiolin and calcic substances. The layer is smooth and lustrous:

   The periostractum and ostracum are secreted by the mantle edge; soft and horny in colour at the beginning, which hardens with subsequent deposition of materials. The nacreous layer is secreted by the outer surface of the mantle. Pearl is secreted by this layer.

Muscles

The muscles are unstriated. The fibres of the muscles run from body to the shell. The foot muscles are not attached to the shell.

1. **Adductor muscles.** Two, anterior and posterior: large transverse bands control the movements of the valves. Contraction closes the shells.
2. **Retractor muscles.** Two, anterior and posterior; run from the foot to the shell. Contraction causes retraction of the foot.
3. **Protractor muscles.** Two, anterior and posterior; contraction compresses the visceral mass.
4. **Pedal muscle.** The foot is made of a complex mass of intrinsic muscles.

Locomotion

The foot is the locomotor organ.

1. The foot is a muscular projection (Fig. 29.2) from the anteroventral end of the body and protrusible through the gape between the shell valves.
2. It is flattened from side to side and terminates in a more or less elongated keel, the anterior end being narrow, resembling a ploughshare (Fig. 29.3) which remains buried in the substratum.
3. A large sinus, the **pedal sinus** in the foot receives haemolymph from the pedal artery. Pressure on the haemolymph exerted by the contraction of foot muscles, forces the foot to extend forward.
4. In the next step, the narrow anterior end of the foot expands laterally and acts as an anchor in the substratum.
5. The foot muscles contract, the haemo lymph from the sinus flows out through

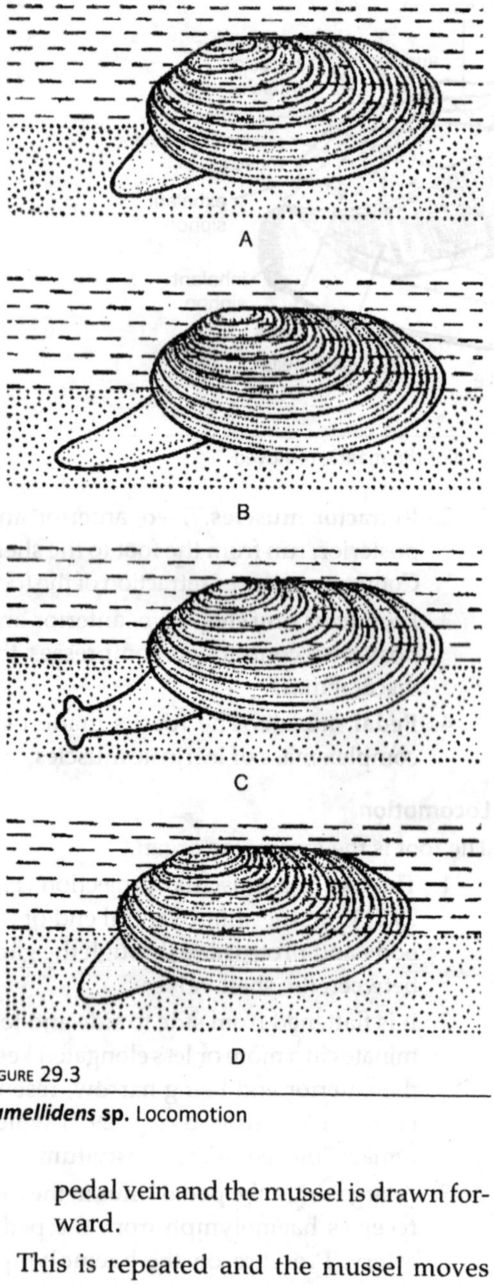

FIGURE 29.3

***Lamellidens* sp.** Locomotion

pedal vein and the mussel is drawn forward.

This is repeated and the mussel moves forward, polughing its foot through the substratum.

Body cavity

In the adult, the coelom is greatly reduced and restricted to pericardium (pericardial chamber), gonad (gonocoel) and lumen of the kidney (urocoel). The haemocoel acts as channels for haemolymph circulation.

☒ Digestive System

The digestive system consists of a coiled alimentary tract, two pairs of labial palps and a pair of digestive glands (Fig. 29.4).

Alimentary canal

Mouth. Immediately below the anterior adductor muscles lies the mouth, which is a transverse slit. On each side of the mouth are two ciliated triangular flaps, the external and internal **labial palps**.

The external palps unite in front of the mouth to form an upper lip and the internal palps unite behind the mouth to form a lower lip. The palps bear cilia.

Oesophagus. A short tube, runs dorsally from the mouth to open into the stomach.

Stomach. A large, oval sac embedded in the digestive gland. The epithelium is thrown into folds. Two narrow ducts from the digestive glands open in the stomach. Occasionally, a mucous rod, the **crystalline style**, which plays an important role in digestion is found in the stomach. It contains diastatic enzymes.

Intestine. The intestine is a narrow tube and, after its origin from the ventral surface of the stomach near its posterior end, descends into the visceral mass above the foot, form two incomplete loops, ascends to the level of the stomach, anterior to the pericardium, turns sharply backward and proceeds straight as the rectum.

Rectum. The rectum pierces the pericardium and the ventricle of the heart, runs parallel to the dorsal surface of the mussel and opens into the exhalant siphon by the anus. The internal wall of the rectum is thrown into longitudinal folds to form a ridge, known as **typhlosole**.

☒ Digestive glands

The digestive glands are a pair of irregular light grey structures situated around the stomach. The secretion of the glands is discharged into the stomach through small ducts. Each gland is made up of numerous diverticula.

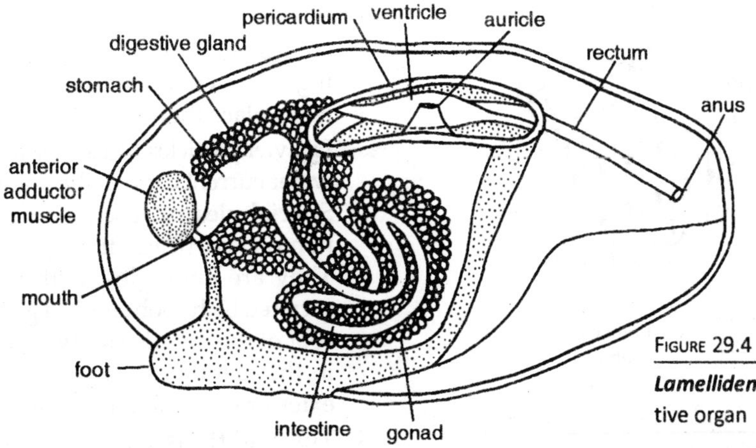

FIGURE 29.4
***Lamellidens* sp.** Digestive organ

☒ Feeding

The fresh water mussel is a ciliary feeder, i.e. it collects food with the help of cilia. The food consists of **detritus** (partially decom-posed organic particles) suspended in water, and microorganisms like bacteria, protozoa, etc. Detritus constitutes four-fifths of the total amount of food.

1. The respiratory, water current created by the movement of cilia on the gills carries food particles and other matters.
2. The heavy particles drop in the mantle cavity and are expelled out through exhalant siphon by the movement of cilia in the mantle.
3. The gill laminae are covered with a thin sheet of mucus and the food particles are entagled in it.
4. The movement of **lateral cilia** on the gills drives mucous sheet to the ventral edge of the gills, bearing **frontal cilia**.
5. The frontal cilia drives the mucous string forward to the **ciliary tracts** of the labial palps.
6. Further assorting takes place here, the larger particles are carried to the tips of the labial palps, drop off in the mantle cavity and removed to the exterior.
7. The mucous string with the food matters are carried to the deep groove, the **food groove** between the two palps, leading to the mouth and driven to the oesophagus.

☒ Digestion

Not much is known about digestion in lamellidens. Intracellular digestion plays an important role in nutrition. Carbohydrases are only enzymes in the stomach and carbohydrate digestion takes place there. Intracellular digestion in the digestive glands hydrolyses protein and lipid to their constitutents.

☒ Respiratory System

The respiratory organs consist of a pair of gills, one on each side between the body and the mantle (Figs. 29.2, 29.5).

1. Each gill consists of two **laminae**, the outer and the inner, and each lamina consists of two **lamellae**, outer and inner. The two lamellae of each lamina are united along the anterior, ventral and posterior edges, but free dorsally. The cavity inside the lamina is divided into several chambers or **water tubes** by means of vertical **interlameller junctions**. Each lamella is made up of many gill **filaments** connected by **interfilamentar junctions** (Fig. 29.5). Minute pores or **ostia** are present between the filaments.
2. The outer lamellà of the outer lamina is attached to the mantle; the inner lamella

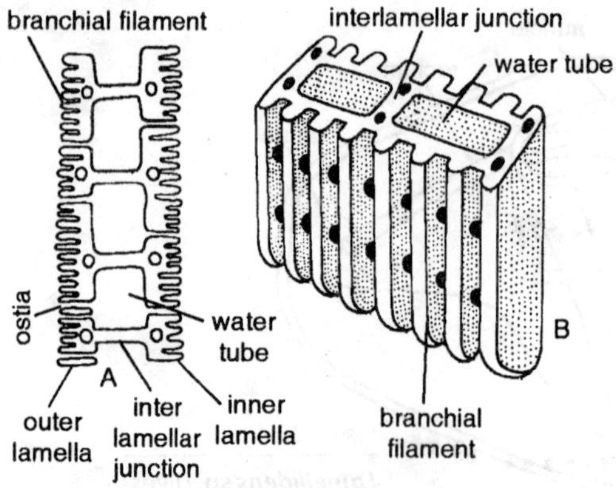

FIGURE 29.5

***Lamellidens* sp.** A. Transverse section of an outer gill lamina. B. Structure of a gill (diagrammatic)

of the outer and the outer lamella of the inner lamina are attached together to the visceral mass; the inner lamella of the inner lamina is attached to the visceral mass. All water tubes of the gill open dorsally into a **suprabranchial chamber**, which, in turn, opens to the exterior through the exhalant siphon.

3. The cells of the gill filaments are provided with cilia and, by their action, a current of water inhalant current, enters the mantle cavity through the inhalant siphon. From there it passes to the water tubes through ostia and goes to the suprabranchial chamber to pass out through the exhalant siphon as exhalant water current. The gill filaments are highly vascular and gaseous exchange takes place there.

   Heavier particles carried by the inhalant water current drop in the mantle cavity. Special cleansing cilia sweep small, nonfood particles off the gill filaments. They are entangled with the mucus secreted by hypobranchial glands in the roof of the mantle cavity. The particles in the mantle cavity are swept to the exterior with exhalant water current.
4. The mantle is richly supplied with haemolymph and gaseous exchange occurs there also.

⊠ Circulatory System

A heart, arteries with their branches, sinuses and veins constitute the open circulatory system of lamellidens (Fig. 12.2, ch. 11). Leucocytes are few and suspended in the colourless haemolymph.

Heart and pericardium

The heart consists of a single ventricle and two auricles, enclosed in a thin-walled pericardium (Fig. 29.6).

1. **Ventricle.** A muscular chamber, surrounding the rectum.
2. **Auricles.** Two thin-walled chambers, communicating with the ventricle by

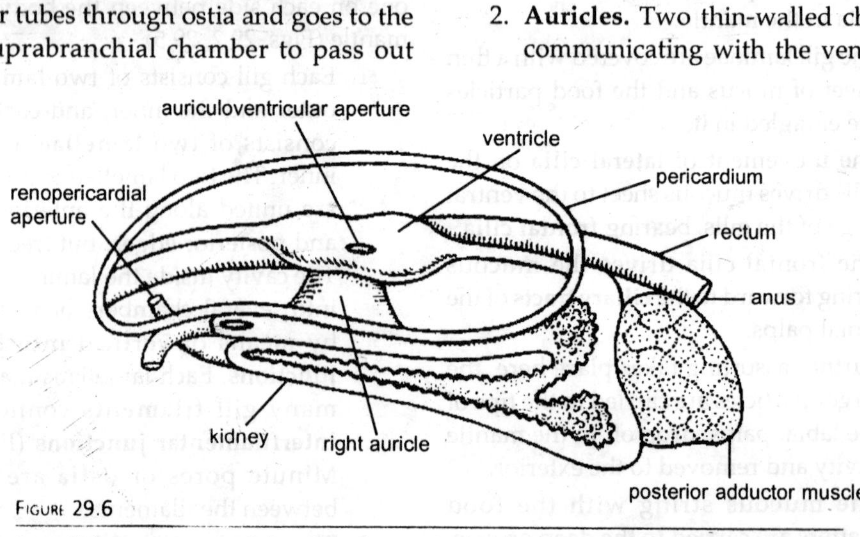

FIGURE 29.6

***Lamellidens* sp.** A part of the visceral mass. Pericardiom with heart, rectum and kidney exposed

auriculo-ventricular apertures guarded by valves, opening only towards the ventricle.

Aortae

An aorta arises from each end of the ventricle from the point of origin they are named as anterior and posterior aorta.

1. **Anterior aorta.** It arises from the anterior end of the ventricle and runs anteriorly above the rectum along the middle line.
   a. At some distance from origin it bifurcates into two—the right and left terminal arteries.
   i. **Terminal artery.** Each sends branches to the mantle lobe, stomach, intestine, digestive gland, palp, foot and other organs at the anterior part of the body and finally terminates as the pallial artery.
2. **Posterior aorta.** The aorta arises from the posterior end of the ventricle and runs posteriorly, ventral to the rectum, along the middle line.

The posterior aorta sends branches to the pericardium, heart, kidney, rectum and other organs at the posterior part of the body.

Sinuses

The returning haemolymph from various parts of the body passes through many sinuses, smaller veins and finally collected in a large, longitudinal **vena cava,** lying between the kidneys.

Veins

Haemolymph is returned to the heart through three circuits—mantle circuit, vena cava circuit and ctenidial circuit.

1. **Mantle circuit.** Small veins arising from the mantle lobe join to form a large vein and returns oxygenated haemolymph directly to the auricle.
2. **Vena cava circuit.** Prior to the origin of renal veins for supplying deoxygenated haemolymph to the kidney, a small vein arising from the vena cava, carries certain amount of deoxygenated haemolymph directly to the auricle.
3. **Ctenidial circuit.** The major amount of haemolymph is returned to the auricle through this circuit.
   a. **Renal veins.** Carry deoxygenated haemolymph from vena cava to kidney.
   b. Afferent branchial veins. Carry haemolymph from kidney to the gills.
   c. Efferent branchial veins. Carry oxygenated haemolymph from gills to auricles.

☒ Excretory System

A pair of kidney or **organs of Bojanus** and a **pericardial gland** or **Keber's organ** constitute the excretory system.

Kidney

1. The kidneys are more or less U-shaped tubes (Figs. 29.2, 29.6) located one on each side of the visceral mass, ventral to the pericardium.
2. The brownish, **spongy**, and **glandular** ventral arm is **kidney** proper, while the thin-walled, **nonglandular, ciliated** dorsal arm is the **urinary bladder**. The cilia produce an outward current.
3. The kidney proper opens in the pericardium through a **renopericardial aperture** and the bladder communicates anteriorly with its fellow of the other side by a large, oval aperture.
4. Each bladder opens in the mantle cavity through a minute **renal aperture** (nephridiopore) between the visceral mass and the inner gill lamina of the side.

Pericardial gland

It is a reddish brown glandular mass, just in front of the pericardium and discharges the excretory matters in the pericardial cavity.

☒ Nervous System

The nervous system consists of paired cerebral, pedal, visceral ganglia; commissure and

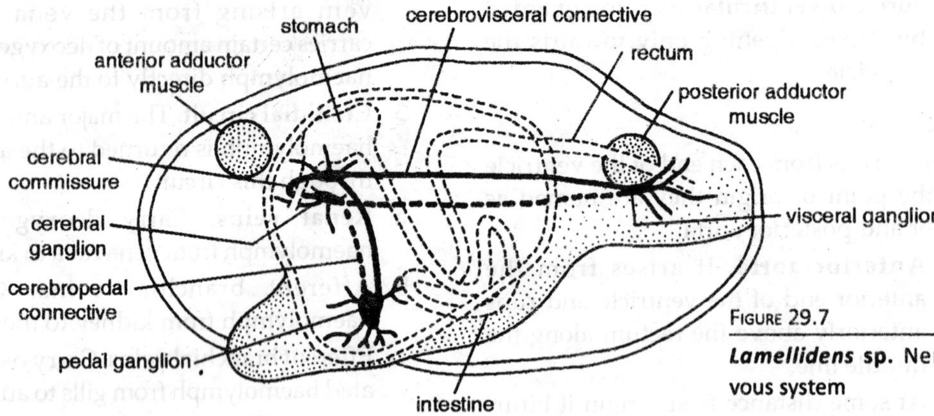

FIGURE 29.7 *Lamellidens* sp. Nervous system

connectives and nerves emanating from the ganglia and connectives (Fig. 29.7).

Cerebral ganglia

Two, triangular in shape, yellow in colour, located at a distance from each other, lying at the base of the labial palps, just under the skin.

Cerebral Commissure

A slender nerve, runs around the oeso-phagus in the anterior part of it and connects the two cerebral ganglia at their anterodorsal border.

Pedal ganglia

The pedal ganglia are two, embedded deeply in the muscles of the foot at about 1 cm above the ventral border of the foot. The ganglia are oval in shape, pale yellow in colour and the two fuse medially to form a bilobed mass.

Cerbropedal connectives

Arising from the posterior border of cerebral ganglia, the two nerves run downward, backward through the visceral mass and muscles of the foot to join the pedal ganglia at the dorsal border.

Visceral ganglia

Two large ganglia, partially fused and located ventral to the posterior adductor muscle, covered by a thin peritoneum. The ganglia are flattened, triangular in shape and pale yellow in colour. The two in *vivo* assumes X-shape.

Cerebrovisceral connectives

Two, arising from the posterolateral border of cerebral ganglion, each runs backward, ventral to the attachment of the gill lamina with the visceral mass. It courses through the mass of the kidney and emerging from its posterior border runs further backwards to join the visceral ganglion of the side at its anterior border.

Innervation

a. Nerves arising from cerebral ganglia innervate labial palps, anterior part of the mantle and adjacent structures.
b. Pedal ganglia send nerves to foot and anterior retractor muscle.
c. Visceral ganglia send following nerves.
   i. Dorsal and posterior pallial nerves to the mantle lobes.
   ii. Posterior renal nerves to kidneys.
   iii. Branchial nerves to gills.
   iv. Nerves to alimentary canal.

Fine nerves* from cerebropedal and cerebrovisceral connectives innervate the statocysts and structures in the visceral mass, respectively.

Receptors and sense organs

The sense organs are poorly developed. Paired osphradia, statocysts and scattered sensory cells constitute sense organs.

* These nerves presumably arise from the cerebral and visceral ganglia.

Osphradium

A patch of brown pigmented sensory epithelium in association with each visceral ganglion. It is a **chemoreceptor** and tests the purity of the water entering mantle cavity.

Statocyst

The statocyst lies close to the pedal gan-glion. It is a round sac, enclosing a space lined by a layer of **sensory cells,** supported by other cell layers (Fig. 29.8). A granule, the **statolith** is present in the space. The statocyst is innervated by the statocyst nerve arising from the cerebropedal connective. It is a balance organ.

Sensory cells occur on the outer epithelim of the body and inner border of the mantle. These act as tactile sense organs.

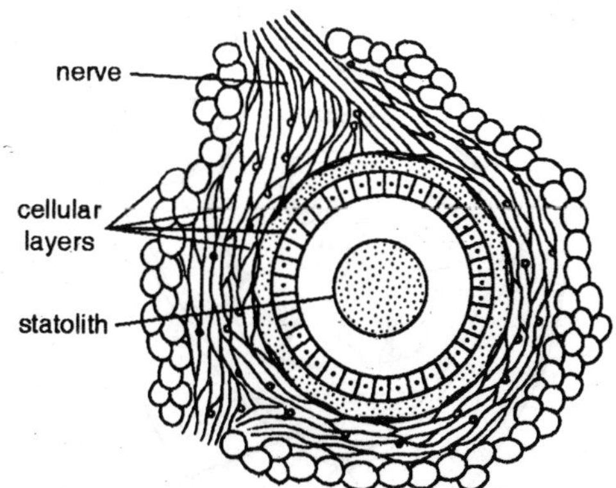

FIGURE 29.8

***Lamellidens* sp.** Statocyst with surrounding cell layers

☒ Reproductive System

The sexes are separate; gonads are paired and occupy the major portion of the visceral mass in which the loops of the intestine are lodged. Both the testes and ovaries are racemose glands; the gonoducts arising from the gonads are short, narrow and open in the front of the renal aperture through a round genital pore (Fig. 29.9).

☒ Male Reproductive System

1. Testes white in colour.
2. A vas deferens arises from the postero-lateral border of each testis.

☒ Female Reproductive System

1. Ovaries reddish in colour.
2. An oviduct arises from the posterolateral border of each ovary.

☒ Fertilization and Development

Fertilization is internal.

1. The eggs discharged in the mantle cavity from the ovaries through female genital pores pass into the suprabranchial chamber.
2. The sperms discharged from the testes in the mantle cavity through male gonopores pass out with water current and reach the suprabranchial chamber of the female.

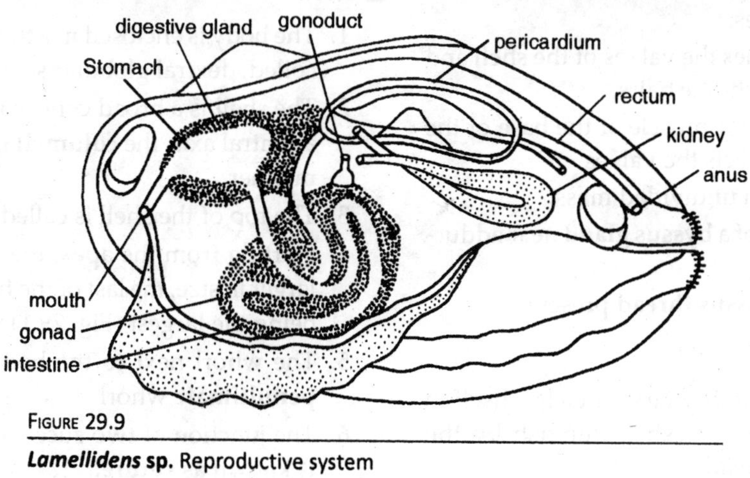

FIGURE 29.9

***Lamellidens* sp.** Reproductive system

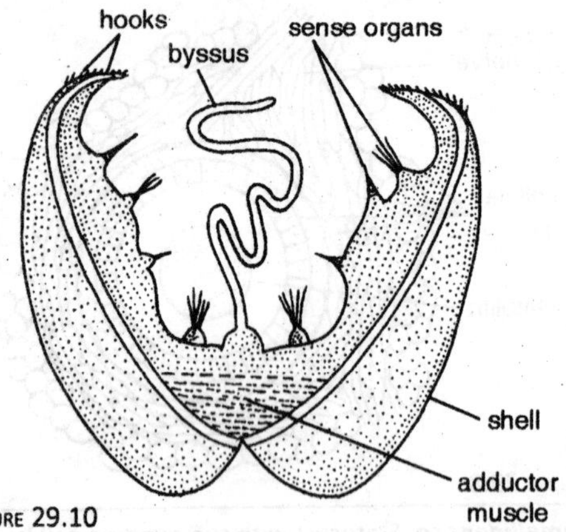

FIGURE 29.10
Glochidium larva

3. Fertilization takes place there and the fertilized eggs are passed to the water tubes.
4. In the water tubes, a number of fertilized eggs are enclosed in a thick, soft, somewhat triangular, whitish **capsule**, the **egg capsule**.
5. Development continues in the egg capsule and the capsules with glochidium larvae are discharged outside.
6. The glochidium larvae (Fig. 29.10) escape from the capsule.

Glochidium larva

a. Presence of a **bivalve shell**.
b. Valves united dorsally and flat ventrally.
c. Ventral ends of the valves curve and bear spines.
d. Mantle lines the valves of the shell and bears brush-like hairs.
e. An adductor muscle at the base of the shell connects the valves.
f. Body is an undivided mass.
g. Presence of a **byssus gland** near adductor muscle.
h. A long **byssus thread** present.

Metamorphosis

1. The glochidium larva attaches itself to the gill of a fresh water fish by the byssus thread.
2. The larva penetrates the host tissue, becomes enclosed by gill epithelium, live temporarily as ectoparasites and undergoes metamorphosis.
3. The byssus thread and sense organs disappear.
4. A **stomodaeum** is formed by ectodermal invagination, immediately posterior to the roof of the byssus thread and soon communicates with the **archenteron**. The opening of the stomodaeum forms the mouth.
5. **Proctodaeum** is not formed. The posterior end of the archenteron is in contact with the ectoderm and the **anus** is formed by a simple process of rupture.
6. **Foot** originates as a median, ventral elevation behind the mouth.
7. **Gill rudiments** appear as two papillae on each side of the mouth.
8. The larva fit for free life, drops from the host, sinks to the bottom and gradually attains adult form.

PILA sp. (Apple snail)

The *Pila* is commonly called apple snail. It lives in ponds, lakes and slow moving streams; found in abundance in paddy fields during monsoon.

⊠ Shell

1. The body is enclosed in a thick, spirally coiled, dextral globular shell.
2. The shell is a broad cone, coiled round a central axis, the **columella**, in a spiral manner.
3. The top of the shell is called **apex**.
4. Starting from the apex, the whorls are larger in size, the last or the **body whorl** being the largest (Fig. 29.11).
5. The whorl next to the body whorl is **penultimate** whorl.
6. The junction of two successive whorls is known as **suture**.

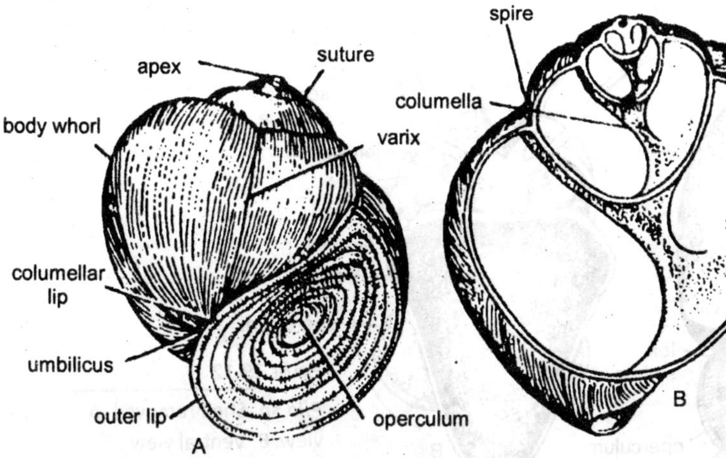

FIGURE 29.11

Pila globasa. A. An entire shell, B. Half of the shell ground off

7. A large aperture called **mouth** is present in the body whorl. The margin of the aperture is smooth and known as **peristome.**
8. The aperture can be closed by an **operculum** which is a flat calcareous plate, **lunate-oblong** in outline.
9. Concentric growth **rings** around the centre or nucleus are present in the operculum.

☒ External features

1. The body is soft, slimy and lodged within the shell.
2. It is attached to the **columella**, the spirally twisted axis of the shell, by a **columellar muscle** arising from the foot.
3. The muscle prevents the animal from extending out of the shell beyond a certain limit and also helps to withdraw it into the shell.
4. The body is divisible into **head, foot** and **visceral mass** (Fig. 29.12).
5. The head is well-marked and drawn anteriorly into a **snout.**
6. The head carries two pairs of tentacles.
   i. The shorter pair are anterolateral in position and called **labial palps** or first tentacles.
   ii. The longer pair or second tentacles are thread-like and bear **stalked eyes** at the base.
7. Two fleshy projections called **nuchal lobes** or **pseudopipodia** are seen on the two sides of the head. The left nuchal lobe is highly developed and form the **respiratory siphon.**
8. The **mouth** is a vertical slit at the end of the snout. True lips are absent. The plicate edges serve as lips.
9. The **anus** is a small round aperture. It is anteriorly placed at about 6 mm distance from the right pseudopipodium.
10. The **foot** is highly muscular and more or less triangular in shape, if viewed ventrally.
    i. The sole or the creeping surface is flat.
    ii. The operculum is attached to the dorsal surface of the posterior part of the foot.
11. The skin covering the visceral mass is known as the **pallium** or **mantle.**
12. In a retracted state, the mantle forms a cloak over the anterior part of the body including the head and its appendages.

☒ Torsion

The twisting of the visceral mass on headfoot axis is termed torsion, a phenomenon peculiar to gastropad molluscs. The gastropod in the early embyonic stage is **bilaterally symmetrical.** In the process of torsion, the headfoot axis remains unchanged, the visceral hump of the embryo rotates 180°, shifting the

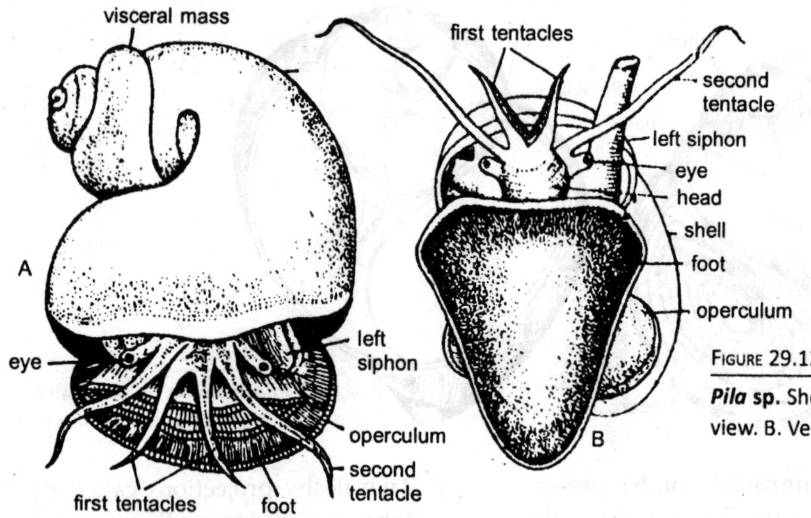

FIGURE 29.12

Pila sp. Shell removed. A. Dorsal view. B. Ventral view

mantle cavity and the posterior part of the visceral mass forward and upwardly, bringing the anus close to the mouth.

The torsion is usually due to unequal growth of the two halves of the visceral mass associated with muscular contraction or rarely by muscular contraction alone. The active growth of one side shifts the structures towards the opposite side, the growth of the structures in the latter half being suspended, viz. the active growth of the left side shifts the structures towards the right and vice versa. The clockwise twisting is called **dextral** and the anticlockwise **sinistral**. In dextral forms, the shell mouth (opening of the shell) is to the left and the twisting of the shell is clockwise, when the shell is held with the apex away from the observer and the mouth ventral. In sinistral forms the situation is reverse. The majority of the gastropods are dextral e.g. Pila, Achatina, etc.

⊠ Locomotion

Locomotion in *Pila* is slow and effected by crawling on the surface with the help of a more or less triangular foot with a flat sole.

1. The foot is provided with vertical, longitudinal and transverse muscles.
2. Glands in the foot secrete a copious amount of slimy mucus, making the surface on which the snail moves, slippery.
3. The foot is protruded through the mouth of the shell due to influx of haemolymph in the pedal sinus.
4. Contraction of the transverse muscles forces the haemolymph forward resulting in the forward elongation of the foot.
5. Contraction of vertical muscles creates wave-like movement on the sole of the foot, helping it to move forward.
6. Contraction of longitudinal muscles pull the posterior end of the foot forward and the snail creeps slowly on the substratum.

⊠ Pallial Complex

The structures present in the mantle or pallial cavity are collectively called pallial complex. The body is enclosed in a **mantle** or **pallium**. In the body whorl it forms a **cloak** over the body, enclosing the **mantle** or **pallial cavity** (Fig. 29.13). In the rest, the pallium is thin and adherent to the visceral mass. The opening of the mantle cavity is anterior and wide. A longitudinal ridge, the **epitaenium**, incompletely divides the cavity into a right and left chamber. The right or **banchial chamber** contains ctenidium, rectum and genital duct. The left or **pulmonary chamber** contains the pulmonary sac.

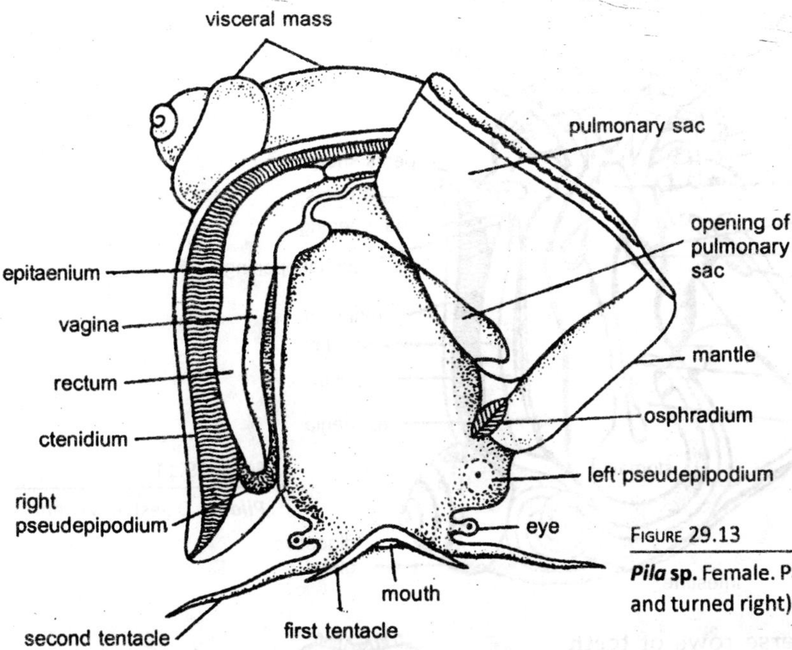

FIGURE 29.13

***Pila* sp.** Female. Pallial complex. (The mantle cut and turned right)

⊠ Body cavity

In the adult, the coelom is greatly reduced and restricted to pericardium (pericardial cavity) and lumen of the kidney (urocoel). The haemocoel acts as channels for haemolymph circulation.

⊠ Digestive System

The digestive system consists of a coiled alimentary canal, a pair of salivary glands, a pair of oesophageal pouches and a bilobed digestive gland (Fig. 29.14).

Alimentary canal

It is divided into three zones—the foregut, midgut and hindgut. The fore and hindgut are of stomodaeal and proctodaeal origin, respectively and lined by ectoderm cells, while the midgut is archenteric and lined by endoderm cells.

Foregut

It includes mouth, buccal mass and oesophagus.

1. **Mouth.** The mouth is a slightly ventral, vertical slit at the tip of the snout. The first pair of tentacles, short in length, are anterolateral projections of the snout and considered as **labial palps**. The mouth leads into a thick-walled buccal cavity.
2. **Bucal mass.** It is a fairly large pyriform muscular structure. The buccal cavity is a spacious chamber, major portion of which is occupied by an **odontophore**. Anteriorly, the buccal mass bears two dorsolaterally placed horny jaws, the anterior cutting edge of which bears two or three teeth. The buccal cavity opens into the oesophagus.

Odontophore. A thick-walled muscular structure. It consists of a radula, radular muscles, a pair of subradular cartilages and a subradular organ with a radular sac.

a. **Radula.** It is a thin, spoon-shaped, **chitinoid ribbon** (Fig. 29.15), the narrow median strip of the broad anterior part is exposed to the floor of the buccal cavity and the narrow posterior portion is lodged in the radular sac.

i. The radula is movably placed on a pair of **radular cartilages** operated by large radular muscles, the two constituting major part of the odontophore.

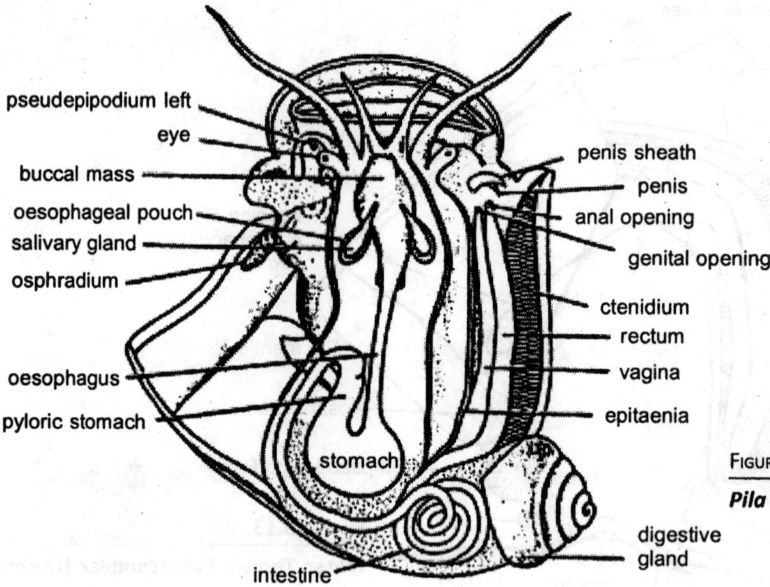

FIGURE 29.14
***Pila* sp.** Digestive system

ii. Numerous transverse rows of **teeth** are present on the radular ribbon. The teeth in a transverse row are one median or **rachidian**, one **lateral** and two **marginals**, the radular formula being 2. 1. 1. 1. 2.

iii. Distally, the radular teeth bear broad cutting edge or **cusp**. A rachidian tooth has a median and two or three lateral cusps on each side.

3. **Oesophagus.** A long median tube, narrows posteriorly, runs medially backwards, turns left and enters the visceral mass to join the stomach.

The salivary glands open into the dorsal aspect of the buccal cavity close to the origin of the oesophagus and the oesophageal pouches (glands) open close to the openings of the salivary glands.

Midgut

The stomach constitutes the midgut.

Stomach. A somewhat oval-shaped, dark red, muscular sac embedded partially in the digestive gland. It has two chambers*. The stomach with the oesophagus and the intestine is a U-shaped body. The base of the U is formed by the stomach and the arms by the oesophagus and the intestine.

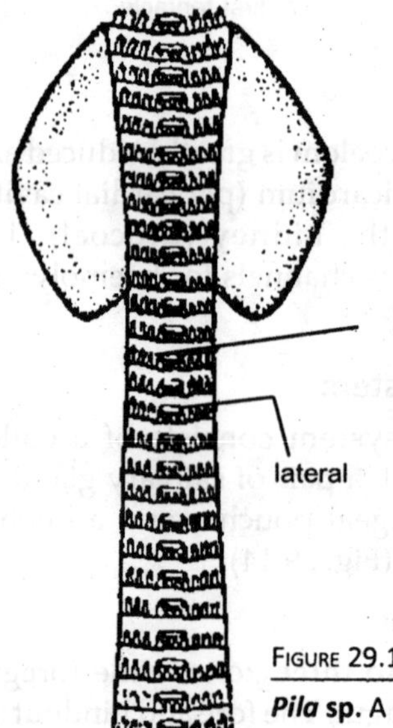

FIGURE 29.15
***Pila* sp.** A radula

Hindgut

It is a long, narrow tube and divisible into intestine and rectum.

Caecum. A small diverticulum at the origin of the intestine, not clearly discernible from outside.

* The division of the stomach into a **cardiac** and a **pyloric chamber** has been rejected by present-day workers.

Intestine. A narrow tube, forms $2\frac{1}{2}$ to 3 coils in the mass of the digestive gland. Emerging from the gland it enters the mantle cavity and joins the rectum.

Rectum. A wide tube, exposed to the mantle cavity and adherent to the inner surface of the mantle on the right side, parallel to the ctenidium. It runs anteriorly to open through the anus in the right side of the mantle cavity near its opening.

⊠ Digestive glands

The salivary glands, oesophageal pouches (glands) and digestive gland* constitute the digestive glands.

1. **Salivary glands.** A pair of whitish glands, one on either side of the oesophagus. The glands are alveolar type and open into the dorsal aspect of the buccal cavity through narrow ducts.
2. **Oesophageal pouches** (glands). A pair of sac-like glands, placed laterally at the origin of the oesophagus and open into the buccal cavity close to the opening of the salivary glands.
3. **Digestive gland.** A large, bilobed, brownish or dull green mass. Roughly, it is a spirally coiled, triangular plate. Two ducts from two lobes of the gland unite to form a common duct before opening in the stomach.

⊠ Feeding and digestion

Pila feeds upon plant matters.

1. The labial palps help in bringing the food close to mouth.
2. Small pieces of plant materials may be cut by the horny jaws and passed to the buccal cavity through the mouth.
3. The major part of the food is procured by rasping with radula. The food is held tightly by the labial palps and the anterior margin of the musculture forming the floor of the buccal mass.
4. The radula is protruded out through the mouth by the action of protractor muscles. The alternate action of upper and lower sets of muscles cause forward and backward, a chain-saw like movement of the radula and the food matter is **rasped** by the cusps of the radular teeth and passed to the oesophagus.
5. The cellulosic cell walls of a considerable portion of the plant food are broken mechanically during rasping and the contents of the cells released, which facilitates digestion.
6. The food gets mixed with saliva and secretion of the oesophageal pouches in the buccal cavity, before it is passed to the oesophagus. In the stomach the food is mixed with the secretion of the digestive gland. Extracellular digestion is partial in the stomach.
7. Small particles of food pass to the digestive gland through the digestive duct and intracellular digestion completes the hydrolysis of food matters in the digestive gland.
8. The undigested part of the food pass out in the form of strings through the anus.

⊠ Respiratory System

Pila are remarkable amphibious inhabi-tants of fresh waters in tropical countries. It is noted for its ability to respire both under water and in air, being provided with a **gill** and a **pulmonary sac**. When under water, Pila maintains the usual respiratory current through the right side of the mantle cavity which contains the gill; at intervals it crawls to the surface or swims up by lashing the snout and tentacles, and forms a funnel leading into the pulmonary sac by rolling up the left nuchal lobe. *Pila globosa* is able to make long excursions on land.

The respiratory organs consist of a gill, a pulmonary sac and the right and left nuchal lobes and afford both terrestrial and aquatic respiration.

* The term **liver**, used by previous workers, is misnomer and has been replaced by digestive gland.

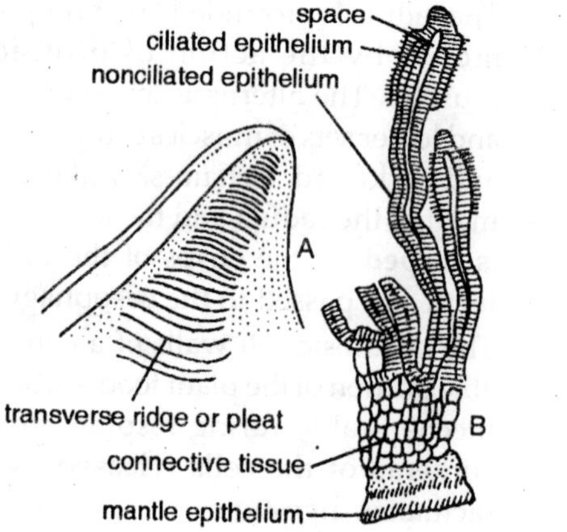

FIGURE 29.16

***Pila* sp.** A. Branchial lamellae. B. Transverse section of two branchial lamellae with mantle

The pallial cavity is incompletely partitioned by a fleshy fold, the **epitaenia** into two chambers, the right one **branchial chamber** containing the gill and the left one **pulmonary chamber** containing the pulmonary sac.

Gill respiration

1. The **gill** or **ctenidium** consists of a series of thin triangular leaflets or **lamellae**. The lamellae are largest in the middle but become smaller towards the extremities (Fig. 29.16) and are lodged in the right portion of the mantle cavity. Each lamella bears a varying number of transverse ridges or pleats on its both outer and posterior surfaces.
2. The lamellae are attached to the mantle along an axis known as **ctenidial axis**. The attachment is effected by the broad bases of the lamellae, the narrow apices hanging freely in the mantle cavity.

⊠ Mechanism of Respiration

Due to extension of the mantle cavity and nuchal lobes water enters the mantle cavity through the left lobe or siphon, which forms a distinct gutter. In its course, water comes in contact with the osphradium. Bathing the gill, the water goes out through the right nuchal lobe or siphon due to the contraction of the mantle cavity. The fill filaments are highly vascular and gaseous exchange takes place there.

Pulmonary respiration

The pulmonary sac hangs down like a sac from the roof of the pulmonary chamber (Fig. 29.17). The sac opens to the exterior at the extreme posterior limit of the mantle cavity.

Mechanism of respiration

When out of water, air enters the pulmo-nary chamber of the mantle cavity through its open-

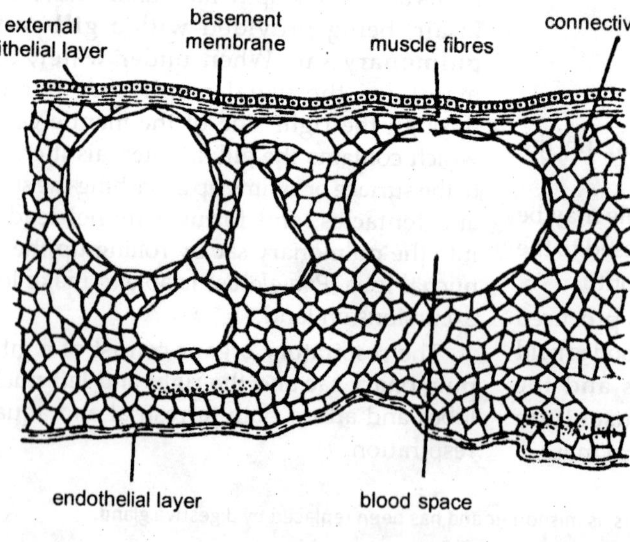

FIGURE 29.17

***Pila* sp.** Outer wall of the pulmonary sac (ventical section)

ing. With the contraction of the muscles in the wall, the cavity of the pulmonary sac enlarges and air enters into it. Gaseous exchange takes place in the highly vascularised wall of the pulmonary sac. Siphon formed by the left nuchal lobe does not play any role in it.

The inspired air is forced out by the compression of the pulmonary sac due to relaxation muscles in its wall.

⊠ Circulatory system

The circularoty system in Pila is open type and consists of heart, pericardium, arteries, sinuses and veins (Fig. 29.18). Phagocytic amoebocytes are present in the haemolymph. The haemolymph contains dissolved haemocyanin, the respiratory pigment, a compound of copper and protein.

Pericardium

1. A thin-walled, ovoid sac lying dorsally on the left side of the body whorl, behind the mantle cavity and extending anteriorly up to the stomach and digestive gland.
2. The pericardium encloses the heart, main aortic arches and aortic ampulla.

Heart

The heart has a single auricle and a ventricle (Fig. 29.18).

1. **Auricle.** A thin-walled, highly contractile, triangular sac. It receives:
   a. An **efferent ctenidial vein** from ctenidium.
   b. An **efferent renal vein** at its apex, from the posterior chamber of the renal organ.
   c. A **pulmonary vein** at a slightly lower level at the anterior end, from the pulmonary sac.*
2. The auricle communicates with the ventricle by an **auriculoventricular opening**, guarded by two semilunar valves, opening towards the ventricle.
3. **Ventricle.** A thick, spongy, ovoid chamber. It sends an **aortic trunk** from its lower end, which divides immediately into two branches—cephalic and visceral aorta. The opening between the aortic trunk and ventricle is guarded by two semilunar valves, which prevent backflow of haemolymph to the ventricle.

Arteries

a. **Cephalic aorta.** It bears a bulbous outgrowth, the **aortic ampulla**, which controls haemolymph pressure. The aorta proceeds anteriorly and sends branches from both the outer and inner sides.

Branches from the **outer side**

i. Cutaneous artery to skin.
ii. Oesophageal artery to oesophagus.
iii. Pallial artery to organs of the left half of the mantle cavity and osphradium.

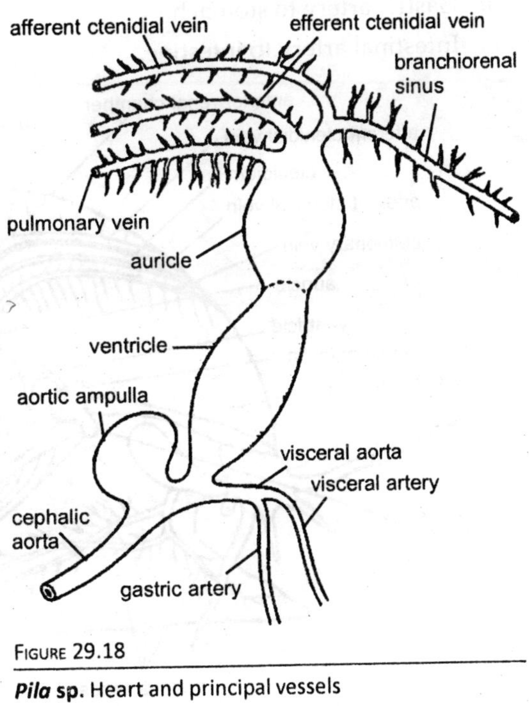

FIGURE 29.18

Pila sp. Heart and principal vessels

* The pulmonary sac has no separate origin. It is formed simply by vascularisation of a part of the mantle.

Branches from the **inner side**

iv. Pericardial artery to pericardium.

v. Renal artery to posterior renal chamber.

The cephalic aorta proceeds anteriorly along the left side of the oesophagus for some distance, then crosses to the right and sends many branches, the important ones are:

vi. Right pallial artery to right side of the mantle.

vii. Right siphonal artery to right siphon.

viii. Penial artery to penis.

ix. Optic arteries to eyes.

x. Tentacular arteries to tentacles.

xi. Radular artery to radular sac.

xii. Pedal artery to foot.

b. **Visceral aorta.** It runs backwards into the visceral mass and supplies haemolymph to visceral organs. The important branches are:

i. Visceral artery to digestive gland.

ii. Gastric artery to stomach.

iii. Intestinal artery to intestine.

iv. Hepatic artery to digestive gland and gonad.

The aorta terminates in the rectum.

Sinuses

The haemolymph from the finer branches of the arteries passes into small **lacunae**. Many small lacunae unite to form a large sinus. There are four large sinuses.

a. **Anterior perivisceral sinus.** It lies above the foot and beneath the floor of the pallial cavity.

b. **Anterior periintestinal sinus.** Located in the columellar axis, runs along the coil of the intestine up to the junction of the anterior and posterior renal chambers.

c. **Branchiorenal sinus.** It lies along the right side of the anterior renal chamber.

d. **Pulmonary sinus.** It is adherent to the walls of the pulmonary sac.

The haemolymph finally passes to the perivisceral and periintestinal sinuses and carried to auricle by veins directly or through gills, mantle and kidney.

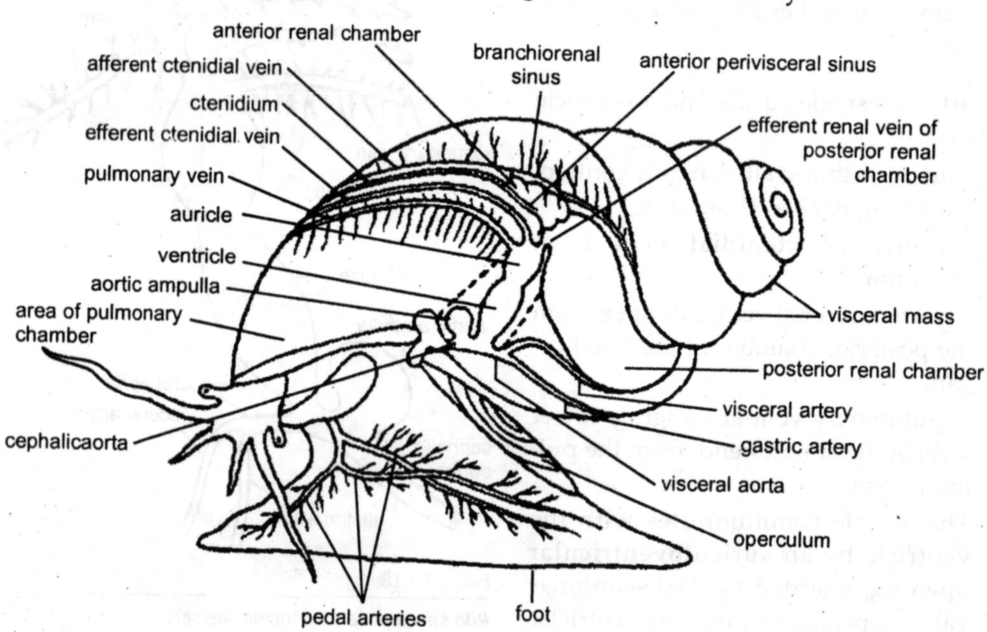

FIGURE 29.19

***Pila* sp.** Heart and principal arteries (diagrammatic). Seen from left side

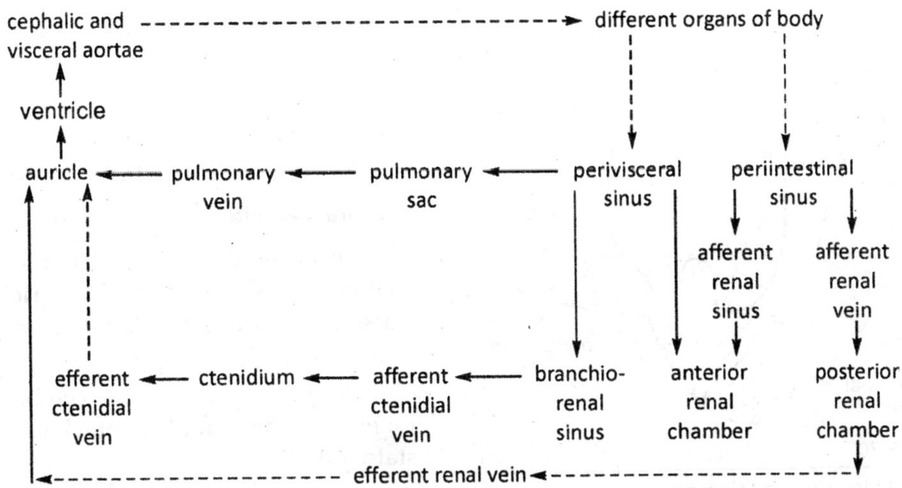

FIGURE 29.19A

***Pila* sp.** Haemolymph circulation. (schematic)

Veins

a. **Afferent ctenidial vein.** It lies above the rectum and receives branches from rectum, terminal part of the genital duct, perivisceral sinus and branchiorenal sinus and sends haemolymph to the gill lamellae.
b. **Efferent ctenidial vein.** Runs along the roof of the anterior renal chamber and drains haemolymph from gill lamellae, mantle and sends it to auricle.
c. **Afferent renal vein.** Located on the roof of the posterior renal chamber. It originates from the periintestinal sinus and sends haemolymph to the posterior renal chamber.
d. **Efferent renal vein.** Carries haemolymph from posterior renal chamber to the auricle.
e. **Pulmonary vein.** Collects haemolymph from the pulmonary sac and open into the auricle.

The efferent ctenidial vein and pulmonary vein carry oxygenated haemolymph, the afferent renal vein carries deoxygenated haemolymph to the auricle.

☒ Excretory System

The kidney is the chief excretory or renal organ.

Kidney

The kidney is a large, thin-walled sac lying behind the pericardium. It consists of two distinct renal chambers, a right anterior and a left posterior (Figs. 29.20, 29.21).

1. **Anterior chamber**

a. It is an ovoid, reddish organ situated in front of the pericardium and posterior renal chamber.
b. The anterior renal chamber projects into the mantle cavity and opens into it in a deep crypt through a slit-like opening, the **renal pore**, lying to the right of the epitaenium.
c. The lamellae on the roof of the chamber are arranged on either side of a median longitudinal axis, the efferent renal sinus.
d. The lamellae on the floor are similarly arranged on either side of a median axis, the afferent renal sinus.

2. **Posterior chamber**

a. It is a broad, brownish, hook-shaped chamber situated behind the anterior renal chamber in between rectum on the left and pericardium and digestive gland on the right. It possesses a large cavity.

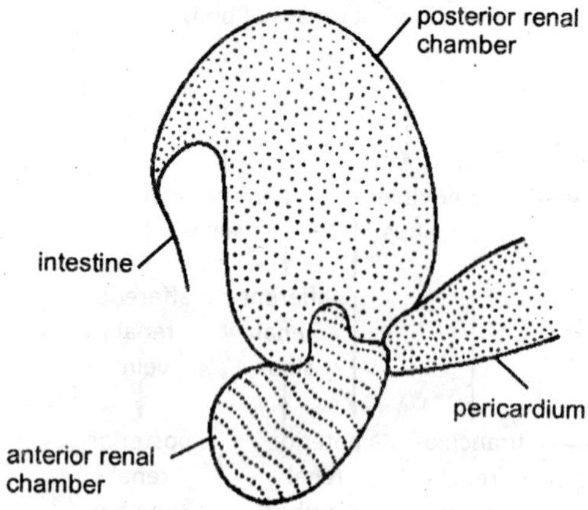

FIGURE 29.20

***Pila* sp.** Kidney (viewed from above)

b. The roof of the chamber is richly supplied with haemolymph.
c. The posterior chamber communicates with the pericardium by a **renopericardial aperture** at one end and with the anterior renal chamber at the other end.

Digestive gland also palys a role in excretion. The lobules contain excretory cells which eliminate waste materials.

The nature of nitrogenous wastes in Pila varies with season and habitat. **Ammonia, urea** and **uric acid** are major constituents of nitrogenous wastes. Pila living in water during aquatic phase produces ammonia, while living out of water during terrestrial phase it excretes urea and uric acid, depending on the period of land life.

⊠ Nervous System

A central nervous system, a peripheral nervous system and a visceral nervous system constitute the nervous system (Fig. 29.22).

The central nervous system comprises paired cerebral, buccal, pleural, pedal, visceral ganglia and an unpaired supraintestinal ganglion; cerebral, buccal, pedal commissures; paired cerebrobuccal, cerebropleural, cerebropedal connectives; supra and infraintestinal visceral connectives and supra and infraintestinal nerves.

Nerves arising from the ganglia and connectives constitute the peripheral nervous system.

Nerves innervating some of the visceral organs form visceral nervous system.

Cerebral ganglia

The cerebral ganglia are paired roughly triangular in shape, situated anteriorly on the dorsolateral sides of the buccal mass at its anterior end.

Nerves arising from the cerebral ganglia innervate the snout, tentacles, eyes and statocysts.

Cerebral commissure

The two cerebral ganglia are connected by a thick, band-like commissure medially, located dorsal to the buccal mass.

Labial commissure

A fine commissure, connects the two cerebral ganglia ventrally.

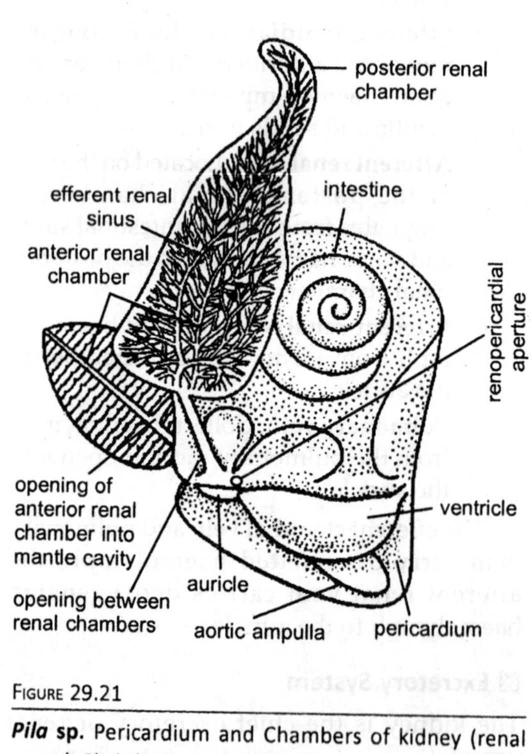

FIGURE 29.21

***Pila* sp.** Pericardium and Chambers of kidney (renal organ). Slightly diagrammatic

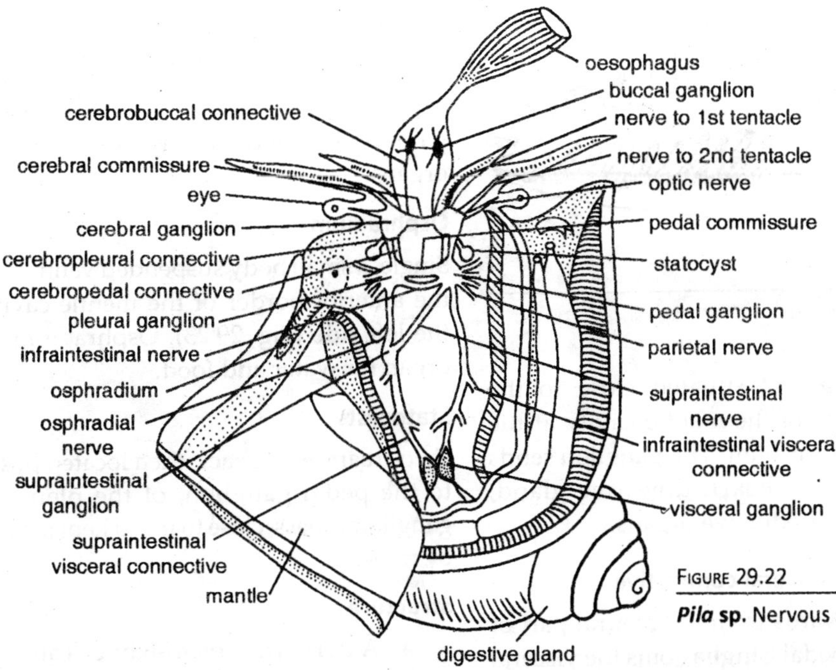

FIGURE 29.22

***Pila* sp.** Nervous system

Buccal ganglia

Two small, oval, roughly triangular, ganglia embedded in the dorsolateral sides of the buccal mass at its junction with the oesophagus, near the radular sac.

Nerves arising from the buccal ganglia innervate the buccal mass, salivary glands and oesophagus.

Buccal commissure

A slender nerve connecting the two buccal ganglia, located ventral to the oesophagus.

Cerebrobuccal connectives

Each buccal ganglion is connected with the cerebral ganglion of the side by a slender cerebrobuccal connective.

Pleuropedal ganglionic mass

A pair of ganglionic masses, one on either lateral side, posteroventral to the buccal mass. The pleural and pedal ganglia of each side are more or less fused, only a slight notch separating the two. The subintestinal ganglion fuses with the right pleurepedal ganglionic mass.

Pleural ganglia send nerves to the mantle complex and nuchal lobes and adjacent organs and the pedals to the foot.

Cerebropleural connectives

Two. Each runs posteroventrally from the cerebral ganglion, around the oesophagus and joins the pleural region of the pleuropedal ganglion of the side. Its origin is medial to that of the cerebropedal connective.

Cerebropedal connectives

A pair. Each runs posteroventrally from the cerebral ganglion, around the oesophagus and joins the pedal region of the pleuropedal ganglion of the side.

⊠ Pedal Commissure

A thick nerve connecting the pedal regions of the two pleuropedal ganglia.

Infraintestinal nerve

A slender nerve, close to the pedal commissure, joining the pleural regions of the two pleuropedal ganglia.

Visceral ganglion

It is formed by the fusion of two spindle-

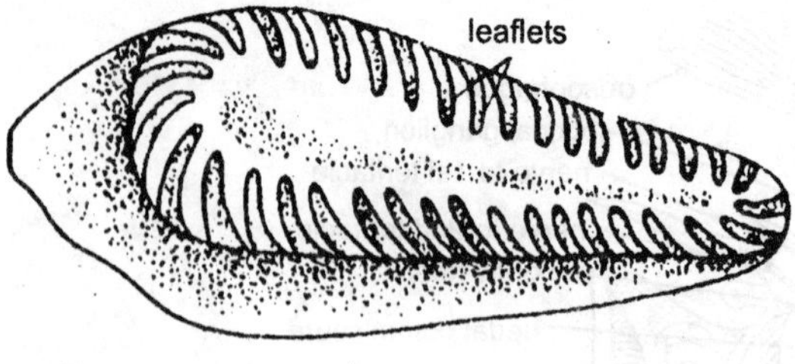

FIGURE 29.23

Pila sp. An osphradium

shaped ganglia and situated close to the posterior border of the mantle cavity at the right of the pericardium. The ganglion sends nerves to kideny, heart, digestive gland, intestine and reproductive organs.

Supraintestinal visceral connective

Running backwards from the pleural part of the left pleuropedal ganglia, joins the visceral ganglion. It bears a **supraintestinal ganglion** at about its middle.

Infraintestinal visceral connective

Runs backward from the pleural part of the right pleuropedal ganglia and joins the visceral ganglion.

The supra and infraintestinal visceral connectives are also designated as left and right pleurovisceral connective, respectively. Both of them send nerves to adjacent organs.

Supraintestinal nerve

A slender nerve runs obliquely backwards and connects the right pleuropedal ganglia with the supraintestinal ganglion.

Supraintestinal ganglion

It is almost fusiform and situated slightly posterior to the pleuropedal mass at some distance from it on the supraintestinal visceral or left pleurovisceral connective. The ganglion sends nerves to gill, pulmonary sac, etc.

Receptors and sense organs

The sense organs consist of tactile organs, osphradium, statocysts and eyes.

Tactile organs

Tactile sensory cells are scattered all over the exposed body surface. In addition, the labial palps and second pair of tentacles are tactile organs.

Osphradium

Single, a small body suspended ventrally from the anterior border of the mentle cavity, on the left side (Fig. 29.23). Osphradium is test organ for water and food.

Statocysts

Two cream-white sacs, each located posterior to the pedal ganglion, of the pleuropedal ganglionic mass, close to it and kept in position by muscles.

Structure

a. A pyriform, cream-white, leathery sac.
b. The internal lining epithelium contains sense cells (Fig. 15.3).
c. A statolith is present in the cavity.

The statocysts are balance organs and innervated by cerebral ganglia.

Eyes

Two, borne on short stalks at the base of the second pair of tentacles.

a. The outermost layer of an eye is an **optic vesicle**, formed by a thick layer of fibrous connective tissue (Fig. 29.24).
b. The optic vesicle is open at the tip and covered by transparent outer and inner cornea.
   i. The **outer cornea** is a continuation of the epithelium of the eye stalk, the cells are small, squarish and trans-parent.
   ii. The **inner cornea** is also transparent and continuation of the connective tissue around the optic vesicle.
c. The inner lining of the optic vesicle is **retina**, made of two types of cells.
   i. **Visual cells.** Large, broad and with brush border.
   ii. **Packing cells.** Small, wedged between visual cells.

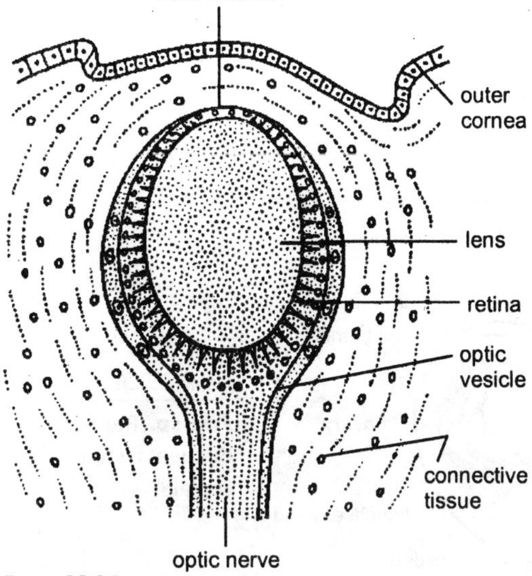

FIGURE 29.24

***Pila* sp.** Eye. Vertical section

d. The space enclosed by the retina is occupied by an ovoid lens.
e. The optic nerve enters the optic vesicle at the posterior end.

The eyes of Pila are poor photoreceptors.

⊠ Reproductive System

The sexes are separate, sexual dimorphism is not pronounced.

The gonad, testis or ovary is single and occupies the same position in male and female. It is restricted to about half the height of the inner concave surface of the first two to three whorls of the digestive gland. The gonad is covered by the mantle, which is very thin and transparent.

Male reproductive system

It consists of a testis, a vas deferens and a penis.

Testis

Much branched, cream coloured and clearly visible through the transparent mantle. It enlarges posteriorly. The lobes of the testis are exposed after the removal of the mantle (29.25).

Vas deferens

Many fine ducts, the **vasa efferentia**, emerging from the lobes of the testis join to form the vas deferens, which leaves the testis at the posterior end and runs along the inner border of the posterior renal chamber. The vas deferens has three distinct zones.

a. **Tubular part.** The slender first part runs along the inner edge of the posteior renal chamber.
b. **Glandular part.** Terminal part of the vas deferens. It is fairly large, swollen

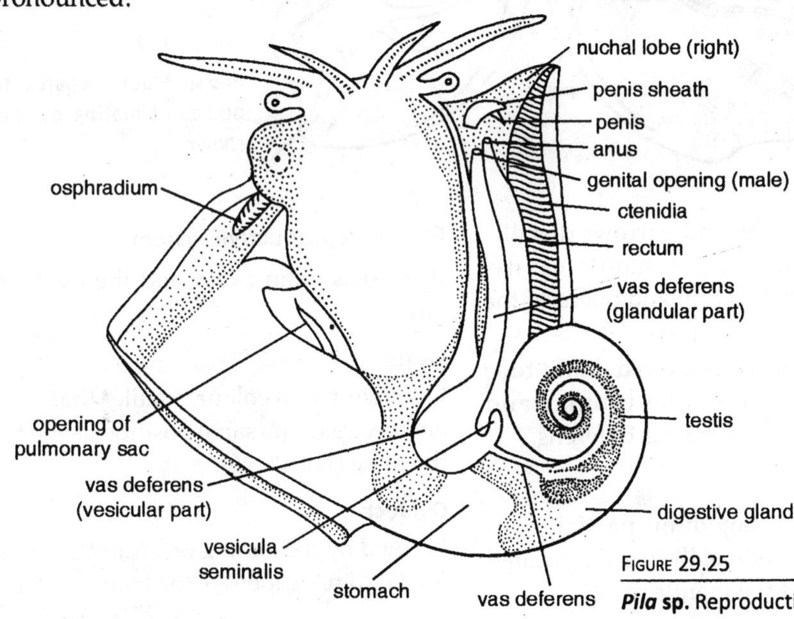

FIGURE 29.25

***Pila* sp.** Reproductive system. Male

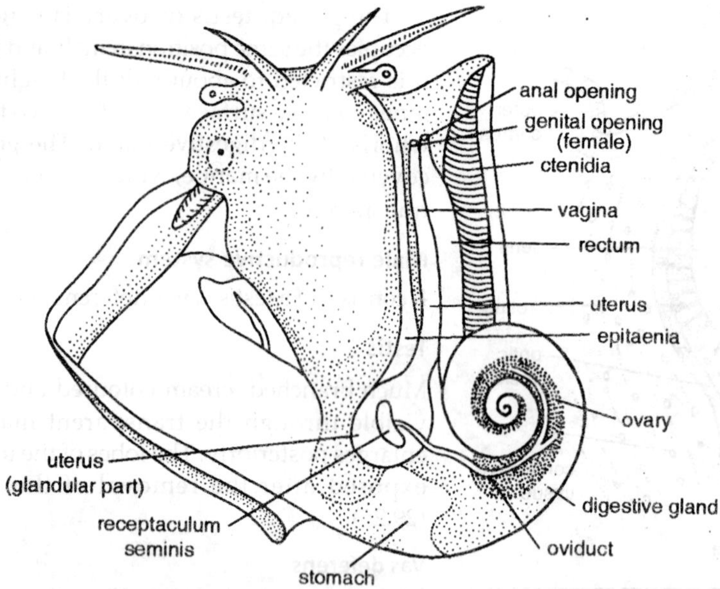

FIGURE 29.26
Pila sp. Reproductive system. Female

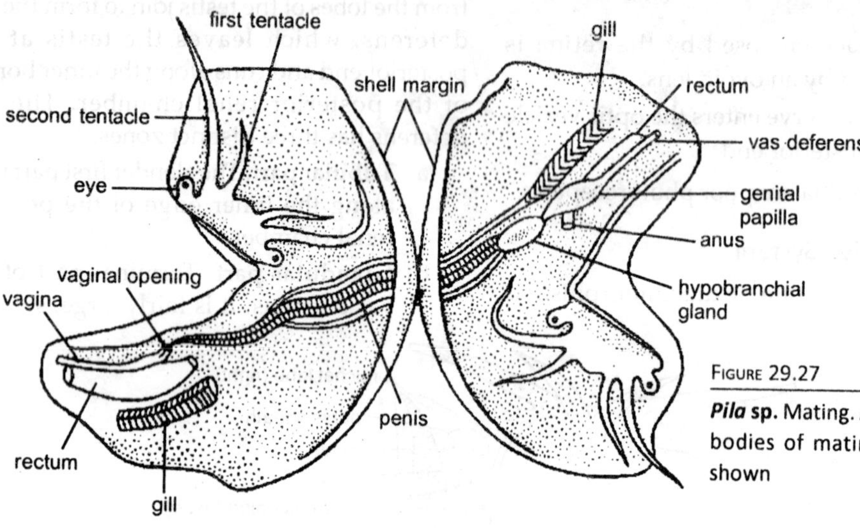

FIGURE 29.27
Pila sp. Mating. A part of the bodies of mating partners shown

at the beginning and narrows distally, being lodged in the mantle cavity, parallel to the rectum and opens at the tip of the genital papilla.

c. **Seminal vesicle.** A curved, blind tube at the junction of the tubular and glandular part of the vas deferens.

Penis

A long, whip-like flagellum, partially ensheathed; an outgrowth of the mantle, located on the right side of the mantle cavity.

Female reproductive system

It consists of an ovary and the oviduct with vagina.

Ovary

Organge-red in colour, smaller than the testis and occupies the same position as the testis in the male (Fig. 29.26).

Oviduct

Formed by the joining of a number of narrow ducts from the lobes of the ovary and leaves it at the posterior end. The oviduct is divisible into two distinct zones.

a. **Tubular part.** The slender first part running along the border of the digestive gland, turns downwards above the level of the kidney and joins the bean-shaped **seminal receptacle** at the junction of the tubular and glandular part. The seminal receptacle receives sperms from the male during mating.
b. **Glandular part.** it lies in the mantle cavity, parallel to the rectum and divisible into two zones.

i. **Uterus.** A bluntly rounded yellowish sac, lying at the junction of the two renal chambers and narrows down to form the vagina.
ii. **Vagina.** Tubular, light cream coloured, narrows down anteriorly and opens through a slit-like aperture on the genital papilla.

⊠ Fertilization and egg laying

Pila breeds in monsoon.

1. Copulation lasts for 2-3 hours. The tip of the penis is inserted into the vagina (Fig. 29.27) and the spermatic fluid is transferred to seminal receptacle, where fertilization occurs.
2. The sperms are of two types (Fig. 29.28 A-B), **Eupyrene** and **oligopyrene**. The eupyrene type fertilize the ova.
3. The **zygote** receives albumen and becomes enclosed in a **double shell**

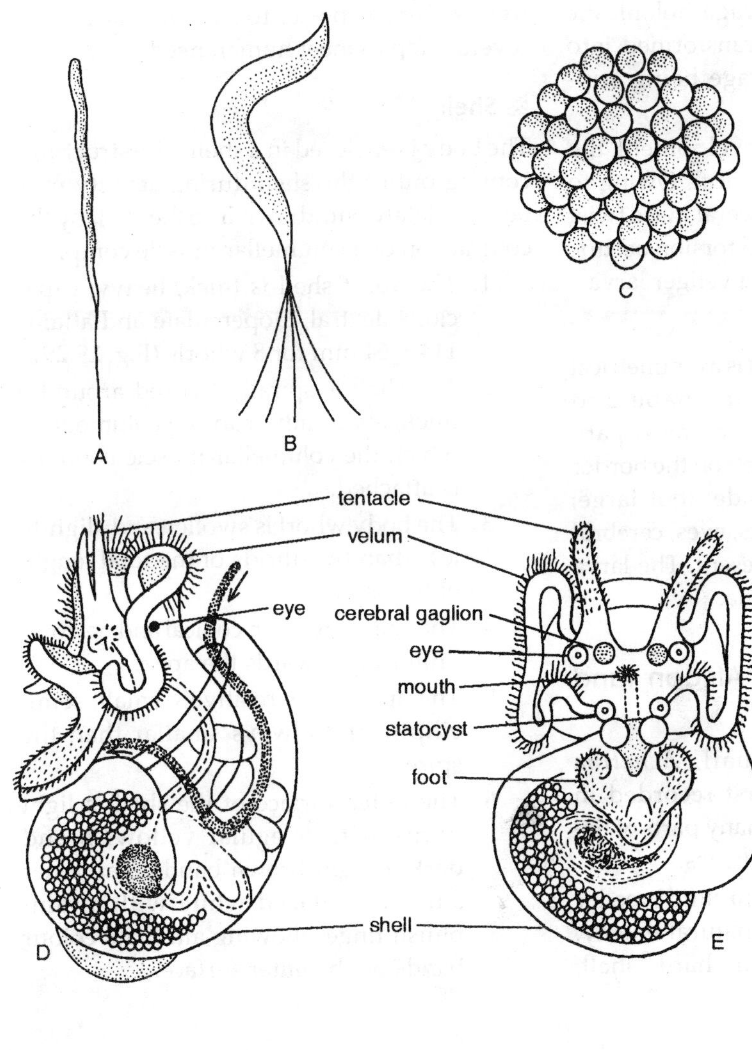

FIGURE 29.28

***Pila* sp.** Gametes and development. A. Eupyrene sperm. B. Oligopyrene sperm. C.-A. cluster of eggs (laid in a hole). D.-E. Veliger larva

membrane, surrounded externally by a white, calcareous (Fig. 29.28C). **egg shell**.

4. The embryo remains suspended in the core of fluid albumen in the egg.
5. 200 to 800 eggs, the number being extremely variable in a brood, are laid in sheltered crevices or holes in the soft soil, made by the snout of the mother, above water level. The eggs are held together by a transparent cementing material.

☒ Development

Development starts immediately after fertilization. Development indirect and passes through a veliger stage, which is considered as an advanced larva. Cleavage holoblastic and spiral. The gastrula is transformed into veliger, the trochophore stage being suppressed.

In the early stages of development the embryo is bilaterally symmetrical. Owing to asymmetrical growth, the visceral mass twist 180° on the head-foot axis and torsion occurs, leading to the appearance of a veliger larva.

Veliger larva

The veliger (Fig. 29.28D.-E) is asymmetrical transparent elongateoval in shape and enclosed in a coiled shell for the most part; mantle cavity rudiment is a cleft on the border of the mantle in the right side; foot large; paired tentacles, coelomic sacs, eyes, cerebral ganglia and statocysts are present. The larva metamorphoses to a young pila.

ACHATINA FULICA (Giant African land snail)

The giant African land snail, *Achatina* (*Lissachatina*) *fulica fulica*, first recorded in Madagascar, have spread to many parts of the globe in the last couple of centuries.

They are voracious eater, prolific breeder and practically without any natural enemy. These, in addition to their hard shell, capability to withstand long dry period and ability to secrete copious thick mucus, making them almost immune to chemicals like pesticides, have led to their population explosion in the countries they have invaded.

Today, their population is quite high in tropical and semitropical countries with high rainfall and humidity. In major part of Indian subcontinent, the whole of Indopacific region and adjacent areas they are serious threat to agrihorticulature. All attempts to manage them have failed.

Snail meat is a delicacy in Continent and U.S.A. The shortage of this delicacy due to depletion of *Helix* population, is being made up with *Achatina* meat, though it is a poor substitute. Of late attention has been shifted from *Achatina* control to *Achatina* farming, a reverse step to meet human need.

☒ Shell

The body is enclosed in a shell. The structures coming out of the shell during activity may be completely withdrawn into the shell by the contraction of a columellar muscle complex.

1. The adult shell is thick, heavy, capacious dextral, inoperculate and attains 114 × 64 mm for 8 whorls (Fig. 29.29).
2. The shell is spirally twisted around a thick, abapically truncate columella to which the columellar muscle complex is attached.
3. The body whorl is swollen and slightly less than two-thrids of the total length of the shell.
4. The spire is regular, conical and narrows uniformly towards the apex.
5. The aperture is relatively small, semi-elliptic and always smallar than the spire.
6. The outer surface of the shell is light horny with irregular yellowish and dark to light brown bands. The inner surface is glazed white with a faint bluish tinge, showing alternate colour bands on the outer surface.

External features

The body is divisible into a head, a foot, a visceral stalk, a neck and a visceral mass. The visceral mass ramains within the shell, the head, visceral stalk, foot and neck and protruded out of the shell during activity.

Head

1. The **head** is small (Fig. 29.29), light brown in colour, placed at the anterior end of the visceral stalk and bears a ventral **mouth** anteriorly and a pair of labial palps, dorsolateral to the mouth.
   a. The **labial palps** are two nearly semicircular in shape, and arise from very near the bases of the ventral tentacles and overhang the mouth.
   b. During feeding the palps act as upper lips and help in bringing food nearer the mouth.
2. The soft, dirty-cream ventral surface of the head extends beyond the anterior end of the foot and is separated from it by a narrow groove.
3. A pair of highly retractile **ventral tentacles**—about 12 mm long at full extension—arise just posterior to the labial palps. They are highly sensitive and are withdrawn completely and quickly by introversion on slight touch.
4. A pair of **ocular tentacles**, about 25 mm long, dorsolateral in position and strictly forming the most anterodorsal structures of the head arise posterior to the ventral tentacles and each bears a small, round, black eye, placed slightly dorsally at its tip.

Visceral Stalk

5. The **visceral stalk** is semicircular in cross section broad at the base and imperceptibly tapers forward.
6. A small, whitish, round **genital area** with the genital aperture at the centre lies at a distance of about 25 mm from the anterior end of the snout and posteroventral to the right ocular tantacle.

Foot

7. The **foot** is elongateoval, about 120 × 50 mm at full extension in an adult snail. It is very thin in the central portion but thick and muscular along the margins.
   a. It narrows anteriorly and is about 12 mm wide at the tip, on which rests the posterior part of the head, but the two are separated by a narrow groove.
   b. Posteriorly, the foot tapers suddenly and is roughly triangular with a blunt and round apex.
   c. The foot is soft and spongy at the anterior end but fairly compact at the posterior end and lateral margins.
   d. During locomotion, narrow transverse waves, limited to the median longitudinal zone, appear at the posterior end

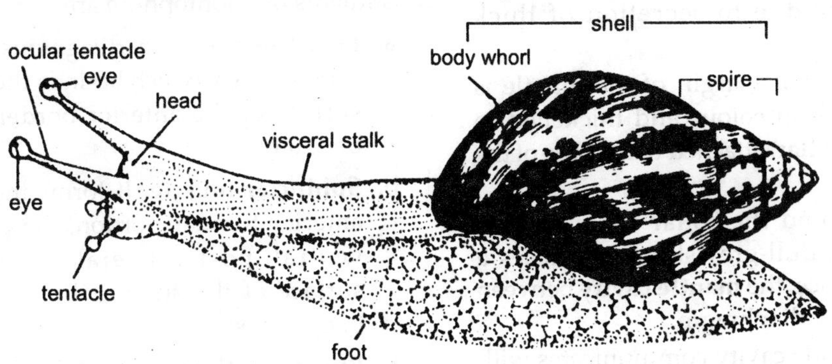

FIGURE 29.29

Achatina fulica. (Giant African land snail)

of the sole and gradually progress forward to disappear at the anterior end. The number of waves at any time is twelve.

Neck

The neck is short and narrow and joins the visceral mass with the foot.

Visceral mass

8. The **visceral mass** is spirally coiled and the number of spirals usually corresponds to the number of whorls of the shell.
9. In the adult, the visceral mass does not extend up to the first whorl of the shell and one or two apical whorls are plug up with calcareous matter. The visceral mass is divisible into two regions, the pleuropedal and apical.
10. The **pleuropedal visecral mass** occupies the body whorl. The heart-kidney complex, major portion of columellar muscles, genital ducts and a part of crop constitute it.
11. The **apical visceral complex** accommodates a large, spirally twisted, dull-blackish-brown digestive gland. The stomach and intestine are embedded in the digestive gland; the albumen gland, ovotestis, ovotestis duct in its spirals, being covered by the mantle.
12. The pleuropedal visceral mass is very soft, enclosed in a **mantle** and extremely slippery due to secretion of thick mucus.
13. The anterior margin of the mantle is brownish in colour and thickened to form a **collar**, attached to the base of the neck.
14. Just behind the collar the mantle is thinner, dull-greyish and encloses a large space, the **mantle cavity** between it and the diaphragm.
15. The mantle cavity communicates with the exterior by a fairly large elongate-oval **pneumostome** at the right side, close to the anterior margin of the mantle.

⊠ Digestive System

The digestive system consists of an alimentary canal, salivary and digestive glands (Fig. 29.30).

⊠ Alimentary Canal

The zones of the alimentary canal (Fig. 29.30A) are buccal mass, oesophagus, crop, stomach, intestine and rectum.

Labial palps. Two, slightly dorsal to the corners of the mouth.

Mouth. A semicircular opening with a plicate margin enclosing a small vestibule posteriorly. It is placed ventrally at the anterior end of the snout and leads to the buccal cavity.

Buccal mass

It is a fairly large, pyriform muscular structure. The buccal cavity is a spacious chamber, major portion of which is occupied by an odontophore. The buccal cavity is lined with a thin layer of cuticle and is divisible into an anterior tubular, and a posterior dorsally compressed arched chamber due to the presence of odontophore.

Odontophore. A thick, highly muscular body, narrower anteriorly and broader posteriorly. It is protracted and rectracted by sets of **protractor** and **retractor muscles**. The constituents of odontophore are:

a. **Jaw.** A semilunar, cartilaginous cuticle-capped body which bounds the dorsal surface of the anterior border of the buccal mass.
b. **Sphincter muscle.** It forms the entire anterior tubular portion, the postero-ventral, posterolateral and ventral surface of the anterior border of the buccal mass.
c. **Buccal cartilage.** Horseshoe-shaped, supports the radula and forms the main mass of the **odontophore**.

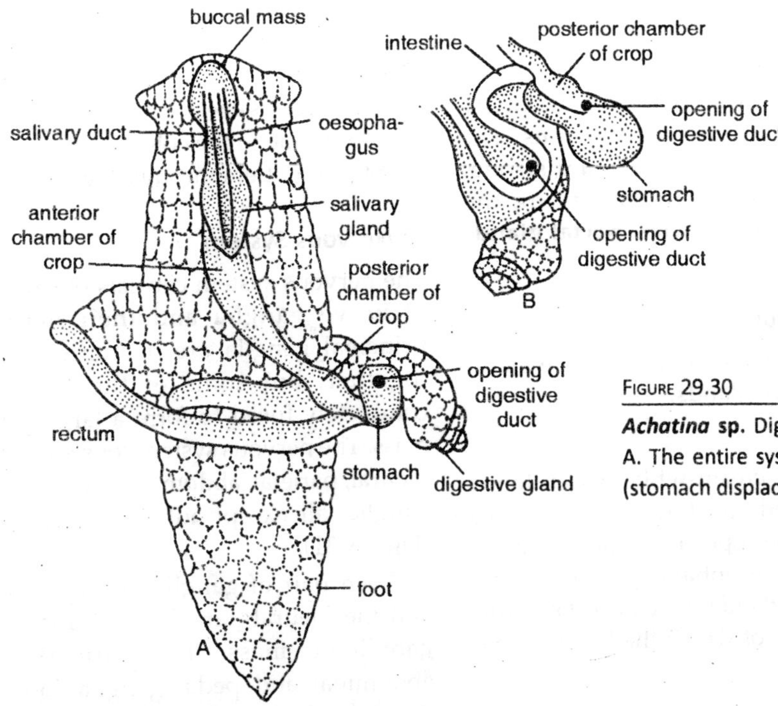

FIGURE 29.30

***Achatina* sp.** Digestive system. A. The entire system, B. A part (stomach displaced)

d. **Radula.** A ribbon-like cartilaginous strip with a narrow anterior and a blunt posterior end. It is posteroventral in position in the buccal mass.

Teeth. Arranged transversely on the radular ribbon, the formula being 64 : 1 : 64 and 140 such rows are usually present.

Oesophagus

A narrow, thick-walled tube bearing longitudinal folds internally. It runs backwards and downwards from the buccal mass and ends posteriorly in the crop. Externally it is covered with a flattened epithelium while internally it is lined with a ciliated epithelium. Oesophageal glands are lodged in the wall.

Crop

A large, thin-walled, elongated sac, divided into an anterior and a posterior chamber by a constriction. The anterior chamber is broadest at its origin and posteriorly it narrows, becomes thick-walled and the constriction acts as a sphincter to inhibit free flow of substances between the two chambers. The posterior chamber narrows to open in the right side of the stomach by a round aperture. A **digestive duct** opens on the dorsal surface of the posterior chamber of the crop near its opening into the stomach. Glands are present in the wall of the crop.

Stomach

A thick-walled, heart-shaped sac. The openings of the crop and intestine are placed side by side, and the crop and the intestine form the limbs of a 'U', while the stomach itself forms the base. A large **digestive duct** opens on the left side of the midventral surface, towards the posterior end of the stomach. The wall is muscular. The inner epithelium is ciliated and bears numerous **gastric glands**.

Intestine

A thin-walled, fairly large, round tube, forms loop, remains fully or almost fully embedded in the mass of the digestive gland and ends in the rectum near the arch of the adrectal ureter. The wall is thin but strengthened with patches of longitudinal fibres. **Intestinal glands** are present.

Rectum

A more or less straight tube, adherent to the thickened posterior wall of the mantle cavity and opens to the exterior by the **anus**. The posterior end of the rectum bears strong, longitudinal muscular folds serving the purpose of sphincter muscles. **Rectal glands** are present.

☒ Digestive Glands

Paired salivary glands and a digestive gland constitute digestive glands.

Salivary glands

Paired, elongated, cream-white, much-lobed structures, united posteriorly but free anteriorly. The ducts open in the buccal cavity by the sides of the oesophagus near its origin. Each lobule of the gland is made up of **mucous** and **serous alveoli** of which the latter type is more numerous.

Digestive gland

A large, spirally twisted, blackish-brown gland, divisible into an anterior and a posterior lobe. The duct from the anterior lobe opens into the posterior chamber of the crop, and that from the posterior lobe opens into the stomach (Fig. 29.30A.-B). The digestive gland consists of a large number of lobules made up of **storage, vesicular, calcic** and **pseudopodial cells**. The lumina of several lobules join to form a narrow interlobular duct, which receives other ducts. The digestive gland acts as a store-house of food and reservoir of calcium.

Feeding and digestion

Feeding in *Achatina* is similar to that in Pila, but the rate of food consmption is much fast. The young snails prefer soft plant matters while the older ones like slightly coarser materials. They are often found to feed upon dead insects, molluscs and other small animals.

Digestion is both extracellular and intracellular, the former playing the major role. All the hydrolysing enzymes, viz., carbohydrases, proteases, lipases and cellulase capable to digest 1-4β glucosides have been recorded from the gut.

The faeces is egested as faceal pellets, green in colour, due to the colour of the undigested vegetable matters through the anus.

☒ Nervous System

The nervous system consists of several pairs of nerve ganglia and their nerves. The principal ganglia are not only aggregated together, but some of them are almost completely fused to form a large ganglionic mass. The nervous system consists of cerebral, pleural, parietal, abdominal, pedal and buccal ganglia, all except adbominal being paried (Fig. 29.31).

Each pair of ganglia—except the buccal and the pleural—are fused to form a single ganglionic mass. The parietal, pleural, abdominal and pedal ganglia form a ring through which the cephalic aorta runs.

Connectives extend between the cerebral and buccal, cerebral and pleural and cerebral and pedal. A single buccal commissure is present.

Cerebral ganglia

The cerebral ganglia are two, almost round, fused, and the mass lies dorsal to the oesophagus. It is surrounded by a thin membranous tissue. From each ganglion arise several nerves which may be arranged in two groups according to their origin, viz., dorsal anteromedian and ventral anterolateral.

The **dorsal anteromedian nerves** include (1) **ocular nerve** to ocular retractor muscle, (2) **ommatophoral nerve** to ocular retractor muscle, eye and the tip of the ocular tentacle, (3) **superior frontal nerve** to the dorsal wall of the snout and (4) **inferior frontal nerve** to the dorsal and dorsolateral wall of the snout.

The **ventral anterolateral nerves** include (1) **posterior oesophageal nerve** to the oesophagus, (2) **cerebrobuccal connective** to buccal ganglion, (3) **nuchal nerve** to antero-ventral wall of the snout and buccal mass, (4) **nuchal nerve proper** to the wall of the

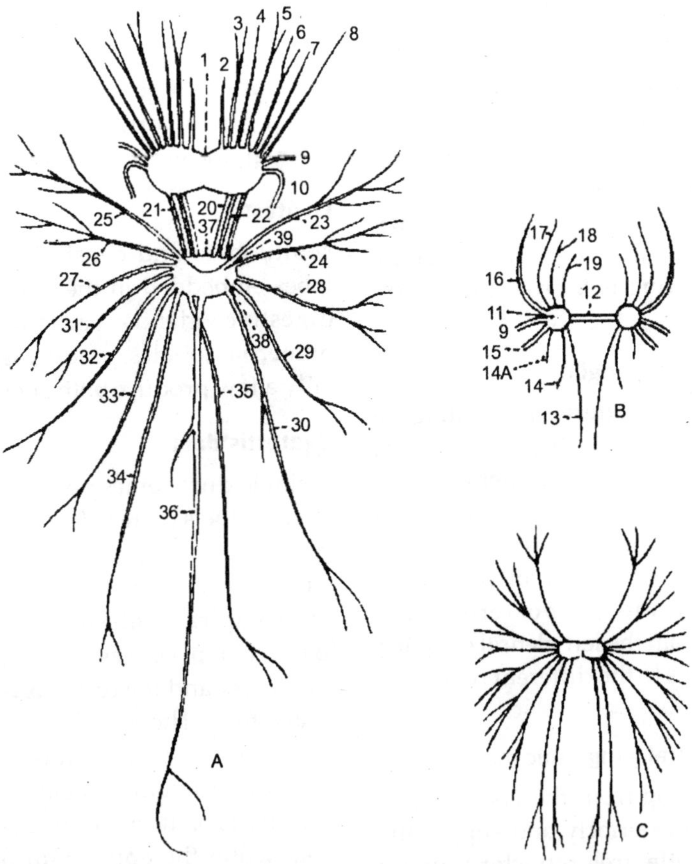

FIGURE 29.31

***Achatina* sp.** Nervous system. A. Nervous system (buccal ganglia not shown), B. Buccal ganglia with commissure and nerves, C. Pedal ganglia with nerves. 1. cerebral gn. 2. ocular n. 3. ommatophoral n. 4. superior frontal n. 5. inferior frontal n. 6. labiotentacular n. 7. nuchal proper n. 8. nuchal n. 9. cerebrobuccal con. 10. posterior oesophageal n. 11. buccal gn. 12. buccal com. 13. oesophageal-salivary n. 14. oesophageal-salivary duct n. 14A. odontophoral n. 15. posterior buccal n. 16. anterior oesophageal n. 17. anterior buccal n. 18. dorsolateral buccal n. 19. lateral buccal n. 20. salivary gland n. 21. cerebropedal con. 22. cerebropleural con. 23. anterogenitonuchal n. 24. posterogenitonuchal n. 25. anteronuchal n. 26. left medionuchal n. 27. posteronuchal n. 28. right retractor n. 29. right lateroparietal n. 30. right medioparietal n. 31. left retractor n. 32. left lateroparietal n. 33. left palliopariet al n. 34. left medioparietal n. 35. right palliopariet al n. 36. visceral n. 37. pedal gn. 38. parietal gn. 39. pleural gn.

snout and (5) **labiotentacular nerve** to the wall of the snout around the base of the ventral tentacle sending branches to its tip.

Buccal ganglia

The buccal ganglia are paired, small, round and placed on the posterior face of the buccal mass between the base of the oesophagus and the radular sac. Several nerves arise from each buccal ganglion both anteriorly and posteriorly to innervate the organs of the side. According to their distribution, nerves may be arranged into an anterior series and a posterior series.

The nerves of the **anterior series** are (1) **lateral buccal nerve**, (2) **dorsolateral buccal nerve**, (3) **anterior buccal nerves** and (4) **anterior oesophageal nerve**.

The nerves of the **posterior series** include (1) **posterior buccal nerve** (2) **odontophoral nerve** and (3) **oesophageal** and **salivary duct nerve**.

Visceral and the pedal ganglionic ring

The paired **parietal, pleural** and **pedal ganglia** and the single **abdominal ganglion** are united together to form a stout ring. The two parietal ganglia and the abdominal ganglion form a heart-shaped visceral ganglionic mass constituting the dorsal part of the nerve ring. The pedal ganglionic mass forming the ventral part of the nerve ring is a broad transverse mass.

Nerves of the pleural ganglion

(1) **Anterogenitonuchal nerve** innervating the visceral stalk, vas deferens, basal vagina and penis, and (2) **posterogenitonuchal** nerve innervating apical vagina, vas deferens and visceral stalk, emerge from the right pleural ganglion, while (3) **anteronuchal,** (4) **medionuchal** and (5) **posteronuchal nerves** originate from the left pleural ganglion supplying branches to the different regions of the visceral stalk.

Nerves of the visceral ganglionic mass

All the nerves arising from the visceral ganglionic mass run posteriorly and supply the visceral stalk, mantle, free retractor muscles and the visceral organs. **Right retractor, lateroparietal** and **medioparietal** nerves arise from the right side while **left retractor, latero-parietal, left pallioparietal, medioparietal and right pallioparietal** nerves originate from the left side of the visceral ganglionic mass. A single stout **visceral** nerve arises from the posterior region.

Nerves of the pedal ganglion

From the lateral sides of each ganglion emanate—one after another—nine nerves, some of which run anterolaterally, others only laterally, some others posterolaterally, and a few posteriorly. All the pedal nerves divide repeatedly and form regular plexuses in the sole of the foot along the borders.

☒ Reproductive System

The genital system is remarkably simplified but functions most efficiently. The system consists of an ovotestis, an ovotestis duct, a spermoviduct with prostatic acini, a vas deferens with penis, an oviduct with vagina and a genital atrium opening to the exterior through the genital aperture (Fig. 29.32).

Ovotestis

Multilobate and consists of 4-5 finger-like lobes embedded in the posterior lobe of the digestive gland. Each lobe consists of numerous lobules, producing sperm cells and only a few produce both sperm cells and ova.

Ovotestis duct

A single duct comes out from each lobe, and all such ducts unite to form the **ovotestis duct,** opening into the **spermoviduct.** The ovotestis duct is divisible into three distinct regions: the narrow **apical ovotestis duct,** the thick much-folded **ovisperm vesicle** where spermatozoa are stored, and the rest, **basal duct,** ending in a **carrefour.** The carrefour is connected with the ovotestis duct, spermoviduct (sulcus and uterus) and albumen gland. The apical and the basal parts of the ovotestis duct are lined with cilia, while the epithelium of the ovisperm vesicle consists of phagocytic cells.

Spermoviduct

All along the length the permoviduct (sulcus and uterus), the **sulcus** is an incom-pletely separate channel through which the sperms travel. It receives secretions from the prostatic acini.

Uterus

Divisible into an **apical** and a **basal** portion. The walls of both the uteri are muchfolded and bear numerous calcic glands, in addition to the mucous ones.

Albumen gland

Cream-white, and enlarges considerably during breeding season. It is slightly curved, flat ventrally and convex dorsally. It is enclosed in a capsule and bears an **albumen canal** at the centre. The gland is made up of a

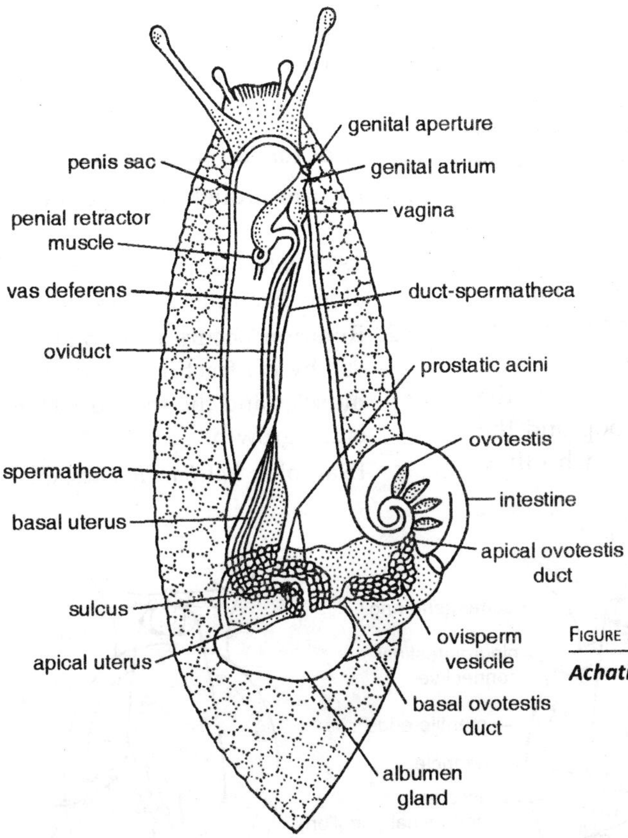

FIGURE 29.32

***Achatina* sp.** Reproductive system

large number of **vesicles** consisting of small columnar cells arranged in a single layer. The vesicles open into the ciliated albumen canal, which leaves the gland as a short albumen duct to end in **carrefour** (Oötype).

Prostatic acini

More than one hundred. Each acinus is made up of a large number of alveoli, opening into a central prostatic canal.

Basal uterus

Ends in a tubular oviduct, which receives the **spermathecal duct** at its anteior end. The mucous membrane of the duct is thrown into folds.

Oviduct

Ends in the vagina, divisible into a tubular **apical portion**, and an anterior dialated muscular **basal portion**, opening in the **genital** atrium.

Vas deferens

It is in continuation with the sulcus. The mucous membrane of the first part is highly. pleated and is transformed into a few folds towards the anterior end.

Penis

A curved muscular structure which continues anteriorly as the **penial prepuce** and ends in the genital atrium. This in turn opens to the exterior through the genital aperture.

TORSION

The anticlockwise twisting of the visceropallium to 180° on the head-foot long axis during early development in gastropod molluscs is **torsion**. It is an early and fundamental event in gastropod ontogeny (the event in phylogeny, which brought the gastropod in being). Most of the organs are affected by torsion and certain structures initially lodged posteriorly are brought to the front.

A narrow neck behind the head connects the head-foot and visceropallium, through which runs the oesophagus, rectum, anterior aorta, visceral loop and muscles attached to the shell. The short, narrow neck is the site of torsion (Fig. 29.33).

⊠ Course of torsion

The pretorsional larva is bilaterally symmetrical; the mantle cavity is posterior and ventral; gut straight, opening posteriorly.

A ventral flexure of the body changes the straight alimentary canal to a loop and the mouth and anus come close to each other. Initially saucer-shaped visceral mass and shell assume a cone-shape and later become spirally coiled.

⊠ Ways of torsion (Mechanism)

Torsion can be brought about in the following ways:

1. 180° rotation by muscle contraction alone. Example: *Acmaea.*
2. Rotation occurs in two stages—first 90° by contraction of larval retractor muscles and the second 90° by differential growth.

   Examples: *Patella, Haliotes,* etc.

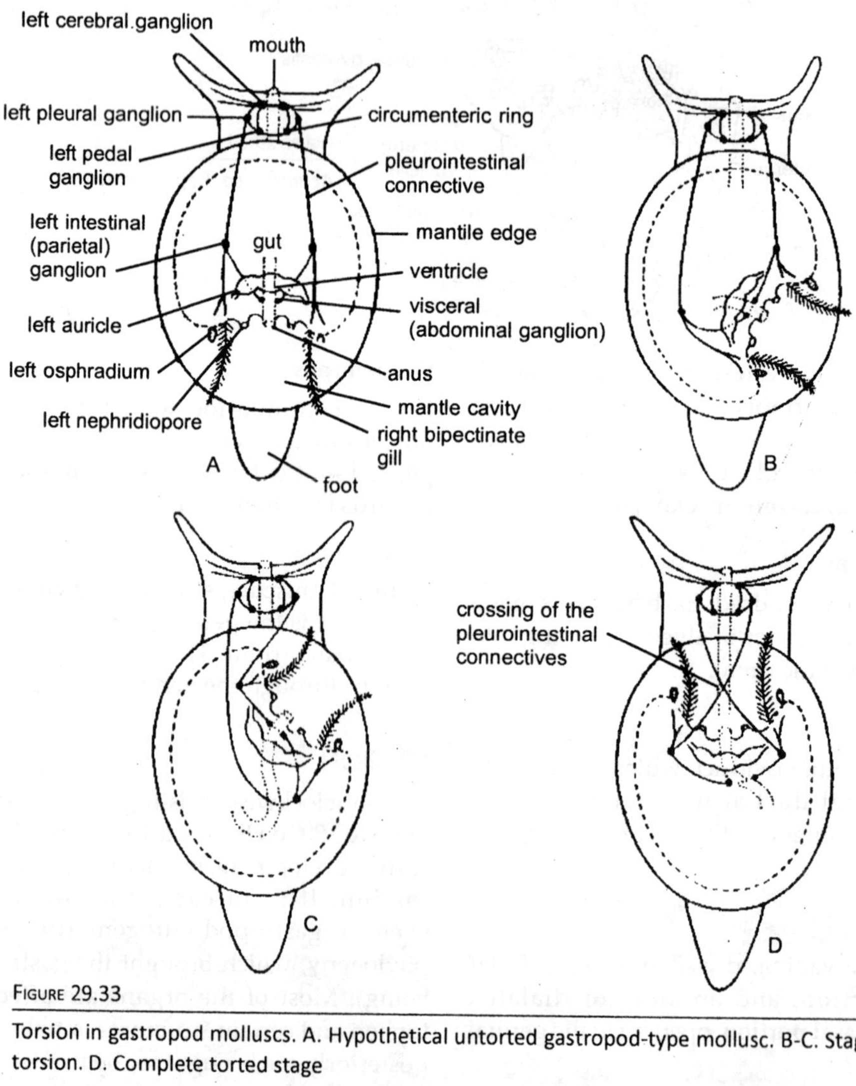

FIGURE 29.33

Torsion in gastropod molluscs. A. Hypothetical untorted gastropod-type mollusc. B-C. Stages of torsion. D. Complete torted stage

3. Rotation by differential growth alone. Example: *Vivipara*.
4. Rotation by differential growth, the anus coming to a position as in the adult.

   Example: *Aplysia*.

☒ Effect of torsion

Position of organs in pretorsional stage

Posteriorly located mantle cavity opens backward; mouth and anus at opposite ends; ctenidia pointed backwards; auricles behind the ventricle; nervous system bilaterally symmetrical; coilled visceral sac and shell dorsal or exogastric.

Position of organs in posttorsional stage

Mouth cavity opens in front just behind the head; gut assumes U-shape; anus lies in front near mouth; ctenidia shifted to front, directed backwards, the right one shifted to the left and vice versa, the auricles become anterior to the ventricle and the position changed—right to left and vice versa; nervous system becomes twisted into a figure of '8' by crossing of the pleurovisceral connectives; coil of visceral sac and shell become ventral or endogastric.

Asymmetry and loss of organs

In specialised forms the visceral organs of one side; viz. the ctenidium, auricle and kidney fail to develop. These organs of the original right side become topographically placed on the left side.

Significance of torsion

Authors are in agreement that torsion is the result of a larval mutation but they do not agree on the advantage of such mutation restricted only to the larva or extends to the adult.

1. Due to torsion the mantle cavity becomes anterior and sensitive parts of the anterior end including the head and also the foot of the larva could be withdrawn into the mantle cavity at the approach of danger. Garstang (1928) holds that torsion represents a larval mutation, which is of advantage to the larva and not the adult, a concept shared by Yonge (1947). By rotation the anteriorly placed mantle cavity provides a refuge for the delicate head and velar lobes of the larva, and it is not directly advantageous to the adult.
2. Morton (1958) believes the final effects of torsion due to larval mutation are profound in the adult snails though in a different way.

a. Torsion promotes stability in the adult by bringing the body of the snail nearer to substratum.
b. The free flow of respiratory water current in the posteriorly located mantle cavity containing the gills is obstructed by the backflow of water current during upstream movement of the snail. In reverse situtation, anteriorly placed mantle cavity is continuously flushed with fresh water current.
c. The forward position of the receptors enables the snail to scan the surroundings and also test the quality of water and mud.

☒ Coiling

The twisting of the visceropallium and shell cone together, the dorsoventral axis forming a helicoid spiral is **coiling**.

Spiral coiling and pallial asymmetry

Visceral coiling is totally distinct from torsion. Coiling probably preceded torsion. Coiling of the shell in a compact spiral ensures rigidity, protection, aggregation of organs, reduced diameter, better balance and a provision for uninterrupted growth. Coiling are of two types.

a. **Planospiral** (found in fossil forms). The shell was bilaterally symmetrical with each spiral completely outside the preceding one and in the same plane.

b. **Conispiral.** The shell is asymmetrical with helical coiling around a conical axis, the columella, with each coil below the preceding one.

Detorsion

Reversal of torsion is detorsion. Detorsion might be the result of some disadvantages met with in the adult life. It occurs in Euthyneura. The pallial complex travels back towards the posterior along the right side and a secondary symmetry is established. The degree of detorsion varies. In opisthobranchs and pulmonates (*Acteon, Helix*) detorsion is incomplete, retaining partially twisted visceral loop and the anus and the ctenidium laterally directed. Total detorsion occurs in *Aplysia*, an opisthobranch in which the shell is lost*.

SEPIA sp. (Cuttle fish)

The *Sepia*, popularly known as cuttle fish, are marine carnivorous molluscs, belong to the class Cephalopoda. They are active swimmers, abundantly found in coasts of a number of countries including Indian subcontinent.

External features

1. Body divisible into a head and a trunk (Fig. 29.34) joined by a short, narrow, neck.
2. Ten arms are present in the head region surrounding the mouth, of which eight are comparatively shorter and stout, provided with suckers, arranged in longitudinal rows.
3. The remaining pair of arms (the fourth pair) are known as **tentacles**. They are also provided with suckers. The tentacles may be retracted in large basal depressions.
4. The fifth arm of the left side, the **hectocotylysed** arm, has suppressed suckers and acts as intromittent organ in males.
5. A pair of large **eyes** are present on the lateral sides of the head.
6. The free extremity of the head bears the **mouth**.
7. The shield-shaped trunk is flat below and slightly convex above with fringed fin on either side. The fin extends from the oral to the aboral end where a notch separates them.
8. The trunk is covered by the thick mantle, forming a ridge at the oral end and ventrally it forms the **oral lip** of the opening of the mantle cavity.
9. The transverse slit-like **pallial aperture** is located at the anterior end of the trunk, on the ventral surface communicating the pallial or mantle cavity with the exterior.
10. A ventral **funnel** is present that opens externally behind the neck and internally in the mantle cavity by a wide aperture, the pallial aperture and serves as the main outlet of the mantle cavity.
11. The skin is provided with chromatophores containing three kinds of pigments and iridocytes.
12. A shell is present that serves as an internal skeleton and float, and is placed within a sac. Calcareous particles are arranged in the form of lamellae in the shell enclosing spaces for gas.

Mantle

1. The mantle forms a thick integument covering the trunk. Towards the oral end it terminates in a **ridge** around the neck.
2. Dorsally, the ridge projects forward as a prominent, round lobe, partially covering the head. Ventrally, it forms the **oral lip, or collar,** which surrounds the neck, between the head and visceral mass.

* Molluscs retaining bilateral symmetry are termed **untorted**, those undergoing torsion **torted** and those with some degree of detorsion **detorted**.

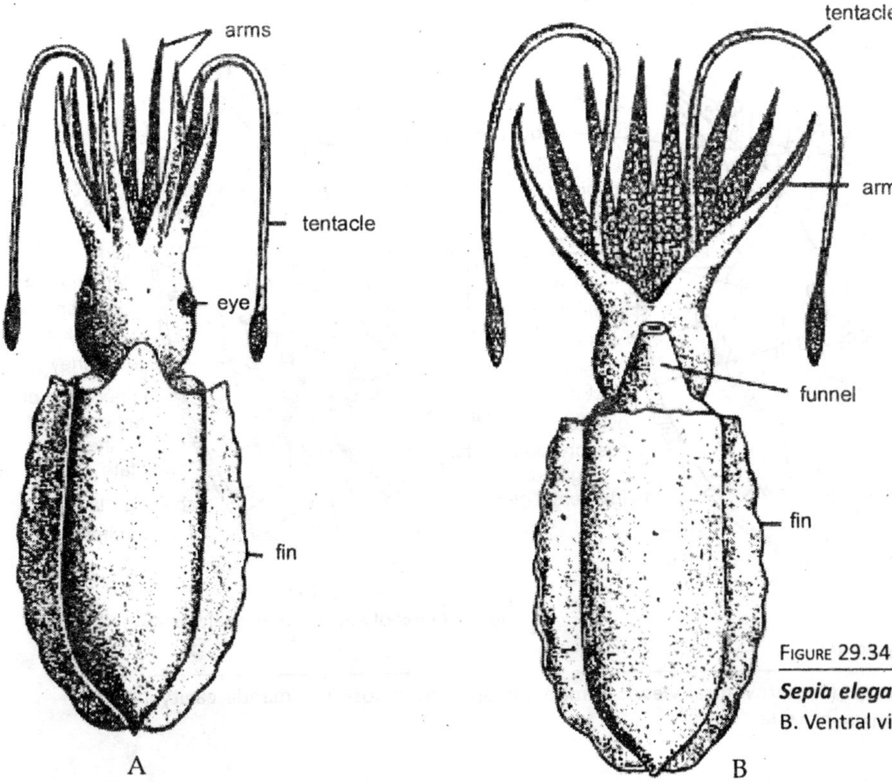

FIGURE 29.34

Sepia elegans. A. Dorsal view. B. Ventral view

3. A pair of **fins**, extending from the oral to aboral end, are formed by lateral extensions of the mantle.
4. The mantle encloses a **mantle** or **pallial cavity**.

⊠ Pallial Complex

The mantle cavity or pallial cavity is a spacious chamber (Fig. 29.35) communicating with the exterior anteriorly through the **funnel** and the pallial aperture between the neck and mantle collar. It encloses:

1. A pair of plume-shaped **ctenidia** or **gill** on either side.
2. The **rectum** in the middle line of the ventral surface, the **anus** opening close to the internal opening of the funnel.
3. A narrow tube, the **renal sac** with a terminal opening is present on each side of the rectum.
4. The genital duct, **spermiduct** or **oviduct** opens on the left side.

⊠ Endoskeleton

The internal shell, the cranial cartilage and other cartilages constitute internal skeleton.

Shell

A calcareous shell enclosed in a sac is present in the dorsal wall of the mantle.

1. The shell is leaf-like, bilaterally symmetrical, broad at the oral and narrower at the distal end, drawn into a sharp, dorsally projected spine (Fig. 29.36).
2. The ventral surface of the shell is convex.
3. The dorsal surface is convex anteriorly and bounded laterally by thin wing-like ridges, converging to meet at the aboral end, which is deeply concave.
4. The body of the shell consists of numerous closely arranged, thin calcareous laminae, enclosing spaces for gas.

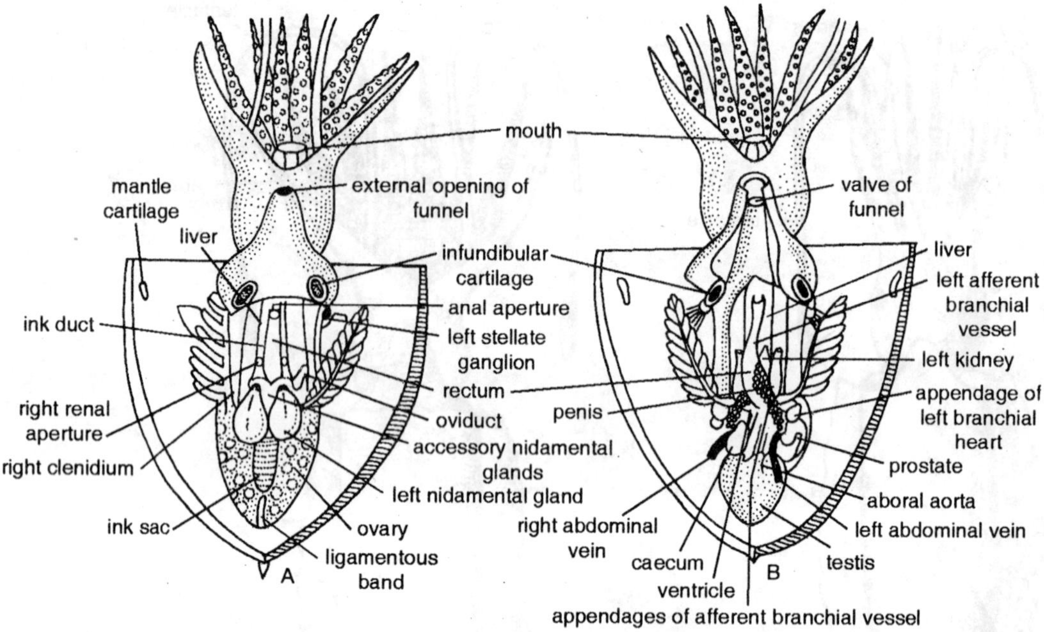

FIGURE 29.35

***Sepia* sp.** (Posteroventral view). Mantle cut open to expose the mantle cavity; A. Female. B. Male

5. A chitinoid layer, thin on the surfaces and slightly thicker along the margins encloses the shell.

Cranial and other cartilages

The cranial cartilage encloses closely aggregated nerve ganglia and statocyst.

In addition, the eyes are protected by **cartilaginous cups**; a thin, shield-shaped **nuchal cartilage** lies in the ventral wall of the neck; a thin **plate** of cartilage in each of the paired elevations of the ventral wall of the funnel and the corresponding depressions on the dorsal surface of the body; thin cartilages supporting the fins and bases of the arms.

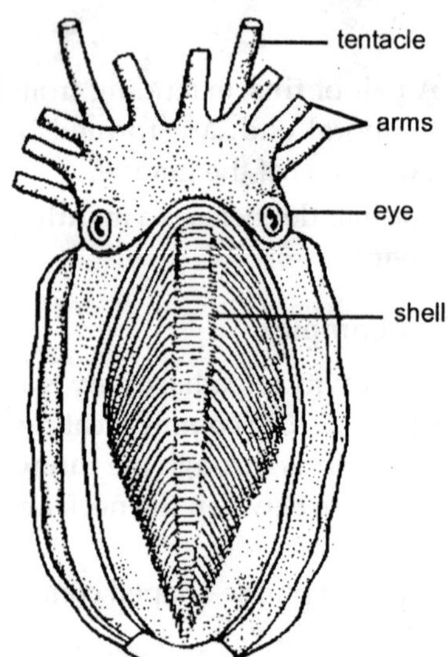

FIGURE 29.36

***Sepia* sp.** Mantle removed to expose internal shell

Locomotion

The locomotor organs are arms* and the funnel, which are modifications of foot, and the lateral fins. The mantle also helps in locomotion.

1. Undulating movement of arms and fins help in slow movement. The fins also help in changing direction.

* Some authors believe the arms are not modified foot but appendages of the head.

2. Darting rapidly through water is effected by rhythmic contraction of the muscular mantle, causing jets of water forced out through the opening of the funnel.

The funnel is used to expel water only from the mantle cavity.

a. With the relaxation of the mantle, water enters the mantle cavity through the cleft, the pallial aperture, between the neck and mantle collar.
b. The cleft is tightly closed, the mantle contracts and water from the mantle cavity is forced out in the form of a jet through the opening of the funnel.
c. The thrust of water shoots the animal backwards when the tip of the funnel is directed forward. By bending the tip backwards, Sepia can dart forward. Forward movement, which helps in catching prey, is slower than that of the backward movement helping to escape from predators.

⊠ Digestive System

The digestive system consists of an alimentary canal, a pair of salivary glands, a digestive gland and a pancreas (Fig. 29.37).

Alimentary canal

The alimentary canal is divisible into a foregut consisting of a buccal mass with an odontophore and an eosophagusl; the midgut constituted by the stomach and the hindgut by intestine and rectum.

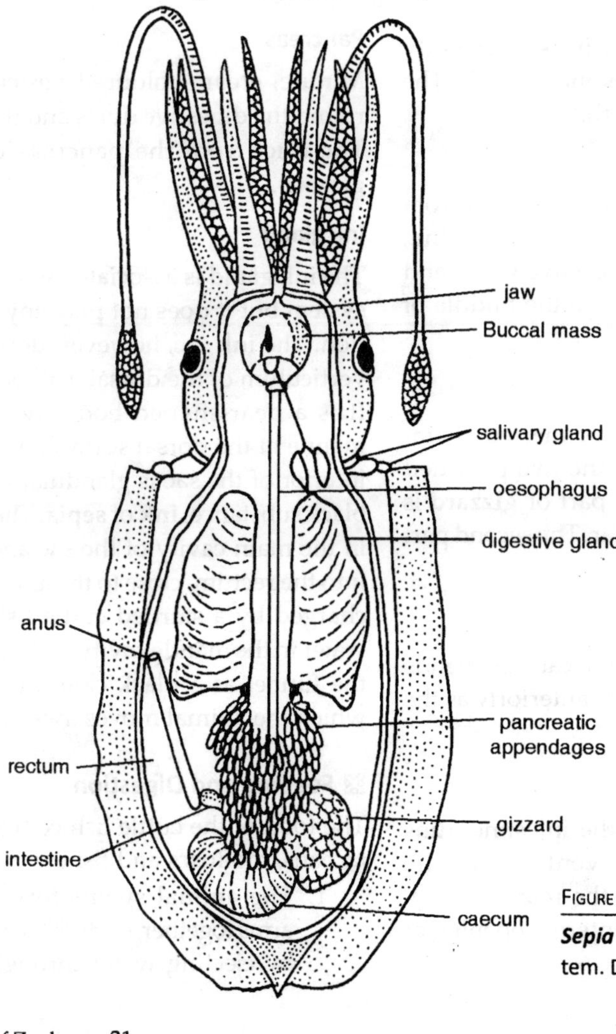

FIGURE 29.37

***Sepia* sp.** Digestive system. Dorsal view

Mouth

A round aperture at the anterior end surrounded by oral arms. A thin, lobed, **peristomial membrane** surrounds the mouth. The lip is circular and beset with many papillae. A pair of jaws are lodged in the lip. The mouth leads to the buccal cavity. The jaws somewhat resemble the beak of a parrot, the ventral one larger and strongly bent than the dorsal one, which it encloses partly.

Buccal mass

It is a fairly large, pyrifom muscular structure containing a spacious chamber, the buccal cavity. A large odontophore is present in the buccal cavity.

Odontophore

A fairly large, pyriform, muscular body containing buccal cartilages and a radula. The radula bears numerous teeth.

Oesophagus

A straight, narrow tube, runs posteriorly from the buccal mass, along the middle line, between the lobes of the digestive gland and joins the stomach posterior to the middle of the trunk.

Stomach

The stomach has two interconnected chambers, the opening between the two provided with a sphincter. The first part or **gizzard** is somewhat oval and muscular. The second part or **caecum** is round.

Intestine

A round tube, arises from the caecum, bends sharply upon itself and runs anteriorly as the rectum.

Rectum

It is of same diameter as the intestine, runs forward, parallel and ventral to the oesophagus and opens in the mantle cavity through **anus**, close to the internal opening of the funnel.

☒ Digestive glands

Digestive glands are of three types:

Salivary glands

A pair of glands located behind the cranial cartilage, one on either side of the oesophagus. Salivary ducts run inward and the two join in the middle. The common duct opens dorsally into the posterior part of, the buccal cavity.

Digestive gland

Commonly called **liver**. A large, brownish gland, divided into right and left halves, partly connected, and lie by the sides of the oesophagus. The ducts from two lobes open separately at the junction of the stomach and intestine.

Pancreas

Minute, cream-coloured vesicles, clustered round the digestive ducts and their openings. The duct from the pancreas opens in the gizzard.

Ink gland

The ink gland is associated with the digestive system but it does not play any role in nutrition. The ink sac, however, develops as a diverticulum of the dorsal wall of the intestine. It is a pear-shaped body, lying below the mantle of the dorsal surface. A portion of the interior of the sac is glandular and secretes a black substance, **ink of sepia**. The ink is stored in the main cavity of the sac and discharged into the rectum, close to the anus. The ink discharged by a startled cuttle fish mixes with water in the mantle cavity and ejected through the funnel as a black cloud, under cover of which the animal moves away unnoticed.

☒ Feeding and Digestion

The food of the cuttle fish comprises of crustaceans, molluscs and fishes.

1. The animal swims forward with the arms together, darts at the prey suddenly by ejecting water through the funnel,

FIGURE 29.38
Sepia sp. A gill

the arms are spread and the prey grasped. All the small eight arms play active role in capture of prey.

2. The suckers are pressed against the body of the victim and a partial vacuum thus created, makes the suckers firmly adhere to the body of the prey and the latter cannot escape.
3. The victim is brought to the mouth by the arms, torn to pieces by the jaws and swallowed. The radula is seldom used in feeding.
4. Small, bottom dwelling animals are quietly covered by the arms and captured.

Digestion is both extracellular and intracellular.

⊠ Respiratory System

The respiratory organs consist of a pair of gills or ctenidia, one on each side, situated in the mantle cavity. The gills are plume-shaped (Figs. 29.35, 29.38).

1. Each **gill** consists of a series of paired delicate **lamellae** along an exis. The lamellae are largest in the middle but become smaller towards the extremities. Each lamella bears a complex system of foldings by which the surfaces are increased. Internally, the lamellae are not in complete contact, and an axial canal is present.
2. The lamellae of each gill are attached to a long axis known as **ctenidial axis**. Along the greater part of its length the gill is attached to the wall of the mantle cavity by a thin muscular fold.

Mechanism of respiration

By sudden contraction of the body, water within the mantle cavity goes out through the opening of the funnel and with the relaxation of the body a fresh current of water enters the mantle cavity through the pallial aperture. The water penetrates freely to all parts of the gills through axial canal. The blood flows to the gills through afferent branchial vein and reaches gill lamellae through minute branches. After gaseous exchange in the gill lamellae, oxygenated blood is returned to the auricle through the main efferent vein formed by the union of minute vessels in the gill lamellae.

⊠ Circulatory System

The circulatory system in Sepia (in all cephalopods) is highly developed and almost the whole of the blood is carried through vessels arteries and veins.

The system consists of a ventricle, two auricles, arteries and veins. The heart is somewhat obliquely placed but the rest of the system is bilateral (Fig. 29.39).

Heart and pericardium

The heart is enclosed in a pericardium and lies superficially in the middle of the ventral surface.

1. **Ventricle.** Median, somewhat asymmetrical in position and divided into two lobes by a median constriction.
2. **Auricles.** Two, lateral and symmetrical; they are simple contractile expansions of the efferent branchial veins.

Arteries

Two aortae arise from the two ends of the ventricle, divide into branches and capillaries in tissues.

1. **Oral aorta.** A large vessel, arises from the oral end of the ventricle and sends

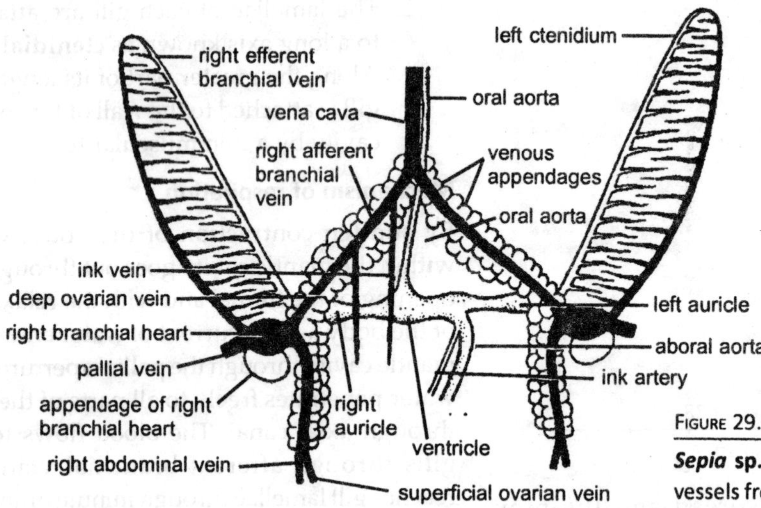

FIGURE 29.39

***Sepia* sp.** Heart and main blood vessels from the posterior aspect

branches to the organs of the anterior part of the body.

2. **Aboral aorta.** A small vessel, arises from the aboral end of the ventricle. It bends over the ink sac, runs aborally and sends branches to the posterior part of the body.

Veins

The veins are formed in the tissues by the joining of the capillaires and they unite to form a system of veins.

Vena cava

A large median vein runs from head to near the rectum, in front of which it bifurcates to form the right and left afferent branchial veins.

Afferent branchial vein

Two. Each runs through the renal organ of the side and receives veins from the aboral organs at the base of the gill.

a. At the base of the gill, the afferent branchial vein dilates to form a contractile sac, the **branchial heart**, to which a round, glandular body, known as **appendix** is appended.
b. The afferent branchial vein runs through the axis of the gill and sends branches to the gill lamellae.
c. The branchial heart pumps blood to the gills through the afferent branchial vein.

Efferent branchial vein

Two. Each arises from one gill and oxygenated blood is returned to the ventricle through auricles, the dilated contractile base of the efferent branchial veins.

☒ Excretory System

A pair of thin-walled, **renal sacs**—right and left—constitute the excretory system. The afferent branchial vein of the side runs through the renal sac. Each sac opens in the mantle cavity through a conspicuous **renal aperture** on either side of the middle line. Each sac is in communication with the pericardium through a renopericardial aperture. The two sacs communicate with one another both orally and aborally. Masses of glandular tissues (Fig. 29.40) surrounding the renal sacs excrete nitrogenous wastes in the form of **guanine** and released in the renal sac. The wastes discharged into the mantle cavity through the renal apertures are thrown outside through respiratory water current.

☒ Nervous System

The nervous system is highly developed; the nerve ganglia are large. An extreme condensation of the nervous system has led to the aggregation of all the nerve ganglia in the head region, around the oesophagus and protected by a cartilaginous cranium. The

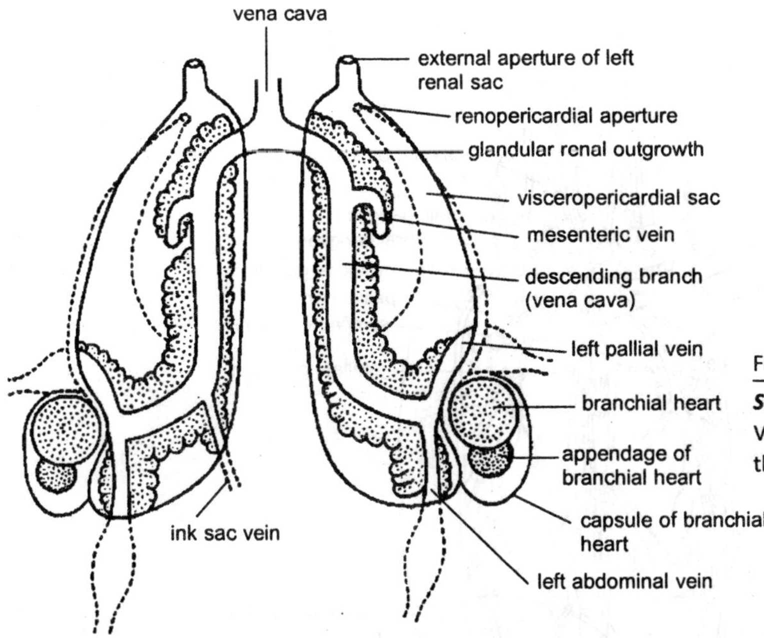

FIGURE 29.40

***Sepia* sp.** Excretory organs. Ventral view. Upper walls of the renal sac removed

cerebrals, superior buccals, inferior buccals, pleurovisceral and pedal genglia and their nerves constitute the nervous system (Fig. 29.41).

Cerebral ganglia

Paired, fused to form a large round mass, dorsal to the oesophagus. Anteriorly, they are covered by a tough, fibrous membrane. A pair of very thick band-like nerves around the oesophagus connect the cerebral ganglia with the pleurovisceral ganglia at the sides and posterior border. From each cerebral ganglion arise an optic nerve going to the eye, a statocyst nerve innervating the statocyst and a cerebrosuperior buccal connective.

Superior buccal ganglia

A small pair, closely united and located dorsal to the oesophagus, close to the buccal mass. The are connected with the cerebral ganglia by cerebrosuperior buccal connectives.

Inferior buccal ganglia

A pair of small, closely united ganglia on the dorsal surface of the oesophagus, posterior to the superior buccal ganglia, with which they are joined by slender connectives.

A loop of sympathetic nerves arising from the ganglia courses along the oesophagus to the stomach and each ends in gastric ganglion.

Pleurovisceral ganglia

The visceral ganglion is below and the pleural ganglia on the lateral sides of the oesophagus. The pleurals and the visceral fuse to form a pleurovisceral ganglionic mass. It gives off:

a. Two **visceral nerves.** Each sends branches to visceral organs and continue as branchial nerve, bearing a **branchial ganglion** at the base of the gill and runs along the axis to its extremity.
b. Two stout **pallial nerves.** Each runs through the neck to the inner surface of the mantle cavity and forms a large, flat **pallial** or **stellate ganglion** in front of the gill. Nerves emanating from the stellate ganglion innervate different parts of the mantle.

Pedal ganglia

The two ganglia are fused and located below the oesophagus. It sends:

a. Ten **branchial** nerves to the arms. The branchial nerves are connected by a ring commissure.

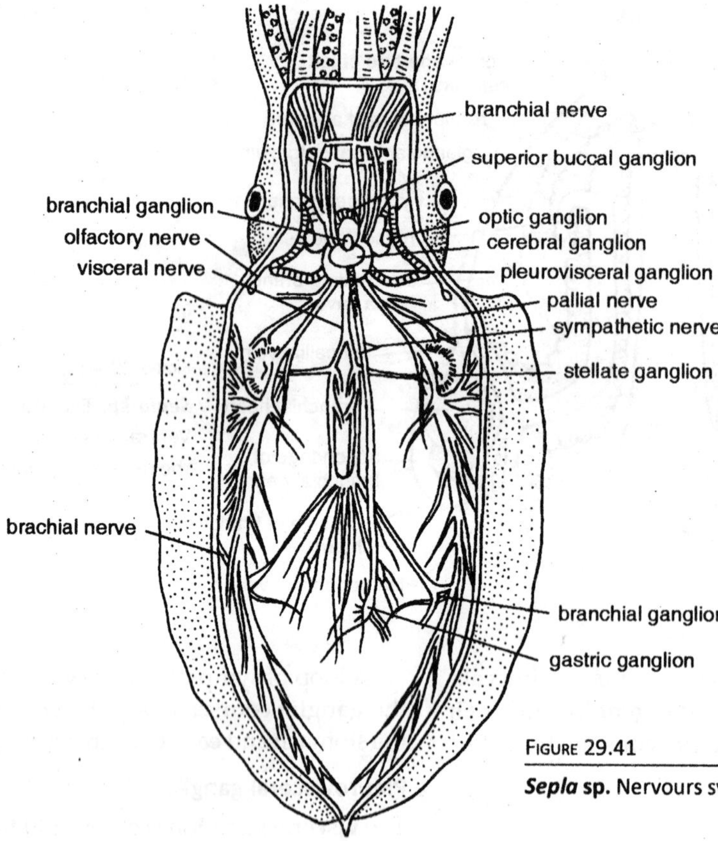

FIGURE 29.41

***Sepla* sp. Nervours system. Dorsal view**

b. A pair of nerves to the funnel.

Receptors and sense organs

The sense organs are highly developed and consists of a pair of eyes, a pair of statocysts, a pair of ciliary pits and a gustatory organ.

Eye

The eyes are simple (Fig. 29.42), located laterally on the head and lodged in some sort of orbit made of curved plates of cartilage, connected with the cranial cartilage.

Structure

1. **False Cornea.** A transparent portion of the integument covering the exposed part of the eye.
2. **Sclerotic.** It is strengthened by plates of cartilage and forms a firm wall of the eye ball.
3. **Pupil.** A large opening of the sclerotic on the outer side towards the head.
4. **Iris.** The part of the sclerotic surrounding the pupil. Muscle fibres in the iris can regulate the diameter of the pupil to a limited extent.
5. **Lens.** A dense, glassy spherical body just internal to the iris and projecting slightly through the pupil. It consists of two planoconcave lenses in close apposition.
6. **Cornea.** A thin layer of cells between two parts of the lens.
7. **Ciliary process.** An annular process, projecting inwards from the sclerotic, supporting the lens.
8. **Cavities.** The lens with the ciliary process divides the cavity of the eye ball into two portions—a smaller outer cavity containing **aqueous humor** and a larger inner cavity containing **vitreous humor.**

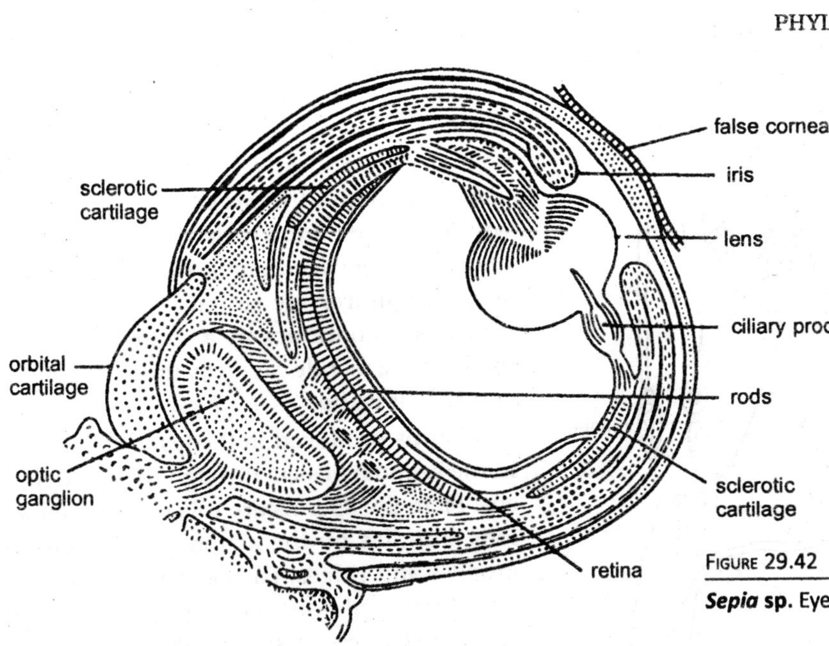

FIGURE 29.42

***Sepia* sp.** Eye (Vertical section)

9. **Retina.** The light sensitive layer lining the wall of the inner chamber. It is composed of two layers:

a. An inner layer of close set parallel **rods.**

b. An outer layer of **retinal cells** in communication with the rods internally and optic nerve fibres externally.

The movement of the eye is very limited and effected by muscles.

Statocysts

The statocysts are highly developed, closely placed organs (Fig. 29.43), separted by a median cartilaginous septum in the posterior portion of the cranial cartilage, close to the pleurovisceral ganglion.

1. The cavity of a statocyst is about 3 mm in diameter.
2. The inner surface is lined with a flattened epithelium and is thrown into a numer of round and pear-shaped elevations.
3. On the posterior surface of the ridge a **crista statica** and a **macula statica** are present. These are composed of large cells with hair-like processes on the surface and processes of the base continuous with the fibres of the statocsyst nerve.
4. A large **statolith** attached to the macula is present in the cavity.

The statocysts are organs of equilibrium.

Olfactory organs

A pair of **ciliated pits** opening by slits on the surface behind eyes. The pits are lined by narrow sensory cells connected to the nerve fibres from a small ganglion close to the optic ganglion.

Gustatory organ

The organ of taste is a small elevation bearing papillae, on the buccal floor, just in front of the odontophore.

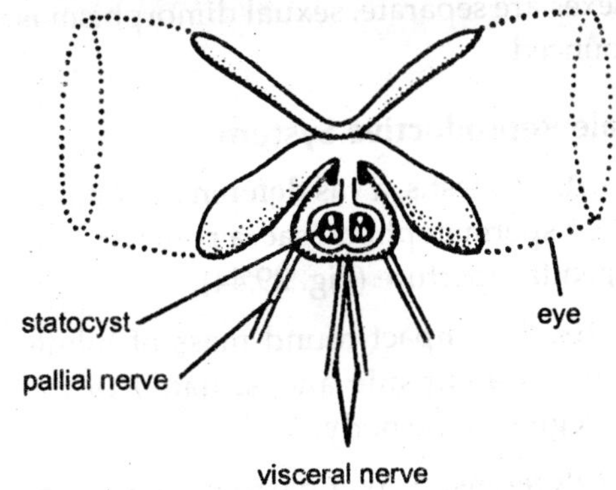

FIGURE 29.43

***Sepia* sp.** Cranial cartilage (Posterior view). The cavities of the statocyst exposed

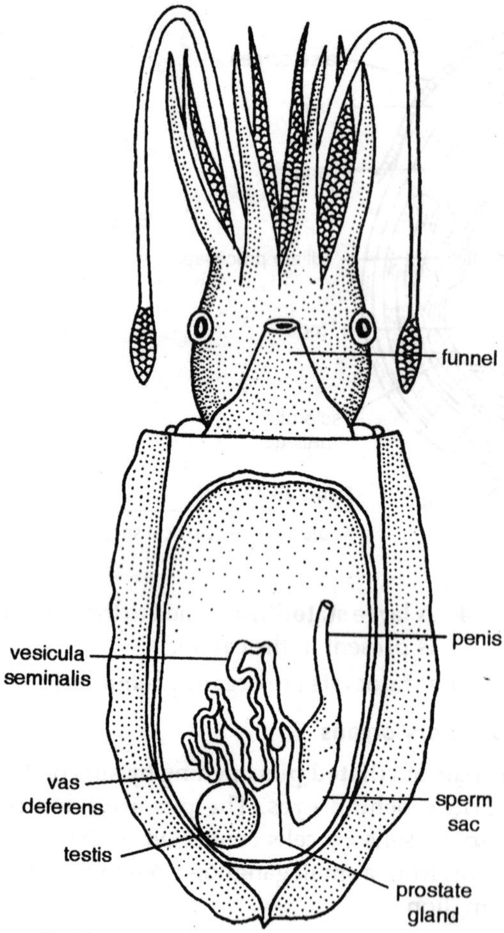

FIGURE 29.44

***Sepia* sp.** Reproductive system. Male. Dorsal view

⊠ Reproductive System

The sexes are separate, sexual dimorphism is pronounced.

⊠ Male Reproductive System

It consists of a testis, a vas deferens, a seminal vesicle, a spermatophoral sac, a penis and the male genital aperture (Fig. 29.44).

Testis. A compact round mass of tubuls enclosed in a cupsule and situated in the aboral region of the body.

Vas deferens. A long narrow tube, arises from the capsule of the testis. It is much coiled, moves to the left and ends in the seminal vesicle.

Seminal vesicle. An elongated tubular structure, the interior of which is provided with grooves and ridges. Appended to it is a glandular body, the **prostate.** The seminal vesicle expands, into a wide sac, the **sprematophoral sac** or **Needham's sac** in which the **spermatophores** are stored. The spermatophoral sac continues as a stout **penis** and opens in the mantle cavity by the male genital aperture at its tip.

Sperms produced in the testis are carried to the seminal vesicle through the vas deferens. Here the sperms are rolled up into thin, long bundles by the action of a system of grooves and ridges. Each bundle is enclosed in a narrow cylindrical chitinoid capsule, the spermatophore. A spring-like structure is present at one end, which causes the rupture of the spermatophore. Spermatophores are stored temporarily in the spermatophoral sac and discharged in the mantle cavity through the genital aperture.

⊠ Female Reproductive System

It consists of an ovary, an oviduct, the female genital aperture and a pair of nidamental glands, in association with an accessory nidamental gland (Fig. 29.45).

Ovrary. A rounded body, enclosed in a capsule and lodged in the aboral region of the body. An axial swelling bears numerous follicles, each having a single ovum.

Oviduct. A short wide tube, arises from the capsule enclosing the ovary and opens in the mantle cavity by the genital aperture at its tip.

Nidamental glands. a pair of large, flattened pear-shaped glands, situated on the anterior wall of the mantle cavity and to the right and left of the ink duct. A median canal is present in the long axis of each gland, on either side of which are present closely set delicate lamellae. The median canal opens into the mantle cavity by a slit. The nidamental glands secrete a viscid material by means of which eggs adhere together after deposition. Another glandular mass, **accessory nidamental gland,**

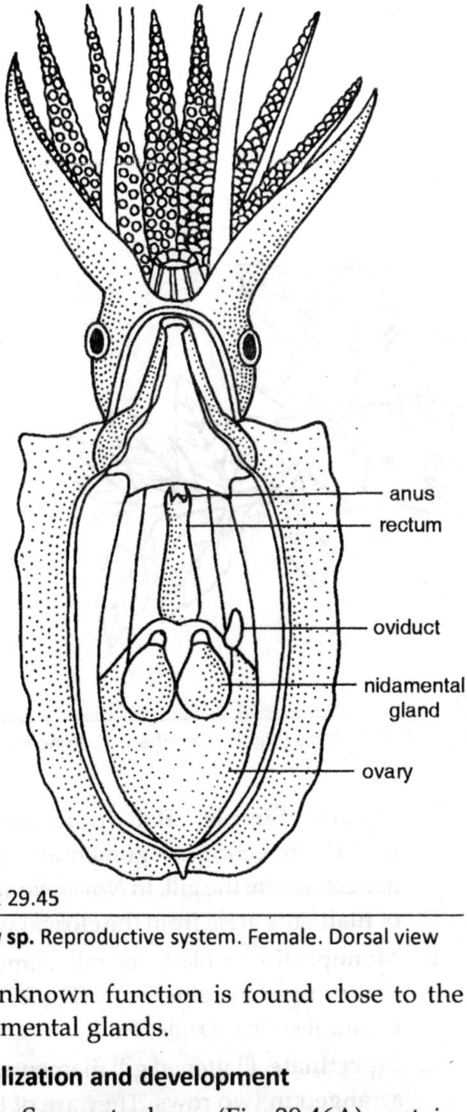

FIGURE 29.45

***Sepia* sp.** Reproductive system. Female. Dorsal view

of unknown function is found close to the nidamental glands.

Fertilization and development

1. Spermatophores (Fig. 29.46A) containing sperms (Fig. 29.46B) are transferred to female during copulation.
2. The hectocotylised arm receives spermatopheres in its specialised tip. The arm is inserted into the mantle cavity of the female and the tip is detached there.
3. Fertilization internal; eggs are large sized, pear-shaped and heavily yolked.
4. The eggs are laid shortly after copulation in masses in a soft, gelatinous substance which becomes attached to some aquatic support (Fig. 29.46C).
5. Development direct; young resembling the adult hatches out and starts independent life.

Accessory reproductive organ

The fifth left arm in male **Dacapoda** and the third right arm in male **Octopoda** became hectocotylysed—an intromittent organ, and transfer spermatophores to female.

RESPIRATORY STRUCTURES AND RESPIRATION IN MOLLUSCA

Respiration, in its broad sense, is a biological phenomenon in which oxygen is absorbed by the organism from its surroundings and carbon dioxide is released. It has two phases—organ respiration and cell respiration. The first one involves uptake of O_2 by respiratory structures and releasing CO_2 following simple physical law. The second one involves all cells and biological oxidation of carbohydrates releasing energy, CO_2 and water (p. 97).

☒ Types of respiration

Depending on the medium, respiration in molluscs are of two types—aquatic and terrestrial.

Aquatic respiration

In this, O_2 dissolved in water is used. The surface of the respiratory structure is greatly increased to allow rapid absorption of O_2. The aquatic respiratory structures may be placed under two heads—primary and secondary.

Primary respiratory structures

The primary respiratory structures are **ctenidia** or **gills**. The gills are located in the mantle cavity, attached to the body by membranes.

Structure of ctenidia

1. Ctenidia are paird, symmetrical, ciliated and two rows of flattened gill filament, arranged one on either side of a long, flattened axis, traversed by afferent and efferent vessels through which haemolymph flows.

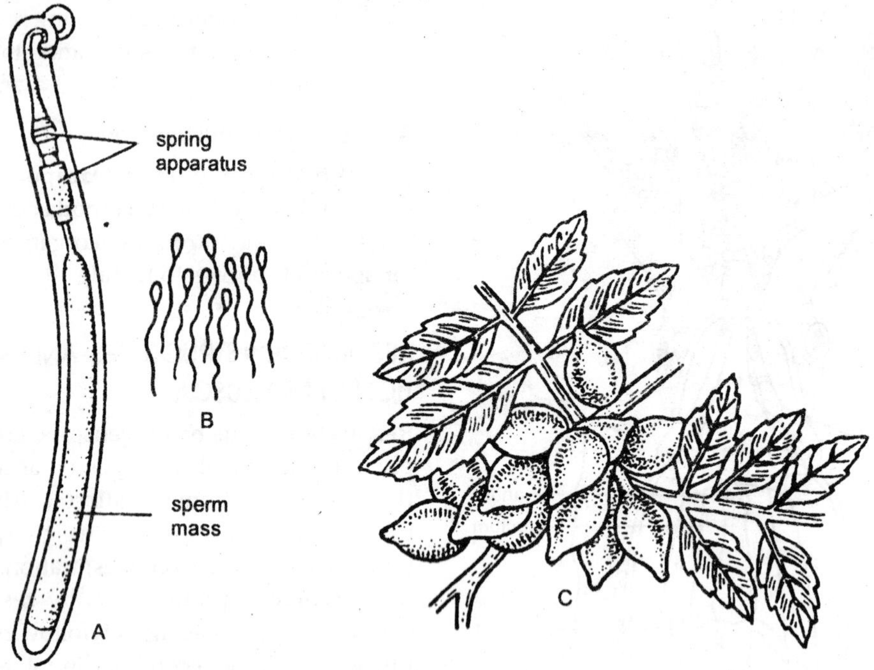

FIGURE 29.46

***Sepia* sp.** Spermatophore and gametes. A. Spermatophore. B. Sperms (highly magnified). C. Bunch of eggs attached to a twig

2. Narrow spaces are left between the gill filaments to permit free water flow between them but close enough so that cilia on adjacent filaments may operate together in generating water current.
3. Haemolymph circulates through the gill filaments.
4. Skeletal rods support the gill filaments.
5. Cilia generate inhalant water current drawing water below the ctenidia and exhalant current expelling water out from them.

Types of ctenidia

Based on topography, the ctenidia in molluscs are placed in following categories:

1. **Holobranchiate.** The ctenidia extend all over the body. The number varies from 14-80 pairs. Example: Polyplacophora.
2. **Merobranchiate.** The ctenidia are restricted to a particular area of the body. Depending on the arrangement of leaflets the gills have been subdivided into four types:

A. **Plicate.** Simple, flat, transversely folded, projecting integumentary laminae constitute the gill. In *Neomania* a tuft of filaments arise from the cloacal wall.

B. **Monopectinate.** Flattened gill filaments are arranged in a single row (Fig. 29.16). Examples: *Pila, Triton,* etc.

C. **Bipectinate.** Flattened gill filaments are arranged in two rows. They are of two types:

a. **Unequal.** Both rows of filaments are present, the right one being smaller in size. Example: *Fissurella.*

b. **Equal.** Both the rows of filaments are of same size. This type is present in Bivalvia, but they are variously modified.

i. The leaflets are short and flat. Example: *Nucula.*

ii. The leaflets are filamentous and long. They may be free (*Arca*) or joined by ciliary connectives (*Mytilus*).

iii. The ciliary junctions are replaced by membrane. Example: *Unio Lamellidens* etc. (Figs. 29.2, 29.5).

iv. The gill is degenerated and represented as transverse partitions. Example: *Poromya*.

D. **Feathered.** Assumes the shape of a feather (Figs. 29.35, 29.38). Examples: *Cephalopods*.

Ctenidia in different groups of molluscs

1. **Monoplacophora.** The pallial groove contains five to six pairs of unipectinate gills. Example: *Neopalina*.
2. **Polyplacophora.** Bipectinate ctenidia located in the mantle groove. (a) Number of ctenidia varies from 14 (Lepidopleura) to 80 (Acanthopleura). (b) Gill rows may be holobranchiate with two exceptions bearing merobranchial type. Examples: *Chiton laeves, Chiton pallasii*.
3. **Aplachophora.** The gills are reduced to a paired, feather-shaped structure situated near the cloacal cavity, one on each side. The gills are merobranchiate. Example: *Chaetoderma*.
4. **Gastropoda.** The gills vary widely in number and position in this group.

A. **Prosobranchia.** Ctenidia lie in front of the heart.

a. **Diatocardia.** The arrangement is most primitive. The gills are two, long, feathered, on each side and lie symmetrically to the middle line. Example: *Fissurella*.

b. **Monotocardia.** The arrangement of the gills is remarkably uniform. A single gill, feathered on one side and united to the mantle along its whole length. Example: *Triton*.

B. **Opisthobranchia.** The gills are partially enclosed in the mantle cavity (Tectibranchiata). The true ctenidium, when present, is little developed and located on the right side of the body. Example: *Aplysia*.

5. **Bivalvia.** The gills are bipectinate, equal on either side, usually very large, having assumed food collecting function in most species in addition to gas exchange.

A. **Protobranchiata.** Gills are smallest and lie behind the foot at the back of the mantle cavity. In *Nucula*, the gill filaments are triangular.

B. **Filibranchiata.** Each gill separately forms a 'W' in section with long, narrow limbs. The gill axis lies at the middle angle of the W. Examples: *Arca, Mytilus, Anomia*.

C. **Pseudolamellibranchiata.** The reflected dorsal tips of the gill filaments have coalesced laterally with the mantle and mesially with the base of the foot. The gill has a greater cohesion than that in the filibranchs. The gills are more complex, with the surface plicate or thrown into folds and grooves. Examples: Ostreidae, Pectinidae, Pleriacea.

D. **Eulamellibranchiata.** The adjacent filaments of gills are united by vascular cross connections, leaving narrow openings, the ostia between them. The two lamellae of each demibranch are attached back to back in the same way. Examples: Cardiacea, Myacea.

E. **Septibranchiata.** The ctenidia do not exist as such, being converted into a horizontal muscular septum running from the base of the foot to the mantle, and extending right back to the siphons. Example: *Poromya*.

6. **Cephalopoda.** The gills are large, paired, bipectinate and one suspended on either side of the rectum by their afferent edges, not by efferent, as in gastropods, and the water current is driven from afferent to efferent side.

A. The gill filaments are firm and fleshy, nonciliated, and thrown into primary and secondary folds to increase respiratory surface. In *Nautilus* the gills are two pairs.

B. The haemolymph flows through each gill is assisted by a pulsatile accessory branchial heart, located at the base of the gill in the course of the afferent ctenidial vessel and placed in an annexe of the pericardium.

C. Pallial contractions drive water between the gill filaments at great pressure.

Evolution of ctenidia

The most primitive ctenidia are found in zeugobranchiate prosobranchs, which are simple outgrowths of the body. In bivalves the trend was

filibranch
eulamellibranch
septibranch

with an increase in length of the ctenidium. In this group three kinds of new tissue connections evolved.

1. Interfilamentar junction between adjacent filaments.
2. Interlamellar junction between two lamellae.
3. Connections between the tips of the filaments and mantle or foot.

In bivalves, the branchial apparatus in addition to respiratory function acts as a food collecting organ. Large, non-food particles are trapped and strained out by the mucus secreted by hypobranchial gland. Osphradium tests water quality.

Adaptive or secondary gills and integument

The gills develop from aberrant sites:

1. **Anal gills.** Delicate leaflets form a rosette around the anus. Example: *Doris*. In *Chaetoderma* a pair of symme-trical lateral gills are present on each side of cloaca.
2. **Cerata** or **dorsal appendages**. In many opisthobranchs the dorsum bears highly vascular appendages called cerata. They may be simple and club-shaped (*Aeolis*), dendritic (*Dendronotus*) or multilobed resembling a bunch of grapes (*Doto chica*).
3. **Pleural gills.** In Pleurophyllida lateral rows of branchial leaflets are situated beneath the mantle.
4. **Pallial gills.** In certain basommatophore pulmonate secondary external gills develop by the enlargement of the pallial lobe, just outside the pneumostome. They, however, lack cilia. Examples: Planorbidae, Ancylidae.
5. **Integument.** Specialised respiratory structures are lacking in some Scaphopoda. Respiration is carried by the internal surface of the mantle, particularly the anteroventral side in *Dentalium*, *Antalis*. In nudibranchs (Gastropoda) the entire dorsum of the body acts as the site of gas exchange. Integumentary gas exchange occurs in parasitic *Entoconcha*, *Conia*, *Limpontia* etc.

Terrestrial respiration

Amphibious adaptation and adaptation to terrestrial life have introduced remarkable modification in the respiratory structures of molluscs. The structures associated with terrestrial respiration are:

1. **Nuchal lobe.** The left nuchal lobe is better developed and forms a long respiratory siphon. Example: Monotocardia.
2. **Plumonary sac.** In some amphibious prosobranchs—*Pila*. (Fig. 29.17), *Ampullarius*, *Siphonaria*, etc. the pallial cavity is incompletely partitioned by a fleshy fold, the **epitaenia**, into a right branchial chamber and a left **pulmonary chamber**. The highly vascularised roof of the pulmonary chamber forms a **pulmonary sac**, the aerial respiratory structure, with a small aperture opening in the pulmonary chambers. Air enters the pulmonary chamber through

an opening at the tip of the left siphon. Aerial respiration takes place during aestivation and when there occurs a depletion of O_2 concentration in water. The pulmonary epithelium comprises cuboidal or squamous cells bearing short microvilli and many mucous cells.

3. **Lung.** In pulmonate—both stylommatophores and basommatophores—respiratory structure is a true lung* of independent origin, not vascularised mantle, which occupies the major part of the roof of the pallial cavity (p. 103). In some it extends to the walls and floor. In slugs Athoracophoridae the lung is nonvascular and its wall gives off a number of delicate branched tubules, named **tracheae** (Hiscock, 1972), which function as respiratory structures.

   The pallial cavity opens to the exterior by a large, oval aperture, the **pneumostome**, the opening and closure of which are controlled by contraction and relaxation of muscles. Large pulmonary veins with extensive ramification drain oxygenated blood from lung to auricle.

⊠ Respiratory pigment

The respiratory pigment is an extra-corpuscular haemocyanin (p. 103). It is more concentrated in gastropods and cephalopods. In *Arca* and *Solen* the pigment is haemoglobin and present in some special corpuscles in the haemolymph.

⊠ Mechanism of Respiration and Oxygen uptake

A. Aquatic respiration

Water is driven over the ctenidial surface by ciliary beating or by muscular pumping.

a. Cephalopods have branchial hearts at the gill bases, the pulsation of which drives haemolymph through afferent vessels.
b. Lack of such additional pump and less efficient funnel contraction are compensated with the development of two pairs of ctenidia in *Nautilus*.
c. Oxygen utilisation from pallial water is high in gastropods and cephalopods. It is 56% in *Haliotis* 79% in *Triton* and 63% in *Octopus*, as compared with 5-9% in sedentary lamellibranchs.

B. Terrestrial respiration

Contraction and relaxation of muscles in pulmonates cause in and outflow of air into and from the pallial cavity.

a. In a relaxed state of muscles the floor of the pallial cavity (diaphragm) arches upward.
b. During inspiration the pneumostome opens, muscular contraction flattens the floor causing fall of pressure in the pallial cavity and air rushes in. This is followed by closure of the pneumostome.
c. A slight delay in the opening of the pneumostome facilitates O_2 uptake by haemolymph.

---

* For a long time the pulmonary sac of prosobranchs and lung of pulmonates were considered homologous structures and formed by vascularisation of a part of the mantle cavity roof. Since the studies were restricted to adult specimens, there was no way to differentiate the two except in prosobranchs the pulmonary sac is less vascularised and occupies a small part of the roof of the pulmonary chamber, while in pulmonates vascularization is very high and the lung covers a large part of the mantle roof, extending to the walls and floor in some cases.

In the early twentieth century it was reported that the respiratory structure of Pulmonata is a true lung, developing independent of the mantle cavity. In early embryonic stage, the mantle cavity and lung develop by separate invaginations from a point at the posterior end of the body, the latter following the former. With growth of the embryo, both the invaginations enlarge in size, the smaller lung rudiment extending in the larger mantle cavity rudiment. Towards the close of embryonic life, the lung shifts dorsally, becomes flattened, the dorsal and ventral walls coming together, and they adhere to the root of the mantle cavity (Ghose, 1963). This led to the erroneous conclusion made from studies in adult specimens that the two structures—pulmonary sac in prosobranchs and lung in pulmonates—are homologous.

d. Respiratory movements increase with lowering of the O_2 concentration and at higher temperature. Such movements have been reported from *Arion*, a slug with a large pneumostome (Runhon and Hunter, 1970).

e. In prosbranchs alternate dilatation and contraction of the mantle and pulmonary chamber wall help in and outflow of air in the pulmonary chamber through left siphon, which enlarges to reach above water leval when the snail remains submerged.

Molluscs are sluggish creatures except cephalopods. The lower respiratory activity in the majority, however, is sufficient to meet with the need of energy of the animals. In cephalopods, a rapid flow of haemolymph through the gills is effected with the development of accessory branchial hearts, helping higher uptake of oxygen by the gills. The presence of diverse but well-organised respiratory structures—gills and lung—and certain other features led some authors to believe that molluscs are diphyletic in origin.

MODIFICATION OF FOOT IN MOLLUSCA

In molluscs the foot is a muscular projection from the ventral surface and acts as a locomotory organ. Molluscs are exposed to diverse habitats—aquatic, semiaquatic and terrestrial. Accordingly they are aquatic, amphibious and terrestrial. Foot in molluscs exhibit a remarkable degree of modifications in response to their habittat and modes of living.

It is believed that in molluscs the primitive and simplest foot was elongateoval with a broad flattened ventral surface suitable for creeping. The structure has been variously modified in response to the life conditions to which they were exposed.

A broad-soled muscular foot is prevalent in gastropods. The foot has three zones- anterior propodium, middle mesopodium and posterior metapodium. Waves of contraction sweep from the anterior to the posterior and of the foot and be lead to forward movement of the animal. A pedal mucous gland is present in the foot, the secretion of which helps in gliding movement and a trail of mucus is left behind.

Foot in molluscs have been placed in different groups on the basis of function.

1. In Chaetaderma a longitudinal ciliated ridge on the ventral surface is the organ of locomotion. A ventral foot—a highly developed organ is absent in this group.
2. **Creeping or crawling organ** (Figs. 29.12, 29.29). In the majority of the gastropods (*Aclean, Achatina, Pila,* etc.) and placophora and broad flat foot with a creeping sole is present. The sole of the foot may be triangular, elongateoval, broad at the middle; narrow, elongateoval and round at both the ends and of other forms. The creeping sole is contractile in *Triton;* in *Bullia* the foot encircles the whole body; in *Atlanta* the posterior part of the foot forms a sucker; in *Murex* the sole is highly glandular and bears a long backwardly bent siphon.
3. **Anchoring organ** (Fig. 29.47). In *Chiton* the foot is narrow and in addition to functioning as a creeping organ acts as a sucker for firm attachment to the support; in *Patella* the foot acts as a powerful organ to firmly adhere to the support; in *Fissurella* the foot is large and used as a sucker for anchoring to rack.
4. **Burrowing organ.** In *Dentalium* the foot is conical and protrusible; in **Unio, Anodonta,** *Lamellidens* (Fig. 29.3) the foot is like a ploughshare and burrows in mud or sand and acts as a creeping organ; in **Mya** the foot is highly developed and acts as a burrowing organ; in **Solen** the foot burrows actively and rapidly into the sea mud.
5. **Looping organ.** A groove is present between the propodium and the rest of the foot and the foot acts as a leaping organ in *Podipus.*

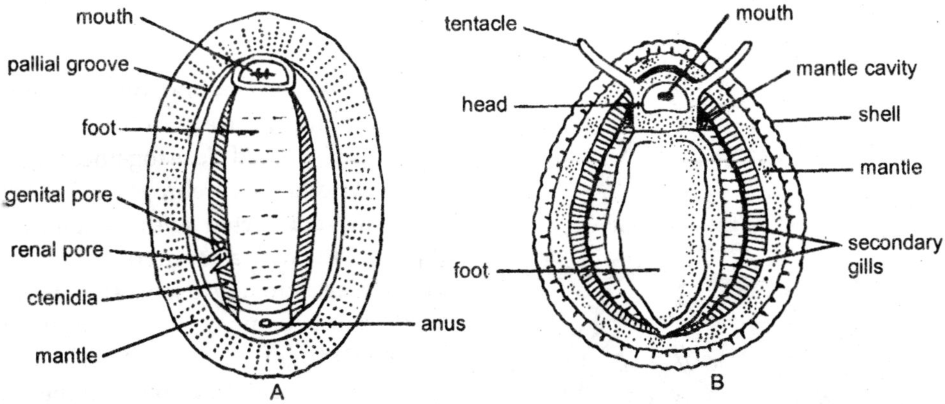

FIGURE 29.47

Modification of foot as anchoring organ. A. *Chiton barnesi*. B. *Patella rustica*. Ventral view

6. **Leaping organ** (Fig. 29.49). Modification of the foot to make it an efficient leaping organ the foot is bent upon itself in *Cardium*; elongated and compressed anteroposteriorly in *Trigonia*; curved with well developed protractor and retractor muscles in *Poromya*; in *Mytilus* the cylindrical foot acts as a springtail.
7. **Swimming organ** (Fig. 29.50). The foot is modified as a swimming organ in a number of molluscs. It is modified as a flat in ventral fin in *Carinaria*; a pair of fleshy lateral outgrowths, the parapodia, which act like fins in *Aplysia*; the largest part of the foot is modified as a pair of fins for active swimming in *Pteropods*; the parapodia are hollow and bear finlike outgrowths in *Oxygurus*; the metapodium is transformed into laterally flattened swimming lobe in *Atlanta*; oral arms* of **cephalopods** (Fig. 29.34) are adapted for swimming in addition to other functions.

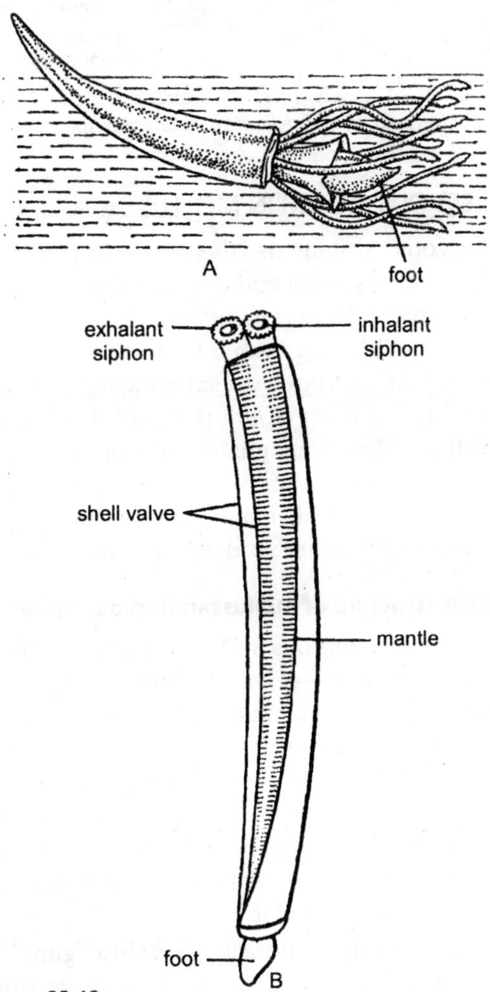

FIGURE 29.48

Foot for burrowing. A. *Dentalium variable*. B. *Solen radiata*

* Some authors do not consider the arms in cephalopods as modified foot and believe these are simply appendages of the head. According to them, only the funnel or siphon is modified foot. The funnel in certain circumstances acts as a rapid locomotory organ and the animal darts rapidly by forceful expulsion of water from the mantle cavity through siphon.

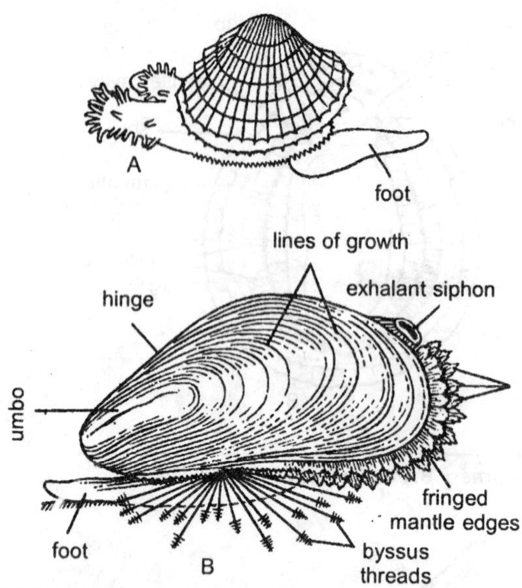

FIGURE 29.49

Foot for leaping. A *Cardium* sp. B. *Mytilus* sp.

NERVOUS SYSTEM IN MOLLUSCA

Nervous system is the structure which respond to stimuli and conduct the waves of excitement to nerve centres. The nervous system of Mollusca exhibit a graded two way evolution, (i) from a dorsal nonganglionated cerebral portion (*Chiton*) to eight pairs of well developed ganglia (Cephalopoda), and (ii) condensation of nerve ganglia to close aggregation of all major ganglia anteriorly indicating a higher grade of organisation.

Basic structure of molluscan nervous system

The central nervous system consists of three pairs of ganglia—cerebral, pleural and pedal. They are located far apart and joined by commissures and connectives. The basic plan with certain modification and better development is retained unaltered in the phylum, except gastropods where torsion, detorsion, exaggeration and elimination of certain organs have greatly altered it.

In **Polyplacephora** the cerebral ganglia are absent and the oesophageal nerve ring is with a thicker dorsal cerebral portion and a thinner ventral commissure (Fig. 29.51) in Aplacophora it is ladder-like with a single or double cerebral ganglia (Fig. 29.52); in **Scaphopoda** the cerebral, pleuals, pedal and visceral ganglia are paired. The cerebral ganglia are situated close together (Fig. 29.53); in **Pelecypoda** the cerebropleural ganglia are fused and connected with its fellow of the opposite side with a nerve cord, the two pedal ganglia are closely associated to form a single bilobed mass, the visceral ganglia are fused together (Fig. 29.54). A higher garde of organisation is found in the nervous system of **Cephalopoda**. The cerebral, pedal, pleurovisceral are of relatively large size and are closely aggregated together around the oesophagus, supported and protected by cranial cartilage (Fig. 29.34).

Streptoneury

In Gastropoda the nervous system exhibit a great deal of deviation from the general plan is Mollusca. The effect of 180° degree twisting of the visceropallium on the head-foot axis or torsion has profound influence on the

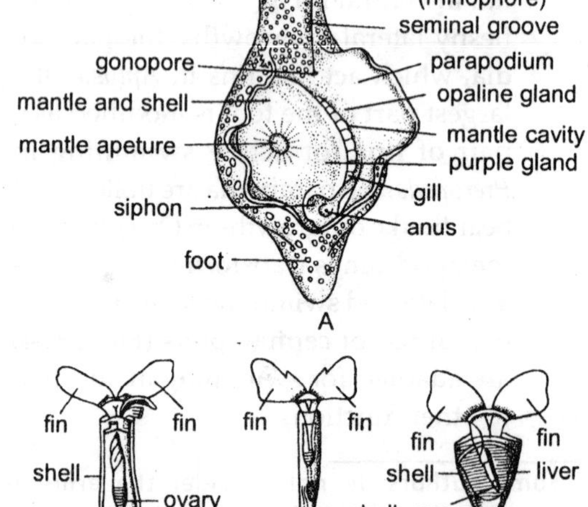

FIGURE 29.50

Foot for swimming. A. *Aplysia depilans. B-D. Teropods*

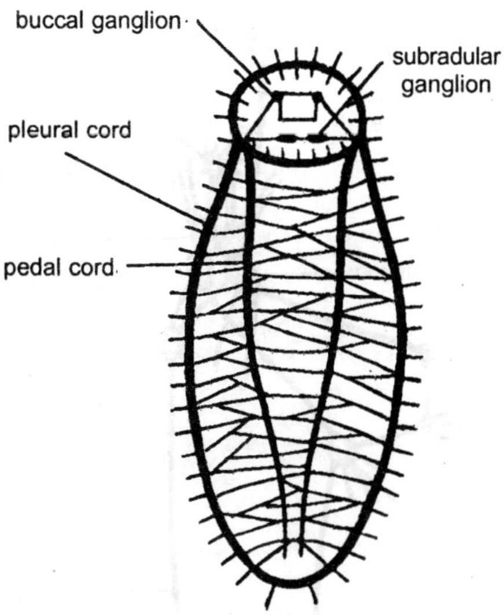

FIGURE 29.51

***Chiton* sp.** Nervous system

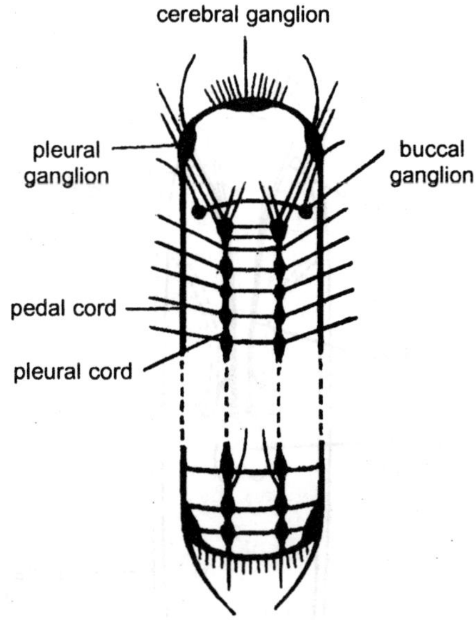

FIGURE 29.52

***Neomenia* sp.** Nervous system

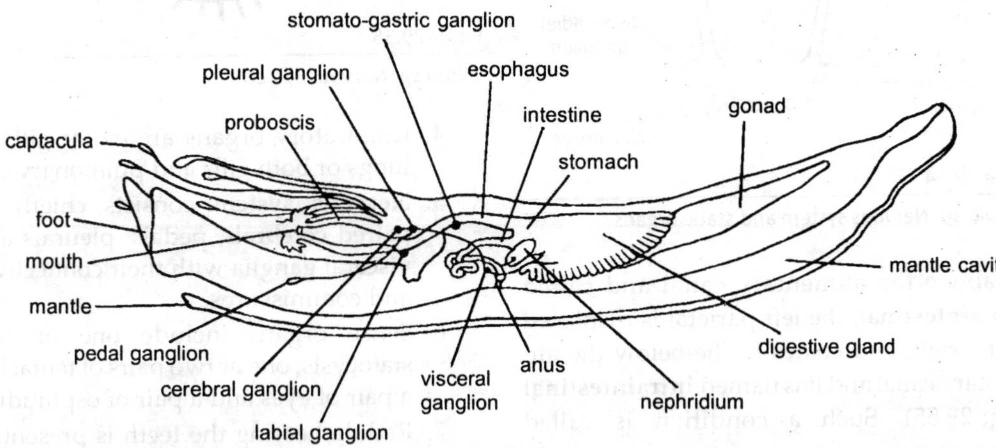

FIGURE 29.53

***Dentalium* sp.** Nervous system

nervous system in gastropods, particularly in prosobranch.

As in other groups of molluscs the paired, closely united cerebral ganglia are located over the gullet, a pair of nerve cords connecting them with a pair of laterally placed pleural ganglia and which are united together behind a median (or paired) visceral ganglion. In the course of the cords there is also a pair of parietal ganglia. A pair of pedal ganglia joined by a transverse commissure and joined to the cerebral and pleural ganglia by connectives. Buccal ganglia and a pair of ospharodial ganglia are present.

In most prosobranchs due to the torsion of the visceral mass, the palliovisceral connectives are twisted into a figure of '8' and the nervous system assumes a secondary asymmetry. The right parietal ganglion becomes displaced to the left and comes to

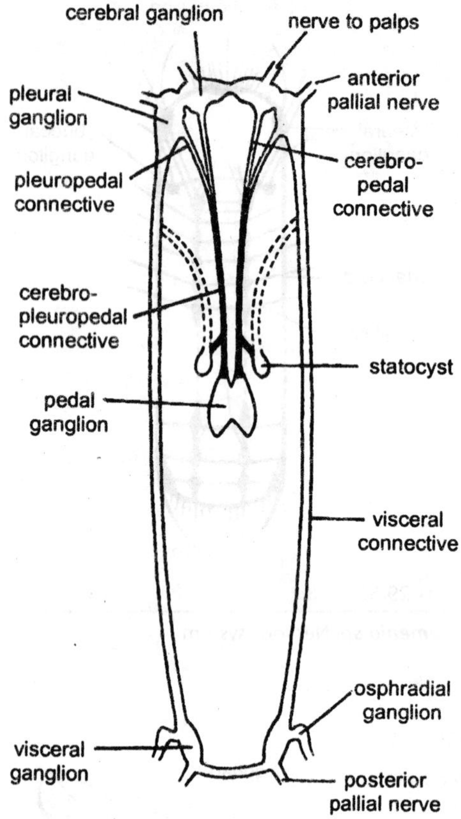

FIGURE 29.54

***Nucula* sp.** Nervous system and static organs

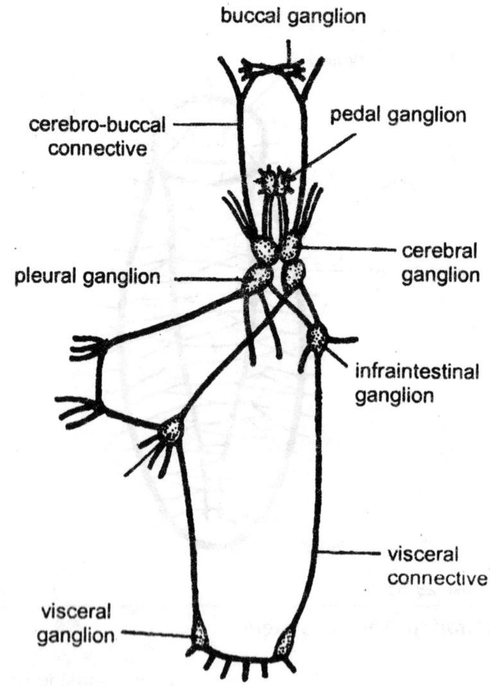

FIGURE 29.55

***Triton* sp.** Nervous system

lie above the alimentary canal and called **supraintestinal**; the left parietal is displaced to the right and comes to flie below the alimentary canal and it is named **infraintestinal** (Fig. 29.55). Such a condition is called *streptoneury*.

CLASSIFICATION: PHYLUM MOLLUSCA

Despite a basic similarity in morphology, molluscs display a bewildering diversity of forms.

1. The body is bilaterally symmetrical, at least in the early stages.
2. A muscular foot is usually present on the ventral surface.
3. Body wall forms a mantle which encloses a space, the mantle cavity and secretes a shell.
4. Respiratory organs are either gills or lungs or both gills and pulmonary sac.
5. Nervous system consists chiefly of paired cerebrals, pedals, pleurals and visceral ganglia with their connectives and commissures.
6. Sense organs include one or two statocysts, one or two pairs of tentacles, a pair of eyes and a pair of osphradia.
7. Radula bearing the teeth is present in the buccal cavity, except a few.
8. Muscles are usually unstriped.
9. Coelom restricted to pericardium.
10. The vascular system almost always well-developed. It is open and the heart lies within pericardial coelom—consists of a pumping ventricle and paired auricles receiving haemolymph from the gills or pulmonary sac (prosobranchs) or lung (stylommatophores).
11. Development with a metamorphosis and the larval stages are either trochopore or veliger.

The classification followed here is that of

Thiele (1929-1935), Wenz (1938-1960), and Morton (1958).

The phylum Mollusca has been divided into following classes.

Class 1. **Monoplacophora**

1. The body is bilaterally symmetrical and covered by a mantle.
2. The single dorsal shell is thin and subcircular.
3. The metamerism is apparent and not represented externally.
4. The coelom is well formed.
5. A flat creeping ventral foot is present.
6. The mouth and anus are situated at the anteromedian and posteromedian ends of the foot.

Distribution—Pacific coast.

Example: *Neopilina galatheae*.

Class 2. **Amphineura**

It is a small class of marine molluscs dating back from the Cambrain period.

1. The typically elongated form, retain bilateral symmetry having terminal mouth and anus.
2. The nervous system is primitive, with longitudinal pallial and pedal cords with cross anastomoses.

The two subclasses of the class Amphineura sharing a similar primitive nervous system are very different in appearance, habitat and development.

Subclass i. **Aplacophora** or **Solenogastres**

1. The body is elongated and worm-like and enveloped by the mantle.
2. The shell absent and the body is covered with a cuticular mantle enclosing spicules of calcified material.
3. A ventral longitudinal groove present, which is connected anteriorly with a ciliated groove and posteriorly with the cavity of the cloaca. In *Chaetoderma* the ventral "foot-groove" is absent and cloaca is a discrete bell-like mantle cavity which also contains a pair or a series of ctenidia.
4. Digestive glands are not distinct; salivary gland is usually present.
5. A ventral foot is absent.
6. The nervous system is ladder-like.
7. *Neomenia* is hermaphaoditic whereas *Chaetoderma* is dioecious.
8. Marine, found in shallow (*Neomenia*) and deep sea (*Chaetoderma*), feeding upon hydroids, corals, etc.

Examples: *Neomenia*, *Proneomenia*, *Chaetoderma*, etc.

Subclass ii. **Polyplacophora**

1. Marine molluscs with an elongated, bilaterally symmetrical flattened body and a broad foot ventrally.
2. The shell consists of eight separate pieces on dorsal surface and studded laterally with scales or spines to form a girdle.
3. The mantle bearing cuticular spicules covers at least a great part of the body.
4. The mouth lies centrally in a reduced head, the anal and excretory apertures are posterior.
5. Between the edge of the foot and the girdle on each side is a mantle groove into which project a number of gills or ctenidia.
6. The nervous system is ladder-like, with slightly developed ganglia.

Occur most commonly in the littoral and sublittoral rocky shores.

Examples: *Chiton*, *Lepidopleurina* etc.

Class 3. **Gastropoda**

1. Asymmetrical molluscs with a spirally coiled mantle and shell of one piece enclosing a corresponding visceral mass. The asymmetry of visceropallium is its fundamental features.
2. A well-developed head bearing eyes and tentacles are usually present.
3. The foot is behind the head and is typically a flat, creeping organ.
4. The buccal cavity contains an odontophore with a radula bearing rows of chitinoid teeth.

5. The respiratory organs consist either of one or two gills, or a gill and a pulmonary sac or a lung.
6. The nervous system consists of cerebral, pleural, buccal, pedal, visceral and abdominal ganglia with their connectives and commissures.
7. The kidney is usually single.
8. The sexes are either separate or united.
9. The larva passes through trochophore and veliger stages.

Subclass 1. **Prosobranchia**

1. The mouth of the shell can be closed by an operculum borne on the foot.
2. Head with a single pair of tentacles.
3. Mantle cavity opens anteriorly and contains two ctenidia anterior to heart.
4. Sexes separate, gonad solitary, opening in the right.
5. Larval stages are trochophore and veliger.
6. Primarily aquatic gastropods, retaining larval torsion and nervous asymmetry.

Upper Cambrain to recent, 25,000 living and 10,000 fossil species.

Examples: *Haliotis, Pila, Bellamya, Viviparus, Patella, Acmaea, Fissurella, Cypraea, Littorina, Murex, Buccinum, Conus*, etc.

Subclass ii. **Opisthobranchia**

1. Marine gastropods displaying reduction or loss of shell (shell sometimes internal).
2. Visceral complex usually incorporated into head-foot complex, leading to secondary external bilateral symmetry.
3. Respiratory structures—seconday gills, never ctenidia.
4. Untwisting and shortening of visceral loop and pleuroparietal nerve connectives due to detorsion is pronounced.
5. Hermaphrodite; larva is a veliger, Carboniferous to recent; 3,000 living and 300 fossil species.

Examples: *Aplysia* (sea hare); *Actean, Gastropteron* (sea slug), *Cliona, Bertholimia, Polycera, Doris, Dendonotus*, etc.

Subclass iii. **Pulmonata**

1. Shell single piece with a simple spiral or none.
2. Head with one or two pairs of tentacles and one pair of eyes.
3. Mantle cavity communicates with the exterior through an oval, closeable pneumostome.
4. Respiratory organ a lung, fused with the mantle in the adult. The mantle cavity acting as the respiratory chamber.
5. Hermaphrodite; gonad single, mostly ovoviviparous; development direct or with suppressed larval stages.

Upper Cambrian to recent; more than 5,000 living and 1,000 fossil species.

Examples: *Achatina, Helix, Limax, Artalimax, Agriolimax* (slug) *Gonaxis, Lymnaea, Planorbis, Ferrisia, Physa*, etc.

Class 4. **Scaphopoda**

1. Marine molluscs with an elongated worm-like body enclosed in a bilaterally cylindrical shell.
2. The shell is a curved tube open at both the ends, the opening being wider at the oral and narrower at the other end.
3. The foot is narrow and trilobed or with a terminal disc which can be protruded through the oral opening of the shell.
4. The mouth opens at the end of a short buccal tube, at the base of which is a circlet of tentacles.
5. Gills absent.
6. The odontophore is with a simple radula.
7. The sexes are separate; the reproductive elements pass out through the right excretory aperture.
8. A free-swimming trochophore larva is succeeded by a veliger.

Devonian to recent; 300 fossil and 100 living species.

Examples: *Dentalium, Entalina*, etc.

Class 5. **Pelecypoda** or **Bivalvia** or **Lamellibranchiata**

1. Bilaterally symmetrical molluscs with a laterally compressed body enclosed in a bivalved calcareous shell.
2. Shell usually symmetrical, with dorsal hinge and ligament, and closed by 1 or 2 adductor muscles.
3. Head is not distinct and without eyes and tentacles.
4. The foot is ventral and plough-shaped.
5. The gills one or two pairs (ctenidia or branchia), commonly plate-like.
6. Jaws and radula absent.
7. The mouth is bounded by two pairs of labial palps acting as lips.
8. The rectum pierces the pericardium and the ventricle, and opens in the exhalant siphon.
9. The heart consists of two auricles and a ventricle.
10. The nervous system consists of paired cerebral, pedal and visceral ganglia with their connectives.
11. Sexes are separate or united; development with a veliger or glochidiam larva.

Ordovician to recent, 11,000 living and 15,000 fossil species.

Examples: *Lamellidens, Unio, Anodonta, Mytilus, Teredo, Crassostrea, Ostrea* etc.

Class 6. **Cephalopoda**

1. Bilaterally symmetrical, free-swimming, marine molluscs.
2. The head is surrounded by a series of arms bearing suckers, which are modified part of the foot.
3. The rest of the foot forms a siphon on the ventral surface of the body.
4. The mantle encloses a large mantle cavity in which are situated the gills, the renal, reproductive and the anal apertures.
5. The shell may be absent or rudimentary; when present may be internal or external, undivided or divided internally into a series of chambers.
6. The head bears a pair of large, simple eyes.
7. A pair of horny jaws and an odontophore with a radula are present.
8. In majority, there is an ink gland opening in the rectum.
9. The nervous system is highly developed; the principal nerve ganglia are aggregated around the oesophagus and a cartilaginous skeleton supports and protects the nerve centre.
10. The sexes are separate and development is direct.

Subclass i. **Nautiloidea** (Tetrabranchia)

1. Shell external, many-chambered, siphunculate and coiled or straight.
2. Head with numerous retractile tentacular appendages which lack suckers.
3. Siphonal funnel of two separate folds.
4. Two pairs of ctenidia, osphradia, renal organs, and auricles.
5. Ink gland absent.
6. Eyes are open vesicles without cornea or lens.

Only one living genus with three species of *Nautilus* in eastern Pacific and Indian Oceans at depths to 560 metres; 2,500 fossil species.

Examples: *Nautilus pompilius*.

Subclass ii. **Coleoidea** (Dibranchia)

1. Body cylindrical or globose, often with fins.
2. Shell internal and reduced or none.
3. Tubular siphon.
4. The head bears eight nonretractile tentacles which are provided with suckers in two rows.
5. One pair of ctenidia, osphradia, auricles and renal organs present.
6. Eyes with lens.
7. Ink gland present.

Examples: *Loligo, Sepia, Octopus, Argonauta,* etc.

30 PHYLUM ECHINODERMATA

The Phylum Echinodermata (*Gr. Echinos* = *spiny* + *derma* = dermis, spiny-skinned) is a very distinct group, which includes exclusively marine animals of pentamerous biradially or radially symmetrical forms like star fishes or sea stars, sea urchins, sea cucumbers, sea lilies having a typical water vascular system with tube feet. They are older groups of animals known from fossils of Cambrian period.

Due to their radial symmetry once they were grouped with Coelenterata as the 'Radiata' but due to their higher well-up organization amongst invertebrates, they are now placed at the end point of invertebrates.

Echinoderms show some chordate-like characteristics, such as (1) the coelom forms as outpushings of the primitive gut cavity or enteron (enterocoel) not by a split in the mesoderm (schizocoel) as in annelids and molluscs; (2) the embryonic mouth is formed as an ectodermal inpocketing (stomodaeum) connecting the endodermal stomach, and the blastopore becomes the adult anus; (3) the egg cleavage is radial (not spiral) as in many invertebrates and usually indeterminate; (4) presence of a mesodermal endoskeleton in addition to an ectodermal one as in other invertebrates; (5) the larva resembles that of some hemichordates (Tornaria larva) and for this reason echinoderms, hemichordates and chordates are often grouped together as the **deuterostomes**.

Some authors believe that echinoderms have regressed from a more advanced and active form, since the larvae are bilaterally symmertical and free-living, whereas the adults are radially symmetrical and of sedentray habit with reduced circulatory and nervous system. Radial symmetry and lack of a 'head' are the adaptation for sedentary habit.

About 6,000 living species and 13,000 fossilised species are known in all depths of the sea.

ASTERIAS sp. (Star fish)

The starfish is a slow moving, marine animal living in shallow sea. It is a predator and also a scavenger. *Asterias* belongs to subclass Asteroidea, class Stelloroidea, subphylum Asterozoa, phylum Echinodermata.

⊠ External features

1. The body is star-shaped, consisting of a central part, the **central disc** and five symmertically arranged **arms or rays**.
2. The arms are broad at the base and taper slightly towards the extremities.

The two arms between which the madreporite opens are **bivium** and the other three are known as **trivium.**

3. The body has two surfaces—the dorsal or **aboral** and the ventral or **oral.** The aboral surface is convex while the oral surface is flat (Fig. 30.1).
4. The body is covered by an **exoskeleton.** The hard integument contains numerous calcareous plates or **ossicles.**
5. A five-rayed aperture called **actionostome** is present at the centre of the oral surface.
6. Five grooves, the **ambulacral grooves** run out radially from the actinostome, one along the middle of each arm, up to its extremity.
7. Two or three rows of movable calcareous spines, the **ambulacral spines** border the ambulacral groove.
8. A number of short spines, supported by ossicles buried in the integument, are arranged in irregular rows on the dorsal surface, parallel to the long axes of the rays.
9. A number of minute pores, the **dermal pores,** are present in the soft interspaces between the ossicles.
10. A very small, soft, filiform process, the **dermal branchus** or papulla, respiratory in function, projects through the dermal pore. They are retractile.
11. The **anus** is a narrow aperture, close to the centre of the aboral surface.
12. The **madreporite** is a flat, circular plate situated midway between the bases of two of the arms (bivium).
13. **Pedicellariae** are present on both the oral and aboral surfaces. They are of both large and small sizes.
   a. A pedicellaria consists of a flexible stalk and three calcareous components—**one basilar plate** and two **jaws.**
   b. Jaws are attached to the basilar plate and directed forward.
   c. Pedicellariae are of two types (Fig. 30.2).
   i. **Straight pedicellaria.** The jaws are more or less straight and, when closed, meet almost throughout the whole length.
   ii. **Crossed pedicellaria.** Basal portion of the jaws curved and cross each other.

   Pedicellariae are organs of offence and defense.

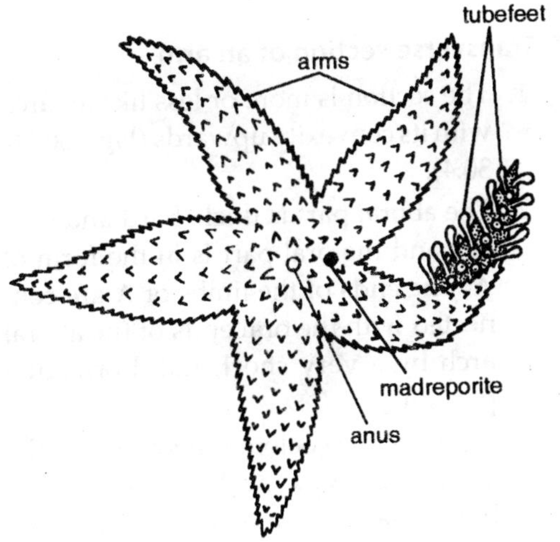

FIGURE 30.1A

Asterias rubens. Aboral (Dorsal view)

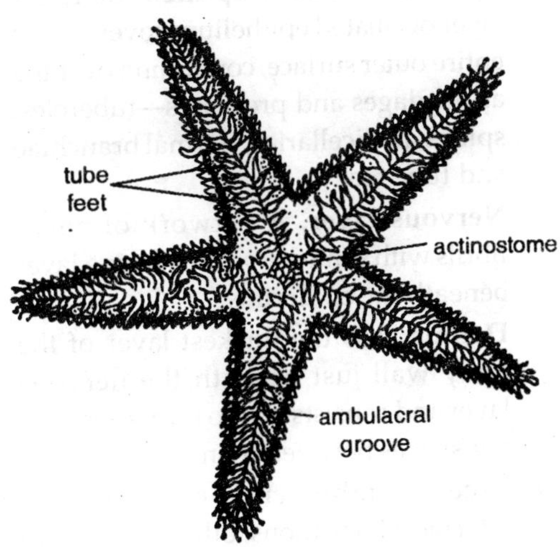

FIGURE 30.1B

Asterias rubens. Oral (Ventral view)

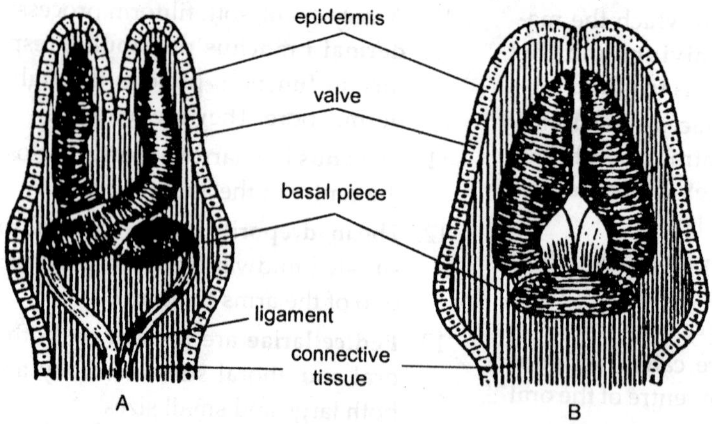

FIGURE 30.2
***Asterias* sp.** Pedicellaria. A. Crossed pedicellaria, B. Straight pedicellaria

14. The tube feet or **podia** are soft, tubular bodies ending in sucker-like extremities. They are present in two double rows in each of the ambulacral groove.
15. A hollow, noncontractile **terminal tentacle** is present at the tip of each arm. A small, bright red spot called the **eye** is located at the base of each tentacle.
16. A median process known as **tentacle**, olfactory in function, is present dorsal to the eye.

☒ Body wall

Following structures take part in the formation of the body wall (Fig. 30.3).

1. **Epidermis** or **deric epithelium.** It is a layer of ciliated epithelium covering the entire outer surface, continuing over the appendages and processes—tubercles, spines, pedicellariae, dermal branchiae and tube feet.
2. **Nervous layer.** A network of nerve fibrils with some nerve cells form a layer beneath the epidermis.
3. **Dermis.** It is the thickest layer of the body wall just beneath the nervous layer and consists of two layers containing spaces between them.

a. Except ambulacral ossicles and interradial partitions, other ossicles are located in the outer layer.

b. The inner layer consists of smooth muscle fibres, arranged in two layers—an outer circular and an inner longitudinal muscle layer.

4. **Coelomic epithelium or parietal layer.** Made of cuboidal, flagellated cells of mesodermal origin and forms the innermost layer of the body wall.

☒ Coelom

A large space lined by coelomic epithelium and filled with **coelomic fluid**.

1. Alimentary canal with its appendages, madreporite canal, ampullae and gonads are lodged in the coelom.
2. The coelomic epithelium covering the organs lodged in the coelom is known as **visceral** or **peritonial layer**.

☒ Transverse section of an arm

1. The section is more or less like an arch with its convexity upwards (Figs. 30.3B, 30.4).
2. The aboral part is thick, hard and convex and the oral part is in the form of Λ. The ends of the limbs of Λ are connected with the oral ends of the aboral arch by a very short, flat, horizontal portion.
3. A part of the coelom is enclosed by the aboral convex arch and the flat horizontal portion on the oral surface.
4. Below this horizontal portion lies the ambulacral groove, supported by two rows of elongated ossicles, the ambu-

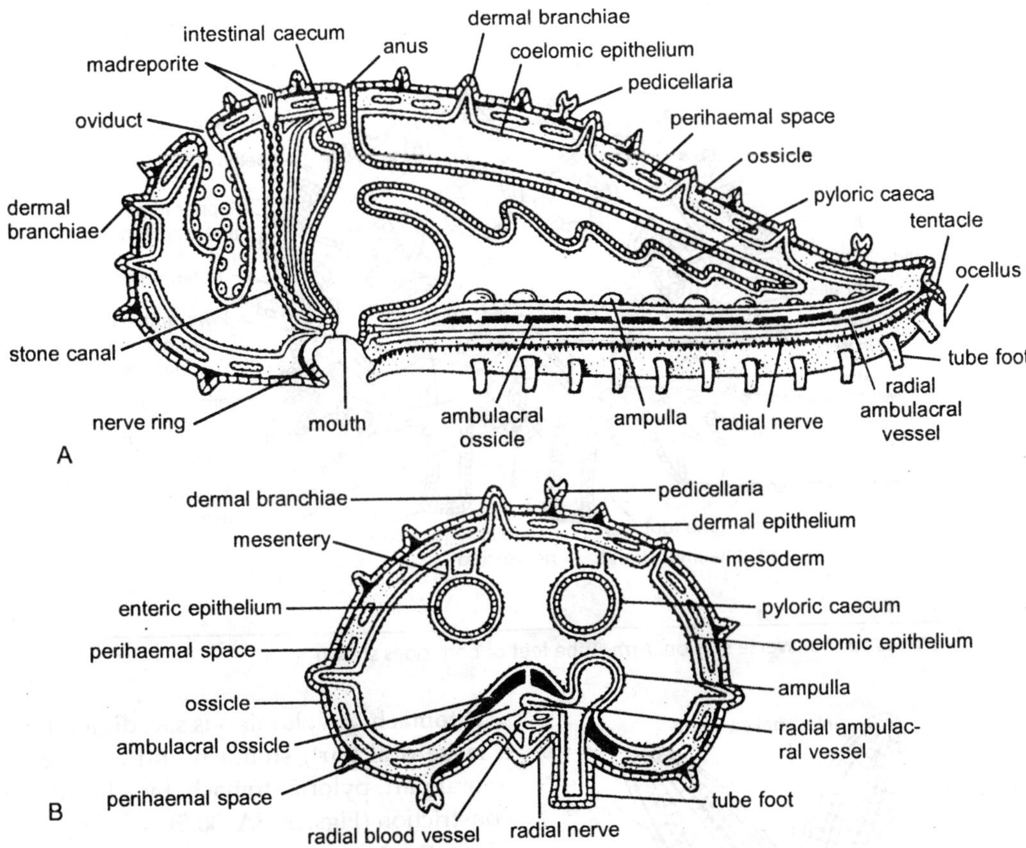

FIGURE 30.3

***Asterias* sp.** A. Vertical section. Right side passing through a **radius**, left side through an **interradius**. B. Transverse section of an arm

lacral ossicles which meet at the summit of the groove.

5. At the edges of the groove a row of ossicles support ambulacral spines and tubercles.
6. Between the ambulacral ossicles of each row are a series of oval openings, the ambulacral pores, through which the tube feet project out.
7. The tube feet are cylindrical structures with suckers at their free ends and ending in swellings or the ampullae at the other end, lodged in the coelom.
8. The tube feet are connected with the transverse vessels, which are branches of a radial ambulacral vessel.
9. Just above the epidermis is the radial nerve of the epidermal system.
10. Perihaemal vessel is present just above the epidermal nervous system.
11. The aboral arch is supported by a number of irregular ossicles and is perforated by numerous small dermal pores through which the dermal branchiae project.
12. The aboral wall is thick and provided with narrow spaces in the mesoderm.

⊠ Digestive System

The digestive system consists of an alimentary canal and digestive glands (Fig. 30.5).

Alimentary canal

It extends from oral to aboral surface and divisible into oesophagus, cardiac stomach, pyloric stomach and intestine.

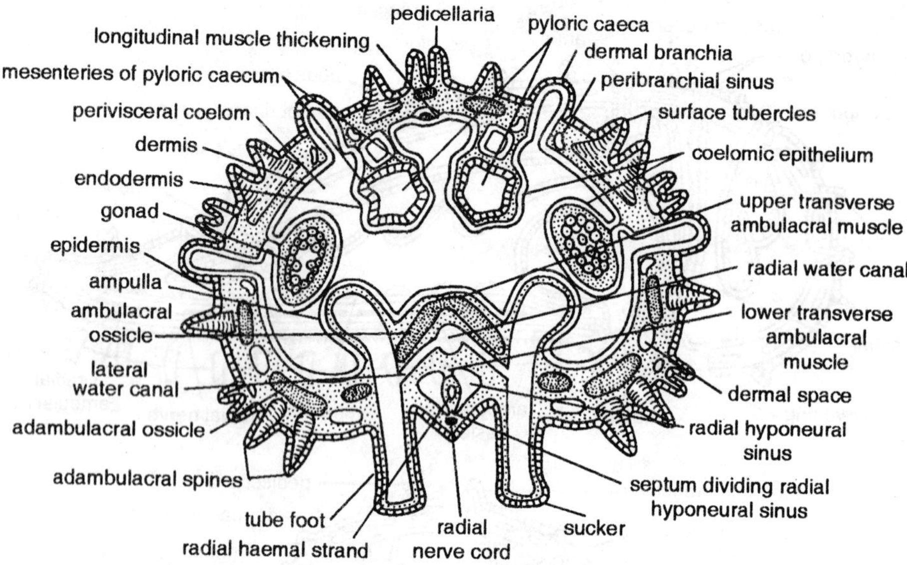

FIGURE 30.4

***Asterias* sp.** Transverse section. Arm. Tube feet of both sides shown

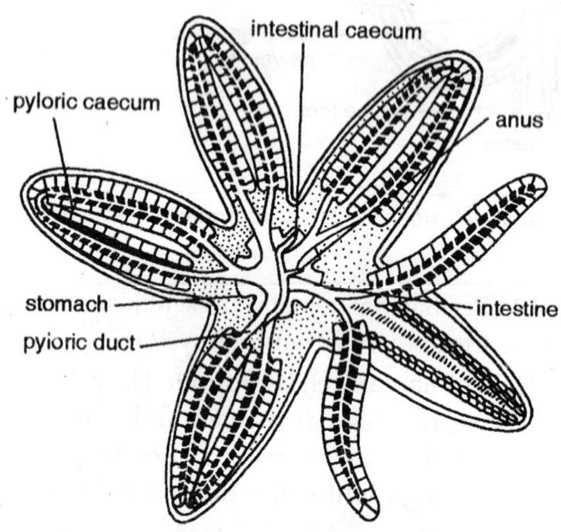

FIGURE 30.5

***Asterias* sp.** Digestive system. Aboral view

Mouth. A five-rayed aperture, also known as **actinostome**, at the centre of the peristomial membrane of the oral surface. It is provided with sphincter muscles and radial muscle fibres, which help in retraction and dilatation. Mouth opens into the oesophagus.

Oesophagus. A very short, vertical tube, opening aborally in the stomach.

Stomach. A voluminous sac, divided into a large oral part, stomach and a flattened aboral part, pyloric stomach, by a horizontal constriction (Figs. 30.3A, 30.5).

a. **Cardiac stomach.** A muscular, five-lobed sac, each of the lobes is opposite to one of the five arms. The wall is greatly folded. The cardiac stomach can be everted through the mouth opening and retracted by contraction of special retractor muscles.

b. **Pyloric stomach.** A small, flat, pentagonal chamber, each angle being drawn out to form a pair of large appendages, the **pyloric caeca.**

i. A cylindrical canal or **duct** connected to the pyloric stomach, bifurcates to form two hollow stems extending to near the extremity of the arm.

ii. Each hollow stem sends two series of short, lateral branches, each bearing a number of bladder-like, glandular pouches.

Intestine. A short, narrow, five-sided tube, runs aborally to open in the **anus**, close to the **madreporite** on the aboral surface.

Two hollow, branched **intestinal caeca** arise from the intestine.

⊠ Digestive glands

The pyloric caeca are considered as digestive glands. The walls of the caeca are glandular and secrete digestive fluid, similar to the secretion of vertebrate pancreas.

Feeding and digestion

Asterias feeds on small animals and bivalve mollucs.

Feeding on small animals

1. By the contraction of the muscles of the body wall, the whole of the cardiac stomach is everted through mouth and the prey is wrapped by it.
2. The cardiac stomach with the prey is retracted into the interior by the contraction of special retractor muscles, arising from the sides of the ambulacral ridges.

Feeding on bivalve molluscs

1. Asterias creeps over the clam, hold it with tube feet and turn the prey to bring the free margins of the shell valves, close to its mouth.
2. The body is arched, holds itself strongly with the substratum by distal tube feet, the more proximal tube feet firmly hold both the shell valves and the cardiac stomach is everted.
3. With the contraction of the muscles in the arms, the tube feet gripping the valves maintain a steady pull for a long time.
4. Since muscles of the clam are unable to maintain a continuous state of contraction, the adductor muscles become fatigued and finally relax and the shell opens.
5. The everted cardiac stomach is inserted into the mantle cavity and digestive enzymes are poured, which partially digest the visceral organs of the clam.
6. The stomach with the partially digested molluscs body is retracted and digestion is completed there.
7. The undigested food is egested directly from the mouth or passed through the intestine and thrown out through the anus.

⊠ Water Vascular (Ambulacral) System

A system of tubes that courses along the inner surface of the ambulacra from the oral region and extends to their tips is collectively known as water vascular system, which is peculiar to the echinoderms. The tubes are filled up with a watery fluid and the system helps in locomotion.

The water vascular system first appeared possibly as hollow tentacles, in each of which the coelom is extended subsequently (Fig. 30.6). It consists of:

1. The **actinostome**, a five-rayed aperture is situated at the centre of the oral surface and from it radiate out five narrow **ambulacral grooves**, one along the middle of the oral surface of each arm.
2. Situated near the anus on the aboral surface, between the bases of two arms is a nearly circular, flat, calcareous plate—the **madreporite**—the surface of which is provided with numerous narrow radiating grooves. A large number of narrow pores are present in the grooves. There may be 200 pores and pore canals. Each pore leads into a pore canal in the madreporite. These canals unite to form collecting canals, which open into an **ampulla** beneath the madreportie.
3. From the ampulla runs downward near the border of the mouth a S-shaped cylinder, the **madreporic** or **stone canal**. The walls of the canal are supported by a series of calcareous rings.
4. The stone canal opens into a five-sided ring-like canal, the **ring vessel** or water

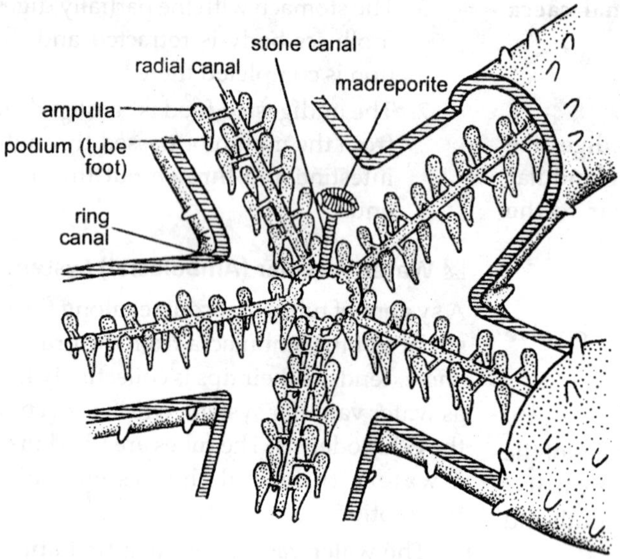

FIGURE 30.6

***Asterias* sp.** Ambulacral system. Aboral view

ring of the ambulcral system, round the mouth (Fig. 30.7).

5. The ring vessel gives off five **radial canals**—one for each arm—which runs along the arm, and lies within the arm beneath the shelter of the rafter-like ossicles of the ambulacral groove.
6. From each radial canal arise paired, short, transverse tubes arranged in two rows, and the two ends of a transverse tube end at the bases of two opposite **tube feet**.
7. To its inner side, the ring canal gives off irregular glandular bodies, the **Tiedemann's vesicles**, interradial in position. A pair of vesicles are located by the sides of each radial canal but only one by the side of that radial canal near which the madreporic canal opens. The Tiedemann's vesicles are supposed to give rise to the amoeboid cells of the water vascular system.
8. Each tube foot consists of a swelling or **ampulla** and a tubular portion, the terminal end of which bears a **sucker**.
9. The walls of the tube feet are muscular and valves to regulate the flow of water are present at the junctions of tubes and ampulae.
10. When extended, the tubes come out through the ambulacral pores.
11. The various parts of the vessels forming the ambulacral system have developed fairly strong muscular walls.

⊠ Mechanism of locomotion

Due to contraction of ampullae the water is forced to the tube feet, which become stretched and come out through the ambulacral pores. The suckers of a set of tube feet are firmly fixed on the support and the muscular walls being to contract. The water flows back to the ampullae and the animal is slightly drawn forward. The suckers are released and the process is repeated. All the tube feet do not perform the same work at a time. When one set is drawn, another set remains fixed, and vice versa.

The backflow of water to the radial canal is prevented by valves. The water which is lost by diffusion, etc., is compensated by drawing water from outside through the porous madreporite.

⊠ Respiratory System

The respiratory organs are papullae or gills and tube feet.

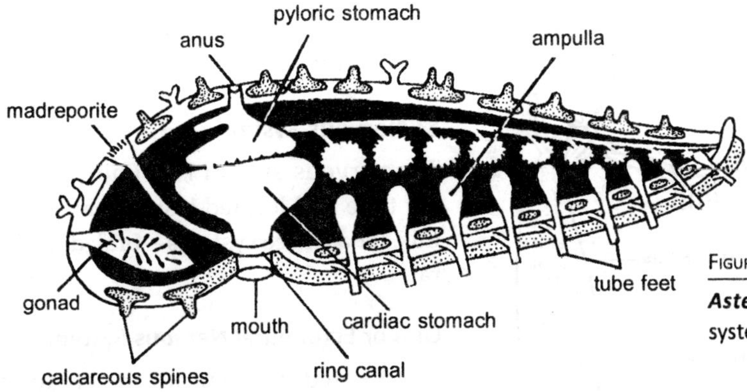

FIGURE 30.7

***Asterias* sp.** Ambulacral system. Aboral view

Papullae

These are the chief respiratory organs.

1. A papulla is a simple contractile, transparent, hollow outgrowth of the body wall on the aboral surface..
2. The epithelium over both the outer and inner surfaces are ciliated.
3. The lumen of a papulla is in direct communication with the coelom.
4. The cilia help in the movement of coelomic fluid and creating respiratory current.
5. Gaseous exchange takes place between the sea water and body fluid in the gills.

Tube feet

The tube feet are thin-walled and richly vascularised. A considerable amount of gaseous exchange occurs in the tube feet.

☒ Haemal System or Blood Vascular System

A system of intercommunicating channels enclosing coelomic spaces consitute haemal system (Fig. 30.8). It consists of:

1. **Oral haemal ring.** A fine channel in the form of a ring around the mouth, running in the septum dividing the hyponeural ring sinus.
2. **Radial haemal sinus.** Five in number, arise from the oral haemal ring and one runs into each arm. It gives off branches to the tube feet.
3. **Aboral haemal ring.** Located on the aboral side and sends branches to the gonads.

   The oral and aboral haemal rings are communicated by an ascending **haemal plexus** in the axial gland.
4. **Gastric haemal tuft.** Two, open in the plexus of the axial gland.
5. **Pyloric haemal channels.** About twenty in number. They are connected with the axial gland and haemal plexus through gastric haemal tufts. The pyloric caeca receive pyloric haemal channels.

Coelomocytes are present in the coelomic fluid of the haemal channels.

☒ Axial Complex

A coelomic cavity enclosing the stone canal and axial gland constitute axial complex (Fig. 30.9).

1. The stone canal is enclosed in a wider canal, the axial sinus, which is a part of the coelomic cavity.
2. The **axial sinus** runs nearly vertical, the oral end opening in the internal division of the oral ring sinus and the aboral end approaches close to the aboral ring sinus but does not open into it.
3. The axial sinus communicates aborally with the stone canal and opens to the exterior through certain pores in the madreporite.
4. The **axial organ** is a fusiform body, folded in a complicated manner, in which both the outer and inner surfaces take part.

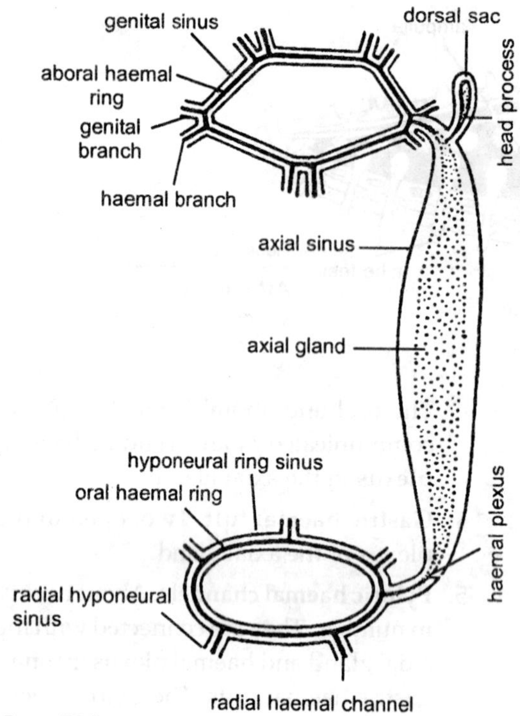

FIGURE 30.8

***Asterias* sp.** Axial complex with haemal system

5. Strands of lacunar tissue are present in the axial organ, which is intimately related with the haemal system.
6. Essentially, the axial organ is a **genital stolon**. Aborally, it is continuous with the **genital rachis**, which is in the form of a ring and gives off branches to gonads.
7. Reproductive cells develop in the aboral end of the axial organ and move to the gonads through genital rachis and its branches.

☒ Exretory System

True excretory organs are absent.

1. The nitrogenous wastes formed in the tissues diffuse into the coelomic fluid and from there to outside through the walls of the gills, tube feet and intestinal caeca.
2. Coelomocytes also extract nitrogenous wastes from the coelomic fluid and pass the same out of the body through papullae or gills.

☒ Nervous System

The nervous system is primitive and nerve-net type and located just beneath the epidermis. It has three systems—oral, deep and aboral.

Oral or Ectoneural Nervous System

It is located just beneath the epidermis and constitute the main part of the nervous system. It consists of:

1. **Nerve ring.** A circumoral pentagonal ring around the mouth in the peristomial membrane. It gives off five radial nerves and also innervate the peristomial membrane and oesophagus.
2. **Radial nerve.** A radial nerve arises at each radius of the nerve ring, runs the whole length of the arm in the ambulacral groove, below the radial canal and terminates in a sensory cushion on the aboral side of the terminal tentacle. The nerve is a thick band and V-shaped in cross-section. It sends branches to the tube feet and is continuous with the subepidermal nerve plexus of the body wall.

Deep or Hyponeural Nervous System

It occurs in the form of a nervous layer, known as **Langer's nerve**.

1. The deep nervous system is located in the lateral part of the oral wall of the hyponeural sinus, beneath the coelomic epithelium, lining the sinus.
2. Langer's nerve extends as five interradial nerve thickenings which innervate the muscles of the corresponding arms.

Aboral or Coelomic Nervous System

At the outer ends of the ambulacral ossicles, the subepithelial nerve plexus gives rise to **marginal nerve** cords on two sides of the arm.

1. **lateral motor nerves.** Arising from the marginal nerve cord proceed aborally to the coelomic epithelium.
2. The nerves from **plexus** beneath the entire coelomic lining and innervate muscles of the body wall and gonads.

⊠ Receptors and Sense Organs

The sense organs are eyes, tentacles and neurosecratory cells.

Eyes

Five pigmented eye spots, one at the base of each, terminal tentacle, towards the oral side. Each eye consists of a varying number of ocelli (Fig. 29.9B) ranging from 80 to 200.

1. Each ocellus is a cup-shaped structure, covered externally by the cuticle.
2. The epidermis beneath the cuticle may form a lens, which may be absent in some cases.
3. The wall of the cup consists of short and stout epidermal cells, bearing **pigment** distally.
4. Elongated **retinal cells** with a bulbous enlargement at the free ends are present between the pigment cells.
5. The basal end of the retinal cells are in continuation with the underlying radial nerve.

The ocelli are light sensitive and the star fish can detect changes in the light intensity.

Terminal tentacles

Tactile organs (Fig. 30.9A), help to survey surroundings during locomotion and also in detecting food.

Neurosecretory cells

The cells are numerous and scattered all over the body, being crowded in the epidermis, ambulacral regions, suckers of the tube feet, pedicellariae, terminal tentacles, spines, etc.

1. A neurosecretory cell is slender and fusiform in shape.
2. The proximal end is connected with the subepidermal nerve plexus and the distal thread-like end goes upto the cuticle.

The cells are tactile and **chemoreceptors**.

⊠ Reproductive System

The starfish is unisexual but sexual dimorphism is absent. The gonads—**testes** or **ovaries**—are five pairs and macroscopically similar.

1. A pair of gonads are placed laterally in the proximal part of each arm, being attached to the aboral body wall, near the interbranchial septum.

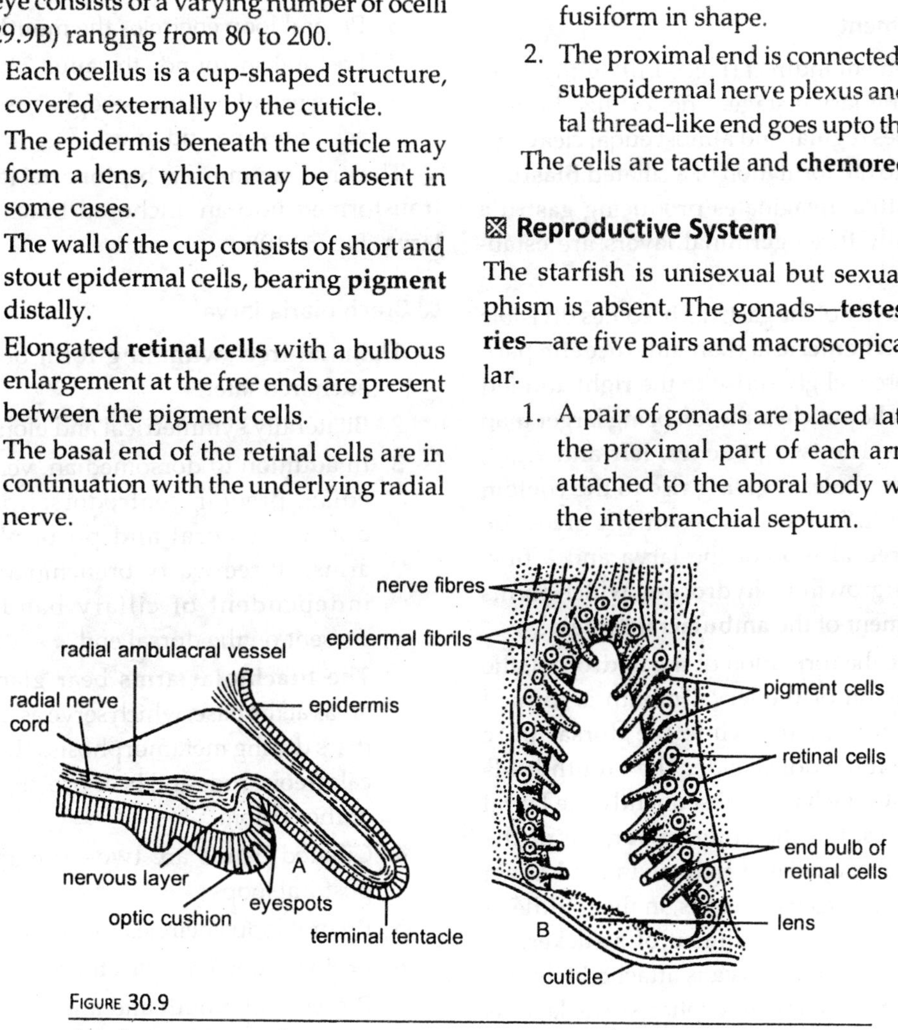

FIGURE 30.9

***Asterias* sp.** A Terminal tentacle and eye spots (section). B. A single eye spot or ocellus (vertical section)

2. The gonad is enclosed in a coelomic sac. The germ cells are located in a connective tissue matrix of the germinal epithelium, lining the gonad.
3. **Gonoduct** very short, ciliated, arises from the gonad, close to its attachment with the interbranchial septum. It opens to the exterior through a **gonopore** on the aboral surface, at the angle of the arms.

Fertilization

Fertilization is external. Sperms and ova are released in sea water.

Development

Development indirect (Fig. 30.10) with a free-swimming larval stage. The fertilized ovum undergoes regular and almost euqal cleavage, leading to the formation of a ciliated **blastula.** The blastula invaginates producing **gastrula** and finally three germinal layers are established.

The cavity of the gastrula becomes divisible into an **archenteric** and an **enterocoelic** part. The enterocoel gives rise to the right and left enterocoelic pouches, the left being larger than the right. The two pouches, left and right, ultimately coalesce giving rise to the **coelom** of the adult. The coelom eventually gives rise to the **preoral lobe** of the larva and a fine, lobed outgrowth or **hydrocoel**, which forms the rudiment of the **ambulacral system**.

Before the formation of the hydrocoel, the **stomodaeum** or the larval mouth is formed on the ventral side, while the **dorsal pore** appears on the dorsal side. The **mouth** leads into the **stomach** and subsequently the larval anus closes. On the anterior surface of the preoral lobe, a ciliated flattened area known as the larval organ develops, in the middle of which appears the rudiment of a sucker, by means of which the larva is attached to some support, prior to metamorphosis. The larva is able to move by this time and known as bipinnaria larva (Fig. 30.10).

⊠ Bipinnaria larva

1. Free swimming, develops in 2-7 days.
2. Bilaterally symmetrical and somewhat angular in shape.
3. Arms or projections of the body are 7 in number—dorsomedian, ventro-median, preoral, centrodorsal, posterodorsal, postoral and posterolateral, (only in *Bipinnaria papillata* the arms are ciliated).
4. Ciliated bands two, **preoral** and **postoral** and they are continuous.
5. **Anus** bordered by ciliary bands.
6. Preoral loop encircles the **mouth**.
7. Postoral surrounds the **anus**.
8. Presence of mouth, oesophagus, stomach, intestine and anus.

The free-swimming bipinnaria larva is transformed into an anchored brachiolaria larva (Fig. 30.10).

⊠ Brachiolaria larva

1. Initial free-swimming form becomes anchored later.
2. Bilaterally symmetrical and elongated.
3. In addition to dorsomedian, ventromedian, preoral, centrodorsal postero dorsal, postoral and posterolateral arms, three warty branchiolar arms independent of ciliary bands, are present on the dorsal end.

   The **brachiolar arms** bear glandular areas at the base, which serve as fixation discs during metamorphosis. They lack calcareous rods and bear prolongations of the coelom.
4. Ciliated bands are two—preoral and postoral loops.
5. Preoral loop encircles the mouth.
6. Postoral loop encircles the anus.
7. Presence of mouth and anus.

The brachiolaria larva in most Asteroidea are formed from auricularia larva.

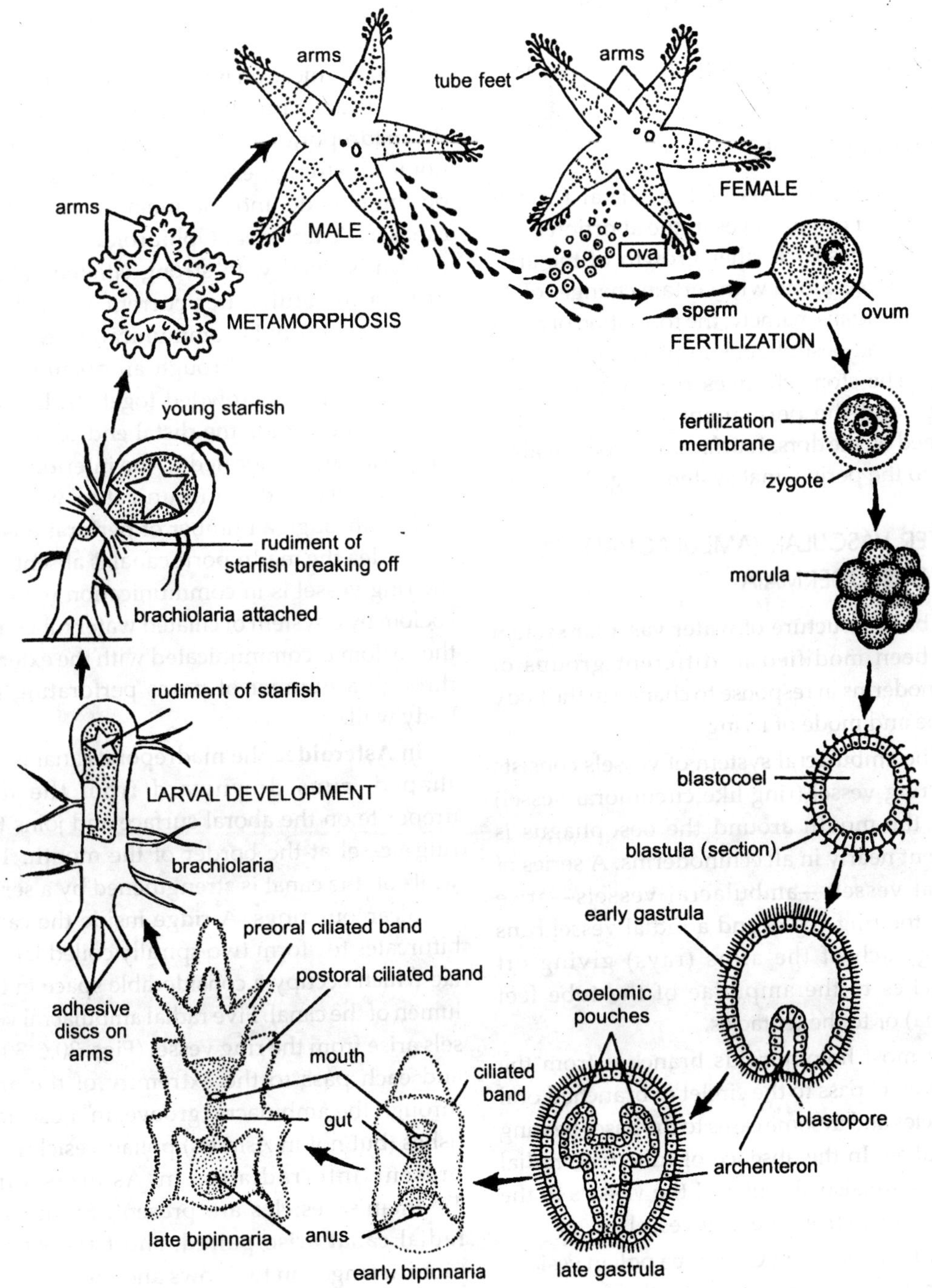

FIGURE 30.10

***Asterias* sp.** Life history stages

The hydrocoel ultimately continues in the radial canals and ring vessel of the ambulacral system, while the madreporic canal develops from the coelom.

Metamorphosis

The metamorphosis of the larval form to young star fish involves in the abortion of the preoral lobes by further growth of the arms and tube feet along with certain changes of the internal organs, namely, the formation of a new mouth and oesophagus in place of the larval ones. The stomach gives rise to diverticular caeca and the permanent anal opening is formed on the dorsal surface. The coelom gives rise to the perihaemal system (Fig. 30.10).

WATER VASCULAR (AMBULACRAL) SYSTEM IN ECHINODERMATA

The basic structure of water vascular system has been modified in different groups of echinoderms in response to change in the body shape and mode of living.

The ambulacral system of vessels consists of a ring vessel (ring-like circumoral vessel) near the mouth around the oesophagus is present nearly in all echinoderms. A series of radial vessels—ambulacral vessels—arise from the ring vessel and a radial vessel runs along each of the arms (rays) giving off branches to the ampullae of the tube feet (podia) or to the tentacles.

In most holothuroids branches from the ring vessel pass to the circlet of branched oral tentacles and in some cases to the bases bearing ampullae. In the absence of tube feet, radial vessels are also absent and the vessels to the tentacles run from the ring vessel.

In all cases except Crinoidea polian vesicles are present and connected with the ring vessel. The Tiedman's vesicles are characteristic of the Asteroidea.

In all cases except the Crinoidea and the majority of Holothuroidea a communication exists between the ring vessel and the surrounding water through a madreporic canal. In Asteroidea, Ophiuroidea and Echinoidea the communication is through a madreporite, which bears a number of pores. The walls of the madreporic canal in Asteroidea and Echinoidea is strengthened by calcareous ossicles on the outside and the inner surface of the canal in Asteroidea, Ophiuroidea and Echinoidea is lined with strong ciliated cells to drive a forceful water current inwards. In some holothuroids communication with the exterior is simply through an opening or a number of pores crowded together. In some other holothuroids the distal end of the madreporic canal is located in the interior of the body cavity and communicated with the coelom through a number of perforations. In Crinoidea the madreporic canal is absent and the ring vessel is in communication with the coelom by a system of ciliated water tubes and the coelom is communicated with the exterior through a number of pores perforating the body wall.

In **Asteroidea** the madreporic canal is 'S'-shaped, runs downward from the madreporite on the aboral surface and joins the ring vessel at the border of the mouth. The walls of the canal is strengthened by a series of calcareous rings. A ridge inside the canal bifurcates to form two spirally coiled lamellae which occupy a considerable space in the lumen of the canal. Five radial ambulacral vessels arise from the ring vessel (Figs. 30.6, 30.7) and each pass to the extremity of the arm through the ambulacral groove. In most starfishes (but not in *Asterias*) polian vesicles are present interradially. In Asterias nine Tiedman's vesicles are present. From each radial canal arise paired, short transverse tubes arranged in two rows and the two ends of a transverse tube terminate at the two opposite tube feet. A tube foot bears a muscular ampulla and a tubular portion, the terminal end of which bears a sucker. The tube feet are organs of locomotion, anchoring and also help in feeding in Asteroidea and Ophiuroidea.

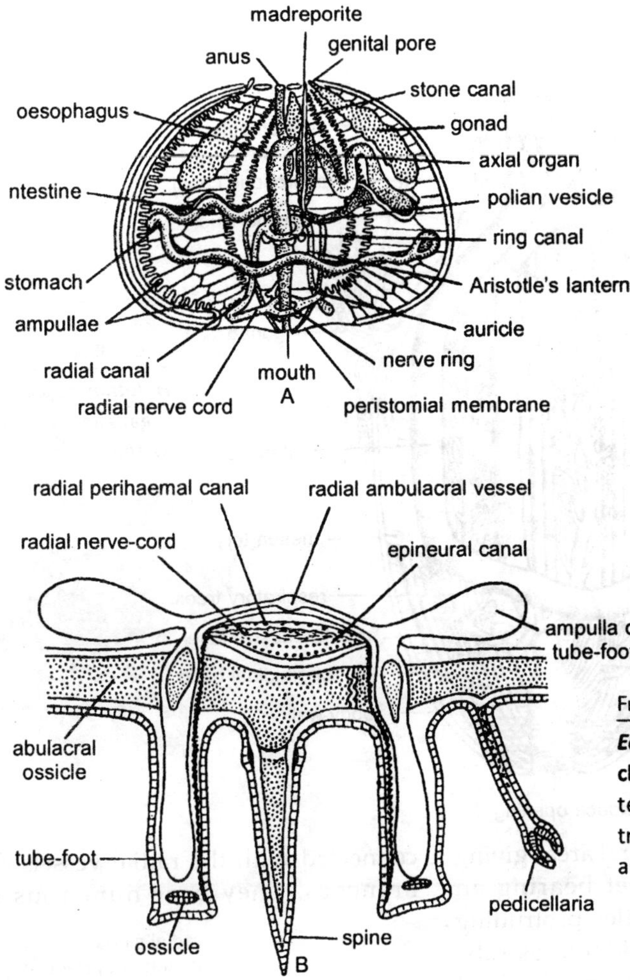

FIGURE 30.11

Echinoid species **(sea urchin).** A. Ambulacral system. B. Diagrammatic transverse section of the ambulacral zone

In **Echinoidea** the ring vessel surrounds the oesophagus at some distance dorsal to the mouth (Fig. 30.11). The five ambulacral vessels radiate from the ring canal and proceed along the inner surface of the corona along the ambulacral zones perforated towards the outer end by two minute pores, the ambulacral pores for protrusion of the tube feet. Appended to the ring vessel are five small mammillated bodies, the Tiedman's bodies. A madreporic canal with soft membranous walls run from the ring canal to the madreporite on the aboral surface. From each side of the radial canal run a series of short branches to the tube feet. Each tube foot has a muscular sac-like ampulla and a tubular portion, the terminal end of which bears a sucker. Ossicles are present in the sucker of the tube foot (Fig. 30.11B). The tube feet help in anchoring and locomotion.

In **Holothuroidea** the ambulacral vessel (ring vessel) is circular and located just behind the bases of the tentacles, surrounding beginning of the Oesophagus.

A large pear-shaped polian vesicle and one or more short sinuous canals, the madreporic canals arise from the ring vessel. The canals end in a perforated extremity in the interior of the coelom. Five radial vessels (Fig. 30.12) run out from the ring vessel. For the first part the radial vessels run towards the tentacles and give off branches to the tentacles and then

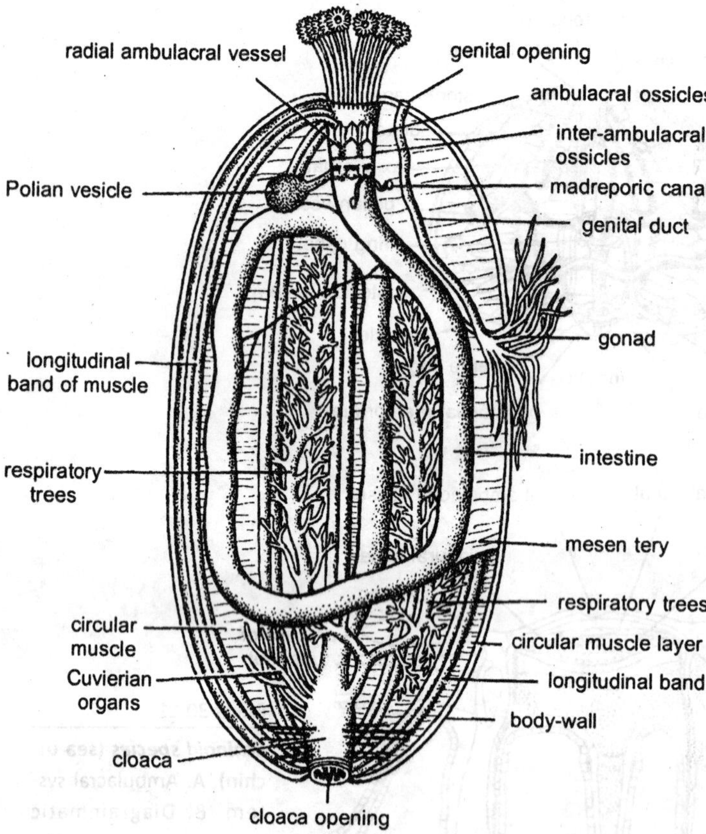

FIGURE 30.12

Holothuria sp. Internal organs and Ambulacral system

backwards along the ambulacral areas giving of branches to the tube feet bearing an ampulla. The circlet of tentacles protruding through the mouth are considered as tube feet. The tube feet help in locomotion and progression.

In **Crinoidea** the ring vessel surrounds the mouth. A series of radial vessels from the ring vessel run in the ambulacral grooves giving of branches to the pinnules (Fig. 30.13) A number of ciliated tubular diverticula, the water tubes are connected with the radial vessels. The water tubes are open at their extremities and remain suspended in the coelom. A large number of vessels with minute ciliated opening, the water pores open through the actinal (oral) wall of the disc. The openings and the ciliated tubes together represent the madreporic canal in the sea urchin.

Minute tubular ciliated appendages the podia or tentacles—homologous with the tube feet of Asteroidea and Echinoidea—are connected with the radial vessels and their branches. They bear numerous sensory

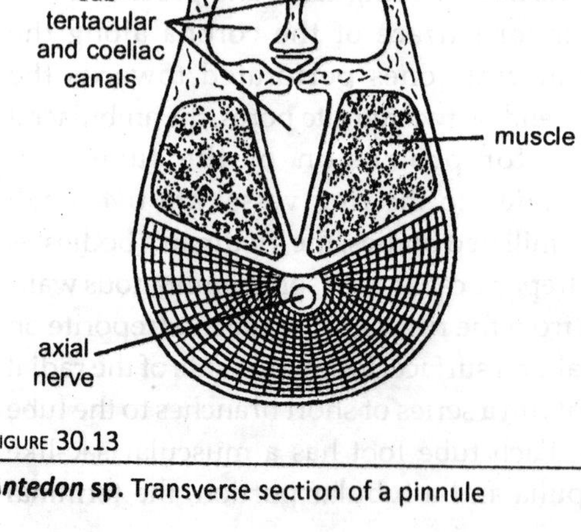

FIGURE 30.13

***Antedon* sp.** Transverse section of a pinnule

papillae and are considered as tactile organs. The podia are locomotory organs.

LARVAL FORMS OF ECHINODERMS

Echinoderms exhibit great diversity in their body forms, habitat and mode of life. The diversity is also expressed in their larval forms.

Echinoderms are dioecious, with a few exceptions. Fertilization takes place in sea water. Development may be direct or indirect through larval forms. Polar or deep sea forms are viviparous.

In addition to bipinnaria and brachiolaria following larval forms occur in echinoderms.

Pluteus larva

1. Five to six pairs of arms supported by calcareous rods and with pigmented tips.
2. Presence of four ciliated bands forming **epaulettes** at the base of the postoral and posterodorsal arms.
3. Arms are preoral, anterolateral, antero-dorsal, postoral, posterodorsal and pos-terolateral.

Two types of pluteus larvae, ophiopluteus and echinopluteus, are present.

Ophiopluteus larva

1. Free-swimming.
2. Arms are four pairs, slender and supported by calcareous skeleton.
3. Posterolateral arms are longest and directed forward, giving the larva V-shaped appearance.
4. Ciliated bands are present on the edges of arms.
5. The alimentary canal is divisible into mouth, oesophagus, stomach and intestine opening through the anus (Fig. 30.14A).

Ophiopluteus is the larva of **Ophiuroidea**. In viviporous forms, *Amphiura vivipara*, the pluteus stage is omitted. In *Ophionotus hexactis*, the development takes place in ovary and pluteus larva (Fig. 30.14B) is devoid of arms and anus.

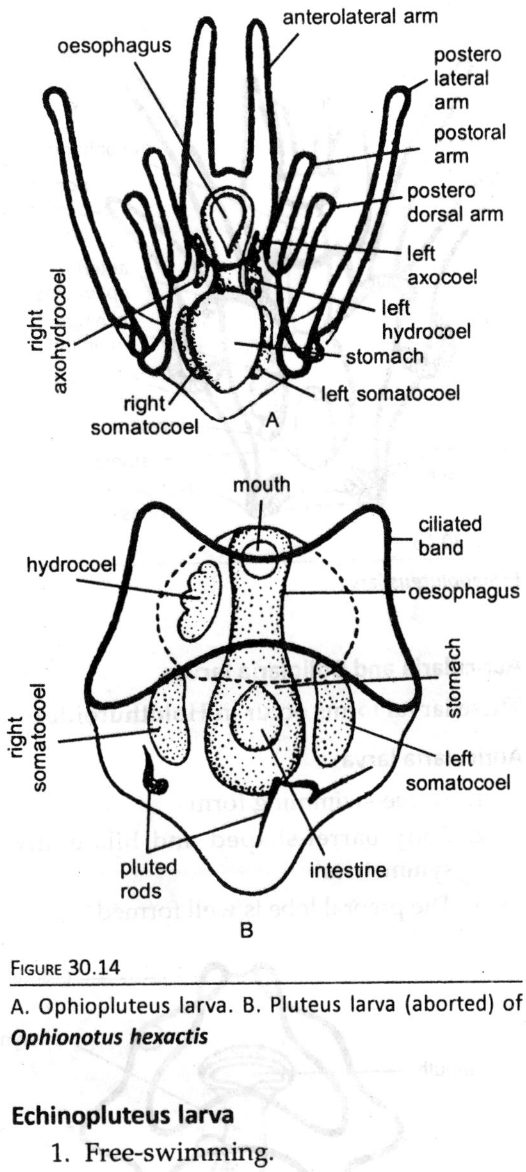

FIGURE 30.14

A. Ophiopluteus larva. B. Pluteus larva (aborted) of ***Ophionotus hexactis***

Echinopluteus larva

1. Free-swimming.
2. Arms five or six pairs, pigmented and supported by calcareous skeleton.
3. The posterolateral arms are very short and directed outward or backward.
4. The skeletal rods simple or thorny or fenestrated or branched.
5. The zones of the alimentary canal are mouth, oesophagus, stomach and intestine opening through the anus.

Echinopluteus (Fig. 30.15) is the larva of **Echinoidea**.

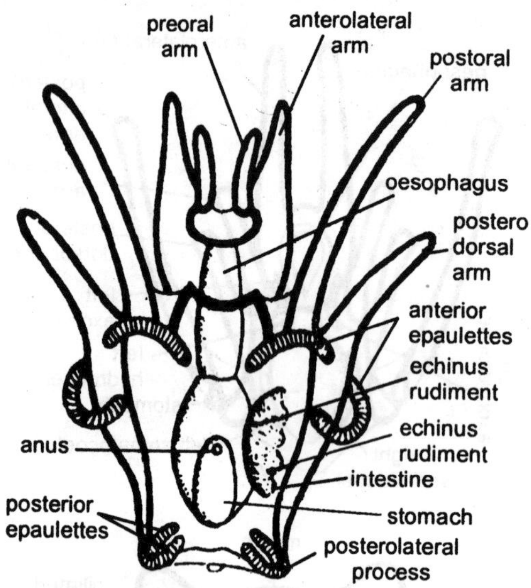

FIGURE 30.15

Echinopluteus larva

Auricularia and Doliolaria larvae

These larval forms occur in **Holothuroidea**.

Auricularia larva

1. A free-swimming form.
2. Body barrel-shaped and bilaterally symmetrical.
3. The preoral lobe is well formed.
4. A single winding ciliated band, which may be produced into lobes (Fig. 30.16A).
5. Gut with mouth, sacciform stomach, hydrocoel and right and left stomocoels and anus.
6. The hydrocoel becomes lobulated forming primary tentacles and communicates with the hydropore by a canal.
7. Calcareous rods replaced by spheroid or star-shaped or wheel-like bodies. The auricularia larva is transformed into a Doliolaria larva similar to that of **Crinoidea**.

Doliolaria larva

1. A free-swimming form.
2. Body barrel-shaped and bilaterally symmetrical.
3. Preoral lobe well-developed.
4. Wavy, continuous band break into 3-5 flagellated, transverse rings (Fig. 30.16B).
5. The gut with distinct zones.

☒ Phylogenetic relationship

Due to presence of enterocoelic coelom and some other minor resemblances, attempts

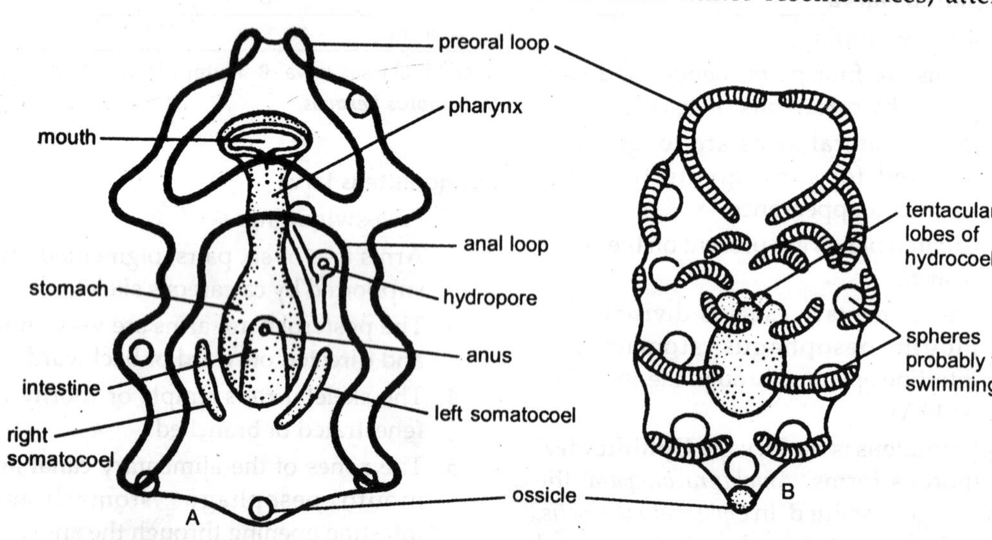

FIGURE 30.16

A. **Auricularia larva.** B. Transitional stage from Auricularia to **Doliolaria larva**

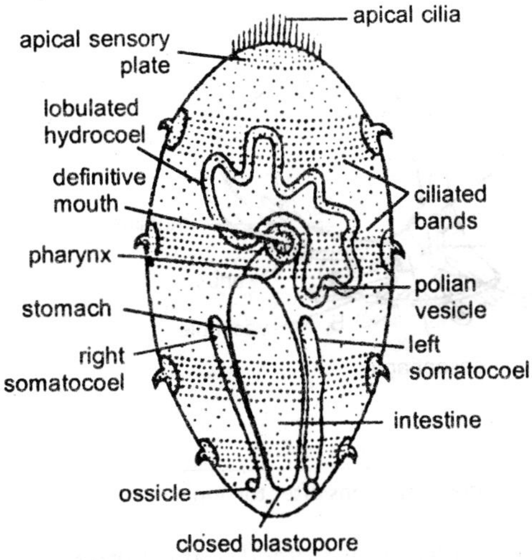

FIGURE 30.17

***Leptosynapta inkaerens*. A mature Dolioloria larva**

have been made to establish a relationship between auricularia and tornaria larva of Hemichor-data. This theory is now in dispute. Garstang (1894) propounded that the tadpole larva of Ascidia probably evolved from auricularia larva.

⊠ Doliolaria larva of Crinoidea

1. A free-swimming form.
2. Body elongateoval, a little narrower posteriorly (Fig. 30.17).
3. Presence of 4-5 transverse ciliated bands around the body.
4. An apical sensory plate with a bunch of cilia at the anterior end.
5. An adhesive pit over the first ciliary band in the midventral line close to the apical plate.
6. Gut with distinct zones and the stomodaeum between the second and third ciliated bands.
7. The skeletal structures present.

The larva attaches itslef to some support and the internal organs rotate at 90° angle from ventral to posterior position. A stalk develops and the larva turns to a **cystidian larva**, with metamorphoses to a young individual.

⊠ Homology and phylogeny of echinoderm larvae

Except for the crinoids, a sedentary group, the larvae of Asteroidea, Holothuroidea, Echinoidea and Ophiuroidea exhibit some fundamental resemblances.

1. Preoral and postoral loops.
2. Ciliated bands V-shaped.
3. Presence of gut with its divisions and openings.
4. Coelom enterocoelic.

These and other common features indicate they had a common ancestor.

Two theories have been forwarded to trace the common ancestry of different groups of echinoderms.

⊠ Dipleurula theory

Bather (1900) suggests that different forms of echinoderms possibly evolved from a common ancestor resembling a hypothetical dipleurula larva

Dipleurula larva

1. Body oval, bilaterally symmetrical with a flat, ventral surface (Fig. 30.18).
2. Gut straight with a stomodaeum, stomach and proctodaeum. Anus is supposed to be formed by atriopore.
3. Three paired coelomic sacs—axocoel, hydrocoel and somatocoel, with water pores on the dorsal side.

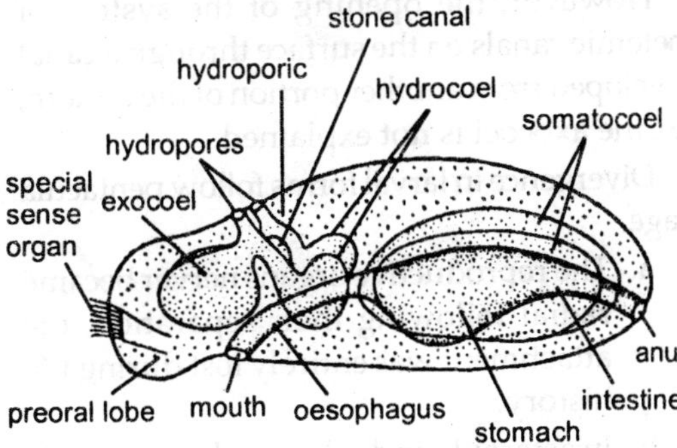

FIGURE 30.18

Dipleurula larva. Diagrammatic reconstruction of hypothetical ancestor

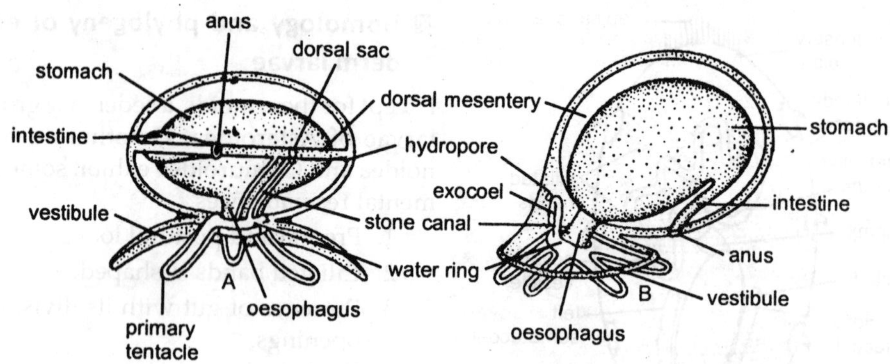

FIGURE 30.19

Pentactula (ancestor larva). A. Bilateral form. B. Radial conditions after torsion

4. Absence of skeletal structures.

The drawback of the concept is that it fails to explain the derivation of water vascular system, the most distinctive feature of echinoderms.

Pentactula theory

Advocated by Samon (1888), developed by Bury (1895), Hyman (1955). The concept advocates evolution of different groups of echinoderms from a common ancestor resembling pentactula larva.

Pentactula larva

1. Free, hollow tentacles around the mouth with coelomic branches from hydrocoel (Fig. 30.19).
2. Hydrocoel separated from the coelom to form water vascular system.

However, the opening of the system of coelomic canals on the surface through a canal developed from another portion of the coelom, i.e., the axocoel is not explained.

Divergence in larval forms follow pentacula stage.

a. The reproductive organs nerver became radial following their separation, the attachment was entirely lost during life history.
b. In asteroids and echionoids the gonads separated after the radial arrangement.
c. In crinoids the attachment of the gonads is retained.

It may be presumed that echinoderms are descendants of a common free-swimming ancestor, possibly pentactula.

⊠ Regeneration in echinoderms

Echinoderms have the remarkable ability to regenerate lost body parts following **autotomy**. In case, an arm is injured or grabbed by some predator or held up in some object, it is usually cast off (autotomy) near the base of the nearest ambulacral ossicle. The broken end is immediately closed by the contraction of adjacent muscles of the body wall for protection of internal organs, and a new arm starts forming from the broken point.

CLASSIFICATION: PHYLUM ECHINODERMATA

1. The echinoderms are noncolonial, coelomate, marine animals, the larval bilateral symmetry, which is replaced by radial symmetry in the adult.
2. The surface is covered with an exoskeleton of calcareous plates or ossicles, which usually support a system of movable or immovable calcareous spines.
3. A large body cavity or coelom, and a part of the coelom (hydrocoel) is converted into a system of vessels, containing water. The hydrocoel is a multipurpose organ-system, capable of meeting

the needs of respiration, locmotion, nutrition, sensory perception, or a combination of these. The alimentary, nervous and vascular systems are well-developed.

4. A characteristic system of vessels, the ambulacral system is connected with locomotion and other functions.
5. In most cases, locomotory organs are tube feet.
6. Almost all the systems exhibit radial symmetry.
7. Abundant amoebocytes are present in the coelom, haemal vessels and water vessels.
8. Reproduction sexual; free-swimming, larvae are bipinnaria, brachiolaria, pluteus, auricularia and doliolaria.

The phylum Echinodermata has been divided into four subphyla.

Subphylum 1. ECHINOZOA

1. Globoid echinoderms that never develop arms.
2. Mouth the anus at opposite poles of the body; in some, due to secondary adaptation for feeding these organs may be displaced.
3. Hydrocoel forms a ring around the mouth, giving rise to meridional water vessels traversing the body wall in the direction of anus.
4. Skeleton, nervous system, reproductive organs and muscular system tend to differentiate in meridional pattern.

The subphylum comprises five classes of which only two are represented by surviving members.

Class 1. **Helicoplacoidea** (extinct)

Lower Cambrian of California

Example: *Helicoplacus.*

Class 2. **Holothuroidea**

1. Free-living, elongated in the adoral-aboral axis, through which passes a dorsoventral plane of bilateral symmetry containing hydropores and gonoduct; one end (anterior) is thicker containing the mouth opening and the opposite end with anal opening.
2. The body is usually five-sided and along each side there extends a double row of suctorial tube feet.
3. No conspicuous external calcareous plates, but the integument is tolerably hard due to the presence of innumerable microscopic calcareous spicules.
4. Surrounding the mouth a circlet of tree-like tentacles present.
5. Madreporite situated in the interior of coelom.
6. A pair of tubular respiratory trees open into the cloaca; besides serving respiration and excretion of waste materials, these organs also have a hydrostatic function.
7. The sexes are separate; ovaries and testes are similar and consist of bunches of tubular follicles which opens on the dorsal surface, at a short distance behind the oral end.

About 500 living species; Ordovician to recent.

Subclass i. **Dendrochirotacea**

1. Pharyngeal retractor muscles connecting the anterior part of the body wall with calcareous ring and capable of retracting it within the oral aperture.
2. Tentacles usually 10 (up to 30), with tree-like branches (dendroid).
3. Tube feet many.
4. Respiratory trees present.

Examples: *Cucumaria, Psolus, Thyone,* etc.

Subclass ii. **Aspidochirotacea**

1. Pharyngeal retractor muscles absent.
2. The oral tentacles usually 20 (15 to 30), each branched from a shield-shaped central stalk.

3. Locomotor tube feet normally present on the ventral side and those on the dorsal side usually converted into sensory organs.
4. Respiratory trees present.

Examples: *Holothuria, Stichopus, Elpidia, Bathyplotes*, etc.

Subclass iii. **Apodacea**

1. Pharyngeal retractor muscle absent.
2. Test vestigeal, reduced to spicules scattered in the dermis.
3. Tentacles digitate or simple.
4. Tube feet reduced or lacking altogether.

Examples: *Molpadia, Leptosynapta, Synapta, Chiridota*, etc.

Class 3. **Edrieasteroidea** (Extinct)

Evolved in mid-Cambrian period and became extinct before the close of the Carboniferous.

Example: *Isorophus*

Class 4. **Echinoidea**

1. Free-living echinozoans with well developed test, composed of meridionally arranged columns of plates.
2. Ambulacral meridians are correlated with the water vascular system (ambulacral plates), and intervening segments comprise interambulacral plates.
3. Tube feet meridionally arranged, primitively suctorial and ampullate.
4. Outer surface of shell plate bears tubercles with which articulate movable spines of varying thickness and length.
5. Oral surface bears the central mouth surrounded by a membranous peristome.
6. Aboral surface bears a well-developed system of apical plates, with anus surrounded by a membranous periproct.
7. Madreporite lies close to anus; pedicellariae are stalked and 3-jawed.

The class includes about 850 living and 5,000 fossil species.

Subclass i. **Perischoechinoidea**

1. Echinoids usually with a flexible test, showing marked instability in the number of columns of plates comprising the ambulacrals and the interambulacrals.
2. Well-defined ambulacral system, test plates present.

Examples: *Eothuria, Melonechinus, Archaeocidaris, Goniocidaris, Ogmocidaris*, etc.

Subclass ii. **Enechinoidea**

1. Enormous variety of body forms, spherical, oval, disc-like, or heart-shaped, bilaterally symmetrical.
2. Lantern of Aristotle modified differently.

Examples: *Echinus, Laganum, Echinosigra, Echinocardium, Spatangus*, etc.

Class 5. **Ophiocistioidea**

Extinct. Ordvician to Devonian period.

Example: *Volchovia*.

Subphylum II. HOMALOZOA

Extinct. Mid-Cambrian to Devonian.

Example: *Enoploura*.

Subphylum III. CRINOZOA

1. Fundamentally globoid echinoderms with a partial meridional symmetry tending to produce an aboral cup-shaped body (calyx) on which the ambulacra are restricted.
2. Usually fixed with a stalk composed of a row or rows of ossicles.
3. The mouth is on the free surface, near or in the centre, and the anus usually on the oral surface.
4. A system of narrow ciliated ambulacral grooves, functioning as food grooves, extends out radially from the mouth on the oral surface.
5. Tube feet, when present, are smaller, ciliated and without suckers.

Class 1. **Crinoidea**

1. Calyx more or less globular, cap-like, strongly pentamerous, with oral surface

directed upwards and aboral surface downward.

2. Mouth usually lies near the centre; anus also present orally, being situated on an anal tube.
3. The five arms—each usually subdivide into one or more forks, producing from ten to thirty branches or, in extreme cases, up to 200; in existing crinoids. The arms usually bears lateral appendages, the pinnules.
4. The stem is attached to the aboral pole of the calyx; it may be long and slender with root-like organs attaching it to the sea floor. Feather-stars discard the stem early in life, assuming a free-living mode of life.
5. The sexes are separate but seldom differ in external features. Gonads usually lie in the dilated bases of pinnules; development with metamorphosis; developing into a barrel-shaped, free-swimming doliolaria larva.
6. Madreporite, spines and pedicellariae absent.
7. Crinoidea includes about 630 existing and 5,000 extinct species.

Subclass 1. **Inadunata**

Extinct. Ranging from upper Cambrian to Permian.

Examples: *Dendrocrinus, Anartiocrinus*, etc.

Subclass ii. **Camerata**

Extinct. Palaeozoic stalked genera, ranging from Ordovician to Permian period.

Example: *Xenocrinus.*

Subclass iii. **Flexibilia**

Extinct. Palaeozoic stalked genera, ranging from Ordovician to Permian times.

Examples: *Forbesiocrinus.*

Subclass iv. **Articulata**

1. Calyx pentamerous, flexible, fundamentally dicyclic.
2. The uniserial arms bear pinnules and are usually branched; the arm retains its movable articulation with the radial plate, despite the incorporation of the lower brachial ossicles into the clayx.
3. In some forms stalk exist with nodal ring of cirri (Metacrinus) and some discard the stem while young and thereafter remain free-swimming (feather stars).
4. Mouth and ambulacral grooves remain exposed.

Examples: *Antedon, Metacrinus, Comatula, Rhizocrinus*, etc.

Subphylum iv. ASTEROZOA

1. Radially symmetrical, free-living echinoderm with a star-shaped body, without stem.
2. The arms (usually five) develop after metamorphosis by the outward growth of the radial water vessels in the horizontal plane, carrying the perivisceral coelom and body wall with them. The gut sends caeca into the arms.
3. The nervous system is external, in the floor of the radial grooves, carried on the adoral surface of the arms.
4. Respiration and excretion by dermal gills called papulae.
5. Pedicellariae possess movable, pincer-like spines.
6. From peristome or mouth, an open, narrow and median ambulacral groove runs along the oral surface of each arm to its tip.
7. Sexes usually separate; testes and ovaries are alike and radially arranged. The larva is either bipinnaria or a brachiolaria.

Class 1. **Stelleroidea**

Open ambulacral groove, carried in long slender arms arising from a small central disc.

Subclass i. **Somasteroidea**

Exhibit characters intermediate between crinoids and other asterozoans.

1. Body flattened; the arm skeleton exhibits the pinnate structure of a biserial crinoid; the arms are petaloid, contracted at the bases and the disc is small.

2. The gut is blind, no anus; madreporite marginal; tube feet small, without suckers.

Examples: *Chinianaster, Ampullaster*, etc.

Subclass ii. **Asteroidea**

1. Body strongly flattened, star-shaped, pentagonal, with 5-40 simple unbranched long or short triangular arms not sharply demarcated from the central disc.
2. Mouth opens in the centre of the oral surface surrounded by a membranous peristome and guarded by spines.
3. A small eccentric anus is situated on the aboral surface and adjacent to it, in an interradius. The madreporite or sieve plate is present.
4. Arms are hollow, each containing a prolongation of coelom and viscera; running along the radial midline in the ambulacral furrow, is the radial ambulacral vessel, from which short side branches open into the bases of the tube feet.
5. About 1,700 living and 300 fossil species; the largest starfish, *Linckia*, is one metre in diameter.

Examples: *Asterias, Luidea, Solaster, Linckia, Astropecten, Ctenodiscus, Pentagonaster, Astrostole*, etc.

Subclass iii. **Ophiuroidea**

1. Body stellate and flattened with distinct oral-aboral surfaces.
2. Arms usually five, sharply demarcated from the body disc; slender, jointed, solid, cylindrical, muscular, flexible, rarely branched and useful for clambering and wriggling; autotomy of arms is frequent and regeneration occurs rapidly, so those are called brittle star.
3. Ambulacral grooves are not covered with ossicles, forming epineural canals.
4. Visceral organs do not extend into the arms. Arms contain muscle skeleton, branches of circulatory canals, nerves and a narrow prolongation of coelom.
5. Alimentary canal is confined to the disc and terminates blindly; anus and caeca absent.
6. Pedicellariae, skin gills and special sense organs absent.
7. Madreporite opens on the oral surface.
8. Tube feet arise ventrolaterally from arms in two rows, usually very small, have no ampullae and terminal suckers, and not locomotory.
9. Gonads lie interradially in pairs in the central disc; do not open directly to the exterior but open into genital bursae, one of which opens on each side at the base of each arm on the oral surface.
10. Sexes separate; the free-swimming larva is a pluteus.
11. There are about 1,800 living and 180 fossil species; large species may have a disc 10 cm across and an arm-spread of 50 cm. In the smallest species the disc and arm together may be no more than a fiew millimetres.

Examples: *Ophiura, Astrothorax, Astrophyton, Asteronyx, Gorgonocephalus*, etc.

31 PHYLUM HEMICHORDATA (ADELOCHORDATA)

The hemichordates are a small group of worm-like marine animals, that once were considered a subphylum of Chordata. This is a lesser deutrostome group providing some additional evidence linking echinoderms and chordates. They are found beneath the stones and shells, in burrows made in sand and mud, and some are tube-dwellers.

The body is divided into three regions. An anterior elongation of the body is the **proboscis**. The middle is a band-like **collar** region followed by a greater part of the body, the **trunk**. Mouth is located at the posteroventral margin of proboscis and just in front of collar. There is a line of perforations. the gill clefts, on either side of the anterior end of the trunk which do not bear gills. The pharyngeal clefts are the respiratory structures.

Hemichordates ingest large quantity of sand and extract the organic detritus from it. They have neither **notochord** nor a true **dorsal nerve cord**. A dense concentration of dorsal epidermal nerve fibres sink inwards within the collar region in some, and form a tube-like arrangement. The sexes are separate. Numerous paired lateral gonads, each with a separate gonopre, are located on either side of the trunk. Development is typical deuterostome* pattern and leads to development of **Tornaria larva**, which is strikingly similar to some echinoderm larvae.

The common pattern of early development (embryonic) of echinoderms, hemichordates and chordates is an important basis for recognizing their deuterostome line of evolution. The strikingly similar larval stages clearly indicate a relationship between hemichordates and echinoderms. On the other hand, the presence of gill clefts and dorsal collar nerve cord relates hemichordates to the chordates. From this it is apparent that these three phyla—Echinodermata, Hemichordata and Chordata—have a common ancestry. The very nature of all the three alleged characters of hemichordates—gill slits, notochord and dorsal tubular nerve cord—previously considered for their inclusion in the chordates has been questioned and they have been placed in a separate phylum Hemichordata.

BALANGLOSSUS sp. Acorn worm; tongue worm

Balanoglossus belongs to class Enteropneusta, phylum Hemichordata. It is a marine animal, widely distributed in all seas of the world. It

* Deuterostome: Blastopore forms the anus in adult and the mouth appears as opening later.

lives in burrows (Fig. 31.1) in inter-tidal zone and shallow water, though some species may live in somewhat deeper zones. The presence of Balanoglossus can be detected by the odour emitted by them, which resembles that of idoform. An individual lives in a U-shaped burrow excavated by the proboscis. Sand grains adhere to a viscid matter secreted by numerous glands in the integument to form a fragile temporary tube. During low tide, the proboscis is protruded out of the burrow in search of food. The posterior end projects out through the other opening to eject faeces.

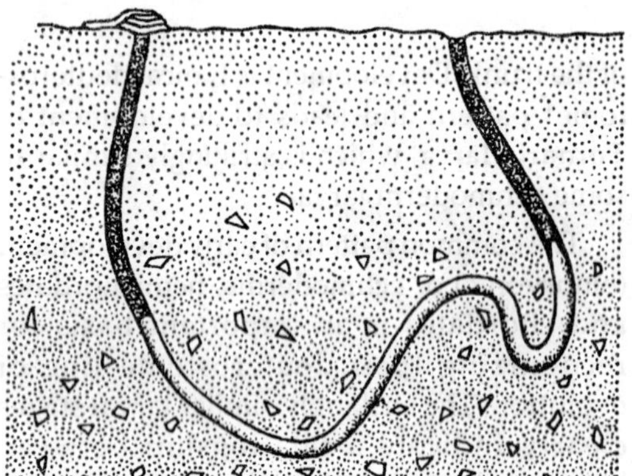

FIGURE 31.1
***Balanoglossus* sp. The animal in its burrow**

External features

Balanoglossus is a soft-bodied, cylindrical, worm-like animal (Fig. 31.2). The size in different species varies from 2 to 3 mm to 2.5 metres. Most of the surface is ciliated.

The body is divisible into three regions—an anterior muscular, ovoid probosis; the collar, just behind the proboscis; and a long, nearly cylindrical but somewhat depressed trunk.

Proboscis

The tongue-shaped proboscis has a thick, muscular wall, enclosing a cavity, the **proboscis coelom**, opening to the exterior by a minute **proboscis pore**. The narrow posterior or the **stalk** of the proboscis is strengthened by a layer of cartilage-like **chondroid tissue**.

Collar

The collar is a prominent muscular fold encircling the proboscis stalk. It contains two coelomic cavities, the **collar cavities** opening to the exterior by a pair of **collar pores**. The mouth (Fig. 31.3) lies on the ventral surface of the collar.

Trunk

The trunk is long and constituted by branchiogenital region, hepatic region and caudal region. A double row of small pores, the **branchial apertures**, are present on the

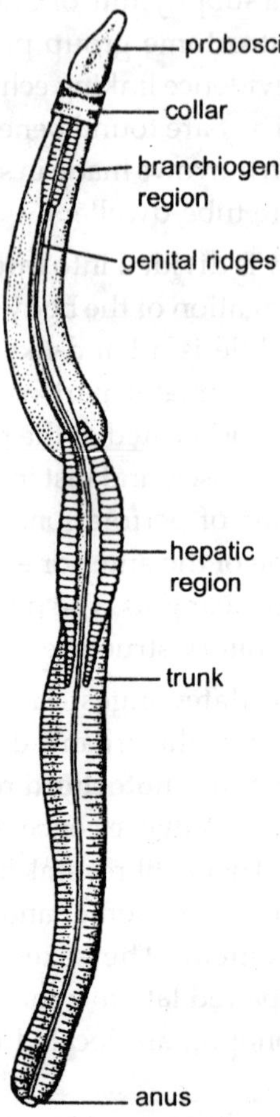

FIGURE 31.2
***Balanoglossus* sp.**

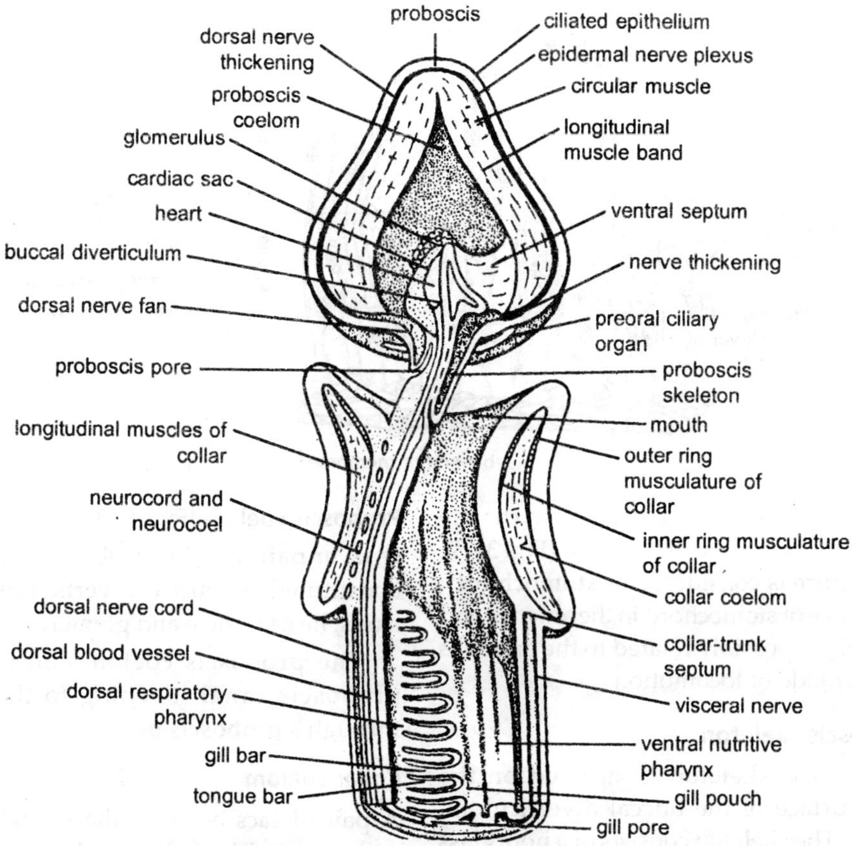

FIGURE 31.3

***Balanoglossus* sp.** Median sagittal section of the anterior end

dorsal surface of the anterior end of the trunk, a row situated in a long furrow. During growth the apertures increase in number. A pair of longitudinal **genital ridges** containing gonads are present in a considerable part of the trunk in the branchial region and behind. Two rows of prominences formed by the hepatic caeca constitute **hepatic region**. The rest of the trunk or the **caudal** region is nearly uniform in diameter and marked externally by irregular annulation.

☒ Body wall

The outermost layer epidermis is made of tall and narrow ciliated cells. Three types of gland cells—**reticulate gland cells, mulberry gland cells** and **goblet cells,** are present in the epidermis. A thick **nervous layer** lies beneath the epidermis, the two being separated by a thin layer of connective tissue, the **muscular sheath** (Fig. 31.4) supported by a basement membrane. The proboscis and collar contain circular muscle fibres, while longitudinal muscle fibres are present in the trunk region. The muscle fibres are **smooth**.

☒ Notochord: Buccal diverticulum: Stomochord

A diverticulum from the dorsal wall of the alimentary canal just behind the mouth, runs forward into the basal part of the proboscis, after sending out a small branch.

The wall of the diverticulum is made of a single layer of very narrow, vacuolated cells of endodermal origin and encloses a narrow lumen. The cells constituting the wall of the

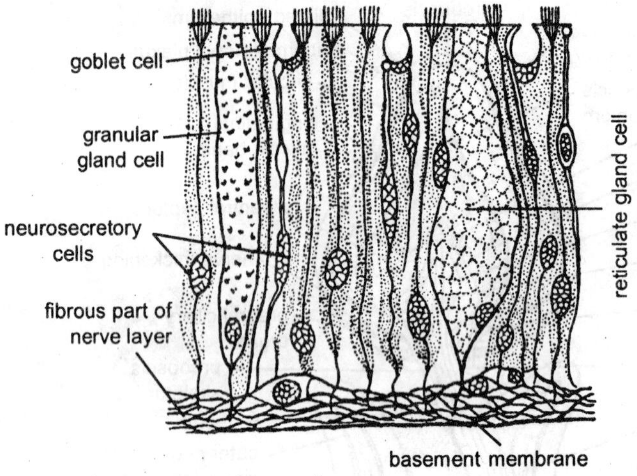

FIGURE 31.4

***Balanoglossus* sp.** Cross section of body wall

diverticulum are continuous with the epithelium of the alimentary canal (Fig. 31.3). This structure is considered as **stomochord***. The presence of stomochord in the anterior end of the body is possibly related to the animal's principal mode of locomotion.

⊠ Proboscis skeleton

The proboscis skeleton is situated on the ventral surface of the buccal diverticulum, close to it. The skeleton consists of a hourglass shaped **median plate** with two backwardly directed diverging limbs. A midventral keel is present on the median plate (Fig. 31.5).

⊠ Locomotion

Locomotion is effected chiefly by the epidermal cilia which are capable of synchronised reversal. Ciliary movement is helped by peristaltic contraction of the longitudinal muscle fibres leading to alternate shortening and lengthening of the body. Burrowing by the proboscis is essentially due to peristalsis of longitudinal muscle fibres.

⊠ Coleom

The coelom is spacious and enterocoelous in origin. It is lined by coelomic peritoneum. The coelom forms a number of sacs.

Proboscis coelom (Fig. 31.3)

It is unpaired. The proboscis complex is constituted by buccal diverticulum, central sinus, heart vesicle and glomerulus.

The proboscis coelom continues as a **proboscis canal**, opening to the exterior through a **proboscis pore**.

Collar coelom

A pair of sacs between the buccal tube and collar wall. The collar coelom open to the exterior through **collar tubes** and **collar pores**.

Trunk coelom

It is divided into two lateral lobes by dorsal and ventral mesenteries. The collar coelom and trunk coelom are separated by a **collar-trunk septum**. Composition of coelomic fluid is different in different coelomic spaces.

⊠ Digestive System

The digestive system consists of a straight alimentary canal starting in mouth and ending in anus and divisible into pharynx, hepatic region and intestine. The whole of the alimentary canal is lined with cilia.

The **mouth** is situated ventrally at the base of the proboscis, within the collar (Fig. 31.3).

---

* Previously the structure was considered as notochord and due to its forward position and being restricted to a small area, the animals were considered as chordates and placed in a separate group Hemichordata, under phylum Chordata.

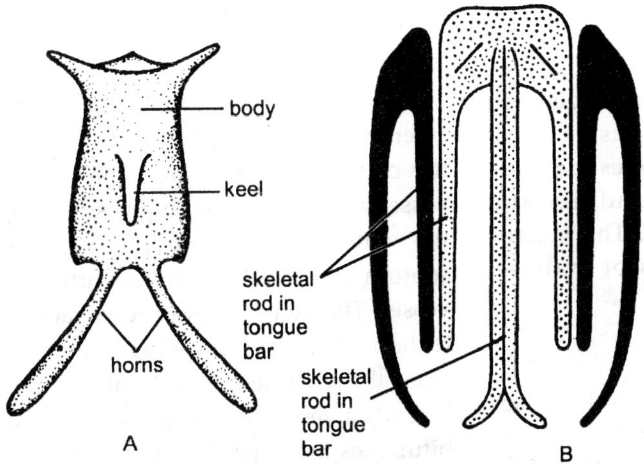

FIGURE 31.5

***Balanoglossus* sp.** Skeleton. A. Proboscis skeleton, B. Skeletal support of the gill apparatus

The mouth leads into a small **buccal tube**, from the roof of which a buccal diverticulum protrudes into the proboscis coelom. The buccal tube opens into the **pharynx**.

A **peripharyngeal** ridge on each side divides the pharyngeal cavity into two incomplete halves—dorsal and ventral (Fig. 31.6). The dorsal chamber, the respiratory part, is pierced by gill slits and the ventral chamber is concerned with food collection.

Internal gill slits open on the dorsal surface of the pharynx. Each gill slit is in the form of a long, narrow U, the two limbs separated by a narrow process, the **tongue bar** containing a prolongation of the coelom. The internal gill slits lead into **gill pouches**, which in turn, open to the exterior through gill pores.

The bars between the gill slits have skeletal supports, each of which is made of a dorsal basal portion and three long narrow rods. One rod is median and two are lateral. The median rod is bifurcated at the end and lies in the septum between the two adjacent gill sacs. The lateral rods lie in the adjoining tongue bars.

The pharynx opens posteriorly in the **intestine**. In the hepatic region, a series of paired **hepatic caeca** formed by the sacculation of the dorsal wall of the intestine are present. The post-hepatic region of the intestine is narrow and terminates in an **anal aperture** at the posterior extremity of the body.

☒ Feeding and respiration

Balanoglossus is a ciliary feeder. Cilia on the proboscis and lateral cilia on the sides of the gill bars play important role in feeding.

Beating of the **proboscis cilia** drives food particles towards the mouth. A respiratory current is created by the beating of **lateral cilia** on the sides of the gill bars. Water entering pharynx through the mouth is driven outwards through the branchial apertures. Food are swept directly into the mouth with respiratory current. Some food particles, silt and sand get entangled with the mucus secreted by the proboscis. The mucus, with its contents, is driven into the alimentary canal partly by the respiratory current and partly by the action of cilia in the buccal tube and gut.

☒ Ciliary organ

A **ciliary organ** bearing sensory cells, immediately anterior to the mouth, on the posteroventral surface of the proboscis is considered as an organ for testing water and matters drawn towards the mouth.

☒ Respiration

The gill slits are highly vascular and the dissolved oxygen in the inward respiratory current is absorbed there. The carbon dioxide released passes out with outward water current.

⊠ Digestion

Nothing is definitely known about digestion in Balanoglossus. Amylase has been detected in proboscis. The oesophagus is most glandular and presumably secretes enzymes. The oesophagus is most glandular and presumably secretes enzymes. The hepatic caeca secrete small amounts of maltase, proteinase and lipase.

⊠ Blood Vascular System

A heart vesicle, blood vessels and sinuses constitute blood vascular system.

Main longitudinal vessels are two; the **dorsal vessal** lies just below the dorsal nerve cord and the **ventral vessel** in the ventral mesentery. Blood flows forward in the dorsal vessel but backward in the ventral vessel.

The dorsal vessel runs forward from the region of anus to end anteriorly in a sinus, the **dorsal sinus** or **heart**, situated in the posterior part of the stomochord. A vessel arising from the posterior part of the sinus bifurcates to supply the proboscis. A number of vessels arising from the anterior part of the sinus forms a plexus, the **glomerulus**, possibly excretory in function. An **efferent vessel** runs backward from each side of the posterior end of the glomerulus and breaks up into a plexus. The two plexus communicate with the ventral vessel and oxygenated blood is received by the ventral vessel.

The ventral vessel supplies the collar with a **ventral collar vessel**. Two distinct plexus in the collar tissue communicate with a ring vessel posteriorly. The ring vessel arises from the ventral vessel, located in the collar-trunk septum and it is connected with the dorsal vessel. The ventral vessel runs backward up to the anal region. It gives off lacunar networks along the whole length of the alimentary canal. It sends a vessel to each gill septum, where it bifurcates to supply the two adjacent tongue bars. Each tongue bar receives two **afferent branchial vessels**, breaking up into a plexus. An **efferent branchial vessel** is formed from the plexus. The vessel runs dorsally, joins with the efferent branchial vessel of the adjacent tongue bar and opens into the dorsal vessel. Blood from the intestinal plexus is drained to the dorsal vessel.

A close **cardiac sac** encloses the dorsal surface of the heart. Muscle fibres present in the ventral wall of the sac contract rhythmically. Similar fibres are present in the dorsal and ventral vessels throughout the length of the trunk. Presumably all the muscles are responsible for driving blood in Balanoglossus.

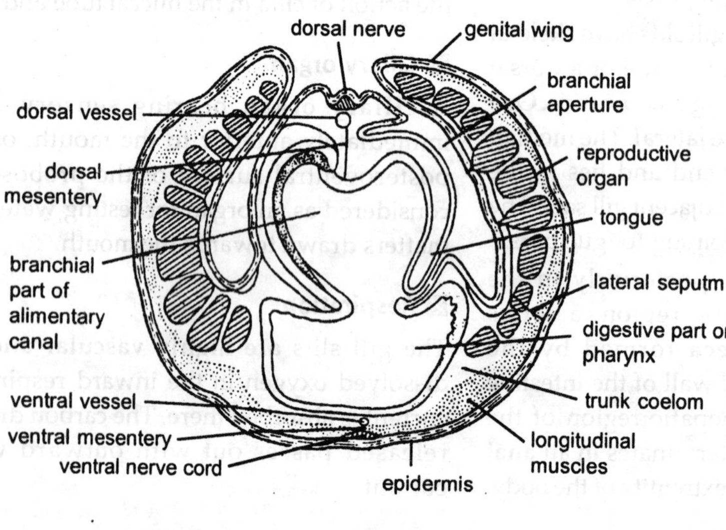

FIGURE 31.6 ***Balanoglossus* sp.** Transverse section of the branchial region

☒ Nervous System

It consists of dorsal and ventral **nerve strands** extending throughout the length of the body, at the base of the epidermis. These are merely thickenings of a layer of nerve fibres. **Giant nerve cells** are scattered in the nerve strands. The nerve strands thicken to form two **nerve cords**, one in the middorsal and the other in the midventral line. The two are connected by a **peribranchial nerve ring** in the collar-trunk septum. The dorsal cord projects into the collar coelom as the **collar cord** or **neurocord**. Nerve fibres run out from the ring. The dorsal and ventral nerve cords are conduction pathways.

☒ Receptors and sense organs

Preoral ciliary organ (see Feeding) is the only organised receptor. Certain cells of the proboscis epidermis, on the anterior edge of the collar, might be sensory in nature.

☒ Reproductive System

The sexes are separate and often differ in shape and colour. Simple or branched saccular organs constitute **testes** or **ovaries**, arranged in a double row along the branchial region of the trunk and further behind. Each gonad opens to the exterior by a **single pore**.

☒ Fertilization and development

Fertilization is external. Development is associated with metamorphosis.

Early development

Segmentation is complete, fairly regular and a round **blastula** is produced. Later, it becomes flattened. **Gastrula** is formed by **invagination**. At this stage the embryo is covered with short cilia and a ring of stronger cilia. Separation of ectoderm and endoderm is followed by hatching, which occurs at about 24-36 hours since fertilization. The larva begins to swim. A diverticulum is separated from the anterior part of the archenteron. **Proboscis coelom** develops from the cavity of the diverticulum. Paired coelomic cavities of the collar and trunk develop either from the diverticulum of the posterior wall of the proboscis coelom or directly from the wall of the archenteron.

The free-swimming larva elongates and a transverse groove appears, posterior to the middle. **Mouth** appears by invagination in the position of the groove. **Anus** appears in the position of the blastopore. A second transverse groove divides the embryo into anterior, middle and posterior regions, the beginning of proboscis, collar and trunk. A tuft of apical cilia develop. The embryo is transformed into a **Tornaria** larva.

TORNARIA LARVA

The size varies from 1 to 3 mm. Body bell-shaped, bilaterally symmetrical and variously folded (Fig. 31.7). The **mouth** is ventral and slightly posterior to the equatorial plane of the body. The **anus** is median and at the posterior end of the body. The portion anterior to the mouth forms the **preoral lobe**. The anterior end bears an **apical plate** with nerve fibres, ganglionic nerve cells, a tuft of cilia and a pair of eye spots. Ciliated bands are three and distinct. The **preoral** and **postoral bands** join for a short distance at the apical plate. The **circumanal band** surrounds the anus. The preoral and postoral bands collect food; the postoral and circumoral band or **telotroch** are main locomotory organs. The **alimentary canal** is short and bent at right angle and

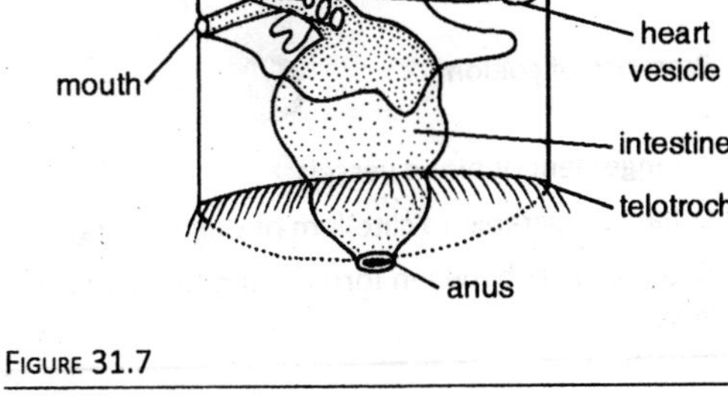

FIGURE 31.7

***Balanoglossus* sp.** Tornaria larva

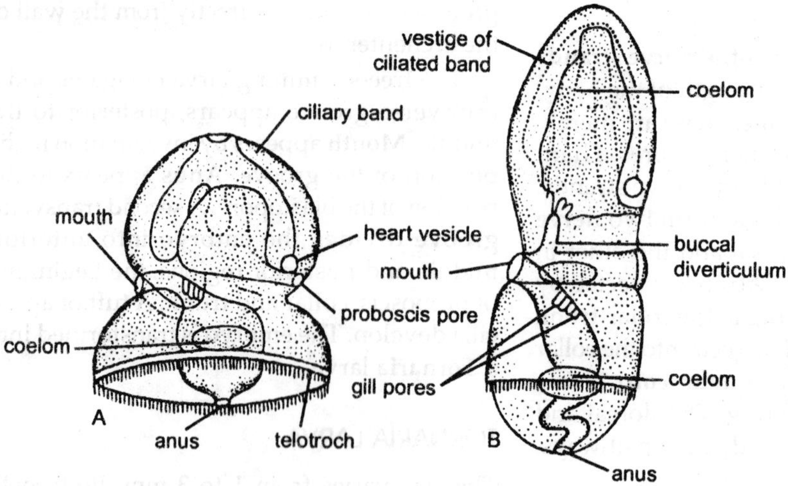

FIGURE 31.8
***Balanoglossus* sp.** Metamorphosis of **Tornaria** larva

divisible into oesophagus, stomach and intestine.

☒ Metamorphosis

Tornaria is **pelagic**. It sinks to bottom and undergoes a series of changes. The apical plate with its structures and ciliated bands are lost (Fig. 31.8). The preoral part of the body becomes ventrally ciliated and form proboscis, being separated from the collar by a constriction. Gill-slits begin to appear, trunk elongates and the larva assumes adult form.

BIOLOGICAL STATUS (SYSTEMATIC POSITION) OF BALANOGLOSSUS

For a long time the thick-walled, hollow, anterodorsal buccal diverticulum, the **stomodaeum** of Balanoglossus, was considered as half a true **notochord** and the animal was placed in Hemichordata under the phylum Chordata. Later, Hemichordata was assigned the status of a nonchordate phylum.

☒ Affinities of Balanoglossus

Based on similarities and differences with nonchordates and chordates, attempts have been made by many workers to ascertain the systematic position of Balanoglossus.

☒ Affinities with Phoronis

Similarities (very few)

1. Disposition of the coelom.
2. A ring of cilia surrounding the anus.
3. Proboscis pore comparable to water pore.

Affinities with Annelida

Similarities	Dissimilarities
1. Vermiform body.	1. Nephridia present in Trochophore but absent in Tornaria.
2. Presence of coelom.	2. Gastrulation opposite—blastopore forms mouth in Trochophore but forms anus in Tornaria.
3. Arrangement of blood vessels.	3. Developmental aspects differ in two.
4. Collar comparable to clitellum of Oligochaeta.	
5. Resemblances between Tornaria and Trochophore larve.	

Affinities with Pogonophora

Similarities	Dissimilarities
1. Coelom enterocoelous with similar divisions. 2. Nervous system intraepidermal.	1. Pogonophores lack alimentary canal. 2. Excretory structures—nephrocytes in Balaroglossus and coelomoducts in Pogonophora. 3. Gonads in the trunk region. 4. Presence of pericardial sinus in some species. 5. Septum between meso and metasome.

Affinities with Echinodermata

Similarities	Dissimilarities
1. Ciliated bands twisted in Tornaria, Auricularia and Bipinnaria. 2. Position of mouth and anus. 3. Divisions of larval alimentary canal. 4. Origin and development of coelom (most convincing resemblance).	1. Eye spots in Tornaria larva wanting in Trochopore larva. 2. Protocoel paired in echinoderms, unpaired in Tornaria. The two are closely related groups and might have evolved from a common ancestor.

Affinities with Chordata

Similarities	Dissimilarities
1. Presence of gill clefts. 2. Presence of lumen only in collar nerve cord. 3. Feeding mechanism in Ammocoete larva and some amphibian larvae.	1. Notochord absent. 2. Nervous system solid except collar nerve cord.

Affinities with Urochordata

Similarities	Dissimilarities
1. Pharynx and branchial apparatus. 2. Development of central part of nervous system. 3. Feeding mechanism.	The two groups differ in all other aspects.

Affinities with Cephalochordata

Similarities	Dissimilarities
1. Branchial apparatus. 2. Arrangement of coelomic sacs.	1. Unsegmented muscle fibres in Balanoglossus. 2. Mode of development.

Zoologists today not only disagree with the inclusion of Balangolossus in Chordata, but even object to the use of the term Hemichordata assigned to it, their preference being for **Adelochordata**, as the buccal diverticulum is not considered by many to be the notochord, some believing it to be a structure similar to the pituitary body. However, the presence of a dorsal, tubular nerve cord and pharyngeal gill slits are two undisputed chordate characters of Blanoglossus. All efforts to assign a definite systematic position of Balanoglossus have failed. Still it can be assumed that the group Hemichordata (Adelochordata) is probably at least a greatly modified branch from the base of the chordate tree.

CLASSIFICATION: PHYLUM HEMICHORDATA

1. Soft-bodied, worm-like, marine, burrowing animals.
2. Body divisible into three regions—proboscis, collar and trunk, the anterior end of which bears gill slits.
3. The nerve cord is dorsal, but not a true nerve cord, lies along the middle line.
4. Coelomic pouches are present.

The phylum Hemichordata has been divided into two classes.

Class **Enteropneusta**

1. Worm-like, sand-burrowing forms, range in length from 25 to 2,500 mm.
2. They are found in shallow waters, although some live at a depth of 4,500 metre.
3. Body divided into proboscis, collar and trunk, the latter being further divided into an anterior branchial region, a middle genital region and a posterior abdominal region terminating into anus.
4. The mouth is always wide open and capable of closing completely, lies on the ventral surface and its lips are the ventral edges of the collar region.
5. Both proboscis and collar are hollow and their cavities communicate with the exterior by pores that permit water to be taken in or expelled out. The coelom in the trunk region is divided into two halves and without any pore.
6. Series of gill slits, even up to 700 pairs in some species, present on both lateral sides of pharynx extending far backwards in the trunk.
7. A thick-walled, hollow diverticulum of the dorsal wall of the mouth cavity extends forwards into the hollow proboscis.
8. The nervous system is constructed on the nerve-net principle and consists of a plexus of cells and fibres from the so-called dorsal tubular nerve cord in the collar region.
9. Larva Tornaria, similar to the **Bipinnaria** larva of starfish.

Example: *Balanoglossus, Dolichoglossus, Glossobalanus, Ptychodera, Saccoglossus*, etc.

Class **Pterobranchia**

1. Nearly all members are marine, sedentary, colonial, forms of minute size.
2. Body divided into three regions— a probosics with proboscis cavity, a collar with two collar cavities and a trunk with two distinct lateral cavities.
3. Buccal diverticulum and so-called nerve cord similar to those of Enteropneusta.
4. The alimentary canal 'U'-shaped, anal opening lies near the mouth.
5. Presence of arms bearing tentacles arising from collar.
6. Gill pores single pair in Cephalodiscus, absent in Rhabdopleura (except a pair of ciliated grooves).
7. In most, sexes are separate, but Cephalodiscus is hermaphrodite.
8. Development is direct.

Examples: *Cephalodiscus, Rhabdopleura, Atubaria*, etc.

32 PHYLUM CHORDATA

CHORDATA: Diagnostic features

Phylum Chordata, the largest of the deuterostome phyla, is a diverse assemblage of marine, fresh water and terrestrial animals. The chordates are triploblastic, bilaterally symmetrical, metameric, coelomate organisms with anteroposterior and dorsoventral axes. It includes sea squirts, amphioxus, cyclostomes, fishes, amphibians, reptilians, birds and mammals. All possess three unique chordate

Nonchordates	Chordates
1. Body with bilateral or radial symmetry.	1. Body bilaterally symmetrical.
2. The paired appendages vary in number and may even correspond to the number of body segments.	2. The appendages are two pairs.
3. Tail absent.	3. A postanal tail is usually present.
4. Branchial clefts absent.	4. The branchial clefts are always present, being restricted to larval stages only in some.
5. The vascular system is open and consists of a heart, arteries and sinuses, the last ones are continuation of primary body cavity.	5. The vascular system is a closed one and consists of a heart, arteries and veins.
6. The heart is less muscular, dorsal and always above the alimentary canal.	6. The heart is highly vascular, ventral and lies below the alimentary canal.
7. No portal veins.	7. Portal veins present in the majority.
8. The respiratory pigment haemocyanin is a prosthetic group of copper or haemoglobin in some cases. Both are dissolved in plasma.	8. The respiratory pigment haemoglobin is an iron compound and present in the erythrocytes.
9. Notochord absent.	9. A solid, unsegmented, flexible axial rod, the notochord is universally present and extends along the whole body length.
10. The main nerve cord is a solid, double one, running midventrally immediately dorsal to the ventral body wall.	10. The spinal cord is a hollow, middorsal tube, placed immediately below the body wall and above the notochord.

features : a dorsal rod-like notochord, a dorsal tubular nerve cord and pharyngeal clefts or slits through the pharyngeal wall. In many chordates some of these characteristics are found only in developmental stages.

The notochord is a primitive internal skeleton consisting of an elastic rod, covered by one or two sheaths of tough connective tissue acting as a fulcrum for the segmental muscles and extends forward up to the middle of the brain where lies the hypophysis. It typically lies beneath the nerve cord and above the alimentary canal and is composed of large vacuolated cells. The gill slits or branchial clefts primitively serve for the passage of water from the pharynx to the outside. In vertebrates several pairs of gill slits are always functional in the adults of fully aquatic classes, and in the larvae of most amphibians.

In many aquatic forms, gills are present within this cleft. In animals equipped with lungs, branchial clefts or grooves are always found in the embryo.

In chordates, there is a single, unpaired, dorsal, hollow, fluid-filled nerve cord, without distinct ganglionic enlargements, but anteriorly differentiated into a brain in advanced forms. The dorsal origin and position of the nerve cord are regarded as related to the original mode of life of ancestral chordates. The concentration of nerve cells to form a central nervous system takes place in relation to the direction from which the greatest stimulation comes. The dorsal position of nerve cord in chordates is regarded as the evidence that the ancestral chordates were free-swimming pelagic forms, receiving their chief stimulation from the sea surface above them, the phenomenon called **neurobiotaxis**.

All vertebrates possess characteristics of chordates. Only two subphyla of chordates—urochordata and Acrania—lack vertebral column but retains notochord and large pharynx with branchial clefts. All vertebrates are chordates but all chordates are not vertebrates.

CLASSIFICATION: PHYLUM CHORDATA

1. The chordates are bilaterally symmetrical, triploblastic, coelomate animals.
2. The notochord or chorda dorsalis is present either in the young stage or throughout life. In the higher chordates, this is replaced by a bony vertebral column at a later stage.
3. The wall of the pharynx is perforated by a system of clefts, the branchial or gill clefts, either in the embryonic or larval condition only or throughout life.
4. A hollow dorsal nerve cord or spinal cord is always present. A cavity or a system of cavities, the neurocoel, is present within the central nervous system.

The phylum Chordata has been divided into three subphyla.

Subphylum UROCHORDATA

1. The adult animals are invested in thick cuticular tests and are mostly sedentary. They exhibit a divergence from the chordate type.
2. The dorsal nervous system, notochord, gill slits and brain-eye develop in the free-swimming larval stage.
3. The larval structures degenerate during retrogressive metamorphosis and are not found in the adult.
4. The ventral heart is tubular.

Examples: *Ascidia*, *Ciona*, etc.

Subphylum ACRANIA (CEPHALOCHORDATA)

1. Lack distinct head, cranium, brain, eyes, auditory organs and jaws.
2. The nerve cord is dorsal and tubular.
3. Notochord is persistent and unsegmented and extends along the whole length of the body.
4. Gill slits are numerous and open in the atrium.
5. The body wall consists of over fifty myotomes.

6. Digestive, blood vascular and nervous system resemble those of the vertebrates.
7. The larval form is asymmetrical and the larval period is long.

Example: *Branchiostoma*.

Subphylum CRANIATA (VERTEBRATA)

1. Notochord does not proceed anteriorly beyond forebrain.
2. Pharynx perforated by not more than seven pairs of gill slits, except in some cyclostomes.
3. Heart muscular and chambered.
4. Renal tubules unite to form paired kidneys opening in ducts.
5. The complex brain is enclosed in a cranium.
6. Brain well-developed, cranial nerves at least ten pairs.
7. Sense organs paired and well-developed.

The subphylum Craniata has been divided into two superclasses.

Superclass AGNATHA

1. The mouth is round and suctorial and without jaws.
2. Paired fins and scales are absent Unpaired fins are without fin rays.
3. Skeleton cartilaginous and the notochord persists.
4. Nasal organ unpaired, no sympathetic nervous system, no conus arteriosus, no distinct pancreas, no spleen, no genital ducts.

Examples: *Petromyzon* (Lampreys), *Myxine*, (Hag fishes) etc.

Superclass GNATHOSTOMATA

1. The mouth is provided with two jaws, the upper and lower.
2. Two pairs of limbs are present.

This superclass has been subdivided into following major classes:

Class **Placodermi (Aphetohyoidea)** (Extinct)

1. Primitive Gnathostomata in which the hyoid arch persists as an effective branchial arch, preceded by a hyoid gill cleft with functional gill.
2. Jaws are attached directly to the cranium by ligaments.

Examples: *Climatius*, *Dinichthys*, etc.

Class **Chondrichthyes** (Elasmobranchii)

1. Scales are placoid.
2. Skeleton entirely cartilaginous with no true bones.
3. The operculum is absent and the gill slits open directly to the exterior.
4. The venous system is expanded in places into large sinuses.
5. A spiral valve is present in the intestine.

Examples: *Scoliodon*, Skates, Rays, etc.

Class **Osteichthyes** (Teleostomi)

Fishes with bony skeleton.

Subclass **Actinopterygii**

1. The skull is hyostylic, the jaws being supported by the hyomandibular arch.
2. Scales in the majority are cycloid and fins with fin rays.
3. Operculum always present, gill filaments project from the gill arches.

Examples: *Channa*, *Labeo*, *Puntius*, *Clarias*, etc.

Subclass **Crossopterygii**

1. Lung partially used as respiratory organ.
2. Nostrils are placed within the mouth.
3. Scales cosmoid.

Examples: *Latimeria*, *Dipnoi* (lung fishes).

Class **Amphibia**

1. The larva possesses gills and the adult bears lungs.
2. The skull has two occipital condyles.
3. Heart with one sinus venosus, two auricles, one ventricle and a conus arteriosus.
4. Metamorphosis is usually present.

Examples: *Rana*, *Bufo*, etc.

Class **Reptilia**

1. The body is covered with epidermal scales.
2. The skull bears a single occipital condyle.
3. Heart with two auricles, one partly divided ventricle and a sinus venosus.
4. The eggs are large, land eggs with calcareous shell; embryo protected by amnion and allantois.

Examples: *Lacerta, Calotes, Naja,* etc.

Class **Aves**

1. Warm-blooded, oviparous, feathered bipeds, forelimbs of which are modified as wings.
2. The skull is light with obliterated sutures and without teeth. A single occipital condyle is present.
3. The pectoral muscles are large and used in flight.
4. Air sacs are present, which continue to air spaces in some bones.
5. Heart with two auricles and two ventricles. The single aortic arch curves to the right side.

Examples: *Columba, Gallus,* etc.

Class **Mammalia**

1. The body is covered with hairs at least in the early stage.
2. Mammary glands are universally present.
3. A diaphragm separates the thoracic cavity from the abdomen.
4. The skull bears two occipital condyles; sutures are present.
5. Heterodont and thecodont dentition.
6. The single aortic arch curves to the left side.
7. Viviparous, and the embryos are nourished through placenta (except monotremes).

Examples: *Cavia, Homo erectus* (extinct), *Homo sapiens,* etc.

33 SUBPHYLUM UROCHORDATA

The members of the subphylum Urochordata are commonly called tunicates or ascidians or sea squirts. The urochordates comprise marine, sessile sea squirts and a few pelagic forms and constitute the largest group of protochordates*. The body is covered by a thick envelope or tunic, containing tunicine, a form of cellulose. The tunic may be soft, delicate or very tough and rigid. One end of the body remains attached to the substratum. The opposite end contains two openings—the oral and atrial siphons. The oral or buccal siphon opens into a large pharyngeal chamber, which occupies the upper half or two-thirds of the body. The pharyngeal wall is perforated by a vast number of slits, which open into an outer surrounding chamber, the atrium, connected to the exterior through the atrial siphon. The body below the pharynx is the abdomen which contains the stomach, intestine, heart and other internal organs. These animals are filter feeders, removing plankton from the water stream passing through the pharyngeal chamber. A deep groove called endostyle, extends down the length of the pharyngeal chamber. Definite excretory organs are lacking. The nitrogenous wastes mostly escape by diffusion. They have an open internal transport system but the blood flows in distinct channels, which course through the pharyngeal basket and abdominal organs. The heart periodically reverses its beat and the flow of blood is also reversed. Many common tunicates are solitary but a considerable numbers are colonial. The colonial tunicates have the individuals symmetrically arranged and the atrial siphons open into a common cloacal chamber. Some colonial and solitary urochordates are unattached and swim using the feeding current as a water jet for locomotion; hence they are also humorously called sea squirts.

Most tunicates are capable of asexual reproduction by budding. All species are hermaphroditic with a single abdominal ovary and a testis. The oviduct and sperm duct open into the atrium. Development is associated with a tadpole larva.

ASCIDIA sp. (Sea squirt)

Ascidia belongs to class Ascidiacea, subphylum Urochordata, phylum Chordata.

Ascidia are widely distributed, solitary, marine animals, sedentary in adult stage, remaining attached to some submerged hard

* Urochordates and cephalochordates together form the protochordates.

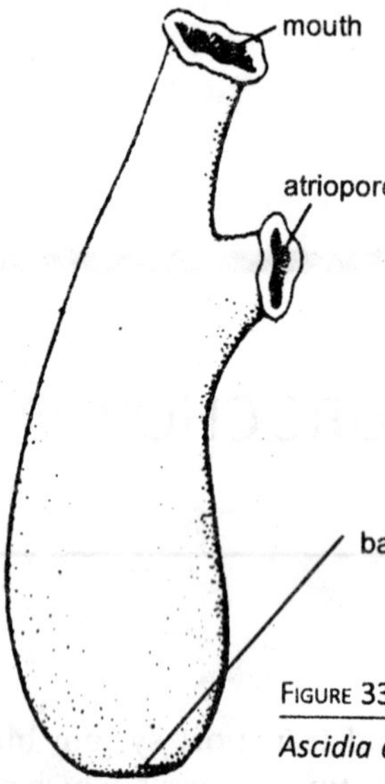

FIGURE 33.1

Ascidia andamanensis

structure, and occur in various depths of ocean.

⊠ External features

Body bag-like, with a basal end attached to some support and the free end directed upwards (Fig. 33.1). Two apertures are present; an eight-lobed **oral funnel** or **oral siphon** at the top and a six-lobed **atrial funnel** or **artial siphon** placed laterally at some distance from the free end, towards the base. The **atripore** is present in the middle of the atrial funnel. The body is enclosed in a tough but translucent **tunic** or **test**. Water enters the body through oral funnel and goes out through atrial funnel.

⊠ Body wall and atrial cavity

The body wall is composed of an outer tunic and an inner mantle (Fig. 33.2). The **tunic** is secreted by epidermis, made of **tunicin** (cellulose and glycoprotein) and contains scattered cells. The cells in the tunic are mesodermal in origin and migrate through the ectoderm to the outer surface. Outgrowths of the body wall with prolongations of blood channel enter the test and ramify in all direction. A longitudinal septum divides a channel into two and each ends in a small bulb. Blood moves along one passage, circulates in the bulb and returns through the second passage.

The soft body wall or **mantle** underlies the test. It remains suspended in the test, being firmly attached to it only round the oral and

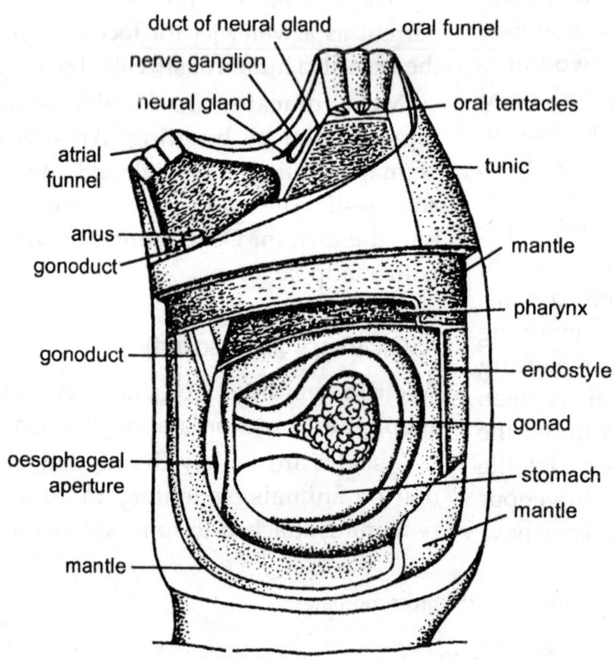

FIGURE 33.2

***Ascidia* sp.** Visceral structures. The greater part of tunic and mantle removed

atrial funnels. Ectoderm cells with underlying layers of connective tissue containing muscle fibres constitute the mantle. The mantle is produced into two short but wide tubular projections—the oral and atrial funnel. The prolongations are continuous with the margins of the apertures of the test and guarded by strong sphincter muscles, capable of closing the openings. The mantle encloses an **atrial** or **peribranchial cavity, atrium**, opening to the exterior through **atriopore**. The atrial cavity is formed by the involution from the outer surface.

⊠ Coelom

The coelom is greatly reduced and restricted to pericardial and gonadal cavities.

⊠ Digestive and Respiratory System

Mouth, stomodaeum, pharynx, oesopha-gus, stomach, intestine and anus constitute digestive system.

Mouth and Stomodaeum

The mouth is surrounded by **velum**, located in the posterior end of the oral funnel. The velum bears **velar tentacles**. Sphincter muscles regulate opening and closure of the mouth. The mouth leads to a short but wide oral passage, the stomodaeum, which communicates with a large chamber, the pharynx or branchial chamber (Fig. 33.3).

Pharynx

The pharynx extends posteriorly near to the base of body and remains attached to the mantle along the ventral side. The space between the mantle and pharynx is called **atrium**. Vascular mesodermal tissue, the **connectives** or **trabeculae** run from the body wall to the wall of the pharynx across the atrium. The wall of pharynx bears numerous slit-like vertically disposed apertures, the **stigmata**, arranged in transverse rows. The edges of the stigmata bear numerous strong cilia. By the end of metamorphosis, two primary gills, **protostigmata** appear on each side of the pharynx. The number increases to six on each side, by the development of independent perforations. With growth, the six protostigmata elongate and subdivide to form a row of definitive stigmata; additional rows continue to form as growth proceeds, and the perforated pharynx is produced.

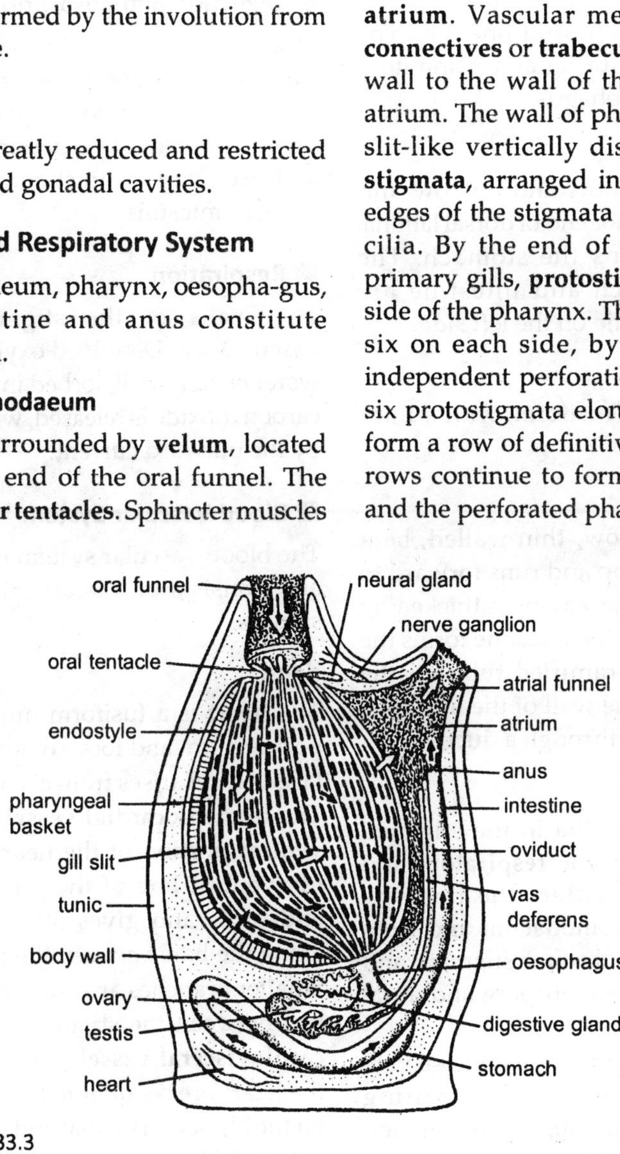

FIGURE 33.3

***Ascidia* sp.** Tunic and mantle of the side removed (lateral view, diagrammatic)

Along the line of adhesion of the pharynx with the mantle, a thickening in the form of a pair of ventral longitudinal folds, separated by a groove, are present on the inner surface of the pharynx. The grooved thickening is called **endostyle**. Two rows of **mucous cells** separated by a row of **ciliated cells** are present on each side of the endostyle. In the median groove of the endostyle, a group of cells with long cilia are present. A ciliated fold, **dorsal lamina** is present on the inner surface of the pharynx along the middorsal line. Pharynx acts as both food capturing and respiratory organ. It leads to oesophagus.

Oesophagus

The oesophagus is short and narrow and located near the posterior end of dorsal lamina. The oesophagus joins the stomach. The oesophagus, stomach and intestine are embedded in the mantle on the left side.

Stomach

The **stomach** is large, fusiform, with a thick glandular wall.

Intestine

The **intestine** is narrow, thin-walled, bent around in a double loop and runs forward to open in **anus** in the atrial cavity. A thickening along the inner wall of the intestine forms the **typhlosole**. Delicate ramified tubules, the **pyloric glands** over the wall of the intestine, open into the stomach through a duct.

Feeding

Beating movement of cilia in the stigmata create a water current, the **respiratory** and **food current**, which enters the pharynx through mouth and stomodaeum and from there goes to the atrium through stigmata, and thence to the exterior via atripore and atrial funnel.

The gland cells of the endostyle secrete viscid mucus. Active, lateral beating movement of the cilia, distributes the secretion on the side wall of the pharynx; the median ventral cilia assist to deflect the mucus to either side of the groove. On reaching the wall of the pharynx, the mucus is driven upwards and towards the middorsal line as a sheet. Additional mucus follows the trail. Food particles carried by water current are entangled in the mucous sheet. The dorsal lamina along the middorsal line of the pharynx collect the food-laden mucus and conducts it backward, towards the oesophagus.

⊠ Digestion and absorption

Digestion occurs in stomach. Strong carbohydrases and weak protease and lipase are secreted by two glands in the stomach. They pyloric gland is possibly both digestive and excretory in function. Absorption takes place in intestine.

⊠ Respiration

The walls of the stigmata are highly vascularised. Dissolved oxygen in the inward water current is absorbed in the stigmata and carbon dioxide is released, which is moved out by the outward current.

⊠ Blood Vascular System

The blood vascular system is well-developed and consists of a heart, blood vessels and sinuses.

Heart

The **heart** is a fusiform muscular sac in the pericardium and located near the stomach. A large vessel arises from each end of the heart. The **branchiocardial vessel** arising from the ventral portion of the heart runs along the midventral line of the pharynx, below the endostyle and gives off branches running along the bars between the rows of stigmata. Smaller branches arising from these branches run between the stigmata of each row. The **cardiovisceral vessel** from the dorsal end of the heart breaks up into branches and ramify on the alimentary canal and other organs. The vessels or lacunae open into a large sinus, the **viscerobranchial sinus**. The sinus runs along the middorsal wall of the pharynx and communicates with the dorsal ends of the series of transverse branchial vessels. In addition to the principal vessels, numerous lacunae extend throughout the body including the test.

The contractions of the heart are **peristaltic**. The contraction follows from one end of the heart to the other end for a certain period, which is followed by a pause. The next contraction takes place in opposite direction. In this way, the course of blood in the heart is reversed at regular intervals.

Blood

The plasma is colourelss and the corpuscles are few. Lymphocytes, phagocytes, macrophages, vacuolated compartmental cells and coloured and colourless cells are present.

⊠ Excretory System

Nephrocytes present in blood are concerned with excretion. Particles of **urates** and **xanthine** in the cells are disposed of by storage in the form of concretions in special **excretory vesicles** or **renal organs**.

⊠ Nervous System

The nervous system is extremely simple. A **neural ganglion** (Fig. 33.4) lies between the oral and atrial funnels, embedded in the mantle. It is dorsoventrally elongated and gives off nerves at each end, which innervate various parts of the body.

⊠ Neural gland

The neural gland lies on the ventral side of the neural ganglion and considered homologous with at least a part of the hypophysis of vertebrates by some workers, which is controversial. A duct runs forward from the neural gland and opens into the cavity of the pharynx through a **ciliated funnel**. The duct is folded upon itself to form a prominent **dorsal tubercle** projecting into the pharyngeal cavity.

⊠ Reproductive System

The sexes are united. The ovary and testis are closely situated on the left side of the body in the intestinal loop (Figs. 33.2, 33.3). The gonoduct—**oviduct** or **spermduct** is long, continuous with the gonad and opens in the atrial cavity close to the anus. Self-fertilization is prevented by production of ova and sperms at different times. Fertilization is external. Self-fertilization has been reported in some species of Ascidia.

⊠ Fertilization and development

Small, yolkless eggs float on water and in the majority, fertilization takes place there.

Development is indirect and associated with metamorphosis.

Segmentation is **holoblastic** and nearly equal. Only a few cells form the **blastula**. Gastrulation is by **invagination**. The **gastrula** elongates and, about three days after fertilization, a fish-like **tadpole larva** hatches out.

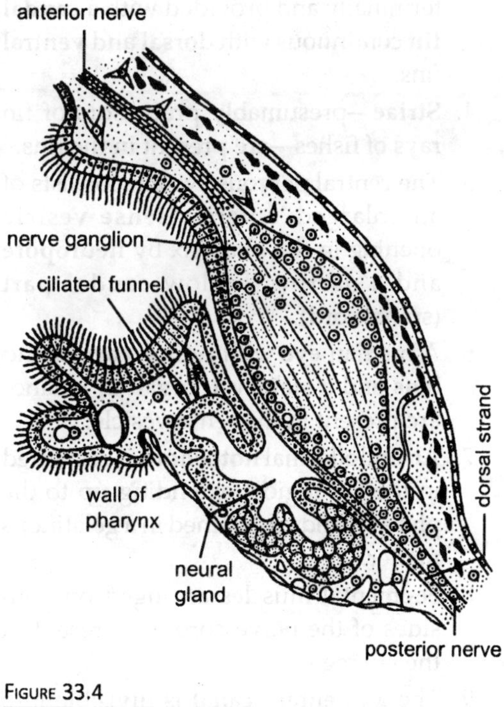

FIGURE 33.4

***Ascidia* sp.** Longitudinal section of the neural gland and nerve ganglion

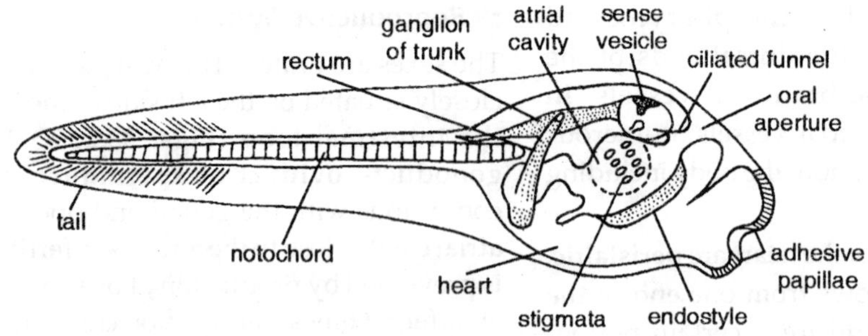

FIGURE 33.5 ***Ascidia* sp.** A free swimming tadpole larva

☒ Larval form

The tadpole larva (Fig. 33.5) is highly motile and moves with the help of tail. It does not take food for some time. The stage is known as **nonfeeding** form.

1. The body is covered by tunic and divisible into an oval **head** and a long **tail.**
2. Three **adhesive papillae** or **chin warts**; one middorsal and two ventrolateral are present in front of the head.
3. The tail is laterally compressed, pointed terminally and provided with a **caudal fin** continuous with **dorsal** and **ventral** fins.
4. **Striae**—presumably precursors of fin rays of fishes—are present in the fins.
5. The central nervous system consists of an enlarged, anterior **sense vesicle** opening in the pharynx by **neuropore** and a narrow, hollow **caudal part** (spinal cord).
6. A median **eye** with **retina, cornea** and **lens** and a **statocyst**, organ of balance are present in the sense vesicle.
7. An endodermal **notochord** is restricted in the tail region, extending up to the pharynx and ensheathed in a gelatinous matter.
8. Segmental muscles arranged on both sides of the nerve cord are present in the tail region.
9. The **alimentary canal** is divisible into two regions. The anterior region consists of a **mouth** and a well-developed **pharynx** with two pairs of **gill slits** and an **endostyle**, and the posterior region gives rise to **oesopha-gus**, stomach and **intestine**.
10. **Atrial sacs** are paired.

The tadpole larva is positively phototactic and negatively geotactic. After a brief active period, the larva becomes sluggish and starts searching for a suitable substratum for fixing itself. The preference is always for hard or rocky one. The larva fixes itself to the support with adhesive papillae, turns negatively phototactic and positively geotactic and metamorphosis begins.

☒ Metamorphosis

Metamorphosis (Fig. 33.6) of tadpole larva in Ascidia is a two-way process. As the majority of the organs of the larva are lost, the process has been termed as **retrogressive**. But at the same time some of the structures further develop, which may be considered as **Progressive**.

☒ Changes in the larva following fixation :

Retrogressive metamorphosis

1. The length of the tail is considerably shortened, the number of striae reduced and finally the tail disappears.
2. The notochord becomes coiled, disorganized and disappears with the tail.
3. The nerve cord becomes restricted to the trunk region and gradually reduced to a solid ganglion.
4. 90° shifting of the mouth occurs from the point of attachment due to rapid

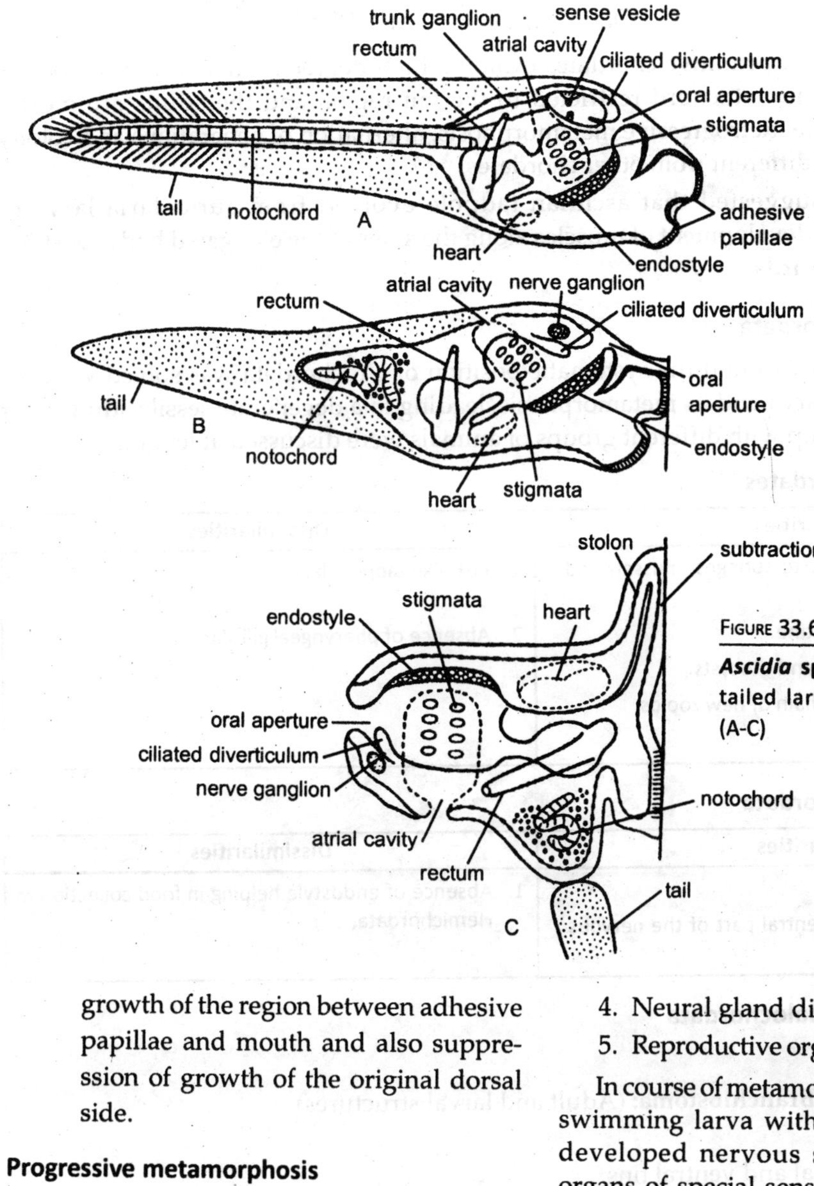

FIGURE 33.6

***Ascidia* sp.** Metamorphosis—free tailed larva into a fixed ascidian (A-C)

growth of the region between adhesive papillae and mouth and also suppression of growth of the original dorsal side.

Progressive metamorphosis

1. The pharynx enlarges in size and the number of stigmata increases.
2. Atrium enlarges.
3. Velum appears.
4. Neural gland differentiates.
5. Reproductive organs gradually develop.

In course of metamorphosis the active, free-swimming larva with a notochord, a well-developed nervous system and complex organs of special senses is degenerated to a fixed, nonmotile adult, in which all the chordate characters disappear, except gill slits, endostyle and structures associated with feeding.

BIOLOGICAL STATUS (SYSTEMATIC POSITION) OF UROCHORDATA

Urochordata, a group of primitive chordates exhibit a high degree of diversity. The body is enclosed in a tunic, the notochord restricted to the tail region and the adults are mostly degenerated and sessile. Retrogressive metamorphosis, alternation of generations and colony formation make them different from other chordates.

Garstang (1894) suggested that ascidian tadpole evolved from auricularia larva of echinoderms with the development of muscles along the sides of the elongated body, pushing the ciliated bands upwards.

☒ Affinities of Urochordata

Uncertainties revolving around the systematic position of Urochordata was solved with the discovery of tadpole larva and its metamorphosis, leading to a degenerate sessile adult. Prior to this, their relationship with different groups of animals were discussed at length.

Affinities with nonchordates

Similarities	Dissimilarities
1. Food collection similar to sponges, mussels and lophophores.	1. Fish-like tadpole larva.
2. Typhlosole in the intestine.	2. Absence of pharyngeal gill slits.
3. Presence of larval eye and otocysts.	
4. Budding leading to a chain of new zooids.	
5. Formation of colony.	

Affinities with Hemichordata

Similarities	Dissimilarities
1. Structure of pharynx.	1. Absence of endostyle helping in food collection in Hemichordata.
2. Development of the central part of the nervous system.	

☒ Affinities with cephalochordata

Similarities

Tadpole larva and Branchiostoma: (Adult and larval structures)

1. Body shape.
2. Presence of dorsal and ventral fins.
3. Presence of velum and velar tentacles.
4. Filter feeders.
5. Cleavage and gastrulation.
6. Mode of development of notochord and nervous system.
7. Presence of notochord.
8. Similar atrial complex with endostyle, pharyngeal groove and epipharyngeal bands.

☒ Affinities with Chordates

Similarities

1. Tadpole larva possesses major basic features of chordates.

2. Presence of gill slits.
3. Presence of notochord.
4. Dorsal, tubular, hollow nerve cord.

☒ Affinities with vertebrates

Similarities

1. Tadpole larva similar to fish larva.
2. Feeding mechanism similar to Ammocoete larva and some larval amphibians.
3. Nasal glands similar to hypophysis.
4. Gonadotrophin stimulates ascidian neural gland.

Scientists are unanimous that Urochordata occupies a position in the lower rung of the hierarchy of Chordata, but they are divided in ascertaining the exact pathway of their origin. Some suggest that the development of sexual maturity along with other complexities in the larval form led to the elimination of ancestral adult phases and the degenerated forms are the descendants of ancestors having all chordate features. Others hold that they were lowly organized forms, which never reached the level of the simplest form of chordates. A third group believes that the appendicular larva is a primitive one giving rise to a stationary adult having degenerated vertebrate characters. Another view is that the doliolids (urochordates) have undergone neoteny to produce present day urochordates.

Regarding the interrelationship within the urochordates, it is believed that larvacea is most primitive. Thaliacea developed from fixed forms close to ancestral compound ascidians, which gave rise to Pyrosomida.

CLASSIFICATION : SUBPHYLUM UROCHORDATA

1. Mostly sessile, marine, nonmetameric, largest group of protochordates with a cellulosic body covering, the tunic.
2. The pharynx is greatly enlarged, saclike, perforated with numerous lengthwise slits, stigmata.
3. A deep groove, the endostyle, below the pharynx extends down the length of the pharyngeal chamber.
4. A dorsal nervous system, notochord, gill slits and brain-eye develop in the free-swimming tadpole larval stage.
5. The larval structures degenerate during retrogressive metamorphosis and are not found in adult. The urochordata have been grouped in three classes.

Class ASCIDIACEA

1. Sessile, marine ascidians inhabitat from rocky shore to ocean depths, but most species live in shallow coastal waters. Distribution polar to the equator seas.
2. The pharynx is large with persistent gill-slits.
3. The compound ascidians have a common external aperture, the cloaca, into which their excurrent siphons discharge.
4. The compound ascidians reproduce asexually by gemmation and also produce eggs and sperms for sexual reproduction.

5. Development in most species proceeds up to the larval stage within the atrium, then the young emerge.
6. Most ascidians are attached (*Botrylus, Ciona*), but some are free-living (*Polycarpa*) on sand or mud.

Order **Enterogona**

1. The body may be divided into two regions.
2. Single gonad located in or behind the intestinal loop.
3. The neural gland usually situated ventral to the nerve ganglion.
4. The larva bears two sense organs.

Examples: *Ascidia, Ciona, Phallusia, Clavelina*, etc.

Class THALIACEA

1. Free-swimming, transparent, tunicates, sometimes simple, sometimes colonial, without caudal appendage in the adult.
2. The test is a permanent structure.
3. The muscular fibres of the body wall are arranged in complete or interrupted ring-like bonds or held diffusely.
4. The mouth and atriopore are situated at opposite ends of the body.
5. The pharynx has either two large—or many small—stigmata leading into atrial cavity which communicates with the exterior by the atrial aperture.
6. There is usually an alternation of generations.
7. Tailed larval stage may or may not be present.

Order 1. **Doliolida (Cyclomyaria)**

1. Hermaphrodite, generally found in phytoplanktonic zone in tropical and subtropical seas, usually up to a depth of 200 metres.
2. The solitary sexual individual or gonozooid is barrel-shaped, with transparent and delicate test, and eight completely encircling muscle bands (hence the name Cyclomyaria).
3. The oral and atrial apertures are at opposite extremities.
4. The oral aperture leads into a wide pharynx occupying at least the anterior half of the body.
5. Only the posterior wall of the pharynx is usually perforated by stigmata.
6. An endostyle and peripharyngeal band occurs but there is no dorsal lamina.
7. The heart lies immediately posterior to the endostyle in the midventral line.
8. The nervous system composed of dorsal ganglion, lying in the third or fourth intermuscular zone.
9. The larva has a tail with notochord, in the asexual stage or 'nurse'. Gradually the tail aborts and two processes, one posterodorsal known as cadophore and the ventral one stolon, grow out from the body of the larva. Zooids develop from cadophore.

Examples: *Doliolum, Doliolopsis, Dolioletta, Dolioloides*, etc.

Order 2. **Pyrosomida**

1. Luminescent, colonial tunicates of warmer seas found up to a depth of 500 metres. The colony consisting of numerous individuals forming a gelatinous tube.
2. The zooids of a colony are seemingly independent of one another.
3. Body cavities two, the pharynx and atrium.
4. The gut and gonads lie near the midline; the heart and stolon are situated directly behind the endostyle.
5. Musculature weak.
6. The pharynx is divided. The prebranchial buccal cavity lies anterior to peripharyngeal bands. The branchial sac has the endostyle extending throughout its length.
7. Development is direct, without tailed larval stage, and takes place within the body of parent.

Examples: *Pyrosoma, Pyrostremma*, etc.

Order 3. **Salpida (Desmomyaria)**

1. Planktonic, prism-shaped or cylindrical, transparent tunicates, 8-190 mm in size, occurring abundantly in almost every sea, down to 200 metres.
2. Forward propulsion is by contraction of muscle bands ejecting water posteriorly.
3. The pharynx is reduced to its median, dorsal and ventral walls, the lateral wall being absent.
4. Occur in two forms—oozooid and blastozooid.
5. The oozooid or solitary form develops from egg. It is bilaterally symmetrical and has a budding stolon but without gonad. Both branchial and atrial apertures are wide and situated at the opposite ends of the body; test is relatively thin and delicate. Gill slits are represented only by a pair of large apertures on either side of a single median gill bar. In the dorsoanterior region occurs a horseshoe-shaped eye. A single large tentacle, the so-called languet, extends dorsally into the buccal cavity in front of the peripharyngeal bands. Reproduction asexually.
6. The blastozooid or aggregate symmetrical form develops from a bud; gonad —single ovary and single testis; no stolon. Blastozooids remain connected in groups initially, but become separate later in life. Development is direct and takes place within the body of parent. A placenta is formed.

Examples: *Salpa, Cyclosalpa, Octacnemus,* etc.

Class LARVACEA (APPENDICULARIA)

1. Transparent, free swimming, oceanic tunicates with a globular body and permanent tail supported by a skeletal notochord.
2. Most are very small in size, with a trunk, rarely more than 1 mm long and a tail generally not more than three times the length of the trunk.
3. The test is a relatively large envelope, formed from the secretion of ectoderm and frequently thrown off and renewed.
4. The pharynx has only two stigmata leading directly to exterior.
5. Peribranchial or atrial cavity absent.
6. Development without metamorphosis and budding never occurs.

Order **Copelata**

1. Pelagic small tunicates with persistent notochord and only two gill apertures.
2. The gill apertures and anus open directly to the exterior.
3. The mouth is at the front end and the anus on the ventral surface.
4. Lacks dermis, and muscles are usually absent in the region of the body.
5. Hermaphrodite, single or double ovary between two testes, the male organ maturing first (protandrous).

Examples: *Oikopleura, Fritillaria, Stegosoma, Megalocercus, Kowalevskia,* etc.

34 SUBPHYLUM ACRANIA

Cephalochordates, members of subphylum Acrania, class Cephalochordata are marine, sand-burrowing creatures, confined to shallow waters all over the world. Lack head, brain, eyes, auditory apparatus and jaws. They are known as Acrania since they have no cranium or brain box. Nervous and blood vascular systems are clearly of the same general pattern as those in fishes and higher vertebrates.

The digestive system is much in common with the digestive arrangements, of vertebrates but their apparatus of ingestion is strikingly similar to that of the tunicates. The excretory system is much more similar to flatworms, annelids, and molluscs. Notochord extends along the whole length of the body in median plane. A hollow dorsal nervous system and gill slits present.

BRANCHIOSTOMA (Amphioxus) sp.

Branchiostoma belongs to class Cephalochordata, subphylum Acrania, phylum Chordata.

Branchiostoma dwells in the shallow waters of tropical and semitropical seas and has apparently retained a great many relatively simple features, despite the fact that it has probably evolved a long way from the actual ancestors of chordates. This small animal shows the chordate plan reduced to almost its lowest terms and its development and structure point the way to the rise of complicated conditions found in higher vertebrates.

It is a sand-burrowing animal, the greater part of the body remains (Fig. 34.1) in sand, only with the anterior end protruding. It

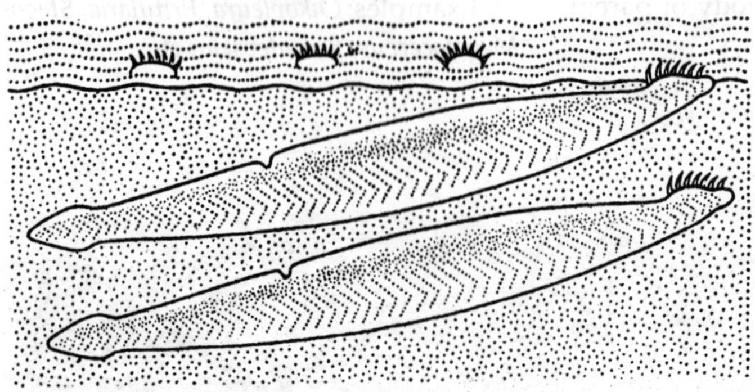

FIGURE 34.1

***Branchiostoma* (Amphioxus) sp.** in their natural habitat (in burrowing state)

burrows rapidly in sand with head or tail foremost. It is capable of swimming but erratically.

Branchiostoma are distributed from Mediterranean to North Sea. The Indian species are *Branchiostoma caribbaeum, B. indicum and B. pelagicum.*

☒ External features

1. *Branchiostoma* is a transparent, fish-like animal occurring near the shore, burrowing in sand.
2. The body is narrow, 2.5-5.8 cm long, laterally compressed and pointed at both the ends. The anterior two-thrids of the body is triangular in cross-section.
3. Along the middorsal line a **dorsal fin** extends the whole body length. It is joined posteriorly to a somewhat broader **caudal fin** around the tail (Fig. 34.2).
4. A **ventral fin** is situated midventrally running from the caudal fin to a median opening, the **atriopore**.
5. The muscles are arranged in sixty-two V-shaped segments or **myotomes**.
6. Three unpaired apertures are present:

a. The **mouth** overarched by the median, ventral oral hood, and fringed with tentacle-like cirri.

b. The **atriopore** in myotome thirtysix is to expel water from the pharynx which enters through the mouth.

c. The **anus** is ventral and slightly to the left behind the atriopore, at some distance from the posterior end of the body.

7. The **notochord** is a flexible, unsegmented rod, pointed at both the ends and runs from one to the other end of the body.
8. The **pharynx** is supported by gill rods, which border the numerous gill slits.
9. The **intestine** is straight and without any loop.
10. The **hepatic diverticulum** is simple.
11. The **circulatory system** is ill-developed. A definite heart is absent but the ventral or the branchial artery is rhythmically contractile.
12. The blood is colourless. A few amoeboid cells are present in it.
13. About ninety pairs of segmentally arranged **nephridia** are present on the dorsolateral walls of the pharynx.
14. The **dorsal nerve cord** is shorter than the notochord, above which it lies.

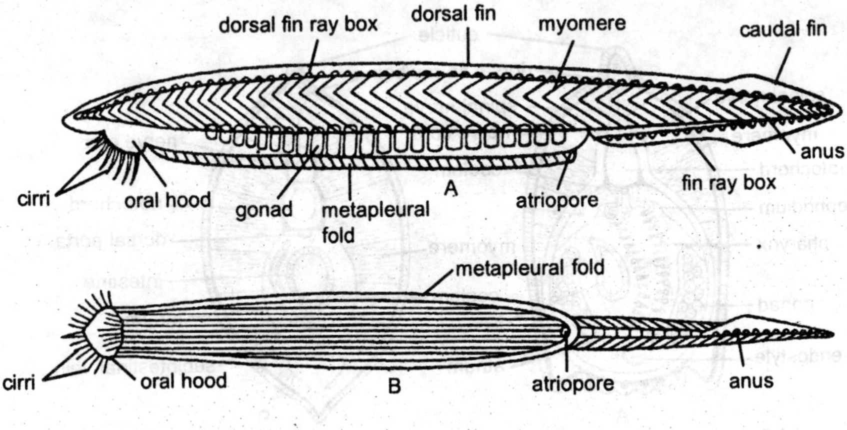

FIGURE 34.2

***Branchiostoma (Amphioxus)* sp.** A. Lateral view, B. Ventral view

15. A definite **brain** is absent although anteriorly the central canal of the nerve cord widens to form the so-called cerebral vesicle, also called brain.
16. Two sets of nerve roots are present. The two sets of roots do not unite as in vertebrates. Spinal and sympathetic ganglia absent.
17. The sexes are separate but similar. The gonads are segmental and without duct.

☒ Body wall

1. The body is covered with a **cuticle** bearing numerous pores.
2. An **epidermis** of a single layer of columnar cells underlines the cuticle. Some of the epidermal cells form **unicellular glands** and some are epidermal **nerve cells.**
3. The dermis beneath the epidermis is formed chiefly of connective tissue.
4. The muscular layer is metamerically segmented and consists of about sixty muscle-segments or **myomeres** (myotomes) (Fig. 34.2) separated from one another by **myocommas** or partitions of dense connective tissue.
5. In surface view, the myotomes appear as a series of open 'V's with forwardly directed apices.
6. Flat, striated **muscle-plates** are arranged longitudinally in each myotome, enabling its attachment to the successive myocommas.
7. The myotomes of the right and left sides are arranged not in opposite but alternately.
8. In the floor of the atrial cavity a set of **transverse muscles** extends across the ventral surface of the anterior two-thirds of the body.
9. The dorsal portion of the muscular layer of the body wall (Fig. 34.3) is greatly thickened.

☒ Skeletal System

Notochord, oral ring, gill rods and fin rays constitute skeletal system.

Notochord

Notochord, the main supporting structure is a cylindrical rod tapering at both ends and extends the whole length of the body (Fig. 34.4) along middorsal line, above alimentary canal. **Notochordal tissue** formed of large vacuolated cells constitute the notochord. It is enclosed in a sheath—the **notochordal sheath**. A structureless layer secreted by the cells of notochord occupies the space between the two. The principal function of the notochord is to prevent shortening of the body during contraction of longitudinal muscles.

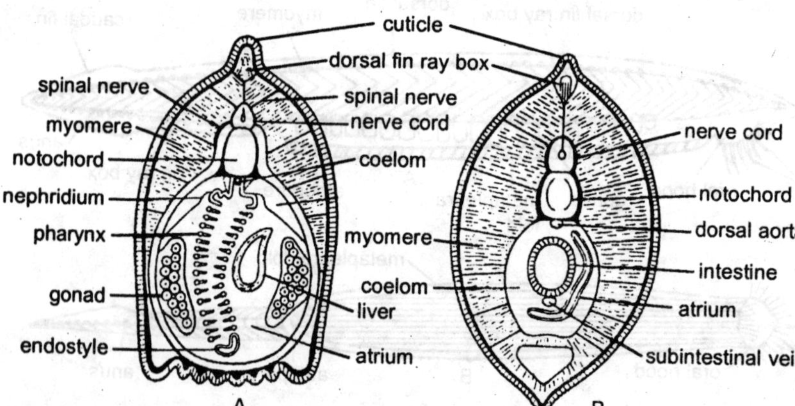

FIGURE 34.3

***Branchiostoma* sp.** Cross section. A. Pharyngeal region, B. Intestinal region

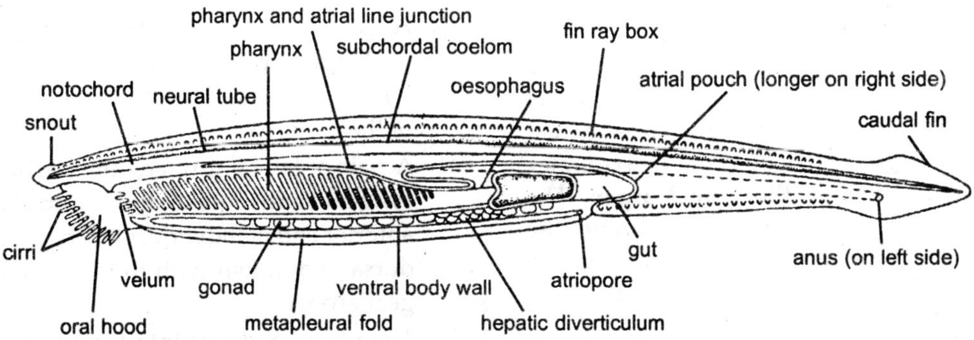

FIGURE 34.4

***Branchiostoma* sp.** Gross internal structures (lateral view). Body wall of a side removed

Oral ring

A ring made of cartilage-like matter; the oral ring, supports the oral hood. The ring consists of separate rod-like pieces arranged end to end corresponding with the number of cirri. From each piece an offshoot runs into the cirrus and supports it.

Gill rods

The pharynx is supported by branchial rods made of elastic fibres (see Pharynx).

Fin rays

Short rods of connective tissue, the fin rays support the fin. The rays are in a single series in the dorsal fin but of double series in the ventral fin.

Locomotion

Branchiostoma is a sluggish animal but it can swim freely. Swimming is effected by contraction of longitudinal muscles of the body. Transverse motion of the body at different angles due to contraction of the myotomes helps the animal to move forward. The myotomes contract anteroposteriorly to produce waves of contraction from the anterior to the Posterior. Myocommas are attached to the notochordal sheath and the notochord helps in efficient contraction of myotomes.

Atrium

The atrium is a wide space between the body wall and pharynx. In cross section it is crescentic (Fig. 34.3A), surrounding the ventral and lateral regions of the pharynx.

The atrium is closed in front, opens outward by an **atriopore** behind the level of the pharynx, and extends backward as a blind pouch. The cavity is lined with atrial epithelium of ectodermal origin.

Coelom

The coelom (Fig. 34.3A.-B) is divided into a number of compartments. The alimentary canal is surrounded by the coelom. In the pharyngeal region, the coelom is restricted in paired cavities, one on either side of the dorsal region of the pharynx, above the atrium. The cavities extend as narrow canals in the primary branchial lamellae and continuous with a median elongated space beneath the endostyle. The coelom surrounding the intestine is much reduced on the right side. The hepatic diverticulum is lodged in a forward extension of the coelom on the left side. In the metaplaures the coelom exists as separate cavities. The spaces in the gonadal pouches are coelomic in origin.

☒ Digestive System

Mouth, foregut, midgut, hindgut, anus and hepatic diverticulum constitute digestive system (Fig. 34.4). The pharynx and oesophagus comprise the foregut, the portion between oesophagus and ileocolon ring is midgut and the rest or the intestine is hindgut.

Mouth

The mouth is situated at the base of a **vestibule** overarched by a median ventral **oral hood** containing tentacle-like structures, called **buccal cirri**. The oral hood contains a complex set of ciliated tract, the **wheel organ**. The mouth is surrounded by a membrane, the **velum**, and is fringed by twelve velar tentacles and leads posteriorly into the pharynx.

Pharynx

The **pharynx** is a high, compressed chamber and occupies anterior half of the body (Fig. 34.4).

1. The walls of the pharynx are perforated by nearly 200 oblique vertical slits, the gill slits or **branchial apertures**.
2. The portions of the pharyngeal wall separating the clefts are known as **branchial lamellae**.
3. Each lamella is supported towards its outer edge by a rod made of elastic fibres, the **branchial rod**.
4. The rods are united with one another dorsally by loops but free ventrally.
5. The rods at their free ends are alternately forked and simple and known as **primary** and **secondary branchial rods**, respectively.
6. The branchial lamellae bearing the primary rods are **primary lamellae** and those lodging the secondary rods are **secondary lamellae**.
7. Transverse branchial lamellae supported by branchial rods connect the primary lamellae at regular intervals.
8. The inner and lateral faces of each lamella are provided with long cilia.
9. On the ventral wall of the pharynx is a longitudinal groove, the **endostyle**. The groove is composed of four tracts of mucous glands, each separated from the other by tracts of ciliated cells.
10. A longitudinal groove, the **epipharyngeal** or **hyperpharyngeal groove**, lined with cilia is present on the dorsal surface of the pharynx.
11. A pair of lateral grooves, the **peripharyngeal bands** encircle the mouth at the anterior end of the endosyle. They meet dorsally and join with the epipharyngeal groove.
12. The pharynx opens into the **oesophagus.**

Oesophagus

1. The oesophagus joins the **midgut** posteriorly. At the junction of the two a large pouch, the **hepatic diverticulum** or **liver** or **midgut diverticulum** is present.
   a. The diverticulum runs forward for some distance along the pharynx on its right side.
   b. Large cilia are present on the inner surface of the dorsal and ventral walls.
   c. **Zymogen cells** in the diverticulum secrete **hydrolyzing (digestive)** enzymes.
2. The densely ciliated posterior end of the midgut is called **ileocolon ring.**

Hindgut

The narrow **hindgut** (intestine) runs straight backward from the ileocolon ring and opens in the ventral **anus**, slightly to the left.

☒ Respiratory and feeding mechanism

Cilia of the wheel organ and pharyngeal wall maintain a continuous flow of water, which enters the pharynx through the mouth and passes to the exterior through the gill slits, atrium and atriopore. The branchial lamellae are highly vascular and a gaseous exchange* takes place there.

---

* Many authors doubt the respiratory function of pharynx. The amount of oxygen required by *Branchiostoma* is very little and oxygenation of blood in superficial lacunae is sufficient to meet the need.

Minute organic particles, chiefly microscopic organisms, are brought into the pharynx by the water current. During feeding, a continuous stream of water enters the pharynx through the mouth and passes to the atrium through the pharyngeal slits, from where it is expelled to the exterior through the atriopore.. The cilia oreinted along the anterior and posterior faces of each gill lamella, known as **lateral cilia**, work in unison with the cilia distributed over the inner (pharyngeal) surface of the gill lamella, known as **frontal cilia**, and thus form an efficient ciliary feeding machine (Fig. 34.5), and, as such, the *Branchiostoma* is known as a **ciliary feeder** (Grove and Newell). The endostyle as well as the pharyngeal epithelium secrete a cord of mucus in which food particles are entangled. By the beating movement of the cilia, food particles entangled in the twisted rope of mucus in the endostyle are driven forwards, then upwards through the peripharyngeal bands and backwards along the epipharyngeal groove and finally into the opening of the oesophagus.

Summary of the feeding mechanism

The cilia in the gill lamellae set up an water current, which enters the pharynx → gill slit → atrium. Food particles are entangled in the mucous layer of the pharyngeal surface, the frontal cilia help in driving the food into the oesophagus.

⊠ Digestion and absorption

Digestion is both extracellular and intracellular. Intracellular digestion occurs in the hindgut. Extracellular digestion begins in the midgut. Carbohydrases, proteinase and lipase are secreted by the epithelium of the hepatic diverticulum and midgut. The hydrolyzing enzymes are mixed with food by ciliary activity. Most of the digestion occurs in hind end of the midgut beyond iliocolon ring, where the mucous cord containing food is rotated by the movement of the cilia and mixing of enzymes and food is ensured.

The chief site for absorption is intestine. Certain amount of absorption possibly takes place in the hinder part of the midgut.

⊠ Blood Vascular System

Branchiostoma does not posses **heart** (Fig. 34.6). The principal arteries have muscular walls but an endothelial lining is present only in the dorsal aorta.

1. A voluminous sac, **sinus venosus** is lodged ventral to the posterior part of the pharynx.
2. A large median artery, **ventral aorta** or **endostylar artery** arises from the sinus venosus and extends forward, ventral to the pharynx.
3. The ventral aorta sends branches, the **branchial vessels**, to the gill bars.
4. The branchial vessel dilates to form a small **branchial bulb or bulbule** at the base of each primary gill bar.
5. A pair of **dorsal aortae**, one on each dorsolateral side of the pharynx receive blood from the gill bars.

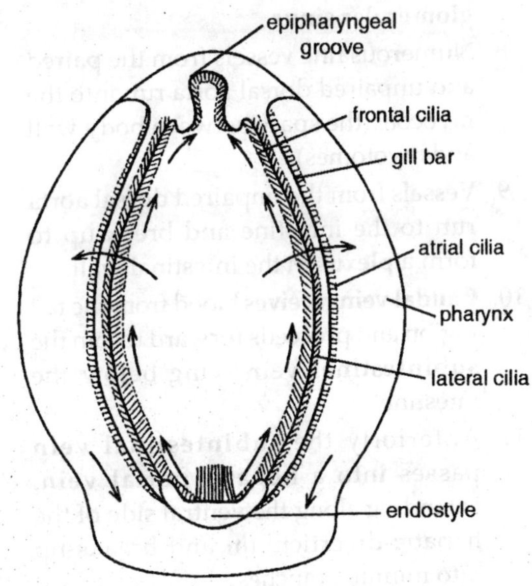

FIGURE 34.5

***Branchiostoma*. sp.** Transverse section of pharynx showing course of water current and movement of cilia

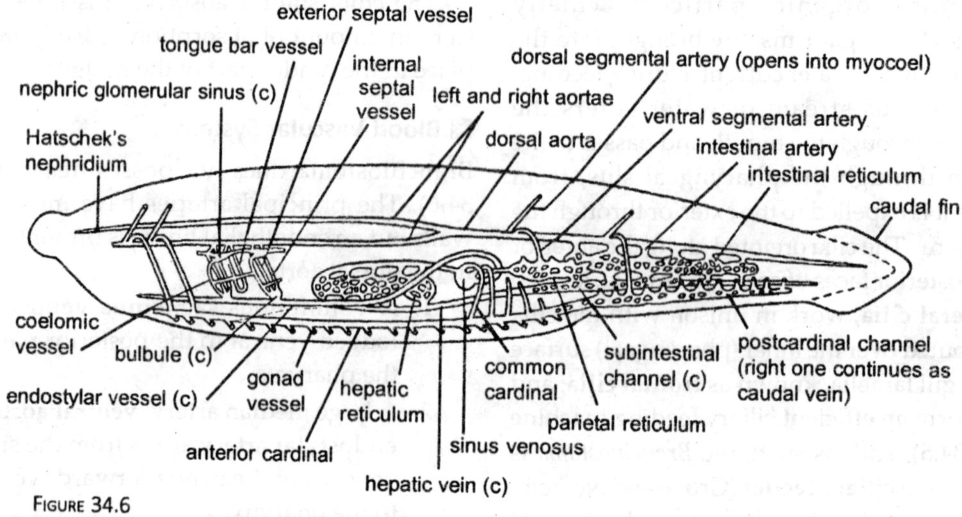

FIGURE 34.6

***Branchiostoma*. sp.** Blood vascular system (diagrammatic) (c) indicates contractile vessel

6. Posteriorly the two dorsal aortae join to form a **median dorsal aorta** extending up to the tail, where it is called **caudal artery**.
7. Small vessels arising from the paired dorsal aortae run to the nephridia, where they form a network, the **nephric glomerular sinus**.
8. Numerous fine vessels from the paired and unpaired dorsal aorta run into the myocoel (the space between body wall and myotomes).
9. Vessels from the unpaired dorsal aorta run to the intestine and break-up to form a **plexus** in the intestinal wall.
10. **Caudal vein** receives blood from the tail region and proceeds forward to join the **subintestinal vein** lying below the intestine.
11. Anteriorly the **subintestinal vein** passes into a **hepatic portal vein**, extending along the ventral side of the hepatic diverticulum and breaks-up into minute branches.
12. The **hepatic** vein arising from the hepatic diverticulum extends along the dorsal part of the latter and joins the posterior end of the **sinus venosus**.
13. Two pairs of veins, **anterior** and **posterior cardinal veins** collect blood from the ventrolateral sides of the body wall.
14. **Common cardinal vein** or **ductus cuvieri** is formed by the joining of two cardinals of each side and passes ventrally to join with sinus venosus.

Blood is colourless and contains no **leucocytes**. Oxygenation of blood occurs in the lacunae, close to integument. Most of these spaces are in the dorsal and ventral fins and paired canals in the metapleures. The sinus venosus and principal vessels contract at a very slow rate.

Blood flows forward from the posterior end through ventral vessel, subintestinal vein and the posterior cardinal veins. Blood flows backward through paired and unpaired dorsal aortae and anterior cardinal veins.

⊠ Excretory System

The excretory organs of *Branchiostoma* consist of about 90 pairs of true nephridia and a pair of brown funnels. Some parts of the atrial wall also serve as excretory organ. Of the different excretory structures, the nephridia are supposed to play the most important role. The nephridium does not obtain wastes from the coelom, instead, it collects the same from the blood vessels that surround it.

☒ Structure of a nephridium

1. The nephridia are situated between the atrial epithelium and the epithelium of the dorsopharyngeal coelom, on the dorsolateral wall of the pharynx.
2. Each nephridium is a bent tube (Fig. 34.7) lined with cilia, and consists of an **anterior vertical** and a **posterior horizontal** limb in relation with the primary gill lamella.
3. The nephridium opens into the atrium opposite the dorsal end of a branchial rod, in one of the dorsal pouches.
4. Close to the opening, it is divided into two canals, the horizontal limb passes forward and then turns ventral and the vertical limb passes backwards.
5. The vertical limb terminates in a large group of solenocytes. Several smaller groups are present on the horizontal limb. A **solenocyte** is a long tubular cell, closed above by a knob containing a nucleus, from which hangs down a long

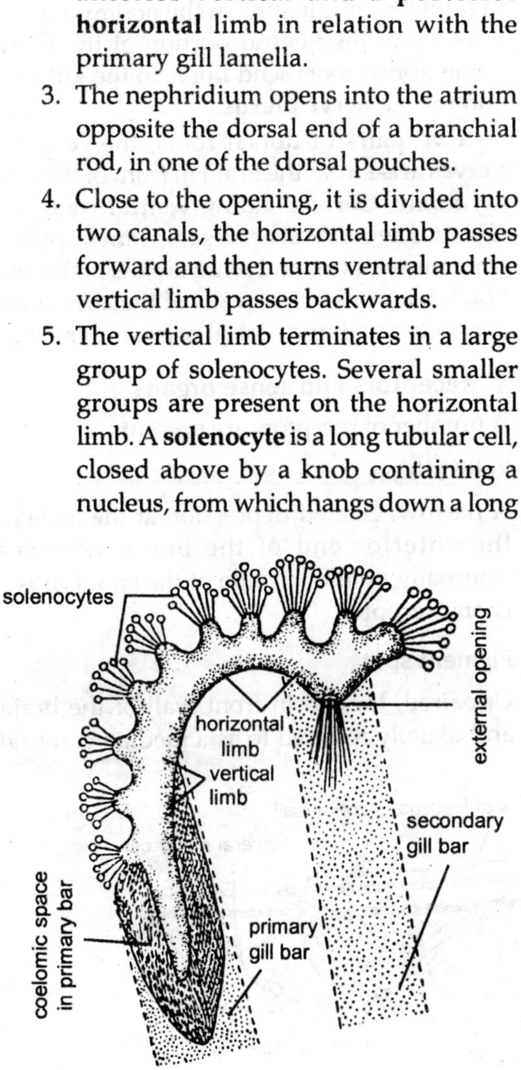

FIGURE 34.7

***Branchiostoma* sp.** A nephridium

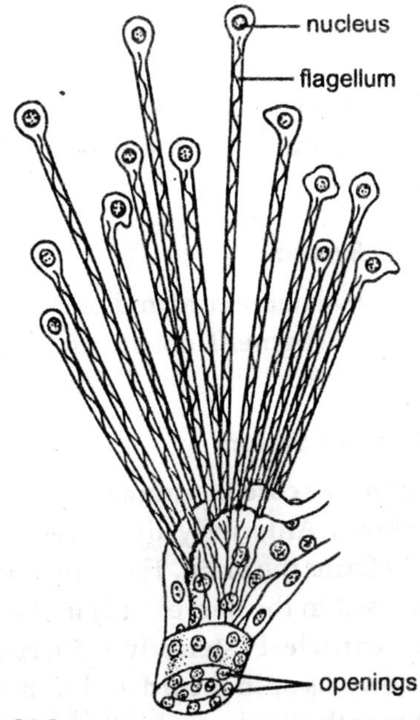

FIGURE 34.8

***Branchiostoma* sp.** Solenocytes. (showing opening into the main excretory canal)

flagellum (Fig. 34.8). The movement of the flagellum flushes out the wastes from the cell-stalk. These solenocytes have a close similarity with those found in polychaete annelids.

6. A tuft of specially long cilia project into the atrium through the renal opening.
7. The nephridia receive special blood supply from the **paired** dorsal aortae and the extracted wastes are discharged into the atrium. Branchiostoma is the only chordate with true nephridia as excretory organs.

Brown funnels

A pair of funnel-shaped structures known as brown tubes, are present on the hinder part of the coelom and open into the atrium at their wide ends. The brown funnels are presumably of excretory function.

Atrial wall

Some parts of the atrial wall are provided with groups of columnar cells, which are supposed to be excretory in function.

The development of the nephridia in Branchiostoma indicates that they are ectodermal in origin and, as such, they are in no way comparable to the pronephros of vertebrates.

⊠ Nervous System

A hollow dorsal nerve cord and paired nerves in each body segment constitute nervous system (Fig. 34.9).

Central Nervous System

The **dorsal nerve cord** lies immediately above the notochord. Anteriorly the nerve cord is dilated and forms the **brain**. The neurocoel (the space enclosed in the nerve cord) in the brain forms the **ventricle**. Posteriorly the nerve cord tapers, known as **spinal cord** and terminates at a point over the hinder end of the notochord. The whole nerve tube constituting **central nervous system** is ensheathed by a delicate connective tissue.

The anteriormost region of the brain bears a small, hollow pouch, the **median olfactory lobe**. The **infundibular organ**, possibly an organ of special sense, consists of small ciliated cells is embedded in the posterior and ventral walls of the cerebral ventricle. In young Branchiostoma the cerebral ventricle opens to the exterior by a **neuropore**, which subsequently closes leaving a depression, the **Kolliker's pit**.

The nerve cord is chiefly made of long nerve fibres, with numerous nerve cells, mostly grouped around the neurocoel. Giant nerve cells are also present.

Peripheral nervous system

Consists of paired nerves given off from the dorsal nerve cord in each body segment. The paired nerves are actually **dorsal** and **ventral roots** not united. The dorsal root is not ganglionated as in vertebrates. The ventral root carries motor fibres ending in a myotome. The dorsal root is mixed, possesses both sensory and motor fibres and passes between the myotomes, terminating in the nonmyo-tomal muscles of the ventral portion of the body. Some dorsal roots send fibres to the gut wall and form a **nerve plexus**.

Two pairs of dorsal roots, the **cerebral nerves** arise from the brain in front of the first myotome. Corresponding ventral roots are absent. The cerebral nerves transmit impulses from the tentacles and sense organs of the oral hood. Neurons in the walls of the atrium are connected with the central nervous system.

⊠ Receptors and sense organs

A number of receptors are present.

Kolliker's pit

A narrow, ciliated depression at the level of the anterior end of the brain. It opens externally on the left side of the snout. It is a **chemoreceptor**.

Pigment spot

Unpaired, located in front wall of the brain and usually referred to as a median **cerebral**

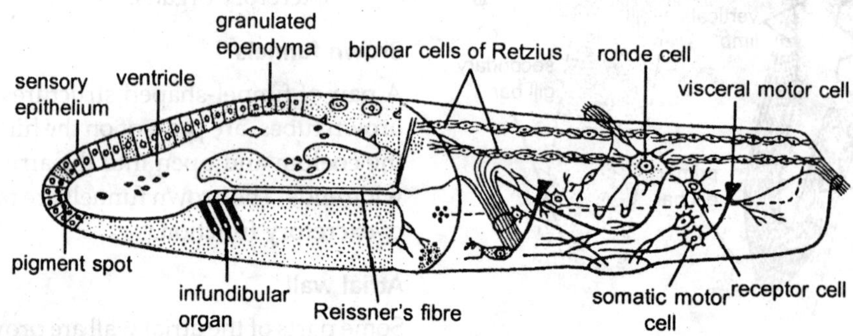

FIGURE 34.9

***Branchiostoma* sp.** Nervous system (brain shown in longitudinal section)

eye. Lens and other accessory structures of an eye are lacking and it is not sensitive to light.

Eye spots

Cups of pigment granules enclosing **photosensitive** cells are present on the spinal cord. They are oriented in different directions.

Sensory papillae

Sensory cells in the oral cirri and velar tentacles are **thermo** and **chemoreceptors**.

Infundibular organ

The infundibular organ (see nervous system) in the posterior and ventral walls of the cerebral ventricle is probably a **photoreceptor**.

Epidermal sensory cells

Sensory cells are scattered on the body surface, particularly on the dorsal body wall.

☒ Reproductive System

The sexes are separate but the reproductive organs are similar in both male and female.

The **gonads** are simple pouch-like segmental organs (Fig. 34.3A), located in the ventrolateral sides of the pharyngeal region between 10 and 30 segments. They project into the atrium and largely fill its cavity. The inner or medial face of each pouch is covered by atrial epithelium.

The **ova** or **sperms** escape with the rupture of the **gonadal pouches**. The gametes reach the atrium and from there escape to the exterior through the atriopore. Fertilization external, segmentation (Figs. 43.5A, 43.6) complete. An active larval stage is present. The active form is transformed into a sluggish adult following **metamorphosis**.

BIOLOGICAL STATUS (SYSTEMATIC POSITION) OF BRANCHIOSTOMA (AMPHIOXUS)

The miniature fish-like lancelet, *Branchiostoma* (*Amphioxus*), a lowly organised animal is a simplified form of chordate, which exhibits important features of the group, viz. presence of a notochord, pharyngeal gill slits and a dorsal, hollow, tubular nerve cord.

☒ Systematic position of Branchiostoma

The Lancelet has had a chequered zoological history. Initially, in the eighteenth century, it was thought to be a slug (gastropod mollusc), later as a fish. Still later, due to lack of important vertebrate characters it was separated from fish and for its resemblances with tunicates, larval form of Agnatha and other features, a special division Acrania (Cephalopoda) was created for it.

Affinities with Annelida

Similarities	Dissimilarities		
	Structure	**Annelida**	**Branchiostoma**
1. Body bilaterally symmetrical and segmented	1. Segmentation	1. Metameric	1. Restricted to myotomal region
2. Nephridia segmentally arranged	2. Coelom	2. Schizocoelic	2. Enterocoelic
3. Coelom well developed	3. Nerve cord	3. Solid, ventral	3. Hollow, tabular, dorsal
4. Disposition of blood vascular system	4. Notochord	4. Absent	4. Present
	5. Gill slits	5. Absent	5. Present

☒ Affinities with Mollusca

Molluscs and Branchiostoma differ widely. Superficial resemblances in ciliary feeding and respiratory mechanism (no gill slits in molluscs) have no phylogenetic significance.

☒ Affinities with Echinodermata

Similarities

1. Early development cleavage, gastrulation and mesoderm formation.
2. Three separate enterocoelic cavities.
3. The pores of the calyx of some fossil carpoid echinoderms are similar to the gills slits of Branchiostoma.
4. Presence of creatin phosphate and phosphagens.

At present, the echinoderms are not considered as an ancestor of chordates. The similarity is presumably due to a remote common origin of both the groups.

☒ Affinities with Hemichordata

Similarities

1. Pharyngeal apparatus.
2. Feeding and respiratory mechanism.
3. Origin and development of enterocoelic coelomic pouches.

☒ Affinities with Urochordata

Similarities

(Tadpole larva and Branchiostoma)

1. Body shape and a fin-bearing tail.
2. Notochord extends nearly whole body length.
3. A hollow nerve cord above notochord.
4. Pharynx with endostyle, pharyngeal and epipharyngeal grooves.
5. Feeding and respiratory mechanism.

☒ Affinities with Cyclostomata

Similarities

(Branchiostoma and Ammocoete larva)

1. Body elongated bearing a continuous dorsal fin.
2. Presence of an oral hood.
3. Velum guards the mouth.
4. Presence of endostyle.

☒ Adult Branchiostoma and Cyclostomata

Similarities	Dissimilarities
1. Presence of myotomes. 2. Presence of gill slits.	1. Absence of vertebral column and paired sense organs in Branchiostoma.

⊠ Affinities with Vertebrata

Similarities

1. Presence of pharyngeal gill slits.
2. Presence of notochord.
3. Presence of dorsal, hollow, tubular nerve cord.
4. Presence of phosphocreatin and phosphoarginine, the sources of energy supply for muscular activity.

The old concept that Branchiostoma is a recent derivative of ascidians has been abandoned. Alternatively, Acrania may be a specialised and degenerate descendant of the Agnatha. The evidence as to the real position of Acrania is still insufficient, but investigations tend to bring them nearer to the craniates. It is probable that the Branchiostoma and its allies, fish-like creatures existed in Silurian seas, and the vertebrates evolved from animals of that kind.

CLASSIFICATION: SUBPHYLUM ACRANIA

1. Lack distinct head, cranium, brain, eyes, auditory organs and jaws.
2. The nerve cord is dorsal and hollow.
3. Notochord persistent and extends along the whole length of the body.
4. Gill slits numerous and open in the atrium.
5. Blood vascular system and nervous system resemble those of vertebrates.
6. Body wall consists of over fifty myotomes.
7. The larval form is asymmetrical and larval period is long.

Subphylum Acrania has only one class.

Class CEPHALOCHORDATA

1. Small, almost transparent about 5.8 cm long, lancelate organisms.
2. The anterior two-thirds of the body is roughly triangular in transverse section, and posterior third is nerarly oval in section.
3. The dorsal fin is a median longitudinal fold along the whole of the dorsal border and continued round the posterior end of the body extending forwards as the ventral fin.
4. Below the pointed anterior extremity is a large median aperture, the oral hood, surrounded by a frill-like membrane, the edge of which is beset with numerous tentacles or cirri. The oral hood encloses a cup-shaped vestibule which communicates with the mouth.
5. Finger-like lobes called "wheel organs" are present on the wall of the oral hood.
6. Along the roof of the vestibule runs a ciliated groove—the groove of Hatschek.
7. The notochord is persistent and unsegmented.
8. The nerve cord is dorsal and tubular.
9. Gill slits are numerous and open in the atrium.
10. The body wall consists of about sixty 'V-shaped' myotomes.
11. The larval form is asymmetrical and the larval period is long.

Examples: *Branchiostoma, Asymmetron.*

SUBPHYLUM CRANIATA (Vertebrata)

The craniates are vertebrates with a cranium or brain box, enclosing a highly complex brain; a muscular two-three or four-chambered heart and blood with red blood corpuscles. Notochord is not continued up to the end of snout, but stops beneath the forebrain. The pharynx is perforated by not more than seven paris of gill slits except in some cyclostomes. The gill pouches never open into an atrium. The liver is massive. The renal tubules unite to form large paired kidneys opening into ducts. The brain is complex and cranial nerves are at least ten pairs. Eyes well developed, complex and paired; auditory organs paired. A single pair of gonads are present. The coelom is always developed as a schizocoele.

The craniata includes a diverse group of vertebrate animals like agnathans, fishes, amphibians, reptiles, birds and mammals. Although they are diverse in their physical appearance, they have general agreement in the basic arrangement of various organ-systems and in the structure of the organs themselves, with a gradual sequential development.

Two superclasses Agnatha and Gnathostomata constitute the Subphylum Craniata or Vertebrata. Agnatha has two classes while Gnathostomata embraces seven classes from the true fishes to the mammals, including man.

35 SUPERCLASS AGNATHA

Agnatha is the first vertebrate group, probably evolved from protochordates by paedomorphosis, appear to have been ancestor of other fish groups. Most of these are extinct, but it is represented today by about 50 species of jawless hagfishes and lampreys. There is no fossil representative of cyclostomes to indicate their course of evolution. The primitive agnathans originated probably in marine environment, subsequently entered fresh water and underwent an extensive adaptive radiation. In the first part of Palaeozoic era they flourished but were replaced by more progressive fishes in Devonian period. They are arranged in six classes—Euphanerida, Heterostraci, Anaspida, Osteostraci, Petromyzontia and Myxinoidea. The first four classes are extinct. All extinct forms had an extensive development of heavy bony scales and plates in their skin and are collectively known as **Ostracoderms**. The existing agnathans, hags and lampreys are often grouped together in class Cyclostomata (Gr. Cyklos, Circulars + stoma, mouth) being distinct from all other living vertebrates by the possession of pouched gills, a single nostril, a single gonad, two semicircular canals in each ear and the absence jaws and paired fins. The hags and lampreys are placed under separate classes probably due to their separate origin.

⊠ Petromyzon and Myxine: Salient features

Petromyzon (Lamprey) belongs to family Petromyzontidae, order Petromyzontia and *Myxine* (Hag) is placed under family Myxinoididae, order Myxinoidea. Both are included in class Cyclostomata, superclass Agnatha, subphylum Craniata (Vertebrata), phylum Chordata.

PETROMYZON (Fig. 35.1)	MYXINE (Fig. 35.15)
1. Live in rivers and seas.	1. Exclusively marine.
2. A median, lobed dorsal fin and a tail fin.	2. The fin is confined to tail.
3. Sensory structures in the complex, glandular, slimy, pigmented skin.	3. Numerous large glands present in the complex slimy skin.
4. Buccal funnel present, surrounded with papillae.	4. Buccal funnel absent. The mouth is edged with tentacles supported by cartilage.
5. Teeth many, yellow and horny. The tongue also bears horny teeth.	5. A single medial tooth and two rows of smaller teeth present on the tongue.

contd.

PETROMYZON (Fig. 35.1)	MYXINE (Fig. 35.15)
6. Eyes hidden and retarded in the larva, exposed and complete in the adult.	6. Eyes hidden and rudimentary.
7. Seven pairs of gill pouches, opening directly to the exterior and less directly into the adult gullet.	7. Six pairs of gill pouches opening directly into the gullet and less directly to the exterior.
8. A slight constriction indicates separation of utricle and saccule, semicircular canals two.	8. Internal ear with a saccule and one semicircular canal.
9. The nostril unpaired, dorsal and ends blindly in the pituitary sac.	9. The nostril large, unpaired, dorsal and opens into the pharynx through a passage, the pituitary sac.
10. A velum guards the respiratory tube from the buccal cavity.	10. A velum separates the buccal cavity from the pharynx.
11. The intestine is without convolutions. Typhlosole in the form of a spiral valve present.	11. The intestine is very wide and bears longitudinal ridges.
12. A urinogenital sinus and two genital pores.	12. No urinogenital sinus, only one genital pore.
13. Skull very imperfectly roofed.	13. Skull without any roof.
14. Hints of vertebral arches, cartilaginous branchial basket round gill pouches.	14. Skeletal system less developed, only a hint of a branchial basket.
15. All the components of the brain are distinct and well-marked.	15. Cerebrum and cerebellum rudimentary.
16. Lateral line sense organs are not enclosed in a canal and exposed.	16. Lateral line sense organs absent.
17. Sexes are separate.	17. Protandrous hermaphrodite.
18. Ova small and spherical.	18. Ova large and oval with attaching threads.
19. Cleavage holoblastic.	19. Cleavage meroblastic in *Bdellostoma*.
20. Development with a metamorphosis.	20. Development direct in *Bdellostoma*.

PETROMYZONE sp. (Lamprey)

Lamprey, also called lamper eels are most common in the European river system. The marine form of sea-lamprey (*P. marinus*) is about one metre long; the fresh water lamprey (*Lampetra fluviatilis*) is about 90 cm long and *L. planeri*, a lesser fresh water form is about 45 cm in length.

Two distinct phases are present in the life-history of lampreys. They spend major part of their adult life in sea and are provided with a suctorial mouth. The larval phase, the **Ammocoetes** lives in fresh water and feed on microscopic food trapped in endostyle.

⊠ External features

The head and trunk are nearly cylindrical and the tail is laterally compressed (Fig. 35.1). The smooth, slimy body surface is devoid of scales. The upper surface is usually dark and the

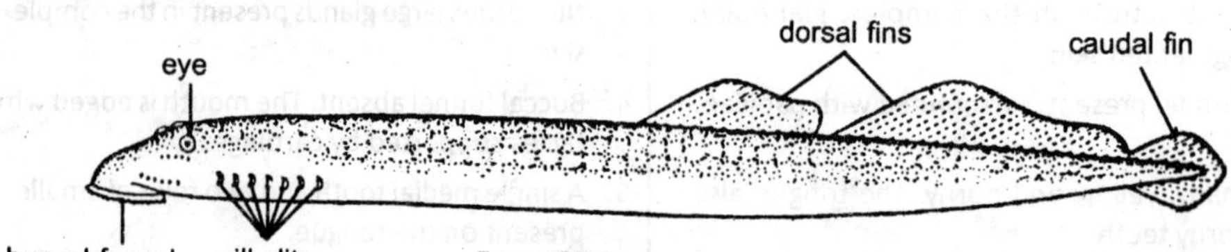

FIGURE 35.1

Petromyzon sp.

lower surface white. A median **dorsal fin** divided into two unequal parts and a **caudal fin** continuous with posterior dorsal fin are present. The fins are supported by fine cartilaginous fin rays.

At the anterior end a large downwardly directed basin-like depression, the **buccal funnel** is present (Fig. 35.2A). It is surrounded by a **marginal membrane**, overlapping the **oral fimbrae** and **sensory cirri** arising from outside the marginal membrane. Radiating rows of **yellow, horny teeth** on cartilaginous pads are present in the buccal funnel. A prominence, the **tongue** bearing large, horny teeth projects from the bottom of the buccal funnel. The narrow **mouth** lies just above the tongue.

A single median **nostril** and the **pineal organ**, closely behind it are present on the dorsal surface of the head. **Eyes** are paired, lateral, without **eyelids** but covered with a transparent skin (Fig. 35.2B, C).

Gills slits or **gill pores** are seven pairs of small openings on the sides of the head.

The **anus**, a narrow opening lies in a small depression, the **vent** on the ventral surface at the junction of trunk and tail.

A small papilla bearing urinogenital aperture lies immediately behind the anus.

Segmental sense organs, the **lateral line sense organs** on either side of the body and head are exposed to the exterior, not enclosed in lateral line canals.

Body wall and muscles

The skin is soft. The epidermis bears numerous mucus secreting unicellular glands. In addition to these, two other types of secretory cells are present (a) **granular cells** of unknown function and (b) elongated **club cells** with hyaline cytoplasm, probably secreting slime. A layer of **collagen fibres** form the **cutis** beneath the epidermis. Pigment cells or **chromatophores** are present in this layer. The pigment cells are star-shaped and change their position under different conditions of illumination, imparting change of colour from pallid to dusky and vice versa.

The muscles of the trunk and the tail are arranged in **myotomes** in a zigzag course.

Locomotion

The lamprey swims by producing waves of curvature from anterior to posterior of the

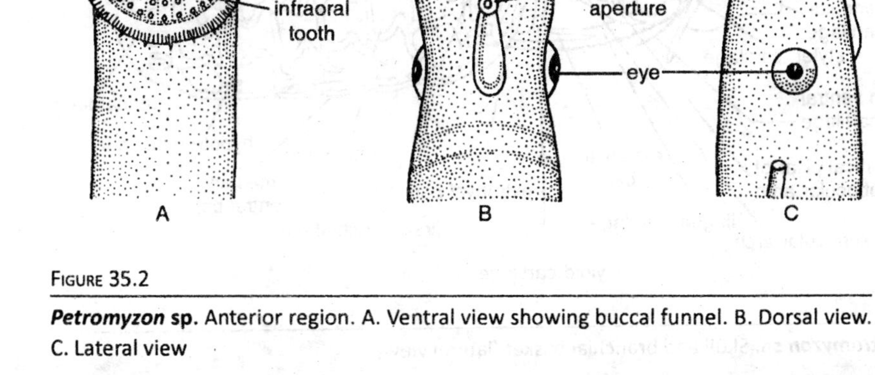

FIGURE 35.2

***Petromyzon* sp.** Anterior region. A. Ventral view showing buccal funnel. B. Dorsal view. C. Lateral view

body, alternately on each side, due to contraction of myotomal longitudinal muscle fibres. Since the waves are short, progress by swimming is slow.

Endoskeleton

The skeleton is wholly cartilaginous. The notochord, ill-developed skull and branchial basket constitute axial skeleton.

Notochord

The notochord (Fig. 35.4) is persistent, unsegmented and enclosed in a tough sheath of an inner fibrous and an outer elastic layer. Small, segmentally arranged, vertical cartilaginous rods are attached to the sides of the notochord bounding the spinal canal. In the caudal region the rods fuse to form a single plate bearing foramina for the spinal nerves and sends processes to the base of the fin.

Cranium

The cranium is primitive. The floor is occupied by a basal plate formed by paired **parachordals** and **trabeculae**. An incomplete box enclosing the brain and special sense organs is formed by pieces of cartilage attached to the basal plate (Fig. 35.3). A large aperture, the **basicranial fontanelle** is present in front of the cranium. The roof of the cranium is made of membranous fibrocar-tilage, supported by a transverse bar. Auditory capsules are united with the posterior end of the basal plate and form the end of the **neurocranium**. The **olfactory capsule** is an imperfectly paired concavo-convex plate, supporting the posterior wall of the olfactory sac. Fibrous tissue joins the capsule with the cranium. A subocular arch extends downward and outward from each side of the nasal plate to support the eye (Fig. 35.3). A slender styloid process hangs from the **subocular arch** and a small **cornual cartilage** is connected to its lower end. The skeleton of the buccal funnel and branchial basket are attached to the cranium

A ring-like **annular cartilage** supports the buccal funnel. The tongue is supported by a long, unpaired **lingual cartilage**.

Visceral skeleton

The visceral skeleton (Fig. 35.3) consists of a **branchial basket** of nine irregular, vertical cartilaginous bars or rods on each side. The first rod is located immediately posterior to the styloid process, the second in front of the first gill slit and the rest posterior to the

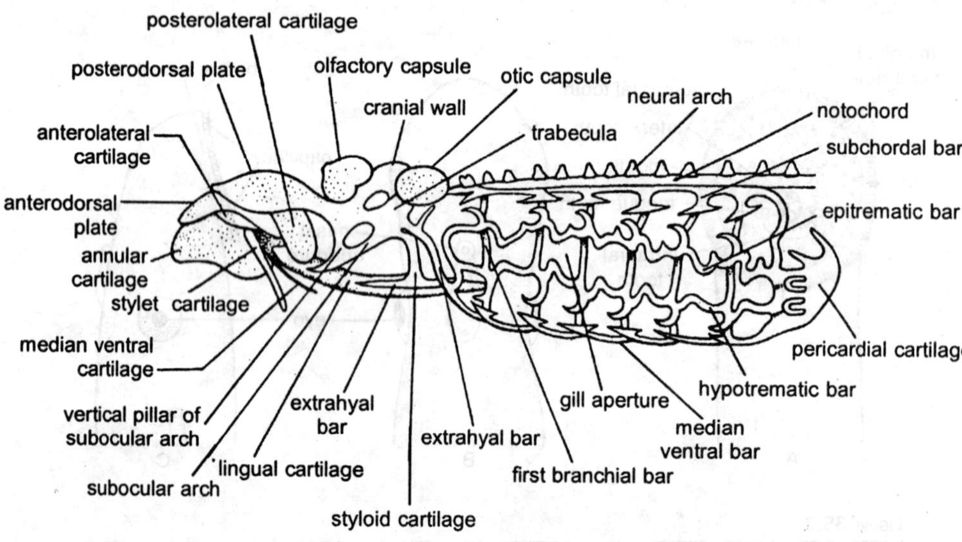

FIGURE 35.3

***Petromyzon* sp.** Skull and branchial basket (lateral view)

remaining gill slits. The vertical bars are joined by longitudinal bars. Posteriorly the branchial basket extends to form a cup-like **pericardial cartilage** accommodating the heart.

Coelom

The coelom is represented in two cavities, a small, anterior **pericardial cavity** containing heart and a large posterior **pleuroperitoneal cavity** containing visceral organs.

☒ Digestive System

An alimentary canal, a pair of buccal glands, a liver and patches of secretory cells constitute the digestive system.

Alimentary canal

It is a nearly straight tube starting at mouth and ending in anus.

Buccal funnel

A large, downwardly directed basin-like depression (Fig. 35.4) at the anterior end of the body. The **mouth** is a narrow opening above the **tongue**, projecting from the bottom of the buccal funnel.

Radiating rows of horny, yellow **teeth** on cartilaginous pad are present in the buccal funnel. The teeth, when worn out, are succeeded by others developing at their bases. The tongue also bears large, horny teeth.

Buccal cavity, oesophagus and respiratory tube

The mouth leads into a **buccal cavity**, communicating behind with two tubes, placed one above the other (Fig. 35.4). The dorsal one is the **oesophagus** and the ventral one, **respiratory tube**, which is closed behind. The opening of the respiratory tube is guarded by a curtain-like **Velum** bearing velar tentacles. The **oesophagus** leads into an uncon-voluted **intestine** through a valvular aperture.

Intestine

The intestine is slightly dilated anteriorly. It bears a **typhlosole, which takes a slightly** spiral course lengthwise and known as **spiral valve.**

The intestine widens posteriorly to form the rectum ending in **anus,** in a small depression, the **vent**.

☒ Digestive glands

Paired buccal or salivary glands, a voluminous liver and patches of secretory cells in the epithelium of the anterior part of intestine constitute digestive glands.

Salivary or buccal glands

They are paired and embedded in the hypobranchial muscles and open separately by ducts in the buccal funnel, below the

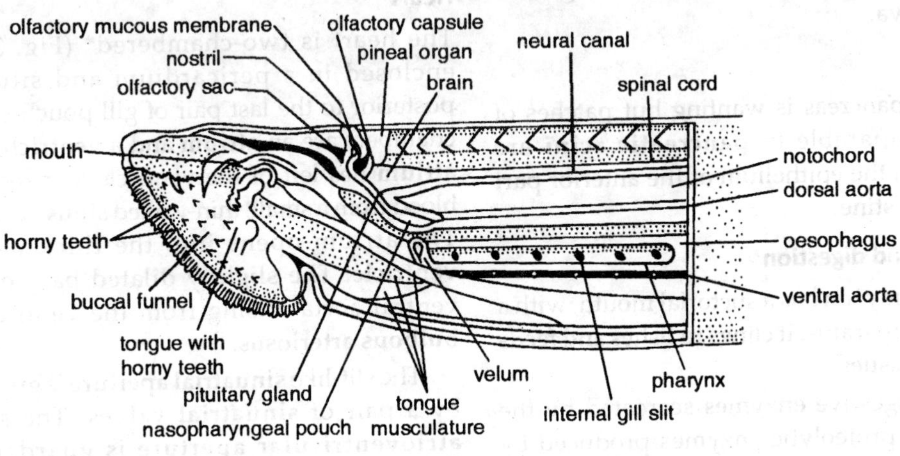

FIGURE 35.4

***Petromyzon* sp.** Anterior end (sagital section)

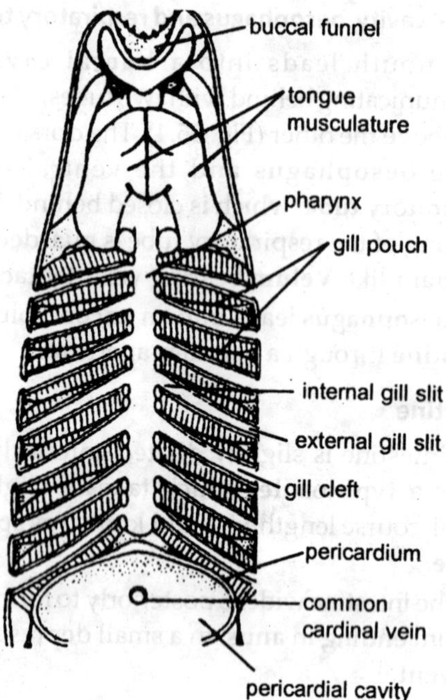

FIGURE 35.5

Petromyzon **sp.** Frontal section of anterior end showing arrangement of gills (seen from below)

tongue. The walls of the glands are folded and the secretion containing **anticoagulant,** presumably prevent coagulation of blood sucked from the host.

Liver

The liver is voluminous, bilobed and lacks **bile duct** in the adult, though the ducts are present in the larva.

Pancreas

Definite pancreas is wanting but patches of cells comparable to **pancreatic acini** are present in the epithelium of the anterior part of the intestine.

Feeding and digestion

The Petromyzon has a suctorial mouth, with a rasping apparatus. It clings to fishes and feeds on their tissues.

The digestive enzymes secreted by the liver and proteolytic enzymes produced by zymogen granules in the so-called pancreatic acini help in digestion of carbohydrates and proteins.

⊠ Respiratory System

The respiratory system consists of seven pairs of **gill pouches** or **branchial pouches** (Fig. 35.5). In the adult connection between the respiratory tube and oesophagus is lost.

A gill pouch has the shape of a biconvex lens with numerous gill lamellae on the inner surface. The gill pouches are separated from one another by **interbranchial septa**. Each gill pouch opens to the exterior separately through **gill pores** or gill slits.

⊠ Mechanism of respiration

Both incurrent and excurrent streams of water pass through gill pouches. The branchial skeleton is elastic. With the expansion of the branchial basket water enters the respiratory tube through gill pores. The water is expelled through the same route by contraction of the branchial basket. Gaseous exchange takes place in the gill lamellae.

⊠ Blood Vascular System

The circulatory system is much advanced and consists of a heart, arterial and venous systems.

Heart

The heart is two-chambered* (Fig. 35.6), enclosed in a pericardium and situated posterior to the last pair of gill pouches. The chambers are an **atrium** and a **ventricle**. The **atrium** lies left of the ventricle and receives blood from a small, thin-walled **sinus venosus**. The atrium opens into the thick-walled ventricle. The slightly dilated base of the ventral aorta arising from the ventricle is **bulbous arteriosus**.

The slit-like **sinuatrial aperture** is guarded by a pair of **sinuatrial valves**. The small **atrioventricular aperture** is guarded by

* The Petromyzon heart has often been described as four-chambered by some authors, which is confusing.

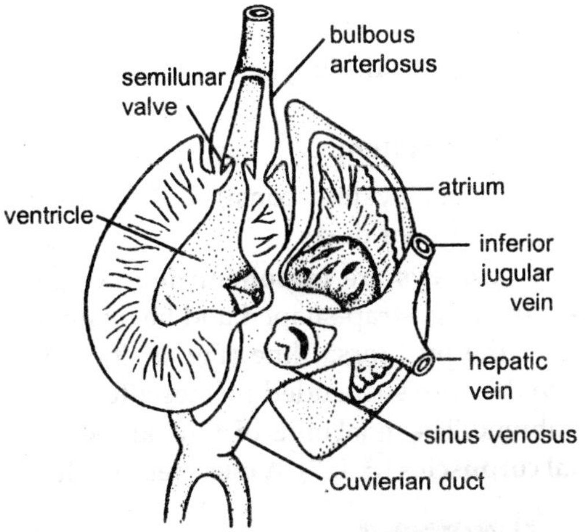

FIGURE 35.6

***Petromyzon* sp.** Heart. Chambers exposed

atrioventricular valves. A single set of **semilunar valves** at the base of the bulbous prevents backflow of blood to the ventricle.

Blood

The colour of blood is red due to the presence of **haemoglobin** in **erythrocytes**, which is intermediate between the haemoglobin of invertebrates and gnathostomata. The red blood corpuscles are nucleated and circular. The white blood corpuscles are almost like the **lymphocytes** and **polymorphs** of higher vertebrates. The blood cells are formed in the tissues present in spiral valve, kidneys and spinal cord.

⊠ Arterial System

The ventral aorta with a slightly dilated base, the **bulbous arteriosus** arises from the ventricle between the last pair of gill pouches. Eight pairs of **afferent branchial arteries** are given off from the ventral aorta, ending in capillaries in the gills.

Afferent branchial arteries

The ventral aorta leaves the heart as a single vessel, which bifurcates at the level of the fourth gill pouch (Fig. 35.7). From each of the paired anterior branch arise four afferent branchial arteries and four pairs are given off from the unpaired portion. The first of the eight vessels on each side supplies the gill lamellae of the **anterior hemibranch** of the first gill pouch, while the last ends in most **posterior hemibranch**. Each of the six remaining afferent branchial arteries divides and supply the lamellae on either side of the septum, i.e. the posterior hemibranch of a gill pouch and anterior hemibranch of the next pouch.

Efferent branchial arteries

These vessels collect blood from the gill lamellae and join to form a single median **dorsal aorta** (Fig. 35.8). The anterior end of the dorsal aorta is paired for a short distance and again join to form the **cephalic circle**. Arteries supplying brain, eyes, tongue and parts of the head arise from the cephalic circle. The dorsal aorta gives off a series of **segmental arteries** to the myotomes and sends branches to gut, kidneys and gonads.

⊠ Venous System

The venous system consists of true **veins** and network of **venous sinuses**. The **caudal vein** bringing back blood from the tail region and

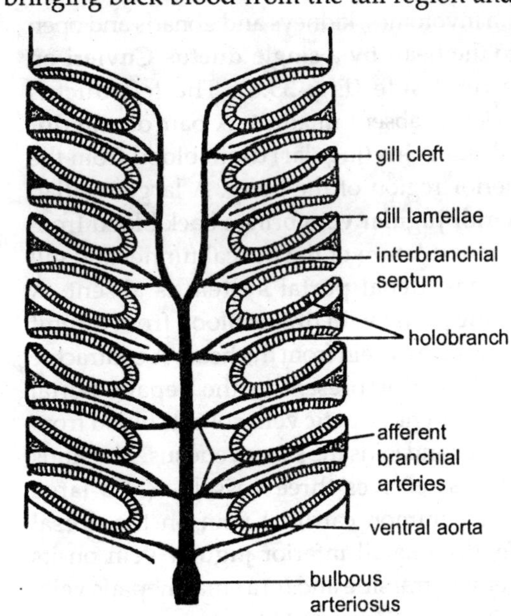

FIGURE 35.7

***Petromyzon* sp.** Afferent branchial system

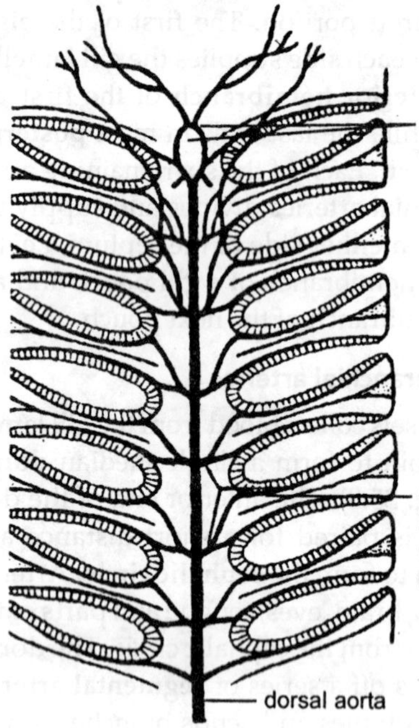

FIGURE 35.8

***Petromyzon* sp.** Efferent branchial system

running anteriorly divides into two **posterior cardinal veins** in the posterior end of the abdominal cavity. The cardinals receive blood from myotomes, kidneys and gonads and open into the heart by a single **ductus Cuvieri** on the right side (Fig. 35.6). The left ductus Cuvieri is absent in adult. A pair of **anterior cardinal veins** (jugular) drain blood from the anterior region of the body. A large median **inferior jugular vein** brings back blood from the musculature of the buccal funnel and gill pouches. Renal portal system is absent. A **hepatic portal vein** drains blood from the gut and receives a vein from the head. A contractile **portal heart** is present in the hepatic portal vein. A single **hepatic vein** collects blood from the liver and joins the sinus venosus. The sinus venosus receives three vessels—(a) a large single common cardinal vein on the dorsal side, (b) a small inferior jugular vein on its anteroventral side and (c) a small hepatic vein.

Only deoxygenated blood passes through the heart. Blood received from different parts of the body by sinus venosus is sent to atrium → ventricle → ventral aorta → gills → body. It is a **single type circulation**.

☒ Excretory System

The excretory system consists of a pair of kidneys and their ducts.

The **kidneys** are **mesonephric** in origin. They are long, strap-shaped and lie on either side of the middorsal line (Fig. 35.9), from which they are suspended by mesentery-like membrane. The renal units of adult kidney are **renal corpuscles** (p. 127). A mesonephric duct,

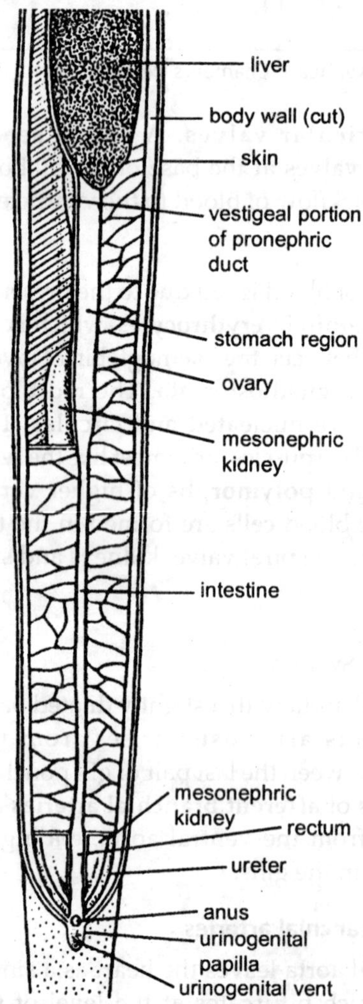

FIGURE 35.9

***Petromyzon* sp.** Urinogenital system (ventral view). Female. Portion of ovary removed to expose kidney

the **ureter**, runs along the free edge of each kidney. In the adult, vestiges of **pronephric ducts** of larva are present anterior to the functional part, or mesonephric kidney. The two ureters open into **urinogenital sinus**, a cavity within the **uringenital papilla**, just behind the rectum. The urinogenital papilla opens to the exterior by a urinogenital aperture, behind the anus in the vent. The urinogenital sinus is in communication with the coelom by **genital pores**, one on each side.

☒ Nervous System

The nervous system is well-developed and similar to that found in the vertebrate series though some of the structures are in a relatively primitive condition.

The central nervous system and peripheral nervous system constitute the nervous system.

☒ Central Nervous System

It consists of a brain and a spinal cord enclosed in a single membrane, the **menix**. The nerves are **nonmyelinated**.

Brain

The brain (Fig. 35.10A, B) is divided into three distinct parts.

a. **Forebrain** (anterior portion) or **prosencephalon**.
b. **Midbrain** (middle portion) or **mesencephalon**.
c. **Hindbrain** (posterior portion) or **rhombencephalon**.

Prosencephalon

It consists of an anterior **telencephalon** made of two **cerebral hemispheres**, the two together is known as **cerebrum**, and a posterior **diencephalon**.

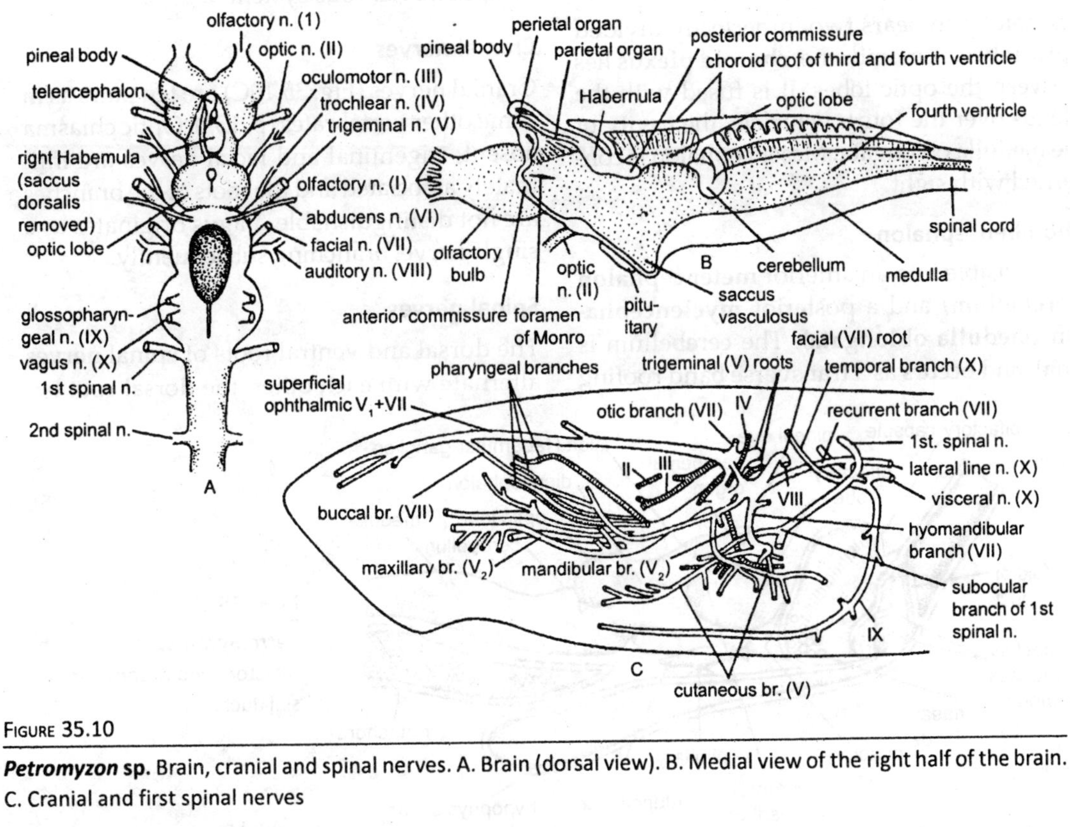

FIGURE 35.10

***Petromyzon* sp.** Brain, cranial and spinal nerves. A. Brain (dorsal view). B. Medial view of the right half of the brain. C. Cranial and first spinal nerves

The cerebral hemispheres are large and lobed. A large **olfactory bulb**, organ of smell is attached to each cerebral hemisphere anterolaterially.

The diencephalon has a membranous roof, projecting outwards as **saccus dorsalis**. The lateral sides are known as **thalamus** and the ventral part is formed by a well-developed **hypothalamus** bearing inconspicuous **inferior lobes**. The **hypophysis** or **pituitary gland** is closely applied to the hypothalamus. Two **ganglia Habernulae**, the right being much larger than the left are present on the dorsal border of the lateral walls. These constitute **epithalamus** and are connected to the **pineal apparatus**. The pineal apparatus consists of a **pineal eye** lying against the transparent area of the cranial roof and the smaller **parapineal body**, below and behind it. A simple crossing of optic nerves, not forming a **chiasma**, is present on the ventral surface.

Mesencephalon

The midbrain bears two imperfectly divided **optic lobes**. A membrane, **choroid plexus** lies between the optic lobes. It is fused with the plexus over the fourh ventricle, the cavity of the medulla oblongata. Mesencephalon is concerned with sight.

Rhombencephalon

It is divisible into an anterior **metencephalon (cerebellum)** and a posterior **myelencephalon (medulla oblongata)**. The cerebellum is small and occurs as a transverse band roofing over the anterior end of the fourth ventricle. The medulla oblongata is largest and highly developed. It controls the powerful sucking structures through trigeminal nerve.

Cavities of the brain

The cavities of the brain are **ventricles** and that of the spinal cord is central canal (p. 134).

Spinal cord

The spinal cord is tubular, of fairly uniform diameter throughout its length, tapering gradually towards the posterior, and encloses the **central canal**. It is flattened, convex dorsally and concave ventrally. **Grey matter** is in the form of a longitudinal band surrounded by **white matter**, the two not sharply differentiated.

⊠ Peripheral Nervous System

Cranial nerves and spinal nerves constitute peripheral nervous system.

Cranial nerves

Cranial nerves (Fig. 35.10C) are ten pairs as in some other vertebrates (p. 137). Optic chiasma absent; trigeminal and facial nerves are intimately associated and the roots of the branches are not distinguishable. Vagus originates as a single nerve, branching subsequently.

Spinal nerves

The dorsal and ventral roots of spinal nerves alternate with each other, the dorsal root be-

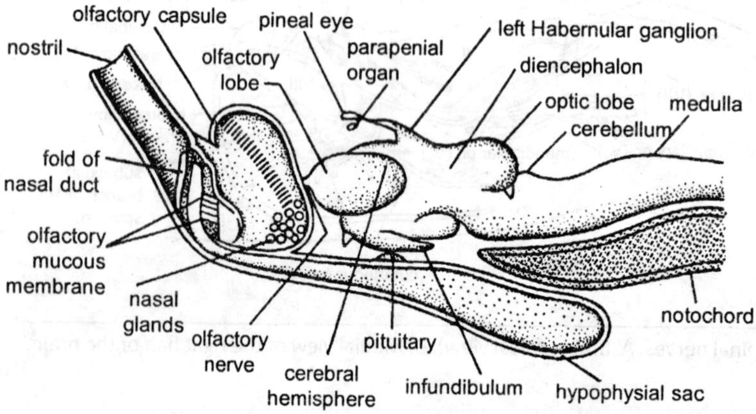

FIGURE 35.11

***Petromyzon* sp.** Brain with olfactory and nasohypophysial ducts

ing somewhat posterior to the ventral root and they do not join. The dorsal root bears a **ganglion**. In the gill region, fibres from adjacent dorsal roots unite to form a nerve (sensory). A nerve (motor) is formed by the union of fibres of adjacent ventral roots. Thus two nerves, one sensory and one motor are formed. These together form the **hypobranchial nerve**, innervating ventral part of the gill region.

☒ Autonomic Nervous System

Segmental **sympathetic ganglia** are present. They are not connected with one another and **sympathetic** trunks are lacking. Preganglionic sympathetic fibres emerge from the spinal cord through dorsal spinal nerves.

☒ Receptors and Sense organs

Olfactory sac (chemoreceptor), eyes (photoreceptors), pineal eye (photoreceptor), ears (equilibrium and phonoreceptor), lateral line sense organs (rheoreceptors), integumentary photoreceptors and gustatory receptors constitute receptor organs.

Olfactory sac (Nasopharyngeal sac)

The external nostril communicates with a round **nasal** or olfactory sac by a short passage. Unlike other vertebrates, the lumen of the hypophysial rudiment, which develops from evagination of buccal epithelium, is continuous with the nasal sac, through a duct, the **nasopharyngeal canal** (Fig. 35.11). Posteriorly, the tube extends to a closed **hypophysial sac** or **nasal sac** below the brain and notochord between the first pair of gill pouches. The internal lining of the nasal sac is folded and contains **olfactory receptor cells**, joined by olfactory nerve.

Eye

Eyes (Fig. 35.12) are almost similar to vertebrate eyes (p. 151). The eye is attached to the rim of the orbit; the eye ball nearly spherical (Fig. 35.12A), cornea somewhat flattened, two-layered, separated by a gelatinous matrix;

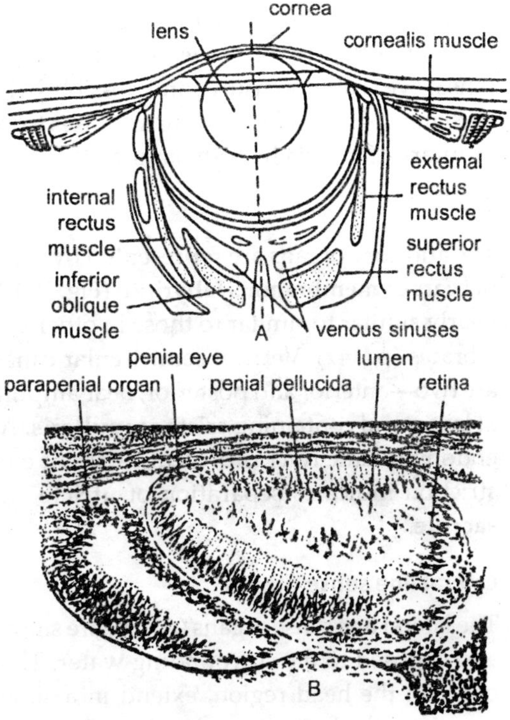

FIGURE 35.12

Petromyzon sp. Geotria sp. Receptors and sense organs. Vertical section passing through different plain. A. **Petromyzon sp.** Eye. B. **Geotria sp.** Penial apparatus

pupil round and capable of a little change in diameter. Extrinsic muscles move the eye-ball. **Cornualis muscle** attached to the outer layer of cornea helps in flattening the cornea and also in accommodation of the eye.

Pineal apparatus

The pineal apparatus consists of a larger, dorsally placed **pineal eye** and a ventral, smaller **parapineal body** (Fig. 35.12B). Both develop from evagination of the roof of the brain and are connected with epithalamus. The pineal eye is joined with the right Habenular ganglion by a stalk containing nerve fibres and the parapineal body to the left Habenular ganglion with a similar stalk. Pigmented **retina** with sensory cells and a flattened imperfect **lens** are present in the pineal eye. The structures are extremely reduced in parapineal body.

In lamprey, the pineal eye is better developed than that in any other vertebrate. The pineal eye is photosensitive and possibly helps in change of colour by sending impulses to the pituitary gland via hypothalamus.

Ear

The auditory organs are represented by internal **ears** or **membranous labyrinth** (Fig. 35.13), nearly similar to similar to those in other vertebrates (p. 147). Vertical semicircular canals are two—anterior and posterior; bear ampulla at lower end containing **crista ampullaris**. An endolymphatic duct is present. A slight constriction indicates separation of **utricle** and **saccule**.

Lateral line sense organs

The lateral line sense organs (p. 150) are simple and exposed to the surrounding water. They occur in the head region, extend in a single line along the body on each side and consists of small patches of sensory cells innervated by branches of fifth and tenth nerves.

Integumentary photoreceptors

Occur in the tail of the ammocoete larva. They are supposed to be present in the skin of other parts of the body.

Gustatory receptors

Organs of taste occur in the wall of the pharynx between gill pouches.

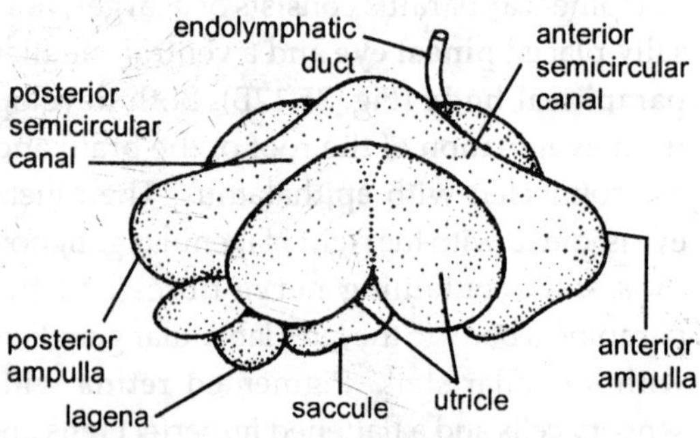

FIGURE 35.13

***Petromyzon* sp.** Membranous labyrinth

⊠ Reproductive System

The sexes are separate in the adult. Both spermatocytes and oocytes may be found in a single gonad in young lamprey. Sexual differentiation occurs at a late stage. The single gonad, **testis**, made of **sperm folicles** or **ovary** (Fig. 35.9) extends the length of the pleuroperitoneal cavity, remaining suspended from the middorsal body wall by a membrane, the **mesorchium** or **mesovarium** in male and female, respectively. Mature **sperms** or **ova** escape to the coelom with the rupture of the gonad and find their way through the **genital pores** to the **urinogenital sinus** and, thence, to the exterior.

Fertilization and development

In the breeding season the adults migrate from sea to river. During spawning the male wraps the posterior part of the female body and sheds sperms over the eggs released by the female. **Fertilization** is external.

The egg is **telolecithal** with considerable amount of yolk in the **vegetal pole**. At about twenty-one days, young Ammocoete larva hatches out.

AMMOCOETE LARVA

It is dissimilar to adult.

1. Body small, about 10 cm long transparent, with a **median**, continuous (Fig. 35.14) fin.
2. Instead of a suctorial buccal funnel of the adult a hood-shaped **upper lip** overhangs the mouth.
3. **Teeth** absent.
4. **Velum**, a pair of cup-shaped structures are attached to the anterior wall of the **pharynx**.
5. Entrance of the alimentary canal is guarded by a set of branched **buccal tentacles**.
6. Rudimentary **paired eyes** are hidden below skin and muscles.
7. Unpaired **pineal eye** well-developed.
8. **Photoreceptors** are abundant in the tail region.

9. The **endostyle** opens into the pharynx by a slit-like aperture and connected with **hyperpharyngeal groove** by a pair of **peripharyngeal grooves**.
10. The endostyle secretes a profuse quantity of muscus to which food particles are entagled. Cilia in the pharynx drive strands of food-laden mucus to the oesophagus and from there to **intestine** by **oesophageal cilia**.
11. The larva remains buried in mud in a U-shaped burrow, coming out only in night in search of food. It swims head downward and burrows again rapidly.

☒ Metamorphosis

The larval period is long and may extend for 3 to 4 years. Metamorphosis usually occurs in winter and the larva assumes the adult form.

1. Endostyle transforms into endocrine **thyroid gland** below the pharynx.
2. Suctorial **buccal funnel** with **tongue**, **teeth** and musculature surrounds the **mouth**.
3. The **eyes** complete their development and move to the surface.
4. The continuous dorsal fin changes to form two **dorsal** and a single **caudal fin**.
5. The oesophagus and respiratory tube turn to independent canals.
6. The velum reduced to guard the opening of the respiratory tube only.
7. The gills develop in sacs, the gill **pouches** opening in respiratory tube.
8. Nasal sac becomes folded internally and retains connection with enlarged **olfactory sac**.
9. The nasopharyngeal sac becomes enlarged posteriorly.
10. The larval cranium and branchial skeleton transform into the adult condition.
11. The skeleton of the upper lip and tongue develop.
12. **Pronephros** degenerates and **mesonephros** form the adult kidneys.
13. The spinal cord turns flattened.
14. The skin colour changes from muddy-brown to metallic on the dorsal surface.

MYXINE sp. (Hagfish)

Myxine also called Hag, Hagfish or slime-eel, are marine, inhabit muddy bottoms and rest in temporary burrows. Myxines are noctural. Body length varies from 45 cm (*Myxine*) to 100 cm (*Eptatretus*).

☒ External features

Body cylindrical (Fig. 35.16); **mouth** terminal, not extensible into a funnel and surrounded by **tentacles** supported by cartilages. The buccal cavity is more extensive. A large, median **palatine tooth** lies above the mouth.

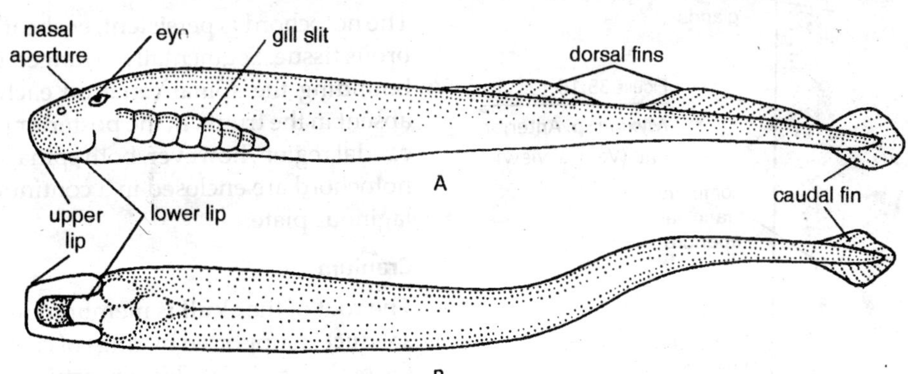

FIGURE 35.14

Ammocoete larva. *Petromyzon* sp. A. Lateral view, B. Ventral view

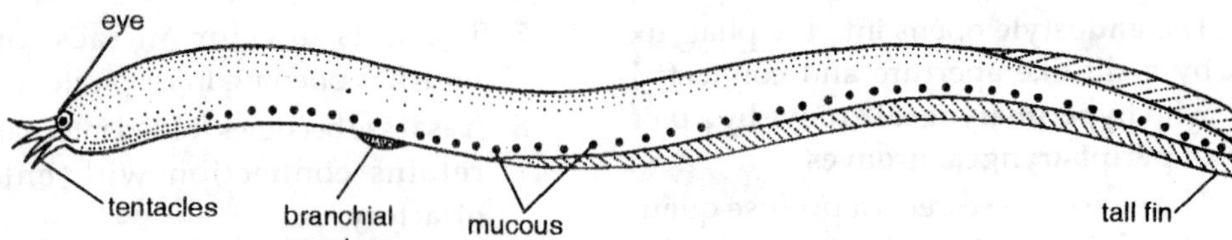

FIGURE 35.15
Myxine glutinosa

Two rows of **smaller teeth** arising from the **dental plate** are present on the **tongue**. The single median **nostril** is terminal. It opens into a tubular nasal duct, leading to the **olfactory organ**, continuous with the **nasohypophysial duct** opening into the buccal cavity, just anterior to the velum. The eyes are paired but rudimentary. External openings of gill pouches, **branchial apertures** or **gill slits** vary from 1 to 15 in different species. Only one pair (Fig. 32.16) are present in Myxine; 13-15 pairs in Eptatretus. Rows of minute openings of **mucous glands**—more crowded on the ventral side—are present. Only a narrow **caudal fin** surrounds the end of the tail.

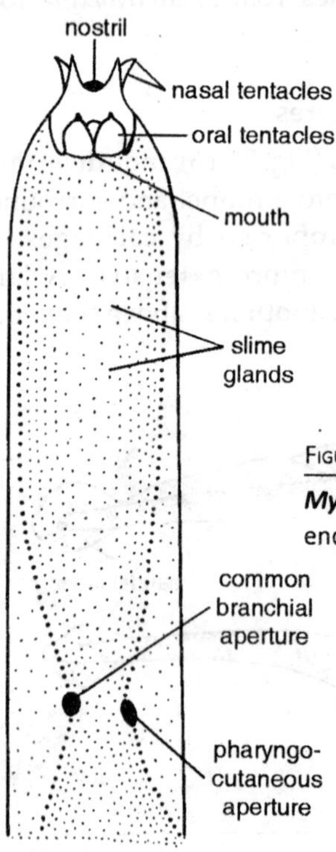

FIGURE 35.16
***Myxine* sp.** Anterior end (Ventral view)

Body wall and muscles

The skin is soft and slimy. Numerous **thread cells**, singly or in the form of **slime glands**, are embedded in the epidermis. A continuous thread of fibrous protein is present in a thread cell. In an agitated myxine, a large number of these cells burst, releasing their threads, 1 to 3 μm in diametre and several centimetres long. The threads absorb water and form a **slimy gel** around the body.

The muscles are arranged in **myotomes**. The myotomes of one side alternate with those of the other side.

Locomotion

Myxines are sluggish animals with poor locomotory power owing to feeble development of body musculature. The mechanism of locomotion is similar to that in lampreys.

Endoskeleton

The skeleton is wholly cartilaginous. The ill-developed notochord, skull and visceral skeleton constitute axial skeleton (Fig. 35.17).

Notochord

The notochord is persistent, ensheathed by fibrous tissue. Segmentally arranged cartilages bounding the spinal canal on each side are absent in the trunk. In the posterior part of the caudal region, however, both spinal canal and notochord are enclosed in a continuous cartilaginous plate.

Cranium

The roof of the **skull** is membranous. The **nasal duct** is supported by rings of cartilages and the oral tentacles by flexible cartilaginous rods. The **cornual cartilage, subneural cartilage** and **basal plates** are present.

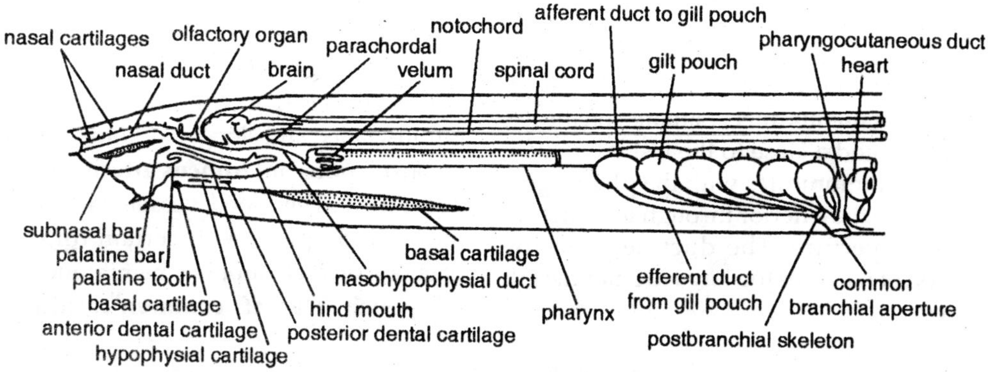

FIGURE 35.17

***Myxine* sp.** Sagittal section (diagrammatic). Skeletal and visceral structures

Visceral skeleton

The visceral skeleton of the **mandibular arch** is formed by the anterior portion of the **subocular arch**, while the posterior portion forms the **hyoid arch**. Only first two branchial arches, the **first extrabranchial cartilage** and the **first branchial cartilage,** and the **second extrabranchial cartilage** are present. A pair of asymmetrical cartilages represent a posterior remnant of the **branchial skeleton** in the region of common branchial aperture.

☒ Digestive System

An alimentary canal, liver and a pancreas-like gland constitute the digestive system.

The alimentary canal is a straight tube (Fig. 35.17); the **mouth** is ventral; the **buccal cavity** long. A single, median **palatine tooth** is located above the mouth. Two rows of small teeth arising from the dental plate are present on the well-developed **tongue**. The **stomach** is slightly dilated; the **intestine** is wide. The **liver** consists of two separate portions, an anterior and a posterior, their ducts opening separately into a **gall bladder**. A **pancreas-like gland** is formed by the aggregation of secretory tubules in the gut epithelium.

☒ Feeding

Myxine feed upon dead or dying fishes and perhaps devour living polychaetes and preapulids. With the powerful sucking device they suck out all the soft structures of the body of dead or dying prey, except skin and bones.

☒ Respiratory System

Respiratory structures are six pairs of **gill pouches** restricted in the branchial region. The gill pouches lack separate external openings. An **efferent duct** arises from each gill pouch and all the ducts from six gill pouches of one side unite together before opening to the exterior, and a single duct thus

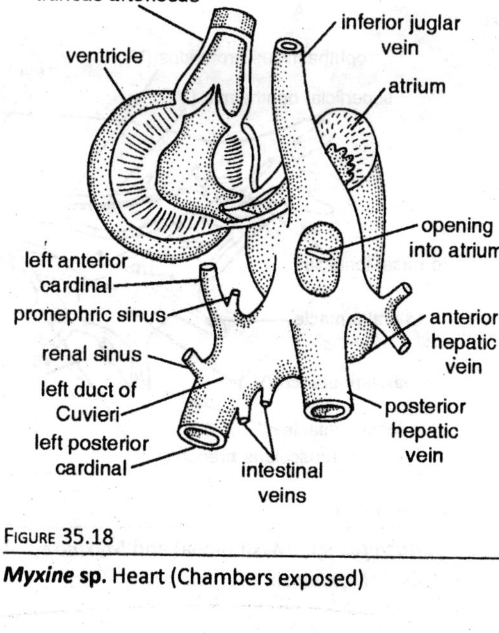

FIGURE 35.18

***Myxine* sp.** Heart (Chambers exposed)

formed opens to the exterior by an **external branchial aperture** (Fig. 35.17). So only one pair of branchial apertures are present. A **pharyngocutaneous duct** is situated on the left side behind and a close to the last gill pouch. The duct communicates with the **pharyngeal cavity** and to the exterior through left external branchial aperture. The duct helps in the expulsion of particles too large to enter the gill pouches.

⊠ Blood Vascular System

The heart is S-shaped, consists of an **atrium** and a **ventricle** (Fig. 35.18). The **sinus venosus** is thin-walled and elongated, **conus arteriosus** inconspicuous but **truncus arteriosus** is well-developed. The blood vessels are similar to those in lamprey, with certain differences. Each **aortic arch** supplies hemibranch of one gill pouch. The **left common cardinal vein**

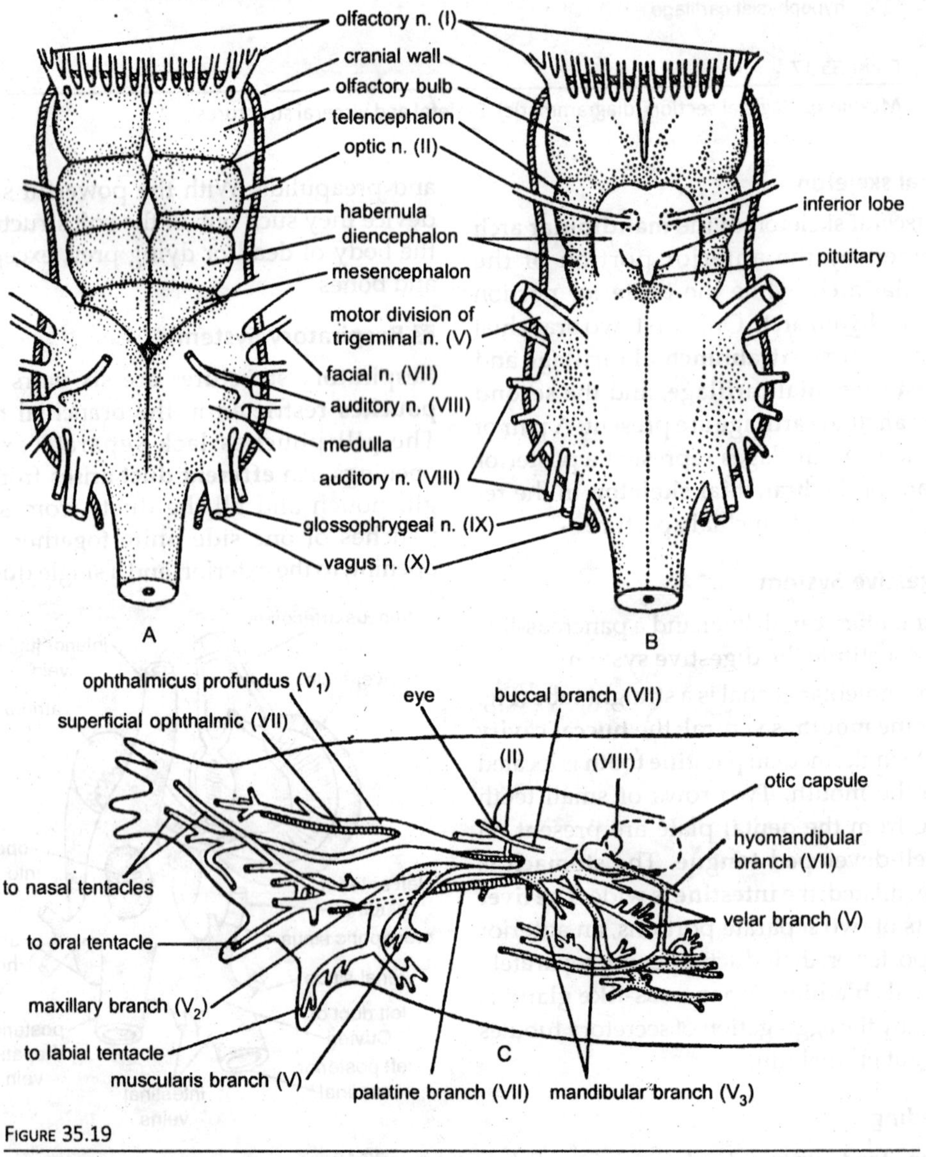

FIGURE 35.19

***Epitatretus* sp.** (Myxinoidea) and ***Myxine* sp.** A-B. ***Epitatretus* sp.** Brain. A. Dorsal view. B. Ventral view, C. ***Myxine* sp.** Cranial nerves

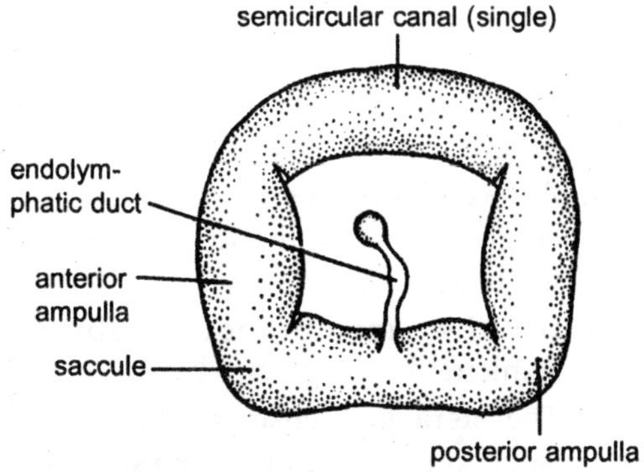

FIGURE 35.20

***Myxine* sp.** Membranous labyrinth

(ductus Cuvieri) persists. The right one forms a **portal heart**. Blood cells are formed in tissues scattered in the submucous layer of the alimentary canal and in pronephros.

☒ Excretory System

The kidneys are paired structures and consists of both pro and mesonephros. The **pronephros** is represented by a pair of small bodies dorsal to the pericardial cavity into which the **nephrostomes** open. A pronephric duct is absent and the tubules end blindly. The functional kidney is **mesonephros**, retaining primitive segmental arrangement. In each segment, the **ureter** sends off a coiled tubule with a single **renal corpuscle** bearing a **glomerulus**.

☒ Nervous System

The nervous system differs from that of lamprey. In the brain (Fig. 35.19A, B) the **olfactory lobes** are large, **choroid plexus** absent, **myelencephalon** large but the rest of the brain is ill-developed. The **ventricles** are reduced. The spinal cord is ensheathed by fibrous tissue only.

Of the cranial nerves (Fig. 35.19C), the **oculomotor** (III), **trochlear** (IV) and **abducens** (VI) are extremely reduced or absent. The **dorsal** and **ventral roots** of the spinal nerves unite but the function is different from that of a typical Gnathostomata.

☒ Receptors and sense organs

The **eyes** are vestigeal, sunk beneath the skin but they are photosensitive. In **membranous labyrinth** (Fig. 35.20) the **semicular canal** is one, vertical and bears an **ampulla** at each end. **Saccule** and **utricle** are not discernible. **Endolymphatic duct** small, with an enlarged sinus at its dorsal end. A **pineal organ** absent. The **nasopharyngeal tube** opens internally into the pharynx.

☒ Reproductive System

Hagfishes are **protandric hermaphrodite**. Eggs are produced from the anterior portion, while sperms are formed in the posterior part of the body. Male gametes are formed prior to the maturation of female gametes (protandry). The **eggs** (Fig. 35.21) are telolecithel, large, cylindrical in shape, enclosed in a horny shell bearing hooked filaments at either pole.

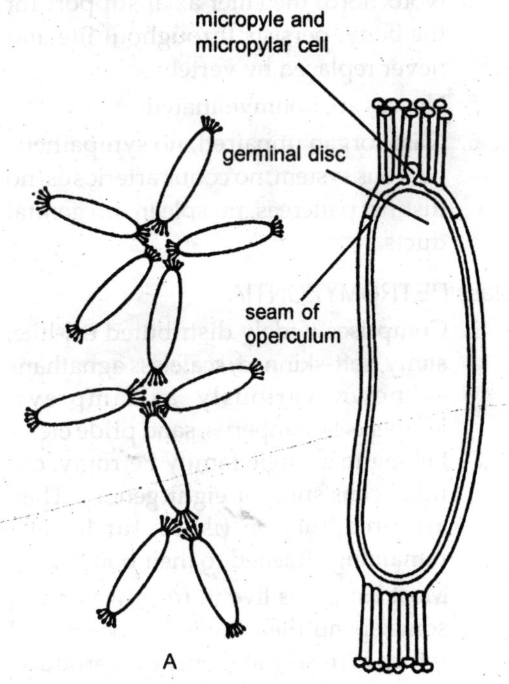

FIGURE 35.21

***Epitatretus* sp.** (Myxinoidea). Eggs. A. A cluster of eggs attached together by anchor filaments. B. An egg with anchor filaments

☒ Development

Development in **Myxine** is unknown. In **Epitatretus** segmentation is meroblastic, being restricted to the animal pole of the elongated egg. The blastoderm extends over the yolk, which is completely enclosed at a late stage. Subsequent development is nearly similar to that of lamprey but development is direct and there is no larval stage and metamorphosis.

CLASSIFICATION: SUPER CLASS AGNATHA

1. Eel-like animals with circular mouth, slimy, scaleless skin and without jaw and unpaired fins with slender cartilaginous fin rays.
2. The brain is encased in a cartilaginous cranium.
3. Pouched gills supported by cartilaginous branchial basket.
4. Notochord, the chief axial support for the body, persists throughout life and never replaced by vertebrae.
5. Nerves are nonmyelinated.
6. Nasal organ unpaired, no sympathetic nervous system, no conus arteriosus, no distinct pancreas, no spleen, no genital ducts.

Class PETROMYZONTIA

1. Comprises widely distributed eel-like, slimy, soft-skinned, scaleless agnathans —known variously as lampreys, lamper eels, lamperus, sand pride etc.— belong to a single family Petromyzontidae consisting of eight genera. They are predators on fishes, turtles etc. remaining attached to their body.
2. Most lampreys live in fresh water, but some spend their life in the ocean and return to fresh water only to reproduce.
3. The head and trunk are nearly cylindrical, with the side to side compressed tail region. Mouth is round, encircled by a large basin-like depressed buccal funnel. Inside the buccal funnel are radiating rows of yellow, horny teeth. At the bottom of the buccal funnel, just below the mouth, is a small tongue, bearing horny teeth.
4. A nasohypophysial sac opens by a single nostril on the dorsal surface of the head.
5. In the membranous labyrinth semicircular canals are two.
6. The lateral line sense organs consists of isolated pits opening at the surface.
7. The paired eyes without eyelids are covered by a transparent area of skin.
8. Pharynx ends blindly, and from a branchial basket little behind the eye. On the sides of the basket are seven pairs of small apertures, the gill pouches opening separately. Numerous gill lamellae are present on the inner surfaces of the gill pouches.
9. Paired appendages are absent, but two approximately equal dorsal fins, and a caudal fin present; the second dorsal fin is continuous with the caudal fin.
10. The stomach is absent; a typhlosole present in the intestine.
11. Liver is a large, bilobed organ; gall bladder and bile duct almost entirely absent in adult but present in larval stage.
12. Spleen absent.
13. Pancreas is represented only by cells embedded in the wall of intestine and in liver.
14. Heart two-chambered.
15. The mesonephric kidneys are drained by pronephric ducts which carry only urine.
16. Genital ducts absent and the gametes pass from the gonads to the coelom.
17. Fertilization external, the adults die after spawning.
18. Ammocoete larva is eel-shaped, lacks specialised feeding mechanism of adult and possesses endostyle for food trapping.

19. Rudimentary vertebrae are present in the form of small cartilages on each side of the notochord, two pairs to each metamere.
20. Metameres are 'W'-shaped.
21. Ninth and tenth cranial nerves not enclosed in cranium.

Examples: *Petromyzon, lampetra, Entosphenus, Ichthyomyzon, Geotria, Mordacia,* etc.

Class MYXINOIDEA

1. Exclusively marine, 'slime eels' or 'hag fishes' descend up to a depth of more than 300 fathoms (540 metres). Some adults are nonparasitic, and most remain attached to muddy bottom where they rest temporarily. They feed on dead or dying animals or prey upon living polychaetes and priapulids.
2. Body cylindrical, 40 to 80 cm long, producing immense amount of slime from the slime glands.
3. Buccal cavity large but the mouth is not expanded into a funnel.
4. Mouth and nostril are surrounded by tentacles.
5. Single median palatine tooth above the oral aperture and two rows of smaller teeth arising from the dental plate on the tongue.
6. The terminal nostril opens into a tubular nasal duct leading to the olfactory organ and is continuous with the naso-hypophysial duct opening into the buccal cavity, just anterior to the velum.
7. A narrow caudal fin surrounds the end of tail.
8. External branchial aperture may be one to fourteen pairs. Each branchial aperture communicates by a short efferent duct with one of the gill pouches which is again connected with the pharynx by another tube.
9. There is no trace of neural arch in trunk but in the posterior part of the caudal region both neural canal and notochord are enclosed in a continuous cartilaginous plate. The roof of the skull is entirely membranous. Spinal cord is covered with fibrous tissue only.
10. The intestine is wide. Liver consists of separate anterior and posterior portions, the ducts of which open separately in gall bladder. A pancreas-like organ, 'the insular organ' present. No definite spleen, but blood forming lymphoid tissue is found scattered in the submucosa of gut.
11. Pronephros is retained in adult myxinoids but mesonephros is the functional kidney retaining primitive segmental arrangement.
12. Olfactory lobes large, ventricle reduced and midbrain smaller than that in lamprey.
13. The ventral and dorsal roots of spinal nerves united.
14. The eyes are vestigeal, sunk beneath the skin.
15. Single semicircular canal having an ampulla at each end in the membranous labyrinth.
16. Lateral line sense organ lacking and the pineal organ is absent.
17. Sexes are separate.
18. Larval stage and development unknown.
19. They are the only vertebrates to resemble marine invertebrates in having blood that is isosmotic to sea water.

Examples: *Myxine, Paramyxine, Eptatretus* (= *Bdellostoma*), etc.

The whole group of Cyclostomes show primitive characters like those of amphioxus or in lacking advanced characters of true fishes, specialized characters and degenerate characters. A comparison of hags and lampreys also shows their relative primitiveness or specialization in various systems.

SUPERCLASS GNATHOSTOMATA

Gnathostomes include all vertebrates with upper and lower jaws. The group embraces a wide range of animals starting from fish at the lowest rung of the evolutionary ladder, the last rung being occupied by anthropoid apes.

The jaws were probably derived from the paired anterior gill bars and the gnathostomes evolved from a fish-like ancestor.

Gnathostomes have been divided into seven classes. The primitive Placodermi are extinct. The classes Chondrichthyes, and Osteichthyes are grouped as Pisces or fishes having similar characters due to their primary aquatic adaptation. The last four classes—Amphibia, Reptilia, Aves and Mammalia—are tetrapods having common features acquired for terrestrial life.

The paired appendages, the fins supported by dermal fin rays in fishes, an adaptation for aquatic life have been replaced by pentadactyl limbs to carry weight of the body and also to function as locomotory organs in the tetrapods. Gills, the aquatic respiratory organs, lost their utility in terrestrial forms and lungs capable of utilizing atmospheric oxygen, evolved.

Shifting from aquatic to terrestrial life exposed the migrants to varied life conditions, the end result of which is the appearance of numerous kinds of animals of today, both terrestrial and aquatic, the latter secondarily adopted in their present environment.

36 CLASS PLACODERMI (APHETOHYOIDEA)

Placodermi are the earliest jawed vertebrates, appeared in the Silurian periods, flourished in the Devonian and Carboniferous and became extinct in the Permian. They had heavy defensive armour and hinged jaws. The hyoid arch did not support the jaw (the aphetohyoidean condition) indicating that autostylic jaw suspension is more primitive than that in any other jawed animals. Their paired fins were essentially adaptive organs.

CLASSIFICATION: CLASS PLACODERMI

Subclass **Acanthodii**

Acanthodii or "needle-finned sharks" are oldest and least specialized known gnathostomes, which became extinct in the Lower Permian. They were fresh water, superficially shark-like, elongated fish with blunt snout, heterocercal tail and series of paired fins between the pectoral and pelvic fins.

Examples: *Climatius, Acanthods,* etc.

Subclass **Arthrodira (Coccostei)**

Arthrodira are allied to elasmobranchs with the head protected by bony plates which are united to form a strong cranial roof. The remaining free part of the body tapered to a whip-like tail and was either naked or in some covered with scales. They were fast swimmers and predatory forms with an operculum covering the branchial arches.

Examples: *Arctolepis, Coccosteus, Mylostoma, Dinichthys,* (Fig. 20.18A), etc.

Subclass **Petalichthyida**

Devonian forms allied to elasmobranchs. The structure of shoulder girdle, the tribasal pectoral fin, is similar to earlier elasmobranchs. Ventral position of mouth and flattening of neurocranium were the result of adaptation to bottom dwelling.

Examples: *Macropetalichthys,* (Fig. 36.1) *Lunaspis,* etc.

Subclass **Antiarchi (Pterichthyomorphi)**

Devonain, small, freshwater forms, sufficiently well preserved specimens reveal certain internal details. Spiral valve, backwardly directed sacs-like lungs, communicating with the pharynx, were present. Mouth ventrally placed, eyes closely placed on the dorsal side of head and tail covered with scales.

Examples: *Bothriolepis* (Fig. 20.18B) *Microbrachius, Ceraspis,* etc.

Subclass **Rhenanida**

Devonian placoderms with dorsoventrally flattened body and broad head, like rays. Bony plates constitute the body armour which are provided with small denticles. The skull fully ossified. The gills are covered by operculum. Spiracle absent. Pectoral fins large and are supported by elongated radials.

Examples: *Gemuendina* (Fig. 36.2) *Asterostens, gagorina*, etc.

FIGURE 36.1

***Macropetalichthys* sp.** (Extinct)

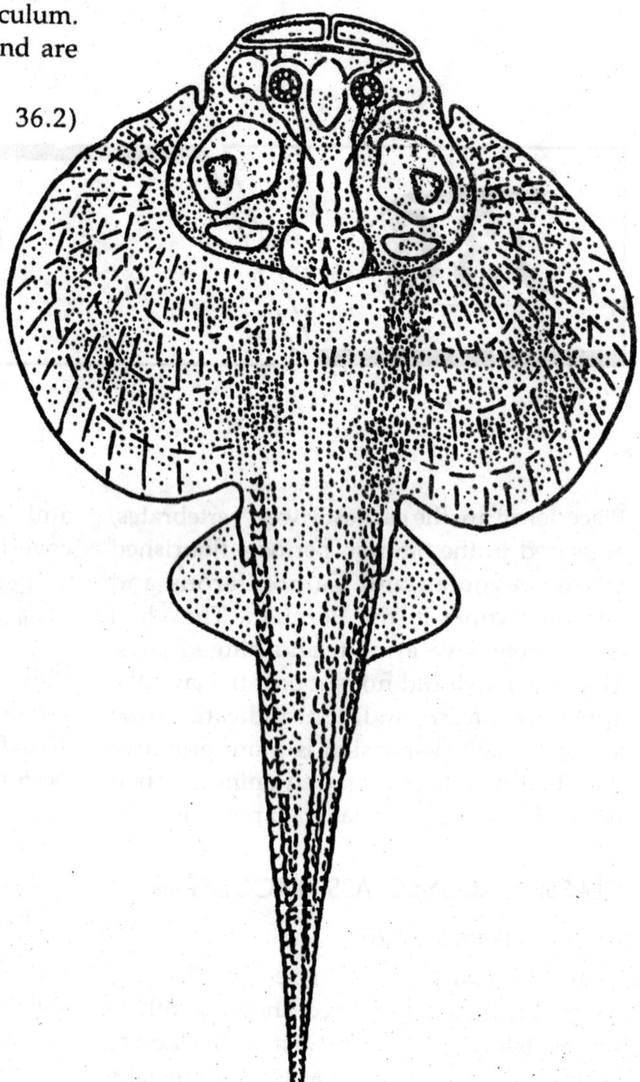

FIGURE 36.2

***Gamuendina* sp.** (Extinct)

Subclass Palaeospondylia

The group having dorsoventrally depressed skull and well-ossified vertebral column. The vertebrae on the posterior side possess strong centra with neural arch and haemal spine. Tail heterocercal. Auditory capsule large. Traces of paired fins present. The hyoid arch complete but never takes part in jaw suspension. The external armour is complete.

Example: *Palaeospondylus.*

37 CLASS CHONDRICHTHYES (ELASMOBRANCHII)

Elasmobranchs are largest known fishes. They are generally marine, predacious, comprising the extant sharks, rays and chimaeras together with a number of extinct orders. The class includes about 700 species of contemporary sharks, rays and similar fishes with a cartilaginous endoskeleton. Although the cartilage is not replaced by bone, calcium salts are sometimes deposited within it. The extensive ancestral dermal armours are replaced by minute placoid scales embedded in the skin. The teeth are enlarged modified placoid sacles. Oviparous. The earliest known species were marine and few contemporary species have secondarily adapted to fresh water. The largest living shark, the oceanic tropical, oviparous whale shark *Rhincodon typus*, attaining nearly 21.5 m length, and weighing 70 tons lives on plankton and is harmless to man. The man eating great white shark, *Carcharodon*, attains a length of about 11m and has flat, triangular serrated teeth. Now-a-days it is considered that elasmobranches evolved from a bony ancestor and perhaps by neoteny.

SCOLIODON LATICAUDAS (Dog fish)

The Scoliodon* is a marine, predatory carnivorous fish widely distributed in different seas of the world. They are swift swimmers and inhabit open seas. Of the nine species of scoliodon, four are present in abundance in Indian seas. *Scoliodon laticaudas* belongs to class

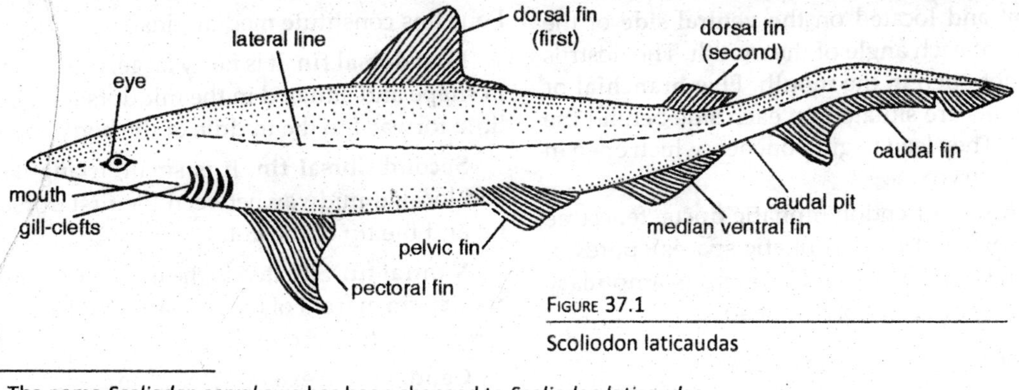

FIGURE 37.1
Scoliodon laticaudas

* The name *Scoliodon sorrakowa* has been changed to *Scoliodon laticaudas*

Chondricthyes (Elasmobranchii), subphylum Vertebrata, phylum Chordata.

External features

The body of Scoliodon is about 60 cm long in the adult. It is laterally compressed, spindle-shaped, tapering at both the ends (Fig. 37.1). The entire body is covered with **placoid scales (odontoids)** arranged in regular, oblique rows. The colour of the dorsal surface is dark grey, the under surface is pale white and the posterior portion of the caudal fin is dark. The body surface in fresh specimen is slippery due to secretion of mucus by the glands in the multilayered epithelium. (Fig. 7.9A). The body is divisible into three regions—the anterior head, middle trunk and posterior tail but distinct boundaries are lacking.

Head

The portion of the body in front of pectoral fins is head. The head is strongly compressed dorsoventrally and ending in a wedge-shaped snout anteriorly. The ventral **mouth** is a large, crescentic opening guarded by upper and lower jaws bearing one or two rows of backwardly directed, **pointed teeth**.

The **eyes** are large, round, one on either side of the head. Two movable **eye-lids**, upper and lower, and a movable **nictitating membrane** in front, are present in the eye. The **pupil** is narrow and vertical. **Nostrils** are two obliquely placed, crescentic openings, partially covered by a small fold of skin in front and located on the ventral side of the head at each angle of the mouth. The nostrils do not open in the mouth. Five **branchial** or gill slits are situated on each side behind the eyes. They lead to gill pouches, which open in the pharynx.

A pair of endolymphatic ducts, fenestrae (openings of perilymphatic spaces), pores of lateral line canals and numerous ampullary pores are present on the head.

Trunk

The portion of the body between the head and cloacal aperture, a longitudinal slit, is trunk. The trunk is almost elliptical in transverse section. It is thickest in front of the middle of the body and tapers behind to end in tail. The trunk bears an unpaired first dorsal fin and paired pectoral and pelvic fins.

Two **abdominal pores** connecting the coelom with the exterior are present on elevations, one on either side of the cloacal opening.

Tail

The postanal part of the body is tail. It is laterally compressed, tapers posteriorly and bent upwards, at a small angle. The tail bears unpaired second dorsal fin, ventral fin and caudal fin. Two shallow pits, the **caudal pits**, one dorsal and the other ventral, are present at the root of the tail.

Lateral lines

Two, one on each side. It is a faint line, on either side of the body, running from behind the head up to the tail, at the middle of the body. The lateral lines indicate the position of the underlying lateral canals.

Fins

Fins are appendages in fishes. These are membranous extensions of the skin supported by dermic **fin rays** (dermotrichia). In Scoliodon the fin rays are cartilaginous (pterygiophores).

Median fins

Four fins constitute median fins.

First dorsal fin. It is fairly large, triangular in shape and situated in the middorsal line, a little forward to the middle of the body.

Second dorsal fin. It is small, triangular and located midway between the first dorsal fin and the tip of the tail.

Ventral fin. Situated in the midventral line about 5 cm in front of the caudal fin and a little forward to the second dorsal fin.

Caudal fin. It extends along the dorsal and ventral surface of the tail in the median line. It has one dorsal and a ventral lobe. The dorsal lobe is narrow and forms a low ridge along the greater part of the upper surface. The ventral lobe is large and divided into two parts—the anterior one is much larger and the posterior one small. Such a tail is known as **heterocercal tail.***

Paired fins

They are two pairs, spread horizontally.

Pectoral fins. Large, triangular, located at the ventrolateral margins of the body, immediately behind the gill clefts.

Pelvic fins. These are smaller than the pectorals, arise close together from the ventral surface at the junction of the trunk and tail and enclose the cloacal aperture between them. In male, a **clasper** or **myxipterygium**, the copulatory organ, is connected with the inner edge of each pelvic fin.

⊠ Skeletal System

The skeletal system consists of exoskeleton and endoskeleton.

Exoskeleton

In Scoliodon the **placoid scales** or **odontoids**, arranged in regular oblique rows covering the whole body and even basal parts of the fins constitute exoskeleton.

Skin

The skin consists of an outer **epidermis** and an inner **dermis**. The epidermis is made of many layers of epithelial cells, a large number of which secrete **mucus**. The **dermis** is a dense connective tissue containing smooth muscle fibres (Fig. 7.9A).

The placoid scales are partly dermal and partly epidermal. The basal plate and the dentine of the spine are dermal, while the enamel covering the dentine is epidermal.

Placoid scale: Structure

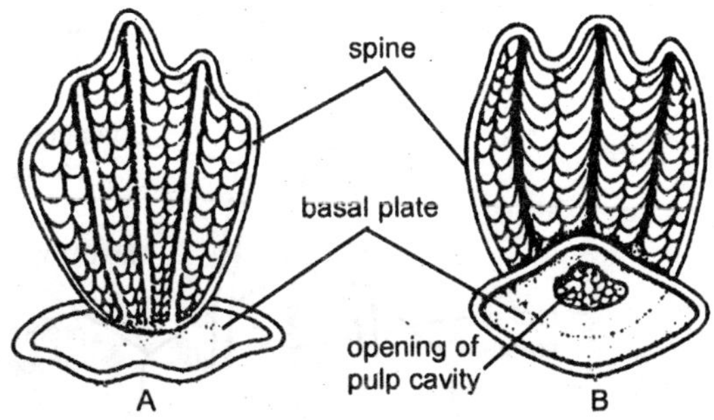

FIGURE 37.2

***Scoliodon* sp. Placoid scale** A. Dorsal view, B. Ventral view

1. A placoid scale consists of a diamond-shaped **basal plate** and a backwardly directed, flat, triradiate **spine** projecting out of the skin (Fig. 7.9A, 37.2).
2. The basal plate is held in the dermis by strong fibres of connective tissue.
3. The surface of the spine is not smooth but stratified.
4. The spine is composed of **dentine**, covered externally by **enamel**.
5. **Pulp cavities** are present both in the spine and basal plate, which are connected by a perforation at the base of the spine.
6. The scales continue in jaws, where they act as teeth.

Endoskeleton

The endoskeleton is entirely cartilaginous, hardened at places by calcification but true bone is absent. It has two components—axial and appendicular skeleton.

Axial skeleton

Consists of skull, visceral arches and vertebral column.

Skull

The skull consists of (a) a cranium enclosing the brain, (b) sense capsules enclosing olfactory and auditory organs (Fig. 37.3) and (c) visceral skeleton including the jaws and skeletal supports of the pharynx and gills.

---

* **Heterocercal tail.** The vertebral column bends upward to enter the dorsal flange of the tail, which, thus, becomes prominent than the ventral flange.

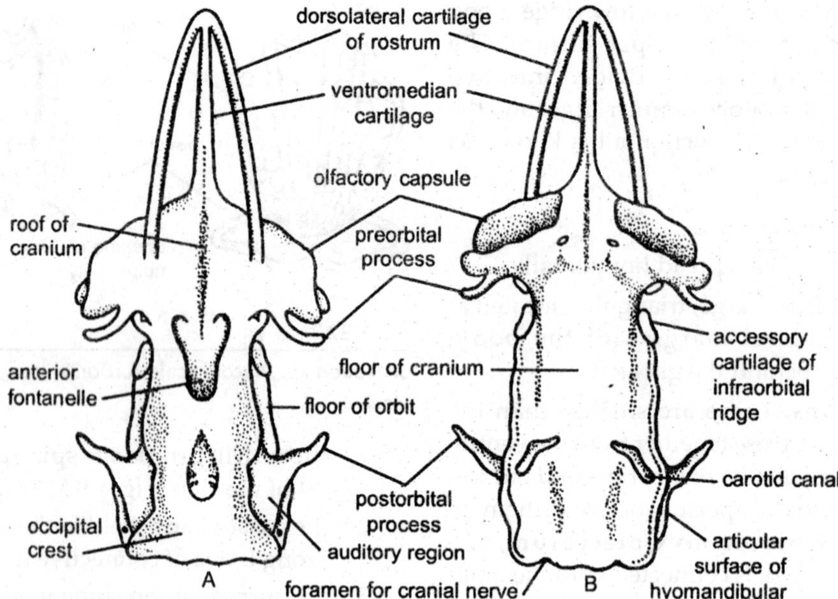

FIGURE 37.3

***Scoliodon* sp.** Skull. A. Dorsal view. B. Ventral view

Cranium

It has four distinct regions

Occipital region constitutes the posterior portion of the skull bearing a large **foramen magnum**. Two **occipital condyles,** one on either side of the foramen magnum are present. **Occipital ridge**, a median crest is present on the roof above the foramen magnum.

Sense capsules

Auditory region comprises the two **auditory capsules** firmly united with the cranium. **Parietal fossa**, an oval depression is present between the two capsules.

Orbital region. A **supraorbital ridge** marks the junction between the roof of the cranium and the orbit.

Preorbital process, a slender cartilage partly encircles the orbit. A **postorbital process** arises from the side of each auditory capsule and runs forward along the upper border of the orbit.

Ethmoid region includes the anterior region of the cranium, olfactory capsules and the rostrum. **Intranasal septum**, a thin median cartilage separates the olfactory capsules. The cranial roof behind and between the olfactory capsules is incomplete, leaving a large **anterior fontanelle**, covered by a tough, fibrous membrane. A pair of **dorsolateral cartilages** arise in front of the anterior fontanelle, run forward to converge and meet in front with a **median ventral cartilage**. The three cartilages together form the skeleton of the **rostrum**.

Visceral skeleton

Seven pairs of half-hooped cartilages encircling the buccal cavity and the pharynx constitute visceral skeleton. The first pair gives rise to **jaws**, the second, the **hyoid arch** and the last five pairs support the gills.

Jaw apparatus

Jaws are two—upper and lower. Each half of the first or **mandibular arch** divides into two segments—an upper palatoptery-goquadrate and a lower Meckel's cartilage.

Uppar jaw. The two stout and curved, rod-like **palatopterygoquadrates** unite with each other in front to form the upper jaw. Just behind the anterior end each palatopterygoquadrate gives off an **orbital process** to which is attached the **ethmopalatine ligament** connecting the upper jaw with the cranium. Several rows of teeth are embedded in the mucons membrane lining the deep grooves of the upper jaw.

Lower jaw. The two **Meckel's cartilages** extend along the posterior margin of the mouth and unite to form the lower jaw. The posterior ends of the palatopterygoquadrates fit on the articular surfaces of the Meckel's cartilages. Each cartilage at its posterior end bears an articular surface for the upper jaw. The Meckel's cartilage bears teeth in shallow grooves along its dorsal border.

Vertebral column

The vertebral column consists of a series of cartilaginous vertebrae. The posterior end of the vertebral column is bent dorsally.

The vertebrae develop around an elastic axis of vacuolated cells, the **notochord**. The centra are **amphicoelous**, with deep concavities at both the ends. The notochord is constricted within the centrum, but greatly dilated in the intervertebral spaces and presents a beaded appearance in sagittal section.

The precaudal vertebrae or the vertebrae of the trunk region differ from the caudal vertebrae or those of the tail region to a certain extent.

Structure of a vertebra

The body or **centrum** is short, cylindrical and hollow. A dorsal **neural arch** encloses the spinal cord. The side of the neural arch is formed by a **vertebral neural plate**, projecting upwards from the dorsal surface of the centrum on each side (Fig. 37.4). **Intervertebral neural plates** fill up the space between the successive vertebral neural plates (Fig. 37.5). The neural arch is completed above by **neural spine**. A pair of transverse processes project ventrolaterally from the centrum.

The transverse processes of the anterior precaudal vertebrae are short and extend outward, while in posterior precaudal vertebrae they bend downward and enclose an open groove. In the caudal region the transverse processes close up to form **haemal canal**. In caudal vertebrae both the neural and haemal spines are long, narrow and pointed. Caudal artery and vein run through the haemal canal formed by successive caudal vertebrae.

Appendicular skeleton

The skeleton of the pectoral and pelvic girdles, paired and median fins constitute appendicular skeletor.

Pectoral girdle

The pectoral girdle lies just behind the branchial arches and consists of two half-hoops of cartilages fused in the midventral line

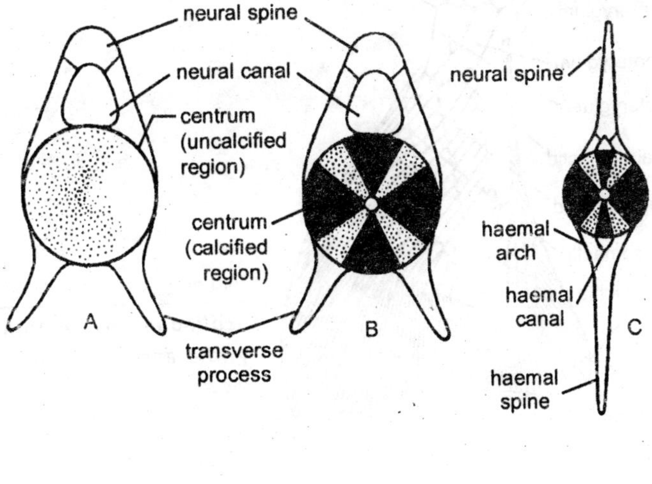

FIGURE 37.4

***Scoliodon* sp. Vertebrae.** A. Precaudal vertebra (front view). B. Section through the middle of the centrum of A. C. Section through the middle of the centrum of a caudal vertebra.

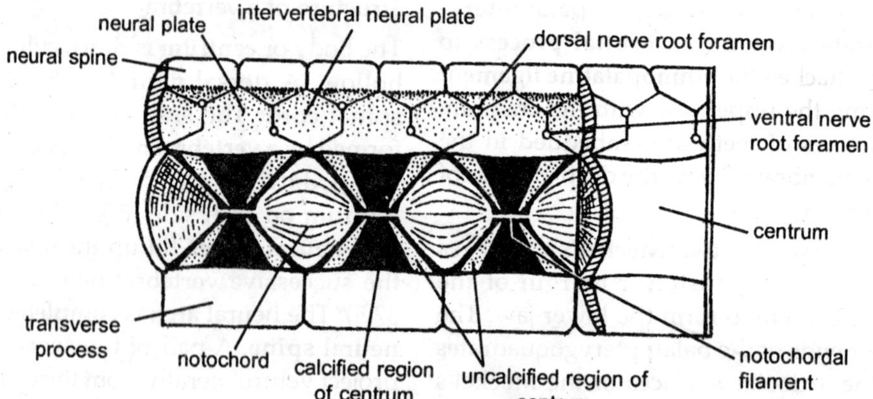

FIGURE 37.5

***Scoliodon* sp.** A portion of the trunk (precaudal) vertebral column (greater part removed)

but free dorsally (Fig. 37.6). **Coracoid** is thin and flattened and forms the ventral portion of the half-hoop, while the dorsal portion is formed by a thick rod-like **scapula**. A triple facet for the articulation of basal cartilages of the fin is present on the outer border of the coracoid and scapula.

Pectoral fin

It consists of three basal cartilages—**propterygium, mesopterygium** and **metapterygium** and a large number of segmented radial cartilages. Each pro- and mesopterygium bears two segmented radials, while the metapterygium, an elongated cartilage, bears eighteen radials with long distal pieces and support the **ceratotrichia** (Fig. 37.6).

Pelvic girdle

It consists of a transversely oriented (Fig. 37.7), flattened cartilaginous rod in front of the cloaca. A basipterygium is attached to each end of the transverse cartilage.

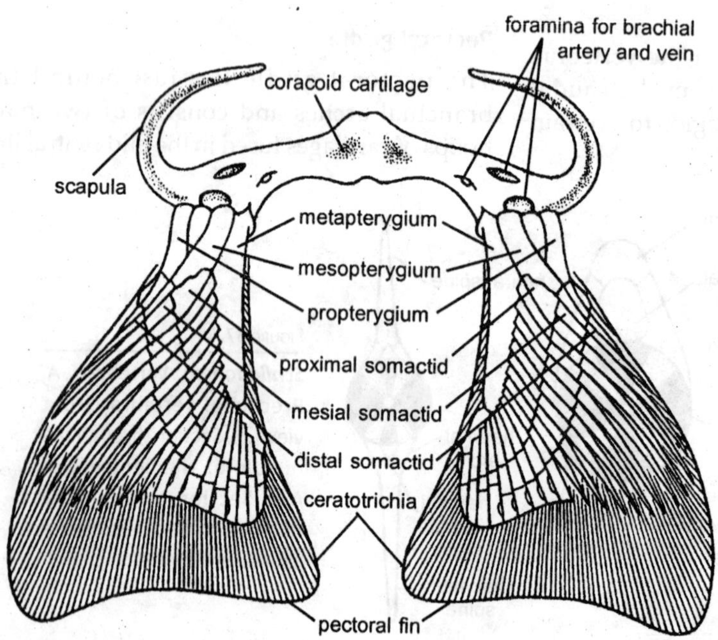

FIGURE 37.6

***Scoliodon* sp.** Pectoral girdle and fins (ventral view)

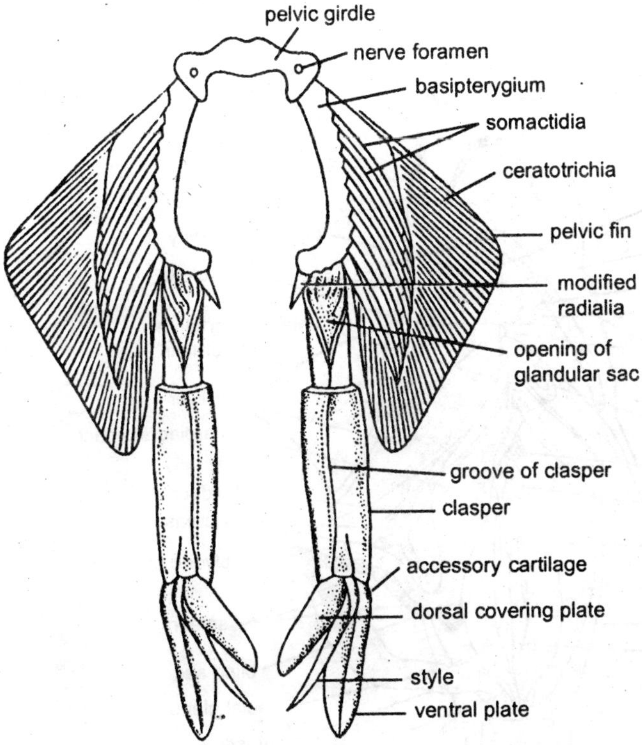

FIGURE 37.7

***Scoliodon* sp.** Pelvic girdle, fins and claspers (ventral view)

Pelvic fin

A curved cartilage, the **basipterygium**, constitutes the skeleton of each pelvic fin. Fifteen or more slender radials are attached to the outer border of the basipterygium. Small, polygonal cartilages bearing ceratotrichia are attached to the distal ends of the radials (Fig. 37.7).

Median fins

Each of the two dorsals and the ventral fin consists of a series of rod-like cartilages—**somactidia** or **pterygiophores**—bearing a double series of numerous horny fin rays or **ceratotrichia**. Each somactid is divided into three segments—proximal, mesial and distal.

Caudal fin

Somactidia are absent, the neural and haemal spines of the vertebrae are elongated and flattened to support the dorsal and ventral lobes of the fin.

Clasper

It consists of a dorsally grooved, tubular cartilage. Distally the groove ends in a sharp style enclosed in cartilaginous **sheathing plates**.

⊠ Muscular System

The muscles of the trunk and tail are segmentally arranged, consisting of a paired series of muscle-segments or **myotomes** (Fig. 7.6), separated from one another by tough connective tissue, the **myocommata**. Four sharp forward and backward bends in the muscle from its dorsal to ventral course give a zigzag appearance to it. Myocommata are inserted on the vertebral column internally and the skin externally. Muscle fibres in myotomes run parallel to the body length. Contraction of muscle fibres exert a pull on the myocommata and this leads to a flexion both of the backbone and the body.

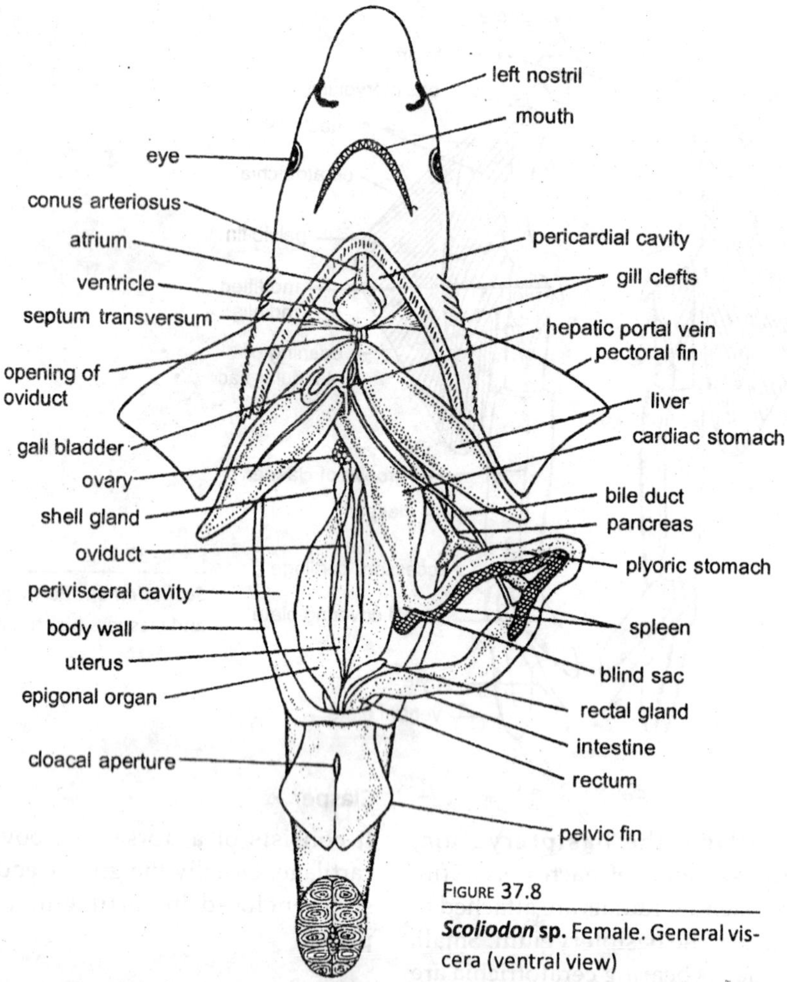

FIGURE 37.8

***Scoliodon* sp.** Female. General viscera (ventral view)

Muscles of the head and branchial region are not segmented and are concerned with the movement of associated structures.

☒ Locomotion

Succession of longitudinal waves of contraction on alternate sides of the body produce waves of curvature of the body (Fig. 7.7). The pressure of the body of the fish against the surrounding water is exerted always by the leading surface. The water is obliquely pushed backward on both sides of the axis of forward movement. The oblique backward thrust alternately by each side of the body give the fish a forward thrust pushing it to the front. The strong lateral movement of the heterocercal tail, in addition to giving a forward thrust, forces the anterior end downwards.* On the other hand, movement of the pectoral and pelvic fins tend to force the anterior end upward with nearly the same amount of force. The median and paired fins maintain body equilibrium in a swimming fish.

☒ Coelom

The coelom is large and divided into a small pericardial and a large perivisceral cavity (Fig. 37.8) separated by a membranous **septum transversum**.

* Heterocercal tail is always associated with a ventral mouth.

Pericardial cavity

Small, triangular, situated beneath the pharynx and encloses the heart. A colourless **pericardial fluid** fills up the cavity. The pericardial cavity communicates with the perivisceral cavity through a **pericardial-peritoneal canal** in septum transversum.

Perivisceral cavity (abdominal cavity)

The cavity is large, accommodates the **visceral organs** (viscera) and is in communi-cation with the exterior through a pair of adbominal pores. The cavity is lined with **peritoneum**, adherent to the inner body wall, but free in the middorsal line to form a double fold of membrane, the **mesentery**, ventral to the vertebral column, by which the gut is suspended in the coelom.

⊠ Digestive System

The digestive system (Figs. 37.9, 37.10) consists of a digestive tract or alimentary canal and its appendages, the digestive glands and a caecal (rectal) gland.

⊠ Alimentary canal

The alimentary canal is divisible into mouth, buccal cavity, pharynx, oesophagus, stomach, intestine and rectum.

Mouth

The mouth is a ventral crescentic opening bounded by folds of integument, the **upper** and **lower lips**, supported by **jaws**. The mouth leads into the buccal cavity.

Buccal cavity

A spacious, dorsoventrally compressed chamber, lined with a thick mucous membrane, raised ventrally at the middle line to form a nonmuscular, nonglandular **tongue**. The mucous membrane bears denticles, which are modifed into **teeth** in the region of jaws. The teeth are homodont, oblique, sharply pointed and arranged in several parallel rows on the inner margin of the jaws, but not attached to the jaw cartilage. The buccal cavity passes imperceptibly into the pharynx.

Pharynx

The pharynx is lined with mucous membrane bearing numerous dermal denticles, crowded in the region of gill slits. On either side of the pharynx an oval pit, the **vestigial spiracles**, are present. They have neither external opening, nor they bear gill lamellae. Five pairs of **gill clefts** are present in the lateral walls of the pharynx. The pharynx narrows posteriorly to open in the oesophagus.

Oesophagus

A short, thick muscular tube, the internal lining membrane bearing series of longitudinal folds. The oesophagus opens in the stomach.

Stomach

It is a large, muscular sac. A crescentic fold of the mucous membrane at the junction of oesophagus and stomach acts as a valve. The stomach is a J-shaped organ. The proximal limb, the **cardiac stomach**, is large and extends to very near the posterior end of the abdominal cavity. The shorter and narrow dorsal limb is the **pyloric stomach**. A small outgrowth, the **blind sac** is present at the junction of cardiac and pyloric stomach. The mucous membrane of the cardiac stomach bears longitudinal folds some of which form circular folds and act as a valve between the cardiac and pyloric stomach. The mucous membrane of the pyloric stomach is smooth proximally but bears small folds distally. **Pyloric valve**, a band of circular muscle fibres, guards the junction of pyloric stomach and a thick-walled small chamber **bursa entiana**, which leads to intestine.

Intestine

The intestine is a wide tube, runs straight backward and narrows posteriorly to form rectum. The internal surface of the intestine bears a characteristic fold of mucous membrane called **scroll valve**, an anticlockwise spiral of about two and a half turns. Scroll valve increases absorptive surface and also checks rapid flow of digested food through the intestine.

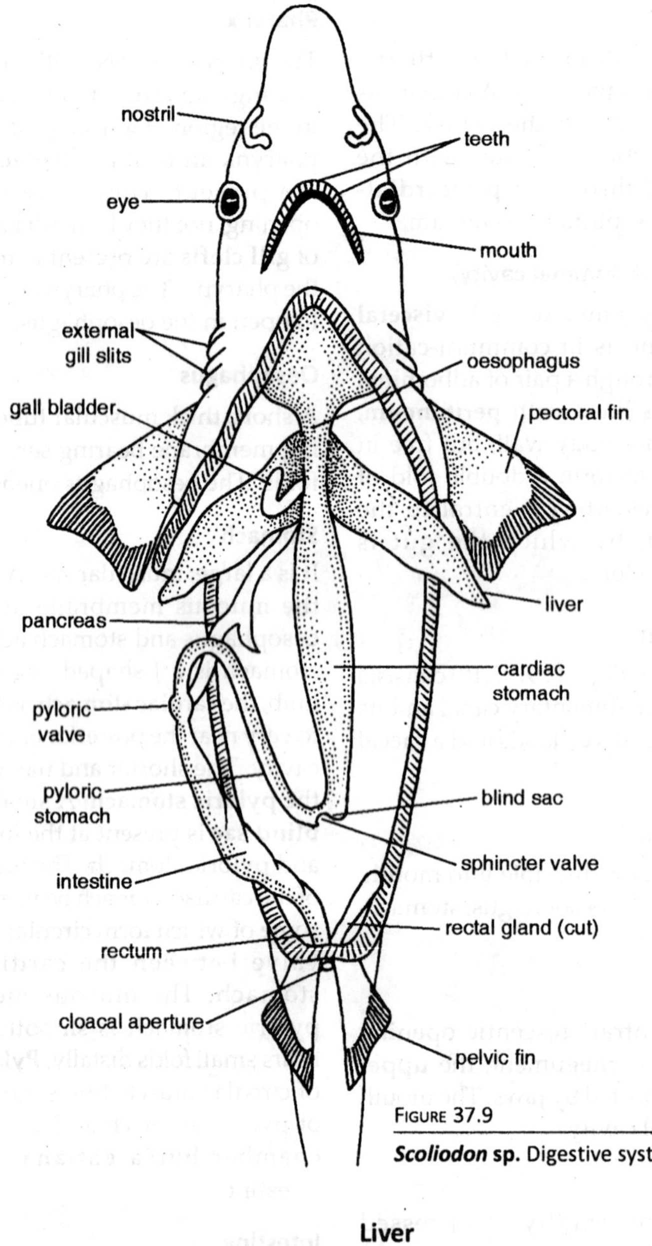

FIGURE 37.9

***Scoliodon* sp.** Digestive system

Rectum

A short, narrow tube in which a tubular digitiform **rectal** or **caecal gland**, opens dorsally. The rectum opens into the cloaca.*

⊠ Digestive glands

Liver and pancreas constitute digestive glands.

Liver

The liver is a massive, bilobed, yellowish gland, the lobes being united anteriorly. **Gall bladder**, a thin-walled V-shaped sac embedded in the anterior portion of the right lobe of the liver receives bile from the liver through a number of small ducts. The **bile duct**, after its

* Cloaca is a common chamber at the posterior end of the abdominal cavity in which rectum, urinary and genital ducts open. Cloaca opens to the exterior through a cloacal aperture.

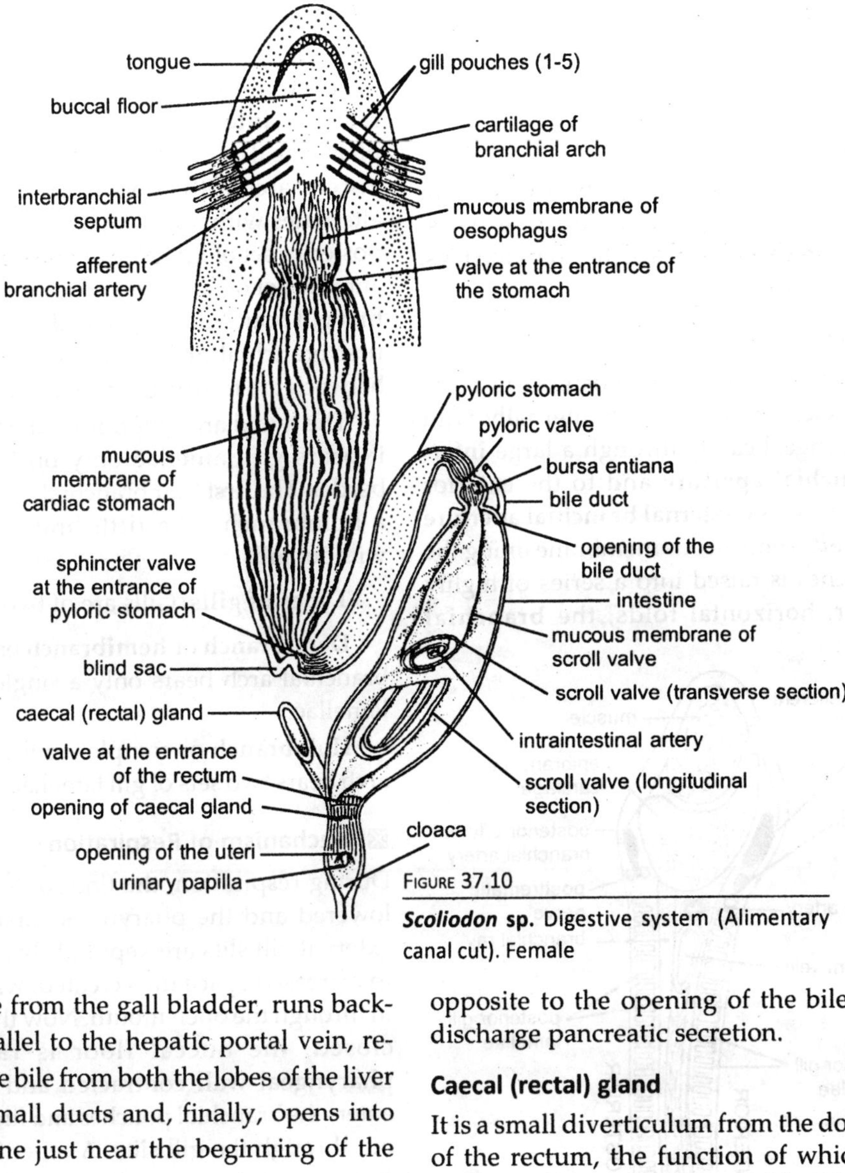

FIGURE 37.10

***Scoliodon* sp.** Digestive system (Alimentary canal cut). Female

emergence from the gall bladder, runs backward, parallel to the hepatic portal vein, receives more bile from both the lobes of the liver through small ducts and, finally, opens into the intestine just near the beginning of the scroll valve.

Pancreas

The pancreas is a compact, pale yellowish, bilobed gland, located in the triangular space between the two limbs of the stomach. The dorsal lobe runs parallel to the posterior part of the cardiac stomach, while the ventral lobe lies against the pyloric stomach. The **pancreatic duct** runs along the whole length of the bile duct and opens into the intestine opposite to the opening of the bile duct to discharge pancreatic secretion.

Caecal (rectal) gland

It is a small diverticulum from the dorsal wall of the rectum, the function of which is not clearly understood. Presumably it has some blood-related function.

⊠ Digestion and Absorption

Scoliodon is carnivorous. Buccal cavity and oesophagus have no digestive function. Stomach secretes hydrochloric acid and proteolytic enzymes. Pancreatic secretion is the main source of hydrolysing enzymes. Bile helps in emulsification of fat and also makes the partly digested food alkaline, which promotes action of the proteolytic enzymes.

Respiratory System

Scoliodon utilises dissolved oxygen in water in respiration.

Respiratory structures

The respiratory structures **gills**, are borne in gill pouches. Gill pouches are five pairs, located in the lateral walls of the pharynx (Figs. 9.3A and 37.10).

Gill pouch

A gill pouch is an anteroposteriorly compressed sac, communicating internally with the pharyngeal cavity through a large **internal branchial aperture** and to the exterior through a narrow **external branchial aperture** or **gill cleft**. The mucous membrane lining the gill pouches is raised into a series of highly vascular, horizontal folds, the **branchial lamellae**. Each gill pouch has two sets of gill lamellae, one on the **anterior** and the other on the **posterior wall**. The gill pouches are separated by **interbranchial septa**, the inner or pharyngeal border of which is supported by a visceral arch with its branchial rays. Each arch alternates with a gill pouch and supports the posterior set of lamellae of one pouch and the anterior set of lamellae of the next pouch. The first gill pouch lies between the hyoid and the first branchial arch. The last gill pouch lies between the fourth and fifth branchial arches.

The hyoid arch supports a demibranch, as it bears gill lamellae only on its posterior border. The first four branchial arches support a holobranch. The fifth branchial arch is without gills.

Types of gills. Gills are of two types.

Demibranch or **hemibranch** or half gill. A branchial arch bears only a single set of gill lamellae.

Holobranch or complete gill. A branchial arch bears two sets of gill lamellae (Fig. 37.11).

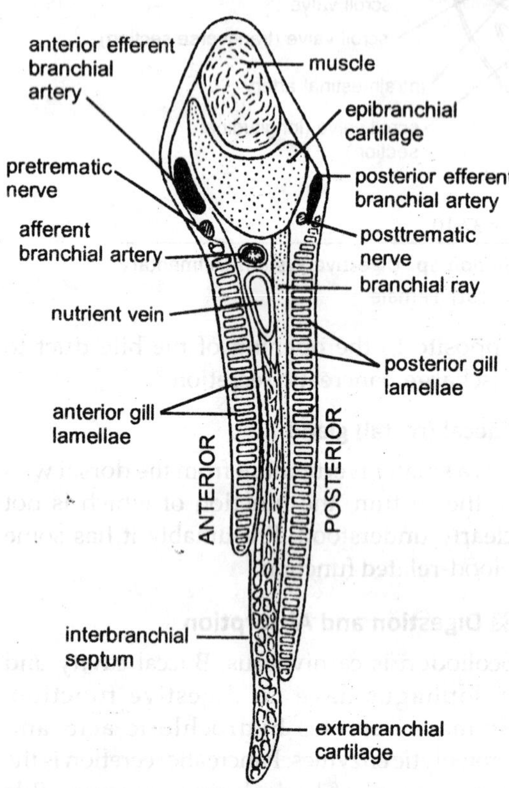

FIGURE 37.11

***Scoliodon* sp.** Horizontal section of a gill (holobranch) parallel to the branchial cartiage

Mechanism of Respiration

During respiration, the floor of the mouth is lowered and the pharynx is expanded. The external gill slits are kept tightly closed. Due to increased space thus created, water rushes in through the open mouth. Now the mouth is closed, the buccal floor is raised, the pharyngeal wall contracted and the water enters the branchial pouches and finally forced out through the gill slits. A repetition of this process maintains a continuous water current through the gill slits which carries dissolved O_2 and food particles with it.

Blood Vascular System

The blood vascular system consists of a heart, an arterial or afferent and efferent branchial system and a venous system (see *Lates*).

Heart

The heart is a muscular, reddish structure situated ventrally in the **pericardial cavity** and enclosed within a thin membrane, the **pericardium**. It consists of three chambers—a dorsally situated **sinus venosus**, a single **auricle** or **atrium** and a **ventricle**, most ventral in position (Fig. 37.12). Deoxygenated blood from different parts of the body is received by sinus venosus through two large, transverse veins, **ducti Cuvieri**. Blood from sinus venosus passes to the atrium through the **sinuatrial aperture** and from the atrium to the ventricle through an **atrioventricular aperture**. The sinuatrial aperture is guarded by a pair of **membrane valves** and the atrioventricular aperture by a pair of **bilabiate** valves. The sinus venosus and the atrium constitute the receiving part of the heart.

The ventricle has a thick muscular wall, the inner surface bearing numerous muscular strands. The **conus arteriosus** is a stout muscular tube, extending from the ventricle to the anterior end of the pericardial cavity. The conus continues forward as **ventral aorta**. Two transverse rows of **semilunar valves** guard the opening of the conus into the ventral aorta. Each row contains three valves—one dorsal and two ventrolaterals. A pair of small, additional **accessory valves** are present, one on either side of each dorsal aorta. The ventricle and the conus constitute the forwarding part of the heart.

⊠ Arterial System

The arterial system consists of two components. The afferent branchial or afferent system sending deoxygenated blood from heart to the gills for oxidation and efferent branchial or efferent system, draining oxygenated blood from gills (Fig. 37.13).

⊠ Afferent Branchial System

The **ventral aorta** runs forward up to the posterior border of the hyoid arch and sends out a series of paired **afferent branchial arteries**, one for each branchial pouch (Fig. 37.13). Anteriorly, the ventral aorta bifurcates into two **innominate arteries**, each of which again divides into first and second afferent branchial arteries. The first afferent runs along

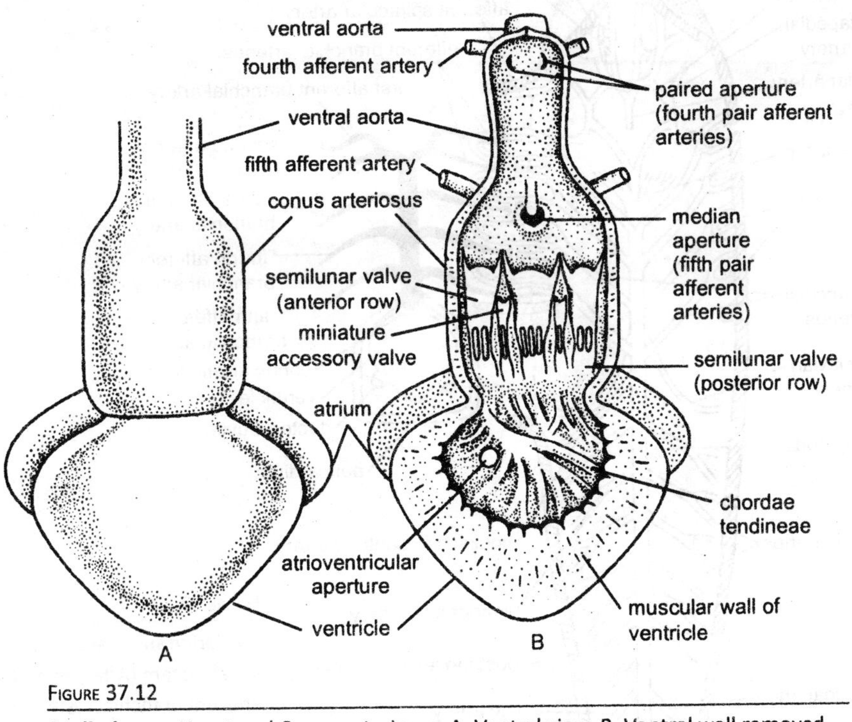

FIGURE 37.12

***Scoliodon* sp.** Heart and Conus arteriosus. A. Ventral view, B. Ventral wall removed the posterior border of the hyoid arch and supplies blood to the hyoidean demibranch. The second one courses along the first branchial arch, supplying blood to both anterior and posterior gill lamellae. Origin of the third, fourth and fifth pairs of afferent branchials from the ventral aorta are almost equidistant and they course along the outer border of the second, third and fourth branchial arches. Each afferent branchial artery has its separate opening in the ventral aorta, except the fifth pair which communicates with the ventral aorta through a common opening.

☒ Efferent Branchial System

Running along the anterior and posterior border of the gill clefts of each side are nine **efferent branchial** arteries. Of these, the first four pairs join to form four complete loops around the first four gill clefts, while the ninth efferent artery arising from the demibranch is connected to the fourth loop by a connective. The four loops are also connected by short connectives. In addition, a network of longitudinal commissural vessels connect all the loops. The four efferent branchial loops continue medially and back ward as **epibranchial arteries** and the four pairs of epibranchial arteries join to form a median **dorsal aorta.** The half-loop of the ninth efferent branchial joins the eighth efferent branchial artery.

Anterior arteries

The arteries of the head—external carotid, afferent spiracular and hyoidean epibranchial —arise from the first or **hyoidean efferent** artery.

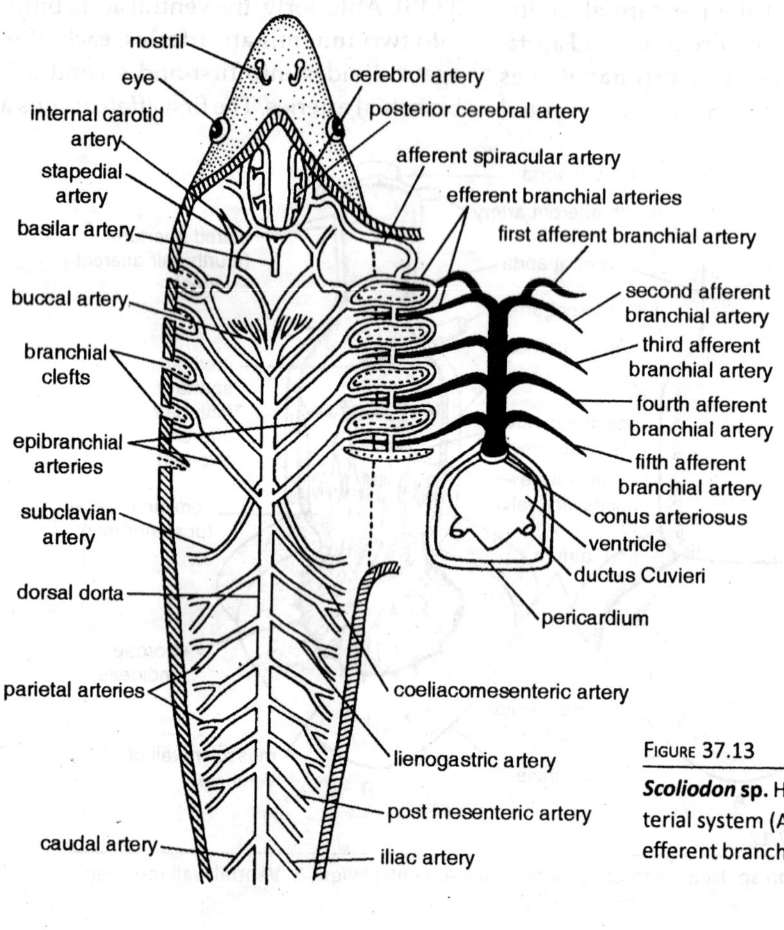

FIGURE 37.13

***Scoliodon* sp.** Heart and arterial system (Afferent and efferent branchial systems)

a. **External carotid artery.** Arises from the anteroventral corner of the first collector loop and divides anteriorly into—

i. **Ventral mandibular**, sending branches to coracomandibular muscles and muscles of the lower jaw.

ii. **Superficial hyoid** sending branches to second ventral constrictor muscle, the skin and subcutaneous tissue of the hyoid arch.

b. **Afferent spiracular artery.** Originates from the middle of the hyoidean afferent, proceeds forward as **spiracular epibranchial artery**. Sending a great **ophthalmic artery** to the eye-ball it enters the cranium, where it joins with a branch of the internal carotid artery to form the cerebral artery. The cerebral artery divides into an anterior and posterior cerebral arteries to supply the brain.

c. **Hyoidean artery.** Arises close to the dorsal end of the hyoidean efferent, runs forward and inward, joined by an anterior branch of dorsal aorta and divides immediately into two branches—the **stapedial** and **internal carotid**. The stapedial artery enters the orbit, sends **inferior orbital artery** to six eye muscles and superficial tissues and runs forward as **superior orbital artery** sending a large branch, the **buccal artery** supplying muscles of the lower jaw and continues forward to the snout as **maxillonasal artery**. The maxillonasal sends **nasal artery** to olfactory sac and several maxillary arteries to upper jaw and continues as **rostral artery** to the rostrum. The internal carotid enters the cranial cavity and divides into two branches, one uniting with its fellow of the opposite side and the other with the stapedial artery to form the anterior and posterior **cerebral arteries**.

Dorsal aorta and its branches

The four pairs of epibranchial arteries join medially to form the dorsal aorta. The dorsal aorta runs backward along the whole length of the body cavity, lying below the vertebral column in the trunk. In the tail region it is known as **caudal artery** and runs through the haemal canal of the tail vertebrae.

In its way backward dorsal aorta sends:

a. **Buccal arteries.** Several small branches to the roof of the buccal cavity.

b. **Vertebral arteries.** Small arteries supplying vertebral column.

c. **Two branches** join the hyoidean epibranchial arteries.

d. **Subclavian arteries.** Paired, origin anterior to the union of fourth epibranchial arteries, run outward and backward and join with the epicoracoid artery. Each divides into:

i. **Brachial artery** supplying pectoral girdle and fin.

ii. **Anterior lateral artery** supplying muscles of the body wall.

iii. **Dorsolateral artery** supplying dorsal body muscles.

e. **Coeliacomesenteric artery.** Large, unpaired, arises slightly behind the junction of the fourth pair of epibranchial arteries. It divides into:

i. **Coeliac artery.** A small vessel, supplying dorsal and ventral wall of the cardiac and pyloric stomach and both lobes of the liver.

ii. **Anterior mesenteric artery.** A large artery supplying pancreas and intestine.

f. **Lienogastric artery.** Unpaired, arises behind the coeliacomesenteric and sends branches:

i. **Spermatic** or **ovarian artery**. Supplying testes or ovaries.

ii. **Posterior intestinal artery.** Supplying the posterior part of the intestine.

iii. **Posterior gastric artery.** Supplying posterior part of cardiac stomach.

iv. **Posterior mesenteric artery.** Supplying posterior end of gonad and rectal gland.

g. **Parietal arteries.** A series of paired arteries arise from the dorsal aorta behind subclavian. Each parietal divides into a dorsal and a ventral branch:

i. **Dorsal parietal artery.** Supplies dorsolateral muscles, vertebral column, spinal cord, lateral line canal and dorsal fin.

ii. **Ventral parietal artery.** Supplies peritoneum and ventral muscles and sends **renal** branch to the kidney.

h. **Iliac arteries.** Paried and extend to the pelvic fins as **femoral arteries.**

☒ Venous System

Veins are thin-walled vessels bringing back deoxygenated blood from different parts of the body. The veins form wide irregular spaces in their course and called **sinuses.** The cardinal system, renal portal system, hepatic portal system, ventral veins and the cutaneous system constitute venous system (Fig. 37.14).

☒ Cardinal System

It consists of anterior and posterior cardinal systems.

☒ Anterior Cardinal System

A pair of large **internal jugular veins** drain blood from the dorsal region of the head and branchial region. Each internal jugular is formed by the union of **olfactory sinus, orbital**

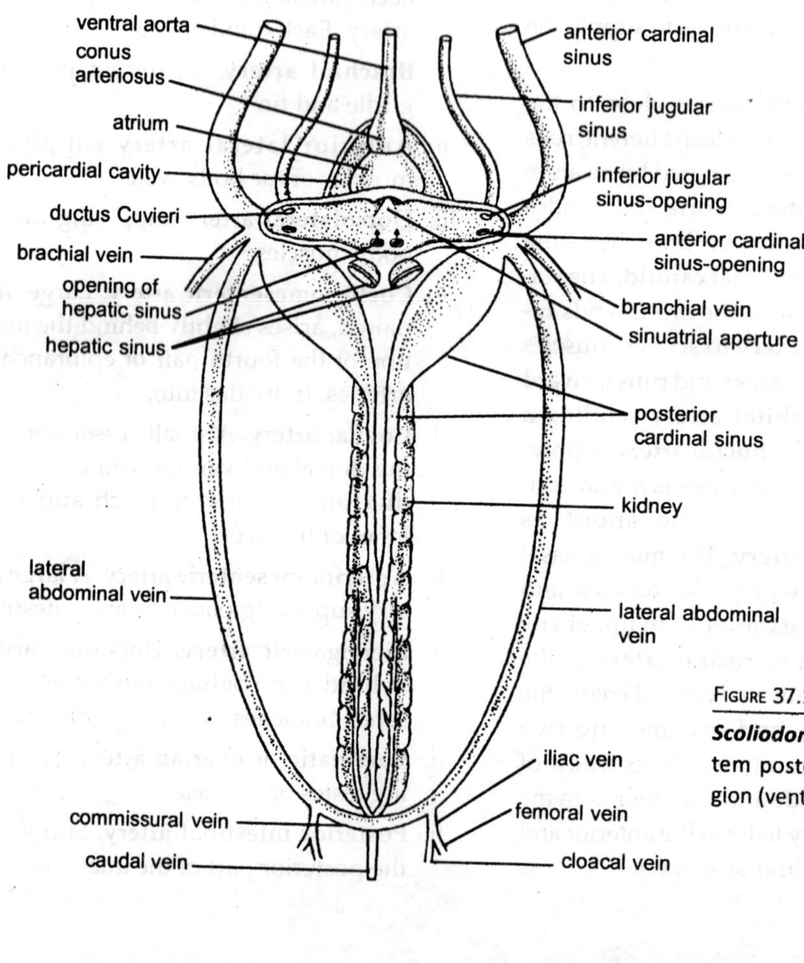

FIGURE 37.14

***Scoliodon* sp.** Venous system posterior to head region (ventral view)

sinus, **postorbital sinus** and the **anterior cardinal sinus**. **Anterior facial vein** drains blood from the rostral region and joins the **orbital sinus**, which opens into the anterior cardinal sinus through postorbital sinus. The **hyoidean sinus** and the five **dorsal nutrient branchial sinuses** join the anterior cardinal sinus which opens into the ductus Cuvieri.

☒ Posterior Cardinal System

Draining blood from the tail, the **caudal vein** runs forward through the haemal canal. In the abdominal cavity the caudal vein bifurcates into right and left **renal portal veins.**

☒ Renal Portal System

Each renal portal vein branch into capillaries within the kidneys. Small parietal veins join the renal portal veins along their course. Series of **renal viens** collect blood from the kidneys, join together and form the **posterior cardinal sinuses,** each of which opens into a Cuvierian duct through a small aperture opposite to the opening of the anterior cardinal sinus. The renal portal system is complete.

☒ Hepatic Portal System

A number of veins collect blood from alimentary canal and digestive glands. The **hepatic portal vein** formed by the joining of **anterior** and **posterior intestinal veins**, near the posterior end of the pyloric stomach, a **lienogastric vein** a little behind the pancreas, two large veins—**anterior** and **posterior** gastric at the level of the cardiac stomach. The vein runs forward, divides into two branches and enters the right and left lobes of the liver, where they branch into capillaries. Blood from the liver is collected in two large, thin-walled sinuses, the **hepatic sinuses**, which open separately into the sinus venosus.

Ventral veins

Two groups of veins anterior and posterior constitute ventral veins.

i. **Anterior veins.** Each **inferior jugular sinus** is formed by joining of **submental sinus** from the lower jaw, the **hyoidean sinus** and **ventral nutrients** from the gills and open into the ductus Cuvieri of the side.

ii. **Posterior veins.** The posterior veins consist of two large, **lateral abdominal veins**, each of which is formed by the joining of a small **cloacal vein** from the wall of the cloaca and a **femoral vein** from the pelvic fin. The two lateral abdominal veins are connected by a commissural vein posteriorly. Anteriorly, in the pectoral region, the lateral vein joins the **brachial vein** and form the large **subclavian vein** opening into the ductus Cuvieri.

Cutaneous system

The cutaneous system consists of a **dorsal**, a **ventral** and two paired **lateral veins.** The **inferior lateral cutaneous vein** joins the lateral cutaneous vein, close to the anterior end of the pectoral fin. Each lateral cutaneous opens into the **brachial vein.**

☒ Nevous System

The central nervous system, peripheral nervous system and autonomic nervous system constitute the nervous system. Both brain and spinal cord are surrounded by a single membrane, the **meninx.**

☒ Central Nervous System

The components of central nervous system are brain (Fig. 37.15, 37.16) and spinal cord.

Brain

The brain is enclosed in a cartilaginous cranium. It is divisible into forebrain (anterior portion) or **prosencephalon**, midbrain (middle portion) or **mesencephalon**, hind brain (posterior) or **rhombencephalon.**

The brain is horizontal and its different parts are in a straight line.

Forebrain

The forebrain consists of **cerebrum** and **diencephalon.**

The cerebrum is a massive mass not divided into **cerebral hemispheres**. An **olfactory tract** or **peduncle** arises from each anterolateral end, runs forward and slightly lateral to end in a large, bilobed mass, the **olfactory lobe**, lying against the **olfactory sac.** The dorsal surface of the cerebrum is smooth and nonnervous. The ventral surface bears a **neuropore**, through which a pair of **terminal nerves** emerge.

The **diencephalon** is narrow and short and completely covered by the forward extension of the cerebellum. The roof of the diencephalon is membranous, thin, nonnervous and bears the **anterior choroid plexus** formed by numerous blood vessels. A slender **pineal body** arises from the posterior portion of the roof.

The **infundibulum** projects downwards and backwards from the floor and bears **pituitary body** or **hypophysis** at its tip. Two thin-walled sacs, **lobi inferiores** bearing glandular sacs, the **sacci vasculosi** at the tip lie by the sides of the infundibulum.

Midbrain

The midbrain consists of two large **optic lobes,** almost completely covered dorsally by the cerebellum and ventrally by infun-dibulum. Longitudinal nerve tracts are present on the sides.

Hindbrain

The hindbrain is divisible into a large, anterior **cerebellum** and a posterior, somewhat triangular **medulla oblongata.**

The cerebellum is rhomboidal in shape, extends up to the cerebrum in front and overhangs the medulla behind. The dorsal surface bears numerous convolutions. Two deep furrows divide it into three lobes.

The medulla oblongata forms the hindmost part of the brain and continues backward as **spinal cord**. A pair of hollow outgrowths, **corpora restiformia** arise from the anterior end of the medulla and lie on its dorsolateral sides, being slightly overlapped by the cerebellum. Corpora restiformia of the two sides are connected by a band of nervous tissue just

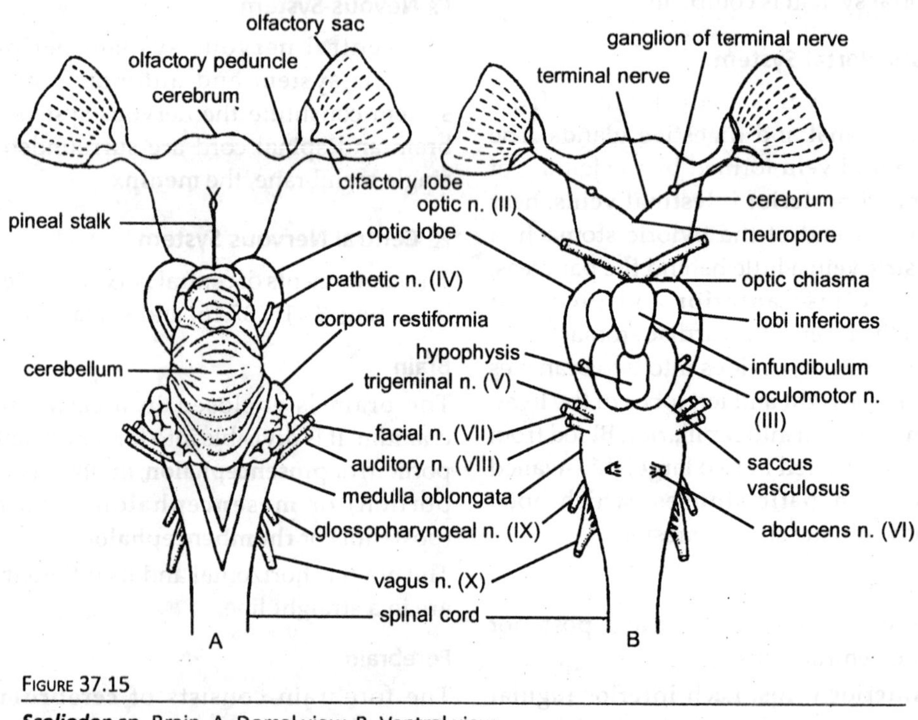

FIGURE 37.15

***Scoliodon* sp.** Brain. A. Dorsal view, B. Ventral view

behind the cerebellum. The floor and sides of the medulla are thick. The roof is nonnervous, extremely thin and bears the **posterior choroid plexus**.

Cavities of the brain

The cavities of the brain are **ventricles** (Fig. 37.16) and that of the spinal cord is **central canal**. The lateral ventricles or the cavities of the cerebrum—also known as first and second ventricles—are narrow and communicate behind with the third ventricle through **foramen of Monro**. The floor of the ventricles are thick and form the bulging **corpora striata**. A tubular olfactory tract connects each lateral ventricle with **rhinocoel**, the cavity of the olfactory lobe. The floor of the **diocoele** or third ventricle, the cavity of the diencephalon, is tubular, bears two longitudinal bands, the **crura cerebri** on the floor. The thin roof bears a **Habenular ganglion**, from which the pineal body develops, while an opening on the floor leads to the cavity of the infundibulum. The fourth ventricle or the cavity of the medulla oblongata is large, continues in front as a wide **iter** or **aqueduct of Sylvius** to communicate with the **epicoele** or the cavity of the cerebellum above, the **optocoeles** or the cavities of the optic lobes on the sides, and diocoele of the diencephalon.

Spinal cord

From the posterior end of the medulla oblongata the spinal cord extends almost to the end of the tail running through the neural canal formed by the vertebrae.

⊠ Peripheral Nervous System

The peripheral nervous system consists of cranial and spinal nerves (p. 128).

Cranial nerves

Nerves arising from the brain are cranial nerves. In addition to ten pairs (Figs. 37.17, 37.18) of cranial nerves, a pair of cranial nerve 'O' are present. Proceeding from the anteriormost to the posteriormost in order, the cranial nerves of each side are:

Cranial nerve 'O' (Terminal nerve). Arises from the ventral surface of the cerebrum; bears a ganglion, **ganglion terminale** near its base and innervates the nasal septum and external nostril. Function: sensory.

I. **Olfactory nerve.** Origin by a number of fibres from olfactory lobe, innervates the olfactory sac. Function: sensory.

II. **Optic nerve.** Arises from the optic thalamus. Nerves of the two sides cross each other to form **optic chiasma**, enters the orbit of the opposite side and innervates the retina. Function: sensory.

III. **Oculomotor nerve.** Origin ventral surface of the midbrain, enters the orbit and dividing into branches innervate anterior, superior, inferior rectus and inferior oblique muscles of the eye-ball. Function: motor.

IV. **Trochlear nerve.** Origin dorsolateral surface of the midbrain between the optic lobe and the cerebellum. It runs

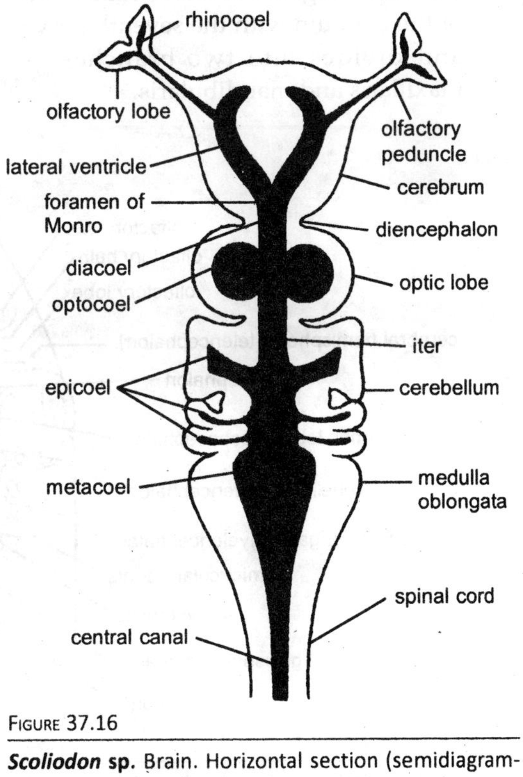

FIGURE 37.16

***Scoliodon* sp.** Brain. Horizontal section (semidiagrammatic)

forward, enters the orbit and innervates the superior oblique muscle of the eye-ball. **Function.** motor.

The fifth, seventh and eighth nerves arise close together from the side of the medulla oblongata just below the corpora restiformia.

V. **Trigeminal nerve.** It has four branches—ophthalmic superficialis, ophthalmic profundus, maxillaris and mandibularis.

a. **Ophthalmic superficialis.** A small branch, It enters the orbit, runs along its upper border, passes to the cranium and supplies the skin of the snout in front of olfactory capsule. Function: sensory.

b. **Ophthalmic profundus.** Origin anterior border of the trigeminal trunk; sends a **ciliary** branch to eye-ball and finally divides into a number of small branches to innervate the snout.

The main trigeminal trunk comes out of the cranium with the seventh nerve and divides into two branches — maxillaris and mandibularis.

c. **Maxillaris.** It proceeds in two groups (i) **maxillaris superior,** a flat ribbon-like nerve innervating skin of the upper jaw and (ii) **maxillaris inferior** innervating the posterior part of the upper lip.

d. **Mandibularis.** Runs obliquely backward in the posterior angle of the orbit, curves round the angle of the mouth and divides into small branches to innervate muscles of the lower jaw. Function: mixed.

VI. **Abducens nerve.** A slender nerve, arises near the midventral line of medulla and innervates posterior rectus muscle of the eye-ball. Function: motor.

VII. **Facial nerve.** Leaves the cranium in two bundles—ophthalmic superficialis and a bundle of mixed nerves.

a. **Ophthalmic superficialis.** A large flat nerve, gives off branches to supraorbital canal, runs forward, divides into branches and inner-vates the supraorbital canal and all the ampullary organs of the snout.

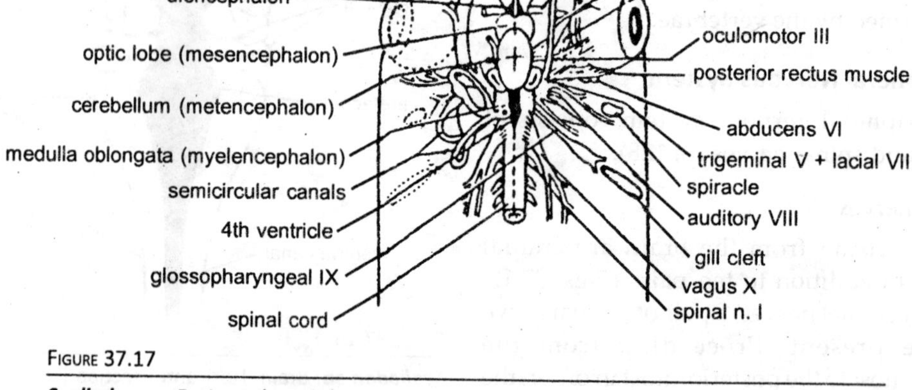

FIGURE 37.17

***Scoliodon* sp.** Brain and cranial nerves. Dorsal view

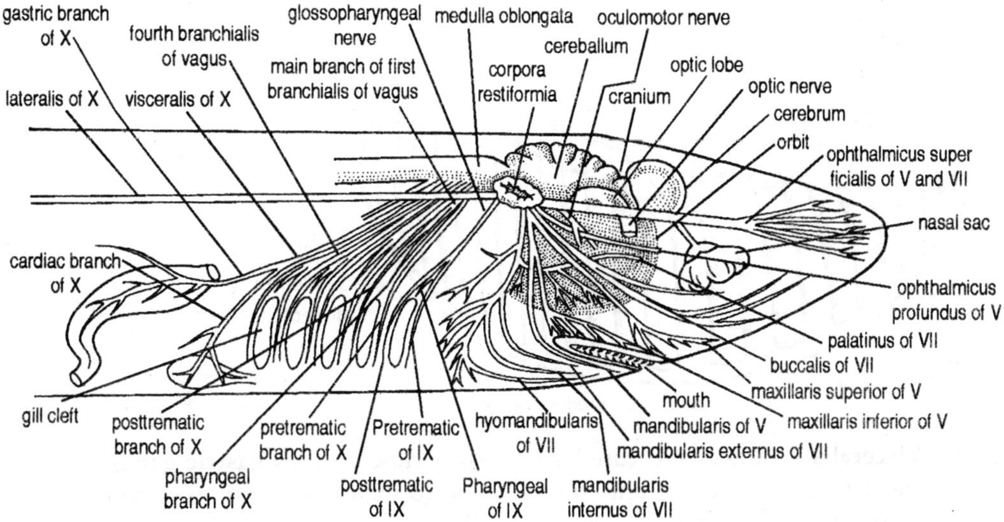

FIGURE 37.18

***Scoliodon* sp.** Brain and cranial nerves. Lateral view

b. The **bundle of mixed nerve** has three branches.

i. **Ramus buccalis.** Sends branches to infraorbital canal below and behind the eye and the trunk, runs forward to innervate the infraorbital canal of the snout and associated ampullae.

ii. **Ramus hyomandibularis.** It runs backward and outward, sends a slender **optic branch** innervating the postorbital canal and an **anterior branch** to the jugal canal. The main hyomandibularis nerve divides into three branches (a) **mandibularis externus** supplying the mandibular canal, (b) **mandibularis internus** innervating mucous membrane of the buccal floor and, (c) **hyoidean** to the muscles of the throat.

iii. **Ramus palatines.** Immediately after leaving the cranium it runs across the floor of the orbit and sending a few small branches to the roof of the pharynx, finally ends in the roof of the buccal cavity. Function: mixed.

VIII. **Auditory nerve.** Origin close to those of the fifth and seventh nerves, divides immediately into two branches—**vestibular** and **saccular**, innervating membranous labyrinth or internal ear. Function: sensory.

IX. **Glossopharyngeal nerve.** Origin ventrolateral side of the medulla, behind the root of the sixth nerve. It runs obliquely forward along the floor of the auditory capsule. Emerging from the capsule, the nerve sends a branch to commissural occipital canal, runs backward and downward and divides into a small **pretrematic** and a large **posttrematic** branch to innervate mucous membrane around the first gill cleft and pharynx and muscles of the pharynx. Function: mixed.

X. **Vagus nerve.** Arises by a number of roots behind the ninth nerve. Leaving the cranium it runs backward and sends many branches. The major branches are branchialis, visceralis and lateralis.

a. **Branchialis.** Consists of four main nerves running around the second, third, fourth and fifth gill clefts. Each nerve divides into a pretrematic and a posttrematic branch innervating mucous membrane around the gill clefts and pharynx and branchial muscles.

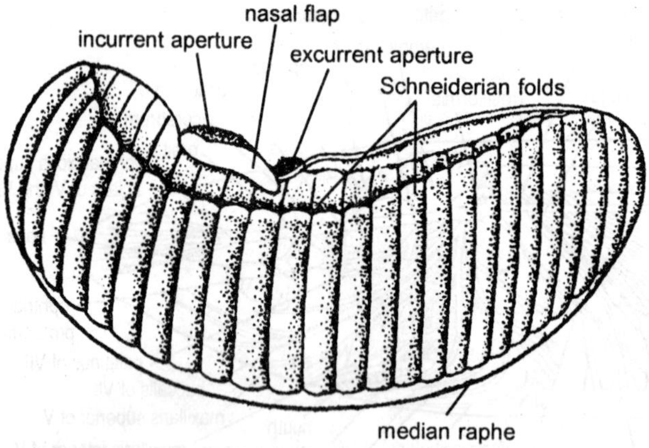

FIGURE 37.19

***Scoliodon* sp.** Right olfactory sac, dissected to expose nasal flaps

b. **Visceralis.** A large nerve, runs backward into the coelom and sends branches to the viscera including alimentary canal, liver, heart and lungs.

c. **Lateralis.** Dorsal to other branches of the vagus, runs backward in the deep fascia between the dorsal and ventral segments of body muscles, parallel to the lateral line canal. Anteriorly it sends a **supratemporal branch** to the commissural occipital canal and sends nerves to neuromast organs of the lateral line canal. Function: mixed.

Spinal nerves

Spinal nerves (Fig. 37.17) are paired. Each nerve arises by two roots, dorsal and ventral (p. 138). The dorsal root passes out through a foramen at the upper margin of the neural plate, while the ventral root passes through a foramen at the lower margin of the intervertebral plate.

☒ Autonomic Nervous System

The autonomic nervous system consists of a paired series of ganglia (p. 139) arranged irregularly. Their joining by longitudinal connectives are also not continuous. Anteriorly, the ganglia are located on the dorsal wall of the posterior cardinal sinus and posteriorly on the dorsal surface of the kidneys. The first ganglion is small and the second, known as **gastric ganglion**, is large and formed by the fusion of several ganglia. The succeeding ganglia are small.

☒ Receptors and Sense organs

Paired olfactory organs, eyes, ears and neuromast organs and ampullae constitute receptors.

Olfactory organs (Chemoreceptors)

The olfactory organs are a pair of blind sacs located ventrally in front of the mouth. The organ, known as **olfactory sac** (Fig. 37.19), opens only to the exterior. It is lodged in the olfactory capsule. The mucous membrane of the inner wall of the sac bears a double series of folds, the **Schneiderian folds** held by a **median raphe**. **Olfactory receptor cells** and **supporting cells** form the epithelium of the folds. The olfactory cells are joined with fibres of olfactory nerve. Each **narial opening** is functionally divided into a lateral **incurrent** siphon and a median **excurrent siphon** by three nasal flaps. Water enters the olfactory sac through incurrent siphon and goes out via excurrent siphon.

Eyes (photoreceptors)

The eyes are large, circular, lodged in the orbits located on the sides of the head. The eye-ball is held in position by six **eye muscles** and a cartilaginous stalk, the **peduncle** (Fig. 37.20). The upper and lower **eyelids** are merely outgrowths of the skin and movable. A

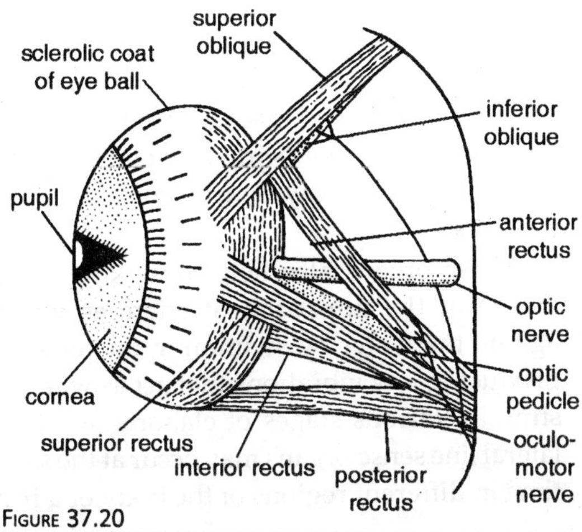

FIGURE 37.20

***Scoliodon* sp.** Eye and eye muscles

movable **nictitating membrane** lies along the anteroventral border and can be drawn to cover the eye. The **pupil** is narrow and vertical.

The eyes of Scoliodon are nearly similar to those in other vertebrates (pp. 151-154).

In Scoliodon the pupil cannot be enlarged and contracted. The lens is **spherical** and **crystalline**. The **retina** consists only of **rods** innervated by the fibres of the optic nerve. The inner surface of the **choroid** is covered by a series of **guanine plates** forming **laptum**, which reflects back incident light to the retina.

The eyes are proportionally large and being set apart, the vision is not binocular. Each eye has its own range of vision. The lack of adjustment capability of the lens is compensated by projecting the lens through the pupil against the round cornea, an arrangement by which light entering at a very wide angle can be focussed on the retina. **Cones** being absent, Scoliodon is **colour-blind**.

Statoacoustic organs (Equilibrium and phonoreceptors)

The **membranous labyrinth** (Fig. 37.21) enclosed in auditory capsules constitute statoacoustic organs. The membranous labyrinth of all vertebrates are similar in basic plan (pp. 148, 149) except in some minor details.

The cavity of the **saccule*** communicates with the exterior through a long ductus **endolymphaticus**. The duct runs upward and opens to the exterior through a minute pore in the parietal fossa of the cranium. The perilymphatic space opens outside through a large aperture, the **fenestra** located immediately behind the opening of the ductus endolymphaticus.

The membranous labyrinth is innervated by auditory nerve and helps in maintaining equilibrium.

Lagena, a simple, blind diverticulum of saccule is the organ of hearing (acoustic organ).

Neuromast organs (Rheoreceptors)

The neuromast organs consist of lateral line receptors and pit organs.

Lateral line receptors. Lateral line receptors are situated within the **lateral line canal**, which may be divided into (i) a trunk and tail canal, and (ii) a head or cephalic canal.

They are innervated by the branches of the cranial nerves—facial and the lateralis branch of the vagus. The lateral line organs or

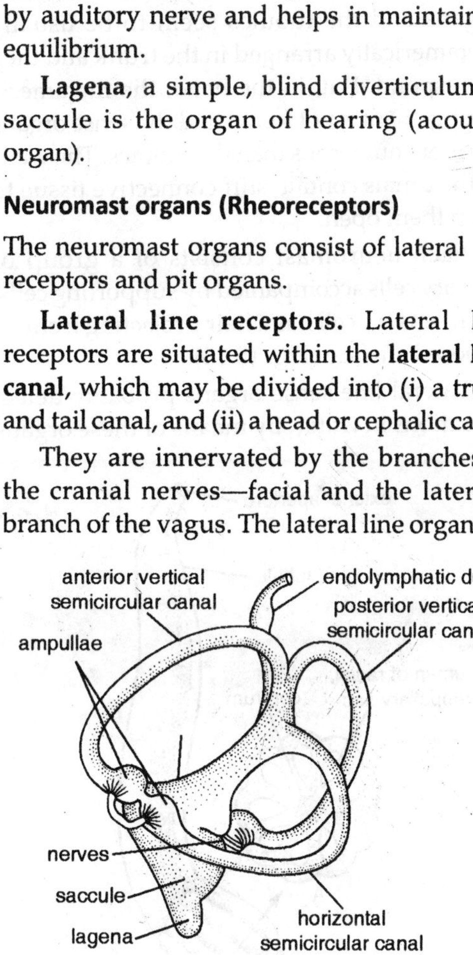

FIGURE 37.21

***Scoliodon* sp.** Membranous labyninth

* The terminology utriculus and sacculus have been changed to utricle and saccule.

neuromasts arranged in rows extending from one end of the body to the other.

The neuromasts lie in the dermis or deeper in the subcutaneous tissue. During development, the neuromasts sink down to form grooves, which then close over to form canals embedded in the skin but with numerous pores opening to the outside.

The lateral line canal runs from the head to the posterior end of the body along the outer edge of the horizontal septum, which divides epaxial musculature from hypaxial. The sense organs and the tubules seem to be usually metamerically arranged in the trunk and they correspond. But, in the head, the metameric arrangement is not found and the sense organs are more numerous than the tubules. The walls of the canals contain stiff connective tissue to keep them open.

Each neuromast consists of a group of sensory cells accompanied by supporting cells. Each sensory cell has a hair projecting from its exposed surface (Fig. 15.13).

Lateral line sense organs probably detect vibrations in water. By means of these organs the fish is enable to face against the water flow, even in darkness or in turbid waters, when no landmarks are visible, and so to maintain its position without being constantly carried downstream or away with flowing tides and currents.

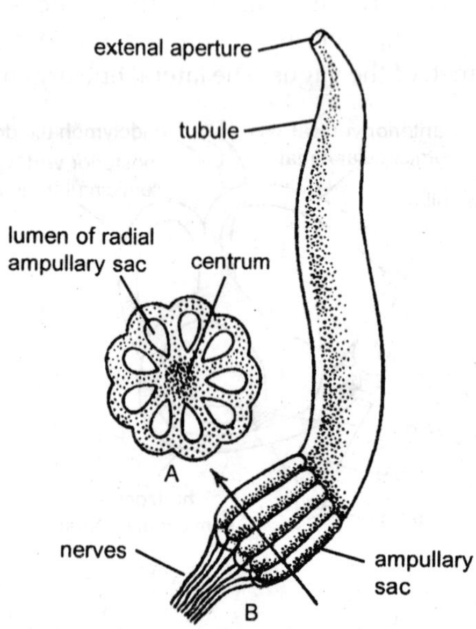

FIGURE 37.22

***Scoliodon* sp.** A. Section through an ampullary sac. B. An ampulla of Lorenzini

These organs also help in making a fish aware of the approach of enemies which agitate the water by swimming movements, producing low vibrations that act as warning stimuli. Various stages of elaboration of the lateral line sense organs may occur at the same time in different regions of the body of a fish.

Pit organs

The pit organs are present on the dorsal and lateral sides of the head. Each organ is sunk in a pit in which groups of receptor cells innervated by nerve fibres are present.

Ampulla of Lorenzini (Thermoreceptor)

Numerous ampullae of Lorenzini are present on the dorsal and ventral surfaces of the head, opening to exterior by narrow pores (Fig. 37.22). According to their location in the head they are named **Supraophthalmic**—lying around the supraorbital canal, **outer buccal**—between supraorbital and infraorbital canals and **inner buccal**—beneath the infraorbital canal. The ampullae are innervated by fibres of the facial nerve.

An ampulla consists of a sac containing eight or nine radially dilated chambers arranged around a central core, the **centrum**. Two types of cells—**pear-shaped gland cells**, which secrete jelly filling up the ampullary tubule and the **pyramidal cells** with hairs, the **receptor cells**, are present in the sac. The ampullary sac lies deep beneath the skin and continues as an elongated tube opening to the exterior through a narrow pore. Ampulae help the fish to respond to change in water temperature.

☒ Urinogenital System

The excretory or urinary system and the reproductive or genital system are collectively called urinogenital system. In male Scoliodon the excretory and reproductive structures are intimately associated (Fig. 37.23), while in female they are completely separate (Fig. 37.24).

☒ Excretory System

Excretory system consists of paired kidneys, ureters, single urinogenital sinus and a urinary aperture. The kidneys are **mesonephric** in origin.

The kidneys are long, ribbon-like and extend from the base of the liver to the sides of the cloaca. The anterior region narrow while the posterior region is greatly thickened and laterally compressed. The posterior region is made up of **nephrons** (p. 118) and excretory in function.

☒ Male Excretory System

In the anterior region of the kidney the collecting tubules formed by the joining of renal tubules are exceedingly small and open in the **Wolffian** duct. In the posterior region the collecting tubules open independently into thin walled **ureters**. The two ureters finally open separately into a wide **urinogenital sinus** opening in turn into the cloaca through a **urinary pore** on **urinary papilla** (Fig. 37.23).

☒ Female Excretory System

The anterior region of the kidney is greatly reduced and forms a long, narrow strand. It does not function as the genital duct. The nephrons are restricted to the thick posterior region of the kidney. The ureters are a pair of long, thin-walled tubes, receiving collecting tubules along their length. Posteriorly, the two ureters unite to form a **common ureter** (Fig. 37.24), opening into a triangular chamber, the urinary sinus, through a median urinary aperture. The urinary sinus opens into the cloaca through a urinary pore on the urinary papilla.

A part of the renal tubules of elasmobranch are capable of reabsorption of urea from the glomerular filtrate. Urine is **hypotonic** to blood.

☒ Reproductive System

The sexes are separate. In male a pair of claspers are present.

Male reproductive (genital) system

The male genital system consists of a pair of testes, vasa efferentia, vasa deferentia, vesiculae seminalis, a urinogenital sinus and a urinogenital aperture (Fig. 37.23).

The **testes** are paired, elongated bodies with a median central canal, located ventral

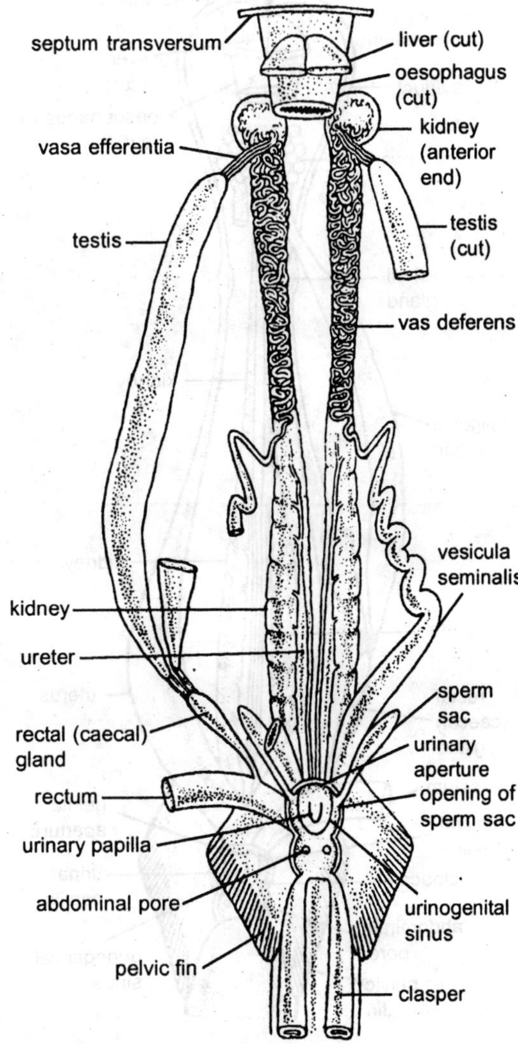

FIGURE 37.23

***Scoliodon* sp.** Male. urinogenital system (Ventral view). Ventral wall of cloaca removed. Cloacal region slightly diagrammatic

to the dorsal abdominal wall and extend from the base of the liver to the caecal gland.

The anterior region of the kidney conveys sperms to the **vesicula seminalis** and known as **epididymis** or organ of **Leydig**.

The **spermatozoa** from the testes pass through minute tubes, the **vasa efferentia**, into the anterior end of the Wolffian duct in the anterior part of the kidney. The Wolffian duct, now termed as vas deferens, is a narrow, closely coiled tube occupying a major part of the anterior section of the kidney. Posteriorly, the vas deferens enlarges greatly to form **vesicula seminalis** running by the side of the kidney, and the two vesiculae seminalis open separately into a triangular urinogenital sinus, which, in turn, opens into the cloaca through a urinogenital aperture in urinogenital papilla.

A pair of elongated, glandular sacs, the **siphons**, are present just beneath the skin of the ventral surface of the body and extend behind from the level of the posterior margin of the pectoral fins. Posteriorly, they continue as **siphon tubes** and each opens into the proximal portion of the groove of the clasper of the side. The sacs contain sea water and it is believed that they force the sperms from the cloaca into the clasper grooves by a current produced by sea water in the sacs.

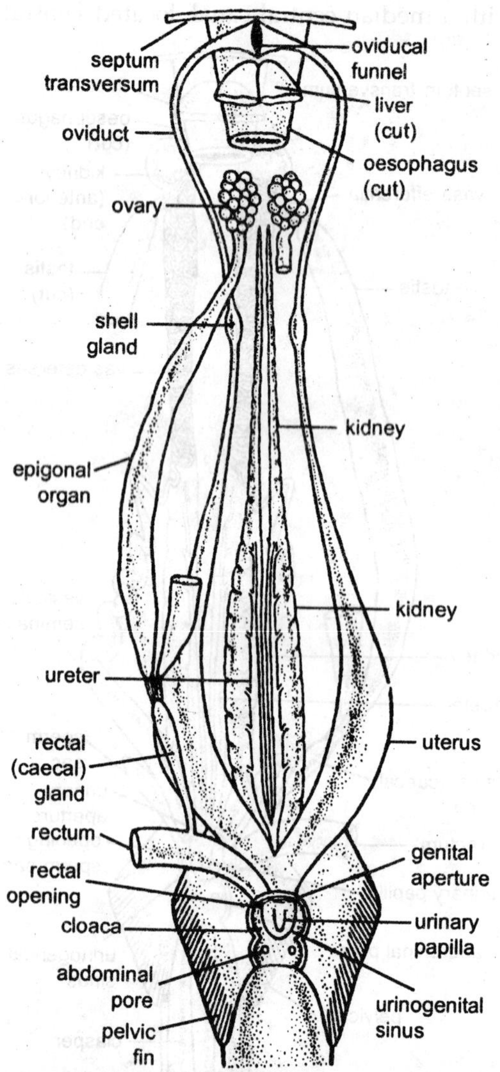

FIGURE 37.24

***Scoliodon* sp.** Female. Urinoganital system (Ventral view). Ventral wall of cloaca removed. Cloacal region slightly diagrammatic

Female Reproductive (genital) System

Paired ovaries, epigonal organs, oviducts and a vagina opening into the cloaca constitute female genital system (Fig. 37.24).

Ovaries with irregular surfaces are located one on either side of the vertebral column behind the base of the liver. A long strand of tubular tissue, the **epigonal organ** is present between the ovary in front and the caecal gland behind. The **oviducts** or **Mullerian ducts** run backward from the pericardia-peritoneal septum anteriorly to the posterior end of the abdominal cavity. Anteriorly, the oviducts meet in the midventral line, close to the base of the liver and open into the coelom through a large, median longitudinal slit, the **oviducal funnel.** Behind the funnel each oviduct curves medially and enlarges to form a small **shell gland**. Posteriorly the oviduct enlarges greatly, which is now termed **uterus.** The two uteri join to form the **vagina**, opening into the cloaca. The mucous membrane of the uterus is divided into a number of compartments, each of which accommodates an embryo. A fold of the mucous membrane at the junction of the vagina and the cloaca acts as a valve, which closes the vaginocloacal opening during development of embryos within the uteri.

The mature ova are released into the abdominal cavity, from where they pass to the oviducts through the single oviducal funnel.

☒ Fertilization and Development

Scoliodon is **viviparous**. During mating one or both the claspers of the male are inserted into the cloaca of the female and the spermatic fluid is forced through the grooves of the claspers into the oviducts through the vagina. Fertilization occurs in the oviduct in front of the shell gland.

In the early development stage, the embryo bears a **yolk sac**. A tubular **yolk stalk** connects the gut of the embryo with the yolk sac. With the progress of development, the yolk in the yolk sac is used up, the sac becomes greatly folded and embedded in the uterine wall and forms **yolk sac placenta**. From this stage onwards, the embryo obtains its nourishment from the uterine tissue. With the formation of placenta, blood vessels develop in the yolk stalk, which now becomes the **placental cord**, one end being attached to the midventral line of the embryo, while the other end terminates in the folds of the yolk sac placenta, attached close to the vaginal end of the uterus. Each embryo has its own placental cord and placenta. An embryo occupies one compartment of the uterus, filled up with a fluid, which surrounds the embryo and protects it. Development is completed within the uterine compartment.

CLASSIFICATION: CLASS CHONDRICHTHYES (ELASMOBRANCHII)

1. The class includes sharks, rays and chimaeras. They are mostly predaceous and marine.
2. The skin is tough, possesses minute placoid scales and mucous glands.
3. Skeleton cartilaginous and notochord persistent. The vertebrae are complete and separate.
4. The median and paired fins are supported by cartilaginous fin rays. The pelvic fins in male are provided with claspers.
5. The tail is heterocercal.
6. Ventrally placed mouth with rows of teeth. Spiral valve is present in the intestine.
7. The swim bladder absent.
8. The nostril one or two, never open in the mouth cavity.
9. The heart is with a contractile conus arteriosus and bears rows of valves.
10. Five to seven pairs of gills, each being placed in a separate cleft. No operculum over the gills except in Holocephali.
11. Sexes are separate. Gonads typically paired and gonoducts open into the cloaca.
12. Eggs large and yolky. Fertilization internal.

The Class Chondrichthyes (Elasmobranchii) is divided into two subclasses, Selachii and Bradyodonti.

Subclass SELACHII

1. Modern sharks. The gills are placed in separate clefts and a spiracle is present behind each eye.

Order 1. **Cladoselachii (Pleuropterygii):** Devonian-Permian

1. Shark-like, fusiform Chondrichthyes became extinct in Devonian to Permian periods. It is the ancestor of all modern sharks.
2. Tail heterocercal with a large lower lobe, and a horizontal keel-like fin on each side of the caudal peduncle.
3. Pectoral, pelvic fins and two dorsal fins were very prominent, sometimes with a spine in front of the first skeleton of fins consist of parallel cartilages of simple structure.
4. The large subterminal mouth, long sharp teeth, relatively enormous forwardly directed eyes, body covered with small denticles, amphistylic jaw suspension, a well-developed postor-

bital process, paired nostrils and absence of claspers are prominent features.

Examples: *Cladoselache, Cladodus, Symmorium, Diademonus*, etc.

Order 2. **Pleuracanthodii (Ichthyotómi):** Devonian-Triassic.

1. Slender, fresh water, shark-like fishes having dorsal fin extending along the body and tail. It became extinct in Triassic period.
2. The pectoral fins are paddle-like and 'archipterygium' type, having long axis with pre- and post-axial rays. This type of fin is supported to be ancestral type of all fins.
3. The tail is diphycercal with equally developed blades of fin above and below the notochord.
4. A long spine on the head and claspers are present in male.

Example: *Pleurocanthus*.

Order 3. **Protoselachii:** Devonian-Cretaceous

1. The pectoral fin is notched at the posterior margin.
2. Claspers in male present.
3. Two types of teeth, one sharply pointed, located anteriorly and the other flat type placed behind for crushing molluscan shells.
4. They floursihed in Palaeozoic seas and became extinct in Permian to Triassic period. Some belonging to the genus Heterodontus are still surviving. From Protoselachii all the euselachians arose in the Mesozoic period.

Examples: *Hybodus, Heterodontus*, etc.

Order 4. **Euselachii:** Jurassic-Recent

1. Sharks and rays constituting this group are abundant in equatorial and temperate seas.
2. Teeth numerous, developed in continual succession.
3. Pectoral fins have three basal pieces, the pro-, meso- and metapterygium, from which a number of preaxial radials spread out.

This order is divided into two suborders.

Suborder i. **Pleurotremata (Squaliformes)**

1. It includes about 225 living species of sharks with spindle-shaped body and heterocercal tail.
2. Gill-slits 5 to 7 pairs—laterally placed.
3. The anterior margin of the pectoral fin is free from the body. The pectoral radials are simple and of a few segments, and, as a rule, only the anterior ones reach the free edge of the fin. The two halves of the pectoral arch are well-separated above.
4. Paired preorbital cartilages attached to the olfactory capsules are often well-developed.
5. The pterygoquadrate has no articulating process to the skull.
6. The hyomandibular and ceratohyal both bear cartilaginous rays, and both take part in supporting the first gill. The ceratohyal is a single cartilage and attached to the lower end of the hyomandibular.

The suborder includes four superfamilies:

a. **heterodontoidea.** Example: *Heterodontus*.

b. **Notidanoidea.** Examples: *Chlamydoselachus, Heptranchius, Hexanchus*.

c. **Gaboidea.** Examples: *Lamna, Carcharodon, Carcharinus, Sphyrna, Scyliorhinus*, etc.

d. **Squaloidea.** Examples: *Squalus, Squantina, Scoliodon*, etc.

Suborder ii. **Hypotremata (Batoidea, Rajiformes)**

1. About 300 extant species including rays and skates with depressed body; chiefly bottom-dwellers in shore waters.

2. Pectoral fins enlarged, joined to the sides of the body or to the head. The pectoral radials are numerous, multisegmented and bifurcated at the ends, all of them reaching the free edge of the fin. Swim by flapping.
3. The halves of the pectoral girdle either fuse with each other or else both fuse to the vertebral column.
4. Gills ventral, five pairs.
5. Spiracle highly functional.
6. Paired preorbital cartilages attached to the olfactory capsules, which are often very well-developed.
7. The pterygoquadrate has no articulating process to the skull.
8. The hyomandibular lacks rays, and takes no part in the support of the gill; ceratohyal is segmented and attached to the hyomandibular—either high up or not at all.

The hypotremata is divided into two groups—the **Narcobatoidea** including electric rays and the **Batoidea** including all other skates and rays.

Examples: *Pristis, Raja, Rhinobatos, Torpedo, Dasyatis, Aetobatus, Manta*, etc.

Subclass BRADYODONTI

1. This aberrant group diverged from the main cartilaginous stock at the Carboniferous period.
2. Vertebrae are reduced to nodules. Teeth are firmly attached to the jaws and modified into crushing plates.
3. Notochord remains unconstricted.

Order 1. **Eubradyodonti:**

Devonian-Palaeozoic

1. Primitive bradyodonts, remarkable for their dentition. The teeth are few and fuse to form flattened crushing teeth.
2. The nature of vertebrae, mode of jaw suspension and cleavage pattern are primitive.
3. They became extinct at the end of Palaeozoic era.

Examples: *Cochliodus, Psammodus, Helodus*, etc.

Order 2. **Holocephali:**

Jurassic-Recent

1. The surviving species are mollusc-eating, deep sea forms with claspers in male and large egg cases in females like those in elasmobranchs.
2. The general anatomy is similar to that of Euselachii.
3. The skin naked except for a few denticles restricted to the head claspers.
4. Jaw suspension holostylic; the upper jaw being immovably united with cranium. Neither skull nor jaw are attached to hyoid arch.
5. The extra claspers are on the head and in front of the ordinary pair on the abdomen.
6. Cloaca and spiracle absent.
7. An operculum covers the gills.
8. The dentition is highly peculiar in the composition of the tooth plates as well as in the shape.
9. The teeth are in the form of plates. The arrangement of these plates, producing the 'parrot-like' beak, has contributed largely to the modification of the whole skull.
10. The first dorsal fin is far forward with strong spine.
11. The tail is slender, whip-like.
12. About 25 species are found in cold ocean waters down to 1,800 meter.

Examples: *Chimaera, Callorhynchus, Harriotta*, etc.

CHONDRICHTHYES and OSTEICHTHYES: Salient feature

Chondrichthyes	Osteichthyes
1. The mouth is ventral in position.	1. The mouth is apical in position.
2. The body is covered with minute and microscopic placoid scales, the skin is tough.	2. The body is covered with cycloid or ctenoid or ganoid scales. The skin is soft. Complete absence of scales in cat fish group.
3. Paired pelvic fins are provided with claspers in males.	3. Claspers are absent.
4. The tail is heterocercal, the vertebral column extending into the dorsal part of the caudal fin.	4. The tail is homocercal, the vertebral column ends at the middle of the base of the caudal fin.
5. The head is dorsoventrally flattened.	5. The head may be dorsoventrally or laterally flattened.
6. Gill slits, usually five pairs, open separately and directly to the exterior.	6. Gill slits, usually four pairs, open in a gill chamber.
7. The gills are hemibranch/holobranch.	7. The gills are holobranch.
8. Operculum always absent.	8. Operculum with branchiostegal membrane always present.
9. A spiracle usually present.	9. A spiracle is absent.
10. The endoskeleton is cartilaginous, with amphicoelous vertebrae.	10. The endoskeleton is bony with amphicoelous vertebrae.
11. A conus arteriosus is present. Aortic arches five pairs.	11. A conus arteriosus is absent. Aortic arches four pairs.
12. A spiral or scroll valve is present in the intestine.	12. A spiral valve in the intestine is absent.
13. Stomach with distinct cardiac and pyloric regions. Intestinal caecum absent.	13. Stomach not differentiated into cardiac and pyloric portion. Intestinal caecum present.
14. Generally carnivorous, with a short intestine.	14. Generally omnivorous, with a short or long intestine.
15. Gut with valves at all the junctions.	15. Valves few and far between.
16. Cloaca present.	16. Cloaca absent.
17. An optic chiasma is present.	17. An optic chiasma is absent.
18. An air bladder is absent.	18. An air bladder is usually present.
19. Gills with intrabranchial septum.	19. Intrabranchial septum absent or rudimentary.
20. Testes are connected with the kidneys.	20. Testes are not connected with the kidneys.
21. Kidneys elongated and narrow, with anterior nonrenal and posterior renal portion.	21. Kidney renal, fused, with a broad anterior and a narrow posterior portion.
22. The eggs are large and except one species, provided with horny case.	22. The eggs are small but heavily yolked and without horny cases.
23. In some cases, development takes place within the oviducts and the young-ones are born alive.	23. With a few exceptions, the young are hatched in an immature condition and undergo larval development.
24. In the embryo, gills project from the gill clefts as filaments.	24. Gill filaments remain covered by operculum.
25. Marine.	25. Marine and fresh water forms.

38 CLASS OSTEICHTHYES

The class Osteichthyes includes fishes having bony skeleton, scales and plates; swim-bladder and advanced type of brain, inhabiting all sorts of water—fresh, brackish or salt; warm or cold; the size being 10 mm to 4 m. They are staple protein food for mankind since antiquity.

The body is spindle-shaped, with tail, which ranges from the heterocercal of sharks to the homocercal and diphycercal form. Most modern fishes have less bony armour in the scales than that in ancient types.

The head is provided with a true skull comprised of both cartilage or replacement bones and membrane or dermal bones. The fins in most bony fishes are supported by many parallel dermal rays, but in crossopterygians each paired fin has a single stout central element or lobe that articulates with the limb girdle (this probably transformed to the limb skeleton in land vertebrates). The fins aid in maintaining equilibrium, in steering and in swimming.

The body is enveloped with soft mucous producing epidermis that facilitates easy movement in water and is also protective. The trunk and tail bear thin rounded dermal scales—lengthwise and diagonal rows. The lateral line sense organ, along either side of the body, is a row of small pores connected to a lengthwise tubular canal under the scales. The canal possesses sensory organs responsive to vibrations of low frequency, such as those from agitation of the surrounding water.

The gills are four pairs. Four gills are lodged in a common gill chamber on each side of the pharynx beneath the operculum; cranial nerves are ten pairs.

Many fishes perform seasonal migration. The shad, hilsa, salmon and some trout travel from salt to fresh water for spawning and are termed **anadromous**, while fresh water eel follow the reverse course and are called **catadromous**.

Some species of fishes are active at all hours, many are quiet at night, and a few are nocturnal. The fishes of fresh water streams habitually head against the current to maintain the position, facilitate respiration and catch food. Marine species are active at all seasons, but many fresh water species become inactive during winter. Some fishes are solitary whereas others are gregarious.

LATES CALCARIFER (Bhetki fish)

Lates is a common carnivorous bony fish living in estuaries. It is tasty edible fish, abundantly available during winter. Bhetki belongs to subclass Actinopterygii, class Osteichthyes, subphylum Vertebrata, phylum Chordata.

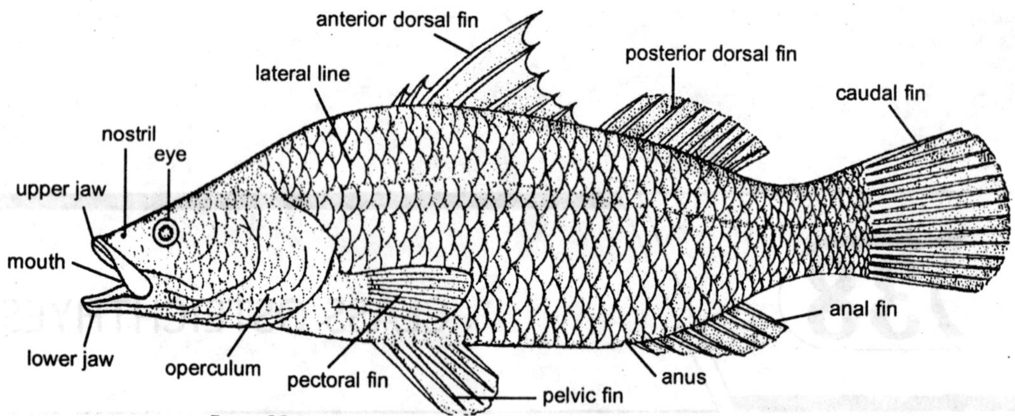

FIGURE 38.1

Lates calcarifer (Bhetki fish)

⊠ External features

The body is spindle-shaped, laterally flat and covered with dermal scales. The body (Fig. 38.1) is divisible into three distinct regions; the anterior head, middle trunk and posterior tail.

A thin, transparent epidermis overlines the scales. The scales are arranged in such a way that the anterior one covers a part of the posterior scale. The free margins of the scales are raised into spines, the **ctenes**. Such scales are known as **ctenoid** scales* (Fig. 38.13). A slimy substance is secreted by the epidermis (outermost layer) of the skin, which makes the fish slippery.

Head

The portion of the body starting from the anteriormost to a point slightly anterior to the pectoral fin is known as head. The mouth is situated at the anterior end of the head and is guarded by two movable jaws, the upper and lower. Anteriormost in position on the dorsal surface, a pair of apertures, the **nostrils** are present, one on each side. The nostrils do not open in the buccal cavity and are used only for smell. The **eyes** are two in number, one on each side, and situated posterior to the nostrils. The eyes are without eyelids but covered with a thin, transparent, protective membrane in front. A movable bony flap, oval on the posterior margin, situated on each side of the head is known as **operculum**. The operculum covers a chamber inside, the **gill chamber**, containing bright red **gills** or respiratory structures.

Trunk

The trunk starts from the posterior end of the head and ends in a line where the anus is situated. The trunk bears the anterior dorsal fin and paired pectoral and pelvic fins.

On the ventral surface at the posteriormost margin of the trunk a depression, the **vent** is present. It bears following apertures . From anterior to posterior they are **anus**, the opening of the rectum, the genital apertures or gonopores, opening of the male or female ducts and an urinary aperture, the opening of urinary bladder.

Tail

The postanal part of the body is known as tail. It gradually narrows towards the posterior end and ends in a broad fin known as caudal fin. The tail bears posterior dorsal, anal and caudal fins.

---

* Cycloid scales are similar to ctenoid scales both in origin and shape but they lack ctenes and the free surface is smooth. Cycloid scales are present in carps.

A distinct line, the **lateral line**, is present on each side of the body. This line is slightly dorsal to the middle line of the body and continues up to the tail from behind the head.

Fins

Fins are membranous extensions of the skin supported by bony rays. The bony fin rays* are knonw as **dermotrichia**. Fins are considered as appendages in fishes.

According to their position on the body, the fins have been placed in two groups—median (unpaired) and paired fins.

Median fins

Anterior dorsal fin. It is situated dorsally and supported by seven bony spines, the third is the largest. It is located in the trunk.

Posterior dorsal fin. It is situated behind the anterior dorsal fin and supported by one spinous ray in front and eleven bony rays behind. The major part is located in the tail, extending anteriorly in the trunk.

Anal fin. It is situated behind the anus and supported by three short spinous rays in front and eight bony rays behind. It is located in the tail.

Caudal fin. The posterior region of the tail continues as caudal fin. It is more or less rounded and supported by nineteen bony rays. The fin rays radiate regularly from the rounded end of the tail proper. The tail with such a symmetrical tail fin is known as **homocercal**.

Paired fins

These are located in the trunk.

Pectoral fins. One is placed on each side of the trunk just behind the operculum and form the first pair of fins. Each fin is supported by about fourteen bony rays. The fins are smaller in size.

Pelivc fins. The second pair of fins situated behind and below the pectoral fins are pelvic fins. One pelvic fin is situated on each side of the body and is provided with a spine in front and four bony rays behind.

The fins are used mainly for maintaining the proper position (balance), steering through water and checking the speed. The tail (caudal) fin being in the same long axis with the body takes part in undulations of the body and helps in locomotion. Other fins are used as locomotory organs only during slow movement.

☒ Skeletal System

The skeletal system consists of exoskeleton and endoskeleton.

Exoskeleton

Hard structures covering and protecting the body from external violence, injury, attack of parasites and other objects are collectively called exoskeleton.

In *Lates*, the ctenoid scales constitute exoskeleton. The scales are **dermal** in origin.

Skin

The skin has two distinct layers. The **epidermis** or the outermost layer is a thin membrane, made of a single layer of flattened cells, most of which secrete mucus. The underlying layer or **dermis** is a thick layer of connective tissue (Fig. 7.9B), from which the scales develop. The scales are thin, nearly rectangular bony plates with semicircular, free border, bearing concentric rings of annual growth. The free margin bears spines, the ctenes (Fig. 38.13). The scales form a complete covering of the body except a small area of the head and the fins. They are arranged on the body in a special way—the posterior part of the scales of the anterior row overlap the anterior part of the scales of the next row.

Endoskeleton

The hard structures or skeleton present inside the body is endoskeleton. The endoskeleton can be divided into two sets, the axial and appendicular skeleton.

* Fins in other groups of vertebrates (except Chondrichthyes and Osteichthyes) are not supported by fin rays.

Axial skeleton

The axial skeleton consists of skull, backbone or vertebral column with ribs, operculum and the visceral arches.

Skull

Prominent **sutures** (lines of union of bones) are present in the skull. The skull may be divided into three zones, (a) posterior or basal or occipital zone, (b) middle or sphenoidal zone and (c) an anterior zone.

Basal zone (occipital region)

The floor is formed by **basioccipital bones,** sides by two **exoccipitals** and the roof by **Supraoccipital bones**. Posteriorly, the occipitals surround a large aperture, the **foramen magnum,** through which the medulla oblongata continues as spinal cord. The basioccipital forms a single concave facet, **occipital condyle** for articulation with the vertebral column. The auditory capsule of each side is formed by five separate bones—the **epiotic, opisthotic, prootic, periotic** and **sphenotic** and is incorporated with the cranium.

Middle zone (sphenoidal region)

The floor is made by **basisphenoid** which continues forward as **parasphenoid.** The sides are made by **alisphenoids** and **orbitosphenoids** near the orbit. The roof is made of two pairs of bones: the first pair, anterior to supraoccipital, are **parietals** articulating with the alisphenoids by its side, and the second pair **frontals,** connected with the orbitosphenoids lie in front of parietals. The orbit is surrounded by bones, the **jugals.**

Anterior region

The roof of the anterior region is formed by a pair of bones, **nasals,** partially overlapping the **ethmoid.** The floor is formed by **vomer** lying in front of the parasphenoid.

Visceral skeleton

The visceral skeleton consists of seven pairs of visceral arches. The first or **mandibular arch** forms the **jaws,** the second is the **hyoid arch** and the rest five arches, the **branchial** arches, support the gills.

Jaw apparatus

The jaws are two, upper and lower.

Upper jaw. The **palatine** forms the palate or roof of the buccal cavity. The **pterygoid** lies in the middle region and the **quadrate** articulates with the lower jaw. The primary cartilaginous jaw is covered by toothed investing bones—the anterior one, **premaxilla** and the posterior one, **maxilla.**

Lower jaw. A large bone, the **articular,** articulates with the quadrate. It continues anteriorly as a cartilaginous rod, the **meckel's cartilage.** The distal end of the cartilage is covered by a large toothed investing bone, the **dentary.** A small bone, the **angular** is attached to the lower and hinder end of the articular.

Hyoid arch

It is made of several pieces of bony and cartilaginous elements. Each **operculum** is made of four bony pieces attached with the hyoid arch.

Five **branchial** arches are present on each side of the pharynx.

Vertebral column

The vertebral column runs through the trunk in the middorsal line above the gut. Anteriorly it articulates with the skull and posteriorly ends in tail. The vertebral column consists of a number of vertebrae, (Fig. 38.2), which are of two types, precaudal and caudal.

Precaudal vertebrae

Vertebrae of the trunk region. Each precaudal vertebra is made of a disc, the **centrum** and bears a dorsal arch on the centrum, the **neural arch.** The neural arch encloses a cavity through which the spinal cord passes. The vertebra is **amphicoelous,** i.e., the centrum is concave on both the anterior and posterior faces.

Successive vertebrae are articulated by pre- and postzygapophyes, the processes of neural

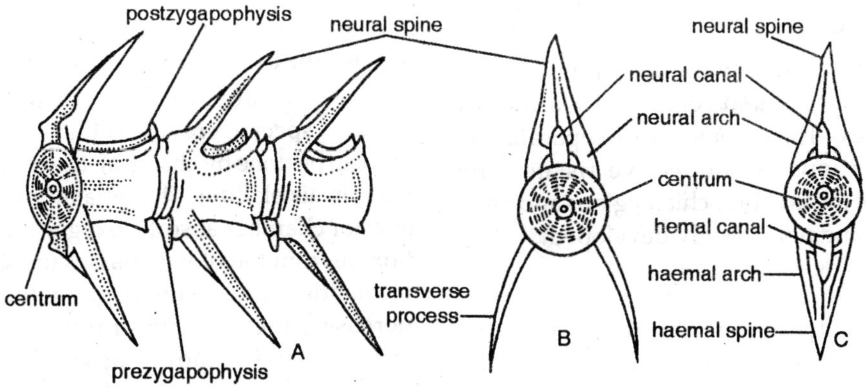

FIGURE 38.2

***Lates* sp.** Vertebrae. A. Caudal vertebrae (3) in natural position. B. Precaudal vertebra. C. Caudal vertebra

arches. Transverse processes develop from the sides of the centrum of precaudal vertebra on each side. Slender **ribs** are attached to the transverse processes and these are directed downwards.

Caudal vertebrae

Vertebrae of the tail region, the caudal vertebra bears a ventral arch corresponding to the dorsal neural arch and is known as **haemal** arch. A spine, **haemal spine** develops from the haemal arch and is directed downwards. Successive haemal arches enclose a canal in which run the caudal artery and vein.

At the posterior end of the vertebral column a few bony plates are present, which support the base of the caudal fin.

Appendicular skeleton

The appendicular skeleton is represented by pectoral girdle, pelvic girdle and bony structures supporting fin rays.

Pectoral girdle

Each half of the girdle (Fig. 38.3) consists of a coracoid and a scapula which are ill-developed and attached to the inner sides of the well-developed clavicle or **cleithrum**. A slender **postclavicle** extends backward from the clavicle. The two clavicles meet ventrally and a **supraclavicle** is present on the dorsal end of

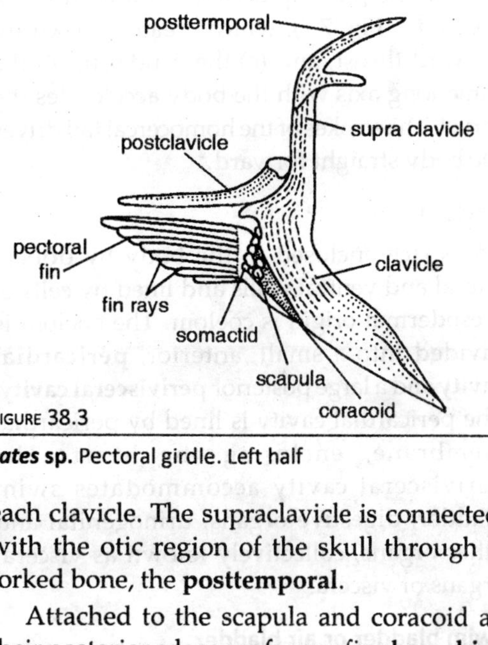

FIGURE 38.3

***Lates* sp.** Pectoral girdle. Left half

each clavicle. The supraclavicle is connected with the otic region of the skull through a forked bone, the **posttemporal**.

Attached to the scapula and coracoid at their posterior edges are four or five **branchial ossicles** also called **somactids** or **pterygiophores**, which support the fin rays or **dermotrichia** of the pectoral girdle.

Pelvic girdle

The position of the pelvic girdle is occupied by two large, flat bones, the **pterygiophores**. To the posterior border of each are attached three small nodules, the distal pterygiophores, to which the dermotrichia or fin rays are attached.

⊠ Muscular System

The somatic muscles of the trunk (Fig. 7.6) and tail are powerful and segmentally arranged (**myotomes**). The myotomes are separated by thin partitions of connective tissue. The muscles of the fins, branchial regions and head are small and not strongly developed.

Locomotion

Water is a dense medium. Lates is a fast swimmer. The ability to swim rapidly is the cumulative effects of certain factors. (a) The body is laterally compressed and spindle-shaped to minimise resistance. (b) The myotomal arrangement of trunk and tail muscles help in rapid lateral undulations of the body (Fig. 7.7), which create a powerful forward thrust, and (c) the caudal fin in the same long axis with the body accelerates the thrust. The strokes of the homocercal tail drives the body straight forward.*

Coelom

The space enclosed in the body by dorsal, lateral and ventral walls and lined by cells of mesodermal origin is coelom. The coelom is divided into a small, anterior, **pericardial cavity** and a large posterior **perivisceral cavity**. The pericardial cavity is lined by pericardial membrane, enclosing the heart. The perivisceral cavity accommodates swim bladder, digestive organs, urinogenital and other organs, collectively known as visceral organs or viscera.

Swim bladder or air bladder

It is a thin-walled, closed, elongated sac, along the whole length of the body cavity and lies below the kidney, close to the dorsal wall of the coelom. It is whitish in colour, tough and consists of two chambers, the anterior one short and wide and the posterior one narrower long and tapering posteriorly. The inner wall of the air bladder is highly vascularised by the capillaries of mesenteric artery, which form **rete mirabile** or **retia mirabilia**.

The air bladder serves to keep the fish at the same specific gravity as that of the water. This can be done by changing the compression of gas in air bladder. The retia mirabilia of the anterior chamber absorbs oxygen and carbon dioxide from the blood, passes the same into the bladder; the amount of gas in the bladder increases, the specific gravity of the fish decreases and it moves upward. The retia mirabilia of the posterior chamber of the air bladder absorbs the gases from the bladder, the pressure inside the chamber falls, the specific gravity of the fish increases and the fish sinks downward. The swim bladder acts as a hydrostatic organ.

⊠ Digestive System

The digestive system (Fig. 38.4) consists of a digestive tract or alimentary canal and its appendages, the digestive glands.

Alimentary canal

The alimentary canal starts from the buccal cavity and ends in the anus, receiving secretions from various digestive glands in its way. Digestion of food and absorption of nutrients take place within the canal. The digestive tract can be divided into following zones.

Buccal cavity

The mouth is bounded by two jaws, upper and lower, provided with sharp denticles. The mouth leads into a wide cavity, the buccal cavity, bearing a small, nonmuscular tongue on the floor and mucous glands in the wall. Posteriorly, the buccal cavity continues as pharynx.

Pharynx

The pharynx is a dilated chamber perforated on each side by five vertically elongated slits, the gill slits, communicating it with the

* A forward mouth is associated with the straight forward movement of the body due to strokes of homocercal tail.

exterior. The pharynx leads to oesophagus posteriorly. Laterally, the pharynx is supported by branchial or gill arches.

Oesophagus

The oesphagus is a short tube which narrows posteriorly and opens into stomach. The opening is guarded by sphincter muscle to prevent entry of water into the stomach.

Stomach

The stomach is an elongated, sac-like structure, broader at the middle and divisible into two parts—the cardiac and pyloric stomach. The sac-like portion to which the oesophagus opens is the cardiac stomach. It is oriented anteroposteriorly, slightly obliquely, the hind end being narrow. The pyloric stomach is tubular, separates from the cardiac stomach, near its anterior end.

Intestine

From the pyloric stomach at a short distance from the oesophagus arises a narrow, long tube, the **small intestine.** The junction of the two is well marked by a pyloric constriction. After its origin the intestine passes forwards and then bends upon itself and runs backward. The intestine continues posteriorly with bends forming loop and ends in rectum. A small portion of the intestine adjacent to anus is designated as **rectum**, but the two are not sharply distinguishable. The rectum opens to the exterior by an aperture, the **anus** at the middle line of the vent on the ventral side of the body anterior to the anal fin. The mesentery supporting the gut is heavily laden with fat.

⊠ Pyloric Caeca

At a short distance from its origin five elongated, blind pouches arise from the intestine. These are known as pyloric caeca. The function of pyloric caeca is not definitely known. They are either secretory of absorptive.

⊠ Digestive Glands

The appendages of the digestive tract, the liver and pancreas, are digestive glands.

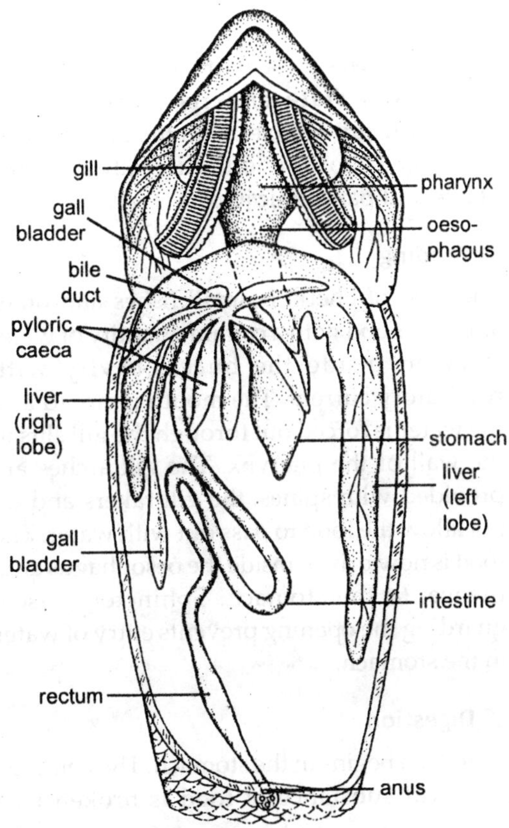

FIGURE 38.4

***Lates* sp.** Digestive system

Liver

The liver is a large, bilobed, yellowish brown mass. The two lobes are much elongated and continue posteriorly in the coelom. The left lobe is longer than the right one. The two lobes of the liver are connected together anteriorly by a short median portion. Slightly right to the median portion a small pear-shaped, thin-walled, greenish sac, the gall bladder is present. The narrow neck of the gall bladder continues behind along the right lobe of the liver and ends in another elongated sac close to the tip of the liver. The secretion of the liver, bile is stored in the gall bladder. Ducts from the liver are connected with the gall bladder and the duct from the gall bladder, the **bile duct**, opens near the commencement of the intestine.

Pancreas

The pancreas is present in the mass of liver in a **diffused state** and its secretion, the **pancreatic juice**, mixes with the bile. The bile discharged into the intestine is really a mixture of bile and pancreatic juice.

⊠ Feeding

The food of bhetki is small fishes and other animals. Food with a certain quantity of water is taken inside the buccal cavity with respiratory current. The mouth is closed and the water is forced out through the gill slits in the wall of the pharynx. The gill arches are provided with spines, the gill rakers and do not allow the food to pass out with water. The food is now driven inside the oesophagus and, thence, to the stomach. Sphincter muscle guarding the opening prevents entry of water in the stomach.

⊠ Digestion

Digestion begins in the stomach. The complex chemical substance of food is broken into simpler forms by the action of enzymes (p. 86) which are secreted by various glands. The food is mixed with the bile secreted by the liver. The partly digested food passes to the intestine, where further digestion takes place and the digested food, which is in a liquid state, is absorbed by the cells of the intestine. The undigested portion of the food passes to the rectum and is finally thrown out through the anus.

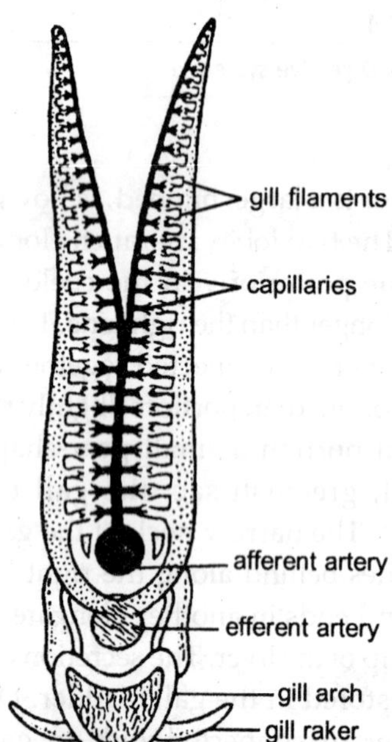

FIGURE 38.5

***Lates* sp.** A gill (holobranch). Sectional view (diagrammatic)

⊠ Respiratory System

The respiratory organs are gills and, being aquatic in habit, the oxygen dissolved in water is absorbed by the gills and carbon dioxide is given out.

Respiratory structures

The pharynx is perforated by five vertical slit-like openings, the gill slits, on each lateral side. Two gill slits are separated from one another by a bony arch, **branchial arch** or gill arch. Each arch bears a double row of filaments (Fig. 38.5), the **gill filaments.** Gills bearing a double set of branchial filaments are known as holobranch. Of the two rows, one is anterior and the other posterior. Each gill arch on its inner concave surface bears a number of bony **projections, gill rakers**. Altogether four branchial arches are present on each side of the pharynx (Fig. 9.3). All the gills are lodged in a gill chamber had covered externally by **operculum**, the posterior margin of which bears a thin **branchiostegal membrane.**

The afferent branchial vessels break up into capillaries inside the gill filaments. Blood brought to the gills through the afferent branchial vessels is deoxygenated and rich in carbon dioxide. Water taken inside the buccal cavity passes out through the gill slits and the gill filaments remain bathed in water. At this stage the oxygen dissolved in water is absorbed by blood and carbon dioxide given out (p. 94). The oxygenated blood passes out of the gills through the efferent branchial vessels.

Mechanism of respiration

During respiration, the branchiostegal membranes are pressed against the body wall. The floor of the mouth is lowered and the pharynx is expanded. Due to increased space thus created water rushes in through the open mouth. Now the mouth is closed, the buccal floor is raised, the pharyngeal wall contracted, the opercula released, branchiostegal membranes released and the water is forced out through the gill slits to the branchial chamber and from there to the exterior (Fig. 9.4). A repetition of this process maintains a continuous water current through the gill slits which carries dissoved oxygen and food particles with it.

⊠ Blood Vascular System

The blood vascular system consists of a heart, an arterial or afferent and efferent branchial system, and a venous system. To complete one circulation through the whole body, the blood passes only once through the heart and such a circulation is known as **single circulation**. The flow of blood from heart to the gills, from there to different body parts and back to heart, is known as **circulation**, and the organs which take part in it are collectively known as circulatory organs or blood vascular system.

Heart

The heart is a reddish structure situated ventrally in the pericardial cavity and enclosed within a thin membrane, the **pericardium**. The heart consists of three chambers; a dorsally situated **sinus venosus**, a single **auricle** or **atrium**, and a **ventricle** most ventral in position (Fig. 38.6). Deoxygenated blood from different parts of the body is received by sinus venosus through two large, transverse veins the **ductus Cuvieri**. Blood from sinus venosus passes to the atrium through an opening, the **sinuatrial (sinuauricular) aperture** and from the atrium to ventricle through an aperture, the atrioventricular (auriculoventricular) aperture.

From the anterior end of the ventricle springs a stout vessel, the **ventral aorta**, the base of which is swollen and known as **bulbous aorta**. The passages communicating the various chambers of the heart are guarded by **valves** and the flow of blood is allowed in one direction only.

⊠ Arterial System

All vessels carrying blood away from the heart are arteries; the arteries and the heart are collectively known as arterial system. The arterial system consists of two components (a) The afferent system or afferent branchial system carrying deoxygenated blood from heart to the gills and (b) the efferent system or efferent branchial system carrying oxygenated blood from gills to different parts of the body.

Afferent branchial system

The ventral aorta runs forward and gives out paired branches, the **afferent branchial arteries** (Fig. 38.7). The number of afferent branchial arteries are four on each side and correspond to the gills of the side. At the anteriormost region, the ventral aorta divides into two branches, one on each side, and enters the first gill of the same side. These are first afferent branchial arteries. Posterior to the first afferent branchial arteries arise two vessels—

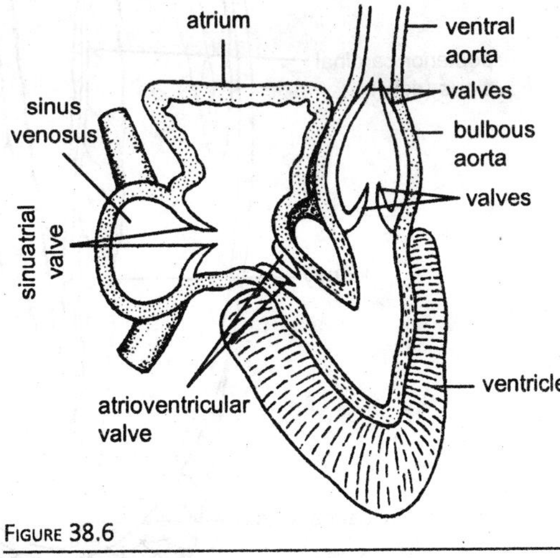

FIGURE 38.6

***Lates* sp.** Heart (showing chambers, diagrammatic)

the second afferent branchial arteries from the ventral aorta, one on each side, and enters the second gill of the side. At about the middle of the length from the bulbous aorta and the origin of second afferent branchial arteries arise two vessels from the ventral aorta, one on each side. Each of these two divides into two, the anterior one the third afferent branchial artery, enters the third gill and the posterior one entering the fourth gill is the fourth afferent branchial artery. The afferent branchial arteries break up into capillaries in the gills and the blood becomes oxygenated there.

Efferent branchial system

Four pairs of **efferent branchial arteries** carry blood from the gills to different body parts. They are four in each side corresponding to the gills. The first and second vessels from the first and second gills are known as first and second efferent branchial arteries.

A **cephalic aorta** arises from each first efferent branchial artery and divides into two branches.

i. **External carotid** supplying orbital region, tongue and hyoid arch.
ii. **Internal carotid** supplying blood to the brain.

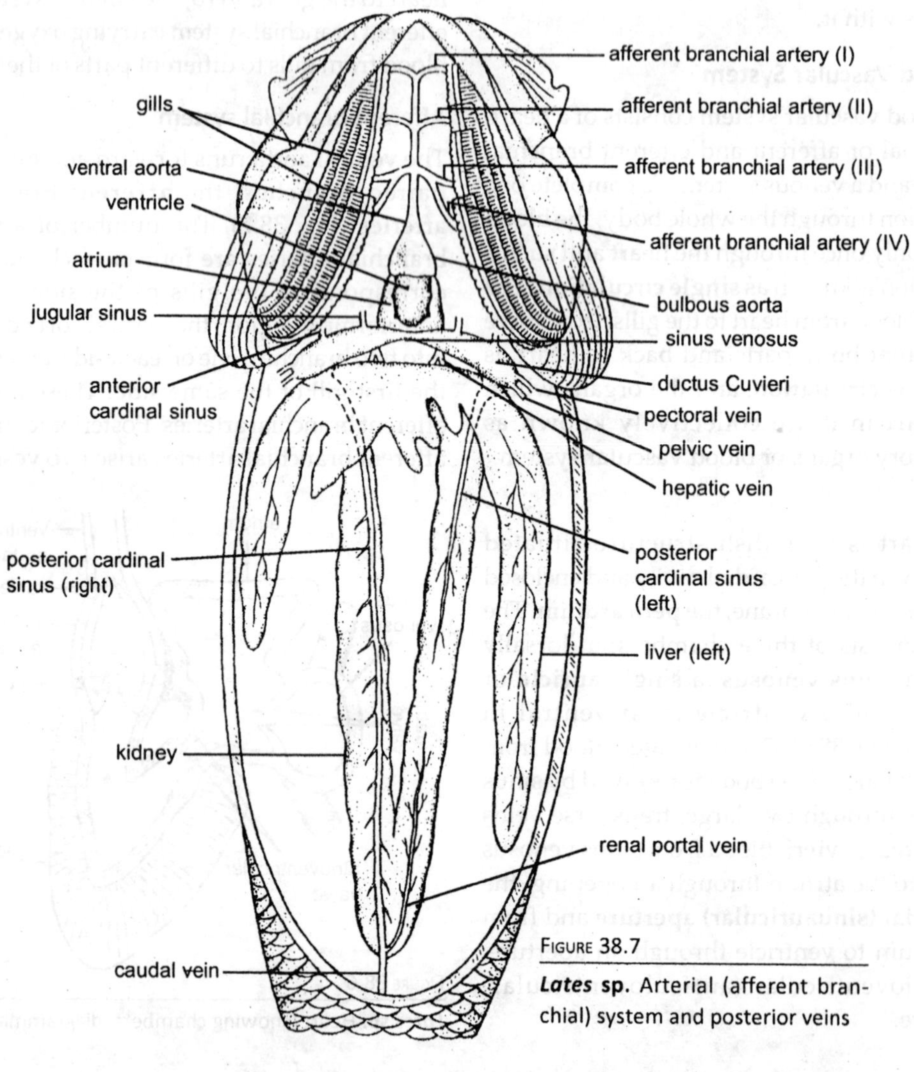

FIGURE 38.7

***Lates* sp.** Arterial (afferent branchial) system and posterior veins

The first and second efferent branchial arteries from each side unite to form an **anterior epibranchial artery**. The anterior epibranchial arteries of both the sides unite posteriorly to form a median **dorsal aorta**. The dorsal aorta runs posteriorly along the ventral line of the vertebral column (Fig. 38.8).

The dorsal aorta in its way backwards receives, at a short distance from its origin, one vessel, on each side, the **posterior epibranchial arteries**. The posterior epibran-chial artery of each side is formed by the union of the third and fourth efferent branchial arteries coming from the third and fourth gills, respectively.

Dorsal aorta

A stout vessel, the **coeliacomesenteric artery**, springs from the dorsal aorta from the point of union of the posterior epibranchial arteries with the aorta. The coeliacomesenteric artery sends branches to swim bladder, gut and other visceral organs.

Posterior to coeliacomesenteric artery, at short distance, the dorsal aorta gives off a pair of branches, one on each side, the **subclavian artery** supplying blood to pectoral and pelvic fins of the side.

In its way backwards between the two kidneys the dorsal aorta sends a number of paired arteries the **renals** to the kidneys, **genitals** to reproductive organs and **segmental arteries** to the myotomes.

Posteriorly, behind the anal region the dorsal aorta continues as **caudal artery** and passes through the haemal canal formed by the heamal arches of the caudal vertebrae.

☒ Venous System

Systemic veins and portal veins constitutes the venous system.

The principal veins, right and left **ductus Cuvieri**, receiving deoxygenated blood from the body open in the sinus venosus. Each ductus Cuvieri is formed by an anterior cardinal sinus, a jugular sinus, a hepatic vein and a posterior cardinal sinus.

Systemic veins

Deoxygenated blood from each side of the anterior region of the body is received by the Cuvierian duct of the side by two veins, the outer, anterior **cardinal sinus** and the inner, **jugular sinus**. The anterior cardinal sinus is stouter and drains blood from the dorsal region. The jugular sinus is narrower and drains blood from the ventral region.

Blood from the middle region of the body is received by the Cuvierian duct of the side through **pectoral** and **pelvic veins** from the pectoral and pelvic region of the side and **hepatic vein** from the liver. Blood from the

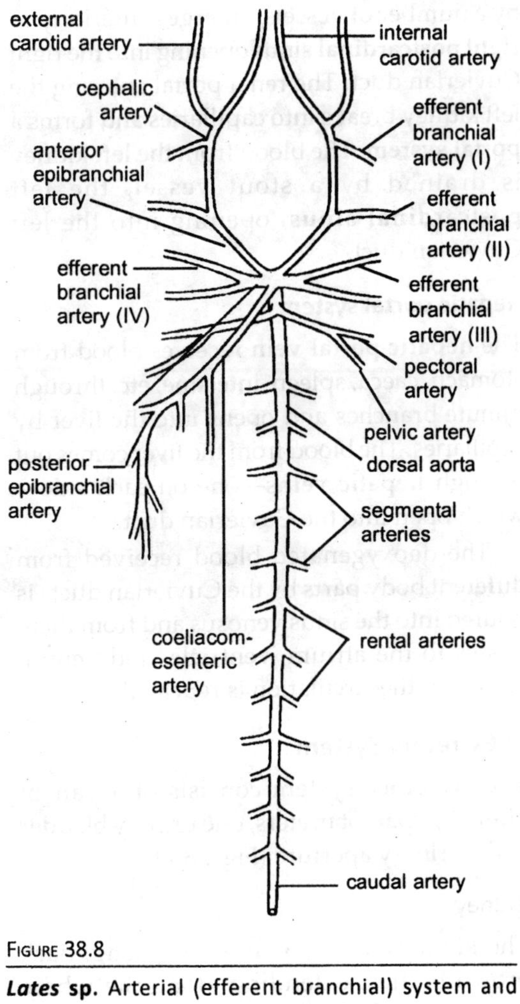

FIGURE 38.8

***Lates* sp.** Arterial (efferent branchial) system and arteries posterior region enters Cuvierian ducts through **postcardinal sinuses.**

Portal system

Both renal and hepatic portal systems are present. The renal portal system is, however, incomplete.

Renal portal system

Blood from the posterior region of the body is returned by a **caudal vein**, which runs through the haemal canal. On reaching the coelom the caudal vein divides into two—the right one **renal vein** and the left one **renal portal vein**. The renal entering the right kidney continues as such and receiving blood from the kidney by a number of vessels, emerges anteriorly as **right postcardinal** sinus opening into the right Cuvierian duct. The renal portal entering the left kidney breaks into capillaries and forms a **portal system**. The blood from the left kidney is drained by a stout vessel, the left **postcardinal sinus**, opening into the left Cuvierian duct.

Hepatic portal system

The **hepatic portal vein** receives blood from stomach, caeca, spleen, intestine, etc. through minute branches and opens into the liver by capillaries. The blood from the liver comes out through **hepatic veins**—one on each side—which open into the Cuvierian duct.

The deoxygenated blood received from different body parts by the Cuvierian ducts is poured into the sinus venosus and from there passes to the atrium, ventricle, and ventral aorta and the circulation is repeated.

⊠ Excretory System

The excretory system consists of a pair of kidneys, a pair of ureters, one urinary bladder and a urinary aperture (Fig. 38.11).

Kidney

The kidneys are two large, flat, long, dark brown bodies, extending along the whole length of coelom, dorsal to the air bladder. They are separate in the anterior part. In this region they lie on either side of the vertebral column and are attached to the dorsal wall of the coelom. The anterior end bears an obliquely oriented, transverse lobe, the outer end of which is directed posterolaterally and the inner end anteromedially. The kidneys gradually narrow down posteriorly, where they are partially fused and this part lies ventral to the vertebral column.

The kidney is formed by the aggregation of a large number of **uriniferous tubules** held together by connective tissue. The tubules are in communication with the coelom through minute pores, the **nephrostomes**. A norrow duct, the ureter, runs backward along the outer border of the posterior part of each kidney to open on the dorsal surface of the **urinary bladder** separately, but very close to each other. The bladder is a thin, elongateoval sac. It narrows posteriorly and opens to the exterior through a **urinary aperture**, the last aperture in the vent.

The kidneys are mesonephric (p. 127) in origin and the bladder is a dilatation of the mesonephric duct.

⊠ Nervous System

The central nervous system, peripheral nervous system and autonomic nervous system constitute the nervous system. Both brain and spinal cord are surrounded by a single membrane, the **meninx**.

Central nervous system

The components of central nervous system are brain (Fig. 38.9) and spinal cord.

Brain

The brain is enclosed in a bony cranium. It is divisible into three regions—forebrain (anterior portion) or prosencephalon, midbrain (middle portion) or mesencephalon and hindbrain (posterior portion) or rhombencephalon.

The brain is horizontal and the different parts are situated in a straight line.

Prosencephalon

It consists of an anterior telencephalon and a posterior diencephalon or thalamen-cephalon.

Telencephalon

Two anteriorly placed **olfactory lobes** and two **cerebral hemispheres** from the telencephalon. The floor of the cerebral hemispheres forms a thick band called **corpora striata** and the roof is smooth and nonnervous.

Diencephalon

It is a small diamond-shaped area between the optic lobes and cerebral hemispheres with a thin roof. The sides are produced into optic thalami. A process called the **epiphysis** or **pineal body** is present on its dorsal surface. The floor is depressed downward and bears a median outgrowth, the **infundibulum**. Posteriorly, the infundibulum is attached with an outgrowth known as **pituitary body**. Each side of the pituitary is provided with a lobed structure, the **lobi inferiores**, which bears a glandular sac, the **saccus vasculosus**. Two **optic nerves** arise from the ventral surface of diencephalon and pass to the eyes. In their way the optic nerves cross each other, the right nerve going to the left and the left going to the right eye but they do not form **optic chiasma**.

Mesencephalon

The midbrain is fairly large and formed by two large, round lobes, the **optic lobes**, dorsally and a longitudinal band, **crura cerebri**—ventrally. The two optic lobes form **corpora bigemina**.

Rhombencephalon

The hindbrain is divisible into two regions. The anterior large portion, **metencephalon** or the **cerebellum** pushes forward under the roof of the mesencephalon to form the **valvula cerebelli** and overlaps medulla oblongata behind. The posterior narrow tubular portion, the **medulla oblongata** or **myelencephalon**, proceeds backward as the spinal cord and runs up to the tail region through the neural canal

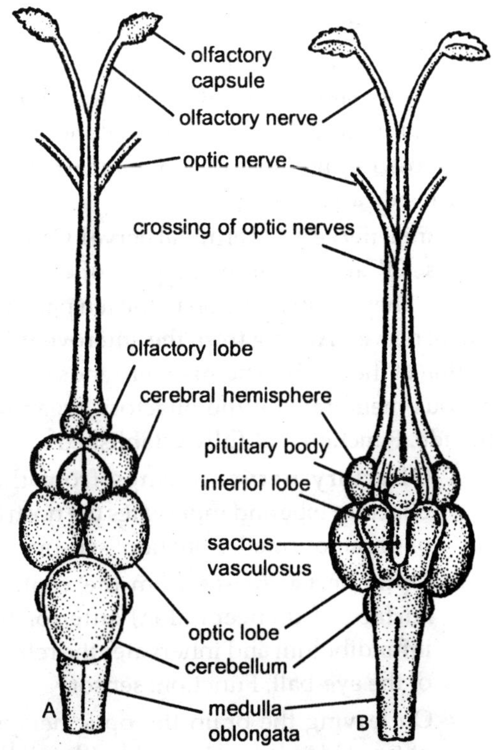

FIGURE 38.9

***Lates* sp.** Brain. A. Dorsal view, B. Ventral view

formed by the vertebral column.

Cavities of the Brain

The cavities of the brain are **ventricles** and that of the spinal cord is **central canal** (p. 135), the latter continuing in the spinal cord.

Functions of different zones

Olfactory lobes concerned with smell, cerebral hemispheres are the seat of memory, intelligence and consciousness. Pituitary body (not a part of the brain) secretes a number of hormones. Optic thalami is the seat of vision and possibly balance. Optic lobes control vision. Metencephalon, is the centre for coordination of muscular movement and partly balance.

⊠ Peripheral Nervous System

The cranial and spinal nerves constitute peripheral nervous system (p. 137).

Cranial nerves

Nerves arising from the brain are cranial nerves. Ten pairs of cranial nerves are present in bhetki. Proceeding from the anteriormost to the posteriormost in order, the cranial nerves of each side are:

Cranial nerve 'O' (Terminal nerve). Cranial nerves 'O' are the anterior pair of cranial nerves. They are in addition to the ten pairs of cranial nerves. Arising from the anteroventral portion of the cerebral hemisphere ends in the mucous membrane of the olfactory organ of the side. Function: not fully established.

i. **Olfactory nerve.** Springs from the olfactory lobe and innervates the nostril or organ of smell. Function: sensory.

ii. **Optic nerve.** Arises from the ventral surface of diencephalon, anterior to infundibulum and innervates the retina of the eye-ball. Function: sensory.

On leaving the brain the optic nerves of two sides cross each other; the right nerve going to the left and the left going to the right eye. (No optic chiasma or crossing of nerve fibres take place).

iii. **Oculomotor nerve.** Emanates from the ventral region of the midbrain and innervates four out of six muscles of the eyeball. Function: motor.

iv. **Trochlear or pathetic nerve.** Springs from the anterolateral part at the junction of the mid and hindbrain and innervates the superior oblique muscle of the eye-ball. Function: motor.

v. **Trigeminal nerve.** Arises from the side of the anterior region of medulla oblongata. A swelling or ganglion, the **gasserian ganglion**, is present at its base and the nerve divides into three branches.

Ophthalmic. Divides to superficial ophthalmic and deep ophthalmic and innervates the skin above the eye-ball.
Maxillary. Innervates the upper jaw.
Mandibular. Innervate the lower jaw.

Function: both motor and sensory, hence a mixed nerve.

vi. **Abducens nerve.** Springs from the ventral surface of the medulla oblongata and innervates the posterior rectus muscle of the eye-ball. Function: motor.

vii. **Facial nerve.** Arises from the side of the medulla oblongata and, proceeding anteriorly, fuses with the gasserian ganglion of the fifth or trigeminal nerve. It divides into two branches after its emergence from the ganglion.

Palatine. Innervates the roof of the mouth cavity.

Hyomandibular. Innervates the hyoid and the lower jaw. Function: mixed.

viii. **Auditory (Statoacoustic nerve).** Emanates from the side of the medulla oblongata behind the facial nerve and innervates the inner ear. Function: sensory.

ix. **Glossopharyngeal.** Springs from the side of the medulla just posterior to the auditory nerve and divides into two branches.

The anterior branch innervates the wall of the suprabranchial chamber. The posterior branch innervates the first gill arch and ventral branchial muscles. Function: mixed.

x. **Pneumogastric** or **vagus.** It is the posteriormost cranial nerve and springs from the side of the hindermost part of the medulla oblongata. A ganglion, the **vagus ganglion**, is present at the base. It divides into two main branches.

Lateralis. A stout branch just after the origin it runs backward up to the tail along the median line supplying branches to lateral line sense organs.

Visceralis. In its backward course it sends three branches.

i. **Branchialis.** Innervating, second, third and fourth gill arches.

ii. **Cardiac.** Innervating the heart.
iii. **Gastric.** Innervating the alimentary canal and fourth gill arch. Function: mixed.

Spinal nerves

The spinal nerves are paired. Each nerve arises from the spinal cord by two roots, dorsal and ventral. The nerves pass out between the neural arches of successive vertabrae and innervate the trunk and tail region (p. 137).

Autonomic nervous system

The sympathetic trunks are devoid of ganglia.

Receptors and sense organs

The receptors and sense organs are mechanoreceptors, rheoreceptors, chemoreceptors, photoreceptors, phonoreceptors and equilibrium receptors.

Mechanoreceptors

Distinct mechanoreceptors are absent, certain tactile cells on the body surface serve as organs of touch. They are scattered all over the body, being crowded in the anterior region, most crowded on the lips.

Rheoreceptors (Lateral line sense organs): **Neuromasts** present along the lateral lines of the body enable the fish to detect low vibration (p. 150).

Chemoreceptors (olfactory receptors)

Two blind sacs represent **olfactory organs**. Each nostril leads internally into a small sac, the **olfactory sac**, lined with sensory cells and innervated by the olfactory nerve. It receives the stimulus of smell (p. 143).

Photoreceptors (eyes)

The eye is constructed on the same principle as in higher vertebrates (p. 143) with the following differences.

The cornea is flat. The lens is globular. The anterior chamber is small. A silvery layer known as **argentea** is present between the sclera and choroid (Fig. 38.10). Choroid processes are absent. A vascular fold of the choroid, the **falciform process**, pierces the retina and continues up to the back of the lens to which it is attached by a retractor **lentis muscle.** The eyes of fishes are made to see near objects, but distant vision is adjusted possibly by moving the lens towards and away from the retina with the help of falciform process. Vision is **monocular** and each eye has different visual field.

Equilibrium and Phonoreceptors (Statoacoustic organs)

The ears are audioequilibrating organs and serve the dual purpose of hearing and balance. External and middle ears are absent. Only the

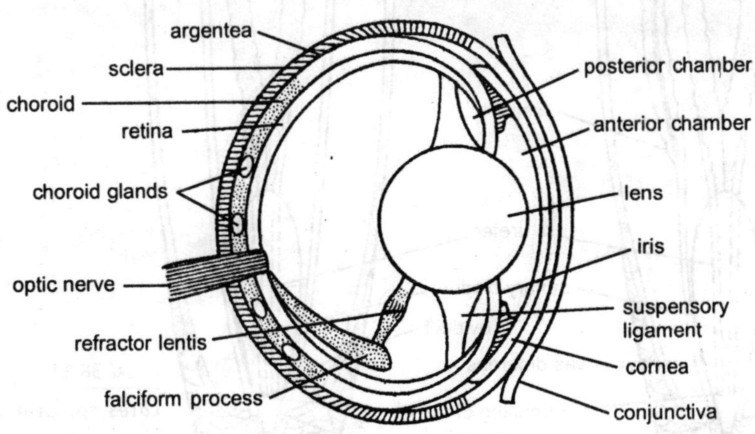

FIGURE 38.10

***Lates* sp.** Eye (vertical section). Diagrammatic

internal ears or membranous labyrinth are present.

Membranous labyrinth (p. 147) is formed by two chambers, the upper **utricle** and the lower **saccule**. The three **semicircular canals** open in the utricle. The saccule is saclike, with an ill-developed **lagena**, on the floor (Fig. 15.10D). The membranous labyrinth is filled up with **endolymph**. Sensory hairs are present in utricle and saccule and in the ampullae of the semicircular canals. **Otoliths** are present in the utricle and saccule. The semicircular canals and utricle play important role in maintaining equilibrium.

The auditory organs are simple blind diverticula, the lagena, at the posterior end of the saccules. Sound waves moving through a dense medium as water have greater impact. Transmission of vibrations through body surface helps in the process of hearing.

⊠ Endocrine System

Only two endocrine glands (p. 156) have been clearly identified in bhetki.

Thyriod

A small, reddish mass along the ventral aorta.

Thymus

Paired structures, one on each side, near the last branchial arches.

⊠ Reproductive System

The sexes are separate. Sexual dimorphism is not pronounced. The pelvic fins are reddish in colour in live male fishes.

Male reproductive system

The testes are two, one on each side (Fig. 38.11A). Each **testis** is a narrow elongated body lying ventral to the kidney. A duct, the **vas deferens**, is continuous from the posterior end of each testis and runs backward. The two vasa deferentia open to the exterior by separate genital pores, situated on either side of the anal depression slightly anterior to the urinary aperture.

During breeding season the testes become greatly enlarged in sexually mature specimens.

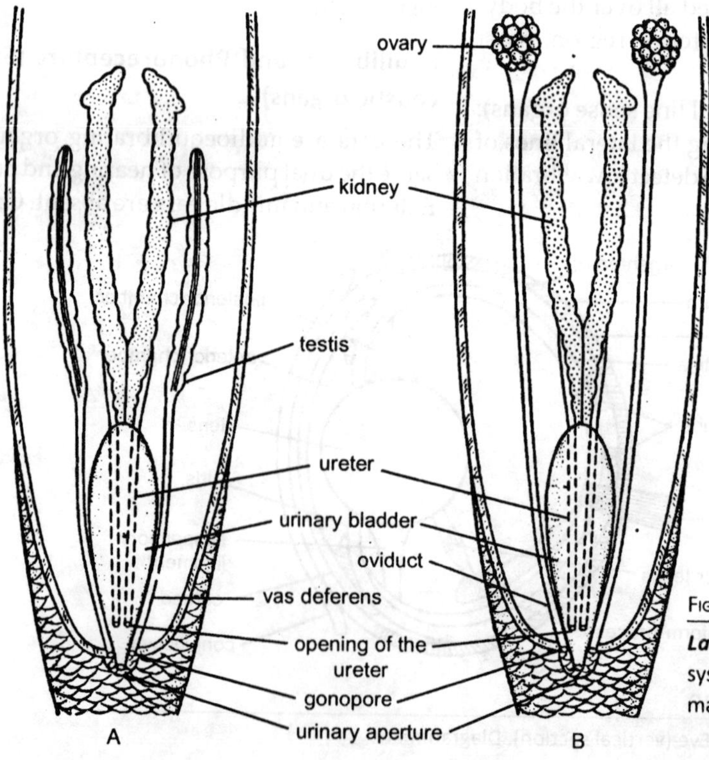

FIGURE 38.11

Lates **sp.** Urinogenital system. A. Male. B. Female

Female reproductive system

The ovaries are two (Fig. 38.11B), one on each side, situated anteriorly very close to the side of the oesophagus. Ovaries are very small in off-season but increase greatly in the breeding season and extend along the whole length of the body cavity. Each **ovary** is wrapped by peritoneum which fuses to form a duct, the **oviduct**. The two oviducts run posteriorly and open to the exterior by the genital pores situated on either side of the anal depression slightly anterior to the urinary aperture.

Breeding

The females are of larger size than the males. Bhetki breeds in winter. At the time of breeding the **ova** are discharged by the female in water and the male discharges the **seminal fluid** (milt) over them. Fertilisation takes place in water. The zygote undergoes segmentation and a tiny fry is formed. The fry is nourished at the early stage of life by the yolk present in the **yolk sac**, attahced to its ventral surface. The yolk is exhausted, the yolk sac disppears, the fry enlarges in size and starts feeding on small, aquatic organisms.

CHANNA PUNCTATUS* (Lata fish; goroi; snake-headed murrel)

Channa is a most common fresh water fish, widely distributed in Indian subcontinent. It belongs to subclass Actinopterygii, class Osteichthyes, subphylum Craniata (Vertebrata), phylum Chordata.

External features

Channa attains a length of about 10-18 cm in adult stage. The body is elongated, sub-cylindrical anteriorly. The abdomen is round and the tail laterally compressed (Fig. 38.12). The colour is blackish grey dorsally while the ventral surface is yellowish grey. Alternate black patches are present above and below the lateral lines. The body is protected with systematically arranged **cycloid** and **prectenoid** scales (Fig. 38.13), covered with a thin, transparent epidermis, which secretes mucus.

The cycloid scale is a thin, nearly rectangular plate of bone with a semicircular, free border. Concentric rings of annual growth are distinct in a cycloid scale.

The body of channa is divisible into three distinct regions—the anterior head, middle trunk and posterior tail.

Head

The head is depressed and bears plate-like scales. The snout is nearly obtuse. The anterior mouth is guarded by two equal jaws, upper and lower, covered by moderate lips. The jaws bear pointed denticles. Nostrils are two, anterior and one on each side. They do not

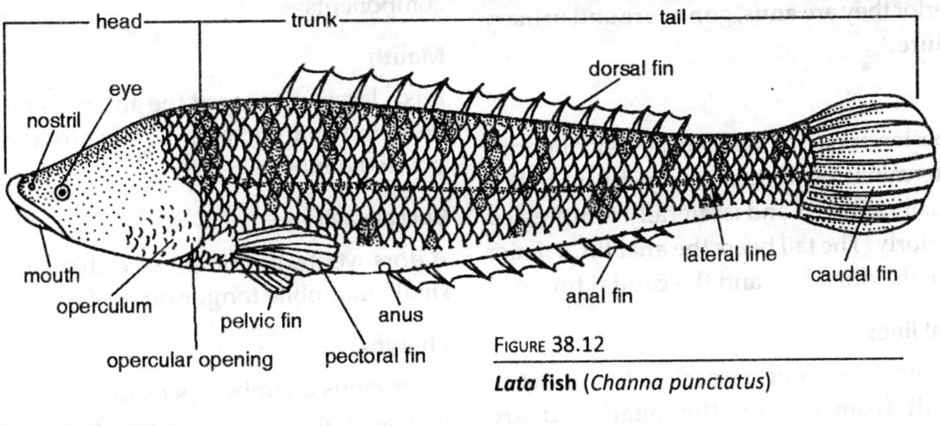

FIGURE 38.12

***Lata* fish (*Channa punctatus*)**

* Description of organ-system similar to those in **Lates**, is drastically abridged or omitted, except the nervous system.

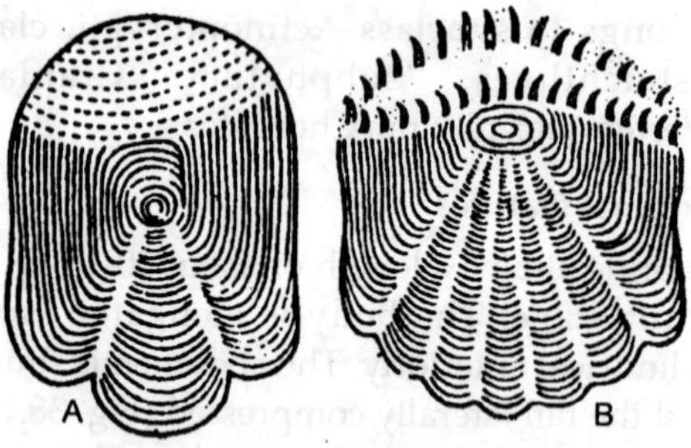

FIGURE 38.13

***Channa* sp. Scales.** A. Cycloid scale, B. Ctenoid scale

open in the buccal cavity. The eyes are two, one on each side, posterior to the nostrils. Eyelids are absent but the eyes are protected by a transparent membrane in front. A movable bony operculum, enclosing a chamber containing gills, are present on each side. Gill openings are wide; membranes of two sides connected beneath the isthmus. An accessory respiratory organ in the form of air chamber is present in each gill chamber.

Trunk

The part of the body posterior to the head and extending up to the anus is the trunk. The anterior portion of the unpaired dorsal fin and paired pectoral and pelvic fins are present in the trunk. A depression, the **vent**, on the ventral surface, at the posterior margin of the trunk, bears three apertures. From anterior to posterior they are **anus, gonopore** and **urinary aperture**.

Tail

The posterior part of the body beyond the anus and terminating in the caudal fin is tail. It gradually narrows and is laterally compressed posteriorly. The tail bears the anal fin, a major part of the dorsal fin and the caudal fin.

Lateral lines

The lateral lines, one on each side, run up to the tail from behind the head and are interrupted at frequent intervals.

Fins

The fins are unpaired median fins and paired fins supported by bony fin rays (dermortichia).

Median fins

Dorsal fin. Long, narrow, without spines and extends from above the pectoral fin to a short distance from the caudal fin.

Anal fin. Extends from anus to a point slightly anterior to the posterior end of the dorsal fin. It is also without spines.

Caudal fin. More or less rounded and extends from the posterior end of the tail, supported by radiating fin rays. Such a tail is known as **homocercal** tail.

Paired fins

Pectoral fins. First pair of fins; small, located on each side of the trunk, just behind the operculum.

Pelvic fins. Second pair of fins, about double the size of the pectoral fins, located behind and below the pectoral fins.

☒ Digestive System

The digestive system consists of a digestive tract or alimentary canal and its appendages, the liver and the pancreas (Fig. 38.14).

Alimentary canal

The alimentary canal consists of the following components:

Mouth

It is a large opening at the anterior end of the head and guarded by two jaws provided with pointed denticles.

Buccal cavity

A dorsoventrally compressed chamber with a small, immobile tongue on the floor.

Pharynx

A spacious chamber, perforated laterally by five pairs of vertical gill slits. The roof bears a pair of dental plates.

Oesophagus

A thick-walled tube, slightly narrows down posteriorly to open in the stomach.

Stomach

The **cardiac stomach** is a fairly large, somewhat oval sac, slightly narrowed at the free end. The **pyloric stomach** is a narrow tube, close to the anterior end of the cardiac stomach.

Intestine

A long narrow tube, folded upon itself for a short distance near the posterior end.

Pyloric caeca

A pair of blind diverticula, one long and the other short, close to the junction of the pyloric stomach and the intestine.

Rectum

Slightly wider than the intestine and bulged at the middle. It opens through the anus.

Liver

A large bilobed, brownish mass, the left lobe larger than the right one. The spherical **gall bladder** is located in the right side, close to the middle of the liver. The bile duct opens in the intestine, near its origin.

Pancreas

The pancreas is present in the mass of the liver in a diffused state. The pancreatic juice mixes with the bile, the secretion of the liver and discharged into the intestine.

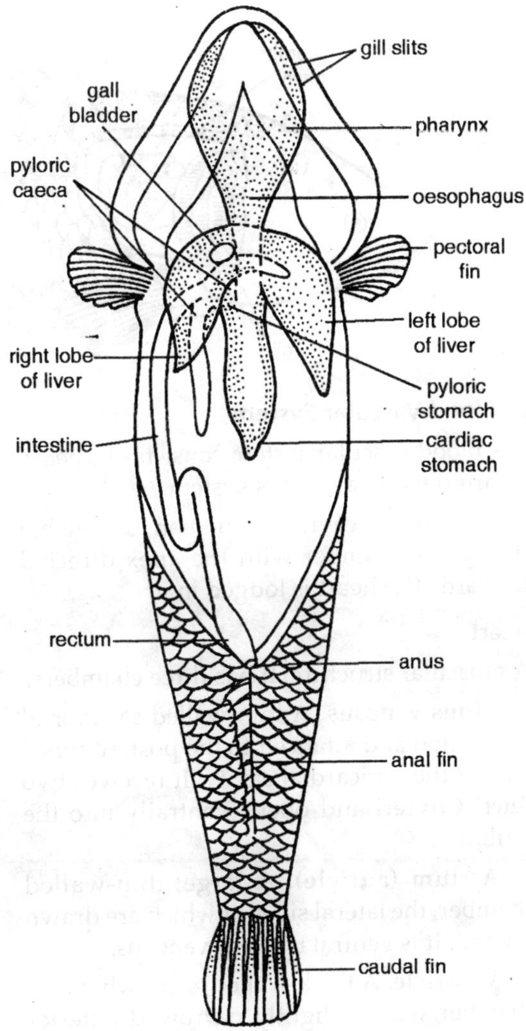

FIGURE 38.14

***Channa* sp.** Digestive system

☒ Respiratory System

Channa has the capability to utilise dissolved oxygen in water with the help of gills and also atmospheric oxygen through accessory respiratory structures.

Gills

The pharyngeal gills are four pairs and holobranch. Deoxygenated blood supplied by afferent arteries is oxygenated in the gill filaments and oxygenated blood is drained from the gills by efferent branchial arteries.

Accessory respiratory structures

Two simple, highly vascularised **air chambers** constitute accessory respiratory structures. They are formed as outgrowths of pharynx, separated from it later and located one in each gill chamber (Fig. 38.15). The air chambers function somewhat like lungs and utilise atmospheric oxygen in respiration. The accessory respiratory structures help the fish to survive in water with a low percentage of dissolved oxygen and also out of water for a considerable period.

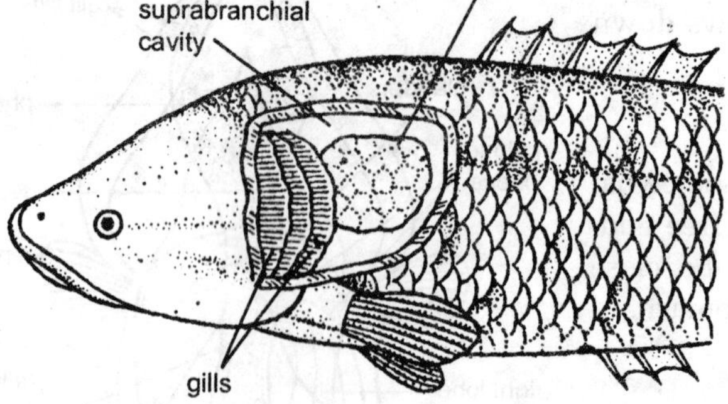

FIGURE 38.15

***Channa* sp.** Accessory respiratory organ

Blood Vascular System

The blood vascular system consists of a heart, an arterial and a venous system.

Pericardial cavity. A median, somewhat triangular chamber with the apex directed forward. The heart is lodged in it.

Heart

A muscular structure. It has three chambers:

Sinus venosus. A thin-walled sac, dorsal in position and attahced to the posterodorsal wall of the pericardial cavity. It receives two **ducti Cuvieri** and opens ventrally into the atrium.

Atrium (auricle). A large, thin-walled chamber, the lateral sides of which are drawn foward. It is ventral to sinus venosus.

Ventricle. A thick-walled, somewhat oval chamber with a slightly narrowed anterior end. It is most ventral in position.

Arterial system

It has two components—the afferent and efferent branchial systems.

Afferent branchial system

Its components are:

Ventral aorta. It arises from the antero-dorsal end of the ventricle, the swollen base of which forms the **bulbous aorta**. The ventral aorta sends laterally four pairs of **afferent branchial arteries** to four paris of gills (Fig. 38.16).

Afferent branchial arteries. They supply deoxygenated blood to the gills for oxygenation.

The posterior or the pair nearest to bulbous aorta are the fourth and the farthest pair are the first afferent branchial arteries.

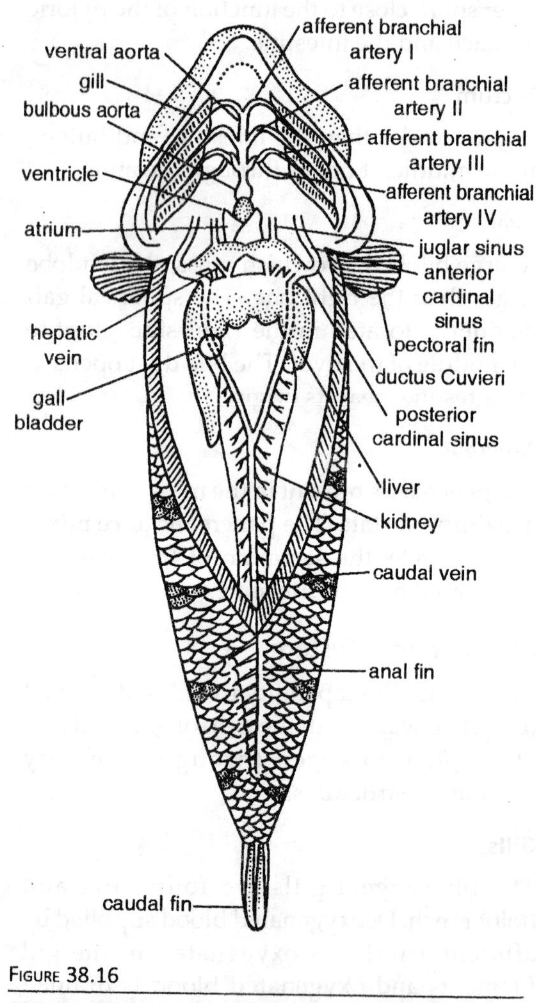

FIGURE 38.16

***Channa* sp.** Heart and afferent branchial system

The fourth and the third pair arise from one point in the ventral aorta. The fourth pair arises from the ventrolateral sides of the ventral aorta. Immediately after its origin the artery forms a loop around the third afferent branchial artery of the side, becomes dorsal to it and courses posterolaterally to the fourth gill.

The third pair is lateral in origin and runs to the third pair of gills.

The origin and course of the second pair are similar to those of the third pair. They supply the second pair of gills.

The first pair is formed by the bifurcation of the ventral aorta and supply the first pair of gills.

Efferent branchial system

The efferent branchial arteries (Fig. 38.17) collect oxygenated blood from the gills. The arteries arising from the first and second pairs of gills, instead of joining the efferent branchial system, supply oxygenated blood directly to the anterior region of the body.

Each of the third pair of efferent branchial arteries forms the **common carotid artery** of the side. Sending a branch, the **external carotid** to the hyoid and ophthalmic region, the

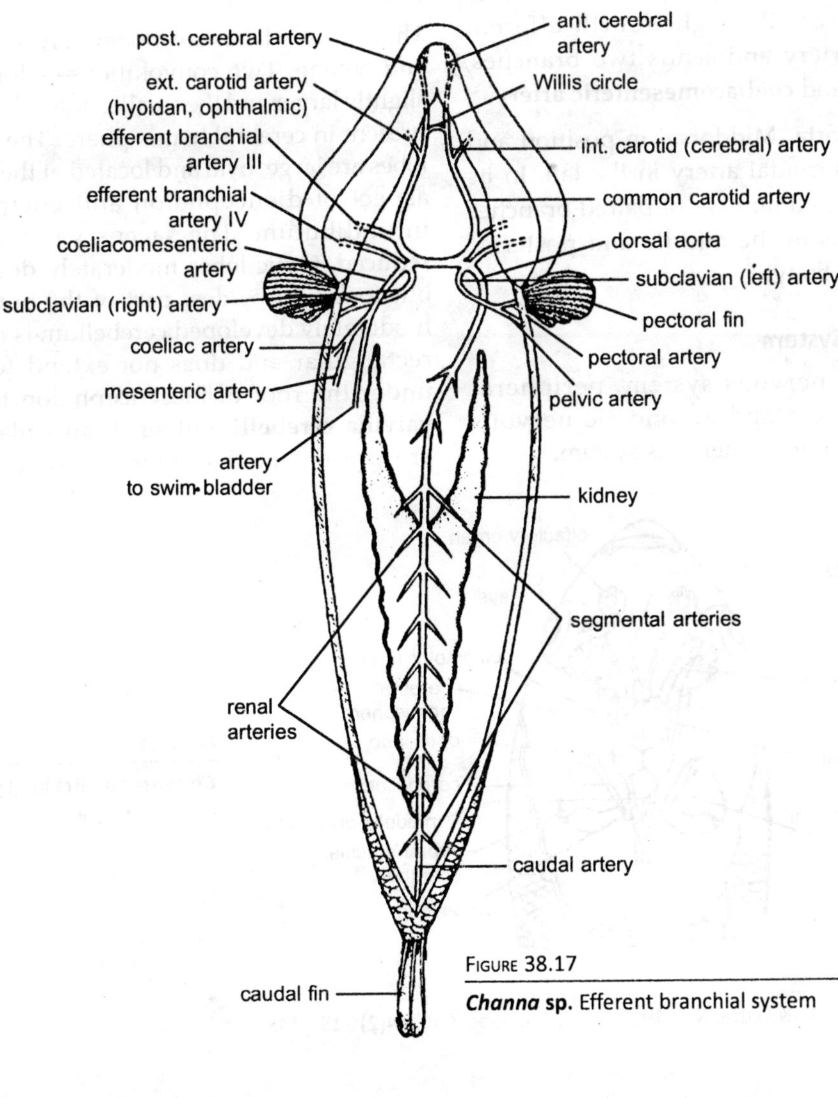

FIGURE 38.17

***Channa* sp.** Efferent branchial system

common carotid proceeds anteriorly as **internal carotid artery**. It sends a short median branch and the two branches join to form **willis cricle**. Anteriorly the main vessel divides into **anterior** and **posterior cerebral** arteries supplying blood to brain and adjacent region. Posteriorly, the two carotids form a transverse vessel which receives the fourth pair of efferent branchial arteries.

A branch on the left side of the transverse vessel receives the left fourth efferent branchial artery and sending a branch, the **subclavian** artery, it runs backward as dorsal aorta.

A branch on the right side of the transverse vessel receives the right fourth efferent branchial artery and sends two branches, **subclavian** and **coeliacomesenteric artery**.

Dorsal aorta. Middorsal in position and continues as caudal artery in the tail. In its course it sends a number of paired branches to the organs in the middle and posterior region of the body.

☒ Nervous System

The central nervous system, peripheral nervous system and autonomic nervous system constitute the nervous system.

Central nervous system

The components of the central nervous system are brain and spinal cord. The brain has three distinct zones—prosencephalon, mesencephalon and rhombenecphalon. The brain (Figs. 38.18, 38.19) is enclosed in a bony cranium.

The brain of *Channa** is nearly similar to that of lates.

Meninx primitive

The membrane covering the brain is quite distinct. A good amount of adipose tissue surrounds the brain.

Brain

The olfactory lobes are moderately developed and oblong. Two convolutions— dorsal one slightly larger and the smaller anterolateral are present in cerebral hemis-phere. The inferior lobes are large, oval and located at the ventral aspect of diencephalon and encircle the infundibulum. The saccus vasculosus is reduced. Optic lobes moderately developed but forms the widest part of the brain. The moderately developed cerebellum is roughly rectangular and does not extend forward under the roof of mesencephalon to form **valvula cerebelli**, but small auricular lobes are present at the anterior aspect of the

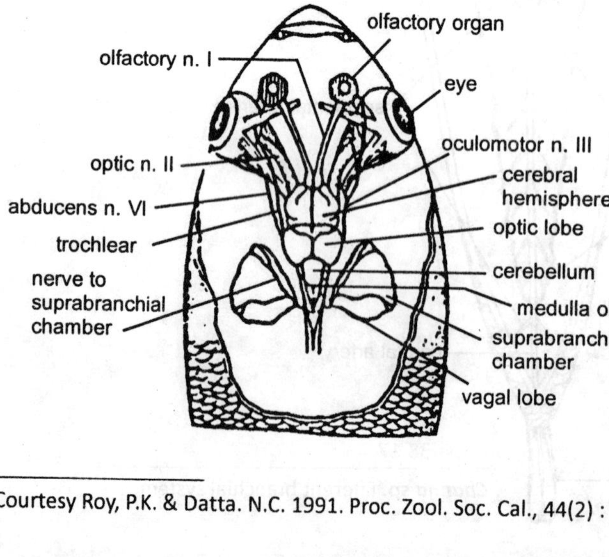

FIGURE 38.18
***Channa* sp.** Brain (Dorsal view) and nerves

* Courtesy Roy, P.K. & Datta. N.C. 1991. Proc. Zool. Soc. Cal., 44(2) : 137-148

cerebellum. The medulla oblongata bears two large **vagal lobes**.

Peripheral nervous system

The cranial and spinal nerves constitute peripheral nervous system (p. 136).

Cranial nerves

Ten pairs of nerves arising from the brain are cranial nerves (Fig. 38.20). Proceeding from the anteriormost to the posteriormost in order, the cranial nerves of each side are:

Cranial nerve O (nervous terminalis) absent.

i. **Olfactory nerve.** Arising from the anterior part of the olfactory lobe as a thick cylindrical nerve, runs forward and outward to innervate olfactory lamella. Function: sensory.

ii. **Optic nerve.** A thick cylindrical nerve, arises from the ventral surface of diencephalon, anterior to infundibulum and innervates the retina of the eye-ball. In their course, the right and left optic nerves cross each other but no optic chiasma is formed. Function: sensory.

iii. **oculomotor nerve.** Arising from the ventral region of the midbrain it splits into four branches and innervates anterior rectus, superior rectus and inferior rectus, inferior oblique muscles of the eye-ball. Function: motor.

iv. **Trochlear nerve.** Arises from the anterolateral part of the junction of mid- and hindbrain and innervates the superior oblique muscle of the eye-ball. Function: motor.

vi. **Abducens nerve.** The thinnest cranial nerve; arises from the ventral surface of the medulla oblongata and innervates the posterior rectus muscle of the eye-ball. Function: motor.

v and vii. **Trigeminal and facial nerves.** The fifth or **trigeminal** is the first nerve of the medulla oblongata. It arises from the ventrolateral side of the medulla, at its anterior

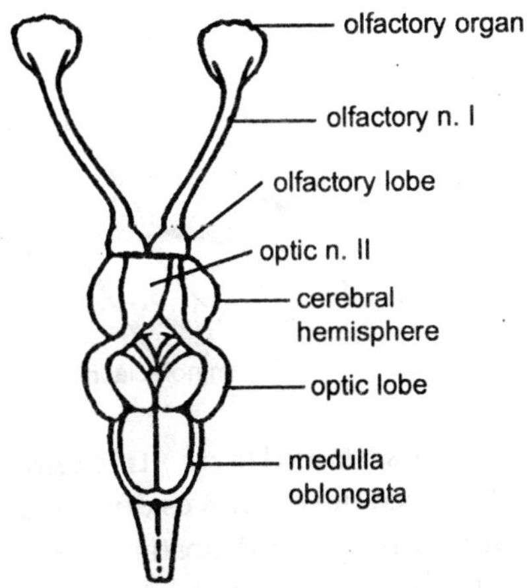

FIGURE 38.19

***Channa* sp.** Brain, Ventral view

end in the form of two laterally compressed, broad bundles of nerves. Four more bundles of nerves arise from the posterior aspect of the trigeminal roots, which are actually roots of the **facial nerve**. The three roots of the facial nerves along with two of the trigeminal coalesce to form a complex mass. The mass receives a short connective from the remaining facial nerve bundle and form the **Trigeminal ganglionic complex.**

Three main nerve trunks arise from the trigeminofacial ganglionic complex—the supraorbital, the infraorbital and the hyomandibular, in the addition to ramus palatinus.

a. **Supraorbital trunk (V).** The dorsolateral branch with a swollen ganglionic base. It divides into two branches.

   i. **Ramus ophthalmicus superficialis facialis**, a thick dorsal branch, running straight forward to innervate the snout area.

   ii. **Ramus ophthalmic superficialis trigemini**, a thin ventral branch, innervating the orbital surface bordering its rim and the skin.

b. **Infraorbital trunk (V).** A solitary branch from the ganglionic base and running posterior to orbit divides into three branches.

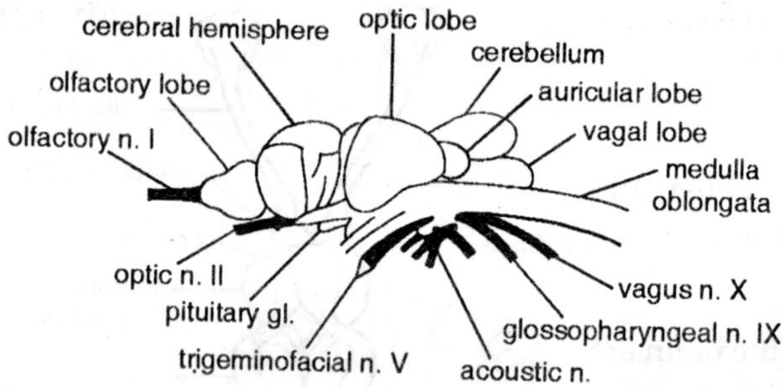

FIGURE 38.20

***Channa* sp.** Brain, Lateral view

i. **Ramus maxillaris trigemini.** A flattened nerve, moves downward and forward along the postero-ventral part of orbit and innervates the upper lip.

ii. **Ramus mandibularis trigemini.** A thick, compressed nerve, turns backward making a sharp angle at the maxillomandibular junction, courses forward to the quadratomandibular joint and innervates mandibular muscle.

iii. **Ramus buccalis facialis.** Emerging from the base of the infraorbital trunk it moves outward and then downward to innervate the posterior section of infundibular sensory canal.

c. **Hyomandibularis trunk (viii).** Leaving the cranium through a foramen of the hyomandibular bone, it divides into:

i. **Ramus hyoideus.** It moves postero-ventral and then turns forward to innervate the operculi and hyoideus adductor muscles, preopercular muscle and mandibular sensory canal.

ii. **Ramus mandibularis facialis** innervates the intramandibular muscle.

d. **Ramus palatinus (vii).** A slender nerve, arises from the ventromesial aspect of the trigeminofacial ganglionic complex and innervates the roof of the buccal cavity. Function: mixed.

viii. **Auditory (Statoacoustic) nerve.** Arises from the side of the medulla oblongata behind the facial nerve and innervates the internal ear. Function: sensory.

ix. **Glossopharyngeal nerve.** Arises from the lateral side of the medulla oblongata behind the auditory nerve, through a single root. Turning backward, it creeps upward along the dorsal plane of the suprabranchial organ. The ganglion, **pretorsal ganglion**, is formed over the suprabranchial chamber from which arise two branches, one anterior and the other lateral.

a. the **anterior branch** innervates the wall of the suprabranchial chamber.

b. The **lateral branch** descends along the lateral part of the first gill arch and finally terminates in the ventral branchial muscles. Function: mixed.

x. **Vagus or pneumogastric nerve.** It arises by two roots from the lateral surface of the medulla oblongata, posterior to the origin of the glossopharyngeal nerve. The roots meet outside the cranium to form **vagus ganglion.** Three groups of nerves, the ramus lateralis, the ramus visceralis and the ramus branchialis arise from the ganglion.

a. **Ramus lateralis.** The outermost group of nerves, arise from the lateral part of vagus ganglion. The main trunk is a broad, flat nerve. It enters deep into the body to reach the vertebral column and runs straight up to the caudal end.

i. **Ramus supratemporalis.** Arises from the base of the lateralis trunk and innervates temporal and supratemporal sensory canals.

b. **Ramus visceralis.** A single stout nerve, arises from the vagus ganglion, moves downward and divides into three branches:

i. The **dorsal branch** innervates the retractor dorsalis muscle.

ii. The **middle branch** enters deep into the visceral cavity and bifurcates. The dorsal branch moves straight backward for a long distance along the ventral surface of the kidney and innervates the air bladder. The ventral branch sends a tributary, the **gastric** to the stomach and continues as the **cardiac** branch, which is thin and most ventral. It bifurcates and sends branches to sphincter oesophagi muscle and heart.

c. **Ramus branchialis.** A group of four branchial nerves.

i. **First branchialis** arises independently from the vagus ganglion, moves outward and bifurcates.

The thinner **pretrematic** branch innervates the first gill arch. The stouter **posttrematic** branch innervates the ventral trough of the second gill arch.

ii. **Second branchialis** arising independently bifurcates as **pre** and **posttrematic** branches supplying second and third gill arches, respectively.

The third and fourth branchialis arise from a common trunk at the base of the visceralis branch.

iii. **Third branchialis.** The pre and **posttrematic** branches innervate the third and fourth gill arches, respectively.

iv. **Fourth branchialis.** At the very onset, it sends a fine **pretrematic** branch to innervate the fourth gill arch and runs anteriorly along the posteroventral part of the branchial basket as a flat nerve, innervating the adjoining structures. Function: mixed.

⊠ Breeding

Channa breeds in stagnant water in monsoon. Nesting has been reported in some species. The fry hatches out with a yolk sac containing yolk. The fries are bright red and all those of one brood move from one place to another in a swarm, jealously guarded by parents. With growth, the red colour of young fishes turns to greyish and finally the adult coloration appears.

DIPNOI

The Dipnoi is a well defined group, a side branch of the bony fishes by the middle of the Devonian period, possibly as an offshoot of the primitive Crossopterygia, the lobe finned fishes.

⊠ Evolution of Dipnoi

The earliest fossil of lung fishes go back over 350 million years. Diabolichthyes (Lower Devonian), the earliest known lung fish, shared some characteristics of crossoptery- gians as well. At times they flourished well all over the globe in Permian and early Triassic period and become rare after Triassic. Today only 6 species survive in three corners of southern Hemisphere. At an early period they began to undergo several specializations that led them off at a tangent from the main line of vertebrate evolution and made it possible for them to give rise to true terrestrial descendants.

⊠ Distribution of surviving Dipnoi

All the three surviving genera of dipnoans- *Neoceratodus, Lepidosiren,* and *Protopterus* are isolated in three continents of southern Hemisphere. The *Neoceratodus fosteri* is the most primitive of (Fig. 38.21) the existing lung fishes. Originally it was found only in the Burnett and Mary rivers in Queensland, Australia and is called Burnett salmon. Until recently, this species was an inhabitant of the Darling river system.

The South American lung fish *Lepidosiren paradoxa* is widely distributed over the Amazan basin.

The *Protopterus* has a wide range of distribution over the continent of Tropical Africa and consists of four species *P. annectans, P. aethiopicus. P. dolloi* and *P. amphibius.*

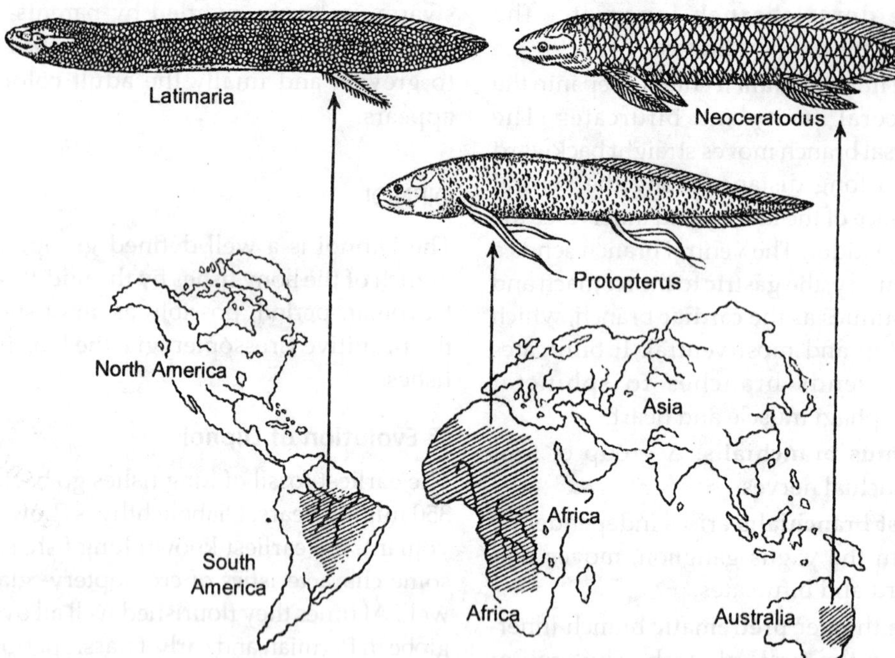

FIGURE 38.21

Dipnoi. Geographical distribution

P. aethiopicus lives in the basins of the Nile, Congo, and in the Tropical Lake Victoria and Edward and Albert of Eastern Africa. *P. annectans* lives in West Africa. *P. dolloi* in the *Ogone* and *P. amphibius* in the lakes of East Africa.

⊠ Characteristics and Natural history of existing lung fishes

A. Neoceratodus forsteri

1. Generalised fish-shaped attaining a length of about 1.5 m. The body covered with large, thin, spine-covered scales (Fig. 20.20D, page 197).
2. The lateral line system somewhat degenerated.
3. The paired fins are archipterygial, leaf-like, or paddle-like. Tail fin symmetrical, pointed and is of diphycercal category.
4. The skull is largely cartilagenous. Rest of the skeleton has been slightly replaced by bone. The membrane bones on top of the skull are rather deeply sunk beneath the skin and are overlaid by scales.
5. The notochord continuous and not constricted by vertebrae. The latter are represented by paired basidorsal and basiventral cartilages which form respectively the neural arch in the trunk and both neural and haemal arches in tail.
6. There are five gill-clefts covered by an operculum, but no external gills.
7. The glottis lies on the floor of the oesophagus and opening into trachea, which passes around right side of the oesophagus, open in lung, a large sac lying above the gut.
8. The cerebral hemispheres are rather large, thin-roofed, and the cerebellum poorly developed.
9. No stomach, no hepatic caeca, but well developed spiral valve present.

10. The 6th embryonic right afferent branchial arch acts as a pulmonary artery (cp. Amphibia).
11. The auricle is partly divided by a septum, the left half receiving oxygenated blood from the lung (cp. higher vertebrates).

Natural history

1. **Neoceratodus** frequents comparatively stagnant ponds or water-holes, still places of river bed, filled with rich vegetation all the year round, lies sluggishly at the bottom, occasionally rising to the surface to gulp in air.
2. It crops the luxuriant vegetation of the river banks to obtain the associated molluscs, crustaceans, worms, insects and fish larvae on which it lives.
3. Apparently, the air bladder is a functional lung at all times, acting in conjunction with gills. At irregular intervals the fish rises to the surface only when the oxygen tension falls below 83 mm Hg in water and protrudes its snout in order to empty the lung and take fresh air through mouth making a peculiar grunting noise. It is unable to live out of water but can survive in fowl water.
4. In spite of its pulmonary respiration, *Neoceratodus* more closely resembles the typical fishes in its habit than any other Dipnoi. It lives all the years round in water, and never passes into summer sleep (aestivation).

B. Lepidosiren paradoxa

1. It is an eel-like, attaining a length of about 1.2 m. The scales are small and deeply buried in the skin (Fig. 20.20B, page 197).
2. The lateral line prominent.
3. There are four gill-clefts bearing much reduced gills.
4. The external gills of larval stage are not retained in adult life.
5. The lung is double, lie dorsally, but their opening is to the oesophagus, ventrally.
6. The olfactory region of forebrain is much reduced.
7. Its diet is partly vegetarian and partly carnivorous *Ampullaria*, a large fresh water snail, being a favourite food article.

Natural history

1. **Lepidosiren** lives in swamps, breaths air more largely, taking several breaths at a time when it comes to the surface. It obtain 98% of its O_2 from the air.
2. In dry season it digs burrow in which it aestivates. The entrance to the burrow is plugged with a mud stopper, ventilated through several round holes.
3. In rainy season, with accumulation of water in the swamps, the mud plug is pushed out and the water enters the burrow. It lays up to 5000 eggs of 6-7 mm diameter in the burrow.
4. The male remains in the burrow guarding the eggs till they hatch out into tadpole-like larvae.
5. While guarding the eggs, the pelvic fins of the male bear numerous vascular filaments, which act as accessory gills, and are waived in water to supply oxygen to the developing eggs.
6. The larva bears four pairs of external gills.

C. Protopterus spp

1. These are eel-like fishes with small scales embedded in the skin, usually 0.6 to 0.9 m long, may attain up to 2.2. m (Fig. 20.20C).
2. The median fin system is a continuous fin fold with dorsal and caudal fins united. Paired fins are long, slender, with a skeleton composed of a central jointed axis and a few delicate lateral rays. These are used as tactile organs, though they may help in locomotion along the bottom.

3. There are five gill-clefts covered by an operculum. Vestiges of the external gills of larval stage sometimes remain on into adult life.
4. The lung is paired.

Natural history

1. The **Protopterus** inhabit marshes near rivers, is a voracious feeder, mainly but not exclusively carnivorous, feeding upon frogs, worms, insects etc. and from time to time rising to the surface of the marsh to take mouthfuls of air.
2. It has extraordinary vitality, can withstand severe wounds, long fasting a desiccation.
3. It is most active at night, swimming rapidly with powerful tail strokes, or walking slowly along the bottom with its slender fins moving alternately on each side, like the legs of a newt.
4. In the dry season, the marshes dry up, the fish makes its way into the mud to a depth of about 45 cm and coil itself into a flask-shaped enlargement at the bottom of the burrow, which is lined by a capsule of hardened mucus secreted by the glands of the skin and undergoes summer sleep or aestivation.
5. The mouth of the flask is closed by a capsular wall or lid which is perforated by a small aperture.
6. During aestivation lung is the sole breathing organ and nutrition is effected by the absorption of fat stored in the kidneys, and gonads.

ACCESSORY RESPIRATORY ORGAN IN FISHES: BIMODAL GAS EXCHANGE

Adult fishes depend chiefly on pharyngeal gill respiration. In some, other devices, however, are present to supplement or replace gill respiration. All such additional respiratory structures are known as accessory respiratory organs.

Air breathing fishes represent one of the most highly adapted group living in swampy environment and inhabit both freshwater and intertidal region. Following structures in air breathing fishes act as accessory respiratory organ.

a. Skin or integument

Common eels ***Anguilla anguilla*** and ***Amphipnous cuchia*** respire through integument both in air and in water. They often can make overland journeys through damp vegetation.

b. Buccopharyngeal epithelium

The highly vascularised buccopharyngeal epithelium of mudskipper, *Periophthalmus* and *Boleophthalmus* helps in absorption of oxygen directly from the atmosphere. These tropical fishes leave water and spend most of the times skipping in the dumpy areas particularly round the roots of the mangrove plants.

c. Pharyngeal diverticula

In some fishes a pair of sac-like diverticula from pharynx, lined by vascular epithelium extends above the gills. The diverticula are small in *Periophthalmus*, small and smooth in *Amphipnous* (Fig. 38.22A) and somewhat folded to form some alveoli in *Channa sp.* (Fig. 38.22B). The diverticula of *Amphipnous* open anteriorly through the midventral gill slit. The fishes have poorly developed gill-filments and the pharyngeal diverticula serve to breathe atmospheric air during aestivation or while coming out of water for some time.

d. Gut epithelium

In some fishes gut epithelium becomes vascular and modified to serve as a respiratory organ. The gut epithelium just behind the stomach in *Misgurus fossilis;* intestine in *Lepidocephalus guntea* and *Gobitus,* a giant European loach; rectum in *Callichthyes, Hypostomus* and *Doras* takes part in respiration. Fresh air is drawn through mouth or anus and void through anus after gaseous exchange.

e. Pelvic fins

During breeding season the pelvictins of American lung fish *Lepidosiren* enlarge,

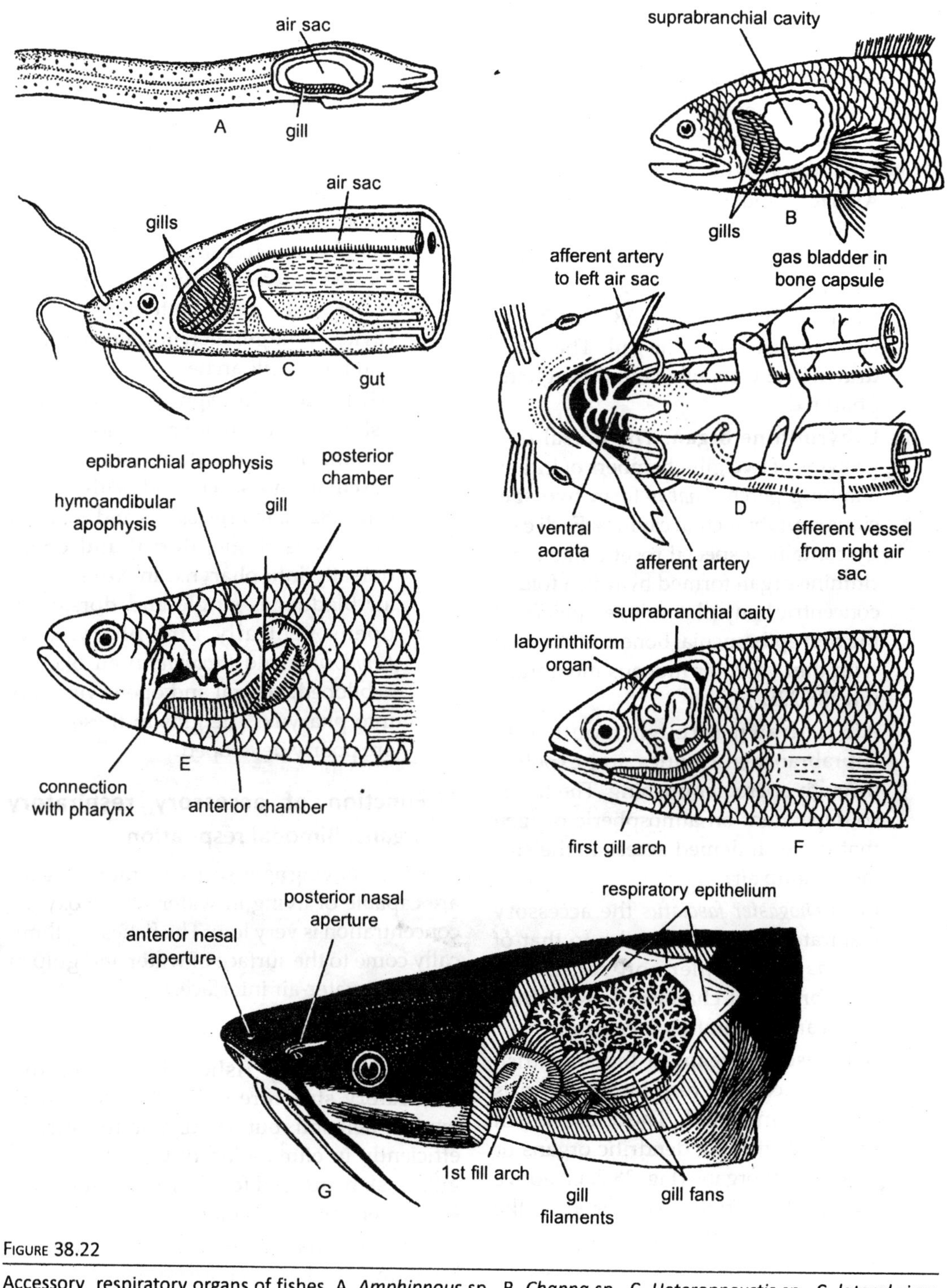

FIGURE 38.22

Accessory respiratory organs of fishes. A. *Amphipnous* sp., B. *Channa* sp., C. *Heteropneustis* sp., C. *lateral* view, D. *ventral* view, E. *Trichogaster* sp., F. *Anabas* sp., G. *Clarias* sp.

develop vascular filamentous outgrowths and provide fresh oxygen to the guarded eggs.

f. Branchial diverticula

The more complicated accessory respiratory organs develop as outgrowths from the gill chambers in the following fishes.

i. **Air sac.** In Indian catfish *Heteropneustes* (= *Saccobranchus*) has a pair of highly vascular long tubular dorsolaterally situated air sacs, arising posteriorly from gill chambers (Fig. 38.22C, D), extend almost up to the tail. The air is drawn in and expelled out through the pharynx.

ii. **Labyrinthine organ.** The dorsal outgrowths of two gill-chambers of Indian climbing perch *Anabas* form two specious suprabranchial cavities. Each cavity contains a special rosette-like labyrinthine organ formed by much folded concentric bony plates developed from the first epibranchial bone and covered with thin vascular mucous membrane (Fig. 38.22E, F).

   Air is drawn through mouth into suprabranchial cavities and expelled through opercular opening. The fish is so dependent on atmospheric oxygen that it dies if denied access to the surface to gulp air.

   In *Trichogaster fasciatus* the accessory respiratory organ is similar to that of *Anabas* but simpler and consists of suprabranchial chamber, labyrinthine organ and respiratory membrane.

iii. **Arborescent organ.** In Indian catfish *Clarias* accessory respiratory organ is more complicated. These are much branched tree-like dendritic organs or arborescent organs (Fig. 38.22G) above gill chambers, from the upper end of the second and fourth gill arches.

   The arborescent organ is composed of numerous terminal knobs, each with a core of cartilage covered by vascular membrane. The entrance of the suprabranchial chamber is guarded by fan-like structures, developed by the fusion of the adjacent gill-filaments of dorsal side of gill-arches. The dendritic organ and suprabranchial chambers functions as lungs.

   Atmospheric air from the pharyngeal cavity is drawn into the suprabranchial chamber through the inhalant aperture located between second and third gill arches and expelled into the opercular cavity by the gill slit lying between third and fourth gill-arches.

iv. **Air bladder.** In Dipnoans and ganoid fishes the air bladder acts as lung to breathe air. The wall of the bladder is vascular and sacculated with alveoli (Fig. 38.23). In *Lepidosteus* and *Amia* air bladder is single, dorsal and opens dorsally into pharynx. In *Neoceratodus* air bladder is single and dorsal but opens ventrally into pharynx. In *Lepidosiren* and *Protopterus* air bladder is bilobed, ventral and opens ventrally into pharynx. The gills in these fishes are poorly developed.

⊠ Function of accessory respiratory organs: Bimodal respiration

The fishes having accessory respiratory organs are capable of living in water where oxygen concentration is very low. The fishes rhythmically come to the surface of water and gulp in air in the water-air interface.

Bimodal gas exchange

In air-breathing fishes the accessory respiratory structure can easily cope with oxygen demand, but are unable to function efficiently in elimination of CO_2. In general gills or skin are used for CO_2 elimination and ion, water and pH regulation.

At the initial stages of evolution of air breathing, the bimodal breathing is required where the air breathing organs are used for O_2 uptake, and the gill respiratory surface is used for CO_2 elimination and pH regulation.

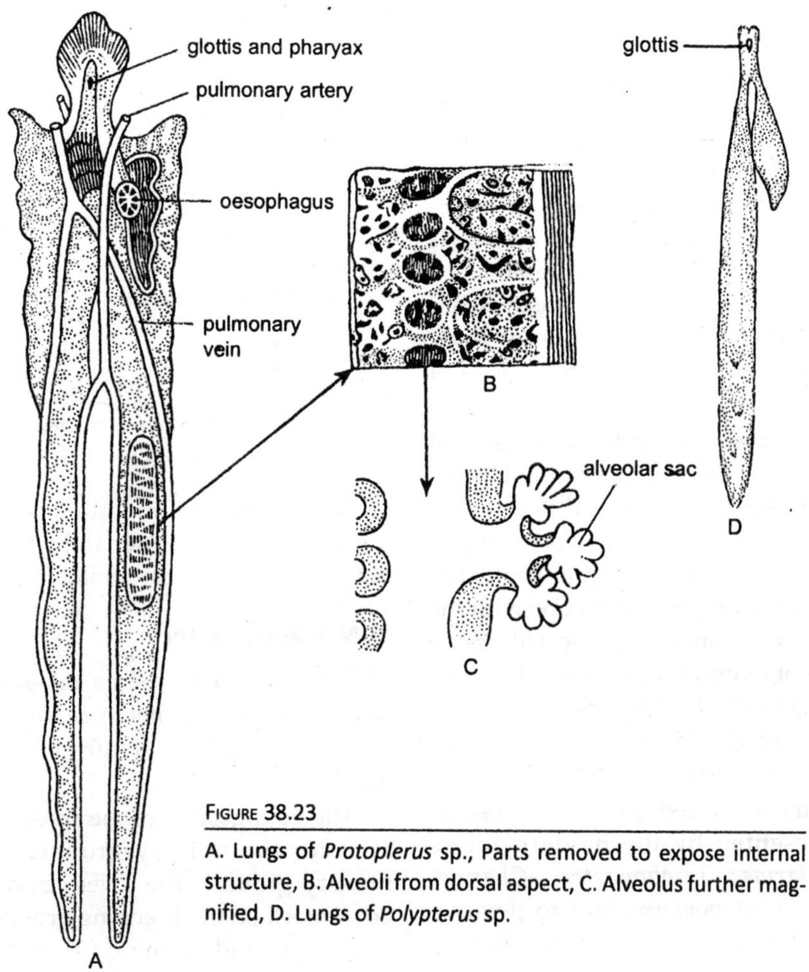

FIGURE 38.23

A. Lungs of *Protoplerus* sp., Parts removed to expose internal structure, B. Alveoli from dorsal aspect, C. Alveolus further magnified, D. Lungs of *Polypterus* sp.

Transitional forms went through stages of bimodal gas exchange with a large exchange ratio initially in the gill-skin system and a low ratio initially in the accessory respiratory organs of primitive lung.

⊠ Evolution of air breathing habit and organs

The air breathing teleosts of the modern times inhabit both freshwater and intertidal environments. Aquatic hypoxia in fresh water is the most important selective force in the evolution of air breathing vertebrates. The first step in this direction was the habit of taking air into the pharynx when at the surface in the air water interface.

The fish ventilates its air breathing pharyngeal or suprabranchial organs by using buccal pump, which is used for gill ventillation. The aspiratory models of ventillation, which have developed in a variety of groups of air breathing animals allow greater control over tidal volume and separation of ventillatory mechanisms of air breathing organs from those associated with feeding. Aspiration has evolved in aquatic as well as terrestrial vertebrates as a mode of ventillating their air breathing organs.

During early Devonian period the atmospheric oxygen was more or less like that in the present period which started declining in the late Devonian with drying up of lagoons, swamps and fresh water bodies.

These adverse ecological conditions led to the development of lungs from pharyngeal/

branchial pouches in dipnoi, crossopterygians and amphibians.

The extinct fish *Cheirolepis* which was the ancestor of the group Palaeoniscoidea, aquired lungs besides the gills. The palaeoniscids later gave rise to Chondrostei group of fishes represented by the modern forms *Polypterus* and *Polyodon*, and the Holostei represented by *Amia* and *Lepisosteus*. The teleostean fishes became the master of fresh as well as sea water during Tertiary and Quarternary periods of Cenozoic era.

The lungs transformed into the specialised hydrostatic organ the swim bladder. During Tertiary and Quarternary period the oxygen level of atmosphere fell down considerably affecting the oxygen concentration in the water. The gills were unable to cope with the depletion of oxygen in water, especially in the fresh waters of rivers and swamps. The lungs were no longer available for aerial respiration as they were modified into the specialised organs of swim bladder.

As such the advanced group of teleostean fishes represented by the modern forms *Anabas, Clarias, Trichogaster, Channa, Heteropneustes, Monopterus* had to develop some other types of accessory respiratory organs. These air breathing organs are essentially modification of the gills. The swim bladder in these fishes is either absent or very much reduced.

The lining of the respiratory organs are essentially mucoid in nature, while the lungs have a surfactant lining of phospholipids.

MIGRATION OF FISHES

The migration is the mass movement of animals from one habitat to another. Fishes migrate to the places where they find the conditions, they require at the particular phase of their life history which sets in toward the end of the migration. The migration, like other properties of the species, have some adaptive significance, ensuring the favourable conditions for the existence and reproduction

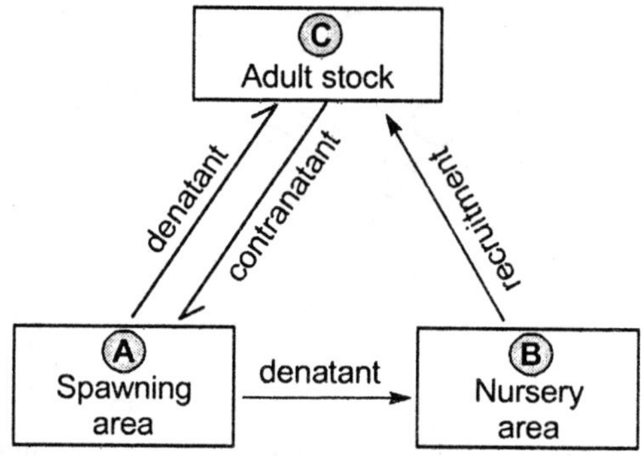

FIGURE 38.24

Migration pattern of fishes from spawning (A), nursery (B) and Adult stock area (C)

of the species. The migration is a link in the life cycle which is inseparably connected to both the preceding and following link.

☒ Migration Pattern

The migration of fishes usually follow a pattern of triangle between adult stock (A) spawning area (B) and nursery area (C). (Fig. 38.24)

The youngest leave the spawning ground at A for the nursery ground B; from the nursery grounds the juveniles move to the adult stock on their feeding ground at C; and the mature and ripening fish move from the feeding ground back to the spawning ground at A. Then as spent, they return to the feeding ground (Fig. 38.25).

☒ Migration Cycle

The cycle of migration of fish thus consists of the following:

1. **Spawning migration**—a movement form the feeding or overwintering grounds to the spawning ground.
2. **Feeding migration**—a movement away from the spawning or overwintering grounds to the feeding grounds.
3. **Wintering migrations**—a movement away from the spawning or feeding grounds to the overwintering grounds.

The migration pattern is believed to be related to that of the water current also. The

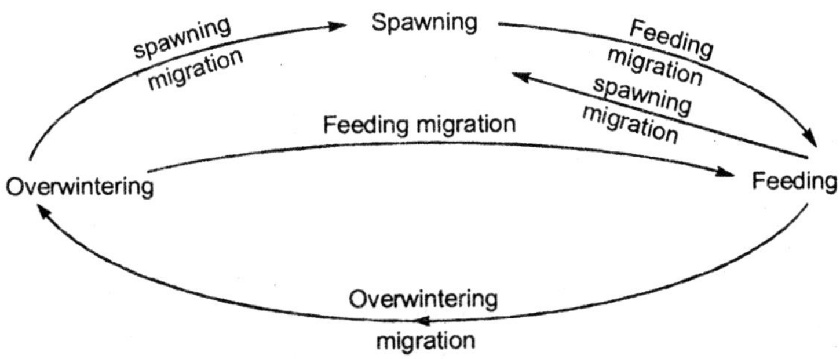

FIGURE 38.25

A scheme of migration cycle of various fishes

young stages drift with the current to the nursery ground; the spawning migration B to C is against the current and the species return to the feeding ground with the current (Fig. 38.25).

☒ Types of fish migration

The migration of fishes may be classified in four categories, (i) Catadromous, (ii) Anadromous, (iii) Potamodromous, (iv) Oceanodromous.

☒ Catadromous migration

The journey of fresh water fishes from the rivers to the seas for spawning is the catadromous migration. The best example is fresh water **eel**, ***Anguila*** spp. The European fresh water eel ***Anguila anguila*** and American species **A.** ***rostata*** change their colour from yellow to metallic silver in Autumn and prepare to migrate to the spawning ground.

These silver eels migrate down the rivers to enter the sea and migrate about 5600 km Westwards from Europe and eastwards from America to reach their spawning ground in the Surgasso sea off Bermuda. The adults die after spawning in deep waters.

The fertilized eggs hatch into little transparent, leaf-like flattened, about 6 mm long, pelagic larvae, called Leptocephalus*

The larvae start their homeward journey, undergo metamorphosis and turn into glass eel, the Elver. The elvers are cylindrical, about 8 cm long, pinkish translucent and takes about three years to attain this stage. The elvers start ascending the rivers in soals to reach a suitable resting place where their parents lived, feed and grow for some years to become the yellow eels.

During ascending the rivers the males remain behind in the brackish waters near coast, while the females ascend fresh water streams and rivers.

☒ Anadromous migration

The journey of fishes from the seas to the river system for spawning is known as **anadromous migration**. Examples: Hilsa shad (*Hilsa* [*Tenuolsa*] *ilisha*), Tolishad (*H. toli*), Allis shad (*Clupea alosa*). Sturgeon (*Acipencer* sp.), Atlantic salmon (*Salmo salar*), etc.

Atlantic salmon (*Salmo salar*) and five species of Pacific salmon (*Oncorhynchus* spp) and *Tenuolosa* spp (hilsa) of Indian ocean are best examples of anadromous migration.

These fishes produce huge numbers of eggs to compensate the risk to which the eggs are subjected during their development. Hilsa breeds in May-June, Salmons spawn in November-December. They leave their feeding grounds at sea ascend to the freshwater mountain streams, reaching the identical spot where they originally grew some years back.

After selection of suitable spawning ground, they seggregate in pairs and produce shallow saucer-shaped nests and spawn there. Following spawning they return to sea.

After hatching the larvae feed and grow for some period in streams before going out

* Until their real identity the Leptocephali Larvae were called glass fishes and placed in the genus **Leptocephalus**

to sea. The young grow faster in the ocean, after getting abundant food.

The ***Tenuolosa*** (= *Hilsa*) spp. of Indian ocean also show similar behaviour ascending river systems of the Ganges and Padma Complex of eastern India reaching upto the upstreams at the foot-hils of the Himalayas in the rainy season (May-June).

The strong olfactory sense of these fishes determines their homing into the original birth place, for different streams have different odours.

⊠ Potamodromous migration

The long journey of fishes within the fresh-water rivers for breeding and resting is known as **potamodromous migration**. The major carps and trouts breeding in the upstream hills migrate for feeding and resting to the lower course in the plains. In the breeding season, in monsoon, they again ascend the high stream following the same course. The spawns in the upper course of the river feed and grow and gradually come down to the lower course of rivers like their parents.

⊠ Oceanodromous migration

The long journey of fihses within the vast sea and oceans for breeding, feeding, growing and resting is known as **oceanodromous migration**.

Mackerels, Tunas and Herrings show such type of migration. Atlanto-Scandinavian herrings, *Clupea harengus* in the Norwegian and Barrents seas lives in the water of Atlantic origin and penetrates through the Faeroes-Shetland Bay between Iceland and Greenland, during spawning in April-May.

BIOLOGICAL SIGNIFICANCE OF FISH MIGRATION

The biological significance of migratory movements in fishes is an adaptation towards abundance of food; there may not be enough food in the spawning or nursery grounds to maintain both the immature and adult members of a large population. So it could be an advantage to have separate spawning, nursery and feeding areas. Accepting migration as an adaptation towards abundance, it would explain why the important commercial species are migratory; they are of commercial interest because they are abundant, and abundant because they are migratory. Furthermore, there would appear to be something to be gained by a species whose adults returned to spawn in an area where the environmental conditions were similar to those under which they themselves survived when young.

CLASSIFICATION: CLASS OSTHEICHTHEYES (TELEOSTOMI)

1. The skeleton is chiefly made of bones (cartilage in sturgeons and some others); tail usually homocercal; pelvic girdle absent.
2. Mouth usually terminal with teeth; jaws well-developed and articulated to the skull; eyes large without eyelids.
3. Skin with many glands; usually with embedded dermal scales of ganoid, cycloid or ctenoid type; some are scaleless a few with enamel covered scale.
4. Both median and paired fins rays of cartilage or bone.
5. Heart two-chambered, one atrium (auricle), one centricle and with sinus venosus, and conus arteriosus containing only venous blood. Four pairs of aortic arches; erythrocytes oval and nucleated, containing haemoglobin.
6. Respiration by pairs of gills on bony gill arches present in a common chamber at each side of pharynx covered by operculum.
7. Usually a swim-bladder with a duct opening into the pharynx present.
8. Cranial nerves ten pairs.
9. Gonads paired, usually oviparous; fertilization external.

10. Body temperature variable, depends upon the enviornmental temperature (poikilotherms). About 17,000 species of extant bony fishes inhabit salt, brackish and fresh water. The class is divided in two subclasses Actinopterygii and Crossopterygi.

Subclass ACTINOPTERYGII

1. Includes ray-finned bony fishes, most common of all the marine species as well as numerous fresh water species and other archaic survivors like sturgeons, garpike, bowfin, etc.
2. In the paired fins the bony or cartilaginous supports reduced to almost vanishing point. The fin-web lying outside the bodywall is supported by fin-rays alone (in the representatives of palaeoniscoids like Polypterini and Chondrostei fin structures are different).
3. The scales of modern forms have gradually become modified by the loss of many elements and may be reduced to thin bony structures or even may be absent.
4. The tail is initially heterocecal where lower fin-lobe is disproportionately developed, and supported only by hypurals. Subsequently, it transfromed into semiheterocercal and finally homocercal or more rarely diphycercal.
5. The primitive groups (strugeons Lepisosteus, Amia) possessed spiracles, abdominal pores and spiral valve in **gut**. Advanced group lost these structures. The respiratory arrange-ments, have changed from a gill-condition near the laminar type to the fully filamentar.
6. A prominent elastic bulbous arteriosus occurs in the heart, the conus being reduced to one or at most two rows of valves.
7. The brain is characterised by large corpora striata, cerebellum and medulla, presence of valvula cerebelli, and absence of cerebral lobes.
8. No trace of internal nostrils (in few forms it is present).
9. The dorsal fin single.
10. A pineal aperture, common in early crossopterygians rarely occurs.

The large number of actinopterygian species are arranged in three superorders Chondrostei, Holostei, and Teleostei.

Superorder **Chondrostei**

Primitive ray-finned fishes with usually ganoid scales, mouth opening large and tail usually heterocercal.

Order 1. **Palaeoniscoidea** (Palaeonisciformes)

Primitive ray-finned Chondrostei probably arose from some ancestors common to themselves and the Crossopterygii in the Devonian period. The order is a collection of families of separate lines of evolution whose actual relations, one to another is still uncertain.

1. Eyes large, sclerotic plates not more than four.
2. The maxilla had a wide postorbital extension which runs far backward.
3. The cheek extended dorsally as far as the upper surface of the skull and helped to obliterate the spiracular cleft.
4. In some palaeoniscoids a series of divergent radials in the fin formed a short reduced lobe, with no defined central axis.
5. The scales were ganiod not cosmoid. On the tail, there was an apparent break of pattern in the lines of scales on the upper tail-lobe.

Examples: *Palaeoniscus*, *Cheirolepis*, *Platysomeus*, *Cheirodus*, *Belonorhynchus*, *Helichthys*, etc.

Order 2. **Acipenseroidei** (Acipenseriformes)

Represented by strugeons of the present time. Twenty species of Acipenser inhabit rivers of Europe, North America and Asia, shovel-nosed *Scaphirhynchus* of North

America and central Asia, spoon-billed *Polydon* of the Mississipi and spoon-billed *Psephurus* of China.

1. Snout long, mouth ventrally placed, small and provided with barbels; tail heterocercal.
2. Skeleton cartilaginous; vertebrae lack centra.
3. Jaws weakly developed and teeth absent.
4. Scales reduced; tail covered with small rhomboidal scales, bony plates bearing spines arranged in five lines on the rest of the body. The skin between the lines is beset with denticles.
5. Spiral valve, conus arteriosus and an open spiracle present.
6. Roof of cerebral hemisphere thin.
7. Jurassic to recent.

Examples: *Acipenser, Scaphirhynchys, Polydon, Psephurus*, etc.

Order 3. **Polypterini** (Cladistia, Polypteriformes, Bachiopterygii)

1. Body slender, covered with thick ganoid, rhomboid scales.
2. Eight or more dorsal finlets, each preceded by spine and caudal fin arrow-shaped.
3. Bony skeleton, vertebral centra biconcave.
4. Paired fins with long posterior axis and preaxial radials and muscular lobe lying outside the body wall.
5. Spiracle helps in expulsion of excess water from the mouth. Swim bladder lung-like, possibly helps in respiration.
6. Internal nostril absent.
7. Spiral valve present in the intestine.

Examples: *Polypterus, Erpetoichthys* (*Calamoichthys*), etc.

Superorder **Holostei**

A polymorphic group indicating a position between the Chondrostei and teleosts.

1. An abbreviated heterocercal tail, superficially resembling the homocercal.
2. The scales have lost the middle layer of cosmine, while the outer ganoine layer retained.
3. The spiracle disppeared; no trace of a clavicle; considerable loss of bony fin rays and a single dorsal swim bladder.
4. A considerable degree of uniformity in the reduction, modification and freedom of maxilla together with striking changes in the preopercular and orbital regions are observed which leads to the development of more advance teleost architecture in future.

Order 1. **Semiontoformes** (Semiontoidea, Ginglimodi, Lepidosteiformes)

The sole representative of the order are garpikes or alligator gars, *Lepisosteus* of the fresh water of North and Central America.

1. **Primitive features**
   - i. A hemiheterocercal tail.
   - ii. Thick rhombic scales in oblique rows with a complete covering of ganoine.
   - iii. Elongated body with long snout.
   - iv. The central jugal plate has been lost and the spiracle is closed.
2. Pyloric caeca present and the structure of generative organs are teleostean.
3. **Specialised features**
   - i. Maxilla reduced and its functional replacement by a number of tooth bearing infraorbital elements.
   - ii. Numerous small cheek plates and opisthocoelus vertebrae.

Upper permian to recent.

Example: *Lepisosteus*.

Order 2. **Amiiformes** (Amioidea, Halecomorphi, Protospondyli)

A single North American bowfin species *Amia* is the sole extant representative of the order with a mixture of primitive and advanced features.

1. The scales are thin, cycloid, overlapping and have lost ganoine layer.
2. The fins have lost all fulcra, the tail is homocercal.
3. The vertebrae solid, amphicoelous, except in the tail region.
4. One jugal plate between the rami of lower jaw and well-developed branchiostegal rays present as in teleost.
5. The lower jaw retains the full number of elements.
6. Lung dorsal but bilobed and cellular, showing possibility of air breathing.
7. A vestigeal spiral valve in the intestine; no pyloric caeca.
8. A small valvular conus in the heart but the bulbous arteriosus is developed.

Triassic to Recent

Example: *Amia calva*.

Order 3. **Pycnodontoidea** (Pycnodontiformes)

Fossil forms; Upper Triassic to Eocene; body laterally compressed, covered by rhomboid scales; reduced opercular apertures.

Examples: *Gyrodus*, *Pycnodus*, etc.

Order 4. **Aspidorhynchoidia**

Fossil forms; Upper Jurassic to Cretaceous; the vertebrae amphicoelous, ring-haped; tail homoceral.

Example: *Belonostomus*, *Aspidorhynchus*, etc.

Order 5. **Pholidophorodiea**

Fossil forms which gave rise to Leptolepidae, considered to be the ancestors to teleosts.

Example: *Leptolepis*

Superorder **Teleostei**

The most successful group comprising more than 25,000 species of fishes inhabiting every zone of waters in all depths of ocean and even coldest polar seas; in every fresh water habitat except in hot springs. They are the offshoot of holostean evolution.

1. **Degenerated or lost characters**

i. The spiracle and spiral valve absent.

ii. Conus arteriosus reduced to one, or two rows of valves; bulbous arteriosus greatly enlarged to compensate the loss.

iii. The swim-bladder, where present, is essentially hydrostatic in function, may be adapted to sound production or respiration.

iv. Exoskeleton and scales lack ganoine layer.

v. Dermal bones tend more and more to sink beneath the skin.

vi. Bones of lower jaw are reduced to three elements dentary, angular and articular.

vii. Pectoral girdle lost clavicle.

2. **Highly modified characters**

i. Gills fully filamentar.

ii. Tail externally homocercal or even gephyrocercal.

iii. Endoskeleton fully ossified. Vertebrae amphicoelous. Additional ribs known as intermuscular bones are present.

iv. A single prevomer and single supraoccipital bone present in the skull.

v. The upper crushing teeth appear to shift to the parasphenoid and corresponding lower ones are attached to elements of the lower pharyngeal floor.

vi. The radials in the paired fins are reduced to a few branchial ossicles, the fin is completely fan-like and without any trace os axis.

vii. Highly specialised urinogenital organs with ducts in both ovaries and testes. No connection between mesonephros and testis.

viii. The brain is highly developed. Upper wall of cerebrum very thin, mid- and hindbrain often very large and the optic nerves cross-over outside the brain without mixing i.e. absence of optic chiasma.

The teleosts are arranged (after Berg) in the following orders.

Order 1. **Clupeiformes** (Isospondyli)

The herring, salmon, trout, etc. are included in this order,

1. The tail is homocercal; single dorsal fin; ventral fin may be absent; fins without spiny rays; pelvic fins abdominal.
2. Well-developed cycloid scales cover the body.
3. Air bladder with duct, opens in pharynx.
4. Auditory ossicles absent.
5. Pyloric caeca usually numerous.

Examples: *Tarpon, Clupea, Sardinops, Sardinia, Albula, Alsoa, Oncorhynchus, Salmo, Salvelinus, Notopterus, Hilsa, Chirocentrus, Megalops*, etc.

Order 2. **Esociformes** (Haplomi)

1. Includes pikes, slender bodied with large mouth and conspicuous teeth. Feeding on fishes.
2. Pyloric caeca absent.
3. Fins are provided with soft fin rays.
4. Body length up to 2.5 m and weight about 50 kg.
5. Mieocene to recent.

Example: *Esox, Dallia, Umbra.*

Order 3. **Myctophiformes** (Iniomi or Scopeliformes)

1. Deep sea fishes called lantern fishes having cylindrical body and luminescent organs.
2. The mouth is wide, jaw of long premaxilla armed with numerous minute teeth.
3. The dorsal and ventral fins lack spines.
4. Swim bladder absent or filled with fat to withstand pressure differences.

Examples: *Harpodon, Myctophum, Synodus*, etc.

Order 4. **Saccopharyngiformes** (Lyomeri)

1. Oceanic gulpers with eel-shaped body without scales and soft fin rays, found at a considerable depth.
2. Mouth is very large.
3. No pelvic and caudal fins and ribs.
4. Tail long and slender.

Example: *Saccopharynx, Eupharynx.*

Order 5. **Cypriniformes** (Ostariophysi)

Includes over 5,000 species of varied fishes, like suckers, loaches, cat fishes, carps etc. Except some catfishes all are fresh water forms.

1. Anterior vertebrae fused, and lateral elements of four vertebrae detached to form Weberian ossicles between swim bladder and membranous labyrinth.
2. Swim bladder of 2 to 3 parts, usually connected with the gut by a duct.
3. Fins are usually provided with spine. The ventral fin is abdominal.

Cypriniformes are arranged in two divisions, Ciprini and Siluri.

1. Division **Ciprini**

The body may be without scales. The scales, when present, are without bony plates. The third and fourth vertebrae never fuse with each other.

Examples: *Labeo rohita, L. calbasu, Catla, Cyprinus, Cirhina, Puntius, Rasbora, Carassius*, etc.

2. Division **Siluri**

The body is scaleless. The maxillary bone reduced and support the barbels. The second, third, fourth and fifth vertebrae are usually fused.

Examples: *Clarias, Heteropneustes, Wallago, Mystus, Rita, Chaca, Bagrius*, etc.

Order 6. **Miripinnati**

1. Scaleless body.
2. The lateral line sense organs are located on projection.

Example: *Miripinna.*

Order 7. **Anguilliformes** (Apodes)

1. Marine eels, with long slender body, usually coloured.

2. Dorsal, caudal and anal fins continuous; caudal sometimes lacking. Pelvic fins usually absent, when present, adbominal in position.
3. Gill opening small.
4. Scales minute or absent.
5. Swim bladder with duct.

Examples: *Anguilla, Muraena*, etc.

Order 8. **Giganturiformes** (Giganturoidea)

Body elongated; fishes without pelvic fins and swim bladder. Lower lobe of the tail fin elongated.

Example: *Gigantura*.

Order 9. **Notocanthiformes** (Heteromi, Halosauriformes)

1. Deep sea spiny eels with slender body, long tail and without caudal fin.
2. Fins with spine.
3. Oviduct absent.

Example: *Natocanthus*.

Order 10. **Beloniformes** (Synentognathi)

Mostly marine needle fishes, gar fishes, halfbeaks and flying fishes.

1. Body covered with cycloid scales.
2. The pectoral fins shifted to the dorsal side of the body and greatly elongated. The ventral fins shifted to abdominal side and dorsal fin above the anal.

Example: *Xenentodon, Strongylura, Exocoetus, Cypselurus, Hemirhamphus, Belone*, etc.

Order 11. **Cyprinodontiformes** (Microcyprini)

Very small or diminutive sized top minnows and cave fishes.

1. Mouth protractile.
2. The lateral line weak and pelvic fins abdominal or absent.
3. Dorsal fin single, fins are spineless.

Examples: *Cyprinodon, Gambusia, Typhlichthyes, Fundulus, Lebistes, Xiphophorus, Molliensia*, etc.

Order 12. **Percopsiformes** (Salmopercae)

North American trout-perches, dorsal and ventral fins are preceded by one to four spines.

Example: *percopsis*.

Order 13. **Syngnathiformes** (Solenichthyes)

Marine sea horses and pipe fishes.

1. Body covered with bony rings of scales.
2. The male bears brood pouch under tail and shelters eggs until hatching.
3. The suctorial mouth at the tip of the tubular snout.
4. Fins are minute and tail slender and prehensile.

Examples: *Hippocampus, Syngnathus, Trachyrhanphus*.

Order 14. **Gadiformes** (Anacanthini, Macruriformes)

Mostly marine, bottom-dwellers of Northern hemisphere; cods, pollacks, hakes, burbots, etc.

1. Fins are soft-rayed, lack spines; pelvic fins often many-rayed and thoracic or jugular in position.
2. The caudal fin when present, formed mainly of dorsal and anal rays and without expanded hypural.
3. The swim bladder ductless
4. Body covered with cycloid scales.

Examples: *Gadus, Merluccius, Macrourus* etc.

Order 15. **Lampridiformes** (Allotriognathi)

1. Mostly oceanic, ribbon-like fishes with large eyes.
2. Fins with soft rays or dorsal with 1-2 spines; pelvic fins with 1-17 rays.
3. Teeth feebly developed or absent.

Examples: *Lampris, Tachypterus, Regalicus*, etc.

Order 16. **Beryciformes** (Barycomorphi)

Deep water marine squirrel fishes.

1. The pelvic fins bear more than five soft fin rays.

2. The caudal fin with 19 principal rays.

Example: *Holocentrus, Beryx.*

Order 17. **Zeiformes** (Zeomorphi) (Atlantic 'John Dory')

1. The anal fin preceded by 1-4 spines forming separate fins.

Examples: *Zenopsis, Zeus.*

Order 18. **Perciformes** (Percomorphi, Acanthopteri)

Includes large number of varieties of fishes commonly called perches, inhabiting all aquatiç habitats, salt and fresh waters and is largest order of vertebrates including more than 8,000 species of diverse form and habits.

1. Fin spines are usually present; pectoral fins well up on sides; pelvic fins far forward, each usually with one spine and five soft rays.
2. Upper jaw bordered usually only by premaxilla.
3. Swim bladder ductless.

Examples: Freshwater basses and sunfishes (*Micropterus, Lepomis, Pomoxis*), fresh water perches (*Perca, Etheostoma, Hadopterus, Cichlids*), Viviparous perches (*Cymatogaster*), Mackerels (*Scomber, Rustrelliger*), Swordfish (*Xiphius*), Marlin (*Makaira*), Gobies, rockfishes (*Sebastodes*), Sculpins (*Cottus*), *Barracudas* (*Sphyraena*), *Lates, Chaetodon Toxofes, Oreochromis* (= *Tilapia*), Silver pomfret (*Stomateus*), *Anabas*, etc.

Order 19. **Gobimorphi** (= Suborder Gobioidea)

1. Carnivorous fishes having suborbital ring devoid of a bony support for the operculum.
2. Gills are four pairs and gill membrane is attached to isthmus.
3. The first dorsal fin may contain few weak spines; the pelvic fins lie below the pectorals and each pectoral contains one spine and 4 or 5 soft rays.
4. Swim bladder and pyloric caeca absent.

Examples: *Gobius, Boleophthalmus, Periophthalmus.*

Order 20. **Gasterosteiformes** (Thoracostei)

1. Fresh or brackish waters stickleback; males prepare nest of sticks and secretions, then guard the eggs.
2. Free spines before dorsal fin.

Example: *Gasterosteus.*

Order 21. **Scleroparei** (= Suborder Cottoidei)

Extra spines are present over the head and body and a bone may be present across the cheek.

Examples: *Cottus, Sebastes*, etc.

Order 22. **Pegasiformes** (Hypostomides)

Small dragon fishes of tropical seas of Old World, body enclosed in a broad bony box and tail in bony rings.

Example: *Pegasus.*

Order 23. **Pleuronectiformes** (Heterosomata)

Usually marine flat fishes on bottom in coastal waters.

1. Asymmetrical, with both movable eyes on one side of head; early young is bilaterally symmetrical, but as the fish soon lies on one side, the lower eye migrates around and both the eyes are on upper surface.
2. Body strongly compressed with dorsal and anal fins fringing the body. Fins lack spine.
3. The swim bladder is absent.

Examples: *Cynoglossus, Solea, Psettodes, Hippoglossus, Paralichthyes, Achirus Symphurus*, etc.

Order 24. **Echeniformes** (Discocephali)

Sucker fishes of temperate seas.

1. Upper surface of head with flat, oval adhesive disc with transverse furrows, which is the modified spinous dorsal fin, by which it attaches to sharks or other fishes.
2. The anal fin and other dorsal fin spineless.

3. Body covered with cycloid scales.
4. Swim bladder absent.

Examples: *Remora, Echeneis.*

Order 25. **Tetraodontiformes** (Plectognathi)

Puffers, head fishes, porcupine fishes, trunk fishes, trigger fishes of both fresh and marine waters.

1. Body forms variable, some globose, some inflate by swallowing water.
2. Jaws short, powerful and teeth usually strong incisors or forming a sharp-edged beak.
3. Scales generally spiny or bony.
4. Gill openings small.

Examples: Globe fish (*Tetrodon*), Porcupine fish (*Diodon*) *Mola, Ostacion*, etc.

Order 26. **Icosteiformes** (Malacichthyes)

1. Endoskeleton less ossified.
2. No spiny rays in the fins.

Example: *Icosteus.*

Order 27. **Gobiesociformes** (Xenopterygii)

Cling fishes remaining attached to the stones or shells with the help of large, sub-circular adhesive disc on abdomen, in the shore waters of tropic and temperate zones. Body naked.

Example: *Gobiesox.*

Order 28. **Batrachoidiformes** (Haplodoci)

Toad fishes of the bottom of tropical and subtropical seas.

1. Head depressed and mouth wide.
2. A spinous dorsal fin and long softrayed dorsal and anal fins present.

Examples: *Porichthyes, Opsanus.*

Order 29. **Lophiiformes** (Pediculati)

1. Deep sea angler fishes usually with luminescent organs.
2. The 'spinous' dorsal fin of a few flexible rays, the first being located on the head and with dilated bulb at tip.

Examples: *Lophius, Ceratias, Metanocetus, Antennarius*, etc.

Order 30. **Mastacembaliformes** (Opisthomi)

Spiny eels of fresh water of Africa and South Asia.

1. Laterally flat body with tubular nostril on fleshy tentacles at the end of snout.
2. The dorsal, caudal and anal fins are continuous, except in *Macrognathus*, where caudal fin has separate entity.
3. Dorsal fin is preceded by separate spines.
4. Pectoral fins present.

Examples: *Mastacembelus, Macrognathus*, etc.

Order 31. **Channiformes** (Ophiocephaliformes)

Common freshwater snake-headed (Ophicephalus) fishes popularly, known as chang, lata, shole, shal in Bengali.

1. Head sepressed and covered with large plate-like scales.
2. The fins are devoid of spines.
3. Swim bladder ductless.
4. The symbranchial organ or accessory respiratory organ is present.

Examples: *Channa punctatus, C. gachua, C. triatus, C. Marulius*, etc.

Order 32. **Synbranchiformes** (Synbranchii)

Synbranchoid eels of fresh water streams of Africa, India and tropical America.

1. Eel-shaped body with dorsal anal fins without fin rays and continuous with caudal fin.
2. No pectoral fins.
3. Gill opening fused as transverse ventral slit.

Examples: *Amphipnous, Synbranchus*, etc.

Order 33. **Mugiliformes**

1. Brackish water fishes.
2. Scales cycloid or ctenoid.
3. Head covered by scales.
4. Pelvic fin abdominal.

Examples: *Mugil parsia, M. spaigleri, Rhinomugil corsula, Polynemus Paradiseus,* etc.

Subclass CROSSOPTERYGII

(Choanichthyes : Sarcopterygii)

The crossopterygians are lung fishes known as Dipnoi, and their allies, have some important tetrapod characters, though they have not attend the status of the latter group. They are the lobe-finned fishes, possibly the ancestors of first land vertebrates. During the whole Devonian Period the lobe fin fishes were the most abundant of the bony fishes; declined during Carboniferous period and only few of them survived in the Mesozoic era. These fishes were pushed furthest up the rivers and lived in inland swamps and ponds, that drained into the rivers. The waters of ponds and swamps were subjected to seasonal changes and dried up mostly during long rainless summers. For survival in this harsh condition the lobe-finned fishes got advantage of having lobed fins which were used to crawl out on land to get access to the nearest, suitable, good pond or pool, and kept on moving as the water dried up more and more. The most severe challange they had to face was the problem of respiration. Ordinary gill-breathers could not survive in this condition but the lobe-finned fishes developed a fairly good air-breathing lung-like equipment. Carbon dioxide tolerance, i.e. ability to retain certain amount of carbon dioxide in blood, without disturbing normal metabolism, is a prerequiste to aerial respiration. To meet the need, they must have developed CO_2 tolerance prior to the functioning of lung-like breathing organs. In addition to this lung-like organ, the lobe finned were able to inhale air without opening the mouth by taking it directly through the already developed internal nares (choanae) connected with external nostrils, as in Dipnoi. These two important adaptations, lungs with internal nares opening in external nostrils and lobed fins made them fit to survive in such an unstable environment. The existing important crossopterygians are dipnoan fishes and coelacanths.

1. The nostrils connect the mouth cavity.
2. The paired fins are provided with a large median lobe containing jointed skeleton and muscles fringed with dermal rays and are covered with scales.
3. The pectoral fins are supported by jointed median axis, with radials arranged on its two sides.
4. The scales are covered with thick cosmine.
5. The jaw suspension is autostylic.
6. Caudal fin with epichordal lobe.
7. The subclass Crossopterygii includes two orders Rhipidistia and Dipnoi.

Order 1. **Rhipidistia**

The order is very much important in that it provides some clues to vertebrate evolution. The members of this group are specialised to be themselves ancestral to the coelacanths, lung fishes, or tetrapods and prototetrapods evolved from a generalised rhipidistian crossopterygian in late Devonian period.

1. Skull ossified and teeth on both premaxilla and maxilla.
2. Tail partly heterocercal.

Suborder i. **Osteolepidoti**

Became extinct during Permian period.

1. Elongated body covered with thick pitted rhomboidal cosmoid scales. Dorsal fin two. Tail heterocercal. Paired fins scaly, lobed form.
2. The skull well-ossified and composed of series of bony plates and the top of the skull is divided into two parts by the presence of a transverse joint between the parietal and post-parietal. An unossified gap is present at the base of the skull.
3. Jaw suspension autostylic.

4. Eyes small and supported by many sclerotic plates.

Examples: *Osteolepis, Rhizodus, Eusthenopteron* etc.

Suborder ii. **Coelacanthini** (Actinistia: Coelacanthiformes)

Mostly extinct rhipidistians with a good fossil records and with only one **extant species**, *Latimaria chalumnae* of South Eastern Africa.

1. Head short, deep and skull with much cartilage.
2. Teeth only at tips of premaxilla and dentary.
3. Tail trilobed, diphycercal.
4. Internal nostril absent in fossil forms.
5. The paired fins paddle-like, short pedunculate, lobed, covered with scales and the fin rays are attached in a fan-shaped fashion. Two dorsal fins.
6. Notochord persists as a massive unconstricted rod.
7. Vertebral column mostly unossified.
8. A calcified swim bladder present.

Examples: *Coelacanthus, Latimeria* (Fig. 20.21A), *Macropoma, Undina*, etc.

Order 2. **Dipnoi** (Dipneusti)

1. Dipnoi the lung fishes are slender eel-like or fish-like animals, 1 to 2 m long. Nostrils two; both lung and gill present.
2. Body covered with thin overlapping cycloid scales.
3. Snout depressed bearing external nostrils enclosed in upper lip.
4. Eyes small
5. Paired pectoral and pelvic fins are narrow lobe-like or filamentous with a central axis of flesh and bone.
6. Dorsal anal and caudal fins fused into a continuous symmetrical diphycercal tail, supported by partly calcified fin rays.
7. Mouth subterminal and ventral.
8. Gill covered on either side by an operculum leaving a single gill slit. Gills may be well developed (*Neoceratodus*) or reduced (*Protopterus* and *Lepidosiren*).
9. Urinary and genital aperture, one or two; abdominal pores and anus lie within a small midventral cloaca.
10. Lateral line system well developed, especially on head region.
11. Glottis may be protected by epiglottis.
12. Internal nares may open into mouth cavity.
13. Swim bladder or air bladder single (*Neoceratodus*) or double (cf. *Protopterus* and *Lepidosiren*), lungs with internal alveoli.
14. Spiral valve present in the intestine.
15. Primary cranium persists with little ossification and with or without fontanelles.
16. Premaxillae and maxillae absent and teeth modified as crushing plates.
17. Notochord persistent with cartilagenous vertebral arches but lacking vertebral centra. Anterior vertebrae fused with the skull. Endoskeleton mostly cartilaginous.
18. Jaw suspension autostylic.
19. Cartilaginous branchial arches, 4 to 6.
20. Pectoral girdle a stout cartilage with a pair of investing bones on either side.
21. Pelvic girdle well developed and cartilaginous.
22. Mid brain is almost undivided into optic lobe. Pallium of cerebral hemispheres has a layer of nerve cells.
23. Auricle and ventricle of heart and sinus are imperfectly divided. The large atrioventricular opening is filled up by a fibrous plug, there is no true valve.
24. Ventral aorta is very small; conus

arteriosus partly or fully partitioned; a spiral valve is present. A pulmonary artery is given off on either side from efferent branchial system. A pulmonary vein opens into left side of auricle.

25. Precaval two (ductus Cuvieri) and a large posterior vena cava (posterior cardinal vein) cf. tetrapods.
26. Paired oviducts open anteriorly into coelom in female. Vasa efferentia carry sperms through excretory part of kidney in male. Fertilization external.
27. Development includes metamorphosis; larva may develop suckers and external gills.

The dipnoans are arranged into two sub-orders.

Suborder i. **Monopneumona (= ceratodontiformes)**

Lateral rays of paired fins jointed. Lung single and well developed.

Examples: *Ceratodus, Neoceratodus.*

Suborder ii. **Dipneumona (= Lepidosireniformes)**

Lateral rays of the paired fins extremely reduced or absent. Lungs double.

Examples: *Lepidosiren, Protopterus, Ctenodus* etc.

39 CLASS AMPHIBIA

The amphibians are the lowest and earliest land vertebrates derived from fish-like ancestors in the Devonian times. These cold-blooded vertebrates having naked skin rich in glands, constitute a true transitional group, neither fully aquatic nor fully terrestrial. They have struck a sort of compromise between the two habitats though not particularly well adapted to either. Amphibians stand between the fishes and reptiles and are the first group among the chordates to live out of water.

Amphibia includes the present day toads and frogs, salamanders, worm-like caecilians and various fossil forms from Devonian time onwards. Most of the species live partly in fresh water and partly on land. To conquer land—several new changes evolved in the architectural plan of their body. Like, modification of body for travel on land and also retaining the swimming ability; development of limbs in place of paired fins; change of the skin to permit exposure to air; replacement of gills by lungs; changes in the circulatory system to provide for respiration by lungs and skin, and aquisition of the sense organs that function in both air and water.

Thus, to be an amphibian (i) the gills of fish have been eliminated, and the swimbladder changed to a lung; (ii) the nostrils, which in the fish end blindly in nasal sacs, are continued through the roof of mouth and are glandular, for keeping them moist; (iii) the median fin and tail reduced and finally disappeared; (iv) the paired fan-like fins modified to legs; (v) scales lost for making the skin soft and glandular for cutaneous respiration; (vi) the eyes—kept moist by water in the fish—are modified and provided with eye-lids, tear-glands and ducts for keeping them moist on land. The eyes are also somewhat elevated from the surface and adapted for vision in the air; (vii) the lateral-line sense organs of fish being useless on land are eliminated; (viii) the ear, largely an equilibrium organ with some ability to pick up vibrations in water in fish, changed to pick up sound waves from the air and thus, becomes a true organ of hearing as well as an organ of equilibrium and an ear drum with bony connection between it and inner ear has been introduced.

The amphibians, fundamentally, are aquatic, because their developmental processes are aquatic. The amphibians never attained the heights of success and of dominance in nature like fishes, reptiles, birds or mammals. Because, they were forced to occupy a narrow strip of territory between the sweet waters and dry land, due to their necessity for amphibious environment. As a result, they were prey to the fishes on one hand and, on the other to the land vertebrates, evolved later. However, at one time, the amphibians occupied an honourable place in nature, reached in some cases giant size and were active and predaceous creatures. Their fall began with the rapid rise of reptiles, the true land vertebrates.

BUFO MELANOSTICTUS (Toad)

The common Indian toads are nocturnal, carnivorous and found in large number in damp and shady places, in dark conrners and water outlets of houses, by the sides of ditches, ponds and in many other places. They are adapted for land life but can survive in water for some time. They are independent of water, except breeding, as the eggs are always laid on water and fertilized eggs develop into young toads in water. Being **poikilothermous** (cold-blooded), they undergo **hibernation** under some cover, holes, cracks, crevices and spend a dormant life during winter. Toad belongs to order Salientia (Anura), class Amphibia, subphylum Vertebrata, phylum Chordata.

⊠ External Features

The body is divisible into two parts—the posterior, short and stout trunk and the anterior, broad, depressed head. There is no neck between the head and the trunk. Tail is absent (Fig. 39.1) Two pairs of limbs, one at the anterior and another at the posterior end of the trunk, are present. The skin is rough and greyish in colour and large number of warts are present on the skin of the dorsal surface. Two elevated ridges are present on the two sides of the head and posterior to these ridges on the dorsal surface of the anterior end of the trunk, two elevated longitudinal glandular structures, the **parotid glands**, are present. In general, the toad is very ugly.

Head

The head is more or less triangular with a nearly semicircular anterior border, drawn

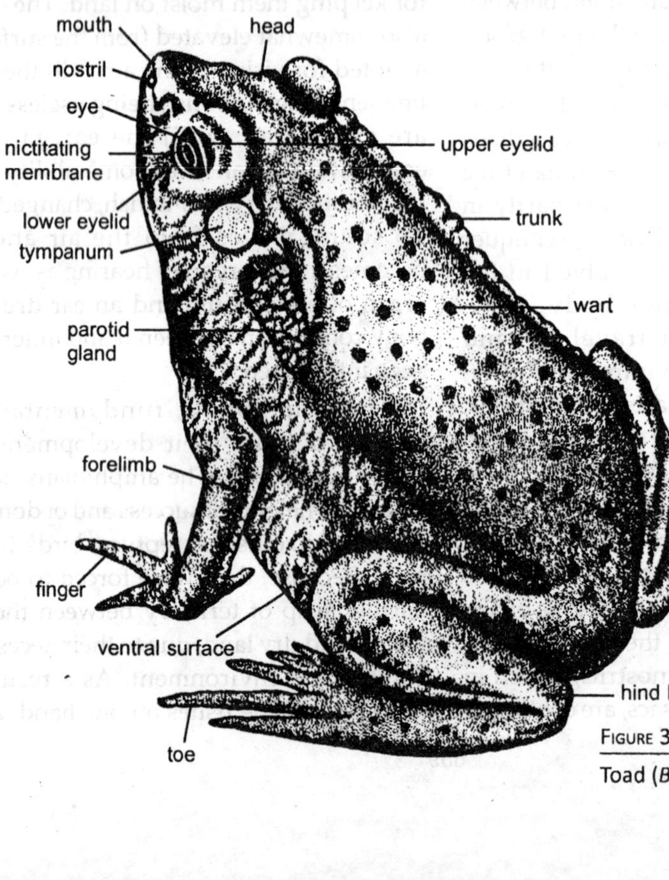

FIGURE 39.1

Toad (*Bufo melanostictus*)

into a blunt **snout**. The **mouth** is large and situated slightly ventral to the snout. The opening of the mouth is transverse, bounded by two jaws and extends on the sides behind the eyes. Two small apertures, the **nostrils**, one on each side, are present on the dorsal surface of the snout. The nostrils open into the buccal cavity and the internal openings are known as **internal nares**. The **eyes** are rounded, large, prominent and located on each side of the head. Each eye is protected by an immovable thick fold of skin on the upper surface, the **upper eyelid**; a movable **lower eyelid** and in continuation with the latter is a semitransparent, thin, membranous **nictitating membrane**, which can be drawn over the eyes. The nictitating membrane is also known as the **third eye-lid**. Just behind and close to each eye a circular area of stretched skin—the **tympanic membrane**—is present.

Trunk

Two pairs of limbs of unequal size are present on two ends of the trunk.

The anterior pair is known as forelimb. The **forelimb** is short and consists of an upper arm or **brachium**, a forearm or **antibrachium** and a hand. The hand is divisible into three parts—the **wrist (carpus)** or the portion next to the forearm, the **palm (manus)** or the broad middle portion and the **digits or fingers**. The digits are short, tapering, four in number and directed forwards. During breeding season, the cushion-like **thum pads** are well-developed at the basal part of two fingers (inner finger of each forelimb) in male.

The posterior pair of limbs, the hindlimbs are much larger in size. Each **hindlimb** consists of a **thigh** or **femur**, a **shank** or **crus** and a **foot**. The foot is divisible into three parts—an **ankle (tarsus)** or the portion next to the shank, the **instep (pes)** or the middle portion and the **digits** or **toes**. The digits are five in number, long and directed forward and united by thin folds of skin, **webs**. The **cloacal aperture** is situated at the posterior end of the trunk. The faeces, urine and the products of reproductive organs come out through this aperture. Sexual differences (**dimorphism**) is present. The male is provided with a **vocal sac** on the ventral surface of the buccal cavity, in addition to thumb pads.

Skin

The skin is water-tight, an adaptation to prevent dehydration or loss of water by evaporation, a characteristic of all land tetrapods.

The skin consists of two layers (Fig. 39.2) outer epidermis and inner dermis.

Epidermis

The epidermis consists of several layers of cells. The outermost layer, **stratum corneum**

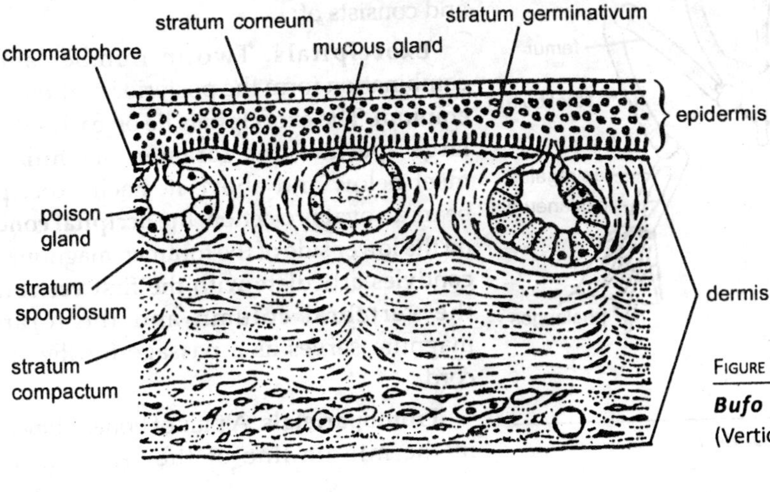

FIGURE 39.2

***Bufo* sp.** Integument (Vertical section)

is thin, scaly and made of dead, cornified cells. During ecdysis the cornified layer is shed in fragments. The deeper layers of cells constituting **stratum germinativum** of the epidermis consists of columnar cells with prominent nucleus.

Dermis

The dermis is composed of two layers—an outer loose layer, stratum spongiosum and an inner, more compact layer, stratum campactum. The outer part of **stratum spongiosum** is more compact than the rest. Blood vessels, lymph spaces, nerves and glands are abundant in stratum spongiosum. Unusually large number of blood vessels help the skin to serve as an important structure for respiration. The glands are of two types—**mucous glands** and **poison glands**, present in the region of warts. The **stratum compactum** is composed of dense

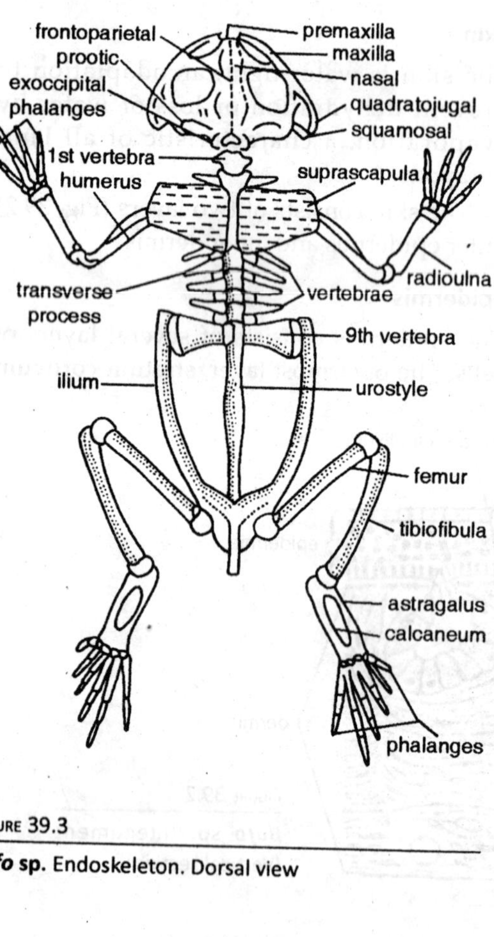

FIGURE 39.3

***Bufo* sp.** Endoskeleton. Dorsal view

connective tissue, smooth muscle fibres, nerves and blood vessels.

Chromatophores or pigment cells present, for the most part, between the epidermis and dermis, brings about change of colour of the integument to blend with the surroundings.

Skeletal System

Exoskeleton being absent, skeletal system is constituted by endoskeleton only (Fig. 39.3).

Endoskeleton

The endoskeleton consists of hard structures as bones and cartilages and functions to support the body and protect internal structures. The skeleton has two components—axial and appendicular skeleton.

Axial skeleton

The axial skeleton consists of the skull, hyoid apparatus and spinal column made of vertebrae lying in a line along the length of the body.

Skull

The skull is a bony structure (Fig. 39.4A, B) which gives the characteristic shape of the head. A cavity is present inside the skull in which the brain is lodged. The skull has three components—cranium, sense capsules and jaw apparatus.

Cranium

It forms the narrow central part of the skull and consists of:

Exoccipitals. Two in number and in combination form the posterior portion of the cranium. The exoccipitals lie on each side of a large aperture, the **foramen magnum** and meet below and above it. Each exoccipital bears an oval projection, the **occipital condyle** on the lower side of the foramen magnum. The condyles articulate with the first vertebra of the vertebral column and the foramen magnum serves as a passage for the spinal cord.

Frontoparietals. A pair of bones; they meet in the middle line of the cranium. Each frontoparietal is formed by the fusion of two bones—a **frontal** and a **parietal**—which remain distinct in early stage.

Parasphenoid. A large ⊥-shaped bone, forming the ventral surface of the cranium. The stem is more or less dagger-shaped and lines the cranium and the two arms extend laterally beneath the auditory capsules.

Sphenethmoid. It is a ventrally placed tubular bone forming the main framework of the skull. Anteriorly it is divided into two compartments by a vertical partition. Each chamber is communicated with the olfactory capsule of the side.

Sense capsules

Sense capsules are three pairs—the olfactory capsules, optic capsules and auditory capsules.

Olfactory capsules. The olfactory capsules are large, situated anteriorly, one on each side. A capsule consists of two paired bones, the dorsal one **nasal** and the ventral one **vomer.** The nasal and vomer in combination enclose a small piece of cartilage. Each nasal bone is triangular in shape and the two meet in the middle line of the skull. The nasal capsules are fused with the cranium.

Optic capsules. Paired cartilaginous chambers in which the eyes rest. Optic capsules do not fuse with the cranium.

Auditory capsules. Paired, abnormally large capsules lodging the auditory organs or internal ears. Each auditory capsule is provided with a bone, the **preotic**. In adult, the preotic fuses with the exoccipital and the basal portion is covered by the extension of the parasphenoid.

Jaw apparatus

The jaw apparatus consists of two components—maxillary arch forming the upper jaw and mandibular arch forming the lower jaw.

Maxillary arch. It is divisible into inner and outer regions. The inner region of each side again is divided into three regions.

a. **Posterior quadrate region or suspensorium.** A small piece of cartilage, the **quadrate**, on either side extends backward, outward and downward from the auditory capsule. It is connected with the pterygoid and ventrally furnish two articular facets, one for the lower jaw and the other for a T-shaped bone, the **squamosal**. The lower process of squamosal is attached with the angle of the mouth and the upper process with the auditory capsule.

b. **Intermediate pterygoid region.** It consists of a triradiate bone, the **pterygoid**, on each side. The anterior end of pterygoid joins with the palatine, the inner limb with the auditory capsule and the outer limb extends to the angle of the mouth.

c. **Anterior palatine region.** A thin rod-like bone, **palatine**, on each side extending outwards from the ventral side of the anterior end of the cranium constitute palatine region.

The outer region of each side can also be divided into three regions.

i. **Premaxillary region.** A narrow, small bone, the **premaxilla**, is placed anteriorly, on each side of the median line, form the premaxillary region.

ii. **Maxillary region.** It consists of a long narrow bone, the **maxilla**, posterior to the premaxilla on each side.

iii. **Quadratojugal region.** It is formed by a small, narrow, posteriorly placed bone, the **quadratojugal**, behind the maxilla on either side.

Mandibular arch. It forms the lower jaw and consists of two identical halves united in front (Figs. 39.4). Each half is made of a core of cartilage, the **Meckel's cartilage**, the outer side of which is formed by two bones—a short one, the **dentary** and a long one, the **angulosplenial.** The dentary extends more

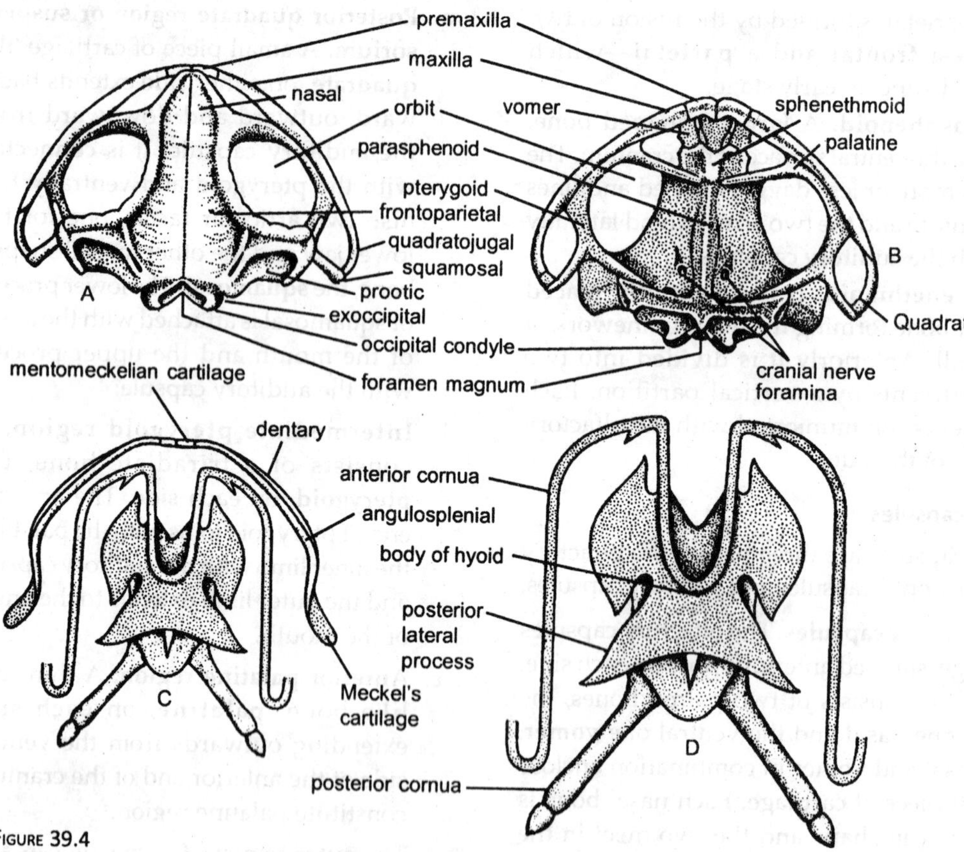

FIGURE 39.4

***Bufo* sp.** Skull, mandible and hyoid apparatus. A. Skull (dorsal view). B. Skull (Ventral view). C. Mandible and hyoid apparatus. D. Hyoid apparatus

than half the length and is situated on the outer face and the angulosplenial on the inner face. The distal end of the Meckel's cartilage is ossified into a small bone, the **mentomeckelian.**

⊠ Hyoid apparatus

The main body of hyoid apparatus is situated in the floor of the buccal cavity. It consists of a shield-shaped, cartilaginous plate (Figs. 39.4 C, D), known as the body of the hyoid. The body is produced at its anterior angles into two slender, curved rods, the **anterior cornua,** which fuse with auditory capsules. The posterior angles of the body are produced into a pair of diverging, partly ossified rods, the **posterior cornua,** extending backward, one on either side and embrace the **glottis.** The glottis is an aperture leading from the buccal cavity to the lungs.

⊠ Vertebral column

The vertebral column is short and consists of nine **vertebrae** anteriorly and a slender, long rod, the **urostyle** posteriorly (Fig. 39.5). The structures of the second to the eighth vertebrae are similar and they are known as typical vertebrae.

Structure of a typical vertebra

The body or the **centrum** is ventrally placed and concave at the anterior and convex at the posterior face. An arch, **neural arch,** is placed on the centrum dorsally, which forms a ring enclosing the spinal cord. A ridge—the **neural spine**—is present on the roof of the arch. Two

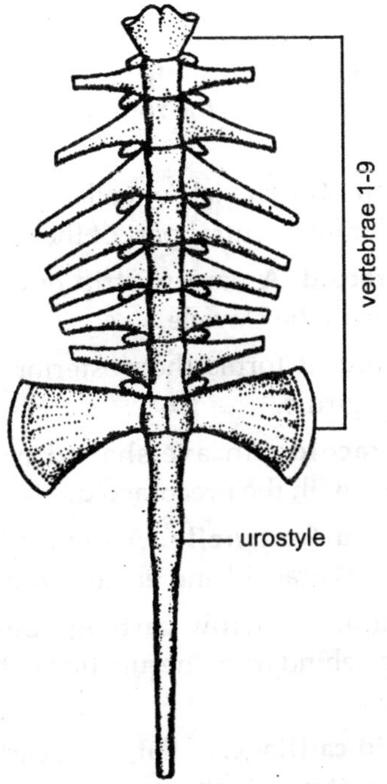

FIGURE 39.5

***Bufo* sp.** Vertebral column (Ventral view)

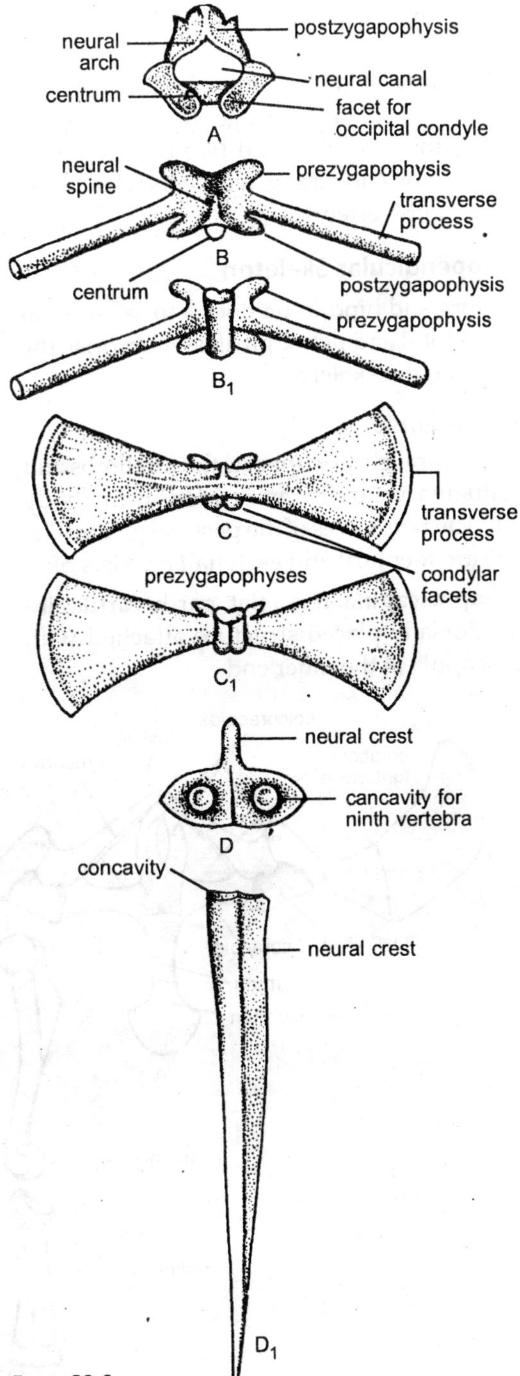

FIGURE 39.6

***Bufo* sp.** Vertebrae and Urostyle. A. Atlas (first vertebra), anterior end on view, B-B_1. Typical vertebra. B. Dorsal view, B_1. Ventral view, C-C_1. Ninth vertebra, C. Dorsal view, C_1. Ventral view, D-D_1 Urostyle, D. Anterior end on view, D_1. Lateral view

lateral processes, **transverse processes**, spring from the sides of the arch. Articulating processes or **zygapophyses** are present on the lateral sides of both the anterior and posterior faces of the arch. The anterior pair is **prezygapophyses** and the posterior pair **postzygapophyses.** Each vertebra articulates with its anterior and posterior vertebrae by the centra and articular processes (Figs. 39.6B, B_1). This type of vertebra with an anterior concave and posterior convex face of the centrum is called **procoelous**.

First vertebra. The first vertebra or **atlas** is a ring-shaped structure (Fig. 39.6A). The centrum is small and without any transverse process; prezygapophyses absent. At the junctions of the centrum and neural arch two oval concave **facets** for articulation with the condyles of the skull are present.

Ninth vertebra. The transverse processes are large, wide and flattened (Figs. 39.6C, C_1). The postzygapophyses are absent but the centrum bears two **condylar facets** which fit into the two concavities at the anterior end of the urostyle.

Urostyle. A long slender bony rod (Figs. 39.6D, D_1) with a low ridge, the **neural crest** on the dorsal surface and two concavities at the anterior end fitting with condylar facets of the ninth vertebra.

Appendicular Skeleton

The fore and hindlimb bones—together with the pectoral and pelvic girdles—comprise the appendicular skeleton.

Pectoral girdle

The pectoral girdle supports the forelimbs and is situated ventrally, embedded in the body wall. The two halves of the pectoral girdle (Fig. 39.7) are identical and each half consists of:

Suprascapula. It is a flat, partly cartilaginous, dorsally placed structure attached with the scapula at the inner end.

Scapula. A bone ventral to suprascapula and provided with a cup-shaped **glenoid cavity**, into which fits the rounded head of the humerus.

Clavicle. The clavicle or collar bone forms the anterior and ventral part of the girdle.

Precoracoid. A narrow strip of cartilage, lying beneath the clavicle.

Coracoid. It forms the posterior ventral part of the girdle.

Epicoracoid. An axe-shaped cartilage continuous with the precoracoid.

Coracoid fontanelle. An empty space between the coracoid and epicoracoid.

Sternum. A narrow cartilaginous plate extending behind from the junction of the two coracoids.

Xiphoid cartilage. A flat, thin, cartilaginous plate posterior to the sternum.

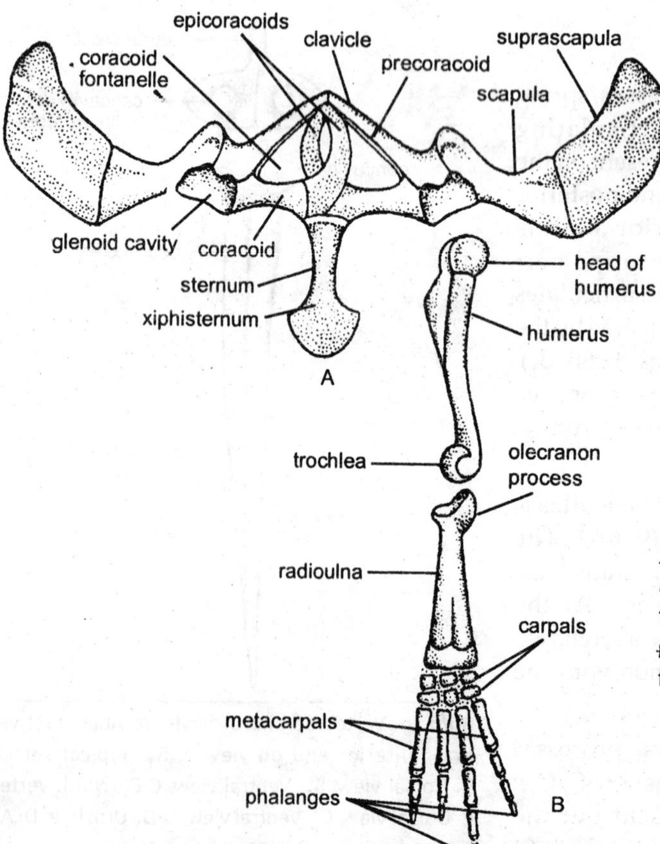

FIGURE 39.7

***Bufo* sp.** Pectoral girdle and forelimb bones. A. Pectoral girdle (Ventral view), B. Left forelimb bones

Pelvic girdle

The pelvic girdle supports the hindlimbs and is situated dorsally at the posterior region of the trunk. The two halves of the pelvic girdle (Fig. 39.8) are identical and united together

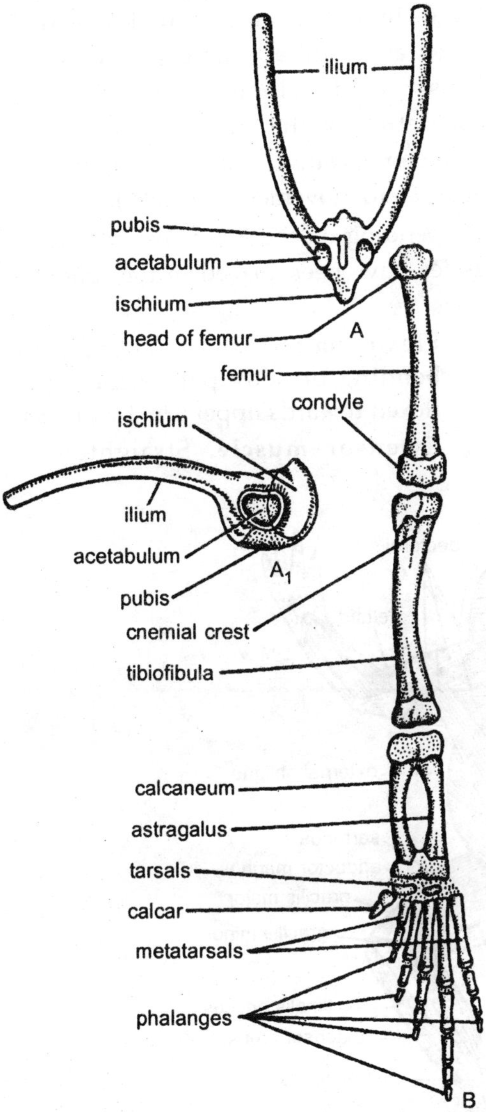

FIGURE 39.8

***Bufo* sp.** Pelvic girdle and hindlimb bones. A. Pectoral girdle (Ventral view). A_1. Pectoral girdle (lateral view). B. Left hindlimb bones

firmly in the middle line forming a disc-like structure beyond the urostyle. Each half consists of:

Ilium. A long bone, articulates anteriorly with the transverse process of the ninth vertebra.

Ischium. It is a small bone making up the posterior ventral portion of the disc.

Pubis. A cartilage comprising the anterior ventral part of the girdle, lying between the ischium and ilium.

Acetabulum. It is a socket at the junction of the above three bones. The head of the femur or thigh bone fits into this socket.

Forelimb

The skeleton of forelimb (Fig. 39.7) consists of several pieces of bones:

Hemerus. The bone of the upper arm. It is a short, rounded bone with a small crest, the **deltoid ridge** on the front and two knob-like structures, one at each end. The knob at the upper end or the **head** fits into the glenoid cavity of the pectoral girdle. The knob at the lower end is the **trochlea**, and articulates with the bones of the forearm.

Radioulna. The forearm bone is a composite one. It is formed by the fusion of two separate bones, **radius** and **ulna**. The bone bears a projection at the upper end, the **olecranon process**, with a cup-like depression to receive the trochlear end of the humerus.

Carpal bones. The bones of the wrist or carpus are carpal bones and consist of six small pieces arranged in two rows of three.

Metacarpal bones. The bones of the palm or manus are metacarpal bones and consist of four long pieces.

Phalanges. The bones of the digits are phalanges. The digits are four in number and the number of segments in each phalange from inner to outer are 2, 2, 3 and 3.

Hindlimb

The skeleton of the hindlimb (Fig. 39.8) consists of following bones:

Femur. The bone of the thigh—the femur—is a long bone with a knob-like **head** at the upper end which fits into the acetabulum of the pelvic girdle and a process at the opposite end to articulate with the shank bone.

Tibiofibula. The bone of the shank or crus is a composite one. It is formed by the fusion of two separate bones—**tibia** and **fibula**. The **cnemial crest**, a slender ridge, is present on the front side of the tibial portion.

Tarsal bones. The bones of the tarsus or ankel are four in number. Next to tibiofibula corresponding with them are two elongated bones, the **astragalus** and **calcaneum** which are fused at either end by cartilage. The second row of the ankle bones are two and much reduced.

Metatarsal bones. The bones of the foot consist of five long pieces:

Calcar. It is a small structure of two cartilaginous pieces situated on the inner side of the first toe.

Phalanges. The bones of the digits are phalanges. The digits are five in number and the number of segments in each phalange from inner to outer are 2, 2, 3, 4 and 3.

⊠ Muscular System

All movements of the body or parts of the body are brought about by muscles (p. 61). In most tetrapods the appendicular muscles are extensively developed, while the axial musculature of the trunk are reduced in volume and relatively of minor significance in relation to movement (Fig. 39.9).

Depending on the functions, the skeletal muscles have been placed in the following groups:

1. **Flexor muscle.** Associated with bending of body parts—forearm is flexed towards upper arm by **biceps**.
2. **Extensor muscle.** Straightens or

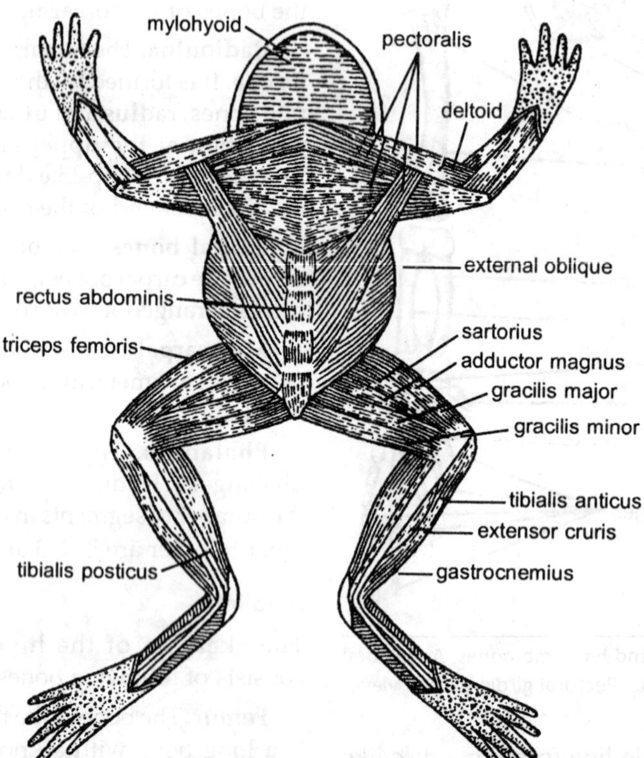

FIGURE 39.9

***Bufo* sp.** Important skeletal muscles. Muscles viewed from the ventral aspect

extends a part—forearm extended away from upper arm by **triceps**.

3. **Abductor muscle.** Draws a part of the body or of a limb away from the axis of the body—arm drawn forward by **deltoid muscle**.
4. **Adductor muscle.** Draws a part of the body or of a limb towards the axis of the body—arm drawn up and moved back by **latissimus dorsi**.
5. **Depressor muscle.** Lowers a part of the body—lower jaw moved down to open the mouth by **depressor mandibular**.
6. **Levator muscle.** Raises a part of the body—lower jaw moved upward to close the mouth by **masseter muscle**.
7. **Rotator muscle.** A part of the body is rotated—femur is raised and rotated by **pyriform muscle**.

☒ Locomotion

Locomotion or change of place takes place in three ways—walking, jumping and swimming.

Walking

In walking movement forelimb of one side is lifted and moved to a forward position, which is followed by forelimb of the other side. This step is followed by forward movement of the hindlimbs in a similar way. Toad can walk only slowly.

Jumping

It is the most active movement. In a sitting posture, the anterior end of the body is supported by forelimbs and directed upward at about 45° angle. The hindlimbs are much long, remain folded in the shape of 'Z' and support maximum weight of the body. During jumping, the toad suddenly extends its hindlimbs and a thrust against ground helps it to move upward and forward. The forelimbs possibly manipulate the direction of jump. All the four limbs touch the ground at a time in each jump.

☒ Swimming

All the four limbs act as propellers during swimming, but the major role is played by the hindlimbs due to their length and webs between long toes.

☒ Coelom

The coelom is a large space enclosed within the body in the trunk region. It is lined by **peritoneum**, a thin membrane of mesodermal origin. The coelom is divided into two compartments—an anterior, small **pericardial cavity** containing heart, lined by a detached portion of the peritoneum, the **pericardium**, and a large posterior **pleuroperitoneal** or **perivisceral cavity**, accommodating **viscera** or visceral organs. The peritoneum leaves the dorsal wall, forms a double membrane, the **mesentery**, which partly or fully encloses visceral organs, viz., alimentary canal, liver, pancreas, lungs, blood and lymph vessels, excretory and reproductive organs. The organs are kept in position by the mesentery.

☒ Digestive System

The digestive system consists of an alimentary canal and its appendages, the digestive glands (Fig. 39.10B). Digestion and absorption of food take place within the canal.

Alimentary canal

The digestive tract can be divided into the following zones:

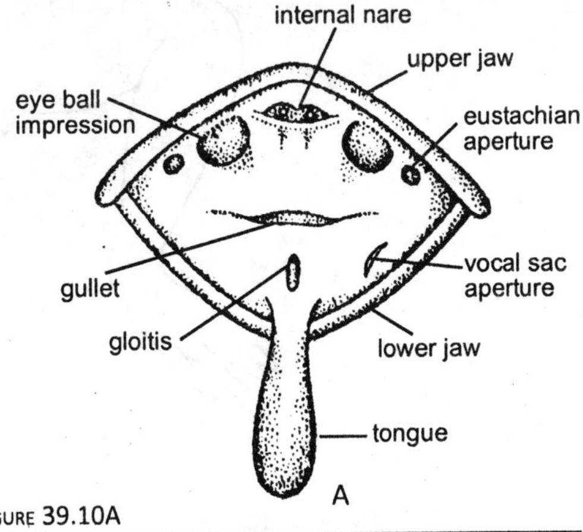

FIGURE 39.10A

***Bufo* sp.** Digestive system. Buccal cavity (male)

Buccal Cavity.

The mouth is bounded by two toothless jaws—the upper and lower. The mouth leads into a wide buccal cavity.

The structures (Fig. 39.10A) present in the buccal cavity are:

a. Two apertures—**internal nares**—the openings of the nostrils on the roof of the cavity towards anterior end.

b. A pair of projections on the roof, one on each side, due to downward bulging of the eyes.

c. Two apertures—the **eustachian apertures**—at the angle of the jaws, one on each side. The eustachian tube is in communication with the tympanic cavity and mouth cavity.

d. A large fleshy **tongue** on the floor. The tongue is attached with the front end, the posterior rounded end remaining free.

e. In male toad, an additional structure, the opening of the vocal sac, is present on the floor towards left side. The **vocal**

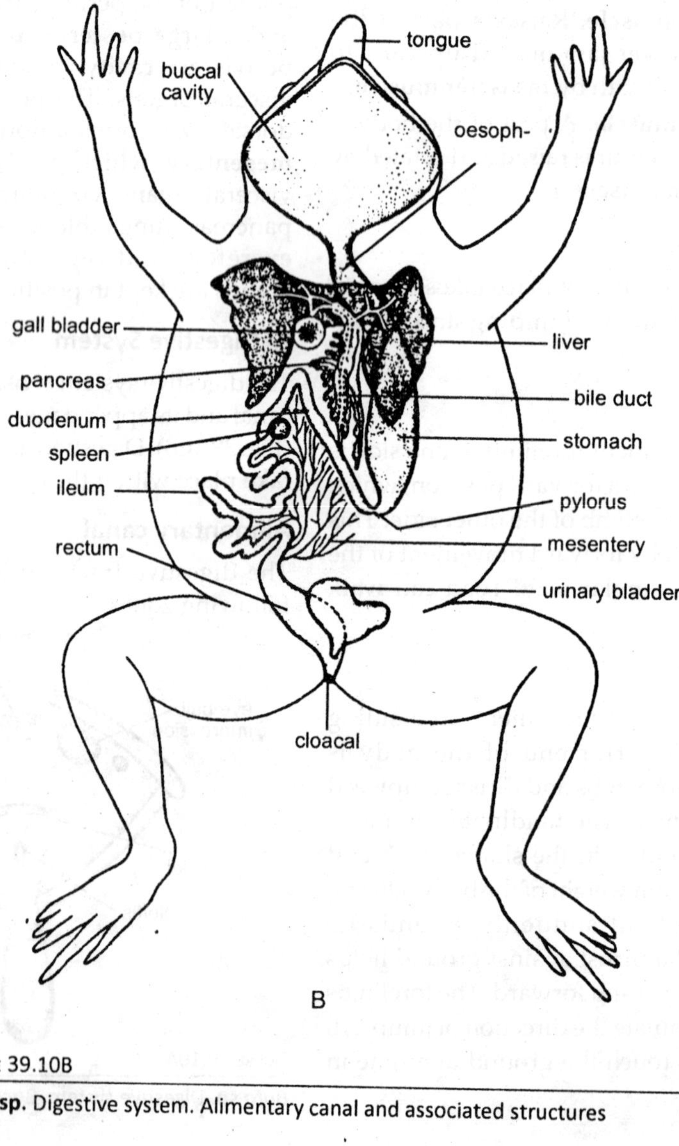

FIGURE 39.10B

***Bufo* sp.** Digestive system. Alimentary canal and associated structures

sac is black-pigmented and situated beneath the floor of the mouth cavity, just covered by the skin.

f. A slit-like aperture, the **glottis**, just behind the tongue. Glottis communicates the mouth cavity with the lungs.

Phaynx

The buccal cavity narrows posteriorly, and this region is known as pharynx.

Oesophagus

The pharynx leads posteriorly to a short, narrow tube, the oesophagus or gullet. The oesophagus opens into a sac-like stomach.

Stomach

The stomach is divisible into an anterior wide chamber towards the oesophagus, the **cardiac stomach** and a posterior narrow portion, the **pyloric stomach**.

Small intestine

The pyloric stomach leads to a narrow tube, the small intestine. The junction of the stomach and small intestine is marked by a sharp constriction. Circular sphincter muscles forming the **pyloric valve** at the constriction regulate the flow of food from the stomach. The first portion of the small intestine lying parallel to the stomach is **duodenum** and the rest long, slender, much coiled portion is **ileum**.

Rectum

The small intestine continues posteriorly as a wide tube, the **rectum** or large intestine. The rectum opens posteriorly into a short chamber, the **cloaca***, which opens to the exterior by an aperture, the **cloacal aperture**.

☒ Digestive glands

Liver and pancreas are digestive glands.

Liver

The liver is a large, reddish-brown, bilobed mass. The left lobe is larger than the right one and sublobed. The two main lobes are connected by a narrow, short, median portion. Between the two lobes, median in position, lies a small, thin-walled, greenish, rounded sac, the **gall bladder**. The secretion of the liver, the **bile**, is stored within the gall bladder. Fine ducts from the liver, the **hepatic ducts**, receive bile from the liver. Some of the hepatic ducts enter the gall bladder and the others unite to form a duct, the bile duct. **Cystic duct** springs from the gall bladder and joins the bile duct. The duct passes through the substance of the pancreas and in its way receives ducts from the latter. The duct, which is now called **common bile duct**, receives secrections from both the liver and the pancreas and opens into the duodenum.

Pancreas

The pancreas is a large, creamy-white, irregular gland, surrounding the bile duct, into which it pours its secretion, the **pancreatic juice**. It lies between the duodenum and the stomach.

☒ Feeding and Digestion

The adult toad is carnivorous and lives on various types of animals, both large and small. Digestion is extracellular.

Food capture

Large-sized animals are captured by the jaws. To capture small animals, like insects the tongue is used. When any food organism comes near a toad, it suddenly throws out the free end of its sticky tongue on the prey. The prey gets attached to the tongue and is taken into the buccal cavity (Fig. 39.11).

---

* Cloaca is a chamber situated in the posterior end of the abdominal cavity. It receives rectum, urinary bladder, urinary and genital ducts. The faecal matters, nitrogenous wastes and reproductive units (sperms or ova) are discharged into the cloaca. These are finally thrown to the exterior through cloacal aperture.

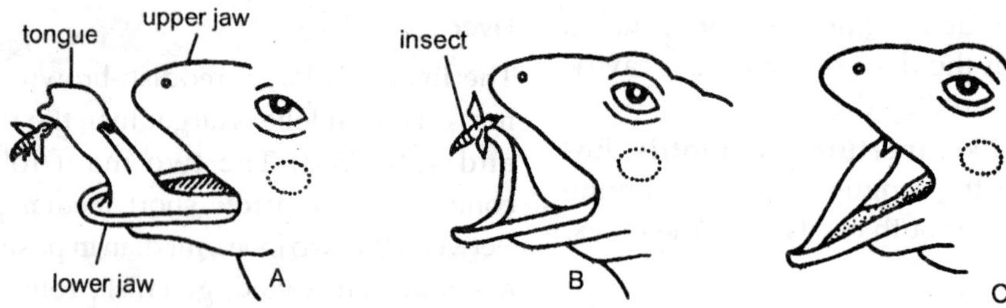

FIGURE 39.11

***Bufo* sp.** Food capture (successive stages)

⊠ Digestion

The toad is toothless and cannot chew the food. The food taken into the mouth cavity is swallowed as a whole and passes down to the stomach via the oesophagus. Digestion begins in stomach. The purpose of digestion is to break down complex food substances like proteins, fats and carbohydrates into simpler compounds (p. 86) so that the cells of the alimentary tract may absorb them.

The breaking up of various food substances is done by specific ferments known as **enzymes**, which are secreted by various glands. Protein is partly broken down in the stomach into peptone by pepsin, in acid medium.

The partly digested food in the stomach passes to the duodenum through the pyloric valve, and is further digested by the action of pancreatic juice, which also neutralises stomach acid. Bile helps in emulsification of lipids. After the completion of digestion, the nutritive substances which are now in a liquid state are absorbed by the intestinal cells and passed to the blood. The undigested portion of the food, the excreta, is stored temporarily in the rectum, and finally thrown out to the exterior through the cloaca.

The internal lining of the digestive tract is different in structures in different regions, and is thrown into folds. The folds are irregular in the stomach, transverse and few in the duodenum, longitudinal and numerous in ileum. The purpose of the folds is to increase the surface area. It is maximum in intestine, the chief site for absorption of nutrients.

The whole alimentary tract is slung to the dorsal body wall by sheets of tissues, the mesenteries.

⊠ Respiratory system

Respiration is a process in which a gaseous exchange takes place between the organism and the surrounding medium. The different modes of respiration in a toad can be placed under two heads—adult respiration and larval respiration.

A. Adult respiration

An adult toad respires by three means:

Pulmonary respiration

From the evolutionary point of view, this mode of respiration in the adult is a step further towards terrestiral adaptation. Pulmonary respiration involves three stages—aspiration, expiration and inspiration (Fig. 39.12).

Aspiration. During aspiration, the floor of the buccal cavity is lowered by the contraction of the muscles and the cavity is increased. The mouth and glottis remain closed but the nostrils remain open. Due to fall of pressure in the buccal cavity, air rushes in through the open nares.

Expiration. Expiration follows aspiration. The trunk muscles contract, the elastic walls of the lung recoil, the glottis opens and the air pumped into the lungs in previous inspiration, which is now rich in carbon dioxide, is forced out to the buccal cavity. The buccal cavity

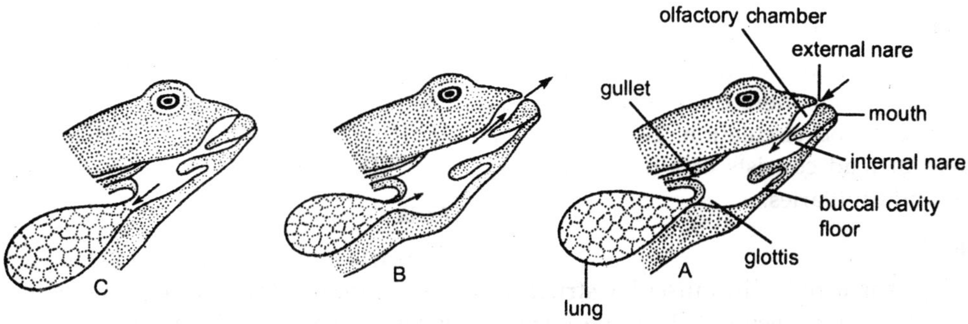

FIGURE 39.12

***Bufo* sp.** Pulmonary respiration. A. Aspiration. B. Expiration. C. Inspiration

already full with atmospheric air during aspiration, becomes very much distended and a part of the mixed air of the buccal cavity is forced out through the nares, which open by this time.

Inspiration. It follows expiration. The nares are closed, the floor of the buccal cavity raised and the mixed air in the buccal cavity is forced into the lungs through glottis. The lungs become distended. The alveoli of the lungs are richly supplied with blood by numerous capillaries and a gaseous exchange takes place in the distended lungs, in which oxygen of the air is absorbed by blood through capillaries and carbon dioxide of the blood is released in the pulmonary cavity.

Since the lungs receive mixed air, in which oxygen concentration is less than that in the atmospheric air, pulmonary respiration is not that effective as in higher tetrapods. Pulmonary respiration, however, is not the usual mode of adult respiration. Whenever extra energy is required, that is made for by pulmonary respiration.

Buccopharyngeal respiration

The whole air in the buccal cavity is never expelled during pulmonary respiration. The walls of the buccal cavity and the pharynx are highly vascular and during pulmonary respiration gaseous exchange takes place to a certain extent in this region. In normal phase of life, this type of respiration is sufficient to meet the demand.

Cutaneous respiration

The skin is highly vascular and a gaseous exchange takes place in the skin. Under normal condition, this exchange is very slight. During hibernation, this is considered as the primary mode of respiration*.

B. Larval respiration

A tadpole larva respires by three means:

External gill respiration

Three pairs of branched filamentous gills are present by the sides of the head of the tadpole. These gills are highly vascular and can absorb oxygen dissolved in water.

Internal gill respiration

At late tadpole stage, gill slits appear in the pharyngeal wall and outgrowths develop on the walls of the slits. These are internal gills and being richly supplied with blood, gaseous exchange takes place in the gills.

Cutaneous respiration

The skin of tadpole larva is highly vascular and capable of absorbing oxygen dissolved in water.

* In frogs, the skin is moist, and cutaneous respiration occurs there. In toads, the skin is dry and it is doubtful if dry skin can act as a respiratory organ, however vascular it may be.

⊠ Blood Vascular System

The flow of blood from heart to the body and lungs and from there back to the heart is known as blood circulation. The circulatory system of toad consists of a heart, arteries, veins and capillaries.

Heart

The heart is a reddish, muscular structure (Fig. 39.13) situated ventrally in the pericardial chamber of the coelom. It is enclosed in a thin, semitransparent membrane, the **pericardium**. The heart consists of five components—a sinus venosus, two atria—the right and left, a ventricle and a conus arteriosus. Such a heart is called three-chambered* heart.

Sinus venosus. A dorsally placed, thin-walled, large triangular sac opening into the right atrium by a **sinuatrial aperture**. The aperture is guarded by a pair of membranous **sinuatrial valves**. Sinus venosus receives three veins. Deoxygenated blood from sinus venosus flows into the right atrium through sinuatrial aperture.

Atria

The atria are two thin-walled chambers situated in front of the ventricle. The right atrium is larger than the left and the two are separated by a partition, **interatrial septum**. The two atria open posteriorly into the single ventricle by a common **atrioventricular aperture** guarded by membranous **atrioventricular valves**, the free ends of which are attached to the wall of the ventricle by fine thread-like **chorda tendineae**. The valves regulate blood flow and allow passage only from atria to ventricle. The pulmonary veins open into the left

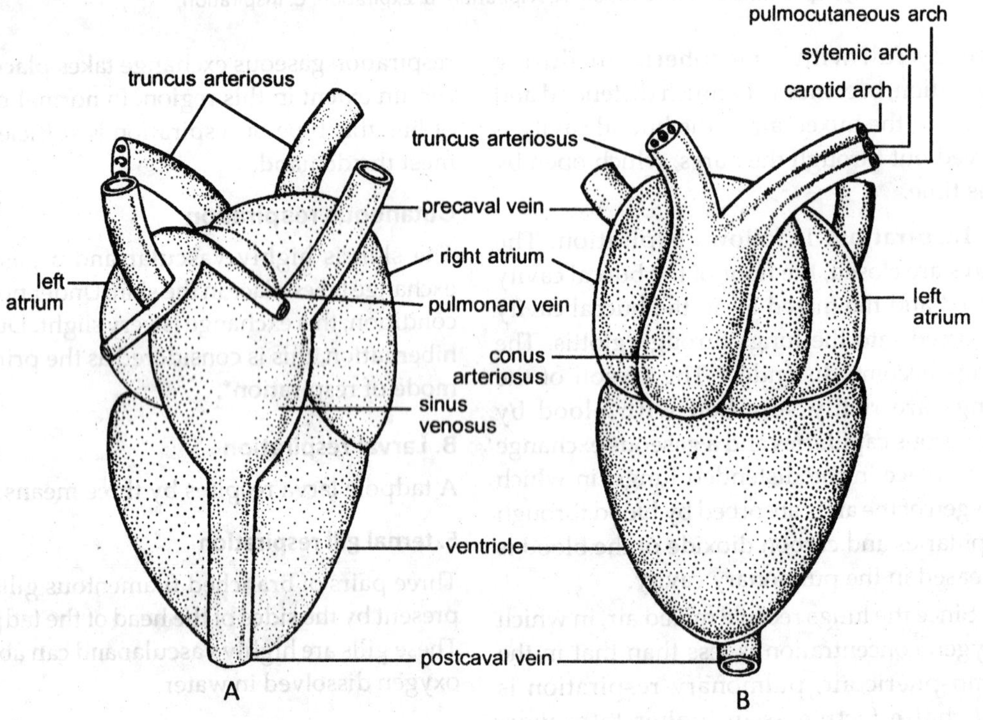

FIGURE 39.13
***Bufo* sp.** Heart. A. Dorsal view. B. Ventral view

* Sinus venosus and conus arteriosus are not considered as chambers of the heart.

atrium. A transverse groove, the **coronary sulcus** marks the separation of atria and the ventricle.

Ventricle

The ventricle is a conical structure with thick muscular walls and the internal wall is provided with ridges of strong interlacing muscles, **columnae carnae.**

Conus arteriosus. Very short, springs from the right side of the base of the ventricle and separated from the latter by three small semilunar valves. The conus proceeds forward obliquely as **truncus arteriosus**. The truncus is not a part of the heart. A twisted longitudinal **spiral valve** divides the cavity of the conus into a right **cavum aorticum** and a left **cavum pulmocutaneum**. The truncus, which is the basal stem of the three main arteries, divides into three vessels by two longitudinal partitions. The vessels are—an inner **carotid arch**, carrying blood to the anterior region of the body, a middle **systemic arch** carrying blood to the middle and posterior region and an outer **pulmocutaneous arch** carrying blood to the lungs and skin.

Circulation through heart

The heart is a sort of machine, which receives from and the drives blood to different parts of the body (Fig. 39.14). The sinus venosus receives blood through three main systemic veins. In the first step, the sinus venosus contracts, driving blood into the right atrium. By this time the left atrium is filled with blood coming from the lungs through pulmonary vein. Flow of blood back in the veins is prevented by valves at the apertures. The two atria contract simultaneously and blood in them is driven into the ventricle. The sinus begins to relax but the backflow of blood from atrium is prevented by the sinuatrial valves.

The right side of the ventricular cavity receives deoxygenated blood from the right atrium and the left side receives oxygenated blood from left atrium. The ventricle contracts and, the passage between atria and ventricle now being blocked by valves, the blood is forced into the conus arteriosus. The deoxygenated and oxygenated blood received by the ventricle from the right and left atria begin to get mixed in the middle region. The nature of blood inside the ventricle is deoxygenated in the right portion, mixed in the central region, and oxygenated in the left side portion*. Conus, being placed on the right side of the ventricle, first receives deoxygenated blood from the right side of the ventricle, then the somewhat mixed blood and finally the oxygenated blood.

The blood proceeds from conus to truncus arteriosus. The truncus, which is divided into three arches, offer different resistances in different arches. The resistance in the way to pulmocutaneous being least, the first portion of the blood, i.e., the most deoxygenated blood, is received by the pulmocutaneous arch.

The next in resistance is the systemic arch, which receives the mixed blood, and the carotid arch which offers highest resistance receives the last portion of blood, i.e., the most oxygenated blood. When it is emptied, the ventricle relaxes but the backflow of blood from conus to ventricle is prevented by three semilunar valves.

The contraction of the ventricle is immediately followed by the contraction of the conus and the latter helps the ventricle to send blood into different arches.

The heart receives blood from the veins and pumps out the same to the arteries, thus acting as a force pump. The contraction and dilation together is known as **heart-beat.**

---

* It has been suggested that oxygenated and deoxygenated blood are not separated in the ventricle. The skin being an effective respiratory organ, the right atrium receives a good amount of oxygenated blood (Foxon, 1995).

☒ Arterial system

all vessels carrying blood away from heart are **arteries**. Arteries are thick-walled vessels and appear red due to the oxygenated blood they contain. The arteries and the heart collectively form arterial system (Fig. 39.15).

The conus arteriosus springing from the right side of the ventricle continues forward as truncus arteriosus. The truncus divides into two, and each branch again divides into three to form the three main arterial arches—the anterior carotid arch, the middle systemic arch and the posterior pulmocutaneous arch (Fig. 39.14).

Carotid arches

Carotid arches are two. At a short distance from its origin each arch forms a swelling, the **carotid labyrinth**. At the point of labyrinth the carotid divides into two, the external and internal carotid, the base of the latter being formed by the labyrinth. Inside the labyrinth, the carotid arch breaks up into capillaries which reunite at the other end. It is a device to regulate blood pressure in internal carotid artery.

External carotid. Supplies blood to the outer side of the head, thoracic musculature, throat and tongue.

Internal carotid. Sends blood to the internal part of the head, meninges and brain. In its backward course, the internal carotid comes very close to the systemic arch to which it is joined by a fibrous **carotid ligament**.

Systemic arches

The systemic arches are two. Each arch passes round the oesophagus and meets its fellow from the other side, dorsal to the oesophagus to form a median vessel, the **dorsal aorta**. The dorsal aorta runs posteriorly between the two kidneys and ventral to the vertebral column.

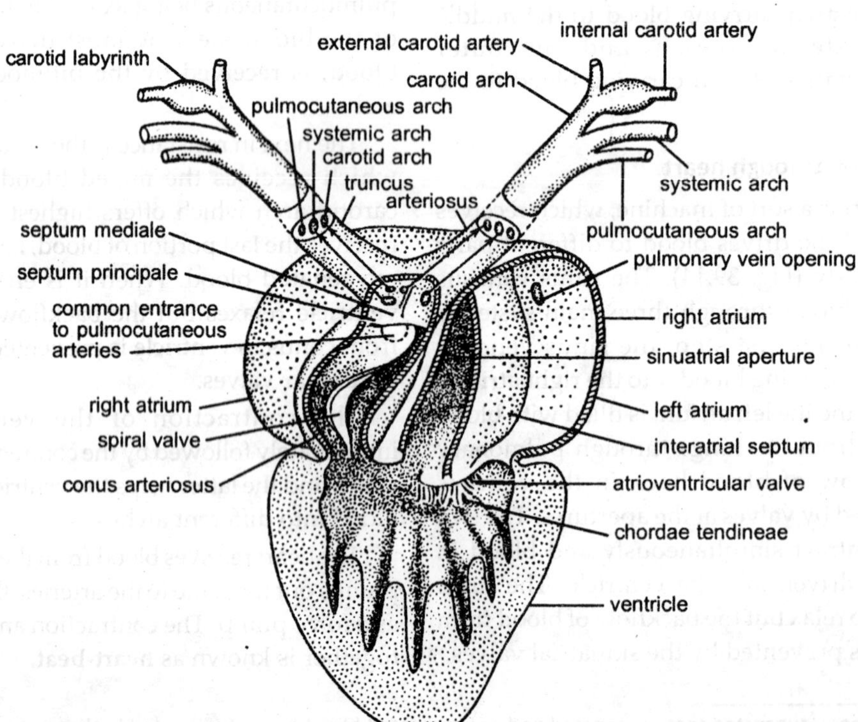

FIGURE 39.14

***Bufo* sp.** Heart. Chambers exposed (semidiagrammatic) and arterial arches

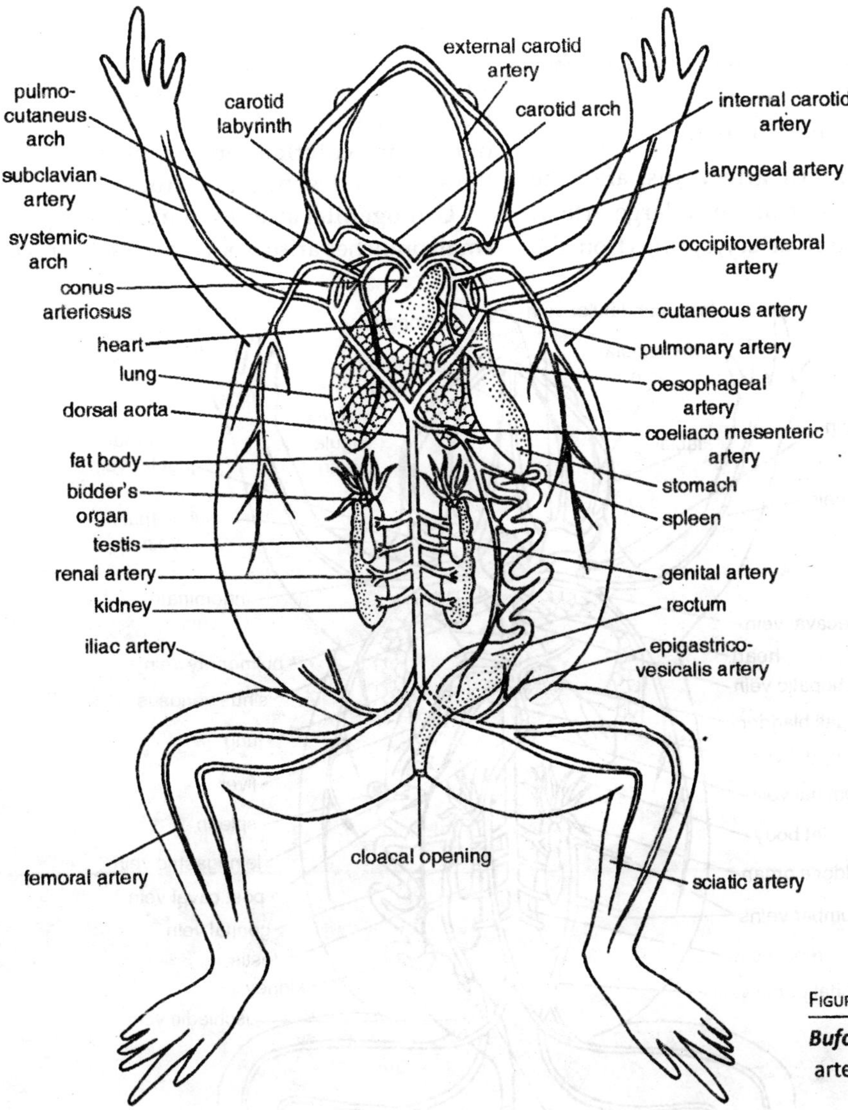

FIGURE 39.15

***Bufo* sp.** Heart and arterial system. Male

Branches of the systemic arch

Following branches are given off by each systemic arch:

a. **Laryngeal artery.** A short vessel springs at a short distance from the origin of the systemic arch and supplies blood to the larynx.

b. **Occipitovertebral artery.** It springs from about the middle of the arch. Sending a branch to the pharyngeal region, it divides into two, one anterior and another posterior. The anterior one supplies blood to the posterior region of the skull and the posterior one further subdivides and sends blood to the vertebral column.

c. **Subclavian artery.** Arises slightly posterior to occipitovertebral and supplies blood to the forelimb.

d. **Oesophageal artery.** It arises from the left systemic arch, slightly anterior to the union of arches, and supplies blood to the oesophagus.

Dorsal aorta

The dorsal aorta sends a number of branches supplying blood to the structures in the middle and posterior region of the body.

Coeliacomesenteric artery. Just after the union of two systemic arches a large artery, the coeliacomesenteric artery, springs from the left side of the dorsal aorta. It divides into two—the anterior branch, **coeliac artery**, supplying blood to the stomach, pancreas, liver and gall bladder and the posterior branch, **mesenteric artery**, supplying blood to the spleen, intestine and rectum.

Urinogenital arteries. A number of paired small branches arise from the dorsal aorta and

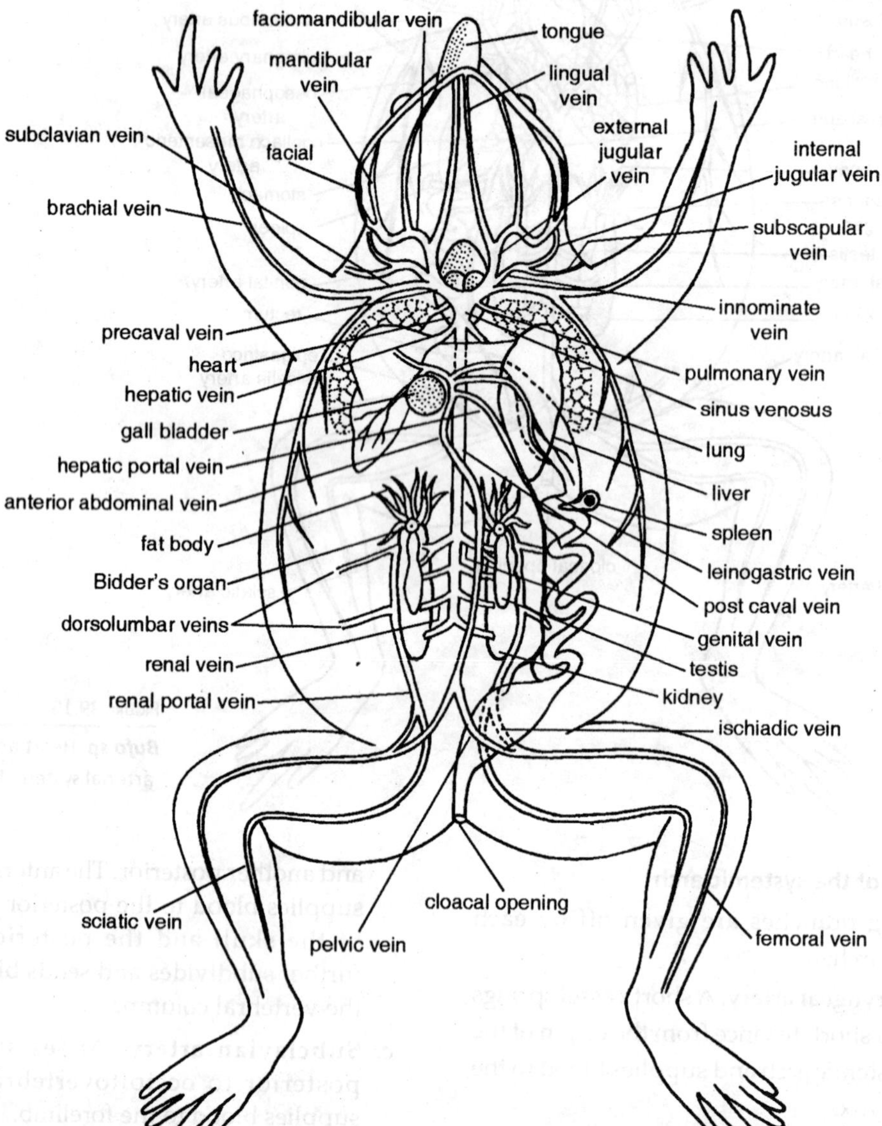

FIGURE 39.16

***Bufo* sp.** Heart and venous system. Male (Lungs displaced, heart overturned)

supply blood to the kidneys and reproductive organs.

The dorsal aorta runs posteriorly and finally divides into two **iliac arteries**. Each iliac sends two branches—the **epigastric artery** supplies blood to the ventral body wall, and the **femoral artery** supplies the skin and thigh muscles. Sending these two branches the iliac continues as **sciatic artery** and supplies blood to the hindlimb.

Pulmocutaneous arches

Each arch divides into two branches. Each branch subdivides into:

a. **Cutaneous artery** supplies blood to skin of the back.
b. **Pulmonary artery** divides repeatedly and supplies blood to the lung. Unlike other arteries, the pulmonary arteries carry deoxygenated blood from the heart to lungs.

☒ Blood

The blood is red. The red blood corpuscles are oval, biconvex and nucleated and the white blood corpuscles are irregular in shape and nucleated (p. 63).

Venous system

All vessels carrying blood towards the heart are **veins**. Veins are thin-walled vessels and appear bluish due to the deoxygenated blood they contain. The veins collectively constitute the venous system. With reference to the region of the body from where the veins bring back blood, the venous system is divided into four main sections—systemic, hepatic portal, renal portal and pulmonary (Fig. 39.16).

Systemic system

The systemic system consists of two anterior **precavals** and one large, median, medioposterior **postcaval** vein. The sinus venosus receives deoxygenated blood from the anterior portion of the body through the precavals and from the median and posterior part through the postcaval.

Precaval vein

Each precaval is identical with the other, receives blood from three branches—external jugular, innominate and subclavian, each of which, in turn, receives two vessels.

a. **External jugular** receives blood from:
   i. **Lingual** from the head region.
   ii. **Faciomandibular** from the snout and lower jaw.
b. **Innominate** receives blood from:
   i. **Internal jugular** from the head region.
   ii. **Subscapular** from the shoulder.
c. **Subclavian** receives blood from:
   i. **Brachial** from the forelimb.
   ii. **Musculocutaneous** from the muscles and the skin.

Postcaval vein

It springs at the level of the kidneys, runs forward and receives blood from the following vessels:

a. **Renal veins** from the kidneys, four or five pairs of vessels.
b. **Genital veins.** A pair from the reproductive organs—testes or ovaries.
c. **Hepatic veins.** A pair from two lobes of the liver.

The two precavals and the postcaval open into the sinus venosus.

Portal system

In its usual course blood from the tissues of the body returns directly to the heart through veins. But, in some cases, instead of ending directly in heart or in some other vein, it enters some organ, divides there into capillaries, which join to form another vein, and this, in turn, ends in a vein, which goes to the heart. This sort of circulation is called portal circulation. When a vein starts in capillaries and ends in capillaries—such a vein is known as **portal vein**. Two sets of portal systems—the hepatic portal and renal portal—are present in toad.

Hepatic portal system

Hepatic portal vein and anterior abdominal vein constitute hepatic portal system. The **hepatic portal** vein is formed by the union of veins from the stomach, intestine, pancreas and spleen. Near the region of the liver it joins with the **anterior abdominal vein** which is formed by the union of two **pelvic veins**; one on either side. The pelvic vein brings back blood from the hindlimb. After the union of the hepatic portal and the anterior abdominal vein, the common vein enters two lobes of the liver by two main branches where they end in capillaries. The two hepatic veins arising from the two lobes of the liver drain blood from the latter and pour it into the postcaval vein.

Renal portal system

It consists of veins opening into the kidneys. Blood from each hindlimb is returned by the **femoral vein**, which, on reaching the coelom, divides into two. The ventral branch, the **pelvic**, goes to form the anterior adbominal and, finally, the hepatic portal system. The dorsal branch of the femoral, which is now known as **renal portal vein**, continues forward along the dorsal body wall and receives a vein, the **sciatic vein**, returning blood from the inner side of the leg. On receiving the sciatic, the renal portal continues anteriorly along the outer border of the kidney and finally breaks up into capillaries in the substance of the kidney. In its way along the side of the kidney each renal portal receives 2-4 veins, the **dorsolumbar** veins, bringing back blood from dorsal body wall. Renal veins drain blood from the kidneys.

Pulmonary vein

The pulmonary vein is very short. It is formed by the union of two pulmonary veins, one from each lung and opens into the left atrium. The blood becomes oxygenated in the lungs and the pulmonary vein, unlike other veins, carry oxygenated blood from lungs to heart.

⊠ Spleen

It is a small, dark-red, spherical body located in the mesentery near the junction of ileum and rectum. It is the only ductless gland that does not secrete hormone. Rather, it is related to blood vascular system and immunity. The spleen contains **T** and **B-lymphocytes**. From the latter are derived plasma cells, which secrete **antibodies**. Wornout red blood corpuscles, platelets and foreign particles are destroyed in spleen. During foetal life, spleen forms red and white blood cells.

⊠ Nervous System

The nervous system may be divided into three components—the central, the peripheral and the autonomic nervous sytem.

Central nervous system

It consists of **brain** and **spinal cord**. The brain and the spinal cord lie along the long axis of the body. The brain is situated dorsally at the anterior region of the body and continues posteriorly up to the anal region as a tubular spinal cord. The brain is enclosed in a bony brain case or **cranium** and the spinal cord lies in the tubular space formed by the union of the vertebrae. Both the brain and the spinal cord are hollow and the space inside them is filled with a fluid the **cerebrospinal fluid**. The brain and the spinal cord are enclosed in two protective coverings or **meninges**—the outer and thicker, tough membrane the **dura mater** and a thin inner membrane, the **pia mater** (p. 135).

Brain

The brain (Fig. 39.17) is divided into:

(a) Forebrain (anterior portion) or prosencephalon, (b) Midbrain (middle portion) or mesencephalon and (c) Hind brain (posterior portion) or rhombencephalon.

Prosencephalon

An anterior **telencephalon** and a posterior **diencephalon** constitute prosencephalon. Anteriorly, the telencephalon is constituted of

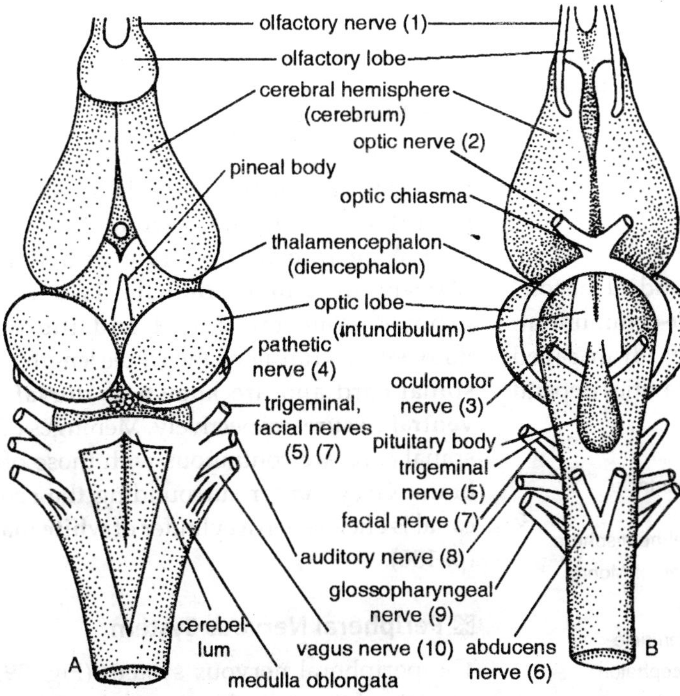

FIGURE 39.17

***Bafo* sp.** Brain. A. Dorsal view, B. Ventral view

a pair of small **olfactory lobes** concerned with smell. The posterior part is formed by a pair of oblongoval bodies, the **cerebral hemispheres**. The two cerebral hemispheres, in combination, from the **cerebrum** with a thin membranous roof and a thick floor, the **corpora striatum**. The two cerebral hemispheres are connected by an **anterior commissure**. An ill-developed **pallial commissure** is present above the anterior commissure.

The cerebral hemispheres are the seat of intelligence, consciousness and voluntary motion.

Posterior to cerebrum, the transversely elongated portion is known as **diencephalon** or **thalamencephalon**. An **epiphysis**, a small projection, is present on the dorsal surface of the diencephalon, in front of which lies a projection, **anterior choroid plexus**. A small process, the **pineal body**, is attached to the epiphysis. The ventral surface of the diencephalon is depressed and bears **optic chiasma**. A hollow structure, **infundibulum**, is present behind the optic chiasma. The infundibulum projects posteriorly downwards and meets with a mass, the **pituitary body**, which is formed as a dorsal diverticulum of the pharynx and is glandular in nature. The pituitary body secretes a number of hormones. The lateral sides of the diencephalon are thickened to form **optic thalami**.

Mesencephalon

It consists of two rounded large bodies on the dorsal side, the **optic lobes** and two longitudinal bands, the **crura cerebri** on the ventral surface. The crura cerebri connects the forebrain with the hindbrain. The optic lobes control vision.

Rhombencephalon

It has two parts. The anterior portion, **metencephalon** or **cerebellum**, is a transverse band-like structure.

The cerebellum is the centre for coordination of muscular movement and partly for balance also.

The posterior swollen portion with a thick floor and sides and a thin, vascular membrane, the **posterior choroid plexus**, forming the roof constitutes **myelencephalon** or **medulla oblongata**. The medulla continues posteriorly as spinal cord.

Medulla regulates respiration, heart beat and metabolism.

Cavities of the brain

The brain is a hollow structure and all the fore, mid and hindbrain contain inter-communicating cavities. The cavities (Fig. 39.18) are called **venteicles** and the sizes of the cavities depend

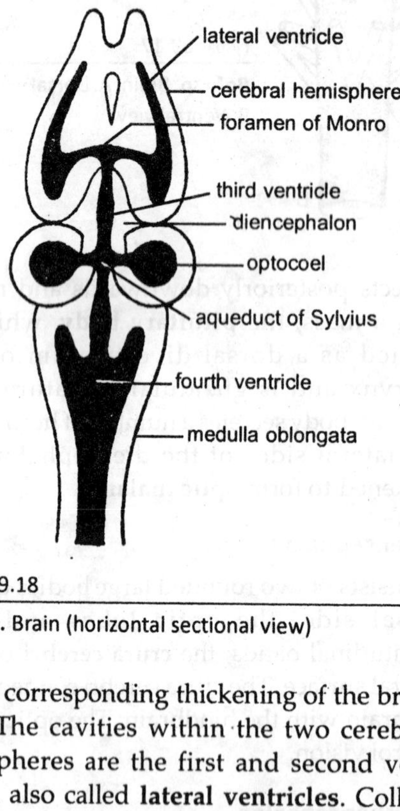

FIGURE 39.18

***Bufo* sp.** Brain (horizontal sectional view)

on the corresponding thickening of the brain wall. The cavities within the two cerebral hemispheres are the first and second ventricles, also called **lateral ventricles**. Collectively, they are termed **prosocoele**. The cavity inside the diencephalon is the third ventricle or **diocoele**. The first and second ventricle communicate with the third ventricle by an aperture, **foramen of Monro**. The third ventricle continues posteriorly as a narrow passage, the **iter or aqueduct of Sylvius**. Two cavities within the two optic lobes are **optocoele**. The cavity inside the medulla oblongata is comparatively large and known as fourth ventricle or **metacoele**.

Spinal cord

The spinal cord runs backward from the posterior end of the medulla through the neural canal of the vertebral column, extending into the urostyle as a slender filament, the **filum terminale**. Two median grooves—one middorsal and the other midventral—run along the whole length of the spinal cord and are known as **dorsal** and **ventral fissure**, respectively. Meninges of the spinal cord are continuous with those of the brain. **Grey matter** surrounding the central canal is enclosed in a cylinder of **white matter** (p. 136).

⊠ Peripheral Nervous System

The peripheral nervous system (Fig. 39.19) consists of the cranial and spinal nerves. The **autonomic nervous system** is a part of the peripheral nervous system.

Cranial nerves

Ten pairs of nerves springing from the brain are cranial nerves (Fig. 39.19). Proceeding from anteriormost to the posteriormost in serial order, the nerves of each side are:

i. **Olfactory nerve.** Springs from the olfactory lobe and innervates the nostril or organ of smell. Function: sensory.

ii. **Optic nerve.** Springs from the ventral surface of the diencephalon anterior to infundibulum and innervates the retina of the eye-ball. Function: sensory.
The two optic nerves of two sides cross each other on the ventral surface and, at the point of crossing, some fibres of each nerve pass to the opposite nerve and a chiasma, the **optic chiasma**, is formed.

iii. **Oculomotor nerve.** Arises from the ventral region of the midbrain, divides into branches to innervate superior, inferior, internal rectus and inferior

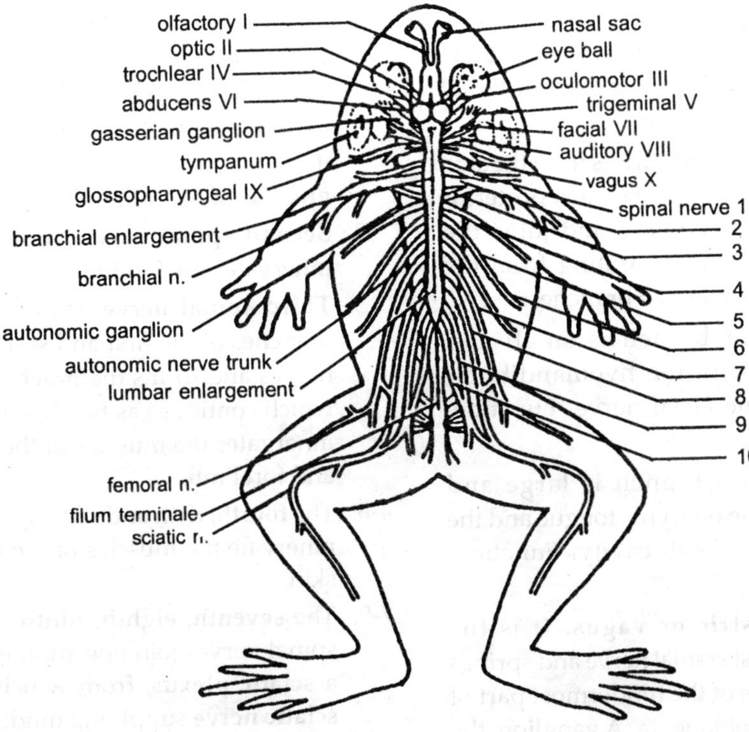

FIGURE 39.19

***Bufo* sp.** Cranial, Spinal and sympathetic nerves. Dorsal view

oblique muscles of the eyel-ball. Function: motor.

iv. **Trochlear or Pathetic nerve.** Springs from the dorsal surface at the junction of the mid and hindbrain and innervates the superior oblique muscle of the eye-ball. Function: motor.

v. **Trigeminal nerve.** Arises from the side, at the anterior region of medulla oblongata. A swelling or ganglion, the **gasserian ganglion**, is present at its base, and the nerve divides into three branches:

a. **Ophthalmic** sends sensory fibres to the skin above the eye-ball, anterior part of the head and mucous membrane of the nose. Function: sensory.

b. **Maxillary** sends sensory fibres to upper lip, mucous membrane of the upper jaw and lower eyelid. Function: sensory.

c. **Mandibular** sends motor fibres to the buccal cavity floor and sensory fibres to the mucous membrane of the buccal cavity. Function: sensory and motor (mixed).

Function of trigeminal nerve is both motor and sensory and it is a mixed nerve.

vi. **Abducens nerve.** Springs from the ventral surface of the medulla and innervates the external rectus muscle of eye-ball. Function: motor.

vii. **Facial nerve.** Arises from the side of medulla oblongata and, proceeding anteriorly, fuses with the gasserian ganglion of the fifth or trigeminal nerve. It divides into two branches after its emergence from the ganglion.

a. **Palatine** innervating the roof of the mouth cavity. Function: sensory.

b. **Hyomandibular** receives a branch from glossopharyngeal and innervates tympanum, mandibular joint and buccal floor. Function: mixed.

viii. **Auditory nerve.** Springs from the side of the medulla oblongata behind the facial nerve and innervates the internal ear. Function: sensory.

ix. **Glossopharyngeal.** Arises from the side of the medulla oblongata just posterior to the auditory nerve and joins the vagus ganglion. Separating from the ganglion it sends two branches:

a. The anterior branch is small and connected with the hyomandibular branch of the facial nerve. Function: sensory.

b. The posterior branch is large and innervates the pharynx, tongue and the floor of the mouth cavity. Function: mixed.

x. **Pneumogastric or vagus.** It is the posteriormost cranial nerve and springs from the side of the hindermost part of the medulla oblongata. A ganglion, the **vagus ganglion**, is present at the base. It divides into four main branches:

a. **Laryngeal** innervating the larynx. Function: motor.

b. **Pulmonary** innervating the lung. Function: motor.

c. **Cardiac** innervating the heart. Function: motor.

d. **Gastric** innervating alimentary canal. Function: mixed.

Spinal nerves

Ten pairs of spinal nerves (Fig. 39.19) spring from the spinal cord. Each spinal nerve arises by two roots—one **dorsal** and one **ventral**. The dorsal root bears a ganglion, the **dorsal root ganglion**. The fibres of the dorsal root are **sensory** or **afferent** and those of the ventral root are **motor** or **efferent**, and the spinal nerve performs **mixed function**. The spinal nerves come out from the vertebral column through the intervertebral foramina of the side.

Following are the spinal nerves of toad:

1. **Hypoglossal.** It is the first spinal nerve. Coming out through the foramen between the first and second vertebrae it runs anteriorly and innervates the muscles of the tongue.
2. **Second spinal nerve.** A stout nerve, innervates the forelimb.
3. **Third spinal nerve.** It joins with the branches of the first and second spinal nerves and forms the **brachial plexus**, which continues as brachial nerve and innervates the muscles of the shoulder and forelimb.

4-6. The **fourth, fifth** and **sixth** spinal nerves innervate the muscles of the body and skin.

7-9. The **seventh, eighth, ninth** and **tenth** spinal nerves join one another to form a **sciatic plexus**, from which arises a **sciatic nerve** supplying hindlimb.

10. The **tenth** spinal nerve comes out through an aperture on the side of the urostyle. It is very small and insignificant.

Autonomic nervous system

Autonomic nervous system consits of two sympathetic trunks (Fig. 39.18), each containing ten ganglia.

Sympathetic trunks. Thin, lie one on each side of the dorsal aorta, close to the vertebral column. The posterior ends of the trunks are near the end of the spinal cord. Anteriorly, they accompany systemic arches. Further forward, each trunk splits to encircle the subclavian artery, enters the cranium, sends a twig to the vagus ganglion and ends by joining the gasserian ganglion of the fifth cranial nerve.

Each sympathetic trunk bears ten ganglia in its course and each sympathetic ganglion is connected to corresponding spinal nerve by a short **ramus communicans**.

Solar plexus. Three slender nerves arising from fifth, sixth and seventh sympathetic

CRANIAL NERVES

Serial No.	Name	Origin	Branches	Distribution	Function
I	Olfactory	Anterior end of olfactory lobe		Olfactory or organ of smell	Sensory
II	Optic	Ventral side of diencephalon, anterior to infundibulum		Retina of the eye-ball	Sensory
III	Oculomotor	Ventral region of midbrain		Four muscles—superior rectus, inferior rectus, internal rectus and inferior oblique of the eye-ball	Motor
IV	Trochlear or Pathetic	Dorsally at the junction of mid and hind-brain		Superior oblique muscle of the eye-ball	Motor
V	Trigeminal	Side of medulla oblongata	1. Ophthalmic	Skin above the eye-ball (upper eye lid)	Sensory (mixed)
			2. Maxillary	Upper jaw	Sensory (mixed)
			3. Mandibular	Lower jaw	Sensory and motor (mixed)
VI	Abducens	Ventral surface of medulla oblongata		External rectus muscle of the eye-ball	Motor
VII	Facial	Side of medulla	1. Palatine	Roof of the buccal cavity	Sensory (mixed)
			2. Hyomandibular	Hyoid and lower jaw	Sensory and motor (mixed)
VIII	Auditory	Side of medulla		Internal ear (organ of hearing)	Sensory
IX	Glossopharyngeal	Side of medulla	1. Anterior	Connected with hyomandibular	Sensory (mixed)
			2. Posterior	Pharynx, tongue, floor of the mouth cavity	Mixed (mixed)
X	Pneumogastric or vagus	Side of medulla	1. Laryngeal	Larynx	Motor (mixed)
			2. Pulmonary	Lung	Motor (mixed)
			3. Cardiac	Heart	Motor (mixed)
			4. Gastric	Alimentary canal	Motor and sensory (mixed)

ganglia of each side join to form a solar plexus around coeliacomesenteric artery and innervates stomach, intestine and related organs as **splanchnic nerve**.

Nerve fibres from the ganglia of the sympathetic chain innervate heart, stomach, liver, kidney and other visceral organs. Autonomic nervous system is associated with the control of involuntary activities, such as digestion, circulation etc.

⊠ Receptors and Sense organs

Sense organs are structures which receive stimuli and transfer the same to the nervous system. The sense organs (p. 142) of toad are—skin, olfactory sacs, taste buds, eyes and ears.

Cutaneous (tactile) Receptors

Receptors for heat, cold, touch and pain are microscopic and present beneath the epidermis.

Olfactory sac or nose (Chemoreceptor)

It is the organ of **smell**. The nostrils are two, situated at the anterior end of the head, one on each side. Each **nasal cavity** is lined with sense hairs which receive chemical stimuli and has the capacity to detect the odour of different gases. The sac is provided with two openings, one external, known as **anterior nares**—opening to the outside and the other internal, the **posterior nares**—opening into the mouth.

Taste buds: Gustatory receptor

Taste buds are similar, both in structure and function in all vertebrates (p. 151). These are chemoreceptors and scattered on the tongue and lining of the buccal cavity.

Eye (Photoreceptor)

Eyes of all tetrapods are structurally similar (p. 143) with minor variations in different groups.

In Bufo, the eyes are large and partly protrude out of the optic capsules. All the three eye-lids are present. The crystalline lens is nearly globular (Fig. 39.20). The eyes being placed on the lateral sides of the head, the vision is monocular. Each eye has its own range of vision. The amount of light received by the eye is regulated possibly by backward and forward movement of the eye.

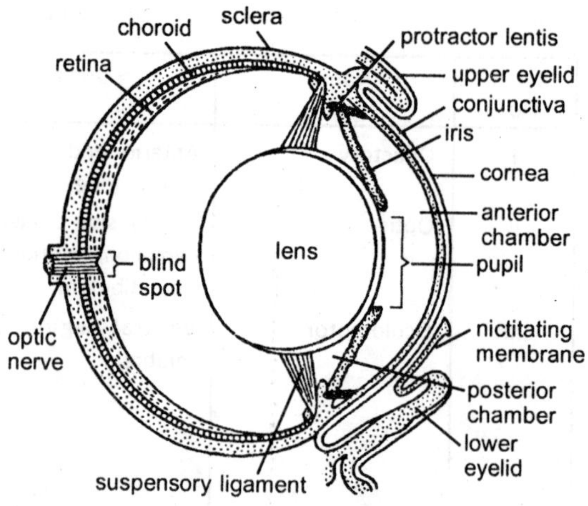

FIGURE 39.20

***Bufo* sp.** Eye (Ventrical section)

Ear (Equilibrium and Phonoreceptor)

The ear or auditory organ is the organ of equilibrium and hearing (p. 145) and situated on either side of the head. In toad, the ear is externally represented by a stretched membrane, the **tympanic membrane** (tympanum) or ear drum. Inside the tympanic membrane, a bony cavity, the **tympanic cavity** is present. On one side, the tympanic cavity leads to a passage, the **eustachin tube**, which communicates with the buccal cavity. The eustachin tube equalises atmospheric pressure on two sides of the tympanum. The tympanic cavity with its contents is known as **middle ear**. A long rod with rounded ends, the **columella**, is present in the tympanic cavity. One end of the columella is fixed with the tympanic membrane and the other with the capsule enclosing the **internal ear** or **membranous labyrinth** (Fig. 39.21). The capsule enclosing the membranous labyrinth is a bony structure known as **periotic** or **auditory capsule**. The periotic capsule bears an aperture, the **fenestra ovalis**, in which fits the rounded end of the columella. The periotic capsule is filled with a fluid, the **perilymph**, in which the membranous labyrinth remains bathed. The membranous labyrinth (Fig. 14.10E) consists of two chambers—a dorsal **utricle** and a ventral **saccule**. Three semicircular canals, two in vertical and one in horizontal postition, arise from the utricle. Each canal opens at both of its ends into the utricle. The lower end of each canal is dilated and known as **ampulla**. Patches of sensory cells are present in the ampullae, utricle and saccule. The saccule gives off a blind pouch posteriorly, the **lagena**, the organ of hearing. The cavity of the membranous labyrinth is filled with a fluid, the **endolymph**. The membranous labyrinth is innervated by auditory nerve.

Equilibrium

Small calcareous particles, **otoliths**, are present in the utricle and saccule. Flow of endolymph in the semicircular canals stimulates sensory cells in ampulla, utricle and saccule, which helps the animal to detect change of position in relation to gravity.

Hearing

Sound waves set the tympanic membrane in vibration, which are transmitted to the perilymph within the periotic or auditory capsule by the columella. The perilymph transfers the same to the endolymph within the membranous labyrinth; the sense hairs receive the vibrations and the impulse is carried to brain by the auditory nerve.

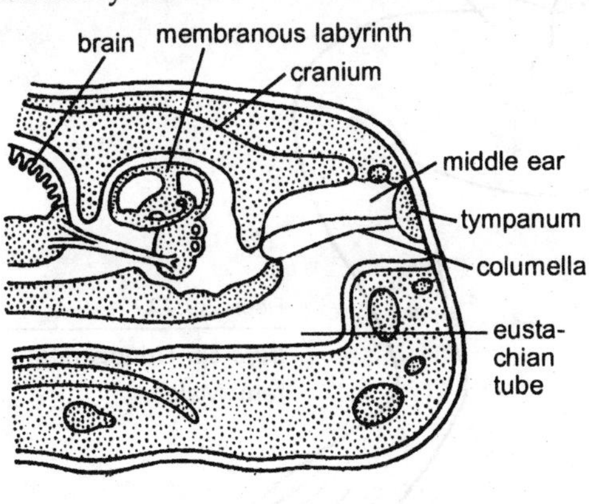

FIGURE 39.21

***Bufo* sp.** Ear (sectional view). Semidiagrammatic

⊠ Endocrine System

The endocrine system is constituted by a number of endocrine or ductless glands secreting hormones (p. 156). Endocrine glands are structurally and functionally similar in all vertebrates. Certain hormone-secreting glands, viz. islet of Langerhans in pancreas and those in gonads are part of other structures and they do not form distinct glands as in the majority of the group.

Thyriod

A pair of small, oval bodies on the floor of the buccal cavity, one on each side of the hyoid apparatus.

Parathyroid

Two yellowish ovoid bodies, close to thyroid and attached to external jugular vein.

Thymus

Small, yellowish bodies on either side of the head, just behind the tympanum.

Adrenal

Narrow, band-like orange-coloured, on the ventral side of the kidneys.

Pituitary

A bilobed gland attached to the tip of the infundibulum, ventral to the floor of the diencephalon (forebrain).

⊠ Urinogenital System

The urinary or excretory and genital or reproductive systems (Figs. 39.22, 39.23) are not completely separated in Bufo and both the systems are described under the common heading 'urinogenital system'. In male, the exrcretory and reproductive organs are not separate but in female they are independent structures. The urinogenital system consists of paired kidneys, a urinary bladder and the male or female reproductive organs.

⊠ Urinary (Excretory) System

The kidneys are paired, flat, somewhat oval, dark-reddish bodies situated in the posterior region of the coelom, one on each side of the vertebral column. Finger-like fat bodies are attached to the anterior end of the kidneys. The kidneys are wrapped by the folds of peritoneum. On the ventral surface of each kidney, an elongated, yellow structure the **adrenal body**, is present. A thin-walled duct, the urinary duct or **ureter**, runs posteriorly from the outer border of the kidney. The ureter of both the sides finally join to form a common duct wich opens by a single aperture into the cloaca on the dorsal surface. The kidneys are mesonephric in origin (p. 127). At the posterior region of the coelom, ventral to the cloaca, a bilobed, thin-walled delicate sac, the **urinary bladder**, is present. The urinary bladder opens into the cloaca on the ventral surface.

☒ Reproductive System

Sexes are separate, sexual dimorphism is present. Males possess thumb pads and a vocal sac and they are usually smaller in size.

Male reproductive system

A pair of testes, a pair of urinogenital ducts and a urinogenital opening constitute male reproductive system (Fig. 39.22).

The **testes** are slightly elongated bodies, each attached on the ventral face of the kidney near the anterior end by the folds of peritoneum. The testis is formed by numerous **seminiferous tubules**, which produce **spermatozoa**. The **Bidder's organs** or ovotestes are rounded structures, one situated anterior to each testis.

Leading from each testis are several small sperm ducts—the **vasa efferentia** which open into the tubules of the kidney. Spermatozoa from the testes passing through the vasa efferentia are carried by the ureter into the cloaca through the urinogenital aperture and thence to the exterior. The ureter, also known as **urinogenital duct** is both **ureter** and **vas deferens** in function.

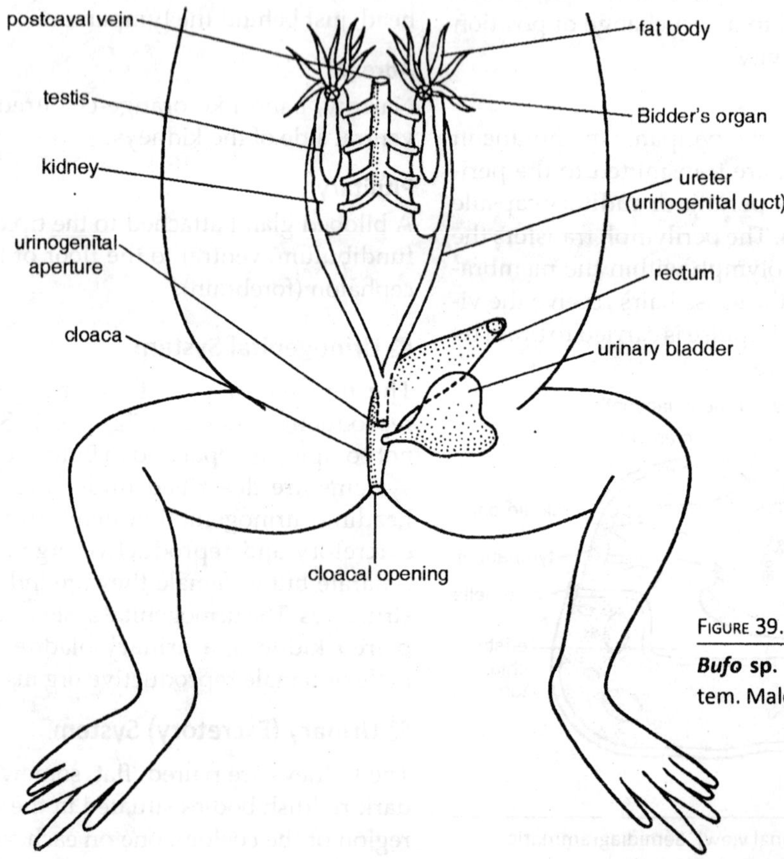

FIGURE 39.22

***Bufo* sp.** Urinogenital system. Male

Female reproductive system

The female reproductive system is constituted by a pair of ovaries, a pair of oviducts and a common genital aperture (Fig. 39.23). **The ovaries** are paired, much folded irregular bodies, one on each side of the coelom. Each ovary is slung by mesentery (mesovarium), from the dorsal wall of the coelom. During breeding season they enlarge greatly, dark in colour and contain numerous developing eggs.

The oviducts are large, convoluted tubes, one on each side, with a funnel-shaped anterior, **oviducal funnel** opening, in the coelom. The oviducts run forward to near the base of the lungs. The oviduct is differentiated into three zones—a narrow, anterior part bearing **funnel**, a median convoluted part bearing **glands** and a posterior thin-walled dilated part, the **uterus**. The two uteri join posteriorly and open into the cloaca by a common genital aperture.

The **ova** or mature eggs are shed into the body cavity. Ova from the body cavity enter the oviducts through the oviducal funnels and are carried to the cloaca and, thence, to the exterior.

The kidneys and ureters of females are structurally similar to those of the males, but the oviducts act as genital ducts and the ureters function as excretory ducts. The opening of the ureter in the cloaca is dorsal to genital opening.

Breeding

Breeding takes place in rainy season. The males move to a shallow water body and produce a croacking sound with vocal sacs to invite females for mating. The male sits on the female firmly grasping the latter with the well-

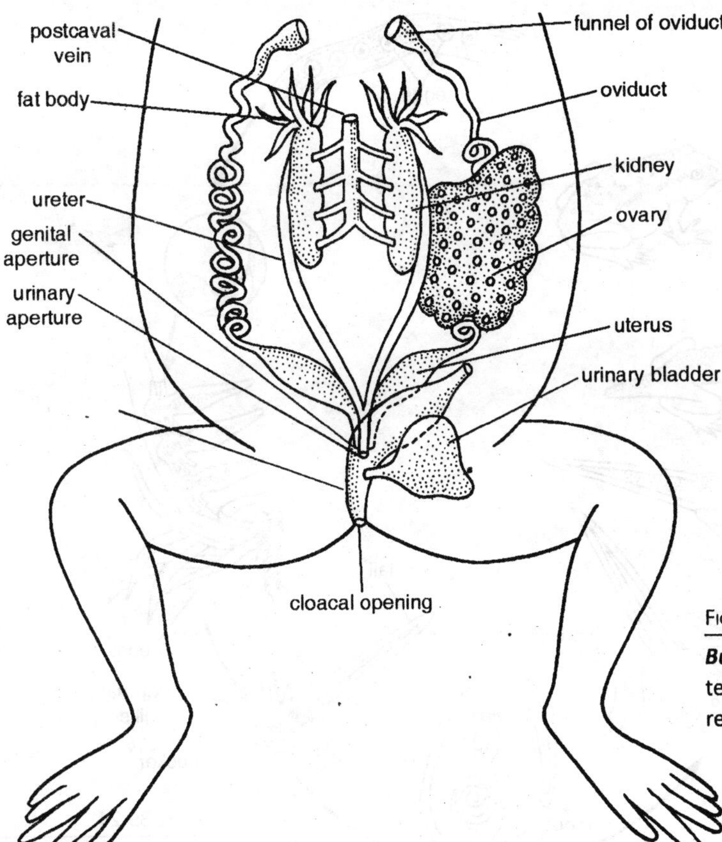

FIGURE 39.23

***Bufo* sp.** Urinogenital system. Female (Right ovary removed)

developed thumb pads of forelimbs and this may continue for more than twenty four hours. Both the partners get excited in this process and the female sheds eggs on water. The eggs are arranged in a single row in a convoluted ribbon. The male discharges **milt (seminal fluid)** simultaneously and the sperms fertilize the ova. The eggs are partly black and partly white, the black portion representing the cytoplasmic mass the white portion the yolk.

Development

The zygote or fertilized egg begins to divide **mitotically** and the cleavage is **holoblastic** (complete). The division continues and a large number of cells are produced.

Undergoing a series of complicated changes a tiny fish-like young individual is formed which remains coiled inside the egg and begins to move.

Metamorphosis

Metamorphosis is a series of changes by which the larva assumes the adult form. The larva is considered as a postembryonic stage in the development of an animal which does not resemble the parents and leads a more or less independent life. The metamorphosis may be progressive or retrogressive. In Bufo, the metamorphosis (Fig. 39.24) is a progressive one.

☒ Tadpole Larva

1. About fourteen hours after fertilization the fish-like tadpole larva hatches out from the egg.
2. It is devoid of mouth, eyes and limbs, but bears a sucker on the ventral surface of the head.

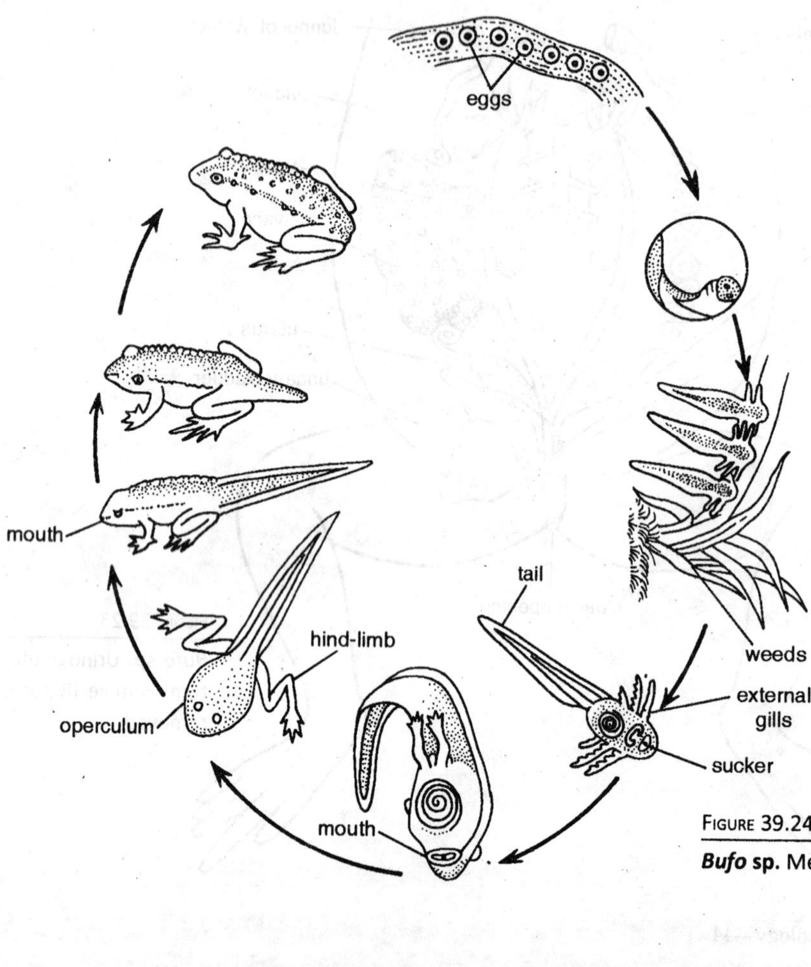

FIGURE 39.24
***Bufo* sp.** Metamorphosis

3. The tadpole attaches itself to some aquatic weed by the sucker and remains quiet without taking any food. Nourishment at this stage is obtained from the yolk granules.

New structures added

1. The mouth and vent (cloacal aperture) are formed and the eyes appear. The mouth is bounded by a pair of horny jaws, the gut is long and remains coiled like a watch spring.
2. Three pairs of branched filamentous gills appear by the sides of the head. These are known as external gills and respiratory in function.
3. The tadpole now swims with the help of its tail, feeds on water weeds and respires by external gills.

Changes to young toad

1. The external gills begin to disappear. Gill slits and internal gills appear in the pharyngeal wall.
2. A fold of skin, the operculum, appears on each side and covers the gill slits except an opening, the branchial aperture on the left side, which acts as an outlet of water.
3. Lungs appear as outgrowths of the anterior end of the gut, and the branchial aperture closes. At this stage, the tadpole begins to respire air.
4. The hindlimbs are formed by this time. The forelimbs, which were covered by the operculum, come out piercing the skin.
5. The gut becomes short and more or less straight.
6. The limbs increase in size and the tail diminishes.
7. The young toad comes on land with a short tail. The tail is gradually absorbed and the tadpole ultimately turns to a young toad (Fig. 39.24). The metamorphosis from the tadpole to young toad is completed in about three months.

PARENTAL CARE IN AMPHIBIA

Taking care of eggs or young until they can take care of themselves to defend from predator is parental care. With parental care continuation of a species can be ensured with a meagre number of eggs.

Different types of parental care in Amphibia can be grouped under two broad heads (a) protection by nests, nurseries or shelters (b) direct caring by parents.

☒ Protection of Eggs

1. **Frothing of water.** Just after laying the eggs some anurans make the surrounding water frothy by the wriggling movement of hind limbs by both the parents—male and female. This helps in escaping the eggs from the sight of the predator and at the same time prevents desiccation.
2. **Foam nests.** Many amphibians secrete copious mucus in their nests to protect the eggs and young.

   Polypedatas *(= Rhacaphorus)* ***maculatus*** secrete jelley-like substance with eggs, which is beaten by the crosswise movement of the hindlimbs of the female partner, leading to the formation of a frothy mass. Within a few hours the auter surface of the ball-like frothy mass hardens. Foam nests are made in the vicinity of water, on the leafy branches of the bushes and other materials overhanging water.

   Some anurans lay eggs in the nests of foam floating on water (Fig. 39.25A).

   Rhacophorus schlegeli lay eggs in a frothy mass in a hole or tunnel dug by them (Fig. 39.25A). The tadpoles are washed down into river or pond during rains.
3. **Mud nests.** The Brazilian tree frog *Hyla fabre* digs a little crater-like hole (Fig. 39.25B) in mud in shallow water in which the female lays eggs. Tadpoles hatch within this relatively safer barrier and develop.

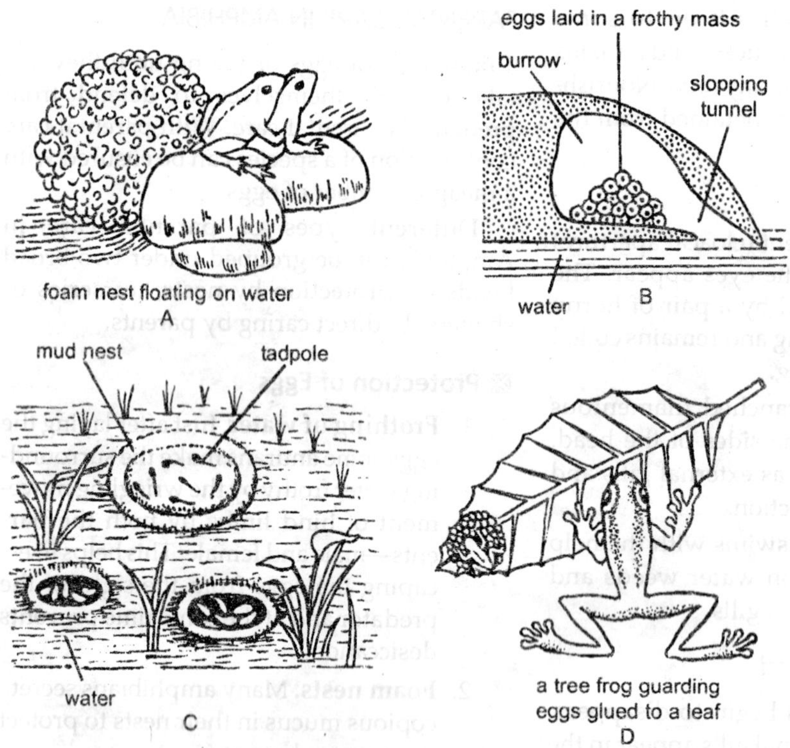

FIGURE 39.25

Parental care in Amphibia. **Nests. A. Floating nest, B. Sloping burrow nest, C. Mud nest, D. Tree nest**

4. **Tree nests.** ***Phyllomedusa hypochondrales,*** south American tree frog make a folded nest by glueing together the margins of the leaf with cloacal secretion and lays eggs within it. The tadpoles after hatching fall straight into water below. ***Hylodes*** sp. a tree frog glues cluster of eggs on the leaf of a tree branch overhanging a water body and guard the brood (Fig. 39.25C).

   Hyla resinfictrix lines a shallow tree cavity with bee's wax collected from the hives of some stingless bees. The female lays eggs in the cavity when it is filled with rain water.

5. **Gelatinous bag nests.** The female ***Phrynixalus bivori*** secretes a large saucer-shaped transparent, membranous, gelatinous bag in which she lays her eggs. The bag is left in mountain streams. A small aquatic salamander ***Salamandrella keyserlingi*** deposit eggs in a gelatinous bag fastened to aquatic plants.

6. **Shoot nests.** *Triton* prepares a short nest by fixing the shoots of the trees in which the eggs are deposited. The whole nest remain covered by a gelatinous secretion.

⊠ Direct Caring by Parents

1. **Placement of eggs in safest part.** The oviparous caecilian ***Ichthyophis*** coils round the egg mass till the larvae hatch out (Fig. 39.26A).
2. **Placed inside vocal sac.** Male ***Rhinoderna darwini*** pick up the eggs by the mouth and place those inside the inflated vocal sac, where they remain until hatching, even up to the completion of metamorphosis (Fig. 39.26B).
3. **Carrying eggs over the body**

a. In ***Alytes obstetricans*** the eggs are fartilized during expulsion at the time of mating by the sperms of the grasping male. The fertilized eggs are

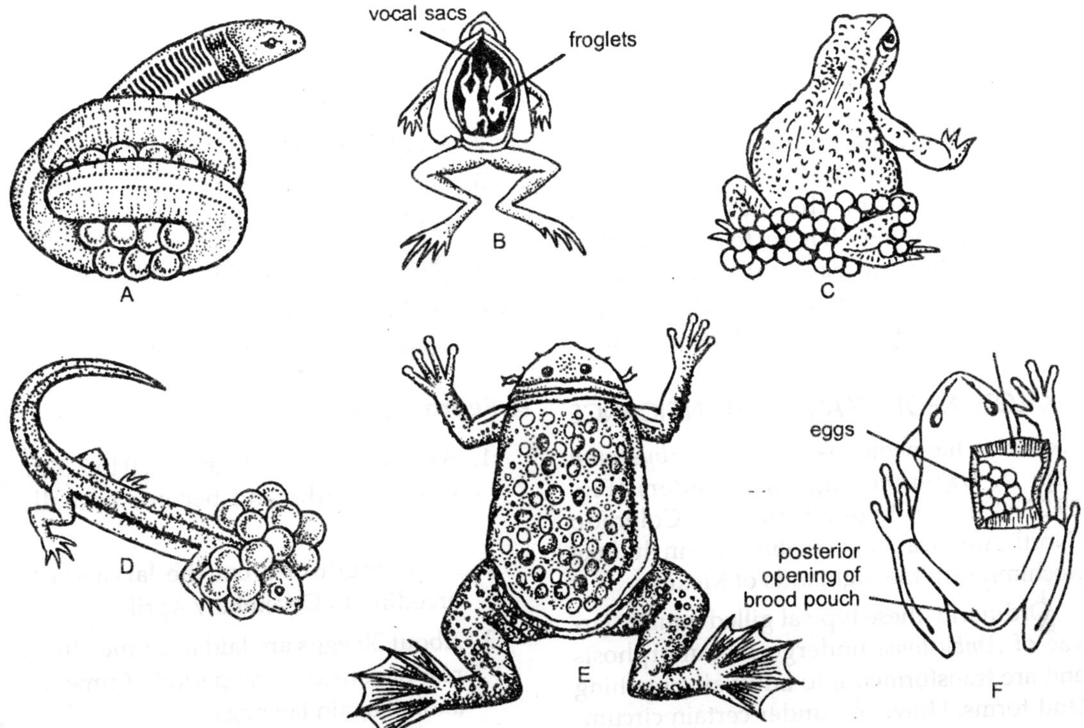

FIGURE 39.26
Parental care in Amphibia. Direct caring by parents. A. *Ichthyophis*. Coiling round the eggs, B. *Rhinoderma*. Froglets inside vocal sac of male (vocal sac cut open to expose froglets), C. *Alytes*. Eggs raped round the thigh, D. *Desmognathus*. With eggs in an underground hole, E. *Pipa*. Carrying eggs on the back, F. *Nototrema*. Flap of dorsal brood pouch cut open to expose eggs

wrapped round the back of the thigh of the male (Fig. 39.26C). The male moves into a hole near a pond. On the point of hatching out of the larvae the male moves to water and the larvae come out.

b. The females of ***Desmognathus*** carry the eggs and live in an underground hole (Fig. 39.26D).

c. The females of ***Pipa pipa*** carries the eggs on their back (Fig. 39.26E).

d. The female ***Hyla goeldi*** carry eggs in their inciplient brood pouch on their back. Tadpoles come out of the pouch.

e. In ***Cryptobatrachus evansi, Pipa dorsigera*** and same other species the eggs are ladged in small pockets on the soft spongy skin of the dorsal side. During breeding season the dorsal skin of the female becomes soft, spongy and gelatinous. The eggs are placed on the back of the female by the male and each egg sinks into a small pit. The pit is covered by an operculum. The partitions between the skin become highly vascular. Development is campleted inside the pits.

f. In ***Nototrema* spp.** the eggs develop in the brood pouch (Fig. 39.26E) bearing a small slit-like opening in the posterior side.

4. **Carrying larvae from one place to another.** Small frogs-***Phyllobates, Arthroleptis, Palobates, Dendrobates*** and others lay eggs on the ground. After hatching the tadpoles fasten them to the back of their parents with sucker-like mouth and transported to water.

Experimental evidences on parental care in Amphibia to establish any parallelism with that in Mammalia are meagre. The parental care in Amphibia probably evolved independently in response to their habitat to protect young from the hazards of the surroundings. It is yet to be established whether the parental care is purely instinctive or due to reflex action developing in the breading season.

A XALOTL, NEOTENY AND ITS SIGNIFICANCE

Axolotl is the neotenus larva of ***Ambystoma*** spp. *A. **triginum*** (tiger salamader) (Fig. 39.27A) live in high altitudes of Colorado (North America). ***A maxicanus*** live in the lake Xochimileo in the highlands of Mexico.

Ordinarily these typical gilled aquatic larvae of *Ambystoma* undergo metamorphosis and are transformed into adult air-breathing land forms. However, under certain circumstances the larvae do not metamorphose, retain external gills, aquatic habitat and reproduce morphologically similar to other members of the group. This stage is called **Axolotl*** (Fig. 39.27B).

⊠ Structure: Axolotl larva

1. Body elongated with a laterally compressed tail.
2. Mouth wide, anterior and semicircular.
3. Eyes well developed.
4. Lateral lines present.
5. Dorsal and caudal fin continuous.
6. Limbs poorly developed.
7. Vomerine teeth present.

⊠ Reproduction

1. Axolotl attains a length of 2.0 to 2.5 cm in six months and become sexually mature.
2. At a length of 8 cm the larva starts breeding in December/April.
3. About 30 eggs are laid at a time after a rest of considerable period of time the axolotl again lay eggs.
4. The larva hatches out within 15 days at optimum temperature 18 to 20°C, feed on *Daphnia, Infusoria*, etc. and metamorphose to adult. The adult measures 12-14 cm.

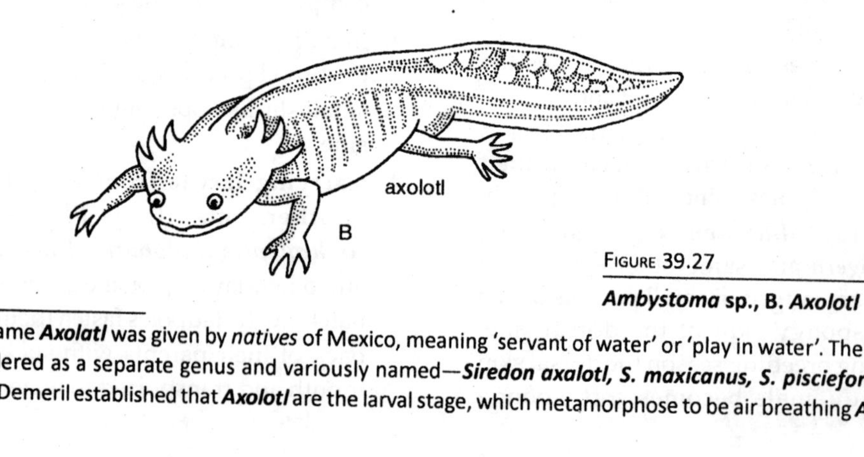

FIGURE 39.27

Ambystoma* sp., B. *Axolotl

* The name ***Axolatl*** was given by *natives* of Mexico, meaning 'servant of water' or 'play in water'. The ***Axolotl*** was causidered as a separate genus and variously named—***Siredon axalotl, S. maxicanus, S. piscieformes***, etc. In 1966. Demeril established that ***Axolotl*** are the larval stage, which metamorphose to be air breathing ***Ambystoma***.

NEOTENY AND AXOLOTL

The phenomenon of attainment of sexual maturity in an arrested larval (**proadult**) stage is known as **Neoteny** or **Progenesis**. Such retardation of metamorphosis of larva to adult and retention of larval characters beyond the normal limited span of time is termed **Paedormorphosis** or **Paedogenesis**.

Paedomorphic forms are found mainly in urodeles. The extensive studied animal is ***Ambystoma triginum***. The axolotl of Ambystoma metamorphose into adult under favourable circumstances.

Some extrinsic and intrinsic factors very much influence the process of neoteny.

Extrinsic factors

a. Abundance of food, cold temperature, sufficient quantity of iodine (a component of thyroxine hormone) may cause failure of metamorphosis and retention of larval features.
b. Drying up of swamps, rise of temperature in the surrounding water and lack of food induce axolotl to metamorphose.

Intrinsic factors

a. The metamorphosis is influenced by varying threshold levels of thyroxine and its analogues.
b. The degree of responsiveness of the larval tissue to hormones.

The level of prolactin, an inhibitor, plays an important role in the overall control of metamorphosis. During early metamorphic changes the level of prolactin remains high. As the process progresses the level of prolactin suddenly falls and the metamorphosis proceeds at a faster rate.

It is suggested that the action for guiding the synthesis of thyroxine is controlled by some operator genes. To produce prolactin the genes are switched on and in reverse case the genes are 'switched off'. In early metamorphic stage the genes are switched on and the hypothalamus become selective to available level of thyroid hormone in blood streams and produces a substance, thyrotropic releasing factor (TRF). The TRF stimulates the anterior pituitary to produce thyroid stimulating hormone (TSH), which in turn enhances the secretion of thyroxine. With the increasing level of TSH during premetamorphosis the genes are switched off and the level of prolactin suddenly falls and the metamorphosis starts. Poor secretion of thyroid glands and irresponsiveness of the larval tissue to the hormones are responsible for neoteny.

☒ Significance of Neoteny

Neoteny is considered to be the case of **atavism** or retarded evolution, or reversion to ancestral condition by Weissmann (1875). At present neoteny is regarded to be a consequence of physiological adaptation to the surrounding environment, where retention of larval gills and other larval features may be advantageous. Existence of gills in neotenous form is secondary, not an atavistic feature. The tail with fin and epidermal sense organs of the neotenous larvae are secondary acquisition rather than ancestral reminiscences. Selection can obviously produce the condition of return to a fully aquatic cycle, which may be advantageous, especially if water is abundent and terrestrial conditions are unfavourable.

CLASSIFICATION: CLASS AMPHIBIA

About 2,500 extant species of three divergent groups—frogs and toads, salamanders, and caecilians are included in Amphibia.

1. The skin is moist, glandular and without scales.
2. A distinct neck is absent.
3. Typically, two pairs of limbs, no paired fins. Forelimbs with four and hindlimbs with five clawless digits.
4. Two nostrils connected with the mouth cavity.
5. Skull with two occipital condyles.
6. Heart 3-chambered; two atria (auricles) and one ventricle; two other chambers —sinus venosus and conus arteriosus, present.

7. Respiration by gills, lungs, skin or lining of buccal cavity.
8. Tympanum present; a rod-like collumela connects the tympanum and inner ear.
9. Eggs with gelatinous coverings, usually laid in water. Larvae usually aquatic.

The class Amphibia is divided into two subclasses—Apsidospondyli and Lepospondyli.

Subclass APSIDOSPONDYLI

The centra of vertebrae formed from cartilage blocks in units of two, which are ossified in varying degrees as anterior intercentra and posterior pleurocentra. Apsidospondyli are placed into two superorders—Labyrinthodontia and Salientia.

Superorder **Labyrinthodontia (Stegocephalia)**

1. The cross section of the teeth exhibits a prominent radiating infolding of the enamel surface, hence the name.
2. Tail present; legs all of about same size.
3. Cranium and cheeks completely roofed with bony plates.
4. Often armoured ventrally with overlapping scales.

Order 1. **Ichthyostegalia**

The earliest known amphibians, perhaps the nearest to true prototetrapods, were prevalent from upper Devonian to upper Carboniferous.

The skull is about 20 cm long with three interesting features:

i. A preopercular and possibly subopercular persisted on the back of squamosal and quadratojugal bones.
ii. The septomaxilla formed part of the dermal covering of the skull.
iii. Like Dipnoi and some early crossopterygians, the nose bears a pit on the under side of the skull, bridged by a process of the maxilla and divided into anterior and posterior part.

Examples: *Ichthystega*, *Ichthystegopsis*, etc.

Order 2. **Rhachitomi**

Typical labyrinthodonts flourished during the Permian and Triassic, became subsequently extinct.

1. About 150 cm long with short powerful legs and five toes.
2. Large depressed skull with large interpterygoid cavities and otic notch.
3. The rhachitomous vertebrae had a semilunar wedge-like intercentrum and one or two posterior pleurocentra in addition to a vertical neural arch.

Example: *Eryops*.

Order 3. **Stereospondyli**

Transitional types arose from rhachitome stock, lived in the early Triassic waters, became extinct towards the end of Triassic.

1. Heavy-headed, shallow-water, bottom-dwellers of large size.
2. Vertebrae with pleurocentra removed, the intercentra joining almost the whole structure.
3. Skull flattened, ossification much reduced, except in the exoccipitals, which formed a double condyle.
4. Parasphenoid highly developed and apparently had firm union with pterygoids.
5. Eyes dorsally placed.
6. Massive supporting pectoral plate formed from clavicles and interclavicles.

Examples: *Capitosaurus*, *Cyclotosaurus*, *Paracyclotosaurus*, *Brachyops*, etc.

Order 4. **Embolomeri**

Fish-like labyrinthodonts appeared in the Carboniferous, became abundant during the Permian and extinct at the end of Triassic. Some were superficially crocodile-like.

1. The teeth had the dentine infolded at the base into labyrinthine grooves.
2. The vertebrae were embolomerous, consisting of a neural arch resting on

two notochordal centra—an anterior intercentrum and a posterior pleurocentrum.

3. The shoulder girdle with increased size.
4. The pelvic girdle not directly attached to the vertebral column but held in place by ligament.

Examples: *Eogyrinus, Palaeogyrinus*, etc.

Order 5. **Seymouriamorpha**

The group represents a stem apparently developed astride the transition line between Amphibia and Reptilia at the lower Permian.

1. Head relatively small, pointed; dorsal nostrils; a comparatively thick body and a short tail.
2. All its limbs were pentadactyl, short, muscular and thrust from the midline.
3. The atlas and axis were formed from anterior cervicals.
4. Interclavicle long and a ventral coracoid ossification in the shoulder girdle.
5. The pelvic girdle was anchored by means of two sacral vertebrae.
6. The ilium was dorsally expanded for attachment of strong walking muscles.
7. The shoulder girdle lay beneath the jaw.
8. The ribs were larger and double-headed.

Example: *Seymouria*

Superorder **Salientia (Anura, Batrachia)**

1. Tailless frogs and toads with short and broad bodies and long hindlimbs.
2. The earliest fossil records are found as early as Jurassic period.

The Salientia is divided into three orders-Eoanura, Proanura and Anura.

Order 1. **Eoanura**

1. These are "dawn frogs"—ancestral to the modern frogs and were prominent in upper Carboniferous.
2. Cranial, vertebral and other features are labyrinthodontine.

Examples: *Amphibamus, Miobatrachus*, etc.

Order 2. **Proanura**

They were ancestral form of Anura in the Triassic period in Madagascar.

1. The trunk was longer than that of the modern frog.
2. The skull unlike that of modern frog.
3. The postcranial skeleton, including caudal vertebrae, was not otherwise anuran; the ilia were elongated.

Example: *Protobatrachus.*

Order 3. **Anura (Salientia)**

Tailless modern toads and frogs with more than 2,200 extant species arranged in ten families.

1. Skull thin without solid roof; bones few.
2. Vertebrae few, and last is a slender urostyle.
3. Ribs reduced or none.
4. Egg deposition and fertilization usually external.

Examples: *Bufo, Rana, Rhacophorus, Xenopus, Ascaphus, Hyla*, etc.

Subclass LEPOSPONDYLI

1. These are small late Palaeozoic amphibians, ranging from a few inches to a few feet in size.
2. The vertebrae lepospondylous, i.e., centra formed directly of bone around notochord, not preceded by cartilage.
3. A few of these animals are superficially snake-like.

Order 1. **Aistopoda**

Small sized, body snake-like with as many as 100 vertebrae and with forked ribs, found in Carboniferous deposits.

Examples: *Ophiderpeton.*

Order 2. **Nectridia**

1. One type was superficially snake-like, limbless and another type with small limbs and a flattened skull, occurred in the Upper Carboniferous.
2. The skull bore an extraordinary armament fromed by the outgrowth of the tabulars.

3. Cranium was largely unossified.

Examples: *Sauropleura, Diplocaulus,* etc.

Order 3. **Microsauria (Adelospondyli)**

1. Permian urodeles but retained too many labyrinthodont features to be closely related to caudata.
2. The neural arches were loosely attached to centrum.
3. Skull incompletely roofed; circumorbital bones lost, which left the orbit open below.
4. Gill breathing in adult continued.

Example: *Lysorophus.*

Order 4. **Phyllospondyli**

1. The broad, heavily roofed skull with a highly diagrammatic arrangement of the investing bones.
2. Two occipital condyles were present in the skull.
3. The notochord and nerve cord remained in the same cavity.

Example: *Branchiosaurus.*

Order 5. **Urodela (Caudata)**

1. Newts and salamanders; degenerate animals with a determinable ancestry from the Cretaceous period. More than 250 species are existing today.
2. The head, trunk and tail usually distinct.
3. Limbs are about equal-sized.
4. Limb girdles largely cartilaginous.
5. No dermal shoulder girdle.
6. Larvae resemble adults in form and have teeth in both jaws.

Modern urodels are placed in eight families.

Examples: *Cryptobranchus, Necturus, Amphiuma, Siren, Typhlomolge, Ambystoma, Trilotoriton, Salamandra,* etc.

Order 6. **Apoda (Gymnophiona, Caecilia)**

1. Aberrant, blind, worm-like, limbless, tropical amphibians with a very short tail and intromittant organ in the male.
2. About 55 species are existing today and without fossil record.
3. The body is usually grooved transversely, and in the grooves are often series of minute scales, a legacy from carboniferous ancestors.
4. The eyes are lidless and sometimes covered by cranial bones.
5. A protrusible sensory tentacle between nostril and orbit present.
6. The vent is almost terminal.
7. The skull compact, roofed with bones.
8. The vertebrae amphicoelous and many.
9. The limb girdles absent.
10. The eggs large and yolky, laid on land.
11. Cleavage meroblastic.

Examples: *Typhlonectes, Rhinotrema, Coecilia, Ichthyophis, Caymnopis, Ureotyphlus,* etc.

40 CLASS REPTILIA

The reptiles are the first true land vertebrates that were adapted in dry places. To colonise on land, their habitat, involves a number of fundamental adaptive changes. The chief changes are those associated with waterproof skin and loss of aquatic adaptation, type of respiration, nature of eggs and fertilization. Originating from a Carboniferous amphibian stock the reptiles underwent fairly extensive radiation during the Permian—that for a hundred million years dominated the earth, air and waters in the Mesozoic era, the age of reptiles. Dinosaurs, pterosaurs were the ruling reptiles, which at the height of their career had a rapid decline due to overspecialisation and at the same time evolution of smaller, swifter, more adaptable, warm-blooded mammals at about the beginning of the Jurassic or even earlier period.

Only four orders comprising about 6,000 species, out of sixteen thousand are existing today. The familiar snakes, lizards, alligators and crocodiles, turtles and tortoises and the New Zealand *Tuatara* are maintaining their existence.

The reptiles form a heterogeneous group, advance over the amphibians in every respect. The keratinous, horny, ectodermal scale is a characteristic and unique feature of all reptiles. The reptilian epidermis is always hardened and cornified sometimes forms substantial plates of horny materials such as scutes of tortoise-shell to protect the soft parts from injury and desiccation. The skin covered by horny epidermal scales (corneoscutes) quite different from dermal osteoscutes of fishes, and the body surface of reptiles is dry and water-proof. The skin has lost most of the glands except some scent glands.

The eggs of reptiles are laid on land. Each egg is covered with a protective porous shell, and has a thick layer of albumen around the yolk and develop embryonic membranes like amnion, allantois and yolk sac.

Amnion preserve watery medium like private aquarium, allantois is the embryonic respiratory organ as well as excretory reservoir and yolk sac with vitelline circulation provides nutrition to the developing embryo. To ensure insemination on land, system of internal fertilization with intromittant organ develops.

The cutaneous respiration of amphibians is replaced by improved lung respiration. The more efficient lung breathing is ensured by the movement of ribes and muscles. In addition to an improved pulmonary circulation with separation of oxygenated and deoxygenated blood; metanephric kidney with ureter developed. The brain became highly developed, associated with much greater activity and complexity of behaviour on land. Thus, the cerebral hemispheres enlarged greatly with introduction of superficial grey matter in the roof of the hemisphere which is the sign of development of neopallium. Two additional pairs of cranial nerves, spinal accessory and hypoglossal are added to the brain region which is regarded as a forward step in cephalization. Well-developed pineal eye and changes in the auditory organs are directly associated with the adaptation on land. The complete ossification of skeleton, quadrate articulation of lower jaw, skull articulation with the first vertebra by a single median condyle, and other changes in the appendicular skeleton are in direct response to survival on land. The presence of insulated body covering, mode of regulation of body temperatures and comparatively well-developed brain regulated their mastery over the land in Mesozoic period. After evolution of more swift, adaptable, warm-blooded, smaller carnivorous mammals which were successful competitors in Cretaceous period, the decline of such monstrous reptiles began.

CALOTES VERSICOLOR (Garden lizard)

The garden lizard is a common reptile in the Indian subcontinent and South China, inhabiting dry, sandy areas with sparse vegetation though they are found in considerable numbers in different parts of India with different climate. The garden lizard belongs to suborder Lacertilia, order Squamata, subclass Lepidosauria, class Reptilia, superclass Gnathostomata, subphylum Vertebrata, phylum Chordata.

☒ External features

1. The body is elongated and slightly compressed laterally. A full-grown lizard is about 25 cm long (Fig. 40.1).
2. The dorsal surface is dark grey and the ventral side pale yellow in colour.
3. The colour of the skin on the ventral surface of the neck of male may change from red to green in breeding season.
4. The body is covered with **scales** overlapping posteriorly.
   a. The scales on the dorsal surface of the head are large and unequal.
   b. The middorsal spiny scales or **frills** are arranged in a row and also along the posterolateral border of the head.
5. The body is divisible into head, trunk and tail. The neck is short.
6. The somewhat triangular **head** is broader posteriorly and narrowing anteriorly ending in a snout.

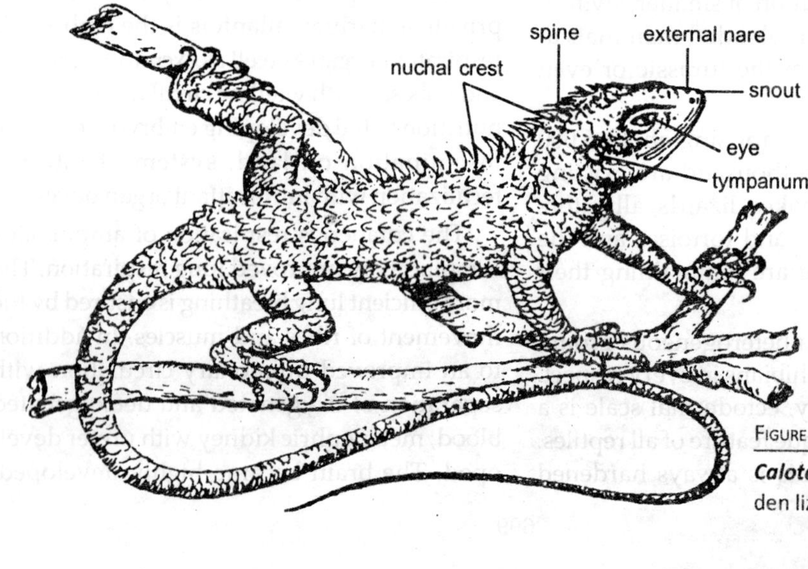

FIGURE 40.1 ***Calotes versicolor*** (Garden lizard)

7. The **mouth** is broad and appears as a transverse slit, bounded by **upper** and **lower** jaws.
8. The jaws bear **homodont teeth**.
9. A pair of **nostrils** open just behind the tip of the snout.
10. The **eyes** are dorsolateral in position and protected by a thick **upper**, a thin **lower** and a transparent **third eyelid** or **nictitating membrane**.
   a. The monocular type of vision is due to the lateral location of the eyes.
11. The **tympanum** is circular, semitransparent, in a sunken area, a little posterior to each eye.
12. Posterodorsal to each tympanum is a **temporal spine** directed backwards.
13. The appendages are two pairs, fore and hind limbs.
   a. The shorter **forelimb** is divisible into upper arm, forearm and hand. The hand consists of wrist, palm and digits.
   b. The **hindlimb** is divided into thigh, shank and foot. The foot consists of ankle, palm and digits.
   c. The limbs are **pentadactyl** and the digits are clawed.
   d. The third and fourth digits of the forelimb and the fourth digit of the hindlimb are longer.
14. The **cloacal aperture** is a transverse slit bounded by a pair of anterior and posterior lips.
15. In the male, a pair of **hemipenes** situated normally in a pair of ventrolateral penial sacs are present at the base of the tail, behind the cloaca.
16. The **tail** is round, long and gradually tapers posteriorly to a point. It is more than twice the length of the trunk.

⊠ Skin

The skin is dry, scaly, nonglandular and made of two layers—outer ectodermal **epider-mis** and inner mesodermal **dermis**.

The epidermis is thin, outer layer cornified to form scales and the deeper layer is known as **germinal epithelium**. The scaly upper layer of the epidermis **moult** occasionally.

The dermis is formed by loose fibrous connective tissue with blood capillaries and nerve endings. **Pigment cells** are present in the outer layer of dermis, while fat bodies are found in the deeper layer (Fig. 7.10A). The dermis is loosely attached to the underlying muscles of the body wall.

⊠ Skelatal System

Exoskeleton and endoskeleton constitute skeletal system.

Exoskeleton

The exoskeleton (p. 79) is well-developed and makes the body impervious to water to prevent desiccation. It also acts as defensive and offensive structures.

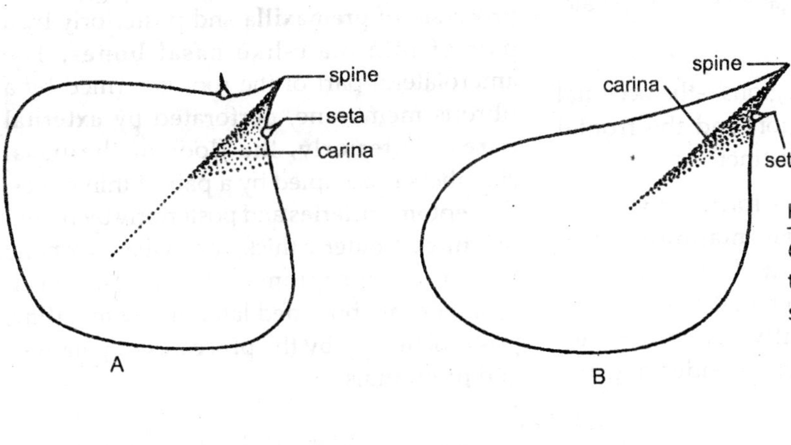

FIGURE 40.2

***Calotes* sp.** Body scales. A. A typical scale. B. A mucronate scale

Scales

The scales covering the body are **cornified epidermis**, the base planted in the dermis and are free along the posterolateral margins (Fig. 7.10A). The scales (Fig. 40.2) form a continuous covering of the body; the dorsal ones being larger than the ventral ones. The scales are of two types—large and small, the latter situated between the rows of the former. Sensory bristles, the **prototrichus** (singular prototrix) are present in some of the scale of both types. The head region is covered by **head shields**, large scales not overlapping one another. The scales along the middorsal line and the posterolateral border of the head are pointed and termed **frills**.

Claws

All the digits of both fore and hindlimbs end in sharply pointed, curved, horny claws and cover the terminal phalanx of digit.

⊠ Endoskeleton

Hard structures inside the body and supporting it constitute endoskeleton. It has two components—the axial skeleton and appendicular skeleton.

Axial skeleton

The axial skeleton includes skull, vertebral column and visceral skeleton.

Skull

The skull (Fig. 40.3A, B) is composed of cranium enclosing the brain, the sense capsules accommodating sense organs, jaws and palate.

A. Cranium

The cranium has three regions—the occipital region, the parietal region and the frontal region. The sutures are distinct.

Occipital region. It is formed by bones bounding the **foramen magnum. The supraoccipital**, and a quadrangular bone forming the roof of the cranial cavity in the occipital region is slightly convex. Paired **exoccipitals** form the lateral sides and the plate-like **basioccipital** constitutes the floor. The posterior margin of the basioccipital forms the **occipital condyle** in conjunction with exoccipitals.

Parietal region. Parietal region is composed of two bones. The flat, squarish, plate-like **parietal bone** forms the roof of the cranial cavity and situated in front of the supraoccipital. The thick, quadrangular **basisphenoid** forms the floor of the cavity and placed in front of basioccipital. The lateral walls are made partly by **prootics** and partly by **fibrocartilaginous membrane**.

Frontal region. The frontal region consists of four bones. The **frontal bone** is flat, somewhat inverted and T-shaped forming most of the roof of the interorbital septum in front of the cranial cavity. The **presphenoid** is long, rod-like, projecting anteriorly from the basisphenoid and forms the bottom of the interorbital septum. The paired **orbitosphenoids** are rod-like, slightly curved, vertically placed in the fibrocartilaginous membranous walls, closing the cranial cavity anteriorly.

B. Sense capsules

Sense capsules include paired olfactory, optic and auditory capsules located in the corresponding region.

Olfactory region. It is formed by a pair of closely placed **nasal chambers** in front of the orbits. A nasal septum in continuation with the **interorbital septum** separates the two nasal chambers. Anteromesially the nasal chambers are roofed by two median dagger-like processes of **premaxilla** and posteriorly by a pair of thin plate-like **nasal bones**. The anterolateral part of the roof is formed by a fibrous membrane, perforated by **external nares**. Anteriorly, the floor of the nasal chambers is occupied by a pair of thin bones, the **septomaxillaries** and posteriorly by paired **palatines. Vomer**, a thick, somewhat oval bone supports the septomaxillaries. The nasal chambers are bounded laterally by **maxillae**, posterolaterally by the processes of palatines and prefrontals.

Optic region. It consists of two orbits accommodating eye-balls on either side of the frontal bone. An orbit is bounded by **frontal** dorsally, by **prefrontal**, slightly curved and with a pair of anterior processes anterodorsally, by **maxilla** anteroventrally, by **jugal**, rhomboidal in shape and bearing a narrow anteroventral process posteroventrally and a slightly curved **postorbital** bone posteriorly. **Interorbital septum**, a median, longitudinal, vertical fibrocartilaginous membrane separates the two orbits.

Auditory region. Paired auditory capsules associated with occipital and parietal regions in the posterior part of the skull form the auditory region. Each capsule is formed mainly by a triradiate **prootic** bone. Excavations of prootic, supraoccipital and exoccipital take part in the formation of the capsule.

A large, oval **supratemporal fossa** is present on each side in the temporal region. A slender, splint-like bone, the **tabular** borders supratemporal fossa. The **squamosal** is a short, curved bone lining the outer boundary of the supratemporal fossa.

Palate. The palate (Fig. 40.3B) forms the roof of the buccal cavity, just below the cranium and sense capsules. It is formed by an unpaired **vomer**, paried **palatines, pterygoids, epipterygoids** and **ectopterygoids**.

C. Jaw apparatus

The jaw apparatus is formed of two parts—upper and lower jaws. The upper jaw is fixed to the skull along its ventral side and forms its anterior and lateral boundaries (Fig. 40.3B), while the lower jaw is movable.

The **upper jaw** consists of a **premaxilla**, paired **maxillae, jugals** and **quadrates**. The **mandible** or lower jaw is made of two similar halves (rami) joined anteriorly by Meckel's cartilage. Each half or **ramus** is formed by six bones (Fig. 40.3C),—**dentary, angular, supraangular, splenial, coronoid and articular**.

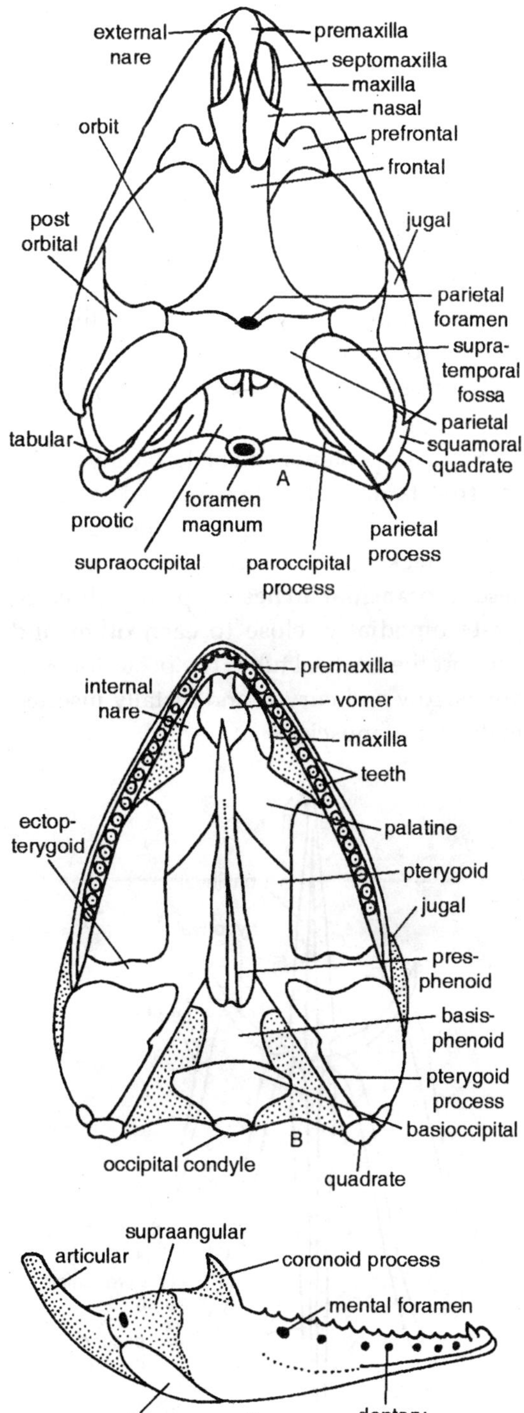

FIGURE 40.3

***Calotes* sp.** Skull and mandible. A. Skull. Dorsal view. B. Skull. Ventral view. C. Mandible (right half). Outer view

⊠ Hyoid apparatus

The hyoid apparatus consists of a hyoid arch and two branchial arches (Fig. 40.4).

The **hyoid arch** is formed by: (a) a **basihyal**, a short, median cartilage, grooved lengthwise, with a tapering and curved anterior end supporting the tongue, (b) **hypohyals**, paired, slender, projected anterolaterally from the base of the basihyal and support the gullet and (c) **ceratohyals**, paired, slender, directed posterolaterally from the hypohyals end support the gullet.

From the posterior end of the basihyal arise the **first branchial arches**—paired, ossified, long and slender, directed posterolaterally in the neck region to the tympanum. The **second branchial arches** are paired, directed posteromedially, close to each other and support the tracheal tube. The posterior ends are narrow and curved and distally inserted in the throat muscles.

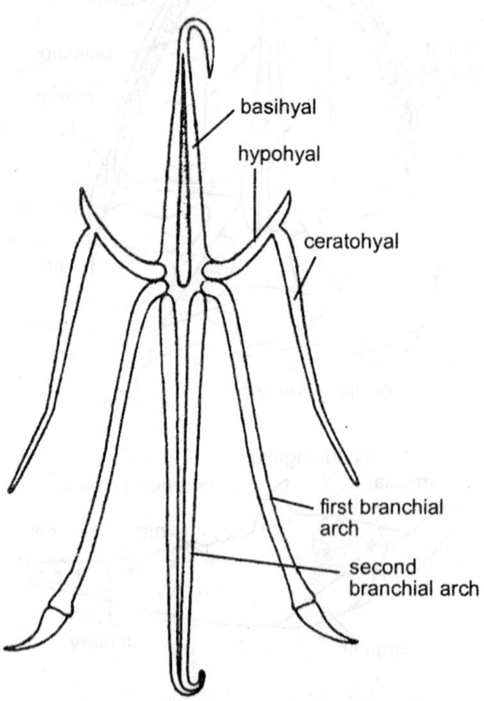

FIGURE 40.4

***Calotes* sp.** Hyoid apparatus

⊠ Vertebral column

The vertebral column formed of a fixed number of **procoelous** vertebrae (p. 667), runs along the neck, trunk and continues in the tail. It is slightly arched in the trunk region between the pectoral and pelvic girdles, which is possibly of some functional importance. The number of vertebrae in the vertebral column is: cervical-8, thoracolumbar-15, sacral-2 and caudal-38-50.

The basic components of a typical vertebra are—a ventral centrum and two lateral neural arches enclosing a neural canal. Pre- and post zygapophyses are present in all vertebrae from axis to caudal.

Atlas (First cervical vertebra). Shape ring-like (Fig. 40.5); **hypocentrum** (incomplete centrum) occupies ventral position; the cavity enclosed by neural arches is horizontally divided into an upper neural canal and a lower **odontoid fossa** by a ligament; a concave **articular facet** ventrally on the anterior face; **hypapophysis** rudimentary; **neural spine** and **prezygapophyses** absent; a pair of **postzygapophyses** and blunt posterolateral processes present.

Axis (Second cervical vertebra). **Odontoid process** (centrum of the first vertebra) bears hypapophysis (Fig. 40.6) in the form of a spine; neural spine crest-like* and directed posteriorly; rudiments of **diapophyses** in form of anterolateral tuberosities.

Cervical vertebra (Typical). Procoelous, centrum well-developed (Fig. 40.7); neural spine posteriorly directed; two anterior **capitular facets** (in the last four [5-8] cervical vertebrae for attachment of ribs); a ventral **hypapophysis**.

Thoracolumbar vertebra. Two **capitular facets** anteriorly; hypapophysis absent (Fig. 40.8).

Sacral vertebra. Two transverse **processes**; capitular facets absent (Fig. 40.9).

Caudal vertebra (anterior). Two transverse processes; a ventral y-shaped chevron bone (Fig. 40.10) encloses **haemal canal**.

Ribs

The ribs are movably attached to the capitular facets of the last four (5-8) cervical and all the thorocolumbar vertebrae. The ribs are single-headed, slender, slightly curved bony rods, usually with cartilage at the tip. Thoracic ribs (1-4 thoracolumbars) are complete, i.e., they curve around and reach the sternum. Other ribs (5-8 cervicals and 5-15 thoracolumbars) are incomplete, i.e., they do not reach the sternum.

Sternum

It is a small, rhomboidal cartilaginous plate (Fig. 40.11), placed ventrally in the thorax. Two large openings, the **sternal fenestrae**, one on each side of the middle line are present. Anterolaterally, the sternum is movably attached to the pectoral girdle and partly covered by interclavicles. Posteriorly it forms xiphisternum.

Appendicular skeleton

It consists of two gridles and limb bones.

Girdles

Girdles are two—the pectoral girdle supporting the forelimb bones and pelvic girdle supporting the hindlimb bones.

Pectoral girdle

The two halves of the pectoral girdle are similar except the median episternum and sternum (Fig. 40.11). The two halves meet along the midventral line. Each half consists of—

Suprascapula. A thin dorsal plate of partly ossified cartilage, attached to the dorsal end of scapula.

Scapula. A short bone attached to the outer end of coracoid.

Coracoid. A flat bone, articulates with sternum.

Precoracoid. Cartilaginous and forms the anterior portion of the coracoid.

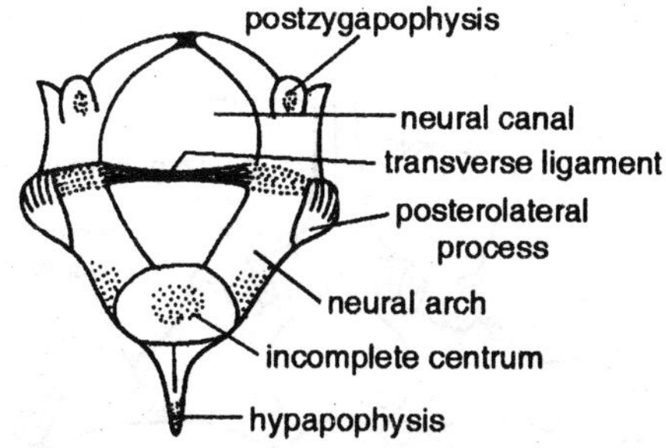

FIGURE 40.5

***Caltoes* sp.** Atlas (First vertebra) Posterior view

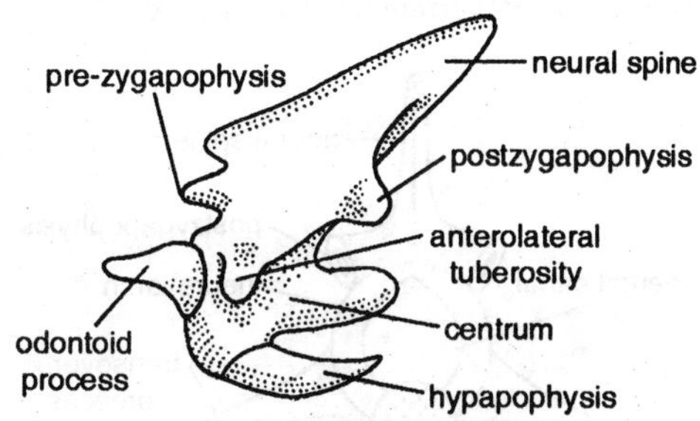

FIGURE 40.6

***Calotes* sp.** Axis (Second vertebra). Lateral view

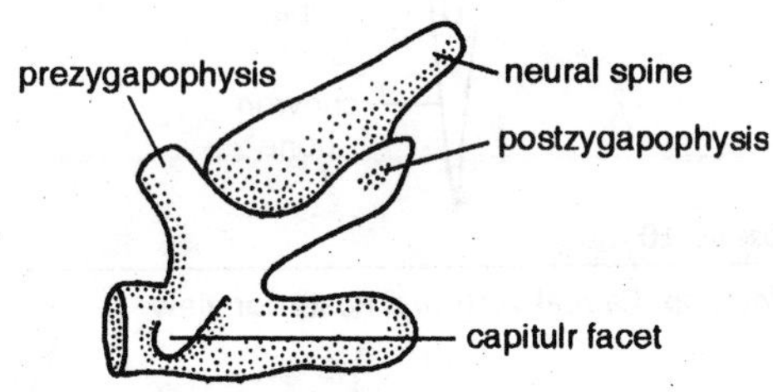

FIGURE 40.7

***Calotes* sp.** Cervical (Fourth vertebra). Lateral view

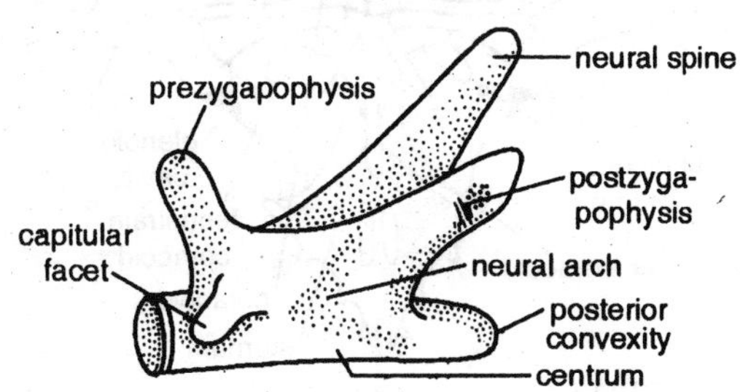

FIGURE 40.8

***Calotes* sp.** Thoracolumbar vertebra. Lateral view

* In all the vertebrae, axis to caudal, the neural spine is crest-like

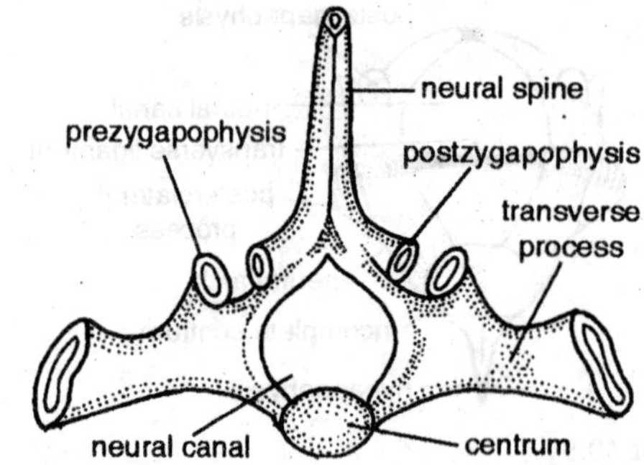

FIGURE 40.9

***Calotes* sp.** Sacral vertebra. Posterior view

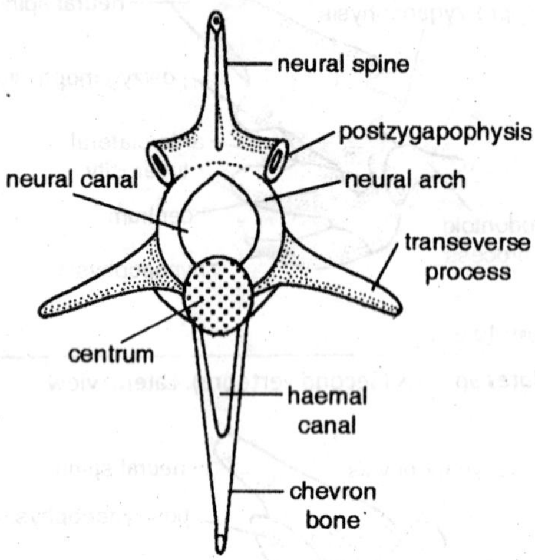

FIGURE 40.10

***Calotes* sp.** Caudal vertebra. Posterior view

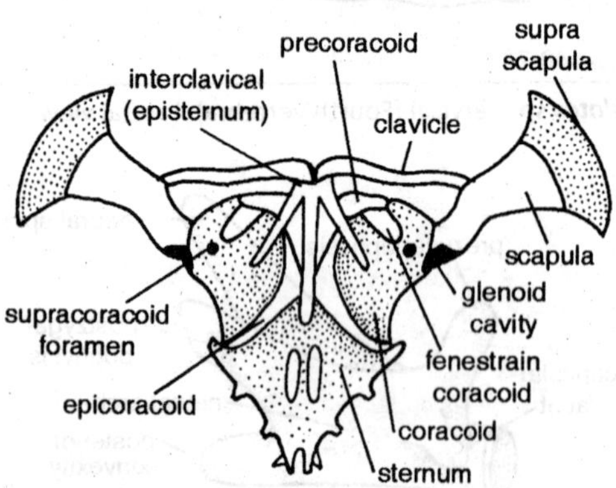

FIGURE 40.11

***Calotes* sp.** Pectoral girdle. Ventral view

Epicoracoid. Cartilaginous and located at the inner side of the precoracoid and coracoid.

Clavicle. A curved bone, joins the lateral arm of the episternum (interclavicle) and the scapula.

Episternum. A somewhat cross-shaped bone, ventral to epicoracoid and sternum and attached to them.

Sternum. A medial rhomboidal bone and bears ribs at posterolateral margins.

Glenoid cavity. A cavity at the junction of coracoid and scapula in which head of the humerus fits.

Pelvic girdle

The pelvic girdle is formed by the fusion of two similar halves except the unpaired hypoischium cartilage (Fig. 40.12). Each half is constituted by the following:

Ilium. Large, rod-like, runs forward and upward from the acetabulum.

Ischium. Flattened, slightly curved, ventral and posterior to the pubis. The two ischia join in the ischiatic symphysis.

Hypoischium. Calcified cartilage, forked at the tip and extends posteriorly from the junction of the two ischia.

Pubis. Slightly curved, extends posterior and outward and the two pubes meet at the pubic symphysis.

Epipubis. a small nodule of calcified cartilage between the anterior ends of the two pubes.

Acetabulum. A cavity, on either side at the junction of ilium, ischium and pubis in which head of the femur fits.

Limb bones. Bones supporting the limbs are limb bones.

Forelimb

The forelimb bones (Fig. 40.13) are—

Humerus (upper arm bone). A strong stout, elongated bone, slightly narrow at the middle. Both ends are swollen. The swelling of the proximal end is the head. The distal end bicondylar. Epiphysis of calcified cartilage present at both ends. Deltoid ridge absent.

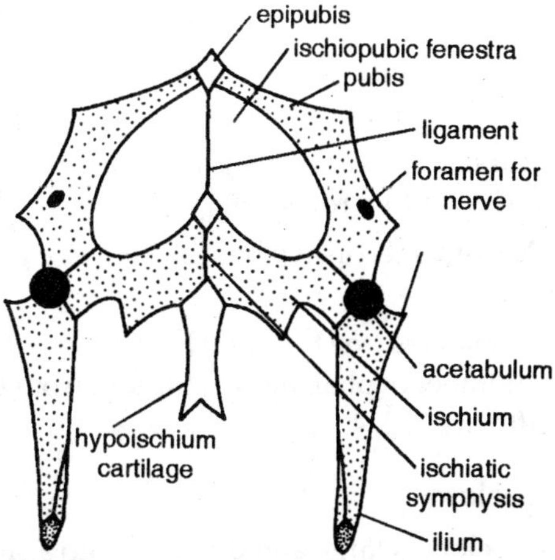

FIGURE 40.12

***Calotes* sp.** Pelvic girdle. Ventral view

Radius and ulna (fore arm bone). Two elongated, separate bones, a little narrow at the middle. Radius slightly narrow—proximal end round, distal end bears a styloid process and a concave facet. Ulna slightly stout—proximal end bears olecranon process, distal end provided with a convex articular facet.

Carpals (wrist bones). Small nodular bones, nine in number, arranged in two rows. Proximal row consists of four bones and the distal row bears five bones.

Metacarpals (palm bones). Five elongated bones.

Phalanges (finger bones). The digits are five, each formed by a number of phalanges. From first to fifth the number of phalanges are 2, 3, 4, 5, 3.

Hindlimb

The hindlimb bones (Fig. 40.14) are:

Femur (thigh bone). A strong, stout, slightly curved elongated bone. Epiphysis present at both the ends. Proximal end bears a round head. A rough-surfaced trochanter present near the head. Distal end bears two ball-shaped condyles.

Tibia and fibula (shank bone). Two elongated, separate bones narrow at the middle. Tibia stout, slightly curved end bears a cnemial crest, proximal end bears two concave articular facets; distal end possesses one concave articular facet. Fibula slender—proximal end bears a concave articular surface; distal end concave.

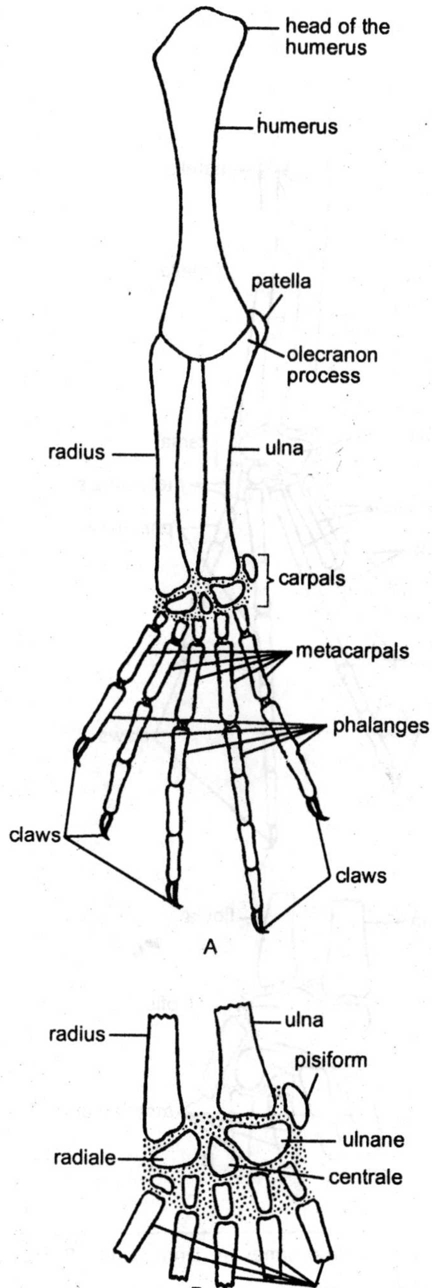

FIGURE 40.13

***Calotes* sp.** Forelimb bones. A. Limb bones. B. Distal bones (part)

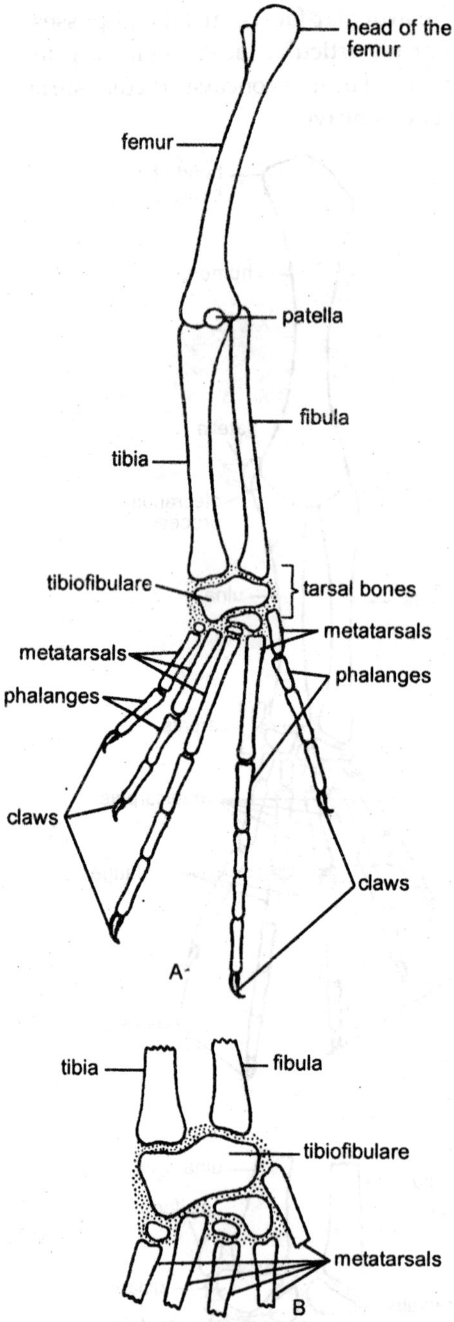

FIGURE 40.14

***Calotes* sp.** Hindlimb bones. A. Limb bones. B. Distal bones (part)

A sesamoid bone (a bone developed in a tendon), **patella** is present at the junction of the femur and tibia and fibula.

Tarsals (ankle bones). Four small bones arranged in two rows.

Metatarsals (instep bones). Five elongated bones.

Phalanges (toe bones). The digits are five. The number of phalanges in the toes from first to fifth are 2, 3, 4, 5, 4.

⊠ Coelom

The coelom is large and divisible into a small **pericardial cavity** containing heart and a large **pleuroperitoneal (pleurovisceral)** cavity in which visceral organs are lodged. The cavities are separated by a thin oblique septum.

The coelom is lined by an epithelium of mesodermal origin. The membrane lining the coelom is called **peritoneum**. It leaves the dorsal wall and forms a double membrane, the **mesentery**, which supports visceral organs.

⊠ Digestive system

The digestive system consists of a long alimentary canal and digestive glands.

Alimentary canal

The alimentary canal (Fig. 40.15A) begins with mouth and ends in cloaca. It is divided into buccopharynx*, oesophagus, stomach, intestine and rectum.

Mouth

The mouth is terminal. It is a large, somewhat inverted V-shaped gape, round in front. The mouth is bounded by a pair of jaws covered with a thin lip-iike skin. The jaws bear homodont teeth.

Buccopharynx

The mouth leads to a spacious buccoph-arynx, the anterior region is the **buccal cavity** and the posterior region, **pharynx** (Fig. 40.15B). In the

* The buccal cavity and pharynx together form a single undifferentiated chamber. The terminology 'buccopharynx' is quite appro-priate

buccal region a short, broad, fleshy, sagittate-shaped **tongue** free at the anterior end is lodged on the floor and two slit-like openings, the **internal nares** are present in the roof. Numerous papillae with goblet cells are present on the free surface of the tongue except the tip. Mucous membrane of the pharynx is thrown into longitudinal folds. The pharynx communicates with the oesophagus through **gullet**. A vertical slit-like opening—the **glottis,** just behind the tongue leads to the **larynx** of the respiratory system. The glottis is supported by cartilages.

Oesophagus

It is tubular, runs in the neck dorsal to trachea and ends in stomach. The mucous membrane is thrown into longitudinal folds.

Stomach

The stomach is thick-walled, swollen, tubular and lies on the left side, somewhat behind the heart and dorsal to the liver. It has three regions—the anterior **cardiac stomach** into which the oesophagus opens, the middle **sac-like portion** and the posterior **pyloric stomach** opening in small intestine. A **pyloric sphincter** is present at the junction of the pyloric stomach and doudenum. The mucous membrane is thrown into longitudinal folds, some of which are prominent.

Small intestine

A long, thin-walled tube thrown into a few loops in the middle and opens posteriorly in the large intestine. The small intestine is divisible into two zones—the slightly curved, anterior region, next to pyloric stomach is **duodenum** and the rest long, looped part **ileum**. The bile ducts and the pancreatic duct open in the duodenum. The mucous membrane is thrown into long, delicate folds.

Large intestine

It is a short tube with an anterior dilated portion, the **colon,** passing imperceptibly into a short, narrow, posterior tubular portion the **rectum** opening in the cloaca. A small, blind, finger-like appendage, the **caecum** arises from near the junction of small intestine and colon.

The wall of the alimentary canal is histologically similar to that in other vertebrates. It consists of a mucosa thrown into folds and bears glands, a submucosa, a muscularis externa and a serosa, the outer most coat.

☒ Digestive glands

Liver and pancreas (Fig. 40.15A) constitute digestive glands.

Liver

The liver is a large, dark red, bilobed gland situated for the most part, ventral to the stomach. The right lobe is longer than the left lobe and extends behind the intestine close to the right gonad. **Gall bladder** is a greenish (due to the colour of the bile) sac embedded in the posterior margin of the right lobe at the point of separation of two lobes of the liver. A **bile duct** arises from each lobe of the liver and the two ducts from two lobes run close to each other to open separately in the duodenum. A small **cystic duct** connects the gall bladder and the right bile duct.

Pancreas

It is a compact, ribbon-like gland, closely applied to the outer surface of the pyloric stomach and duodenum, being lodged in the loop formed by them. A part of the gland extends along some length of the bile ducts. The pancreatic duct opens into the duodenum close to the openings of the bile ducts.

☒ Food and Feeding

The lizards are chiefly insectivores. It has been reported that they feed on earthworms, small eggs of birds, tiny frogs, etc.

Small animals are captured by tongue. During feeding the tongue is protruded and the food organisms get stuck to the mucus, covering the anterior part of the tongue. The larger animals like cockroaches, beetles, grasshoppers and earthworms are captured and crushed by snapping of toothed jaws. The teeth also prevent escape of the prey. The food is swallowed and passed to the oesophagus through gullet; the hind part of the tongue closes the glottis during feeding.

☒ Digestion and absorption

Not much is known about the secretion of hydrolysing enzymes in the first part of the alimentary canal. Digestion is similar in all vertebrates (p. 86). Hydrochloric acid and pepsin are secreted in the stomach. HCl kills living organisms and also helps in acid hydrolysis. Pepsin reacts with protein in the food. Partly digested food passes to the duodenum. In this region bile emulsifies fat. Pancreatic juice makes the food alkaline and enzymes there hydrolyse both protein and carbohydrate and to some extent lipid.

The nutrients, amino acids, monosaccharides and certain amount of mono- and diglycerides and undigested triglyceride (lipid), now in a fluid state pass to ileum with undigested portion of the food. The nutrients and water are absorbed in ileum, the faeces stored temporarily in the rectum and finally thrown out through cloacal aperture, via cloaca.

☒ Respiraory System

The respiratory structures consist of external nares, nasal chambers, internal nares, pharynx, larynx, glottis, trachea and lungs.

The **external nares** are a pair of small anterolateral openings on the snout near its tip. An external nare leads into a **nasal chamber** opening anteriorly into the buccopharynx through an **internal nare** in the roof. Glottis is situated on the floor of the buccopharynx in the median impression of the posterior margin of the tongue. Glottis leads into larynx and trachea. During feeding, the hind portion of the tongue closes the glottis to prevent entrance of food into the latter. The **larynx** is a

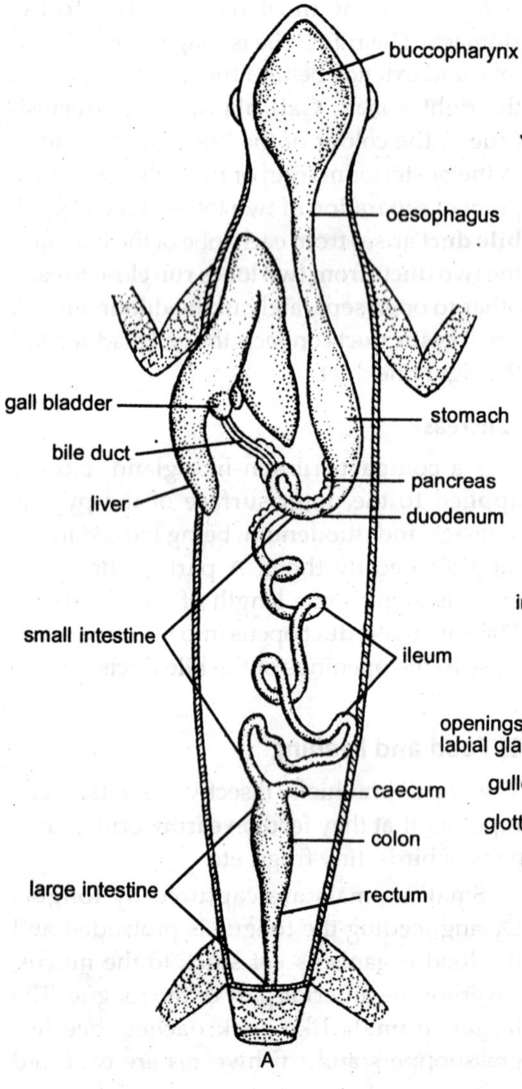

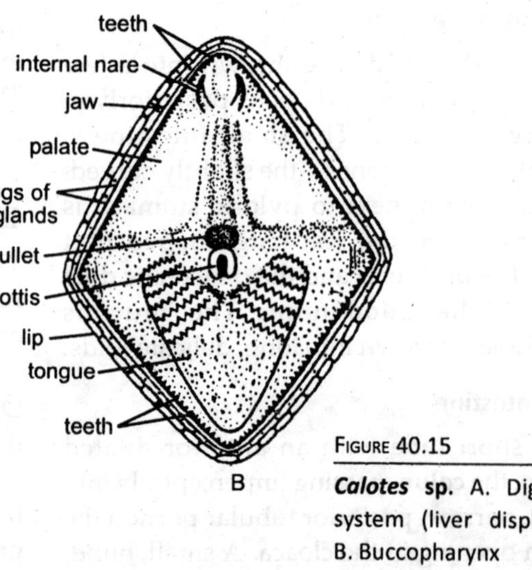

FIGURE 40.15

***Calotes* sp.** A. Digestive system (liver displaced). B. Buccopharynx slightly swollen structure bounded by three cartilages. The trachea (Fig. 40.16) is along tube in the neck region, below the oesophagus, supported by cartilaginous rings. The trachea bifurcates into two bronchi close to the lung and each opens into a large cavity in the lung.

The lungs are a pair of spindle-shaped, brownish sacs, lying somewhat dorsolaterally in the pleuroperitoneal cavity, extending posteriorly up to the gonads. The lung cavity is narrower anteriorly and wider posteriorly. The wall is spongy anteriorly and sac-like posteriorly.

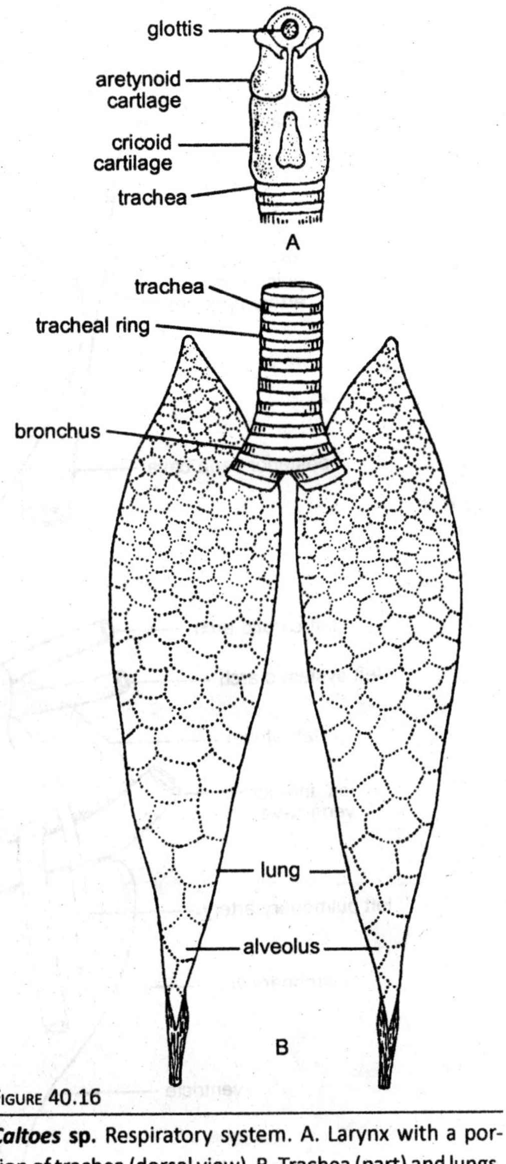

FIGURE 40.16

***Caltoes* sp.** Respiratory system. A. Larynx with a portion of trachea (dorsal view). B. Trachea (part) and lungs.

☒ Mechanism of respiration

The mechanism of respiration is similar* in all vertebrates possessing ribs (p. 96). The respiratory movement is controlled by thoracic muscles. The contraction of intercostal muscles raises the ribs, the size of the pluroperitoneal cavity increases and the pressure within the cavity is reduced. This in turn reduces pressure in the lungs and causes an inflow of air, **inspiration**, through nasal passages, larynx and trachea. Conversely, relaxation of muscles causes lowering of the ribs, the pleuroperitoneal cavity returns to original volume, resulting in an increased pressure in the lungs, leading to forcing out air, **expiration**, through the same route. Inspiration is passive, expiration is active.

☒ Blood Vascular System

A heart, arteries, veins and capillaries constitute blood vascular system.

The heart is situated anteromedially and ventrally in the pericardial cavity. The pericardium enclosing the heart is thin but tough and white in colour. The heart is broadly conical (Fig. 40.17A, B) with a broad anterior end and a narrow posterior end. It consists of a **sinus venosus**, right and left atria (auricles) and a ventricle.

The **sinus venosus** is thin-walled and situated on the dorsolateral side of the heart overlapping the major part of the right atrium and a part of the right side of the ventricle.

The **atria** are sac-like, broad, placed anteriorly and transversely.

The **ventricle** is conical, highly muscular, incompletely divided and located posterior to the atria.

* In mammals, diaphragm plays an important role in respiration.

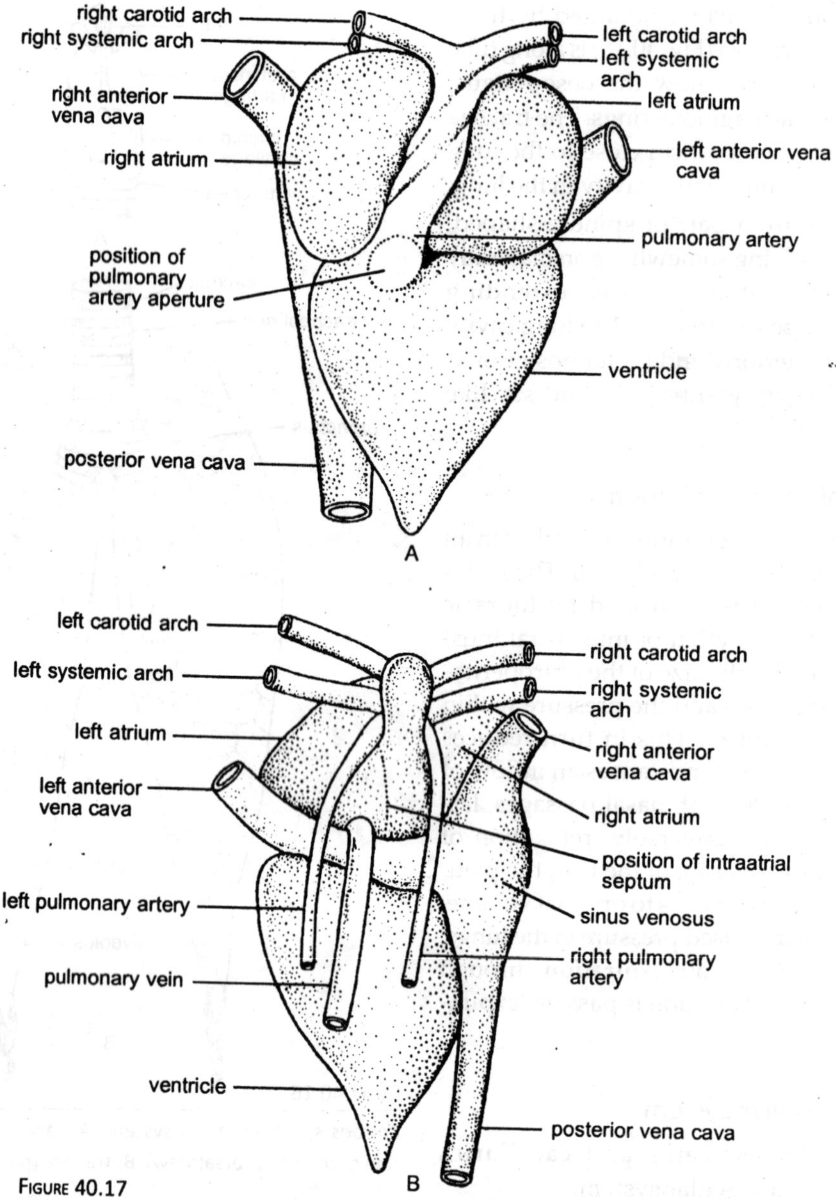

FIGURE 40.17

***Calotes* sp.** Heart and principal blood vessels. A. Ventral view. B. Dorsal view.

A shallow **interatrial fissure** is present at the junction of the atria. A deep atrioventricular fissure, the **coronary sulcus** runs transversely at the juncton between the atria and the ventricle. A short **diverticulum** is present at the anterior mesial corner of the right atrium.

The three venae cavae open at the three corners of the sinus venosus without any vulvular arrangement. The sinus venosus communicates with the right atrium through **sinuatrial aperture** guarded by the folds of the atrial wall. The left atrium, which is smaller than the right one, receives the pulmonary vein dorsally near the interatrial septum. The interatrial septum is complete and extends on to a flap of **atrioventricular valve**, resulting in the separate passages for blood from two atria to the ventricle through two openings guarded by a pair of thick, concave valves opening into the ventricle. The inner walls of the atria are raised into muscular ridges the **musculi pectinati**.

Internally, the ventricle is provided with an almost complete, oblique **interventricular septum**, separating the right ventral part of the ventricle into **cavum ventrale**. Being adjoining to the base of the pulmonary artery, it is also called **cavum pulmonale**. The rest of the ventricle constitutes the **cavum dorsale**. The cavum dorsale is partially subdivided by myocardial trabeculae. A prominent ridge is formed by the fusion of some of these myocardial trabeculae, resulting in a partial division of the cavum dorsale—**cavum arteriosum** on the left and **cavum venosum** on the right. So the ventricle presents three incompletely separated cavities (Fig. 40.18).

Three principal aortae arise from this tripartite ventricle. Each principal artery is guarded basally by a pair of thin, membranous, pocket-like valves. Pulmonary artery comes up from the right side (cavum ventrale). Dorsal to it, the left systemic arch arises. From the more dorsal side and to the left of the left systemic arch, the right systemic arch originates.

Circulation through heart

Venous blood from sinus venosus enters the right atrium. Oxygenated blood comes to the left atrium from lungs by pulmonary veins. The atria contract and the blood from both empties into the cavum dorsale of the ventricle. Blood enters the cavum ventrale from the cavum dorsale over the edge of the incomplete septum. The contraction of the atrium is followed by the contraction of the ventricle. The interventricular septum is so placed that its free edge is brought against the opposite walls of the ventricle resulting in a functional division of the ventricle in such a way as to connect the cavum ventrale with the base of pulmonary artery.

The difference in the size of the two atria, the left being smaller and the right larger and the differential capacity of the cavum dorsale and cavum ventrale stand in the way of sending equal amount of blood to the systemic and pulmonary circulation; resulting in a somewaht mixture of arterial and venous blood. Oxygenated blood from the left atrium passes to the left part of the cavum dorsale and from there to the body through both carotid arches and the right systemic arch. The blood from the right part of the cavum dorsale move out through left systemic arch and probably into the left carotid through ductus caroticus. Deoxygenated blood from the right atrium enters the right portion of cavum ventrale and from there to the lungs through pulmonary arteries.

Arterial system

The arteries end in capillaries in the tissues. All the arteries and the heart constitute the arterial system.

Three principal arches—right and left systemic and pulmonary—arise directly from the tripartite ventricle. Each principal artery is guarded basally by a valve. The pulmonary arch takes its origin from the cavum ventrale which is ventralmost in position. The right and the left systemic arches take their origin from the left and right side of the cavum dorsale of the ventricle, respectively. These two trunks cross each other in course of their forward passage and have a change in their dorsoventral course, as evident by twist of the truncus (Fig. 40.19).

Systemic trunks

The origin of the left systemic is a little anterior and right to that of the pulmonary trunk. The right systemic trunk is slightly mesial and anterior to that of the left one. The systemic trunks loop round the alimentary canal, turn dorsal and run posteromedially. Posteriorly, they meet each other to form the dorsal aorta. The left systemic trunk twists around itself at the anterior end of the arch and forms a small

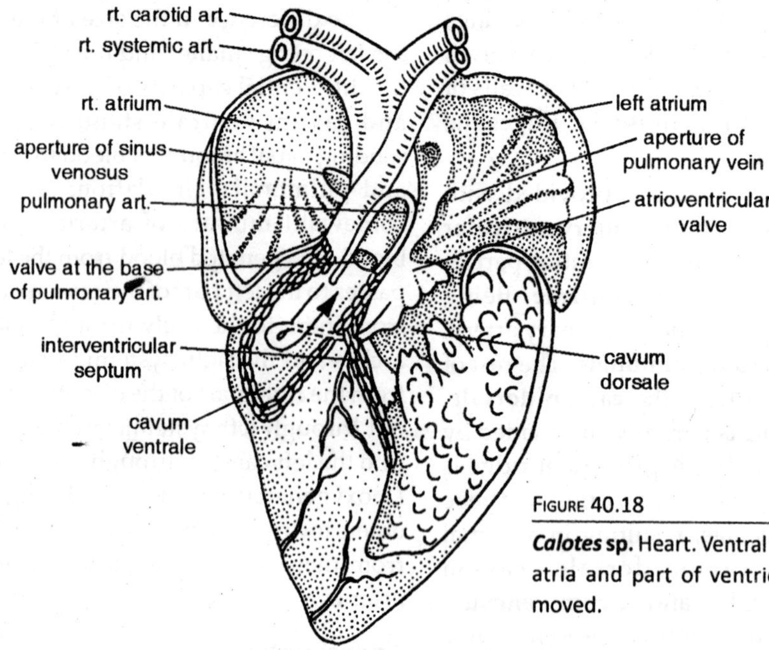

FIGURE 40.18

Calotes **sp.** Heart. Ventral wall of atria and part of ventricle removed.

loop. The branches of the systemic trunks are not similar. The right systemic carries almost fully oxygenated blood and, unlike the left, sends branches to the cephalic and anterior region of the head.

Right systemic trunk (arch)

It sends the following branches.

Coronary. A small vessel, arises close to the origin of the trunk, turns backward and supplies blood to the heart.

Carotids. The arteries arise close to each other and run by the side of the oesophagus, known as right and left carotids. Near the throat each carotid divides into a pair of branches, the **external** and **internal carotid**. At the junction of the branches, the **ductus caroticus** joins the carotid artery.

External carotid. Superficial in location and gives off branches to buccopharynx, tongue, larynx and trachea.

Internal carotid. Deep seated, divides into two branches—palatine and stapedial. The **palatine** supplies the palate region and sends a median **hypophysial**, which enters the cranial cavity and supplies the brain. The **stapedial** artery enters the cranial region, divides into two branches—**mandibular** supplying the lower jaw region and **temporal** sending branches, the **superior** and **inferior orbitals** to orbital areas and **maxillary** to the upper jaw.

Ductus caroticus. Arises from a point little behind the anteriormost end of the arch formed by the systemic trunk and runs forward to join the carotid artery.

Oesophageal. Three to four small vessels, posterior to the root of the ductus caroticus supply the oesophagus.

Vertebral. Posterior to the oesophageals, and ends in the region of cervical vertebrae through a number of branches.

Subclavian. A pair, the right and left. They have a common origin and supply the right and left forelimbs and the shoulder. Each subclavian sends branches, the **scapular** and **coracoid** to the shoulder region. The main artery proceeds laterally into the forelimb as **brachial artery**.

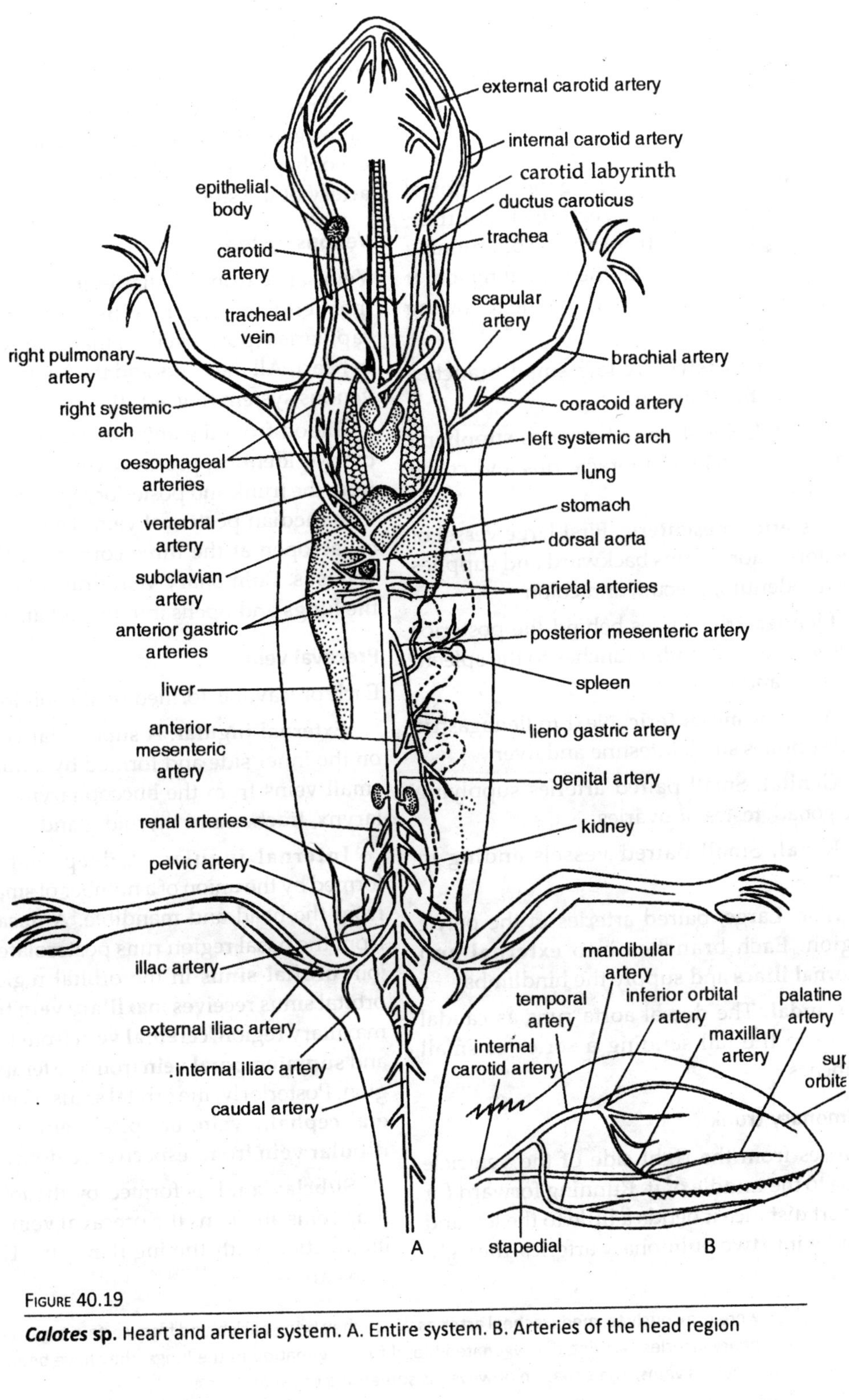

FIGURE 40.19

***Calotes* sp.** Heart and arterial system. A. Entire system. B. Arteries of the head region

Left systemic trunks

It sends only the **ductus caroticus** and **oesopageal arteries**, which are similar to those arising from the right systemic trunk.

Dorsal aorta

It runs posteromedially along the mid-dorsal line just below the vertebral column, remaining free in the trunk region but running through haemal canal in the caudal region. It sends many branches of which the following are important:

Anterior gastric. A few small branches supplying the stomach.

Parietal. Small paired arteries supplying the trunk vertebrae, thoracic ribs and costal muscles.

Posterior mesenteric. First large vessel of the dorsal aorta, runs backward and supplies the duodenum, caecum and colon.

Lienogastric. Arises behind the posterior mesenteric and sends branches to the spleen, intestine and liver.

Anterior mesenteric. Next to lienogastric and supplies small intestine and liver.

Genital. Small paired arteries supplying the gonad, testes or ovaries.

Renal. Small paired vessels ending in kidneys.

Iliac. Large, paired arteries in the pelvic region. Each branches into **external** and **internal iliacs** and supply the hindlimbs.

Caudal. The dorsal aorta runs as caudal artery in the tail sending a series of small branches.

Pulmonary trunk

It arises from the right side of the ventricle close to the middle of it. Running forward for a short distance it bends a little to the left and divides into two pulmonary arteries, the right and left.

Pulmonary artery. It loops over the heart anterodorsally, proceeds backward to ramify in the lung of the side.

Tracheal vein*. A vessel bringing deoxygenated blood from the side of trachea, oesophagus and larynx joins the pulmonary artery of the side.

Venous system

Veins arise from capillaries in the tissues and may end in larger veins or terminate in capillaries (portal vein). Finally, all veins end in heart. All the veins and the heart constitute venous system (Fig. 40.20).

Blood from the anterior region is drained by two identical precaval veins, while blood from the trunk and posterior region is collected by a median postcaval vein. The three venae cavae open at the three corners of the sinus venosus. Pulmonary vein drains blood from the lungs and opens into the left atrium.

Precaval vein

Each precaval is formed by the joining of—

External jugular. A superficial vein lying on the inner side and formed by a number of small veins from the buccopharynx, tongue, larynx, trachea and thyroid gland.

Internal jugular. A deep-seated vein formed by the union of a number of small veins from the head and mandible. A **nasal sinus** from the nasal region runs posterolaterally to join **orbital sinus** in the orbital region. The orbital sinus receives **maxillary vein** from the maxillary region, **cerebral vein** from the brain and **supratemporal vein** from the temporal region. Posteriorly, the orbital sinus receives **lateral cephalic vein, occipital vein** and **mandibular vein** from respective regions.

Subclavian. It is formed by the joining of four veins and joins the precaval vein close to its junction with the jugular vein. The four veins are—

* The vessels were previously termed **tracheal arteries**. Since they bring back blood from certain organs and open in the pulmonary arteries carrying deoxygenated blood for oxygenation in the lungs, they have been properly renamed as tracheal veins. The situation however, is something unusual.

Brachial from the forelimb, **Subscapular** from the scapular region and skin and muscles of the back, **Pectoral** from the pectoral region and **Anterior vertebral** from the anterior portion of the vertebral column.

Coronary venous sinus. Collects blood from the heart and opens into the left precaval vein.

Postcaval vein or Vena cava

It is a large vein situated a little to the right of the middle line. The major part of it is embedded in the right lobe of the liver. It arises in front of the kidneys and receives the following veins:

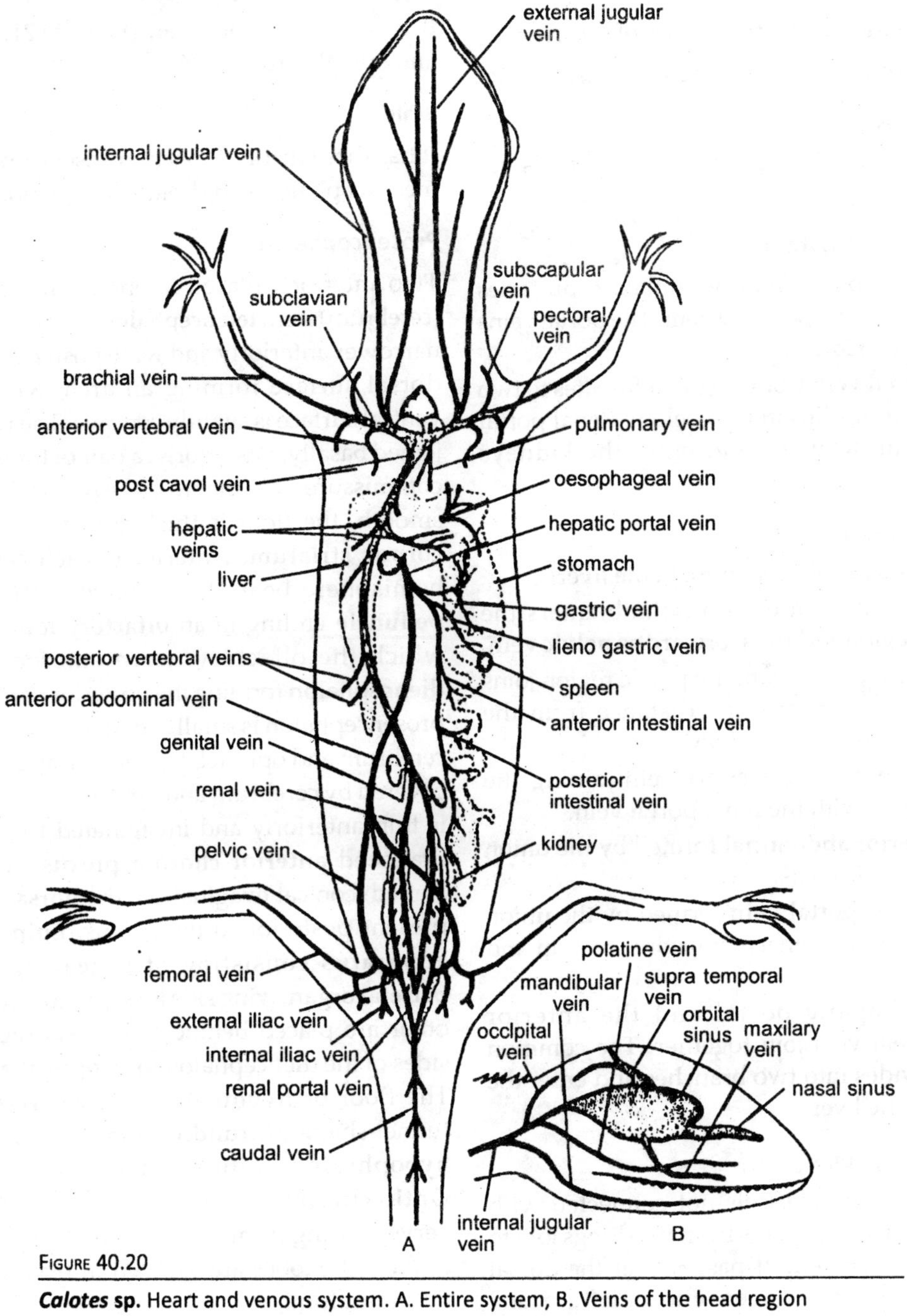

FIGURE 40.20

***Calotes* sp.** Heart and venous system. A. Entire system, B. Veins of the head region

Renal. Paired, spring from the kidneys. The two renals are continuous at the posterior extremity, where they form a narrow loop.

Genital. Paired arise from the gonads, testes or ovaries.

Posterior vertebral. One or two from the vertebral column.

Hepatic. A few, from lobes of the liver.

Oesophageal. A vein from the oesophagus.

Portal systems

Both renal and hepatic portal systems are present.

Renal portal system

Two renal portal veins with their tribu-taries constitute renal portal system. The portal veins end in the kidneys.

Caudal vein from the tail bifurcates to two renal portal veins in the coelom. Renal portal veins run forward and enter the kidneys laterally.

Hepatic portal system

It consists of veins opening in the liver.

Femoral from each hindlimb enters the pelvic region and runs forward as **pelvic vein**.

Internal iliac, from the pelvic region joins the pelvic vein at some distance from the posterior end of the kidney.

External iliac, a short vein joining the pelvic vein with the renal portal vein.

Anterior abdominal formed by the union of two pelvic veins.

Hepatic portal vein formed by the union of veins from stomach, intestine, rectum and spleen.

The hepatic portal and the anterior abdominal vein join together. The common vein divides into two branches and enter the lobes of the liver.

Pulmonary veins

Small but wide vessels not connected with major veins. Two veins from two lungs join to form a single vein. It passes over the dorsal side of the heart and opens into the left artrium.

⊠ Nervous System

The central nervous system and peripheral nervous system constitute the nervous system. The basic structures of nervous system are similar in all vertebrates (p. 133).

Central nervous system

It is formed by the brain (Figs. 40.21, 40.22) and spinal cord (p. 136).

Brain

It has three distinct zones—prosence-phalon, mesencephalon and rhombence-phalon.

Prosencephalon

Two anteriorly placed cerebral hemis-pheres (cerebrum) form **telencephalon**. These are large narrower anteriorly and wide posteriorly, the dorsal surface forming an arch, while the ventral surface is slightly concave. The two are joined basally. Posteriorly, a pair of transverse commissures join the two. The roof is thin and smooth, the floor is thick and occupied by **corpus striatum**. Anteriorly, each cerebral hemisphere bears long, narrow **olfactory peduncle** ending in an **olfactory lobe**, from which the olfactory nerve emerges. The **diencephalon** forming the posterior part of the prosencephalon is small, situated between the cerebrum and optic lobes. Dorsally, it is almost covered by cerebrum and optic lobes. The roof is thin anteriorly and invaginated to form a branched **anterior choroid plexus**. A thin-walled, conical evagination, the **dorsal sac** is present posterior to the plexus. **Epiphysial apparatus** consisting of **pineal eye** and **parietal organ**, lying side by side, close to each other are placed behind the dorsal sac. The sides of the diencephalon form **optic thalami**. The floor or **hypothalamus** gives rise to a funnel-shaped **infundibulum**, to which the **hypophysis** or **pituitary body** is attached. **Optic chiasma** formed by the crossing of nerves arising from the anteroventral corners of the optic lobes are situated in front of the infundibulum.

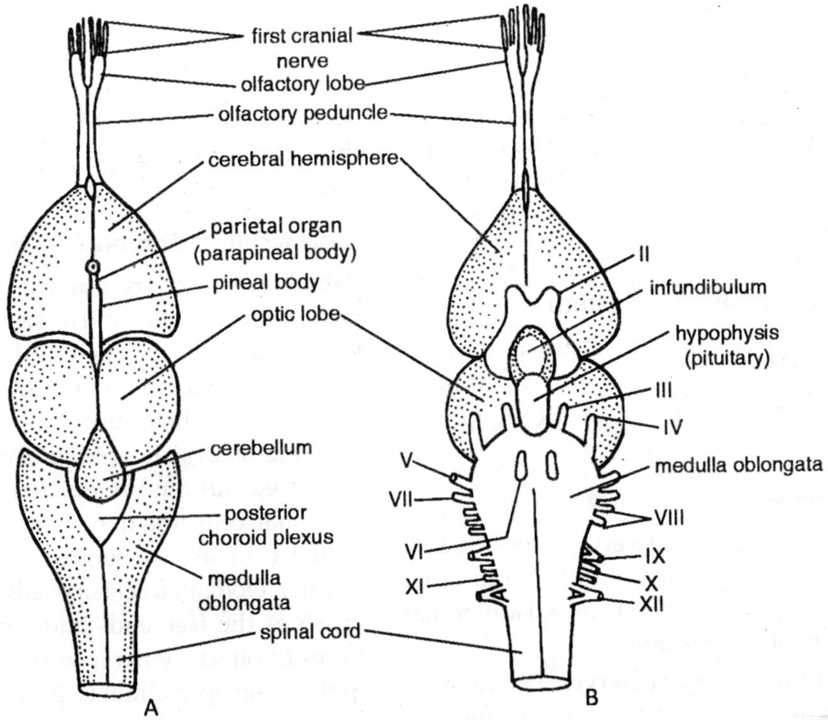

FIGURE 40.21

Calotes sp. Brain. A. Dorsal veiw, B. Ventral view

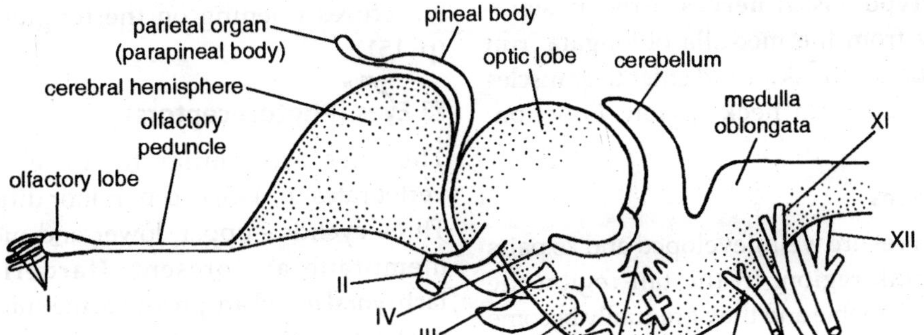

FIGURE 40.22

***Calotes* sp.** Brain.Lateral view I-XII Cranial nerves

Mesencephalon

A pair of large, oval **optic lobes** or **corpora bigemina** occupy the dorsolateral position of the mesencephalon. **Crura cerebri**, a pair of longitudinal bands, ventral to the optic lobes join the hind brain to the mid brain. The floor serves mainly as a track for transmission.

Rhombencephalon

The anterior or **metencephalon** or **cerebellum** is small and covers the optic lobes medially from behind. The posterior **myelencephalon** or **medulla oblongata** is ventrally flexed and continues behind as spinal cord. The roof of the myelencephalon is membranous, deeply vascular and known as **posterior choroid plexus.**

Spinal cord

A thick-walled, cylindrical nerve tube situated in the neural canal of the vertebral column. In the tail region it gradually becomes slender and tapers to a blind end.

⊠ Peripheral Nervous System

Cranial nerves and the spinal nerves form the peripheral nervous system. Autonomic nervous system is a part of the peripheral nervous system (p. 137).

Cranial nerves

The cranial nerves are twelve pairs. The first ten pairs (I-X) are similar to those in other vertebrates (p. 137). The XI and XII are similar to those in other **amniotes.**

XI. **Spinal accessory nerves.** Paired, arise close to the roots of the vagus in medulla oblongata, runs posteriorly, enter vagus ganglia and merge with vagus nerves. Function: motor.

XII. **Hypoglossal nerves.** Arise postero-ventrally from the medulla oblongata, run forward below the skull and enter the muscles of the tongue and neck region. Function: motor.

Spinal nerves

Spinal nerves are well-developed and 26 pairs in the trunk region. In the tail region the number is variable and they gradually become short and slender.

Receptors and sense organs

The sense organs are skin, olfactory sacs, Jacobson's organs, taste buds, eyes and ears.

Tactile (Touch sense organs: Mechanoreceptors)

These are present in skin scales. The sense organ is very minute, knob-like epidermal outgrowth consists of 2-3 centrally situated, spindle-shaped tactile receptor cells, connected with nerve fibres and surrounded by 5-6 supporting cells (Fig. 40.23). The structure is situated below the base of a tactile bristle of the skin and projects into it. Any pressure to the tactile bristles stimates the tactile receptors.

Nasal sacs (Olfactory organs: Chemoreceptors)

Paired nasal sacs are situated in the anterior part of the head region. The sac is narrow and tubular anteriorly, opening to the exterior through external nare but large posteriorly, which opens into the buccal cavity by internal nare (Fig. 40.24). The anterior part is lined with stratified squamous epithelium; ciliated columnar cells line the posterior part. A few goblet cells are present in the epithelium. Slender, elongated sensory cells bearing hairy brush at the free ends and connected with fibres of olfactory nerve at bases are present in the nasal epithelium (Fig. 40.25).

⊠ Taste Buds (Gustatory Receptors)

Taste buds are present on the broad papillae of the tongue. The buds are elongateoval structures opening on the tongue surface (p. 151).

⊠ Eyes (Photoreceptors)

The eyes are similar to those in other vertebrates (p. 143) except minor differences. Three eyelids—upper, lower and nictitating membrane are present. **Harderian** and **lachrymal** gland are present at the anterior and posterior corner of the eye-ball, respectively. The watery secretions of the glands keep the eye-ball moist. A vascular, pigmented, cushion-like projection, the **pecten** or **ectodermal conus** (Fig. 40.26A) projects into vitrous body from the blind spot of the retina. Accommodation for near vision is effected by changing the lens curvature with the contraction of ciliary muscles. The dorsolateral position of the eyes in the head region indicates that the vision is more of the monocular type than of biocular type.

⊠ Pineal Eye (Parietal organ)

The pineal eye is a dorsal outgrowth of the diencephalon. Structureally (Fig. 40.26B) it resembles the eye to a certain extent. It is saccular, transversely oval in vertical section and lies just below the parietal foramen. The walls are made of simple epithelium. The skin over the organ is thin, depigmented and transparent and resembles cornea. A lens-like structure is present below this layer. The rest of the wall resembles retina. It is innervated by Habenular nerve arising from the diencephalon. The function of pineal eye as an organ of vision is, however, doubtful.

⊠ Ears (Equilibrium and Phonoreceptors)

Ears are paired, located posterolaterally behind the eyes in the head region. The components of an ear are—an external ear, a tympanum, a middle ear and an internal ear or membranous labyrinth (p. 143).

External ear. A shallow, circular depression bounded by fibrous connective tissue covered with skin.

Tympanum. A tough, circular semitransparent membrane at the bottom of the external ear. A thin, protective circular scale, **tympanic scale**, covers outer surface of the tympanum.

Middle ear. The middle ear is a long. narrow cavity, posteroventrally connected with the **eustachian tube** by a narrow elongated slit. A partially ossified rod-like **columella auris** is present in the middle ear (Fig. 40.27). It is made of two pieces—the columella proper is long, bony and applied to the **foramen ovale** of the otic capsule and the small cartilaginous **end piece** attached to the inner surface of the tympanic membrane acentrically.

Membranous labyrinth. It consists of three **semicircular canals, a urtricle** and **a saccule** (Fig. 40.28). Utricle, saccule and semicircular canals help in maintaining equilibrium.

Legena, the structure for hearing is a short, blunt, terminal projection of the saccule. A pair

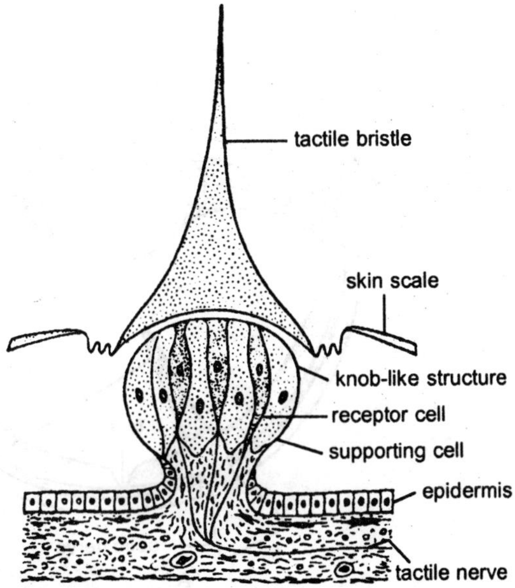

FIGURE 40.23

***Calotes* sp.** Tactile sense organ (semidiagrammatic)

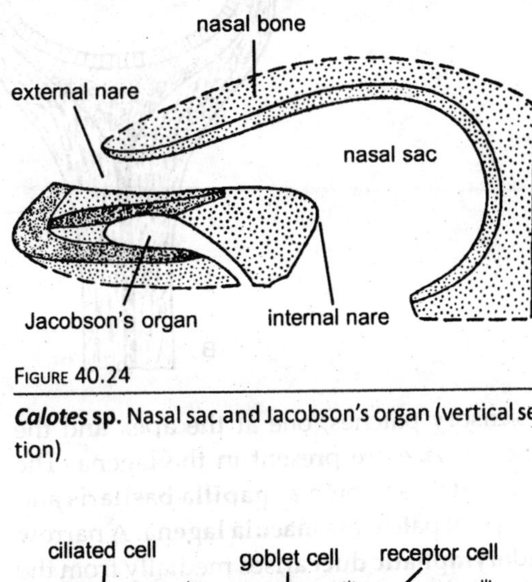

FIGURE 40.24

***Calotes* sp.** Nasal sac and Jacobson's organ (vertical section)

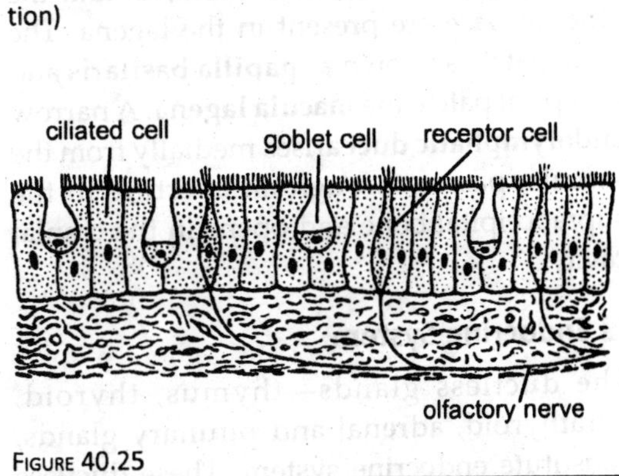

FIGURE 40.25

***Calotes* sp.** Olfactory mucous membrane (nasal sac)

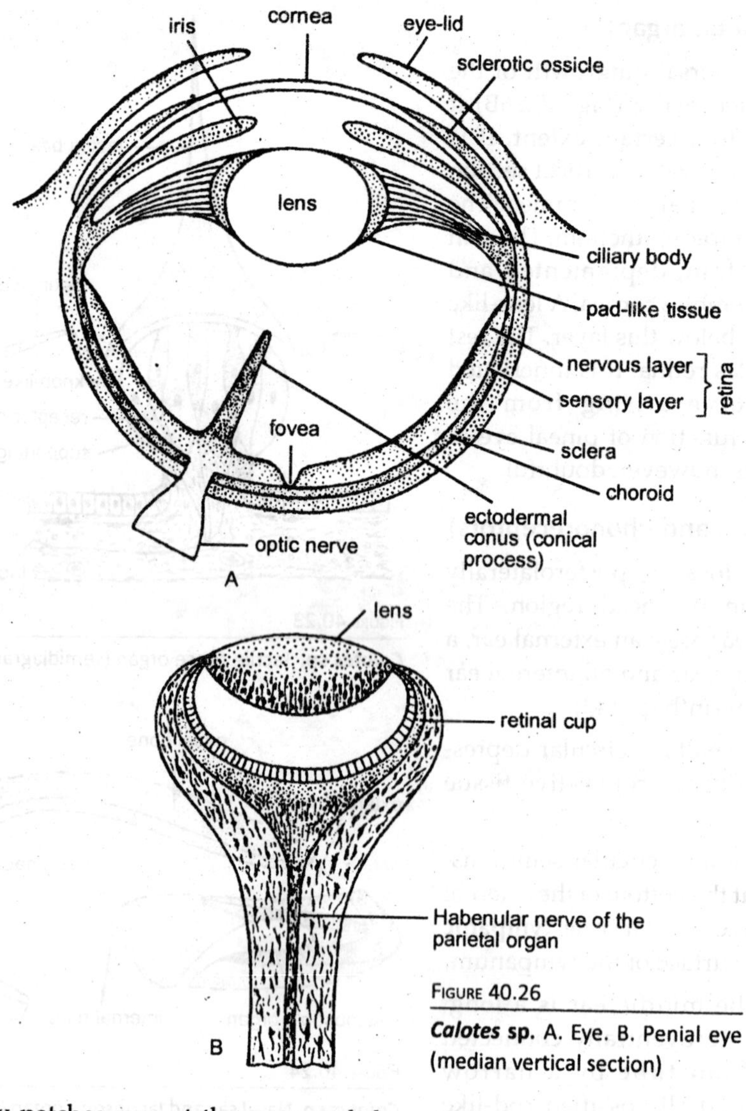

FIGURE 40.26

***Calotes* sp.** A. Eye. B. Penial eye (median vertical section)

of sensory patches, one at the apex and the other at base are present in the lagena. The basal patch is known as **papilla basilaris** and the apical patch, the **macula lagena**. A narrow **endolymphatic duct** arises medially from the saccule, passes through an aperture in the supraoccipital bone and opens in the cranial cavity.

⊠ Endocrine System

The ductless glands—thymus, thyroid, parathyroid, adrenal and pituitary glands, constitute endocrine system. These ductless glands secrete **hormones** (p. 156).

Thymus

A pair of small-whitish glands in the neck region in association with the internal carotid arteries.

Thyroid

A pair of club-shaped, cream coloured glands in the neck region, just in front of the pectoral girdle.

Parathyroid

A pair of small, nodule-like reddish glands at the junction of the ductus caroticus with the carotid artery.

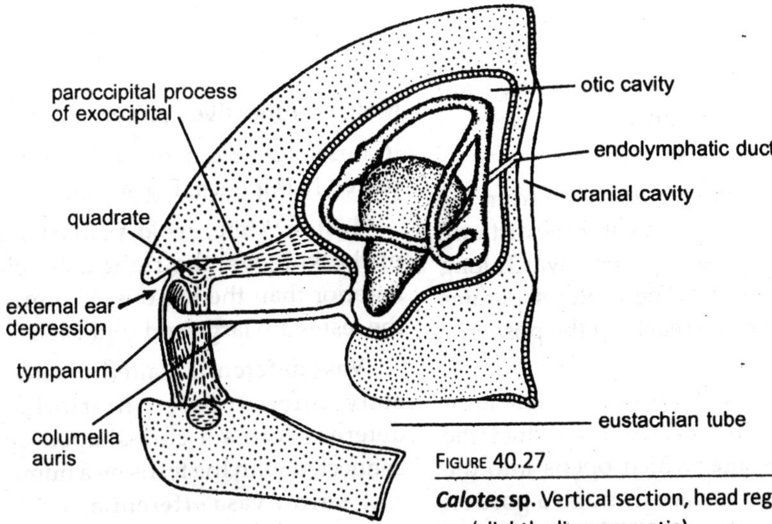

FIGURE 40.27

***Calotes* sp.** Vertical section, head region with ear (slightly diagrammatic)

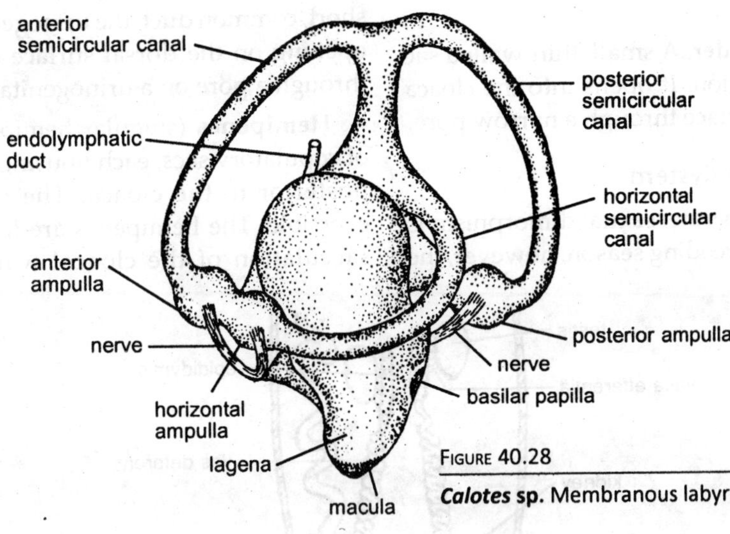

FIGURE 40.28

***Calotes* sp.** Membranous labyrinth

Adrenal
A pair of elongted whitish bodies along the inner side of the gonads.

Pituitary
It is situated on the ventral side of the diencephalon (forebrain), attached to the tip of the infundibulum.

Pancreas (islet of Langerhans), **ovaries** and testes scerete hormones.

☒ Urinogenital System

The urinary and genital systems in Calotes are not completely separate and it is customary to describe the two systems together as urinogential system. The interdependence of the two systems are limited only to posteriormost part of the genital and renal ducts in male.

Urinary (Excretory) system

It consists of paired kidneys, ureters and a urinary bladder.

Kidneys. The kidneys are metanepharic (p. 127). These are brownish, elongated, fleshy, slightly lobulated bodies dorsomedian in position, situated in the posterior region of the abdomen, distally extended into the tail region. The kidneys are attached to the dorsal wall of the body cavity. Anteriorly they are free and tapering. Posteriorly the two are joined and form a ploughshare-like vertical ridge at the end (Figs. 40.29, 40.30).

Ureters. The urinary ducts arise from the ventral surface of the kidneys at some distance from the anterior end and run backward along the midventral surface. The ureter separates from the kidney before reaching the posterior end of the latter.

In male, the vas deferens of the side joins the ureter to form, a short, common duct, the **urinogenital passage,** which opens into the cloaca on its dorsal surface through a pore on a **urinogenital papilla.** In females, the ureters open directly and independently into the cloaca.

Urinary bladder. A small, thin-walled sac in the pelvic region. It opens into the cloaca on its ventral surface through a narrow pore.

⊠ Reproductive System

The sexes are separate, sexual dimorphism is absent. During breeding season, however, the colour in males turns to pinkish-red, particularly on the ventral side.

Male reproductive system

It consists of paired testes, vasa deferentia and paired hemipenes (Fig. 40.29).

Testes. White, ovoid, paired bodies in front of the kidneys, the right one being a little anterior than the left one. In a mature lizard the testes are large and pinkish in colour.

Vasa deferentia. Paired, greatly elongated, convoluted tubes. Anteriorly each vas deferens forms a club-shaped **epididymis** connected with the testis by a number of small and narrow **vasa efferentia.**

Urinogenital passage. Posteriorly the vas deferens joins the ureter and the two form a short, common duct, the urinogenital passage, opening on the dorsal surface of the cloaca through a pore on a urinogenital papilla.

Hemipenes (singular hemipenis). A pair of copulatory sacs, each housing a hemipenis posterior to the cloaca. The hemipenis is eversible. The hemipenes are formed by the sacculation of the cloacal wall extending

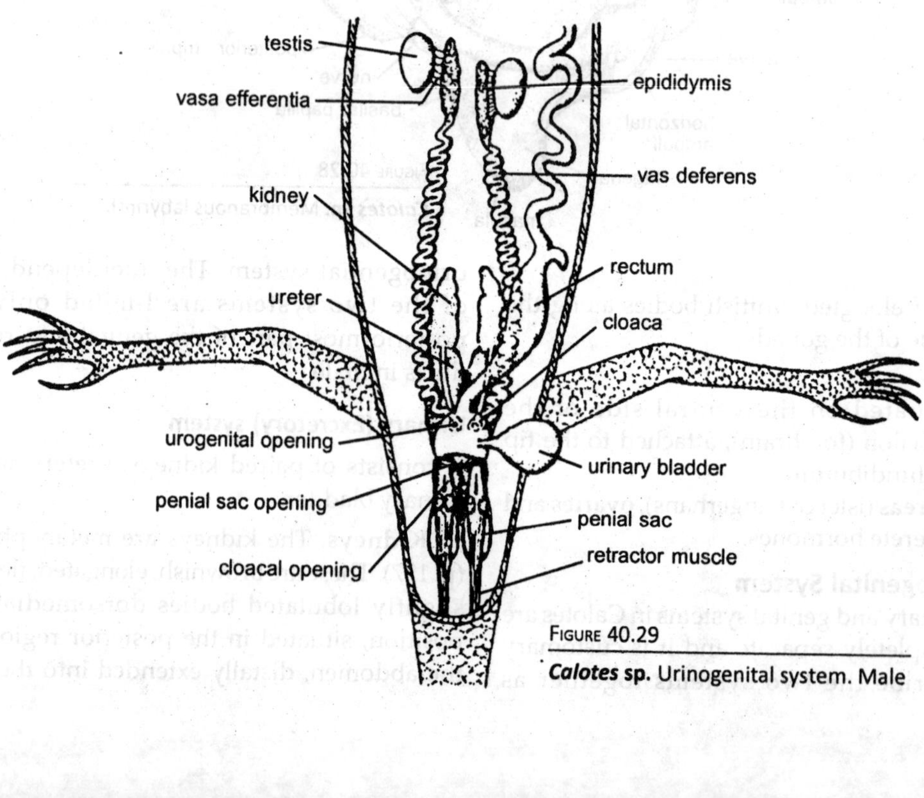

FIGURE 40.29
***Calotes* sp.** Urinogenital system. Male

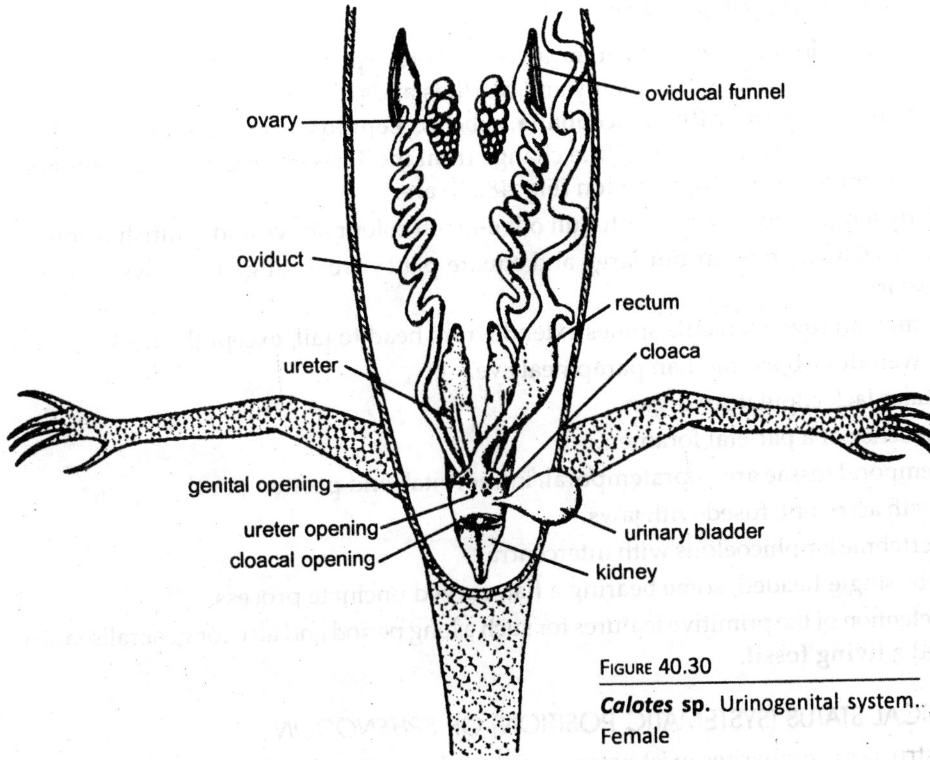

FIGURE 40.30

***Calotes* sp.** Urinogenital system.. Female

posteriorly beneath the skin. When everted the hemipenis assumes the shape of a cylindrical stalk, with a large swollen apex or **head**. The head is four-lobed. A median elongated groove runs along the stalk and communicates with the opening of the urinogenital passage. The groove serves as a passage for the spermatic fluid from the cloacal cavity of the male to that of the female.

Female reproductive system

It consists of paired ovaries and oviducts (Fig. 40.30).

Ovaries. A pair of clusters of round germ cells of different sizes in front of the kidneys constitute ovaries. They are slung to the dorsal wall of the body cavity by mesentery.

Oviducts. Two large convoluted tubes, lateral to the ovaries. The oviduct ends in a funnel at the anterior end, opening in the coelom in the vicinity of the ovary. Posteriorly, the oviducts are wide, less convoluted and open separately into the cloaca on its dorsal surface through pores on genital papillae.

⊠ Fertilization and shell formation

Fertilization takes place in the anterior region, **vitellogenesis** (deposition of albumen) in the mid region and **shell formation** occurs in the posterior region of the oviduct.

Egg laying

Eggs are enclosed in a calcareous, porous shell and laid on land under protective coverings or holes dug out in the soil by the mother and covered with the dug out materials to keep out of sight of the predators. Young ones hatch out breaking the shell and start independent life.

SPHENODON: The living fossil

The **Sphenodon**, locally named **Tuatara**, an endangered species, is restricted today within limited areas—some islets of Bay of Plenty and North Islands of New Zealand. It is the only living representative of the order **Rhyncocephalia**, subclass **Lepidosauria**, class Reptilia. The sphenodon is a relic and surviving with little change from the Triassic, beginning 200 million years ago. The salient features of sphenodon (Fig. 48.4B) are:

1. Body length about 60 cm with dull olive-green colour above and whitish below.
2. Scales small, granular but large and square plates are arranged in rows on the ventral surface.
3. A median row of erectile spines extends from head to tail, except the neck.
4. A well-developed median parapineal eye.
5. Males lack copulatory organ.
6. Presence of a parietal foramen.
7. Temporal fossae are supratemporal, infraorbital, and posttemporal.
8. Teeth acrodont, fused with jaws.
9. Vertebrae amphicoelous with intercentra.
10. Ribs single-headed, some bearing a flat, curved uncinate process.

The retention of the primitive features for such a long period and also for generalised structure it is called a **living fossil**.

BIOLOGICAL STATUS (SYSTEMATIC POSITION) OF *SPHENODON*

Certain structural similarites exist between *Sphenodon* and Amphibia and members of several subclasses of Reptilia.

Affinities with Amphibia

Similarities	Dissimilarities
1. Main arterial trunks arise from a common vessel—conus arteriosus. 2. Courses of carotid, laryngotracheal arteries and abdominal vein are similar to those of Caudata. 3. Course of blood through arteria interossea is similar.	1. Presence of moderately developed ductus arteriosus and ductus caroticus. 2. No other similarities are present in the two groups.

Both Caudata (Amphibia) and Rhyncocephalia are primitive groups. The resemblances are not of any phylogenetic significance.

Affinities with Dinosaurs

Similarities	Dissimilarities		
	Structure	Sphenodon	Dinosaurs
1. Skull diapsid. 2. Presence of abdominal ribs and uncinate process	1. Ribs 2. Teeth 3. Clavicle, interclavicle and parietal organs	single headed acrodont present	double headed thecodont absent

The Rhyncocephelia and Protosauria are most generalised and closely allied groups.

Affinities with Crocodilia

Similarities	Dissimilarities		
	Structure	Sphenodon	Crocodilia
1. Skull diapsid 2. Quadrate immovable 3. Cochlear process tubular 4. Proatlas present 5. Abdominal ribs present 6. Caudal ribs fused with vertebra 7. Ribs with uncinate process 8. Chevron bones present	1. Nasal opening 2. Teeth 3. Vertebrae 4. Clavicle 5. Pecten 6. Penis	double acrodont amphicoelous present absent absent	single thecodont procoelous absent present present

There are certain similarities between Sphenodon and Crocodilia but the differences are wide

Affinities with Chelonia

Similarities	Dissimilarities		
	Structure	Sphenodon	Chelonia
1. Quadrate movable 2. A process of parietal bone reaches the squamosal 3. Caudal ribs fused with vertebra 4. Pecten absent 5. Urinary bladder present	1. Vomer 2. Sternum 3. Cloacal opening 4. Penis	paired present transverse absent	unpaired absent longitudinal present

Resemblances exist between Sphenodon and Chelonia but the differences are quite pronounced.

Affinities with Lacertilia

Similarities	Dissimilarities		
	Structure	Sphenodon	Chelonia
1. General body feature 2. Presence of spiny scales 3. Proatlas present 4. Caudal vertebrae separated by septum 5. Amphicoelous vertebrae in some geckos 6. Ribs single-headed, extensively developed in sphenodon, being present even in the caudal region 7. Chevron bone present 8. Remnants of notochord present between the vertebrae 9. Cloacal glands present 10. Presence of parapenial organ	1. Quadrate 2. Union of rami (mandible) 3. Vertebrae 4. Clavicle and inter clavicle 5. Uncinate process 6. Parapineal organ 7. Conus arteriosus 8. Pecten 9. Copulatory organ	Immovable ligament amphicoelous present present well-developed present absent absént	movable symphysis procoelous except in some Geckos absent in limbless lizards absent less developed absent present present

Some authors suggested inclusion of sphenodon in Lacertilia but the majority are against the idea.

The biological status of *Sphenodon* is a much discussed topic. The similarities and dissimilarities with other groups of vertebrates; particularly Reptilia, are some-what balanced, being very close to Lacertilia. For primitive features not present in other reptiles, a separate order **Rhyncocephalia** has been created, which includes several extinct forms and the only living genus *Sphenodon*

POISONOUS AND NONPOISONOUS SNAKES

The poisonous and nonpoisonous snakes differ morphologically and are easily identified.

Poisonous	Nonpoisonous
1. Body colour generally bright.	1. Body colour not so bright.
2. Neck constricted.	2. Neck unconstricted.
3. Head long, triangular, and wide due to the presence of poison glands on two sides of the head.	3. Head usually narrow and elongated.
4. In majority, a hood is present.	4. Hood usually absent.
5. Tail abruptly tapering except in sea snakes.	5. Tail gradually tapering.
6. Head shield very large.	6. Head shield small.
7. The scales on the top of the head are small.	7. The scales are large and usually nine in number.
8. The scales on the dorsal surface of the trunk are smaller but spinal scales are large and hexagonal.	8. The dorsal scales are large but spinals are smaller and not hexagonal.
9. The ventral surface is covered with large transverse plates. No small scales are usually visible from below. All poisonous snakes must have broad plates on the belly but snakes having broad plates are not always poisonous.	9. Ventral surface is covered either with small scales or small scales are visible on both sides of the transverse plate below.
10. Teeth are not uniform (Fig. 40.31A). The maxillary teeth are large and called 'fangs.' Fangs are grooved or with a canal.	10. The teeth are uniform and solid (Fig. 40.31B).
11. Two poison glands are present.	11. Poison glands absent.

POISON APPARATUS AND BITING MECHANISM OF SNAKES

⊠ Poison apparatus

All poisonous snakes* have poison apparatus which includes a pair of poison glands with ducts, muscles and ligaments and fangs associated with glands.

A. Poison glands

1. Paired sacs, situated on either sides of upper jaw, below the eyes, and somewhat behind. These are the modified superior **labial** or **parotid glands.**
2. The glands, may be small and oval (sea snakes) or large, tubular (Viper). Each gland is encapsulated with fibrous

* Poisonous snakes: Corba, *Naja* spp., *Vipere* spp. etc.

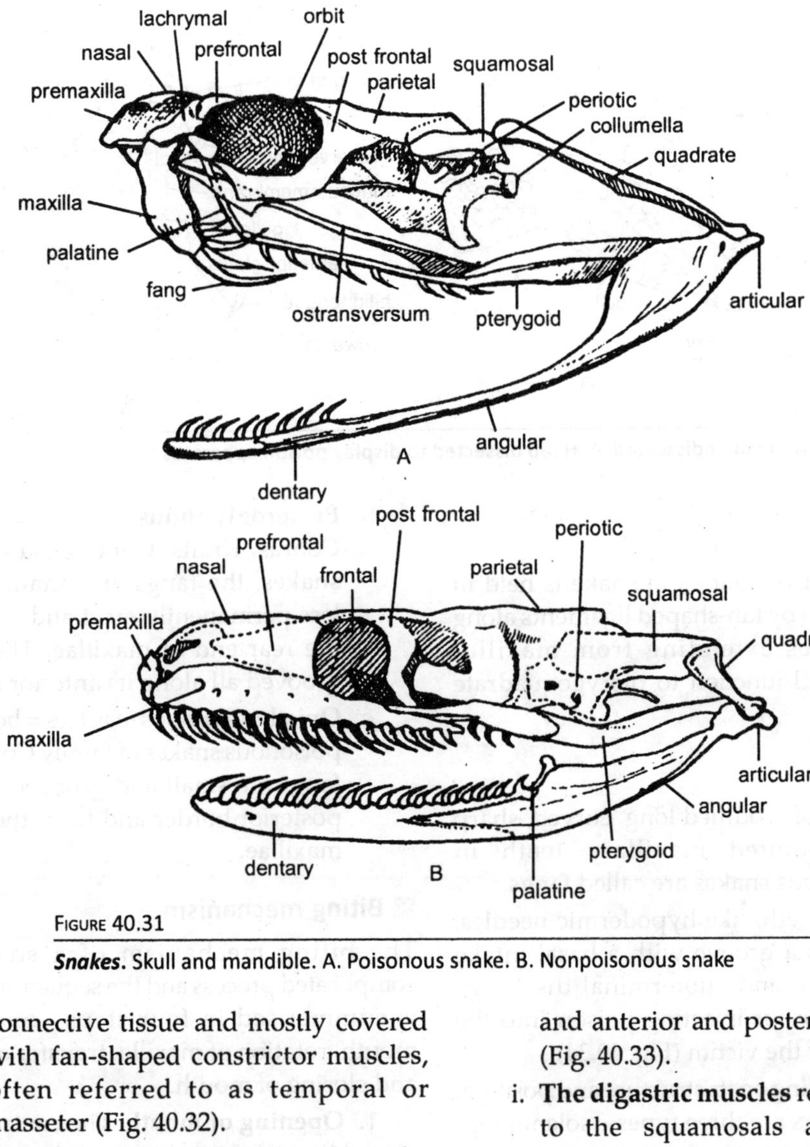

FIGURE 40.31

Snakes. Skull and mandible. A. Poisonous snake. B. Nonpoisonous snake

connective tissue and mostly covered with fan-shaped constrictor muscles, often referred to as temporal or masseter (Fig. 40.32).

3. A narrow poison duct arises from the anterior end of each poison gland, which passes forward along the side of the upper jaw and loops over itself just in front of the fang and opens either at the base of the fang or at the base of the tunnel of the fang, which remains covered by muscle sheath.

B. **Muscles and ligaments**

1. The poison apparatus is associated with bands of three types of specialised muscles—digastric, sphenopterygoid and anterior and posterior temporalis (Fig. 40.33).

   i. **The digastric muscles** remain attached to the squamosals and quadrate junction of the skull at one end and articular of the lower jaw at the other end.

   ii. The **sphenopterygoid or protractor-pterygoid** muscles remain attached anteriorly to the basal orbitosphenoidal region and the dorsal surface of the pterygoid posteriorly.

   iii. **Anterior and posterior temporalis (massetor)** or **capito-mandibularis superficialis** muscles are fan-shaped remain attached to the side walls of post

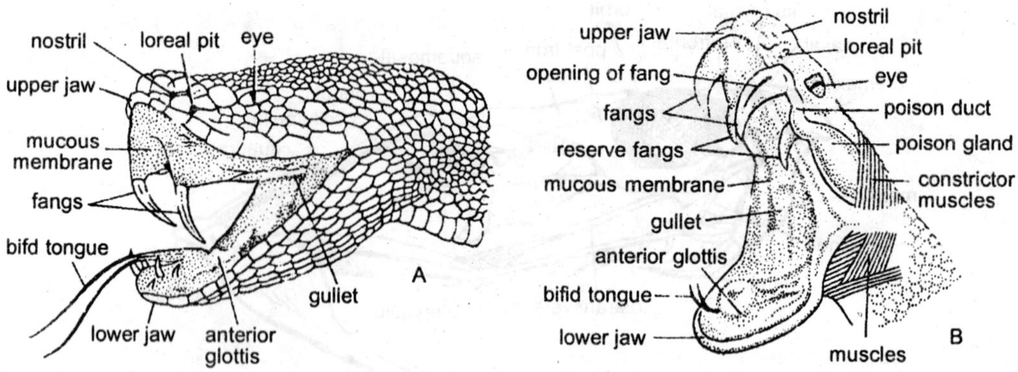

FIGURE 40.32

***Pit viper*. A. Head undissected, B. Head dissected to display poison apparatus**

frontal and parietal ridges of cranium and the lower jaw.

2. The poison glands in snake is held in position by fan-shaped ligaments along its sides extending from maxillo-lacrymal junction to pterygoquadrate bone.

☒ Fangs

1. A pair of modified long, curved, sharp and pointed maxillary teeth in poisonous snakes are called fangs.
2. These teeth, like hypodermic needles, possess a groove with a basal intake aperture and subterminal discharge aperture for injecting poison into the body of the victim (Fig. 40.34).
3. Depending on its structure and position, the fangs are three types—solenoglyphous, proteroglyphous and opisthoglyphous.

a. **Solenoglyphous** (*Solen* = pipe, *glyph* = hollow). In vipers and rattle snakes, a pair of large functional fangs are present at the rear end of maxillae followed by another pair of reserve teeth. These are long, with hollowed poison canal, lined with enamel, capable of vertical movement, thus turn inside to lie close to the roof of mouth when it is closed.

b. **Proteroglyphous** (*Protero* = first). In Cobras, Kraits, Coral snakes and sea snakes, the fangs are small, more or less permanently erect and situated at the rear end of maxillae. The fang is grooved all along its anterior face.

c. **Opisthoglyphous** (*opisthos* = behind). In poisonous snakes of family Colubridea, fangs are small and grooved along its posterior border and lie at the back of maxillae.

☒ Biting mechanism

The biting mechanism of a snake is a complicated process and the sequence of biting is summerised in four steps—opening of mouth, rotation of maxilla, transfer of venom and closing of mouth.

1. **Opening of mouth.** The contraction of digastric muscles result in opening of mouth by depressing the lower jaw (Figs. 40.32, 40.33).
2. **Rotation of maxilla.** As the mouth opens the spheanopterygoid or protractor-pterygoid muscles contract. This contraction results in the forward movement of pterygoid and uppushing of ectopterygoid. This upward movement of ectopterygoid brings about a rotation of the maxilla on its own axis

* The functional fang may be replaced when it is lost or damaged.

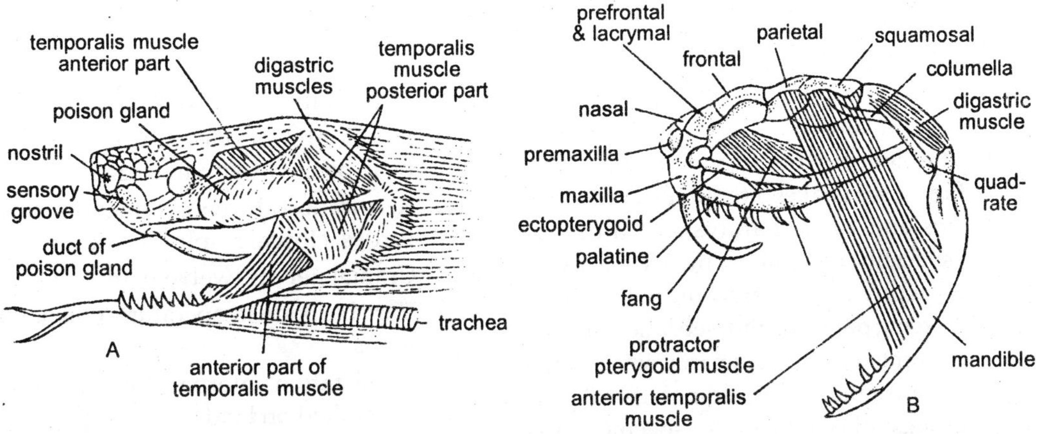

FIGURE 40.33

skull of piver showing biting mechanism. A. Mouth at rest (closed), B. Mouth opened when striking (only involved muscle and skull bones diagrammatically shown)

round the lacrimal and as a result the fangs are raised and become directed forward. The fangs remain nearly horizontal when mouth is closed but the fangs assume almost vertical during opening of mouth to bite.

3. **Transfer of venom.** During contraction of the digastric muscle the posterior ligament is relaxed and during rotation of the squamosal bone the fan-shaped ligaments are stretched to squeeze the wall of the poison gland. This makes the poison to come out of the poison gland through the ducts and fangs.
4. **Closing of mouth.** The contraction of the temporalis and sphenopterygoid muscles close the mouth and the tip of the fang is indirected backward horizontally. Longer time is needed to open the mouth than to close it.

☒ Poison or snake venom

1. Snake venom is a clear, transparent, straw-coloured or greenish, acidic (pH 5.8 to 6.6), thermolabile fluid with specific gravity of 1.02-1.07.
2. It is soluble in water and glycerine, precipitable and destroyed when treated with silver nitrate, potassium permanganet, alcohol, sodium and potassium hydroxide. It becomes thermostable and crystalised by drying and lyophilisation.
3. It is a tasteless (cobra venom is slightly bitter), odourless complex mixture of enzymes and specific toxins and is a good digestive juice secreted and stored by the poison glands.
4. The important enzymes present in the snake venom are hyaluronidase, L-arginine hydrolases, L-amino acid oxidase, phospholipase-A, B, C, phisphodiestererase, transaminase, cholinesterase, alkaline and acid phosphatase, deoxyribonuclease, ribonuclease, etc.
5. It also contains metals like zinc, sulphur, copper, etc.
6. Venom of particular species of snake contains one or more enzymes and several active nonenzymatic proteins having at least two to three peptides.

Mode of action : Signs and symptoms of venom poisoning

1. On snake biting the venom is introduced into the subcutaneous tissue which subsequently reaches general cir-

culation or direct into the vein. The venom of different snakes acts differently either as neurotoxin or haemotoxin.

2. The neurotoxins in venom of cobra, kraits, sea snakes, etc. attack nerve centres and cause paralysis of muscles, especially respiratory muscles leading to asphyxiation. Symptoms include piercing pain, burning sensation ending in numbness of bitten parts, which turns bluish or blackish; giddiness, weakness of legs, high pulse rate, spechlessness droping of saliva and eye lids, contraction of pupils, vomiting and ultimately death due to failure of respiratory (asphyxia) or cardiac failure.
3. In case of krait bite, the symptoms in addition to the above are severe abdominal pain due to internal haemorrhage; destruction of RBC (haemolysis) and paralysis of trunk and limbs followed by death within 6 to 24 hrs.
4. The haemotoxins of typical viper venom affect the circulatory and nervous system more severely. Local swelling, discolouration of the bitten part with acute burning pain, oozing out of red fluid or blood from the wound due to massive necrosis; dilation of pupils, increase of pulse rate profuse vomiting and gradual unconciousness occurs. Death results due to paralysis of vasomotor centres and exhaustion from profuse bleeding.

CLASSIFICATION: CLASS REPTILIA

1. Body covered with dry cornified skin, usually with scales or scutes. Few surface glands are present.
2. Two pairs of limbs, each typically with 5 digits, ending in horny claws and suited to running, crawling or climbing. Limbs reduced or absent in Ophidia and paddle-like in marine turtles.
3. Skull with one occipital condyle. Skeleton completely ossified.
4. Heart imperfectly four-chambered.
5. Aortic arches one pair, erythrocytes oval, biconvex and nucleated.
6. Respiration always by lungs. In marine turtles, cloacal respiration occurs.
7. Cranial nerves twelve pairs.
8. Variable body temperature, poikilothermous.
9. Fertilization internal. Eggs are much yolky, laid on land.
10. Segmentation is meroblastic. Embryonic membranes amnion, allantois, yolk sac develop.
11. Metamorphosis absent.

The class Reptilia is divided into five subclasses—Anapsida, Ichtyopterygia, Lepidosauria, Archosauria, and Synapsida.

Subclass ANAPSIDA

1. Skull with solid roof, without opening behind the eye. It has two orders—Cotylosauria and Chelonia.

Order 1. **Cotylosauria**

Oldest reptiles ranging from upper Carboniferous to upper Triassic.

1. Skull roofed completely. Pelvis flattened, plate-like.
2. Body size varies from 30 to 60 cm.

Two suborders—Captorhinomorpha and Diadictomorpha are recognized.

Examples: *Captorhinus*, *Seymouria*, *Diadectes*, etc.

Order 2. **Chelonia (Testudinata)**

1. Includes turtles, terrapins, and tortoises. Body box-like, broad, with rounded dorsal carapace and ventral plastron; joined at sides, and covered by leathery skin or polygonal scutes. The head and limbs are retractile within the box. About 335 extant species live on land, fresh or marine waters.
2. The nasal opening single and anteriorly disposed.

3. Anus longitudinal. Copulatory organ single. Oviparous.
4. Teeth absent in adult and jaws covered with horny sheaths.
5. The ventricle is completely divided by the interventricular septum.
6. The skull is secondarily anapsid without temporal vacuities and monimostylic. Quadrate immovably articulated. Skull bones are firmly united to each other. A bony secondary palate separates the nasal cavity from the mouth.
7. Vertebrae are procoelous or amphicoelous. The vertebrae and ribs of thoracic region are firmly fused with the body carapace. The cervical vertebrae are eight, lumbar vertebrae absent, sacral two and caudal are a few in number.
8. Sternum absent. The pectoral girdle consists of ventral coracoid and dorsal scapula attached to the carapace.

The order is subdivided into four suborders Eunotosauria, Amphichelydia, Pleurodira and Cryptodira. Eunotosauria became extinct in Permian and Amphichelydia in the Lower Cretaceous. The modern groups are placed under last two suborders.

Suborder i. **Pleurodira**

Side-necked turtles. Neck folds sideways under shell. Pelvis fused to shell.

Examples: *Chelys, Chelodina, Pelomedusa, Podocnemis*, etc.

Suborder ii. **Cryptodira**

Neck bends in vertical S-curve, if retractile. Pelvis not fused to plastron.

Examples: *Chelydra, Chelonia, Chrysemys, Dermatemys, Dermochelys, Emys, Macrochelys, Lepidochelys, Trionyx, Testudo, Terrapene, Sternotherms*, etc.

Subclass ICHTHYOPTERYGIA (PARAPSIDA)

The roof of the skull with upper opening behind eye and bounded below by supratemporal and postfrontal.

Order 1. **Mesosauria**

Fresh water lake-dwellers, slender-bodied, not more than a metre long. The hindlimbs were more powerful than the forelimbs and tail laterally compressed. They were found in the late Carboniferous or early Permian beds.

Examples: *Mesosaurus.*

Order 2. **Ichthyosauria**

Marine fish-like reptiles occurred between mid-Triassic and upper Cretaceous. Some were of considerable length, even up to 10-15 metres. Skull with a single temporal vacuity. Large head produced into elongated snout. Tail long and limbs were in the form of paddles. The vertebrae were amphicoelous and cervical ribs with double head. Abdominal ribs were present. No sacrum, the ilia had lost connection with vertebral column.

Examples: *Ichthyosaurus, Ophthalmosaurus* etc.

Subclass SYNAPTOSAURIA (SYNAPSIDA)

Late Palaeozoic or Mesozoic aquatic reptiles having a single temporal (parapsid) vacuity high in the skull bounded above by postorbital and squamosal. It has two orders:

Order 1. **Protosauria**

Lizard-like terrestrial reptiles not exceeding 30 cm in length existed during lower Permian. The vertebrae were amphicoelous.

Example: *Araeoscelis.*

Order 2. **Sauropterygia**

Aquatic forms with single temporal vacuity in the skull bounded by a postorbital squamosal arch. The coracoids were single. Feet webbed. They existed between Permian and Triassic period.

The order includes three suborders Nothosauria, Placodontia, and Plesiosauria.

Examples: *Placodus, Cyamodus, Elasmosaurus* etc.

Subclass LEPIDOSAURIA

Lizard-like reptiles with two temporal vacuities in the skull, although in some more specialized forms the vacuities reduced. The subclass comprises three orders.

Order 1. **Eosuchia**

The eosuchians are considered to be the ancestors of the squamata, and appeared in the upper Permian. They had interparietal and tubular bones and parietal foramen in the skull.

Example: *Youngina*.

Order 2. **Rhynchocephalia**

the widespread representatives of this order are known from the mid-Triassic, but the sole present-day representative of this order is *Sphenodon punctatum* of coastal islands of New Zealand.

Sphenodon is the oldest surviving lepidosaurian and a Mesozoic fossil (*Homoeosaurus*) points to the continuity of the race.

1. Scales granular; a middorsal row of low spines present.
2. Quadrate immovable and rami of mandible joined by ligament.
3. Teeth acrodont.
4. Vertebrae amphicoelous and intercentra are present.
5. The ribs are single-headed and have uncinate process.
6. Sternum and abdominal ribs present.
7. Anal opening transverse and hemipenis absent.
8. Lungs, heart, and brain similar to Squamata.

Examples: *Hyperodapedon, Sphenodon Rhynchosaurus*, etc.

Order 3. **Squamata**

The existing widely distributed lizards and snakes and some extinct forms are included in this order.

1. Skin with horny epidermal scales or shields.
2. Skull with temporal fossa and no inferior temporal fossa.
3. Quadrate bone movable. The maxilla, palatine and pterygoid are immovably articulated with the skull.
4. Lower jaw composed of several pieces of bones.
5. Teeth either acrodont or pleurodont.
6. The vertebrae and procoelous. Chevron bones present.
7. The ribs are single-headed.
8. The cloacal aperture transverse.
9. A pair of eversible hemipenes present.

The order is subdivided into two suborders —Lacertilia and Ophidia.

Suborder i. **Lacertilia (Sauria)**

1. Lizards with usually slender body and moderate mouth gape. Existing number of species 3,140.
2. Limbs are often well-developed, typically four, sometimes reduced or absent.
3. Mandible fused anteriorly.
4. Pterygoid in contact with quadrate.
5. Eyelids and a tympanum present.
6. Pectoral girdle well-developed or vestigeal in some.
7. Tongue usually entire.

Of the 30 families 26 are existing.

Examples: *Hemidactylus, Calotes, Varanus, Phrynosoma, Anolis, Coleonyx, Chamaeleo; Crotaphytus, Iguana, Lacerta, Heloderma, Anguis, Ophisaurus*, etc.

Suborder ii. **Ophidia (Serpentes)**

Surviving species of snakes are about 2,500.

1. Elongated body and wide mouth opening.
2. Limbs, feet, ear openings, sternum, episternum and urinary bladder absent. The middle ear is degenerated.
3. The rami of the mandible joined anteriorly by ligaments at the symphysis. Maxillae, palatine and pterygoids are freely movable.
4. Teeth slender, conical, usually on jaws and on roof of mouth (palatines and pterygoids).
5. No separate supratemporal ossification.

6. Eyelids absent, eyes covered with transparent spectacles.
7. Tongue slender, bifid and protrusible.
8. Left lung reduced.

Examples: *Phython, Boa, Naja, Uropeltis, Anilins, Xenopeltis, Natrix, Bungarus, Crotalus, Vipera, Hydrophis*, etc.

Subclass ARCHOSAURIA

It is a great and widely varying assembly of reptiles probably all descendants of Thecodontia.

1. The skull is diapsid without interparietal, tabular bones or a parietal foramen.
2. The palatal teeth were lost in almost all except few primitive forms.
3. The lower jaw with a vacuity on the outside between dentary, subangular and angular.
4. Antorbital vacuity in front of the orbit almost always present.
5. Some were bipedal and exhibited correspondingly striking modification in the limb girdles.

Archosaurians are arranged in two orders—Thecodontia and Crocodilia.

Order 1. **Thecodontia**

1. Carnivorous small reptiles arose in the Triassic period.
2. The hind legs were longer, slim skulls, and sharp teeth in sockets along the jaw edges.
3. Some were aquatic with features like modern crocodiles.
4. The external nostrils were sometimes set above the skull level, posteriorly placed near the eyes, enabling the animal to breathe while in water.

Examples: *Ornithosuchus, Aetosaurus*, etc.

Order 2. **Crocodilia**

The amphibian, carnivorous, extant crocodiles, alligators, caimans, gavialis and their allies are included in this order and are the largest existing reptiles, the only remnants of a once widespread Mesozoic archosaurian stock.

1. The elongated body with powerful long heavy compressed tail, four limbs, five digits in the forelimbs and four digits in the hindlimbs and forelimbs webbed.
2. Body covered with rows of sculptured bony scutes. Epidermal scales present.
3. The head large and long with powerful long jaws, rimmed with numerous bluntly conical teeth.
4. The external nostrils occur at the end of the snout which can be closed or opened by muscular system during submersion.
5. Tympanic membrane is protected by two scaly flaps which close the external meatus at the time of diving. Eustachian system present.
6. Cloacal opening longitudinal. A median erectile grooved penis in male and a clitoris in female is present.
7. The quadrate is immovable.
8. The secondary palate is formed by grown out portions of maxillae-palatines and pterygoids and internal nares open into the throat.
9. The centra of vertebrae are either amphicoelous, falt at each end or procoelous. The anterior theracic vertebrae have elongated, bifid transverse processes. The sacral vertebrae are two. The first caudal vertebra with convex ends.
10. The ribs are bifid at vertebral end.
11. A sternum and abdominal ribs present.
12. The pelvis has pubis extended from the acetabulum.
13. The ventricle of the heart is completely divided, but the aortic arches two. The roots of the left and right aortic arches are communicated by an aperture, the foramen of Panizzae.

Examples: *Crocodylus, Gavialis, Alligator, Caiman, Tonistoma*, etc.

Order 3. **Pterosauria** (Pterodactyla)

Toothed, tailed flying fossil reptiles with pnenmatic skeleton, like birds. Forelimbs were converted to wings; fourth finger highly enlarged to support membranous wing, other fingers were small. The sternum was keeled. The external nares placed far back. The eyes bordered with sclerotic plates.

Examples: *Pteranodon, Pterodactylus. Orinothodesmuns*, etc.

Order 4. **Saurischia**

Pelvic girdle was triradiate, teeth were on the premaxillae. There were two or one antorbital vacuities in the skull.

Examples: *Tyrannosaurus, Branchiosaurus, Gorgosaurus*, etc.

Order 5. **Ornithischia**

1. Extinct bird-like herbivorous dinosaurs with a pelvis, essentially like that in modern birds, in which the pubis lay parallel with an extended ischium.
2. The presence of predentary bone in mandible might have supported the beak.

Examples: *Iguanodon, Demetrodon, Nodosaurus, Ankylosaurus, Stegosaurus, Anatosaurus*, etc.

Subclass SYNAPSIDA (SYNAPTOSAURIA)

1. Mammal-like replites, skull roof with lower opening and lateral temporal vacuity behind eye, bounded above by postorbital and squamosal.
2. The lower jaw flat with heterodont teeth.
3. Precoracoids and coracoids were present in shoulder girdles.

The sub-class has three orders—**Pelycosauria, Therapsida and Ictidosauria**.

Order 1. **Pelycosauria**

1. Herbivorous or carnivorous forms existed during upper Carboniferous and Lower Permian period.
2. Skull with temporal vacuity. Neural spines much elongated. Limbs small.

Examples: *Dimetodon, Varauosaurus, Edaphosaurus*, etc.

Order 2. **Therapsida**

1. Large, heavy, mammal-like reptiles existed between Permian and lower Triassic, with double occipital condyle in some, big temporal vacuity in skull, reduced quadrate and quadratojugal bones.
2. In some forms secondary palate was present.
3. Teeth distincly differentiated into incisors, canines and molars.

Examples: *Lycosaurus, Cynognathus Dicynodon, Endothiodon*, etc.

Order 3. **Ictidosauria**

1. Mosaic reptilian group with most of the promammalian features. Existed between upper Traissic and mid Jurassic time.
2. The large temporal opening confluent with the orbit.
3. Quadrate and articular bones reduced.
4. Dentition of heterodont type.

Examples: *Tritylodon, Bienotherium*, etc.

41 CLASS AVES

The birds constitute an extraordinary successful, highly diversified, climax group at the height of their evolutionary career. They have undergone very extensive adaptive radiation. Some are specialised as expert divers, waders, soarers, runners, ground burrowers, cave-dwellers, tree-dwellers and distance fliers.

Almost every part of the body is modified to suit their aerial mode of life. All birds possess feathers, beak and feet modified in relation to their diverse habitat. The forelimbs are modified as wings. The sternum and shoulder girdles are altered strikingly to give mechanical framework for suporting the flight muscles. The pelvic girdle and hindlimbs have changed to support the entire weight of the body on the ground. To make the body light, the entire skeleton has become pneumatic and fully ossified. The left aortic arch and right ovary and oviduct disappeared. Brain is highly developed. In most species degenerated olfactory apparatus is accompanied by an extraordinary enlar-gement of eyes with enhanced efficient vision. With the evolution of incomparable power of flight the respiratory system became more refined, enabling a more complete absorption of oxygen for generation of greater amount of heat and energy to maintain a constant body temperature and also to lead a most active life.

Birds are the best known, very common, active and most easily recognised of all the animals. They are unique in having feathers that clothe and insulate their bodies to make possible a regulated body temperature, and aid in flight, which no other animal possesses. The ability to fly enables the birds not only to occupy spaces denied to other animals but also to move rapidly.

The birds represent a significant advance-ment over all other vertebrates except mammals in having an insulated body covering, regulated body temperature, complete separation of venous and arterial circulation in the heart, high rate of metabolism, highly developed mechanism for voice production, hearing and sight, specialised care for the young. The ability of the mammals to exploit the environment more successfully is the result of evolution in a direction different from that of the birds in many respects.

ORIGIN OF BIRDS

Birds, warm-blooded, feathery animals, prominently creatures of the air, are equally at ease in all three media—air, land and water. They occupy varied habitats, ranging from ice belts to scorching deserts, high treetops to underground burrows.

Fossil remains to trace ancestry of birds are few but strong enough to establish their origin from reptiles, though the exact stock of origin is still disputed. Birds are so aligned with reptiles in their anatomy and embryology that Huxley (1880) termed them "**glorified reptiles**" and placed both reptiles and birds under the superclass Sauropsida.

☒ Salient features of Aves

1. Presence of feathers.
2. Forelimbs modified into wings.
3. Skull monocondylar with large brain case and orbits.
4. Jaw suspension autostylic; jaws beaked; quadrate single or double-headed.
5. Coracoid stout.
6. Cervical vertebrae heterocoelous.
7. Pelvic girdle opisthopubic (pubis placed below ischium and behind acetabulum).
8. Heart large, four-chambered.
9. Single right aortic arch.
10. Brain large and microsmatic (with small olfactory part); optic lobes and cerebellum well-developed.
11. Oviparous, eggs telolecithal and cleidic (box-like).
12. Foetal membranes protect developing embryos.

☒ Reptilian Ancestry of Birds: Evidences Supporting Reptilian Origin

Anatomical

1. Dry epidermal scale present on shanks and feet of birds.
2. Skin glands relatively few.
3. Bones of *Pterosaurs* (flying reptiles of Mesozoic era) were pneumatic.
4. Skull monocondylar.
5. Mandible with many bones (no postdentary reduction).
6. Jaw articulation (quadrate-articular) autostylic.
7. Single middle ear ossicle—columella auris.
8. Uncinate process in *Sphenodon* and crocodiles.
9. Intertarsal ankle joint in birds and Archosaurs.
10. Air sacs present in *Chamaeleo.*
11. R.B.C. nucleated.
12. Presence of pecten in some lizards.
13. Genital organs and cloaca similar.

Embryological

1. Eggs large, telolecithal and cleidic.
2. Cleavage meroblastic and discoidal.
3. Morphogenesis similar; presence of foetal membranes.
4. Development of feathers and scales similar. Transformation of scales into feathers occurs in ostrich chick and penguins.

Serological

Seroprotein identical.

Palaeontological

Discovery of the fossil of *Archaeopteryx lithographica* (Fig. 41.5) of the upper Jurassic era in Bavaria (Germany) by Wagner in 1861 laid the foundation of reptilian ancestry of bird. Chatterjee (1986) discovered another fossil bird *Protoavis* in North America, which is 7.5 million years older than *Archaeopteryx.* McFarland *et al.,* (1979) tried to establish phylogenetic relationship between structures of *Archaeopteryx* and reptiles.

The Jurassic fossil bird named *Archaeornis* was found to be a synonym of *Archaeopteryx.* Unfortunately, any fossil that might suggest origin of Archaeopteryx from a definite reptilian progenitor is lacking. Several suggestions regarding origin of birds have been made by different workers:

1. Galton (1970). Ornithischian dinosaurs (bird-footed) are much related to *Archaeopteryx* than saurischian dinosaurs (beast pelvis). In *Ornithischia* and modern birds, the pelvis is opisthopubic but in saurischian including coelurosaurs, the pelvis is propubic.

Table: *Archeopteryx* characters and Phylogenetic relationship

Archaeopteryx characters	Phylogenetic relationship
1. Presence of feather.	1. Avian trait.
2. Hallux (great toes) opposable.	2. Present in many reptiles and modern birds.
3. Bones nonpneumatic.	3. Bones pneumatic in fliers.
4. Large orbit with sclerotic rings.	4. Found in modern birds and thecodont reptiles like *Euparkeria*.
5. Antorbital foramen.	5. Archosaurian feature.
6. Postdentary fenestra.	6. Archosaurian feature.
7. Teeth socketed.	7. Thecodont feature.
8. Vertebrae amphicoelous.	8. Nonavian feature.
9. Cervical vertebrae six and bear ribs.	9. Reptilian character. Modern birds have 13-25 cervical vertebrae and none bears rib.
10. 19-20 trunk vertebrae; 6 sacral vertebrae, only 5 fused to form synsacrum.	10. Reptilian feature; more vertebrae fused in modern birds.
11. Free caudal vertebrae 20, no pygostyle.	11. Primitive feature, modern flying birds bear a pygostyle.
12. Sternum lacks a keel.	12. Found in ratite (flightless birds).
13. Abdominal ribs.	13. Reptilian character, present in sphenodon and crocodiles.
14. Clavicles fused to form a furcula.	14. Avian character.
15. None of the wing bones fused.	15. Reptilian feature; modern birds have fused carpometacorpus.
16. Digits of the manus reduced to 3 with external and functional claws.	16. Common in bipedal Mesozoic reptiles; vestigeal external claws present in some modern birds, e.g. Hoatzin of the Amazon forest.

Ancestral stock of birds

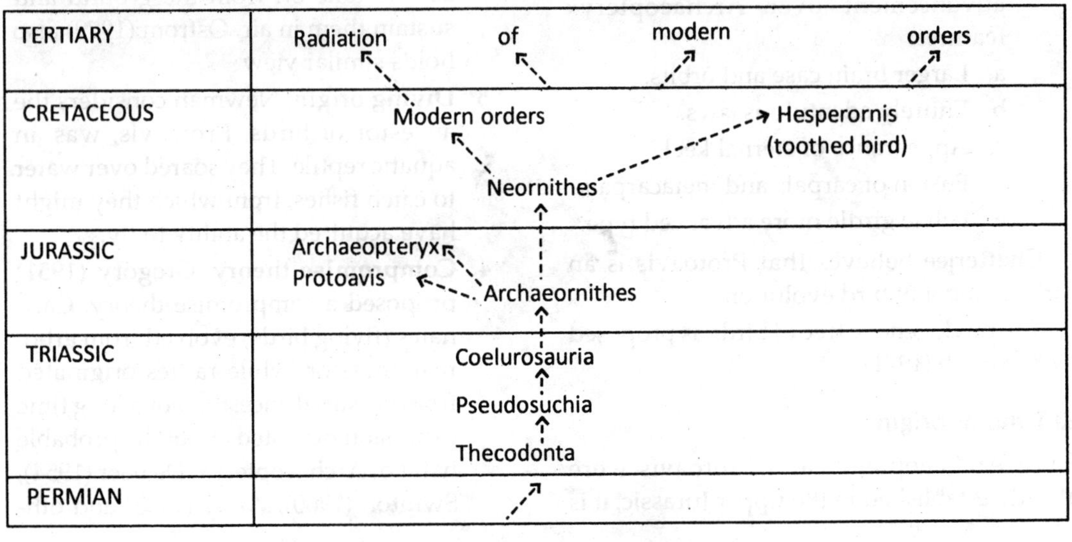

2. Ostrom (1973). Birds evolved from some unknown coelurosaurs. He considers that homology in the skeleton of fore-and hindlimbs of coelurosaurs and Archaeopteryx are due to phylogenetic relationship. The concept is supported by Baker and Galton (1974).
3. Heilman (1926), Lowe (1935, 1944), Holmgren (1955) and Ostrom (1976) believe that the possible phylogeny of birds is —Pseudosuchia→Coelurosauria→Archaeopteryx→modern birds. Martin *et al.* (1980) support the view.
4. Ostrom (1979) suggests that there was perhaps an intermediate stage between Coelurosaurs and *Archaeopteryx,* which he named *Protoarchaeopteryx.* Protoarchaeopteryx was a biped cursor-like Archaeopteryx. In course of time its primary remiges—carpals and metacarpals of the forelimb—elongated. Finally, the pectoral muscles became more developed. The reconstructed phylogeny appears as—Pseudosuchia→ Coelurasauria→Protoarchaeopteryx →Archaeopteryx→modern birds.
5. Chatterjee (1986). *Protoávis,* older than Archaeopteryx, shows following advancement over Archaeopteryx features:
   a. Larger brain case and orbits.
   b. Entirely edentulous jaws.
   c. Appearance of sternal keel.
   d. Fusion of carpals and metacarpals.
   e. Pelvic girdle more advanced type.

Chatterjee believes that Protoavis is an early offshoot of bird evolution.

The phylogenetic tree of birds as proposed by Brodkarb (1971).

☒ Time of origin

Since Archaeopteryx and Protoavis were already established in the upper Jurassic, it is assumed that evolution of birds started sometimes in middle Jurassic. The initial impetus was presumably to avoid and escape from Mesozoic predators.

☒ Place of origin

All five fossil Archaeopteryx were discovered from Europe and Protoavis from North America. It appears that the ancient Northern Supercontinent **Laurasia** was perhaps the place of origin of birds.

☒ Origin of feathers and flight Feathers

Authors are unanimous that feathers are transformed reptilian scales. Primary function of feather is insulation, while flight is only secondary, which is well documented in the ratite.

☒ Flight

1. **Arboreal origin.** Heilman, Osborn and Beebe hold that ancestors of the birds were arboreal. They used their extended forelimbs as parachutes while jumping from one tree branch to another, a view supported by De Beer (1954).
2. **Cursorial origin.** Nopsca believes that the ancestral Protoavis, was a biped cursor. It used to flap air with forelimbs to gain speed. Eventually they were able to take off from the ground and sustain them in air. Ostrom (1979) also holds similar view.
3. **Diving origin.** Newman considers the ancestor of birds, Protoavis, was an aquatic reptile. They soared over water to catch fishes, from which they might have acquired the ability to fly.
4. **Compromise theory.** Gregory (1951) proposed a compromise theory. Carinates (flying birds) evolved from arboreal ancestor, while ratites originated from cursorial ancestor. For a long time confusion persisted about the probable habit of Archaeopteryx. De Beer (1954), Swinton (1960), Welty (1963) and oth-

ers believe that Archaeopteryx was arboreal and a poor flier. A flat, keelless sternum, primitive condition of pectoral girdle elements indicate its weak flying ability. The idea is supported by Yalden (1961). From the structure of pedal phalanges Ostrom (1974) suggests that the bird was a cursor. He attaches a new function to clawed digits on its wings—to knock down the insect-prey to ground. Conical and recurved, socketed teeth helped them to pick up insects quickly from the ground. The climbing function of the clawed wing digits, advocated by early workers, however, was not supported by him.

☒ Monophyletic/Diphyletic origin

Lowe (1928, 1944), Bellaris and Attridge (1975) believe that birds are diphyletic in origin. They hold that ratites differing from the carinates in several important features, represent a natural primitive group. They diverged from the ancestral stock quite early.

On the other hand majority of the workers believe that the birds are monophyletic. A fossil bird *Eleutherornis* is regarded as the probable ancestor of ostrich. Curiously enough, the fossil has remarkable affinity to the ancestor of flying bird (McFarland, 1979).

COLUMBA LIVIA DOMESTICA (Domestic pigeon)

The pigeon belongs to order Columbiformis, superorder Neognathae, subclass Neornithes, class Aves, superclass Gnathostomata, subphylum Vertebrata, phylum Chordata. Distribution of pigeon is worldwide. The domestic pigeons have many varieties, viz., panter, fantail, tumbler and others differing in size, coluration and details in feather arrangement. All of them are, however, descendants of rock pigeon *Columba livia*, produced by selective breeding.

☒ External features

Body spindle-shaped, size varying from 20-25 cm and covered by coloured feathers except beak and a small portion of the hindlimbs (Fig. 41.1).

The body is divisible into head, neck, trunk and a very small, conical tail.

Head

The head is round and drawn anteriorly into a strong, hard, pointed beak.

a. The **mouth** is a terminal wide gape, guarded by elongated upper and lower **beaks** covered with a horny sheath or **rhampotheca**.

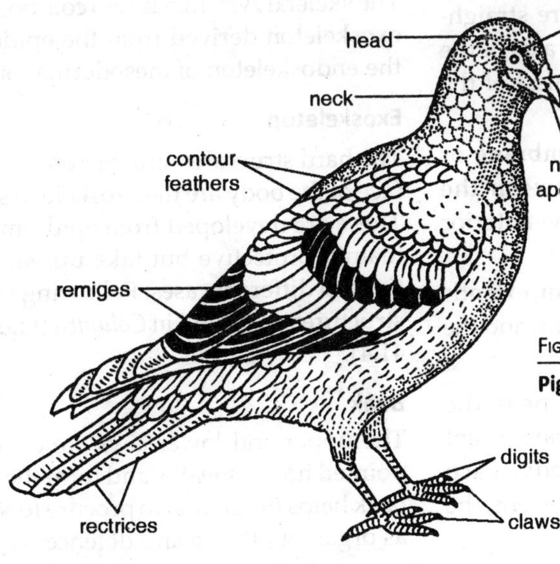

FIGURE 41.1

Pigeon (*Columba livia domestica*)

b. A swollen area of soft skin, the cere, surrounding the **nostril**, is present on each side of the upper beak.
c. The **eyes** are large and guarded by **upper** and **lower eye-lids** and a transparent third eye lid, the **nictitating membrane**.
d. Small, paired openings, the **external ears** are situated at a short distance behind the eyes, concealed by **auricular feathers**. Each opening leads into a short external **auditory meatus**, closed below by the **tympanic membrane**.

Neck

The neck is long and motile and helps in movement of the head in various directions.

Trunk

The trunk is compact, heavy and covered by wings on the dorsal and dorsolateral sides.

Forelimbs. Paired, located in the anterior region of the trunk and modified as wings. The wing has three typical divisions of the forelimb—**upper arms, forearm** and **hand**. The hand has only three imperfectly marked **digits** (fingers), the second being much longer than the first.

i. At rest, the three divisions of the wing are bent upon one another in the form of a Z; during flight they are straightened and extended and the axis of the wing is at right angle with that of the trunk.
ii. A fold of skin, the **alar membrane** or **prepatagium**, streches between the upper and forearm along the anterior or **preaxial border** of the limb.
iii. A small fold, the **postpatagium**, extends **postaxially** between upper arm and the trunk.

Hindlimbs. Paired, situated near the posterolateral sides of the trunk and are meant for carrying the weight of the body. Each hindlimb or leg has three typical divisions, the **thigh, shank** and **foot**.

i. At rest the hindlimb is in a plane parallel with the sagittal plane of the trunk.
ii. The thigh and shank are covered with feathers.
iii. The foot is divisible into a proximal **tarsometatersus** and four **digits (toes)**, of which one, the **hallux**, is directed backwards, the others—the second, third and fourth, directed forwards. The tarsometatersus and the toes are covered with scales (rhamotheca), each toe ending in a pointed, curved claw.

Tail

The tail or **uropygium** is a small, conical structure posterior to the trunk. On the dorsal surface of the uropygium lies a papilla bearing the opening of the **preen** or **coccygeal** or **oil gland**. The cloacal aperture is a large, transversely elongated aperture on the ventral surface at the junction of the uropygium and trunk.

Skin

The skin is thin, dry and devoid of any gland. It is capable of secreting **keratin**. Feathers, scales and horny sheath covering the beak are dirved from epidermis.

Skeletal system

The skeletal system has two components—the exoskeleton derived from the epidermis and the endoskeleton of mesodermal origin.

Exoskeleton

The hard structures situated on the outer surface of the body are the exoskeletal structures. These are developed from epidermis and are mainly protective but take up various functions in different cases. Following exoskeletal structures are found in *Columba* (Figs. 41.2 and 41.4).

Beak

The upper and lower jaws are covered with pointed horny sheaths and form the beak. The beak helps the animal to procure food and also as organ of offence and defence.

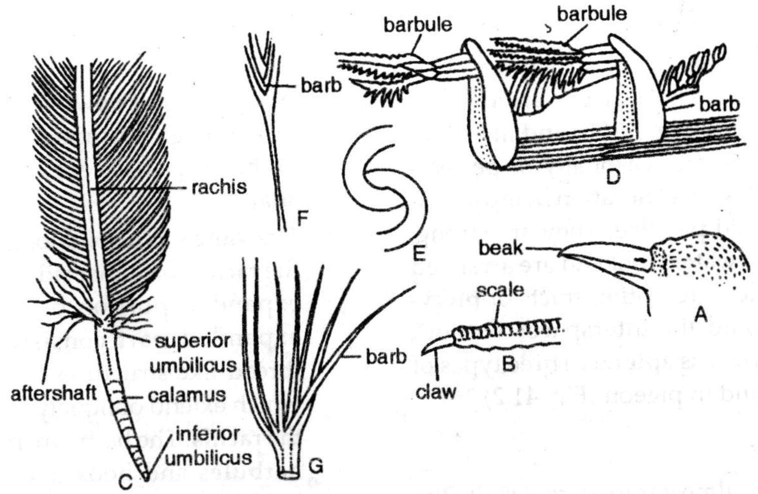

FIGURE 41.2

***Columba* sp.** Exoskeleton. A Beak. B. Claw and scale. C. Contour feather. D. Vane (a part magnified). E. Hooklet. F. Filoplume. G. Nestling down feather

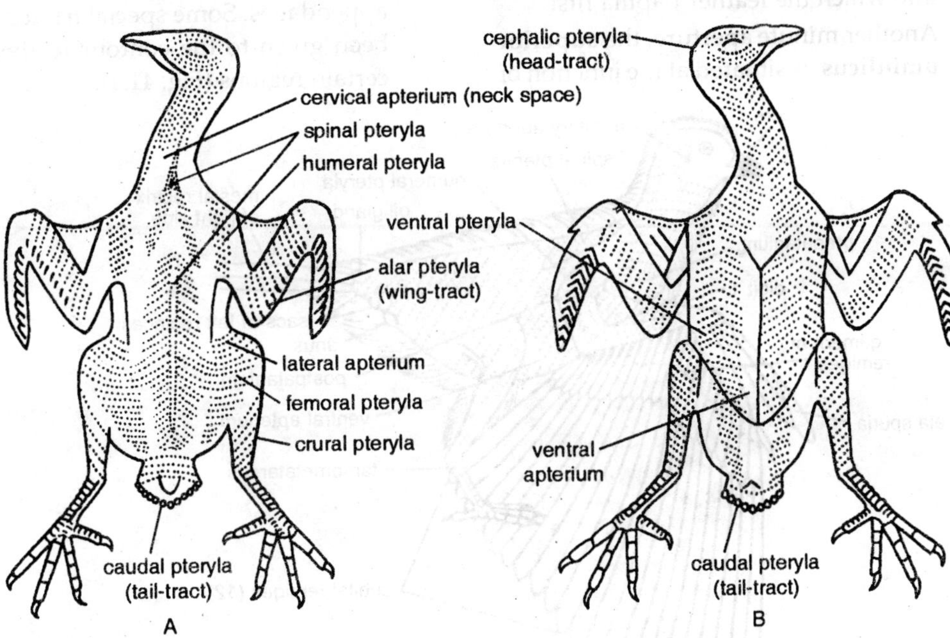

FIGURE 41.3

***Columba* sp.** Pterylosis (feather tracts). A. Dorsal surface. B. Ventral surface

Claws

The digits of the foot (toes) are provided with pointed horny structures or claws at their tip. The claws are organs of offence and defence.

Scales

The shanks and digits are provided with thickened and cornified epidermis resembling reptilian scales and are protective in function.

Feathers

Feathers are integumentary structures, characteristic of birds. The feathers are derived from the epidermis (Fig. 7.10B) and like other similar structures, are seasonally or periodically cast off, the new generation being developed from the old papillae. They are strong, light, elastic and waterproof and are arranged on the skin in definite feather tracts or **pterylae** (Fig. 41.3) and the interspaces without feathers are known as **apterae**. Three types of feathers are found in pigeon (Fig. 41.2).

Contour feather

1. A hollow, almost transparent stalk, the **quill** or **calamus**, and an expanded distal portion, the **vexillum** or **vane**.
2. A small aperture, the **inferior umbilicus** is situated at the base of the quill into which the feather papilla fits.
3. Another minute aperture, the **superior umbilicus**, is situated at the junction of the quill and the vane on the ventral surface.
4. A small tuft of down feather is present in the neighbourhood of the superior umbilicus. These are known as **after-shaft**.
5. The **vane** has a solid longitudinal axis, the **rachis**, on either side of which is the expanded portion or the vane. The expanded part consists of delicate, thread-like structures known as **barbs** which extend obliquely outwards from the **rachis**. The barbs are provided with **barbules** and hook-like structures or **hooklets** on both the sides, by which the barbs remain attached with one another and form the expanded portion. Contour feather forms the general covering of the whole body and appendages. Some special names have been given to the contour feathers in certain regions (Fig. 41.4).

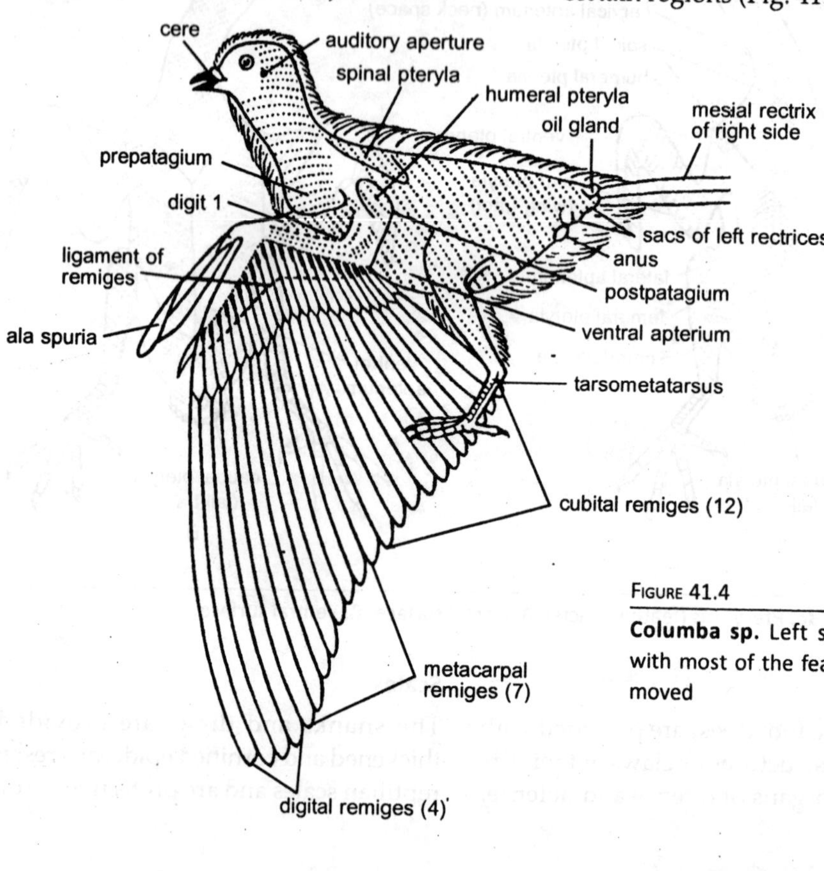

FIGURE 41.4

Columba sp. Left side view with most of the feathers removed

i. **Remiges or wing quills.** The large feathers on the hinder border of the wing. Twelve of them attached to the ulna are known as **cubitalis** and the rest are called **primaries.** Of the 11 primaries,—7 called **metacarpals**, are attached to the metacarpal region and 4 digitals, attached to phalanges.
ii. **Rectrices.** Twelve long quills spring in a semicircle from the uropygium.
iii. **Coverts.** Several rows of feathers cover the bases of large feathers of remiges and rectrices are called **upper** and **under wing** and **tail coverts**, respectively.

Filoplumes

These feathers have delicate hair-like long axis and a few **barbs** devoid of barbules. These are situated in between the contour feathers.

Down feathers

These feathers have small axis and a few barbs devoid of locking apparatus at the distal end. Nestling pigeons are covered with a temporary investment of down feathers.

Endoskeleton

Hard structures within the body and supporting it are endoskeleton. The skeleton of pigeon is strong but lightly built. The texture of the bone is often very spongy and air spaces from the lungs may continue into many of the bones, making them light and more or less devoid of marrow. These are called **pneumatic bones.** Most of the bones—except those of the tail, forearm, hand and hindlimb contain air spaces. The components of the endoskeleton are axial skeleton and appendicular skeleton (Fig. 41.5).

Axial skeleton

The skull, hyoid apparatus, vertebral column and thoracic basket constitute axial skeleton.

Skull

The skull is **schizognathous** (Fig. 41.6) with a round cranial cavity, large orbits and a narrow, pointed beak, chiefly composed of premaxillae. The sutures are obliterated and the skull is light with a highly polished surface. All the bones are fixed except the quadrate, lower jaw, columella and hyoid. Teeth absent.

Cranium

The **foramen magnum** is directed down-ward and backward, the anterior margin being formed by a single, small, rounded **occipital condyle.** The foramen magnum is surrounded by **basioccipital**, two **exoccipitals** and the **supraoccipital**. The basioccipital forms the **occipital condyle.**

The roof of the cranial cavity is formed by paired **parietals** and **frontals.** The frontals are small and partly overlapped by the upward extensions of premaxillae. The posterior part of the floor is formed by **basioccipital**, while the anterior part by **basisphenoid**, the latter being drawn into a rostrum, which represents the anterior part of the **parasphenoid.** A pair of membrane bones, the **basitemporals**, are present on the ventral aspect of basisphenoid. In the adult, these bones probably represent the posterior portion of the parasphenoid.

Sense capsules

The tympanic cavity is bounded by the **squamosal** firmly united with other cranial bones. The main part of the **auditory capsule** is occupied by a large ossified **prootic.**

Embryonic **opisthotic** unites with the exoccipital; the **epiotic** with supraoccipital. The vertical **interorbital septum** is formed by **parasphenoid** and **mesethmoid.** The septum is partly ossified and partly cartilaginous.

The **ectoethmoids** of the olfactory organs are poorly developed. The **lacrymals** are large; the **nasals** are forked, each forming both the inner and outer boundary of the corresponding nostril.

Jaw apparatus

Practically the whole of the **upper beak** is formed by the union of two **premaxillae** forming a large triradiate bone. The **maxillae** are small, their anterior ends produced inward

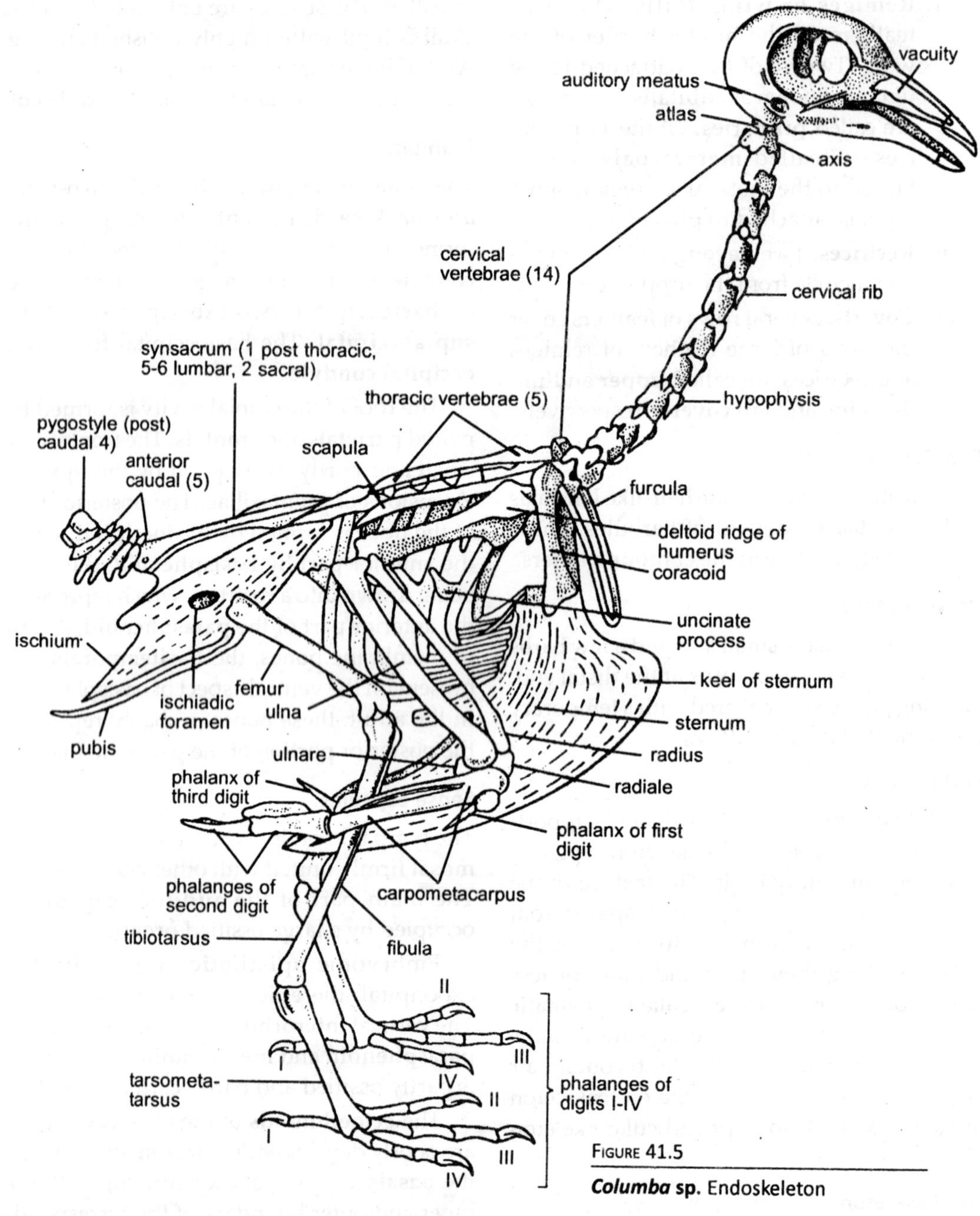

FIGURE 41.5

***Columba* sp.** Endoskeleton

into spongy **maxillopalatine processes.** The slender posterior end of the maxilla continues backwards by a slender **jugal** and **quadrato-jugal** to the **quadrate.** The quardrate is stout, triradiate and articulates with two facets on the **otic process** with the roof of the tympanic cavity. The quadrate sends an orbital process anteriorly and bears a condyle below for articulation with the **mandible.**

The slender anterior end of the palatines are ankylosed with the maxillae and the posterior ends with the pterygoids and the rostrum. The **pterygoids** are rod-shaped and articulate behind with the quadrate and with the **basipterygoid process** at about the middle of its length.

The mandible (Fig. 41.6C) consists of two similar halves or **rami**. Each **ramus** is formed by five bones, **dentary, splenial, supra-angular, angular** and **articular**.

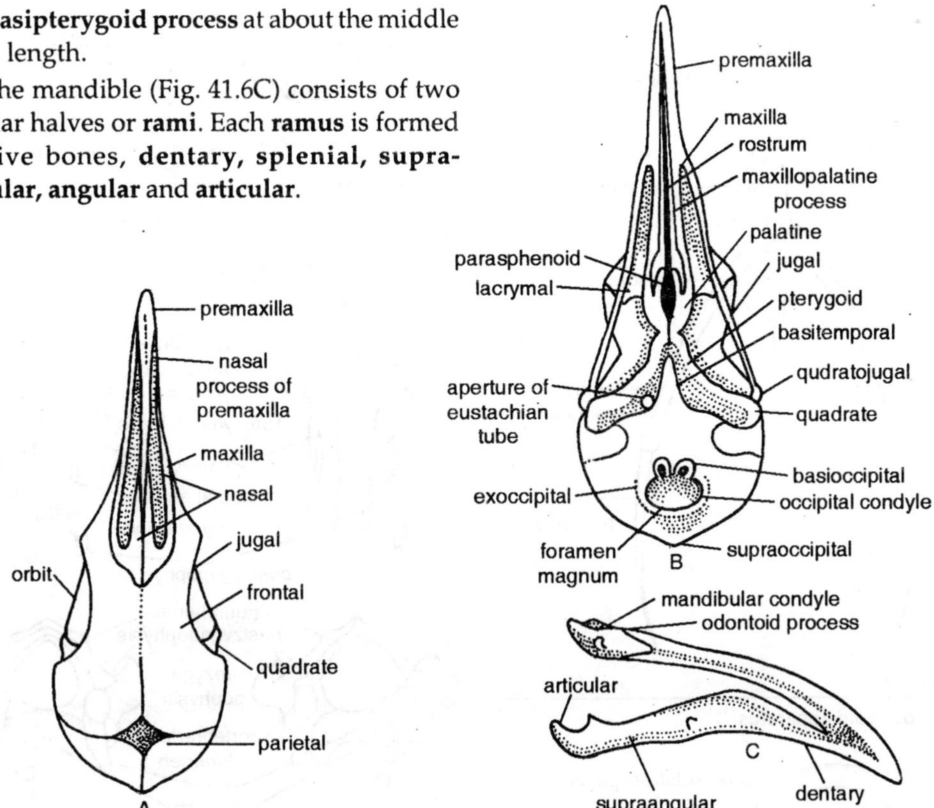

FIGURE 41.6

***Columba* sp.** Skull and mandible. A. Skull. Dorsal view. B. Skull, Ventral view. C. Mandible. Dorsolateral view

☒ Hyoid apparatus

The hyoid apparatus (Fig. 41.7) is formed by (a) **basihyal**, median, arrow-shaped, bony, except the tip, (b) **ceratohyals (hyoid cornua)** short paired bones with cartilaginous tip and directed posteroventrally from the base of the basihyal, (c) **first basibranchial**, a median bone posterior to basihyal, (d) **second basibranchial**, a median bone with a narrow, cartilaginous tip, posterior to the first basibranchial, (e) **ceratobranchials**, paired bones running posterolaterally from the junction of the two basibranchials, and (f) **epibranchials**, paired, continuous with the ceratobranchials, curved posteriorly and gradually narrow down, cartilaginous at the tip.

☒ Vertebral column

The vertebral column is divided into five regions—cervical, thoracic, lumbar, sacral and caudal. The numbers of vertebrae are—cervical 14, thoracic 5, lumbar 5-6 sacral 2, anterior caudal 5 and posterior caudal 4 (pygostyle). One posterior thoracic, 5-6 lumbars, and 2 sacrals fuse to form the synsacrum.

The centrum of the vertebra is **heterocoelous**, i.e., surface saddle-shaped. Anterior face concave from side to side and convex from above downwards; posterior face convex from side to side and concave from above downwards.

Atlas (first vertebra). Small, ring-like*, bones completely fused and without any suture (Fig. 41.8); a distinct centrum lacking;

* A neural canal is present in all vertebrae from atlas to free caudal.

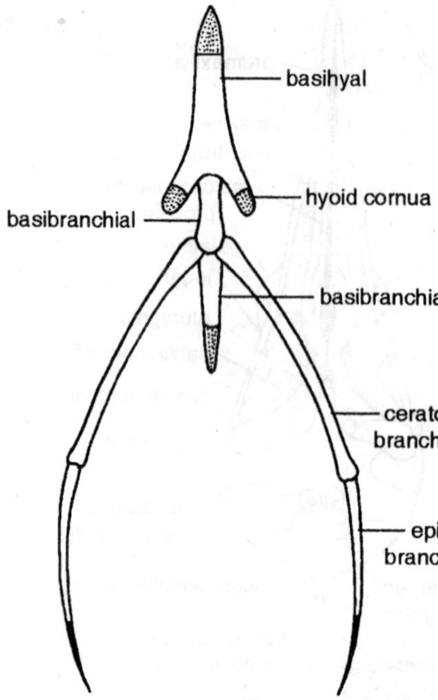

FIGURE 41.7

***Columba* sp.** Hyoid apparatus

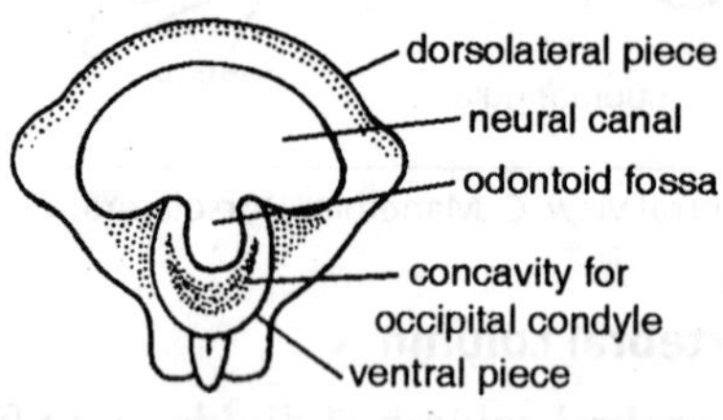

FIGURE 41.8

***Columba* sp.** Atlas (First vertebra). Anterior view

odontoid fossa, a notch, present at the posterior side of the thickened ventral portion; a cup-shaped facet for articulation with occipital condyle present anteriorly; postzygapophyses present, prezygapophyses absent.

Axis (second vertebra). Centrum heterocoelous; odontoid process (centrum of first vertebra) peg-like (Fig. 41.9) neural **arch*** encloses neural canal; neural spine blunt; pre- and postzygapophyses** well-developed; hypapophysis small and flattened.

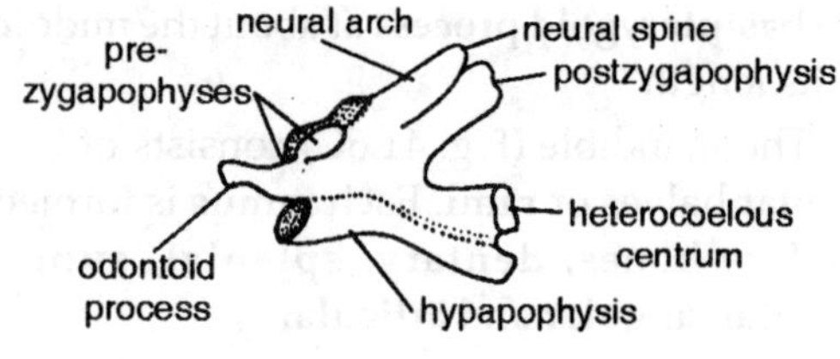

FIGURE 41.9

***Columba* sp.** Axis (Second vertebra). Lateral view

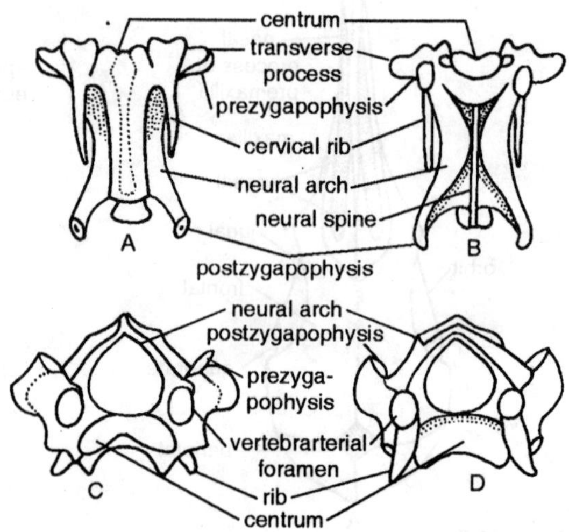

FIGURE 41.10

***Columba* sp. Cervical vertebra.** A. Ventral view. B. Dorsal view. C. Anterior view. D. Posterior view

Typical cervical vertebra. Neural spine (Fig. 41.10) crest-shaped (short in anterior vertebrae; longer in posterior ones); transverse processes present; vestigeal, double-headed ribs fused with the vertebra and directed posteriorly; vertebrarterial foramen present; articulating surfaces of the centrum contain synovial capsules, meniscus and supplementary ligament.

N.B.: Posterior cervicals bear hypapophysis.

Free thorocic vertebra. Neural spine large and crest-shaped (Fig. 41.11); transverse processes present; presence of capitular facets on centrum and tubercular facets on transverse processes for articulation with

* Neural arch is present in all vertebrae from axis to free caudal.

** Pre-and postzygapophyses are present in 2-14 cervical and anterior four thoracic vertebrae.

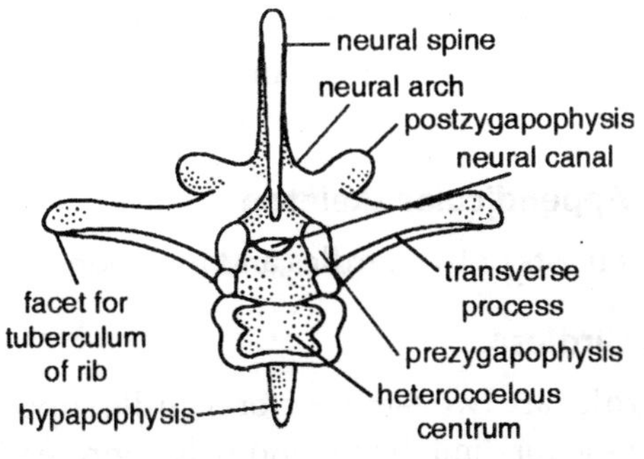

FIGURE 41.11

***Columba* sp.** Free thoracic vertebra. Anterior view

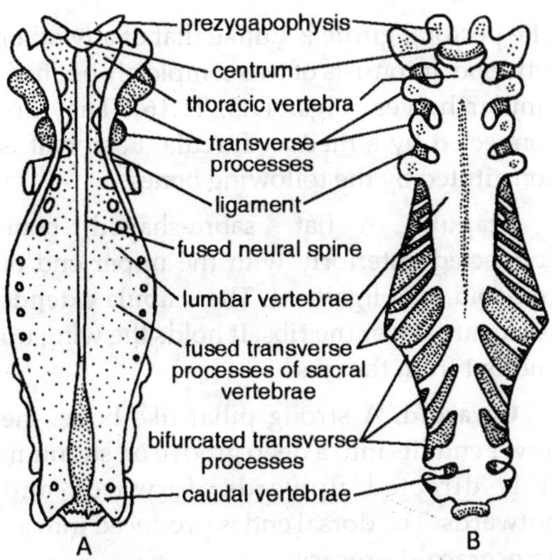

FIGURE 41.12

***Columba* sp.** Synsacrum. A. Dorsal view, B. Ventral view

double-headed ribs; vertebrarterial foramen present; articular surfaces of centrum similar to those in cervical vertebra; hypapophysis present.

Synsacrum. Elongated, more or less triangular in shape (Fig. 41.12), formed by the fusion of 8 to 9 vertebrae.

Free caudal vertebra. Centrum short but prominent and appears as acoelous (flat on both the faces); neural arch quadrangular (Fig. 41.13); neural canal narrow; neural spine moderately developed and bifid.

Pygostyle or ploughshare bone. Laterally compressed, upturned posteriorly (Fig. 41.14); formed by fusion of last four caudals; centrum heterocoelous but appears as acoelous; facet at the anterior end for articulation with the free caudal vertebra.

⊠ Ribs

A pair of ribs remain attached with a vertebra. The ribs are of two types—incomplete, not reaching the sternum from the vertebral column and complete—joining the sternum with vertebral column and forming the thoracic cage. A complete rib is a slightly curved structure (Fig. 41.5) with two heads—a **capitulum**, articulating with the centrum and a **tubercle**, articulating with the transverse process. The ventral part of the rib adjoining the sternum is called **sternal rib**, and joined at an angle to the dorsal part, the **vertebra rib**, articulating with a vertebra. A bony **uncinate** process arises from the posterior edge of the vertebral rib of the first four pairs and ankylosed with the ribs. Complete ribs are five pairs attached to the five thoracic vertebrae. Rest of the ribs are incomplete. **Vestigeal ribs** with the head fused with the centrum and tubercle with transverse process are present in third to twelfth cervical vertebrae. The last, or last two cervical vertebrae possess **cervical ribs**, which are free and do not unite with the sternum.

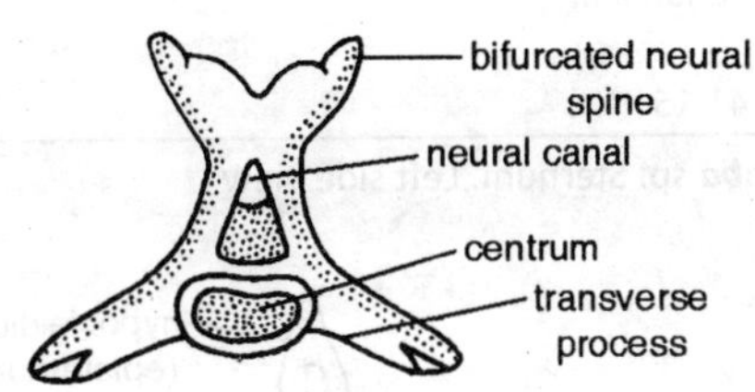

FIGURE 41.13

***Columba* sp.** Free caudal vertebra. Anterior view

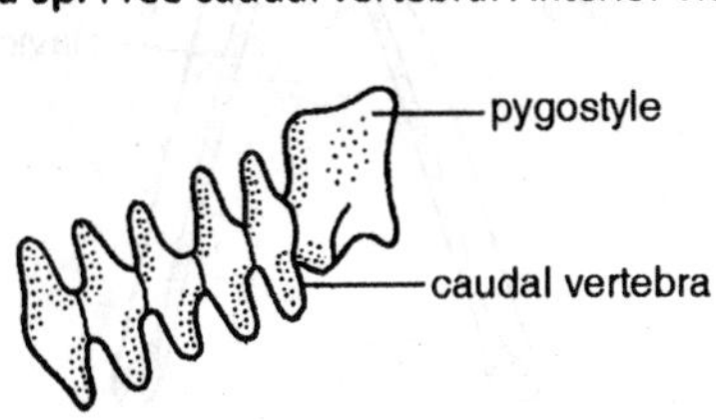

FIGURE 41.14

***Columba* sp.** Pygostyle. Side view

⊠ Sternum

The sternum is most characteristic of a flying bird, which provides greater surface for attachment of large and powerful flight muscles. It is a boat-shaped structure (Fig. 41.15) made up of thin bone— anterior end broad and convex laterally, posterior end narrow. A deep keel or **carina sterni** is present in the sagittal plane, on the ventral surface of the bony plate. Two pairs of notches are present on the posterior border. Anterior edge bears a pair of deep grooves for articulation with coracoids. Dorsal concave surface bears five pairs of costal facets for articulation with ribs; posterior to costal facets arise a pair of short, stout, oblique lateral xiphoid processes, directed posterodorsally. The pair of metasternal processes are slender, long and directed posteriorly.

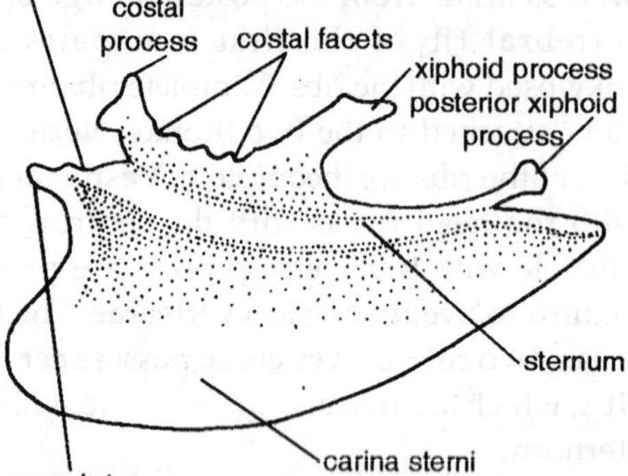

FIGURE 41.15

***Columba* sp.** Sternum. Left side view

⊠ Appendicular skeleton

It consists of two girdles and limb bones.

⊠ Girdles

Girdles are two—the pectoral girdle supporting the forelimb bones and pelvic girdle supporting the hindlimb bones.

Pectrol girdle

The pectoral girdle is unlike that of any other tetropod. It consists of two completely ossified, similar halves (Figs. 41.5, 41.16). They are connected by a median furcula. Each half is constituted by the following bones.

Scapula. A flat, sabre-shaped bone connected anteriorly with the upper end of coracoid by a ligament. The scapula extends backwards over the ribs. It holds the wing up and out from the ribs.

Coracoid. A strong pillar-like bone, the lower end fits into a deep groove on sternum. It is directed upwards, forwards and outwards. The dorsal end is produced into an **acrocoracoid process**.

The **coracoscapular** angle is less than 90°.

Clavicle. A slender bone, the upper end meets the junctions of coracoid and scapula and the other ends of the two clavicles fuse to

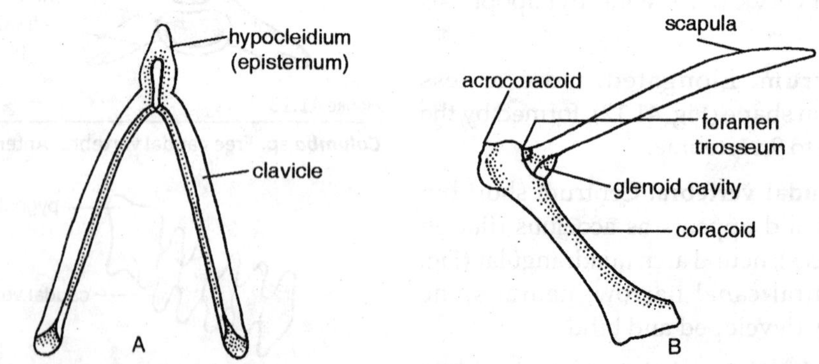

FIGURE 41.16

***Columba* sp.** Pectoral girdle. A. Furcula, B. Girdle (left half)

form a **furcula**. It braces the mobile wing out from the sternum.

Interclavicle. A small bone, fused with the furcula and connects it with the sternum.

Foramen triosseum present at the junction of coracoid, scapula and clavicle.

Glenoid cavity. At the junction of coracoid and scapula.

Pelvic girdle

The pelvic girdle does not resemble that of any other living tetrapod. It is formed by two separate halves. The pelvic girdle is compact and ankylosed with the synsacrum.

Each half (Fig. 41.17) is constituted by the following bones.

Ilium. Thin, broad, large and flat, concave anteriorly and convex posteriorly and with an **antitrochanter** process. The ilium is divisible into **pre-** and posttrochanter portions.

Ischium. Broad, fused with the posterior end of the ilium and anteriorly separated from the **ischiadic foramen.**

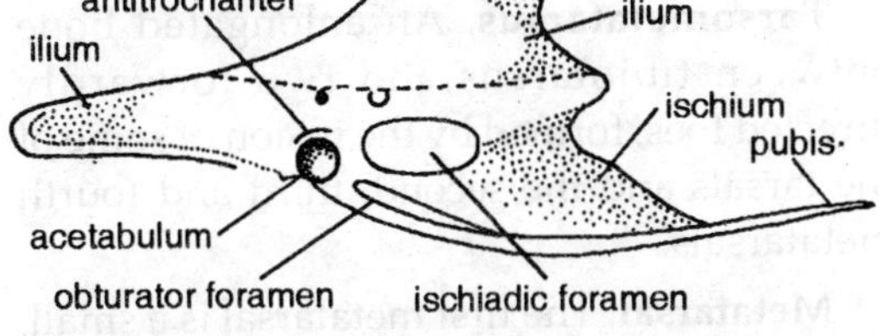

FIGURE 41.17

***Columba* sp.** Pelvic girdle. Left half. Lateral view

Pubis. Slender, long curved and parallel to the ischium. Both pubis and ischium are directed sharply backwards from the acetabular ends.

Obturator foramen. Separates ischium from pubis.

Acetabulum. Present at the junction of the ilium, ischium and pubis.

Limb bones

Bones supporting the limbs are limb bones.

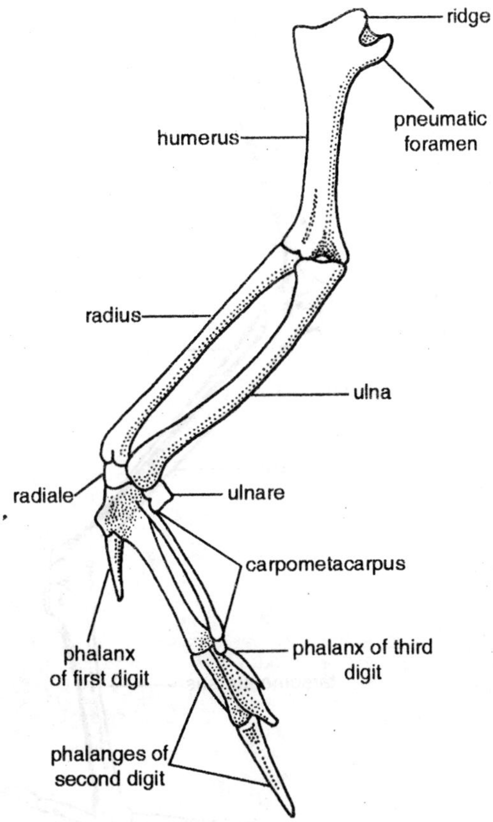

FIGURE 41.18

***Columba* sp.** Left forelimb bones

Forelimb

The forelimbs are modified as wings and consequently have been modified in a different direction from other tetrapods, which is more pronounced at the distal end. The forelimb bones (Fig. 41.18) are:

Humerus. (a) The bone is light but strong. (b) Shaft long, slightly flat and bent. (c) Proximal end greatly expanded and bears head. (d) A prominent ridge and a pneumatic foramen present very near the head. (e) Presence of two condyles separated by an intercondylar foramen at the distal end. Expanded proximal end and the ridge provide surfaces for attachment of pectoral muscle.

Radius and ulna. (a) Two unequal, light, long bones joined at both ends. (b) Radius slender and nearly straight, (i) a concave facet

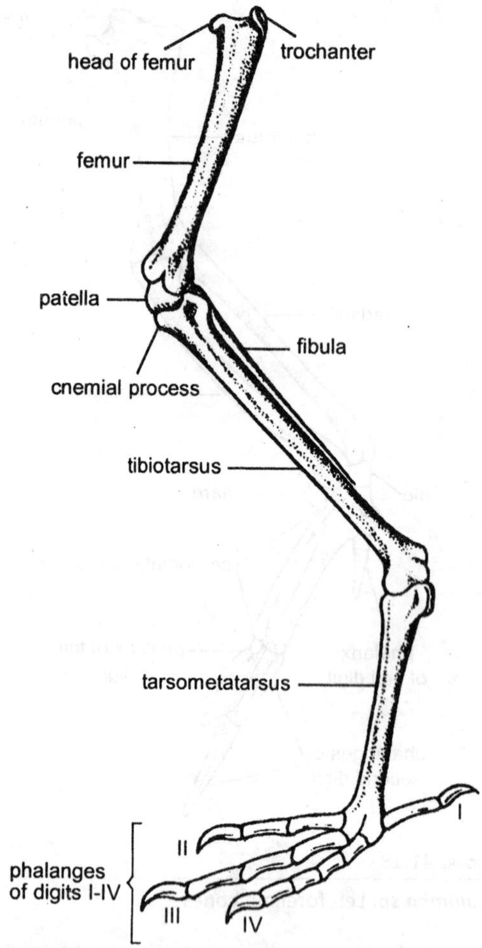

FIGURE 41.19

***Columba* sp. Left hindlimb bones**

present at the proximal end, (ii) distal end convex. (c) Ulna stout and gently curved, (i) proximal end bears an olecranon process and a concave sigmoid notch, (ii) distal end convex, (iii) presence of nutrient foramen.

Carpals. Two free and three united with metacarpals. The two free carpal bones are **radiale** and **ulnare.**

Carpometacarpus. Formed by the union of distal carpals with three metacarpals. The first is a step-like projection; the second and third are two rod-like bones, fused with one another at both the proximal and distal ends.

Phalanges. Phalanx articulated with the first metacarpal is single and pointed. The second metacarpal bears two phalanges; the third bears a single, pointed phalanx.

Hindlimb

The hindlimb bones (Fig. 41.19) are:

Femur. (a) A long, stout cylindrical bone with broadened ends. (b) Proximal end bears a round head at right angle to the shaft. (c) A trochanter present near the head. (d) The distal end pulley-like and the condyles separated by an intercondylar fossa.

Tibiotarsus and fibula. (a) Two long bones, tibiotarsus and fibula, partially fused. (b) Tibiotarsus stout, straight, long and formed by the fusion of tibia and two proximal tarsals (astragalus and calcaneum), (i) proximally a cnemial crest present, (ii) proximal end bears a concave articular surface, (iii) distal end pulley-like. (c) Fibula, a small, slender, distally pointed bone, located on the lateral surface of the tibiotarsus, (i) proximally articulates with the tibiotarsus.

Patella. A sesamoid bone (a bone developed in a tendon) is present on the exterior side of the knee-joint.

Tarsometatarsus. An enlongated bone between tibiotarsus and two forwardly directed toes, formed by the fusion of some of the tarsals and the second, third and fourth metatarsals.

Metatarsal. The first metatarsal is a small, irregular bone, attached to the inner side of the tarsometatarsus near its distal end.

Phalanges. The phalanges in the toes are two in the backwardly directed hallux. In the three others from inner to outer the phalanges are three, four and five. All the phalanges end in a pointed, curved, claw.

Muscles of flight and perching

Flight is perhaps the most distingushing characteristic that separates birds from other groups of vertebrates, that has made them the technical marvel of the animal world. Flight in the air involves a typical physical phenomenon, in that it depends on the generation of two kinds of forces, a lift-force acting upwards and a drag-force tending to stop the motion. These lift and drag-forces are proportional to the square of the speed.

To maintain buoyancy in air, various types of modifications have taken place in the body of a bird, of which muscles concerned with flight are most highly modified. The birds cannot remain in air for an indefinite period and they must return to some support for rest after a flight. In percher birds, the muscles and tendons of legs (Figs. 41.20, 41.21, 41.22) are so modified that they not only help in perching but also enable the bird to keep the body in position during rest or even in sleep.

Flight muscles

Wings, the modified forelimbs are organs for flight. The musculature of the forelimbs are greatly modified in response to the function they are to perform (Figs. 41.20, 41.21, 41.22). Flight is the coordinated effect of a number of paired muscles, of which the following are most important.

Pectoralis major (Depressor muscles). These are the largest breast muscles which arise from the whole of the keel of the sternum and from the clavicle, and are inserted into the ventral surface of the humerus by a complicated tendon of insertion. The two muscles fill the space between the body and the keel of the sternum forming the breast of the bird and are about onefifth of the body weight. These muscles depress the wings. The bird balances on its wings mainly by the action of the pectoralis major.

Pectoralis minor (Supracoracoideus or Elevator muscle). These are smaller but longer than pectoralis major and arise from the anterior part of the body of the sternum dorsal to the pectoralis major. Its tendon passes through the foramen triosseum between furcula, scapula and coracoid and inserted into the dorsal aspect of the humerus. The foramen acts as a pulley and the direction of action of the muscle is changed; the backward pull of the tendon raises the humerus. The elevation of the wing is assisted by other two small muscles, **latissimus dorsi** and **deltoid muscle**.

Latissimus dorsi. Arise from the spines of thoracic vertebrae and inserted on the humerus between the attachment of triceps and deltoid muscles. It pulls the humerus medially.

Deltoid muscles. These are complex muscles, each with three **tensores patagialis**. These muscles keep the prepatagium tensely stretched in extended wings. A small muscle, **tensor membrane posterioris alae** keeps the postpatagium tense.

Coracohumerals and scapulohumerals. These are two pairs of small muscles. The posterior one, arises chiefly from the coracoid but in part from the sternum, and its tendon is inserted into the ventral surface of the humerus. It helps a little in depressing the wings. The anterior muscle arises from the outer surface of the scapula and produces adduction and lateral rotation of the humerus in the glenoid cavity, resulting in a slight elevation of the wing (Fig. 41.21).

Biceps. Attached to scapuloclavicular joint and head of the humerus by two heads and inserted to anteroproximal surface of the radius.

Triceps. A large, fleshy muscle attached to humerus by two heads and inserted to the olecranon process of ulna along with the tendon of **anconeus muscle**.

The biceps and triceps operate the elbow and perform adjustments during flight.

Extensor carpi radialis and extensor carpi ulnaris. The forearm muscles, act in conjunction with triceps and biceps in stretching and folding the wing.

Brachioradialis muscles. Two, they rotate the radius medially.

Digit muscularis. Inspite of being greatly reduced they perform certain important functions in disposition of parts of wings and even individual feathers during flight.

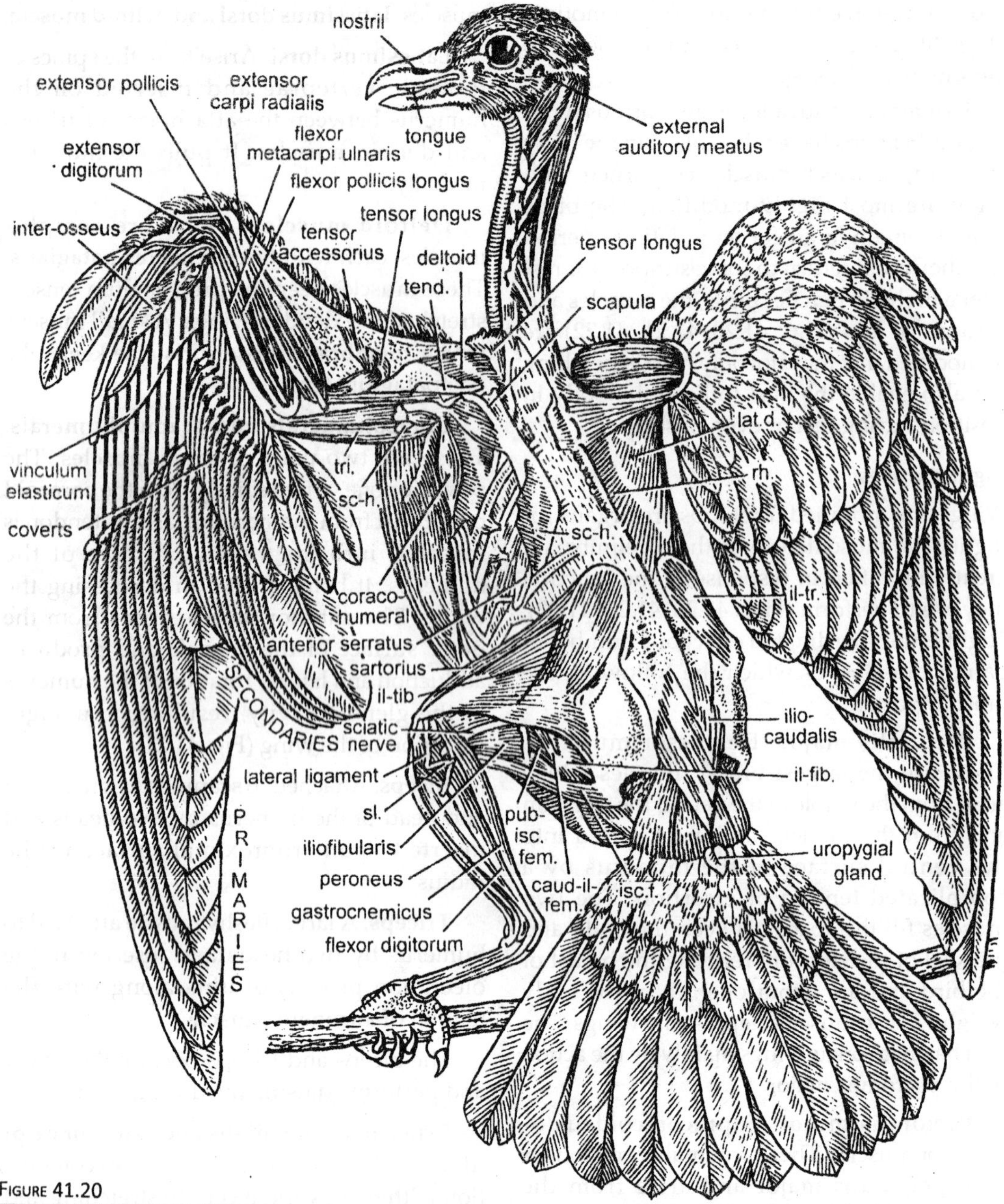

FIGURE 41.20

***Columba* sp.** Flight muscles and leg muscles, (dorsal-side view)

Pelcis muscles. Move the bastard wings (first digit), possibly controlling the front edge of the wing and to produce extra forward momentum during take-off.

There are several smaller muscles also, which help in the elevation and rotation of the wings. The muscles of the neck, tail and legs also help the bird to a considerable extent in its flight. An elaborate system of tendons controls the position of the individual feathers.

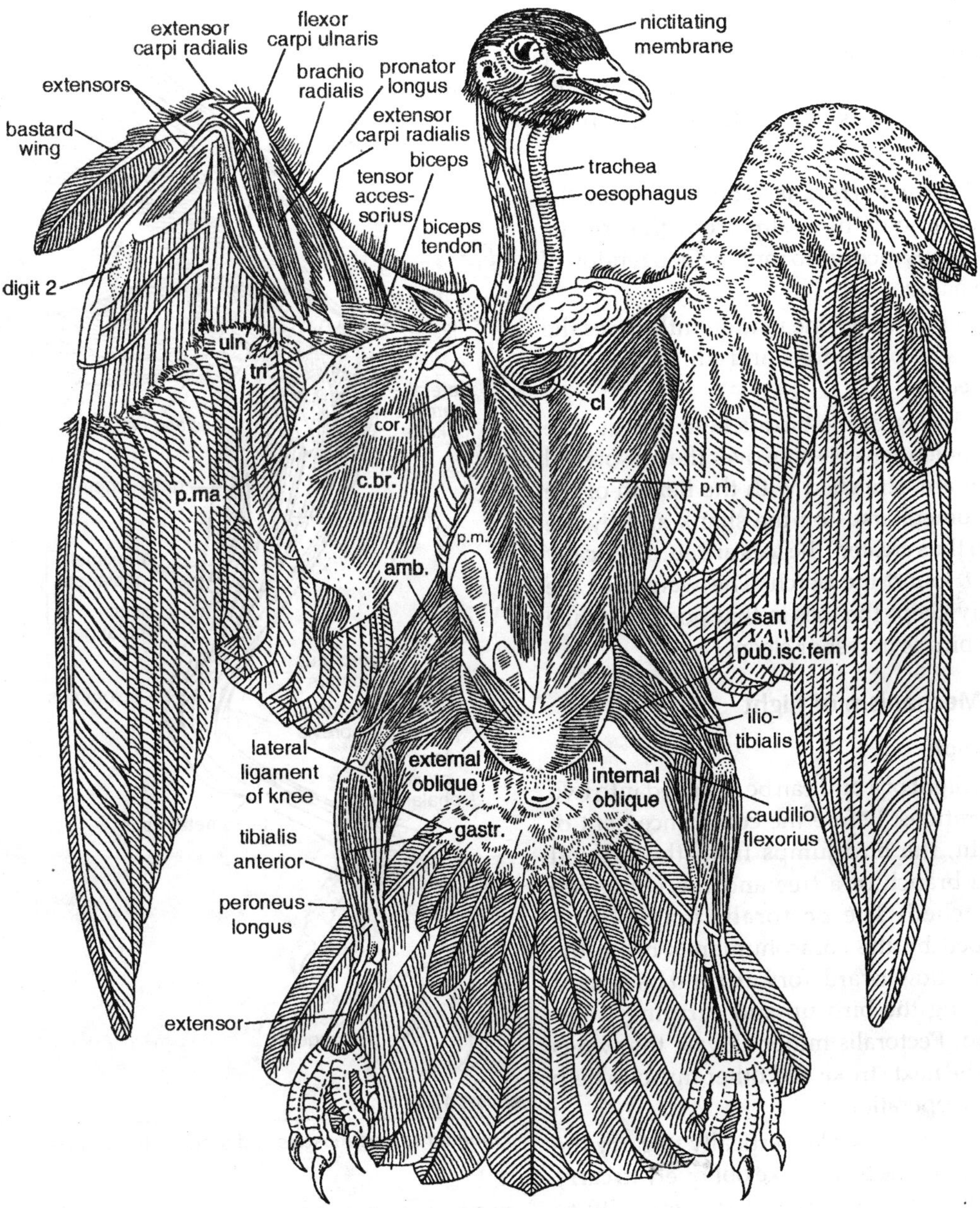

FIGURE 41.21

***Columba* sp.** Fligut muscles (viewed from ventral surface) (S/S) Dissected from ventral surface *amb.* ambiens; *c.br.* coracobrachialis; *cl.* clavicle; *cor.* coracoid; *gastr.* gastrocnemius; *p.ma.* pectoralis major; *p.am.* pectoralis minor (supracoracoideus); *pub-ise-fem.* pubischiofemoralis; *sart.* sartorius; *tri.* triceps; *ubt.* ulna

Leg muscles

Peroneus medius. The tendon of the muscle (Fig. 41.22) trifurcates and form the deep tendons of the three forwardly directed toes.

Flexor perforans. The tendon of the muscle joins by a slip from that of the peroneus medius to form a deep tendon of the hallux.

A pull upon one tendon influences the two sets of tendons and all the toes are flexed.

The leg of the bird is bent at rest. The flexor tendons passing over the mesotarsal joint are stretched with the flexion of the tarsometatarsus on the shank and by the pull thus created, the toes are automatically bent round the perch. The weight of the body keeps the toes in that position while the bird is in rest or sleep.

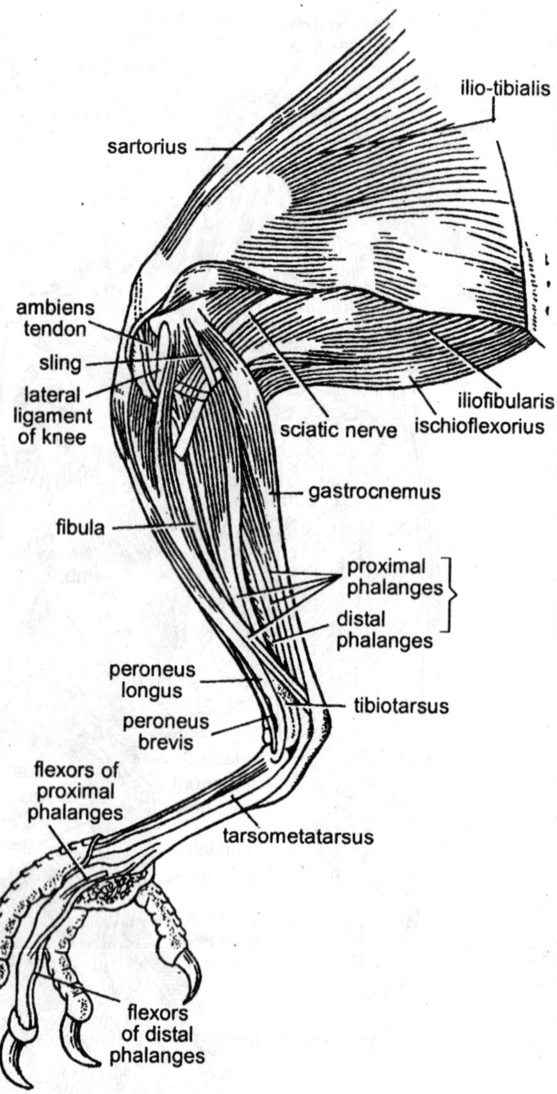

FIGURE 41.22

***Columba* sp.** Leg muscles and tendons (Lateral view)

⊠ Mechanism of flight

Take-off

The flight of a bird can be resolved into two operations. Before the commencement of flight, the bird jumps from the ground or a branch of a tree and the wings are stretched. The pectoralis major being helped by the coracohumerels bring the wings downward, forward and backward keeping the bird up and carrying it onward. Pectoralis minor elevate the wings for the next stroke and other muscles help in the operation.

The wings strike vertically to carry the weight of the bird; strike obliquely to carry the bird onward and more vertically to push the bird upward.

Landing

It is a very delicate operation, specially since it often involves coming to rest suddenly. This is achieved by lowering and fanning out the tail feathers which thus act as a flap. Legs are then lowered, one further wing-stroke is given to bring the bird forward to drop on the perch.

⊠ Modes of flight

1. **Common mode of flight.** The strokes are active, in which the wings move downward and forward, backward and upward.
2. **Gliding flight.** The wings remain stretched but do not flap, the curvature of the surface may, of course, be changed. This takes place after an active flight in which velocity has been acquired.

3. **Sailing flight.** The bird is motionless with spread wings. In such cases the weight of the bird is supported upon upward currents of air.

Digestive system

The digestive system consists of an alimentary canal and digestive glands (Fig. 41.23).

Alimentary canal

It starts from mouth and ends in rectum opening in cloaca.

Mouth

A fairly large opening bounded by two horny **beaks, Teeth** absent.

Buccal cavity

Mouth leads to a buccal cavity (Fig. 41.23B). A large **tongue**, pointed anteriorly is present on the floor. **Taste buds** and **mucous glands** are present on the tongue. **Salivary glands**—a median **sublingual** below the tongue and paired **angle glands** of the gape are present.

Pharynx

A small chamber, posterior to the mouth cavity. The internal nare communicating with the external nares open on the roof. A dorsally situated aperture, the **glottis**, leading to trachea is situated near the base of the tongue. Posteriorly the pharynx leads into oesophagus through gullet.

Oesophagus

A fairly long, thick-walled tube extending from pharynx to the **proventriculus**. The major part of the oesophagus lies in the neck running by the side of the trachea.

Crop

The posterior part of the oesophagus is dilated into a large, thin-walled sac at the base of the neck, between the skin and the muscles, and

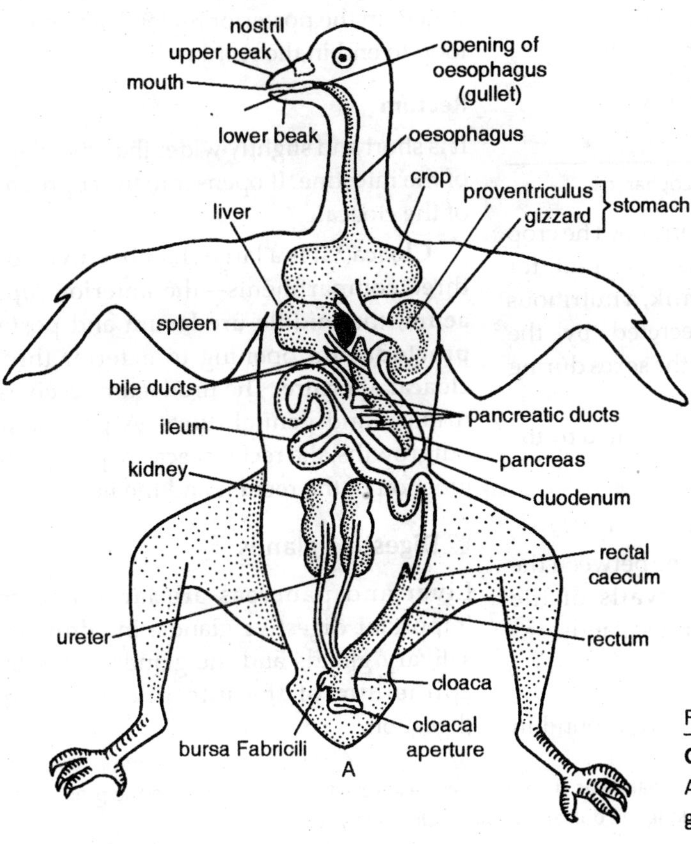

FIGURE 41.23A

Columba sp. Digestive system. A. Alimentary canal and digestive glands

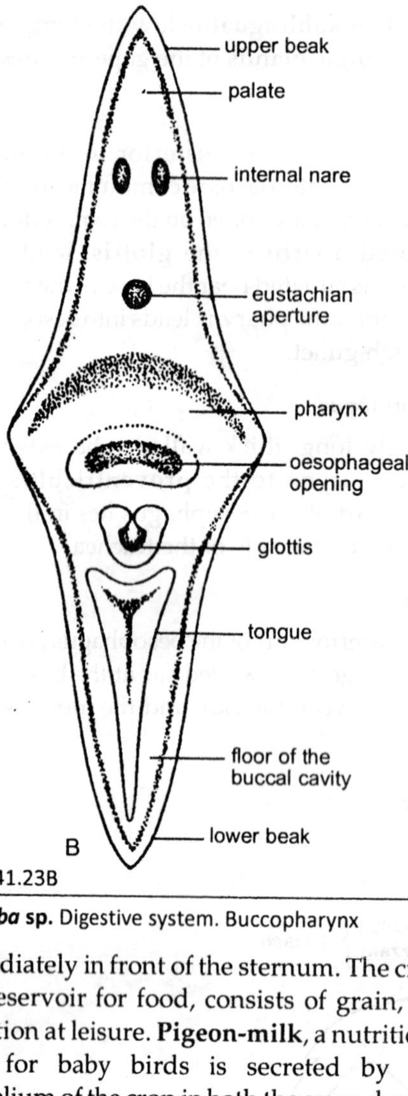

FIGURE 41.23B

***Columba* sp.** Digestive system. Buccopharynx

immediately in front of the sternum. The crop is a reservoir for food, consists of grain, for digestion at leisure. **Pigeon-milk**, a nutritious food for baby birds is secreted by the epithelium of the crop in both the sexes during breeding season.

In pigeon the stomach is represented by the **proventriculus** and **gizzard**.

Proventriculus

It is a slightly dilated, small tube between the crop and the gizzard. The walls of the proventriculus contain numerous glands.

Gizzard

A highly muscular, biconvex body, round in outline, with very thick walls and a small lumen. The wall is formed by immense development of muscles, which radiate from two tendons, one on each of the convex surface. Narrow, minute, tubular glands in the epithelium of the gizzard secrete a yellow or green-coloured fluid, which becomes thick and horny and forms the lining of the gizzard. The cavity always contains small stones which are swallowed by the bird for grinding food in the gizzard.

Duodenum

The first part of the intestine. It leaves the gizzard close to the entrance of the proventriculus and forms a U-shaped loop enclosing the pancreas. Three pancreatic ducts and two bile ducts open in the duodenum.

Ileum

The rest of the intestine. It forms a loop at the beginning. The major part of the rest is spirally coiled. In the posterior region it again forms a loop to end in the rectum.

Rectum

It is short and slightly wider than the diamater of the intestine. It opens into the coprodaeum of the cloaca.

Cloaca.* It is a large chamber divided into three compartments—the anterior **coprodaeum**, the middle **urodaeum** and posterior **proctodaeum** opening to exterior through **cloacal aperture**. The urodaeum receives the urinary and genital ducts. A pair of small outgrowths, the **rectal caeca** are present at the junction of the rectum and the ileum.

⊠ Digestive glands

Liver and pancreas are large and most important digestive glands. In addition, the salivary glands and the glands in the lining epithelium of the intestine also help in digestion.

* **Bursa fabrici.** A small, thick-walled glandular chamber in close association and opening in the proctodaeum is absent in the adult but present along the dorsal wall of the cloaca in young birds.

Liver

A large, reddish brown, bilobed gland, ventral to the gizzard. Gall bladder is absent. Two ducts from the two lobes of the liver open directly into the two arms of duodenum.

Pancreas

A compact reddish gland lying in the loop of the duodenum. It is a combined **exocrine** and **endocrine** gland. The secretion of the exocrine part, the pancreatic juice, is discharged into the duodenum by three ducts.

Feeding

The pigeon is a granivore. The beaks are modified to pick up grains efficiently. The **saliva** contains a **diastatic enzyme**. The swallowed food is moistened by mucus and a large amount can be stored in the crop.

⊠ Digestion

The ingested food in the crop is macerated. The transfer of food from crop to proventriculus is regulated. The food gets mixed with HCl and pepsin in the proventriculus. In the gizzard, the grains are completely churned by the action of gizzard walls, the stones helping in the operation. The churned food is mixed with hydrolysing enzymes in the pancreatic juice, which also neutralises stomach acid; the bile helping in emulsification of fat.

⊠ Absorption and egestion

The nutrients, the products of digestion, are absorbed in ileum; water and minerals absorbed in the rectum and the faeces is discharged to the exterior through cloacal aperture.

⊠ Respiratory structures and vocal organ

Very large supply of oxygen is essential for the active metabolism of birds and this is provided by special arrangement, the air sacs, placed beyond the lungs.

The external nares open into the pharynx on the roof. **Glottis** is a narrow, slit-like aperture situated just behind the root of the tongue and leads into larynx, supported by cartilaginous **cricoid** consisting of four pieces and a paired **arytenoids**. The larynx leads to **trachea**, an elongated tube ventral to the oesophagus and supported by bony rings.

The vocal organ, characteristic of birds, occurs at the junction of the trachea and paired bronchi. The last three or four rings of trachea and first or bony half-ring of each broanchus are modified to form a slightly dilated chamber, the **syrinx** (Fig. 41.24).

The mucous membrane of the syrinx forms a cushion-like thickening on each side. **Pessulus**, a cartilaginous bar extends dorsoventrally at the junction of the bronchi.

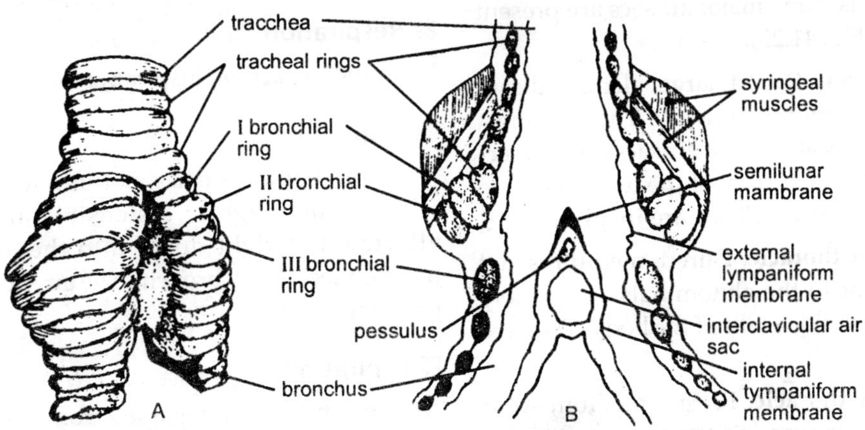

FIGURE 41.24

***Columba* sp.** Syrinx. A. Syrinx with trachea and bronchi, B. Longitudinal section of A

Pessulus supports a small **membrana semilunaris** (semilunar membrane). Paired, intrinsic syringeal muscles run from the trachea to the syrinx. Sound is produced by the vibration of membrana semilunaris.

The trachea bifurcates into right and left bronchi, the **primary bronchi** in the trunk. A primary **bronchus** enters a lung (Fig. 41.25) through a small space, the vestibulum and extends up to the posterior extremity of the lung. Within the lung, the primary brounchus divides into paired **secondary** or **lateral bronchi** and branch again forming a system of **tertiary** or **parabronchi**, ramifying into fine branches and anastomosing tubules, which come into intimate contact with the blood capillaries of the lung.

The lungs are small, pink-coloured, only slightly distensible, spongy organs, their dorsal surfaces fitting closely into the spaces between the ribs and are without peritoneal covering. The ventral surface is covered with a strong sheet of fibrous tissue, the **pleura**, to which small **costopulmonary muscles** arising from the junction of the vertebral and sternal ribs are inserted.

☒ Air sacs

The secondary bronchus continues in the lung and enters a thin-walled, bladder-like air sac, formed as a dilation of mucous membrane of the bronchus. Nine major air sacs are present in pigeon (Fig. 41.25).

Abdominal. Paired, large, elongated sacs lying in the abdominal cavity. They are most posterior in location. The air sacs are connected with ventral surface of the lungs, close to the posterior border. They are **inspiratory**.

Posterior thoracic. Paired, medium-sized sacs, anterior to the abdominals and closely applied to the side walls of the body. They are **inspiratory**.

Anterior thoracic. Paired, medium-sized sacs, slightly smaller than the posterior thoracics. They are anterior to the posterior thoracic and closely applied to the body walls.

Interclavicular. A median sac, dorsal to the bronchi and anteromedian to the anterior thoracics. It is connected to the ventral surface of both the lungs. On each side it sends two branches, the **clavicular**, in the armpit, and the **axillary (extraclavicular)** continuing in the humerus.

The anterior thoracics, interclavicular and its branches are expiratory.

☒ Functions of air sacs

1. The air sacs do not take part in gaseous exchange but act as reservoirs during respiratory movements.
2. The arrangement of air sacs is such that a proper centre of gravity is established during flight. They also reduce the weight of the bird.
3. The air sacs act like pads between the muscles to reduce friction during flight.
4. Being filled with warm, moist air, the air sacs help to maintain and regulate the body temperature and aid mechanically in keeping the wings stretched during soaring.
5. Help in the efficient circulation of air through the lungs by acting as bellows and make the respiratory arrangement highly specialised by eliminating dead spaces.

☒ Respiration

Expiration is **active**, inspiration is **passive.**

☒ Inspiration

At inspiration air rushes, not into the lungs, which being slightly distensible are little affected, but right down into the air sacs through secondary bronchi. The air sacs act as reservoirs.

☒ Expiration

At expiration the air passes from the more posterior air sacs through the lungs by special recurrent bronchi into the anterior air sacs. From these the air may be expelled to the

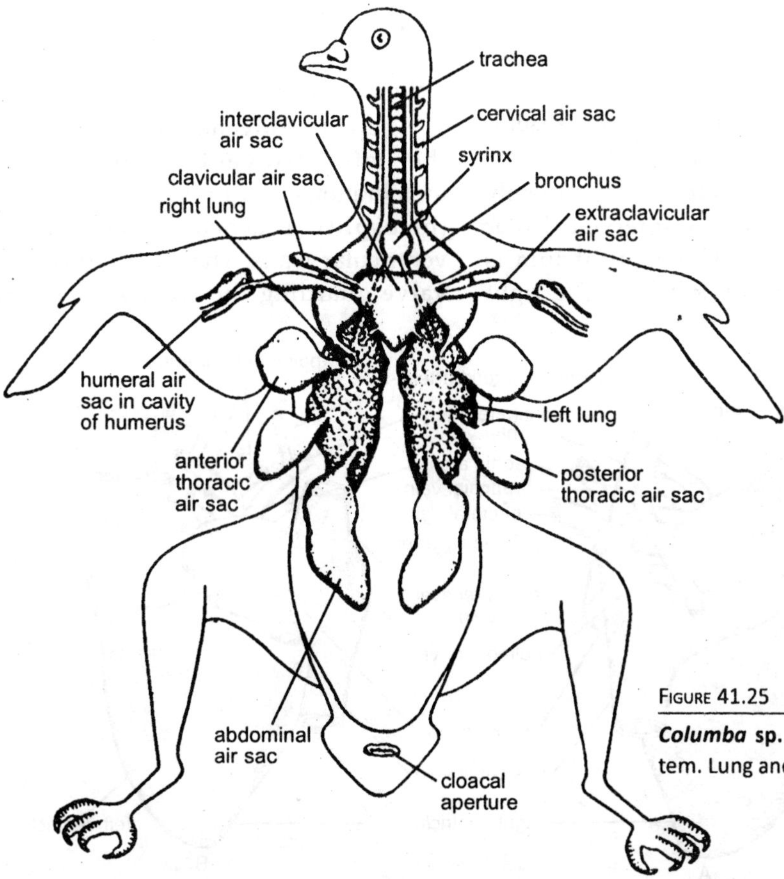

FIGURE 41.25

***Columba* sp.** Respiratory system. Lung and air sacs

exterior but cannot come back to the lungs due to closure of sphincters.

Mechanism of Respiration

The mechanisms are different in rest and in flight.

In sitting

While the bird is sitting, the expiration and inspiration are effected by the rise and fall of the sternum due to the contraction and relaxation of abdominal and intercostal muscles.

In flight

While the bird is in flight the inspiration and expiration are effected by the sternum moving away from and towards the vertebral column, due to the movements of the pectoral muscles.

In either case, the expiration is more active than inspiration. The more rapid is the flight, greater is the rate of respiration in birds.

⊠ Blood Vascular System

The components of the blood vascular system are the heart, arteries, veins and capillaries. The heart is a part of both the arterial and venous systems.

Heart

The heart is located at the anterior end of the coelom, oriented obliquely dorsoventrally, the anterior end being directed dorsally. It is protected by pericardium, a thin, whitish membrane in which serous fluid is present. The heart is reddish brown, massive and four-chambered, the apex being directed posteroventrally. A groove, the **coronary sulcus**, is present between the atria and ventricles.

The chambers of the heart are right and left atria and highly muscular right and left ventricles, the chambers being lined with **endocardium**. The **sinus venosus** absent, being absorbed in the right atrium.

The **atria** and **ventricles** are completely separated by **interatrial** and **interventricular septum**, respectively. The right atrium is slightly larger than the left one. The right ventricle partly encircles the left (Fig. 41.26) the former having a crescentic, and the latter a circular cavity in transverse section.

The left **atrioventricular valve** guarding the left **atrioventricular aperture** is membranous and provided with two cusps (**bicuspid**), attached to the ridges of the ventricular walls. The right **atrioventricular valve** guarding the right **atrioventricular**

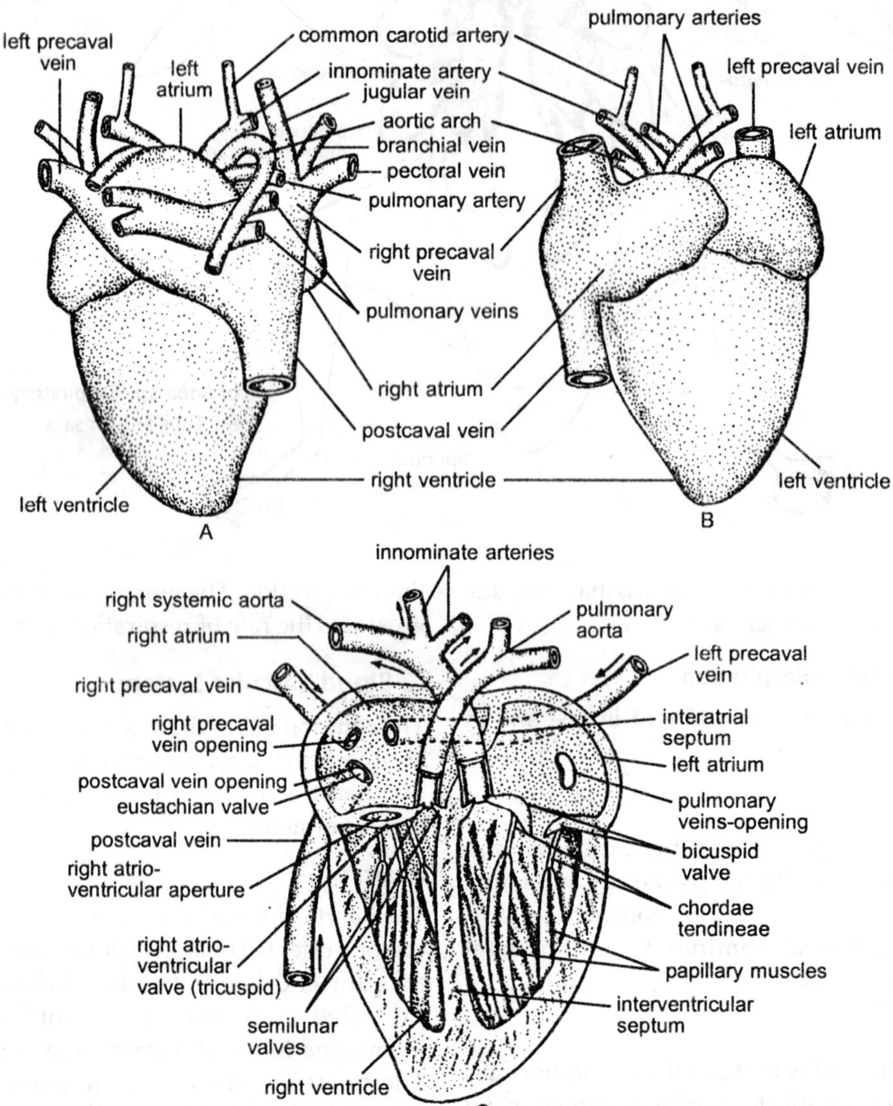

FIGURE 41.26

***Columba* sp.** Heart. A. Dorsal view. B. Ventral view. C. Ventral wall removed to expose chambers and principal blood vessels (slightly diagrammatic)

aperture is a large, muscular fold, very characteristic of birds. **Chorda tendinae**, chord-like fibres are attached to the margins of the atrioventricular valves and to the walls of the ventricle by **papillary muscles**, the latter controlling the activities of the atrioventricular valves. The openings of the aortic arches in the ventricle are guarded by three cusp-like (**tricuspid**) semilunar valves.

An elaborate **coronary system** ensures circulation of blood to the heart. There occur a **pace-maker** or **sinuatrial node** in the atrial wall; an **atrioventricular node** in the atrial septum and a right **atrioventricular ring** of **Purkinje fibres**, which constitute the intrinsic nervous system of the heart. The wave of contraction starts at the SA node, proceeds through the atria to the AV node and from there to the whole of the heart through branches.

Circulation through heart

The right atrium receives the right and left precaval veins and the postcaval vein; the left receives four large pulmonary veins. The right ventricle gives off one aortic trunk, the pulmonary arch, and the left ventricle gives off one aortic trunk, the right aortic arch, the left being absent.

The heart relaxes during **diastole**, the right atrium receives deoxygenated blood through three caval veins and the left atrium oxygenated blood through pulmonary veins. The **systole** starts at the SA node and passes to the AV node and thence to the remaining part of the heart. With contraction of the atria blood passes to the respective ventricles through atrioventricular apertures. Contraction of ventricles drives blood to the aortic arches, the back flow being prevented by valves. The pulmonary arch receives deoxygenated blood from right ventricle; the aortic arch receives oxygenated blood from left ventricle, back flow being prevented by semilunar valves. There is no mixing of blood in the whole of the circulatory system, except in the middle portion of the capillaries between arteriole and venule.

Arterial system

The pulmonary arch arises from the right ventricle, and the right aortic arch leaves the left ventricle. The left aortic arch is absent. The arteries appear as white vessels (Fig. 41.27).

Pulmonary arch

Soon after its origin from the right ventricle it divides into two, the right and left **pulmonary arteries**. Each artery loops over the heart anterodorsally, runs backward to end in the lung of the side.

Aortic arch

It curves over the right bronchus, reaches the dorsal body wall and runs backward as **dorsal aorta** along the middorsal line.

Branches of the aortic arch: Innominate

The innominate arteries are actually larger than the aorta itself owing to the immense size of the pectoral muscles, which they supply. The right and left innominate arteries arise from the aortic arch, close to each other, before it turns backward. Each innominate divides into two, the **subclavian** running laterally and the **common carotid** running anteriorly.

Subclavian. Immediately after its origin it divides into two—the **axillary** and **pectoral**.

Axillary. Curves forward and runs dorsolaterally to enter the forelimb. It proceeds outwards as **brachial**.

Pectoral. Sends two slender **internal mammary arteries** to the outer wall of the thoracic cavity and itself divides into three branches in the pectoral muscle.

Common carotid. The two common carotids run parallel to each other in the neck. After giving out a short mesial branch supplying oesophagus, trachea, syrinx and bronchus, each common carotid divides into three—a slender **comes nervi vagi** and two stout vessels, the **vertebral** supplying the cervical vertebrae and **internal carotid**.

Ascending oesophageal. Arises mesialy from the comes nervi vagi. The latter sends

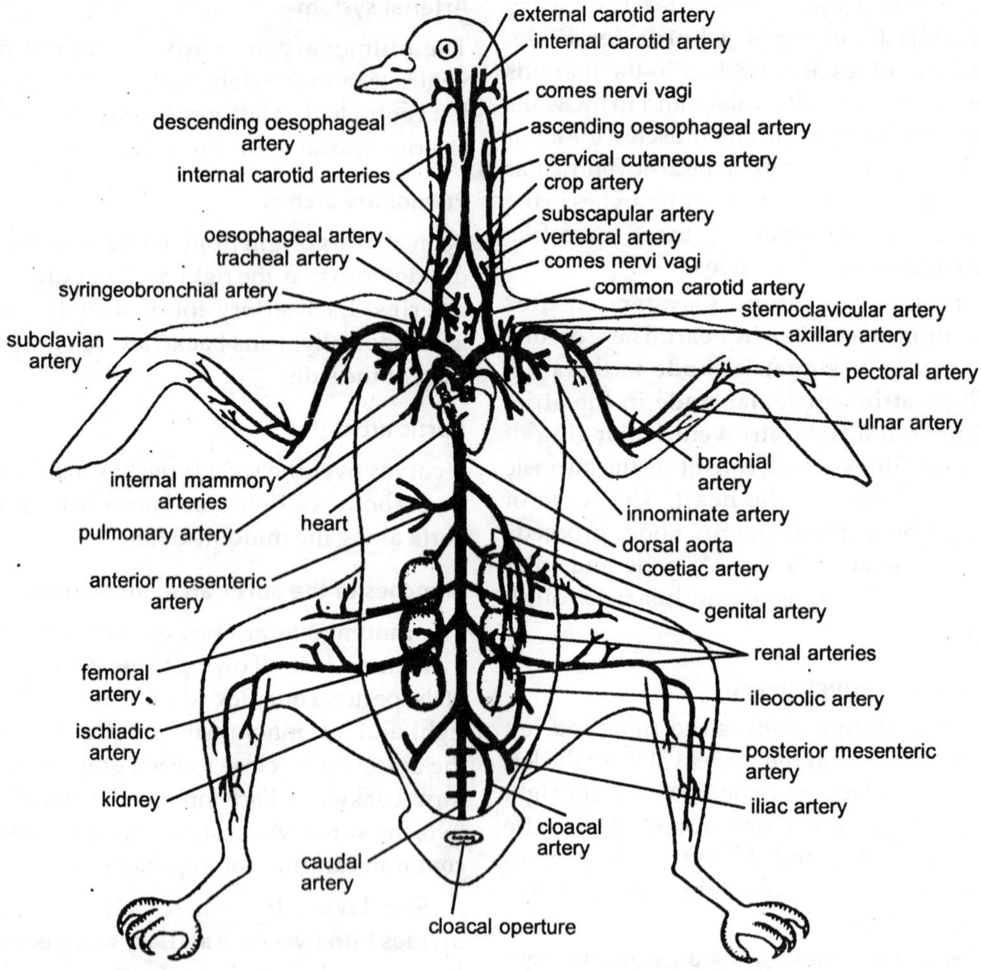

FIGURE 41.27
Columba sp. Heart and arterial system. Female

branches also to thyroid, oesophagus, skin of the neck and adjoining structures.

Internal carotid. The two internal carotids converge anteromesially, overlap for some distance anteriorly and then run forward side by side. Further anteriorly, they move away, each gives off an external carotid and proceeds forward. Internal carotid supplies blood to palate, brain, orbital areas, upper and lower jaw.

External carotid. Close to its origin it receives the comes nervi vagi and immediately sends out the **descending oesophageal branch**. The external carotid gives out a branch, the **hyomandibular** and proceeds forward as **facial artery**, sending branches to adjoining structures.

Dorsal aorta

It runs posteromedially sending following arteries to different organs.

Dorsal intercostal. Supplying intercostal muscles.

Coeliac. Unpaired, runs to the stomach and liver.

Anterior mesenteric. Unpaired, ends in the intestine.

In correspondence with the position of the legs, the femoral and sciatic arteries arise far forward.

Femoral. Paired, end in the anterior part of the thigh.

Sciatic (ischiadic). Paired, running to the posterior portion of the thigh.

Renal. Three pairs, one pair arise from the dorsal aorta and the other two pairs from the sciatics. The renals of each side end in the three lobes of the kidney of the side.

Gonadial (genital). Paired in males. Arise from the anterior pair of renal arteries and end in the testes and their ducts. In female, only the left one is present.

Iliac. Paired, end in pelvis.

Posterior mesenteric. Unpaired, supplies the rectum and the cloaca.

Caudal. A small vessel supplying the caudal vertebrae.

☒ Venous system

The veins in pigeon are placed in three groups—the pulmonary, the systemic and portal veins (Fig. 41.28). Two pulmonary veins from each lung open into the left atrium; two precavals draining blood from the anterior region and one postcaval from the middle and posterior regions, the three forming the systemic veins open into the right atrium.

Precaval vein

The vessels forming the precaval veins are similar. Each precaval vein receives the following veins from one side of the anterior part of the body.

Jugular. Arises from the head and runs backward along the neck close to the vertebral column and dorsal to the oesophagus. Anteriorly, the two jugulars, right and left, are joined by a **transverse vessel**. Posteriorly it receives:

Vertebral from vertebral column and spinal cord.

Small veins from crop and shoulder.

Facial from skin and muscles of the head.

Tracheal from trachea.

Cervical cutaneous from a plexus in the skin and neck.

Oesophageal from oesophagus.

The jugular joins the precaval, which receives:

Brachial from the wing.

Pectoral from the pectoral muscles.

Internal mammary from the inner surface of the thoracic wall, coracoid region and ribs.

The right precaval vein opens into the right anterior angle and the left precaval into the left border of the right atrium.

Postcaval vein

A large vein situated a little to the right of the heart. It is formed by the union of veins coming from the posterior half and middle of the body.

Caudal. Brings back blood from the tail region. In the posterior part of the abdomen it bifurcates to two **renal portal** veins.

Coccygeomesenteric. A large vein returning blood from the rectal region. It joins the caudal vein, close to the point of its bifurcation.

Renal portal. Each receives a few small renal veins from the kidney and the main veins runs forward through the substance of the kidney, sending a few small branches to it.

Sciatic. Brings back blood from the thigh and joins the renal portal vein of the side.

Femoral. Brings blood from the leg of the side and joins renal portal vein to form the **iliac** vein.

Gonadial (genital). Short, runs from gonad to end in the iliac vein. Two in male; only left one present in female.

Postcaval. Two iliacs unite to form the postcaval vein.

Epigastric. Returns blood from the great omentum and ends in one of the hepatic veins of the left side.

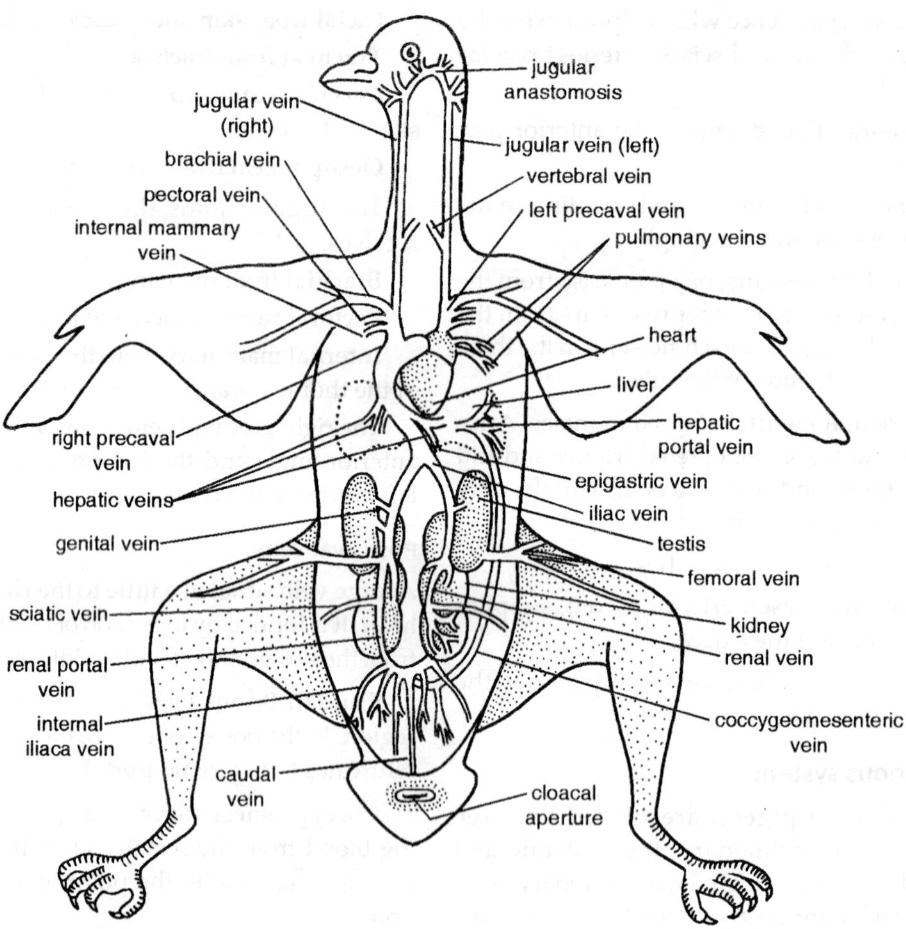

FIGURE 41.28

***Columba* sp. Heart and venous system. Male**

Hepatic. Three pairs of veins from the lobes of the liver.

The postcaval vein runs forward through the right lobe of the liver for some distance and opens on the posterior surface of the right atrium.

Hepatic portal system

It consists of veins opening in the liver. The **hepatic portal vein** is formed by the joining of:

Inferior mesenteric from the rectal region.

Gastroduodenal from the intestine and right side of the gizzard.

Anterior mesenteric from the greater part of the intestine.

The hepatic portal vein opens in the lobes of the liver through branches.

Pulmonary veins

Small but wide veins not connected with the major veins. Two veins from each lung turn dorsal to the heart and open separately into the left atrium on its dorsal surface.

⊠ Blood

Red blood corpuscles oval and nucleated. **Leucocytes** include lymphocytes, neutrophils, polymorphonuclear-pseudoeosinophilic granulocytes, basophils, eosinophils and monocytes.

⊠ Excretory (urinary) System

The excretory system is situated in the posterior part of the abdominal cavity. It consists of two kidneys and two ureters opening separately into the cloaca (Figs. 41.32, 41.33).

Kidneys. Two dark red, elongated, flat, three-lobed bodies, dorsal in position and fit closely into the hollow of the pelvis. They are metanephric in origin. The loop of Henle of each nephron or kidney tubule is long, which helps in water conservation by reabsorbing most of the water from glomerular filtrate.

Ureters. The urinary ducts are narrow tubes, arise from the ventral surface of the kidneys at some distance from the anterior end, and run backward along their midventral surface. Posteriorly, they open separately on the dorsal surface of the **urodaeum** or middle chamber of the cloaca, the urine is discharged with faeces through cloacal aperture.

⊠ Nervous system

The nervous system is basically similar to other vertebrates.

Brain

The five divisions of the brain are not horizontal in disposition as in lower tetrapods due to unequal growth of the components. The ventral flexure is well-marked in medulla oblongata.

The anteriormost part of the brain is formed by poorly developed **olfactory lobes.** The **cerebral hemispheres** (Fig. 41.29) are large, consists of massive and complex **corpora striata**, which form greater part of the hemispheres, concerned with reflex behaviourism. **Cortex** in the pallial region is smooth. The backward growth of the cerebral hemispheres has pushed the optic lobes laterally.

The backward and forward growth of the cerebal hemispheres and cerebellum, respectively obscure the dorsal surface of **diencephalon** from view and the **third ventricle** is a narrow, vertical cavity (Fig. 41.29D), bounded laterally by large thalami and communicating on each side by **foramen of Monro** with the **lateral** ventricles. The thalami are directly connected with the forebrain and spinal cord through sensory fibres. A **hypophysial portal system** join the **hypothalamus** with well-developed **pars distalis** of the **anterior pituitary. Pineal body** small and projects from the dorsal surface between the cerebral hemispheres and cerebellum.

The optic lobes are large and spherical and pushed laterally by the backward growth of cerebral hemispheres. The optic nerves and tracts are highly developed, the latter are crossed and terminate in the **mesencephalon** and in the thalami. The **optocoels** are extension of aqueduct of Sylvius.

The **cerebellum** is large, solid and consists of a large median portion, the **vermis** and two small lobes or flocculi. The surface of the vermis bears transverse grooves resulting in a great increase in the superficial **grey matter**.

The fourth ventricle is a small cavity between the cerebellum and medulla oblongata and does not extend into the cerebellum. The cerebellum possibly controls movement during flight.

The **medulla oblongata** has a well-marked ventral flexure and continues posteriorly as spinal cord.

Spinal cord

The spinal cord bears two enlargements in the **brachial** and **lumbar** regions, from which the nerves of the fore and hindlimbs, respectively, are given off. The dorsal columns of the cord diverge in the lumbar enlargement, as a result the **central canal** becomes open forming a diamond-shaped cavity, the forming the **rhomboidalis**, bound above only by the membranes of the cord.

Peripheral nervous system

The cranial nerves are twelve pairs (p. 137).

Spinal nerves (p. 137) are thirty-eight pairs.

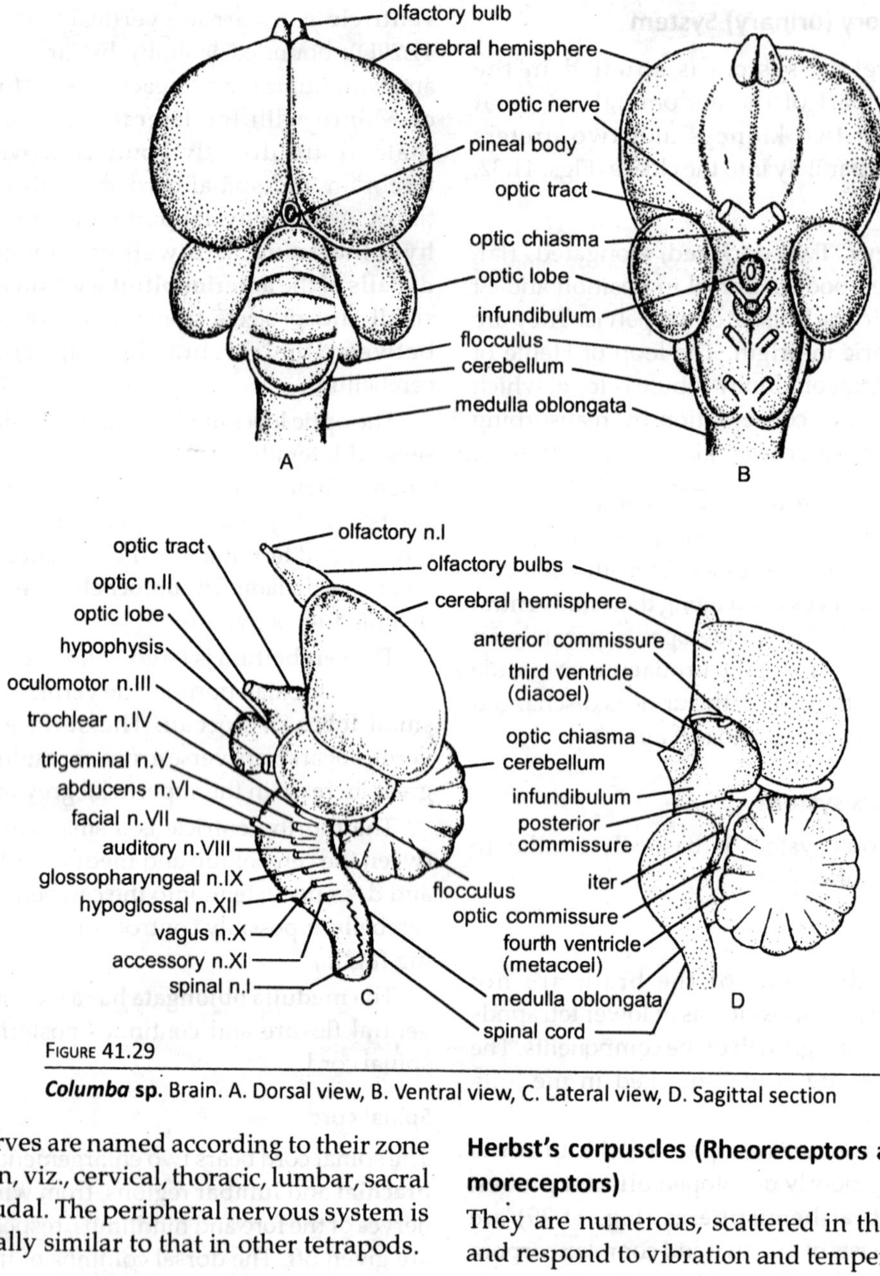

FIGURE 41.29

Columba **sp.** Brain. A. Dorsal view, B. Ventral view, C. Lateral view, D. Sagittal section

The nerves are named according to their zone of origin, viz., cervical, thoracic, lumbar, sacral and caudal. The peripheral nervous system is essentially similar to that in other tetrapods.

⊠ Receptors and sense organs

Grandy's corpuscles, Herbst's corpuscles, olfactory organs, eyes, ears and taste buds constitute sense organs.

Grandy's corpuscles (Mechanoreceptors)

These are tactile sense organs (touch), present in the beak.

Herbst's corpuscles (Rheoreceptors and thermoreceptors)

They are numerous, scattered in the dermis and respond to vibration and temperature.

Olfactory organs (Chemoreceptors)

Paired chambers at the base of the beak, separated from each other by mesethmoid bone and bounded externally by ectoethmoid. The ectoethmoid is produced inwards into three scroll-like processes, the **turbinals**, which greatly increases the surface of the mucous membrane. The anterior portion of the nasal chamber—including the anterior turbinal covered by a laminated epithelium—acts as **vestibule**. An one-layered epithelium, the **Schneidarian membrane** covers the posterior portion of the chamber including the middle and posterior turbinals. Fibres of the olfactory nerves terminate in the membrane. Sense of smell is poor.

Eyes (photoreceptors)

The eyes are large and biconvex (Fig. 41.30), the shape being maintained by bony **sclerotic plates**. The **nictitating membrane** slides over the eye-ball and presumably protects the cornea by closing it, during flight. The eye is essentially similar (p. 143) to that in other amniotes. A **pecten** is present at the point of entry of the optic nerve into the eyeball. The pecten is nearly square in outline; a membranous structure, with close set longitudinal folds and heavily black pigmented. It possibly helps in long distance vision. Accommodation is effected by striated ciliary muscles.

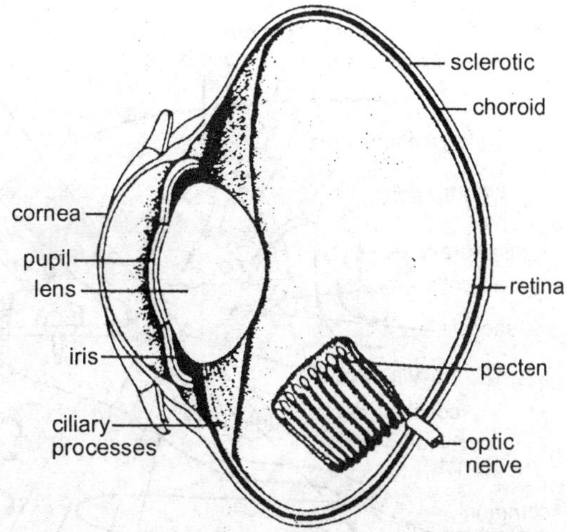

FIGURE 41.30

***Columba* sp.** Pecten. Sagittal section of an eye

Ears (Equilibrium receptors and Phonoreceptor)

The ear is highly developed. It has three compartments (Fig. 41.31A) the outer, middle and internal ear (p. 148).

External ears are small openings, covered by auricular feathers. Each opening leads into a short **external auditory meatus** closed below by the **tympanic membrane**. The **middle ear** is a long, narrow cavity, **postero-ventrally** connected with the **eustachian tube** by a narrow elongated slit. The **columella** is a long, narrow rod, at the inner end of which is attached the **stapes**, applied to **foramen ovale**. At its outer end the columella bears a three-rayed cartilage, the **extracolumella**, fixed to the tympanic membrane.

The **membranous labyrinth** (internal ear) is closely invested by a layer of dense, ivory-like bones, the **bony labyrinth**. **Utricle, saccule** and **cochlea** are well-developed (Fig. 41.31B). The anterior semicircular canal is of large size. There are prominent connections between the cerebellum and vestibular apparatus, suggestive of its importance in flight.

The cochlea encloses a **basilary membrane** bearing **hair cells**. Aggregation of hair cells and calcareous particles occur at the apical end of the cochlea. The cochlea bears a **lagena**. The legena possibly responds to low, and basilar membrane to high frequency. vibrations.

Taste buds (Gustatory receptors)

Taste buds are only a few in the tongue. Strucurally they are similar to taste buds of other Gnathostomata (p. 151).

⊠ Endocrine System

The endocrine organs (p. 156) are well-developed.

Thyroid

Paired, one occurs on each side of the trachea near its division into two bronchi. The secretion, thyroxine contains iodine and important in general metabolism. They are also concerned with periodic moulting.

Parathyroid

Paired, situated close to thyroids, control calcium and phosphate metabolism.

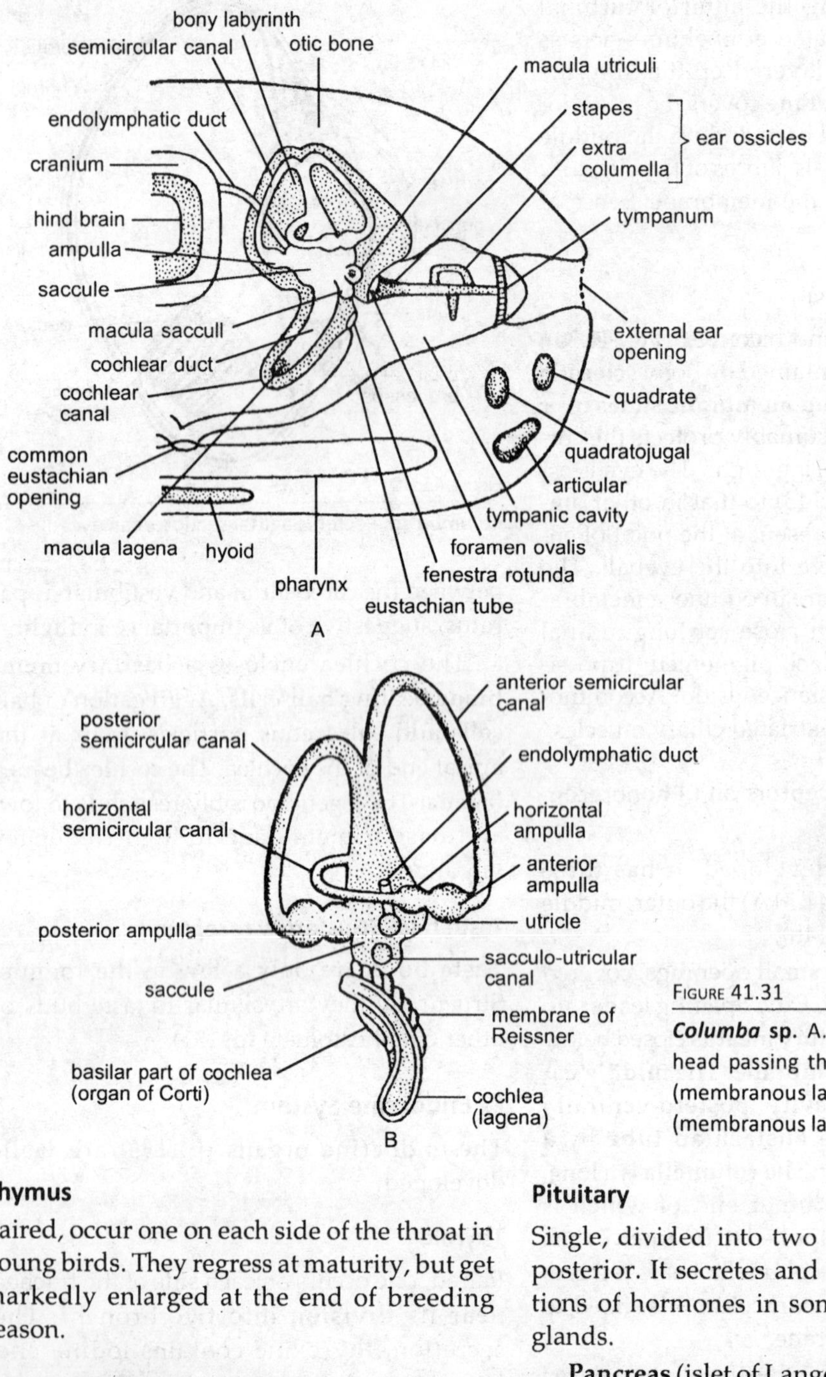

FIGURE 41.31

***Columba* sp.** A. Transverse section of head passing through the internal ear (membranous labyrinth). B. Internal ear (membranous labyrinth)

Thymus

Paired, occur one on each side of the throat in young birds. They regress at maturity, but get markedly enlarged at the end of breeding season.

Adrenal or suprarenal

Paired, yellowish, situated anterior to the kidneys. The secretion, adrenalin, in addition to many functions increases cardiac rate and vasoconstriction.

Pituitary

Single, divided into two lobes, anterior and posterior. It secretes and also induces secretions of hormones in some other endocrine glands.

Pancreas (islet of Langerhans) and gonads also secrete hormones.

⊠ Reproductive System

The sexes are separate but sexual dimor-phism is not always well-marked.

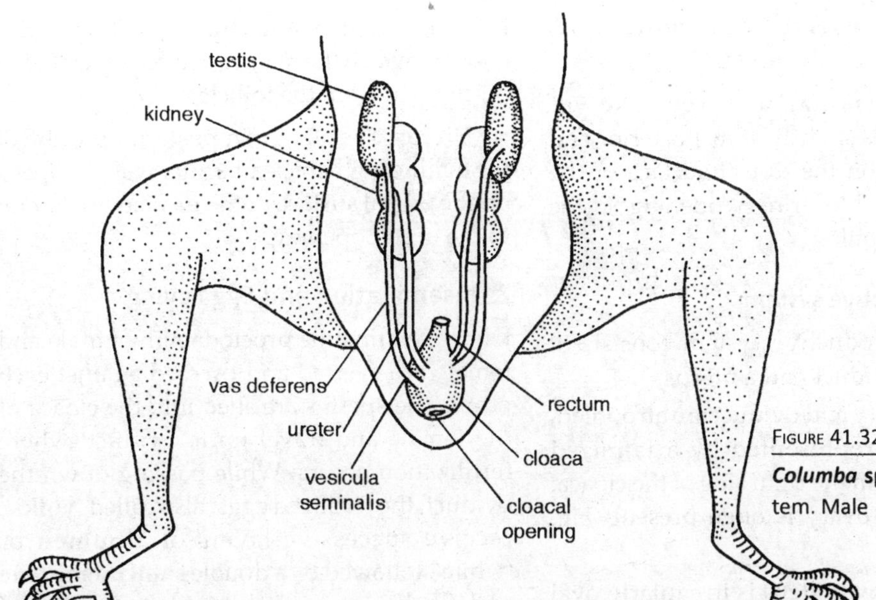

FIGURE 41.32

***Columba* sp.** Urinogenital system. Male

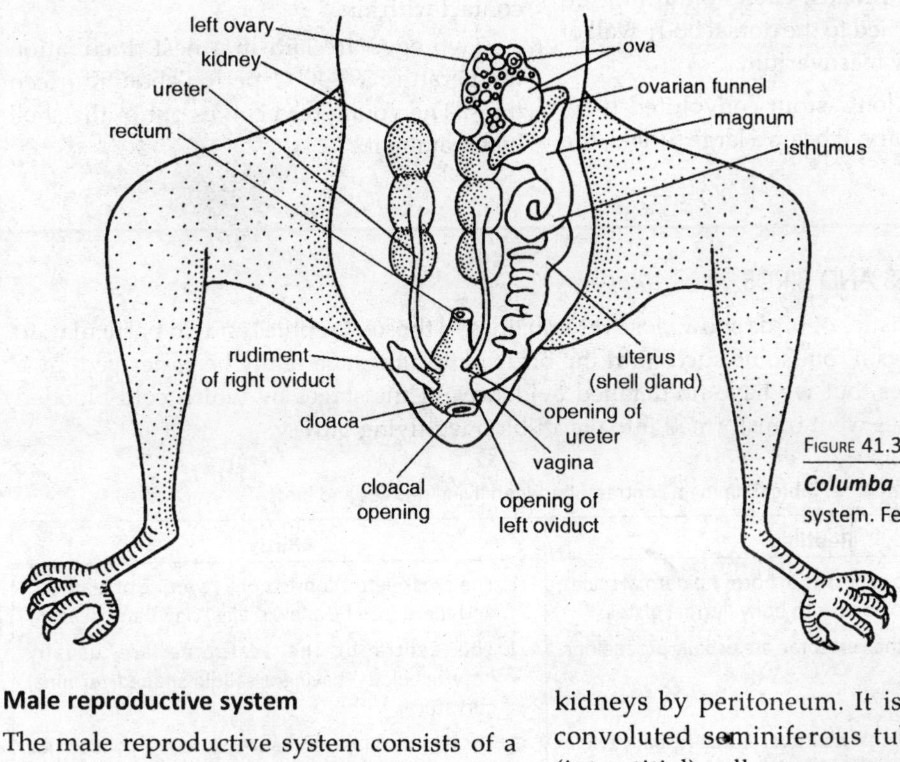

FIGURE 41.33

***Columba* sp.** Urinogenital system. Female

Male reproductive system

The male reproductive system consists of a pair of testes, a pair of vasa deferentia and two vesiculae seminalis (Fig. 41.32).

Testes. White, ovoid bodies, attached to the ventral surface of the anterior end of the kidneys by peritoneum. It is composed of convoluted seminiferous tubules. Leydig (interstitial) cells are present between the tubules.

Vasa deferentia. Paired, narrow, convoluted tubes, arise from the inner border of the testes. The vas deferens runs posteriorly, parallel and close to the ureter.

Vesiculae seminalis. Paired. The vas deferens enlarges slightly at its posterior end before opening on the dorsal surface of the urodaeum, lateral to urinary opening, at the tip of a small papilla.

Female reproductive system

The female reproductive system consists of one ovary, an oviduct and a uterus.

The right ovary and oviduct are atrophied. The oviduct is represented by a fairsized vestige, opening in the right side of the cloaca. Vestige of right ovary is often present (Fig. 41.33).

Ovary. The ovary (left) is irregularly oval with the surface raised up into rounded elevations or follicles, each containing an ovum. It is attached to the dorsal body wall of the abdomen by mesovarium.

Oviduct. A long, stout convoluted tube, lateral to the ovary. It bears a large funnel, the **infundibulum** (ovarian funnel) at the anterior end into which the mature ova pass after being discharged from the follicles.

Uterus. The glandular posterior portion of the oviduct with thick muscular walls. It opens on the dorsal surface of the urodaeum, lateral to the ureter opening.

⊠ Insemination and Egg Laying

During mating the proctodaeum of male and female are everted and pressed against each other. The sperms are shed into the cloaca of the female and travel up to oviduct, where fertilization occurs. While passing down the oviduct, the fertilized eggs, also called '**yolks**', receive successive layers of **albumen** or 'white', followed by a double **shell membrane** and, finally, a calcareous **shell**. At the time of laying, the shell is soft, which later hardens in contact with air.

Two eggs are laid in a nest. Incubation temperature 39°-40°C; period about fourteen days. The young bird comes out of the shell by breaking it.

LIVING REPTILES AND BIRDS

Many characteristics of birds show close resembalnce to those of reptiles and, in particular, to the bipedal types. From some such form the birds have almost certainly been derived, by a series of changes, but we have no detailed evidences of the stages by which cold-blooded terrestrial reptiles were transformed into warm-blooded flying birds.

Table: The main contrasts between living reptiles and birds

Reptiles	Birds
1. The exoskeleton consists of horny epidermal scales, sometimes provided with bony dermal plates.	1. The exoskeleton consists of a covering of feathers, and there may be a few scales, claws and beak.
2. The centra of the vertebrae are usually procoelous.	2. The centra of the vertebrae are usually heterocoelous, having a saddle-shaped terminal curvature.
3. Sacrum when present, its vertebrae (usually two in number) bear large expanded ribs, with the ends of which the ilia articulate.	3. The two sacral vertebrae are without expanded ribs and a long synsacrum is formed by their fusion with others vertebrae.
4. The cartilaginous sternum may become long, but it is never replaced by a membrane bone.	4. The cartilaginous sternum is replaced by a membrane bone from several centres.

Contd.

Reptiles (*Contd.*)	Birds (*Contd.*)
5. Episternum or interclavicle, when present, remains distinct from the clavicle and sternum.	5. Interclavicle, when present, is fused with the clavicles.
6. The forelimb bears more than three digits and at least three are provided with claws.	6. The forelimb never bears more than three digits and more than two digits are never clawed. The forelimbs are modified as wings, carpals fuse with the fused metacarpals.
7. In the pelvic girdle, the ilium is elongated further behind than in front of the acetabulum. The pubis slopes downward and forward. Pubic and ischiatic symphyses are usually present.	7. The ilium is elongated in front of the acetabulum. The pubis slopes backward parallel to the ischium. Pubic and ischiatic symphyses are very rarely present.
8. In the hindlimb, five digits are often present. The tarsals and metatarsals remain distinct.	8. In the hindlimb, digits are not more than four. The proximal tarsal unites with the tibia forming a tibiotarsus. The first metatarsal, when present, remains free. The other three are fused and fuse with the distal tarsals forming a tarsometatarsus.
9. Sutures in the skull are prominent.	9. The sutures in the skull are obliterated.
10. The aortic arches are at least two. The heart is four-chambered only in Crocodilia. The posterior part of the body always gets a supply of mixed blood.	10. Only the right aortic arch is present. The heart is four-chambered and the body always gets a supply of oxygenated blood.
11. The body temperature varies with the temperature of the surrounding medium (poikilothermal).	11. The body temperature is high and almost constant (homeothermal).
12. The optic lobes are placed on the upper surface of the brain.	12. The optic lobes are placed on the sides of the brain.
13. Air sac-like structures are present only in *Geckos* and *Chameleons*.	13. Air sacs are present.
14. Both the ovaries and their ducts are present and functional.	14. The right ovary and its duct almost atrophies.

RATITAE AND CARINATAE

Birds are characteristically flying animals but some of them lack flying ability. The former are placed in Carinatae and the latter in Ratitae.

Table: Contrast between Ratitae and Carinatae

Ratitae	Carinatae
1. Running birds. The wings are more or less degenerated and not used in flight.	1. Flying birds. The wings are well-developed and almost always used in flight.
2. The sternum or breast bone is keelless and raft-like.	2. The sternum or breast bone is provided with a keel.
3. The skull is dromaeognathous, i.e., the vomer is interposed between the palatines, the pterygoids and the basisphenoidal rostrum.	3. The skull is not dromaeognathous, i.e., the vomer is not fused with the bones of the palate and the palatines articulate with the basisphenoidal rostrum.
4. The sutures or the lines of union in the skull remain distinct for a long time.	4. The sutures or the lines of union in the skull almost always disappear at a very early stage.
5. The quadrate articulates with the skull by a single head.	5. The quadrate articulates with the skull by a double head.

Contd.

Ratitae (*Contd.*)	**Carinatae** (*Contd.*)
6. The scapula and coracoid are fused and their long axes near the point of union lie almost in the same line or form a very obtuse angle.	6. The scapula and coracoid are connected with each other by ligament only and the two bones meet almost at right angles.
7. The clavicles are small or absent.	7. The clavicles are well-developed in most cases.
8. Except in old *Rheas, Emus,* the ilium and ischium are not united posteriorly.	8. The ilium and ischium unite and enclose a sciatic foramen.
9. A pygostyle is never present.	9. A pygostyle is usually present.
10. The feathers have free barbs and the barbules are without hooks.	10. The barbs of the feathers are usually united and the barbules are with hooks.
11. Except a few, the oil gland is absent.	11. The oil gland is usually present.
12. Pterylae are not regularly arranged.	12. Pterylae are regularly arranged.
13. The male has penis.	13. The male rarely has a penis.
14. The young birds are always precocious (independent).	14. The young birds may be precocious or altrices (dependent).

PARENTAL CARE IN BIRDS

The parental care in birds is one of the important behaviourial aspects in bringing together the two sexes for breeding, brooding and caring of the young.

In birds the eggs and young are not left uncared for, and consequently the breeding habits involve the development of elaborate systems of mutual association and cooperation, which not only make the parents to stay together but also extending mutual help in building nests, incubating eggs, feeding young since hatching and protecting them till they can fend for themselves.

⊠ Nest Building

Damage of eggs, mortality of newly hatched and young are quite high in egg laying animals. This is more true for tree dwellers. Nest, to a considerable extent, help to minimize the damage by protecting them from biotic and abiotic hazards.

Nest building is an intricate process in which both male and female birds cooperate. In addition, they together zealously guard the territory around the nest from both intruders and predators, and ensure safety to the eggs and growing young birds in the nest.

1. The sex-fidgeting in birds is common before, during and after mating. Ritual offering of nest material is an important element in many courtship displays. In some birds, however, (magpie) male birds build the nest.
2. The site and location of the nest is usually chosen by male. Sometimes the male brings, the materials and the female uses it. She may do the fetching as well, the male do not take active part in the process.
3. The nest-building and copulation occur at about the same stage of reproductive cycle and both are induced by oestrogen.

⊠ Types of Nests

Nests are of different types with protective devices and are built in a way to ward off invasion by enemies and also to withstand natural forces (Fig. 41.35).

The materials used and the method of building nests varies with species of birds. Many are quite simple, while others involve most elaborate tying of creeper, grass, wire, twines, etc. collected from surroundings and adjacent households, to branches of trees,

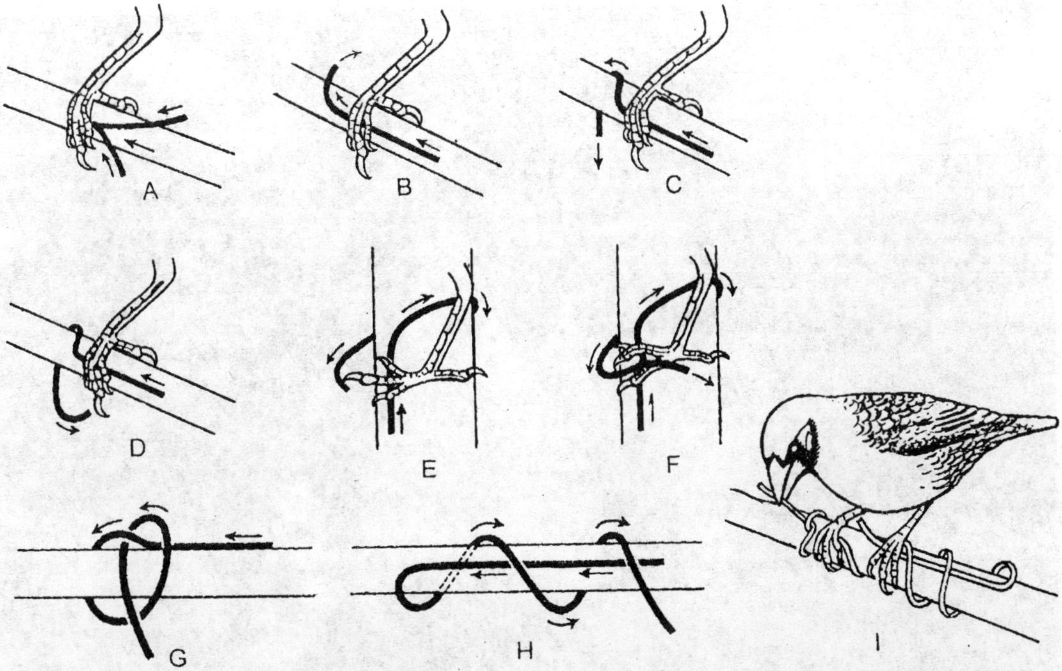

FIGURE 41.34

Nest building by a weaver bird (***Quelea* sp.**) A-H—the arrows show directions in which the piece of grass is pulled. I cooperative weaving by the foot and beak and point of holding by the beak

which is done by beak or feet or both (Fig. 41.34). The binding is strong enough to withstand shaking of trees and other physical forces.

1. In some birds the nest is simply a hollow in the ground.
2. Owls use existing holes in trees as nests.
3. Wood packers drill holes in the trees, which are used as nests.
4. The swifts build nearly the whole nest with saliva.
5. In storks and cranes the nest is just a heap of sticks neatly arranged with concavity at the centre for laying eggs at the junction of high branches of trees to prevent disintegrating or felling down with storm or other physical forces.
6. The most complicated form of nests (pitcher-like) are built mainly by passerine birds.
   a. The nests are open above where predators frequent the ground.
   b. In region having smokes and other tree-climbing predators the nest is mostly dome-shaped and hang from branches of trees (weaver bird) (Fig. 41.35).

The space within the nest meant for laying eggs and accommodation of young after hatching is lined with hair, pieces of cloth, breast feather (duck), etc. Thrush lines the nest with mud moistened by saliva.

⊠ Brooding and Care of Young

Brooding is the complicated act of sitting on the eggs, associated with development of vascular and oedematous response with a local moult of feathers in a part of the skin (brood spot), which is induced by oestrogen and prolactin hormones.

In fowl usually the hen brood the eggs and cocks assist. In some species the male does brooding.

FIGURE 41.35

Nests of weaver bird on the top branches

FIGURE 41.36

A mother Tailor bird feeding her baby

At hatching the chick breaks the shell with egg tooth (caruncle), the parents may assist in the process. The care of the nestling is a very elaborate process in most birds, involving many separate actions.

a. Some hatchlings are precocial. They are feathered and capable of taking care of themselves (Fowl, Duck) and may even fly (magapodes).
b. Some hatchlings are altrices. They are naked and depend for food on their parents (pigeon). Their food requirement is quite high, which is provided by the parents.

Parents collect and carry the food to the nestling in their beak and feed them. Some species swallow food and regurgitate, mixed with crop milk to feed the young (pigeon). Gaping of brightly coloured mouth of the hatchlings stimulate parents to feed them (Fig. 41.36).

The nest is kept clean by careful removal of faeces and other materials.

The period of caring for young birds is variable:

a. The young passerines fledge in about two weeks. They are after cared by their parents for additional one to three weeks.
b. The larger species of birds require longer period of parental care.
c. The young of twany owl fledge in a month, but they receive parental care for three months after fledging.
d. The young albatross requires a year to lead an independent life.

MIGRATION OF BIRDS

Mass movement of a population from one place to another and return to the original site is migration. Broadly birds are of two types, Resident or non migratory and migratory.

⊠ Resident or non migratory

Sparrows, pigeons, sand grouses, etc. are non migratory and remain it throughout the year in a country.

⊠ Migratory

Certain birds exhibit a regular, periodic to and fro seasonal movements between their summer and winter homes or from a breeding and nesting place or to a feeding and resting place (non breeding). They cover thousands kilometers in their periodic journey.

⊠ Types of migration

Several distinct types of migration are found in birds.

1. **Latitudinal migration.** Movement from north to south hemisphere and vice-versa is latitudinal migration.

   Several North American and Eurasian birds cross the equator to spend winter in deeper parts of South America and Africa. The American golden plover *Pluvials dominica* spend nine winter months in the Pampus plains of Argentina covering about 12,872 km south and enjoy two summers in a year.

   Most migratory birds live in the land masses of the northern hemisphere and subarctic zones where facilities for nesting and feeding in summer exist. In winter the northern hemisphere is covered with ice and the birds move towards south for shelter.
2. **Longitudinal migration.** Some birds migrate east to west and vice versa. The starling migrates from breeding area in East Europe and Asia towards the Atlantic coast to avoid continental winter. This type of migration is longitudinal migration.
3. **Altitudinal migration.** Migration from high mountains to low valleys in winter and from low valley's to mountains in summer is altitudinal migration. In India, wood cock, scolopa migrate from

planes to slopes of Himalayas ascending thousands of meters above sea level in summer and return to planes with the commencement of winter. Such vertical migration takes place in coots and grebes of Andes in Argentina; violet and green swallows in Great Britain.

4. **Seasonal migration.** Summer visitors, winter visitors and short time visitors are seasonal migrants. In Great Britain summer visitors—swifts, swallows, nightangles, cuckoos, etc. arrive in spring from south, breed and leave for south in autumn. The field fare, snow bounting, red wing, etc. are winter visitors. They arrive in autumn chiefly from south, stay through the winter and fly northwards in spring. Birds like sand pipers, snipes, etc. are **bird of passage**. They stay for a short time, twice in a year on their way to colder and warmer countries in autumn and spring.

5. **Partial migration.** Partial migration is the influx of new individuals of the same species in an area of resident birds for a short period. The barn owl (*Tyto alba*), blue birds, many jays of Canada and Northern United States travel southwards to mingle with the resident population of the southern states. Song thrush, red breast, finch, etc. found throughout the year, actually represent partial migrants. The birds found in summer are not same as in winter.

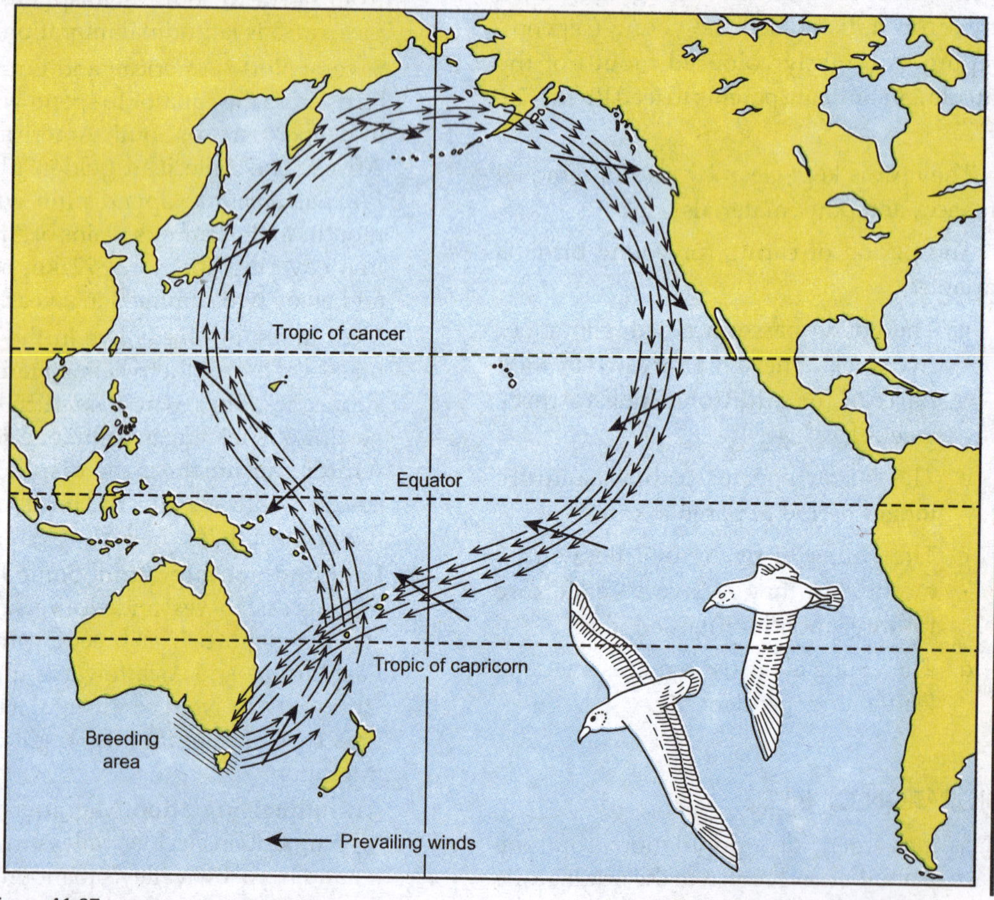

FIGURE 41.37

The migratory path of the short tailed shearwater from Australian breeding area to northern range

⊠ Significant features of migration

1. **Diurnal migration.** Diurnal migrants fly in daytime in well organised flocks. The ducks, geese, swans, etc step to forage in suitable sites.

   Swallows, swifts, etc. capture their insect food in the air in course of their travel.
2. **Nocturnal migration.** Nocturnal migrants fly at night under cover of darkness to escape enemies. These are small birds like sparrow, warblers, thrushes, etc. they rest in daytime procure food and resume their flight again with the advent of night.

⊠ Range of migration

The distance covered by the migratory birds depend upon the species and the prevailing conditions. The Himalayan snow patridges descend a few hundred metres and only cover hardly 2 to 3 kilometre, while the chicades descend nearly 3,000 metre. The arctic tern spends summer and breed along the northern most ice free coasts of Labrador, far inside the artic circle, then travels a distance of about 6,800 km to reach their destination to the edge of Antarctica in winter and return again through the same distance in summer. Short tailed shearwaters, make an annual migration between their breeding range in Southern Australia and the North Pacific, that requires a round trip of more the 30,000 km (Fig. 41.36).

⊠ Altitude of flight

Most routine migration of birds is restricted within 915 metre altitude. Some small birds migrate at night flying 1,500 to 4,300 metre altitude. Some even cross the Andes and the Himalayas at the altitude of 6,096 metre or more.

⊠ Speed and distance of migration

The average flight velocity of most small birds seldom exceeds 48 km hour. The two Indian species of swifts cover 275 to 320 km per hour.

Several hundred km may be covered in nonstop flight in a day or in a night. The golden plover holds the world record with the longest nonstop flight from Hudson Bay and Alaska in South America, distance of about 3,861 km.

⊠ Routes of migration

The migratory birds usually follow a definite route for flight. Noctural migration of small land birds proceed with the general air flow.

In spring they travel northwards from the south along warm air currents from the south, and in autumn more southwards with the cold winds of north. Occasional deviation of route may be due to configuration of land, coastline, course of great rivers or intervening mountain chains, etc.

⊠ Order of migration

As a rule the migratory birds almost strictly follow a definite order. The maturity of gonads in the adult induces migration. During migration towards breeding ground, normally the adults are followed by youngs. The adult males take the lead, adult females are next in order, birds of the year follow them; weak and wounded ones are at the end. During back flight the order is reversed. The new born followed by their parents in the same route. Inspite of hazards of long distance and vagaries of weather, several species of migratory birds maintain a striving regularity year after year in their timings of arrival and departure.

⊠ Navigation of bird migration

The migration of birds and binding their way in the almost endless horizon elicit one of the greatest wonders. Following suggestion and theories on navigation have been forwarded.

a. **Visual landmark.** The landmarks—great rivers, valleys, coastal lines, mountain ranges, Oceanic islands, etc. attribute to the sense of direction in the daytime. But a vast majority of birds migrating at night cannot use these landmarks for direction. Birds crossing great streches of seas also have no seamarks to follow.

b. **Celestial bodies.** The celestial bodies are sun and constellation of stars.

i. **Sun.** The birds travelling by day use the sun as compass for orientation. Based on experiments with starling in cages, Kramer (1949) suggests that birds possess an internal time sense or time clock with which they make necessary adjustments in their course according to changes in the angle or position of the sun as the day progresses. Kramer proved this by altering the direction of flight with the false apparent direction of the sun by using mirror (Fig. 41.38). The young birds that never migrated before make similar navigational orientation to sun, while travelling independently of their parents. It is suggested that the instinct is inherited.

ii. **Stars.** The nocturnal migrants navigate by the constellation of the stars, as seen in night flying old world warblers (Saner, 1957) (Fig. 41.39).

c. **Earth's magnetic field.** It has been suggested that the magnetic field of the Earth plays an important role in migration of birds. The inner ear reacts to the mechanical Coriolis effect produced by the rotation of the earth.

d. **Experience.** Some older members with the experience of the route of migration act as leaders to guide the younger generations.

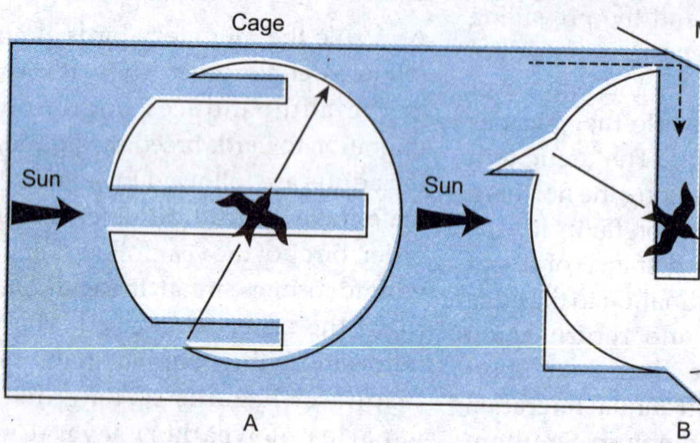

FIGURE 41.38

Stars used by night migrants as guides (Franz Sauez's experiment)

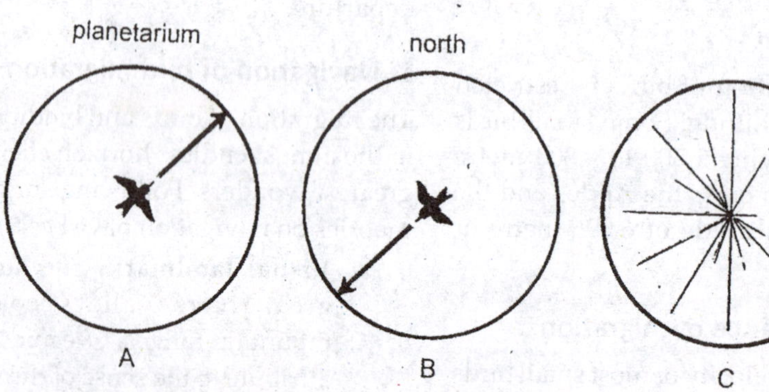

FIGURE 41.39

Stars used by night migrants as guides (Franz Sauez's experiment)

Birds certainly do not learn their route of migration from the elders. Some birds do not fly in flocks at all. In many cases, young birds make their first journey independently without guidance of their parents. They are guided by instinct impressed in their nervous system in same way through countless generations.

Advantages of migration. Migration helps birds to avoid harsh climatic extremes. Migration from higher altitudes and latitudes (breeding areas of north) during winter affords protection from cold and stormy weather. The migrants get more food and better living condition in their winter quarters in temperate and tropical regions. Return in summer to the breeding areas in north once again provides suitable and unconqured breeding grounds, abundance of food with minimum efforts, and long hour of daylight for searching food. In this way the birds utilise the food supply in both the hemispheres and exploit new areas for nesting.

CLASSIFICATION: CLASS AVES

1. Body has three distinct divisions with a flexible neck and covered with feathers of different kinds for the most part.
2. Limbs two pairs, the anterior pair modified as wings for flight, and the posterior pair adapted for perching, walking or swimming.
3. Each foot is usually with 4 toes; shanks and toes are sheathed with cornified skin.
4. Mouth opens through a projecting beak or bill provided with horny sheath.
5. The skeleton is strong, bones fully ossified and pneumatic.
6. Skull with a single occipital condyle; sutures obliterated.
7. Pelvis fused with a number of vertebrae but open ventrally forming synsacrum.
8. Tail vertebrae few and compressed posteriorly forming pygostyle in modern birds.
9. Sternum enlarged, usually with a median keel.
10. Syrinx, the voice box, at the base of the trachea.
11. The lungs are compact, attached to the ribs and connected to thin-walled air sacs extending between internal organs.
12. Heart four-chambered with two atria and two ventricles. Only the right aortic (systemic) arch persists. Red blood corpuscles nucleated.
13. No urinary bladder, excreta semi-solid.
14. Females usually with only left ovary and oviduct.
15. Fertilization internal. Oviparous. Eggs with large amout of yolk, telolacithal; cleaveage meroblastic.

About 8,600 species of birds are arranged under two subclasses—Archaeornithes and Neornithes.

Subclass ARCHIORNITHES

Mesozoic ancestral or lizard birds with long tail and teeth in both the jaws.

Order 1. **Archaeopterygiformes**

So far two genera have been described, each with a single species from the upper Jurassic lime stones of Solenhofen in Bavaria.

1. Forelimbs bear remiges.
2. Tail long with many vertebrae, which gradually taper to the distal end. The rectrices are arranged in two lateral rows on each side of the caudal vertebrae.
3. The carpals and metacarpals free and hand with three clawed digits.
4. The eyes were large.
5. The skull proportionately large, with a rounded bird-like brain case and strong jaws, in each of which bearing a series of conical teeth lodged in socket.
6. Enamelled crowned teeth on both the jaws.
7. The sacrum made up of only about six vertebrae.

8. The cervical and thoracic ribs probably devoid of uncinates. Thin abdominal ribs or gastralia or so-called ventral ribs were on ventral wall of abdomen.
9. The cerebral hemispheres smooth, long and narrow and cerebellum small.

Examples: *Archaeopteryx lithographica, Archaeornis siemensi.*

Subclass NEORNITHES

These birds first appeared in Cretaceous period and includes few extinct and all other birds known up to date.

1. Tail greatly shortened usually ending in a pygostyle, around which the rectrices, when present, are arranged in a semicircle.
2. Except a few extinct forms all are without teeth.
3. The sternum is well-developed and usually keeled or carinate.

The neornithine birds are arranged in four superorders—Odontognathae, Palaeognathae, Impennae and Neognathae.

Superorder **Odontognathae**

New World toothed birds found in Upper Cretaceous period.

1. Clavicle not fused. Sternum without keel. Wing of vestigeal humerus only.
2. Both the jaws were with teeth.
3. Flightless, specialised for swimming, about 150 cm in length.

Order 1. **Hesperornithiformes**

1. Flightless, about 90 cm, adapted for swimming.
2. Forelimbs with vestigeal humerus only.
3. The hindlimbs laterally directed with webbed feet.
4. The pygostyle absent.
5. Teeth pointed, arranged in grooves in both the jaws. Premaxilla without teeth.
6. The quadrate single, basipterygoid process absent.
7. Sternum without keel.
8. Long neck provided with heterocoelous vertebrae.
9. Pectoral girdle much reduced, and clavicles not fused.

Examples: *Hesperornis, Hargeria, Neogaeornis, Baptornis, Enaliornis,* etc.

Order 2. **Ichthyornithiformes**

Small marine flying birds, of Upper Cretaceous of North America.

1. The neck vertebrae amphicoelous.
2. The sternum had a well-developed keel, and the wings were unlike modern birds.
3. Clavicles fused.

Examples: *Ichthyornis, Apatornis,* etc.

Superorder **Palaeognathae**

Large-sized flightless birds, descended from volant ancestors at the end of the Mesozoic period.

1. The feathers are primitive, without hooked barbules. Barbs remain fee. Apterae are usually absent in adult. The rectrices are absent or irregularly arranged.
2. The wings reduced in size or vestigeal or absent.
3. Pygostyle small or undeveloped.
4. The sternum is flat, raft-like. The coracoid and scapula are comparatively small and completely ankylosed.
5. The coracoscapular angle approaches to right angle and the acrocoracoid process is vestigeal.
6. Large basipterygoid process developed from the basisphenoid. The vomer is large and broad and separates the palatines; which do not come in contact with the cranium.
7. The quadrate articulates with the skull by a single or partially divided facets.
8. The males are with erectile penis and the females have clitoris. The young are precocious.

The superorder has seven orders of which two are extinct.

Order 1. **Struthioniformes**

1. Largest existing running birds of Africa and Arabia, about 210 cm tall, 170 kilogram in weight, inhabit arid lands in flocks of 3 to 20.
2. Flightless, terrestrial omnivores with small wings, usually kept folded during running but may be expanded to act as a steering.
3. Head, neck and leg sparsely feathered.
4. Feathers are without aftershaft. Tail feathers are replaced by tail coverts.
5. Hindlimbs are strongly built and only 2 toes (3rd and 4th) in each foot are present.
6. The sternum lacks keel.
7. Pubic symphysis present. Pygostyle absent.
8. At nesting, a male defends 4 to 5 females laying up to 30 eggs in a single nest.

Example: *Struthio camelus (Ostrich).*

Order 2 **Rheiformes**

1. Large flightless running birds of South America, live in groups, about 120 cm tall, feeding on vegetation to insects.
2. Each foot with 3 front toes, webbed at the base.
3. Sternum unkeeled.
4. At nesting a male defends 3-7 females, digs nest and all females lay about 50 lemon-yellow eggs in the same nest; male incubate them for about 40 days.

Examples: *Rhea americana, Pteronemia pennata.*

Order 3. **Casuariformes**

1. The emus and cassowaries of Australia, Papua New Guinea, and East Indies.
2. Flightless, terrestrial, running herbivore birds with comparatively small wings supported by a single digit.
3. The neck and body densely feathered. Feathers with aftershaft nearly equal to shaft.
4. The plain and savannah-dwelling emus are nearly 180 cm tall with incomplete lining of feathers on neck; inner one of the three toes is a defensive structure, provided with claw.
5. Emu is monogamous. The male digs nest. Female lays 15 or more green eggs. The male incubates first brood while the female incubates the 2nd brood of eggs for about 60 days.
6. The fruit-eating nocturnal cassowaries are confined to the dense tropical rain forests; over 150 cm tall with bare skin on head and neck and with a helmet-like horny casque on head which may help in advancing through dense forest.

Examples: *Dromicieus* (Emu), *Casuarius*.

Order 4. **Apterygiformes**

1. Relatively small, almost wingless, henlike, nocturnal, omnivorous running birds confined to New Zealand.
2. Bills long and slender with nostril at the tip.
3. Wings degenerated with vestigeal humerus, only one digit and no flight feathers.
4. Each leg bears four-clawed toes.
5. Body plumage fluffy, hair-like, without aftershafts.
6. Sternum unkeeled. The eyes are small. Lays only one white egg at a time. Eggs are largest of that in all living birds.

Example: *Apteryx* (Kiwi).

Order 5. **Dinornithiformes**

1. Turkey-like running birds called Moas of New Zealand, about 3 m tall. Became extinct few centuries back.
2. Feathers with large aftershafts and without barbicels.
3. Beak short.
4. Hindlimbs massive with four toes.
5. Sternum reduced and unkeeled. Cora-

coid, scapula and wing bones reduced or absent.

Example: *Dinornis* (Moa).

Order 6. **Aepyornithiformes**

1. Giant flightless Madagascar elephant birds, about 3 m tall, became extinct few centuries back after the extinction of Moas.
2. Wings were relatively tiny but legs stout and powerful and with four toes.
3. Egg size 32 × 24 cm; largest of all known animal eggs; capacity about 10 litres.

Example: *Aepyornis* (elephant bird).

Order 7. **Tinamiformes**

1. Partridge or quail-like, almost tailless, herbivores, essentially cursorial birds, known as Tinamon, can fly clumsily over a short distance, confined to Southern Mexico and Central and South America.
2. Sternum is keeled, and the palate palaeognathus.
3. Pygostyle reduced.
4. The eggs with glossy white shell and incubated by males.

Examples: *Crypturellus, Eudromea, Tinamus,* etc.

Superorder **Impennae**

Evolved from ancestors having wings used for both swimming and flight and gradually changed to form an effective paddle for swimming. The superorder has only one order.

Order **Spheniseiformes**

Gregarious, penguins, wonderfully adapted to swimming and diving, confined to the southern part of South America, South Africa, Australia, Antarctic regions and Galapagos Islands.

1. Body sleek, streamlined, and offers least possible resistance to diving and submarine activities.
2. Closely packed plumage; feathers small, scale-like, dense, over entire body without apteria.
3. Hindlimbs modified for swimming; metatarsal, unlike those of other birds only partly fused. The feet are strongly webbed.
4. The integument is provided with thick, fatty, insulating layer.
5. The bones, excluding certain skull bones, are solid. Air sacs absent.
6. The bones of featherless wings flattened and united to form a powerful, resistant paddle or flipper moving only at the shoulder point.
7. Monogamous and one egg is laid at a time. Feed on fishes, crustaceans, squids, etc.

Examples: *Apenodytes* (King penguin), *Pygoscelis, Eudyptes, Eudyptula, Spheniscus* (Penguin).

Superorder **Neognathae**

1. Modern birds with well-developed wings, keeled sternum, and fully adapted for flight.
2. Beaks without teeth.
3. Slender vomer separates the palatines imperfectly.
4. The palatines are protruded posteriorly and in contact with the base of cranium and remain movably articulated with small pterygoids.
5. Tail vertebrae 5 or 6. Pygostyle absent.
6. Forelimbs with metacarpals joined, and fingers included in wings.

Order 1 **Gaviiformes**

1. Piscivorous sea or lake-birds of North America, Europe and Arctic, adapted for diving and swimming. Capable of flight also.
2. The neck is long. Legs are posteriorly placed and completely encased with skin.
3. The digits webbed; wings short.
4. Eggs are laid in nests among piles of vegetation.

Example: *Gavia* (diver).

Order 2 **Podicipediformes**

1. Fresh water, agile, compact-bodied divers or grebes, almost cosmopolitan in distribution.
2. Toes lobate.
3. The reduced tail bears degenerated tail feathers.
4. Lungs are placed far back.

Examples: *Podiceps* (grebe), *Aechmophorus, Podilymbus, Centropelara,* etc.

Order 3 **Procellariiformes**

1. Tube-nosed, long-winged sea birds, including albatrosses, fulmars, petrels etc, enjoying cosmopolitan distribution. Come to land only in breeding season and build-up nest in holes.
2. Nostril tubular, and horny sheath of hooked bill composed of several plates.
3. Hind toe vestigeal or none (as in diving petrels).
4. Plumage compact and oily.
5. Wings long, narrow, span may be over 3 metre.

Examples: *Diomedea* (albatross), *Puffinus* (Sheerwater), *Oceanodroma, Pachyptila, Pelecanoides,* etc.

Order 4. **Pelecaniformes**

1. Voracious fish-eaters, diving pelicans, cormorants, boobies, gannets, etc. with all 4 toes included in the foot-web.
2. Nostril vestigeal or absent.
3. A gular Pouch or throat, except in tropical birds.
4. In pelicans; body is heavy, bill up to 45 cm. long, pouch used to scoop fishes from water.

Examples: *Pelecanus, Phalacrocorax, Sula, Anhinga, Fregata, Phaethon,* etc.

Order 5. **Ciconiiformes**

1. Long-necked, long-billed, herons, storks, ibises, flemingos, feeding principally in tropical and subtropical marshes and mudflats.
2. They are either with decorative plumages, bare areas on head (storks); bill abruptly decurved at middle (flamingos).
3. The pattern of beak varies and with sharp cutting edges.
4. Except flemingos web between the toes are absent.
5. Speedy fliers; legs adapted for quick walking on mudflats.

Examples: *Ardea, Egretta* (heron), *Nycticorax, Balaeniceps, Scopus, Ciconia* (stork), *Threskiornis, Plegadis, Phoenicopterus* (flamingo), etc.

Order 6. **Anseriformes**

1. About 200 species of swans, geese and ducks, all are efficient fliers and almost cosmopolitan in distribution.
2. Bills broad, covered with soft cornified epidermis containing numerous sense pits with harder cap at tip. The margins of bill with many transverse horny lamellae (ridges).
3. Tongue fleshy, legs short. Feet webbed. The tail is usually short and many feathered.

Examples: *Ansera, Coscoroba, Anas, Branta, Cairina, Somateria, Polysticta, Mergus, Oxyura,* etc.

Order 7. **Falconiformes**

1. The powerful, predaceous birds like vultures, kites, hawks, falcons, eagles etc. are distributed almost everywhere except Antarctica.
2. They are notable for the spread of their sharp claws for catching and holding prey and their stout hooked beaks for cutting and tearing.
3. Mandible sharp-edged.
4. Feathers are stiff.
5. Eyes laterally placed and with extremely sharp power of vision.
6. They feed on living vertebrates including fish and, in some cases, even insects.

7. Mostly monogamous. Females incubate eggs. The males feed the females during incubation.

Examples: *Gypes, Gymnogypes, Aegypices, Haliaster, Aquila* (eagle), *Buteo, Circus, Falco, Accipiter, Nesphron* (Vulture), *Milvus* (kite).

Order 8. **Galliformes**

1. Gregarious, nonmigratory, ground dwelling game-binds, grouse, quail, pheasants, turkeys, etc., cosmopolitan in distribution.
2. The head is small and body compact.
3. The legs are massive, clawed and used for scratching soil for searching food. Three toes are anteriorly directed.
4. Wings, in the majority, short and round.
5. The beaks are strong built, arched, suitable for picking up grains or seeds.
6. Males are aggressive polygamous.

Examples: *Crax, Lagopus* (grouse), *Alectoris, Coturnix, Frauedinus, Phasianus* (pheasant), *Pavo, Melagris* (turkey), *Opisthocomus, Gallus*.

Order 9. **Gruiformes**

1. The order includes polymorphic forms, either of large size and strong flight, inhabiting open marshes or prairies (cranes) or of medium to small size, of weak flight, or flightless, and shy inhabitant of marshes enjoying cosmopolitan distribution except in poles.
2. Toes are not webbed. Nest may be built on the ground.

Examples: *Grus (Cranes), Rallus, Fulica, Gallinula, Otis, Rhynochetos, Turnix, Monias*, etc.

Order 10. **Diatrymiformes**

Large flightless European and North American terrestrial birds with massive head, beak and neck; lived in the Eocene period. They were perhaps allied to the cranes.

Examples: *Diatryma*.

Order 11. **Charadriiformes**

Shore birds, gulls, auks etc. are universally distributed, except in poles.

1. The order comprises of variety of forms, which live near water, some in shore, some in aquatic leaves and some are aquatic.
2. Plumage dense and firm.
3. Toes usually webbed, at least at the base.
4. More or less long-legged (shore birds), strong-winged (gulls), or with only 3 toes and legs far back in position.
5. Eggs heavily spotted.

Shore birds probe for food in sand, mud or shallow waters.

Examples: *Charadrius, Jacana, Rostratula, Haemantopus, Vanellus (lapwing), Gallinago, Calidris (Sandpiper), Numenius, Dromas, Glareola, Larus (Gull), Fratercula, Alca*, etc.

Order 12. **Columbiformes**

Pigeons and doves, distributed world-wide except in poles.

1. Herbivorous, size varies from swan to pigeon.
2. The wings are long, pointed and enable the birds to make sustained flight at a great speed.
3. Bills short and slender; the base of beak covered by a soft skin bearing longitudinal slit-like nostril called cere.
4. The head and neck small and compact.
5. Legs are small. All the toes lie in same plane and are provided with straight tarsus usually shorter than toes.
6. The crop is large producing 'pigeon milk.'
7. Monogamous but lives in groups.
8. Young are helpless and both male and female nourish the young with the help of 'pigeon milk' produced by both sexes during breeding season.
9. Usually lay 2-3 eggs; both male and female incubate the eggs.

One family of this order is extinct which includes *Dodo*.

Examples: *Columba, (pigeons), Pterocles, Syrrhypta, Streptopelia, Cabonas, Treron, Coura, Didunculus, Raphus (dodo)*, etc.

Order 13. **Psittaciformes**

1. These are parrots distributed in forests of tropics and subtropics; concentrated in Australia, New Zealand, and South America. Adapted for arboreal life and can also climb.
2. Loud-voiced birds, chiefly frugivorous with brilliant plumage or green, blue, yellow or red.
3. Beak stout, narrow, sharp-edged, and hooked at the tip used for climbing.
4. The upper mandible is movably articulated with the frontal bone of the skull.
5. Bill with soft cere, often feathered.
6. The toes two in front and two behind, the outer hind toe is not reversible. Used for grasping.
7. Live in flocks, but strictly monogamous. The nest is built in the holes of trees.
8. The young are nourished till they attain maturity.

Examples: *Ara (Macaw), Kakatoe, Psittacus (parrot), Conuropsis, Melapsittacus, Agapornis, Psittacula (tota), Nestor, Rhynchopsitta, Probasciger* (cockatoo) *etc.*.

Order 14. **Cuculiformes**

Tropical and subtropical cuckoos, road-runners, anis, etc., common in temperate zones also.

1. Toes are two in front, and two behind. Outer toe reversible; the feet are not adapted for grasping.
2. Bill moderate, tail long.
3. Many old-world cuckoos are parasitic, the females lay eggs in the nest of other birds like crow, for incubation and rearing of young.

Examples: *Cuculus, Chalecites, Geococcyx,* etc.

Order 15. **Strigiformes**

The order includes nocturnal birds, the owls, which prey upon small animals at night.

1. Plumage soft-textured. The outer vane of each outermost wing feather is narrow to constitute a loose fringe, making a soundless flight and approach possible.
2. Head large and round. Owl rotates its neck to look sideways.
3. The beak is short.
4. Feet adapted for grasping, claws sharp.
5. Ear openings large, often with flap-like cover, sometimes asymmetrical.
6. Eyes large, round, directed forward and each in a disc of radial feathers.
7. Retina contains principally rods to perceive low intensity light.
8. The prey is swallowed whole.
9. Eggs are white. Nests are built in holes of trees or buildings.

Examples: *Bubo (eagle owl) Tyto, (barn owl), Asio, Otus, Glaucidium, Speotyto, Nyctea,* etc.

Order 16. **Caprimulgiformes**

1. Tropical, subtropical and temperate; nocturnal night hawks, whippoorwills, etc. living in woods or bush, forage near ground or high in air, and prey on insects at night or dusk.
2. Colouration protective, helps the bird in concealing.
3. Plumage soft and arranged in the fashion of that of owl.
4. Legs and feet small and weak, not adapted for grasping.
5. Bills are small, delicate, but the mouth opening wide and margined with long bristle-like feathers with specialised sensory receptors.

Examples: *Podargus, Nyctibius, Chordeiles, Caprimulgus (nightjar) Nyctidramus,* etc.

Order 17. **Apodiformes (Micropodiformes)**

Small swifts and humming birds, mostly tropical.

1. Legs very short, feet very small, wings pointed and bills small and weak (swifts) or slender with long tubular tongue (humming birds).
2. In swifts the mouth is broad and eyes are large.

3. In humming birds, plumage is brilliant, iridescent, specially on head and neck of males.
4. Feed on nectar, small insects and spiders from the blossom by tubular protrusible tongue and needle-like bill.
5. Eggs are white.

Examples: *Apus (swift), Cypsiurns, Chaetura, Patagona, Loddigeria, Trochilus* (humming bird), etc.

Order 18. **Coliiformes**

1. Small (27.5-35 cm), passerine-like, colies or mouse birds of Sourthern Africa with short legs, sharp curved claws. Live in a group of 20 to 30.
2. The 1st and 4th toes reversible, which help in creeping on trees.
3. The feathers hair-like and soft; tail very long.
4. Usually a crest is present on the head.

Example: *Colius* (mouse bird).

Order 19. **Trogoniformes**

1. Small Trogons and Quatjals of Asia, Africa and Tropical America, lead a solitary life in dense forest and capture insects in the early morning or during twilight.
2. The bills are short, and stout with bristles at the base. The tip of the beak hooked and, in some, indented.
3. Feet small and weak. The 1st and 2nd toes are directed backwards and 3rd and 4th toes anteriorly placed.
4. Feathers soft and lax, help in noiseless flight.
5. Tail long and stiff, supports against vertical surface at the time of digging.

Examples: *Herpaetes, Hypodermes, Calures, Pharomechrus*, etc.

Order 20. **Coraciiformes**

1. Brilliantly-coloured kingfishers, hornbills, motmots, bee-eaters, rollers, and hoopoes, found world-wide, especially in tropical climates, show varied structures.
2. In most, wings and legs are short and beak long.
3. In some, anteriorly directed toes are fused at the base (syndactylous) and in others one of the toes is reversed (zygodactylous).
4. Nests are built in holes of trees, etc.

Examples: *Alcedo* (kingfisher), *Ispidina, Dacelo, Chlorocerle, Merops* (bee eater), *Coracis, Upupa, Buceros, Anthracoceros*, etc.

Order 21. **Piciformes**

1. Woodpeckers, jacanas, puffbirds, barbets, honey guides, toucans, enjoying world-wide distribution, except in Australia and poles; live in forest, cling to the tree trunk, dig insect and its larvae out of wood and excavate nest cavities in trees.
2. The feet is zygodactylous. 2nd and 3rd toes are directed forward and 1st and 4th pointed backward.
3. Tail feather stiff with pointed tips.
4. Bill stout, tongue protrusible, roughed, or with barbs near tip.

Examples: *Picus (woodpecker), Sphyrapicus, Fynx, Indicator, Ramphastos, Notharcus, Megalaima*, etc.

Order 22. **Passeriformes**

1. Small perching birds or passerines, adapted to various habitats and distributed widely. The order includes about 5,100 known species.
2. All are adapted for land life and the four digits remain in the same level.

Examples: *Muscivora, Tyrannus, Xenicus, Acanthisitta, Oxyruncus, Menura, Pycnonotus, Aegithnia, Lanius* (shrike), *Dulus, Corvus, Pica, Sturnus*, (sterling), *Passer* (house sparrow), *Fringilla, Aethopyga, Moho*, etc.

42 CLASS MAMMALIA

The mammals are the highest evolved group in the Animal Kingdom and sharply differentiated from other vertebrate classes. It includes animals ranging from small shrews to the existing largest animals like whales and elephants. The shrews and mice are less than 5 cm in head-and-body length and weighing only a few grammes while a blue whale grows up to 30.5 m weighing 119 tons. Others diverse groups are moles, bats, kangaroos, rodents, cats, deers, monkeys and also man.

The group is enriched by myriad forms of animals—reptile-like egg-laying mammals, the Monotremata of Australia; some pouch or marsupium bearing with half gestation period; the Marsupialia of Australia, and others with full gestation period developing a filtering placenta within uterus and giving birth to young, the Eutheria.

Due to their thermoregulatory capacity and hyperdevelopment of cerebrum, mammals, particularly the eutherians, with capacity to produce and carry in her body perfectly suitable food for the unborn young, have exploited diverse habitats ranging from polar regions to tropics, from the sea to the densest forest and inhospitable deserts to the aerial mode of life with ecolocation capacity.

Of more than 8,500 extant species, many are nocturnal and escape notice of man. Holes in the trees are used by mice, raccoons, chipmunks, opossums, and black bears. Bats resort to caves, trees or buildings. Tree squirrels, some wood rats and mice build nests amid tree foliage. Rabbits, ground squirrels, coyotes dig burrows in the ground. Whales, seals and walrus exploit the vast ocean.

The mammals are derived from the Synapsida, the mammal-like reptiles of an ancient stem which appeared in the Carboniferous and subsequently diversified, and also glorified, as in man.

CAVIA PORCELLUS (Guinea pig)

Cavia belongs to suborder Hystricomorpha, order Rodentia, infraclass Eutheria, subclass Theria, class Mammalia, superclass Gnathostomata, subphylum Craniata (vertebrata), phylum Chordata.

Guinea pigs are cosmopolitan in distribution. The wild forms are gregarious, usually living within the limits of a territory, preferably close to marshy lands. They are poor burrowers and take shelter in crevices, burrows dug by other animals or in bushes, hiding them in thick vegetation and avoid strong light. The guinea pigs are herbivores and feed upon fresh plant matters and fruits. They are coprophagus* (eating faeces).

☒ External features

The guinea pig is a quadruped animal about 20 cm in length. The whole body surface is covered with variously coloured soft hairs, except the tip of the noses, foot pads and teats. The body is divided into two distinct regions, the **head** and **trunk** joined by a short **neck** (Fig. 42.1).

Head

The head is elongated and produced into a **snout** anteriorly located. **Mouth** is a transverse slit, bounded by two soft **lips** and situated slightly ventral to the anterior extremity of the head. The **upper lip** is cleft at the middle exposing chisel-shaped front teeth or incisors. A pair of oblique slits, the **nostrils**, are situated just above the mouth. The **eyes** are large, paired and situated at the sides of the sloping head. Each eye is provided with three **eyelids**; the upper and lower movable eyelids and third eyelid, the nictitating membrane, at the inner corner of the eye. Some long, stiff hairs known as **vibrissae** are scattered above and below the eyes and on the snout. Behind the eyes and on the top of the head a pair of external ears or **pinnae** supported by cartilage are present, one on each side. Each pinna on its concave side leads to the external auditory opening, which continues as **external auditory meatus**, a short tube closed behind by **tympanic membrane**.

Neck

The neck is distinct constriction, but very short and connects the head and the trunk.

Trunk

The trunk is divided into two regions, the anterior **thorax** or chest protected by ribs and sternum and the posterior **abdomen** or belly without any bony support. The trunk bears two pairs of **pentadactyl** limbs. The anterior pair **forelimbs** are shorter and consist of upper arm, forearm and hand; the posterior pair, **hindlimbs** are longer and consist of thigh, shank and foot. The forelimb carries four digits or fingers and the hindlimb three digits or **toes** each ending in a **claw**.

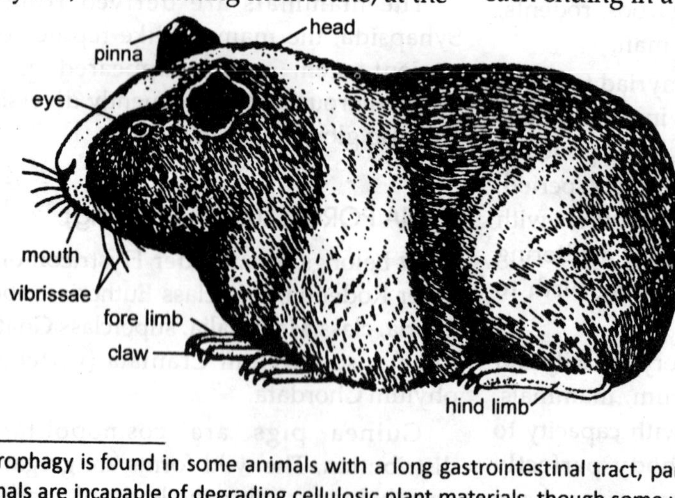

FIGURE 42.1
Cavia porcellus

* Coprophagy is found in some animals with a long gastrointestinal tract, particularly in herbivores. As a rule, animals are incapable of degrading cellulosic plant materials, though some workers believe that certain lower animals have that capacity. Cellulolytic bacteria harbour the alimentary canal of many animals. Due to rapid multiplication, population density of the bacteria, moving with the food, becomes higher towards the posterior part of the tract, reaching a peak in the large intestine. Only some of the products of biological degradation caused by the bacteria are used by them while in rest, which are useful to the host, pass out with the faeces. By eating its own faeces these substances are taken back into the alimentary canal and the nutrients absorbed in the small intestine, while passing through it.

The teats are present on the ventral surface of the abdomen. They are well-developed and longer in females and the mammary or milk glands open in them.

In male, the perineum* consists of prominent, paired, lateral scrotal swellings, indicating the presence of scrotal pouches, in which tests are lodged. Of the openings in the perineum, the **urethal opening** is anteriormost and covered by folded skin of the prepuce of the penis, which forms the **preputial ostium**. The **prepuce** is a loose sheath of skin in which the penis is retracted. Anterior to the urethral orifice is a longitudinal, median cylindrical ridge representing the elongated **glans** of the **penis**. Posterior to the preputial ostium is a longitudinal cleft covering the opening of the large perineal sac and the **anus**, which is a corrugated opening.

In female, the perineum bears one pair of slightly raised, lateral labia. The anteriorly placed urethral orifice is bounded ventrally by a short **clitoris**. The clitoris is a vestige of the penis of male. The centrally placed vaginal orifice is U-shaped, separated from the anus by perineal sac. The anus is posterior to the vagina.

Skin

The skin consists of two layers, an outer epidermis and a deeper dermis (Fig. 42.2). A subcutaneous layer of loose connective tissue is present beneath the dermis.

Epidermis

A stratified, keratinized epithelium made of two layers.

Stratum corneum, the superficial keratinized layer of flat cells. The cells are dead and scale-like. In parts subject to pressure and friction, viz., palm of the hands and sole of the feet, the horny layer thickens and hardens to form **tori** (singular torus). The claws are hardened with modified pieces of horny layers.

Stratum germinativum, the deep layers of columnar cells. The cells divide actively to replace the superficial layers of keratinized

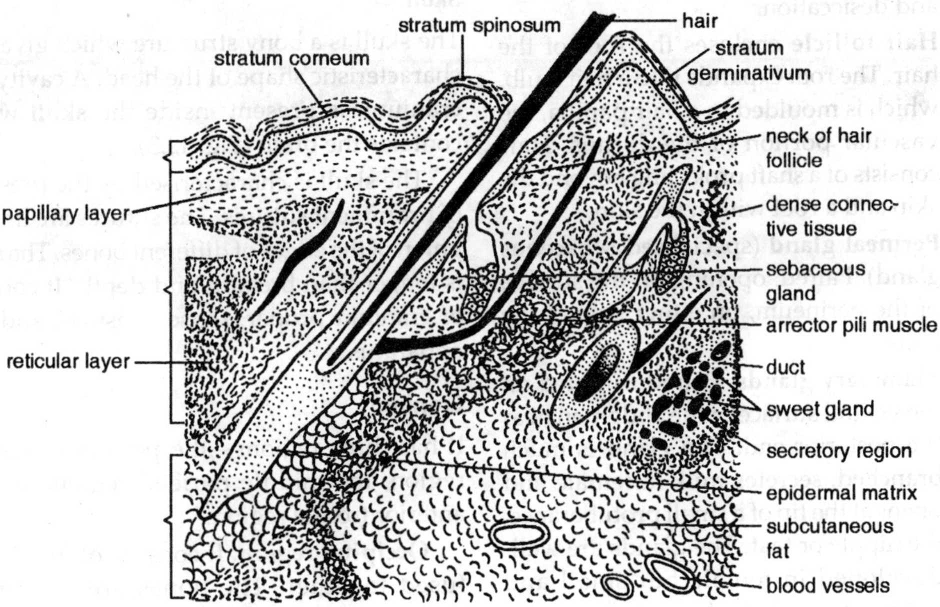

FIGURE 42.2

Skin. Mammalia. (*Cavia*) Vertical section

* A part of the abdomen between the anus and genital organs.

cells. Melanin pigment is present in the deeper cells of stratum germinativums.

Dermis

It consists of connective tissue. (a) The outer part dense and bears papillae projecting into the epidermis. (b) The deeper part consists of loose connective tissue in which blood vessels, nerve fibres, glands and hair follicles are present.

i. **Sweat gland.** A coiled tube, the duct, runs straight outward and passing between two dermal papillae and epidermis opens on the surface as a sweat pore. Sweat secreted by sweat gland excrete certain minerals and keeps the body cool by evaporation.
ii. **Sebaceous gland.** Small, sac-like, alveolar or branched and opens in the hair follicle. The **sebum**—secreted by the sebaceous gland lubricates the adjacent hair-shaft and skin. It is bactericidal and probably also helps render the skin impervious to moisture and desiccation.
iii. **Hair follicle** encloses the root of the hair. The root expands into a hair-bulb, which is moulded in a hair papilla, the vascular portion of the dermis. Hair consists of a shaft projecting beyond the skin and a root within the skin.
iv. **Perineal gland** (specialised sebaceous gland). Paired, opening on the surface of the perineum, on either side of the anus.
v. **Mammary glands.** Paired, situated on the ventral surface of the abdomen near the posterior end. The gland is much branched, secretes milk in female and opens at the tip of a small protuberance, the nipple or **teat**. The glands are well-developed in female but small and nonfunctional in male.

☒ Skeletal system

The hard parts or structures which protect and support the body form the skeletal system. The skeletal system has two components—the hard structures outside the body and of protective function, the **exoskeleton** and those inside the body which support different parts of the body—the endoskeleton.

☒ Exoskeleton

The hairs forming a coating on the body and the claws on the digits constitute the exoskeleton. These are formed from the cells of the outer layer of the skin.

☒ Endoskeleton

The hard parts inside the body which form a supporting framework is known as endoskeleton. The endoskeleton can be divided into two sets—the axial skeleton and the appendicular skeleton.

Axial skeleton

The skull, hyoid apparatus, vertebral column and the thoracic basket constitute axial skeleton.

Skull

The skull is a bony structure which gives the characteristic shape of the head. A cavity, the cranium, is present inside the skull which contains the brain (Fig. 42.3).

The skull is characterised by the presence of prominent sutures. The sutures are lines of union of the edges of different bones. The skull is longer than breadth and depth. It consists of cranium proper, sense capsules, and jaw apparatus.

Cranium

It has three regions—the posterior occipital region, the middle parietal region and the anterior frontal region.

Occipital region. It consists of four bones fused together. The bones are—a pair of **exoccipital** on two sides, a **supraoccipital** above and a **basioccipital** below. These bones surround a large aperture, **foramen magnum**, which serves as a passage for the spinal cord. The two exoccipitals bear two projections, the

occipital condyles, which articulate with the first vertebra. Each exoccipital is drawn to a long curved process ventrally, the **paroccipital process** for attachment of muscles.

Parietal region. Five pieces of bones form the parietal region; a single piece, **basisphenoid** below, a pair of **alisphenoids** ventrolaterally and two **parietals** on the roof. A small bone, the **interparietal**, is situated between the two parietals. The posterior margin of the interparietal rests against the supraoccipital.

Frontal region. Two **frontals** on the dorsal surface form the roof and **sphenoid** forms the floor. The anterior and posterior end of sphenoid is called **presphenoid** and **basisphenoid**, respectively. The lateral walls are formed by **alisphenoid** and **orbitosphenoid**, and two **squamosals**. The squamosals take part in the formation of zygomatic arches. The inner anterior wall of the cranium is formed by a perforated **cribriform plate**, separating the cranium from the olfactory capsule.

Sense capsules

The sense capsules are olfactory, optic and otic or auditory.

Olfactory capsules. The olfactory capsules house nostrils. The roof is formed by a pair of **nasals**, the floor and lateral walls by the inward projection of the **premaxilla** anteriorly and **maxilla** posteriorly. A vomer is situated along the midventral line, posteriorly articulating with the **presphenoid**. A vertical cartilaginous plate, the **mesethmoid** separates the nasal cavities. Small, spongy bones, the **turbinals**, support the olfactory epithelium.

Optic capsules. A small, perforated bone, the **lacrimal bone**, bearing lacrimal foramen forms the anterior wall of the optic capsule. The zygomatic arch formed by **maxilla, jugal** and **squamosal** forms the ventral wall of the orbit. The **interorbital septum** forms the inner wall and also separates the orbits from each other. **Orbitosphenoid** forms the upper part and **presphenoid**, the lower part of the septum.

Otic (auditory) capsules. The otic capsule is composed of a periotic bone, a tympanic bone and auditory ossicles. The **periotic bone** is irregular in shape, formed by fusion of three bones and encloses the **membranous labyrinth**. The **tympanic bone** is firmly applied to the outer surface of the periotic. It is flask-shaped, the open end is **external auditory meatus** and the swollen base, the **tympanic bulla** encloses the tympanic cavity and the auditory ossicles. The **pinna** (external

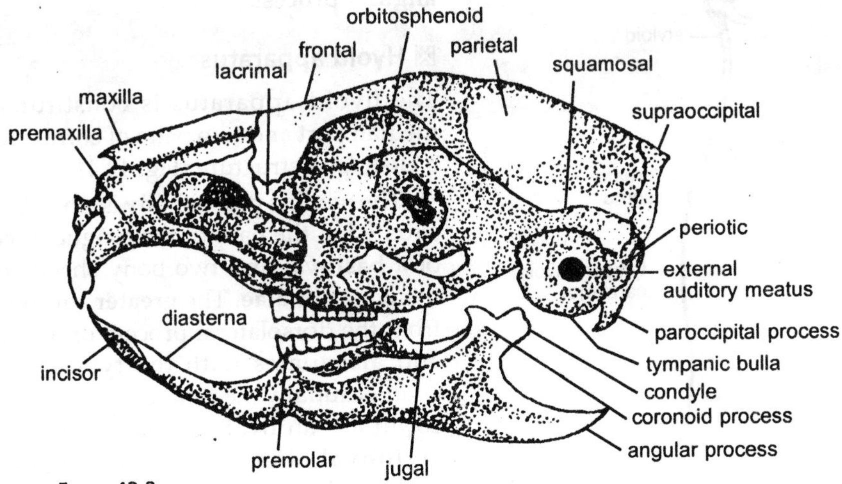

FIGURE 42.3

***Cavia* sp.** Skull and mandible (lateral view)

ear) is attached to the rim of the external auditory meatus. The auditory ossicles form a chain of three small bones—**malleus, incus** and **stapes**.

Jaw apparatus

The jaw apparatus consists of two parts—the maxillary arch, forming the upper jaw and the mandibular arch, forming the lower jaw.

The **maxillary arch** of each side consists of a **premaxilla** anteriorly and a **maxilla** posteriorly. The premaxilla of one side unites with the premaxilla of the other side and bears the **incisors** or cutting teeth anteriorly. The maxilla bears **premolars** and **molars** or crushing teeth and gives off a horizontal process, the **palatine**, from its inner surface. The palatines of two sides articulate in the middle and form the anterior portion of the **bony palate**. The maxilla gives off another process from the outer surface which forms the anterior portion of the cheek arch or **zygomatic arch**. A process arises from the anterior region of each palatine, and the two processes from the two palatines articulate in the middle line to form the posterior portion of the **bony palate**. The **pterygoid** is a small bone which articulates with the palatine in front and the pterygoid process of the alisphenoid behind.

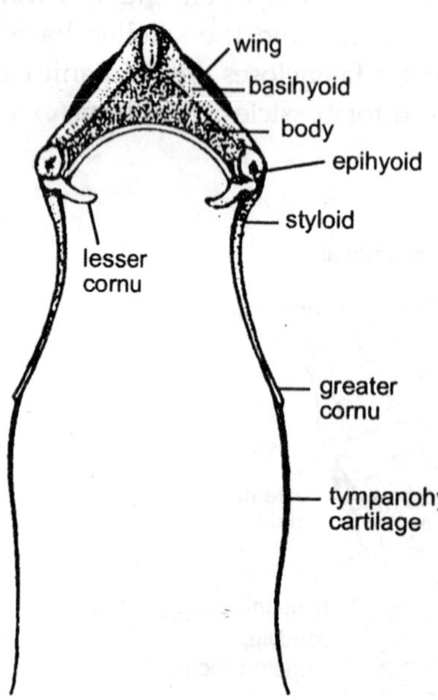

FIGURE 42.4

***Cavia* sp.** Hyoid apparatus (dorsal view)

In front of each perietal is a broad bone, the **squamosal**, bounded below by alisphenoid; the parietal and frontal above and the frontal and alisphenoid in front. Anteriorly the squamosal sends off a process—the **zygomatic process**, which forms the posterior portion of the zygomatic arch. The **jugal**, which forms the middle portion of the zygomatic arch, articulates with the zygomatic process of the squamosal posteriorly and with the process of maxilla anteriorly. The zygomatic process of squamosal is provided with a longitudinal facet for articulation of the lower jaw.

The **mandibular arch** consists of two halves, each known as ramus. The two rami are connected together in front by a rough articular surface or **symphysis**. Each **ramus** is formed by a single bone, the **dentary**, consists of a broad vertical portion posteriorly and a tapering horizontal portion anteriorly.. The tapering portion bears incisor anteriorly and premolars and molars on the upper surface of the posterior portion. The broad vertical portion bears a facet for articulation with the zygomatic process of the squamosal. The angle of union of the horizontal and vertical portion of the ramus gives off a projection known as **angular process**.

⊠ Hyoid apparatus

The hyoid apparatus is constituted by a median part and two pairs of cornua forming a U-shaped structure (Fig. 42.4).

The median part or **body** is stout, bony and situated at the base of the tongue. It consists of a **basihyal** and two bony sheets forming the wings or **alae**. The **greater cornu** extends from the dorsolateral process of the basihyal and articulates with the tympanic bulla. Craniocaudally the components of the greater cornu are an oval **epihyal**, a long, slender **stylohyal** and a narrow cartilaginous **tympanohyal**. The **lesser cornu** is slender, cartilaginous and extends dorsomedially from the basihyal, as **thyrohyal** and articulates with the thyroid cartilage of the larynx.

Vertebral column

The vertebral column usually consists of 37 vertebrae articulated together. The vertebral column can be divided into five regions—the cervical or neck region, the thoracic or chest region, the lumbar or abdominal region, the sacral or pelvic region and the coccygeal or rudiment tail region. A cartilaginous pad, the **intervertebral disc** is present between two vertebrae.

Strucutre of a typical vertebra (cervical)

The body of the vertebra or **centrum** is ventrally placed and both the anterior and posterior faces are flat. This type of vertebra is known **acoelous.** A **neural arch** enclosing the neural canal is present on the centrum. The spinal cord runs through the neural canal. The roof of the nerual arch is known as **lamina** and the part joining the centrum is **pedicel**. A **neural spine** runs dorsally from the summit of the neural arch. The notched margins of the pedicel leave **intravertebral foramina** between the successive vertebrae, through

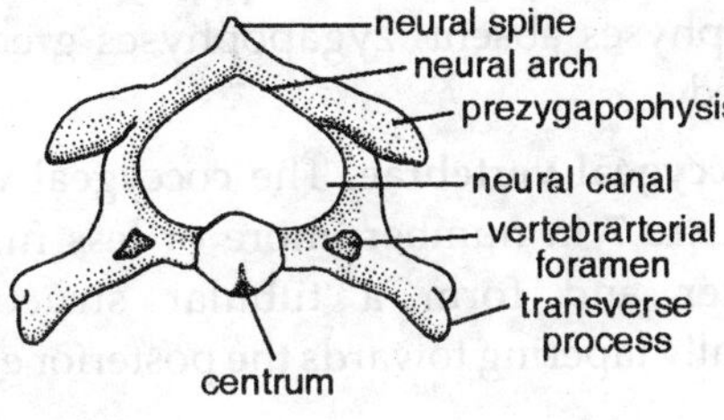

FIGURE 42.5

***Cavia* sp.** A typical cervical vertebra. Anterior view

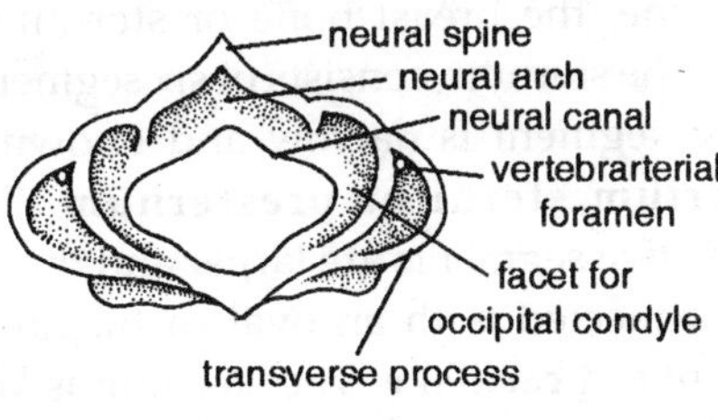

FIGURE 42.6

***Cavia* sp.** Atlas (First vertebra). Anterior view

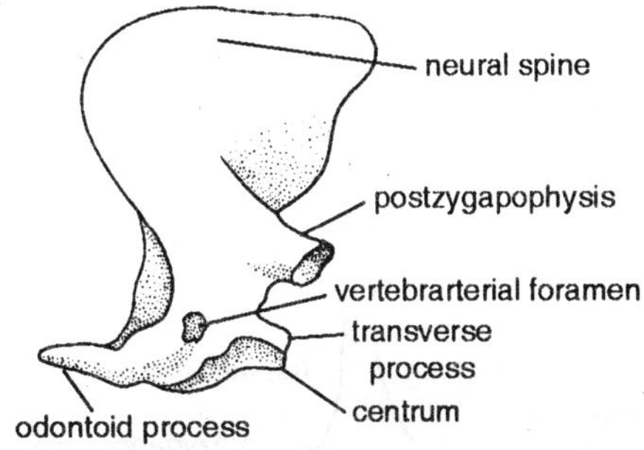

FIGURE 42.7

***Cavia* sp.** Axis (Second vertebra). Side view

which the spinal nerves emerge. A pair of **transverse processes**, one on each side, run outward and downward from the junction of the centrum and neural arch. **Prezygapophyses**, a pair of anterior articular surfaces are present on the inner side of a pair of large, forwardly directed **metapophyses**. A pair of posterior articulating processes, the **postzygapophyses**, are borne on a pair of large, backwardly directed **anapophyses** (Fig. 42.5).

Cervical bertebrae. 7 in number; centrum small; neural spine short; the craniocaudal base of transverse process pierced by a **vertebrarterial foramen.**

Atlas (first cervical bertebra). Ring-shaped (Fig. 42.6). without any distinct centrum; two concave facets on the anterior face for articulation with occipital condyles of the skull; neural spine very small; transverse processes small and flat; the neural canal divided horizontally into an upper and a lower compartment by a fibrous ligament.

Axis (second cervical vertebra). A peglike **odontoid process** on the anterior face of the centrum fits on the ventral surface of atlas; neural spine large, laterally compressed forming a vertical plate exending anteroposteriorly; transverse processes short; prezygapophyses absent; postzygapophyses present (Fig. 42.7).

Thoracic vertebrae. 12 pairs; each bears a pair of lateral ribs. They have been grouped into—(a) an anterior series of first nine vertebrae, and (b) a posterior series of the last three.

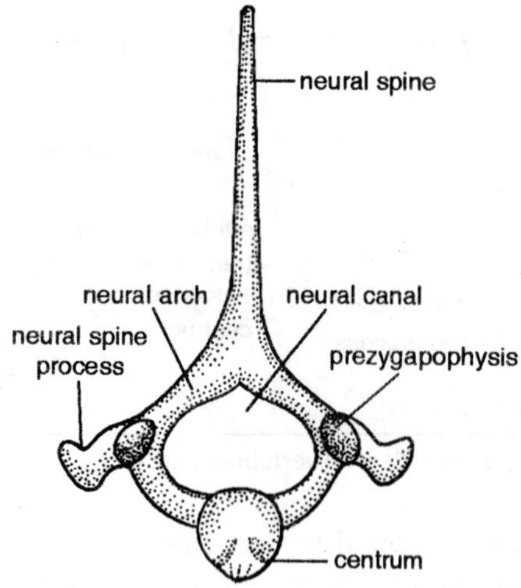

FIGURE 42.8

Cavia sp. Anterior thoracic vertebra. Anterior view

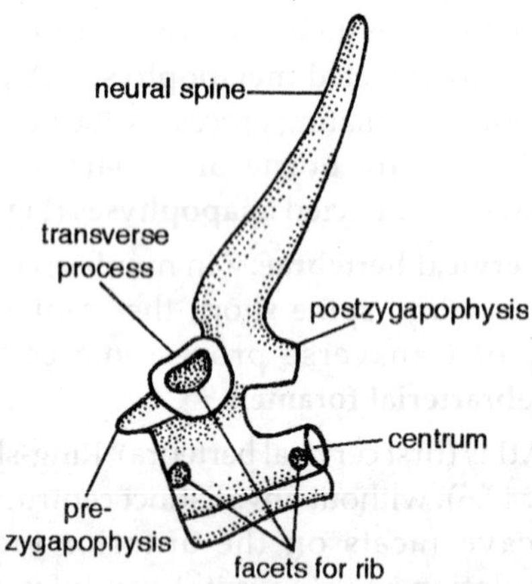

FIGURE 42.9

Cavia sp. Posterior thoracic vertebra. Anterior view

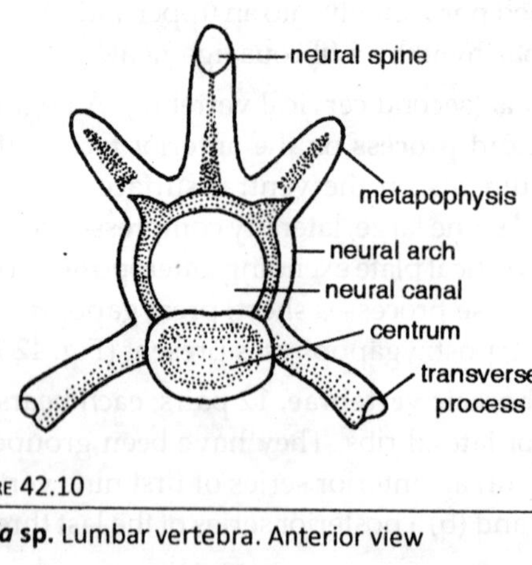

FIGURE 42.10

Cavia sp. Lumbar vertebra. Anterior view

Anterior thoracic vertebrae (Fig. 42.8). Centrum short and thick; neural spine tall, slender and backwardly directed; transverse processes stout and horizontal; articular facets for capitulum and tuberculum of rib on centrum and transverse processes, respectively.

Posterior thoracic vertebrae. Well-developed articular facets for capitulum on centrum; transverse processes lack articular facets for ribs; neural spine flattened from side to side and inclined forward (Fig. 42.9).

Lumbar vertebrae. 7 in number; largest vertebrae; size increases from anterior to posterior end; neural spine flat, directed forward; transverse processes large and expanded (Fig. 42.10); centrum of first two bears a ventral **hypapophysis** for attachment of diaphragm muscles.

Sacral vertebrae. 3-4 in number; fused to form a single bone, the **sacrum** (Fig. 42.11) wedged between the two halves of the pelvic girdle; first one is largest; gradully reducing in size posteriorly; neural spine of first sacral upright; transverse processes expanded and attached to pelvic girdle; metapophyses small, anapophyses absent; zygapophyses greatly reduced.

Coccygeal vertebrae. The coccygeal vertebrae are 7 in number; more or less fused together and form a tubular structure gradually tapering towards the posterior end.

⊠ Sternum

On the ventral surface of the thorax at the middle line, the breast bone or sternum is present. The sternum consists of six segments. The first segment is narrow and known as **manubrium sterni** or **presternum**. The posterior five segments are larger and the last one is connected with an oval cartilaginous plate, **xiphoid cartilage**. The sternum is kept in position by means of ribs, the other ends of which are connected with the vertebral column.

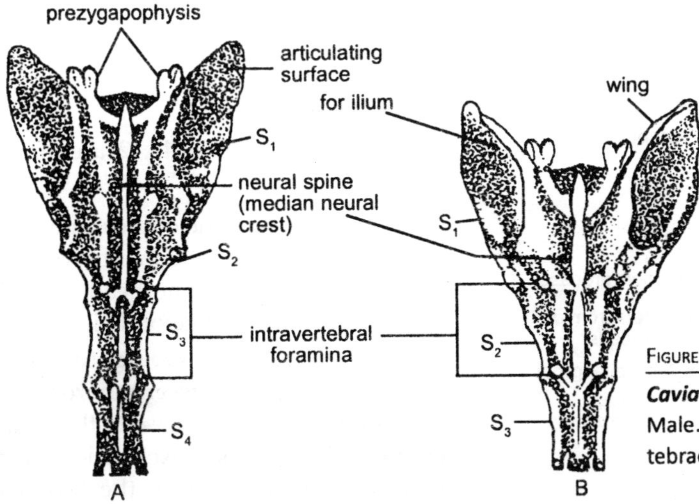

FIGURE 42.11

***Cavia* sp.** Sacrum (dorsal view) A. Male. B. Female, S_1-S_4. Secral vertebrae

☒ Ribs

Twelve pairs or ribs are present in the thoracic region attached to the sides of the 12 thoracic vertebrae. Of these, the first seven pairs are **true ribs**, i.e., they are connected with the sternum and the last five pairs are **false ribs**, as they do not reach the sternum. The portion of the rib articulated with the vertebral column is bony and known as **vertebral portion** and the portion articulated with the sternum is cartilaginous and known as **sternal portion** (**costal cartilage**). Each rib is articulated with the vertebra by two heads—(a) the **capitulum** or the head, that articulates with the centrum and (b) the **tuberculum** or the head, that articulates with transverse process. The costal cartilage of the eighth and ninth are attached to the seventh costal cartilage in front and, thus, indirectly connected with the sternum. The ventral ends of last three ribs remain free and they are known as **floating ribs**.

☒ Appendicular skeleton

The fore- and hindlimb bones together with the pectoral and pelvic girdle comprise the appendicular skeleton.

Pectoral girdle

The pectoral girdle consists of two separate halves, termed **scapula**, not joined by any structure. Each half is made of a scapula, suprascapula and clavicle. The **scapula** is a broad, thin, triangular bone (Fig. 42.12). The broad end has a thin strip of cartilage, **suprascapula**. The scapula bears a **spine**. The spine ends below in a long process, the **acromion process**. The acromion process bears another process, the **metacromion**. A concave cavity, the **glenoid cavity** is present at the narrow lower end of the scapula in which the head of the humerus fits. Internal to glenoid cavity a small inwardly curved process, the **coracoid process is present**. The **clavicle** is ill-developed and extends a part of the distance between the presternum and scapula.

Pelvic girdle

The two halves of the pelvic girdle are identical and attached with the sacral vertebrae anteriorly and also with each other in the midventral region along a line, the **pubic symphysis** (Fig. 42.13). Each half of the pelvic girdle is known as **osinnominatum** and consists of 3 bones—the ilium, ischium and pubis fused together. On the outer surface of osinnominatum at the point of union of the three bones a concave cavity, the **acetabulum**, is present. The head of the femur or thigh bone fits in the acetabulum. The **ilium** is a long bone projected forward and expanded to articulate with the laterally expanded sacral vertebrae. The **ischium** is projected behind and the

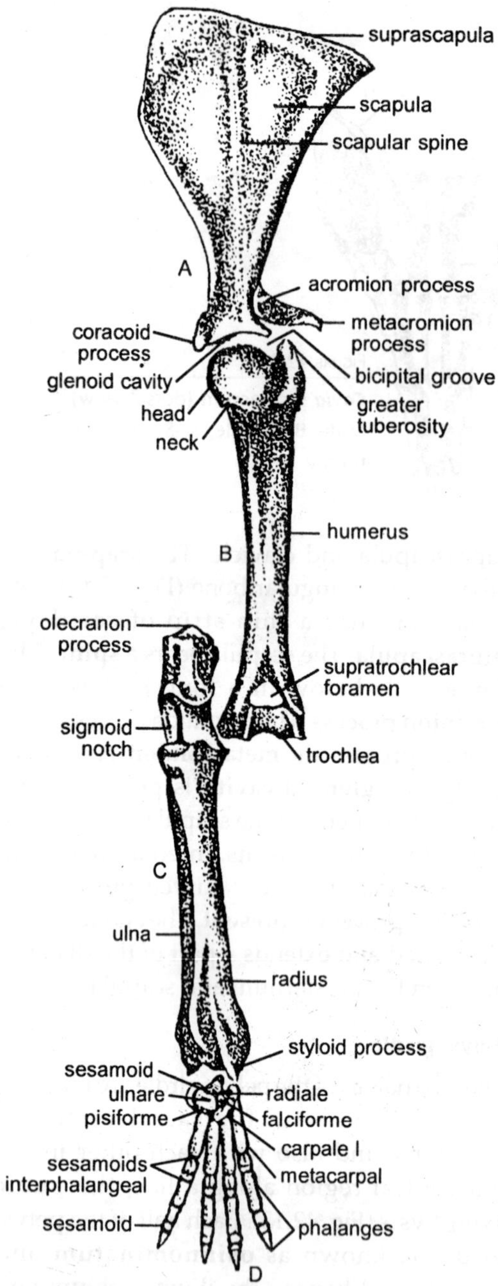

FIGURE 42.12

***Cavia* sp.** Scapula and forelimb bones. A Pectoral girdle (left half). B. Humerus. C. Radius and ulna. D. Distal bones

posterior margin is thickened. The **pubis** starts from the acetabulum and projects backward and downward and the two pubes of two sides unite along the posteroventral borders. The pubis and ischium enclose a large foramen, the **obturator foramen.**

Forelimb bones

The forelimb bones (Fig. 42.12) are constituted by:

Humerus. The bone of the upper arm, the humerus, is long, rounded with a round **head** at the proximal end. The head fits into the glenoid cavity of the pectoral girdle.

Two tuberosities, a **greater tuberosity** on the outer side, and a **lesser tuberosity** on the inner side, are present close to the head. The tendon of the biceps muscle is lodged in the bicipital groove between the tuberosities. Anteriorly, the humerus bears a **deltoid ridge**. At the distal end is a pulleylike structure, the **trochlea** for articulation with radius and ulna of the forearm. In the two depressions above the trochlea, the **coronoid fossa** in front and **olecranon fossa** behind fits the radius and the **olecranon process** of the ulna, respectively. A small tube, the **supratrochlear foramen** joins the two fossae.

Radius and ulna. The forearm bone is a composite one and known as radius and ulna, firmly fixed together. The radius is shorter than ulna and located on the inner side. Near its proximal end the ulna is provided with a deep notch, the **sigmoid notch** for articulation with the humerus. A prominent process, the **olecranon process**, extends beyond the sigmoid notch. Articulating surfaces for carpal bones are present at the distal ends of both radius and ulna.

Carpal bones. The bones of the wrist are carpal bones and consist of seven pieces arranged in two rows. The **radiale, intermedium** and **ulnare** constitute the proximal row. Four small bones constitute the distal row.

Metacarpal bones. The bones of the palm are metacarpal bones and consist of four pieces corresponding to the four digits.

Phalanges. The bones of the digits (fingers) are phalanges. The digits are four in number and each digit bears three phalanges, the terminal one ending in a horny claw.

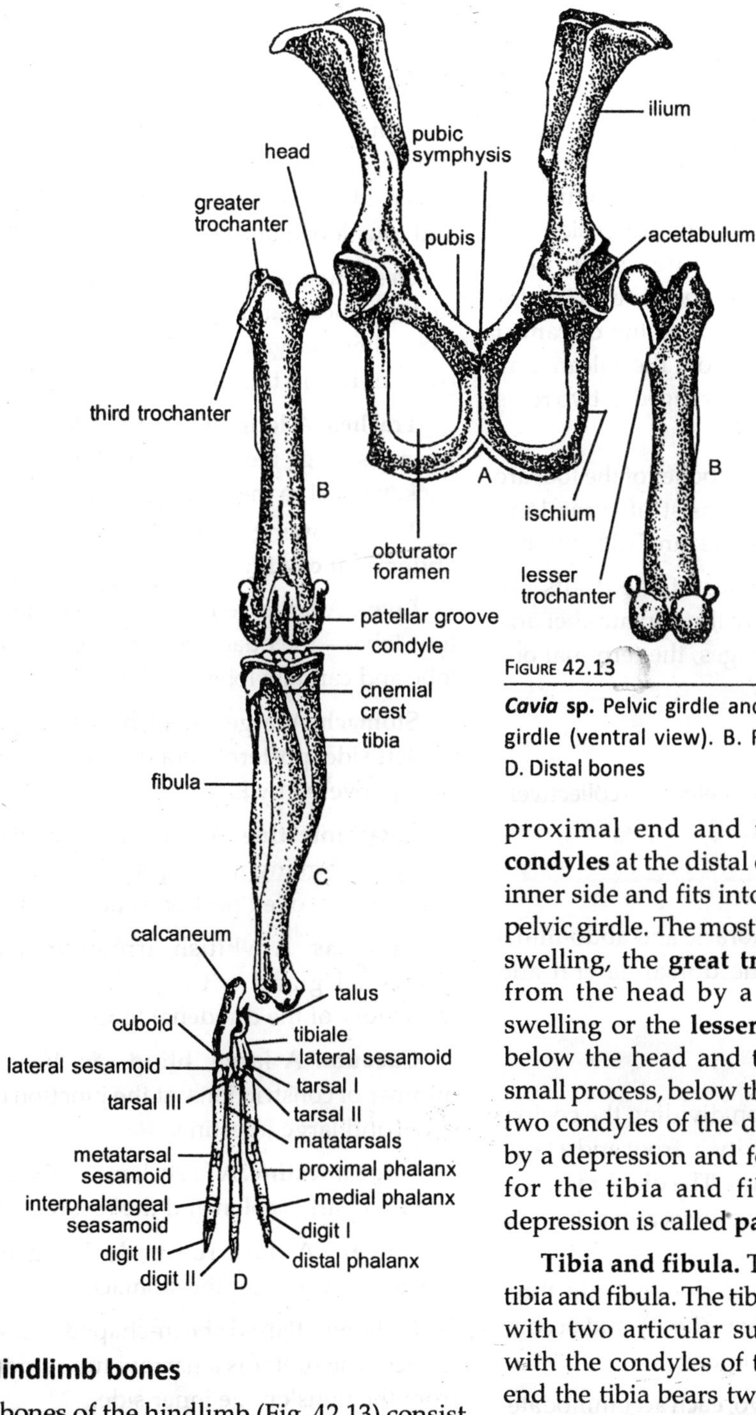

FIGURE 42.13

***Cavia* sp.** Pelvic girdle and hindlimb bones. A. Pelvic girdle (ventral view). B. Femur. C. Tibia and fibula. D. Distal bones

⊠ Hindlimb bones

The bones of the hindlimb (Fig. 42.13) consist of following pieces:

Femur. The bone of the thigh, the femur, is long rounded, with a prominent head and three other swelling, the **trochanters**, at the proximal end and two prominences or **condyles** at the distal end. The **head** is on the inner side and fits into the acetabulum of the pelvic girdle. The most anterior and prominent swelling, the **great trochanter** is separated from the head by a groove. The smaller swelling or the **lesser trochanter** is situated below the head and the **third trochanter**, a small process, below the great trochanter. The two condyles of the distal end are separated by a depression and forms articular surfaces for the tibia and fibula of the leg. The depression is called **patellar groove.**

Tibia and fibula. The bones of the leg are tibia and fibula. The tibia is stout and provided with two articular surfaces for articulation with the condyles of the femur. At its distal end the tibia bears two articular surfaces for the astragalus and calcaneum of the ankle. The tibia bears a prominent ridge, the **cnemial crest**, on the front surface of its anterior end. The fibula is a slender bone and remains distinct from the tibia.

A small cup-lke bone known as knee-cap or **patella** is situated in front of the knee-joint for articulation between the femur and tibia.

Tarsal bones. The bones of the tarsus or ankle are tarsal bones and consist of six pieces. The first two, **astragalus** and **calcaneum**, are large. The astragalus lies on the inner side and the calcaneum on the outer side, both articulating with the respective articular surfaces of the tibia. A process the **calcaneal process** extends behind from the calcaneum. The other four bones are arranged in two rows and, much smaller in size.

Metatarsal bones. The bones of the foot are metatarsal bones and consist of three long pieces corresponding to the three digits (toes).

Phalanges. The bones of the digits are phalanges. The digits are three in number and each bears three phalanges, the terminal one ending in a horny claw.

⊠ General viscera

The organs lodged in the coelom are collectively termed **viscera** (Fig. 42.14).

⊠ Coelom

A large cavity in the thoracic and abdo-minal region bounded by the dorsal, lateral and ventral body walls.

⊠ Diaphragm

It is a muscular partition dividing the coelom into a small thoracic cavity in front and a large abdominal cavity behind. The diaphragm is a dome-shapped muscle, convex anteriorly and concave posteriorly.

The thoracic cavity is divided into three chambers by membranous tissue and accommodate the following organs:

Pleural cavities. Two, each accommodates a lung.

Mediastinum. The space between the two pleural cavities in which the pericardial cavity, the oesophagus and blood vessels are lodged.

Pericardial cavity. A small chamber in which the heart, enclosed by a double-layered pericardium, is lodged.

Lungs. Paired, spongy, bright pink in colour, enclosed in pleural sheath, located on either side of the heart. The right one is longer and four-lobed and the left one three-lobed.

Oesophagus. A narrow, straight tube, dorsal to the heart and median to the lungs, runs from the pharynx, passes through the diaphragm to join the stomach.

Trachea. A long tube, supported by cartilaginous rings along the middle line of the neck, ventral to the oesophagus.

The following organs are lodged in the **abdominal cavity**:

Liver. A large, reddish-brown gland with four lobes—the quadrate lobe, right lobe, left lobe and caudate lobe.

Stomach. A large, broad, bent sac, lying on the left side, posterolateral to the liver, being partly covered by it.

Small intestine. A long, narrow tube, the first part forming the loop is the duodenum and the rest coiled portion is ileum.

Pancreas. A whitish, irregular-shaped, elongated gland with small lobes in the mesentery of the duodenal loop.

Caecum. A large, blind sac, bearing a number of constrictions, at the junction of the small and large intestine.

Large intestine. A fairly long, narrow tube. The first part is colon and the rest rectum.

Spleen. A dark red body lying in the greater curvature of the stomach.

Kidneys. Paired, bean-shaped, dark red bodies. The ureter is a narrow tube emerging from the hilus on the inner side.

Adrenal glands. Paired, ovoid bodies, anterior to the kidneys.

Urinary bladder. A pear-shaped muscular sac with a narrow neck.

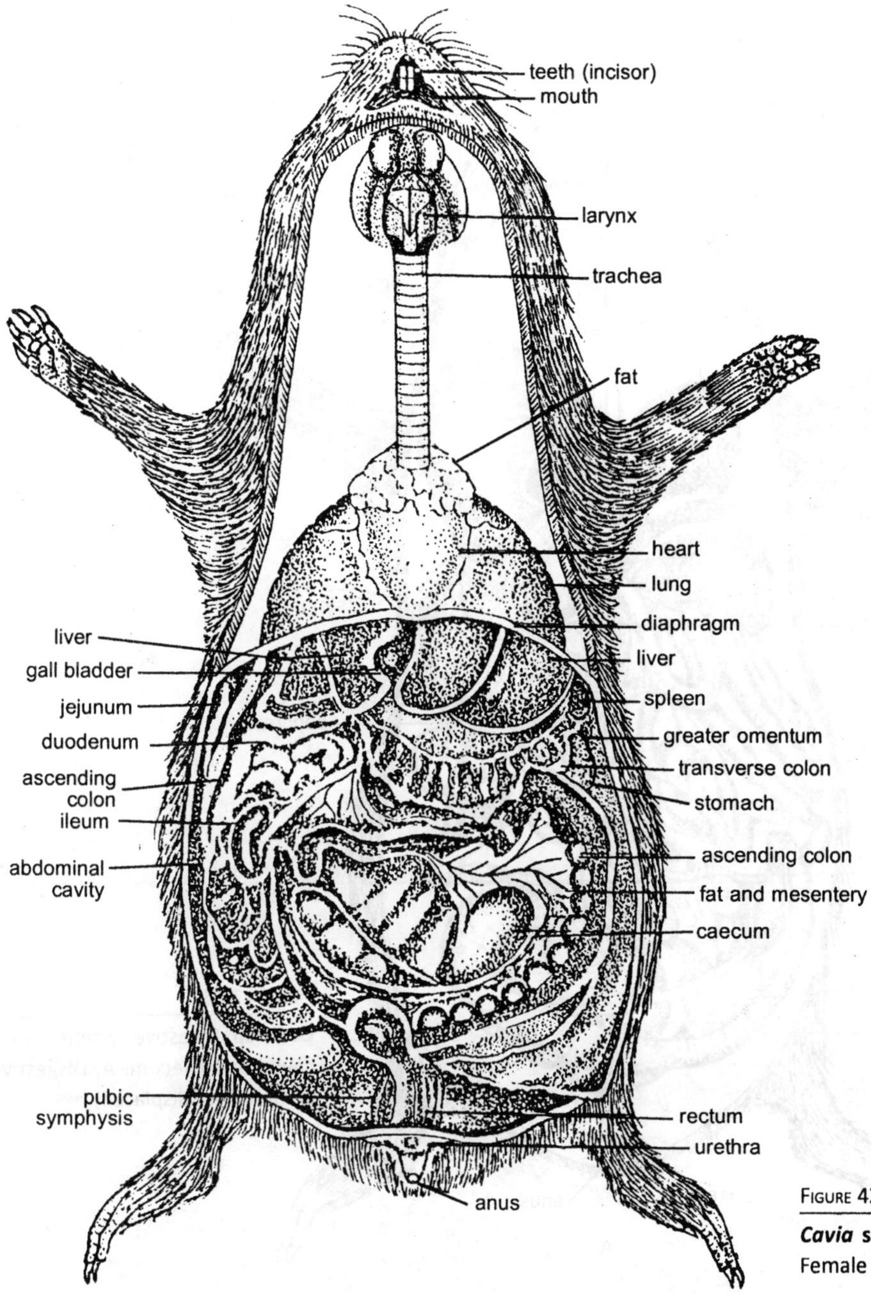

FIGURE 42.14

***Cavia* sp.** General viscera. Female

Ovaries. Paired, small ovoid bodies ventrolateral to the kidneys in female.

⊠ Digestive System

The digestive system consists of a long alimentary canal and its appendages (Fig. 42.15).

Alimentary Canal

The alimentary canal starts from the buccal cavity and ends in anus receiving secretions from various appendages or digestive glands in its way. Digestion of food and absorption take place within the canal. The alimentary canal is divided into the following zones:

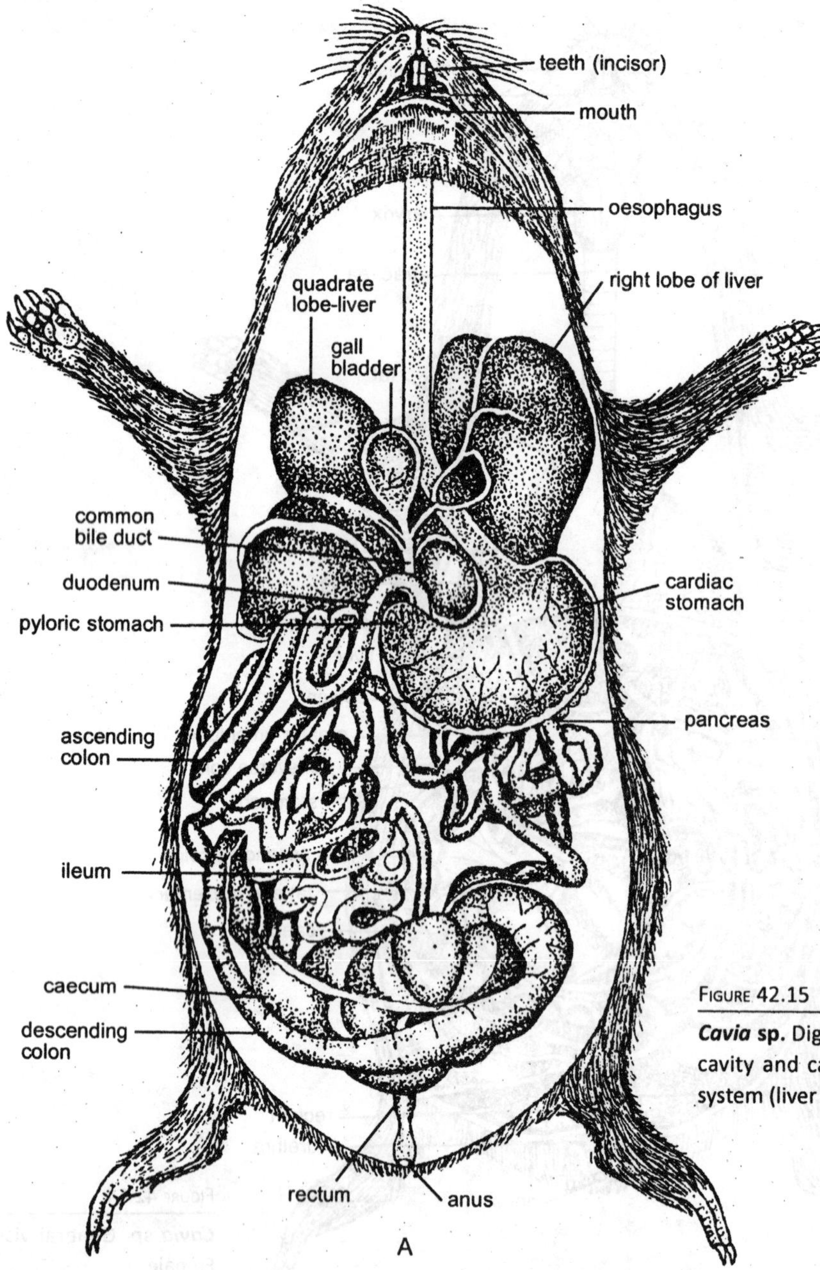

FIGURE 42.15

***Cavia* sp.** Digestive system, buccal cavity and caecum. A. Disgestive system (liver displaced)

Buccal cavity

The mouth is bounded by two soft **lips**, the upper one of which is cleft in the middle. The mouth leads into a cavity, the buccal cavity (Fig. 42.15B). The roof of the mouth is called **palate**. Anteriorly, the palate is supported by bone and known as **hard palate** and, posteriorly, it has no bony support and known as **soft palate**. The palate separates the buccal cavity from the nasal cavities. On the floor of the buccal cavity is present a muscular **tongue** covered with mucous membrane and bears small outgrowths or papillae, some of which contain **taste buds** or organs of taste.

Each side of the buccal cavity receives **saliva**, secretions from four **salivary glands**

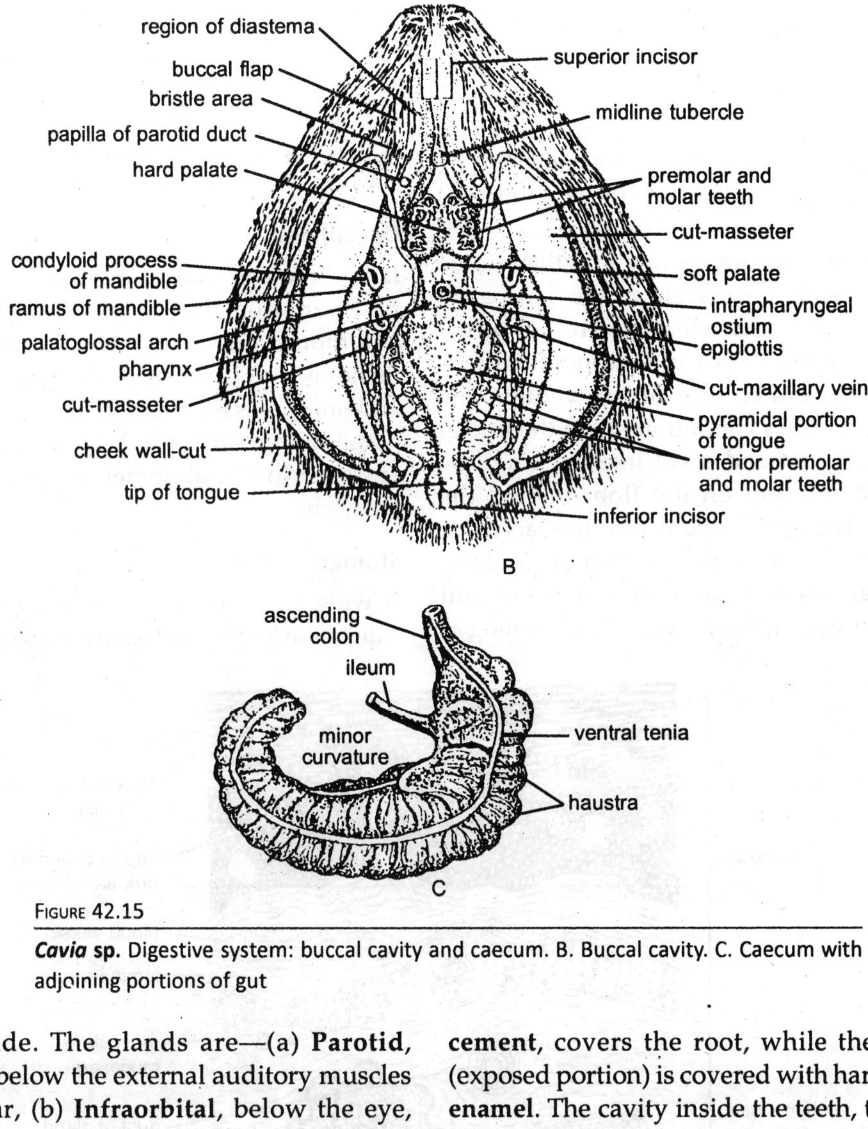

FIGURE 42.15

***Cavia* sp.** Digestive system: buccal cavity and caecum. B. Buccal cavity. C. Caecum with adjoining portions of gut

of the side. The glands are—(a) **Parotid**, situated below the external auditory muscles of the ear, (b) **Infraorbital**, below the eye, (c) **Sub-maxillary**, at the angle of the mandible, and (d) **Sublingual** on the inner side of the lower jaw.

The buccal cavity contains **teeth**, which are provided with roots, implanted deep in the sockets of the jaws, the **thecodont dentition**.

The sets of teeth—the **milk set** and the **permanent set** are present. At an early stage of life the milk set is replaced by the permanent set which continues throughout life, the **diphyodont** dentition. The teeth are of different types, the **heterodont dentition**. The teeth are made of **dentine**. A substance, **cement**, covers the root, while the crown (exposed portion) is covered with hard, shiny **enamel**. The cavity inside the teeth, the pulp cavity is filled up with jelly-like pulp, blood vessels and nerves. Similar teeth are present in the upper and lower jaws.

The teeth on each side of a jaw are—one curved, chisel-shaped, cutting tooth, the **incisor** and four cheek teeth or grinding teeth with transverse ridges on the upper surface. Of these four grinding teeth, the first one is **premolar** and the rest three are **molars**. The toothless gap in the jaw between incisor and premolar is **diastema**. The dental formula is

$$\frac{1.0.1.3}{1.0.1.3}.$$

Pharynx

It is posterior to the buccal cavity and divided into two parts—dorsal or nasal division, the **nasopharynx** and a ventral or buccal division, the **buccopharynx**. In front, the nasopharynx is continuous with the nose through internal nostrils and on each side bears a small **eustachian aperture** leading into the middle ear. The posterior margin of the soft palate extends into the nasopharynx and forms the **velum**. **Tonsil**, a mass of lymphoid tissue is present on either side of the velum. The buccopharynx is continuous with the buccal cavity in front and a slit-like opening, the **glottis** is present on the floor behind the tongue. The glottis opens into the **larynx** or voice box. A movable, thin flap of cartilage, the **epiglottis** overhangs the glottis and shut the latter during swallowing. The opening of the oesophagus or the **gullet** is placed dorsal to the glottis.

During swallowing, the posterior nasal openings become closed by the soft palate, the epiglottis protects the trachea or windpipe and the food pass into the oesophagus.

Oesophagus

It is a narrow elastic muscular tube, runs backward from the pharynx through the neck and thorax, dorsal to the larynx and trachea in the neck and to heart and lungs in the thorax and finally passing through an aperture in the diaphragm reach the adbdomen. Just after its entrance in the abdomen it opens into the stomach.

Stomach

A wide, bent, muscular sac lying on the left side in the abdominal cavity, close to the pos-

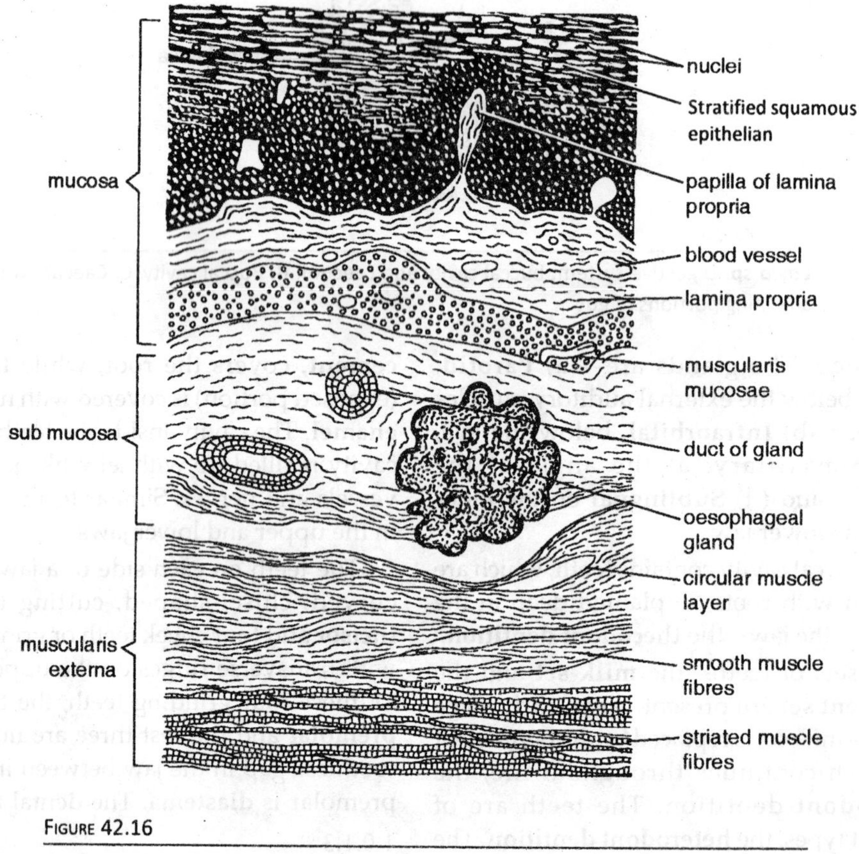

FIGURE 42.16

***Cavia* sp. Oesophagus.** Mammalia. Transverse section

terior face of the diaphragm. The concave anterior border is known as **lesser curvature** and the convex posterior border, the **greater curvature**. The anterior wider portion receiving the oesophagus is called **cardiac end** and the posterior narrower portion giving out the small intestine is **pyloric end**. The mucous membrane of the stomach bears a large number of glands secreting mucus and gastric juice. A constriction, **pyloric constriction**, is present at the junction of the stomach and small intestine.

Small intestine

An elongated, narrow, coiled tube divided into two portions. The first portion, just after its origin, runs anteriorly to form an U-shaped loop with the stomach. It is known as **duodenum**. The rest of the much coiled small intestine is **ileum**, the loops being held together by mesentery. Posteriorly, the small intestine leads into the large intestine. At the junction of the small and large intestine a wide blunt diverticulum, the **caecum**, is present. The caecum (Fig. 42.15C) is marked with spiral constrictions externally, indicating the presence of a narrow spiral valve inside.

Large intestine

A wide, long tube. The first and major part is **colon**, the inner wall of which is sacculated. The smooth-walled narrow posterior portion is termed **rectum**. It opens in the anus.

The inner wall of the whole alimentary canal is provided with longitudinal folds; the sizes and structures of the folds vary in different regions.

The alimentary canal is kept in position by sheets of tissues, the **mesenteries** attached to the dorsal wall of the coelom.

Alimentary canal (microanatomy)

Four distinct layers constitute the wall of the alimentary canal.

Oesophagus (Fig. 42.16). **External fibrous coat**, a narrow layer of loose connective tissue. **Muscular coat** (muscularis externa) consists of an outer longitudinal and an inner circular

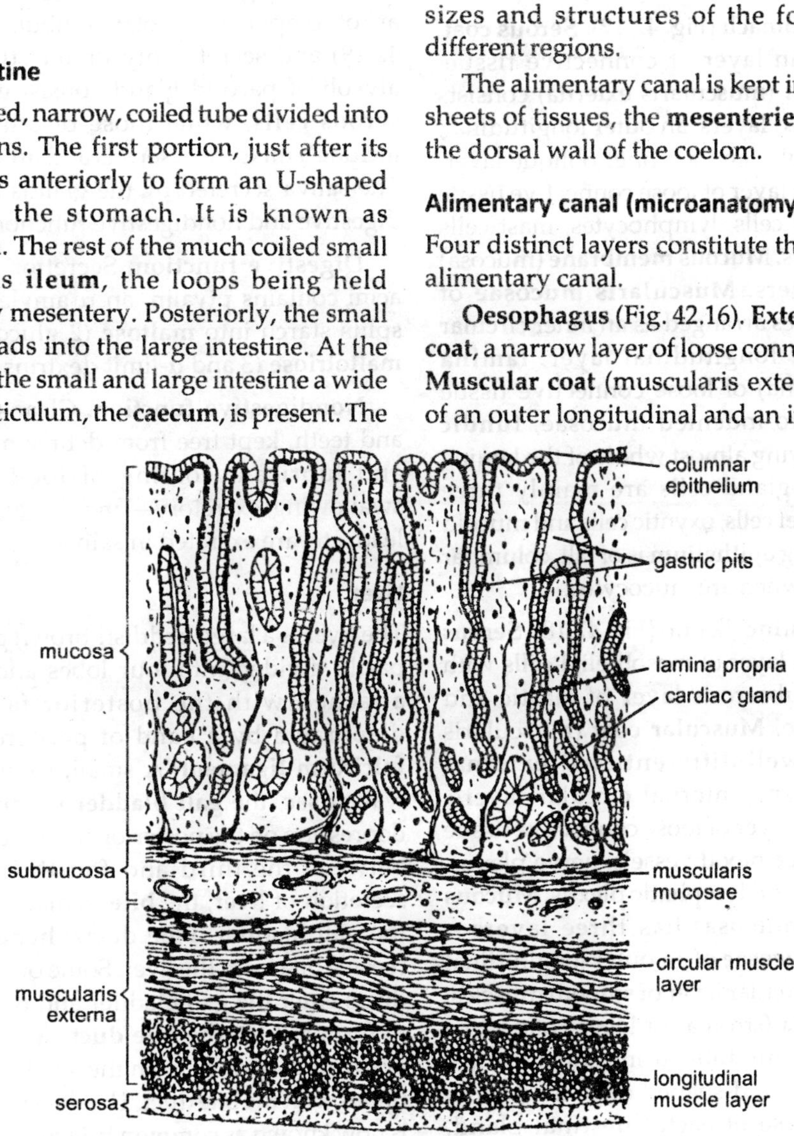

FIGURE 42.17

***Cavia* sp., Stomach.** Cardiac region. Mammalia. Transverse section layer of smooth fibres containing many bundles of oblique fibres. **Submucosa**, loose connective tissue containing large blood vessels and lymphatics. **Mucous membrane** (mucosa) has three layers. **Muscularis mucosae**, a thin layer of smooth muscles with fibres arranged longitudinally; **lamina propria** (stroma) or loose connective tisue, which projects into the indented epithelial layer as long papillae; **nonkeratinized stratified epithelium**, indented on its deep surface.

Cardiac stomach (Fig. 42.17). **Serous coat**, (serosa) a thin layer of connective tissue. **Muscular coat** (muscularis externa) consists of ill-developed layers, an outer longitudinal, a middle circular and an inner oblique layer. **Submucosa**, a layer of loose connective tissue containing fat cells, lymphocytes, mast cells and lymphatics. **Mucous membrane** (mucosa) has three layers. **Muscularis mucosae** of smooth muscles arranged as an inner circular and an outer longitudinal layer; **lamina propria** (stroma) of loose connective tissue projecting into indented mucosae; **fundic glands** occupying almost whole of the lamina propria. The gland cells are mainly three types—the chief cells, oxyntic cells and mucoid cells. The **lining epithelium** is of tall, columnar cells, most of whch are mucocytes.

Small intestine (ileum) [Fig. 42.18]. **Serous coat** (serosa) a layer of mesothelial cells with underlying subserous layer of elastic and collagen tissue. **Muscular coat** (muscularis mucosa) of well-differentiated external longitudinal and internal circular layers. **Submucosa**, a layer of loose connective tissue containing large blood vessels, nerve plexus, adipose tissue and lymphatic plexus. **Mucous membrane** (mucosa) has three layers—**muscularis mucosae** of an outer longitudinal and an inner circular layer of smooth muscles; **lamina propria** (stroma) of loose connective tissue packed with tubular intestinal glands; a group of pyramid-shaped cells, the paneth cells at the base of each intestinal gland. Muscularis mucosae and stroma together are thrown up into minute projections, the **villi**. **Epithelium** covers the villi and consists of single layer of columnar cells containing many goblet cells.

☒ Digestive glands

The large glands associated with feeding and digestion are salivary glands, liver and pancreas.

Salivary glands

The salivary glands are four pairs. The glands are of compound alveolar or tubular type (Fig. 42.19) and secrete only under stimuli. The alveoli of parotid gland consist entirely of **serous cells**, while those of other glands include a mixture of serous and **mucous cells**. The saliva secreted by the glands have both digestive and nondigestive function.

Digestive function. Secretion of serous acini contains ptyalin, an α amylase, which splits starch into **maltose** (2 glucose units), **maltotriose** (3 and α-limit dextrins 5-9).

Nondigestive function. Cleanses mouth and teeth, kept free from debris; moistening and lubricant coating of food to assist swallowing; excretory—urea, sugar, mercury, lead, etc. are excreted in saliva.

Liver

The liver is a large, reddish-brown gland. It is partly divided into four lobes and remains attached with the posterior face of the diaphragm by a band of peritoneum, the **falciform ligament**. A small, rounded thin-walled sac, the **gall bladder** is situated in a depression on the posterior surface of the liver between the third and fourth lobes. The secretion of liver, the **bile**, remains stored in the gall bladder. Fine ducts, **hepatic ducts** receive bile from the liver. Some of the hepatic ducts enter the gall bladder and others unite to form a duct, the **bile duct**. A duct, called **cystic duct** springs from the gall bladder and unites with the bile duct. The duct thus formed is now known as **common bile duct** and opens into the anterior end of the duodenum near the pyloric end of the stomach.

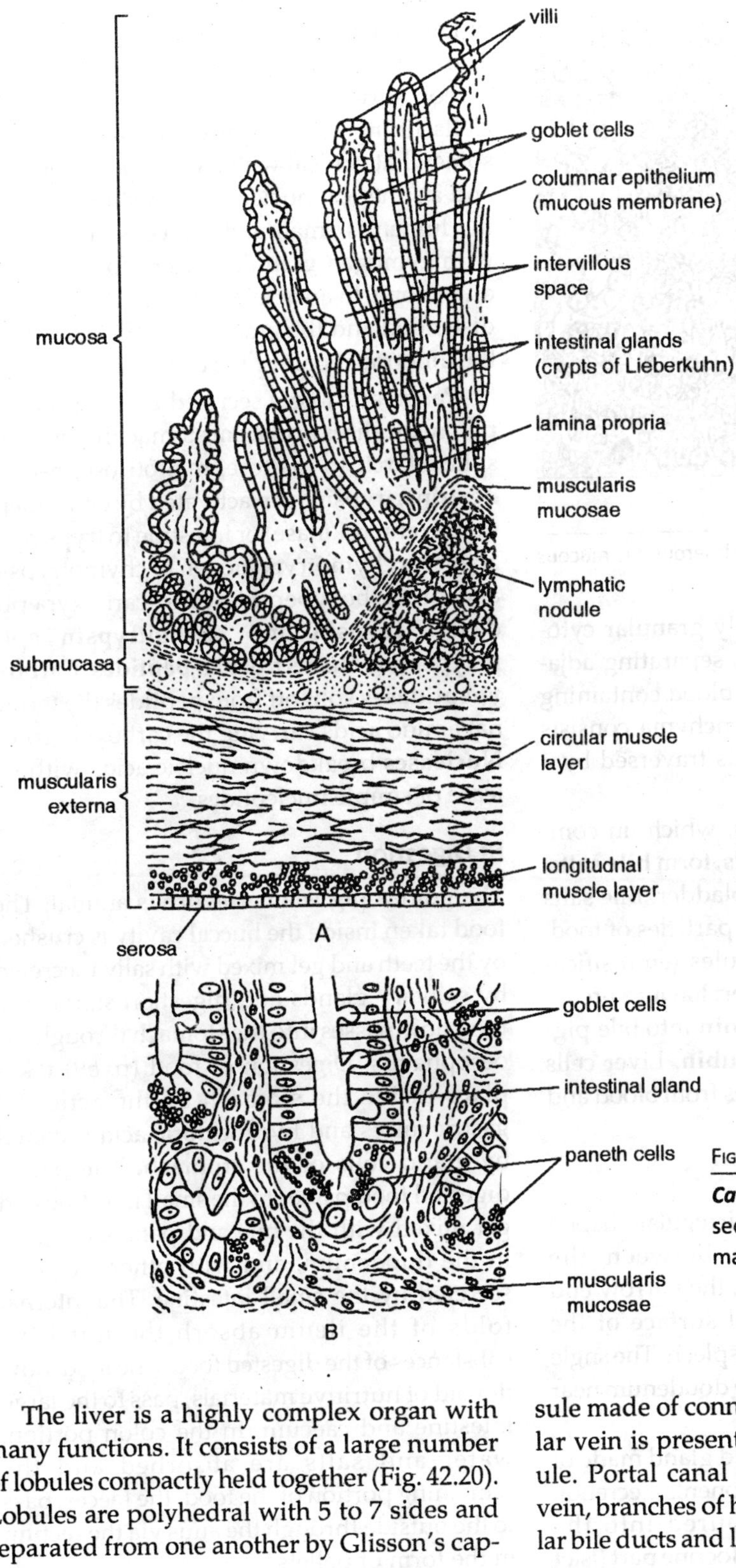

FIGURE 42.18

***Cavia* sp.** Ileum. A. Transverse section, B. A part of A, further magnified

The liver is a highly complex organ with many functions. It consists of a large number of lobules compactly held together (Fig. 42.20). Lobules are polyhedral with 5 to 7 sides and separated from one another by Glisson's capsule made of connective tissue. An intralobular vein is present in the middle of each lobule. Portal canal consists of an interlobular vein, branches of hepatic artery, the interlobular bile ducts and lymphatic vessels. Liver cells

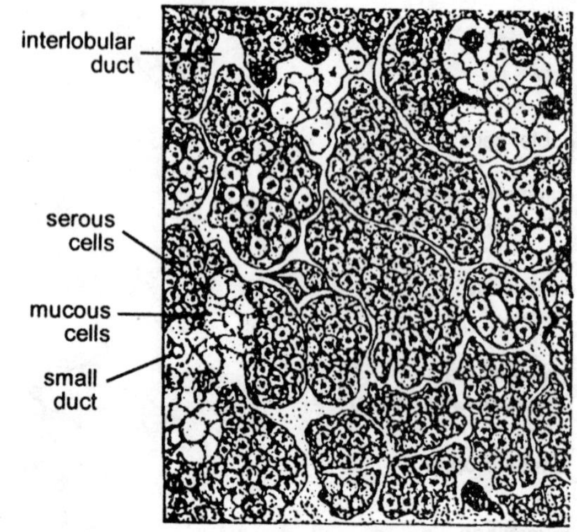

FIGURE 42.19

***Cavia* sp.** Submandibular gland with serous and mucous acini. Section

are polyhedral with coarsely granular cytoplasm or empty. The plates separating adjacent lacunae are formed by blood containing sinusoids. The hepatic parenchyma consists of a continuous mass of cells traversed by a complex system of lacunae.

Liver secretes **bile** acids, which, in combination with other chemicals, form **bile salts**. Bile is concentrated in gall bladder. Bile salts have detergent action on fat particles of food, breaking it into small globules (emulsification). Liver stores iron, convert **haem**; a breakdown product of **haemoglobin** into bile pigments, **biliverdin** and **bilirubin**. Liver cells can take up certain substances from blood and excrete those in the bile.

Pancreas

The pancreas is an elongated, irregular-shaped gland on the mesentery between the duodenum and the stomach, the narrow end extending under the dorsal surface of the stomach and ending near the spleen. The sngle pancreatic duct opens into the doudenum near its posterior extremity.

The pancreas is a combined gland made of exocrine and endocrine components. Secretion of the exocrine part is poured into the duodenum and that of the endocrine part (islet of Langerhans) passes into the blood stream. The exocrine part consists mostly of tubular alveoli surrounded by a lamina (Fig. 42.21). Cells lining the lumen of the alveolus are somewhat conical with a granular cytoplasm and a spherical nucleus. In a resting alveolus the lumen is small and the cells are packed with zymogen granules except for a narrow outer zone. In an active alveolus, the lumen is distended and the granules restricted only to the innermost part of the cell.

Pancreatic juices secreted by the exocrine part neutralises acid from stomach; α **amylase** splits starch to maltose, maltotriose, α-limit dextrins; trypsinogen activated by enteropeptidase (enterokinase) or intestine to trypsin → activates chymotrypsinogen to chymotrypsin and procarboxypeptidases to carboxypeptidases. **Trypsin** and **chymotrypsin** split polypeptide chains to small **peptides. Carboxypeptidases** split off from peptides the terminal amino acids carrying the carboxyl group. **Nucleases** breakdown nucleic acids with release of simple nucleotides.

⊠ Digestion

The guinea pig is a herbivorous animal. The food taken inside the buccal cavity is crushed by the teeth and get mixed with saliva secreted by salivary glands and digestion starts. The food now passes into the stomach through the oesophagus. Digestion of food (p. 59) takes place inside the stomach by the action of gastric juices and hydrochloric acid secreted by gland cells of the stomach. The partly digested food passes to the small intestine and get mixed with the bile and pancreatic juice in the duodenum, and with other enzymes secreted by the small intestine. The internal folds of the ileum absorb the nutritive substances of the digested food. The food, now devoid of nutritive materials, pass to the large intestine and caecum. In the colon portion, water and salts are absorbed and the remaining portion of the food, the faeces, pass to the outside through the anus via the rectum in the form of pellets.

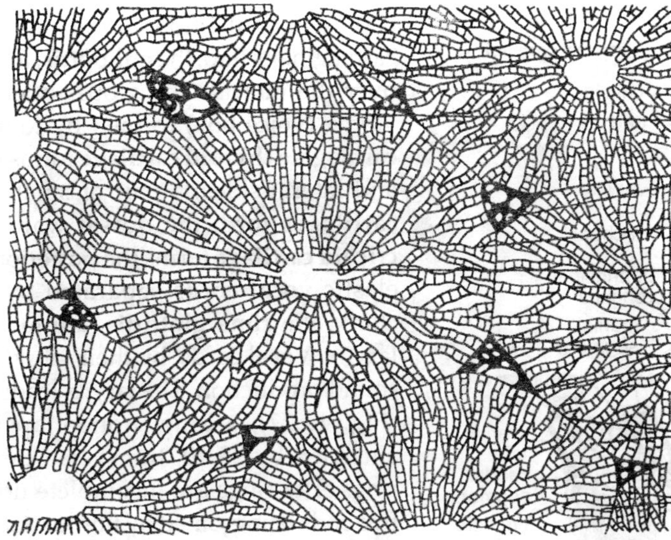

FIGURE 42.20

***Cavia* sp., Liver.** Section

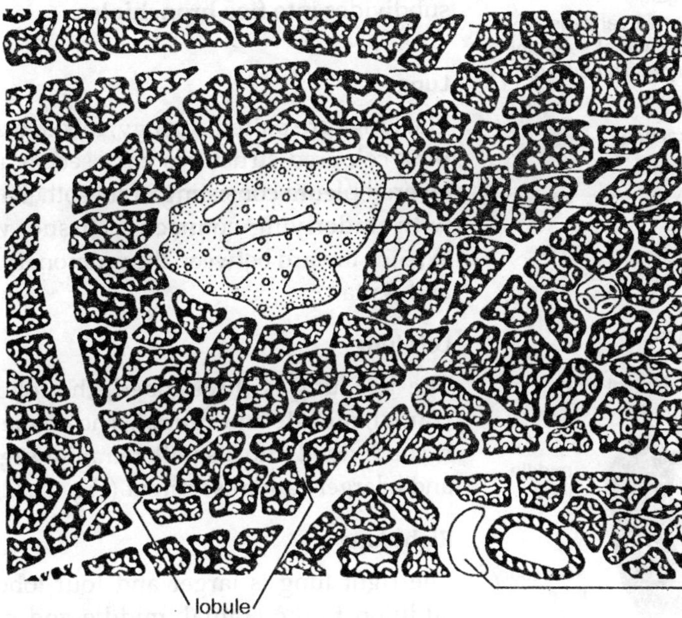

FIGURE 42.21

***Cavia* sp. Pancreas.** Section

☒ Respiratory system

The lungs are respiratory structures, in communication with the exterior through a number of passages. **External nares** are paired openings at the tip of the snout leading to paired **nasal chambers**, separated by a **nasal septum**. The nasal cavities are separated from the buccal cavity by the palate and open into the **nasopharynx** through **internal nares**. **Glottis**, a small opening for the passage of air,

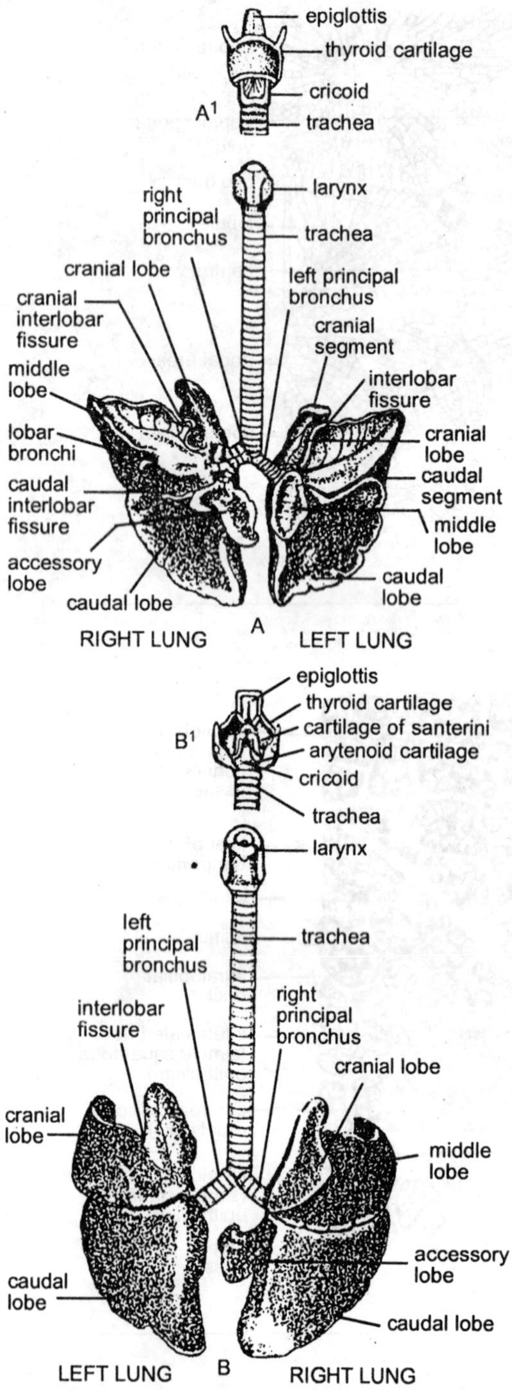

FIGURE 42.22

Cavia **sp. and Oryctolagus sp.** (A, B). Larynx and respiratory organs with a part of trachea, Ventral view (A), Dorsal view (B), Larynx dorsal view (A^1), Ventral view (B^1)

is present on the floor of the buccal cavity. A cartilaginous flap, the **epiglottis**, forms a cover over the glottis. The glottis opens into the larynx (voice box) formed by four pieces of cartilage, a **thyroid cartilage** forms the ventral and lateral sides; a pair of small **aretynoid cartilages** form the dorsal wall, a circular **cricoid cartilage** forms the posterior part. Two elastic **vocal cords** separated by a narrow chink, the **rima glottidis**, are lodged in the larynx. Vocal cords, by their vibraton, help in sound production. The larynx leads to **trachea** (Fig. 42.22), a long, noncollapsible tube, encircled by 35-40, transverse cartilaginous rings. The rings are incomplete dorsally, the free ends being joined by fibrous and muscular tissue. The trachea runs midventrally through the neck and, on entering the thorax, bifurcates into two bronchi. Each **bronchus** enters into the lung of the corresponding side, divides and subdivides into fine **bronchioles**.

Lungs

The lungs are paired, spongy, lobed, brightly pink coloured structures enclosed in a two-layered pleura consisting of mesothelial cells with underlying connective tissue, which extends inward between the pulmonary lobes and septa.

Left lung

The left lung is three-lobed, the lobes are **cranial, middle** and **caudal**. The cranial lobe is subdivided into a smaller **cranial segment** and a larger **caudal segment** (Fig. 42.22).

Right lung

The right lung is larger and four-lobed. In addition to the cranial, middle and caudal lobes, it has an extra **azygos lobe** (Fig. 42.22).

Within the lung the bronchioles divide into finer branches, 2-11 **alveolar ducts**, each bearing several dilated culs-de-sacs called **alveoli**. The alveoli are lined by a single layer of flattened epithelial cells, identified only in those places where nuclei occur (Fig. 42.23). The alveolar cells are about 200 μm thick and with a dual basement membrane separating them from the endothelial cells of the lung capillaries. Small fenestrae, the **alveolar pores**, are scattered on the alveolar wall. Gaseous exchange takes place across the alveolar wall. The capillaries are formed by the pulmonary vein on one side and the pulmonary artery on the other.

Mechanism of respiration

During **inspiration**, the chest muscles contract and the diaphragm is lowered, resulting in an expansion of the thoracic cavity. The air pressure falls and the air is sucked into the lungs from outside through external nares → nasal cavities → glottis → larynx → trachea → bronchi → bronchioles → alveoli. After gaseous exchange in the lungs the **expiration** takes place. During expiration the chest muscles relax and the diaphragm is raised, resulting in a contraction of the thoracic cavity. The air pressure increases and the air is forced out of the lungs through the same route. The lungs are also of elastic nature and greatly help in expiration. A certain amount of residual air always remain in the lungs (p. 78).

⊠ Blood vascular system

The blood vascular system consists of a heart, arterial and venous system.

Heart

The heart is a muscular, reddish-brown structure, situated ventrally in the thoracic cavity, slightly left to the middle line, in the space between the pleural sacs enclosing the lungs. The base of the heart is formed by two atria directed forward and the apex by the **ventricles directed backward** (Fig. 42.24). The heart is enclosed in a two-layered pericardium, the outer **parietal wall** and the inner **visceral**, investing the heart. A narrow cavity enclosed by the two layers contains a little lubricating pericardial fluid. The heart is four-chambered, the cavities of the right and left sides are completely separated from one another by **interatrial** and **interventricular** partitions. The walls of both the atrias and ventricles are thick, the latter being thicker. The muscles of the wall, **cardiac muscles**, are involuntary in function but histologically resemble voluntary (skeletal) muscles (striated). Unlike skeletal muscles they bear **anastomosis**.

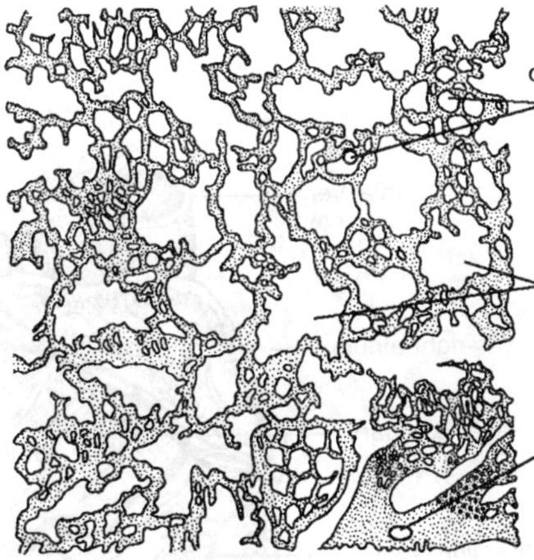

FIGURE 42.23

***Cavia* sp., Lung.** Mammalia. Transverse section

Three **sulcus** or grooves are present in the heart. The transverse groove demarcating the atria and the ventricle is **coronary sulcus**. Two **interventricular grooves**, one dorsal and one ventral are located between the two ventricles.

The sinus venosus is absent. A small node, the **sinuatrial** (SA) **node** or pacemaker, is present in the wall of the atrium. The heart beat originates in the SA node. The excitation wave is conducted through the **atrioventricular** (AV) **node** → **atrioventricular** (AV) **bundle** and **Purkinje network**.

Atria

The walls of the atria are thick. Numerous cords of muscle fibres, **musculi pectinati**, arise from the inner surface of the atria.

The **right atrium** is larger than the left. It receives deoxygenated blood from different parts of the body. The **interatrial septum**, at its right atrial surface, bears a small, oval depression, the **fossa ovalis**, where the partition is thinner than the rest. This marks the position of the **foramen ovale**, that occurs

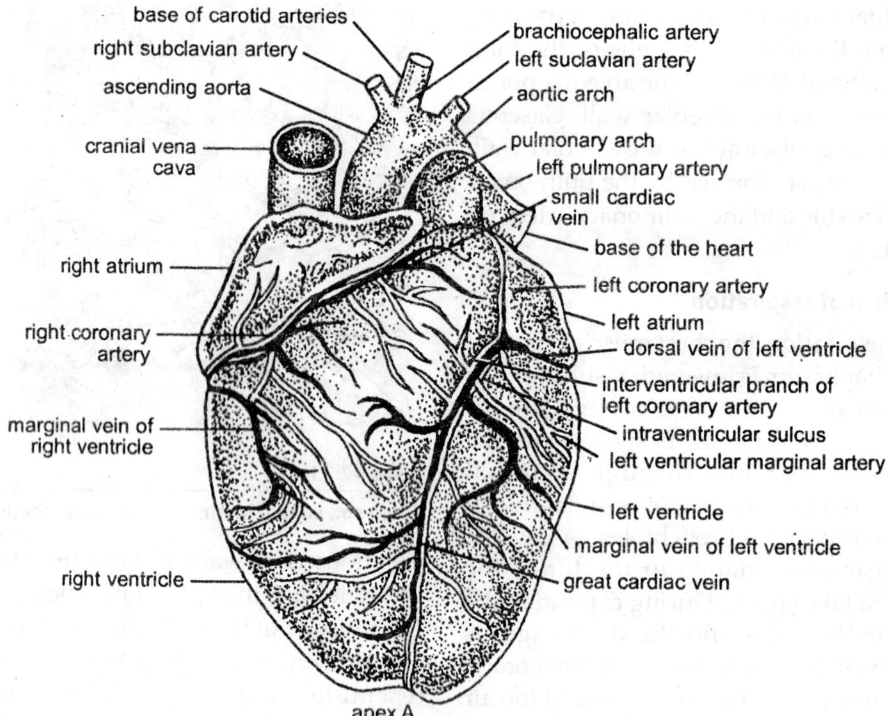

FIGURE 42.24A

***Cavia* sp.** Heart. A. Ventral view. B. Dorsal view. C. Internal structure

in the foetus. The atrium receives three vessels—**coronary sinus** draining blood from heart and opening in the posterodorsal aspect; **cranial vena cava** draining blood from the anterior part and opening in the craniodorsal aspect; **caudal vena cava** bringing blood from the middle and posterior part of the body and opening in the caudodorsal aspect. The **atrioventricular opening** communicates the cavity of the atrium with that of the ventricle.

The opening of the cranial vena cava is bounded by a crescentic fold, the valve of **Thebesius**. The right atrioventricular opening is guarded by **tricuspid valve**, formed by three delicate and transparent cusps. According to their position they are named **ventral angular cusp**, **dorsal parietal cusp** and **medial septal cusp** (Fig. 42.24C). The bases of the valves are continuous with the border of the atrioventricular opening and the free ends project into the opening. The cusps are held in position by **chordae tendineae** and **papillary muscles**.

The **left atrium** is smaller than the right, receive oxygenated blood from the lungs through a large **pulmonary vein** opening on its dorsal surface. The left **atrioventricular opening** is guarded by a **mitral valve** (bicuspid) consisting of two membranous cusps with chordae tendineae and musculi papillaris.

Ventricles

The **right ventricle** is much thicker than the atrium and forms the right side of the conical apical portion of the heart but does not extend up to the apex. The walls are raised up into muscular ridges, **columnae carnae**. The **pulmonary artery**, the base of which is guarded by three pouch-like **semilunar valves** arises from the left anterior angle.

The **left ventricle** is larger than the right one, the wall is thick, columnae carnae more developed than that of the right ventricle. The **aortic arch** (left) arises from the basal end, being guarded by three semilunar valves.

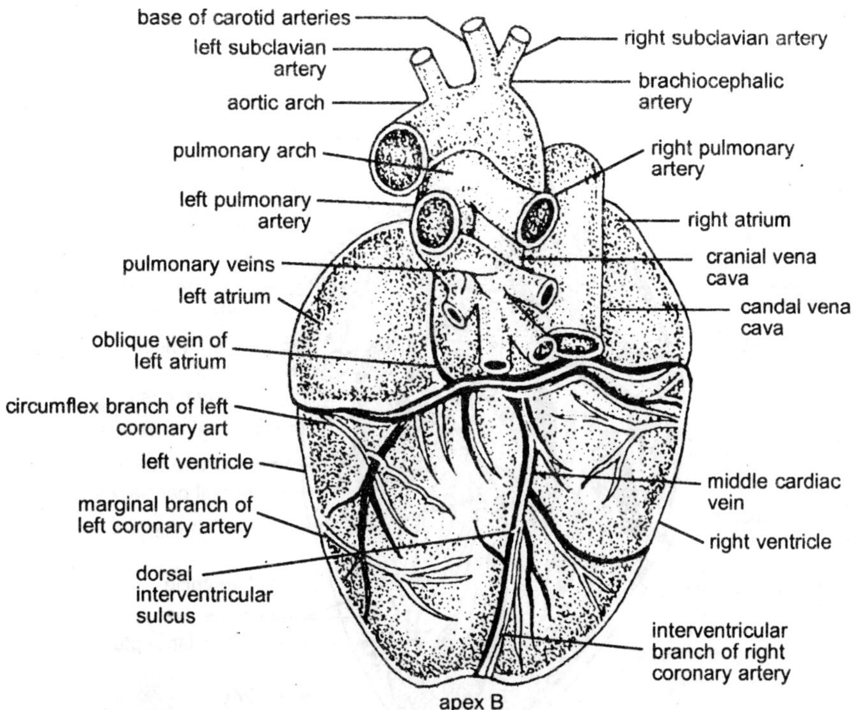

FIGURE 42.24B

***Cavia* sp.** Heart (continued)

☒ Circulation through heart

The heart contracts and relaxes rhythmi-cally. One **systole** (contraction) and one **diastole** (relaxation) constitute one heart beat.

The right atrium receives deoxygenated blood from heart and anterior part of the body through coronary sinus and cranial vena cava, respectively. Blood from the middle and posterior part is received through caudal vena cava. The left atrium receives oxygenated blood from the lungs.

Systole starts at SA node. The atria contract simultaneously driving their contents to the respective ventricles through atrioventricular openings. Backflow of blood is prevented by atrioventricular valves (tricuspid in the right and bicuspid in left). The contraction of the atria is immediately followed by the contraction of the ventricles. Deoxygenated blood from right ventricle is forced into the pulmonary trunk and from there to the lungs; oxygenated blood is driven into aortic arch (systemic aorta) and thence to different parts of the body. The backflow of blood is prevented by semilunar valves. Diastole follows systole and the atria are refilled with blood.

The great veins do not have valves guarding their entrance into the heart. The thickening and contraction of muscles around their mouth prevent backflow of blood from heart.

The heart being four-chambered, no mixing of deoxygenated and oxygenated blood is allowed in the heart and the blood pumped to body is fully charged with oxygen.*

* Body temperature and efficiency of the tissues increase as a result of increased oxidation and the animals like mammals and birds can maintain a constant body temperature to a certain degree, irrespective of the atomospheric temperature

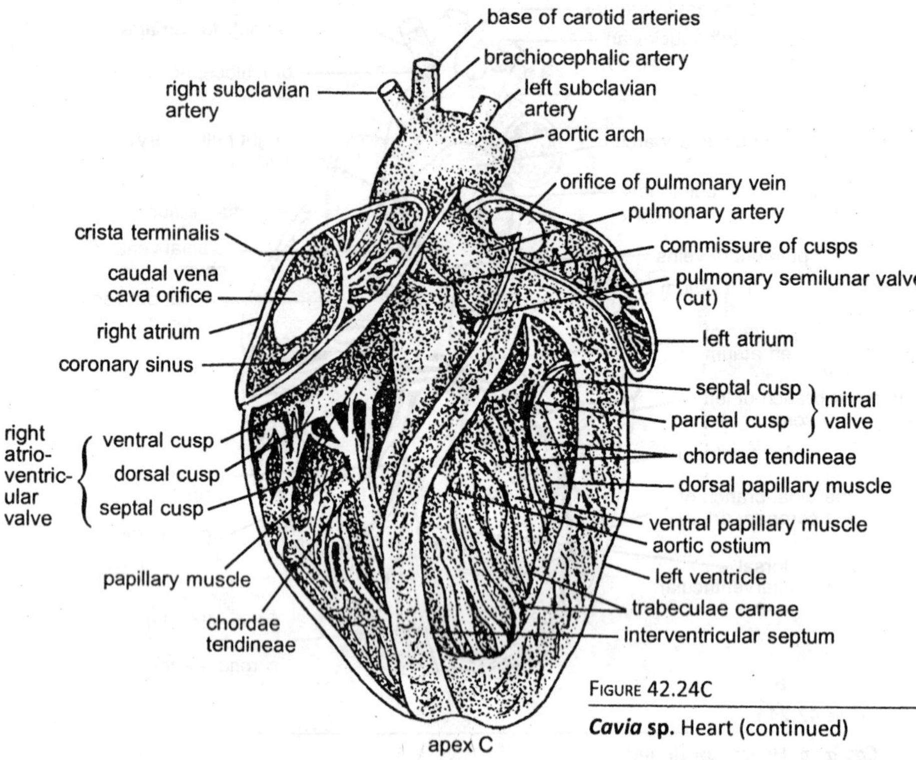

FIGURE 42.24C

Cavia sp. Heart (continued)

☒ Arterial system

Two main vessels—the aortic trunk and pulmonary trunk and their branches constitute arterial system (Fig. 42.25).

Aortic trunk (left)

The aortic trunk (right is absent) arises from the base of the left ventricle (Fig. 42.25A). It turns left, forms an arch round the left bronchus and runs posterodorsally and medially as dorsal aorta along the middorsal line of the thorax in close contact with and ventral to the vertebral column. From the thorax the dorsal aorta passes to the abdomen through an aperture in the diaphragm. The aortic arch has been divided into three portions. The basal part going anteriorly is the **ascending aorta**, the next part curving dorsally, the **arch of the aorta** (aortic arch) and the portion turning medially to reach the middorsal line of the body, the **descending aorta**, which runs backward and named **dorsal aorta**.

Coronary arteries

Two or three small vessels arise close to the base of the ascending aorta and supply blood to the heart.

Brachiocephalic artery

A short but wide vessel, the first large branch of the aortic arch. Through a number of branches it sends blood to the head, neck and right forelimb. The brachiocephalic artery divides into a lateral branch, the right subclavian artery and a medial short, stout branch, forming the base of the right and left common carotids.

Right subclavian

Arising from the brachiocephalic artery, it runs outward between the first rib and clavicle and sends the following branches:

Vertebral courses forward and enters cranial cavity. It sends branches to the paravertebral musculature and spinal cord.

Caudal thyroid. It runs forward obliquely to the left, divides into right and left caudal thyroid, which proceed anteriorly by the sides of the trachea and supply the thyroid gland.

Superficial cervical parallel to external jugular vein and supplies muscles of the neck region.

Internal thoracic supplies sternal region.

Costocervical supplies neck and upper part of the thoracic wall and the diaphragm through **musculophrenic** branch.

Axillary. The portion of the artery beyond the origin of costocervical up to the brachial artery. It sends the following branches:

a. **External thoracic** supplies pectoral and trapezius muscles and axillary lymph nodes.
b. **Lateral thoracic** supplies pectoralis profundus and cutaneous trunk muscles through a number of branches.
c. **Subscapular** supplies upper arm, scapulla and thorax.

Brachial continues in the extremity of the forelimb and divides into a number of branches.

Common carotids

The right and left common carotids arise from a common base.

Right common carotid. It runs to the head lying parallel and right to the trachea. Sending a branch to the lymph node, the artery divides into two, the internal and external carotids and supplies the structures of the right side in the anterior region.

Internal carotid. Lateral in position, enters the skull and supplies blood to the brain and deep structures of the head.

External carotid. Medial and sends blood to head, tongue, jaw, parotid gland, pharynx, oesophagus and thyroid gland through a number of branches.

Left common carotid. It proceeds forward, lying left to the trachea and supplies structures of the left side of the anterior region. Its branches and course of the branches are similar to those of the right common carotid.

Left subclavian

Arises directly from the aortic arch at a short distance from the brachiocephalic artery. Branches are similar to those of the right subclavian, except the caudal thyroid, which it lacks.

Dorsal aorta*

The Aortic arch curves medially and runs posteromedially as dorsal aorta sending branches to the organs of the middle and posterior part of the body (Fig. 42.25B).

Intercostals. A few pairs of small arteries supplying the thoracic region.

The dorsal aorta pierces the diaphragm and enters the abdominal cavity. The branches in the adbominal region are:

Coeliac. A large, unpaired vessel—sends branches to oesophagus, stomach, spleen and pancreas (**gastropancreatic**), transverse and proximal descending colon (**middle colic**), liver (**hepatic**), major part of intestine, caecum, ascending, transverse and descending colon (**cranial mesenteric**) and other associated structures.

Renal. Two pairs. The anterior pair supply kidneys and send branches to (a) **adrenal** (adrenal arteries), and (b) to diaphragm (**caudal phrenic** arteries). The posterior pair ends in the caudal end of the kidney.

Genital. Paired, supply gonads and their ducts. In male, the **testicular** sends branches to the testis and in female the **ovarian**, which is a short vessel, supplies the ovary. The testicular arteries are long, run posteroventrally to supply testes and associated structures (Fig. 42.26) lodged in the scrotum.

---

* Some authors prefer zonations of the dorsal aorta. (a) **Thoracic aorta** from the aortic arch at the level of the second and third thoracic vertebrae to the diaphragm. (b) **Abdominal aorta** from the termination of the thoracic aorta to the bifurcation to two common iliacs. The term '**dorsal aorta**' is retained in the description.

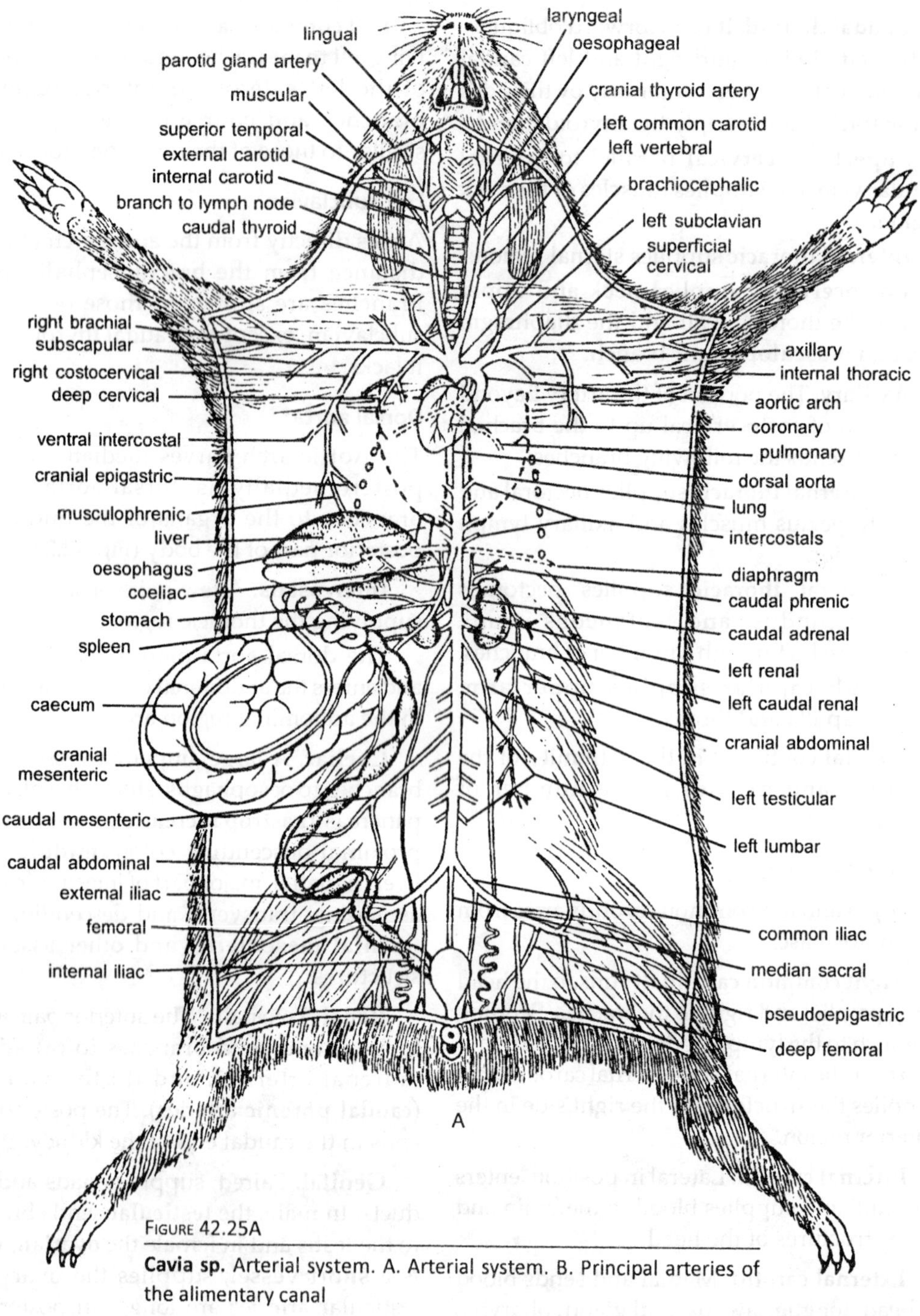

FIGURE 42.25A

Cavia sp. Arterial system. A. Arterial system. B. Principal arteries of the alimentary canal

Lumbar. A few paired arteries posterior to the genitals supply dorsomedian muscles of the lumbar region.

Caudal mesenteric. Unpaired, arises from the dorsal aorta in between the second and third pairs of lumbars and sends branches to descending colon, rectum and anal region.

Median sacral. A slender vessel arises at the junction of the two common iliacs and sends blood to the sacral and coccygeal region.

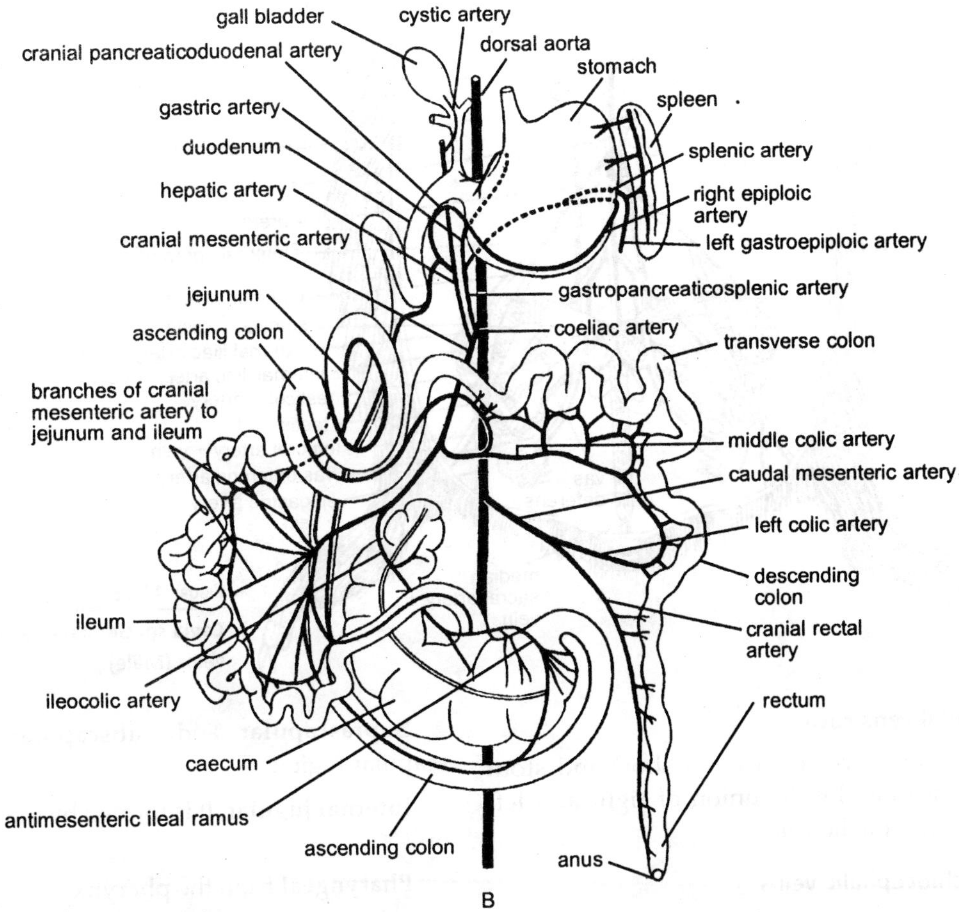

FIGURE 42.25B

***Cavia* sp.** *Arterial system.* (Continued)

Common iliac. Paired, the right and left common iliacs are formed by the bifurcation of the dorsal aorta. Each divides into two major branches. In addition, a few small branches arise from the main vessel to supply the adjacent region.

External iliac. Supplies abdominal wall and urinogenital organs through **pseudoepigastric** branch and continues as **femoral** artery in the hindlimb.

Caudal abdominal. Two. arise from the external iliacs and supply the structures in the posterior abdomen.

Internal iliac. It runs backward and supplies the pelvic region and urinary bladder.

Pulmonary trunk

Soon after its origin from the right ventricle it courses towards the dorsal side and bifurcates into two. The right pulmonary artery passes dorsal to the aortic arch, divides into four branches and supply four lobes of the right lung. The left pulmonary artery passes ventral to the aortic arch, divides into three branches and supply three lobes of the left lung.

⊠ Venous system

The cranial and caudal vena cava open into the right atrium, the former in the craniocaudal and the latter in the caudodorsal region (Fig. 42.27). Only hepatic portal system is present.

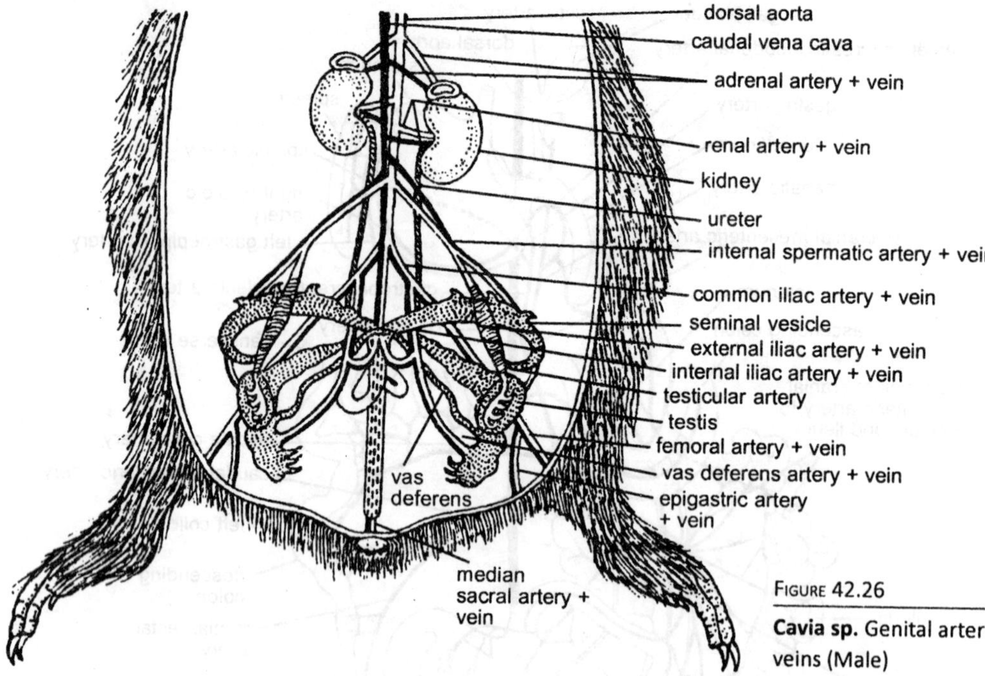

FIGURE 42.26

Cavia sp. Genital arteries and veins (Male)

Cranial vena cava

The cranial vena cava is a short and stout vein formed by the union of right and left brachiocephalic veins.

Brachiocephalic veins

Two brachiocephalic veins bring back blood from the head and anterior region of the body. A brachiocephalic vein is formed by the joining of external jugular, internal jugular and subclavian. The jugular veins bring back blood from the head, cervical and scapular region and subclavian drains blood from the thoracic region and forelimb of the side.

External jugular

It is formed by the union of following vessels.

Linguofacial formed by the joining of:

a. **Lingual** bringing blood from the tongue and adjacent region.
b. **Facial** draining blood from the facial region.

Maxillary from maxillary region.

Deep cervical from deep pectoral muscles.

Suprascapular and **subscapular** from scapular region.

Internal jugular. It is formed by the union of:

Pharyngeal from the pharynx.

Oesophageal and tracheal. A number of small veins from oesophagus and trachea.

A stout vein formed by the joining of **occipital, muscular** and **vertebral** vein opens in the internal jugular vein at about its middle.

Posteriorly, the external and internal jugular veins unite and open into the brachiocephalic vein of the side.

Right subclavian. It is formed by the joining of:

Brachial. Drains blood from the extremity of the forelimb through a number of small veins. It continues inward as axillary vein.

Axillary. It receives:

Costocervical vein from the upper part of the thoracic wall, deep neck muscles and diaphragm.

First intercostal from anterior intercostal space.

Supreme intercostal drains blood from the first two intercostal spaces.

Left subclavian. It is similar to the right subclavian.

The **cranial vena cava** directly receives:

Vertebral. A small vein bringing back blood from the vertebral column.

Internal thoracic draining blood from the fifth and sixth intercostal spaces.

Azygos. Unpaired, formed by the union of **right intercostal** from the lower part of the thoracic wall and **phrenic** from the diaphragm*.

Caudal vena cava

It is a large median vein formed by the joining of a number of veins bringing blood from the organs in the posterior and middle region of the body. The caudal vena cava is formed by the joining of two common iliacs—right and left, and the median sacral vein.

Median sacral. A slender vein from the sacral and coccygeal region joins the caudal vena cava at the junction of the two common iliacs.

External iliac formed by the union of:

a. **Femoral** from the hindlimb.
b. **Pseudoepigastric** from abdominal muscles and urinogenital organs.

Internal iliac from the urinary bladder and pelvic region.

The two iliacs join to form a **common iliac**.

Caudal abdominal. Single. Drains blood from the posterior region of the abdomen and rectum and joins the common iliac of the right side.

In its forward course the caudal vena cava receives the following veins:

Lumbar. Six pairs from the lumbar region.

Genital. Paired, arise from the testes (**testicular**) or ovaries (**ovarian**) and their associated structures. In male, the testicular veins are long, since the testes are lodged in the scrotum (Fig. 42.26). Anteriorly each opens in the renal vein of the side. In female, the ovarian veins are short.

Renal. Paired, drain kidney, formed by the union of two veins. It receives a small **adrenal vein** from the adrenal body.

Caudal phrenic. Paired, drain blood from the diaphragm.

Cranial abdominal vein. Paired, drain blood from the abdominal region. The veins open either in the caudal vena cava or in the renal veins.

Hepatic. Unpaired, formed by the union of four or five branches from the lobes of the liver.

The caudal vena cava pierces the diaphragm and proceeds forward following a middle course in the thoracic cavity. On reaching the apex of the heart it forms a shallow arch to the right and opens in the right atrium.

Hepatic portal system

A number of veins draining blood from the majority of the visceral organs join to form the large **hepatic portal vein**, which branches in the lobes of the liver.

Cystic vein from the gall bladder.

Cranial mesenteric. Largest and most important vein of the system. It receives blood from distal portion of duodenum, jejunum, ileum, caecum and pancreas.

Caudal mesenteric from the rectum.

Splenic from spleen, distal pancreas, stomach and caudal portion of oesophagus.

Gastroduodenal from pylorus, proximal duodenum, great curvature of stomach and pancreas.

---

* In some cases the azygos veins are paired.

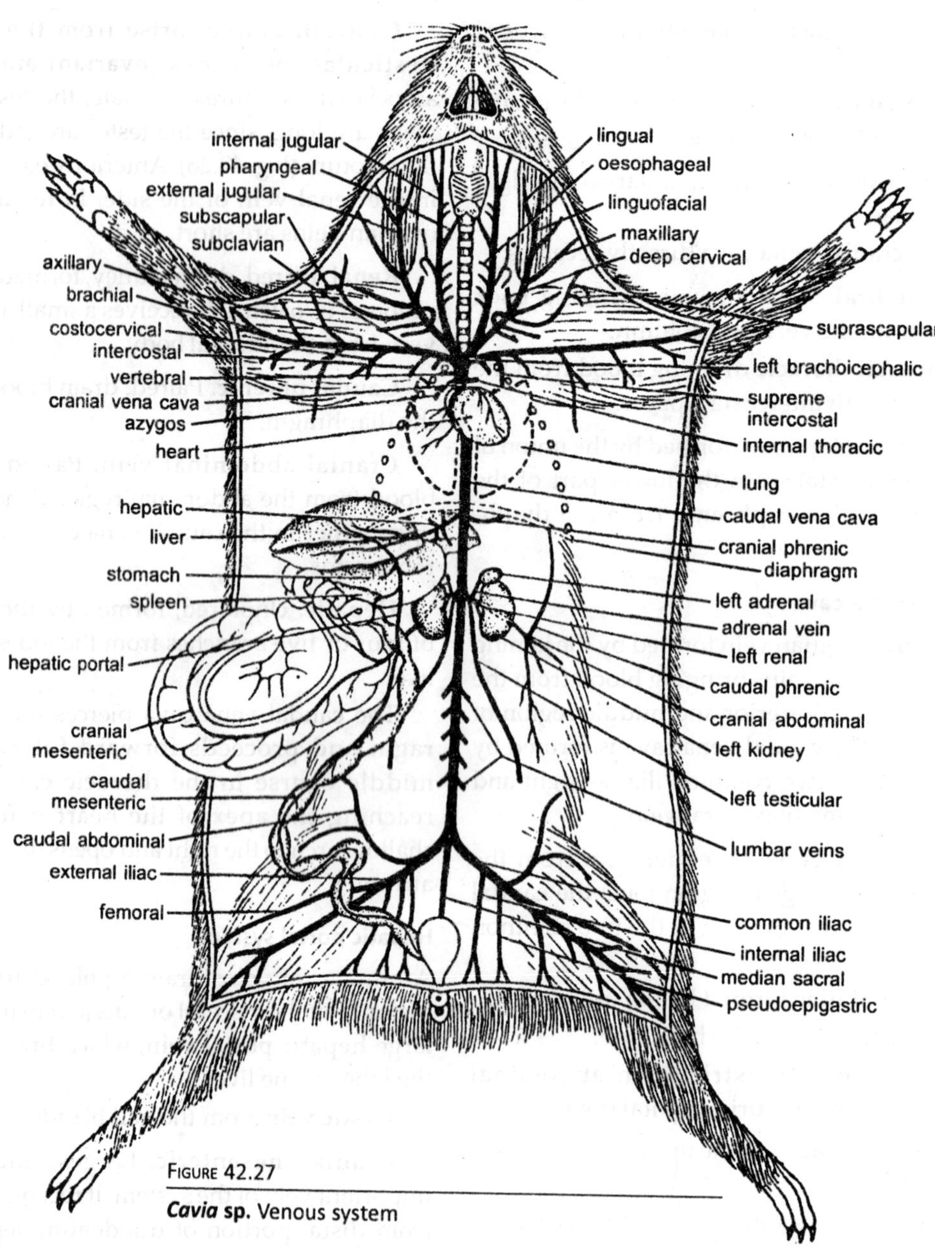

FIGURE 42.27

***Cavia* sp.** Venous system

From the liver a single hepatic vein carries blood to the caudal vena cava.

Pulmonary veins

Blood is brought back by a single vein from the right lung and two veins, which join to form one, from the left lung. The two join to form a common pulmonary vein and opens in the left atrium.

⊠ Blood

Blood is red due to presence of haemo-globin in **erythrocytes** (RBC). The red blood corpuscles are circular, biconcave and nonnucleated (p. 62-64).

⊠ Lymphatic System

Lymph (p. 64, 102) is a clear watery fluid containing a few leucocytes. Slender lymphatic vessels are associated with veins of every organ. The vessels ultimately unite to form two large trunks—the thoracic duct and right lymphatic duct. The **thoracic duct** on the left side of the dorsal aorta joins the left subclavian vein at its junction with the jugular vein. The **right lymphatic duct**, on the right side of the dorsal aorta joins the right subclavian vein.

⊠ Spleen

The spleen is a ductless gland but it does not secrete hormone. Rather it is concerned with blood. The spleen is a dark red, elongated body attached to the dorsal surface of the stomach by a fold of mesentery.

A single layer of mesothelial cells covers the capsule enclosing the spleen. The capsule is made of a dense connective tissue with elastic and smooth muscle fibres. **Lymphoid follicles** separated by splenic pulp; the white and red pulps are scattered throughout the spleen (Fig. 42.28). The **white pulp** is fusiform or cylindrical in shape; consists of reticular fibres, reticular cells and lymphocytes; large type lymphocytes in dividing stages are often present in a germinal centre. The **red pulp** is a reticular network, honeycombed by sinuses; nongranular leucocytes are numerous; splenic cells (macrophages) present; granular leucocytes and erythrocytes found in the meshes of stroma.

Blood is filtered in red pulp. Plasma cells, which secrete antibodies, are derived from the T and B lymphocytes of white pulp. Destroy abnormal red blood cells and other (foreign) particles.

⊠ Nervous System

The nervous system is basically similar to that of other vertebrates (p. 126-130). It consists of central nervous system and peripheral nervous system. The autonomic nervous system is a part of the peripheral nervous system.

⊠ Central Nervous System

The brain and spinal cord constitute central nervous system.

Brain

The brain has undergone marked specialization and the horizontal disposition of the five parts of the brain is lost (Fig. 42.29). Due to large size of the cerebrum and cerebellum some of the components of the dorsal surface of the brain are obliterated from view.

Meninges or membranes covering the brain and spinal cord have three layers—the tough outer **dura mater**, the highly vascular inner **pia mater** and a spongy layer, the **arachnoid**, in between the two.

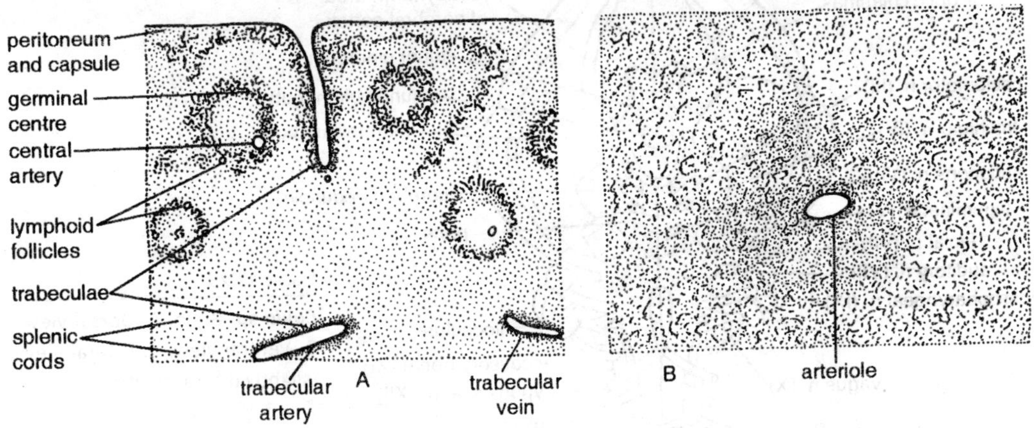

FIGURE 42.28

***Cavia* sp. Spleen.** Transverse section. A. Peripheral zone, B. A part of A, further magnified

Prosencephalon

The anterior part of the brain is formed by a pair of small, club-shaped **olfactory lobes** attached to the cerebral hemispheres. The two **cerebral hemispheres** (cerebrum) constituting **telencephalon** are large, narrower anteriorly and wider posteriorly. A deep groove, the **median fissure** demarcates the two. A slight depression, the **Sylvian fissure**, at the side of each hemisphere, separates off a lateral portion into a **temporal lobe**. The rest of the hemisphere is divided into a **frontal lobe** anteriorly, a **parietal lobe** in the middle region and an **occipital lobe** posteriorly.

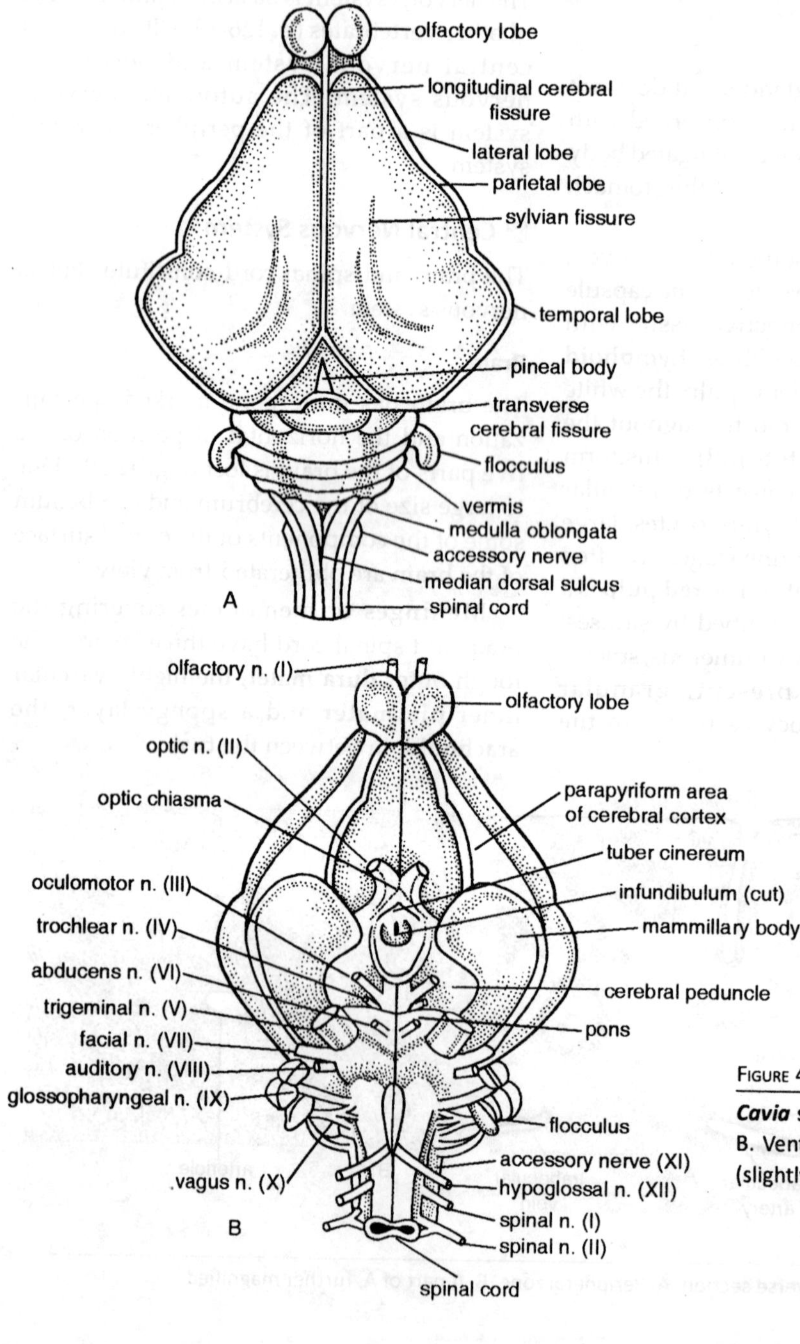

FIGURE 42.29

***Cavia* sp.** Brain. A. Dorsal view. B. Ventral view. C. Lateral view (slightly diagrammatic)

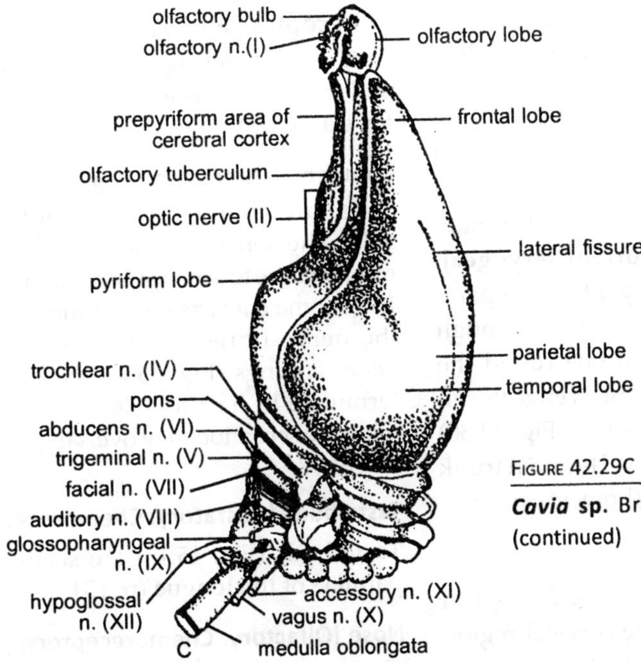

FIGURE 42.29C

***Cavia* sp.** Brain. (lateral view) (continued)

The superficial part of the **cerebral cortex** or **pallium** is formed by nerve cells. The ventral and lateral walls are thickened to form **corpus striatum**. A transverse band of nerve fibres, the **corpus callosum** on the floor of the median fissure join the two hemispheres.

The **diencephalon** is overlapped by cerebral hemispheres on the dorsal surface. The roof is nonnervous and bears **anterior choroid plexus**. Its thick side walls form optic thalami; epiphysis ending in **pineal body** on the dorsal surface. **Optic chiasma** and **hypophysis** (infundibulum) bearing pituitary body are present on the ventral surface. A pair of small, round **mammillary bodies** are located behind the pituitary.

Mesencephalon

The thick roof of **mesencephalon** bears four rounded **optic lobes** or **corpus quadrigemina**. The thickened **corpora peduncles** or **crura cerebri**, ventral to the optic lobes connect forebrain with hindbrain.

Rhombencephalon

The **metencephalon** (cerebellum) is large and subdivided into a median **vermis** bearing transverse grooves and two lateral lobes, each ending in a small **flocculus** bearing transverse folds dorsally. Ventrally, the two halves of the cerebellum are joined by a broad band, **pons varolii.**

The **myelencephalon** or **medulla oblongata** narrows behind and continues posteriorly as spinal cord. The roof of the medulla is nonnervous and bears **posterior choroid plexus.**

Spinal cord

It is basically similar to other tetrapods. A deep longitudinal groove, the **ventral fissure**, runs along the midventral line. The cord is slightly swollen in the brachial and lumbar regions. Posteriorly it ends in a narrow cone, **the conus terminalis**, from which arise a tuft of nerves, forming **filium terminale**.

☒ Peripheral Nervous System

The cranial and spinal nerves constitute peripheral nervous system.

Cranial nerves

Cranial nerves are twelve pairs (p. 127). The

distribution of vagus (X) is a little different from others. It bears a large **vagus ganglion,** immediately after its emergence from the cranium. Close to the ganglion arise two branches, **cardiac depressor** going to heart and **anterior laryngeal** to the larynx. Running posteriorly through the neck region, the main trunk sends a branch, **recurrent laryngeal,** near the thorax. Immediately after its separation from the vagus, the left one runs through the aortic arch and the right one round the subclavian artery. The recurrent laryngeals run forward on either side of trachea (Fig. 42.30) close to it and enter the larynx. The main trunk of the vagus enters into the thoracic cavity.

Spinal nerves

Spinal nerves (p. 128) are 32 pairs. The fourth and fifth spinal nerves of the cervical region of each side unite to form a **phrenic nerve** innervating the diaphragm. The last four cervical and first thoracic of each side unite to form **brachial plexus** to innervate the forelimb of the side. The last two lumbar and the sacral nerves of each side form **sciatic plexus** supplying the hindlimb of the side.

⊠ Receptors and Sense Organs

Cutaneous sense organs, taste buds, nose, eyes and ears constitute **exteroceptors** or receptors for external stimuli.

Cutaneous (Mechanoreceptors)

These are abundantly distributed over the skin. The sensations which stimulate these receptors are touch, heat, cold and pain (p. 135-138). Some consists only of the free ending of the nerve fibres, others are encapsulated nerve endings (Fig. 15.9). In mammals they terminate in hair follicles. Areas concerned with one sensation may overlap others to some extent.

Taste buds (Gustatory: Chemoreceptors)

Taste buds (Fig. 15.15) are scattered on the papillae of the tongue (p. 151).

Nose (Olfactory: Chemoreceptors)

The mucous membrane (Fig. 15.14) of the nasal cavities are concerned with smell (p. 151).

Eyes (Photoreceptors)

Eyes (Fig. 15.17) are similar to those of other tetrapods (p. 151-154). The upper and lower eyelids bear fine, small hairs. Eyelashes are absent.

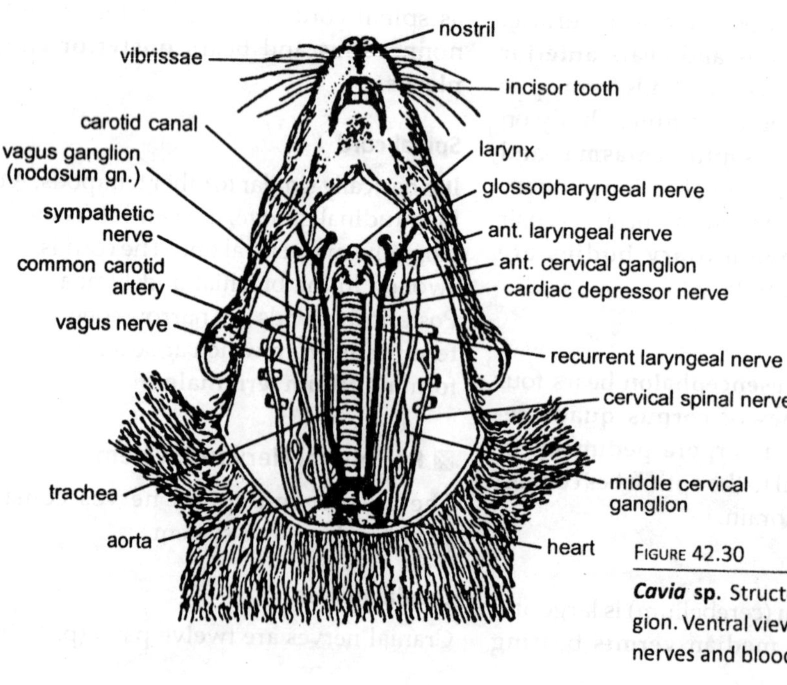

FIGURE 42.30

***Cavia* sp.** Structures in neck region. Ventral view. Distribution of nerves and blood vessels.

Ears (Equilibrium and Phonoreceptors)

The pinnae are flap-like. Structurally (Fig. 15.11) the ears are similar to those of other tetrapods (p. 147, 148).

⊠ Endocrine System

The principal ductless glands constituting the endocrine system the thyroid, parathyroid, thymus, adrenals, islet of Langerhans (pancreas), pituitary and gonads. The hormones secreted by the endocrine (ductless) glands are complex chemicals, specific in action and act on targets, usually at a distance from the source (p. 156, 158).

Thyroid

A bilobed gland situated at the anterior end of the trachea behind the thyroid cartilage and larynx. The gland is enclosed in a two-layered capsule. Septa of fibroelastic tissue extend inwards from the inner capsule and divide the gland into lobules. The glandular tissue consists of groups of irregularly spherical follicles. The follicles are supported by connective tissue carrying blood vessels, lymphocytes and nerves. The follicles are lined by cuboidal cells with spherical nucleus and supported by a thin basement membrane (Fig. 42.31). Microvilli are present on the apex of the cells. Cytoplasm contains granules and droplets. Golgi apparatus well-developed and the apical cytoplasm contains numerous membrane bound dense granules.

Thyroid secretion, **thyroxine** and **tri-iodothyroxine** directly influence the rate of cell metabolism with an increase in oxygen consumption and heat production. Both are also concerned with growth and maturation.

Parathyroid

Two pairs of globular glands closely associated with thyroid. The gland is com-posed of two types of cells, **principal (chief)** and **oxyphil**, arranged in clumps, cords and follicles. The cytoplasm of the principal (chief) cells is completely clear and the nucleus large and

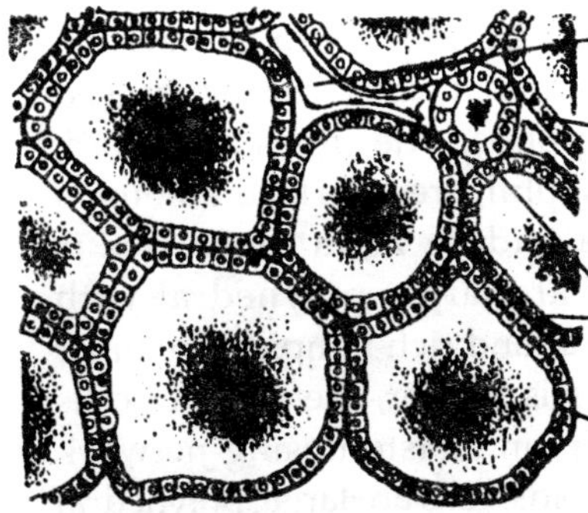

FIGURE 42.31

***Cavia* sp. Thyroid gland.** Transverse section

granular. The oxyphil cells have granular acidophilic cytoplasm and a deeply staining nucleus. A framework of reticular cells and a network of sinusoids are present between the epithelial cells.

The hormones, parathormone, **1, 25-dihydroxycholecalciferol** (1, 25 DHCC) and calcitonin act on kidney and gut to keep blood ionized calcium constant. Parathormone plays important role in calcium and phosphate metabolism.

Thymus

A fairly large, pinkish gland in the thorax enclosing the base of the aorta. It is relatively large in young. The gland closely resembles a lymphoid node. **Nodules** or follicles contain large number of **lymphocytes**, closely packed in the outer cortex; less densely packed in the central medulla where occasional concentric **capsules of Hassal** are found. Some lymphocytes undergo part maturation in the thymus and so they are called **T-lymphocytes**. The thymus is important in immunity.

Adrenal

Oval, yellowish glands attached to the anterior end of the kidneys. The gland is enclosed in a thick capsule of fibrous connective tissue containing a few muscle fibres. The gland consists of a firm, outer cortex and a soft, central medulla.

Adrenal cortex consists of three layers, somewhat vaguely differentiated (Fig. 42.32).

The outer **zona glomerulosa** is narrow, with small, columnar cells closely packed in rounded groups, the nucleus round, cytoplasm basophil, free surface adjoins a capillary. The middle **zona faciculata** is thickest, cells are large, polyhedral with vesicular nucleus and rich in lipid. The inner **zona reticularis** is thicker, the cords of cells are arranged in an anastomosing network round the sinusoids; cells are large, polyhedral with vesicular nucleus and less lipid.

Adrenal medulla consists of polyhedral cells arranged in groups and in contact with venous sinuses; cells are supported by a meshwork of fine fibres and contain small granules.

Adrenal cortex secretes **steroid hormones** derived from **cholesterol**. The hormones are three types. (a) **Mineralo corticoids,** specially **aldesterone**, but also **deoxycorticosterone**, act on kidney tubules and promote retention of Na^+ in exchange for K^+ and H^+. (b) **Gluco-corticoids**, specially **cortisol** but also **corticosterone** causes protein metabolism. (c) **Androgens** and **oestrogen** promotes protein metabolism and growth; have minor effects on reproductive function.

Adrenal medulla secretes **adrenaline** and **noradrenaline**. The hormone adrenaline reinforces action of sympathetic nervous system; noradrenaline causes vasoconstriction and consequent rise in blood pressure.

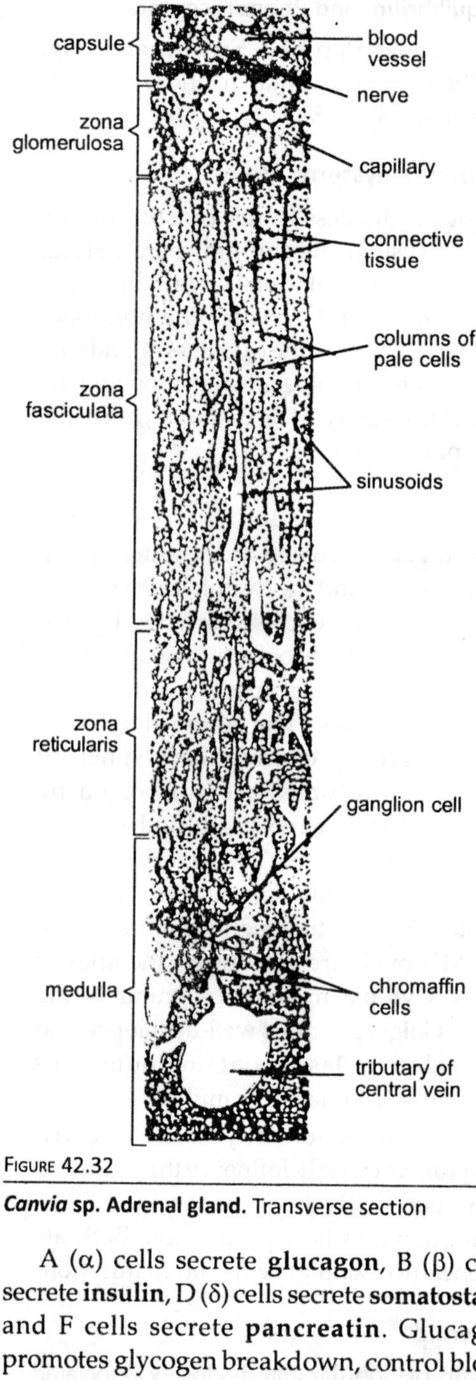

FIGURE 42.32

***Canvia* sp. Adrenal gland.** Transverse section

⊠ Islet of Langerhans

Scattered in the pancreas and constitute 1 to 2% of the pancreatic tissue. An islet of Langerhans consists of solid groups of lightly staining cells marked off from the alveolar portion by a thin reticular membrane (Fig. 42.21). Blood supply is extremely rich. The cells are granulated and of four types. (a) Coarsely granular A (α) cells, less in number; (b) B (β) cells containing much more numerous small granules; (D) (δ) cells and F cells are less granulated*.

A (α) cells secrete **glucagon**, B (β) cells secrete **insulin**, D (δ) cells secrete **somatostatin** and F cells secrete **pancreatin**. Glucagon promotes glycogen breakdown, control blood sugar; insulin enhances uptake and utilization of glucose and amino acids by muscle and glucose by adipose tissue. Control is by negative feedback with blood glucose.

* Functions of D (a) and F cells are not known.

Pituitary

The pituitary gland consists of anterior, intemediate and posterior lobes. The gland is composed chiefly by irregularly branched cords of cells, separated by fine reticular fibres and sinusoids. Occasionally, the cells form acinus-like bodies. The cells are classified by staining reactions of thier cytoplasm.

(a) **Chromophobe cells**. Smaller in size, cytoplasm stains poorly, nuclei clumped more closely together. The cells are clustered near the centre of the chord.

(b) **Chromophil cells**. Cytoplasm stain red (acidophils) or purple blue (basophils); the nuclei of each cell type vary in size and staining intensity.

Pituitary is the master gland of the endocrine system. It regulates the activity of other endocrine glands including the gonads and influences all metabolic activities including growth.

Gonads

Testes and ovaries secrete hormones and are considered as part of the endocrine system.

☒ Urinogenital System

The urinary (excretory) and genital (reproductive) organs are collectively described as urinogenital system. The urinary and genital system in *Cavia* are not completely separate. The interdependence of the two systems is restricted to the urethra, only in male, which acts as a common passage for both urine and sperms.

☒ Urinary (excretory system)

The excretory system (Figs. 42.33, 42.34) consists of a pair of **metanephric** kidneys (p. 118-122), two ureters, urinary bladder and urethra.

Kidneys. Two dark red, bean-shaped bodies with a notch, the **hilus**, on the inner side. The right kidney is slightly anterior to the left one. The kidney is divisible into two distinct zones—the outer **cortex** and inner medulla (Fig. 13.2).

Ureters. The urinary ducts arise from the hilus. They are long, thin, tubular and lie against the dorsal abdominal wall; run backward and open separately on the dorsal surface of the narrow neck of the urinary bladder.

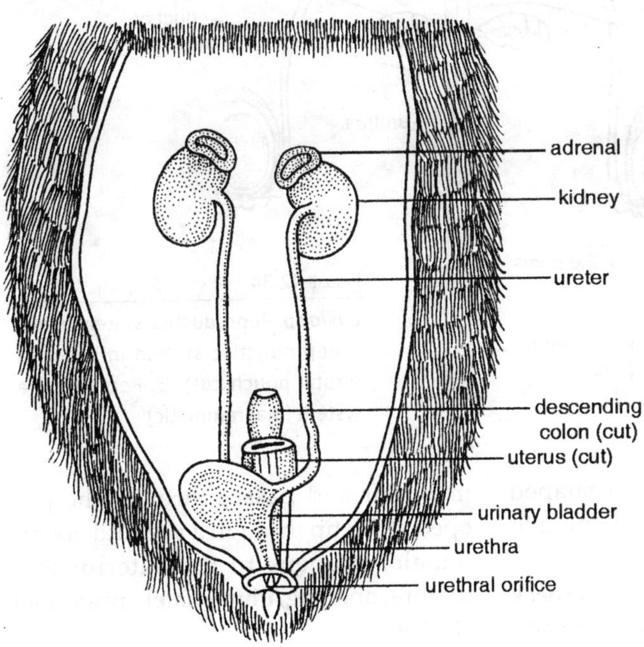

FIGURE 42.33

***Cavia* sp.** Urinary (excretory) system, female. Ventral view

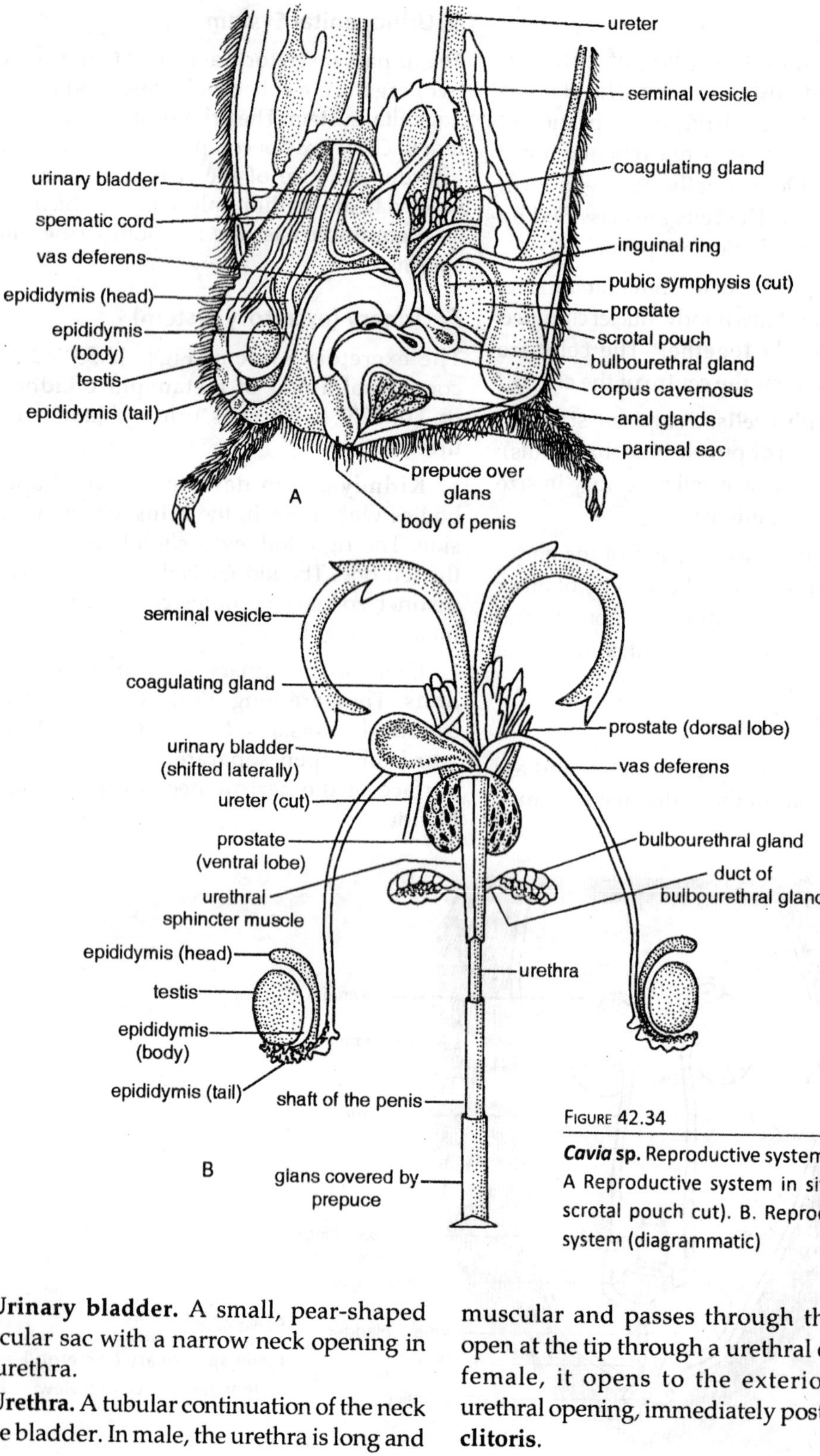

FIGURE 42.34

***Cavia* sp.** Reproductive system, male. A Reproductive system in situ (left scrotal pouch cut). B. Reproductive system (diagrammatic)

Urinary bladder. A small, pear-shaped muscular sac with a narrow neck opening in the urethra.

Urethra. A tubular continuation of the neck of the bladder. In male, the urethra is long and muscular and passes through the penis to open at the tip through a urethral opening. In female, it opens to the exterior through urethral opening, immediately posterior to the **clitoris**.

☒ Reproductive System

The sexes are separate.

Male reproductive system

The paired scrotal pouches are externally visible swellings on either side of the perineum, extending a little behind the posterior end of the body. The testes are lodged in scrotal pouches.

Testes. White, ovoid, paired bodies lodged in the cushion-like muscular portion of the scrotal sacs (Fig. 42.34), being attached to them at their posterior end by a band-like tissue, the **gubernaculum**. The cavity of the scrotum is continuous with the abdominal cavity through **inguinal canals**, which serve as passages for blood vessels and nerves from the abdomen to the testes.

Epididymis. Attached to the inner border of the testis. An epididymis is formed by the convolutions of numerous narrow tubules the **vasa efferentia**.

Vasa deferentia. Paired tubes. The vas deferens arises from the postrior end of the epididymis, passes forward through the inguinal canal and reaches the adbominal cavity. It loops round the ureter and opens in the urinary bladder near the base of the urethra.

Seminal vesicles. A pair of large, curved and sacculated organs associated with the vasa deferentia.

Urethra. A common passage for urine and sperms. It runs through the penis.

Coagulating glands. A pair of fairly large, lobed glands attached to the inner concave surface of the seminal vesicles.

Prostate glands. Two small, lobed bodies. The glands are attached to the junction of the bladder and urethra by stalk.

Penis. A long, median muscular organ, ensheathed distally by a fold of skin, the **prepuce**. The terminal region of the penis is called **glans**.

Cowper's glands. Two small, flask-shaped glands at the base of the penis.

Preputial glands. Flat bodies lying between the prepuce and the skin on either side of the penis.

Female reproductive system

The female reproductive organs (Fig. 42.35) are restricted to the abdominal cavity.

Ovaries. Paired, compact, oval bodies in the abdominal cavity, ventrolateral to the kidneys. They are suspended with dorsal body wall by **mesovarium**.

Oviducts. Two tubes corresponding to the ovaries. The oviduct is differentiated into two distinct zones:

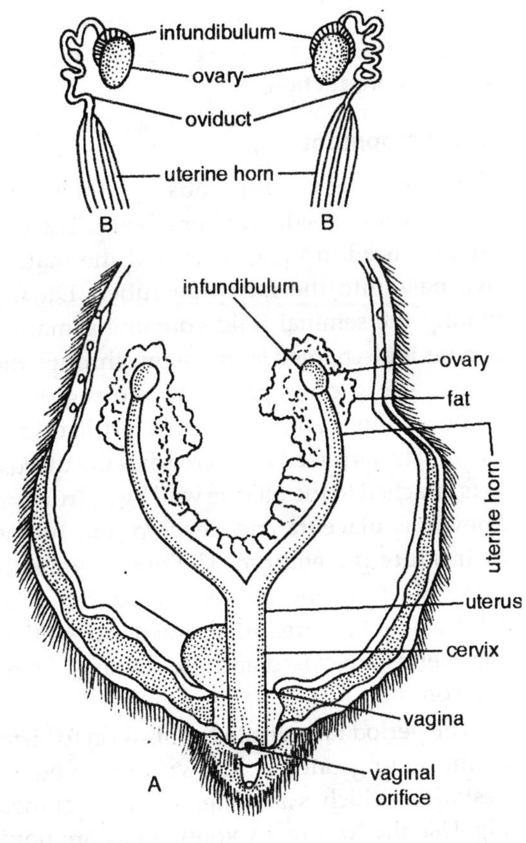

FIGURE 42.35

Cavia **sp.** Reproductive system, female (slightly diagrammatic). A. The reproductive system. B. Anterior part of A.

(a) **Fallopian tube.** The anterior, narrow, looped region opening in the body cavity through a ciliated funnel, facing the ovary.

(b) **Uterus.** The thick-walled, dilated portion following the fallopian tube.

Vagina. The two uteri join to form a median vagina. It is a muscular passage, ventral to the rectum but dorsal to the urethra. It opens to the exterior by **vulva**, immediately posterior to the urethral opening. The vulva includes the clitoris, vaginal orifice and the labia.

Preputial glands. Similar to those in male.

Clitoris. A small, rod-like body anterior to the vulva.

The **mammary glands** provide milk to young but these are really components of reproductive organs, since they do not take part in reproduction.

⊠ Development

The guinea pig is **viviparous**, i.e., the young ones develop inside mother's body. The **ova** are produced in the ovaries and the mature ova pass into the fallopian tubes. During mating the seminal fluid containing mature sperms is discharged in the vagina through the penis. The sperms pass into the region of fallopian tubes and one sperm fertilizes one ovum. The fertilized ovum reaches the uterus, gets attached to the uterine wall by a structure known as **placenta** and develop into **foetus** or **intrauterine embryo**. The young animals come out from the uterus in a condition in which all organs are fully formed, except that the eyelids are closed and the coating of hair is incomplete.

The period taken from fertilisation to birth of the young ones is known as period of **gestation**, which is about ten weeks in guinea pig. Usually four to six young ones are born in a litter and they are capable to pick up their own food within 2 to 3 days. Sexual maturity occurs in about one year. Life span is 5 to 8 years.

DENTITION IN MAMMALIA

The hard structures implanted in the upper and lower jaw for collecting food, capturing prey, cutting and crushing food-stuffs are teeth. Teeth are present in all mammals with rare exceptions. Mammalian dentition has been classified in the following way.

Homodont. Teeth are all alike in shape and size. Examples: All odontocete whales—dolphins, porpasises, etc.

Heterodont. The teeth are of different shape, size and function. Example: Majority of the mammals including human.

Thecodont. The teeth are lodged in the bony sockets of the jaws. (Fig. 42.36). Examples: Majority of the mammals.

Monophyodont. Only one set of teeth develops in the life time. Examples; Marsupials, Odontocete, etc.

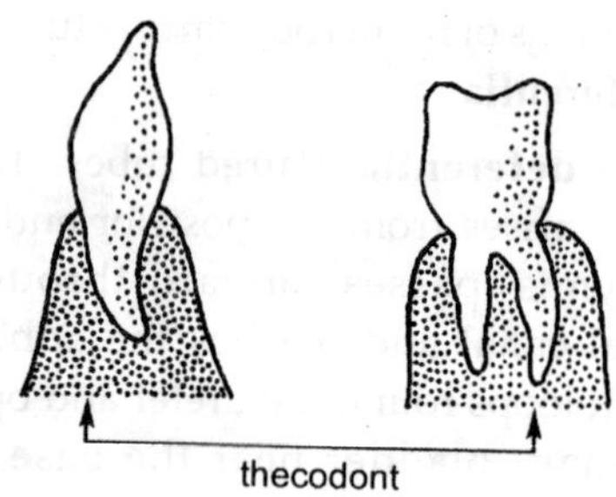

FIGURE 42.36

Mammalia. Teeth. *Thecodont*. diphyodont.

Diphyodont. The first temporary set of teeth called deciduous teeth—milk teeth, lacteal teeth. They are replaced by permanent teeth. Examples: Majority of mammals.

⊠ Occurrence of Teeth

Teeth develop in all mammals. In some cases the teeth are lost afterwards.

Echidna (*Tachyglossus*). Teeth are absent in all stages of life. **Baleen whales** (*Balaena*) and in others of the group the rudimentary embryonic teeth are replaced by baleen in the adult.

⊠ Structure of a mammalian tooth

A mammalian tooth has two parts—**crown** or the portion above the jaw and the **root**—the hidden part in the gum, which is anchored to the socket of the jaw bone. The junction of the crown and the root is called neck. A pulp cavity containing blood vessels, nerves and connective tissue is enclosed in the middle along the length of the tooth (Fig. 42.37). The bony outer part of the tooth is made of enamel, dentine and cement.

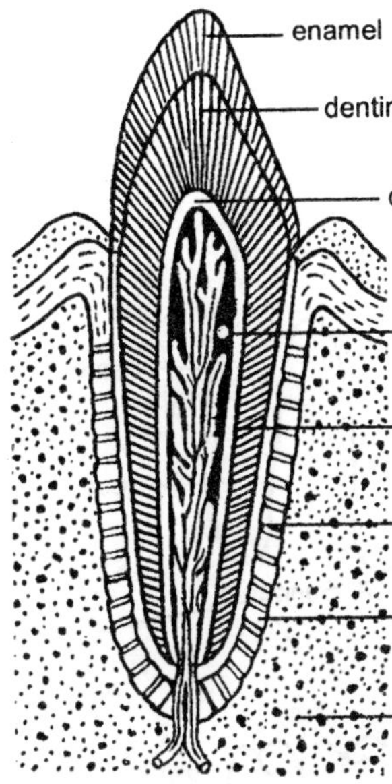

FIGURE 42.37

Mammalian typical tooth (Longitudinal section, diagrammatic

Crown

The crown consists of an outer enamel and central dentine.

Enamel. A thin, very hard glistening layer covering the teeth. It is hardest and heaviest tissue in vertebrates and composed of crystals of hydroxyapatite [$3(Ca_3PO_4)\ Ca(OH)_2$]. It is acellular and ectodermal in origin.

Dentine. A hard dermal bony layer inside enamel. It is harder than bone but softer than enamel, and mesodermal in origin. The dentine continues in the root in between the periodontal membrane and odontoblast.

Root. The root is covered by a thin layer of cement and a vascular periodental membrane.

Cement. A thin layer of nonmuscular bone, usually acellular and softer than dentine. It is rich in collagenous fibres and mesodermal in origin.

Periodontal membrane. It is made of strong connective tissue fibre.

Odontoblast. The pulp cavity is ensheathed by a layer of bone cells—odontoblast.

⊠ Types of teeth

Four types of teeth are found in mammals.

Incisor. Located anteriorly on the premaxilla in upper jaw and tips of dentaries in lower jaw; monocuspid; root single and conical. Function-cutting or cropping food.

Canine. Located next to incisor; one in each half of the jaw; large, pointed, long-crowned; root single. In rodents canines are absent leaving a space—**diastema** between incisor and premolar. Function—piercing and tearing flesh of the prey.

Premolar. Located next to canine; crown bicuspid and roots two. Function—grinding of food. In some carnivores last upper premolar is chisel-shaped with short cusps—the carnassial teeth for cracking bone and shearing flesh and tendon.

Molar. Located next to premolar and last teeth after series. Cusps several; roots two or more. Function—grinding of food. In some carnivores the first lower molar is modified to **carnassial** tooth.

Premolars and molars are collectively called **cheek teeth**. In man the eruption of last molar—**wisdom tooth** is delayed.

⊠ Cusp patterns of cheek teeth in living eutherians (Fig. 42.38)

Bunodont. Cusps separate and rounded, used for grinding food. Examples: Man and some omnivore.

Lophodont. Cusps are joined to form ridges or lophs, enamel and dentine with

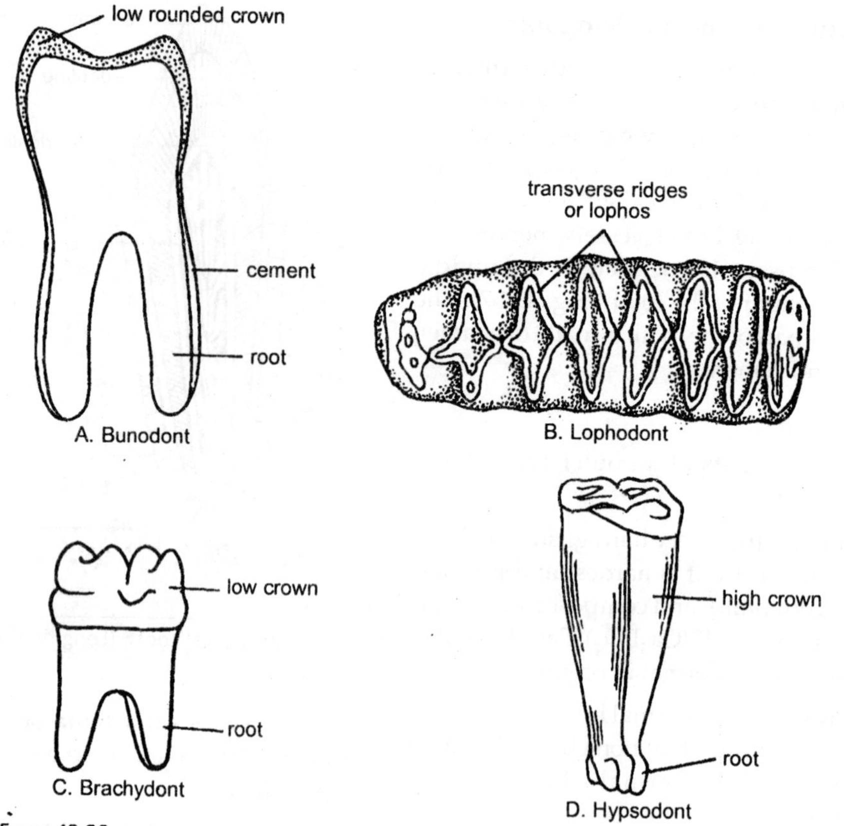

FIGURE 42.38

Mammalian teetch Different types (A-D) Cusp pattern

intricate foldings. Used for grinding plant food and grasses. Examples: Elephant.

Secodont. Crown sharp for cutting, used for cutting and shearing flesh. Examples: Terrestrial carnivores.

Selenodont. Cusps crescent-shaped. Used for grinding coarse plant materials. Example: Horse.

Brachydont. Crown low; roots comparatively long. Used for cutting different types of food. Example: Man.

Hypsodont. Crown high; roots short and open. Used for grinding food. Examples: Horse, elephant, etc.

Dental formula

Mathematical expression of types and number of teeth in a species is known as dental formula.

Eutherian mammals possess 44 teeth, the number varies in some cases.

$$\frac{\text{Half of the upper jaw}}{\text{Half of the lower jaw}} = \frac{\text{i.c.pm.m}}{\text{i.c.pm.m}} : \frac{3.1.4.3}{3.1.4.3}$$

$$= 22 \times 2 = 44$$

Dental formula of some mammals

Metatheria (Kangaroo) $\frac{3.1.2.4}{1.0.2.4} = 34$

(Didelphis) $\frac{5.1.3.4}{4.1.3.4} = 50$

Horse and pig $\frac{3.1.4.3}{3.1.4.3} = 44$

Bat $\frac{2.1.0.4}{3.1.05} = 32$

Sheep, cow, goat $\frac{0.0.3.3}{3.1.3.3} = 32$

Cat $\frac{3.1.3.1}{3.1.2.1} = 30$

Seal $\frac{3.1.4.1}{2.1.4.1} = 34$

Walrus $\frac{1.1.3.0}{0.1.3.0} = 18$

Rat $\frac{1.0.0.3}{1.0.0.3} = 16$

Guinea pig $\frac{1.0.1.3}{1.0.1.3} = 20$

Hare and rabbit $\frac{2.0.3.3}{2.0.2.3} = \frac{8}{7} = 30$

Hyrax $\frac{1.0.4.3}{2.0.4.3} = 34$

Elephant $\frac{1.0.3.3}{0.0.3.3} = 26$

Old world monkey $\frac{2.1.2.3}{2.1.2.3} = 32$

New world monkey $\frac{2.1.3.3}{2.1.3.3} = 36$

Modification of Teeth

Elephant's tusk. Second pair upper incisors
Pig's tusk. Upper canines
Barking deer's tusk. Upper canines
Walrus's tusk. Upper canines.

LIMB BONES: EQUUS (HORSE)

Limbs of horse are greatly modified in response to cursorial adaptation, the most important being elongation due to lengthening of the distal segments, associated with loss of universal movement, which is found in other tetrapods. The restriction of movement of the limbs to one plain only (pendulum-like) has ensured safety to the animal while it is in gallop.

Forelimb

1. The bones of the forelimb consist of a humerus, a radius and an ulna, carpal and metacarpal bones and phalanges (Fig. 42.39).
2. **Humerus.** It has a short, rounded and slightly bent shaft. The proximal end bears a tuberosity and a head to fit into the glenoid cavity. The head is separated from the tuberosity by a bicipital groove. The distal end bears a pulley-like structure and a depression just anterior to it.
3. **Radius and ulna.** It consists of two narrow, elongated, separate bones—the radius and the ulna. The radius is slender and smaller than the ulna and bears a concave facet at the proximal end. The distal end is almost flat. The ulna is long and bears an olecranon process and a sigmoid notch for the humerus.

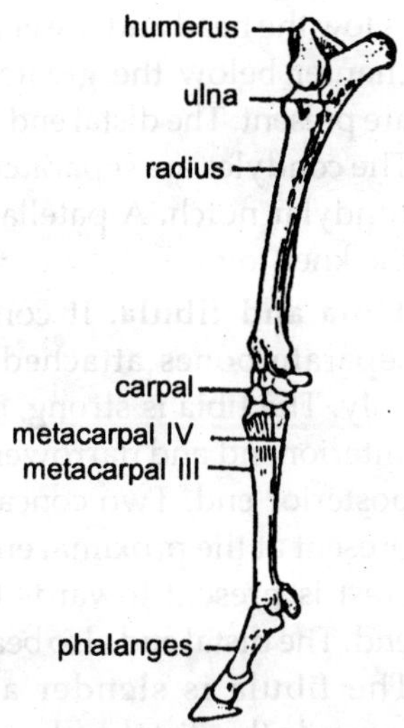

FIGURE 42.39

***Equus* sp. Forelimb bones**

4. **Carpal bones.** These are seven bones of the wrist arranged in two rows. The proximal row contains pisiform, cuneiform, lunar and scaphoid; and the distal row contains trapezoid, magnum and unciform.
5. **Metacarpal bones.** These are the bones of the palm. The first and fifth are completely degenerated the 2nd and 4th are represented by narrow splint bones. The third one is a large, elongated, round bone, known as **cannon bone**.

6. **Phalanges.** These are the bones of the fingers. The 2nd and 4th fingers are represented by small bones, the sesamoids. The third finger is elongated and consists of three phalanges, the last one bearing a hoof.

☒ Hindlimb

1. The bones of the hindlimb consist of a femur, a tibia and fibula, tarsal and metatarsal bones and phalanges (Fig. 48.2D).
2. **Femur.** It has a long, stout, slightly curved shaft. The proximal end bears a prominent head to fit into the acetabulum of the pelvic girdle. Three elevated areas—greater trochanter immediately below the head and lesser and third trochanter below the greater trochanter, are present. The distal end is pulley-like. The condyles are separated by an intercondylar notch. A patella is present at the knee joint.
3. **Tibia and fibula.** It consists of two separate bones attached at the ends only. The tibia is strong, stouter at the anterior end and narrower towards the posterior end. Two concave facets are present at the proximal end. A cnemial crest is present towards the proximal end. The distal end also bears concavity. The fibula is slender and narrows towards the distal end.
4. **Tarsal bones.** These are the bones of the ankle and are six in number. The astragalus has a pulley-like surface above for articulation with the tibia. Its distal surface is flattened and articulates more with the navicular than with the cuboid. The calcaneum does not articulate with the fibula. The fused meso-, ento- and ecto-cuneiform articulate distally with the metatarsals.
5. **Metatarsal bones.** These are the bones of the foot. The first and fifth are completely degenerated. The second and fourth are represented by narrow splint bones. The third one is a large, elongated, round bone known as **cannon bone.**
6. **Phalanges.** These are the bones of the fingers. The 2nd and 4th figners are represented by small bones, the sesamoids. The third finger is elongated and consists of three phalanges, the last one bearing a hoof.

RUMINANT STOMACH

In same animals the food is regurgitated from stomach to mouth cavity for mastication. They are known as **ruminants**. The stomach of ruminants differ structurally from that of the others and named ruminant stomach. Ruminant stomach is found in the members of the group **Ruminentia**.

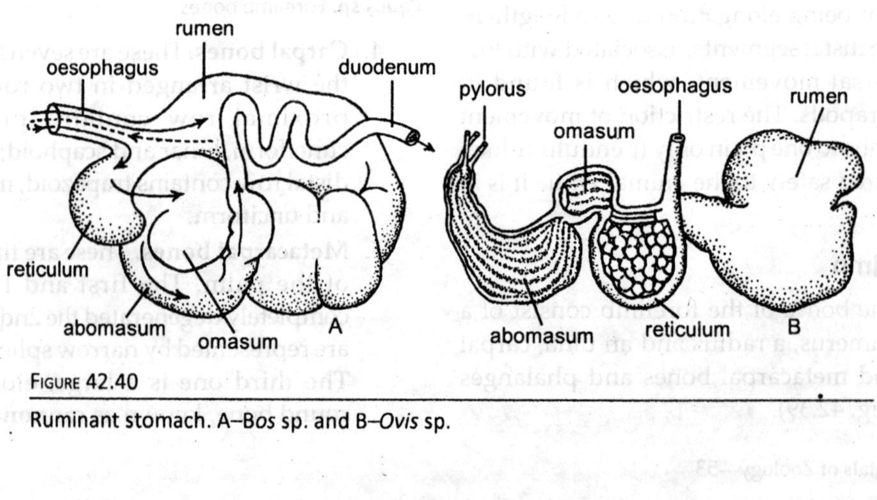

FIGURE 42.40

Ruminant stomach. A–*Bos* sp. and B–*Ovis* sp.

⊠ Structure of Ruminant Stomach

The ruminant stomach has four intercommunicating chambers—**rumen** (paunch), **reticulum** (honeycomb). **omasum** (psalterium or manyplies) and **abomasum** (rennet) (Fig. 42.40).

The rumen is large, sac-like, and a site for temporary storage of food. The herbivores eat rapidly and load the rumen with food. The food is churned in the rumen by muscular contraction of the walls of the rumen. Biodegradation of food also occurs in rumen due to the presence microorganisms.

Food from the rumen passes by degrees into the reticulum or directly to the oesophagus and from there back to the oral cavity by a process of **retroperistalsis**. The regurgitated food is **cud** or **bolus**.

In the buccal cavity the cud is thoroughly masticated, mixed with saliva and passed back to rumen. The process is repeated for several times, till the food is well masticated. From rumen the masticated food passes to the reticulum → omasum → abomasum. Gastric glands are present in the abomasum.

ECHOLOCATION IN MAMMALIA

Return of sound back to the sound producer after striking an object at a distance is **echo**. Determination of location or movement of an object at a distance with the help of echo is called echolocation. In such a situation the path of the sound waves must be unhampered. The radar system in some animals is considered as a unique adoptation.

⊠ Echolocation in Bat

Bats are nocturnal. They belong to order Chiroptera. It has two suborders—Megachiroptera (fruit eater) and Microchiroptera (insect eater). Microchiroptera capture insects with the help of a radar system (echolocation).

During flight towards a stationary object the bisonar pulses emitted by a bat strike the object, become compressed or Doppler shifted and reflected. The echo received by the bat is of higher frequency than the emitted pulses. When the bat flies to a moving object, the movements of the object introduce oscillating frequency shifts, which are superposed on the Doppler shift.

Sound production

While in flight the bats produce ultrasonic sound. The high discrete pulses, whose frequency is up to 150 kilohertz (kHz) is produced from the mouth and nose. The larynx is large and the cartilaginous rings are ossified to from a strong framework. Some horseshoe bats (*Rhinolophus*) passes a special resonating chamber and the face is modified to a nose-leaf to beam the sound forwards (Fig. 42.41). About 800 species of microchiroptera live in various habitats and vary in their behaviour and physical features. They use various types of signals to receive information about their environment—position, distance and path of the potential prey. The bisonar pulses of microchiroptera may differ in different species. Three different pulses have been recorded.

1. **Constant frequency (CF) type.** These consist of a single frequency or tone. The pulses are omitted through nostrils and are longer (40 to 1000 millisecond duration) than frequency modulated pulses. Repetition rates are less than 10 per second.
2. **Frequency modulated (FM) type.** FM pulses are emitted by vespertilionids and some others. The pulses are emitted through mouth, sound like chirps, sweeps downward and are of short duration. The pulse repetition rate varies from less than ten per second at rest and above 100 when hunting.
3. **Constant frequency and frequency modulated type (CF-FM)** . CF-FM pulses are of a long constant tone

* Doppler = changes in the frequency of echo in relation to original signal.

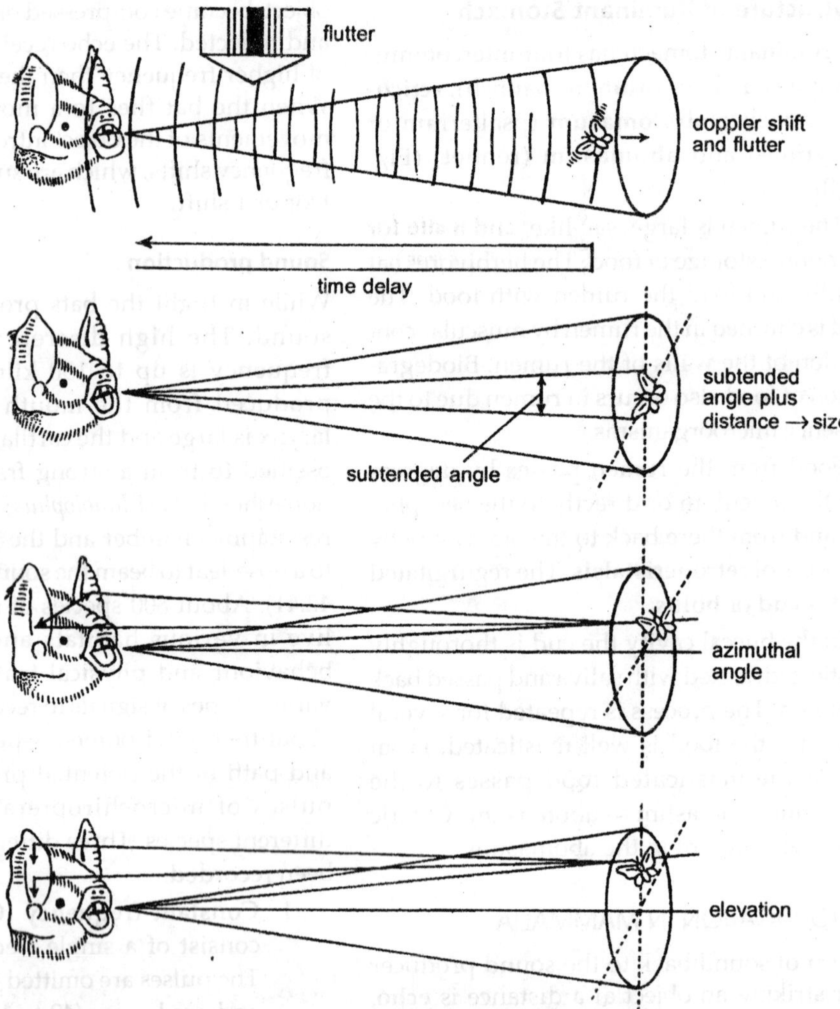

FIGURE 42.41
Echolocation in bat. Varied information extracted from bisonar signals

followed by a downward chirp. The tones are a mixture of first harmonic and second higher harmonics.

In addition to the information about target, bisonar signals provide other information —relative velocity of the prey, wing beat and size of the target, etc.

Pinna

The shape and size of the pinnae (external ear) varies in microchiroptera and play an important role in echolocation. Pinnae receive the echo, which are transmitted to **cochlea** via middle ear and membranous labyrinth.

The small-eared fast flying bats emit loud sound and prey upon flying objects. The large-eared bats emit faint pulses and prey upon insects on ground or vegetation.

The behaviour and hunting method of a species of bat are directly related to the bisonar pulses they emit.

⊠ Echolocation in Cetacea

Cetaceans are fully aquatic animals. There are two lineages among extant cetaceans—Mysticeti and Odontoceti. Ballen whales belong to Mysticeti; toothed whales—dolphins, porpoises and others belong to Odontoceti.

☒ Mysticeti

Echolocation in true sense is absent in this group. The sound produced is below 3 kilohertz in the majority. It is higher in some but pulses emitted are always of low frequencies, in most cases below 35 Kilohertz (kHz). The calls of baleens have been given different names, of which **Moans** (12-500 Hertz) are most common. Vocal songs produced by baleens are associated with their social behaviour and communications in distress and excitement. Baleens feed on small animals and there is no need to locate the position of the prey.

☒ Odontoceti

Odontoceti are of smaller size, inhabit coastal areas of the sea and fresh water in rivers. They feed upon squids, fish and other animals.

The water of the coastal areas is turbid due to extreme tides and waves. River water is also murky and the odontocetes are to capture prey in limited vision. In such a situation the sonar system (echolocation) is only means for navigation, communication and capture of prey (hunting)

Echolocation

In echolocation clicks of sounds or structural pulses emitted by the animal are reflected back to the sound producer as echoes by an object in their path. The toothed whales can detect the shape, size and movement of the prey and other objects from the form of echos including loudness. The features of the surroundings are detected by low frequency clicks –0.25 to 1 kilohertz. An object is identified by clicks 2 to 220 kilohertz. The distance between the sound producer and the object is determined by the time lapse between emitted signal and receiving its echo. Equipped with such an arrangement the odontocetes detect, chase and catch the moving prey. **Tursiops**, a bottle-necked dolphin can avoid obstacle in its way in total darkness, with the help of the system.

Till today it is not clear which particular organ produces sound in whales. Some species of dolphin produce two types of sounds of different frequencies at a time. Presumably there are two different sites of sound production. Most likely the nares connecting the larynx with the blow hole are sound producing organs. In between whales, however, the sound is produced by larynx.

Echolocation clicks of dolphins differ from the calls of the baleen whales. The clicks of dolphins are short, less than one millisecond duration and repeated many times in a second. In sperm whales the clicks are long and repetition rate is less.

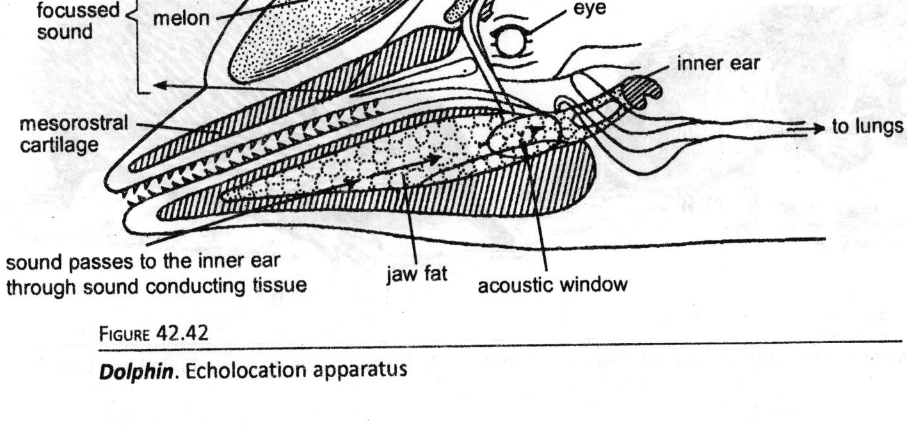

FIGURE 42.42

Dolphin. Echolocation apparatus

Sound production

Currently there are two opinions on the sources of sound in cetaceans. Some authors believe that the nasal plug produces the sound, while others consider larynx is the sound producing organ.

Nasal plug concept. The sound is produced by the nasal plug around the blow hole and focussed by **melon** (Fig. 42.42) a fatty tissue located dorsal to the upper jaw. The shape of the melon changes in some during echolocation. The focal length depends on the change of the muscles in the head.

Larynx concept. The sound is produced by larynx located far forward. Complex folds and muscles are present in the larynx, but vocal cords are absent. Focussing of sound is done by bouncing the parts of the skull.

BIOLOGICAL STATUS (SYSTEMATIC POSITION) OF MONOTREMATA

Monotremata (Prototheria) is an endangered group of living representative of the greatest antiquity. The past history of the group is practically unknown. The group is represented by three genera belonging to two families, differ widely from one another in appearance and some other features. The genus *Ornithorhynchus anatinus* (Duckbilled platypus, Duckmole (Fig. 42.43) is the sole representative of the family **Ornithorhynchidae** restricted to Eastern Australia and Tasmania. The **Tachyglossidae** (**Echidnidae**) are represented by two closely allied genera—*Zyglossus bruijni* (long-nosed spiny ant eater) restricted to New Guinea and *Tachyglossus* (*Echidna*) *aculeata* (Australian spiny ant eater) and *T. solosus* (Tasmanian spiny ant eater) ranges from New Guinea to Tasmania.

☒ Salient features of Monotremata (Fig. 42.43)

1. Body covered with a dense, soft fur (*Ornithorhynchus*), and with strong pointed spines between which are coarse hairs on the dorsal surface and only hairs on the lower surface (Tachyglossus).
2. The upper jaw is produced into a depressed muzzle like the beak of a duck in Duckmole. The jaws are produced into a rostrum in Echidna.
3. Teeth present only in young, replaced by horny plates in the adult Duckmole; entirely absent in Echidna in any stage of life.
4. Digits webbed in Duckmole; ending in sharp claws in Echidna.
5. Mammary glands devoid of teats, open in shallow depression on the ventral side.
6. Brain lacks corpus callosum.
7. Eggs large segmentation meroblastic.

FIGURE 42.43

Monotreme. A. *Ornithorhynchus anatinus (Platypus),* B. *Tachyglossus aculeata* (Spiny and eater)

☒ Affinities with Reptilia

Similarities	Dissimilarities
1. Presence of ectopterygoid	1. Skin covered with soft hairs or with hairs among spines.
2. Presence of vomer	2. Presence of mammary glands.
3. Epipterygoid homologus with Echidna pterygoid.	3. Skull smooth and polished with the disappearance of sutures.
4. Zygomatic arch perforated by a temporal canal.	4. Two halves (rami) of the lower jaw joined anteriorly by Meekel's cartilage.
5. Ribs single-headed.	5. Presence of true mammalian teeth at least in young.
6. Cervical ribs present.	6. Epiphysis in vertebral column weak or absent.
7. Precoracoid well developed.	7. Optic lobes four.
8. Coracoid platelike	8. Presence of diaphragm.
9. Interclavicle plate-like	9. Body temperature 25°C-28°C.
10. Epipubic bones present.	
11. Acetabulum perforated.	
12. Epiphysis absent.	
13. Corpus callosum absent.	
14. Presence of cloaca	
15. Viviparous, eggs large, meroblastic.	

The presence of reptilian characters suggests that the group separated from the main line of evolution of mammals from reptiles and presumably due to their restricted habitat with primitive features, and inability to migrate to different habitat, their primitiveness has been preserved; while others underwent transformation, diversified and higher mammals appeared.

☒ Affinities with Aves

Similarities	Dissimilarities
1. Shape of the back (Platypus) similar to that of duck.	1. Absence of feathers or scales.
2. Feet webbed (Platypus).	2. Absence of beak.
3. Skull smooth, sutures obliterated.	3. Cranial sutures present in Echidna.
4. Teeth absent (Echidna)	4. Presence of true mammalian teeth in Duck mole (young).
5. Presence of a hollow tarsal spur.	5. Cervical vertebrae bear ribs.
6. Presence of oil gland.	6. Absence of flight muscles.
7. The right atrioventricular valve (Duckmole) partly muscular.	7. Absence of air sacs.
8. Left ovary larger than right.	8. Presence of milk glands.
9. Urinogenital canal communicates with both cloaca and penial canal.	9. Syrinx absent.
10. Ova large with abundant yolk.	
11. Cleavage meroblastic.	

The avian features of monotremes are rather superficial and the similarities are presumably due to the origin of birds and mammals from a common reptilian ancestor.

Affinities with Allotheria (Mammalia)

Similarities	Dissimilarities
1. Presence of hairs.	1. Pinna absent.
2. Presence of mammary glands.	2. Cloace present.
3. Presence of sweat gland.	3. Skull bones transitional or reptilian.
4. Presence of oil gland.	4. Lacrimal bone absent.
5. Presence of diaphragm.	5. Auditory bulla absent.
6. Double occipital condyle.	6. Vertebrae for most part with epiphysis.
7. Jaw bone single-dentary.	7. Ribs one-headed.
8. Jaw attachment craniostylic.	8. Cervical ribs present.
9. True mammalian teeth in young Duckmole.	9. Pectoral girdle with large separate coracoid bones, and interclavicle.
10. Palate present.	10. Corpus callosum absent.
11. Cervical vertebrae seven.	11. Cochlea simple, not coiled.
12. Sternum segmented.	12. Paired oviducts open into the cloaca.
13. Heart four-chambered.	13. Oviparous, lay large eggs.
14. Cerebellum well developed, optic lobes four.	14. Segmentation meroblastic.

The two families of Monotremata—Ornithorhynchidae and Tachyglossidae differ from one another in appearance and in some minor anatomical characters. The differences are presumably the result of adaptation to different modes of life. The *Ornithorhynchus* lives in the ponds, rivers and creeks in the bank of which they make burrows and nest. The *Tachyglossus* are land living nocturnal, insect eaters and are capable burrowers.

There are two views regarding the phylogeny of the monotremes. Monotremes evolved from reptiles; in the making of mammals they separated from the main line and lived in isolation in the areas, where geophysical changes were almost nil. The other view advocates that monotremes evolved from early marsupials and their differences with present day marsupials is the result of degeneration and reversion of characters due to living in an area where the primitive climate has been maintained.

CLASSIFICATION: CLASS MAMMALIA

The mammals are endothermic tetrapods.

1. The skin is clothed with hairs and provided with sweat and sebaceous glands. Sweat evaporation keeps the body cool and sebum keeps the hair and skin soft and pliable.
2. Dermal milk gland or mammary gland present.
3. The teeth are heterodont with differentiation into incisors, canines and molars; thecodont, each tooth lying embedded in an alveolar pocket of jaw bones; and diphyodont, having two sets, a milk set replaced by a permanent set.

4. Fleshy cheeks and lips cover the edges of jaws and teeth.
5. A muscular diaphragm separates the abdominal and thoracic cavities.
6. The skull articulates with the atlas by two exoccipital condyles.
7. The lower jaw is composed of one pair of bones, the dentaries. Each dentary articulates with squamosal bone of the skull.
8. The vertebrae are gastrocentrous. Each vertebra consists of a centrum, a neural arch and two epiphyses. A cartilaginous intervertebral disc separates the centra of adjacent vertebrae.
9. The ribs articulate with vertebrae by two heads, capitulum and tuberculum.
10. Cervical vertebrae are seven.
11. The coracoid bone of pectoral girdle is reduced as a coracoid process on the scapula.
12. The heart four-chambered, A single left aortic arch; enucleate erythrocytes.
13. The opening of the trachea is guarded by an epiglottis.
14. The cerebral hemispheres, especially the cerebral cortex or neopallium, highly developed; corpus callosum, a massive bundle of fibre tracts, connect the two cerebral hemispheres (absent in monotremes); cerebellum solid, large and complex.
15. Pinna present (except some aquatic mammals); tympanic membrane greatly depressed at the bottom of the external passage or auditory meatus; cochlea, the auditory sense organ, is spirally coiled (not coiled in mono-tremes).
16. The middle ear contains three minute auditory ossicles or bonelets—malleus, incus and stapes—which serve to transmit sound vibrations from tympanum to perilymph. The bonelets are modified articulate, quadrate and hyomandibular bones, respectively, of reptiles.
17. The nasal passages are provided with turbinal bones, which serve to warm the inbreathed air.
18. Part of oviducts are specialised to form uterus, uterine gestation (except monotremes) occur; eggs microscopic (except monotremes, laying large eggs); placenta allantoic (not in monotremes and marsupials); viviparous (except monotremes).

The mammals are placed in three subclasses—Prototheria, Allotheria and Theria.

Subclass PROTOTHERIA

Primitive mammals with both reptilian and mammalian characters, the former predominating. The living prototherians are included in the order Monotremata.

Order **Monotremata**

In many respects monotremes are transitional between reptiles and mammals and sometimes are referred to as "living fossil".

1. Pinna absent.
2. Cloaca present.
3. Skull bones in many respect transitional or reptilian, lacrimal bone absent; jugal reduced or absent; auditory bulla absent; traces of reptilian prevomer present.
4. Vertebrae for most part are without epiphyses.
5. The ribs, except the cervicals, are one-headed, the tuberculum being absent; cervical ribs present (absent in higher mammals).
6. Pectoral girdle possesses large, separate coracoid bones and an interclavicle.
7. The pelvic girdle possesses epipubic or marsupial bones, extending from the front of pelvis.
8. Corpus callosum absent.
9. Cochlea is a simple process, not coiled as in higher mammals.
10. Paired oviducts open separately into the shallow cloaca.

11. Lays 1 to 3 eggs, measuring about 12 × 18 mm.
12. Embryonic development reptilian.
13. The typical mammalian characters are (i) milk glands, (ii) hair, (iii) a complete diaphragm.

Examples: *Orinthorhynchus, Tachyglossus.*

Subclass ALLOTHERIA

Extinct forms. Flourished between Upper Jurassic and Eocene.

Example: *Ctenacodon.*

Subclass THERIA

Marsupials and Placental mammals.

1. Pinna present.
2. Cloaca absent.
3. Ureters open into the base of urinary bladder.
4. Testes of males usually in a scrotal sac; vasa deferentia and bladder opening through a common urethra running through the penis.
5. Oviduct differentiated into upper fallopian and lower uterine region opening into a distinct vagina.
6. Mammary glands with nipples.
7. Viviparous (produce living young).

The existing therians are placed in two infraclasses.

Infraclass METATHERIA (Pouched mammals).

Order **Marsupialia** or **Didelphia**

1. Epipubic bone usually present.
2. Incisor teeth not equal in 2 jaws. Dental formula is $\frac{5.1.3.4}{4.1.3.4}$
3. Female usually with marsupium (ventral pouch) or marsupial folds surrounding nipples or abdomen.
4. Uterus and vagina double.
5. Usually no placenta.
6. Fertilization internal; development begins in uterus, after a few days premature baby leaves and crawls to marsupium and attaches closely to a mammary nipple by its mouth and remains as "mammary foetus" until fully formed.
7. All existing marsupials, except Didelphidae and Caenolestidae, are restricted and dominant mammals in Australia.

Examples: *Didelphys, Myrmecobius, Dasyurus, Thylacines, Peragole, Notoryctes, Petaurus, Phalanger, Macropus, Phascolomys, Caenotestes, Phascolarctos,* etc.

Infraclss EUTHERIA OR PLACENTALIA OR MONODELPHIA

The placental mammals include about 95 per cent of all living mammalian species.

1. Marsupial bones and pouch absent.
2. Dental formula $\frac{3.1.4.3}{3.1.3.4}$.
3. Internal carotid artery enters the skull through the foramen lacrum medius or through the bulla.
4. Corpus callosum connects the large cerebral hemispheres; brain highly organised by remarkable development of neopallial region of the hemispheres.
5. Urinary ducts pass to the bladder directly.
6. Vagina single.
7. Single uterus (uterus simplex), placenta develops for providing nourishment throughout long gestatation period till young are born; placenta allantoic and highly organised.

Classification of Eutheria proposed by Simpson.

Cohort **Unguiculata**

Includes extinct and livng orders in which archaic mammalian characters are retained.

Order 1. **Lipotyphla** (Upper Cretacoeus-Recent)

1. Small mammals with plantigrade manus and pes; all five digits present, except in a few.

2. The snout prolonged to form a mobile proboscis.
3. The mammary glands are usually many, 3-6 pairs (only 2 pairs in Chrysochloroidea).
4. The frontal part of orbital wall is formed by extension of the maxilla and palatine does not reach the lacrimal.
5. The tympanic bone is wing-shaped.
6. The zygomatic arch is incomplete, when present, it is usually slender and the jugal greatly reduced.
7. Nodular structures, the remains of chevron bones are present in the vertebral column between the vertebrae on the ventral side.
8. The pubic symphysis is reduced to a short contact between the ischia (completely absent in Soricoidea).
9. The incisors are frequently enlarged to function as forceps (in Soricidae) and the canines are small or absent.
10. Predominantly insectivorous with sharp molar cusps and interlocking teeth.
11. The tympanic bone usually ring-shaped.
12. No caecum.
13. The cerebral hemispheres are small and smooth; olfactory lobes large.
14. Testes are frequently abdominal; fully developed scrotum absent.
15. Uterus bicornuate.
16. The number of young given birth usually large, even up to 21 in a litter.

Examples: *Sorex* (shrews), *Suncus*, *Talpa* (moles), *Erinaceus* (hedgehog), *Solenodon*, *Blarina*, *Chrysochloris*, etc.

Order 2. **Monotyphla** (Upper Cretaceous—Recent)

Small mammals with two distinct existing families—the elephant shrews of Africa and tree shrews of Orient.

1. The eyes are relatively large.
2. One to three pairs of mammary glands.
3. Caecum present.
4. The cranial cavity larger and extends forward between the large orbits.
5. A vertical plate of palatine extends up to the anterior medial wall of the orbit to meet the lacrimal.
6. Jugal well-developed.
7. Pubic symphysis not reduced.
8. The number of youngs in a litter never more than three.

Examples: *Rhynchocyon*, *Elephantubus*, *Tupaia*, *Ptilocercus*, etc.

Order 3. **Primates** (Palaeocene-Recent) (Lemurs, Monkeys, Apes, Man)

1. Primarily arboreal with prehensile digits.
2. Plantigrade, walking position of the feet.
3. Hair covers entire body except palms, soles and part of the face.
4. Presence of thumb, the greater toe shorter than other digits and more or loss opposable to the latter.
5. Terminal flattened nail instead of claws.
6. A single pair of usually pectoral mammae.
7. Eyes directed anteriorly instead of laterally.
8. The orbit completely surrounded with bone.
9. Clavicle always present.
10. The stomach simple.
11. The cerebrum usually large and well-convoluted.

Suborder i. **Prosomii**

More primitive primates; low brain case, uncovered cerebellum, non-deciduous placenta.

Examples: *Lemur*, Chiromys (Aye-aye), *Galago*, *Perodicticus*, *Loris*, *Tarsius*, etc.

Suborder ii. **Anthropoidea**

1. Pinnae reduced and generally pressed against the head.

2. The upper lip lacks a ligament and freely protrusible.
3. The eyes are forwardly directed and vision is binocular. The retina is provided with cones and macula lutea is generally present.
4. The cranium is expanded as the cerebral hemispheres reach their highest development and overhang the cerebellum and the medulla as well and the neopallium is generally markedly convoluted with the development of the Sylvian fissure, and central and other sulci. The occipital visual areas well-developed.
5. The nostrils are completely ringed round by naked skin.
6. The enhancement of visual sense is correlated with a corresponding reduction in olfactory reception.

Examples: Cebid monkey (*Cebus*), spider monkey (*Ateles*); howler monkey (*Alouatta*), woolly monkey (*Lagothrix*), rhesus monkey (*Macaca*), mandrill (*Mandrillus*), langur (*Semnopithecus*), baboon (*Papio*), gorilla, Gibbon (*Hylobates*), orang-utan (*Pongo*), [*Homoerectus* [Upright Man (extinct)]. *Homo sapiens* (Wise Man), etc.

Order 4. **Dermoptera** (Palaeocene—Recent)

1. Bat-like, nearly as large as cat, with hairy, muscular parachuting membrane, patagium. The membranes stretched between the head and the hand, hand and feet and feet and tail.
2. The brain is primitive, macrosmatic and the corpora quadrigemina uncovered by the cerebrum.
3. The bulla is formed by the tympanic ring, which also form the lower border of the spout-like external auditory meatus.
4. The lower incisors are procumbent and comb-like. The molars multicuspidate.

Example: *Cynocephalus* (*Galeopithecus*). The only genus named flying lemur.

Order 5. **Chiroptera** (Eocene-Recent)

The chiropterans or bats are capable of true flight.

1. The forelimbs have greatly elongated segments, especially the forearm and the fore-ulnar digits. These support a thin membranous fold of the integument stretching up to the hindlimbs to form the wing.
2. The sternum is keeled for attachment of the large pectoral flight muscles.
3. A cartilaginous rod, calcar, is often attached to the inner side of the ankle, which helps to support a second membranous fold, the interfemoral membrane, extending between hindlimbs and sometimes involving tail.
4. The ulna is vestigeal.
5. The pollex much shorter than other digits, directed forwards and terminates in a well-developed curved claw.

The chiropterans are arranged into two suborders.

Suborder i. **Megachiroptera**

1. Frugivorous bats with large eyes, long snout, without foliaceous nasal or auricular appendages.
2. The muzzle is nearly always elongated and pinna of the ear simple.
3. The second digit of manus usually end in claw.
4. The tail, when present, is not enclosed in the interfemoral membrane, but lies below it.
5. The crowns of the molar teeth are devoid of sharp cusps.

Examples: *Pteropus, Xantharpyia*, etc.

Suborder ii. **Microchiropera**

1. Insectivorous bats; the hindlimbs rotate outward, the knee is directed backward instead of forward.
2. The five pedal digits are all provided with claws and the limbs are so adapted for flight and suspension that bats can only shuffle with difficulty on ground.

3. The second digits of the manus are always clawless.
4. Usually with short snout and frequently foliaceous nasal and auricular appendages.
5. The muzzle is generally short, the pinna is usually complicated by the presence of an inner lobe or tragus, and often produced into arborescent appendages.
6. The tail, when present, is at least partially enclosed in the interfemoral membrane.
7. The body is covered with soft fur, except in a group of microchiropteran in which integument is practically naked.
8. The crowns of molar teeth have typically sharp cusps. The dentition complete, heterodont, diphiodont.
9. The cerebral hemispheres are smooth and do not overlap the cerebellum.
10. The olfactory centres are reduced.
11. The penis is large and pendent.
12. The testes abdominal, descend seasonally to a prominent situation in the groin.
13. The uterus is simple or bicornuate.
14. The placenta is deciduous and discoidal.

Examples: *Vespertilio*, *Rhinolophus*, *Desmodus* (blood-sucking vampire bat) *Diphylla*, etc.

Order 6. **Edentata** (Paleocene—Recent)

1. The name of the order implies a total lack of teeth (except ant bears).
2. The dentition of toothed edentates is peculiar in that incisors or canines are absent and the teeth in the definitive condition are without enamel.
3. The clavicle is always present.
4. An additional pair of zygapophyses on the posterior dorsal and lumbar vertebrae.
5. The testes are abdominal.
6. Distribution restricted to America since their origin.

Suborder i. **Palaeanodonta**

1. The North American species appears to represent a primitive Palaeocene offshoot from the ancestry.
2. Reduction and degeneration of teeth and enamel-covered canines.
3. No extrazygapophyses and ischio-caudal symphysis.

Examples: *Palaeanodon*, *Metacheiromys*, *Epoicotherium*, etc.

Suborder ii. **Xenarthra**

The armoured armadillos, ant-eaters, sloth, etc. of almost entirely South American forms constitute this suborder.

1. Feet heavily clawed.
2. Enamel is entirely absent from teeth in living groups.
3. The teeth are transformed into pillar-like persitently growing stumps (the arma-dillos) or reduced in number (the sloths) or may be altogether absent (ant-eaters).
4. An extra pair of zygapophyses upon the posterior dorsal and lumbar vertebrae present (like snakes or Iguana).
5. The frequent reduction or even absence of the zygomatic arch.
6. Fusion of coracoid process to the front border of the scapula so as to enclose a large foramen and articulation of several of the anterior caudal vertebrae to the ischium, producing an unusually long ischiosacral region.
7. Brain simple type.
8. Persistently abdominal testes.
9. Heavily clawed feet.

Examples: *Myrmecophaga*, *Tamandua*, *Bradypus*, *Chaelopus*, *Tolypeutes*, *Dasypus*, *Glyptodon*, *Mylodon*, etc.

Order 7. **Pholidota** (Oligocene—Recent)

1. The scaly ant-eater or pangolin, with body covered by imbricating horny scales over the head, body and tail and few scattered hairs. Lower surface is hair-covered.

2. The head is produced into a short pointed muzzle.
3. The tongue is long and teeth are entirely absent.
4. The skull is long and cylindrical.
5. No bony separation of orbit and temporal fossa.
6. Jugal arch and clavicle absent.
7. About five feet long from head to tail end.

Examples: *Manis*.

Cohort **Glires**

Includes rodents, rabbits, pikas, etc. which show early departures from the primitive eutherian insectivoran stem and have many adaptive parallel morphology, physiology and behaviour.

The cohort has two orders—Rodentia and Lagomorpha.

Order I. **Rodentia** (Palaeocene—Recent)

1. The premaxillae and maxillae enlarged to provide greater surface for attachment to enlarged masseter jaw muscle.
2. Antorbital vacuity enormously enlarged.
3. A single pair of chisel-shaped incisors in the upper and lower jaw with persistently open roots; posterior surface without enamel; grow throughout life and are always of great size.
4. No symphysis between the mandibles, special modification of the mylohyoids and masseters permit the separation and return of the prominent lower incisors causing scissor-action in feeding.
5. Canines absent and diastema between the incisors and cheek-teeth present.
6. The glenoid cavities for the mandibular condyles are elongated in a fore-and-aft direction, enabling the jaws to move forward.
7. The jaw muscles are highly enlarged.
8. A clavicle is usually present.
9. The caecum tends to be large.
10. The brain is little convoluted and the cerebellum is not covered by the cerebral lobes.
11. 3,000 living species, the largest of all the mammalian groups showing adaptive character for gnawing.

Suborder i. **Sciuromorpha**

1. Primitive, two upper and one lower premolars retained.
2. The antorbital foramen not enlarged.

Examples: *Funambulus*, squirrels, gophers, chipmunks, (*Petaurista* flying squirrels), etc.

Suborder ii. **Myomorpha**

1. The infraorbital opening is large.

Examples: *Dipus*, *Rattus* (rats), *Mus* (mice), jarboas, hamsters, etc.

Suborder iii. **Hystricomorpha**

1. The tibia and fibula remain separated.
2. The infraorbital canal is very enlarged.

Examples: *Cavia* (guinea pigs), *Hystrix*, (true porcupines), *Lagostomus*, etc.

Order 2. **Lagomorpha** (Palaeocene—Recent)

1. Maxillae laterally fenestrated.
2. The orbital foramina small.
3. A diastema present.
4. Incisors with persistent pulps and with additional small pair of upper incisors.
5. Three upper, two lower premolars and three molars are present in each jaw.
6. Cheek pouch present.
7. Masseter muscle enormous and temporal muscle weak.
8. Tail is invariably reduced and the hindlimbs are highly specialised for a saltatorial mode of progression.

Examples: *Lepus* (hares), *Oryctolagus* (rabbits), pika, etc.

Cohort **Mutica**

Constitutes highly specialised, widely divergent mammals like whales. They are perfectly adapted to aquatic life.

Order **Cetacea** (Eocene—Recent)

Assemblage of highly specialised aquatic mammals profoundly modified for marine life.

1. The large torpedo-shaped body with naked smooth skin.
2. The forelimbs or flippers and dorsal fin, when present, act as radders and stabilizers.
3. The dorsal fin and tail flukes are specialised skin folds, devoid of any bony skeleton and supported by underlying fibrous tissue.
4. Extra digits (hyperdactyly) and extra phalanges (hyperphalangy) serve to enlarge the surface of the flipper.
5. A thick insulating layer of dermal fat, the blubber in the hypodermis, constitute as much as 27 per cent of total weight.
6. The proximal elements of forelimbs are enclosed inside the body and the skeleton is secondarily specialised by the addition of extra joints in the digits.
7. The skull has greatly elongated facial and dorsoventrally compressed cranial portions. Nasal openings often asymmetrical, placed far back on the upper surface of head; tympanic bulla is loosely attached; the cranial bones, in general, are much modified in shape.
8. The cervical vertebrae shortened, compressed and fused together to a greater or lesser extent (complete fusion in *Balaena*).
9. The sternum is reduced.
10. Scapula with extremely reduced prescapular fossa, forwardly directed acromion process and absence of spine.
11. Bilobed liver, relatively simple, and nonlobulated lungs.
12. The stomach is subdivided into several digestive regions.

The features which have been lost are:

i. Hair, except for a very few facial bristles.
ii. Finger nails, except for traces in foetus.
iii. All skin glands.
iv. Lacrimal glands.
v. Nictitating membrane in the eye.
vi. The heads of ribs in some.
vii. The odontoid process in most.
viii. Hindlimbs, except traces in some.

Modern Cetaceans are divided into two suborders—Odontoceti and Mysticeti.

Suborder i. Odontoceti

1. Skull with a varying degree of asymmetry; the external nostril single and usually crescentic in outline.
2. Teeth present and whale bone absent.
3. The upper jaw bones widen posteriorly.
4. The mandibular rami are wide at their articular end, gradually narrowed to the symphysis. It may be of considerable length (dolphins and sperm whale) or greatly shortened (porpoise and pilot whale).
5. The anterior ribs usually have capitulum and tuberculum.
6. The sternum is formed of two or more sternebrae.

Examples: *Physeter* (sperm whale), bottle-nosed whale, beaked whale, *Phocaena* (porpoises), *Dolphinus* (dolphin), white whale, *Orca* (killer whale), etc.

Suborder ii. Mysticeti

1. Baleen whales having numerous horny baleen (or whale bone) plates arranged in two series, one on each side of the upper jaw. The baleen is a horny material developed from the epithelial lining at the upper jaws acting as a strainer.
2. The teeth are rudimentary, functionless, present in young but in adult replaced by baleen.
3. Skull symmetrical.
4. Nostrils paired.
5. Sternum single.
6. The ribs are one-headed, articulating only with transverse processes of the vertebrae.

Examples: *Balaena, Caperea, Eschrichtius, Balaenoptera, Megaptera*, etc.

Cohort **Ferungulata**

A large group, divided into five super orders.

Superorder **Ferae**

The sole representative of the superorder is order Carnivora.

Order **Carnivora** (Palaeocene—Recent)

1. Universal retention of the full number of incisors and well-developed canines in these flesh-eaters.
2. The post-canine dentition always with a certain amount of reduction and a rather wide range of adaptive variation.
3. The pedal digits are usually five and never less than four.
4. The toes are armed with sharp claws which are ofen retractile.
5. The brain is well-developed.

The living carnivores are arranged in two suborders.

Suborder i. **Fissipedia**

1. Terrestrial carnivores with six incisors of small size in each jaw; canines are large and strong, the last premolar and first molar "carnassial" or cutting teeth and the last two molars are crushing teeth.
2. The carnivores are distributed in all the continents except Australia.

Examples: *Vulpes* (foxes), *Viverricula* (civet), *Felis* (cat), *Panthera* (lion, tiger), *Herpestes* (mongooses), Hyaena, *Canis* (dogs), *Procyon* (racoons), weasels and *Ursus* (bears).

Suborder ii. **Pinnipedia**

1. Marine forms in which there has been a secondary adaptation of the whole body for aquativ life.
2. The greater part of limbs are enclosed within the body skin.
3. The claws are reduced and the digits increased in number.
4. The milk dentition is feeble and shed early.
5. The cranial cavity is large as compared with face.
6. No carnassial teeth; never more than five pairs of lower post-canines, four premolars and one molar. The teeth are basically conical, with a single crown.

Examples: *Otter* (sea lion), *Odobenus*, (walrus), *Phoca* (seals).

Superorder **Protungulata**

1. Hoofed grazing animals with specialised grinding teeth of various kinds as well as numerous muscular, articular and alimentary adaptations and with long necks and legs.

Order **Tubulidentata**

1. The body is thickset and covered with thick skin bearing scattered hairs; the head produced into a long muzzle.
2. The forelimbs are short and stout with four clawed digits, the palmar surfaces are placed on the ground for walking.
3. The hindlimb has five clawed digits.
4. The tongue long and mouth small.
5. 4-5 peg-like teeth of peculiar structure on each jaw present, Teeth lack enamel but possess a coating of cement and the body of vasodentine is perforated by large number of fine tubules (hence the order name).

Examples: *Orycteropus*.

Superorder **Paenungulata** (Subungulata)

A widely diversified "near" or "subungulate" group forming a loose assembly of related animals like hyrax, elephants, dugongs and other extinct forms.

Order 1. **Hyracoidea** (Oligocene—Recent)

Small, hooved, rabbit-like animals of Africa and South-West Asia, included in a single family, Procaviidae (= Hyracidae) with genera *Procavia* (= Hyrax) and *Dendrohyrax*.

1. Manus and pes are plantigrade. The former bears four digits and a vestige of a pollex; the latter bears three digits of which the third is the longest.

2. All digits terminate in small flat hooves except the second pedal digit with a curved claw.
3. The centrale persists in the carpus.
4. The radius and ulna partially crossed.
5. The astragalus differs greatly in shape from that of any other mammal in that the malleolus of the tibia has a large flat bearing on it.
6. Clavicle absent.
7. The postorbital process arises largely from the parietal.
8. The jugal is a very stout bone extending backwards to take a large share of glenoid cavity.
9. The periotic and tympanic are ankylosed together, but not to the squamosal.
10. The tympanic bulla and a spout-like auditory meatus present.
11. The scapula triangular and lacks an acromion.
12. Dorsolumbar vertebrae as many as twenty two.
13. A single pair of large curved upper incisors grow from persistent roots like those of rodents; canine absent; four premolars and three molars which are lophodont.
14. Lower jaw carries two pairs of spatulate, comb-like incisors.
15. Coronoid foramen just behind the last molar present.
16. No gall bladder.
17. Unlike other mammals, the caecum has a pair of caecal pouches.
18. Testes abdominal.
19. A dorsal gland on the back present.

Examples: *Procavia*, *Dendrohyrax*, etc.

Order 2. **Proboscidea** (Pliocene—Recent)

Comprises the largest and in many respects the most highly specialised terrestrial mammals.

1. The elongation of nose and upper lip form a very long trunk.
2. The facial region shortened; the nasal opening recedes towards the tip of the head and the nasal bones reduced in size.
3. Both fore-and hind feet possess five functional digits.
4. In forelimb the ulna and radius are permanently crossed in pronation and the ulna is unique in having its distal articulation with the carpus larger than that of the radius.
5. The skull is immensely big and its bones contain large air cavities.
6. The incisor teeth of upper jaw are specialised into great tusks. The ivory of the tusks is solid, specialised dentine. Enamel absent.
7. Clavicle absent.
8. The pelvis has broad ilia and is vertical in direction.
9. Paired precaval veins.
10. The cerebral hemispheres are much convoluted but do not cover the cerebellum.
11. The testes are permanently abdominal.

Examples: *Loxodonta africanus* (African elephant), *Elephas maximus* (Indian elephant).

Order 3 **Sirenia** (Eocene—Recent)

1. Fully aquatic and present a number of structural features convergent with those of the Cetacea.
2. Thick, almost hairless, skin with an underlying layer of blubber.
3. External hindlimbs, external ear, and sacrum absent.
4. The forelimbs are paddle-like.
5. The tail is flattened and either rounded or rhomboidal.
6. The diaphragm, like in the cetacean, is oblique and very muscular.
7. The skull and dentition are specialised.
8. The tympanic bone is ring-shaped.
9. The cervical vertebrae are not ankylosed; the whole skeleton is strong and made of very strong bones.

10. The scapula is long and narrow.

Examples: *Trichechus* (manatee), *Dugong* (sea cow).

Suborder **Mesaxonia**

1. "Odd-toed" hooved animals including horses, tapirs and rhinoceroses and certain extinct forms.
2. In the mesaxonic foot, exemplified by the horse, the skeletal axis passes down the third digit and other digits are reduced and do not support any weight.

Order **Perissodactyla** (Eocene—Recent)

1. The second and fourth digits are represented in horses by the splint bone (metacarpals and metatarsals).
2. Odontoid process peg-like.
3. Dorsolumbar vertebrae twenty-two to twenty-three.
4. Almost complete molarisation of the premolar teeth.
5. Presence of a third trochanter.
6. No ruminant stomach and true horns with a boney horn-core.

Suborder i. **Hippomorpha**

1. The second and fourth toes are represented by vestigeal remnants called "splint bones".
2. The molar teeth are highly complex and wear down through most of the life of individual.

Examples: *Equus cabalus* (horses), *E. hemionus* (asses), and *E. zebra* (zebras).

Suborder ii. **Ceratomorpha**

1. Least specialised perissodactyls; feet have stout digits, four in the manus and three in the pes, the third digit in each case being somewhat enlarged.
2. The anterior nasal openings are enlarged and recessed. The nasal bones are small and pointed upwards in correlation with the proboscis.
3. The teeth are persistently brachydont and without cement. The molar teeth are bilophodont.

Examples: *Tapirus indicus* (Tapirs) and *T. terrestris*.

Suborder iii. **Ruminentia**

1. Total loss of upper incisors. The upper canine is also usually lost but, in rare cases where it persists, it is much enlarged.
2. Flat spatulate lower incisors. The lower canine has the same shape as the incisors and is closely pressed against them.
3. The feet have always a cannon bone. The lateral toes are usually lost and the feet absolutely didactyle. In some cases the second and fifth digits remain, but are nerver complete.
4. The ruminating stomach attains its highest development with four complete chambers.
5. Horns of several types are usually present.

Examples: deers, giraffes, okapi, antelope, gazelles, impala, eland, yak, *Bos* (cow), *Capra* (goat), *Ovis* (sheep), etc.

PART THREE

ORIGIN AND CONTINUITY OF ANIMAL LIFE:

ENVIRONMENT: HUMAN INTERFERENCE: DISEASES AND ANIMALS

43 GAMETE FORMATION: EARLY DEVELOPMENT

GAMETOGENESIS

The series of changes through which the germinal epithelial cells pass to produce germ cells or **gametes** is called **gametogenesis** (Fig. 43.1). The development of the sperm is **spermatogenesis** and that of the egg is **oogenesis**. The phases of development in both the gametes are essentially similar.

• Spermatogenesis

1. It occurs in the testes. The walls of the seminiferous tubules of vertebrates contain **primordial germ cells** (Fig. 43.1).
2. The primordial germ cells begin to enlarge with increased metabolism and undergo mitotic division, and produce **spermatogonia**.

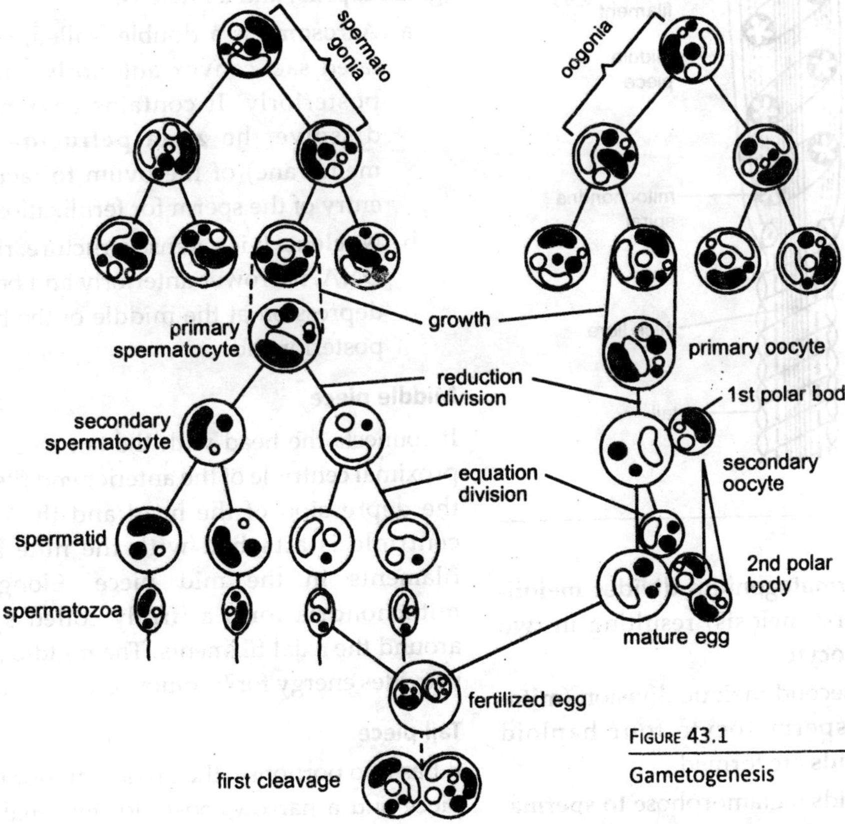

FIGURE 43.1
Gametogenesis

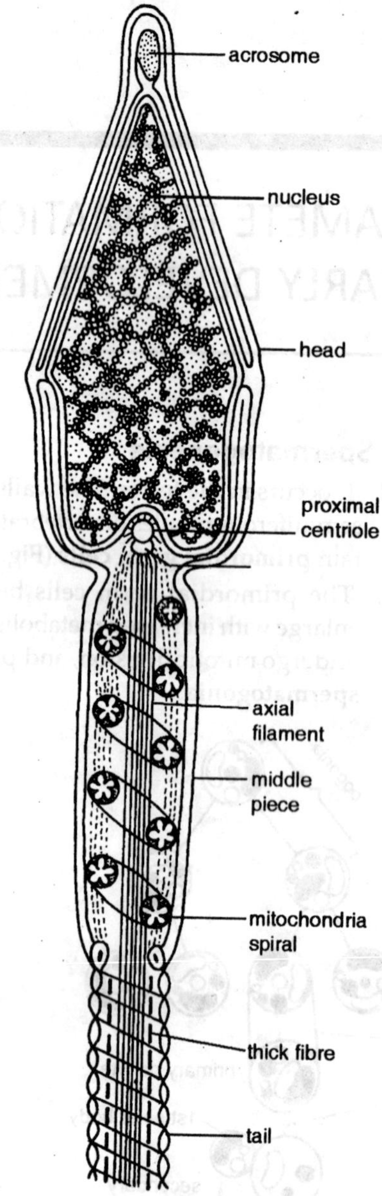

FIGURE 43.2

A human sperm

3. The spermatogonium divides **meiotically** (first meiosis) resulting in two **spermatocytes**.
4. Due to second meiotic division (**mitosis**) of spermatocyte four haploid **spermatids** are formed.
5. Spermatids metamorphose to **spermatozoa**. In the process spermatid increases in size; centriole divides into two; distal centriole forms the main axis of the tail; the axial filament is surrounded by a fibre coat; mitochondria form a spiral sheath; Golgi complex form acroblast and give rise to acrosome.
6. The mature spermatozoon has three distinct zones—**head, trunk** and **tail**. It can move actively in fluid.

☒ Structure of human spermatozoa

In human, the spermatozoon (Fig. 43.2 measures 60 μm long and 2.5 to 3.5 μm in diameter at the broadest point in the head. The spermatozoon is rich in deoxyribonucleic acid (DNA).

Head

The head is constitued by an anterior **acrosome** (**galea capitis**) and a nucleus.

a. **Acrosome** is a double-walled, granulated sac, convex anteriorly and flat posteriorly. It contains enzymes to dissolve the **zona pellucida** (egg membrane) of the ovum to facilitate entry of the sperm for fertilization.
b. **Nucleus.** It is a dense structure, rich in DNA, narrower anteriorly and bears a depression at the middle of the broad posterior end.

Middle piece

It connects the head with the tail piece. The proximal centriole of the anterior end fits into the depression of the head and the distal centriole is attached with the nine axial filaments in the mid piece. Elongated mitochondria form a firmly coiled spiral around the axial filaments. The middle piece provides energy for locomotion of sperms.

Tail piece

It has two portions—the broad anterior main piece and a narrow, posterior terminal end piece.

a. The main piece of the tail consists of a spiral mitochondrial sheath surrounding a group of eleven fibrils. Two fibrils are centrally placed and the rest form a ring around them. Contraction and relaxation of peripheral fibrils help the spermatozoa to move forward. The enzyme **adenosine triphosphatase** (ATP ase) is present in the whole length of the main piece.

b. **Terminal end piece** (flagella). It is the slender, last portion of the tail piece. The fibrils are naked and resemble cilia.

⊠ Oogenesis

1. It occurs in the ovary. The phases of division in multiplication are similar to those in spermatogenesis (Fig. 43.1).
2. The primary germ cells of the ovary undergo several mitotic divisions to produce **oogonia.**
3. The oogonium gives rise to **primary oocytes** mitotically.
4. A primary oocyte undergoes first meiotic division and produces one large **secondary oocytes** and one small first **polar body** or **polocyte.**
5. The oocyte and the polocyte divide mitotically (second meiotic division). The oocyte produces one large **ootid** and one small second polar body and the polocyte two polar bodies.
6. In oogenesis one haploid ootid and three polar bodies are formed. On maturity, the ootid is transformed into **ovum**. The ova are of one type only. The nucleus with haploid chromosome number is termed **pronucleus.**

⊠ Structure of ova

The ovum is a large, round structure. In human, according to the stage of development, it measures 117 to 142 μm in diameter.

The cytoplasm of the ovum is termed **yolk** or **ooplasm**, the nucleus as the **germinal**

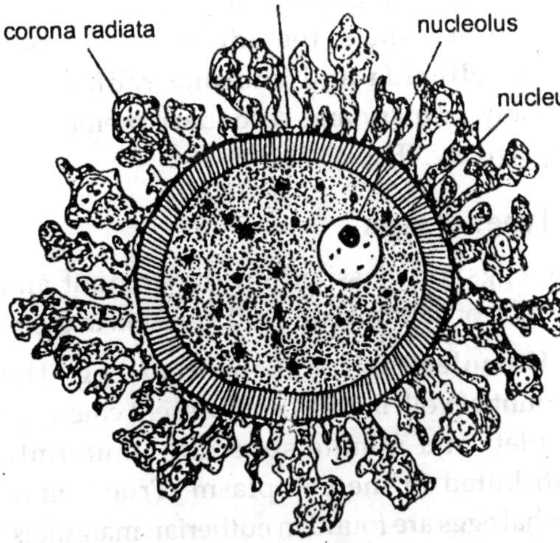

FIGURE 43.3

A human ovum

vesicle and the nucleolus as **germinal spot.** The yolk comprises (a) **cytoplasm** similar to that of other cells and (b) **deutoplasm** or **nutritive yolk,** consists of fatty droplets containing lecithin, a phospholipid. The mode of distribution of deutoplasm in the ovum differ in different animals. During oogenesis mitochondria increase in numbers and become uniformly distributed in the cytoplasm. The Golgi apparatus spreads out, being restricted usually near the periphery. The amount of yolk in the ovum varies in different groups of animals.

The mature ovum or egg is surrounded by two membranes, the **plasma membrane** (egg membrane) and the **vitelline membrane.** The vitelline membrane is thin and surrounds the plasma membrane. In addition to these primary membranes, secondary membrane secreted by the ovary and tertiary membrane secreted by the glands in the oviduct may surround the egg.

In *Branchiostoma*, only the plasma membrane and vitelline membrane are present. In frogs and toads a sphere of jelly around the egg is enclosed in a thin membrane. In birds, a two-layered shell membrane and a calcareous shell are secreted by the oviduct and, in mammals, the additional layers—**zona pellucida** and **chorona radiata** are possibly primary and secondary membrane, respectively, (Fig. 43.3).

☒ Types of eggs

The eggs are classifed on the amount and position of **deutoplasm** (yolk) present in them.

Homolecithal or **isolecithal egg.** The amount of yolk is small and present chiefly in droplets and minute spherules, uniformly distributed in the cytoplasm. True homolecithal eggs are found in eutherian mammals.

Microlecithal egg. The amount of yolk is small. Example: Cnidarians.

Megalecithal egg. The amount of yolk is large. Examples: Some invertebrates, reptiles, birds prototherian mammals.

Centrolecithal egg. The yolk is present around the nucleus in the form of a large sphere. Example: Arthropod eggs.

Telolecithal egg. Eggs are large, megalecithal and the yolk occupies almost the entire egg, except a minute area in the animal pole. In chordates, telolecithality occurs in varying degrees.

a. In *Branchiostoma*, the yolk is concentrated at one pole, the region of future endoderm cells.
b. In cyclostomes and amphibians, the quite high amount of yolk is concentrated in the vegetal pole.
c. In reptiles and birds, the amount of yolk is large and occupies almost the entire egg and the active cytoplasm. The egg nucleus forms a small cap on the yolk in the animal pole.

☒ Fertilization

Fertilization is a complete fusion of sperm and ovum except the chromosomes, resulting in a single diploid cell, the **zygote**. The union of the nuclei of male and female gametes is **karyogamy** and that of their cytoplasm is **plasmogamy**. The steps in fertilization (Fig. 43.4) are:

I. Approach of the spermatozoon to the egg

a. By chemotactic movement the sperm swims towards the egg.
b. The jelly coat of the egg produces **fertilizin** and the sperms produce **antifertilizin**.
c. The fertilizin and the antifertilizin combine to form an initial bond to facilitate penetration of spermatozoon into the egg.

II. Penetration of the sperm

a. The bound or agglutinated spermatozoon produces an enzyme, **lysine**, which dissolves egg membranes in the local area and clears the path for spermatozoon to reach the egg surface.
b. The acrosomal filament is pushed through the jelly and vitelline membrane, and touches the surface membrane of the egg cytoplasm.

III. Reaction in egg during sperm penetration

a. Coming in contact with the acrosomal filament the egg cytoplasm bulges forward to produce a conical projection, the **fertilization cone**.
b. It gradually engulfs the spermatozoon and carries it inward.
c. Simultaneously profound cortical reaction occurs in the cytoplasm.
d. Immediately after penetration of the spermatozoon, the vitelline membrane separates from the plasma membrane. The vitelline membrane thickens and forms the **fertilization membrane** which prevents entry of other sperms and thus ensures entrance of only one sperm into the egg.

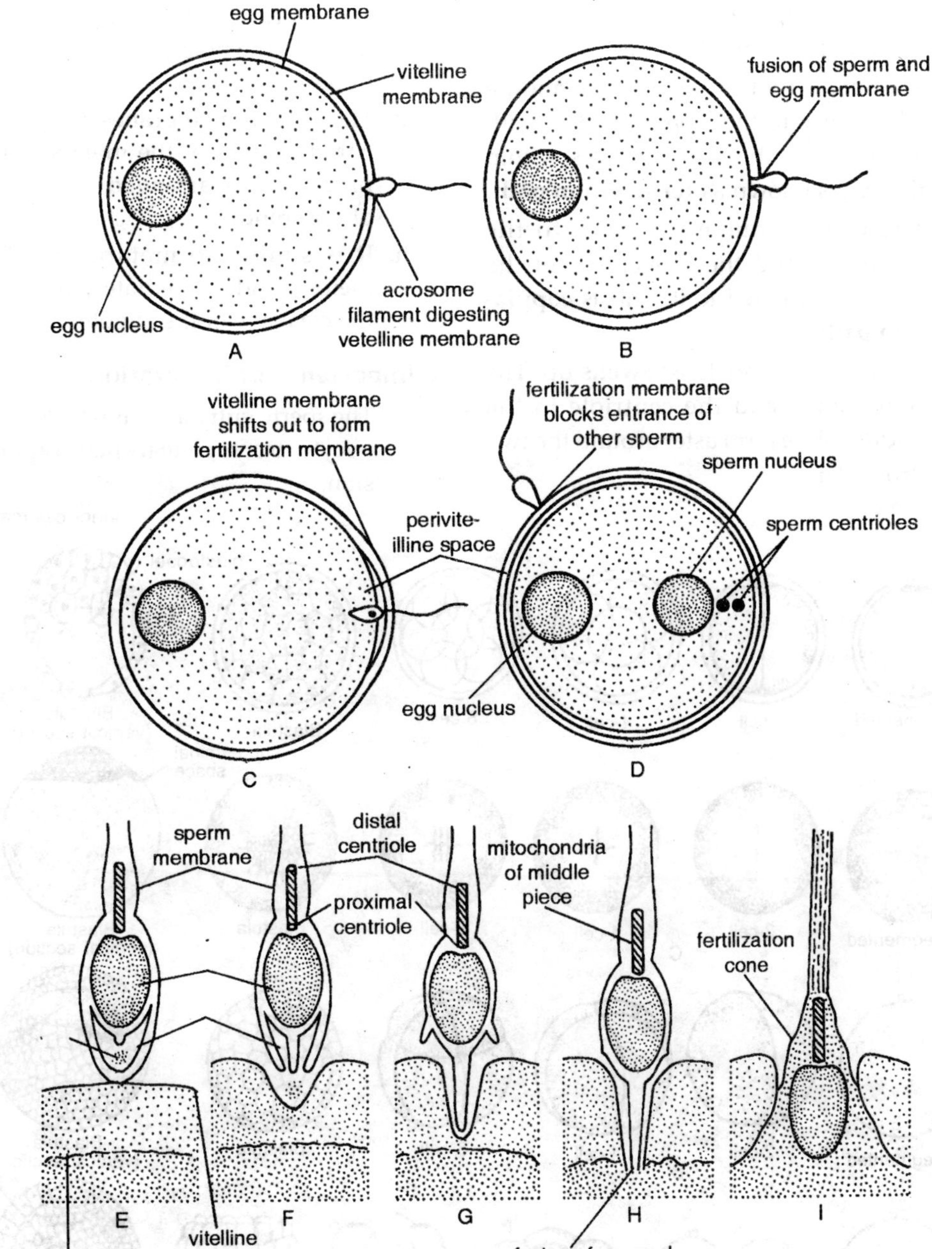

FIGURE 43.4

Fertilization. A-D. Union of a spermatozoon and an ovum. E-1. Insemination

e. The cortical granules swell, become liquefied and the fluid is released in the space between egg cytoplasm and fertilization membrane.

IV. Behaviour of sperm within the ovum

a. The head and middle piece enter the egg cytoplasm but the tail is left outside.

b. The egg completes its meiosis immediately and loses the centriole.

c. The sperm rotates 180º and its middle part comes to the front end. The path along which the sperm moves within the egg is known as **fertilization path**.

d. The egg pronucleus moves to the periphery and the movement of the sperm is oriented to it. This is **copulation path**.

e. The sperm pronucleus* swells up. The centrosome and the centriole in the middle piece form aster around the two pronuclei.

f. The sperm centriole divides into two, move to the poles, being connected by fibres.

g. The nuclear membranes disappear, the chromosomes of the male and female pronuclei are arranged in the equator of the spindle.

h. It is similar to metaphase plate of meiosis and immediately followed by the cleavage of the zygote.

⊠ Importance of fertilization

a. The sperm entry activates the secondary oocyte to complete its maturation division.

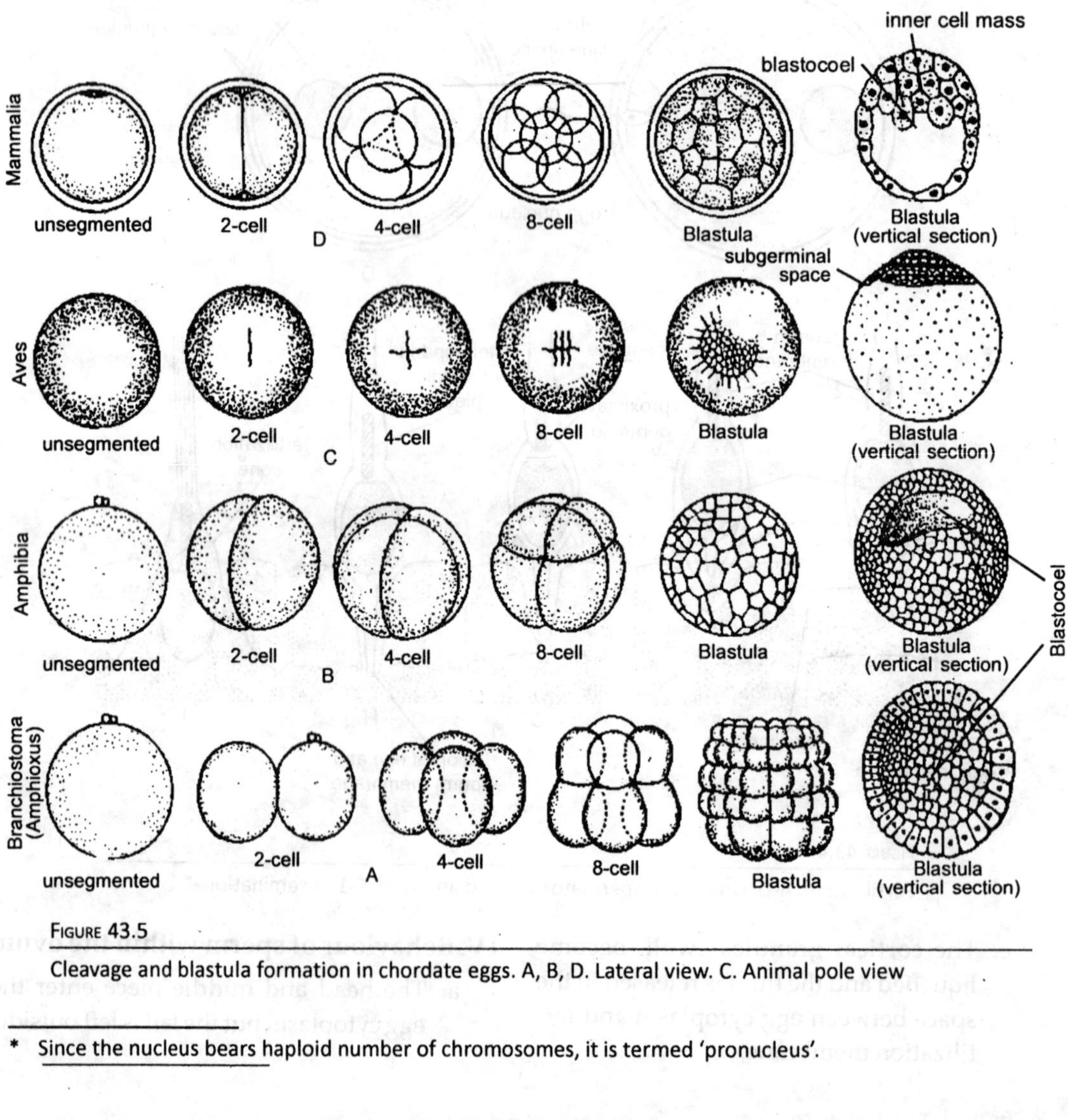

FIGURE 43.5

Cleavage and blastula formation in chordate eggs. A, B, D. Lateral view. C. Animal pole view

* Since the nucleus bears haploid number of chromosomes, it is termed 'pronucleus'.

b. Restores diploid number of chromosomes and recombines the maternal and paternal genetic materials.
c. Evokes development and cleavage of egg to form a complete organism.

SEGMENTATION

The life of a muticellular individual starts with the zygote. The zygote undergoes repeated mitotic divisions, **cleavage** or **segmentation** following a particular pattern, which varies to some extent in different groups of animals (Fig. 43.5). The amount and distribution of yolk in the egg also influence the mode of division. The daughter cells formed by the divisions of the zygote are known as **blastomeres**. The small blastomeres are called **micromeres** and the large ones **macromeres**. As a rule, the micromeres are formed in the animal pole and the macromeres in the vegetal pole.

☒ The plane of division of the zygote

The early cleavage takes place along a predestined plane, which differs in different groups of animals.

Meridional cleavage

The cleavage furrow starts at the middle of the animal pole and ends in the vegetal pole dividing the zygote into two equal blastomeres. In the majority of the animals the first cleavage is meridional.

Vertical cleavage

The cleavage plane is similar to that of the meridional cleavage but the cleavage furrow proceeds along a line lateral to the axis cutting both the poles. The blastomeres are of unequal size. In rabbit, the first cleavage is vertical.

Equatorial cleavage

The cleavage furrow proceeds along the equator of the zygote, producing blastomeres of equal size.

Latitudinal cleavage

The cleavage occurs in a plane parallel to the equator of the zygote latitude, nearer to the animal pole producing micro-and macromeres.

☒ The types of cleavage

The amount and distribution of the inert yolk influence the types of cleavage.

Holoblastic or complete cleavage

The cleavage is complete involving the whole of the zygote and distinctly separate (but not separated from one another) blasto-meres are formed. This occurs in homolecithal eggs. Holoblastic segmentation may be:

Equal. The blastomeres produced are of equal sizes. The cleavage may be lateral, radial and spiral.

Unequal. The blastomeres produced are of unequal sizes.

Meroblastic or incomplete cleavage

The division is incomplete and occurs in telolecithal eggs. The inert yolk occupying the vegetal pole does not divide and the division is restricted to the cytoplasmic part of the zygote in the animal pole.

Transitional cleavage

The cleavage is transitional between holo-and meroblastic cleavage.

SEGMENTATION: Formation of germinal layers : *Branchiostoma, Rana* (frog), *Gallus* (chick) and *Oryctolagus* (rabbit): A Comparative chart

The three layers of cells which are produced by the arrangement of cells or blastomeres formed due to cleavage or repeated mitotic divisions of the zygote are known as the three germinal layers. Presence of three germinal layers is characteristic of all triploblastic animals and all the different organs of an animal are produced form these three germinal layers, by a process known as **organogeny** or organ formation.

Branchiostoma (Fig. 43.6)	**Rana** (Fig. 43.7)	**Gallus** (Fig. 43.8-43.10)	**Oryctolagus** (Fig. 43.11)
1. The egg is telolecithal, the small amount of yolk is restricted to one pole. **Impregnation** is external.	1. The egg is telolecithal with an upwardly directed, pigmented protoplasmic mass and a lower white portion loaded with yolk. **Impregnation** is external.	1. The egg is large and telolecithal. The germinal disc is situated on the upper surface of the heavy yolk-mass. **Impregnation** is internal.	1. The egg is small and isolectihal. **Impregnation** takes place while the ova are still in the fallopian tube. Development starts while the zygote comes down to the uterus, and rest of the embryonic period is spent in the uterus.
2. **Segmentation.** Segmentation is holoblastic but slightly unequal. The first **cleavage** takes place along the meridional line, the second is also meridional but at right angle to the first, the third is latitudinal. Further divisions take place and a **blastula** enclosing a large **blastocoel** or primary body cavity filled with a fluid is formed. The cells of the lower or vegetal pole are larger than the rest.	2. **Segmentation.** Segmentation is holoblastic but slightly unequal. The first **cleavage** takes place along the meridional line, the second is also meridional but at right angle to the first, the third is latitudinal, and four small black and four large white cells are formed. Further divisions take place and a **blastula** enclosing a small **blastocoel** is formed. The divisions of the upper pigmented cells are rapid than the lower yolk cells.	2. **Segmentation.** Segmentation starts prior to laying of eggs. It is meroblastic and **discoidal**. The first **cleavage** is vertical and does not reach the edge of the germinal disc. The second one is also vertical but at right angle to the first. After several cleavages a single layer of cells is produced. Now a horizontal cleavage takes place and the cells are separated from the yolk, except the border of the germinal disc. A cavity, the	1. **Segmentation.** Segmentation holoblastic but irregular. The blastomeres produced are approximately of equal sizes and a spherical mass of cells or **morula** is formed. Vacuoles appear in one side of the morula and a fluid begins to accumulate within it. The **morula** becomes a **blastocyst**, corresponding to the blastula of lower forms. The blastocyst consists of an outer cell layer, the **trophoblast**, and attached to it is a group of

Branchiostoma (Contd.)	**Rana** (Contd.)	**Gallus** (Contd.)	**Oryctolagus** (Contd.)
		blastocoel or **subgerminal cavity** appears between the cells and the yolk. It is now known as **blastula** and the cellular region, the **blastoderm.**	cells, the **inner cell mass**. The cavity of the blastocyst is **blastocoel** or **subgerminal cavity**. From the inner cell mass arise the embryonic ectoderm and the entire endoderm. The trophoblastic ectoderm is **extraembryonic,** becuase it is concerned with the nourishment of the embryo.
3. **Gastrula.** The single-layered blastula is transformed into a two-layered gastrula enclosing an archenteron, by a process known as gastrulation. Gastrulation takes place by **invagination, involution** and a kind of **epiboly**. The rapid divisions of the micromeres lead to the invagination of the blastula. Gradually, a two-layered embryo is formed and the blastocoel is completely obliterated. The **gastrula** is now like a basin. It gradually bends, the **archenteron** is formed, and the **blastopre** destined to form the **anus** becomes narrower. The dorsal surface of the	3. **Gastrula.** The blastula is transformed into a **gastrula** enclosing an archenteron by a process known as gastrulation. Gastrulation takes place by **epiboly, involution, invagination** and **convergence**. The pigmented cells grow down over the yolk cells. Cells produced from the yolk cells acquire pigment, develop the character of ectoderm and gradually extend. The whole embryo is now covered by **ectoderm** cells except a small patch known as **yolk plug**. The margin of the ectoderm layer surrounding the yolk plug represents the **lip of the blastopore**. The	3. **Gastrula.** The blastula is transformed into a two-layered **gastrula** enclosing an archenteron, by a process known as gastrulation. Gastrulation takes place by **invagination** and **delamination**. In invagination, the cells move from the surface layer and detach within the subgerminal cavity, and finally arrange themselves to form a continuous layer. The original layer is termed **ectoderm** and the new layer, **endoderm**. (Fig. 43.8). In delamination, scattered cells are separated off the under surface of the blastoderm and being arranged in a row help	3. **Gastrula.** The blastocyst is transformed into a two-layered gastrula enclosing an archenteron, by a process known as gastrulation. Gastrulation takes place by delamination and infiltration of cells. The cells are produced by **delamination** and **infiltration** from the inner cell-mass on the ventral side and arrange themselves in a single row. This is the **endoderm** layer. The rest of the cells of the inner cell mass give rise to the **embryonic ectoderm** which is situated outside the endoderm. The cavity enclosed by the endoderm is the **archenteron**. The ectoderm and endoderm are in contact at one point only.

Branchiostoma (Contd.)	**Rana** (Contd.)	**Gallus** (Contd.)	**Oryctolagus** (Contd.)
embryo is flattened, the ventral surface convex and the blastopore is posterodorsal. Cilia develop from the ectoderm. Now the outer cell layer is known as **ectoderm** and the inner layer, the **endoderm**.	whole phenomenon is **epiboly**. A splitting takes place among the yolk cells, beginning at the edge of the blastopore and gradually extends forward. The process is aided by invagination of the small cells of the upper hemisphere (ectoderm) and the archenteron is formed. Archenteron does not communicate with the exterior for a long time, the blastopore being closed by the yolk plug. The **archenteron** extends forward and the blastocoel disappears. The yolk cells surrounding the archenteron become differentiated and form the definitive **endoderm**.	in the formation of the endoderm. The blastocoel which becomes enclosed by the endoderm layer above and the yolk below is the **archenteron**.	
4. The ectoderm of the median portion sinks below and form the **medullary plate**. The medullary folds gradually cover the plate, which forms the **neural tube**. The endoderm cells, lining the archenteron, develop a pair of dorsolateral longitudinal	4. The endoderm cells surrounding the archenteron undergo splitting and the **mesoderm** cells appear. These form several layers deep surrounding the archenteron. Mesoderm exists throughout the greater part of the embryo between the ectoderm and the endoderm.	4. The blastoderm is now distinguishable into a central clearer area, the **area pellucida** which overlies the subgerminal cavity, and a surrounding opaque area, the **area opaca** which is placed, directly on the yolk. By the thickening of the ectoderm	4. At the point where the ectoderm and endoderm are in contact, the **primitive streak** appears by the thickening of the ectoderm cells. From either side of the primitive **streak**, cells are produced rapidly and they arrange themselves in a single layer,

Branchiostoma (Contd.)	**Rana** (Contd.)	**Gallus** (Contd.)	**Oryctolagus** (Contd.)
folds. Transverse folds appear and divide the longitudinal folds into a series of coelomic pouches and this is the beginning of the formation of **mesoderm**. Gradually the endoderm cells which become cut off from the cells lining the archenteron give rise to the **definitive mesoderm** and the endoderm cells, which are now drawn to line the archenteron completely form the **definitive endoderm**.		cells, at the point where the two layers are continuous, the **primitive streak** appears (Fig. 43.9). The primitive streak has a central groove, and the groove runs along the centre of the area pellucida. From either side of the primitive streak a layer of cells is produced between the ectoderm and the endoderm. This is the **mesoderm** (Fig. 43.10). All the three cell layers are connected at the sides of the primitive streak.	the **mesoderm**, in between the ectoderm and the endoderm. The mesoderm soon splits into two, the outer layer is known as **somatic mesoderm**; the inner layer, **splanchnic mesoderm** and the space between the two is the **coelom**.

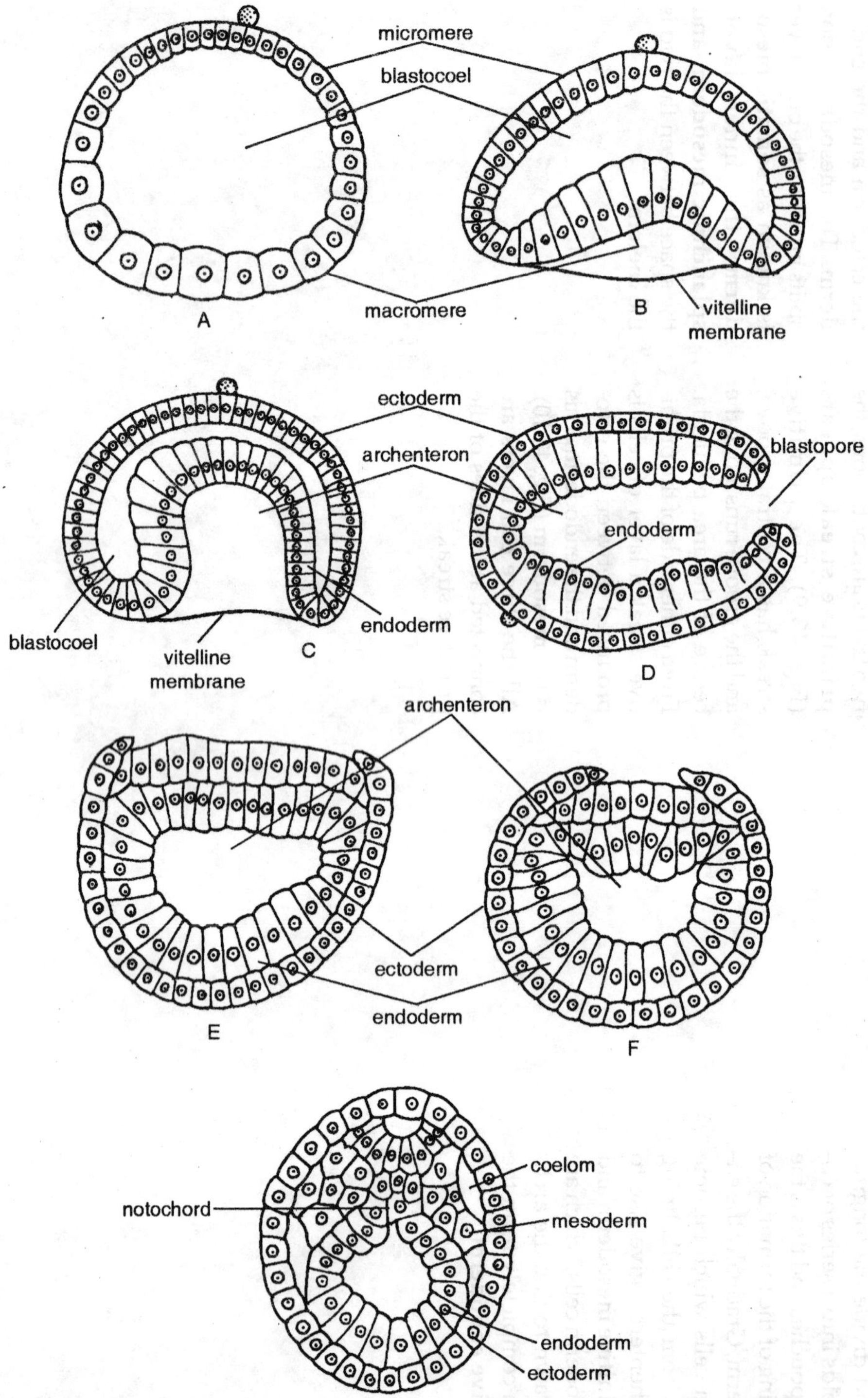

FIGURE 43.6

Branchiostoma sp. Formation of germinal layers (Sectional view). A. Blastula. B. Beginning of gastrulation. C-D. Gastrulation and formation of archenteron. E. Formation of neural plate. F. Formation of neural plate and mesoderm folds. G. Formation of mesoderm and notochord

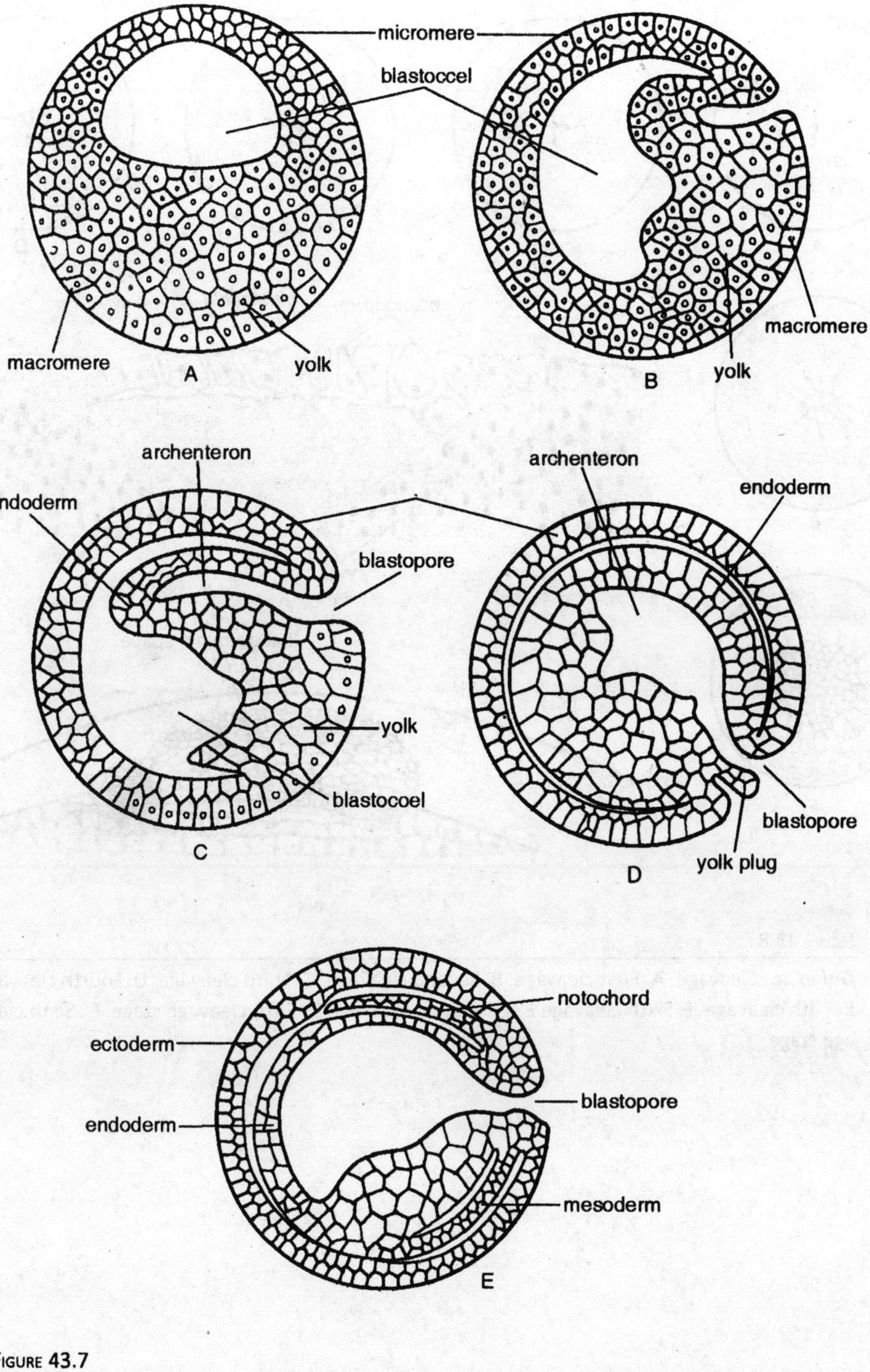

FIGURE 43.7

***Rana* sp.** Formation of germinal layers (sectional view). A. Blastula. B. Beginning of gastrulation. C-D. Gastrulation and formation of archenteron. E. Formation of mesoderm and notochord

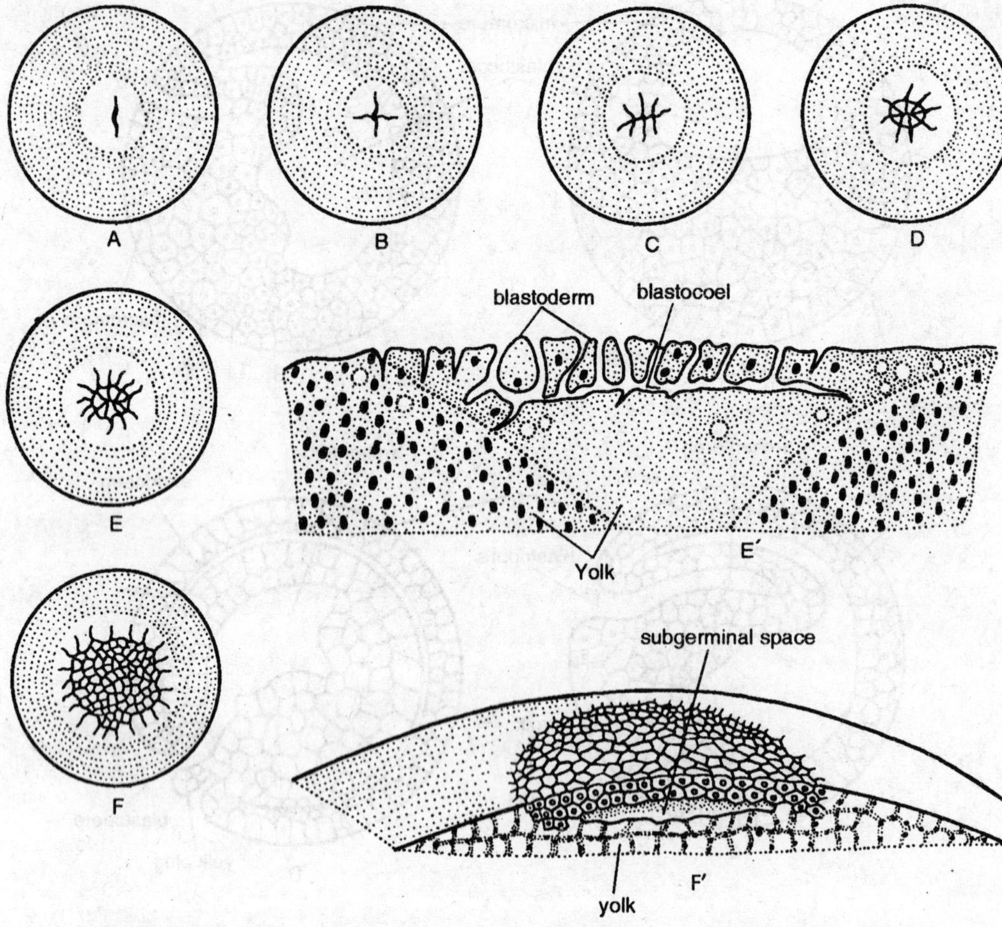

FIGURE 43.8

***Gallus* sp.** Cleavage. A. First cleavage. B. Second cleavage. C. Third cleavage. D. Fourth cleavage. E. Fifth cleavage. F. Sixth cleavage E′-F′. Vertical sections. E′. Fifth cleavage stage. F′. Sixth cleavage stage

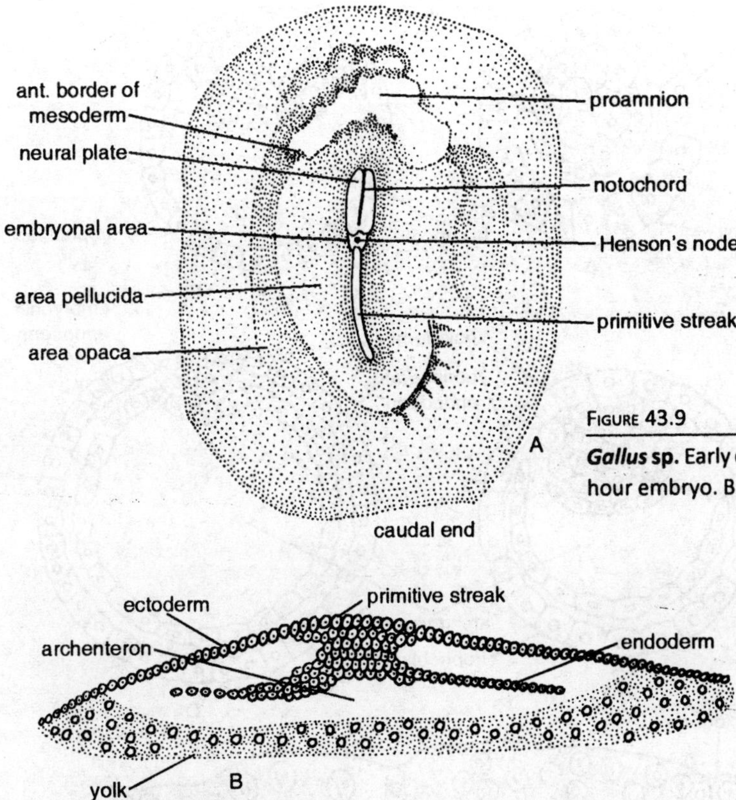

FIGURE 43.9

***Gallus* sp.** Early development. A. An 18-hour embryo. B. Ventrical section of A

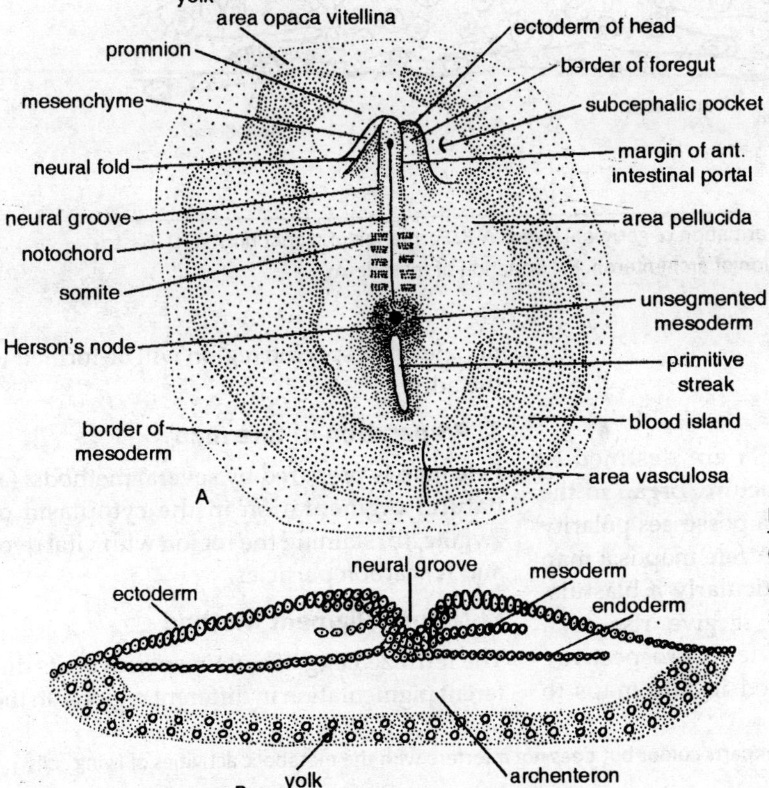

FIGURE 43.10

Gallus sp. Early development. A. 24-hour embryo. B. Vertical section of A

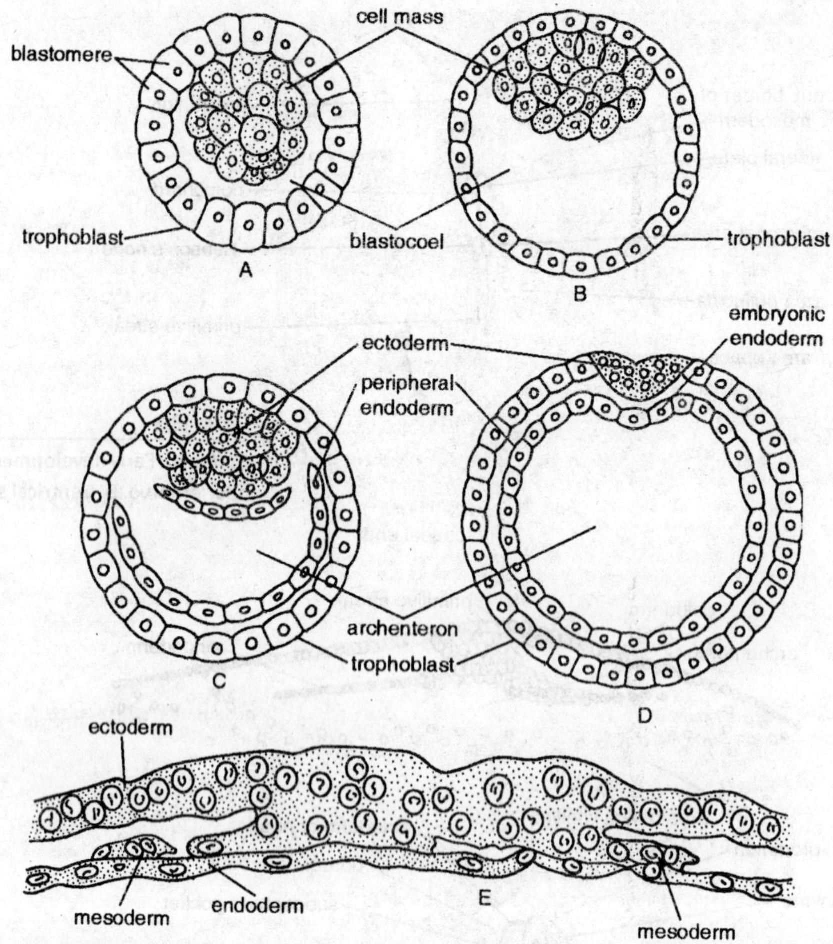

FIGURE 43.11

***Oryctolagus* sp.** Formation of germinal layers. A-B. Formation of trophoblast. C-D. Gastrulation and formation of archenteron. E. Formation of mesoderm

FATE MAP

⊠ Fate Map

Certain cells in a blastula are destined to produce a particular structure/organ in the adult animal. The blastula possesses polarity and bilateral symmetry. A fate map is a map of an early embryo, particularly a blastula, showing which part of it give rise to a particular structure. The term prospective/presumptive is often used in fate maps to donate what specific region will be formed in the adult.

⊠ Preparation of fate map

Fate map is prepared by several methods: (a) natural pigmentation in the cytoplasm of Zygote, (b) staining the region with vital dye* and (c) carbon particles.

⊠ Natural Pigment Method

The fertilized egg of certain species bears different pigmentation in different regions. In the

* **Vital dye.** The stain which imparts colour but does not interfere with the metabolic activities of living cells.

ascidian *Cynthia partita* the cytoplasm of the upper hemisphere is **light yellow crescent** posterventrally. **Gray crescent** anterodorsally and **slaty gray** in vegetal area. From the development of the pigmented regions it has been concluded that the upper light hemisphere represent prospective epidermal ectoderm, the yellow crescent prospective mesoderm, the gray crescent prospective neural ectoderm and notochord and the slaty gray perspective endoderm.

⊠ Vital Stain Method

A piece of thin agar or cellophane sheet* of a particular shape and size soaked in vital dye-Nile blue sulphate, neutral red, Bismarck brown— is placed on the surface of the ambryo (blastula) in the desired position, without removing the vitelline membrane, since the membrane is permeable to vital dyes. The dye diffuse from the sheet in minutes and the cells of the embryo to which it has been applied take up sufficient dye to produce a stain on the surface of the embryo. The embryo continues to develop normally. The vital dye remain restricted to the cells which had originally taken up the dye and to their descendants. The stained cells in the differentiated embryo clearly show the fate of the stained area.

⊠ Carbon Particle Method

Tiny carbon particles are applied to the surface of the embryo. The particles stick to the surface and are used as markers to follow the movement of cells and draw the fate map.

⊠ Fate map: Amphibia

The whole surface of an amphibian blastula may be divided into three main regions—(a) Animal region, an area around the animal pole, (b) an intermediate or marginal zone and (c) vegetal region, an area around the vegetal pole, the areas corresponding to the future organ of the animal (Fig. 43.12).

⊠ Animal Region

The animal region is deeply pigmented. The region has two major areas—fate of one is to develop nervous system and that of the other, skin epidermis. Material for sense organs is present in both the areas.

A. **Nervous system area.** It is a small sub area which forms eyes of the embryo.

B. **Epidermis area.** Material for nose and the ears and the ectodermal part of the mouth are present in the area.

C. **Intermediate or marginal region.** The zone is much broader on one side than the other and less pigmented than the animal region. A number of structures develop from this region. Material for notochord occupy a large area on the dorsal side of the blastula. Materials for the prechordal connective tissue or the prechordal plate lies nearer the vegetal pole. Further down towards the vegetal pole, the intermediate zone contains material for the anterior region of the alimentary canal endodermal lining of the mouth, pharynx and gill region. Material for the segmental muscles occupy the region on both sides of the notochord area. The lateral and ventral part of the intermediate zone produce the mesodermal lining of the body cavity, the kidneys, etc.

⊠ Vegetal region

The vegetal region is practically devoid of any pigment. The cells of the region produce the mid and hindgut.

MORPHOGENETIC MOVEMENT—CELL MIGRATION

The term morphogenesis is used to describe the process of formation of embryonic structures. The morphogenetic movements are the movement of the large parts of the whole embryo. These include streching, folding, contraction and expansion of the parts of the

* Currently thin cellophane sheet instead of agar is used as it produces better results.

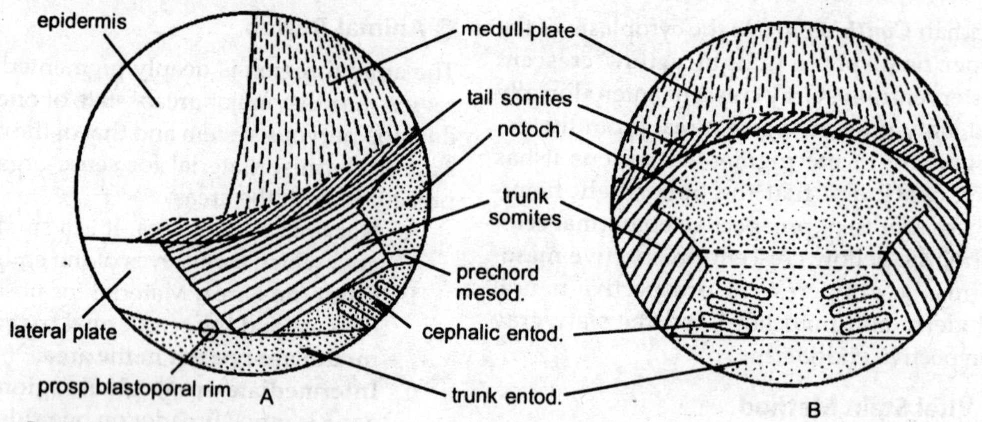

FIGURE 43.12
Fate map of amphibian embryo (Axolotl). A, Lateral view, B. Dorsal view

embryo. Each moving part consists numerous cells, and the movement of the whole would be an integrated result of the movements of the individual cells.

⊠ Movement of individual cells (migration)

Ingression, egression. Individual cells move into the embryo or move out towards the surface.

Delamination. Splitting of an epithelium into two or more layers. In the process individual cells move out of the epithelium.

Intercalation. In the merger of two or more epithelial sheets the individual cells move into one epithelial sheet.

Condensation. Dispersed mesenchyme cells congregate to form an epithelium.

Dispersal. Epithelium cells bręak up to form dispersed mesenchyme cells.

⊠ En masse movement of cell sheet

Invagination, evagination. Folding of a sheet of cells and formation of an in pushed or out pushed structure. In invagination the fold is inwards, in evagination the fold is outwards.

Involution. The sheet of cells expand inwards spread over the internal and a second layer is formed.

Epiboly. Spreading of a sheet of cells due to proliferation of cells.

Convergent extension. Cells converge at the organizer and then extend along the future anteroposterior axis of the embryo.

ORGANIZER

The ability of the dorsal lip of the blastophore to cause when transplanted, the development of a complete whole that led to the dorsal lip of the blastopore is called organizer or primary organizer (Spemann, 1938).

Currently, it is held that any region of an embryo with ability to direct the development of surrounding tissues with inductive signals is an organizer. The axes of a developing structure is generated by the organizer and thus initiate pattern formation. An organizer when grafted, a tissue having its own organizer leads to the duplication of the developing structure. The organizer is a highly specialised structure having multiple role in axis determination, pattern formation and morphogenesis.

In a broad sense the term organizer—referred to as **primary embryonic organizer,** has been given different names.

Embryonic shield—Zebra fish
Spemann organizer—Amphibia
Henson's node—Birds and mammals

As a rule developmental systems have their own organizers. From the studies on *Xenopus* it has been ascertained that:

a. organizer for hindlimb development—posterior margin of the limb bud.
b. organizer for midbrain development—isthmus of mid and hind brain.

⊠ Organizer's role in development

A number of genes coding transcription factors and secreted proteins are specifically expressed in the organizer. Several organizer-specific transcription factors have been recorded.

a. Initiation of cell movement in gastrulation.
b. Formation of anterior endoderm, dorsal mesoderm, somittes, neural floor plate and dorsal mesoderm of the gut during gastrulation.
c. Generates a range of intermediate mesodermal cell types along the dorsoventral axis by dorsalizing the adjacent ventrolateral mesoderm.
d. The neural plate is induced by dorsalizing the overlying ectoderm.
e. Induces head development from the anterior dorsal mesoderm and anterior endoderm.
f. Induces anteroposterior positional information on mesoderm and endoderm cells.
g. Introduce left-right symmetry.

INDUCTION AND COMPETENCE

The ability of one cell or a group of cells to change developmental fate of another is induction. The ability to respond to a specific signal is termed competence.

⊠ Induction

The inductive signal produced by a cell (inducer) may be a protein or another molecule. The signal may interact with the surface receptor of the responding cell or may pass through the plasma membrane to react with the cytosol receptor. The signal alters the functioning of the transcription factors or proteins, causing change in the pattern of gene interactions.

Induction can take place in different ranges—locally or at some distance. Local induction takes place in early development with the release of diffusable substances (paracrine signaling) or by deposition of a substance in the matrix. The range of signals in most cases is restricted to a few cells but it may be extended by a relay mechanism from one cell to another and so on. In the event of induction at a distance as in endocrine signaling, the inductive signals—hormones, released by the cells of endocrine glands are moded to a distant target—induction receiving cell through blood.

In addition to·the inductive signals that induces a single response in responding cells, there exist signals that elicit different responses in different concentrations. Such inductive signal is known as **morphogen**. Such a situation is found in neural induction.

⊠ Competence

The capability of a cell to respond to inductive signal by undergoing molecular changes that allow it to follow the induced development pathway is competence. The cells follow a different pathway in the absence of induction and it leads to loss of competence in response to induction. In the case of endocrine or paracrine signaling competence depends upon several factors that link inductive signal to the target. If any of the factors is missed the cell loses competence. The cells become irreversibly committed to a particular development pathway, is loss of competence.

DIFFERENTIATION AND GROWTH

The main features of the organisation of an animal is established with the formation of the rudiments. The organ rudiments are, however, incapable of performing specific functions, on which depends the ability of the animal to lead an independent life. The establishment of the morphological plan is followed by differentiation and growth to bring the animal to its functional state.

☒ Differentiation

The process by which the cells become structurally and functionally specialized is called differentiation.

The term differentiation is used at least in two sanses.

1. In the broad sense, the cells or other parts of an organism become differentiated from one another and also differ from their previous state.
2. In the narrower sense, the term differentiation is histological differen-tiation.

The special functions of cells are different from the basic functions of life, which are common to all living cells. Every cell—undifferentiated or differentiated is capable of performing the processes of metabolism.

Differentiation leads to the formation of distinct cell types— myocytes, erythrocytes, leucocytes, keratinocytes, neurons, etc. to carry out different functions. The structural and functional specificity of a cell depends on the proteins it synthesizes.

Synthesis of different proteins may be achieved by DNA rearrangement. Since most cells contain the same DNA, only selective expression of certain gene products can lead to the formation of different sets of proteins. A series of steps—transcription, RNA processing, protein synthesis and post translational protein modifications can lead to the synthesis of a functional protein.

Any or all steps of functional protein synthesis can be regulated. Activation of a particular regulatory molecule may result in differentiation.

☒ Growth

Growth is the increase in the size of an organism or its parts due to synthesis of protoplasm or of apoplasmatic substances. Substances produced by cells such as fibres of the connective tissue, the matrix of bone and cartilage are apoplasmatic substances.

Growth occurs when **anabolism** (synthesis) is greater than **catabolism** (destruction). There is **no growth** in the bulk of the organism when synthesis and destruction are at the same level. In certain cases catabolism exceeds anabolism as in prolonged starvation—when synthetic process cannot take place due to lack of food supply, but catabolism continues to meet the energy requirements of the body. In the process the internal food reserves—the fat in the adipose tissue is used and finally exhausted. At this stage, energy is produced at the expense of protein in the protoplasm, which causes decrease in the mass of the living matter, called **degrowth.**

FOETAL MEMBRANES OF GALLUS (chick) and ORYCTOLAGUS (Rabbit)

Foetal membranes are specialised embryonic tissues, develop as membranes round the growing embryo during organogesis in amniotic vertebrates to carry certain definite functions necessary for the embryo. These membranes are the yolk sac, amnion, chorion (serosa) and allantois.

A. Comparative of foetal membranes of *gallus and Oryctolagus*

Gallus (Fig. 43.13)

A. **Formation of Amnion**

The amnion begins to form on the second day of incubation and is completed on the fourth day.

1. A fold of blastoderm occurs just in front of the head of the embryo. It consists of ectoderm and endoderm. With the appearance of mesoderm, it takes part in the same and finally the **head-fold** of the amnion is constituted by an extra embryonic ectoderm and a somatic mesoderm. The fold gradually grows back over the embryo.

   At the beginning of the thrid day, another fold, the amniotic **tail-fold**, appears at the posterior end of the body and it gradually grows towards the head-fold. It consists of ectoderm and somatic mesoderm from the beginning.
2. At the beginning of the fourth day the amniotic folds meet and fuse at the centre and the embryo is surrounded by a cavity (Fig. 43.13B), the **amniotic cavity**, which is bounded by two membranes—the outer and the inner.
3. The amniotic cavity is filled up with a fluid, the amniotic fluid. The fluid acts as a shock absorber and protects the embryo. The outer layer is known as the **chorion** and the inner layer as the **amnion**. The space between the amnion and chorion is known as the **extraembryonic coelomic cavity**. Both the layers consist of an outer ectoderm and an inner mesoderm facing the amniotic cavity.
4. Muscle fibres in the amniotic wall contract rhythmically which rocks the embryo in the amniotic fluid and the adhesion of the embryo to the amniotic wall is prevented.

Oryctolagus (Fig. 43.14)

A. **Formation of Amnion**

The amnion begins to form after the appearance of the coelom as a result of splitting of the mesoderm into two layers—the outer somatic and the inner splanchnic layer.

1. The amnion arises by two folds, one in front of the embryo, the **head-fold**, and the other at the posterior end of the embryo, the **tail-fold**. Both the folds grow upward and inward. Each fold consists of an outer trophoblastic ectoderm and an inner somatic mesoderm. These constitute a part of the extraembryonic layers.
2. The two amniotic folds meet at the middle over the embryo and unite to form two layers. The embryo is surrounded by a cavity, the **amniotic cavity** (Fig. 43.14B), which is bounded by two membranes—the outer and the inner.
3. The amniotic cavity is filled up with a fluid, the amniotic fluid. The fluid acts as a shock absorber and protects the embryo. The outer layer is known as **chorion** and the inner layer is **amnion**. Both the layers consist of an outer trophoblastic ectoderm and an inner somatic mesoderm facing the amniotic cavity. The chorion becomes closely associated with the mucous membrane of the uterine wall and sends out processes or **primary villi** by which attachment and nourishment are effected.
4. The rhythmic contractions of the muscle fibers in the amniotic wall rock the embryo in the amniotic fluid and prevent the adhesion of the embryo to the amniotic wall.

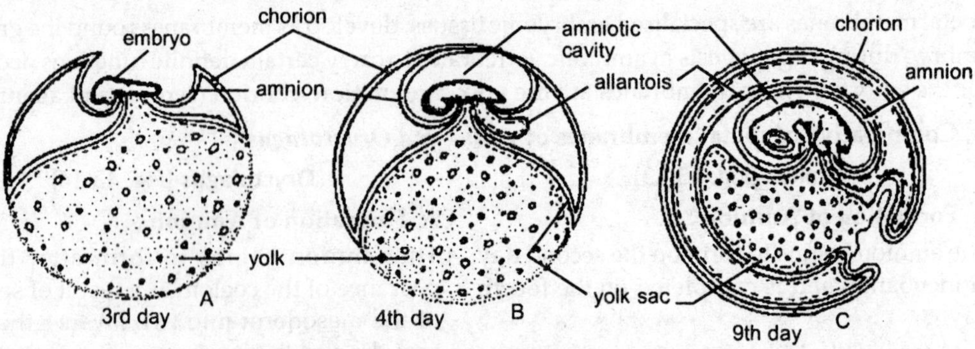

FIGURE 43.13

***Gallus* sp.** Development of foetal membranes

B. Formation of Allantois

The allantois begins to form on the third day and becomes well-developed by the end of the fifth day.

B. Formation of Allantois

The allantois begins to appear at the time of establishment of the amnion and grows out rapidly.

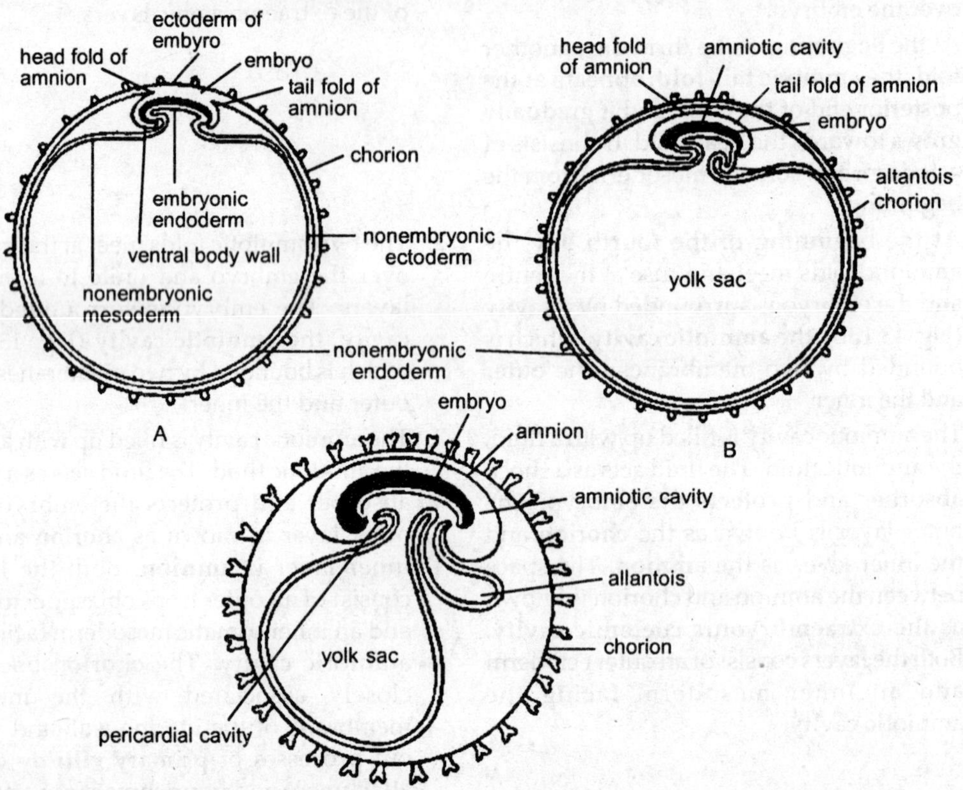

FIGURE 43.14

Oryctolagus sp. A. Early stage of the formation of amnion and yolk sac (Vertical section). B. Amnion about to close and beginning of allantois formation (Vertical section). C. Embryo with completed amnion, larger allantois with villi on the chorion (Vertical section)

Gallus (Contd.)

1. The allantois arises as an outpushing from the ventral wall of the hindgut and consists of an inner endodermal and an outer mesodermal layer.
2. The allantois gradually pushes out through the coelomic space between the somatic umbilicus and the yolk-stalk, and spreads out in the extra embryonic coelom. The neck of the allantois is known as **allantoic stalk** and two arteries and one vein pass through it.
3. It grows rapidly and covers the amnion within two days and coming very close to the shell acts as a respiratory organ due to the presence of ramified blood vessels. By this time it fuses with the chorion.
4. The allantois finally encloses the whole embryo and yolk-sac with the remains of the albumen. It absorbs the white of the egg and acts as an organ of nutrition. It is also excretory and acts as the urinary bladder of the embryo, receiving wastes from the kidney.

Oryctolagus (Contd.)

1. The allantois arises as an outpushing from the ventral wall of the hindgut, and consists of an inner endodermal and an outer splanchnic layer of mesoderm.
2. The allantois gradually pushes into the extra-embryonic coelom between the chorion and amnion. The neck of the allantois is known as **allantoic stalk** and two arteries and one vein pass through it.
3. It grows rapidly, covers the amnion and fuses with the chorion and trophoblast in a small area. The wall of the allantois is highly vascular.
4. The allantois finally encloses the whole embryo and the yolk-sac. The region where the chorion, trophoblast and allantois unite, sends vascular outgrowths or **villi** which burrow into the corresponding crypts or depressions on the uterine wall and an **allantoic placenta** is formed, through which nutrition, excretion and respiration are performed. Prior to the formation of allantoic placenta, nourishment is temporarily drawn by yolk-sac placenta. Part of the yolk-sac wall unites with the chorion and trophoblast and forms the foetal part of the **yolk-sac placenta**.

FORMATION OF PLACENTA IN ORYCTOLAGUS (Rabbit)

The placenta is a double vascularised organ (Fig. 43.15) in the formation of which both maternal and foetal tissues take part. Placenta brings the maternal blood in close contact with the foetus and thereby supplies the different requirements of the growing foetus and at the same time removes metabolic wastes formed in the foetus.

Changes in the embryo

1. After fertilization the zygote descends in the oviduct, undergoes cleavage, and a blastocyst, corresponding to the blastula of lower forms is produced.
2. The epiblast formed during gastrulation is differentiated into an inner cell mass and a subzonal layer, the **trophoblast**.
3. The trophoblast is rapidly differentiated into an outer layer—the **plasmotrophoblast**, and an inner layer or layers of **cytotrophoblast**.

Implantation of the ovum

1. A pair of folds arise upon the mesometric side of the uterus and the blastocyst lies in between the folds and becomes attached by the trophoblast.

2. The enzymic activity of the trophoblast erodes the uterine epithelium, and the embryo becomes surrounded by a cushion of liquefied cells and blood called **histotroph**.

Later development

1. In the beginning, the chorionic villi surround the whole embryo but later all except those between the uterine wall and the chorion atrophy.

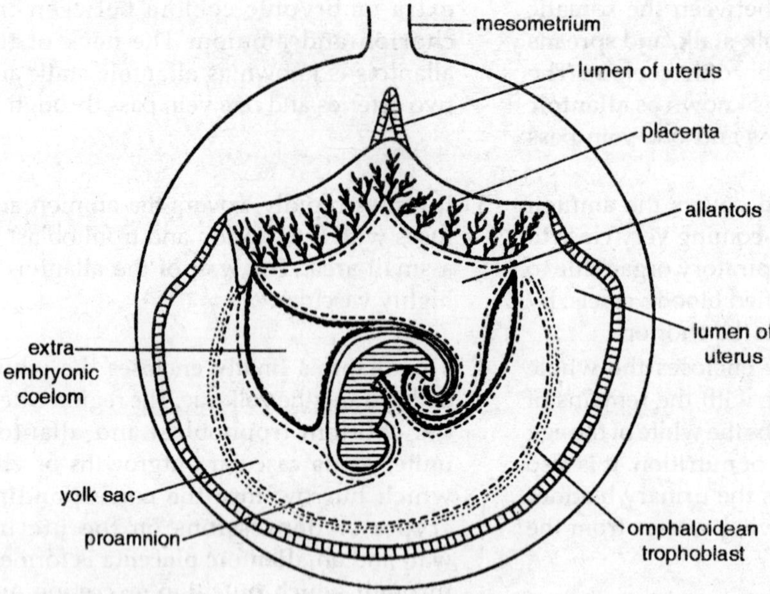

FIGURE 43.15 *Oryctolagus* **sp.** Foetal membrane and placenta

3. The embryo sends finger-like projections or **villi** in the uterine tissue.

Appearance of primary villi

1. The plasmotrophoblast soon turns to a solid mass of multinucleated and vaculated protoplasm.
2. The vacuoles are termed primary **lacunae.**
3. The plasmotrophoblast sends projections into the uterine mucosa and pushing out the somatic cells approaching the capillaries, establishes communication between the lacunae and the capillaries.
4. The lacunae break up the trophoblast into trabeculae giving the entire chorion a villus appearance, the **primary villi**.

Secondary or true chorionic villi

The cytotrophoblast also forms villi communicating with the chorion and these are called **true villi**.

2. The region where the chorionic villi are restricted is known as **chorion frondosum**.
3. The placenta now consists of the villi of the chorion frondosum and part of the uterus in which the villi are lost.

Nutrition in placenta

1. **Histotrophic nutrition.** Nutrition is supplied from the mother to the embryo.
2. **Haemotrophic nutrition.** Trophoblast penetrates the trophospongia facilitating the osmotic interchange of fluid and diffusion of gases between the foetal and maternal blood.

⊠ Functions of Placenta

1. Anchoring of the foetus to the uterine wall of mother.
2. Supply of nutrients and other requirements to the foctus and removal of metabolic wastes from factus to maternal circulation. Act as a foetal

lung and kidney and prevent bacterial infection.

☒ Physiology of the Placenta

A semipermeable membrane acting as a barrier is present between the maternal and foetal blood, and the blood of mother and embryo do not mix.

1. Small malecules pan through the semi-permeable membrane by simple diffusion.
2. Water, oxygen, simple salts of sodium, potassium, magnesium, monosaccharides, etc. pass from maternal to foetal blood.
3. Penetration of some complex substances is effected by active transport and in some cases by pinocytosis.
4. Vitamins, hormones, antibodies, etc. pass through placental barrier.
5. Substances like carbon dioxide, urea and other metabolic wastes pass out from foetal to maternal blood by diffusion.

PLACENTATION IN MAMMALIA

Placenta is an organ built up of maternal and foetal tissues jointly for transport of nutrient from the tissues of the mother into those of the embryo and at the same time passing out of metabolic wastes and other substances from foetus to maternal blood.

☒ Types of placenta

Two types of placentae are present in mammals.

1. **Chorio-vitelline placenta.** The connection is established between uterine wall of the mother and the part of the chorion of the foetus lined by the yolk sac with its network of vitelline blood vessels. Examples: Marsupials (*Didelphys, Macropus*, etc.).
2. **Chorio-allantoic placenta.** Connection is established between the chorion lined by the allantois with the allantoic blood vessel system. Examples: Eutheria, some marsupials (*Parameles, Dasyrus*), etc.

☒ Placenta formation

After the embryo enters the uterus, the zona pellucida surrounding it dissolves and the embryo, now in the **blastocyst stage** is bathed in the uterine fluid containing organic substances produced by the tubular glands of the uterine wall.

The embryo absorbs some of these substances till an intimate connection is not established with the uterine wall.

To maintain a steady and abundant flow of organic substances a close relationship is established between the foetal and maternal tissues.

Nondeciduous placenta

Fingerlike outgrowths develop from the surface of the blastocyst, which penetrate into the depressions of the uterine wall. The outgrowths are known as **chorionic villi**, initially formed by the trophoblast—the epithelium covering the blastocyst. Later connective tissue and blood vessels enter the outgrowths (Fig. 43.16).

In **chorioallantoic placenta** the blood vessels are formed by the ramification of allantoic blood vessels. In primitive **choriovitelline placenta** blood vessels are branches of vitelline blood vessels.

At the time of parturition the chorionic villi are easily drawn out from the depressions of the uterine wall and material and foetal tissues are separated practically without any damage. Such a placenta is known as **nondeciduous placenta**. Examples: Pig, Cattle, etc.

Deciduous placenta

The degree of connection between maternal and foetal tissues is intimate. Through the action of trophoblast the uterine wall becomes

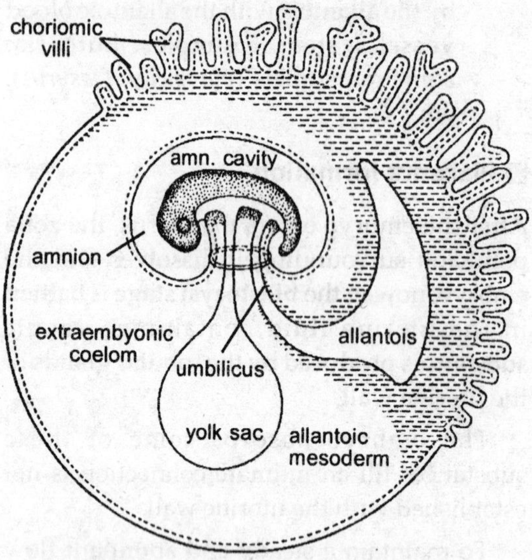

FIGURE 43.16

Embryo of placental mammal, surrounded by embryonic membrane (diagrammatic)

eroded and the embryonic tissue penetrates into the uterine wall. This facilitates the passage of materials from mother to foetus and from foetus to mother. At the end of the gestation period the uterine wall is no more intact. During delivery, the removal of the foetus with its membranes including chorion causes more or less extensive haemorhage from the uterine wall. The uterine wall participating in the formation of such a placenta is called the **decidua**. Example: Anthropoid apes, human.

Interstitial implantation

Following the establishment of contact of the blastocyst with the uterine wall, the adjacent epithelium of the latter is destroyed and the blastocysts sinks into the underlying connective tissues. Finally the uterine wall closes over the implanted blastocyst. The trophoblast proliferates profusely and sends out growth, which penetrate deeper in the uterine wall. The cell boundaries of the external lining part of the trophoblast disappear and the mass of the trophoblast becomes syncytium—the **syncytiotrophoblast**. The portion of the trophoblast nearer the embryo retains its cell boundaries and termed **cytotrophoblast**. The trophoblast becomes permeated by a system of cavities—the **trophoblastic lacunae** (Fig. 43.17).

The whole mass of trophoblast, which is a meshwork of irregular strands with interstices in between, increases and penetrates deeper into the uterine wall. It reaches the small blood capillaries and breaks up the endothelium of the capillaries. Thus a communication is established between the lumen of the capillar-

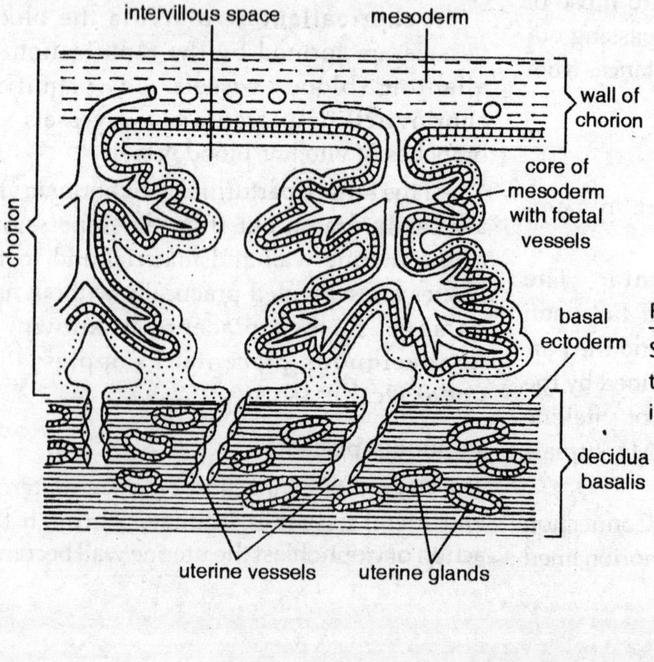

FIGURE 43.17

Schematic diagram of interrelations of foetal and maternal tissues in formation of placenta

ies and the trophoblastic lacunae. Both the arterial and venous branches of the blood vessels become communicated with the trophoblastic lacunae and blood circulation is established.

In course of development the cytotrophoblast penetrates into the strands of syncytiotrophoblast and form a cellularized core within the syncytial mass. Following this, the extraembryonic tissue underlying the trophoblast penetrates the cytotrophoblastic strands. This is soon followed by blood vessels extensions and ramifications of the allantoic blood vessels cellularized (Fig. 43.18).

The shape and arrangement of trophoblastic strands change and form a dendrite system, with the main stem arising from the surface of the blastocyst and fingerlike branches—the villi spread out into the blood-filled trophoblastic lacunae. Some of the villi reach the maternal tissue and anchor the whole foetal system to the uterine wall.

PLACENTAE IN DIFFERENT GROUPS OF MAMMALS

Differences exist in placentae of various groups of mammals due to (a) structure and arrangement of villi and also (b) the degree of connection between the maternal and foetal tissues.

Diffuse placenta. The villi may be scattered all over the surface of the chorion. Example: pig.

Cotyledon placenta. The villi are arranged in patches or groups and the rest of the chorion surface is smooth. The patches are termed cotyledon. Example: cattle.

Zonary placenta. The villi develop in the form of a belt around the middle of the more or less elliptical blastocyst. Example: Carnivores.

Discoidal placenta. In the beginning the whole chorion is covered with villi but later the villi continue to develop only on the side turned away from the lumen of the uterus. The villi in other parts of the chorion are reduced and the functional placenta is disc-shaped. Examples: Anthropoids apes, human.

Bidiscoidal placenta. The placenta consists of two discs. Example: Monkey.

ORGANOGENESIS

With the begining of differentiation in an early embryo—the primordia—structures destined to produce organs appear first. In the course of development the primordia form rudiments, the beginning of organs. The development of the organs in the embryo is organo genesis.

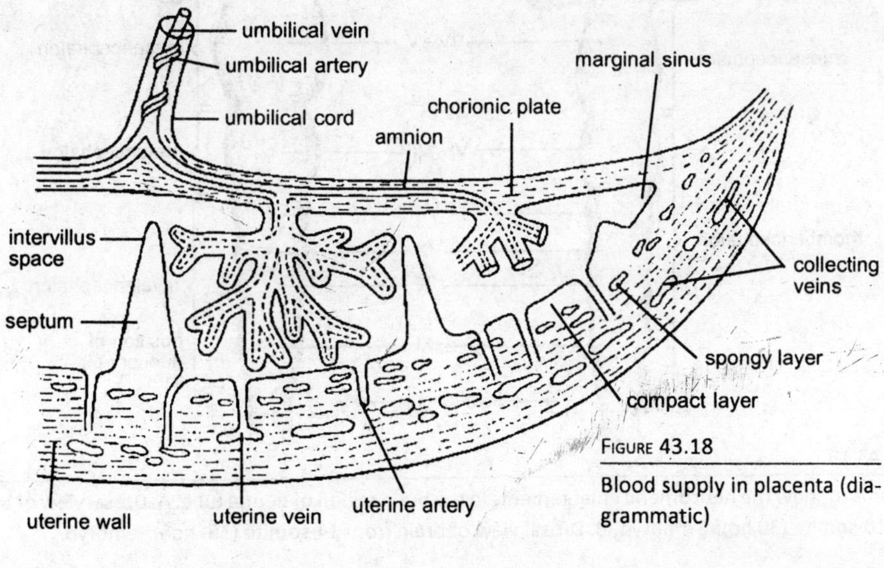

FIGURE 43.18

Blood supply in placenta (diagrammatic)

CHICK: DEVELOPMENT OF BRAIN

The earliest indication of the development of the central nervous system is the formation of the **neural plate** by the local thickening of the ectoderm in 16 to 18 hours of development.

1. **Neural tube.** The neural groove is formed by longitudinal folding of the neural plate. The margins of the neural folds fuse to form the neural tube and it separates from the body ectoderm.
2. **Brain and spinal cord formation.** The cephalic portion of the tube dilates to form the brain and the rest gives rise to the spinal cord.
3. **Formation of brain vesicles.** A series of enlargements in the ventral and lateral walls establish the three-vesicle condition of the brain. The demarcation lines between **prosencephalon, mesencephalon and rhombencephalon** (Fig. 43.19) are formed by the exaggeration of some interneuronic constrictions and the obliteration of others. **Metencephalon** and **myelencephalon** are formed by the division of the rhombencephalon by 40 hours.

⊠ Prosencephalon

By 55 hours the prosencephalon enlarges considerably and a slight constriction on its dorsal wall indicates the demarcetion of telencephalon and diencephalon.

⊠ Telencephalon

It is formed by two lateral evagination, telencephalon vesicles of the brain—one on either side of the middle, by the end of the third day. The openings connecting the cavities of the vesicles with the lumen of the middle portion are future **foramina of Manro**. The telencephalon divides into three components—two lateral vesicles and the median portion of brain. So the lumen of the telencephalon has three divisions—a median **telocoele** continuing posteriorly with the **diocoele** and the **lateral telencephalic vesicles** (Fig. 43.20).

Lamina terminalis. The most rostral part of the brain lying in the middle line is known as lamina terminals.

Cerebral hemispheres. The telencephalic vesicles become the cerebral hemispheres and their cavities—**lateral ventricles**. the hemispheres enlarge greatly, extend rostrally, dor-

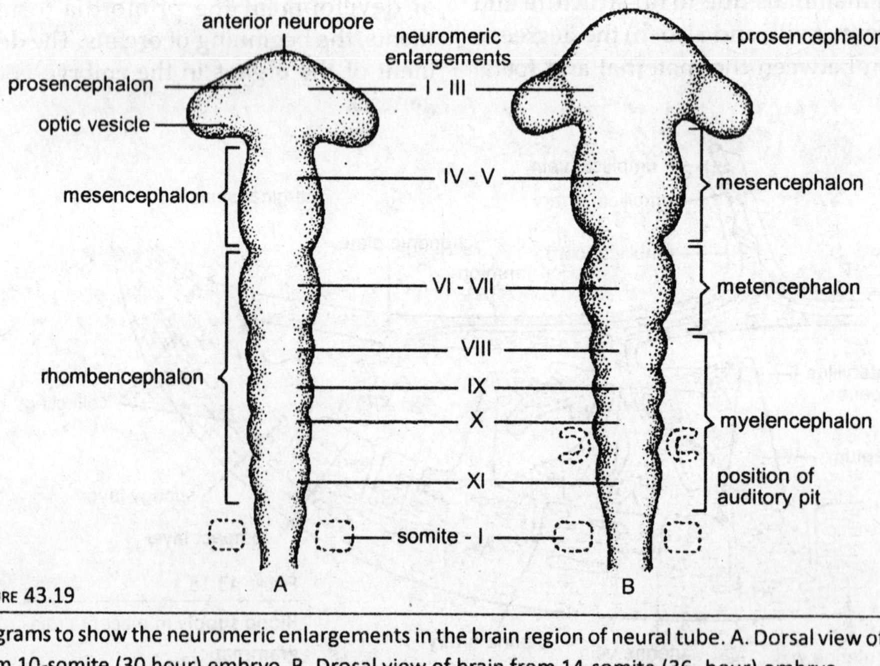

FIGURE 43.19

diagrams to show the neuromeric enlargements in the brain region of neural tube. A. Dorsal view of brain from 10-somite (30 hour) embryo. B. Drosal view of brain from 14-somite (36- hour) embryo.

sally and posteriorly and cover the entire diencephalon and mesencephalon posteriorly.

☒ Diencephalon

The diencephalon produces a great variety of structures with different functions. Except in the ventral portion where the optic stalks merge into the wall, the differentiation of the lateral walls is very little. **Epiphysis** arises as a small evagination of the middorsal wall of the diencephalon. In adult it is differentiated into the **pineal body**. The infundibular depression on the floor of the diencephalon appearing at about 35 hours of development deepened and lies close to a newly formed ectodermal invagination. The Rathke's pocket, with which it fuses in the formation of **hypophysis**. Later the dorsal wall of the diencephalon thickens greatly and form the **optic thalami**—reducing the size and changing the shape of the **diocoele**, the third ventricle. The anterior part of the roof of the

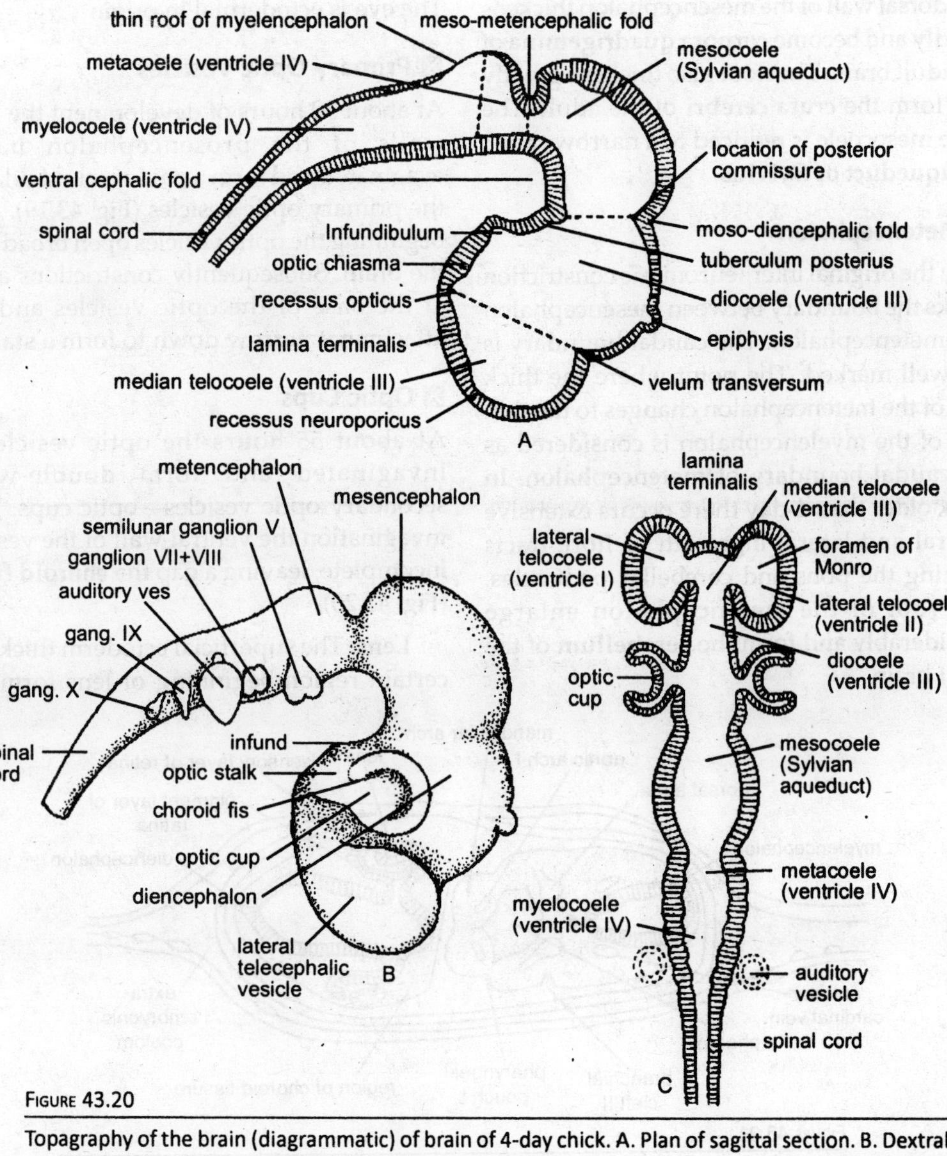

FIGURE 43.20

Topagraphy of the brain (diagrammatic) of brain of 4-day chick. A. Plan of sagittal section. B. Dextral view of a brain. C. Schematic frontal section plan of brain

diencephalon remains thin and becomes richly vascular. Later these blood vessels push into the third ventricle and form the anterior **choroid plexus**. The original dorsal constriction between the primary forebrain and midbrain to the rounded elevation, tuberculum posterius in the floor of the brain is considered as the boundary between the diencephalon and mesencephalon.

☒ Mesencephalon

The dorsal wall of the mesencephalon thickens rapidly and become **corpora quadrigemina** of the adult brain. The floor also thickens greatly and form the **crura cerebri** of the adult. The large mesocoele is reduced to a narrow canal, the **aqueduct of Sylvius**.

☒ Metencephalon

Only the original interneuromeric constriction marks the boundary between mesencephalon and metencephalon. The caudal boundary is not well marked. The point where the thick roof of the metencephalon changes to the thin roof of the myelencephalon is considered as the caudal boundary of metencephalon. In chick older than 4-day there occurs extensive ventral and lateral ingrowth of fibre tracts forming the pons and cerebellar peduncles. The roof of the metencephalon enlarge considerably and form the **cerebellum** of the adult brain.

☒ Myelencephalon

The dorsal myelencephalic wall is reduced to the thin roof of the **medulla**. It is richly supplied with small blood vessels which, carrying the roof with them grow into the **myelocoel** to form the **posterior choroid plexus**. The ventral and lateral walls become the floor and side walls of the medulla.

CHICK : DEVELOPMENT OF EYE

The eye is ectodermal in origin

☒ Primary Optic Vesicles

At about 30 hours of development the lateral walls of the prosencephalon become evaginated and form a pair of round dilations, the primary optic vesicles (Fig. 43.19). At the beginning the optic vesicles open broadly into the brain. Subsequently constrictions appear at the base of the optic vesicles and their attachment narrow down to form a stalk.

☒ Optic Cups

At about 55 hours the optic vesicles are invaginated and form double-walled secondary optic vesicles—optic cups. In the invagination the ventral wall of the vesicle is incomplete, leaving a gap the **choroid fissure** (Fig. 43.20).

Lens. The superficial ectoderm thickens at certain region-beginning of lens formation.

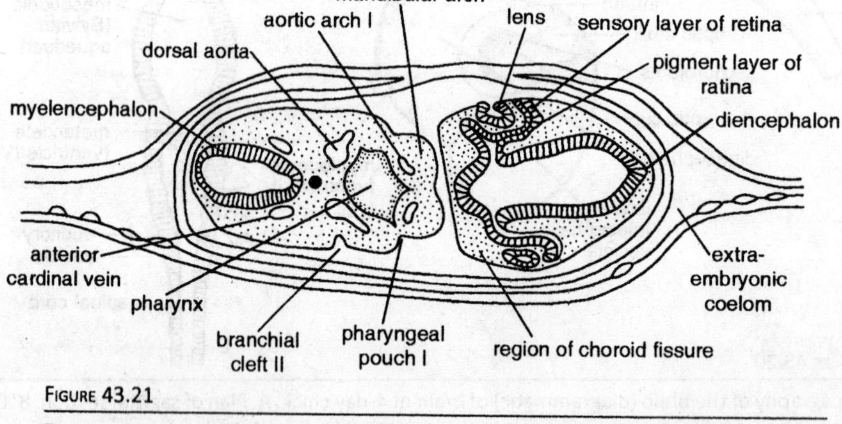

FIGURE 43.21
Transverse section of 55 hour (30-somite) chick embryo (diagrammatic)

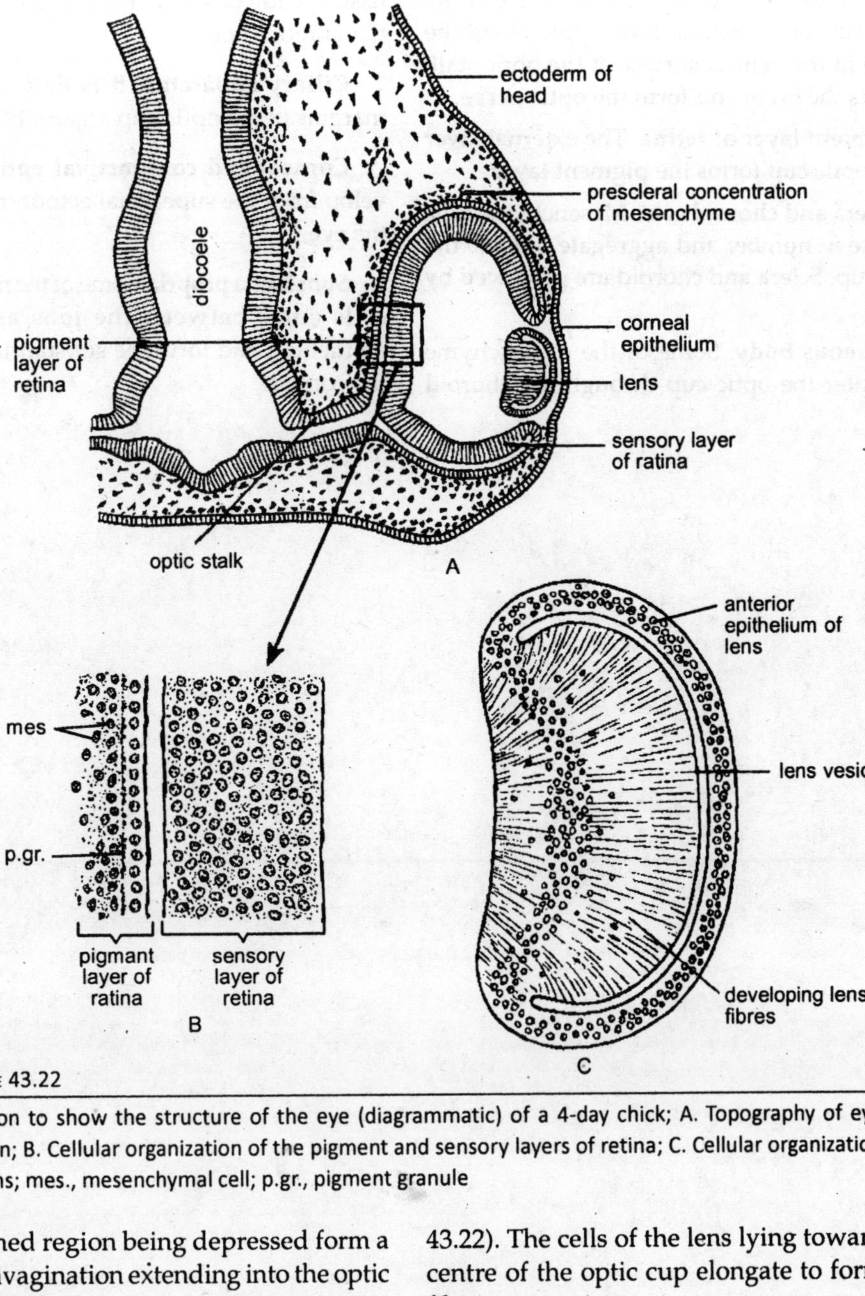

FIGURE 43.22

Section to show the structure of the eye (diagrammatic) of a 4-day chick; A. Topography of eye region; B. Cellular organization of the pigment and sensory layers of retina; C. Cellular organization of lens; mes., mesenchymal cell; p.gr., pigment granule

The thickened region being depressed form a vesicular invagination extending into the optic cup (Fig. 43.21).

In a 4-day embryo the choroid fissure becomes narrow due to the growth of the optic cup walls on either side. The orifice of the optic cup also narrows by convergence of its margins towards the lens. By this time the lens becomes free from the superficial ectoderm and forms a completely closed vesicle (Fig. 43.22). The cells of the lens lying towards the centre of the optic cup elongate to form lens fibres.

☒ Adult eye

In a 4-day embryo differentiation of the structures of the adult eye starts.

Sensory layer of retina. The thickened internal layer of the optic cup forms the sensory layer (Fig. 43.22).

Optic nerve. The fibres arising from the nerve cells of the retinal layer grow along the groove in the ventral surface at the optic stalk towards the brain and form the optic nerve.

Pigment layer of retina. The external layer of the optic cup forms the pigment layer.

Sclera and choroid coat. Mesenchyme cells increase in number and aggregate outside the optic cup. Sclera and choroid are produced by them.

Vitreous body. Some of the mesenchyme cells enter the optic cup through the choroid fissure and constitute the cellular elements of the vitreous body.

Ciliary apparatus. It is derived from the margins of the optic cup adjacent to the lens.

Corneal and conjunctival epithelium develop from the superficial ectoderm overlying the eye.

Substantia propria. Some of the mesenchyme cells enter between the lens and corneal epithelium and form the substantia propria of the cornea.

44 ORGANIC EVOLUTION

ORIGIN OF LIFE : CHEMICAL HYPOTHESIS

It is universally accepted that life arises from life and a continuity is maintained. It is equally true that there was no trace of life on the earth at the time of its origin and naturally life had a beginning on this planet* at a later date and that too from the nonliving. The issue is a complex one and we can only make vague guesses.

All life, including the viruses, has as its physical abode the nucleoproteins, highly complex organic compounds. Except some amino acids, synthetically prepared in the laboratory, the majority of the amino acids, constituents of nucleoproteins, are synthesized exclusively in the living organisms. Spontaneous origin of these energy-rich compounds is highly improbable, still attempts have been made to trace the origin of these substances without intervention of life. This is known as **chemical hypothesis** of the origin of life.

For this we are to visualize a kind of simple virus capable of utilizing organic substances and multiply in an inorganic environment as a free-living organism, unlike the parasitic viruses of the present day. The conditions which were prevalent on the earth when it was young, were entirely different from those of the present age, and they cannot be duplicated. The high temperature permitted water to exist only in the form of superheated steam. There was no free oxygen or ozone in the atmosphere and high energy ultraviolet radiation could reach the earth's surface freely. With the cooling of the earth, liquid water appeared and conditions were set for random compound formation and reformation.

⊠ Chemical Evolution of Life

The most important elements—carbon, hydrogen, nitrogen and oxygen reacted with each other and with many other elements and countless compounds could be formed. Methane (CH_4), the simplest organic compound, was probably formed in such a situation. (Methane has been detected in the atmosphee of some cooler stars). Ammonia (NH_3) was probably formed in such a way. High energy ultraviolet radiations reached the earth's surface and possibly simple aldehydes, such as formaldehyde (CH_2O), appeared

* Some people believe that the first life was capable to withstand the hazards of space journey and might have reached earth from some extraterritorial source.

which could slowly condense into sugars. Organic acids also could be formed, and reacting with ammonia could have rise to other compounds including simple amino acids, such as glycine (CH_2NH_2COOH). Protein could be formed by linking together several amino acids.

These chemical compounds accumulated in large quantities in sea water, as there was no living organisms to consume those.

The sea water was still hot and it has been described as **'The hot diluted soup.'** Given enormous length of time, interaction between the simple organic molecules resulted in complex compounds like polysaccharides, fats, purines, pyrimidines, proteins, nucleosides and nucleotides. A large number of nucleotides were linked together in different sequences resulting in nucleic acids. Nucleoproteins were formed by the union of nucleic acid and protein molecules. The nucleoprotein molecules were presumably comparable to free-living genes. A number of them possibly aggregated to form **protovirus.**

☒ Biological Evolution of Life

The large molecules combined in various sequences and precipitated as colloid particles, termed **Coacervates**. These distinct bodies suspended in water possessed several types of proteins, nucleoproteins and other organic and inorganic molecules. Substances like lecithin and cephalin organised to form the limiting or surface membrane around the coacervates. Some of the proteins were transformed into enzymes, which could lead to new reactions. Gradually, the coacervates were surrounded by nutrient shell and living membrane, and the first living unit or **cell** was formed.

Of the early cell types, two are considered significant.

Moneran type

The nucleoproteins are arranged loosely in the cell substance. The bacteria and blue green algae evolved from them.

Protistan type

The nucleoproteins condensed into a mass, surrounded by a membrane and separated from the cell substance. They are ancestors of other organisms.

It may be emphasized that no direct evidences are available and any or none of these steps may be correct.

☒ Evidences of Organic Evolution

Evolution is a process by which an organism changes itself to adapt best with the environment. It, in fact, signifies that animals, after their origin, have descended from their ancestors with modifications, resulting in the variety of today. The complex animals and plants of today are the result of a series of slow and steady changes of simpler forms of the bygone days through generations which have taken place during millions of years.

The principle of evolution is supported by numerous facts or evidences, which are apparently indirect.

1. Evidences from Comparative Anatomy or Morphology

Basic relationship

A basic relationship exists between all animals from lowly organized to most highly evolved, in the structure and composition of protoplasm and cells. The cells, however, exhibit difference in shape and size and in response to functions but their fundamental constitution remains unchanged. Throughout the animal kingdom the protoplasm contains very nearly the same elements almost in the same proportions of proteins, carbohydrates, fats, water and other substances.

Homologous organs

Homologous organs are those which are developmentally alike and may or may not perform similar functions. The limbs, brains, hearts, etc. of different vertebrates are homologous organs.

Thus the skeletal elements in the wings of pterodactyls (extinct flying reptiles), birds and

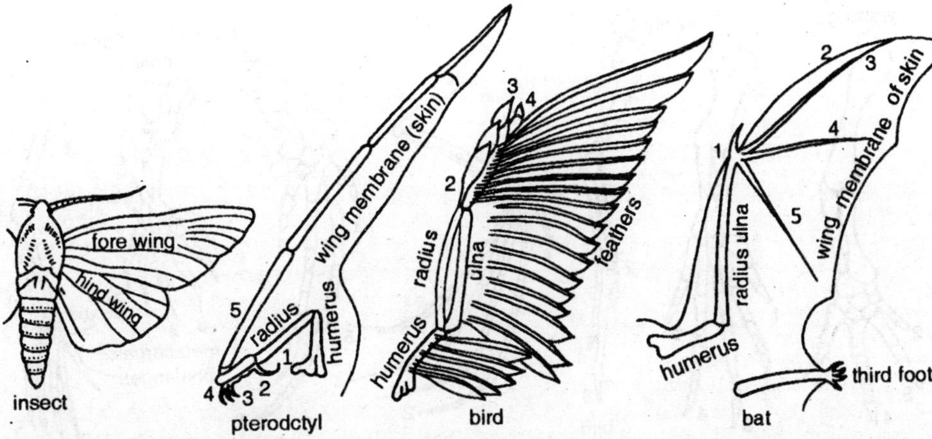

FIGURE 44.1

Analogy between wings of insects and vertebrates

bats are homologous in that all are modifications for flight of the common pattern of forelimb in land vertebrates.

The wings of insects are analogous to those of vertebrates, although used for flight, they are derived not from limbs, but presumably as extension of the body wall (Fig. 44.1).

Homologous structures are present in different groups of animals. The fore- and hindlimbs of different vertebrates are pentadactyl (Fig. 44.2). A comparison of the forelimbs of salamander, crocodile, bird, bat, whale, mole and man shows that the limbs are homologous in being composed of comparable bones, which in each group of animals are adapted for special uses by differences in the length, shape and bulk of various bones. Forelimbs of salamander and crocodile are used for walking, those of birds and bats help in flying, and that of whale for swimming, mole for digging and man for handling objects.

In all vertebrates, the heart is ventrally placed and circulation takes place through a closed system of vessels, arteries and veins. A gradual evolution towards perfection is found in different classes of vertebrates. It is two-chambered in fishes, three-chambered in amphibians and reptiles and four-chambered in birds and mammals, completely separating the arterial and venous blood.

The presence of identical organs in different groups of vertebrates can only be explained by accepting that all of them are descendants of a common ancestor, from whom they inherited these structures, but as they are variously modifed, the minute structural differences are due to various life conditions to which they have been subsequently adapted.

Connecting links

It is proved that the mammals, like the birds, have evolved from reptiles. But, these two groups of animals are widely separated by striking differences, which, however, may be thought to be bridged over, if some intermediate animals with some common characters are found.

Monotremes are animals found in Australia, in which most important characters of reptiles and mammals are combined and serve as a link between these two groups. They possess hairs and mammary glands like the mammals but lay eggs like reptiles and again suckle the young after hatching. It is presumed that monotremes are in the making of mammals from reptiles.

The fossil bird *Archaeopteryx* (Fig. 44.5) connects the reptiles with the birds. This bird had teeth and long tail like reptiles and wings with feathers similar to birds.

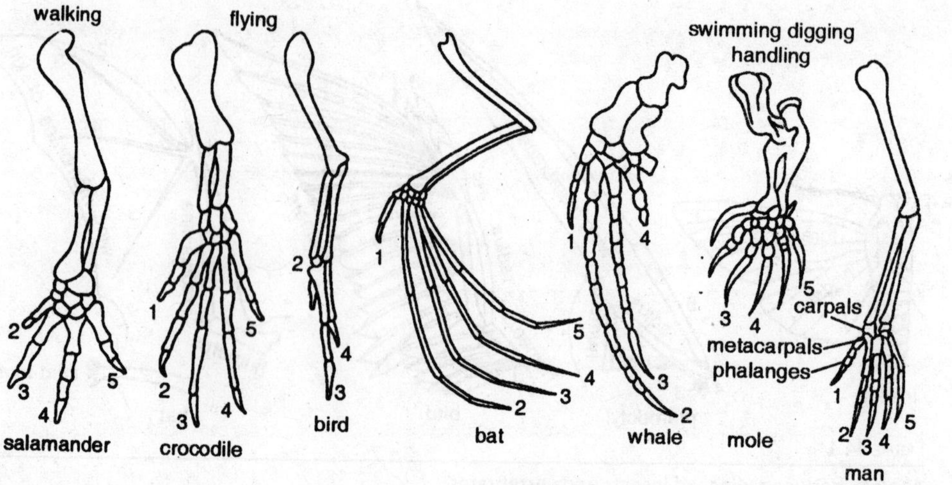

FIGURE 44.2

Homology and adaptation in bones of left forelimbs in land vertebrates

Vestigeal organs

Organs which were once well-developed and functional but have been rendered functionless due to change of environment and are in the way of degeneration, are known as vestigeal organs. The handful of bones in the pelvic girdle of whales, the small amount of flesh (rudimentary nictitating membrane) at the inner corners of our eyes, the caecum, and the coccyx in man, the pair of vestigeal wings in the flightless New Zealand kiwi are the examples of vestigeal organs. Their presence indicates that the possessors have originated from a group of animals in which these structures were functional.

2. Evidences from embryology

The law of embryogenic development states, (a) General characters appear in development prior to special characters, (b) From the general, the less general, and ultimately the special characters develop. (c) During development an animal gradually drifts progressively from other animals. (d) Young stages of an animal resemble the embryonic or young stages of lower animals but are unlike the adult of those animals.

A common basic plan of development with similar embryonic stages exists in different classes of vertebrates (Fig. 44.3). All of them in their foetal stages possess visceral clefts and a visceral skeleton, though not used for respiration in true land vertebrates, viz., the reptiles, birds and mammals.

Certain larval stages are common to some groups of invertebrates also. The trochophore larva is found both in Annelida and Mollusca; the nauplius larva is common to most of the orders of the class Crustacea.

All vertebrate embryos develop a series of aortic arches (usually six) running from the ventral aorta to the dorsal aorta similar to Branchiostoma (Amphioxus). In course of development the number of aortic arches are reduced to five in sharks, four in teleosts, three in amphibians and reptiles and two in birds and mammals (Fig. 12.7).

From the the development of some organisms it is revealed that in some cases the embryos do not resemble the parents but resemble the adults of lowly evolved animals. The tadople larval stage of frogs and toads is exactly like a fish. It is believed that the larval stages of an animal repeat the structures of the

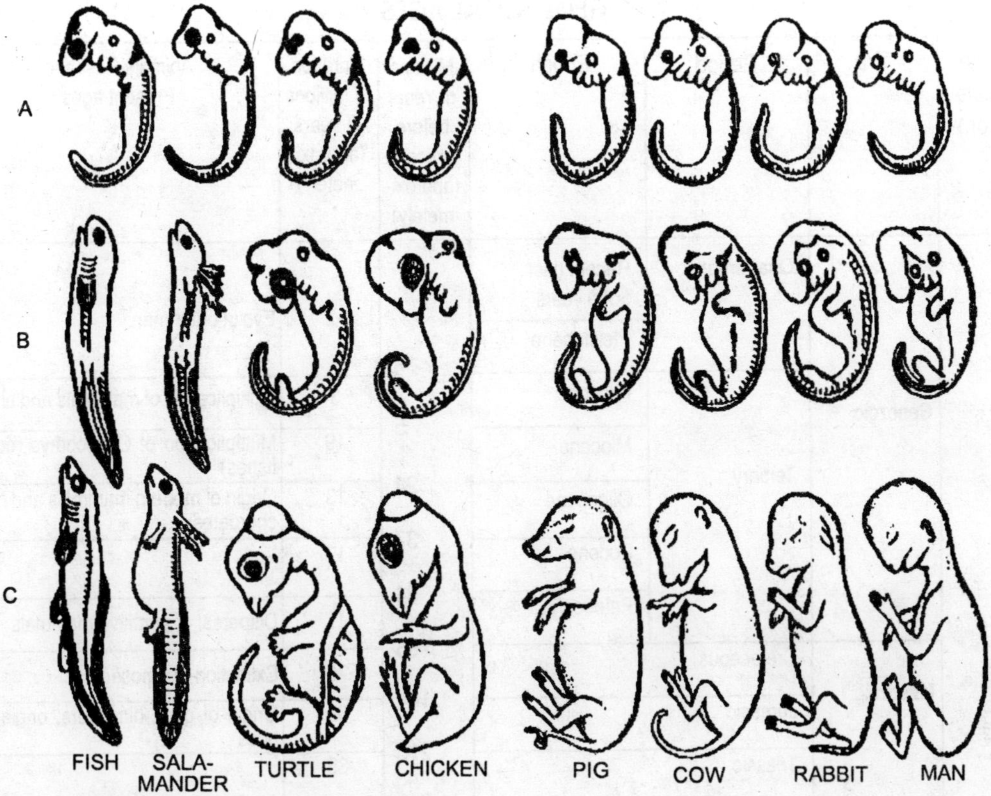

FIGURE 44.3

Series of vertebrate embryos in three successive and comparative stages of development. A. All are much alike in the earliest stage, B. Differentiation has set in but the four mammals (right) are similar, C. Late stage; each with its distinctive features

ancestor, from which it has evolved. This is also known as '**Ontogeny recapitulates Phylogeny***, or the development of an individual repeats the evolution of the race in a condensed form. Nowadays this principle is used in a very limited sense, because the pattern of development is liable to alterations to some degree, due to the influence of the environment and also the amount of yolk present in the egg.

3. Evidences from palaeontology

Palaeontology is the science of fossils. Fossils are the remains of animals and plants of the bygone days and are of six types. Fossils are present in different layers of earth's crusts. Lower the layer, older is the age of the organism and the reverse. The age of a geologic strata may be accurately determined by either lead method or radiocarbon method. The oldest era with animal fossils for certain is the

* Structural similarities in early developmental stages of vertebrates led van Baer to postulate **Recapitulation theory**. Haeckel elaborated it further and advanced his **Biogenetic Law** 'Ontogeny recapitulates phylogeny', which was rejected by experimental embryologists of early twentieth century. Some workers, however, suggest that the law is acceptable only in a redefined form—**ontogeny repeats the ontogeny of phylogeny**. Today the theory is only of historic importance.

GEOLOGICAL AGES

Time Scale (eon)	Era	Period	Epoch	Millions of Years before present (approxi-metely)	Duration in millions of years (approxi-metely)	Animals-Past and Present ages
Phanerozoic	Cenozoic	Quarternary	Recent (last 5000 years		2	Evolution of man
			Pleistocene	2		
		Tertiary	Pliocene	5	3	Multiplication of mammals and birds
			Miocene	24	19	Multiplication of Osteichthys (bony fishes)
			Oligocene	37	13	Origin of modern mammals and non chordates
			Eocene	54	17	
			Paleocene	65	11	Dispersal of primitive mammals
	Mesozoic	Cretaceous		144	79	Extinction of dinosaurs
		Jurassic		213	69	Origin of giant dinosaurs; origin of birds
		Triassic		248	35	
	Paleozoic	Permian		286	38	Multiplication of reptiles, extinction of many early non chordates
		Carboniferous	Pennsylvanian	320	34	Origin of reptiles
			Mississippian	360	40	Origin of amphibians and insects
		Devonian		408	48	Rise and multiplication of fishes, origin of land vertebrates
		Silurian		438	30	Origin of terrestrial nonchordates
		Ordovician		505	67	Multiplication of nonchordates, origin of vertebrates
		Cambrian		590	85	Large number of fossils of non-chordate groups
Precambrian	Proterozoic	Upper		900	310	Origin of multicellular organism
		Middle		1600	700	Appearance of Eukaryotic cell
		Lower		2500	900	Appearance of Prokeryotic cell
	Archean			3000	1400	
	Hadean			4500	600	

Geological Ages: Origin, Multiplication and Extinction of animals

Proterozoic (2,000 million years). Beginning with invertebrates (505 million years), reptiles (255 million years) appeared in the Paleozoic era (505-230 million years). Birds and small mammals evolved in the Mesozoic era (205-135 million years) or the **age of reptiles**, when dinosaurs reached peak and became extinct later. The last or Cenozoic era (75-0 million years) has been marked by the rise to dominance of mammals and man, and is still in progress (Table–Geological ages).

Fossil records definitely indicate trend of evolution through different era.

Mollusca

The trends of evolution of fossil ammonites (Cephalopoda) reveal coiling at the initial stage, and uncoiling of the shell at a later period. Structures between successive chambers were simpler in the older stocks and more complex in those of later periods. The first chambers formed by some of the later ammonites were simple, like those of their ancestors. Then, surprisingly, some of the last ammonites had simpler sutures again and their shells were coiled in the young but straight in the adult stages. Finally, the entire line of ammonites ended in the Cretaceous period (Fig. 44.4).

Archaeopteryx

An extinct bird, possessed feather like most birds but unlike them had homodont reptilian teeth and a less developed forelimb (Fig. 44.5, see page 892). These explain that *Archaeopteryx* represents a stage in the evolution of birds from reptilian ancestors. Birds appeared later than mammals but very soon they became specialized and static.

The mammals were in the process of continuous change, the primitive egg-laying forms gave rise to more specialised ones, first the marsupials, then the placentals. From the study of the fossils, complete evolutionary histories of horse, camel and elephant have been revealed and we can read the story step by step.

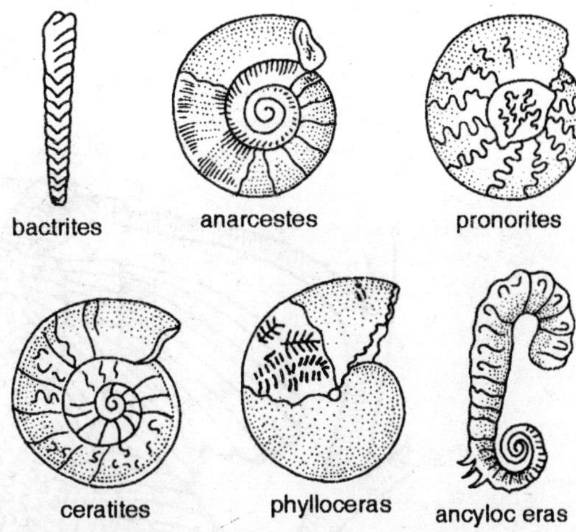

FIGURE 44.4

Stages in evolution of the fossil ammonites. Coiling and later uncoiling of the shell and changes in the forms of the sutures

Evolution of the horse

The history of the horse covers a period of 50 million years beginning in the Eocene period. The known ancestor of the present day horse, *Equus*, was *Hyracotherium* (including *Eohippus* and others). The browsing, less than 30 cm high ancestral horse changed through *Mesohippus* (Oligocene, 39 million years), *Miohippus, Parahippus, Merychippus* (Miocene, 28 million years), and *Pliohippus* (Pliocene, 12 million years) to *Equus* (Pleistocene to recent 1-0 million years). The evolution of *Equus* from *Eohippus* is a history of the enlargement of the body both in length and girth, loss of digits in both pairs of limbs, transformation of low crowned cheek teeth to high-crowned and highly specialised for grinding coarse, siliceous grasses, associated with many other changes (Fig. 44.6, see page 893).

Fossil records also suggest that a number of animals have become extinct for various reasons. This is very conspicuous in reptiles. Dinosaurs (large lizard-like), pterodactyls (flying reptiles) and ichthyosaurs (fish-like reptiles) are all extinct.

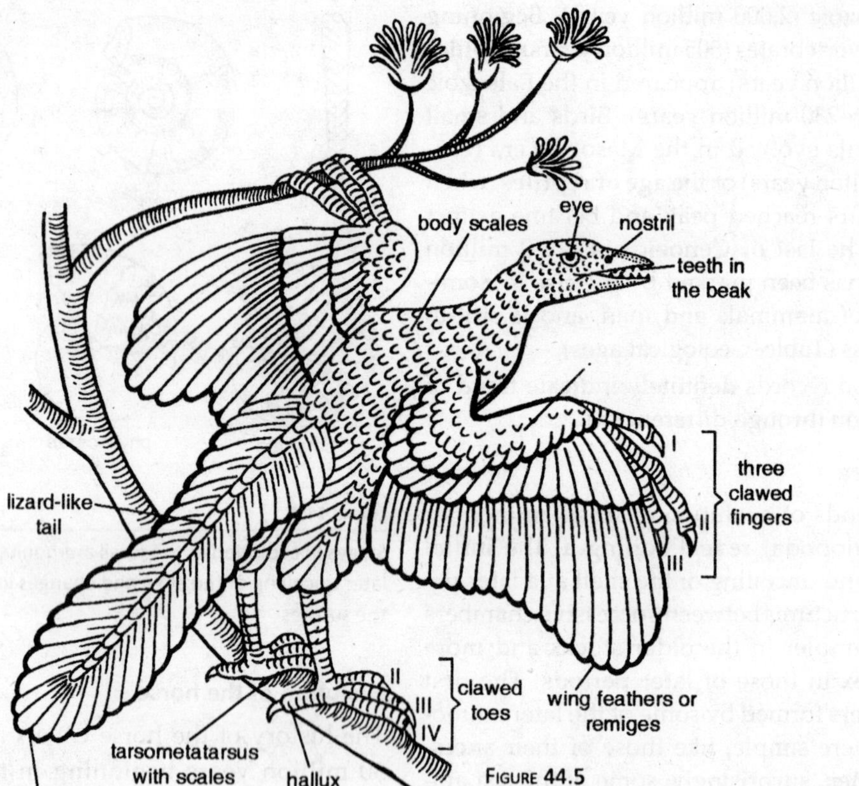

FIGURE 44.5

Archaeopteryx (Restored from fossil)

☒ Fossilization and Types of Fossils

a. **Preservation of hard parts.** The endoskeleton of vertebrates are preserved unaltered. From such fossils it is possible to get some idea about the shape, size and posture of animals.

b. **Footprints.** The footprints left on soft mud are preserved by subsequent hardening of the mud into rock. These provide clues for body structure, locomotory organs and mode of locomotion of animals.

c. **Petrified fossils.** The organic substances of the body are replaced by inorganic substances, particle by particle, in minute detail and an exact duplicate is left. The replacement is so perfect that muscle fibres with cross striations are recognised.

For the formation of petrified fossil the organism or part of it must be embedded and buried in some soft medium. This may happen either at the bottom of a body of water or on land by accumulation of windblown sand, soil or volcanic ash. The dead body is subsequently covered with layers of mud. The organic substances of the body are slowly replaced by minerals from the surroundings and gradually turn to hard rock.

Sedimentary rock containing fossils may be carried by the rivers to the sea, where it sinks to the bottom, and deposited in subsequent layers. Entire body or parts of organisms are often included in these layers.

d. **Moulds and Casts.** The natural moulds of organisms are formed by hardening

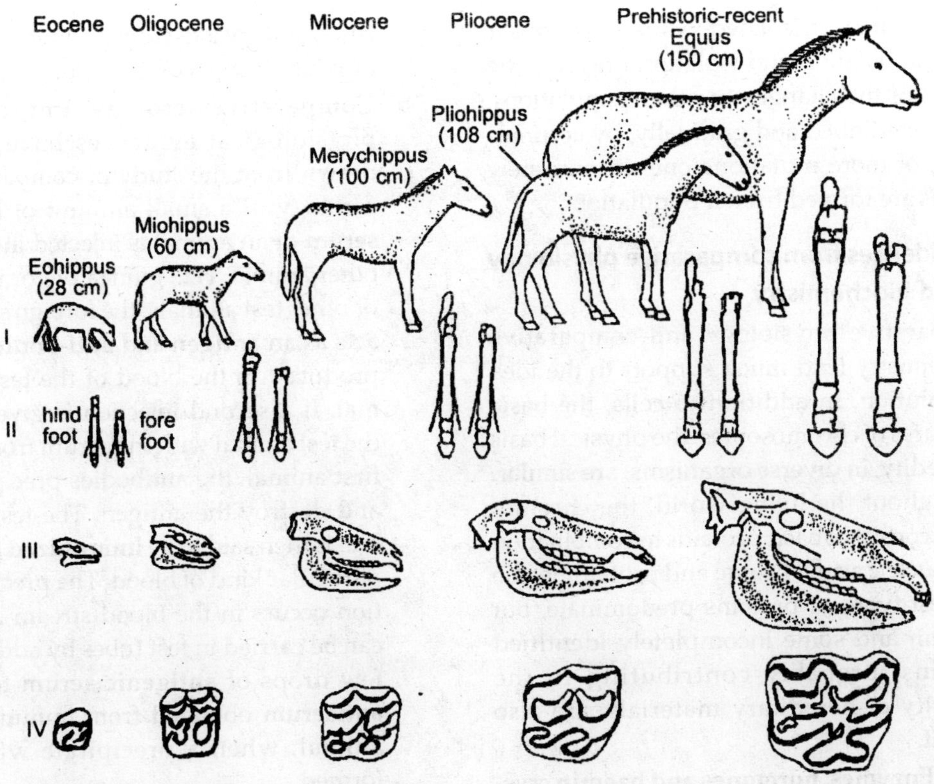

FIGURE 44.6

Evolution of horse. Row I. Progressive changes in size and conformation from the forest dwelling **Hyracotherium (Eohippus)** of the Eocene period to the large, modern, plain dwelling Equus. Row II. Reduction in bones of lateral toes of hind and forefoot from Hyracotherium to Equus. Only the third toe is functional, the second and fourth reduced to splints. Row III. Changes in size and contour and closing of parorbital processes in skull. RowIV. Increasing complexity of enamel pattern of the grinding surfaces of second upper molar teeth.

of materials surrounding buried bodies, which later disintegrate and are removed by seepage of the ground, leaving a hollow cavity. It presents the exact external features of the organism buried. The soft animals like jelly fish, worms, etc, are preserved in this way.

e. **Amber preservation.** Sometimes delicate animals, viz., crustaceans, insects, spiders, etc. are preserved in oil shales, asphalt, amber, etc.

f. **Ice preservation.** The carcass of many animals are preserved, as such, within ice at subzero temperature. Wooly mammoth, rhinoceros of past have been recovered from Siberian ice.

4. Evidences from geographical distribution

The distribution of animals and plants on the earth's surface is not uniform. Every country has its peculiar flora and fauna. Elephants are found in Africa and Asia; Tigers in parts of Asia, Rhinoceros in Africa and India, Marsupialia, Kangaroos in Australia and South America, Lions in India and Africa, and so on.

Particular species of organism evolves at a particular place and with the increase in number they migrate in all available directions. Climatic conditions in no two places are exactly alike and the organisms exhibit modifications. Due to different barriers, populations are separated or isolated. Mutations appear in the population and isolation prevents the spread of the changes to other populations. This is continued and gradually, by accumulation, of more mutations one or more new species are formed from a population.

5. Evidences from comparative physiology and biochemistry

Comparative physiology and comparative biochemistry lend much support to the idea of evolution. In addition to cells, the basic structures of chromosomes, the physical basis of heredity, in diverse organisms, are similar. Throughout the living world, the chromosomes consist of basic proteins in combination with nucleic acid. Histone and protamine, the simplest types of proteins predominate, but globulin and some incompletely identified proteins, probably contributing to the diversity of hereditary materials, are also present.

a. **Enzymes, hormones and haemin crystals.** Identical or closely similar enzymes and hormones are common to large groups of animals. **Digestive enzyme**, viz., trypsin (proteolytic enzyme) is common in animals, Protozoa to Mammalia. Amylase (starch-splitting enzyme) has been recorded from sponges to man.

   The **thyroid hormone** is found in all vertebrates and it is interchangeable, i.e. beef thyroid can be used to meet thyroid deficiency in man. Metamorphosis of a tadpole larva continues if it is fed with mammalian thyroid after the removal of its own thyroid gland.

   Various classes of vertebrates can be distinguished on the basis of **haemin crystals** alone prepared from the haemoglobin. The crystals differ from species to species in structural details and they parallel the present classification. Crystals from different species of a genus share many common characteristics, and crystals from different classes also have some common structures.

b. **Comparative serology.** Impressive physiological evidences have been drawn from the study of comparative serology. If a small amount of blood serum of an animal is injected into another animal, viz., guinea pig or rabbit or other test animals, the foreign serum acts as an **antigen** and **anti-bodies** are produced in the blood of the test animal. If a second injection is given to the test animal with the serum from the first animal, the antibodies precipitate and destroy the antigen. The test animal is then said to be **immunized** to the particular kind of blood. The precipitation occurs in the blood stream but it can be carried in test tubes by adding a few drops of **antigenic serum** to the **antiserum** obtained from immunized animal, when a precipitate will be formed.

Antigen-antibody reactions are highly specific, still the reactions take place in related species, but in ever-decreasing degrees as the relationship grows more distant. A rabbit may be immunized with the blood from *Necturus*, the common salamander. If a few drops of antigenic serum from *Necturus, Amphiuma, Siren* and *Cryptobranchus* are added to different test tubes containing antiserum from the immunized rabbit, the heaviest precipitation will occur in the first tube, the second and third will give about equal precipitates but less than the first, and the fourth will give only a slight precipitate. These suggest *Necturus, Amphiuma* and *Siren* are closely related while *Cryptobranchus* is a much more primitive salamander.

This method is being applied in the problems of plant taxonomy also. Here, plant proteins are antigen and rabbit is the test animal. It has been possible to ascertain the relationship between plants with antibody containing serum from immunised rabbit.

THEORIES OF ORGANIC EVOLUTION

The organisms of different ages are products of evolutionary changes. Various theories have been put forward to explain the factors or mechanism of organic evolution.

☒ Darwinism

The theory "The Origin of Species by means of Natural Section", also called Darwinism, was postulated by the English biologist Charles Darwin (1809-1882).

All individuals and species in nature vary in their characters and all the different characters are hereditary; tendency of the organisms is to overproduce but the food and space are limited. Organisms are subjected to natural selection. Results the death of many individuals.

The individuals with favourable inheritable variations survive and hand over these variations to a few of the next generation. The favourable variations increase from generation to generation and new species are formed.

Variations in nature

Sexually reproducing individuals show variations of all degrees in size, proportions, colouration, external and internal structures, physiology and habits in nature.

Overproduction

The number of offsprings produced by a pair of parents is always much more than the number that ultimately survives. One female toad may lay as many as 12,000 eggs but actually we never get such a big population as there are some limiting factors or checks.

Limiting factors

Increase of population in geometric ratio faces the more slowly increased food supply and constant space restriction. Predatory animals constitute another check on population size. If the lions on the plains of Africa were exterminated, the zebra population would increase rapidly for a time, until a new limit is imposed. Disease is another limiting factor; epidemics occur among animals, particularly if they become overcrowded.

Climate, seasonal changes, drought, severe heat in summer, severe cold in winter are also limiting factors.

The struggle for existence

The number of individuals increases sharply as a result of overproduction. This, in turn, leads to a competition among the off-springs for food, home territory and mate. Besides there are check factors, viz., predators, diseases and the severities of natural catastrophe. Striving to survive against the odds was termed by Darwin as 'struggle for existence.'

Darwin identified three types of struggles:

Intraspecific struggle. Struggle amongst the members of same species. When a large number of seedlings grow under a tree from the seeds of the same plant, many of them die due to want of water and minerals, light and space. This is severest of all the struggles.

Interspecific struggle. Struggle against the members of different species. In a tank, carps and other fishes are devoured by carnivorous fishes and there is a decrease in the number of fishes.

Environmental struggle. Struggle against the physical or abiotic environment, like earthquake, volcanic eruption, flood, etc., which kill many organisms suddenly.

A constant struggle is going on in the living world for their very existence. Variations are of common occurrence and are of three types—beneficial, harmful and indifferent.

Natural selection

Individuals with slightly favourable variations are able to meet the conditions of life more successfully and survive and propagate their kind. Herbert Spencer termed this process "**the survival of the fittest** ." Under natural selection those lacking such variations would perish or fail to breed so that the characters which they possessed would be eliminated from the population; in other words **elimination of the**

unfit. Thus the species in a changing environment or one that had migrated to some new environment would be gradually altered to suit the new environment.

Origin of Races and Species

Being favoured by nature the favourable variations gradually increase in size and are handed over to a few of the many offsprings. These increase from generation to generation which may manifest first by small differences as varieties or races or subspecies. Later, being isolated from one another they cannot interbreed and the advantageous variations become permanent and a species is established. A continuation of such divergence would lead, in time, to the production of still other species and, in turn, to wider differences (genus, family, etc.). Thus a large number of species and larger categories of the animal kingdom have been established through the long duration of geological time.

☒ Criticism of Natural Selection

1. It accounts for the survival of the fittest but not the arrival.
2. Due to natural selection, organisms become overspecialized and overspecialization leads to destruction.
3. It recognizes widespread occurrence of variations and its effect but does not explain their origin. Today the mechanism of inheritance is well-known and a number of theories try to explain the causes of origin of variations.
4. Causes of degeneracy are not explained by natural selection.
5. It could not distinguish heritable and nonheritable variations. Heritable variations alone are important in evolution.
6. Slight variations are considered more important than the abrupt ones; abrupt changes also play an important role in species formation.

☒ NeoDarwinism

Darwin's theory of natural selection could not satisfactorily explain the origin of species. Since then, a large mass of evidences has been discovered, and a theory acceptable to majority of modern evolutionists has been derived from Darwin's theory. This is known as neoDarwinism. NeoDarwinism is really a synthesis of the findings of genetics, systematics, comparative anatomy, morphology, palaeontology, ecology, etc., and it clearly differs from Darwin's original theory.

Different types of variability occur within a species. These are not uniformly distributed in the whole population of a species, but somewhat restricted to local population, commonly more or less isolated from their neighbours. The population with a distinctive pattern of variable characters is given the status of a **subspecies**. A subspecies is considered as an **incipient species**, which becomes ultimately converted into a good species through natural selection.

Each species may have one or more subspecies. A **complex of gene-determined** characters is associated with each subspecies, and these various genes are derived by mutation from a more uniform ancestor. The genes of a particular subspecies are not peculiar to it but are present in different sub-species in varying degrees. Individual genes are not very important. A particular combination of genes is associated with a subspecies. These combinations are screened out by ecological factors. **Agents of selection** act on this gene combination, not on individual gene. Random mutation continues within each subspecies, as a result different alleles arise in different subspecies. Subspecies are at least partially geographically isolated from each other. In addition, there might have ecological and physiological isolation also. New mutations appearing in an isolated subspecies cannot be distributed in the whole population of a species. In this way mutation gradually accumulates and new species are formed ultimately. NeoDarwinians believe that mutations which serve as raw materials for natural slection to be acted upon are very small and are likely to be overlooked in the usual genetical experiments. Large mutation, being disadvantageous, cannot serve as a raw material for natural selection.

Goldschmidt and his followers believe that only subspecies within a species could be formed by neoDarwinian hypothesis, and by considering only very minor mutations significant for evolution, which are subject to very slight selection pressure, too great a burden is placed upon natural selection. He suggests that a decisive step, the **systemic mutation**, is necessary for the evolution of species or higher ranks. A change in the chromosome structure might induce such mutation and new species could be formed by one or a few steps. They would then immediately be subjected to natural selection, and to subspeciation by NeoDarwinian evolution.

⊠ Present status of Natural Selection

Nonrandom or differential reproduction of genotypes in a population is termed natural selection,if the differential reproductive success cannot be explained by mutation or drift. In the study of a population, if a population differs from that of the past to such a degree that statistical analysis indicates it is highly improbable that the differences are due to drift or mutation, it can be said that selection has occurred.

In a genetically heterogeneous population the chances of reproductive success of some genotypes will be usually higher than the probability of success of others. As a result, certain kinds of genetic information will be more and more common in the gene pool of the population while other kinds of information will become less and less common. The gene frequencies will also change.

When a nonrandom set of genotypes leaves more offspring than other, selection has occurred. And it is not correct to suggest that natural selection is '**acting**' or '**operating**' in nature. Various factors of the environment or **selective agents** act or operate in ways leading to differential reproduction in genotypes. The result is natural selection.

Natural selection does not create the species. Rather it creates some sort of order in nature, by reducing the diversity of living organisms. Individuals, containing certain types of combination of genetic information, are inviable and do not persist. It only combats the trend towards increasing entropy (disorder) in nature.

INHERITANCE OF ACQUIRED CHARACTERS

The theory of inheritance of acquired characters, also known as **lamarckism**, was postulated by the French biologist Jean Baptiste de Lamarck (1744-1829).

Living organisms and their component parts tend to increase in size. Environmental conditions are always changing. The new environment creates a new need and new changes or variations appear to adjust with the environment. These are **acquired characters.** Development of an organ is directly proportional to its use. If the organs or parts thereof are constantly used, they develop to perfection, but if not used, they gradually decrease in size and finally disappear. This is known as the **Law of use and disuse**. Lamarckism states that these somatic changes are transmitted to the next generation. Modifications produced by the above principles during the lifetime of an individual will be inherited to the offsprings with the result that changes are cumulative over a period of time.

1. The giraffe got long necks as a result of constant stretching of their neck muscles through generations to collect leaves from high branches.
2. The muscles on the upper part of the body of a blacksmith are well-developed due to constant use.

3. The limbs of snakes degenerated and finally disappeared due to lack of use.
4. Absence of light in caves led to the eyelessness or degeneration of organs of vision in cave animals.

The acquired characters are inherited to the offsprings. The offsprings acquire new characters and this is repeated. In this way a new species appears after many generations.

⊠ Criticism of Lamarckism

The first part of theory "use and disuse of parts" is accepted by all but the main criticism centres round whether the characteristics developed by an individual during its lifetime are inherited by its offsprings or not.

Through the years many experiment of varying kinds have been performed in attempts to demonstrate inheritance of acquired characters. Pavlov trained mice to come for food at the sound of a bell and claimed that fewer and fewer trials were needed to teach mice for succeeding generations, and MacDougall has claimed similar results in the training of rats. Lysenko and his followers in the Soveit Union strongly favour Lamarckism.

Apart from these neither have produced convincing results. A new organism develops from the germ cells of its parents, not from the somatic cells. The germ cells are set aside early in the growth of an individual and are subjected to little or no effect from the body cells or environment. This was demonstrated by Weismann by mutilation of tail of mice and penis of dog. Neither mice in the next generation developed cut tail character nor the dog with cut penis. Castle and Philips replaced ovaries of a white guinea pig with those from a black female. The former was then bred twice to a black male. All the six young produced were black and homozygous. These experiments are criticised on the ground that mutilation is something done to an animal, something in which the animal does not participate actively. The muscles of an athlete increase in strength and bulk with extensive use but recede if exercises are discontinued; children nerver inherit such acquired characters of father. Geneticists are of the opinion that acquired characters are not hereditary and Lamarck's position as an evolutionist is still uncertain.

⊠ NeoLamarckism

The "Theory of inheritance of acquired characters" has been interpreted in a somewhat different way by many workers. This is known as **neoLamarckism**. It holds that environment has profound influence on the organisms, and characters appearing in response to it are hereditary, at least to a certain extent.

Studies on behaviourism of animals suggest that organism and environment are not separate things. Organism modifies the environment and the environment, in turn, influences the organism. If the stress due to the influences of the environment is strong it is likely to modify the **pathways of development** of the organism in the ontogenetic stages and this will lead to some changes, however little those might be, in the genetic constitution or the hereditary materials.

1. White mice reared at 20°-$30^{o}C$ developed larger body, larger hindlimbs and a long tail. These characters were transmitted to their offsprings in subsequent generations.
2. Marked physical changes were recorded in potato beetles exposed to extreme moisture and temperature. The offsprings inherited the characters in successive generations.
3. Larvae of *Drosophila* were grown in a medium containing a high amount of salt. A large number of fly grubs died. The two anal papillae on either side of the anus enlarged in those who survived the high salt food. The anal papillae presumably regulate the salt content of the body fluid. The offsprings of those *Drosophila* were subjected to high salt food, and this was repeated

for about twenty generations. The enlarged anal papillae remained unchanged even after withdrawal of high salt food.

4. The eggs of *Drosophila* were subjected to a treatment with ether vapour. In response to the environmental stress the third body segment had been modifed into a replica of the second segment (bithoracic) in some of the adults. After twenty or more generations, this response was generically assimilated and there was no reversal to the original state following withdrawal of the stress.

EVOLUTION OF MAN

Homo habilis, the earliest fossil of the genus *Homo,* the known ancestor of man, found both in East and South Africa dates between 2.2 and 1.8 million years ago at the Pliocene-Pleistocene period. *Homo habilis* was bipedal, short stature, about 140 cm like ***Australopithecus*** (Southern ape). The cranial capacity of *Homo* was 600-700 ml, higher than that of *A. afarensis* with a cranial capacity about 450 ml. The hominids were stone tool makers. The tools were presumably used for hunting and butchering animals.

Following the *H. habilis* fossil period—about 1.8 million years ago, new groups of hominid fossils, named *Homo erectus* were found in Java, Africa, China and Europe. They were taller than *H. habilis,* about 170 cm long. Skull bones were thicker, brow ridges heavier, teeth smaller and brain capacity 750 to more than 1000 ml. They used improved stone tools with large hand axes and were probably full scale hunters, and could handle large animals bear, elephant, wild boar and others. They knew use of fire. From fossil records distinction between different groups of *H. erectus* could not be established. The distinction between *H. erectus* and *H. sapiens,* our own species is blurred. It is not clear at which points separation or transitions of the two species of *Homo* occurred. The fossil from near solo river in Java, dated less than 250,000 years ago and from Europe (England and Germany) dated to about 200,000 years ago had prominent brow ridges and some other features like older fossils, but the cranial volume was significantly larger—1,100 to 1,200 ml in both Java fossil and the European fossil. The increased brain volume and anatomical traits, as found in fossils indicate that there were several intermediate forms between *H. erectus* and *H. sapiens.*

Though not very clear, it is accepted that *H. sapiens* evolved from *H. erectus,* through gradual changes in various intermediate populations. The European Neanderthals (*H. sapiens neanderthalensis*) along with similar types found in Zimbabwe (Africa) and Shanidar (Iraq) between 50,000 to 100,000 years ago had an average brain volume 1,300 to 1,500 ml similar to that in present time man. However, they were shorter than modern humans, with distinctive brow ridges, large jaws and small chins. They were skilled hunt-

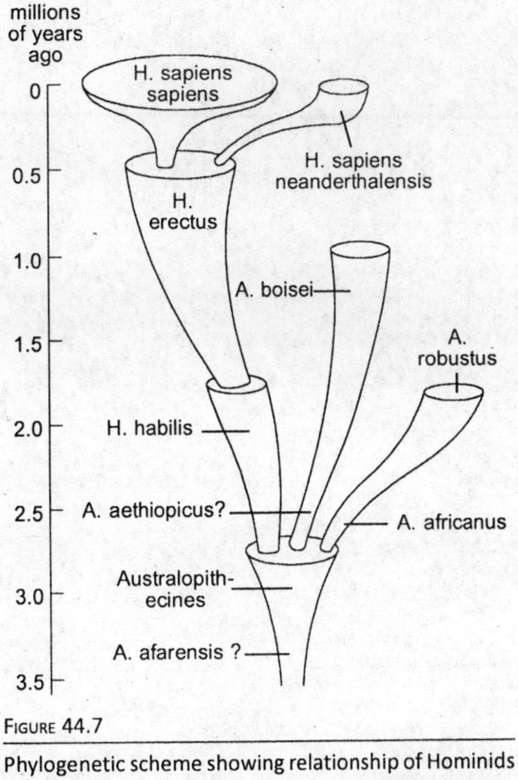

FIGURE 44.7

Phylogenetic scheme showing relationship of Hominids

ers of large animals like cave beer, mammoth and others. The *H. sapiens sapiens* began to replace the neanderthals in different parts of the world about 40,000 years ago.

By about 35,000 years ago in the upper paleolithic (last part of the old stone age), the Neanderthals were apparently replaced by Cro-Magnon types with reduced brow bridges, higher skull vault, smaller and less prognathous face. The Cro-Magnon had an advanced culture and was like modern man in appearance.

Beginning with *Australopithecus afarensis* a possible phylogenetic scheme showing relationship (Fig. 44.7) among various known hominid fossil groups can be traced with the help of molecular technique. The divergence of mitochondrial DNA (mt DNA) sequence rate is likely between 2 and 4 per cent nucleotide change per million years in vertebrates. The end point of the phylogeny—relations among existing human races can be analysed by following the technique. Differences between mt DNA indicate that the ancestral DNA of African origin between 140,000 and 200,000 years ago is the source of modern human mt DNA sequences. It may be concluded that the *H. sapiens* evolved in Africa at about that period and spread worldwide without mixing with other extant species.

45 BIOGEOGRAPHICAL DISTRIBUTION AND BARRIER

The distribution of animals over the earth's surface is known as zoogeographical distribution and that of both plants and animals as Biogeographical distribution. It is presumed that a species evolves at a particular place and, with the increase in number, the members of the group migrate in all possible directions with subsequent modifications. Present distribution of a species is generally maximal, and indicative of the limits imposed by physiographic or climatic barriers.

DISTRIBUTION OF ANIMALS AND PLANTS

Distribution of animals and plants reveals that the same of closely related species sometimes exist in widely separated places, with no representatives in the intermediate territory. This may be explained by accepting that the animals and plants in the intermediate territory have become extinct or the land masses with their flora and fauna have separated from one another.

Every species is expected to be present in every place on the earth unless (a) they were unable to reach there, (b) on reaching there they could not survive, (c) after establishing themselves they were modified to another species. Distribution is the result of two opposite forces—**expansion** and **repression**.

Multiplication in number and the consequent scarcity of food and space, and the change in the environment (soil, vegetation, light, temperature, food, parasite, predator, etc.) compel animals to migrate.

BARRIERS TO DISPERSAL OF ANIMALS

The factors which act as check to the distribution of animals are known as barriers. There is almost no geographical feature which is not a barrier to the dispersal of some species. Again, a barrier to one species may be the highway of dispersal for another.

☒ Physical Barrier

The land is a barrier to aquatic animals and water is barrier to most land animals.

Large bodies of water

This is the most effective barrier to almost all land animals including even the birds. The opposite banks of Amazon river are inhabited by different species of birds.

Salt water

It is a nearly absolute barrier to amphi-bians. (i) The Pacific Islands are usually uninhabited by amphibians, except that they have been introduced by man. (ii) In the Pacific coast many freshwater rivers follow almost parallel courses to the ocean. Each river has its own subspecies, even species. The expanse of salt water separating the mouths of rivers, even very small, is not crossed by fishes. If the streams are joined by flood in rainy season, the fishes move from one to the other river. (iii) The marine animals living in open sea—where the salinity is about 3.5 per cent—do not move to areas of low salinity like brackish and estuarine water, and vice versa.

⊠ Mountains and Lowlands

Mountains are described as islands in the sea of low lands. Mountains serve as barriers to low-land animals and the low lands are barrier to animals inhabiting mountain ranges. The mountain goat and big horn sheep are restricted to mountains as they cannot cross low lands. The American opossums cannot cross high mountains. They thrive well in the low coastal range of California but could not invade Sierras. White-footed American mice are restricted to low lands.

⊠ Forests and Prairies

Extensive forests serve as barriers to the dispersal of grassland animals, prairies serve as barriers to forest animals. Here, adaptation to a particular type of food and also the distance are most important.

Forest animals do not usually cross the grassland, while grassland animals do not cross large forests.

Temperature

The temperature is an effective barrier to dispersal of animals. (i) Animals in the surrounding areas are different from those living in deserts. Only those, specially adapted, cross the deserts and the desert animals do not move out of their habitat. (ii) Arctic fox, polar bear, penguine and other animals of cold region are restricted within a definite temperature range and the same is true for animals of the tropical and subtropical zones. (iii) Fishes can detect small differences in water temperature, as low as 0.1°C and possibly 0.03°C. The temperature difference sets a limit to the distribution of many fishes.

Biological barrier

The effectiveness of such barriers in the dispersal of animals is, however, somewhat doubtful. A few such facts may be mentioned.

Sendentary habit

A number of marine invertebrates, viz., cnidarians, echinoderms and others are sessile, but their free-living larvae can travel long distances. Certain vertebrates with powerful locomotor organs often keep them restricted within a limited area of distribution.

Homing instinct

In birds and other vertebrates, homing instinct is an effective barrier and limits distribution. In the geese, the family groups migrate together resulting in a high degree interbreeding.

ZOOGEOGRAPHICAL REALMS

The zoogeography or the knowledge of the distribution of animals on the earth has led to the division of the world into a number of distinct zoogeographical (biogeographical also) realms (Fig. 45.1). Originally these were based on the mammalian faunas of the various parts of the world, but their validity is general.

1. **The Holarctic realm.** This realm is often broken up into the Palearctic realm (Old World) and the Nearctic realm (North America), and are separated by the Atlantic and pacific Oceans.

A. **The Palearctic realm.** Includes whole of Europe, Asia—north of Himalayas and Ming Ling Mountains and Africa—north of the Sahara desert.

Amphibia—Necturus and Siren.

Mammalia—Rodents, oxes, sheep, goat, deer, antelope, bear, etc.

B. **The Nearctic realm.** The realm consists of North America and Greenland. In addition to many species of palearctic realm it possesses:

Reptilia—iguanas and rattle snakes.

Mammalia—American opossum, racoons, and some rodents.

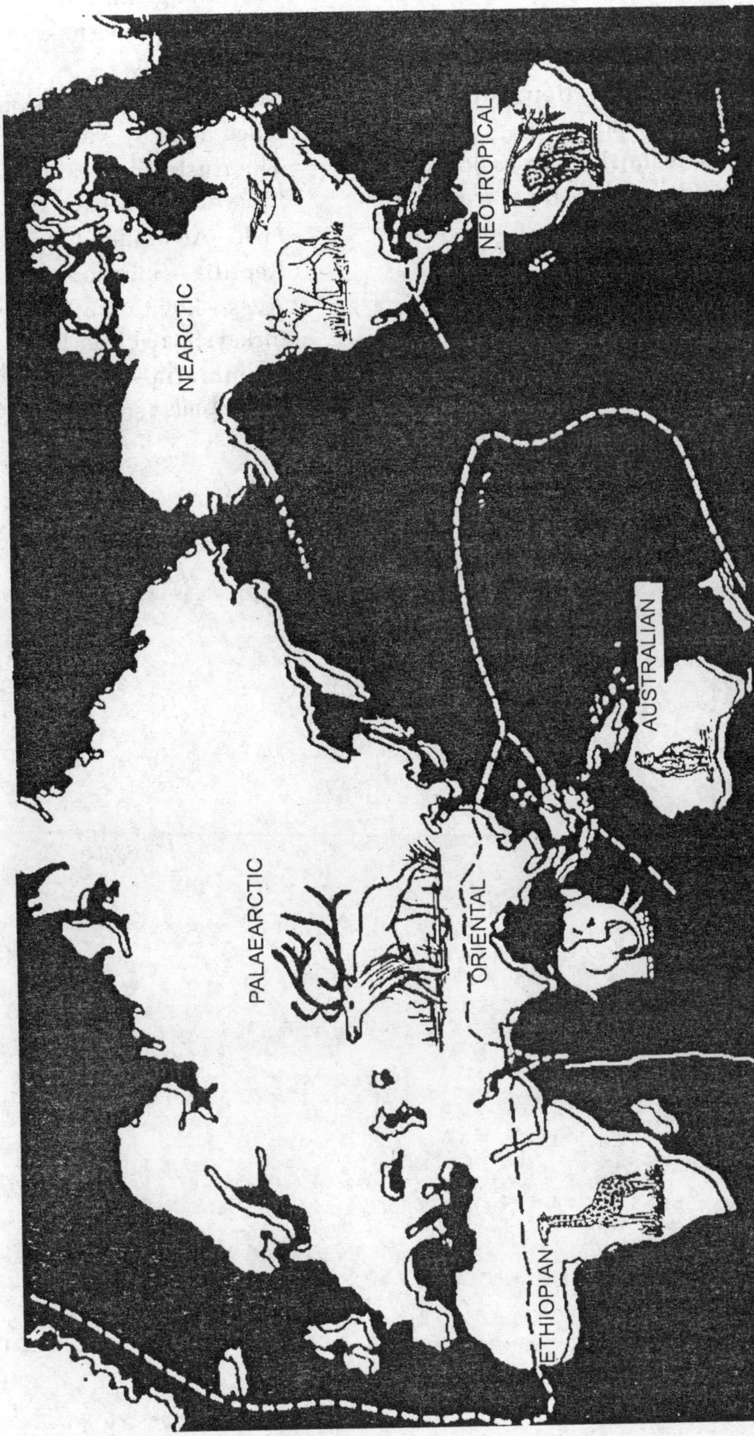

FIGURE 45.1

The Biogeographical regions of the world with characteristic animal

2. **The Ethiopian realm.** It comprises Africa—south of the Sahara desert.

   Aves—ostrich, secretary bird.

   Mammalia—giraffe, lion, hippopotamus, elephant, rhinoceros.

3. **The Oriental realm.** It includes portions of Asia—south of the Himalayas and Ming Ling Mountains.

   Aves—peacock.

   Mammalia—tarsiers, orang-utang, Indian elephant, rhinoceros, flying fox.

4. **The Neotropical realm.** It comprises South and Central America.

   Fish—lungfish.

   Reptilia—rattle snake, alligator.

   Aves—Rheas.

   Mammalia—tapirs, sloth, prehensile-tailed monkey, vampire bats.

5. **The Australian realm.** It includes Australia and associated islands.

   Fish—Australian lungfish;

   Reptilia—Sphenodon.

   Aves—kiwi, emu, cassowaries, brush turkeys, lyre birds, birds of paradise.

   Mammalia—Monotremata, most marsupials.

46 ORIGIN OF SPECIES

Aspecies is a distinct biological unit consisting of individuals united by the common bonds of sexual union (reproductive community), as well as of common descent and common parenthood. Genetically distinct allopatric populations of a species are called local varieties, subspecies, geographic races or simply **races**.

The origin of a number of distinct reproductive communities from a single interbreeding population (species) is referred to as **speciation** or origin of species. The present discussion is restricted to the mechanism promoting the differentiation leading to the formation of new species.

ORIGIN OF SPECIES

Individual does not evolve; population evolves. The evolution of a species is a continuous process, but it can be divided into a few logical steps, which often overlap one another.

1. The origin and accumulation of genetic diversity in existing species.
2. Environmental selection leading to a higher proportion of advantageous genes in the whole population or to that part which has been subjected to selection pressure.
3. Genetic assimilation.
4. The isolation of genetically diverse groups.
5. The crystallisation of new species (harmonisation of the genetic constitution of the isolated groups and the formation of a new species).

☒ Origin and accumulation of genetic diversity

It is presumed that gene mutation caused by cosmic radiation and other mutagens, modifier genes, inversion of parts of chromo-somes, translocation of genes from one chromosome to another, elimination or reduplication of chromosomes, introduce genetic diversity in natural populations.

The accumulation and preservation of genetic diversity are believed to be the effects of different types of evolution in different kinds of populations (Simpson, 1944; Dobzhansky, 1951). Perfection of the genotype and a refined adaptation towards the environment occurs in large and continuous populations through '**centripetal evolution,**' while '**centrifugal evolution**' leads to an increase in genetic diversity in small and partially divided populations.

☒ Environmental selection

Selection is not the indiscriminate mass mortality of populations but the higher mortality of individuals with one genetic constitution than with another. The **agents of selection** are both **abiotic** (physical) and **biotic** (biological). Light, temperature, oxygen concentration, salinity, pH and others belong to the first category, while the second category includes predator, parasite, competitor, food and others. The agents of selection affect the fecundity of individuals or cause their death before attaining sexual maturity. Selection leads to a greater number of individuals of a genotype (with advantageous genes) than another. (In a genetically heterogenous population, the possibility of reproductive success of some genotypes will be higher than the probability of success of the other. Thus certain kinds of genetic information will be more and more common in the gene pool of the population and other kinds will be less and less common).

⊠ Genetic Assimilation

The pathways of development of organs in an individual are different and they tend to end in normal state. In cases, developmental hazards force a deviation in one or more pathways; the organs in the phenotype become somewhat different from that in the normal. Selection pressure stabilises the changes in the genotype and makes those irreversible.

The processes of development leading to the appearance of functioning adult organism from a zygote are interrelated and form a system, which has been termed **epigenotype** (Waddington). It is a branching system of developmental pathways, each of which leads to one of the components of the adult. Each individual in a population in a given environment will have its own genotype and, therefore, its own epigenotype, which will eventually result in the adult phenotype. Selection to preserve fitness in the given environment may eliminate gentypes producing imperfect phenotype leading to the selection for a well-canalised **epigenetic system.**

With the change in the environment some individuals may produce fit phenotypes without immediate genotypic change. In course of time, genotype change occurs in the new environment and developmental paths are stabilised by selection. If the organisms are returned to their original environment, original phenotypes are not produced due to the change in the genotype. The original phenotypic (epigenetic) response to changed environment has become incorporated in the genotype. This is genetic assimilation.

⊠ The isolation of genetically diverse groups

Isolation of interbreeding population bearing advantageous genes from the rest of the populations is a prerequisite for further progress towards speciation. The term **'isolating mechanism'** is used to include all the agents that curtail or stop the gene exchange between populations. The chief isolating mechanisms are:

1. **Geographic or spatial isolation.** The populations inhabit different territories. The gene exchange in allopatric populations is prevented because they live in different territories. Sympatric sexual populations remain distinct, only if the gene exchange between them is hindered or prevented by their intrinsic, genetically conditioned properties.
2. **Reproductive isolation.** Restriction or suppression of gene exchange between populations due to genetically conditioned differences.

a. **Ecological isolation.** Due to different habits, potential mates from different populations do not meet and interbreeding is prevented, though they live in the same general area. (i) In the water snake, *Natix sipedon*, the fresh water and salt water races are kept separate by their habitat, though they may come very close together.

b. **Seasonal or temporal isolation.** Mating period occurs at different seasons and prevent mating. (i) In North Eastern United States three species of *Rana* breed in the same pond. *R. sylvaticus* breeds first at water temperature 44°F (6.67°C), *R. pipiens* at 55°F (12.78°C) and *R. clamitans* at above 60°F (15.56°C). If crossing takes place, embryos fail to develop.

c. **Sexual, physiological and ethological isolation.** Lack of mutual attraction and behaviour differences —courtship, song, scent, etc. play a major role to prevent random mating. (i) The mating dances of salamanders and turtles, courtship motion in crabs, sexual behaviour in drosophila are selective and reduce the chance of mating between different populations. (ii) Songs of birds, scents of butterflies lead to the same result. (iii) In some oysters, the chemical substance released in water with gametes stimulate other individuals to discharge masses of sex cells. (iv) The swarming of pololo worm *Eunice vixidis* in the south Pacific, the maturation of eggs and sperms of *Echinus esculentus* in the Mediteranean Sea are related to the phases of the moon. (v) The jelly coat around the egg containing fertilizin and also the vitelline membrane and cortical layer of the egg promote penetration of a sperm of the same species or inhibit cross-fertilization with other species. The mechanism ensures effective fertilization within a local population.

d. **Mechanical isolation.** Physical non-correspondence of the genital organs. In many animals, specially in insects, the male and female genitalia do not match for mating and are considered as mismatching sex organs.

e. **Gametic isolation.** Spermatozoa are not attracted by the eggs.(i) In the cross-insemination between the related species of *Dorsophila viridis*, *D. americana*, the mobility of the sperm in the sperm receptacle of females of foreign species is lost rapidly.

f. **Hybrid inviability.** The zygotes are inviable or inferior to those of the parental species. (i) The hybrid inviability is total in North American species of *Rana pipiens*, *R. palustris*, *R. clamitans* and *R. catesbeiana*. (ii) Sheep and goat hybrids appear to be normal in the early embryonic stages but die much before birth.

g. **Hybrid sterility.** Failure of hybrids to produce F_2 generation. (i) The F_1 hybrids produced by crossing related species of *Drosophila pseudoobscura* and *D. similis* are fully vigorous like paternal species but the hybrid males are sterile.

h. **Hybrid breakdown.** Inviability or adaptive inferiority, of all, or a part of the F_2 or back cross hybrids.(i) The robust and vigorous hybrids of F_1 in many species produce degenerate hybrids in F_2.

☒ The crystallisation of new species

Every race of a species is not an incipient species. With the selection and isolation of a new genotype with advantageous characters, a new species is in the process of formation. The phenotypic characters can be fully exploited only after harmonisation of the genetic constitution with the environment. The harmonisation is not a separate step, but continuous with genetic mutation, selection and isolation, finally leading to the formation of a new species.

SPECIES CONCEPT

From the point of view of the origin, geographic area inhabiting and morphological features, three kinds of species have been described.

☒ Allopatric species and allopatric speciation

Geographically isolated species are allopatric species. Isolation of a population followed by speciation is characterised as **allopatric speciation**. In allopatric speciation, the daughter populations are separated (isolated) from parent populations. Sufficient subsequent genetic divergence prevents interbreeding and complete mixing (introgre-ssion) when the two isolated populations are subsequently brought into contact. Allopatric speciation is the most important cause of organic diversity.

☒ Sympatric species and sympatric speciation

A number of species living in a common geographic area are sympatric species. The process of origin of sympatric species is of enduring controversies. One argument against **sympatric speciation** is that hybridization will swamp out differences produced by selection pressure. It is generally believed that allopatric species formed by isolation return to a common geographic area and they are considered as sympatric species. A number of workers hold that sympatric species do not owe their origin to allopatric species. Factors like host specificity, living in the same locality, mate selection, lead to the origin of new races. (i) The initial differences, when reinforced by factors like different times of emergence on plants with different fruiting time, results in the sympatric evolution of the races of *Rhagoletis pomonella*, an insect, widespread on apple trees in North Eastern and coastal United States. (ii) The ground finches *Goespiza magnirostris*, *G. fortis* and *G. fuliginosa* of Galapagos finches are widely sympatric. The three forms occur together, the individuals recognise and mate only with the members of the proper form.

☒ Sibling species and speciation

Diverse species with superficial similarity are termed sibling species. The holarctic butterflies (Alpine butterflies) of the *Erbia tyndarus* group are similar in outward appearance and in the morphology of male and female genital structures but their chromosome number is widely divergent (n = 8, 10, 11, 15, 24, 25, 51 and perhaps 52). Not much is known about the origin of sibling species. It is presumed that closely related species living in a geographic area but not competing with one another have developed similar morphological adaptations under selection pressure, leading to the origin of the sibling species.

47 COLOURATION AND MIMICRY

Striations and refractive granules on the surface, pigments in epihelial cells and blood supply in the skin produce colouration. Mimicry is imitation of another. The adaptive values of both colouration* and mimicry are great. The former is of immense help to both the aggressor and the victim, while the role of the latter is mostly defensive, only a few aggressors having this adaptation.

COLOURATION

Colours directly helping the animals in their survival are **valuable colours.** On the basis of their uses they have been placed under several heads.

⊠ Sympathetic, cryptic or concealing colouration

The colour of the animal blends with the surroundings and the animal cannot be detected easily. It may be protective or aggressive.

⊠ Protective and aggressive colouration

It is of value both to the prey and the predator. To the first, it is protective coloura-tion, e.g. the Arctic hare and to the second, it is aggressive colouration, e.g. the Arctic fox.

⊠ Local and seasonal colouration

The same species vary in colour in two ways.

1. The caterpillar of the hawk moth, *Sphinx convolvuli*, when full grown, is either green, like the food plant or brown, resembling the ground colour.
2. The varying hare, *Lepus variabilis*, retains brown colour in southern Sweden during winter; the same species remains white for six or seven months in the higher Alps, for nine months in the south of Norway and white throughout the year in northern Greenland.

Standard faunal colour

The surrounding of an animal influences its colouration to a considerable (Fig. 47.1) extent;

---

* The discussion is restricted only to the colours of biological significance or valuable colouration. Besides this, several types of colouration are found in animals, which presumably are not of biological importance.

1. **Indifferent colours.** The brilliant colours of some deep sea animals, living in darkness belong to this category. It is believed that such colouration was of importance to their ancestors.
2. **Albinism.** Total absence of colour in skin, hair, feather and iris of the eye.
3. **Melanism.** A pigment melanin imparts black colour to the whole or parts of the body of animal.

Both albinose and melanose arise from mutation. They are heritable and follow Mendel's law in the ratio of their appearance.

the deserts, the plains, the jungles and other habitats, all play some roles in the life of their inhabitants.

Concealment colours

1. The camel, gazelle, lion and other desert animals are grey, matching the colour of the sand.
2. The buckskin horse simulate the colour of dry grass.
3. The tiger and zebra are striped to blend with the shades of trees or long grasses.
4. Arctic creatures are white.
5. Many sea birds are steel grey or blue above and white ventrally.
6. Nocturnal animals are mottled brown or grey.

All perfectly match with the surrounding colours.

Alluring colours

The colouration which attracts other animals for one or more reasons.

1. Certain spiders allure insects by closely resembling an orchid blossom both in colour and form.
2. Some African lizards have protective colouration all over the body, except for a brilliant patch of colour at the corner of the mouth to attract the prey.
3. The orchid mantis with its anterior portion resembling an orchid flower attracts insects on which it feeds (Fig. 47.2).

Warning or revealing colours

The colour resembling that of poisonous or distasteful animals like stinging wasps, poisonous snakes, etc. serves as warning and the animals are immune from attack. The colours are usually conspicuous reds or yellows. Many moths, butterflies and their caterpillars protect them by such adaptation. Sometimes the attitude of dangerous animals is associated with warning colouration.

A small moth, *Smerinthus*, has large eye-like spots on the hind-wings, which are concealed under the fore-wings. When the fore-wings are raised, the eye-like markings, resembling the markings on the back of the head of a cobra, creates terror in the assailant and it retreats.

Mimetic colouration

The colour resembling an animal or even an inanimate object to conceal the animal or to warn others is called mimetic colouration (Fig. 47.1). The walking-stick insect (*Carausius*), the walking-leaf insect (*Phyllium*), the non-poisonous snakes imitating colour of poisonous coral snakes are examples of mimetic colouration.

Signal marks

It is an adaptation in gregarious animals to send warning signals to fellow animals heralding some danger.

1. A conspicuous white area in the posterior part of the body of the virginian deer remains covered with lowered tail. Sensing danger when one starts off, the tail goes up, exposing the white area. This acts as a warning and other members of the herd are instantly in full retreat.
2. The signal flag of the cotton tail rabbit *Lepus sylvaticus* gleams in the dark and directs the young to safety.

Recognition colouration

Colouration helping members of the same species to recognise one another is called recognition mark or sympathetic colouration.

1. The dorsal surface of a brooktrout (*Salvelinus fontinalis*) is dusky, while the ventral surface is light-coloured. Red and orange spots are pesent on the sides of the body. In a shadowy pool the colours of the dorsal and ventral surfaces harmonize with the surroundings when viewed from either above or below. But the characteristic spots on the sides are clearly visible to its fellow members at the same level.

FIGURE 47.1

Concealment, **Protective** and **Mimetic colouration** and **Warning mimicry**. A. Great green bush cricket (*Tettigonia viridissina*). B. **Goat moth** (*Cossus cossus*). C. **Thorn tree hopper** (*Thelia maculata*). D. **Angle shade moth** (*Phlogophora meticulosa*). E. **Leaf butterfly** (*Kalima poralekta*). F. **Dark brown beauty moth**. G. **Stick insect** (*Carausius morosus*). **Warning mimicry**—H. **Eyed hawk moth** (*Deilaphila elpenor*)

FIGURE 47.2

Concealment, Protective and Mimetic colouration. A **Cinnabar moth** (*Callimorpha jacobaeae*). B_1 **Puss moth caterpillar** (*Cerura vinula*). B_2 **Frightened** Puss moth caterpillar. C. **Hornet moth** (*Sesia apiformis*). D. **Lady bird beetle** (*Cocciniella variabilis*). **Alluring colouration**—E. **Orchid mantis** (*Mantis religosa*)

2. Many moths, butterflies and other insects possess sympathetic colours on the exposed portions, chiefly wings.

Confusing colouration

Sudden disappearance of the sympathetic colours and harmonization of the body colour with the surroundings confuse the pursuer or the viewer. It is known as confusing colouration.

1. The eyes of a viewer or a pursuer follow the brilliant recognition colouring of a moth or a butterfly on the surface exposed in flight but not at rest. The colour disappears with the settling of the insect and the viewer loses sight of it.
2. Brilliant sympathetic colour is present on the underside of the tail of certain sombre-coloured lizards. The colour is exposed in running lizards and disappears when it stops and the tail is lowered. The body colour harmonizes with the sand and the lizard escapes sight of the pursuer.

Sexual colouration

The colouration with the attainment of sexual maturity is sexual colouration. The males of many species of insects, birds, mammals attain beautiful colouration at sexual maturity, while the females have only protective colouration.

1. The protective colouration of female birds help them to escape notice of predators during nesting.
2. The spotted coat of young deer, lion and puma is of protective value under the conditions surrounding babyhood. The infantile colouration increases the percentage of survival in young animals.

MIMICRY

Imitating another is mimicry. The resemblances of an organism to another or even to an inanimate object in form, colour, attitude and action to escape notice of others or signalling apparent harmfulness or dangerous nature fall under mimicry. Mimicry is probably the most effective adaptation for protection of animals. Mimicry may be **passive** or unconscious and **active** or conscious. Most of the mimics fall under passive category.

☒ Protective Mimicry

Protective mimicry has two aspects— concealment (Fig. 47.1) and warning (Fig. 47.2).

Concealment mimicry

1. The smooth rounded form and texture of the crab, *Cryptolithodes* harmonize perfectly with the white quartz pebbles of the beach.
2. The walking-stick insect, *Carusius* (Fig. 47.1) with its slender body, attenuated limbs, sympathetic colouration, slow movement oftens stiffening into rigidity, perfectly harmonizes with the twig on which it rests.
3. In walking-leaf insect *Phyllium*, the expanded body, flattened wings and appendages are green except for small, yellowish, irregular spots resembling the fungus that grows upon a dead leaf. It is a unique specimen in which resemblance is precise.
4. In the dead-leaf butterfly, *Kallima paraleeta*, the hind-wings elongate to resemble the stem on which it rests and also bear other leaf-like markings with extreme perfection. At rest, it perfectly resembles a dead and sere leaf in colour. The midrib, lateral ribs of leaf and black and mouldly spots also occur. The wings bear striking recognition colouration above and the under surface has protective colouration. The recognition colouration, seen in flight, disappears on alighting, with the folding of the wings and the protective colour of the undersurface is exposed.

Effectiveness of protective mimicry

Protective mimicry is a successful tool for defence, only under certain conditions.

1. Both imitatives and imitators must occur in the same area.
2. The imitators are more defenceless.
3. The imitators are always less in number in an area.
4. The imitation is always external.

Warning mimicry

Mimicry of distasteful, dangerous and poisonous forms are considered warning mimicry.

1. Harmless snakes of different genera mimic the poisonous coral snakes in colouration and are immune from attack of other animals.
2. The viceroy butterfly, *Basilarchia archippus* (= *Limenitis disippus*), which is palatable, protects itself from predators by mimicking inedible monarch butterfly *Anasia plexippus* (= *Danais archippus*).

Aggressive mimicry

Mimicry which renders a predator invisible and helps to capture prey is aggressive mimicry.

1. The yellow bodies of certain carnivorous spiders harmonize with golden red and other flowers on which they rest. The insects visiting the flowers fail to detect them and become spider's prey.
2. Spiders may resemble oak galls, droppings of birds and other forms and pounce upon unsuspected preys.

Simulation of death

Feigning as a dead animal is conscious mimicry.

1. Simulation of death is common in American opossums, *Didelphis virginiana*. Some believe that it is not actually feigning death, the animal faints in fright. However, the adaptation is of much importance for its survival.
2. Many hard-bodied beetles feign death and drop down like pebbles when they apprehend some danger.

48 ADAPTATION

The changes in the organism by which it suits best in the environment and leave behind offsprings are known as adaptation. Had there been no adaptation, evolution would have been impossible. The earth and all the earthly things are under constant change. The capability of an organism to change itself with the changing world determines its survival. Greater the degree of adaptation, higher is the ability to exploit the environment resulting in the production of greater number of offspings and consequent higher chances of survival of the species.

ADAPTATION

The adaptation* may be of two types—Primary and Secondary (acquired).

☒ Primary Adaptation

The adaptations remain in the original direction and the subsequent changes remain more or less in the same direction from the very appearance of the animal.

Examples: Aquatic adaptation of fish.

☒ Secondary (acquired) adaptation

The adaptive changes deviate from the original direction. The changes are diverted along different lines due to change of the original home of the organisms and they subsequently adapt themselves to some new environment.

Example: Aquatic adaptation of whales.

☒ Convergence and divergence

Animals belonging to different groups living under similar life conditions show similar adaptation. This is known as convergence or parallelism.

Animals belonging to same stock migrate to different geographical areas to meet their need for food and safety and are subjected to different life conditions. This leads to varied and diverse adaptations. This is called divergence.

ADAPTIVE RADIATION

From a primitive stem the animals move out in four different directions and are adapted to aquatic, amphibious, terrestrial and arboreal life. Adaptation in a particular environment may again be of different types.

Terrestrial < Fossorial → Subterranean
Terrestrial < Semiarboreal → Arboreal ↓ volant

From environmental point of view following and other adaptations are found in the animals.

* Figs. 48.1 to 48.7 at the end of the chapter.

AQUATIC ADAPTATION

Adaptation to water medium. Water is the densest homogeneous medium and adaptations to aquatic life (Fig. 48.1) are necessarily directed to minimise the resistance during movement.

⊠ Modifications

Modifications are much less in primary aquatic forms but extensive in secondary aquatic ones.

Primary aquatic adaptation

1. Head, body and tail compressed into a spindle-like form (Fig. 48.1A, A').
2. The head is wedge-shaped, the edges of the jaws and gill covers fit perfectly.
3. Locomotion by lateral undulation of the body aided by the fins.
4. Hydrostatical air bladder to maintain a plane of flotation.

Examples: All fishes.

Secondary (acquired) aquatic adaptation

Modifications of structures are extensive in secondarily adapted aquatic forms. Ancestors of these animals were terrestrial lung-breathers. Search for safety and competition for food presumably forced them to move to water, to which they have been subsequently adapted with a remarkable degree of success.

1. The **body contour** spindle-shaped, neck constriction disappears and the tail enlarges.
2. The neck becomes shorter.
3. The chest tends to become cylindrical.
4. External ears lost.
5. External nostrils shift towards the apex of the head.
6. The eyes shift higher on the face.
7. Limbs are fan-like, tail flattened like the caudal fin of a fish but horizontal in most cases.
8. Loss of hair, skin glands and salivary glands reduced.
9. Development of a layer of fat beneath the skin.
10. Mouth armament. Teeth simplified with an increase in number or even totally lost (baleen whale).
11. Speed usually high, about 39-40 knot per hour in porpoises.
12. Attain a large size and exceed all other animals.

Reptilia. Sea turtles (Fig. 48.1B), Plesiosaurs, Proganosauria, crocodiles (amphibious).

Aves. Hesperonis, penguins (Fig. 48.1C), great auk, albatross, pelican (amphibious).

Mammalia. Pinnipedia, Cetacea (Fig. 48.1D), Sirenia; hippopotamus (amphibious ungulata).

CURSORIAL ADAPTATION

Adaptation for speed (Fig. 48.2, 48.3). In addition to the appearance of some most beautiful and perfect animals, cursorial adaptation played a significant role in the origin of birds and mammals. Some primitive dinosaurs started walking on hindlegs and their greatly increased activity resulted in the evolution of birds. Unlike birds, which came from reptiles active on hindlegs, mammals evolved from reptiles in which all the four limbs were active.

⊠ Modifications

1. **Body contour.** Body spindle-shaped, with a streamline contour to minimise resistance to the medium.
2. Loss of general utility of the limbs, specially the hindlimbs.
3. **Change in foot posture** (Fig. 48.2A-C). Plantigrade (walking on sole) to digitigrade (walking on finger toe).

a. Loss of digits. In transformation from planti-to digitigrade, the animals became four-, three-, two-, or even one-toed.

b. Metacarpals and metatarsals fuse to form canon bone.

c. The limbs are lengthened without elongation of the bones in the digitigrade forms.
d. The fibula and ulna are greatly reduced or even vestigeal.

4. The universal movement of the limbs lost and restricted to one plain only (pen-dulum-like).
5. Elongation of limbs due to lengthening of the distal segments, the foot and the shin (Fig. 48.2D) and the hand and forearm (Fig. 42.36).

Reptilia. Lizards (Fig. 48.3A) and snakes, specially the frilled lizards of Australia and desert lizards.

Aves. Flightless birds, emu, (Fig. 48.3B), ostrich, rhea, etc.

Mammalia. Tasmanian wolf, hares and rabbits, dogs, cats, hunting leopards, horse, deer, giraffe, (Fig. 48.3C), antelope (Fig. 48.3D), stoat (Fig. 48.3E) etc.

FOSSORIAL ADAPTATION

Adaptation to subterranean life. Fossorial animals belong to two categories—partial, living above ground but dig for food and wholly fossorial, dig burrows to live underground (Fig. 48.4).

☒ Partial Fossorial

1. Presence of digging structures like snout, tusks, etc.
2. Skull and jaws are modified.

Examples: Swine, elephant, etc.

☒ Wholly Fossorial

The majority of the wholly fossorial animals are small, pentadactyl, plantigrade with moderate to large claws, defenceless and more primitive members of their groups.

☒ Modifications

1. **Body contour.** The body is spindle-shaped or fusiform to minimise resistance to subterranean passage. The body is cylindrical in some cases (*Caecilia*).
   a. The diameter of the shoulder region is greatest.
   b. Jaw musculature feeble.
2. The tail is short and probably acts as a tactile organ.
3. The eyes are reduced or vestigeal and in some cases covered with skin.
4. The external ears are very small, minute or even lost.
5. **Digging mechanism.** The digging organs are snout, nose, teeth and limbs.
   a. The snout is truncate and upturned at the tip in swine and hog-nosed snakes (*Heterodon*).
   b. A prenasal bone is present at the tip of the nose cartilage in swine and the mole (*Talpa*).
   c. The incisor (elephant) and canine (swine) teeth are directed forward and protrude out.
   d. All the limbs are very short, stout and clawed. The forelimbs are broad and with long claws.
   e. The forelimbs loosen the earth, the hindlimbs throw it further backward and also resist the occasional backward thrust recived from the forelimbs.
6. Hibernation occurs in fossorial animals beyond the limits of the tropics.

Amphibia. Caecilians (Fig. 48.4A).

Reptilia. *Sphenodon* (Fig. 48.4B), *Uromastix*, legless lizards, desert snakes.

Aves. Burrowing owls (Fig. 48.4C), cliff swallow.

Mammalia. Monotremata, kangaroo rat, marsupial mole, mole (*Talpa*), mole rats, otter (*Lutra*), porcupine (48.4D), mongoose (48.4E), prairie dog (rodent) (Fig. 49.4F).

SCANSORIAL ADAPTATION

Adaptation to tree life. In the beginning, relatively feeble animals climbed up trees for safety, retreat and easily available food. For the early mammals, arboreal life was a safe retreat to escape the assaults of reptiles and that paved the way for further mammalian evolution.

Based on their adaptation, scansorial animals (Fig. 48.5) have been divided into three groups.

☒ Wall and Rock Climbers

These animals are not tree-living and adapted to climb walls of buildings and vertical rocks. Geckoes and flying squirrels belong to this group. It is believed they are descendants of arboreal forms.

☒ Terrestrio-arboreal Forms

They are equally adapted for tree and ground life. These forms are expert climbers, can rest on the trees and at the same time are equally at home on ground. The climbing adaptations are, however, not very marked. Insectivore and carnivore rodents belong to this group.

☒ Arboreal Forms

The group has made trees their home. They may come down occasionally but are not at ease on the ground. On the mode of locomotion arboreal animals have been put under three groups.

Branch runners

The vast majority of the group exhibit perfect arboreal adaptation. All the four limbs are used for moving on branch surface and living there. Chameleons, tree marsupials, squirrels, lemurs belong to this group.

Branch suspended forms

They cannot walk on the branches but move and rest in a suspended, inverted position (belly upward) as in sloth with the help of powerful curved claws of all the four limbs. On land they move with great difficulty.

Swinging forms

The forelimbs are modified for swinging and are practically only organs for rapid locomotion. The limbs swing with great speed and accuracy and the animal moves from limb to limb and from tree to tree. The hindlimbs are used in walking and rest on the branches of trees. Many primates belong to this group.

☒ Modifications: Arboreal forms

1. **Body contour.** There is not much change in the shape of the body but certain organs are greatly modified.
   a. The thorax is subcircular anteriorly.
   b. The ribs are much curved and numerous to support the viscera in an inverted position.
   c. The dorsolumbar vertebrae increase in number and are often elongated.
2. The shoulder girdle is strong, both the clavicles and scapullae well-developed.
3. The ilia of the pelvic girdle are strong and broad.
4. Proximal limb segments elongated. The elongation is great in suspended (sloths) and fast-leaping (gibbons) forms.
5. The feet may be prehensile with opposing digits of nonprehensile.
6. The claws well-developed in nonprehensile forms, squirrels, cats, etc.
7. Presence of adhesive pads on the tips of the digits. In tree frogs a sticky substance is secreted in the pads.
8. The tail usually prehensile.

Fishes. Climbing perchs (Fig. 48.5A), mud skippers, etc.

Amphibia. Tree frogs and toads (Fig. 48.5B).

Reptilia. Geckos (Fig. 48.5C), chameleon, tree snakes, etc.

Aves. Passerine birds, parrots, woodpeckers (Fig. 48.5D), guanas, snake birds, etc.

Mammalia. Opossums, tree sloths (Fig. 48.5E), tree hyrax, civets, squirrels, flying lemurs, monkeys (Fig. 48.5F, F_1), etc.

VOLANT ADAPTATION

Adaptation to aerial life (Fig. 48.6). Air is a homogenous medium which offers the least possible resistance to the attainment of speed. Aerial creatures cannot live in the air for an indefinite period of time and must come to land or water or branches of trees for rest. Hence their adaptation is always a double one. Practically all flying vertebrates, except fishes, evolved from ancestors adapted to tree life (scansorial adaptation).

Flight is of two types—passive or gliding flight and true flight.

Passive or gliding flight

In passive fliers (Fig. 48.6A-C) only the sustaining surface is present and no muscular force is applied. Fishes differ from other fliers in their flight; they jump out of water and cover some distance before dropping. Other passive fliers move to a high point and take a leap. The animal is held up by the sustaining surface and, being pulled down by gravity, glides to a lower level covering a considerable distance. Gravity is the only locomotive force. Passive flight is found in fishes, amphibians and reptiles.

True flight

In addition to the sustaining surface, muscular power is applied for the movement of the sustaining surface. True flight involves sustained movement through the air (Fig. 48.6D-E).

☒ Modifications

1. **Body contour** streamlined to minimise resistance through the air.
2. Sustaining surface may be simply patagium or folds of skin and wings.
   a. **Patagium.** It is a fold or series of folds of the skin between the fore- and hindlimbs and sometimes extend to the neck and between the hindlimbs and tail.
   b. **Wing.** The wing has independently evolved three times. In the **bat wing**, the humerus is well-developed, the radius is long and curved, the ulna is vestigeal. The pollex is always free and clawed for crawling and climbing. In the **pterodactyl wing**, the radius is a little shorter than ulna, the single wing finger, presumably the fourth is huge, the wing being flexed between metacarpal and proximal phalanx. In the **bird wing**, the digits are reduced to three and more or less fused together, patagium is vestigeal and the sustaining surface is formed by feathers.
3. Pneumatic bones. The bones are hollow and air-filled.
4. Air sacs. Sacs filled with air not only render the body light, but help in maintaining a sort of double respiration.
5. Sternum well-developed and bears a median keel. The keel is lacking in running birds.
6. In birds, shoulder girdle is rigid by the development of clavicles and the pillar-like coracoids.
7. Scapula and coracoid are united by an articulation in Carinatae, but firmly ossified in Ratitae.
8. Brain well-developed with enlarged cerebrum and large optic lobes.
9. The eyes are large with so-called sclerotic plates.

Fishes. The fins are usually enlarged and some flying fishes either soar in the air or skip over water for some distance.

Examples: *Exocoetus* (Fig. 48.6A), *Dactylopterus*, *Pantodon*, etc.

Amphibia. *Rhacophorus* (Fig. 48.6B) is the only Amphibia having volant adaptation. The digits bear terminal adhesive pads and are connected by web-like expansion of the skin. Rudiments of patagia present in front and behind the arms. The webbed feet sustain it in prolong leaps.

Reptilia. In flying dragon, *Draco*, (Fig. 48.6C), the patagium is supported by a number of extended ribs. In the flying gecko, *Ptychozoön*, the skin expands laterally along the sides of the neck, body, tail, limbs and between the toes. The extinct pterodactyls is or flying dragons were a remarkable group of reptiles.

Aves. All flying birds (Fig. 48.6D) exhibit volant adaptation. Running birds, previously thought to be degenerated flying birds, are really not so. It is suggested that during their early experiments for flight some birds took to air while a group remained on the ground and was adapted for running on land.

Mammalia. More than thirteen separate volant adaptations are found in mammals, of which only the bats attained the power of true flight. In Microchiroptera (Fig. 48.6E), the patagium extends from the rear of the head along the front of the arm, between the fingers to the base of the claws, between the fore- and hindlimb, webbing the toes as well as the fingers, and between the hindlimbs and the tail.

DESERT ADAPTATION

Adaptation to desert life (Fig. 48.7). A vast stretch of drifting sand with a little or no rains, devoid of vegetation except in the rare oases, and scarcely populated by rare animals is known as desert. The struggle for existence is most severe in the desert. Moisture getting and moisture conservation, and defence against both physical and biotic environments are most acute.

☒ Modifications

1. **Moisture getting.** The sources of moisture are plant sap, dews collected on plants and blood of the prey. The desert lizard, *Moloch*, absorbs water through the skin like a blotting paper when it comes in contact with it.

   The problem of water conservation in desert animals is met with in several ways.

2. **Moisture conservation**

   a. **Nonperspiring.** Perspiration has a physiologic value and helps in excretion and to maintain standard body temperature. But the loss of moisture in desert animals has greater ill effect and, to conserve the scarce and precious water, many desert animals have become nonperspiring. Kangaroo rats lack sweat glands.

   b. **Economising of sweat.** A fluctua-tion of 7°C body temperature helps camels to minimise water loss through perspiration. The body temperature drops to 34°C at cool dawn and much of the day elapses before the temperature rises to 41°C, at which sweating begins*.

   c. **Highly concentrated urine and solid faeces.** The urine is highly concentrated and the volume drasti-cally reduced. A 450 kilogram camel excrete only about 1,200 ml urine in a 24 hour period in summer. The urine of kangaroo rat contains about 23 per cent urea and 7 per cent salt. The faeces of kangaroo rat is almost dry.

   d. **Body insulation.** The body is often covered with hard plates, spines, etc. in arthropods and reptiles and a thick covering of furs in mammals. Fur greatly reduces the formation of sweat, which may reach 40 per cent.

   e. **Conversion of urea to protein.** Micro-organisms in the gut convert urea back to protein and water loss from the kidney to excrete urea is saved.

---

* The rumen chambers of the stomach in camel does not contain water cells as described by previous workers. Nor the hump play any role in providing water through oxidation of fat present in it. The camel possesses no special reserve of water. (Knut Schmidt-Nielsen, The Physiology of Camel **Scientific Americana**, Dec. 1959, 201, 140-151.)

f. **Drinking of enormous quantity of water.** A camel can drink 150 litres water—one-third of its body weight—in ten minutes and compensate the water loss. A desert donkey quenches his thirst in two minutes.

g. **Tolerance of water loss.** Tolerance of water loss is quite high. It may go up to 25 per cent in camels.

3. **Defence.** To protect themselves against rigorous physical conditions, viz., extreme heat, cold, sand storm, scorching sun rays, etc. and also against predators, the desert animals have adapted themselves in different ways.

a. The nostrils are directed upwards in digging species. In snakes, they are protected by complicated valves or reduced to pinholes.

b. The eyes are overhung by the head shields, or margins of the lids are broadened into plates and furnished with scales. The upper eyelid may be enlarged, the lower enlarged eyelid provided with a window.

c. The ear openings are reduced, or protected by fringes of scales or lost.

d. The colour of the animals generally harmonises with the colour of the sand or rock. Sometimes the same species show double colour adaptation to suit local conditions. Warning colouration is most common.

e. Hard surface and spinescence skin protect insects and reptiles.

f. Venom is a characteristic of the desert animals. Venomous spiders, insects and snakes are most common in deserts.

g. Speed. Most of the animals can move with great speed.

h. The senses of sight, hearing and smell are well-developed.

i. Intelligence is highly developed, particularly in defenceless animals.

Arthropoda. Wasps, spiders, *Tarantula* (spider) (Fig. 48.7A), etc.

Reptilia. *Moloch* (Fig. 48.7B), *Agama*, *Phrynocephalus*, *Mabuya*, (lizards). *Typhlops* (burrowing snake), rattle snake (Fig. 48.7C), etc.

Aves. Ostrich (Fig. 48.7D).

Mammalia. *Saiga*, *Antelocapra* (antelopes) gazelle, kangaroo rat (Fig. 48.7E), camel (48.7F), etc.

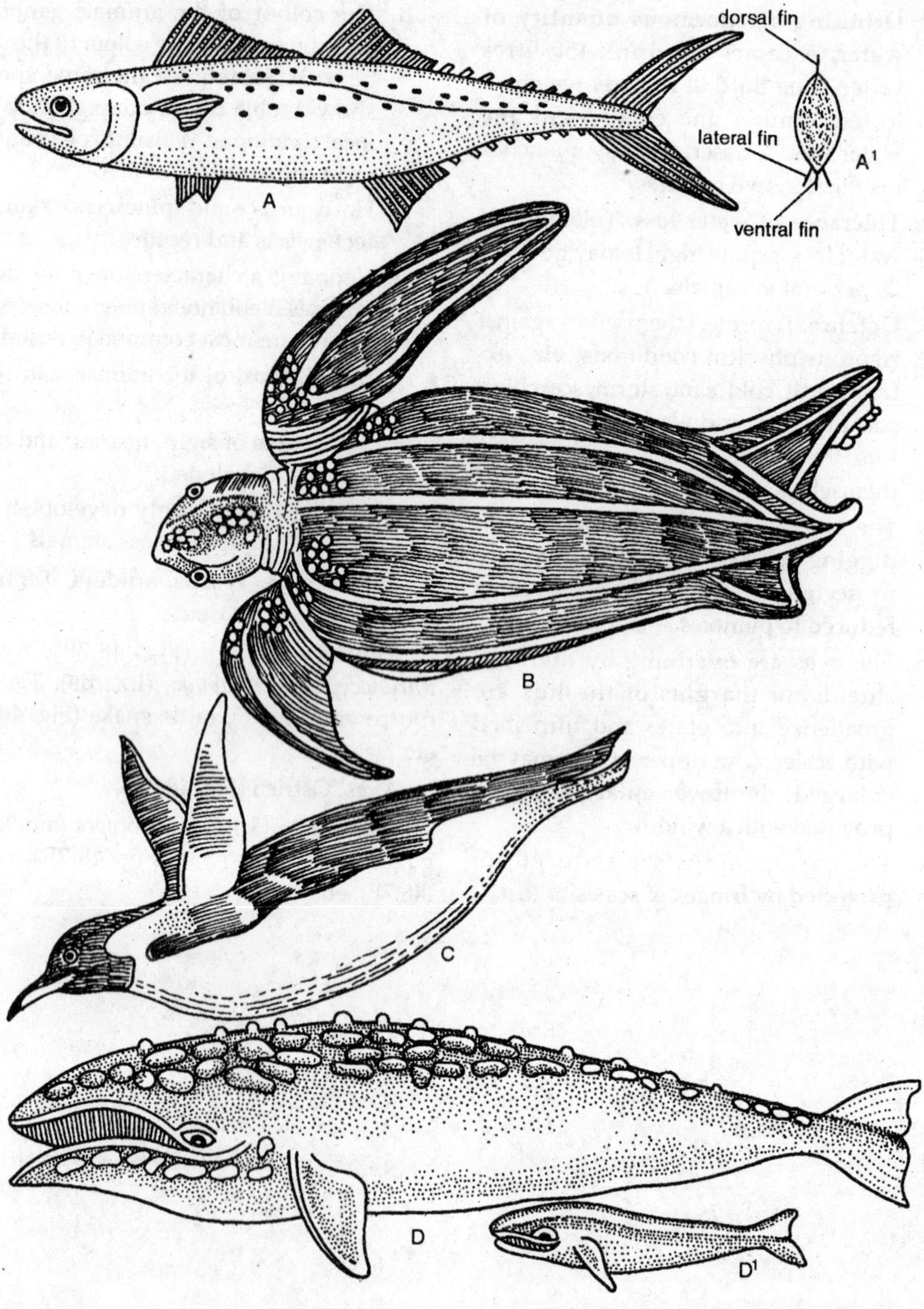

FIGURE 48.1

AQUATIC ADAPTATION. A. **OSTEICHTHYES.** Spanish Mackerel (***Scomberomorus* sp.**) A[1]. Cross section of the body. B. **REPTILIA.** Leather-back turtle (***Chelone* sp.**) C. **AVES.** Emperor penguine (***Aptenodytes* sp.**). D. **MAMMALIA.** Grey whale (***Megaptera* sp.**). Barnacles and whale lice growing on back. Adult length about 12 m. D[1]. New born calf about 3.6 m.

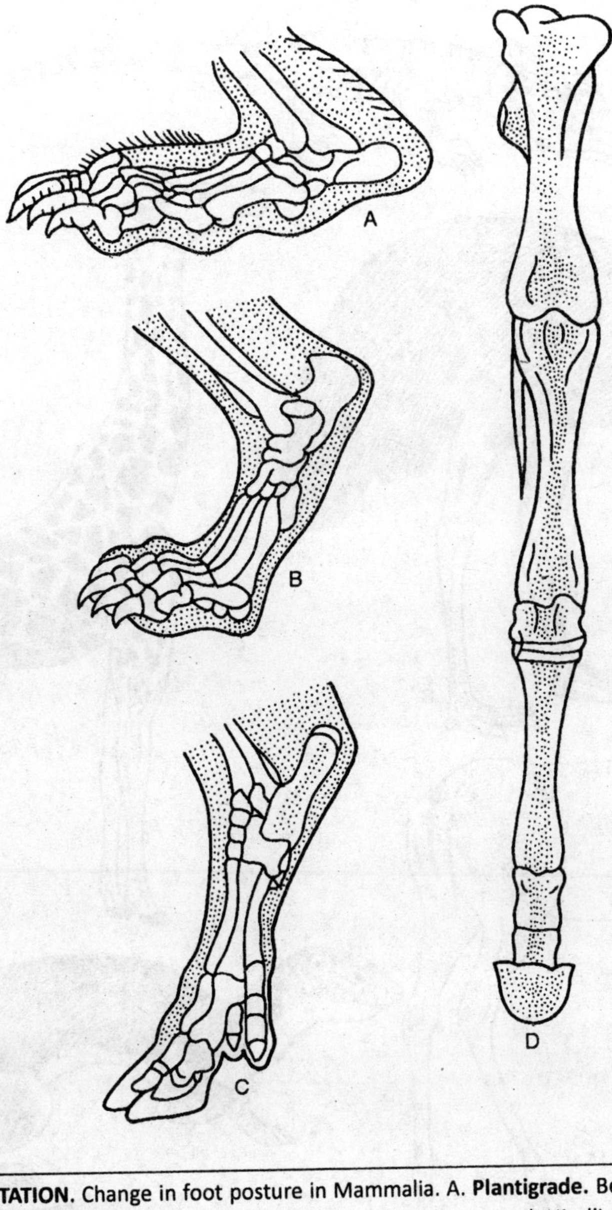

FIGURE 48.2

CURSORIAL ADAPTATION. Change in foot posture in Mammalia. A. **Plantigrade.** Bear (***Ursus* sp.**). B. Digitigrade. Hyaena (***Hyaena* sp.**) C-D. Unguligrade. C. Pig (***Sus* sp.**). D. Horse (***Equus* sp.**). Hindlimb bones pulley-like to restrict movement in one plane

FIGURE 48.3

CURSORIAL ADAPTATION. A. **REPTILIA.** Lizard (***Goanna* sp.**). B. **AVES.** Emu (***Dromaeus* sp.**). C-E. **MAMMALIA.** C. GIRAFFE (***Giraffe* sp.**). D. Antelope (***Apeceros* sp.**). E. Stoat (***Mustella* sp.**)

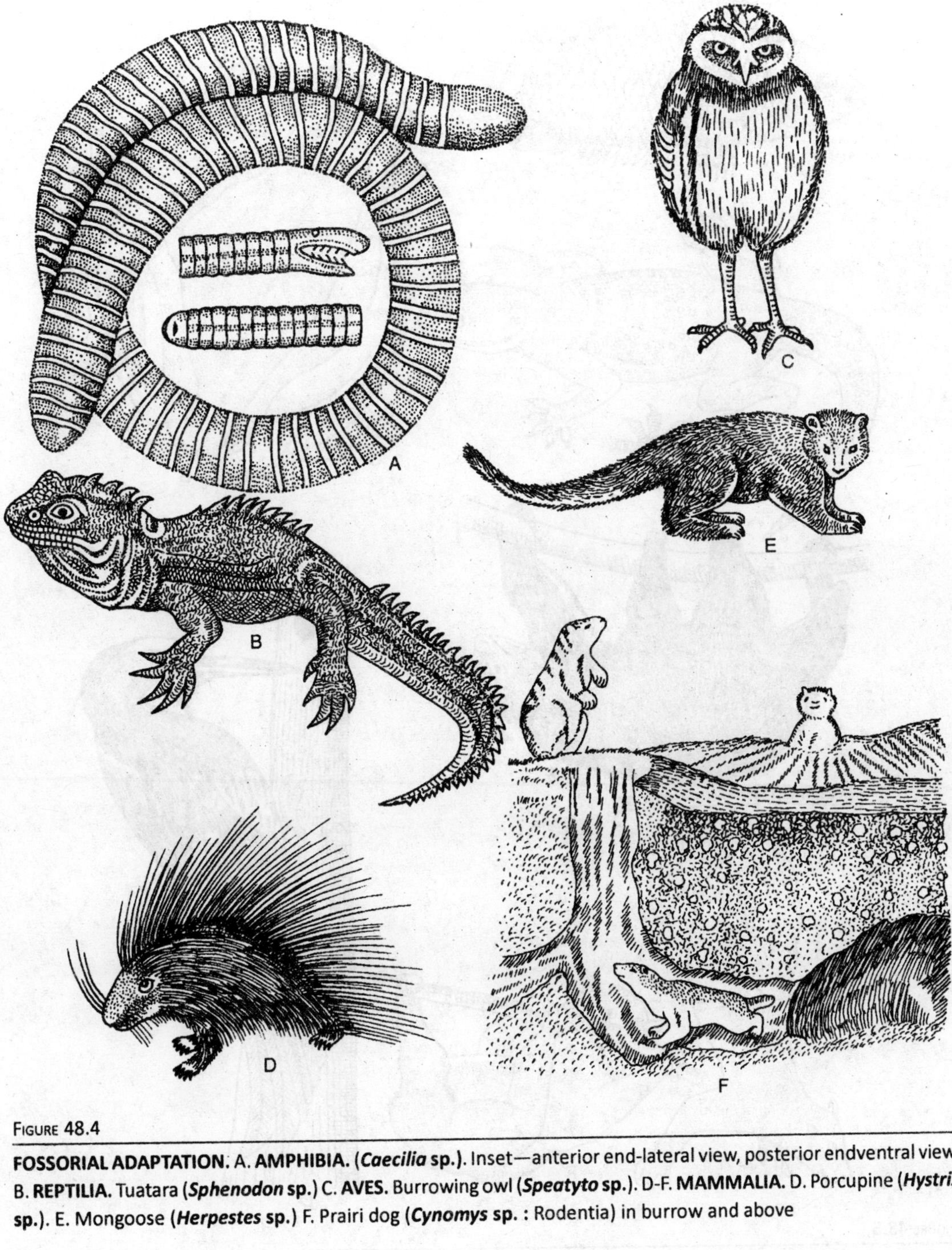

FIGURE 48.4

FOSSORIAL ADAPTATION. A. **AMPHIBIA.** (***Caecilia* sp.**). Inset—anterior end-lateral view, posterior endventral view. B. **REPTILIA.** Tuatara (***Sphenodon* sp.**) C. **AVES.** Burrowing owl (***Speatyto* sp.**). D-F. **MAMMALIA.** D. Porcupine (***Hystrix* sp.**). E. Mongoose (***Herpestes* sp.**) F. Prairi dog (***Cynomys* sp.** : Rodentia) in burrow and above

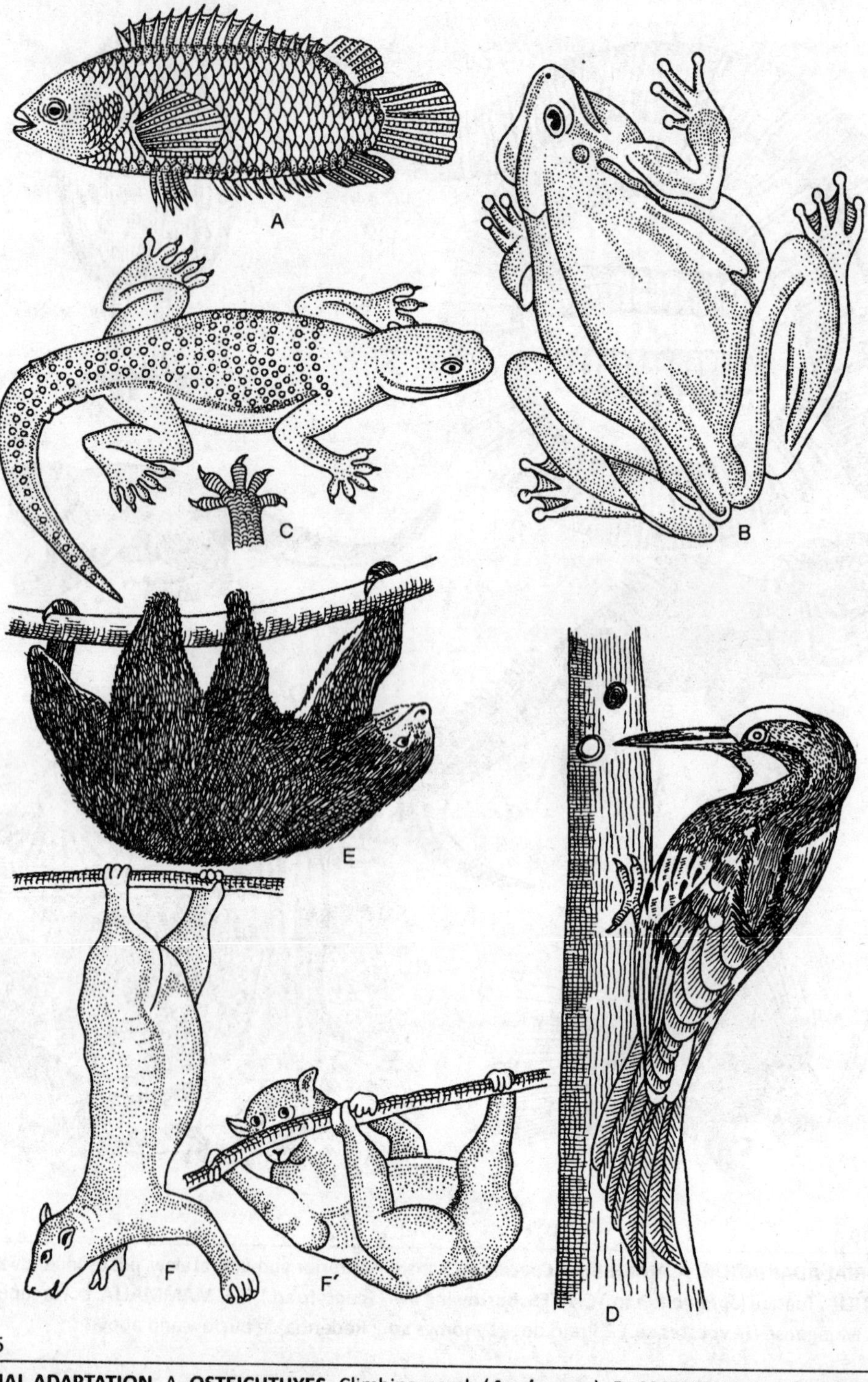

FIGURE 48.5

SCANSORIAL ADAPTATION. A. **OSTEICHTHYES.** Climbing perch (***Anabas* sp.**). B. **AMPHIBIA.** Tree frog (***Hyla* sp.**). C. **REPTILIA.** Lizard (***Gecko* sp.**). D. **AVES.** Wood-pecker (***Melanerpes* sp.**). E-F_1. **MAMMALIA.** E. Two-toed sloth (***Choloepus* sp.**). F-F_1 Spider mankey (***Angwantibo = Arctocabus* sp.**). F. Two limbs, F′ all four limbs in use

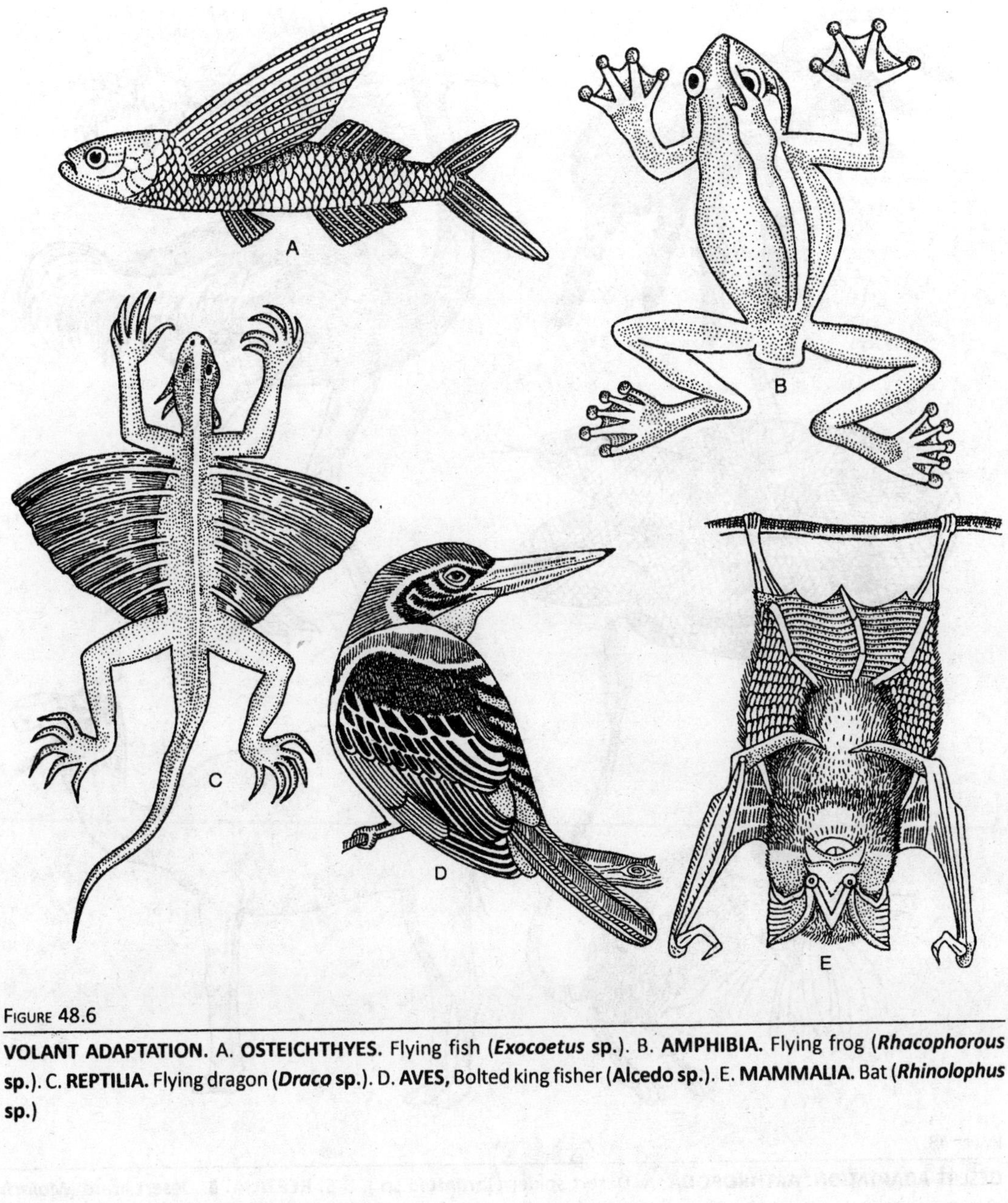

FIGURE 48.6

VOLANT ADAPTATION. A. **OSTEICHTHYES.** Flying fish (***Exocoetus* sp.**). B. **AMPHIBIA.** Flying frog (***Rhacophorous* sp.**). C. **REPTILIA.** Flying dragon (***Draco* sp.**). D. **AVES,** Bolted king fisher (**Alcedo sp.**). E. **MAMMALIA.** Bat (***Rhinolophus* sp.**)

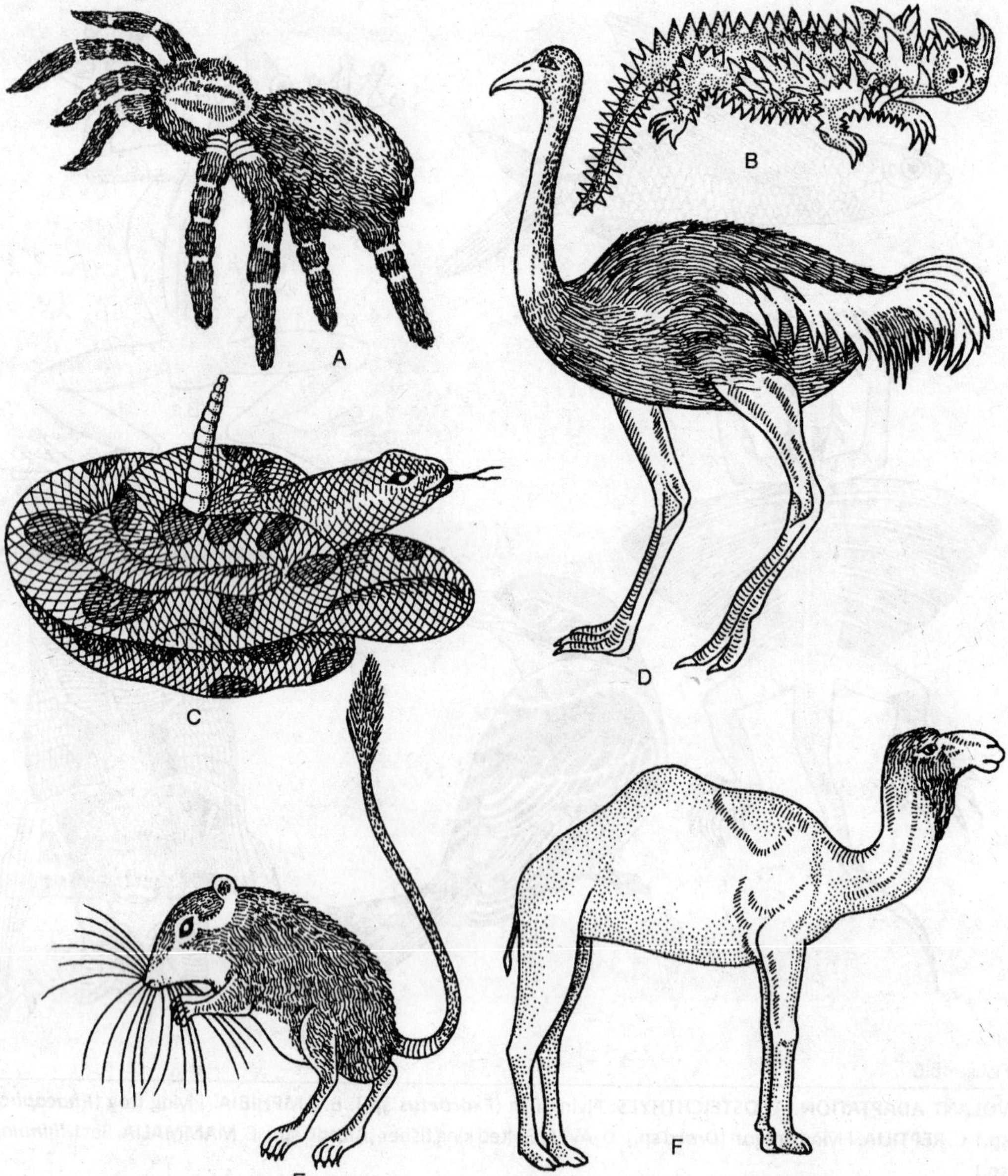

FIGURE 48.7

DESERT ADAPTATION. ARTHROPODA. A. Desert spider (***Tarantula*** **sp.**). B-C. **REPTILIA.** B. Desert lizard (***Moloch*** **sp.**). C. Rattle snake (***Crotalus*** **sp.**). D. **AVES**. Ostrich (***Struthio*** **sp.**). E-F. **MAMMALIA.** E. Kangaroo rat (***Dipodymis*** **sp.**). F. Camel (***Camelus*** **sp.**).

49 GENETICS

Gregor Johann Mendel (1822-1884), an Austrian monk, was the first person to conduct experiments on heredity. From his experiments on garden peas, Mendel postulated simple laws of heredity, which were published in 1866. From a humble beginning in the hands of 'Father of Genetics,' genetics has progressed remarkably and unfolded many mysteries of life.

The advancement of our knowledge in the field within a century is simply stupendous. Today not only the mystery of resemblances and dissimilarities between parents and offsprings is known, we are also aware of which particular gene is responsible for which particular character and that has opened a new field—genetic engineering—with the application of which scientists are dreaming of life free from many hereditary disorders and creating new varieties of animals and plants to meet the ever-increasing demands of man.

⊠ Monohybrid Cross

When individuals differing in one pair of contrasting characters are crossed, such an experiment is known as a monohybrid cross.

Monohybrid cross with pure tall and dwarf pea plant

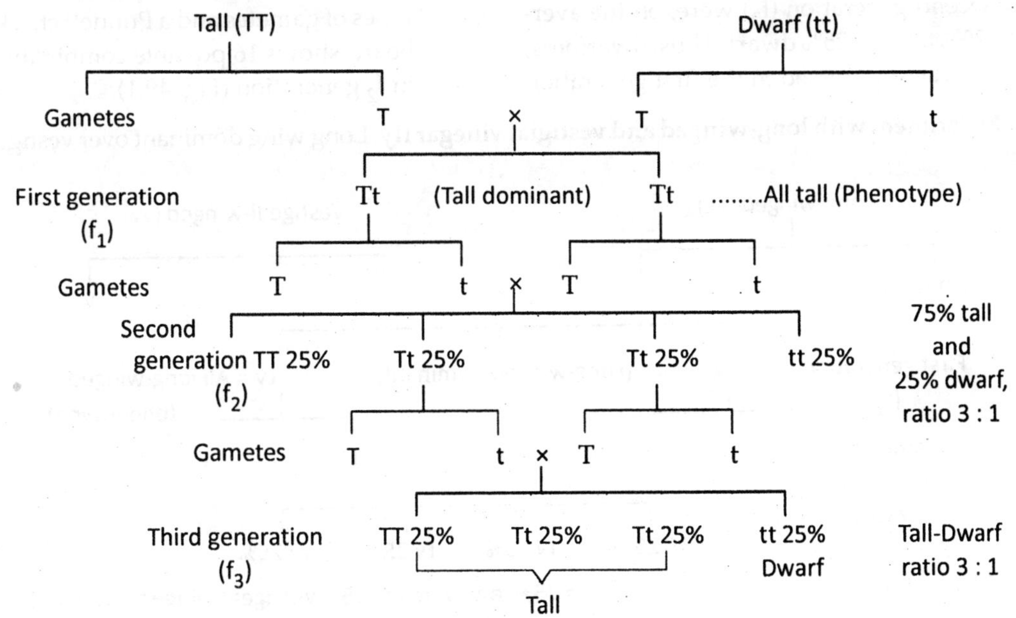

MENDEL'S LAWS OF HEREDITY

From his experiments on plant crossing Mendel established the basic laws of genetics. He carried his plant crossing experiments with ordinary garden peas (*Pisum* sp.). Mendel selected pure tall and pure dwarf plants for crossing. These are contrasting characters. It was found that when two contrasting characters are brought together in a cross, one is **dominant** (expressed or evident) in the next filial (f_1) generation, and the other remains **recessive** (latent or recedes from view). Likewise, tallness is dominant over dwarfness. The factors responsible for a pair of alternative or contrasted characters are termed **alleles** (allelomorph). A class of offspring as determined by genetic composition is referred to as a **genotype** and one determined by its appearance is referred to as **phenotype**.

⊠ Description

Pure tall and pure dwarf plants were crossed and all the seeds obtained produced hybrid tall in the first generation (f_1), the dwarfness remained suppressed. As a result of random crossing in the first generation, the plants of the second generation (f_2) were, on the average, 75% tall and 25% dwarf. Thus, dwarfness, which was suppressed in the first generation became expressive, in a quarter of the second generation. When the f_2 dwarf plants were self-pollinated, only dwarf plants resulted (f_3), whereas among the tall f_2 plants, 25% produced pure tall and 50% produced tall and dwarf plants in the ratio 3 : 1, showing that in 50% plants of f_2 the introduced characters were not completely separated. Thus, of the tall and dwarf characters, the tall became dominant in f_1 generation but in the following f_2 generation these two characters were segregated or separated, in an average 3 to 1 ratio.

DIHYBRID CROSS

When individuals differing in two pairs of contrasting characters are crossed, such an experiment is kwon as dihybrid cross.

1. Mendel crossed a variety of peas producing yellow and round seeds with one producing green and wrinkled seeds. He knew that yellow colour (Y) is dominant over green (g) and that wrinkled seed-surface (w) is recessive to round (R). The f_1 generation hybrid seeds were, as expected, yellow and round. This f_1 hybrid produced four types of gametes and a Punnett checker board shows 16 possible combinations in f_2 generation (Fig. 49.1).

Experiment with long-winged and vestigial vinegar fly. Long wing dominant over vestigeal wing.

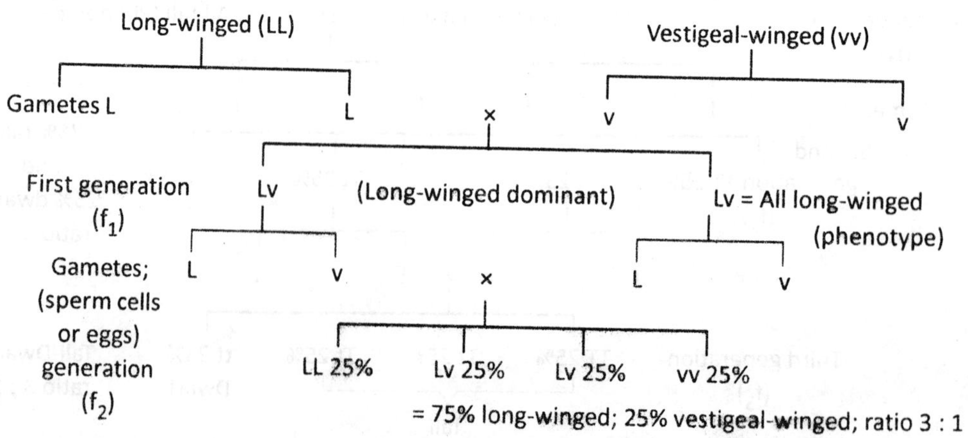

Monohybrid cross in *Drosophila*

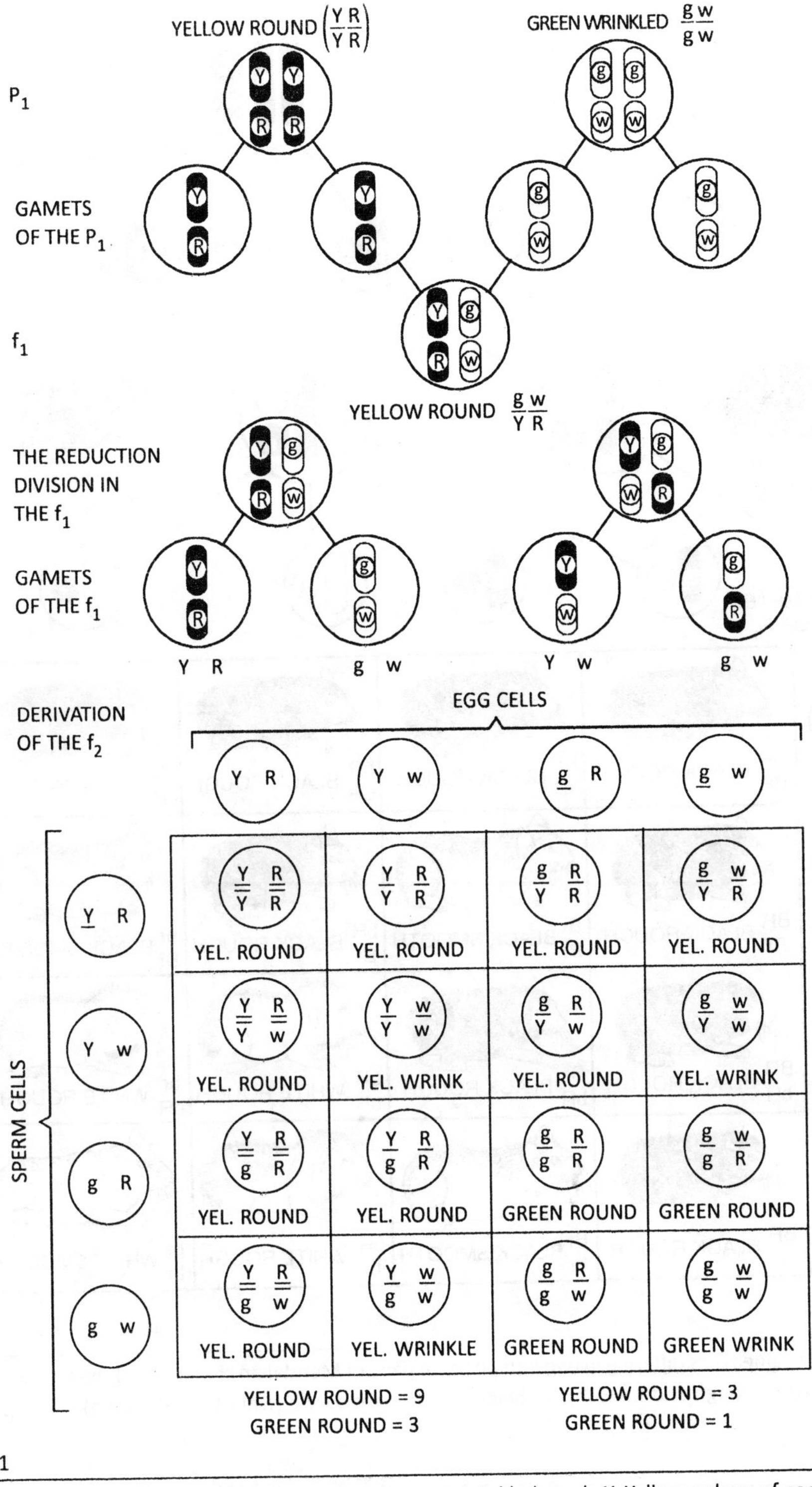

FIGURE 49.1

Dihybrid cross with peas having yellow-round and green-wrinkled seeds Y. Yellow colour of seed coat, R. Round seed, g. Green colour of seed coat, w. Wrinkled seed; Yellow (Y) is dominant over green (g) and round (R) over wrinkled (w)

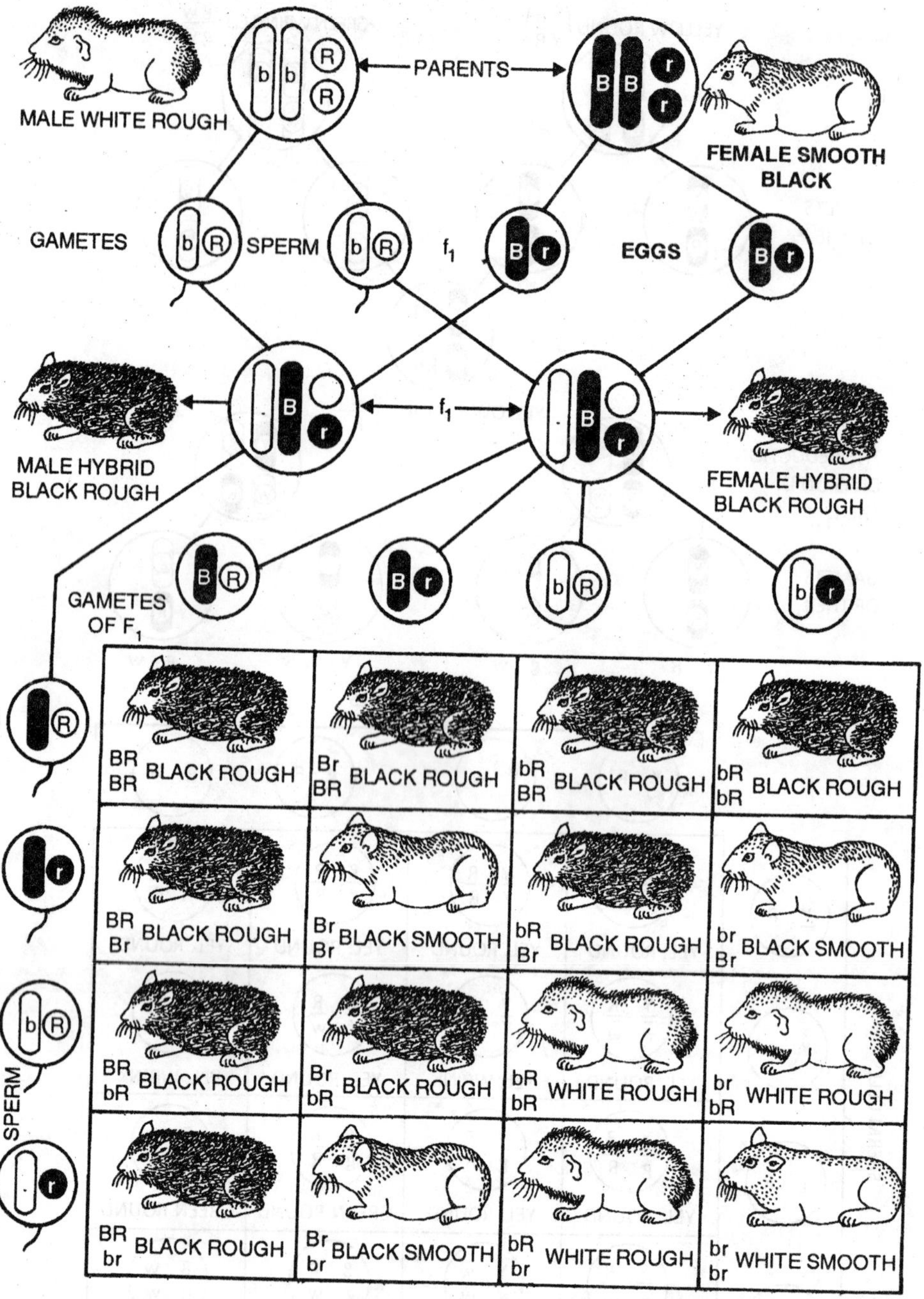

FIGURE 49.2

Dihybrid cross with guinea pigs differing in two independent pairs of Mendelian characters of the hair colour (black and white) and texture (rough and smooth). B. (black) is dominant over b (white) and R (rough) over r (smooth)

The f_2 phenotypes include 9 yellow-round (YR) like the dominant parents, and 1 green-wrinkled (gw) like the recessive parents.

Two new combinations appeared of which 3 were green-round (gR) and 3 yellow-wrinkled (Yw). The 9 : 3 : 3 : 1 ratio is characteristic of a dihybrid cross.

Considering one pair of contrasting characters, there are 9 + 3 = 12 yellow to 3 + 1 = 4 green and 9 + 3 = 12 round to 3 + 1 = 4 wrinkled seed-producing individuals. Thus, the ratio of 3 dominant to 1 recessive in each case follows Mendel's first law, as in a monohybrid cross. Thus Mendel found that each pair of character **is inherited independently of the other.**

2. The dihybrid cross may be illustrated in the case of guinea pig where black coat (B) is dominant over white and (b) rough coat (R) is dominant over smooth (r). The mating of white-rough with black-smooth results black-rough hybrid. The black-rough of f_1 generation of each sex produces four kinds of gametes. When the members of f_1 generation are intercrossed, sixteen possible combinations appear in the f_2 generation. The f_2 phenotypes include 9 black-rough (BR) like the dominant parents, and 1 white-smooth (br) like recessive parents. Two new combinations have appeared, of which there are 3 black-smooth (Br) and 3 white-rough (bR). As in dihybrid crosses with pea, 9:3:3:1 ratio is obtained (Fig. 49.2). It is the characteristic of a dihybrid cross.

Mendel formulated the following laws from these experiments.

1. **Law of segregation** (from monohybrid cross).

   Whatever unit characters may be introduced into the zygote at fertilization, such factors are not completely blended, but are separated out in the gametes it produces.

2. **Law of independent assortment** (from dihybrid cross)

   When considering two or more pairs of unit characters, the inheritance of each pair of unit character is independent of the other.

LINKAGE AND CROSSING OVER

In Mendelian experiment the common sweet peas with two or more "**elements**" (now called **genes**) are considered heterozygous. In a hybrid, the segregation of any one is completely independent of the rest. That means, all the genes are distributed to the gametes at random.

Mendel considered inheritance of seven characters and all the characters showed random assortment. It was later demonstrated cytologically that peas have seven pairs of chromosomes. It was genetically shown that the genes for the characters studied by Mendel are each located in a different member of the seven pairs of chromosomes. The independent assortment between two gene pairs, each with complete dominance and affecting different characters, results in 9 : 3 : 3 : 1 ratio among the offspring of heterozygotes for both pairs (AaBb × AaBb).

But Bateson and Punnet in England in 1906 found exception to this law when they bred a dihybrid strain of sweet pea. They crossed purple (P) flowered-long pollen grained (L) sweet peas with red flowered (p) round pollen grained (l) ones. The f_1 generation was purple flowered-long pollen grained (P-L).

But in f_2 generation purple-long, purple-round, red-long and red-round appeared in the ratio of 11 : 1 : 1 : 3 with too many parental types. Bateson and Punnet explained that the **gametic coupling** between P and L and p and l lead to such high ratio of parental combination. Again, in the next generation, Pl and pL characters tend to come together more often. This is called **repulsion** where the parental associations did not form but new combinations tend to persist.

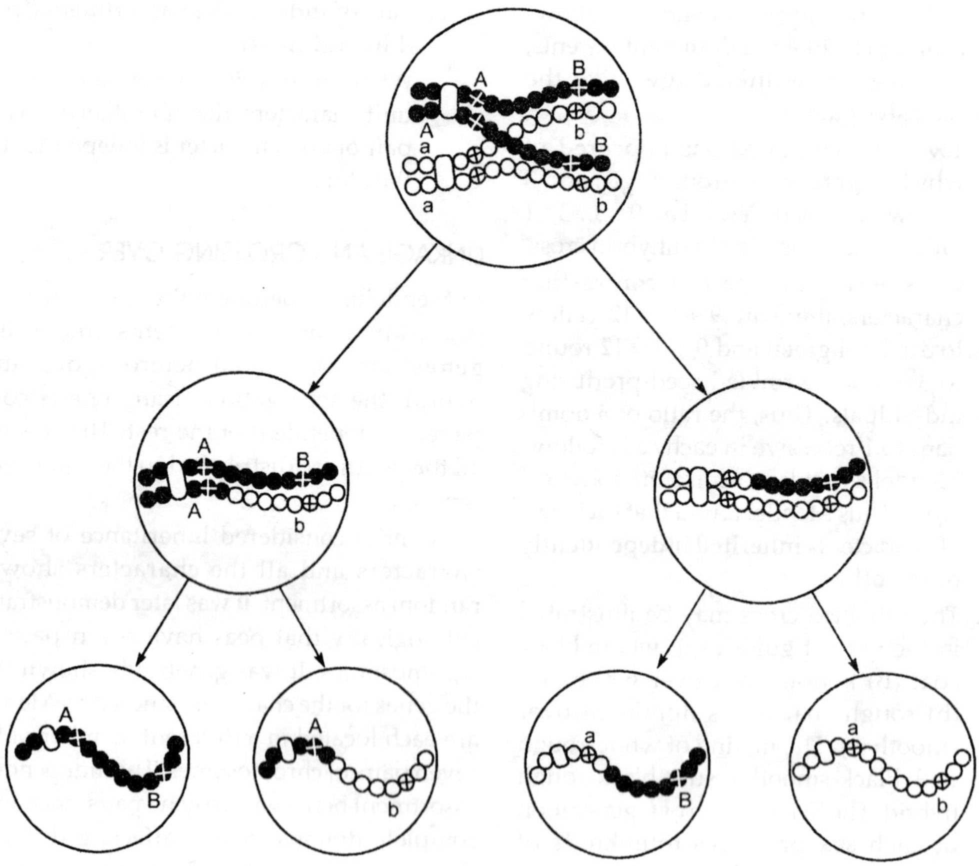

FIGURE 49.3

Crossing over giving rise to Ba and bA recombinations

Both coupling and repulsion are possible when genes are arranged linearly along the length of chromosomes. When two genes remain in the same chromosome they tend to be inherited together to the offspring. This is called **linkage** and two genes are **linked**. But, during reduction division, exchange of segments of a pair of chromosomes occur, as a result linked genes are separated and recombination occurs. This is called **crossing over** (Fig. 49.3). T.H. Morgan (1911) showed in his experiment with *Drosophila*, the linear arrangement of genes in a chromosome and gave explanation of both coupling and repulsion.

So linkage may be defined **as the tendency of two or more genes in the same chromosome to remain in their original combination in the process of inheritance**. Thus, (i) the linked genes remain in the same chromosome and tend to be transmitted together generation after generation, (ii) linked genes never show Mendel's Principle of Independent Assortment, (iii) less distance between the genes show stronger linkage bond.

⊠ Example of Linkage

In *Drosophila*, grey body (b^+) and long wing (v^+) characters are dominant over black body (b) and vestigeal wing (v). When a pure grey-long is crossed $\left(\frac{b^+v^+}{b^+v^+}\right)$ to a black -vestigeal $\left(\frac{bv}{bv}\right)$ the offspring produced in f_1 generation are grey-long hybrid $\left(\frac{b^+v^+}{bv}\right)$ (Fig. 49.4).

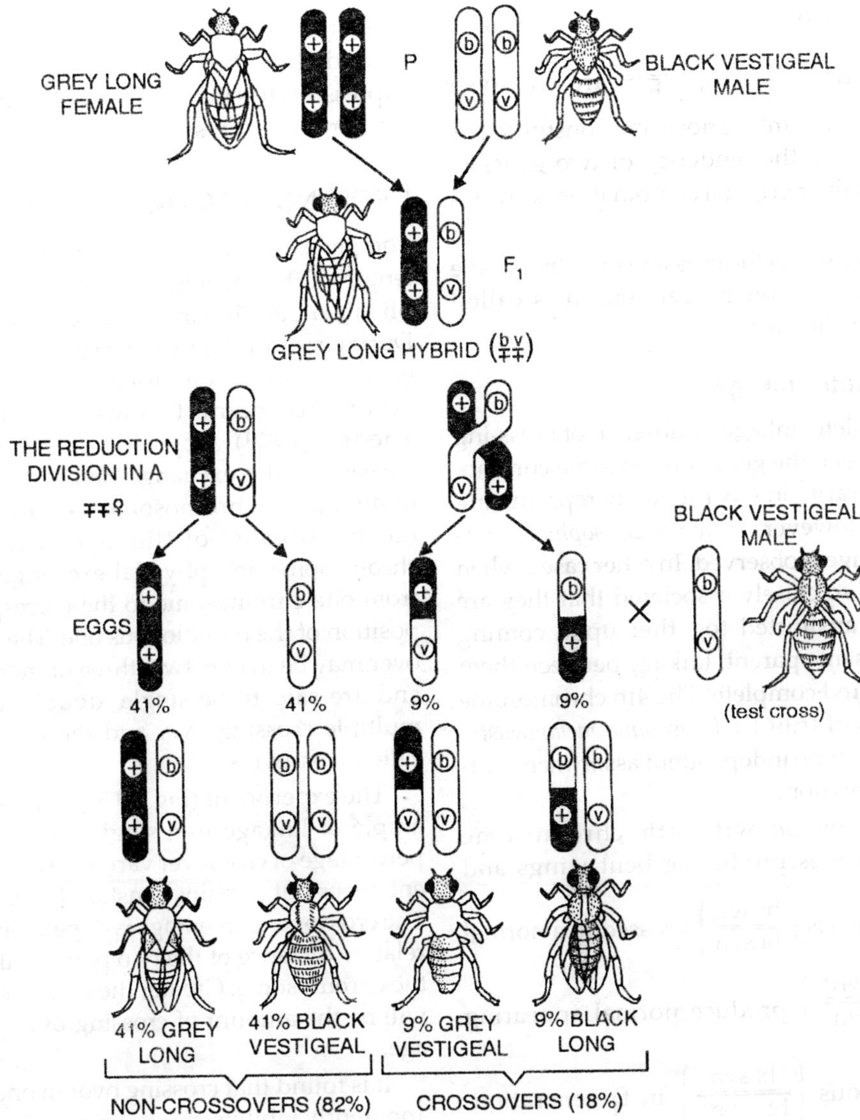

FIGURE 49.4

Linkage in *Drosophila*: grey-long × black-vestigeal (grey is dominant over black colour and long is dominant over vestigeal wing)

In the f_1 female genotype, the chromosome might come together and separate without crossing over between $\underline{b}^+$ and $\underline{v}^+$, giving eggs of two noncrossover classes $\underline{b^+v^+}$ and $\underline{b\ v}$. The second possibility is that crossing over might take place between $\underline{b}^+$ and $\underline{v}^+$ giving eggs of two crossover classes $\underline{b^+\ v}$ and $\underline{b\ v^+}$.

If a cross is made between f_1 female and a black-vestigeal male (double recessive) four categories of offsprings will be produced (Fig. 49.4). The grey-long $\left(\frac{b^+v^+}{bv}\right)$ and black-vestigeal $\left(\frac{bv}{bv}\right)$ are the noncrossovers (parental combination) and constitute more than 50% of the total offsprings, while the grey-vestigeal $\left(\frac{b^+v}{bv}\right)$ and black-long $\left(\frac{bv^+}{bv}\right)$ are crossover classes (new combination) and constitute less than 50%. So the tendency of two genes to remain in their original combination is always greater.

In this case, as the crossing over breaks the linkage in certain percentage, it is called incomplete linkage.

☒ Complete Linkage

The complete linkage or absence of cross-ing over between the genes on the same chromosome is a rarity in most sexually reproducing species. However, in male *Drosophila*, complete linkage is observed. In other cases, when genes are so closely associated that they are always transmitted together upon coming from the same parent, linkage between them is considered complete. The 4th chromosome mutations of fruit fly *Drosophila melanogaster* show little or no independent assortment during transmission.

A *Drosophila* with 4th chromosome carrying genes producing bent wings and shaven bristles $\left(\frac{bt\ svn}{bt\ svn}\right)$ crossed to a normal fly $\left(\frac{bt^+\ svn^+}{bt^+\ svn^+}\right)$ produce normal appearing heterozygous $\left(\frac{bt\ svn}{bt^+\ svn^+}\right)$ in f_1 generation. When such f_1 flies are crossed to the homozygous bent-shaven $\left(\frac{bt\ svn}{bt\ svn}\right)$ stock, the offsprings are almost all phenotypically either bent and shaven or normal for both characters. The complete absence of new combination is the evidence of very strong and complete linkage between the two genes on the same homologue.

Test cross

The cross of f_1 hybrids with a double recessive male or female is known as test cross or **back cross**. In this process, only recessive genes will be added so that all the mutations will be expressive. It is the simplest way to determine the per cent of crossing over.

CROSSING OVER AND LINKAGE

The genes remain in linear order along the length of the chromosome and linkage is the physical relationship between the genes. During meiosis a physical crossover between gene pairs on homologous chromosomes occur. According to Jonssen's crossover Theory (1909) a cytologically observed chiasma is the exchange point between the homologous chromosomes. Crossing over means breaks of linkage within the chromosome and physical exchange of gene from one chromosome to the corresponding position of the homologous one. The crossing over may be in one, two, three or more points and are said to be single, double, triple or multiple crossing over and the gametes are called crossovers.

The experiment (Fig. 49.5) to show the example of linkage in a single crossover. The percentage of crossover varies between different genes but remains constant for each pair. The crossover percentage is dependent on the relative distance of the two pairs of alleles on the chromosome. Greater the distance, higher will be the amount of crossing over percentage.

It is found that crossing over in one region apparently inhibits or interferes with crossing over in a neighbouring region. Muller termed this as '**interference**.' There are only few double crossovers within a 10 unit or less long section of chromosome due to interference. When the distance between two genes increases, the interference becomes less or even nil.

The double crossover is the coincidence or coming together of two single crossovers and involves three genes on the same chromosome. When double crossovers occur in expected numbers, the coincidence is considered as 100 per cent and interference is 0.

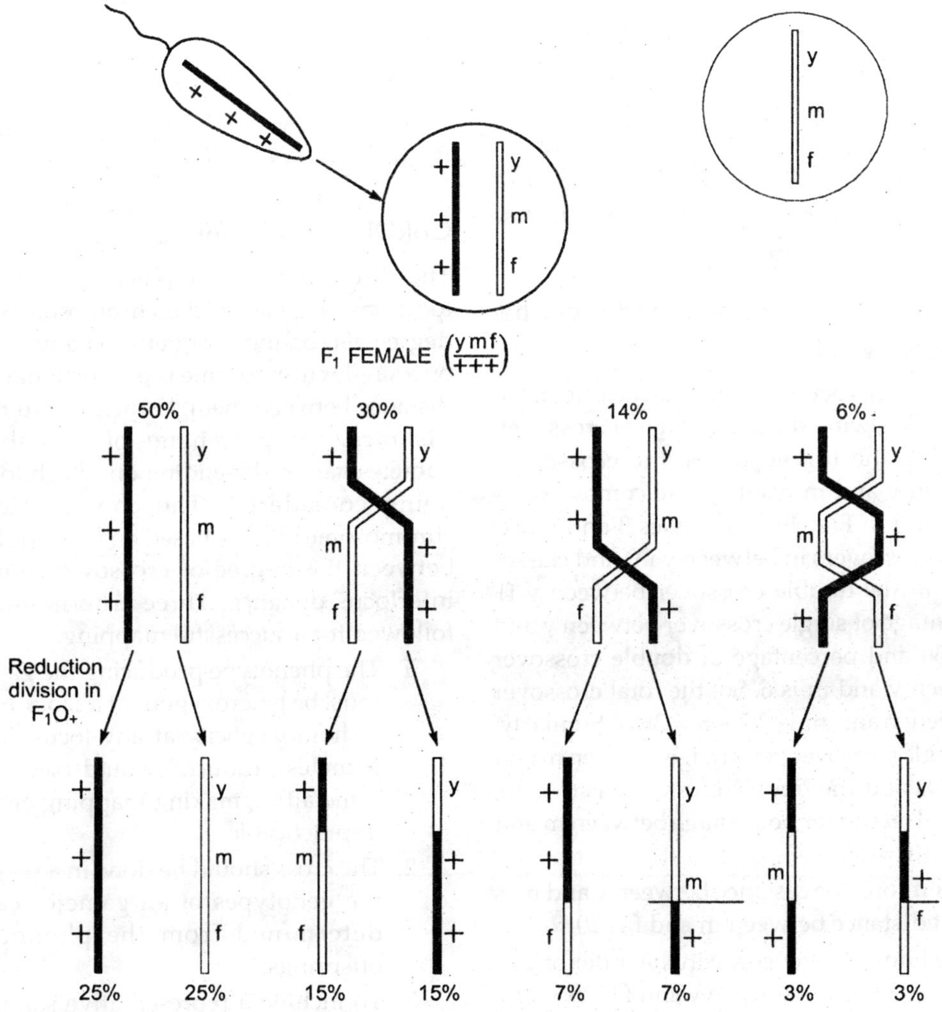

FIGURE 49.5

Linkage involving three pairs of genes

Alternatively, when there is no double crossover, the interference is 100% and coincidence is 0. The ratio of observed frequency to the expected frequency of double crossovers is known as "**coincidence coefficient.**"

☒ Double crossover in Drosophila

In *Drosophila*, yellow body (y), miniature wing (m) and forked bristles (f) are three recessive mutations in the X-chromosome. The normal fly possesses grey body (y^+), long wings (m^+) and straight bristles (f^+). A cross between yellow-miniature-forked female $\left(\frac{ymf}{ymf}\right)$ and a normal male ($\underline{y^+m^+f^+}$) pro-duces in f_1 female with genotype $\left(\frac{ymf}{y^+m^+f^+}\right)$. During ovulation in this female, the charo-mosome might pair in four possible ways, resulting eight classes of combinations (Fig. 49.5). The first and 2nd classes of combinations ($\underline{y^+m^+f^+}$ and $\underline{ymf}$) are the crossover, the third and fourth classes ($\underline{y^+m^+f^+}$, $\underline{ym^+f^+}$) are single crossovers between y and m, the 5th and 6th classes (y^+m^+f, ymf^+) are single cross-overs between m and f and 7th and 8th classes (y^+mf^+, ym^+f) are double crossovers between y and f. If these are test crossed with triple recessive male (ymf), then all these eight classes of offspring will be produced.

Calculation of distance and order between y m f

The distances between y m f can be calculated now by knowing the percentage of crossover. Ignoring the forked locus, the crossovers between y and m result in the combinations are found in Fig. 49.6 as classes 3 and 4 (the single crossovers in between y m) and classes 7 and 8 (the double crossover between y f). Percentage of single crossover between y and m is 30 and percentage of double crossover between y and m is 6. So, the total crossover between y and m is 30 + 6 = 36%. Similarly, the single crossover percentage between m and f is 14% and the double crossover is 6%. So, the total crossover percentage between m and f is 14 + 6 = 20.

Therefore, the distance between y and m is 36 and distance between m and f is 20.

We know the genes are in the order of y m f and the distance between y and f is 36 + 20 = 56 (Fig. 49.6). The double crossovers are counted twice, because a double crossover is equivalent to two single crossovers, one between y and m and another between m and f.

With the test cross the actual percentage of crossovers between y and m and f will give the real order of y m f.

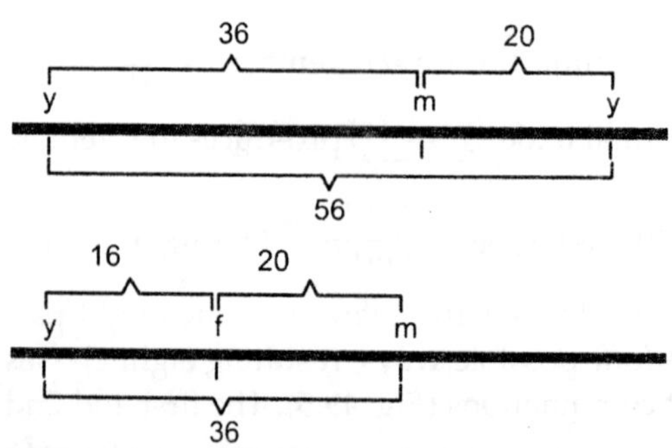

FIGURE 49.6

Possible arrangements of three genes when only two of the distances are known.

CHROMOSOME MAP

The chromosome map indicates the relative locations of genes in the chromosomes. The degree of crossing over between any two loci on a single chromosome is proportional to the distance between them, known as **interlocus distance.** The percentage of recombinant gametes varies, depending on which loci are being considered. The construction of chromosome maps is based on the correlation between the degree of crossovers and the interlocus distance. Three criteria must be followed for a successful mapping.

T. The phenotype producing the gametes must be heterozygous at all loci. In case of homozygocity at any locus, all the gametes produced would contain the same allele, making mapping analysis impracticable.
2. The cross should be done in a way that the genotypes of all gametes can be determined from the phenotypes offsprings.
3. To include a representative sample of all crossovers classes, production of a sufficient number of offsprings is essential.

GENE MAP

Genetic map is based on recombination frequency. The mapping is concerned with identifying the positions of the genetic loci, which are fixed and lie in a linear order. The extent of recombination between two genes on the same chromosome is the **map distance.** The map distance is used to measure their relative location.

$$\text{Map distance} = \frac{\text{Number of recombinants} \times 100}{\text{Total number of progeny}}$$

one map unit [Centi-Morgan (cM)] equals 1% crossover.

Genetic map does not give an accurate measurement of distance between the loci, because the frequency of cross over is not constant throughout the genome.

HARDY-WEINBURG EQUILIBRIUM

Population genetics is the study of alleles of genes in a population and the forces changing or maintaining the frequencies of certain alleles, and genotypes in a population. Evolution is the change of frequencies of alleles in the gene pool. The gene pool is the total genetic stock of a population.

Assuming random mating, equal reproductive ability and no effects of selection pressure or migration affecting particular genotypes, the expected genotype frequencies from allele frequencies can be measured with the Hardy-Weinburg equilibrium.

It is presumed that a population differing from Hardy-Weinburg predictions is subjected to some sort of selection pressure operating to disturb it.

GENETIC DRIFT

Genetic drift is the random change in allele frequencies in a population. Large interbreeding populations are essential to the maintenance of Hardy-Weinburg equilibrium. Significant random fluctuations in allele frequencies are possible by chance deviation in small populations. With the decrease of population size the degree of fluctuation in allele frequencies increase. This is known as **genetic drift**.

MUTATION

Mutation is the random, minute and irreversible, measureable change in the gene of an organism or a cell, with the necessary implications that any phenotypic manifes-tation of the change may potentially or actually be shown by all descendant organisms or cells. Mutation in somatic cells is of the same general quality as in germ cells. Somatic mutation is associated with the functional activities of a cell type while germ cell mutation is responsible for introducing variations in the offsprings.

Possible causes of mutation are cosmic radiation, atomic radiation, and different stresses, viz., chemicals, heat, ultraviolet rediation, etc. are **mutagenic** (mutation inducing).

Mutations are always random and nondirectional. They are mostly lethal or deleterious but some mutations are useful when organisms are placed in a changed environment. Environmental changes disrupt the adaptedness of an organism and the harmony between the environment and the organism can be restored only through a genotypic reconstruction by mutation.

Mutation and selection occur simultaneously. Most frequently selection works against mutation and the rate of change, if the mutations are not lethal, will be very slow. But if a particular mutation is advantageous to the organism, and mutation rate is appreciable, the combined action of mutation and selection might well cause a rapid change.

It is believed that mutations are the raw materials for evolution.

☒ Types of mutation

Mutations are of two types

a. **Spontaneous mutation.** These occur in nature in animals or plants not subjected to any artificial stress.
b. **Induced mutation.** These are produced when the organism is subjected to different mutagens, viz., X-rays, gamma-rays, chemicals, etc.

☒ Stage of occurrence of mutation

Mutations may take place in different stages in the life of an organism and also at different locations.

a. **Gametic.** It occurs prior to the maturation of gametes.

b. **Zygotic.** It appears immediately after the fertilization of egg.

c. **Somatic.** It occurs in somatic cells as a result of which the cells produced by the multiplication of the mutant cells become genotypically different from the nonmutant cells.

Classifications of Mutation

Mutational changes can be grouped into two major classes. The mutation proper is caused by chemical alteration of individual genes, and chromosomal aberrations leading to a grosser structural change involving destruction, multiplication or spatial arrangement of the genes in the chromosomes. The chromosomal aberrations may be of the following types:

I. Numerical changes. Affecting the number of chromosomes.

a. **Haploidy.** Higher organisms for the major part of their life are mostly diploid. Under experimental conditions, some diploid organisms have produced haploid aberrants with a single set of chromosomes in the cells instead of the normal double sets.

b. **Polyploidy.** Normally diploid organisms in some cases may produce forms having more than two sets of homologous chromosomes.

II. Structural changes. Affecting the arrangement of the genes in the chromosomes.

A. Changes due to reduplication or loss of some genes.

a. **Duplication.** In addition to its normal location a section of the chromosomes may be present elsewhere. In the salivary gland cells of *Drosophila* certain sections of the normal chromosomes are represented two or more times in the haploid set.

b. **Deletion** (deficiency). A section of the chromosome containing one gene or a block of genes is lost resulting in a deficient chromosome.

B. Changes due to altered arrangement of genes.

a. **Translocation.** Two chromosomes may exchange parts giving rise to a new chromosome.

b. **Inversion.** Due to a rotation of 180° the arrangement of the genes in the chromosome may be altered.

Point Mutations

In nature different types of mutation occurs. In point mutation each mutation involves the change of a single base, influencing the protein encoded by the gene.

a. **Missense mutation.** It alters a single coded amino acid. Such mutations usually occur in one of the first two bases of a **codon***. It produces a mutant phenotype.

b. **Nonsense mutation.** It changes a codon for an amino acid into a terminal codon and produces shortened polypeptides.

c. **Frameshift mutation.** It involves insertion of extra bases or deletion of existing bases and produces an altered reading frame.

d. **Silent mutation.** It occurs at the third base of a codon and due to degeneracy of the genetic code the amino acid is not altered. It has no effect on encoded protein.

SEX DIFFERENTIATION IN DROSOPHILA

In the fruit fly (*Drosophila sp.*) sex of the individual depends on the ratio of X-chromosomes to sets of autosomes. Flies are male when the ratio is 0.5 or less; flies are female when the ratio is 1.0 or higher (Table).

* **Codon.** The DNA sequence of a gene consists of a series of units of three bases. Each set of three bases is known as codon.

Table. Ratio of X chromosomes to sets of autosomes and sex differentiation in Drosophila

No. of X chromosome (X)	No. of sets of autosomes (A)	X : A ratio	Sex
3	2	1.5	Female
3	3	1.0	Female
2	2	1.0	Female
2	3	0.67	Intersex
1	2	0.5	Male
1	3	0.33	Male

Abnormal individuals known as **gynandromorphs** are present in *Drosophila*. This is due to nondisjunction in the somatic cells. Gynandromorphs are with male and female sex phenotypes in different parts of the body.

SEX DETERMINATION IN MAN

Sexes, male and female are distinct in all higher animals including human. Hereditary differences are quite pronounced in male and female individuals.

1. Sex of an organism is associated with certain chromosomes, known as **sex chromosomes**. In human the sex chromosomes are XY in male and XX in female.
2. The male which is heterozygous, produces two types of gametes or **sperm** X and Y. The homozygous female produces only one type of gametes or **ovum** X.
3. An ovum fertilized with Y bearing sperm gives rise to a male offspring (XY) while an ovum fertilized with X bearing sperm produces a female (XX).
4. The maleness and femaleness of the offsprings is dependent on the type of sex chromosomes in the zygote.
5. The only difference between male and female sex is that the male has a haploid set and the female a diploid set of sex chromosomes*.

* X is the female sex chromosome, Y is the male sex chromosome.

CONGENITAL ABNORMALITIES IN MAN

Congenital means present since birth. Congenital disorders are not always heredi-tary. It might me due to some extrinsic factors also. The congenital abnormalities in human are (a) Down's syndrome, (b) Colour blindness, (c) Albinism and (d) Sickle-cell trait.

☒ Down's Syndrome

It is hereditary abnormality due to gentic disorder, autosomal abnormality. It is also known as **Mongolism** or **Mongoloid idiocy**.

The individual possesses a small extra chromosome No. 21, the total number of chromosomes being 47.

1. The extra chromosome is presumably due to nondisjunction or failure of homologous chromosome in the first meiosis, or of the two chromatids of a chromosome in the second meiosis to separate during the formation of ovum.
2. It may also be caused by translocation between chromosome No. 21 and 14 or 15 (translocation interchange).

About 0.15 per cent people suffer from Down's syndrome. It is believed that the disease is related to the age of the mother and greatly increases with age. It is independent of the age of the father and preceding pregnancy of the mother.

Symptoms. Stature short; stubby hands; hand and feet bear simian folds; a broad face; flat or stubby nose, obliquely set eyes with mongoloid eye-lid fold (though not similar); open mouth; fat and soft skin; flaccid muscles; mentally retarded and malformed heart.

☒ Colour Blindness

The cone cells of the retina of human eye are of three different types and respond to three primary colours—blue, green and red. The development of cone cells is controlled by colour gene. It is difficult or impossible for a person to distinguish between the colours of longer wave lengths due to lack of either the red or the green cones. This is known as **red-green colour blindness.** The colour genes are sex linked and present in the female sex chromosome. 'C', the gene for normal vision is dominant over 'c' (recessive allele), gene for colour blindness colour blindness is almost absent in females, as they posses two female sex chromosomes (XX). Males have only one female sex chromosome (X).

Colour blindness is about 8 per cent in men but only about 0.4 per cent in females.

Genotypes

The genotype for a colour blind male is 'c', the recessive allele (no colour gene is present in male or Y chromosome).

The genotype of a normal visioned male is 'C' (dominant).

The genotype of a colour blind female is 'cc' (recessive).

The genotype of a normal visioned female is 'CC' (dominant).

The offsprings of both the sexes receive x chromosome from a normal visioned or a colour blind mother. The father transmits x chromosome only to the daughter.

Transmission of colour blindness

A. Offsprings of a normal man and a colour blind woman.

Normal male Colour **xy** (C–) × Colour blind female **xx** (cc)

Colour blind sons **xy** (c–) | Normal daughter (carrier) **xx** (Cc)

All the sons will be colour blind.

All the daughters will be normal but carriers.

B. Offsprings of a colour blind son and a normal carrier daughter.

Colour blind son **xy** (c–) × Normal carrier daughter **xx** (Cc)

Normal son **xy**(C) | Colour blind son **xy** (c–) | Normal carrier daughter **xx**(Cc) | Colour blind daughter **xx**(cc)

50% of the offsprings (male and female) will be normal, 50% of the offsprings (male and female) will be colour blind.

C. Offsprings of a colour blind male and a normal female (homozygous).

Colour blind male **xy** (c–) × Normal homozygous female **xx** (CC)

Normal son **xy**(C–) × Normal daughter (carrier) **xx**(Cc)

All the males will be normal.

All the daughters will be normal but carriers.

Albinism

It is a congenital absence of pigment from the skin, hair, choroid coat and iris. It may be partial or localised and total or generalised.

1. **Melanin,** a black pigment is synthesised in melanocytes following a series of enzymatic reactions. The amino acid tyrosine plays an important role in it.
2. One-gene one-enzyme hypothesis advocates that a gene controls synthesis of one enzyme. The enzyme is responsible for a specific metabolic reaction.
3. Albinism is the result of a **genetic block** involving a step in the pathway of melanin formation from tyrosine, possibly due to gene mutation.

☒ Sickle-cell Trait

In certain persons the erythrocytes (red blood cells) instead of being round as in normal human are sickle or crescent-shaped, and this is known as sickle-cell trait.

1. The disease sickle-cell trait is caused by mutation.
2. The amino acid, **glutamic acid** of the haemoglobin of a normal person is replaced by **valine** in sickle-cell haemoglobin.
3. As a result the normal haemoglobin and consequently the normal RBC is converted into abnormal haemoglobin causing sickling RBC.
4. The blood of persons heterozygous for sickle-cell gene contains both normal haemoglobin (Hb^A) and mutant (Hb^S) alleles and the haemoglobin is a mixture of both kinds.
5. The persons homozygous for sickle-cell gene contain only mutant haemoglobin.
6. Sickle-cell trait occurs in Africans. The persons homozygous for sickle-cell gene suffer from anaemia, skin lesions, kidney damage and spleen enlargement, and die early.
7. The persons heterozygous for sickle-cell gene suffer from anaemia but they are immune to *Plasmodium*, causing malignant malaria.

HAEMOPHILLIA

Haemophillia is a single gene disorder of blood coagulation characterised by a life long tendency to excessive haemorrhage and greatly prolonged coagulation time. The disease is due to X-linked recessive mutation. A haemophilic man do not transmits the trait to sons. The trait is transmitted to the daughters. There is a 50% chance that sons of female carriers will suffer and a 50% chance that daughters will be carriers.

Haemorage into the joints is the common manifestation of haemophillia associated with swelling, warmth and muscle spasm. There are two types of haemophilia. A and B. (Table).

⊠ Klinefelter and Turner Syndrome

The abnormalities caused by the syndrome: genitalia and internal ducts are usually male but testes underdevoped and incapable of producing sperm. In the patients certain musculine development occur but female sexual development is not entirely suppressed. Enlargement of breast is slight. This ambiguity in sexual development is termed **intersexuality**.

Klinefelter syndrome. Individuals have more than one X-chromosome, most often they have XXY copplement in addition to 44 autosomes. Individuals with Karyotype are designated 47 XXY. Klinefelter syndrome is found in about 2 individuals in every 1000 male births.

Turner syndrome. Individuals have female external genitalia and internal ducts and rudimentary ovaries. Stature short (usually under five feet), neck webbed; chest broad and shield-like. Individuals with Turner syndrome are most often monosomic with only 45 chromosomes including only a single X-chromosome. They are designated 45, X. Turner syndrome occurs in about 1 individual in 3,000 female births.

⊠ Cri-du-chat Syndrome

Cri-du-chat syndrome (cry of the cat) has not been reported beyond birth. Individuals with the syndrome are conceived but do not survive embryonic and foetal development.

Table. Haemophilia A and B compared

Disorder	Frequency per 1000 birth	Pattern of inheritance	Mutated gene	Characteristics
Haemophillia A	0.1	X-linked	Factor VIII	Abnormal bleeding
Haemophillia B	0.03	X-linked	Factor IX	Abnormal bleeding

Survivors with partial monosomy, where only part of one chromosome is lost however, is on record. The genetic constitution may be designated as 46, –5p– i.e., the individual has 46 chromosomes but some or all of the p arm (the peptite or short arm) of chromosome 5 is missing.

Cri-du-chat syndrome has been reported 1 in 50,000 live births.

Infants are born with anatomic malformations, glottis and larynx abnormally develope, with gastrointestinal and cardiac complication and often mentally retarded.

50 MICROBIOLOGY

Microbiology deals with living organisms ordinarily too small to be seen without magnification. Such microscopic organisms are collectively referred to as micro organisms or microbes. Bacteria, viruses, fungi, protozoans, algae and helminths are the major groups of microbes.

Microbiology is one of the largest and most complex of the biological sciences because, it deals with many diverse biological disciplines. In addition to studying the natural history of microbes, it also deals with every aspect of microbe-human and microbeenvironmental interaction. These interactions include genetics, metabolism, infection, disease, drug therapy, immunology, genetic engineering, industry, agriculture and ecology. Micro-biology has numerous practical uses in industry and medicine. Some important areas those are based on applications in microbiology are—immunology, public health microbiology, epidemiology, food micro-biology, dairy microbiology, aquatic microbiology, agricultural microbiology, biotechnology, industrial microbiology, recombinant technology etc.

⊠ Morphology and Fine Structure

Viruses

Viruses are unique group of non cellular biological entities known to infect every type of cell, including bacteria, algae, fungi, protozoa, plants and animals. Viral replication inside a cell causes death or disease of that cell. (i) Viruses are ultramicroscopic in size, ranging from 20 to 450 nm; (ii) it is not a cell; its structure is very compact; (iii) it is inactive macromolecule outside of the host cell and active only inside hosts cells; (iv) geometric, can form crystal-like masses. (v) basic structure consists of protein capsid and nucleic acid; (vi) capside is made of repeating subunits encloses and protects nucleic acid; (vii) nucleic acid can be either DNA or RNA but not both, (viii) nucleic acid can be double-stranded DNA, single stranded DNA, single-stranded RNA or double-stranded RNA; (ix) Molecules on virus surface import high specificity for attachment to host cell; (x) multiply by assembly-line method; do not divide. The cycle includes attachment of viruses to host cell, penetration by genetic materials, production of virus components by the cell, assembly of new viruses and release from host cell; (xi) lack enzymes for most metabolic processes; (xii) lack machinery for synthesizing proteins; (xiii) obligate intracellular parasites of bacteria, protozoa, fungi, algae, plants and animals, (xiv) resistant to antibiotics.

Morphology

1. Viruses are of different shape and sizes. They may be spheroid, cuboid (adenoviruses), elongated (potato vi-

ruses), coiled (beet yellow), bullet-shaped (rabis virus), filamentous (bacteriophage M_{13}), pleomorphic (alfalfa mosaic) etc. Size vary from 20 to 400 nm in diameter, the largest being small pox viruses (400 nm).

2. A virion* consists of a nucleic acid core surrounded by a protein coat or **capsid.** The complete set of virion is known as **nucleocapsid.** The nucleocapsid may be naked or enveloped by a loose covering. The envelope is composed of a large number of subunits known as **Capsomeres.**

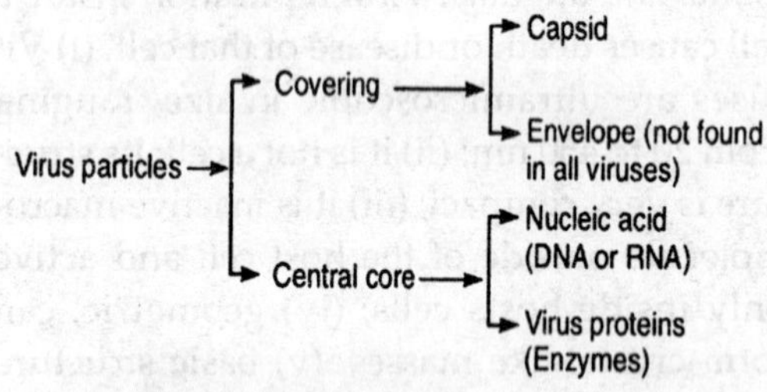

3. Chemically the envelope is made up of proteins and glycoproteins. Envelope is composed of both host and viral components. Protein is virus specific and carbohydrates and lipids are host specific.
4. The nucleic acids, either single or double-stranded DNA or RNA molecules, are in linear or circular form and have plus or minus polarity. The segmented nucleic acid is also found in wood-tumour virus, influenza virus and reovirus.
5. Proteins found in viruses may be grouped into four categories.
   i. **Envelope protein.** contain glycoprotein which is specified by both virus and host cell. Glycoprotein also differs from virus to virus. In influenza virus the main protein is the carbohydrate free protein which comprises of 50% of evelope protein and 35-40% of virion protein as a whole. Herpes virus, pox virus and leukoviruses do not contain protein in the envelope.
   ii. **Nucleocapsid protein.** The viral capsids are made up totally of identical subunits. The helical capsids contains single type of protein and icosahadral capsid contains several types of protein. For example TMV contains single type protein, adenovirus contain 14 types, T_4 bacteriophage contains 30 types of protein.
   iii. **Core protein.** Protein found in nucleic acid is known as core protein; The nucleoproteins of influenza virus and protein of V and VI of adenovirus are examples.
   iv. **Viral enzymes.** In animal virus, especially in the enveloped viruses and many virions specific enzymes have been characterised. RNAse and reverse transcription in retrovirus, protein kinase in herpes and adenoviruses, DNA dependant RNA polymerase in pox viruses are examples.
6. **Carbohydrates** specified by either host cell (e.g. arbovirus) or viral genome (e.g. vaccinia virus) is found in viral envelope. The carbohydrates are hexoses, hexosamines which are present in the form of glycoprotein and glycolipids. galactose, mannose, glucose, fucose, glucosamines, galactosamines are found in influenza virus.
7. There are certain projections on the envelope known as spikes which are arranged into distinct units. There are haemoglutinin spikes. The viral haemaglutinin (glycoprotein) has a special affinity for another but different glycoprotein located in the receptor areas in the red cell envelope.

* The complete assembly of the infections particles is known as virion.

☒ Types of Viruses

The morphological types of viruses observed through electron microscope and crystallography have been categorised into three groups.

1. **Helical viruses.** These are elongated, rod-shaped, rigid or flexible. Their capsid is a hollow cylinder with a helical structure. Capsid consists of monomer arranged helically in a rotational axis. The helical capsids may be naked or enveloped.

a. **Naked viruses.** These are rod-shaped measuring 280 × 150–180 nm. consists of a protein tube with a lumen of 20Å enclosing a single stranded helix of coiled RNA (Fig. 50.1 TMV).

i. Protein coat contains a number of identical subunit (monomers) which are arranged in a helical manner.

ii. A capsid consists of several capsomers each composed of a few monomers. Example-Tobacco mosaic virus (TMV).

b. **Enveloped viruses.** When helical viruses are enclosed within an envelope they are called enveloped helical viruses. Example influenza virus.

2. **Polyhedral (icosahedral) viruses.**

i. These are several animal, plant and bacterial viruses which have either naked enveloped icosahedral shape.

ii. Polyhedral structure has three possible symmetry, such as tetrahedral, octahedral and icosahedral. Icosahedral symmetry is the best one for packaging and bonding of subunits. An icosahedral is a regular polyhedron with 20 triagular faces and 12 corners. The capsomere of each face form an equatorial triangles and 12 intersepting points or corners (Fig. 50.2). The number of capsomers are 12 in poliovirus, 92 in reovirus, 162 in herpesvirus, 252 in adenovirus. Icosahedral virus are of two types.

a. **Naked icosahedral virus.** Poliovirus, adenovirus, bacteriophage etc. belong to this group. Adenovirus are large icosahedral containing dsDNA. The capsid has the ring-like capsomers each containing pentamers and hexamers.

b. **Enveloped icosahedral virus.** Herpes virus where the capsid is enclosed inside an envelope of 30 nm thickness, made up of a glycoprotein-lipid complex. The envelope consists of spikes on its surface.

3. **Complex viruses.** The viruses which have the unidentifiable capsids or have the capsids with additional structures are called complex viruses. For ex-

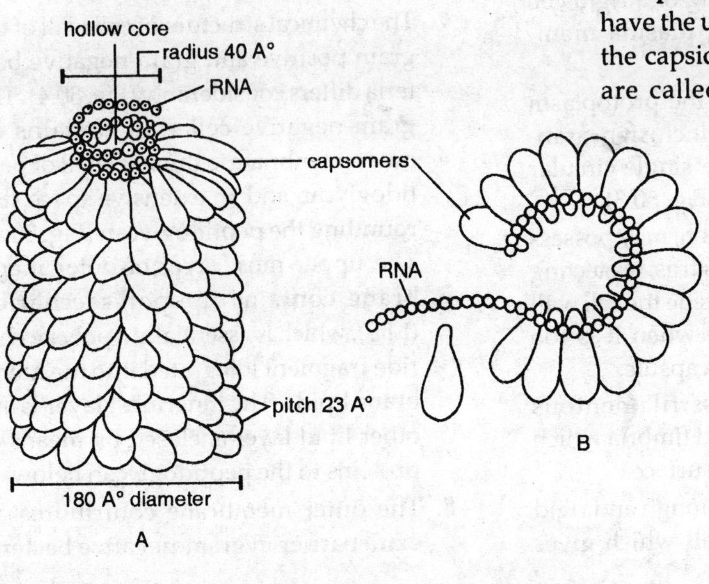

FIGURE 50.1
Structure of tobacco mosaic virus. A. helical model of TMV showing structure of RNA containing capsomers; B. arrangement or capsomers on RNA in one turn

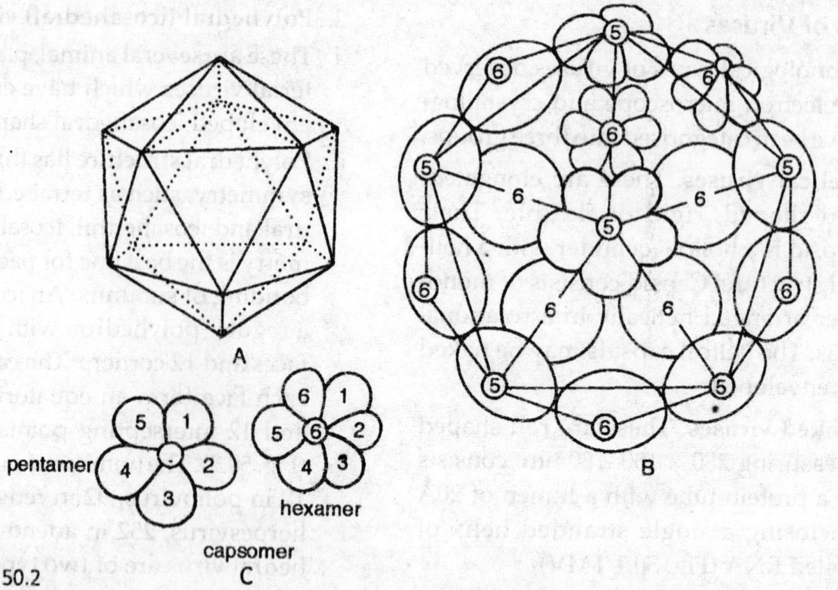

FIGURE 50.2

Icosahedral symmetry (A), architecture (B) and subunits of capsomers (C)

ample, vaccinia virus and T-even bacteriophages (infecting bacteria).

Morphology of Bacteria

1. Bacteria are prokaryotic micro-organisms, divide by binary fission and do not possess chlorophyll and true branching (except actinomycetes). Their size varies from 0.5 to 15 μm in length and 0.2 to 2 μm in breadth.
2. Bacteria is covered with outer layer or cell envelope, consisting of a rigid cell wall and underlying plasma membrane.
3. The cell wall encloses the protoplasm including ribosomes, inclusion granules, mesosomes and a single circular chromosome of DNA (Fig. 50.3).
4. Some bacteria, in addition, may possess additional structures such as, protecting gelatinous covering outside the cell wall known as capsule and when it is too thin is known as microcapsule.
5. Some bacteria possess filamentous appendages-flagella and fimbria which protrude from the cell surface.
6. Most bacteria possess a tough and rigid 10-20 nm thick cell wall which gives rigidity to bacteria in addition to giving protection against osmotic damage. It is elastic, porous, freely permeable to solute molecules of less than 10,000 molecular weight. It weight 20 to 25% of the dry weight of bacteria cell. Cell wall takes part in cell division by forming an ingrowth and mucopoly peptide component of cell wall is the terget site for antibiotics, lysozymes and bacteriophages.
7. The chemical structure of cell wall of the gram positive and gram-negative bacteria differs considerably (Fig. 50.4). The gram-negative cell wall contains an outer membrane, a thinner shell of peptidoglycan and an extensive space surrounding the peptidoglycan (Fig. 50.4). The upper most layer of outer membrane contains lipopolysaccharide (LPS), which is essentially a polysaccharide fragment integrated into the membrane lipids. The innermost layer is another lipid layer enclosed by means of proteins to the peptidoglycan below.
8. The outer membrane contributes an extra barrier in gram-negative bacteria

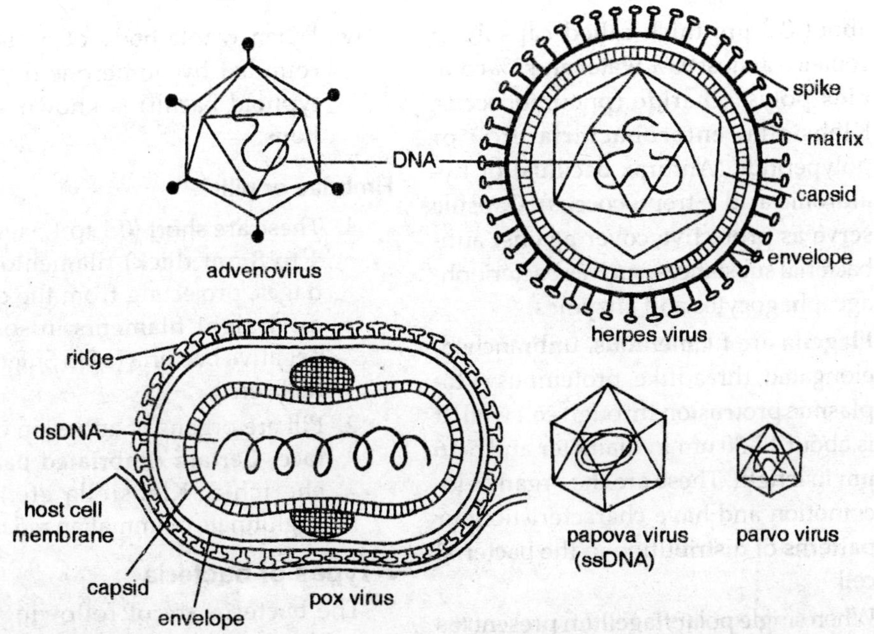

FIGURE 50.3

DNA containing animal viruses

that makes them more impervious to some antimicrobic chemicals, dyes and disinfectants.

9. The gram-negative lipopolysaccharides called endotoxins are a complicating factor causing fever and sock in meningitis and typhoid fever.
10. The gram-positive cell wall is a thick homogeneous sheath of peptidoglycan ranging from 20 to 80 nm in thickness. It contains tightly bound acidic polysaccharides including techoic acid and lipotechoic acid. Techoic acid is a polymer of ribitol or glycerol and phosphate embedded in the peptidoglycan sheath.

Plasma membrane. (i) It is a thin, 5-10 nm thick, elastic, semipermeable layer and lies beneath the cell wall separating it from the cell cytoplasm. (ii) It has a lipid bilayer structure and is the physical and metabolic barrier between the interior and exterior of the bacterial cell.

Cytoplasmic Components. (i) Cytoplasm is a colloidal, viscous watery solution containing chromatin body, ribosomes, mesosome and storage granules. The bacterial cells lack a nuclear membrane, instead the haploid double stranded DNA is concentrated in the cytoplasm as a nucleoid. (ii) If undergoes semiconservative replication by simple fission and maintains genetic characteristics. (iii) Some bacteria may contain circular, much smaller, extranuclear genetic material in the cytoplasm consisting of DNA, known as **plasmids**. (iv) Plasmids are not essential to bacterial growth and metabolism, but they often confer protective traits such as resisting drugs and producing toxins and enzymes.

External structures

1. Capsule, flagella and pili are found outside the cell envelope. Some bacteria secrete amorphous viscid polysaccharide materials on its surrounding and remains as a loose slime layer.
2. Surrounding the cell wall there is an outer thick jelly-like outer covering of

about 0.2 μm thick called capsule. It contains almost 98% water and 2% complex polysaccharide (pneumococcus, Klebsiella, enterobacteria etc.) or polypeptide (Anthrax bacillus) or hyaluronic acid (Streptococcus). Capsule serve as protective cover against antibacterial substances, such as bacteriophage, phagocytes and enzymes.

3. **Flagella** are filamentous, unbranched, elongated, threadlike, proteinous cytoplasmic protrusion through cell wall. It is about 12-30 nm in diameter and 5-16 μm in length. These are the organ of locomotion and have characteristic four patterns of distribution in the bacterial cell.

i. When single polar flagellum present (as in *Vibrio cholerae*) is called **monotrichous.**

ii. When both ends remain attached to flagellum (as in **Alkaligenes faecalis**) is called **amphitrichous.**

iii. When tufts of flagella are attached at one or both ends (as in spirilla) is called **lophotrichous.**

iv. When whole body of bacteria is surrounded by numerous flagella (as in typhoid bacilli) is known as **peritrichous.**

Fimbriae or pili

1. These are short (0.1 to 1.5 μm long and 4 to 8 nm thick) filamentous appendages, projecting from the cell surface as straight filaments in some Gram-negative bacteria (as in Shigella Salmonella etc.).
2. Pili are organs of adhesion on cell surface. Certain fimbriated bacteria (Escherichia, Klebsiella etc.) strongly agglutinate mammalian red blood cells.

• Types of bacteria

The bacteria are of following different types depending on the arrangement and shape of cells (Fig. 50.5).

1. **Coccus form**

a. **Diplococcus.** Cells divide in one plane and remain attached permanently in pairs.

b. **Streptococcus.** Cell divide in one plane and remain attached to form a linear chain of cells.

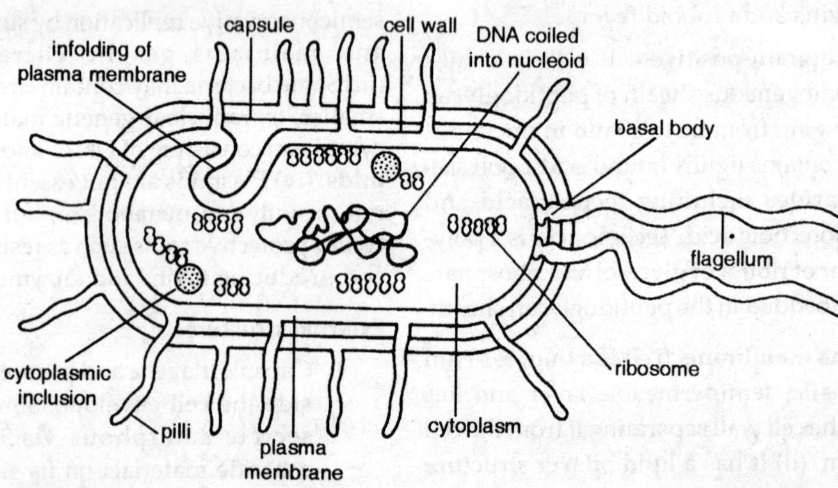

FIGURE 50.4

Section of a bacterial cell

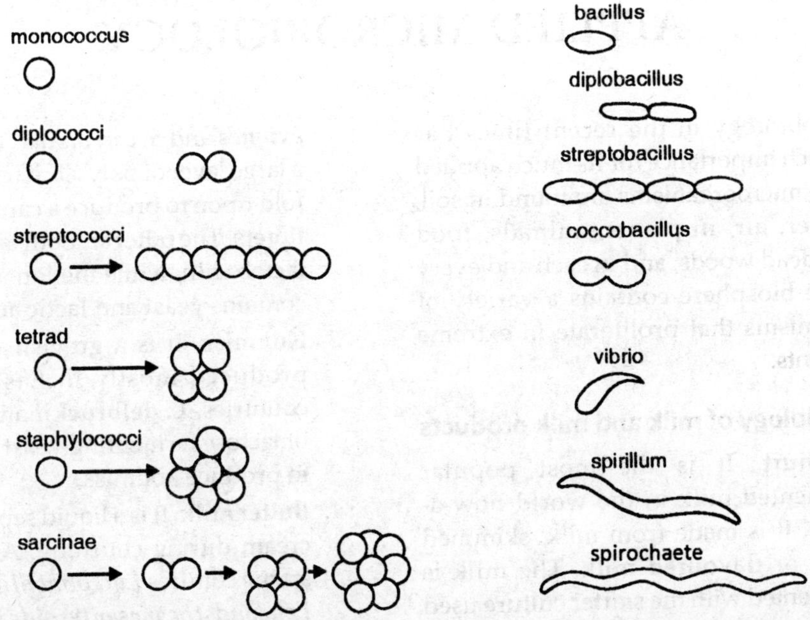

FIGURE 50.5

Diagram of cocci. bacilli and spiral bacteria

c. **Tetracoccus.** Cells divide in two planes, form groups of four cells.

d. **Staphyllococcus.** Cells divide in three planes in an irregular pattern producing bunches of cocci.

e. **Sarcinae.** cells divide in three planes in regular pattern producing bunches of cocci.

2. **Bacillus form**

a. **Monobacillus.** Single, elongated cell freely present in nature.

b. **Diplobacillus.** Cells after division remain adhered and appear paired form.

c. **Streptobacillus.** Cells after division remain attached as chain and appear like strand.

d. **Coccobacillus.** The oval cells in the chain looking like cocci.

3. **Forms of spirilli**

a. **Vibroid.** Bacterial cells having less than one complete twist form-vibroid shape (*Vibrio cholerae*).

b. **Helical form.** Cells that have more than one twist forming a distinct helical shape (e.g. spirillum, spirochaete).

APPLIED MICROBIOLOGY

The microbiology in the recent times has gained much importance for its much applied value. The microorganisms are found in soil, mud, water, air, in plants, animals, food products, dead woods, and in each and every space. The biosphere contains a variety of microorganisms that proliferate in extreme environments.

☒ Microbiology of milk and milk products

1. **Yoghurt.** It is the most popular fermented milk in the world now-a-days. It is made from milk, skimmed milk or flavoured milk. The milk is fermented with the starter culture used containing ***Lactobacillus delbruckii*** and ***Streptococcus thermophilus***. Within four hours the lactose is converted into lactic acid, pH decreases to 6.3 to 4.7.
   Flavour of acetaldehyde develop due to its fermentation at the rate 23 to 41 mg/kg. It is then cooled to 5 °C which helps in keeping up to 3 weeks.
2. **Kafir.** It is fermented milk produced by a mixed ***Lactobacillus kafir*** and alcoholic yeast like ***Saccharomyces exiques*** and ***S. cerevisiae***. The Kafirian, a large layer of polysaccharide materials fold upon to produce a cauliflower-like florets. The outer smooth layer contains lactobacilli, while the inner rough side contains yeast and lactic acid bacteria.
3. **Koumiss.** It is a greyish white drink produced mostly in East-European countries. **L. delbruckii** and a number of lactose fermenting yeast responsible to produce koumiss.
4. **Butter milk.** It is a liquid separated from cream during churning. A mixture of startic cultures ***Lactobacillus lactis*** and ***Leuconostoc mesenteroids*** is used. The final products contains 0.7 to 0.9% lactic acid.
5. **Butter.** The organism like ***Leuconostoc citrovorum, Streptococcus cremoris, S. lactis*** produce a small amount of acetoin which is spontaneously oxidised to diacetyl. The butter is produced by the action of enzymes that attack the citrates in the milk to produce diacetyl.
6. **Cheese.** There are about 2000 varieties of cheese which are of three categories—soft, hard and semihard. The soft

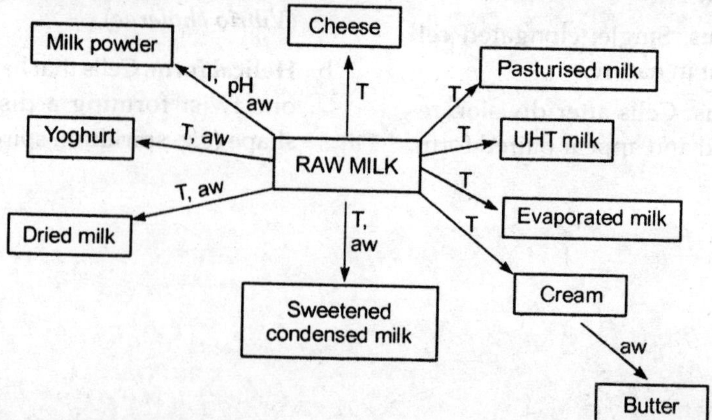

FIGURE 50.6

Milk and milk product; T-elevated temperature, pH-reduced pH, aw-reduced water pressure

cheeses are ripend by enzymes from yeast and other fungi that grow on the surface. Hard cheese is ripened by lactic acid bacteria which grow throughout the cheese, die, autolyze and release hydrolytic enzymes. The semisoft cheese are ripended by proteolytic and lipolytic organisms which soften the curd and give it flavour. Quality of cheese depends on the microorganisms used.

⊠ Spoilage of Poultry, Fish and Seafood

Meat of chicken, goose, duck, quail etc include in poultry products. Meats are usually contaminated from the skin, lining of body cavity during washing, plucking and evisceration. Enterobactor, Escherichia, Bacillus, Flavobacterium, Micrococcus, Pseudomonas, Staphylococcus, Corynebacterium and Salmonella are usually prevalent in poultry products. These bacteria are the main causative agents of spoilage of poultry products, meats and fishes. Use of heat (canning), freezing (frozen storage), irradiation, drying and curing by salt helps in preservation.

⊠ Microbiology of Endemic Diseases

The diseases which are caused by viruses, bacteria, fungi, protozoans and helminths after being transmitted from one host or reservoir to other hosts is known as infections diseases. A disease maintaining a steady, low level frequency at a regular interval, like common cold, is known as endemic disease. When a disease, like typhoid, occurs occasionally and at irregular intervals in a human population is known as sporadic disease. A sudden increase in occurrence of a disease beyond limit is known as **epidemic**. The infectious organisms perpetuates in hosts and are transmitted through vectors and is known as **disease cycle**. The cycle of disease includes sources of disease, reservoirs and vectors. Source is the site or location from where the pathogens spreads to a new host either through environmental factors (e.g. soil, water, air, food) or indirectly through animals or humans. Reservoir are the natural environment (soil, water or air) or susceptible animal hosts where pathogens survive. Carriers are the individuals already infected with pathogens.

Perpetuation of disease and survival of pathogen transmission from one host to other occurs through air contract, vehicle and vector. A living organisms that transmit a pathogen by biting or injecting one host to other is the vector. Examples dog, cat, bat, fleas, flies, ticks, mites, insects etc.

⊠ Pathogen and Disease of Man

A. **Viral diseases**

a. **Air-borne diseases**

1. Influenza (flue)—by orthomyxo virus (A,B,C variants)
2. Measles (Rubeola)—by Morbillivirus
3. Mumps—by Paramyxovirus
4. Rubella (German measles)—Rubella virus.
5. Small pox—by variola

b. **Insect-borne diseases**

6. Yellow fever—by Flavivirus transmitted by Aedes mosquitoes.
7. Dengue—by RNA virus, transmitted by Aedes mosquito.
8. Encephalitis, like california Encephalitis, Colorado tick fever, Eastern Equine Encephalities (EEE), St. Louis encephalitis, (SLE), venezuelan encephalitis (VEE), western equine encephalitis (WEE)—caused by arboviruses are transmitted by Aedes mosquitoes.

c. **Food and water borne diseases**

9. Polio (Polio myelitis)—by Polio virus.

d. **Direct contact**

10. Viral hepatitis—nine viruses are known to cause hepatitis.

11. Rabis—by different strains of Lysa virus.
12. Cold sores (fever blisters)—by herpes virus.
13. AIDS (Acquired immune deficiency syndrome)—by Human immunodeficiency virus (HIV).

B. **Bacterial diseases**

a. **Air-borne disease**

1. Tuberculosis—by *Mycobacterian tuberculosis* (gram+)
2. Diphtheriae—by *Corynebacterium diphtheriae* (gram+)
3. Maningitis—by *Streptococcus pneumoniae,* Mycobacterium and many fungi, amoeba etc.
4. Pertusis (Whooping cough)—*Bordetella pertussis* (gram–)
5. Pneumonia—by *Streptococcus pneumoniae.*

b. **Food-borne disease**

6. Cholera—by *Vibrio cholerae* (Gram–)
7. Botulism—by *Clostridium botulinum* (Gram–)
8. Shigellosis—by *shigella* sp (Gram–)
9. Typhoid fever—by *Salmonella typhi* (Gram–).

c. **Soil-borne diseases**

10. Tetanus—by *Clostridium tetani* (Gram+)
11. Anthrax—*Bacillus anthracis* (Gram+)

d. **Sexually transmitted diseases**

12. Gonorrhea—by *Neisseria gonorrhae* (Gram–)
13. Syphilis (Lyme disease)—by *Treponema pallidum, Sporellia* spp. (Spirochete).
14. Leprosy—by *Mycobacterium leprae*

C. **Fungal diseases**

1. Mycotoxicoses—by ingestion of fungal toxins (mushroom poisoning etc.)
2. Phycomycoses—by Absidia, Mucor, Rhizopas.
3. Candidiasis—by *Candida albicans* and about 200 other species.
4. Actinomycosis, Dermatophytosis, Aspergillosis, Otomycosis, Penicilinoses etc. are other fungal diseases.

51 BIOTECHNOLOGY

The technique by which a new individual of a species—animal or plant can be created by a process other than natural, and also alteration of the structure of a chromosome of living organisms leading to a genetically modified form to benefit man.

GENETIC ENGINEERING

It involves using recombinant DNA technique to transfer genes from one organism to another. The term recombinant DNA means the creation of a new combination of DNA molecules that do not occur together naturally. It involves DNA molecules produced by joining molecules from different biological sources. Unlimited quantity of genes can be isolated by using recombinant DNA technology. Large amount of biologically useful proteins can be produced by gene transfer between two organisms. The procedure is carried in several steps.

a. Using **restriction enzymes** that recognise and cut DNA molecules, DNA fragments with specific nucleotide sequences are produced.

b. The fragments are joined to other DNA molecules that serve as **vectors** or carrier molecules. A **recombinant DNA molecule** is a vector with an inserted DNA.

c. The recombinant DNA molecule is transferred to a host cell. The recombinant DNA molecule produces many identical molecules by replication within the host cells. The identical molecules are called **clones**.

d. With the replication of the host cells, the recombinant DNA molecules pass to the daughter cells and a population of host cells is established. Each daughter cell (host cell) contains copies of cloned DNA sequence.

e. The cloned DNA can be recovered from host cells, purified and analysed.

f. The cloned DNA can be transcribed, its mRNA translated and gene products isolated.

Variant forms of natural proteins can be produced by using recombinant DNA technology. This is known as **protein engineering**. By *in vitro* **mutagenesis** cloned gene sequences can be altered and protein with modified amino acid sequences having altered characteristics and properties can be produced. Protein engineering can produce variants of normal proteins. This rapidly emerging field is call **proteomics**.

The important application of genetic engineering are:

To produce proteins used as drugs to treat diseases; proteins used in industrial processes and also to produce plants resistant to draught, diseases, pests and high-yielding crops (transgenectic crop plants).

Insulin is a human gene product using recombinant DNA. Insulin is a protein hormone that regulates sugar metabolism.

As for transgenetic crops—controversies are around the consumption of vegetables and fruits of plants grown from seeds of genetically altered forms. In general, human society is, however, reluctant to accept genetically altered disease and pest resistant crops. On the other hand people include genetically altered high-yielding, draught resistant, and higher nutritional value crops, viz., Soybean, corn and other in their food list.

Recombinant human protein as drugs

Protein	Use
α-1 antitrypsin	Emphysema
Calcitronin	Rickets
Chorionic gonadatropin	Infertility
Erythropoietin	Anaemia
Factor VIII	Haemophillia
Growth hormone	Growth retardation
Interleukins	Cancer
Insulin	Diabetes
Interferons (λ, β, γ)	Cancer, viral infection
Tissue plasminogen activator	Blood clot

Recombinant enzymes used in industry

Protein	Use
α-amylase	Beer manufacture
Bromelain	Meat softening, juice clarification
Catalase	Antioxidant in food
Cellulase	Alcohol and glucose manufacture
Lipase	Cheese making
Protease	Detergents
Renin	Cheese making

52 ECONOMIC ZOOLOGY

The discipline of zoology dealing with animals directly or indirectly related to human economy is termed '**economic zoology**'. This includes a vast number of animals embracing those destroying the products of our labour or natural resources (harmful) and also those providing us with materials of commercial value (beneficial). They are also called **enemies** and **friends**, respectively.

PESTS

The organisms damaging plants and animals and their products of economic value are pests*. The damage caused by the pests are neither occasional nor accidental, but of regular occurrence, since their very survival depends on it. Those damaging standing crops, cereals, fruits and other plant products of commercial importance are designated as **crop pests**; those destroying stored grains are **stored grain pests**; those damaging household articles are **household pests**.

The moth *Tryporiza* is a crop pest; the rodent *Bandicota* is a stored grain pest; the bookworm, wood borer are household pests.

Thousands of economically destractive species of plants and animals classified as bacteria, viruses, fungi, weeds, some vertebrates and invertebrates including insects are pests.

PLANT PARASITIC NEMATODE PEST

1. ROOT-KNOT NEMATODE, *MELOIDOGYNE INCOGNITA*

Behaviour, ecology and control

1. Meloidogyne incognita, a root-knot nematode, is widely distributed and attack a wide range of mostly dicot and some monocot crops and largely associated with jute and vegetables.
2. The juveniles feed on the underground plant parts and penetrates the newly formed tissue. The hydrolytic enzymes secreted by the nematodes induce excessive conversion of tryptophan into indole acetic acid and certain other changes and form giant cells or **syncytia**.
3. This results in cortical hypertrophy and extensive pericycle hyperplasia (excessive multiplication) giving rise to tiny swellings on the roots or primary

* If the yield loss is above 15 per cent of the total production the causative organism is categorised as major pest and if loss is below 10 per cent of total production is minor pest.

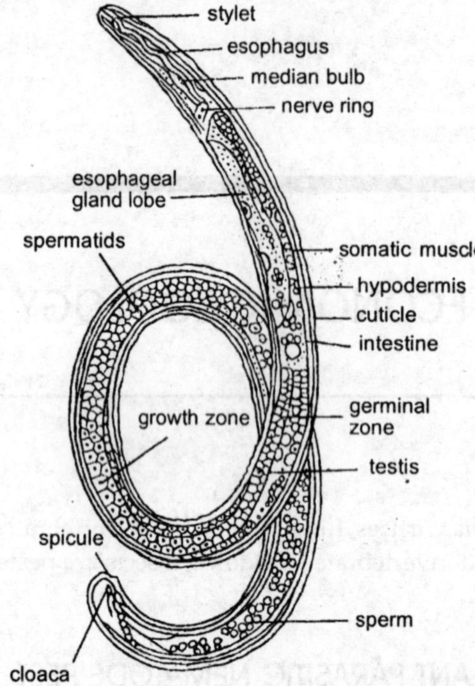

FIGURE 52.1

Meloidogyne sp., male, Root knot nematode.

galls (knot), several of which may merge into big multiple gall. Formation of gall cause heavy loss of food, fibre and timber plants.

External features

1. Meloidogyne is an elongated, cylindrical, unsegmented, bilaterally symmetrical, eel-like (or saccular in female), nematode.
2. Body covered with tough transparent cuticle.
3. A tail or post-anal elongation of the body is present in all the stages.
4. Distinct dimorphism present. Male (Fig. 52.1) is slender, long, vermiform with a very short, rounded tail. The average diameter and length of the male is 30 microns and 1400 microns respectively.
5. Female is pyriform and its average diameter is 400 microns and length 700 microns (Fig. 52.2).
6. Mouth is surrounded by six lips, a buccal stylet, a cuticular spear of less than 0.5 microns diameter.

Life Cycle (Fig. 52.3)

1. Female lays about 400-500 eggs in a gelatinous matrix secreted by the rectal glands within the host's tissue (gall). The egg laying is completed in a period of one week. The females degenerate after producing eggs.
2. The eggs are oval, elongated, 40 microns in thickness and 90 microns in length.
3. Eggs generally protrude out of the host tissue, lent some may be embedded in it.
4. The embryo and first stage juveniles are highly resistant to moisture stress. The first moult occurs within the egg (J_1). The second stage juvenile is ready to hatch out. The gelatinous egg matrix assists in the survival of the second stage juvenile (J_2).
5. The 2nd stage juvenile is vermiform with an elongated conoid tail. The overage diameter at this stage is 15 micron, and length 400 microns.

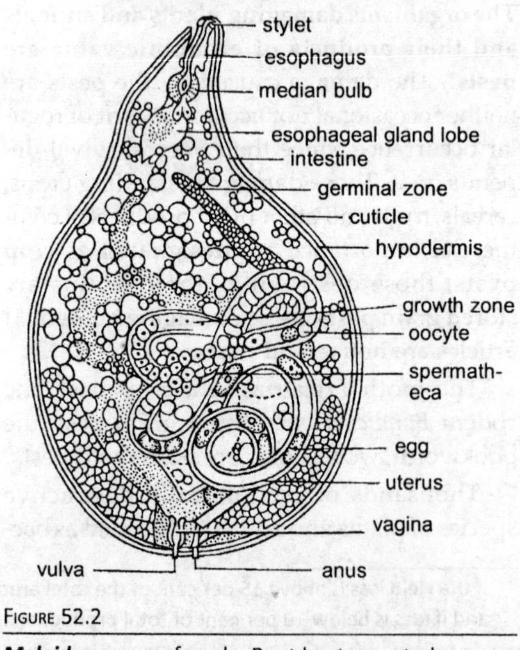

FIGURE 52.2

Meloidogyne sp., female, Root knot nematode

6. The juveniles escape into the soil and attack fresh plants or reinfect the uninfected parts of the same plant, feed on the underground parts and penetrate the newly formed tissue where hypertrophy and hyperplasia of cells develop galls; the nematodes develop to sexual maturity within the gall. They may live for months without feeding on and attacking fresh plants.
7. The filiform juveniles change in shape. The females become more plump, incapable of movement within the roots, assume a pyriform or saccular shape (Fig. 52.3), remain confined within the galls, the outer tissue of which disintegrate at the top through which they often protrude.
8. The males, after three moults, remain vermiform, leave parent gall, wander through the soil to find and fertilize females on other plants or in the same plant. The males die after fertilizing the females.
9. The total duration of life cycle under optimum conditions is 3-4 weeks.

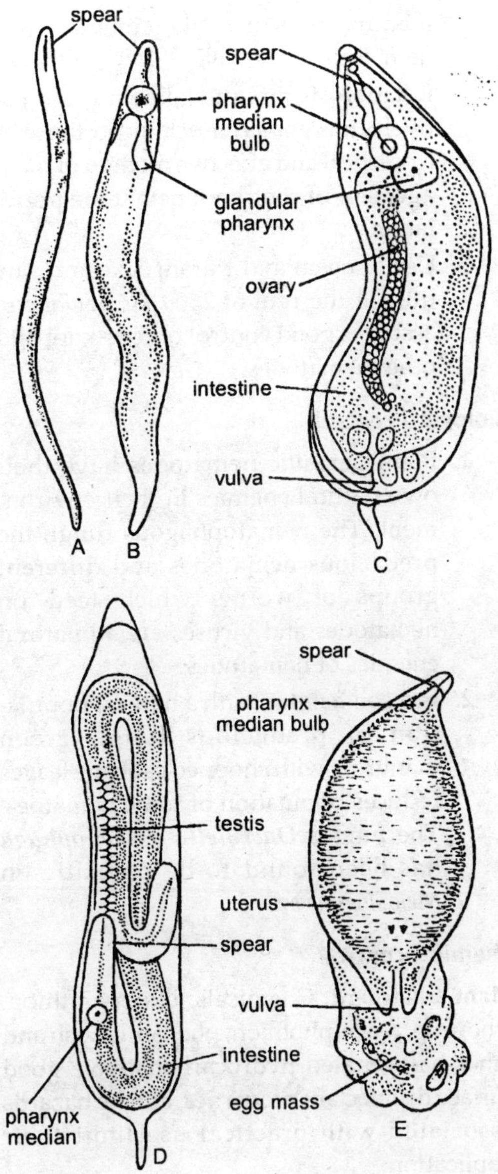

FIGURE 52.3

Meloidogyne sp. Root knot nematode Development stages. A-C. Second, third and fourth stage females; D. Fourth stage male inside third cuticle; E. Mature female

Management

The management depends on regulating the density of the nematode in time to minimise damage and crops can be raised profitably.

Physical

Heat treatment

1. Treatment of soil with heat for a short period kills most nematodes. Heat treatment of plant materials, like bulbs infected with endoparasitic nematode, are suitable method in a limited area.
2. Burning of trash materials or dry litter on the soil causes limited control of certain nematodes, including the root-knot nematodes up to 10 to 20 cm depth of soil.

Flooding. Flooding is a very effective method for nematode control, as nematodes can not withstand submergence of soil. But this may affect adversely by destroying beneficial aerobic microorganisms.

Cultural

1. Ploughing during hot summer months, planting in right time, mulching, manuring, and irrigation and use of green

manure help in management of this nematode in the field.

2. Rotation of crop, and fallowing alternatively, has gained much importance as a practical and effective method of management of root-knot nematode population.
3. Use of neem and Karanj cake and saw dust at the rate of 2500 kg per hactre assumes good control of root-knot and other nematodes.

Biological control

1. Plant parasitic nematodes have their own natural enemies in their environment. The nematophagous fungi, the predacious nematodes and different groups of worms which feed on nematodes and viruses are all natural enemies of nematodes.
2. In a soil habitat, with a natural population of predacious fungi, green manuring with chopped cabbage leaves reduces population of some nematoes. The fungus ***Dactylella asthenophaga*** has been found to be parasitic on *Meloidogyne* sp.

Chemical control

Plant treatment. Chemicals, like, parathion, ethoprop, fensulphothion, phenamiphosh and other halogenated hydrocarbons have good surface disinfectant properties, but the hazards associated with practical use limit their application.

Soil treatment

1. Complete or partial destruction of nematodes in soil before cropping is caused by the fumigation with methyl bromide, chloropicrin, etc.
2. Systemic application of soil nematicide chemicals, like, tetrahydrodizenthion (Dazomat), aldicarb and phorate, phenamiphos, etc. are effective.
3. Application of foliar spray with plant growth retardants, like, maleic hydrazide reduce nematode infection on host plants.

PADDY PEST

2. BROWN PLANT HOPPER: *NILAPARVATA LUGENS*

Behaviour, ecology and control

1. The brown plant hopper, once a minor pest in traditional local varieties of paddy, has now turned to a serious pest, causing epidemic in high yielding varieties in rice growing areas.
2. The reasons for increased activity of the pest:
   - i. high tillering dwarf plants in close spacing,
   - ii. application of high doses of nitrogen fertilizers,
   - iii. good water management, and
   - iv. double and continuous cropping.
3. Adult and nymphs exhibit negative phototaxis and prefer high humidity.
4. They confine their activity in the basal region under protected conditions provided by close luxuriant plant growth.
5. They damage the plants by sucking plant sap and plugging vascular bundles with their feeding sheaths.
6. The infested plants start drying up and a condition, commonly known as "**hopper burn**" appears in patches. With increase in hopper population the number of patches increase and the crop of the whole field is destroyed.
7. In addition to damage by sucking, the pest has also found to be responsible for spreading viral disease called "**grassy stunt**" or "**ragged stunt**".
8. They are distributed in India, Sri Lanka, china, Fiji, Indonesia, Malaysia, Philippines, Taiwan, Thailand, Korea, and New Ginea.

External features

1. Adults are uniformly light or dark brown and about 3 mm long. There are two distinct winged forms (Fig. 52.4).
   i. The micropterous with normal wings well adapted for migration.
   ii. The brachypterous with very much reduced wings, the hind wings being rudimentary. The brachypterous are relatively larger in size, has longer legs and ovipositor.

Life Cycle

1. Flight dispersal from one area to another generally takes place before the egg deposition stage in the evening of hot humid days.
2. The females lay eggs by lacerating the leaf mid ribs with their sharp ovipositors. Each female may lay 300 to 350 eggs.
3. Eggs are cylindrical and transparent, laid in clusters of 4 to 10 with anterior ends attached to one another.
4. Eggs hatch in 8 to 9 days.
5. Young nymphs are dark brown in colour, 0.5-0.6 mm long when freshly hatched and growing up to 2 mm by fifth and final moult.
6. Nymphs, take 12 to 14 days to reach the adult stage.

Management

1. The spraying of systemic insecticide, carbofuran applied as granules gives satisfactory management of the pest.
2. A number of resistant varietis of paddy have been identified and developed in which incidence of damage is very low.

3. TRYPORIZA (*SCHOENOBIUS*) *INCERTULUS*

Behaviour, ecology and control

1. The lepidopteran insect (moth) *Tryporiza*, locally named '**Mazra poka**' is a common paddy pest. Its larvae bore into the stem tissue, hence it is called **stem borer** (Fig. 52.5).
2. It has a wide range of distribution and is found in all the paddy-producing states of India, viz., West Bengal, Bihar, Orissa, Uttar Pradesh, Madhya Pradesh, Karnataka, Andhra Pradesh, Kerala, Assam and others.
3. The larva (**caterpillar**) bores into the stem at a lower point of the plant.
4. It cuts the central shoot, which turns yellow and dies, and no grains are produced.
5. Infestation occurs throughout the year, the peak is in September-October period.

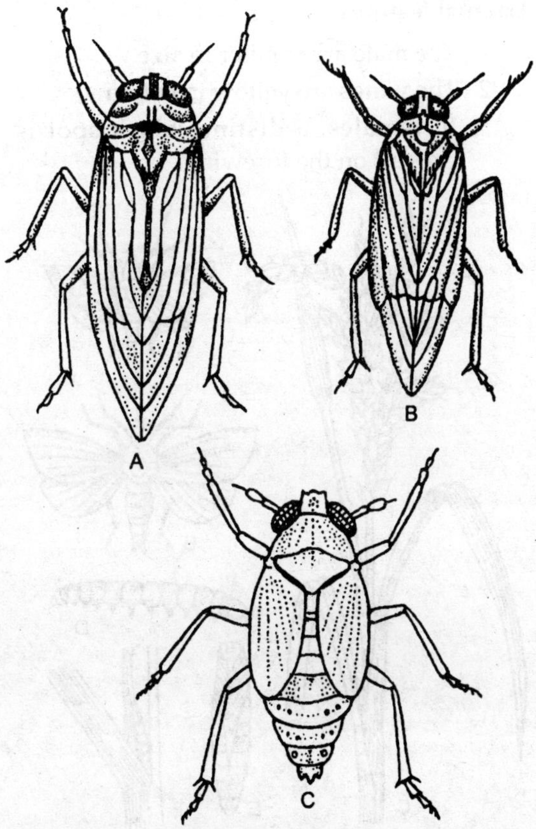

FIGURE 52.4

***Nilaparvata* sp.** Brown plant hopper. A. Female adult, B. Male adult, C. Nymph

External features

1. The male are smaller in size.
2. The wings are yellow in colour.
3. In females, a distinct black spot is present on the forewing.

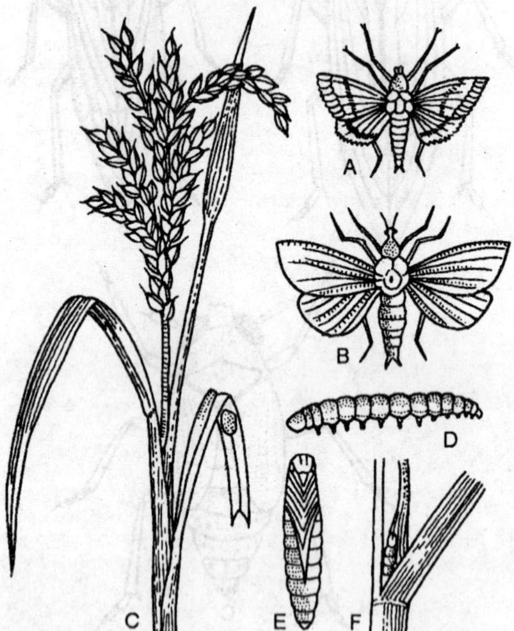

FIGURE 52.5

***Tryporiza* sp.** Life cycle. A. Male. B. Female. C. Egg mass. D. Larva. E. Pupa. F. Pupa inside paddy plant

Life cycle of Tryporiza

1. The female moth lays about 150-300 eggs in phases on the lower surface of paddy leaves.
2. Usually the egg mass is covered by buff colour hairs.
3. The eggs are greyish-white whith sculptured surfaces.
4. The larva hatches out within 6-8 days depending on temperature, moisture etc.
5. The caterpillar larva makes a small bore in the stem devouring the tissue.
6. In 4-5 weeks within the stem it is about 2 cm long.
7. A full-grown larva is faint yellow, with a yellowish-orange chitinoid head, three thoracic and twelve abdominal segments. The larva may leave the original plant and attack a few more plants before pupation.
8. The larva pupates within the stem. Before pupation it makes another hole on the plant for escape after emergence from pupa.

Management

1. **Physical**

(a) Light trap with persistent insecticides, (b) Rotation of crops, (c) Hand-picking and destruction of infected plants are somewhat effective.

2. **Chemical**

Laying of eggs on the undersurface of the leaves keep the eggs out of sight. Immediately after hatching, the larvae enter paddy plants and remain protected there. But the larva often leaves the original plant and attacks a number of plants one after another.

Advantage of this behaviour is taken for chemical control.

a. Spraying of water soluble BHC and other insecticides are of limited use since the chemicals usually do not reach the eggs on the undersurface of leaves.
b. Poisoning of water in the field with systemic insecticide is quite effective as the larvae come in contact with water in their way from plant to plant.

4. BRINJAL SHOOT AND FRUIT BORER: *LEUCINODES ORBONALIS*

Behaviour, ecology and control

1. *Leucinodes* is a pest of brinjal shoots and fruits. Its caterpiller causes damage by boring into the petiole and midribs of large leaves, young shoots, fruit-buds and fruits.
2. Besides brinjal it feeds on many other solanaceous plants.
3. It is a widely distributed pest in India, Sri Lanka, Pakistan, Malaysia, Myanmar, East Africa.

4. The caterpillars hibernate in winter and pupate early in spring. Moths appear in March-April.

External features

1. The moth is white, with pale brown or black spots on the dorsum of the thorax and abdomen. The head and thorax blackish-brown (Fig. 52.6A).
2. The wings are white with a pinkish or bluish tinge and are ringed with small hairs along the apical and anal margins.
3. The forewings are ornamented with a number of pale black and light brown spots.
4. The moth measures about 20-22 mm across spread wings.

Life Cycle (Fig. 52.6)

1. During the life span of 2-5 days, a female lays about 80 to 150 creamy white elongated eggs, singly or in batches of 2-4 on the underside of leaves, on green stem, flower-buds or the calyses of fruits.
2. The eggs hatch in 3 to 6 days at 27°C.
3. The young larvae immediately after hatching enter the tender plant tissue near the growing-points, into flower-buds, or into the fruits. When available, it bores in other parts and one caterpillar may destroy as many as 4 to 6 fruits. Even healthy looking fruits and thick stem may harbour a number of caterpillars (Fig. 52.6).
4. While boring into the leaf petioles or young shoot, the larvae eat the internal tissues. The infested portion wilts and droops.
5. The caterpillars are creamy white when young, but light pink in full-grown stage. A full-grown caterpillar measures about 18-23 mm in length.
6. The larvae grow through five stages and the larval period varies 9 to 28 days, being about 9 days at 27°C.

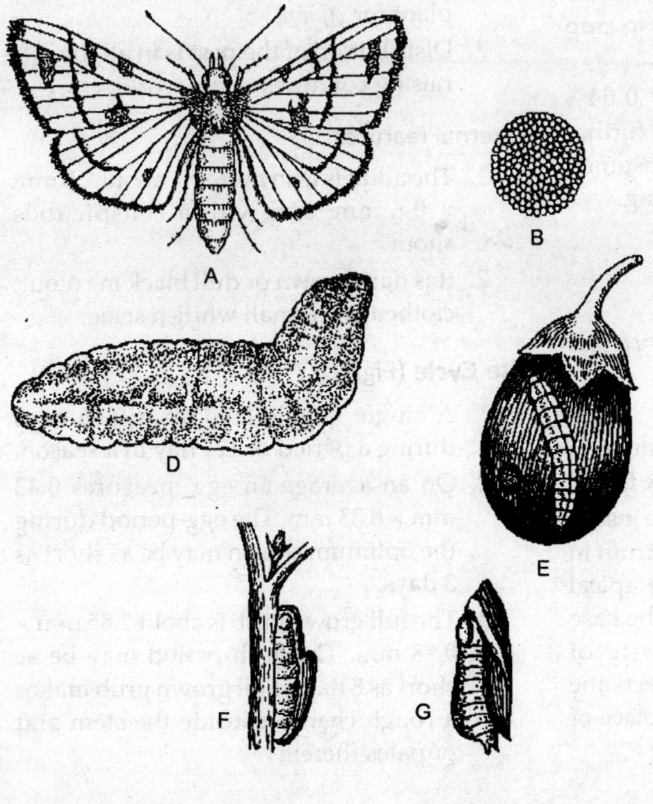

FIGURE 52.6

Leucinodes sp. Brinjal fruit-borer *Life cycle stages*. A. Adult moth, B. egg cluster, C. Young larva, D. Full grown larva, E. Larva emerging through the hole on the brinjal, crawl to make another bore, F. Larva pupating on the twig, G. Pupa inside Cocoon

7. The mature larvae come out of their feeding tunnels and pupate in tough silken cocoons among the fallen leaves. The pupal stage lasts 6-17 days and life cycle is completed in 20-43 days during active season.
8. There are five overlapping generations in a year.

Management

1. For proper management, close watch on the early stage of the crop is necessary, when the attack on the young shoot can easily be spotted. Eggs on the leaf, affected shoots, fruits along with caterpillar should be removed and destroyed. Thus rest of the plants can be saved.
2. Before plantation, special precaution is necessary not only to avoid ratooning of the brinjal crop but also to uproot and burn the plants well before planting new crop, because the old brinjal plants harbour the borers and carry the infestation from crop to crop and season to season.
3. Spraying 0.02% endrine, or 0.04% parathion helps in preventing further attack. Fruits should not be consumed before three weeks after spraying.

JUTE PEST

5. JUTE STEM-WEEVIL: *APION CORCHORI*

Behaviour and ecology

1. The weevil (snouted beetle), a jute stem borer, damages the quality of jute fibres.
2. The female weevil a very minute insect, bores a hole by means of its rostrum in the jute stem, generally in the apical region and lays an egg there. The base of petiole is the favoured site of oviposition. This adversely affects the development of fibre near the place of injury.
3. The larva on hatching, begins to feed on the surrounding tissues and gradually damages the fibre. Where such damage is excessive, the shoot above this point withers and a number of side branches develop. Such plants yield short fibres of poor quality.
4. Whole plant may succumb when attacked at the seedling stage. The quality of jute fibres deteriorates very much due to a number of chemical reaction changes brought about at the site of injury. A mucilagenous substance oozes out at the site of damage and it firmly cement the adjacent tissues with the insect excreta into a hard structure.
5. A corky tissue develops round the wound, which later in the season shifts from apical portion to the basal region, resulting in a hard compact mass with the skin sticking to the fibre.
6. There are a number of alternate host-plants but jute is the most preferred host plant for *Apion*.
7. Distribution of the pest is in all the jute raising countries of the world.

External features

1. The adult is a tiny weevil, about 1.8 mm × 0.8 mm, and with a conspicuous snout.
2. It is dark brown or dull black in colour, clothed with small whitish setae.

Life Cycle (Fig. 52.7)

1. A single female lays up to 675 eggs during a period of 124 day in a season.
2. On an average an egg measures 0.43 mm × 0.33 mm. The egg-period during the optimum season may be as short as 3 days.
3. The full grown grub is about 2.85 mm × 0.98 mm. The grub period may be as short as 8 days. Full grown grub makes a rough chamber inside the stem and pupates therein.

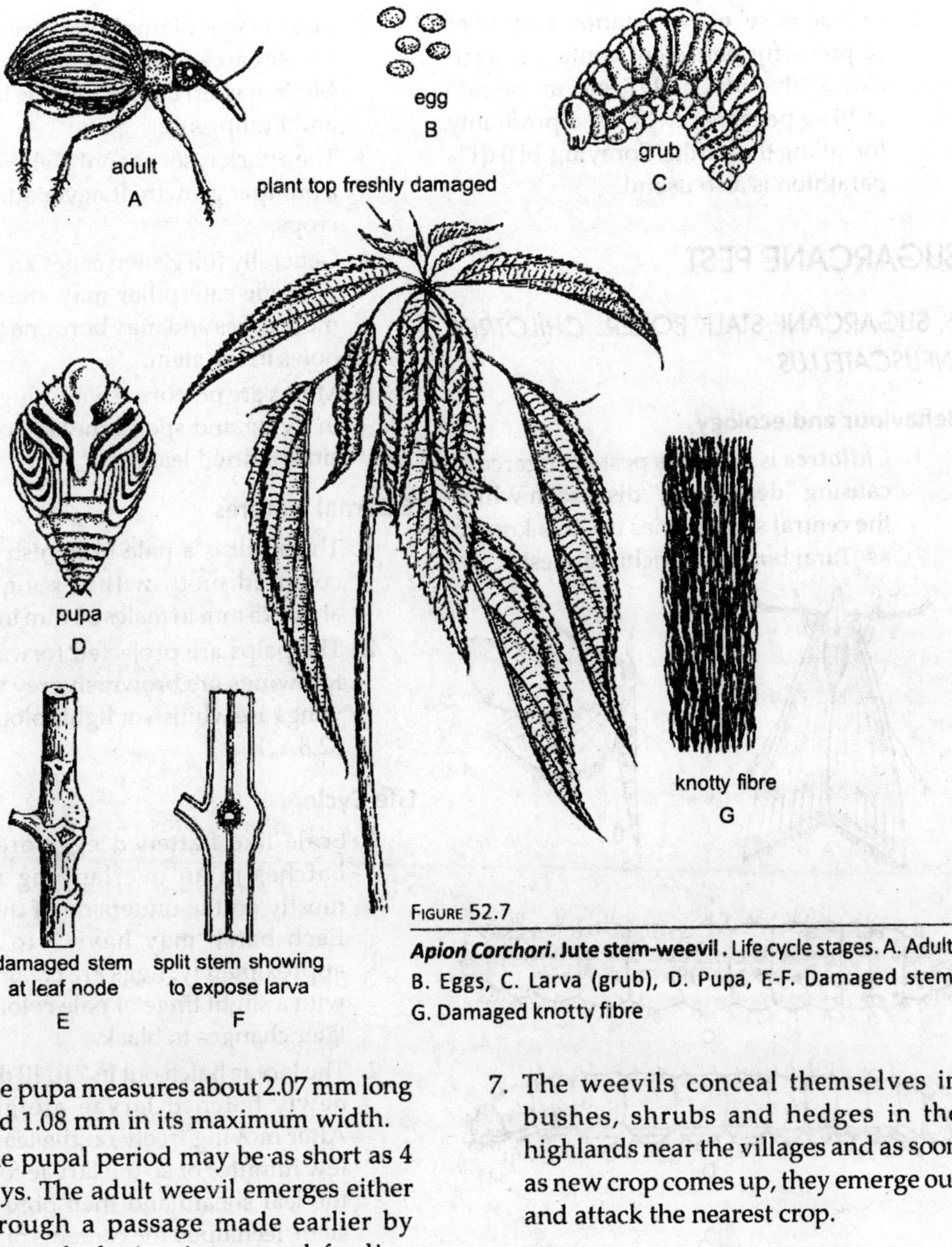

FIGURE 52.7

***Apion Corchori*. Jute stem-weevil**. Life cycle stages. A. Adult, B. Eggs, C. Larva (grub), D. Pupa, E-F. Damaged stem, G. Damaged knotty fibre

4. The pupa measures about 2.07 mm long and 1.08 mm in its maximum width.
5. The pupal period may be as short as 4 days. The adult weevil emerges either through a passage made earlier by the grub during its normal feeding activity or by making a fresh hole for its emergence.
6. The whole life-cycle is completed in about 15 days; the maximum longevity of adult has been reported to be as much as 209 days. As a result there are a large number of overlapping generations in a year. The adults are generally found in winter.
7. The weevils conceal themselves in bushes, shrubs and hedges in the highlands near the villages and as soon as new crop comes up, they emerge out and attack the nearest crop.

Management

1. Removal and destruction of infested plants in the first stage is very effective. Burning of stubbles after harvest, quick steeping of harvested jute are quite effective for management.
2. Use of nitrogenous manures increases infestation but potash and phosphate fertilizers decrease the same.

3. In case of severe infestation a number of powerful persistent contact insecticide 5-10%. BHC, DDT, etc. at the rate of 10 kg per acre can be used profitably for killing the adults. Spraying of 0.01% parathion is also useful.

SUGARCANE PEST

6. SUGARCANE STALK BORER: *CHILOTREA INFUSCATELLUS*

Behaviour and ecology

1. ***Chilotrea*** is a serious pest of sugarcane causing "**dead heart**" disease in which the central shoots dries up. It is known as 'Tarai borer' or "richman pest".

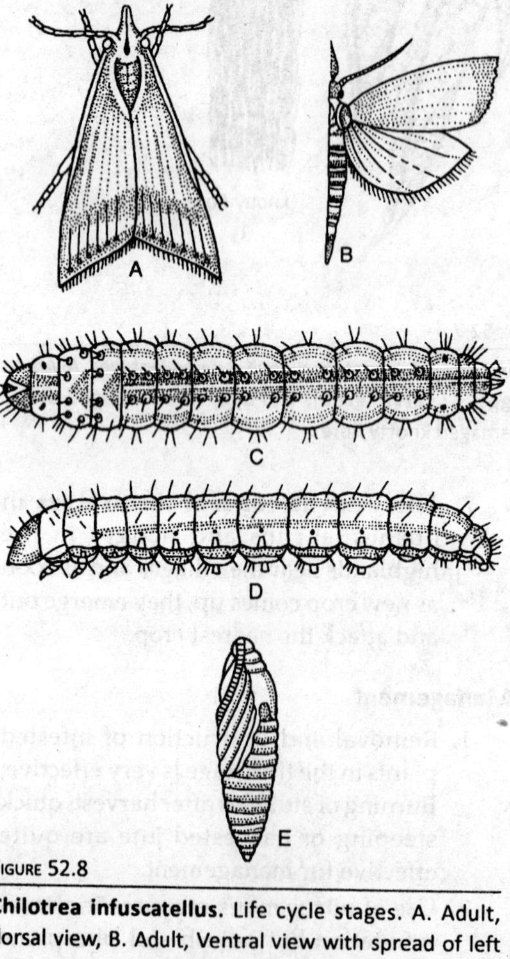

FIGURE 52.8

Chilotrea infuscatellus. Life cycle stages. A. Adult, dorsal view, B. Adult, Ventral view with spread of left wing, C-D. Lava, full grown, dorsal and side view, E. Pupa

2. The ravages of this borer is restricted to the sugarcane raising areas of India, Myanmar, Sri Lanka, Indonesia, Taiwan and Philippines.
3. The attack is severe only when crop has a bumper growth. It never attacks poor crops.
4. Generally full grown canes are attacked. A single caterpillar may attack one or more canes and may bore one or several holes in the stem.
5. Moths are nocturnal, very sluggish, shy in flying and spend the day on underside of dried leaves.

External features

1. The adult is a pale brownish or straw coloured moth with a wing span of about 25 mm in males 30 mm in females.
2. The palps are projected forwards. The forewings are brownish grey and hind wings are whitish or light coloured (Fig. 52.8A, B)

Life Cycle

1. Scale-like flattened eggs are laid in batches in an overlapping manner, mostly on the underside of the leaves. Each batch may have 7 to 70 eggs approximately. Eggs are creamy-white with a slight tinge of pale colour which later changes to black.
2. The larvae hatch out in 7 to 10 days. The newly hatched larvae are greenish. After moving freely on the leaves for a few minutes or so the larvae feed upon the leaf sheath and then bore into the stem, feed upon the contents of the stem and make a tunnel extending through several nodes and internodes of the cane.
3. The caterpillars become full grown in about 3 to 4 weeks and measures about 25 to 30 mm in length. There are five violet longitudinal stripes on the dorsal and lateral sides of a full grown larva. The head is reddish brown, the prothoracic tergite dirty brown.

Crockets are presents in complete circles on the legs.

4. It pupates in the tunnel made in the cane. Pupal period lasts for 6 to 9 days after which the moth emerges through the exit hole made by larva before pupation.
5. There are five generations in a year between spring and fall. The winter is passed in hibernating larval stage.

Management

1. The rejected older shoots, if left unharvasted, serve as source of infection to the newly planted cane. As a preventive measure the stubbles, uprooted plants and dead-heart bearing plants should be burnt.
2. Spraying 0.35% DDT or 0.03 to 0.1% endrine at 15 to 18 litres per acre when moths and egg masses are observed reduces the infection. Spray of 0.08% telodrin 6 times at fortnight intervals in the months of August to October is very effective.
3. Natural enemies of chilotreal in India are two species of insect hyper parasites attacking the egg-stage, about half a dozen attacking larval stage and one species attacking the pupal stage. But effective solution is yet to be known.

STORE-GRAIN PEST

7. RICE WEEVIL: *SITOPHILUS ORYZAE*

Behaviour and ecology

1. **Sitophilus oryzae** popularly known as rice weevil or black weevil, is one of the most destructive pest of stored grains, particularly rice (both husked and unhusked), maize, wheat, jowar, bajra, etc.
2. It frequently causes complete destruction of grains in granaries, leaving behind only the skin of the grain, which is unsuitable for seed or human consumption.
3. The adults and larvae are found working inside the grains. They bore the grains with the projected snout on the head and feed voraciously on the karnel.
4. The adults do not eat much. The females cause much damage by laying eggs in the grains. Most of the damage is done by grubs feeding on the karnel.
5. The insect is almost cosmopolitan in distribution. The damage of grains however, is very high in warm countries where it breeds continuously and rapidly destroy grains.

External features

1. The adult is brownish in colour, about 3.13 mm long with a cylindrical body and a curved snout (Fig. 52.9).
2. Females are comparatively bigger in size than males.
3. Adults are capable of flight. In hot periods they migrate from areas of higher temperature to that of lower temperature.

Life Cycle

1. A female can lay eggs up to 250, singly, in small holes made on the surface of the grain. The hole is plugged after oviposition with a starchy material. Generally it lays one egg in one grain, and only one egg in a day.
2. The eggs are translucent, oval, and measure on the average 0.7 mm long and 0.3 mm broad. The larva hatches out in about 4 days during summer but takes 6 to 7 days in winter.
3. The larva or grub is small, legless, white, fat bodied, with small yellowish brown head, lives within the grain feeding on its starchy contents (Fig. 52.9).
4. The grub is full grown in 3 to 4 weeks, and measures 3 mm long. Following total consumption of the karnel of the grain, it pupates within it after passing 1 or 2 days as a prepupa.

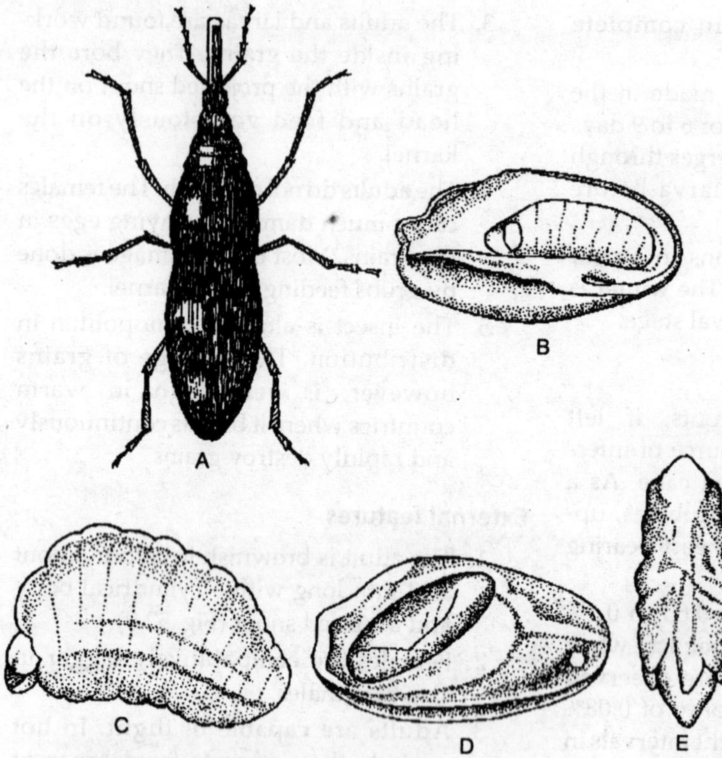

FIGURE 52.9

***Sitophilus oryzae*.** Rice weevil, A. Adult, B. Larva, early stage, C. Full grown larva, exposed; D. Pupa inside grain, E. Pupa, exposed, ventral side

5. A pupa is yellowish-white, measures about 45 mm long. The pupal stage lasts from 5 to 7 days depending on prevailing temperature.
6. The young weevil does not come out of the grain until complete chitinisation occurs. It can withstand starvation even for 2 to 3 weeks.
7. The adult weevil starts breeding as soon it emerges from the grain. It often lives from 7 to 8 months and may survive over 2 yrs in optimum conditions.
8. Commonly 5 to 6 broods are passed in a year in Indian climate.

Management

1. The grains should be thoroughly sun-dried before storing.
2. The godowns should be clean, damp-proof, insect proof and rat-proof, with no holes or cracks.
3. The godowns may be heated to 65°C for 12 hours. Spraying with 0.4% BHC, or DDT at the rate of 4.5 litre per 90 m^2 before the grains are stored.
4. The grains in new or sterilized gunnies are stacked, with a 60 cm gap between the stacks. The insecticide –5% BHC may be sprayed in the gap.
5. Fumigation with methyl bromide, at a dose rate of 450 ml per 90m^2 under atomospheric pressure with an exposure of about 24 hours is helpful. Great care should be taken in handling fumigants and application should be done under supervision of an expert.
6. Fumigation with phostoxin, at a dosage rate of 3 tablets per ton is very much effective.

MAMMALIAN GRAIN PEST

8. *BANDICOTA BENGALENSIS; RATTUS NORVEGICUS*

Behaviour and ecology

The major field and stored grain mammalian pests are *Bandicota bengalensis* (Fig. 52.10) and

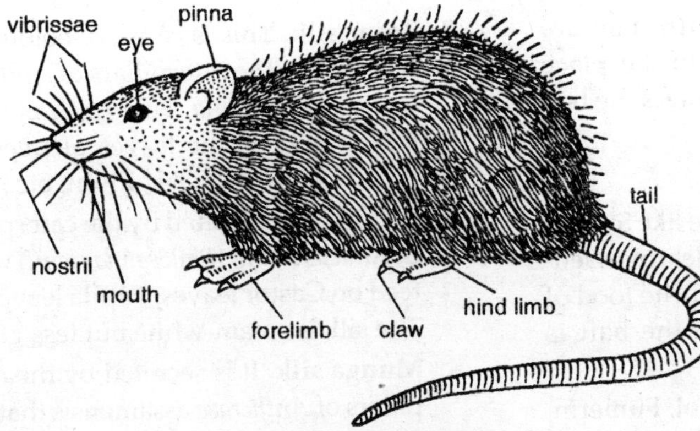

FIGURE 52.10

Mole Rat. (*Bandicota bengalensis*)

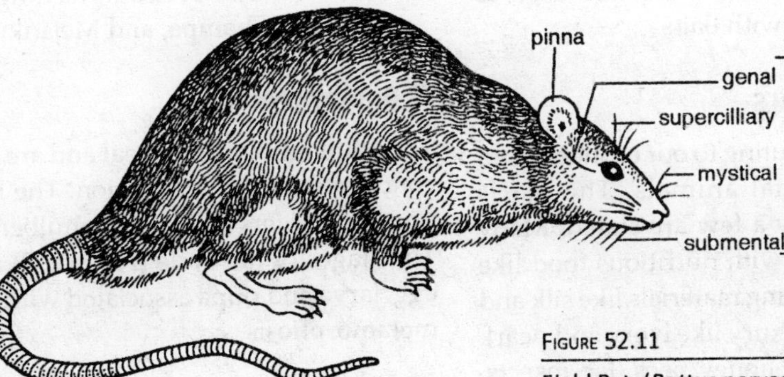

FIGURE 52.11

Field Rat. (*Rattus norvegicus*)

Rattus norvegicus (Fig. 52.11). The Bandicoot rat is commonly known as Indian mole rat and *R. norvegicus* as field rat and rural rat.

1. They are widely distributed in the Indian subcontinent, South-east Asia and other countries.
2. A mature Bandicoot weighs about 200 gms and lives for nearly 200 days.
3. They are good climbers and swimmers.
4. They inhabit fields, grasslands and houses. These fossorial and nocturnal animals are most active immediately after dusk.
5. They burrow in soil. The burrow extends to a depth of 1.0 to 1.5 m and ramifies laterally up to 12 m. From a central chamber, lateral canals extend to open outside by separate openings.
6. The rats store grain inside the burrow and a single burrow may have 5 to 6 kgs of grain.
7. Usually they prefer wheat and paddy. Each rat consumes 11 to 12 gms of grain daily and spoils at least 3 times of it with urine and faeces.
8. The males mature at about 58 days while females require 90 to 100 days.
9. They reproduce throughout the year and all the life.
10. The gestation period is 23 days. 5 to 14 offsprings are born in a litter.

Management

1. **Physical**

a. **Protectors.** The rats usually live outside near the house. At night they enter the store or dwelling house through doors, windows, holes and drains. Regular maintenance of sanitation, blocking of holes, closing doors, etc. may prevent their entrance.

b. **Traps.** Various traps are in use. The most effective is box traps of 30 × 17.5 ×

12.5 cm size. Baits with attractant are put in the traps and set in the place haunted by rats. The rat enters the box for food and is trapped.

2. **Chemical**

a. A single dose of rodenticide like Strychnine, Zinc phosphide, Calcium arsenate, etc. is mixed with favourite food of rat and left. One bite on the bait is enough to kill a rat.

b. Several applications of Pivol, Fumerin, Rozol etc. kill rats. These are used as fumigants or with baits.

BENEFICIAL ANIMALS

The animals contributing to our economy and health are beneficial animals. The list is exhaustive and only a few are stated. Many animals provide us with nutritious food like milk and meat; clothing materials like silk and wool and items of luxury like ivory and pearl. The silkworms, the honey bees, lac insects, fowls, fishes, prawns, bivalve molluscs belong to this category.

⊠ Sericulture

Sericulture is the scientific management of production and marketing of natural silk threads of quality. It involves (a) Selection and cultivation of food plants for silk moth larvae, (b) Maintenance, improvement and rearing of silk moth stock, (c) Protection against pests and diseases, (d) Extraction and finishing of silk threads and (e) Marketing.

Types of silk and kinds of silkworms

Chiefly based on the species of silkmoth and food, four major types of silk have been identified.

1. **Mulberry silk.** It is secreted by the caterpillars of *Bombyx mori* which feed on mulberry (*Morus indica*) leaves. The silk is highly valued for its smoothness, glaze and cream and cream-white colour.
2. **Tasar silk.** This is coppery colour and secreted by the caterpillars of *Antheraea mylitta*, *A. paphia*, *A. royali*, etc. The larvae feed on the leaves of trees like Arjun, Asan, Sal, Oak, Plum, etc.
3. **Eri silk.** It is secreted by the caterpillars of *Attacus ricini*, (*Philosamia ricini*) which feed on Castor leaves, Papila leaves etc. The silk is cream-white but less glossy.
4. **Munga silk.** It is secreted by the caterpillars of *Antheraea* assamensis that ever exist only in Assam, feeding on Som, Soalu, Champa, and Mejankuri leaves.

⊠ Bombyx Mori

The adult moths seldom eat and are primarily concerned with reproduction. The larvae are voracious eaters and feed on mulberry leaves. The stages of life cycle (Fig. 52.12) are adult, egg, larva and pupa associated with complete metamorphosis.

Life Cycle

Adult

The adult moth is whitish in colour, 25 mm long with 40 to 50 mm wing span. The female is larger than the male. The body has three divisions—head with a pair of eyes and a pair of antennae, thorax with three pairs of legs and two pairs of wings, and one large abdomen.

Egg

Immediately after mating the female starts laying eggs on mulberry leaves. The eggs are oval and covered by a hard, smooth chitinoid shell. At about 24°C a moth lays 300 to 500 eggs within 24 hours. In favourable temperature (18-24°C) the larvae hatch out within 10 to 11 days.

Larva

1. A newly hatched larva is about 3 mm long. The cylindrical body is covered with a chitinous skin and divided into 13 (or 14) segments, with a head at the anterior end and a caudal horn near the posterior end of the body.

2. It has three pairs of thoracic, four pairs of abdominal and a pair of caudal legs.
3. The larva goes to dormancy or **diapause** for four times. These are called 1st, 2nd, 3rd and 4th '**sleeps**'.
4. Each 'sleep' is followed by **ecdysis**. The part of life in between hatching and first ecdysis is 1st stage, and between the first and second ecdysis the 2nd stage and so on.
5. Towards the end of 5th stage, the larva is mature and starts spinning cocoon. The larval age is about 20 days.
6. The first **instar** (newly hatched larva) is a voracious eater, and is fed with finely cut, young mulberry leaves. Advance larvae are fed with matured and entire leaves.
7. The 1st instar eats for three days, stops eating and moults to 2nd instar.
8. The 2nd instar eats for $2\frac{1}{2}$ days and moults to 3rd instar.
9. After eating for three days the 3rd instar moults to 4th instar.
10. The 5th instar is formed after eating for 4 days.
11. The 5th instar eats for eight days, ceases to eat and starts spinning silk around it from outside to inside.
12. For a complete covering the larva rotates its head about 60,000 to 300,000 times and the silk thread is liberated at the rate of 15 cm per minute.
13. The protective covering is called cocoon which is formed by an unbroken silk thread 400 to 1,500 metres.
14. A caterpillar larva takes about 4 days to complete a cocoon and then turns to a completely immobile larva, the **pupa**.
15. The pupa is transformed into a full grown adult or **imago** after ten days.

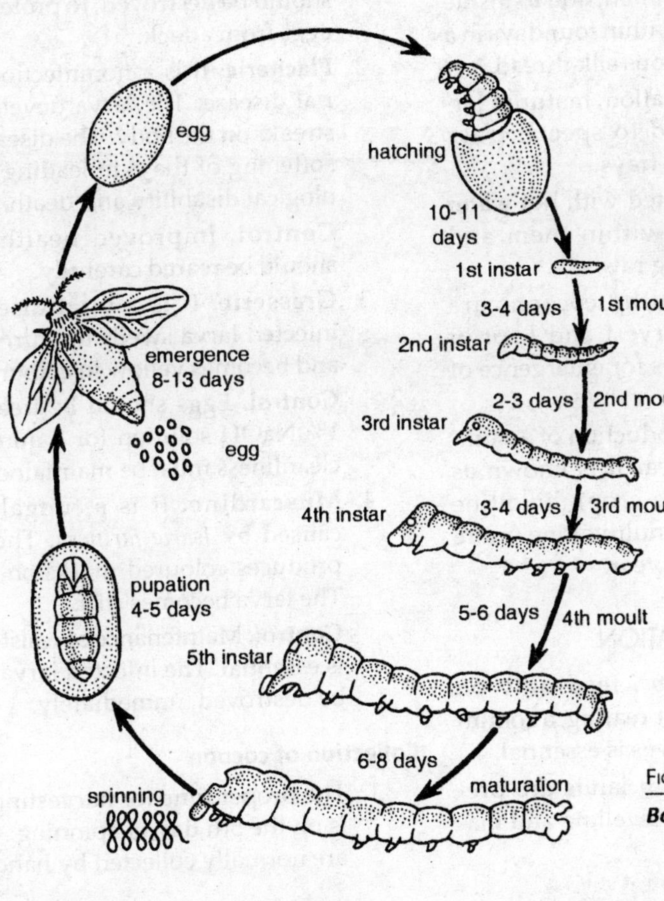

FIGURE 52.12
***Bombyx* sp.** Life Cycle stages

16. The imago secretes a fluid which dissolves the cocoon at one end and the adult emerges through it.

☒ Culture of *Bombyx mori*

1. The moth is made to lay eggs, on a blotting paper. The eggs are collected and treated with 2% formaldehyde.
2. The larvae hatch out within 10 to 11 days. They are reared in bamboo trays and strict hygiene is maintained. The larva is about 3 mm long and called 1st instar.
3. The larvae are voracious easters. Chopped, young mulberry leaves are given as food in the beginning, but with advanced age, entire and matured leaves are supplied.
4. The 1st, 2nd, 3rd, 4th and 5th instars eat for 3, $2^1/_2$, 3, 5 and 8 days, respectively.
5. The 5th instar starts spinning silk thread around its body from outside to inside and encloses itself within four days in a cocoon of a continuous silk thread.
6. Before cocoon formation, matured larvae are transferred to special trays known as **spinning trays.**
7. The cocoon are treated with hot water to kill the pupae within them and processed for getting raw silk.
8. For next crop, healthy cocoons are selected during harvest and kept in well-ventilated cages for emergence of adults.
9. According to the production of a number of crops in a year it is known as monovoltine (single crop), bivoltine (double crops) and multivoltine (more than two) crops in a year.

MULBERRY PLANT CULTIVATION

The larvae of *Bombyx mori* feed only on mulberry leaves. For regular rearing, a continuous supply of healthy leaves is essential.

1. Selected plots of high lands are prepared by ploughing, levelling and manuring.
2. Saplings are implanted there after sufficient rainfall.
3. Watering, manuring and tilling of the soil around the saplings are done regularly.
4. The fields are protected from herbivores and sufficient pest management is needed.

Common diseases of silkworm and their management

Viral, fungal, bacterial and protozoon infection cause much loss to silk industry. Following are the most common diseases of silkworm.

1 **Pabrine.** Caused by protozoan parasite *Nosema bombycis*. This is contagious and transovarian transmission also occur. The infected larva becomes thin and elongated with black spots over the body.

   Control. Infected female and their eggs should be destroyed, to protect healthy eggs from attack.

2. **Flacherie.** It is a noninfectious, bacterial disease. The larva develops black streaks on the body. The disease causes softening of the skin leading to physiological disability and death.

   Control. Improved healthy stocks should be reared carefully.

3. **Grasserie.** It is a viral disease. The infected larva swells up, turns restless and becomes yellow in colour.

   Control. Eggs should be treated with 1% NaOH solution for 2 minutes and cleanliness must be maintained.

4. **Muscardine.** It is a fungal disease caused by *Isaria farinesa*. The disease produces coloured marks on the skin. The larva becomes stiff.

   Control. Maintenance of resistant stock is essential. The infected larvae should be destroyed, immediately.

Collection of cocoon

1. The proper time for harvesting cocoon is on the 5th day of spinning. Cocoons are normally collected by hand.

2. A simple device using a stick is used to separate cocoons from the cavities in the blocks of the cocooning frames or Chandraki (a circular bamboo mountage).
3. The collected cocoons are sorted as good, defective, double, pierced stained, etc. The good cocoons are cleaned by removing any feacal matter on the surface. Cocoons are assorted on the basis of their colour and texture.

Reeling of silk

To obtain standard variety of raw silk latest technology is employed for speedy and economical reeling of raw silk.

1. The cocoons are cooked dipping in a container of hoot water at about 70 to 80 °C or with steaming arrangement for 10 minutes.
2. The cocoons are continuously stirred in the hot water with a stick.
3. In the process the outer gum portion (sericin) of the cocoon is loosened and removed in the form of long tape and the end of the continuous silk thread is obtained.
4. The threads of about 8 to 10 cocoons are hold together and passed through the "glass eye" on to the reel.
5. The thread is skillfully unreeled in a wheel. Thread drawn in this way from several cocoons form the raw silk.
6. The silk thread is treated variously with soap and soda to make it suitable for spinning.
7. Nearly 3,500 cocoons are necessary to get about 1 Kg of raw silk, depending on the quality of cocoon.

APICULTURE

Bees are economically important social insects. They not only provide us with honey and wax, they are also responsible for pollination of flowers of the majority of commercially important plants. The common Indian honey bees are *Apis (Megapis) dorsata* (Fig. 52.13), *A. (Microapis) florea* and *A. indica*.

Apiculture or bee-keeping is the technique of rearing honey bees for honey and wax from their comb or beehives. Selection of sites for quality honey and protection of bees and combs from pests and diseases are part of apiculture.

⊠ Honey Bee : Apis Cerana Indica

Life history (Fig. 52.15)

1. The queen bee is the only functional female (Fig. 52.14) in the colony. Queen is impregnated by the male or drone once in their life time during nuptial flight.
2. The spermatozoa are received in the spermatheca and the drone dies after copulation.
3. The spermatozoa thus received may live for many years and fertilize all the eggs produced by the queen.
4. The ovaries of female begin to enlarge to fill the long abdomen and in a day or two queen begins to lay eggs.
5. The queen can control fertilization of eggs. Unfertilized eggs develop into genetically haploid drones and fertilized eggs produce diploid females (worker).

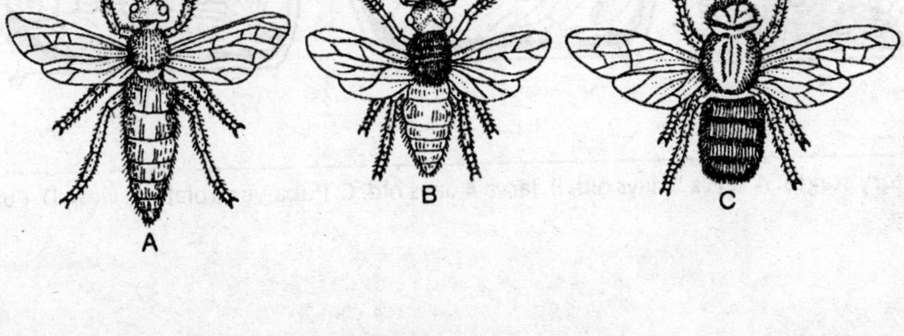

FIGURE 52.13

Apis dorsata. A. Queen, B. Worker, C. Drone

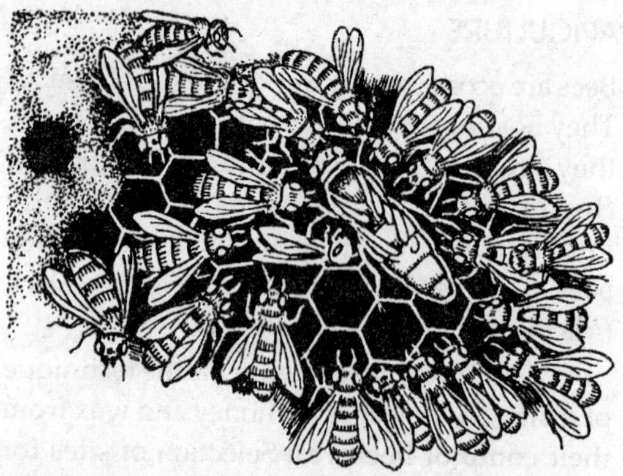

FIGURE 52.14

A part of honey bee colony; Queen bee (centre) surrounded by her retinue of young nurse-bee

6. Queen lays one egg in each cell of brood chamber at the rate of 1500 to 2000 per day. For generation of new queen, separate egg is laid in the queen cell, a specially prepared large cell at the sides of the brood chamber.
7. The eggs are pearly white, about 1.5 mm long, weighing less then 0.1 mg, attached by the queen to the cell bases with an adhesive substance.
8. A tiny annulated, white larva hatches out within 3 days and immediately starts consuming the royal jelly, already kept by the nurse bees at the bottom of the cell.
9. From the third day of their larval life, the queen, drone and workers are fed differently. The queen receives almost exclusively royal jelly. From the 4th day of larval life, the drone and workers are fed with honey and bee bread (a mixture of honey and pollen). The curled larva filling the basal end of the cell begging to straighten on the fourth day (Fig. 52.15)
10. During 6 days of life the larva ensheathed itself with a soft chitinous skin. It is less than a milligram in weight and consumes about 200 mg food daily.
11. On 6th day the larva is straightened and stops eating. It lies in its cell with its head towards the opening. Its stomach, filled up with food is connected with the rectum.
12. The larva starts spinning a cocoon, covering the cell walls with "bee-silk" and the worker bees seal the "pupa house" with a porous wax cap. A queen larva does not enclose itself completely with the cocoon, the last segment remains out of the cocoon. It takes a queen larva one day to spin its cocoon, a worker two days and a drone three days.
13. In their houses of silk and wax, sleeping pupae spend two days (queen), three days (workers) and four days (drone). In general drones develop at a slow rate through all the stages of development.
14. During this period the larval organs transform and tissues and organs of the future bee are built anew from antenna to sting, from head to legs. Nothing of the larva is left in a mature pupa.
15. Pupating larva weighs only 141 mg, and in pupating period it loses 21 mg. Freed from its skin, the true pupa is colourless and almost wingless. Pigmentation and

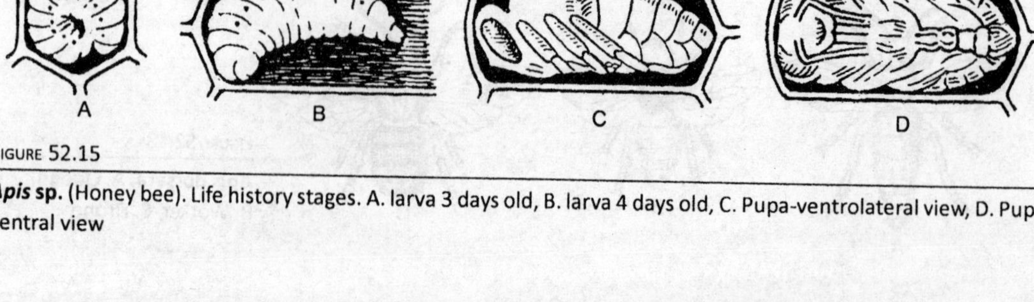

FIGURE 52.15

***Apis* sp.** (Honey bee). Life history stages. A. larva 3 days old, B. larva 4 days old, C. Pupa-ventrolateral view, D. Pupa-ventral view

wings appear at the final stage before emergence.

16. The imago's body is soft but gradually becomes more elastic. The head of imago is slightly bent and presses the unfolded long proboscis against the thorax with the ends of the antennae forming a triangle under the proboscis. The first pair of legs is folded on the thorax and seems to hold the end of the proboscis.
17. After complete metamorphosis the adult bee wakens, shed the pupal skin, and creeps out of the cell. In a queen cell this usually happens sixteen days after laying of egg, in a worker cell twenty one days and in a drone cell twenty four days.
18. The bee's busy life begins at the moment of emergence.

Methods of bee-keeping in India

1. Indigenous methods

a. **Immovable structures.** It is practised in villages from time immemorial. Small structures are made in secluded and protected places. During construction of dwelling houses, small permanent chambers are made in the outer wall of the house for bees to build combs.

Sometimes mud chambers are constructed. On the outerside of the chamber a horizontal slit is made for the entry of bees, while on the inside wall a large opening is left for removal of comb.

b. **Movable structures.** Bee chambers are made up of hollow bags, empty wooden boxes, earthen pots, etc. which can be moved from place to place, and put in a suitable location for the bees.

These methods are not much satisfactory, as the comb is lost in the process of extraction of honey. The quality of honey is inferior due to presence of dust, tissues of damaged bee larvae, pollen grains, etc. in it.

2. Modern methods: Beehive

In modern apirary, Longstroth's frame hive is most suitable and used commercially for production of honey.

1. It is a two-tier structure. The chambers can be removed from or added to, as required.
2. The hive is made up of wooden box.
   a. It has a basal plate or bottom board on which is placed a wooden box called **brood chamber**.
   b. A small opening at the bottom of brood chamber permits passage for bees.
   c. Inside the brood chamber several frames hang vertically from the top. These frames can be removed independently. For this arrange-ment, a modern hive is also called movable frame hive.
   d. The distance between the two frames, the **bee space** is narrow and serves as a passage for the workers but small for building a comb.
   e. Above the brood chamber is placed another similar chamber, but of lesser height.
   f. It is meant for storage of honey only and known as **honey chamber**. The queen is never allowed to enter the chamber. In some cases two honey chambers are used.
3. Above the honey chamber an inner covering is placed over which lies the roof.

Tools for bee keeping

1. **Comb foundation.** A small piece of comb is necessary to tie with one of the frames from where the bees will start comb-building.
2. **Bee gloves.** Leather gloves are used to prevent bees from stinging during handling of the comb and bees.
3. **Bee veil.** A bee veil is required to cover neck, face and head of the keeper during handling. Usually it is made of linen.

4. **Smoker.** A smoker must be used while capturing bees in a hive. Smoke from paper, wood and coconut cover makes the bees inactive. There is fire box in a smoker in which smoke-producing materials and fire are put. A bellow system is fitted to blow the smoke.
5. **Hive tool.** It is a long, narrow and flat piece of steel with a slightly bent head to scrap away dirty materials deposited by bees or some other factors.
6. **Honey extractor.** It is used for extraction of honey from the frames without damaging the comb. It consists of a metal drum with several pockets around a rotating wheel. The frames are hanged from the pockets and the pockets are made to rotate round a central axis. The centrifugal force created by rotation separates honey from the comb which is collected in the drum. The honey is taken out from the drum through a hole at the bottom. The combs and frames are again placed in the hive.

Typical location of apiary

A locality for apiary must have different varieties of pollen and nector-producing plants in sufficient number within a distance of 1.5 to 2.5 km. area. Neem, Rita, Mango, Tamarind, Cheery, Apple, Citrus, Lily, Lotus, various wild plants and crops are good sources for both nectar and honey.

Diseases of the bees

Bees suffer from different contagious diseases and are very often subjected to curious organic disorders. The organisms for contagious diseases are *Aspergillus*, viruses, mites and Protozoa.

a. *Aspergillus*, a fungus, causes paralysis in worker bees.
b. Virus cause paralysis and high percentage of mortality to all categories of bees.
c. Gut protozoa, particularly *Nosema apis*, cause death of bees.

Control of disease

1. The colony should be completely destroyed by burning if larvae are infected with virus, and bacterial diseases.
2. Larvae infected with fungus *Pericystis apis* recover if proper arrangement for sufficient light and air is made.
3. Adults are usually infected by protozoa *Nosema apis*. Cotton soaked with 150 ml 80% acetic acid and kept in the windows of hive cures the disease.
4. Bees infested with mite *Acarapis woodi* recover if cotton soaked in a mixture of nitrobenzen 6 parts, methyl salisylate 2 parts, and petrol 5 parts kept in the food passage of the hive at every alternate day for seven days.

LAC CULTURE

Lac culture is the scientific management of lac insects to obtain a high amount of quality lac. This involves selection and maintenance of host plants, inoculation of host plants with healthy lac insects, collection and processing of lac and protection against enemies.

Lac is the resinous secretion of lac insects. Two species of lac insects *Tachardia lacca* and *T. chinensis* are common, of which the former one is predominant in India. India is the highest lac-producing country. Thailand is next in order.

Host plants. Black siris (Enterolobium) Kusum, Ranjeeni (Khair) and Ber (Plum) trees. The insects live upon plant juice.

Inoculation. 20 to 30 cm long twigs of host plant with old lac crusts are cut and tied with branches of new plants for propagation. The crust contains eggs laid by female insects (Fig. 52.16).

Types of lac

1. **Kusmi lac.** It grows on Kusum trees. Inoculation is done in January-February and harvested in June-July.

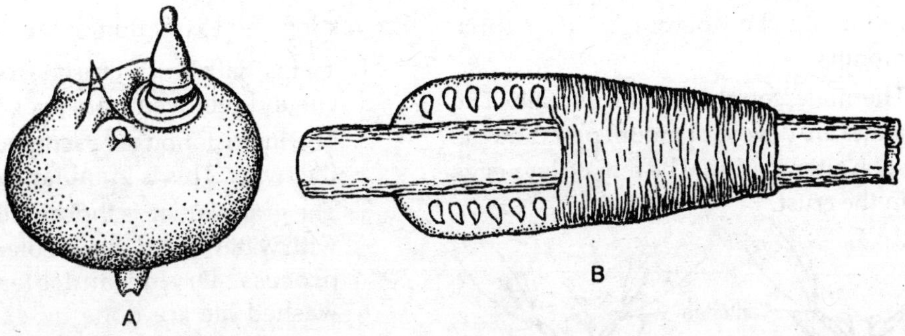

FIGURE 52.16

Tachardia sp., A. female, exposed. B. Lac incrustation around host plant (a part removed)

2. **Ranjeeni lac.** It grows on trees other than Kusum. Inoculated in October-November and harvested in next May-June.

Life cycle of lac insects

A resinous exudate comes out of the insect body and forms a crust around it, which gets attached to the host plant. Gradually a thick crust of the resinous substance or lac surrounds the twig.

1. Following inoculation, the larvae hatch out of the eggs present in the old crust. They are **nymps** and come out of the crust. This is **swarming** (Fig. 52.17).
2. The abandoned old crust is called **Phunki**.
3. The phunki is removed within three weeks of inoculation.
4. The nymphs are boat-shaped, reddish in colour and possess one pair of antennae, three pairs of thoracic walking legs and a pair of caudal setae.
5. Some of the nymphs are males—both winged and wingless, and others are females.
6. The nymphs move to new branches of the host plant. With the help of maxillae and mandibles modified for sucking, they suck cell sap from the branches of the host plant to thrive. The nymphs settle on the branch and liberate the exudate.
7. Most of the body structures are lost and the nymphs moult several times.
8. The rejected skin and the exudate form a crust around it containing a pair of branchial pores and a large anal tubercular opening.
9. An operculum is present at the tubercular opening. The male comes out

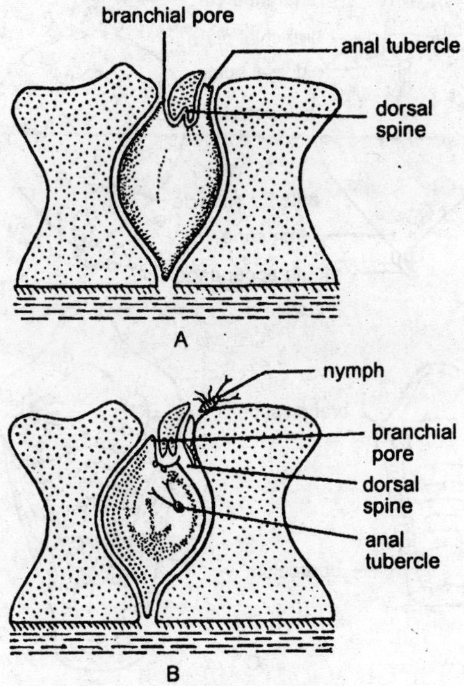

FIGURE 52.17

Tachardia sp., Cross section, A. Position of egg within incubating cell; B. Nymph coming out after metamorphosis

removing the operculum after three months.

10. The male copulates with female. The males are devoid of mouth parts and die soon after mating. The female lays eggs in the crust.

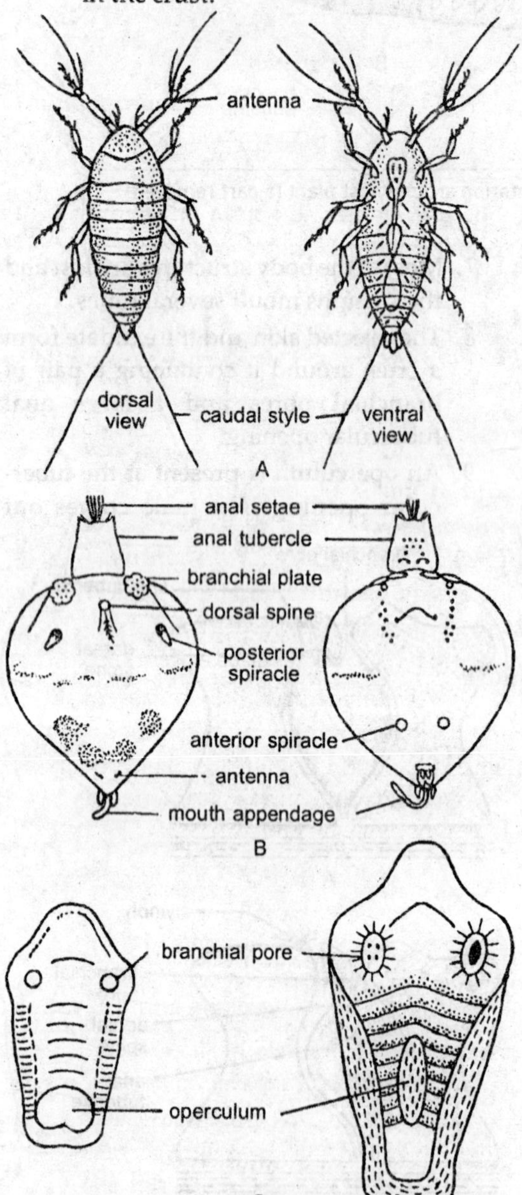

FIGURE 52.18

***Tachardia* sp.**, different stages: A. Male, dorsal and ventral view; B. Female, dorsal and ventral view; C. Immature male in incrustation

Harvesting and extraction of lac

1. Twigs with thick crusts around it are cut and removed from sites.
2. The incrustations are scraped out from the twigs. This is granular lac.
3. The granular lac is thoroughly washed with water. A red dye is obtained in the process. Drying and bleaching of washed lac are done by exposure to sunlight.
4. The granules are melted in pot over open charcoal fire.
5. The lac melts and comes out of the crust. Thin sheets are prepared from the molten lac. The sheets are dried and broken into pieces and marketed as flakes.

Chemical Composition of lac

1. Lac is composed of resin, carbohydrate, protein, water soluble salts, silica, volatile oil and three types of pigments.
2. Resin contains monodibasic hydroxy aldehydic acid. Of the acids 30% is trihydroxy pamitic acid which is known as aleuritic acid.

Uses of lac

1. Lac has been used for the welfare of mankind since antiquity.
2. Lac is used in making toys, bracelets, scaling wax, etc.
3. It is also used in making grinding stones, for filing ornaments, for manufacturing varnishes and paints, for silvering the mirror, for encashing cable wires, for preservation of water and oil paintings water proof ink, nitrocellulose lacars, nail police, shoe police dental plate, photographic negative varnish and many other cosmetics.
4. Stick lac, seed lac, button lac, sellac etc. are also used differently.

Enemies of lac

Some chalcidoid and lepidopteran insects lay eggs inside the crusts. The nymphs of lac insects are eaten up by their larvae.

Proper selection of seed lac, killing of predator eggs in harvested lac, and use of insecticides greatly minimise damage to the crop.

POULTRY

Management of domestic birds providing meat and eggs, viz., fowl, ducks, geese, turkeys in a commercial scale, including disease control and marketing, is known as poultry. Emphasis, however, is given on fowl.

Types of breeds of fowls

The domestic fowl, *Gallus domesticus*, are descendants of wild fowl *Gallus gallus*. The following breeds of fowls are common.

1. **Asiatic breeds.** Large-sized birds with feathered shanks and heavy bones; ear lobes small and red; wattles minute. Eggs small and light brown. Aseel, Brahma, Chittagong and Cochin breeds are common.
2. **American breeds.** Large-sized; shank yellow and unfeathered; skin yellowish; ear lobes red; eggs large and light brown. Examples: Bucueye, Delwares, Jersy black giant, Rhode Island red, etc.
3. **English breeds.** Large-sized; shank white and unfeathered; skin white; ear lobes red; eggs large and light brown. Examples: Australorp, Cornish, Dorking Sussex, etc.
4. **European breeds.** These breeds are named after their country of origin. They are of very little economic importance. Examples. Polish class (Poland), Hamburg class (Holland), French class (France), Continental class (Belgium and Germany), Oriental class (South-eastern Asia).
5. **Mediterranean breeds.** Medium-sized; shank yellowish and unfeathered: skin yellow; comb single or rose-like; eggs medium-sized and white. Examples: Anena, Minorea, White leghorn, etc.

Methods of rearing

Newly hatched chicks are transferred to **Brooder** and kept there in controlled environment for four weeks. On the termination of this period, young chicks are transferred to **deep litters** or **battery cages** or **slatted floors**. Here the birds are properly fed and preventive measures for diseases are taken.

Feeding

It is done with the help of feeders or containers in which food is supplied. The arrangement in a feeder is such that the birds can reach the food easily but cannot sit on it. This prevents spoiling of food. The number of birds receiving food from a feeder is reduced with the age of birds.

Water supply

Water supply to birds is done through containers called **waterers**. Demand for water by the birds is quite heavy and a constant water supply is essential. The waterer is of fairly large size and is heavy so that it cannot be upturned by birds and at the same time the birds should have easy access to water. The waterer is covered to prevent spoiling of water with bird droppings and other materials.

Nest boxes

On the average one bird requires 0.15-0.3 sq m floor space in semi-intensive system and 2.5-4.0 sq m in a shed. The ideal size of a community nest is 3.65 × 1.25 m with a depth 0.6 m 0.2 m openings at a distance of 1.5 m are provided on the back of the cage for entry of birds. Openings are provided on the front for collection of eggs. In most cases the floor of the cage slopes towards the front to facilitate egg collection. A nest must have the following arrangements:

1. It must be strongly built and well-ventilated.
2. It should allow easy passage for birds.
3. Collection of eggs and cleaning can be done easily.
4. The inner side is less lighted.
5. It can be closed at night.

Diseases and their management

Diseases of broods

1. **Brooder pneumonia** or **Aspergillosis** is caused by a fungus *Aspergillus* sp. Clean brooder management is somewhat preventive.
2. **Pullorum disease** or **Salmonellosis** is caused by the bacteria *Salmonella pullorum*. White diarrhoea is the symptom. The disease is transmitted through eggs (transovarian) and appears in the chick 2-20 days old.

   Prevention and treatment. (a) Prevention through pullorum control programme and by not exposing the chick to cold for the first few days. (b) In treatment, strict hygiene is to be maintained. 0.4% Furazolidene (NF 180) is effective.
3. **Coccidiosis** is a protozoan disease infecting caecum and intestine of 3-9 weeks chick. Loss of appetite and colour of comb and wattle and ruffled feathers indicate infection by coccidian *Eimeria tenella*. There may be blood diarrhoea.

   Prevention and treatment. (a) Free movement of chicks in open and dry cages are preventive. (b) Coccidiostat should be administered.
4. **Ranikhet disease** is a highly infectious viral disease. The mortality rate is 25 to 100%. Respiratory difficulty causing shivering and gasping. The head is placed between the legs or at the back of the shoulder. Marked salivation and foetid diarrhoea occur.

   Prevention and treatment. (a) Nasal drops of Fi strain vaccine for one day and again at 4 to 6 weeks, and injection of R_2B vaccine at 10 to 12 weeks are preventive. (b) No treatment is known.

Diseases of growing flocks

1. **Worms.** Both round and flat worms cause much harm to the birds. The worms are endoparasites. Symptoms—general weakness, loss of appetite, retarded growth, less egg-laying.

   Prevention and treatment. (a) Proper hygienic conditions for round worm and this, alongwith not allowing to eat intermediate hosts are preventive for tapeworms. (b) Phenothiazine fed with food or water for roundworms and Dicestal or Nilverm for tapeworm are effective.
2. **Marek's disease.** A serious viral disease causing paralysis and high mortality.

   Prevention and treatment. (a) One-day chick vaccinated against this disease ensures life long immunity. (b) No treatment.

Diseases of egg-laying birds

1. **Chronic respiratory disease.** A transovarian viral disease causing cough, respiratory trouble and nasal discharge.

   Prevention and treatment. (a) Identification of infected birds through serological test and their elimination. (b) Application of antibiotics with feed.
2. **Laryngotracheitis.** A very serious viral disease. Watery discharge from mouth and eyes, violent cough, forced breathing are symptoms.

   Prevention and treatment. (a) Vaccination of six weeks' bird is preventive. (b) No treatment.
3. **Lymphomatosis.** It is a viral disease. The disease is named after the symptoms in the infected bird. (a) **Fowl paralysis.** Partial paralysis of one or more organs, (b) **Grey eye disease.** Iris turns grey and the contour changes. (c) **Visceral lymphomatosis**. Weight and appetite loss, pale combs. (d) **Thick bone disease.** Bones thickened.

   Prevention and treatment. (a) Hygienic condition, nourishing food and organism-free water are preventive. (b) No treatment.

Table: Principal diseases of chicken and their percentage distribution

Disease	Age of birds					
	1-4 weeks	5-12 weeks	13-24 weeks	7-12 months	1-2 years	over 2 years
Colds and Coryza	2.5	6.8	10.1	11.5	7.4	7.7
Coccidiosis	20.5	49.7	24.5	8.6	4.9	1.4
Laryngotracheites	1.4	2.6	2.7	2.6	2.8	4.6
Paralysis	—	3.8	14.6	3.9	2.5	1.1
Pullorum	27.9	0.7	0.6	0.7	1.8	1.6
Round worm (Intestine)	—	1.7	7.9	11.3	10.8	7.2
Round worm (Gizzard)	10.9	4.9	0.8	2.0	3.2	2.0
Tape worm	—	0.2	2.9	2.7	1.4	1.8
Fowl cholera	—	—	—	6.3	7.7	9.4
Fowl pox	—	—	—	2.8	3.2	4.9
Chick bronchitis	3.1	2.6	0.7	—	—	—
Pneumonia	3.6	0.4	0.2	—	—	—

4. **Chicken pox or Fowl pox.** A viral disease transmitted through contact or mosquito, affecting skin, comb and oral mucous membrane. Black blisters on comb and wattles, appetite loss and reduction of egg production are the symptoms.

   Prevention and treatment. (a) Vaccination at 6-8 days is preventive. (b) No treatment.

5. **Fowl cholera.** *Pasteurella aviseptica*, a bacteria, is the causative organism. It is highly infectious and causes high mortality. Inactivity, droopiness, fever, dark head colour and green diarrhoea are symptoms.

   Prevention and treatment. (a) Hygienic environment and vaccine are best preventive. (b) Treatment with Sulphamethazine is effective.

6. **Tuberculosis.** The *Mycobacterium tuberculosis* cause the disease. Dry skin, pale colour, shrinkage of comb and wattles, weight loss and cessation of egg laying are symptoms.

   Prevention and treatment. (a) Hygienic condition and nutritious food are preventive. (b) Streptopenicillin treatment is effective.

FISHERIES

The cultivation and management of fishes, prawns and pearl oysters in a scientific way to make it commercially viable and, at the same time, making harvest of quality food animals and pearls possible, is known as fishery.

Types of fisheries

In its broad concept fishery may be divided into two categories.

1. **Culture fishery.** The fishes, prawns and pearl oysters are grown, managed and harvested in a scientific way.
2. **Capture fishery.** The fishes, prawn and pearl oysters are simply collected from their natural habitats and processed or straight way despatched to the market.

CARP CULTURE

Carp culture involves cultivation, propagation and harvest of edible and marketable carps including Indian and exotic forms or both.

The practice of carp culture in India is unscientific in most cases as the ratio of major carps required to fully utilise the space and available food in a water body is not maintained. As a result the yield is poor and varies from 300 to 600 kilograms per hectare.

Composite Culture

Indian major carps are:

A. **Catla** (*Catla catla*)
B. **Rohu** (*Labeo rohita*)
C. **Mrigal** (*Cirrhinus mrigala*)

The popular exotic major carps are:

A. **Cyprinus carp** (*Cyprinus carpio*)
B. **Silver carp** (*Hypophthalmichthys molitrics*)
C. Grass carp (*Ctenopharyagodon idella*)

In present day carp culture, species of carps (Fig. 52.19) with different feeding habits and habitats are grown at a definite ratio in a pond to fully occupy the space and utilise available food. This is known as '**composite fish culture**' or '**mixed fish farming**' or '**polyculture**' or '**intensive fish farming**'. An average yield of 3,000 kilogram/hectare/year has been obtained.

1. Catla and silver carp occupy the upper stratum of pond water. Catla feeds on zooplankton while silver carp is predominantly a phytoplankton feeder.
2. Rohu and grass carp occupy middle stratum. Rohu feeds on small aquatic vegetation and grass carps mainly on soft aquatic macrophytes.
3. Mrigal and cyprinus carp occupy lower stratum of water and feed upon benthonic organisms and faeces of

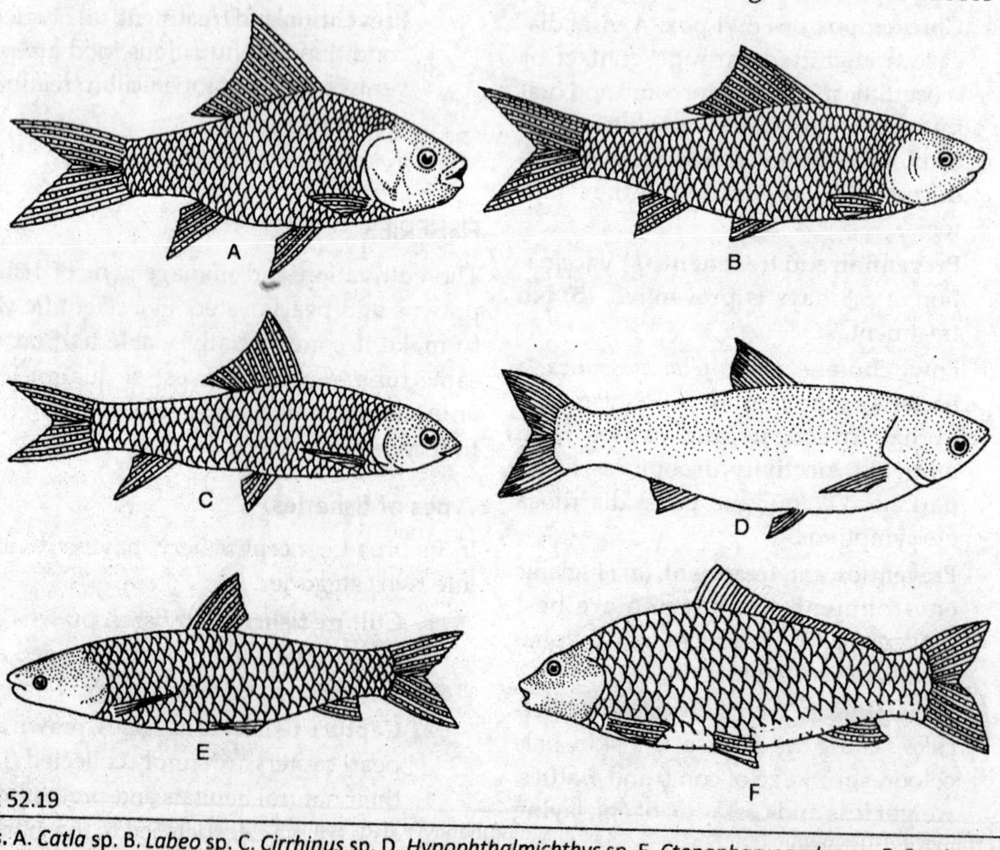

FIGURE 52.19

Carps. A. *Catla* sp. B. *Labeo* sp. C. *Cirrhinus* sp. D. *Hypophthalmichthys* sp. E. *Ctenopharyngadon* sp. F. *Cyprinus* sp.

upper dwellers and other organic matters raining down.

Six species of carp, cultured at a time, may yield 3,000 to 9,000 kg/ha/yr.

Basic principles of composite culture

1. The use of fast-growing nonpredatory food fishes, that effectively utilize both natural and supplementary fish food.
2. Use of compatible species—not competitors— for food and space.
3. Stocking of requisite number of each species with a total adequate population to properly utilize the whole water body and all food resources.
4. Increasing carrying capacity of the pond on its fish productive potential through the application of suitable manure and also increasing the stocking density by providing supplementary food.
5. Manipulation of the number of fishes in the course of culture through intermediate harvest.

Criteria for selection of fishes

1. Plankton feeders; food chain being short, energy output is higher.
2. Rapid growth rate.
3. Compatible, i.e., capable to live with other species of fishes.
4. Good looking and good to taste, i.e., marketable.
5. Nonpredatory.

Composite fish farming in India

Three combinations are adopted in composite fish culture in India. The ideal combinations are:

A. **Indian major carps alone**

a. Ratio—Catla 4: Rohu 3: Mrigal 3.
b. Density 3,750 fingerlings/ha.
c. The average weight attained 700gm/yr.
d. Yield varies with ratios and density and ranges from 1,400 to 3,000 kg/ha/yr.

B. **Exotic carps alone.** Successfully introduced in 1957.

a. Ratio—Silver carp 3: Grass carp 1: Cyprinus carp 2.
b. Density 5,000/ha.
c. Yield 2,900-3,000 kg/ha/yr.

C. **Indian and exotic carps together**

a. Ratio—Catla 1 : Rohu 3 : Mrigal 1.5 : Silver carp 2 : Grass carp 1 : Cyprinus carp 1.5.
b. Density 5,000/ha.
c. Yield 6,000 kg/ha/yr.

Composite culture management: Improved technique

a. Use of 'Mohua' (*Madhuka latifolia*) oil cake to eradicate unwanted fishes and other aquatic organisms. It is initially a fish poison but turns to a good fertiliser after proper treatment.
b. Use of both organic and inorganic manures, viz., cowdung and triple superphosphate.
c. Application of lime prior to the use of fertilizer.
d. Use of quality supplementary feed like ground nut oil cake, rice bran and aquatic weeds.
e. High stocking density 5,000 to 7,500 and even 10,000 fingerlings/ha.
f. Correct stocking ratio of fingerlings of different species.
g. Yield 8,000 to 9,000 kg/ha/yr.

NATURAL BREEDING

The Indian major carps do not breed in confined water. Natural spawning also has many hazards.

a. Factors required for breeding are beyond human control.
b. Spawns are not of good quality and often mixed with those of unwanted fishes, particularly predators.
c. High mortality in transit from collection point to culture pond.

INDUCED BREEDING

Present day wide spread **hypophysation** or induced breeding is the result of contributions of a number of workers. Pioneers in this field are: Houssay (1930) of Argentina suggested the idea; Brazil introduced it in 1934; in India, Khan (1937) made the first attempt; Chaudhuri (1955) was successful with *Esomus danricus;* Chaudhuri and Ali Kunhi (1957) introduced large scale spawning in *Labeo rohita, L. bata, Cirrhinus mrigala, C. reba, Puntius sarana*, etc.

In induced breeding, the potential breeders, i.e., sexually mature male and female are made or induced to spawn in confined water at desired time, of course with certain limitations, by injecting them with pituitary gland extract. It ensures:

a. Spawning more than once a year.
b. Production of quality seeds.

Technique adopted

1. **Collection and preparation of gland extracts.** The pituitary gland is located on the ventral side of the diencephalon and lodged in a cavity in the base of the skull.

Collection of gland

a. Intact glands are collected in May to July from fully ripe, fresh or ice-preserved (5-7 days maximum) carps of both sexes of same or closely related species, and stored in absolute methanol.
b. After 24 hrs. glands are transferred to fresh ethanol in another vial.

Preparation of gland extract

a. The required amount of glands are taken out and placed on a filter paper for partial drying.
b. The glands are macerated in a tissue homogenizer with a small amount of distilled water or 0.3% saline solution.
c. The tissue debris is centrifuged off and the aliquot (supernatant) containing the hormones is drawn into a syringe for injection.

2. **Preservation of gland extract**

a. Distilled water is added to **homogenised glands** to make 1/3rd of the desired volume to be used and stored in a refrigerator for 24 hours.
b. Two volumes of glycerine (double the volume of the extract) is added and again stored in a refrigerator for another 24 hours.
c. The tissue debris in the mixture is centrifuged off and the aliquot transferred to ampules, sealed and stored for future use.

3. **Selection of breeders for injection**

a. A mature male is distinguished by roughness of the dorsal surface of the pectoral fin, which is smooth in female.
b. Milt oozes out freely in ripe males with gentle pressure on the belly near the vent.
c. The belly of a mature female is soft, rounded and bulging and the genital opening is pinkish.
d. One set of breeders, one male and two females are released in a **breeding hapa** after recording their individual body weight.

4. **Rearing of breeders prior to selection.** The potential breeders, each weighing above 1 kg, are stored 5 to 6 months piror to breeding season and are reared properly. Health and maturity status are periodically examined.

5. **Determination of dose for injection**

a. It is calculated in terms of mg/pituitary/kg body weight of the recipient fish.
b. The female is given 2 injections. The first one at a dose of 2-3 mg/kg and the second one 5-8 mg/kg after 6 hours.
c. The male is given only one injection, 2-3 mg/kg, simultaneously, with the second injection to the female.
d. Very ripe breeders need small dose for spawning.

6. **Site of injection**

a. Intermuscular injection at 45° is given in the region of caudal peduncle or in the maximum fleshy portion of the body dorsal to the lateral line.

b. During injection the breeders should be held with sponge mattress.

7. **Breeding**

a. Injected male and female breeders are released in a breeding hapa prepared at a corner of the pond.

b. Breeding starts 3-6 hours after the second injection.

c. Sexual activity is manifested by jumping and darting movements. The female spawns and the male sheds milt.

d. The released eggs are immediately fertilized in water.

e. Fertilized eggs are transparent, swell up to a diameter of 3.5 to 5.5 mm. Unfertilized eggs are whitish in colour.

f. A female catla yields about 0.15-02, a rohu 0.2-0.3 and mrigal about 0.7 million eggs in a single spawn.

8. **Hatching and management of spawns**

a. Eggs are transferred to hapa for hatching. During transfer, mortality is very high due to sudden change of temperature and other factors.

b. A hatching hapa is really two rectangular hapas—made of net—one placed inside the other. The inner hapa measures 1.8 × 0.9 × 0.9 m and its meshes are larger than that of the outer one.

c. The eggs are uniformly spread on the bottom of the inner hapa. At an optimum temperature 27°-31°C the eggs hatch out by 15 to 18 hours.

d. The hatchling escape to the outer hapa. Unfertilised eggs and egg shells left are removed with the inner hapa.

e. On the third day, the spawns are collected and transferred to nursery ponds.

Abiotic factors conducive to spawning

a. Rain and rain water
b. Water current range
c. Temperature range 27-31°C
d. pH of water
e. Light—soft or shady.

STRIPPING TECHNIQUE FOR EXOTIC CARPS

Stripping or artificial insemination is adopted if spawning with injection fails.

a. 4-6 hours following the second injection the belly of the female is pressed and the eggs are collected in a large enamel tray.

b. Milt from male is pressed out on the eggs in the same way.

c. The milt and the eggs are thoroughly mixed with a feather to ensure fertilization of the eggs.

d. Stripping is usually practised in silver carp and grass carp but can be successfully applied for many other fishes.

Recent advances in induced breeding

Since 1957 intensive research is being carried for augmenting seed production of cultured fishes. The important gains are:

a. Pituitary bank.
b. Multiple breeding.
c. Induced breeding with human chorionic gonadotropin (HCG).
d. Improved breeding in tank and hatchery system throughout the year.
e. Stripping technique.

In 1997-98 West Bengal produced 8,500 millions fry, which is more than the requirements of the state. The excess fries are exported to other states or even abroad.

PRAWN CULTURE

Crustacean fishing is an important aspect of Indian fishery. The prawn fishery has a major share in it. Both capture and culture fisheries are practiced.

A. Important cultural prawns and their importance in fishery

A good number of prawn species are of commercial value. The following are important.

a. *Penaeus indicus* (Fig. 53.20). Maximum size 230 mm. One of the most important commercial species. Young ones are fished in estuary.
b. *Penaeus monodon*. Maximum size 320 mm. An important commercial species. Good fishery in West Bengal and Orissa.
c. *Metapenaeus monoceros*. Maximum size 180 mm. Very important commercially.
d. *Metapenaeus brevicornis*. Maximum size 125 mm. Good fishery in northern region of south and east coasts. Juveniles are also fished in estuary.
e. *Palaemon styliferus*. Maximum size 90 mm. One of the most important fishery in Gangetic delta.
f. *Macrobrachium rude*. Maximum size 130 mm. Good seasonal fishery in West Bengal and Orissa.
g. *Macrobrachium rosenbergii* (Fig. 25.1). Maximum size 320 mm. Very good fishery in monsoon and post-monsoon months in Kerala.
h. *Macrobrachium malcolmsonii*. Maximum size 230 mm. Fairly good fishery in North-east coast in monsoon months.

B. Prawn seed collection: sites and time

The creeks, canals and shallow waters, preferably with sloping sides, are suitable places for seed collection. The lunar phases are the best time period for collection. Different types of gears are used for collection.

a. A battery of nets are operated in longitudinal rows in shallow water.
b. The shooting net is suitable for the pelagic forms.
c. Scoop nets and drag nets are used in intertidal pools and shallow inundations.

Different species of prawn do not breed in confined water. Induced breeding with **hypophysation** has been a success.

C. Prawn culture in brackish water

Four types of prawn culture are prevalent in West Bengal.

a. Only stocking and harvesting without any care to prawn population and food supply.
b. Stocking of known quantity and cultured with natural food.

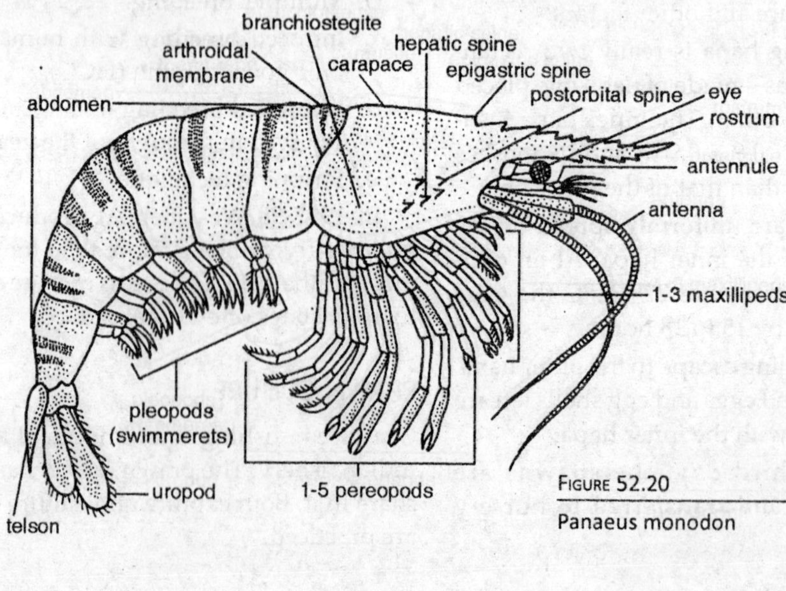

FIGURE 52.20
Panaeus monodon

c. Stocking at higher concentration and extra nutrient added. In this method, production can be increased by 500 per cent.

d. Stocking selected prawns at higher concentration with fertilisation, supplemental feeding and stock manipulation.

1. **Pond preparation**

   a. Prior to stocking, the culture pond is dewatered and its bed is exposed to sun to dry. It helps in high soil fertility.

   b. Mohua oil cake, an organic toxicant, is spread over the bed at a rate of 100 to 150 kg/ha and water is drawn to a depth of 4 to 5 cm.

   c. The toxin dissolves in water and the unwanted organisms are killed.

   d. The water is again drained out after 2-3 days and the pond is flushed with fresh tide water to wash away rest of the toxicants.

2. **Fertilisation**

   a. For a better growth of prawn food organisms, manuring of ponds is essential.

   b. This is repeated monthly with oil cake @ 50 to 100 kg urea and superphosphate each 50 kg/ha.

3. **Water management**

   a. Water is periodically drawn through inlets guarded with close mesh screen.

   b. This prevents entry of undesirable organisms and also escape of prawn.

   c. The optimum ranges are—temperature 25-30°C; dissolved oxygen 4-8 PPM; pH 8.0.

4. **Stocking**

Fry and fingerlings are directly stocked in ponds.

   a. Stocking density is high in brackish water than in fresh water as the growth rate is slow in the first.

   b. Supplementary feeding is done depending on the stocking density and productivity of the pond.

   c. Stocking density largely depends on the culture practised.

   i. Single stocking and single harvest.

   ii. Single stocking and multiple harvest.

   iii. Multiple stocking and multiple harvest.

5. **Supplementary feeding**

Shortage of natural food needs supplementary feeding. It leads to a better growth in a pond where prawns are stocked at a high concentration.

6. **Harvest**

   a. Partial harvesting is done by setting bag nets at the outlet during low tide.

   b. Total harvest is possible only with complete dewatering.

⊠ Prawn culture in Kerala

Prawns are cultured in backwaters along with paddy crops. This is known as **paddy cum prawn culture**. Juveniles of all commercially important species are trapped in paddy field.

a. The variety of paddy 'Pokali' is harvested in September.

b. From now onwards the salinity of paddy fields gradually increases and prawn juveniles are allowed to enter with water through controlled sluice gates.

c. In and outflow of water are allowed daily during high and low tides but prawns are prevented to escape with bamboo mats placed inside the gates.

d. In April, the culture comes to an end. Prawns are completely harvested by repeated netting and dewatering of the field.

⊠ Problems of prawn culture

The hazards of brackish water prawn culture are:

a. **Mechanical**

i. Structural failure of water gates.

ii. Landslide from dykes.
iii. Erosion and silting up of farm ponds.
iv. Overflowing of ponds.

b. **Physical**
i. temperature
ii. Turbidity of water.

c. **Chemical**
i. Salinity.
ii. Organic pollution of farm ponds.
iii. Oxygen deficiency.

d. **Biological**
i. Parasites and diseases.
ii. Pests.

These, however, can be controlled with proper care.

PEARL CULTURE

Pearl is secreted by the mantle of marine bivalve molluscs, the pearl oysters (Fig. 52.21). The secretion is known as **nacre** or mother of pearl. It is deposited in extremely thin layers.

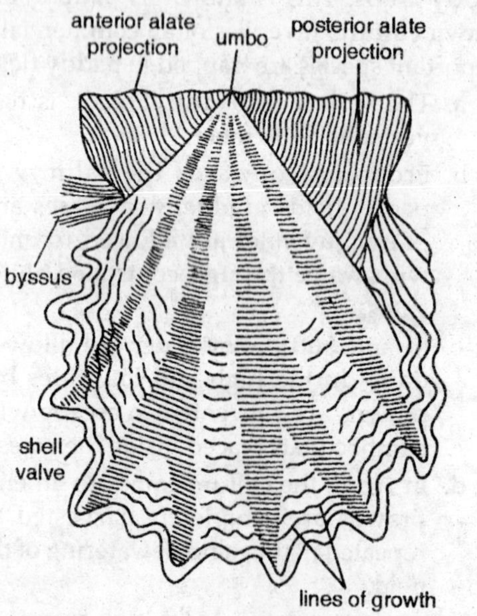

FIGURE 52.21

Pearl oyster. *Pinctada margaritifera*

Collection of oyster from the sea bed, opening the shell (valves) and getting the pearl from them is collectively termed **Pearl fishery**.

Important pearl-producing oysters

a. *Pinctada fucata*
b. *Pinctada chemnitzi*
c. *Pinctada margaritifora* (Fig. 52.21)
d. *Pinctada anomioides*
e. *Pinctada atrapurpurea*

The first one occurs in Indian seas while the others are found in other Oriental seas.

Distribution in Indian waters

Two zones in Indian coast are well-known for pearl oysters.

a. The Gulf of Kutch along the northen coast of Gujarat.
b. The Gulf of Mannar adjoining Tamil Nadu coast.
c. The beds are vast in the Gulf of Mannar. Oyster beds are present at a depth of 10 to 20 metres and at a distance of 11 to 16 kilometer from the coast.
d. The area of continental shelf is maximum at the eastern coast and these beds yield high-quality pearl.

Biology of pearl oyster

a. The oyster usually breeds several times in succession.
b. The larvae are free-swimming and feed upon microscopic organisms.
c. They undergo metamorphosis and finally settle, known as **Spatfall**, at the bottom, in large numbers.
d. The young oyster has a fast growth rate and attains about 50 mm size in a year.

Fromation of pearl

a. If a small foreign particle, an inert material, e.g. sand particle, hard object, etc.; living object, e.g. parasite, penetrates the mantle of the oyster, the cells of the mantle get attached to it.
b. The foreign particle forms the nucleus over which more or less concentric layers of **mother of pearl** is secreted.
c. In case the particle penetrates deep enough, a sac or cyst is first formed around the particle and pearl formation occurs within the sac.

d. With time, due to more secretion, it enlarges in size and the pearl is formed.

e. The natural pearl is rarely a sphere, as it assumes the shape of the nucleus.

f. The process is very slow and continues for the whole life of an oyster.

g. The value of a natural pearl depends on its size, shape, colour and lusture.

Artificial pearl culture

Japan was the first to develop scientific pearl culture in 1907. Introduction of large particle of desired shape and size made it possible to get pearls of different shapes and sizes.

☒ Large-scale pearl culture

a. Sandwich type nets made of iron rods are used. A net measures 60 × 40 cm and is partitioned into 5 equal sections. Each section is woven with 2 mm thick nylon twine to get 2 cm^2 meshes. The iron frame is heavily painted to prevent rusting.

b. The young oysters collected from their habitat are placed in the nets and firmly secured by a twine running all along the four sides of the frame.

c. The nets are suspended from a wooden raft 6.0 × 4.5 m in suitable area in sea.

d. Planks are fitted to the raft for working space and also for periodical examination of the oysters. Each net accommodates 150 to 250 oysters and each raft holds up to 30 such nets. The raft is fixed to empty wooden barrels or metallic drums.

e. After the attainment of desired size, the oysters are cleaned, taken to laboratory and kept in sea water tanks for conditioning.

f. Healthy oysters are anaesthesized in menthol solution in sea water.

g. The specimen is firmly fixed on a special stand and the two valves of the shell are opened by a pair of tongs.

h. An incision is made in the pallial ridge and a piece of mantle from another oyster is introduced in the incision and implanted in the tissue.

i. The oyster is put back in a basin containing sea water, which is changed for several times a day.

j. After recovery the oysters are transferred to tanks with well-aerated water. After 7 days they are placed in pearl nets in sea water.

k. Deposition of 'nacre' starts after a month of implantation and pearl can be obtained within 108-190 days.

The rate of nacre deposit is faster in warm tropical waters. India has knowhow for warm water pearl culture but much more sophistication including modernisation is necessary.

DAIRY BREED, ARTIFICIAL INSEMINATION AND FARM MANAGEMENT

DAIRY BREED

The breed is a group of animals related by descent and similarity in most characters like general appearance, features, size, configuration, etc.

Most important breeds of cow in India are:

1. **Milch breeds** (producer of more milk)
   i. **Deoni.** Origin and distribution is North-West and Western portion of Hyderabad (Sind).
   ii. **Gir.** Origin and distribution is Gir forest, Junagarh, Gujarat.
   iii. **Sahiwal.** Distributed in Panjab, Delhi, Uttar Pradesh, Bihar, Madhya Pradesh.
   iv. **Sindhi.** Distributed in Karachi, and portion of Sind.
2. **Dual purpose breed** (milk, producer as well as hilly drought living)
   i. **Hariana.** Distributed in Rohtak, Hissar, Gurgaon, Karnal, Delhi.
   ii. **Ongole.** Distributed in Andhra Pradesh.

iii. **Tharparkar.** Distributed, in south-east Sind, Amarkot, Kutch desert, Marwar and Palampur of Bombay.

iv. **Kankrej.** Distributed in south-east of Runn of Kutch, from north-west corner of Tharpar district in Sind to Dholka in Ahmedabad.

3. **Draught breed** (in hilly draught area)

i. **Siri.** Distributed in Darjeeling hill tracts, Sikkim and Bhutan.

ii. **Malvi.** Distributed in Malwa tract in Madhya Pradesh and Rajasthan.

iii. **Kangayam.** Distributed in Kangayam division of Dharampuram taluk of Coimbatore district, Udamalpet, Palladam and in other parts of South India.

iv. **Amrit mahal.** Distributed in Mysore.

ARTIFICIAL INSEMINATION

It is the method of breeding involving artifical collection of semen from the bull, preservation of semen and insemination of cow in proper time under hygienic conditions.

☒ Requirement

A healthy bull
A Dummy Cow
An artificial vagina
A single service plastic tube.

☒ Artificial vagina

The artificial vagina is almost similar to natural vagina and different kinds of artificial vagina are used for different classes of animals.

An artificial vagina consists of following components (Fig. 52.22):

1. An outer heavy rubber cylinder fitted with an water and air outlet, which communicates inner side of the cylinder with outside. The air pressure in between the outer cylinder and inner sleeve can be regulated with screws fitted at the water and air outlets.
2. An inner sleeve made of rubber.
3. A semen receiving cone.
4. A semen collecting graduated vial made of glass or plastic.

☒ Assembling of the components

The components are thoroughly cleaned, sterilized and assembled into the artificial vagina.

a. The inner rubber sleeve is put into the outer rubber cylinder and both the ends of the sleeve are reflected over the cylinder. A water tight space is enclosed between the two, which is in communication with the exterior through the water and air outlets.

b. The semen collecting vial is fitted with the semen receiving at the narrow end of the latter.

c. The cone along with the vial is slipped at one end of the barrel (rubber cylinder + rubber sleeve) and lightly secured with rubber bands. (Fig. 52.22)

d. The space between the outer cylinder and the rubber sleeve is filled with warm water of 40-45 °C.

☒ Collection of semen

Three to four inches at the open end of the artificial vagina is lubricated with sterilized jelley. The bull is made to mount a dummy cow. Just prior to mounting the artificial vagina is held at 45° angle with the horizontal plane of the dummy, because the penis of the bull enters the vagina of the cow at that angle. When the bull mounts the penis is quickly guided by a trained hand into the artificial vagina. Following dismount, the artificial vagina is removed from the penis and kept in an upright position after releasing the pressure by draining out water and air, out of it. The ejaculate flows from the cone to the vial, which is detached from the cone for subsequent proceeding.

1. The semen is a suspension of spermatozoa in seminal fluid. It is a white

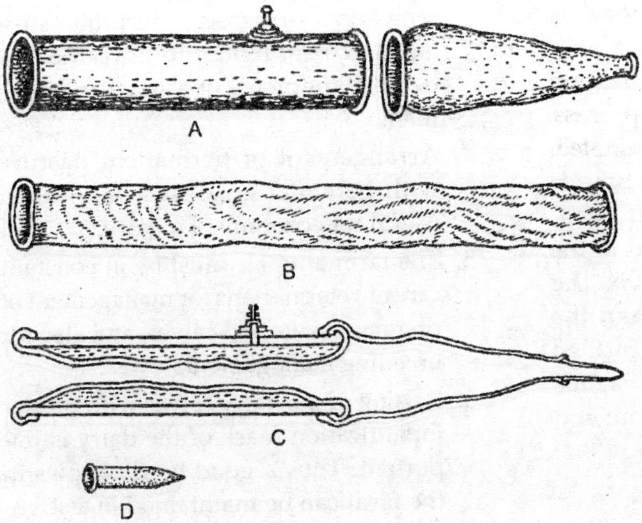

FIGURE 52.22

Artificial vagina; different parts; A. Cylinder (heavy rubber), B. Inner sleeva of rubber, C. Section of whole (assembled instrument, D. Semen collecting graduated vial

to light cream-coloured opaque and viscous fluid.

2. The semen is collected in an artificial vagina. It consists of an elongated rubber tube blind at one end and with a flexible rubber ring at the other end to keep it open (Fig. 52.22).
3. After proper sterilization and lubrication the artificial vagina is inserted into the vagina of the female to copulation with the open end outside.
4. When the bull dismounts after coitus it is taken out and the semen is collected.
5. The semen is a suspension of spermatozoa in seminal fluid. It is a white to light cream-coloured apaque viscous fluid.
6. The spermatozoa are generated in the testes and stored in the epididymis. The seminal plasma is contributed by he secretory fluids produced in the accessory organ like epididymis, prostate gland, seminal vesicle, and Cowper's gland.

☒ Processing of Semen

1. Immediately after collection, the semen is diluted with 50% by volume with 2.94% sodium citrate (dihydride) sol^n.
2. The diluted semen is cooled in a refrigerator at 5 °C for 3 to 5 hrs. Again diluted with 16% glycerol by volume in 2.94% sodium citrate sol^n at 5°C.
3. Finally this diluted semen is mixed in proportion of 5 : 3 : 1 with 8% glycerol, 25% egg yolk, and 2.2% sodium citrate, mixed over a period of several hours, distributed in different vials and preserved in deep freeze.

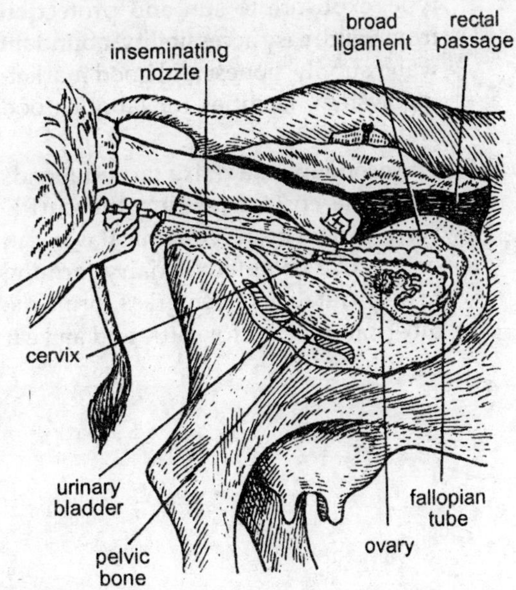

FIGURE 52.23

Recto-Vaginal method of Insemination (diagrammatic)

Insemination. Insemination is done by cervical insemination and recto-vaginal deposition methods. Recto-vaginal method (Fig. 52.23) is the most popularly used process. Within the heat period the cow is inseminated. Single unbreakable service plastic tube is inserted in the vaginal passage. Left hand, maintaing proper hygiene, is inserted in the rectum to guide the tube up to the cervix. The preserved vial of semen is broken and the content is taken in a syringe which is pushed through the tube already put in the vagina. After insemination the tube is taken out and discarded.

DAIRY FARM MANAGEMENT

1. The farm management is the act and science of combining ideas, facilities, processes, materials and labour to produce and market a worthwhile product or service succefully.
2. The dairy management depends on
   i. good farm records
   ii. proper annual and partial budgetting.
   iii. well planned and adequate housing for cattle with proper drainage, good soil type, exposure to sun and protection from wind; easy accessibility, abundant water supply, honest and good marketing system, facilities for labour, food and electricity.
3. The economy of a dairy farm depends upon the cost of feeding the cuttle. Feeds grown in an economical way can only prove commercial dairy farming to be profitable. Green grass forms the most natural food for cattle and an year long supply of green, nutritious, lush and sacculant fodder to the cattle in a dairy is the mark of a good management.
4. Arrangement of permanent pasture with peremial grasses is very much useful for economy of a dairy.
5. The farm animals must be in constant care of veterinarians for management of pregnant cows and calves, and also for breeding management.
6. Raising of the dairy calves with proper identification mark of the dairy is important. Thus a good herd of cows in the farm can be maintained in a scientific manner.
7. Records in the dairy farm and its preservation is necessary.
8. The registers in a dairy farm are—
   i. milk record register,
   ii. cattle feed register,
   iii. cattle history and pedigree sheet register,
   iv. lactation record
   v. cattle service register
   vi. health record
   vii. calf register
   viii. financial register
   ix. feed indent register
   x. stock register of cattle.
9. Most essential for dairy management are daily inspection, regularity of care, kindness in handling, exercise, grooming dairy cows, drying of cows, and proper watering.

53 ECOLOGY

The study of animals and plants in their relation to each other and to their environment is called ecology.

SCOPE

It deals with a number of factors governing the life of organisms.

1. The local and geographic distribution and abundance of organisms.
2. The temporal change in the occurrence, abundance, and activities of organisms.
3. The interrelations between organisms in populations and communities.
4. The structural adaptation and functional adjustments of organisms to their physical environment.
5. The behaviour of organisms in natural conditions.
6. The evolutionary development of all these interrelations.
7. The biological productivity of nature and how this may best serve mankind.
8. The development of mathematical models to relate interaction of parameters and predict effects.

ENVIRONMENT CONCEPT

⊠ Atmosphere

The space, 960 kilometres above sea laval is considered as atmosphere. It is made up of gases and vapour and receives solar energy from the sun. Terrestrial animals and plants live at the bottom of the atmosphere, where the air is densest. The air thins out higher up.

⊠ Troposhere

The lowest layer of atmosphere with confined weather is known as troposphere. It extends to a height of about 9.6 kilometres from the surface of the earth. The temperature falls with increasing altitude. In general the lower part of the atmosphere contains a constant proportion of certain gases—Nitrogen 78%. Oxygen 21%, Carbon dioxide 0.03% and minute traces of argon, hilium and some rare gases, in addition to solid particles, dust and smoke. The amount of water as water vapour is extremely variable.

⊠ Stratosphere

It lies above the atmosphere and extends upwards for 80 kilometres or more. It is very cold, cloudless and with extremely thin air and without any gas or vapour or particulate matters.

⊠ Ionosphere

It extends several kilometres up beyond stratosphere. It has electrically conducted layers, which make short wave transmission possible over long distance.

☒ Hydrosphere

The waters of the earth dislinguished from lithosphere and atmosphere.

☒ Lithosphere

The solid outer layer of the earth above the semisolid asthenosphere, and includes the crust and solid upper part of the mantle down to about 75 kilometres.

DIVISIONS OF ECOLOGY

Animal ecology. Concerned with animals.

Plant ecology. Concerned with plants.

Bioecology. Dealing with both animals and plants.

Synecology. Study of communities.

Autecology. Study of species.

Based on the fundamental and basic facts, ecology can again by classified:

Community ecology. The local distribution of animals in various habitats, the recognition and composition of community units and succession.

Ecosystem dynamics. The process of soil formation, nutrient cycling, energy flow and productivity.

Population ecology. The manner of population growth, structure and regulation.

Evolutionary ecology. The problems of niche segregation and speciation.

Geographic ecology. Concerned with distribution, paleoecology and biomass.

Referring to special consideration of the ecology of taxonomic group, we speak of **mammalian ecology, avian ecology, insect ecology, parasite ecology, human ecology**.

With emphasis on habitat the terminology are **oceanography**, the study of marine ecology; **limnology**, fresh water ecology, **terrestrial ecology**, ecology of land organisms and so on.

ECOLOGICAL NICHE

The interaction of a species with all the factors of the environment, both biotic and abiotic, in a particular area is known as ecological niche. There might be more than one niche in an ecosystem but only one species can exist in a niche.

Types of niche

Broadly three types of niche have been identified.

Spatial or habitat

An area restricted to only one species. Several species of ants living in restricted areas in different zones, viz., bark, wood, leaf, etc. of the same plant.

Trophic niche

Habitat related to food supply. Consumers feeding on different food can live in the same area. Fishes eating different species of plankton inhabit the same water body. Herbivore species eating different land plants are found in the same area.

Multidimensional or hypervolume niche

A large area where a number of biotic and abiotic factors influence the life of animals. All the factors, however, are not essential for the survival of a particular species. Due to the various factors present, a large number of spatial niche are formed and many species may live there. A large number of species belonging to different groups may live in the same water body under different condition, viz., in the mud of the bottom, at different depths of water, at different temperatures, eating different food, etc. Such large niche on land supports many terrestrial species.

ECOLOGICAL FACTORS

Of the different factors present in the environment of organisms, temperature, light and oxygen are most important.

Temperature

A definite range of body temperature is essential for the maintenance of optimum metabolic activities in animals. The temperature is maintained either by producing sufficient heat by respiration or getting it from the surroundings.

Homiotherms

Animals capable of maintaining a near constant body temperature, irrespective of the surrounding temperature.

Poikilotherms

Animals unable to maintain a constant body temperature and it varies with that of the environment. Temperature control to a certain degree is present in some poikilotherms.

a. Python coils round the eggs to raise the temperature at least for a short time.
b. Gregarious behaviour in cold weather. Examples: Lackey moth; social insects, honey bee, etc.
c. Fanning with wings to lower the temperature in hive. Example: Honey bee.

In all cold-blooded and many small warm-blooded animals prolonged low temperature reduces metabolism and they pass the period through **aestivation** or **hibernation**. Many invertebrates survive the winter in egg stage.

Temperature vs. Size

Besides influencing structure and behavi-our, temperature influences the size of the animal. This is governed largely by the balance between the heat they lose to the atmosphere and the amount they generate or absorb from the surroundings.

The surface area of an animal increases as the sequence of its linear dimension, while the volume increases as their cube.

a. Small animals present a greater surface area than large ones.
b. Homiotherms tend to attain larger size in cold regions. The reverse is true for poikilotherms.

Light

Light influences organisms in two ways.

a. **Physiological.** Governing some of the vital processes, viz., synthesis of vitamin D (calciferol) in mammals. Vitamin D regulates absorption of calcium and phosphorus. It is found in food only in small quantities and the major amount is synthesized, in which the ultraviolet component of light takes part.
b. **Visual.** Enabling animals to see or be seen. It plays a vital role in promoting recognition, concealment, protection and deceit in animals.
   i. External colours have evolved.
   ii. Nocturnal animals avoid light.
   iii. Diurnals are light-lovers but some prefer shade.

Oxygen

Oxygen is essential for respiration and the consequent ATP release.

a. Terrestrial forms obtain oxygen from atmosphere.
b. Aquatic animals use oxygen dissolved in water.
c. Parasites and animals having no access or adaptation to utlise oxygen directly, depend largely on anaerobic respiration in which glycogen is broken down to CO_2 and fatty acids.

ECOSYSTEM

There are two different types of ecosystems—**terrestrial** and **aquatic**. The terrestrial ecosystem based on predominant types of vegetation includes **grassland**, **forest** and **tundra** ecosystem. The aquatic ecosystem based on major chemical differences covers **fresh water**, **estuarine** and **marine** ecosystems.

Living organisms and their nonliving environment are inseparably interrelated and interact upon each other. Any unit that includes all the organisms in a given area interacting with the physical environment resulting in a flow of energy which leads to clearly defined trophic structure, biotic diversity and the material cycles within the system, is an ecological system or ecosystem.

The material cycle implies the cyclic change of materials between the living and nonliving components of the system. From the trophic angle, two constituents—autotrophic and heterotrophic—comprise an ecosystem. The autotrophic part includes the use of radiant energy in synthesizing complex organic molecules from simple inorganic molecules and thereby storing energy and the heterotrophic part covers rearrangement, degradation and utilization of complex organic molecules and thereby releasing energy for work.

In an ecosystem the energy circuits are two.

a. Transfer of radiant energy fixed by autotrophs to heterotrophs.
b. Release of the fixed energy with the decomposition of the dead bodies of heterotrophs by the decomposers the microorganisms.

COMPONENTS OF ECOSYSTEM

An ecosystem has three components—structure, energy flow and biogeochemical cycles.

I. STRUCTURE

It has two factors—Abiotic and Biotic.

A. Abiotic factors

These include the following components:

a. **Inorganic substances.** Vital elements like C, H, O, N and P and other mineral salts that are found both inside and outside the organism. They participate in material cycle, i.e., exchange of materials between living and nonliving and as such remain in a constant state of flux.
b. **Climate regime.** It comprises physical factors like solar energy, rainfall, humidity, etc.

B. Biotic factors

These include:

a. **Producers.** Only the green plants can manufacture food from inorganic salts through photosynthesis. In deep sea, where light does not reach, the diatoms are replaced by bacteria.
b. **Consumers or Phagotrophs.** Animals cannot manufacture their food, and as they feed upon producers, they are called consumers. Consumers are both **microphagous** and **macrophagus.**

Different categories or tiers of consumers are:

a. **Primary consumer.** Herbivorous animals or herbivores feeding upon producers. Examples: copepods, oysters, periwinkles, etc.
b. **Secondary consumer.** Those living on primary consumers. Examples: herring, hydroid, etc.
c. **Tertiary consumer.** Large carnivores feeding upon secondary consumers. Examples: sharks, carnivore fishes, big cats, etc.
d. **Scavengers.** Living on matters left by secondary and tertiary consumers. Examples: crab, crow, etc.
e. **Detritus feeders.** Organisms on the surface of the bottom of a water body or partially submerged in the mud feed on decomposing organic matters (detritus) raining from above. Examples: Mussels, some gastropods, etc.

C. Decomposers or reducers

The decomposers derive their nutrition from dead organisms and in this process some organic constituents of the dead body are released to the soil. Examples: bacteria and fungi.

Ecosystem of a small water body: a small lake or a pond (Fig. 53.1)

The components of the system are two—abiotic (nonliving) and biotic (living).

The **nonliving components** include water, carbon dioxide, dissolved oxygen, inorganic salts and organic compounds.

The **living components** include producers, consumers and decomposers.

A. **Producers are of two types**

a. **Microscopic phytoplankton (algae).** In small water bodies the phytoplankton are distributed throughout the water as the light penetrate the water to reach the bottom.

b. **Macroscopic or larger plants.** Floating on water or growing in shallow water; border of the water body and adjacent areas are included in this group.

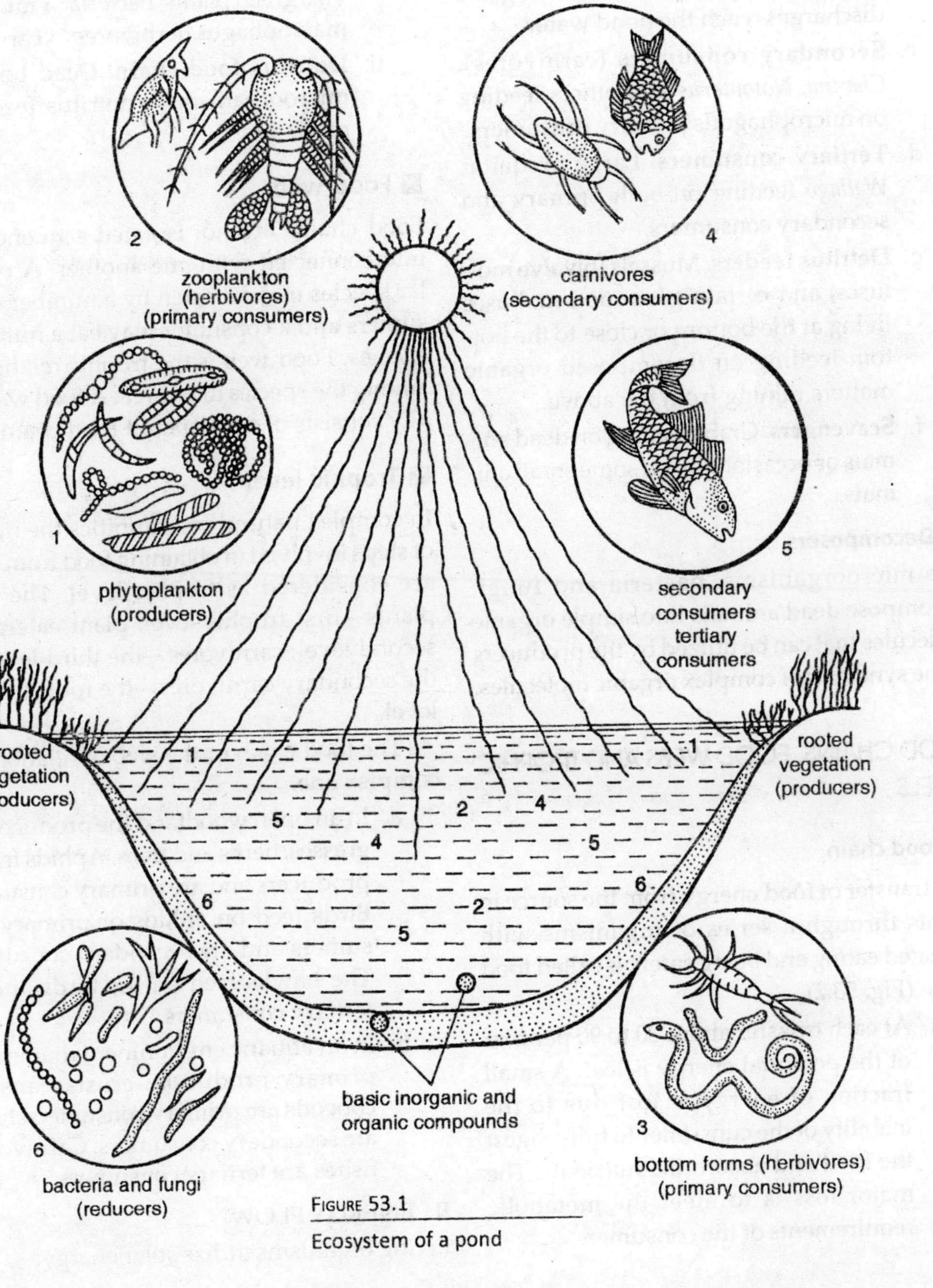

FIGURE 53.1
Ecosystem of a pond

B. Comsumers are primary, secondary and tertiary and scavengers

a. **Primary consumers (Microphagous herbivores).** Crustaceans, insect larvae and fishes feeding on phytoplankton.

b. **Primary consumers (Macrophagous herbivores).** Goat, sheep, cattle feeding upon larger plants and their organic discharges reach the pond water.

c. **Secondary consumers** (carnivores). *Channa, Notopterus* and others feeding on microphagous primary consumers.

d. **Tertiary consumers.** Large predator. *Wallago* feeding on both primary and secondary consumers.

e. **Detritus feeders.** Mussels (bivalve molluscs) and certain gastropod molluscs living at the bottom or close to the bottom feeding on decomposed organic matters raining from the above.

f. **Scavengers.** Crabs feeding on dead animals or occasionally on some small animals.

C. Decomposers

The microorganisms, bacteria and fungi, decompose dead animals into simple organic molecules that can be utlized by the producers in the synthesis of complex organic molecules.

FOOD CHAINS, FOOD WEBS AND TROPHIC LEVELS

⊠ Food chain

The transfer of food energy from the source in plants through a series of organisms with repeated eating and being eaten, is called food chain (Fig. 53.2).

a. At each trnasfer about 80 to 90 per cent of the potential energy is lost. A small fraction of energy is lost due to the inability of the cunsumer to fully digest the food and absorb the nutrients. The major loss is to meet the metabolic requirements of the consumer.

b. To minimise the loss, the number of steps or **links** in a sequence is limited, usually to four or five.

c. The shorter the chain, the greater is the available energy.

Food chains are of two basic types:

i. **Living food chain.** Starting from a base, viz., green plants, bacteria → micro- and macrophagus herbivores → carnivores.

ii. **Detritus food chain.** Dead bodies → microorganisms → detritus feeders → predators.

⊠ Food web

Food chains are not isolated sequences but interconnected with one another. A particular species may be eaten by a number of consumers and a consumer may eat a number of species. Food web is the trophic relationship among the species in an area: A food web usually consists of a number of food chains.

⊠ Trophic level

In complex natural communities the number of steps involved in obtaining food from plants are considered as trophic level. The green plants—first trophic level; plant eaters—the second level; carnivores—the third level and the secondary carnivores—the fourth trophic level.

The food chain or the web of relations is a complex one.

a. In an open woodland the producers are grasses, herbs and trees. Aphids feed on producers and are primary consumers. Birds feed on aphids or primary consumers and are secondary consumers. The hawks feed on the birds and are tertiary consumers.

b. In an aquatic environment diatoms are primary producers, crustaceans like copeods are primary consumers. Fishes are secondary consumers. Carnivorous fishes are tertiary consumers.

II. ENERGY FLOW

Living organisms utilize solar energy.

a. The radiant energy* is transformed into chemical energy.
b. The chemical energy is transformed into mechanical and heat energy in cellular metabolism.
c. The solar flux is a constant and its value is 2 calories per square centimetre per minute (2 cal/cm^2/min).
d. Plants absorb about 45% of the solar energy reaching an ecosystem, a part of which is lost as heat and the rest is used in photosynthesis.

⊠ Energy fixation by autotrophs

This is best expressed by the equation of photosynthesis.

$$6CO_2 + 12H_2O \xrightarrow[\text{Chlorophyll}]{\text{673 kilo cal.}} C_6H_{12}O_6 + 6CO_2 + 6H_2O$$

Gross primary production

The amount of carbohydrate synthesized by plants.

Net primary production

Certain amount of carbohydrate produced is used in respiration of plants and the rest is net primary production.

Gross primary production – loss in respiration = net primary production.

Net community production

A part of the net primary production is consumed by animals. The sum total of the part left is known as net community production.

FIGURE 53.2

Food Chain. Interrelationship of organisms on land

* About one-fifty millionth of the solar energy reaches the outer atmosphere of the earth at a constant rate.

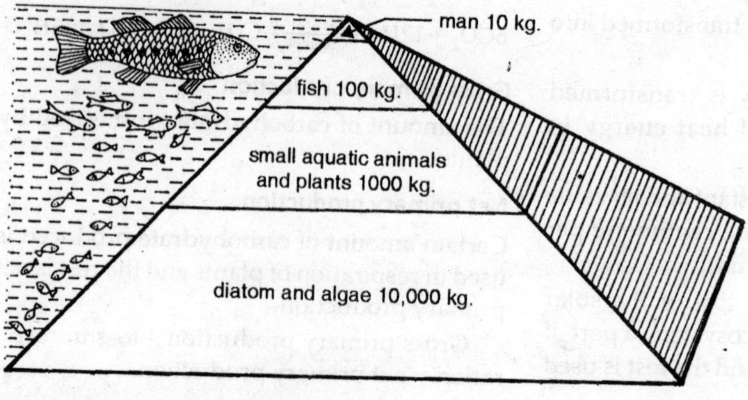

FIGURE 53.3

Biomass pyramid. Relationship between size of the organisms and biomass at different trophic levels

☒ Measurement of primary producers

A good number of techniques, viz., Harvest method, Oxygen and Carbon dioxide measurement, pH method, Radio-active method and, disappearance of raw materials, have been developed to measure primary production, i.e., the amount and rate of energy fixation by autotrophs.

Energy flow beyond producer level

1. Energy flow is unidirectional, i.e., producer to consumer.
2. The loss of energy is about 80 to 90% in each step of energy transfer.

Fate of energy beyond the primary consumers

The herbivores are primary consumers. In a rough estimate the herbivores incorporate about 15 g cal/cm^2/yr of energy. Of this they lose 4.5 g cal/cm/yr or 30% of energy due to metabolic activities. The rest 10.5 g cal/cm^2/yr of energy is available to the carnivores. But the actual amount utilised is only 3 g cal/cm^2/yr which passes to the carnivores.

In the carnivore level again, 60% of the energy is spent in metabolism and the rest 40% becomes part of the nonutilised sediments.

PYRAMID OF BIOMASS

1. The interactions of the food chain phenomena and the size-metabolism relationship results in communities having a definite trophic structure, which is often characteristic of a particular type of ecosystem.
2. Trophic structure and the trophic function may be shown graphically by means of ecological pyramids (Fig. 53.3), in which the first or the producer level forms the base, and the successive levels or the tiers make up the apex.
3. The pyramid of biomass is based on the total dry weight, caloric value, or other measure of the total amount of living materials.
4. In general, the biomass pyramid gives a rough picture of the overall effect of the food chain relationships for the ecological group as a whole.
5. When the total weight of individuals at successive trophic levels is plotted, a gradually sloping pyramid may be expected as long as the size of the organisms dose not differ greatly.

III. BIOGEOCHEMICAL CYCLE OF THE ECOSYSTEM

The elements, including all the essential elements of protoplasm, circulate in the biosphere in characteristic paths from environment to organisms and back to the environment. This is biogeochemical cycle. The elements and dissolved substances essential to life have been termed **biogenic salts** or **nutrients**. These are divided into two groups. **Macronutrients**—carbon, hydrogen, oxygen, nitrogen, potassium, calcium, magnesium, sulphur and phosphorus, and **Micronutrients**—iron, manganese, copper, zinc, boron, sodium, molybdenum, chlorine and cobalt. The movements of nutrients are known as **nutrient cycling**.

Three major **nutrient cycles** have been identified.

Hydrologic cycle

Hydrological cycle (Fig. 53.4) is concerned with the movement of a compound, the water. Water is an agent of geologic change, eroding in one place and depositing in another. It distributes many chemicals including salts and dissolved gases. Water prevents great fluctuation of temperature and helps in metabolic activity of the organisms.

1. The cycle involves an interchange of components between the earth's surface and the atmosphere via precipitation and evaporation.
2. The world precipitation is about 5.20×10^{20}g (geogram) annually. Of this about 1.08×10^{20}g falls on land surfaces, the remainder 4.12×10^{20}g on ocean surfaces. The distribution of particular types of eco systems is dependent on precipitation and evaporation.
3. More water evaporates from ocean than that returns to it via rainfall and, conversely, less water evaporates from land and more returns to it through rainfall.
4. That means a part of the rainfall which supports land ecosytem comes from the water evaporated from the ocean. In Mississippi valley as much as 90% of the rainfall comes from the ocean.
5. The amount of fresh water in ponds, lakes and rivers is about 0.25 geogram*. The annual rainfall on land is 1.08 geogram and the run-off is 0.37 geogram per year. Thus the annual recharge rate of ground water is 1.08 – 0.37 = 0.71 geogram.
6. The circulation pattern of atmosphere determines the pattern of precipitation.

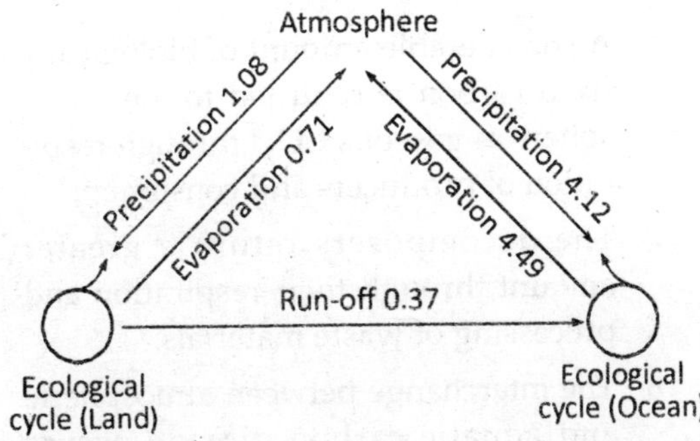

FIGURE 53.4

The central pattern of hydrologic cycle and distribution of water on earth

Gaseous nutrient cycle

The atmosphere constitutes the major reservoir of elements that exist there in gaseous state. Carbon and nitrogen are chief constituents of the gaseous nutrient cycle.

Carbon cycle

The carbon cycle is simplest of all the nurient cycles.

1. The cycling of carbon from atmosphere to producers → consumers → decomposers and back to atmosphere is known as carbon cycle (Fig. 53.5).
2. Ocean is the chief reservoir of carbon. It is stored as bicarbonate and the amount is fifty times greater than that in the air. This carbon regulates the amount of carbon in the atmosphere.
3. Atmosphere is the minor reservoir, but the basic movement occurs through it.
4. The CO_2 concentration in atmosphere ranges from 0.03 to 0.04 per cent.
5. The annual fixation of CO_2 by green plants is about 4 to 9×10^{13} kg.

* One geogram (g) = 10^{20} gram.

6. A considerable amount of biologically fixed carbon is returned to the atmosphere as gaseous CO_2* through respiration of producers and consumers.
7. The decomposers return a greater amount through their respiration and processing of waste materials.
8. The interchange between atmospheric and aquatic carbon dioxide occurs through diffusion, the direction of flow governed by the concentration.
9. The aquatic phase occurs through precipitation.
10. A litre of rain water contains about 0.3 ml CO_2. The dissolved CO_2 combines with the water of the ecosystem and forms carbonic acid, the reaction being always reversible.
11. The carbonic acid in its turn dissociates in a reversible reaction into H^+ and HCO_3^- ions. The bicarbonate ions again dissociate reversibly into H^+ and CO_3^- ions. CO_2 of atmosphere

$$\downarrow \quad \uparrow$$

$$CO_2 \text{ dissolved} + H_2O \rightleftharpoons H_2CO_3 \rightleftharpoons H^+ + HCO_3^-$$

$$HCO_3 \rightleftharpoons H^+ + CO_3^-$$

⊠ Nitrogen Cycle

The movement of nitrogen from atmo-sphere to earth and back to the atmosphere is called nitrogen cycle (Fig. 53.6). It is a complete and perfect cycle.

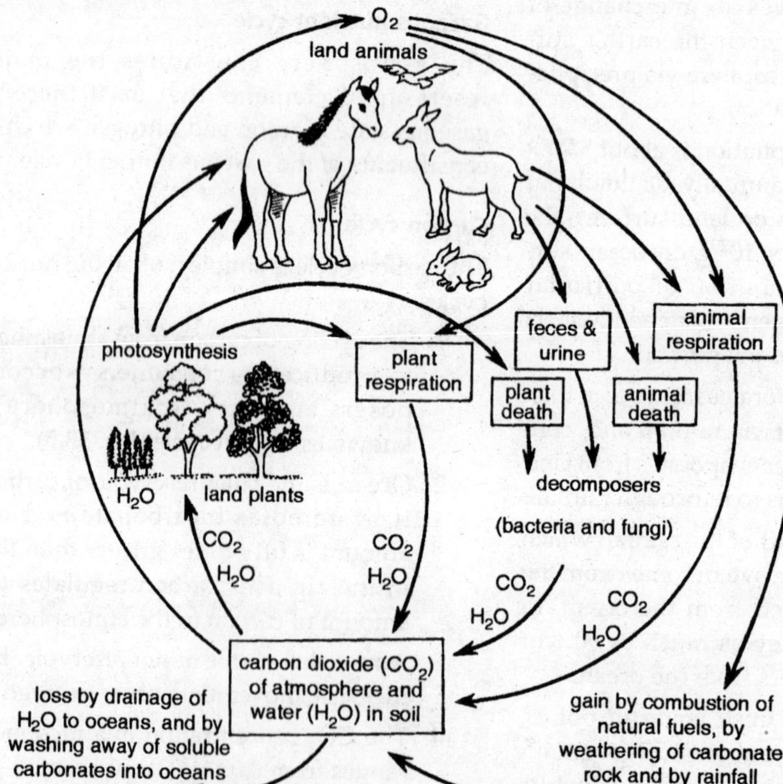

FIGURE 53.5
Carbon cycle

* It has been estimated that 8 billion tons of CO_2 is injected annually into the atmosphere. Of this 6 billion tons come from burning of fuel or accidental fire and 2 billion tons from cultivation of land or agriculture. Most of this quickly passes into the sea and kept there stored in the form of carbonate.

1. The atmosphere contains about 79% nitrogen but most organisms are unable to use this nitrogen.
2. Nitrogen for the use of organisms comes from inorganic substances like ammonia, nitrite and nitrate and also in organic forms like urea, protein and nucleic acids.

☒ Nitrogen fixation

The nitrogen of the air is transformed into nitrates by (A) electrification and photochemical means and (B) nitrogen fixing organisms, chiefly microorganisms.

A. The amount of nitrogen fixed by electrification and photochemical processes is about 35 mg/m^2/yr.

B. The amount of nitrogen fixed by biological means is about 140-700 mg/m^2/yr.

a. **Nitrogen fixing bacteria**

i. Free living. Examples: *Azetobactor* (aerobic), *Clostridium* (anaerobic).

ii. Symbiotic nodule bacteria on leguminous plants. Example: *Rhizobium*.

iii. Purple bacteria. Example: *Rhodospirillum*.

b. **Blue green algae.** Examples: *Anabaena*, *Nostoc*.

c. **Some lichens.** Nitrates and other simpler nitrogen compounds are used by plants for the synthesis of amino acid and protein, which, in turn, reach the animals.

Course of events in nitrogen cycle Ammonification

The protoplasm of dead animals and plants is acted upon by bacteria, actinomycetes and fungi, both on land and in water. While utilising the organic nitrogenrich materials, they convert those to and release the organic

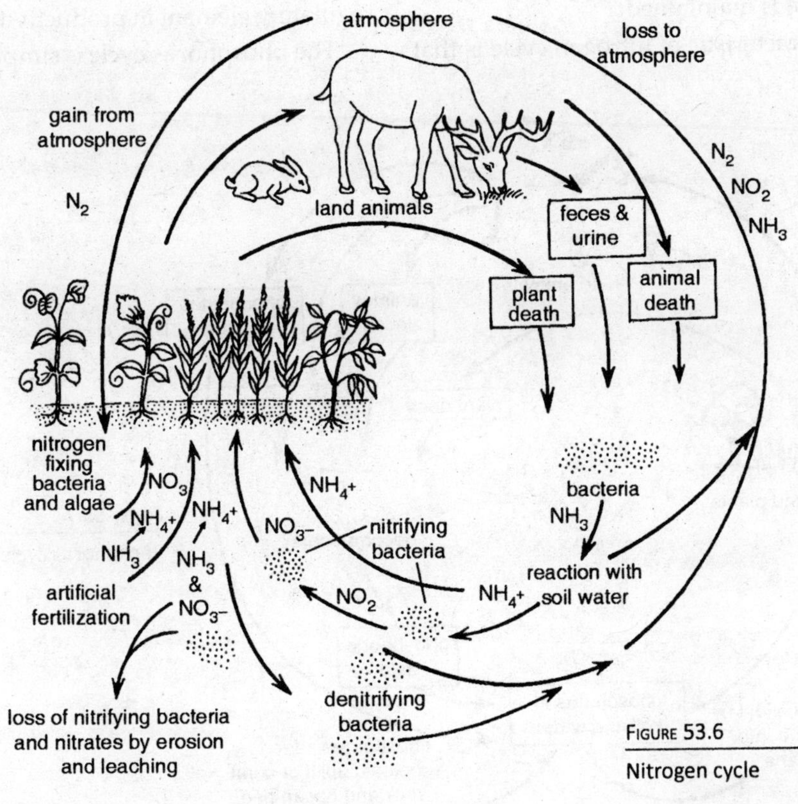

FIGURE 53.6

Nitrogen cycle

form, ammonia. The process is ammonification or mineralization. A good amount of ammonia is also gained from volcanic action.

Nitrification

The process by which ammonia or ammonium salts are converted to nitrates is known as nitrification. It is slow, pH-dependent and occurs in acid condition. The process has two steps—(a) The bacteria *Nitrosomonous* convert ammonia to nitrite (NO_2) and (b) other bacteria, viz., *Nitrobactor* convert nitrite into nitrate (NO_3). Some organisms can use nitrogen in ammonia, but the majority are incapable of doing this. Nitrate is the accessible form to them.

Denitrification

Release of nitrogen from nitrogenous compounds is denitrification. By the action of denitrifying bacteria *Pseudomonas, Bacillus denitrificans, Micrococcus,* which can free the nitrogen, it enters the air as a gas and the nitrogen cycle is maintained.

The characteristic of nitrogen cycle is that the steps in conversion of portein to nitrates provide energy to the organisms which accomplish the breakdown. But the return steps from nitrite to protoplasm require energy from other sources, sunlight or organic matter.

⊠ Sedimentary nutrient cycle

The lithosphere constitutes the major reservoir of the elements—phosphorus and sulphur. Phosphorus and sulphur play important role in the metabolism and growth of organisms. In the pathway of these essential nutrients short or long term stagnation occurs. Sedimentation in the oceans and deep continental lakes are most significant stagnation stages.

Phosphorus cycle

The ratio of phosphorus to other elements in organisms is considerably greater than the ratio of phosphorus in the available and primary sources. Ecologically, phosphorus is very significant as it is the limiting or regulating element in productivity.

The phosphorus cycle is simple (Fig. 52.7).

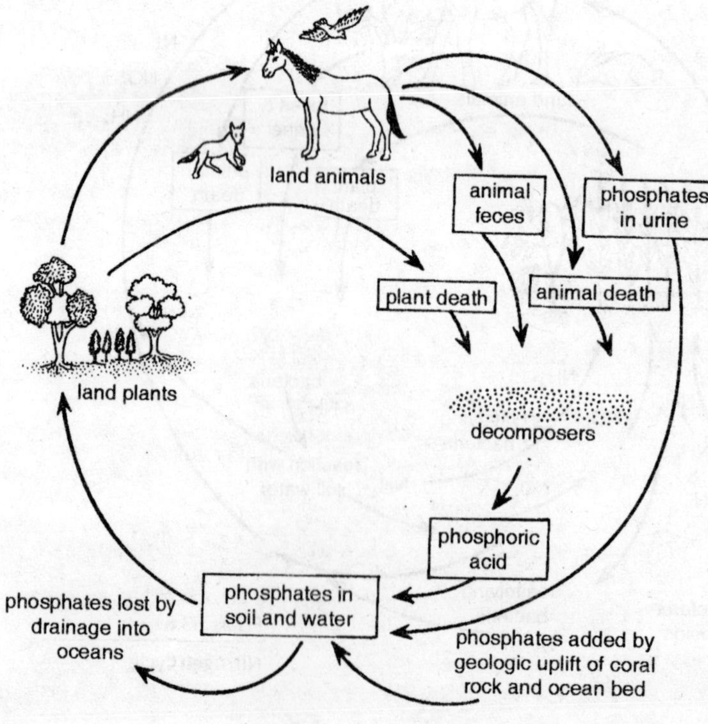

FIGURE 53.7
Phosphorus cycle

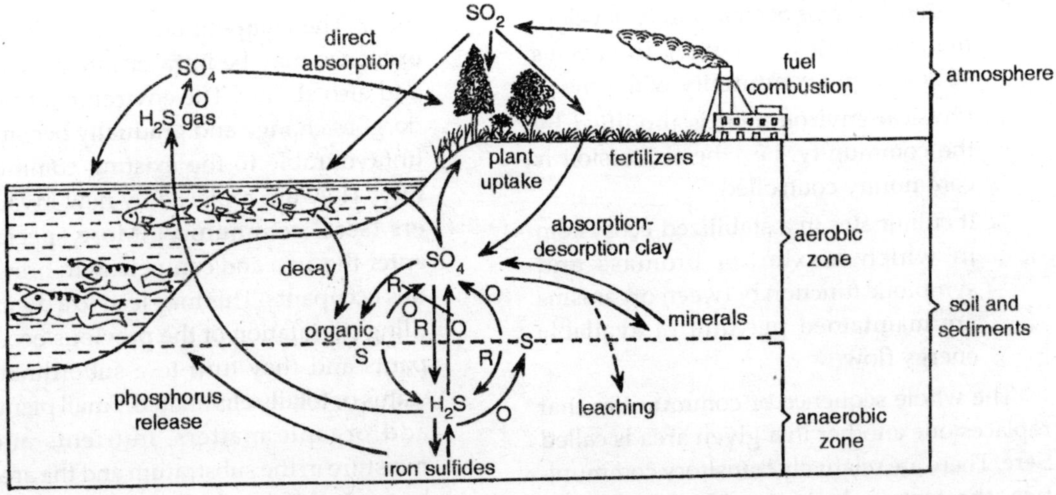

FIGURE 53.8
Sulphur cycle

Release of phosphorus

1. The rocks or other deposits formed in geological ages are the reservoirs of phosphorus.
2. Phosphorus is released to the ecosystem with the erosion of the reservoirs.
3. Much of the released phosphorus reaches sea, where a part of it is deposited in the shallow sediments, and a part is lost in the deep sediments.

Return of phosphorus

The return to the cycle is inadequate.

1. Sea birds bring back phosphorus through their '**guano**' deposits.
2. Marine fish harvested by man is also a good source of phosphorus.

Plants take inorganic phosphate as orthophosphate ions. The phosphorus is transferred to consumers. After death, dissolved phosphate is released with the action of microorganisms, the decomposers. The excreta of animals also return phosphorus to the cycle.

☒ Sulphur Cycle

The reservoirs of sulphur are soil and sedimentary rocks. The atmosphere is a minor reservoir formed from fuel combustion.

1. The sulphur cycle (Fig. 53.8) is regulated by a group of specialised microorganism, each carrying out a particular oxidation or reduction.
2. The sedimentary aspect of the cycle involves precipitation of sulphur in the presence of iron in anaerobic condition.
3. Ferrous sulphate is insoluble in neutral or alkaline water and as a result the sulphur has the potential for being bound up under these conditions to the limit of the amount of iron present.
4. The biologically incorporated sulphur is mineralized by bacteria and fungi in decomposition. Some such sulphur is also directly reduced to sulphides, including hydrogen sulphide, by bacteria.

The sulphur cycle is regulated by the interaction of geochemical, meteorological and biological processes. Inorganic sulphate SO_4 is the source of elemental sulphur in the ecosystems. The sulphate is reduced to elemental sulphur or to hydrogen sulphide by anaerobic bacteria.

ECOLOGICAL SUCCESSION

Ecosystem development or ecological succession involves:

1. Orderly process of community development leading to changes in species structure and community with time.
2. Physical environment is modified by the community, i.e., the succession is community controlled.
3. It culminates in a stabilized ecosystem in which maximum biomass and symbiotic function between organisms are maintained per unit of available energy flow.

The whole sequence of communities that replaces one another in a given area is called **Sere**. There are relatively transitory communities, the terminal; the terminal stabilized system is known as **climax community**.

⊠ Process of succession

The process of succession involves several steps (Fig. 53.9).

1. **Nudation.** It is the basic process of succession. These are new lifeless bare areas caused by both abiotic and biotic factors.
2. **Invasion.** Reproducing organisms (plants and animals) invade the bare area. The invasion occurs usually in three ways.
   a. **Migration.** Organisms, seeds, spores, etc., reach the bare area through air, water or animals.
   b. **Ecesis.** The organisms are established in the new area and propagate. In the primitive unfavourable condition only a few are able to propagate and the majority do not survive.
   c. **Aggregation.** It is the final stage of invasion. The successful immigrants reproduce, their numbers increase and a large population is established. Individuals of different species come close to each other and interact.
3. **Competition and reaction.** With the increase in number of individuals competition for food and space starts among them. The competition or struggle for existence may be intra or interspecific and also abiotic. The environment undergoes change and gradually become unfavourable to the existing community. They are replaced by new invaders (seral community). More species enter the area and compete with previous occupants. This may lead to a dwindling population of the previous occupants and they turn to a subordinate status or totally eliminated. Small plants add organic matters, nutrients and moisture in the substratum and the area become suitable for growth of larger forms. Availability of food increases and various kinds of animals join the community. Interactions between them further modify the environment and more species of plants and animals move there in the process of succession.
4. **Stabilization or climax.** The end product of succession is the **climax community** or **final community**. The climax community does not alter the environment and its members can resist invasion of near species. It is believed that climax community does not change but in reality the community may change, being influenced by various biotic and abiotic factors.

WETLAND ECOSYSTEM

Lands perennially under shallow water or saturated by surface or ground water and in a state to maintain flora and fauna adapted for saturated soil conditions are **wet lands**. Wetlands include areas water logged or saturated with water for the major part of the year. The role of biotic and abiotic factors and their influence both on the organisms living in wetland and their relationship is termed wetland ecosystem. Total wetlands of the world including tropical and boreal regions is 8.36 million square kilometres.

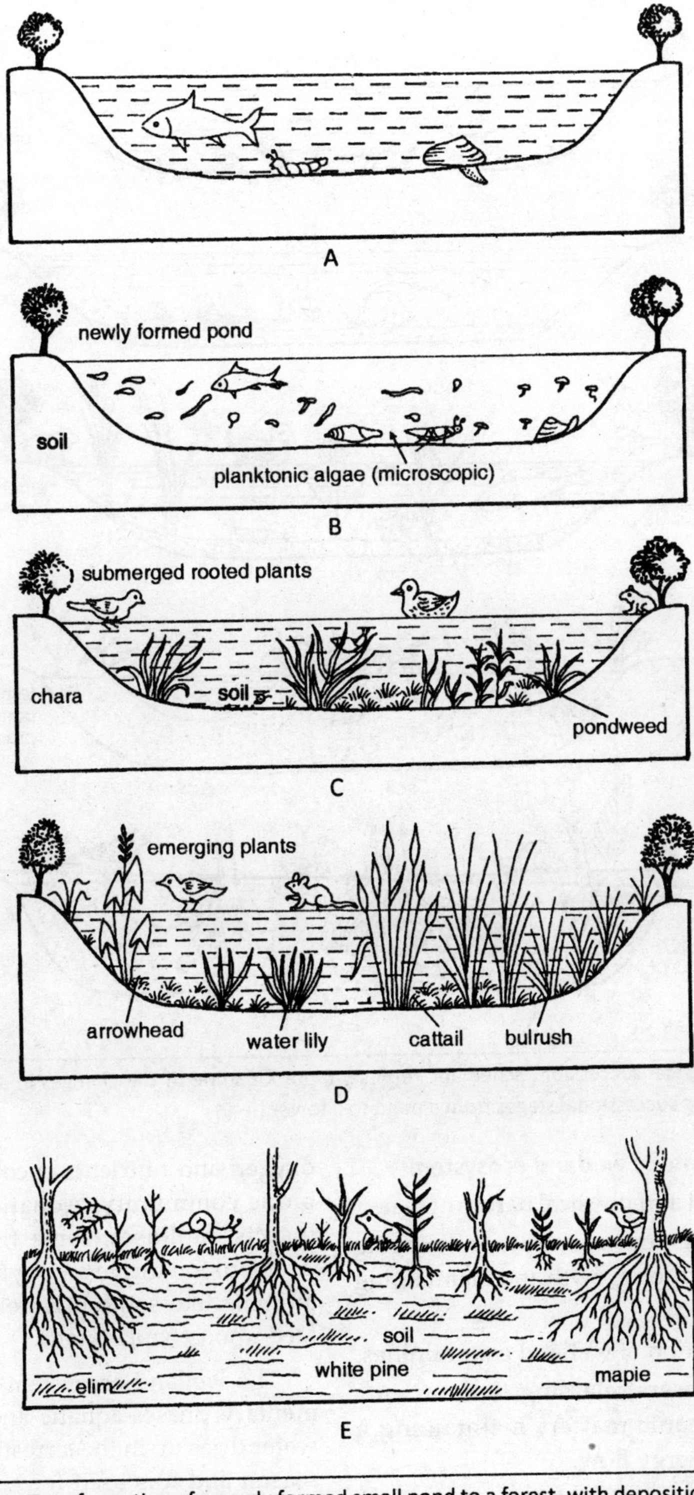

FIGURE 53.9

Ecological succession. Transformation of a newly formed small pond to a forest, with deposition of silt and debris, A. Freshly dug pond, B. Appearance of planktonic algae (microscopic), C. Appearance of submerged rooted plants, D. Appearance of emerging plants, E. Appearance of trees replaced by meadow or forest.

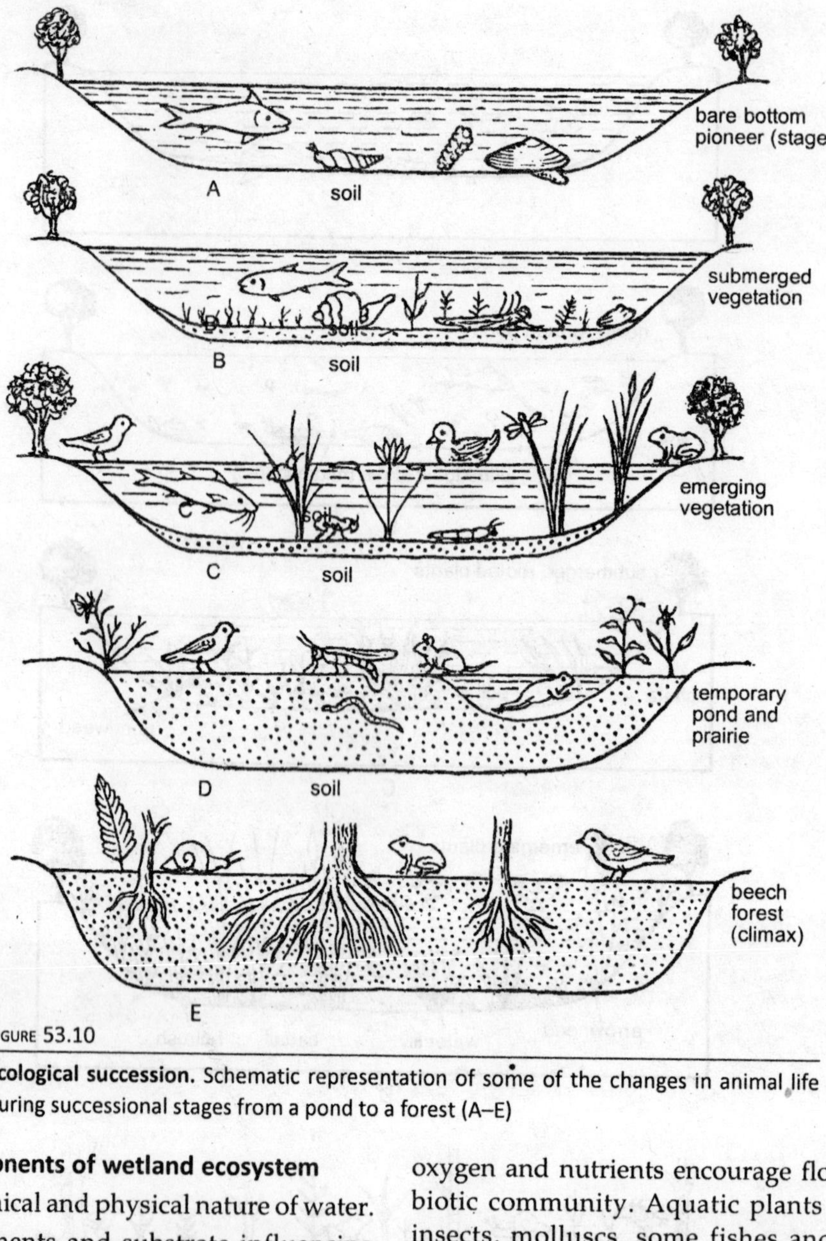

FIGURE 53.10

Ecological succession. Schematic representation of some of the changes in animal life during successional stages from a pond to a forest (A–E)

Basic components of wetland ecosystem

1. Chemical and physical nature of water.
2. Sodiments and substrate influencing the vegetation growth and maintaining vegetation.
3. Composition of wetland communities.
4. Decomposers releasing the nutrients from organic matters maintaining a steady energy flow.

As a rule wetlands exhibit great diversity of flora and fauna. High amounts of dissolved oxygen and nutrients encourage flourishing biotic community. Aquatic plants support insects, molluscs, some fishes and others; zooplankton thrive on phytoplankton, larger animals like some invertebrates and fishes feed upon zooplanktons.

The wetland ecosystem has two complementary phases-aquatic and terrestrial. The water dries up in the terrestrial system in dry season and it is used for grazing of animals and cultivation, which benefit growth of aquatic flora and fauna in wet season.

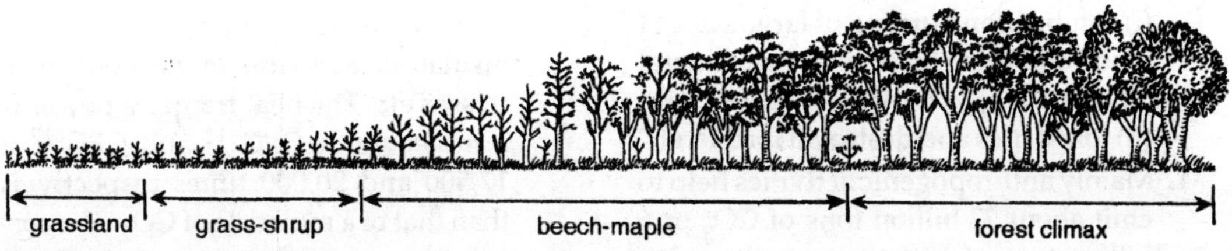

FIGURE 53.11

Ecological succession. An abandoned, plowed field first colonized by annual plants, which are gradually replaced by perrenial plants such as grasses, gradually the field become suitable for shrubs, beach maple sapling then forest (climax)

Legislation for protection of wetlands

The wildlife Protection Act, 1972.

The Environment Protection Act, 1986

The Environment impact Assessment Notification, 1994.

The Tamil Nadu Aquaculture (Regulation) Act.

The West Bengal Fishery Act

The West Bengal Town and Country Planning Act.

Wetland management and sustainable development

Various organic and inorganic compounds remain in a dissolved state in water. Suspended particles are present in rainy season and the period following it. Various kinds of flora and fauna from micro to macro live there. The aquatic environment is a mixture of all these factors.

The nature of basin—shapes, size, depth; surrounding soil, pH, amount of radiant energy received and the biotic community collectively determine the nature of a wetland. Excess or deficit of one or more components disturb the balance. If the disturbance is due to changed natural conditions, the disturbing factors are gradually corrected by nature and the homeostatic condition is restored. Recovary is difficult if the disturbance is man made and this may lead to an ecological disaster.

Importance of wetlands

Wetlands played a major role in human history. Some of the prehistoric cultures developed in glacial lake borders and coastal countries of the world.

1. Welands act as sponges—store rain water, reduce flooding and run off and slowly release the water afterwards.
2. Wetlands recharge ground water. Water seeps downwards to aquifer rock or impervious layer of clay below.
3. Wetland helps in stabilization of banks and shores by the binding effect on vegetation in the area. Acretion of sediments prevents erosion, subsidence and raising of sea level.
4. Deposition of nutrient rich silt helps in better growth of crop plants.
5. Wetlands prevent nitrate build up and lead to eutrophication, which encourage vegetative growth.
6. Wetland microflora biologically degrade organic matters and provide nutrients for plants.

Green house gases

1. Of the green house gases CO_2 is responsible for almost 60%, methane 2.0%, CFC 15%, and NO 5% effects.
2. **Carbon-di-oxide** (CO_2) in air is larger than other green house gases (at present about 0.035%), except moisture. CO_2 plays the most vital role in trapping the heat radiated from the earth. CO_2 does not obstruct visible light but it absorbs infrared heat radiated from the earth's surface.
3. The amount of CO_2 in atmosphere is rapidly increasing mainly due to

(i) combustion burning of large scale of wood and fossil fuels, like coal, mineral oil, natural gas, hydrocarbon felt, and (ii) forest fires and destroyal forests, etc.

4. Mainly anthropogenic activities help to emit about 22 billion tons of CO_2 or 6 billion tons of carbon annually on an average in the atmosphere. Plants and oceans absorb only 50% of this amount and the remaining portion is accumulated in the atmosphere. As a result CO_2 is increasing in the atmosphere at the rate of 0.4% per year. If the rate of increase of atmospheric CO_2 continues at such rate it will be doubled (0.07%) by the middle of the next century. From pre-industrial revolution time to the present time the amount of CO_2 in air has been already increased by about 25-26%.
5. **Methane (CH_4)** is produced in the marshy lands, paddy fields by the anaerobic decomposition of plants bodies, fermentation of urine, faeces of animals and petrification of dead bodies. In the air methane gas is about 1.7 ppm. (parts per million) and the rate of increase on an average 1.1% per year. Methane traps heat 25 times than that of CO_2 molecule wise and its contribution to the green house effect is next to CO_2.
6. **Nitrous oxide** (NO) is produced in nature by burning of biomass, like forest fire, slow oxidation of crop residues, addition of nitrogenous fertilizers to the soil and large scale combustion of fossil fuels (in motor vehicles). This gas increasing at a rate of about 0.3% per year in an average. Its combustion to the green house effect is about 4-5%. A molecule of NO can trap heat 250 times compared to CO_2.
7. **Chlorofluorocarbons** (CFC) is artificially prepared compound used in refrigeration (cold storage, motor cars, small refrigerators etc.) as refrigerants, propellants, for spraying cosmetics, as insulators, as foams in cushions, mattresses etc. The heat trapping power of a molecule of CFC^{-11} and CFC^{-12} is 17,500 and 20,000 times respectively than that of a molecule of CO_2. The contribution of CFCs in the green house effect is about 10-12%.
8. **Ozone** (O_3) in the ambient air (troposphere) is harmful for public health but ozone in layer covering the earth as umbrella in the stratosphere absorbs almost 99% of harmful ultraviolet rays coming from the sun, thus serving the flora and fauna from the ultraviolet rays of the sun.

Ozone layer of atmosphere is stretched over a distance of 10-35 kilometers from the ground. The tropospheric ozone is very must effective as green house gas. The fossil fuel combustion in the motor cars produces oxides of nitrogen (NO) which react with O_2 to produce this gas. The average rate of increase of this gas per year at present in the atmosphere is about 0.5-2.0%.

Scientists now observe that the CFC compound combines with ozone in the stratosphere causing depletion or thinning of ozone layer and in some places over Antarctica already formed a hole. This is called **ozone hole**. In 1992 a survey by the National Aeronautics and Space Administration, USA observed this hole to grow about 23 million sq. km. and is 80% thinner than usual. This hole will allow much greater penetration of dangerous ultraviolet B (UV-B) radiation from the sunrays. This UV rays cause various ill effect on all the living organisms in different ways, including genetic changes.

GLOBAL WARMING

The raising of mean temperature of the lower atmosphere (troposphere and stratosphere) of the planet's surface is known as global warming.

1. The global warming occurs due to green house effect of the increasing green house gases by the human activity.
2. Due to an increase of average global temperature, water from verious sources on earth will evaporate more rapidly. Overall rainfall will increase but rainfall may be uneven, some parts will get heavy rainfall and other parts will face drought. In tropical zone summer may be prolonged, temperature of polar region in winter will be higher.
3. The ice-caps and glaciers of the polar region will melt and the sea level will rise. In the last 100 years the sea level has risen by about 0.15 metre. If the global temperature increases by 3-5°C on an overage the sea level will rise 0.5 to 1.5 metre. As a result, a vast populous low lands of coastal areas will be flooded, fertile agricultural lands will be submerged with salt waters.

☒ Green house and green house gases

1. In cold countries a room made up of glass is used for saving the plants from severe cold is known as **green house.** The green house can retain heat within it which helps plant to survive. The sun rays entering the green house causing heating effect within green house and the heat can not come out of the green house through the glass. So heating effect is retained.
2. The sun rays with different visible range of spectrum (light) and electromagnetic spectrum (heat) enter the atmosphere. The heat is absorbed by the earth's surface and radiated back at a longer wave length than the incoming radiation. The long wave length radiation is incapable of travelling back through the atmosphere, as is partly absorbed by some natural gases like CO_2 Ozone (O_3), water vapour, chlorofluorocarbon (CFC), methane, Nitrogen oxide (NO) etc. As a result the atmospheric temperature is raised and thus global warming is caused. The gases responsible for raising temperature are known as green house gases and the effect of warming is the green house effect.

BIODIVERSITY AND CONSERVATION

1. Biodiversity or Biological diversity is the total variety of life forms on earth's surface. It refers to the sum total of varieties of species of microbes, plants and animals found in the biosphere and associated with ecosystem. It may be (i) diversity of species within community, (ii) diversity of biotic communities within an ecosystem and (iii) genetic diversity.
2. The biotic community or species diversity is the variability of population with respect to the reproductive distinctness of each of the individuals. It is the taxonomic unit of biological world.
3. The ecosystem diversity is the variety of habitats, biotic communities and ecological processes in the biosphere as well as diversity within ecosystem.
4. The genetic diversity is the total genetic information contained in the genes of individuals of flora, fauna and microrganisms. Each species is the repository of an immense amount of genetic information.
5. The population diversity originate due to constantly cropping up of new genotypes in a population through genetic phenomena like mutation, recombination etc. and through immi-gration of individuals. Diversity in population is eliminated by natural selection acting through differential reproduction and mortality and lost through emigration of individuals.

☒ Community and Ecosystem Diversity

Diversity at the levels of community and ecosystem has three perspectives—alpha, beta and gamma diversity.

1. **Alpha diversity** (within community diversity) refers to the diversity of organisms sharing the same community/ habitat.
2. The rate of replacement of species along a gradient habitats or communities is called **beta diversity** (between community diversity) Higher the heterogeneity in the habitats in a region or greater the dissimilarity between communities.
3. Diversity of the habitats over the total landscape or geographical area is called **gamma diversity**.
4. Ecosystem diversity describes the number of niches, trophic levels and various ecological processes that sustain energy flow, food webs and the recycling of nutrients.
5. Among the biogeographical regions, Deccan Peninsula has the most extensive coverage of the Indian landmass (about 42%).

   The most biodiversity-rich zones, western ghats and north-east, account for 4 and 5.2% of the geographical area, respectively.
6. About 33% of the flowering plants recorded in India are endemic to the country.
7. Indian region is also notable for endemic fauna. For example, out of the recorded vertebrates, 60% amphibians, 53% freshwater fish, 36% reptiles and 10% mammalian fauna are endemic.

☒ Gradients of Biodiversity

Change in **latitude** and **altitude** causes variation in biodiversity. As we move from high to low latitudes (i.e., from the poles to the equator), the biodiversity increases.

In temperate region the climate is severe.

☒ Importance of Biodiversity

1. Biodiversity is of immense value to human life. All foods ultimately come from plants, supply energy to sustain human life and all other life forms on earth.
2. Biological diversity helps for the preservation of ecosystem. All species in a particular ecosystem are interde-pendant. Loss of a particular species will break the food chain and food web and the life supporting system of other dependent species will be disturbed.
3. 25% of drugs in pharmacy are derived from 120 species of plants but throughout the globe traditional plants used by the natives are around 25000 species. Even the medicines prepared now synthetically were originally discovered in nature mostly from different plants, fungi, bacteria and animals.
4. Biodiversity is a pool of variable genetic resource. The genetic organization of traditional and wild varieties of organisms are time tested evolution of species for millions of years and remain as the collection of traits and gene-pool for future. The reduction of biological diversity results in adverse effect upon future breeding and improvement of species.
5. For tourism purpose for visiting, viewing and learning about natural communities, ecological integrity value, maintaining all functionally critical aspects of an ecosystem, ethical value, moral obligation the sustenance of biodiverity is very much needed.
6. The species diversity is essential for supporting and proper functioning of communities and for emergence of community level properties.

India is rich in biodiversity due to its diverse physiographic and climatic condition. More then 45000 plant species are known from India which accounts 7% of the flora of the world. Of these 15000 are flowering plants, 64 gymnosperms, 2843 bryophytes, 1012 pterydophytes, 1940 lichens, 12480 algae, and 23000 fungi. The animal species are estimated to be around 81000. This accounts for 6.4% of the world fauna. More than 5000 molluscs. 57000 insects, 2546 fishes, 204 amphibian, 428 reptiles, 1228 birds, and 372 mammal species are known from India.

Considering such importance of biological richness for our life supporting system, the United Nation's Conference on Environment and Development (UNCED) Earth summit held at Rio de Janeiro, Brazil in June 1992, and discussed about bio-diversity and took resolution on

i. the intrinsic value of biodiversity.
ii. the sovereign rights of states over their biological resources.
iii. the fundamental requirements of *in situ* conservation of ecosystems and natural habitats,
iv. the supporting role of *ex situ* conservation,
v. the vital role of local communities and women in the conservation and sustainable use of biodiversity,
vi. desirability of shaving equitably the benefits arising from the use of traditional knowledge, skills, innova-tions and practice,
vii. importance and need to promote regional and global cooperation for conservation, and
viii. requirement of substantial investments to conserve biological diversity.

Loss of biodiversity

1. The biodiversity of earth's surface has been continuously facing mass destruction. Many of the animal species have already been lost, out of which mammals include 60, reptiles 30, and birds 75. Now about 210 reptiles, 1100 birds and 650 mammals are threatened.
2. Destruction of rain forests leading to degradation of environment and habitat of many species, the industrial production and release of wastes also act as pollutants and destroys habitat of many species. Such habitat lost for various reasons are the prime cause for extinction of species.
3. Between 1975 to 2000 one million species become extinct. The major causes of extinction of species are—

i. habitat destruction.
ii. hunting of wild life.
iii. introduction of new species in the habitat.
iv. international trades of biotic resources.
v. pollution of habitats, and
vi. natural catastrophy.

Conservation of Biodiversity

Conservation is the management and maintenance of the biodiversity in its proper natural habitats called *in situ* conservation for the sustainable development. Conservation in zoo gardens, parks, and some localised spaces outside the habitats for experimental or tourism purposes are called *ex situ* conservation.

***In situ* Conservation Strategies**

In situ strategies of conservation emphasise protection of total ecosystem. It includes protection of a group of typical ecosystems through a network of protected areas.

Protected Areas

These are areas of land and / or sea, especially dedicated to the protection and maintenance of biological diversity and of natural and associated cultural resources.

Examples of protected areas are National Parks and Wildlife Sanctuaries.

World Conservation Monitoring Centre has recognized 37000 protected areas around the world.

As of September 2002, India has 581 protected areas. (89 National Parks and 492 Wildlife Sanctuaries), covering 4.7% of the land surface, as against to 10% international suggested norm.

Jim Corbett National Park was the first National Park established in India.

Benefits of Protected Areas

- Maintains viable populations of all native species and subspecies.
- Makes it possible for species/habitats to shift in response to environmental changes.
- Maintains the number and distribution of communities and habitats, and conserves the genetic diversity of all the present species.
- Prevents human caused introductions of alien species.

Biosphere Reserves

Biosphere reserves are a special category of protected areas of land and/or coastal environments, wherein people are an integral component of the system.

Biosphere reserves are representative examples of natural biomes and contain unique biological communities.

The concept of biosphere reserves was launched in 1975 as a part of UNESCO's Man and Biosphere Programme, dealing with the conservation of ecosystems and the genetic resources contained therein.

Till 27th May 2002, there were 408 biosphere reserves located in 94 countries.

There are 13 biosphere reserves in India. In India, biosphere reserves are also notified as National Parks.

A biosphere reserve consists of core, buffer and transition zones.

Natural or **core zone** comprises an undisturbed and legally protected ecosystem.

Buffer zone surrounds the core area and is managed to accommodate a greater variety of resource use strategies and research and educational activities.

Transition zone, the outermost part of the biosphere reserve, is an area of active, cooperation between biosphere reserve management and the local people, wherein activities like settlements, cropping, forestry, recreation and other economic uses continue in harmony with conservation goals.

Chief functions of Biosphere Reserves

Conservation. Ensure the conservation of landscapes, ecosystems, species and genetic resources. It also encourages traditional resource use.

Development. Promote economic development which is culturally, socially and ecologically sustainable.

Scientific research, monitoring and education. Provide support for research, monitoring, education and information exchange related to local, national and global issues of conservation.

Ex situ Conservation Strategies

- The *ex situ* conservation strategies include botanical gardens, zoos, conservation stands and gene, pollen, seed, seedling tissue culture and DNA banks.
- Conservation of biological diversity in botanical gardens is already in practice. There are more than 1500 botanical gardens and arboreta (botanical gardens where specific tree and shrub species are cultivated) in the world containing more than 80000 species.

  Similarly, there are more than 800 professionally managed zoos around the world with about 3000 species of mammals, reptiles, birds and amphibians.

  Plants and animals conserved in botanical gardens, arboreta, zoos and aquaria can be used to restore degraded lands, reintroduce species into wild and restock depleted populations.

☒ Hot Spots of Biodiversity

1. Biodiversity is not uniformly distributed across the geographical regions of the earth. Certain regions of the world are mega diversity zones where a very large number of species are found. For example, India accounts for only 2.4% of the land area of the world, but it contributes about 8% species to the global diversity.
2. In 1988, **Norman Myers** developed **the hot spots concept** to designate priority areas for *in situ* conservation.

   The hot spots are the richest and the most threatened reservoirs of plant and animal life on the earth.
3. 25 terrestrial hot spots for conservation of biodiversity have been identified worldwide. Among the 25 hot spots of the world, two (Western Ghats and Eastern Ghats) are found in India and these extend into the neighbouring countries also. These areas are rich in flowering plants, also in reptiles, amphibians, swallow-tailed butterflies and some mammals; and also show a high degree of endemism.
4. The Eastern Himalayas hot spot extends to the north east India and Bhutan. The temperate forests are found at altitudes of 1780 to 3500 metres. Many deep and semi-isolated valleys found in this region are exceptionally rich in endemic plant species.
5. Tropical forests appear in 15 hot spots, Mediterramean-type zone in 5 and 9 hot spots are mainly or completely made up of islands. As many as 16 hot spots are in the tropics.

   About 20% of the human population lives in the hot spot regions.
6. The key criteria for determining a hot spot are—(i) Number of endemic species, *i.e.*, the species which are found nowhere else and (ii) degree of threat, which is measured in terms of habitat loss.

54 WILD LIFE: ACT AND CONSERVATION

Wild animals refer to all undomesticated animals living in their natural habitats or in the habitats to which they are subsequently adapted.

The earth is the only known planet where life not only exists but has diversified itself in myriad forms. The space on earth and the life-sustaining materials available are limited. These two and a number of factors collectively led to the evolution of an extensively rich flora and fauna on the earth. Their relationship is of a very delicate interdependence and is finely balanced. Any interference is bound to snap the invisible chain binding them together and upset the balance of nature.

Man, the cruellest animal, is destroying nature's balance ruthlessly, inspite of paying heavy penalties for his misdeeds. With the growth of population at an alarming rate man is bound to destroy global environment encouraging life, inviting his own destruction, if drastic measures are not taken without further delay.

Fortunately, realisation that exploitation of natural resources cannot continue for a long time and the wanton destruction of fabrics of life would lead to the annihilation of mankind, have compelled us to device strategies, at least to minimise the damage; a complete stop to it is still a far cry.

The importance of wild life and its conservation were clearly understood in ancient India. This was reflected in the protection of forests and wild animals by Royal Charters. The '**Arthashastra**' by Kautilya (300 B.C.) and the fifth pillar Edict of Emperor Asoke (242 B.C.) are evidences of the noble attempts. Since then nothin was done for about 2,000 years. For the conservation of wild life in the national level (Indian subcontinent) the British Government convened a meeting in 1935 to make a blue print for wild life conservation. Only in 1952, the Indian Board of Wild Life was formed. Still later, State Wild Life Boards were constituted in the international level. The World Wild Life Fund (WWF) was established in 1961 with its headquarter in Switzerland to conserve wild life throughout the world. The principal functions of WWF are to help and advice.

Since then quite a number of **Sanctuaries** and **National Parks** were established in many countries. In India the '**game sanctuaries**' established by the British Government for the propagation of games and '**game preserves**', for shooting games by princely states before independence of India, were restored as National Parks and Sanctuaries in 1947.

Following this there has been manifold increase in the number of National Parks and

Sanctuaries. By 1989, 65 National Pakrs and 386 Sanctuaries were established in India (Fig. 49.1) under the Wild Life Protection Act 1972. The total area covered by the Act is about 90,000 square kilometres, which is nearly three per cent of the total land area of India.

NATIONAL PARK

It is a well-defined area meant for protection of wild life, conserve natural and historical objects and other features, for all times. It is under statute by the Government of that country and subject to modifications as required by local conditions for its maintenance.

SANCTUARY

The word sanctuary is equivalent to **reserve.** It is an area constituted by competent authority in which killing, hunting, shooting or capture of any species of birds, mammals, reptiles or any other form of wild life are prohibited by law except by or under the control of the highest authority in the department responsible for its management. Saving of adequate areas of natural habitats of wild life provide 'living space' for wild animals. A sanctuary is a **living space** of that type.

In India, administration of National Parks and Sanctuaries vests with the State governments. By the 42nd Amendment of Indian Constitution in 1976, the preservation and protection of wild life and forests have been placed in the Concurrent List. This authorises the central government to control the preservation and protection of National Parks and Sanctuaries and also to acquire the areas, if necessary.

CONCEPT OF THREATENED SPECIES

The rare species of flora and fauna are recognised and classified by the International Union of Conservation of Nature and Natural Resources (IUCN) for conservation purpose on the basis of (a) the present and past distribution of the species or taxa, (b) abundance and quality of natural habitat, (c) decline of population in course of time, and (d) the biology and ecological value of species.

1. **Endangered species.** These are the species or taxa whose numbers are reduced to a critical level or the habitats are drastically reduced in such a way that they are in instant danger of extinction.
2. **Vulnerable species.** These are the species or taxa whose population are still abundant but are under threat through out their range. These species are likely to move into the endangered category in the near future, if the causal adverse factors continue to operate.
3. **Rare species.** The species with very small population in the world, usually localised within restricted geographical areas or habitat or are thinely scattered over a more extensive range, not endangered or vulnerable at present, but are at risk.

UTILITY OF WILD LIFE STUDY

☒ Biological Importance

The wild life is an integral part of nature and helps in maintaining balance in an ecosystem.

a. In an ecosystem the biotic factors are related to each other through food chain.
b. Killing of too many predators like tiger, leopard, etc. helps in the rise in number of primary consumers like deer, antelope, pig, etc. This, in turn, may pose as serious crop pests in areas adjacent to forests.
c. Conversely, depletion of preys induces the carnivores to turn man-eaters or killer of domestic animals.

Cultural and aesthetic importance

Mythology, art and literature of all nations are saturated with the mention of wild life forms.

a. Creative art of man would have been dull had there been no reference to wild life.

b. Beauties of nature help our very existence, and wild life is an important part of nature's beauty.

Economic importance

Wild life is an important resource. Preserved, encouraged and judiciously exploited, it would yield good dividends for ever.

Preservation of genome

The genome of a species differs from all other genomes. It does not appear for a **second time in nature** nor can it be **recreated** by man. Extinction of a species means loss of a precious global wealth for ever.

SOME IMPORTANT SANCTUARIES AND NATIONAL PARKS OF INDIA

Jaldapara

The sanctuary was established in 1941 to protect the Indian Rhino. It is located in Jalpaiguri District, West Bengal. The total area is 115.00 sq. km. It is a riverine forest criss-crossed with water courses and dense tall grasses. The wild life are rhesus monkey, tiger, leopard, jungle cat, sloth bear, elephant, one-horned rhinoceros, sambar, barking deer, hog deer, gaur.

Jim Corbet

It is the first National Park in India established in 1935 and named Hailey National Park. In 1957 it was renamed Jim Corbett National Park after the famous hunter and wild life lover, Jim Corbett. It is located at the foothill of Himalayas in Garhwal District, Uttar Prodesh. The area is 525 sq. km. The vegetation is a luxurious growth of grass dotted with trees like Samul, Dhan and Sal. It is a home for tigers. The other wild life are bear, leopard, jungle cat, hyaena, sambar, chital, barking deer, hog deer, nilgai, peafowl, partridges, mugger, gharial, python.

Bandipur

It was established in 1941 as a game preserve of the Maharaja of Mysore. The National Park is located in Mysore District, Karnataka. The area is 874.20 sq. km. The rainfall is poor; forest dry and deciduous and blaze into colour in spring. It is famous for gaur. The other wild life are wild dog, sloth bear, jungle cat, tiger, leopard, hyaena, palm civet, elephant, sambar, chital, Malabar squirrel, jungle fowl, partridges.

Gir

The National Park with an area of 140.00 sq. km. is located in Junagadh District, Gujarat. It was started as a reserve forest and later upgraded to National Park. It is a dry, deciduous forest of teak, with patches of acacia bushes. It is the last hold of the famous Asiatic lion *Panthera leo persica*. The other wild life are hyaena, jungle cat, leopard, sambar, chital, chinkara, four-horned antelope mugger.

Sunderbans

The National Park* is a vast unbroken swamp formed by the great Ganga Brahmaputra river system in the South 24 Parganas District, West Bengal. The area covers 3,978.00 sq. km. Vegetation—dense mangrove and evergreen plants. It is famous for Royal Bengal Tiger. Certain area of Sunderbans have been set aside as Sanctuaries.

The Sunderbans National Park has two distinct zones.

A. **Wilderness zone**

1,330 sq. km. covering the forest areas of Mayadwip, Chotahardi, Gosaba, Gona, Matla, Chempta and Bagmara.

B. **Buffer zone**

The remaining area in the north and east of wilderness zone.

1. **Tiger Sanctuary**

2,585.00 sq. km. comprising 119 small islands for conservation of Royal Bengal

* Formerly designated as Reserve forest.

NATIONAL PARKS AND SANCTUARIES IN INDIA

(As on March 1, 1989)*

Name of State/ Union Territory	No. of National Parks	No. of Sanctuaries	Total
Andaman & Nicobar Islands	6	94	100
Andhra Pradesh	0	16	16
Arunachal Pradesh	2	4	6
Assam	2	7	9
Bihar	1	15	16
Chandigarh	0	1	1
Delhi	0	1	1
Goa	1	3	4
Gujarat	4	12	16
Haryana	0	7	7
Himachal Pradesh	2	29	31
Jammu & Kashmir	4	13	17
Karnataka	5	18	23
Kerala	3	11	14
Madhya Pradesh	11	31	42
Maharashtra	4	25	29
Manipur	2	0	2
Meghalaya	2	3	5
Mizoram	0	1	1
Nagaland	0	3	3
Orissa	1	14	15
Punjab	0	5	5
Rajasthan	4	18	22
Sikkim	1	3	4
Tamil Nadu	2	10	12
Tripura	0	4	4
Uttar Pradesh	5	16	21
West Bengal	3	17	20
Total	**65**	**386**	**451**

* 7 Sanctuaries—1 in each of Goa, Jammu and Kashmir, Nagaland, Sikkim and 3 in Rajasthan in the list prepared by the Ministry of Environment and Forests, Government of India, not enlisted by respective State Governments, till March 1, 1989.

Tiger. The sanctuary includes the entire Basirhat range and has been divided into 15 blocks. This is one of the nine san-ctuaries founded in collaboration with World Wild Life Fund. This area offers brightest promise for tiger conservation. Other wild life are rhesus macaque, jackal, Indian fox, jungle cat, leopard cat, fishing cat, chital, wild boar, herons, egrets, pelican, esturine crocodiles, python, Russel's viper, king cobra, common cobra, banded krait.

2. **Bird Sanctuary : Sajnakhali**

   The water bird sanctuary covers 362.33.00 sq. km. The nesting birds are spot-billed pelican, little cormorant, darter, white ibis, black-necked stork, cattle egret, little green bittern. Other wild life are tiger, chital.

3. **Halliday island**

   5.96 sq. km. Wild life are tiger, chital, wild pig, water birds.

4. **Lothian island**

   37.96 sq. km. Wild life are tiger, otter, chital, wild boar, Gangetic dolphin, estuarine crocodile, water birds.

WAYS AND MEANS OF WILD LIFE CONSERVATION

The World Wild Life Fund (WWF) was established in 1961 with its headquarter in Switzerland to conserve wild life throughout the world. The principal functions of WWF are to help and advice.

a. Collection of fund.
b. Education and training in wild life conservation.
c. To finance projects for conservation of near-extinct wild life forms and to send experts, if necessary.
d. To arrange seminars and conferences to impress upon the general public the importance of wild life.
e. To train guards for reserve forests or sanctuaries, and to provide them with transport like land rover and gadgets like walkietalkies to prevent operation of poachers.

Project Tiger

The project tiger is launched as per the recommandations of a special task force of the Indian Board of Wild life. Its main objectives are—(1) to ensure maintainance of available population of tigers in India for scientific, economic, esthetic, cultural and ecological values; (2) to preserve, for all times the areas of such biological importance as a national heritage for the benefit, education and enjoyment of the people. In India at present 27 Tiger Reserves spreading over 14 states and covering about 37,761 sq. km. areas are present.

In that protected areas a number of traditional and nontraditional management practices are undertaken for maintaining adequate wildlife protection, where threatend wild life population not only increased but also help to get back for consumptive uses (Wild life cropping).

METHODS OF WILD LIFE CONSERVATION

The following are proved, fruitful methods in wild life conservation.

1. Eastablishment of (a) National parks, (b) Wild life sanctuaries, (c) Protected areas, (d) Reserved forests.
2. Promulgation of laws.
3. Mass education on the importance of wild life.

PRINCIPAL MAMMALIAN TYPES TO BE PRESERVED

Following are the **endangered** mammalian fauna of India.

1. **Blue whale** (*Balaenoptera musculus*)
2. **Dugong** (*Dugong dugon*)

3. **Gangetic dolphin** (*Platanista gangetica*)
4. **Humpback whale** (*Megaptera nevacangliae*)
5. **Caracal** (*Felis caracal*)
6. **Clouded leopard** (*Neofelis nebulosa*)
7. **Golden cat** (*Felis temmincki*)
8. **Indian lion** (*Panthera leo persica*)
9. **Indian wolf** (*Canis lupes*)
10. **Himalayan brown bear** (*Ursus arctos*)
11. **Leopard** (*Panthera pardus*)
12. **Leopard cat** (*Felis bengalensis*)
13. **Red Panda** (*Ailursus fulgens*)
14. **Lynx** (*Felis lynx*)
15. **Malay or Sun bear** (*Belarctos malayanus*)
16. **Marbled cat** (*Felis mormorata*)
17. **Otter** (*Lutra lutra*)
18. **Rust spotted cat** (*Felis rubiginosa*)
19. **Snow leopard** (*Panthera uncia*)
20. **Tiger** (*Panthera tigris*)
21. **Wild dog** (*Quon alpinus*)
22. **Pallasa's cat** (*Felis manul*)
23. **Gaur** (*Bos gaurus*)
24. **Thamir** (*Cervus ebai*)
25. **Indian wild ass** (*Equus hemionus*)
26. **Kashmir** stag (*Cervus elaphus hanglu*)
27. **Markohor** (*Capra folconeri*)
28. **Musk deer** (*Maschus moschiferus*)
29. **Nyan** (*Ovis ammon hodgsoni*)
30. **Pigmy hog** (*Sus salvanius*)
31. **Rhinoceros** (*Rhinoceros unicornis*)
32. **Serow** (*Capricornis sumatraonsis*)
33. **Swamp deer** (*Cervus duvancali*)
34. **Wild yak** (*Bos grunniens*)
35. **Crab-eating macaque** (*Macaca irus umbnosa*)
36. **Golden langur** (*Presbytis geei*)
37. **Hoolock** (*Hylobates hoolock*)
38. **Nilgiri langur** (*Presbytis johnii*)
39. **Hanuman langur** (*Presbytis entellus*)
40. **Chinese pangolin** (*Manis pentadactyla*)
41. **Indian Pangolin** (*Manis crassicaudata*)
42. **Giant squirrel** (*Ratiya macroura*)
43. **Hispid hare** (*Caprolagus hispidus*)
44. **Indian elephant** (*Elephas maximus*)
45. **Pig-tailed macaque** (*Macaca nemestrina*)
46. **Stump-tailed macaque** (*Macaca arfoides*)
47. **Urial** (*Ovis vignei*)

55 ENVIRONMENT POLLUTION

POLLUTION

Incorporation of any substance, harmful to living forms, in the medium where the organisms live, is termed pollution.

The substances causing pollution are **pollutants.** All pollutants may be placed under two categories. **Chemical pollutants**—affect the physiological and biochemical processes of plants and animals and **physical pollutants**—causing physical damages, which may finally interfere with physiological processes.

Every organism can withstand a certain degree of pollution, with apparently no damage. The level is termed **tolerance level.** Organisms survive in pullated environment to some extent beyond tolerance limit but their normal body functioning is impaired. Beyond this it is **lethal.**

Pollution is largely man-made. It occurs in nature only in a very limited scale and that too restricted to localized areas. Arsenic and fluoride poisoning in subsoil water; pollution due to mass mortality of living forms in a catastrophe, fall under this group.

Global pollution of today is the result of chiefly two factors—**human population explosion** and tremendous **growth of industry** and **agriculture** to meet the need of mankind.

All forms of pollution have been placed broadly under three categories—water pollution, air pollution and sound pollution.

WATER POLLUTION

Incorporation of pollutants in natural water causes water pollution.

Sources

Chief sources of water pollution are industrial wastes, chemical fertilizers and toxic substances used in agriculture; sewage in towns and cities, and waste products of man and animals and their decomposing carcasses.

Industrial wastes

Discharge of affluent from industries usually contain a number of hazardous sub-stance, viz., acid, alkali, cyanide, detergent, pigments, starch, tanin, hydrocarbon, ammonia, phenol, heavy metals, suspended particles, oil droplets, etc.

Management

Affluents should be discharged only after proper treatment. The techniques used are—neutralization of toxic chemicals; clarification, sedimentation of heavy metals and suspended particles; filtration and biological treatment.

Agricultural inputs

The toxic substances present in chemicals, viz., pesticides, fungicides and weedicides used, and excess chemical fertilizer applied cause pollution in two ways. (1) Direct con-sumption of agricultural products, e.g. green fruits and vegetables containing toxic materials. (2) Consumption of aquatic forms—both plants and animals—exposed to pollutants. Pollution with agricultural inputs is extensive and countrywide.

A part of the materials applied in fields finally reach water bodies—tanks, canals, rivers, through irrigation system or being washed by rain. The aquatic organisms are thus exposed to toxic materials*.

Management

The following procedure may be adopted to minimise pollution. (a) Total ban on the use of compounds with long residual effect; (b) application of optimum amount of chemicals and fertilizers to ensure complete absorption of the materials by plants; (c) creation of barriers to prevent flow of chemicals to water bodies; and (d) plant protection by biological control, wherever possible.

☒ Sewage

Water containing liquid and solid wastes is termed sewage. The waste water of towns and cities contains a large amount of organic substances, viz., faecal matters, waste food, kitchen and house rejects and other matters in addition to some inorganic compounds. Presence of organic matter in sewage encourages rapid multiplication of micro-organisms. The drainage system carrying sewage to another place is known as **sewerage**, through which sewage is released in large water bodies, viz., canals, rivers and seas.

Management

The sewage is treated in sewage treatment plants following clarification and sedimentation processes in which residual organic matters are degraded (broken down) by microorganisms (biodegradation) and, finally, the clear water is released. The sediment, the **sludge**, is a good fertilizer.

Organic wastes

Organic wastes include metabolic wastes of animals and dead bodies of man and animal.

In underdeveloped countries defecation and urinating in the open is a common practice. These finally reach water bodies. Bathing of man and animals, cleansing of clothes and other items of daily use pollute the water, which is used for other purposes, even drinking. The carcasses of animals and man are often thrown into rivers, causing widespread water pollution.

Management

It is a gigantic task to minimise such pollution. Construction of community toilets in villages with proper water supply, and arrangement for quick disposal of carcasses may reduce the pollution level to a considera-ble extent. Most important measure is creating health consciousness in rural people.

AIR POLLUTION

Release of harmful gases, dust, coal dust, droplets of oil and pollen grains of some plants in the atmosphere cause air pollution. Air pollution is chiefly restricted to heavily populated areas and industrial zones. Villages are,

* The chemicals used for crop protection belong to three groups—organochlorines, organophosphates and carbamates. The organochlorine compounds Endrin. Aldrin, DDT etc. cause serious hazards to living forms for their long residual effect (degradation requires long time, even years).

Producers absorb the toxic materials from water and these pass through consumers to supercarnivores. In each step of the food chain, the toxic materials, which are not excreted, increase in amount (concentration), reaching a peak in the top carnivores. This is known as **Biological magnification**.

however, not entirely free from air pollution, but the degree of pollution is negligible.

Sources

Major pollutants are smoke from burning fuel (incomplete combustion); hydrogen, carbon monoxide, carbon dioxide, methane, etc. from engines; sulphur dioxide, coal dust from thermal power plants and iron and steel industries, from burning coal; fluoro-carbon from cosmetic industries and refrigerators.

Reaction between sulphur dioxide and moisture in the atmosphere causes **acid** (sulphuric acid) **rain**, which damages plants and organ-systems of animals. Accumulated gases in the atmosphere cause **greenhouse effect**, resulting in rise of earth atmospheric temperature. **Ozonosphere**, a layer of ozone gas above the atmosphere prevents cosmic radiation reacting the earth. Nitrous oxide, fluorocarbon punch holes in the ozone layer and the barrier is destroyed, allowing free flow of radiation to earth.

Management

Air pollution can be greatly reduced by adopting the following measures: (a) Ensuring complete combustion of fuel or use of nonpolluting fuel in the engines, (b) preventing obnoxious emission from factories, (c) absorption or burning of poisonous gases, (d) fitting of mechanical particle collector like **scrubber** or **cyclone** in chimneys to prevent escape of harmful substances. Fitting of long chimneys does not prevent air pollution; it only helps in distribution of small particles, smoke and gas in a large area and, thereby, concentration of pollutants in the vicinity of the factory is reduced.

SOUND POLLUTION

Any disturbance in air produces sound. Sound is transmitted through air from its point of origin to the tympanum or eardrum of man and other animals. In the internal ear the sound energy is transformed into electrical energy and passes to the auditory centre of the brain through auditory nerve. Exposure to loud sound, even with some intervals causes physical and mental strain, particularly in cardiac patients. Sound pollution is chiefly restricted to large cities and industrial areas.

Sources

Sound is produced by bursting of crackers, talking men in crowds; slogan-shouting in processions and meetings; loud speakers, rock music, oil engines, jet planes, machines, etc. Continuous exposure to loud sounds (90 decibel) for 8 hours a day and for a long period damages auditory organs and even can make a man deaf.

Management

Ban on bursting of crackers, use of loud speakers above permissible limit and that also restricted to a limited period and removal of air horn from automobiles. Fitting of silencer in oil engines and shifting of airports to far away places from cities will reduce noise pollution considerably.

DECIBEL

The unit of sound is decibel (dB). Absolute silence— 0 dB (10^{-10} microwatt/cm^2 or 0.0002 dyne/cm^2). (Absolute silence does not exist in the world). Pin-drop silence— 2 dB; heart-beat sound— 13 db; ticking of a watch— 20 dB; whisper— 30 dB; normal conversation— 60–65 dB; sound produced by automobiles— 70–90 dB; heavy traffic—80 dB; loud radio music— 85 dB; rock music 110 dB; diesel train 110-115 dB; discomfort pneumatic hammer at 1 metre—120 dB; jet plane—140 dB.

Threshold of human hearing

Sound up to 60 dB is not harmful and is considered as permissible limit.

56 EXPLANATORY NOTES

Allopatric species

Different species living in geographically isolated territories are allopatric species. They may have intermediate forms.

Ambulacral groove

In the centre of the oral surface of *Asterias*, a five-rayed aperture, the actinostome, is present. Five narrow grooves, the ambulacral grooves, run out from the actinostome, each extending along the middle of the oral surface of one arm up to its extremity. The ambulacral groove is bordered by two or three rows of movable, calcareous spines, the ambulacral spines. Two double rows of tube feet are present in each ambulacral groove.

Atlas

The first vertebra of a vertebrate is known as atlas. It is usually ring-like, with a small centrum and bears no spines or processes. One or two articular facets are present anteriorly for articulation with the condyles of the skull. The number of facets correspond to the number of condyles. Two postzygapophyses are present for articulation with the prezygapophyses of the second vertebra.

Axolotl

During the early development of salaman-der, *Ambystoma tigrinum* (Amphibia) the larvae exhibit a precocious growth. While it is still in the larval stage and possesses gills, the larva attains sexual maturity and reproduces. This larva is known as Axolotl larva. After breeding, the metamorphosis starts and the larva turns to a terrestrial salamandrine form.

The reproduction in larval stage is known as **Paedogenesis** and the phenomon is known as Neoteny.

Barriers

The factors which act as an obstruction to the migration of animals in all possible directions is known as barrier, viz. sea, river mountain, desert, etc. Which is barrier to an animal may be the passage of migration to other animal. Sea and rivers are barriers to land animals but not to aquatic animals.

Bipinnaria

It is a free-living larva of Echinodermata. The body is elongated in which the postoral region is broad and the preoral region is narrower. Three pairs of soft bilateral arms are present. Two ciliated bands, preoral and postoral are present and these are thrown into folds peripherally due to the development of arms. A bent alimentary tract with distinct mouth and anus is present.

Biramous appendage

The appendages of Crustacea bear two rami

or branches—hence called biramous. It is jointed. The basal portion articulating with the body is known as protopodite. The proximal portion of the protopodite is coxopodite and the distal portion, basipodite. At its distal end the protopodite (basipodite) bears two branches—endopodite, the inner and exopodite, the outer.

Blastula

The zygote undergoes repeated divisions and a large number of rounded cells or blastomeres are produced. These cells are arranged in a single or several layers, enclosing a space within. This is blastula. In most cases, the wall of the blastula is made of a single layer of cells. The cavity within it is known as blastocoel or segmentation cavity.

Botryoidal tissue

Situated just beneath the longitudinal layers of muscles and within the connective tissue of the body wall of leech. It bears heavy dark-brown pigment. It consists of a network of branched canals which anastomose and their free ends communicate with the haemocoelomic fluid. Botryoidal tissue has an excretory function.

Bursa fabricii

A thick-walled glandular sac, located on the dorsal surface of the proctodaeum in the young birds, is known as Bursa fabricii. It opens into the proctodaeum dorsally. Initially it functions as a blood forming organ but it often becomes a fibrous connective tissue mass, which usually disappears during adolescence.

Carapace

The cephalothorax in Crustacea is covered by a hard chitinoid shell known as carapace. It is drawn anteriorly into a sawlike median process, pointed at the tip. This is rostrum. The carapace is in contact with the cephalothorax on its dorsal surface, but it hangs freely on the sides and encloses on either side a narrow space, the gill chamber. The chamber contains gills attached to the cephalothorax.

Cephalothorax

The cephalothorax in Crustacea is a more or less cylindrical, rigid, anterior region of the body. It is formed by the fusion of 5 cephalic and 8 thoracic segments. The cephalothorax bears two pairs of antennae, one pair of jaws, two pairs of maxillae, three pairs of maxillipeds and five pairs of walking legs, the pereopods of which first two pairs are chelate. Eight pairs of gills are attached to the sides of the cephalothorax.

Chiton

It is a marine mollusc, belonging to the class Polyplacophora. The body is dorsoventrally flattened, with a convex dorsal and a flat ventral surface. The head is not distinct and the eyes and tentacles are absent. The dorsal surface bears a calcareous shell made up of eight separate pieces. The foot is ventral and flat. Along the margins, horny or calcified tubercles and spicules are present. A series of small gills are present under the cover of small lateral folds of the mantle. The mouth and the anus are median, terminal and situated at the opposite ends.

Choanocyte

The radial canals of *Scypha* are lined with peculiar cells of endodermal origin known as choanocytes. These are large, columnar cells and each has a prominent nucleus, one or more vacuoles and a single, long, whip-like flagellum at the inner end, the base of which is surrounded by a transparent, collar-like upgrowth. These cells resemble the choanoflagellate protozoa.

Clitellum

The body cavity of the pheretima is divided into a large number of segments by means of thin septa. These divisions are marked externally by rings or segments. The skin of the segments 14-16 is swollen and forms a band-like structure, the clitellum. The intersegmental grooves on the skin are absent in these segments. The female gonopore is present on the ventral surface of the clitellum. During mating, the clitellar glands secrete a membranous girdle around the animals which gradually hardens to form a cocoon. The fertilization of the eggs and development of the worm takes place within the cocoon.

Cloaca

It is a chamber at the posterior end of the abdominal cavity of vertebrates in which the genital ducts, the urinary ducts and the rectum open. The urinary bladder, when present, opens into the cloaca. The cloaca opens to the exterior through an aperture, the cloacal aperture.

Coelomic fluid

Coelom is a cavity within the body of an organism and is completely lined with cells of mesodermal origin. The fluid which fills up the coelomic cavity is known as coelomic fluid. The animals which bear coelom are coelomata and the others are acoelomata. The coelomic fluid performs different functions in different animals.

Columella

The shell of a gastropod mollusc is spirally twisted due to head-foot torsion. It consists of a number of whorls or varying sizes. The whorls of the shell coil round a central axis known as columella. It is hollow and opens to the exterior by an opening called umbilicus. The body of the snail is attached to the columella by columellar muscles arising from the foot. By contraction of these muscles, the body can be retracted into the shell.

Crossing Over

At the end of synapsis (pairing of homologous chromosomes), towards the close of pachytene stage in meiosis, when chromosomes separate, they sometimes break, and a part of one is joined to another. This is an exchange of parts between paternal and maternal chromosomes. This phenomenon is known as crossing over or genetic crossover (breakage and reciprocal fusion). This break of chromosomes leads to an exchange of whole block of genes. The exchange of parts of homologous chromo-somes is always equal, and the position of the genes remains the same. If the break of the chromosomes occurs at one point, it is called single crossover; if there are two breaks, a double crossover. 2, 3 and 4 breaks are common, but break above 4 is rare. Ordinarily the distance between the genes in a chromosome has some relation with the amount of crossing over. Genes far apart will crossover often than those close together. Crossover values are dependent on linkage. If the genes are so closely linked that they break apart only once in a hundred times, they might still be more closely linked and break apart only once in a thousand times.

During synapsis each of the homologous chromosomes splits into two chromatids forming tetrad. Crossing over takes place between the two chromatids, one paternal and the other maternal. The other two do not take part in this process.

At the end of the pachytene stage the synaptic attraction between the homologous chromosomes seems to disappear suddenly. The separation would have been complete except for the fact that there are certain places where two out of four tetrads remain attached to each other and this is known as a chiasma (plural : chiasmata). The shorter chromosomes (Fig. 56.1A,B) show a single chiasma and the bivalent looks like a cross (× or +). In case of several chiasmata, the bivalent looks like a chain of short length, with loops or links between the successive chiasmata (Figs. 56.1C,D). A paternal chromatid is always associated with another paternal one and a maternal with another maternal on each side of the chiasma, hence a rotation of the bivalent takes place after a genetic crossover.

Ctenidium

Usually the gills of invertebrates are called ctenidia. The gills are made up of thin filamen-

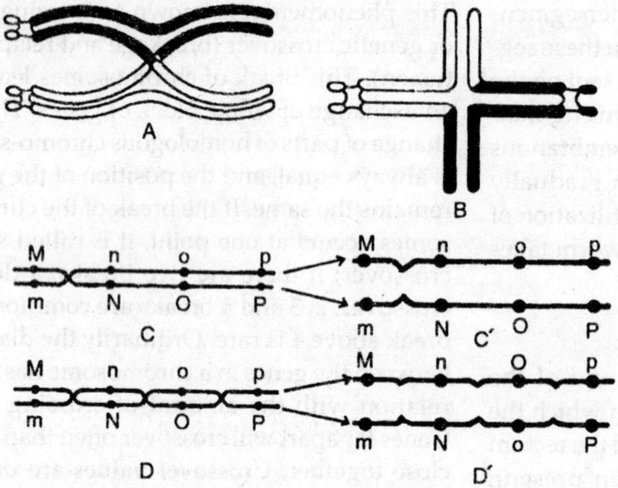

FIGURE 56.1

Crossover. A. Single crossover before rotation. B. Same, after rotation. C-D. Crossover between chromosomes having four pairs of unlike genes. C-D. Pairing chromosomes. C′-D′. Resulting chromosomes

tous structures, gill filaments, arranged in series. The filaments are richly supplied with blood which gives up carbon dioxide and absorbs oxygen dissolved in water and the blood becomes oxygenated. The oxygenated blood goes back to different parts of the body. The ctenidia are respiratory organs, but they differ widely in structure in different invertebrates.

Ctenoid scale

The flat horny structures which cover the body of an animal are known as scales. In fishes, the scales are dermal in origin. Ctenoid scales are found in *Lates*, *Anabas*, etc. Concen-tric or eccentric rings are present in the scale and its exposed posterior half bears pigment. The free posterior margin of the scale is provided with numerous spines. Scales are organs of defence in fishes.

Cycloid scale

It resembles a ctenoid scale, but the rings are concentric and the posterior free margin is entire and without spine.

Endomixis

It is a process of periodic nuclear reorganisation of an organism. It is found in *Paramoecium*. After the degeneration of the macronucleus, the micronucleus divides into fragments and from these fragments new micro- and macronuclei are formed. It is not supposed to be a normal condition in protozoa. After endomixis the activities of the individual are accelerated. Endomixis may be considered as a process which assists in rejuvenation.

Flame cell

These are specialised cells concerned with excretion in the helminthes. A flame cell is a large nucleated cell with a space in the cytoplasm which is in continuation with the excretory vessel. In this space lies a bundle of vibratile cilia resembling a flickering candle flame. Certain processes of the cytoplasm the cell are also present. Wastes absorbed by the cytoplasm and discharged into the lumen are driven to the exretory canals by the movement of cilia.

Gametocyte

During sexual reproduction certain acellular animals undergo changes to produce a large number of gametes. When an individual is in a stage to produce the gametes, that stage is called gametocyte. The nucleus of the gametocyte undergoes repeated divisions and produces a large number of daughter nuclei. The cytoplasm breaks up a small amout aggregates around each daughter nucleus and many gametes are formed.

Gastrula

The blastula is a hollow ball-like structure in the early stage of development of an embryo. The wall of the blastula is usually made of a single layer of cells and encloses a cavity or blastocoel within it. The hollow blastula undergoes invagination at a point, as a result of which a two-layered embryo or gastrula is formed. In a gastrula, the blastocoel is usually obliterated. A new cavity, the archenteron appears in the gastrula, which communicates with the exterior through a narrow pore, the blastopore. Gastrula may also be formed by the overgrowth of cells from a particular point of the embryo.

Gemmule

Gemmules are internal buds. In some sponges, asexual reproduction takes place by the formation of internal buds. Just outside the lining epithelium of the radial canal, by repeated divisions of some of the amoe-bocytes, a spherical group of cells are produced. The bud thus formed remains enclosed in an envelope, often with peculiarly-shaped siliceous spicules. The gemmules are set free only when the part of the parent body, which bears the gemmules, gets detached and undergoes disintegration. After liberation, each gemmule gives rise to a new sponge.

Gizzard

It is a part of the alimentary tract modified for crushing food substances. Usually, the walls of the gizzard are thick and muscular and the internal folds are provided with chitinoid caps or the inner surface may be formed of a thick tough structure. The hard structures of the gizzard operate against the food substances and they are crushed to fine particles. In case of *Columba*, hard structures are not present in the gizzard but the animals takes small pieces of stones which act as hard structures to crush the food.

Glochidium larva

The larva of Bivalvia is known as Glochidium larva, the body of which has acquired a bivalve shell of triangular form. The ventral angle of the shell is produced into incurved hooks beset with spines. The body of the larva, an undivided mass, projecting between the valves of the shell, becomes cleft from below upwards, and thus divided into a single dorsally placed body proper and paired, right and left mantle lobes. Peculiar brush-like sense organs are located on the mantle lobes. A glandular pouch is formed on the ventral surface of the body, which secretes a long thread, the provisional byssus. An immense adductor muscle, formed out of the mesoderm cells is present, the fibres of which extend from valve to valve.

Haemocoel

These are narrow passages or channels and act as blood vessels in certain animals. When the coelom is obliterated by the development of muscles, narrow interspaces are left between them in such an intercommunicating way that these passages may act as blood vessels. These are not true vessels and are without any wall. Being connected with the heart, as in prawns, they function as veins.

Larva

It is a postembryonic stage in the development of an animal. The stages in the development of an animal from zygote to young are known as embryos. Larva is a peculiar stage in the development, which does not resemble the parents and is more or less independent. A larva undergoes metamorphosis or a series of complicated changes by which the dissimilar larva is transformed to resemble the parents.

Laurer's canal

In the female reproductive system of *Fasciola* (liver fluke) and some other platyhel-minthes a central round structure or oötype is present in which the oviduct/uterus, shell gland and duct of the vitelline glands open. From this oötype runs dorsally a narrow duct, the Laurer's canal, to open to the exterior. This is supposed to be the passage by which spermatozoa are received by a fluke from another fluke during mating.

Linkage

It is assumed that the genes are arranged in a linear order in a chromosome, something like beads on a loose string. The genes always maintain their position of linear order. All the genes in a chromosome form a group, and this is known as linkage. It is the opposite of free assortment and the genes in each chromosome form a natural linkage group. The number of such groups is variable in different animals. The linkage of genes located in a chromosome need not be absolute. In a number of cases certain percentage of recombination appears in the offspring of hybrids. Recombination of linked genes takes place due to the occurrence of chiasmata, of the exchange of sections between paired paternal and maternal chromosomes at meiosis. Neighbouring genes in a chromosome are strongly linked; those further apart are exchanged more frequently.

The idea that linked characters tend to stay together in a cross is supported by the results of a cross between grey-vestigeal and black-long fruit flies. In f_1 all flies are grey-long as the grey and long are dominant characters. Four types of gametes—grey-vestigeal, black-long and a small number of two crossover types—would be produced by the f_1 female (there is no crossover in the male germ cells of fruit fly). The f_1 female black crossed to black-vestigeal male would produce 41 per cent grey-vestigeal; 41 per cent black-long; 9 per cent grey-long and 9 per cent black-vestigeal. In each of the two cases, the majority are the original combinations of parents, and new combinations are comparatively few.

Meiosis

It is a type of cell division in which the number of chromosomes in daughter cells is reduced to half the number present in the mother cell. Meiosis is strictly confined to germ cells and takes place only once in a spermatocyte or oöcyte, prior to the formation of spermatozoon or ovum, respectively. The chromosome number is again restored in the zygote at fertilization. The whole process has been divided into four phases—Prophase, Metaphase, Anaphase and Telophase. The prophase is a long-drawn one and has again been subdivided into Leptotene, Zygotene, Pachytene, Diplotene and Diakinensis.

Mesogloea

It is a noncellular gelatinous layer, protein in nature and found in between the ectoder-mal and endodermal cell layers of Porifera and Cnidaria. To some extent, it adds to the rigidity of the animal body. Different cells and spicules are often found in the mesogleal layer but these are not the components of the same and have possibly migrated there.

Metabolism

The sum total of the vital activities of an organism is known as metabolism. It is divisible into two aspects—catabolism or destructive process in which complex organic molecules are broken down and energy is released, e.g. respiration, and anabolism or constructive process in which complex organic molecules and protoplasm are formed by synthesis from simpler elements, e.g. photosynthesis, glycogenesis.

Mimicry

To protect themselves from enemies, certain weak and helpless animals imitate the colour and to some extent the shape of dangerous or poisonous animals. Simple imitation of other objects to avoid the sight of enemy is also usually called mimicry. The stick insects are often mistaken as small dead twigs of plants. The grasshoppers and butterflies appear like green leaves with their venation and are often mistaken as leaves. Mimicry also occurs in some predators to hoodwink their preys.

Mushroom gland

It is a part of the male reproductive system of *Periplaneta*. A pair of delicate tubes, the vasa deferentia, run from the two testes and each ends in a vesicula seminalis. Each vesicula seminalis assumes a more or less branched glandular structure and the two in combination are called the mushroom gland. These open into the ejaculatory duct.

Mutation

It is a random and irreversible measureable change in the gene of an organism or a cell, with the necessary implications that any phenotypic manifestation of the change may potentially or actually be shown by all descendant organisms or cells. Mutation in somatic cells is of the same general quality as in germ cells. Somatic mutation is associated with the functional activities of a cell type while germ cell mutation is responsible for introducing variations in the offspring.

Causes of mutation are not definitely known but possible atomic radiation and different stresses, viz., chemicals, heat, etc. are mutagenic (mutation inducing).

Mutations are always random and may take place in any direction. They are mostly lethal or deleterious but some mutations are useful when organisms are placed in environments other than those in which they usually occur. Environmental changes disrupt the adaptedness of an organism and the harmony between the environment and the organism can be restored only through a genotypic reconstruction by mutation.

Mutation and selection occur simultaneously. Most frequently selection works against mutation and the rate of change, if the mutations are not lethal, will be very slow. But if a particular mutation, advantageous to the organism, stands selection pressure and if the mutation rate is appreciable, the combined action of mutation and selection might well cause a rapid change.

Modern theories of evolution consider the process of mutation to be the source of raw materials from which evolutionary changes may be constructed.

Notochord

It is a supporting structure, found in chordates. It is a cylindrical rod, pointed at both ends and usually extends from anterior to posterior end of the body in median plane, just above the gut. It is formed of large vacuolated cells and is covered by a cellular tissue known as notochordal tissue. It is persistent throughout life in lower chordates but replaced by vertebral column in the vertebrates.

Osculum

The sponge may be in a solitary, cylindrical form or in a colony consisting of a number of cylindrical branches joined at the base. At the free end of each branch a small but distinct openig, the osculum, surrounded by what appears like a delicate fringe, is present. The osculum leads to the paragastric cavity or spongocoel, running at the middle of a branch along its whole length. The osculum is also known as exhalant pore, since water entering through the ostia or inhalant pores of a branch is finally forced out through it.

Otolith

Patches of elongated sensory cells produced into hairs are present in the ampullae, utricle and saccule of the internal ear or membranous labyrinth of vertebrates. The sensory cells are richly innervated by the auditory nerve. The membranous labyrinth is filled up with a fluid known as endolymph. The utricle and saccule contain small calcareous granules, the otoliths. Change of position affects the gravitational pull on the otoliths and hence the pressure on the sensory hair cells. This helps the animal to understand its balance and it can reorient itself.

Ovum

Ovum or female gamete is an unicellular structure produced from the ovum mother cell or oöcyte by reduction division and maturation. Ovum is usually passive, large and contains half the original number of chromosomes. It has the capacity to unite with a male gamete and being activated can produce a new individual. In certain cases, the ovum can produce a new individual by parthenogenesis, i.e. without being united with a sperm cell or male gamete.

Parapodium

Parapodia are small, lateral processes of the body wall bearing a number of bristles or setae, which, in the living animal (*Neanthes*), are used in active movement, aiding in creeping or acting as a series of oars for propelling through water. Anatomically, it is a biramous structure consisting of two distinct divisions, a dorsal notopodium, and ventral neuropodium. Each of these is further subdivided into two lobes and bears a bundle of setae, lodged in setigerous sac, which can be protruded or retracted and turned in various directions by strands of muscular fibres in the interior of the parapodium. The setae are nearly always chitinous and generally solid. Usually each bundle contains a stouter, straight, dark-coloured seta, the pointed apex of which rarely projects on the surface and is known as aciculum. Parapodium bears a short, cylin-drical tentacle-like appendage on the dorsal side, known as dorsal cirrus, and a similar but somewhat shorter appendage on the ventral side, the ventral cirrus. Opening of the nephridium, a fine aperture, is located on the ventral surface near the base of the parapodium.

Pecten

It is present mostly in birds. It is a small, black, corrugated, fan-shaped structure, projects into the vitreous humour from the blind spot, where the optic nerve enters the eye. It has a remarkable capacity to accommodate light in birds and makes the eye a most powerful organ of vision. In some reptiles, pecten is present and helps to stabilise the pressure of the fluid within the eye.

Pelagic

The aquatic animals, which remain floating on the surface of water and can go from place to place by swimming throughout their life, are known as pelagic animals. Water current or waves very often play an important role in their movements. In pelagic animals, colonies are never formed. Examples; Ctenophora, Phyto and Zooplankton.

Phagocytes

The blood of animals consists of a liquid portion, the plasma and a solid portion, the corpuscles, platelets, etc. In the blood of vertebrates the corpuscles are of two types—the red corpuscles or erythrocytes which are coloured red due to the presence of haemoglobin, an iron compound and white corpuscles or leucocytes, which can produce pseudopodia and devour bacteria. These are known as phagocytes.

Placoid scale

It is found in elasmobranch fishes and forms a covering on the surface of the body. Each scale is based in bone, cored with ivory or dentine and tipped with enamel. The enamel is formed partly from ectoderm (epidermis) and partly from mesoderm (dermis). Usually the base is diamond-shaped, the body is palm-shaped, from which project several finger-like spines. The spines are directed backward.

Polian vesicle

Attached to the ring vessel of ambulacral system of many echinoderms are interradially placed, five pairs of thin-walled bladders with long narrow necks. These are known as Polian vesicles and act as reservoirs of water for the ambulacral system. Polian vesicles are absent in *asterias*.

Processus falciformis

It is a sickle-shaped organ containing blood vessels, nerves and muscles, found in the posterior chamber of the eye of teleostean fishes. It is absent in elasmobranch fish. Close to the entrance of optic nerve, it extends from the choroid through the retina to the back of the lens where it ends in a knob—the campanula Halleri. The processus falciformis with the campanula Halleri takes an important part in changing the position of lens by which the eye becomes adapted to forming and receiving images of objects at various distances.

Pseudopipodium

These are simple projections of the mantle. In *Pila* two fleshy nuchal lobes or parapodia project anteriorly over the foot, on either side of the head. The left one is better developed and forms a long respiratory siphon. For aerial respiration the left nuchal lobe elongates and forms a tubular structure by rolling up its edges. The respiratory siphon thus formed projects above the water level. Air enters through it and finally reach the pulmonary sac in the pulmonary chamber.

Radula

It is found in the buccal mass of Mollusca. It is a chitinoid ribbon-shaped structure on the floor of the buccal cavity and is placed on an elevation called odontophore. It bears numerous horny teeth arranged in rows and is continued posteriorly into the radular sac. The radula helps the animal in rasping food.

Rectum

Rectum is the posteriormost part of the alimentary canal. It is wider than the intestine and opens either in the cloaca or to the exterior by anus. The wall of the rectum is either smooth or bear very slight flods. Undigested food substances are temporarily stored in the rectum and finally discharged to the exterior through the anus or the cloacal aperture.

Semicircular canals

These are narrow curved canals associated with the membranous labyrinth or the inner ear of the vertebrates. At their both ends they open into the utricle. One of the semicircular canals is horizontal, the other two are vertical but at right angles to each other. Each canal bears a dilated portion, the ampulla, at one end. The canals are filled up with a fluid known as endolymph. The semicircular canals help the animal to maintain equilibrium. The semicircular canals are one in Myxine, two in Petromyzon and three in Gnathostomata.

Sibling species

Such species are very similar in their morphological characters and they cannot be distinguished from external features. But they are good species as they do not interbreed.

Species

A species is a group of genetically closed animals or plants, the individuals of which differ from each other only in minor morphological characters, except sex, and sharply separated from all other species in their features. The members of a species are mutually fertile, but at least partially sterile when crossed to other species.

Spermatheca

These are thin-walled sac-like structures associated with the female reproductive organs of certain animals. The sacs may or may not be in direct communication with the reproductive organs. Spermatozoa received from the male during mating are stored temporarily in the spermathecae and are subsequently used to fertilize the eggs. The number of spermathecae may be one (cockroach) or many (earthworm).

Statocyst

It is an organ of balance. It is an almost circular cuticular sac within the basal segment of the antennule of prawn and is communi-cated with the exterior by a small pore. Within the sac are a large number of delicate setae or sense hairs arranged in the form of an oval ring. Each seta consists of a swollen basal portion and a shaft and is innervated by the branches of the antennulary nerve. The free ends of the setae are directed towards the centre, where particles of sand grains are present. When the position of the animal is changed the weight of the sand grains fall more heavily on certain setae and, being aware of its changed position, the prawn can reorient itself. Statocyst is common in molluscs and lodged close to the pedal ganglion. The centre of the statocyst may be occupied by small particles, the statoconia (arthropods) or by a large particle—the statolith (molluscs).

Stigmata

These are narrow openings for inhalation and exhalation of air in respiration. In Periplaneta, these are ten pairs situated laterally between the terga and sterna. Two pairs are located in the thorax, one between pro- and mesothorax and the other between meso- and metathorax. Eight pairs occur in the abdomen. Each spiracle leads internally to the main air tube or trachea with an elastic ring or spiral, acting as a valve for closing the passage. Stigmata are present in many air-breathing arthropods.

Subspecies

A subspecies is a population subordinate to a species. A species may have one or more subspecies at any time. If the subspecies are isolated, diversities accumulate in different subspecies, and when reproductive isolation appear in them, they are considered as species.

Swim bladder

The swim bladder is found in fishes. It is a sac-like structure, develops as an outgrowth from the foregut. With a few exceptions it arises from the dorsal aspect of the gut. In certain cases, it maintains connection with the pharynx, and in others the connection is lost. It acts as a hydrostatic organ and helps the fish to maintain the same specific gravity as that of water in any depth. This is done by increasing or decreasing the pressure of gas in the bladder. Red glands (rete mirabile), abundantly supplied with blood, are present on the bladder and they open into the cavity of the bladder. These glands can secrete or absorb gases. In some cases, the bladder acts as an accessory respiratory organ.

Symbiosis

When two organisms live in intimate association and both the partners are mutually benefited, such an association is known as symbiosis. It takes place usually between an animal and a plant. Numerous rounded green algae, Zoochlorellae, are found within the nutritive epithelial cells of *Hydra*. *Hydra* gives protection to the algae and supplies nitrogenous wastes to the plant which the latter uses as its food. On the other hand, the plant supplies carbohydrate food to the *Hydra* which it manufactures by photosynthesis.

Sympatric species

Different species living in the same geographic territory are sympatric species. They have no intermediate forms.

Syngamy

The fusion of two gametes is known as syngamy. In certain protozoa, the whole animal behaves as a gamete and two such unite to form the zygote. In some other protozoa a large number of gametes are produced from each individual and, after liberation, these gametes unite in pairs with similar other gametes produced from different individuals. In higher animals, the whole body does not take part in the formation of gamates but only special cells, known as germ cells, produce gametes. The gametes are dissimilar and their union is called fertilization. Usually, the term 'syngamy' is used to mean complete union of similar gametes produced in lower animals.

Syrinx

It is a characteristic vocal organ in birds not found in any other class. Birds have evolved an additional secondary 'larynx' called syrinx which is located at the lower end of the trachea at its junction with the bronchi. The last 3 to 4 tracheal rings and the first or bony half-ring of each bronchus are modified to form a dilated chamber, the tympanum, the mucous membrane of which forms a cushion-like thickening on each side. A bony ridge, the pessulus, extends dorsoventrally and supports a fold of mucous membrane, the membrana semilunaris. Internal tympaniform membranes form the inner walls of bronchi. In connection with this organ a pair of intrinsic syringeal muscles and sternotracheal muscles are found. The cartilages of the syrinx, which are modifications of tracheal and bronchial rings, are combined with intercartilages, membranous walls, and a variety of muscles, into an efficient vocal mechanism. By means of this apparatus, the bird is capable to change the shape of the tracheobronchial chamber, thus producing a variety of different sounds.

Telson

The last abdominal region of the prawn is called telson. It is somewhat horizontally flattened, broader and thick anteriorly and pointed at the free end. The broader portion is attached to the sixth abdominal segment, the pointed end bears two spines and a number of setae. The telson is not considered a segment, as appendages and nerve ganglia are absent in it.

Tornaria

It is the free-living larval stage of hemichordates. The larva bell-shaped. The gut is bent at right angle; the mouth is ventral and the anus posterior and terminal. An apical sensory plate with two eye spots and sense hairs is present at the centre of the narrow anterior end. Three circlets of cilia are present. The body of the larva is variously folded into lobes and the margin of the lobes is bordered by a wavy band of oral circlet of cilia. Posterior to the oral band is another ring-like band of cilia, the anal ring of cilia. A small ring of cilia is present very near the posteriormost end; this is known as the secondary anal ring of cilia.

Trochophore

The free-living larva of Annelida is known as Trochophore. It is pear-shaped with a broad anterior and a narrower posterior end. The mouth lies in the middle of the body and the anus at the posterior end. The gut is bent at a right angle. A thickening, known as apical plate, bearing an eye spot and a tuft of sensory hairs, are present at the centre of the broad end. A preoral and a postoral circlet of cilia pass round the body in front and back of the mouth. An anal circlet of cilia is present at the narrow posterior end. Excretory organs and mesodermal bands are also present.

Weberian ossicle

It is an accessory equilibrium apparatus found in fishes. Usually, it is represented by a chain of small bones, which connect the air bladder with the internal ear. It controls the passage of gas to and from the bladder by reflex action. Any change in volume of the air bladder is communicated to the auditory organ through the Weberian ossicle and ensures equilibrium at any depth.

Zoea

It is a larval stage of crab. The body is divisible into a large cephalothorax bearing a pair of large eyes and an abdomen. The cephalothoracic shield is drawn into a long, gradually tapering spine dorsally. The appendages of the cephalothorax are typically biramous. The abdomen is slender, segmented and without appendages. It metamorphoses to a megalopa stage and finally resembles the adult crab.

57 ENGLISH NAMES: ZOOLOGICAL NAMES: LARVAL FORMS: DISEASES AND ANIMALS

ENGLISH NAMES: ZOOLOGICAL NAMES

Majority of the macroscopic and also a few microscopic animals of a region are identified by the people of an area through their local names. Such names vary from one to another country and even in the different regions of the same country.

Knowing local names of animals (plants too) is advantageous in field studies. To locate the habitat of animals in an uncharted area, cooperation and assistance of local people is a must. To meet this end, the people there have to be made familiar with the animals looking for. This is possible either by telling the local name or making a sketch of the animal.

Being the world language, English has an edge over other languages. It is the link language between monolinguistic countries and also between different regions of a multilingual country. India belongs to the second category.

With the aim of helping the beginners to identify animals properly from their English names, a few individual have been picked up at random from different groups and their names are arranged alphabetically. In the process, the evolutionary hierarchy had to be ignored, which can be ascertained from the classification in the text.

Major division	English name(s)	Zoological name
PHYLUM APICOMPLEXA	Malaria parasite	*Plasmodium*
PHYLUM CILIOPHORA	Bell animalcule	*Vorticella*
	Slipper animalcule	*Paramoecium*
PHYLUM	Slime mold; slime fungi	*Myxomycetes*
	Amoeba	*Amoeba*
SARCOMASTIGOPHORA	Chicken liver	*Chondrilla*
PHYLUM PORIFERA	Crumb of bread	*Halichondria*
	Fire sponge	*Tedania*
	Glass-rope sponge	*Hyalonema*
	Neptune's cup	*Poterion*
	Venus flower basket	*Euplectella*

Contd.

Major division	English name(s)	Zoological name
PHYLUM CNIDARIA	Brain coral	*Meandrina*
	Common sea anemone	*Metridium*
	Dead man's fingers (coral)	*Alcyonium*
	Jelly fishes	*Aurelia*
	Horn coral	*Madreporadrina*
	Lettuce coral	*Agaricia*
	Little sail	*Vellela*
	Mushroom coral	*Fungia*
	Organ pipe coral	*Tubipora*
	Portuguese man-of-war	*Physalia*
	Red coral	*Corallium*
	Sea anemone	*Adamsia*
	Sea fan	*Gorgonia*
	Sea nettle	*Chrysaera*
	Sea pancy	*Renilla*
	Sea pen	*Pennatula*
	Sea wasp	*Chironex*
	Stag horn coral	*Acropora*
PHYLUM PLATYHELMINTHES	Blood fluke	*Schistosoma*
	Chinese liver fluke	*Opisthorchis*
	Liver fluke	*Fasciola*
	Tape worm	*Taenia*
PHYLUM ASCHELMINTHES	Eye worm	*Loa*
	Guinea worm	*Dracunculus*
	Hook fluke	*Ancylostoma*
	Round worm	*Ascaris*
PHYLUM ANNELIDA	Bamboo worm	*Axiothella*
	Blood worm	*Glycera*
	Christmas tree worm	*Spirobranchus*
	Earthworm	*Pheretima*
	Fan worm	*Sabella pavenina*
	Leech	*Hirudinaria*
	Lug worm	*Arenicola*
	Peacock worm	*Sabella melanostigma*
	Quil worm	*Hyalinoecia*
	Rag worm/Clan worm	*Neanthes*
	Samon palolo worm	*Polola*

Contd.

Major division	English name(s)	Zoological name
PHYLUM ANNELIDA (Contd.)	Scale worm	*Lagisca*
	Sea mouse	*Aphrodite*
	Thread worm	*Drilonereis*
PHYLUM ARTHROPODA SUBPHYLUM MANDIBULATA	Ant lion	*Myrmelean*
	Bed bug	*Cimex*
	Bird lice	*Menopan*
	Blow fly	*Calliphora*
	Book worm; Silver fish	*Lepisma*
	Brine shrimp	*Branchipus*
	Bristle tail	*Campodoa*
	Cray fish	*Astacus*
	Carp louse	*Argulus*
	Cockroach	*Periplaneta*
	Chinese mitten crab	*Eriocheir*
	Dead leaf butterfly	*Kallima*
	Death's head hawk moth	*Acherontia*
	Death's watch trap	*Xastobium; Anobium*
	Devils conch horse	*Acoleoptera*
	Dragon fly	*Aeshna*
	Ear-wigs	*Forficula*
	Fairy shrimp	*Branchinecta*
	Goose barnacle	*Lepas*
	Grasshopper	*Melanopus*
	Glowworm	*Lampyrus*
	Ghost crab	*Ocypoda*
	Gandi poka	*Leptocorisa*
	Hawk moth	*Protoparce*
	Honey bee	*Apis*
	House cricket	*Gryllus*
	House fly	*Musca*
	Human head louse	*Pediculus*
	Jute stem borer	*Apion*
	King crab	*Lithodes*
	Lady bird beetle	*Coccinella*
	Leaf insect	*Phyllium*
	Locust	*Schistocerca*
	Mole crab	*Hippidea*
	Mole cricket	*Gryllotalpa*

Contd.

Major division	English name(s)	Zoological name
MANDIBULATA (Contd.)	Pea crab	*Dissodactylus*
	Peppermint shrimp	*Hippolysmata*
	Plant lice	*Aphid; Aphis*
	Prawn; Lobster	*Macrobrachium; Penaeus*
	Rat flea	*Xenopsylla*
	Rice stem borer	*Tryporyza*
	Rock barnacle, Acorn barnacle	*Balanus*
	Root-headed barnacle	*Sacculina*
	Rust red; Flour beetle	*Tribolium*
	Sand fly	*Phlebotomus*
	Soldier crab	*Myctyris*
	Spider crab (Largest living arthropod)	*Macrocheira*
	Spring tail	*Podura*
	Stick insect	*Carusius*
	Stone crab	*Menippe*
	Tadpole shrimp	*Triops*
	Water boatman	*Notonecta*
	Water flea	*Daphnia*
	Wire worm	*Cylindroiulus (Julus)*
	Wood louse, Saw bug, Pillbug	*Oniscus*
SUBPHYLUM ARACHNOMORPHA (CHELICERATA)	Black widow	*Laterodectus*
	Desert spider	*Tarantula*
	Garden spider	*Araneus*
	Horseshoe crab	*Limulus*
	Menge mite	*Sarcoptes*
	Orb weaver	*Argiope*
	Scorpion	*Palamnaeus*
	Soft tick	*Argas*
	Trap door spider	*Nephila*
	Whip scorpion	*Mastigoproctus*
PHYLUM MOLLUSCA CLASS AMPHINEURA	Sea mouse	*Chiton*
CLASS SCAPHOPODA	Elephant's tusk shell	*Dentalium*
CLASS BIVALVIA	Fresh water mussel	*Lamellidens*
	Giant clam	*Tridacna*
	Indian pearl oyster	*Pinctada*

Contd.

Major division	English name(s)	Zoological name
CLASS BIVALVIA (Contd.)	Jewel Box	*Chama*
	Oyster	*Ostrea*
	Pen shell	*Atrina*
	Razor fish	*Solen; Tagelus*
	Sea mussel	*Mytilus*
	Ship worm	*Teredo*
	Winged oyster	*Pteria*
CLASS GASTROPODA	Apple snail	*Pila*
	Bubble shell	*Hydatina*
	Buckie carrier shell	*Xenophora*
	Cockle	*Cardium*
	Cowrie	*Cypraea*
	Dog whelk	*Nucella*
	Ear shell	*Haliotis*
	Keyhole limpet	*Fissurella*
	Limpet	*Acmaea*
	Moon shell	*Lunatia*
	Periwinkle	*Littorina*
	Scallop	*Pecten*
	Sea butterfly	*Globa*
	Sea hare	*Aplysia*
	Sea lemon	*Doridium*
	Slipper shell	*Crepidula*
	Slug	*Laevicaulis*
	Violet shell	*Janthina*
	Worm shell	*Spiroglyphus*
CLASS CEPHALOPODA	Cuttle fish	*Sepia*
	Devil fish	*Octopus*
	Giant squid	*Architenthis*
	Sea grapes	Bunch of *Sepia* eggs
	Squid	*Loligo*
PHYLUM ECHINODERMATA	Basket star	*Astrophyton*
	Brittle star	*Ophiura; Ophicoma*
	Feather star	*Antedon*
	Pencil urchin	*Heterocentrotus*
	Sand dollar	*Leodia*
	Sea biscuit	*Clypeaster*
	Sea cucumber	*Cucumaria; Holothuria*

Contd.

Major division	English name(s)	Zoological name
PHYLUM ECHINODERMATA (Contd.)	Sea lily	*Ptilocrinus*
	Sea star	*Astropecten*
	Sea urchin	*Echinus*
	Star fish	*Asterias*
	Sun star	*Crossaster*
	'Trepang'/beche-de-mer	Pacific Holothurians
PHYLUM HEMICHORDATA	Tongue worm	*Bolanoglossus*
	Acorn worm	*Saccoglossus*
ABERRANT GROUPS		
PHYLUM CHAETOGNATHA	Arrow worm	*Sagitta*
PHYLUM BRACHIOPODA	Lamp shells	*Magellania*
PHYLUM ROTIFERA	Wheel animalcule	*Brachionus*
PHYLUM CHORDATA		
SUBPHYLUM UROCHORDATA	Sea squirts	*Ascidians*
	Tunicates	*Caleolus; Polycarpa*
SUBPHYLUM ACRANIA (CEPHALOCHORDATA)	Lancelet	*Branchiostoma*
CLASS CYCLOSTOMATA	Hag fish	*Myxine*
	Lamprey; Tempereel	*Petromyzon*
	Nine eyes	Ammocoete larva (*Petromyzon*)
CLASS CHONDRICHTHYES (ELASMOBRANCHII)	Banja ray	*Rhinobatis*
	Basking shark	*Cetorhinus*
	Common rays and skates	*Raja*
	Dog fish	*Scoliodon*
	Eagle ray	*Mylobatis*
	Electric ray	*Torpedo*
	Great white shark	*Cercharodon; Carcharias*
	Hammer-headed shark	*Sphyrna*
	Saw fish	*Pristis*
	Sea cat	*Chimaera*
	Sea devil	*Mobula*
	Shark (Indian)	*Scoliodon*
	Spiny dog fish	*Squalus*
	Sting ray	*Dasyatis*

Contd.

Major division	English name(s)	Zoological name
CLASS OSTEICHTHYES (TELEOSTOMI)	American koi (Tilapiya)	*Oreochromis*
	Anchor fish	*Toxotes*
	Angle fish	*Chaetodon*
	Angler fish	*Lophius*
	Arapina	*Arapima*
	Atlantic 'John Dory'	*Zens*
	Bat fish	*Platax*
	Bhangon fish	*Rhinomugil*
	Bhetki fish	*Lates*
	Boal fish	*Wallago*
	Bombay duck; Nehari	*Harpadon*
	Bow fish	*Amia*
	Brook trout	*Salvelinus*
	Bull head	*Cottus*
	Butter fish	*Pholis*
	Butterfly fish	*Chaetodon*
	Cat fish	*Heteropneustes; Clarias, etc.*
	Snake head	*Channa;*
	Chittal fish	*Notopterus*
	Climbing perch (Koi fish)	*Anabas*
	Cling fish	*Gobiesox*
	Cod fish	*Gadus*
	Cuchia	*Monopterus*
	Dragonet	*Callionymus*
	Dragon fish	*Pegasus*
	Eel	*Amphipnous*
	European carp	*Ciprinus*
	Flying fish	*Exocoetus*
	Globe fish	*Tetrodon*
	Gold fish	*Carasius*
	Gulper eel	*Anguilla*
	Half-beak fish	*Hemirhamphus*
	Hatchet fish	*Gasteropalecus*
	Herring	*Clupea*
	Lantern fish	*Harpodon*
	Mackerel	*Rastrelliger*
	Magur fish	*Clarias*
	Major carps	*Catla; Labeo; Mirgal; Cirrhinus*

Contd.

Major division	English name(s)	Zoological name
CLASS OSTEICHTHYES (Contd.)	Minor carps	*Puntius*
	Meedle fish	*Xenentodon*
	Oar fish	*Regalacus*
	Paddle fish	*Polyodon*
	Parse fish	*Mugil parsia*
	Pike	*Esox*
	Pipe fish	*Syngnathus*
	Pomfret	*Pampus*
	Porcupine fish	*Diodon*
	Ribbon fish	*Trichiurus*
	Sail fish	*Istiophorus*
	Sand goby	*Pomataschistus*
	Shanny	*Lophrus*
	Scorpion fish	*Pterois*
	Sea horse	*Hippocampus*
	Shovel-shaped strugeon	*Scaphirhynchus*
	Singi fish	*Heteropneustes*
	Snake-headed murrel	*Channa*
	Spiny eel	*Mastacembelus*
	Squirrel fish	*Holocentrus*
	Stickle back	*Gasterosteus*
	Sucker fish	*Echeneis*
	Tangra fish	*Mystus*
	Thread fin	*Polynemus*
	Tripod fish	*Bathypterois*
	Toad fish	*Porichthyes*
	Trout	*Salmo*
	Tuna	*Thunnus*
	Viper fish	*Chauliodus*
CLASS CROSSOPTERYGII	Fringe-finned fish	*Latimeria*
	Lung fish, mud fish (Dipnoi)	*Lepidosiren; Neoceratodus; Protopterus*
CLASS AMPHIBIA	Bull frog	*Rana catesbeiona*
	Darjeeling salamander	*Tylototriton*
	Edible frog	*Rana esculenta*
	Flying frog	*Rhacophorous*
	Grass frog	*Rana temporaria*
	Green tree toad; Tree frog	*Hyla*
	Hell bender	*Cryptobranchus*
	Horned frog	*Ceratophrys*

Contd.

Major division	English name(s)	Zoological name
CLASS AMPHIBIA	Midwife toad	*Alytes*
	Mud puppy	*Necturus*
	Newts	*Ambystoma*
	Salamander	*Salamandra*
	Surinam toad	*Pipa*
	Toad	*Duttaphrynus melanostictus*
CLASS REPTILIA	Blind snake; Burrowing snake	*Typhlops*
	Chain viper	*Vipera*
	Chameleon	*Chamaeleon*
	Cobra	*Naja*
	Common krait	*Bangurus*
	Flying dragon	*Draco*
	Garden lizard; Girgit	*Calotes*
	Gilla monster	*Heloderma*
	Glass snake	*Ophisurus*
	Golden Lizard	*Mabuya*
	Green turtle	*Chelone*
	Horned toad	*Phrynosoma*
	House lizard	*Hemidactylus*
	Iguana	*Iguana*
	Monitor	*Varanus*
	Marine iguana	*Amblyrhynchus*
	Mud tortoise	*Trionyx*
	Ornate box turlle	*Geocoelamys*
	Slow worm	*Angus*
	Spectacle cayman	*Cayman*
	Tuatara; Newzealand lizard	*Sphenodon*
	Turtle	*Chelone*
CLASS AVES	Albatross	*Diomadea*
	Barn owl	*Tyto*
	Black bird	*Turdus*
	Bolted kingfisher	*Alcedo*
	Bulbul	*Pycnonatus*
	Bull flinch	*Pyrrhula*
	Burrowing owl	*Speotyto*
	Bustard	*Otis*
	Cassowories	*Casuarius*
	Cockatoo	*Kakatoe*
	Cow bird	*Molothrus*
	Crab plover	*Dromas*
	Cranes	*Grus*

Contd.

Major division	English name(s)	Zoological name
CLASS AVES (Contd)	Crow	*Corvus*
	Dipper	*Cinclus*
	Egrets	*Egretta*
	Elephant bird	*Aepyornis*
	Emu	*Dromoeus*
	Falcon	*Falco*
	Flemingo	*Phonicopte*
	Finches	*Loxia*
	Fowl	*Gallus*
	Frogmouth	*Pedargus*
	Golden eagle	*Aquila*
	Goose	*Anser*
	Guinea fowl	*Acryllium*
	Herons	*Ardea*
	Hornbill	*Buceros*
	House sparrow	*Passer*
	Humming bird	*Archilochus*
	Kingfisher	*Aleodo*
	King penguin	*Aptenodytes*
	Kites	*Haliastur*
	Kiwi	*Apteryx*
	Little owl	*Asio*
	Lyre bird	*Memura*
	Macaw	*Ara*
	Moa	*Dinormis*
	Mocking bird	*Mimus*
	Mouse bird	*Colius*
	Ostrich	*Stuthio*
	Parakeet; Tota	*Psittacula*
	Passanger pigeon	*Ectopistes*
	Peacock	*Pavo*
	Pelican	*Pelecanus*
	Penguin	*Eudyptes*
	Pigeon	*Columba*
	Sandpiper	*Calidris*
	Secretary bird	*Sagittarius*
	Shoe bill	*Balaeniceps*
	Sky lark	*Alanda*
	Snag thrush	*Hyloeichla*
	Snipe	*Gallinago*
	Snowy owl	*Nyctea*
	Stork	*Cioconia*

Contd.

Major division	English name(s)	Zoological name
CLASS AVES (Contd)	Sun bird	*Aethopya*
	Swan	*Coscorbea*
	Swallow	*Hirunda*
	Swifts	*Apus*
	Vulture	*Gyps*
	Weaver bird	*Ploceus*
	Whistling duck	*Dendrocygna*
	Woodpecker	*Picus*
CLASS MAMMALIA		
PROTOTHERIA*		
THERIA (METATHERIA)+		
THERIA (EUTHERIA)	Aard-varks	*Orycteropus*
	Agoutis	*Dasyprocta*
	Ant eater	*Myrmecophaga*
	Armadillo	*Dasypus*
	Aye aye	*Dauben tonia*
	Azara's opossum+	*Didelphis*
	Badger+	*Moles*
	Banded ant eater	*Myrmecobius*
	Bear	*Ursus*
	Beaver	*Castor*
	Big-eared bat	*Plecotus*
	Brazilian bat	*Vampyrus*
	Chimpanzee	*Pan*
	Common seal	*Phoca*
	Coyote	*Canis*
	Duck bill Platypus*	*Ornithorhynchus*
	Elephant seal	*Mirounga*
	Elephant shrew	*Macroscelides*
	Flying fox	*Pteropus*
	Flying lemur	*Galiopithecus*
	Flying phalanger+	*Petaurus*
	Flying squirrel	*Petaurista*
	Giant otter	*Lutra lutra*
	Gibbon	*Hylobates*
	Golden mole	*Chrysochloris*
	Gorilla	*Gorilla*
	Great Panda	*Ailuropoda*
	Grey seal	*Halichoerus*
	Himalayan Panda	*Ailurus*
	Hippopotamus	*Hippopotamus*
	Hedgehog	*Erinaceus*

Contd.

Major division	English name(s)	Zoological name
CLASS MAMMALIA (Contd.)	Horseshoe bat	*Rhinolophus*
	Hyaena	*Hyaena*
	Indian palm squirrel	*Funambulus*
	Jarboas	*Jaculus*
	Kangaroo+	*Macropus*
	Killer whale	*Orcinus*
	Koala bear+	*Phascolarctos*
	Lemings	*Lemmus*
	Long ear hedgehog	*Hemiechinus*
	Manatee	*Trichechus*
	Marsupial mole+	*Notoryctes*
	Mermots	*Murmata*
	Mole (Common)	*Talpa*
	Monk seal	*Monachus*
	Orang-utan	*Pongo*
	Otter	*Luthra*
	Pig	*Sus*
	Pole cat	*Putonus*
	Porcupine	*Hystrix*
	Pouched mouse	*Sminthopsis*
	Prairie dog	*Cynomys*
	Puma	*Puma*
	Rhesus monkey	*Macaca*
	Scaly ant eater	*Manis*
	Sea cow	*Halicose (Dugong)*
	Sea lion	*Otaria*
	Sea otter	*Lata*
	Skunk	*Mephitis*
	Sloth (two-toed)	*Choleopus*
	Shrew	*Sorex*
	spiny ant eater*	*Tachyglossus (=Echidna)*
	Squirrel	*Sciurus*
	Stoat and weasal	*Mustela*
	Tapir	*Tapirus*
	Tasmanian devil+	*Sarchophilus*
	Tasmanian tiger cat+	*Dasyurus*
	Tasmanian wolf+	*Thylacinus*
	Tree shrew	*Tupaia*
	Vampire bat	*Desmodus*
	Voles	*Microlus*
	Walrus	*Odobenus*
	Wombat+	*Vombatus*

Contd.

LARVAL FORMS

The larva is an off-the-track stage in the embryogeny of an animal. The zygote, the starting point of multicellular life, is provided with an immune system; certain degree of protection against onslaught of biotic and abiotic environment; in growth and organ-system development, leading to the appearance of a young one capable to start an independent life.

The developmental pathway of an animal is predetermined and genetically fixed. As a rule, the embryos are helpless, almost lacking physical activities in the majority. In exceptionally rare cases the situation is different, only in a limited scale as in larvae (often defined as postembryonic stage) and they can act independently, almost like fully-formed individuals. They have locomotory powers; they can defend themselves against both abiotic and biotic hazards; procure their own food for growth and further development and store energy-rich materials for vital activities in the nonfeeding stage following the completion of development.

The larval adaptation, viz., a free-living state associated with asexual multiplication in sedentary animals (Cnidaria) to ensure wide dispersal; enormous increase in number through repeated asexual multiplication in parasites, where many perish in transmission from host to host and even sexual reproduction (Axolotl) to avoid the danger of elimination due to drying up of water in their breeding sites, are of immense help in the successful continuation of the species.

These are extreme cases and occur rarely in nature but their rare occurrences point to the enormous capability of life to adust itself with the changes in the environment where they live.

LARVAE

Phylum	**Class**	**Larva**	**Free living/Sedentary/ Parasitic**
PORIFERA		Parenchymula	Free-living
		Stomoblastula	Free-living
		Amphiblastula	Free-living
CNIDARIA	Hydrozoa	Actinula	Free-living
		Planula	Free-living
		Hydrula	Sedentary
		Hydratuba	Sedentary
	Scyphozoa	Planula	Free-living
		Scyphistoma	Sedentary
		Strobila	Sedentary
		Ephyra	Free-living
PLATYHELMINTHES	Turbellaria	Muller's larva	Free-living
	Trematoda	Gyrodactylid	Free-living
		Miracidium	Free-living
		Sporocyst	Parasitic
		Redia	Parasitic
		Cercaria	Free-living (non-feeding)
		Metacercaria	Parasitic (non-feeding)

Contd.

Phylum	Class	Larva	Free living/Sedentary/ Parasitic
	Cestoda	Oncosphere	Free-living
		Cysticercus	Parasitic
NEMERTINA		Pilidium	Free-living
ASCHELMINTHES		Filariform	Free-living
		Microfilaria (embryo)	Parasitic
		Rhabditiform	Free-living
ANNELIDA		Trochophore	Free-living
		Metatrochophore	Free-living
ARTHROPODA	Crustacea	Nauplius	Free-living
		Zoea	Free-living
		Megalopa	Free-living
		Cyclippid	Free-living
	Insecta	Caterpillar	Free-living
		Maggot	Free-living
		Grub	Parasitic
MOLLUSCA	Placophora	Trochophore	Free-living
	Bivalvia	Glochidium	Free-living/Parasitic
	Gastropoda	Trochophore	Free-living
		Veliger	Free-living
ECHINODERMATA	Asteroidea	Bipinnaria	Free-living
		Brachiolaria	Free-living
	Holothuroidea	Auricularia	Free-living
		Doliolaria	Free-living
		Vitellaria	Free-living
	Crinoidea	Crinoid	Free-living
		Vitellaria	Free-living
	Ophiuroidea	Pluteus	Free-living
		Ophiopluteus	Free-living
		Echinopluteus	Free-living
HEMICHORDATA		Tornaria	Free-living
SUBPHYLUM			
UROCHORDATA	Tadpole	Tadpole	Free-living
PHYLUM	Cyclostomata	Ammocoete	Free-living
CHORDATA	Amphibia	Tadpole	Free-living
		Axolotl	Free-living

DISEASES AND ANIMALS

A number of animals are causative organisms for diseases in other animals. The endoparasites causing diseases, mostly fatal, to human, domesticated and game animals are recorded.

Phylum SARÇOMASTIGOPHORA

Disease. **Entamoebiasis** (amoebic dysentery); hepatitis and liver abcess

Parasite. *Entamoeba histolytica.*

Host. Human.

Habitat. Mucous and submucous layers of large intestine; In heavy infestation—liver, lungs and brain are damaged.

Geographical distribution. Worldwide.

Transmission. Direct—man to man; faecal contamination of water, vegetables and food. Indirect—through house fly; use of human excreta in vegetable cultivation.

Disease. **Giardiasis** (Intestinal disorder)

Parasite. *Giardia intestinalis.*

Host. Human.

Habitat. Surface epithelium of duodenum.

Geographical distribution. Worldwide; incidence is higher in tropical, subtropical and temperate zones.

Transmission. Direct—man to man; faecal contamination of water, vegetables and food; Indirect—through house fly.

Phylum APICOMPLEXA

Disease. **Eimeriasis** (Intestinal disorder)

Parasite. *Eimeria sp.*

Host. Calves, chicken, turkey, fishes, etc. (Presence in human faeces is the result of eating infected fishes).

Habitat. Epithelial cells of intestinal (lower portion of ileum) mucosa.

Geographical distribution. Worldwide.

Transmission. Faecal contamination of food; eating of infected food.

DISEASES CAUSED BY BLOOD PARASITES

TRYPANOSOMA (TRYPANOSOMIASIS)

Disease. **Sleeping sickness** (Mid- and West Africa)

Parasite. *Trypanosoma gambiense.*

Host. Human.

Geographical distribution. Western and Central equatorial Africa.

Transmission. Tse-tse fly (vector), *Glossina palpalis; G. tachinoides* (blood- sucking insects).

Disease. **Sleeping sickness** (East Africa)

Parasite. *Trypanosoma rhodesiense.*

Host. Human.

Geographical distribution. South-East and Central Africa.

Transmission. Tse-tse fly (vector), *Glossina morsitans.*

Disease. **Chagas disease**

Parasite. *Trypanosoma cruzi.*

Host. Human—children and adolescent.

Geographical distribution. Central and South America.

Transmission. Reduviid bug (vector), *Panstrongylus magistus.*

Disease. **Nagana**

Parasite. *Trypanosoma brucei*

Hosts. Domesticated and game animals.

Geographical distribution. Tropical Africa.

Transmission. Tse-tse fly (vector), *Glossina* sp.

Disease. **Congo disease** (Bovine trypanosomiasis; Nagana)

Parasite. *Trypanosoma congolense.*

Host. Domestic animal (fatal disease).

Geographical distribution. Tropical Africa.

Transmission. Tse-tse fly (vector), *Glossina morsitans*

Disease. 'Mal-de-caderas' (a fatal contaminative disease of horse in South America)

Parasite. *Trypanosoma equinum.*

Host. Horse.

Geographical distribution. South America

Transmission. Mechanical. Tabanid—biting insect.

Disease. Dourine (Stallion's disease)

Parasite. *Trypanosoma equiperdum*.

Hosts. Horse and ass.

Geographical distribution. Europe, America, North Africa, Indian subcontinent.

Transmission. Mechanical—during coitus.

Disease. Surra

Parasite. *Trypanosoma evansi*.

Hosts. Horses and mules in India.

Geographical distribution. Indian subcontinent.

Transmission. Mechanical—Tabanid fly

Disease. Souma

Parasite. *Trypanosoma vivax*

Host. Horse and cattle.

Geographical distribution. Tropical Africa.

Transmission. Tse-tse fly (vector), *Glossina* sp., Tabanid fly.

Disease. Strumisch disease (violent ungulate and primate disease)

Parasite. *Trypanosoma simiae*

Hosts. Goat (chronic disease), pigs and monkeys (fatal).

Geographical distribution. Tropical Africa.

Transmission. Tse-tse fly (vector), *Glossina* sp.

Disease. Rat blood disease (nonpathogenic)

Parasite. *Trypanosoma lewisi*.

Host. Rat.

Geographical distribution. Worldwide.

Transmission. Contaminative rat flea (vector), *Xenopsylla* sp and others.

OTHER BLOOD PARASITES

Disease. Kala-azar (Dum-Dum fever in Calcutta)

Parasite. *Leishmania donovani*

Host. Human.

Geographical distribution. India, China, Africa, South Europe.

Transmission. Sand fly (vector), *Phlebotomus argentipes*.

Disease. Oriental sore (limited to a local lesion of the skin and subcutaneous tissue)

Parasite. *Leishmania tropica*

Host. Human.

Habitat. Intracellular parasite (reticuloendothelial system).

Geographical distribution. West to Central Asia, parts of Central and Western India and places in Central Africa.

Transmission. Sandfly (vector), *Phlebotomus* sp.

Disease. Texas fever (Red water fever of cattle).

Parasite. *Babesia bigemina*

Host. Cattle.

Habitat. Blood.

Geographical distribution. Worldwide.

Transmission. *Tick, Boophilus* (vector).

☒ Malaria

Occurrence. Malaria fever occurs in human, higher apes, monkeys, bats, birds, canaries, domestic fowls, ducks, pigeons, sparrows, lizards, etc.

Geographical distribution. Restricted to 60° N and 40° S. Rate of incidence varying with temperature.

Habitat. Blood and reticuloendothelial cells of liver; parenchymatous cells of liver in monkeys.

Fever. High body temperature at certain intervals* associated with shivering.

Disease. Benign tertian (B.T.) malaria (vivax malaria). Fever recurring at 48 hour interval, i.e., fever every third day.

* **Quotidian malaria.** In double infection with *P. vivax* and *P. malariae* and also double infection with *P. falciparum* and *P. ovale*, fever recurs at 24 hours interval, i.e., fever every day.

Parasite. *Plasmodium vivax.*
Host. Human
Transmission. Female *Anopheles* mosquito (vector).

Disease. Quartan malaria. Fever recurring at intervals of 72 hours, i.e., fever every fouth day.
Parasite. *Plasmodium malariae*
Host. Human
Transmission. Female *Anopheles* mosquito (vector).

Disease. Ovale tertian malaria. Fever recurring at 48 hour interval, i.e., fever every third day.
Parasite. *Plasmodium ovale.*
Host. Human.
Transmission. Female *Anopheles* mosquito (vector).

Disease. Malignant tertian (M.T.) malaria. Fever recurring at 48 hour interval, i.e., fever every third day.
Parasite. *Plasmodium falciparum.*
Host. Human.
Transmission. Female *Anopheles* mosquito (vector).

Pernicious malaria (Cerebral malaria, Malignant malaria) occurs due to the infection of *P. falciparum*; the patient dies within 1 to 3 days.

Blackwater fever (Haemoglobinuric fever)

It is a pernicious manifestation of falciparum malaria occurring in previously infected patients and associated with sudden intravascular haemolysis followed by fever and passage of haemoglobin in the urine.

Disease. Simian malaria
Parasite. *Plasmodium cynomolgi*
Host. Monkey. *Macaca* sp.
Transmission. Female *Anopheles maculipennis* (vector).

Disease. Bird (Fowl) malaria
Parasite. *Plasmodium gallinaceum.*
Host. *Gallus* sp.
Transmission. Female *Anopheles* mosquito (vector).

Disease. Bird (Duck) malaria
Parasite. *Plasmodium lophurae.*
Host. Ducks.
Transmission. Female *Anopheles* mosquito (vector).

Disease. Bird (Sparrow) malaria
Parasite. *Plasmodium praecox.*
Host. Sparrow.
Transmission. *Culex papiens* female mosquito (vector).

Phylum PLATYHELMINTHES

Disease. Liver rot (facioliasis hepatica)
Parasite. *Fasciola hepatica* (Liver fluke).
Hosts. Sheep, goat, cattle, rarely man.
Habitat. Biliary passages of liver.
Geographical distribution. Worldwide.
Transmission. Amphibious snail (Pulmonata) *Lymnaea* (vector).

Disease. Fasciolopsiasis in man
Parasite. *Fasciolopsis buski* (largest trematode parasite in man)
Hosts. Man, pig.
Habitat. Small intestine.
Geographical distribution. Asiatic-China, (Malaya, Federation of), Bengal (undivided), Assam and other Oriental regions.
Transmission. Aquatic snail, *Segmentina* (vector).

Disease. Endemic jaundice (clonorchiasis)
Parasite. *Clonorchis sinensis* (Chinese/Oriental liver fluke).
Hosts. Man, dog, cat, pig.
Habitat. Biliary tract of liver, common bile duct and pancreatic duct.
Geographical distribution. Far East and the endemic areas in Japan, Korea, Formosa, Indo-China and Southern China.

Transmission. Two intermediate hosts: aquatic snail *Bithynia* and cyprinoid fishes. Infection in man through eating prickled or insufficiently cooked fishes.

Disease. Schistosomiasis haematobia (endemic haematuria).

Parasite. *Schistosoma haematobium* (vesical blood fluke).

Host. Human.

Habitat. Pelvic venous plexus—particularly vesical, prostatic and uterine plexuses. The adult worms live in copula.

Geographical distribution. Endemic in the Nile valley, Egypt; parts of Africa and West Asia, spreading with extended irrigation canals.

Transmission. Direct, intermediate host fresh water snail, *Bulinus*.

Disease. Taeniasis saginata (Beef tapeworm infection)

Parasite. *Taenia saginata* (Beef tapeworm).

Hosts. Adult host—man (definitive host); larval host—cattle.

Habitat. Larval stage, *Cysticercus bovis* in adipose tissue and striated muscles of cattle. Adult in small intestine of man, attached to mucosa.

Geographical distribution. Worldwide, restricted to beef-eating people.

Transmission. Man to man via cattle through undercooked beef containing *Cysticercus bovis*.

Disease. Taeniasis solium (Pork tapeworm infection)

Parasite. *Taenia solium* (Pork tapeworm)

Hosts. Adult host man; larval host pig, occasionally man.

Habitat. Larval stage, *Cysticercus cellulosae* in striated muscles of tongue, neck, shoulder, ham and cardiac muscle of pig. Adult in anterior third of the small intestine of man, attached to mucosa.

Geographical distribution. Worldwide, restricted to pork-eating people.

Transmission. Man to man via pig through undercooked 'measly pork' containing *Cysticercus cellulosae*.

Phylum ASCHELMINTHES (Nemathelminthes)

Disease. Trichinelliasis or Tricinosis

Parasite. *Trichinella spiralis*.

Hosts. Man, pig, rat.

Habitat. Small intestine—duodenum or jejunal mucosa.

Geographical distribution. Europe, United States of America, Syria, China, Africa.

Transmission. Normally from pig to pig and rat, and from rat to rat. The same animal harbours both larvae and adults. Infection spreads through eating striated muscles infected with encysted larva.

Disease. Ancylostomiasis (Hookworm disease)

Parasite. *Ancylostoma duodenale* (old world hookworm).

Host. Primarily man.

Habitat. Small intestine, particularly jejunum, less often in duodenum and rarely in ileum.

Geographical distribution. Tropical and subtropical countries from 36°N to 30°S.

Transmission. Direct—the filariform larva in the soil, penetrates the skin of the host and move inside.

Disease. Ascariasis

Parasite. *Ascaris lumbricoides*.

Host. Man.

Habitat. Small intestine.

Geographical distribution. Worldwide, particularly the tropics—India, China and Pacific Islands.

Transmission. Direct—through food, drink and raw vegetables contaminated with infective embryo in the egg.

Disease. Elephantiasis (Wuchereriasis bancrofti)

Parasite. *Wuchereria bancrofti.*

Host. Man.

Habitat. Adults—lymphatic nodes and lymph vessels; embryos—circulating blood.

Geographical distribution. Tropical and subtropical contries, viz., India, South China, Japan, Pacific Islands, North Australia, West Indies, Puerto Rico.

Transmission. Mosquitoes (vector). *Culex fatigans* in India and China; *Aedes scutellaris* in Pacific Islands.

Periodicity of the larva. *Microfilaria bancrofti*

a. In India and China, the larvae appear in peripheral blood at night from 22.00 hours to 04.00 hours and transmitted by night-feeding *C. fatigans.*

b. In Pacific Islands, the larvae are present in the peripheral blood both in day and night and transmitted by *Aopheles scutellaris female.*

Disease. Loaiasis (fugitive swellings)

Parasite. *Loa loa.*

Host. Man.

Habitat. Connective tissue; often found in the conjunctiva of the eyes.

Geographical distribution. West Africa, particularly along the coasts of Congo river.

Transmission. Blood-sucking mangrove fly *Chrysops* (vector).

Periodicity of the larva. *Microfilaria loa*

The larvae are diurnal, i.e., appear in the peripheral blood in day-time only.

ANNEXURE I

GLOSSARY

Abiogenesis. Origin of life from nonliving matters.

Aboral. Opposite to mouth.

Absorption. Taking up of fluid or other substances.

Abyssal. The deep ocean, 4,000 to 6,000 m.

Accessory. Extra and subordinate.

Acclimation. A change of a single adaptation, a readjustment of the physiology of organism reaching a new state.

Acclimatization. An adjustment of the physiological state of an organism in response to a series of intercorrealated parameters changing in concert.

Acentric fragment. A fragment of a chromosome lacking a centromere.

Acetylcholine. The acetic acid ester of the organic base choline, normally secreted at the ends of many neurons.

Acid. A substance which releases hydrogen ions (H^+) in water.

Acoelomate. Animals having no body cavity.

Acrocentric Chromosome. A chromosome with a terminal or near terminal centromere.

Acron. The anterior nonsegmented part of the body of a metameric animal.

Acrosome. A cap-like structure covering the head of the spermatozoa.

Acting. Minute protein filament found in most contractile cells.

Action potential. Generation of a slight current on activation of a tissue.

Active transport. The transfer of a substance across the plasma membrane against a concentration gradient, a process requiring energy.

Adaptation. The modification of the organs of an animal to suit best with the environment.

Adaptive radiation. Evolutionary diversification of a group of animals or their structures, following their entry into a new adaptive zone.

Adaptive value. The survival value and the reproductive capacity of a given genotype relative to other genotypes in the population.

Adaptive zone. The 'way of life' of a single species or a group of organism.

Adenine. One of the two purines (amino purine) involved in the structure of DNA and RNA.

Aestivation. Dormancy during dry seasons.

Aerobe. Growing only in presence of air.

Afferent. Conducting inward to an organ; leading inwards.

Agglutination. The collection of cells or particles of a fluid.

Agnatha. A group of vertebrates without jaws.

Air bladder. An elongated, thin-walled sac containing gas; acts as a hydrostatic organ in fishes.

Alimentary canal (gut). The tube concerned with ingestion and digestion of food and absorption of the products of digestion (nutrients).

Alkaloids. Nitrogenous plant products, some of which have physiological activity in animals, e.g., morphine, caffeine and nicotene.

Allantois. An extraembryonic membrane in mammals, which carries requirements of the embryo from mother and removes wastes from the embryo.

Allele. One of the several alternative states of a functional gene unit.

Allergy. A hypersensitivity to some substances causing disorder in the body.

Allopatric. Animals living in nonoverlapping geographical areas.

Allopatric speciation. The differences of geographically isolated populations into distinct species.

alveolus (i). A small, thin-walled, sac-like dilation, the terminal chamber of bronchi in vertebrate lung.

Amoeboid movement. Movement of a cell by the slow flow of the cellular contents, sol to gel.

Amino acids. The chemical building blocks of proteins; proteins yield amino acids on hydrolysis.

Amitosis. Division of a nucleus into two halves by a simple constriction.

Amnion. A fluid filled extraembryonic membrane around the embryo in amniota (reptiles, birds and mammals).

Amylase. An enzyme capable of catalyzing the hydrolysis of starch, α-1 $\rightarrow$ 4 glucocidic bonds of polysaccharides.

Anabolism. Chemical reactions in which smaller molecules are joined together to form large, complex molecules.

Anaerobic. Metabolism occurring in absence of oxygen.

Analogy. Likeness in function among nonhomologous structures.

Anaphase. The stage of nuclear division in which the chromatids of the chromosomes separate and move towards opposite poles of the spindle.

Anion. An ion containing negative charge.

Anoxia. Lacking oxygen.

Antibody. (aggutinin). A protein produced in response to a foreign substance in the blood or tissue.

Antigen (agglutinogen). A foreign substance, usually a protein or proteinpolysaccharide complex, which induces formation of specific antibody in an organism.

Anadromous fish. Fishes spawning in fresh water (river) but spending major part of their life in sea

Aorta. A primary artery distributing blood to different parts of the body through branches.

Archenteron. The rudiment of the gut in the gastrula lined by endoderm cells.

Arteriole. A minute branch of an artery, which branches into a number of narrow vessels, the capillaries.

Artery. A thick-walled vessel carrying blood away from the heart.

Artificial selection. The choosing of the genotypes, within a limitation, by man, contributing to the gene pool of successive generations.

Asexual reproduction. Reproduction of the individual without production of gametes and zygote.

Atoll. A series of islands arranged in a horseshoe or circular form capping a coral reef system around, as oceanic volcanic mountains.

Back-cross. The mating of an offspring with one of its parents or the parental type.

Bacteriophage. Virus which attacks bacteria.

Barrier. A physical, physiological or behavioural condition that obstructs or prevent the distribution of animals.

Baroreceptor. A change of pressure detecting receptor.

Base. A compound that releases hydroxyl ions (OH^-) in solution with water.

Benthos. The animals or plants living at the bottom of lakes or oceans.

Binary fission. Division of a cell into two parts; a mode of asexual reproduction.

Biological clock. An internal timing mechanism in animals governing and regulating circadian rhythm and other cycles.

Biological oxidation. Removal of electrons from one atom or molecule and transfer of the same through electron transmitter system in mitochondrion (respiration).

Bioluminescence. Light emission by organisms.

Biomass. The total weight of all living organisms per unit area or volume, usually expressed as grams of carbon; total dry weight and so on.

Biome. Large community unit formed in similar physical and biotic factors.

Biotic potential. The capability of a species to increase in number under optical condition.

Bivalent. The synapsed pair of chromosomes in the meiotic prophase.

Bisexual (hermaphrodite). Having both male and female organs in an individual (Monoecious in plants)

Blastocoel. The cavity of a blastula.

Blastocyst. The modified blastula stage in a mammalian embryo.

Blastopore. The opening of the archenteron in a gastrula.

Blastula. An early stage of embryonic development, usually a hollow sphere of a single layer of cells (blastomeres), enclosing a cavity, the blastocoel.

Blood pigment. A molecule capable of carrying oxygen in animals, present in blood; haemoglobin, haemocyanin are chief blood pigments in chordates and nonchordates, respectively.

Boring. Capability to penetrate a substratum by scraping or chemical dissolution.

Bowman's capsule. A hollow double-walled sphere of flattened cells enclosing a glomerulus in vertebrate kidney.

Brackish sea. Large semienclosed bodies of water having its own circulation pattern inspite of tidal stirring and flow of freshwater from rivers.

Brain stem. The portion of the brain left after the removal of cerebral hemispheres and the cerebrum.

Breeding size. The number of individuals in a population that are involved in reproduction in a generation.

Browsers. Organisms feeding by scraping thin layers of living organisms from the surface of a substratum, e.g., periwinkles, sea urchins, etc.

Buccal. Pertaining to mouth.

Budding. Asexual reproduction; a small part of the body separates from the parent body to become an independent individual or remains attached to the parent animal to form a colony.

Buffers. Substances in a solution reducing to some extent the change in hydrogen ion level (pH), which otherwise would affect in adding acids or bases.

Bulk flow. The transport of materials, gas or solid, present in the extracellular fluid.

Caecum. A cavity with an opening at one end.

Calcareous. Being made of calcium carbonate.

Capillary. A thin-walled minute vessel connecting an arteriole with a venule.

Carbohydrate. An organic compound of a chain of carbon atoms in which hydrogen and oxygen are bound in a ratio 2 : 1.

Carnivore. Animal feeding upon freshly killed animals.

Carrying capacity. The total number of individuals in a population that can be supported in a given area.

Catadromous fish. Fishes spawning in sea but spending major part of their life in estuarine or fresh water.

Caudal. Concerning tail.

Cations. Ions bearing a positive charge.

Cellulose. An insoluble carbohydrate, the microfibrils of which are formed by polysaccharide chains of monasaccharides (glucose).

Cell wall. The thin structure enclosing a plant cell, containing primarily of cellulose and other polysaccharides.

Centriole. An organelle generally found in pairs and functioning in nuclear control of cell division and also in the origin and development of cilia and flagella.

Centromere (Kinetochore). The portion of a chromosome to which the spindle fibres are attached in cell division.

Centrosome. The substance surrounding the centriole, presumably associated with organization of protein fibres.

Cephalic. Pertaining to head.

Cephalization. (Head formation; concentration of the nervous tissue and sense organs at the anterior end of the body.

Cercaria. The last free swimming larval stage of a trematode parasite.

Cerebellum. Muscular contraction coordination centre of vertebrates, a part of the posterior brain.

Cerebrum. The two cerebral hemispheres unite to form the cerebrum, the largest part of the vertebrate central nervous system, (brain) occupying the upper part of the cranium.

Character. Features peculiar to a group and that vary in different groups.

Chelate. The terminal part of an appendage in the form of pincers (chelate legs of arthropods).

Chelicera. A pair of pincer-like cephalic appendages (present in arachnids).

Chemosynthesis. Primary production of organic matters using various substances without light as energy source.

Chiasma (chiasmata). A cross-shaped appearance of the chromosomes in a bivalent in the first meiotic prophase.

Chitin. An insoluble, horny glycoprotein, abundant in many arthropods and other animals.

Chlorophyll. Magnesium complexes of polyparphyrins present in green pigments in the cells of majority of plants.

Chloroplast. Membrane bound, chlorophyll containing organelle, present in a most eukaryotic plant cells.

Choanocyte. A cell with a flagellum surrounded by a protoplasmic collar, present in protozoan and sponges.

Chorion. An extraembryonic membrane in amniotes forming an outer covering of the embryo; contributes to placenta formation in mammals.

Chromatid. A half chromosome (subdivision of a chromosome) connected with the other half by a centromere.

Chromatin. The stainable portion of a nucleus forming a network of fibrils, made of protein and DNA.

Chromatography. The separation of different chemical compounds with gels or paper, based on their different migration through or on various substances.

Chromatophore. Pigment producing and pigment retaining cell.

Chromonema. The first visible longitudinal subdivisions of a chromosome.

Chromosome. Filamentous or rod-shaped bodies in the nucleus of a cell bearing genetic information and contain centromere or spindle attachment point.

Cilium (a). A small protoplasmic extension of a cell and moving in a particular pattern; usually a large number of cilia are present in an individual.

Cleavage (segmentation). The early divisions of the zygote producing blastomeres.

Clitellum. Segments of earthworm and leech having genital pores and glands, the latter secreting cocoon.

Cloaca. The common chamber at the posterior end of the gut, in which rectum, genital and urinary ducts open. The cloaca opens to exterior.

Clone. A population produced by asexual reproduction of a single individual.

Cnidoblast (Cnidocyte). A stinging cell of cnidarians having a sac-like nematocyst with a long, barbed thread and poisonous fluid and a trigger hair, the cnidocil.

Coacervetes. Aggregate of varying degrees of complexity, resulting from the interaction of two or more colloids.

Cochlea. The coiled portion of the membranous labyrinth (inner ear) having sound receptor cells.

Coelom. Body cavity of triplobastic animals lined with mesoderm.

Coenzyme. Substances in combination with specific proteins form complexes that are catalytically active.

Colloid. Dispersal of a solute in a solvent having properties of both a solution and a suspension.

Colony. Association of a number of individuals of the same species in which a division of labour is present.

Commensal. Living together; an association of two species, beneficial to one but having no effect on the other.

Community. Groups of organisms living in a particular area or habitat, the members of which interact with one another in various ways.

Competition. An interaction between or among two or more individuals or species for exploitation of resources which affect others negatively.

Cones. Thecone-shaped photoreceptor cells in the retina of the eye, sensitive to bright light.

Conformer. An organism whose physiological state is identical to or varies with that of the surroundings.

Conjugation. Union of two cells or organisms during which all or part of the genetic information of one is transferred to the other.

Conservation of energy. The amount of energy is constant in any given system; energy is neither created nor destroyed but transformed from one form to another.

Consumer organism. Organisms eating other organisms; the term is used in ecology.

Continental shelf. A broad expanse of ocean bottom sloping to the sea from the shore line up to a depth of 100 to 200 m.

Contraclile vacuole. Temporary osmoregulatory structure in protozoa.

Convergence. Superficial resemblance in animals of different groups due to their living in the same adaptive zone.

Coral reef. A wave resistant structure formed from cementation process of the skeletal elements of corals, calcareous algae and other calcium carbonate secreting organisms.

Corpus luteum. A glandular mass of cells in mammalian ovary around the ovum.

Corpus striatum. A subcortical mass of neurons in the base of the cerebral hemispheres.

Countercurrent. Flow of two fluids in opposite directions.

Cranium. The part of the skull holding the brain.

Crop. A sac-like structure in the gut, anterior to the stomach.

Crossing over. The synapsis and exchange of segments of homologous chromosomes in meiosis.

Cyst. The protective covering formed around an organism in unfavourable conditions.

Cytokinesis. The division of the cytoplasm during cell division.

Cytopharynx. Funnel-like organelle leading from surface to endoplasm in certain ciliates and other protozoans.

Cytosome. The mouth opening of certain ciliates and other protozoans.

Deamination. Removal of amino group (—NH_2) from an amino acid or some other organic molecule.

Dehydrogenation. Removal of hydrogen atoms from a molecule, a type of oxidation.

Deletion. The loss of a segment of a chromosome.

Denaturation. Change in the three dimensional structure of a protein or any macromolecule with the change in physical properties.

Density dependent factor. Factor, as resource availability, that varies with population density.

Density dependent selection. Fitness of a genotype depending upon the population density.

Deoxyribonucleic acid (DNA). Molecules containing the genetic inforamtion, which are replicated and equally distributed to daughter nuclei.

Deposit feeders. Organisms both aquatic and terrestrial feeding on detritus (decomposing organic matters) mixed with the soil.

Dermal bone (membrane bone). Bone that develops from direct deposit of calcic materials in the connective tissue.

Dermis. The layer of the skin beneath the epidermis.

Deuterostomia* (secondary mouth). Blastopore forming the anus and the mouth is a new opening. The animals are termed deuterostomes. Examples: echinoderms, branchiopods, polyzoans, phoronis, hemichordates, tunicates and chordates.

Diaphragm. A fibromuscular partition separating the body cavity into an anterior thoracic and a posterior abdominal cavity in mammals.

Diastole. Relaxation of heart muscles causing enlargement of the chambers of the heart.

Diatom. Planktonic algae with silicious test occuring in single cells or chains of cells.

differentiation. Transformation of a developing embryo in which diverse kinds of tissues and organs are formed.

Diffusion. The movement of the molecules of a substance from a higher to the region of lower concentration.

Diploid. In which the number of chromosomes is twice the basic (gamete) number of chromosomes.

Dominance. The capability of a character to express itself in a generation by total suppression of its allele, the opposite character.

Ductus arteriosus. A short vessel emerging from the dorsal part of the sixth aortic arch and connecting the pulmonary artery and dorsal aorta in the embryo.

Duodenum. The first part of the small intestine (near the stomach).

Duplication. The presence of a section of chromosome more than once in a genome.

Ecdysis. The moulting or shedding of exoskeleton.

Ecological efficiency. The amount of energy obtained from a trophic level, divided by the amount of energy supplied to that level.

Ecological niche. The interaction of a species with the biotic and abiotic factors of the environment.

Ecology. The study of interactions of organisms and the environment, both abiotic and biotic.

Ecosystem. An unit of living and non-living components that interact to produce a stable system in an area.

Ectoderm. The outer germinal layer of a gastrula.

Effector. The structures with which an organism reacts to stimuli.

Efferent. Leading away from.

Electrolyte(s). Substance(s) that dissociates into ions in a solution.

Electrophoresis. A technique of separating molecules with different charges by their different rates of migration in an electric field.

Elements(s). A natural or man made matter, the atoms of which are similar in every respect.

Embryo. The early stage of development of a multicellular organism.

Endocrine. Ductless gland, secretion of which has a specific effect on an organ—the target.

Endoderm. The inner germinal layer of a gastrula lining the archenteron.

Endopeptidase. A proteolytic enzyme that cleaves peptide bond within a peptide chain by hydrolysis.

---

* Actually the mouth and anus are formed in various ways and such division is unfortunate.

Endoplasm. The cytoplasm in a sol state, occupying the space of a cell inside the ectoplasm and contains nucleus.

Endoskeleton. Hard structures within supporting the body and giving it a definite shape.

Endostyle. A longitudinal groove lined by ciliated mucous cells on the floor of the pharynx in some lower chordates and cyclostome larva.

Endotherm. The ability to maintain a near constant body temperature due to higher metabolism; warm blooded animals.

Enterocoel. The cavities in the pouches formed by outgrowths of the primitive gut.

Entomology. A branch of Zoology dealing with insects.

Environment. Sum total of the factors—both abiotic and biotic—surrounding an organism including the organims itself.

Environmental stress. Environmental change to which an organism cannot adjust or environmental change that increase the probability of death.

Enzyme. An organic catalyst produced by cells that can accelarate specific chemical reaction.

Epiboly. A type of gastrulation in which the micromeres of the animal pole grow over and enclose the macromeres of the vegetal hemisphere.

Epidermis. The outermost covering cell layer of an organism.

Epiglottis. A lid-like structure in pharynx, which prevents the entry of food into the glottis.

Erythrocytes(s). Haemoglobin containing cells in the blood of vertebrates.

Estrogen. A female sex hormone secreted by ovarian follicle, promoting the development of secondary sex characters.

Estuary. A semi-enclosed water body having a free communication with open sea and in which sea water is measurably diluted with fresh water received from the rivers.

Eukaryotic cells. Cells having a well-defined nucleus and cellular organelles.

Euryhaline animals. Animals capable of tolerating a wide range of salinities.

Eutheria. A group of mammals in which the placenta is well-developed and the young ones born are fully formed.

Excretion. Getting rid of metabolic wastes.

Exergonic. A chemical reaction in which energy is released.

Exopeptidase. A proteolytic enzyme that cleaves only the terminal bonds of a peptide chain.

Exoskeleton. Hard covering on the surface protecting and supporting the body.

Faeces. Wastes, undigested portion of the food, finally egested through anus.

Faecal pellet(s). Compacted aggregation of undigested portion of food, finally egested through anus.

Fallopian tube(s). The oviducts in mammals, which receive ova after discharge from the ovaries.

Fat. An organic compound containing carbon, hydrogen and oxygen; the basic substances of fat are glycerol and fatty acid.

Fatty acid. An organic acid of long chain carbon.

Fecundity. The total number of eggs produced by a female per unit time (per spawning season).

Feedback. The influence of the products of a reaction on the process of reaction.

Fenestra. A moderate size opening in a structure, e.g. fenestra ovalis (an oval opening) in the foetal heart of mammals.

Fermentation. Anaerobic degradation of an organic compound by an enzyme, e.g. the splitting of sugar into alcohol and carbon dioxide with the enzyme produced by yeast.

Fertilization. The fusion of a sperm with an ovum producing a zygote.

Foetus. The embryonic stages in which development has progressed considerably.

Filtration. The passing of a liquid through a filter following a pressure gradient, e.g. in the glomerulus of the kidney.

Fission. Splitting; in asexual reproduction a unicellular organism divides into two equal halves; binary fission.

Fluke. Common English name for some parasitic flat worms, liver fluke.

Food(s). Substances yielding organic compounds from which biomolecules are synthesized; used as fuel; in building and repairing the body.

Food chain. A sequence of organisms in a community starting from primary producers ending in tertiary carnivore; energy is transferred from one to another in each step.

Food web. The trophic relationship or feeding interactions of the species in an area.

Foramen. An opening in a structure; it may be small (foramen of Monro—brain) or large (foramen magnum—skull).

Fossil. Remains of animals and plants of bygone days preserved in different geologic starta.

Gamete. A mature germ cell, sperm or ovum.

Gametocyte. A cell, which, through divisions, produces gametes.

Gametogenesis. The continued process of cell division and differentiation resulting in the formation of gaametes.

Ganglion. An aggregation of cell bodies of a number of neurons, outside the central nervous system.

Gastrula. A two-layered developmental stage of an embryo enclosing two cavities; the cell layers are ectoderm and endoderm. Ectoderm encloses the blastocoel and the endoderm encloses the archenteron.

Gene. A self-duplicating hereditary unit having more than one state and whose different states produce differences in the phenotype.

Gene flow. The movement of genetic information within and between populations.

Gene frequency. The number of loci at which a given allele is found within a population divided by the total number of loci at which it could occur.

Gene pool. The total genetic information present in a population.

Generation time. The time period from birth to average age of reproduction.

Genetic assimilation. The incorporation in a genotype of the characteristics appearing in ontogeny through the operation of agents of selection in the environment.

Genetic drift. Random fluctuation of gene frequency, due to sampling error, in populations; the effects are pronounced in very small populations.

Genome. A complete set of hereditary potentialities in the haploid set of chromosomes.

Genotype. The genetic make up of an organism.

Genus (pl. genera). The level of taxonomic hierarchy above the species but below the family level.

Gill(s). Highly vascularised respiratory organ of aquatic animals, the surface of which increases with the appearance of folds, lamellae, filaments, etc.

Gizzard. A specialised part of the gut, usually stomach, meant for mastication of food.

Gland. A group of cells, capable of secreting materials, which play important role in the metabolism of the animal.

Globulin. A protein; in some cases the gamma globulin factor of which acts as antibodies.

Glossary. A collection of specialised terms accompanying definition.

Glottis. Slit-like opening in the pharynx leading to the trachea (wind pipe).

Golgi complex. A secretary cell organelle in close proximity of the nucleus, comprising a series of concentrically arranged cisternae.

Glycolysis. Conversion of complex sugars into simple sugars.

Glucose. A six-carbon sugar, a primary source of energy for most organisms.

Gnathostomata. Vertebrates with jaws.

Gonad(s). A reproductive structure, testis or ovary, producing gametes, sperms or ova.

Gonoduct(s). The ducts carrying gametes to a particular site of the body or to the exterior; the ducts may or may not be connected with the gonads.

Gonopore(s). An opening through which gametes escape to the exterior.

Grab. A sampling device made of two or more curved metal plates designed to converge and grab (collect) a specified volume of bottom sediment.

Grazer. A predator, feeding upon animals far smaller than it in size, e.g., copepods, periwinkles, etc.

Green gland(s) (antennary gland). Excretory organs of certain crustaceans; paired structures located at the base of the antennae.

Habitat. The natural abode of a species of animals.

Haemocoel (Hemo). The narrow, intercommunicating spaces between tissues, lined with an epithelium and filled with blood (haemolymph) in invertebrates.

Haemocyanin. A respiratory pigment containing prosthetic group of copper, found in the majority of the invertebrates, it is shining blue in oxidised state and colourless in deoxidised state.

Haemolymph. A circulating fluid in invertebrates with open circulatory system, having the properties of blood and lymph fluid.

Haemostatis. The mechanism stopping out flow of blood from an injured blood vessel.

Haploid. Having the gametic (n) number of chromosomes, i.e. half the diploid (2n) number or that of the somatic cells.

Hardy-Weinburg law. It determines the frequencies of genotypes in a population at a locus by random mating and allelie frequency.

Heparin. Secretion by basiphilic leucocytes and mast cells, that inhibit blood clotting.

Hepatic. Concerning the liver.

Herbivore. An animal feeding upon plants.

Hermaphrodite. Having both the male and female sex organs in an individual; individual possessing the phenotype of both sexes with male-female sex differentiation.

Heterotroph(s). Organisms incapable of synthesizing its own food from inorganic substances and entirely depend on substances produced by autotrophs.

Heterozygous. An organism in which the alleles of a gene are different.

Hibernation. A greatly depressed state of metabolism in which certain animals pass the winter.

Homeostasis. A state of equilibrium in the internal environment of an animal.

Homeotherm(s). Animal capable of maintaining a near constant body temperature irrespective of the temperature of the environment.

Homologous organs. Organas developmentally alike, may be functionally unlike.

Homology. Structural similarity of organs in different groups of animals due to their inheritance from a common ancestor.

Homozygous. An organism in which the alleles of a gene are similar.

Hormones. Extremely complex chemicals secreted by endocrine (ductless) glands, which induce specific reaction in targets (effector).

Humus. Decomposed plant materials that impart a brown or black colour to the soil.

Hybridization. Gene flow between distinct populations.

Hydrogen bond(s). A weak bond between two molecules by sharing a hydrogen atom between two atoms, usually oxygen.

Hydrographic. Arrangement and movement of bodies of water, viz., current and water masses.

Hydrolysis. Splitting of a molecle by addition of a water molecule between certain bonds.

Hydroskeleton. A turgid column of fluid within the body of an organism giving support or rigidity to it.

Hyoid. A cartilaginous or partly cartilaginous and partly bony structure at the base of the tongue supporting the adjacent structures.

Hypertonic. A solution containing greater number of solute molecules when compared with the concentration of another solution.

Hypothalamus. The floor of the third ventricle in a vertebrate brain containing centres controlling temperature, water balance, visceral activities, sleep, etc.

Hypothesis. A refutable statement in science or in any other subject about one or more abiotic and biotic phenomena.

Hypotonic. A solution containing less number of solute molecules when compared with concentration of another solution.

Ileum. The last division of the small intestin—duodenum, jejunum and ileum.

Ilium (a). The bones comprising the dorsal part of the pelvic girdle.

Inducer. A chemical which brings upon the appearance and development of a structure.

Induction. Determination of the developmental fate of one tissue of an embryo by another tissue or an organizer.

Ingestion. The process by which a food particle is taken inside the cell in an acellular animal or passing the food to the alimentary canal through mouth in metazoan.

Immune response. The synthesis of antibodies in response to antigen.

Immunity. The capability to resist the attack of a disease causative organism due to presence of antibodies (both active and passive immunity).

Instar. The developmental stage of an arthropod between successive moults.

Integument. The covering of the body, the skin.

Interspecific competition. Condition arising out of the exploitation of limited resource of one species negatively affecting another species.

Interstitial. Spaces between the cells or small structures like sedimentary grains in a soft sediment.

Interstitial fluid. The fluid in which the cells of a tissue are bathed, the fluid in the spaces between the cells.

Invagination. Infloding of one cell layer into another in a single cell-layered blastula; invagination leads to the development of a two-layered gastrula.

Inversion. Reversal of the sequence of a portion of the chromosome.

Ion(s). An atom or a group of atoms having electric charge : positive (cation) or negative (anion).

Isolating mechanism. The factors that prevent interbreeding between two populations living in a geographical area.

Isomer. Compounds with same chemical formula but different chemical structures, e.g., glucose, fructose.

Isotonic. The number of solute molecules in a given volume being same in two solutions.

Isotope(s). Different forms of an element having same atomic number but different mass number (the number of neutrons vary).

Isozyme. Enzymes with similar catalytic functions but with structurally different molecules.

Jugular. Pertaining to neck.

Karyotype. The characteristic phenotype of the chromosomes of an organism.

Keratin. A horny, water-insoluble substance, the constituent of feathers, horns, nails, hairs, etc.

Labial. Pertaining to lips.

Larva. An embryonic stage of animals, which can carry life independently and morphologically dissimilar to the parents; it undergoes metamorphosis to assume the shape of the parents.

Meiosis. Two successive cell divisions producing four cells, each containing one-half of the number of chromosomes in parental nucleus.

Melanin. A dark brown or black pigment present in the integument of many animals.

Mendelian population. A reproductive group, sharing a common gene pool.

Meroplankton. Organisms spending a part of the life as plankton and a part as benthos.

Merozoites. Young parasites formed by the splitting of schizonts in the human cycle of malaria parasite (*Plasmodium*).

Mesenchyme. A group of cells derived from mesoderm, which are destined to form connective tissues in different organs.

Mesentery(ies). A two layered epithelium containing connective tissues, blood vessels and nerves run from body wall to visceral organs and from one organ to another.

Mesoderm. The third primary germinal layer of an embryo, present between ectoderm and endoderm.

Mesoglea. A gelatinous layer between the ectoderm and endoderm in cnidarians.

Mesopelagic. Organisms living in 150 to 2,000m depth zone.

Messenger RNA (m RNA). A ribonucleic acid synthesized in the nucleus, move out to cytoplasm, combines with ribosome RNA and forms a template for the synthesis of protein (Polypoptide).

Mesozoic. The period between 200 to 700 million year ago, the age of reptiles.

Myelin. A fatty material deposited in the form of a sheath around the axons of the neurons of the peripheral nervous system.

Myofibrils. The contractile fibres present in a muscle cell.

Myomeres. The muscle segments in an adult animal.

Myopia. Short or near sightedness.

Myosin. Minute protein filaments associated with actin, found in muscle cells.

Natural selection. Nonrandom (differential) reproduction of a genotype in natural populations.

Nauplius larva. The earliest stage of a crustacean larva with three pairs of appendages.

Nekton. Somewhat large forms of plankton that are active swimmers.

Nematocyst. A sac-like structure filled with a poisonous fluid and continuing into a fine, hollow thread, the stinging structure, lodged in specialized cells, the cnidoblasts (cnidocytes) in cnidarians.

Nephridium (a). The excretory tubules, collecting and also absorbing nitrogenous wastes from the coelom or haemocoelomic fluid discharging the same to the exterior directly or indirectly, found in many invertebrates.

Nephron. The functional unit of vertebrate kidney concerned with removal of nitrogenous wastes and osmoregulation.

Nerve net. A net-like primitive nervous system in some cnidarians, in the neurons, of which axon and dendrites are not differentiated.

Neuroglia. Connecting and supporting cells, surrounding the neurons, present in the central nervous system.

Neuron(s). The structural unit of the nervous system; a neuron consists of a cell body, dendrites and an axon.

Neurosecretion. The hormones produced by certain nerve cells.

Niche. A way of life; the range of environmental space occupied by a species.

Nitrogen fixation. The conversion of gaseous nitrogen to nitrate by bacteria and some physical factor like lightning.

Nondisjunction. Failure of the chromosomes in a bivalent or a monovalent to separate at the first meiotic anaphase.

Notochord. A dorsally placed rod of turgid cells, ventral to the nerve cord, functioning as a hydroskeleton, present in lower chordates; its place is occupied by vertebral column in vertebrates.

Nucleic acid. Complex acid composed of nucleotides, present both in the nucleus and cytoplasm.

Nucleolus. A spherical body, the site of ribosome production, present within the nucleus and composed primarily of RNA and protein.

Nucleotide. The chemical building block of a nucleic acid composed of a phosphate ester of N-glycoside of a nitrogenous base.

Nucleus. A membrane bound organelle of eukaryotic cells, the primary site of DNA and controlling centre of cell activities.

Nutrients. Constituents required by animals to synthesize biomolecules, organic matters obtained from the hydrolysis of foodstuffs—carbohydrates, proteins and lipids.

Ntrient cycling. The pattern of transfer of nutrients among the constituents of a food web.

Nymph. A juvenile insect resembling the adult; pupal stage absent.

Oceanic ridge. A tortuous ridge rising from the ocean floor.

Occipital. The base of the skull.

Ocellus (i). A simple, primitive light receptor.

Olfactory. Pertaining to the sense of smell.

Ommatidium. One of the many complete visual elements of a compound eye having lens and retina.

Omnivore. Animals feeding on more than one type of food, viz., plant food, animal food.

Oocyte. Cells having the potentialities to develop to a mature ovum.

Oogenesis. The origin and maturation of ovum.

Oogonium. The primordial germ cell from which the ovum is produced through division, growth and maturation.

Organelle(s). A structure with distinct morphology and function, within cytoplasm of a cell; the unicell analog of an organ in a multicellular animal.

Organic. Derived from living organism.

Organic detritus. Particles of decomposing organic materials.

Organic nutrients. Nutrients in the form of biomolecules synthesized by an organism or obtained from the synthesis of other organisms.

Organiser. A portion of an embryo that determines the course of development of another cell mass in its contact.

Osmoconformer. An animal whose osomotic concentration in the body change with the concentration of ions in the surrounding medium; incapable of maintaining a definite osmotic concentration.

Osmoregulator. An animal capable of maintaining a constant osmotic concentration in the body irrespective of ionic concentration in the surrounding medium.

Osmosis. The movement of solvent molecules across a membrane from the lesser to greater concentration of solute following concentration gradient.

Osphradium (a). Sense organs capable of testing water quality, present in molluscs.

Ovary(ies). The female gonad which produces eggs.

Ovulation. The discharge of mature ovum from the ovary.

Ovum (a). The female reproductive cell, which, being fertilized by a sperm, develops to a new individual.

Oxidation. A chemical reaction in which oxygen is gained, hydrogen is lost.

Parapodium. Paired appendages bearing setae and projecting laterally from each body segment; polychaete.

Parasite. An organism living within or on some organism (host) from which it obtains its food.

Parasitism. An intimate association between two organisms of two different species in which one—the parasite—obtains its food and protection from the other—the host.

Parallelism. Similar adaptation in closely related forms.

Parthenogenesis. The development of an individual from an unfertilized egg.

Pecten. A heavily black pigmented with close

longitudinal folds, light sensitive organ in the eyes of birds, helping in long distance vision.

Pectine. Comb-like tactile sense organs on the ventral surface of the second mesosomatic segment of scorpion.

Pedal. Related to foot.

Pelagic. Minute organisms with feeble moving capacity, living in open water.

Penis. The capulatory organ in male.

Pentadactyl. Limb having five fingers or toes.

Peptides. Chains of amino acids, components of a protein molecule.

Pericardial cavity. A chamber in the coelom containing heart.

Pericardial sinus. An enlarged portion of the haemocoel containing the heart; associated with open circulatory system.

Peristalsis. Rhythmic waves of muscular contraction and relaxation of the walls of tubular organs to move its contents, e.g. in intestine.

Peritoneum. Coelomic epithelium and underlying connective tissue lining the coelomic cavity.

pH. The acidity of water (–log 10), of the activity of hydrogen ions in water.

Phagocytosis. The engulfing of organisms or particles by a cell.

Pharynx. The muscular anterior part of the gut for ingestion of food.

Phenotype. The product of interaction between genetic information with the environment.

Pheromone. The secretion of an individual in the environment influencing the behaviour of another individual.

Phosphate bond (high energy). The anhidride linkage between phosphate ions.

Photolysis. Splitting water molecules with light energy.

Photoperiodism. The physiologic response of organisms to variation in light intensities and darkness.

Photosynthesis. The process of carbohydrate synthesis from inorganic materials, carbon dioxide and water with sunlight as energy source.

Phylogeny. The course of evolution of a species.

Phytoplankton. The microscopic, photosynthesizing organisms, usually algae, present throughout the water in open sea or lakes or ponds.

Pituitary. An endocrine gland located just below the hypothalamus of the brain and attached to it; secretes a large number of hormones and termed as 'master gland'.

Placenta. A structure connecting the foetus with the uterine wall of the mother, formed by contribution of both maternal and foetal tissues.

Plankton. Organisms suspended in water and incapable of moving against water current.

Planula. The planktonic larval forms of cnidarians.

Plasma. The liquid portion of blood containing fibrinogen in which blood cells are suspended.

Plasma membrane. The outer limiting membrane of an animal cell, separating it from other cells.

Platelets. Small, colourless blood corpuscles helping in clotting of blood present in mammals.

Plastids. Organelles of plant cells, usually containing pigments and those with green pigment synthesize carbohydrate molecules from inorganic molecules.

Pleistocene. The major part of the quaternary period, covering the period one million to 10,000 years ago.

Poikiloosmotic. Same as osmoconformer.

Poikilotherm. An organism whose body temperature is nearly same as that of the external environment.

Polar body(ies). The small products of meiosis in oogenesis that do not develop into eggs.

Polymorphism. Members of a species morphologically different : distinct phenotypes in a population.

Polyp. A single cnidarian or one of the members of a cnidaria colony.

Polypeptide. A long amino acid chain, the acid group of one molecule joining with the base of another molecule by losing a molecule of water.

Polypoid. Having more than two sets of haploid chromosome complements in the reproductive cells other than gametes.

Polysaccharide. Large carbohydrate molecule formed by the joining of a large number of monosaccharide molecules.

Pons. The central portion of the metencephalon of vertebrate brain.

Population. A number of individuals of a species living in a geographic area.

Population biology. Relationship and interactions of the members of a population in relation to space and time, functioning, taxonomy, population genetics, ecology and behaviour.

Population structure. The factors governing the pattern of union of gametes produced by the individual of a population..

Portal vein. A vein which arises from a set of capillaries and again divides into another set of capillaries.

Preadaptation. Adaptation to a situation apparently absent; in reality, the situation is present but not pronounced; the appearance of carbon dioxide tolerance (prerequisite for aerial respiration) in ancestral fishes of amphibia living in water bodies with less amount of dissolved oxygen.

Precursor. a chemical producing another chemical.

Predator. An organism that kills animals and feed upon them.

Primary production. The production of living matters by organisms through photosynthesis or chemosynthesis.

Primitive. A character or an organism less changed in quite a long period of time in comparison with others.

Primitive streak. A longitudinal thickening in the post gastrula embryo of birds and mammals, developing from migration of cells and formation of mesoderm.

Proboscis. A narrow, tubular outgrowth in the oral region used for feeding.

Proctodaeum. An ectodermal invagination at the posterior end of an early embryo destined to form the hind gut and communicates with the achenteron; the blastopore closes in many invertebrates.

Producer. An organism capable of using radiant energy or chemical substance to synthesize energy-rich organic molecules.

Proglottis (tids). The segments of the body of a tape worm.

Prokaryotic. Cells lacking a well-defined nucleus, plastids, Golgi bodies and mitochondria.

Pronephros. The kideny tubules developing from archinephric duct in the anterior end of the coelom; present in some vertebrate emebryo and adult cyclostomes.

Prophase. The first phase of mitosis in which the chromosomes become visible, shorten and move towards equatorial plate.

Prostomia (Protostomia, Primary mouth). Blastopore forming the mouth and the anus in a new opening. The animals are called Prostomes. Examples: platyhelminthes, annelids, arthropods and molluscs (see Deuterostomia).

Protease. A protein-splitting enzyme.

Protein. High molecular weight, large molecules formed by the joining of a very large number of amino acids.

Pseudocoel (pseudocoelom). A persistent blastocoel, a cavity between mesoderm and endoderm.

Pseudopodium(a). Temporary cytoplasmic protrusion of an amoeboid cell functioning as locomotory and feeding organ.

Pupa. An inactive stage of insect larva between larva and adult (imago).

Purine. An organic base of two condensed heterocyclic rings, a pyrimidine ring and an imidazole ring.

Pyrimidine. An organic base having a six membered ring system with nitrogen atoms in 1 and 3 positions.

Putrefaction. Anaerobic degradation of protein and amino acid with enzymes.

Pygostyle (ploughshare bone). The few last vertebrae in birds fused to form it, which is turned upwards; the number of vertebrae is four in **columba**.

Race. A small subgroup of a species, a population having certain morphological and physiological traits.

Radula. A chitinoid ribon with a large number of cusped teeth arranged in rows for rasping food in the majority of molluscs.

Range. The geographic area occupied by the members of a species.

Reabsorption. Selective uptake of certain substances from the glomerular filtrate by the tubules of the kidney; also in the excretory tubules in invertebrates.

Recapitulation. The appearance of stages, similar to the adults of presumptive ancestors, during development of an individual, 'ontogeny recapitulates phylogeny'.

Receptor(s). A sensory cell or a nerve-ending capable to respond to stimulus.

Recessive genes. The genes that cannot express their phenotype in the presence of dominant genes; the genes can express their phenotype only in homozygous state.

Recombination. Appearance of genes combination not inherited from parents.

Redia. The second larval stage in the life cycle of flukes, capable to produce generation of larvae by asexual reproduction.

Reduction. Opposite to oxidation; a chemical reaction in which oxygen is lost or hydrogen is added.

Regulator. An organism capable of maintaining some aspects of its physiologic state despite different and changing state in the environment.

Regeneration. Reconstruction of a lost part of the body.

Renal. Pertaining to kidney or excretory organ.

Renal corpuscle. The glomerulus and the Bowman's capsule of a renal tubule together, in a vertebrate kidney.

Reproduction. Appearance of new individuals from a single individual or through gametes produced by male and female, the offsprings resembling the parents.

Reproductive cells. The gametes and the cells from which they are produced by division.

Reptiles. A group of terrestrial amphibious and aquatic vertebrates in which the body is covered with horny epidermal scales or plates.

Respiration. The act of breathing or inspiration and expiration of air; the chemical reaction in which oxygen is used and carbon dioxide, water and energy are produced by oxidation of foodstuffs.

Respiratory pigment. Metal containing proteins capable of binding oxygen and carbon dioxide, forming an unstable compound, a reaction which is reversible.

Retina. The innermost of the three layers of the wall of the eye-ball containing light receptor cells, rods and cones and innervated by the optic nerve; an inverted image of the object is formed in the retina.

Rh factor. An agglutinogen present in the erythrocytes of about 85 per cent white population, orginally discovered in rhesus monkey, hence so named.

Riboncucleic acid (RNA). Nucleic acid having ribose sugar; found both in the nucleus and the cytoplasm; participates in portein synthesis.

Rods. The rod-shaped photoreceptor cells in the retina of eye, senstitive to dim light and colours.

Saccule. A hollow sac in the inner ear, lined with sensitive hairs and containing endolymph and otoliths; organ for static balance.

Salinity. Grams of dissolved salts in one liter of solution with water.

Scavenger. An animal feeding upon dead or decomposing animals.

Sclera. The fibrous outer supporting membrane of the eye-ball in the posterior five-sixths, the anterior part forming the cornea.

Scrotum. A thin pouch consists of integument, muscular and connective tissue, which contains the testes in most mammals.

Secretion. Materials elaborated by glandular tissue having some specific physiological activities.

Segmentation. Division of an animal body into some nearly similar parts; cleavage of zygote in the early development.

Selection. Nonrandom (differential) production of genotypes.

Semicircular canals. The three canals in the membranous labyrinth containing sense hairs and endolymph, capable of detecting changes in the angular acceleration of the body.

Seminar receptacle(s). A part of the male genital duct, usually dilated to store sperms in premating period.

Serum. The light-yellow fluid portion of the blood after removal of its other constituents.

Sessile. Permanently attached to some substratum and immobile for life.

Setae. Small, needle-like chitinoid structures, projecting from the body surface; locomotory structures in annelids.

Soma. Body tissue.

Sex chromosome. Heterochromatic chromosomes, determining sex in individual, at least in part; Y is the maleness determining chromosome, X is the femaleness determining chromosome.

Sexual reproduction. Reproduction by the union of male and female gametes.

Sinuatrial node. The junction between sinus venosus or superior vena cava and right atrium, made of modified cardiac muscle, initiates heart beat and controls contraction rate of heart.

Sinus venosus. A chamber in the heart of lower vertebrates, which receives deoxygenated blood and opens into the right atrium.

Solute. A substance that dissolves in a solution.

Solvent. The liquid in which a substance (solute) dissolves.

Slope. A sharp downward bottom extending seaward from the continental shelf to abyssal depth.

Somatic growth. Growth of the body excluding gonads.

Spatial distribution. The distribution of individuals in a space.

Speciation. The origin of new species from a preexisting species: evolution of higher taxa from species.

Species. A small group of similar individuals, capable of interbreeding but unable to interbreed with members of other species (reproductively isolated) in natural population.

Spermatophore(s). A capsule enclosing numerous sperms packed inside; secreted by accessory male reproductive organ in some invertebrates.

Spicule. Needle or rod-like or variously-shaped mineral structure, forming skeletal elements in sponges and rarely in some other animals.

Spindle. The ellipsoid mass of protein fibers appearing during cell division, to which chromatins are attached.

Spiral cleavage. The cleavage planes are oblique in relation to polar axis of the zygote.

Spongin. A skeletal substance made of flexible protein fibres, found in sponges.

Stabilizing selection. Selection in which genotypes close to the mean for a character have some advantage over others at a distance.

Statocyst(s). A hollow, oval or round organ with sense hairs and some particles, helping the animal to detect change in body position in relation of gravity.

Stenohaline. Animals restricted to a narrow range of salinity; unable to withstand a wide range of salinity.

Steroid. Closely related to sterols, a class of solid alcohol having cyclopentano perhydrophenanthrene ring, found in the sex hormones, hormones of adrenal cortex, bile acids etc.

Stomodaeum. An ectodermal invagination at the anterior end of an early embryo destined to form the foregut, communicates with the archenteron; occur in many invertebrates.

Subspecies (incipient species). A geographic subdivision of a species, considered as a distinct group, which would attend the status of a species in due course.

Subtropical. The part of the temperate zone close to the equatorial region.

Swim bladder. An elongated, thin-walled sac, filled with a gas and lying in the dorsal part of the abdominal cavity in teleosts; a hydrostatic organ.

Symbiosis. Intimate beneficial association between two dissimilar species.

Sympathetic system. One of the two constituents of the autonomic nervous system which functions automatically; innervates most of the visceral organs and antagonistic to parasympathetic system.

Sympatric. Occuring in the common geographic area.

Sympatric speciation. Speciation without geographical isolation.

Synapse. The junction between the axon of one neuron and dendrites of another neuron.

Synapsis. The association of homologous chromosomes in the prophase of first meiotic division.

Syncitium. A multinucleate mass of cytoplasm not divided into cells.

Syngamy. Union of two gametes to form a zygote.

Systole. The rhythmic contraction of the heart.

Taxon (Pl. Taxa). An unit in classification; at the lowest level it is species, at the highest, kingdom.

Taxonomy. Describing, naming and assigning proper place of an animal in classification.

Telencephalon. The anterior part of the brain comprisd of olfactory lobes and cerebral hemisphères.

Teleost(s). The group of bony fishes.

Telophase. The last phase of nuclear division in which chromosomes lose their identity and daughter nuclei are formed.

Tetrad. Four homologous chromosomes together at the end of the first meiotic prophase.

Tetrapod. Four-legged vertebrates, amphibia to mammalia.

Thorax. Insect's middle part of the three divisions of the body; mammal's chest cavity.

Thrombin. The enzyme that converts fibrinogen to fibrin.

Thymus. A gland at the base of the neck in young mammals, produces T lymphocytes; atrophies in the adult.

Thymine. 5-methyl-2, 4-deoxypyrimidine, the pyrimidine involved in the structure of DNA, not RNA.

Tidal current. A water current produced by regularly varying tidal forces.

Tides. Periodic movements of oceanic water caused by gravitational pull between earth, sun and moon.

Trachea(e). Insect's narrow air tubes conducting air from atmosphere to the tissues through its branches (tracheoles); land vertebrates—a tube conducting atmospheric air from outside to lungs through its branches (bronchi).

Transfer RNA (t-RNA). A form of RNA acting as an adaptor in the synthesis of protein.

Trochophore. A free-ling, top-shaped larva with girdles of cilia in polychaetes.

Trophic level. A level in food chain containing organisms of similar feeding habits with respect to the chain, e.g. herbivores.

Tropical. The latitudinal zone bounded by two tropics, 23-50 north and south.

Turbidity. The weight of particulate matters per unit volume of water.

Umbilical cord. The cord connecting the foetus with the placenta, umbilical artery and vein run through it.

Ungulata. Mammals walking on tips of the digits, protected by a hoof.

Univalent. A chromosome lacking a homolog.

Urea. Water soluble end product of protein metabolism.

Ureter(s). A tube carrying urine from the kidney to the exterior through cloaca or to the urinary bladder.

Urethra. The passage carrying urine from the bladder to the exterior in mammals.

Uric acid. A sparingly water soluble end product of nucleic acid and protein metabolism.

Urinogenital. Related to reproductive and urinary system, collectively.

Uterus. A hollow muscular organ in female mammals for carrying and nourishing the young during development, till birth.

Vaccination. Treatment of a person with an antigen to stimulate antibody production; a precautionary measure to prevent occurrence of diseases.

Vacuole. A space within a cell, usually filled with intracellular fluid.

Vagina. The terminal portion of the female genital duct, running from uterus to exterior (mammals).

Vein(s). Vessel carrying blood from tissues to the heart.

Veliger. A free-living larval stage of gastropod molluscs.

Viability. The capability of living or developing.

Ventricle(s). A muscular chamber of the heart (two-chambered in birds and mammals) to receive blood and pump it out through arteries.

Vertebrata. A subphylum of chordata; members possessing a vertebral column or backbone.

Vestigeal organ(s). Organ, which was once well-developed and functional, is in the process of degeneration due to loss of function.

Villus(i). Small, narrow, finger-like projection from the surface to increase the surface area; small intestine of vertebrates.

Visceral arch(es). A narrow, cartilaginous or bony arch-like structure in the wall of the pharynx between two gill slits, supporting the gills in fishes.

Vitamin(s). Chemicals required in trace concentration, acting as cofactors with enzymes in catalyzing bio-chemical reactions.

Viviparous. Giving birth to young; the embryos develop within the body of the mother and young ones are born.

Water mass. A distinctly separate body of water with its own characteristics like temperature, salinity, pH, etc.

Weberian ossicle (apparatus). A chain of small bones in carps and siluroids connecting the air bladder with the inner ear, helping in maintaining equilibrium while moving from shallow to deeper water.

Yolk. Food material stored within the egg.

Yolk sac. A pouch-like outgrowth from the gut of amniotes that envelops the yolk mass and digest it to provide nourishment to the developing embryo.

Zooid. An animal, an individual in a colony carrying a particular function is usually termed zooid.

Zoology. The branch of Biology that deals with animals.

Zooplankton. Animal members of the plankton community.

Zygote. The product of the union of male and female gametes.

Zygotene. A stage of the first meiotic prophase during which the pairing of homologous chromosomes occur.

ANNEXURE II

TERMINOLOGY

A—, (La.) prefix—without, separation
Abdomen. (La.) to hide
Abducens. (La.) *ab* = away + *ducens* = leading
Abductor. (La.) *ab* = away + *ducere* = to lead
Accessory. (La.) *accessorius* = supplementary
Acelous. (Gr.) *a* = without, not, + *koilos* = hollow.
Acetabulum. (La.) *acetabulum* = vinegar cup
Acrania. (Gr.) *a* = without + *krania* = heads
Ad—, (La.) prefix—to, toward, at, near
Adductor. (La.) *ad* = to + *ducere* = leading
Adrenal. (La.) *ad* = near + *renes* = kidneys
Allantois. (Gr.) *allas* = sausage + *eldos* = form
Alveolus. (La.) *alveolus* = small cavity
Amnion. (Gr.) *amnion* = foetal membrane
Amphi—, (Gr.) prefix—on both sides, double
Amphibia. (Gr.) *amphi* = double + *bios* = life
Amphicoelous. (Gr.) *amphi* = both + *koilos* = hollow.
Amphioxus. (Gr.) *amphi* = both + *oxys* = sharp
Ampulla. (La.) *ampulla* = a vessel with swelling in the middle.
An–, (Gr.) prefix—without, not
Anamniota. (Gr.) *an* = without + *amnion* = foetal membrane
Anapsid. (Gr.) *an* = not + *apsis* = arch
Anastomosis. (Gr.) *anastomocin* = cause to communicate
Annulus. (La.) *anulus* = a ring
Anura. (Gr.) *an* = without + *oura* = tail
Anus. (La.) *anus* = fundament
Apo—, (Gr.) prefix—form
Apoda. (Gr.) *a* = without + *poda* = foot
Appendicular. (La.) *appendere* = to hang upon
Arachnoid. (Gr.) *arachnes* = spider + *eidos* = shape of like
Arch—, (Gr.) prefix—first, chief
Archenteron. (Gr.) *arch* = first + *enteron* = intestine, gut.
Astragalus. (Gr.) *astragalos* = ankle bone
Atrium. (La.) *atrium* = entrance hall, court
Auditory. (La.) *auricula* = external ear.
Auricle. (La.) *auricula* = external ear
Auto—, (Gr.) prefix—*autos* = self
Autonomic. (Gr.) *autos* = self + *nomas* = law
Autostylic. (Gr.) *autos* = self + *stylos* = pillar
Axial. (La.) *axis* = the line about which body turns
Azygos. (Gr.) *a* = not + *zygos* = unpaired
Basal. (La.) *basis* = base or footing
Basi—, (La.) prefix—pertaining to the base
Basibranchial (La.) *basis* = base + *branchiae* = gills
Basihyoid. (La.) *basis* = base + (Gr.) *hyoeides* = Y-shaped
Bi—, (La.) prefix—two, twice, double
Biceps. (La.) *bis* = twice + *caput* = head
Bilateral (La.) *bi* = two + *latus* = side
Blasto—, (Gr.) prefix—bud, germ
Blastocoel. (Gr.) *blastos* = germ + *Koilos* = hollow
Blastoderm. (Gr.) *blastos* = germ + *diskos* = a round plate
Blastomere. (Gr.) *blastos* = germ + *meros* = part
Blastopore. (Gr.) *blastos* = germ + *poros* = opening
Blastula. (La.) *blastos* = germ

Abbreviations used: Fr.—French; Gr.—Greek; La.—Latin; NLA.—New Latin; Sp.—Spanish.

Brachial. (La.) *branchialis* = related to arm
Branchial. (La.) *branchiae* = gills
Bronchus. (Gr.) *bronchia* = end of windpipe
Buccal. (La.) *bucca* = cheek
Bulbous. (La.) *bulbus* = bulb, swollen root

Caecum. (La.) *caecus* = blind
Calcaneum. (La.) *calcaneum* = heel
Callosum. (La.) *callosus* = thick skinned
Capillary. (La.) *capillaris* = pertaining to the hair
Capitulum. (La.) *caput* = small head
Carapace. (NLa.) *carapax* = bony or chitinous covering
Cardiac. (Ga.) *kardiakos* = pertaining to the breast
Carnassial. (Fr.) *carnissier* = carnivorous
Carnivorous. (La.) *caro* = flesh + *vorare* = to devour
Carpus. (Gr.) *karpos* = wrist
Caudal. (La.) *cauda* = tail
Cava. (La.) *cavus* = hollow
Cephalic. (Gr.) *Kephale* = head
Cerebellum. (La.) *cerebrum* = brain
Cervical. (La.) *cervix* = neck
Chiasma. (Gr.) *chiasma* = figure of X
Chondrichthyes. (Gr.) *chondros* = cartilage + *iethys* = fish
Chorda. (Gr.) *chorde* = cord, string of gut
Choroid. (Gr.) *chorion* = skin + *eides* = like
Chorion. (Gr.) *chorion* = skin
Chromatophore. (Gr.) *chroma* = colour + *pherein* = to bear
Chromosome. (Gr.) *chroma* = colour + *soma* = body
Chyme. (Gr.) *chymos* = juice
Ciliary. (La.) *cilium* = eyelash
Circum—, prefix—around
Cloaca. (La.) *cloaca* = drain, sewer
Cnemial. (Gr.) *kneme* =lower leg
Cochlea. (La.) *cochlea* = spiral, snail shell
Coeliac. (Gr.) *koilia* = body
Coelom. (Gr.) *koiloma* = a hollow
Colon. (La.) *colon* = great gut
Columella. (Gr.) *columna* = little pillar
Commissure. (La.) *commissura* = connection
Condyle. (Gr.) *kondylos* = knuckle
Conjunctiva. (Gr.) *conjunctivus* = connecting
Constrictor. (La.) *constringere* = to draw together
Cornea. (La.) *corneus* = horny
Coronary. (La.) *coronarius* = encircling
Cortex. (Gr.) *cortex* = bark, rind
Costa. (La.) *costa* = rib
Cranial. (Gr.) *kranion* = skull
Cribriform. (La.) *cribrium* = sieve + *forma* = form
Crista. (La.) *crista* = crest
Ctenoid. (Gr.) *kteis* = a comb + *eidos* = form
Cuneiform. (La.) *cuneus* = wedge + *forma* = shape
Cutis. (La.) *cuits* = skin
Cycloid. (Gr.) *Kyklos = Circle + eidos = form*
Cytoplasm. (Gr.) (*kytos* = cell + *plasma* = plasma

De—, (La.) *prefix* = down, away from, deprived of
Deferens. (La.) *de* = away + *ferens* = carrying
Depressor. (Gr.) *de* = down + *premere* = to press
Dermal. (Gr.) *derma* = skin
Di—, (Gr.) prefix—twice, double, twofold
Diaphragm. (Gr.) *diaphragma* = partition
Diaphysis. (Gr.) *dia* = between + *physis* = growth
Diapophysis. (Gr.) *dia* = apart + *apophysis* = outgrowth
Diapsid. (Gr.) *di* = double + *apsis* = arch
Diastema. (Gr.) *diastema* = interval
Digit. (La.) *digitus* = finger
Diphycercal. (Gr.) *diphyes* = twofold + *kerkos* = tail
Dipnoi. (Gr.) *di* = double + *pnein* = to breathe
Distal. (La.) *distare* = to stand apart
Dorsal. (La.) *dorsum* = back
Duct. (La.) *ducere* = to lead or draw
Duodenum. (La.) *duodeni* = twelve (twelve finger breadths)

E—, Ex—, (La.) *prefix*—out, out of, from
Ectoderm. (Gr.) *ektos* = outside + *derma* = skin
Effector. (La.) *effecere* = to bring, to pass
Efferent. (La.) *ex* = out + *ferre* = to bear
Ejaculatory. (La.) *e* = out + *jacere* = to throw

Ek—, Ekto—, (Gr.) prefix—out of, outside, from

En—, Endo—, (Gr.) prefix—in, within

Endocardium. (Gr.) *endon* = within + *Kardia* = heart

Endocrine. (Gr.) *endon* = within + *krinen* = to separate

Endoderm. (Gr.) *endon* = within + *derma* = skin

Endolymph. (Gr.) *endon* = within + (La.) *lympha* = water

Endometrium. (Gr.) *endon* = within + *metra* = womb

Endoskeleton. (Gr.) *endon* = within + *skeletos* = dried up

Endostyle. (Gr.) *endon* = within + *stylos* = pillar

Endothelium. (Gr.) *endon* = within + *thele* = nipple

Enzyme. (Gr.) *en* = in + *zyme* = leaven

Epiaxial. (Gr.) *epi* = on + *axis* = centre

Epibranchial. (Gr.) *epi* = on + *branchia* = gills

Epicardium. (Gr.) *epi* = on + *kardia* = heart

Epicondyle. (Gr.) *epi* = on + *kondylos* = knuckle

Epidermis. (Gr.) *epi* = on + *derma* = skin

Epididymis. (Gr.) *epi* = on + *didymoi* = testicles

Epiglottis. (Gr.) *epi* = on + *glotta* = tongue

Epiphysis. (Gr.) *epi* = on + physis = growth

Epithalium. (Gr.) *epi* = on + *thele* = nipple

Erythrocyte. (Gr.) *erythros* = red + *kytos* = cell

Ethmoid. (Gr.) *ethmos* = sieve + *eidos* = form

Excretion. (Gr.) *ex* = out + *cernere* = to shift

Exocrine. (Gr.) *ex* = out + *krinein* = to separate

Extensor. (La.) *ex* = out + *tendere* = to stretch

Extrinsic. (La.) *extrinsecus* = on the outside

Facialis. (Gr.) *facies* = face

Falciform. (La.) *falx* = sickle + *forma* = shape

Fascia. (La.) *fascia* = band

Fibre. (La.) *fibra* = string, thread

Fibril. (NLa.) *fibrilla* = a little thread

Filoplume. (La.) *filum* = thread + *pluma* = soft leather, down

Fissure. (La.) *fissura* = a cleft

Flagellum. (La.) *flagellum* = a little whip

Flexor. (La.) *flexus* = bend

Flocculus. (NLa.) *floccus* = a tuft of wool

Follicle. (La.) *folliculus* = a small bag

Fovea. (La.) *fovea* = small pil

Frontal. (La.) *frons, frontis* = forehead, brow

Fundus. (La.) *fundus* = bottom

Gametes. (Gr.) *gametes* = spouse

Ganglion. (Gr.) *ganglion* = a swelling under the skin

Gastrula. (NLa.) *gaster* = stomach

Genital. (La.) *genitalis* = pertaining to birth

Germinal. (La.) *germen* = germ, bud

Glans. (La.) *glans* = acorn

Glenoid (Gr.) *glene* = socket + *eidos* = form

Glomerulus. (La.) *glomus* = a ball

Glossopharyngeus. (Gr.) *glossa* = tongue + *pharynx* = throat

Glottis. (Gr.) *glottis* = mouth of wind pipe

Gnathos. (Gr.) *gnathos* = jaw

Gnathostomata. (Gr.) *gnathos* = jaw + *stoma* = mouth

Gonad. (Gr.) *gone* = seed

Granulocytes. (La.) *granulum* = small grain + (Gr.) *kytos* = cell

Guanin. (Sp.) *guano* = dung of sea fowl

Gular. (La.) *gula* = throat

Haemal. (Gr.) *haima* = blood

Haemoglobin. (Gr.) haima = blood + (La.) globus = Sphere

Hemi. (Gr.) prefix—half

Hemibranch. (Gr.) *hemi* = half + *branchia* = gills

Hemichordata. (Gr.) *hemi* = half + *chorde* = string

Hemipenis. (Gr.) *hemi* = half + (La.) *penis* = penis

Hemisphere. (Gr.) *hemi* = half + *sphaira* = ball

Hepatic. (La.) *hepar* = liver

Hetero—, (Gr.) prefix—other, differnt

Heterocoelous. (Gr.) *heteros* = other + *koilus* = hollow

Heterocercal. (Gr.) *heteros* = other + *kerkos* = *tail*

Heterodont. (Gr.) *heteros* = other + *odons* = teeth

Hippocampus. (Gr.) *hippos* = horse + *kampos* = sea monster

Histology. (Gr.) *histos* = web + *logos* = account, discourse

Holo—, (Gr.) prefix—whole
Holoblastic, (Gr.) *holos* = whole + *blastos* = *germ*
Holobranch (Gr.) *holos* = whole + *branchia* = *gill*
Holocephali. (Gr.) *holos* = whole + *kephale* = head
Holostei. (Gr.) *holos* = whole + *osteon* = bone
Homo—, (Gr.) prefix—one and the same
Homocercal. (Gr.) *homos* = same + *kerkos* = tail
Homoeothermus. (Gr.) *homos* = same + *thermos* = hot
Homolecithel. (Gr.) *homos* = same + *lekithos* = yolk
Homology. (Gr.) *homos* = same + *logos* = ratio
Hormone. (Gr.) *hormaein* = to excite
Humor. (La.) *humor* = fluid, moisture
Hyaline. (Gr.) *hyalos* = glass
Hyoid. (Gr.) *hyoeides* = U-shaped
Hyomandibular. (Gr.) *upsilon* = Y=-shaped + (La.) *mandibula* = jaw
Hyostylic. (Gr.) *upsilon* = Y-shaped + *stylos* = pillar
Hypaxial. (Gr.) *hypo* = under + (La.) *axis* = centre line, axis
Hypo—, (Gr.) prefix—under, below
Hypobranchial. (Gr.) *hypo* = below + *branchia* = gills
Hypoglossal. (Gr.) *hypo* = under + *glossa* = tongue
Hypophysis. (Gr.) *hypo* = under + *physis* = growth
Hypothalamus. (Gr.) *hypo* = under + *thalamus* = chamber
Ileum. (Gr.) *eilein* = to wind or turn
In—, (La.) prefix—not, in into, within, toward, on
Incisor. (La.) *incisus* = cut
Incus. (La.) *incus* = anvil
Inductor. (La.) *inducere* = excite, to lead on
Infra—, (La.) prefix—below, lower than
Infundibulum. (La.) *infundibulum* = a funnel
Inguinal. (La.) *inguina* = groin
Integument. (La.) *in* = over + *tegere* = to cover
Inter—, (La.) prefix—between, among
Intercostal. (La.) *inter* = between + *costa* + rib
Interstitial. (La.) *inter* = between + *sistere* = to set
Intervertebral. (La.) *inter* = between + *vertebra* = joint
Intestine. (La.) *intus* = within
Intrinsic. (La.) *intrinsecus* = inward
Invagination. (La.) *in* = in + *vagina* = sheath
Invertebrate. (La.) *in* = not + *vertebratus* = jointed
Iris. (Gr.) *iris* = rainbow
Ischiofemoral. (Gr.) *ischion* = lip + (La.) *femur* = thigh
Ischium. (Gr.) *ischium* = hip
Iso—, (Gr.) Prefix—equal
Isolecithal. (Gr.) *isos* = equal + *lekithos* = yolk
Isomer. (Gr.) *isos* = equal + *merox* = part
Jejunum. (La.) *jujunus* = empty (of food)
Jugal. (La.) *jugum* = yoke
Jugular. (La.) *jugularis* = pertaining to the neck
Labial. (La.) *labialis* = pertaining to the lips
Lacrimal. (La.) *lacrina* = tear
Lagena. (La.) *lagena* = flask
Lamina. (La.) *lamina* = thin plate
Larva. (La.) *larva* = ghost, musk
Larynx. (Gr.) *larynx* = upper part of wind pipe
Lateral. (La.) *lateralis* = pertaining to a side
Leukocyte. (Gr.) *leukos* = white + *Kytos* = cell
Ligamentum. (La.) *ligamentum* = a bandage
Lipid. (Gr.) *lipos* = fat + eidos = alike
Lphodont. (Gr.) *lophos* = ridge + *odons* = teeth
Lumbar. (La.) *lumbare* = apron for the loins
Luteum. (La.) *luteus* = yellow
Lymphocyte. (La.) *lympha* = water + *kytos* = cell
Macrophage. (Gr.) *makros* = large + *phagein* = to eat
Malleus. (La.) *malleus* = hammer
Mammillary. (La.) *mammillaris* = of or in the breast
Marginal. (La.) *marginalis* = bordering
Marsupium (La.) *marsupium* = pouch
Mastoid. (Gr.) *mastos* = bread + *eidos* = like
Matrix. (La.) *matrix* = mold, groundwork
Meatus. (La.) *meatus* = passage
Medial. (La.) *medialis* = pertaining to the middle

Mediastinum. (La.) *mediastinus* = servant, drudge

Medulla, (La.) *medulla* = marrow, pith

Melanin. (Gr.) *melas, melanos* = black

Membrane. (La.) *membrana* = skin

Mes—, Meso—, (Gr.) prefix = middle

Mesencephalon. (Gr.) *mesos* = middle + en = in + *kephate* = head

Mesenchyme. (Gr.) *mesos* = middle + en = in + *chymos* = juice.

Mesentery. (Gr.) *mesos* = midway between + *enteron* = gut

Mesoderm. (Gr.) *mesos* = middle + *derma* = skin

Mesolecithal. (Gr.) *mesos* = middle + *lekithos* = yolk

Mesonephros. (Gr.) *mesos* = middle + *nephros* = kidney

Mesorchium. (Gr.) *mesos* = middle + *orchis* = testis

Mesovarium. (Gr.) *mesos* = middle + *ovarium* = ovary

Meta—, (Gr.) prefix—after, next, indicating change of time or situation

Metabolic. (Gr.) *metabole* = change

Metacarpus. (Gr.) *meta* = after + (La.) *carpus* = wrist

Metamere. (Gr.) *meta* = after + *meros* = part

Metamorphosis. (Gr.) *meta* = change + *morphe* = form

Metanephors. (Gr.) *meta* = after + *nephros* = kidney

Metatarsus. (Gr.) *meta* = after + *tarsus* = ankle

Metencephalon. (Gr.) *meta* = after + *enkephalos* = brain

Molar. (La.) *molaris* = millstone

Monocyte. (Gr.) *monos* = single + *kytos* = cell

Mucus. (La.) *mucus* = slippery secretion

Myelencephalon. (Gr.) *myeles* = marrow + *enkephalos* = brain

Myelin. (Gr.) *myelos* = marrow

Myo—, (Gr.) prefix—muscle

Myocardium. (Gr.) *myos* = muscle + *kardia* = heart

Myomere. (Gr.) *myos* = muscle + *meros* = part

Myotome. (Gr.) *myos* = muscle + *tome* = a cutting

Naris. (La.) *naris* = nostril

Neopallium. (Gr.) *neos* = new, youthful + (La.) *pallium* = cloak

Nephridia. (Gr.) *nephridias* = kidney

Neural. (Gr.) *neuron* = nerve

Neurilemma. (Gr.) *neuron* = nerve + *lemma* = sheath

Neuro—, (Gr.) prefix—nerve

Neuroglia. (Gr.) *neuron* = nerve + *gloia* = glue

Neuromast. (Gr.) *neuron* = nerve + *mastos* = around hill

Neuron. (Gr.) *neuron* = sinew, tendon, nerve

Nictitating. (Gr.) *Nictare* = to wink

Nidamental. (La.) *nidamentum* = materials for a nest

Node. (La.) *nodas* = knot

Notochord. (Gr.) *noton* = back + *chorde* = cord

Nuchal. (La.) *nucha* = nape of the neck

Oculomotor. (La.) *oculus* = eye + *motor* = mover

Odontoblast. (Gr.) *odous* = tooth + *blastos* = shoot, germ

Oesophagus. (Gr.) *oisein* = to carry + *phagein* = to eat

Olecranan. (Gr.) *olekranon* = point of the elbow

Olfactory. (La.) *olere* = to smell + *facere* = to make

Omentum. (Gr.) *omentum* = fatty membrane enclosing the bowels

Ontogeny. (Gr.) *onto* = existing + *gennan* = to beget

Operculum. (La.) *operculum* = lid

Ophthalmic. (Gr.) *opthalmos* = eye

Opistho—, (Gr.) prefix—backward, behind

Opisthocelous. (Gr.) *opisthe* = behind + *koilos* = hollow

Optic. (Gr.) *opsis* = sight

Osseous. (La.) *ossa* = bone

Ossicle. (La.) *ossiculum* = small bone

Osteoblast. (Gr.) *osteon* = bone + *blastos* = germ

Osteocyte. (Gr.) *osteon* = bone + *kytos* = cell

Otic. (Gr.) *otikos* = in the ear
Otolith. (Gr.) *otos* = ear + *lithos* = stone
Ovum. (La.) ovum, egg

Palaeontology. (Gr.) *palios* = ancient, Onto = existing matter + logos = science
Pallium. (La.) *palma* = the hand
Pancreas. (Gr.) *pan* = all + *kreas* = flesh
Papilla. (La.) *papilla* = pimple
Para—, (Gr.) prefix— near, alongside
Parathyroid. (Gr.) *para* = near + *thyreas* = oblong shield + *eidos* = form
Parietal. (La.) *paries* = wall
Parotid. (Gr.) *para* = near + *otos* = ear
Pecten. (La.) *pecten* = comb
Pectoral. (La.) *pectoralis* = in the breast
Pelvic. (La.) *pelvis* = basin
Peri—, (Gr.) prefix—around
Pericardial. (Gr.) *peri* = around + *kardia* = heart
Perilymph. (Gr.) *peri* = around + (La.) *lympha* = fluid
Periosteum. (Gr.) *peri* = around + *osteon* = bone
Pharynx. (Gr.) *phanynx* = throat
Phylogeny. (Gr.) *phylon* = race + *gennan* = to beget
Pineal. (La.) *pinea* = pine cone
Pinna. (La.) *penna* = feather, wing
Pituitary. (La.) *pituita* = slime
Placenta. (La.) *placenta* = a flat cake
Plastron. (Fr.) *plastron* = breast plate
Pleuro—, (Gr.) prefix—side
Plexus. (La.) *plexus* = braid
Pneumatic. (Gr.) *pneumatikos* = breath
Poikilothermous. (Gr.) *poikilos* = changeful + *thermos* = heat
Pons. (La.) *pons* = bridge
Portal. (La.) *porta* = gate
Porus. (Gr.) *poros* = passage
Pous. (Gr.) *pous (pod)* = foot
Pre—, (La.) prefix—in front, before
Premolar. (La.) *pre* = in front + *molaris* = molar
Prepuce. (La.) *praeputium* = foreskin
Primordial. (La.) *primordium* = beginning
Pro—, (Gr.) prefix—prior, before, in front of
Procoelous. (Gr.) *pro* = in front + *koilos* = hollow
Proctodaeum. (Gr.) *proktos* = anus + *hodaios* = a way
Pronephros. (Gr.) *pro* = before + *nephros* = kidney
Proprioreceptor. (La.) *proprus* = special + *capere* = to take
Prosencephalon. (Gr.) *pros* = before + *enkephalos* = brain
Prostrate. (Gr.) *pro* = in front + *stare* = to stand
Protonephros. (Gr.) *protos* = first + *nephnos* = kidney
Protoplasm. (Gr.) *protos* = first + *plasma* = form
Proximal. (La.) *proximus* = next
Pseudobranch. (Gr.) *pseudo* = false + *branchion* = gills
Pterygoid. (Gr.) *pteryx* = wing + *eidos* = resemblance
Pterylae. (Gr.) *pteron* = feather + *hyle* = matter, wood
Pubis. (La.) *pubis* = mature
Pulmonary. (La.) *pulmo* = lung
Pygostyle. (Gr.) *pyge* = rump + *stylos* = pillar
Pylorus. (Gr.) *pylouros* = gate-keeper
Pyriform. (La.) *pirum* = prior + *forma* = shape

Radial. (La.) *radius* = rod, spoke
Receptor. (La.) *recipere* = receive
Rectus. (La.) *rectus* = straight
Remiges. (La.) *remex* = rower
Renal. (Gr.) *renes* = kidney
Rete. (La.) *rete* = network
Reticulum. (La.) *reticulum* = a small net
Retina. (La.) *rete* = network
Retractor. (La.) *retrahera* = to draw back
Retrices. (La.) *retro* = back + *cedere* = to go
Rhombencephalon. (Gr.) *rhombas* = a type of parallelogram + *enkephalos* = brain
Rostrum. (La.) *rostrum* = back

Saccule (La.) *sacculus* = a small bag
Sacrum. (La.) *sacer* = sacred

Sagittal. (La.) *sagitta* = sagitta = arrow
Sarcolemma. (Gr.) *sarx* = flesh + *lemma* = skin
Sclera. (Gr.) *skleros* = hard
Scrotum. (La.) *scrotum* = skin
Sebaceous. (La.) *sebum* = tallow; grease
Seminiferous. (La.) *semen* = seed + *ferre* = to bear
Septum. (La.) *saeptum* = fence
Sinus. (La.) *sinus* = cavity, curve
Somatic. (Gr.) *soma* = body
Somatopleure. (Gr.) *soma* = body + *pleura* = side
Somatic. (Gr.) *soma* = body + *ite* = origin
Spermatozoon. (Gr.) *sperms* = seed + *zoon* = animal
Spheniod. (Gr.) *sphen* = wedge + *eidos* = like
Sphincter. (Gr.) *sphingein* = to bind tight
Spiracle. (La.) *spiraculum* = air hole
Splanchnic. (Gr.) *splanchna* = viscera
Stapes. (La.) *stapes* = stirrup
Stomodaeum. (Gr.) *stoma* = mouth + *hodaios* = a way
Striatum. (La.) *striatus* = streaked
Styloid. (Gr.) *stylos* = pillar + *eidos* = resemblance, likeness
Sub—, (La.) prefix— under, beneath, near
Subcostal. (La.) *sub* = under + *costa* = rib
Sublingual. (La.) *sub* = under + *lingus* = tongue
Sulcus. (La.) *sulcus* = furrow
Sym—, Syn—, (Gr.) prefix—with, together
Sympathetic. (Gr.) *sym* = together + *patherin* = to suffer
Symphysis. (Gr.) *sym* = together + *physis* = growth
Synapse. (Gr.) *syn* = together + *haptein* = to fasten
Synsacrum. (Gr.) *syn* = together + (La.) *sacer* = sacrum
Syrinx. (Gr.) *syrinx* = pipe

Tarsus. (Gr.) *tarsos* = sole of the foot
Tegumentum. (La.) *tegumentum* = covering
Telencephalon. (Gr.) *telos* = end + *enkephalos* = brain
Telolecithal. (Gr.) *telas* = end + *lakithos* = yolk
Temporal. (La.) *tempora* = the temples
Tendon. (La.) *tendere* = stretch
Terminalis. (La.) *terminare* = to limit
Testis. (La.) *testis* = testicle
Tetropod. (Gr.) *tetra* = four + *podos* = foot
Thalamus. (Gr.) *thalamos* = chamber
Thecodont. (Gr.) *thake* = sheath + *odous* = teeth
Thorax. (Gr.) *thorax* = breast plate, the chest
Thrombocyte. (Gr.) *thrombos* = clot + *kytos* = cell
Thymus. (Gr.) *thymus* = sweet bread
Thyroid. (Gr.) *thyreos* = shield + *eidos* = form
Trachea. (Gr.) *tracheia* = windipipe
Triceps. (La.) *tres* = three + *coput* = head
Trigeminus. (La.) *trigeminus* = three born together
Trochanter. (Gr.) *troches* = pulley
Trochlea. (Gr.) *Trochilia* = pulley
Trophoblast. (Gr.) *trophe* = nourishment + *blastos* = germ
Tunica. (La.) *tunica* = under garment
Tympanic. (La.) *tympanum* = drum

Umbilical. (La.) *umbilicus* = navel
Unciform. (La.) *uncus* = hook + *forma* = shape
Uncinate. (La.) *uncinatus* = with a hook
Urea. (Gr.) *ouron* = urine
Urodaeum. (Gr.) *ovron* = urine + *hodaios* = pertaining away
Urodela. (Gr.) *aura* = tail + *delos* = evident
Urinogenital. (Gr.) *ouron* = urine + (La.) *genitalis* = genital
Uropygial. (Gr.) *orros* = end of sacrum + *pyge* = rump
Urostyle. (Gr.) *oura* = tail + *stylos* = pillar
Uterus. (La.) *uterus* = womb
Utricle. (La.) *Utriculus* = leather bottle

Vagus. (La.) *vagus* = wandening
Vas. (La.) *vas* = vessel
Vascular. (La.) *vasculum* = small vessel
Ventral. (La.) *Venter* = belly
Ventricle. (La.) *Ventriculus* = small cavity
Vermiform. (La.) *Vermis* = worm + *forma* = shape
Vesicle. (La.) *vesicula* = a small bladder

Vestibulum. (La.) *vestibulum* = entrance court
Vibrissa. (La.) *vibrissa* = hair in the nostril
Villus. (La.) *villus* = shagy hair
Visceral. (La.) *viscera* = bowels
Vitelline. (La.) *vitellus* = egg yolk
Vitreous. (La.) *vitreus* = of glass
Viviparous. (La.) *vivus* = living + *parere* = to beget
Vomer. (La.) *vomer* = ploughshare

Xiphiplastron. (Gr.) *xiphos* = sword + *plastron* = breast plate

Zygapophysis. (Gr.) *zygon* = yoke + *apophysis* = process of a bone
Zygomatic. (Gr.) *zygoma* = cheek bone

INDEX

All numbers refer to pages; Numbers in **thick** types are number of pages with figures; Words in *italics* are names of genus and species; Words with asterix are names of higher divisions.

ABBREVIATIONS

A1. Ascaris, A2. Achatina, A3. Asterias, A4. Ascidia, B1. Balanoglossus, B2. Branchiostoma, B3. Bufo, C1. Channa, C2. Calotes, C3. Columba, C4, Cavia, F. Fasciola, H. Hirudinaria, L1. Lamellidens. L2. Lates, M1. Macrobrachium, M2. Myxine, N. Neanthes, P1. Pheretima, P2. Periplaneta, P3. Palamnaeus. P4. Pila, P5. Petromyzon, S1. Scoliodon, S2. Scolopendra, S3. Sepia, T. Taenia.

A

D

G

M

P